CARNIVOROUS MARSUPIALS

CARNIVOROUS MARSUPIALS

Volume 1

Edited by

Michael Archer

Published by

Royal Zoological Society of New South Wales

P.O. Box 20, Mosman, 2088, N.S.W., Australia

ISBN—0 9599951 3 7
National Library Canberra

Published February, 1982

Printed in Australia by
SURREY BEATTY & SONS PTY. LIMITED
43 Rickard Road, Chipping Norton, 2170, N.S.W.

FOREWORD

As Michael Archer has pointed out in the Introduction to these Volumes, the American carnivorous marsupials had an auspicious beginning in European history, under Royal patronage. The Australian carnivorous marsupials had a much more mundane entry into European knowledge. The first one was observed by the savants on the *Endeavour* in 1770; it was described by Parkinson as "a spotted [quadruped] of the viverra kind" and by Banks as "about the size and something like a polecat". The Aboriginal name was recorded variously as Je–quoll, Taquol, Jaquol or Dekol, incidentally rather different words to the misbegotten abbreviation "Quoll" currently being imposed as the common name for all of the spotted dasyures. Even when further species and specimens of dasyurids came to light with the first European settlement at Sydney Cove (1788), their special nature was usually not appreciated and they were often classified with mustelids or viverrids. Apparently only females with pouches were identified with certainty as "opossums". Thus, in *The Voyage of Governor Phillip to Botany Bay* (1789), the Native Cat *(Dasyurus viverrinus)* figured, a female, is called the Spotted Opossum, whereas the Tiger Cat *(D. maculatus)* illustrated, of unrecorded sex, is listed as a Spotted Martin or Martin Cat. In any case these creatures intruded into European affairs mainly as raiders of hen houses or, when the Thylacine was discovered in Tasmania, as sheep killers.

As European exploration and settlement of Australia progressed, the study of the Dasyuridae, in common with that of other faunal groups, settled down to a long descriptive period during which they were separated from other marsupials as a distinct family. Numerous genera and species were discovered and described, and superficial observations were recorded on their "habits and economy".

During this long discovery period the first Australian fossil marsupials were collected by the Surveyor–general and explorer T. L. Mitchell. In his narrative (1838), fragments of a dasyurid *(Sarcophilus laniarius)* and a *Thylacoleo carnifex* are figured. In Mitchell's text, Richard Owen described the former but failed to mention the latter. He described it later (1858) from new material and pronounced it carnivorous — "one of the fellest and most destructive of predatory beasts". This generated a lively controversy on whether the creature was carnivorous or herbivorous, an argument which remains highly interesting as shown in the pages of Volume 2.

The modern scientific period began in the 1880s after the general acceptance of Darwin's views on evolution and the publication of Huxley's paper (1880) in which the three major groups of mammals were placed in a linear sequence, with the Metatheria being considered but a stage in the evolutionary development of the "higher" mammals. It became important to study the monotremes and marsupials in order to have a better understanding of the evolution of the higher mammals. This kind of science is well exemplified by the elegant works on embryology and microanatomy published by Richard Semon and his collaborators following his visit to Australia to collect material in 1891–92. The first of the native schools that specialized in research on monotremes and marsupials developed in Sydney University following the appointment of J. T. Wilson as Professor of Anatomy in 1890. Some of his colleagues and associates made notable contributions to the study of the Dasyuridae. In 1902, C. J. Martin published the first paper on thermoregulation and respiratory physiology, including experiments with a *Dasyurus maculatus*. J. P. Hill was the outstanding pioneer worker on marsupial reproduction. Among other things he carried out detailed investigations of the embryology of *D. viverrinus* and, with his student, C. H. O'Donoghue, worked out the basic reproductive cycle of that species. T. T. Flynn, a Sydney graduate and later Professor of Biology in Tasmania, included reproduction in *Sarcophilus harrisii* among his studies. To most people he is chiefly remembered as the father of the famous film star Errol, whose life history perhaps parallels in some respects that of the males of certain small dasyurids that have a short, wild love life and then drop dead.

The study of Australian marsupials, most of it in the nature of comparative anatomy, continued in a few institutions with slow but steady progress into the 1950s, although little that was new was published on the Dasyuridae. The first descriptions of chromosomes of dasyurids were published in 1923, and in the 1940s and 1950s Pearson made detailed investigations of the female reproductive systems of the carnivorous marsupials native to Tasmania. J. P. Hill's last work on *D. viverrinus* appeared in 1955.

As we all know there was an enormous upsurge in research on Australian marsupials beginning in the early 1950s, and happily this has accelerated with time. Many research centres and people are now involved and most fields represented, including systematics and phylogeny of modern and fossil species, ecology, behaviour, genetics, biochemistry, reproductive biology, physiology and so on.

Until very recent times the study of the Dasyuridae has lagged behind that of other marsupials, largely, I think, because most people found the taxonomy too difficult, and the animals would not readily breed in captivity. *Antechinus stuartii* and one or two other small dasyurids must have been common around or close to Sydney in the early days of European settlement. In the Watling drawings painted in Sydney between 1788 and 1794 and now housed in the library of the Zoology Department, British Museum (Natural History), No. 93, here identified for the first time, is a representation of *A. stuartii* and bears the following legend — "Native name Mirrin. The Penis and Testacles (sic) made visible. The former placed between the latter and the anus. This Animal I cut out of a fallen Rotten Tree, into which my dog drove it." Yet the species was not described scientifically until 1841. Only one small dasyurid was recognised before 1838 — *Antechinus minimus* from Tasmania, described in 1803 by the great French zoologist Étienne Geoffroy Saint-Hilaire. Only within the last ten years has the systematics and phylogeny of the fossil and modern Dasyuridae been finally placed on a sound basis, largely by Michael Archer. Pat Woolley deserves the credit for initiating modern biological studies of dasyurids in the early 1960s with her work on reproduction in *A. stuartii* and other *Antechinus* species, and her discovery of post–mating die–off of males. This stimulated much new work on ecology, endocrinology and parasitology in attempts to explain this remarkable phenomenon.

The carnivorous marsupial symposium and these two volumes arising in part from it are the first major stocktaking of the current state of our knowledge of the carnivorous marsupials. Because of the geographical location and the nationality of the majority of contributors, the symposium was naturally heavily weighted towards the Australian fauna. However, South American groups intrude significantly in questions relating to phylogeny. The 64 contributions in these volumes contain an enormous amount of new information on the Dasyuridae, Thylacinidae and Thylacoleonidae in virtually all fields of endeavour, and much that was previously known is comprehensively reviewed.

We all owe a great debt to the Royal Zoological Society of New South Wales for holding the symposium and publishing "Carnivorous marsupials", and we are especially indebted to Michael Archer who organized the symposium, edited the volumes and was himself a major contributor.

John Calaby

Canberra
January 1, 1982

CONTENTS

Volume 1

PHYSIOLOGY

BEHAVIOUR

Volume 2

PALAEONTOLOGY

MORPHOLOGY, CYTOLOGY, GENETICS, ELECTROPHORESIS AND PHYLOGENETIC RELATIONSHIPS

1. GENESIS: AND IN THE BEGINNING THERE WAS AN INCREDIBLE CARNIVOROUS MOTHER

MICHAEL ARCHER[1]

The first marsupial brought to the attention of Europeans was a Brazilian opossum with pouch young presented by the Spanish explorer Vincente Yáñez Pinzón to Queen Isabella and King Ferdinand of Spain in 1500. Her Majesty's curiosity about this most "incredible mother", as she put it, led her to poke her Royal Fingers into its swollen pouch. This Royal Intrusion was the first European research project carried out on a carnivorous marsupial.

Fig. 1. The Royal Intrusion of 1500, upon which occasion Her Majesty Queen Isabella of Spain placed her Royal Fingers into the pouch of a carnivorous marsupial mother.

(Drawing by Pamela Woods)

DURING the subsequent centuries, hundreds of living and extinct marsupials found themselves the objects of intense scientific curiosity, and of these, the carnivorous marsupials of South America and Australia attracted a major share of the attention. This probably reflects both the mild fear and yet sense of kinship we seem to feel towards these alert, bright–eyed killers of the night. This special interest gripped a long and venerable succession of twentieth century mammalogists including Oldfield Thomas, George Tate, Ellis Troughton, Basil Marlow, David Ride, John Calaby and Pat Woolley.

David Ride's keen interest in marsupial carnivores frequently infected colleagues. Shortly after my arrival in Perth in 1967 as a student of vertebrate palaeontology, he suggested it would be interesting to examine Pleistocene assemblages of small mammals from Western Australian caves. But after an abortive first effort to identify the small dasyurids present in these fossil deposits, it became evident that the basic species–level systematics of small dasyurid genera were badly in need of revision. So, simply in order to identify the cave fossils, I committed what I assumed would be no more than a year to sort out the necessary species–level taxonomy. That was in 1968. Fourteen years, three generic revisions, eight new genera and ten species descriptions later, my extraordinary naivete appalls me. Further, only now realizing the extent of David Ride's determination to have the systematics of the small dasyurid genera overhauled, I suspect him of gross subterfuge when he originally suggested the fossil project to me in 1967.

Whatever his real motives, this diversion into the world of living beasts made me realize there was a great deal more to dasyurids than a set of fossilized teeth. For example, there is simply nothing like having a large pet research subject fang you on the nose in the middle of the night to make you fully appreciate form–function complexes in mammalian teeth. Similarly, one's first look at the erect male intromittent organs of a healthy Western Quoll instantly dooms all previous limericks about such subjects to mediocrity (in this regard, see Pat Woolley's article on useful taxonomic tools). Clearly these fascinating beasts deserved attention into all aspects of their being.

ZOOLOGICAL SYMPOSIA OF THE ROYAL ZOOLOGICAL SOCIETY OF NEW SOUTH WALES

In July, 1970, The Royal Zoological Society held a symposium on Kangaroos at the Australian Museum. The results, carefully edited by

[1] School of Zoology, University of New South Wales, P.O. Box 1, Kensington, New South Wales, Australia, 2033

John Simons, were published as a special issue of the *Australian Zoologist* (Vol. 16, Pt 1) entitled "Kangaroos and men". In May, 1978, the Society held a symposium on monotremes at the University of New South Wales and, with the skillful editorship of Mike Augee, produced another special issue of the *Australian Zoologist* (Vol. 20, Pt 1), this time entitled "Monotreme biology". The obvious success of these ventures encouraged the Society to offer its support for yet another zoological symposium. Being clearly of unsound mind, but by this time anxious to have a broad–spectrum look at carnivorous marsupials, I tossed in the idea. It was accepted by the Society and, with the continued backing of the University of New South Wales, well over 200 Australian and overseas registrants participated in the Carnivorous Marsupial Symposium held in May, 1980. For three days, participants listened to the results of research into all aspects of carnivorous marsupial biology.

"CARNIVOROUS MARSUPIALS", FROM THE NICE ALTRUISTIC SOCIETY THAT BROUGHT YOU "MONOTREME BIOLOGY"

Birth Pangs

By August, 1980, sixty–nine manuscripts had been submitted for consideration in the anticipated publication. Of these, about half had been invited. The others were volunteered during or immediately following the 1980 Symposium. This response was about double that expected, and even allowing that some would not survive refereeing, the size and production cost of the eventual publication was clearly too great to be printed (as originally planned) as an issue of the *Australian Zoologist*. As a result, I did the rounds of the potential scientific book publishing companies. To make a long and iconoclastic story short, several companies said they would be happy to publish it but only if we allowed a retail price that was often as much as five times their production cost, and then only if the Society subsidized about half the production costs. For this academic "prostitution", the Society would receive 10% of the retail price of each volume. Because this seemed outrageous to the Society (they could not hope to recover their investment) and certainly not in the Society's interests of making the published results readily available to students and professionals alike, we decided to go it alone. Surrey Beatty and Sons, the printers of *Australian Zoologist* and *Australian Mammalogy,* offered to print "Carnivorous marsupials" as two volumes and at a price that would make them less expensive to buy than to photocopy. This offer enabled the Society to make the volumes financially available to students as well as professionals and, although the low retail price charged would not enable the Society to make a profit, there was at least a good chance that the investment could be recovered if all the volumes were sold. Clearly, there was little reason for further deliberation. But this raised in my mind a general call to arms for academics.

It is, we have found, economically feasible to produce high quality books such as this at prices much lower than those charged by academic book publishing companies. These companies are not altruistic. They are looking for short–term recovery of investments and can usually rely on libraries to purchase enough copies at the inflated prices to recover costs and make a profit. This trend (witness, for example, the important three volume "Ecological biogeography of Australia" ed. by A. Keast, 1981, Junk Press, which retails at about $350.00, and is thus, for all practical purposes, unavailable to students) has placed many academic books beyond the financial reach of professional academics let alone aspiring students. From our experience, there is no need to accept this as the fixed order of things. The alternative is to have books published by scientific societies using outside printers, such as Surrey Beatty and Sons. The trade–off for producing inexpensive academic books is the need to overcome three (not insurmountable) problems. First, the scientific society needs capital. In our case, capital required by the printer was about the same as that quoted by several scientific book publishing companies, but of course the latter were only intending to return 10% of the sales value to the Society (as opposed to the whole of it if it is produced by a printer) and intending to charge a retail price that was about five times as much as we needed to charge to recover the costs of production using a printer. Second, someone must shoulder the load as editor. Third, means must be found for drawing the academic community's attention to existence of the publication. In our case, we printed 3000 pre–publication announcements and distributed them at appropriate scientific meetings in Australia and overseas; sent advertisements to relevant scientific journals; and intend to send copies for review. So far, the reaction has been encouraging. As of October, 1981, about a third of the production cost had been recovered through pre–publication sales and hopefully sales will increase substantially when the volumes are actually available for distribution in February, 1982. Although this is an unusual topic to take up in an introduction, to anyone concerned with the recent trend in costs of academic books, I think it is an important diversion.

Editorial Midwifery

All manuscripts were refereed by at least two referees, then rejected or returned for revision. The generally high standards of the original papers, and the excellent critical refereeing, considerably simplified the editing. Unless they requested otherwise, these referees have been noted in the acknowledgements of each paper.

Following the recommendations of the Vernacular Names Committee of the Australian Mammal Society (Strahan, R., 1980. Recommended common names of Australian mammals. *Aust. Mamm. Soc. Bull.* **6** *(2)*: 13–23), standard vernacular names for native and introduced species have (where possible) been used. Also, following the same recommendations, these names are capitalized if they refer to a *particular* species. For example: "The Cinnamon Antechinus (*Antechinus leo*) is one of many antechinuses with thin tails".

Particular scientific names used for the marsupial carnivores is a common arena for inconsistency. Although consistent species level taxonomy has been maintained (I hope) throughout these Volumes, generic level taxonomy varies. Despite this, I would like to recommend to the reader the higher level taxonomic conclusions proposed in articles 37, 38, 39, 43, 44, 45, 50, and 51. Inconsistencies between these papers may remain at the suprageneric level. As far as I am aware, other authors who have used different generic concepts have not done so because of contradictory taxonomic convictions. Rather, because of the simultaneous presentation of these papers, most authors have had little or no time to assimilate and assess the changes proposed by colleagues. Two principal exceptions to this were the reviews of dasyurid and thylacinid evolution and classification (articles 37 and 38) which were deliberately written after consideration of all other contributions in order to attempt syncretic resolutions of conflict in phylogenetic interpretation.

One species level amendment should be noted here. Several articles (e.g. 4, 12, 30 and 52) refer to *Ningaui* sp. from the Ayers Rock area of the Northern Territory. Until about a month ago, the taxonomic status of this beast was uncertain. As our work with the enzymes of dasyurids continues (see the summary presented in article 52), it has become evident that the Northern Territory beast is *Ningaui ridei.*

As an aid to readers uncertain about the most up–to–date generic and species level thylacinid, myrmecobiid, notoryctid, thylacoleonid and dasyurid systematics, the following list is provided. Innovations presented here are based on the results reported in all of the papers presented in these Volumes. The age of extinct species is indicated.

Thylacinidae: *Thylacinus potens* (Miocene); *T. cynocephalus* (Pliocene to Recent). Myrmecobiidae: *Myrmecobius fasciatus*. Notoryctidae: *Notoryctes typhlops*. Thylacoleonidae: *Wakaleo oldfieldi* (Miocene); *W. vanderleuri* (Miocene); *W. alcootaensis* (Miocene); *Thylacoleo hilli* (Pliocene); *T. crassidentatus* (Pliocene); *T. carnifex* (Pleistocene). Dasyuridae: *Ankotarinja tirarensis* (Miocene); *Keeuna woodburnei* (Miocene); *Dasylurinja kokuminola* (Miocene); *Antechinus flavipes; A. stuartii; A. swainsonii; A. leo; A. minimus; A. godmani; A. bellus; A. mayeri* (and other New Guinean species, a still unresolved blob of species); *Phascogale tapoatafa; P. calura; Murexia longicaudata, M. rothschildi; Myoictis melas; Myoictis wallacei; Phascolosorex doriae; P. dorsalis; Neophascogale lorentzii, Wakamatha tasselli* (? Miocene); *Sminthopsis murina; S. ooldea; S. leucopus; S. virginiae; S. douglasi; S. butleri; S. floravillensis* (*Plio–Pleistocene*) *S. macroura; S. hirtipes; S. granulipes; S. psammophila; S. crassicaudata; S. longicaudata; S. laniger; Planigale ingrami; P. tenuirostris; P. maculata; P. gilesi; P. novaeguineae; Dasycercus cristicauda; Dasyuroides byrnei; Dasyuroides achilpatna* (Pleistocene); *Dasyurus viverrinus; D. maculatus; D. dunmalli* (Pliocene); *Satanellus hallucatus; S. albopunctatus; Glaucodon ballaratensis* (Pliocene); *Sarcophilus harrisii; S. moornaensis* (Pleistocene); *Parantechinus apicalis; Parantechinus bilarni; Pseudantechinus macdonellensis; Pseudantechinus* sp. (= the "Ningbing Antechinus"); and *Dasykaluta rosamondae.*

Of course, this is not all. Future work with New Guinean dasyurids will clarify species and generic level uncertainties. Further, descriptions of four additional modern species of *Sminthopsis*, one extinct *Antechinus* and five additional extinct dasyurid genera are underway. For those seeking some kind of permanent taxonomic stability, my advice is not to hold your breath. Yet, despite this council of despair, I hope that the level of phylogenetic understanding achieved in "Carnivorous marsupials" will serve as a solid stepping stone for future research.

ACKNOWLEDGEMENTS

The appearance of these volumes is a measure of the value of teamwork. In the first place, none of it would have happened without the Royal Zoological Society of New South Wales and the backing of its Council. As the then President of Council, Mike Augee deserves special acknowledgement. The Utah Foundation aided the Society with a greatly appreciated financial contribution towards publication of "Carnivorous marsupials". Rosendo Pascual (University of La Plata, Argentina) provided, through herculean effort, the Spanish translations of nearly every paper. Both of us hope that this will be of considerable value to zoologists in South America, the other half of the marsupial world. The *Resumens* for articles 15, 55 and 56 were kindly provided by Hector Sosa. Any mistakes in Spanish grammar that remain in these summaries, however, are my fault. The Queensland Museum deserves very special thanks for allowing access to Margaret Oakden's and Mary McKenzie's superb scraper–board drawings of carnivorous marsupials. Some of these were used in the brochure announcing the Volumes and as miniatures adjacent to the abstract of each paper. The drawings will appear with about 100 others in "Marsupials and monotremes of eastern Australia" (by M. Archer) to be published by the Queensland Museum, hopefully in late 1982. Identifications of the particular miniatures used in the Volumes are given below. Margy Oakden also provided, by special request, the scraper–board drawing of *Planigale gilesi* used with several of the abstracts (see below) and the drawing of *Sminthopsis* (*Antechinomys*) *laniger* produced on the spines of the Volumes. Elizabeth Archer carefully prepared the index to mammal species. Georgina Clayton and Sue Hand helped with the tedious job of marking up the final manuscripts in preparation for printing. The photographs of the Tiger Quoll (*Dasyurus maculatus*) on the cover of Volume I and of the Mulgara (*Dasycercus cristicauda*) on the cover of Volume II were taken by Barry Baker. Deanne Thomas and Penny Lawry, as brilliant typists in the School of Zoology, University of New South Wales, saved many an author from ignominy by converting their chicken scrawls into English.

Particular thanks go to Ivor Beatty and his team including Bruce Lyons, Robert Wales, Harry Gleeson, Laurie Fernandes, Jim Johnson and Kevin Kearns for their superb effort in organizing, compositing, typesetting and printing the Volumes. Any annoying little errors that still remain (and, oh woe, there are some) despite their attentions, are entirely my fault. Additionally, I most sincerely want to thank Ivor Beatty for his determinations to help the Society sidestep the otherwise unacceptable publication "offers" of the academic publishing companies.

The scraper–board drawings from "Marsupials and monotremes of eastern Australia" (to be published by the Queensland Museum) were placed as miniatures adjacent to the abstracts of the articles as follows. The drawings of *Dasyuroides byrnei, Antechinus stuartii* and *Satanellus hallucatus* were by Mary McKenzie. All others were by Margaret Oakden.

Antechinus flavipes (58, 62)
Antechinus leo (61, 64)
Antechinus stuartii (7, 15, 16, 17, 19, 28,
Antechinus swainsonii (10, 14)
Dasycercus cristicauda (41)
Dasyuroides byrnei (34, 52)
Dasyurus geoffroii (20, 47)
Dasyurus viverrinus (18)
Ningaui ridei (4)
Parantechinus apicalis (2)
Phascogale tapoatafa (3, 32, 53)
Planigale gilesi (8, 21, 24, 51)
Planigale ingrami (31)
Planigale maculata (9, 13, 30)
Pseudantechinus macdonnellensis (63)
Sarcophilus harrisii (6, 25, 27, 43)
33, 35, 36, 56)
Satanellus hallucatus (26, 49)
Sminthopsis (*Antechinomys*) *laniger* (11, 50)
Sminthopsis crassicaudata (12 ,57, 60)
Sminthopsis macroura (29)
Sminthopsis murina (5, 54)
Sminthopsis virginiae (37)
Thylacinus cynocephalus (22, 23, 38, 44, 55, 59)

For use of these drawings, we would again like to express sincerest thanks to the Queensland Museum.

2. LIFE HISTORY STRATEGIES OF DASYURID MARSUPIALS

A. K. LEE[1], P. WOOLLEY[2] and R. W. BRAITHWAITE[3]

A review of the life histories of 30 dasyurid species reveals six distinct life history strategies. These are distinguished on the basis of frequency of reproduction in females, the duration and timing of male reproductive effort, the seasonality of breeding and the age at maturity.

The Dasyuridae show a high incidence of species with monoestrous females, species in which males die after a single short breeding season, and species which time reproduction so that lactation coincides with the period of maximum insect accessibility. Few dasyurids are capable of breeding in the season of birth. Circumstances favouring the selection and persistence of these strategies are discussed.

Pages 1–11 *in* CARNIVOROUS MARSUPIALS, ed. by M. Archer, February, 1982. Roy. Zool. Soc. New South Wales: Sydney, Australia.

RECOGNITION of the potential of the Dasyuridae for studies of life history strategies developed from the description of the unusual life history of *Antechinus stuartii* (Woolley 1966a, Wood 1970). Although the life histories of only a few dasyurid species are known in the same detail, sufficient information is available on the reproduction and longevity of 30 of the 50 or so dasyurid species for us to portray the variety of life history strategies evident in this family. A classification of these life history strategies is presented. The classification is necessarily preliminary, but it does point to opportunities for empirical comparisons between strategies, and is necessary before more realistic comparisons between the eutherian and marsupial "grand strategies" are possible (c.f. Low 1978).

TERMINOLOGY

STRATEGY AND TACTICS. Many ecologists appear to use the terms "strategy" and "tactics" interchangeably, or arbitrarily use one term to the exclusion of the other. Consistent use of the terms in a military context is long established. For example, Wright (1942) defined "tactics" as ". . . the management of military operations in direct contact with the enemy in order to win battles . . ." and "strategy" as ". . . the management of operations so as to effect such contact under maximum advantage in order to win campaigns . . .".

If we equate a species with a nation and compare the survival of the species with the conduct of war, tactics and strategy may be considered in an ecological context. It is proposed that use of the term "tactics" should be reserved for the various combinations of the more flexible components of the biology of a species, that is, those components which exhibit considerable variation within or between populations (e.g. litter size in *A. stuartii*). "Strategy", on the other hand, should be used to convey the combination of tactics available to a species. Such a combination is conservative and likely to be subject to only slow evolutionary change. It should give a description of what appears to be the "approach to existence" of the species. To this end, it should not be restricted to demographic characteristics but also include other characteristics, including behavioural ones. Diet is likely to be of fundamental importance in determining both strategy and tactics. As Fleming (1979: 2) points out, "To discuss the evolution of demogaphic strategies without consideration of a species' physical, energetic and social environment would be a futile exercise indeed".

It is further suggested that the epithet "life history" should be used when only traditional demographic traits are considered, but when the broader concept is applied, "ecological" is a better epithet. While we will endeavour to discuss the ecological strategies of the Dasyuridae, often the only useful information available is on reproduction from laboratory studies from which we can only infer demographic consequences. This limitation in the information has forced us to confine ourselves primarily to life history strategies.

SEMELPARITY AND ITEROPARITY. Lee *et al.* (1977) and Braithwaite and Lee (1979) adopted the term *semelparity* for the strategy characterized by males which die after their first short breeding season, and *iteroparity* for strategies in which males participate in one long breeding season or several breeding seasons. Pine (in prep.) has objected to this application of semelparity on the grounds that a percentage of females, albeit small, survive to produce a second litter. Pine observed that Cole (1954: 105) defined semelparous forms as ". . . those

[1] Dept. of Zoology, Monash University, Clayton, Victoria, Australia, 3168.
[2] Dept. of Zoology, La Trobe University, Bundoora, Victoria, Australia, 3083.
[3] National Museum of Victoria, Melbourne, Australia, 3000 (present address — C.S.I.R.O. Division of Wildlife Research, Private Mail Bag 44, Winnellie, Northern Territory, Australia, 5789).

Figure 1. Key factors in the classification of dasyurid life histories.

Oestrous pattern	No. of seasons per ♂	Duration of breeding season	Seasonality of breeding	Age of sexual maturity	Type of strategy	No. of species
monoestrous	annual	restricted	seasonal	11 mo.	I	9
monoestrous	perennial	restricted	seasonal	11 mo.	II	7 or 8
monoestrous or polyoestrous	perennial	restricted	seasonal	11 mo.	III	3 or 4
polyoestrous	annual	extended	seasonal	6 mo.	IV	3
polyoestrous	annual	extended	seasonal	8–11 mo.	V	4
polyoestrous	annual	extended	aseasonal	?	VI	3

which reproduce only once in a lifetime . . ." and iteroparous forms as ". . . those which reproduce more than once in a lifetime . . .". Semelparity (*sensu stricto*) means giving birth once, and in view of the concern of population ecologists with the consequences of producing one, or more than one, litter, on population growth, strict application of the terms is desirable. Pine (in prep.) proposed the terms *semelseminant* for species in which males have a single period of rut and *iteroseminant* for those in which there are two or more periods of rut. As Pine has defined the term, semelseminant implies nothing about the duration of the rut or the number of females inseminated. Thus a species is semelseminant if males breed throughout the year and for successive years. Pine's terminology does not distinguish between the vastly different male strategies of *Sminthopsis crassicaudata,* in which males may survive through a six to eight month breeding season, and *A. stuartii* in which males die after a two to three week breeding season. To avoid further confusion, we observe that all dasyurids are iteroparous *sensu stricto,* and have used familiar terms to describe each strategy.

LIFE HISTORY STRATEGIES

At least six distinctive life history strategies are evident in the Dasyuridae (Fig. 1, Table 1). These are distinguished on the basis of five components: the frequency of oestrus in females, the duration and timing of male reproductive effort, the seasonality of breeding and the age at maturity (Fig. 1).

Sixteen of the 30 species we are considering are monoestrous, and show two distinct strategies:

STRATEGY I. Nine species are characterized by the mortality of all males at the conclusion of their first short breeding season. The breeding season is shorter than the gestation period (i.e. restricted) in those species for which we have adequate data (*A. stuartii,* Lee *et al.* 1977; *A. swainsonii,* Lee and Martin unpublished observations). Some females survive to breed in a second season and very rarely in a third.

Details of this strategy are best known for *Antechinus stuartii.* In a population at Powelltown, Victoria, mating commenced in early August and all males died by the last week of that month. Births occurred in mid- to late-September, and lactation continued until January. Only at this time did males re-enter the trappable population. These males became sexually mature and commenced mating in the August after their birth (Barnett 1972, Lee *et al.* 1977).

There is some variation in the timing of these events even between adjacent populations of the same species (Dickman, this volume), but the overall pattern of events is the same in each species which has been studied in detail (*A. stuartii* — Woolley 1966a, Wood 1970, Leonard 1972, Lee *et al.* 1977, Braithwaite 1979; *A. minimus* — Wainer 1976; *A. flavipes* — Woolley 1966a, Inns 1976; and *A. swainsonii* — Leonard 1972, Reed and Wallis 1975, Lee and Martin unpublished observations). All breed in late winter or early spring and lactation coincides with spring and summer. The punctuality of these events in the life histories in dense populations is remarkable. In a population of *A. swainsonii* in the Otway Ranges, Victoria, only 1 of 22 males surviving to 16 July was captured on 25 July, 1978, and only 3 of 23 males surviving to 20 July, 1979 were captured on 27 July, 1979, and none were captured the following night. They may be less punctual in low density populations.

The post mating mortality of males has been shown to be stress induced in the three species that have been examined, *A. stuartii* (Lee *et al.* 1977, Bradley *et al.* 1980), *A. swainsonii* and *A. flavipes* (McDonald *et al.* in prep.). This stress appears to be a consequence of attempts by males to maximize their reproductive fitness. The specific nature of male behaviour during the breeding season is not precisely defined, but male *A. stuartii* show increased movements (Wood 1970, Braithwaite 1979) and increased levels of aggressiveness (Braithwaite 1974). Male *A. stuartii* given access to females in the laboratory copulate for more than 5 h and up to 12 h at a time (Marlow 1961, Woolley 1966b). These females often show loss of fur from the neck and flanks after repeated copulations, but this has only rarely been observed in female *A. stuartii* in natural populations. This led Braithwaite (1979) to conclude that females may avoid successive matings in nature. However, Lee and Martin (unpublished observations) have found severely debilitated females with fur loss in a dense population of *A. swainsonii* at the conclusion of the breeding season, suggesting that females of this species may mate repeatedly.

Dispersal of juvenile *A. stuartii* from their maternal home ranges occurred in May/June at Mt. Glorious, Queensland, after which individuals nested alone. Individual ranges overlapped considerably, but each year, the set of males could be arranged into a dominance hierarchy, in which the most dominant males occupied the same area of the grid in two successive years (Braithwaite 1979). Braithwaite argued that the abundance of logs and soil moisture in this area contributed to a greater insect abundance than elsewhere. As a consequence, these males grew throughout winter, were larger than subordinates at the onset of mating, copulated with females in their own area and in the areas occupied by subordinate males, and tended to die earlier than subordinate males. If *A. stuartii* females do not copulate repeatedly in nature, subordinate males may make little if any genetic contribution (Braithwaite 1979).

The mating system has yet to be established for other species with this strategy, but all are noticeably sexually dimorphic in size (Table 1). Dasyurids showing this strategy are generally small (20 to 240g as adults), mature at about 11 months of age, and have relatively high teat numbers and therefore, the capacity for large litters (Table 1).

The single annual litter and the attachment of neonates to the teat limit the tactical options open to species with this strategy. Unlike most small mammals, variation in fecundity in these dasyurids results solely from variation in litter size, itself effectively governed by teat number. Woolley (1966b) has shown that female *A. stuartii* and *A. flavipes* produce more eggs (and young) than there are available teats, and this probably holds for other species with this strategy. In those species in which litter size has been studied in detail (*A. stuartii, A. flavipes, A. minimus, A. rosamondae* and *A. swainsonii*, Woolley 1966a, Wood 1970, Leonard 1972, Wainer 1976, and Woolley, Lee and Martin unpublished observations), litters where all or all but one teat are occupied are most frequent.

Generally teat number and litter size are remarkably constant within populations, but vary geographically within species. Variation in teat number is greatest in those species with the widest distribution (*A. stuartii* and *A. swainsonii*). Teat number is least in Tasmanian populations of *A. minimus* and *A. swainsonii,* and in *A. stuartii* and *A. swainsonii* in parts of their range nearest the southern Australian coast. Teat number is highest in alpine and inland populations of *A. stuartii,* in alpine populations of *A. swainsonii,* and overall in inland populations of *A. flavipes* (Lee and Martin, unpublished observations).

STRATEGY II. In the seven remaining monoestrous species, some males and females survive to breed in at least a second year. This strategy is exemplified by *Antechinus bilarni* and a description of the demography of a population of this species is provided by Begg (1981). The following observations are relevant to this review. Breeding in *A. bilarni* occurs in late June and early July, and births occur in August. Neonates remain in the pouch until mid-September, and juveniles first enter the trappable population in December. These juveniles attain maturity in the following June. Thus the annual cycle of events resembles that of the Strategy I species in all but one respect: male *A. bilarni* survive to a second breeding season, and show spermatogenesis at that time (Begg 1981, Woolley in press a). These males disappear from the population before their third breeding season. Female longevity appears similar, although a few individuals survive to a third breeding season but do not reproduce.

There is circumstantial evidence that reproductive effort in each breeding season may be less intense than in Strategy I species. Begg observed that 12 and 35 per cent of females did not cary pouch young in the two breeding seasons of this study. In comparison all females (n = 47) in three breeding seasons carried pouch young in a Victorian population of the Strategy I species *A. swainsonii* (Lee and Martin, unpublished observations).

Sexual dimorphism in body weight is slight in species with small body size but is appreciable in the species with large body size (Table 1).

Both small and large dasyurids show this strategy. The range of teat numbers (4 to 10) is generally lower than in Strategy I species. There is no information available on demographic tactics within species.

STRATEGY III. This strategy occurs in three species — *Dasyurus viverrinus, D. maculatus* and *Dasycercus cristicauda.* In these species females are usually monoestrous, but some females may undergo a second oestrus if unmated or after premature loss of the first litter (Fletcher 1977, Settle 1978, Woolley 1971b). They therefore have some insurance against reproductive failure. A bimodal distribution in birth dates of *D. viverrinus* in New South Wales suggests that this has natural significance (Tyndale Biscoe, in press).

These are among the larger dasyurids and show marked sexual dimorphism in body weight (Table 1). Births occur in winter, and teat numbers are generally towards the lower end of the range for dasyurids (4 to 8). Once again, demographic tactics have not been described for these species.

Association of the Tasmanian Devil, *Sarcophilus harrisii,* with one or other of Strategies II or III, or a distinct strategy, in which maturation requires two years, is uncertain at this stage. Guiler (1970) found that weaning occurs at seven months, which is substantially longer than the three to four months typical of other dasyurids. He also found that females do not breed until their second year, and so differ from other dasyurids which mature in 11 months or less. It is not clear from that paper whether males mate first in their second year. Nor is it clear whether females can produce a second litter in one season. Buchmann and Guiler (1977) suggest that individuals survive to at least six years of age in the field, but do not comment on the distribution of reproductive effort. Births occur in early winter rather than spring, and litter size is relatively small. Males are conspicuously heavier than females.

The remaining ten species are polyoestrous and show three strategies. The information available does not permit us to define these strategies with the same certainty as Strategies I and II.

STRATEGY IV. In this strategy the breeding season is extended but seasonal. Litters are produced in winter, spring and summer. Females of these species attain sexual maturity in about six months under laboratory conditions (Godfrey and Crowcroft 1971), which should permit mating within the breeding season of birth. Although Morton (1978a) found no evidence of reproduction in the season of birth in the three natural populations of *S. crassicaudata* he examined, we consider this capacity of such potential signficance that we have recognized it as a criterion for a distinct strategy. *Sminthopsis murina* is tentatively placed with species showing this strategy. Age at maturity is not known but age to weaning (60 to 65 days; Fox and Whitford, this volume) is similar to other species within this strategy.

This strategy is exemplified by *Sminthopsis crassicaudata,* in which litters are produced over a six to eight month breeding season between June and February. Females attempt to raise two litters during the breeding season, one before October and the second after that month. Some females survive to breed in two seasons. Some males born in the previous breeding season survive to the end of the next, but it is unlikely that any survive beyond this (Morton 1978a). In laboratory colonies greater than 90 per cent of litters are produced between July and February (Godfrey and Crowcroft 1971)

In contrast to *A. stuartii, S. crassicaudata* showed no evidence of offensive aggression in contrived encounters and little change in the levels of agonistic and amicable behaviour from the non-breeding to breeding periods; nor were there behavioural differences between females without litters and males (Morton 1978b). In view of this it is surprising that individuals begin to nest singly with the onset of breeding. In captivity, Ewer (1968) observed that males only fought in the presence of oestrous females, and most nest groups found by Morton during the breeding season consisted of a male with a pre-oestrous or oestrous female. Males of *S. murina* captured during the breeding season show wounds consistent with fighting (Fox and Whitford, this volume). Ewer also noted that *S. crassicaudata* females with young became increasingly hostile towards other animals, and again Morton's observations suggest that the females behave most aggressively when the young are about 40 days old.

Small dasyurids (12 to 25g) show this strategy. Our index showed little or no evidence of dimorphism in body weight (Table 1), but Morton (1978a) observed a slightly higher mean weight in breeding males compared with breeding females (see also Fox and Whitford, this volume). Litter sizes (6 to 10) are towards the middle of the dasyurid range. Litter size does not vary between the first and second litter in *S. crassicaudata* (Morton 1978a).

Morton (1978a) compared reproduction in three populations, one from coastal Victoria and the others from central and western New South Wales. Despite marked differences in the predictability of the climate, there were only minor differences in the life history tactics at these sites. Neither teat number nor litter size varied consistently between populations. Although most females appeared to produce two litters per season, there were differences in timing and duration of breeding seasons between sites and years.

STRATEGY V. In contrast to Strategy IV, *Planigale maculata maculata, Antechinomys laniger* and *Dasyuroides byrnei* mature in 8 to 12 months (Aslin 1975, Woolley, unpublished observations), and are therefore unlikely to breed in the season of birth. Again we consider this capacity so important that we have tentatively placed these species in a distinct strategy. These are small to medium sized dasyurids (7 to 150g), and show considerable variation in teat number and litter sizes, although these are smaller and less variable in the larger species (*A. laniger* and *D. byrnei*) (Table 1).

Planigale tenuirostris and *P. gilesi* (not included in Table 1), and *P. ingrami,* are tentatively grouped under this strategy. Data summarized by Archer (1976) and Morton (this volume) suggest that there may be conspicuous differences in the duration of the breeding season and in teat number and litter sizes within and between these species.

STRATEGY VI. A third polyoestrous strategy may occur in *Antechinus melanurus,* and possibly *A. naso,* from the highlands of New Guinea, and *Planigale maculata sinualis* from the north of the Northern Territory. Unlike most Australian

Table 1. Some life history correlates of dasyurid marsupials. Field weight have been used wherever possible. * Dimorphic ratio = ratio of maximum male to maximum female body weight. See text for full names of species.

Strategy	Species	Adult body weight (g) ♂♂	♀♀	Dimorphic ratio*	Month of births	Usual litter size	Teat number	Source
I	*A. bellus*	55-70			Sept.			8,44
	A. flavipes	35-55	20-35	1.6	July-Sept.	8-12	10-12	15,39
	A. godmani	90-105			Aug.-Sept.		6	35,46
	A. leo	75					10	34
	A. minimus	46-100	30-58	1.7	July-Aug.			28,36,37
	A. rosamondae	25-40	20-30	1.3	Oct.-Nov.	6-8	8	45,46
	A. stuartii	28-40	19-30	1.4	Sept.-Oct.	6-10	6-10	28,38,39
	A. swainsonii	50-90	40-60	1.5	July-Aug.	6-8	6-10	28,42,46
	P. tapoatafa	180-240	120-200	1.2	Aug.	7-8	8	9,12
II	*A. apicalis*	60-100	40-75	1.3	April	8	8	40
	A. bilarni	20-36	22-28	1.1	Aug.-Sept.	5	4-6	6,8,45
	A. macdonnellensis	25-45	20-40	1.1	Aug.-Sept.	6	6	42,46
	S. leucopus	15-25	15-25	1.0	Aug.-Sept.	10	10	47
	Ninbing Antechinus	20-25	15-20		July	1-4	4	46
	D. hallucatus	600-800	400-450	1.8	July-Aug.	6-8	6,7,8	17
	D. geoffroii				May-June	6		1,3
III	*D. cristicauda*	75-130	60-95	1.4	June-July	5-7	4,6,8	16,27,30,41
	D. viverrinus	1010-1800	700-1225	1.5	May-Aug.	6	6	19,24,25
	D. maculatus	7000	4000	1.8	July-Aug.	5	6	14,32
II or III?	*S. harrisii*	9000	6500	1.4	April-Sept.	2-4	4	7,13,23
IV	*S. crassicaudata*	16-22	12-21	1.0	June-Feb.	6-8	6-11	31,42
	S. macroura	14-21	15-21	1.0	July-Feb.	6	6	22,46
	S. murina	15-25	15-20	1.25	Sept.-Jan.	4-10	10	20
V	*P. m. maculata*	13-14	7-11	1.3	Spring	5-10		2,18
	P. ingrami				Dec.-Mar.	4-12		2,43
	A. laniger	25-35	25-35	1.0	July-Nov.	4	6,8	42,46
	D. byrnei	85-150	70-120	1.3	May-Jan.	5-6	5-7	4,26,29,42,43
VI	*A. melanurus*	31-39	24-32	1.2	All months	3-4	4	11
	A. naso	44-63	37-48	1.3	All months(?)	4	4	11
	P. m. sinualis	8-22	7-15	1.5	All months (?)	8-12	12	2,4,10,33

Sources: [1]Archer (1974); [2]Archer (1976a); [3]Arnold and Shield (1970a and b); [4]Aslin (1974); [5]Aslin (1975); [6]Begg (1981); [7]Buchmann and Guiler (1977); [8]Calaby and Taylor (in press); [9]Cuttle (1978); [10]Davies (1960); [11]Dwyer (1977); [12]Fleay (1934); [13]Fleay (1935) [14]Fleay (1940); [15] Fleay (1949); [16]Fleay (1960); [17]Fleay (1962); [18]Fleay (1965); [19]Fletcher (1977); [20]Fox and Whitford (this symposium); [21]Godfrey (1969); [22]Godfrey and Crowcroft (1971); [23]Guiler (1970); [24]Hill and Hill (1955); [25]Hill and O'Donoghue (1913); [26]Hutson (1976); [27]Jones (1923); [28]Lee and Martin (unpublished observations); [29]Mack (1961); [30]Michener (1969); [31]Morton (1978a); [32]Settle (1978); [33]Taylor, Calaby and Redhead (this symposium); [34]Van Dyck (1980); [35]Van Dyck (in prep.); [36]Wainer (1976); [37]Wilson (unpublished observations); [38]Wood (1970); [39]Woolley (1966a); [40]Woolley (1971a); [41]Woolley (1971b); [42]Woolley (1973); [43]Woolley (1974); [44]Woolley (in press a); [45]Woolley (in press b); [46]Woolley (unpublished observations); [47]Woolley and Ahern (in prep.)

dasyurids which breed seasonally in nature, these dasyurids produce litters in most, if not all, months of the year (Dwyer 1977, Taylor *et al.*, this volume). These are small dasyurids, and litter sizes are usually towards the middle to lower end of the dasyurid range (3-8). Other aspects of the strategy have not been defined, but it is highly probable that females are polyoestrous and males survive to mate over an extended period (see Whitford, Fanning and White, this volume).

DISCUSSION

The forcgoing description points to a surprising diversity of life history strategies in the Dasyuridae. Notable features are, first, the high incidence of species with monoestrous females, and in which, as a consequence, variation in litter size assumes great importance as a reproductive tactic; second, the relatively high incidence of species in which males participate in a single short breeding season then die abruptly; third, the low incidence of species which are capable of reproducing in the season of birth; and fourth, the high frequency of species which mate in winter and lactate in spring and summer.

Sixteen of the 30 species included in our analysis are monoestrous, and although this proportion will undoubtedly fall as data becomes available for the remaining twenty or so species (predominantly small desert and New Guinea dasyurids), this proportion contrasts with the high frequency of polyoestrous species in other Australian marsupial groups (Tyndale Biscoe, in press) and in small, similar-sized placentals. It follows that variation in fertility (see Krebs 1978) may only be achieved by variation in litter size in these species. Dasyurids appear to produce more young than can be accommodated on the teats (Woolley 1966b, 1971b, Guiler 1970) and teats tend to be fully utilized in most species (Table 1). Consequently adjustments in fecundity may only be achieved through variations in teat number. If there is any relaxation of the relationship between litter size and teat number, it occurs among the polyoestrous *Sminthopsis* spp. (see Morton 1978a) and *Planigale* spp. (Taylor *et al.*, this volume). Among Australian marsupials only the Burramyidae resemble the Dasyuridae in the close association between litter size and teat number.

The phenomenon of mortality of all adult males after a short mating period (< gestation) from which females produce only a single litter, is found in approximately 20 per cent of all dasyurid species. We know of no other examples of this reproductive phenomenon outside of the Dasyuridae. Although many small mammals may only survive to breed in one season, females usually produce several litters in this time. Circumstances favouring the origin and existence of this phenomenon are considered below.

The potential for breeding in the season of birth has been shown to occur in only two dasyurids, both *Sminthopsis* spp., and both from environments where the food supply is seasonally predictable but may be temporarily in short supply (Morton 1978a). The majority of dasyurids require eight months or more to attain maturity and most breed at 11 months of age. Their inability to breed in the season of birth reflects the short period suitable for their reproduction in most Australian environments, and the long period of maternal dependence. For example, age to weaning is in excess of 100 days in most small dasyurids, and even requires 60 to 70 days in the *Sminthopsis* spp. (Godfrey 1969, Godfrey and Crowcroft 1971, Fox and Whitford, this volume, Woolley 1973).

Parker (1977) suggested that maternal investment by marsupuials is minimal during gestation. Although eutherian maternal investment is greatest during late lactation (e.g. Millar 1975), Parker contends that marsupial investment at this time is of greater importance relative to that of eutherians. This conclusion is supported by the coincidence of lactation with the time of maximum abundance of food resources in a variety of marsupials (Tyndale Biscoe 1979). In most small dasyurids, which are primarily insectivorous this relationship is strong. All but a few dasyurids mate in late winter and lactation occurs in spring or summer (Table 1), when insect prey is likely to be most accessible. Further, slight differences in timing of reproduction between populations and species can be related to likely peaks in insect accessibility in the feeding niches (Wainer 1976, but see also Dickman, this volume). For example, where they occur together, mating in the soil-fossicking insectivore *A. swainsonii* precedes mating in the scansorial insectivore *A. stuartii* by three weeks to a month, and in the order predicted from peaks in insect emergence from the soil and flight. Another advantage in this timing is that juveniles approach adult weight before insect abundance falls in winter. Relaxation of this seasonality, particularly in the New Guinea antechinuses and northern Australian planigales, occurs where insects are probably continuously available.

Evolution of life history strategies. Braithwaite and Lee (1979) suggested that an inability to successfully raise two litters during the period of the year when insects were reliably both abundant and accessible, was the circumstance favouring the evolution of Strategy I. In *A. stuartii,* for example, a second litter conceived immediately following weaning of the first would be weaned in late autumn when insect abundance declines markedly. These young would face winter with an energetically disadvantageous small mass. Significant failure at this second reproduction should favour selection of monoestrous females. The reproductive success of those females successful in raising their first litter is presumably enhanced by withholding their residual reproductive effort until the spring of their second year. The advantage of coinciding lactation and weaning, the period of maximum maternal investment, with peak insect abundance will, in a highly predictable environment, favour synchronous oestrus. The strategy of males will then depend upon their probability of survival to reproduce in a second year. Where this is low, as it presumably is in Strategy I species, males should maximize their reproductive effort at first reproduction, even though longevity may be reduced as a consequence. The conspicuous sexual dimorphism (Table 1), male aggressiveness and the appetite of males for prolonger copulation, as seen in *A. stuartii,* are consistent with intense reproductive effort.

It follows that the probability of survival of males exhibiting Strategy II to a second reproduction must be higher, and their reproductive effort during any mating period less than Strategy I species. Evidence of reduced reproductive effort is largely wanting; sexual dimorphism in some Strategy II species is less, but it is pronounced in others.

Morton (1978a) argued that since *D. byrnei,* a polyoestrous species, has a similar 120-day period between conception and weaning, as *A. stuartii,* an element of extended maternal care may be involved in the evolution of monoestry. Braithwaite and Lee (1979) proposed that the conditions necessary for the evolution of Strategy I are that food resources are *seasonally restricted but reliably timed.* Long developmental times of small marsupials generally, make them prime candidates for only being able to reliably raise one litter per season under such circumstances. We feel that an increase in litter size and an increase in duration of maternal care occurred secondarily and as consequences of monoestry. Indeed, in the likely absence of parent-offspring conflict (after Trivers 1974) in a monoestrous species, it is probable that continuance of maternal investment would serve to increase female reproductive success. This is particularly likely in the case of species of small body size where individuals have a less than even chance of breeding in the next year (e.g. *A. stuartii,* Wood 1970).

However, available evidence suggests that female reproductive success in one year does not influence success in the following year in *A. stuartii* (see Braithwaite 1979). Therefore, as female residual reproductive value (after Williams 1966a and b) is apparently unaffected by reproductive effort in the first year, there seems to be little to inhibit prolonging maternal care. At Mt. Glorious mothers and offspring appeared to remain together for as long as 4 to 5 months after weaning (see Braithwaite 1979), but at Cockatoo, Victoria, all animals were nesting alone within two months of weaning (n = 4; Braithwaite and Lumsden, unpublished observations). Dispersal of groups at Mt. Glorious appeared to be related to the decline in insect abundance with the onset of winter, whereas at Cockatoo dispersal may be associated with the dry summer conditions and their presumed effect on insect abundance. The diagnosis of what constitutes parental care and when it is occurring is difficult with these cryptic species, but it should not be assumed that maternal investment finishes with the cessation of lactation. Certainly high juvenile survival (a consequence of a predictable environment) is a necessary condition for the evolution of Strategy I (see Stearns 1976) and prolonged maternal care should enhance this.

In view of the lack of insurance against reproductive failure, Strategy I species should be confined to habitats where the availability of resources is predictable. The majority of these species are found in forests, woodlands and heaths of northern, eastern, southeastern and southwestern Australia, and only *A. rosamondae* from the sand plain desert of northwestern Australia appears to contravene this prediction.

Species exhibiting Strategy II should also occur where resources are predictable and highly seasonal, but also where adult survival is enhanced. They in fact occur in a surprising variety of habitats including forest, heath and rock outcrops, which show few attributes in common. The rock outcrops inhabited by *A. bilarni* and *A. macdonnellensis,* and the large body size of some of these species, conceivably reduce predation and enhance adult survival.

Body size appears to significantly influence the life history strategies of dasyurids, and it is likely that this relationship is often a compromise with other ecological benefits of body size. Thus, the selective advantage of body size in the Tasmanian Devil may lie primarily with its feeding strategy, and the slower rate of maturation of females may be consequent upon large body size.

In the Strategy III species, the polyoestrous condition in some females appears to be an insurance against an unsuccessful mating or premature loss of the first litter (Fletcher 1977, Tyndale Biscoe, in press). It is curious that this insurance against reproductive failure is not found in all females.

The polyoestrous condition occurs under at least two different circumstances in dasyurids, where resources are relatively abundant all year around as in topical rainforest and grasslands in New Guinea, and where resources are seasonal but may be temporarily in short supply as in temperate grassland and habitats in arid Australia (Morton 1978a, Braithwaite and Lee 1979). Strategy VI species occur where resources are relatively continuously abundant. Strategy IV and V species occur where resources may be temporarily in short supply, and where the probability of reproductive failure is greater than in the monoestrous species. Strategies IV and V represent two solutions to this problem. In Strategy IV the frequency with which reproduction may occur is increased, with probable although unmeasured reduction in maternal care; litter size is not markedly reduced. On the other hand, maternal investment may be greater in Strategy V sacrificing the potential for increase in litter frequency. Litter size is small in the larger Strategy V species, but potentially large in the very small planigales.

Fox (this volume) associated the polyoestrous *S. murina* with early seral stages of heathland following disturbance. Later seral stages at the same site were occupied by the monoestrous *A. stuartii*. *Sminthopsis murina* also has a relatively large litter size (10) and it is tempting to suggest that this, in association with the polyoestrous condition, is related to this species exploiting short-lived, spatially unpredictable habitats.

These polyoestrous species show little sexual dimorphism and, judging from observations on *S. crassicaudata* by Morton (1978a), little intermale aggression or overt copulatory behaviour (although see Fox and Whitford, this volume). This is consistent with the reduction in (current) reproductive effort and an increase in residual reproductive value in species which breed more than once (after Williams 1966a and b).

Ecological strategies may occasionally be subject to surprisingly rapid change, accompanied by genetic divergence (e.g. Braithwaite 1980), but the evolution of certain strategies may trap a species at an adaptive peak. For example, Braithwaite (1979) has argued that this may have happened to Strategy I species. The general lack of correlation between life history strategy and systematics in dasyurids, however, is evidence of a general theme of flexibility and relatively rapid change. If Braithwaite (1979) is correct, it means that Strategy I has arisen more than once according to Archer's (1976b) phenogram based on cranial characters.

BIBLIOGRAPHY

Archer, M., 1974. Some aspects of reproductive behaviour and the male erectile organs of *Dasyurus geoffroii* and *D. hallucatus* (Dasyuridae: Marsupialia). *Mem. Qd Mus.* **17:** 63–7.

Archer, M., 1976a. Revision of the marsupial genus *Planigale* Troughton (Dasyuridae). *Mem. Qd Mus.* **17:** 341–65.

Archer, M., 1976b. The Dasyurid dentition and its relationships to that of Didelphids, Thylacinids, Borhyaenids (Marsupicarnivora) and Peramelids (Peramelina: Marsupialia). *Aust. J. Zool. Supp. Ser.* **39**.

Arnold, J. E. and Shield, J., 1970a. Growth and development of the chuditch *(Dasyurus geoffroii)*. *Bull. Aust. Mammal. Soc.* **2:** 198.

Arnold, J. E. and Shield, J., 1970b. Oxygen consumption and body temperature of the chuditch *(Dasyurus geoffroii)*. *J. Zool. (Lond.)* **160:** 391–404.

Aslin, H., 1974. The behaviour of *Dasyuroides byrnei* in captivity. *Z. Tierpsychol.* **35:** 187–208.

Aslin, H. J., 1975. Reproduction in *Antechinus maculatus* Gould (Dasyuridae). *Aust. Wildl. Res.* **2:** 77–80.

Barnett, J. L., 1972. Stress and ageing in *Antechinus stuartii*. Ph.D. thesis, Monash University, Victoria.

Begg, R. J., 1981. The small mammals of Little Nourlangie Rock, N.T. 2. Ecology of *Antechinus bilarni*, the Sandstone Antechinus (Marsupialia: Dasyuridae). *Aust. Wild. Res.* **8:** 57–72.

Bradley, A. J., McDonald, I. R., and Lee, A. K., 1980. Stress and mortality in a small marsupial (*Antechinus stuartii* MacLeay). *Gen. Comp. Endocrinol.* **40:** 188–200.

Braithwaite, R. W. 1974. Behavioural changes associated with the population cycle of *Antechinus stuartii* (Masupialia). *Aust. J. Zool.* **22:** 45–62.

Braithwaite, R. W., 1979. Social dominance and habitat utilization in *Antechinus stuartii* (Marsupialia). *Aust. J. Zool.* **27:** 517–28.

Braithwaite, R. W., 1980. The ecology of *Rattus lutreolus*. III. The rise and fall of a commensal population. *Aust. Wildl. Res.* **7:** 119–215.

Braithwaite, R. W., and Lee, A. K., 1979. A mammalian example of semelparity. *Amer. Nat.* **113:** 151–5

Buchmann, O. L. K., and Guiler, E. R., 1977. Behaviour and ecology of the Tasmanian Devil, *Sarcophilis harrisii*. Pp. 155–68, *in* "The biology of marsupials" ed. by B. Stonehouse and D. Gilmore. MacMillan: London.

Calaby, J. H. and Taylor, J. M. (In press). Reproduction in two marsupial mice, *Antechinus bellus* and *Antechinus bilarni* (Dasyuridae), from tropical Australia. *J. Mammal.*

Cole, L. C., 1954. The population consequences of life history phenomena. *Q. Rev. Biol.* **29:** 103–37.

Cuttle, P., 1978. The behaviour of *Phascogale tapoatafa*. MSc. thesis: Monash University, Clayton, Victoria.

Davies, S. J. J. F., 1960. A note on two small mammals of the Darwin area. *J. Proc. R. Soc. West. Aust.* **43:** 63–6.

Dwyer, P. D., 1977. Notes on *Antechinus* and *Cercartetus* (Marsupialia) in the New Guinea highlands. *Proc. R. Soc. Qd* **88:** 69–73.

EWER, R. F., 1968. A preliminary survey of the behaviour in captivity of the dasyurid marsupial, *Sminthopsis crassicaudata* (Gould). *Z. Tierpsychol.* **25:** 319–65.

FLEAY, D., 1934. The Brush-tailed Phascogale. First record of breeding habits. *Vict. Nat.* **51:** 88–100.

FLEAY, D., 1935. Notes on the breeding of Tasmanian Devils. *Vict. Nat.* **52:** 100–5.

FLEAY, D., 1940. Breeding of the Tiger-cat. *Vict. Nat.* **56:** 159–163.

FLEAY, D., 1949. The Yellow-footed Marsupial Mouse. *Vict. Nat.* **65:** 273–7.

FLEAY, D., 1960. Breeding the Mulgara. *Vict. Nat.* **78:** 160–7.

FLEAY, D., 1962. The Northern Quoll *Satanellus hallucatus*. *Vict. Nat.* **78:** 288–293.

FLEAY, D., 1965. Australia's 'needle-in-a-haystack' marsupial. *Vict. Nat.* **82:** 195–204.

FLEMING, T. H., 1979. Life-history strategies. Pp. 1–61 *in* "Ecology of small mammals" ed. by D. M. Stoddart. Chapman & Hall: London.

FLETCHER, T. P., 1977. "Reproduction in the Native cat, *Dasyurus viverrinus* (Shaw)." Hons thesis: University of Tasmania, Hobart.

GODFREY, G. K., 1969. Reproduction in a laboratory colony of the marsupial mouse *Sminthopsis larapinta* (Marsupialia: Dasyuridae). *Aust. J. Zool.* **17:** 637–54.

GODFREY, G. K., AND CROWCROFT, P., 1971. Breeding the Fat-tailed Marsupial Mouse *Sminthopsis crassicaudata* in captivity. *Int. Zoo Yearb.* **11:** 33–8.

GUILER, E .R., 1970. Observations on the Tasmanian Devil, *Sarcophilus harrisii* (Marsupialia: Dasyuridae). II. Reproduction, breeding and growth of young. *Aust. J. Zool.* **18:** 63-70.

HILL, J. P., AND HILL, W. C. O., 1955. The growth stages of the pouch young of the Native Cat *(Dasyurus viverrinus)* together with observations on the anatomy of the new born young. *Trans. Zool. Soc. Lond.* **28:** 349–453.

HILL, J. P., AND O'DONOGHUE, C. H., 1913. The reproductive cycle in the marsupial *Dasyurus viverrinus*. *Q. J. Microsc. Sci.* **59:** 133–74.

HUTSON, G. D., 1976. Grooming behaviour and birth in the dasyurid marsupial *Dasyuroides byrnei*. *Aust. J. Zool.* **24:** 277–82.

INNS, R. W., 1976. Some seasonal changes in *Antechinus flavipes* (Marsupialia: Dasyuridae). *Aust. J. Zool.* **24:** 523–31.

JONES, F. W., 1923. "The mammals of South Australia. Part 1. The monotremes and carnivorous marsupials". Government Printer: Adelaide.

KREBS, C. J., 1978. "Ecology: The experimental analysis of distribution and abundance". Harper: New York.

LEE, A. K., BRADLEY, A. J., AND BRAITHWAITE, R. W., 1977. Corticosteroid levels and male mortality in *Antechinus stuartii*. Pp. 209–20. *in* "The biology of marsupials" ed. by B. Stonehouse and D. Gilmore. MacMillan: London.

LEONARD, B. V., 1972. "The effect of fire upon selected small mammals and leaf litter fauna in sclerophyll forest in southern Australia". M.Sc. thesis: Monash University, Clayton, Victoria.

LOW, B. S., 1978. Environmental uncertainty and the parental strategies of marsupials and placentals. *Amer. Nat.* **112:** 197–213.

MACK, G., 1961. Mammals from southwestern Queensland. *Mem. Qd Mus.* **13:** 213–29.

MARLOW, B. J., 1961. Reproductive behaviour of the marsupial mouse *Antechinus flavipes* (Waterhouse) (Marsupialia) and the development of the pouch young. *Aust. J. Zool.* **9:** 203–18.

MICHENER, G. R., 1969. Notes on the breeding and young of the Crest-tailed Marsupial Mouse *Dasycercus cristicauda*. *J. Mammal.* **50:** 633.

MILLAR, J. S., 1975. Tactics of energy partitioning in breeding *Peromyscus*. *Can. J. Zool.* **53:** 967–76.

MORTON, S. R., 1978a. An ecological study of *Sminthopsis crassicaudata* (Marsupialia: Dasyuridae). III. Reproduction and life history. *Aust. Wildl. Res.* **5:** 183–211.

MORTON, S. R., 1978b. An ecological study of *Sminthopsis crassicaudata* (Marsupialia: Dasyuridae). II. Behaviour and social organization. *Aust. Wild. Res.* **5:** 163–82.

PARKER, P., 1977. An ecological comparison of marsupial and placental patterns of reproduction. Pp. 273–86 *in* "The biology of marsupials". ed. by B. Stonehouse and D. Gilmore. MacMillan: London.

REED, G. F., AND WALLIS, R. L., 1975. Studies of *Antechinus swainsonii* and other small mammals in an area of Sherbrooke Forest Park. *Vict. Nat.* **92:** 84–90.

SETTLE, G. A., 1978. The quiddity of Tiger Quolls. *Aust. Nat. Hist.* **19:** 164–9.

STEARNS, S. C., 1976. Life history tactics: a review of the ideas. *Q. Rev. Biol.* **51:** 3–47.

TRIVERS, R. L., 1974. Parent-offspring conflict. *Amer. Zool.* **14:** 249–64.

TYNDALE BISCOE, C. H., 1979. Ecology of small marsupials. Pp. 343–79 *in* "Ecology of small mammals". ed. by M. Stoddart. Chapman and Hall: London.

TYNDALE-BISCOE, C. H. (in press). Mammalia — Marsupialia in Marshall's "Physiology of reproduction", 4th Edition, Volume A, Chapter 5. Ed. by E. E. Lamming. Churchill Livingston: Edinborough.

VAN DYCK, S., 1980. The Cinnamon Antechinus, *Antechinus leo* (Marsupialia: Dasyuridae), a new species from the vine-forests of Cape York Peninsula. *Aust. Mammal.* **3:** 5–17.

WAINER, J. W., 1976. Studies of an island population of *Antechinus minimus* (Marsupialia: Dasyuridae). *Aust. Zool.* **19:** 1–7.

WILLIAMS, G. C., 1966a. "Adaptation and natural selection". Princeton University Press: Princeton.

WILLIAMS, G. C., 1966b. Natural selection, the costs of reproducton, and a refinement of Lack's principle. *Amer. Nat.* **100:** 687–90.

WOOD, D. H., 1970. An ecological study of *Antechinus stuartii* (Marsupialia) in a southeast Queensland rainforest. *Aust. J. Zool.* **18:** 185–207.

WOOLLEY, P., 1966a. Reproduction in *Antechinus* spp. and other dasyurid marsupials. *Symp. Zool. Soc. Lond.* **15:** 281–94.

WOOLLEY, P., 1966b. "Reproductive biology of *Antechinus stuartii* Macleay (Marsupialia: Dasyuridae)". Ph.D. thesis: Australian National University, Canberra.

WOOLLEY, P., 1971a. Observations on the reproductive biology of the Dibbler, *Antechinus apicalis* (Marsupialia: Dasyuridae). *J. Proc. R. Soc. West. Aust.* **54:** 99–102.

WOOLLEY, P., 1971b. Maintenance and breeding of laboratory colonies of *Dasyuroides byrnei* and *Dasycercus cristicauda*. *Int. Zool. Yearb.* **11:** 351–4.

Woolley, P. 1973. Breeding patterns and the breeding and laboratory maintenance of dasyurid marsupials. *Exp. Anim.* **22** Suppl.: 161–72.

Woolley, P., 1974. The pouch of *Planigale subtilissima* and other dasyurid marsupials. *J. R. Soc. West. Aust.* **57:** 11–5.

Woolley, P. (in press a). *Antechinus bellus*, another dasyurid marsupial with a post mating mortality of males. *J. Mammal.*

Woolley, P. (in press b). Laboratory maintenance of dasyurid marsupials in captivity. Management of Australian mammals in captivity. *R. Zool. Soc. Vict.*

Wright, Q., 1942. "A study of war". University of Chicago Press: Chicago.

RESUMEN

2. Estrategias de la vida de loss marsupiales dasiúridos. Por A. K. Lee, P. Woolley y R. W. Braithwaite.

Una revisión de las historias de la vida de 30 especies de dasiúridos revelaron seis distintas estrategias. Se distinguen éstas sobre la base de la frecuencia de reproducción de las hembras, la duración y los tiempos de los esfuerzos reproductivos de los machos, la estacionalidad de la reproducción y la edad de la madurez.

Los Dasyuridae muestran una alta incidencia de especies con hembras monoestras, en las que los machos mueren después de una corta estación de fecundación, y especies que "programan" su reproducción para que la lactación coincida con el período de máxima existencia de insectos. Pocos dasiúridos pueden continuar la crianza en la estación del nacimiento. Se discuten las circunstancias que favorecen la selección y la persistencia de esas estrategias.

3. LIFE HISTORY STRATEGY OF THE DASYURID MARSUPIAL

Phascogale tapoatafa

PETER CUTTLE[1]

Phascogale tapoatafa is monoestrous and has a restricted breeding season. Ovulation is synchronized and supernumerary young are produced. Males show increasing vagility and aggressiveness as the mating period approaches, and disappear from the population before parturition occurs. Lactation occupies the period August to November, during which food resources are relatively abundant. Litter mortality is high during early lactation and just prior to weaning when the young begin to appear in the population. Some females survive to a second breeding season. Males may survive the breeding season in captivity, but afterwards show reproductive senescence. There is no redevelopment of sexual characters in subsequent breeding seasons in captivity. Some captive animals died during the breeding season of a stress-related syndrome.

This life-history strategy is very similar to that of *Antechinus stuartii* in which a total post-mating male mortality is well documented. The implications of this strategy in *P. tapoatafa* are discussed in terms of adaptedness to the highly predictable seasonal environment and other factors. The advantages to polytocous marsupials of producing supernumerary young are discussed.

Pages 13–22 *in* CARNIVOROUS MARSUPIALS, ed. by M. Archer, February, 1982. Roy. Zool. Soc. New South Wales: Sydney, Australia.

THE REPRODUCTIVE biology of dasyurid marsupials has attracted keen interest since the discovery of the spectacular life–history strategy adopted by certain members of the genus *Antechinus*. The case of *A. stuartii* has been particularly well documented (Marlow 1961, Woolley 1966, Wood 1970, Braithwaite 1974, Braithwaite and Lee 1979). In this monoestrous species, breeding is synchronous and the mating period is very restricted. Males breed only once, and die shortly after mating, before the end of their first year. Females may survive to a second breeding season.

One dasyurid which is widespread in Australia, yet which remains little known, is the brush-tailed phascogale *Phascogale tapoatafa* (Meyer). Considered to be closely related to the *Antechinus* group (Tate 1947), this species parallels *A. stuartii* in several aspects of its life–history strategy. Some of the most important similarities concern the occurrence of, and the inferred causes of an annual post–mating male mortality. That such a large mammal (males — 220 g) should live less than one year and breed only once is of immediate interest. Reproductive strategy involves a compromise between various factors, including the time of breeding, litter size and the amount of parental investment in each offspring, the partitioning of resources between male, female and offspring, and the amount of energy invested in growth to sexual maturity. The disadvantages of a short life with only one attempt at breeding are counterbalanced by advantages conferred by other aspects of the life–history strategy.

This study was conducted to examine the reproductive strategy of *P. tapoatafa,* in particular its behavioural and ecological aspects, as a basis for understanding the behaviour of this species in captivity (Cuttle 1978). The implications of the various aspects of the life–history strategy are considered in terms of the adaptive value of the strategy as a whole.

MATERIALS AND METHODS

Phascogale tapoatafa occurs in many widely scattered areas of mainland Austrialia (Fig. 1). The nominate subspecies *P. t. tapoatafa* occurs in a broad band around the coast from Eyre Peninsula in South Australia (34°30′S 135°30′E) to near Maryborough in Queensland (25°31′S 152°37′E). A break in the distribution occurs in New South Wales between latitudes 33°S and 35°S. An isolated population occurs in the vicinity of Perth, Western Australia (31°57′S 115°52′E). There are also isolated records from near Rockhampton (23°22′S 150°32′E) and Coen (13°52′S 143°12′E) in Queensland. All known localities[2] of *P. tapoatafa* lie between the 500 mm and the 2000 mm isohyets. Most localities lie in areas of low annual variability of either summer or winter rainfall.

[2] Locality data are compiled from published reports and from records of the following museums: Queensland Museum, Brisbane; Australian Museum, Sydney; National Museum, Melbourne; South Australian Museum, Adelaide.

[1] Dept. of Zoology, Monash University, Clayton, Victoria. Present Address: Natureland Zoological Gardens, P.O. Box 109, Coolangatta, Queensland, Australia, 4225.

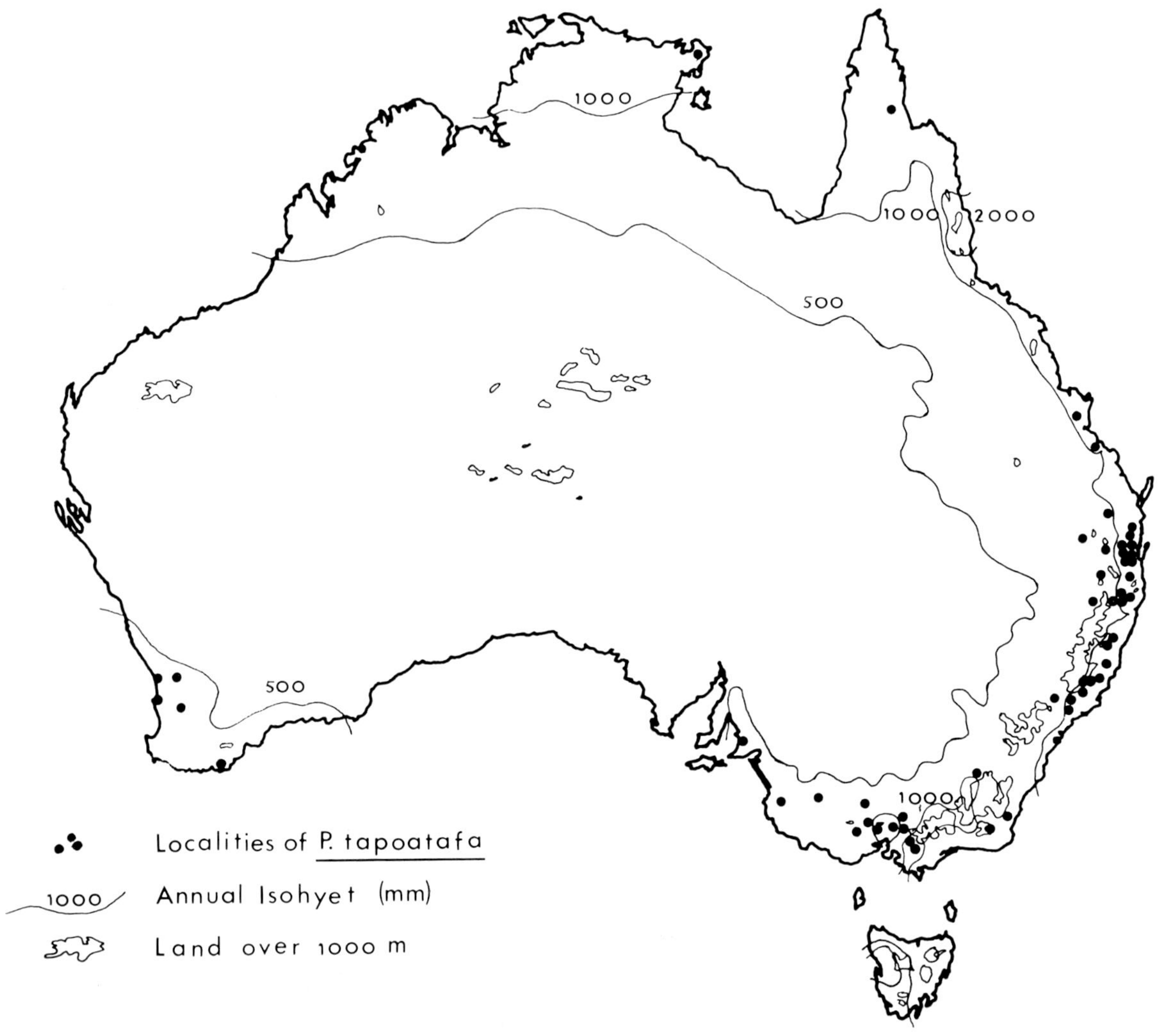

Fig. 1. Localities of recent records of *P. tapoatafa.*

The habitat occupied by *P. tapoatafa* in Victoria is open dry sclerophyll forest, predominantly an association of *Eucalyptus* spp. of the box and stringybark groups. Messmate *E. obliqua* and red stringybark *E. macrorhyncha* commonly occur either separately or together, mixed with red box *E. polyanthemus,* yellow box *E. melliodora* or long–leaved box *E. goniocalyx*. Isolated specimens or patches of red ironbark *E. sideroxylon* may also occur. The dominant understorey species is either black wattle *Acacia mearnsii* or spreading wattle *A. diffusa.* The ground cover is composed of sparse low herbaceous plants such as bush pea *Pseudanthus* spp. and common correa *Correa reflexa,* with creepers such as sarsaparilla *Hardenbergia violacea* and *Pultenea pedunculata,* and ground orchids such as the green-comb spider orchid *Caladenia deletata.* Grasses are very sparse or absent. Much of the ground is covered with only leaf litter.

This vegetation association occurs chiefly on stony ridges, especially on the drier north– and east–facing slopes, in foothills and up to about 600 m above sea level. The forest is open, with easy visibility to about 200 m, and canopy cover of 30-70%. The habitat of *P. tapoatafa* in southeast Queensland is similar to that in Victoria (R. Braithwaite, pers. comm.). *Trichosurus vulpecula* and *Sminthopsis murina* were trapped in the same areas as *P. tapoatafa. Antechinus stuartii* and *Rattus rattus* were trapped in gullies adjacent to capture sites of *P. tapoatafa,* where other *Eucalyptus* spp. become dominant and the understorey is much less open. *P. tapoatafa* were not trapped in such situations. *Macropus giganteus, Wallabia bicolor, Pseudocheirus peregrinus* and *Petaurus breviceps* were observed to range through both habitat types.

Phascogale tapoatafa were studied on a grid of 13.5 ha at Christmas Hills, Victoria (37°35′S 145°20′E). Trap sites were located at 50 m intervals on the grid. Each alternate row of sites was trapped for three consecutive nights. Animals were captured in wirecage live traps baited with a mixture of equal parts rolled oats, peanut butter and honey. Each animal captured was marked and uniquely numbered with a metal fingerling tag inserted in the pinna. For each capture, or recapture, the identity, sex, body weight and breeding condition of the individual were noted, together with the trap location and the date. The number of young in the pouch area of females and the number of nipples suckled were also noted. Trapping extended over the entire breeding seasons of 1975 and 1976 and the intervening period. Trapping was conducted on at least six consecutive nights each month.

Further information on reproduction was gathered from close observation of pregnant and lactating females isolated in indoor cages 60 cm x 60 cm x 45 cm high. Development of the pouch area, number of young born and the rate of growth of the young were observed. Regular observations were made on body weight, size and stage of sexual development. Captive animals were routinely housed in large indoor enclosures approximately 3.6 m x 3.0 m x 3.0 m high.

RESULTS

CHANGES IN THE POPULATION STRUCTURE: Trapping success for *P. tapoatafa* changed throughout the year. From January to June, up to five animals were trapped per 100 trap nights. After June, trap success fell, with success as low as one animal per 100 trap nights. Success began to increase again between November and January. Changes occurred also in the proportions of the different sex and age classes trapped throughout the year (Fig. 2). Approximately 70 percent of captures between January and June were of males. From July, only pregnant or lactating females were trapped. Juveniles of both sexes began to appear in the population in November, but juvenile females trapped in November and December were not trapped again. This was probably due to the small numbers initially trapped, rather than to a change in the behaviour of juvenile females. Small numbers of females, judged to be parous on the basis of pouch area characters (see below), were trapped between November and June. Some females were known to breed successfully a second time, but none was known to survive to a third breeding season. All males trapped at the same time of year were judged to be of the same age class on the basis of body weight and stage of sexual development (see below).

CHANGES IN THE MALE: Male *P. tapoatafa* show a rapid increase in body weight between November and April, when they attain adult size (Fig. 3). By April the penis has reached its maximum length of 15 mm extruded, and the scrotum its maximum width or 16 mm. The scrotum is then well–furred and and held close to the body. The sternal gland takes on a granular appearance between February and April and commences to secrete a substance which stains the covering fur a dark yellow colour. The data in Fig. 3 indicate that male *P. tapoatafa* grow rapidly and achieve maturity well in advance of the breeding season. Males are relatively easy to trap prior to the mating period. Trap results indicate that the vagility of males is consistently higher than that of females of the same age class, and increases towards the mating period. The available data are insufficient to indicate the existence of male territories in *P. tapoatafa*. No evidence was found for a dispersal of males at any stage of the breeding cycle. All males disappeared from the field population after mating and are presumed to have died (Fig. 2).

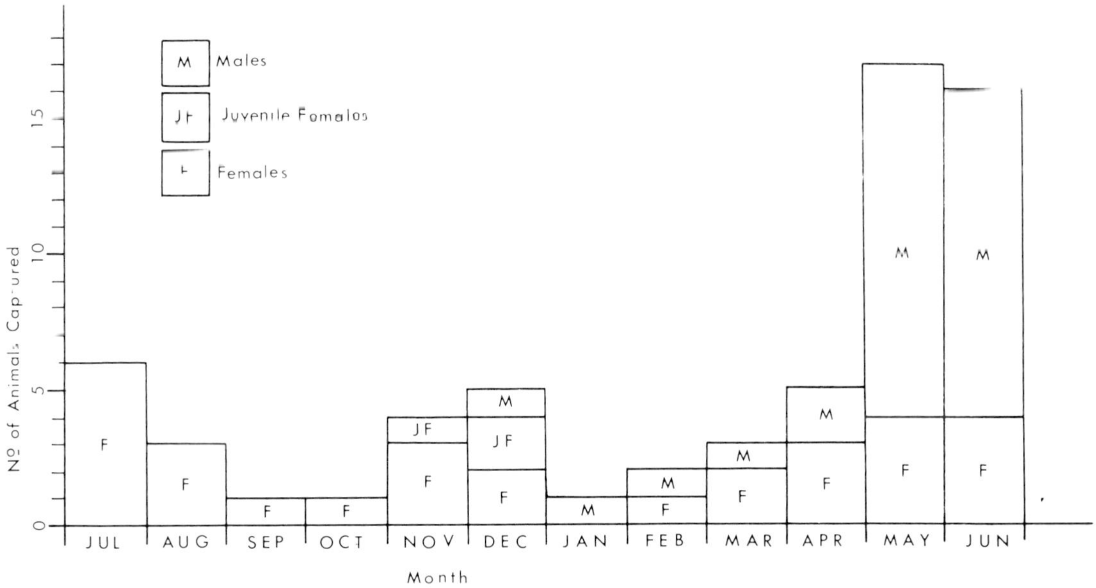

Fig. 2. Numbers of *P. tapoatafa* in three sex–age classes trapped in each month.

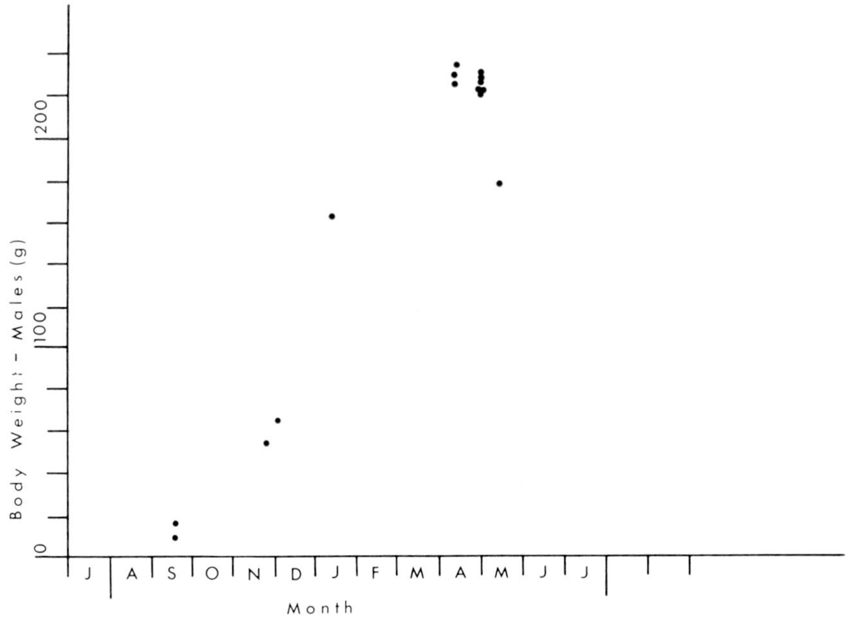

Fig. 3. Body weight of male *P. tapoatafa* trapped at different times of the year.

In captivity, male *P. tapoatafa* lived for over three years but showed signs of reproductive senescence after the first year. Although the scrotum did not become flaccid, pendulous and darkly pigmented as does that of *Antechinus stuartii* (Woolley 1966), spermatogenesis ceased completely (Presidente and Beveridge, pers. comm.). The penis became flaccid and non-retractile, but did not decrease in size. The sternal gland ceased to secrete, but the granular appearance of the covering skin remained throughout life. The fur stain gradually faded, presumably as fur was shed and replaced. Males that survived to a second breeding season showed no signs of redevelopment of sexual characters or of sexual behaviour. Throughout life they remained aggressive towards other males while active at night, but often shared a nest box with another male during the day.

Mortality in the Captive Colony: Several *P. tapoatafa* of both sexes died in captivity during the mating season. Two males and two females housed together, and five males housed together, in large indoor enclosures, all died during the period 18-24 June, 1974. A gram-positive bacillus *Erysipelothrix rhusiopathiae* was cultured from the heart blood of two males (K. Cuttle, pers. comm.), and one male was found to have a cytomegalo viral infection of the prostate (P. Carbonell, pers. comm.). All were in their first year and one female was found *post mortem* to have been pregnant.

One male and five females housed together in a large indoor enclosure died during the period 8-12 June, 1977. Stress–related symptoms such as splenic lymphoid depletion and hepatic centrolobular necrosis were found in all six animals, while gastric and duodenal ulceration and haemorrhage were found in one female (Beveridge and Presidente, pers. comm.). One female was in her fourth year, while the other five animals were in their third year. In this, as in the previous case of multiple mortality in the colony, symptoms preceding death were loss of appetite, dramatic loss of weight, diarrhoea, lassitude and loss of motor coordination. Time from onset of visible symptoms to death was 24-72 hours.

Changes in the Female: A true pouch is lacking in *P. tapoatafa* as in related *Antechinus* spp. and in many other dasyurids (Woolley 1966, 1974). The pouch of *P. tapoatafa* is an extreme form of Woolley's (1974) type 1 pouch, where no part of the pouch area is ever covered by a fold of abdominal skin (Fig. 4). The term "pouch area" is used here to describe the area surrounding the nipples, including the skin covering the mammary glands and forming the anterior and lateral "ridges" which develop during lactation (Cuttle 1978). The pouch area in a nulliparous female is discernable only as an area on the ventrum, not more than 10 mm in diameter, covered with sparse pale hair. The skin is pale and smooth and the nipples are minute, about 0.5 mm in length and in diameter. Parous females can be identified as the pouch area is somewhat larger — about 15 mm in diameter — than in a nulliparous female and the covering fur is stained a dark rufous to yellow colour. The skin of the pouch area is pale and smooth, but the nipples are easily discernable and are about 1.0 mm in diameter and in length.

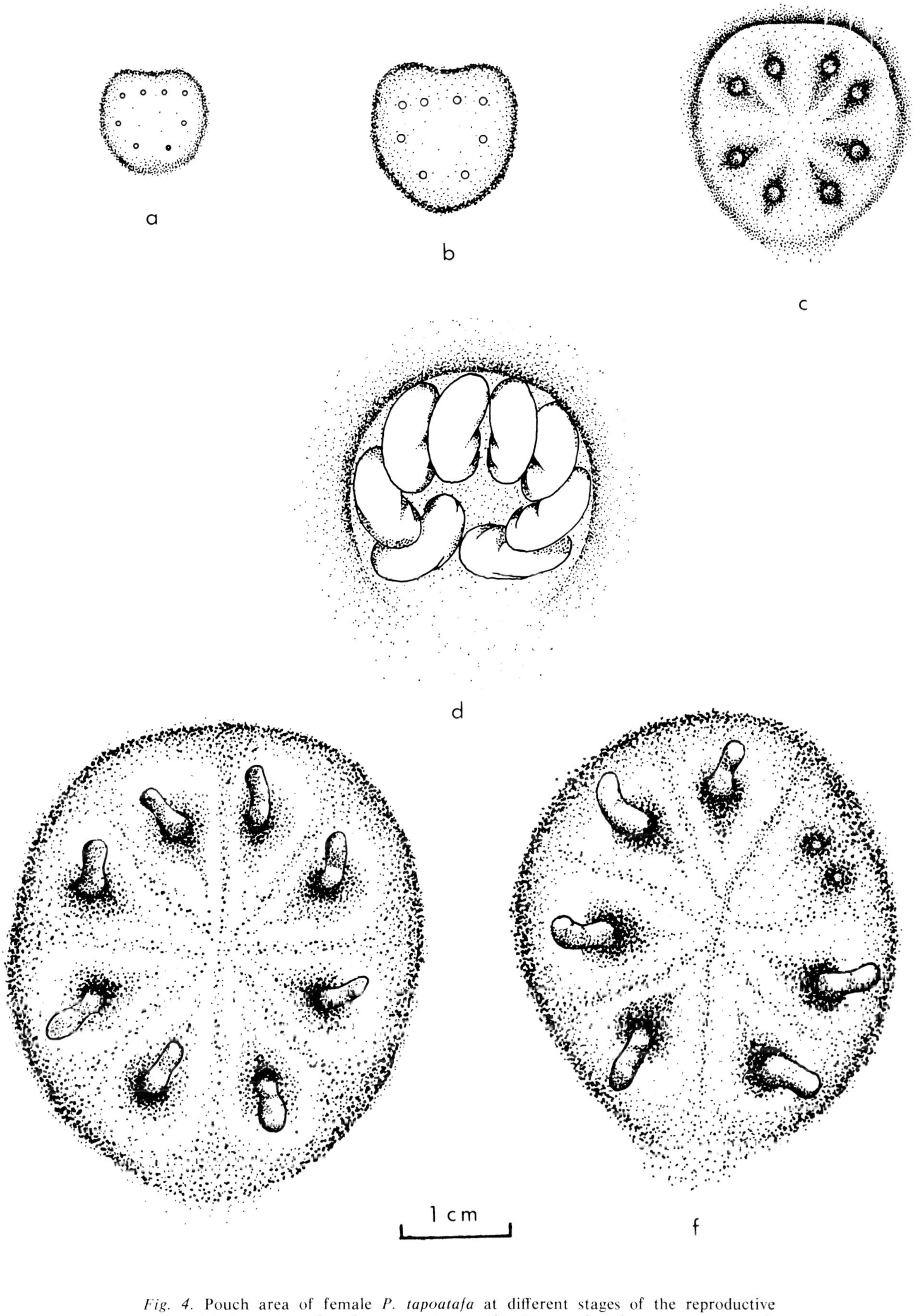

Fig. 4. Pouch area of female *P. tapoatafa* at different stages of the reproductive cycle: (a) Immature female, pre-oestrous; (b) Parous female, pre-oestrous; (c) immediately prior to parturition; (d) early fixed lactation, with young attached to the nipples; (e) prior to weaning; (f) prior to weaning, showing regression of nipples and distortion of pouch area due to litter mortality during fixed lactation.

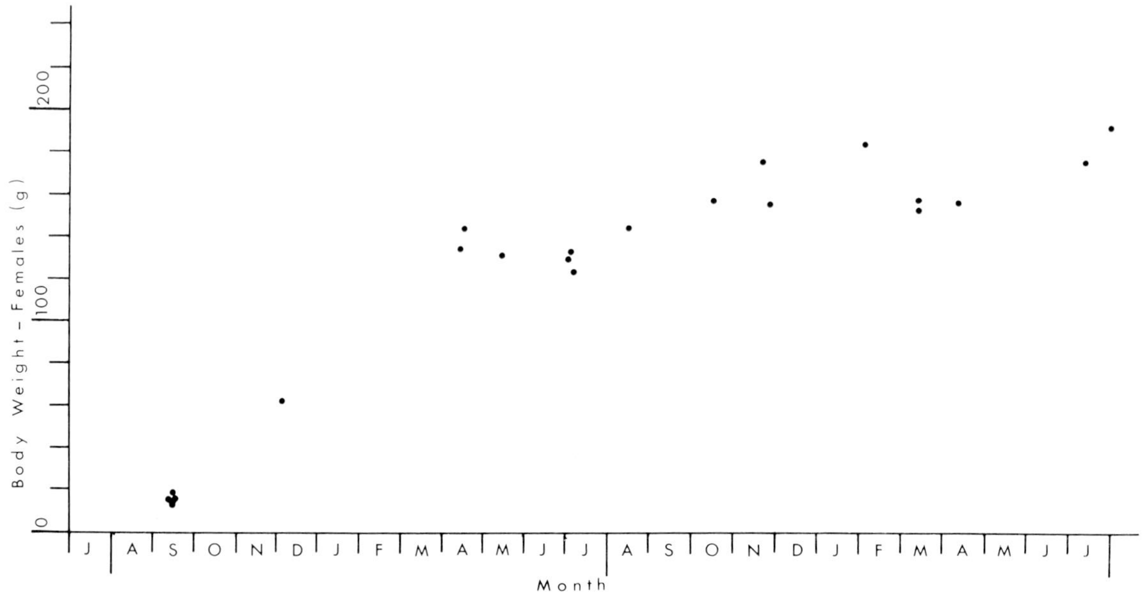

Fig. 5. Body weight of female *P. tapoatafa* trapped at different times of the year.

Female *P. tapoatafa* show a steady increase in body weight between November and April with a slight further increase until the following November (Fig. 5). Thereafter, body mass changes little and is not a useful indicator of age in females over one year of age. The rate of increase of body mass between November and April is much greater in males than in females, and adult males achieve a much greater mass at maturity (Fig. 3) than do females. Females are relatively easy to trap prior to the mating period, but are difficult to trap during the entire period of lactation (Fig. 2). Trap results suggest that females show less movement about the environment during lactation than they do before the mating period. This is especially true during the period of fixed lactation (see below) when the female's movement is hampered by the presence of young in the pouch, and her foraging radius must be reduced.

Trap results indicate that litters containing young females may be trapped in the vicinity of their maternal nest until the mating period. Two pregnant females were captured as they fled with two others from the same nest in a newly felled tree. It is inferred that these animals had not yet dispersed. During lactation, females appear to occupy exclusive but adjoining ranges of about four to five hectares each. Two females were trapped at the same sites only along the common border of their inferred ranges. Thus it is surmised that female dispersal occurs during the lactation period. Little aggressive behaviour was observed between females (Cuttle 1978) and the mechanism of female dispersal is not fully understood.

Exact information on the length of the gestation period is not available. However, two females captured in early pregnancy and subsequently isolated from males gave birth 26 days and 29 days respectively after capture. Parturition occurs in the wild in late July — early August. No births were observed to result from captive matings, but most captive females showed symptoms of pregnancy at the same time as wild-trapped females. Three captive female *P. tapoatafa* gave birth on 1 August and 4 August in this study and on 4 August, 1970 (J. Kloeden, Super 8 mm film and pers. comm.). Three wild-caught litters were each estimated to have been born during the last week of July or the first week of August. One female that died on 27 June, 1974 carried embryos at early stages of development.

The period from birth to weaning can be clearly divided into two periods based on the two modes of lactation of marsupials. The term "fixed period of lactation" is used here to mean the period from attachment of young to the nipples until the first time the young are free of the nipples. The term "free period of lactation" is used here to mean the period from the end of the fixed period of lactation until the end of weaning. The fixed period of lactation in *P. tapoatafa* lasts seven to eight weeks (N = 2 litters), and the free period a further eight to nine weeks, ending in late November.

Regression of individual nipples and the associated mammary tissue occurs if young are lost during the fixed period. This results in a distortion of the shape of the pouch area (Fig. 4). Regression of nipples during the fixed period gives an estimate of litter mortality at this stage (see below). It is not certain that each young remains faithful to a single nipple during the free period (but see Fleay 1965). Thus the count of regressed nipples during the free period may underestimate mortality to weaning, as it does in *A. stuartii* (Leonard 1972).

Symptoms of pregnancy occur in captive females which fail to produce a litter, with changes in the pouch area occurring exactly as for pregnant females until the expected time for parturition. The pouch area then regresses rapidly until it is indistinguishable from the pouch area of a nulliparous female. If a captive female loses her entire litter at birth, regression of the pouch area occurs exactly as for an unmated female. There is no recurrence of earlier changes in the pouch area during the same breeding season in unmated females or those that lose the entire litter, indicating that *P. tapoatafa,* like *A. stuartii,* is monoestrous (Woolley 1966). In females which raise a litter, the pouch area regresses shortly after weaning. By March of the year following parturition, it resembles the pouch area described above for a parous female prior to the mating period.

At the end of the fixed period of lactation, each young weighs 17-19 grams, giving a possible combined mass of 152 grams, or almost the mass of a female in her first breeding season. By the end of weaning, a full litter may weigh a total of over 400 grams, more than twice the weight of the mother. Thus the female's reproductive effort is large compared to her non–breeding energy budget, and most of it is expended during the later stages of lactation. The neonatus is minute, and the effort invested in gestation is small in comparison with lactation.

In captivity, female *P. tapoatafa* lived for over three years. Captive females were observed to show symptoms of oestrus in their first and second breeding seasons, but not in their third.

LITTER SIZE AND SURVIVAL: Because each young is initially attached to a single nipple, litter size in marsupials cannot exceed the number of nipples available. In this study of *P. tapoatafa* no evidence was found for more than or fewer than eight nipples in any living animal or in any preserved museum specimen. A suggestion that *P. tapoatafa* may have ten nipples may be attributable to Krefft (1864) and has been repeated several times without comment (Thomas 1888, Fleay 1934, Woolley 1966). No evidence of variability in nipple number was found throughout the eastern part of the range of the nominate subspecies *P. t. tapoatafa.* Krefft may have referred to specimens of *P. t. tapoatafa* from Western Australia or to specimens of *P. t. pirata* (Thomas) from northern Australia. Other dasyurid species are known to show variability in nipple number (Woolley 1966, Godfrey and Crowcortt 1971, Leonard 1972).

The number of young found in the pouch area or in the nest, and the number of nipples either fully or partially regressed, were recorded for wild-caught and captive lactating female *P. tapoatafa* (Table 1). One birth in captivity resulted in considerably more than eight young being born for the eight nipples available, but

Table 1. Litter size in *P. tapoatafa.*

(a) Wild–caught Litters

♀ No.	Period of Lactation	No. of Young	No. of Nipples Regressed fully	No. of Nipples Regressed partially	Estimated Litter Size
422	early fixed	8	0	0	8
423	early fixed	7	0	1	7
4279*	fixed	8	0	0	8
426	early free	7	0	0	7
427	late free	–	0	0	8
429	late free	–	0	2	6
451	late free (1976)	–	3	2	3
451	late free (1977)	–	1	0	7

(b) Captive–born Litters

♀ No.	Date of Birth	No. of Young
452	1.8.76	8
453	4.8.76	3
4†	4.8.70	6
11†	.8.70	8
13†	.8.70	7

* A. Heislers, personal communication.
† J. Kloeden, personal communication.

only eight young attached to the nipples (J. Kloeden, Super 8 mm film and pers. comm.). These data suggest that the usual litter size at birth for *P. tapoatafa* in the wild is eight.

Mortality during lactation accounts for a large proportion of overall mortality. Four of the eight wild–caught females had lost from one to five young at various stages of lactation. Of the estimated 14 young of two wild–caught females only four were trapped as juveniles or sub–adults. Four females in captivity reared litters of five, seven, one and one respectively.

DISCUSSION

There are striking similarities between the life–history strategy of *P. tapoatafa* and those of *Antechinus stuartii, A. flavipes* and *A. swainsonii. P. tapoatafa* is monoestrous and has a restricted breeding season. Fleay's (1934) statement to this effect is confirmed by: (1) the appearance of young in the field population during only November of each year, (2) the distribution of body weights of juveniles, (3) the synchoronous circannual oestrus cycle and (4) the rapid regression of the pouch area if a litter is not established. Thus a parallel may be drawn with certain *Antechinus* spp. which are monoestrous and have a restricted breeding season (Marlow 1961, Woolley 1966, Wood 1970, Inns 1976, Lee, Bradley and Braithwaite 1977).

The increase in vagility of male *P. tapoatafa* in the wild and their increasing aggressiveness towards conspecific males in captivity as the mating period approaches (Cuttle 1978) closely parallel behaviour changes that occur in male *A. stuartii* (Wood 1970, Leonard 1972, Braithwaite 1973, 1974). Changes in the behaviour of male *A. stuartii,* coupled with the prodigious reproductive effort of the males (Braithwaite and Lee 1979) are implicated in the aetiology of the adrenocortical response which contributes largely to their death (Lee *et al* 1977).

The disappearance of male *P. tapoatafa* from the wild population, and the pattern of mortality amongst captive animals, strongly reflect these phenomena in *A. stuartii.* Post–mating laboratory and field deaths of *A. stuartii* have been widely investigated (see Arundel *et al* 1977 who summarised probable causes of death). Some of the deaths were attributed to septicaemia associated with focal hepatic necroses containing Gram–positive bacilli *Listeria monocytogenes* (Lee *et al* 1977). This organism, like *E. rhusiopathiae,* is a member of the Corynebacteriaceae. Both are common pathogens in man, domestic animals and wildlife species (Blair *et al* 1970). Other deaths were attributed to an anaemic state (Cheal, Lee and Barnett 1976) which was related to chronic gastric and duodenal haemorrhage in moribund captive animals (Lee *et al* 1977).

In *A. stuartii* a link was found between post–mating mortality and an increase in plasma corticosteroid concentration, coupled with a decrease in cortisol binding activity of the plasma. Incidence of infection of the prostate by cyto–megalo virus, and incidence of depletion of splenic lymphoid elements were related to increasing concentration of unbound plasma corticosteroids (Lee *et al* 1977). The cortisol binding capacity of the blood of *P. tapoatafa* was investigated and the result indicated that plasma corticosteroid concentrations may rise in *P. tapoatafa* in the mating period, but not to the same marked extent as in *A. stuartii* (A. Bradley, pers. comm.). These results strongly suggest that there is a total post–mating male mortality in *P. tapoatafa* as in *A. stuartii,* and that the mechanisms underlying the phenomena in the two species are very similar.

Braithwaite and Lee (1979) discussed theoretical considerations for the evolution of "semelparity" in small mammals, with particular reference to *A. stuartii.* For *P. tapoatafa,* the open forest habitat in southeastern Australia represents a highly predictable seasonal environment, where breeding occurs at such a time that food will be abundant during lactation and weaning. The breeding period is long enough for a female to successfully rear one litter, but not two. These conditions, and the added advantage of producing supernumerary young, may have favoured selection of a monoestrous reproductive pattern, synchrony of oestrus and restricted mating period (Braithwaite and Lee 1979). With these constraints on reproduction, and the generally low chance of survival of small mammals to reproduce in a second year (French, Stoddart and Bobek 1975), selection should strongly favour males which maximise reproductive effort during their first breeding season, even at the expense of longevity (Braithwaite and Lee 1979). Changes in the behaviour of male *P. tapoatafa* are consistent with this hypothesis.

A constant feature of dasyurid reproduction which is shared with polytocous species in other marsupial families is the production of large numbers of ova, well in excess of the number of nipples (Hill 1910, Hartman 1919, Enders 1966, Woolley 1966, 1971, Godfrey 1969). The production of supernumerary young may be adaptive for some eutherian mammals where inbreeding is part of the life-history strategy and cooperative suckling occurs, e.g. *Spermophilus tridecemlineatus* (Zimmerman 1974). However, as cooperative suckling is precluded in marsupials by the exclusive occupancy of a single nipple by each young, supernumerary young may be seen as apparent wastage. Prenatal mortality claims a large proportion of ova — many do not survive cleavage — and disintegrating blastocysts and dead embroyos may be found in the uteri (Godfrey 1969).

Reproductive effort involves a trade-off between number of young born and amount of parental investment in each young. An increase in either may increase the Darwinian fitness of the parent, but both cannot be increased simultaneously. The strategy of producing large numbers of supernumerary ova can be summed up thus: a very large number of potential offspring (ova) receive very little parental care, and have little chance of survival; a smaller number of offspring receive enough care to survive to parturition, while a very small and strictly limited number of offspring — those that attach to nipples — are given a great deal of parental care, and have a strong chance of survival. In *P. tapoatafa* a potentially still smaller number survive the rigours of the open pouch to become virtually assured of survival to weaning in the safety of the nest.

Marsupial young are born at an earlier stage of development than eutherian young, so the cost of producing supernumerary young is smaller for a marsupial than for a comparable sized eutherian. The actual effort required to bring a fertilised ovum to birth in a dasyurid is not known, but it must certainly be very much smaller than the effort of raising a single young to weaning. The lactation period involves a far greater energy turnover than does the gestation period. Thus the cost of producing supernumerary young is minimised in marsupials, where gestation is comparatively brief, especially in the polytocous species such as the Dasyuridae. The larger the potential litter size, the smaller is the marginal cost of each additional supernumerary foetus.

Not all marsupials produce supernumerary young. The Macropodidae, for example, normally give birth to a single young. They are larger and less susceptible to disturbance during birth than the Dasyuridae. They live longer and breed repeatedly. If one reproductive attempt is unsuccessful, a new blastocyst may develop to parturition within 30 days. Many dasyurids are monoestrous with a restricted breeding season. They have only one chance per year to reproduce, and little certainty of survival to another breeding season. With these constraints on reproduction, it is vital that reproduction should succeed at the first opportunity. Thus the advantage of producing supernumerary young, to *P. tapoatafa* and other marsupials which exhibit the phenomenon, can be seen in its being a cheap and reliable strategy to ensure that the nipples are fully occupied by young at birth, thus minimising the potential impact of the high litter mortality during certain stages of lactation.

ACKNOWLEDGEMENTS

I thank J. E. Nelson for his supervision of the study, and A. K. Lee for his valued comments on an earlier draft. I am indebted to the Melbourne and Metropolitan Board of Works and to E. H. M. Ealey for permission to study animals on their respective properties. I am deeply grateful to Mr. and Mrs. W. R. Maughan, of Natureland Zoological Gardens, Coolangatta, Queensland, for generous financial support throughout the study. Jo-ann Hyndman kindly typed the manuscript.

REFERENCES

Arundel, J. H., Barker, I. K., and Beveridge, I., 1977. Diseases of marsupials. *In* "The biology of marsupials" ed. by B. Stonehouse and D. Gilmore. Macmillan: London.

Blair, J. E., Lennette, E. H., and Truant, J. P., 1970. "Manual of clinical microbiology". American Socety for Microbiology: Bethesda, Maryland.

Braithwaite, R. W., 1973. "An ecological study of *Antechinus stuartii* (Marsupialia: Dasyuridae)". MSc. thesis: University of Queensland, Brisbane.

Braithwaite, R. W., 1974. Behavioural changes associated with the population cycle of *Antechinus stuartii* (Marsupialia). *Aust. J. Zool.* **22:** 45–62.

Braithwaite, R. W., and Lee, A. K., 1979. A mammalian example of semelparity. *Amer. Nat.* **113:** 151–5.

Cheal, P. D., Lee, A. K., and Barnett, J. L., 1976. Changes in the haematology of *Antechinus stuartii* and their association with male mortality. *Aust. J. Zool.* **24:** 299–311.

Cuttle, P., 1978. "The behaviour in captivity of the dasyurid marsupial *Phascogale tapoatafa* (Meyer)". M. Sc. thesis: Monash University, Melbourne.

Enders, R. K., 1966. Attachment, nursing and survival of young in some didelphids. *In* "Comparative biology of reproduction in mammals" ed. by I. W. Rowlands. Academic Press: London. (*Symp. Zool. Soc. Lond.* **15:** 195–203).

Fleay, D., 1934. The Brush–tailed Phascogale — first record of breeding habits. *Victorian Nat.* **51:** 88–100.

Fleay, D., 1965. Australia's "needle–in–a–haystack" marsupial. *Victorian Nat.* **82:** 195–204.

French, N. R., Stoddart, D.M., and Bobek, B., 1975. Patterns of demography in small mammal populations. *In* "Small mammals: their productivity and population dynamics" ed. by F. B. Golley, K. Petrusewicz and L. Ryszkowski. Cambridge University Press. Cambridge.

Godfrey, G. K., 1969. Reproduction in a laboratory colony of the marsupial mouse *Sminthopsis larapinta* (Marsupialia: Dasyuridae). *Aust. J. Zool.* **17:** 637–54.

Godfrey, G. K., and Crowcroft, P., 1971. Breeding the Fat–tailed Marsupial Mouse *Sminthopsis crassicaudata* in captivity. *Int. Zoo Yb.* **11:** 33–8.

Hartman, C. G., 1919. Studies in the development of the opossum *Didelphis virginiana* L. III. Description of new material on maturation, cleavage and entoderm formation. IV. The bilaminar blastocyst. *J. Morph.* **32:** 1–142.

Hill, J .P., 1910. The early development of the Marsupialia, with special reference to the Native Cat (*Dasyurus viverrinus*). *Q. Jl. microsc. Sci.* **56:** 1–134.

Inns, R. W., 1976. Some seasonal changes in *Antechinus flavipes* (Marsupialia: Dasyuridae). *Aust. J. Zool.* **24:** 523–31.

Krefft, G., 1864. "Catalogue of the Mammalia in the collecton of the Australian Museum". Australian Museum: Sydney.

Lee, A. K., Bradley, A. J., and Braithwaite, R. W., 1977. Corticosteroid levels and male mortality in *Antechinus stuartii*. *In* "The biology of marsupials" ed by B. Stonehouse and D. Gilmore. Macmillan: London.

Leonard, B. V., 1972. "The effect of fire upon selected small mammals and leaf litter fauna in sclerophyll forest in southern Australia". M.Sc. thesis: Monash University, Melbourne.

Marlow, B. J., 1961. Reproductive behaviour of the marsupial mouse, *Antechinus flavipes* (Waterhouse) (Marsupialia) and the development of the pouch young. *Aust. J. Zool.* **9:** 203–18.

Tate, G. H. H., 1947. On the anatomy and classification of the Dasyuridae (Marsupialia). *Bull. Am. Mus. Nat. Hist.* **88:** 97–156. (Results of the Archbold Expeditions. No. 56).

Thomas, O., 1888. "Catalogue of the Marsupialia and the Monotremata in the collection of the British Museum (Natural History)". British Museum: London.

Wood, D. H., 1970. An ecological study of *Antechinus stuartii* (Marsupialia) in a southeast Queensland rainforest. *Aust. J. Zool.* **18:** 185–207.

Woolley, P., 1966. Reproduction in *Antechinus* spp. and other dasyurid marsupials. *In* "Comparative biology of reproduction in mammals" ed. by I. W. Rowlands. Academic Press: London. (*Symp. Zool. Soc. Lond.* **15:** 281–94).

Woolley, P., 1971. Maintenance and breeding of laboratory colonies: *Dasyuroides byrnei* and *Dasycercus cristicauda*. *Int. Zoo Yb.* **11:** 351–4.

Woolley, P., 1974. The pouch of *Planigale subtilissima* and other dasyurid marsupials. *J. Proc. R. Soc. West. Aust.* **57:** 11–5.

Zimmerman, G. D., 1974. Co-operative nursing behaviour in *Spermophilus tridecemlineatus* (Mitchell). *J. Mammal.* **55:** 680–1.

RESUMEN

3. Estrategia de la vida del marsupial dasiúrido. *Phascogale tapoatafa*. Por Peter Cuttle.

Phascogale tapoatafa es monoestro y tiene una restringida estación de apareo. La ovulación está sincronizada y produce hijos supernumerarios. Los machos muestran un incremento de la movilidad y de la agresividad a medida que se aproxima el período de apareo, y desaparecen de la población antes de que ocurra el parto. La lactación ocurre en el perído Agosto–Noviembre, durante el cual los recursos alimentarios son relativamente abundantes. La mortalidad más alta ocurre durante las primeras etapas de la lactación y un poco antes del destete, cuando los jóvenes comienzan a aparecer en la población. Algunas hembras sobreviven hasta una segunda estación de apareo. Los machos pueden sobrevivir la estación de apareo en cautividad, pero poco después muestran una senectud reproductiva. No se produce un nuevo desarrollo de los caracteres sexuales en las subsiguientes estaciones de apareo mientras estánen cautividad. Algunos animales cautivos murieron durante la estación de apareo de un síndrome tensional relacionado.

Esta estrategia biológica es muy similar a la de *Antechinus stuartii*, en el que se produce una total mortalidad de los machos después del apareo. Las implicancias de esta estrategia en *P. tapoatafa* son discutidas en términos de adaptación al altamente pronosticable ambiente estacional y a otros factores. Se discuten las ventajas para los marsupiales politocus (productores de huevos extra) de producir jóvenes supernumerarios.

4. REPRODUCTION, GROWTH AND DEVELOPMENT IN *Ningaui* sp. (DASYURIDAE, MARSUPIALIA) FROM THE NORTHERN TERRITORY

F. DOMINIC FANNING[1]

The breeding season for *Ningaui* sp. was from early September until late February. Females proved to be polyoestrus. Oestrus females called in response to mate-seeking calls uttered by males and, when permitted, mated on a number of occasions over extended periods. Mating was typically dasyurid in fashion with an active and passive phase, and lasted from 20 to 310 minutes. Gestation periods, timed from the last mating, ranged from 13½ to 21¼ days (± ¼ day). Observations were made on changes in the females' pouches, changes in females' weights and the behaviour of females and young. Body measurements were taken from young Ningauis and the weights of attached young were estimated from the females' weights. The Ningaui neonatus is typically dasyurid in appearance and develops gradually, with an apparent increase in the rate of growth at about 30 days. Young were first seen free of the teats at 42–44 days and were left in the nest at 48–49 days, at which time the eyes began to open. The teeth were also beginning to erupt at about 48 days and the auditory orifice was open by day 54. Young were independent between 76 and 81 days, when they were able to eat fully–grown mealworms and were uttering threat calls.

Ningauis possess an extensive vocal repertoire, with vocalizations evident during all types of social interaction. A large vocal repertoire and a reproductive strategy involving polyoestry and repeated matings during oestrus may be of considerable value to a small, socially intolerant, arid zone dasyurid.

Pages 23–37 *in* CARNIVOROUS MARSUPIALS, ed. by M. Archer, February, 1982. Roy. Zool. Soc. New South Wales: Sydney, Australia.

NINGAUI is a genus of tiny dasyurids which was first described from Western Australia by Archer (1975). Representatives of the genus have subsequently been recorded from all the mainland states of Australia (Baverstock and Aslin 1975, Archer 1979, Fleming and Cockburn 1979, Johnson and Roff 1980) and are confined to arid and semi-arid habitats. The animals on which this investigation was undertaken were trapped in open-hummock grassland near Ayers Rock in the Northern Territory (Johnson and Roff 1980) and their specific status is still unknown. *Triodia basedowii* was the dominant plant in the area, with scattered trees and shrubs. Virtually nothing is known of the ecology of this genus of dasyurids.

METHODS

Three adult Ningauis were received from Dr. Ken Johnson of the Territory Parks and Wildlife Service, Alice Springs, in January, 1979. These animals had been captured near Ayers Rock and included two adult males and a female with seven pouch young. The young were first seen detached from the nipples eight days later and were removed and housed separately 47 days after their arrival. During the study, this wild caught female (No. 4) and three females of her litter (No. 1, No. 2 and No. 3) were used. A male from this litter and the wild caught males were also involved.

Animals were housed separately in a variety of cages ranging from glass aquaria 60 x 30 x 30 cm to glass–fronted wooden boxes measuring 90 x 60 x 45 cm. All cages were covered with fly screen tops. The floors of the cages were covered with fine red soil from Fowlers Gap Station (31°05′S, 141°45′E), western New South Wales, and pieces of bark were included for shelter. After females had produced young, wooden nest boxes 18 x 14 x 14 cm were included in the cages to facilitate observation and trapping of animals for measurement. The Ningauis were housed in a room with natural lighting and were exposed to the normal photoperiod for Sydney. The cages were placed so that they were in direct sunlight for at least a short time each day.

Animals were fed on a fillet mince, containing beef, mutton and kangaroo. To this was added dry ground dog food, a vitamin and mineral supplement and dicalcium phosphate powder. Fly pupae and diced cheese were added to the mince and the diet was occasionally supplemented with mealworms, *Tenebrio* sp. Females with young were given 5–10 mealworms every day. The animals were usually fed a few mealworms between 12 and 2 p.m. and were given the meat mixture at about 5 p.m. Water was supplied *ad libitum*.

Vocalizations made by Ningauis were recorded at 7.5 i.p.s. using a Uher 4000 Report L tape

[1] School of Zoology, University of New South Wales, P.O. Box 1, Kensington, New South Wales, Australia, 2033.

recorder and an AKG directional microphone, and the circumstances relevant to the calls were noted. The calls were analyzed on a Kay Electric Co. Sonograph at half their recorded speed, thus increasing the frequency range of the printout. No attempt was made to determine whether ultrasonic frequencies were present in Ningaui vocalizations. Observations of behaviour were made throughout the day and evening at irregular periods. As animals were housed separately, mating only occurred when a female which had been calling was placed in a male's cage. Interactions between animals were noted, and the duration of some activities and the developmental stages of females and of young were recorded.

The weights of females were monitored for various periods before parturition and at intervals thereafter. Weights of females and of young were recorded throughout the period of growth of the young. From the pattern of weight changes for each female, it was possible to obtain an estimate of the weights of the attached young. Body measurements of the young Ningauis were

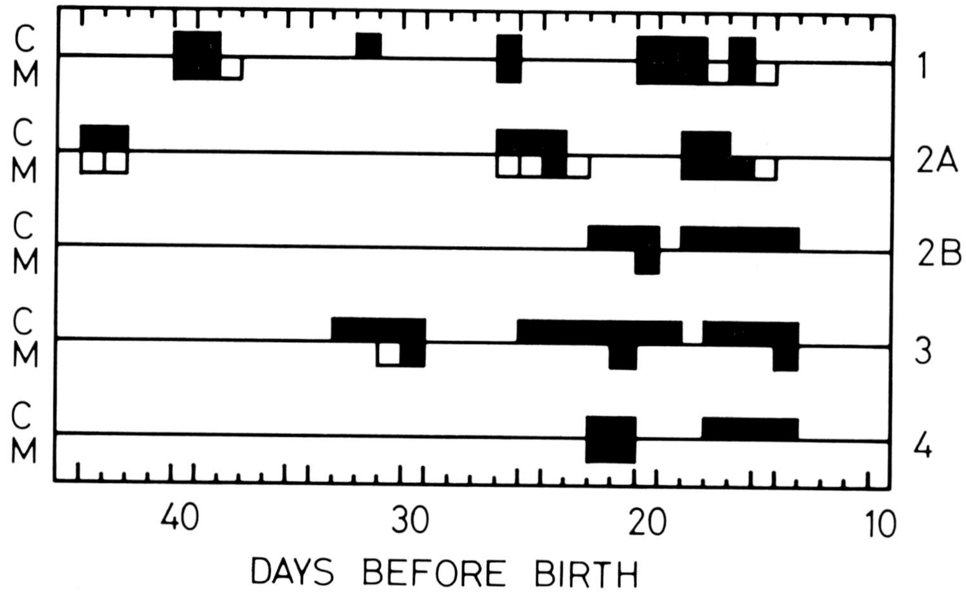

Fig. 1. The incidence of calling (C) and mating (M), by female Ningauis. Black squares indicate the days on which these activities occurred. Open squares indicate when females were placed with males but mating did not occur.

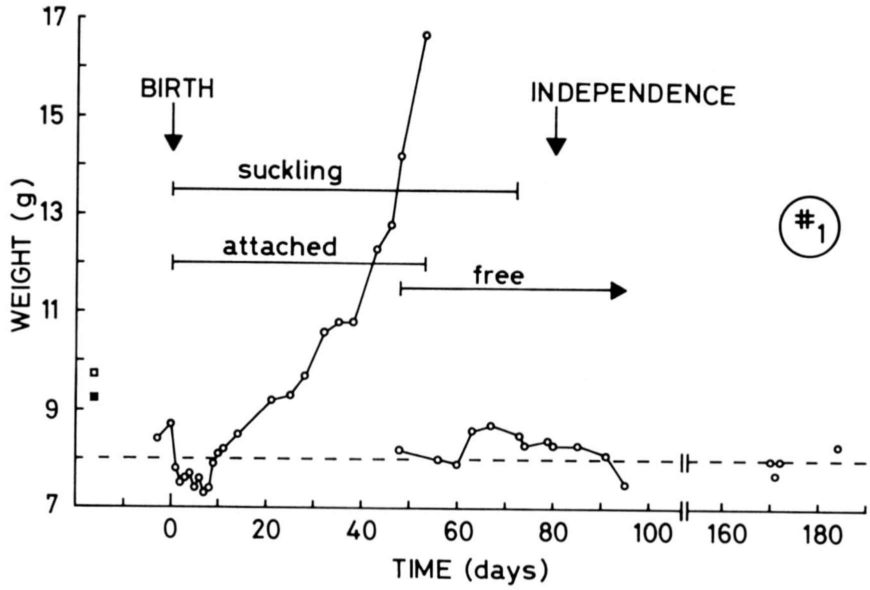

Fig. 2. Changes in the weight of female No. 1 from before her litter was born until the young were 184 days old. The period during which the young are dependent on suckling, and during which they are attached and free are indicated. The dashed line is the reference weight for this female, the open square indicates the last day of calling and the closed square indicates the last day of mating.

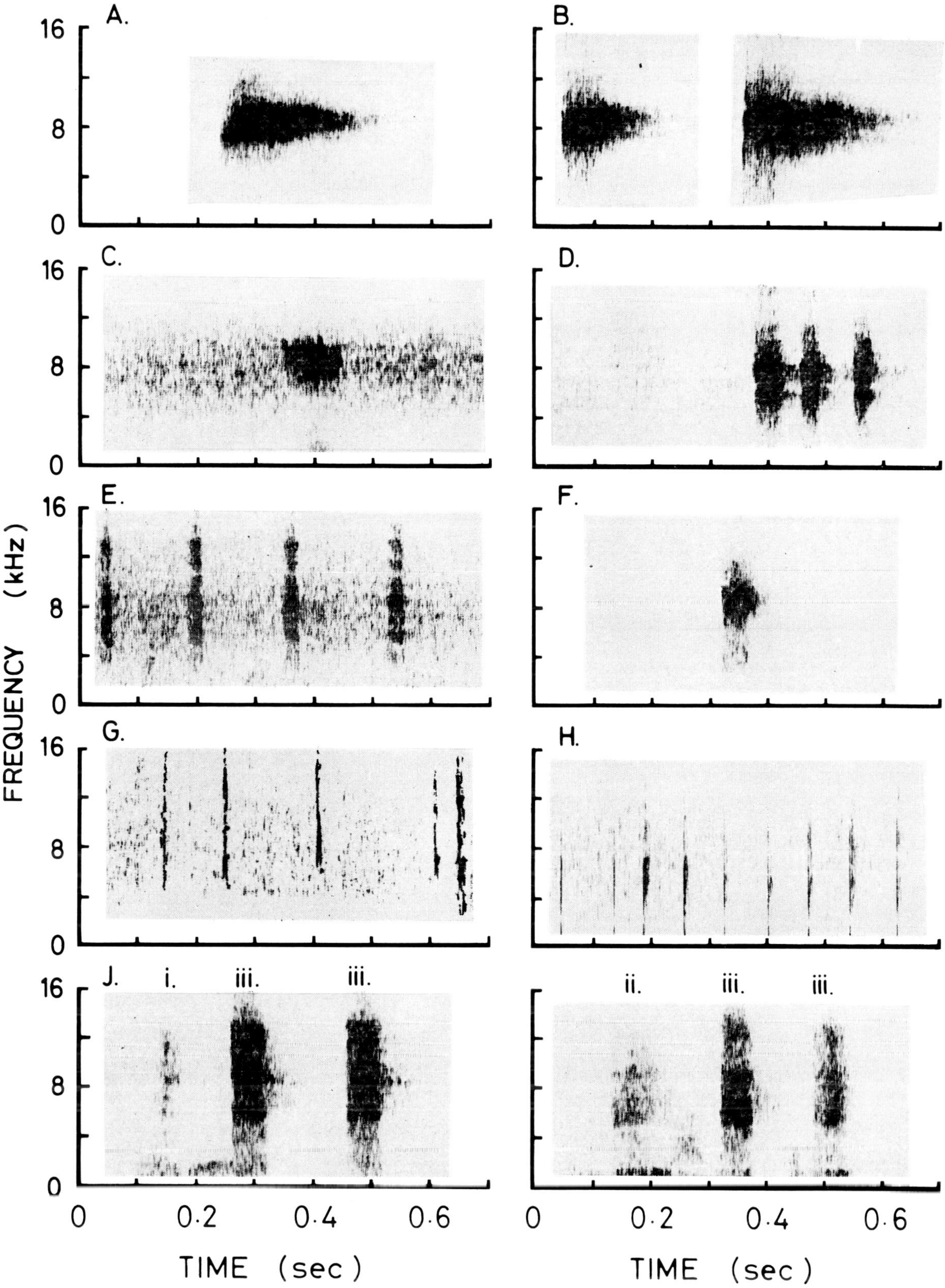

Fig. 3. Sonograms of the calls uttered by Ningauis.
A — the "male call". B — part of a "female call". C — the copulation call. D — the triple call. E — the sibilant call. F — the monosyllabic call. G — investigatory twittering. H — soft chattering. J — threat calls. i — least intense response. ii — medium intensity response. iii — strongest response.

taken using vernier calipers. In young up to about 40 days, crown-rump length and head width measurements were taken. From about 40 days records were kept of head–vent, tail–vent, hindfoot (heel to distal end of longest toe, excluding claw) and head lengths. Once the young became detached from the nipples, individual weights were recorded. Developmental changes in the young were also noted.

RESULTS

1. REPRODUCTION

The Breeding Season: The first indication that the Ningauis were entering a reproductively active period was the onset of mate–seeking calls by adult males and females. Males began calling in early September and continued until late January, with females beginning a few weeks later. Females subsequently called, mated and gave birth within this period. Including the period during which females were lactating, the breeding season extended from early September to late February. Captive adults were intolerant of one another throughout the rest of the year and had to be housed separately. The young of the wild–caught female (No. 4) were weaned in mid to late February. It was therefore estimated, using the data obtained throughout this study, that she had mated in mid–November and given birth in early December. These dates correlate well with the breeding season observed in the captive animals of this study.

Calling and Marking: Throughout the breeding season animals became very active around dawn and dusk, and typically called from high vantage points in their cages. When responding to calls made by the opposite sex, however, they would sometimes pause in their activity and call from the floor of the cage. Males generally called while standing with one forepaw slightly raised in the "indecision–alert" posture (Ewer 1968). Females began to call in this position but finished in a sitting position with both forepaws off the ground. Calls were made with the mouth wide open and the head was inflected upwards as each syllable was uttered. After calling, the animals paused as if listening, and then resumed their activity. Calling occurred predominantly around dawn and dusk, and neither males nor females were heard to call after dark.

Males made a distinctive mate–seeking call, designated the "male call", which usually consisted of only one syllable (Fig. 3A). Females initially called in resonse to a "male call". The vocalization of the female sounded similar to that of the male, but females repeated the syllable four or five times in rapid succession (the "female call", Fig. 3B). If females remained unmated over four to six days, the number of syllables increased to about twenty. The frequency and intensity of calling also increased, with females calling earlier in the day and without a male having called. A similar behaviour has been reported for *Sminthopsis murina* (Van Dyck 1979). Females called until between fourteen and seventeen days before parturition (Fig. 1). Both males and females called and mated in the breeding season which followed their birth. One male began calling at an age of approximately 275 days, and a female began at approximately 290 days.

In association with the onset of calling there was a considerable increase in marking behaviour by male Ningauis. Particularly around dusk, when males were active and calling, they often marked specific pieces of bark, which became noticeably stained. Marking consisted of an animal rubbing his chest and chin on the bark as he passed over it, and was generally preceded by a short period of intensive sniffing of the area. The male Ningaui possesses a sternal gland on his chest and the hair in this region becomes stained a ginger colour during the breeding season. Scent may be transferred to the animal's feet as an animal passes over the marked area, and thereby be spread over a larger area.

Males were also observed marking with the cloaca, which was briefly pushed down onto the substrate while the animals were active. The tail was held up in the air during cloacal marking and on several occasions males were seen to leave cloacal drag marks 10–12 cm long in the loose soil. When females in oestrus called, males which had been running about in their cages became more active and increased their rate of marking, or stopped and called in reply.

Mating: Females were placed in a male's cage within three days of their starting to call. Initially the female slowly moved around the male's cage with her ears erect and pointing forwards, uttering a soft chattering or a high–pitched investigatory twittering. Mouth and nose sniffing, and the sniffing of the female's cloaca by the male, as reported by Ewer (1968), Aslin (1974) and Van Dyck (1979) for other small dasyurids, did not occur, but males did smell the substrate where females had temporarily stopped. Within several minutes a short fast chase ensued with the male generally catching the female quickly. A scuffle followed and both animals tumbled over, during which the male managed to grasp the hair on the back of the female's neck in his teeth. The male then manoeuvered himself into position on the female's back, with his forepaws clasped tightly around the female's abdomen and in front of her hindlimbs. Eventually they attained an upright position with the male mounted from above and behind in typical dasyurid fashion. Sometimes, mating did not occur despite a great deal of chasing by the males. In these instances, males were rarely able to catch the females, but if they did, the females quickly escaped.

Once mounted, the male released his grip on the female's neck and pushed his snout into the hair on the back of her neck, just behind the ears. A neck grip was only used hereafter when the animals fell over or when the female was attempting to escape. Intromission took place very quickly, as observed in pairs that toppled over soon after mating began, and on one occasion intromission had occurred just twenty seconds after the normal mating position had been attained. During the first part of the mating sequence, the active phase, the animals moved about and pelvic thrusting was alternated with short periods of quiescence. The male wriggled his tail at times throughout the active phase and frequently bent over to alternate sides to lick at the genital regions.

Two other actions were frequently seen during the active phase. One of these consisted of the male rubbing his chin and throat laterally across the back of the females neck and head, as reported for *Sminthopsis crassicaudata* (Ewer 1968). This occurred soon after copulation had started and often followed a series of five or six thrusts. The other action observed involved the male raising a hind leg as if to scratch the female's flank. Scratching, however, was never observed. A scratching reflex has been reported for *A. stuartii* by Marlow (1961) and vibrating of, or scratching with, the hindfoot has been recorded from other dasyurids (Ewer 1968, Aslin 1974, Van Dyck 1979). Male *Planigale gilesi* alternate the hindlegs in scratching against the female's rump (Whitford *et al.* 1982). Throughout the active phase, the animals often toppled over onto their sides and males were noted thrusting even while in this position. It was not clear as to what had promoted the temporary loss of balance in these instances. Marlow (1961) records that after a single coital thrust by male *A. stuartii,* the copulating animals rolled over onto their sides, and Van Dyck (1979) records *P. maculata* toppling over after a series of powerful thrusts. Thrusting was not observed, however, to initiate the loss of balance in Ningauis.

The second part of the mating sequence I have termed the passive phase. It was marked by long periods in which the female lay quietly on her ventral surface with her neck and chin on the ground or with her head tucked underneath her body with the crown resting on the ground. The female often looked as if she was asleep and her eyes were closed or partly closed. The male remained in the mounted position, periodically bending over to lick the genital area. Thrusting occurred infrequently but more deeply than during the active phase, and the male's eyes remained closed for most of the time. In one closely observed mating, the passive phase lasted for approximately 3½ hours and the earlier active phase lasted about 1½–2 hours. At intervals during the passive phase, the female appeared to wake with a start, the pair leapt several centimetres into the air and fell over onto their sides. On these occasions the male renewed his neck grip and appeared to try to right the pair, thrashing his tail about.

Table 1. Matings and gestation periods for female Ningauis. The values in brackets under duration indicate periods in which females had escaped from males between two matings.

female		mating duration (minutes)	mating occurrence (days before birth)	gestation period (days)	birth date
No. 1		20	40		
		85	39		
		160+	26		
		150+	20		
		60 (5) 180	19		
		85	18		
		210	16	16	29.10.79
No. 2	A.	15 (5) 240	24		
		120+	18		
		120+	17		
		90+	16	15½	4.11.79
	B.	310	20	19½	8.12.79
No. 3		—	30		
		30 (10) 15	21		
		197	14	13½	13.11.79
No. 4		270	22		
		30 (75) 120	21	21¼	8.11.79

Throughout the passive phase male Ningauis uttered a low call (Fig. 3C), which was also heard when males were chasing females in oestrus. This was frequently repeated and may have a placatory effect on the female or may simply reflect the level of excitement of the male (Ewer 1968). Matings were concluded by the female breaking away from the male, and he subsequently spent a considerable period grooming himself. Matings generally lasted for considerable periods (Table 1), as is commonly the case for dasyurids (Croft 1982). On four occasions females broke away from males after only short matings, but were caught again and mating resumed. Matings occurred throughout the day and appeared to be independent of the time of day at which the animals were put together.

Three of the females produced only one litter during the breeding season of this study, however female No. 2 produced two litters (Table 1). Her first litter (No. 2A) consisted of two young, which disappeared from the pouch between days 10 and 12, when the female was heard uttering the mate-seeking call. She subsequently called for seven of the following eight days, and mated two days after first resuming her calling. This mating occurred 20 days before the birth of her second litter (No. 2B) and suggests that she had returned to oestrus very soon after losing her first litter. A rapid return to oestrus has been previously reported for *Sminthopsis macroura* (Godfrey 1969) and *Planigale maculata* (Aslin 1975). It would therefore appear that *Ningaui* sp. is polyoestrus, but there is no indication as to whether females could rear two litters in the one breeding season.

GESTATION: Gestation periods were calculated from the beginning of matings to the mid-point of the period during which the young were born. As the time for parturition approached (indicated by changes in females' pouches), they were checked often for the presence of young, so that birth could be assigned to a specific 7–12 hour period. On this basis, the gestation period obtained for the five litters born in captivity ranged from 13½ to 21¼ days (Table 1). In only two of these five instances, females had not mated at intervals prior to the last mating (Fig. 1). The second litter of female No. 2 resulted from a single mating and yielded a gestation period of 19½ days. Female No. 4 mated on consecutive days only, yielding a gestation period of 21¼ days. In apparent corroboration of these values, females in the other three cases (No. 1, No. 2A, No. 3) had mated 18 to 21 days before parturition (Fig. 1).

However, all females continued to call during the 4–7 days following these matings. In the three cases mentioned above (No. 1, No. 2A, No. 3) the females were subsequently placed with males, and readily mated (Fig. 1). These instances yielded gestation periods of 13½, 15½ and 16 days. In the two cases which yielded the longer gestation periods (No. 2B, No. 4) the females were not permitted to mate. Both females called until 14 days before parturition, and in the last 3–4 days their calling assumed the intensity and frequency of females that had not mated. The possible implications of this situation will be considered in the discussion.

CHANGES IN THE POUCH: The pouch of immature female Ningauis is present only as a small area on the ventral surface of the abdomen. It is sparsely covered with short white hairs and the nipples are minute. However, in the females that raised litters in captivity, the pouch area and the nipples remained slightly larger. The pouch hair returned to its normal colour, but the nipples of these females showed a varying degree of pigmentation. Additionally, a dark ring, situated outside the area of the nipples, developed during the later stages of pregnancy. This ring was of a uniform width, and appeared to lie in the upper layers of skin, perhaps being a deposit of melanin (Fig. 9). It was retained by females subsequent to their raising a litter and was present to varying degrees throughout the rest of the year. Consideration of the above characteristics, along with the weights of females, may enable differentiation in the field between females that have and those that have not reared a litter.

Following Woolley's (1974) classification of dasyurid pouches, *Ningaui* sp. has a type 1 pouch, and all females in this study possessed seven nipples. However, it was not possible to diagnose oestrus from the condition of the pouch. A dark, dry secretion was evident in the pouch from the time of mating until several days before birth. The pouch area increased markedly during the later stages of pregnancy and ridges formed across the pouch floor, causing the nipples to look countersunk. The skin in the anterior portion of the pouch (in front of the nipples) became greatly enlarged and red, and the whole pouch area assumed a granular appearance. Approximately one day before birth, droplets of a clear fluid accumulated around the nipples. As the young suckled, and particularly after they had become detached, the nipples became enlarged and the floor of the pouch became inflated. This swelling, and the overall size of the pouch, did not begin to decrease until about 90–95 days after parturition. Female No. 1 continued to lactate until at least day 98, when milk was observed oozing from the nipples. She had been separated from her young 5 days earlier.

WEIGHT CHANGES OF FEMALES: Throughout the breeding season, there were considerable changes in the body weights of the female Ningauis. A reference weight was established for each female, to which her weight changes could be compared, in an attempt to estimate the weights of the attached young. The reference weight for each female was determined from body weights taken: before pregnancy, immediately after parturition, as soon as the young had become detached from the nipples, and well after the breeding season. The weight of one female (No. 1) was recorded as frequently as possible, from three days before parturition until the young were 184 days old (Fig. 2). A similar determination was made for the other females.

During the six to eight days prior to parturition, females gained from 1.0 to 1.5 g in weight. This pre-birth weight gain has been observed in other dasyurids (Woolley 1966). At birth, this excess weight is lost, presumably due to the loss of fluids during parturition. For the next six to nine days, females tended to lose weight and remained below their reference weights. By the tenth day after parturition, however, all females had regained their normal weight and thereafter the weight of female plus her attached young increased. The return to normal weight of the female by day 10 is attributed largely to an increase in the weight of the female rather than to the increasing weights of the young. This assumption is supported by data for female No. 2, whose weight ten days after the birth of her first litter had returned to her reference weight, even though there were only two young in the litter.

As the young grew, the weights of females with their attached litters obviously increased. This increase is attributed to the growth of the young and not to any increase in the weight of the female. Female (No. 2) died when her second litter was 41 days old and was exactly the same weight as her reference value. The two females that successfully raised their litters, were within 0.2 g of their reference weights when they were first weighed after the young became detached. It would appear therefore that there is little nett change in the weight of females while the young are permanently attached. Following the detachment of the young the weights of females increased by about 0.8 g (Fig. 2). This is apparently associated with an increase in mammary tissue, indicated by a swelling of the pouch area. As the young began taking solid food (72–75 days) and the intensity of suckling decreased, the weights of the females gradually declined, returning close to their reference weights by 85–89 days.

SAND BATHING: During the breeding season, female Ningauis were observed to engage in two types of sand bathing. The first consisted of flattening the ventral surface onto the loose red soil and scurrying along using the hindlimbs. This activity lasted for several minutes and was interspersed with brief pauses for vigorous scratching under the chin. It was seen immediately following mating. *Dasyuroides byrnei* sandbathe in a similar fashion and Aslin (1974) interprets this as being functional both in maintenance of the fur and in marking.

The second sand bathing activity involved the animal twisting the rear part of the body over onto its side and dragging itself along using only its forepaws. This was often repeated, and the animal then jumped up, shook itself, and resumed running about in the cage. Ewer (1968) describes a somewhat different sand rolling in *Sminthopsis crassicaudata,* which she reports as a common post–copulatory behaviour, and suggests that it functions primarily in maintenance of the fur. The precise functions of sand bathing behaviour in female Ningauis remain unclear.

NEST BUILDING: Nest boxes, which were provided after the birth of the young, were only occasionally used at first as the females preferred to shelter under favourite pieces of bark. Nests were constructed by females only after the young had reached 40 days old. Although plenty of nesting material was available, the nests were rudimentary constructions, consisting of a few pieces of dried grass and leaves lining a shallow depression in the soil, or a shallow bed in the shredded tissues provided. These nests were simpler and contained less material than those constructed by some other dasyurids, including *Antechinus stuartii adustus, A. godmani, A. flavipes* and *Sminthopsis murina* (D. Whitford, pers. obs.). All females engaged in some digging activities when their young were between 30 and 40 days old. The digging behaviour and the small quantity of material used in nests suggest that in the wild, animals may use a burrow in which to leave their young once they cease to be attached to the nipples. Captive Ningauis in the Northern Territory spent considerable time digging the soil of their cages, and Johnson and Roff (1980) suggested that wild individuals may dig burrows to avoid high diurnal temperatures. *Dasyuroides byrnei* also constructs simple nests and is thought to dig and use burrows for shelter (Aslin 1974).

2. CALLING

In the course of this study, the Ningauis displayed an extensive repertoire, with vocalizations being evident during all types of social interaction. Some calls were therefore limited to specific periods in the year, e.g. the mate–seeking calls, whilst others were uttered throughout the year, e.g. threat calls. Eisenberg *et al.* (1975)

have described four syllable types commonly produced by marsupials, these being noisy, tonal, mixed (tonal in structure but with noise superimposed) and click. The vocal repertoire of Ningauis appeared to contain representatives of each of these syllable types, but a detailed investigation of the sonic characteristics of the calls was neither intended nor attempted.

"MALE CALL": uttered by males as a mate-seeking call throughout the breeding season. It is a distinct sound, like "tsitt" and is confined to a specific frequency range (Fig. 3A). The call generally consists of a single syllable.

"FEMALE CALL": a mate-seeking call uttered by females during the breeding season when approaching, or during, oestrus. Each syllable is similar to the "male call", although they may occasionally be somewhat attenuated (Fig. 3B). Females repeated the syllable, usually four or five times, but if they remained unmated, the number of syllables increased to about twenty, and the frequency and intensity of calling also increased.

COPULATION CALL: a low repeated "tssz" uttered by males during copulation (Fig. 3C) and perhaps having a placatory effect on the female. While males were chasing females, they often uttered a call which sounded almost the same. A similar call has been recorded from *Sminthopsis crassicaudata* and *Antechinus flavipes* (Ewer 1968) and also from female *S. murina* during copulation (D. Whitford, pers. obs.).

TRIPLE CALL: uttered by females at various times throughout the breeding season. A trisyllabic call (Fig. 3D) repeated two or three times in quick succession, or sometimes heard at the beginning or end of a "female call". It's significance remains uncertain, although it did not appear to be specifically associated with mate-seeking.

SIBILANT CALL: uttered by young Ningauis from the time they first became detached from the nipples, whenever they were left in the nest or were disturbed. It was a rapid stanza of syllables (Fig. 3E), sounding like "ss-ss-ss-", and became louder as the young grew older. This call generally elicited retrieval or other maternal behaviour from the mother. A similar call has been reported for many dasyurids (Croft 1982). Settle and Croft (1982) suggest that this type of vocalization is probably important in ensuring the continuance of maternal behaviour.

MATERNAL CALL: a low "tssz–tssz" used by females to reply to the sibilant calls of their young. The call was similar to that uttered by males during copulation and may have a placatory or reassuring effect on the young.

INVESTIGATORY TWITTERING: a high-pitched twittering (Fig. 3G), covering a broad frequency range, uttered by all animals when investigating a new environment or a new food. This sound was very low in volume and often difficult to hear as the animals were generally moving about while making it. A similar sound has been reported for *Planigale maculata* (Van Dyck 1979) and heard from some *Antechinus* spp. (D. Whitford, pers. obs.).

SOFT CHATTERING: heard from young when disturbed and from females when placed in a male's cage. It was a soft call, made with the mouth almost or fully closed, and sounded like a distant typewriter (Fig. 3H). This call appears to be associated with stress (D. Croft, pers. comm.).

MONOSYLLABIC CALL: uttered by one female when her young were 22, 70 and 85 days old, and also well after the breeding season, during May and June. The call was of short duration and restricted to a small frequency range (Fig. 3F), and its function remains obscure.

THREAT CALLS: made when animals were handled or disturbed, when adults were placed together outside the breeding season and when females did not want the young to suckle. The threat calls consisted of three sounds which were made under different levels of stimulation and were generally preceded by an open–mouth threat. The least intense response was a soft exhalation like a very gentle "shh" (Fig. 3Ji). This was usually uttered only once and the animal then attempted to escape, although it was sometimes heard at the beginning of a threat call sequence. Continued or more intense stimulation evoked a sharper exhalation, "rhit", which was often repeated or interspersed with the third threat call (Fig. 3Jii). This last sound was much harsher and lower, sounding like "ch–ch–" (Fig. 3Jiii) and was usually repeated a number of times. It was uttered under the most intense provocation and was often heard in long stanzas. The young are regarded as being capable of independence at the age they were heard making threat calls.

3. GROWTH AND DEVELOPMENT

Six litters were involved in this study, five of which were bred in captivity and the original litter which was received with the wild caught female (No. 4). This female subsequently produced a litter of six young, and the first litter of female No. 2 contained only two young. The remaining litters contained seven young, corresponding to the number of nipples possessed by each female, and there was no evidence of supernumerary young. Of the 31 young which were sexed, 15 were females and 16 were males, and all litters contained young of both sexes.

Only two of the five litters born during this study survived to independence. Female No. 1 initially had seven young but one was found dead in the nest box and partly eaten at an age of 67 days. Female No. 3 gave birth to seven young, but she was disturbed during parturition, and one neonate was found on a piece of bark. Her remaining six young were successfully raised.

WEIGHTS OF YOUNG: Until the young were left in the nest, an estimate of their mean weight (Fig. 4) was obtained from the differences between the weight of mother and young, and her reference weight. Although this method of obtaining mean weights for the pouch young may contain errors, it does provide estimates that are in agreement with weights taken when the young were first weighed, and with values obtained when one female with 41-day-old young died. Subsequently, individual young or litters that had been left in the nest, were weighed and measured. The weights of the young in two litters were monitored long after they were weaned and the resultant curve is illustrated in Fig. 5. This curve includes the estimated values of the young from Fig. 4.

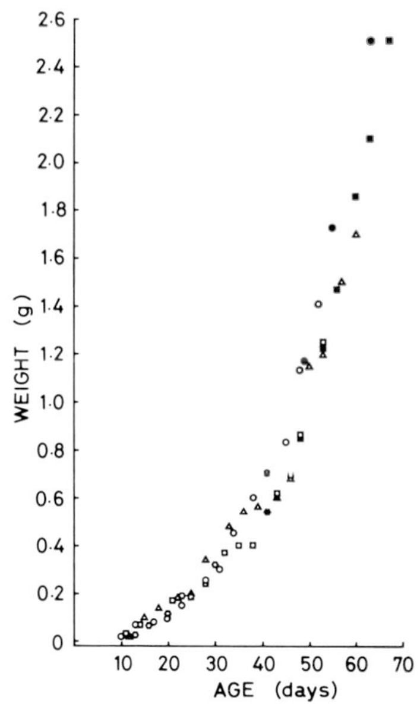

Fig. 4. The growth in weight of young Ningauis from 10 to 70 days. Open symbols indicate weights estimated from the weight of females, closed symbols indicate weights obtained from free young.

Figure 5 indicates an increasing rate of growth of the young Ningauis over their first fifty to sixty days. The rapid rate of growth, particularly after 30 days, correlated with the increased rate of growth in crown–rump lengths after this age (Fig. 6). A similar pattern of growth rates can be inferred from the photographs of young *Antechinus stuartii* (= *flavipes*, Marlow 1961). After 60 days, the rate of growth declined. By 180 days, the young were still approximately 2 g. lighter than their mothers. The mean weights for both litters fell immediately after the young were housed individually. This loss in weight may have partly resulted from the removal of the energetic benefits derived from huddling.

MEASUREMENTS OF YOUNG: The initial measurements taken from young Ningauis were crown-rump length and head width (Fig. 6). The small amount of data available for head width indicates a relatively linear increase from day 9 to day 63, although there may have been a slightly lower rate of increase prior to day

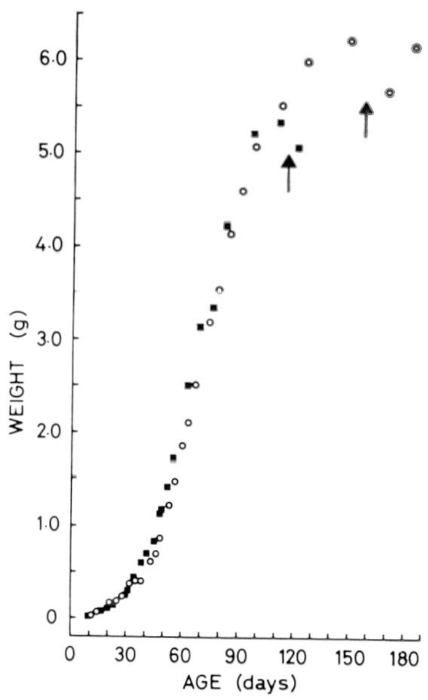

Fig. 5. The weights of young from two litters of Ningauis, including both estimated and measured weights. The arrows indicate when the young of each litter were housed individually.

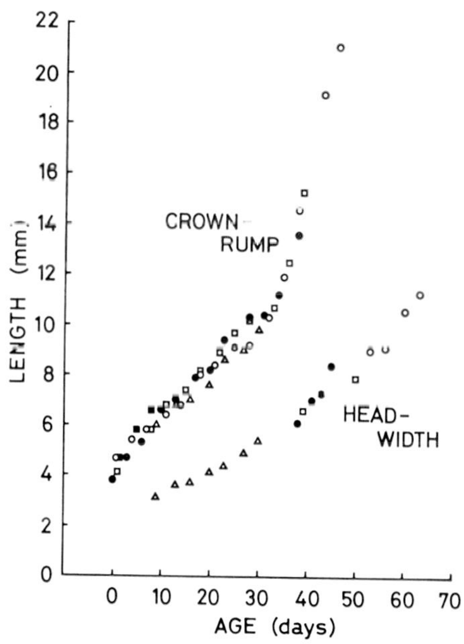

Fig. 6. Crown–rump length and head width of attached Ningauis. The mean values for each litter are represented by different symbols.

30. With more investigation, head width may be found to be a useful measurement for aging wild caught young, particularly when they are small. The crown-rump length of Ningauis increased by about 1 mm every 5 days until about 30 days of age, after which the rate of increase became approximately 1 mm every 1½ days. This increase in the growth rate coincides with the increase in the calculated weights of young Ningauis obtained from the female's weight (Fig. 4).

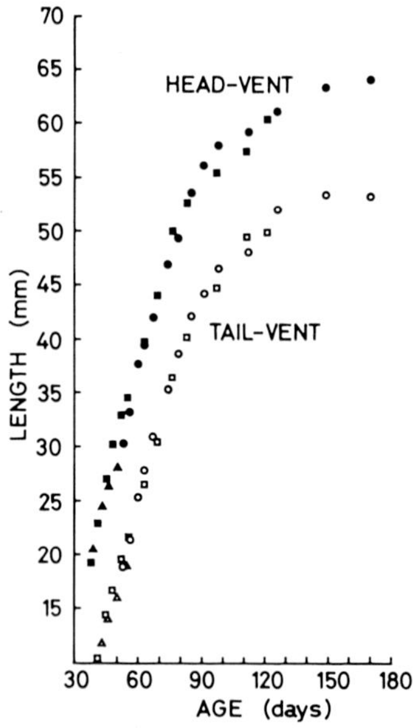

Fig. 7. Head–vent length (closed symbols) and tail–vent length (open symbols). Mean values for each litter are represented by different symbols.

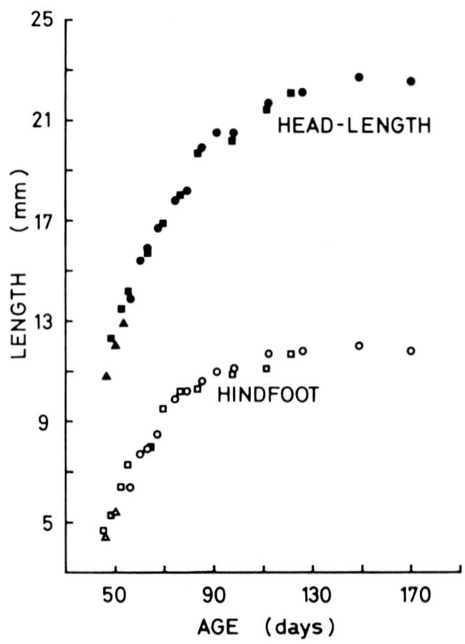

Fig. 8. Hindfoot length (open symbols) and head–length (closed symbols) from individual litters. Mean values for each litter are represented by different symbols.

From 37 days of age, head–vent measurements were taken from the young, although they were still attached to the nipples (Fig. 7). The rate of increase from day 38 to day 90 was essentially linear, at approximately 1 mm every 2 days, enabling accurate determinations of age. From day 90 onwards, the rate of increase in this measurement gradually fell. The values within each litter were very consistent, with standard deviations being generally considerably less than 5% of the mean value for each litter. Mean values for tail–vent measurements for the various litters also displayed a great deal of similarity (Fig. 7), although the variation within each litter was considerably greater than for head–vent measurements (standard deviations being up to 10% of means).

Hindfoot measurements displayed a rapid rate of increase to about 80 days with a gradual decrease in the growth rate thereafter (Fig. 8). The variation within each litter was remarkably small, being generally less than 3% of the mean value. The most consistent measurement obtained from the young was head length (Fig. 8). Values for the two litters from which extensive data were obtainable indicate a very high degree of correlation, and standard deviations for each of the litters was less than 3% of the mean values. There appeared to be a general decline in the rate of growth of the head length from about day 48, but reliable estimates of young would be obtainable up to day 90 at least.

Development of the Young: Newborn Ningauis have a crown–rump length of 3.5–4.5 mm and weigh approximately 5 mg. Although considerably smaller, the Ningaui neonate is essentially the same as described for *Antechinus stuartii* (= *flavipes,* Marlow 1961). The sequence of events shown below, with Figs 10-14, detail the growth and development of the young.

Day 0. birth; pink skin; eye is a faint dark ring.

Day 10. ear buds first appear.

Day 13. creases appear on hindlimbs, delineating digits.

Day 16. claws on forefeet shed; head and dorsum light grey; patches of papillae for whiskers on snout.

Day 20. tiny claws appear on forefeet.

Day 23. eye is a large dark spot; crease for eyelids in skin; head and dorsum dark grey.

Day 27. ears are flat triangular pads, folded forward tightly on sides of head; digits on hindfeet well developed and separating; claws on hindfeet; foot pads appear; fine hairs appear on body; whiskers first appear from papillae; body dark except for ears and lower forelimbs.

Day 33. ears thinner, directed backwards; lips formed but fused; body well-haired, a few tiny hairs on limbs and tail.

Day 38. ears fully formed, dark; tail and hindlimbs dark; vibrissae well developed.

Day 41. lips no longer fused; sparse white hairs on belly and ventral tail; other hair is dark grey.

Day 43. eyelids separable; tiny hairs on ears; brown hairs appearing on head and shoulders.

Fig. 9. The pouch of a Ningaui, 97 days after birth, displaying enlarged teats and the uniform black ring surrounding the pouch (photo: D. Whitford).

Fig. 10. The young of Ningauis at 3 days (photo by D. Whitford).

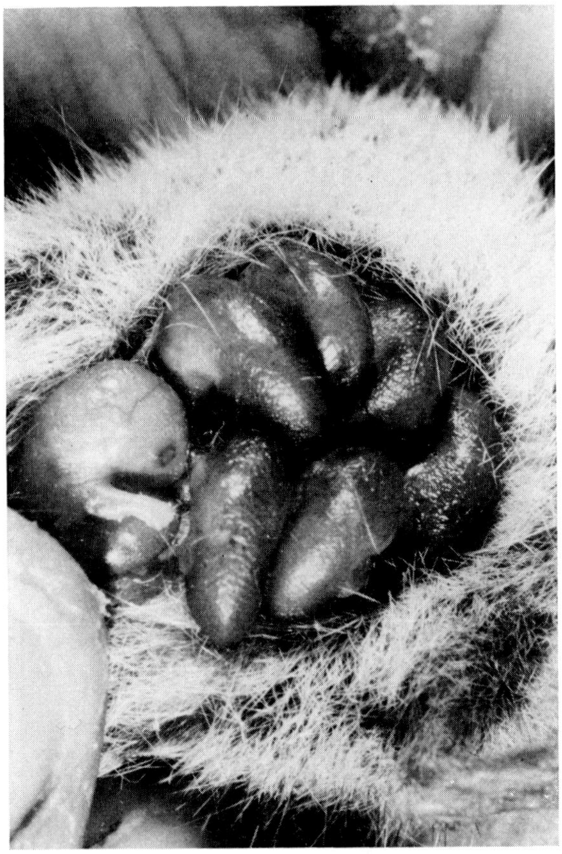

Fig. 11. The young of Ningauis at 20 days (photo by D. Whitford).

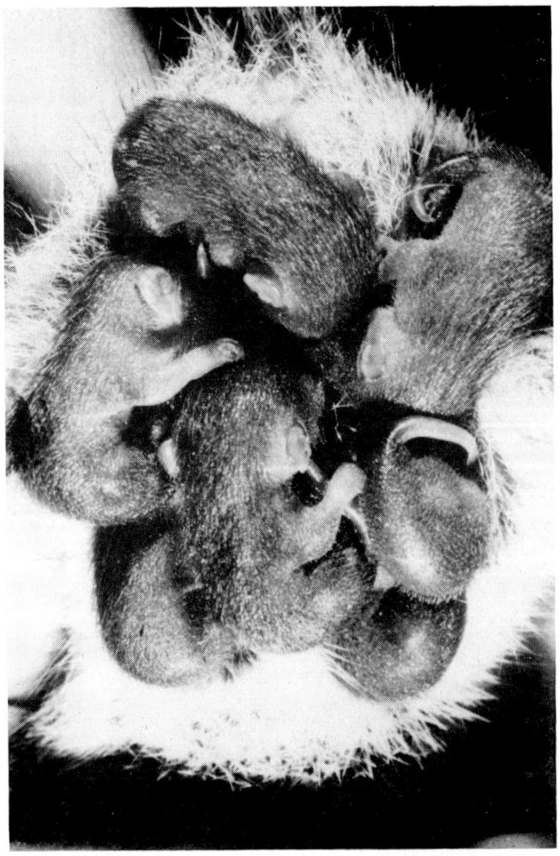

Fig. 12. The young of Ningauis at 38 days (photo by D. Whitford).

Day 48. eyes partly open for first time; teeth erupting.

Day 54. all young with open eyes; auditory orifice open.

Day 63. head, ears, shoulders, pes, lower forelimbs, ventral tail surface are light brown, flecked with grey; hindquarters and dorsal tail are dark grey with flecks of brown; belly is white to very light brown.

Day 70. hindquarters and dorsal tail surface are brown/grey; young attain adult colouration.

At 42-44 days young were first seen detached and the two litters that survived were left in the nests at 48 and 49 days, although they continued to suckle for a further 43 days. The young initially uttered sibilant calls when disturbed or distressed, to which the female often replied. Females also retrieved calling young by pulling them underneath their bodies so that the young could attach to a nipple. Until 60 days, young were being dragged about the cage attached to the nipples, but were subsequently seen clinging to the females' backs and flanks. From 54 days onwards, females often emerged with one or two young, either on their nipples or clinging to their backs, and dislodged them by rolling over. The young were left to explore the the cage and find their own way back to the nest. By 70 days, the young were noticed eating small pieces of mealworm discarded by the females. They were independent at 76-81 days, when they could eat fully grown mealworms and had been heard uttering threat calls. By 86 days the young displayed all the cleaning and maintenance behaviour of the adults, including sandbathing. Investigatory twittering was heard from this age, and soft chattering was heard at 97 days. After removal of the mothers at 93 and 97 days, the young rested by day in various groupings, often being found loosely huddled together. They did not, however, adopt a typical huddling pattern as seen in *Planigale tenuirostris* and *P. gilesi* (pers. obs.) and as reported for *P. maculata* (Van Dyck 1979). They were housed separately from 119 to 157 days, when individuals were first noticed with chewed or damaged tails.

DISCUSSION

The individuals of *Ningaui* sp. used in this study were generally intolerant of conspecifics. The only exceptions to this observation were mating pairs, females with offspring and littermates until 120–150 days old. An effective means of communication, which does not require physical contact, would therefore be beneficial for the transmission of information between individuals. Additionally, this species of *Ningaui* inhabits arid to semi–arid environments which tend to consist of scattered vegetation rather than of a continuous cover. Thus, a mode of non–tactile communication, by minimizing the animals' exposure to possible predation, may be a distinct advantage. Throughout the investigation, the animals produced a large number of calls,

Fig. 13. The young of Ningauis at 52 days (photo by D. Whitford).

Fig. 14. The young of Ningauis at 63 days (photo by D. Whitford).

some of which would enable them to transmit information over considerable distances. Animals could thereby advertize their reproductive state, for instance, without the need for physical contact.

Ningauis are very small and are insectivorous. They are therefore likely to be subject to large fluctuations in population, related to environmental conditions and to the availability of food. The possession of an efficient means of long-distance communication would enhance population survival during and after a period of adverse conditions, when the population density would be low. A large vocal repertoire may therefore be of considerable adaptive significance for *Ningaui* sp. In addition, the instances of marking by male and female Ningauis may assist in communication between these socially intolerant animals.

Females in captivity were observed to call for considerable periods, and to mate frequently. In three instances, in which the gestation periods were from 13½ to 16 days, females ceased all calling after the last mating. On two other occasions however, which yielded the longer gestation periods of 19½ to 21¼ days, the females continued to call vigorously after their last mating until 14 days before parturition. In these two instances, their calling assumed the increasing intensity displayed by females that had not mated. The question arises as to why these two females continued to call so vigorously if fertilization had occurred near to mating, or why the remaining three females ceased calling after last mating. It is possible that calling by females and their receptivity to males may simply extend beyond the time of ovulation and fertilization. Thus the three females which provided the shorter gestation periods and which did not call after their last matings, may have been mating during the early stages of embryogenesis.

Perhaps, however, the two females which continued to call provide instances in which either embryogenesis or fertilization had been delayed. Females were observed in torpor on some mornings throughout their pregnancies, and the low body temperatures associated with this activity may impede the rate or timing of embryogenesis. Alternatively, ovulation and therefore fertilization may not have occurred until some time after mating in these two females. Hill and O'Donoghue (1913) reported that in *Dasyurus viverrinus* copulation almost always preceded ovulation by about 5-6 days and that spermatozoa could remain alive in the gland lumina, opening into the fallopian tubes, for at least fourteen days. This would appear to involve delayed fertilization in *D. viverrinus,* and a similar mechanism may occur in *Ningaui* sp. It must be stressed at this point that these alternative possibilities are largely conjecture and that the data obtained during this investigation are insufficient to determine which strategem has been adopted by *Ningaui* sp.

Much of the reproductive behaviour and the growth and development of young Ningauis is similar to that displayed by other small dasyurids (e.g. Marlow 1961, Woolley 1966, Aslin 1974, Van Dyck 1979) and has been covered in detail in earlier sections of this paper.

Ningaui sp. may be polyoestrus, as is another arid zone dasyurid, *Sminthopsis crassicaudata* (Morton 1978). Although the young took slightly longer to reach independence than those of *Planigale gilesi* (Whitford *et al.* 1982) and *Sminthopsis murina* (Fox and Whitford 1982) it is possible that Ningauis may be capable of rearing more than one litter in a breeding season. Additionally, if a female is unable to find and mate with a male during her first oestrus period, or loses her first litter, she may still have the opportunity of producing young. This again is a strategy which appears to be well-adapted for a socially intolerant, widely dispersed semi-arid species.

ACKNOWLEDGEMENTS

Dr. Ken Johnson of the Territory Parks and Wildlife Service kindly supplied the original wild-caught animals. Mr. Dick Whitford contributed greatly to this paper in carrying out many of the observations and measurements. The author would like to thank Dr. Barry Fox for assistance at various times with measurements, and Drs. David Croft, Michael Archer and Barry Fox for their discussions and helpful suggestions. Professor B. K. Martin and Dr. Roy Caddy are thanked for their assistance with the sonograms; Ms. Jenny Taylor for the photographs of the sonograms. Ms. Jessica Johansen and Ms. Deanne Thomas very patiently dealt with the typing.

REFERENCES

Archer, M., 1975. *Ningaui*, a new genus of tiny dasyurids (Marsupialia) and two new species, *N. timealeyi* and *N. ridei*, from arid Western Australia. *Mem. Qd Mus.* **17:** 237–49.

Archer, M., 1979. The status of Australian dasyurids, thylacinids and myrmecobiids. Pp. 29–43 *in* "The status of endangered Australasian wildlife" ed. by M. J. Tyler. Proc. Roy. Zool. Soc. S.A.: Adelaide.

Aslin, H., 1974. The behaviour of *Dasyuroides byrnei* (Marsupialia) in captivity. *Z. Tierpsychol.* **35:** 187–208.

Aslin, H., 1975. Reproduction in *Antechinus maculatus* Gould (Dasyuridae). *Aust. Wildl. Res.,* **2:** 77–80.

Baverstock, P., and Aslin, H., 1975. A new marsupial. *Hemisphere* **19:** 20–3.

Croft, D. B., 1982. Communication in the Dasyuridae: A review. This volume.

EISENBERG, J. F., COLLINS, L. R., AND WERNMER, C., 1975. Communication in the Tasmanian Devil *(Sarcophilus harisii)* and a survey of auditory communication in the Marsupialia. *Z. Tierpsychol.* **37:**

EWER, R. F., 1968. A preliminary survey of the behaviour of the dasyurid marsupial, *Sminthopsis crassicaudata* (Gould). *Z. Tierpsychol.* **25:** 319–65.

FLEMING, M. R., AND COCKBURN, A., 1979. Ningaui: a new genus of dasyurid for Victoria. *Vict. Nat.* **96:** 142–5.

FOX, B. J., AND WHITFORD, D. 1982. Polyoestry in a predictable coastal environment: reproduction, growth and development in *Sminthopsis murina* (Dasyuridae, Marsupialia). This volume.

GODFREY, G. K., 1969. Reproduction in a laboratory colony of the marsupial mouse *Sminthopsis larapinta* (Marsupialia: Dasyuridae). *Aust. J. Zool.* **17:** 637–54.

HAPPOLD, M., 1972. Maternal and juvenile behaviour in the marsupial jerboa *Antechinomys spenceri* (Dasyuridae). *Aust. Mammal.* **1:** 27–37.

HILL, J. P., AND O'DONOGHUE, 1913. The reproductive cycle in the marsupial *Dasyurus viverrinus*. *Quart. J. Microscop. Sci.* **59:** 133–74.

JOHNSON, K. A., AND ROFF, A. D., 1980. Discovery of Ningauis (*Ningaui* sp: Dasyuridae: Marsupialia) in the Northern Territory, Australia. *Aust. Mammal.* **3:** 127–9.

MARLOW, B. J., 1961. Reproductive behaviour of the marsupial mouse, *Antechinus flavipes* (Waterhouse) (Marsupialia) and the development of the pouch young. *Aust. J. Zool.* **9:** 203–18.

MORTON, S., 1978. An ecological study of *Sminthopsis crassicaudata* (Marsupialia: Dasyuridae). III. Reproduction and life history. *Aust. Wildl. Res.* **5:** 183–211.

SETTLE, G. A., AND CROFT, D. B., 1982. Maternal behaviour of *Antechinus stuartii* (Dasyuridae, Marsupialia) in captivity. This volume.

SORENSON, M. W., 1970. Observations on the behaviour of *Dasycercus cristicauda* and *Dasyuroides byrnei* in captivity. *J. Mammal.* **51:** 123–30.

VAN DYCK, S., 1979. Behaviour in captive individuals of the dasyurid marsupial *Planigale maculata* (Gould 1851). *Mem. Qd Mus.* **19:** 413–39.

WHITFORD, D., FANNING, F. D., AND WHITE, A., 1982. Some information on reproduction, growth and development in *Planigale gilesi* (Dasyuridae: Marsupialia). This volume.

WOOLLEY, P., 1966. Reproduction in *Antechinus* spp. and other dasyurid marsupials. *Symp. Zool. Soc. Lond.* **15:** 281–94.

WOOLLEY, P., 1974. The pouch of *Planigale subtilissima* and other dasyurid marsupials. *J. Roy. Soc. W. Aust.* **57:** 11–5.

RESUMEN

4. Reproducción, crecimiento y desparpollo del *Ningaui* sp. (Marsupialia, Dasyuridae) del Northern Territory. Por F. Dominic Fanning.

La estación de apareamiento de *Ningaui* sp. fue desde comienzos de Septiembre hasta fines de Febrero. Las hembras demostraron ser poliestras. Las hembras en celo respondieron a los llamados de apareamiento proferidos por los machos y, cuando se les permitió, se aparearon en numerosas ocasiones durante extensos períodos. El apareamiento fue típicamente dasiúrido en su forma, con una fase activa y otra pasiva, y duró desde 20 hasta 310 minutos. Los períodos de gestación, medidos desde el último apareamiento, variaron desde 13½ a 21¼ días (± ¼ día). Se hicieron observaciones sobre el cambio de los marsupios de las hembras, sobre sus pesos y el comportamiento de las hembras y los hijos. Se tomaron medidas corporales desde los estados juveniles, y los pesos de los jóvenes fijados a la madre fueron deducidos del peso de ésta. El neonato Ningaui es típicamente dasiúrido en su apariencia, y se desarrolla gradualmente, con un aparente incremento del ritmo de crecimiento en aproximadamente los 30 días. Los primeros jóvenes destetados fueron vistos entre los 42 y los 44 días, y fueron dejados en el nido a los 48–49 días, en cuyo tiempo los ojos comenzaron a abrirse. También los dientes comenzaron a salir en aproximadamente 48 días y el orificio auditivo se abrió por el día 54. Los jóvenes fueron independientes entre los 76 y los 81 días, siendo entonces capaces de comer los gusanos de las comidas ya bien crecidos y de proferir amenazantes llamados.

Los Ningauis poseen un extenso repertorio vocal, con vocalizaciones evidentes durante cierta interacción social de hostigamiento. El variado repertorio vocal y la estrategia reproductiva, que incluye la poliestría y el repetido apareamiento durante el celo, pueden tener un considerable valor para un dasiúrido pequeno, socialmente itolerante y de zona árida.

5. POLYOESTRY IN A PREDICTABLE COASTAL ENVIRONMENT: REPRODUCTION, GROWTH AND DEVELOPMENT IN *Sminthopsis murina* (DASYURIDAE, MARSUPIALIA)

BARRY J. FOX[1] AND DICK WHITFORD[1]

This paper presents information on the reproduction of *Sminthopsis murina* (Waterhouse 1838), based on five females and on the growth and development of a litter of 10 young conceived and raised in the laboratory. The species is polyoestrous with a cycle length of 24 ± 1 (s.e.m.) days. Mean gestation length was taken as 12.5 ± 0.3 days. A maximum of 10 young can be raised and the young remain attached to the teats for 34 days and are weaned by 65 days. Field and laboratory data suggest that mating may occur from late August to late January with lactation extending to the end of March, giving a seven month breeding season.

Sminthopsis murina occurs early in the succession following fire and reaches its greatest abundance in early and mid seral stages. It is suggested that *S. murina,* as an opportunist species, evolved polyoestry as a means of increasing fecundity in response to the need for rapid population increase. This species has a high plasticity which has enabled adaptations to shorten gestation length, accelerate development of the young and reduce parental care, facilitating the production and survival of a second litter in the breeding season. This contrasts sharply with the co-existing *Antechinus stuartii* which reaches maximum abundance in a late seral stage of succession following fire, and has persisted with monoestry.

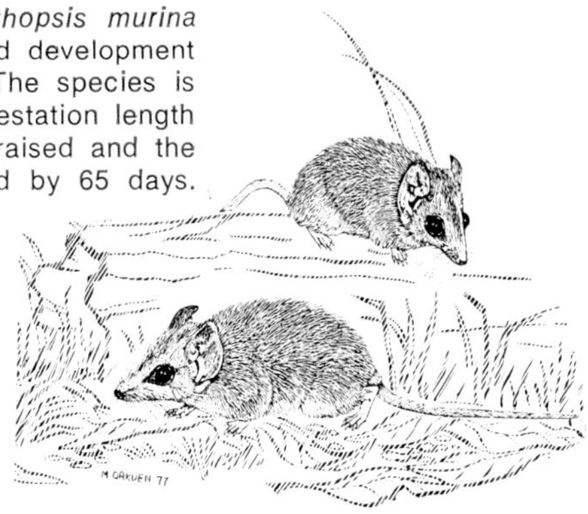

Keywords: Breeding season; Dasyuridae; development; growth; monoestry; polyoestrous; reproduction; reproductive strategy; *Sminthopsis murina.*

Pages 39-48 *in* CARNIVOROUS MARSUPIALS, ed. by M. Archer, February, 1982. Roy. Zool. Soc. New South Wales: Sydney, Australia.

IN THIS PAPER we present information on the growth and development of a litter of 10 *Sminthopsis murina* (Waterhouse), the Common Dunnart, together with aspects of this species' reproduction, based on five females. We then use this information to interpret field observations and establish the breeding season for this species in the Myall Lakes area of N.S.W.

Sminthopsis murina is a species that most commonly inhabits the open forests, woodlands and heathlands of southeastern and southwestern Australia: areas with seasonal, predictable rainfall usually exceeding 500 mm annually. The species has a wide distribution and may also be found extending into more arid areas but seldom in areas with less than 250 mm annual rainfall (Archer in press, Morton 1978c; Morton, Wainer and Thwaites 1980). This allows comparisons to be made between *S. murina* and its more arid–adapted, polyoestrous congeners, such as *S. crassicaudata* and *S. macroura;* as well as with monoestrous *Antechinus stuartii,* a species of eastern Australian forests, with which it coexists in N.S.W. (Fox 1982a).

Morton (1978a) has reviewed what is known of the biology of *S. crassicaudata.* Its reproduction has been well studied both in the laboratory (Godfrey 1969a, Godfrey and Crowcroft 1971 and Ewer 1968) and in the field (Morton 1978b and c). *Sminthopsis crassicaudata* breeds readily in the laboratory and this has facilitated studies such as Ewer's (1968) survey of its behaviour. The only information we can find on growth in *Sminthopsis* is that presented by Morton (1978c) for *S. crassicaudata.* The only other species for which data is available in *S. macroura,* its laboratory reproduction (as *S. larapinta*) is reported by Godfrey (1969b). The biology of other members of the genus *Sminthopsis* Thomas is in general poorly known and *S. murina* is no exception. We know of no published work on the growth, development, reproduction or life history of this species.

Several unsuccessful attempts to breed *Sminthopsis murina* were made between 1977 and 1979 (B.J.F.). The first success came with a male and female obtained from Myall Lakes area (32°24′S, 152°29′E) in August, 1979. The animals were housed separately, the male in a glass aquarium (85 x 40 x 55 cm) and the female in a glass fronted wooden cage (90 x 60 x 40 cm) with a wire gauze lid. A small wooden nest box (25 x 15 x 10 cm) was placed in the female's cage. Abundant leaf litter was provided as floor covering and also for nesting material. Pieces of bark were introduced as shelter sites.

---

[1] School of Zoology, University of New South Wales, P.O. Box 1, Kensington, New South Wales, Australia, 2033.

Water was provided *ad libitum* and the animals were fed on a diet of minced meat (Pets Pack) and dry ground dog food (Lucky Dog Dinner) to which a vitamin and mineral supplement (Petvite) and D.C.P. powder (Bassett's) were added. The animals were maintained in a room with natural lighting. The female was not placed with the male until she showed signs of oestrus.

To avoid distressing the female and so minimise the risk to the young, the female and litter were examined at about weekly intervals. Measurements could only be taken for crown–rump length, tail length (tip to cloaca) and hind foot length (without claw). When the young were left in the nest, head length and weight were also measured, while head and body length (nose tip to cloaca) replaced crown–rump length. All measurements on the young were initially taken to the nearest 0.1 mm. After day 60, tail and head and body lengths were measured to the nearest 0.5 mm only. Weight was measured to the nearest 0.1 g. Measurements were taken at the same time of day. Growth phases were determined as described in Fox (1979) using the method of Brody (1945).

Observation of behaviour during the dark hours was made possible by illuminating the cages with a dim red light that did not appear to disturb the animals.

During 1980 four females were maintained following the same procedures. From the first week in November the females were checked for the presence of cornified epithelial cells in the urine, using the method described by Godfrey (1969). She successfully used this method to determine the oestrous cycle in *S. macroura* (as *S. larapinta*). Urine samples were collected between 0900 and 1000 daily and examined under X100 magnification, the highly refractive epithelial cells were clearly visible without staining.

OBSERVATIONS AND RESULTS

Calling and Mating: Many calling and mating sequences were observed, they all contained the same elements, one such sequence will be described in detail. The female first called after dark on 31 August, 1979, nine days after capture. The call was a repetitive "chit–chit–chit–chit–chit" comprising five apparently identical syllables; the whole call was then repeated at intervals. The call was recorded and detailed descriptions will be given elsewhere. We have interpreted the call as mate attracting; the females readily mated the night they began calling and calling ceased after mating, in addition cornified epithelial cells were found in the urine of calling females, but not at other times. The male call was usually a single syllable "chit", but occasionally the call comprised two syllables. These syllables were indistinguishable from those made by the female. The male call was also mate seeking and males called each night of the breeding season, but not at other times.

The female was placed in the male's cage from 2000 to 2300 hr on 3 September, and again for several hours on 4 September. Although no mating was observed on the second occasion, on the first occasion mating lasted for 2 hrs (2150–2350 hr) during which time the female often emitted a low "tssz–tssz–tssz" with pauses between syllables. The animals first approached one another slowly with ears erect and forward, nose and mouth sniffing occurred briefly, after which the male smelt the female's cloaca. Then the male grasped the female around the waist region with his forelegs before mounting, after which the male buried his snout in the hair immediately behind the female's ears. It was not possible to see if the neck was gripped. Thrusting by the male soon followed, the female continued to vocalize throughout this period. The male frequently rubbed the chin and throat across the back of the female's neck

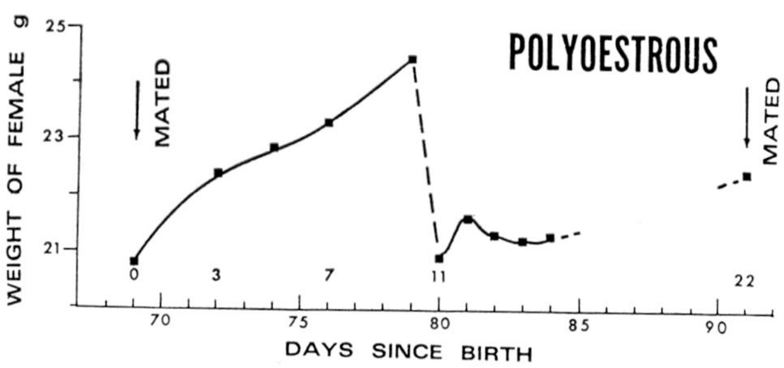

Fig. 1. Weight gain and matings for the adult female *Sminthopsis murina* from 69 to 91 days after the birth of the litter.

The animals were active for the first hour, with bouts of thrusting by the male. Then they became quiescent and the female sat with her head curled underneath her body and resting on the ground. The male occasionally bent over to lick his genital area. At various times throughout the active and quiescent phases the female would give a start and both animals would leap up to 2 cm in the air, often rolling onto their side. The male would then grasp the back of the female's neck with his mouth and struggle to remain mounted. This would sometimes happen several times in quick succession. Mating ended with the female breaking away from the male who sat cleaning his genital area.

The female was observed to have a litter of 10 young on the 14th day after mating. If parturition occurred in the daylight hours as might be expected for a nocturnal animal, the gestation period in this case would have to be greater than 13 days 6 hr but less than 13 days 20 hr

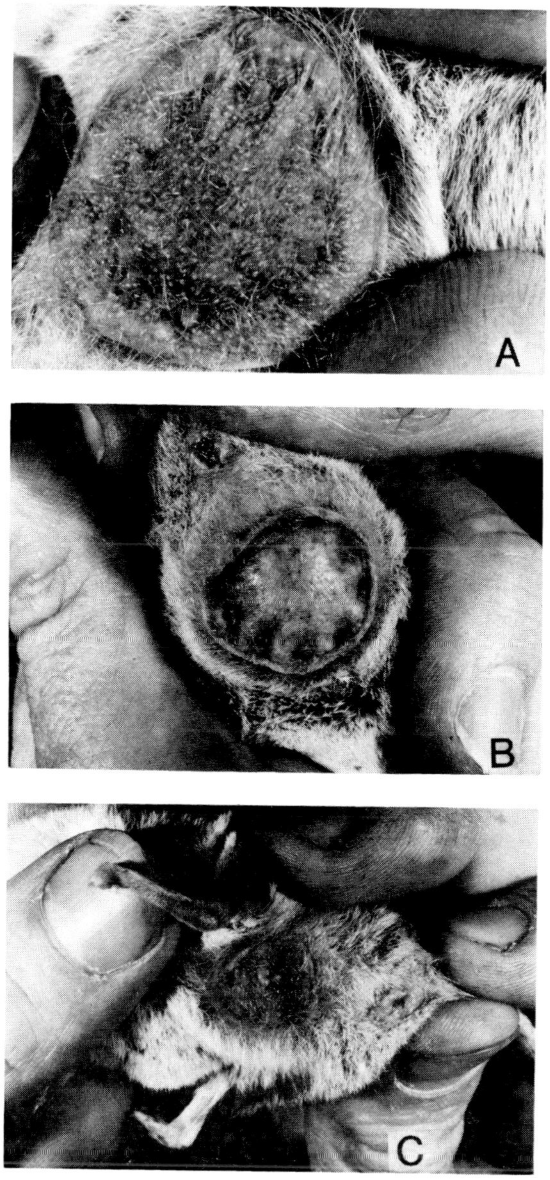

Fig. 2. Pouch development: (A) 38 days, 4 days after young first left in nest; (B) 53 days showing development of mammary tissue when young are growing at their fastest rate; (C) 70 days showing regressed pouch after young are weaned.

On 23 November, 67 days after the birth of the litter, the female signified oestrus by calling again. She mated two days later. For ten days the female gained weight steadily but by the 11th day had lost 4 g and discharged a cloudy discoloured fluid from the cloaca. No young were observed. This represented the first indication that *S. murina* was not monoestrous. On 7 December (91 days after the birth of the litter), the female called at 1430 hr. and mated for 4½ hr between 2130 and 0145 hr. This was 24 days after the female first called at the previous mating (see Fig. 1 for weight change). No young were produced from this mating. It was not clear whether the latter two matings produced true pregnancies, or pseudo–pregnancies as has been described for *Dasyurus viverrinus* (Hill and O'Donoghue 1913).

Young males were observed to call at 90 to 100 days of age but we were not able to determine the function of these calls. The call was similar to the mate–attracting call of the female but with a longer interval between syllables. The adult male called on most nights over the whole period from the initial mating until January. Males and females appear to use very similar sounds, but differ in the number of syllables and time between syllables.

POUCH DEVELOPMENT: The pouch first showed signs of development 10 days after mating, and was clean and moist on day 13. The pouch completely enclosed the 10 young so that it would be a type 3 pouch as defined by Woolley (1974). This type of pouch has been described for only two other members of this genus, *S. crassicaudata* and *S. macroura* (as *S. larapinta*) (Woolley 1974) and so appears to be confined to *Sminthopsis*. There was a remarkable development of mammary tissue (Fig. 2) during lactation, though it may have been a function of suckling ten young. The pouch regressed after the young were weaned but the teats remained clearly visible; the pouch retained a reddish stain on the skin and hair for several months. Using pouch characteristics it is possible to distinguish between parous and nulliparous females. This has been used to identify 2nd year females in the field.

DEVELOPMENT OF THE YOUNG: The developing young are illustrated in Fig. 3 and a catalogue of development as a function of days since birth is given below:

17 days — eyes visible; hind foot digits free; pinnae fused; no claws on manus.

25 days — eyes still covered; pinnae free.

28 days — ears open.

31 days — fine fuzz of white hair.

33 days — pigmentation on tail and dorsal surface; viens visible through skin; female has difficulty walking.

34 days — young first left in nest as female feeds; young furred dorsally; no pigmentation or hair ventrally; crawl with body off ground; young can burrow out of the nest into leaf litter.

44 days — eyes closed; very active crawling; both forefeet together then both hindfeet together.

49 days — eyes fully open; fully furred ventrally (white); climb actively and hang on female's back.

58 days — young first seen out of nest alone.

63 days — young observed eating; solid faeces; threat displays; hissing; biting; know way around cage.

69 days — young separated from female.

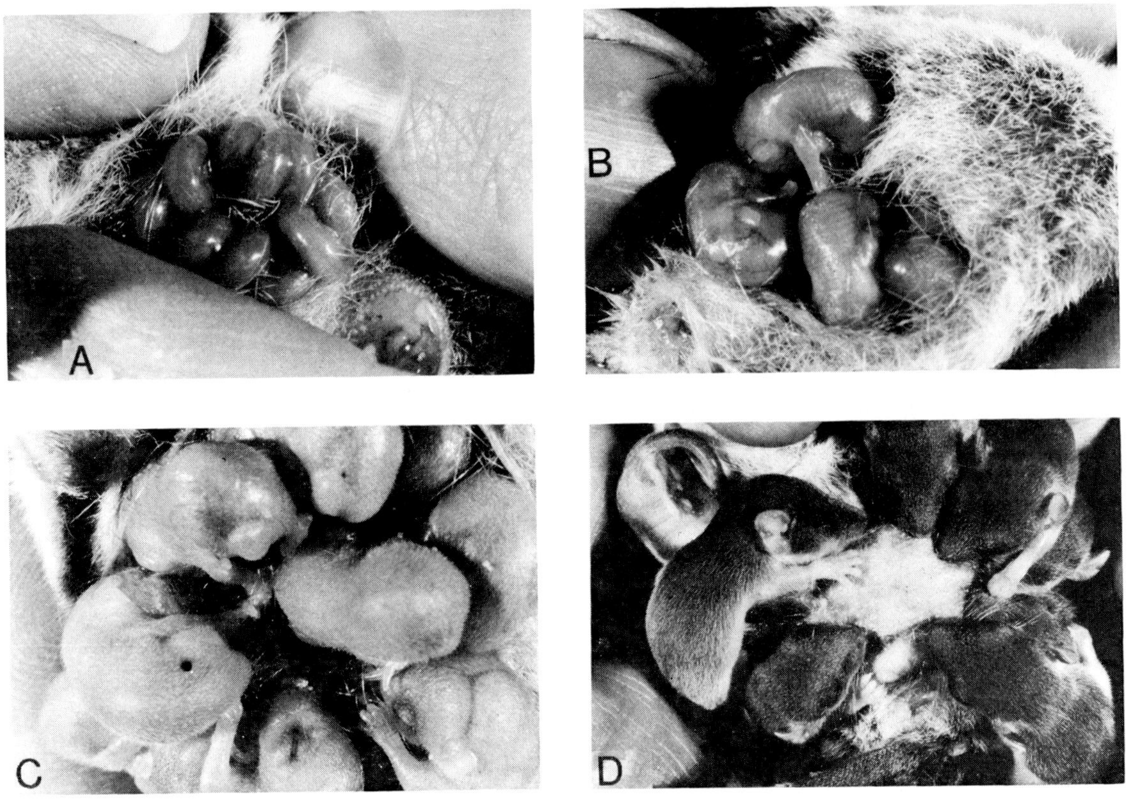

Fig. 3. Development of pouch young: (A) 3 days, crown–rump length 5.4 mm; (B) 17 days, C–R length 9.3 mm; (C) 31 days, eyes visible now, C–R length 17.3 mm; (D) 42 days, H + B length 39 mm, fully furred but eyes still closed.

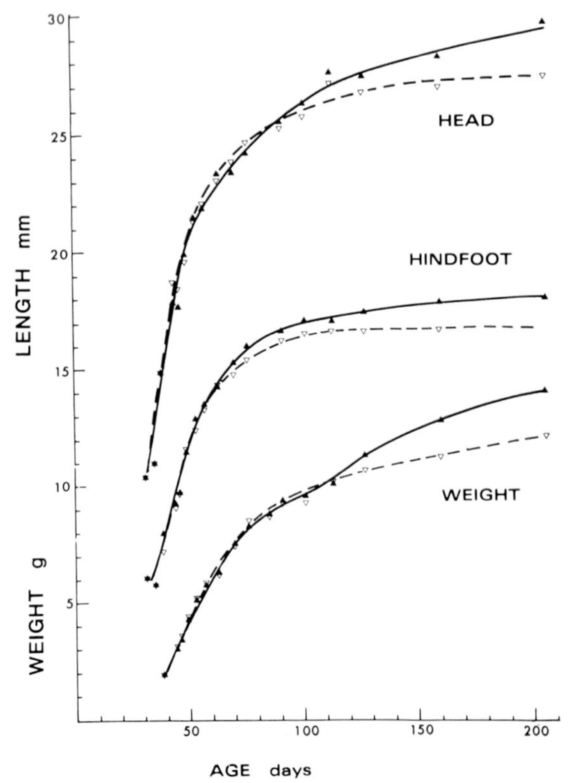

Fig. 4. Growth in body parameters as a function of age for *Sminthopsis murina* pouch young. Mean values for six males are shown as solid symbols while mean values for four females are shown as open symbols, curves are fitted by eye.

GROWTH: Adult *S. murina* exhibit a significant sexual dimorphism. For this reason males and females have been indicated separately although the sexes do not usually show differences until after weaning, and do not become significantly different until about 150 days for most parameters.

Increases in head length, hind foot length and body weight as functions of time are shown in Fig. 4. Hind foot length was first to approach the adult value at about 100 days for females and about 150 days for males. Head length attained the adult value at about 130 days for females, but not until after 200 days for males. Body weight continued to increase steadily for almost a year to adult values of 18 to 20 g for females and 20 to 25 g for males.

Figure 5 shows the growth of tail and head + body with time. Female tail length reached the adult value at about 100 days while female head + body length did not plateau until 200 days. Male tail length appeared to plateau after 150 days, but head + body was still increasing at 200 days.

Growth appears to occur in four phases, as can be seen in Fig. 6. There is some evidence that the stages occur at different ages for males and females, but more litters and individual's growth patterns need to be studied before this can be confirmed.

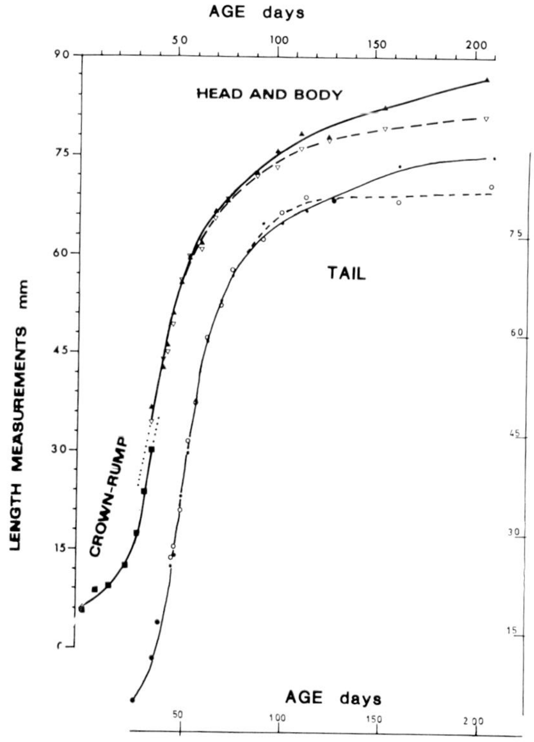

Fig. 5. Growth in body parameters as a function of age for *Sminthopsis murina* pouch young. Symbols are as for Fig. 4. Head and body length with crown–rump length up to 38 days and Tail length.

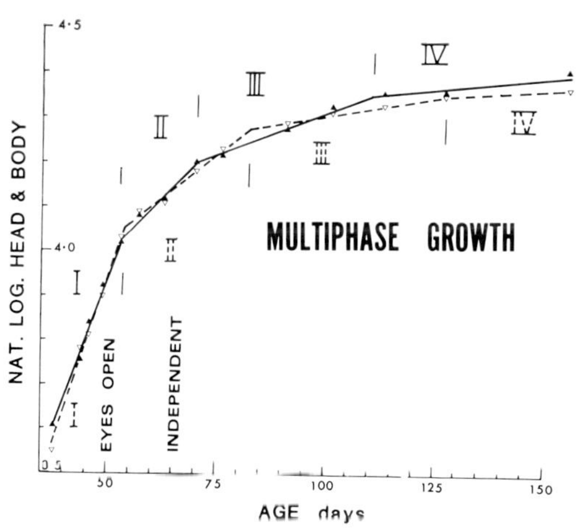

Fig. 6. Multiphase growth shown as the natural logarithm of head and body length as a function of age. Symbols as for Fig. 4. Growth phases shown with roman numerals.

Gestation: Observations on four females during the 1980 breeding season are shown from mid August until the end of December (Fig. 7). Changes in the weight of each female are related to calling, presence of a male, matings and births. For the latter part the presence, in the urine, of cornified epithelial cells and/or sperm was noted. Female 1 (Fig. 7A) produced two litters of 5 and 6 young with weight drops of 4.0 and 3.2 g respectively. Following this the female called for 22 days continuously, a male was placed in the cage, mating occurred and the calling ceased. No further young were produced.

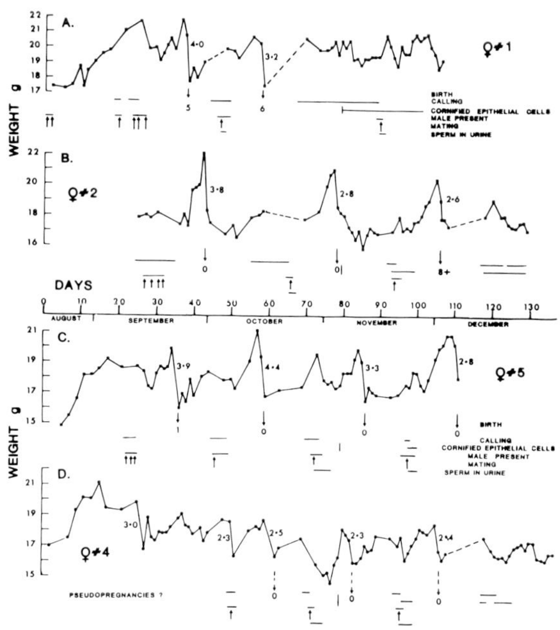

Fig. 7. Weight changes shown for four female *Sminthopsis murina.* (A) ♀ 1, (B) ♀ 2, (C) ♀ 5, (D) ♀ 4. Also shown on the figure are periods when: the females called; cornified epithelial cells were found in their urine; a male was present; mating occurred; sperm was found in ♀ urine. Weight drops associated with, birth of young, inferred birth of young, and pseudopregnancy are indicated.

Female 2 (Fig. 7B) showed three periods with marked weight gains, each followed by a very rapid drop in weight, (3.8, 2.8, 2.6 g). No young were observed from the first two, but more than eight were produced from the third. Female 5 showed four consecutive gains in weight, each followed an abrupt drop in weight (3.9, 4.4, 3.3, 2.8 g). All appeared well synchronised with the calling and mating cycles observed, however young were observed in only one case (the first, a single pouch young).

No young were observed at any time for female 5 (Fig. 7D). Three sudden drops in weight (2.5, 2.3, 2.4 g) were synchronised with mating. Two other weight drops (3.0 and 2.3 g) occurred over a period in which no male had been in the cage for almost a month prior to the first drop. These drops in weight are less than those recorded for other animals that produced live young, and the female did not show the pattern common to other females; of a rapid gain prior to the loss of weight.

No female successfully raised a litter during the 1980 breeding period. Three of the five young produced by female 1 were missing two days after birth and all were missing by the third day. This female was left undisturbed after the birth of her second litter, but all six were missing when examined seven days after birth. Of eight young observed to be delivered by female 2, none remained 1½ hr later.

In the light of these observations on the poor survival of litters we have interpreted sudden weight drops (greater than 2.5g), that follow a marked increase over the preceding week, as representing births. Three 1980 females (1, 2, 5) meet these criteria, as does the 1979 female. On the other hand we have interpreted female 4 to have undergone pseudopregnancies only. This female showed many of the signs associated with a pregnant female; such as pouch development with moistness on day 13, and tightness in the abdomen. The drop in weight however often extended over 3 days. Hill and O'Donaghue (1913) described pseudopregnancy in *Dasyurus viverrinus*, noting that very many of the characteristics of pregnancy were also observed in pseudopregnancy. Similar pseudopregnancies were recorded for *S. macroura* (Godfrey 1969b).

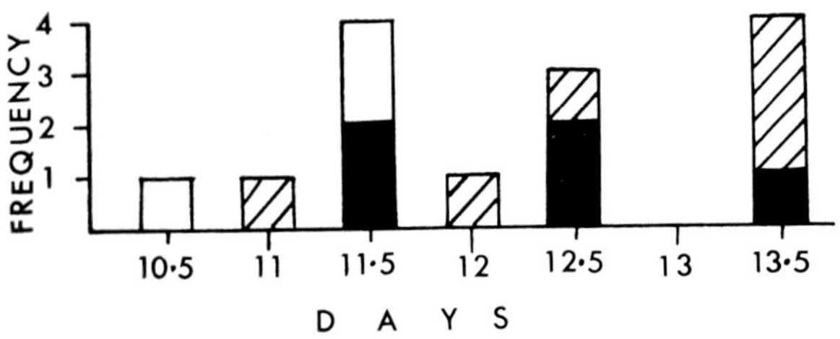

Fig. 8. A histogram of observed gestation lengths. Shading indicates those for which young were actually observed, hatching indicates those for which birth was inferred from weight loss, and clear columns indicate pseudopregnancies.

Gestation length can be estimated from Fig. 7 and from the 1979 female, these values are shown in Fig. 8. We have differentiated between time from mating to (a) actual birth of young, (b) inferred birth of young, and (c) weight drop only. There is no significant difference between categories (a) mean = 12.3 ± .37 (s.e.) and (b) mean = 12.7 ± .42 (t=0.7) but each is significantly greater (1–tailed t–test) than the pseudogregnancy value (c) mean = 11.2 ± 0.33 (t_a = 2.28, $p<.05$, t_b = 2.81, $p<.025$).

The best estimate of gestation length is 12.5 ± 0.28 (s.e.) days, which is also the median value for the distribution, using births and inferred births ((a) + (b)). The half day is to be expected given that most matings observed occurred late in the night, and observed births occurred in the late morning.

Oestrous Cycle: Estimates of the length of the oestrous cycle can also be obtained from the data contained in Fig. 7. We attempted to do this using several methods, the results can be seen in Table 1. Methods employed with their median values were: (a) the interval between the commencement of successive periods of calling (24d); (b) the interval between successive first matings (24d); (c) the interval between successive final matings (24d); (d) the interval between successive births (24d); and (e) the interval between the first appearance of cornified epithelial cells, in the urine, in successive cycles (23d). Unfortunately data from cornfield epithelial cells could only be used from two cycles. The best estimate, based on median values is 24 days, while taken as the mean of means for all methods it is 23.8 ± 0.3 days.

Litter Size: Information from these laboratory litters provide values of 10, 8, 6, 5 and 1; mean = 6. Given the high neonate mortality this should be regarded as an underestimate. Data from three litters observed in the field (10,10,4) give a mean estimate of 8 young/litter.

Table 1. Oestrous cycle length values estimated in five different ways, based on data contained in Fig. 7. Best estimate is 24 days, within a range of 20 to 30 days. Based on observations from five females.

Interval	Sample size	Range	Median	Mean ± Standard error
First Call to first call	11	20–30	24	24.6 ± 0.7
First Mating to first mating	9	20–28	24	24.0 ± 0.9
Last Mating to last mating	7	20–26	24	23.3 ± 1.0
Birth to birth	8	20–28	24	24.1 ± 1.0
First appearance of cornifield epithelial cells	2	21–25	23	23.0 ± 2.0
		All methods	24	23.8 ± 0.3

BREEDING SEASON: By combining information from field notebooks with what has been observed in the laboratory, it is possible to assemble a description of the breeding season for this species in the Myall Lakes area (Fig. 9). Initial field observations by one of us (B.J.F.) from 1976 onwards showed that females with pouch young could be found in September and December, indicating that *S. murina* was at least asynchronous, if not polyoestrous in southeastern Australian forests and heathlands. However as no female has been observed with two successive litters the possibility of different females breeding at different times could not be discounted. Attempts to breed *S. murina* in the laboratory were unsuccessful for two years, and it was not until the successful breeding reported here that polyoestry was confirmed.

The earliest observed mating for a field animal was on 22 August from a 2nd year female captured on 17 August, 1977, and placed with a male on 21 August; however no young were produced. This would indicate the female was about to come into oestrus in the field. We report a female in the laboratory calling on 31 August and mating on 3 September, 1979; while the first of the 1980 laboratory females called and mated on 18 August.

Calaby (1966) reports active sperm in the testes of male *S. murina* in August, 1960 in the Clarence River region of northern N.S.W., and males caught at Myall Lakes (B.J.F.) on 6 and 7 September, 1977 show wounds consistent with fighting. A male found dead in a trap on 20 August, 1975, had severe wounds around the snout, consistent with having been grasped by a similar sized animal. Observations on our laboratory colony showed that males live together happily in the one cage, without fighting, until the breeding season approaches, at which time we had to separate males due to wounds caused by fighting.

A female caught during the last week of November, 1978, had a regressing pouch and had apparently raised 10 young as all 10 teats were uniformly enlarged. Nine immature but independent young were trapped in the vicinity. During the first week of December three adult males were caught at the same site in a single pitfall trap, one was dead and the other two were alive but had wounds on the snout.

The end to mating is inferred by estimating the age of juveniles at first capture (from the growth curves) and working backwards to a date for mating. The latest matings estimated for two separate litters are 28 January, 1977 (Myall Lakes) and 29 January, 1980 (at Wyee, 2 specimens from D. Andrew (pers. comm.)). Laboratory males ceased calling in January for both 1979 and 1980 breeding periods. This gives a mating perid of 145 to 150 days from mid to late August until the end of January. Lactation would then continue from the second week in September until the beginning of April, so that breeding activities occupy at least seven months for this species. Observed events and inferred events from field data are shown separately in Fig. 9. Seven West Australian *S. murina* females with pouch young were taken in August (1), October (5), November (1) (Archer in press).

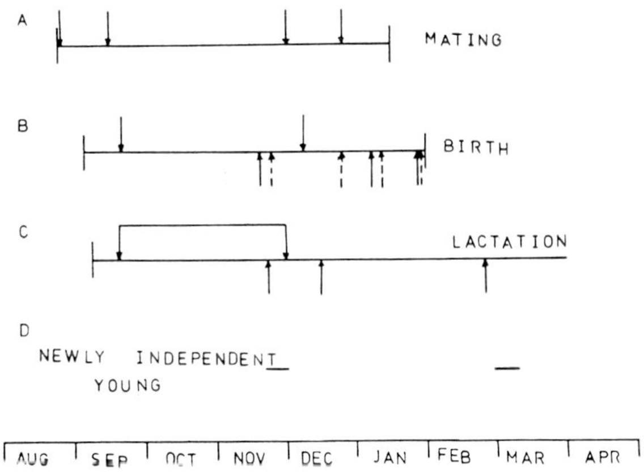

Fig. 9. Breeding season for *Sminthopsis murina* in the Myall Lakes area of N.S.W. Observed laboratory events are shown as arrows pointing downwards. Field observations and births estimated by ageing field juveniles are shown as arrows pointing upwards. The lines represent the periods over which each activity occurs, estimated from both field and laboratory observations. (A) Mating, (B) Births, (C) Lactation, (D) Newly independent young caught. Dotted arrows from Wyee, N.S.W.

DISCUSSION

The mating behaviour shown by *Sminthopsis murina* is similar to that described for other dasyurids: mounting was preceded by mouth smelling and cloacal smelling by the male. Ewer (1968) regarded mouth smelling in *Sminthopsis crassicaudata* as the main form of individual recognition. Van Dyck (1979) reported that *Planigale maculata* appeared to recognize each other by mouth smell, while Aslin (1974) reported naso-nasal contact in *Dasyuroides byrnei*. Both Ewer and Van Dyck reported cloacal smelling of oestrous females by males. The starting and jumping by both animals during the period the male was mounted, has also been reported for *Planigale gilesi* by Whitford *et al.* (1982).

Call monitoring has proved to be a simple and effective aid to obtaining successful breeding in dasyurids (Whitford, unpublished data; Whitford *et al.* (1982)). We have shown that calling by females is a reliable indicator of oestrus, based on the close correlation between cycle length estimates based on calling, mating and the appearance of cornified epithelial cells. The female does not call if caged with a male and this is why we use the term, mate attracting call. On occasions when females were not allowed access to males, they continued to call for long periods, up to 22 days in one case. Cornified epithelial cells were found in the urine of this female for more than 20 successive days (Fig. 7A). This female mated when a male was added on the 22nd day and calling ceased, no pregnancy resulted and this calling may have been abnormal. A second female mated, when a male was introduced, after ten successive days of calling (Fig. 7B). In this case a pregnancy was inferred to have occurred. Female *S. murina,* when unmated, had been noted to call for approximately ten days by Van Dyck (1979). This limited information indicates that females appear to remain in oestrus for lengthy periods if they remain unmated. However it will require further more detailed data to confirm what is in fact taking place. Godfrey (1969b) recorded epithelial cells for up to nine days (with a mean of 5.2 days) for *S. macroura,* and found no difference in oestrus cycle length between mated and unmated females. She reported mean cycle lengths for *S. macroura* (as *larapinta*) of 26.2 days (n=32) based on appearance of epithelial cells, and 26.6 days (n=9) based on the interval between successive matings.

The minimum time between copulation and the appearance of pouch young was 11½ days, but both the median and mean values, for births plus inferred births (n=11), were 12.5 days (Fig. 8). As both of the 11½ day values came from the one animal, we have taken the more conservative view, that gestation length in this species is 12½ days, until a larger sample size (that might confirm 11½ days) is available. Values of 12½ days for *S. macroura* (Godfrey 1969b) and *ca.* 13 days for *S. crassicaudata* (Godfrey and Crowcroft 1971) have been reported. These values are less than half that found for *Antechinus stuartii* (31.5 days) by Marlow (1961) (as *A. flavipes*).

Growth parameters showed a modified exponential form with the rate of increase becoming smaller with time. Tail length indicated an initial slow rate of increase. Head and body length, with the incorporation of crown–rump length to cover the first 40 days, clearly illustrates a full logistic type curve. Growth curves of this form for eutherians are found only when embryonic measurements are included.

Given the continued increase in body weight, this parameter would appear most useful for ageing field animals, with head + body (or crown–rump) providing an alternative for younger animals. It should, however, be kept in mind that the data is for a single litter. Morton (1978c) was able to use a linear regression of scrotal width on time to age *S. crassicaudata* up to 160 days. This was not successful for *S. murina,* possibly because of our small sample size (n=6 ♂♂). Given the length of the mating season (Fig. 7) and the time to weaning, it would be possible for a female to produce three litters within one breeding season but this appears unlikely from field observations. Our estimate of 60 to 65 days (mean = 63) to weaning would seem to be appropriate as the earliest independent young trapped in the field have been aged (from Figs. 4 and 5, taken as the mean of five estimates from the five growth curves), at 57 ± 3 days, 64 ± 3 days (two animals) and 75 days. Godfrey (1969b) reported weaning for *S. macroura* at 70 days while Ewer (1968) reported *S. crassicaudata* females mating a second time at 65, 67 and 69 days respectively, at which time the young were weaned. This time to second mating is also in agreement with our observed value (69 days). Polyoestry appears to be a reproductive strategy common to all *Sminthopsis* studied: *S. crassicaudata* (Ewer 1968); *S. macroura* (Godfrey 1969b); *S. murina* (this study); and *S. virginiae* (L. Taplin pers. comm.).

The development of young *S. murina* is much faster than for *A. stuartii* (Marlow 1961, as *A. flavipes*). Young *S. murina* are first left in the nest at 34 days (permanent attachment for 40 days for *A. stuartii*); eyes open at 49 days (62); the young are fully furred by 49 days (70) and are independent by 69 days (90). This rapid development means that *S. murina* are independent 20 days earlier than *A. stuartii,* giving a much reduced time from mating to weaning of 75-80 days rather than 120 days (Marlow 1961).

In N.S.W. *Sminthopsis murina* may occur sympatrically with *Antechinus stuartii,* although there is a marked ecological separation between them (Fox 1982a). This has important implications. Morton (1978c) interpreted the adoption of polyoestry as a response to reduced predictability of seasonal fluctuations in food supply. Braithwaite and Lee (1979) suggest that monoestry will only persist where the food supply is predictable and highly seasonal. They also argue that the production of a second litter by *A. stuartii* in southeastern Australia would not be possible, as it would coincide with the hot dry summer conditions. Yet *S. murina* contravenes this model, so there are two similar species of dasyurid in the same environment with two markedly different reproductive strategies. Compared to *Antechinus,* the shorter mating–weaning cycle in *S. murina* may facilitate survival of a second litter, as would the reduced parental care of *Sminthopsis,* already noted by Morton (1978c).

In the Smith Lake–Myall Lakes area *S. murina* begins mating in the last week of August, 30–40 days later than *A. stuartii* (mid July), reducing this apparent advantage in accelerated development and reduced parental care. Dickman (1982) has advanced the idea that separation in the breeding time of two different sized *Antechinus* plays an important role in maximising the size difference between juveniles of each species. The smaller species (*S. murina*) also breeds later in this case, the size difference between these two species is illustrated in Fox (1982b.)

Another possible explanation that might be put forward would be for *Sminthopsis* to have evolved from a basically arid adapted stock with species invading humid coastal environments retaining polyoestry. However Archer (in press) regards *Sminthopsis* as derived from humid adapted stock, with the arid adapted species each having a closely related coastally distributed sister species from which they most probably speciated. Relevant examples of this would be *S. ooldea* as the arid adapted version of *S. murina* and *S. macroura* as the arid adapted version of *S. virginiae*. This explanation must then be rejected.

The adoption of polyoestry or monoestry as a strategy would appear more closely related to phylogenetic similarity rather than predictability of seasonal fluctuations in food supply. Previously this distinction could not be made, because of a confounding effect, polyoestrous *Sminthopsis* were confined to unpredictable environments, while monoestrous *Antechinus* were confined to predictable environments. Polyoestry in a predictable coastal environment, demonstrated by *S. murina,* allows the distinction to be made. *Sminthopsis murina* shows that a predictable, highly seasonal environment is not, of itself, sufficient to ensure the persistence of a monoestrous life history strategy.

There is, however, one further factor that should be considered, this is the opportunism component associated with unpredictability. *Sminthopsis murina* reaches maximum densities in early to mid seral stages following fire in heath and open forest (Fox 1982a; Fox and McKay 1981). Although in the long term, the probability of the occurrence of fire may be determined; at any one site it must be regarded as an unpredictable event, within the context of time spans from several to several tens of years. A population of *Sminthopsis murina* derives considerable benefit from rapid, opportunistic increase during early seral stages, as it is able to maximise its use of resources, before the later seral stage *A. stuartii* begins to increase (Fox and McKay 1981). This increase is facilitated by the ability to raise two litters in a breeding season. *Sminthopsis murina* is able to achieve this by more rapid development of the young and a reduction in the length of parental care.

Although polyoestry in *S. murina* may be seen as a response to unpredictability; the reason appears to be to increase fecundity, without the incentive to reduce the risk of reproductive failure, as has been suggested for arid adapted *Sminthopsis* (Morton 1978c). The opportunism exhibited by *S. murina* indicates a high degree of plasticity in this species, in terms of its role as an early seral stage species, that extends to its ability to evolve towards rapid development and reduced parental care, making polyoestry possible. This is in marked contrast to the inability of *A. stuartii* to alter its strategy (Braithwaite 1979), either in terms of response to disturbance or adherence to monoestry. Three species of *Sminthopsis* with confirmed polyoestry (*crassicaudata, macroura* and *murina*) have by far the widest distribution among *Sminthopsis,* both geographically and in terms of habitats occupied. This is probably correlated with high plasticity, the ability to adapt rapidly, enabling them to occupy a wide range of habitats, including those in unpredictable environments. We suggest it may be this ability to adapt in *Sminthopsis,* and the lack of it in *Antechinus* that has produced the phylogenetic dichotomy of *Sminthopsis* evolving polyoestry, while *Antechinus persists* with monoestry.

ACKNOWLEDGEMENTS

We would like to thank Dr. George McKay for supplying one female *Sminthopsis murina.* Debbie Andrew and Laurie Taplin kindly allowed us access to unpublished work while Mike Archer provided a great deal of useful discussion with reference to his wide knowledge of the genus *Sminthopsis*. The work was supported by a Faculty Research Grant from the University of N.S.W. to one of us (B.J.F.).

REFERENCES

Archer, M. A revision of the dasyurid marsupial genus *Sminthopsis* Thomas. *Bull. Am. Mus. Nat. Hist* (in press).

Aslin, H., 1974. The behaviour of *Dasyuroides byrnei* (Marsupialia) in captivity. *Z. Tierpsychol.* **35:** 187–208.

Braithwaite, R. W., 1979. Social dominance and habitat utilization in *Antechinus stuartii* (Marsupialia). *Aust. J. Zool.* **27:** 517–28.

Braithwaite, R. W. and Lee, A. K., 1979. A mammalian example of semelparity. *Am. Nat.* **113:** 151–5.

Brody, S., 1945. "Bioenergetics and growth". Reinhold Publishing Corporation: New York.

CALABY, J. H., 1966. Mammals of the Upper Richmond and Clarence Rivers, New South Wales. *C.S.I.R.O. Division of Wildlife Res. Tech. Paper No. 10*: Pp. 1–55.

DICKMAN, C., 1982. Some ecological aspects of seasonal breeding in *Antechinus* (Dasyuridae, Marsupialia). Pp.139–50 *in* "Carnivorous marsupials" ed. by M. Archer. Royal Zoological Society of New South Wales: Sydney.

EWER, R. F., 1968. A preliminary survey of the behaviour in captivity of the dasyurid marsupial, *Sminthopsis crassicaudata* (Gould). *Z. Tierpsychol.* **25:** 319–65.

FOX, B. J., 1979. Growth and development of *Rattus lutreolus* in the laboratory. *Aust. J. Zool.* **27:** 945–58.

FOX, B. J., 1982a. Ecological separation and coexistence in *Sminthopsis murina* and *Antechinus stuartii* (Dasyuridae, Marsupialia). A regeneration niche? Pp.187–97 *in* "Carnivorous marsupials" ed. by M. Archer. Royal Zoological Society of New South Wales: Sydney.

FOX, B. J., 1982b. A review of dasyurid ecology and speculation on the role of limiting similarity in community organization. Pp.97–116 *in* "Carnivorous marsupials" ed. by M. Archer. Royal Zoological Society of New South Wales: Sydney.

FOX, B. J. AND MCKAY, G. M., 1981. Small mammal responses to pyric successional change in a eucalypt forest. *Aust. J. Ecol.* **6:** 29–42.

GODFREY, G. K., 1969a. The influence of increased photoperiod on reproduction in the dasyurid marsupial *Sminthopsis crassicaudata. J. Mamm.* **50:** 132–3.

GODFREY, G. K., 1969b. Reproduction in a laboratory colony of the marsupial mouse *Sminthopsis larapinta* (Marsupialia: Dasyuridae). *Aust. J. Zool.* **17:** 637–54.

GODFREY, G. K. AND CROWCROFT, P., 1971. Breeding the Fat–tailed Marsupial Mouse *Sminthopsis crassicaudata* in captivity. *Int. Zool. Yb.* **11:** 33–8.

HILL, J. P. AND O'DONOGHUE, C. H., 1913. The reproductive cycle in the marsupial *Dasyurus viverrinus. Qd. J. Microsc. Sci.* **35:** 37–44.

MARLOW, B. J., 1961. Reproductive behaviour of the marsupial mouse *Antechinus flavipes* (Waterhouse) (Marsupialia) and the development of the pouch young. *Aust. J. Zool.* **9:** 203–17.

MORTON, S. R., 1978a. An ecological study of *Sminthopsis crassicaudata* (Marsupialia: Dasyuridae). I. Distribution, study area and methods. *Aust. Wild. Res.* **5:** 151–62.

MORTON, S. R., 1978b. An ecological study of *Sminthopsis crassicaudata* (Marsupialia: Dasyuridae). II. Behaviour and social organization. *Aust. Wildl. Res.* **5:** 163–82.

MORTON, S. R., 1978c. An ecological study of *Sminthopsis crassicaudata* (Marsupialia: Dasyuridae). III. Reproduction and life history. *Aust. Wildl. Res.* **5:** 183–211.

VAN DYCK, S., 1979. Behaviour in captive individuals of the dasyurid marsupial *Planigale maculata* (Gould 1851). *Mem. Qd. Mus.* **19:** 413–29.

WHITFORD, D., FANNING, F. D., AND WHITE, A. W., 1982. Some observations on reproduction, growth and development in *Planigale gilesi* (Dasyuridae, Marsupialia). Pp.77–81 *in* "Carnivorous marsupials" ed. by M. Archer. Royal Zoological Society of New South Wales: Sydney.

WOOLLEY, P., 1974. The pouch of *Planigale subtilissima* and other dasyurid marsupials. *J. Roy. Soc. W. Aust.* **57:** 11–5.

RESUMEN

5. Poliestría en un ambiente costanero predecible: reproducción, crecimiento y desarrollo de *Sminthopsis murina* (Dasyuridae, Marsupialia). Por Barry J. Fox y Dick Whitford.

Este trabajo presenta información de la reproducción de *Sminthopsis murina* (Waterhouse 1838) basada en cinco hembras y en el crecimiento y desarrollo de un parto concebido y sostenido en el laboratorio. La especie es poliestra, con una longitud del ciclo de 24 ± 1 (error standard medio) días. La longitud de la gestación media fue tomada como de 12.5 ± 0.3 días. Un máximo de 10 hijos pueden ser criados, el joven permanece fijado al pezón por 34 días y se desteta hacia los 65 días. Los datos de campo y de laboratorio sugieren que el apareo puede ocurrir desde fines de Agosto hasta fines de Enero, con la lactación extendida hasta fines de Marzo, lo cual da una estación de crianza de siete meses.

Sminthopsis murina aparece tempranamente en la sucesión que sigue al fuego y alcanza su mayor abundancia en los estados tempranos y medios de la sucesión. Se sugiere que *S. murina,* como una especie oportunista, desarrolló la poliestría como medio para incrementar su fecundidad en respuesta a la necesidad de un rápido aumento de la población. Esta especie tiene una alta plasticidad que le ha permitido lograr adaptaciones para acortar la extensiòn de la gestación, acelerar el desarrollo de los jóvenes y reducir el cuidado parental, facilitando el cumplimiento y sobrevivencia de un segundo parto en la estación de crianza. Esto contrasta marcadamente con el co–existente *Antechinus stuartii,* que alcanza su máxima abundancia en los últimos estados de la sucesión que sigue al fuego, y que ha persistido como monoestra.

6. REPRODUCTION IN THE TASMANIAN DEVIL *Sarcophilus harrisii* (DASYURIDAE, MARSUPIALIA)

R. L. HUGHES[1]

Sexually mature female Tasmanian Devils usually weigh more than 4.5 kg. and in the wild the majority of pregnancies occur during the months of March, April and May. The onset of the breeding season is marked by a striking hypertrophy of both the pouch and the reproductive tract. After ovulation of up to 56 eggs, the uterus exhibits extreme enlargement. At least one–fifth of the eggs fail to cleave and less than half attain the unilaminar blastocyst stage. All four teats are commonly occupied by newborn young.

Spermatogenesis is active at least during the months of April and August in males with body weights that range between 3.5 kg and 8.25 kg. The male reproductive system is without gross specializations to distinguish it from the typical marsupial pattern. The glans penis is not forked and the shaft of the penis lacks keratinous spines. Spermatozoa have an overall length of about 213 μm and a maximum diameter of 2 μm. Ultrastructural specializations of the sperm are characteristic of other dasyurid marsupials.

Pages 49-63 *in* CARNIVOROUS MARSUPIALS, ed. by M. Archer, February, 1982. Roy Zool. Soc. New South Wales, Sydney, Australia.

THE TASMANIAN DEVIL (Fig. 1) was believed by Flynn (1939) to be confined to the unsettled mountainous parts of Tasmania and was considered to be rapidly disappearing. Despite repeated efforts Flynn was unable to obtain a single pregnant female during the whole of his sojourn in Tasmania (22 years). However the few animals he collected enabled him to describe the reproductive system (Flynn 1910, 1911) and to give an account of the microscopic structure of one set of 22 unfertilized eggs that were recovered from the much enlarged uteri of a single female captured during the early winter of 1920 (Flynn 1922, 1939). During about the past 15 years the Tasmanian Devil has exhibited a population explosion that has been associated with an increasingly widespread distribution first noted in 1960 (Guiler 1970a).

The present observations were made during the years from 1974 to 1980 inclusive, and are based on data obtained from 63 females and 24 males. The approximate duration of each of the seven field trips involved was about two weeks and these took place in the months of April (N=3), May (N=3) and August (N=1). The primary purpose of this study was to collect data on the embryology of the Devil. However considerable amount of data has also accumulated on other aspects of reproduction.

Fig. 1. Sexually mature male Tasmanian Devil *Sarcophilus harrisii.*

[1] Department of Anatomy, University of Queensland, St. Lucia, Queensland, Australia.

FEMALE REPRODUCTION

AGE AT SEXUAL MATURITY AND BREEDING SEASON: A relatively small proportion (4/27) of the pregnant or lactating females had a body weight of less than 5 kg. The body weights of the four females concerned ranged between 4.4 kg and 4.7 kg. Guiler (1970a) found that the weight at which breeding first occurred at Granville was 4.5 kg for a lactating female. In the present study observation on the degree of fusion of the tibial epiphysis at the knee joint showed that fusion was in progress or rarely completed in animals with a body weight range between 4.5 kg to 5.0 kg. Guiler (1970b) reported that the majority of females wean their pouch young in November-December and that the body weight gain of a few recaptured juvenile Devils was about 0.5 kg per month. Fleay (1952) reported that a litter raised in captivity had almost ceased to suckle in middle November when the young were estimated to be about 22 weeks of age. This would suggest that a high proportion of the animals captured in the months of April and May with a body weight of less than 5 kg and with unfused tibial epiphyses might be expected to be young from the preceeding breeding season and consequently belonging to the one year old age group.

As the body weight of 23 of the 27 pregnant or lactating females captured in the present study was greater than 5 kg and none of these animals when examined exhibited unfused tibial epiphyses it is considered highly probably that the greater proportion of the breeding females had an age of more than one year. Although both Guiler (1970b) and Fleay (1952) considered that female Devils did not breed until they were two years of age, confirmation of this awaits a more detailed study of the relationship between body weight and puberty.

In animals with a body weight of less than 4.0 kg the pouch folds were poorly developed or virtually absent and the nipples and mammary glands were small. The juvenile condition of these animals was readily confirmed by the lack of fusion of the tibial epiphyses. The onset of breeding was marked by a spectacular pre-ovulatory enlargement of the pouch to a moist "fist-sized" structure with four well developed mammary glands. Fleay (1935 and 1952) reported that the pouch developed from its resting condition in April and he considered that this was a sign of the onset of the pairing season. The pouch when relaxed was observed to open in a backward direction. However when the sphincter muscles were contracted to close the pouch its orifice shifted towards the centre of the pouch area and resulted in the formation of a caudal pouch fold. Pocock (1926) reported that the pouch of the Tasmanian Devil was structurally identical with that of *Thylacinus cynocephalus*. Both opened backwards and downwards and contained four teats.

MATING: None of the female reproductive tracts examined during the course of the present study gave evidence of a recent mating. However, in five females captured between the 9th April and the 9th May, and with body weights that ranged between 4.8 kg and 5.8 kg, the uteri exhibited marked hypertrophy in conjunction with a pre-ovulatory enlargement of the Graafian follicles. Guiler (1970b) reported that at oestrus the male remained close to the female and grasped her with his forelimbs during copulation which was observed to take place both by day and night. In three females guiler (1970b) reported the date of matings to be on about 10th, 5th and 22nd March in the years 1966, 1967 and 1968 respectively. These observations together with data to be presented later in the text, that estimates the date of birth suggests that the onset of breeding exhibits some degrees of synchronisation throughout the population.

OVULATION: In six females captured between 21st April and 10th May, the number of uterine eggs or blastocysts (or both) recovered from each animal ranged between 11 and 56 (mean 39) Earlier reports that *S. harrisii* was polyovular were those of Flynn (1922, 1939) who recovered a total of 22 unfertilized uterine eggs from a single female, and also the record of 15 uterine embryos recovered from a female by Guiler (1970b). The reported out-of-phase breeding based on the size of five sets of pouch young recorded between August and October by Guiler (1970b) and Green (1967) would suggest the possibility that the Tasmanian Devil may not be monoestrus and that lactational anoestrus might occur during pouch life. Currently pouch-young are being removed from females in a natural population at Avoca, Tasmania in order to determine whether any of these females give birth to more than one litter before entering anoestrus.

PREGNANCY AND DATE OF BIRTH: The gestation period of the Tasmanian Devil was reported by Asdell (1964) to be 31 days. However Mr. Geoff Wood, who has observed matings at his private zoo near Hobart, considers the gestation period to be about three weeks.

The date of birth of 21 litters examined during the months of April and May in the present study was estimated from the growth data of Guiler (1970b). Sixteen of these litters were estimated to be born in the first half of April while the remaining five litters were born in the second half of April. Two early pregnancy stages were obtained in April and a further four stages in

May. These data suggest that a small proportion of births might be expected to continue until at least early June. Fleay (1935) reported a birth in captivity in May and estimated the birth of an earlier litter in this female to have occurred in the middle of June of the preceding year.

LITTER SIZE AND SEX RATIO: The number of pouch young per litter in the present study ranged between one and four. However a modal litter size of four occurred in 15/21 litters examined. Contrary to the statement of Guiler (1970b) the sex of pouch young could not be determined by external examination at birth. Rudiments of neither the pouch nor scrotum were present in six litters of pouch young with head lengths that ranged between 6.0 mm and 11.0 mm. However these rudiments were visible in pouch young with a head length greater than 11.0 mm. The growth data of Guiler (1970b) indicate that the approximate post partum age of these young would be greater than 10 days.

Guiler (1970b) reported that females were present in larger numbers than males in the pouch young and that this ratio continued throughout the adult life until old age, when males became more numerous. The present study indicated that the sex ratio of pouch young also favours females (27/48). However the small sample size was considered inadequate for this to be firmly established.

FEMALE REPRODUCTIVE SYSTEM: As in other marsupials, the female reproductive system was characterised by a uterus duplex and a vaginal complex where the ureters pass between the lateral vaginal canals before entering the urinary bladder. The cervical canals of the uteri exhibited marked elongation (Fig. 3) and each os-cervix was located terminally on a conspicuous papilla with lobed edges. MacKenzie and Owen (1919) wrongly considered the cervical canals to be part of the vaginal complex. According to Flynn (1911) each cervix opened separately into a "median vaginal neck", a unique but rudimentary anterior extension of each of the separated vaginal cul-de-sacs. MacKenzie and Owen (1919) reported the absence of a median vaginal septum in some specimens, however, in no instance was communication found between the median vagina and the urogenital sinus. The lateral vaginal canals were of diminutive size and became slightly enlarged during the periovulatory stage of the reproductive cycle. The clitoris consisted of a small median longitudinal element invested on each side by a well developed lateral fold.

The reproductive tract increased in size and vascularisation with onset of the breeding season. However the most striking change was in the degree of enlargement of the corpus uteri. In prepubertal or anoestrous females the uterine bodies weighed approximately 0.4 g. However with the onset of the breeding season about a tenfold increase in weight occurred before ovulation. After ovulation the uterine endometrium exhibited an accelerated degree of proliferation. During the early cleavage stages of embryogenesis and before the expansion of the unilaminar blastocyst the uterine body weighed about

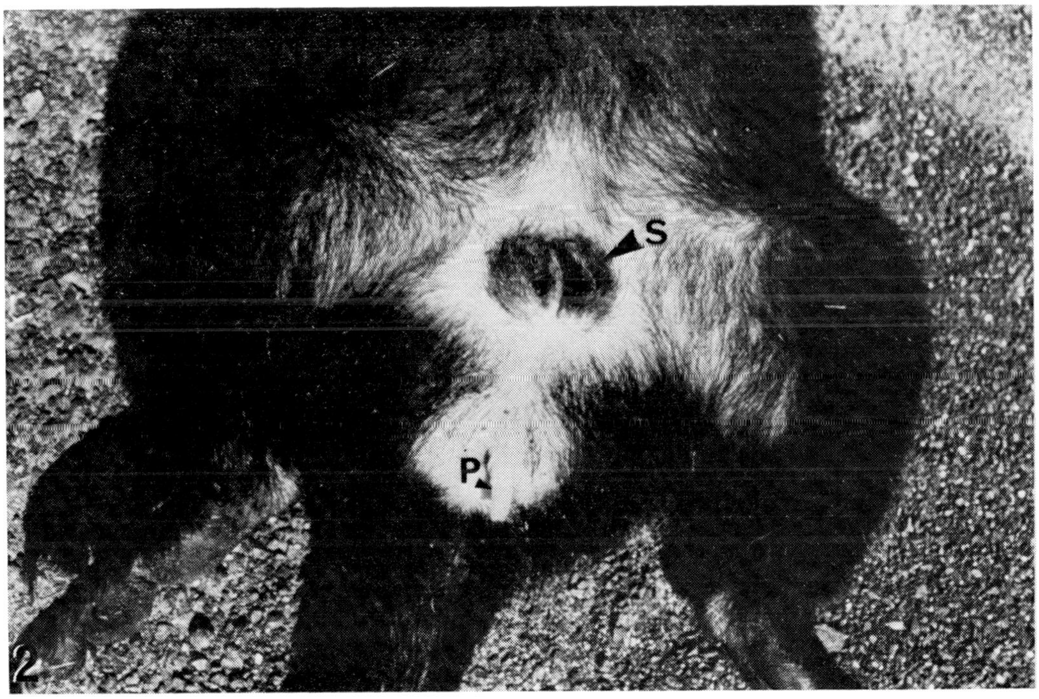

Fig. 2. The inguinal region of a male Tasmanian Devil showing the pendulous scrotum (S) located in front of the penis (P).

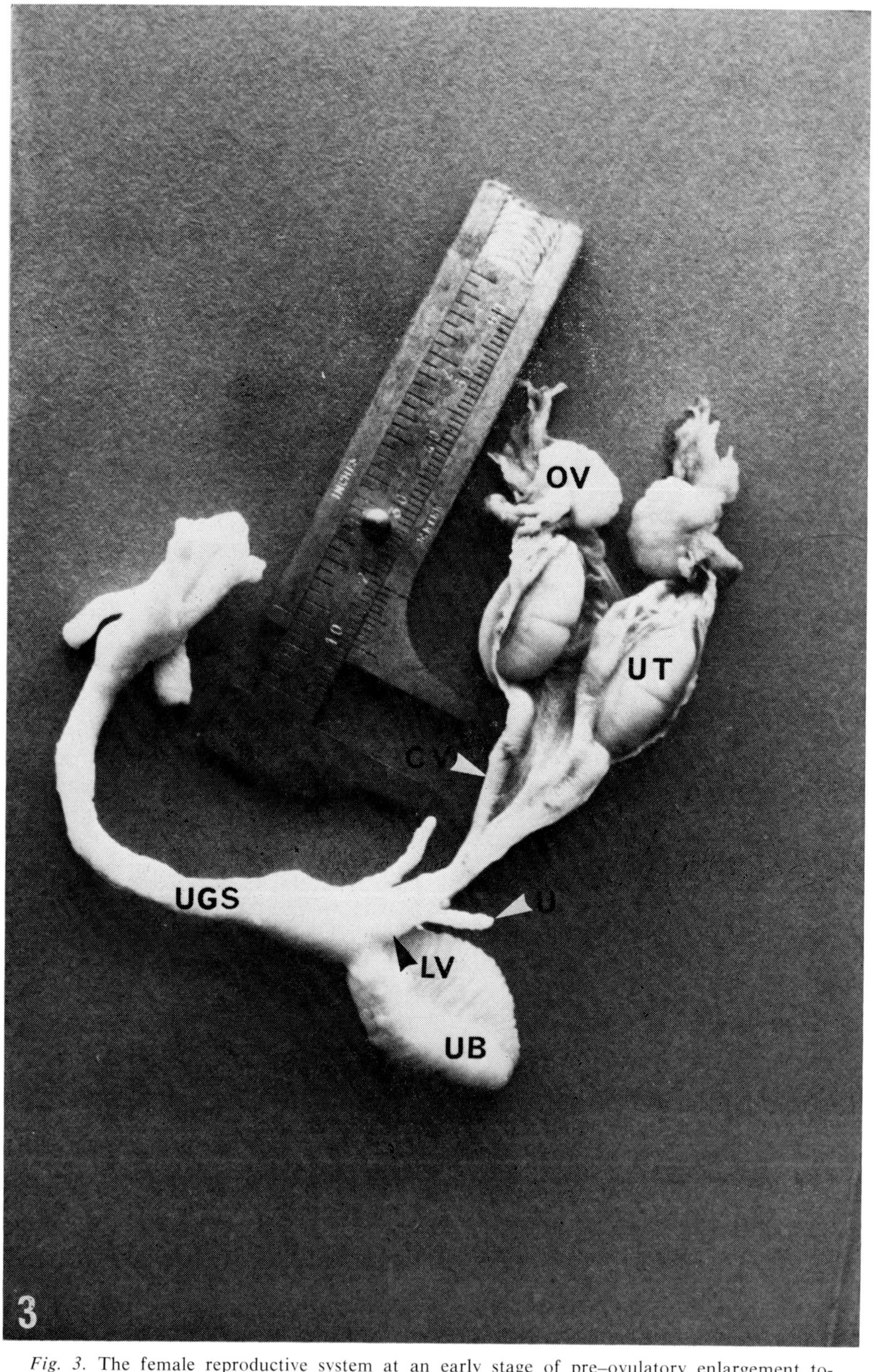

Fig. 3. The female reproductive system at an early stage of pre–ovulatory enlargement towards the onset of the breeding season. The small metric scale divisionse are in mm. CV, cervix; LV, diminutive lateral vaginal canal; OV, ovary; UB, urinary bladder; UGS, urogenital sinus; UT, uterus; U, ureter.

5.0 g. In an animal with cystic endometrium of both uteri and early cleavage stages the combined weight of both uterine bodies was 72 g. In pregnant females at an early stage of blastocyst expansion (overall blastocyst diameters 2.0 mm to 2.5 mm) the uterine bodies each weighed about 20 g and measured about 36 mm x 36 mm x 24 mm. The cervix then had a length of about 40 mm. Pearson and De Bavay (1953) have compared the morphology of the female urogenital system of *Sarcophilus* with that of other Dasyuridae.

OOGENESIS: As in other mammalian species the early growth of the primary oocytes was accompanied by the formation of primary ovarian follicles of increasing complexity (Fig. 4). The enlarged primary oocytes located within secondary ovarian follicles (zona pellucida present) were characterised by a marked degree of cytoplasmic vacuolation (Fig. 5) of unknown significance. Both ovaries were equally functional and were characterised by Graafian follicles with small antral cavities (Fig. 6). The fully formed Graafian follicles had an overall diameter of about 1.3 mm and did not project appreciably from the ovarian surface. The small size of the mature Graafian follicles is believed to be casually related to the large number of eggs (up to 56) ovulated from relatively small ovaries. Newly formed corpora lutea were also small and did not project markedly from the ovarian surface (Fig. 7). The diameter of young corpora lutea was 1.3 mm and at later stages in their develop-

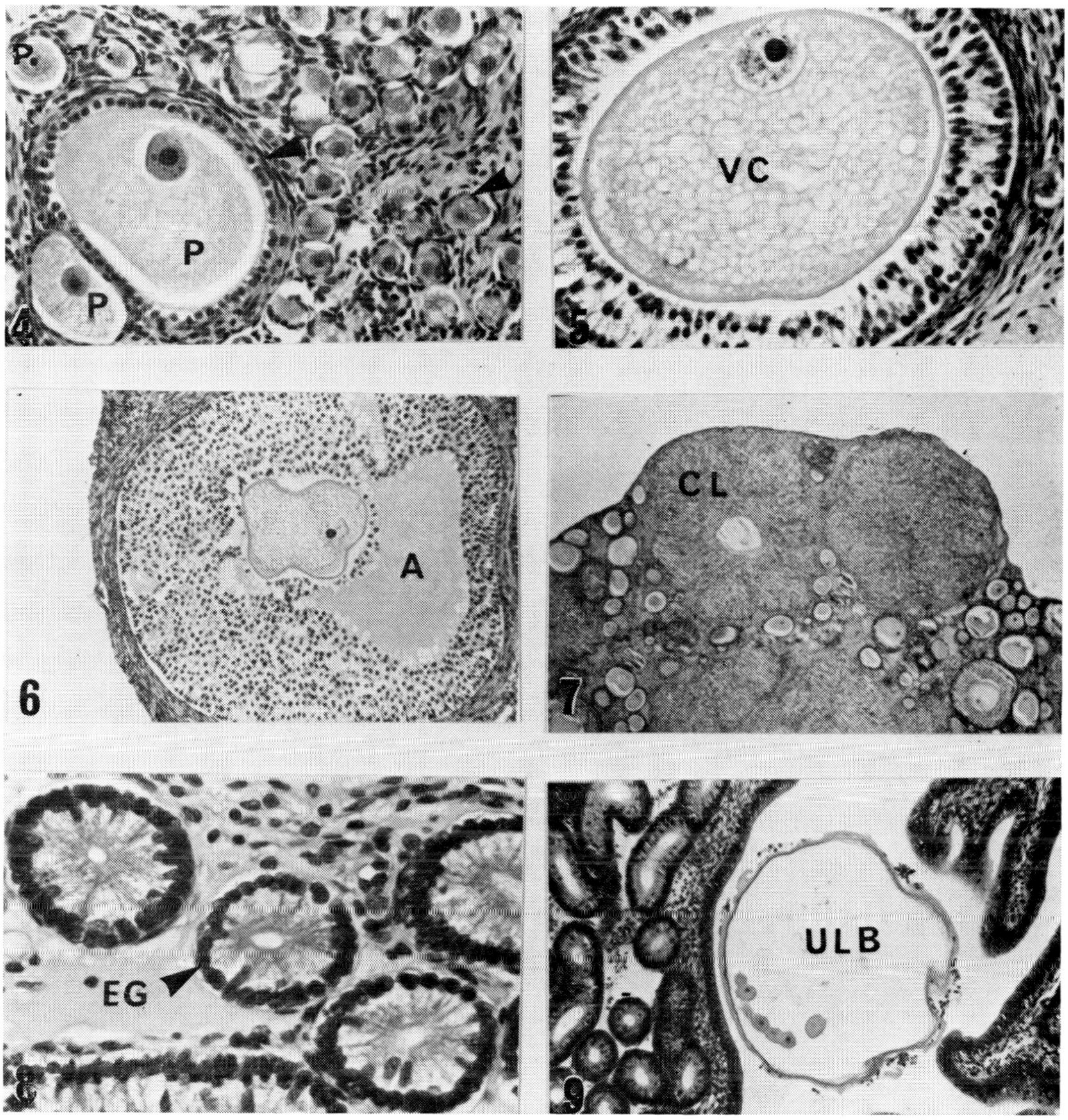

Fig. 4. Histological section of the ovary showing the early growth phase of primary oocytes (P) invested by primary ovarian follicles (arrowed) of increasing complexity. *Fig. 5.* An enlarged primary oocyte located within a secondary ovarian follicle. Note the marked degree of cytoplasmic vacuolation (VC). *Fig. 6.* A Graafian follicle with a small antral cavity (A). *Fig. 7.* Section of the ovary showing newly formed corpora lutea (CL). *Fig. 8.* Section of the uterine endometrium at an early unilaminar blastocyst stage. Note the luteal enlargement of the endometrial glands (EG). *Fig. 9.* Section through the uterine lumen showing a unilaminar blastocyst (ULB). Same uterus as Figure 8.

ment these increase in diameter to at least 1.7 mm.

UTERINE HISTOLOGY: The uterine luteal phase as in other marsupials was characterised by enlarged endometrial glands lined by tall columnar cells with basally located nuclei. A well developed luteal phase occurred at a relatively early stage of gestation when embryonic development was at the early unilaminar blastocyst stage (Fig. 8 and Fig. 9).

EARLY EMBRYOLOGY: The overall diameter of the vitellus of newly ovulated eggs was approximately 210 μm (Hughes 1976a). In twenty single celled uterine eggs the vitellus was invested successively by three egg membranes (Fig. 16); the zona pellucida, the mucoid coat, and the shell membrane with mean thicknesses of 4 μm, 28 μm and 5 μm respectively. The overall diameter of single celled uterine eggs was either about 300 μm where the zona pellucida closely invested the vitellus or approximately 435 μm where the zona pellucida had elevated from the vitellus to form a perivitelline space.

The unilaminar blastocyst stage, as in other marsupials, formed directly from the cleaving eggs without an intervening morula stage. The sequence of developmental events in this process is shown in Figs 16 to 21 inclusive. Both the zona pellucida and the mucoid coat containing numerous trapped sperm were present at the unilaminar blastocyst stage. The shell membrane also remained intact at the unilaminar blastocyst stage.

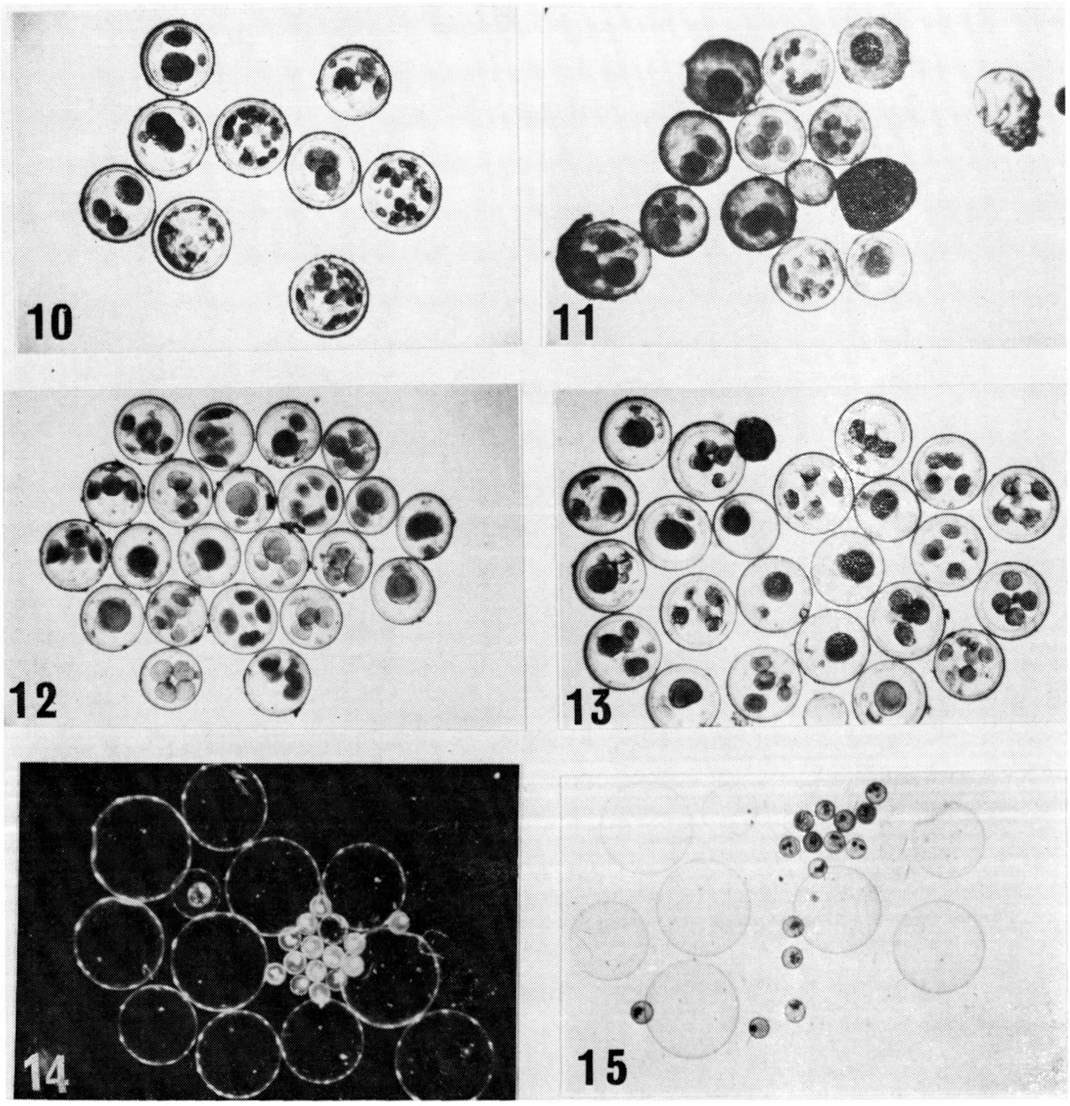

Fig. 10. Eggs and early blastocysts recovered from the right uterus of animal number SH75–9. *Fig. 11.* Eggs and early blastocysts recovered from the left uterus of animal number SH75–9. *Fig. 12.* Eggs recovered from the right uterus of animal number SH77–2. *Fig. 13.* Eggs recovered from the left uterus of animal number SH77–2. *Fig. 14.* Degenerating eggs (N=15) and expanding blastocysts (N=12). recovered from the right uterus of animal number SH79–2. *Fig. 15.* Degenerating eggs (N=15) and expanding blastocysts (N=8) recovered from the left uterus of animal number SH79–2.

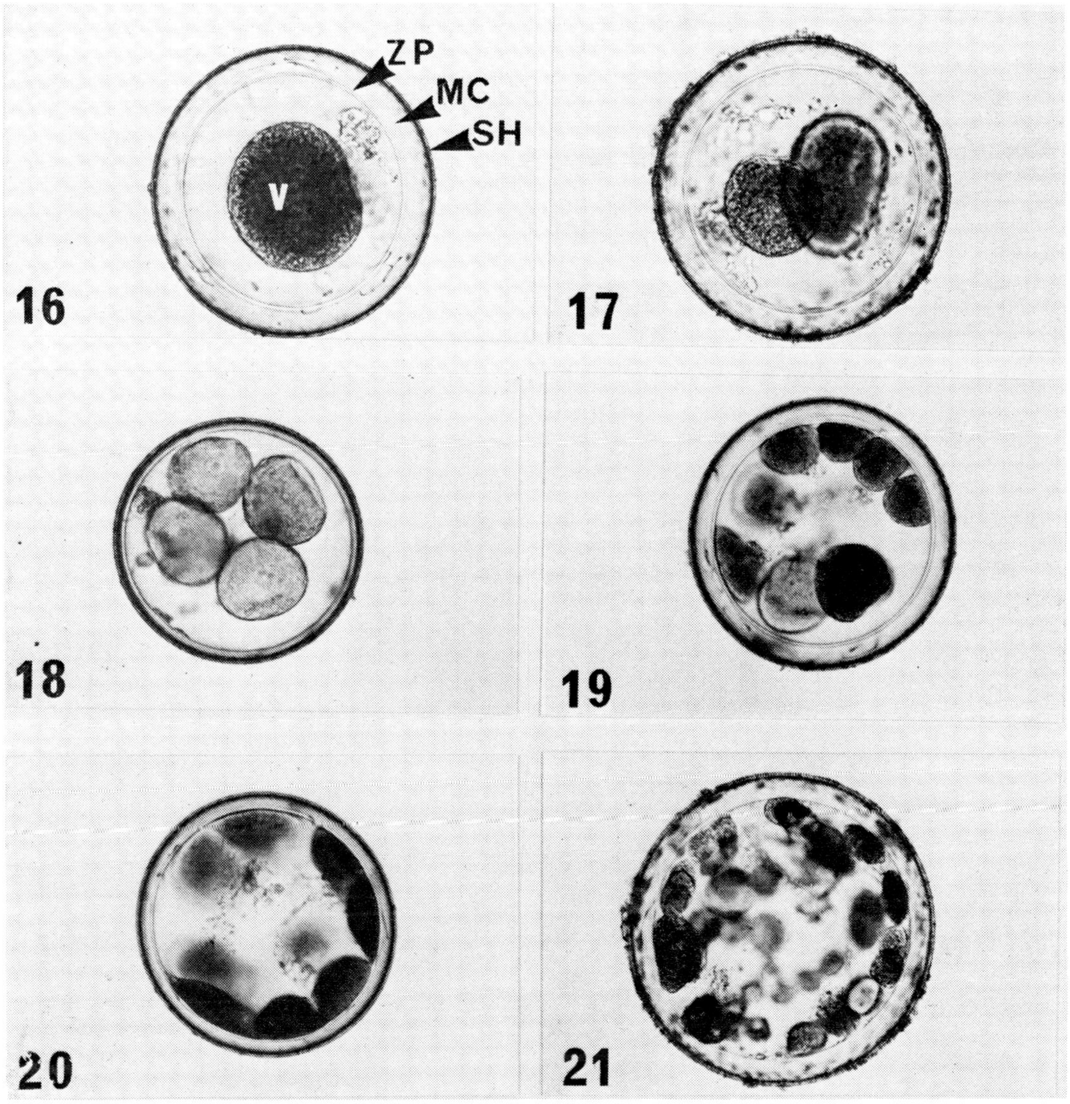

Fig. 16. Single celled uterine egg. V, vitellus; ZP, zona pellucida; MC, mucoid coat; SH, shell membrane. *Fig. 17.* Two celled uterine egg. Note the slightly unequal cleavage of the blastomeres. *Fig. 18.* Four celled uterine egg. The blastomeres were rounded and separated. *Fig. 19.* Stage in the formation of the unilaminar blastocyst. The blastomeres of unequal size adhered to the inner surface of the zona pellucida. *Fig. 20.* An early unilaminar blastocyst stage formed directly from the cleaved blastomeres without an intervening morula stage. *Fig. 21.* An early unilaminar blastocyst. Note that the zona pellucida, mucoid coat and shell membrane are still intact.

EMBRYONIC LOSS DURING EARLY PREGNANCY: A total of 233 uterine eggs or embryos (or both) was obtained from six individual pregnancies. The number of eggs shed at each of the ovulations ranged from 11 to 56 with a mean number of 39 eggs per set. All the pregnancy stages obtained from 3/6 females are shown in Figures 10 to 15 inclusive. Feulgen preparations of whole–mounted eggs showed that cleavage failed to occur in at least one–fifth of the ovulated eggs. The degenerating vitellus in one of these uncleaved single–celled eggs is shown in Fig. 22. More than one–third of the cleaving eggs from individual animals failed to reach the 8–celled stage. However normal development to the 8–celled stage, as indicated by either undegenerated interphase nuclei (Fig. 23) or the presence of mitotic activity, occurred in at least half of the cleaving eggs. Figure 24 shows ten normally developing blastomeres of unequal size from a 24–celled early unilaminar blastocyst stage. In two females less than one half of the ovulated eggs (15/46 and 18/50 respectively) had developed to the expanded blastocyst stage with diameters that ranged between 2.0 mm to 2.5 mm. In these females the remaining eggs were all at an earlier stage of development and exhibited conspicuous degenerative changes including fragmentation of the blastomeres.

The pouch of the Tasmanian Devil contains only 4 nipples so it is inevitable that considerable wastage must occur before young attach to the teats. However both the nature of this wastage as well as its proportionate distribution between the terminal stages of pregnancy and the immediate postnatal period remain to be established.

New Born Young: Four new born young of the Tasmanian Devil (Fig. 25) each weighed approximately 244 mg and had a head length of about 6.0 mm. As in other marsupials the forelimbs were well developed and the manus was furnished with conspicuous functional claws. The hindlimbs were rudimentary immobile paddles that clasped the median recurved tail.

MALE REPRODUCTION

Very little is known of male reproduction in this species. MacKenzie and Owen (1919) described the urinogenital system. Light microscope observations have been made on the morphology

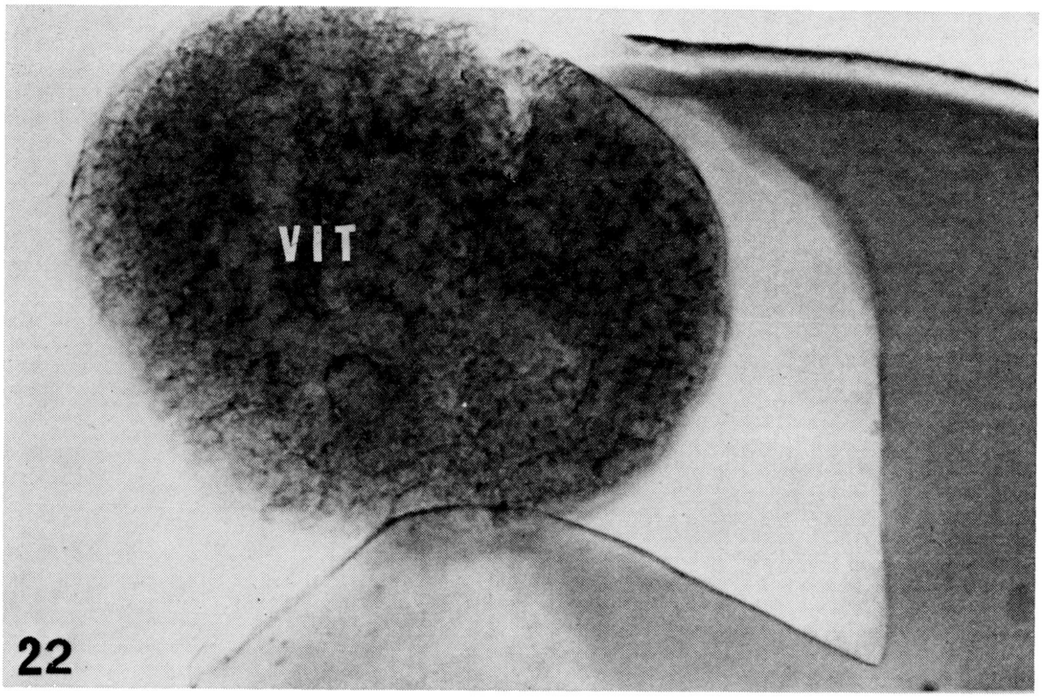

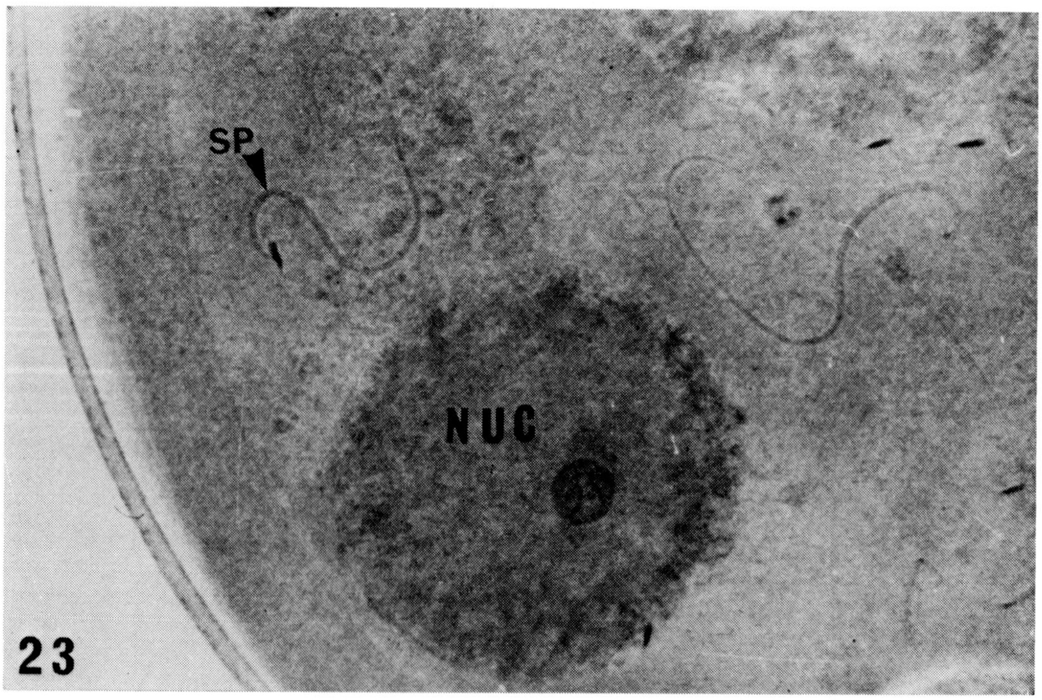

Fig. 22. A Feulgen whole-mount preparation of a single celled uncleaved uterine egg showing the degenerating vitellus (VIT). *Fig. 23.* A Feulgen whole-mount preparation of an 8-celled uterine egg showing an undegenerated interphase nucleus (NUC) of one of the blastomeres. The mucoid coat contained many sperm remnants. (SP).

and dimensions of epididymal spermatozoa (Hughes 1965). Guiler and Heddle (1970) described the anatomy of the scrotum, the cremaster muscles and the blood vessels of the rete mirable and made observations on both the testicular and the body temperatures.

The material examined consisted of 24 males taken in the wild. Eighteen of the animals were caught during either April or May during the years from 1974 to 1980 inclusive. The remaining six animals were captured in August, 1975. The body weights of the animals ranged from 3.5 kg to 9.8 kg. Spermatogenesis was active in all animals. The terminal stages of sperma-

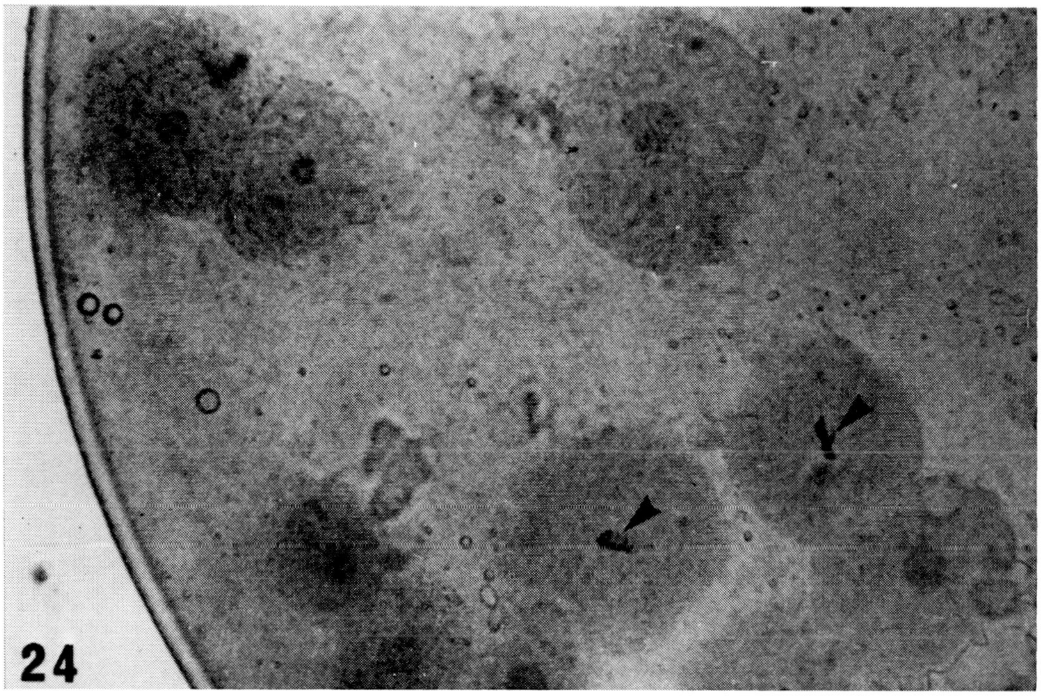

Fig. 24. A 24–celled early unilaminar blastocyst showing ten normally developing blastomeres of unequal size. Note that the nuclei of two of the blastomeres (arrowed) are at the metaphase plate stage.

Fig. 25. Four new born young. Each weighed about 244 mg and had a head length of 6.0 mm. The small metric scale divisions are in mm. The manus was furnished with conspicuous functional claws. The hindlimbs were rudimentary immobile paddles that clasped the median recurved tail.

togenesis in males with a body weight of over 5 kg was numerous morphologically mature sperm. In three males with body weights of between 3.5 kg and 4.0 kg. captured in August, 1975 the testis tubules exhibited very active spermatogenesis. However, the morphological maturation of a high proportion of the sperm was incomplete. This was particularly evident in the structure of the sperm head. The body weight and age at the onset of puberty in males requires further investigation. The present study would suggest a prolonged period of active spermatogenesis from at least April to August. However, it is not known whether spermiogenesis ceases during any portion of the year in adult males.

The Tasmanian Devil exhibits sexual dimorphism with respect to body weight. Guiler (1970a) reported that the average weight of 91 males was a little more than 2 kg greater than that found in 103 females. Preliminary observations from the present study suggested that the fusion of the tibial epiphyses of males occurred in animals with a body weight of between one to two kg heavier than that found in females.

MALE REPRODUCTIVE SYSTEM: The scrotum was pendulous and located in the inguinal region of the abdomen in front of the penis (Fig. 2). The scrotal skin was darkly pigmented and a slight pigmentation was also evident in the tunica vaginalis. As in other marsupials the dissected reproductive tract showed no evidence of either seminal vesicles or ampullary enlargements of the vas deferens (Fig. 26). A single carrot–shaped disseminate prostate gland completely surrounded the prostatic portion of the urethra. Two pairs of Cowper's glands were present. The Cowper's glands were wrongly interpreted by MacKenzie and Owen (1919) to be ductless glands. Two pairs of erectile bulbs (the crura of the penis and the urethral bulbs) were located at the base of the penis and were invested by thick coats of striated muscle. The penis was retracted within a pocket when not erect, and its shaft was not ornamented with hooks. The glans of the penis was slightly furrowed but was not forked.

Guiler and Heddle (1970) reported that in all male Devils captured in the field the scrotum was withdrawn by the action of the cremaster muscles into a pseudo–pouch consisting of ventrolateral folds of abdominal skin. This was believed to be the "normal position" of the scrotum and the pseudo-pouch was considered to play an important part in the thermoregulatory mechanisms of the region as well as providing protection for the testis. The present study showed that in mature males the cremaster muscles relaxed immediately after death and resulted in the complete disappearance of the pseudo–pouch. Further studies are required in order to determine the functional significance of contraction of the cremaster muscles. It is of interest that Pocock (1926) who made observations on the external reproductive organs of the male Tasmanian Devil did not describe the presence of a pseudo–pouch investing the scrotum. However he reported that the scrotum of *Thylacinus cynocephalus* was located within a permanent scrotal pouch that consisted of a continuous anterior and lateral fold of skin. The scrotal pouch fold was absent posteriorally and although no dimensions were given, Pocock's figure shows that the greater part of the ventral aspect of the scrotum was not invested by the folds of the scrotal pouch. Pocock (1926) considered the scrotal pouch of *Thylacinus* to be unique among marsupials and he reported that it was better developed in younger animals than in the adult.

A characteristic feature of the histology of the testis was the high proportion of interstitial tissue (Fig. 27). Whole mounts of sperm were prepared from both the native cat *Dasyurus viverrinus* and the Tasmanian Devil in a manner previously described by Hughes (1965). The gross sperm morphology of both species was indistinguishable. In the proximal third of the epididymis sperm tails were inserted at about the mid-point of the under surface of the head plate. The long axis of the head was supported at an angle of form 60 to 90 degrees to the axis of the tail by a conical cytoplasmic droplet (Fig. 28). Sperm taken from the distal two–thirds of the epididymis lacked the cytoplasmic cone and the head plate had rotated to parallel the axis of the tail (Fig. 29).

Specimens for ultrastructural examination were prepared from the testis and the head, body and tail of the epididymis of three male Devils captured in April, 1975. Observations on the sperm have been previously published by Hughes (1976b). The sperm head was a narrow plate (Fig. 30) of about 11.1 μm in length. It was folded in a median longitudinal ventral plane (Fig. 30) to form a finely tapered stylus. The sperm tail of approximately 207 μm in length was inserted near the midpoint of the concave ventral surface of the nuclear plate by way of a ball socket joint. The socket consisted of a deep implantation fossa within the nuclear plate and was lined by a basal plate, a slightly thickened portion of the nuclear membrane. The ball component of the joint was the proximal junction body of the neck region which was fused to the lateral subterminal surface of the sperm tail.

The acrosome did not form a head cap as in eutherian mammals but consisted of an acrosome vesicle, comprising an inner and an outer acrosome membrane. The acrosome occupied most of the convex upper surface of the head

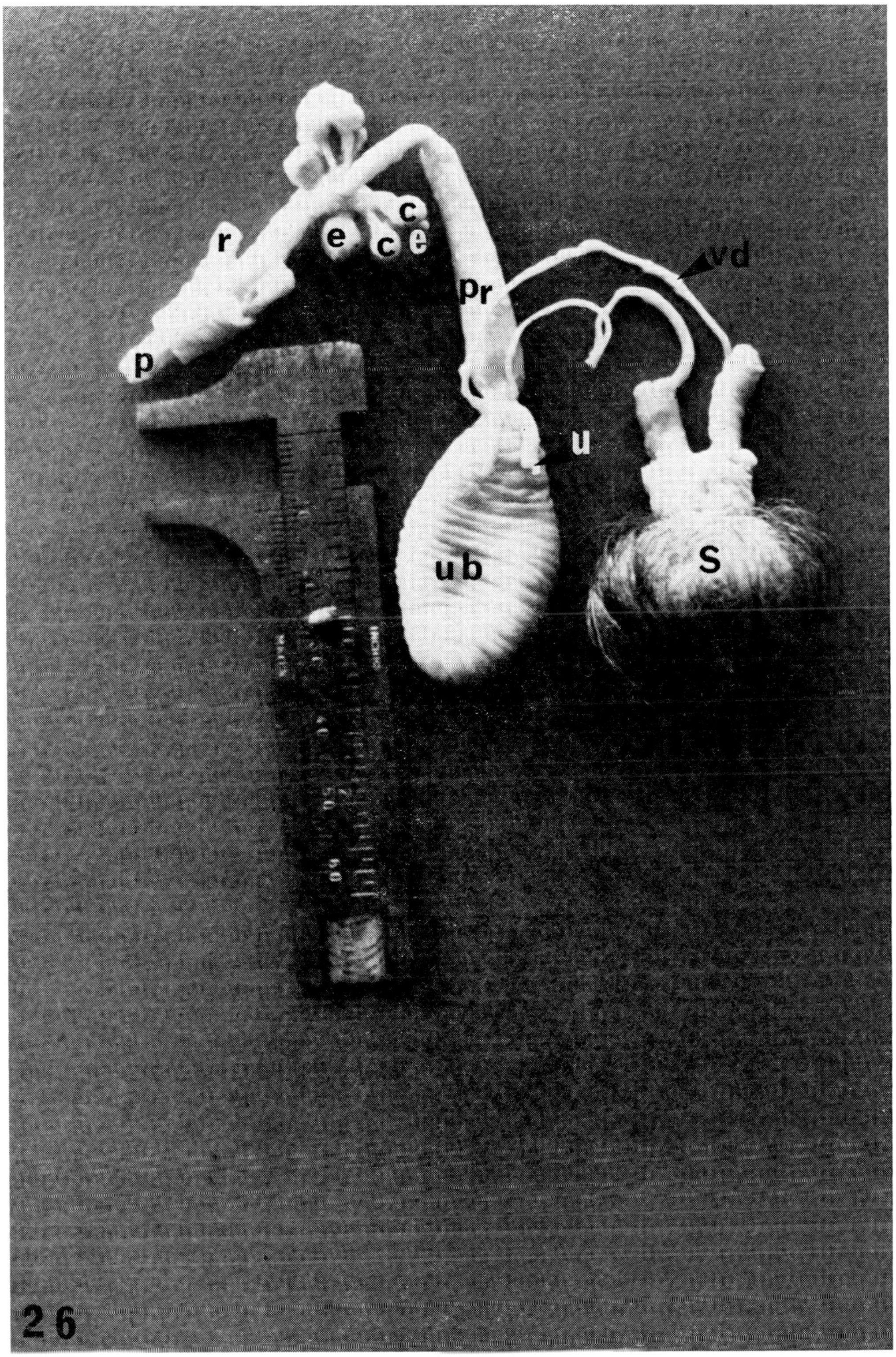

Fig. 26. The reproductive system of a sexually mature male. p, penis; r, retractor muscle of penis pocket; e, erectile bulbs of the penis; c, Cowper's glands; pr, disseminate prostate gland; u, ureter; ub, urinary bladder, vd, vas deferens; S, scrotum.

plate and its proximal portion lay within a shallow depression of the nuclear plate. A periacrosomal area of the overlying plasma membrane was thickened and marked a point of membrane fusion, involving the membrane of the acrosomal vesicle and possibly also the nuclear membrane. This area of membrane fusion is suggested to be equivalent to the postacrosomal nuclear ring of eutherian mammals and as such might be expected to play in important role in fertilization of the egg of this marsupial.

A neck sheath with up to 20 striations enclosed the relatively undegenerated proximal and distal centrioles.

The middle piece had an overall length of 34.4 μm and consisted of approximately 115 tightly wound mitochondrial gyres. The cristae of the mitochondria, as in other marsupial sperm, were concentrically arranged about a central matrix. The inner surface of the mitochondrial sheath was variably indented by two or more of the nine coarse peripheral longitudinal fibres of the axial filament complex (Fig. 31). A non–mitochondrial helical sheath of filaments regularly spaced at about 70 nm, and not represented in the sperm of eutherian mammals, was closely applied to almost the entire inner surface of the plasma membrane that invested the middle piece of sperm from the tail region of the epididymis (Fig. 31). The formation of this distinctive midpiece fibre network which occurs as part of epididymal sperm maturation changes has been reported for a wide variety of Australian marsupial species, including the Dasyuridae (Harding *et al.* 1979).

The axonemal complex consisted of a cylindrical core of 20 microtubules with an overall diameter of about 210 nm. The microtubules had the typical 2 + 9 × 2 arrangement of other mammalian sperm (Fig. 32). Outside the axonemal core, nine coarse longitudinal peripheral fibres each initially with a cross sectional area equivalent to that of the entire axonemal

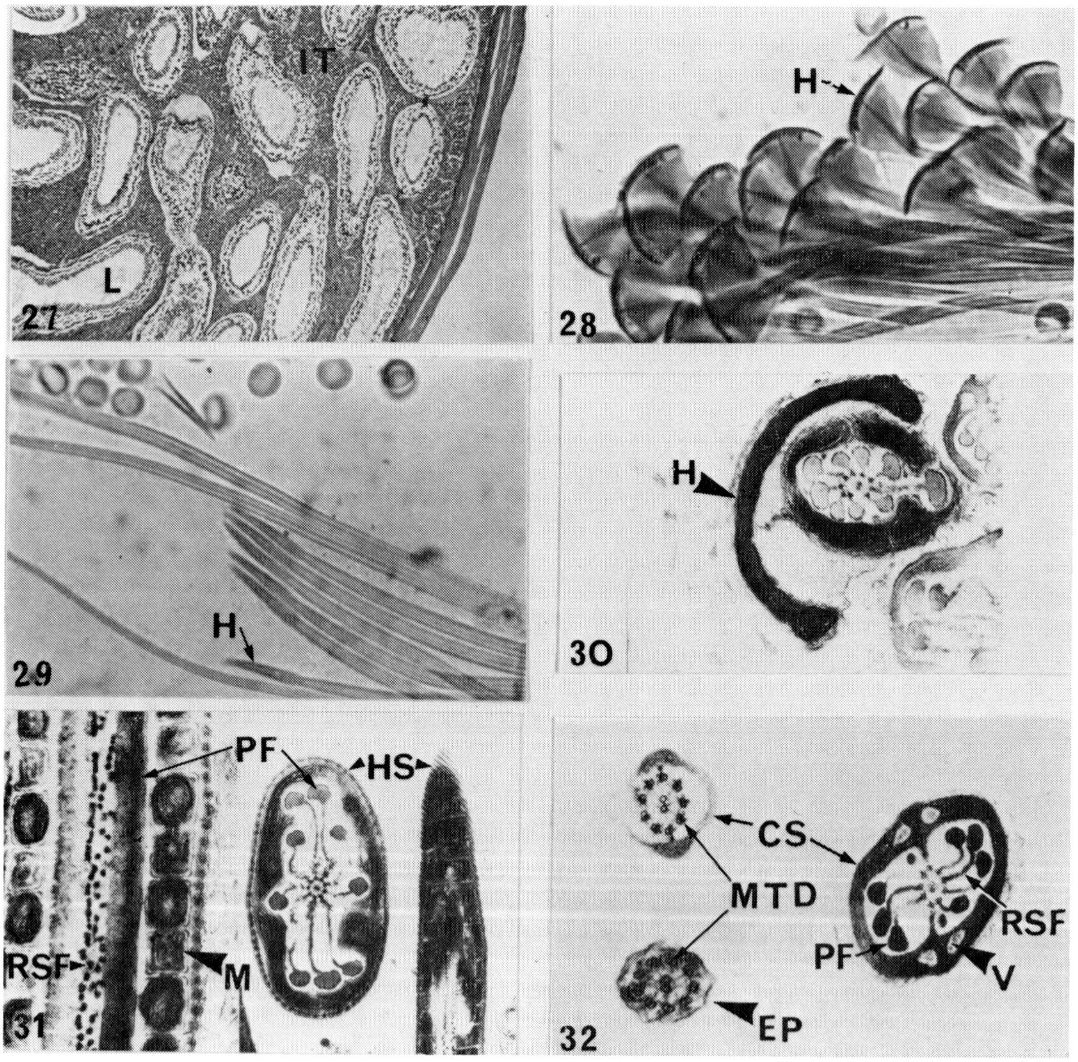

Fig. 27. Section of a testis of a sexually mature male. IT, interstitial tissue; L, lumen of semeniferous tubule. *Fig. 28.* Sperm from the proximal third of the epididymis of *Dasyurus viverrinus*. The long axis of the sperm head (H) was supported at an angle of from 60 to 90 degrees to the axis of the tail by a conical cytoplasmic droplet. *Fig. 29.* Sperm from the distal two–thirds of the epididymis of *Dasyurus viverrinus*. The sperm heads had rotated to parallel the axis of the tail and cytoplasmic cones were absent. *Fig. 30.* Electron micrograph showing transverse section of the narrow sperm head plate (H). *Fig. 31.* Electron micrographs of the midpiece region of the sperm. M, mitochondria; HS, non–mitochondrial helical sheath; PF, longitudinal peripheral fibres; RSF, radial satellite fibrils. *Fig. 32.* Electron micrographs showing transverse sections of the main piece and end–piece (EP) of the sperm tail. CS, cortical fibrous sheath; MTD, microtubule doublets of the axonemal complex; PF, longitaudinal peripheral fibres; RSF, radial satellite fibrils; V, vacuoles within the longitudinal columns of the cortical fibrous sheath.

core (Fig. 31 and Fig. 32), extended from the neck region to the subterminal extremity of the main piece of the tail. Numerous radial satellite fibrils, not found in eutherian mammals, arose at regular intervals along the inner surface of the coarse peripheral fibres (Fig. 31 and Fig. 32) and after varying degrees of curvature fused with the outer surface of the pair of microtubules of the axoneme, that corresponded in numbering to the peripheral fibres of their origin. The system of numbering follows that of Fawcett (1975). The radial satellite fibrils are identical with the narrow fibrous bands or connecting laminae which Harding *et al.* (1979) considered were characteristic of the sperm of the Dasyuridae and Peramelidae.

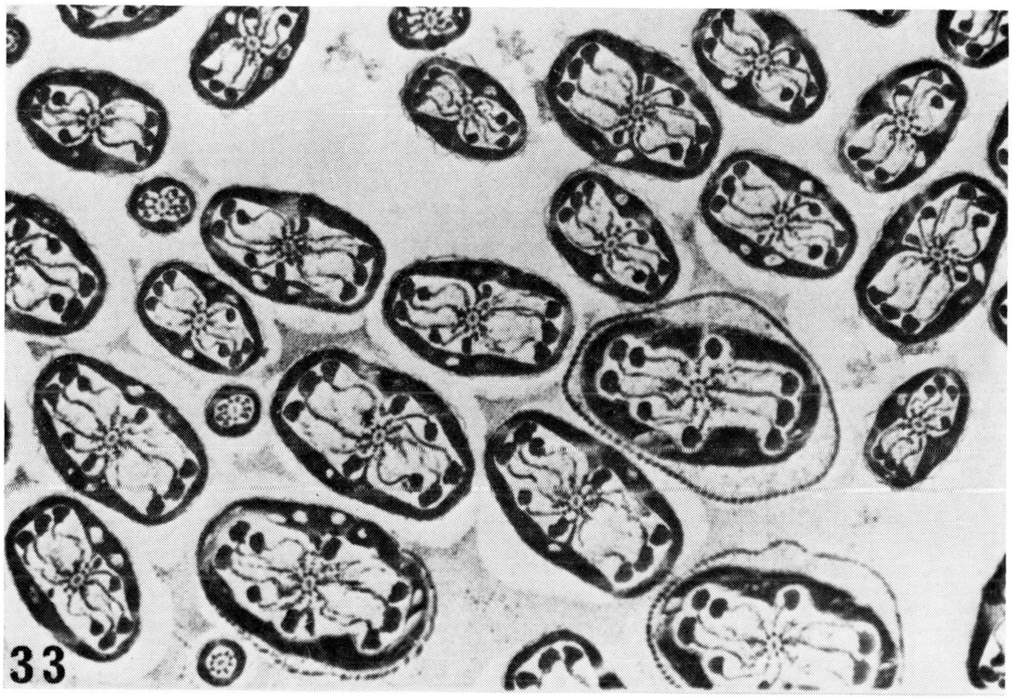

33

Fig. 33. Electron micrograph showing transverse sections of the sperm tail. Note the progressive simplification of the axonemal complex and the cortical fibrous sheath towards the distal extremity of the sperm tail.

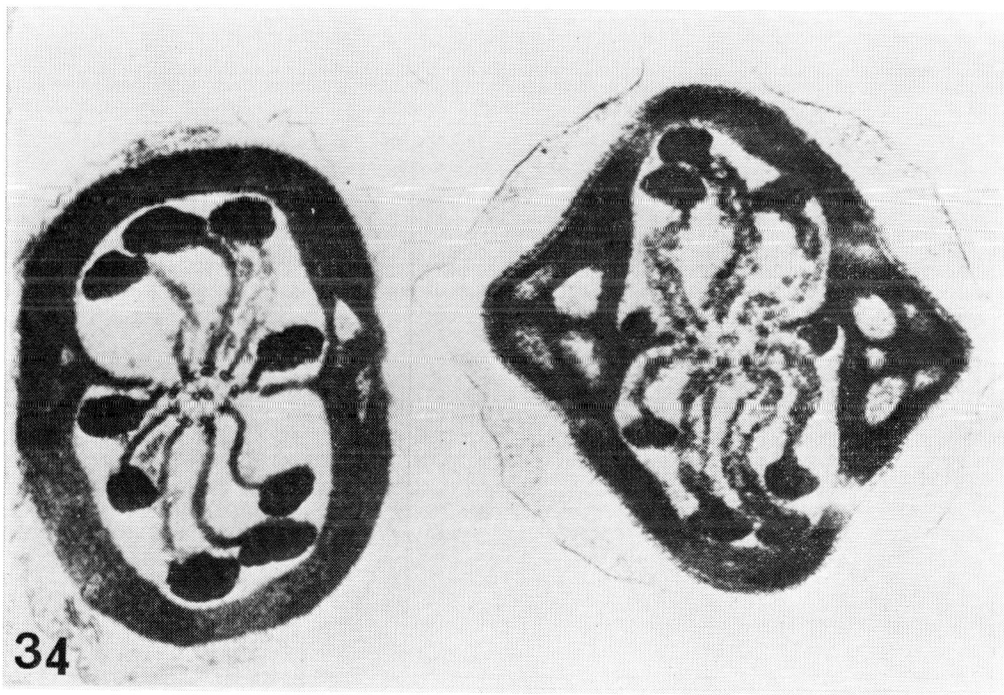

34

Fig. 34. Electron micrograph of transverse sections of the sperm tail main piece of *Sarcophilus harrisii* (left) and *Perameles nasuta* (right).

The junction between the middle piece and the main piece was marked by the annulus. Here the main piece had its maximum diameter of about 2 μm and from this point it tapered distally for about 170 μm where it had a terminal diameter of about 0.4 μm. An extensively developed fibrous sheath was situated beneath the plasma membrane of the main piece (Fig. 32 and Fig. 34). Throughout the greater part of the main piece the fibrous sheath consisted of two longitudinal columns on either side of the sperm tail. Two longitudinal rows of vacuoles occurred in each of the longitudinal columns which were themselves situated in the plane of the two central microtubules of the axoneme. An inwardly directed spur from each longitudinal column of the sheath fused with the outer surface of numbers 3 and 8 of the axonemal microtubule pairs respectively (Fig. 32 and Fig. 34). The longitudinal columns of the fibrous sheath were united on each side by a closely spaced series of semi–circular band–like ribs. The ribs occasionally branched and anastomosed with neighbouring ribs. In the distal extremely of the main piece of the tail the fibrous sheath was simplified to a relatively thin concentric structure (Fig. 32 and Fig. 33). The cross sectional area of the coarse peripheral fibres diminished distally after they reached the main piece and this was accompanied by the shortening and eventual loss of the radial satellite fibrils (Fig. 32 and Fig. 33). Coarse fibres numbers 3 and 8 disappeared relatively proximally in the main piece. However, the remaining peripheral fibres persisted in the sub–terminal region of the main piece, as small remnants adhering to their corresponding pair of axonemal microtubules. In the terminal region of the tail the microtubules of axoneme were surrounded by the greatly reduced concentric fibrous sheath and coarse peripheral fibres were absent (Fig. 32).

An end-piece (Fig. 32) was located at the distal extremity of the main piece. In this region of the flagellum the axoneme was invested only by the plasma membrane of the spermatozoan.

The overall morphology of the Tasmanian Devil sperm is typical of that of most other marsupials (Hughes 1965). As fertilization has not been reported for any marsupial species it is not possible to relate the unique structure and location of the acrosome to its role in the penetration of the egg.

The large size of the Tasmanian Devil sperm presents the possibility of manipulating individual spermatozoa. A high proportion of the mass of the giant sperm tail is occupied by the longitudinal coarse peripheral fibres of the axial filament complex and the fibrous cortical sheath is also extremely well developed. This presents a ready opportunity for the chemical dissection and identification of these elements.

Non-mitochondrial helically arranged fibrils have also been reported to occur beneath the plasma membrane in the region of the middle piece in the Brush–tailed Possum, *Trichosurus vulpecula* (Harding *et al.* 1975).

In ultrastructure the spermatozoan of the Eastern Quoll, *Dasyurus viverrinus,* is virtually identical with that of the Tasmanian Devil (Hughes, unpublished). The giant sperm of these two species are also characteristic of other dasyurid marsupials and are equalled in size by the marsupial bandicoots (peramelid group). The length of the Tasmanian Devil sperm is exceeded by *Tarsipes spencerae* (Harding *et al.* Carnivorous Marsupials Symposium). The ultrastructure of the main piece of bandicoot sperm more closely resembles that of the dasyurid group than any other. The affinities of the sperm of *Tarsipes spencerae* with the Dasyuridae are separately discussed by Harding *et al.* (This volume). In the bandicoots *Perameles nasuta* and *Isoodon macrourus,* both the fibrous sheath and coarse peripheral fibres of the axial filament complex exhibit an equivalent development and similar distribution to that of the Tasmanian Devil and the Eastern Quoll (Fig. 34). The coarse peripheral fibres in the bandicoot species exhibit a double series of radial satellite fibrils attached to their inner surface (Hughes, unpublished), and these are obviously homologous to the single series found in the Tasmanian Devil and the Eastern Quoll. Sapsford *et al.* (1970) mistakenly believed that the radial satellite fibres of *Perameles nasuta* to be a double lamina. In both *Perameles nasuta* and *Isoodon macrourus,* each of the longitudinal columns of the fibrous sheath exhibit two pairs of longitudinally arranged vacoules (Hughes, unpublished) and not a single pair on each side as occurs in the Tasmanian Devil (Fig. 34). Despite these minor differences the overall ultrastructure of the main piece of the bandicoot species strikingly resembles that of the Tasmanian Devil and the Eastern Quoll. This is considered to indicate an evolutionary affinity rather than functional convergence of unrelated marsupial groups.

ACKNOWLEDGEMENTS

I wish to express my sincere thanks to Dr. Ian Eberhard and the staff of the Tasmanian National Parks and Wildlife Service for help in obtaining the animals; to Dr. E. R. Guiler for making available holding yards and laboratory facilities at the Department of Zoology, The University of Tasmania; to B. V. Harmon, G. N. Rogers, D. J. Harbrow and R. Coleman of the electron microscope unit of the School of Anatomy, University of Queensland for assistance with the ultrastructural processing of the material.

ADDED IN PRESS

Fourteen litters of small pouch young and seven prefoetal pregnancy stages were obtained from Tasmanian Devils captured in the Avoca district between 26 and 31 March, 1981. The mean head length of individual litters ranged from 3.2 mm to 9.0 mm. A further sixteen young litters and two pregnancy stages were obtained during the first half of April, 1981. These observations indicate that the breeding season can commence up to at least one month earlier than previously reported and that the size of new born young were only approximately half that of the previous specimens earlier described. A total of 44 females were captured during March and April, 1981, that either carried litters or were pregnant. None of these animals had a body weight of less than 4 kg. Nineteen of the animals weighed between 4 kg. to 4.9 kg. while the remaining twenty-five animals were all over 5 kg. These data contrast with the previous observations which indicated that only a relatively low proportion of females reached sexual maturity before a body weight of 5 kg. was attained.

REFERENCES

ASDELL, S. A., 1964. "Patterns of mammalian reproduction". 2nd ed. p.15. Cornell University Press.

FAWCETT, D. W., 1975. The mammalian spermatozoon. *Developmen. Biol.* **44:** 394–436.

FLEAY, D., 1935. Notes on the breeding of Tasmanian Devils. *Vic. Nat.* **52:** 100–5.

FLEAY, D., 1952. The Tasmanian or marsupial Devil — its habits and family life. *Aust. Mus. Mag.* **10:** 275–80.

FLYNN, T. T., 1910. Contributions to a knowledge of the anatomy and development of the marsupialia. I. The genitalia of *Sarcophilus satanicus*. *Proc. Linn. Soc. N.S.W.* **35:** 873–86.

FLYNN, T. T., 1911. Notes on marsupial anatomy. II. On the female genital organs of a virgin *Sarcophilus satanicus*. *Proc. Roy. Soc. Tasmania* **1911:** 144–56.

FLYNN, T. T., 1922. Notes on certain reproductive phenomena in some Tasmanian marsupials. *Ann. Mag. nat. Hist.* **10:** 225–31.

FLYNN, T. T., 1939. Note on the egg of the rare marsupial, *Sarcophilus*. *Proc. Roy. Irish Acad.* **1938–39:** 16–19.

GREEN, R. H., 1967. Notes on the Devil *(Sarcophilus harrisii)* and the Quoll *(Dasyurus viverrinus)* in northeastern Tasmania. *Rec. Queen Vic. Mus.* **27:** 1–12.

GUILER, E. R., 1970a. Observations on the Tasmanian Devil, *Sarcophilus harrisii* (Marsupialia: Dasyuridae). I. Numbers, home range, movements, and food in two populations. *Aust. J. Zool.* **18:** 49–62.

GUILER, E. R., 1970b. Observations on the Tasmanian Devil, *Sarcophilus harrisii* (Marsupialia: Dasyuridae). II. Reproduction, breeding and growth of pouch young. *Aust. J. Zool.* **18:** 63–70.

GUILER, E. R. AND HEDDLE, R. W. L., 1970. Testicular and body temperatures in the Tasmanian Devil and three other species of marsupials. *Comp. Biochem. Physiol.* **33:** 881–91.

HARDING, H. R., CARRICK, F. N. AND SHOREY, C. D., 1975. Ultrastructural changes in spermatozoa of the Brush-tailed Possum *Trichosurus vulpecula* (Marsupialia) during epididymal transit. Part I: The flagellum. *Cell. Tiss. Res.* **164:** 121–32.

HARDING, H. R., CARRICK, F. N. AND SHOREY, C. D., 1979. Special features of sperm structure and function in marsupials. *In* "The spermatozoon–maturation, motility, surface properties and comparative aspects" ed. by D. W. Fawcett and J. M. Belford. Urban and Schwarzenberg: Baltimore, Munich.

HUGHES, R. L., 1965. Comparative morphology of spermatozoa from five marsupial families. *Aust. J. Zool.* **13:** 533–43.

HUGHES, R. L., 1976a. Preliminary observations on reproduction and embryonic development in the marsupial Devil, *Sarcophilus harrisii. J. Reprod. Fert.* **46:** 504–5.

HUGHES, R. L., 1976b. Reproduction in the male marsupial Devil *Sarcophilus harrisii* with particular reference to the ultrastructure of epididymal spermatozoa. *Proc. VIIIth Int. Congr. Anim. Reprod. and A.I. Krakow* **4:** 908–10.

MACKENZIE, W. C. AND OWEN, W. J., 1919. The genito–urinary system in monotremes and marsupials. *In* "The comparative anatomy of Australian mammals. Part IV". Jenkin, Buxton and Co.: Melbourne.

PEARSON, J. AND DE BAVAY, J. M., 1953. The urogenital system of the Dasyurinae and the Thylacininae (Marsupialia, Dasyuridae). *Pap. Proc. Roy. Soc. Tas.* **87:** 175–99.

POCOCK, R. I., 1926. The external characters of *Thylacius, Sarcophilus* and some related marsupials. *Proc. Zool. Soc. Lond.* **1926:** 1037–84.

SAPSFORD, C. S., RAE, CLARE A. AND CLELAND, K. W., 1970. Ultrastructural studies on the development and form of the principal piece sheath of the bandicoot spermatozoon. *Aust. J. Zool.* **18:** 21–48.

RESUMEN

6. Reproducción del "diablo de Tasmania", *Sarcophilus harrisii* (Marsupialia, Dasyuridae). Por R. L. Hughes.

La hembras sexualmente maduras del "diablo de Tasmania" pesaron más de 4.5 kg, y en la vida silvestre la mayoría de las preñeces ocurrieron durante los meses de Marzo y Mayo. El advenimiento del período de reproducción esta señalado por una notable hipertrofia del marsupio y del tracto reproductivo. Después de una ovalación de hasta 56 huevos, el útero exhibe un agrandamiento extremo. Al menos una quinta parte de los huevos fallan en clivarse y menos de la mitad alcanzan el estado unilaminar del blastocito. Las cuatro tetas son comúnmente ocupadas por neonatos

La espermatogénesis es activa al menos durante los meses de Abril y Agosto en machos de pesos corporales que oscilan entre 3.5 kg y 8.25. El sistema reproductivo del macho carece de grandes especializaciones que lo distingan del típico de los marsupiales. El glande no está dividido y el cuerpo del pene carece de espinas queratinosas. Los estpermatozoides tienen un longitud total de approximadamente 213μ y un diámetro máximo de 2μ Las especializaciones ultraestructurales del esperma son las características de otros marsupiales dasiúridos.

7. A REVIEW OF MATURATION AND FERTILIZATION IN MARSUPIALS WITH SPECIAL REFERENCE TO THE DASYURID: *Antechinus stuartii*

LYNNE SELWOOD[1]

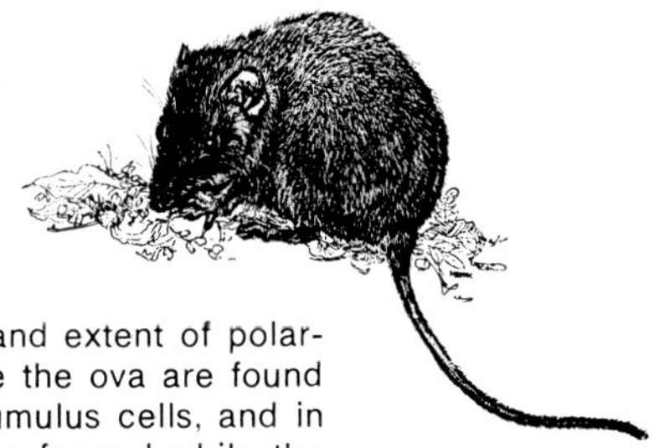

In *A. stuartii,* meiosis 1 is halted at the dictyate stage of late diplonema. Maturation starts in the ovary, towards the end of behavioural oestrus, and is completed in the ampulla following fertilization. The first polar body, which sometimes later divides, is eliminated in the ovary and the second is given off after sperm entry. The ovum is ovulated as a secondary oocyte with the chromosomes arrested in metaphase 2. Polarisation of the yolky cytoplasm in the ovum of *A. stuartii* does not occur until activation following fertilization. The size and degree of yolkiness of other marsupial ova are related to the time and extent of polarisation of the yolky cytoplasm. Fertilization occurs in the ampulla, where the ova are found to be 93.2μm $\pm$ 2.8 (S.E.) by 118.0μm $\pm$ 2.4 (S.E.) (N = 12), without cumulus cells, and in the early stages of mucin deposition. The male and female pronuclei are formed while the ovum is still in the ampulla and the ova enter the uterus in the pronuclear stage. It is suggested that possible mechanisms to prevent polyspermy in *A. stuartii* include the very small numbers of sperm which reach the ovum and the entrapment of sperm by mucin.

Pages 65–76 *in* CARNIVOROUS MARSUPIALS, ed. by M. Archer, February, 1982. Roy. Zool. Soc. New South Wales: Sydney, Australia.

MATURATION of the egg in this study is interpreted to extend from the resumption of the first meiotic division, which results in the formation of the secondary oocyte and the first polar body, to the completion of the second, which results in the formation of the mature egg and the second polar body. In the case of marsupials, as in many mammals (Austin 1961), maturation is normally not completed unless fertilization occurs. This interpretation is the same as that used by Austin (1961) and Masui and Clarke (1979) in reviews on mammalian eggs and oocyte maturation, respectively, and is very commonly used in reference to non-mammalian eggs (Balinsky 1975).

While a number of stages in maturation and fertilization have been reported in studies on marsupials (Table 1), the events of maturation and fertilization and the histology of the primary and secondary oocytes and fertilization have been studied in greatest detail in two species of marsupials, *Didelphis virginiana* (Hartman 1916, 1919; McCrady 1938; Guyaya 1968) and *Dasyurus viverrinus* (Sandes 1903; Hill 1910; Hill and O'Donoghue 1913). In addition, in *D. virginiana,* the timing of ovulation relative to oestrus has been clarified (Hartman 1923, 1928).

Of those other species where oocytes in the process of maturation in the Graafian follicle in the ovary have been reported (Table 1) most histological information is available on the polyprotodont *S. macroura* (Godfrey 1969), and the diprotodonts *Phascolarctos cinereus* (Caldwell 1887), *Pseudocheirus peregrinus* (Hughes, Thomson and Owen 1965), *Setonix brachyurus* (Waring *et al.* 1955; Sharman 1955a) and *Macropus eugenii* (Sharman 1955c; Renfree and Tyndale–Biscoe 1978).

Because ovulation is spontaneous in all marsupials so far studied and the interval of time between ovulation and the entry of the ova into the uterus is relatively short (see Tyndale–Biscoe 1973; Lyne and Hollis 1977; Renfree and Tyndale–Biscoe 1978 for review), secondary oocytes in the process of ovulation have only been obtained in two species, *Sminthopsis macroura* (Godfrey 1969), *Macropus eugenii* (Sharman 1955c).

Didelphis virginiana (Table 1) is the only marsupial species in which both the unfertilized and fertilized tubal eggs and the unfertilized and fertilized uncleaved uterine egg have been described. Penetration of the egg by the sperm and the subsequent activation of the egg have not previously been described for any marsupial.

In dasyurid marsupials, because behavioural oestrus is relatively extensive and ovulation is spontaneous (Woolley 1973), collection of early stages is difficult, and relatively little is known of maturation and fertilization. In Hill's (1910) study of development in *D. viverrinus,* the emphasis is on development prior to ovulation and post fertilization *in utero,* as no normal material was obtained during the intervening period. In *S. macroura,* where behavioural oestrus is relatively short, Godfrey (1969), obtained a comprehensive range of material from

[1] Department of Zoology, Monash University, Clayton, Victoria, Australia, 3168. (Present address: Department of Zoology, La Trobe University, Bundoora, Victoria, Australia, 3083).

the primary oocyte to the unfertilized tubal egg, but stages during fertilization and activation were not obtained. In this study maturation and fertilization in marsupials are reviewed and these events are described in *Antechinus stuartii*.

Table 1. Marsupial species in which stages during maturation and fertilization have been described.

Species	Author	Preovulation: Primary oocyte	Preovulation: Secondary oocyte	Postovulation: Unfertilized tubal egg	Postovulation: Fertilized tubal egg	Postovulation: Fertilized uncleaved uterine egg
Antechinus stuartii	Selwood 1980	—	—	+	+	+
	This study	+	+	+	+	+
Dasyurus viverrinus	Sandes (1903)	+	—	—	—	—
	Hill (1910)	+	+	—	—	+
	O'Donoghue (1912)	+	—	—	—	—
	Hill and O'Donoghue (1913)	+	+	—	—	—
Sminthopsis macroura (= *larapinta*)	Godfrey (1969)	+	+	+	—	+
Didelphis aurita	Hill (1918)	+	+	—	—	—
Didelphis virginiana	Hartman (1916)	+	+	+	—	+
	Hartman (1919)	+	+	+	—	+
	McCrady (1938)	+	+	+	+	+
Marmosa robinsoni	Godfrey (1975)	+	—	+	—	+
Phascolarctos cinereus	Caldwell (1887)	+	—	—	—	+
Trichosurus vulpecula	Pilton and Sharman (1962)	+	—	—	—	—
	Hughes and Shorey (1973)	—	—	+	—	—
Pseudocheirus peregrinus	Hughes, Thomson and Owen (1975)	+	+	—	—	—
Macropus giganteus	Clark and Poole (1967)	+	—	—	—	—
Setonix brachyurus	Waring, Sharman, Lovat and Kahan (1954)	+	+	—	+	—
	Sharman (1955a)	—	+	—	—	—
	Sharman (1955b)	—	—	—	+	—
Macropus eugenii	Sharman (1955)	—	+	—	—	—
	Renfree and Tyndale-Biscoe (1978)	—	—	—	+	+
	Tyndale-Biscoe (1979)	—	—	—	+	+
	Tyndale-Biscoe and Rodger (1975)	—	—	+	—	—

MATERIALS AND METHODS

Specimens of *A. stuartii* were trapped in the localities of Kinglake, Warburton, Donna Buang and Lake Mountain around Melbourne from 1975 to 1978. The females were either mated in the field before capture or were mated in a laboratory colony, which was maintained according to the methods outlined by Woolley (1966a, 1973) and Selwood (1980) for this species. In 1977, females 7 to 11 were obtained from a colony maintained by Mr. G. Young, Dept. of Genetics, La Trobe University.

Selwood (1980) established that ovulation in *A. stuartii,* occurred around the time when the number of epithelial cells in the urine of the female started to decrease. Using this method, reproductive tracts were removed from females during maturation, fertilization and early cleavage. For females which were trapped after mating in the field, material was collected at the end of the breeding season and in the week following the male "die–off" in that area (pers. comm. Assoc. Prof. A. K. Lee, Dr. A. Bradley).

Following laparotomy the ovaries, oviducts and uteri were examined to assess the stage of development of the eggs. Where the uteri were deemed to contain eggs they were flushed with phosphate buffered saline and the freed eggs were fixed with Karnovsky's fixature for 1–2 hours. In one case (female 10, Table 2), the uterus and enclosed eggs were fixed and sectioned without prior flushing. The remainder of the reproductive tracts were fixed in aqueous Bouin for 24 hours and the right hand side, and in some cases both sides (Nos. 1, 7, 11, 12, Table 2), were dehydrated, embedded in paraffin and serially sectioned at 8μm. Sections were stained in haematoxylin and eosin.

The number and state of the Graafian follicles and oocytes in the ovarian sections and the number and stage of development of the uterine eggs were assessed. The number and stage of development of the eggs and the presence or absence of sperm were determined in sections of the oviducts while the level of sperm and eggs were recorded in reconstructed oviducts.

Additional information on the eggs *in utero* was obtained from slides of 1–4 cell stages of *A. stuartii* lent by Dr. P. Woolley, Department of Zoology, La Trobe University.

RESULTS

PREOVULATION

Prior to ovulation during late behavioural oestrus, the Graafian follicles were obvious as blisters about 0.5mm in diameter on the surface of the ovary. By transmitted light, the ova could be seen lying within the follicles. At this time the reproductive tract, apart from the infundibulum of the oviduct, which had large, ciliated fimbriae folded around the ovary, was small and inconspicuous. In some females the uterus was slightly larger and pinker with more conspicuous blood vessels. The results from females sampled during this stage of development are recorded in Table 2. In one female (4), in which cell numbers had been high in the urine for 2 days, early stages in antral formation were seen as isolated patches of antral fluid collected between the follicle granulosa cells. In female 6 (Table 2), in which a high cell number had been recorded in the urine for 4 days, antral formation was more advanced and droplets of antral fluid had begun to coalesce in the centre of the follicle. For the 12 follicles in these two females, the mean diameters of the ovoid follicles were 438.0μm ± 15.2 (S.E.) by 501.7μm ± 14.7 (S.E.) and the enclosed ovoid primary oocytes had mean diameters of 101.3μm ± 3.73 (S.E.) by 122.6μm ± 2.7 (S.E.) which are of similar dimensions to the diameter reported by Lintern–Moore *et al.* (1976).

Specimens 3, 7, 7L, 9 and 17 all had large antral cavities and the follicles measured 593.4 μm ± 22.2 (S.E.) by 681.1μm ± 22.1 (S.E.) N=11). The enclosed primary oocytes measured 115.3μm ± 7.06 (S.E.) by 131.1μm ± 1.92 (S.E.) (N = 11). About 2–5 cell layers comprised the narrowest part of the membrana granulosa and the cumulus oophorus was composed of about 2–3 cell layers at its narrowest point. The primary oocyte was attached to the wall of the follicle by a broad discus proligerus which was sometimes broken up into a retinaculum. Cell numbers were high in the urine of females 3 and 17 but no information is available on cell numbers in the urine of females 7 and 9. Specimen 3 was killed 4 days before the male "die–off" in the area of capture (Dr. A. Bradley pers. comm.). During these stages of follicle formation, the primary oocyte (Fig. 1 a) was surrounded by a zona pellucida 1.2μm ± 0.08 (S.E.) wide and consisted of an internal vaculated cytoplasmic region surrounded by a narrow cortical zone, 3–8μm wide, composed of finely granular and fibrous eosinophilic material. No zona radiata was visible in these oocytes.

The peripherally situated, ovoid nucleus, 22.6μm ± 2.6 (S.E.) by 38.3μm ± 1.9 (S.E.) (N = 11) was in the dictyate or resting stage of late diplonema stage of prophase 1. It was in the form of a germinal vesicle in which narrow threads of chromatin with occasional granular material could be seen (Fig. 1a). A single nucleolus, 5.8μm ± (0.47) which contained a single vacuole of variable size or several smaller vacuoles was present. For all these females, where the ova were at the primary oocyte stage (Table 2), numerous sperm were found lying in clusters in an area 1–1.5 mm long in the isthmus of the oviduct anterior to the uterotubal junction. Their heads were closely associated with the columnar epithelium. No sperm were found anterior to this region in the ampulla or infundibulum of the oviduct.

In 6 females, oocytes in various stages of the first meiotic division and up to metaphase of the second meiotic division were found (Table 2). In each ovary, the meiotic divisions were not synchronized and two or three consecutive stages of meiosis were found. Where both ovaries from the one female were examined (1, 11, 12) the oocytes from each ovary were in approximately the same stages of meiosis. No record is available for the number of cells in the urine of these 6 females. The tracts from females 1 and 3 were obtained post-mortem, as these females had died in the traps in the field. Female 2 was trapped 4 days before, and female, 1 was trapped during the first day after the male "die–off" for that area (Assoc. Prof. A. K. Lee, Dr. A. Bradley, pers. comm.).

The Graafian follicles during these meiotic divisions were 475.8μm ± 14.75μm (S.E.) by 581.5μm ± 12.9 (S.E.) and the oocytes measured 116.0μm ± 3.3 (S.E.) by 145.9μm ± 3.3 (S.E.). The cumulus was composed of 1 to 2 layers of cells and the membrana granulosa had 2–3 layers of cells. The oocyte was still attached by a wide discus proligerus. The width of the zona pellucida had not increased and no zona radiata was visible, but in some specimens, a narrow clear cytoplasmic zone lay beneath the zona pellucida. However, the amount of cortical material increased during meiosis 1, so that by the time of formation of the secondary oocyte and the first polar body, the cortical cytoplasm was up to 24μm wide in the region of the nuclear material and 8μm at the opposite pole or formed an even layer about 14μm wide. Occasional patches of cortical material were found in the inner regions of the oocyte.

In 2 females (2 and 11) Table 2, Fig. 1b) the nucleus was passing out of the dictyate stage and entering the diakinesis of prophase 1, where the nucleolus had disappeared and condensation of the chromosomes was apparent. At metaphase 1, the chromosomes, as short rods,

Table 2. Stage of development of the ova from a unilateral sample of the reproductive tract of *Antechinus stuartii.* S, polyspermic ovum.

Serial No. of ♀♀	Organ	Description	No. of ova
4	ovary	primary oocyte-dictyate stage of diplonema — germinal vesicle with nucleolus	6
6	ovary	primary oocyte-dictyate stage of diplonema — germinal vesicle with nucleolus	6
3	ovary	primary oocyte-dictyate stage of diplonema — germinal vesicle with nucleolus	5
7	ovary	primary oocyte-dictyate stage of diplonema — germinal vesicle with nucleolus	5
7L	ovary	primary oocyte-dictyate stage of diplonema — germinal vesicle with nucleolus	8
9	ovary	primary oocyte-dictyate stage of diplonema — germinal vesicle with nucleolus	7
17	ovary	primary oocyte-dictyate stage of diplonema — germinal vesicle with nucleolus	7
1	ovary	Primary oocyte-metaphase 1	1
		-elimination of 1st Polar body	3
		Secondary oocyte-1st Polar body, metaphase 2	2
		Sections with nuclear material missing	1
1L		Primary oocyte-metaphase 1	3
		-telophase 1	1
		-elimination of 1st Polar body	1
		Secondary oocyte-Polar body, metaphase 2	4
2	ovary	Primary oocyte-diakinesis of prophase 1	1
		-metaphase 1	2
		-anaphase 1	1
		Sections with nuclear material missing	1
8	ovary	Primary oocyte-metaphase 1	2
		-anaphase 1	2
		Sections with nuclear material missing	3
11	ovary	Primary oocyte-metaphase 1	3
		-anaphase 1	3
		-late telophase	1
11L	ovary	Primary oocyte-diakinesis of prophase 1	1
		-metaphase 1	2
		-anaphase 1	1
		Sections with nuclear material missing	2
12	ovary	Primary oocyte-metaphase 1	3
		-telophase 1	1
12L	ovary	Primary oocyte-metaphase 1	6
13	oviduct	Secondary oocyte-telophase 2, sperm penetration	$2+1^S$
		-telophase 2, ♂ pronucleus	1
		2nd Polar body, sperm penetration	$1+1^S$
		-2nd Polar body ♂ pronucleus	1^S
14	oviduct	Secondary oocyte-metaphase 2	1
		-2nd Polar body, ♂ pronucleus	1
		-2nd Polar body, ♀ + ♂ pronuclei	1
		2nd Polar body, ♀ + ♂ pronuclei adjacent	2
15	uterus	1 cell	2
		2 cell	3
5	uterus	2 cell	1
		3 cell	1
		4 cell	2
10	uterus	4 cell	2
17	uterus	3 cell	2
		4 cell	2
18	uterus	2 cell	3
		3 cell	1

were lined up on the equater of the spindle. The long axis of the spindle was usually tangential to the circumference of the oocyte. During anaphase, the more extended chromosomes were pulled apart (Fig. 1c) and then became more condensed and compacted together at the poles of the spindle at telophase. The chromosome number of 14 which has been established for this species (Sharman 1961) could be counted in some oocytes during metaphase and anaphase.

Near the outer pole of the spindle the plasma membrane became wrinkled and eventually a small polar body, containing very condensed chromosomal material was given off (Fig. 3 and Fig. 1d). The polar body was variable in size and ranged from 8μm x 3μm to 16μm x 4μm. It lay flattened against the zona pellucida and formed a slight depression in the secondary oocyte. No interphase or prophase of the second meiotic division were seen and the condensed

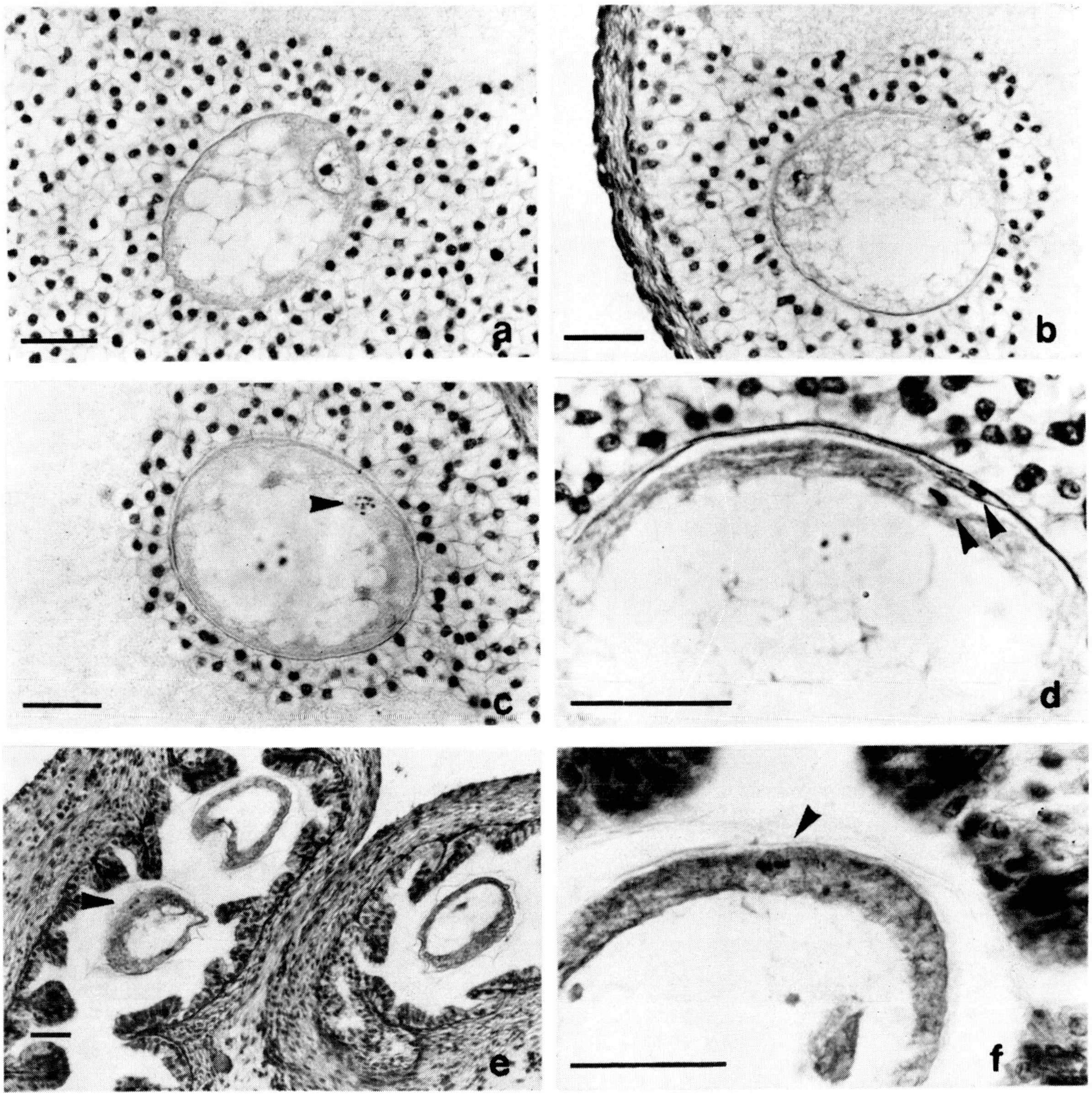

Fig. 1. Length of bar equals 50 μm. a. Primary oocyte in the dictyate stage of prophase 1 surrounded by cumulus cells in the Graafian follicle of *Antechinus stuartii*. b. Primary oocyte without nucleolus, in late prophase 1, surrounded by cumulus cells in the Graafian follicle of *A. stuartii*. c. Anaphase chromosomes (arrow) in a primary oocyte of *A. stuartii*. d. Secondary oocyte showing the first polar body and egg chromosomes (arrows) in the Graafian follicle of *A. stuartii*. e. Ova in the ampulla of the oviduct in *A. stuartii*. Telophase 2 chromosomes are visible in the fertilized ovum on the left (arrow). f. Unfertilized egg of *A. stuartii*, showing the chromosomes in metaphase 2 (arrow) in the even band of cortical cytoplasm.

chromosomes appeared to pass straight on to metaphase 2 (Table 2, 1 and 1L). These processes which occurred during maturation in the ovary are summarised in Fig. 3.

The ova appeared to be randomly distributed around the periphery of the follicle in relation to the ovary and the position of the nuclear material in each ovum appeared to be random in relation to the position of the ovum in the follicle. Some abnormal follicles were found in the ovaries of two females (Table 2, 1 and 2). In these abnormal smaller follicles, no antrum had developed, the ova were smaller than in normal follicles and the nucleus was frequently found in a central position thus altering the arrangement of the cytoplasmic material. Oocyte nuclei were found in the dictyate stage of prophase 1 although some had progressed to metaphase 1. Polynuclear oocytes and polyovular follicles were not found in this study although polyovular follicles were found in the ovary of an *A. stuartii* from the 1976 colony (Selwood, unpublished) and by Woolley (1966b) in her study of *A. stuartii*. Numerous sperm were encountered in the isthmus of the oviduct during these stages but no sperm were found in the ampulla or infundibulum of the oviduct.

POST OVULATION

TUBAL OVA: Ova which were completely free of cumulus cells were found in the ampulla of the oviduct (Fig. 1e) in two females (Table 2) in which the number of epithelial cells had begun to fall in the urine (Selwood 1980). At this time the uterus was relatively small, obviously pinker and with prominent blood vessels. The infundibulum of the oviduct was large and conspicuously folded around the ovary. In the ovary, no ova could be seen in the follicles, some of which had small amounts of blood associated with them. No ova were found in the infundibulum or isthmus of the oviduct. The epithelium lining the oviduct was ciliated in the infundibulum, ampulla (Fig. 2c) and isthmus.

Not all eggs were in the same stage of maturation or fertilization in each oviduct (Table 2) and the eggs which were lowest in the oviduct tended to be at the most advanced stages of development. Most stages in the second meiotic division and fertilization, from metaphase 2 to formation of the male and female pronuclei, were obtained. Maturation and fertilization in the oviduct of *A. stuartii* are summarised in Fig. 3. All eggs found in this study were in the very early stages of deposition of the mucoid

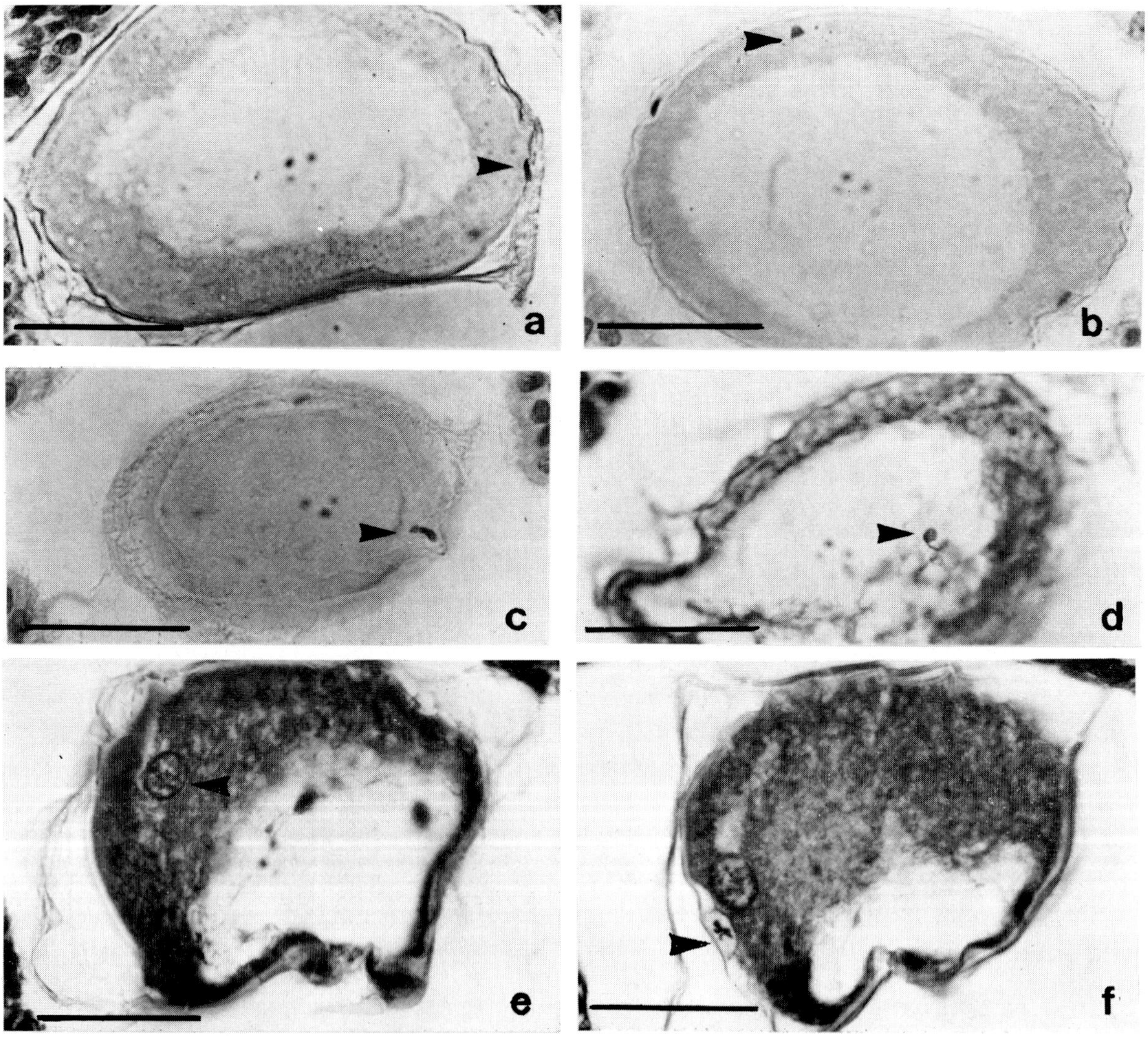

Fig. 2. Length of bar equals 50 μm. a. Polyspermic ovum of *A. stuartii* showing sperm head lying against zona (arrow). b. Section from the same polyspermic egg as in Fig. 2a showing three sperm heads of which one (arrow) has completely penetrated the zona. c. An early stage in fertilization in the late telophase 2 ovum of *A. stuartii*. The sperm head (arrow) has penetrated the zona and has started to swell. d. The sperm head, attached to portion of the tail, has swollen to form a small pronucleus (arrow), in a fertilized egg of *A. stuartii*. e. Male pronucleus (arrow) in the fertilized egg of *A. stuartii*. f. Female pronucleus and the second polar body (arrow) in the same fertilized egg of *A. stuartii* as in Fig. 2e.

layer and strands of mucoid material were found around, but not closely applied to the surface of each egg (Fig. 1e, f and Fig. 2a–f). The eggs in the oviduct had an irregular ovoid shape and were reduced in size to 93.2μm ± 2.8 (S.E.) by 118.0μm ± 2.4 (S.E.) and were each surrounded by a narrow zona pellucida, of similar dimensions to that found around eggs in the ovary. No zona radiata was visible.

One unfertilized secondary oocyte (14, Table 2) in which the chromosomes were in metaphase 2 with the long axis of the spindle radially oriented was found (Fig. 1f). The internal structure of the oocyte was similar to that found in late secondary oocytes in the ovary, in that the central core of vacuolated cytoplasm was surrounded by an even layer of cortical cytoplasm, 13μm wide. Some clumps of cortical material were found in the internal region. No perivitelline space was present. The first polar body, in which the chromatin formed a compact mass 3 to 6μm in size, was about 4μm by 10μm in size and lay in a slight depression beneath the zona pellucida. No sperm were found on the surface of the oocyte but several strands of mucin were found associated with it.

Large clusters of sperm were found associated with the epithelium of the isthmus of the oviduct, anterior to the uterotubal junction. However, separate sperm were also found anterior to this area in the ampulla of the oviduct, below and around the eggs. No sperm were found in the infundibulum.

The earliest stages in fertilization were found in the ampulla in female 13 (Table 2). The number of sperm in the lumen of the ampulla around each egg ranged from 0–7 and they were mostly close to or flattened against the zona pellucida (Fig. 3). Sperm were found in the zona in three oocytes. Of the sperm found lying against the zona pellucida, eight were lying with the dorso–proximal or acrosomal surface flattened against the zona. The sperm appeared to penetrate the egg at an oblique angle (Fig. 3). Swelling of the sperm head appeared to occur very soon after penetration of the zona (Fig. 2b and 2c). In two oocytes, where the sperm head was still in contact with the inside of the zona, the sperm head had started to swell. The expanding sperm was at first triangular, 1.5x4x4μm then 4x4x4μm, then many small granules appeared in the nucleoplasm and then it became a spherical pronucleus about 3–4μm in diameter containing many small granules (Fig. 3). Pronuclei during these stages of formation were found in the cytoplasm, mostly in the cortical regions of the oocyte. The sperm tail was associated with some pronuclei (Fig. 2d). Once the sperm pronucleus had migrated to the region of the female chromosomes and the polar bodies, the granular material was no longer visible and many strands of chromatin could be seen in the nucleoplasm (Fig. 2e). No nucleoli were present and the pronucleus measured 8μm x 10μm by 12μm.

Early stages of sperm penetration of the secondary oocyte (i.e. when the sperm was penetrating the zona) were found when the chromosomes were in late telophase 2 (Fig. 1e) and when the second polar body had been eliminated (Table 2). Oocytes with supernumerary sperm were found in three cases in female 13 (Table 2), and these oocytes each had 2 or 3 sperm present inside the zona in the vitellus of the egg. The second polar body was somewhat variable in size, 6μm x 10μm to 4μm by 16μm but was usually large and contained more distinct chromosomes (Fig. 2f) than the first polar body which measured 3μm x 10μm to 5μm x 12μm. In four cases, the first polar body appeared to be in the process of division. The female pronucleus formed soon after the elimination of the second polar body and measured 6μm x 10μm to 12μm x 15μm. A nucleolus, 2μm in diameter, was seen in the smallest female pronucleus. All female pronuclei contained densely staining strands of chromatin material (Fig. 2f).

At the end of telophase 2 and during the formation and migration of the male pronucleus, the cortical material became less evenly distributed in the oocyte so that by the time the female pronucleus was formed, the cortical material was about 50μm wide in the hemisphere containing the pronuclei and polar bodies and about 2μm wide in the opposite one (Fig. 2f, Fig. 3). In the most advanced tubal oocyte found, the male and female pronuclei were about 20μm apart and the cortical and vacuolated cytoplasmic material had become distinctly polarised so that most of the eosinophilic granular fibrous cytoplasm was found in the hemisphere which contained the pronuclei and polar bodies and most of the vacuolated cytoplasm was confined to the opposite pole (Fig. 3). No zona radiata was visible and in ten ova a narrow perivitelline space was visible beneath the zona pellucida.

UTERINE OVA: Reproductive tracts were examined in 5 females in which the oocytes had entered the uterus (Table 2). In all the females, the uterus was larger than in previous specimens and had a pinkish colour and well developed vasculature. The infundibulum was relatively smaller and less conspicuous in those females in which the uteri contained 4 cell stages. Very much reduced numbers of sperm were found in the isthmus of the oviduct and those sperm which were present were not lying in close proximity to the epithelium nor were they lying

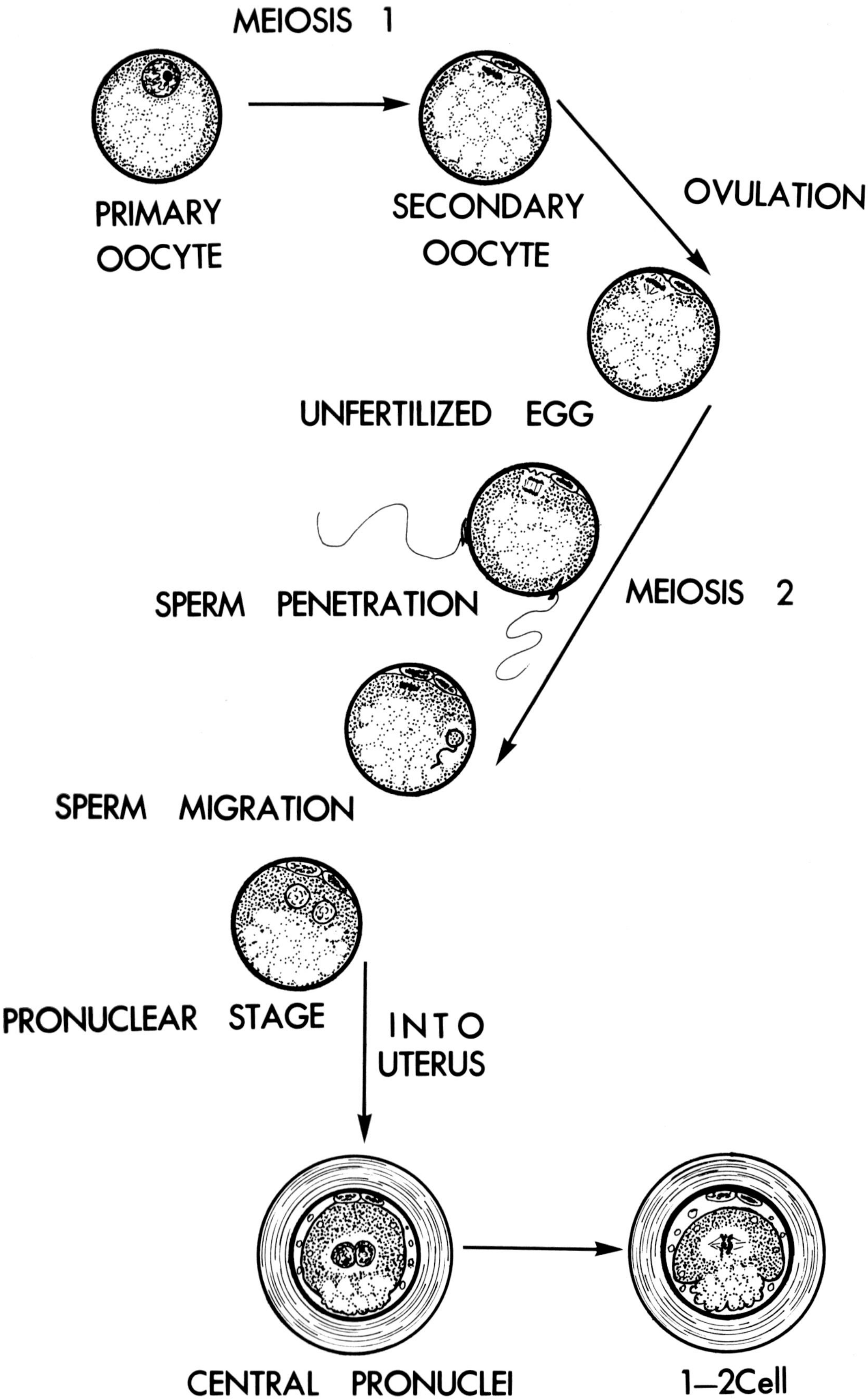

Fig. 3. A diagram of maturation and fertilization in *A. stuartii*. Maturation begins in the ovary and is completed in the ampulla of the oviduct, where fertilization occurs. The eggs receive the mucoid layer and shell after fertilization and enter the uterus in the pronuclear stage. Cleavage occurs in the uterus. Not drawn to scale. Based on material obtained in this study and from Dr. P. Woolley, Zoology Dept., La Trobe University.

in dense clusters. In females where the embryos were at the 4 cell stage degenerating sperm were found in the oviduct. Epithelial cell numbers had started to decline in the urine of females 5 and 15 but no record is available of the epithelial cell numbers in the urine of the other females.

For live early uterine stages the zona was 7μm, the mucoid layer was 18μm wide and the total diameter was 242μm (Selwood 1980). The shell was about 4μm wide in the earliest specimens and up to 6μm wide in the 4 cell material. In sectioned material of female 10 it was not possible to determine most dimensions because of fragmentation of the membranes but the zona was about 1.2μm wide and the shell was about 5μm wide. Many sperm were embedded in the mucoid layer. In the uncleaved uterine ova, the vitellus before fixation measured 140μm x 112μm. While the whole egg was spherical, the vitellus was distinctly flattened on one side where the 'yolky' material was accumulating. In addition, globules of 'yolky' material were extruded from the rounded surface of the vitellus and isolated globules could be seen in the perivitelline space.

By transmitted light, in live whole specimens, the pronuclei could be seen in a central area above the flattened side and when the first cleavage had commenced, a spindle which had the long axis parallel to this flattened side was visible. In sectioned material (from Dr. P. A. Woolley, pers. comm.) this yolky material was seen to be the same as the vacuolated cytoplasmic material which was accumulated at one pole of the egg in the tubal ova. The pronuclei ,each with one or two nucleoli, lay in the centre surrounded by clear cytoplasm. A spindle formed in this central area and the chromosomes were situated on the equator for the metaphase of the first mitotic division of cleavage. Fusion of the male and female pronuclei to form a single pronucleus was not observed .

The final phases of fertilization in the uterus are summarised in Fig. 1.

DISCUSSION AND REVIEW

Because ovulation is spontaneous and behavioural oestrus is relatively long, in many dasyurids (Woolley 1966a, 1973) the timing of ovulation in relation to oestrus is unknown. However, in a number of species, the presence of epithelial cells in the urine of the female have been shown to be characteristic of the period of behavioural oestrus (Godfrey 1969, Woolley 1971).

Development of the antrum and Graafian follicle enlargement have been described in both polyprotodont (Sandes 1903, O'Donoghue 1912, 1916, Hill and O'Donoghue 1913, Martinez–Esteve 1942, Godfrey 1969) and diprotodont marsupials (Caldwell 1887, Waring *et al.* 1955, Sharman 1955a, Hughes *et al.* 1965). Antral enlargement occurs during oestrus when epithelial cells reach high numbers in the urine of the female in *A. stuartii.* Unfortunately, because no records were available, it was not possible to say whether resumption of the first maturation division occurred before or after the fall in cell numbers in the urine. Tubal ova and early cleavage stages in the uterus were obtained once cell numbers began to decline (Selwood 1980).

The elaborate development of the fimbriae of the infundibulum of the oviduct which occurs about a week before ovulation in *A. stuartii* presumably facilitates transport of the eggs to the oviduct. Marked regression of the fimbriae occurs by the time the uterine ova have completed the second cleavage which is approximately 1–2 days after fertilization (Selwood 1980).

Sperm Storage and Movement: Sperm storage has been reported to occur in the isthmus of the oviduct in *A. stuartii* (Woolley 1966, Selwood 1980), *S. macroura* (Godfrey 1969), *M. robinsoni* (Godfrey 1975) and *D. viverrinus* (Hill and O'Donoghue 1913) for periods of up to 15 days (O'Donoghue 1913, Woolley 1966, Selwood 1980) and in the vaginal caeca of bandicoots (Lyne and Hollis 1977). Such long term storage has not been found in other marsupials, although Tyndale–Biscoe and Rodger (1978) have demonstrated that the cervical canal on the ovulation side of *M. eugenii* acts as a reservoir for and possibly a selective barrier to sperm migration for 24 hours after copulation. As many sperm remain in the isthmus above the uterotubal junction during fertilization in *A. stuartii,* it seems likely that this area also acts as a selective barrier to further sperm migration. Similar functions have been assigned to the cervix and uterotubal junction in eutherian mammals (Thibault 1973). The role of the anterior vaginal canals was not investigated in *A. stuartii* and they appear to provide not more than transient storage in those marsupials which have been investigated (Hartman 1924, Hughes and Rodger 1971, Tyndale–Biscoe and Rodger 1978). Passage of ova through the isthmus removes most of the sperm and the oviducts are free of intact sperm by the time the embryos have reached the 4 cell stage, which accounts for the final disappearance of sperm from the urine of recently mated females a day or so after ovulation (Selwood 1980).

Primary and Secondary Oocytes: Primary and secondary oocytes have been reported in a large number of marsupials (Table 1) but their histology has only been described in detail in *D. virginiana* (Hartman 1916, 1919, McCrady 1938), *D. viverrinus* (Hill 1910), *P. cinereus* (Caldwell 1887). In these four species, as in *A. stuartii* meiosis is arrested at the dictyate stage of late diplonema in the primary oocyte, at a similar stage to that which occurs in eutherians (Austin 1961, McGaughey 1978). As in many eutherians (Austin 1961, McGaughey 1978), after formation of the secondary oocyte the chromosomes pass straight into metaphase 2, and the oocyte is ovulated in this state in those marsupials where this stage has been obtained (Table 1). Shedding of the cumulus starts in the follicle in the ovary in *D. viverrinus* (Hill 1910), is completed in the ovary in *S. macroura* (Godfrey 1969) and is completed by the time the oocytes enter the oviduct in *D. virginiana* (Hartman 1912, 1916) and *A. stuartii*. Accidental retention of some cumulus cells in the mucoid layer have been reported in *P. cinereus* (Caldwell 1887) and *D. aurita* (Hill 1919). Primordial follicles and oocytes in *Isoodon macrourus* have been described by Ullmann (1978).

FERTILIZATION: The unfertilized tubal egg has been reported in a number of species (Table 1) and has been described in some detail as a secondary oocyte with chromosomes in metaphase 2 by Hartman (1916, 1919) in *D. virginiana*. Penetration of the egg by the sperm has not been described in any marsupial, but has been assumed to occur in the ampulla of the oviduct (Hill 1910, Hartman 1916) as it does in eutherians (Austin 1961, Masui and Clarke 1979) and in *A. stuartii*. As the unfertilized egg in *A. stuartii* was a secondary oocyte with chromosomes in metaphase 2, and early stages in sperm penetration were found in oocytes in early telophase 2, it is assumed that penetration of the egg triggers completion of the second maturation division. In those marsupials where they have been described, the ripe secondary oocyte and the unfertilized tubal or uterine egg (Table 1) are in metaphase 2, the stage in which the second meiotic division is arrested until fertilization in most mammals (Austin 1965).

Swelling of the sperm nucleus to form a male pronucleus appears to be very similar to that described by Austin (1961) for the rat. Unlike in eutherian mammals, the male pronucleus, in *A. stuartii* before reaching its maximum size, moves towards the female pronucleus which remains in the peripheral cytoplasm near the polar bodies. After the ova have descended the oviduct and entered the uterus, the two pronuclei are found in a central position in those species where this stage has been described. The pronuclear stage has been reported in the lower oviduct in *S. brachyurus* (Sharman 1955) and in *D. virginiana* (McCrady 1938).

The unusual distribution of the acrosomal material along the proximodorsal surface in *A. stuartii* and other dasyurids (Hughes 1965, Harding *et al.* 1975) may be the reason for the flattened orientation of the sperm against the zona in *A. stuartii* in the early stages of penetration. The progressive secretion of mucoid material around the egg may assist in binding the sperm to the surface of the zona. The oblique passage of sperm through the zona which was found in *A. stuartii,* has also been reported in some eutherian mammals (Austin 1965).

POLYSPERMY: In tubal eggs found in *A. stuartii,* the incidence of polyspermy was relatively high (25% compared to the 1–2% incidence which occurs in eutherians (Austin 1961). This high incidence in *A. stuartii* may be a reflection of the small size of the sample and a high incidence of polyspermy in some individuals. Polyspermy occurs in another dasyurid, *D. viverrinus* (Hill 1910) but no information is available on polyspermy in other marsupials.

Examination of the *Dasyurus viverrinus* slides and notes in the Hill Collection at the Hubrecht Laboratory shows that polyspermic eggs were found in three of the four samples of fertilized uncleaved eggs (catalogue nos. MS57, MS59, MS61, MS62).

There appear to be a number of possible mechanisms to reduce the incidence of polyspermy in *A. stuartii.* Of those few sperm which get past the reservoir/barrier area in the isthmus of the oviduct, only those which reach the egg before it is covered by too much mucin and at the appropriate angle so that the acrosome is applied to the zona will be able to penetrate it. Hartman (1916) suggested that the mucin may trap excess sperm in *D. virginiana* and it is likely that such a mechanism is of general occurrence in marsupials as sperm have been reported in the mucoid material of all groups which have been studied so far, although the thickness of the mucoid material is variable (Lyne and Hollis 1977).

The presence of supernumerary sperm in the zona pellucida in tubal eggs of *A. stuartii* and occasionally in the perivitelline space in uterine material suggests that while a vitelline reaction or change in the plasma membrane occurs following entry of one sperm, that the zona reaction (Austin 1961) is absent or inefficient in this species. It may be that the entrapment of sperm by the deposition of mucin partially or entirely supplements the zona reaction in this and other marsupials. In species where the histology has been studied, sperm have been reported in the zona as well as the mucoid layer as in *S. brachyurus* (Sharman 1955), but have not been reported in *D. virginana* (Hartman 1916, 1919, McCrady 1938) and *D. viverrinus* (Hill 1910). Presumably in marsupials as in eutherians, the efficiency of the zona reaction varies from species to species (Austin 1965).

Ultrastructural studies are required to demonstrate the presence of a cortical reaction in marsupials. At the light microscope level the lack of obvious cortical granules in unfertilized eggs and the narrowness of the perivitelline space in fertilized tubal eggs suggests that the cortical reaction is minimal or delayed in *A. stuartii*. This may also occur in *D. virginiana,* where a perivitelline space is not found until deutoplasmolysis commences in fertilized uterine eggs (Hartman 1916, McCrady 1938). A perivitelline space has been shown in tubal eggs in the tammar (Renfree and Tyndale–Biscoe 1978, Tyndale–Biscoe 1979).

Polarity and Egg Structure: As the pronuclei and polar bodies are found in the opposite hemisphere to the yolky material in tubal eggs in *A. stuartii,* the figures are shown with this hemisphere at the top of the page, in accordance with normal embryological conventions (Balinsky 1975). Hartman (1916, 1919) used the angle of the cleavage plans in *D. virginiana* and Hill (1910) used the presumptive fate of the blastomeres in *D. viverrinus* to determine the orientation of their figures.

In marsupials, polarity due to the unequal accumulation of 'yolk' material in one hemisphere of the egg occurs while the eggs are still in the ovary in the primary oocyte in *P. cinereus* (Caldwell 1887) and in the secondary oocyte in *D. viverrinus* (Hill 1910) but has not been demonstrated in ovarian, tubal or uterine ova of *D. virginiana* (Hartman 1916, 1919, McCrady 1938) or *D. aurita* (Hill 1918). In contrast, in *A. stuartii,* a marked unequal accumulation of yolky cytoplasm occurs during activation of the tubal eggs following fertilization.

This suggests that in those marsupials with large yolky eggs, polarisation of the yolk occurs early in the ovary (e.g. *P. cinereus, D. viverrinus*), and in those with smaller eggs and a relatively large amount of "yolk", polarisation occurs during the cytoplasmic movements of activation (e.g. *A. stuartii*) while in those with relatively small amounts of yolk, no polarisation occurs (e.g. *D. aurita* and *D. virginiana*). A complete histological study of these stages is not available for any other marsupial but photographs of sectioned material support this theme. For example, the larger eggs of *S. macroura* show polarisation in the ovary, and ampulla before fertilization (Figs. 4, 8, 9 and 10 of Godfrey 1969) while the small fertilized tubal egg of *S. brachyurus* (pl.1, Fig. 1 of Sharman 1955c) does not.

The degree of yolkiness and the extent of polarisation, in turn affects the type of deutoplasmolysis which occurs, from a generalised elimination in *D. virginiana* (Hartman 1916, 1919) to the localised extrusion which occurs in *D. viverrinus* (Hill 1910). In contrast, in the intermediate type of egg as typified by *A. stuartii,* deutoplasmolysis occurs by generalised elimination in addition to localised extrusion.

ACKNOWLEDGEMENTS

This work was supported by a grant from the M. A. Ingram Trust and the Australian Research Grants Committee. Many thanks are due to Dr. P. Woolley and Mr. G. Young of La Trobe University for supply of some specimens and especially to Dr. Woolley for many helpful discussions and advice. I would also like to thank the Hubrecht Laboratory and Dr. E. C. Boterenbrood, Curator of the Central Embryological Collection for the opportunity to examine the Hill Collection.

REFERENCES

Austin, C. R., 1961. "The mammalian egg". Blackwell Scientific Publications: Oxford.

Austin, C. R., 1965. "Fertilization". Prentice Hall Inc. Englewood Cliffs: New Jersey.

Balinsky, B. J., 1975. "An introduction to embryology". Saunders: Philadelphia.

Caldwell, M. A., 1887. The embryology of Monotremata and Marsupialia — Part 1. *Phil. Trans. Roy. Soc. B.* **178:** 463–86.

Clark, M. J. and Poole, W. E., 1967. The reproductive system and embryonic diapause in the female grey kangaroo, *Macropus giganteus*. *Aust. J. Zool.* **15:** 441–59.

Godfrey, G. K., 1969. Reproduction in a laboratory colony of the marsupial mouse *Sminthopsis larapinta* (Marsupialia: Dasyuridae). *Aust. J. Zool.* **17:** 637–54.

Godfrey, G. K., 1975. A study of oestrus and fecundity in a laboratory colony of mouse opossums (*Marmosa robinsoni*). *J. Zool.* **175:** 541–55.

Guraya, S. S., 1968. Histochemical study of developing ovarian oocyte of the American opossum. *Acta embryologiae et morphologiae experimentalis* **10:** 181–91.

Hartman, C. G., 1916. Studies in the development of the opossum *Didelphys virginiana* L. I. History of the early cleavage. II. Formation of the blastocyst. *J. Morph.* **27:** 1–84.

Hartman, C. G., 1919. Studies in the development of the opossum *Didelphys virginiana* L. III. Description of new material on maturation, cleavage and entoderm formation. IV. The bilaminar blastocyst. *J. Morphol.* **32:** 1–144.

Hartman, C. G., 1923. The oestrus cycle in the opossum. *Am. J. Anat.* **32:** 353–421.

Hartman, C. G., 1924. Observations on the motility of the opossum genital tract and the vaginal plug. *Anat. Rec.* **27:** 293–303.

HARTMAN, C. G., 1928. The breeding season of the opossum *(Didelphys virginiana)* and the rate of intra uterine and postnatal development. *J. Morphol.* **46:** 143–215.

HILL, J. P., 1910. The early development of the Marsupialia, with special reference to the Native Cat (*Dasyurus viverrinus*). *Quart. J. Microsc. Sci.* **56:** 1–134.

HILL, J. P., 1918. Some observations on the early development of *Didelphys aurita. Quart. J. Microsc. Sci.* **63:** 91–140.

HILL, J. P. AND O'DONOGHUE, C. H., 1913. The reproductive cycle in the marsupial *Dasyurus viverrinus. Quart. J. Microsc. Sci.* **59:** 133–74.

HUGHES, A. L., THOMPSON, J. A., AND OWEN, W. H., 1965. Reproduction in natural populations of the Australian ringtail possum *Pseudocheirus peregrinus* (Marsupialia: Phalangeridae) in Victoria. *Aust. J. Zool.* **13:** 383–406.

HUGHES, R. L. AND RODGER, J. C., 1971. Studies on the vaginal mucous of the marsupial *Trichosurus vulpecula. Aust. J. Zool.* **19:** 19–33.

LINTERN–MOORE, S., MOORE, G. P. M., TYNDALE–BISCOE, C. H. AND POOLE, W. E., 1976. The growth of the oocyte and follicle in the ovaries of monotremes and marsupials. *Anat. Rec.* **185:** 325–32.

LYNE, A. G. AND HOLLIS, D. E., 1977. The early development of marsupials with special reference to bandicoots. Pp. 293–302 *in* "Reproduction and evolution" ed. by J. H. Calaby and C. H. Tyndale–Biscoe. Australian Academy of Science: Canberra.

MCCRADY, E. JR., 1938. The embryology of the opossum. *Am. Anat. Mem.* **16:** 1–233.

MCGAUGHEY, R. W., 1978. *In vitro* oocyte maturation. Pp. 1–19 *in* "Methods in mammalian reproduction". Academic Press: New York.

MARTINEZ–ESTEVE, P., 1942. Observations on the histology of the opossum ovary. *Contr. Embryol.* **30:** 19–26.

MASUI, Y.,AND CLARK, H. J., 1979. Oocyte maturation. *Intern. Rev. Cytol.* **57:** 185–282.

O'DONOGHUE, C. H., 1912. The corpus luteum in the non-pregnant *Dasyurus* and polyovular follicles in *Dasyurus. Anat. Anz.* **41:** 353–68.

O'DONOGHUE, C. H., 1916. On the corpora lutea and interstitial tissue of the ovary in the marsupialia. *Q. Jl. Microsc. Sci.* **61:** 433–79.

PILTON, P. E. AND SHARMAN, G. B., 1962. Reproduction in the marsupial *Trichosurus vulpecula. J. Endocrinol.* **25:** 119–36.

RENFREE, M. B. AND TYNDALE–BISCOE, C. H., 1978. Manipulation of marsupial embryos and pouch young. Pp. 307–331 *in* "Methods in mammalian reproduction". Academic Press: New York.

SANDES, F. P., 1903. The corpus luteum of *Dasyurus viverrinus* with observations on the growth and atrophy of the Graafian follicle. *Proc. Linn. Soc. N.S.W.* **2:** 364–405.

SELWOOD, L., 1980. A timetable of embryonic development of the dasyurid marsupial *Antechinus stuartii* (Macleay). *Aust. J. Zool.* **28:** 649–68.

SHARMAN, G. B., 1955a. Studies on marsupial reproduction. II. The oestrus cycle of *Setonix brachyurus. Aust. J. Zool.* **3:** 44–55.

SHARMAN, G. B., 1955b. Studies on marsupial reproduction. III .Normal and delayed pregnancy in *Setonix brachyurus. Aust. J. Zool.* **3:** 56–70.

SHARMAN, G. B., 1955c. Studies on marsupial reproduction. IV. Delayed birth in *Protemnodon eugenii* Desmarest. *Aust. J. Zool.* **3:** 156–61.

SHARMAN, G. B., 1961. The mitotic chromosomes of marsupials and their bearing on taxonomy and phylogeny. *Aust. J. Zool.* **9:** 38–60.

THIBAULT, C., 1973. Sperm transport and storage in vertebrates. *J. Reprod. Fert. Suppl.* **18:** 39–53.

TYNDALE–BISCOE, H., 1973. "Life of marsupials". Edward Arnold: London.

TYNDALE–BISCOE, C. H., 1979. Hormonal control of embryonic diapause and reactivation in the tammar wallaby. Pp. 173–191 *in* "Maternal recognition of pregnancy. Ciba Foundation Series 63 (New Series)". Excepta Medica: North Holland.

TYNDALE–BISCOE, C. H. AND RODGER, J. C., 1978. Differential transport of spermatozoa into the two sides of the genital tract of a monovular marsupial, the Tammar wallaby *(Macropus eugenii). J. Reprod. Fertil.* **52:** 37–43.

ULLMAN, S. L., 1978. Observations on the primordial oocyte of the bandicoot *Isoodon macrourus* (Peramelidae, Marsupialia) *J. Anat.* **128:** 619–631.

WARING, H., SHARMAN, G. B., LOVAT, D. AND KAHAN, M., 1955. Studies on marsupial reproduction. I. General features and techniques. *Aust. J. Zool.* **3:** 34–43.

WOOLLEY, P., 1966a. Reproduction in *Antechinus* spp. and other dasyurid marsupials. *Symp. Zool. Soc. Lond.* **15:** 281–94.

WOOLLEY, P., 1966b. "Reproductive biology of *Antechinus stuartii.* Macleay (Marsupialia: Dasyuridae)". Ph.D. thesis: Australian National University, Canberra.

WOOLLEY, P., 1971. Maintenance and breeding of laboratory colonies of *Dasyuroides byrnei* and *Dasycercus cristicauda. Int. Zool. Yb.* **54:** 351–4.

WOOLLEY, P., 1973. "Breeding patterns, and the breeding and laboratory maintenance of dasyurid marsupials. *Exp. Animals* **22**, Suppl." "Proceedings of the ICLA Asian Pacific meeting on laboratory animals" 1971, Tokoyo and Inuyama 161–72.

RESUMEN

7. Revisión de la maduración y fertilización de los marsupiales, con especial referencia al dasiúrido *Antechinus stuartii.* Por Lynne Selwood.

En *A. stuartii* la meiosis 1 es detenida en la etapa *dictiotena* de fines de la fase diplonema. La maduración comienza en el ovario, hacia el fin del proceso estral, y es completada en la ampulla que sucede a la fertilización. El primer cuerpo polar, que ε veces se divide más tarde, es eliminado en el ovario y el segundo es despedido con la entrada del esperma. El huevo es ovulado como un oocito secundario con los cromosomas detenidos en la metafase 2. La polarización del citoplasma vitelado en el huevo de *A. stuartii* no ocurre hasta la activación que sigue a la fertilización. El tamaño y el grado del vitelo de los huevos de otros marsupiales están relacionados al tiempo y extensión de la polarización del citoplasma vitelado. La fertilización ocurre durante la ampulla, en cuyo momento los huevos son de $93.2\mu \pm 2.8$ (error standard) por $118.0\mu \pm 2.4$ (error standard) (N=12) sin células cumulus, y en los primeros estados de la depositación de mucina. Los pronúcleos de los machos y de las hembras se forman mientras el huevo está todavía en la fase ampulla, de manera que los huevos entran al útero en el estado prenuclear. Se sugiere que los posibles mecanismos para prevenir la polispermia en *A. stuartii* incluyen el bajo núero de espermas que alcanzan el huevo y el entrampamiento del esperma por la mucina.

8. SOME INFORMATION ON REPRODUCTION, GROWTH AND DEVELOPMENT IN *Planigale gilesi* (DASYURIDAE, MARSUPIALIA)

Dick Whitford[1], F. Dominic Fanning[1] and
A. W. White[2]

Specimens of *Planigale gilesi* from Western New South Wales were maintained and bred in captivity. In the laboratory, matings occurred between late November and late December. During this period, oestrous females made high pitched clicking calls for approximately 5 days. Copulation lasted about two hours. During copulation, the males used a neck grip and scratched with their hindfeet against the female's rump. Males made a low clicking sound during copulation. The gestation period was 14½ to 16½ days. The young were first detached from the nipples at 37 days. Their eyes were open by 43 days. They were independent by 65 days. Changes in the weights of females during pregnancy and lactation were observed. Growth rates, developmental changes and weights of the young are given.

Pages 77–81 *in* CARNIVOROUS MARSUPIALS, ed. by M. Archer, February, 1982. Roy. Zool. Soc. New South Wales: Sydney, Australia.

PLANIGALE GILESI is a robust planigale and the only member of this genus with 2 premolar teeth (Aitken 1972). The tail is, depending on the animal's condition, either thin or variably incrassated for the proximal two-third of its length. *Planigale gilesi* has been recorded from northeastern South Australia, western New South Wales and southwestern Queensland (Aitken 1972, Denny 1975, Archer 1976).

The animals used in the present study were caught at Fowlers Gap Station (31°05′S., 141°45′E) in western New South Wales where they occur sympatrically with 5 other small mammals: *Planigale tenuirostris; Sminthopsis crassicaudata; S. macroura; Leggadina forresti;* and the introduced *Mus musculus* (Morton 1978).

Denny (1975) has recorded information about the ecology of *P. gilesi* in northwestern New South Wales, but very little is known of their breeding biology.

METHODS

The *Planigale gilesi* were trapped during 1979. One immature female was caught on 24 January. The other 4 (2 adult ♂♂ and 2 adult ♀♀) were caught on 5 October.

Animals were housed in glass aquaria measuring 60 x 30 x 30 centimetres. The floor of each cage was covered with fine red soil from Fowlers Gap. Pieces of bark were provided for shelter. They were exposed to the natural photoperiod of Sydney. The animals were housed individually. Females were only placed in a male's cage when they showed signs of oestrous.

Animals were fed with minced meat containing beef, mutton and kangaroo. The mince was mixed with dry ground dog food, a vitamin mineral supplement and calcium powder. Small pieces of cheese and fly pupae were sprinkled on top of the mince mixture. This diet was supplemented three days a week with mealworms. Females with young older than 17 days refused the mince mixture and were fed on mealworms. Water was supplied *ad libitum*.

Females were weighed from the time of mating until they were separated from their young. Measurements of the young included crown-rump length, head width (across the broadest part of the head), head–vent (nose tip to cloaca), tail length (tip to cloaca), head length and hind foot length (excluding the claws). Once the young became detached from the nipples it was also possible to record body weights of individuals.

RESULTS AND DISCUSSION

CALLING BEHAVIOUR ASSOCIATED WITH MATING: The two male *Planigale gilesi* were heard calling after dusk on 18 October, thirteen days after their arrival in Sydney, and they continued calling for the next several nights. Unfortunately, the two males and the three females were used in physiology experiments for the next several weeks and calling by both males abruptly ceased. After these experiments the males did not call again. However, the three females called and mated during late November and late December. Each female was first observed calling respectively on 25 November, 13 December, and 29 December. Each female called every night for 4 to 6 consecutive nights.

[1] School of Zoology, University of New South Wales, P.O. Box 1, Kensington, New South Wales, Australia, 2033.
[2] Department of Zoology, University of Sydney, Sydney, New South Wales, Australia, 2006.

The calls made by both sexes consisted of a repeated high pitched clicking that was very similar to the sound made by small insectivorous bats in flight. Given that each animal was housed separately, the calls of the males and females were interpreted to be mate attracting calls (see Fox and Whitford 1982).

In the laboratory, males of several dasyurid species call for a specific 5 to 6 month period of the year (e.g. *Sminthopsis murina,* Fox and Whitford 1982, and Ayers Rock *Ningaui,* D.W. pers. obs. as noted in Fanning 1982). There is some evidence from field data that the period in captivity during which the males of these two species call corresponds with the breeding season in wild populations. We could not establish whether male *P. gilesi* called for a specific period of the year corresponding with the breeding season in the wild, because they stopped calling when they were used in the physiology experiments.

More information is available about calling in female dasyurids. The females of both *Sminthopsis murina* and Ayers Rock *Ningaui* called and mated within the same 5-6 months during which their respective males called. Female calls associated with oestrous and mating have been recorded for *Dasyurus maculatus* (Settle 1978, D.W. pers. obs), *Planigale maculata* (Van Dyck 1979), *Sminthopsis virginiae* (Taplin, pers. comm.). Godfrey (1969) diagnosed oestrus and oestrous cycle length for *Sminthopsis macroura* (as *larapinta*) by the cyclic presence or absence of cornified epithelial cells in the urine of females. Woolley (1971) has also successfully used this method to diagnose oestrus in *Dasyuroides byrnei.*

Fox and Whitford (1982) used this method to show that calling by female *Sminthopsis murina* was a reliable indication of oestrus and was synchronized with the presence of epithelial cells in the urine.

Similarly, calling by female *P. gilesi* was also a reliable indicator of oestrus. Each of the three females mated almost immediately when they were placed in a male's cage.

A change in behaviour coincides with calling in female *P. gilesi.* Besides calling throughout the night, they frequently rub the cloaca and sternal area on specific objects in their cage. The same behaviour has been reported for female *P. maculata* (Van Dyck 1979).

COPULATION: A typical mating in *P. gilesi* was as follows. At first the animals approached one another and touched noses briefly. The male then moved behind the female and sniffed her cloaca. He clasped the female around the thoracic area with his forepaws, gripping her hair in this region. Thrusting by the male was observed several seconds later, followed by vigorous lateral movements of his chin and throat across the back of the female's neck. Soon after thrusting started, the male raised both hindlimbs and placed his hindfeet against the female's rump. The male scratched the female's rump 5 or 6 times on one side. This was immediately followed by the same action on the other side. Scratching continued throughout almost all of the copulation sequence. Intermittently during copulation, the male made a low clicking call. About fifteen minutes after copulation had started, the male used his mouth to grip the hair on the back of the female's neck. Both animals became much more active, turning in circles several times. Occasionally the animals jumped several centimetres into the air landing on their sides, but quickly regained an upright position. Throughout this period the female's body moved from side to side, with alternate raising and lowering of her forefeet. Her eyes were closed and the anterior part of her body trembled for much of the time. The male retained the neck grip, and frequently lifted the female's forefeet off the ground while thrusting. This was followed by a quiescent period during which the female curled her head underneath her body so that the top of her head was resting against the substrate. After several minutes of intense activity, during which both animals jumped up in the air and toppled over several times, the female suddenly struggled and broke away from the male, thus terminating the mating. Males, unlike females, always groomed their genital region immediately following copulation.

The mating behaviour of *P. gilesi* was very similar to matings described for *P. maculata* (Van Dyck 1979). A notable difference was the low clicking sound made by male *P. gilesi* during part of the mating period. Behavioural elements and sequences in dasyurid matings for other species have not been discussed here because these are reviewed by Croft (1982).

Each female mated only once and was returned to her cage immediately after the mating was concluded. Two copulations were observed to last 130 and 135 minutes respectively. The third copulation, observed only in part, lasted at least 140 minutes.

GESTATION PERIODS, BIRTH, AND MATERNAL WEIGHT CHANGES RELATED TO BIRTH: Gestation periods, timed from the end of mating until birth, were 14½, 15½ and 16½ days. The maximum number of young observed was 7. Although the three females had twelve nipples each, a number consistent with that recorded by Aitken (1972). One female, picked up just after birth, had three young attached to her nipples and four unattached but in the pouch. These four young failed to attach to nipples and were recovered from the floor of the cage ½ h later.

With each of the other two females, only four young were found attached to the nipples. Two births occurred during the daylight hours at 1030 h. and the other at about 1510 h. The three litters were as follows: 4 ♀♀; 3 ♂♂ and 1 ♀ and 2 ♂♂ and 1 ♀.

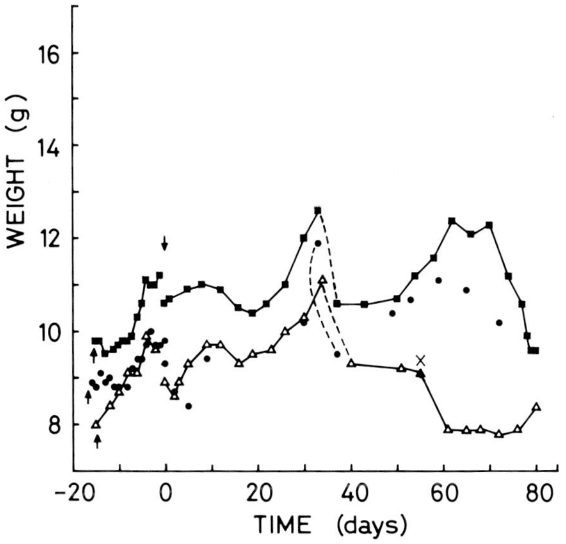

Fig. 1 Weights for three female *P. gilesi* with attached young until 37–40 days, and then females alone until the end of lactation. Symbols: Arrow pointing up, mating; arrow pointing down, birth; - - -, equals interval during which young became detached from female; X, time when the female killed two of her three young.

Changes in body weight of females occurred during pregnancy, after birth and during lactation. In the last 8–10 days of the gestation period, female body weight increased by up to 1½ grams. This weight gain was partially lost at birth. Maternal weight loss at birth may be attributed to several factors: loss of fluid during birth; failure of neonates (including possible supernumaries) to attach to the nipples; and failure to eat before giving birth. This last factor has been a good predictor of imminent births in the laboratory.

After birth, the body weight of the females plus attached young decreased for several days. Following this initial decrease, the weight of each female plus attached young increased until about 12 days after birth. A second weight loss then occurred between 12 and 16 days, possibly because the females refused the usual food of mince and were, in consequence, fed on mealworms for the rest of the lactation period. Between about 16 and 37 days, the weight of each female plus attached young once again increased. The overall increase in weight from about 3 to 37 days was attributed to an increase in the weight of the growing young (Fig. 1). Similar weight changes during pregnancy, birth and lactation have been recorded for other dasyurids (e.g. *Antechinus stuartii*, Woolley 1966, and *Sminthopsis murina*, Fox and Whitford 1982). Van Dyck (1979) recorded a change in the diet of *P. maculata* 17 days after birth and noted that, after that time, only live insect food was eaten. During the later stages of lactation in *P. gilesi* the weight of two of the three females increased until the young became independent at 65 to 70 days. We attribute this weight increase to development of mammary tissue, because after the young became independent, the decline in the females' body weight corresponded with regression of the pouch area. The third female killed two of her three young at 55 days and increase in her body weight was not observed. This may be attributed to reduction in demand for milk resulting in atrophy of some of the mammary tissue.

Early Development of the Young: Observations of development of young *P. gilesi*, based on the 3 litters born in captivity, are as follows (weights given in this sequence are based on the individuals of one litter only):

Birth (day 0): skin pink; veins visible through skin; claws on manus; ears not visible; one neonate just after birth weighed 9.7 mg; young completely enclosed by pouch.

Day 8: eye ring visible; ear buds visible; forelimb movements; hind limbs paddle-shaped.

Day 12: skin pigmentation starting (dorsally); digits of manus clawed; mysticial and genal papillae visible.

Day 22: ears free, directed laterally; young darkly pigmented; mysticial vibrissae starting; veins no longer visible through skin; head sparsely haired; hair starting on body.

Day 37: young first detached from nipples and left in nest; eyes closed; weight (♀ 0.77), (♂ 0.80), (♂ 0.74), (♂ 0.78) g; vocalize when handled.

Day 43: eyes open; hair on head; hindlimbs and forelimbs grey; long white hairs on rump; nails, manus and pes nearly black.

Day 48: body well haired all over; ventrallly white; anterior (head and shoulders) adult coloured; posterior still dark.

Day 59: young have now assumed adult colour all over.

Day 65: young observed eating; solid black faeces; presumably independent; weight (♀ 3.55), (♂ 3.94), (♂ 4.14), (♂ 4.37)

As indicated in Fig. 2, crown–rump length showed a steady increase while the young were permanently attached to the nipples (until 37 days). Between 26 and 40 days, there was a marked increase in crown-rump length. Head width showed a linear relationship with age at least up to the time when the young became independent. For each of these body parametres, there was little difference between litters.

Growth of head length (Fig. 3) began to plateau at about 70 days. At this time, a sexual dimorphism became apparent, with males being slightly larger than females. By 150 days, the head length of males was 3 mm longer than females in the same litter. The increase in hind foot length began to plateau at approximately 60 days and there was a noticeable difference in hind foot length between the sexes up to 170 days. In each of these parametres, differences between litters were not apparent.

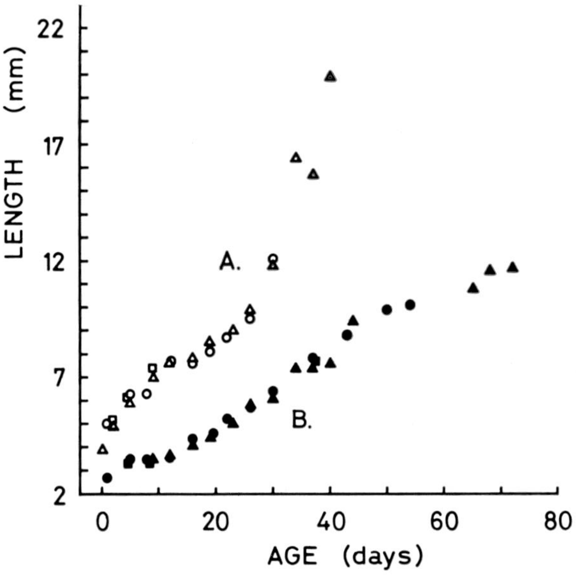

Fig. 2. Mean values for measurements taken from each of the three *P. gilesi* litters. Each litter is represented by a different symbol. A, crown–rump length (open symbols). B, head width (closed symbols).

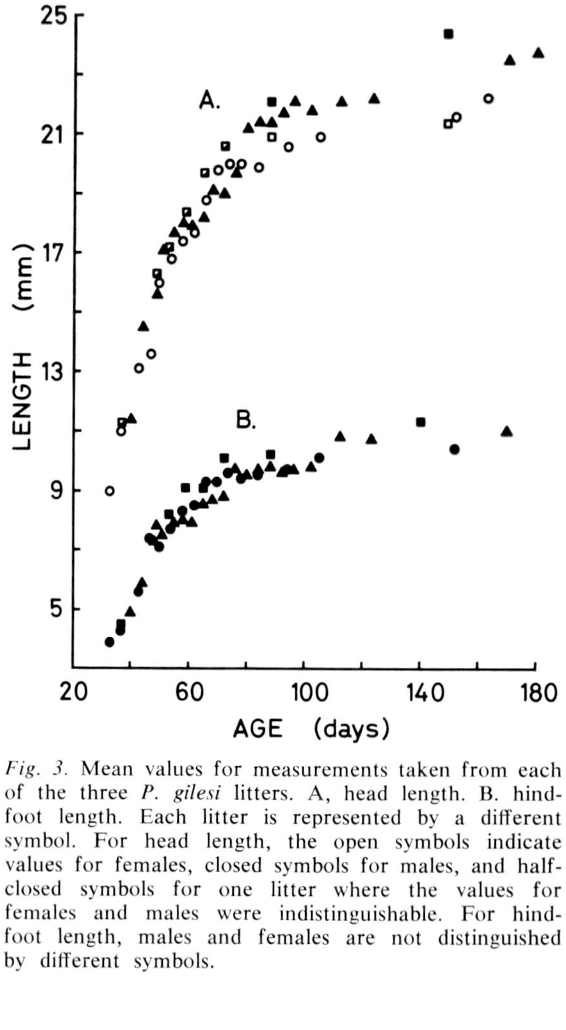

Fig. 3. Mean values for measurements taken from each of the three *P. gilesi* litters. A, head length. B. hind-foot length. Each litter is represented by a different symbol. For head length, the open symbols indicate values for females, closed symbols for males, and half-closed symbols for one litter where the values for females and males were indistinguishable. For hind-foot length, males and females are not distinguished by different symbols.

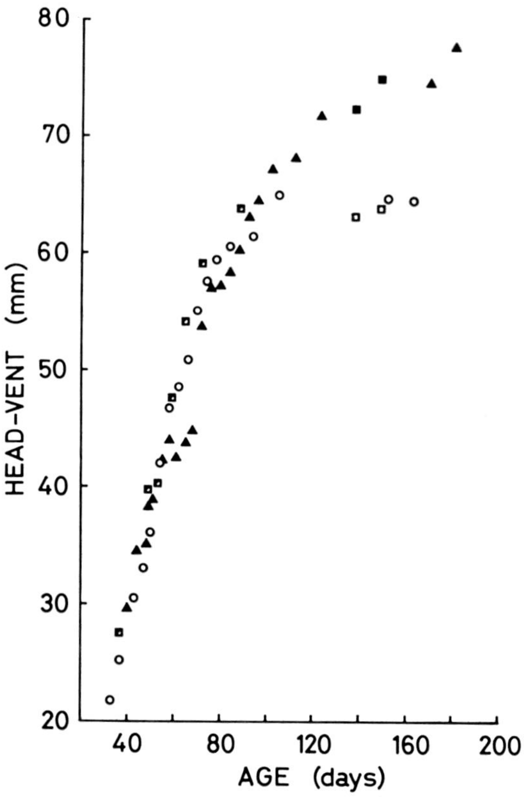

Fig. 4. Mean values for head–vent measurements of young *P. gilesi*. Symbols as for Fig. 3.

The head–vent length (Fig. 4) continued to increase during the period of measurement, although the rate of growth declined from about 120 days. From about 100 days, the head–vent lengths of males were noticeably longer than those of females. Little difference was seen between litters.

Body weights showed a linear relationship up to 90 days (Fig. 6). After 90 days, the growth rate declined in both males and females. Males were also noticeably heavier than females. As was the case in the other parameters noted above, there was little difference between the litters during the observed period of growth.

Later Development of the Young: Young *P. gilesi* approximate adult proportions in head width by 80 days, and in head length by 150 days for females and 180 days for males. Hind foot length (Fig. 3) reached adult values at 120 days for females and 140 days for males. At 170 days, head–vent length (Fig. 4) of both sexes were approximately 95% of adult values (mean of 73 mm, O.R. (observed range) 60–80mm, pers. comm. D. Read). At 160 days, tail–vent lengths of the young (Fig. 5) were close to adult values (mean of 62mm, O.R. 55–70mm, pers. comm. D. Read) for both sexes. By 170 days (Fig. 6), females had reached adult weights (mean of 6.9g, O.R. 5.0–9.0g, pers. comm. D. Read), and males 90% of adult weight (mean of 11.5g, O.R. 9.5–14.0g, pers. comm. D. Read).

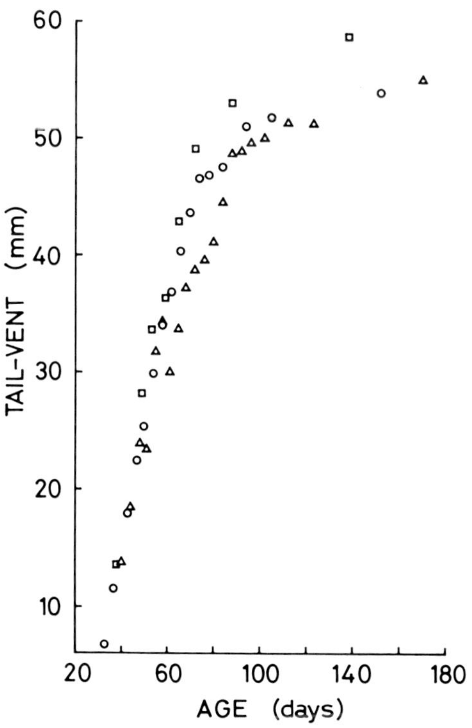

Fig. 5. Mean values for tail–vent measurements of young *P. gilesi*. Each litter is represented by a different symbol. Sexes are not distinguished.

In the parameters measured, young female *P. gilesi* reached adult proportions earlier than males, by 20 days in some cases. Adult male *P. gilesi* are larger than females. This sexual difference in size is noticeable primarily in head length, nose-vent length and body weight.

ACKNOWLEDGEMENTS

The authors would like to thank Mr. David Read for capturing the breeding stock. Dr. B. J. Fox, Dr. M. Archer and Dr. B. Porter read the manuscript and their criticism and advice in preparation of this paper is greatly appreciated.

REFERENCES

AITKEN, P. F., 1972. *Planigale gilesi* (Marsupialia, Dasyuridae); a new species from the interior of south eastern Australia. *Rec. S. Aust. Mus.* **16**: 1–14.

ARCHER, M., 1976. Revision of the marsupial genus Planigale Troughton (Dasyuridae). *Mem. Qd Mus.* **17**: 341–65.

CROFT, D. B., 1982. Communication in the Dasyuridae: A review. *In* "Carnivorous marsupials", ed. by M Archer. Royal Zoological Society of New South Wales: Sydney.

FANNING, F. D., 1982. Reproduction, growth and development in *Ningaui* sp. (Dasyuridae, Marsupialia) from the Northern Territory. *In* "Carnivorous marsupials", ed. by M. Archer. Royal Zoological Society of New South Wales: Sydney.

DENNY, M. J. S., 1975. Mammals of Sturt National Park, Tibooburra, New South Wales. *Aust. Zool.* **18**: 179–95.

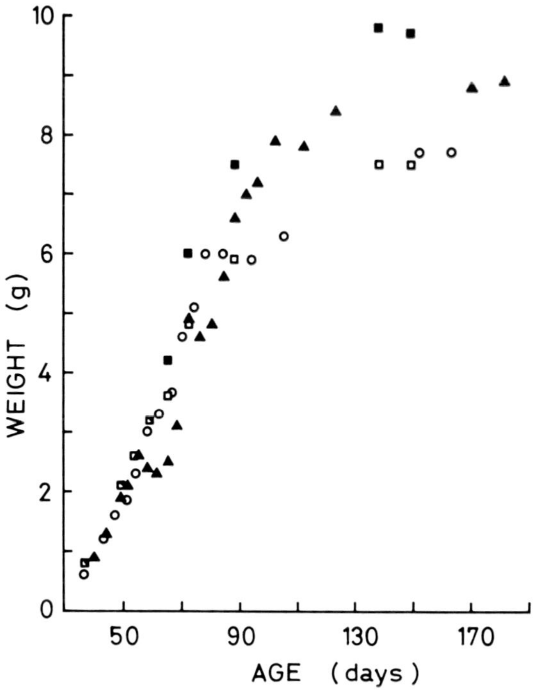

Fig. 6. Mean weights for young *P. gilesi* from age of detachment. Symbols as for Fig. 3.

FOX, B. J. and WHITFORD, D., 1982. Polyoestry in a predictable coastal environment: reproduction, growth and development in *Sminthopsis murina* (Marsupialia: Dasyuridae). *In* "Carnivorous marsupials", ed. M. Archer. Royal Zoological Society of New South Wales: Sydney.

GODFREY, G. G., 1969. Reproduction in a laboratory colony of the marsupial mouse *Sminthopsis larapinta* (Marsupialia: Dasyuridae). *Aust. J. Zool.* **17**: 637–54.

MORTON, S. R., 1978. An ecological study of *Sminthopsis crassicaudata* (Marsupialia: Dasyuridae) I. Distribution, study areas and methods. *Aust. Wildl. Res.* **5**: 151–62.

SETTLE, G. A., 1978. The quiddity of Tiger Quolls. *Aust. Nat. Hist.* **19**: 164–9.

VAN DYCK, S., 1979. Behaviour in captive individuals of the dasyurid marsupial *Planigale maculata* (Gould, 1851). *Mem. Qd Mus.* **19**: 413–29.

WOOLLEY, P., 1966. Reproduction in *Antechinus* spp. and other dasyurid marsupials. Pp. 281–94 *in* "Comparative biology of reproduction in mammals", ed. I. W. Rowlands. Academic Press, London.

WOOLLEY, P. 1971. Maintenance and breeding of laboratory colonies of *Dasyuroides byrnei* and *Dasycercus cristicauda*. *Int. Zoo. Yb.* **11**: 351–14.

RESUMEN

8. Alguna informacion sobre la reproduccion, el crecimiento y el desarrollo de *Planigale gilesi* (Dasyuridae: Marsupialia). Por Dick Whitford, F. Dominic Fanning Y A. W. White.

Se describre la copulacion de especimenes de *Planigale gilesi* en cautividad, cuyo periodo de gestacion es de 14½ a 16½ dias, y cuyos jovenes destetan a los 37 dias, abren sus ojos a los 43 y devienen independientes a los 65.

9. BREEDING IN WILD POPULATIONS OF THE MARSUPIAL-MOUSE *Planigale maculata sinualis* (DASYURIDAE, MARSUPIALIA)

J. MARY TAYLOR[1], J. H. CALABY[2] and T. D. REDHEAD[2]

Planigale maculata sinualis occupies flood plain sedgeland of rivers of the coastal plain in the monsoonal Northern Territory. When this habitat floods during the wet season, *Planigale* moves to the fringing savannah woodland on higher ground.

Our observations indicate that the species breeds in all months. Teats varied in number from 8 to 13. The sizes of 21 litters varied from 4 to 12 (average 8). Sexually mature females averaged 10.0 g (range 7–15) in weight and males 13.5 g (range 8.7–22).

The gonads and associated tissues of some animals were examined histologically. Two records of ovulation rate exceeded the number of teats which in turn was usually greater than the number of pouch young. Seminiferous tubules of sexually active males are of large diameter as is usual in dasyurids.

The breeding pattern is in sharp contrast to those of other common sympatric dasyurids which are monoestrous and have restricted breeding seasons during the dry season.

Pages 83-87 *in* CARNIVOROUS MARSUPIALS, ed. by M. Archer, February, 1982. Roy. Zool. Soc. New South Wales: Sydney, Australia.

THE GENUS *PLANIGALE* contains five species of small insectivorous marsupial–mice, among which are the smallest of all marsupials. Except for one species found in southern New Guinea, *Planigale* occurs only in the monsoonal northern and eastern parts or the semi–arid interior of Australia (Archer 1976). Prior to Archer's revision, identification of specimens was difficult, resulting in erroneous determinations in the literature.

Only a few isolated observations of breeding in wild populations have been published on any species of *Planigale*. In the case of *Planigale maculata sinualis,* Davies (1960) (as *Phascogale ingrami*) and Aslin (1975) (as *Antechinus maculatus*) recorded a few observations on breeding in the wild. Aslin bred the subspecies in captivity and showed that it is polyoestrous and produces young in most months. Morrison (1975) published observations on reproductive behaviour in captivity.

The present study is based on specimens obtained during 1972–74 in the northern part of the Northern Territory, between 12°20′ and 12°35′S latitude. They were collected incidental to a study of the population ecology of *Rattus sordidus colletti* on the Adelaide River flood plain (Redhead 1979) or as part of a fauna survey of the Alligator Rivers region (Calaby and T. O. Wolfe, unpublished data). Davies' (1960) and Aslin's (1975) field observations were made in the Adelaide River study area.

Most animals were caught in 33 x 10 x 9 cm folding live-traps baited for rodents with small pieces of leather soaked in raw linseed oil or with rolled oats. It is unlikely that *Planigale* were attracted by either bait. Some specimens sheltering under rock or fuel drums or in long grass, were caught by hand.

ENVIRONMENT

The region has a monsoon climate with high temperatures the year round. On the Adelaide River the mean maxima and minima for the hottest and coolest months are 35.5°, 23.8°C (November) and 30.9°, 15.1°C (July); the corresponding figures for the East Alligator River are 37.1°, 23.4°, and 31.5°, 17.8°C. The average annual rainfall for the two areas is 1407 and 1351 mm, respectively. The rainfall is highly seasonal and reliable and about 90 per cent falls between November and March. The winter months are very dry and somewhat cooler than the humid late spring and summer.

In the region, *P.m. sinualis* is found in two habitats on the coastal plains, flood plain sedgeland and the adjacent myrtle–pandanus savannah. The coastal plains are low and flat and are dissected by a number of wide meandering rivers that are tidal for 60 km or more from the sea. The rivers are bordered by flood plains from one to ten km wide. The flood plain soil is a dark heavy clay and it is covered with dense sedges and grasses. In the wet season much of the flood

[1] Department of Zoology, The University of British Columbia, Vancouver, British Columbia, Canada, V6T 1W5.
[2] Division of Wildlife Research, CSIRO, P.O. Box 84, Lyneham, A.C.T., Australia, 2602.

plain is flooded from heavy rain and runoff, and vegetation growth is rapid, but during the dry season the soil dries out and becomes intersected by large cracks up to 10 cm wide and 50 cm deep. Much of the tall sedge and grass dries off in the dry season and forms a mat on the soil surface. Fringing the flood plain is the myrtle-pandanus savannah, a woodland with a variety of trees, including species of *Acacia, Eucalyptus, Melaleuca* and particularly *Pandanus*, and usually a long grass understorey. For further details on the environment see Story *et al.* (1969, 1976), Redhead (1979) and Calaby and Taylor (1981).

The flood plain sedgeland is the preferred habitat of only two small mammals, *P.m. sinualis* and *R.s. colletti*. Both utilize the large cracks in the soil but the rat is also a burrower. The sedgeland is uninhabitable for small mammals in the wet season and both species move to higher ground, particularly to the myrtle–pandanus savannah. Both habitats are considerably damaged by feral water buffalo (*Bubalus bubalis*) but in spite of this both small mammals are common species. Davies (1960), the only previous author to discuss the habitat of *P.m. sinualis,* described only the fringing savannah.

MATERIALS AND METHODS

Animals collected were sexed, and most were weighed and the head–and–body and tail lengths recorded. Scrotal or pouch size and condition were noted, teats were counted if visible, and number of pouch young, if any, recorded. The majority were released after examination but some were retained for museum specimens. The gonads and associated tissues of those retained were preserved in Bouins fluid at field temperatures. They were stored in 70 per cent alcohol, serially sectioned at 5μ, and stained with eosin and Delafield's haematoxylin.

The total sample was 121 males and 79 females of which some were collected in all months. Histological examination was based on 12 males and eight females collected between January and March, and one female and one male taken in July and September, respectively.

RESULTS AND DISCUSSION

Annual Breeding Regime: As there appeared to be no obvious seasonal influence on breeding, the specimens examined during the three years of the study have been lumped into monthly samples.

Table 1 is a summary of animals captured and litters recorded. In all cases the young were naked and hence were less than four weeks old (Aslin 1975), and would have been born in the month of capture or the previous one. Single females without pouch young trapped in September, November and December had large elongated teats and were possibly still suckling young. Aslin (1975) records four females trapped in June that were carrying young less than a month old, and Davies (1960) lists females trapped in January (1), February (1) and March (3), that had young in their pouches or were pregnant. We conclude that *P.m. sinualis* breeds in all months, although there may be some decrease of breeding activity during the wet summer months.

The breeding pattern of *P.m. sinualis* contrasts sharply with those of the other three common dasyurids in the region, all of which have restricted breeding seasons during the dry months. In *Dasyurus hallucatus* small pouch young were recorded as early as mid–July. Females trapped in late September to November were without pouch young but were lactating (Calaby and Wolfe, unpublished data). *Antechinus bilarni* gives birth from mid–August into Setember and *A. bellus* from late September into October (Calaby and Taylor 1981). Two further dasyurids are found uncommonly in the region, *Phascogale tapoatafa* and *Sminthopsis virginiae*. Information obtained from five females and three males of *P. tapoatafa* suggests that it also is monoestrous and has a restricted breeding season, giving birth in July. Little can be said about the breeding regime of *S. virginiae* as only three breeding females were recorded, one each in late July, early September and late October. All had small pouch young.

Rattus s. colletti, the only small mammal that shares the habitat and seasonal distribution of *P.m. sinualis,* breeds seasonally, chiefly in the early part of the dry season, but breeding may be prolonged by unseasonal rain later in the dry season (Redhead 1979).

Table 1. Numbers of animals captured in each month, and litters in pouches.

	J	F	M	A	M	J	J	A	S	O	N	D
Males	2	9	7	18	9	11	5	7	4	15	30	4
Females	4	4	6	17	5	7	4	4	2	7	12	4
Litters	—	—	2	5	—	2	1	4	—	3	1	—

Sexual Maturity: The species is sexually dimorphic, and there is no evidence of a sternal gland in either sex. The pouch is of Woolley's (1974) Type 2. In nulliparous females it is shallow, hairy and unpigmented. In females carrying pouch young the pouch is deep, sparsely haired and stained with a red–brown pigment. In non–breeders that have had litters the pouch regresses but some yellowish–brown staining of of the pouch hairs remains.

External features provide no visible indication as to when young animals begin to breed. Among six young nulliparous females ranging in weight from 5.5 to 7.3 g, one weighing 6.2 g had slight development and staining of the pouch but the others showed none, and the ovaries of all but one were histologically immature. The exception was a reproductively active female in early pregnancy that weighed 7.2 g. Histology of the testes of a male weighing 8.7 g indicated that it was in breeding condition while that of another male weighing 7.2 g was classed as immature. These data, although limited, suggest that attainment of body weights of approximately 8.5 g for males and 7.0 g for females can be associated with onset of sexual maturity. The mean weight of 38 sexually mature females was 10.0 g ± 2.40 S.D. (range 7–15) and that of 46 males was 13.5 g ± 3.87 S.D. (range 8.7–22).This large variation suggests that these animals may continue to grow throughout life. The largest females and males had head–and–body lengths of about 85 mm and 100 mm respectively. The smallest sizes of both sexes trapped with any consistency was 4 g, and one specimen was caught that weighed only 3 g. The smallest independent young are presumably too light to spring the treadle. Some measurements of specimens examined histologically are given in Table 2.

genus among Australian marsupials, and needs investigation. Some of the variation may be accounted for by the fact that teats of nulliparous females are very small and difficult to locate in the shallow hairy pouches, as are non-suckled teats of post-breeding females with regressed pouches. The most commonly reported maximum number in *Planigale* is 12 (Archer 1976) and this figure may well be the true teat number for most forms of *Planigale*. The female reported in this paper as having 13 teats presumably had a supernumerary one.

Fleay (1967) reported a wild female *Planigale* with 15 "well-furred" young, and this figure is usually quoted as the maximum litter size in the genus. The number of teats was not given. The specimen was identified as *P. ingrami* but was apparently an example of *P.m. maculata,* a form in which the maximum teat number recorded is ten (Archer 1976). There can be little doubt that two litters were involved. Females of small dasyurids are known to adopt, or at least tolerate, the small young of their conspecifics. For example, Dr. P. Woolley (pers. comm.) had two captive female *Antechinus macdonnellensis* with litters, separated by a partition that prevented adults but not small mobile young from changing cage compartments. Some of the young transferred from one female to the other.

Table 2. Mean measurements for *Planigale maculata* examined histologically.

Months	January	February	March	July	September
		FEMALES			
Weight (g)	7.3	6.2	8.2	10.3	
Hind foot length (mm)	11.5	10.8	10.7	10.6	
Head and body length (mm)	70.0	67.0	75.8	72.0	
Follicle maximum diameter (μ)	250	320, 265, 273 (mean = 286)	273, 208 260, 260 (mean = 235)	106	
Corpora lutea maximum diameter (μ)			455, 477 (mean = 466; Sample = 2)		
Sample size	1	3	4	1	
		MALES			
Weight (g)	12.0	10.1	10.4		11.2
Hind foot length (mm)	11.5	11.4	11.9		11.2
Head and body length (mm)	78	74	80		79
Epididymal tubule diameter (μ)		71.5	78.5		49.8
Seminiferous tubule diameter (μ)	361.4	309.6	297.0		239.0
Average testis length (mm)	7.4	6.2	6.6		5.5
Average testis width (mm)	4.0	4.4	4.6		4.0
Sample size	1	6	5		1

NUMBER OF TEATS: Teats ranged in number from 8 to 13. The great variability reported for this species, and indeed all forms of *Planigale* in which an adequate number of specimens has been examined (Archer 1976), is unique to this

Lee (1974) has discussed variation in teat number in two species of *Antechinus* but in those cases the variation has a geographical basis, and the number is consistent within populations. The number of teats in *Antechinomys laniger* also varies geographically although there is some doubt about the numbers within populations (Archer 1977). It is well established that teat number in *Didelphis virginiana* is very variable and not all teats are functional (Enders 1966, Hunsaker 1977).

LITTER SIZES The sizes of 21 pouch litters were as follows: 4 (2 litters), 5 (1), 6 (1), 7 (5), 8 (4), 9 (3), 10 (1), 11 (2), 12 (2). The average litter size was 8. Davies (1960) noted a litter of 8 and another of 12, and Aslin (1975) recorded three litters of wild–caught females, of 8, 10 and 11 respectively.

The smallest naked young of one female averaged 5.7 mm in crown–rump length.

REPRODUCTIVE HISTOLOGY OF FEMALES: Among females examined histologically, four taken in late January and early February were all non–breeding. The lack of development of the pouch area together with the small size of the reproductive tract, suggest that they were nulliparous. A mid–March sample included a pregnancy at blastocyst stage, a female with relatively new pouch young and two non–breeders, one of which had a well–developed pouch and conspicuous teats of uniform size, indicative of a weaned litter some weeks earlier. A female taken in late July had naked, but not newborn, young.

Ovaries and production of young. The ovarian follicles in our sample are small and lack antra. Increment in both follicular number and in size of the largest follicles among the four females sampled in early February is perhaps indicative of a forthcoming breeding surge after the wet summer months. This becomes more evident when comparison of follicular size is made with the mid–March sample. In the latter group, follicular size was largest (273μ) in the female that had recently mated. She had corpora lutea, measuring to 455μ, in which luteal cells were filling in the ruptured follicles. The pregnancy was at unilaminar blastocyst stage. In the two March females with follicles to 260μ, one had corpora lutea to 477μ in diameter and relatively new pouch young, while the other had no corpora lutea but an enlarged pouch with visible teats that suggested recent lactation. The reproductively inactive female taken this month had a maximum follicular size of only 208μ. Measurements of follicles and corpora lutea are given in Table 2.

Corpora lutea are short lived following parturition, for the female obtained in July with young of 9.2 mm crown–rump length had no trace of corpora lutea and the uteri were reduced to a size approaching that of non–breeders. It is reasonable to judge, from the rapid postnatal growth of this species (Aslin 1975), that pouch young of this size are less than a week old. If this assumption is correct, corpora lutea disappear within a week postpartum. Such rapid degeneration among dasyurids has previously been described only in *Antechinus bilarni* and *A. bellus* (Calaby and Taylor 1981).

The ovulation rate, determined by corpora lutea counts of one pregnant and one postpartum female was 17 and 14 (average 15.5). The pregnant female had 17 corpora lutea and 14 blastocysts. In these instances the ovulation rate exceeds the number of teats by about one–third, yet the teat number usually exceeds the litter size. These limited data suggest that high mortality before pouch life is characteristic for this species, as it is for some other polyoestrous marsupials (Godfrey 1969).

Seasonal changes of the endometrium. The histology of the endometrium in pregnant, lactating and non–breeding females is very similar to that of *A. bilarni* and *A. bellus* (Calaby and Taylor 1981), however, two aspects deserve comment. The first is that, although the endometrial condition of the non–breeding females captured in January and February is similar to that of other non–breeders, some proliferation of the uterine glands indicates that these females are not fully quiescent in reproductive state. The second aspect concerns the simultaneous occurrence in March of non–breeders along with breeders. In the non–breeders the uterine glands are minimal in extent and the reproductive systems are very small. Lack of histological evidence of impending breeding in these females from the same population in which breeding females occur, indicates that in March at least females may have no reproductive synchrony with their conspecifics under similar field conditions.

HISTOLOGY OF MALES: Among the 12 males taken from late January to March, all but one were in breeding condition (Table 2). The exception was a sexually maturing subadult (7.4 g) with small testes (4.9 x 3.4 mm). An adult male trapped in September was also non–breeding, and had small, recrudescing testes (5.5 x 4.0 mm).

Changes in the testis and epididymis. The testes of the two non–breeding males provide some information on impending sexual maturation. The subadult specimen had numerous primary spermatocytes lining and in the seminiferous tubule lumina but the testes had not progressed to the spermatid stage. The seminiferous tubules average 230μ in diameter and the interstitial tissue was highly cytoplasmic and abundant. In the non–breeding adult, the larger testes were in advanced spermatogenic activity through the spermatid stage, and the seminiferous tubule diameter averaged 239μ. The fully adult status of this male with activating, but still aspermic, testes suggests that males may fluctuate between breeding and non-breeding activity in the field. Among the testes sampled, those measuring 5.8 x 4.6 mm and above, with seminiferous tubules averaging 280μ in diameter, or larger, were fully spermatogenic. The largest testis recorded of a breeding male was 7.1 x 5.0 mm. Diameters of the seminiferous tubules in fully spermatogenic testes are recorded to 377μ. This large tubular size is comparable with those of other species of dasyurids, described by Woolley (1975). Mean measurements of testes, and epididymal and seminiferous tubules are given in Table 2.

ACKNOWLEDGEMENTS

We thank the former Forestry, Fisheries, Wildlife, Environment and National Parks Branch, Department of the Northern Territory, for permission to collect and export protected fauna. The skilled technical assistance in the field of Tony Wolfe and John Wombey was crucial to the success of the project. Our thanks go also to Wendy Hindmarsh Jones for her help in the laboratory and to Sarah C. Smith for assistance in library searches. Financial support for the laboratory part of this project came from a National Research Council of Canada grant A3462 awarded to J. M. Taylor.

REFERENCES

Archer, M., 1976. Revision of the marsupial genus *Planigale* Troughton (Dasyuridae). *Mem. Qd Mus.* **17:** 341–65.

Archer, M., 1977. Revision of the dasyurid marsupial genus *Antechinomys* Krefft. *Mem. Qd Mus.* **18**: 17–29.

Aslin, H., 1975. Reproduction in *Antechinus maculatus* Gould (Dasyuridae). *Aust. Wildl. Res.* **2**: 77–80.

Calaby, J. H. and Taylor, J. M. 1981. Reproduction in two marsupial–mice *Antechinus bellus* and *Antechinus bilarni* (Dasyuridae), of tropical Australia. *J. Mammal.* **62:** 329–41.

Davies, S. J. J. F., 1960. A note on two small mammals of the Darwin area. *J. Roy. Soc. W. Aust.* **43**: 63–6.

Enders, R. K., 1966. Attachment, nursing and survival of young in some didelphids. *Symp. Zool. Soc. Lond.* **15**: 195–203.

Fleay, D., 1967. Planigale holds record family number. *Vic. Nat.* **84**: 202.

Godfrey, G. K., 1969. Reproduction in a laboratory colony of the marsupial mouse *Sminthopsis larapinta* (Marsupialia: Dasyuridae). *Aust. J. Zool.* **17**: 637–54.

Hunsaker, D., 1977. Ecology of New World marsupials. Pp. 95–156 *in* "The biology of marsupials", ed. by D. Hunsaker. Academic Press, New York.

Lee, A. K., 1974. Variations in nipple number in female *Antechinus*. *Aust. Mammal.* **1**: 300. (Abstract).

Morrison, R. G. B., 1975. Emergence of the Pygmy Antechinus. *Aust. Nat. Hist.* **18:** 164–7.

Redhead, T. D., 1979. On the demography of *Rattus sordidus colletti* in monsoonal Australia. *Aust. J. Ecol* **4**: 115-36.

Story, R., Galloway, R. W., McAlpine, J. R., Aldrick, J. M. and Williams, M. A. J., 1976. Lands of the Alligator Rivers area, Northern Territory. *CSIRO Land Res. Ser. No. 38.*

Story, R., Williams, M. A. J., Hooper, A. D. L., O'Ferrall, R. E. and McAlpine, J. R., 1969. Lands of the Adelaide–Aligator area, Northern Territory. *CSIRO Land Res. Ser. No. 25.*

Woolley, P., 1974. The pouch of *Planigale subtilissima* and other dasyurid marsupials. *J. Roy. Soc. W. Aust.* **57:** 11–15.

Woolley, P., 1975. The seminiferous tubules in dasyurid marsupials. *J. Reprod. Fert.* **45**: 255–61.

RESUMEN

9. La fecundación en las poblaciones naturales de la rata marsupial *Planigale maculata sinualis* (Marsupialia: Dasyuridae). Por J. Mary Taylor, J. H. Calaby y T. D. Redhead.

Planigale maculata sinualis ocupa los juncales de las llanuras de inundación de los ríos de las llanuras costeras monzónicas del Northern Territory. Cuando este biótopo se inunda durante la estación lluviosa, *Planigale* de desplaza hacia las sabanas de los bosques marginales de las tierras más altas.

Nuestras observaciones indican que la especie se reproduce en todos los meses. Las tetas varían en número de 8 a 13. El número de individuos en 21 pariciones varió desde 4 hasta 12 (promedio 8). El peso promedio de hembras sexualmente maduras fue de 10.0 gr. (variación 7–15), y el de los machos de 13.5 gr. (variación 8.7–22).

Se examinaron histológicamente las gónadas y tejidos asociados de algunos animales. Dos registros del porcentaje de ovulación excedieron el número de tetas, y a la vez fueron usualmente mayores que el número de hijos en el marsupio. Los tubos seminíferos de machos sexualmente activos son de gran diámetro, como ocurre usualmente en los dasiúridos.

El tipo de reproducción contrasta marcadamente con las de otros dasiúridos simpátridos que son monoestros y que tienen restingidas estaciones de reproducción durante las estaciones secas.

10. THE LIFE CYCLE OF *Antechinus swainsonii* (DASYURIDAE, MARSUPIALIA)

RAY WILLIAMS[1] and ANNE WILLIAMS

Little information has been published on the biology of *Antechinus swainsonii* (Dasyuridae). This paper describes the growth and development of two laboratory conceived litters and two laboratory raised but wild born litters.

In the wild, mating occurs from early May through to September on the mainland with later breeding records in Tasmania. From records collected it appears that altitude plays an important role in stimulating mating; as coastal and low altitude colonies breed earlier regardless of latitude. Males usually die off about 3 weeks after mating. Females appear to be monoestrous.

The gestation period was estimated to be 29-35 days. When born the young are about 4.5 mm and are bright pink in colour. The pouch usually contains a full complement of young (8). The fur starts developing at 4–5 weeks. At 8 weeks old the young were left in the nest and shortly afterwards their eyes open. They are well furred at this time. At 12 weeks the young were observed eating solid food and appear to be weaned at about 14 weeks. The young then slowly develop, reaching adult size just before the next breeding season.

Pages 89-95 *in* CARNIVOROUS MARSUPIALS, ed. by M. Archer, February, 1982. Roy. Zool. Soc. New South Wales: Sydney, Australia.

ANTECHINUS SWAINSONII is found in south-east Australia from Southern Queensland to eastern South Australia and in Tasmania. It is one of the largest Antechinus with males often weighing over 120 g and females to about 70 g. The abundance of food may limit the weight (Recher *et al.* 1975) so that some populations may not reach these weights.

Antechinus swainsonii is most commonly found in wet sclerophyll and rainforest gullies but specimens have been found in other habitats where ground cover is abundant. In Tasmania the species is restricted to beech (*Nothofagus cunninghamii*) rainforest and its periphery (Green 1972). In N.S.W. and Victoria heathlands, coastal sand dunes and swamps are utilised (Wakefield and Warneke 1963, Recher *et al.* 1975). The species is also found in sub-alpine woodland with a dense shrubby understorey in Kosciusko National Park (Ann Jelinek, pers. comm.).

Very little research has been carried out on this species. Wakefield and Warneke (1963) and Green (1972) give detailed accounts of the habitat and description of the species but very little information on the life cycle, although there are some detailed descriptions of pouch conditions, the pouch young and breeding seasons. Some limited information on the life cycle has been forthcoming from field studies (Press (1976), Reed & Wallis (1975), Van Dyck & Ogilvie (1977), Recher *et al.* (1975) and Leonard (1970). Population levels are usually low and capture rates are low and unpredictable making it difficult to obtain information from field studies.

Efforts to keep the species in captivity have not been very successful and to our knowledge the only record of this species being bred in captivity before is by Fleay (1932). He makes some notes on the life cycle of the species. In this study 2 litters were born in captivity and their growth and development were followed. Some observations were also made on two litters present in the pouch of two wild caught females.

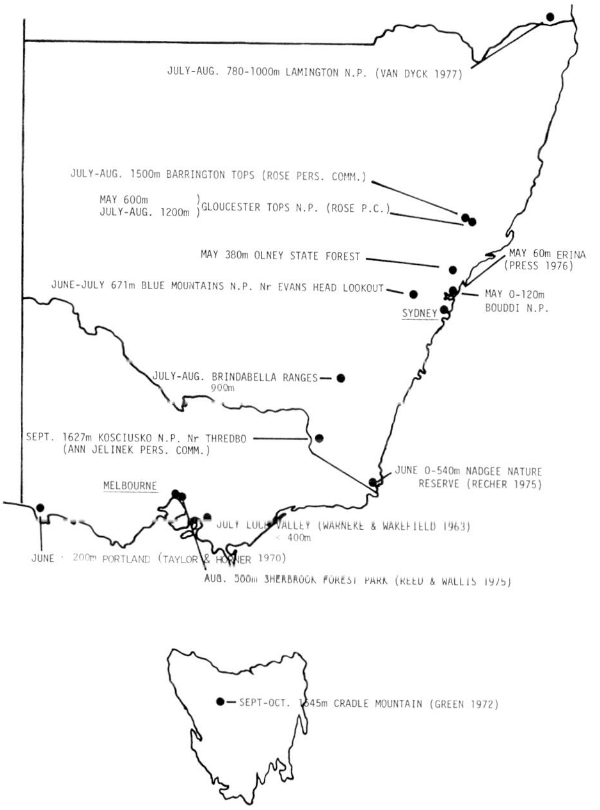

Fig. 1. Mating seasons in *Antechinus Swainsonii.*

[1] School of Zoology, University of New South Wales, P.O. Box 1, Kensington, New South Wales, Australia, 2033.

METHODS

The animals were kept in pairs prior to mating and the males were removed before the predicted date of birth. They were kept in two glass fronted cages 1.2m x 1.2m x 1.4m high and 0.9m x 1.8m x 0.75m high. Separated males were kept in smaller cages. Nest boxes were built on the sides of the cages, as high off the cage floor as possible. The cage floor was covered with soil up to 5 cms deep and this was covered with a layer of dry leaves. Moss covered logs and rocks were also provided.

Live food was given as much as possible, and mealworms, grasshoppers, earthworms and beetle larvae were offered. Kangaroo mince with calcium powder added was given initially but because of this being the possible cause of enteritis infection a change was made to the proprietory dog food, "Luv" mince (with egg added).

The crown-rump measurement was taken with vernier calipers at weekly intervals until the young were 7 weeks old, several young were measured each time and an average taken. From seven weeks onwards the following measurements were taken weekly with vernier calipers, head and body (tip of the nose to the vent), tail (vent to tip, excluding any fur), head (tip of the nose to the back of the skull) and the hind foot (heel to the tip of the longest toe — minus the claw), Weights were taken to 0.1 g using a balance. Visual observations were made of external changes in the young.

RESULTS

Breeding Season and Mating: The breeding season of *A. swainsonii* varies with locality. It has been suggested that photoperiod, combined with temperature, availability of food and other factors may be a controlling factor on the breeding season of some species (Recher *et al.* 1975, Press 1976). In the case of *A. swainsonii* altitude and climate seem to play an important part in triggering mating, with the low altitude colonies breeding earlier.

The earliest evidence of breeding is in May in the Central Coast district of N.S.W. A male was caught by the authors in Olney State Forest (Alt. 380m) N.W. of Wyong N.S.W. at the end of May. It was dead in the trap and had large bald patches over the body, a typical symptom of male "die off" (Van Dyck 1977). Another specimen near death was found at the same time by a Forestry Ranger, it also had bald patches on the body. In wild populations it is known that there is usually a total male "die off" (Recher *et al.* 1975, Reed & Wallis 1975 and Leonard 1970). In captivity the bald patches and loss of condition occurred in males about 3 weeks after mating, these survived and recovered their condition so that mating could have occurred in early May at Olney State Forest. Press (1976) reports young in the pouch 8.5 mm long at Erina N.S.W. (Alt. 60 m) in June, this indicates mating occurred in early May. The authors received a female with pouch young 11 mm long from Bouddi National Park, N.S.W. (0-120m) early in July, also suggesting mating in May. At Lamington National Park, Queensland, the most northerly range of the species, Van Dyck & Ogilvie (1977) found breeding occurred in July-August at an altitude of 780-1000 m. Recher *et al.* (1975) found mating to occur in late June for Nadgee Nature Reserve N.S.W. (0-540m) and Warneke & Wakefield (1963) found births in mid-August at Loch Valley Vic. (< 400m) suggesting mating in July. At Cradle Mountain Tasmania (1545m) Green (1972) recorded births in October and lactating females in February suggesting a longer breeding period. A similar trend was found in populations in Kosciusko N.P. near Thredbo (1627m) (Ann Jelinek, pers. comm.).

The principal animals used in this study came from the Brindabella Ranges on the border of the A.C.T. and N.S.W. (900m). Mating was observed by Tony Stewart (A.N.U.) from the 31/7/77-2/8/77 after being in captivity for 2 weeks. The offspring from this pair mated in Sydney in June, 1978, but although the females pouches developed, no young were born, suggesting a possible lack of synchronisation between male and female. These matings were 6 weeks earlier than the wild populations in the Brindabella Ranges and not much later than wild populations near Sydney.

Attempted matings lasted for a period of a week with a definite copulation observed two days after the commencement of mating activities. The male would rush the female and seize her with his jaws by the scruff of the neck. Sometimes the female would resist and roll on her back kicking vigorously forcing the male to release his grip. A hissing open-mouthed threat was used by the female when resisting. Eventually the male, maintaining his grip, moved into the mounting position, while being dragged around by the female, who eventually succumbed. Copulation sometimes took place for an hour or more (the exact time was not recorded).

Pouch Development: About 19 days after mating the nipples became more obvious, at 21 days the pouch area was slightly enlarged and the teats were very obvious, and 23 days after mating the pouch had deepened and the edge was ridged. A central ridge divides the pouch into left and right halves. This was well concealed by body fur. Five days before birth there was no more obvious development, but 2 days before birth the pouch became fleshy, hung down and although covered by fur the position of the pouch was clearly visible as the animal moved about. The day before birth the pouch appeared moist.

Birth: The birth was observed by Densey Clyne and Jim Frazier while they were attempting to film the event. Their observations are as follows:

"On arrival home we saw one of the female *A. swainsonii* gathering leaves and pulling them around herself. After a few moments of this she adopted what seemed to be a strange attitude, on all fours with the rear end somewhat raised off the substrate, and perfectly motionless.

I moved forward to take a closer look at what struck me as an unusual posture, and the female, startled by my approach, moved back a few inches. Her movement revealed a wet patch on the leaves and a struggling neonate, which she began to lick.

The female settled down quickly, seeming too intent on what she was doing to be further interested in the lights and movement. I managed to film what turned out then to be two neonates. Later the pouch was found to contain the full complement of 8 neonates, which did not include the two referred to above.

This observation was excellent preparation for our later attempts — finally successful — to film the birth of *Dasyuroides byrnei*, although this was not realized at the time. Five attempts were made to get this film sequence, and it was the amazing speed of the neonates in their movement to the pouch that caused our difficulties.

The important points to emerge from our experience were that: (1) both *A. swainsonii* and *D. byrnei* adopt the same basic birth position, i.e. on all fours with the hindquarters somewhat raised; (2) They both expel large amounts of fluid; (2) Although *A. swainsonii* was not seen from underneath, we believe that the journey from the cloaca to the pouch is both rapid and, because of the mother's posture, slightly downhill as in *D. byrnei;* (2) The supernumerary young were not eaten but discarded."

From this account it is obvious that birth is a rapid process and that as recorded for this and other dasyurid species more young are produced than the number of teats available (Green 1972).

Gestation: The gestation period was estimated to be between 28–35 days. Only limited information could be obtained — as only one pair mated in captivity and gave birth. The gestation for this pair was 28–30 days; but the other pair was collected on 28 July, 1977, and no matings were observed. Birth was on the 31 August, 1977, giving a gestation of at least 35 days.

Growth and Development of the Young: Two litters, one of 8 and the other 9, were observed from birth. At birth, the crown rump measurements were 4.5–4.7 mm (see Fig. 2). They were bright pink in colour and hard to count as they were pressed together tightly in two rows. A specimen of a new born young was examined under a dissecting microscope. Claws on the fore legs were well developed. The hind legs were poorly formed with no toes or claws. There were large nostrils and the mouth was just a circular hole. There was no sign of ears and the position of the eyes could just be seen. A short tail was present.

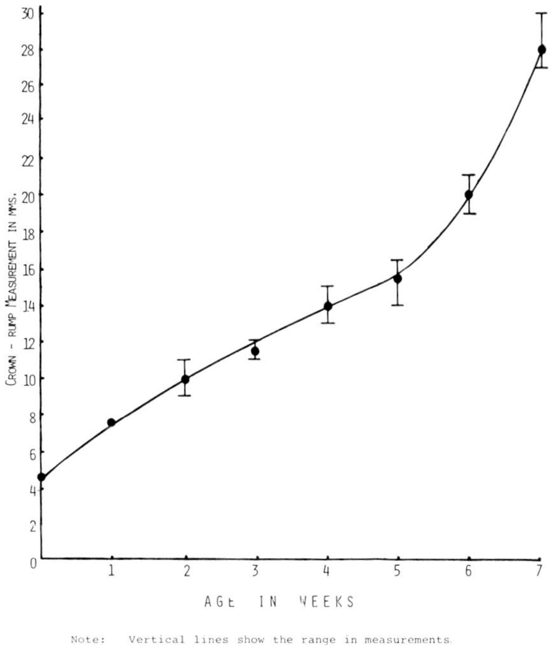

Fig. 2. Growth of pouch young up to seven weeks.

Marlow's (1961) description of the new born young of *A. stuartii* shows the two species to be similar.

After a week the young had increased in size and were pale pink. By two weeks the young were making some writhing movements and were visible in the pouch without parting the fur. At 3 weeks it was harder to make the crown–rump measurements because of the bodies being stretched.

At 4 weeks the toes were developed on the hind foot and examination of a specimen approximately this age under a dissecting microscope showed the claws on the hind feet to be developing as buds. The ear buds were not free from the head and dark eye spots were clearly visible. There was also a slight darkening on the head as the fur had started to grow. The position of the lips can be seen.

At 5 weeks the young were still pink in appearance although there was a layer of fine dorsal fur, which was slightly longer on the head. The ears were still not free.

The ears were free of the head at 6 weeks and the eyelids were well defined. The young were now well covered with fine grey dorsal fur.

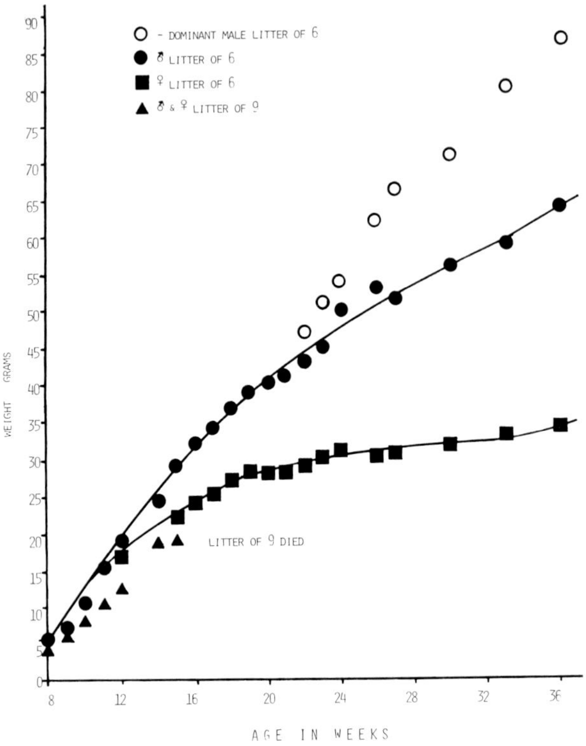

Fig. 3. Average weight of young from eight weeks.

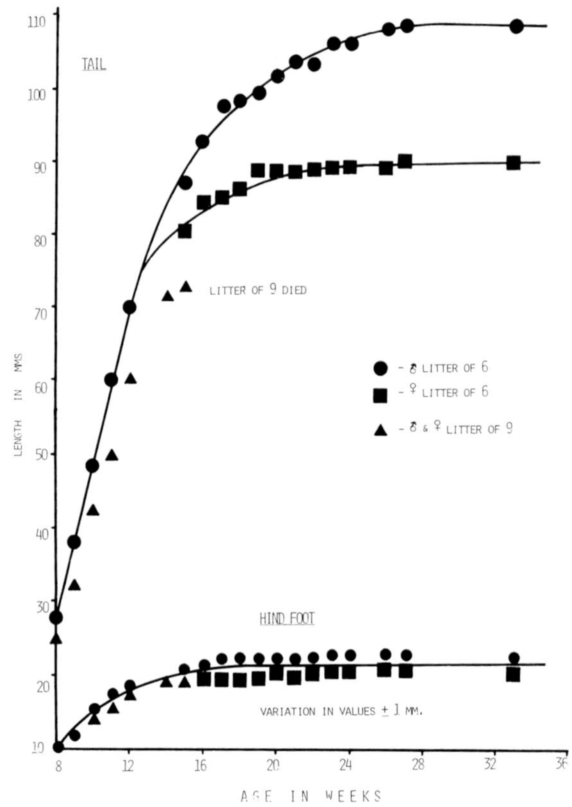

Fig. 5. Tail and hind foot measurements from eight weeks.

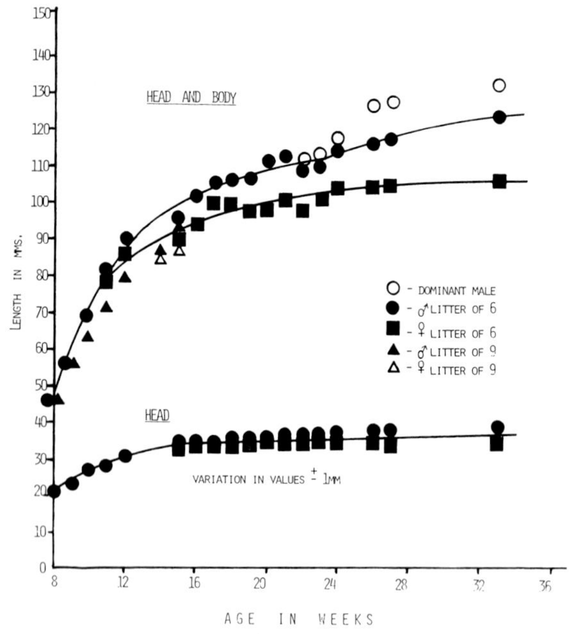

Fig. 4. Head and body measurements from eight weeks.

At 7 weeks it was impossible to measure the crown–rump measurements so head and body, head and hindfoot measurements were taken (see Figs. 4 and 5). The dorsum was well furred but the venter was still pink. The lips were separating and one young was heard calling. The young were still attached to the teats and hung well out of the pouch.

The young were left in the nest for the first time at 8 weeks old. The eyes opened at 58 days. The dorsal fur was grey with longer brown fur on the head. The venter was covered with fine light grey fur. Some teeth were present. This was the first chance to observe the teeth so it was not possible to determine when the first teeth erupted. One litter was reduced to 7 whilst the other still had 9 young. The litter of 7 were larger in all measurements and weighed 5.2 g compared with 4.1 g – 4.2 g for the litter of 9 (see Fig. 3).

By 9 weeks the young were well furred and still carried on the teats by the female if she was disturbed. The young showed signs of venturing from the nest box themselves. Some aggression by the young was shown for the first time when caught by hand. Another of the litter of 8 disappeared leaving 6.

At 11 weeks the young were still clinging to the mothers back when disturbed and their heads still appeared out of proportion with their bodies. The young were still observed suckling at 12 weeks and they started to eat a limited amount of solid food. The litter of 6 was still outgrowing the litter of 9.

Unfortunately at 13 weeks the mother with the litter of 9 and one young were found dead. The remaining 8 young were fed on milk, earthworms, mealworms and "Luv" mince but they all died when 15½ weeks old.

The mother of the litter of 6 died at 14 weeks and the post mortem could not determine the cause. These young were also fed milk, earthworms, mealworms and "Luv" mince and survived to maturity. Possibly the fact that they were more robust than the litter of 9 helped them survive on their own.

The weights of these survivors gradually increased up to the breeding season in mid June 1978. At this stage the weights were quite variable. The males weighed 66g, 71.5g and 96g. The females weighed 43.5g and 27g (abnormally low). The third female died just before the breeding season.

Captive Observations: Female *A. swainsonii* appear to be excellent mothers, as up until weaning the mortality rate was low. It was observed on many occasions that if young fell from the pouch they would make a "hissing" distress call and the female would immediately return and push the young towards the pouch area with the fore paws and squat until the young reattached itself. While the young were in the pouch they were groomed regularly. When left in the nest the female only stayed out for short periods of 10–15 minutes and if disturbed she would immediately return to the nest and collect the young and move them to another nest. Sometimes the young were left in another nest closer to where the female was feeding or the young stayed on the teats and were dragged about.

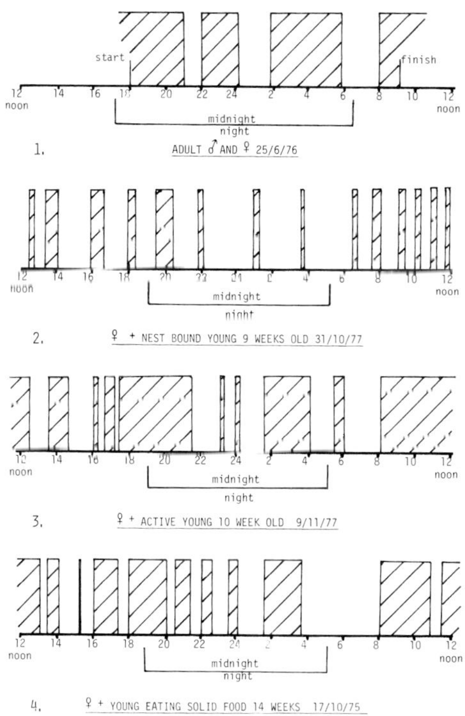

Fig. 6. Activity periods in different social groups.

When the young were about 10 weeks old they started to venture from the nest themselves. The female returned them to the nest after they climbed on her back. If, when she left the nest, young were still hanging on the teats she would return until she could leave on her own.

Young were observed to be carried on the female's back until they were 12–13 weeks old.

ACTIVITY PATTERN: *A. swainsonii* have been reported to be diurnal and have frequently been sighted during the day in the wild (personal observation, Brian Wilson, Harry Recher, pers. comm., S. Hall 1980).

Some limited information was obtained on the activity of the species in captivity. This was carried out by observing the animals for 24 hour periods where possible. Four different social groups were observed. One was a female with active young eating solids; the second, was a female with 10 week old young which were active but not eating solid food; the third, a female with young nearly 9 weeks old which were nest bound; and the last group, an adult pair with no young. Only 15 hours of observation was done on the adult pair.

Animals showed periods of activity over the whole 24 hours although there appeared to be two periods of reduced activity. One occurred either side of 12 midnight (10.00 p.m.–2.00 a.m.) and lasted for 2–4 hours. Some short periods of activity sometimes occurred during this time, the main activity being defaecation. The second period was either side of dawn (4.00–8.00 a.m.) for about 4 hours and also comprised of short periods of activity. During the rest of the daylight hours the time was spent fairly evenly between activity and rest.

The exception was the female with young in the nest who, understandably, spent more time in the nest box and only came out to feed and defaecate, etc., for short periods of 10–15 minutes. Even so there was a lack of activity during the major periods of rest mentioned above. The adult pair did not remain inactive for such long periods but their longest rests were also around midnight and dawn.

NEST BUILDING: Nests were constructed of a ball of Eucalypt leaves which were tightly packed and completely filled the nest boxes. Nest boxes were preferred to nests made on the ground under bark and logs. Leaves were carried singularly to the nest and placed in position. The shape of the interior of the nest was maintained by the occupant turning head to tail several times in the nest.

FOOD PREFERENCES: Live food was preferred. Earthworms were eaten readily by *A. swainsonii* but were refused by *A. stuartii*. Freshly killed mice were refused by *A. swainsonii* although readily eaten by *A. stuartii*. Mealworms, grasshoppers, beetle larvae and cockroaches were readily taken by both species. The "Luv" mince was taken readily and was given ad. lib. and used as the base food.

DISEASES: It was interesting to note that the mothers in each of the four litters observed died when the young were approximately at the same stage of development (14 weeks old). Three of the litters died about 2 weeks after their mother.

The mothers died from different causes. One died from an enteritis infection, another showed no obvious cause except a possible viral infection; another had a large infestation of carpet python worms *Amplocaecum robertsii* and one died of pneumonia. A post mortem of a male showed degeneration of major organs, a symptom of male "die off". One litter of young was killed by an enteritis outbreak whereas another litter showed no obvious reasons for death except a possible viral outbreak.

DISCUSSION

The life cycle of *A. swainsonii* appears to be almost identical to its close relative *A. stuartii*. Where the species occur sympatrically *A. swainsonii* usually breeds earlier. In the Central Coast district of N.S.W. breeding may be 3 months earlier. In most other localities breeding is about 1 month earlier. The only exception to be found of this was at Sherbrooke Forest Park where Read & Wallis (1975) found that both species breed at the same time.

From Table 1 it can be seen that the only other point in the life cycles that appeared to differ greatly was the point at which the young were left in the nest while the female foraged for food. This extended length of time may be a factor of being in captivity although field work by Leonard (1970) and observations by Fleay (1932) agree with our findings. Marlow (1961) did not mention the young of *A. stuartii* being left in the nest but noted that they first released the teats at 40 days. Wood (1970) reported that young were left in the nest at 35 days in the wild. This compared with 56 days in *A. swainsonii*.

An attempt was made to estimate the ages from some juvenile records (Tony Rose, pers. comm.) using the growth curves obtained from this study. It was found to be almost impossible because of the great variation in measurements between populations and individuals within populations. Once the young were over 16 weeks old the head and hind foot measurement stabilised; the tail was fairly reliable, but varied considerably between individuals. Body weight and head and body length were too variable to be of much use. Therefore it should only be possible to age animals under 16 weeks old with reasonable accuracy especially when the young are still in the pouch.

ACKNOWLEDGEMENTS

We would like to thank Tony Stewart and Chris Tidemann (ANU), Tony Rose (NSW N.P. & W.S.) and Barry Baker, for supplying the animals used in this project, and Drs. Mike Archer, Barry Fox and Martin Denny for their reading and criticising of the manuscript. We would also like to thank Jim Frazier and Densey Clyne for the use of their observations on the birth of the species and Peter Jensen for assisting with the maintenance of the animals in captivity.

Table 1. Comparison of the life cycles of *A. swainsonii* and *A. stuartii.*

	A. swainsonii	*A. stuartii*
Breeding Season (mating and births) Gestation period	May–October 29–35 days	July–October 30–33 days (Marlow 1961)
Maximum litter size at birth equals number of teats available	Usual 8 Var. 6–9	Usual 8. Var. 8–10 (Rose pers. comm.)
Size of young at birth	4.5–4.7mm	4.9 mm (Marlow 1961)
Young first left in nest	56 days	35 days (Wood 1970)
Eyes open	58 days	62 days (Marlow 1961)
First venture from nest	63–70 days	——
Eating solid food	84 days	——
Weaned	98 days	90 days (Marlow 1961)
Sexually mature	10–11 months	11 months (Marlow 1961)
Life span of ♂s in the wild	350 days (Reed & Wallis 1975)	350 days (Reed & Wallis 1975)

We are also grateful to the National Parks & Wildlife Service of N.S.W. for granting a permit to carry out this work and also to Ted Finney of Taronga Zoo for carrying out the post mortems.

REFERENCES

FLEAY, D., 1932. Swainson's Phascogale (The "Bush Mouse") *Vic. Nat.* **49:** 132–4.

GREEN, R. H., 1972. The murids and small dasyurids in Tasmania. Part 5. *Rec. of Queen Victoria Museum, Launceston.* No. **46**.

HALL, S., 1980. Diel activity of three small mammals coexisting in forest in southern Victoria. *Aust. Mammal.* **3**: 67–79.

LEONARD, B. V., 1970. Effects of control burning on the ecology of small mammal populations. *Second Symposium on Fire Ecology, Forests Commission of Victoria*: Pp. 1–4.

MARLOW, B. J., 1961. Reproductive behaviour of the marsupial mouse, *Antechinus flavipes* (Waterhouse) (Marsupialia) and the development of the pouch young. *Aust. J. Zool.* **9**: 203-18.

PRESS, A., 1976. "Some aspects of the ecology and habitat utilisation of four species of native small mammals." B.Sc. Hons. thesis: University of Sydney, Sydney.

RECHER, H. F., LUNNEY, D. AND POSAMENTIER, H., 1975. A grand natural experiment, the Nadgee wildfire. *Aust. Nat. Hist.* **18**: 154–163.

REED, G. F. AND WALLIS, R. L., 1975. Studies of *Antechinus swainsonii* and other small mammals in an area of Sherbrooke Forest Park. *Vic. Nat.* **92**: 84–90.

TAYLOR, J. M. AND HORNER, B. E., 1970. Gonadal activity in the marsupial mouse *Antechinus bellus* with notes on other species of the genus (Marsupialia: Dasyuridae). *J. Mammal.* **51:** 659–68.

VAN DYCK, S. AND OGILVIE, P., 1977. *Antechinus swainsonii* (Waterhouse, 1840). The Dusky Marsupial Mouse, an addition to the mammal fauna of Queensland. *Mem. Qld Mus.* **18:** 69–73.

WAKEFIELD, N. A. AND WARNEKE, R. M., 1963. Some revision in *Antechinus* (Marsupialia) 1. *Vic. Nat.* **80**: 194–219.

WOOD, D. H., 1970. An ecological study of *Antechinus Stuartii* (Marsupialia) in a south-east Queensland rain forest. *Aust. J. Zool.* **18**: 185–207.

RESUMEN

10. El ciclo biológico de *Antechinus swainsonii* (Dasyuridae, Marsupialia). Por Ray y Anne Williams.

Se ha publicado poca información sobre la biología de *Antechinus swainsonii.* En este trabajo se describen el crecimiento y desarrollo embriológico de dos pariciones concibedas en el laboratorio y dos más criadas en el laboratorio pero nacidas en el ambiente natural.

En la naturaleza el apareo ocurre desde comienzos de Mayo hasta Septiembre en Australia, con apareos más tardíos registrados en Tasmania. De la información obtenida parece que la altitud juega un rol importante en la estimulación del apareo; las colonias de las costas y de bajas altitudes se aparean más temprano, caulquiera sea la latitud. Usualmente los machos mueren aproximadamente 3 semanas después de apareo. Las hembras parecen ser monoestras.

El período de gestación se estimó entre 29 y 35 días. Cuando recién nacidos los jóvenes son de aproximadamente 4.5 mm, y presentan un color rosado brillante. El marsupio usualmente contiene un total complemento de jóvenes (8). La piel comienza a desarrollarse a las 4–5 semanas. A la edad de 8 semanas los jóvenes fueron dejados en el nido y muy poco después abrieron sus ojos. En este tiempo están bien cubiertos de pelo. Se observó que a las 12 semanas los jóvenes comían alimentos sólidos y parecieron destetarse aproximadamente a las 14 semanas. Desde ontonces los jóvenes se desarrollaron lentamente, alcanzando el tamaño de adultos poco antes de la próxima estación de reproducción.

11. A REVIEW OF DASYURID ECOLOGY AND SPECULATION ON THE ROLE OF LIMITING SIMILARITY IN COMMUNITY ORGANIZATION

BARRY J. FOX[1]

The dasyurids are a particularly successful group that have colonized the entire range of terrestrial habitats in Australasia. Despite a marked increase in ecological studies over the last eight years, information is available for less than a quarter of the species. The majority of studies have been conducted in southeastern Australia in forest or heath. There is a particular need for studies on species inhabiting the arid and semi-arid zones and also for studies in tropical northern Australia. *Antechinus stuartii* has been the subject of many studies so that its ecology is relatively well known, however data on many of the other species is derived from single studies. There remains a great deal to be done on population ecology, habitat requirements and fire ecology of many species.

Dietary information indicates that with the exception of a few of the larger dasyurids, arthropods are most likely to be the major diet component. This lack of feeding specialization is consistent with the overall lack of specialization in the Dasyuridae. This may have contributed to the reduced number of species of dasyurid found at any site (α – diversity) in comparison to the high β – diversity (species turnover between sites). Dasyurid assemblages do not appear to exhibit close packing of species. This may be influenced by the strong sexual dimorphism present in many dasyurids, and the resulting expanded niche width for species.

Body size plays an important role in the organization of dasyurid species assemblages. Sympatric species show marked separation in body weights, often greater than minimum ratios predicted by the theory of limiting similarity. When considered over the broad biogeographical region of eastern Australia, congeners show a linear relationship between the logarithm of body weight and head plus body length. A similarly distinct relationship exists between the logarithm of body weight and species rank. The differences found for each genus reflect the fundamental differences between genera in the relationship between body shape and body weight. A speculative argument is presented to explain how constant weight and size ratios might occur for species that are allopatric or parapatric. This relationship is not restricted to the Dasyuridae, and other examples are given. The logarithmic size sequences of species within genera do not appear to result from chance, the use of ranking or the classification process. Limiting similarity is an insufficient explanation at a regional level where species are not coexisting. Latitudinal gradients in productivity and prey size, and the degree of overlap in the distribution of species appear as factors more likely to be influencing the observed pattern.

Key words: Community structure, competition, Dasyuridae, diet, ecology, fire habitat, limiting similarity, population, review.

Pages 97–116 *in* CARNIVOROUS MARSUPIALS, ed. by M. Archer, February, 1982. Roy. Zool. Soc. New South Wales: Sydney, Australia.

IN 1972 Eleanor Russell reviewed marsupial ecology as part of a status report on Australian Mammalogy. She noted that 13 years before in 1959, Main, Shield and Waring in a similar review made the statement" . . . there has been no comprehensive ecology of Australian marsupials". She went on to say that the position had changed little in the intervening period. Of the studies she reviewed only three could be regarded as dealing specifically with dasyurid ecology: Calaby's 1960 study of *Myrmecobius fasciatus;* Guiler's 1970 study of *Sarcophilus harrisii;* and Wood's 1970 study of *Antechinus stuartii.* A further seven studies contributed information on dasyurid ecology (see Russell 1972).

For this review the position is much better; although there is a tremendous amount still unknown, our knowledge has rapidly increased over the last few years. From this symposium alone 19 papers deal directly with dasyurid ecology, and a further 15 cover related topics. Although this represents a welcome increase in research, it is important to realise how the research effort (since 1972) has been distributed over the 50 or so species of dasyurids and between major fields of ecology. Several aspects of dasyurid ecology have been included in a recent review of the ecology of small marsupials (Tyndale–Biscoe 1979).

POPULATION ECOLOGY

Understanding population processes in dasyurids has been made more difficult by the variety of life history strategies exhibited (Lee *et al.*

[1] School of Zoology, University of New South Wales, Kensington, New South Wales, Australia, 2033.

Table 1. Dasyurid species for which some information on population ecology is available.

Species	Habitat	Location	Reference
Antechinus stuartii	Tall Closed Forest (Rainforest)	Mt. Glorious, Qld.	Wood 1970
A. stuartii	Open Forest (Dry Sclerophyll)	Daylesford, Vic.	Leonard 1972
A. stuartii	Tall Closed Forest (Rainforest)	Mt. Glorious, Qld.	Braithwaite 1973
A. stuartii	Tall Woodland Open Forest (Dry Sclerophyll) Closed Forest (Wet Sclerophyll) Tall Closed Forest (Rain-forest)	Petroi, N.S.W.	Fletcher 1977 Statham 1982 Statham & Harden 1982
A. stuartii	Closed and Open Forest	Central Coast, N.S.W.	Press 1976
A. stuartii	Closed Forest (Wet Sclerophyll)	Myall Lakes, N.S.W.	Lim 1979
A. stuartii *A. swainsonii*	Closed Forest (Wet Sclerophyll)	Powelltown, Vic.	Leonard 1972
A. swainsonii	Closed Forest (Wet Sclerophyll)	Sherbrooke Forest, Vic.	Reed & Wallis 1975
A. minimus	Closed Heath	Glennie Island, Bass Strait	Wainer 1976
A. bilarni	Scree Slope	Kakadu National Park, N.T.	Begg 1981
Sarcophilus harrisii	Heath, Open Forest and Farmland	Cape Portland, Tasmania.	Guiler 1970
Sminthopsis crassicaudata	Low Shrubland and Grassland Low Shrubland and Grassland Grassland	Fowlers Gap, N.S.W. Willandra, N.S.W. Werribee, Vic.	Morton 1978c
Dasyurus viverrinus	Open Forest and Farmland	Tasmania	Goodsell 1982
Phascoagale tapoatafa	Open Forest (Dry Sclerophyll)	Christmas Hills, Victoria	Cuttle 1982
Planigale	Arid zone habitats	Sturt National Park, N.S.W.	Denny 1982

1982). The phenomenon of post–mating male mortality was first documented for *A. stuartii* by Wood (1970). During the mating season male *A. stuartii* become more aggressive (Braithwaite 1974) and show increased movements (Wood 1970), Braithwaite 1979). Lee *et al.* (1977) have shown a stress induced increase in free corticosteroids produces ulceration of the gastrointestinal tract and suppresses the immune reaction, leaving males vulnerable to pathogens and resulting in their death. No increase in aggression or marked increase in free corticosteriod levels has been observed in females. Similar stress induced mortality of males has been observed in *A. flavipes* and *A. swainsonii* and Lee *et al.* (1982) have allocated a number of other species of *Antechinus* and *Phascogale* to this life history strategy. Wood's (1970) study of *A. stuartii* in a rainforest habitat has served as a model for a growing number of ecological studies on small dasyurids, and much of the work has continued on this common species. Braithwaite (1973, 1974 and 1979) followed with a study in the same area, including detailed examination of behavioural changes and habitat ultilization. He concluded that dominant males occupied the best habitats with access to most food, grew faster and as a result of increased size had greater mating success, improving their genetic fitness over subordinate males that may not mate during the short mating season. This has given a better understanding of population processes in *A. stuartii,* particularly the reduced longevity in males.

Fletcher (1977) and Statham (1982a) showed the order of abundance of *A. stuartii* to be: closed forest (wet sclerophyll forest) > tall closed forest (cool temperate rainforest) > tall woodland or open forest (dry sclerophyll forest) in northeastern New South Wales. Marked differences in trappability (number of captures/individual) occur between vegetation types, and often there were marked differences in trappability between sexes. Using tracking plots she was able to show that up to 18% of individuals may not be trapped at all. Seasonal fluctuations in *A. stuartii* abundance were greatest in the wet sclerophyll forest as well (Statham and Harden 1982) with a higher proportion of males being trapped there.

Antechinus stuartii showed a trend of decreasing winter abundance, from 1969–72, in open forest at Nadgee during two dry winters, following wetter years, but preceding the 1972 fire (Recher *et al.* 1978). In the same study *A. swainsonii* shows no trend in abundance, but body weights show a marked decreasing trend. Shorter term prefire demographic changes for these species at Nadgee are provided by Newsome *et al.* (1975) in heathland as well as open forest. Two additional studies on the central coast of New South Wales have also examined demographic changes and habitat use in *A. stuartii* (Press 1976 and Lim 1979).

Other species of *Antechnius* have been less well studied. A very high density population of *A. minimus* (80 ha^{-1}), in closed heath on Glennie Island (Victoria) has been studied by Wainer (1976). Limited information on *A. swainsonii* is available from a study by Reed and Wallis (1975) in Sherbrooke forest (Victorian closed forest). Both studies show a demographic pattern similar to that of *A. stuartii,* but shifted in time. *A. minimus* breeds in winter, 2–3 months earlier than the late August breeding for *A. stuartii* in Southern Victoria, and Wainer (1975) has suggested that *A. minimus* breeding may be synchronised with the winter peak of insect larvae, as these form a large component of its diet (Wainer 1976). *Antechinus minimus* fills the niche of a soil fossicking insectivore in Victorian heathland (Braithwaite *et al.* 1978). Dickman (1982) has made a strong case to support his suggestion that differences in mating time between sympatric populations of *A. swainsonii* (also a soil fossicking insectivore) and *A. stuartii* increase the niche separation between these species by maximising the difference in body size of juveniles. Such reasoning requires the species to be competing, this is supported by the experimental confirmation of interspecific interaction between these species (Dickman 1980). These hypotheses are not mutually exclusive and both may be contributing to the observed differences.

Antechinus bilarni is one species of *Antechinus* studied (Begg 1981), that does not exhibit postmating male mortality. Both males and females survive to breed in a second breeding season, although reproductive effort may be less intense in each season than for the monoestrous *Antechinus* spp. (Lee *et al.* 1982). Longevity in *A. bilarni* inhabiting a scree slope in Kakadu National Park (N.T.) seems greater than for other species of *Antechinus* and breeding is early, occurring in late June or early July (Begg 1981).

The ecology of two Tasmanian dasyurids have been studied in detail, *Sarcophilus harrisii* (Guiler 1970, Buchman and Guiler 1977) and *Dasyurus viverrinus* (Godsell 1982). Guiler concluded that man exerted the greatest influence on population regulation in *S. harrisii,* detrimental in terms of direct predation and incremental in provision of excess food. However, greatly reduced juvenile recruitment was observed at the site with higher adult abundance. Guiler found *S. harrisii* to be most often trapped in heathlands and open forest. Goodsell (1982) concluded that her *D. viverrinus* population appeared to be regulated by intense intra–specific competition, and that differential dispersal of the sexes was an important factor in understanding the mechanisms involved in population regulation.

Morton (1978a,b,c) carried out a very comprehensive study of *Sminthopsis crassicaudata* despite the difficulties inherent in trapping this species. This represents the first detailed study of a polyoestrus dasyurid and shows several important differences. Although breeding was seasonal it extended over six months with females producing two litters during the breeding season. Intra–specific competition was not important, there was little aggression between individuals, and during the colder months (including the early part of the breeding season) nest sharing was common, with up to eight individuals found in the one nest (mean > 3 in April–June) (Morton 1978b). Both sexes are mobile and have weak site attachment. Morton (1978c) concludes that these adaptations are responses to a high variability in food supply (see also Morton 1982).

Our knowledge of population processes in dasyurids may be most conveniently summarised into a generalised pattern of demographic change based on small dasyurid species (largely *A. stuartii*) but using other information where available. Mating is usually tightly synchronised (as are births), usually occurring in winter or spring (Lee *et al.* 1982). This period usually corresponds with the population's minimum, because of the male mortality that follows mating in many species. Obvious exceptions are the polyoestous, and larger more long lived species. Population peaks occur in summer with juvenile recruitment. The peak occurs later for polyoestous species as juveniles from second litters become independent (Morton 1978c). Because of differences in seasonal trappability (Fletcher 1982a) a "trappability peak" might be observed at the time of juvenile dispersal and male territory formation (Braithwaite 1979) or more likely at mating time, when an abnormally large proportion of males may be captured, reflecting the increased movement and aggression of males at this time (Wood 1970, Braithwaite 1974).

Species abundance may differ in different habitats within the same area (Statham and Harden, 1982) and may be related directly to food supply (Guiler 1970), although year to year differences in abundance (Braithwaite 1973), or body weight (Recher *et al.* 1978), may be indirectly related to rainfall. Climate may play an important role in determining breeding times for species, via its regulation of insect emergence times (Van Dyck 1982, Wainer 1976). Inter–specific competition may play an important additional role in determining mating times for sympatric species of dasyurids, to maximise niche separation, and may be important in population regulation (Braithwaite 1973, Dickman 1980, 1982). Intra–specific competition is intense in many species (Braithwaite 1973, 1979, Lee *et al.* 1977, Godsell 1982) and plays an important role in population regulation, in contrast to natural predation which has a minor role only (Guiler 1970, Braithwaite 1973). There may be differential dispersal between sexes, (Goodsell 1982) influencing population structure, which may also be affected by abundance–induced decreases in juvenile recruitment in areas of high food availability (Guiler 1970).

The extremely high density encountered for an island population of *A. minimus* (80 ha^{-1}), is consistent with the above model, given the absence of dispersal and reduced competition and predation on the island. Mainland densities for other small species are an order of magnitude lower (*A. stuartii* — 4.4 ha^{-1}, Statham and Harden this volume, N.S.W.; *A. flavipes* – 3.5 ha^{-1}, Dwyer *et al.* 1979, Qld., and 6 ha^{-1}, Christensen and Kimber 1975, W.A.; *S. murina* – 6 ha^{-1}, Fox and McKay 1981, N.S.W.). Wood (1970) found densities for *A. stuartii* of 45 ha^{-1}, but recognised the Mt. Glorious area as a 60 ha "island" of subtropical rainforest surrounded by pasture and open forest.

A species' ecology embraces all factors in its environment capable of influencing population processes. In particular, life histories, reproductive parameters, habitat requirements, behaviour and interactions with other species. I will confine myself to aspects not covered by reviews in this volume: behaviour (Croft); life history strategies (Lee *et al*); arid zone ecology (Morton); and *Planigale* ecology (Denny).

HABITAT

Here I am considering studies primarily aimed at measuring habitat usage or requirements of one or more species rather than incidental notes on habitat. The last three or four years has seen a number of these studies and information now exists for *A. stuartii, A. swainsonii, A. flavipes, A. minimus, S. murina, S. leucopus* and *S. crassicaudata*. Again we know most about *A. stuartii* (Table 2).

Barnett, How and Humphreys (1978) used structural characteristics to describe *A. stuartii* habitat use in pine and eucalypt forests in northern N.S.W. and Stewart (1979) used structural parameters in his analysis of habitat use by *A. stuartii* and *A. swainsonii*. Statham and Harden (1982) carried out an extensive study of *A. stuartii* habitat use at Petroi in northern N.S.W. for tall woodland, closed forest (wet sclerophyll), tall closed forest (cool temperate rainforest), logged tall closed forest and open forest (dry sclerophyll). A large number of parameters were employed, such as tree counts, log counts

Table 2. Dasyurid species for which information on habitat use is available.

	HABITAT USE		
Species	Habitat	Locality	Reference
Antechinus stuartii	Woodland to rainforest (see Table 1)	Petroi, N.S.W.	Statham & Harden 1982
A. stuartii	Open forest and pine forest	Clouds Creek, N.S.W.	Barnett, How and Humphreys 1978
A. stuartii	Pine forest and Open forests	Myrtleford, Vic.	Suckling and Heislers 1978
A. stuartii *A. flavipes* *A. minimus*	Open forests and Heathlands	Grampians Frankston Cape Liptrap, Vic.	Braithwaite, Cockburn and Lee 1978
A. stuartii *A. swainsonii*	Lowland Heath and Highland Heath	Nadgee and Kosciusko, N.S.W.	Newsome and Catling 1979
A. stuartii *A. swainsonii*	Closed forest (Creek margins)	Brindabella Range, A.C.T.	Stewart 1979
A. stuartii *A. swainsonii*	Closed forest	Sherbrooke Forest, Vic.	Hall and Lee 1982
A stuartii *Sminthopsis murina*	Open forest and Heathland	Myall Lakes, N.S.W.	Fox and Fox (in press)
S. murina *S. leucopus*	Woodland and Heath	Victoria	Morton, Wainer and Thwaites 1980
S. leucopus	Coastal Heath	Westernport, Vic.	Ahern 1974

and description, shrub, vine, sedge and fern counts, ground cover, horizontal obstruction, light penetration and vegetation biomass below 1 m. *Antechinus stuartii* was only found in areas with a relatively thick shrub layer, detailed habitat requirements were complex and abundance correlated with: the number of dwarf shrubs; the number of high stems and log length; and horizontal obstruction; respectively, in each of the three years studied. Predictive equations using these variables were tested on an open forest not previously trapped, but results were satisfactory for only one out of the three years. The year to year changes in predictive equations may result from many of the independent variables being highly internally correlated. Newsome and Catling (1979) in a study of heath habitats over an altitudinal gradient, devised an additive habitat index that considers aspects of the physical environment as well as vegetation. They used 4 point scales to describe tree canopy, shrub canopy, ground herbage, ground cover (as logs, rocks, debris etc), and soil moisture. Fox (1982) was able to use structural variables and plant species diversity to quantify habitat niche separation between *A. stuartii* and *S. murina*.

Braithwaite and Gullan (1978) developed a method of habitat analysis that included both structural and floristic characters that has been used by Braithwaite *et al.* (1978) to examine the habitat requirements for *A. stuartii, A. flavipes* and *A. minimus* in Victoria. *Antechinus stuartii* at Frankston used no floristic group markedly more than any other group, but structural groups that were layered (i.e. shrubs and trees), with high cover up to 1.5 m, were used most often. In the Grampians *A. flavipes* showed greatest use of structurally complex dry heath sites with small trees of *Eucalyptus aromaphloia*. At Cape Liptrap a single floristic and a single structural group received greatest use, with *A. minimus* often caught in short heath areas (up to 1 metre tall) dominated by *Melaleuca squarrosa, Casuarina paludosa, Leptospermum juniperinum, Leptocarpus tenax and Calorophus lateriflorus*. Gullan and Robinson (1980) used floristic classification methods to describe groups of trapping sites which they could relate to *A. stuartii and A. swainsonii* distributions. They were able to extrapolate from their trapping grid densities, to total forest abundance using vegetation maps of the total forest, illustrating the occurrence of the floristic groups recognised. However, the critical test of this predictive method has yet to be made. Given that *Antechinus* spp. are insectivorous, Gullan and Robinson (1980)) make the point that the relationship to floristics is necessarily indirect, whereas for omnivorous species, such as *Rattus fuscipes,* there is a more direct relation to food resources.

Fox and Fox (1981) compared several methods of vegetation classification in terms of their effectiveness in describing and matching the small mammal classification of sites. Using a method which weighted floristic and structural characters as equally as possible, they partitioned the total variance associated with the distributions of *A. stuartii* and *S. murina*. The variance was partitioned into that portion contributed by floristic and structural components, when both were used, as well as the proportion of the total variance explained if each set of components was used alone. The relevant values are tabulated in table 3 together with results for *R. fuscipes* and *Pseudomys novaehollandiae,* for comparison. For the omnivorous *R. fuscipes,* floristic components account for all of the variance explained and nothing extra is added by structural information. For both dasyurids, structural components provide the better description and although a reasonable description can be made from floristic data, it explains less of the variance and does not add to the amount of variance explained if included with structural data. This contrasts markedly with granivorous/omnivorous murid. *P. novaehollandiae* for which the variance is equally partitioned.

From the above information it would seem that the contribution of the various components to a species' habitat requirements is closely related to its dietary requirements, and good habi-

Table 3. The proportion of variance accounted for by structural and/or floristic variables. The significance of the simple and multiple regressions used is shown as: * $P < 0.05$, ** $P < 0.01$; *** $P < 0.001$. (Data taken from Fox and Fox in press).

DEPENDENT VARIABLES	Structure Only	Floristics Only	INDEPENDENT VARIABLES Both = Structure + Floristics
R. fuscipes	0.06 N.S.	*** 0.40	*** 0.40 = 0 + 0.40
A. stuartii	** 0.39	* 0.19	*** 0.40 = 0.29 + 0.11
S. murina	*** 0.46	** 0.34	*** 0.47 = 0.47 + 0
P. novaehollandiae	** 0.23	** 0.26	** 0.36 = 0.19 + 0.17

tat characterization requires complete structural and floristic analysis of the vegetation (Cheal 1979).

There remain two clear needs in this area: first, the obvious one is to obtain similar habitat information on those species for which it is currently lacking; the second, and not so obvious need, is for an experimental approach that manipulates these habitat variables to test the hypotheses formed from descriptive studies. Such an approach would improve confidence in predictions made on the basis of descriptive studies and have obvious applications to improving our knowledge of the distribution of habitats suitable for dasyurids. I think we can see from the proliferation of studies over the last few years, for which results are only now becoming available, that both of these needs are being, or will be, met.

DIET

Dasyurids have been thought of as insectivores, yet there has been very little published information on diet (see review in Statham 1982b). Leonard (1972) has shown the diet of *A. stuartii* and *A. swainsonii* at Powelltown and Daylesford in Victoria to be composed of arthropods. A study by Hindmarsh and Majer (1977) has shown *A. flavipes* from karri forests in Western Australia to be an opportunistic feeder, taking all but the smallest arthropods in roughly the same proportions in which they occur. A very complete study by Statham (1982b) has shown similar opportunistic feeding by *A. stuartii* at Petroi in northern N.S.W. with Coleoptera, Araneae and Blattodea occurring most often in a diet of ground–dwelling arthropods (0.5–2.0 cm length).

An interesting addition to the diet of *A. stuartii* was vegetation: some plant material was eaten throughout the year, and flowers of the shrub *Acrotriche aggregata* were eaten by all specimens taken from closed (wet sclerophyll) forest during August and September (Statham 1982b). Statham has shown abundance of arthropods to be lowest at this time, and may not meet the energy requirements of *A. stuartii. Acrotriche aggregata* may serve as a source of nectar, and hence energy. Carpenter (1978) has shown that some mammals such as *Rattus fuscipes* are able to play an important role in the pollination of plants such as Banksia, implying that nectar may form part of their diet. It would seem more likely that these pollination mechanisms coevolved with small marsupials rather than the more recently arrived rodents. Evidence of fruits of blackberries and native raspberries have been found in the faeces of *A. swainsonii* (A.K. Lee *pers. comm*).

Hall (1980) has shown that although there is much overlap in diet between sympatric *A. stuartii* and *A. swainsonii*, there are important differences in both prey type and prey size for arthropods taken, but that both are opportunistic feeders. *Sminthopsis murina* also appears to be an opportunistic feeder on arthropods (B. J. Fox, unpublished data).

Blackall (1980) has shown that insects also form an important component of the diet of *Dasyurus viverrinus*, being present in 97% of scats, while birds and small mammals form a minor component (17% of scats). Plant material was also important in the diet of this dasyurid, and was found in 100% of scats, with grasses and herbs making up 40% of the volume of scats, on average. The diet of *Sarcophilus harrisii* has almost no insect component, and very little plant material, with *Vombatus ursinus, Macropus rufogriseus* and Sheep, taken as carrion making up the major components (Guiler 1970).

These results emphasise the need for dietary studies on many of the remaining dasyurids and reinforce the importance of taking floristic as well as structural habitat data.

FIRE ECOLOGY

There has been an imbalance in the aspects of ecology studied, with considerable resources directed towards the study of the fire ecology of dasyurids and other groups. Studies of the effects of fire on *A. stuartii and A. swainsonii,* in both open forest and closed (wet sclerophyll) forest in Victoria, have shown that where much of the forest understorey remains unburnt (30–50%), these dasyurid populations quickly recover (Cowley, Heislers and Ealey 1969, Leonard 1970, 1972, Heislers 1978). The extensive studies on the 1972 Nadgee Wildfire (Newsome *et al.* 1975, Recher *et al.* 1975, 1978) have shown *A. swainsonii* to suffer local extinction. *Antechinus stuartii* survived on some grids (Newsome *et al.* 1975) where unburnt open forest patches remain, but on others (including heathlands) this species steadily declined to local extinction, as also occurred in the study by Recher *et al.* (1978). Christensen and Kimber (1975) have shown *A. flavipes* to be very detrimentally effected by fire in Western Australia where it was found to be most abundant in an area of Jarrah forest unburnt for 40 years, this has been supported by Schmidt and Mason (1975). Fox and McKay (1981) found *S. murina* most abundant in areas of open forest burnt two to four years previously, while *A. stuartii* in the same area of Myall Lakes National Park, recovered more slowly. Similar results were found in nearby heathland (Fox 1980).

The differences in the regeneration niche of these two species is examined in Fox (1982).

This underlines how important fires and fire regimes are as components of Australian heath and forest habitats. The dynamics of fire–perturbed populations has contributed much to our understanding of the ecology of these species, and has illustrated how species replacement can result in a change in the structure of community (Fox 1980, Fox and McKay 1981).

A closely related field of research is the study of disturbance effects other than fire. These studies have usually been done to answer specific questions, often related to management of forest. Barnett, How and Humphreys (1977) found that in northern N.S.W., populations of *A. stuartii* were restricted to native eucalypt forest, and only scattered individuals are found in nearby pine forests. Suckling and Heislers (1978) studied *A. stuartii and A. swainsonii* in different aged Victorian pine plantations and adjacent eucalypt forests. From a value of 5.8 captures/100 trap nights in mature eucalypt forest, *A. stuartii* showed an almost linear decrease in abundance in pine forests falling from 3.5/100 T.N. in 8 year old pine to 1.9/100 T.N. in a 42 year old pine plantation. *Antechinus swainsonii* was three times more abundant in 8 year old pine plantation (1.5/100 T.N.) than in mature eucalypt forest (0.5/100 T.N.), but in mature 42 year old pine forest the abundance had fallen to 1/5th of this value (0.1/100 T.N.). The effects of clear felling eucalypt forests followed by windrow burning prior to the planting of pine plantations in Victoria has been examined by Friend (1979), but the populations of *A. stuartii* and *A. swainsonii* were very low and results were inconclusive.

INTERACTIONS

Dasyurid communities are subject to many external influences apart from fire, and one which should be considered is the effect of the murids, peramelids and burramyids, which together with dasyurids comprise the total small mammal community. Inter–family, and indeed, inter–order competition may have an important bearing on how the dasyurid members of the community function. This has been well illustrated by Braithwaite (1973) who found that year to year variation in *A. stuartii* abundance was negatively correlated with small mammal species diversity. The largest component of the effect appeared due to *R. fuscipes* interacting aggressively with, and inflicting injuries on *A. stuartii*, an observation supported by laboratory trials. From this one can see the need to consider the whole community as part of the environment of the dasyurids and not consider dasyurids independently of other groups.

Interactions within the dasyurids have not received a great deal of attention in the past. Van Dyck (1982) considers possible interactions between *A. stuartii* and *A. flavipes*. Fox (1982) presents evidence of competitive interactions between *A. stuartii* and *S. murina* and illustrates that niche parameters in this community depart significantly from those expected in a non–interacting community (Fox, in press). However the most conclusive study to date has been that of Dickman (1980), who shows experimentally that *A. swainsonii* has the ability to displace and out compete *A. stuartii* in "islands" represented by closed forest (wet sclerophyll) gullies in the midst of open forest (dry sclerophyll) habitat in the Brindabella Ranges, A.C.T.

The need for additional experimental studies like this one is not only obvious, but is essential to unequivocally establish the roles played by intra–specific, inter–specific and inter–generic interactions in determining the structure of assemblages of coexisting species.

COMMUNITY ORGANIZATION

The Dasyuridae show a marked lack of specialization when compared to other marsupial families, such as the Phalangeridae. The diets of all small dasyurids so far examined, up to and including *Dasyurus viverrinus*, contain arthropods as the major component (see Leonard 1972, Hindmarsh and Majer 1977, Blackall 1980, Hall 1980, Statham 1982b). All appear to be opportunistic feeders and Hall (1980) has shown a high degree of overlap in both the size and type of prey taken by sympatric *A. stuartii* and *A. swainsonii*. But the larger species (*A. swainsonii*), takes a significantly greater proportion of larger prey.

This leads to an interesting hypothesis in relation to community organisation within the Dasyuridae : the lack of feeding specialization means that partitioning of the food resource (arthropods) appears to occur on prey size alone; this should accentuate the associated affect on body size differences. It should follow from this that within any one habitat there should be a reduced number of species segregated on body size, but there should also be a concommitant relative increase in the amount of habitat and microhabitat segregation by species. The latter is based on the theoretical conclusions of McArthur and Pianka (1966) and MacArthur and Wilson (1967), that partitioning space is easier than partitioning food on size.

The idea that differences in body size might produce differences in the food niches of species was first expressed by Hutchinson (1959), although Brown and Wilson (1956) had considered changes in body size between species in allopatry and sympatry (character displacement). Hutchinson's important contribution was the assembly of empirical evidence for a limiting similarity between sympatric species, expressed as the ratio of the linear dimensions of their trophic appendages (this value lies in the range 1.2 to 1.4).

For insectivorous birds, body weight is a better predictor of prey size than any single bill character, and further, the ratio of two species body weights is proportional to the overlap in food preference (Hespenheide 1971). Diamond (1975) used body weights to examine community structure in the guild of fruit pigeons in New Guinea. Eight species encountered at one locality formed a graded size series with body weight ratios between adjacent species, ranging from 1.3 to 1.6 with a mean of 1.5 (average weights 49, 76, 123, 163, 245, 414, 592, and 802g). Body weights were also used by Brown (1975), studying community structure among seed–eating rodents in the (a) Sonoran and (b) Great Basin deserts. Brown found body weight ratios ranging from 1.4 to 2.7 with an overall mean of 1.9 ((a) 7.2, 11.4, 17.1, 24.3, 45.3, and 120g; (b) 7.1, 12.5, 18.1, 37.6 and 101g; average weights). Some species showed a shift in weight between desert areas when an additional species is supported. Brown interpreted this as character displacement, to maximise the ecological separation between species in the community.

Much of the evidence for a limiting similarity has come from demonstrating character displacement. The subject has been thoroughly reviewed by Grant (1972). Although many well documented examples of differences between sympatric and allopatric populations do exist, it was not possible to unequivocally implicate natural selection as the necessary mechanism for the shift in character states in any single study. Grant concluded that the present evidence for the ecological aspect of morphological character displacement is weak and that this should be recognised and rectified by more detailed studies. Evidence for the behaviourial and ecological analogues of character displacement (habitat selection) is more satisfactory, as is that for character release, because it is easier to obtain (Grant 1972). Schoener (1974) has reviewed the closely related field of resource partitioning, again documenting the existence of niche differences in a great many cases but without proving causality.

The ecological aspect important here, is the implication that interspecific competition is operating to prevent the coexistence of species that are too similar. This is known as the competitive exclusion principle, attributed to Gause (1934) and most concisely expressed by Hardin (1960) as "complete competitors cannot coexist". May and MacArthur (1972) have provided a theoretical basis to limit the degree of similarity that can be tolerated (a separation of one standard deviation along the resource axis, between normal resource utilization curves). Hespenheide (1971) has shown the distribution of prey sizes for some insectivorous birds to be log–normal. Abrams (1975, 1976) has questioned the assumptions made in the derivation of the May and MacArthur model, but the empirical evidence does point to some limit to the similarity of coexisting species (see Schoener 1974). Two common strands run throughtout: one is the necessity for species to be using the same resource, hence the need to deal with guilds (insectivores, seed–eaters, etc), but not necessarily congeners; the second is sympatry, the species must be attempting to make use of the same resource in the same place (and at the same time). A closer examination of sympatric, insect–eating dasyurids would seem warranted to test this hypothesis on community structure.

The six species of dasyurid in Tasmania can be used to test this hypothesis. The species, the body weights and habitats most utilized (Green 1973) are shown in Table 4. The two species most similar in weight, (*Antechinus*) occur most frequently in different habitats. Each broad habitat category supports three species only, that are well separated on body weight (Table 4); in

Table 4. Tasmanian dasyurids showing mean male body weight and preferred habitat (from Green 1973). Weight ratios are shown in parentheses.

Species	Mean Male Weight	Preferred Habitat: Heath	Preferred Habitat: Forest
Sarcophilus harrisii	8.3 Kg		*
			(2.1)
Dasyurus maculatus	3.9 Kg		*
D. viverrinus	1.2 Kg	*	(59)
Antechinus swainsonii	66 g	(21.1)	*
A. minimus	57 g	*	
		(1.9)	
Sminthopsis leucopus	30 g	*	

some cases by an order of magnitude more than would be predicted for limiting similarity.

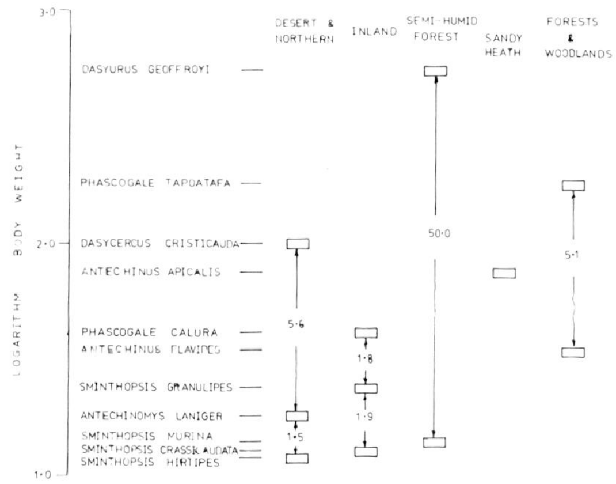

Fig. 1. Body weight separation shown for dasyurid species assemblages found in southwestern Australia. Assemblages taken from Baynes (1975) and body weights provided by A. Baynes *pers. comm.* Values are shown on a logarithmic scale.

Baynes (1975) lists a total of 11 dasyurids in all habitats in the south–west of Western Australia and allocates the members of this fauna to a series of suites occupying different habitats. These are shown in Fig.1 with the weights of the species depicted on a logarithmic scale and weight ratios between adjacent species shown. In any one habitat there occurs only one, two or three species, all well separated on body weight, and in some cases with weight ratios greatly exceeding those predicted for limiting similarity.

Victoria also has a rich fauna of small dasyurids, Braithwaite *et al.* (1978) report six species in heathland (*A. stuartii, A. flavipes, A. swainsonii, A. minimus, S. murina and S. leucopus*). However their heath sites do not support more than two species of small dasyurid each, again indicating a segration of species in space with only limited segregation on prey size. Morton, Wainer and Thwaites (1980) have examined the distribution of two *Sminthopsis* species (*murina and leucopus*), both of which can be found in the heathlands as well as woodlands and open forest; they found no site where both species were coexisting so they must be considered to be parapatric. These *Sminthopsis* inhabited similar habitats, but *S. leucopus* occupied areas with more dense understorey than those occupied by *S. murina.* Wakefield and Warneke (1967) similarly found *A. flavipes* and *A. stuartii* to be allopatric in Victoria although their distributations appear more consistent with parapatry.

Braithwaite *et al.* (1978) recognised two foraging niches for insectivores in southeastern Australia: one small and scansorial; the other larger and soil fossicking; these are usually filled by *Antechinus* spp. Hall and Lee (1980) found that both of the species they studied foraged in leaf litter but *A. swainsonii* had larger prey size diversity (11.3 *v* 8.6), because of access to larger prey. *Antechinus stuartii* had a larger prey type diversity (11.2 *v* 9.2), because by foraging up trees they had access to bark and foliage dwelling arthropods as well.

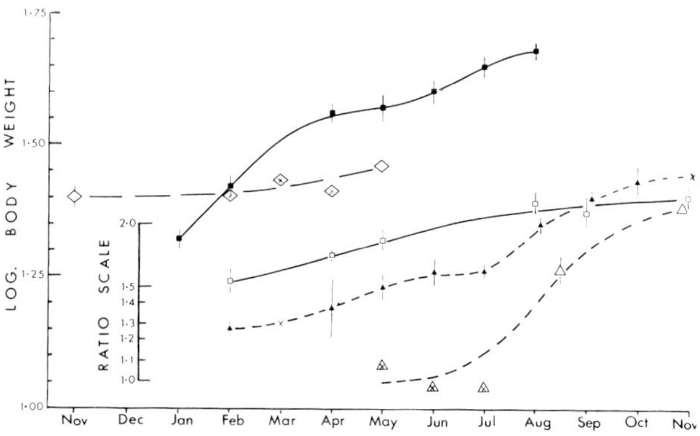

Fig. 2. Body weight changes for *Antechinus stuartii* and *Sminthopsis murina* for Myall Lakes National Park. Mean values, as logarithm of body weight during the year, are shown for *A. stuartii* ♂♂, ■; 1st Yr. ♀♀, □; 2nd Yr. ♀♀, open diamond; and *S. murina* ♂♂, closed triangle; ♀♀, open triangle. Single values are shown (x), error bars are ± s.e. and a ratio scale for body weights is included.

Seasonal and sexual weight differences are shown for two sympatric, similar species : *A. stuartii* and *S. murina* at Myall Lakes (Fig. 2). Differences between sex and age classes for *A. stuartii* (the larger species) increase during the period of lowest arthropod abundance, to be greatest at mating time before the males die. The small *S. murina* does not increase in body weight prior to breeding until the demise of male *A. stuartii* begins, and differences between age, sex and species classes disappear during the late spring–early summer arthropod flush. This is consistent with food availability influencing class–size differences. A similiar pattern is exhibited by *A. stuartii* at Mt. Glorious (Braithwaite 1973).

The weight differences between species at Myall Lakes, are distinct, however what is interesting are the differences maintained between the five age–sex classes throughout much of the winter period. The sexes behave as "species" with body weight differences during May ranging from 1.25 to 1.4, between successive members of the graded size series (Fig. 2). Hall (1980) 1978c, Lee *et al.* 1982) may limit this alternative (Fig. 3) is illustrated in an analysis of *Antechi-* during May. However if sexual dimorphism is considered the body weight ratios for the four "species" series (*A. stuartii* ♀♀, ♂♂, *A. swainsonii* ♀♀, ♂♂) become 1.21:2.22;1.35. A more complete picture for these two species can be obtained from Dickman (1982, table 5), who shows a remarkably constant head and body length ratio (1.30 to 1.35) between males of each species, and also between females of each species for seven different localities. When considered as four "species" the mean and range for size ratios in the above series become, 1.11 (1.09–1.12) : 1.20(1.17–1.23) : 1.12(1.09–1.13), over six localities. However, in the seventh locality the greatest difference was between *A. stuartii* sexes (1.22:1.08:1.11). The extremely narrow ranges illustrate a degree of uniformity that strongly supports the idea of "sex–species". These examples illustrate: interspecies length ratios at limiting similarity (1.2) while intra-species length ratios are constant, but less than this; interspecies weight ratios are greater than limiting similarity (1.5) while intra–specific weight ratios are less.

The reduced ratios probably occur because of behavioural responses between sexes in each species enabling them to reduce their spatial overlap (Braithwaite 1973), thus increasing the degree of prey overlap and presumably body size overlap that can be tolerated. This is particularly noticeable where additional niche separation has evolved (e.g. habitat selection).

The importance of sexual dimorphism in producing differential niche utilization in woodpecker sexes, and so facilitating coexistence between species, has been thoroughly examined by Selander (1966). Schoener (1967) related reduced prey size overlap and microhabitat overlap in the lizard *Anolis conspersus,* to the extreme sexual dimorphism in body size exhibited by this species.

This introduces a further hypothesis to explain the low number of dasyurid species to be found at any one site: "sex–species" effectively double the number of species present and are able to utilise a broader range of resources than could monomorphic species. The fact that not all dasyurids are sexually dimorphic (Morton 1978c, Lee *et al.* 1982) may limit this alternative explanation but it cannot be dismissed out of hand.

A further difficulty associated with determining body weight is the geographical variation found in many species. An example of this (Fig. 3) is illustrated in an analysis of *Antechinus* by Van Dyck (1982). Contrary to Bergmann's rule, *Antechinus stuartii* shows an increasing trend in body weight from Victoria to southeastern Queensland, while *A. flavipes* remains relatively constant over the same range where it is allopatric with *A. stuartii.* However in northeastern Queensland, Van Dyck reports the two species to be sympatric. *A. flavipes* shows a marked increase in weight while *A. stuartii* reverses its trend to show a marked reduction in weight. Such changes from allopatry to sympatry are consistent with character shifts to facilitate coexistence. *Antechinus flavipes* is the more advanced form (Van Dyck 1982), that has been derived from a rainforest form so its character shift represents a character release.

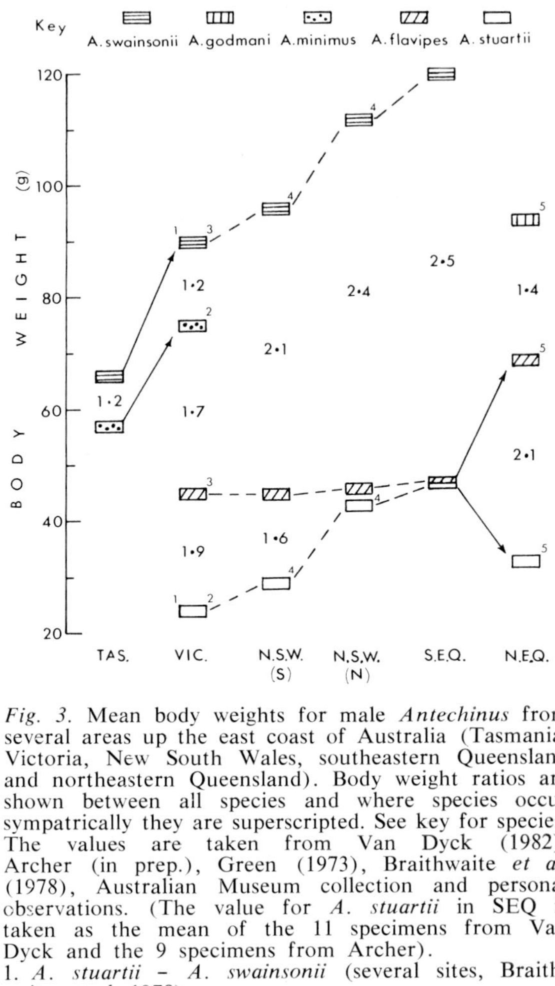

Fig. 3. Mean body weights for male *Antechinus* from several areas up the east coast of Australia (Tasmania, Victoria, New South Wales, southeastern Queensland and northeastern Queensland). Body weight ratios are shown between all species and where species occur sympatrically they are superscripted. See key for species. The values are taken from Van Dyck (1982), Archer (in prep.), Green (1973), Braithwaite *et al.* (1978), Australian Museum collection and personal observations. (The value for *A. stuartii* in SEQ is taken as the mean of the 11 specimens from Van Dyck and the 9 specimens from Archer).
1. *A. stuartii* – *A. swainsonii* (several sites, Braithwaite *et al.* 1978).
2. *A. stuartii* – *A. minimus* (Kentbruck & Cape Liptrap, Braithwaite *et al.* 1978).
3. *A. flavipes* – *A. swainsonii* (Mirranatwa Gap, Braithwaite *et al.* 1978).
4. *A. stuartii* – *A. swainsonii* (N.S.W., Personal observation).
5. *A. stuartii* – *A. flavipes* – *A. godmani* (N.E.Q., Van Dyck, this volume).

Antechinus swainsonii also shows a strong latitudinal increase in weight, contrary to Bergmann's rule, and is the heaviest *Antechinus* in N.S W. but is replaced in northeastern Queensland by *A. godmani* which is sufficiently large to be able to coexist with *A. flavipes* and *A. stuartii.* The body weight ratios for pairs of sympatric species all clearly demonstrate their marked separation. The abrupt changes in body weight for *A. stuartii and A. flavipes* from southeastern to northeastern Queensland both coincide with subspecific replacements that may represent speciation events in progress. Such a speciation event may also have occurred further north between *A. leo* and *A. flavipes* (Van Dyck 1980). A replacement of *A. swainsonii* with *A. godmani* as the largest *Antechinus* in some communities, has already been mentioned, while the relationship between the subspecies of both southern species, *A. swainsonii and A. minimus* is not clear. The distribution of these *Antechinus* then presents a series of overlaps and replacements among congeneric species utilising the same food resource, but not necessarily in contact (sympatry) at present.

Southwestern Australia supports a much higher dasyurid diversity than does eastern Australia. The increase is due to the increased β–diversity resulting from the higher degree of habitat specialization observed (Baynes 1975, 1982). The differences may be related to the much steeper rainfall gradient in the South–west which can be associated with more rapid changes in habitat type along the gradient, that may have assisted speciation.

Weight and size ratios for sympatric dasyurids may be much greater than those predicted for limiting similarity (see Fig. 1 and Table 4), exhibiting an absence of close species packing. This may stem from the similarity of diet among dasyurids (Hall 1980, Blackhall 1980) which may in turn result from their lack of feeding specialisations. In contrast the Australian *Rattus* show clear dietary separation within the genus (Braithwaite *et al.* 1978, Watts and Braithwaite 1978). The restriction on dasyurid diet may have had the result that Australia had no small mammals exploiting the substantial plant food resources until the arrival of the conilurine and later the murine rodents. The lack of species filling this large group of related niches facilitated the radiation of these rodents so that they occupy all available habitats and fill most of the food niches available to small mammals (see Braithwaite *et al.* 1978).

The effect of the dasyurid's lack of diet specialization is well illustrated by comparing the number of small dasyurids and murid rodents captured at study sites in south–eastern Australia (Fig. 4). There are most commonly one or two species of dasyurid, to which might be added one species of *Dasyurus* while there are most commonly three or four species of rodents. This further supports the community organisation hypothesis. From this and examples from dasyurid communities in other areas several conclusions emerge:

1. Dasyurid species richness at any one site (α–diversity) is low (no more than 2 or 3), while there may be high β–diversity (species turnover between sites).

2. Dasyurid assemblages do not exhibit close packing so that species are seldom near to limiting similarity.

3. This may be related to the lack of specialization in dasyurids.

4. An alternative explanation for low α–diversity is that sexual dimorphism increases the niche width of each species making less of the resource spectrum available to additional species.

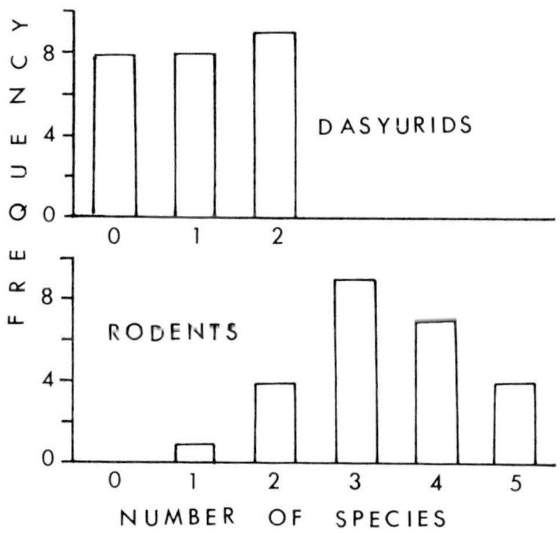

Fig. 4. The number of species of dasyurids and rodents that occur in a range of southeastern Australian heath habitats. Taken from a Table in Fox (1980), based on information collated from Braithwaite *et al.* (1978), Posamentier (1976) and unpublished data (B. Fox).

It is very difficult to draw overall conclusions for dasyurids on the topics discussed, because too few species have been studied and an overwhelming proportion of the information is concentrated at the small end of the size spectrum. In all fields there is a great need to obtain information covering a wider range of dasyurid species, to be followed by more intensive studies using an experimental approach. One glaring deficiency is in the area of the dietary preference. It is most important to test whether in fact different sized species in each genus do partition prey on size. The obvious areas to test this are in situations amenable to experiment, similar to those examined by Dickman (1982). It is also most important to test the hypothesis of character displacement implicit in Van Dyck's (1982) paper and discussed here (see Fig. 3) together with other similar cases. It is only with this sort of experimental approach, and the type of information it can provide about species interactions, that we can hope to improve our understanding of the ecology of the Dasyuridae.

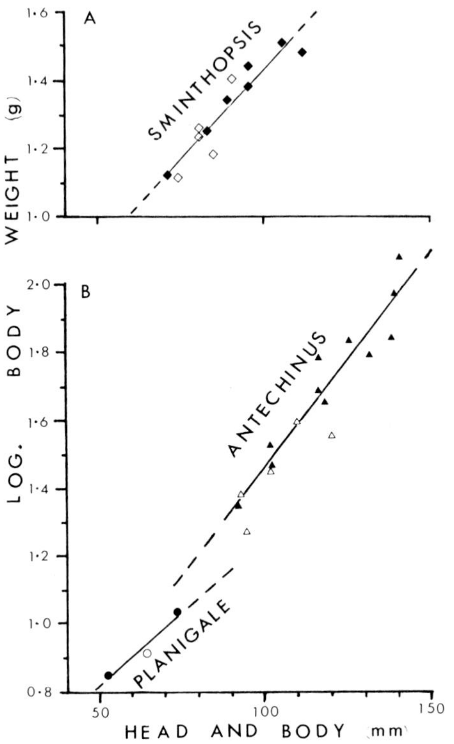

Fig. 5. Values of the logarithm of body weight as a function of head and body length for different populations of species and subspecies of small dasyurids. Regression equations are (A) *Sminthopsis* (bold diamond, ♂♂ and open diamond, ♀♀) $y=0.3729 + (0.0106 \pm 0.0002)x$, $r=0.928$, $P<0.01$, $n=12$; (B) *Antechinus* (closed triangle, ♂♂ and open triangle, ♀♀) $y=0.2052 + (0.0125 \pm 0.0001)x$, $r=0.939$, $P<0.001$, $n=24$; *Planigale* (●, ♂♂ and o, ♀), $y=0.4016 + (0.0084 \pm 0.0010)x$, $r= 0.9511$, $P<0.05$, $n=3$. All slopes differ significantly from one another; *S.* v *A.*, $t=8.94$, $P<0.005$; *S.* v *P.*, $t=2.16$, $P<0.05$. Values shown for ♂♂ and ♀♀ for *S. crassicaudata, S. macroura, S. murina, S. virginiae, S. leucopus* (Vic.). *A. macdonnellensis, A. stuartii stuartii* (N.S.W., S.E.Q.), *A. stuartii adustus* (N.E.Q.), *A. flavipes* (S.E.Q., N.E.Q.), *A. flavipes rubeculus* (N.E.Q.), *A. leo, A. godmani, A. swainsonii* (Vic.), *P. maculata* and ♂♂ only for *S. murina* (Vic.), *S. leucopus* (Tas.), *A. minimus* (Vic., Tas.), *A. swainsonii* (Q.). Values taken from Archer (in press): Van Dyck, this volume; Green (1973).

SPECULATION ON BODY WEIGHT PATTERNS WITHIN GENERA

The pattern of body sizes to be found in species of *Antechinus* that are not necessarily sympatric (Fig. 3), poses some very interesting questions. I have presented this information (Fig. 5) as a plot of the logarithm of body weight as a function of head and body length, using males and females (separately), from the populations available to me. Data from *Sminthopsis* and *Planigale* are treated similarly. Within each genus there is a common pattern reflecting a basic intrageneric similarity in shape, differences between genera represent a fundamental difference in body shape. The species of both *Sminthopsis* and *Antechinus* show significant positive intrageneric correlations, and there are significantly different regression slopes between all three genera (see caption Fig. 5).

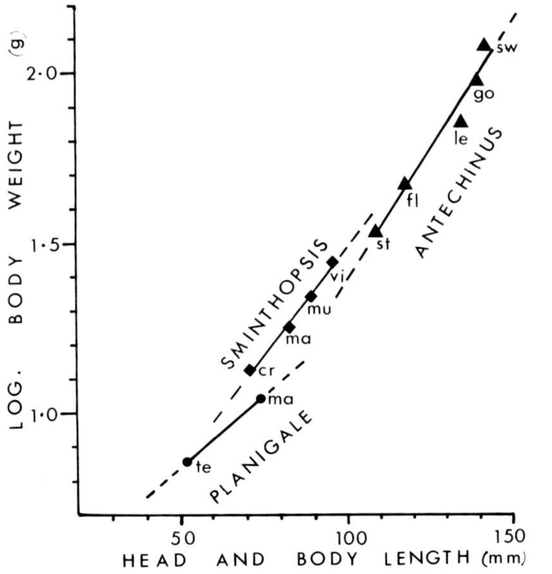

Fig. 6. Mean values of the logarithm of body weight as a function of head and body length for male dasyurids from three genera: (in order) *Planigale* (●), *P. tenuirostris, P. maculata; Sminthopsis* (bold diamond), *S. crassicaudata, S. macroura, S. murina, S. virginiae; Antechinus* (▲), *A. stuartii, A. flavipes, A. leo, A. godmani, A. swainsonii.* Values taken from Archer (in prep.).

A clearer picture emerges when a mean value is used for each species (Fig. 6). Where possible, I have chosen mean male body weights and head + body lengths from a single source, (Archer, manuscript in preparation). This minimises any bias that I might introduce and helps to overcome some of the problems already discussed, in determining mean body weights. For south eastern Australian species of dasyurids not represented, I have taken weights and measurements from the literature.

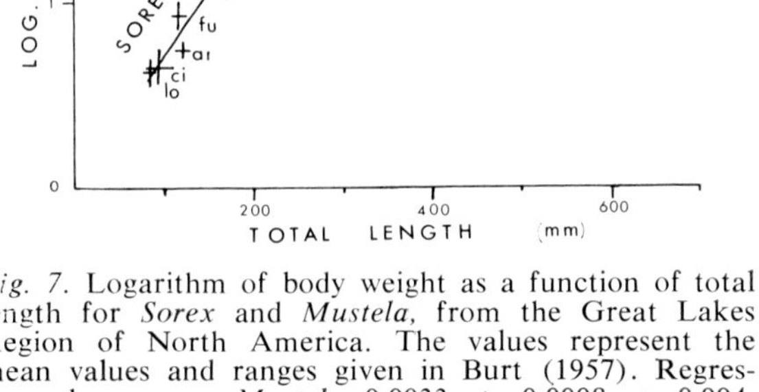

Fig. 7. Logarithm of body weight as a function of total length for *Sorex* and *Mustela*, from the Great Lakes Region of North America. The values represent the mean values and ranges given in Burt (1957). Regression slopes are: *Mustela* 0.0033 ± 0.0008, $r=0.994$, $P<0.05$, 2d.f.; *Sorex* 0.0073 ± 0.0004, $r=0.879$, $P<0.05$, 4d.f. The slopes are significantly different $t=4.96$, $P<0.05$. The species used are, in order: *Sorex longirostris, S. cinereus, S. arcticus, S. fumeus, S. palustris; Mustela rixosa, M. erminae, M. frenata, M. vison.*

Such patterns of body size increase are not restricted to the dasyurids and a similar picture is presented by the small carnivores of the Great Lakes region of North America (Fig. 7). The genera *Mustela* and *Sorex* show similar linear relationships. The reduced slope of the *Mustela* line, reflecting the thin elongate shape of weasels, is significantly less than that for *Sorex*.

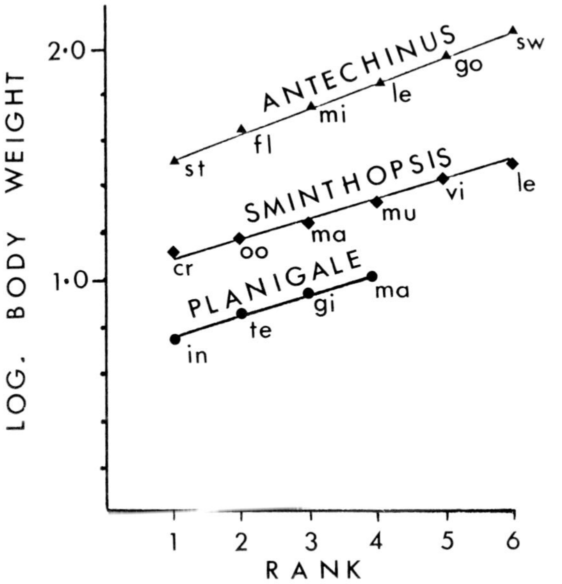

Fig. 8. The logarithm of male body weight shown as a function of rank for three genera of dasyurids. In order they are: *Planigale ingrami, P. tenuirostris, P. gilesi* and *P. maculata* (r=0.995, P<.01, mean body weight ratio = 1.4); *Sminthopsis crassicaudata, S. ooldea, S. macroura, S. murina, S. virginiae, S. leucopus* (r=0.997, P<0.001, mean body weight ratio = 1.2); *Antechinus stuartii, A. flavipes, A. minimus, A. leo, A. godmani, A. swainsonii* (r=0.997, P<0.001, mean body weight ratio = 1.3). Data taken from Archer (in prep).

To examine separation on body size within genera I have graphed the logarithm of body weight against the rank of each species in the genus (Fig. 8). I used only species that could be considered to form an overlapping or replacement series, so that they belong to the same geographic zone (eastern Australia) but are not necessarily sympatric. A straight line relationship on such a graph represents a constant ratio of body weights between successive species.

The partitioning of the food resource and the limiting similarity between sympatric species competing for that resource has already been discussed. However, the above result indicates that there may also be some limit to the degree of similarity tolerated between congeners, that are not all sympatric but do use the same resource (arthropods in this case). The possible importance of limiting similarity in this context, in addition to its already documented importance in an ecological context should be considered. Species of *Mustela* (Fig. 7) also illustrate equal spacing of body weights on a logarithmic scale, although there is some overlap in species of *Sorex*. Rudd (1955) also shows overlap in body size for three species of Californian shrews (*Sorex*), all taken in the San Francisco Bay Region. Rudd found similarily large seasonal changes in body weight as has been shown for *A. stuartii* (Fig. 2) and also commented on the extreme intraspecific variation and strong habitat selection that was found in these shrews.

Rosenzweig (1966: 604) in a study of community structure in sympatric North American carnivores has shown clear increases in body size with prey size, within each hunting set or group of species with similar body build and hunting strategy. No similar pattern was found between sets, as different hunting strategies were used. Least prey size difference was found among weasels (*Mustela*). This may be due to the coarseness of prey classes used (la = 0 to 50 g, lb = 50 to 150g, 2 = 150 – 700g, etc.), so that almost all weasel prey fell into one class (la).

Wilson (1975) has examined the adequacy of body size as a niche difference and concludes: that larger predators utilize larger prey sizes unavailable to smaller predators, while both utilize small prey sizes; such asymmetrical resource utilization gives larger animals a competitive advantage; but this advantage can be enhanced, diminished or completely negated by changes in the prey size distribution or in the predator's ability to utilize the lowest levels of resource. Both *A. stuartii* and *A. swainsonii* have the predicted asymmetrical success of capture distribution, as a function of prey size (see Hall 1980), so that the outcome of competition (Dickman 1980) and character shifts (Dickman 1982) are both as would be predicted by Wilson's model.

Rosenzweig's (1968) analysis of the overall strategy of body size in mammalian carnivores examines the nature and relative magnitude of evolutionary pressures effecting body size, in particular, clines covered by Bergmann's rule. He concluded that with few exceptions, biotic variables of competitive pressure and size were not important predictors of body size at a continental scale. Actual annual evapotranspiration, as a measure of total primary productivity was the best predictor: food availability sets upper limits to body size.

The observed patterns within the Dasyuridae are found within genera which can be identified (Fig. 6) with Rosenzweig's "hunting sets". Based on Rosenzeig's (1968) findings one would expect to find an increase in *Antechinus* body size with decreasing latitude, in response to increasing productivity (arthropods) as measured by actual annual evapotranspiration. A comparable measure of productivity (Growth Index) has been used by Fitzpatrick and Nix (1970) and when correlated with latitude it shows the expected negative correlation ($r = -0.980$, $P < 0.0005$, $n = 7$, 1–tailed), using data from Fig. 3 plus *A. leo* from Cape York (Van Dyck 1980), it is possible to correlate the mean weight, for *Antechinus* species in each area, against latitude. The expected negative correlation is significant ($r = -0.692$, $n = 7$, $P < 0.05$, 1–tailed) and the regression equation predicts an increase in mean weight of 3.7g for each 10° of latitude toward the equator. These weight changes parallel similar changes in the size of many arthropods. The mean body weight value increases from 60g in Tasmania to 70g for the single species (*A. leo*) on Cape York. For much of this range *A. swainsonii* lies well above the mean with *A. stuartii* and *flavipes* well below. These species are replaced by others at the northern and southern ends of their ranges. The two smaller species are parapatric or allopatric (except for S.E.Q.) so that at any one site one large, one small (or possibly a single small) *Antechinus* may be supported (Van Dyck 1982). Northeastern Queensland offers an exception with three species sympatric, one close to the mean, one above and one below (see Fig. 3). At any one site the species present occupy the two (or three) body sizes best suited to utilize the distribution of arthropod prey sizes at that site (see Wilson 1975). The combination of this latitudinal increase in body weight and prey size, and overlapping species distributions maintain a body weight separation between replacement species at either end of the range, despite the fact that they are allopatric, and can have no direct interaction (see Fig. 3). This would explain why the patterns are more strongly expressed within zoogeographic regions, and less so where there is an absence of overlapping distributions. However several possible alternate explanations of the observed phenomenon should also be considered.

LIMITING SIMILARITY IN AN EVOLUTIONARY CONTEXT?

Diamond (1973) has provided a thorough explanation of his reconstruction of the speciation of the altitudinally segregated New Guinea birds, drawing examples from present day distributions. Such speciation becomes possible when an initially continuous population breaks into geographically isolated units, which diverge but then re–establish geographical contact (parapatric) when effective reproductive isolating mechanisms have evolved, so that they are able to reinvade each other's geographical range if their ecological separation is sufficient to allow coexistence. Examples of other spatial segregating mechanisms (vertical stratification and habitat segregation) are also cited, along with several nonspatial segregating mechanisms, including body size differences (Diamond 1973).

The *Antechinus* example (Fig. 3) may represent an example of this type of speciation, with body size differences, encountered along a latitudinal prey size gradient, acting as the segregating mechanism. They represent regular body weight ratios for a monophyletic group of species, occupying one zoogeographic region, that reflect their past interaction as well as any present day interaction. This represents a somewhat speculative extension of the previous argument.

The mean body weight ratios for the three genera (Fig. 7) are *Planigale*, 1.4; *Sminthopsis*, 1.2; and *Antechinus*, 1.3. This is less than for sympatric species (> 1.5), as might be expected where the influence is indirect, between many species. Greater prey and body size overlap could be tolerated by congeners only indirectly competing, particularly where additional niche separation has evolved (e.g. habitat selection, temporal separation).

Another possible explanation that should be considered, is that the observations are a chance occurrence resulting from a fortuitous data set that is not representative of the real world. To test this I have examined two further data sets, the most comprehensive being taken from the skull measurements for all species of *Sminthopsis*, presented in a review of that genus (Archer 1981). I selected basicranial length as being representative of the size of the animal, as well as a measure of the maximum size of the arthropod prey it was capable of handling. The second set was the *Antechinus* body weight data used to construct Fig. 3, as an alternative to that from Archer (in prep.).

For the *Sminthopsis* case, I chose data from the eastern region, with its significant latitudinal gradients, in productivity and mean *Antechinus* body weight. As a test of the latitudinal effect I chose the northern region of the continent (< 21° latitude) where there is no signicant productivity gradient. I divided each region into a series of small areas, and mean basicranial length measurements are presented for each of these areas (Table 5), together with mean values for each region (from Archer 1981).

For each area with three or more (n) species I calculated a correlation coefficient (r) between the logarithm of basicranial length and its rank, together with a mean value for the ratio of basicranial lengths for successive species (antilogarithm of regression slope). Mean values for the areas in each region and values for the whole region were also calculated and all values presented in Table 6. For comparison a mean

Table 5. Mean basicranial lengths (mm) from seven species of *Sminthopsis* from (A) eastern and (B) northern regions of Australia, together with mean overall values for each region. Mean values shown for each species in each area are taken directly from Archer (1981).

A. Eastern Region	Tas	Vic.	N.S.W.	S.Q.	N.Q.		Mean Overall
					28.3	*S. douglasi*	28.3
					26.2	*S. virginiae*	26.2
	24.6	25.6				*S. leucopus*	25.1
		22.1	23.3	22.1	23.7	*S. murina*	22.8
		22.4	22.4	22.4	22.4	*S. crassicaudata*	22.4
			20.9	20.9	22.2	*S. macroura*	21.3
Number of species	1	3	3	3	5		6

B. Northern Region	N.W. Aust.	N.T.	N.W.Q	N.E.Q.		
			28.3		*S. douglasi*	28.3
	26.0	26.0	26.2	26.2	*S. virginiae*	26.1
				23.7	*S. murina*	23.7
			22.4		*S. crassicaudata*	22.4
	22.7	22.4	22.2		*S. macroura*	22.4
	21.1				*S. butleri*	21.1
Number of species	3	2	4	2		6

Table 6. Correlation coefficients (r) for the logarithm of mean basicranial length as a function of rank for n species of *Sminthopsis*. The mean size ratio between successive species is also shown. (A) Areas in the eastern region with strong latitudinal gradient in productivity (B) Areas in the northern region with no productivity gradient. Mean values of areas and regions are shown for comparison. Mean values significantly higher than expected for random rankings are shown in bold type. Statistically significant correlations are indicated: * P<O.05; ** P<0.01; *** P<0.001. Data from Table 5.

REGIONS	n	r	Mean Ratio	n	r	Mean Ratio	AREAS
A. Eastern Region	6	*** **0.986**	1.06	5	* **0.967**	1.07	N. Queensland
				3	0.943	1.04	S. Queensland
				3	**0.988**	1.06	N.S.W.
				3	0.905	1.08	Victoria
				1	–	–	Tasmania
Mean of Regions (East, Central, West)		*** **0.981**	1.06		0.951	1.06	Mean of Eastern Areas
B. Northern Region	6	*** **0.965**	1.06	3	**0.985**	1.11	N.W. Australia
				2	–	1.16	N. Territory
				4	0.955	1.09	N.W. Queensland
				2	–	1.11	N.E. Queensland
Mean of Regions (North, Central, South)		*** **0.963**	1.07		0.970	1.12	Mean of Northern Areas

value for regions (east, central, west; and north, central, south) are also provided.

Antechinus body lengths for the eastern region were analysed in the same way using the same areas as Fig. 3 and the results presented in Table 7.

For both of the eastern region analyses (*Sminthopsis,* Table 6A and *Antechinus* Table 7) the area correlation means are markedly lower than regional values. For the northern region (*Sminthopsis,* Table 6B) the area correlation mean is higher than the regional value. The difference in trend appears to result from the absence of the productivity gradient across the northern region. The area correlation mean for *Antechinus* is much less than the value for *Sminthopsis* in the same eastern region. This difference may result from the different distribution patterns found in each genus. *Sminthopsis* has an extremely widespread distribution covering almost the entire continent (Archer 1981), with several species having continent-wide distributions (e.g. *S. crassicaudata* and *macroura*), *Antechinus* distribution is very largely restricted to the less arid habitats, resulting in a series of very disjunct distributions (Tyndale–Biscoe 1979) with a much reduced overlap of distributions. For *Antechinus* the mean body weight ratio for areas (1.51) is markedly reduced at the regional level (1.22). There is little change in *Sminthopsis* size ratios.

Table 7. Correlation coefficients (r) for the logarithm of mean male body weight as a function of rank for n species of *Antechinus*. The mean ratio of body weights between successive species is also shown. Values are calculated for areas with three or more species, and for the whole region. Data derived from Fig. 3.

	n	r	Ratio	n	r	Ratio
N.E.Q.	3	**0.977**	1.69	REGION		
S.E.Q.	3	0.871	1.60			
N.N.S.W.	3	0.896	1.61	6	** **0.973**	1.22
S.N.S.W.	3	**0.988**	1.82			
VIC.	4	0.831	1.21			
TAS.	2	–	1.16			
Mean Value		0.913	1.51		** **0.973**	1.22

I think these analyses allow rejection of the possibility that the observed patterns result from a fortuitous data base. Rather the patterns appear related to a latitudinal gradient in productivity (and probably prey size), and less strongly to overlap in the distribution of species.

One other obvious possible explanation should also be considered: the pattern observed may be an artifact of the ranking procedure. To examine this I selected 30 sets of random numbers, with 6 numbers in each set, these numbers were ranked and then correlated against rank, and the distribution of correlation coefficients examined. The mean correlation was 0.933, with a 95% confidence interval for the mean value of 0.912 to 0.951. The procedure was repeated for 30 sets of 3 numbers, which gave a mean value of 0.961, with a 95% confidence for the mean of 0.944 to 0.974 (upper limits for the 99% interval were 0.957 (n=6) and 0.983 (n=3)).

The high correlation for sets of random numbers is most surprising, but it should be noted that although the mean correlations for individual areas are not significantly greater than expected by chance, those for regions are. Region means lie above the 99% confidence interval (0.957, n=6) for the mean of random correlations, and the mean of regions differ significantly from random values ($P<.01$). It is also interesting to note that as the sample size rises there is a decrease in mean values and confidence intervals for the random correlations, whereas the observed correlations increase, in the eastern region. These results allow the rejection of an explanation that the pattern is an artifact of the ranking process.

These analyses do however illustrate one important point, they test equal spacing between species rather than a limit to the similarity between species on that character. Equal spacing can result from a limit to similarity, but is not a necessary consequence (see Fig. 1), and species may differ by much more than the limiting ratio. Equality of spacing for weight and size ratios ensures the maximum utilization of morphological space by a genus, with a reduction in the ratio as more species are included in that space. The observed phenomenon that speciation tends to more often add larger species (Gingerich 1977) may stem from two related factors: it may be easier to expand morphological space than to compress size ratios; because the larger end of any prey size distribution will be less heavily exploited by other species (Wilson 1975, Hall 1980) thus allowing for an increase in morphological space. This is supported by field observations that when only one species of *Antechinus* is found it is almost always a small species; the larger species will usually be found together with a smaller species.

The notion of equality of weight or size ratios does introduce one further possible explanation that should be considered. The logarithmic sequence of species in a genus may be an inherent component of the classifications erected by taxonomists. Williams (1976; 128, quoting Zipf 1946) notes that such logarithmic sequences are not functions of the material being classified, but rather of the human mind doing the classification. This may be tested on the *Sminthopsis* data. For two species groups the mean correlation of ranked basicranial lengths is 0.949 (*S. murina* group n=4, *S. macroura* group n=5; Archer 1981), not significantly different from random rankings and below mean values observed for *Sminthopsis* in regions. Hence I can reject the hypothesis that the pattern of equal spacing on a logarithmic scale is an artifact of the human taxonomists classifying these groups.

There is a much reduced mean size ratio for *Sminthopsis,* resulting from the similar size of *S. macroura* and *S. crassicaudata* (Table 5). Archer (1981) indicates that *S. crassicaudata* shows greater dental specialization than *S. macroura.* This may be consistent with habitat separation between these two which has been reported by Spencer (1896), Parker (1973) and Morton (1978a). The close body weight similarity for south western *Sminthopsis* resulted from habitat selection (Fig. 1 and Baynes (1975).

Hespenheide (1975: 174) made the following comments, "Size differences are important and interpretable within single taxonomic groups which are also similar in food; . . . They become more difficult to interpret at the community or faunal level where many taxonomic groups are included, each with its own prey size/predator size relationships".

CONCLUSIONS

The regional pattern of body weight and size differences within genera is similar to that found between coexisting species in a guild, but ratios between successive species are less. The logarithmic size sequences of species within genera do not appear to result from chance, the use of ranking or the classification process. Limiting similarity is an insufficient explanation at a regional level, where species are not coexisting. Latitudinal gradients in productivity and prey size, and the degree of overlap in the distribution of the species appear as factors more likely to be influencing the observed patterns. The lower size ratios exhibited by *Sminthopsis,* probably result from their strong habitat selection.

ACKNOWLEDGEMENTS

I would like to thank Michael Archer for his continued encouragement to write this paper and also for providing access to his manuscript "Marsupials and monotremes of eastern Australia." Thanks must also go to Alex Baynes for his comments and access to his data for Western Australia species. The papers of Stephen Van Dyck and Chris Dickman stimulated my thinking. I would also like to express my appreciation to Marilyn Fox, Tony Lee and Dick Braithwaite for valuable, critical comments on the manuscript, and to Graeme Pyke and Mark Westoby for discussion of some points. However, errors or misinterpretations remain my own. My thanks also go to Ann Lonergan for typing several drafts of the manuscript.

REFERENCES

ABRAMS, P. A., 1975. Limiting similarity and the form of the competition coefficient. *Theor. Popul. Biol.* **8:** 356–75.

ABRAMS, P. A., 1976. Niche overlap and environmental variability. *Math. Biosci.* **28:** 357–72.

AHERN, L. D., 1974. A trapping study to determine habitat utilization and biology of small mammals at Sand Point, Westernport with particular emphasis upon *Sminthopsis leucopus* (Gray). Hons. thesis: LaTrobe University, Melbourne.

ARCHER, M., 1981. Results of the Archibold Expeditions No. 104. Systematic revision of the marsupial dasyurid genus *Sminthopsis* Thomas. *Bull. Am. Mus. Nat. Hist.* **168:** 65–223.

ARCHER, M. (in prep.). "Marsupials and monotremes of eastern Australia". Queensland Museum: Brisbane.

BARNETT, J. L., HOW, R. A. AND HUMPHREYS, W. F., 1977. Small mammal populations in pine and native forests in northeastern New South Wales. *Aust. Wildl. Res.* **4:** 233–40.

BARNETT, J. L., HOW, R. A. AND HUMPHREYS, W. F., 1978. The use of habitat components by small mammals in eastern Australia. *Aust. Wildl. Res.* **5:** 277–85.

BAYNES, A., 1975. The distribution of the native mammals of southwestern Australia. *Aust. Mamm.* **1:** 404–5.

BAYNES, A., 1982. Dasyurids in Late Quaternary communities in southwestern Australia. *In* "Carnivorous marsupials" ed. by M. Archer. Royal Zoological Society of New South Wales: Sydney.

BEGG, R., 1981. The small mammals of Little Nourlangie Rock, N.T. II. Ecology of *Antechinus bilarni,* the Sandstone Antechinus (Marsupialia: Dasyuridae). *Aust. Wildl. Res.* **8:** 57–72.

BLACKALL, S., 1980. Diet of the Eastern Native-cat, *Dasyurus viverrinus* (Shaw), in southern Tasmania. *Aust. Wildl. Res.* **7:** 191–8.

BRAITHWAITE, R. W., 1973. An ecological study of *Antechinus stuartii* (Marsupialia: Dasyuridae). M.Sc. thesis: University of Queensland, Brisbane.

BRAITHWAITE, R. W., 1974. Behavioural changes associated with the population cycle of *Antechinus stuartii* (Marsupialia). *Aust. J. Zool.* **22:** 45–62.

BRAITHWAITE, R. W., 1979. Social dominance and habitat ultilization in *Antechinus stuartii* (Marsupialia). *Aust. J. Zool.* **27:** 517–28.

BRAITHWAITE, R. W. AND GULLAN, P., 1978. Habitat selection by small mammals in a Victorian heathland. *Aust. J. Ecol.* **3:** 109–27.

BRAITHWAITE, R. W., COCKBURN, A. AND LEE, A. K., 1978. Resource partitioning by small mammals in lowland heath communities of southeastern Australia. *Aust. J. Ecol.* **3:** 423–45.

BROWN, J. H., 1975. Geographical ecology of desert rodents. Pp.315–41 *in* "Ecology and evolution of communities" ed. by M. L. Cody and J. M. Diamond. Belknap Press: Cambridge, Mass.

BROWN, W. L. AND WILSON, E. O., 1956. Character displacement. *Syst. Zool.* **5:** 49–64.

BUCHMANN, O. L. K. AND GUILER, E. R., 1977. Behaviour and ecology of the Tasmanian Devil, *Sarcophilus harrisii.* Pp.155–68 *in* "The biology of marsupials" ed. by B. Stonehouse and D. Gilmore. Macmillan Press: London.

Burt, W. H., 1957. "Mammals of the Great Lakes Region". 246pp. University of Michigan Press: Ann Arbour, Mich., U.S.A.

Calaby, J., 1960. Observations on the banded ant-eater *Myrmecobius f. fasciatus* Waterhouse (Marsupialia), with particular reference to its food habits. *Proc. Zool. Soc. Lond.* **135:** 183–207.

Carpenter, L., 1978. Hooks for mammal pollination. *Oecologia (Berl.)* **35:** 123–32.

Cheal, P. D., 1979. Vegetation analysis — an essential part of (small) mammal studies. *Bull. Aust. Mamm. Soc.* **6:** 46–7.

Christensen, P. E. and Kimber, P. C., 1975. Effect of prescribed burning on the flora and fauna of S.W. Australian forests. *Proc. Ecol. Soc. Aust.* **9:** 85–106.

Cowley, R. D., Heislers, A. and Ealey, E. H. M., 1969. Effects of fire on wildlife. *Victoria's Resources.* **11:** 18–22.

Croft, D., 1982. Behaviour in dasyurids: A review. *In* "Carnivorous marsupials", ed. by M. Archer. Royal Zoological Society of New South Wales: Sydney.

Cuttle, P., 1982. Life–history strategy of the dasyurid marsupial *Phascogale tapoatafa* (Dasyuridae, Marsupialia). *In* "Carnivorous marsupials", ed. by M. Archer. Royal Zoological Society of New South Wales: Sydney.

Denny, M. D. S., 1982. Review of planigale (Dasyuridae, Marsupialia) ecology. Pp.131–38 *in* "Carnivorous marsupials" ed. by M. Archer. Royal Zoological Society of New South Wales: Sydney.

Diamond, J. M., 1973. Distributional ecology of New Guinea birds. *Science* **179:** 759–69.

Diamond, J. M., 1975. Assembly of species communities. Pp.342–444, *in* "Ecology and evolution of communities" ed. by M. L. Cody and J. M. Diamond. Belknap: Cambridge, Mass.

Dickman, C., 1980. Niche shift and interaction among *Antechinus* (Dasyuridae). *Bull. Aust. Mamm. Soc.* **6:** 32.

Dickman, C., 1982. Some ecological aspects of seasonal breeding in *Antechinus* (Dasyuridae, Marsupialia). Pp.139–50 *in* "Carnivorous marsupials" ed. by M. Archer. Royal Zoological Society of New South Wales: Sydney.

Dwyer, P., Hocking, M. and Willmer, J., 1979. Mammals of Cooloola and Beerwah. *Proc. R. Soc. Qd* **90:** 65–84.

Fitzpatrick, E. A. and Nix, H. A., 1970. The climatic factor in Australian grassland ecology. Pp.3–26 *in* "Australian grasslands" ed. by R. M. Moore. Aust. Nat. Univ. Press: Canberra.

Fletcher, H. L., 1977. Habitat relationships among small mammals at Petroi, northeastern New South Wales. M. Nat. Res. Thesis. University of New England.

Fox, B. J., 1980. The ecology of small mammal community: secondary succession, niche dynamics, habitat partitioning, community structure and species diversity. Ph.D. thesis: Macquarie University, Sydney.

Fox, B. J., 1982. Ecological separation and coexistence in *Sminthopsis murina* and *Antechinus stuartii* (Dasyuridae, Marsupialia): A regeneration niche? Pp.187–97 *in* "Carnivorous marsupials", ed. by M. Archer. Royal Zoological Society of New South Wales: Sydney.

Fox, B. J. Niche parameters and species richness. *Ecology* (In press).

Fox, B. J. and McKay, G. M., 1981. Small mammal responses to pyric successional change in a eucalypt forest. *Aust. J. Ecol.* **6:** 29–42.

Fox, B. J. and Fox, M. D., 1981. A comparison of vegetation classifications as descriptors of small mammal habitat preference. Pp.166–80 *in* "Vegetation classification in the Australian region" ed. by A. N. Gillison and D. J. Anderson. CSIRO and Australian National University Press: Canberra.

Friend, G. R., 1979. The response of small mammals to clearing and burning of eucalypt forest in southeastern Australia. *Aust. Wildl. Res.* **6:** 151–63.

Gause, G. F., 1934. "The struggle for existence". Hafner: New York. (Reprinted 1964).

Gingerich, P. D., 1977. Patterns of evolution in the mammalian fossil record. Pp.469–500 *in* "Patterns of evolution" ed. by A. Hallam. Elsevier Scientific Publishing Co.: Amsterdam.

Godsell, J., 1982. Aspects of the population ecology of the Eastern Quoll, *Dasyurus viverrinus* (Dasyuridae, Marsupialia), in southern Tasmania. Pp.199–207 *in* "Carnivorous marsupials", ed. by M. Archer. Royal Zoological Society of New South Wales: Sydney.

Grant, P. R., 1972. Convergent and divergent character displacement. *Biol. J. Linn. Soc.* **4:** 39–68.

Green, R. H., 1973. "The mammals of Tasmania". Green: Launceston.

Guiler, E. R., 1970. Observations on the Tasmanian devil, *Sarcophilus harrisii* (Marsupialia: Dasyuridae) I. Numbers, home range, movements and food in two populations. *Aust. J. Zool.* **18:** 49–62.

Gullan, P. and Robinson, A. C., 1980. Vegetation and small mammals of a Victorian forest. *Aust. Mamm.* **3:** 87–96.

Hall, S., 1980. The diets of two coexisting species of *Antechinus* (Marsupialia: Dasyuridae). *Aust. Wildl. Res.* **7:** 365–78.

Hall, S. and Lee, A. K., 1982. Habitat use by two *Antechinus* spp. and *Rattus fuscipes* in tall open forest in southern Victoria. Pp.209–20 *in* "Carnivorous marsupials", ed. by M. Archer. Royal Zoological Society of New South Wales: Sydney.

Hardin, G., 1960. The competitive exclusion principle. *Science* **131:** 1292–7.

Heislers, A., 1978. Effects of wildlife on mammals in a Victorian eucalypt forest Pp.37–42. *Third Fire Ecol. Symp. Papers.* Forests Commission of Victoria: Melbourne.

Hespenheide, H. A., 1972. Food preference and the extent of overlap in some insectivorous birds, with special reference to the Tyrannidae. *Ibis* **113:** 59–72.

Hespenheide, H. A., 1975. Prey characteristics and predator niche width. Pp.158–80 *in* "Ecology and evolution of communities" ed. by M. L. Cody and J. M. Diamond. Belknap: Cambridge, Mass.

Hindmarsh, R. and Majer, J. D., 1977. Food requirements of Mardo *(Antechinus flavipes)* (Waterhouse) and the effect of fire on Mardo abundance. *Forest Department of W. Aust. Research Paper No.* **31.**

Hutchinson, G. E., 1959. Homage to Santa Rosalia or Why are there so many kinds of animals?. *Am. Nat.* **93:** 145–59.

Lee, A. K., Bradley, A. and Braithwaite, R. W., 1977. Corticosteroid levels and male mortality in *Antechinus stuartii.* Pp.209–20 *in* "The biology of marsupials" ed. by B. Stonehouse and D. Gilmore. Macmillan Press: London.

Lee, A. K., Woolley, P. and Braithwaite, R. W., 1982 Life history strategies of dasyurid marsupials. Pp.1–11 *in* "Carnivorous marsupials" ed. by M. Archer. Royal Zoological Society of New South Wales: Sydney.

LEONARD, B. V., 1970. Effects of control burning on the ecology of small mammal population. Symposium Papers 2nd Fire Ecology Symposium. Forest Commission Victoria.

LEONARD, B. V., 1972. "The effect of fire upon small mammals and leaf litter fauna in sclerophyll forest in southern Australia". M.Sc. thesis: Monash University, Melbourne.

LIM, L. T., 1979. Some aspects of the ecology of small mammals and a small mammal survey of Wallingat State Forest, New South Wales. M.Sc. thesis: Macquarie University, Sydney.

MACARTHUR, R. H. AND PIANKA, E. R., 1966. On optional use of a patchy environment. *Amer. Nat.* **100:** 603–9.

MACARTHUR, R. H. AND WILSON, E. O., 1967. "The theory of island biogeography". Princeton University Press: Princeton.

MAIN, A. R., SHIELD, J. W. AND WARING, H., 1959. Recent studies on marsupial ecology. Pp.315–31 *in* "Biogeography and ecology in Australia". *Monogr. Biol.* **8:** Den Haag: W. Junk.

MAY, R. M. AND MACARTHUR, R. H., 1972. Niche overlap as a function of environmental variability. *Proc. Nat. Acad. Sci.* **69:** 1109–13.

MORTON, S. R., 1978a. An ecological study of *Sminthopsis crassicaudata* (Marsupialia: Dasyuridae). I. Distribution, study areas an dmethods. *Aust. Wildl. Res.* **5:** 151–62.

MORTON, S. R., 1978b. An ecological study of *Sminthopsis crassicaudata* (Marsupialia: Dasyuridae) II. Behaviour and social organisation. *Aust. Wildl. Res.* **5:** 163–82.

MORTON, S. R., 1978c. An ecological study of *Sminthopsis crassicaudata* (Marsupialia: Dasyuridae) III. Reproduction and life history. *Aust. Wildl. Res.* **5:** 183–212.

MORTON, S. R., 1982. Dasyurid marsupials of the Australian arid zone: an ecological review. Pp.117–30 *in* "Carnivorous marsupials" ed. by M. Archer. Royal Zoological Society of New South Wales, Sydney.

MORTON, S. R., WAINER, J. W. AND THWAITES, T. P., 1980. Distributions and habitats of *Sminthopsis leucopus* and *S. murina* (Marsupialia: Dasyuridae) in southeastern Australia. *Aust. Mamm.* **3:** 19–30.

NEWSOME, A. E. AND CATLING, P., 1979. Habitat preferences of vertebrates inhabiting heathlands of coastal mountain and alpine regions of southeast Australia. Pp.301–16 *in* "Heathlands and related shrublands of the world. A descriptive study" ed. by R. Specht. Elsevier Scientific Publishing Co.: Amsterdam.

NEWSOME, A. E., MCILROY, J. AND CATLING, P., 1975. The effects of an extensive wildlife on populations of 20 ground vertebrates in S.E. Australia. *Proc. Ecol. Soc.* **9:** 107–23.

PARKER, S. A., 1973. An annotated check list of the native land mammals of the Northern Territory. *Rec. South Australian Mus.* **16:** 1–57.

POSAMENTIER, H. G., 1976. "Habitat requirements of small mammals in coastal heathlands of New South Wales". M.Sc. thesis: University of Sydney, Sydney.

PRESS, A., 1976. "Some aspects of the ecology and habitat utilisation of four species of native small mammals". B.Sc. (Hons.) thesis: University of Sydney, Sydney.

RECHER, H. F., LUNNEY, D. AND POSAMENTIER, H. G., 1975. A grand natural experiment. *Aust. Nat. Hist.* **18:** 154–63.

RECHER, H. F., LUNNEY, D. AND POSAMENTIER, H. G., 1978. Effects of wildlife on small mammals at Nadgee Nature Reserve. Symposium Papers, Forestry Commission of Victoria, 3rd Fire Ecology, Symposium.

REED, G. F. AND WALLIS, R. L., 1975. Studies of *Antechinus swainsonii* and other small mammals in an area of Sherbrooke Forest Park. *Vic. Nat.* **92:** 84–90.

ROSENZWEIG, M. L., 1966. Community structure in sympatric carnivora. *J. Mamm.* **47:** 602–12.

ROSENZWEIG, M. L., 1968. The strategy of body size in mammalian carnivores. *Amer. Midl. Nat.* **80:** 299–315.

RUDD, R. L., 1955. Age, sex and weight comparisons in three species of shrews. *J. Mamm.* **36:** 323–39.

RUSSELL, E. M., 1972. Recent ecological studies on Australian marsupials. *Aust. Mamm.* **1:** 189–211.

SCHMIDT, W. AND MASON, M., 1973. Fire and fauna in the northern Jarrah forest of Western Australia. *West. Aust. Nat.* **12:** 162–4.

SCHOENER, T. W., 1967. The ecological significance of sexual dimorphism in size in the Lizard *Anolis conspersus*. *Science*. **155:** 474–6.

SCHOENER, T. W., 1974. Resource partitioning in ecological communities. *Science* **185:** 27–39.

SELANDER, R. K., 1966. Sexual dimorphism and differential niche utilization in birds. *Condor* **68:** 113–51.

SPENCER, W. B., 1896. Mammalia. Pp.1–52 *in* "Report on the work of the Horn Scientific Expedition to Central Australia Pt. 2. Zoology". Melville, Mullin and Slade: Melbourne.

STATHAM, H. L., 1982a. *Antechinus stuartii* (Dasyuridae, Marsupialia) diet and food availability at Petroi, northeastern New South Wales. Pp.151–63 *in* "Carnivorous marsupials", ed. by M. Archer. Royal Zoological Society of New South Wales: Sydney.

STATHAM, H. L., 1982b. Trappability of *Antechinus stuartii* (Dasyuridae, Marsupialia) at Petroi, northeastern New South Wales. Pp.333–44 *in* "Carnivorous marsupials" ed. by M. Archer, Royal Zoological Society of New South Wales: Sydney.

STATHAM, H. L., AND HARDEN, R., 1982. Habitat preferences of *Antechinus stuartii* (Dasyuridae, Marsupialia at Petroi, northeastern New South Wales. *In* "Carnivorous marsupials", ed. by M. Archer. Royal Zoological Society of New South Wales: Sydney.

STEWART, A. P., 1979. Trapping success in relation to trap placement with three species of small mammals, *Rattus fuscipes, Antechinus swainsonii* and *A. stuartii*. *Aust. Wildl. Res.* **6:** 165–72.

SUCKLING, G. C. AND HEISLERS, A., 1978. Populations of four small mammals in radiata pine plantations and eucalypt forests in northeastern Victoria. *Aust. Wildl. Res.* **5:** 305–15.

TYNDALE–BISCOE, C. H., 1979. Ecology of small marsupials. Pp.343–79. *in* "Ecology of small mammals" ed. by D. M. Stoddart. Chapman and Hall: London.

VAN DYCK, S., 1980. The Cinnamon Antechinus, *Antechinus leo* (Marsupialia: Dasyuridae), a new species from the vine–forests of Cape York Peninsula. *Aust. Mammal.* **3:** 5–18.

VAN DYCK, S., 1982. The relationships of the Australian marsupials *Antechinus stuartii* and *A. flavipes* (Dasyuridae, Marsupialia) in Queensland. *In* "Carnivorous marsupials" ed. by M. Archer. Royal Zoological Society of New South Wales: Sydney.

Watts, C. H. S. and Braithwaite, R. W., 1978. The diet of *Rattus lutreolus* and five other rodents in Victoria. *Aust. Wildl. Res.* **5:** 47–58.

Williams, W. T., 1976. "Pattern analysis in agricultural science". C.S.I.R.O. and Elsevier Scientific Publishing Co.: Amsterdam.

Wilson, D. S., 1975. The adequacy of body size as a niche difference. *Amer. Nat.* **109:** 769–84.

Wood, D. H., 1970. An ecological study of *Antechinus stuartii* in a southeast Queensland rain forest. *Aust. J. Zool.* **18:** 185–207.

Zipf, G. K., 1946. "Human behaviour". Addison–Wesley Press: Cambridge, Mass.

RESUMEN

11. Revisión de la ecologia de los dasiúridos y especulación sobre el rol de la similitud limitante en la organización de una communidad. Por Barry J. Fox.

Los dasiúridos son un grupo particularmente exitoso que ha colonizado la total variedad de hábitats terrestres en Australasia. A pesar de un marcado incre-- mento de los estudios ecológicos en los últimos ocho años, sólo disponemos de información para menos de un cuarto de las especies. La mayoría de los estudios han sido realizados en los bosques o matorrales del Sureste de Australia. Existe una particular necesidad de estudios sobre las especies que habitan las zonas áridas o semi–áridas, como también de aquellas del norte tropical de Australia. *Antechinus stuartii* ha sido objeto de muchos estudios por lo que su ecología es relativamente bien conocida; sin embargo, los datos sobre muchas de las otras especies provienen de estudios aislados. Queda todavía mucho por hacer sobre la ecología de las poblaciones, requiremientos de hábitats y ecología de los incendios de muchas especies.

La información alimentaria disponible indica que, con la excepción de unos pocos de los más grandes dasiúridos, los artrópodos representan los más probables componentes de sus dietas principales. Esta falta de especialización dietaria es consistente con la falta total de especialización que caracteriza a los Dasyuridae. Esto puede haber contribuido al reducido número de especies que los dasiúridos manifiestan en ualquier lugar (diversididad α) en comparación a la alta diversidad β (diferencias en la composición específica entre sitios). Los grupos de dasiúridos no parecen exhibir un estrecho agrupamiento por especies. Esto parece estar influenciado por el marcado dimorfismo sexual presente en muchos de ellos, y por el expandido nicho de las especies resultante.

El tamaño corporal juega un rol importante en la organización del agrupamiento de las especies. Especies simpátridas muestrans una marcada separación por peso corporal, a menudo mayor que los radios mínimos pronosticados por la teoría de la limitante similitud. Cuando se considera el problema al nivel de la amplia region biogeográfica del Este autraliano, los congéneres muestran una relación linear entre los logaritmos del peso corporal y la longitud de la cabeza y el cuerpo. Una similar relación distintiva existe entre el logaritmo del peso del cuerpo y el rango de la especie. Las diferencias de cada género reflejan las diferencias fundamentales entre géneros según la relación de la forma y el peso del cuerpo. Se presenta un argumento especulativo para explicar cuán constantes radios del peso y tamaño podrían ocurrir en especies que son alopátridas o parapátridas. Esta relación no está específicamente restringida a los Dasyuridae, por lo que se proveen otros ejemplos. Las secuencias de tamaños logarítmicos de especies dentro de géneros no parecen ser el resultado del azar, de la jerarquización o de los procesos clasificatorios. La limitante similitud es una explicación insuficiente en un nivel regional donde las especies no coexisten. Los gradientes latitudinales de la productividad y del tamaño de las presas, y el grado de superposición en la distribución de especies aparecen como los factores que más probablemente influenciaron el modelo estudiado.

12. DASYURID MARSUPIALS OF THE AUSTRALIAN ARID ZONE: AN ECOLOGICAL REVIEW

S. R. MORTON[1]

The dasyurid marsupials of the Australian arid zone are tabulated, and their distributions within the broad habitat types of the region are examined. The analysis shows that dasyurids are particularly rich in species in hummock grassland, but much poorer in species in *Acacia* shrubland than expected. Our sparse knowledge of their behaviour, life-history, and ecology is reviewed, and it is concluded from the available information that selective pressures operating upon these animals are dominated by the considerable variability of the invertebrate populations which constitute their food supply. The data necessary to establish the characteristics of short–term variability will be relatively easy to collect, but it is argued that long–term variability may also have important consequences for dasyurids, yet be less easily quantified. Methods of inference concerning long–term variability are investigated, therefore, and the likely effects of climatic fluctuations examined. Future studies of desert–dwelling dasyurids will be most fruitful if they concentrate upon species inhabiting hummock grasslands, and upon the measurement and interpretation of variability in food supply.

Pages 117–30 *in* CARNIVOROUS MARSUPIALS, ed. by M. Archer, February, 1982. Roy. Zool. Soc. New South Wales: Sydney, Australia.

"Away to the south of Amadeus, [we] had an exciting chase after a little, but extremely active, mouse–like creature. It was finally captured as the result of a well–directed shot from a boot, hastily snatched for the purpose from the foot of our energetic leader whilst he was on horseback. It turned out to be a new species of the genus *Sminthopsis* . . . We never had the good fortune to see, or secure, another specimen and . . . it is likely to remain for long the sole representative of its special kind known to science, because there is little to tempt ordinary travellers into this desolate, forsaken part of Australia. . ." (Spencer 1928: p.164).

THE DASYURID specimen discussed here by Spencer, *Sminthopsis psammophila,* indeed remained unique for many years, and it was not until comparatively recently that further individuals were collected (Aitken 1971). Paucity of knowledge, or even of specimens, has not been confined to this particular species, and new species of desert-dwelling dasyurids continue to be discovered (Aitken 1972, Archer 1975, Johnson and Roff 1980); consequently, these dasyurids constitute one of the least–known groups of Australian mammals. Early scientific visitors to the Australian arid zone did what they could, but the limited information which emerged was confined to taxonomic explication, with some notes on reproduction and behaviour (Spencer 1896, Finlayson 1933, 1961). Ecological work on these animals has been hampered not only by the remoteness of much of the region, as Spencer (1928) suggested, but also by the apparent sparseness of most species and the difficulties involved in capturing them. Recently, the advent of alternative capture techniques such as pit-trapping and spotlighting has provided better means of conducting ecological studies of desert–dwelling dasyurids, and it is work resulting from such methods that forms a major part of this review. In addition, information concerning the habitat preferences of dasyurids is assembled here so that the pattern of species diversity across the arid zone can be discerned. Finally, the problems involved in quantifying and identifying the effects of the substantial variability of arid Australian environments, and the importance of these factors for dasyurids, are examined and discussed. The aim of the review, in brief, is to indicate which areas of the biology of the desert–dwelling dasyurids, and which species, would be most fruitful for future investigation. For these purposes, the arid zone (Fig. 1) is delimited by the 250 mm isohyet in the south and southeast and the 550 mm isohyet in the northeast and north (Newsome and Corbett 1975).

THE DESERT–DWELLING DASYURIDS AND THEIR HABITATS

Of the 19 species of dasyurids occurring in the arid zone (Table 1), ten are apparently confined therein. The remaining nine, in their ranges outside the arid zone, occur primarily in the semi–arid belt surrounding the region, most of them to the south–east and south–west. The proportion of desert–dwelling dasyurids actually confined to the arid zone (53%) is greater than that of other terrestrial desert–dwelling mammals similarly confined (41%; Table 2), and

[1] School of Biological Sciences, Zoology Building, University of Sydney, Sydney, New South Wales, Australia, 2006.

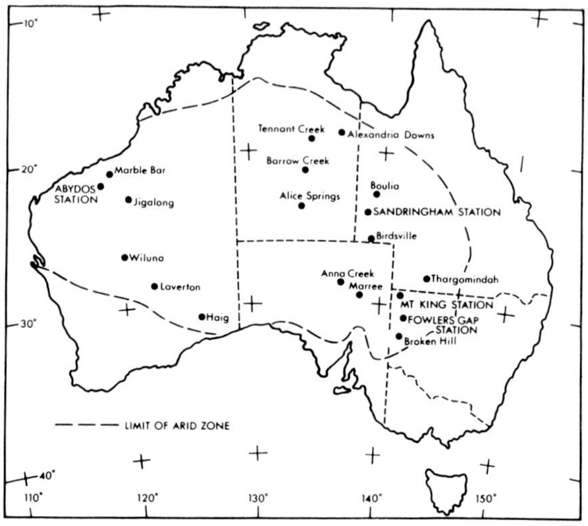

Fig. 1. Map of Australia showing the extent of the arid zone, the location of four sites at which communities of dasyurids have been investigated (large print), and localities for which rainfall data are presented in Table 3.

further differences are evident in the proportions of different groups of desert–dwelling mammals from the Australian mainland that occur in the arid zone (45% of dasyurids compared to 30% of other mammals; Table 2). These increased proportions suggest that the arid zone is of some importance to the pattern of species diversity in dasyurids, and that this group has found it easier to radiate into arid habitats than have other groups. The significance of this result is emphasized in Fig. 2, which shows the species density of dasyurids throughout Australia. The density is very constant relative to other groups of vertebrates; in general, mammals have distinctly greater species density along the east coast, while, at the other extreme, lizards show pronounced increases in species density in the arid centre of the continent (Schall and Pianka 1978, Pianka and Schall 1981, Cogger and Heatwole 1981). The lack of distinct gradients of species density in dasyurids supports the suggestion that dasyurids have been successful relative to most other mammals occupying arid habitats. Nevertheless, dasyurids have not radiated in the arid zone to anywhere near the extent that insectivorous lizards, in particular, have done.

Fig. 2. Contour map of the species density of Australian dasyurid marsupials. The contour lines are isograms for numbers of species occurring in quadrats 240 × 240 km. (From an unpublished study by R. W. Braithwaite, G. D. Sanson and S. R. Morton.)

Examination of the patterns of habitat selection by dasyurids throws some light on these patterns. In Table 1, each species has been assigned to the broad habitat types in which it is known to occur. These habitat types, together with their approximate contributions to the arid zone in terms of area (in brackets; modified from Leigh and Noble, 1969), are: low shrubland (8%); *Eucalyptus* shrubland (8%); *Acacia* shrubland (33%); tussock grassland (9%); and hummock grassland (31%). In addition, three restricted but geomorphologically distinct formations are recognised: stony desert (gibber); rock outcrops; and sandridges. It is clear

Table 2. Proportions of different groups of native mammals from the Australian mainland occurring in the arid zone, and of desert–dwelling species occurring in the two arid habitats of broadest distribution*.

	1. Dasyurids	2. Other Marsupials, Monotremes and Carnivores	3. Rodents	4. Combined totals of columns 2 and 3
Proportions of species that occur in the arid zone	45%	26%	36%	30%
Proportions of desert–dwelling species confined to the arid zone	53%	38%	44%	41%
Proportions of desert–dwelling species that occur in hummock grassland	40%	29%	17%	24%
Proportions of desert–dwelling species that occur in *Acacia* shrubland	21%	50%	28%	41%

* Chiroptera excluded; calculated from Ride (1970), Archer (1979), Morton (1979) and Watts (1979).

Table 1. Dasyurids of the Australian arid zone, and references to knowledge of ecological aspects of their biology.

SPECIES	WEIGHT (g)	HABITAT	REPRODUCTION – LABORATORY	REPRODUCTION – FIELD	BEHAVIOUR – LABORATORY	BEHAVIOUR – FIELD	DIET	ENERGY METABOLISM	WATER METABOLISM
Planigale gilesi	10	Low shrubland; tussock grassland	—	Read, 1982	—	Read, 1982	Read, 1982	Dawson & Wolfers, 1978	—
P. tenuirostris	7	Low shrubland; tussock grassland	—	Read, 1982	Andrew & Settle, 1982	Read, 1982	Read, 1982	Dawson & Wolfers, 1978	—
*Ningaui ridei**	10	Hummock grassland	—	—	—	—	—	—	—
*N. timealeyi**	10	Hummock grassland	—	—	—	—	—	—	—
N. sp.*	10	Hummock grassland	Fanning 1982	—	—	—	—	—	—
Sminthopsis crassicaudata	15	Low shrubland; tussock grassland; stony desert	Martin 1965; Ewer 1968; Godfrey 1969a; Smith & Godfrey 1970; Smith *et al.* 1978	Godfrey & Crowcroft 1971; Morton 1978c	Crowcroft & Godfrey 1968; Ewer 1968	Morton 1978b, 1978d	Morton 1978c	Godfrey 1968; MacMillen & Nelson 1969; Dawson & Hulbert 1970; Kennedy & Macfarlane 1971; Baudinette *et al.* 1976b; Morton 1978d, 1980a	Kennedy & Macfarlane 1971; Haines *et al.* 1974; Morton 1980a
*S. hirtipes**	20	Hummock grassland; sand ridges	—	—	—	—	—	—	—
*S. longicaudata**	20	Hummock grassland	—	—	—	—	—	—	—
S. macroura	20	Low shrubland; tussock grassland	Godfrey 1969b; Woolley 1973	—	Crowcroft & Godfrey 1968	—	—	—	—
S. murina	15	*Eucalyptus* shrubland	—	—	—	—	—	—	—
*S. ooldea**	10	*Acacia* shrubland; hummock grassland	—	—	—	—	—	—	—
S. psammophila	25	Hummock grassland	—	—	—	—	—	—	—
Antechinomys laniger	30	Low shrubland; *Acacia* shrubland; tussock grassland; stony desert	Woolley 1973, & unpublished	—	Happold 1972	—	—	MacMillen & Nelson 1969; Macfarlane 1975; Baudinette *et al.* 1976a, 1976b	Macfarlane 1975
*Antechinus macdonnellensis**	35	Rock outcrops	Woolley 1973, & unpublished	—	—	—	—	MacMillen & Nelson 1969	—
*A. rosamondae**	30	Hummock grassland	Woolley, unpublished	—	—	—	—	—	—
*Dasycercus cristicauda**	100	Hummock grassland; sand ridges	Jones 1949; Fleay 1961; Mack 1961; Michener 1969; Woolley 1971	—	Ewer 1969; Sorenson 1970	—	—	MacMillen & Nelson 1969; Kennedy & Macfarlane 1971	Schmidt-Nielsen & Newsome 1962; Kennedy & Macfarlane 1974; Haines *et al.*
*Dasyuroides byrnei**	110	Low shrubland; tussock grassland; stony desert	Mack 1961; Woolley 1971	—	Sorenson 1970; Aslin 1974; Hutson 1975, 1976, 1982	—	—	MacMillen & Nelson 1969; Macfarlane & Howard 1972; Baudinette *et al.* 1976b	Macfarlane & Howard 1972; Haines *et al.* 1974
Phascogale calura	160	*Acacia* and *Eucalyptus* shrublands	—	—	—	—	—	—	—
Dasyurus geoffroii	1200	Ubiquitous?	Arnold & Shield 1970a	—	—	—	—	Arnold & Shield 1970b	—

* Denotes species confined to the arid zone

that *Acacia* shrubland and hummock grassland are by far the largest habitat types in the arid zone. Three striking facts emerge from inspection of the habitat preferences indicated in Table 1. First, nine species (40%) occur in hummock grassland, an association dominated by *Triodia* spp. and *Plectrachne* spp. that has no real parallel in desert ecosystems outside Australia (Burbidge 1953). All but one of these dasyurid species appear to be confined to hummock grassland, although two also occur on sandridges. Second, in contrast to the first point, only four species occur in *Acacia* shrubland, even though this association covers an area slightly greater than that of hummock grassland. Further, none of the species appears to be confined to *Acacia* shrubland; the one possible exception, *Sminthopsis ooldea,* has recently been recorded from hummock grassland in south–eastern Western Australia (Burbidge *et al.* 1976). Finally, no dasyurids appear to be specific to sandridges, despite the widespread occurrence of patches of this habitat across much of inland Australia. In summary, it is apparent that the presence of hummock grassland is of great importance in determining the species richness of dasyurids in arid Australia, and is probably a major contributing factor to the even nature of the dasyurid radiation (Fig. 2).

Comparisons with patterns of habitat selection in other vertebrate groups provide further information (Table 2). Other terrestrial mammals have not occupied the hummock grasslands to the same relative extent as have dasyurids; rodents are particularly poor in species in this habitat. In further contrast, these other mammals have much larger proportions of their species inhabiting *Acacia* shrubland than do dasyurids, and the proportions are more like those to be expected in terms of the area involved. Many of these other mammals are apparently confined to *Acacia* shrubland. Thus, dasyurids are unusual among mammals of the arid zone in being rich in species in hummock grassland and poor in *Acacia* shrubland. The pattern of avian diversity is different from either of the groups outlined above. Comparatively few species of birds are confined to either hummock grassland or *Acacia* shrubland (about 14 in each), although slightly more species find *Acacia* shrubland to be suitable habitat than do hummock grassland (Keast 1961, Frith 1973). In desert–dwelling lizards, habitat specificity is highly developed, with many species occurring in either hummock grassland, *Acacia* shrubland, sandridges, or rock outcrops (Pianka 1969, 1972). In summary, the comparative richness of dasyurids in the hummock grasslands is paralleled only in lizards. Possible reasons for this observation are difficult to discern; one is briefly mentioned in the Conclusion.

COMMUNITIES OF DESERT–DWELLING DASYURIDS

No detailed studies of communities of dasyurids have been carried out within the Australian arid zone, but here I attempt to assemble available information concerning the species of dasyurids known to coexist in particular habitats. Sufficient work has been conducted at four sites (Fig. 1) to give some idea of the constitution of such communities.

1. FOWLERS GAP STATION, N.S.W. Information is available for a low shrubland habitat dominated by *Acacia* spp. and *Sclerolaena* spp. (Morton 1978a, Read 1982). Four species of dasyurids exist in this area — *Planigale gilesi, P. tenuirostris, Sminthopsis crassicaudata,* and *S. macroura.* According to Morton (1978a) there is microhabitat separation, with *P. gilesi* and *S. macroura* more likely to be found in thick vegetation along creek beds and gullies, and *P. tenuirostris* and *S. crassicaudata* on the surrounding more open country. Read's (1982) more detailed studies show the situation to be more complex. All four species can be found together, especially in areas of deeply cracked soil; however, it seems probable that not all species are at peak densities in such areas. *S. crassicaudata* appears to be the only species which occurs consistently in areas of very sparse vegetation.

2. MOUNT KING, STURT NATIONAL PARK, N.S.W.: Denny (1975) recorded *P. gilesi, S. crassicaudata,* and *S. macroura* from a trap–line running across a creek and thus across a vegetational gradient. *P. gilesi* was confined to low shrubland of *Leptochloa digitata* and *Chenopodium* sp which occurred along the creek's channel, while *S. macroura* and *S. crassicaudata* were found in low shrubland of *Bassia spp.* and *Atriplex* spp away from the channel, where the soil was less deeply cracked. In this case, therefore, the *P. gilesi* were separated from the *Sminthopsis* spp. by habitat. Later work has shown that *P. gilesi* and also *P. tenuirostris* occur in areas away from creeks where soils again are deeply cracked (Read 1982). The composition of this fauna and the preference of *Planigale* spp. for soils that are deeply cracked indicate a very similar pattern to that seen at Fowlers Gap Station.

3. SANDRINGHAM STATION, QLD: Three general habitats exist in this region: tussock grassland dominated by *Astrebla* spp., which grades into gibber plain; sandridges with *Zygochloa paradoxa;* and small areas of hummock grassland (*Triodia basedowii*) on low stable sandridges and on sandplain. Dasyurids known to occur in the tussock grassland include *P. tenuirostris, S. crassicaudata, S. macroura, Antechinomys lani-*

ger, and *Dasyuroides byrnei* (Morton *et al.* 1977, Morton, unpublished data). *P. gilesi* has been recorded from a claypan habitat (Morton, unpublished data), but probably occurs in other more thickly vegetated areas such as it does at Fowlers Gap and Mount King Stations. Little is yet known about patterns of microhabitat selection in this community. Only one species of dasyurid, *Dasycercus cristicauda,* is known from sandridges and hummock grassland (P. Woolley, personal communication).

4. Abydos station, W.A.: Information is available for hummock grassland of *Triodia* spp. on sandplain. *Antechinus rosamondae* occurs among the *Triodia,* while *A. macdonnellensis* has been captured only at termite mounds that are scattered throughout the grassland (P. Woolley, personal communication). *Ningaui timealeyi* also inhabits the grassland (Archer 1975), and *D. cristicauda* has been recorded in this habitat in the past (P. Woolley, personal communication).

5. Summary: The available data show that up to five species of dasyurids can be found in a particular arid habitat, but microhabitat selection probably results in a restriction of the number of coexisting species to not more than three.

ECOLOGY OF DESERT–DWELLING DASYURIDS

Knowledge of the ecology of these dasyurids is very sparse. Inspection of those columns in Table 1 concerning behaviour, reproduction, and diet in the field shows that detailed data are available for only one species, *S. crassicaudata,* and even this study (Morton 1978a, b, c) relied heavily upon information obtained from a population inhabiting a non-arid environment. However, an as yet uncompleted study of *P. gilesi* and *P. tenuirostris* (Read 1982) will add considerable knowledge. In sections 1 and 2 below, I will summarize major attributes of the ecology of *S. crassicaudata* and, where possible, compare these with attributes of the two species of *Planigale.*

1. Behaviour and social organization: In a non–arid grassland, *S. crassicaudata* displayed a mobility well beyond that observed in forest–dwelling dasyurids, and Morton (1978b) concluded that individuals inhabited overlapping home ranges which changed considerably through time. The length of this "drifting home range" during the life of an individual was estimated to be at least 600 m; there was some evidence to suggest that males were more mobile than females. No indications of territorial behaviour were observed, although females with advanced young were more aggressive than any other individuals and thus may defend their nest–sites. Compared to forest–dwelling dasyurids, *S. crassicaudata* had a very loose form of social organization — there were no apparent long–term bonds between individuals or between an individual and its "range" Both *P. gilesi* and *P. tenuirostris* have shown movements at least as great as those reported for *S. crassicaudata* (Read 1982), and it seems likely that they too will have a comparatively loose form of social organization.

2. Reproduction and life history: Of the dasyurids listed in Table 1 whose reproduction has been investigated, all but two are polyoestrous; the exceptions are *A. rosamondae* and *D. cristicauda* (Lee *et al.* 1982). Information on reproduction in the field, however, is available only for *S. crassicaudata.* This species appears to raise two consecutive litters in the breeding season, which extends from mid or late winter until late summer (Morton 1978c). There is no evidence of opportunistic breeding in response to rainfall outside of this season. Most individuals breed in only one season, and thus live to a maximum of about 18 months. *P. gilesi* and *P. tenuirostris* appear to breed in spring and summer, but the extent of their breeding seasons and of their life–spans are not yet adequately known (Read 1982). It seems reasonable to suggest that the other polyoestrous dasyurids have a similar life-history to that of *S. crassicaudata.* The monoestrous desert–dwelling dasyurids are of great interest in their divergence from the more common polyoestry, and they demand investigation.

Morton (1978c) showed that terrestrial invertebrates forming the food supply of *S. crassicaudata* exhibited a seasonal cycle in abundance and activity, being more common in spring and summer than in other seasons; thus, he postulated that reproduction in *S. crassicaudata* was timed to coincide with this increase. Such interpretations appear realistic, but lack of further information about invertebrate populations prevents me from determining if this relationship holds throughout the arid zone. The site at which invertebrate numbers were assessed, Fowlers Gap Station, is on the southern fringe of the arid zone, where winter temperatures are lowest and where the activity of invertebrates might be expected to show the most pronounced cycling. On the other hand, more northerly parts of the arid zone tend towards a pattern of summer rainfall, and this effect almost certainly favours a coincident cycle among invertebrate populations. Consequently, it seems very likely that most desert–dwelling dasyurids breed in spring and summer.

3. Physiological ecology: Laboratory studies of water and energy metabolism in dasyurids

are relatively common (Table 1), although only one study concerns free–living animals (*S. crassicaudata* in a non–arid environment; Morton 1980a). Here discussion is restricted to three aspects of physiological ecology that have direct ecological import — water usage, torpor, and fat storage.

The food supply of dasyurids, be it invertebrates or small vertebrates, contains at least 60% water. Thus, these animals should face few problems of water supply because they avoid heat stress and because their food and water are obtained simultaneously. Surprisingly, therefore, Kennedy and Macfarlane (1971) reported that *S. crassicaudata* had a very high rate of water turnover and, in contrast to *D. cristicauda* (see also Schmidt–Nielsen and Newsome 1962), could not survive without drinking water for more than about four days. Later, Macfarlane (1975) showed that *A. laniger* had a high rate of water turnover similar to that of *S. crassicaudata* and suggested that it also suffered from an inability to do without drinking water. Morton (1980a) has shown, however, that *S. crassicaudata* is capable of surviving for extended periods without drinking water when eating insects, and Macfarlane's (1975) suggestions are erroneous. It seems probable that, in the wild, desert–dwelling dasyurids rarely or never drink.

Five of the species in Table 1 are known to use torpor in the laboratory — *P. gilesi, P. tenuirostris, S. crassicaudata, S. macroura,* and *D. cristicauda* (references in Table 1; Wallis 1982). With more work the phenomenon will probably be demonstrated in most species weighing less than 100 g, although the ease with which it can be stimulated in the laboratory may vary greatly between species. The function of torpor is the conservation of energy during stressful periods, and Morton (1978d) showed that, in the field, *S. crassicaudata* used torpor infrequently and only during the colder part of the year. He suggested that the major stimulus for torpor was short-term energy shortage resulting from unusually poor foraging conditions combined with relatively cold weather.

Fat storage, similarly, is undoubtedly a safeguard that buffers animals against food shortage. Dasyurids are a little unusual in this regard because caudal fat storage is so prevalent among them, and eight species from Table 1 use such storage — *P. gilesi, S. crassicaudata, S. hirtipes, S. macroura, S. ooldea, A. macdonnellensis* and *D. cristicauda*. Morton (1980b) summarized studies of caudal fat storage in *S. crassicaudata* and examined the occurrence of this storage among other small mammals. He concluded that caudal fat storage was common in two main groups of small mammals: desert–dwelling insectivores; and species with heterogeneous diets but which share the habit of utilizing extended torpor. The dasyurids in general belong to the first group; unusually large fat reserves which spill over into caudal storage appear to be necessary because the food supply of these animals is inherently variable (Morton 1978c, 1980b, Morton and Alexander 1982).

4. Predation: Studies in low shrubland habitats in the eastern half of the arid zone show that most dasyurids occurring in these regions are eaten by the ubiquitous barn owl, *Tyto alba* (Morton *et al.* 1977, Morton and Martin 1979). Barn owls occur in every habitat within the arid zone, and so there is every reason to expect that all species of desert–dwelling dasyurids are preyed upon. Carnivores such as dingoes, *Canis familiaris,* foxes, *Vulpes vulpes,* and cats, *Felis catus,* probably take dasyurids as well (see Newsome and Corbett 1975). The impact of these predators on the behaviour and population dynamics of dasyurids is impossible to determine at present; nevertheless, the relative sparseness of most dasyurids would act against specialization upon them by any predator, and thus the dasyurids have almost certainly escaped all but the most general effects of predatory pressure.

5. Summary: Major Selective Pressures: The foregoing discussion makes possible an attempt at identification of the dominant selective influences that may operate upon dasyurids. First, however, consider aspects of the biology of these animals that do not appear to have resulted in major adaptive changes. The impact of predators has had no unusually dramatic effect on the biology of desert–dwelling dasyurids. Shortage of free water, the characteristic that comes immediately to mind when arid ecosystems are discussed, does not seem to have required significant alterations in physiological or behavioural attributes. In contrast, the two most obvious physiological traits among desert–dwelling dasyurids – torpor and caudal fat storage – appear to be directed towards buffering the animals against variability in food supply. Similarly, the most striking aspects of behaviour and social organization in the species that have been studied can be interpreted as modifications brought about by variability of food supply in space and time. Increased mobility, lack of territoriality, and decreased degree of home range fidelity are best explained as adaptations designed to overcome high levels of short–term unpredictability in the abundance and dispersion of invertebrates. Such effects also provide possible explanations for the prevalence of polyoestry and repeated breeding in desert–dwelling dasyurids, compared with the monoestry evident in many forest–dwelling dasyurids (Morton 1978c, Braithwaite and Lee 1979, Lee *et al.* 1982).

The arguments advanced here face a critical lack of data concerning resources. There is little information on the way in which invertebrate numbers fluctuate, either in the short–term or in the long–term. There are no serious impediments to obtaining data on short–term fluctuations or seasonal changes in invertebrate activity; such studies are relatively simple. However, variability in the long–term is, potentially, as important, because year–to–year fluctuations in the abundance of invtrtebrates will have a direct impact on behaviour, reproductive success, and population dynamics. The lengthy studies necessary for identification of the degree of such changes and their effects on dasyurids probably lie some distance in the future, and until then we will have no rely on conjecture and inference. I wish to examine some methods of inference concerning resource levels in the following section.

LONG–TERM VARIABILITY IN THE ARID ZONE, AND SOME CONSEQUENCES

Many authors have emphasized the importance of periods of drought and heavy rainfall to the animals of inland Australia, and in particular to small mammals (Finlayson 1939, Newsome and Corbett 1975). Because the bouts of plentiful rainfall provide a convenient and apparently important focus, I will investigate these in detail. The questions posed here are: what proportion of generations of dasyurids will, on average, experience unusually heavy rainfall; and what are the probable biological effects of this experience? Any attempt at answering these questions must rely, initially, on patterns of rainfall, for moisture availability is a major determinant of vegetative productivity, which in turn profoundly affects the invertbrates that constitute the diets of dasyurids.

1. Variability in Annual Rainfalls: Representative rainfall data (derived from records provided by the Bureau of Meteorology, Melbourne) are given in Table 3; the localities are depicted in Fig. 1.

First, rainfall in arid Australia is markedly variable, as shown by the very high coefficients of variation at these 15 sites (mean 51.3%); Leeper (1970) demonstrated that most of in-

Table 3. Selected attributes of annual rainfall at fifteen localities in the Australian arid zone.

Locality	Mean Annual Rainfall ± SD (mm)	Median Annual Rainfall (mm)	Discrepancy between Mean and Median Rainfall (% of mean)	Coefficient of variation (% of mean)	Decile 9 of annual rainfalls as a proportion of the mean*	Mean period between incidences of Annual Rainfall exceeding 1.5 times the Mean (years)	Mean period between incidences of Annual Rainfall exceeding twice the Mean (years)	Number of Rainfall Records (years)
Western Australia								
Haig	186 ± 81	166	10.8	43.8	—	5	—+	47
Jigalong	272 ± 151	252	7.4	55.6	—	5	15	59
Laverton	213 ± 94	192	9.9	44.0	—	9	61	61
Marble Bar	335 ± 135	325	3.0	40.3	1.5	9	84	84
Wiluna	241 ± 127	201	12.4	52.9	1.7	7	39	77
South Australia								
Anna Creek	138 ± 76	124	10.1	55.3	1.6	6	30	89
Marree	160 ± 78	140	12.5	48.6	1.5	9	13	93
Northern Territory								
Alexandria Dns.	338 ± 182	346	10.8	47.0	—	10	25	76
Alice Springs	278 ± 144	257	7.6	51.7	1.6	7	26	102
Barrow Creek	312 ± 172	275	11.9	55.3	1.4	7	17	104
Tennant Creek	365 ± 158	343	6.0	43.2	1.6	8	>102	102
Queensland								
Birdsville	161 ± 114	128	20.5	70.7	2.0	5	12	81
Boulia	269 ± 169	210	21.9	62.8	2.0	5	12	92
Thargomindah	281 + 147	265	5.7	52.2	1.7	6	20	100
New South Wales								
Broken Hill	248 ± 113	233	6.0	45.6	1.5	8	90	90
Mean	—	—	10.4	51.3	1.6	7	39	84

* Calculated where possible from Bureau of Meteorology (1968)

+ No records of annual rainfall exceeding twice the mean

land Australia receives a rainfall that is 10 to 20% more variable than the world average for regions of equivalent mean annual rainfall. Second, and of greater importance to the present inquiry, this variability in annual rainfalls is not normally distributed, because mean annual rainfalls are consistently higher than median annual rainfalls by an average of 10.4% and by a maximum of 22% (far western Queensland; Table 3). Thus, annual rainfalls are skewed in the direction of unusually large totals; these appear to result primarily from infrequent incursions of tropical cyclones, and sometimes monsoonal fronts, into inland Australia, the results of which are very heavy and widespread rains (for descriptions see Gentilli 1971, 1972, Newsome and Corbett 1975). Climatologists have noted that daily, weekly, and monthly rainfall totals are not normally distributed (Gibbs and Maher 1967, Gentilli 1971, 1972), but there has been insufficient documentation of the fact that, in arid Australia, annual totals are also skewed (Fleming 1978). To demonstrate briefly that this phenomenon is common to the arid zone and not confined to the localities considered in Table 3, I have compared the proportions of meteorological stations in arid Australia receiving mean annual rainfall greater than median annual rainfalls with the equivalent proportions in non–arid southern and central Victoria (Table 4). These figures show that annual totals are significantly skewed towards high values in arid Australia but are not so in non–arid parts of the continent. The widespread impression about inland Australia that these data contradict is this, that it is a region of low, unpredictable rainfall which experiences periodic, catastrophic drought. This conception is only partially true; it is closer to reality to say that it is a region which normally experiences low, unpredictable rainfall but which has sporadic wet years, as recognized for some localities by Gentilli (1971).

Having established the occurrence of this type of variation, we wish to know how frequently a population of organisms is likely to be exposed to conditions of unusually heavy rainfall and the concomitant changes in resource levels. Gibbs and Maher (1967) recommended the use of deciles for the description of rainfall distributions, a method which avoids arbitrary choice of significant departures from mean rainfall. A decile delimits 10% of the distribution of a particular set of rainfalls; thus the ninth decile is that rainfall amount which is exceeded by only 10% of the totals ("very much above average"), the eight decile is that amount exceeded by 20% of totals ("much above average"), and the seventh decile is that amount exceeded by 30% of totals ("above average") (Gibbs and Maher 1967). To relate the rainfalls of "very much above average" to the mean, I calculated the proportions of the mean represented by decile 9 rainfalls for the localities in Table 3. Decile 9 rainfalls average 1.6 times the mean; in comparison, the coefficients of variation at these localities average 51.3%, indicating that the upper limits of the standard deviations average just over 1.5 times the mean. These calculations emphasize the point that the annual rainfalls we are considering are exceptionally heavy. To provide further information on frequency of heavy rainfall, I determined the incidence of annual rainfalls exceeding 1.5 times the mean (i.e. exceeding the mean upper limit of the standard deviation) from data provided for each of the 15 localities by the Bureau of Meteorology; on average, rainfall of this magnitude occurred every seven years (Table 3). It is worth noting here that the incidence of unusually high numbers of days of vegetative growth per year in central Australia has been estimated at approximately one in 10 years (Fitzpatrick *et al.* 1967), a result which adds weight to the present analysis. Organisms such as dasyurids, whose life–spans are between one and two years, are consequently liable to experience unusually productive conditions once in an average of about seven generations. It seems highly likely that specific adaptations resulting from selective pressures within these periods could be maintained at such a rate of incidence. The frequency of occurrence of annual rainfall totals exceeding twice the mean, however, are so low that it seems

Table 4. The proportions of meteorological stations receiving mean annual rainfall that is greater than median annual rainfall in arid Australia and in non-arid Victoria.

	Number of Rainfall Stations	Number of Stations where Mean > Median Annual Rainfall	Deviation from an even ratio of Mean > Median: Mean < Median among all stations (chi–square test)
Arid Australia*	342	326	Significant ($p < 0.001$)
Southern and central Victoria⁺	299	154	Not significant ($p > 0.5$)

* Meteorological districts 4, 13, 15–17, 37, 38, 46 (Bureau of Meteorology, 1977)

⁺ Meteorological districts 85–88 (Bureau of Meteorology, 1977)

unlikely that rainfalls of this magnitude could have lasting selective impact (one generation in about 40; Table 3). Sites in far western Queensland are exceptional, and may experience sufficiently frequent rainfalls of twice the mean to provide permanent adaptations in dasyurids (one generation in 12).

Table 5. Estimates of the frequency of occurrence of heavy rainfall in two consecutive years within the Australian arid zone.

	Average period between occurrences (years)*	Range of periods (years)*
1. Contiguity of years with rainfalls of deciles 8 or 9	29	10–81
2. Contiguities from row 1, plus contiguity of years of decile 7 rainfall with years of decile 8 or 9 rainfall	14	7–81

* Methods used to produce these estimates are described in the text. The number of years considered is 81 (1885–1965).

It is of further value to gain some idea of the likelihood of one year of above average conditions immediately succeeding another, so as to enable us to determine the frequency with which very extended periods of unusual productivity may occur. Thus, I have used the maps provided by Gibbs and Maher (1976), showing the decile ranges in which annual rainfall has occurred within Australia, to provide some idea of the likelihood of two years of heavy rainfall occurring contiguously over large areas of the arid zone. Estimates were obtained by dividing the arid zone into 16 regions of roughly equal areas, then noting from Gibbs and Maher (1967) the years in which rainfall above deciles 7, 8 or 9 fell over a majority of each particular region[2]. The period between contiguous year of heavy rainfall was then determined for each of the 16 regions, and averages calculated for the entire arid zone (Table 5). Although crude, these estimates show that contiguity of years of rainfall in deciles 8 or 9 is unlikely to be of lasting selective significance (once in 30 years), but inclusion of years of rainfall in decile 7 reduces the period to 14 years. It seems reasonable to suggest that these latter frequencies, in association with those described in the previous paragraphs, could be sufficient to maintain adaptations specific to periods of abundant production, particularly in the more variable regions.

Is there direct evidence that the invertebrates which form the food supply of dasyurids respond to these unusually heavy rainfalls? Data are available only for the plague locust, *Chortoicetes terminifera,* a species of economic importance. Detailed analyses show that rainfall has a markedly positive effect on the densities of locusts, and suggest that there is a non–linear relationship between rainfall and reproductive rate (Clark *et al.* 1969, Clark 1974, 1978, Farrow, 1979). The net result of these effects is that densities of adults vary by a factor of about 50, with rainfall being the driving force (Clark 1978). This species is almost certainly more variable in numbers than most acridids of the region, but others do reach very high densities in parts of the arid zone (Clark 1978). Invertebrates apart from acridid grasshoppers may not reach such dramatic peaks, but it seems inevitable that they fluctuate in response to the same conditions. These fluctuations, in turn, will result from increases in vegetative productivity induced by periods of heavy rain (for a summary see Williams 1979).

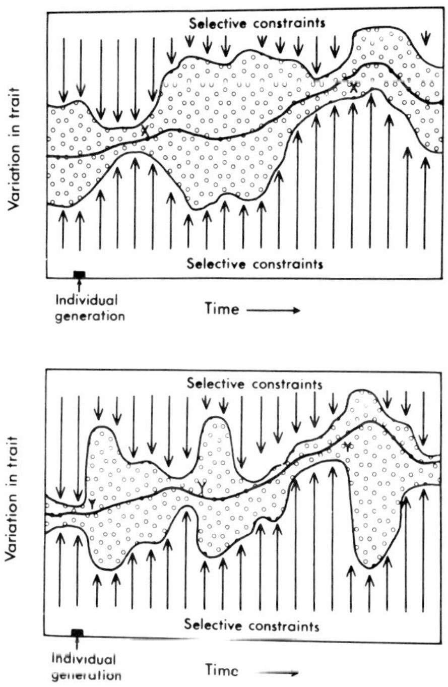

Fig. 3. Changes in the mean phenotypic expression of a trait in a population (solid lines), and in the variation about this mean (shaded areas), under changing selective conditions, the duration of individual generations being indicated on the abcissae. A: Modified from Wiens (1977). X indicates an "ecological crunch" during which the selective constraints are particularly severe. B: An altered version of A representing conditions in the Australian arid zone. Y indicates a period of heavy rainfall, and consequent elevated vegetative production, during which selective constraints are relaxed. Following point Y, the selective constraints return to severity and gradually reduce the variation of the trait; eventually, another Y period intervenes.

2. Effects of Long–term Variability on Dasyurids: The foregoing discussion shows that dasyurids of the Australian arid zone inhabit an environment in which resource levels fluctuate to

[2] Decile 7 averages 1.1 times the mean for the localities in Table 3 (Bureau of Meteorology 1968); decile 8 is not listed in the records, but is probably 1.3 to 1.4 times the mean.

an extraordinary extent and the dominant aspect of this variation is the tendency towards years of unusually heavy rainfall and dramatically increased production. In an effort to express these alterations in terms of changing selective pressures and their effects, I have drawn on an illuminating diagram from Wiens (1977) (Fig. 3). Wiens (1977) noted that continental arid regions are subject to considerable climatic fluctuation and that the effects of these changes upon the evolution of and variation in resource–utilization traits of consumers may be profound. His diagrammatic representation of this phenomenon (Fig. 3A) emphasized variation in phenotypic expression of some relevant trait in a population, resulting from relaxation of selective constraints, and compared this with intensive selection in periods of resource limitation resulting in restriction of variation in the population. In Fig. 3B, I have modified Wiens' (1977) diagram to represent the data on variability presented above. In this situation, I suggest that an overriding consideration is the sudden relaxation in selective constraints springing from elevated resource levels in wet years. Such periods of relaxed selective pressures may then result in greater phenotypic variation, which will gradually become more severe during the drier years that follow. If, subsequently, relaxation of constraints occurs before the variation or the traits favoured during high rainfall periods have disappeared from the population, then high levels of variability and polymorphy will be maintained. In Fig. 3B, the periods of relaxation are depicted at intervals of seven to ten generations in order to represent the frequencies discussed previously.

What does this signify for dasyurids? Two broad areas of their biology seem likely to be affected – reproduction and life history, and morphology. High variability in nipple number within populations of *S. crassicaudata* (six to eleven; Morton 1978c) and of *D. cristicaudata* (four to eight; Spencer 1896) contrasts with the rigid patterns of nipple number in populations of forest–dwelling *Antechinus* spp. (Woolley 1973, Lee 1974). Thus Morton's (1978c) interpretation of the polymorphism in *S. crassicaudata* in terms of environmental variability receives support from the present analysis. Variability of nipple number within populations of other desert–dwelling dasyurids has not been recorded. The prevalence of polyoestry in desert–dwelling dasyurids, especially in large species such as *D. byrnei,* may also result partially from infrequent years of high production. At first sight it seems anomalous that *D. byrnei* should wean young from its second litter in autumn, leaving them to face the critical task of reaching maturity during winter (Woolley 1971, 1973). If, however, this species is able to use small mammals as food in addition to insects, then late breeding may be very successful during the plagues of rodents that tend to follow prolonged vegetative flushes (Finlayson 1939, 1961, Plomley 1972, Newsome and Corbett 1975); thus the polyoestry may persist, in evolutionary terms, so as to allow *D. byrnei* to take advantage of these irregular irruptions. Similarly, the dimorphism in monoestry and polyoestry exhibited by *D. cristicauda* (Lee *et al.* 1982) seems best interpreted as persistence of traits directed towards different conditions of productivity in different generations.

Few studies of morphological variation within populations of dasyurids have been conducted, although early workers noted considerable variability (Spencer 1896, Jones 1949). However, Archer's (1977) study of the genus *Antechinomys* demonstrated considerable variability that was difficult to interpret; perhaps this variation is due to the effects described here. Intra–populational variation in the external morphology of *S. crassicaudata* has been interpreted in terms of year–to–year variability (Morton 1978c, Morton and Alexander 1982), and again, the present analysis supports this hypothesis. It is important to realize that variation in morphology will be directly related to use of resources, and thus will have an important influence on our interpretations of dasyurid ecology. For example, the types of insects most commonly used for food, the amount of energy and water used, and the amount of fat stored, may all be affected by or govern morphological variation. These aspects of variation may make it extremely difficult to determine differences between coexisting species, as Wiens (1977) recognised when developing Fig. 3A.

Periods of unusual vegetative production almost certainly result in elevated population levels in dasyurids, but the few data available suggest that it may be simplistic to expect a one–to–one correspondence between rainfall and population levels. Several authors have noted increases in the densities of dasyurids following periods of extended rainfall (Finlayson 1933, 1961, Denny, 1975, Morton and Martin 1979), and it seems reasonable to assume that dasyurids increase in a similar manner to that of rodents (Finlayson 1939, 1961, Plomley 1972, Newsome and Corbett 1975). However, Newsome and Corbett (1975) noted that numbers of rodents declined from a peak after one year of heavy rainfall and high production, even though rainfall continued to be very heavy during the subsequent collapse in the rodents' numbers; this result was interpreted as evidence of predatory pressure (Newsome and Corbett 1975). Because soils of the Australian arid zone contain very low quantities of nitrogen, however, exhaustion of nitrogenous reserves may prevent vegetation from reaping the full benefit of heavy rainfall following a prior year of high production (Charley and Cowling 1968,

Cowling 1977, West and Skujins 1978). Consequently, consumers may not actually experience continuously favourable conditions during prolonged wet periods in the arid zone. In addition, Davies (1973) postulated that an inverted U–shaped curve described the productivity–rainfall relationship in arid Australia, and suggested that in very wet years large hatchings of certain herbivorous insects could greatly depress the growth of some plants, thereby modifying vegetative productivity. Thus, it is appropriate to conclude this discussion with the cautionary not that exceptionally good rainfall may not automatically result in immediately increased resources and population levels of consumers.

CONCLUSION

The mammalian fauna of the Australian arid zone is characterized by a high species richness of insectivores, most of which are dasyurids. This fauna contains about four times the number of species of insectivorous mammals found in North American deserts (Morton 1979), and although detailed comparisons with other arid regions are lacking, I know of no other such fauna which contains anything like the richness of insectivores occurring in the Australian arid zone. It is most unfortunate, therefore, that our knowledge of these dasyurids in the field is so sparse (Table 1), and it is my hope that this review will show, at the least, which areas of their biology are most likely to provide general insights. First, short–term variability in food supply appears to have had a significant influence on the ecology of those species that have been studied. In addition, it is argued here that long–term variability may have equally important effects, albeit less easily identifiable ones. Consequently, I suggest that future studies of desert–dwelling dasyurids must concentrate on measurement of the fluctuations in food supply and on their effects. Second, a major influence on the species richness of dasyurids is their preference for hummock grassland as a habitat. There are no immediately obvious reasons for this pattern; the only suggestion I can make is that the permanent cover provided by *Triodia* hummocks of this habitat results in a greater stability of invertebrate populations, both in the short–term and the long–term. Whatever the case, further advances in our knowledge of desert–dwelling dasyurids will depend on studies conducted in hummock grassland. In this case, hopefully, we will see the *Triodia*–inhabiting *Sminthopsis psammophila* rescued from the neglect it has suffered since Spencer (1982) wrote of it with such prescience.

ACKNOWLEDGEMENTS

The manuscript was written while I held a University of Sydney Postdoctoral Research Fellowship. Mr. D. Read generously shared with me his impressions of the evology of *Planigale* spp., Dr. R. W. Braithwaite and Dr. G. D. Sanson kindly allowed me to use Fig. 2, and Mr. P. M. Fleming provided advice concerning climatology. I wish also to acknowledge the stimulation I have received from those writings of Dr. A. E. Newsome and of Dr. E. R. Pianka concerning the Australian arid zone. Drs. M. Denny, B. Fox and D. Croft provided useful criticisms of a draft of this paper.

REFERENCES

Aitken, P. F., 1971. Rediscovery of the large desert sminthopsis (*Sminthopsis psammophilus* Spencer) on Eyre Peninsula, South Australia. *Victorian Nat.* **88:** 103–11.

Aitken, P. F., 1972. *Planigale gilesi* (Marsupialia, Dasyuridae); a new species from the interior of southeastern Australia. *Rec. S. Aust. Mus.* **16** (10): 1–14.

Andrew, D. and Settle, G., 1982. Observations on the behaviour of *Planigale* (Marsupialia, Dasyuridae) with particular reference to the narrow-nosed planigale *(P. tenuirostris).* Pp.311–24 *in* "Carnivorous marsupials", ed. by M. Archer. Roy. Zool. Soc. New South Wales: Sydney, Australia.

Archer, M., 1975. *Ningaui,* a new genus of tiny dasyurids (Marsupialia) and two new species, *N. timealeyi* and *N. ridei,* from arid Western Australia. *Mem. Qd Mus.* **17:** 237–49.

Archer, M., 1977. Revision of the dasyurid marsupial genus *Antechinomys* Krefft. *Mem. Qd Mus.* **18:** 17–29.

Archer, M., 1979. The status of Australian dasyurids, thylacinids and myrmecobiids. Pp.29–43 *in* "The status of endangered Australasian wildlife" ed. by M. J. Tyler. Royal Zoological Society of South Australia: Adelaide.

Arnold, J. and Shield, J., 1970a. Growth and development of the Chuditch *(Dasyurus geoffroii). Bull. Aust. Mammal Soc.* **2:** 198.

Arnold, J. and Shield, J., 1970b. Oxygen consumption and body temperature of the Chuditch *(Dasyurus geoffroii). J. Zool., Lond.* **160:** 391–404.

Aslin, H. J., 1974. The behaviour of *Dasyuroides byrnei* (Marsupialia) in captivity. *Z. Tierpsychol.* **35:** 187–208.

Baudinette, R. V., Nagle, K. A. and Scott, R. A. D., 1976a. Locomotory energetics in a marsupial *(Antechinomys spenceri)* and a rodent *(Notomys alexis). Experientia* **32:** 583–4.

Baudinette, R. V., Nagle, K. A. and Scott, R. A. D., 1976b. Locomotory energetics in dasyurid marsupials. *J. comp. Physiol.* **109B:** 159–68.

Braithwaite, R. W. and Lee, A. K., 1979. A mammalian example of semelparity. *Amer. Nat.* **113:** 151–5.

Burbidge, A. A., McKenzie, N. L., Chapman, A. and Lambert, P. M., 1976. The wildlife of some existing and proposed reserves in the Great Victoria and Gibson Deserts, Western Australia. *Wildl. Res. Bull. West. Aust.* **5:** 1–16.

Burbidge, N. T., 1953. The genus *Triodia* R.Br. (Gramineae). *Aust. J. Bot.* **1:** 121–84.

BUREAU OF METEOROLOGY, 1968. "Review of Australia's water resources. Monthly rainfall and evaporation" Government Printer: Melbourne.

BUREAU OF METEOROLOGY, 1977. "Rainfall statistics Australia". Australian Government Publishing Service: Canberra.

CHARLEY, J. L. AND COWLING, S. W., 1968. Changes in soil nutrient status resulting from overgrazing and their consequences in plant communities of semi-arid areas. *Proc. Ecol. Soc. Aust.* **3:** 28–38.

CLARK, D. P., 1974. The influence of rainfall on the densities of adult *Chortoicetes terminifera* (Walker) in central western New South Wales, 1965-73. *Aust. J. Zool.* **22:** 365–86.

CLARK, D. P., 1978. The significance of the availability of water in limiting invertebrate numbers. Pp.198–207 *in* "Studies of the Australian arid zone. III. Water in rangelands" ed. by K. M. W. Howes. C.S.I.R.O.: Melbourne.

CLARK, D. P., ASHALL, C., WALOFF, Z. AND CHINNICK, L., 1969. Field studies on the Australian plague locust (*Chortoicetes terminifera* Walk.) in the Channel Country of Queensland. *Anti-Locust Bull.* **44:** 1–104.

COGGER, H. G. AND HEATWOLE, H., 1981. The Australian reptiles: origins, biogeography, distribution patterns and island evolution. Pages 1331–73 *in* "Ecological biogeography in Australia" ed. by A. Keast. W. Junk: The Hague.

COWLING, S. W., 1977. Effects of herbivores on nutrient cycling and distribution in rangelands ecosystems. Pp.277–98 *in* "The impact of herbivores on arid and semi-arid rangelands". Australian Rangeland Society: Perth.

CROWCROFT, P. AND GODFREY, G. K., 1968. The daily cycle of activity in two species of *Sminthopsis* (Marsupialia: Dasyuridae). *J. Anim. Ecol.* **37:** 63–73.

DAVIES, S. J. J. F., 1973. Environmental variables and the biology of native Australian animals of the mulga lands. *Tropical Grasslands* **7:** 127–34.

DAWSON, T. J. AND HULBERT, A. J., 1970. Standard metabolism, body temperature and surface areas of Australian marsupials. *Amer. J. Physiol.* **218:** 1233–8.

DAWSON, T. J. AND WOLFERS, J. M., 1978. Metabolism, thermoregulation and torpor in shrew sized marsupials of the genus *Planigale*. *Comp. Biochem. Physiol.* **59A:** 305–9.

DENNY, M. J. S., 1975. Mammals of Sturt National Park, Tibooburra, New South Wales. *Aust. Zool.* **18:** 179–95.

EWER, R. F., 1968. A preliminary survey of the behaviour in captivity of the dasyurid marsupial, *Sminthopsis crassicaudata* (Gould). *Z. Tierpsychol.* **25:** 319–65.

EWER, R. F., 1969. Some observations on the killing and eating of prey by two dasyurid marsupials: the Mulgara, *Dasycercus cristicauda*, and the Tasmanian Devil, *Sarcophilus harrisi*. *Z. Tierpsychol.* **26:** 23–38.

FANNING, D., 1982. Reproduction, growth and development in *Ningaui* sp. (Dasyuridae: Marsupialia) from the Northern Territory. Pp.23–37 *in* "Carnivorous marsupials" ed. by M. Archer. Roy. Zool. Soc. New South Wales: Sydney, Australia.

FARROW, R. A., 1979. Population dynamics of the Australian plague locust, *Chortoicetes terminifera* (Walker), in central western New South Wales. I. Reproduction and migration in relation to weather. *Aust. J. Zool.* **27:** 717–45.

FINLAYSON, H. H., 1933. On mammals from the Lake Eyre basin. Part 1. The Dasyuridae. *Trans R. Soc. S. Aust.* **57:** 195–202.

FINLAYSON, H. H., 1939. On mammals from the Lake Eyre basin. Part 5. General remarks on the increase of murids and their population movements in the Lake Eyre basin during the years 1930–1936. *Trans. R. Soc. S. Aust.* **63:** 348–53.

FINLAYSON, H. H., 1961. On central Australian mammals. Part IV — The distribution and status of central Australian species. *Rec. S. Aust. Mus.* **14:** 141–91.

FITZPATRICK, E. A., SLATYER, R. O. AND KRISHNAN, A. I., 1967. Incidence and duration of periods of plant growth in central Australia as estimated from climatic data. *Agric. Meteorol.* **4:** 389–404.

FLEAY, D., 1961. Breeding the Mulgara. *Victorian Nat.* **78:** 160–7.

FLEMING, P. M., 1978. Types of rainfall and local rainfall variability. Pp.18–28 *in* "Studies of the Australian arid zone III. Water in rangelands" ed. by K. M. W. Howes. C.S.I.R.O.: Melbourne.

FRITH, H. J., 1973. "Wildlife conservation". Angus and Robertson: Sydney.

GENTILLI, J., 1971. The main climatological elements. Pp.119–88 *in* "Climates of Australia and New Zealand" ed. by J. Gentilli. Elsevier: Amsterdam.

GENTILLI,, J., 1972. "Australian climate patterns". Nelson: Melbourne.

GIBBS, W. J. AND MAHER, J. V., 1967. Rainfall deciles as drought indicators. *Bull., Bureau of Meteorol.* **48:** 1–33.

GODFREY, G. K., 1968. Body-temperatures and torpor in *Sminthopsis crassicaudata* and *S. larapinta* (Marsupialia — Dasyuridae). *J. Zool., Lond.* **156:** 499–511.

GODFREY, G. K., 1969a. The influence of increased photoperiod on reproduction in the dasyurid marsupial *Sminthopsis crassicaudata*. *J. Mammal.* **50:** 132–3.

GODFREY, G. K., 1969b. Reproduction in a laboratory colony of the marsupial mouse *Sminthopsis larapinta* (Marsupialia: Dasyuridae). *Aust. J. Zool.* **17:** 637–54.

GODFREY, G. K. AND CROWCROFT, P., 1971. Breeding the fat-tailed marsupial mouse *Sminthopsis crassicaudata* in captivity. *Int. Zoo Yb.* **11:** 33–8.

HAINES, H., MACFARLANE, W. V., SETCHELL, C. AND HOWARD, B., 1974. Water turnover and pulmocutaneous evaporation of Australian desert dasyurids and murids. *Amer. J. Physiol.* **227:** 958–63.

HAPPOLD, M., 1972. Maternal and juvenile behaviour in the marsupial jerboa, *Antechinomys spenceri* (Dasyuridae). *Aust. Mammal.* **1:** 27–37.

HUTSON, G., 1975. Sequences of prey-catching behaviour in the Brush-tailed Marsupial Rat (*Dasyuroides byrnei*). *Z. Tierpsychol.* **39:** 39–60.

HUTSON, G. D., 1976. Grooming behaviour and birth in the dasyurid marsupial *Dasyuroides byrnei*. *Aust. J. Zool.* **24:** 277–82.

HUTSON, G. D., 1982. An analysis of offensive and defensive threat displays in *Dasyuroides byrnei* (Dasyuridae, Marsupialia). Pp.345–63 *in* "Carnivorous marsupials", ed. by M. Archer. Roy. Zool. Soc. New South Wales: Sydney, Australia.

JOHNSON, K. A. AND ROFF, A. D., 1980. Discovery of ningauis (*Ningaui* sp.: Dasyuridae: Marsupialia) in the Northern Territory, Australia. *Aust. Mammal.* **3:** 127–9.

JONES, F. WOOD, 1949. The study of a generalized marsupial (*Dasycercus cristicauda* Krefft). *Trans. Zool. Soc. Lond.* **26:** 409–501.

KEAST, A., 1961. Bird speciation on the Australian continent. *Bull. Mus. Comp. Zool. Harv.* **123:** 304–495.

KENNEDY, P. M. AND MACFARLANE, W. V., 1971. Oxygen consumption and water turnover of the fat-tailed marsupials *Dasycercus cristicauda* and *Sminthopsis crassicaudata*. *Comp. Biochem. Physiol.* **40A:** 723–32.

LEE, A. K., 1974. Variations in nipple number in female *Antechinus*. *Aust. Mammal.* **1:** 300.

LEE, A. K., WOOLLEY, P. AND BRAITHWAITE, R. W., 1982. Life-history strategies in dasyurids. Pp.1–11 *in* "Carnivorous marsupials", ed. by M. Archer. Roy. Zool. Soc. New South Wales: Sydney, Australia.

LEEPER, G. W., 1970. Climates. Pp.12–20 *in* "The Australian environment", ed. by G. W. Leeper. C.S.I.R.O.: Melbourne.

LEIGH, J. H. AND NOBLE, J. C., 1969. Vegetation resources. Pp.73–92 *in* "Arid lands of Australia", ed. by R. O. Slatyer and R. A. Perry. Australian National University Press: Canberra.

MACFARLANE, W. V., 1975. Ecophysiology of water and energy in desert marsupials and rodents. Pp.389–96 *in* "Rodents in desert environments", ed. by I. Prakash and P. K. Ghosh. W. Junk: The Hague.

MACFARLANE, W. V. AND HOWARD, B., 1972. Comparative water and energy economy of wild and domestic mammals. Pp.261–96 *in* "Comparative physiology of desert animals", ed. by G. M. O. Maloiy. Academic Press: London. *(Symp. Zool. Soc. Lond.* **31.**)

MACK, G., 1961. Mammals from southwestern Queensland. *Mem. Qd. Mus.* **13:** 213–29.

MACMILLEN, R. E. AND NELSON, J. E., 1969. Bioenergetics and body size in dasyurid marsupials. *Amer. J. Physiol.* **217:** 1246–51.

MARTIN, P. G., 1965. The potentialities of the Fat-tailed Marsupial Mouse, *Sminthopsis crassicaudata* (Gould), as a laboratory animal. *Aust. J. Zool.* **13:** 559–62.

MICHENER, G. R., 1969. Notes on the breeding and young of the Crest-tailed Marsupial Mouse, *Dasycercus cristicauda*. *J. Mammal.* **50:** 633–5.

MORTON, S. R., 1978a. An ecological study of *Sminthopsis crassicaudata* (Marsupialia: Dasyuridae) I. Distribution, study areas and methods. *Aust. Wildl. Res.* **5:** 151–62.

MORTON, S. R., 1978b. An ecological study of *Sminthopsis crassicaudata* (Marsupialia: Dasyuridae) II. Behaviour and social organization. *Aust. Wildl. Res.* **5:** 163–82.

MORTON, S. R., 1978c. An ecological study of *Sminthopsis crassicaudata* (Marsupialia: Dasyuridae) III. Reproduction and life-history. *Aust. Wildl. Res.* **5:** 183–211.

MORTON, S. R., 1978d. Torpor and nest-sharing in free-living *Sminthopsis crassicaudata* (Marsupialia) and *Mus musculus* (Rodentia). *J. Mammal.* **59:** 569–75.

MORTON, S. R., 1979. Diversity of desert-dwelling mammals: a comparison of Australia and North America. *J. Mammal.* **60:** 253–64.

MORTON, S. R., 1980a. Field and laboratory studies of water metabolism in *Sminthopsis crassicaudata* (Marsupialia: Dasyuridae). *Aust. J. Zool.* **28:** 213–27.

MORTON, S. R., 1980b. Ecological correlates of caudal fat storage in small mammals. *Aust. Mammal.* **3:** 81–6.

MORTON, S. R. AND ALEXANDER, F., 1982. Geographic variation in the external morphology of *Sminthopsis crassicaudata* (Marsupialia: Dasyuridae). *In* "Carnivorous marsupials", ed. by M. Archer. Roy. Zool. Soc. New South Wales: Sydney, Australia.

MORTON, S. R., HAPPOLD, M., LEE, A. K. AND MACMILLEN, R. E., 1977. The diet of the barn owl, *Tyto alba*, in southwestern Queensland. *Aust. Wildl. Res.* **4:** 91–7.

MORTON, S. R. AND MARTIN, A. A., 1979. Feeding ecology of the barn owl, *Tyto alba*, in arid southern Australia. *Aust. Wildl. Res.* **6:** 191–204.

NEWSOME, A. E. AND CORBETT, L. K., 1975. Outbreaks of rodents in semi-arid and arid Australia: causes, preventions, and evolutionary considerations. Pp.117–153 *in* "Rodents in desert environments" ed. by I. Prakash and P. K. Ghosh. W. Junk: The Hague.

PIANKA, E. R., 1969. Habitat specificity, speciation, and species density in Australian desert lizards. *Ecology* **50:** 498–502.

PIANKA, E. R., 1972. Zoogeography and speciation of Australian desert lizards: an ecological perspective. *Copeia* **1972:** 127–45.

PIANKA, E. R. AND SCHALL, J. J., 1980. Species densities of Australian vertebrates. Pages 1675–94 *in* "Ecological biogeography of Australia" ed. by A. Keast. W. Junk: The Hague.

PLOMLEY, N. J. B., 1972. Some notes on plagues of small mammals in Australia. *J. nat. Hist.* **6:** 363–84.

READ, D., 1982. The ecology of two arid zone planigales (Dasyuridae, Marsupialia). Pp.227–31 *in* "Carnivorous marsupials", ed. by M. Archer. Roy. Zool. Soc. New South Wales: Sydney, Australia.

RIDE, W. D. L., 1970. "A guide to the native mammals of Australia". Oxford University Press: Melbourne.

SCHALL, J. J. AND PIANKA, E. R., 1978. Geographical trends in numbers of species. *Science* **201:** 679–86.

SCHMIDT-NIELSEN, K. AND NEWSOME, A. E., 1962. Water balance in the Mulgara *(Dasycercus cristicauda)*, a carnivorous desert marsupial. *Aust. J. Biol. Sci.* **15:** 683–9.

SMITH, M. J., BENNETT, J. H. AND CHESSON, C. M., 1978. Photoperiod and some other factors affecting reproduction in female *Sminthopsis crassicaudata* (Gould) (Marsupialia: Dasyuridae) in captivity. *Aust. J. Zool.* **26:** 449–63.

SMITH, M. J., AND GODFREY, G. K., 1970. Ovulation induced by gonadotrophins in the marsupial *Sminthopsis crassicaudata* (Gould). *J. Reprod. Fert.* **22:** 41–7.

SORENSON, M. W., 1970. Observations on the behaviour of *Dasycercus cristicauda* and *Dasyuroides byrnei* in captivity. *J. Mammal.* **51:** 123–30.

SPENCER, B., 1896. Mammalia. Pp.1–52 *in* "Report of the Horn Expedition to central Australia, Part II — Zoology". Dulau and Co.: London.

SPENCER, B., 1928. "Wanderings in wild Australia". Macmillan: London.

WALLIS, R. L., 1982. Adaptations to low environmental temperatures in the carnivorous marsupials. Pp.285–91 *in* "Carnivorous marsupials", ed. by M. Archer. Roy. Zool. Soc. New South Wales: Sydney, Australia.

WATTS, C. H. S., 1979. The status of endangered Australian rodents. Pp.75–83 *in* "The status of endangered Australasian wildlife ed. by M. J. Tyler. Royal — Zoology" led by B. Spencer Dulau and Co.: London.

WEST, N. E., AND SKUJINS, J., 1978. "Nitrogen in desert ecosystems". Dowden, Hutchinson & Ross: Stroudsburg, Pennsylvania.

WIENS, J. A., 1977. On competition and variable environments. *Amer. Sci.* **65:** 590–7.

WILLIAMS, O. B., 1979. Ecosystems of Australia. Pp.145–212 *in* "Arid land ecosystems: their structure, functioning and management, Vol. I" ed. by D. W. Goodall and R. A. Perry. Cambridge University Press: Cambridge.

WOOLLEY, P., 1971. Maintenance and breeding of laboratory colonies of *Dasyuroides byrnei* and *Dasycercus cristicauda. Int. Zoo. Yb.* **11:** 351–4.

WOOLLEY, P., 1973. Breeding patterns, and the breeding and laboratory maintenance of dasyurid marsupials. *Exp. Animals* **22:** (Suppl.): 161–72.

RESUMEN

12. Los marsupiales dasiúridos de la zone árida australiana: una revisión ecológica. Por S. R. Morton.

Se provee una lista de los marsupiales dasiúridos de la zona árida australiana, examinandose sus distribuciones dentro de los grandes tipos de hábitat de esa región. Éste análisis muestra que los dasiúridos son particularmente ricos en especies en los pastizales de las colinas, pero que decrecen en especies en los bosques de *Acacia* más de lo esperado. Nuestro disperso conocimiento de su comportamiento, historia evolutiva y ecología es revisado, concluyéndose – de acuerdo a la información disponible — que las presiones selectivas que operan sobre estos animales están regidas por la considerable variabilidad de las poblaciones de invertebrados que constituyen su alimento. Los necesarios datos para establecer las caraterísticas de la variabilidad a corto plazo serían relativamente fácil de obtener, pero se sostiene que la variabilidad a largo plazo puede tener también importantes consecuencias sobre los dasiúridos, aunque es menos fácil de cuantificar. Por lo tanto se investigaron los métodas de inferencia de la variabilidad a largo plazo, y los probables efectos de las fluctuaciones climáticas. Futuros estudios de los dasiúridos de los desiertos serán mas provechosos si se concentran en las especies que habitan los pastizales de las colinas, y en la medida e interpretación de la variación del suministro del alimento.

13. REVIEW OF PLANIGALE (DASYURIDAE, MARSUPIALIA) ECOLOGY

M. J. S. DENNY[1]

Four species of planigale are known from Australia. These are restricted to the northern parts of the continent. Planigales are the world's smallest marsupials, weighing as little as three grams. Many aspects of their sparsely known ecology reflect this small size.

Most planigales are found in areas associated with water, either near permanent water or in areas periodically flooded. *Planigale tenuirostris* is an exception and can be found in drier habitats. Cracking soil and vegetation cover seem to be a habitat requirement of all planigales. Planigales exhibit periods of high activity followed by relatively quiescent periods. They are ferocious hunters capable of killing and eating animals their own size. They build nests which may be communal, and, in captivity, have a stereotyped behaviour pattern between interacting individuals. Planigales display a range of vocalization patterns and, although they do not appear to establish territories, have a dominance hierarchy.

Planigales have metabolic rates and body temperature similar to other small marsupials. They have the ability to enter torpor when cold and/or deprived of food and use "sunbasking" as a thermoregulatory device. Their low metabolic rate is not reflected in their water usage. The water turnover for *Planigale tenuirostris* is similar to equivalent sized eutherians.

All Australian species appear to be polyoestrous with a slight seasonal effect. Available data on several aspects of the population dynamics of planigales are also given.

Pages 131–8 *in* CARNIVOROUS MARSUPIALS, ed. by M. Archer, February, 1982. Roy. Zool. Soc. New South Wales: Sydney, Australia.

The Shrew
"Strange as it seems, the smallest mammal
is the shrew, and not the camel
And that is all I ever knew,
or wish to know, about the shrew."
(Nash 1943)

IT IS FORTUNATE that not everyone takes the same attitude towards small mammals as Ogden Nash. Otherwise this review would have been even more difficult to write than it is. As relatively new arrivals to the scientific world (the first species of planigale was described in 1928, by Troughton, and the most recent, in 1972, by Aitken), published references to aspects of the ecology of planigales are few. Consequently part of the information used in this review is labelled in prep., in press or as pers. comm. Also, data available from other contributors to "Carnivous Marsupials" will be referred to when applicable. It should be noted that this paper is essentially a review of current knowledge about planigale ecology and does not attempt to reach any broad conclusions. These must wait for further data.

Planigales are a group of dasyurid marsupials that are the smallest marsupials in the world. There are five known species of planigales (Archer 1976). Four of these reside in Australia and one in New Guinea (*Planigale novaeguinae*). The four species in Australia are: *Planigale ingrami* (in three forms, including *brunnea* and *subtilissima*), *P. maculata* (and its *sinualis* form), *P. tenuirostris* and *P. gilesi*. In this review, only the Australian species will be discussed and *P. ingrami* will be its *brunnea* form unless otherwise noted.

PLANIGALE SIZE

Although not as tiny as some of the shrews which can weigh 3 to 5g and have a head–body length of only 2cm (Linstedt and Jones 1980), planigales are still extremely small. A summary of all available information about planigale body size is given in Table 1 and this shows that body weights range from 4 g for *P tenuirostris*, 5.7 g for *P. ingrami*, 9.4g for *P. gilesi*, 9.7g for *P. maculata*, and up to 12g for *P.m. sinualis* (these are means for both sexes). The average head body lengths for both sexes are 5.4cm for *P. ingrami*, 6.8cm for *P. gilesi* and 7.0cm for *P. maculata*.

PLANIGALE DISTRIBUTION

The distribution of planigales shows them to be limited to the northern parts of Australia (Archer 1976). Few specimens are known south of the 30° latitude and none has been found in Victoria or Tasmania, *Planigale maculata* is found on the northeastern and northern coastal areas of Australia, while *P.m. sinualis* is limited to the northern coast. *Planigale ingrami* is dis-

[1] Australian National Parks and Wildlife Service, Canberra, A.C.T., Australia. Present address: Nexus Environmental Studies, 3 West Street, North Sydney, New South Wales, Australia.

Table 1. Flesh measurements from planigales. As far as could be ascertained, all measurements were taken from live animals.

Planigale Species	Body Weight (g) ♂	Body Weight (g) ♀	Head–body Length (cm) ♂	Head–body Length (cm) ♀	Tail Length (cm) ♂	Tail Length (cm) ♀
maculata	6.4–7.9[4] 6.6–9.8[5]	7.8–8.4[5]	7.6[4] 7.2[5]	6.6[4] 6.5[5]	5.1[4] 5.4[5]	5.4[4] 5.5[5]
m. sinualis	14.5[13] 15.3[1] 8.7–22[8]	9.1–16.5[13] 10.9[1] 7.15[8]	7.4–8[8]	6.7–7.6[8]		
ingrami	4.3[3] 4.2[7]	5.6[6]	4.5[12]	6.3[15]	5.0[12]	5.7[12]
tenuirostris	5.8–9.3[10] 3.5[13] 3–6.5[14]		5.5[11] 7.0[12] 4.4–7.1[14]		4.5[11] 5.0[12]	
gilesi	9.17[9] 6.3[14]	9.7[10] 9.2[14] 4–11[14]	6.7[2] 7.1–7.9[9] 5.8–8.2[14] 6.5[11]	7.1[11] 4.4–7.3[14]	6.7[2] 6–7.2[9] 4.8[11]	5.9[9] 5.2[11]

REFERENCES: 1. Morton & Lee 1978; 2. Denny 1975; 3. Heinsohn 1970; 4. Fleay 1965; 5. Archer in press; 6. Woolley 1974; 7. R. Beeton pers. comm.; 8. Taylor *et al.* 1982; 9. Aitken 1972; 10. Dawson & Wolfers 1978; 11. P. Bayliss pers. comm.; 12. Troughton 1965; 13. Aslin 1975; 14. Denny *et al.* in prep.

tributed throughout the farnorth of Australia i.e. overlapping with *P. maculata,* but it is also found further inland. *Planigale gilesi* and *P. tenuirostris* are both found in the inland areas of eastern Australia (see distribution maps in Archer 1976).

PLANIGALE HABITAT PREFERENCES

Habitat preferences by planigales vary within the group and within each species. I have divided the habitat into three categories:

(a) the areas where planigales were found

(b) the vegetation in that area, and

(c) any details of exactly where caught, if by hand. Details of these categories are given in Table 2.

As a very loose generalization, it can be said that individuals of *P. maculata* are mainly found in or near wet swampy areas with a vegetation cover of trees, scrub, sedge and/or grass. The *sinualis* form is found on flood plains and its borders, with a cover of grasses and trees. *Planigale ingrami* is also found in wet, swampy areas such as flood plains, creeks and even in a rice paddy. Their habitats are characterized by black, cracking soil and grass cover. *Planigale tenuirostris* has been found in a variety of habitats ranging from river flood piains and lake edges to gibber plains and stony hills. Cover includes grass, chenopod shrub and woodland. All habitats are characterized by red or grey cracking soils. *Planigale gilesi* appears to be restricted to wet habitats, e.g. bore drains, creek floodouts, swamps' riverplains, and beside lakes. However, they have also been found on sandplains and interdune areas. Vegetation cover includes grasses, sedges, shrub and woodland. Most planigales captured by hand have been sheltering under objects such as logs, rocks, sticks, sheets of iron, fuel drums and wet decomposing grass.

Except *Planigale tenuirostris,* there is a general tendency for planigales to be found in areas associated with water, either near permanent water or in areas periodically flooded. *Planigale tenuirostris* is also found in drier habitats such as stony plains and hills. However, all species share a common requirement of some type of cracking soil and at least some vegetation cover. Two species (*P. ingrami* and *P. tenuirostris*) have been captured in cracks in the soil or in crab–hole depressions (Van Deusen 1969 and Sinclair, pers. comm). Aitken (1972) postulated that dense vegetation cover was important to *P. gilesi* as protection against aerial and terrestrial predators.

PLANIGALE AND CO–EXISTING SPECIES

Planigales, like most small mammals, have overlapping distributions with many other species. Table 3 gives a list of those species known to inhabit the same area as planigales. One study has shown that there was a 61% overlap in the ranges of *P. gilesi* and *Rattus villossisimus* and a 85% overlap in the ranges of *P. gilesi* and *Mus musculus.* The two rodents together nearly covered the whole range of *P. gilesi* (95%, Denny *et al.,* in prep). Although *P. gilesi, R. villossisimus* and *Mus musculus* occupied the same habitat, two other marsupial species found in the area (*Sminthopsis macroura* and *S. crassicaudata*) were mainly caught in open grassland, and did not significantly over-

Table 2. Habitat Preferences of Planigales

	Planigale maculata	*Planigale m. sinualis*	*Planigale ingrami*	*Planigale tenuirostris*	*Planigale gilesi*
Area found	Flooded marsh and rocky areas[1] Coast swamp fringes[2] Wet soil near creek[3] Sandy ridge near island edge[4,5]	Flood–plain borders[10] Sand on edge of creek[11] Flood–plain and fringing areas[9]	Dry swamps, streams[12] Black–soil plains, dry swamps and water courses amongst rocks[3] Dry cracking black–soil in swamp, near streams[13] Flood–plain[14] Rice paddy [15]	River flood–plain[12,26] Stony and Gibber plain[18] Alluvial plain[6] Lake edge with black soil[19] Red gravel ridge, claypan[20] Stony plain, hills and flood–plains[21]	Bore drain[23] Creek floodout, red cracking soil[24] Moist, cracking, clay swamp and creek bed[18] Margins of drainage tracts[6] River flats[25] River plain, beside creeks and lakes, sandplains and interdunes[19] Flood–plains[21,26] and Interdunes[21]
Vegetation cover	Sclerophyl, rainforest[6] Timber country[1] Pine plantation and woodland savanna[7] Sedgeland[2] Open Forest[3] Tall woodland, grassland[3] Wattle scrub[4,5]	Pandanus and Melaleuca[10] Hummock grass[11] Sedgeland and savanna woodland[9]	Tussock grass[12,16] Tall grass[13]	Chenopod shrub, mitchell grass[22] Woodland[19] Grass, chenopod shrub[21,26] and Canegrass[21]	Canegrass, lignum and dense shrub[18] Woodland, bluebush and rosewood[19] Bullrush and sedge[23] Canegrass, sedges and grasses[24]
Where caught by hand	In and under logs, sheets of iron, dead sticks[3] Under, rocks, in long grass under fuel drums[9]		Under wet decomposing grass[17] In soil cracks[13] Under bark[13]		Under iron[25]

REFERENCES: 1. Fleay 1965; 2. Aslin 1975; 3. Calaby 1966; 4. Johnson 1964; 5. Parker 1973; 6. Marlow 1958; 7. Archer in press; 8. Lavery & Grimes 1974; 9. Taylor *et al.* 1982; 10. Davies 1960; 11. Kitchener, in Archer 1976; 12. Troughton 1965; 13. Van Deusen 1969; 14. R. Beeton pers. comm.; 15. Lavery & Johnson 1974; 16. Rudeforth 1950; 17. Rudeforth, in Archer 1976; 18. Aitken pers. comm.; 19. P. Bayliss pers. comm.; 20. A. Rose pers comm.; 21. Denny *et al.* 1979; 22. R. Sinclair pers. comm.; 23. Aitken 1972; 24. Denny 1975; 25. R. Miller pers. comm.; 26. Andrews & Settle 1982.

lap the range of *P. gilesi* (Denny 1975). Many of the animals occupying the same habitat as planigales differ from these marsupials in size or feeding preferences. However, competition between different planigale species and between planigales and other mammals e.g. *Mus musculus* may occur (Denny 1975).

PLANIGALE BEHAVIOUR

There have been several reports on planigale behaviour, but many of these were based on casual observation. Those aspects of planigale behaviour that have been observed are listed in Table 4. Planigales, like shrews, exhibit periods of high activity followed by relatively quiescent periods. They are ferocious hunters and are capable of killing and eating animals their own size. Planigales are known to build nests which may be communal with individuals of their own or other species.

Detailed observations of the behaviour of captured planigales have been undertaken by Van Dyck (1979) and Andrews and Settle (1982). Results of these observations show that agnostic interactions follow a set pattern of threat and appeasement. Contact fighting occurred between individuals, particularly over possession of food items. Andrews and Settle (1982) suggest a dominance hierarchy amongst captured planigales and Van Dyck (1979) and Morrison (1975) record dominance by female

Table 3. Other small mammals found with planigales.

Planigale maculata sinualis:	*Rattus colletti**, *Dasyurus hallucatus*, *Antechinus bilarni*, *A. bellus*, *Phascogale tapoatafa*, *Sminthopsis virginiae* (Taylor *et al.* 1982).
Planigale ingrami:	*Rattus culmorum*, *Leggadina* sp. (Van Deusen 1969).
Planigale tenuirostris:	1. *P. gilesi**, *Sminthopsis macroura*, *S. crassicaudata*, *Mus musculus*, *Rattus villossisimus* (Denny 1975).
	2. *Sminthopsis macroura*, *S. crassicaudata*, *Mus musculus*, *Leggadina forresti* (R. Sinclair, pers. comm.).
Planigale gilesi:	1. *P. tenuirostris**, *Sminthopsis macroura*, *S. crassicaudata*, *Mus musculus*, *Rattus villossisimus* (Denny 1975).
	2. *Sminthopsis macroura*, *Mus musculus*, *Pseudomys desertor*, *Rattus villossisimus* (Aitken 1972).

* Sympatric species.

PLANIGALE POPULATION DYNAMICS

Another aspect of planigale ecology that has not been fully investigated is that concerning population dynamics. Denny (1975) observed changes in numbers of *P. gilesi* trapped over several years. Further study has given an estimate of population density of 7.4 per ha for *P. gilesi* living beside a creek (Denny *et al.* 1979). This value is of the same magnitude as that found for *Sminthopsis crassicaudata* (2.7 per ha, Morton 1978). Home range areas for *P. gilesi* have been estimated at between 1600 and 2250 m^2 (Denny *et al.*, in prep.). Taylor *et al.* (1982) describe how *P. maculata sinualis* move from flooded plain into surrounding areas during the annual "wet" season.

Table 4. Observations of planigale behaviour.

Pattern of Behaviour	*Planigale maculata*	*Planigale ingrami*	*Planigale tenuirostris*	*Planigale gilesi*
Thermo–regulatory	1*,9	2,3	2,3	2,3
Feeding	1,2,9	2,4,5	2	2
Breeding (includes nesting)	1,6,7,9	5	2	2
Parental	1,6,9	—	2	—
Agonistic	6,9	2,8	2	2
Vocalisation	1,9	5	—	—
Activity	6,9	2,8	2	2

*REFERENCES: 1. Fleay 1965; 2. Andrews & Settle 1982; 3. Dawson & Wolfers 1978; 4. Van Deusen 1969; 5. Heinsohn 1970; 6. Morrison 1975; 7. Aslin 1975; 8. McKay 1974; 9. Van Dyck 1979.

P. maculata over males. At least five specific vocal calls were recorded (*P. tenuirostris*, Andrews and Settle 1982; *P. maculata*, Van Dyck 1979). These were most frequently associated with agonistic interactions and copulatory behaviour.

Although there was no evidence for defence of areas, planigales follow specific pathways in enclosures and some areas remained more or less exclusive to particular individuals. Olfactory marking occurred in *P. tenuirostris* and *P. maculata*, both sternal gland and cloacal rubbing were observed.

Van Dyck (1979) suggested that marking behaviour in *P. maculata* was part of a process of familiarization of the planigales habitat. Van Dyck further suggested that *P. maculata* was probably socially intolerant, but this was not supported by Andrews and Settle.

Planigales use basking and torpor as means of thermoregulatory behaviour. Troughton (1965) suggested that the flattened skulls of planigale species (although *P. maculata* does not have a flattened skull; Archer 1976) was advantageous in allowing the animals to move in small cracks. This suggestion is reasonable in view of the preferred habitats and with the sinuous body movement found in three of the four species (again, *P. maculata* is an exception; Andrews and Settle 1982).

PLANIGALE DIET

There have been many reports about the diet of captive planigales. Food items known to be eaten by planigales in captivity include insects meat, pet food and dead birds (see Fleay 1965, Aslin 1975, Morton and Lee 1978, Van Deusen 1969, Heinsohn (1970). However, there is little information available about eating habits in the wild. It is known that planigales will eat other animals captured with them in pit–traps, e.g. insects, house mice and reptiles, but this situation is similar to that of caged planigales. Several authors regard planigales as being at least partially insectivorous. Aitken (1972) and Watts (1972) comment on *P. gilesi* eating insects. McKay (1974) states that *P. ingrami* use crickets and grasshoppers as their main foods. Troughton (1965) describes *P. ingrami* eating crickets and *P. tenuirostris* eating grasshoppers and plague locusts.

PLANIGALE PREDATORS

Because of their small size and their occasional preference for daylight activity, planigales are susceptible to attack from a variety of predators. Remains of *P. gilesi* and *P. tenuirostris* have been found in Barn Owl casts (Morton and Martin 1979), and Parker (1973) found *P. ingrami* in bird of prey pellets.

Cats (*P. maculata*, Archer 1976 and *P. gilesi* Denny, pers. observation), cane toads (*P. maculata*, Covacevich and Archer 1975) and at least two species of snake, are known to prey upon planigales. Sinclair (pers. comm.) believes that the Myall Snake (*Suta suta*) preys upon *P. tenuirostris* and the Eastern Brown Snake (*Pseudonja textiles*) has been observed hunting *P. gilesi* caught on canegrass during flooding (Denny, pers. observation).

PLANIGALES AND FLOODS

Another threat to planigales is that of flooding of their habitats. Because planigales prefer habitats associated with water, flooding may displace these species from low–lying areas. Denny (1975) described the effects of flooding upon *P. gilesi* and noted that many planigales were found marooned upon canegrass and trees during the flooding of the Bulloo Overflow in western N.S.W. A similar situation occurred during the flooding of Cooper's Creek in southwestern Queensland (Denny, pers. observation). Kitchener and Butler (cited in Archer 1976) described how *P. ingrami* were found on clumps of spinifex on isolated hills whilst the water of Lake Argyle were rising. Taylor *et al.* (1982) point out that *P. maculata sinualis* are capable of moving to and from flooded areas. Although *P. gilesi* appear to be used to flooded conditions (they can swim relatively strongly in an "otter like" fashion), occupation of trees and grasses in floodwaters would lead to increased environmental stress as well as vulnerability to predation. As several carcases of *P. gilesi* have been found in trees during flooding (Denny, pers. observation) it may be assumed that flooding is an important factor in the life history of some planigales.

PLANIGALE TEMPERATURE REGULATION, OXYGEN AND WATER REGULATION

Temperature regulation studies have shown planigales to have a standard metabolic rate similar to that of other marsupials. It is not as high as that suggested for shrew–like animals (Dawson and Wolfers 1978, Morton and Lee 1978). All species of planigales showed a daily pattern of torpor, although *P. gilesi* enters torpor less readily than other planigale species (Dawson and Wolfers 1978, Morton and Lee 1978). Differences in the ability to enter torpor may differentiate sympatric planigale species, particularly during food shortage.

This has been suggested for *Sminthopsis* species by Tyndale–Biscoe (1973) but there is no evidence from dietary studies to substantiate it.

Planigales have a low body temperature (about 34°C) compared to larger marsupials· but it is similar to other small marsupials. A low body temperature assists in reducing the gradient for heat loss from the body to the environment (Dawson and Wolfers 1978, Morton and Lee 1978). The pattern of behaviour and activity of planigales indicates a need for heat from the sun during winter. Sunbasking has been observed in all four planigale species (see Table 4) and *P. gilesi* have been found to be active in the wild, during the day, but rarely at night, during the winter months (Denny *et al.* 1979).

Animals with a relatively low standard of metabolism are usually found to have a low water turnover rate e.g., macropodids (Denny and Dawson 1975). However, the only planigale studied, *P. tenuirostris,* has a water turnover rate higher than expected. Water usage for this planigale is similar to an eutherian of equivalent size. According to Nagle (pers. comm) the slope of the equation relating water turnover and body weight for dasyurids shows statistical differences from that found for eutherians. Nagle further suggests that it is habitat rather than phylogenetic status that is the primary determinant of water turnover.

Planigales do show a lower water usage than other dasyurid marsupials studied. A comparative study of metabolism of *Sminthopsis crassicaudata* and *Planigale maculata* by Morton (1980), showed that *P. maculata* were better capable of withstanding water deprivation than *S. crassicaudata.* However, planigales do not appear to have the same ability to exist without water as several other small mammal species, nor do they have the high urine concentration found in some rodents. The highest urine concentration measured in *P. maculata* was 2.95 Osm/1 (Morton 1980).

PLANIGALE REPRODUCTION

There have been sufficient observations of the breeding habitats of planigales to allow some generalisations to be formulated. Details of the reproductive cycles are summarized in Table 5. It would seem that all four species of planigales are polyoestrous. This has been studied in detail for *P. maculata sinualis* by Aslin (1975) and Taylor *et al.* (1982) and casual observations indicate that the other three species are probably the same. However, there appears to be a peak in the breeding of *P. ingrami subtilissima* during summer (Woolley 1973). Taylor *et al.* (1982) noted a decrease in breeding of *P. maculata sinualis* during the wet summer months, when there was a disruption in their occupation of flood plain habitat. Lee *et al.* (1982) have classified planigales into two different life history strategies on the basis of differences in their breeding season length.

Numbers of young born vary within species from 4 up to 15, although some of the higher numbers recorded probably contained supernumerary young. The gestation period of *P. maculata sinualis* is 19-20 days (Aslin 1975). Woolley (1974) described the pouch of *P. ingrami subtilissima* as different from that found in other planigales and suggested that this type of pouch provides complete cover for the young for at least some period of their lactation.

CONCLUSION

Many of the patterns of behaviour and habitat utilization exhibited by planigales reflect their exteremely small size. A very small mammal like a planigale has problems with heat and water loss, because of their high surface area to

Table 5. Reproduction of planigales.

	Planigale maculata	*Planigale m. sinualis*	*Planigale ingrami*	*Planigale tenuirostris*	*Planigale gilesi*
Breeding Season	Oct to Jan[1] Mated Sept[2] July[3] Polyoestrous[18]	All year, in in wild[5]	Feb to April, Nth. Terr[6] Dec. to March, Qld[7], Sept to Aug[8] Mainly Summer[9]	May or June[10] Summer[11]	July with pouch young Aug. onwards Oct. with pouch young[13] Summer
Number of young	5–15[14] 5[3], 7[1], 4[18]	5–11[4] 4–12[5]	4–6[6] 4–12[7]	8[15] 4[16]	
Other information	Uses Strat. 5[17] Gestation period 24–28 days[18]	Polyoestrous, Gestation period 19–20 days[4] Uses Strat. 6[17]	Uses Strat. 5[17]	Uses Strat. 5[17]	Uses Strat. 5[17]

REFERENCES: 1. Fleay 1965; 2. Archer 1976; 3. Morrison 1975; 4. Aslin 1975; 5. Taylor *et al.* 1982; 6. De Lestang, in Archer 1976; 7. Heinsohn 1970; 8. McKay 1974; 9. Woolley 1973; 10. Troughton 1928; 11. Denny *et al.* 1979; 12. P. Aitken, pers. comm.; 13. D. Read, pers. comm.; 14. Archer, in press; 15. A. Rose, pers. comm.; 16. M. Denny, pers. comm.; 17. Lee *et al.* 1982; 18. Van Dyck 1979.

mass ratio. Thus, the voracious feeding behaviour of planigales (they are capable of eating their own weight of insects daily) indicates a need for a continual supply of energy whilst active. Use of basking in the sun and torpor when cold or deprived of food are also related to their need to conserve energy.

Occupation of wet habitats well covered with vegetation reflect planigales' need for an environment which will not be dehydrating. Cracks in the soil are an ideal habitat for such animals. Newsome (1969) pointed out that a crack in the soil provided a cool, micro–environment. The relatively high water turnover by planigales indicates a group of species more suited to a moist environment than a true desert dweller. A moist environment or an ecotone of wet and dry environments may also provide a greater amount of food for planigales than other habitats. Their pattern of reproduction appears to be correlated with the variable nature of the environments in which most of the species live.

ACKNOWLEDGEMENTS

I am grateful to Dr. B. Fox and Dr. S. Morton for useful criticisms of a draft of this paper.

REFERENCES

AITKEN, P. F., 1975. *Planigale gilesi;* a new species from the interior of southeastern Australia. *Rec. S. Aust. Mus.* **16:** 1–14.

ANDREW, D. E. AND SETTLE, G. A., 1982. Observations on the behaviour of *Planigale* with particular reference to the narrow–nosed planigale (*Planigale tenuirostris*). Pp.311–24 *in* "Carnivorous marsupials", ed. by M. Archer. Roy. Zool. Soc. New South Wales: Sydney, Australia.

ARCHER, M., 1976. Revision of the marsupial genus, *Planigale. Mem. Qld Mus.* **17:** 341–65.

ASLIN, H., 1975. Reproduction in *Antechinus maculatus. Aust. Wildl. Res.* **2:** 77–80.

CALABY, J. H., 1966. Mammals of the Upper Richmond and Clarence Rivers, N.S.W. *Div. Wild. Res. Tech. Paper* No. 10 C.S.I.R.O. Aust.

COVACEVICH, J. AND ARCHER, M., 1975. The distribution of the Cane Toad in Australia and its effects on indigenous vertebrates. *Mem. Qld Mus.* **17:** 305–10.

DAVIES, S. J. J. F., 1960. A note on two small mammals of the Darwin area. *J. Roy. Soc. W. Aus.* **43:** 63-6.

DAWSON, T. J., AND WOLFERS, J. M., 1978. Metabolism, thermoregulation and torpor in the shrew sized marsupials of the genus *Planigale. Comp. Biochem. Physiol.* **59*A*:** 305–9.

DENNY, M. J. S., 1975. Mammals of Stuart National Park, Tibooburra, N.S.W. *Aust. Zool.* **18.** 179–95.

DENNY, M. J. S., GIBSON, D. AND READ, D., 1979. Some results of field observations on planigales. (Abstract) *Aust. Mamm. Soc. Bull.* **5:** 27.

DENNY, M. J. S., AND DAWSON, T. J., 1975. Comparative metabolism of tritiated water by macropodid marsupials. *Am. J. Physiol.* **228:** 1794.

FLEAY, D., 1965. Australia's "needle–in–a–haystack" marsupial. *Vict. Nat.* **82:** 195-204.

HEINSOHN, G. E., 1970. World's smallest marsupial, the Flat–headed Marsupial Mouse. *Animals* **13:** 220–2.

JOHNSON, D. H., 1964. Mammals of the Arnhem Land Expedition. In "Records of the American–Australian scientific expedition to Arnhem Land, Vol. 4". Melbourne University Press: Melbourne.

LAVERY, H. J. AND GRIMES, R. J., 1974. Mammals and birds of the Ingham District, north Queensland. *Qld J. Ag. Sc.* **31:** 383–90.

LAVERY, H. J., AND JOHNSON, P. M., 1974. Mammals and birds of the Lower Burdekin River District, north Queensland. *Qld. J. Ag. An. Sc.* **31:** 97–104.

LINSTED, S. L. AND JONES, J. H., 1980. Desert shrew. *Nat Hist.* **89:** 46–55.

MARLOW, B. J., 1958. A survey of marsupials of N.S.W. *C.S.I.R.O. Wildl. Res.* **3:** 71–114.

MCKAY, G. M., 1974. *Planigale ingrami* in captivity. *Koolewong* **3:** 4–5.

MORRISON, R. G. B., 1972. Emergence of the Pigmy Antechinus. *Aust. Nat. Hist.* **18:** 164–7.

MORRISON, S. R., 1978. An ecological study of *Sminthopsis crassicaudata.* II. Behaviour and social organisation. *Aust. Wildl. Res.* **5:** 163–82.

MORTON, S. R., 1980. Field and laboratory studies of water metabolism in *Sminthopsis crassicaudata. Aust. J. Zool.* **28:** 213–27.

MORTON, S. R., AND LEE, A. K., 1978. Thermoregulation and metabolism in *Planigale maculata. J. Thermal Biol.* **3:** 117–20.

MORTON, S. R., AND MARTIN, A. A., 1979. Feeding ecology of the Barn Owl, *Tyto alba,* in arid southern Australia. *Aust. Wildl. Res.* **6:** 191–204.

NASH, O., 1943. "The face is familiar". George Jaboor: Melbourne.

NEWSOME, A. E., 1969. A population study of House Mice permanently inhabiting a reed–bed in South Australia. *J. Anim. Ecol.* **83:** 361–77.

PARKER, S. A., 1973. An annotated checklist of the native land mammals of the Northern Territory. *Rec. S. Aust. Mus.* **16:** 1–57.

RUDEFORTH, B. F., 1950. Some notes on an interesting marsupial. *Scope* **1:** 10–1.

TAYLOR, J. M. CALABY, J. H., AND REDHEAD, T. D., 1982. Breeding in wild populations of the marsupial mouse *Planigale maculata sinualis* (Dasyuridae, Marsupialia). Pp.84–7 *in* "Carnivorous marsupials" ed. by M. Archer. Roy. Zool. Soc. New South Wales: Sydney, Australia.

TROUGHTON, E., LEG., 1928. A new genus, species and subspecies of marsupial mice. *Rec. Aust. Mus.* **16:** 281–8.

TROUGHTON, E., LEG., 1965. "Furred animals of Australia". Angus and Robertson: Sydney.

TYNDALE–BISCOE, H., 1973. "Life of marsupials". Edward Arnold: London.

VAN DEUSEN, H. M., 1969. Feeding habits of planigale. *J. Mamm.* **50:** 616–8.

Van Dyck, S., 1979. Behaviour of the dasyurid marsupial, *Planigale maculata*. *Mem. Qld Mus.* **19:** 413–29.

Watts, C. H. S., 1972. "Handbook of South Australian rodents and small marsupials". Field Naturalists Society of South Aust.: Adelaide.

Woolley, P., 1973. Breeding patterns, and the breeding and laboratory maintenance of dasyurid marsupials. *Exp. Animals* **22:** 161–72.

Woolley, P., 1974. The pouch of *Planigale subtilissima* and other dasyurid marsupials. *J. Roy. Soc. W. A.* **57:** 11–5.

RESUMEN

13. Revisión de la ecología de *Planigale* (Dasyuridae, Marsupialia). Por M. J. S. Denny.

Cuatro especies de planigale son conocidas de Australia, las que están restringidas a las partes septentrionales del continente. Los planigales son los pequénos marsupiales del mun do, pesando tan poco como tres gramos. Muchos aspectos de su ecología, dispersamente conocida, reflejan ese pequéno tamáno.

La mayor parte de los planigales son encontrados en áaras asociadas con el agua, sean éstas permanents como de periódicas inundaciones. *Planigale tenuirostris* es una excepción ya que puede encontrársela en hàbitats más secos. Los suelos cuarteados y la cubierta de vegetación parece ser el hábitat requerido por los planigales. Exhiben períodos de alta actividad seguidos por otros de relativa calma. Son feroces cazadores capaces de mater y comer animals de su propio tamáno. Construyen nidos, que pueden ser comunitarios, y en cautividad muestran un estereotipado tipo de comportamiento entre los individuos. Despliegan un vasto modelo de vocalización y, aunque no parecen establecer territorios, tienen una jerarquía de dominio.

Los planigales tienen ritmos metabólicos y temperatura corporal similares a otros pequénos marsupiales. Tienen la habilidad de entrar en letargo cuando hace frío y/o carecen de alimentos, usando el "asoleado" como medio termo–regulador. El cicio ael agua en *Planigale tenuirostris* es similar al de los euterious de equivalente tamáno.

Todas las especies australianas parecen ser poliestras, con una leve eficiencia estacional. Se dan también todos los datos disponsibles sobre varios aspectos de la dinámica poblaional de los planigales.

14. SOME ECOLOGICAL ASPECTS OF SEASONAL BREEDING IN *Antechinus* (DASYURIDAE, MARSUPIALIA)

C. R. DICKMAN[1]

Geographical variation in the time of mating of *Antechinus stuartii* and *Antechinus swainsonii* was examined in relation to altitude, latitude, longitude and sympatry. Over its entire range, the time of mating of *A. stuartii* varies from late July to late September. Mating begins later at higher altitudes and also at lower latitudes. The findings accord neither with a simple (isophane) model, nor with the expectation that mating should be progressively later from north to south in correspondence with the appearance of invertebrate food. The time of mating of *A. swainsonii* varies from early May to late September, in accordance with the isophane model, and again is later at higher altitudes.

Interspecific differences in time of mating may reflect differential responses to seasonally fluctuating environmental factors, and thus, niche segregation. It is suggested that niche segregation is further facilitated by a shift in the time of mating of *A. stuartii* where it is sympatric with *A. swainsonii*, and by the maintenance of a constant interspecific difference in body size.

Pages 139–50 *in* CARNIVOROUS MARSUPIALS, ed. by M. Archer, February, 1982. Roy. Zool. Soc. New South Wales: Sydney, Australia.

THE unusual life-history of *Antechinus stuartii* (Marsupialia: Dasyuridae) has stimulated a considerable amount of research (Lee *et al.* 1977, Tyndale-Biscoe 1979). Mating is restricted to a short period (about 2 weeks) in winter, and young are weaned four months later in summer. Males die following mating at the age of 11½ months, whereas females may live for up to three years. Mating in any locality is usually synchronous, varying little from year to year (Wood 1970, Braithwaite 1973). Life-history events are similar in the larger *Antechinus swainsonii,* except that they generally occur a month or more earlier. Neither the ultimate nor the proximate factors controlling reproduction are clearly known, though Lee *et al.* (1977) have suggested that mating is so timed that lactation and the weaning of the young coincide with a flush in insect abundance in spring and summer. Wainer (1976) has further suggested that reproductive phase differences in a two-species system (*A. minimus* and *A. stuartii* in Victoria) arise fom dietary preferences for different insect groups that peak at different times, but quantitative information is lacking.

This paper attempts to correlate the times of mating of *A. stuartii* and *A. swainsonii* with altitude, latitude and longitude, and to examine a possible shift in timing arising from interspecific interaction. The roles of seasonal breeding and body size in facilitating coexistence are considered.

MATERIALS AND METHODS

Information on breeding in natural populations was obtained from five sources: published material, unpublished theses, museum specimens, personal observations and field observations kindly provided by many researchers in Australia and overseas (Appendix 1). Published and unpublished material was surveyed for records of the times of i) mating, ii) male die-off, iii) birth of young, iv) vacation of the pouch, v) first appearance of juveniles in the trappable population, and for additional information on the appearance of the external sex organs, and on the size and appearance of pouch young. These records were compared with published information on the appearance of reproductive and post-reproductive *A. stuartii* (Woolley 1966a, Marlow 1961, Moore 1974) and with observations on the growth and development of pouch young (Marlow 1961, unpublished observations). Museum spirit specimens with young in the pouch were examined, and measurements of the crown-rump lengths of the young were made with calipers in the manner outlined by Lyne and Verhagen (1957). Gestation periods of 31 days for *A. stuartii* (Woolley 1966a) and 28 days for *A. swainsonii* (unpublished observations) were used to estimate probable mating times in all cases.

Field observations of breeding were made at 11 localities near Canberra, Australian Capital Territory, during the winters of 1978 and 1979. Trapping was conducted at 10 localities for 1-2 day periods, at intervals of 7-12 days. One locality in the Brindabella ranges was trapped twice a week to obtain detailed reproductive information.

Breeding results obtained from the five sources above were not directly comparable: those from field studies usually reported a mating period of 7 to 14 days, whereas museum specimens with

[1] Dept. of Zoology, Australian National University, Box 4, GPO, Canberra, ACT, 2600.

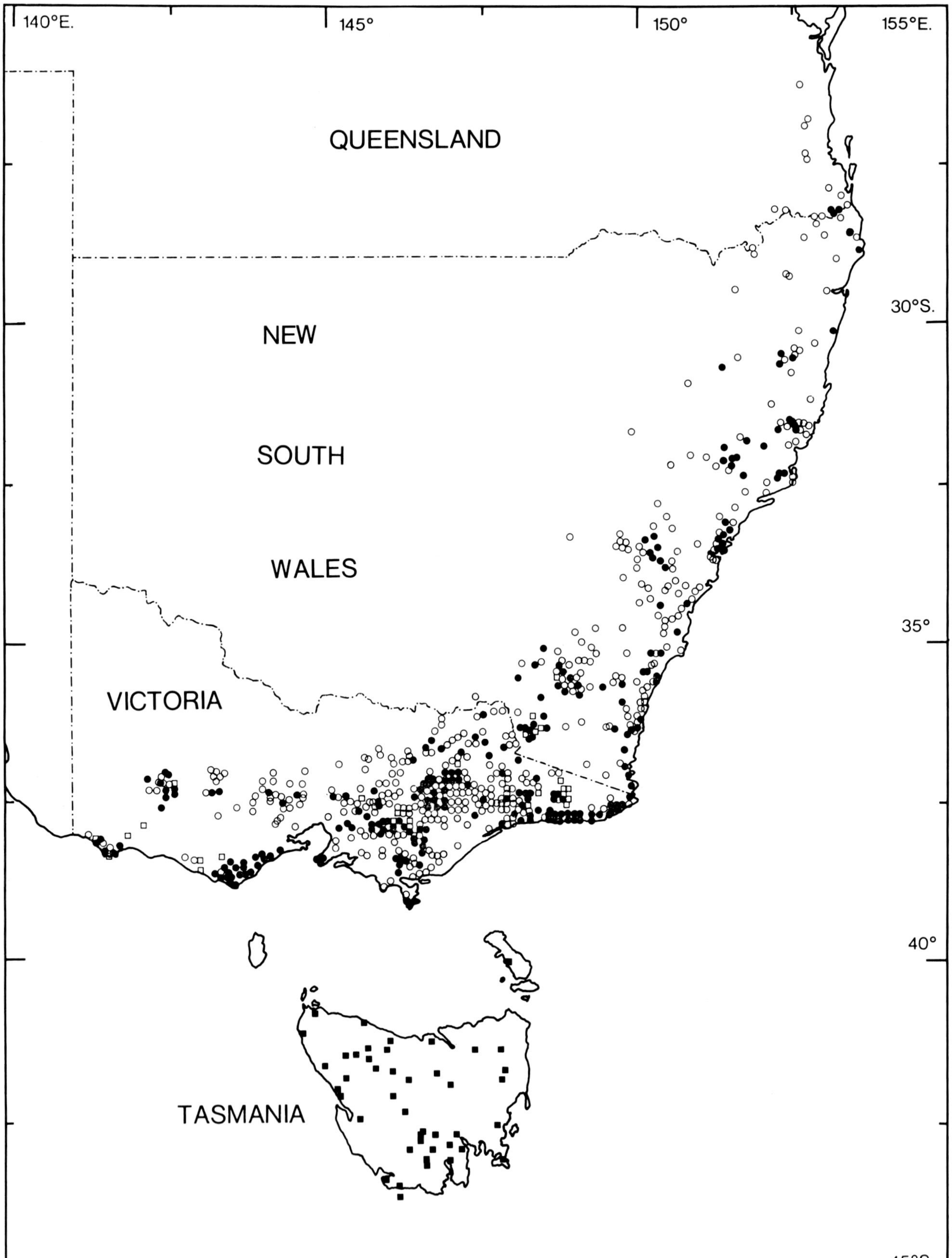

Fig. 1. Distributions of *Antechinus stuartii* and *A. swainsonii.* Open circles represent allopatric *A. stuartii stuartii,* open squares allopatric *A. swainsonii mimetes.* Closed circles represent both forms together, closed squares allopatric *A. swainsonii swainsonii* (Tasmania). *Antechinus stuartii adustus* is not shown. (Distribution records are not listed, but may be obtained on request from the author).

pouch young were often estimated to have mated over a more restricted period. To combine these findings, months were split into quarters. For numerical analysis, mating in a field population in the last two weeks of August was then assigned a value of 8.75, on the assumption that mating was centred about the third quarter of the eighth month. A population mating between the end of June and early July (the sixth and seventh months) was assigned a value of 7. The procedure therefore gives a mean value for time of mating, and ignores the variance. Estimates of time of mating from museum specimens with pouch young were similarly placed in the nearest quarter-month category.

Altitude, latitude and longitude were determined from maps of source localities. The sympatric or allopatric status of populations was sometimes more difficult to resolve, but close examination of habitat type in the source locality and the construction of a detailed distribution map (Fig. 1) ensured fairly certain placement.

Times of matings were initially regressed on isophanes of the source localities. Isophanes are calculated from latitude, longitude and altitude of the localities and therefore reflect gross climatic features (Hopkins 1938, Johnston and Selander 1964). Least squares regressions were plotted for latitude, longitude and altitude, and the R^2 statistic (Zar 1974) used to indicate the proportion of the variation in time of mating accounted for by each. The results of multiple

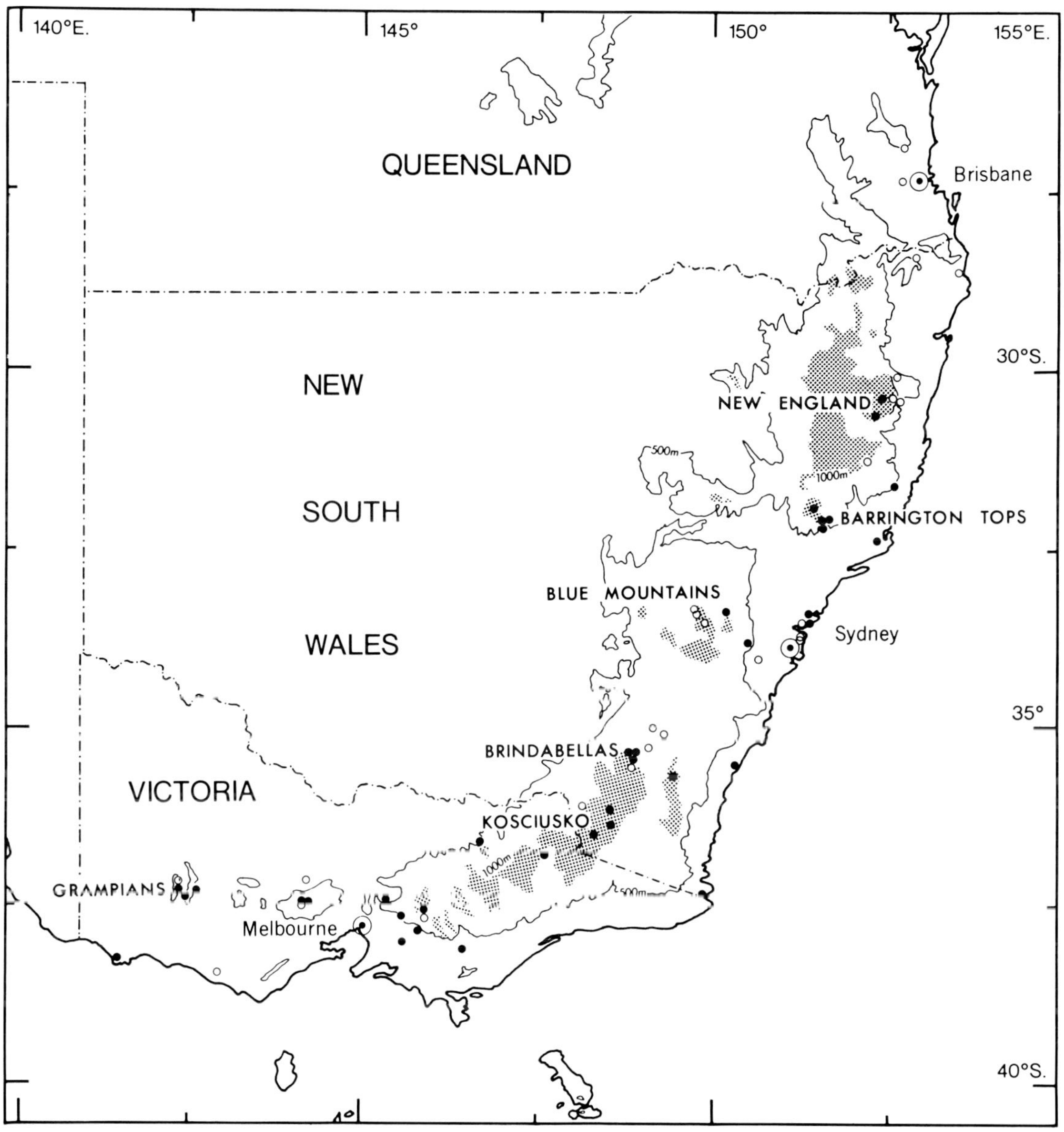

Fig. 2. Localities in which the time of mating of *Antechinus stuartii* is known. Thin contours show land above 500m; shading, land above 1000m. Other symbols are as in Fig. 1. Localities less than 3km. apart are represented by one symbol.

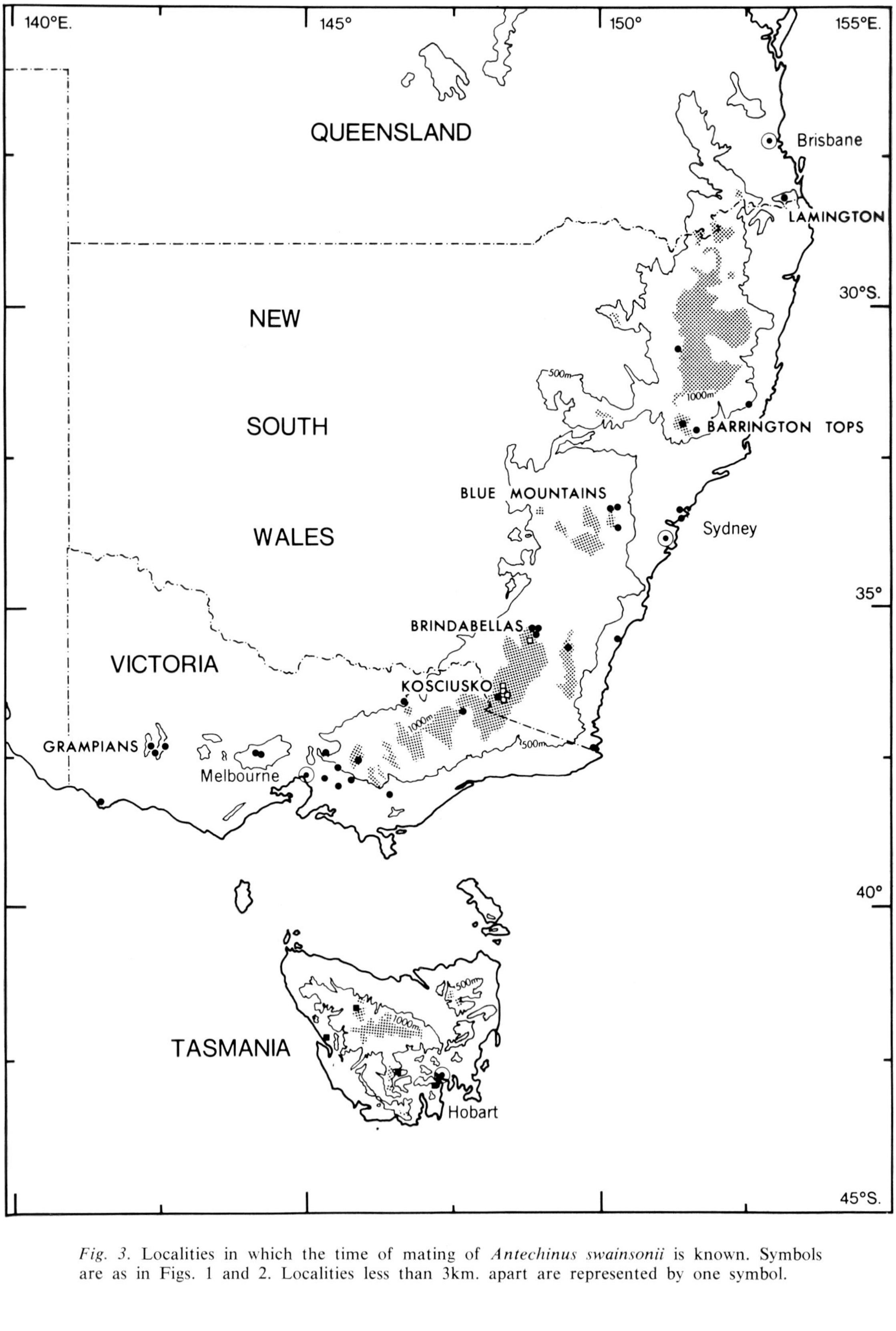

Fig. 3. Localities in which the time of mating of *Antechinus swainsonii* is known. Symbols are as in Figs. 1 and 2. Localities less than 3km. apart are represented by one symbol.

regressions are presented in a similar manner. Beta values (standardised regression coefficients, or measures of slope) are also given.

RESULTS

The distributions of *A. stuartii* and *A. swainsonii* are shown in Fig. 1. The localities in which the time of mating is known are shown in Figs. 2 and 3. Information was obtained on *A. stuartii* in 65 localities and on *A. swainsonii* in 43 localities. In 28 of these there was information on both species. The total distribution of time of mating records is shown in Fig. 4. *Antechinus stuartii* usually mates later than *A. swainsonii*, with a peak around late August. The distribution of mating times of *A. swainsonii* deviates from normality (D'Agostino's test, in Zar 1974: $D = 0.2647$, $P<0.05$). The distribution is skewed to the left, but also shows a tendency towards bimodality. This is probably the result of unequal sampling intensity in all areas. Few records were obtained on breeding either among mainland allopatric populations of *A. swainsonii mimetes*, or among populations of the Tasmanian subspecies *A. swainsonii swainsonii*. These records are therefore combined in all the analyses below.

Time of mating is plotted against isophane in Fig. 5. For *A. swainsonii* 46.5% of the variation is attributable to the linear regression effect ($P<0.001$). For *A. stuartii* less than 1% of the variation is explained ($P>0.1$).

Tables 1 and 2 summarise the simple linear relationships between the component factors that influence time of mating. For *A. swainsonii*, altitude accounts for over 45% of the total variation. There is no evidence that other factors have significant effects. For *A. stuartii*, altitude, latitude and longitude are all correlated with

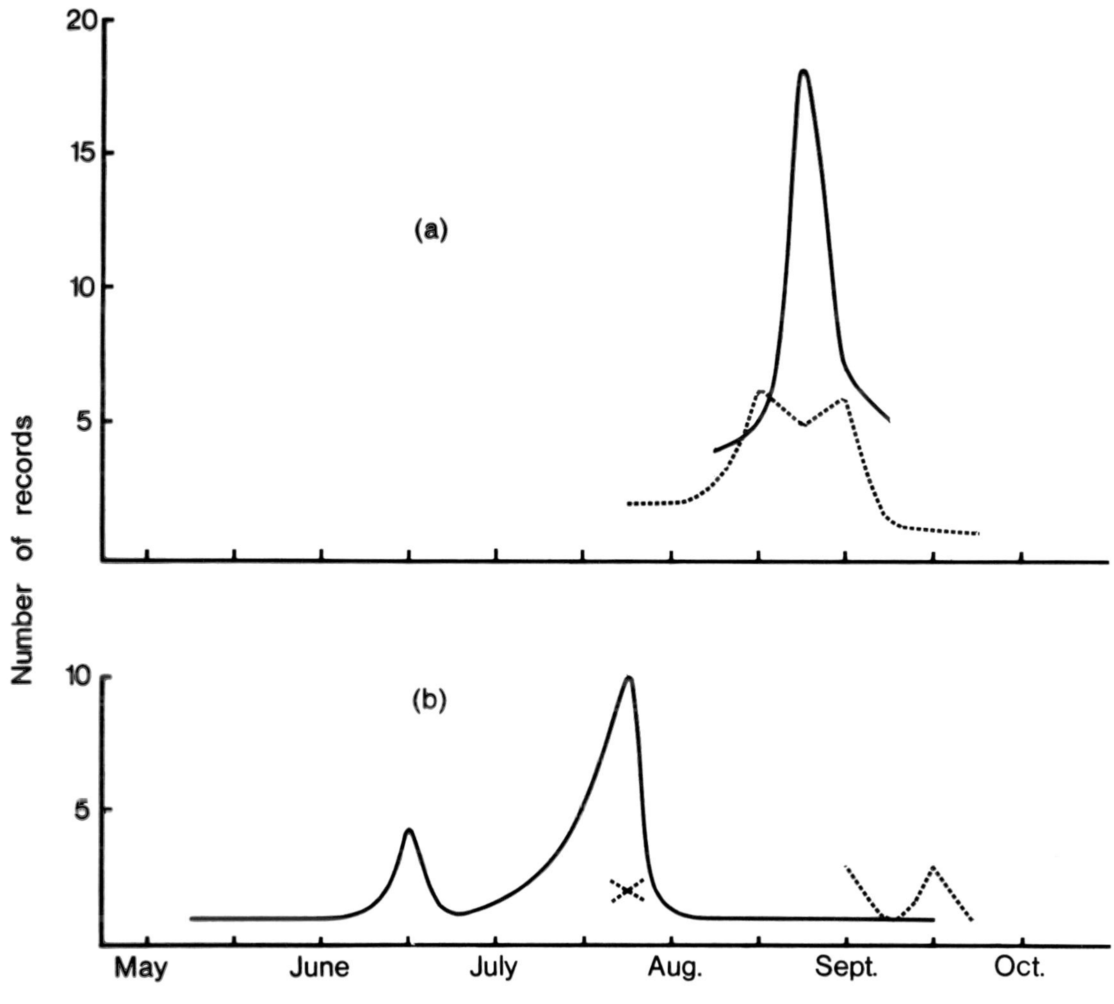

Fig. 4. Distribution of time of mating records of *Antechinus*. (a) *A. stuartii*: dotted line, where allopatric; solid line, where sympatric with *A. swainsonii*. (b) *A. swainsonii*: dotted line, where allopatric; solid line, where sympatric with *A. stuartii*.

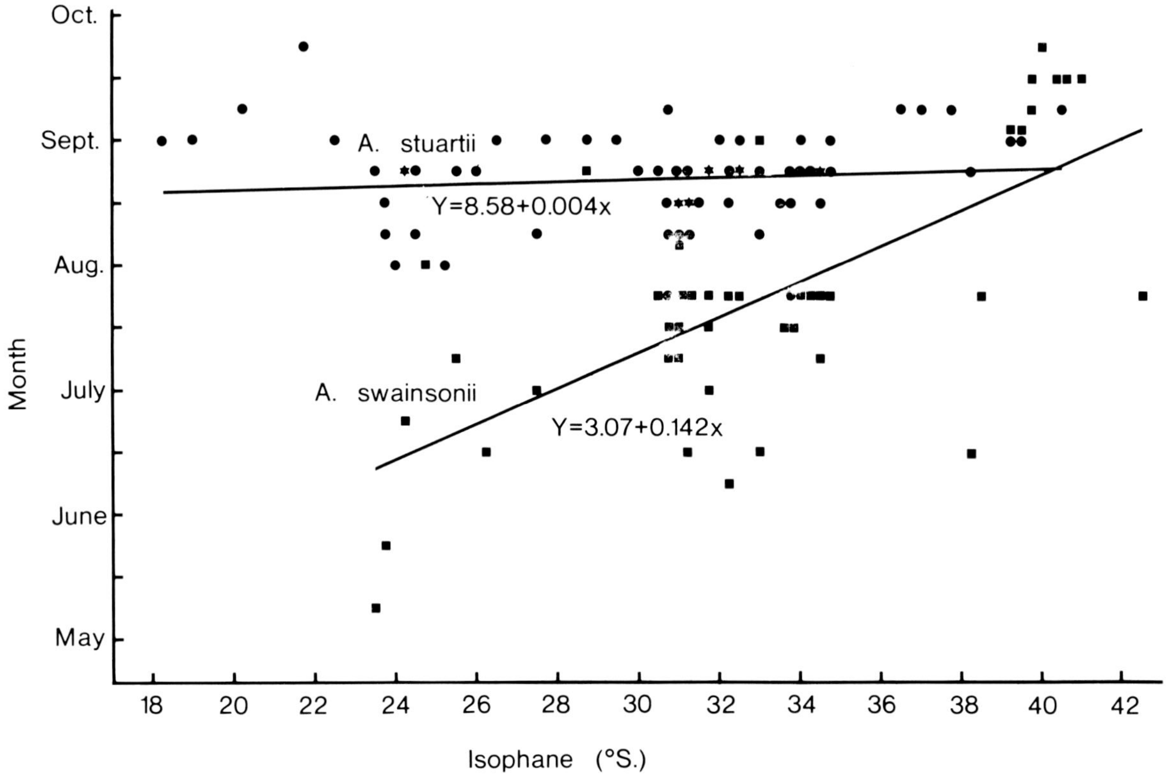

Fig. 5. Regressions of time of mating on isophane. Circles, *Antechinus stuartii;* stars, overlapping *A. stuartii* points; squares, *A. swainsonii.*

time of mating, but there is a difference between allopatric populations and those sympatric with *A. swainsonii.* But use of the *t* distribution to compare regression slopes Zar (1974) yielded a difference only between the longitude regressions ($t = 2.87$, $P<0.01$).

Examination of the raw data revealed some correlations among variables. For example, latitude and longitude are correlated in the 65 localities where the time of mating of *A. stuartii* is known ($r = -0.86$, $P<0.001$). To remove the effects of spurious correlations, multiple regressions were performed (Tables 3 and 4). The total variation in time of mating accounted for was 61% for *A. swainsonii* and 49% for *A. stuartii.* Altitude was again the only factor

Table 1. Linear relationships between factors that affect the time of mating of *Antechinus stuartii.*

		R^2 (%)		
	N	Altitude	Latitude	Longitude
Sympatric	39	44.9***	21.6**	23.4**
Allopatric	26	8.5	29.9**	22.3*
Overall	65	21.5***	16.1***	16.3***

Statistical significance of associated correlation coefficients *(r)*: * $P<0.05$; ** $P<0.005$; *** $P<0.001$.

Table 2. Linear relationships between factors that affect the time of mating of *Antechinus swainsonii.*

		R^2(%)		
	N	Altitude	Latitude	Longitude
Sympatric	33	34.9***	1.1	1.1
Allopatric	10	14.9	34	5.2
Overall	43	45.3***	3.6	0.4

Statistical significance of associated correlation coefficients *(r)*: *** $P<0.001$.

related to the time of mating of *A. swainsonii*, though sympatry (15%) was almost significant at the 5% level. Altitude, latitude and sympatry were related to the time of mating of *A. stuartii*.

The effect of sympatry on breeding was further investigated by comparing the time of mating of one species with that of the other at 28 localities (Fig. 6). There was very little overlap: *A. stuartii* usually mates 4 or more weeks after *A. swainsonii*, though at one locality (Thredbo Valley, Kosciusko) it apparently mates slightly earlier.

Table 3. Multiple regression results for *A. stuartii*.

Variables	R^2(%)	Beta	F	P
Altitude	21.2	0.4331	21.805	<0.001
Latitude	16.0	–0.4777	6.476	<0.05
Longitude	0.1	0.0482	0.071	n.s.
Sympatry	11.5	–0.3552	12.703	<0.002
Overall	48.8		14.267	<0.001

Table 4. Multiple regression results for *A. swainsonii*.

Variables	R^2(%)	Beta	F	P
Altitude	45.3	0.5253	14.865	<0.001
Latitude	0.6	0.2211	1.172	n.s.
Longitude	0.6	0.1293	0.609	n.s.
Sympatry	14.9	0.3147	3.546	n.s.
Overall	61.4		15.122	<0.001

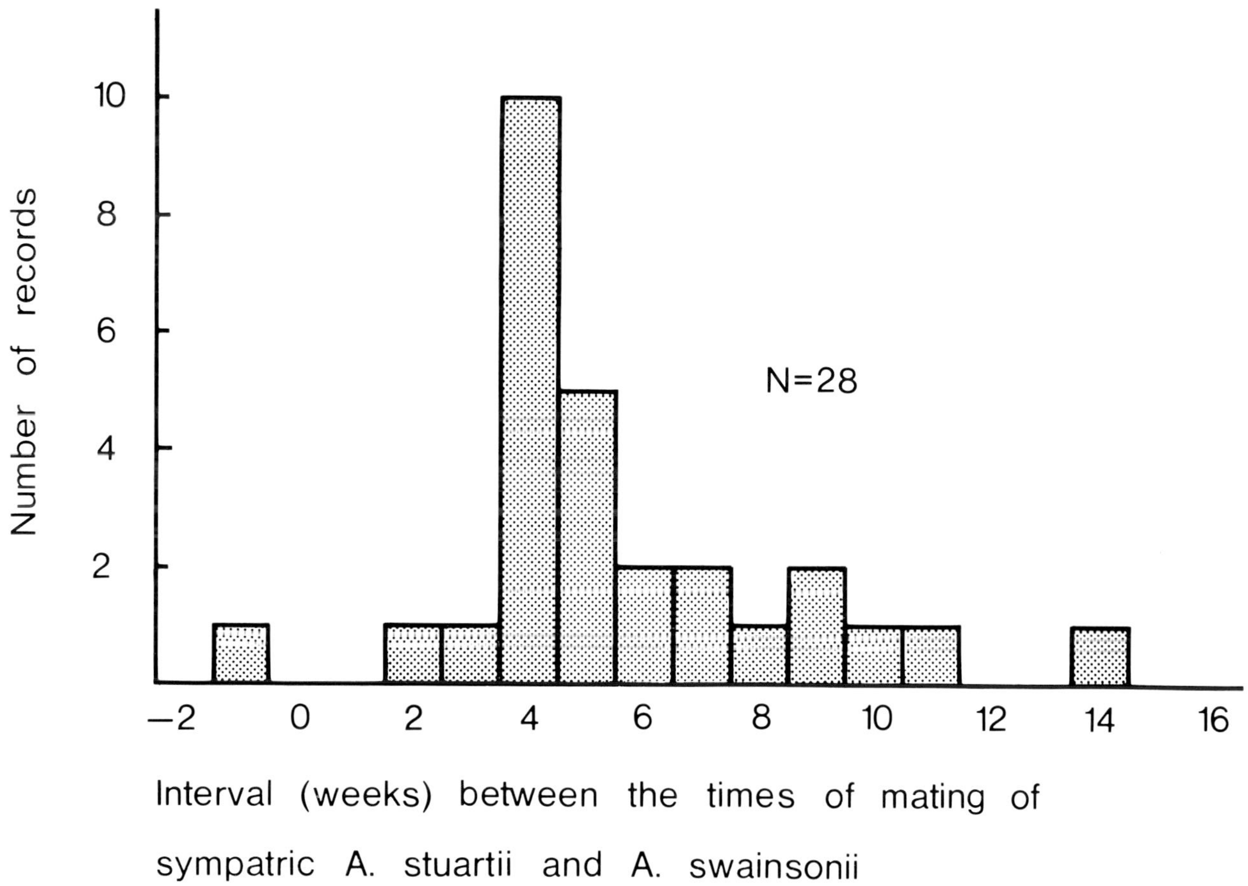

Fig. 6. The time of mating of *Antechinus stuartii* relative to *A. swainsonii*. Positive intervals refer to localities where *A. stuartii* mates after *A. swainsonii*, negative intervals to the converse situation.

DISCUSSION

Recent theoretical work has demonstrated the complex interplay of factors that influence the nature and evolution of life-history strategies (Gadgil and Bossert 1970, May 1977, Knowlton 1979). In this respect the relatively simple and predictable population cycle of some species of *Antechinus* (a highly synchronised breeding season and subsequent sudden death of all males) is of unique interest. The aetiology of die–off of male *A. stuartii* has received much attention (Moore 1974, Lee *et al.* 1977, Bradley *et al.* 1980) and its significance in the life–history has been further explored by Braithwaite and Lee (1979). The analysis above complements previous studies by defining the variation in the mating patterns of *A. stuartii* and *A. swainsonii*. The effects and possible significance of some factors contributing to this variation are discussed below.

ALTITUDE: Increasing altitude delays the onset of reproductive events in both species, but the effect on time of mating of *A. swainsonii* is double that on *A. stuartii*. At high altitudes (>1500m) in Kosciusko National Park, mating in both species begins in early or mid September, often when snow covers the ground. Synchrony of breeding is maintained from year to year under these conditions (P. L. Carron, personal communication), though the role of proximate cues, such as photoperiod, is obscure. At coastal sites, at only slightly lower latitudes, *A. swainsonii* may breed up to 4 months earlier, and *A. stuartii* up to 5 weeks earlier, than in Kosciusko National Park. The magnitude of this variation, and the interspecific difference, are both surprising. The time of the peak in insect abundance certainly varies with altitude, and the time of mating may vary in the same direction. But, unless particular prey species emerge at widely different times according to altitude, or dietary overlap is greater at higher than at lower altitude, food may be of variable importance to the two species as a factor ultimately controlling reproduction. Factors contingent upon altitude, such as habitat, precipitation, temperature and oxygen tension (cf. Dunmire 1960), may also affect time of mating directly, or indirectly by acting on food resources. Moore (1974) has postulated such a role for rainfall in stimulating mating of *A. stuartii* in the Brindabella ranges, near Canberra. Quantitative information, either on arthropod abundance patterns or on their relationship to factors associated with altitude, is, however, lacking.

LATITUDE: The effect of latitude on time of mating of *A. swainsonii* is very small, whereas *A. stuartii* generally breeds later as latitude decreases. The latter finding is unexpected, as spring phenological events (and presumably arthropod peaks) occur earlier at lower latitudes than at higher ones. (This trend forms the basis of robust predictive phenological models, such as those of Hopkins (1938) and of Lieth and Radford (1971). The result accounts for the lack of significance of the isophane regression for *A. stuartii* (Fig. 5) because the effects of altitude and latitude act in opposite directions.

Most literature on latitudinal variation in reproductive activity concerns birds (Cody 1966, Immelmann 1971, Morton 1976). Lack (1948) states that the clutch size of some birds is larger in northern than in southern populations (in the northern hemisphere), and suggests that longer day lengths at higher latitudes allow parents more time to gather food and nourish young. Differences in day length would, however, have little effect on the ability of *A. swainsonii* to obtain food, since activity is continuous through 24 h (Hall 1980, unpublished observations). In contrast, *A. stuartii* is nocturnal, and earlier breeding at higher latitudes could result in the weaning of young in conditions of slightly shorter daylength. Unfortunately, such an effect does not adequately account for the 2-month difference in time of mating between south-eastern Queensland (late September) and western Victoria (late July). Van Dyck (1982) suggests instead that the arrival of food peaks depends on increases in temperature at higher latitudes, on increases in rainfall at lower ones, and on the resultants of both in areas in between Other attempts to account for latitudinal variation in reproductive events, such as "climatic stability" at lower latitudes (Cody 1971), differential mortality (Lord 1960), and avoidance of predation and interspecific competition (Cody 1966), appear to be unsatisfactory.

LONGITUDE: Clutch size of some birds varies along east-west clines (Lack 1954) and may be associated with decreased climatic stability away from the coast. The distributions of *A. stuartii* and *A. swainsonii* are both coastal (Fig. 1), and longitude accounts for an expected small amount of the variation in time of mating.

SYMPATRY: Multiple regression analysis indicates that *A. stuartii* mates later in the presence than in the absence of *A. swainsonii,* and that the latter mates earlier (though $P>0.05$) in the presence than in the absence of *A. stuartii*. These results provide *prima facie* evidence for character displacement. MacArthur (1971) suggests that seasons provide birds with an opportunity to stagger their nesting times, and so to avoid interspecific competition. A rigorous analysis by Ricklefs (1966) failed to support this notion, but showed instead that specific feeding differences and temporal diversity in the availability of different food resources could account for reproductive staggering. Clearly, as Grant (1972) points out, evidence ostensibly supporting character displacement must be carefully scrutinized.

In the present study, allopatric populations of *A. stuartii* were generally found in dry habitats that provide little ground cover, or in Queensland beyond the distribution range of *A. swainsonii* (Wood 1970, S. Van Dyck, personal communication). Allopatric populations of *A. swainsonii* were found only at high altitude in treeless alpine heath, and in Tasmania. Since allopatry therefore appears to be associated with habitat type, it is also possible that habitat-specific factors influence time of mating. The relative importance of habitat, interspecific interactions, and the form of niche shift in sympatry are being investigated (Dickman 1980).

SEASONAL BREEDING AND BODY SIZE: *Antechinus stuartii* mates later than *A. swainsonii* in 27 out of 28 localities where they are known to be sympatric (Fig. 6). Similar differences may ensure reproductive isolation among sympatric populations of *A. stuartii* and the slightly larger congener, *A. flavipes* (Woolley 1966a,b). Attempts to mate male *A. stuartii* with female *A. swainsonii* (and vice versa) in laboratory conditions have, however, consistently failed (unpublished observations). These failures contrast with successful intraspecific pairings among the same individuals both before and after interspecific encounters. I therefore suggest that reproductive isolation between *A. stuartii* and *A. swainsonii* does not depend on differences in reproductive timing.

Variation in the time of mating of one species relative to the other may, alternatively, lead to differences in size. In species that breed once annually, and live, on average, for little more than one year, the species that breeds first gains an advantage in growth over the other species. When the earlier breeding species is also larger (as is *A. swainsonii*), the early growth advantage will tend to enhance and maximise interspecific differences in size. Table 5 presents head-body measurements of adult *A. stuartii* and *A. swainsonii* from sympatric populations, and compares them with the times of matings of one species relative to the other. A surprisingly constant ratio of 1.3–1.35 in the head–body length of the larger to the smaller species emerges. A similar ratio, discovered initially with respect to the lengths of the trophic apparatus of certain sympatric mammals and birds, was considered to represent the minimum size ratio possible for coexistence (Hutchinson 1959).

Table 5. Head–body measurements of sympatric *A. stuartii* and *A. swainsonii* in relation to time of mating.

Locality and reference source	Mating interval (weeks)		Mean head–body length (mm) *A. stuartii*	*A. swainsonii*	Size ratio
Kosciusko region[1,2] (New South Wales)	–1	♂	92.9 (n=9)	125.1 (n=5)	1.35
		♀	84.7 (n=7)	114.6 (n=9)	1.35
Powelltown[3] (Victoria)	3	♂	105 (n=5)	140 (n=6)	1.33
		♀	95 (n=10)	124 (n=6)	1.3
Tidbinbilla[1,4] (Australian Capital Territory)	4	♂	98 (n=4)	129.8 (n=3)	1.32
		♀	88.5 (n=5)	114.9 (n=2)	1.3
Lees Creek[1] (Australian Capital Territory)	4	♂	97.2 (n=11)	128.8 (n=8)	1.33
		♀	89.3 (n=11)	116.2 (n=5)	1.3
Loch Valley[5] (Victoria)	4	♂	95.3 (n=58)	127.9 (n=21)	1.34
		♀	84.9 (n=13)	114.1 (n=37)	1.34
Guy Fawkes region[6,2] (New South Wales)	4	♂	112.1 (n=17)	145.3 (n=4)	1.3
		♀	99.4 (n=5)	130.7 (n=18)	1.31
Kioloa[1,7] (New South Wales)	6	♂	101.6 (n=6)	136.3 (n=2)	1.34
		♀	91.5 (n=6)	121.3 (n=5)	1.33
Nadgee[8] (New South Wales)	5	♀	89 (n=7)	118 (n=2)	1.33
Mean					1.32

Sources: [1] C. R. Dickman (unpublished data). [2] C.S.I.R.O. Museum, Division of Wildlife Research. [3] Leonard (1972). [4] P. Filmer (personal communication). [5] Wakefield & Warneke (1963, 1967). [6] British Museum, American Museum of Natural History. [7] C. R. Tidemann and D. P. Woodside (personal communications). [8] Australian Museum, Recher *et al.* (1975).

As a corollary, we may expect differences in body size to be maximal among sympatric congeneric populations that breed at nearly the same time, and to be minimal among those that breed some weeks apart. The constant size difference of 1.3–1.35, achieved despite differences of up to 7 weeks in the time of mating of the two species of *Antechinus* (Table 5), suggests that this may be so. Table 5 compares measurements of size obtained from populations at a single sampling time. Because of differences in the mating interval, these measurements are of individuals of different ages. If, however, the mating interval is adjusted to 0 by measuring the later breeding species (*A. stuartii*) at the appropriate interval after the earlier breeding species (*A. swainsonii*), it is possible to compare measurements of individuals of the *same* age. At two localities for which data were available (Lees Creek, Australian Capital Territory, and Kioloa, New South Wales), the size ratio of the larger to the smaller species was considerably lower (Table 6).

Table 6. Head–body measurements of sympatric *A. stuartii* and *A. swainsonii* of the same age.

Locality and reference source	Mating interval (weeks)		Mean head–body length (mm) *A. stuartii*	*A. swainsonii*	Size ratio
Lees Creek[1] (Australian Capital Territory)	0	♂	100.7 (n=11)	128.8 (n=8)	1.28
		♀	91.4 (n=11)	116.2 (n=5)	1.27
Kioloa [1,7] (New South Wales)	0	♂	104.6 (n=6)	136.3 (n=2)	1.3
		♀	94.1 (n=4)	121.3 (n=5)	1.29

Sources: See Table 5.

Body size may be an important niche difference in sympatry (Wilson 1975), perhaps permitting differential access to particular resources. Similar hypotheses have been advanced to account for patterns of coexistence among arthropods of which the larger of two species breeds first (Hutchinson 1959, Enders 1976, Uetz 1977). It seems reasonable to suggest that differences in the times of mating of *Antechinus* help to facilitate niche segregation and coexistence both directly and also by maximising differences in body size. The observed shift in the time of mating of *A. stuartii* in the presence of *A. swainsonii* is consistent with this interpretation.

Breeding patterns of *Antechinus* populations may vary considerably with local conditions, and obscure attempts to investigate proximate and ultimate control of reproduction. Future work will therefore need to simultaneously examine the effects of habitat, population density, and an array of physical, climatic and topographical variables on reproduction and growth.

ACKNOWLEDGEMENTS

A large number of people very kindly provided unpublished observations on breeding of *Antechinus*. These are — Mr. A. Bennett, Mr. G. Blair, Mr. R. Cavanah, Ms. P. L. Carron, Drs. A. Cockburn, G. Friend, Mr. P. Filmer, Drs. D. Goldney, R. H. Green, Mr. K. Green, Mr. A. Heislers, Professor B. E. Horner, Mr. C. Moritz, Mr. K. Norris, Mr. W. Osborne, Mr. W. Rohan-Jones, Mr. A. B. Rose, Mr. C. R. Tidemann, Mr. S. Van Dyck, Mr. J. A. Wapstra, Dr. P. Woolley, and Ms. D. P. Woodside. D. Goldney, K. Norris, A. B. Rose and S. Van Dyck were particularly generous in providing unpublished distribution records in addition. I am grateful to all. Dr. I. R. Bishop (British Museum), Dr. J. H. Calaby (CSIRO), Ms. J. M. Dixon (National Museum of Victoria), and Dr. B. J. Marlow (Australian Museum) provided distribution records or allowed me to examine specimens in their care. Others, too numerous to mention individually, generously provided distribution records of *Antechinus*.

The multiple regression analysis (SPSS REGRESSION) was kindly performed by Mr. C. R. Tidemann. I thank Dr. D. C. D. Happold and Professor S. A. Barnett for critically reviewing the manuscript, and Ms. C. McKechnie for her support at all times.

REFERENCES

Bradley, A. J., McDonald, I. R., and Lee, A. K., 1980. Stress and mortality in a small marsupial (*Antechinus stuartii*, Macleay). *Gen. and Comp. Endocrin.* **40:** 188–200.

Braithwaite, R. W., 1973. "An ecological study of *Antechinus stuartii* (Marsupialia: Dasyuridae)". M.Sc. thesis: University of Queensland, Brisbane. 143pp.

Braithwaite, R. W. and Lee, A. K., 1979. A mammalian example of semelparity. *Am. Nat.* **113:** 151–5.

Cody, M. L., 1966. A general theory of clutch size. *Evolution* **20:** 174–84.

Cody, M. L., 1971. Ecological aspects of reproduction. Pp.461–512 *in* "Avian biology", Vol. 1, ed. by D. S. Farner and J. R. King. Academic Press: New York.

Dickman, C. R., 1980. Niche shift and interaction among *Antechinus* (Dasyuridae). *Bull. Aust. Mammal. Soc.* **6:** 32.

Dunmire, W. W., 1960. An altitudinal survey of reproduction in *Peromyscus maniculatus*. *Ecology* **41:** 174–82.

Enders, F., 1976. Size, food finding and Dyar's constant. *Environ. Entomol.* **5:** 1–10.

Gadgil, M. and Bossert, W. H., 1970. Life historical consequences of natural selection. *Am. Nat.* **104:** 1–24.

Grant, P. R., 1972. Convergent and divergent character displacement. *Biol. J. Linn. Soc.* **4:** 39–68.

Hall, S., 1980. Diel activity of three small mammals coexisting in forest in southern Victoria. *Aust. Mammal.* **3:** 67–79.

Hopkins, A. D., 1938. "Bioclimatics: a science of life and climate relations". *U.S. Dept. Agric. Misc. Publ.* **280**. Government Printing Office: Washington. 188pp.

Hutchinson, G. E., 1959. Homage to Santa Rosalia *or* why are there so many kinds of animals? *Am. Nat.* **93:** 145–59.

Immelmann, K., 1971. Ecological aspects of periodic reproduction. Pp.341–389 *in* "Avian biology", Vol. 1. ed. by D. S. Farner and J. R. King. Academic Press: New York.

Johnston, R. F. and Selander, R. K., 1964. House sparrows: rapid evolution of races in North America. *Science, N.Y.* **144:** 548–50.

Knowlton, N., 1979. Reproductive synchrony, parental investment, and the evolutionary dynamics of sexual selection. *Anim. Behav.* **27:** 1022–33.

Lack, D., 1948. The significance of clutch size. *Ibis* **89:** 302–52; **90:** 25–45.

Lack, D., 1954. "The natural regulation of animal numbers". Oxford University Press (Clarendon): London and New York.

Lee, A. K., Bradley, A. J. and Braithwaite, R. W., 1977. Corticosteroid levels and male mortality in *Antechinus stuartii*. Pp.209–220 *in* "Biology of marsupials" ed. by B. Stonehouse and D. Gilmore. MacMillen Press: London and Melbourne.

*Leonard, B. V., 1972. "The effect of fire upon selected small mammals and leaf litter fauna in sclerophyll forest in southern Australia". M.Sc. thesis: Monash University, Melbourne. 142pp.

Lieth, H. and Radford, J. S., 1971. Phenology, resource management, and synagraphic computer mapping. *Bioscience* **21:** 62–70.

Lord, R. D. Jr., 1960. Litter size and latitude in North American mammals. *Am. Midl. Nat.* **64:** 488–99.

Lyne, A. G. and Verhagen, A. M. W., 1957. Growth of the marsupial *Trichosurus vulpecula* and a comparison with some higher mammals. *Growth* **21:** 167–95.

MacArthur, R. H., 1971 Patterns of terrestrial bird communities. Pp.189–221, *in* "Avian biology", Vol. 1 ed. by D. S. Farner and J. R. King. Academic Press: New York.

*Marlow, B. J., 1961. Reproductive behaviour of the marsupial mouse, *Antechinus flavipes* (Waterhouse) (Marsupialia) and the development of the pouch young. *Aust. J. Zool.* **9:** 203–18.

May, R. M., 1977. Optimal life-history strategies. *Nature* **267:** 394–5.

*Moore, G. H., 1974. "Aetiology of the die-off of male *Antechinus stuartii*". Ph.D. thesis: Australian National University, Canberra. 124pp.

Morton, M. L., 1976. Adaptive strategies of *Zonotrichia* breeding at high latitude or high altitude. Pp.322–336 *in* "Proc. 16th International Ornithol. Congr. Canberra, 1974" ed. by H. J. Frith and J. H. Calaby. Australian Academy of Science: Canberra.

*Recher, H. F., Lunney, D., and Posamentier, H., 1975. A grand natural experiment: the Nadgee wildfire. *Aust. Nat. Hist.* **18:** 154–63.

Ricklefs, R. E., 1966. The temporal component of diversity among species of birds. *Evolution* **20:** 235–42.

Tyndale-Biscoe, C. H., 1979. Ecology of small marsupials. Pp.343–379 *in* "Ecology of small mammals" ed. by D. M. Stoddart. Chapman and Hall: London.

Uetz, G. W., 1977. Coexistence in a guild of wandering spiders. *J. Anim. Ecol.* **46:** 531–41.

Van Dyck, S., 1982. The relationships of the dasyurids *Antechinus stuartii* and *A. flavipes* in Queensland. (This volume).

Wainer, J. W., 1976. Studies of an island population of *Antechinus minimus* (Marsupialia, Dasyuridae). *Aust. Zool.* **19:** 1–7.

*Wakefield, N. A., and Warneke, R. M., 1963. Some revision in *Antechinus* (Marsupialia) 1. *Vict. Nat.* **80:** 194–219.

*Wakefield, N. A., and Warneke, R. M., 1967. Some revision in *Antechinus* (Marsupialia) 2. *Vict. Nat.* **84:** 69–99.

Wilson, D. S., 1975. The adequacy of body size as a niche difference. *Am. Nat.* **109:** 769–84.

*Wood, D. H., 1970. An ecological study of *Antechinus stuartii* (Marsupialia) in a southeast Queensland rainforest. *Aust. J. Zool.* **18:** 185–207.

Woolley, P., 1966a. "Reproductive biology of *Antechinus stuartii* Macleay (Marsupialia: Dasyuridae)". Ph.D. thesis: Australian National University, Canberra 136pp.

Woolley, P., 1966b. Reproduction in *Antechinus* spp. and other dasyurid marsupials. *Symp. Zool. Soc. Lond.* **15:** 281–94.

Zar, J. H., 1974. "Biostatistical analysis". Prentice-Hall Inc.: New Jersey.

* Denotes those references also used in Appendix 1.

APPENDIX 1

Sources of breeding data used in the analysis.

Published material —

Barker, I. K., Beveridge, I., Bradley, A. J., and Lee, A. K., 1978. Observations on spontaneous stress-related mortality among males of the Dasyurid marsupial *Antechinus stuartii* Macleay. *Aust. J. Zool.* **26:** 435–47.

Barnett, J. L., How, R. A., and Humphreys, W. F., 1977. Small mammal populations in pine and native forests in northeastern New South Wales. *Aust. Wildl. Res.* **4:** 233–40.

CALABY, J. H., 1966. Mammals of the upper Richmond and Clarence Rivers of New South Wales. CSIRO *Wildl. Res. Tech. Paper No.* **10:** 1–55.

EMISON, W. B., PORTER, J. W., NORRIS, K. C. AND APPS, G. J., 1978. Survey of the vertebrate fauna in the Grampians–Edenhope area of southwestern Victoria. *Mem. Nat. Mus. Vict.* **39:** 281–363.

FLEAY, D., 1932. Swainson's Phascogale (the "bush mouse"). *Vict. Nat.* **49:** 132–4.

GREEN, R. H., 1972. The murids and small dasyurids in Tasmania. Part 5. *Rec. Queen Vict. Mus., Launceston, Tas.* **46:** 1–13.

HORNER, B. E., AND TAYLOR, J. M., 1959. Results of the Archbold Expeditions. No. 80. Observations on the biology of the yellow–footed marsupial mouse, *Antechinus flavipes flavipes. Amer. Mus. Nov. No.* **1972:** 1–24.

REED, G. F., AND WALLIS, R. L., 1975. Studies of *Antechinus swainsonii* and other small mammals in an area of Sherbrooke Forest Park. *Vict. Nat.* **92:** 84–90.

SUCKLING, G. C., BACKEN, E., HEISLERS, A., AND NEUMANN, F. G., 1976. The flora and fauna of radiata pine plantations in northeastern Victoria. *For. Comm. Vict. Bull. No.* **24:** 1–58.

THOMAS, D. J., AND GILMORE, A. M., 1976. An annotated list of the terrestrial vertebrates of the Dartmouth Dam inundation area. *Fish. Wildl. Paper, Vict. No.* **10:** 1–25.

VAN DYCK, S., AND OGILVIE, P., 1977. *Antechinus swainsonii* (Waterhouse, 1840), the Dusky Marsupial Mouse, an addition to the mammal fauna of Queensland. *Mem. Qd. Mus.* **18:** 69–73.

WOOLLARD, P., 1971. Differential mortality of *Antechinus stuartii* (Macleay): nitrogen balance and somatic changes. *Aust. J. Zool.* **19:** 347–53.

Unpublished material —

FLETCHER, H. L., 1977. "Habitat relationships among small mammals at Petroi, northeastern New South Wales". M. Nat. Res. thesis: University of New England, Armidale, New South Wales. 195pp.

GALL, B. C., 1976. "An assessment of the human impact on the Thredbo Valley ecosystem". M.Sc. thesis: Macquarie University, Sydney. 215pp.

HOCKING, G. J., 1975. "Studies on the behaviour and ecology of *Antechinus swainsonii*". Hons. thesis: University of Tasmania, Hobart. 167pp.

MORITZ, C., 1979. "Some aspects of the ecological and genetical structure of a natural population of *Antechinus stuartii* (Dasyuridae: Marsupialia)". Hons. thesis: University of Melbourne, Melbourne. 42pp.

PRESS, A., 1976. "Some aspects of the ecology and habitat utilisation of four species of native small mammals". Hons. thesis: Sydney University, Sydney. 86pp.

REECKMAN, B. M., 1975. "Aspects of the ecology of *Antechinus flavipes* (Waterhouse) (Marsupialia: Dasyuridae)". Hons. thesis: University of Melbourne, Melbourne. 42pp.

Museum specimens —

Specimens examined and measured are listed with the following abbreviations: AM, Australian Museum, Sydney; CM, C.S.I.R.O. Museum, Division of Wildlife Research, Canberra; USNM, National Museum of Natural History, Washington. All specimens are females from New South Wales.

Antechinus stuartii: AM M8906, Byron Bay; AM M4954, Chichester; AM M8145, Cobbity; AM M9730, Bungwahl; AM M8624, Wootton; AM M11084 and AM M11086, Newnes Plateau; CM 6111 and CM 6112, Snowy Plains (Kosciusko); CM 6186, Khancoban.

Antechinus swainsonii: AM M11007 and AM M11008, Mt. Cameron/Newnes Plateau; CM6191, Perisher Valley (Kosciusko); CM6193, Daners Gap (Kosciusko); USNM 221306 and USNM 221307, Tamworth region.

Personal communications and unpublished findings —

Forty–four of the data points used in the analysis were provided by 22 researchers in Australia. Eleven points were from unpublished personal observations.

RESUMEN

14. Algunos aspectos ecológicos de la reproducción de *Antechinus* (Dasyuridae, Marsupialia). Por C. R. Dickman.

Se examinó la variación geográfica del tiempo de apareo de *Antechinus stuartii* y de *Antechinus swainsonii* en relación a la altitud, latitud, longitud y simpatría. En toda su amplia distribución, el tiempo de apareo de *A. stuartii* varía desde fines de Julio hasta fines de Septiembre. El apareo comienza más tarde en altas latitudes, y también en bajas latitudes. Estos descubrimientos no están de acuerdo con un simple (isofano) modelo, ni con lo que podía esperarse de que el apareo ocurriera progresivamente más tarde del norte hacia el sur, en correspondencia con la aparición de los invertebrados que son su alimento. El tiempo de apareo de *A. swainsonii* varía desde comienzos de Mayo hasta fines de Septiembre, de acuerdo con el modelo isofano, y también es más tarde en altas latitudes.

Las diferencias interespecíficas en el tiempo de apareo pueden reflejar respuestas diferentes a los factores ambientales que fluctúan estacionalmente, y por ello una segregación de los nichos. Se sugiere que la segregación de los nichos es además facilitada por una asincronía del tiempo de apareo de *A. stuartii* donde es simpátrida con *A. swainsonii*, y por el mantenimiento de una constante diferencia interespecífica en el tamaño del cuerpo.

15. *Antechinus stuartii* (DASYURIDAE, MARSUPIALIA) DIET AND FOOD AVAILABILITY AT PETROI, NORTHEASTERN NEW SOUTH WALES

H. L. STATHAM[1]

A total of 520 *Antechinus stuartii* scat samples was collected from live traps in tall woodland, dry sclerophyll forest, wet sclerophyll forest, cool temperate rainforest and logged cool temperate rainforest between September, 1974 and August, 1976. Each sample was examined microscopically and fragments were identified using standard keys and photomicrographs of arthropods collected from the area.

The technique of pitfall trapping was used to estimate the relative abundance of ground dwelling arthropods on each of the five trapping grids. It is suggested that *A. stuartii* is an opportunistic feeder because, with the exception of carabid beetles and non–arthropods, the components of its diet followed closely the group composition of the pitfall captures. There was little dietary difference between habitats, with Coleoptera (minus carabs), Araneaeand Blattodea being the most important arthropod groups eaten.

Plant material was eaten occasionally throughout the year in all habitats, possibly unintentionally. However, all animals sampled in wet sclerophyll forest during August and September had eaten *Acrotriche aggregata* flowers. The eating of these flowers in other habitats followed the pattern of this shrub's distribution.

Pages 151–63 *in* CARNIVOROUS MARSUPIALS, ed. by M. Archer, February, 1982. Roy. Zool. Soc. New South Wales: Sydney, Australia.

IT IS generally accepted that insects form the major component of the diet of *A. stuartii* (Troughton 1954, Harrison 1962, Wakefield and Warneke 1967, Leonard 1972 and Frith 1973). However, only four accounts of *Antechinus* spp. dietary analyses are available in the literature. Harrison (1962) examined stomach contents from four *A. godmani*. One stomach contained nothing but insects, the other three were empty. Green (1972) made a qualitative study of *A. swainsonii* stomach contents and concluded that insects form a major part of the diet. Leonard (1972) examined contents of 70 stomachs of male *A. stuartii*. He did not quantify the results but reported that they included beetles, ants, wasps, mites, larvae and cockroaches. Finally Hindmarsh and Majer (1977) from analysis of stomach contents of *A. flavipes* found that the numerically predominant invertebrates Araneae, Blattodea, Coleptera and Hymenoptera are the principal food sources of this species. They found no correlation between the abundance of *A. flavipes* and that of forest invertebrates during the post fire succession and therefore suggested that food availability is not a major factor in population changes.

Woollard (1971), working with *A. stuartii* from Condor Creek, Australian Capital Territory, reported an energy intake for males ranging approximately from 84 kJ/day in March to 145 kJ/day in the breeding season, with that of females increasing from 84 kJ/day to 105 kJ/day over the same period. Cowan *et al.* (1974), working in February on *A. swainsonii* from the Snowy Mountains, New South Wales, gave figures of approximately 100 kJ/day energy intake for 52–57 g males and 96 kJ/day for 45 g females. Nagy *et al.* (1978) reported that in winter, steady state, field *A. stuartii* from wet sclerophyll forest near Healesville, Victoria (mean body weight 25.7g) consumed about 60% of their body weight in arthropods each day. They ingested 807K cal/kg/day and metabolized 670 K cal/kg/day. Figures on basal metabolic rate, but not food intake, for *A. stuartii* and other dasyurids have been given by MacMillen and Nelson (1969).

With little data available on *A. stuartii* diet and food requirements, the lack of information on food availability is understandable. Prior to this study, there had been only one assessment of insect fauna made in conjunction with *A. stuartii* population estimation (Leonard 1972). Unfortunately his study was restricted to separation of litter fauna, most of which is smaller than that eaten by *A. stuartii*. The aims of this study were to investigate the diet and food availability of *A. stuartii* throughout the year in five different habitats and to examine whether food availability is a limiting factor in determining habitat preference and the phenomenon of male die–off.

This research is part of a team study of various aspects of the Petroi ecosystems being made by the National Parks and Wildlife Service and students of the University of New England.

[1] School of Natural Resources, University of New England, Armidale, New South Wales. Present address: Forestry Commission, Tasmania; C/o Department of Agriculture, P.O. Box 407, Launceston, Tasmania, 7250.

METHODS

From March, 1974, until June, 1974, stomach samples of *A. stuartii* were obtained from animals captured in break–back traps set in areas of tall woodland, wet sclerophyll forest, cool temperate rainforest and logged cool temperate rainforest similar to, but geographically isolated from, those of small mammal capture–recapture trapping grids (Statham and Harden 1982). Samples were stored in 70 per cent alcohol. A preliminary examination of the contents revealed the presence of arthropod fragments, in particular cuticle, antennae, leg parts and mouth parts. In this examination a greater proportion of potentially identifiable fragments were found in the intestines than in the stomach. In light of this, break–back trapping was terminated and from August, 1974, until September, 1976, *A. stuartii* scats were collected monthly from live traps in in each grid after the first night's trapping. Each sample consisted of an animal's nightly defecation and was comprised of up to a dozen individual scats. Scats were taken from traps in dry sclerophyll forest from February, 1976, to September, 1976. Scats from each animal were stored separately in 70 per cent alcohol.

In early August, 1974, twelve male *A. stuartii* were collected from a wet sclerophyll area well isolated from any of the trapping grids. An energetics study was made on these animals by G. Wallace, Zoology Department, University of New England. These animals were caged separately and maintained on diets which consisted entirely on scarab larvae, spur throated locusts or lean mince.

The technique of pitfall trapping was used to estimate the relative abundance of ground dwelling arthropods in each of the five trapping grids. Four pitfall sites were located at random within each nine hectare grid. Sixteen 20–ounce jars, the bases of which had been painted white, were used at each site. Thus there were 64 pitfall jars per small mammal trapping grid. The jars were sunk into the ground in holes spaced five metres apart in a square grid formation (4 x 4). When not in use the jars were turned upside down in the holes.

At commencement of trapping (coincident with small mammal trapping for each grid) the jars were turned the right way up and placed so that the top of each jar was at ground level. Thirty ml of beer was placed in each jar to act as attractant and preservative. Daily catches of animals > 3 mm in length were collected from each of the 16 jars per site and stored separately in 70 per cent alcohol. These were identified to order, and usually to family using standard keys (Barnes 1968, C.S.I.R.O. 1973, Tillyard 1926 and York Main 1964). The contents of each jar on each day were oven dried and weighed. Photomicrographs of antennae, mouth parts, tarsal segments and any features of special interest of arthropods captured most abundantly in the pitfall traps were taken. These photomicrographs were then mounted on 8" x 5" cards and filed systematically.

A. stuartii scats were examined under a binocular microscope. Each scat was teased out with a mounted needle and pair of tweezers. Fragments thought to be possibly identifiable were separated out. These were then identified using the photomicrographs and standard keys.

RESULTS

DIET

A total of 520 *A. stuartii* scat samples, collected from the live traps, was analysed microscopically. Each scat contained a mixture of soft (digested) and hard (undigested) material, the majority of which was derived from arthropods. In addition, live and dead plant material, sand and soil particles and even chewed pieces of aluminium were identified.

The identifiable arthropods were divided into ten groups. These were Araneae, Myriapoda, Coleoptera, Blattodea, Hymenoptera, Orthoptera, Diptera, Dermaptera, Larvae and "Others". Because of the predominance of carabid beetles in the summer pitfall trapped samples, Coleptera were subdivided into carabs and non-carabs. The monthly percentage occurrence of these groups, plus plant material and unidentified animal remains in *A. stuartii* scat samples from each habitat are given in Appendix 1. For any group, e.g. Araneae, the monthly percentage occurrence was obtained by determining the percentage of the number of individuals identified in the total number of animals and plants represented in the month's scat samples. Apart from plant material, which was scored as present or absent in each scat, the analyses made allowance for cases where remains of more than one individual per group were found in a single scat.

No statistical analyses were made because of the essentially qualitative nature of the method. Identifiable fragments did, however, suggest that most arthropods ingested were in the 0.5 to 2.0 cm length range.

From Appendix 1, it can be seen that remains of Coleoptera, Araneas and Blattodea (beetles, spiders and cockroaches) were found more abundantly and more frequently than those of the other groups. It is therefore suggested that these animals comprise the major constituents of the diet of *A. stuartii*. The seasonal percentage occurrence of these three groups is given in Fig. 1.

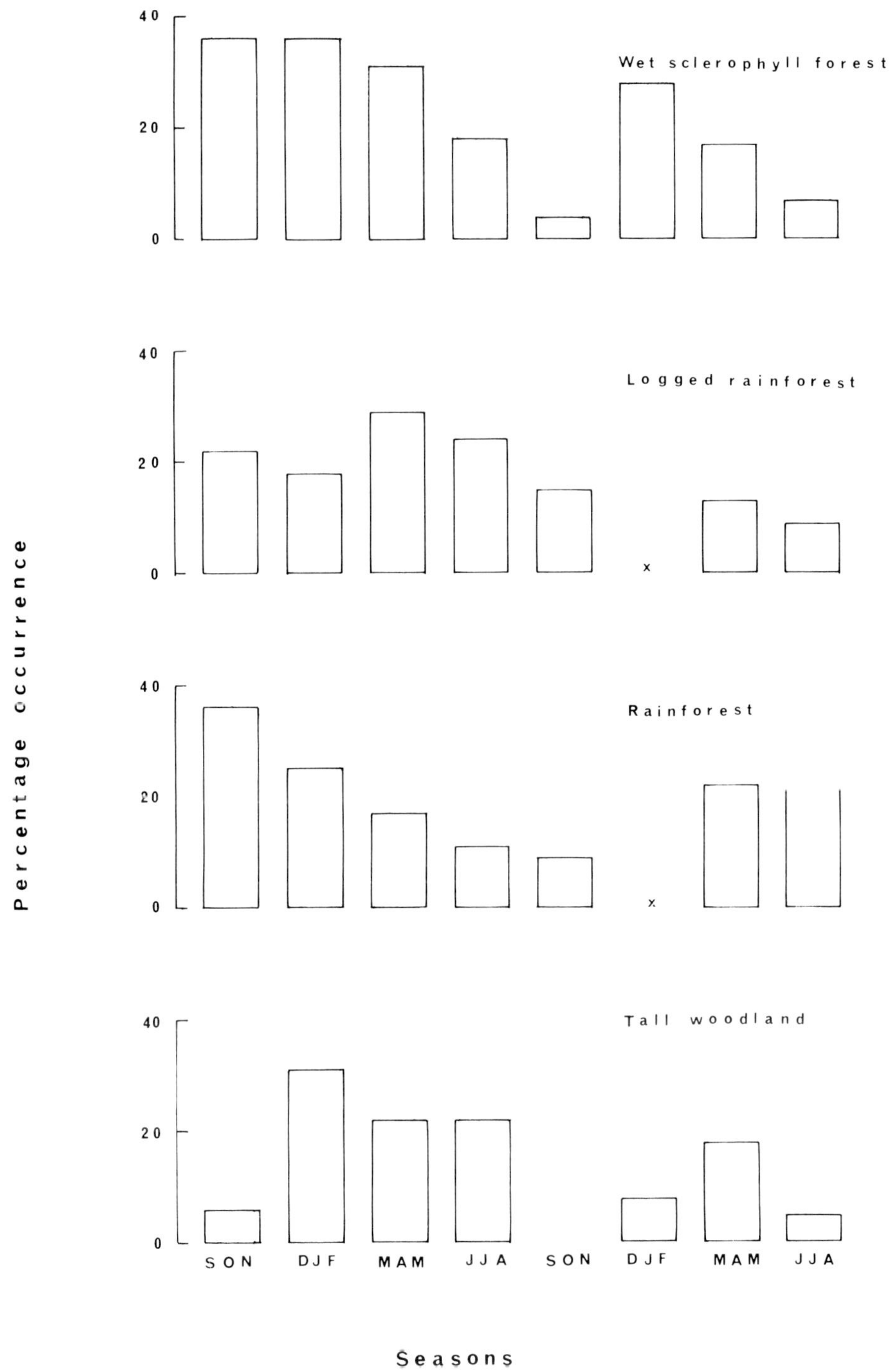

Fig. 1. Seasonal percentage occurrence of the three major arthropod groups found in *A. stuartii* scats from each habitat. Sept. 1974–August 1976; seasons defined by the three monthly groups shown.

1A: Coleoptera (minus carabs). x means less than 5 scats analysed.

In wet sclerophyll forest, Coleoptera, Araneae and Blattodea comprised approximately 50 per cent of the identified scat fragments. The percentage occurrence of Coleoptera was higher in summer and autumn than in winter, that of Araneae remained relatively constant whilst Blattodea were probably more importantin summer. All other groups, except Dermaptera, carabs and "others" were also eaten during most months of the year. The three groups Hymenoptera, Larvae and Orthoptera occurred consistently in the January to August, 1976, samples. An interesting feature of the results from this habitat related to the peak in occurrence of plant material eaten during August and September each year. At this time, plant material comprised up to 41 per cent of the groups recognised and consisted entirely of flowers of the epacrid shrub, *Acrotriche aggregata*.

Coleoptera, Araneae and Blattodea collectively made up approximately 50 per cent of the logged cool temperate rainforest scat remains. The percentage occurrence of Coleoptera was highest in autumn, 1975, and lowest in winter, 1976, but did not show marked seasonal variation as in wet sclerophyll forest. The percentage occurrence of Araneae remained constant whilst that of Blattodea tended to peak in summer. The only other groups recorded in more than 50 per terial were recorded during half of the months samples but showed no seasonal variation in percentage occurrence.

In cool temperate rainforest remains of Coleoptera, Araneae and Blattodea comprised approximately 40 per cent of the scat fragments. Variation in percentage occurrence of Coleoptera was greater between years than between seasons. This was also true for Araneae and Blattodea, both of which peaked in occurrence during summer, 1974–75 and winter, 1975. Orthoptera, unidentified arthropods and plant material were recorded during half of the months samples but showed no seasonal variation in percentage occurrence.

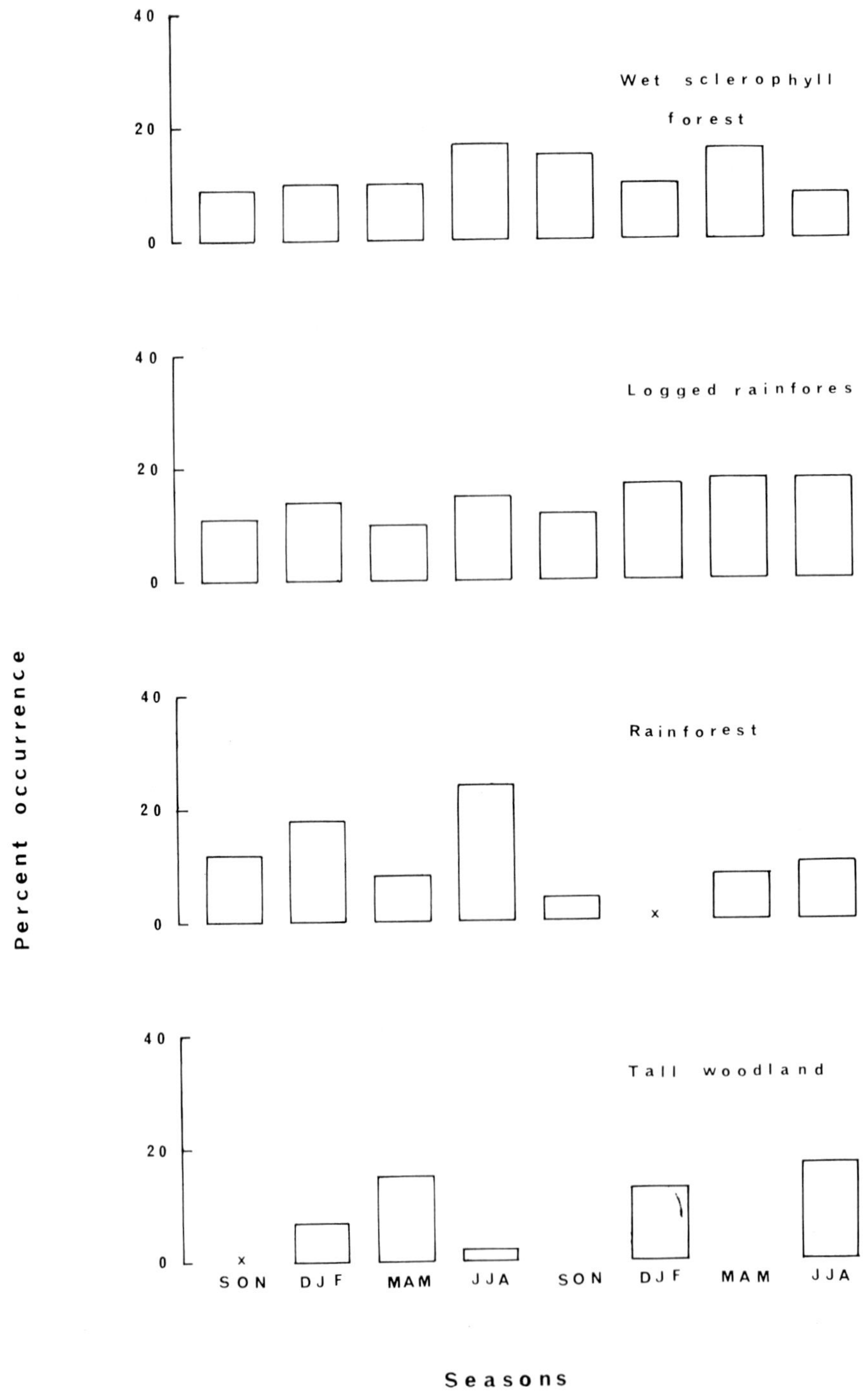

Fig. 1B: Araneae.

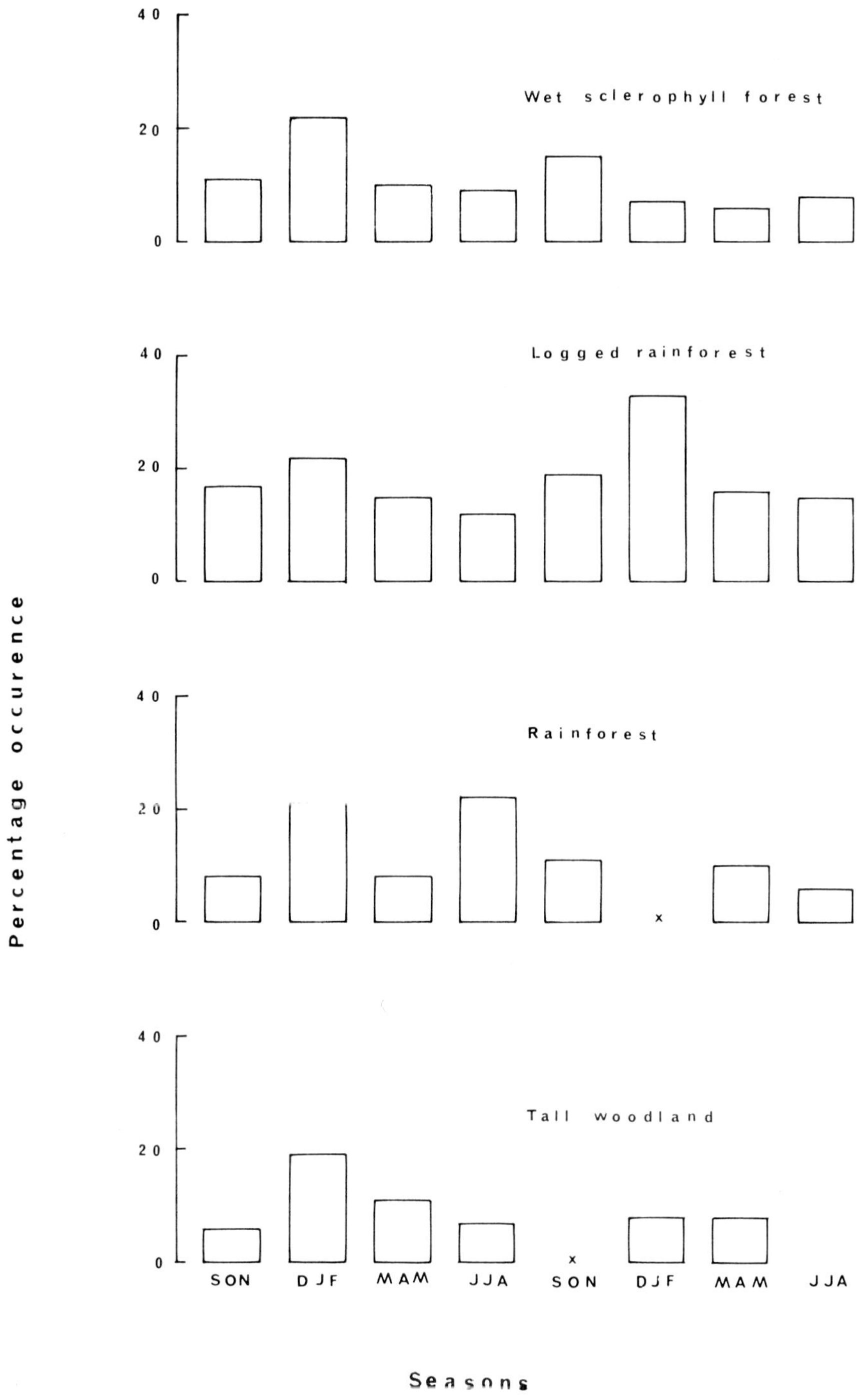

Fig. 1C: Blattodea.

In tall woodland, unidentified arthropod remains occurred during more months than those of any other group. They also occurred more often here than in the other habitats. Coleoptera, Araneae and Blattodea comprised approximately 30 per cent of the tall woodland scat fragments with each group showing as much variation between years as between seasons. There were no peaks in the occurrence of plant material, although some *A. aggregata* flowers were found in the August, 1976 samples.

Araneae and Larvae occurred in scats collected every month from dry sclerophyll forest (February–August, 1976), while Diptera and Dermaptera occurred in all but one of the months. As in tall woodland, *A. aggregata* flowers were found in some of the August samples.

Although no vertebrate remains were found in the scat samples two field observations indicated that *A. stuartii* does eat meat. On one occasion, at the time of male die–off, one and a half male *A. stuartii* were found in one trap. Also, during the break–back trapping several partially eaten

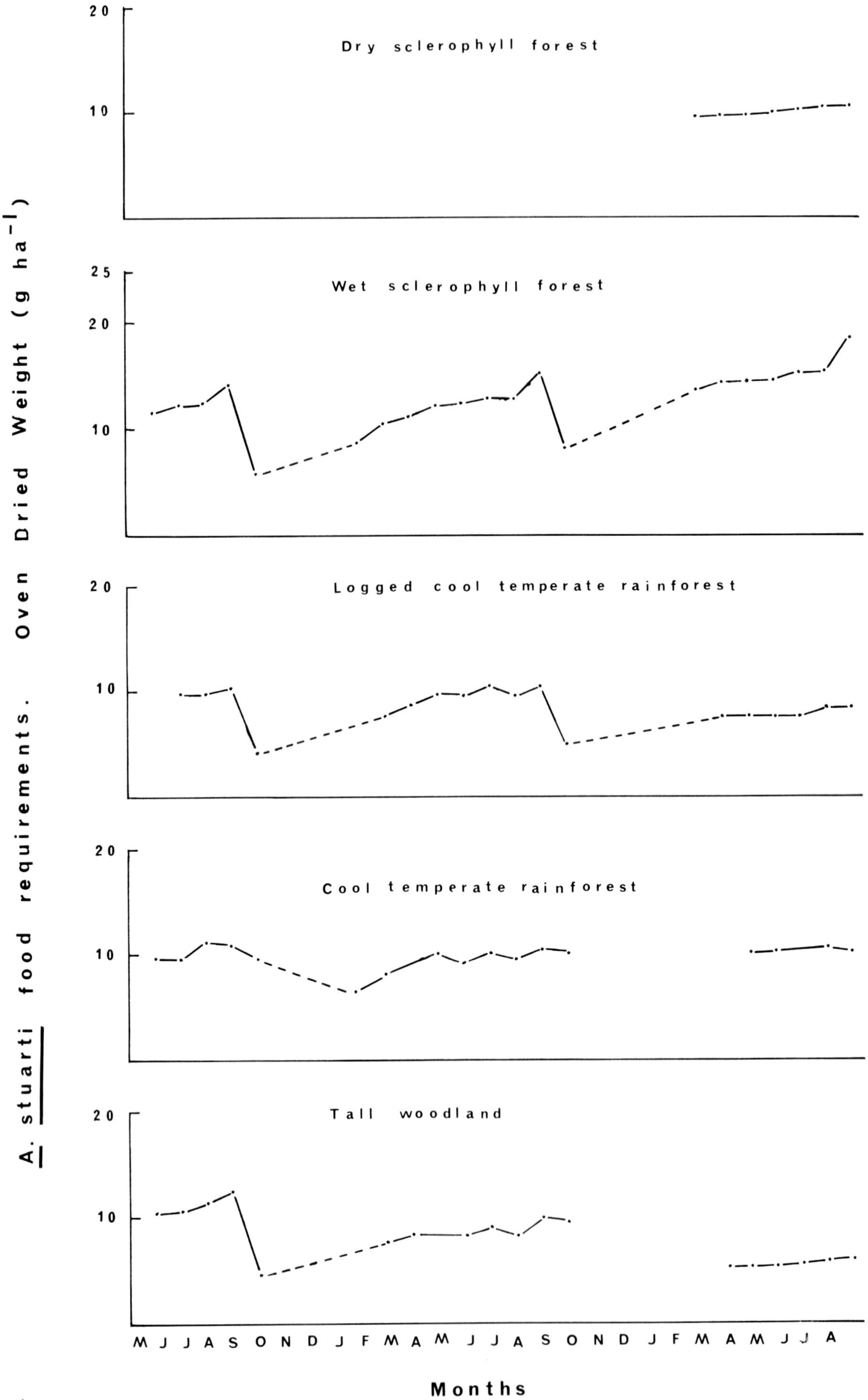

Fig. 2. A. stuartii food requirements, daily requirements of populations within each habitat, May, 1974–August, 1976.

R. fuscipes and *A. stuartii* were captured and examination of bite size and style implicated *A. stuartii.*

Food Requirement

The digestible energy intake/day requirement for 35 g *A. stuartii* for maintenance and growth and some activity has been calculated to be 84 kJ (I. Hume, pers comm.) and digestibility of insect material = 58% (G. Wallace, unpublished data)

∴ gross energy intake requirement/animal/day
= 145 kJ

the gross energy content of ingested arthropods
= 22.0 kJ/g
(oven dry weight) (Cummins & Wuycheck 1971)

then, daily requirement of each 35 g *A. stuartii*
= 6.6 g (oven dry weight) or 0.46 g/g metabolic body weight.

Using the figures of Nagy *et al.* (1978) the daily requirement of a 35 g *A. stuartii* would be 5 g per day. These figures however are based on an intake of beetles and leaf hoppers which are of higher digestibility that the arthropods found to be eaten in this study.

The daily food requirements of *A. stuartii* populations within each habitat (Fig. 2) were determined on a monthly basis. For each sex the annual maximum density was multiplied by the mean monthly metabolic body weight by 0.46. Values for males and females were then pooled. Annual maximum densities were used in these calculations in preference to monthly estimates because of variation in trappability and little indication of immigration throughout the study (Statham 1982).

The estimated biomass of "edible" arthropods required by *A. stuartii* populations in the five habitats ranged approximately from 5.23 to 18.64 g (oven dry wt.)/ha per day.

c) *Food Availability*

A. stuartii food availability was defined as "the food which is easy to find and is being chosen and eaten by these animals" (Grodzinski 1966 cited Drozdz 1967, Grodzinski 1971). Thus, whilst pitfall trapping was used to measure this food availability (Table 1), it was important to consider only the arthropod groups eaten by *A. stuartii* rather than the entire array of pitfall trapped animals.

Most of the animal groups identified in the pitfall samples were eaten by *A. stuartii.* However, during summer months, it was normal for more than 50 per cent of the biomass of pitfall trapped animals to be made up of carabid beetles. Since very few carabs were eaten by *A. stuartii* (Appendix 1), their biomass plus that of snails, frogs, lizards and crayfish was subtracted from each total figure to give the biomass of "edible" arthropods.

Table 1. List of animal groups identified from pitfall samples in each of the habitats.

Animal group		Habitat				
		W.Sc.	L.Rf.	Rf.	T.Wdl.	D.Sc.
Blattodea		+	+	+	+	+
Dermaptera		+	+	+	+	+
Orthoptera	Tetrigidae	+	+	+	+	+
	Gryllidae	+	+	+	+	+
	Stenopelmatidae	+	+	+	+	+
	Tettegoniidae	+	+	+	+	+
Coleoptera	Scarabaeinae	+	+	+	+	+
	Carabidae*	+	+	+	+	+
	Curculoinidae	+	+	+	+	+
	Other	+	+	+	+	+
Diptera		+	+	+	+	+
Hymenoptera		+	+	+	+	+
Lepidoptera			+		+	+
Heteroptera		+				+
Larvae		+	+	+	+	+
Myraipoda		+	+	+	+	+
Amphipoda		+	+	+	+	+
Snails**		+	+	+	+	+
Frogs**		+		+	+	+
Lizards**		+			+	+
Fresh water crayfish**		+	+	+		

* Denotes group occurring in great abundance in pitfall traps but rarely eaten by *A. stuartii.* ** Denotes groups not eaten by *A. stuartii.* W.Sc. = Wet sclerophyll forest; L.Rf. = Logged cool temperate rainforest; Rf. = Cool temperate rainforest; T.Wdl. = Tall woodland; D.Sc. = Dry sclerophyll forest.

The basic theory of pitfall trapping is that ground dwelling animals fall into the traps at random during the course of normal movements. Thus the technique measures activity of animal populations rather than population numbers. This simplistic situation rarely, if ever exists and consequently quantification of pitfall data at the population level is fraught with problems arising from variation in activity between different animal groups. The situation is compounded by fluctuations in climatic conditions, the effects of ground vegetation and the use of bait or preservative (Williams 1959, Greenslade 1964, Greenslade and Greenslade 1971, Davidson 1976).

Two tests were carried out to check pitfall trapping techniques employed in this study. Firstly, in 1976, two extra pitfall sites, of identical design to the others, were set out in the tall

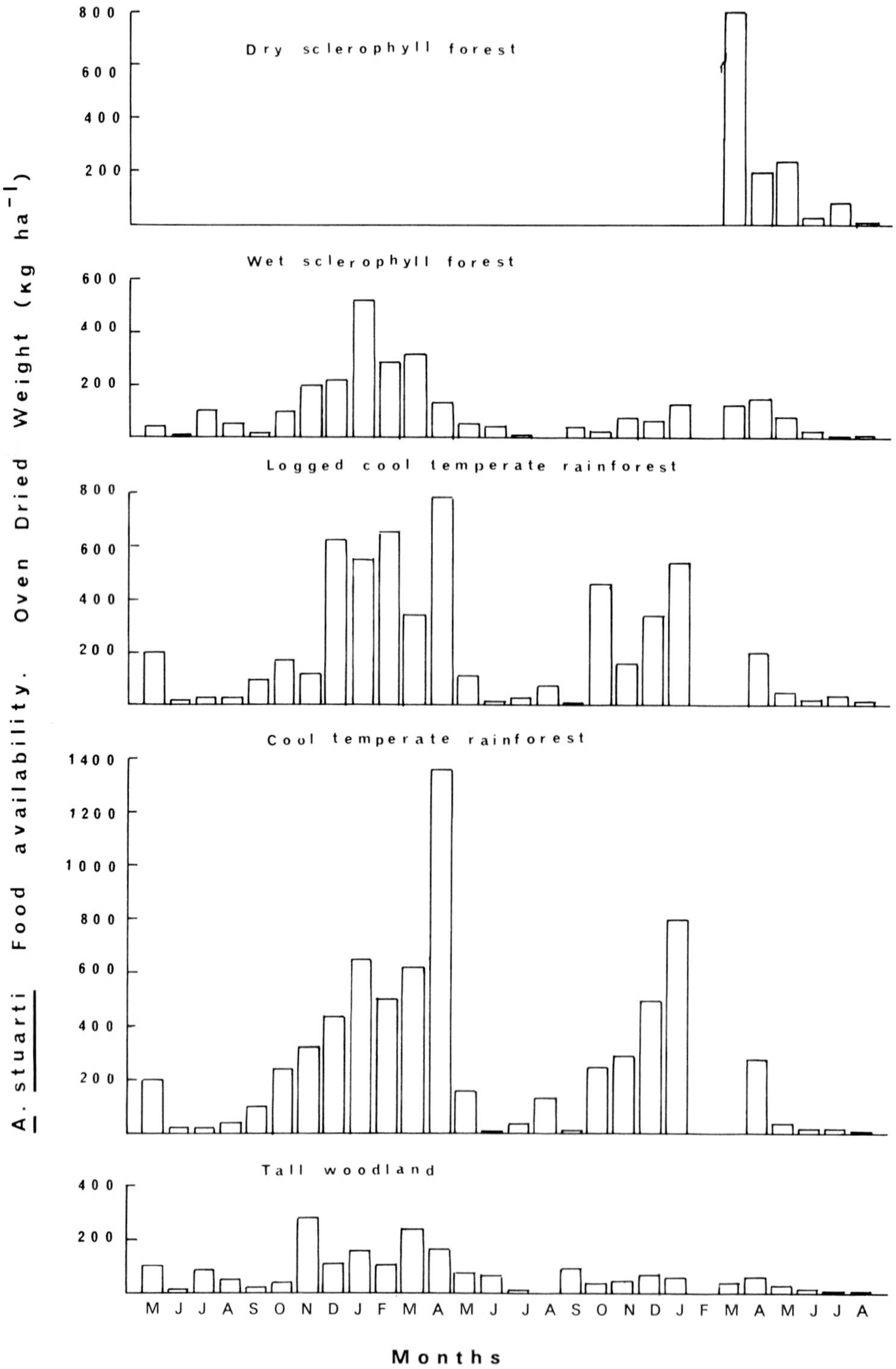

Fig. 3. Pitfall trapping, maximum estimates of *A. stuartii* food availability in each habitat, May, 1974–August, 1976.

woodland and trapped in conjunction with the previously used sites. There were no differences in the group composition and biomass of animals captured at new and old sites, suggesting that the pitfall trapping programme had not depleted local athropod populations. Secondly, analysis of variance comparisons showed no significant variation between biomass of "edible" arthropods captured at interior or peripheral pitfall traps, which suggested that the jars were spaced sufficiently far apart to prevent interference of outer jars on capture rates of inner jars.

The maximum monthly estimates of pitfall trapped "edible" arthropods were taken as a measure of their availability (Fig. 3). It was considered more meaningful to use these maxima rather than mean values for each trapping period because capture of a large number of arthropods on one night suggests that similar numbers were present on the other nights, but were not sufficiently active to be captured. Also, results of the tests described above suggested that depletion of arthropod populations was negligible with trapping system.

Estimates of "edible" biomass peaked over the summer and autumn months and dropped to a low in late winter. The higher peaks in logged cool temperate rainforest and cool temperate rainforest were primarily caused by captures of Orthoptera, in particular large specimens of *Australostoma* sp. These were often longer than 2 cm, and, as such were much larger than most animals eaten by *A. stuartii*. Most of the "edible" arthropods captured in winter measured 0.5–2 cm, corresponding with the size range of animals identified from the scats.

Assuming zero attraction to each pitfall trap, then each jar (diameter 5.5 cm) would have sampled 23.76 cm^2, giving a sampling area per habitat of 0.152 m^2. Under these conditions, the available biomass of "edible" arthropods reached annual peaks ranging approximately from 100 kg/ha in tall woodland during April, 1976, to 1,400 kg/ha in cool temperate rainforest during April, 1975. The winter lows always exceeded 3 kg/ha, suggesting that there was always adequate food available for *A. stuartii* in each of the habitats. If on the other hand, the pitfall sampling area was defined by the external dimensions of each pitfall grid, representing 900 m^2 per habitat (4 x 15 m x 15 m), then the estimated food availability ranged approximately from 0.01 kg/ha to 0.25 kg/ha and thus availability usually exceeded the requirements in summer and autumn but was never adequate in winter. Since the first night's trapping did not deplete the area, and inner trap captures did not differ from outer, the "true" catchment obviously fell between these extremes and was probably closer to the zero attraction situation. For this reason only the zero attraction values have been in Fig. 3.

DISCUSSION

The method of *A. stuartii* diet determination used here was preferable to the more frequently used one of gut analysis because it enabled sampling from known animals within the five trapping grids and eliminated otherwise unnecessary killing. Microscopic separation and identification of arthropod fragments was both tedious and time consuming, although simplified by previous entomological study. It was difficult to determine the size and often number of ingested arthropods from the fragmented remains. No key was made to aid arthropod identification because it seemed impractical due to the fact that several fragments were usually required for positive identification. Instead, I found it more meaningful, if less scientific, to piece together identifiable fragments from each scat sample.

The method was essentially qualitative and simply allowed compilation of tables of arthropod groups eaten by *A. stuartii*. It made no allowance for ingested food items that may have been wholly digested. Also, several ants and larvae were found intact. These may have been superficial to the scat, or, if eaten obviously had not contributed to the digestible fraction of the diet.

Many of the problems encountered in the use of pitfall traps have been considered with the results. It is considered that much of the variation attributable to climatic conditions and differences between animal groups was reduced by the extremely large sample size (320 jars) and length of study 3–4 nights per month, for 28 months. Although I have been unable to determine accurate estimates of *A. stuartii* food availability, I feel this method preferable for my purposes to that of sticky traps (Southwood 1966, Cameron 1975). It was also more practical than the hand collection of ground dwelling arthropods (Gillon and Gillon 1967), or the use of a mechanically operated aspirator (Petrusewicz and MacFadyen 1970). These latter methods would have been difficult to carry out at night.

It was encouraging to find that the pitfall traps were catching basically the same animals as those being eaten by *A. stuartii*. Finally, it is suggested that the relative arthropod activity between habitats and under different seasonal and climatic conditions, may possibly be a meaningful measure of *A. stuartii* food availability. The rationale here is that the more active arthropods would be more frequently encountered by *A. stuartii* and thus they would be more likely to be eaten.

Comparison of food availability estimates with results in the liaterature suggest that the values obtained using the sum of areas of jars exposed more closely approximated ground dwelling arthropod biomass than those based on a sampling area of the whole pitfall grid. For example,

Hutchinson and King (personal communication) estimated invertebrate abundance on *Phalaris/Trifolium* grassland at Armidale to be 178 kg (oven dry weight)/ha (assuming 40 per cent conversion from wet weight (Cummins and Wuycheck 1971) and Nef (1957, cited Ghilarov 1967) estimated 300–1,000 kg soil animals per hectare in Belgian forest.

Throughout the two years, arthropods and plant material made up the diet of *A. stuartii* in wet sclerophyll forest, logged cool temperate rainforest and tall woodland at Petroi. Coleoptera, Araneae and Blattodea occurred more often in the scats than the other arthropod groups, which were Hymenotera, Orthoptera, Diptera, Myriapoda, Dermaptera, Larvae, others and unidentified arthropods. In dry sclerophyll forest Coleoptera, Araneae, Blattodea, Hymenoptera, Orthoptera, Dermaptera and Larvae were identified in scats collected in at least four of the six months studied. The occurrence of Coleoptera in scats from wet sclerophyll forest tended to peak in summer as did that of Blattodea in wet sclerophyll forest and logged cool temperate rainforest. Araneae were found at the same level of occurrence, during most seasons, in these habitats. In cool temperate rainforest and tall woodland there was more variation between years than between seasons in percentage occurrence of these three main groups.

The most striking dietary difference between habitats was in the occurrence of plant material. This was eaten occasionally throughout the year by *A. stuartii* in all habitats, possibly unintentionally. However, all animals sampled in wet sclerophyll forest during August and September had eaten *Acrotriche aggregata* flowers. The eating of these flowers in other habitats followed the pattern of this shrub's distribution. Thus, in dry sclerophyll forest and tall woodland, where the shrub was less abundant than in wet sclerophyll forest, some *A. stuartii* ate them, and since the shrub did not extend to either of the rainforest grids, it was not eaten there. It was interesting to observe that *A. stuartii* in these habitats did not eat other plant material at this time but continued to eat mainly arthropods. Where the flowers were taken they were obviously picked from the shrub, where they occur close to the stem, rather than collected from the ground because they were being eaten before any had fallen to the ground. There seems no obvious reason for the eating of these flowers.

Very similar groups of animals were captured in the pitfall traps located in each of the five habitats. Peaks occurred in the numbers and biomass of catches in the summer and autumn months. In the arthropod groups eaten by *A. stuartii,* summer peaks were greater in logged and unlogged cool temperate rainforest than in wet sclerophyll forest and tall woodland. These were mainly due to large *Australostoma* sp. which I think, were larger than those species eaten by *A. stuartii.* Results in dry sclerophyll forest during 1976 suggested that arthropod activity was probably greater than in the other habitats. During winter arthropod activity was low in all habitats.

Since there was little variation in the relative proportions of the arthropod groups eaten by *A. stuartii* and because the groups eaten were similar in composition to those captured in the pitfall traps, it is considered that *A. stuartii* is an opportunistic feeder, eating ground–dwelling arthropods within a range of approximately 0.5 to 2.0 cm length. Food requirements would therefore be in terms of biomass of "edible" arthropods rather than specific groups.

It is unrealistic to consider the "edible" arthropods estimated as being eaten only by *A. stuartii* since other animals, for example frogs, lizards, birds, bandicoots and Tiger Quolls, are also potential predators. Because of differences in times of activity and feeding, it is unlikely that frogs, lizards and most birds would eat these crepuscular and nocturnal arthopods, unless they dug them out from under bark and logs. Thus food availability estimates would probably only need to include over–lapping requirements for nocturnal birds, bandicoots and Tiger Quolls.

Using even the most conservative estimates, the pitfall trapping results indicated adequate food availability in all habitats for *A. stuartii during* summer and autumn. Thus, if food shortage does occur, it would most likely be in winter when arthropod numbers and activity are lowest.

However males in all habitats, representing more than half of the population, showed a 5–10 g (20 per cent) increase in body weight between July and August each year (Fletcher 1977). It is improbable that this energy expensive event would take place in a food limited situation. The subsequent loss of this weight and male death after breeding may reflect reduced food seeking rather than low availability.

Therefore, I suggest that there was no shortage of food for *A. stuartii* in any of the habitats during this study.

ACKNOWLEDGEMENTS

This research was part of a study on habitat relationships among small mammals at Petroi, undertaken in fulfilment of the requirements for the degree of Master of Natural Resources in the University of New England. I should like to thank Mr R. H. Harden, N.P.W.S. for encouraging me to make the study and for the guidance and assistance. I am grateful to other members of the N.P.W.S. team for assistance with field work. I should also like to thank my supervisor,

Dr. J. Jarman. Sincere thanks are extended to Mr. and Mrs. Joe Davies and Mr. B. Davies for permission to work and camp on Petroi and for their kind hospitality. I am grateful to the late Mr. C. Frazier for assistance in identification of arthropods and to Dr. I. Hume for assistance in determining *A. stuartii* food requirements. Numerous students are thanked for assistance with field work and Mr. J. Woods, Armidale Bowling Club, is thanked for collecting the beer slops used in the pitfall trapping. I am grateful to the N.P.W.S. for the loan of traps plus assistance with travel and other aspects of the study. For two years of the work I was supported by a Commonwealth Postgraduate Research Award. Mr. R. Harden, Dr. B. Fox and Dr. M. Denny constructively criticised the manuscript. Ms. G. Clayton helped letter Figs. 1–3.

REFERENCES

BARNES, R. D., 1968. "Invertebrate zoology". Second Edition. Saunders: Philadelphia.

CAMERON, E. E., 1975. "Resource utilization of three sympatric species of flycatchers with particular reference to feeding ecology". M.Sc. thesis: University of New England, New South Wales.

COWAN, I. MCT., O'RIORDAN, A. M. AND COWAN, J. S. MCT., 1974. Energy requirements of the dasyurid marsupial mouse *Antechinus swainsonii* (Waterhouse). *Can. J. Zool* **52:** 269–75.

C.S.I.R.O., 1973. "The insects of Australia". Melbourne University Press: Melbourne.

CUMMINS, K. W. AND WUYCHECK, J. C., 1971. Caloric equivalents for investigations in ecological energetics. *Mitt. Internat. Verein. Limnol.* No.**18:** 1–151.

DAVIDSON, S. J., 1976. "Mesofaunal response to cattle dung with special reference to Collembola". B.Sc. Hons. thesis: University of New England, New South Wales.

DROZDE, A. 1967. Food preference, food digestibility and natural food supply in small rodents. *In* "Secondary productivity of terrestrial ecosystems. Vol. 1"., ed. by K. Petrusewicz. Polish Scientific Publishers: Warsaw.

FLETCHER, H. L., 1977. "Habitat relationships among small mammals at Petroi, north-eastern New South Wales." M. Nat. Res. thesis: University of New England, Armidale, New South Wales.

FRITH, H. J., 1973. "Wildlife conservation". Angus and Robertson: Sydney.

GILLON, Y. AND GILLON, D., 1967. Methodes d'estimation des nombres et des biomasses d'arthropodes en savanne tropicale. *In* "Secondary productivity of terrestrial ecosystems", ed. by K. Petrusewicz. Polish Scientific Publishers: Warsaw.

GHILAROV, M. S., 1967. Abundance, biomass and vertical distribution of soil animals in different zones. *In* "Secondary productivity of terrestrial ecosystems", ed. by K. Petrusewicz. Polish Scientific Publishers: Warsaw.

GREEN, R. H., 1972. The murids and small dasyurids in Tasmania. Parts 5, 6 and 7. *Records of the Queen Victoria Museum, No.* **46:** 1–34.

GREENSLADE, P. J. M. 1964. Pitfall trapping as a method of studying populations of Carabidae (Coleoptera). *J. Anim. Ecol.* **33:** 301–10.

GREENSLADE, P. AND GREENSLADE, P. J. M., 1971. The use of baits and preservatives in pitfall traps. *J. Aust. Ent. Soc.* **10:** 253–60.

GRODZINSKI, W., 1971. Energy flow through populations of small mammals in the Alaskan Taiga Forest. *Acta Theriol.* **61:** 231–75.

HARRISON, J. L., 1962. The food of some Innisfail mammals. *Proc. Roy. Soc. Qld.* **73:** 37–43.

HINDMARSH, R. AND MAJER, J. D., 1977. Food requirements of Mardo (*Antechinus flavipes*) (Waterhouse) and the effect of fire on Mardo abundance. Forests Department of W.A. Research Paper. **31.**

LEONARD, B. V., 1972. "The effect of fire upon selected small mammals and leaf litter fauna in sclerophyll forest in southern Australia." M.sC. thesis: Monash University, Victoria.

MACMILLEN R. E. AND NELSON, J. E., 1969. Biogenetics and body size in dasyurid marsupials. *Am. J. Physiol.* **217** (4): 1246–51.

NAGY, K. A., SEYMOUR, R. J., LEE, A. K. AND BRAITHWAITE, R., 1978. Energy and water budget in free-living *Antechinus stuartii* (Marsupialia: Dasyuridae). *J. Mammalogy* **59:** 60–8.

PETRUSEWICZ, K. AND MACFADYEN, A., 1970. "Productivity of terrestrial animals, principles and methods". IBP Handbook No. 13. Blackwell Scientific Publishers: Oxford.

SOUTHWOOD, T. R. E., 1966. "Ecological methods with particular reference to the study of insect populations". Methuen. London.

STATHAM, H. L., 1982. Trappability of *Antechinus stuartii* (Dasyuridae, Marsupialia) at Petroi, northeastern New South Wales. Pp.333–44 *in* "Carnivorous marsupials", ed. by M. Archer. Roy. Zool. Soc. New South Wales: Sydney.

STATHAM, H. L. AND HARDEN, R. H., 1982. Habitat utilization of *Antechinus stuartii* (Dasyuridae, Marsupialia) at Petroi, northeastern New South Wales. Pp.165–85 *in* "Carnivorous marsupials", ed. by M. Archer. Roy. Zool. Soc. New South Wales: Sydney.

TILLYARD, R. J. 1926. "The insects of Australia and New Zealand". Angus and Robertson: Sydney.

TROUGHTON, E. LEG., 1954. "Furred animals of Australia". 5th edition. Angus and Robertson: Sydney.

WAKEFIELD, N. A. AND WARNEKE, R. M., 1967. Some revision in *Antechinus* (Marsupialia). *Vict. Nat.* **84:** 69–99.

WILLIAMS, G., 1959. The seasonal and diurnal activity of the fauna sampled by pitfall traps in different habitats. *J. Anim. Ecol.* **28:** 1–13.

WOOLLARD, P., 1971. Differential mortality of *Antechinus stuartii* (Macleay) : Nitrogen balance and somatic changes. *Aust. J. Zool.* **19:** 347–53.

YORK MAIN, B., 1967. "Spiders of Australia". Jacaranda Press: Brisbane.

RESUMEN

15. *Antechinus stuartii* (Dasyuridae, Marsupialia) viveres y disiponibilidad a Petroi, Nordestal New South Wales. Por H. L. Statham.

Un total de 520 *Antechinus stuartii* ovejas ejemplos fueron recolecta–os con vida de trampas en alto arbolado, secos bosques, humedos bosques, agradables temperaturas en bosques lluviosos y cortar agradables temperaturas en bosques lloviosos entre Septiembre, 1974 y Agosto 1976 cada ejemplo fue examinado con micros copios y fragmentos fueron identifica dose usando. Estandar llaves y fotomicrografica de artropodos recolectados de el area.

La tecnica de hoyo–trampa fue usa da para estimar la relativa abundancia de terreno–vida artropodos sobre cada uno de 5 trampas area. Es sugerido que *stuartii*.

Es una oportunidad alimenta dora por que, con la exepcion de carabidos escarabajos y artropodos los componentes de muerte segidos de serca el grupo de composicion de la hoyotrampa captura.

Hay fue la chica dietica diferencia entre habitos, con coleopterous (sin carabids) semejante y blajiodea fueron los mas importantes artropodos. Grupos comiendo.

Plan material fue comido ocacional dentro del ano en todos habitos. Posible no intencional. Como sea todos los animales ejemplos en humedo bosques entre Agosto y Septiembre an comido acrotriche ***aggregata*** flores. La comida de esas flores en otros habitos segidos de area de esta especie distribucion.

APPENDIX 1

A. stuartii scat analyses. Monthly percentage occurrence of the major groups recognised September 1974–August, 1976.

Month	% occurrence of major groups recognised														
	N_1	N_2	Co	Ar	Bl	Hy	La	Or	Di	My	De	Ca	Ot	Un	Pl
							Wet sclerophyll forest								
September, 1974	6	16	25	13	6	—	—	13	6	—	—	—	—	—	38
October	4	12	31	15	8	—	8	8	—	—	—	—	—	8	23
November	7	16	53	—	20	—	—	—	7	20	—	—	—	—	—
December	10	23	48	—	26	—	—	9	9	—	4	4	—	—	—
January, 1975	9	20	30	10	25	10	—	5	5	—	10	—	—	5	—
February	8	16	25	25	13	—	6	6	6	—	—	6	—	13	—
March	6	20	42	11	11	5	16	—	—	—	—	5	—	5	5
April	5	12	17	8	8	8	8	—	8	17	8	—	—	17	—
May	0	0	+	+	+	+	+	+	+	+	+	+	+	+	+
June	6	14	—	20	20	7	7	7	7	7	—	—	7	7	7
July	10	43	23	16	9	19	—	9	3	5	—	—	7	3	2
August	13	20	20	15	—	—	5	—	5	—	5	—	—	15	35
September	0	0	+	+	+	+	+	+	+	+	+	+	+	+	+
October	1	3	—	33	33	—	—	—	33	—	—	—	—	—	—
November	4	23	4	13	13	4	9	4	44	4	—	—	—	4	—
December	0	0	+	+	+	+	+	+	+	+	+	+	+	+	+
January, 1976	8	25	40	16	4	20	4	8	4	—	—	—	4	—	—
February	13	36	19	6	8	33	8	8	3	3	—	3	—	3	6
March	13	35	14	14	8	8	3	11	—	11	11	6	—	3	11
April	7	18	21	5	11	11	26	11	—	—	5	—	—	5	5
May	16	57	18	21	4	14	13	7	2	9	5	2	—	2	4
June	13	35	11	9	14	9	14	9	11	3	—	—	9	9	3
July	8	31	3	7	3	13	48	—	3	7	—	—	—	3	13
August	13	32	6	9	6	—	13	9	3	—	—	—	—	13	41
							Logged rainforest								
September, 1974	4	10	10	20	30	20	—	10	—	—	—	—	—	10	—
October	3	10	20	20	—	10	—	—	20	10	—	—	—	20	—
November	8	16	31	—	19	13	—	6	—	6	—	6	6	6	6
December	9	23	17	4	26	—	9	4	22	—	9	4	—	4	—
January, 1975	1	7	29	29	43	—	—	—	—	—	—	—	—	—	—
February	6	21	14	19	10	29	5	5	5	5	—	5	—	—	5
March	6	21	48	5	14	5	—	5	—	—	5	5	—	10	5
April	5	20	10	15	15	5	—	10	5	5	20	—	—	5	10
May	0	0	+	+	+	+	+	+	+	+	+	+	+	+	+
June	5	16	31	25	7	—	6	6	6	6	—	—	6	—	6
July	4	12	8	8	17	—	8	—	17	33	—	—	8	—	—
August	1	5	40	—	20	20	—	—	—	20	—	—	—	—	—
September	3	18	16	11	16	—	5	11	5	16	—	—	5	—	16
October	2	7	14	14	29	—	—	29	—	—	14	—	—	—	—
November	0	0	+	+	+	+	+	+	+	+	+	+	+	+	+
December	0	0	+	+	+	+	+	+	+	+	+	+	+	+	+
January, 1976	0	0	+	+	+	+	+	+	+	+	+	+	+	+	+
February	3	6	—	17	33	—	17	17	—	—	—	—	—	—	17
March	0	0	+	+	+	+	+	+	+	+	+	+	+	+	+
April	5	15	7	20	20	—	13	7	—	—	20	7	—	7	—
May	6	23	17	17	13	4	9	9	4	—	13	4	—	9	—
June	5	22	14	18	14	—	18	5	23	—	—	—	—	—	9
July	0	0	+	+	+	+	+	+	+	+	+	+	+	+	+
August	4	11	9	18	18	—	9	—	—	—	—	—	9	27	9

Month	N_1	N_2	% occurrence of major groups recognised												
			Co	Ar	Bl	Hy	La	Or	Di	My	De	Ca	Ot	Un	Pl
						Rainforest									
September, 1974	5	18	33	11	22	—	—	6	—	—	—	—	—	—	28
October	1	1	—	—	—	—	—	—	—	—	—	—	—	100	—
November	3	6	50	17	33	—	—	—	—	—	—	—	—	—	—
December	2	2	—	50	—	—	—	—	50	—	—	—	—	—	—
January, 1975	6	11	27	9	36	—	—	18	—	—	9	—	—	—	—
February	12	15	27	20	13	—	—	7	—	—	—	—	—	27	7
March	0	0	+	+	+	+	+	+	+	+	+	+	+	+	+
April	5	12	17	8	8	—	—	17	8	—	—	—	17	17	8
May	0	0	+	+	+	+	+	+	+	+	+	+	+	+	+
June	6	15	13	33	27	—	—	—	—	7	7	—	—	7	7
July	6	22	9	18	18	—	—	9	9	9	—	—	9	14	5
August	0	0	+	+	+	+	+	+	+	+	+	+	+	+	+
September	7	21	19	19	14	—	5	10	5	10	—	—	14	5	—
October	6	24	8	—	8	8	8	8	42	—	—	8	—	4	4
November	0	0	+	+	+	+	+	+	+	+	+	+	+	+	+
December	0	0	+	+	+	+	+	+	+	+	+	+	+	+	+
January, 1976	1	3	—	—	—	—	33	33	—	—	—	—	—	—	33
February	1	1	—	—	—	—	100	—	—	—	—	—	—	—	—
March	0	0	+	+	+	+	+	+	+	+	+	+	+	+	+
April	6	26	8	19	12	31	8	8	—	—	4	—	4	8	—
May	8	22	9	26	9	4	13	—	4	9	9	—	4	9	43
June	8	20	10	30	5	10	10	—	—	—	15	5	—	5	10
July	0	0	+	+	+	+	+	+	+	+	+	+	+	+	+
August	9	31	9	16	7	—	25	6	—	—	—	6	13	16	3
						Tall woodland									
September, 1974	3	4	25	—	25	—	—	—	—	—	—	—	—	50	—
October	4	13	—	—	—	8	—	8	8	—	—	—	69	—	8
November	1	1	—	—	—	—	—	—	—	—	—	—	—	100	—
December	2	4	25	—	25	—	25	—	—	—	—	—	—	25	—
January, 1975	5	9	33	22	11	—	—	—	—	—	—	11	—	11	11
February	4	3	33	—	33	33	—	—	—	—	—	—	—	—	—
March	5	9	22	—	11	—	—	23	11	11	—	11	—	11	—
April	0	0	+	+	+	+	+	+	+	+	+	+	+	+	+
May	0	0	+	+	+	+	+	+	+	+	+	+	+	+	+
June	1	1	—	—	—	—	—	—	—	—	—	—	—	100	—
July	5	14	7	14	7	—	7	—	14	14	14	—	—	7	14
August	11	26	31	19	8	12	8	—	4	4	8	—	—	4	4
September	0	0	+	+	+	+	+	+	+	+	+	+	+	+	+
October	1	1	—	—	—	—	—	—	—	—	—	—	—	100	—
November	0	0	+	+	+	+	+	+	+	+	+	+	+	+	+
December	1	3	—	33	—	—	—	33	—	—	—	—	—	—	33
January, 1976	5	7	—	29	—	—	—	—	43	—	—	—	14	14	—
February	6	16	13	25	13	13	—	13	—	6	—	19	—	—	—
March	12	42	24	19	10	14	12	10	2	—	—	2	5	—	2
April	9	27	12	19	4	4	23	12	4	—	12	—	4	4	4
May	6	9	11	22	11	—	—	11	—	11	—	—	—	22	11
June	4	6	17	—	—	17	—	—	—	—	—	—	—	33	33
July	6	20	—	10	—	—	60	5	5	—	—	—	5	10	5
August	6	21	7	—	—	25	—	—	—	7	—	—	7	19	31
						Dry sclerophyll forest									
February, 1976	8	20	20	10	10	10	5	15	10	—	—	—	—	10	10
March	12	38	32	5	11	29	3	8	—	5	3	5	—	—	—
April	10	17	18	24	6	6	18	12	—	—	12	—	6	—	—
May	12	32	19	6	—	19	16	9	6	13	6	—	—	—	6
June	13	20	—	30	5	10	15	5	—	5	5	10	—	5	10
July	5	9	—	35	—	—	11	—	22	—	22	—	—	11	—
August	11	21	10	14	5	—	14	10	—	5	5	—	5	14	19

N_1 = Number of scats examined; N_2 = Number of identified fragments; Co = Coleoptera — excluding carabs; Ar = Aranaea; Bl = Blattodea; Hy = Hymenoptera; La = Larvae; Or = Orthoptera; Di = Diptera; My = Myriapoda; De = Dermaptera; Ca = Coleoptera (carabs only); Ot = Others; Un = Unidentified; Pl = Plant Material; + = No samples analysed.

16. HABITAT UTILIZATION OF *Antechinus stuartii* (MARSUPIALIA) AT PETROI, NORTHERN NEW SOUTH WALES

H. L. STATHAM[1] AND R. H. HARDEN[2]

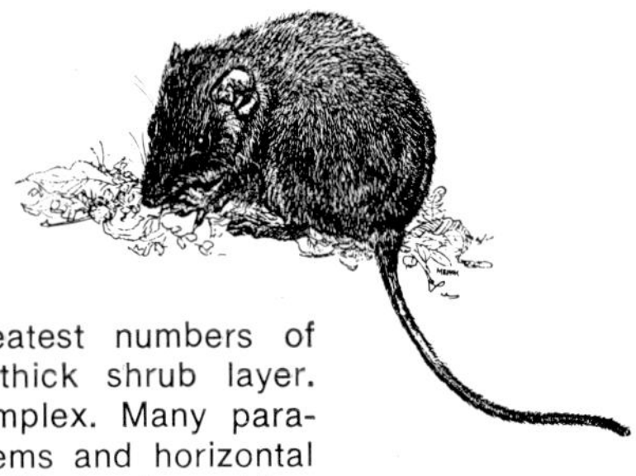

Antechinus stuartii were live trapped monthly on grids of nine hectares in tall woodland, wet sclerophyll forest, cool temperature rainforest, and logged cool temperate rainforest from March, 1974, until September, 1976, and in dry sclerophyll forest from February, 1976, until September, 1976. Habitat was quantified by the technique of multiparameter measurement of variables at fixed circular plots, with quantification of parameters of possible specific significance to small mammals, e.g. number of hollow trees and logs, in conjunction with general habitat quantification. In all habitats, the greatest numbers of *A. stuartii* were captured at stations characterised by a relatively thick shrub layer. However, habitat requirements for this species were found to be complex. Many parameters, including log length, the number of dwarf shrubs and high stems and horizontal visual obstruction at 50 cm appeared important at different times and in differing intensities.

Pages 165–85 *in* CARNIVOROUS MARSUPIALS, ed. by M. Archer, February, 1982. Roy. Zool. Soc. New South Wales: Sydney, Australia.

FEW FIELD STUDIES on *Antechinus stuartii* have been made other than those designed to collect material for reproductive or taxonomic work (e.g. Horner and Taylor 1959, Woolley 1966a, 1966b). Wood (1970) conducted a three year capture–recapture study of *A. stuartii* in southeast Queensland but his study area of 0.8 ha was not large enough to cover the home range of any males. He gave a brief, non-quantitative account of the animals' habitat. Braithwaite (1974), working in the same area, monitored behaviour of field populations of *A. stuartii*. Leonard (1970, 1972) studied the effect of control burning on the ecology of several small mammal species including *A. stuartii*. He also gave qualitative but not quantitative habitat relationships. The effects of fire on small mammal populations including *A. stuartii* in heath and forest areas of New South Wales have been reported by Recher *et al.* (1978) and Fox (1980).

Habitat has been defined as "a physical entity possessing a number of measurable features or dimensions" (Emlen 1956). In this study we wish to make a quantitative assessment of habitat and, by relating this to the distribution and abundance of *A. stuartii,* to identify habitat parameters of importance to *A. stuartii*. Such information can be used in the description and understanding of the habitat preference or requirements of this species.

This research is part of a team study of various aspects of the Petroi ecosystems being made by the National Parks and Wildlife Service and students of the University of New England.

THE STUDY AREA

Petroi is situated in the scarplands of northern New South Wales (30° 37′ S, 152° 20′ E) approximately 70 km east of Armidale and 15 km southwest of Point Lookout, New England National Park. It is a plateau of about 3,500 hectares at an altitude of 1,000 metres.

The parent rock covering most of Petroi, including all areas considered in this study is Botumburra adamellite (Leitch 1972). The soils are shallow and acid with a mean pH of 4.6 (Harden, unpublished data). The mean annual rainfall on Petroi is in excess of 1,500 mm although the annual total rainfall fluctuates widely due to variable easterly rains during summer. Daily temperatures range from 14°C to 32°C in January and from –1°C to 12°C in July.

Vegetation on Petroi consists of a continuum of sub-forms ranging from sod tussock grassland through tall woodland, dry and wet sclerophyll forests to temperate rainforest (Beadle and Costin 1952).

Until the 1930's Petroi was used as an initiation area for the Macleay River Aborigines. The area was taken up for selection in 1889 and since then the grassland and adjacent sclerophyll areas have continuously supported 50–100 head of cattle. The National Parks and Wildlife Service has been carrying out wildlife research in the area since 1969. Intermittent logging of both softwood and hardwood species also commenced at that time.

THE HABITATS

Small mammal trapping grids were located in the most homogeneous nine hectare area of

[1] School of Natural Resources, University of New England, Armidale, New South Wales. Present address: Forestry Commission Tasmania; c/- Department of Agriculture, P.O. Box 407, Launceston, Tasmania, Australia, 7250.
[2] National Parks and Wildlife Service, c/- Zoology Department, University of New England, Armidale, New South Wales, Australia, 2351.

each of tall woodland, dry sclerophyll forest, wet sclerophyll forest and logged and unlogged cool temperate rainforest. No area was completely homogeneous and eight vegetation types, based on the classification of Williams (personal communication) were described:

Tall woodland: overstorey of mature ecualypts characterised by a crown depth to bole length ratio of approximately 1:1 compared to <1:1 for forest, sparse shrub layer.

Dry sclerophyll forest: overstorey of mature eucalypts, well developed xeromorphic shrub layer.

Wet sclerophyll forest type II: overstorey of mature eucalypts with some mesomorphic shrub species, tree ferns present.

Wet sclerophyll forest type III: overstorey of mature eucalypts with more mesomorphic shrub species, vines present in the shrub layer.

Wet sclerophyll forest type IV: scattered eucalypts throughout the overstorey, mature rainforest trees present.

Cool temperate rainforest: eucalypts absent, tree layer dominated by *Nothofagus moorei* with an understorey of *Ceratopetalum apetalum* and other mesmorphic species (this rainforest is classified as *Nothofagus moorei*–alliance by Beadle and Costin 1952 and Baur 1957).

Ecotone: vegetation transitional between wet sclerophyll forest type II and dry sclerophyll forest.

Closed scrub: dominated by *Callicoma serratifolia,* tree ferns present, found in gullies.

TALL WOODLAND: The dominant tree species are *Eucalyptus andrewsii* and *E. cameronii* with *E. saligna* in the more elevated areas and *E.*

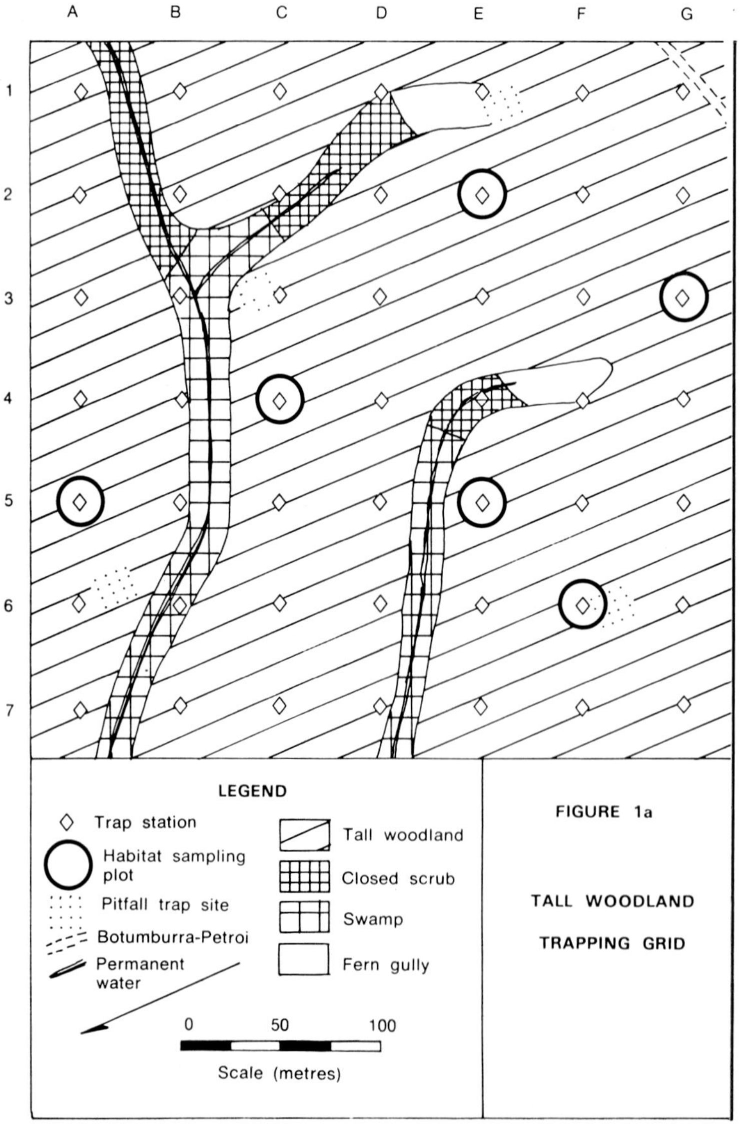

FIGURE 1a

TALL WOODLAND TRAPPING GRID

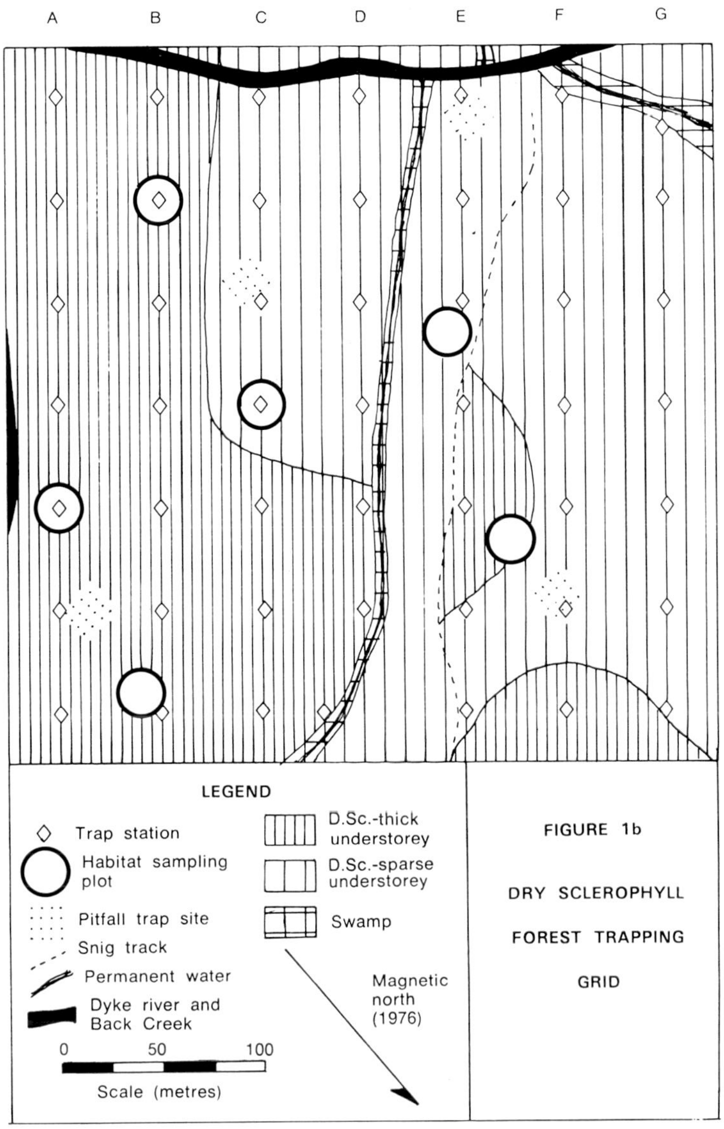

FIGURE 1b

DRY SCLEROPHYLL FOREST TRAPPING GRID

viminalis lining the gullies. Several clumps of *Casuarina rigida* trees and a dense patch of *Acacia filicifolia* suckers are present. Shrubs are sparse on the northern side of the grid and absent on the southern side.

The two gully systems which dissect the grid in an east–west direction contain ferns and closed scrub which are replaced down–stream by *Leptospermum* and reed swamps (Fig. 1a).

Some of the 50–100 head of cattle grazing on Petroi use the area of this grid regularly. The landowner normally lights ground fires through the area each year, but due to this trapping programme he lit it only once during the study. Thus the area to the south and west of the forked gully was burnt in September, 1974.

DRY SCLEROPHYLL FOREST: The dominant tree species are *Casuarina rigida, Eucalyptus andrewsii, E. radiata* and *E. caliginosa.* The grid can be divided into two dry sclerophyll types on the characteristics of the shrub layer one has a well developed shrub layer dominated by *Tasmannia stipitata,* whilst in the other the shrub layer is sparse.

The grid is bordered to the southeast by Back Creek and to the southwest by Dyke River. A thickly vegetated gully runs through the grid separating trap lines A, B, C and D from lines E, F, G (Fig. 1b). Cattle are frequently present throughout the grid.

WET SCLEROPHYLL FOREST: The vegetation ranges from wet sclerophyll forest types II and III in the southern section of the grid through an ecotone to dry sclerophyll forest along the northern edge (Fig. 1c). The dominant tree species are *Eucalyptus andrewsii* and *E. cameronii.*

The shrub layer is dense. Closed scrub occupies two gullies on the western side and the north–west corner of the grid contains a swamp dominated by *Leptospermum flavescens, Baeckea utilis* and *Sphagnum* sp.

The area was selectively logged in 1971. The resultant snig tracks and other disturbed areas were initially colonised by secondary succession species notably *Zieria smithii* and *Haloragus micrantha.* More recently these have been partially replaced by *E. andrewsii, E. cameronii* and *Acacia falciformis* saplings and clumps of *Gahnia erythrocarpa.* Cattle are often present in the northern part of the grid.

COOL TEMPERATE RAINFOREST: Although only the central 60 per cent of this grid is composed of true cool temperate rainforest, the area represents the largest, most uniform patch of this vegetation type remaining on Petroi. This rainforest is dominated by *Nothofagus moorei* with an understorey of *Ceratopetalum apetalum, Ackama paniculata, Doryphora sassafrass* and other mesophyll species. Many epiphytes grow on these trees. The sparse shrub layer consists mostly of climbers which include *Cephalaralia cephalobotrys, Ripogonum discolor* and *Smilax glyciphylla.* Except for a few dense patches of ferns, the forest floor is generally bare of vegetation and is covered by a layer of dead leaves and twigs. Towards the margins of the grid the vegetation grades to wet sclerophyll forest types III and IV.

A permanent creek flows along the eastern side of the grid from G2 to F7 with gullies joining near E4 and F5. A semi-permanent creek flows across the northern side of the grid (Fig. 1d).

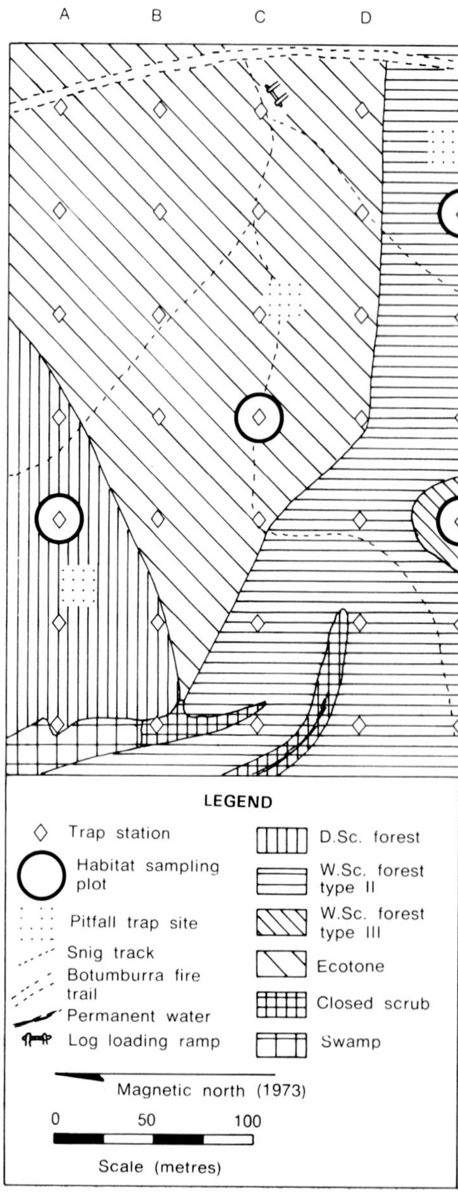

FIGURE 1c

WET SCLEROPHYLL FOREST TRAPPING GRID

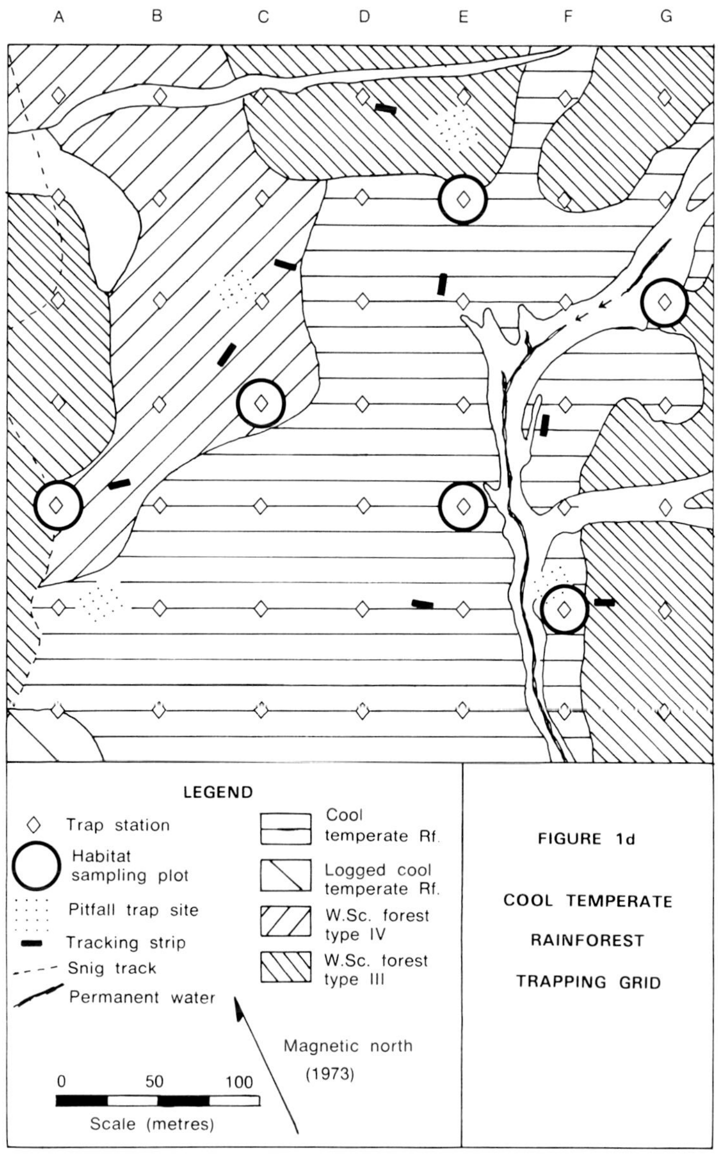

FIGURE 1d

COOL TEMPERATE RAINFOREST TRAPPING GRID

Logged Cool Temperate Rainforest: Until 1971, 95 per cent of this grid contained cool temperate rainforest dominated by *Nothofagus moorei* with an understorey of *Ceratopetalum apetalum* (Coachwood), *Ackama paniculata, Doryphora sassafrass* and other mesophyll tree species. In 1971, all millable Coachwood was logged from this patch of rainforest leaving a maze of snig tracks, large log piles, several clearings and three log loading ramps. Many trees which survived the logging process have subsequently been weakened by exposure and many have died. Thus at the time of study, falling trees were common, especially in windy weather.

Cleared areas were initially colonised by secondary succession species notably *Solanum* sp. More recently large clumps of *Gahnia erythrocarpa* have formed, blocking many of the former snig tracks. *N. moorei* and *C. apetalum* are both regenerating profusely from the seedlings and by coppicing.

A permanent creek flows along the south–east side of the grid beside the G. trap line. It crosses the grid from G5 to A7. A semi–permanent creek flows from a seepage near C1 and joins the main creek at C6 (Fig. 1e).

METHODS

The capture–recapture technique and grid layout were used to study *A. stuartii* population dynamics in five habitats described above. The grid system used was designed to cover as large an area as possible whilst enabling capture of most of the trappable small mammals in the area. The wide trap spacing of 50m was based on 200 foot observed range length for a male *A. stuartii* in south–east Queensland sub–tropical rainforest (Wood 1970). Each grid consisted of seven rows (A–G) of seven trap stations (1–7) and covered an area of nine hectares.

From March, 1974, until January, 1976, the tall woodland, wet sclerophyll and logged and unlogged cool temperate rainforest grids were trapped on four consecutive nights per month. The dry sclerophyll grid was established in February, 1976, to test *A. stuartii* habitat preference models derived from results obtained in the other four trapping grids. From that time until September, 1976, all five grids were trapped for three consecutive nights each month. Throughout the study the number of traps per trap station was adjusted so that there was always at least one spare trap at each trap station.

Elliott Type A live traps (aluminium: 34 cm x 10 cm x 10 cm) were used. Traps from all grids were collected and washed every three to four months. Those from the tall woodland and dry sclerophyll forest grids were collected between trapping periods to prevent trampling by cattle.

Three measures were used to reduce the 25% *A. stuartii* trap mortality which had occurred in 1973 (Fletcher 1974). Firstly, a handful of dry leaves was placed inside each trap behind the trigger plate; secondly, each trap was wrapped in a sheet of clear plastic; and thirdly, traps were cleared at night during winter and on exceptionally cold nights at other times of the year.

For most of the year, traps were cleared of animals at first light and left open for the next

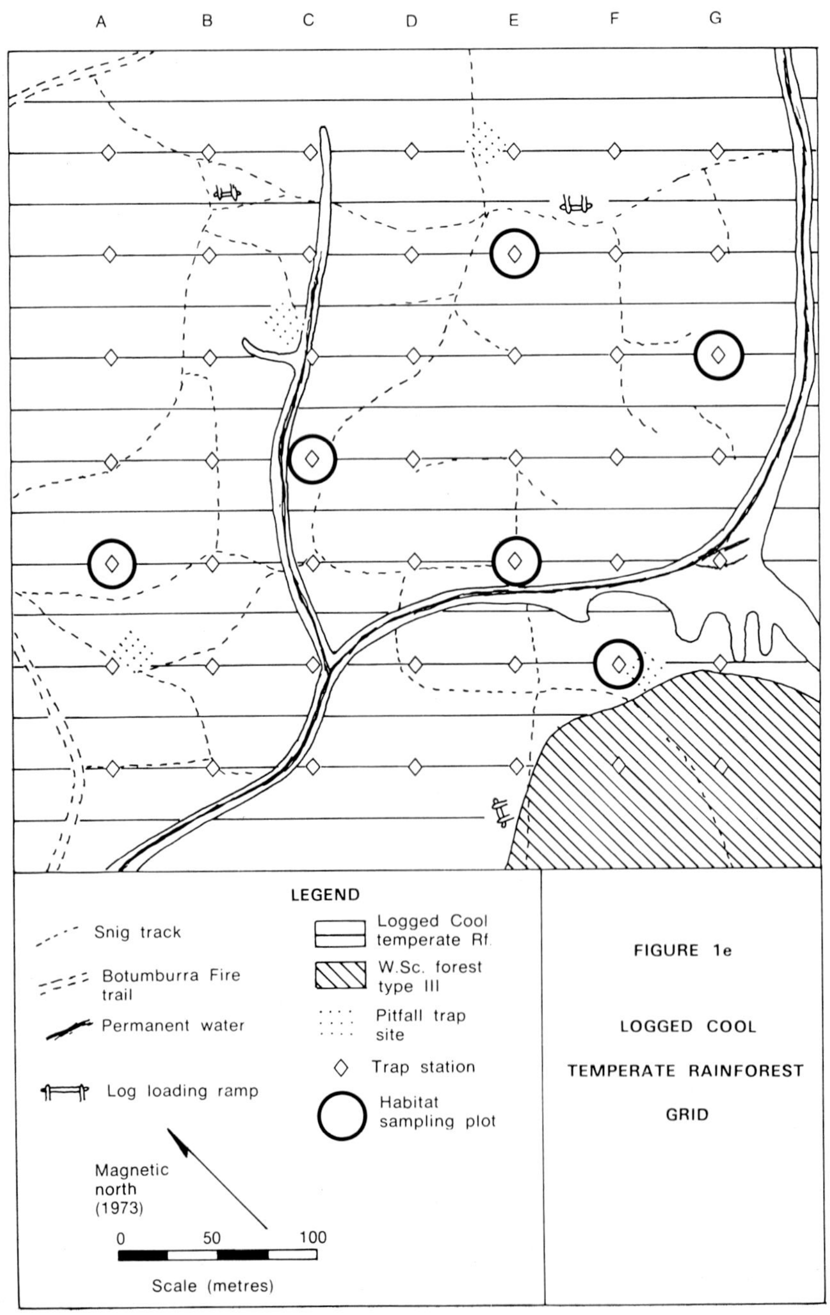

FIGURE 1e

LOGGED COOL TEMPERATE RAINFOREST GRID

night's trapping. During winter months when traps were cleared at night (starting at 2300 hours) they were closed after clearing and reopened the following day.

Captured animals were individually marked by an ear cutting system. Each *A. stuartii* was weighed and its number, sex, reproductive status, head-length and trap site were recorded prior to release.

Habitat was quantified by the measurement of a number of variables at fixed circular plots. This method was chosen for two reasons. Firstly, whilst plotless designs have been used to obtain structural data (Knight 1963), they have been found to be less accurate than areal methods (Daubenmire 1959, James and Shugart 1970 and Posamentier 1975). Secondly, the circular shape had been shown to have better field and computational efficiency than other plot designs (Lindsay *et al.* 1958). The basic methodology, modified from James and Shugart (1970) was developed at Petroi by Harden, Statham and Muir (in prep.) and was described by Fletcher (1974). The method is flexible in that certain basic parameters are usually measured but any other measurements of particular relevance to the study in question may be included. Other recent Australian methods of measuring small mammal habitats have been developed for heath rather than forest habitats (Braithwaite and Gullan 1978, Fox 1979).

Six plots were located at randomly chosen trap stations in each trapping grid. Exact specifications of the plot design are given in Fig. 2. Parameters of possible specific significance to small mammals, e.g. number of hollow logs and trees, were counted in conjunction with general habitat quantification. Some measurements such as tree and log counts were made only once during the three year of study whilst others such as shrub counts were made annually to measure temporal changes.

Most plants were identified using the Petroi reference herbarium and standard texts (Francis 1951, Gray 1961, Anderson 1968, Hall *et al.* 1970, Beadle 1971, 1972 and Beadle *et al.* 1972). Mr. J. B. Williams (Botany Department, University of New England) identified unknown specimens.

i) Tree Counts: A tree was defined as a woody plant of height greater than eight metres and a stump as a tree broken or cut off below eight metres. Stems separate at breast height but

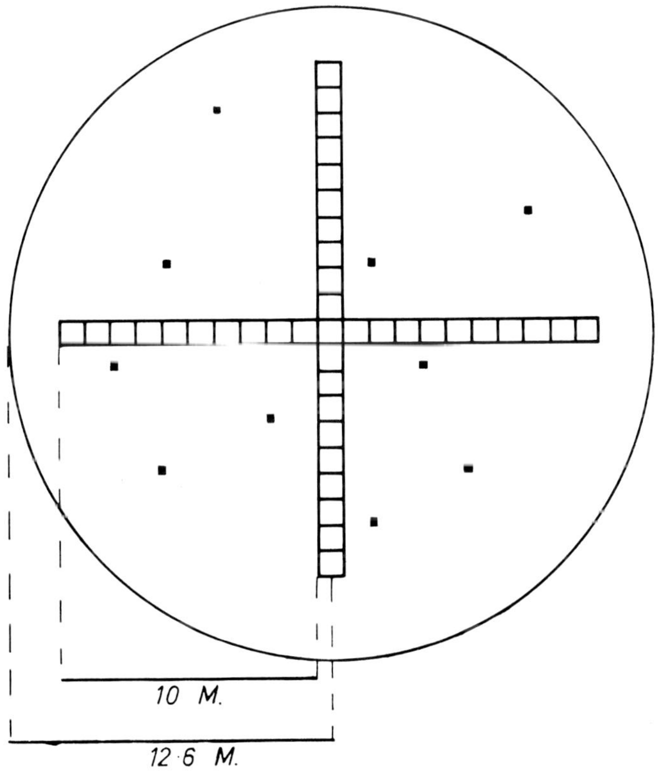

Fig. 2. Habitat analysis, plot design, showing 0.05 ha grid, 40 square metre transects and ten randomly allocated quadrat sites.

sharing a common root system were counted individually. The species, height and diameter at breast height (D.B.H.) of all live and dead trees and stumps within the 0.05 ha plot were recorded. Trees and stumps were examined for the presence of holes of diameter > 2.5 cm occurring up to one metre above ground. Counts were made once during the study.

ii) Log Counts: All logs within the 0.05 ha plot were scored for the following parameters once during the study:

a) diameter class: 1: 6–25 cm; 2: 26–50 cm; 3: 51–75 cm; 4: 76–100 cm; 5: > 100 cm;

b) part of tree: bole or crown;

c) state of decomposition: green, dead or rotten;

d) length of log in the plot: metres;

e) soundness: sound or hollow. A log was classified hollow if it had an internal pipe of diameter > 2.5 cm or if there were at least one hole of diameter > 2.5 cm penetrating more than 30 cm into it.

iii) Shrub, Vine, Sedge and Fern Counts: A shrub was defined as a woody plant of height less than eight metres, and a vine as a climbing plant. The two groups were divided into three height classes, dwarf 0–1 m, low 1–2 m and high 2–8 m. All shrub, vine, sedge and fern stems encountered in the 40 square metres of transect were recorded according to species and height. Counts were made three times, in August of 1974, 1975 and 1976.

iv) Ground Cover: Ground cover was defined as live vegetation or logs. It was scored as a percentage estimate from a height of one metre for each of the 40 square metre sections. Measurements were made twice, in August of 1975 and 1976.

v) Horizontal Visual Obstruction: A sight board made from a piece of timber 100 cm x 10 cm was used to measure horizontal obstruction. The board was divided into alternate red and white squares numbered 1 to 10 with black paint. Horizontal obstructions over five metres distance was measured along the north, south, east and west transects at 10, 50, 100 and 200 cm above ground. Readings were taken with one eye, at the same height as the board. A square was scored positively obstructed if less than half of it could be seen. Measurements were made at all trap sites in the initial four grids in August, 1975 and again in August, 1976, when all trap sites in the dry sclerophyll grid were also measured.

vi) Light Penetration: The vertical incident light intensity (in foot candles) was measured at 0 m, and 2 m above ground level at 25 points in the transect using a Gossen Lunasix three light meter. These readings were taken between 1000 h and 1400 h on clear sunny days. Percent penetration was then calculated by comparison of light intensity at the above positions with light intensity on the grassland at the same time. Measurements were made in August, 1975, but could not be repeated in 1976 due to the cloudy weather.

vii) Biomass in the First Metre: Square quadrats of area 0.1 sq. m were used to measure biomass of live and dead material in the 0–1 m height class. They were cut twice, in August of 1974 and 1976. Two different sets of ten random sites per plot were used for the two samplings. Material for the 10 sites per plot was pooled in the following categories: shrubs, herbs, wood (1–5 cm diameter) and litter (0–1 cm diameter). In August, 1976, the additional categories vines, ferns, grasses and forbs were used. The samples were dried in a forced draught oven at 70°C for 48 hours and then weighed.

RESULTS

A total of 968 individual *A. stuartii* was captured during this study. The cumulative numbers of individuals captured each generation in each habitat are shown in Fig. 3. More individuals were captured in the wet sclerophyll forest each generation than in any other habitat. Numbers in the other habitats were similar to each other with no repeatable trend between generations. The curves usually flattened off except in logged cool temperate rainforest in 1974, tall woodland in 1975, and cool temperate rainforest in 1975 and 1976 when large numbers of new animals were captured in the last trapping period of the season.

The proportions of males, juvenile females and parous females in the monthly known to be alive estimates for each habitat are given in Fig. 4. Juvenile *A. stuartii* first entered the traps in December or January, about three months after their birth in September. Peak capture records and numbers known to be alive occurred in autumn except in cool temperate rainforest in generation 3 when numbers peaked in September. There were usually more males than juvenile females known to be alive, with exceptions in logged cool temperate rainforest during 1974 and in tall woodland in 1975.

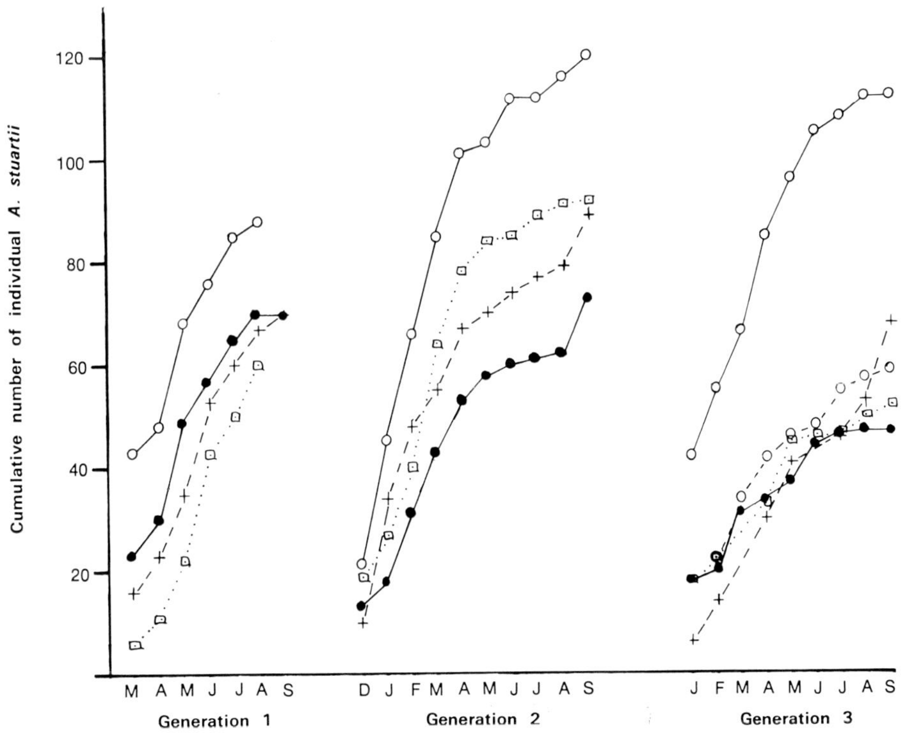

Fig 3. A. stuartii cumulative number of individuals captured each generation in each habitat. Symbols: o———o Wet sclerophyll forest; •———• Tall woodland; o- - -o Dry sclerophyll forest; □ . . . . □ Logged cool temperate forest; + - - - + Cool temperate rainforest.

The maximum number of juvenile *A. stuartii* known to be alive during one month of each generation was always greatest in wet sclerophyll forest. Here the numbers increased over the three generations from 50 to 66. The trend was reversed in each of the other habitats. In tall woodland the maximum number of juveniles known to be alive at any one time decreased from 40 to 18 over the three years, whilst the numbers in logged and unlogged cool temperate rainforest decreased from 32 to 23 and from 37 to 31 respectively.

For the purposes of comparison between habitats and with other studies, the estimation of population density is more valuable than that of population size. Several authors (e.g. Stickell 1954, Pelikan 1967, Hansson 1969 and Faust *et al.* 1971) refer to the problem of obtaining density estimates from capture–recapture and/or removal studies due to the difficulty of estimating the actual sampling area. Stickel (1954) proposed the addition of a buffer zone strip BZ to the area of the grid trapped. He showed that the most accurate estimates were obtained if the width BZ was half the observed range length (O.R.L.) of the species being considered. Pelikan's (1967) method consisted of calculating a central homogenous area and discarding all data obtained from trapping stations outside this area.

These ideas were considered in the calculation of densities of *A. stuartii* in each habitat. The numbers of individual *A. stuartii* captured at interior and peripheral traps were statistically compared by the analysis of varience. In each case, there were no significant differences between the numbers of individuals captured at the 25 interior traps and at the 24 peripheral traps of the 7 x 7, nine hectare grids. That is, there was no edge effect shown in the capture data. Since the O.R.L. for male and female *A. stuartii* in each trapping grid ranged approximately from 86 metres to 160 metres (Fletcher 1977), no precise value for the width of BZ could be given. A buffer zone of 50 metres was therefore assumed for each trapping grid making the effective trapping area 16 hectares. The maximum number of individuals per hectare for each grid during each generation was then determined by dividing the maximum number of *A. stuartii* known to be alive during that generation by 16 (Table 1).

A. stuartii density ranged from 4.4 per hectare in wet sclerophyll during generation 3 to 1.6 per hectare in the logged rainforest also during generation 3. The mean density over the three generations was 3.7 individuals per hectare for wet sclerophyll and about two individuals per hectare for each of tall woodland logged and unlogged rainforest. The density estimate increased each year for wet sclerophyll and decreased each year in tall woodland and rainforest. In the logged rainforest *A. stuartii* density was greatest during the second generation and lowest in the third. Finally, the 1976 density in dry sclerophyll was similar to that in the rainforest in that year.

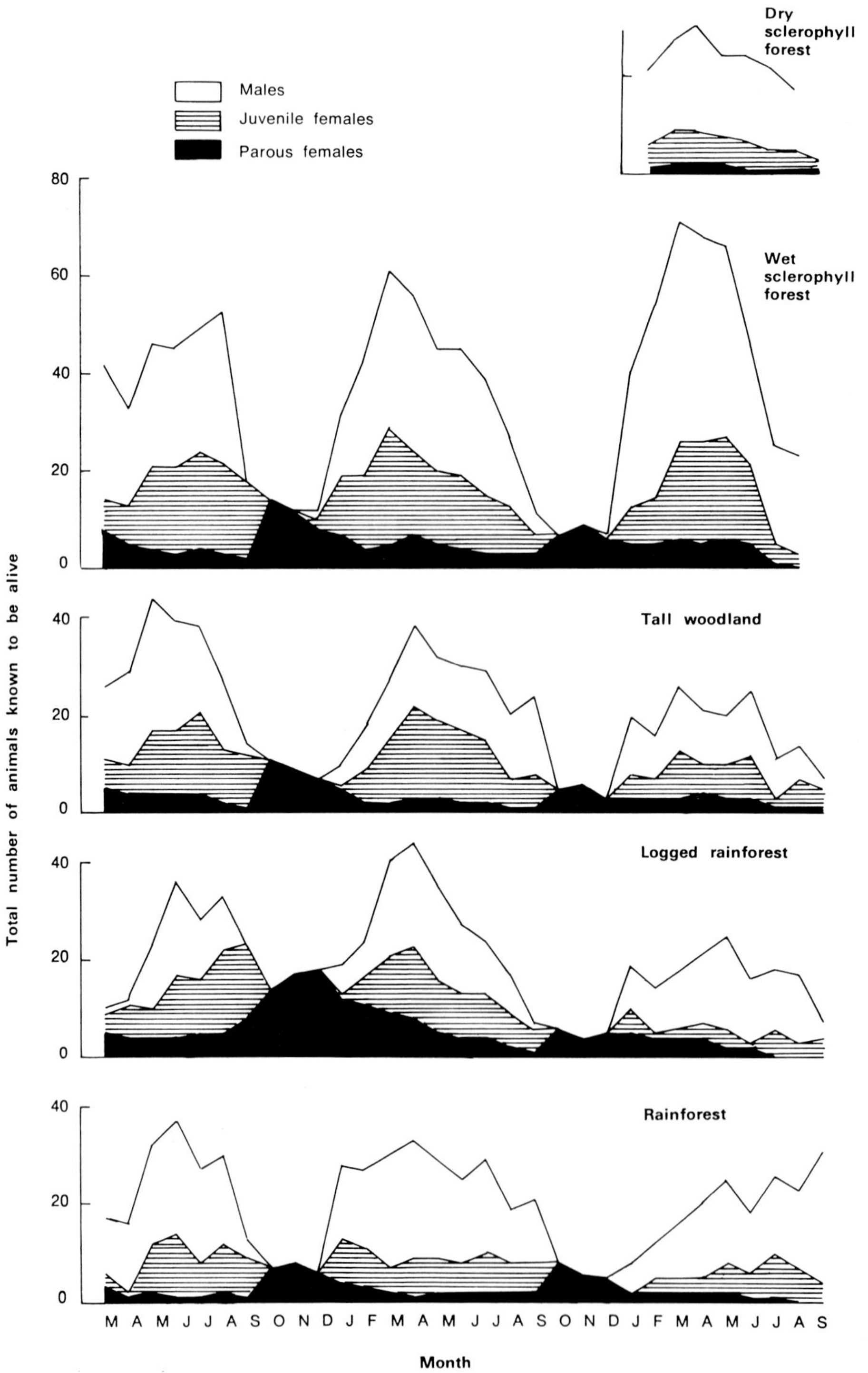

Fig. 4. A. stuartii population structure, number of males, juvenile and parous females known to be alive each month in each habitat, March, 1974–September, 1976.

Table 1. A. stuartii maximum density (number per hectare) in each habitat for each generation.

Generation		Habitat				
		T.Wdl.	D.Sc.	W.Sc.	Rf.	L.Rf.
1.	March–September, 1974	2.8		3.3	2.3	2.3
2.	December, 1974–September, 1975	2.4		3.5	2.1	2.8
3.	December, 1975–September, 1976	1.6	1.9	4.4	1.9	1.6
Mean		2.3	1.9	3.7	2.1	2.2

Abbreviations: T.Wdl. = tall woodland; D.Sc. = dry sclerophyll forest; W.Sc. = wet sclerophyll forest; Rf. = cool temperate rainforest; L.Rf. = logged cool temperate rainforest.

Detailed statistical analyses of the habitat parameters have been given previously (Fletcher 1977). Data obtained from the log counts, 1976 shrub, vine, sedge and fern counts and 1976 ground cover estimates have been used to construct structural diagrams representing vegetation below eight metres in each habitat (Fig. 5).

The correlation matrices given in Tables 2, 3 and 4 show values and significance levels of r for measurements of 10 habitat parameters made in wet sclerophyll forest, logged and unlogged cool temperate rainforest and tall woodland in 1974, 1975 and 1976. Only those correlations significant at the 0.01 level are described here.

In 1974, dwarf shrubs were positively correlated with dwarf total stems, log length, percentage ground cover and horizontal visual obstruction (H.V.O.) (50 cm). Dwarf total stems were positively correlated with log length and H.V.O. (50 cm). Log length was positively correlated with H.V.O. (50 cm) and biomass of litter below 1 m which was itself negatively correlated with light penetration at 0 m.

Only three comparisons differed in significance between 1974 and 1975. Dwarf stems had become positively correlated with biomass of green plant material below 1 m and were no longer highly correlated with log length. Also,

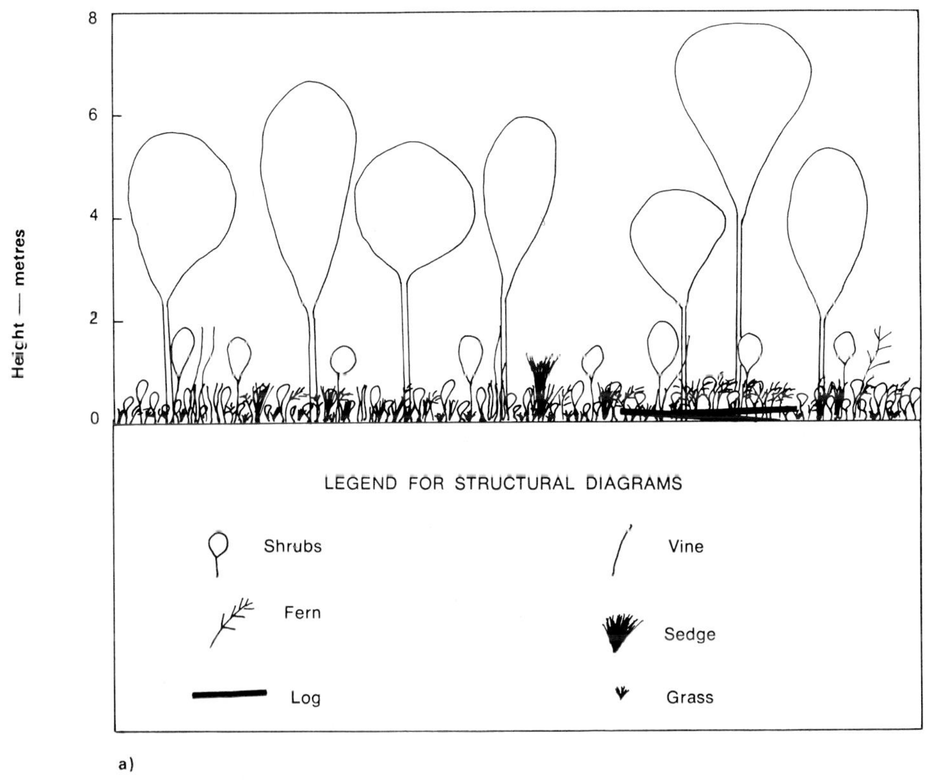

Fig. 5. Diagrammatic representation of vegetation structure in the shrub layer. Each diagram represents a transect strip 20 m × 0.5 m. (a) Wet sclerophyll forest.

dwarf stems were no longer highly correlated with ground cover.

Light penetration estimates were not obtained in 1976. In that year, dwarf shrubs and dwarf total stems were positively correlated with each other and with log length, H.V.O. (50 cm) and biomass of live plant material below 1 m. High stems were positively correlated with H.V.O. biomass of live plant material below 1 m which was itself positively correlated with H.V.O. (50 cm).

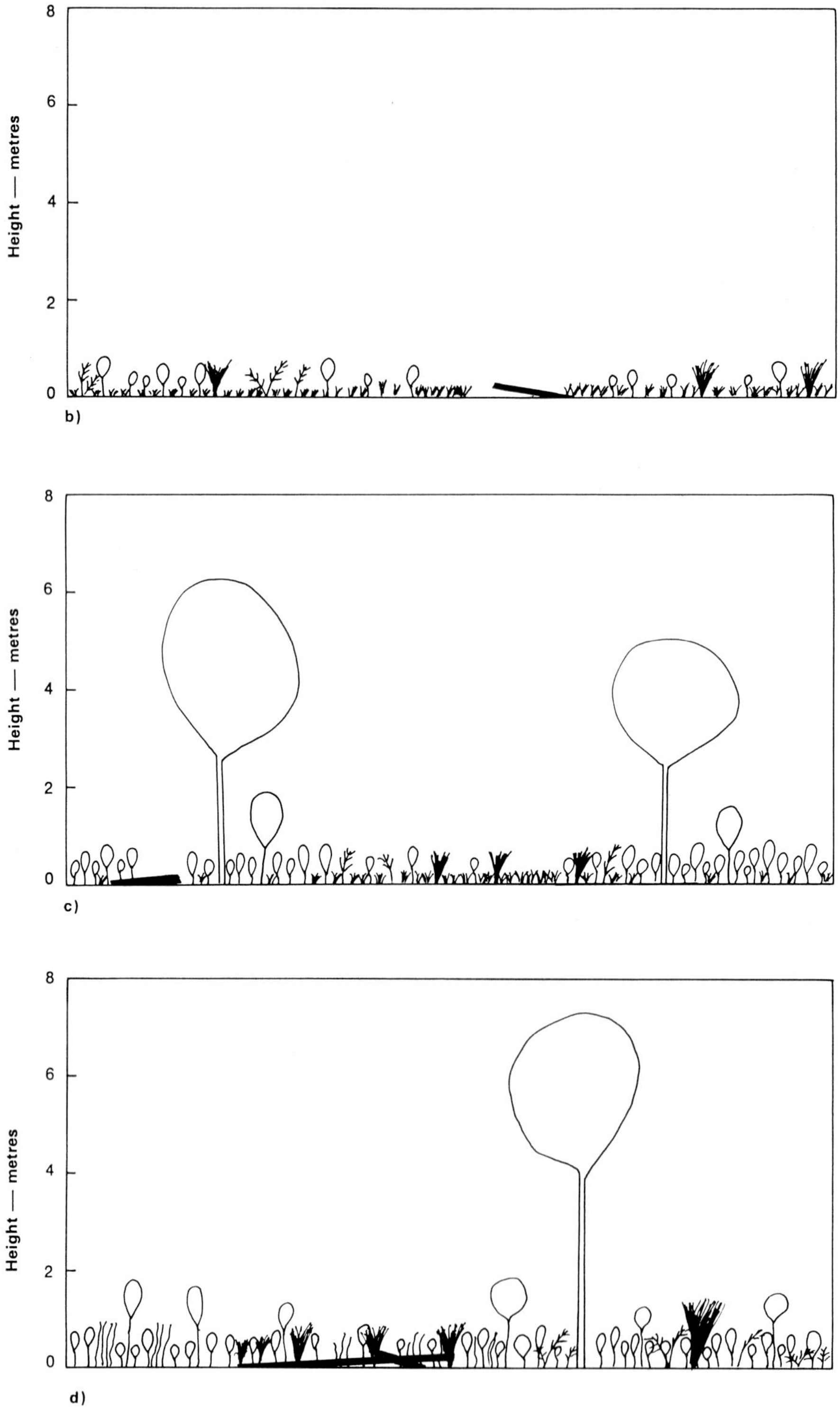

Fig. 5 (continued). (b) Tall woodland, (c) Dry sclerophyll forest. (d) Logged rainforest

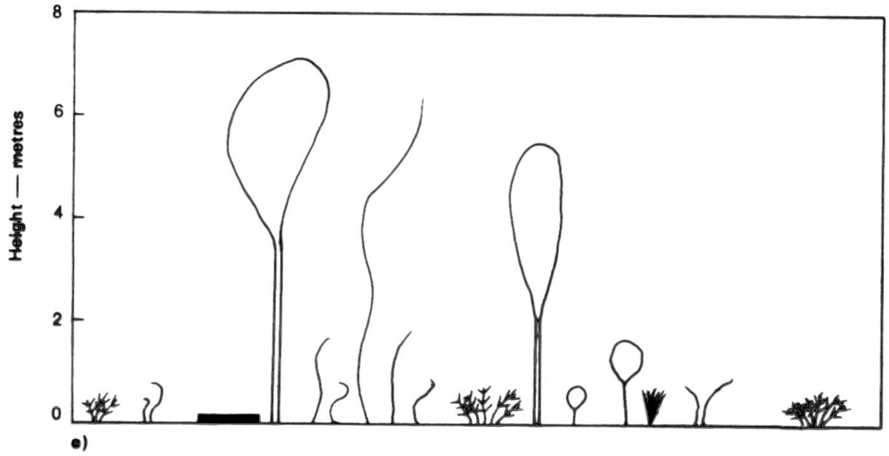

Fig. 5. (continued). (e) Rainforest.

Table 2. Correlation matrix (r) for 10 habitat parameters, N = 24, August, 1974 data

	Dwarf shrubs †	Dwarf stems †	High stems †	Log length !	Nest sites !	Ground cover ‡	H.V.O. 50 cm ‡	Bio-mass green cut †	Bio-mass litter †
Dw. shrubs									
Dw. stems	0.80***								
H. stems	0.41*	0.48*							
Log length	0.70***	0.58**	0.42*						
Nest sites	–0.32	–0.18	–0.20	–0.11					
Ground cover	0.59**	0.37–	–0.09	0.22	–0.39–				
H.V.O. 50 cm	0.53**	0.71***	0.44*	0.75***	0.07	0.27			
Biomass Green cut	0.51*	0.46*	0.27	0.07	–0.23	0.40–	0.14		
Biomass Litter	0.22	0.17	0.28	0.62**	0.26	–0.23	0.47*	0.03	
Light penetr. OM	–0.05	–0.39–	–0.40–	–0.38–	–0.23	0.50*	–0.43*	0.05	–0.60**

Abbreviations as follows: Stars denote level of significance; – = 0.1; * = 0.05; ** = 0.01; *** = 0.001; † measurements made 1974; ‡ measurements made 1975; ! measurements once during whole study.

Table 3. Correlation matrix (r) for 10 habitat parameters, n = 24, August, 1975 data.

	Dwarf shrubs ‡	Dwarf stems ‡	High stems ‡	Log length !	Nest sites !	Ground cover ‡	H.V.O. 50 cm ‡	Bio-mass green cut †	Bio-mass litter †
Dw. shrubs									
Dw. stems	0.82***								
H. stems	0.40–	0.42*							
Log length	0.52**	0.48*	0.43*						
Nest sites	–0.34	–0.19	–0.01	–0.11					
Ground cover	0.47*	0.43*	–0.13	0.22	–0.39–				
H.V.O. 50 cm	0.55**	0.65***	0.49*	0.75***	0.07	0.27			
Biomass green cut	0.39–	0.58**	0.28	0.07	–0.23	0.40–	0.14		
Biomass litter	0.09	0.20	0.24	0.62**	0.26	–0.23	0.47*	0.03	
Light penetr. OM	–0.17	–0.43	–0.44*	–0.37–	–0.23	0.50*	–0.43*	–0.05	–0.60**

Table 4. Correlation matrix (r) for nine habitat parameters, n = 24, August, 1976 data.

	Dwarf shrubs +	Dwarf stems +	High stems +	Log Length +	Nest sites !	Ground cover !	H.V.O. 50 cm +	Bio-mass green cut +
Dw. shrubs								
Dw. stems	0.79***							
H. stems	0.48*	0.48*						
Log length	0.56**	0.54**	0.46*					
Nest sites	−0.36–	−0.21	−0.05	−0.11				
Ground cover	0.33	0.20	−0.04	0.08	−0.34			
H.V.O. 50 cm	0.62**	0.68***	0.64***	0.74***	−0.08	0.02		
Biomass green cut	0.71***	0.78***	0.37–	0.71***	−0.32	0.25	0.62**	
Biomass litter	−0.09	0.05	0.30	0.32	0.40–	−0.22	0.34–	0.04

Abbreviations as follows: Stars denote level of significance: – = 0.1; * = 0.05; ** = 0.01; *** = 0.001; ! measurements made once during whole study; + measurements made in 1976; ‡ measurements made in 1975; † measurements made in 1974.

The numbers of individual *A. stuartii* captured at each trap station of each trapping grid during each generation, are shown in Fig. 6. Major drainage patterns and vegetation types identified in this study are given for each trapping grid.

In wet sclerophyll forest the distribution of *A. stuartii* was continuous over the whole nine hectares. Although the number of individuals captured per trap station within a particular generation ranged from 15 (A5, generation 2) to one (E7, generation 2 and 3 and G4, generation 3), there was no definite pattern in the distribution. More individuals seem to have been captured at traps in the dry sclerophyll and ecotonal areas of the A and B lines during generations 1 and 2, than at most trap stations in the true wet sclerophyll areas. During generation 3, however, this trend was reversed. Thus, only a few trap station (e.g. B4, B5, E5 and F7) consistently captured large numbers of individuals over the three generations. The trap stations at which few individuals were captured (e.g. C1, C7, E7, G4 and G7) were all located in small clearing caused by logging, such as log loading ramps (C1) or snig tracks.

There was a definite pattern in the distribution of *A. stuartii* in tall woodland. The most noticeable feature was the increased number of individuals captured in or adjacent to gullies. B2, C1, C2, D7 and E5 are the only trap stations where 10 or more individuals were captured within one generation. There was also a decrease in number of individuals captured as one goes across the grid from the A line to the G line.

In dry sclerophyll forest between February and September, 1976, *A. stuartii* were captured in 45 of the 49 trap stations. All four trap stations with zero individuals and most with only one or two were located in the areas characterised by a sparse shrub layer. Thus, although *A. stuartii* were distributed over most of the grid, more individuals were captured in the areas that either had a dense shrub layer or were close to water.

A. stuartii were captured in all cool temperate rainforest trap stations, but not during each generation. The peaks in the number of individuals at G5 and G6 in generation 1 were probably influenced by the densely vegetated gully at G5. No such explanation is proposed for the peak at B6, which also occurred in generation 1. As in logged cool temperate rainforest, the number of individuals captured at a particular trap station fluctuated widely between generations. In this habitat, the number of individuals per trap station was generally intermediate between values for wet sclerophyll and logged cool temperate rainforest.

The distribution of *A. stuartii* captures in logged cool temperate rainforest was more uneven that in wet sclerophyll forest, both in terms of different trap sites and different generations. G2 was the only trap site where 10 or more individuals were captured within a generation. Capture records at G2 and movement of individuals when released from there strongly suggested that a litter of *A. stuartii* was reared inside the hollow *Nothogafus moorei* which marked that trapsite. As in the wet sclerophyll forest, the sites of least captures were generally associated with clearings caused by the logging operation.

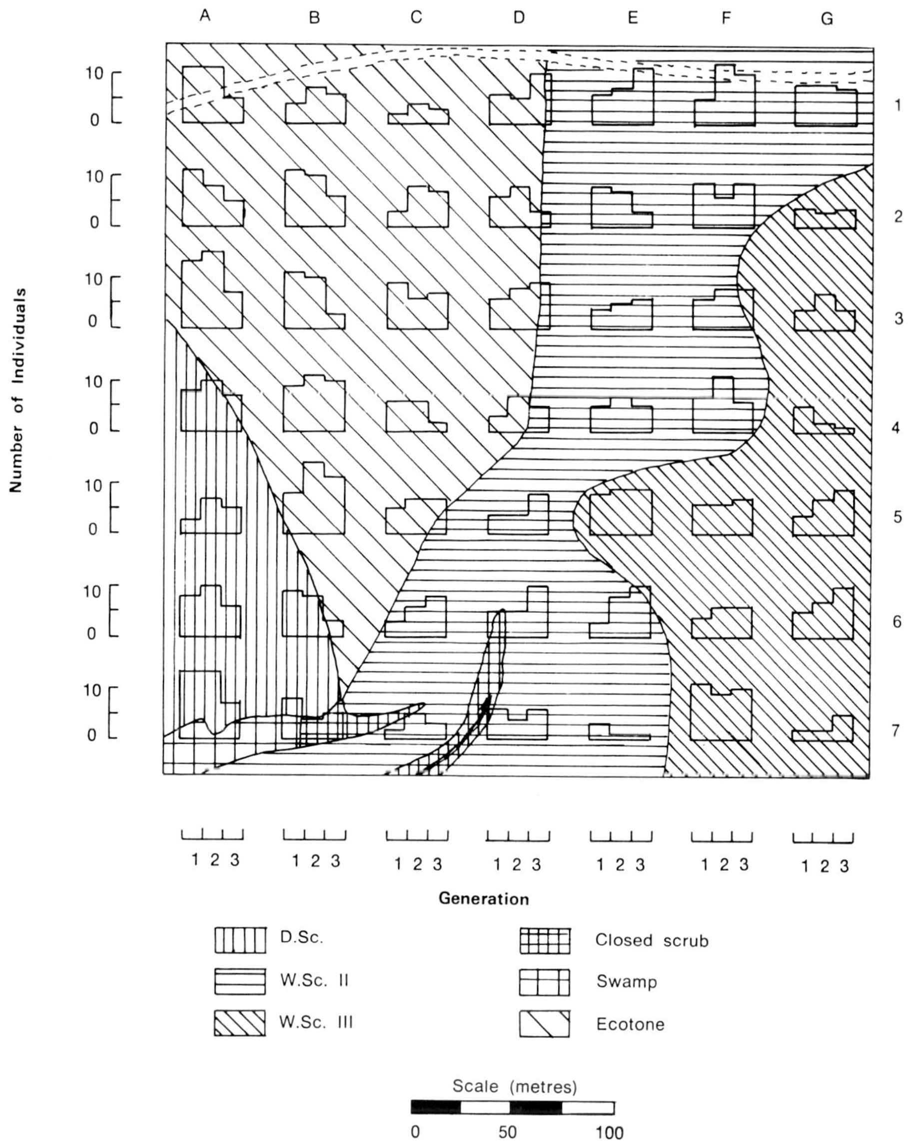

Fig. 6. A. stuartii habitat utilization, the number of individuals captured at each trap station during each generation. (a) Wet sclerophyll forest.

Stepwise regression analysis was carried out using data from all trapping grids except the dry sclerophyll. The numbers of individual *A. stuartii* captured at each of the habitat sampling sites were used as the dependent variable and 10 habitat parameters used in the correlation matrices (Tables 2, 3 and 4) as independent variables. The variables accounting for the greatest proportion of the variation were then introduced into a multiple regression model. Three multiple regression analyses were carried out, one for each of the three generations of animal data and habitat measurements. The predictive equations obtained were:

generation 1 —

No. of *A. stuartii* =

$1.65 + 0.27 \sqrt{\text{no. of dwarf shrubs} + 1}$

$P<0.05 \qquad (r^2 = 0.29)$

generation 2 —

No of *A. stuartii* =

$0.81 + 0.80 \sqrt{\text{no. high stems} + 1} + 0.005 \log \text{length}$

$P<0.05 \qquad (r^2 = 0.56)$

Generation 3 —

No. of *A. stuartii* =

$0.41 + 0.06 \arcsin \sqrt{\%\ \text{H.V.O. 50 cm}}$

$P<0.001 \qquad (r^2 = 0.53)$

These equations were used to obtain expected values of the number of *A. stuartii* captured at

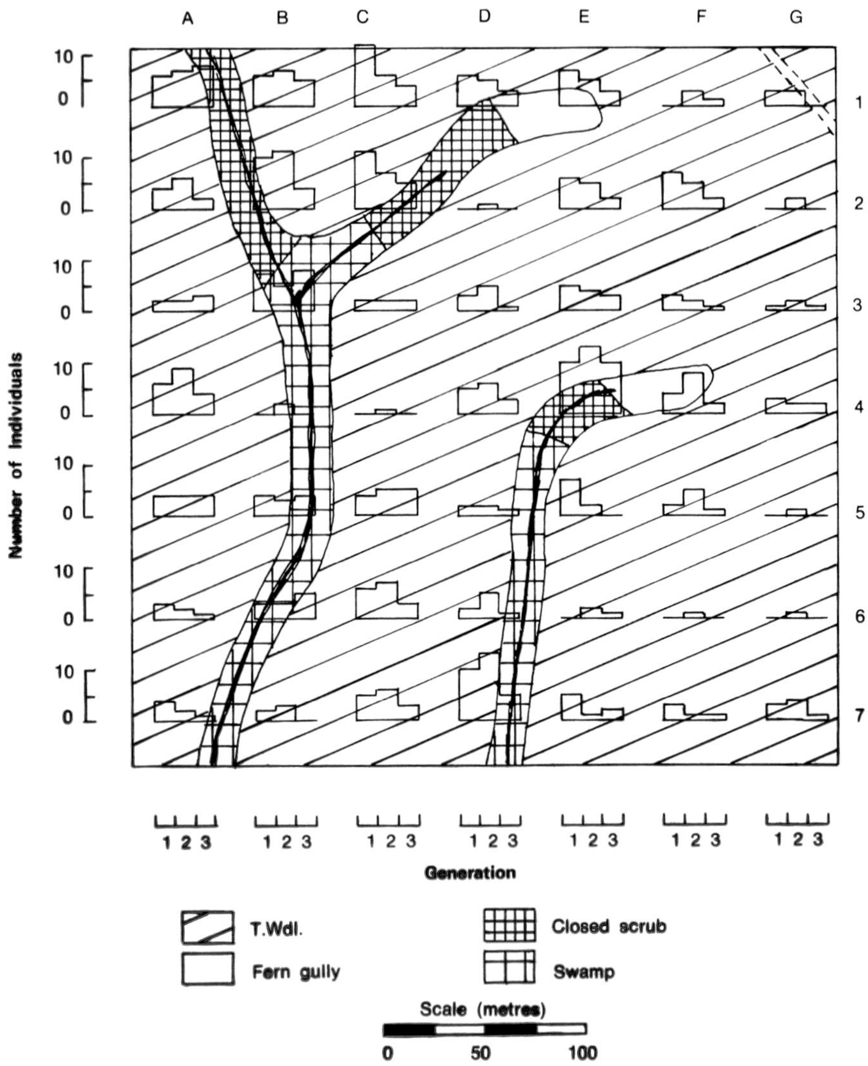

Fig. 6 (continued). (b) Tall woodland.

each of the trap stations in dry sclerophyll forest for which habitat was measured. These expected values were then compared with observed valued using an approximate x_7 test (Table 5). It is realised that the expected values were derived from a linear relationship and not from a set of probabilities as would be the case in the classical χ_2 theory.

Table 5. Results of approximate χ^2–tests comparing the number of individual *A. stuartii* captured at six trap sit in dry sclerophyll forest and the number expected from each predictive equation.

Category	Trap Site					
	A5	B2	B7	C4	E3	E5
No. observed	3	4	7	1	0	2
No. expected (gen. 1)	4.4	4.3	4.6	3.4	4.4	7.1
No. expected (gen. 2)	4.5	4.8	2.6	3.9	3.9	2.3
No. expected (gen. 3)	5.1	2.4	5.1	3.5	2.0	4.1

Generation 1	$\chi^2 = 11.48$	df = 4	P<0.05
Generation 2	$\chi^2 = 14.18$	df = 3	P<0.01
Generation 3	$\chi^2 = 7.50$	df = 4	n.s.

All but two of the expected valued given in Table 3 are less than five and therefore results of the approximate χ_2–tests should be considered with caution. The results suggested that the use of generations 1 and 2 equations to predict numbers of *A. stuartii* in 1976 would not be satisfactory because the expected and observed numbers differed significantly. Using the generation 3 equation, the variation between expected and observed numbers was not significant. It is therefore probable that the use of this equation is more meaningful.

DISCUSSION

At Petroi, *A. stuartii* was the most abundantly captured small mammal in all habitats over the three years of study. The trap records illustrated the definite annual cycling in *A. stuartii* populations previously reported in the literature (Horner and Taylor 1959, Marlow 1961, Woolley 1966b, Wakefield and Warneke 1967, Wood 1970, Recher *et al.* 1978). The number of individuals captured in each habitat reached an annual low after the mating season, with the male die-off occurring in late August and early September. Peaks in trappable populations occured in autumn and early winter. During this study, the trappable population in wet sclerophyll was always largest, with those in other habitats being similar to each other. There was little variation between generations in any of the habitats.

A. stuartii density was found to range from 4.4 per hectare in wet sclerophyll during generation 3 to 1.6 per hectare in tall woodland and logged rainforest also during generation 3. It is difficult to compare these figures with those for *A. stuartii* elsewhere, since neither Wood (1970) nor Leonard (1972) made density estimates and Barnett *et al.* (1976) presented their data in terms of percentage trap success, a factor which must vary according to human trapping effort and animal trap behaviour as well as population density. Fleming (1975) reported an average *A. stuartii* density of 12.4 per hectare, with a range of 3.0 to 24.9 per hectare, based on the date of Wood (1970). However, we consider that his values are invalid. Our reasoning is that since the trapping grid used by Wood (1970) was too small to encompass the home range of any males, the sampling area cannot be calculated. Furthermore, had Wood (1970) been able to determine his sampling area he would presumably have given density estimates himself.

A. stuartii density estimates at Petroi were 20–40 times lower than those of *A. minimus* on

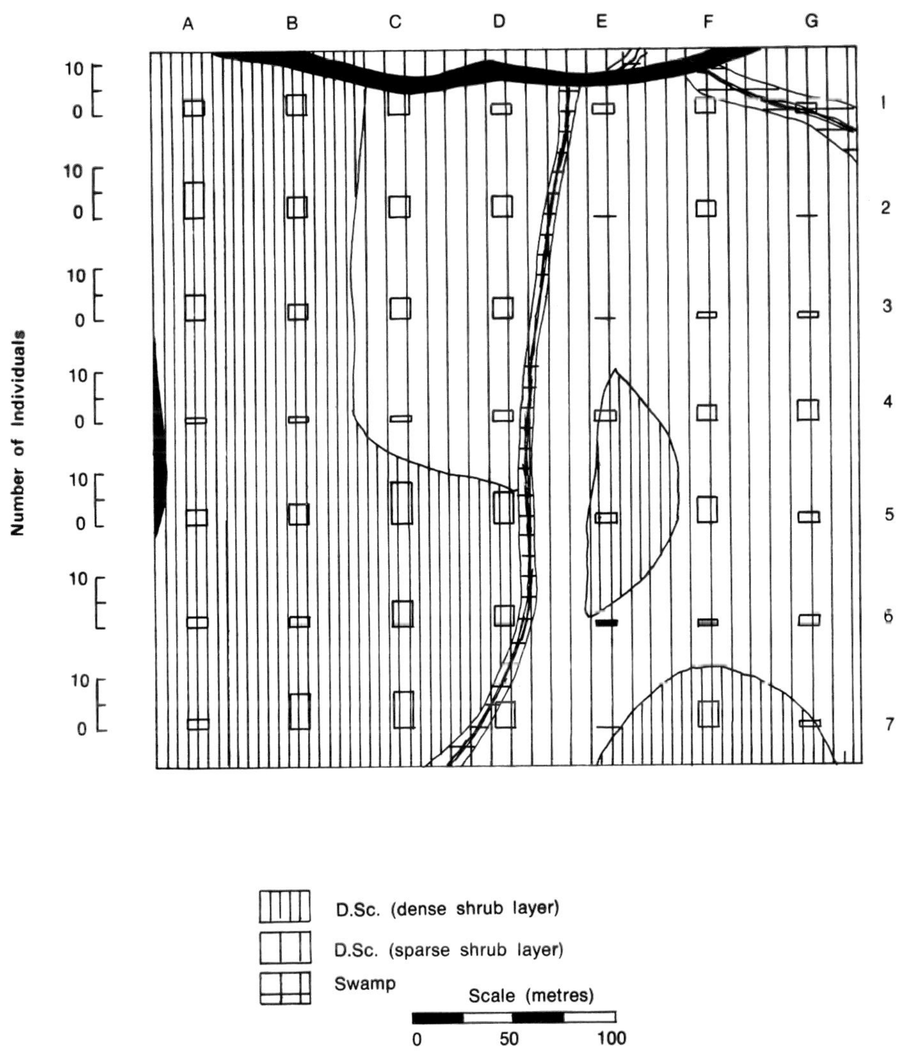

Fig. 6 (continued). (c) Dry sclerophyll forest. N.B. Traps at D7 and G1 were located on the opposite sides of swamps to those shown here, thus distances between C7 and D7 and G1 and G2 were only 25–30 (see Figure 3.10).

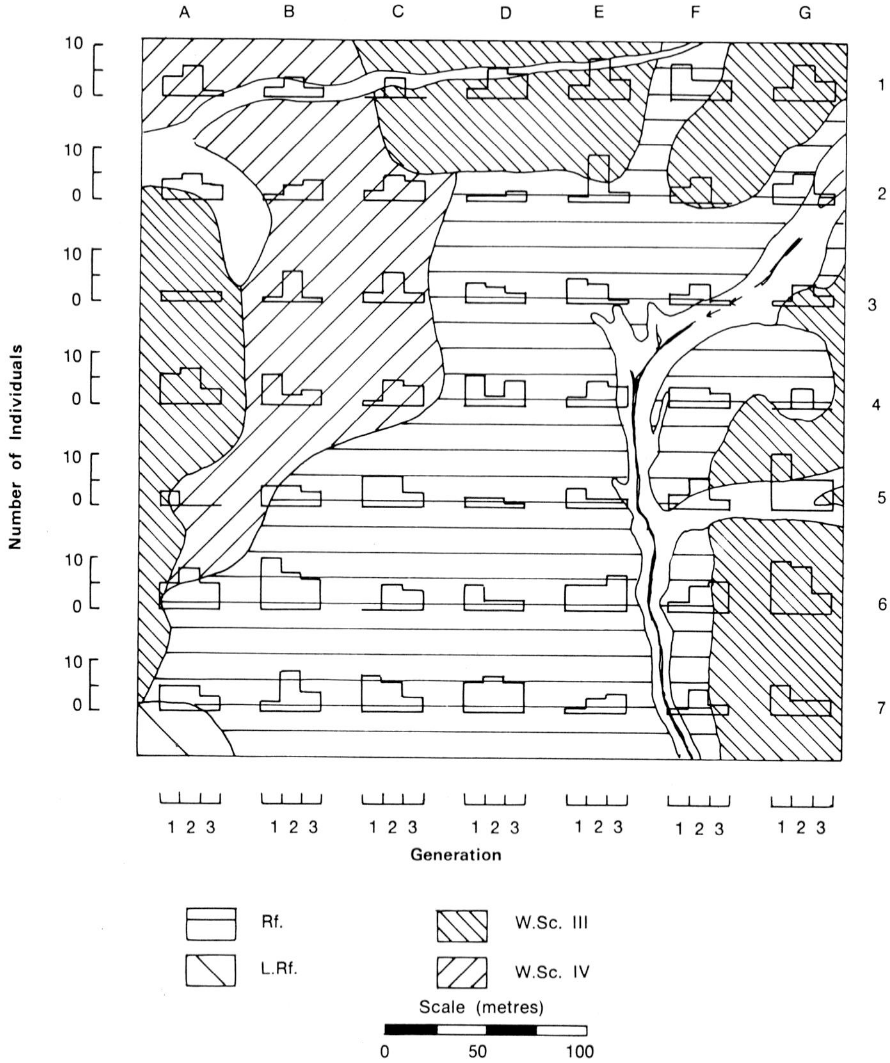

Fig. 6 (continued). (d) Rainforest.

Great Glennie Island, Bass Strait (Wainer 1976). Our densities were much more on par with those for shrews, the Northern Hemisphere eutherian counterparts of this species. Buckner (1969) estimated an annual range of 1.8 to 6.9 *Sorex araneus* per hectare in Wytham Wood, and Berkshire and Randolph (1973) reported 7.7 *Blarina brevicauda* per hectare in a deciduous woodlot in Ontario.

Prior to the pilot study (Fletcher 1974), information on *A. stuartii* habitat utilization was limited to qualitative descriptions. Such information on small mammals has generally come from capture of behaviour studies in different habitats (e.g. Leonard 1972, Kirkland and Griffin 1974, Grant 1971, 1975, Murie 1971, Wecker 1963) or studies of food resource utilization of sympatric species (e.g. Brown and Lieberman 1974, Vaughan 1974, Zimmerman 1965, Meserve 1976).

To be quantitative rather than descriptive, habitat analyses should quantify parameters which effect the distribution of the animals being studied. In practice, the process is limited to measurement of as many parameters as is practicable, with particular emphasis on those thought to be of possible importance to the particular species in question. Such an approach, as used here, has also been made by other workers. For example, Posamentier (1975) has quantified habitat for *Pseudomys novaehollandea, Rattus lutreolus* and *Mus musculus* in coastal heathlands of New South Wales. He used break back trapping and habitat quantification at fixed circular plots similar in design to those of Rosenzweig and Winakur (1969) who studied desert rodent communities. Similar work with birds (Kikkawa 1968, James 1971, Anderson and Shugart 1974) has been based on direct observations of animals and has therefore been unaffected by trap bias. Finally, Rosenzweig (1973), as an extension of

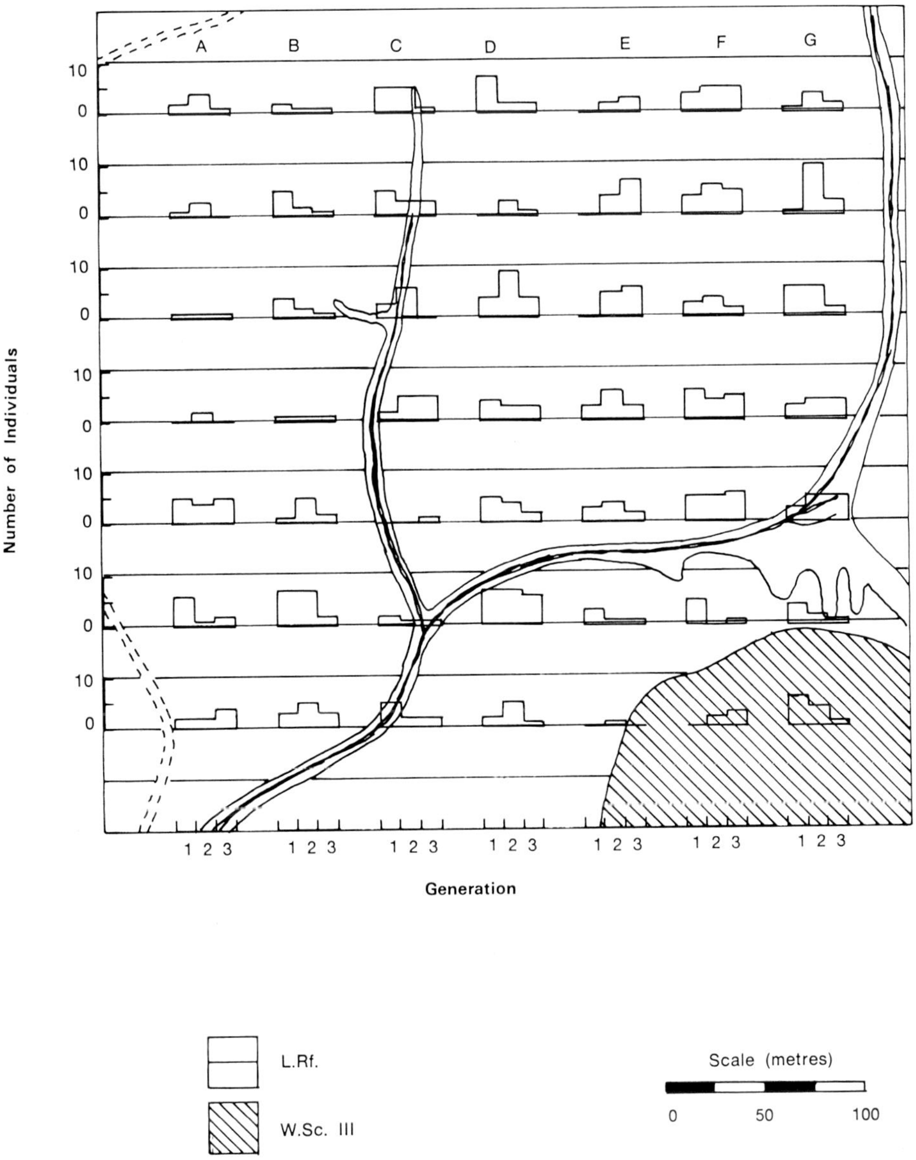

Fig. 6 (continued). (e) Logged rainforest.

his study mentioned above, reported the change in rodent land use patterns following structural alteration of vegetation.

Individual *A. stuartii* were captured continuously throughout each of the five habitats studied with a distribution ranging from the very open tall woodland through dry and wet sclerophyll forest to logged and unlogged cool temperate rainforest. They are thus able to exist under a wide variety of habitats and consequently it seems that their distribution should be considered in terms of habitat preferences rather than habitat requirements.

A. stuartii density was greatest in wet sclerophyll forest during each of the three generations studied at Petroi. It is therefore suggested that conditions in this habitat were more favourable for *A. stuartii* than those in each of the other habitats. In all habitats the greatest numbers of *A. stuartii* were captured at stations characterised by a relatively thick shrub layer, although in tall woodland, where the shrub layer was usually very sparse, one or two *A. stuartii* were captured at most stations.

The habitat parameters most highly correlated with *A. stuartii* capture data were the numbers of dwarf shrubs and high stems, total log length and horizontal visual obstruction at 50 cm. However, none of these parameters explained more than 56 per cent of the total variation in capture data. Also, despite tremendous growth in the shrub layer of logged cool temperate rainforest during this study, there was little change in *A. stuartii* density in that trapping grid. It is therefore suggested that the concept of habitat for *A. stuartii* is complex and that it cannot be expressed in terms of a simple multiple regression equation. Thus, no matter how many more parameters were measured, the situation would not necessarily clarify because different factors would come into play at different times and at differing intensities. A broad preference for dense undergrowth with a large number of logs is suggested. This seems logical since such places would provide adequate nest site availability and maximum protection from predators and climatic extreme.

We consider that further attempts to elucidate the habitat preference of *A. stuartii* should follow an experimental approach rather than the observational approach utilized here. Thus by calculated manipulation of the vegetation structure near ground level the proposed habitat relationship of *A. stuartii* could be tested. Initially two such manipulations are recommended. Firstly, vegetation in the shrub layer should be slashed to predetermined heights and densities. Secondly, the hollow log portion of available nest sites should be removed. Population monitoring of manipulated and control areas in a properly designed experiment would further elucidate *A. stuartii* habitat preference.

ACKNOWLEDGEMENTS

This research was part of a study on habitat relationships among small mammals at Petroi, undertaken in fulfilment of the requirements for the degree of Master of Natural Resources in the University of New England. We are grateful to other members of the N.P.W.S. team and numerous students for their assistance with field work. Sincere thanks are extended to Mr. and Mrs. Joe Davies and Mr. B. Davies for permission to work and camp on Petroi, for their kind hospitality and for going out of their way to exclude fire from the sclerophyll trapping grids during the study.

Mr. J. B. Williams identified unknown plants and Mr. V. Bofinger helped with statistical analyses, Mr. S. Williams programmed the animal capture data and did the regression analyses. Dr. B. Fox and Dr. M. Denny provided useful comments on the manuscript.

REFERENCES

Anderson, R. H., 1968. "The Trees of New South Wales". 4th Edition. Government Printer: Sydney.

Anderson, S. H. and Shugart, H. H. Jr., 1974. Habitat selection of breeding birds in an east Tennessee deciduous forest. *Ecology*. **55:** 828–37.

Barnett, J. L., How, R. A. and Humphreys, W. F., 1976. Mammals of Clouds Creek, Northeastern New South Wales, their distribution in pine and native forests. *Aust. Zool.* **19:** 23–34.

Baur, G. N., 1957. Nature and distribution of rain–forests in N.S.W. *Aust. J., Bot.* **5:** 190–233.

Beadle, N. C. W., 1971. "Students flora of north eastern New South Wales. Part I. Pteridophytes". University of New England, Armidale, N.S.W.

Beadle, N. C. W., 1972. "Students flora of northeastern New South Wales. Part II. Gymnosperms, Key to the angiosperm Families, Angiosperms: Families 37–83". University of New England: Armidale, N.S.W.

Beadle, N. C. W. and Costin, A. B., 1952. Ecological classification and nomenclature, *Proc. Lin. Soc. N.S.W.* **77:** 61–82.

Beadle N. C. W., Evans, O. D. and Carolin, R. C., 1972. "Flora of the Sydney region". Reed: Sydney.

Braithwaite, R. W., 1974. Behavioural changes associated with the population cycle of *Antechinus stuartii* (Marsupialia). *Aust. J. Zool.* **22:** 45–62.

Braithwaite, R. W. and Gullan, P., 1978. Habitat selection by small mammals in a Victorian heathland. *Aust. J. Ecol.* **3:** 109–27.

Brown, J. H., and Lieberman, G. A., 1974. Resource utilization and coexistence of seed–eating desert rodents in sand dune habitats. *Ecology* **54:** 788–97.

Buckner, C. H., 1969. Some aspects of the population ecology of the common shrew, *Sorex araneus*, near Oxford, England. *J. Mammal.* **50:** 326–32.

Daubenmire, R. F., 1959. A canopy coverage method of vegetation analysis. *Northwest Sci.* **33:** 43–64.

Emlin, J. T. Jr., 1956. A method for describing and comparing avian habitats. *Ibis.* **98:** 566–76.

Faust, B. F., Smith, M. H. and Wray, W. B., 1971. Distance moved by small mammals as an apparent function of grid size. *Acta Theriol.* **16:** 161–77.

Flemming, T. H., 1975. Chapter 12: The role of small mammals in tropical ecosystems. In "Small mammals, their productivity and population dynamics" ed. by F. B., Golley, K., Petrusewicz and L. Ryszkowski. IBP 5. Cambridge University Press: Cambridge.

Fletcher, H. L., 1974. "A study of the distribution and abundance of two small mammal species in four different habitats". Dip. Nat. Res. Dissertation: University of New England, New South Wales.

Fletcher, H. L., 1977. "Habitat relationships among small mammals at Petroi, northeastern New South Wales". M. Nat. Res. thesis: University of New England, Armidale.

Fox, B. J., 1979. An objective method of measuring the vegetation structure of animal habitats. *Aust. Wildl. Res.* **6:** 297–303.

Fox, B. J., 1980. Ph. D., thesis: School of Biological Science, Macquarie University, Sydney.

Francis, W. D., 1951. "Australian rain–forest trees". Forestry and Timber Bureau, Halstead Press: Sydney.

Grant, P. R., 1971. Habitat preference of *Microtus pennsylvanicus* and its relevance in the distribution of this species on islands. *J. Mammal.* **52:** 351–61.

Grant, P. R., 1975. Population performance of *Microtus pennslyvanicus* confined to woodland habitat, and a model of habitat occupancy. *Can. J. Zool.* **53:** 1447–65.

GRAY, M., 1961. A list of the vascular plants occurring in the New England Tablelands, New South Wales, with notes on distribution. *Contributions from the N.S.W. National Herbarium*, **3** (1): 1–82.

HALL, N., JOHNSON, R. D. AND CHIPPENDALE, G. M., 1970. "Forest trees of Australia". Australian Government Publishing Service: Canberra.

HANSSON, L., 1969. Home range, population structure and density estimates at removal catches with edge effect. *Acta. Theriol.* **14:** 153–60.

HORNER, B. E. AND TAYLOR, J. M., 1959. Results of Archbold expeditions, No. 80. Observations on the biology of the marsupial mouse *Antechinus flavipes*. *Am. Mus. Novit.* **1972:** 1–24.

JAMES, F. C., 1971. Ordinations of habitat relationships among breeding birds. *The Wilson Bulletin* **83:** 215–35.

JAMES, F. C. AND SHUGART, H. H. JR., 1970. A quantitative method of habitat description. *Audubon Field Notes*. **24**(6): 727–36.

KIKKAWA, J., 1968. Ecological association of bird species and habitats in eastern Australia; similarity analysis. *J. Anim. Ecol.* **37:** 143–65.

KIRKLAND, G. L. JR. AND GRIFFIN, R. J., 1974. Microdistribution of small mammals at the coniferous–deciduous forest ecotone in northern New York. *J. Mammal.* **55:** 417–27.

KNIGHT, D. H., 1963. A distance method for constructing forest profile diagrams and obtaining structural data. *Tropical Ecology* **4:** 89–94.

LEITCH, E. C., 1972. "The geological development of the Bellingen–Macleay Region — a study in the tectonics of the New England fold belt". Ph.D. thesis: University of New England, Armidale.

LEONARD, B. V., 1970. "Effects of control burning on the ecology of small mammal populations." Symposium Papers. Forests Commission: Victoria.

LEONARD, B. V., 1972. "The effect of fire upon selected small mammals and leaf litter fauna in sclerophyll forest in southern Australia". M.Sc. thesis: Monash University, Melbourne.

LINDSAY, A. A., BARTON, J. D. JR. AND MILES, S. R., 1958. Field efficiencies of forest sampling methods. *Ecology* **39:** 428–44.

MARLOW, B. J., 1961. Reproductive behaviour of the marsupial mouse *Antechinus flavipes* (Waterhouse) and the development of the pouch young. *Aust. J. Zool.* **9:** 203–18

MESERVE, P. L., 1976. Habitat and resource utilization by rodents of a California sage scrub community. *J. Anim. Ecol.* **45:** 647–66.

MURIE, J. O., 1971. Behavioural relationships between two sympatric voles (*Microtus*): relevance to habitat segregation. *J. Mammal.* **52:** 181–86.

PELIKAN, J. 1967. The estimation of population density in small mammals *in* "Secondary productivity of terestrial ecosystems. Vol. I". ed. by K. Petrusewicz. Polish Scientific Publishers: Warsaw.

POSAMENTIER, H. G., 1975. "Habitat requirements of small mammals in coastal heathlands of New South Wales". M.Sc. thesis: University of Sydney, Sydney.

RANDOLPH, J. C., 1973. The ecological energetics of a homeothermic predator, the Short–tailed Shrew. *Ecology* **54**: 1166–87.

RECHER, H. F., LUNNEY, D. AND POSAMENTIER, H. G., 1978. "Effects of wildfire on small mammals of Nadgie Nature Reserve, N.S.W." *Third Fire Ecol. Papers*, Pp.30–6. Forests Commission of Victoria: Melbourne.

ROSENZWEIG, M. L., 1973. Habitat selection with a pair of coexisting heteromyid rodent species. *Ecology* **54:** 111–7.

ROSENZWEIG, M. L. AND WINAKUR, J., 1969. Population ecology of desert rodent communities: habitat and environmental complexity. *Ecology*. **50:** 558–72.

STICKEL, L. F., 1954. A comparison of certain methods of measuring ranges of small mammals. *J. Mammal.* **35:** 1–15.

VAUGHAN, T. A., 1974. Resource allocation in some sympatric subalpine rodents. *J. Mammal.* **55:** 764–95.

WAINER, J. W., 1976. Studies of an island population of *Antechinus minimus* (Marsupialia; Dasyuridae). *Aust. Zool.* **19:** 1–7.

WAKEFIELD, N. A., AND WARNEKE R. M., 1967. Some revision in *Antechinus* (Marsupialia). *Victor. Nat.* **84:** 69–99.

WECKER, S. C., 1963. The role of early experience in habitat selection by the Prairie Deer Mouse *Persomyscus maniculatus bairdi. Ecol. Mongr.* **33:** 307–25.

WOOD, D. H., 1970. An ecological study of *Antechinus stuartii* (Marsupialia) in a southeast Queensland rainforest. *Aust. J. Zool.* **18:** 185–207.

WOOLLEY, P., 1966a. "Reproductive biology of *Antechnus stuartii* Macleay (Marsupialia: Dasyuridae)". Ph. D. thesis: Australian National University, Canberra.

WOOLLEY, P., 1966b. Reproduction in *Antechinus* spp. and other dasyurid marsupials. *In* "Comparative biology of reproduction in mammals", ed. by I. W. Rowlands. *Symp. Zool. Soc. Lond.* **15:** 281–94.

ZIMMERMAN, E. G., 1965. A comparision of the habitat and food of two species of *Microtus*. *J. Mammal.* **46:** 605–12.

RESUMEN

16. Apreciación edl hàbitat de *Antechinus stuartii* (Marsupialia) en Petroi, Noreste de New South Wales. Por H. L. Statham y R. H. Harden.

Especímenes de *Antechinus stuartii* fueron entrampados desde Marz de 1974 hasta Septiembre de 1976 en reticulados de nueve hectàreas en altos bosques, bosques esclerófilor húmedos, con lluvias temperaturas frías, y en forestas taladas templadofrias; desde Febrero de 1976 hasta Septiembre de 1976 en los bosques secos escleróftlos. El hábitat fue cuantificado por la técnica de la medida multiparamétrica de variables en sitios circulares prefijados, con la cauntificatión de parámetros de posible significatión para pequénos mamíferos, como por ejemplo número de huscos en los árboles y troncos, en conjunción con la cauntificación general de hábitat. En todos los hàbitats se capturó el mayor número posible de individuos de *A stuartii* en estaciones caracterizadas por una relativamente espesa capa de arbustos. Se determinó que los requerimientos de habitat de esta especie son complejos. Muchos paràmetros, incluyendo la longitud del tronco, el número de arbustos enanos, altos tallos y obstrucción de la visual horizontal hasta 50 cm, resul taron ser importantes en diferentes tiempos y en diferentes intensidades.

17. ECOLOGICAL SEPARATION AND COEXISTENCE OF *Sminthopsis murina* AND *Antechinus stuartii* (DASYURIDAE, MARSUPIALIA): A REGENERATION NICHE ?

BARRY J. FOX[1]

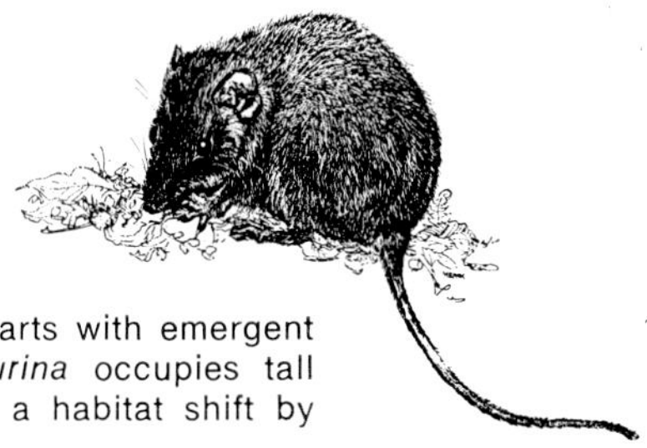

Superficially *Sminthopsis murina* and *Antechinus stuartii* are very similar species of small dasyurid that share the insectivore niche in Myall Lakes area of New South Wales (32°28'S, 152°30'E). Their coexistence is facilitated by their marked ecological separation. In open forest *S. murina* utilises much drier habitats with less dense vegetation below 1m and much reduced leaf litter, while *A. stuartii* is found in denser vegetation with higher plant species diversity.

When both species occur together in heathland, *A. stuartii* occupies parts with emergent mallee-form eucalypts, forest edges or short wet heaths while *S. murina* occupies tall dry heathland and is not found at all in wetter areas. This represents a habitat shift by *S. murina* relative to the habitat occupied in open forest.

Their roles in post–fire succession also differ. *Sminthopsis murina* occupies mid–seral stages where it reaches its maximum abundance before decreasing in later stages. On the other hand, *A. stuartii* increases gradually to reach maximum abundance in late seral stages. Consistent with this the polyoestrous *S. murina* has a much higher reproductive capacity that would facilitate a rapid recovery following disturbance.

Long–term coexistence between these species results from the ability of *S. murina* to exploit a "regeneration niche" that is available in areas with high fire frequencies that produce the required mid–seral stages.

Key words: *Antechinus stuartii,* coexistence, Dasyuridae, ecological separation, habitat shift, interspecific competition, regeneration niche, *Sminthopsis murina,* succession.

Pages 187–97 *in* CARNIVOROUS MARSUPIALS, ed. by M. Archer, February, 1982. Roy. Zool. Soc. New South Wales: Sydney, Australia.

IN THIS PAPER I have attempted to draw together all of the available information relevant to the coexistence of *Sminthopsis murina* and *Antechinus stuartii* in the Myall Lakes National Park. Superficially *S. murina* and *A. stuartii* appear as very similar species of small dasyurid that share the insectivore niche in the Myall Lakes National Park. They offered an opportunity to examine the mechanisms by which species are able to coexist in apparent defiance of the competitive exclusion principle. This is most often attributed to Gause (1934), but has been most concisely stated by Hardin (1960) as "complete competitors cannot coexist". An examination of the degree of ecological separation of these two species would allow comparison with that predicted from the theory of limiting similarity first formulated by Hutchinson (1959). The empirical evidence on which this theory was based, and further evidence in its support, indicate that species differ by at least a factor of 1.2-1.4 in dimensions that can be related to the use of a limiting resource. May and MacArthur (1972) have presented a theoretical basis for limiting similarity but this has been challenged by Abrams (1975, 1976).

Examples of the coexistence of small dasyurids can be readily found in the literature. *Antechinus stuartii* and *A. swainsonii* are most commonly found together (Newsome *et al.* 1975, Braithwaite *et al.* 1978, Hall 1980, Dickman 1982) but the additional pairs of *A. stuartii* — *A. minimus* and *A. swainsonii* — *A. flavipes* are also found to coexist in southeastern Australia (Braithwaite *et al.* 1978). In all of these cases there is a substantial weight difference between coexisting species as is discussed for dasyurids in general, Fox (1982). Dickman (1982) has provided a very good specific example of sympatric *A. swainsonii* — *A. stuartii* populations, and noted that observed differences in mating times between these two species increases the size difference between the developing juveniles. In northeastern Queensland three species exist sympatrically: *A. stuartii, A. flavipes* and *A. godmani,* showing marked weight differences (33g: 68g: 94g) Van Dyck (1982). In southeastern Queensland, where there is much less weight difference between *A. stuartii* and *A. flavipes,* sympatry is uncommon and contact areas are regarded by Van Dyck (1982) to be temporary and examples of distribution boundaries in flux.

METHODS

The field data are derived from two separate studies carried out in the Myall Lakes National Park (32°28'S, 152°30'E). One was a survey of a eucalypt open forest as part of a study on

[1] School of Zoology, University of New South Wales, P.O. Box 1, Kensington, New South Wales, Australia, 2033.

the effects of fire on the forest fauna. Sixteen 1 ha plots were set up in areas of forest with different regeneration ages, these plots were simultaneously trapped five times over a 20-month period during 1976-7. Detailed information on the methods used in this study are given in Fox and McKay (1981). The second investigation was nearby and dealt with one study plot in heathland close to the open forest boundary. This study provided information from the plot prior to burning (1974) and also monitored regeneration for five years after the fire (1974-9). Detailed information on the methods used in this study are given in Fox (1980). Data on the life history and reproduction of *S. murina* are drawn from a study by Fox and Whitford (1982), and details of methods used are provided therein. The use of separate studies in the open sclerophyll forest and the heathlands allowed independent assessment of parameters, such as the position of each species in the mammalian secondary succession in response to vegetation changes following fire.

RESULTS AND ANALYSIS

FOREST PLOTS: The coverage in the forest was extensive with 16 one ha plots over the 2000 ha of forest. The number of individuals captured for each species on each of the 16 study plots

Table 1. The number of individuals of *Antechinus stuartii (A.s.)* and *Sminthopsis murina (S.m.)* captured on each of the study plots over the two years of the study. Replicated plots are linked while sites with asterisks recorded no dasyurids at all on them or a replicate, and may have been unsuitable.

Last Burned		1976 *A.s./S.m.*	1977 *A.s./S.m.*	Total *A.s./S.m.*
January, 1976	(ML4 S *			
	(ML4 N *			
	(ML2 S	/1	/1	/2
	(ML2 N	/1		/1
	ML10		2/	2/
August, 1974	(ML1 S		/1	/1
	(ML1 N		1/	1/
	ML9	/4	/3	/7
February, 1975	(ML7 S		/4	/4
	(ML7 N		/5	/5
October, 1971	ML8	/2	/1	/3
	(ML3 S	1/	3/	4/
	(ML3 N			
November, 1968	(ML5 S		2/	2/
	(ML5 N	1/	2/1	3/1
	ML6 *			

Table 2. Mean Habitat parameters for forest sites occupied by *Sminthopsis murina* and *Antechinus stuartii* respectively, ranked on the results of a one-way analysis of variance to test the difference between the means for each parameter.

Variable	*SMINTHOPSIS MURINA* Mean ± 1S.E.	*ANTECHINUS STUARTII* Mean ± 1S.E.	$F_{1,22}$	P
Leaf Litter (Kg m^{-2})	0.88 ± 0.08	1.38 ± 0.13	11.32	$.001<P<.005$
Plant Species Diversity	3.0 ± 0.5	5.3 ± 0.7	8.18	$.001<P<.005$
Regeneration Age (Yrs.)	3.2 ± 0.4	5.5 ± 1.1	5.67	$.025<P<.05$
Understory Height (m)	1.02 ± 0.11	1.64 ± 0.31	5.42	$.025<P<.05$
% Light at Ground	55 ± 1	48 ± 5	3.95	$.05<P<.1$
Percent Cover	50 ± 3	61 ± 6	3.23	$.05<P<.1$
Vegetation Index 0–200 cm	0.36 ± 0.01	0.49 ± 0.11	2.89	$.1<P<.25$
Vegetation Index 100–150 cm	0.32 ± 0.09	0.66 ± 0.21	2.81	$.1<P<.25$
Forest Species Index	29.5 ± 1.0	32.3 ± 1.2	2.61	$.1<P<.25$
Vegetation Index 20–50 cm	0.08 ± 0.04	0.13 ± 0.04	2.54	$.1<P<.25$

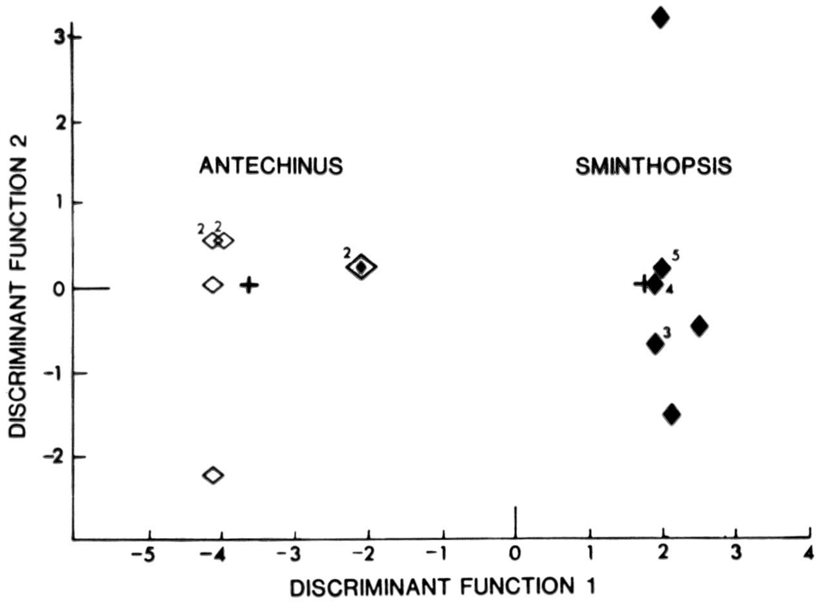

Fig. 1. Sminthopsis murina and *Antechinus stuartii* capture sites displayed on the first two discriminant functions based on the 9 habitat parameters for the forest study (see text).

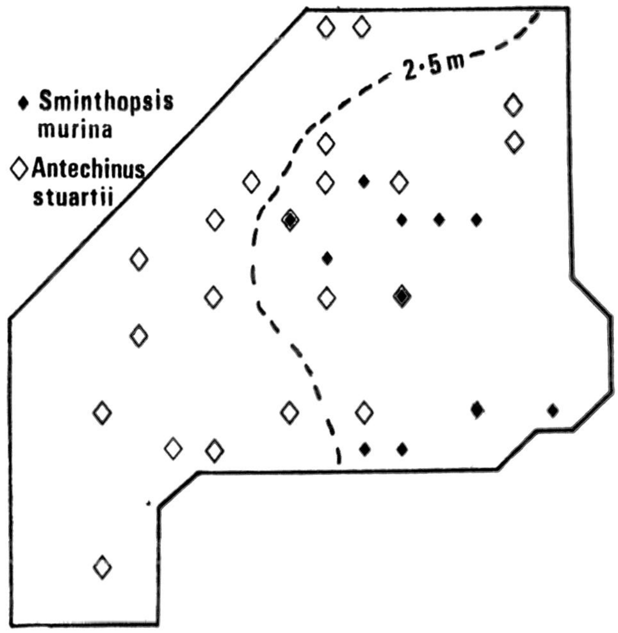

Fig. 2. The distribution of capture sites for *Sminthopsis murina* and *Antechinus stuartii* on the 7ha heath study plot, before the area burned.

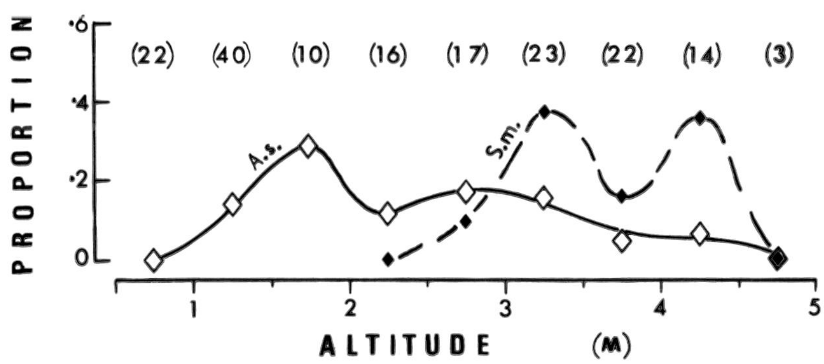

Fig. 3. The distribution of captures for *Sminthopsis murina* and *Antechinus stuartii* over 50 cm altitude bands. Values are shown as the proportion of each species total capture density in each band. The numbers in parentheses are the number of stations in each band.

during 1976 and 1977 are shown in Table 1. Although both species were often caught in the same regeneration stage, there was only one plot (ML5N) where both dasyurids were recorded together. On four plots *A. stuartii* only was recorded and on seven plots *S. murina* only was recorded while there was only one of the plots, shown to be suitable for these small dasyurids, on which neither was recorded. A contingency table then indicated a significant negative association between these species ($G = 6.13$, $P < 0.025$, 1 d.f.).

Mean values of habitat parameters for the sites occupied by each species are shown in Table 2. Values are included only for the parameters having the highest F-values; the first four show significant differences between species. However three of these parameters are correlated with the fourth (regeneration age) as well as with one another. A partial correlation analysis, using stepwise discriminant functions was carried out to overcome this problem. The first variables chosen were, in order: leaf litter (+); a vegetation index for the 100 to 150 cm layer (+); a vegetation index for the 50 to 100 cm layer (—); regeneration age (—); the number of plant species (+); an index of the proportion of true forest plant species present (+); mammal species diversity (—); a vegetation index for the 20 to 50 cm layer (—); and a measure of the patchiness of the total vegetation up to 2m (—). An explanation of these variables is given in Fox and McKay (1981), and the vegetation index used is described in Fox (1979). The results of this analysis clearly illustrate the degree of separation between the two species based on these habitat measures (Fig. 1) and the difference between the multivariate group means is significant ($F_{9,14} = 10.7$, $P < .001$).

HEATH PLOT: Distributional data from the heath is intensive and confined to the one 7 ha plot. The distribution of capture sites is shown in Fig. 2. The *S. murina* capture sites are all in tall dry heath and tree heath (woodland) habitats above the 2.5m contour line. The distribution of captures, for each species, along a topographic gradient is shown in Fig. 3, where the proportion of captures in each 50 cm altitude band is presented graphically. A similar picture is apparent in the distribution of captures among habitat types (Fig. 4) when the habitats are arranged in order of increasing dryness and structural complexity in the vegetation. Detailed descriptions of the vegetation types recognised are given in Fox (1980). To quantify the degree of overlap between these species, I considered each vegetation type or habitat to be a resource state and used Pianka's (1974) modification of MacArthur and Levins (1967) equation:

$$O_{12} = O_{21} = \frac{P_{1k} \cdot P_{2k}}{\left(\sum_k P_{1k}^2 \cdot \sum_k P_{2k}^2 \right)^{1/2}}$$

where P_{1k} and P_{2k} represent the proportional use of the kth resource state by the 1st and 2nd species; O_{12} and O_{21} represent the overlap between the species pair. The between habitat overlap between *A. stuartii* and *S. murina* calculated in this way is 0.63.

If, however, each trap station is considered as a resource state then the overlap value becomes a measure of the degree to which the two species occupy the same trap stations. This within habitat overlap has a value of 0.13, indicating that there is a marked tendency for each species to occupy trap stations from which the other species is absent.

A clearer picture of each species' distribution in relation to both topography and habitat can be obtained by using the mean altitude of each habitat patch as an objective criterion, thus providing an interval scale for the abcissa (Fig. 5). The observed overlap now more closely represents that calculated for within-habitat overlap.

HABITAT REQUIREMENTS: Examination of habitat requirements for each species on the heath plot also indicates differences between them. The background to this analysis is provided in Fox and Fox (1981) and Fox (1980). Contribution coefficients calculated from the multiple regression equations for each species are given in Table 3. The contribution coefficient for each independent variable is an estimate of the proportion of the total variance contributed by that variable (Fox and Fox 1978).

Structural and floristic variables were separately submitted to factor analyses to reduce the variables to a manageable number of interpretable factors (see Fox and Fox, 1981, for full details). As floristic factors were not selected for the regressions they will not be described. The factor loadings for the habitat parameters most highly correlated with each of the structural factors are given in Table 4. Structural factor one represents a vegetation profile with a sparse and patchy distribution of foliage below 20 cm but with substantial vegetation between 1m and 2m. Minor contributions come from foliage above 2m and homogeneity of vegetation from 1m to 1.5m. This description most closely matches a tall dry heath profile. Structural factor two relates to the amount and constancy of vegetation below 1m as well as the total amount of vegetation present. This is an adequate description of the profile to be expected in a short wet heath, or any area with dense vegetation below 1m only. Structural factor three represents a reduced total vegetation and a patchy

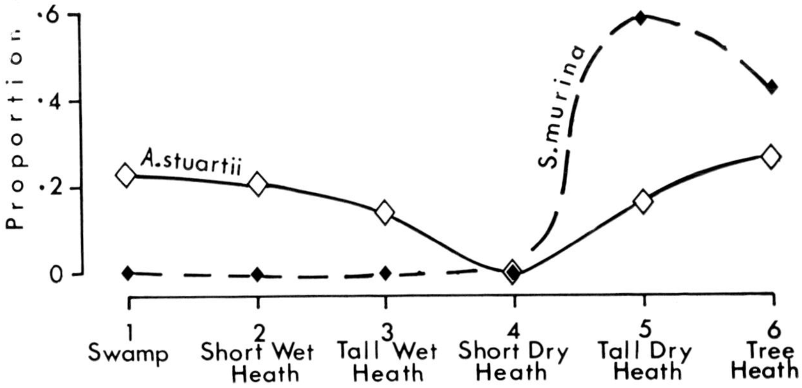

Fig. 4. The distribution of each species captures shown as the proportion of each species total captures in each habitat with the habitats ranked on a gradient of increasing dryness and structural complexity.

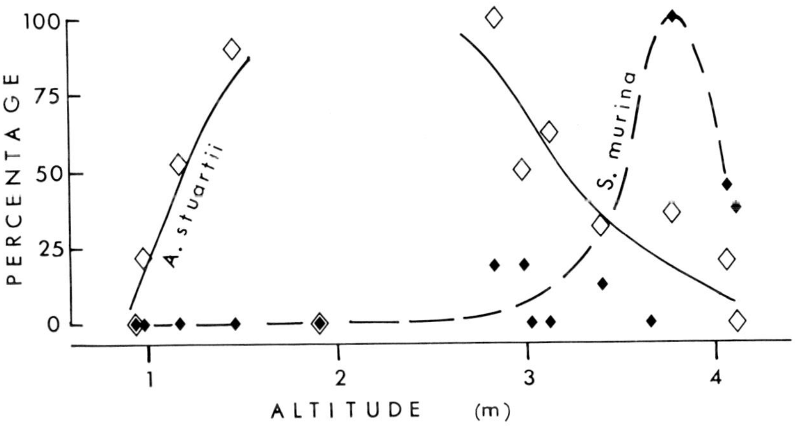

Fig. 5. Each species capture density in habitat patches as a function of the mean altitude of the patch. Values are shown as a percentage of the maximum capture density for that species.

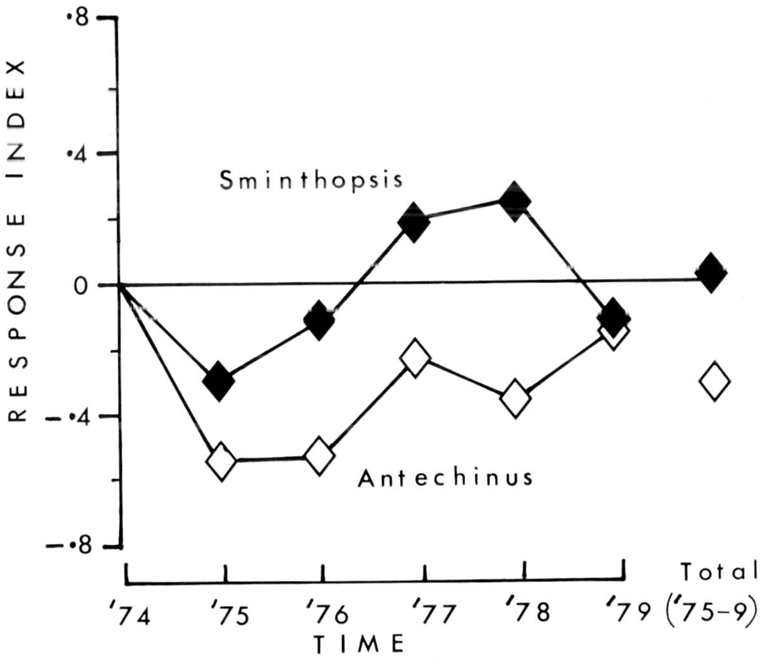

Fig. 6. The fire response index for each species as a function of the time since fire. The response index is the log of the ratio of peak abundance in any post-fire year to the pre-fire value.

Table 3. Contribution coefficients (see Fox and Fox 1978) for independent variables, determined from stepwise multiple regression equations. Significance levels for each contribution and the total equation are shown (* P <0.05, ** P <0.01, *** P <0.001). Data from Fox and Fox (1981).

Dependent Variable	Independent Variables: Floristic Factor 1	Floristic Factor 2	Floristic Factor 3	Floristic Plant Species Diversity	Structural Factor 1	Structural Factor 2	Structural Factor 3	Structural Foliage Height Diversity	Total R^2	Equation F–value
Antechinus stuartii				* –.11	** +.14		* –.15		0.40	*** 7.2
Sminthopsis murina					** +.13	*** –.26	* +.08		0.47	*** 9.8

horizontal distribution of what is present. It is not readily interpretable but might reflect a transition zone between short dry heath and tall dry heath where *Banksia serratifolia* appears as an emergent tall shrub out of a relatively sparse layer of smaller shrubs.

FIRE RESPONSE: As the heath plot has been monitored since 1974, it is possible to chart each species' response to the vegetation changes triggered by the August, 1974, fire. I have done this using a fire response index for each species for each year, calculated as the logarithm of the ratio of the population maximum for that year to the maximum level recorded before the fire. The behaviour of this index for each species is shown in Fig. 6. A total index for each species is calculated from the ratio of the sum of the five post–fire maxima to five times the pre–fire maximum and represents an overall view of the species' post–fire performance. *Sminthopsis murina* appeared to be neutral in this respect, but *A. stuartii* still shows a negative overall response.

HABITAT SHIFT: Although the two species did not occupy identical areas, both did exist in the forest so that they may be considered locally allopatric but regionally sympatric within the forest. *Sminthopsis murina* is the more abundant species, perhaps due to the higher fire frequency in this forest (Fox and McKay 1981). However, when the distribution on the heath plot is examined *A. stuartii* was found in the forested area while *S. murina* was restricted to the tall dry heath areas with occasional captures in the tree heath habitat. It would appear that *S. murina* is exhibiting a habitat shift when the two species are in local as well as regional sympatry. This shift is difficult to quantify but I am able to show quantitative shifts in habitat following the fire (Fig. 7). The proportional use of habitat shown allows the calculation of a median habitat for each species before and after the fire. *Antechinus stuartii* shows a shift towards the more structurally complex habitats (mainly the forest) after the fire, whereas *S. murina* shows an opposite shift away from the more structurally complex habitats.

SUCCESSION: With different response indices it might be expected that these species would also occupy different successional stages. This is true on the heath plot from which the response indices were calculated but more importantly a very similar picture was found in the forest study (Fig. 8). As these studies were independent, in that they used two entirely different experimental designs, the similar results greatly strengthen the conclusions that can be drawn from them. In both studies *S. murina* reached a peak at about four years post–fire, whereas *A. stuartii* peaked around six years or later. *Sminthopsis murina* occupied a mid-seral stage while *A. stuartii* was more abundant in the late-seral stage.

Table 4. Variables contributing to the first three structural factors and their factor loadings as determined from the factor analysis. Vegetation indices for each layer were calculated as described in Fox (1979) and represent the amount of vegetation present in that layer. Data from Fox and Fox (1981).

Structural Factor One	Structural Factor Two	Structural Factor Three
–0.82 Vegetation 0 to 0.2m	+0.91 Vegetation 0.5 to 1.0m	+0.82 Coeff. var. Total vegetation
+0.78 Vegetation 1.5 to 2.0m	–0.76 Coeff. var. 0.5 to 1.0m	
+0.78 Vegetation 1.0 to 1.5m	+0.65 Total vegetation	–0.66 Total vegetation
+0.66 Coeff. var. 0 to 0.2m		

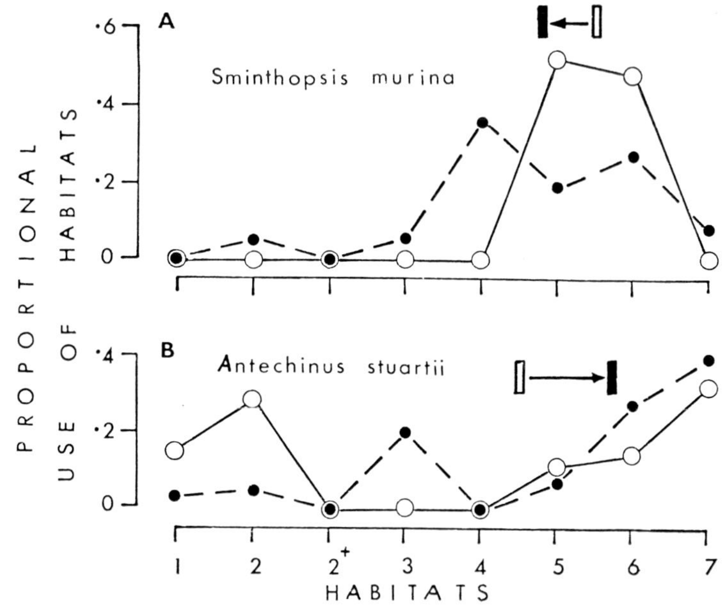

Fig. 7. Proportional use of habitats by (A) *Sminthopsis murina* and (B) *Antechinus stuartii.* Open symbols represent pre–fire distribution and solid symbols post–fire distribution. Bars represent the median habitat used. Habitats are 1 – Swamp, 2 – Short Wet Heath, 2+ – S.W.H. and Hakea thickets, 3 – Tall Wet Heath, 4 – Short Dry Heath, 5 – Tall Dry Heath, 6 – Tree Heath (mallee–form Eucalypts), 7 – Forest.

REPRODUCTION: *Sminthopsis murina* is polyoestrous, and the breeding season may extend from early September until the end of March; females probably produce only two litters in this time (Fox and Whitford 1982). Females have 10 nipples and a field observation of a female with a regressing pouch and all ten teats equally enlarged (Fox and Whitford 1982) indicates that they have the ability to wean ten young. The mating to weaning period of 74–79 days in *S. murina* contrasts sharply with the 120–160 days for *A. stuartii* (Woolley 1973). Mating in *A. stuartii,* at this site, occurs in mid–July (personal observation), six weeks earlier than *S. murina,* but the monoestry exhibited by *A. stuartii* and the smaller number of teats (8; Marlow 1961) means that *A. stuartii* has a considerably lower potential productivity than *S. murina.* The greater productivity resulting from polyoestry and large teat number (10) in *S. murina* is consistent with a species occupying mid–seral stages. Possible reasons for the evolution of these two contrasting reproductive strategies in the same environment are discussed by Fox and Whitford (1982).

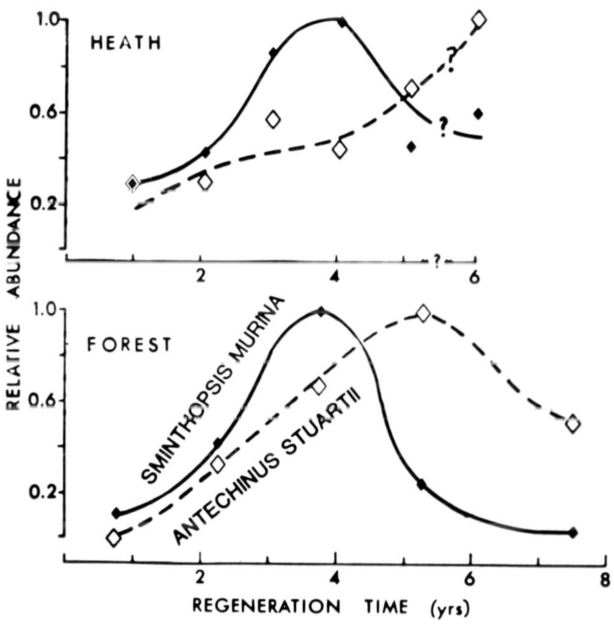

Fig. 8. Relative abundance of *Sminthopsis murina* and *Antechinus stuartii* as a function of regeneration time. Values are shown relative to the maximum value for that species. Heath and forest plots are shown separately.

DISCUSSION

The similarity of these two species is indeed superficial, as my comparison of their ecology has clearly shown. Two of the differences appear to be fundamental and perhaps contribute to some of the other differences observed. The first is the markedly different life histories of these two species. *Antechinus stuartii* is a mono estrous and highly seasonal, synchronous breeder with females investing between four and five months to raise a single litter comprising at most eight young (on the mid north coast of N.S.W.); very few females, and no males, survive to breed in a second year. *Sminthopsis murina* is also seasonal but because it is polyoestrous it may not need to be as tightly synchronised; females invest up to seven months in raising two litters, that could comprise a total of up to 20 young, and can survive to breed in a second year. Although there is no male die-off there is as yet no field confirmation that males breed in a second year.

The second fundamental difference is in habitat requirements, in particular those observ-

Table 5. Body weights (mean ± s.e.) for *A. stuartii* and *S. murina* in Myall Lakes National Park for three months during the year (sample sizes in parentheses). Body weight ratios are shown between the groups indicated.

		MAY		SEPTEMBER		NOVEMBER	
Antechinus stuartii	♂♂	36.9±2.1 (8)		–		–	
			1.29				
Antechinus stuartii	2nd Yr. ♀♀	28.7±1.5 (3)		–		–	
			1.37				
Antechinus stuartii	1st Yr. ♀♀	21.0±1.1 (4)		23.4±1.5 (4)		25.2±1.4 (3)	
			1.23		1.06		1.09
Sminthopsis murina	♂♂	17.1±1.4 (7)		24.9±1.1 (9)		27.5 (1)	
			1.43		1.30		1.05
Sminthopsis murina	♀♀	12.0 (1)		18.0±1.5 (4)		24 (1)	

(In September and November the ratios 1.06 and 1.09 compare *A. stuartii* 1st Yr. ♀♀ with *S. murina* ♂♂; 1.30 and 1.05 compare *A. stuartii* 1st Yr. ♀♀ with *S. murina* ♀♀.)

ed on the heath plot. *Sminthopsis murina* exhibits an ability to utilise much drier habitats than does *A. stuartii*. This is evident in the distribution of captures before the plot burned (Fig. 2); *S. murina* capture sites usually had less dense vegetation below 1m and much reduced leaf litter. Both species relate to structural rather than floristic factors (Table 3) which is not surprising for insectivores. Correlation of density of *A. stuartii* with plant species diversity may be related to Statham's (1982) observation that *A. stuartii* had significant quantities of plant material in its diet.

THE REGENERATION NICHE: *Sminthopsis murina* was most abundant in heath and forest habitats that have regenerated for 2 to 5 years after fire. In a study of recently burned desert grassland, Christian (1977) suggested that the rapid return by some species of small mammals

Table 6. A summary of some of the parameters which illustrate ecological separation between these two species.

Habitat Parameters	Summary of Ecological Separation *Sminthopsis murina*	*Antechinus stuartii*
HEATH		
Altitude and Soil Moisture	Higher and drier	Lower and wetter
Vegetation up to 1m	Less dense	More dense
Tree Presence	Dry Heath	Forest Edge and Tree Heath
FOREST		
Leaf litter	Less	More
Plant Species Diversity	Lower	Higher
Understory height	Shorter	Taller
FIRE		
Successional stage	Mid	Late
Population Peak	4 years	6+ years
REPRODUCTION	Polyoestrous (2 litters/year) 10 teats	Monoestrous (1 litter/year) 8 teats

following the fire was a result of those species' normal ranges encompassing natural areas equally as harsh as the post–fire environment.

This introduces the concept of a regeneration niche: if the vegetation regenerating after a fire is regarded as a time related "resource" axis, then species will have different abilities to take advantage of this "resource". These different abilities will be reflected in the species' fire response indices (Fig. 6), which then define the species' regeneration niche and so determine its position along the regeneration time axis (Fig. 8). This can be seen in the secondary mammal succession in forest (Fox and McKay 1981) and heath (Fox 1980). The concept of a regeneration niche has been used by Grubb (1977), as an explanation for the maintenance of species–richness in plant communities.

Interspecific Interaction: There are four kinds of indirect evidence consistent with some form of competitive interaction between these species. The first is the significant negative association between them in the forest study. However it is not possible on this evidence to exclude the possibility that the species have habitat requirements too different to occur on the same plots.

The apparent displacement of *S. murina* from the forested area of the heath plot is the second kind of evidence. It represents a habitat shift by *S. murina* from the forest study plots where it is caught with *A. stuartii* on only one plot (local allopatry) to the heath plot where it coexists with *A. stuartii* at a macro–habitat level (local sympatry). This evidence is not consistent with habitat selection alone but requires some interaction.

The third type of evidence is the difference in between–habitat overlap (0.63) and within–habitat overlap (0.13). Spatial separation is shown at the microhabitat level within habitat patches, which is consistent with interspecific interaction.

The reciprocal habitat shift shown after the fire on the heath plot strongly indicates interspecific interaction and is the fourth piece of evidence. It represents a further displacement of *S. murina* away from the forest habitat. The direction of the shift, with *A. stuartii* displacing *S. murina*, is that which would be expected given that *A. stuartii* is larger than *S. murina*. The marked sexual dimorphism in *A. stuartii* complicates body weight relationships as does the male "die–off", and there may be three ecological units to be considered. Body weights and ratios for both *A. stuartii* and *S. murina* are presented in Table 5 for May, September and November. During May all five units show differences consistent with those predicted for limiting similarity. However in September *S. murina* males and *A. stuartii* 1st Yr. females show less difference than expected, while in November both male and female *S. murina* are more similar to *A. stuartii* 1st Yr. females than predicted. Insects are most abundant during the late spring–early summer period so that resources may no longer be limiting. The absence of *A. stuartii* males and the absence or the reduced abundance of 2nd Yr. females would also reduce the demand on resources. The earlier breeding of *A. stuartii* maximises the size difference between juveniles of each species.

Coexistence: *Sminthopsis murina* is not a common species in southeastern Australia despite being called the Common Dunnart. Its distribution is very patchy but where it does occur it can be locally abundant with densities as high as 6 ha^{-1} during the mating periods (personal observation). In all cases where *S. murina* has been caught in reasonable numbers it has been in habitats regenerating after a fire (Smith Lake, 3 yrs., Dick Whitford and Mike Archer, pers. comm.; Wyee, N.S.W., 4 yrs., Debbie Andrews, pers comm.; Ourimbah, N.S.W., 4 yrs., Broadbent and Cranwell (1979); Beerwah, Qld., 2 yrs. Dwyer *et al.* (1979)).

Much of the ecological separation observed between the two species (summarised in Table 6) can be directly related to the presence of the appropriate seral stages which provide the required habitat parameters. It is possible that *S. murina* may exist at low abundance in habitats that provide many of the habitat requirements but I would put forward the hypothesis that *S. murina* can only reach peak abundance and coexist with *A. stuartii* in habitats regenerating after fire. And to go further, I would predict that long term coexistence is only possible for *S. murina* in areas with fire histories which ensure that mid–seral stages of regenerating vegetation are always present.

ACKNOWLEDGEMENTS

Much of the original work on which this paper is based was carried out while I was a postgraduate student at the School of Biological Sciences, Macquarie University. I would like to acknowledge the assistance provided to me there, in particular from Dr. George Mckay. This paper benefitted greatly from many conversations with Dr. Michael Archer and critical comments from Dr. Marilyn Fox, Dr. Steven Morton and Mr. Chris Dickman who read the manuscript. Mr. Dick Whitford was responsible for much of the reproduction results while the initial data collection was supported by a Commonwealth Post Graduate Research Award to me and a Department of Environment, Housing and Community Development Grant (76/328) to Dr. G. M. McKay.

REFERENCES

ABRAHAMS, P. A., 1975. Limiting similarity and the form of the competition coefficient. *Theor. Popul. Biol.* **8:** 356–75.

ABRAHAMS, P. A., 1976. Niche overlap and environmental variability. *Math. Biosci.* **28:** 357–72.

BRAITHWAITE, R. W., COCKBURN, A. M., AND LEE, A. K., 1978. Resource partitioning by small mammals in lowland heath communities of southeastern Australia. *Aust. J. Ecol.* **3:** 423–45.

BROADBENT, J. A. AND CRANWELL, I., 1979. Faunal studies for the proposed Mount White–Kariong–Ourimbah sections of the Sydney–Newcastle Freeway (No. 3). Environmental and Urban Studies Report No. 45. December 1979, Macquarie University, North Ryde.

CHRISTIAN, D. P., 1977. Effects of fire on small mammal populations in a desert grassland. *J. Mammal.* **58**: 423–7.

DICKMAN, C., 1982. Some ecological aspects of seasonal breeding in *Antechinus* (Dasyuridae, Marsupialia). Pp.140–50 *in* "Carnivorous marsupials", ed. by M. Archer. Roy. Zool. Soc. New South Wales: Sydney, Australia.

DWYER, P., HOCKINGS, M., AND WILLMER, J., 1979. Mammals of Cooloola and Beerwah. *Proc. R. Soc. Qd.* **90:** 65–84.

FOX B. J., 1979. An objective method of measuring the vegetation structure of animal habitats. *Aust. Wildl. Res.* **6:** 297–303.

FOX, B. J., 1980. "The ecology of a small mammal community: secondary succession, niche dynamics, habitat partitioning, community structure and species diversity." Ph.D. thesis: Macquarie University, Sydney.

FOX, B. J., 1982. A review of dasyurid ecology and speculation on the role of limiting similarity in community organization. Pp.97–116 *in* "Carnivorous marsupials", ed. by M. Archer. Roy. Zool. Soc. New South Wales: Sydney, Australia.

FOX, B. J. AND FOX, M. D., 1978. Recolonization of coastal heath by *Pseudomys novaehollandiae* (Muridae) following sand mining. *Aust. J. Ecol.* **3:** 447–65.

FOX, B. J. AND FOX, M. D., 1981. A comparison of vegetation classifications as descriptors of small mammal habitat preference. Pp.166–80 *in* "Vegetation classification in the Australian region", ed. by A. N. Gillison and D. J. Anderson, C.S.I.R.O. and A.N.U. Press: Canberra.

FOX, B. J. AND MCKAY, G. M., 1981. Small mammal responses to pyric successional changes in eucalypt forest. *Aust. J. Ecol.* **6:** 29–42.

FOX, B. J., AND WHITFORD, R., 1982. Polyoestry in a predictable coastal environment: Reproduction, growth and development in *Sminthopsis murina* (Dasyuridae, Marsupialia). Pp.39–48 *in* "Carnivorous marsupials", ed. by M. Archer. Roy. Zool Soc. New South Wales: Sydney.

GAUSE, G. F., 1934. The struggle for existence". Hafner: New York. (Reprinted 1964).

GRUBB, P., 1977. The maintenance of species–richness in plant communities: the importance of the regeneration niche. *Biol. Rev.* **52:** 107–45.

HALL, S., 1980. Diel activity of three small mammals coexisting in forest in Southern Victoria. *Aust. Mammal.* **3:** 67–79.

HARDIN, G., 1960. The competitive exclusion principle. *Science* **131:** 1292–7.

HUTCHINSON, G. E., 1959. Homage to Santa Rosalia or Why are there so many kinds of animals? *Am. Nat.* **93:** 145–59.

MACARTHUR, R. H., AND LEVINS, R., 1967. The limiting similarity, convergence, and divergence of coexisting species. *Am. Nat.* **101:** 377–85.

MARLOW, B. J., 1961. Reproductive behaviour of the marsupial mouse, *Antechinus flavipes* (Waterhouse) and the development of pouch young. *Aust. J. Zool.* **9:** 203–18.

MAY, R. M., AND MACARTHUR, R. H., 1972. Niche overlap as a function of environmental variability. *Proc. Nat. Acad. Sci.* **69:** 1109–13.

NEWSOME, A. E., MCILROY, J., AND CATLING, P., 1975. The effects of an extensive wildfire on populations of 20 ground vertebrates in S.E. Australia. *Proc. Ecol. Soc. Aust.* **9:** 107–23.

PIANKA, E. R., 1974. Niche overlap and diffuse competition. *Proc. Nat. Acad. Sci.* **71:** 2141–5.

STATHAM H. L., 1982. *Antechinus stuartii* (Dasyuridae, Marsupialia) diet and food availability at Petroi, northeastern New South Wales. Pp.151–63 *in* "Carnivorous marsupials", ed. by M. Archer. Roy. Zool. Soc. New South Wales: Sydney, Australia.

VAN DYCK, S., 1982. The relationships of the Australian marsupials *Antechinus stuartii* and *A. flavipes* (Dasyuridae, Marsupialia) in Queensland. *In* "Carnivorous marsupials", ed. by M. Archer. Roy. Zool. Soc. New South Wales: Sydney.

WOOLLEY, P., 1973. Breeding patterns and the breeding and laboratory maintenance of dasyurid marsupials. *Exp. Animals* **22** (Supp.): 161–72.

RESUMEN

17. Separacion ecologica y coexistencia de *Sminthopsis murina* y *Antechinus stuartii*: regeneracion de un nicho? Por Barry J. Fox.

Superficialmente *Sminthopsis murina* y *Antechinus stuartii* son especies de pequéños dasiúridos muy similares que comparten el nicho ecologico insectívoro en el área de los Lagos Myall de New South Wales (32° 28'S, 152° 30'E). Su coexistencia se ve facilitada por su marcada separación ecológica. En los bosques abiertos *S. murina* utiliza los hábitats más secos, con vegetación menos densa, de menos de 1 m y con más reducida hojarasca. *A. stuartii*, por el contrario, se encuentra en vegetación más densa, con más alta diversidad de especies de plantas.

Cuando ambas especies se encuentran en los matorrales, *A. stuartii* ocupa las partes con los emergentes y achaparrados eucaliptos, los márgenes de los bosques o matorrales de poca humedad. Por su lado, *S. murina* ocupa los altos matorrales secos y falta totalments en todas las áreas húmedas. Representa éste un desplazamiento del área que *S. murina* ocupa en el bosque abierto.

Sus roles en la sucesión posterior a los incendios también difieren. *Sminthopsis murina* ocupa estados intermedios de la sucesión, donde alcanza su máxima abundancia, para decrecer en los estados posteriores. Por otro lado, *A. stuartii* aumenta gradualmente hasta alcanza su máxima abundancia en los últimos estados de la sucesión. Consistente con esto la poliestral *S. murina* tiene una más alta capacidad reproductiva que facilitaría su rápida recuperación después de la alteración ambiental.

La longeva coexistencia entre estas especies resulta de la habilidad de *S. murina* para explotar un "nicho en regeneración" que está disponible en areas con alta frecuencia de incendios, que producen los requeridos estados de sucesión intermedia.

18. THE POPULATION ECOLOGY OF THE EASTERN QUOLL *Dasyurus viverrinus* (DASYURIDAE, MARSUPIALIA), IN SOUTHERN TASMANIA

JANET GODSELL[1]

A population of Eastern Quoll, *Dasyurus viverrinus* (Shaw), studied in southern Tasmania, fluctuated regularly over 20 months. Summer peaks were caused by the influx of juveniles and were followed by heavy mortality of juveniles and adults. There were differences in the times and rates at which adult males and females disappeared from the population. Females were more restricted in movements than males. This difference in dispersion patterns may play an important role in their survival rates.

Pages 199–207 *in* CARNIVOROUS MARSUPIALS, ed. by M. Archer, February, 1982. Roy. Zool. Soc. New South Wales: Sydney, Australia.

THE EASTERN QUOLL is one of five species in the genus *Dasyurus*. Common to all is a dearth of ecological information. Their small geographical ranges and low densities in most places may partly explain this lack of knowledge.

The Eastern Quoll, *Dasyurus vivverrinus* (Shaw), was once distributed over much of eastern Australia, possibly as far north as southern Queensland to South Australia and Tasmania. At the turn of the 20th century an epidemic is believed to have severely reduced its numbers and range (Jones 1923), and today its status is uncertain. It is described as extinct in South Australia and Victoria and as rare in New South Wales (Archer 1978). Only in Tasmania have populations regained their former numbers and, indeed, in the mosaic of eucalypt forest and farmland of parts of the state, it is believed to have surpassed its previous abundance (Green 1967). Hence, Tasmania provides the opportunity to examine the population ecology of a species for which conservation measures on the mainland of Australia may be crucial for its survival.

The Eastern Quoll is reported to occur in a variety of habitats including dry sclerophyll forest (Ride 1970), scrub, heathland and cultivated land (Green 1973). From analysis of the stomach contents of Eastern Quoll from northeastern Tasmania, Green (1967) concluded that invertebrates represented a large proportion of the diet. Vertebrates were recorded in a few stomachs and grass was present in most. Sharland (1962) lists frogs and lizards as part of the diet of the species. Scats analyzed in the present study indicate that the Quoll is an opportunistic carnivore for which seasonally occurring agricultural pests such as army worms, *Persectania ewingi*, and the cockchafer beetle, *Aphodius* sp., represent important food resources. Vertebrates, principally small mammals and birds, occur in low but constant frequencies in diet throughout the year. Plant matter occurs regularly; and fruits such as blackberries, *Rubus fruticosus*, are seasonally popular.

The Eastern Quoll has a short synchronized breeding season in Tasmania; mating usually occurs from mid May until early June. Gestation has been recorded as lasting 16 days (Hill and O'Donoghue 1913) and 19–23 days (Fletcher 1978). Females may have up to six young which are carried in the pouch until mid August. They are then left in the den while the mother forages for food; weaning is complete by the end of October (Fleay 1935, present findings).

This paper describes the changing size and structure of a population of Eastern Quoll in southern Tasmania during 20 months, August, 1978, to March, 1980, and suggests possible explanations of the changes.

METHODS

STUDY SITE: The study site was near Cradoc on the Huon peninsula, about 40 km south of Hobart in a habitat of pastureland and dry sclerophyll forest. The area has been under cultivation for at least 30 years and is much disturbed by grazing cattle and sheep, periodic control burning, ploughing of fields and the growth of blackberries. Within the pastures are copses of eucalypt scrub which consist principally of stringybark *Eucalyptus obliqua*, with an understorey of prickly mimosa *Acacia verticillata,* native gorse *Daviesia ulicifolia,* manuka *Leptospermum scoparium,* blackberries *Rubus fruticosus,* and bracken *Pteridium esculentum.*

PROCEDURE: Twenty–five wire cage traps, 56 x 21 x 21 cm (Mascot Wire Works, Sydney), were baited with beef liver and were set in five areas indicated in Fig. 1. Population estimates

[1] Zoology Department, Australian National University, Canberra, A.C.T., Australia, 2600.

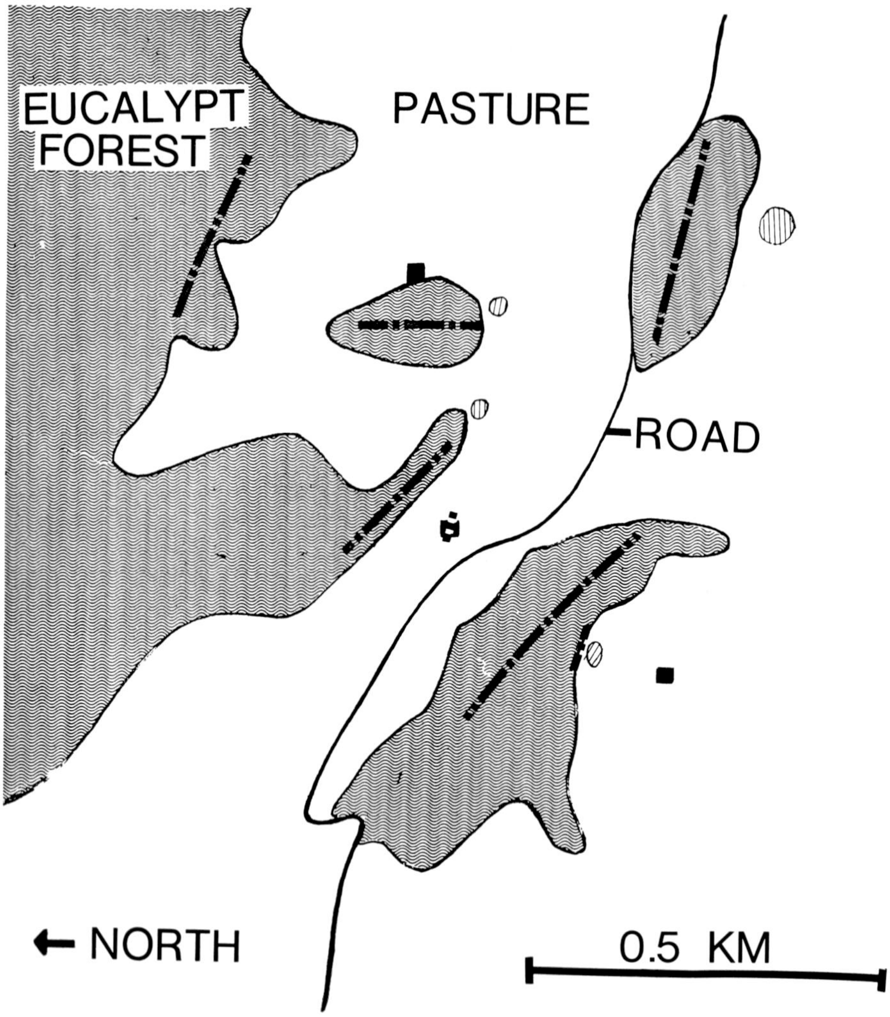

Fig. 1. Map of the study area near Cradoc, Tasmania. Trap lines are represented by the dashed lines. Hatched areas represent ponds and black squares represent buildings now used as hay sheds.

are based on the trapping results of two consecutive nights a month. Animals were ear-tattooed for individual recognition. Weights and body measurements were used to distinguish adults from juveniles.

Population Size and Structure: Monthly estimates of the population size are based on the number of animals known to be alive in each month. Individuals were assumed to be present if, after they had been trapped at least once, they were not trapped during one monthly session but were trapped subsequently.

Classes of animals are defined as follows: (a) resident adults: animals caught in three or more consecutive months; (b) juveniles: born during 1978 and 1979 (these are classified in their respective cohorts); and (c) transient adults: animals caught during one (or, rarely, two) of the successive monthly sessions but not caught subsequently. Although the last class might be suspected of trap–shyness, as most animals were caught on both nights of a monthly session, it is believed that trap–shyness was not a problem. Animals that were caught during one or two consecutive monthly sessions and thereafter not retrapped for at least four months are also defined as transients, though it is believed that these are neighbours from adjoining areas that made occasional exploratory visits into the study area. At present transient and resident juveniles are not differentiated.

Movements: Mobility of resident adults is analyzed in terms of the movements of resident Quolls in the study area. These are recorded as the number of times that an animal is trapped in (a) different localities and (b) its core area. The core area is defined as the locality where the animal is most frequently captured. It is assumed that this is the area most commonly used by that individual.

RESULTS

Population Size: There were annual fluctuations in the numbers of animals known to be alive, with peaks in summer and troughs in wint-

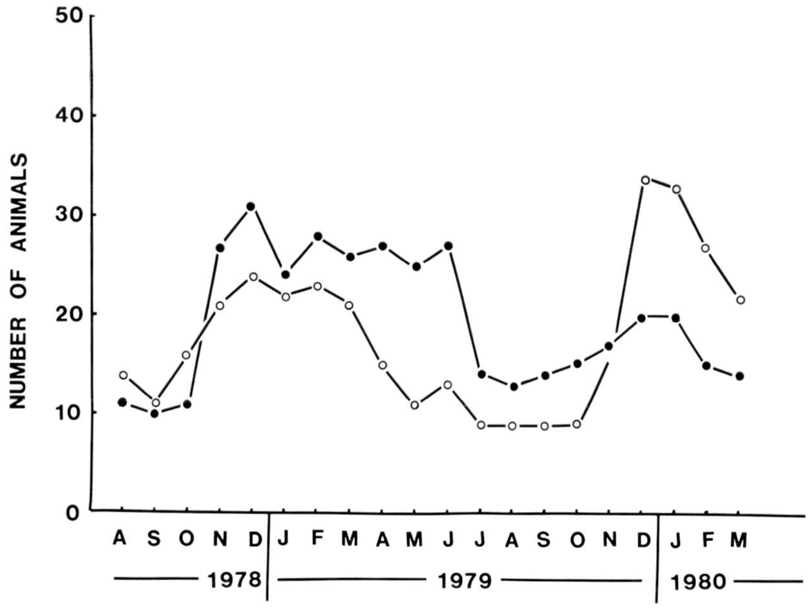

Fig. 2. Number of male and female *D. viverrinus*. Females, open circles; males, closed circles.

er (Fig. 2). In December of 1978 and 1979, the total population was 54 and 52 individuals, respectively. The lowest population numbers, 21 and 22 individuals, were in September, 1978, and August, 1979, respectively.

In total, 171 Eastern Quoll were individually marked. There were 95 males and 76 females, or 1.25 males to a female ($\chi^2=2.10$, $p>0.05$). Males formed the larger proportion of the population in 1979 but the trend was reversed in the early months of 1980 (Fig. 2). Minimum numbers of both sexes occurred during the spring in each year.

POPULATION STRUCTURE: The number of resident adults of the 1978 cohort declined dramatically from 23 in November, 1978, to 2 by February, 1980 (Fig. 3). This represents a 30% mortality rate by the mating season in May, 1979, and a mortality of 91% by February, 1980. Most females of the 1978 adults disappeared in the summer of 1978 (Fig. 4). Of the 12 original resident females, 50% survived until the mating season, and 33% survived until August, 1979, when the young first left the den.

Males remained numerous during the summer of 1979; a few disappeared in the autumn (March — June, 1979) but 83% survived until the mating season. In July, 1979, shortly after the mating season, six males disappeared. This reduced the resident male population to 25% of its original size.

The numbers of transient adults fluctuated annually (Fig. 5). Among the transient adults during the summer there were more females than males (but $\chi^2=2.25$, $p> 0.05$). Transient males were more abundant than transient females during the breeding season ($\chi^2=4.56$, $p<0.05$).

Figure 6 gives two estimates of the numbers of juveniles in the 1978 and 1979 cohorts. The first is based on the number of juveniles known to be alive as determined from the trap records; the assumption is made that the date of first capture indicates when the animal was first trappable. On this basis, the peak number, 30, of the 1978 cohort, was reached in November, 1978; 63% of these juveniles survived to the 1979 mating season and 23% survived until March, 1980. The 1979 cohort showed a peak, 33, in January, 1980. Juveniles gradually disappeared from January until March, 1980.

For the second estimate, it was assumed that all the juveniles that were trapped were born to resident females: that is that all juveniles caught during November through March were present in the population from the date of their becoming independent of the den in November, regardless of the month in which they were first trapped. Such an estimate, however, ignores the effects of immigration and emigration of juveniles. On this assumption, the peaks of juveniles, 62 in 1978 and 60 in 1979, occurred in November of each year. Of the 1978 cohort, 31% survived until May, 1979, and 11% until March, 1980. This is a much lower survival rate than that based on the first estimate. It is, however, probably more realistic, as can be seen by comparing both estimates with the number of young that would have entered the population in November if none died in the pouch or den.

The total number of young born to resident females was determined by counting all young in the pouch in July and the number of extended nipples of resident females whose young were al-

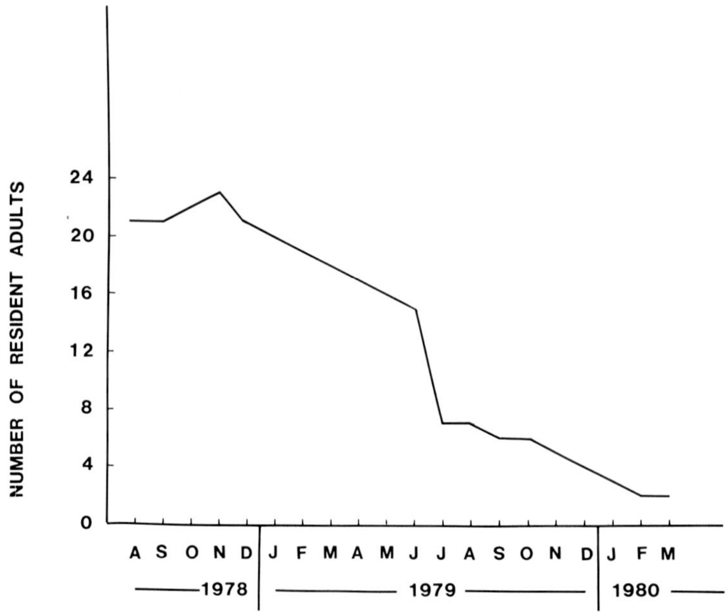

Fig. 3. Number of resident adult *D. viverrinus* of the 1978 cohort.

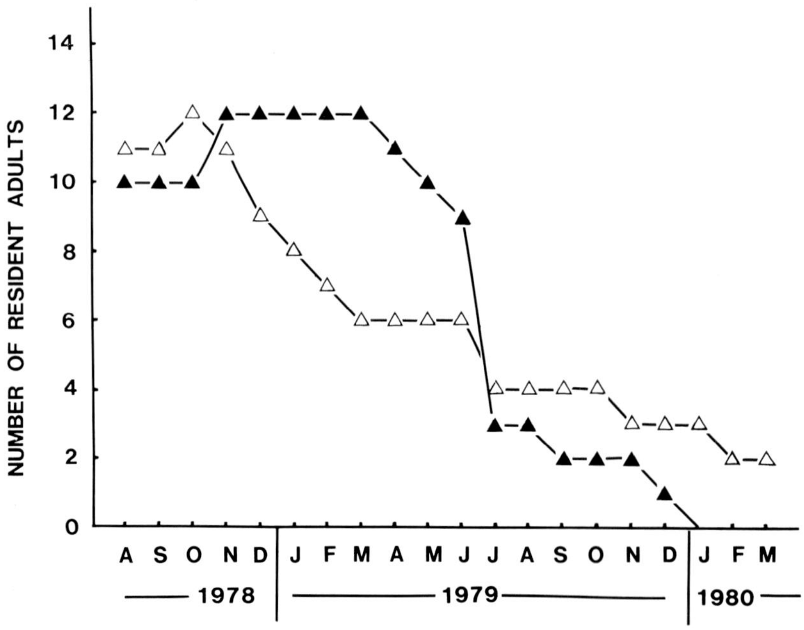

Fig. 4. Number of male and female *D. viverrinus* of the 1978 resident adults. Females, open triangles; males, closed triangles.

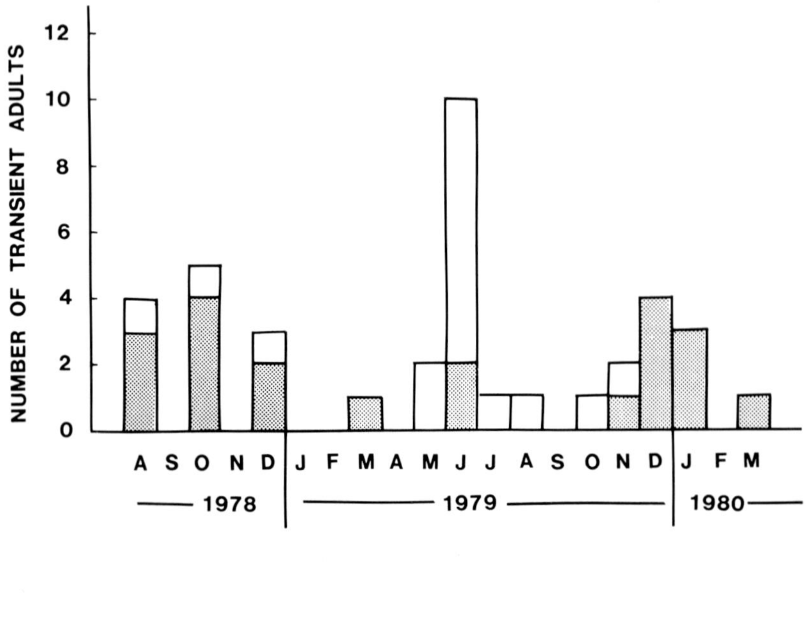

Fig. 5. Number of transient adult *D. viverrinus*. Females, shaded area; males, non–shaded area.

ready in the den. For the 1978 and 1979 cohorts these values were 57 and 54, respectively, close to the November values of 62 and 60 shown in Fig. 6. They greatly exceed the maxima of 30 and 33 for the number of juveniles known to be alive.

The total number of young born to resident and transient females include the number of young born to transient females as determined by their extended nipples. For the 1978 and 1979 cohorts these values were 91 and 78, respectively, both greatly in excess of the other estimates (Fig. 6).

SEX RATIOS OF JUVENILES: Of all young counted in the pouch in 1978 and 1979, the ratio of males to females was 35:23, or 1.52 males to female ($\chi^2 = 2.48$, $p > 0.05$).

The ratio of males to females trapped as juveniles in the 1978 cohort was 41:21, or 1.95 males to a female ($\chi^2 = 6.45$, $p < 0.050$). Figure 7a illustrates the survival rates of both sexes; this analysis assumes that all juveniles were born to resident females. Most losses occurred between November, 1978 and May, 1979. Although males disappeared at a higher rate than females, 32% of the original juvenile males of 1978 survived to the 1979 mating season compared to 24% of the females. The numbers approached parity by September, 1979.

The estimate of the number of juveniles known to be alive (Fig. 7b) indicates that there were about equal numbers of males and females until March, 1979. Compositional changes, especially among males, were, however, high during this period. Female numbers then declined and reached a minimum by the mating season, while male numbers decreased slowly to a minimum by September.

In the 1979 cohort, the ratio of males to females was 29:31, or 0.93 males to a female ($\chi^2 = 0.06$, $p > 0.05$). If one assumes that all juveniles were born to resident females, both sexes had similar mortality rates; but if the values of the juveniles known to be alive each month are used, then a higher proportion of females were present in the summer of 1979. This difference disappeared by March, 1980.

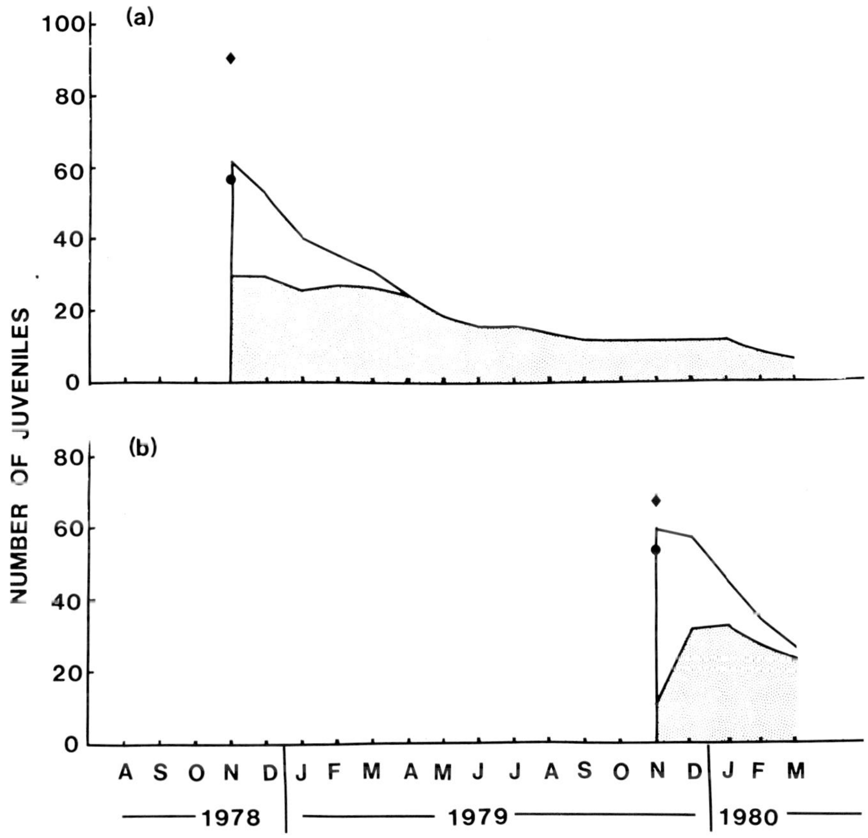

Fig. 6. Number of *D. viverrinus* of the 1978 and 1979 cohorts. (a) 1978 cohort; (b) 1979 cohort. The number of juveniles known to be alive is represented by the shaded area; the estimated number of juveniles, if all juveniles were born to resident females, is represented by the non–shaded area. (●) represents the actual number of young born to resident females and (♦) is the number of young born to resident and transient females.

MOVEMENTS: Figure 8 summarizes the differences between the sexes in mobility between localities in the study area. Fifty percent of females, but only 5% of males, were caught in only one locality. Correspondingly, 47.5% of all males, but only 5.5% of all females, were caught at three or more localities ($\chi^2 = 13.79$, $p < 0.01$).

Figure 9 illustrates the differences between the sexes in the percent frequency with which an individual was trapped in its core area. Of all females, 83% were trapped in one locality on more than 75% of occasions. The corresponding value for males was 47.5% ($\chi^2 = 11.50$, $p < 0.01$).

DISCUSSION

The numbers of Eastern Quoll in the population studied fluctuated seasonally over 20 months. Because of a short, synchronized mating season (May-June), large numbers of young entered the population in early summer (November-December), resulting in remarkably high short–term densities. Heavy losses restored the population to a minimum by August. Between November, 1978, and August, 1979, the number of juveniles declined from 62 to 14 (77% loss), and the number of adults from 23 to 7 (70% loss). Thus while the reduction in the size of the 1978 juvenile cohort accounted for the greatest part of the reduction in population size, the survival rates of the juvenile and adult cohorts were similar.

The months during which most 1978 adults disappeared from the population differed between the sexes. Fifty percent of the adult females disappeared during the summer of 1978–1979. This was the time when the young were newly weaned and the population was most dense. The number of adult males, however, did not decline during this period. Differences in the dispersion patterns of the sexes may have influ-

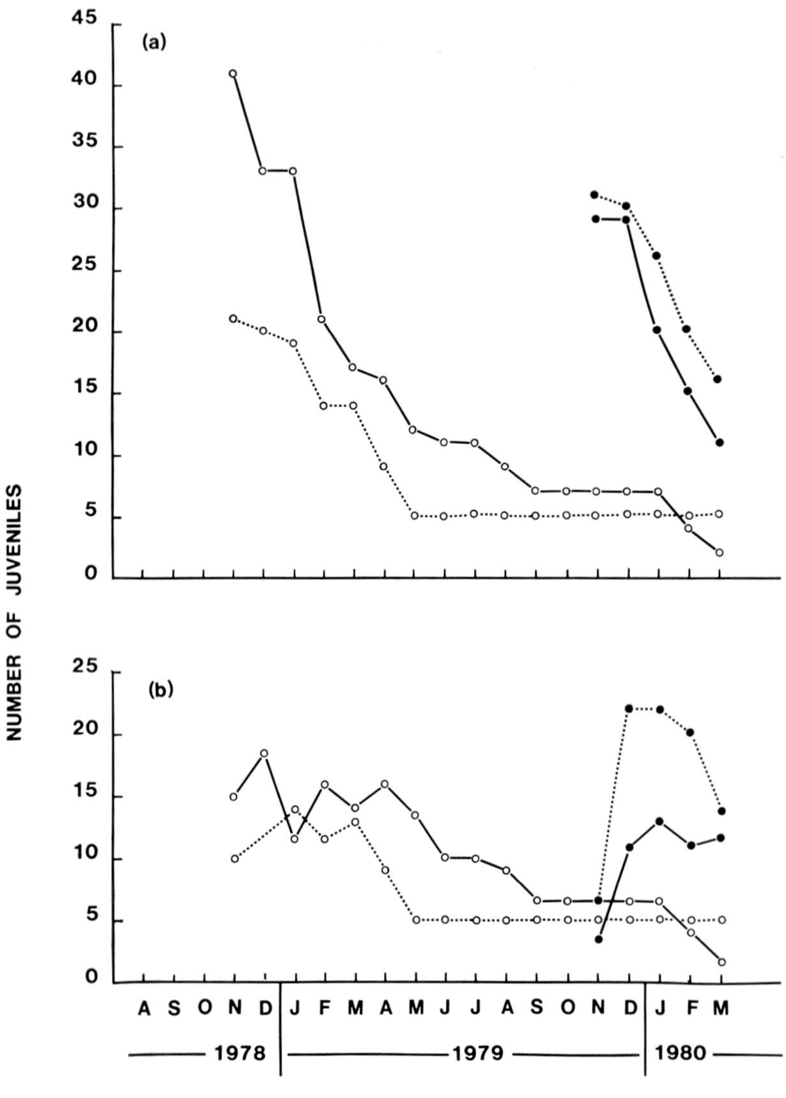

Fig. 7. Number of male and female *D. viverrinus* of the 1978 and 1979 cohorts. a) The estimated number of male and female juveniles of 1978 and 1979 if all juveniles were born to resident females. b) The number of male and female juveniles known to be alive. 1978 cohort: females o - - - - - o, *males* o——o; 1979 cohort: females ● - - - - - ●, males ●——●.

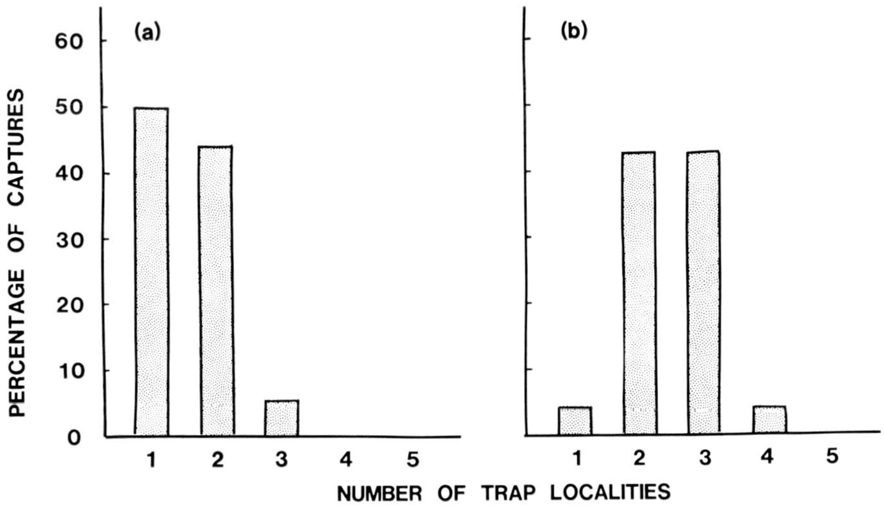

Fig. 8. Percentage of resident *D. viverrinus* trapped at five localities. a) females, b) males.

enced the degree to which adult males and females were in competition for resources with the large new juvenile cohort. Unlike the more widely ranging males, each female was generally restricted in movements to a single copse (Fig. 8 and Fig. 9), and female core areas greatly overlapped. Most juveniles confined their movements to these same areas during early summer. In consequence, competition for food within and around the copses was probably more intense than elsewhere in the study area and may have had a greater effect on adult females than on adult males. The increase in the number of transient females in the population during the summer of 1979 and 1980 supports this theory by suggesting that similar pressures resulted in the dispersal of females from neighbouring areas.

By August, 1979, only 4 (33%) of the original adult females and 5 (24%) of the 1978 juvenile females remained in the population. Although the survival rate of juvenile females of 1978 was slightly lower than that of adult females of 1978, sufficient females of the 1978 cohort survived to represent just over 50% of the breeding females of 1979. Dens were likely to be an important resource for females by August, because the young were then too large to be carried in the pouch. Competition for dens may have contributed to the decline in the number of adult females of 1978 between the mating season and August.

Only two females of 1978 remained in the population by March, 1980. Most of the original resident females were replaced by their offspring in less than two years of observation.

Between November, 1978, and August, 1979, the number of adult males declined from 12 to 3 (75% loss) while the number of juvenile males of 1978 declined from 41 to 9 (78% loss). Juvenile males disappeared at a fairly constant rate from November to August. Adult male losses, however, were greatest at the end of the mating season. Competition for oestrous females, or increased activity affecting the nutritional condition of the males, may have been responsible for the dramatic losses of adult males during the June–July period. During the mating season of 1979, many transient males appeared in the population, perhaps because similar factors resulted in the dispersal of adult males from other localities.

By March, 1980 there were no adult males of 1978 remaining in the population; and the number of males of the 1978 cohort was reduced from 9 in August, 1979, to 2 by March, 1980. Hence, there was a high turnover of both male and female resident adults from one year to the next (Table 1).

Table 1 Composition of the resident populations of 1978 and 1979.

	Aug. 1978	May 1979	Aug. 1979	Mar. 1980
1978 Adults				
Males	10	10	3	0
Females	11	6	4	2
1978 Juvenile Cohort				
Males	0	13	9	2
Females	0	5	5	5

Findings of the present study suggest that there is a difference in the roles of the sexes in population regulation, and that these are expressed differentially with season and with changes in population density. Behavioural differences, reflected in the spacing patterns of the sexes,

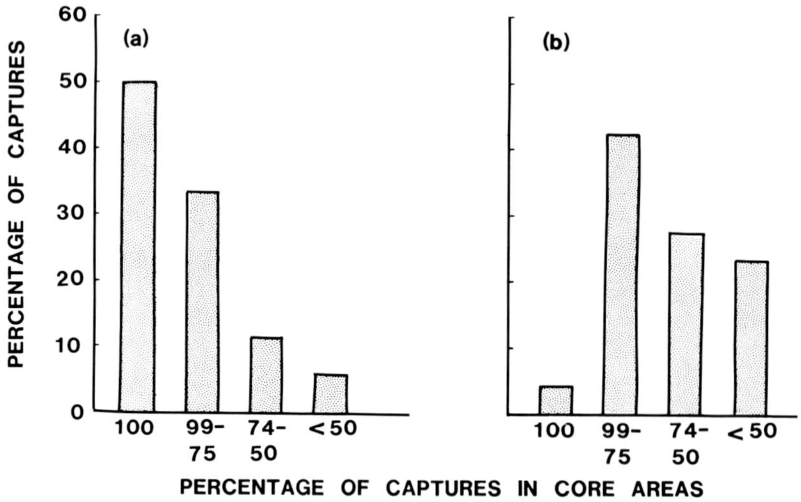

Fig. 9. Percentage of resident *D. viverrinus* trapped in their core areas. The number of captures was determined for each individual and the number of which occurred in the core area was expressed as a percent. a) females b) males.

may result from differences in the requirements of the sexes; mates for males and dens for females. This indicates a situation of greater intrasexual than intersexual competition. Where optimal denning areas are limited resources (in the island copses), it would be selectively advantageous for females to be familiar with the area and in possession of prime den sites. This would account for the restricted movements of females about specific copses. Differences in the possible role of the sexes in population regulation has been suggested by Redfield *et al.* (1978). The results of their experimental manipulations of the sex ratios of *Microtus townsendii* indicated that the negative influence of resident females on juvenile survival and female recruitment could be critical for vole population regulation. Experiments conducted on supplemental feeding of a population of *Peromyscus maniculatus* (Fordham 1971) reported differing numerical responses between males and females. It was suggested that sex–related behavioural mechanisms were responsible for maintaining the constant numbers of males during the breeding season. Boonstra (1977) removed individuals from a declining population of *Microtus townsendii* and found differences in the survival rates of the sexes. He proposed a behavioural model in which females compete for sites to raise young while males compete for females. The current study agrees with this proposal and suggests that the Eastern Quoll exhibits sex specific spacing behaviour and intrasexual competition. This influences both population size and structure.

The Eastern Quoll is capable of producing large numbers of young and, as demonstrated in the present study, can establish dense populations in an environment greatly modified by man. These traits may help in the conservation of the species on the mainland of Australia if populations of Quolls are still to be found there.

ACKNOWLEDGEMENTS

I would like to thank Dr. D. C. D. Happold for his supervision during this study and Professor S. A. Barnett and Dr. B. Fox for their criticism of the manuscript. I am indebted to the Zoology Department of the University of Tasmania for much help and to Mr. R. Rose and Mr. R. H. Green for their help and encouragement. I am grateful to the National Parks and Wildlife Service of Tasmania for both their cooperation and loan of their traps, Mr. N. C. Griggs for permission to trap on his land, Mr. J. Russell for use of his farm as a field station, and Mr. P. Laing for his field assistance on many occassions.

This study was conducted with the support of a graduate scholarship from the Australian National University.

REFERENCES

Archer, M., 1978. The status of Australian dasyurids, thylacinids and myrmecobiids. Pp.29–43 *in* "The status of endangered Australian wildlife" ed. by M. J. Tyler. Royal Zoological Society of South Australia: Adelaide.

Boonstra, R., 1977. Effect of conspecifics on survival during population declines in *Microtus townsendii. J. Anim. Ecol.* **46:** 835–51.

Fleay, D. H., 1935. Breeding of *Dasyurus viverrinus* and general observations on the species. *J. Mammal.* **16:** 10–6.

Fletcher, T. P., 1978. Reproduction of the Native Cat *Dasyurus viverrinus* (Shaw). Honours thesis: University of Tasmania; Hobart.

Fordham, R. A., 1971. Field populations of deermice with supplemental food. *Ecology* **52:** 138–46.

Green, R. H., 1967. Notes on the Devil *(Sarcophilus harrisii)* and the Quoll *(Dasyurus viverrinus)* in northeastern Tasmania. *Rec. Queen Vic. Mus.* **27:** 1–13.

Green, R. H., 1973. "The mammals of Tasmania". Launceston.

Hill, J. P. and O'Donoghue, C. H., 1913. Reproductive cycle of *Dasyurus viverrinus*. *Quart. J. Micros. Sci.* **59:** 133.

Jones, F., Wood, 1923. "The mammals of South Australia. Part 1. The monotremes and carnivorous marsupials". Government Printer; Adelaide.

Redfield, J. A., Taitt, M. J. and Krebs, C. J., 1978. Experimental alteration of sex ratios in populations of *Microtus townsendii*, a field vole. *Can. J. Zool.* **56:** 17–27.

Ride, W. D. L., 1970. "Native mammals of Australia." Oxford University Press: London.

Sharland, M., 1962. "Tasmanian wildlife". Melbourne University Press: Melbourne.

RESUMEN

18. Evologia de la poblacion de *Dasyurus viverrinus* "Eastern Quoll" (Dasyuridae, Marsupialia), en el sur de Tasmania. Por Janet Godsell.

Una población del "Eastern Quoll", *Dasyurus viverrinus* (Shaw), estudiada en el sur de Tasmania mostró regulares fluctuaciones cada 20 meses. Algunos picos fueron causados por la afluencia de jóvenes, seguidos por alta mortalidad de jóvenes y adultos. Se registraron diferencias en el tiempo y en los ritmos de la desaparición de los machos y de las hembras adultas de la población. Las hembras manifestaron una mayor restricción en sus movimientos que los machos. Esta diferencia en la pauta de dispersión pudo haber jugado un rol importante en la extensión de sobrevivencia.

19. HABITAT USE BY TWO SPECIES OF *Antechinus* AND *Rattus Fuscipes* IN TALL OPEN FOREST IN SOUTHERN VICTORIA

S. HALL[1] and A. K. LEE[1]

Antechinus stuartii, A. swainsonii and *Rattus fuscipes,* coexisting in tall mountain ash *(Eucalyptus regnans)* forest, were examined for association with floristics and vegetation structure on a 6.2 ha trapping grid.

In this relatively homogeneous vegetation, *A. stuartii* was not consistently associated with particular floristic or structural groups. In most months *A. swainsonii* also showed no consistent association with particular floristic groups but was most often trapped where vegetation cover was thickest between 62 and 184 cm. *Rattus fuscipes* showed association with particular floristic and structural groups and this changed during the study.

Antechinus stuartii and *R. fuscipes* were negatively associated during colder months, but not over all months. *A. stuartii* and *A. swainsonii* were negatively associated in May, 1975, and in all months combined. Captures of *A. swainsonii* and *R. fuscipes* were not correlated. *A. swainsonii* and *R. fuscipes* appeared to partition the habitat on a moment to moment basis.

Niche dimensions are discussed in light of these data and knowledge of diet, diel activity patterns, and use of trees.

Pages 209–20 *in* CARNIVOROUS MARSUPIALS, ed. by M. Archer, February, 1982. Roy. Zool. Soc. New South Wales: Sydney, Australia.

ANTECHINUS STUARTII Macleay, *A. swainsonii* (Waterhouse), and *Rattus fuscipes* (Waterhouse) coexist over much of the range of *A. swainsonii* in southeastern Australia. Hall (1980b) showed that the diets of the two insectivorous *Antechinus* spp. were very similar, both in relation to prey size and prey types, where the two species occurred together in Sherbrooke Forest Park, Victoria. In the same forest, Robinson (1976) found that arthropods accounted for about 50 per cent of the summer diet of *R. fuscipes* and about 20 per cent of the winter diet. The apparent overlap in the diet of these three species, at least during the summer months, suggests that either this resource exceeds the capacity of the species to harvest it, or there is some ecological separation which otherwise reduces competition for the resource.

Hall (1980a) found that while the peaks in activity of the largely nocturnal *A. stuartii* and *R. fuscipes* were similar, *A. swainsonii* was active across the 24 h day, and so showed some temporal separation from the other two species.

Gullan and Robinson (1980) described habitat preferences of these three species in Sherbrooke Forest Park, Victoria, based on floristic communities, and found that when population numbers were high in 1971, there was no apparent partitioning of the study area. In the next two years, after a drop in sizes of the populations of all species, the floristic community most used by *R. fuscipes* was also most used by one of both of the *Antechinus* spp. However, an analysis of their data shows that there were no significant correlations between the use of habitat made by these species in any year (Kendall Rank Correlations, Siegel 1956).

Others have attempted to define the habitats of one or more of these species (e.g. Warneke 1971, Fletcher 1977, Barnett *et al.* 1978, Cheal 1978, Braithwaite 1979) in terms of vegetation and/or soil type, but the results of these studies do not offer the resolution necessary to see whether the species might show ecological separation within, for example, *Eucalyptus regnans* forest. There have previously been no studies of habitat partitioning of these species incorporating both floristics and vegetation structure.

This study was designed to examine the association of these small, coexisting mammals with floristic groups and vegetation structure in mountain ash (*E. regnans*) forest. An attempt was also made to determine whether the species were segregated in their use of arboreal environments and the forest floor, and in their use of nest sites.

METHODS

GRID TRAPPING: The study area was located in tall mountain ash forest in Sherbrooke Forest Park, Victoria (see Hall 1980A). A 10 x 12 point grid with 25 m intervals was established on a south facing slope (Fig. 1). Animals were live-trapped in May, August, and December, 1975, and in February, May, and September, 1976. In all trapping seasons except December, 1975, Elliott aluminium folding traps were baited with

[1] Dept. of Zoology, Monash University, Clayton, Victoria, Australia, 3168. Correspondence to Assoc. Prof. A. K. Lee.

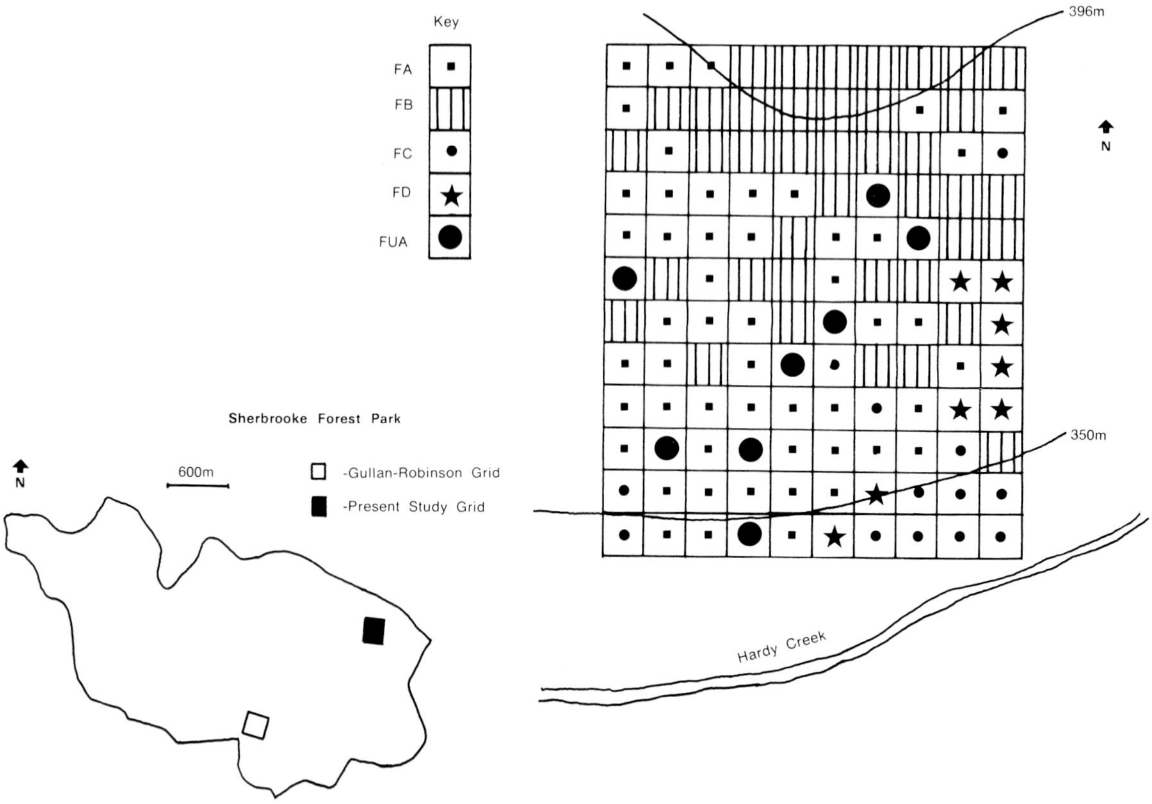

Fig. 1. Map of study area and distribution of floristic groups.

peanut butter and oatmeal and set at alternate grid points in alternate rows, so that one quarter of the grid points were trapped. The traps were set in the late afternoons of two days and were checked on the following mornings. They were then moved to the remaining grid points in the same rows, and again set for two nights. The procedure was repeated for sites in the remaining rows until all trap sites had been trapped for two nights. In December, 1975, two rows on the upper slope and two rows on the lower slope were not trapped due to shortage of traps. In August and September, trapping occurred on alternate nights so that pregnant female *Antechinus* were not confined to traps on consecutive nights. Up to four traps were set at each site to insure that open traps were available at all times. In colder weather, the traps contained shredded newspaper and were placed in plastic bags. Animals were identified, marked by toe clipping, sexed, and weighed, and reproductive status was noted.

FLORISTIC ANALYSIS: The plant species falling with a 10 m radius of each grid point were assigned relative cover/abundance values according to the Braun-Blanquet scale (see Gullan 1978). These data were then analyzed and sorted into floristic groups using a polythetic, agglomerative, non-heriarchical, clustering classification, designed by Gullan (1978). The results were printed in a two-way table (see Hall 1980c, table 16).

MAMMAL-FLORISTIC ASSOCIATION: Trapping data were assembled for each floristic group and entered into a one-way analysis of variance (Zar 1974) to determine whether the frequency of captures of animals varied between floristic groups. Where significant differences occurred· the Students-Newman-Keuls (SNK) multiple range test (Zar 1974) was used to determine the order of association of floristic groups.

STRUCTURE ANALYSIS: Nine point quadrats were taken at each trap site within a 5 m radius, using a 2.5 cm diameter pole calibrated in 30.5 cm intervals. The presence or absence of vegetation touching the pole was noted for each of twelve height intervals. Actual measurements were recorded up to 305 cm. Estimates of contacts were used for heights over 305 cm. Each vegetation height was then assigned a cover value between 0 and 9 at each trap site. The relative cover values for the 12 vegetation levels at each trap site were entered into the clustering programme (Gullan 1978), which sorted the quadrats into similar structural groups (see Hall 1980c, table 17).

MAMMAL-STRUCTURE ASSOCIATION: Trapping data were assembled for each structural group and were analyzed by a one-way analysis of variance (Zar 1974) to determine whether the frequency of captures of animals varied between structural groups. In addition, an SPSS (Nie *et al.* 1975) simple and multiple correlation analysis was run to determine the correlation of reproductive classes with cover at each tier and at combinations of tiers at each trap site.

TREE TRAPPING: Twenty-four stations, each 25 m apart around the perimeter of a 225 x 75 m rectangle, were trapped in April-May, June-July, and October, 1975; and in February, May, and July-August, 1976. Two traps were set on "L" frames (see Wood 1970) at about 165 cm on the most convenient trees (<4 m from the grid point) at alternate sites. At the other sites, 3 traps were set on the ground. Traps were opened for alternate 2-hour periods, over several days, until each 2-hour period in a 24-hour day had been trapped twice (For details of the methods used to record captures, see Hall 1980a.)

NEST SITES: Radiotelemetry was used to locate the nests of *A. swainsonii* and *R. fuscipes*. (*A. stuartii* were not large enough to carry the transmitters.) Silastic-coated, 3 cm, 4 g, 27 MHz Citizen Band Transmitters were fastened to the hair on the napes of 8 *A. swainsonii* and 4 *R. fuscipes* using super glue. The transmitters were located each day using a telcon receiver with a whip aerial.

RESULTS

POPULATION PARAMETERS: The numbers of captures and number of individuals in each reproductive class in each trapping period are shown in Table 1. The number of individuals and capture frequencies of each reproductive class varied considerably from month to month. *A. stuartii* was the most numerous and readily captured species in all months except September, 1976. In 1975, the last male was captured on 19 August. Young were on teats of all females by 17 September, but the exact dates of birth are unknown. Independent juveniles were first trapped in mid December. In 1976, the last male died between 27 August and 3 September. The first young appeared on teats on 20 September. Independent young could not be caught in early January, 1977.

A. swainsonii was the least abundant and least trappable species in all months except May, 1976. In 1975, adult males were not present by 24 August and young were on the teats of all females by 19 September, but the exact dates of the male die-off and birth of young are not known. Juveniles were trappable in the second week of December, 1975, but not again until the first week of March, 1976. In 1976, the last male was captured on 12 August and the first young were born between 27 August and 3 September. Juveniles were first captured in the first week of January, 1977. This species was infrequently trapped in August, December, 1975, and February, 1976. Three females of the 1973 cohort accounted for 9 of the 14 captures of the species between August, 1975 and February, 1976. A fourth female of the 1973 cohort was trapped during activity trapping but was not trapped during grid trapping.

The population of *R. fuscipes* remained between 22 and 27 individuals in all periods except August, 1975 and September, 1976, when a trough (15 individuals) and a peak (38 individuals) occurred in the population size. Although the absolute number of individuals on the grid did not change greatly, there were changes in the composition of the population. For example, the increase of 5 animals between December, 1975 and February, 1976, was a consequence of the gain of 12 females and 8 males (mostly new recruits) and the loss of 2 females and 13 males.

FLORISTIC GROUPS: The study area was composed of five floristic groups and eight unassigned quadrats (Fig. 1). All of the floristic groups shared the lower canopy and scrub species *Coprosma quadrifida, Bedfordia arborescens, Olearia argophylla* and *Pomaderris aspera,* the ground fern *Polystichum prolifera,* and wire grass *Tetrarrhena juncea.*

Group FA: The dominant canopy tree in this group was *E. regnans*. All of the lower canopy, dicotyledonous species found in the study area, and the tree fern *Alsophila australis*, were abun-

Table 1. Numbers of captures and numbers of individuals (in brackets) in each reproductive class (* = castrated by trap in June).

			MAY 1975	AUG. 1975	DEC. 1975	FEB. 1976	MAY 1976	SEPT. 1976	TOTAL CAPTURES
	parous	♀♀	51(13)	27(11)	129(25)	45(11)	22(8)	19(3)	
A. stuartii	non-parous	♀♀	79(24)	54(24)	1(1)	108(44)	127(38)	111(22)	1136
		♂♂	99(24)	0	2(2)	128(35)	134(35)	0	
	parous	♀♀	29(7)	4(3)	0	5(2)	28(6)	61(15)	
A. swainsonii	non-parous	♀♀	8(5)	0	1(1)	0	18(8)	0	269
		♂♂	39(11)	0	2(2)	2(2)	67(17)	5(1)*	
R. fuscipes		♀♀	30(8)	15(7)	17(8)	33(16)	38(13)	86(17)	496
		♂♂	56(17)	25(8)	29(14)	23(11)	28(10)	116(20)	

dant in this group. The sword grass *Lepidosperma elatius* was also common.

Group FB: *Eucalyptus regnans* was the dominant tree in this relatively species-rich group. The area supported species characteristic of disturbed vegetation (e.g. *Clematis aristata, Rubus fruiticosus, Cynoglossum latifolium, Stellaria flaccida,* and *Hydrocotyle geranifolia*) and many minor species uncommon in the other vegetation groups. The lower canopy species, *Pomaderris aspera* and *Olearia argophylla* were relatively less abundant.

Group FC: The canopy in this wet gully group was dominated by the tree ferns *Dicksonia antarctica* and *Alsophila australis*. *E. regnans* and several of the smaller plant species common in groups FA and FB were almost absent here. Ferns were abundant.

Group FD: This group was species poor. While *E. regnans* was the dominant tree, both tree ferns were uncommon, as were those plants missing from group FC.

Group FA is similar to the vegetation group S3 of Gullan and Robinson (1980). Groups FD and FB are subsets of their group S2, and group FC is similar to their group S1.

UNASSIGNED QUADRATS: Unassigned quadrats did not share sufficient numbers of species with any existing vegetation group or with each other to be assigned to a floristic group. However, the quadrats on this grid appeared to share low diversity and a low abundance of small ground plants.

STRUCTURAL GROUPS: Seven structural groups were defined by the sorting programme. The average covers at each level and the distribution of structural groups are represented in Fig. 2.

Group SA: This group was characterized by dense cover up to 92 cm, a dense lower canopy between 6 and 9 m, and a thin upper canopy.

Group SB: In this group the middle layer of vegetation between 93 cm and 614 cm was sparse. The lower canopy (6–9 m) and upper canopy (12 m) were thick.

Group SC: Cover up to 92 cm and between 3 and 6 m was dense in this group. There was very little vegetation between 9 and 15 m.

Group SD: This group was characterized by sparse cover between 30 and 123 cm and between 9 and 15 m. A thick lower canopy existed between 1.5 and 3 m.

Group SE: The ground cover up to 184 cm was very thin in this group, but the lower and

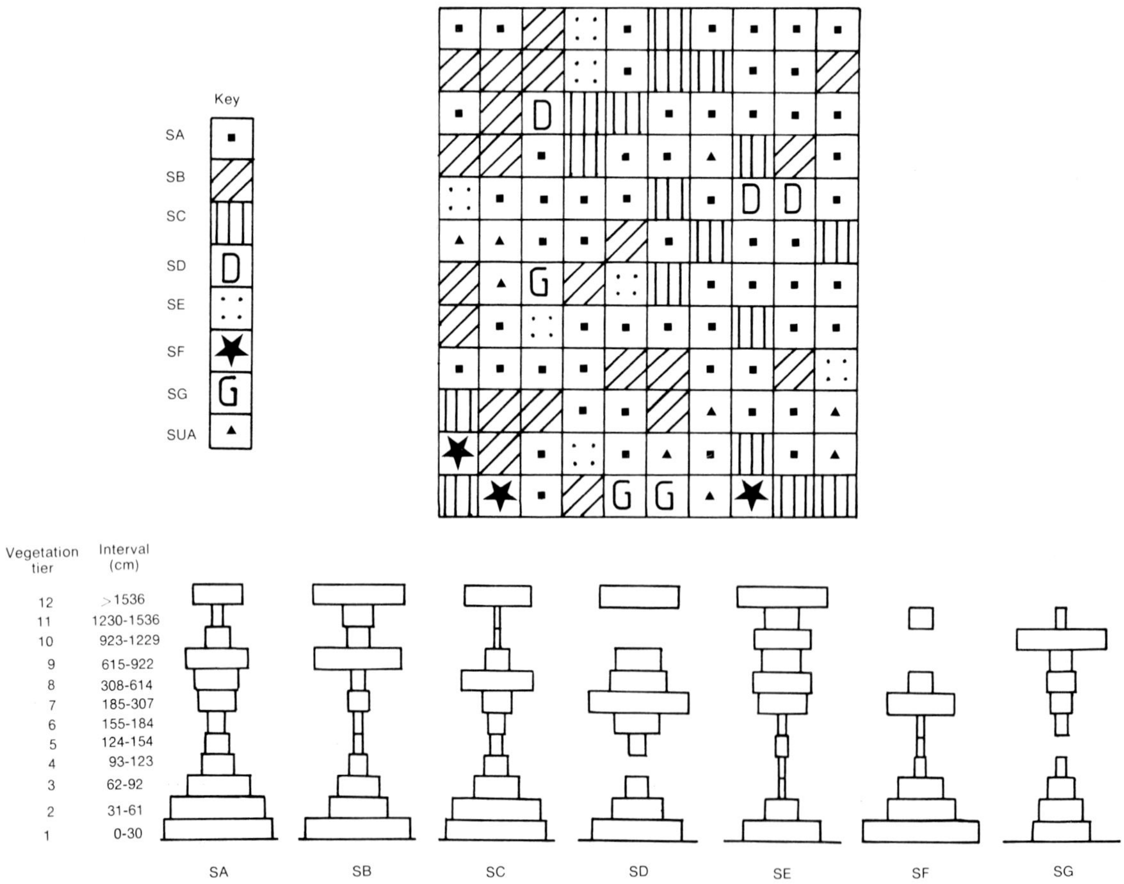

Fig. 2. Distribution of structural groups and representation of cover values at each level. The base line under each set of horizontal bars represents an average cover value of "7" (see Methods: Structure Analysis).

upper canopy (>3 m) was denser than in most other groups.

Group SF: The cover from 0 to 30 cm was densest in this group, but with the exception of moderate cover between 185 and 307 cm, there was no other canopy layer.

Group SG: This group was identified by thick cover between 9 and 12 m and by little cover at any other level. This group had the least ground cover up to 61 cm.

There were no correlations between structural and floristic groups.

MAMMAL-FLORISTIC ASSOCIATION: Although the one-way analysis of variance showed significant differences in capture rates between some species, and between some reproductive classes of each of the three species, and floristic groups, the Students-Newman-Keul test was unable to distinguish between floristic groups in any month (Table 2). This may be due to a relatively small amount of data (Zar 1974).

Antechinus stuartii: This species was captured at all trap sites over the period of the study. In August, 1975, parous females showed higher relative use of floristic group FB, and the unassigned, species–poor quadrats. Otherwise, there were no correlations between the captures of this species and any of the floristic groups.

Antechinus swainsonii: No captures of this species were recorded at thirty-eight trap sites. In May, 1975, males showed greatest relative use of the species–poor floristic group FD and unassigned quadrats. In May, 1976, parous females showed greatest relative use of the wet gully community FC and the species–poor group FD. Otherwise, there were no correlations between captures of this species and floristic groups.

Rattus fuscipes: No captures of this species occurred at nine widely scattered trap sites. In May, 1975, males used the *E. regnans* group FA and the tree fern group FC most. In December, 1975, and February and May, 1976, capture rates were highest in the fern gully and the species–poor groups FC and FD.

MAMMAL–STRUCTURE ASSOCIATION: One–way analysis of variance revealed no preference by any species or any reproductive class of a species for any structural group in any trapping period. This may have been due to variability of structure within groups, the low number of grid points in some groups, and the relatively large number of unassigned quadrats (n=9).

The simple and multiple correlation analysis (Nie *et al.* 1975) revealed correlation between cover at some levels and the occurrence of some reproductive classes of all species (Table 3). In some instances (e.g. *A. swainsonii* over all months), captures of a species or reproductive class were singularly correlated with cover at a vegetation tier (significant simple–r), but cover at that level did not necessarily contribute more to explaining variance of captures when combined in the multiple correlation (non–significant multiple–r). This was sometimes a consequence of high correlation between some vegetation tiers (especially, 2 and 3, r = .725; 5 and 6, r = .504). When a tier in one of these pairs contributed significantly to the multiple correlation, the other did not necessarily do so. Tiers which had either significant simple–r values or significant multiple–r values have been included in Tables 3. However, significant correlations in Table 3 did not account for much of the varience observed in capture rates. In only 6 cases did single vegetation levels account for more than 9% ($r \geq .30$, $r^2 \geq .09$) of the variation in

Table 2. One–way analysis of variance between mammal captures and floristic groups (FA, FB, FC, FD and FUA). Only variances which produced significant F–values have been listed. Horizontal lines under floristic groups indicate floristic groups which were not significantly different, as determined by the Students–Newman–Keuls test (Zar 1974).

Month	Reproductive class	Mean capture rate/trap site					F ratio	P
May 1975	*A. swainsoni* ♂♂	FD=1.12	FUA=0.75	FA=0.28	FB=0.23	FB=0.18	4.97	<0.002
	R. fuscipes ♂♂	FA=0.71	FC=0.69	FD=0.25	FB=0.18	FUA=0	4.88	<0.005
Aug. 1975	*A. stuartii* parous ♀♀	FB=0.40	UA=0.38	FA=0.14	FC=0.08	FD=0	3.88	<0.02
Dec. 1975	All *R. fuscipes*	FC=1.75	FD=0.83	FA=0.54	FUA=0.43	FB=0.42	3.08	<0.05
Feb. 1976	*R. fuscipes* ♀♀	FC=1.00	FD=0.62	FUA=0.38	FB=0.20	FA=0.08	9.11	<0.001
	All *R. fuscipes*	FC=1.15	FD=0.88	FUA=0.62	FB=0.48	FA=0.18	5.95	<0.001
May 1976	*A. swainsonii* parous ♀♀	FC=0.69	FD=0.38	FA=0.18	FB=0.15	FUA=0.12	3.08	<0.05
	All *R. fuscipes*	FC=1.30	FD=1.12	FB=0.55	FA=0.37	FUA=0.12	4.47	<0.005

Table 3. Association of mammals with vegetation cover. Vegetation tiers are shown in brackets. The number to the right of the tier indicates the significant changes in the multiple correlation coefficient of determination ($\triangle r^2$ expressed as a percentage) for each vegetation tier (Zar 1974, Nie *et al.* 1975). The "+" or "–" signs indicate a significantly positive or negative ($p < 0.05$) simple correlation coefficient (simple r). NS = change in the coefficient of determination not significant. (NS) = simple r not significant. NC = not calculated. p♀♀ = parous females. np♀♀ = non-parous females.

		MAY 1975	AUG. 1975	DEC. 1975	FEB. 1976	MAY 1976	SEPT. 1976	ALL MONTHS
A. stuartii	p♀♀		(10) 4.6 – (5) 2.5 +(NS)	(12) 5.8 + (11) 3.9 –(NS)	(9) 5.2 + (11) 3.7 + (2) 2.8 +(NS)		(2) 5.3 – (3) NS –	NC
	np♀♀	(9) 6.9 +	(7) 4.2 +					NC
	♂♂				(9) 5.8 –			NC
	♀♀ + ♂♂	(2) 5.4 – (6) 2.1 –(NS)		(12) 5.8 + (11) 3.9 –(NS)				
A. swainsonii	p♀♀	(6) 5.9 + (3) 5.1 + (12) 3.0 +			(4) 14.1 + (6) 2.1 + (2) 1.9 + (1) NS + (5) NS +		(3) 4.3 + (9) 3.7 – (1) 3.7 –(NS) (2) NS +	
	np♀♀					(7) 5.9 + (3) 2.9 +(NS)		NC
	♂♂					(3) 9.6 + (6) 3.2 +(NS) (2) NS +		NC
	♀♀ + ♂♂				(4) 14.2 + (6) 2.6 + (2) 3.3 + (3) NS +	(3) 7.4 + (7) 2.6 +(NS)	(2) 5.8 + (1) 3.9 –(NS) (9) 3.7 – (3) NS +	(3) 9.5 + (9) 3.1 –(NS) (6) 1.5 +(NS) (2) NS +
R. fuscipes	♀♀	(1) 5.7 + (2) NS+	(1) 4.8 +	(2) 6.8 + (11) 1.4 –(NS)	(8) 7.0 +			NC
	♂♂			(2) 14.0 + (3) NS +			(10) 4.2 – (7) 2.7 +(NS)	NC
	♀♀ + ♂♂			(2) 18.0 + (3) NS +	(8) 8.0 +	(8) 4.7 +	(10) 5.3 – (1) NS + (11) NS –	(2) 5.8 + (7) 4.7 +

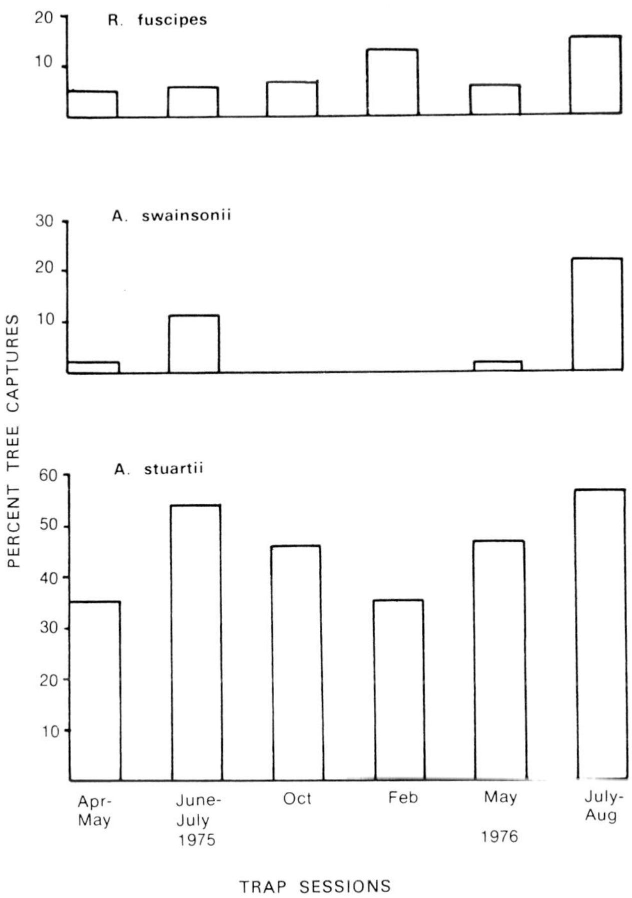

Fig. 3. Arboreal captures as a percentage of total captures.

mammal captures. In most instances, pooling of the effects of at least 2 or 3 vegetation levels was necessary to explain more than 9% of the variation.

Antechinus stuartii — Over all months, *A. stuartii* exhibited no significant correlation with cover at any single or group of vegetation levels. Reproductive classes were not consistently associated with cover in different seasons. For example, parous females showed association with sparse cover at level 10 in August, 1975, but were associated with thick cover at level 12 in December, 1975 and level 9 in February, 1976. In September, 1976, they were associated with sparse cover at levels 2 and 3.

Antechinus swainsonii — Over all months, this species was closely associated with thick cover at levels 2, 3 and 6, and with sparse cover at level 9. Its association with thick cover levels at 2 or 3 was consistent in those months that this species was trappable.

Rattus fuscipes — Over all months, *R. fuscipes* captures were positively correlated with cover at levels 2 and 7. In 1975, females were captured in association with cover at levels 1 and 2, but in 1976, the species was more closely associated with thick cover at levels 8 and 7.

MAMMAL–MAMMAL ASSOCIATION. The correlation between the captures of different species at the same trap sites are shown in Table 4. Captures of *A. stuartii* and *R. fuscipes* were negatively correlated in May and August, 1975 and in May and September, 1976, but not overall. Captures of *A. stuartii* and *A. swainsonii* were negatively correlated in May, 1975 and over all months. There were no correlations between captures of *A. swainsonii* and *R. fuscipes*.

ARBOREAL ACTIVITY. The capture rates of the three species in tree traps in different seasons is shown in Fig. 3. (No correction was made for the fact that there were more ground traps than tree traps, because comparisons of tree captures between species and between months within a species do not depend on proportional frequencies.) *A. stuartii* was the species most commonly captured in trees. Up to 57% of captures of this species in any period, and never less than 35% of captures, were tree captures. The rate of tree captures for this species varied with its overall rate of capture. *A. swainsonii* was regularly captured in trees only in July–August, 1976, while *R. fuscipes* was only occasionally captured in trees in any month.

Twenty–four traps were set on 9 species of trees (*Acacia dealbata*–1, *Acacia melanoxylon*–2, *Alsophila australis*–3, *Bedfordia arborescens*–7, *Coprosma quadrifida*–4, *Dicksonia antarctica*–3, *Eucalyptus regnans*–1, *Olearia argophylla*–1, *Pomaderris aspera*–2). There were not sufficient numbers of each tree species to test for preference, but two trends were evident. Traps on four *Coprosma quadrifida* captured fewer *stuartii* than expected ($\chi^2 = 17$, $p < .001$) and traps one one *Olearia argophylla* captured more *R. fuscipes* than expected ($\chi^2 = 6.69$, $p < .01$). Expected frequencies were calculated using whole–grid trapping capture rates, where all traps were set on the ground. Otherwise, none of the mammal

Table 4. Correlation coefficients of mammal–mammal associations. r significant at: * $p < 0.05$, ** $p < 0.001$.

	MAY 1975	AUGUST 1975	MAY 1976	SEPTEMBER 1976	OVER ALL
A. stuartii–R. fuscipes	–0.224*	–0.217*	–0.202*	–0.223*	
A. stuartii–A. swainsonii	–0.405**				–0.228*

species appeared to favour or reject particular tree species.

NEST LOCATIONS: *Antechinus swainsonii* – Transmitters which had been attached to two lactating females from different cohorts, tagged 5 days and 50 m apart in October, 1976 were both found in the same nest. The nest was located in a burrow in friable, moist soil at the southern end of the observed range of both animals. Two entrances, 40 cm apart, were concealed by thick growth of *Polystichum proliferum*. The entrance tunnel was about 4 cm in diameter and sloped steeply across the hill to a chamber approximately 30 cm below the surface and 30 cm in diameter. The chamber was lined with dried leaves of *E. regnans*.

A second nest was found by tracking a female in January, 1977. The transmitter was later located within the base of the trunk of a small (30 cm diameter) fallen tree. The nest was lined with dried *E. regnans* leaves and *Polystichum* fronds and was about 22 cm in diameter. A third suspected nest was located in January in the centre of the base of the trunk of a large (1.5 m) fallen tree overgrown by wire grass and covered with the remains of a second fallen tree. A female was recorded, alternately present and absent, at this location over several days. The nest was about 0.5 m above the ground, but the entrance was not located nor was the nest examined.

Rattus fuscipes – One rat nest was found in a burrow in dry, packed soil. Two entrances, about 50 cm apart, were surrounded by sparse vegetation. The entrance tunnel was about 3 cm in diameter and sloped steeply to a leaf-lined chamber about 30 cm in diameter and 30 cm below the surface. Two other rats were found deep within large (1.5 m) rotting logs but nests were not located.

Antechinus stuartii – Two nests, possibly used by *A. stuartii,* were found in January in the hollows of dead, standing trees. One was about 2 m, and the other about 1.5 m, above the ground. Both were lined with dried leaves.

DISCUSSION

There have been a number of attempts to define the habitats of the three mammal species studied here. Wakefield and Warneke (1967) observed that *A. stuartii* was found in a wide variety of habitats within wet and dry sclerophyll forest in eastern Australia. *A. stuartii* has been associated with structural complexity (Cheal 1978) and with fallen logs within forest habitat (Barnett *et al.* 1978, Braithwaite 1979). Fletcher (1977) found that *A. stuartii* captures were associated with log length, number of dwarfed shrubs and high stems, and horizontal visual obstructions at 50 cm, at different times. However, Stewart (1979) found no association between this species and logs or vegetation cover. While Leonard (1972) noticed a correlation between captures of female *A. stuartii* with dense creek vegetation, he found no association between males and specific vegetation cover. Leonard captured both sexes in areas where ground cover had been eliminated by fire. Press (1976) detected no correlation between captures of *A. stuartii* with either cover at 25 cm, density of cover, or distance to optimal cover in sclerophyll forest in New South Wales. While this species was most often captured in mature sclerophyll forest, they were often captured in regenerating forest and in adjacent grassland. In Sherbrooke Forest Park, Victoria, Gullan and Robinson (1980) found that the capture rates of *A. stuartii* were higher in species–poor, *E. regnans,* community than in a mixed *E. obliqua* and *E. cypellacarpa* community.

Antechinus swainsonii has been captured in a variety of habitats ranging from woodland and tussock grass to wet *Nothofagus* forest. It has been most commonly associated with dense cover and thick ground litter in fern gullies in wet sclerophyll or *Nothofagus* forests (Wakefield and Warneke 1963, Green 1972, Leonard 1972, Hocking 1975, Stewart 1979). In Sherbrooke Forest Park, Gullan and Robinson (1980) observed that *A. swainsonii* was commonly captured in the species–poor *E. regnans* community and, like *A. stuartii,* least commonly captured in the mixed *E. obliqua* and *E. cypellocarpa* community.

R. fuscipes has often been associated with dense cover and wet, structurally complex vegetation in both forest and heathland (Warneke 1971, Leonard 1972, Braithwaite *et al.* 1978, Cheal 1978, Stewart 1979). Barnett *et al.* (1978) found a similar association in native forest but not in *Pinus taeda* plantations. Press (1976) considered that *R. fuscipes* avoided areas with very thick cover below 25 cm, as this may impede movement across the ground. Fletcher (1977) concluded that *R. fuscipes* was associated with areas containing logs, but Stewart (1979) found no association for logs in his study. In heathland at Cape Liptrap, Victoria, Cheal (1978) found that the highest capture rates of this species were associated with floristic groups which reflected a requirement for moist friable, comparatively productive loamy soil. Gullan and Robinson (1980) observed a diametric change in the use of floristic communities by *R. fuscipes* over a three year period in Sherbrooke Forest Park. In the first year, the species preferred a wet, fern gully community, and in the second and third years, the species preferred a disturbed, grassy, mixed eucalypt community on ridge tops.

The foregoing information suggests that *A. swainsonii* may have the most restricted habitat preferences of these species, while *R. fuscipes* has the broadest habitat preferences. However, it does not provide consistent information for predicting the precise habitat preferences of these mammals when they coexist in relatively homogeneous forest.

The forest in this study was relatively homogeneous in that the differences between floristic groups were much less than the floristic differences between the range of habitats occupied by each of these species, and less than the differences between floristic communities defined in another section of Sherbrooke Forest Park by Gullan and Robinson (1970). There was also little variety in the structure of vegetation.

On the study grid, neither floristics nor vegetation cover correlated with the capture of *A. stuartii*. Correlations of some reproductive classes with either floristics or vegetation cover in some months did not recur in subsequent months, and in several instances, were not readily related to resource use. It was therefore difficult to interpret these correlations.

Again, floristics were not consistently correlated with captures of *A. swainsonii* in this relatively homogeneous habitat. However, captures were correlated with thick cover between 31 and 92 cm and between 155 and 184 cm, and with sparse cover between 6 and 9 m during the months that *A. swainsonii* was trappable. *A. swainsonii* has been observed approaching baited areas (see Hall 1980a) along broad wombat tracks and foraging in natural openings. Stewart (1979) made similar observations. These two observations are not consistent with a strong preference for cover. The apparent requirement for thick cover above 31 cm is logically related to need to move easily across open ground while being sheltered from predators. Yet, the movement and foraging of *A. swainsonii* in the open reveals apparent disregard for exposure to potential predation.

Use of floristic groups by *R. fuscipes* changed from season to season and from year to year, as Gullan and Robinson (1980) found for their nearby population of *R. fuscipes*. Association with vegetation cover also changed. Precipitation was higher in 1975 (1497 mm) than in 1976 (1185 mm; Nicholas Institute, Burnham Beeches, Sherbrooke, Victoria), and in the drier year *R. fuscipes* was more common in the wetter habitats (FC and FD). Gullan and Robinson (1980) observed that *R. fuscipes* had higher capture rates in wetter habitat in wetter years. The dynamic nature of habitat useage in this species in this forest probably resulted from a shift in resource abundance, but this was not measured.

Thus, in this study area the use of habitat by *A. stuartii* was not correlated with either floristic or structure as measured. Although captures of both *R. fuscipes* and *A. swainsonii* were associated with floristic or structure in some seasons, this changed from season to season. The dynamic nature of the observed habitat use suggests that it is not related to easily measured or isolated variables.

The problems of predicting habitat use by small mammals from associations with floristic or structural groups in specific areas are considerable. Roughgarden (1977) developed a model which suggets that patchiness in a population's spatial distribution (and therefore apparent partitioning) may be an inevitable consequence of dispersal in a stochastic environment, and that it is not necessary to assume that the environment exhibits underlying patchiness of resources. Green (1971) noted that a proportion of the environmental factors measured will be correlated invariant or irrelevant (e.g. see Results: Mammal–Structure Association) and Stewart (1979) found that the placement of traps may influence perceived abundance of small mammals. The mere presence of baited traps, may in fact, destroy evidence of habitat partitioning. However, there is good evidence that floristics or structural features do reflect more primary dependence of some small mammals on soil type or non-tracheophytic plans (e.g. *R. fuscipes* – Warneke 1971, Cheal 1978) or on invertebrate fauna (e.g. *A. stuartii* – Braithwaite 1979). The usefulness of using either floristics or structure to predict small abundance must depend on the heterogeneity of the study area, and on the resolution of floristic and structural descriptions relative to the specific requirements of the animals involved.

Rosenzweig *et al.* (1975), Douglass (1976), and Holbrook (1979) have shown experimentally that habitat partitioning by some desert rodents was partially a function of interspecific interaction. Grant (1972) considered that competition was a major factor limiting the distribution of small mammals. Kenner (1972) experimentally demonstrated interspecific avoidance by *A. stuartii, A. swainsonii* and *R. fuscipes* in contrived encounters (see also Braithwaite 1973). The two *Antechinus* spp. avoided each other and both avoided *R. fuscipes*.

The present study supports the idea that interspecific avoidance occurs among these species. In May, 1975, and over all months, captures of *A. swainsonii* and *A. stuartii* were negatively correlated with captures of *R. fuscipes*. Thus, they showed strongest partitioning when insects were relatively scarce, or most inaccessible during winter. The fact that overall, there was no significant negative correlation between *A. stuartii* and *R. fuscipes* captures, suggests that the

Table 5. Summary of habitat and resource use of *A. stuartii, A. swainsonii* and *R. fuscipes* in the Sherbrook Forest area. Sources: (1) Hall 1980a, (2) Hall 1980b, (3) Warneke 1971, Robinson 1976, (4) includes observations by Kenner 1972, (5) Robinson 1976.

Species	Floristic Group	Vegetation cover	Use of trees	Temporal activity[1]	Diet	Nest sites	Interspecific avoidance[4]
A. stuartii	No overall association	No overall association. Inconsistent association between seasons.	Climbs readily. Arboreal activity increase with increased abundance of other species	Largely nocturnal	Beetles, spiders, amphipods, weevils, other arthropods[2]	In tree hollows	Avoided *R. fuscipes* in May, Aug., Sept.; avoided *A. swainsonii* in May 1975 and overall
A. swainsonii	Species–poor *E. regnans* community most used in May of both years; no other overall association	Associated with thick cover between 31 and 92 cm	Did not normally climb, but entered tree traps in colder months	Even diel activity; avoided periods of activity of *A. stuartii and R. fuscipes*	Beetles, spiders, amphipods, other arthropods, perhaps annelids[2]	Burrows or surface hollows	Probably displaced *A. stuartii* in May 1975 and overall; mutual avoidance of *R. fuscipes* on a time to time basis
R. fuscipes	Association changed from species–poor to wet tree fern community	Associated with thick cover between 31 and 61 cm in 1975; between 185 and 614 cm in 1976	Rarely climbs	Strictly nocturnal except in spring	Fungi, fruit, fleshy vegetation; insects in summer[3]	Burrows or surface hollows	Probably displaced *A. stuartii* in May, Aug., Sept.; mutual avoidance of *A. swainsonii* on time to time basis

trap sites which were used by *R. fuscipes,* and less used by *A. stuartii,* changed from season to season.

Only *A. swainsonii* and *R. fuscipes* showed no evidence of habitat partitioning, but the possibility remains that these species may avoid each other from moment to moment at any location. This seems reasonable in view of relatively large home ranges and distances between consecutive captures for each of these species (Hall 1980c), and the following observations. In September, 1976, at least two *A. swainsonii* and one *R. fuscipes* alternately used a baited area for over an hour. At times individuals of both species could be seen at opposite ends of the small clearing sniffing the air. One individual would enter, feed and then leave the area followed by a gap of several seconds before an individual of another species would enter the area. Yet, the mechanism of interspecific avoidance is not clear. On occasions, all three species were captured at the same trap sites on the same nights.

A. stuartii is known to be semi-arboreal (Wood 1970, Kenner 1972), and was the only species trapped in substantial numbers in trees in this study. In the present study, the arboreal activity of *A. stuartii* increased in seasons when the numbers of captures of *A. swainsonii* and *R. fuscipes* increased (Kendall Rank Correlation, r = 0.667, S = 11, p <.028; Siegel, 1956; Figure 3 and Table 1). C. Dickman (pers. comm.) found that *A. stuartii* was trapped more frequently on the ground when *A. swainsonii* was removed from the area. Hall (1980b) concluded that although the diets of the two *Antechinus* spp. were very similar, *A. stuartii* may have taken arboreal weevils which were unavailable to *A. swainsonii.* It seems likely then, that *A. stuartii* may be using trees to avoid confrontations, and perhaps to obtain food resources untapped by the other species.

Both *A. swainsonii* and *R. fuscipes* were found to build nests in burrows or in low hollows. Green (1972) recorded the nest of an *A swainsonii* at 2.5 m in a "stringy bark," but most other nests of this species have been recorded at or below ground level (see Wakefield and Warneke, 1963). Warneke (1971) also described the nests of *R. fuscipes* as subterranean. Although *A. swainsonii* and *R. fuscipes* have similar nest sites, it is difficult to see a limitation on nest sites in this type of forest.

Nests, thought to be those of *A. stuartii* were found in hollows above ground. Others have shown that *A. stuartii* builds nests both near the ground (Wakefield and Warneke 1967) and in hollows and under bark of standing trees (see Leonard 1976). Thus, *A. stuartii* avoids the prospect of competition with the other two species for nest sites.

A summary of use made of the relatively homogenous mountain ash forest in Sherbrooke Forest Park by these three species is shown in Table 5. *A. stuartii* and *R. fuscipes* were both nocturnal and had peaks of activity at almost the same period of the night. They used different parts of the forest floor during the colder months. The basis of horizontal partitioning was not obviously related to recognized floristic or structural characteristics. Unlike *R. fuscipes, A. stuartii* was commonly captured in trees and presumably spent time foraging in trees.

A. swainsonii had the same general requirements for vegetation structure as *R. fuscipes.* Both were terrestrial and used similar nest sites. They may utilize similar food during the summer. Habitat partitioning between these species appears to depend on mutual avoidance on a time–to–time basis, but the frequency of confrontations may have been minimized by *A. swainsonii* extending its activities across the 24 h day.

There was a significant overlap in the diets of the two *Antechinus* spp. *A. stuartii* may reduce the potential for competition for food by foraging in trees, and *A. swainsoni* by foraging over the entire diel cycle. *A. swainsonii* used a restricted part of the study area, but the basis of its choice was not explained by either the floristics or structural parameters used here. *A. stuartii* made less use of the habitat occupied by *A. swainsonii* but was still unbiquitous across the study area. The nests of these species may be sited in different parts of the habitat. *A. swainsonii* mated about 3 weeks before the smaller cogner.

ACKNOWLEDGEMENTS

We wish to thank Dr. Paul Gullan, Dr. Andrew Cockburn, Dr. Graeme Suckling, and especially Mr. Michael Fleming for the guidance and assistance with computing. Assessment of floristics on each quadrat was conducted by Dr. Paul Gullan and Annie Opie. Drs. Barry Fox and Tony Robinson, provided useful criticisms of a draft of this paper.

Gratitude is expressed to the Forests Commission of Victoria for permission to conduct this investigation in Sherbrooke Forest Park.

We wish to thank the trustees of the M. A. Ingram Trust for financial support.

REFERENCES

BARNETT, J. L., HOW, R. A. AND HUMPHREYS, W. F., 1978. The use of habitat components by small mammals in eastern Australia. *Aust. J. Ecol.* **3:** 277–85.

BRAITHWAITE, R. W., 1973. "An ecological study of *Antechinus stuartii* (Marsupialia: Dasyuridae)". M.Sc. thesis: University of Queensland, Brisbane.

BRAITHWAITE, R. W., 1979. Social dominance and habitat utilization in *Antechinus stuartii* (Marsupialia). *Aust. J. Zool.* **27:** 517–28.

BRAITHWAITE, R. W., COCKBURN, A. AND LEE, A. K., 1978. Resource partitioning by small mammals in lowland heath communities of southeastern Australia. *Aust. J. Ecol.* **3:** 423–45.

BRAITHWAITE, R. W. AND GULLAN, P. K., 1978. Habitat selection by small mammals in a Victorian heathland. *Aust. J. Ecol.* **3:** 109–27.

CHEAL, P. D., 1978. "Ecological interrelationships of the small mammals and vegetation (forest and heathland) at Cape Liptrap, Victoria". M.Sc. thesis: Monash University, Clayton, Victoria.

DOUGLASS, R. J., 1976. Spatial interactions and microhabitat selections of two locally sympatric voles *Microtus montanus* and *Microtus pennsylvanicus*. *Ecology* **57:** 346–52.

FLETCHER, H., 1977. Habitat relationships among small mammals at Petroi, northeastern New South Wales. M. Nat. Res. thesis: University of New England, Armidale, New South Wales.

GRANT, P. R., 1972. Interspecific competition among rodents. *Ann. Rev. Ecol. Syst.* **3:** 79–106.

GREEN, R. H., 1971. A multivariate statistical approach to the Hutchinsonian niche: bivalve molluscs of central Canada. *Ecology* **52:** 543–56.

GREEN, R. H., 1972. The murids and small dasyurids in Tasmania, Part 5, *Antechinus swainsonii* (Waterhouse, 1840). *Rec. Queen Vict. Mus.* **46:** 1–34.

GULLAN, P. K., 1978. Vegetation of the Royal Botanic Gardens Annexe at Cranbourne, Victoria. *Proc. Roy. Soc. Vict.* **90:** 225–40.

GULLAN, P. K. AND ROBINSON, A. C., 1980. Vegetation and small mammals of a Victorian forest. *Aust. Mammal.* **3:** 87–95.

HALL, S., 1980a. Diel activity of three small mammals coexisting in forest in southern Victoria. *Aust. Mammal.* **3:** 67–79.

HALL, S., 1980b. The diets of two coexisting species of *Antechinus* (Dasyuridae, Marsupialia). *Aust. Wildl. Res.* **7:** 365–78.

HALL, S. W., 1980c. "Habitat and resource use by three small mammals *(Antechinus stuartii, A. swainsonii* and *Rattus fuscipes)* coexisting in tall, open forest in southern Victoria". M.Sc. thesis: Monash University, Clayton, Victoria.

HOCKING, G. J., 1975. "Studies on the behaviour and ecology of *A. swainsonii*". Hons. thesis: University of Tasmania, Hobart.

HOLBROOKE, S. J., 1979. Habitat utilization, competition interactions, and coexistence of three species of cricetine rodents in east-central Arizona. *Ecology* **60:** 758–69.

KENNER, A. N., 1972. "Aspects of the behavioural ecology of three sympatric species of small mammal, *Antechinus swainsonii, Antechinus stuartii* and *Rattus fuscipes*". Hons. thesis: Monash University, Clayton, Victoria.

LEONARD, B. V., 1972. "The effect of fire upon selected small mammals and leaf litter fauna in sclerophyll forest in southern Australia". M.Sc. thesis: Monash University, Clayton, Victoria.

NIE, N. H., HULL, C. H., JENKINS, J. G., STEINBRENNER, K. AND BENT, D. H., 1975. "Statistical package for the social sciences". McGraw Hill: U.S.A.

PRESS, T., 1976. "Some aspects of the ecology and habitat utilization of four species of native small mammals". Hons. thesis: Sydney University, Sydney, New South Wales.

ROBINSON, A. C., 1976. "Population ecology of *Rattus fuscipes* (Waterhouse) (Muridae: Rodentia)". Ph.D. thesis: Monash University, Clayton, Victoria.

ROSENZWEIG, M. L., SMIGEL, B. W. AND KRAFT, A. 1975. Patterns of food, space and diversity. Pp.241–268 *in* "Rodents in desert environments" ed. by I. Prakash and P. K. Ghosh. Junk: The Hague.

ROUGHGARDEN, J. D., 1977. Patchiness in the spatial distribution of a population caused by stochastic fluctuations in resources. *Oikos* **29:** 52–9.

SIEGEL, S., 1956. "Non parametric statistics for the behavioural sciences". McGraw Hill: Tokyo.

STEWART, A. P., 1979. Trapping success in relation to trap placement with three species of small mammals, *Rattus fuscipes, Antechinus swainsonii* and *A. stuartii. Aust. Wildl. Res.* **6:** 165–72.

WAKEFIELD, N. A. AND WARNEKE, R. M., 1963. Some revision in *Antechinus* (Marsupialia) – 1. *Vict. Nat.* **80:** 194–219.

WAKEFIELD, N. A. AND WARNEKE, R. M., 1967. Some revision in *Antechinus* (Marsupialia) – 2. *Vict. Nat.* **84:** 69–99.

WARNEKE, R. M., 1971. Field study of the Australian Bush Rat, *Ratus fuscipes.* (Rodentia: Muridae). *Wildl. Contr., Vict.* **14:** 1–115.

WOOD, D. H., 1970. Ecological study of *Antechinus stuartii* (Marsupialia) in a southeast Queensland rainforest. *Aust. J. Zool.* **18:** 185–207.

ZAR, J. H., 1974. "Biostatistical analysis". Prentice Hall: U.S.A.

RESUMEN

19. Los habitats usados por dos especies de *Antechinus* y *Rattus fuscipes* en los altos bosques abiertos del sur de Victoria. Por S. Hall y A. K. Lee.

Las especies *Antechinus stuartii, A. swainsoni* y *Rattus fuscipes,* coexistentes en los altos bosques montañosos de "ash" *(Eucalyptus regnana),* fueron examinadas en su asociación con estructuras florísticas y de vegetación en un reticulado de entrampamiento de 6.2 hectáreas. En esta vegetación relativamente homogénea, *A. stuartii* no estuvo consistentemente asociada con particulares grupos florísticos o estructurales. En la mayoría de los meses *A. Swainsonii* tampoco mostró alguna asociación consistente con particulares grupos florísticos, pero fue capturada más a menudo donde la cobertura de la vegetación era más espesa, entre 62 y 184 cm. *Rattus fuscipes* mostró su asociación con particulares grupos florísticos y estructurales, la cual cambió durante el estudio.

Antechinus stuartii y *R. fuscipes* estuvieron negativamente asociadas durante los meses más fríos, aunque nó durante todos los meses. *A. stuartii* y *A. swainsonii* estuvieron negativamente asociadas en Mayo, 1975, y en todos los meses combinados. Las capturas de *A. swainsonii* y *R. fuscipes* no estuvieron correlacionadas. *A. swainsonii* y *R. fuscipes* parecieron compartir el hábitat momentáneamente.

A la luz de estos datos y del conocimiento de la dieta, de los tipos de actividad dietaria y del uso de los árboles se discuten las dimensiones de los nichos.

20. THE WESTERN QUOLL, *Dasyurus geoffroii* (DASYURIDAE, MARSUPIALIA) IN THE NORTHERN TERRITORY: HISTORICAL RECORDS FROM VENERABLE SOURCES

K. A. JOHNSON and A. D. ROFF[1]

The Western Quoll was once common in Central Australia, where it is believed to be now extinct. It is still remembered by elderly Aborigines who once hunted it, and continue to maintain its mythology and ceremony.

Aspects of the Quoll's natural history were recorded during interviews with these people and is presented. These and other data are reviewed and reasons for the Quoll's extinction are discussed.

Pages 221–26 *in* CARNIVOROUS MARSUPIALS, ed. by M. Archer, February, 1982. Roy. Zool. Soc. New South Wales: Sydney, Australia.

THREE SPECIMENS of Western Quoll have been collected from the Northern Territory. These were obtained at the turn of the century from Crown Point (25°30′S : 134°23′E), Barrow Creek (21°31′S : 133°53′E) and Alice Springs (23°42′S : 133°51′E) which occur along the old telegraph line which once linked Darwin with Adelaide (Parker 1973). Finlayson (1961) considered the species to have been formerly widespread and abundant throughout Central Australia.

Regrettably there have been no documented reports of its existence in recent times and it is now either very rare or extinct in the Northern Territory. The species has disappeared from its former range in Queensland, New South Wales, Victoria and South Australia (Archer 1979) and little is known of its ecology in these areas. Fortunately there is a source, also rapidly becoming extinct, of information on the Quoll in the Northern Territory. This reservoir is the old Aboriginal men and women of the Western Desert tribes who valued the animal for its culinary uses and as an important part of their mythology and ceremony.

This paper reports information obtained from Aboriginal people and discusses reasons for the decline of the species.

STUDY AREA AND METHODS

Information was obtained from elderly men of the Western Desert tribes of Central Australia. The chief linguistic groups involved were Walpiri, Pintupi, Aranda and Pitjantjatjara, and informants were spoken with at Yuendumu, Hermannsburg, Ayers Rock, Giles, Jameson and Blackstone (Fig. 1).

Although the area is generally referred to as desert the landscape does not resemble the sandy wastes normally associated with the term. The region is predominantly covered by a complex of grassland and shrubland communities and the otherwise flat, sandy topography is disrupted by scattered rocky hills and ranges. The prominences comprise a small proportion of the total area, but they nevertheless have a significant effect on the surrounding vegetation by acting as watersheds feeding onto nearby run-on areas, or into the usually dry watercourses. The foothills and adjacent plains contain mulga (*Acacia aneura*) and ironwood (*A. estrophiolata*) communities and are generally rich in grasses suitable for stock feed. These areas have naturally been taken up for pastoral purposes, with the exception of the Petermann Range area in the south-west. The watercourses are fringed with River Red Gums (*Eucalyptus camaldulensis*), which in the large systems, become substantial trees with many hollows.

During the infrequent wet periods these river systems channel water to one of two end points: salt lakes or floodouts. Some salt lakes are not fed by river systems and are charged only with flow from the surrounding plains. They are scattered through the western areas and some are immense (Lake McKay covers approximately 3300 square kilometres). Their surrounding vegetation is invariably diverse and forms a rich interface with that of the adjacent plains. Floodouts occur where the water loses momentum at the end of watercourses and eventually percolates down to the groundwater systems, or is evaporated. The silt loads deposited in floodouts make them relatively fertile and productive.

The remaining area consists of sand plains and sand dunes and these form about 50% of the area discussed. The sands are infertile and spinifex (*Triodia spp* and *Plectrachne spp*) is the dominant ground cover. Shrublands together with *Acacia, Casuarina* and *Eucalyptus* woodlands are scattered throughout in favourable areas.

[1] Conservation Commission of the Northern Territory, P.O. Box 1046, Alice Springs, Northern Territory, Australia, 5750.

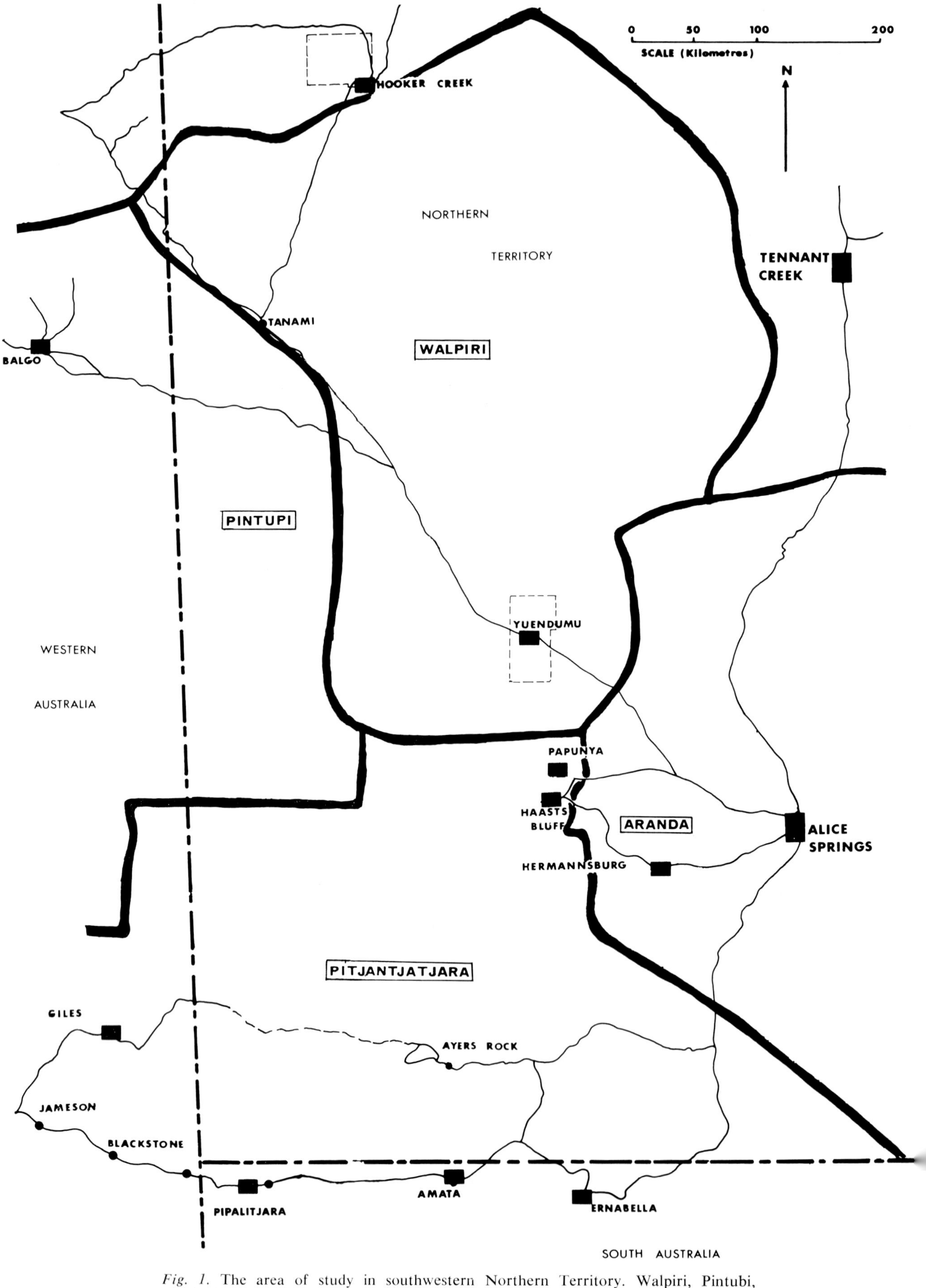

Fig. 1. The area of study in southwestern Northern Territory. Walpiri, Pintubi, Pitjtjanjara and Aranda are the names of the linguistic groups in the area. Heavy lines denote the boundaries between these groups.

Discussion of *D. geoffroii* with tribal elders took place on many occasions but in the initial contact a museum study skin was always used. No *D. geoffroii* material was available but that of *D. hallucatus* proved to be an acceptable substitute because the people did not distinguish between the two. Many would say that they were "all finished" locally but indicated that there were "big mobs" in the Darwin area where the Northern Quoll is common.

Study skins were invaluable as a focus for discussion. They provided a notion of size while the feet could be examined (to establish what tracks would be made), and the fur studied for texture, length, and colour distribution. In contrast, photographs gave a confusing idea of scale to those not familiar with them, and they did not provide the degree of reality imparted by skins, a factor which is so helpful in generating discussion.

Information given by Aboriginal informants can be misleading for two reasons: misunderstanding of the questions put and misinterpretation of the replies. Mindful of this problem, the only information given here is that which has been corroborated by several members, interviewed separately, from one or more linguistic groups.

RESULTS

The Western Quoll has at least one name in each of the various linguistic areas. These names may be changed from time to time for cultural reasons, but those currently in popular use are Jajirdi (Walpiri, Pintupi), Tjilpa (Aranda) and Parrtjata (Pitjantjatjara).

None of the old Aranda men from Hermannsburg had seen Quolls and they were only able to recognise the specimen by recalling the descriptions provided in the dreamtime stories. Elderly members of the other linguistic groups remembered it well. Many Pitjantjatjara men said it lived in burrows in the sand dune country and, where available, in rabbit burrows. Other potentially suitable holes (e.g. those of the Bilby, *Macrotis lagotis,* and Boodie, *Bettongia lesueur*) were not used. Some maintained that like the common Brush–tail Possum (*Trichosurus vulpecula*), it also lived in tree hollows whereas others noted that it only took refuge in trees when frightened.

Dogs were used to chase Quolls up trees from which they were dislodged with spears.

The Walpiri and Pintupi people all said that it lived only in burrows which it made itself, and that it never climbed trees. Some even described how Quolls eluded pursuers by running around trees but not climbing them.

The Walpiris noted how the burrows were sometimes complex and consisted of several entrances. These led down to a maximum depth of about 1.2 m where they joined a large chamber. This was sometimes occupied by up to three adults which "made big mob piccaninny" (young), implying a number greater than five. They were said to have five young which were carried on their backs.

One procedure in catching Quolls from burrows was first to kill them with smoke. This was done by lighting spinifex in the burrows and then fanning the smoke down using a leafy branch. Quolls were then dug from the burrows and eaten. The people respected the power of Quoll jaws so the smoking procedure was evidently to prevent being bitten. A similar technique was also used in excavating Boodies, which were also considered "proper cheeky" because of their inclination to bite and kick.

Little information was obtained about food of the Quoll except that it appeared to eat just about anything. Witchetty grubs (a generic term applied to the larvae of several species of moths and beetles), lizards, hopping mice (*Notomys sp.*) and frogs were specifically mentioned.

All people maintained that Quolls had "all finished up" in their country. This disappearance appears to have occurred rapidly over a period of perhaps 10 years, but at different times in different areas. At Hermannsburg, Aranda men of 60 years and more had never seen Quolls whereas elderly men in other areas had often eaten them.

The informants related the time of the Quolls disappearance to some stage in their life such as boyhood, adolescence and when whiskers began growth. A fairly consistent period of about 40 years B.P. was obtained from Walpiri and Pitjantjatjara people. Prior to this time they were said to be quite common and because they were good tucker, everybody ate them.

MYTHOLOGY

The ancestral Quolls of the dreamtime stories all started their journeys south of the Northern Territory and travelled in a general northerly direction on several roughly parallel tracks (Strehlow 1947).

There were five Quoll groups in the Walpiri stories. One set out from the MacDonnell Ranges to the east of Alice Springs and travelled north through country to the east of the Stuart Highway to Darwin. Another set out from Kings Canyon in the George Gill Range and travelled north on a track passing by Hermannsburg and through Glen Helen Gorge. They arrived at a place just east of Yuendumu where they met a group of possum men. Both groups wanted to occupy the same area so there was a big fight leaving many bodies scattered about. These are signified by a group of bloodwoods (*Eucalyptus terminalis*) still growing in the area. The surviving Quolls continued northwards via the Lander River while the possums returned to their proper home at Glen Helen.

A third group began at Port Augusta and eventually arrived at Haasts Bluff. They proceded to Yuendumu and then north to Arnhem Land.

A fourth group travelled from salt lake country west of Mt. Wedge passing through the Waite Creek area then northwards to the sea. The fifth group began still further west and journeyed north through Balgo Mission country then to the north coast.

DISCUSSION

The meagre record of three *D. geoffroii*, collected at the turn of the century along the old telegraph line, belies what appears to have been a much more abundant and widely distributed species. Finlayson (1961), relying mainly on Aboriginal informants, considered it to have been formerly widespread and abundant over the entire area south of Tennant Creek. The dreaming trails of ancestral quolls are widely dispersed across the western deserts. Because the travels of mythological beings are often associated with areas where the real animals were once present, the quoll dreamings lend support to the notion of the species being widely distributed in the Northern Territory.

The few museum specimens indicate that Quolls occurred in the rocky areas of the central region whilst the Aboriginal people maintain that they also occupied the desert sandhills and sandplains as far west as the Northern Territory – Western Australian border.

We have found no previous reference to burrowing by any of the *Dasyurus* species, the usual den localities being among rocks and crevices, under stumps and fallen limbs, and in hollows. Aborigines said that rabbit burrows were used, and to the Quoll, this would be equivalent of living in the pantry. Other potentially suitable burrows made by Boodies and Bilbies were not used and the type of burrow constructed by Quolls appears singular to the species.

Eucalypts are thinly scattered throughout most of the western desert country and many contain hollows potentially suitable for Quolls. Indeed they were used for shelter by possums (Giles 1889, Finlayson 1943) and their former existence in these desert regions is well known to the Aborigines we spoke with. However, the woodland areas with hollows and rocky areas with den sites form a very small proportion of the total region and the adoption of a fossorial habit by the species would have greatly extended its potential range. Like the Brushtail Possum and the medium sized fossorial marsupials (Boodie, Bilby and Mulgara, *Dasycercus cristicauda*), the Western Quoll most probably occupied areas devoid of permanent surface water.

The confusion among Aborigines regarding the arboreal or terrestrial behaviour of the Quoll is no less conflicting than that found in European literature. Krefft (1866), Thomas (1907) and Gould (1863) all maintained that the Quoll was arboreal whereas Troughton (1965) found these claims questionable on the basis that the foot pads were not morphologically suited for climbing. The evidence favours a flexible combination of arboreal, terrestrial and fossorial life styles.

The Quoll appears to have declined rapidly in the period from about the mid 1930s until the late 1940s or early 1950s. This period coincides with that when many other marsupials also rapidly vanished from central Australia, e.g. Boodie, Golden Bandicoot (*Isoodoon auratus*), Common Brush–tail Possum, and Mala (*Lagorchestes hirsutus*). They apparently survived the protracted drought of 1925 to 1930, which was most severe in 1928 and 1929, but then declined during the good seasons and not so severe droughts which followed in the 1940s and early 1950s. Finlayson (1961), noted anecdotal accounts of Quolls in an area to the north of Alice Springs, and in another to the west of Tennant Creek, prior to 1956. Glauert (1933) reported a specimen in Western Australia near the Northern Territory border along Stuart Creek to the north of Balgo Mission.

The decline of Quolls was undoubtedly caused by a complex of factors and competition for food with Foxes (*Vulpes vulpes*) and feral Cats (*Felis catus*), is an attractive possibility. Certainly the Eastern Quoll (*D. viverrinus*) is still abundant in Tasmania where Foxes have not yet established. Foxes had reached the northern parts of South Australia by 1910, and were north of Alice Springs in 1937 (Finlayson 1961). However, on the basis of their rate of spread in other parts of Australia, they probably followed Rabbits into the southern parts of the Northern Territory well before the 1930s. Foxes are still rare, and have apparently always been so, in the western and northern sandplain and sandhill country, yet Quolls have nevertheless vanished from these areas.

Feral Cats have been present in central Austrialia since before the turn of the century. One was seen by the Elder expedition 1891 at Mt. Squires in the Great Victoria Desert, 640 km. from the nearest European settlement (Finlayson 1961). If Cats were singular in their effect on Quolls it certainly took many decades to have their effect, and then did so in a very rapid and powerful manner.

Dingoes have increased in numbers and expanded their distribution following the expansion of the pastoral industry and the concommitant increase in the number of stock watering points. They may well have competed strongly with Quolls in these areas. However the western desert country remains unused for cattle production yet Quolls have vanished from it.

There have been no published accounts of disease sweeping through the central Australian Quoll population but the possibility cannot be discounted. Disease caused an alarming decline in Koalas (*Phascolarctos cinereus*) in the years 1887–9 and 1900–3 (Troughton 1965) and in the late 1940s an epidemic appears to have been responsible for an immense decline of common Ring-tail Possums (*Pseudocheirus peregrinus*) in Tasmania (Green 1973). Terry (1937) reported an eye disease which was killing large numbers of Rabbits in the Warburton region, 20 years before myxomatosis was first established in Australia. Finlayson (1961) noted large scale mortalities of Rabbits showing symptoms outwardly similar to myxomatosis, but predating its deliberate introduction by 40 years. If these earlier, naturally occurring diseases were not host specific, as is the case with toxoplasmosis, Quolls feeding on dead and dying Rabbits may well have been exposed to the pathogen. Nevertheless, these reported epidemics occurred well before the major decline in Quolls occurred, so unless it took some years for a virulent strain to arise, these particular diseases are unlikely to have been important. Moreover, the normally very observant Aborigines have not reported seeing moribund animals.

Rabbits arrived in central Australia in 1880, and were apparently almost as far north as Tennant Creek by 1905 (Finlayson 1961). This invasion by Rabbits would have greatly increased the potential food supply for Quolls and most probably their numbers increased. The stories from Aborigines of a plentiful supply of Quolls living in Rabbit holes during the early days supports this contention.

Central Australian Rabbit populations exhibit immense oscillations in numbers, rising to very high levels after good seasons, and crashing during droughts. Their ability to progressively denude the landscape of vegetation as drought conditions develop was noted by Terry (1937) who in 1931 observed intense grazing of shrubs by Rabbits near Warburton. Some had become caught up in shrubs at head height to a rider on Camel back. Remnant populations survive during drought in favourable areas (e.g. beside salt lakes and water courses) which most probably formed drought refuge for Quolls, Foxes, Cats, and the native prey species. It is conceiveable that Rabbits indirectly caused the extinction of Quolls in these favourable areas during dry times. Their intensive grazing (which becomes more destructive as a drought develops) most probably removes a substantial portion of the food and shelter resources required by native vertebrate and invertbrate prey species. These species would then decline to very low levels, or extinction, while the Rabbit population persisted for a short period before also crashing to very low numbers. Before the arrival of exotic herbivores the droughts would have been much less destructive to vegetation allowing a suite of native prey species to survive and provide a variable and sufficiently abundant food resource capable of maintaining a Quoll population. By simplifying the prey supply to few species, often in low numbers, Rabbits have amplified the normal feast or famine situation of desert regions. Faced with sudden disappearance of food, together with fierce competition from Cats and Foxes, the Quoll would have slid quietly into oblivion.

However, whereas Rabbits may have been instrumental in the extinction of Quolls in some areas, for general application the timing is wrong. Quolls survived the intense droughts of 1925–30 and 1940–41 when the above effect should have been critical. Moreover, they vanished from the Hermannsburg area long before they did elsewhere.

Aborigines have several explanations for the disappearance including the following three: 1, the animals all moved off to another country; 2, someone, or something, rounded them up "like big mob bullocks"; 3, the people moved from their country to settlements, missions and outstations causing the disruption of totemic links, the decline in ceremonial activities, and the failure to care for the country in the traditional way.

Caring for the country includes the proper use of fire, and it is this aspect, from a European viewpoint, which has greatest potential in disturbing the vegetation and wildlife communities. Regular burning of the country is well documented by early explorers in the Centre, especially Giles (1889). It was used for hunting (Finlayson 1943), signalling (Terry 1937) and to fire particular country to encourage the growth of edible plants (Bolton and Latz 1978). This scattered burning would have produced a mosiac of vegetation in varying seral stages. Such a mosiac is resistant to large scale wildfires (Latz, pers. comm.) and is apparent in the journals of explorers that fires of limited scale were preponderate. Giles (1889) found a burn extending for 60 km worthy of comment, yet now such extensive fires are commonplace, some burning unchecked for hundreds of kilometres (Bolton and Latz 1978). Large fires destroy vast tracts of vegetative cover used by desert animals for food and shelter, and this obliterates the situation where regenerating burned areas can be easily recolonised by animals from adjacent areas in a later seral stage.

A large component of diversity in desert vegetation comes from the effects of fires which produce communities of different ages. For animals the mature communities generally provide shelter and heat refuge, but they are relatively unproductive and provide little food. Many of the short lived, highly productive grasses and herbs are the first to colonise recently burned areas. Mosaic burning patterns therefore produce a large "edge effect" which enables the convenient use of different resources in adjacent habitats. Moreover, regular patchy burning provides a continuous supply of habitats combined in the most useful seral stages. As one community matures and becomes unsuitable there are others available for colonisation. This combination of habitats is important to the herbivorous Mala (Bolton and Latz 1978) and the omnivorous Bilby (Johnson, unpubl.). It is most probably important also in the availability of vertebrate and invertebrate prey items of the Quoll.

The Aranda people began congregating at Hermannsburg Mission soon after the first missionaries arrived in 1877. The availability of European commodities at the Mission reduced their nomadic way of life together with the degree and regularity with which their country was burned. It was not until the 1930s and 1940s that the Pitjantjatjara, Walpiri and Pintupi people began leaving their country, so the effect of reduced burning would have occurred much later in these areas. Old Aranda men in their 60's had never seen a Quoll locally yet similarly aged men in the other tribes had caught and eaten them when they were young men. This gives a consistency in timing between the extinction of the species and the depopulation of traditional country by Aborigines.

We believe that the decline in regular burning of desert areas by Aborigines caused a decline in the vegetation diversity and allowed the immensely destructive and homogenising effects of large scale fires to take place. This in turn has tipped the Quoll into extinction, or best, decline massively in range where only tiny, difficult to locate pockets of species now persist.

In the richer salt lake and foot hills country it is possible that grazing by domestic and feral animals, especially the Rabbit, and competition with Foxes and Cats, may have had significant effects as well.

If the Western Quoll remains in the Northern Territory it will probably be found in the non–pastoral areas between 170°S and 19°30'S, south of the range occupied by *D. hallucatus*. It forms an interzone between the tropical and arid climatic influences and so vegetation diversity is likely to be high. Rabbits and Foxes are rare. The area is now rarely traversed by Aborigines so they are unlikely to know of Quoll populations residing in it. It is also an area which the fauna has not been surveyed using conventional techniques.

ACKNOWLEDGEMENTS

I would first like to thank the Aboriginal men of the western desert tribes for their help and information. I would also like to thank Mr. M. Quinell and Mr. H. Butler for providing comments on a draft of this paper.

REFERENCES

Bolton, B. L., and Latz, P. K., 1978. The Western Hare–wallaby *Lagorchestestes hirsutus* (Gould) Macropodidae), in the Tanami Desert. *Aust. Wildl. Res.* **5:** 285–93.

Finlayson, H. H., 1943. "The red centre." Angus and Robertson: Sydney, London.

Finlayson, H. H., 1961. On central Australian mammals. Pt IV – the distribution and status of central Australian Species. *Rec. S. Aus. Mus.* **41:** 141–91.

Giles, E., 1889. Australia twice traversed, being a narrative compiled from the journals of five expeditions into central South Australia and Western Australia, 1872–76. Vol. 1". Sampson, Low, Marston, Searle and Rivington Ltd: London.

Glauert, L. 1933. The distribution of the marsupials in Western Australia. *J. Roy. Soc. W. Aust.* **19:** 17–32.

Gould, J., 1863. "The mammals of Australia". 3 Vols. The Author: London.

Green, R. H., 1973. "The mammals of Tasmania". Queen Victoria Museum: Launceston.

Krefft, G., 1866. On the vertebrated animals of the Lower Murray and Darling, their habits, economy and geographical distribution. *Trans. Phil. Soc. N.S.W.* **1862–5:** 1–33.

Parker, S. A., 1973. An annotated check list of the native land mammals of the Northern Territory. *Rec. S. Aust. Mus.* **16:** 1–57.

Strehlow, T. G. H., 1947. Aranda traditions. Melbourne University Press: Melbourne.

Terry, M. 1937. "Sand and sun". Michael Joseph Ltd: London.

Thomas, O., 1907. List of further collections of mammals from Western Australia, including a series from Bernier Island, obtained for Mr W. E., Balston; with field notes by the collector, M. G. C., Shortridge. *Proc. Zool. Soc. Lond.* **1906:** 763–77.

Troughton, E., 1965. "Furred animals of Australia". Angus and Robertson Ltd: Sydney, London, Melbourne.

RESUMEN

20. El "Western Quoll" (*Dasyurus geoffroii*) del Northern Territory: registros historicos de fuentes tradicionales. Por. K. A., Johnson y A. D. Roff.

Alguna vez el "Western Quoll" fue común en el centro de Australia. Se cree que no vive màs allí. Todavía es recordado por los viejos aborigenes que alguna vez lo cazaron, preservando todavía sus mitolgias ceremonias.

Algunos aspectos de la historia natural del "Quoll" fueron registrados durante entrevistas con aquellos habitantes, los que sepresentan en este trabajo. Estos y otros datos son revisados, y se discuten las razones de la extinción del "Quoll".

21. OBSERVATIONS ON THE MOVEMENTS OF TWO ARID ZONE PLANIGALES (DASYURIDAE, MARSUPIALIA)

DAVID READ[1]

Populations of *Planigale gilesi* and *P. tenuirostris* in northwestern New South Wales appear to be mobile with individuals moving large distances. Home ranges are stable only for a short period of time. With a capture, mark, release and recapture technique over 4 ha. and 50 ha. trapping grids, movements of individuals were followed during winter, spring and summer.

Males had greater observed ranges than females and movements in excess of 600 m within a few days of first capture were recorded. *Planigale tenuirostris* individuals moved further than *P. gilesi*. It is speculated that drifting home ranges are an adaptation to an arid environment.

Breeding was observed in both species from late winter through summer.

Pages 227–31 *in* CARNIVOROUS MARSUPIALS, ed. by M. Archer, February 1982. Roy Zool Soc. New South Wales: Sydney, Australia.

DISTRIBUTION maps of the two arid zone planigales, *Planigale gilesi and P. tenuirostris* show overlap through northwestern New South Wales. Trapping surveys have shown the two species to be sympatric which provokes the question of how do these two species show ecological separation. Andrew (1977) reports that for these species information on their ecology is poor and largely anecdotal in nature. To differentiate between these two species a four year study to compare their ecology was commenced in March 1979. The results presented in this paper refer to a small aspect of the data collected in the first ten months of this study. As such only preliminary results are reported in this paper.

HABITAT

Study sites have been established on three different land systems: Gum Vale and Olive Downs within the Stuart National Park at Tibooburra, New South South Wales, and Conservation Land System at Fowler's Gap, 110 km north of Broken Hill, New South Walts (N.S.W. Soil Conservation Service 1979). These three land systems differ in habitat but in each occur soil types that form deep cracks which can provide shelter for small mammals. Trapping results contained in this paper come from all three land systems with Gum Vale being trapped in June, Olive Downs in July, and Conservation in May, August, October and January.

METHODS

For the May, June and July trapping, a 10 m grid of 250 Elliott live box traps was used in a rectangle 140 m by 170 m. In May and July this grid was supplemented by four pit trap drift lines (200 m long) placed parallel to each other 50 m apart (Fig. 1A). Animal movements on this 4 ha grid suggested that home range movements may be larger than this area. Consequently a new trapping pattern was designed. This new pattern comprised 13 points arranged with a centre point and two concentric circles of six points each. Placement was such that each point was equidistant (200 m) to the closest points in all directions and so formed a "Star of David" pattern. Each point consisted of a drift fence 95 m to 100 m long and 15 pits placed 6 m apart. The fence was aligned tangential to the circle and 8 Elliott box traps were placed in a 10 m grid pattern at each end of the drift fences. Using this pattern, approximately 50 ha was sampled during August, October and January (Fig. 1B).

Both the 4 ha and 50 ha patterns were trapped for 20 consecutive days at each trapping period. All animals were measured, marked, weighed and released at the site of capture. Traps on the 4 ha plot were cleared each six hours and on the 50 ha pattern at dusk, midnight and dawn.

RESULTS

Trapping data, including male body weights, are in Table 1. Lower total numbers for May, June, and July reflect the smaller area trapped and the fewer traps deployed on the 4 ha grid. From the recapture data, movements and mean observed ranges were calculated for the respective trapping periods. The length of a movement was taken as the distance between one capture and the next capture while the observed range for an animal was taken to be the maximum distance from any one point of capture to any other point of capture of that animal. Table 2 presents the means of observed ranges for the respective animal class and trapping period, and the statistical analysis of the results. Longer distances in

[1] School of Zoology, University of New South Wales, P.O. Box 1, Kensington, New South Wales, Australia, 2033.

August and following months are indicative of the more extensive trapping grid.

During the winter (May, June, July), little was determined about the movements of *P. tenuirostris* because few were recaptured (Table 1) and no drift fences were used in June. These have been the only mechanism effective (for the author) for capturing this species. Movements of *P. gilesi* during winter were similar for both sexes (Table 2). A high percent recapture of *P. gilesi* in these months probably indicates that the animals were not ranging great distances. Low captures in July for both species may reflect less suitable habitat at Olive Downs where the trapping was performed.

Figure 1. Plan of trapping grids

A.

4 Ha grid

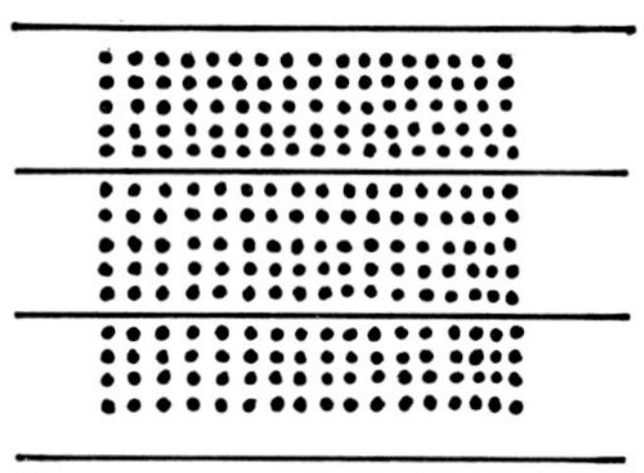

——Drift line

······Elliott livetraps

Scale= 12mm:100m

B.

50 Ha grid

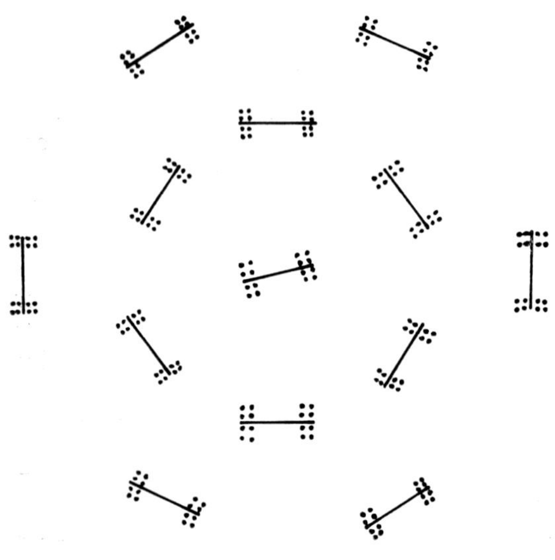

Combining data from August through to January for both species showed males to have greater ranges than females as follows:—

Species Combined

	Number of Animals	Mean Observed Range	Standard Deviation
males	72	198 m	± 84 m
females	43	63 m	± 84 m

The greater range of males is significant at 0.1% level (Student t test). A comparison of all *P. tenuirostris* with all *P. gilesi* from August through January was also significant at the 0.1% level (Student t test) with *P. tenuirostris* having the greater range. With each species, males had larger ranges than females (Table 2).

The 50 ha trapping pattern consisted of 13 drift fences (Fig. 1). If one regards each drift fence as a core area of the home range, then movement between core areas more clearly distinguishes the difference in range movement between males and females (Table 3). Observed range movements of males listed in Table 3 were not significantly different ($p > 0.05$, Student t test) between the species. Most females were observed to stay within the core area but some individuals moved away from the initial core area during any one trapping period.

DISCUSSION

For both species there is an unbalanced sex ratio in the captures for August and October (Table 1). Differences in behaviour patterns between the sexes may well account for this imbalance, particularly in regard to activity and movement in the deep soil cracks. Because the traps have been placed on the soil surface, the range of movement below ground level has not been recorded. Experiments to measure this aspect of animal movement are underway.

Breeding began in late August and pouched young were found in October and January. Less than 10 per cent of the January captures were adults, the majority being free–ranging juveniles. A comparison of male body weights shows this change in age structure (Table 1). However judged from scrotal width measurements, there was no suggestion of male senility in mature males caught in January.

The changes in observed range sites from winter through summer invite speculation about social organization and home range characteristics. In June, before breeding commenced, male and female *P. gilesi* had similar ranges and movement patterns. The small ranges and high recapture rate suggest a sedentary and communal life style.

Table 1. Recapture percentages.

Month of Trapping Period		Mean Body Weight (gms)	Total Caught	Total Recaptured*	% Recaptured of Total Caught
1979	a	5.8	4	2	50
	b	–	9	5	56
May	c	5.1	3	1	33
	d	–	3	1	33
	a	9.2	8	5	63
June	b	–	10	5	50
	c	–	–	–	–
	d	–	–	–	–
	a	8.76	4	1	25
July	b	–	–	–	–
	c	7.84	4	1	25
	d	–	3	1	33
	a	12.2	45	15	33
** August	b	–	22	6	27
	c	7.6	29	13	45
	d	–	11	3	27
	a	11.49	22	15	68
** October	b	–	10	6	60
	c	6.36	17	11	65
	d	–	6	4	67
1980	a	5.4	33	9	27
**	b	–	31	15	48
January	c	4.41	20	3	15
	d	–	16	6	37

Abbreviations: a, *P. gilesi* males; b, *P. gilesi* females; c, *P. tenuirostris* males, d, *P. tenuirostris* females; * Includes animals marked at a previous trapping period; ** Breeding recorded in these months.

Table 2. Mean Observed Ranges (m) of planigales.

Month of Trapping period	Mean range, ± standard deviation (number of animals)					
	P. gilesi males	*P. gilesi* females	Significantly different †	*P. tenuirostris* males	*P. tenuirostris* females	Significantly different †
May 1979	62±18 (2)	33±32 (3)	—	80 (1)	35 (1)	—
June 1979	83±28 (5)	73±27	n.s.	—	—	—
July 1979	60 (1)	—		130 (1)	5 (1)	—
August 1979	191±182 (15)	18±12 (9)	*	222±178 (13)	10± 5 (3)	**
August to October 1979	227±196 (13)	54±73 (6)	n.s.	247±126 (9)	25± 35 (2)	—
October 1979	244±222 (6)	5 (1)	—	120± 95 (4)	227±385 (2)	—
October 1979 to January 1980	400 (1)	275±278 (3)	—	550 (1)	217±258 (2)	—
January 1980	31±68 (8)	13±18 (8)	n.s.	5 (1)	72± 91 (3)	—
August to January 1980 (animals not caught) in October	—	30±38 (4)		200 (1)	35 (1)	—
Total August to Janaury	190±194 (32)	58±122 (25)	**	224±188 (20)	142±199 (9)	***
Total August to January (sexes combined)	128±103 (74)			180±104 (42)		††

Abbreviations: † Results of Mann–Whitney tests of the null hypothesis that male planigales move no further than females; —, insufficient data; n.s., not significant; * $P<0.05$; ** $P<0.01$; *** $P<0.001$; ††, significant $P<0.001$ (student t test).

During the breeding season, August to February, the home range and movement patterns of females did not change significantly but some did move in excess of 200 metres. Males, on the other hand, changed their movement patterns with the breeding season, and the majority of recaptured individuals had moved 200 m or more. With *P. tenuirostris,* the pattern of change in observed range size is similar to *P. gilesi* during the breeding season but there is insufficient data for range sizes prior to breeding.

Males probably move more rapidly than females. One male *P. gilesi* took less than six hours to cover 200 m. The fastest recorded female was a *P. tenuirostris* that took 16 days to cover 550 m. These distances are the minimum distances between capture points.

In both species, the more extensive movement of the males during breeding periods suggests that the males are searching for oestrous females. There is some circumstantial evidence that this is so. In one instance, a pair of *P. gilesi* were found mating in a pit trap, with a male occupying the same pit as the copulating pair, and a further three males in adjacent pits. Such a high concentration of individuals in adjacent pits is extremely uncommon and unlikely to have occurred by chance.

With few exceptions females did not show large range movements (Table 3). This may be due to the requirements of the pouched young and perhaps the necessity of staying close to a nest. Larger movements (over 200 m) may occur between pregnancies, as suggested by *P. gilesi* females from October to January (Table 3). Measurements of pouched young indicate that the litter carried in January was not the same as that carried in October. Detailed work on the life history of these species is not yet complete. Evidence that the same range behaviour occurs in *P. tenuirostris* is not yet clear.

During January, a high proportion of captures were juveniles. The males and females *P. gilesi* showed no significant differences in movement patterns because most recaptures were at the same trap line as the original capture. Very few of the recaptures were animals caught in August and October (Tables 1 and 3), and although almost an equal number of each sex was caught, the recapture of females was nearly double that for males. This higher recapture rate coupled with the smaller observed range, although similar to males, indicates that even at the juvenile stage females are more sedentary than males (assuming that the sexes have an equal trap response). A similar inference can be drawn from the results of *P. tenuirostris* juveniles caught dur-

Table 3. Observed range of planigales recaptured at a second trap line.

Trapping Period Month of and between periods.		*P. gilesi* males	*P. gilesi* females	*P. tenuirostris* males	*P. tenuirostris* females
August	e	15	9	13	3
	f	7	9	4	3
	g	8	—	9	—
	h	335 ± 127	—	303 ± 154	—
August to October	e	13	6	9	2
	f	5	5	1	2
	g	8	1	8	—
	h	353 ± 142	200	272 ± 107	—
October	e	6	1	4	2
	f	2	1	2	1
	g	4	—	2	1
	h	350 ± 191	—	200 ± 35	550
October to January	e	1	3	1	2
	f	—	1	—	1
	g	1	2	1	1
	h	400	400 ± 247	550	400
January	e	8	8	1	3
	f	7	8	1	2
	g	1	—	—	1
	h	200	—	—	175
August to January (Animals not captured in October)	e	—	4	1	1
	f	—	4	—	1
	g	—	—	1	—
	h	—	—	200	—

Abbreviations: e, total recaptured; f, recaptured at same trap line as first capture; g, recaptured at different trap line from first capture; h, Mean observed range (m) and standard deviation of (g) above.

ing January. Some juveniles of both species moved more than 200 m in 24 hours and, with the low recapture rate, these results suggest that juvenile planigales do not maintain a firm attachment to any one location.

The emerging characteristics of social organization show it to one of mobility with perhaps the most sedentary period being the winter months. From late winter there is a dispersal of adults prior to breeding and perhaps the mobility of both sexes is greatest in this period. During breeding periods males have a wide range and show little evidence of being fixed to one area. In contrast, females stay within a small range but its location may shift over a period of months.

Trap lines run in October were at the same sites as the ones run in August. For both species total captures were less in October than August but the recapture rate in October was much higher. This suggests that sedentary animals were a smaller proportion of the August population. Assuming the same trapability in both periods, then either the animals died between August and October or they moved out of the trapping area before October. A similar trend was observed in adult animals between October and January. While these are only initial observations, the above factors suggest that the two arid zone planigales have drifting ranges.

Although *P. tenuirostris* is the smaller species, it ranged over a greater distance than *P. gilesi*. Thus while both species may have drifting home ranges, more subtle differences in social organization may occur and need to be elucidated.

Sminthopsis crassicaudata has been trapped sympatrically with the two Planigale species. Drifting home range is one of the main attributes of the social organization of *S. crassicaudata* (Morton 1978a) and this also appears to be shared by the two planigale species. Morton (1978b) also found that *S. crassicaudata* exhibit nest sharing behaviour during the winter months. The smaller range of *P. gilesi* during winter may indicate nest sharing in this species also. However no conclusive evidence is yet available.

SUMMARY

Data collected so far for these two arid zone planigales indicate that they both occupy home ranges that are stable for only short periods of time. Females are more sedentary than males. Insufficient evidence is available to conclude whether this characteristic of social organization is maintained for the whole year or only occurs during the breeding season. It is interesting to speculate that having a drifting home range may be an adaptation to an arid environment. With the continuation of this study it is hoped to discover more evidence on home ranges and social structure of these tiny carnivores.

ACKNOWLEDGEMENTS

Drs. David Croft and Martin Denny read and constructively criticised earlier drafts of this paper. The staff at Fowler's Gap Research Station helped with much hospitality and general support.

REFERENCES

Andrew, D., 1977. "The behaviour of *Planigale tenuirostris* (Marsupialia: Dasyuridae), with comparative notes on other species". Hons. thesis: School of Zoology, University of New South Wales: Sydney.

Morton, S. R., 1978a. An ecological study of *Sminthopsis crassicaudata* (Marsupialia: Dasyuridae). II Behaviour and social organization. *Aust. Wildl. Res.* **5:** 163–82.

Morton, S. R., 1978b. Torpor and nest–sharing in free–living *Sminthopsis crassicaudata* (Marsupialia) and *Mus musculus* (Rodentia). *J. Mammal.* **59:** 569–75.

Soil Conservation Service of New South Wales, 1979. Land Systems Series Map Sheet 5454–7.

RESUMEN

21. Observacion de los movimientos de dos Planigales (Dasyuridae, Marsupialia) de la zona arida. Por David Read.

Poblaciones de *Planigale gilesi* y *P. tenuirostris* del Noroeste de New South Wales mostraron ser móviles, con los individuos desplazándose largas distancias. El área habitacional fue estable sólo durante un corto período de tiempo. Con la técnica de una captura, marcado, liberación y recaptura, sobre un reticulado de entrampamiento de 4 por 50 hectáreas, se siguieron los movimientos de individuos durante el invierno, la primavera y el verano. Los machos mostraron tener una mayor distribución que las hembras, y se registraron movimientos que excedieron los 600 m en unos pocos dias de la primera captura. Los individuos de *Planigale tenuirostris* se desplazaron más que los de *P. gilesi*. Se infiere que los desplazamientos de las áreas habitacionales responden a la adaptación a un ambiente árido. El apareo de ambas especies fue observado desde fines del invierno hasta todo el verano.

22. RECENT ALLEGED SIGHTINGS OF THE THYLACINE (MARSUPIALIA, THYLACINIDAE) IN TASMANIA

D. E. ROUNSEVELL[1] and S. J. SMITH[1]

There has been a marked increase in the frequency of alleged Thylacine sightings since 1930.

Some features of 104 sightings reported since 1970 are presented. The sightings were often reported after publicity was given to searching for Thylacines. They occurred throughout the State but mainly in northern Tasmania. Most reports are of single animals seen briefly from cars at night, by local people who had not previously reported a sighting. More than half of the sightings occurred between 7 p.m. and midnight.

Although examination of these sightings neither proves nor disproves that the Thylacine still exists, the frequency of reported sightings provides a basis for cautious optimism.

Pages 233–36 *in* CARNIVOROUS MARSUPIALS, ed. by M. Archer, February, 1982. Roy. Zool. Soc. New South Wales: Sydney Australia.

THE STATUS of the Thylacine, (*Thylacinus cynocephalus*), in Tasmania, has not changed during the past 46 years (Pearse 1976). No indisputable proof of its existence has emerged during this period and it could be claimed that the animal is extinct. However, the frequency of alleged sightings of Thylacines reported in the past decade provides a basis for cautious optimism, and is the justification for continued attempts to rediscover the animal. This paper considers sightings reported since 1970 and discusses some of their characteristics.

METHODS

Details of alleged Thylacine sightings are obtained by the Tasmanian National Parks and Wildlife Service (N.P.W.S.), from reports by individuals or the press. Observers are interviewed, if possible, and the information obtained is recorded on standard questionnaires.

Some alleged sightings are quickly reported to the National Parks and Wildlife Service or the press, but many are received belatedly.

This sample of sightings is known to be incomplete. The number of sightings which remain unreported is not known but it is believed that it may be comparable to the number which have been reported. People have given a variety of reasons for maintaining secrecy or not reporting sightings promptly. Some fear that an animal will be threatened if its location is disclosed. Others imagine that changes to the tenure and management of the locality may occur if they report a sighting, and that this could threaten their livelihood or recreation. Fear of personal ridicule is another important reason for people not reporting sightings.

The sightings considered in this paper are therefore from people who openly claim they saw, or have seen, a Thylacine. To be included in the present sample, however, the descriptions supplied by the observers were vetted to a certain standard. These descriptions needed to be sufficient to identify a Thylacine — like animal under the particular circumstances of the sighting. Reports from people who did not disclose the locality were disregarded.

RESULTS

In the past decade (1970 – 1979), 104 alleged sightings of Thylacines were reported and in the previous decade (1960 – 1969), 84 were reported. The total numbers of sightings reported in each decade since 1930 are shown in Fig. 1.

There has been a dramatic increase in the number of sightings reported since 1960.

The number of sightings reported annually has varied from 27 in 1970 to 1 in 1976 (Fig 2). Distinct peaks, occurred in 1970 and 1978. The timing and amplitude of these peaks coincide with the timing and amount of publicity produced by two privately organised Thylacine searches conducted during that decade; one by Brown, Malley and Griffiths in 1972 (Brown 1973), and the other by Tangey in 1977–78 (Anon 1977). It is apparent that the number of reports is directly related to the publicity given to searching for Thylacines.

[1] National Parks and Wildlife Service, P.O. Box 210, Sandy Bay, Tasmania, Australia, 7005.

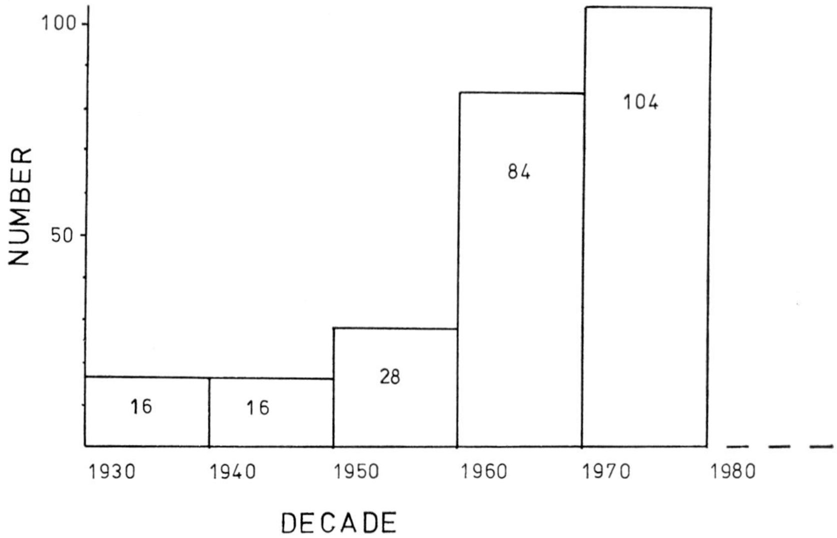

Fig. 1. The number of alleged Thylacine sightings per decade from 1930 to 1980.

DISTRIBUTION: Figure 3 shows the locations of 104 alleged sightings reported since 1970. The sightings occurred in many parts of Tasmania but most were located in the north of the State. Their distribution corresponds approximately to the distribution of government bounty payments made while the Thylacine bounty scheme operated between 1888 and 1909 (Guiler 1961). The distribution of sightings is not closely related to the distribution of roads or to the distribution of the human population in Tasmania.

OBSERVERS: Of the 97 people who reported the alleged Thylacine sightings in the period 1970 to 1979, 91 reported single sightings, five reported two sightings and one person reported three separate sightings. A total of 178 people, in groups of from one to six, witnessed the 104 sightings.

Table 1 shows that 56 people witnessed sightings when alone, 31 had one companion, eleven had two companions and six people had from

Table 1. The number of witnesses to and the duration of alleged Thylacine sightings reported since 1970.

	No. people present						
Duration	1	2	3	4	5	6	
Unspecified	34	20	3	3	0	0	60
0–5 secs.	10	7	2	0	1	1	20
5–10 secs.	5	1	0	0	0	1	7
10–30 secs.	6	2	1	0	0	0	9
30–60 secs.	0	0	1	0	0	0	1
> 1 min.	1	1	4	1	0	0	7
	56	31	11	4	1	1	104

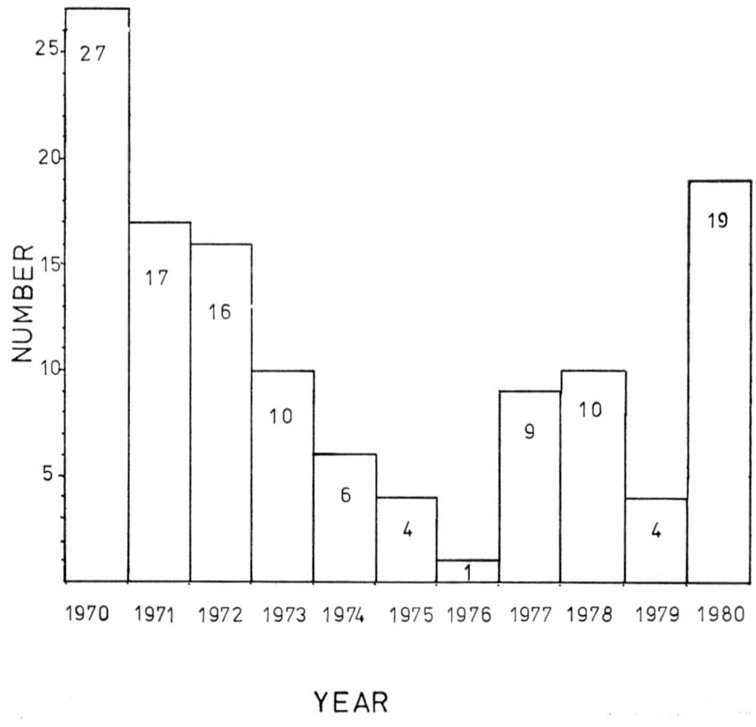

Fig. 2. The number of alleged Thylacine sightings from 1970 to April, 1980.

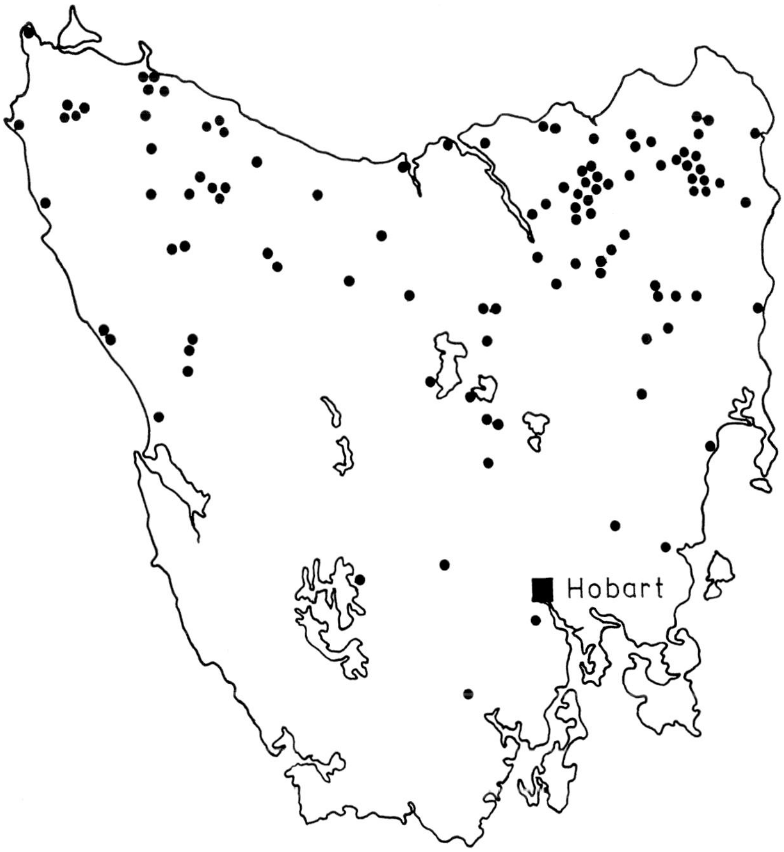

Fig. 3. Alleged Thylacine sightings from 1970 to April, 1980.

three to five companions. Many witnesses did not specify the duration of the sightings, but most of those who did, reported that the sighting occupied less than 30 seconds. Many sightings in this category lasted less than five seconds, and were by observers travelling in motor vehicles. All of the seven people who reported sightings that lasted for more than one minute were on foot during their sightings.

Fifty–eight of the 97 people who reported sightings were local residents or locally employed people and seven were tourists. In addition there were five hunters, two bushwalkers, two motorcyclists, two policeman and an angler. Information on the activities of the remaining 20 people was not available.

Seventy one people reported sightings made whilst they were travelling in cars on roads, and three sightings were made from vehicles not travelling on roads. Sixteen sightings were made whilst the observers were on foot, and one was made from a fishing boat.

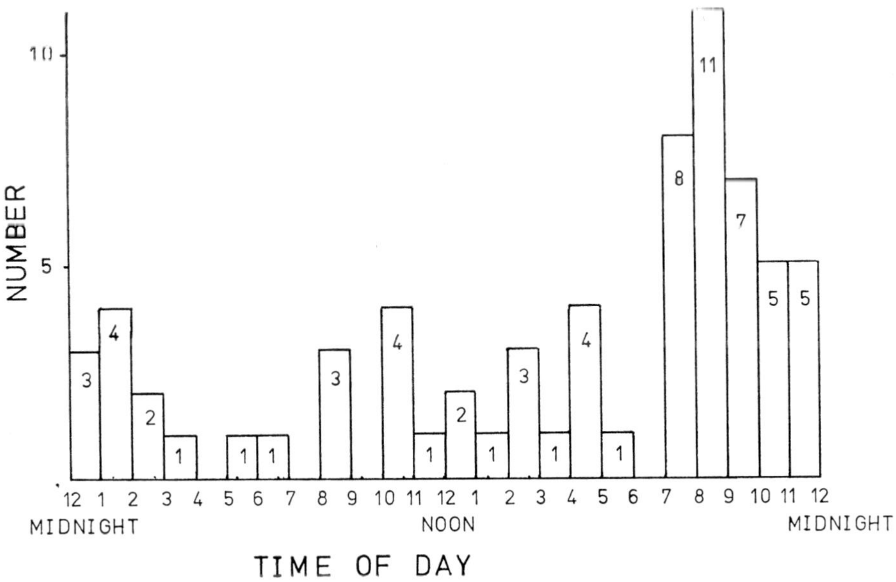

Fig. 4. The times of day when alleged sightings of Thylacines have been made.

SIGHTINGS: All of the alleged sightings were of single live animals. Descriptions of the observed features and the behaviour of the animals varied, but most contained details of an animal with striped hindquarters. Usually no reason could be found to doubt the observers who were certain they had seen a Thylacine. Some, who made sightings of short duration or under difficult conditions, were not certain they had seen a Thylacine but still felt it was important to report their experience.

Figure 4 shows the time of day at which each of 68 sightings occurred. Over half (53%) of these sightings occurred at night between 7.00 p.m. and midnight. The remainder were fairly evenly distributed throughout the rest of the day. The greatest number of sightings in one hour (eleven) was reported to occur between 8.00 p.m. and 9.00 p.m.

The distribution of sighting times throughout a day, shown in Fig. 4, is similar to that which might be expected for a nocturnal animal, which is occasionally active during daylight.

There was little seasonal variation in the frequency of reported sightings: 30% occurred in the Summer months, 28% in Autumn, 21% in Winter and 22% in Spring.

DISCUSSION

There is little to be gained here by further discussion of the similarities between, and characteristics of each of the alleged sightings. They neither prove nor disprove that the Thylacine is extant, however they are phenomena that should not be ignored.

The relationship between publicity afforded to organised Thylacine searches, and the frequency of reported sightings has long been recognised (Guiler 1958). This relationship is being exploited by one of the authors (S.J.S.) who is currently engaged full–time in an investigation of the Thylacine organised by the N.P.W.S. (Tas.), and funded by the World Wildlife Fund (Australia).

The publicity associated with searches certainly encourages people to report recent past sightings. It is not known, however, whether such publicity actually increases the probability of people seeing, or believing that they have seen, Thylacines. The increasing frequency of reports since 1930 (Fig. 1) is marked. It may be related to both the increased ease of mass communication, and changes in interest in the natural environment shown by people generally.

It could be suggested that any 'genuine' sightings reported since 1930 would be increasingly concealed by more 'false' sightings reports, as the above changes in technology and social attitudes have occurred. In an attempt to limit the reporting of 'false' sightings the N.P.W.S. has been deliberately silent about Thylacine sightings reported publicly over the past decade. Instead, efforts were directed towards investigating and documenting the sightings which were reported, and providing unbiassed and, if required, confidential conditions for the reporting of sightings. It is clear that in the absence of publicity fewer reports were received in the mid–1970's (Fig. 2). however, some very impressive reports of sightings were received throughout the decade.

It is important that relevant scientific institutions in Tasmania have an appropriately receptive attitude to alleged Thylacine sightings, in order to be able to collect and evaluate this source of information. If Thylacines still exist in Tasmania, they are few, and the difficult task of rediscovering them may be facilitated by careful analysis of the growing collection of reported sightings.

ACKNOWLEDGEMENTS

We would like to express thanks to Mr. J. Mahoney and Mr. D. Fleay for useful comments on an earlier draft of this manuscript.

REFERENCES

ANONYMOUS, 1979. "The Mercury" (6/12/77) p.3. Hobart.

BROWN, R., 1973. Has the Thylacine really vanished? *Animals.* **15:** 416–9.

GUILER, E. R., 1958. The Thylacine. *The Australian Museum Magazine.* **12:** 352–4.

GUILER, E. R., 1961. The former distribution and decline of the Thylacine *Australian Journal of Science.* **23:** 207–10.

PEARSE, R. J., 1976. Thylacines in Tasmania. *Australian Mammal Society. Bulletin.* **3:** 58. (Abstract of paper presented to the 18th General Meeting, Launceston).

RESUMEN

22. Recientes ale gadas visiones del lobo marsupial (*Thylacinus cynocephalus*) en Tasmania. Por D. E. Rounsevell y S. J. Smith.

Desde 1930 se ha producido una marcada frecuencia de denuncias de haber visto el lobo marsupial *(Thylacinus cynocephalus).*

Se presentan 104 denuncias ocurridas desde 1970. La búsqueda del lobo marsupial se vió impedida porque las denuncias se hicieron luego de su publicidad. Las visiones ocurrieron en todo el Estado, pero principalmente en el norte de Tasmania. La mayoría de esas denuncias fueron de animales solos, vistos brevemente en la noche y desde automóviles, y por lugarénos que no los habían visto antes.

23. REVIEW OF THE THYLACINE (MARSUPIALIA, THYLACINIDAE)

MALCOLM SMITH[1]

The literature on the Thylacine (*Thylacinus cynocephalus,* Harris), is reviewed. At the time of white settlement the species was extant only in Tasmania, where it suffered persecution until its populations collapsed around 1910. Since then a number of expeditions have uncovered much circumstantial evidence for its continued survival. Fossils have been found throughout the mainland, as well as in Papua New Guinea, and there have even been a number of questionable reports of sightings on the mainland. Taxonomically, the thylacine appears to be isolated among the Marsupicarnivores. Its reproductive system is primitive, but the rest of its anatomy has been specialized for cursorial hunting, and it parallels the Canidae in its large size, elongated jaws and sectorial molars. However, its limb proportions are similar to those of *Dasyurus* and its brain, although large and convoluted compared to a hypothetical dasyurid of similar size, is small in comparison to the Canidae. Little is known about the animal's ecology and behaviour, but what can be gleaned from observations on captive animals and anecdotes on wild individuals is presented here.

Pages 237–53 *in* CARNIVOROUS MARSUPIALS, ed. by M. Archer, February, 1982. Roy. Zool. Soc. New South Wales: Sydney, Australia.

"Our cots we fence with firing and slumber while we can
To keep the wolves and tigers from us in Van Diemen's Land."

IT IS UNLIKELY that the author of the above ballad had ever been to Van Diemen's Land, but the fact that it was sung in Ireland until the early days of this century (O Lochlainn 1946) is a credit to the mystique of the animal known as the "Tasmanian Wolf", or "Tiger", and to the more erudite as the "Zebra Opossum" or "Thylacine". The Thylacine is to Tasmania what the Dodo is to Mauritius. It is included in the state coat of arms and is portrayed on the medal of the Royal Society of Tasmania, despite the fact that is is now almost, if not completely, extinct. If still extant, its centres of population are unknown, and although fossil remains are being discovered and the existing collections of skeletons still being examined, our knowledge of the animal's biology is unlikely to be enlarged significantly. It is time, therefore, for the available information to be brought together in a single review.

THE SCIENTIFIC NAME

Didelphis cynocephala Harris, 1808. *Trans. Linn. Soc., London* 9: 174.

Thylacinus Temminck, 1824. *Monographies de mammalogie* 1: xxiii, 23, 60, 267.

The usual translation of the name, "pouched dog with a wolf's head", which appears to have originated with Sharland (1939), is incorrect. The generic name is a Latinized adjective from the Greek *thýlákos,* "a pouch or meal–bag", while the specific epithet comes from the Greek *kyön* (stem *kyno–*), "a dog" and *kephàlè,* "a head".

HISTORY: 1805–1933

The earliest account of the thylacine appears to be one published in the *Sydney Gazette and New South Wales Advertiser* vol. 3. no. 112, 21 April, 1805. It reads:

"An animal of a truly singular and nouvel description was killed by dogs the 30th of March on a hill immediately contiguous to the settlement at Yorkton Port Dalrymple; from the following minute description of which, by Lieutenant Governor PATERSON, it must be considered of a species perfectly distinct from any of the animal creation hitherto known, and certainly the only powerful and terrific of the carniverous and voracious tribe yet discovered on any part of New Holland or its adjacent Islands.

It is very evident his species is destructive, and lives entirely on animal food; as on dissection his stomach was found filled with a quantity of kangaroo, weighing 5 lbs. the weight of the whole animal 45 lbs. From its interior structure it must be a brute peculiarly quick of digestion; the dimensions were, from the nose to the eye 4½ inches: length of the eye, which is remarkably large and black, 1¾ inches; breadth of the eye ¾ of an inch; from the nose to the extent of the mouth in the upper jaw, 6 inches; and to the extent of the under jaw, 4½ inches; breadth of the forehead, 5¾ inches; from the eye to the ear, 3¾ inches; the ear round, diameter 3 inches; from the ear to the shoulder, 1 foot; from the shoulder to the first stripe, 7 inches; from the first stripe to the extent of the body, 2 feet; length of the tail, 1 foot 8 inches; length of the fore leg, 11 inches; and of the fore foot, 5 inches; the fore foot with 5 blunt

[1] No. 7, 23rd Avenue, Brighton, Brisbane, Queensland, Australia, 4017.

claws; height of the animal before, 1 foot 10 inches; stripes across the back 20, on the tail 3; 2 of the stripes extend down each thigh; length of the hind leg from the heel to the thigh, 1 foot; length of the hind foot, 6 inches; the hind foot with 4 blunt claws, the soles of the feet without hair; on each side the mouth are 19 bristles, length of each 4 inches; and 6 bristles on each side under the ear, 9 on the lower jaw upon each side, and 8 under the throat; 8 fore teeth in the upper jaw, and 6 in the under; 4 grinders of a side, in the upper and lower jaw; 3 single teeth also in each; 4 tusks, or canine teeth, length of each 1 inch: circference of the head before the ears, 1 foot 6 inches, and behind the ears, 1 foot 5 inches; smallest part of the neck, 1 foot 4 inches; circumference before the shoulder, 2 feet; the body short hair and smooth, of a greyish colour, the stripes black; the hair on the neck rather longer than that on the body; the hair on the ears of a light brown colour, on the inside rather long. — The form of the animal is that of the hyoena, at the same time strongly reminding the observer of the appearance of a low wolf dog. The lips do not appear to conceal the tusks".

It says a lot for the popular interest in natural history at the time that this description was written by a Lieutenent Governor and published in a newspaper. Some of the definitive descriptions of species to be found in contemporary scholarly journals are less detailed.

The initial wonder soon turned to hostility when unfenced livestock started to fall prey to the animal. Between 1832 and 1849, Surrey Hills Station alone had 147 sheep killed by Thylacines, 299 by dogs, and a further 458 by unidentified predators (Guiler 1961b). Guiler also relates how in 1836 the Van Diemen's Land Co. provided a trapper with a hut and his keep to kill Thylacines at Woolnorth, while in 1840 they placed on the species a bounty of 6/- each for fewer than ten scalps and 10/– each for more. From 28 April 1888 the Government paid £1 each for adult scalps and 10/– each for juvenile ones – this in the days when £2 was considered a reasonable weekly wage. Individual farmers offered equivalent bounties.

The policy had its desired effect. As early as 1863 John Gould was warning:

"When the comparatively small island of Tasmania becomes more densely populated, and its primative for-

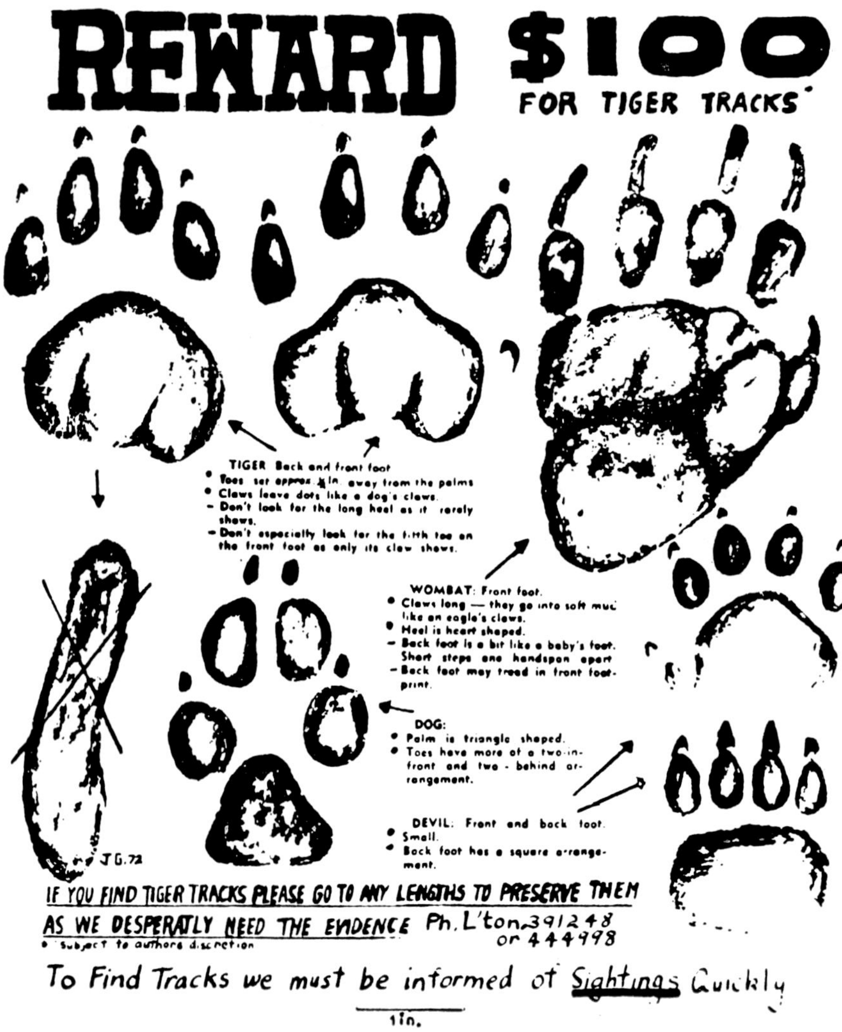

Fig. 1. Footprint identification poster of the Griffith–Malley–Brown team (reprinted with permission). The reward no longer stands, and the present contact is J. F. Malley: Ph. Trowutta 56-5112.

ests are intersected with roads from the eastern to the western coast, the numbers of this singular animal will speedily diminish, extermination will have its full sway, and it will then, like the Wolf in England and Scotland, be recorded as an animal of the past: although this will be a source of much regret, neither the shepherd nor the farmer can be blamed for wishing to rid the island of so troublesome a creature".

Griffith (1972) reports that between 1878 and 1893 a Tasmanian tannery exported 3,482 skins to London to be made into waistcoats, and that as late as 1909 newspaper advertisements could be found offering "tiger shoots for visitors in search of fun." One "old timer" told him how, about the year 1900, a team of two men required just four months to collect the bounty on 300 thylacines. Anecdotes of this nature must however, be treated cautiously. The figure is too high to be consistent with the bounty records examined by Guiler (1961b). The Tasmanian Government received the scalps of 2,040 adults and 144 juveniles, while the Van Diemen's Land Co. took another 84. The number received by independent farmers was probably of the same order of magnitude. Very few came from the dense, wet, sparsely populated rainforest of the west coast.

The peak kill (172) occurred in 1900, after which a rapid decline set in. The Government bounty was dropped in 1910 but the Company bounty lasted till 1914, when only three scalps were obtained. The surprising thing, however, is that the decline was *not* due directly to hunting pressure since, as Guiler pointed out, the scalp harvest suffered a sudden collapse instead of a gradual, steady decline. Several old timers he interviewed claimed that the decline was very rapid and simultaneous throughout the island. One grazier even said that in 1910 all the marsupial carnivores died of something resembling distemper. Throughout the 1920s, Thylacines were still being exported to foreign zoos (Grzimek 1967). According to Guiler (1966), the last confirmed wild Thylacine was shot at Mawbanna in April, 1930, the same year as a closed season was granted for December, the supposed breeding season. In 1933 the last captive died in the Hobart Zoo. In 1936, the species received complete protection, but the day of grace had already passed.

HISTORY SINCE 1933

After 1933 inconclusive expeditions and unverified sightings substituted for hard evidence in regard to the species' continued existence and whereabouts. The problem was not eased by the close resemblance of Thylacines to Dogs, especially under adverse lighting conditions. Furthermore, to the untrained, their tracks can easily be confused with those of Wombats (Fig. 1), particularly if the soil is hard.

In May, 1937, Roy Marthick spent three weeks searching for Thylacines, and claimed to have found tracks of twenty individuals and to have seen some of the makers at dusk (Grzimek 1967). The next two expeditions were sponsored by the Animal and Birds Protection Board. On 28 November, 1937, Trooper A. L. Fleming and prospector Lesley Williams set out into the sparsely populated west coast, much of which was inhospitable to hunters (Fleming 1938). After reaching the Cardigan River, they climbed the hills to south of the Raglan Range and passed down a valley to the Franklin River and Frenchman's Cap. On the way they found several Thylacine tracks[2]. More tracks of a large and a small animal were found west of Frenchman's Cap. South of the southern tip of the Raglan Range, they discovered eleven tracks from at least four different individuals, all within a radius of 16 km.

Next year Fleming led another expedition into the area directly southeast of the first (Sharland 1939). Having crossed the King William Range and the Jane, Franklin and Lodden Rivers, they entered the beech forest of Thirknell's Creek and made plaster casts of Thylacine tracks. They also saw the quite different spoor of a feral dog. Having made their camp in a rough pine cutter's cabin on the Erebus River, they searched an area of 2000 km^2 between the King William Range and the Jane River hills, finding tracks near a tributary of the Jane River and high on the slopes of Frenchman's Cap, as well as around their own cabin, which was, they said, visited by a Thylacine in the night.

In late 1943 or early 1944, a former Thylacine hunter called Charles Spencer was cutting a track when he saw a beautifully marked specimen "at least six feet from tip to tip." He also claimed to have seen a female with young at Adamsfield the previous March (Spencer 1944).

In November 1945, another expedition was led by David Fleay, who wanted an individual for the Healesville Sanctuary. It was based at a slab hut on the Jane River. Traps were set along the Erebus River. The delights of this expedition are best summarized by Fleay himself (1946):

"We lived in a heavy rainfall belt, usually with eight days wet out of ten and not uncommonly summer snow and frost; we struggled with Horizontal Scrub and cling ing Bauera, or walked through dark and silent forests of dripping myrtle sassafras, leatherwood and pine. Floundering journeys with heavy packs and equipment across boggy button–grass plains and strenuous climbing over high ranges were every–day events, not to mention the endless torment from mosquitoes, blow–flies, biting fleas, even rheumatism."

The empty forest was festooned with snares set for possums and wallabies. They were told

[2] It has not been possible to confirm the identity of the Thylacine tracks reported from this, and subsequent, expeditions. The observers were all competent bushmen and unlikely to be mistaken, but it should be pointed out that Guiler and Meldrum (1958) identified some of the footprint casts obtained by the 1945–46 Fleay expedition as belonging to wombats.

how, in a single winter, two brothers had set 2,000 snares. Poison had also been laid for the dasyurids, which would otherwise have attacked the helpless victims. There were no Thylacine tracks anywhere where there were snares. The ranger, A. D. Ferguson, claimed to have seen a Thylacine the previous winter, and a road patrolman, M. Tiffin, told how he had been sawing a tree 2½ years before when a young Thylacine ran out.

An attempt to trap Thylacines by dragging trails of fresh meat around the Collingwood Range failed, but when promising tracks were found on Franklin Hill, Fleay moved camp to Dump Hut, near Calder's Pass, and heard a weird cry, which "suggested the brief, sharp creak of a door and was quite unlike any cry of a mammal or bird I have ever heard" (Fleay 1946). Next month they found some tracks in fresh mud and, on Poverty Plain, a Thylacine was believed to have taken a Bennett's wallaby (*Macropus rufogriseus*) from a trap, leaving a tuft of fur behind. For the next two days it evidently left its tracks all around the traps, but avoided them all. Finally, the expedition was forced to return home.

In October, 1957, Meldrum heard of three sheep killed in a strange manner on a Derwent farm. There was no sign of a struggle, and the only parts eaten were the throat, lower jaw, most of the face, and the turbinate bones. On a neighbouring property, there was a similar killing and, later on the first property, a lamb had its liver consumed but its throat and face were left intact. Guiler and Meldrum (1958) did not think it was the work of a Tasmanian Devil (*Sarcophilus harrisii*) because of the "tidy" method of killing, the lack of a struggle, the fact that the predator did not return to the kill, and the fact that Devils were common in the area but did not kill Sheep. Nor did they think a Dog was responsible, because they would expect a Dog to turn "killer" after its success. The last conclusion is less convincing when one realizes that a Dog was later caught in the vicinity (Griffith *et. al.* 1972). However, footprints resembling a Thylacine's were found, and the owner of a neighbouring property claimed to have seen a "Tiger".

In 1957, an alleged Thylacine was followed by helicopter at Birthday Bay, but examination of the subsequent photograph suggested that the animal was a Dog (Grzimek 1967).

In January, 1958, a small party discovered tracks in the deep mud of the Old Davey Track between Pt. Davey and Muydena. They were bigger than a Devil's with all four toes pointing forward, the claws almost level with one another, and the hind pads much longer than a Dog's. Guiler identified them from a photo as belonging to a Thylacine (Muir 1961). Griffith (1972) referred to brief, unsuccessful expeditions made by a Walt Disney film unit in 1958, and by Sir Edmund Hillary in 1960. I have found no other mention of these expeditions.

In 1961, Charles Reuben took plaster casts of tracks discovered at Mawbanna which Griffith *et. al.* (1972) described as "impressive but not indisputable". They are now owned by Mr I. Stephens of Mole Creek. In August the same year, two fishermen, Bill Morrison and Laurie Thompson, were camping at Sandy Cape when they heard an animal trying to get at the basket of bait outside their tent. Thompson hit it with a stick, and the next morning allegedly found a young Thylacine dead outside. Unfortunately, the carcass was gone when they returned to the camp the following evening (Grzimek 1967, Guiler 1966).

The report did, however, encourage the Government to grant $2000 for a new expedition headed by Eric Guiler. As a result, 700 snares were set along a 16km line at Green's Creek in the Sandy Cape area, followed by 800 snares at the northwest property of Woolnorth, the scene of alleged sightings, footprints and dropping since 1956 (no details supplied). Finally, a trapping program was initiated in 1964 near Balfour. All three programs were unsuccessful (Guiler 1966).

Footprints were discovered at Ormley by Mr Blacklow, and further investigations disclosed a lair, but the footprints, unfortunately, were those of a wombat (Griffith *et al.* 1972). In 1966, another lair was discovered in an abandoned boiler on the White River near Mawbanna. Guiler arrived next year with an electronic eye camera and a large trap, but had no success. In fact, the footprints nearby had been discovered by a witness who subsequently was unable to distinguish Thylacine from wombat tracks. A hair sample taken from the spot in 1967 was identified by Hans Brunner as "definitely not that of a thylacine" (Griffith *et. al.* 1972).

The years 1968 to 1972 saw literally thousands of kilometres covered by a team of very efficient amateurs in the most thorough search to date. Because only a short description of this search has been given by Griffith (1972), I shall cite further details from an unpublished mimeographed report (Griffith, *et. al.* 1972). In 1967, Jeremy Griffith came to Tasmania, unsponsored, and explored parts of the northwest and southwest of the state. In March, 1968, he was joined by James Malley, who had been investigating Thylacine reports for ten years. They examined the site of the White River "lair", crossed the Arthur River to the Balfour Plains in December, 1968, and searched the northwest coast to the Pieman River in February, 1969.

Further excursions were made from December, 1970, to February, 1971, during the first nineteen days of which they averaged five hours' tracking a day, and drove 3,360 km. The area covered by the Fleay expedition was searched in vain. The land was heavily snared and now an even more deadly poison, 1080, was in use. The only large, still untouched area was the land southwest from Macquarie Harbour to Pt. Davey. As a result, in December, 1970, they flew in with Peter Simms' expedition searching for aboriginal carvings, explored down to Moore's Valley and Birch's Inlet, tracked extensively along the beaches, and finally hailed a fishing boat to return to civilization. The discovery of tracks at Beulah in May, 1971, caused Griffith to immediately fly back to Tasmania, only to find the prints too indistinct for positive identification. He and Malley examined the Fingal area and collected records of 40 sightings from the previous decade.

The two investigators were frequently interrupted by requirements of their normal occupations as well as by the urgent need to obtain finance. In 1972, they were joined by a medical doctor, Robert Brown, and a centre was set up to gather sightings from the public. A total of 82 reports were received from northwestern Tasmania alone. An earlier system of trail monitoring by means of a caged fowl decoy and an automatic photograph device had proved unworkable. Now they were aided by Trevor Briggs who constructed six tracking pads which, when triggered by a trap line set too high for a Devil, would transmit a radio signal via a television station to the team's receiver. From July to December 1972, Griffith traveled an average of 800 km a week, but although he thoroughly searched every area where a reasonable sighting had been reported, nothing turned up. The only potential Thylacine areas not examined by the end of the five year search were the King William Range, the Florentine Valley, and the Hartz Mountains. Griffith and Brown were totally disillusioned about the possibility of the species' survival.

Malley was more optimistic. He claimed: "The tiger . . . is not extinct . . . The only evidence I can produce to back this claim is 20 photographs of indistinct footprints found after the sighting of a Tiger at Beulah in May 1971. These tracks were definitely those of a Tiger. However they were not distinct enough for anyone who was not thoroughly familiar with animal pads to recognize as such. A plaster cast taken at Mawbanna in August 1961 is definitely that of a Tiger. In recent years many clear sightings have been made by people whom I know personally. I have no doubts of any sort in their sincerity and honesty. The majority of recent sightings have been in three areas of the State – the central East Coast, the Northern part of the Arthur River Basin, and the northern edge of the Central Plateau" (Griffith *et. al.* 1972).

The team's files are now stored in the Queen Victoria Museum, Launceston, under the custody of the curator of vertebrates. James Malley is still investigating reports, and in a letter to me (2 February, 1973) he tells of a sighting 16 km north of Sandy Cape, of which there are poor photographs and a sketch of footprints which were "spot on in every detail". He also stated: "In figures of sightings I have received virtually hundreds, 70 to 80% you can't discredit in any way".

Every now and then a member of the public claims to have seen a Thylacine. When a Sydney newspaper, the *Sun* (11 February, 1974), published an old photo of the animal, a correspondent wrote: "While on a tourist bus in June, 1969, travelling over the snow fields of the Prince William ranges, two passengers and I observed a small animal near an old logging camp. It appeared to be a greyhound in the puppy stage. It was running along a fence close to the road. It was brown and the stripes as marked as those of the animal shown in 'The Sun' (Fisher 1974).

On the night of 19 August, 1977, two policemen, Tom Banfield and John Wilson, saw a dog-like animal cross the road in front of their car near Derby. It had dark, ring–like markings on its tail, but no distinguishing marks could be seen on its back (Anon. 1977b, Strachan 1977). No doubt reports of this kind will continue to emanate from Tasmania for many years, but it is certain that if the Thylacine is not already extinct its condition is extremely precarious.

PREHISTORIC RANGE

Formerly the Thylacine ranged over most of the mainland, as well as Papua New Guinea, and died out only recently. Bone fragments, associated with a Thylacine humerus found at Tunnel Creek, Western Australia, have been dated at 0 ± 180 radiocarbon years BP, though this date may not apply to the Thylacine bone itself (Archer 1974). Other than that, the most recent finds are the 3,000 year old specimens from Padypadiy, Northern Territory (Calaby and White 1967) and Murray Cave, Western Australia (Archer 1974).

Wright (1972) discovered very naturalistic drawings of Thylacines in caves in the Pilbara near Mt. Edgar Station, Juna Downs and Tom Price Station. The animal he found depicted at Abydos Station could have been either a Thylacine or Numbat (*Myrmecobius fasciatus*). It was older than the others, and produced by an engraving technique rather than by pecking. It is unlikely that the stripes were merely a stylized depiction of body contour because they were absent from the other species portrayed.

The cave art of Arnhem Land and its environs has been little explored. On the Upper East Alligator River, E. J. Brandl found an exposed vertical wall bearing what appeared to be a very attenuated Thylacine (Brandl 1972). Under the shelter of a nine–metre sandstone rock at Deaf Adder Creek was a 35 cm portrait of a Thylacine, painted in red (ibid). Further along the creek was the drawing of a 35 cm striped creature with a long tail and forelegs much longer than its hindlegs. Another red ochre painting, 110 cm long, was discovered in an extremely inaccessible position at the Cadell River Crossing. None of the aborigines interviewed could identify the subjects even from legends (ibid). According to later press reports (Anon. 1974 a,b), ten similar paintings, including one of a female feeding her young, have been discovered in a valley in the Mt Brockman massif of the Kakadu National Park by Messrs. Chaloupka and Woerle, who may have been the first white men to enter the valley.

The classification of fossil Thylacinids has been complicated by the failure of some nineteenth century taxonomists to recognize variation within the species *T. cynocephalus*. When, for instance, DeVis (1894) received the left side of a skull found at Ellagowan, on the Darling Downs, he compared it to just a single skull of the modern species. Finding the fossil skull slightly larger, he pronounced it to be a separate species, *Thylacinus rostralis. Thylacinus breviceps* was supposed to have been a small species, but was in fact based on an immature specimen (Moeller 1968). DeVis (1894), however, recognized that *T. major* from the Wellington Caves was identical to the widespread "cave tiger", *T. spelaeus,* which Ride (1964) showed to be the same as the modern species.

Ride (1964) examined specimens, both fossil and recent, from all over Australia. By taking eight measurements on each specimen he was able to demonstrate that *T. spelaeus* falls within the wide range of variability of *T. cynocephalus*. He was unwilling to pass judgment on *T. rostralis,* and remarked that the specimens from southwestern Australia appeared smaller than average. Lowry (1972) measured the specimens' teeth, which do not grow continuously, to avoid complications caused by measuring immature skulls. The southwestern fossils were 10% smaller than the mean, and Eucla specimens 13% smaller, but still within an acceptable range for a single species. On present evidence, therefore, all Quaternary thylacines should be regarded as belonging to the modern species. However, the description by Woodburne (1967) of *Thylacinus potens,* from the Miocene Alcoota fauna of the Northern Territory, suggests that this species was distinct, and more massive than *T. cynocephalus*. It also differed in several other anatomical details.

Fossils have been found in Tasmania at Scotchtown (Gill 1953), and in Victoria at Keilor (Marshall 1974), in a cave near Gismore (Gill 1953), at Buchan (Hope 1974), and at Mt. Hamilton, 175 km west of Melbourne (Wakefield 1963). Remains were also discovered at Nelson (Colliver 1938), while in nearby South Australia, Pleistocene specimens were unearthed at Victoria Cave, near Naracoorte (Smith 1972). Those discovered at Henschke's Quarry Cave nearby are 32,000 years old (Pledge 1974), and found in association with species typical of the dry sclerophyll and dense heath environments which still prevail in the district. Other South Australian finds were at the Buckalowie Caves near Carrieton (Jones 1923), and a single abnormal tooth from Fromm's Landing, dated at 3,800 B.P. (Mulvaney *et. al.* 1964, and Archer 1971).

In New South Wales, specimens have been recorded from the Wellington Caves (Owen 1877), the Wombeyan Caves (Broom 1896), and Lake Menindee (Tedford 1967), the last of which was found in a stratum between two layers of charcoal, dated at 26,300 and 18,800 B.P. respectively. Queensland sites include Cement Mills, Gore (Bartholomai 1977), as well as Ellangowan, Pilton, Gowrie Creek and Chinchilla (DeVis 1894).

In Western Australia, very recent bones have been unearthed at Tunnel Creek in the Napier Range, and older ones at Murray Cave, 40km north of Perth, and at Horseshoe Cave near Madura (Archer 1974). Undated bones were found with those of a Dingo in the Monajee Cave, Cape Range (Kendrick and Porter 1973). In the Coronation Cave in the southwest a skeleton was found complete to the smallest bones (Glauert 1954), and other specimens were obtained from the Mammoth and Museum Caves (Glauert 1954), the Moondyne Cave, Augusta (Howlett 1960), and the Murra-el-elevyn Cave on the Nullarbor (Partridge 1967). *Sarcophilus* was associated with many of these sites. In Webb's Cave, near Eucla, Devils left not only their bones, but also coprolites, one of which contained the molar of a Thylacine, which had probably been scavenged (Cook 1963). Other remains came from an aboriginal midden at Devil's Lair (Bayne, Merrilees and Porter 1975).

The species once extended to Papua New Guinea. Susan Bulmer excavated part of a Thylacine mandible at Kiowa, in the Eastern Highlands (Van Deusen 1963).

By far the most spectacular find was made in October, 1966, at Thylacine Hole on Mundrabilla Station, 110km west of Eucla, Western Australia (Lowry and Lowry 1966). A near-vertical shaft 11½ metres deep led to a broad cave littered with the remains of native animals. Countless individuals had fallen down the shaft and perished, their bodies mummifying in the atmosphere which, in the main chamber, showed a temperature of 19°C and a relative humidity of 67%. Apparently carnivores had been attracted by the odour and had also fallen in. Their bones were far fewer than those of the herbivores. 140 metres from the entrance lay a completely intact mummy of a Thylacine. Much of the soft tissue had decomposed to a tarry substance with a musty odour, but the tongue and left eyeball were still perfectly recognizable. The skin and hair were perfectly preserved. Even the stripes were clearly visible. At first it was hoped that the specimen was freshly dead, but three separate radiocarbon tests have established an age of 4,600 to 4,700 years (Lowry and Merrilees 1969, Merrilees 1970). So far, no further study has been made of the specimens from Thylacine Hole, nor have the innumerable other holes on the Nullarbor been thoroughly searched.

ALLEGED MAINLAND SIGHTINGS

Every so often the suggestion is made that the Thylacine is not yet extinct on the mainland. "Wild dog" and "Tiger" folklore is rife throughout Australia's rural communities, stories no doubt bolstered by sightings, under conditions of unusual lighting, of feral Dogs bearing abnormal patterns of coloration or mange. North Queensland has been the focus of 'Tiger" reports for decades. Troughton (1967) and Heuvelmans (1962) have reviewed the subject in some detail, with additional information being supplied by Anon (1969), and Makeig (1970). These reports are not discussed further in this paper, because nearly all describe the animal as "cat-like", i.e. with a short face. The footprint described by Scott (1872) from Cardwell is unlike that of any known Australian predator.

However, in January, 1973, the press reported two sightings of a striped, dog-like creature in the Longland Gap area of the Atherton Tableland (Anon. 1973*a,b*). Enquiries by Dr. J. Winter brought the number of reported sightings to four, all between December, 1972, and mid-January, 1973, with no more than a mile separating the furthest sightings. The descriptions were detailed and consistent, and fitted the Thylacine fairly accurately. One sighting occurred in the early morning, suggesting that the stripes were not just the result of nocturnal lighting. John Winter, however, was unable to observe the subject himself (J. Winter, personal communication).

1967 and 1968 produced an outbreak of "Thylacine" sightings in South Australia. According to Harris (1968), 100 people claimed to have seen the animal at 35 different localities between the Coorong and Nelson, Victoria, but especially in the Beachport–Millicent region. Some of the sightings are suggestive of the Thylacine. They concern Dog–like animals with a striped body and a tapering tail like a kangaroo's. One of the witnesses was a National Park Keeper, and another drew a sketch which looked remarkably like a Thylacine. Harris' article included a photo of a Dog-like creature with broad, white stripes on its forequarters. This would rule out a Thylacine, and since the hindquarters were completely white, it was probably a lighting effect. I once saw an unidentified press cutting of the same photograph which, in that case, was alleged to have been taken by Miss Rita Martin in 1964 in a completely different area. The subject was called the "Ozenkadnook Tiger".

One puzzling account is that of S. J. Paramonov (1968) who, in 1949, saw on the Bourke-Wanaaring Road, a Dog-like creature which "was uniformly grey–brown, with short hairs; the strange tail, extremely wide at the base, seemed to be a continuation of the hind quarters; the hind leg was strongly marked with almost horizontal stripes", i.e. the hindquarters were the same as a Thylacine's but the stripes were not. "The most remarkable feature was the strange manner of running: although the animal was swinging regularly sideways, the hind part of the body made a kind of bobbing, up and down movement; the impression was as if the animal was drunk." Paramonov made notes at the time, but was unwilling to publish them until the Thylacine Hole carcass was publicized. The observations were made at a range of 15-20 metres, in broad daylight. However, this barren environment is the last place one would expect to find a Thylacine.

Recently, a group of eight Thylacines, including a female with pouch young, were reported from near the New South Wales-Victorian border. The location was not disclosed, but a number of photographs have become available, one of which was published in the daily press (Anon. 1977*a*). The picture is clear and convincing, and if genuine its importance cannot be overestimated.

Finally, Kolig (1973) reported that the Wolmadjeri people of the Southern Kimberleys have a word, *waldagi,* for a sinister, Dog-like animal quite distinct from the Dingo (*moran*) and the Dog (*gunjar*). It was definitely not a spirit creature, and he speculated that it might be the Thylacine.

EXTERNAL ANATOMY

The Thylacine is the marsupial equivalent of the Coyote — a Dog-like animal of fairly solitary behaviour, the size of a large German Shepherd Dog (Table 1). If Lt.-Gov. Paterson's measurements were correct, the first specimen shot must have been outsize. Even Harris' type

Table 1. External measurements.

Authority	Paterson	Harris (1808)	Thomas (1888)	Australian Museum*		
Sex	unknown	unknown	female	male	male	male
Head–body (mm)	1390	1180	1080	1054	1054	1016
Tail (mm)	508	620	530	521	483	407

* Measured by the author.

specimen was rather large. The 13 skeletons examined by Moeller (1968) may have belonged to a younger sample than the Australian Museum collection, for their average size was: head and body—806mm, tail—492mm. The tail ranges from 40% to 60% of the head-body length.

Paterson's Thylacine weighed about 20kg. Moeller (1968) could find only two other records of weight—15 and 27kg, but he was able to reconstruct from a number of skeletons a mean weight of 25kg, with a range of 15-35kg.

Table 2. Length of limbs as percentage of trunk (after Moeller 1968).

Species	Forelegs	Hindlegs
Thylacinus	76	94
Dasyurus viverrinus	74	102
Dasyurus maculatus	73	93
Sarcophilus	96	101
Canis lupus	99	106
Neofelis	76	94

The species' build is variable enough for bushmen to distinguish two types, the "bulldogs" and the "greyhounds" (Wart 1919), but the body is only superficially canine. Moeller (1968) measured skeletons of a number of species and compared the length of the limbs to that of the trunk i.e. the spinal column minus the tail. As revealed in Table 2, the Thylacine's legs are much shorter than a Wolf's and more like those of the dasyures. Consequently, it would be expected to rely more on stalking than on prolonged chasing to catch game, and would be more at home in dense forests than open plains. In fact, of all eutherian carnivores examined by Moeller, the most similar in limb proportions is the Clouded Leopard, *Neofelis nebulosa,* an inhabitant of the Asian jungles. Moeller also showed that the greatest lengthening of the Wolf's limbs occurred at the wrist and ankle, to preserve the mechanical efficiency of the muscles.

The Thylacine is digitigrade and the feet show many differences from those of dasyurids

Table 3. Proportions of foreleg (%) (After Moeller 1968).

Species	Humerus	Radius	Metacarpals
Thylacinus	44	44	12
D. viverrinus	45	45	10
D. maculatus	46	44	10
Sarcophilus	43	44	13
C. lupus	40	42	18
Neofelis	48	39	13

(Pocock 1926). Striations are completely absent. On the manus, the toes are short and the claws short and blunt. The pollex is especially short, and its pad not clearly distinguished. Unlike the case in the Dog, there is absolutely no webbing to hold the toes together while running. The three interdigital pads have been fused into a single, trilobed plantar pad. Behind it is a long area of naked skin separating it from a carpal pad twice as long as broad. Hair encroaches on the carpal region, but all the skin around the plantar pad is naked, and the creases between toes, 2, 3 and 4 appear granular. Contact with the ground is made by the digital pads on the tips of the toes and the anterior part of the plantar pad. These pads are thickly covered with finely pointed papillae.

The pes is much longer. The hallux is absent, but the four remaining toes are long and unwebbed. There is only a single plantar pad, bilobed posteriorly, but without any sign of a pad behind it. Instead, hair encroaches on the heel, leaving a long, narrow, irregular tract of bare skin stretching to the end of the heel (Fig. 2).

Table 4. Proportions of hindleg (%) (After Moeller 1968).

Species	Femur	Tibia	Metatarsals
Thylacinus	42	44	14
D. viverrinus	38	45	17
D. maculatus	43	44	13
Sarcophilus	42	45	13
C. lupus	39	43	18
Neofelis	42	43	15

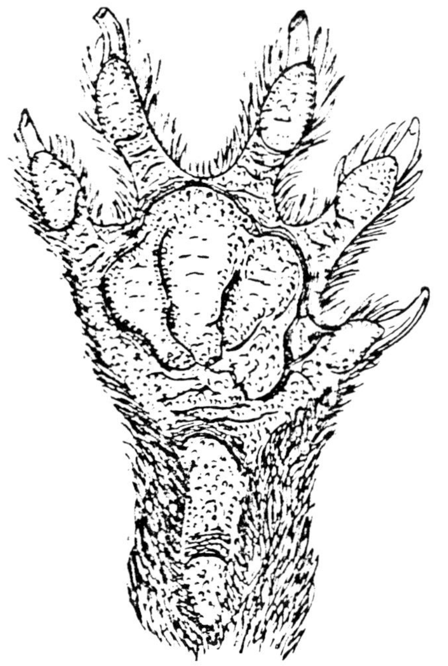

Fig. 2. Pes, with toes extended (from Pocock 1926).

The body hair is dense, short and crisp, with a grizzled, greyish brown colour tinged with tawny. From 15 to 20 black or dark brown stripes extend from just behind the shoulders to near the root of the tail, in a series of unbroken bands. Such a pattern is shared with only three other species: the Banded Hare Wallaby, *Lagostrophus fasciatus,* Hardwick's Civet, *Hemigalus derbyanus,* and the miniature Duiker, *Cephalophus zebra* (Renshaw 1938). As with all striped animals, there is much individual variation (Moeller 1968), but the stripes are always longest over the rump, where they descend the hip almost to the knee, and grow progressively shorter anteriorly until they disappear. The never reach all the way to the abdomen.

The shape of the hindquarters is very characteristic. The body is not separated abruptly from the tail, but tapers strongly like a kangaroo's, so that the tail is really an extension of the hindquarters. It is a pity that observers of alleged Thylacines do not pay more attention to this feature, which is not shared with any feral dog. The tail itself is very thinly haired, except at the base, and has a blackish tip and a slight crest above and below (Thomas 1888).

Superficially, the face is Dog-like. The large nostrils point outwards and forwards, surrounded by a naked rhinarium 1¼ times as broad as long, projecting beyond the edge of the upper lip. From the front it is as wide as it is high, but the lower third is constricted to form a philtrum with a pair of short, vertical grooves reaching up partway to the upper lip (Pocock 1926). The mean number of vibrissae are: 7 genal, 7 supraorbital, and 10 interramal (Lyne 1959). The mystacials are profuse but the medial antebrachials and anconeals are absent, as in the Devil. The ulnar and carpal vibrissae are also completely absent, being of little use to a cursorial animal.

There are whitish markings around the eyes and the bases of the ears (Le Souef and Burrell 1926). The pinnae of the ears are of a basic dasyurid-type (Pocock 1926). The anterior passage of the auditory meatus is a long, vertical cleft strengthened by a short, thick, cartilaginous crest (Fig. 3). The lower edge is formed by a thick, upstanding, curved ridge leading onto the cartilagious antitragus. An anterior ridge, directed outwards and downwards, forms the upper edge, and next to this is the supratragus. The body of the pinna is broad, smooth and hairless, with the edge unfolded. The ear cavity can be closed by folding down the supratragus, thus pushing the anterior ridge onto the posterior.

SKULL

There seem to have been few skull measurements recorded. Table 5 shows the measurements of 16 skulls examined by the author at the Australian Museum, Sydney. Some of the specimens were almost certainly immature, and few were sexed.

Moeller (1973) made an allometric comparison of the Thylacine with a hypothetical *Dasyurus viverrinus* of similar brain-case length. Nearly every dimension was longer in the Thylacine. The most outstanding was the intertemporal breadth, which bulged out 196.6% more than in the dasyure to accommodate a brain 73.82% larger. The posterior palatal foramina

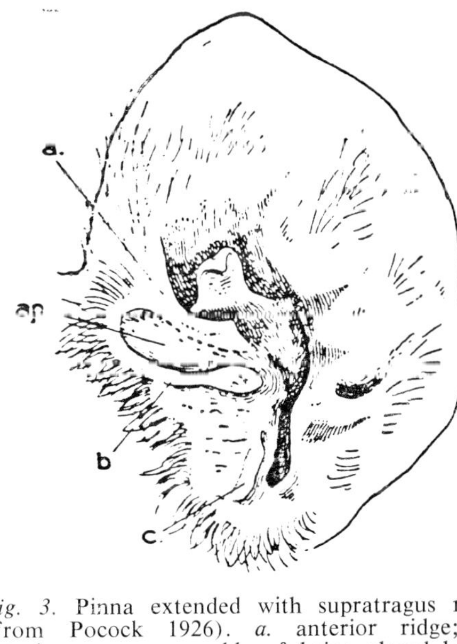

Fig. 3. Pinna extended with supratragus raised (from Pocock 1926). *a.* anterior ridge; *ap.* anterior passage, capable of being closed by the raising of ridge *b* against *a; c.* ridge strengthening border of inferior passage in front of auditory orifice.

Table 5. Measurements of the skull (mm).

Basal Length	212.4	Intertemporal Breadth	30.1*
Palatal Length	55.0*	Interorbital Breadth	16.6
Anterior Palatal Foramen	6.8	Nasal Length	41.4
Posterior Palatal Foramen	12.3*	Nasal Breadth	12.9
Zygomatic Breadth	56.2	Molar Length	46.4^{+}

Measurements taken on 16 specimens from the Australian Museum. * n=14 $^{+}$ n=13.

were twice as large. The nasals were 53.4% larger due to the size of the teeth and, when the tooth–row was compared to the basal skull length, it was found to be 49.2% longer in the Thylacine. The cheekteeth alone were 64.4% longer.

Moeller (1968) also showed how, despite the presence of the typical marsupial cranial features, the facial region is similar to that of a Wolf although, because of the smaller teeth, a better comparison would be with a Coyote or Jackal. The skulls are quite variable both in absolute size and in the proportions of the parts, but no more so than in the Wolf. The young of both species differ considerably, but they grow more similar with age. The sexual differences are due wholly to the slightly larger male skull, with its correspondingly longer face.

The coronoid process of the mandible is much slanted back. The nasals are long and slightly expanded posteriorly. The fenestrations of the palate are large. The anterior one is long and narrow, ending behind the centre of the canine. The posterior one stretches from opposite the second molar to just behind the last (Thomas 1888).

The frontal sinuses are very large, causing the characteristic bulge in the forehead. Their size is reliably correlated with the length of the cranium, and brain capacity (Moeller 1968). The para-occipital processes are much reduced, but the postorbital processes are strong due to the relatively forward directed eyes (Tate 1947). Behind them the skull is markedly constricted. The large, oval infra–orbital foramen has a small bony bar dividing it into two unequal parts (Thomas 1888). The antorbital canal which, in dasyurids, lies wholly within the maxilla, lies so far back in the Thylacine that its posterior margin is bordered by the jugal (Tate 1947).

Archer's (1976b) detailed study of the basicranium can only be touched on here. *Thylacinus* resembles didelphids of the New World in the absense of a mastoid epitympanic sinus and the presence of a well–developed basioccipital, and approximates them in the way the dorsal surface of the internal carotid is enclosed for only a short distance. The pituitary fossa is small or absent. The squamosal epitympanic sinus is of moderate size and, unlike the condition in most marsupicarnivores, the alisphenoid does not contact the parietal. The postglenoid fossa is unusual in that it is split into two branches: one in the squamosal and the other between the squamosal, periotic and, sometimes, the alisphenoid. However, either branch may be absent in individual specimens.

The one thing in which the Thylacine is much inferior to the eutherian Wolf is the size of its brain. The average length of 30 Thylacine skulls measured by Moeller (1968) was 207.1mm, and the average capacity of the braincase only 53.5cc, compared to 134.4cc for the mean of an equal number of wolf skulls, of an average length of 213.4mm. The braincase itself is tubular, and "telescoped" posteriorly, with the

Table 6. Dimensions of the teeth (crown lengths in mm).

Upper Teeth										
Authority	i_1	i_2	i_3	p_1	p_2	p_3	m_1	m_2	m_3	m_4
Tate (1947)	3.0	3.1	5.2	6.3	8.0	10.7				
Lowry (1972)							11.6	15.1	17.8	
Lower Teeth										
Authority	i^2	i^3	i^4	p^1	p^2	p^3	m^1	m^2	m^3	
Tate (1947	3.4	3.2	4.2	6.1	8.4	12.0				
Lowry (1972)				6.2	9.2	10.8	9.6	11.9	14.1	15.8

basisphenoids and basioccipitals broad and flattened. The paraoccipitals are close to the small, almost transparent bullae, and the mastoid processes are compressed close to the glenoid processes (Tate 1947). Because of the small size of the braincase, strong sagittal and lambdoidal crests are developed to hold the jaw muscles.

DENTITION

The jaws are long, and the dental formula is 4/3 1/1 3/3 4/4 = 46 (Thomas 1888), with the teeth of moderate size, as Table 6 shows. Whereas in some of the more specialized dasyurids the tooth row has been shortened, in *Thylacinus* it is elongate. The last upper incisor is almost twice the size of the others. The premolars are narrow and increase in size posteriorly, being separated from one another by wide diastemata. Features of the adult dentition are described by Archer (1976c).

Flower (1867) described a deciduous premolar in a young female with a head-body length of only 23cm and a head length of 7.35cm. No teeth were showing above the surface of the gums, but in the upper jaw the crowns of four developing incisors were calcified, as were the apices of the crowns of the canine, premolars, and the first two molars. The lower jaw told a similar story. From the outside the only evidence of teeth was a bulge in the gum which concealed a deciduous premolar, $2\frac{1}{2}$mm long, triangular and pointed, but hard and white, unlike the brown germs of the permanent teeth. It was only 26-30% of the size of the last permanent premolar.

The molars are adapted to slicing and crushing. In the upper molars the protocone is reduced, with a slightly trenchant posterior border, and the metacone is much enlarged, forming a long cutting edge. In contrast to other dasyurids, there are almost no external stylar cusps. In the lower molars the metaconid is completely absent. The cutting ridge passes from the paraconid, through the protoconid, to the hypoconid, leaving a fairly well-developed, shelf-like entoconid–hypoconid talonid (Bensley 1903). This condition, as Ride (1964) notes, is quite different from that of dasyurids in which there is a large metaconid.

POST–CRANIAL SKELETON

The number of vertebrae is similar to that of *Dasyurus* and *Sarcophilus* — 7 cervical, 13 thoracic, and 6 lumbar. There are only two sacral vertebrae instead of the three in the other two genera, and the tail runs to 25 vertebrae rather than 20 or 21 (Moeller 1968). The two most notable features of the post-cranial skeleton are the vestigal clavicles and epipubic bones. As the species is adapted to running rather than climbing, the clavicles are merely narrow, curved rods 5cm long (Cunningham 1882). All that remains of the epipubic bones are "two small, oblong, flattened fibro-cartilages, imbedded in the internal pillars of the abdominal rings, and [they] appear each as a thickened part of the tendon of the oblique abdominal muscle, which forms the above pillar" (Owen 1843).

SOFT ANATOMY AND REPRODUCTION

Cunningham (1882) published a full description of the species' musculature, of which only the cardinal points can be reviewed here. The reduction of the clavicle has required the rerouting of the muscles normally attached to it. The cleido–mastoid is partly inserted into the inner half of the clavicle and partly joins the clavicular region of the deltoid muscle. Of the latter muscle, the scapular section is enlarged, and there is no spinal section such as occurs in dasyurids. The subclavian muscle cannot be inserted into the upper margin of the clavicle and instead runs under the clavicle, changes direction, and inserts into the fascia covering the supraspinatus. The supinator longus is very poorly developed, and the anconeus externus is absent. However, as an adaption to running, the triceps is very large, and the pectoralis minor is relatively well developed compared to its generally small size in dasyurids. The pectoralis quartus is rudimentary.

Beddard (1891) examined a single brain and noted that, although it was more complex than in dasyurids, there were fewer folds than in a macropodid of the same size. The cerebral hemispheres were not large enough to overlap the cerebellum. They diverged posteriorly, as in other marsupials, partly exposing the *corpora quadrigesima*. The total length from the front of the olfactory lobe to the posterior end of the cerebellum was 76mm. The cerebral hemispheres were 48mm long and 26mm high.

Moeller (1970) made a much more thorough study on six brains, though he was handicapped by very poorly preserved material. He estimated the whole animal weights at 15-35kg. Mean brain–case length was 78.0mm (range 68.0–91.2mm, $n = 32$) and mean capacity 53.4cc (range 46.5–62.00cc, $n = 30$). A *Dasyurus* of similar weight would have had a brain–case of only 28cc. The Thylacine brain is gyrencephalic. The neocortex is large and folded. Despite the importance of the sense of smell, the olfactory bulbs are small compared to those of dasyurids, and are somewhat flattened laterally. The relatively small olfactory tubercle is truncated at the rear. The piriform lobe is bordered by a deep furrow, and the telencephalon arches upwards dorsally and laterally. The roof of the midbrain is completely overlain by the cerebrum. The sulcus rhinalis is deep and slopes backwards and downwards. Behind it are three more or less parallel furrows. The first is the orbital sulcus, separating off the rostral lobe, with two others branching off. The second is composed of two or three furrows combined, and the third divides at the base. Although the cerebellum is large, it does not reach the level of the cerebrum. The vermis bends laterally at the rear. There are no paraflocculi.

The only other soft parts studied have been those concerned with reproduction. The area around the scrotum is bare and insunk to form a cup–like "pouch" composed of a flap of skin in front and on the sides, gradually disappearing posteriorly. Pocock (1926) thought it might prevent the scrotum swinging during the chase. Perhaps it also helps in thermoregulation.

The female system is similar to that of dasyurids. Vestigial Wolffian ducts occur in the young, but the situation in adults is uncertain. The uterine necks run side by side, and are half the length of the uterine body. The culs–de–sac are noteworthy. They are a fifth of the anterior–posterior length of the vaginal complex. To quote Pearson and de Bavay (1953): "It is important to note that even in these extremely immature specimens the culs–de–sac have reached a higher stage of development than that found in mature specimens of *Antechinus*."

The female has a well-developed pouch opening at the rear, with four teats arranged in two rows. The two on each side are close together and well separated from the two on the other (Pocock 1926). The surrounding hair is directed towards the pouch. Owen (1868) measured a pouch which contained three young, each 30cm long including tail. Pouch depth was 20cm, with the width of the aperture 9cm, extending forward to a maximum breadth of 15cm. The teats were each 10 cm long.

The three young observed by Gunn (1863) were so big they could only just fit into the pouch which hung so low it "seemed almost a deformity". When giving suck, the female used to lie on her side like a Dog. The pouch sphincter relaxed so that the young, lying outside, had easy access to the teats. When disturbed, they would crawl in upside down, and the mother used to lower her hindquarters and push them against the side of the cage to allow the young extra purchase with their hind feet.

At this stage they would have been kept in a nest, one of which was reported by Mr. Flinty, who discovered two young "in a dry fern-bed under the drooping and still attached dead fronds of a tree–fern. These reached the ground all around the butt, thus forming a natural tent–like shelter and a perfect camouflage for the youngsters" (Le Souef and Burrel 1926).

When Guiler (1961a) examined the records of scalp bounties he was able to judge the breeding season. The records included 55 "half–grown" animals, 52 "pups" out of the pouch when caught, and 43 other young of unspecified age. Young were taken in all months, but the number of "pups" peaked in May and July-August, followed by peaks of "half-growns" a month later. Therefore, either the breeding season lasted four months or, less likely, there were two breeding seasons separated by a gap of two months. Probably a female would start breeding in autumn and have a second litter after the first had been weaned. In any case, there was some out–of–season breeding.

Boardman (1945) examined some pouch young preserved in museums. One litter consisted of two males and two females of a crown–rump length of 7.5mm. They had no hair except for a tiny amount on the head. The stripes were absent, but the skin contained a greyish pigment. The lips were sealed at the sides, and the eyes were unopened, although eyelashes were present. The females each possessed a small, V-shaped pouch with two nipples. An older female with a head–body length of 288mm and a tail 119mm long had close, coarse hair all over her body, but longest on the head. Stripes were present. The hair radiated in all directions from the pouch, and there were slightly more facial vibrissae. The lips were fully formed and the eyes open. It was probably old enough to have left the pouch.

BEHAVIOUR

The behaviour of the Thylacine can be reconstructed only from bushmen's anecdotes and accounts of captive animals. In January, 1963, Ronald Gunn shipped a female with three pouch young to London Zoo and made some notes on its behaviour (Gunn 1863). Twenty years later a Mr. Crowther of Launceston caught an adult pair which arrived in London on 14 November, 1884 (Renshaw 1938). The male died on 5 February, 1891, and was promptly dissected by Beddard (1891). The female lasted until 2 April, 1893, when she was at least $8\frac{1}{2}$ years old (Flower 1931). This appears to be the record for longevity. Although several were displayed in European zoos, only Renshaw (1938) seems to have made any observatoins. However, a film held by the Tasmanian Government of the last surviving captive allowed Moeller (1968) to observe something of its behaviour.

Gould (1863) claimed that the animal was very shy in captivity, and when alarmed would dash around uttering "a short guttural cry resembling a bark." LeSouef and Burrel (1926) stated:

"When excited it makes a series of husky, coughing barks, the breath being indrawn with a wheeze". Apart from that, they seem to have been rather silent. The one observed by Renshaw "was active during the early part of the day, running to and fro in its exercise yard, the head held low like a hound's on scent. The beast often paused abruptly, as if to reconnoitre, standing motionless with head raised; it took little notice of the attendent entering the yard, as if half-blinded by the sun. When the light became powerful it often retreated to the inner den, where it curled itself up like a dog; but it would also recline to bask in the sun, a strange habit for a semi-nocturnal animal. In sleep it lay on one side fully extended; the upper-most ear remaining erect".

Gunn (1863) once watched a female spring 6 to 8 feet (1.8-2.4m) from the floor of the cage to the top of the wall and from cross–beam to cross–beam with remarkable agility. But the most remarkable aspect of behaviour was recorded on the film examined by Moeller, who recognized two methods of locomotion. The first was the plantar walk shared with most mammals, i.e. diagonally opposite limbs moving alternately, but with the whole pes, even the long heel, touching the ground. (It would not, of course, have been able to run like this.) The second was a bipedal hop. The animal would stand upright, its forelegs in the air, resting on its long hindfeet, and using the tip of its tail as a prop. In such a posture it looked remarkably like a kangaroo, and occasionally it would give a short hop.

For their size they must have been savage fighters. According to Hugh Mackay (quoted by Le Souef and Burrell 1926):

"A bull–terrier was once set upon a wolf (thylacine) and bailed it up in a niche in some rocks. There the wolf stood, with its back to the wall, turning its head from side to side, checking the terrier as it tried to butt in from alternate and opposite directions. Finally, the dog came in close, and the wolf gave one sharp, fox-like bite, tearing a piece of the dog's skull clean off, and it fell with the brain protruding, dead."

There are, however, only two records of attacks on human beings in the wild. The *Scottsdale North–Eastern Advertiser,* early in 1900, told of a Miss Priscilla Murray of South Springfield, near Mt. Scott, who was washing at a bench in the open when a one–eyed Thylacine, possibly mistaking her moving right arm for a mammal, seized it in its jaws. As she tried to push the beast away, her left hand was also bitten, but when she accidently trod on its tail it let go. She was to carry the scars of several of the 32 teeth–marks for years (Troughton 1967).

On 22 May, 1872, the *Sydney Morning Herald* carried the headline: "Attacked by a Tasmanian Tiger" and continued:

"A correspondent writes from Jericho informing the *Hobart Town Mercury* that on Saturday 11th instant, Mr. James Jones was proceeding along the road towards the lakes, and, when on a run known as Poolmarsh, he was attacked by a large tiger which came out of the scrub. Mr. Jones had a large stick in his hand, and succeeded in keeping the animal off, but he was followed by it for a distance of about half a mile. Mr. Jones says that the tiger was the largest he ever saw, standing about five feet high. He had a dark collar, and was of a different species to the common tiger, which has white marks across its back, and a very yellow collar" (Whitley 1973).

The species' preferred habitat was never properly determined. It appears to have ranged throughout Tasmania, but Le Souef and Burrell (1926) claimed that it preferred hilly country, lying up in timber and scrub by day, and hunting on the edge of the thickets in the late afternoon and evening. Lord (1927) said much the same, but in neither case was proof given, and at that time the species had been exterminated from the cleared areas, and was suffering persecution in its last redoubts.

Information on feeding habits is equally scanty. Some individuals must have been fairly adroit, for the earliest accounts refer to the eating of spiny ant-eaters (Troughton 1967). One bushman mentioned the discovery of a den half–full of bones, including those of a half–grown calf. Even if the calf had been scavenged, this implies a considerable strength on the part of the predator (Wart 1919). Gunn (1863) claimed that they invariably eat only what they themselves have killed, and never return to the site of a kill, with the result that farmers found the baiting of sheep kills unproductive. However, this might have reflected merely the vulnerability of the domestic stock combined with harassment by man. Certainly, the animals he shipped back to England readily ate the meat he provided. Because Thylacines did not find it profitable to eat much of a sheep, the legend has arisen that they prefer to drink blood. This has been accepted by such authorities as Lord (1927) and Guiler (1958) who at least admits that they may also eat the nasal tissue, liver and kidney fat, and occasionally some meat. The latter hypothesis was the result of the discovery of the unusual sheep killings mentioned earlier (Guiler and Meldrum 1958), but it was not proved that a Thylacine had been responsible, nor do three such killings constitute evidence for this as a normal practice.

The native macropodids were undoubtedly eaten, but no description of an individual hunt exists. Instead, we are left with generalisations which derive ultimately from the statements of farmers, bushmen, and similar questionable sources. Nevertheless, some of them are plausible. Sharland (1939) claimed that the Thylacine hunts mostly by scent and wears down the prey by chasing it for hours, a method which would be appropriate only in open habitat, and was perhaps suggested by analogy to the Wolf. Le Souef and Burrell claimed that they trot until the quarry shows signs of exhaustion, and then rush it. Finally, Lord (1927) considered that they cover large areas of ground on hunting forays, returning to the lair by day. "During the breeding season" he writes, "a male Thylacine has been known to follow the same route across many miles of country, and one particular animal used to regularly leave a trail of slaughtered sheep along the same line every year, but he was trapped eventually." He also stated that packs of four or five had been noted, although the animals were normally solitary. This no doubt refers to a female with her almost mature young.

One of the most popular photographs of the Thylacine shows a captive of Hobart Zoo in profile with its jaws stretched so wide apart that they stand in almost a single vertical line. David Fleay once described to the author how, right after he had taken the photo, the subject turned and sank its teeth into his thigh. This would suggest a yawn–threat similar to that of *Sarcophilus* (personal observation), but it is impossible to determine anything else from its biting of Mr Fleay.

DISCUSSION

The anatomy of the Thylacine has been recorded almost as well as that for most other Australian mammals, with the possible exception of the macropodids, and it is possible to discuss its position within the Marsupialia. The Thylacine is, or was, an animal with a fairly primitive reproductive system combined with many advanced features (large size, digitigrade stance, enlarged brain, lengthened jaws and sectorial molars) paralleling those of canids. They are all adaptions to cursorial hunting of large prey and represent the end–product of a long period of evolution from a primitive marsupicarnivore stock.

Most earlier authorities (e.g. Le Souef and Burrell 1926, Troughton 1967) tended to group it with the Dasyuridae at least in part because the dasyurids are the only carnivorous, polyprotodont, didactylous group in Australia. Most of the distinctive characteristics of the Thylacine are specializations of features present in all Marsupicarnivore stocks, not to mention eutherian carnivores. Despite its size, the Thylacine's limb proportions are similar to those of the dasyurids, as is the reproductive system. On most scores, *Thylacinus* is no further removed from the Dasyuridae than is *Myrmecobius*. Tate (1947) considered it to be related to the New Guinea species of *Murexia* on the basis of the progressive increase in size of the posterior molars, and the tubular brain–case with post–orbital constriction which is found in some older male specimens of *Murexia*. However, these features are almost the only points of similarity between the genera, (Archer 1976c), and his theory has found few supporters.

On dental grounds, the relationship becomes tenuous. The formation of the shearing crest is different from that of all dasyurids, and the loss of the stylar cusps in a form with enlarged third premolars is also in contrast to the general Dasyurid trend, suggesting that the two groups separated very early. Bensley (1903) saw a closer relationship to the borhyaenids, a view partly upheld by Archer (1976c). On the other hand, basicranial evidence tends to point to didelphid rather than Borhyaenid affinity (Archer 1976b). It would be premature, however, to suppose that Thylacinids originated outside of Australia. Primitive marsupicarnivore fossils have recently been discovered in central Australia (Archer 1976a), and the last word on the subject has yet to be written.

The preferred habitat of the species and its preferred method of hunting are in doubt. The widespread fossil distribution proves that, even allowing for climatic change, its range of habitats was broad, a fact confirmed by the fossils with which it is associated. Similarly, its alleged preference for dense forest is quite likely a result of persecution. The idea that it used to hunt by prolonged chasing is not supported by hard evidence and, in any case, is likely to have been based on the pursuit of relatively helpless livestock. The animal's limb proportions, similar to those of a dasyurid, would be better adapted to stalking under dense cover followed by a sudden burst of speed. There is no reason to doubt, however, that on the open plains a Thylacine could run down peramelids and small macropodids, even if the large kangaroos presented a greater problem. It would certainly have been a better runner than *Sarcophilus,* with which it was often associated. Probably, as one of Australia's largest mammalian predators, it was flexible in both behaviour and habitat, but thick forest may have been the preferred habitat.

Its extinction on the mainland, as well as on New Guinea, can probably be blamed on the Australian Dingo (Archer 1974) and New Guinea wild Dog, both of which are adapted to open country. It is possible that the Thylacine might have survived in dense forest if they were its only competitors, but the additional presence of man was probably critical. To this must be added the decline in natural prey. Rolls (1969) has assembled evidence that at the time of European settlement kangaroo populations were less dense on the mainland than in Tasmania, and only increased once the Dingo and Aboriginals had been removed.

The reasons for the decline in Tasmania must remain conjecture. If hunting pressure was not directly responsible, it at least put such a strain on the species that it was unable to cope with other pressures. Disease may have been the precipitating factor, if Guiler's (1961b) informant was correct. Whatever the cause, the chance to investigate it has been irretrievably lost. The early days of this century were not the age of the field ecologist. Unlike the case with *Sarco-*

philus, which was also persecuted, there appear to have been no sizeable breeding muclei left from which Thylacine populations could grow back to their original size. The extensive campaign of poisoning mentioned by Fleay (1946) and Griffith *et al.* (1972) probably prevented recovery. Its range, in fact, probably shrunk in the post–war years, because the areas where Fleay discovered footprints yielded nothing during Griffith's search.

The Old Davey Track prints (Muir 1961) constitute the best evidence for the species' continued existence. They were fully described by the discoverers and identified by a qualified zoologist. Similarly, some of the many alleged Tasmanian sightings may be genuine, but the reports of similar questionable sightings on the mainland should act as a warning. It can be argued that the latter reveal only the unreliability and suggestibility of amateur witnesses, in which case the same would apply to the Tasmanian reports. Certainly, the existence of mainland Thylacines requires proof before it can be accepted, but a necessary preliminary is the collection of all relevant data, some of which has been presented in this review. Many more alleged sightings are made in Tasmania than are actually published. (Griffith *et. al.* heard of well over 100). The Tasmanian fraternity should give priority to investigating and collating these claims.

REFERENCES

Anonymous, 1969. Tigers, Devils, monsters and things that walk in the night. *Wildl. in Aust.* **6:** 54.

Anonymous, 1973a. Another strange beast sighted on Tableland. *Cairns Post* 17 Jan. 1973.

Anonymous, 1973b. Aloomba woman reports seeing Tableland beast. *Cairns Post* 18 Jan. 1973.

Anonymous, 1974a. Paintings of Tas. Tiger found in NT. *Sydney Morning Herald* 27 April 1974.

Anonymous, 1974b. The Northern Territory's prehistoric tiger. *Sydney Morning Herald* 6 May 1974.

Anonymous, 1977a. 'Tiger' now in baby mystery. *Sunday Telegraph* (Sydney) 27 March 1977.

Anonymous, 1977b. 'Extinct' Tiger seen claim two. *Sunday Telegraph* (Sydney) 21 Aug. 1977.

Archer, M., 1971. A re–evaluation of the Fromm's Landing Thylacine tooth. *Proc. Roy. Soc. Vict.* **84:** 229–34.

Archer, M., 1974. New information about the Quarternary distribution of the Thylacine (Marsupialia, Thylacinidae) in Australia. *J. Roy. Soc. W. Aust.* **57:** 43–50.

Archer, M., 1976a. Miocene marsupicarnivores (Marsupialia) from central South Australia, *Ankotarinja tirarensis* gen. et sp. nov., *Keeuna woodburnei* gen. et sp. nov., and their significance in terms of early marsupial radiations. *Trans. Roy. Soc. S. Aust.* **100:** 54–73.

Archer, M., 1976b. The basicranial region of marsupicarnivores (Marsupialia), inter–relationships of carnivorous marsupials, and the affinities of the insectivorous marsupial peramelids. *Zool. J. Linn. Soc. Lond.* **59:** 217–322.

Archer, M., 1976c. The dasyurid dentition and its relationships to that of didelphids, thylacinids, borhyaenids (Marsupicarnivora) and peramelids (Peramelina: Marsupialia). *Aust. J. Zool., Suppl.* **39.**

Bartholomai, A., 1977. The fossil vertebrate fauna from Pleistocene deposits at Cement Mills, Gove, southeastern Queensland. *Mem. Qd Mus.* **18:** 69–73.

Baynes, A., Merrilees, D. and Porter, J. K., 1975. Mammal remains from the upper level of a late Pleistocene deposit in Devil's Lair, Western Australia. *J. Roy. Soc. W. Aust.* **49:** 88–90.

Beddard, F. E., 1891. On the pouch and brain of the male Thylacine. *Proc. Zool Soc. Lond.* **1891:** 138–45.

Bensley, B. A., 1903. On the evolution of the Australian Marsupialia; with remarks on the relationships of the marsupials in general. *Trans. Linn. Soc. Lond.* **9:** 83–218.

Boardman, F., 1945. Some points on the external morphology of pouch young of the Marsupial, *Thylacinus cynocephalus* Harris. *Proc. Linn. Soc. N.S.W.* **70:** 1.

Brandl, E. J., 1972. Thylacine designs in Arnhem Land rock paintings. *Archaeol. Phys. Anthrop. Oceania* **7:** 24–30.

Broom, R., 1896. Report on a bone brecia near Wombeyan Caves, N.S.W. *Proc. Linn. Soc. N.S.W.* **21:** 48–61.

Calaby, J. H. and White, C., 1967. The Tasmanian Devil (*Sarcophilus harrisii*) in northern Australia in recent times. *Aust. J. Sci.* **29:** 473–5.

Colliver, F. S., 1938. Cave hunting in Victoria. *Vict. Nat.* **54:** 152–5.

Cook, D. L., 1963. *Thylacinus* and *Sarcophilus* from the Nullarbor Plain. *W. Aust. Nat.* **9:** 47–8.

Cunningham, D. J., 1882. Some points on the anatomy of the Thylacine (*Thylacinus cynocephalus*), Cuscus (*phalangista maculata*) and Phascogale (*Phascogale calura*), collected by H.M.S. Challenger, during the years 1873–1876; with an account of the comparative anatomy of the intrinsic muscles and nerves of the mammalian pes. *The Voyage of H.M.S. Challenger. Zoology.* V. Part II. Pp.192.

DeVis, C. W., 1894. A thylacine of the early Nototherium Period in Queensland. *Proc. Linn. Soc. N.S.W.* **8:** 443–7.

Fisher, B. D., 1974. Tasmanian Tigers still exist. *Sun* (Sydney) 25 Feb. 1974.

Fleay, D., 1946. On the tracks of the Marsupial Wolf. *Vict. Nat.* **63:** 129–35, 154–9, 174–7.

Fleming, A. L., 1938. The Thylacine. Reports on two expeditions in search of the Thylacine (Tasmanian Tiger or Tasmanian Wolf). *J. Soc. Preservation Fauna of Empire* **35:** 20–5.

Flower, S. S., 1931. Contribution to our knowledge of the duration of life in vertebrate animals. 5. Mammals. *Proc. Zool. Soc. Lond.* Pp. 145–234.

Flower, W. H., 1867. On the development and succession of the teeth in the Marsupialia. *Phil. Trans. Roy. Soc. Lond.* **157:** 631–41.

Gill, E. D., 1953. Distribution of the Tasmanian Devil, the Tasmanian Wolf, and the Dingo in Southeast Australia in Quanternary times. *Vic. Nat.* **70:** 86–90.

Glauert, L., 1954. The cave fossils of the southwest *W. Aust. Nat.* **1:** 100–4.

Gould, J., 1863. "Mammals of Australia". London.

Griffith, J., 1972. The search for the Tasmanian Tiger. *Nat. Hist.* **81:** 70–7.

GRIFFITH, J., MALLEY, J. AND BROWN, R., 1972. "The report of the search for the Thylacine that was conducted by Jeremy Griffith, James Malley and Robert Brown". Unpublished mimeographed report.

GRZIMEK, B. 1967. "Four–legged Australians. Adventures with animals and men in Australia". Collins: London and Sydney; translated by J. M., Brownjohn.

GUILER, E. R., 1958. The Thylacine. *Aust. Mus. Mag.* **12:** 352–4.

GUILER, E. R., 1961a. Breeding season of the Thylacine. *J. Mammal.* **42:** 396–7.

GUILER, E. R., 1961b. The former distribution and dedecline of the Thylacine. *Aust. J. Sci.* **23:** 207–10.

GUILER, E. R., 1966. In pursuit of the Thylacine. *Oryx* **8:** 307–10.

GUILER, E. R , AND MELDRUM, G. K., 1958. Suspected sheep killing by the Thylacine, *Thylacinus cynocephalus. Aust. J. Sci.* **20:** 214–5.

GUNN, R., 1863. Letter announcing the shipment of living Thylacines, with remarks on their habits. *Proc. Zool. Soc. Lond.* **31:** 103–4.

HARRIS, G. P., 1808. Description of two new species of *Didelphis* from Van Diemen's Land. *Trans. Linn. Soc. Lond.* **9:** 174–8.

HARRIS, SAMELA, 1968. Hold that tiger! *Walkabout* **34 (6):** 28–31.

HEUVELMANS, B. 1962. "On the track of unknown animals." Rupert Hart–Davis: London.

HOPE, JEANETTE, 1974. Palaeoecological reconstruction from small mammal faunas in the Buchan area, Victoria. Paper presented at the 15th General Meeting of the Australian Mammal Society, Monash University: Melbourne. May 1974.

HOWLETT, R. M., 1960. A further discovery of *Thylacinus* at Augusta, Western Australia. *W. Aust. Nat.* **7:** 136.

JONES, F. W., 1923. "The mammals of South Australia. I. Monotremes and the carnivorous marsupials". Government Printer: Adelaide, Sth Aust.

KENDRICK, G. W., AND PORTER, JENNIFER, K., 1973. Remains of a Thylacine (Marsupialia: Dasyuroidae) and other fauna from caves in the Cape Range, Western Australia. *J. Roy. Soc. W. Aust.* **56:** 116–22.

KOLIG, E., 1973. Aboriginal man's best foe? *Mankind* **9:** 122–4.

LESOUEF, A. S., AND BURRELL, H., 1926. "The wild animals of Australasia, embracing the mammalogy of New Guinea and the nearer Pacific Islands. With a chapter on the bats of Australia and New Guinea by Ellis LeG. Troughton". George G. Harrap: London.

LORD, C., 1927. Existing Tasmanian marsupials. *Pap. Proc. Roy. Soc. Tasm.* **1927:** 17–24.

LOWRY, D. C., AND LOWRY, JACOBA, W. J., 1967. Discovery of a Thylacine (Tasmanian tiger) carcass in a cave near Eucla, Western Australia. *Helictite* **5:** 25.

LOWRY, JACOBA, W. J., 1972. The taxonomic status of small forest Thylacines (Marsupialia, Thylacinidae), from Western Australia. *J. Roy. Soc. W. Aust.* **55:** 19–29.

LOWRY, JOCOBA, W. J., AND MERRLIEES, D., 1969. Age of the desiccated carcase of a Thylacine (Marsupialia, Dasyuriodea) from Thylacine Hole, Nullarbor Region, Western Australia. *Helictite.* **7:** 15–6.

LYNE, A. G., 1959. The systematic and adaptive significance of the vibrissae in the Marsupialia. *Proc. Zool Soc. Lond.* **133:** 79–133.

MAKEIG, P., 1970. Is there a Queensland marsupial tiger? *N.Qd Nat.* **37**(152): 6–8

MARSHALL, L. G., 1974. Late Pleistocene mammals from the "Keilor Cranium Site", southern Victoria, Australia. *Mem. Natl. Mus. Vict.* **35:** 63–86.

MERRILEES, D., 1970. A check on the radiocarbon dating of desiccated Thylacine (Marsupial "wolf") and dog tissue from Thylacine Hole, Nullarbor Region Western Australia. *Helictite.* **8:** 39–42.

MOELLER, H., 1968. Zur Frage der Parallelerscheinungen bei Metatheria und Eutheria. Vergleichende Untersuchungen an Beutelwolf und Wolf. *Z. wiss. Zool.* **177:** 283–392.

MOELLER, H., 1970. Comparative studies of the brain of the greater dasyurids (*Thylacinus, Sarcophilus* and *Dasyurus*) with regard to their evolutionary status. *Z. Zool. Syst. Evolutorsch.* **8:** 69–80.

MOELLER, H., 1973 Zur Kenntnis der Schädelgestalt großer Raubbeutler (Dasyuridae *Waterhouse*, 1938). Eine allometrische Formanalyse. *Zool. Jb. Anat.* **91:** 2257–303.

MUIR, E. COLLINE, 1961. Tracks of the Thylacine. *Vic. Nat.* **78:** 69–70.

MULVANEY D. J., LAWTON, G. H., AND TWIDALE, C. R., 1964. Archaeological excavation of Rock Shelter No. 6. Fromm's Landing, South Australia. *Proc. Roy. Soc. Vict.* **77:** 479–516.

O LOCHLAINN, C., 1946. "Irish street ballads." Sign of the Three Candles: Dublin.

OWEN, R., 1843. On the rudimentary marsupial bones in the *Thylacinus. Proc. Zool. Soc. Lond.* **11:** 148–9.

OWEN, R., 1868. "On the anatomy of the vertebrates. III. Mammals".

OWEN, R., 1877. "Extinct mammals of Australia. Vol. 1".

PARAMONOV, S. J. 1968. Is the Tasmanian Tiger (*Thylacinus*) extinct on the Australian mainland? *W. Aust. Nat.* **10:** 171–2.

PARTRIDGE, JEANETTE, 1967. A 3,300 year old Thylacine (Marsupialia, *Thylacinus*) from the Nullarbor Plain. *J. Roy. Soc. W. Aust.* **50:** 57–9.

PEARSON, J., AND DE BAVAY, J. 1953. The urogenital system of the Dasyurinae and Thylacininae (Marsupialia, Dasyuridae). *Pap. Proc. Roy. Soc. Tasm.* **87:** 175–99.

PLEDGE, N., 1974. Excavations in the Henschke's Quarry Cave, Naracoorte – a late Pleistocene fauna. Paper presented at the 15th General Meeting of the Australian Mammal Society, Monash University: Melbourne May, 1974.

POCOCK, R., 1926. The external characters of *Thylacinus, Sarcophilus* and some related marsupials. *Proc. Zool. Soc. Lond.* **1926:** 1037–84.

RENSHAW, G., 1938. The Thylacine. *J. Soc. Preservation Fauna of Empire.* **35:** 47–9.

RIDE, W. D. L., 1964. A review of Australian fossil marsupials. Presidential Address, 1963. *J. Roy. Soc. W. Aust.* **47:** 97–131.

ROLLS, E. C., 1969. "They all ran wild. The story of pests on the land in Australia". Angus and Robertson: Sydney.

SCOTT, W. J., 1872. Letter, addressed to the Secretary, respecting the supposed "native tiger" of Queensland. *Proc. Zool. Soc. Lond.* **1872:** 355.

SHARLAND, M. S. R., 1939. In search of the Thylacine. *Proc. Roy. Zool. Soc. N.S.W.* **1938–39:** 20–36.

SMITH, MEREDITH, 1972. Small fossil vertebrates from Victoria Cave, Naracoorte, South Australia. II. Peramelidae, Thylacinidae and Dasyuridae. *Trans. Roy. Roc. S. Aust.* **96:** 125–38.

SPENCER, C., 1944. Tasmanian Tiger reappears. *Vic. Nat.* **61:** 44.

STRACHAN, L., 1977. Tasmanian Tiger sighted. *Australian* 23 Aug. 1977.

TATE, G. H. H., 1947. Results of the Archbold Expeditions. No. 56. On the anatomy and classification of the Dasyuridae (Marsupialia) *Bull. Amer. Mus. Nat. Hist.* **88:** 101–55.

TEDFORD, R. H., 1967. The fossil Macropodiae from Lake Menindee, New South Wales. *Univ. Calif. Publ. Geol. Sci.* **64:** 1–156.

THOMAS, O., 1888. "Catalogue of the Marsupialia and Monotremata in the collection of the British Museum (Natural History)". British Museum (Natural History). London.

TROUGHTON, E. LEG., 1967. "Furred animals of Australia". 9th ed. Angus and Robertson: Sydney.

VAN DEUSEN, H., 1963. First New Guinea record of *Thylacinus. J. Mammal.* **44:** 279–80.

WAKEFIELD, N. A., 1963. Sub-fossils from Mount Hamilton, Victoria. *Vict. Nat.* **79:** 323–30.

WART (PSEUDONYM), 1919. Untitled letter to the editor. *The Bulletin* 11 Nov. 1919. p.20.

WHITLEY, G. P., 1973. I remember the Thylacine. *Koolewong* **2**(4): 10–1.

WOODBURNE, M. O., 1967. The Alcoota fauna, central Australia. An integrated palaeontological and geological study. *Bur. Min. Res. Geol. Geophys. Bull.* **87:** 1–187.

WRIGHT, B. J., 1972. Rock engravings of striped marsupials: the Pilbara Region, Western Australia. *Archaeol. Phys. Anthrop. Oceania* **7:** 24–30.

RESUMEN

23. Revisión del Tilacino (Thylacinidae, Marsupialia). Por Malcolm Smith.

Se revisa la literatura sobre el Tilacino, *Thylacinus cynocephalus* (Harris). En el momento del establecimiento del hombre blanco la especie vivía solamente en Tasmania, donde sufrió un colapso alrededor de 1910. Desde entonces numerosas expediciones han encontrado evidencias circunstanciales de su sobrevivencia. Sus restos fósiles han sido encontrados en el continente australiano, como también en Nueva Guinea Papua, y hasta ha habido un número de cuestionables denuncias de haberlo visto en el continente. Taxonómicamente el Tilacino parece estar aislado entre los Marsupicarnivora. Su sistema reproductivo es primitivo, pero el resto de su anatomía se ha especializado para la caza cursorial, paralelizando los cánidos por su gran tamaño, sus largas mandíbulas y sus molares sectoriales. Sin embargo, las proporciones de sus miembros son similares a las de *Dasyurus* y su cerebro, aunque grande y con muchas circunvoluciones comparado al de un hipotético dasiúrido del mismo tamaño, es pequeño comparado al de los cánidos. Poco se conoce sobre la ecología y comportamiento de este animal, pero se presenta aquí lo que ha podido obtenerse de los animales en cautividad y de las anécdotas de individuos salvajes.

24. METABOLIC SCOPE IN RESPONSE TO COLD OF SOME DASYURID MARSUPIALS AND AUSTRALIAN RODENTS

T. J. DAWSON[1] and W. R. DAWSON[2]

The metabolic response to cold of two small dasyurid marsupials, *Planigale gilesi* and *Dasyuroides byrnei*, was compared with that of two Australian rodents, *Notomys cervinus* and *Pseudomys gracilicaudatus*. The mean resting body temperatures in the thermoneutral zone were 36.2 ± 0.98 C (mean ± s.d.) for the marsupials and 37.2 ± 0.67°C for the eutherians. The standard metabolic rates (SMR) were 2.22 ± 0.17 W $kg^{-0.75}$ and 3.27 ± 0.27 W $kg^{-0.75}$, respectively for the marsupials and eutherians, these being similar to predicted values. In response to severe cold, the marsupials attained levels of summit metabolism at least matching those of the eutherians. The metabolic scope of the marsupials was 8-9 times SMR compared with 3-5 times SMR for the eutherians, indicating that the lower SMR of marsupials has not restricted their abilities for chemical thermoregulation.

Pages 255–60 *in* CARNIVOROUS MARSUPIALS, ed. by M. Archer, February, 1982. Roy. Zool. Soc. New South Wales: Sydney, Australia.

THE BASIC CONCERN of this study is the metabolic capabilities of marsupials relative to those of other mammals, with the thermogenic responses to cold being the specific topic of investigation. Conjecture has occurred regarding this aspect of the physiology of marsupials since the pioneering work of Martin (1902). Martin, prompted by earlier studies of Sutherland (1897), examined the body temperatures and metabolic rates of marsupials and eutherians, as well as of monotremes. Marsupials were reported to have body temperatures and metabolic rates intermediate between those of eutherians and monotremes; the metabolic rates of the marsupials were less than one third of those of the eutherians. For some five decades Martin's work was cited as demonstrating that marsupials represented an intermediate level of homeothermic development, below the "advanced" eutherians but better than the "very primitive" monotremes. The notion of marsupials as "primitive" mammals became entrenched (Johansen 1962).

The situation has been much clarified over the past decade. The minimal resting metabolic rate or standard metabolic rate (SMR) of marsupials has been demonstrated to vary with the 0.75 power of body mass in the same manner as in eutherians, but, at any mass, the level is about 70% of the corresponding eutherian level (Dawson and Hulbert 1969, 1970, MacMillen and Nelson 1969). The thermoregulatory abilities of a variety of marsupials, especially those of some of the larger macropodids, also have been examined and, generally, have been shown to be similar to those of eutherians (Dawson 1973a, Dawson *et al.* 1974). Smaller species, such as many of the dasyurids, however, do make extensive use of torpor (Dawson and Wolfers 1978, Wallis 1979), as do small eutherians, both thereby avoiding some of the energetic and thermoregulatory difficulties in cold environments.

Small homeothermic animals have problems in the cold because they potentially have high rates of heat loss. This arises because the volume of heat producing tissue and the animal's surface area, through which heat is lost, vary with different exponents of body mass, 1.0 and 0.67 respectively. Therefore as mass decreases the surface available for heat loss declines relatively less rapidly. The problem is compounded by the constraints on the amount of insulation that a small mammal can carry, the level of insulation being largely determined by fur thickness (Scholander *et al.* 1950). Maintenance of similar body temperatures in the face of a high heat loss should require equivalently high heat production by small marsupials or any small mammals. It is here that marsupials may have problems, for many birds and mammals appear unable to sustain metabolic rates in the cold that exceed about 3–4 times SMR (Gelineo 1964). Were small marsupials such as dasyurids to follow this pattern, their lower SMR would be accompanied by absolute metabolic capabilities that would limit their cold tolerance relative to that of their eutherian counterparts. This possibility led us to attempt to define the metabolic scope of these animals.

MATERIAL AND METHODS

The dasyurids examined were the Paucident Planigale, *Planigale gilesi,* and the Kowari, *Dasy-*

[1] School of Zoology, University of New South Wales, Kensington, New South Wales, Australia, 2033.
[2] Division of Biological Sciences, University of Michigan, Ann Arbor, Michigan, U.S.A., 48109.

uroides byrnei. Both these species are found in semiarid and arid regions, *P. gilesi* coming from Western N.S.W. and *D. byrnei* from southwestern Qld. The rodents were the Fawn Hopping–mouse, *Notomys cervinus*, and the Eastern Chestnut Mouse, *Pseudomys gracilicaudatus*. The *N. cervinus* came from southwestern Queensland and the *P. gracilicaudatus* from the central coast of N.S.W. Animals had been kept in an unheated animal house for several months prior to the experiments, which were carried out during winter and early spring (June – September). The only exception involved three planigales, which were wild caught in September. The temperature of the animal house followed outside ambient; usual daily maxima were 16–21°C. with minima of 8-14°C. Nest material or nest boxes were provided. Animals were fasted overnight only in experiments at the highest temperatures, i.e., in the thermoneutral zone. Fasting prior to low temperature experiments occasionally led to torpor in the marsupials.

Metabolic responses to various temperatures were tested in either of two temperature controlled cabinets, which operated down to –3°C and –20°C, respectively. To enable the animals to rest undisturbed in a normal posture, body temperature was not monitored continuously. Temperatures were measured with copper–constantan thermocouples. Body temperatures, taken as deep rectal temperature, were obtained at the beginning and end of the experiments.

The basic gas measuring system used the positive pressure, open circuit technique. Air was passed through a small perspex chamber containing the animal, at a rate measured by a calibrated dry gas meter. Oxygen consumption and carbon dioxide production were determined using a Beckman F3 paramagnetic analyser and a Beckman model 864 infra–red gas analyser, respectively. The oxygen analyser was calibrated manometrically before each experiment and the carbon dioxide analyser was calibrated using standard gases. Oxygen consumption and carbon dioxide production were calculated as dry gas at S.T.P. Heat production was estimated from oxygen consumption and RQ, which varied between 0.84 and 0.71, using the appropriate conversion factors (Brody 1945). Whole body conductances, uncorrected for heat loss by evaporation or changes in T_b, were measured by the formula.

$$\text{Conductance (in W m}^{-2}{}^{\circ}\text{C}^{-1}) = \frac{\text{Heat production}}{(T_b - T_a)\ \text{Surface area}}$$

Surface area was calculated using the equation S.A. (in m^2) = 0.12 $kg^{0.67}$. The Meeh factor adopted, 0.12, was one considered appropriate for small mammals from the work of Dawson and Hulbert (1970).

To obtain good base line readings for SMR measurements, all species were placed in the chambers for some 4–6 hours, starting at 09:00 h. Reported measurements were taken over two, 20 min periods in the latter part of the exposure, but before 15:30 h. Preliminary experiments indicated that after approximately 16:00 h, oxygen consumption could start to increase as the animals became more restless. The same pattern was followed during cold exposures, except experiments were terminated if oxygen consumption noticably declined. In these cases, the oxygen consumption reported was that obtained before the decline. Because of concern about the limited number of planigales, these tiny dasyurids were examined at less extreme temperatures and for shorter periods, the exposure period before data collection being 3 h.

RESULTS

Values for body temperatures and metabolic rates for the four species at thermoneutral temperatures and two low temperatures are summarised in Table 1. The data in thermoneutral conditions for body temperature and metabolic rate of the marsupials and eutherians follow a conventional pattern. The overall mean body temperatures of the marsupials was 36.2 ± 0.98°C as opposed to a group mean for the two eutherians of 37.2 ± 0.67°C ($p<0.02$). The SMR for the marsupials and eutherians in the mass-independent units of W $kg^{-0.75}$, were 2.22 ± 0.17 and 3.27 ± 0.29 respectively ($P < 0.001$). These values are close to the predicted ones for the SMR of 2.35 W $kg^{-0.75}$ (Dawson & Hulbert 1970) and 3.34 W $kg^{-0.75}$ (Kleiber 1961), respectively for marsupials and eutherians. The marsupial values are 68% of the eutherian ones.

The responses to cold environments are interesting, but difficult to interpret, unless the time course of experiments is kept in mind. With the exception of the tiny planigales, all species easily maintained body temperature down to about 0°C, and even some of the larger male planigales (11.5 – 12.4 g) maintained T_b at a T_a close to 0°C for periods in excess of 3 h. Below zero, –10 to –13°C,, the picture changed and only *D. byrnei* maintained T_b for the test period of 4 h. Of the eutherians the small hopping mouse *N. cervinus* performed well and body temperature was reasonably maintained for some 3 h, whereas the larger *P. gracilicaudatus* was unable to maintain T_b more than 1½ h at –10°C. Reasons for differing thermoregulatory capacities are suggested by the data on metabolic responses. *Pseudomys gracilicaudatus* had the lowest maximum metabolic rate both in mass–specific and mass–independent terms.

The pattern of metabolic response to cold varied between the marsupials and eutherians (Table 1). While the marsupials started with a lower basal heat production, their maximum me-

Table 1. Standard metabolic rate and metabolic responses to cold exposure of small marsupials and rodents.

N	Mass g	T_{air} °C	T_{body} °C	O_2 Consumption 1 kg^{-1} h^{-1}	Heat production W $kg^{-0.75}$	MR/ /SMR	Time to data* min.	Conductance W m^{-2} °C^{-1}
Planigale gilesi								
5	9.5±2.2	33.4±0.8	36.8±0.5	1.28±0.12	2.32±0.12	–	245±95	4.1±1.32
5	9.8±2.3	10.7±0.8	35.8±0.9	8.89±0.42	15.4 ±1.46	6.6±0.6	189±53	3.6±0.47
5	10.1±2.3	5.1±1.0	35.2±1.3	10.98±2.00	19.0 ±2.30	8.2±1.0	132±41	3.7±0.57
Dasyuroides byrnei								
5	115±12	31.1±1.1	35.6±1.0	0.65±0.06	2.12±0.17	–	336±64	3.6±1.00
5	118±12	1.1±0.8	36.2±1.6	4.05±0.46	13.2 ±1.66	6.2±0.8	290±35	2.7±0.33
5	118±14	−10.3±0.8	37.1±0.5	5.80±0.53	18.6 ±2.08	8.8±1.0	225±57	2.8±0.34
Notomys cervinus								
5	32.5±3.0	32.0±0.9	37.0±0.7	1.44±0.14	3.42±0.33	–	291±64	4.4±0.83
5	32.5±1.4	−0.2±1.5	36.9±0.8	6.06±0.24	14.1 ±0.58	4.1±0.2	288±58	2.4±0.14
5	33.1±3.2	−13.1±1.0	34.2±2.6	8.85±0.85	21.0 ±0.45	6.1±0.5	150±28	2.8±0.16
Pseudomys gracilicaudatus								
5	79.8±6.9	31.9±1.3	37.4±0.6	1.05±0.05	3.12±0.17	–	255±69	4.0±1.02
5	75.6±8.5	−0.7±2.4	36.2±0.2	3.27±0.16	9.6 ±0.42	3.1±0.2	151±68	1.8±0.18
5	81.2±11.3	−10.6±2.9	29.4±4.0	4.03±0.46	12.0 ±1.54	3.8±0.5	113±59	2.1±0.36

* Time to data indicates the time to the beginning of data collection. Values are means ± s.d.

tabolic responses were similar to that of *N. cervinus,* the eutherian with the better thermogenic capacity of the two species tested. It should also be noted that the value for *D. byrnei* was probably submaximal because T_b was maintained to the end of the experiment. This pattern is best appreciated by the ratio of maximum metabolic rate to SMR. In both marsupials the relative increase, 8–9 times, was significantly higher than in either of the eutherians, and was double that of *P. gracilicaudatus.*

For the maintenance of body temperature in a cold environment, the reduction of heat loss is the other important variable apart from increased heat production. The variation in conductance at different air temperatures is shown in Table I. The general pattern which emerged for all species was consistent with that for homeotherms generally. In the thermoneutral zone, conductance was elevated as thermoregulatory vasomotor and pilomotor adjustment occurred. Conductance was minimal around zero but increased at the lowest temperatures, particularly well below freezing. This effect presumably resulted partially from increased shivering, which may break up the air boundary layer. Possibly more important in these results could be an increased blood flow to the periphery to give protection against freezing of the tissues, i.e., frost bite. No marked pattern was seen between the minimal conductance levels of the two groups of mammals. The planigales, with their very short fur had a high conductance. The *P. gracilicaudatus,* which had the smallest metabolic scope, had the lowest conductance, that is the highest total body insulation.

DISCUSSION

The pattern of difference between marsupials and eutherians which is evident in resting, thermoneutral body temperatures, together with that in standard metabolic rates, is in agreement with that found in many of the studies undertaken in the past decade, following the primary studies of Dawson and Hulbert (1969, 1970) and MacMillen and Nelson (1969). Few studies have actually compared eutherians and marsupials directly, although Dawson and Hulbert (1970) examined rabbits to reinforce their results. This general lack of direct comparative study has enabled the perpetuation of the idea that the difference in SMR between the two groups could be due to a Q_{10} effect associated with differences in body temperature (Kinnear and Shield 1975). When marsupials and eutherians are treated similarly, as in the current experiments, the data do not support this proposition and it is probably time that the idea was laid to rest.

It may be argued that comparisions of the metabolic status of various groups of homeotherms based on the SMR are not particularly valid, because the conditions under which the SMR is determined are rather artificial and rarely related to the animal's activities in the real world. However, there is ample evidence that other levels of physiological activity in marsupials are related to the SMR. These include normal maintenance, nitrogen and energy requirements (Brown and Main 1967, Hume 1978) and water requirements (Denny and Dawson 1975). Also, most importantly, the maximal sustained metabolic responses to cold in some groups of birds and mammals consistently appear to be about quadruple the SMR (Gelineo 1964). Therefore such maximal responses could be limited in absolute terms in those groups with low SMR (Dawson 1973b). Data on bandicoots (Hulbert and Dawson 1974) suggest that this may be also true for marsupials, but other studies have argued against a limiting relationship between metabolic capability and SMR for marsupials. Baudinette *et al.* (1976) have in fact proposed that the metabolic scope of dasyurids in response to exercise may be greater than that of eutherians.

The relationship between metabolic responses to cold and the energy expenditure during exercise is unclear because, for large mammals such as the red kangaroo, an increase in metabolic rate during sustained exercise representing at least 20 times the SMR is possible (Dawson and Taylor 1973). While several workers have dealt with this problem, no concensus as yet exists concerning what is maximum metabolic activity and how it should be assessed. Hart (1971) reviewed much of the initial work, particularly that dealing with small mammals and he recognised at least two different levels of maximum metabolic activity, peak metabolic effort and summit metabolism. The first of these, peak metabolic effort (Hart 1957), is the highest rate of oxygen consumption that can be sustained for a short time, some 3–20 minutes. This may be attained by exposure to extreme cold under conditions where heat loss initially exceeds heat production, or by the addition of exercise to a submaximal cold exposure. This latter test is made on the assumption that the heat production due to shivering and that due to exercise are, in the short term, interchangeable. Some doubt exists about whether peak rates obtained by these two procedures are comparable, but they seem similar, at least in small rodents. Peak metabolic effort appears to be about 7 times SMR in small mammals (Hart 1971), but is obviously higher in exercise in larger cursorial species.

Summit metabolism is the other measure of metabolic scope used for comparative purposes (Gelineo 1964), and this is the highest metabolic rate that can be sustained in the cold by a sedentary animal. Some doubt exists about the appropriate time scale for measurements of summit metabolism, but a period of at least several hours is now accepted. Over a range of homeotherms it seems that most species have a summit metabolism 3–4 times basal (Gelineo 1964) although a few northern hemisphere mammals from cold climates have been reported to exceed this figure in winter (Rosenmann *et al.* 1975). Some small birds which reside in the boreal winter of the northern hemisphere also have exceptional metabolic capabilities but it has also been noted that they have a high SMR (Dawson and Carey 1976).

Our examination of the metabolic capabilities of marsupials was primarily interested in maximum metabolic responses, sustained by homeothermic animals over an extended period in severe cold. We have thus been concerned with an approximation of summit metabolism, rather than in peak effort. In this regard two cold test temperatures were used to characterise the responses more completely. The three larger species were able to maintain T_b for the test time of approximately 5 h at the temperatures near 0°C. However, even planigales, with the exception of a 7.7 g female, maintained T_b for their test period near freezing for 3 hours.

Below freezing, the responses of *D. byrnei* and rodents differed. *Dasyuroides byrnei* maintained T_b for the test period of 4 h, whereas neither of the eutherians could do so for this interval. The summit metabolism of the eutherians therefore develops between the two cold test temperatures and is approximately 3–5 times their SMR; *N. cervinus* reached a level 6 times but could not maintain it. The dasyurids therefore have a higer summit metabolism relative to their SMR, some 8–9 times SMR, one kowari actually maintained a metabolic level 10.4 times SMR. The dasyurids, then, are not restrained in their response to cold stress by their initially low SMR. The true relative metabolic capabilities are appreciated by consideration of the data in the mass independent terms of $W.kg^{-0.75}$ (Table 1). In this comparison the marsupials have rates of heat production approximating that of *N. cervinus* and substantially exceeding that of *P. gracilicaudatus*. However, *N. cervinus* could not maintain this rate whereas *D. byrnei* could.

How then do these small dasyurids manage this metabolic scope? Rodents rely heavily on nonshivering thermogenesis (NST) in their responses to cold (Hart 1971) and since these experiments were carried out in winter and the animals were kept in unheated quarters NST might be functional. One explanation for the large metabolic scope of the dasyurids could be a greater capacity for NST, but this appears unlikely. Reports of NST in marsupials are as yet conflicting. Reynolds and Hulbert (1982) suggest that it is absent and even those studies reporting NST in marsupials are equivocal because the doses of noradrenaline required to elicite a response are far greater than those needed for rodents (Nicol 1978, Wallis 1979). It is thus possible that the dasyurids rely primarily on shivering for their regulatory heat production. In this regard it is interesting that the maximum sustained rates of heat production in exercise and cold thermogenesis are both approximately 8–9 times SMR in dasyurids (Baudinette *et al.* 1976, R. E. MacMillen and T. J. Dawson, unpublished data).

An interesting parallel to this discussion of the metabolic capabilities of dasyurids and other marsupials is their very low resting heart rates (Kinnear and Brown 1967). These are much lower than those of eutherians, lower even than would be expected from their low SMR. On the other hand, marsupials compensate for these low heart rates by having a much larger heart stroke volume (Dawson and Needham 1981). If, as Baudinette (1978) has reported, marsupials and eutherians have similar maximal heart rates, then maximal cardiac outputs may be greater in marsupials. This would be quite consistent with the data indicating the high sustained aerobic capacity of marsupials and it is conceivable that marsupials are superior to eutherians of comparable size in aerobic capacity.

It had been suggested that low metabolic rates in marsupials might be compensated for by a lower rate of heat loss than that seen in eutherians (Dawson 1973b). While this option appears used by larger marsupials (Dawson 1973a, Degabriele and Dawson 1979), it is not available for smaller species. MacMillen and Nelson (1969) failed to find a difference between their data for dasyurid conductance and the predictions for eutherians by Herried and Kessel (1967), and no pattern of difference was noted in the current study. In very small animals the core to skin surface component of insulation can have only a small influence on total insulation, because of the hight rate of heat transfer through tissue (about 10 times that through fur) and the very small distances involved. As animals become larger however, this situation changes and marked adjustments can be made by the utilization of mechanisms such as countercurrent arrangements (Dawson and Fanning 1981). In small animals, however, fur depth largely determines total insulation and there are constraints on the amount of fur a small mammal can carry. Given these circumstances, if a small dasyurid or any other small animal is to remain homeothermic over a wide temperature range, it must have a large metabolic capacity. In marsupials this means a large functional metabolic scope. Whether this is a feature of marsupials in general or a specific adaption to small size such as may occur in some groups of shrews has yet to be determined.

ACKNOWLEDGEMENTS

The authors wish to thank F. D. Fanning for his technical assistance and B. Fox and D. Read for providing animals. The work was supported by a grant from the A.R.G.C. W. R. Dawson was in receipt of a Visiting Professorship from the University of New South Wales and received some support from the Horace H. Rackham School of Graduate Studies, the University of Michigan. Drs. R. T. Wells and A. J. Hulbert read and constructively criticized a draft of this paper.

REFERENCES

Baudinette, R. V., 1978. Scaling of heart rate during locomotion in mammals. *J. Comp. Physiol.* **127:** 337–42.

Baudinette, R. V., Nagle, K., Scott, R. A. D., 1976. Locomotory energetics in dasyurid marsupials. *J. Comp. Physiol.* **109:** 159–69.

Brody, S., 1945. "Bioenergetics and growth". New York Reinhold Publishing Corporation: New York.

Brown, G. D. and Main, A. R., 1967. Studies on marsupial nutrition. 5. The nitrogen requirements of the Euro *Macropus robustus. Aust. J. Zool.* **15:** 7–27.

Dawson, T. J., 1973a. Thermoregulatory responses of the arid zone kangaroos, *Megaleia rufa and Macropus robustus. Comp. Biochem. Physiol.* **46***A***:** 153–69.

Dawson, T. J., 1973b. Primitive mammals. Pp.1–46 *in* "Comparative physiology of thermoregulation. Vol. 3". ed. by G. C. Whittow. Academic Press: New York.

Dawson, T. J. and Fanning, F. D., 1981. Thermal and energetic problems of small semi aquatic mammals: a study of the Australian Water Rat, including comparisons with the Platypus. *Physiol. Zool.* **54:** 285–96.

Dawson, T. J. and Hulbert, A. J., 1970. Standard metabolism, body temperature and surface areas of Australian marsupials. *Am. J. Physiol.* **218:** 1233–8.

Dawson, T. J. and Needham, A. D., 1981. Cardiovascular characteristics of two resting marsupials: an insight into the cardiorespiratory allometry of marsupials. *J. Comp. Physiol.* (in press).

Dawson, T. J., Robertshaw, D. and Taylor, C. R., 1974. Sweating in the kangaroo: a cooling mechanism during exercise but not in the heat. *Am. J. Physiol.* **272:** 494–8.

Dawson, T. J. and Wolfers, J. M., 1978. Metabolism, thermoregulation and torpor in shrew–sized marsupials of the genus *Planigale. Comp. Biochem. Physiol.* **59A:** 305–9.

Dawson, T. J. and Taylor, C. R., 1973. Energetic cost of locomotion in kangaroos. *Nature* **246:** 313.

Dawson, W. R. and Carey, C., 1976. Seasonal acclimatization to temperature in cardueline finches. I. Insulative and metabolic adjustments. *J. Comp. Physiol.* **112:** 317–33.

Degabriele, R. and Dawson, T. J., 1979. Metabolism and heat balance in an arboreal marsupial, the Koala *(Phascolarctos cinereus). J. Comp. Physiol.* **134:** 293–301.

Denny, M. J. S. and Dawson, T. J., 1975. Comparative metabolism of tritiated water by macropodid marsupials. *Am. J. Physiol.* **228:** 1794–9.

Gelineo, S., 1964. Organ systems in adaptation: the temperature regulating system. Pp.259–282 *in* "Handbook of physiology. Sec. 4". ed. by D. B. Dill: Amer. Physiol. Soc. Washington, D.C.

Hart, J. S., 1957. Climatic and temperature induced changes in the energetics of homeotherms. *Rev. Can. Biol.* **16:** 133–74.

Hart, J. S., 1971. Rodents. Pp.1–149 *in* "Comparative physiology of thermoregulation. Vol. 2" ed. by G. C. Whittow. Academic Press: New York.

Herried, C. F. and Kessel, B., 1976. Thermal conductance in birds and mammals. *Comp. Biochem. Physiol.* **21:** 405–14.

Hulbert, A. J. and Dawson, T. J., 1974. Thermoregulation in perameloid marsupials from different environments. *Comp. Biochem. Physiol.* **47A:** 591–616.

Hume, I. D., 1974. Nitrogen and sulphur retention and fibre digestion by Euros, Red Kangaroos and Sheep. *Aust. J. Zool.* **22:** 13–23.

Johansen, K., 1962. Evolution of mammalian temperature regulation. Pp.73–131 *in* "Comparative physiology of temperature regulation" ed. by J. P. Hannon and E. Viereck. Arctic Aeromedical Laboratory: Alaska.

Kinnear, J. E. and Brown, G. D., 1967. Minimal heart rates of marsupials. *Nature Lond.* **215:** 1501.

Kinnear, A. and Shield, J. W., 1975. Metabolism and temperature regulation in marsupsials. *Comp. Biochem. Physiol.* **52A:** 235–45.

Kleiber, M., 1961. "The fire of life". John Wiley: New York.

MacMillen, R. E. and Nelson, J. E., 1969. Bioenergetics and body size in dasyurid marsupials. *Am. J. Physiol.* **217:** 1246–51.

Martin, C. J., 1902. Thermal adjustment and respiratory exchange in monotremes and marsupials. *Phil. Trans. Roy. Soc. Lond. Ser. B.* **195:** 1–37.

Nicol, S., 1978. Non–shivering thermogenesis in the Potoroo, *Potorous tridactylus* (Kerr). *Comp. Biochem. Physiol.* **59C:** 33–7.

Reynolds, W. and Hulbert, A. J., 1982. Cold adaptation in the Brown Marsupial Mouse *Antechinus stuartii.* Pp.279–83 *in* "Carnivorous marsupials", ed. by M. Archer. Roy. Zool. Soc. New South Wales: Sydney.

Rosenmann, M., Morrison, P. R. and Feist, D., 1975. Seasonal changes in the metabolic capacity of Red–backed Voles. *Physiol. Zool.* **48:** 303–10.

Scholander, P. F., Walters, V., Hock, R. and Irving, L., 1950. Body insulation of some arctic and tropical mammals and birds. *Biol. Bull.* **99:** 225–36.

Sutherland, A., 1897. The temperatures of reptiles, monotremes and marsupials. *Proc. Roy. Soc. Victoria* **9:** 57–67.

Wallis, R. L., 1979. Responses to low temperature in small marsupial mammals. *J. Thermal. Biol.* **4:** 105–11.

RESUMEN

24. Cambio metabólico y conducta de algunos marsupiales dasiúridos y roedores australianos en respuesta al frío. Por T. J. Dawson y W. R. Dawson.

Se comparó la respuesta metabólica al frío de dos pequeños marsupiales dasiúridos, *Planigale gilesi* y *Dasyuroides byrnei,* con la de dos roedores australianos, *Notomys cervinus* y *Pseudomys gracilicaudatus.* Las temperaturas medias del cuerpo en descanso en la zona termoneutral fueron 36.2 $\pm$ 0.98°C (media $\pm$ desviación standard) para los marsupiales y 37.2 $\pm$ 0.67°C para los euterios. Las "standard metabolic rates (SMR)" fueron 2.22 $\pm$ 0.17 Peso kg$^{-0.75}$ y 3.27 $\pm$ 0.27 Peso kg$^{-0.75}$, respectivamente para los marsupiales y los euterios, siendo e los similares a lo valores previstos. En respuesta a fríos severos los marsupiales alcanzaron niveles de metabolismo máximo, al menos igualando aquellos de los euterios. El alcance metabólico de los marsupiales fue 8–9 veces el SMR comparado con 3–5 veces el SMR de los euterios, indicando que el más bajo SMR de los marsupiales no había restringido sus habilidades para la termoregulación guímica.

25. THE ENERGETICS OF LOCOMOTION IN DASYURID MARSUPIALS

R. V. BAUDINETTE[1]

The incremental cost of locomotion (energy used per unit body mass per unit distance travelled) is similar in dasyurids and eutherians. Earlier claims of differences are re-examined. Dasyurids show a greater dependence on heat storage during exercise, and heat loss by evaporative routes always exceeds that mediated through non-evaporative routes. This contrasts with the general eutherian strategy of minimal heat storage and evaporative heat loss.

There is little correlation between the low metabolic levels at rest in dasyurids and the physiological response to exercise. Only the dependence on heat storage can be clearly related to this trait, with the lower resting body temperature providing these animals with a greater storage capacity.

Pages 261–65 *in* CARNIVOROUS MARSUPIALS, ed. by M. Archer, February, 1982. Roy Zool. Soc. New South Wales: Sydney, Australia.

MARSUPIALS generally have lower basal or standard rates of metabolism ($\dot{V}o_{2STD}$) and lower body temperatures than eutherians of equivalent body size (Martin 1902, MacMillen and Nelson 1969, Dawson and Hulbert 1970). Whether this lower metabolic level reflects a conservative phylogentic state, or a specialised adaptation to habitat, is the subject of considerable debate (see Kinnear and Shield 1975, McNab 1978, Hulbert 1980).

Levels of $\dot{V}o_{2STD}$ in the family Dasyuridae are about 70% of those found in eutherians (MacMillen and Nelson 1969, Dawson and Hulbert 1970, Arnold and Shield 1970, Morton and Lee 1978), and appear independent of habitat or diet. The same depressed metabolic levels are found in carnivorous dasyurids from moist forest habitats (such as *Sacrophilus harisii* and *Satanellus hallucatus*), as well as in smaller insectivorous forms such as *Sminthopsis crassicaudata* and *Dasyuroides byrnei* from dry habitats. Furthermore, McNabb (1978) has suggested that while the metabolic levels of eutherian and marsupial herbivores and folivores are similar, marsupial carnivores have lower basal rates of metabolism than placentals having an equivalent niche.

These observations of a generally lower level of $\dot{V}o_{2STD}$ in all members of the family, and its apparent independence from the ecology of the animals, suggest a conservative or primitive character state. In this contribution I examine the consequences of this state on the power requirement and the mechanisms of heat dissipation during locomotion.

THE ENERGETIC REQUIREMENTS FOR LOCOMOTION

THE INCREMENTAL COST OF LOCOMOTION: The first published measurements of rates of oxygen consumption ($\dot{V}o_2$) during locomotion were probably those of Edward Smith on English prisoners in 1857. Reports of treadmill exercise in animals appeared around the turn of the century when Zuntz (1897) and Slowtzoff (1903) measured rates of oxygen consumption in dogs and horses. For many years interest in locomotory energetics centered on human subjects, until Taylor *et al.* (1970) studied locomotory energetics in a range of eutherians. These investigators found that the relation between rates of oxygen consumption and running speed was linear, but the slopes of the lines describing this relationship (the incremental cost of locomotion, T_{run}), was greater in small animals. Taylor *et al.* derived an equation for this relationship:

$$T_{run} = 8.46\ m^{-0.40},$$

where T_{run} is the incremental cost of locomotion in units of ml 0_2 (g km)$^{-1}$, and m is body mass in grams. [1 ml 0_2 (g km)$^{-1}$ = 20.1 J (kg m)$^{-1}$].

This elegant model prompted further research on birds and marsupials since it considered only "domesticated" laboratory species. Dawson and Taylor (1973) measured $\dot{V}o_2$ in red kangaroos hopping on a treadmill and found that after the kangaroo changed from a pentapedal "walk" to a hop, its oxygen consumption remained relatively constant and independent of speed. A detailed discussion of the mechanisms underlying this mode of locomotion can be found in Alexander and Vernon (1975).

Fedak *et al.* (1974) measured the energy cost associated with running in birds and derived an equation:

$$T_{run} = 2.64\ m^{-0.20}$$

for the incremental cost of locomotion. The exponent in the equation, a term which reflects the effect of size within the group, was statistically different from that derived for eutherians, and it followed that the energetic demands of bipe-

[1] Biological Sciences, Flinders University, Bedford Park, Adelaide, South Australia, Australia, 5042.

dalism and quadrupedalism show a different body–mass dependence. Baudinette *et al.* (1976 a,b) measured rates of oxygen consumption during locomotion in five species of dasyurid marsupials. The results were qualitatively similar to those for eutherians in that $\bar{V}O_2$ increased linearly with running speed, and the increase was proportionately greater in smaller animals. In similar fashion to Taylor *et al.* (1970) and Fedak *et al.* (1974), an allometric equation was used to describe the incremental cost of locomotion as a function of body size.

$$T_{run} = 4.75\ m^{-0.34}$$

The value of the exponent (–0.34) is statistically similar to that of Taylor *et al.* for eutherians, but differs from the value reported for birds. The constant (4.75) is significantly lower than the eutherian value of 8.46, and suggests that marsupials are not as capable of high power increments. Baudinette (1980) later attributed this difference to greater rates of energy usage in contracting marsupial muscle at lower speeds, or alternatively to behavioural differences in the experimental situation.

In 1979 two papers put an end to the speculation that birds, dasyurids and eutherian mammals differed in their locomotory energetics. Paladino and King (1979) reanalysed all the locomotory data then available, and Fedak and Seeherman (1979) published a similar analysis adding new data for the ostrich and horse. From this analysis of 66 vertebrate species, it is now apparent that there are no significant differences in the incremental energy cost of locomotion for bipeds or quadrupeds of any size category. These new analyses show that for any size class, values of T_{run} may vary by a factor of two, indicating a great deal of "biological noise" in the physiological response to exercise. The dasyurids, like reptiles, birds and monotremes, can now all be grouped with eutherians, and the following equation of Fedak and Seeherman used to describe the relationship between the incremental costs of locomotion and body size during running.

$$T_{run} = 3.89\ m^{-0.28} \quad (r^2 = 0.84)$$

THE "ZERO-SPEED COST OF LOCOMOTION": If the straight line relating $\bar{V}O_2$ to running speed is extrapolated to zero speed, a multiple of basal metabolism can be simply derived. In the earlier eutherian study of Taylor *et al.* (1970), this multiple was around 1.7, however for dasyurids it ranged from 3.9 to 6, with a mean of 5.5 (Baudinette *et al.* 1976b). This difference is statistically valid, but may simply represent stress in a laboratory situation. Paladino and King (1979) have suggested that some of the variation in this parameter may be reduced if the ratio of the extrapolated value to the measured standing value is used rather than the basal value. Further analysis of dasyurids is required to clarify this apparent anomaly.

MAXIMUM LEVELS OF OXYGEN CONSUMPTION

Taylor *et al.* (1978) derived the following equation to describe the maximum levels of oxygen consumption ($\bar{V}O_{2max}$) for treadmill locomotion in wild animals:

$$\bar{V}O_{2max} = 28.8m^{0.79} \quad (\text{where } \bar{V}O_{max} \text{ is units of ml } o_2\ g^{-1}\ h^{-1}).$$

$\bar{V}O_{2max}$ was defined as the point where increased speed no longer caused an increase in oxygen consumption, and blood lactate levels increased throughout the running period. For dasyurids, Baudinette *et al.* (1976b) measured $\bar{V}O_{2max}$ under similar conditions but did not measure lactate levels. Their equation:

$$\bar{V}O_{2max} = 21.2\ m^{0.82}$$

has a similar exponent, and considering the much smaller range of body size used in its derivation, suggests that dasyurids are capable of similar maximal levels to eutherians.

STORAGE AND DISSIPATION OF HEAT GENERATED DURING LOCOMOTION

HEAT STORAGE: Heat storage may be regarded as a means of deferring dissipation of heat, and is seen in animals such as the cheetah which rely on short sprints (Taylor and Rowntree 1973). A tolerance of the resultant higher body temperatures during exercise seems common among animals which appear to be "sprinters" rather than "stayers".

Unpublished data from our laboratory suggest that eutherians and marsupials, (including dasyurids) reach similar peak temperatures during exercise. For example, both the dingo and Tasmanian Devil show a steady state body temperature of between 39–40°C when running, a level usually only exceeded by animals having a carotid rete and a cavenous sinus to aid cooling (see Taylor and Lyman 1972).

Since the resting body temperature of dasyurids is lower than that of eutherians, it follows that for a given exercise regime, the former should store a greater proportion of the heat generated. Baudinette *et al.* (1976b) confirmed this hypothesis in 5 species of small dasyurids, and Bell, Baudinette and Nicol (unpubl.) repeated the finding in the Tasmanian Devel. In these 6 species, about 50% of the metabolic heat produced during treadmill exercise was manifest as a rise in body temperature. Among eutherians, only the cheetah approaches this level of storage and even at high speeds most other animals examined store less than 20 per cent of their heat production (Taylor and Rowntree 1973). Data from dasyurids and representative eutherians is included in Table 1.

EVAPORATIVE AND NON–EVAPORATIVE AVENUES OF HEAT LOSSS: The rate of heat loss through non-evaporative routes (i.e. convection, conduction and radiation) is dependent on the temperature gradient between the animal and its environment, and the relative surface area involved. Small animals with high surface temperatures should therefore favour this method. Data in Table 1 illustrates that small eutherians dissipate about ninety per cent of their heat production by non-evaporative means, and in larger eutherians, this avenue accounts for about one half of the total heat loss. However even in the smaller dasyurids, non-evaporative heat loss during locomotion is always less than in eutherians.

Evaporative heat loss is usually an emergency response to heat stress in small mammals as they lack sweat glands, generally do not pant, and may have counter-current mechanisms to reduce water loss through the nares. Small dasyurids rely proportionately more on evaporative heat loss than do eutherians (Table 1). The only reported levels approaching those of the dasyurids are from chipmunks (Wunder 1970), in which 53 per cent of the total heat produced during exercise is lost evaporatively.

Overall Pattern of Heat Balance During Locomotion: A major reliance on heat storage and evaporative routes of heat loss during locomotion is consistent with low body temperature, low heat conductance across the body surface, and high rates of water turnover. Changes in conductance during locomotion have not been investigated in dasyurids, and have been only briefly considered in eutherians. Both Taylor (1976) and Wunder (1970) report a general increase in rates of heat transfer with increasing speed in eutherians, but an analysis of causative mechanisms such as peripheral blood flow changes in small mammals has not been attempted. There is a little information on rates of water turnover in dasyurids. Green and Eberhard (1979) measured rates of water turnover in the Tasmanian Devil *S. harrisii,* and the native cat *D. viverrinus,* and found levels around 130 ml $m^{-0.82}$ day $^{-1}$. Nicol (1978) reported a value slightly less than this for *S. harrisii.* Among smaller dasyurids, Macfarlane *et al.* (1971) found a level of 233 ml $m^{-0.82}$ day^{-1} in *Sminthopsis crassicaudata,* and in the desert species, *Dasyuroides byrnei,* Haines *et al.* (1974) measured a water turnover rate in the laboratory of 93 ml m $^{-0.82}$ day $^{-1}$. None of these rates indicate an extreme capacity for water economy (see Nicol 1978). Even though the application of these data to running animals may be unjustified, we can anticipate that dasyurids have relatively high rates of water turnover, and that this is consistent with the relatively high water content of a carinivorous diet (see Haines *et al.* 1974).

Table 1. Heat dissipation during locomotion in dasyurids and eutherians.

Species	Body wt. (kg)	Speed (km h^{-1})	Percentage of total heat production: Stored	Lost by nonevaporative routes	Lost by evaporation	Source
Sminthopsis crassicaudata	0.015	0.6	10	28	62	Baudinette *et al.* (1976b)
		1.2	12	26	62	
		1.8	16	24	60	
Antechinomys spenceri	0.030	2.0	28	–[a]	–	Baudinette *et al.* (1976a)
Dasyuroides byrnei	0.115	1.5	29	28	43	Baudinette *et al.* (1976b)
		3.4	31	32	37	
		5.5	29	38	33	
Dasyurus viverrinus	1.120	2.8	35	6	59	Baudinette *et al.* (1976b)
		3.4	41	3	56	
		5.5	38	6	56	
Sarcophilus harrisii	6.8	5.6	39	16	46	Bell, Baudinette and Nicol (unpubl.)
Eutherians						
Dipodomys spectabilis	0.100	1.5	0	92	8	Raab and Schmidt–Nielsen (1972
Rattus norvegicus	0.100	1.5	0	85	15	Raab and Schmidt–Nielsen (1972)
Domestic Dog	3.0	5.0	3	58	39	Taylor *et al.* (1971)
		10.0	14	56	30	
Lycaon pictus	8.7	5.0	13	60	27	Taylor *et al.* (1971)
		10.0	14	56	30	

[a]Dash indicates no data available.

CONCLUSIONS

The initial question posed was whether the low level of metabolism seen in dasyurids is reflected in their locomotory energetics. It appears that only the low resting body temperature associated with depressed metabolism can be closely related to exercise, in that it permits more "heat ballast" for storage. The other major difference seen between dasyurids and eutherians, that of greater dependence on evaporation for heat dissipation, is probably more closely correlated with a carnivorous/insectivorous diet than with overall levels of metabolism. In incremental rates of power input during locomotion, dasyurids fall within the ranges seen for other mammals, and earlier claims of differences in energy requirements for locomotion must be reassessed.

ACKNOWLEDGEMENTS

Dr. T. J. Dawson read and constructively commented on a draft of this manuscript.

REFERENCES

ARNOLD, J., AND SHIELD, J., 1970. Oxygen consumption and body temperature of the Chuditch (*Dasyurus geoffroii*). *J. Zool., Lond.* **160:** 391–404.

ALEXANDER, R. McN. AND VERNON, A., 1975. The mechanics of hopping by kangaroos (Macropodidae). *J. Zool., Lond.* **177:** 265–303.

BAUDINETTE, R. V., NAGLE, K. ADD SCOTT, R. A. D., 1976a. Locomotory energetics in a marsupial (*Antechinomys spenceri*) and a rodent (*Notomys alexis*). *Experientia* **32:** 583–4.

BAUDINETTE, R. V. NAGLE, K. AND SCOTT, R. A. D., 1976b. Locomotory energetics in dasyurid marsupials. *J. Comp. Physiol.* **109:** 159–68.

BAUDINETTE, R. V., 1980. Physiological responses to locomotion in marsupials *in* "Comparative physiology: primitive mammals" ed. by K. Schmidt–Nielsen, L. Bolis and C. R. Taylor. C.U.P.

DAWSON, T. J. AND HULBERT, A. J., 1980. Standard metabolism, body temperature, and surface areas of Australian marsupials. *Amer. J. Physiol.* **218:** 1233–8.

DAWSON, T. J. AND TAYLOR, C. R., 1973. Energetic cost of locomotion in kangaroos. *Nature* **246:** 313–4.

FEDAK, M. A., PINSHOW, B. AND SCHMIDT–NIELSEN, K., 1974. Energy cost of bipedal running. *Amer. J. Physiol.* **227:** 1038–4.

FEDAK, M. A. AND SEEHERMAN, H. J., 1979. Reappraisal of energetics of locomotion shows identical cost in bipeds and quadrupeds including ostrich and horse. *Nature* **282:** 713–6.

GREEN, B. AND EBERHARD, I., 1979. Energy requirements and sodium and water turnovers in two captive marsupial carnivores: the Tasmanian Devil, *Sacrophilus harrisii*, and the Native Cat, *Dasyurus viverrinus*. *Aust. J. Zool.* **72:** 1–8.

HAINES, H., MACFARLANE, W. V., SETCHELL, C. AND HOWARD, B., 1974. Water turnover and pulmocutaneous evaporation of Australian desert dasyurids and murids. *Amer. J. Physiol.* **227:** 958–63.

HULBERT, A. J., 1980. Evolution of energy in mammals. *in* "Comparative physiology: primitive mammals" ed. by K. Schmidt–Nielsen, L. Bolis and C. R. Taylor. C.U.P.

KINNEAR, A. AND SHIELD, J. W., 1975. Metabolism and temperature regulation in marsupials. *Comp. Biochem. Physiol.* **52*A*:** 235–45.

KLEIBER, M., 1961. "The fire of life. An introduction to animal energetics". New York: Wiley.

MCNAB, B. K., 1978. The comparative energetics of neotropical marsupials. *J. Comp. Physiol.* **125:** 115–28.

MACFARLANE, W. V., HOWARD, B., HAINES, H., KENNEDY, P. J. AND SHARPE, C. M., 1971. Hierarchy of water and energy turnover of desert mammals. *Nature, (Lond.)* **234:** 483–4.

MACMILLEN, R. E. AND NELSON, J. E., 1969. Bioenergetics and body size in dasyurid marsupials. *Amer. J. Physiol.* **217:** 1246–51.

MARTIN, C. J., 1902. Thermal adjustment and respiratory exchange in monotremes and marsupials. *Phil. Trans. Roy. Soc. Lond. Ser. B.* **195:** 1–37.

MORTON, S. R. AND LEE, A. K., 1978. Thermoregulation and metabolism in *Planigale maculata* (Marsupialia: Dasyuridae). *J. Thermal. Biol.* **3:** 117–20.

NICOL, S. C., 1978. Rates of water turnover in marsupials and eutherians: A comparative review with new data on the Tasmanian Devil. *Aust. J. Zool.* **26:** 465–73.

PALADINO, F. V. AND KING, J. R., 1979. Energetic cost of terrestrial locomotion: biped and quadruped runners compared. *Rev. Can. Biol.* **38:** 321–3.

RAAB, J. L. AND SCHMIDT–NIELSEN, K., 1972. Effect of running on water balance of the kangaroo rat. *Amer. J. Physiol.* **225:** 1230–5.

SMITH, E., 1857. The influence of the labour of the treadmill over respiration and pulsation; and its relation to the waste of the system and the dietary of prisoners. *Brit. Med. J.* **1:** 591–2.

SLOWTZOFF, B., 1903. Ueber die Beziehungen zwishen Korpergrosse und Stoffverbrauch der Hunde bei Ruhe and Arbeit. *Arch. Ges. Physiol.* **95:** 158–91.

TAYLOR, C. R., SCHMIDT–NIELSEN, K. AND RAAB, J. L., 1970. Scaling of energetic cost of running to body size in mammals. *Amer. J. Physiol.* **219:** 1104–7.

TAYLOR, C. R., SCHMIDT–NIELSEN, K., DMI'EL, R. AND FEDAK, M., 1971. Effect of hyperthermia on heat balance during running in the African Hunting Dog. *Amer. J. Physiol.* **220:** 823–7.

TAYLOR, C. R. AND LYMAN, C. P., 1972. Heat storage in running antelopes, independence of brain and body temperatures. *Amer. J. Physiol.* **224:** 848–51.

TAYLOR, C. R. AND ROWNTREE, V. J., 1973. Temperature regulation and heat balance in running Cheetahs: a strategy for sprinters? *Amer. J. Physiol.* **224:** 848–51.

TAYLOR, C. R., 1976. Exercise and thermoregulation *in* "Environmental physiology II" ed. by D. Robertshaw. University Park Press.

TAYLOR, C. R., SEEHERMAN, H. J., MALOIY, G. M. O., HEGLUND AND KAMAU, J. M. Z., 1978. Scaling maximum aerobic capacity ($\dot{V}_{O_2}max$) to body size in mammals. *Fed. Proc.* **37(3):** 271.

WUNDER, B. A., 1970. Energetics or running activity in Merriam's Chipmunk, *Eutamias merriami*. *Comp. Biochem. Physiol.* **33:** 821–36.

ZUNTZ, N., 1897. Ueber den Stoffverbrauch des Hundes bei Muskelarbeit. *Arch. Ges. Physiol.* **68:** 191–211.

RESUMEN

25. La energía locomotriz de los dasiúrid marsupiales. Por R. V. Baudinette.

El incremento del costo de locomoción (energía uaada por unidad de masa corporal por unidad de distancia recorrida) es similar en los dasiúridos y euterios. Se reexaminan las pretendidas diferencias sostenidas anteriormente. Los dasiúridos muestran una mayor dependencia del calor almacenado durante el ejercicio; en ellos el calor perdido por evaporación excede siempre a aquel perdido por otras vias. Esto contrasta con la estrategia general de los euterios, en los que se produce un mínimo almacenaje de calor y de pérdida por evaporación.

En los dasiúridos existe poca correlación entre los bajos niveles metabólicos en descanso y la respuesta fisiológica en ejercicio. Sólo la dependencia del calor acumulado puede estar claramente relacionada con estte rasgo, pues es la baja temperatura corporal durante el descanso la que provee a estos animales de una mayor capacidad de almacenaje.

26. THE SENSITIVITY OF AUSTRALIAN CARNIVOROUS MAMMALS TO 1080 POISON

J. C. McILROY[1]

Marsupial carnivores appear more tolerant to 1080 poison than eutherian carnivores in Australia. Both groups displayed similar behavioural reactions when poisoned, with wide individual and interspecific variation in the time until onset, sequence, and duration of symptoms. Initial symptoms occurred 0.1-23 hours after ingestion of 1080, tending to appear sooner in animals that ingested higher doses. Most deaths occurred 1.5-147 hours after dosing. All survivors showed signs of recovery after 2-60 hours.

The risk a species faces if it consumes 1080 in baits during Dingo control operations has been assessed. Theoretically, the Dingo faces the greatest risk, followed by the smaller dasyurids and the feral Cat. Of the species studied, the larger dasyurids and the Long–nosed Bandicoot face the least risk. However, from field studies of a simulated aerial baiting programme and a trail baiting campaign against Dingoes, it appears that the actual effect of 1080-poisoning campaigns on carnivore populations may depend upon the pattern of distribution and amount of baits per unit area, the concentration of 1080 in them, the proportion of each population which eat lethal quantities, the time of the year during which the baiting is carried out and the natural resiliency of the populations of species concerned.

Pages 267–71 *in* CARNIVOROUS MARSUPIALS, ed. by M. Archer, February, 1982. Roy. Zool. Soc. New South Wales: Sydney, Australia.

COMPOUND 1080, or sodium fluoroacetate, is a tasteless, odourless poison which inhibits the enzymes responsible for catalysing citrate and succinate metabolism in the body, and thus blocks the Krebs cycle. The resultant lack of energy production causes impairment of cell function either in the central nervous system or cardiac system, depending upon the species involved (Atzert 1971).

1080 is widely used in Australia and overseas against rodents, predators and other vertebrate pests. In Australia it is mainly used against Rabbits, *Oryctololagus cuniculus,* but also against Dingoes, *Canis familiaris dingo,* Pigs, *Sus scrofa,* and in Tasmania, against Brushtail Possums, *Trichosurus vulpecula,* Pademelons, *Thylogale billardierii,* and Bennett's Wallabies, *Macropus rufogriseus.* Eutherian carnivores are recognised as being particularly sensitive to the poison (Chenoweth 1949) but until recently nothing was known about the sensitivity of marsupial carnivores. Consequently, there has been concern about the fate of these non-target animals during poisoning campaigns, particularly those in which meat baits, treated with 1080, are used against Dingoes.

Recently McIlroy (1981b) measured the sensitivity to 1080 of 12 representative species of carnivorous mammals in Australia. These included 10 species of marsupial carnivores. Briefly, the marsupial carnivores tested were markedly more tolerant to 1080 than eutherian carnivores that have been tested both in Australia and overseas (Table 1). The Eastern Native Cat, for example, is nearly four times as tolerant as an eutherian counterpart, the Marten, *Martes americana,* and over nine times more tolerant than the feral Cat, *Felis catus,* while the Tasmanian Devil, *Sarcophilus harrisii,* is 38 times more tolerant than the Dingo. Despite the differences in tolerance both groups of mammals displayed similar behavioural reactions when poisoned, particularly in the time until onset, sequence and duration of symptoms. There was as much variation in response between individuals as between species. Symptoms did not normally appear in poisoned individuals until 0.1 – 23 hours after ingestion of 1080. Typical symptoms, in approximate order of appearance, included depression or increasing sensitivity to stimuli, restlessness, rapid respiration, bouts of trembling, vomiting and convulsions. Most deaths occurred between 1.5 – 147 hours after ingestion of the poison while all survivors showed signs of recovery after 2–60 hours. Inverse relationships generally existed between the amount of 1080 ingested and the time until onset of symptoms, including until onset of convulsions or death. Details of latent periods, times until death or signs of recovery and of symptoms for each species are given by McIlroy (1981b).

Although the smaller mouse–like dasyurids are mainly insectivorous and the Long–nosed Bandicoot, *Perameles nasuta,* is omnivorous, these and other carnivores tested are all known to eat meat baits. All of them except the Tasmanian Devil also occur, or may occur, in areas where meat baits containing 1080 are used for control of Dingoes so they are potentially at risk through eating the baits.

Using LD50s, McIlroy (1981b) assessed the theoretical risk that members of each carnivore

[1] Division of Wildlife Research, CSIRO, P.O. Box 84, Lyneham, A.C.T., Australia, 2602.

Table 1. LD50 values (including 95% confidence limits) for marsupial and eutherian carnivores.

	Species	LD50 and 95% C.L. (mg/kg)	n*
Marsupialia			
Dasyuridae			
Fat-tailed Dunnart[1]	*Sminthopsis crassicaudata*	2.06 (1.58–2.69)	12
Stripe-faced Dunnart[1]	*Sminthopsis macroura*	0.95 (0.57–1.60)	12
Brown Antechinus[1]	*Antechinus stuartii*	1.85 (1.43–2.40)	20
Dusky Antechinus[1]	*Antechinus swainsonii*	3.21 (2.43–4.23)	20
Kowari[1]	*Dasyuroides byrnei*	ca. 2.85	3
Northern Native Cat[1]	*Dasyurus hallucatus*	5.66 (3.91–8.20)	12
Eastern Native Cat[1]	*Dasyurus viverrinus*	3.73 (3.18–4.38)	20
Tiger Cat[1]	*Dasyurus maculatus*	1.85 (1.28–2.68)	12
Tasmanian Devil[1]	*Sarcophilus harrisii*	4.24 2.76–6.60)	20
Peramelidae			
Long-nosed Bandicoot[1]	*Perameles nasuta*	7.70 (5.28–11.23)	12
Carnivora			
Mustelidae			
Marten[4]	*Martes americana*	ca. 1.0	?
Ferret[5,6]	*Mustela putorius*	ca. 1.0–1.4	10
Badger[2]	*Taxidea taxus*	ca. 1.0–1.5	6
Felidae			
Cat[1]	*Felis catus*	0.40 (0.31–0.52)	12
Bobcat[2]	*Lynx rufus*	<0.66	5
Canidae			
Kit Fox[7]	*Vulpes macrotis*	0.20 (0.14–0.31)	8
Grey Fox[2]	*Urocyon cinereoargenteus*	<0.3	4
Coyote[2]	*Canis latrans*	0.10	22
Dog[3]	*Canis familiaris*	0.06	37
Dingo[1]	*Canis familiaris dingo*	0.11 (0.09–0.15)	12

[1]McIlroy (1981b); [2]Ward and Spencer (1947); [3]Tourtellotte and Cooon (1951); [4]Robinson (1953); [5]Marshall (1963); [6]Tucker and Crabtree (1970); [7]Schitoskey (1975). n* = number of animals tested. Data for Dog and Kit Fox recalculated on a 100% 1080 basis to allow comparison with data from McIlroy (1981b); purity of 1080 used for other species not known.

species face from primary poisoning with 1080 during Dingo control operations. Amongst the species studied, the Dingo faces the greatest risk followed in descending order by the smaller dasyurids, the feral Cat and the larger dasyurids and Long-nosed Bandicoot.

A common problem in the assessment of risk is that of extrapolating from the results of experiments on animals in laboratories, generally under highly controlled conditions, to animals living under field conditions. The theoretical assessments of poisoning risk, for example, were based on (i) mean weights of only small samples of animals, (ii) LD50s or measurements of sensitivity to 1080 that are strictly only an indication of the sensitivities that might be expected from repeated trials on the same strain of animal under the same experimental conditions, and (iii) mainly laboratory-dervied data on the potential intake of meat and 1080 by free-living individuals (see McIlroy 1981a,b).

A more practical assessment of risk would be obtained by monitoring the effects of actual poisoning campaigns on field populations of each species. The opportunity to do this with populations of Brown Antechinus, *Antechinus stuartii,* arose during 1977 and 1980. The results obtained have been compared with theoretical assessments of the risk the populations faced during each poisoning campaign.

METHODS

Poisoning campaigns against Dingoes in different regions of Australia consist of either aerial or trail baiting. In aerial baiting the poisoned baits are distributed by dropping them from an aircraft. On some occasions several handfuls of baits or the last dozen or so remaining in a bag are dropped simultaneously. In trail baiting individual baits are laid by hand by a person walking or riding along forest roads or in other places thought to be used by Dingoes. The size of baits and concentrations of 1080 used vary considerably. Sizes range from 6 g manufactured baits to 230 — 250 g ('½ lb') chunks of muscle tissue and ocasionally organs from horses, cattle or kangaroos. Listed concentrations of commercial 1080 (approximately 90% purity) in baits have ranged as high as 1.2 mg/g of bait and amounts of 1080 in a single bait to 50.4 mg. Nowadays the most common concentration of commercial 1080 used in baits is 0.015–0.11 mg/g or the most common amounts per bait, 3.5 – 8.5 mg of 1080.

The first poisoning situation monitored was a simulated aerial baiting in which, in an extreme situation, many baits might land in a small area. The distribution of the baits was far more uniform and their density far higher than would normally be expected. There were two reasons for this. The first was to ensure that there was a high probability that each Brown Antechinus present would find a bait, and consequently that the circumstances would be provided to discover whether this automatically led to their death, as the theoretical assessment suggested. The second reason was simply to find out if such an extreme baiting situation seriously affected the resident Brown Antechinus population. If it did not, then it would be reasonable to assume that Brown Antechinus populations in the same general area would be unlikely to be seriously affected by other more normal aerial baitings at that time of the year.

The second poisoning situation monitored was an actual trail baiting in which the baits were distributed along lines at much lower density than in the simulated aerial baiting, giving every Brown Antechinus far less of an opportunity to discover and eat a bait.

The simulated aerial baiting study was carried out during 1977 in sclerophyll forest in the Brindabella Range, in southeastern New South Wales. Two 6.8 ha rectangular grids each measuring 380 x 180 m were marked out 3 km apart in forest areas with similar habitat conditions. Each area contained 200 trapping stations spaced 20 m apart. Standard Elliott small mammal traps were set at a quarter of the trapping stations in each area for four consecutive nights during each sample period on a rotational system that spread 100 traps (2 per site) throughout each area per night. Trapping bait was a mixture of rolled oats and peanut butter. All Brown Antechinus caught were sexed, weighed, individually marked by toe clipping and released.

On 24 June 1977, immediately after a trapping session, 200 poison baits, marked with rhodamine dye, were distributed by hand so there was one at every grid station in Area A. Another 200 unpoisoned baits marked with rhodamine were simultaneously distributed in a similar fashion in Area B. The size of baits (range 22 104 g, mean 52 g) and concentration of commercial 1080 (0.2 mg/ bait) replicated as closely as possible the type of baits known to be used in aerial baiting campaigns against Dingoes in southern New South Wales. Both areas were retrapped 10 days later.

The trail baiting study was carried out during June and July 1980 in sclerophyll forest, part of Bondo State Forest, near Tumut in New South Wales. Twenty trapping areas, each consisting of 2 lines of 10 traps set 7 m apart, were selected from amongst the areas alongside access roads that were to be baited. The trapping bait used and the treatment of captured animals were similar to those in the simulated aerial baiting study. Each area was trapped for four consecutive nights 19 days before the poisoning and again 7 days afterwards.

On 16 July 275 poison baits were laid alongside 32 km of roads (density of 8.5 baits/km). One poison bait was deliberately laid in the centre of each trapping area but other baits were laid at sites where the bait layer considered Dingoes were most likely to find them. Some of these sites were within 50 m of the trapping areas. The mean size of 95 baits sampled was 212.3 g ± 4.5 g. Each bait had been injected with 1.5 ml of 1080 solution so that it contained 7.5 mg of 1080. All baits were prepared and distributed by personnel of the Forestry Commission of New South Wales.

RESULTS

SIMULATED AERIAL BAITING STUDY: Patchy light snow fell on both areas during the first 24 hours after bait distribution (up to 15 cm on Area B and 5 cm on Area A). Despite this, 92% of the unpoisoned baits disappeared from Area B within 24 hours, 98% within 48 hours and 100% within 4 days. Judging from tracks in the snow, Foxes, *Vulpes vulpes,* appeared to be mainly responsible, followed by Pied Currawongs, *Strepera graculina,* and Ravens, *Corvus* spp. At least 3 of the 13 Brown Antechinus and 2 of the 14 Bush Rats, *Rattus fuscipes,* known to be present immediately before the baiting also fed on the baits because, when subsequently caught, they showed signs of rhodamine dye on their fur and feet, or in their faeces or urine. In contrast, only 7.5% of the poison baits in area A had been taken within 24 hours, 10% within 48 hours, 49% within 4 days and 63% within 10 days. This lower rate of bait removal may well have been due to the immediate poisoning of Foxes and possibly some of the birds. None of the Brown Antechinus or Bush Rats subsequently caught showed any signs of rhodamine dye.

During the two months before the baiting there were no significant differences between the numbers of Brown Antechinus trapped in each area (Table 2). Area A, the area to be poisoned, though, did contain 24–31% *more* antechinus than Area B. Two weeks later, however, after the poisoning, Area A contained 73% *less* antechinus than Area B, a significant difference. It is possible that emigration, trapping success and natural mortality rates differed between each area because of some intrinsic differences in the habitats but it seems unlikely that these could account for such a sudden dramatic change of relative population density indices within 2–3 weeks. During the two months before and two weeks after the baiting there were no significant differences between the sex ratios in each area (the overall 82% reduction in numbers in Area A was spread equally between both sexes). There were also always sufficient empty traps at each trapping station for the trappable animals in the population, the proportion being monitored, to enter and be caught. Instead I think the reduction is likely to have resulted from the poisoning. The species is known to eat the baits and to be theoretically capable of ingesting a lethal dose of 1080. There were also enough baits present within each animal's home range for every individual to experience a high probability of encountering and feeding on a bait.

Table 2. Changes in population numbers of Brown Antechinus before and after a simulated aerial poisoning operation with 1080.

	Trap month	Area A (Poisoned)	Area B (Unpoisoned)
	May	26	21
	June	17	13
Poisoning			
	July	3	11
	Aug	7	6
	Sept	4	6

As might be expected, given the small size of the areas, Brown Antechinus began immigrating into Area A (or at least were trapped for the first time) shortly after the baiting and within seven weeks there was no significant difference again between the trappable populations in both areas (Table 1). By September, as normally occurs after mating, all the males had died in both areas and the females were close to parturition.

TRAIL BAITING STUDY: Patchy snow, up to 8 cm deep, also fell on the trail baiting area before and just after the baiting. The poison baits disappeared at a faster rate from the trails than from the grid area of the simulated aerial baiting study. After one day 28% had disappeared, 46% after two days, 69% after 4 days, 85% after 10 days and 98% after 32 days. All baits disappeared from the trapping areas within the seven days before the traps were reset. Once again, Foxes appeared to be mainly responsible for removal of baits, followed by feral Pigs, Ravens and feral Dogs.

Trapping results indicated that the trail baiting had no significant effect on population numbers of Brown Antechinus; 15 males and 12 females were caught before the poisoning and 16 males and 11 females after the poisoning.

DISCUSSION

Basically, the marsupial carnivores have an advantage over eutherian carnivores during Dingo poisoning operation because of their higher tolerence to 1080. However, this is negated for some species because of their small size. The Brown Antechinus, for example, is nearly 17 times more tolerant to 1080 than the Dingo, but because of its small size (1/450th of that of the Dingo), requires only 0.07 mg of 1080 for a LD50 compared to 1.79 mg for the Dingo. Only the Native Cats, Tasmanian Devil and Long–nosed Bandicoot amongst the marsupial carnivores studied have a greater tolerance ratio than weight ratio (McIlroy 1981b) and because the Dingo and the Tasmanian Devil do not occur in the same area, comparison between their relative susceptibilities to Dingo baits is only of theoretical interest. The feral Cat is in much the same position as the Dingo; what it gains by being slightly more tolerant it loses through being smaller. Similarly, although the marsupial carnivores eat far less than the Dingo (only 1–44% of its intake, see McIlroy 1981b), and thus should ingest far less 1080, the amounts they do consume represent a much higher proportion of their body weight than for the Dingo. Consequently, their small size and disproportionally greater appetites would appear to place some of them at risk during Dingo poisoning campaigns.

In theory, Fat–tailed Dunnarts, *Sminthopsis crassicaudata,* Stripe–faced Dunnarts, *S. macroura,* Brown Antechinus, *Antechinus stuartii,* Dusky Antechinus, *Antechinus swainsonii,* and Kowari, *Dasyuroides byrnei,* could all eat sufficient bait containing 0.015 mg/g of commercial 1080 (a relatively low concentration advocated for some campaigns) during a poisoning campaign for them to face a 73–100% chance of being killed (McIlroy 1981b). Such a concentration though, would pose far less of a risk to the larger marsupial carnivores. The overall situation would obviously be far worse were much higher concentrations of 1080 used.

Narrowing the discussion down to the two field poisoning situations studied, theoretical assessments of risk before each baiting indicated that if each individual in each population found a bait, it would be poisoned. These assessments were based on the LD50 for Brown Antechinus (i.e. 1.85 mg/kg), the mean weight of individuals in each population prior to baiting (i.e. 19.4 g for aerially–baited animals and 20.8 g for trail-baited animals), the concentration of 1080 used (0.2 mg/g and ca. 0.035 mg/g, respectively), plus the amount of bait each individual could possibly eat (McIlroy 1981b). In each case deaths of all individuals in the populations did not occur but it is evident from the results that the higher density of baiting and/or higher concentration of 1080 involved in the simulated aerial baiting had a greater reducing effect on Brown Antechinus numbers than those associated with the trail baiting. Even so, it seems unlikely, given the small areas of actual baiting involved and the high probability of ensuing immigration, that aerial baiting in general during this time of the year would have any long–lasting effects on

Brown Antechinus populations. The situation might, however, be different if the poisoning was carried out during other months of the year, particularly during spring and early summer before the young antechinus become independent.

The field studies revealed, as anticipated, that theoretical assessments cannot allow for all the variables involved and possibly over-estimate the risk carnivores face from 1080 poisoning. For example, for ease of computation, the relative concentrations of 1080 involved were assumed to be evenly distributed throughout each bait. This is not always so (T. Korn, pers. comm. 1979), which means that animals eating similar amounts of bait could ingest different amounts of 1080. The number and distribution of baits in relation to the population density of each non–target species is also likely to influence the effect of the poisoning campaign on different animals as will the length of time the baits remain available to each species before they are eaten by other animals (e.g. insects and birds), or rot away or the 1080 is leached out by dew or rainfall.

Detailed field monitoring of populations before and after poisoning campaigns, however, is not always practical, particularly for those species which are rare or occur naturally at low densities. Even with Brown Antechinus, one of the most common species of small mammals in the Southern Highlands of New South Wales, it was difficult to monitor sufficient numbers for statistical comparisons before and after poisoning. The densities involved, though, were typical of those normally found in the habitats that, in this region, are baited for Dingoes during June or July (unpubl. data). Theoretical assessments based on sensitivity, therefore, might be the only way of estimating the risk faced by individuals in some species. Perhaps the precision of such assessments could be increased if more data were used in the assessment on the proportions of each population that are likely to find and eat lethal amounts of baits, given a certain density of baiting. In the meantime, management of poisoning operations to minimise the concentration of 1080 in baits and to limit the number distributed per unit area will obviously reduce or avert the possibility of primary poisoning of non-target carnivorous mammals.

ACKNOWLEDGEMENTS

The field studies were carried out under permits granted by the National Parks and Wildlife Service of New South Wales and the Forestry Commission of New South Wales. Messrs S. Bishop and D. Ridley of the latter organization cheerfully co–operated over requirements for the study during the trail baiting. Technical assistance was provided by Messrs R. Cooper and E. J. Gifford. Dr. Steven Morton and Mr. R. Warneke read and constructively criticised an earlier draft of this manuscript.

REFERENCES

ATZERT, S. P., 1971. A review of sodium monofluoroacetate (compound 1080). Its properties, toxicology and use in predator and rodent control. *U.S. Fish and Wildl. Serv. Spec. Sci. Rep. Wildl.* **146.**

CHENOWETH, M. B., 1949. Monofluoroacetic acid and related compounds. *J. Pharmac. exp. Ther.* **97**: 383–424.

McILROY, J. C., 1981a. The sensitivity of Australian animals to 1080 poison. I. Intra–specific variation and factors affecting acute toxity. *Aust. Wildl. Res.* **8:** (In press).

McILROY, J. C., 1981b. The sensitivity of Australian animals to 1080 poison. II. Marsupials and eutherian carnivores. *Aust. Wildl. Res.* **8:** (In press).

MARSHALL, W. H., 1963. The ecology of mustelids in New Zealand. *N.Z. Dept. scient. ind. Res. Inf. Ser.* **38:** 1–32.

ROBINSON, W. B., 1953. Coyote control with Compound 1080 stations in national forests. *J. For.* **51:** 880–5.

SCHITOSKEY, F., 1975. Primary and secondary hazards of three rodenticides to Kit Fox. *J. Wildl. Manage.* **39:** 416–8.

TOURTELLOTTE, W. W. AND COON, J. M., 1951. Treatment of fluoroacetate poisoning in mice and dogs. *J. Pharmac. exp. Ther.* **101:** 82–91.

TUCKER, R. K. AND CRABTREE, D. G., 1970. "Handbook of toxicity of pesticides to wildlife". *U.S. Fish and Wildl. Serv. Res. Publ.* **84:**

WARD, J. C. AND SPENCER, D. A., 1947. Notes on the pharmacology of sodium fluoroacetate–compound 1080. *J. Am. Pharm. Assoc.* **36:** 59–62.

RESUMEN

26. La sensibilidad de los mamiferous carnivoros australianos al Veneno 1080. Por J. C. McIlroy.

Los marsupiales carnívoros parecen ser màs tolerantes al Veneno 1080 (fluoracetato de sodio) que los carnívoros euterios en Australia. Ambos grupos desplegaron similares reacciones de comportamiento cuando fueron envenenados, con una amplia variacón individual e interespecífica en la manifestacion, secuencia y duración de los síntomas. Los sintomas iniciales ocurrieron entre 0.1 y 23 horas después de la ingestíon del 1080, con tendencia a aparecer antes en los animales que ingirieron mas altas dosis. La mayor parte de las muertes ocurrieron entre 1.5 y 147 horas después de dosificados. Todos los sobrevivientes mostraron signos de recuperación después de 2 a 60 horas.

El riesgo que enfrentan las especies si consumen el 1080 encebos durante las operaciones de control del dingo ha sido evaluado. Teòricamente el dingo enfreta el màs grande riesgo, seguido por los más pequénos dasiúridos y por el goto salvaje. De las especies estudiadas, los más grandes dasiúridos y el "bandicoot" de hoico largo corren el menor riesgo. Sin embargo, de los estudios de campo de un programa de envenenamiento aereo simulado y de una campána de envenenamiento por rastreo del dingo, parece que el verdadero efecto de las campànas con el 1080 contra las poblaciones de carnívoros puede depender del tipo de distrbuciòn y de la cantidad del cebo por unidad de área, de la concentraciòn del 1080 en ellos, del tiempo del áno durante el que es cumplido y de la natural tolerancia de las poblaciones de especies involucradas.

27. PHYSIOLOGY OF THE TASMANIAN DEVIL *Sarcophilus harrisii* (DASYURIDAE, MARSUPIALIA)

S. C. NICOL[1]

The Tasmanian devil, *Sarcophilus harrisii*, despite its large size shows little evidence of departure from the conservative life strategy of the majority of small dasyurids. Most physiological studies on *S. harrisii* have centred on thermoregulation and energy metabolism. Evaporative heat loss is exclusively by panting, both during exercise and exposure to high temperatures, although a high proportion of heat is stored under these conditions. At neutral temperatures thermoregulation may be relaxed leading to a shallow torpor. Basal heat production is very low (1W.Kg^{-1}).

None of the haematological parameters for *S. harrisii* are unusual but devil blood has been reported to have a low oxygen affinity. Arterial P_{CO_2} is normally low (30 mm Hg) but may rise during sleep, while venous P_{O_2} is high.

Pages 273–78 *in* CARNIVOROUS MARSUPIALS, ed. by M. Archer, February, 1982. Roy. Zool. Soc. New South Wales: Sydney, Australia.

THE TASMANIAN DEVIL is the largest of the extant marsupial carnivores, and this gives an intrinsic interest to studies of the species. But, in addition, the large size of *S. harrisii,* ease of trapping, and the fact that it is surprisingly easy to handle make it possible to carry out physiological investigations which might not be possible with the smaller dasyurids.

In this paper I will attempt to bring together aspects of devil physiology from the literature, and new data from our laboratory. The reproductive physiology of *S. harrisii* is dealt with elsewhere in this volume.

If this review appears unbalanced, it reflects the uneven distribution of research effort in marsupial physiology in general, with work being concentrated on such areas as temperature regulation and energy metabolism while large areas such as digestion in carnivorous marsupials have been nearly totally neglected. This is particularly evident in an animal such as *S. harrisii* to which only a small number of workers have had access.

It is hoped that by bringing together the available data on a single species, this review will point out areas for future study.

THERMOREGULATION AND ENERGY METABOLISM

The thermoregulatory physiology of *S. harrisii* has been recently reviewed by Nicol and Maskrey (1980). Despite a report by Robinson and Morrison (1957) that *S. harrisii* sweats, high air temperatures do not affect cutaneous water loss during rest or exercise (Hulbert and Rose 1972, Bell, Baudinette and Nicol, in preparation). Panting occurs at ambient temperatures above the thermoneutral zone, which extends from 28.5-32°C. At temperatures below neutrality heat production is raised by shivering, there is no indication of non-shivering thermogenesis. These effectors provide good thermoregulation at both low and high environmental temperatures but in intermediate conditions *S. harrisii* has a labile body temperature (Tb). We have measured body temperatures from 31-38°C from animals resting in neutral temperatures. Cold conditions however stimulate increased heat production apparently through skin receptors rather than central thermoreceptors, producing a rise in body temperature. Panting at high temperatures prevents a rise in Tb past 38.5°C, even after several hours at Ta = 40°C, but during exercise Tb may rise further (Bell, Baudinette and Nicol, in preparation). Because of this labile Tb a large proportion of the heat produced during exercise is stored (Table 1). This passive variation in Tb over a wide range must reduce energy expenditure and in fact the Tasmanian Devil appears to possess a number of such energy conserving mechanisms, including shallow daily torpor which probably occurs in all dasyurids.

Table 1. Heat balance of *S. harrisii* running at 6 km/hr (from Bell, Baudinette and Nicol, in preparation).

Respiratory evaporative heat loss	44.5%
Cutaneous evaporative heat loss	1.3%
Non-evaporative heat loss	15.6%
Heat storage	38.6%
Heat production	21.3 ± 1.1 W/kg (n=5)

In our laboratory resting Devils, once accustomed to experimental procedure, nearly always fell asleep, spending a large proportion of this time in paradoxial sleep (PS). Although some authors have suggested that

[1] Physiology Department, University of Tasmania, Hobart, Tasmania, Australia, 7000.

during PS mammals become poikilothermic (Parmeggiani 1977) we observed a weak panting response to high temperatures by *S. harrisii* during PS. There is however a significant relaxation of thermoregulation during PS which will result in further energy savings. MacMillen and Nelson (1969) measured a basal heat production of 1.57 W/kg, but during sleep this may drop to 1 W/kg.

Water turnover of an animal will be related to its metabolic rate, and Nicol (1978) measured basal water turnover rates of 100 ± 7 ml/g $^{0.82}$ for *S. harrisii* at *Ta* = 20°C. Green and Eberhard (1979) measured water turnover rates at Ta = 13°C, and obtained values of 129 ± 11 ml/kg$^{0.82}$. This higher value can be attributed to the higher overall rate of metabolism at the lower air temperature. Green and Eberhard (1979) also investigated food intake and digestibility in *S. harrisii* and *Dasyurus viverrinus,* finding an apparent dry matter digestibility of 80% and apparent energy digestibility of 88% in both species; these digestive efficiencies being similar to those of eutherian carnivores.

CARDIOVASCULAR PHYSIOLOGY

Heart rate is also related to metabolic rate, and Kinnear and Brown (1967) found a minimum heart rate for *S. harrisii* of 54 beats/minute for one individual, and 66 beats/minute for another. During our thermoregulatory study (Nicol and Maskrey, in press), we measured heart rate in some animals using a telemetered ECG signal, and found a mean value of 102 ± 17 beats/min. (n = 20). We observed a large sinus arrhythmia in the heart rate, so that when measured on a beat-to-beat cardiotachometer heart rate often fell to about 50 beats/min. between breaths. During direct measurement of the arterial pressure via a carotid artery we obtained a mean rate of 142 ± 30 beats/min. (mean of 29 observations on 5 animals). This high rate may be in part attributed to a higher mean body temperature of these animals, and an increased cardiac output due to slight anaemia following surgery.

Parsons *et al.* (1970) measured an arterial pressure of 94/42 in a deeply anaesthetized animal. from resting, unanaesthetized animals at thermoneutrality we obtained a systolic pressure

Table 2. Summary of the haematological data for *S. harrisii*, showing mean values and ranges. 1, Blood taken by cardiac puncture into syringes containing sodium edetate; 2, Parsons *et al.* (1971a); 3; Bland and Holland (1977).

	Bartels *et al.* (1966) n=1	Parsons *et al.* (1970) n=3	This report[1] n=4	Average for other marsupials
Red blood cells (10^{12}/1)	6.70		6.33 (5.84 – 6.59)	
Haematocrit (%)	47	33.4 (28 – 38)	40.8 (39.1 – 41.7)	47[2] (43 – 51)
Mean corpuscular volume (fl)	70.1		65.5 (64 – 68)	
Haemoglobin (g/dl)	20.06	12.2 (11 – 13)	14.8 (13.8 – 15.0)	15.8(14.1 – 17.1)[2] 15.7(13.2 – 17.4)[3]
Mean corpuscular haemoglobin (pg)	30		23.2 (22.3 – 24.1)	
Mean corpuscular haemoglobin concentration (g/dl)	42.7		35.2 (34.6 – 35.5)	33.4[2] (32 – 35)
White blood cells (10^9/1)		15 (n = 1)	12.2 (8.6 – 14.6)	7[2] (2.8 – 13)
Polymorphs (%)		74 (73 – 76)	48 (44 – 52)	42[2] (23 – 56)
Lymphocytes (%)		24 (20 – 27)	49 (45 – 51)	51[2] (31 – 69)
Monocytes (%)		2 (0 – 4)	3.5 (2 – 6)	6[2] (1 – 13)

Table 3. Blood gas and acid–base data for *S. harrisii.* Measurements were made at 37°C and converted to body temperature using the method of Kelman and Nunn (1966) and the oxygen dissociation curve of Bartels *et al.* (1966).

		mean	S.D.	No. observations	No. animals
Arterial	pH	7.440	0.04	44	6
	PO_2 (mm Hg)	98.5	15	43	6
	PCO_2 (mm Hg)	30.8	3.95	44	6
	HCO_3^- (mmol/1)	19.3	2.24	44	6
	Tb (°C)	37.3	0.87	43	6
Venous	pH	7.438	0.029	7	1
	PO_2 (mm Hg)	57.4	10	6	1
	PCO_2 (mm Hg)	33.1	3.8	7	1
	HCO_3^- (mmol/1)	21.1	2.9	7	1
	Tb (°C)	31.1	0.58	7	1
Arteriovenous difference	pH	− 0.0226	0.0187	7	1
	PO_2 (mm Hg)	−58.7	16.8	6	1
	PCO_2 (mm Hg)	+ 4.9	1.3	7	1
	HCO_3^- (mmol/1)	+ 2.1	0.74	7	1

of 131 ± 26 mm Hg, and diastolic pressure of 98 ± 23 mm Hg (mean of 28 observations on 5 animals).

PLASMA ENZYMES AND HAEMATOLOGY

While levels of most enzymes in the blood of *S. harrisii* are similar to those in other mammals, there are some exceptions. Levels of transaminases and lactic dehydronase are slightly elevated above normal mammalian levels (Parsons *et al.* 1971b) but the most noteworthy feature of Devil plasma is the extremely high acid phosphatase activity. Of all the other marsupials or eutherian mammals in which it has been examined, only the closely related Tiger Quoll *Dasyurus maculatus* has a comparable level of acid phosphatase, with an activity 2/3 that of *S. harrisii* (see review by Sallis and Guiler 1977). It now appears that the unusually high levels of acid phosphatase in Devil (and presumably Tiger Quoll) plasma may result, not from tissue spillage, but from certain chemical or structural properties that afford the enzyme a long circulating half-life (J. Sallis, personal communication).

White Blood Cells: The two sets of values shown in Table 2 for the differential white cell count differ markedly but the figures obtained in the present study are in closer agreement with the values reported by Parsons *et al.* (1971) for 6 species of marsupial other than *S. harrisii.*

Table 4. Changes in ventilation measured in three experiments in which Devils breathed 10% O_2.

	% change	S.D.
Respiratory frequency	+ 199	135
Tidal volume	— 16	18
Total ventilation	118	86

Red Blood Cells: The red cell count is well within the normal mammalian range (Table 2). Our value for the haematocrit lies between the values obtained by Bartels *et al.* (1966) and Parsons *et al.* (1970). Bartels *et al.,* reported that *S. harrisii* had small cells containing a high concentration of haemoglobin. Our results confirm the small size, but show a normal corpuscular haemoglobin concentration.

RESPIRATION

Respiratory Functions of Blood: Bartels *et al.* (1966) constructed an oxygen dissociation curve for *S. harrisii* using blood from one animal from the Basel Zoo. They obtained a value for P_{50}, the oxygen tension at which half the haemoglobin is saturated with O_2 of 41.2 mm Hg (at 37°C and pH 7.4) and a Bohr effect, $-\Delta \log PO_2/\Delta pH$, of 0.47. The Bohr effect value is well within the normal mamalian range, but the P_{50} is one of the highest reported for any mammal. The mean marsupial P_{50} is 33.6 ± 8.1 mm Hg (Mean of 7 species, data from Bland and Holland 1977 and Hoversland *et al.* 1973) when calculated at 31°C and pH 7.4. Values higher than *S. harrisii* have been reported from *Antechinus stuartii* the only other dasyurid studied, and the opossum, *Didelphis virginiana.* Schmidt-Nielsen and Larimer (1958) and Bartels (1964) proposed that an inverse relationship existed between the body weight of mammals and their P_{50}; a lower oxygen affinity (high P_{50}) for small animals being consistent with their higher metabolic rate per unit body weight, and Bland and Holland (1977) reported such a relationship for a number of marsupials. However Lahiri (1975) demonstrated that no such physiologically signficant correlation exists for eutherian mammals and an examination of the literature, particularly the P_{50} values gathered by Metcalfe and his co-workers, confirms this. The P_{50} obtained for *S. harrisii* by Bartels *et al.* (1966) departs widely from the relationship found by Bland and Holland as does the value for the opossum (Hoversland *et al.* 1973). Parer *et al.* (1970) point out that a high P_{50} if combined with a high oxygen capacity will lead to a high oxygen tension in mixed venous and in tissue capillary blood unless the cardiac output is depressed. Bartels *et al.* (1966) reported a "remarkably high" oxygen capacity for *S. harrisii* of 26.9 $ml O_2$/100 ml blood. This high oxygen capacity was associated with haemoglobin of 20.06 g/dl, however, a comparison with other measurements of Hb in *S. harrisii* and other marsupials (Table 2) suggests that this was not a representative Hb concentration, and the normal oxygen carrying capacity could be expected to be correspondingly lower.

Parsons *et al.* (1971b) measured blood gases in *S. harrisii* and obtained the following values: pH = 7.29; Po_2 : 62 mm Hg; Pco_2 : 39 mm Hg; HCO_3^- 18 mmol/1. However the animal from which the sample was taken was in deep ether anaesthesia which will produce both respiratory and metabolic acidosis. We have measured arterial blood gas values in six Tasmanian Devils by sampling through a carotid catheter, in addition in one of these animals a catheter was inserted in the jugular vein and advanced until in the right ventricle, allowing sampling of the mixed venous blood. Samples were taken from conscious animals resting in a metabolic chamber at least 24 hours after surgery. Results are summarized in Table 3.

Sarcophilus harrisii has a low arterial Pco_2 implying a high ratio of alveolar ventilation to metabolic CO_2 production. This would also explain the high pH and Po_2. A high Po_2 would ensure adequate O_2 loading of the haemoglobin at the lungs despite a high P_{50}.

The mean venous Po_2 of 57 (range 42–69) is very high, and although consistent with a high P_{50}, is well outside the range reported by Dhindsa *et. al.* (1979) for a number of mammalian species. Dhindsa *et al.*, suggested that venous Po_2 is maintained at approximately 40mm Hg in all mammals, species with a high P_{50} having a lower tissue blood flow. In the one Tasmanian Devil for which we have information this does not appear to be the case. Clearly this point needs further investigation as there are important implications for tissue oxygen delivery, particularly during arterial hypoxia and periods of high metabolic rate.

CONTROL OF RESPIRATION: All mammals respond to a raised CO_2 level or low O_2 level in their inspired air by increasing total ventilation (VE). *S. harrisii* responds to hypercapnia by increasing both respiratory frequency (f) and tidal volume (V_T) with the principal effect being on V_T, while hypoxia produces a large increase in frequency and a small fall in V_T (Table 4). This is similar to the response seen by Farber (1972) in adult opossums.

Figure 1 shows the results of one experiment in which a Devil breathing air was then switched to 10% O_2. This figure also illustrated some of the points raised in the previous section.

A fall in arterial Po_2 stimulated a threefold increase in f leading to an increased ventilation despite the slight fall in V_T. The increased VE leads to a fall in arterial and venous Pco_2 and a rise in pH, while the concomitant increased respiratory evaporative water loss produced a fall in Tb. As the fall in venous Po_2 was small compared with the fall in arterial Po_2, oxygen delivery to the tissues had to be maintained by increasing cardiac output. This increased work of the heart and respiratory muscles were presumably responsible for the large increase in

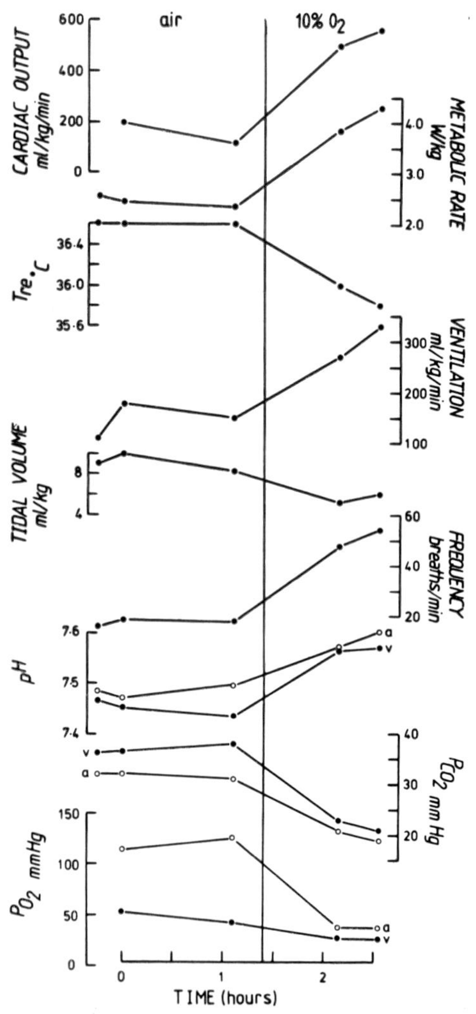

Fig. 1. Effect of exposing a Tasmanian Devil to 10% O_2.

metabolic rate. By contrast Farber *et al.* (1972) observed a fall in metabolic rate in the adult Opossum during hypoxia.

CONCLUSIONS

The data reviewed here suggest that some aspects of Devil physiology, particularly those concerned with gas transport, are unusual by general mammalian standards. It is unlikely however that *S. harrisii* is unique in this regard, but rather this shows the need for further investigation of oxygen transport and haemodynamics in dasyurid marsupials, particularly since information is now available for two diprotodont marsupials (Needham and Dawson, in press) and the American Opossum (Hoversland *et al.* 1975). One measurement which would seem worthwhile making on *S. harrisii* bearing in mind the low oxygen affinity of the blood is the concentration of 2,3-diphosphoglycerate (2,3–DPG) in the red cells. 2,3–DPG regulates the oxygen affinity of haemoglobin in mammals, but a number of eutherians, principally the ruminants and Feloidea, have haemoglobin of low oxygen affinity (high P_{50}'s) and very low concentrations of 2,3–DPG (see review by Bunn, 1980). Kay (1977) has suggested this is associated with a high resting body temperature and a capacity to mobilize energy quickly. Such animals show bursts of activity producing large increases in body temperature and a fall in plasma pH, both of which facilitate oxygen unloading. While *S. harrisii* has a low resting body temperature, it shows greater rise in body temperature during strenuous activity than does the archetypal sprinter, the Cheetah. Hence it would be of considerable interest to know the 2,3–DPG level of Devil red cells, and also the plasma pH changes which occur during exercise, and how these contribute to oxygen delivery.

This perhaps underscores the interrelation of physiological systems and the value of investigating a wide range of physiological parameters in a single species.

ACKNOWLEDGEMENTS

I am indebted to those people who assisted in obtaining the new data presented here.

R. Aitkenhead of Hobart Pathology Services for the haematological analyses, L. McLeod for cannulating the animals, M. Maskrey for assistance in all parts of the investigation, and D. Lovell for technical assistance. Drs. T. Hulbert and L. Hughes read and constructively criticised a draft of this paper.

REFERENCES

BARTELS, H., 1964. Comparative physiology of oxygen transport in mammals. *Lancet* **2**, Part 1: 599–604.

BARTELS, H., RIEGEL, K. AND KLEIHAUER, E., 1966. Comparative studies of the respiratory functions of mammalian blood. II Marsupialia: Great Grey Kangaroo and Tasmanian Devil. *Resp. Physiol.* **1:** 145–50.

BELL, B. C., BAUDINETTE, R. V. AND NICOL, S. C. (In prep.) The role of sweating during exercise in marsupials.

BLAND, D. K. AND HOLLAND, R. A., 1977. Oxygen affinity and 2,3–diphosphogylycerate in blood of Australian marsupials of differing body size. *Resp. Physiol.* **31:** 279–90.

BUNN, H. F., 1980. Regulation of haemoglobin function in mammals. *Amer. Zool.* **20:** 199–211.

DHINDSA, D. S., HOVERSLAND, A. S. AND METCALFE, J., 1979. Resting haemodynamics and oxygen transport in unanaesthetized adult cats. *Resp. Physiol.* **38:** 105–14.

FARBER, J. P., 1972. Development of pulmonary reflexes and pattern of breathing in the Virginia Opossum. *Resp. Physio.* **14:** 278–86.

FARBER, J. P., HULTGREN, H. N. AND TENNEY, S. M., 1972. Development of the chemical control of breathing in the Virginia Opossum. *Resp. Physiol.* **14:** 267 77.

GREEN, B. AND EBERHARD, I., 1979. Energy requirements and sodium and water turnovers in two captive marsupial carnivores: the Tasmanian Devil, *Sarcophilus harrisii*, and the Native Cat *Dasyurus viverrinus*. *Aust. J. Zool.* **27:** 1–8.

HOVERSLAND, A. S., DHINDSA, D. S., MURPHY, W. S., PARER, J. T. AND METCALFE, J., 1973. Respiratory properties and 2,3–diphosphoglycerate (DPG) concentration in blood of the adult Opossum. *Physiologist.* **16:** 348.

HOVERSLAND, A. S., MURPHY, W. S., DHINDSA, D. S., PARER, J. T. AND METCALFE, J., 1975. Oxygen transport and haemodynamics in unanaesthetized American Opossum *(Didelphis virginiana). Comp. Biochem. Physiol.* **50A:** 519–25.

HULBERT, A. J. AND ROSE, R. W., 1972. Does the Devil sweat? *Comp. Biochem. Physiol.* **43A:** 219–22.

KAY, F. R., 1977. 2,3–diphosphoglycerate, blood oxygen dissociation and the biology of mammals. *Comp. Biochem. Physiol.* **57A:** 309–16.

KELMAN, G. R. AND NUNN, J. F. 1966. Nomograms for correction of blood Po_2 Pco_2 pH and base excess for time and temperature *J. appl. Physiol.* **21:** 1484–90.

KINNEAR, J. E. AND BROWN, G. D., 1967. Minimum heart rates of marsupials. *Nature* **215.**

LAHIRI, S., 1975. Blood oxygen affinity and alveolar ventilation in relation to body weight in mammals. *Am. J. Physiol.* **229:** 529–36.

MACMILLEN, R E. AND NELSON, J. E., 1969. Bioenergetics and body size in dasyurid marsupials. *Am. J. Physiol.* **217:** 1246–51.

NEEDHAM, A. D. AND DAWSON, T. J., 1980. Basal cardiovascular parameters in marsupials. *Aust. Mam. Soc. Bull.* **6**(2): 54–5.

NICOL, S. C., 1978. Rates of water turnover in marsupials and eutherians: a comparative review with new data on the Tasmanian Devil. *Aust. J. Zool.* **26:** 465–73.

NICOL, S. C. AND MASKREY, M., 1980. Thermoregulation, respiration and sleep in the Tasmanian Devil *Sarcophilus harrisii* (Marsupialia: Dasyuridae). *J. Comp. Physiol.* **B140:** 241–8.

PARER, J. T., HOVERSLAND, A. S. AND METCALF, J., 1970. Comparative studies of the respiratory functions of mammalian blood VI Young lion and tiger. *Resp. Physiol.* **10:** 30–7.

PARMEGGIANI, P .L., 1977. Interaction between sleep and thermoregulation. Waking and Sleeping. **1:** 123–32.

PARSONS, R. S., HEDDLE, R. W. L., FLUX, W. G. AND GUILER, E. R., 1970. Studies on the blood of the Tasmanian Devil. *Comp. Biochem. Physiol.* **32:** 345–51.

PARSONS, R. S., ATWOOD, J., GUILER, E. R. AND HEDDLE, R. W. L., 1971. Comparative studies on the blood of monotremes and marsupials – I Haematology. *Comp. Biochem. Physiol.* **39*B*:** 203–8.

PARSONS, R. S., GUILER, E. F. AND HEDDLE, R. W. L., 1971b. Comparative studies on the blood of monotremes and marsupials – II Electrolytes organic constituents, proteins, gas analysis and enzymes. *Comp. Biochem. Physiol.* **39*B*:** 209–17.

ROBINSON, K. W. AND MORRISON, P. R., 1957. The reaction to hot atmospheres of various species of Australian marsupial and placental animals. *J. Cell. Comp. Physiol.* **49:** 455–78.

SALLIS, J. D. AND GUILER, E. R., 1977. Unusual plasma acid phosphatase activity in two species of Tasmanian marsupials. *In* "The biology of marsupials" ed. by B. Stonehouse and D. Gilmore. MacMillen Press: London.

SCHMIDT–NIELSEN, K. AND LARIMER, J. L., 1958. Oxygen dissociation curves of mammalian blood in relation to body size. *Am. J. Physiol.* **195:** 424–8..

RESUMEN

27. Fisiología del "Tasmanian Devil", *Sarcophilus harrisii* (Dasyuridae, Marsupialia). Por S. C. Nicol.

El "Tasmanian Devil", *Sarcophilus harrisii,* muestra que a pesar de sugran tamáno poco se desviado de la conservadora estrategia de vida la mayoría de los pequeños dasiúridos. La mayoría de los estudios fisiológicos sobre *S. harrisii* se han centrado en la termoregulación y enelmetabolismo de la energía. La pérdida de color por evaporación es exclusivamente por jadeo, tanto durante actividades físicas como en las exposiciones a altas temperaturas, aunque un alta proporción del calor es tlmacenado bajo esas condiciones. En temperaturas nuetralales la termoregulación puede mitigarse llevando a un incipiente letargo. La producción de calor básico es muy bajo ($1W.Kg^{-1}$).

Ninguno de los parámetros hematológicos de *S. harrissii* es inusual, pero la sangre del "Devil" ha mostrado tener una baja atracciónal oxígeno. El P_{CO_2} arterial es normalmentse bajo (30 mm Hg), pero puede aumentar durante el sueño, mientras que el Po_2 es alto.

NOTE ADDED IN PROOF

Measurements in our laboratory on blood from three Devils gave a whole blood P_{50} of 44 mm Hg at 37°C and $P^{H}=7.4$, slightly higher than that found by Bartels *et al.* (1966), while blood from a smaller dasyurid, the Eastern Quoll (*Dasyurus viverrinus,* weight 1.2 kg), had a P_{50} of 38 mm Hg under the same conditions.

28. COLD ACCLIMATION IN A SMALL DASYURID MARSUPIAL: *Antechinus stuartii*

WERNER REYNOLDS[1] and A. J. HULBERT[1]

Strategy of cold adaptation was compared in the dasyurid marsupial mouse *Antechinus stuartii* and the eutherian mouse *Mus musculus*. Each species consisted of a cold adapted group (6 weeks at 5°C) and a warm adapted group (6 weeks at 25°C). Cold adapted *Antechinus* maintained their body temperature at 5°C by an increased heat production compared to warm adapted *Antechinus* who were unable to maintain a constant body temperature in the cold. Cold adapted *Mus* were also better able to combat cold than were warm-adapted *Mus* due to an increased heat production.

This was due to development of non-shivering thermogenesis (mediated by noradrenaline). The increased heat production in cold adapted *Antechinus* was not due to noradrenaline-mediated non shivering thermogenesis. There was no change in either body conductance or fur insulation during cold adaptation in either *Mus* or *Antechinus*. The marsupial mice were more successful at cold adaptation than the *Mus*, they showed a similar strategy of cold adaptation as the classical eutherian pattern but appeared to use a different heat production mechanism.

Pages 279–83 *in* CARNIVOROUS MARSUPIALS, ed. by M. Archer, February, 1982. Roy. Zool. Soc. New South Wales: Sydney, Australia.

OVER THE LAST 15–20 YEARS, thermoregulation in marsupials has been studied in some detail. Most studies have, however, been concerned with describing the responses of a variety of marsupials during short–term exposure to different temperatures. Virtually no work has been done on any changes that may take place in marsupials following long term exposure to either temperature extreme. Small mammals are particularly affected by cold environments for two reasons; the first is because they have a relatively large surface area through which to lose body heat; the second is that their small size precludes the carrying of a thick insulating layer of fur. Extensive studies on cold acclimation in eutherian mammals (mainly rodents) have shown that they have a distinctive response to long–term cold, they increase their heat production by a non–shivering mechanism that involves brown adipose tissue and is mediated by the release of noradrenaline from nerve terminals (Jansky 1973).

The present study was designed to determine whether (i) small marsupials have an acclimation response to long–term cold exposure as do small eutherians and (ii) whether any such response includes noradrenaline mediated non–shivering thermogenesis as it does in small eutherian mammals. For this reason we have compared the response of the small dasyurid marsupial *Antechinus stuartii* with a similar sized eutherian mammal, the House Mouse *Mus musculus*.

MATERIALS AND METHODS

The *Antechinus stuartii* were caught in the Wollongong area during March – April, 1979. Once caught they were housed in individual cages and allowed to adjust to captivity for at least 2 weeks. They were fed with fresh dog food every day and given *ad libitum* water. The *Mus musculus* were housed individually in identical cages and were fed Allied Foods Rat and Mouse Kubes with water *ad libitum*. Both species were given wood wool with which they could build a nest. Once they had settled, they were randomly divided into two groups; one transferred to a cold room at 5°C (±3°C) whilst the other two groups were housed at 25°C (±3°C). Both the cold–exposed group and the warm–exposed group were exposed to a 12 hr light: 12 hour dark photoperiod. After 6 – 8 weeks exposure to the above conditions the following parameters were measured: (a) body temperature and oxygen consumption at 5°C and 30°C (b) metabolic response to a noradrenaline injection and (c) insulation value of the fur pelt.

All temperatures were measured using 42 S.W.G. copper constantan thermocouples. The rectal thermocouple was inserted 2cm and taped to the animals tail. All thermocouples were calibrated to 0.1°C against a standard thermometer and their outputs were continuously recorded on a Leeds & Northrup Speedomax recording potentiometer.

Oxygen consumption was measured on non-fasting animals using an open flow system. Each animal was weighed, its rectal thermocouple inserted and was placed in a small fibreglass gauze restraining cage, which was then sealed in a perspex metabolic chamber of about 500 cm^3 capacity. Air flow through the chamber varied from 60–140ml/minute and was measured with a Brooks Sho–rate flowmeter. Oxygen content of

[1] Department of Biology, University of Wollongong, Woollongong, New South Wales, Australia, 2500.

inflow and outflow air was measured using a Taylor Servomex type OA272 oxygen analyzer, the output of which was recorded on the Leeds & Northrup recording potentiometer. Heat production was calculated using a conversion factor of 5.59 W.lo_2^{-1}.hr. The animal was placed in a temperature controlled cabinet and the temperature and oxygen consumption were monitored firstly at 5°C and 30°C. Each oxygen consumption value to the average over 20–30 minutes. In most cases the animals were in equilibrium with environmental temperature. However, both warm–exposed groups became hypothermic at 5°C and for these animals it was thus necessary to average their body temperature and oxygen consumption over the measurement period.

Body conductance (W.m^{-2}.°C^{-1}) was calculated using the following formula:

$$\text{Conductance} = \frac{\text{Heat production}}{\text{Surface area (Body temp.–Ambient temp)}}$$

In those cases where body temperature fell throughout the measurement period, the average body temperature was used. The numerator was increased because of this extra loss of body heat using the value of 0.83 cal. g^{-1} for the specific heat of body tissues in the manner of Dawson and Schmidt–Nielsen (1966). Surface area was calculated using a Meeh factor of 12.3 for the *Antechinus* (Dawson and Hulbert 1970) and 11.4 for *Mus* (Rubner 1902).

Non–shivering thermogenesis was assessed by measuring the oxygen consumption at 30°C of anaesthetized animals before and following an injection of noradrenaline. Animals were anaesthetized with Sagatal (May and Baker Ltd.) at the dose of 65 mg/kg body weight for *Antechinus* and 75 mg/kg body weight for the *Mus*. Noradrenaline (Sigma Chemical) was administered by intraperitoneal injection, the dose being 600ug/kg body weight for both species (Mejsnar and Jansky 1971).

At the end of the study the animals were killed and their fur pelts were kept and dried flat. They were hosed with compressed air to remove any debris and ensure an even lie of the fur. Insulation was measured in the mid–back region by the same method and apparatus described in Hulbert and Dawson (1974).

A students t–test was used to test for significant differences between treatments.

Table 1. Effect of long term exposure to cold on fur insulation in *Antechinus stuartii* and *Mus musculus.*

	Mus musculus	*Antechinus stuartii*
warm exposed	0.166 ± 0.004(6)*	0.249 ± 0.007(6)
cold exposed	0.165 ± 0.007(7)	0.242 ± 0.017(5)

* Units are W^{-1}.m^2.°C and values represent mean ± S.E.M. (N=).

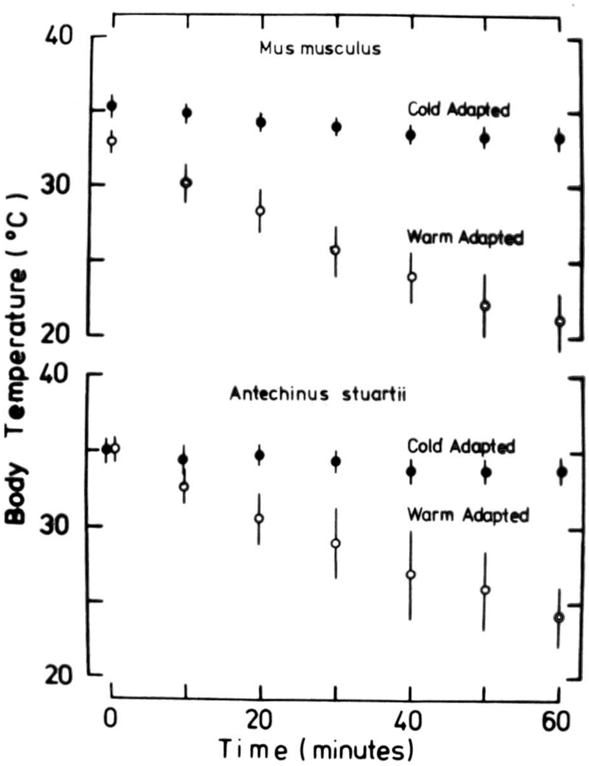

Fig. 1. Effect of short term exposure to 5°C on the body temperature of warm–adapted (N = 6) and cold–adapted (N = 7) *Mus musculus* and warm–adapted (N = 6) and cold–adapted (N = 5) *Antechinus stuartii.* Circle represents mean. Line represents ± S.E.M.

RESULTS

That an acclimation process took place in both species is demonstrated by Fig. 1 where it is obvious that both warm exposed *Mus* and *Antechinus* were unable to maintain a constant body temperature at 5°C whereas the cold exposed *Mus* and *Antechinus* were able to cope with this environmental temperature. Table 1 shows that this increased ability to tolerate cold was not due to an increase in fur insulation after cold exposure. Although *Antechinus* have a much more insulative fur than *Mus* there is no difference in either species between the cold exposed and warm exposed groups. As is shown in Table 2 there were no significant changes in total body conductance following long term cold exposure in either *Mus* or *Antechinus* but the acclimation to cold in both species included an increase in the heat production at 5°C. In the *Mus* the long term exposure to cold resulted in a 31% increase in heat production at 5°C. In *Antechinus* long term exposure to cold resulted in a much higher 81% increase in heat production. As can also be seen from Table 1, at the relatively neutral temperature of 30°C there were no differences in heat production between the cold exposed and warm exposed *Mus,* however in the *Antechinus* the situation was different in that even at 30°C the cold exposed *Antechinus* had a heat production that was significantly greater (60%) than that of

Table 2. Effect of long term exposure to cold on the thermal balance of *Antechinus stuartii* and *Mus musculus* at 5°C and 30°C.

		Mus musculus		*Antechinus stuartii*	
		Warm Exposed	Cold Exposed	Warm Exposed	Cold Exposed
N =		6	7	6	5
Weight (g)		29.2 ± 1.7	31.2 ± 1.9	32.1 ± 1.8	34.2 ± 3.2
5°C	Body temperature (°C)	28.5 ± 1.3	32.3 ± 1.0*	30.8 ± 1.5	34.7 ± 0.6
	Oxygen consumption ($mlO_2.g^{-1}.hr^{-1}$)	7.3 ± 0.5	9.6 ± 0.6**	4.5 ± 0.7	8.2 ± 0.4**
	Body conductance ($W.m^{-2}.°C^{-1}$)	6.7 ± 0.5	5.7 ± 0.4	3.6 ± 0.4	4.1 ± 0.2
30°C	Body temperature (°C)	35.9 ± 0.3	35.6 ± 0.4	33.7 ± 0.5	34.7 ± 0.4
	Oxygen consumption ($mlO_2.g^{-1}.hr^{-1}$)	3.2 ± 0.3	2.7 ± 0.2	1.1 ± 0.1	1.8 ± 0.2**
	Body conductance ($W.m^{-2}.°C^{-1}$)	8.6 ± 1.1	8.0 ± 0.6	5.8 ± 1.1	5.7 ± 0.6

All values are mean ± S.E.M. * represents significant difference ($P<0.05$) from warm exposed group; ** represents highly significant difference ($P<0.01$) from warm exposed group.

the warm exposed *Antechinus*. These differences in heat production between the groups and species are illustrated in Fig. 2. The metabolism values at 30°C should not be treated as standard metabolism for 2 reasons; the first is that the animals were not fasted and the second is that 30°C is probably slightly below the thermoneutral zone for both species.

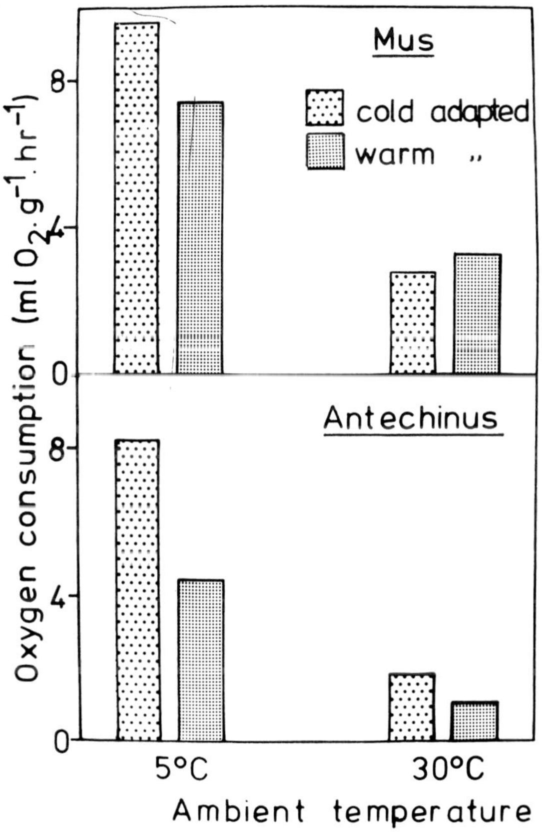

Fig. 2. Effect of cold adaptation on the metabolic rate of *Mus musculus* and *Antechinus stuartii* at 5°C and 30°C.

The increased heat production following cold exposure in *Mus* was probably due to an increase in noradrenaline mediated non–shivering thermogenesis. This is shown in Figure 3 where it can be seen that the cold exposed *Mus* showed a much greater thermogenic response to exogenous noradrenaline than the warm–exposed group. Even the warm–exposed *Mus* would appear to possess some noradrenaline mediated non–shivering thermogenesis. In contrast to the *Mus*, neither the cold exposed or warm exposed *Antechinus* showed any thermogenic response to the injected noradrenaline.

DISCUSSION

Previous studies on *Antechinus stuartii* (Wallis 1976), *Sminthopsis crassicaudata* and *Sminthopsis larapinta* (Godfrey 1968) and the *Planigale* genus (Dawson and Wolfers 1978) have all shown that these small dasyurid marsupials generally lower their body temperature during cold stress, but that most of these species do not enter torpor unless deprived of food. The *Antechinus* used in the present study were not deprived of food and only the warm acclimated marsupials became hypothermic as a physiological strategy or because they were unable to produce any more heat metabolically is unknown. The metabolic rate at 30°C of these warm exposed Antechinus was 1.10 $ml0_2.g^{-1}.hr^{-1}$ which is close to the 1.01 $ml0_2.g^{-1}.hr^{-1}$ reported as the standard metabolism of this species (Dawson and Hulbert 1970). Their metabolic rate at 5°C was 4.51 $ml0_2.g^{-1}.hr^{-1}$ which is 4.5 times their standard metabolism. If this represents their maximum possible heat production whilst resting (i.e. their summit metabolism) then they are similar to perameloid marsupials (Hulbert and Dawson 1974) and many other mammals (Gelineo 1964). This sort of increase has also been found by Wallis (1976) for *Antechinus stuartii.*

Dawson and Dawson (in this volume) report that they have measured metabolic rates in some dasyurid marsupials that are 8 – 9 times their standard metabolism. These animals were kept at winter ambient temperature in an external animal house and could thus be considered to be partly cold acclimated. It was not possible to ascertain the maximum heat production (summit metabolism) of the cold exposed *Antechinus* from this study. However since this group were able to maintain a constant body temperature at 5C° we can say it was at least 4 – 5 fold their heat production at 30°C and at least 8–fold the heat production of the warm–exposed group at 30°C. The body temperatures of the cold acclimated *Antechinus* at both 30°C and 5°C were similar to the 34.4°C previously reported by Dawson and Hulbert (1970) for this species.

Warm–acclimated *Mus* were unable to maintain body temperature at 5°C and thus their metabolic rate at this temperature may have been their summit metabolism. Although this level is 2½ times their 30°C metabolic rate, it is 4½ times the standard metabolism reported for this species (Pennycuik 1967). The cold acclimated group had a metabolic rate at 5°C that was 6 times the standard metabolic rate of *Mus*.

It has been shown in many eutherian mammals that the process of cold acclimation involves an increase in non–shivering thermogenesis. This non–shivering thermogenesis is located largely in two tissues, brown adipose tissue and skeletal muscle, and is controlled by the release of the neurotransmitter noradrenaline (Jansky 1973). One of the tests for non–shivering thermogenesis is to measure the thermogenic response of an animal to either the injection or infusion of noradrenaline, the greater the response the greater the amount of non–shivering thermogenesis. As was expected, cold–acclimation of *Mus* resulted in an increase in noradrenaline mediated non–shivering thermogenesis. The complete lack of response to noradrenaline in either group of *Antechinus* implies they do not possess this classical type of non–shivering thermogenesis.

Classical non–shivering thermogenesis is important in 3 main situations (a) acclimation to cold (b) neonatal heat production and (c) arousal from hibernation. In all these situations there are indications that both monotremes and marsupials may not possess this form of heat production. The arousal of the Echidna from hibernation may take 20 hours (Augee and Ealey 1968) compared to 2–3 hours in the similar sized eutherian hibernator, the marmot (Benedict and Lee 1938). Arousal rates in marsupial hibernators are also lower than that found in similar sized eutherians (Wallis 1978). An extensive comparative survey of neonates in a wide range of mammals failed to find any trace of brown adipose tissue in young marsupials or monotremes whereas it was found in the majority of eutherian groups examined (Rowlatt, Mrosovsky and English 1971). The present study has shown that cold acclimation in a small dasyurid marsupial does not involve classical non–shivering thermogenesis. There is one report of non–shivering thermogenesis in a marsupial in the literature. Nicol (1978) reported a thermogenic response to injected noradrenaline in the macropodid marsupial *Potorous tridactylus*. However, the small response observed and the large dose of noradrenaline required casts doubt on its physiological significance.

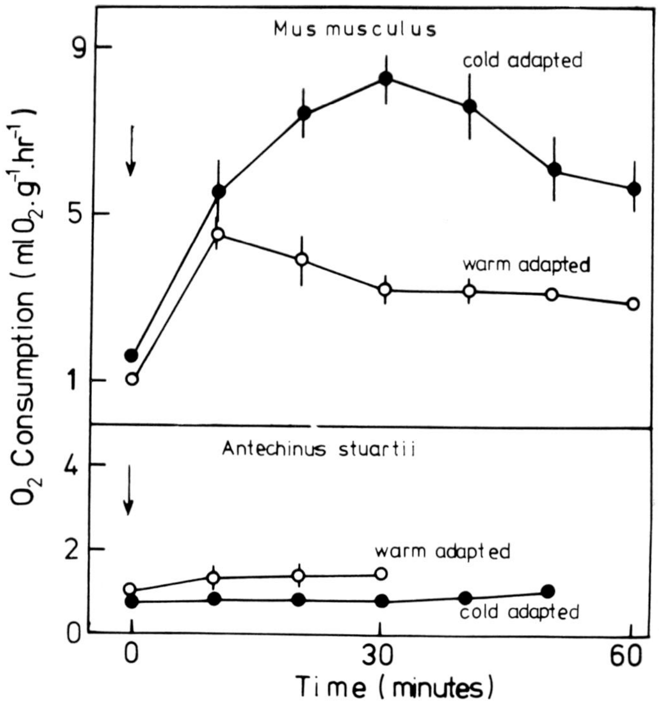

Fig. 3. Effect of cold adaptation on the thermogenic response to an intraperitoneal injection of noradrenaline (600 μg/kg body weight) in *Mus musculus* and *Antechinus stuartii.*

The present study has not shown that non–shivering thermogenesis does not occur in *A. stuartii,* a small marsupial, during acclimation. There was an increase in heat production during cold acclimation but whether this increase was due to increased shivering thermogenesis or a different form of non–shivering thermogenesis (one mediated by something apart from noradrenaline) awaits further study. That cold acclimation resulted in a significant increase in resting metabolism at 30°C in *Antechinus* suggests that an investigation of thyroid gland activity following cold acclimation would be worthwhile.

ACKNOWLEDGEMENTS

We would like to thank the New South Wales National Parks and Wildlife Service for permission to capture the *Antechinus*. This work was supported by a University of Wollongong Research Fund Grant. Thanks are due to Judy Ward for typing of the manuscript. Drs. S. C. Nicol and M. Augee kindly read and constructively criticized an earlier draft of this manuscript.

REFERENCES

AUGEE, M. L. AND EALEY, E. H. M., 1968. Torpor in the Echidna *Tachyglossus aculeatus*. *J. Mammal.* **49:** 446–54.

BENEDICT, F. G. AND LEE, R. C., 1938. Hibernation and Marmot physiology. *Carnegie Inst. Wash. Publ.* No. **497.**

DAWSON, T. J. AND HULBERT, A. J., 1970. Standard metabolism, body temperature and surface areas of Australian marsupials. *Am. J. Physiol.* **218:** 1233–8.

DAWSON, T. J., AND SCHMIDT-NIELSEN, K., 1966. Effect of thermal conductance on water economy in the Antelope Jack Rabbit *Lepus alleni*. *J. Cell Physiol.* **67:** 463–72.

DAWSON, T. J. AND WOLFERS, J. M., 1978. Metabolism, thermoregulation and torpor in shrew-sized marsupials of the genus *Planigale*. *Comp. Biochem. Physiol.* **59A:** 305–9.

GELINEO, S., 1964. Organ systems in adaptation: the temperature regulating system. *In* "Handbook of physiology. Section IV". Amer. Physiol. Soc. Washington: D.C.

GODFREY, G. K., 1968. Body temperatures and torpor in *Sminthopsis crassicaudata* and *S. larapinta* (Marsupialia-Dasyuridae). *J. Zool.* (Lond.) **156:** 499–511.

HULBERT, A. J. AND DAWSON, T. J., 1974. Thermoregulation in perameloid marsupials from different environments. *Comp. Biochem. Physiol.* **47A:** 59–-616.

JANSKY, L., 1973. Non-shivering thermogenesis and its thermoregulatory significance. *Biol. Rev.* **48:** 85–132.

MEJSNAR, J. AND JANSKY, L., 1971. Non shivering thermogenesis and calorigenic action of catecholamines in the white mouse. *Physiol. Bohemoslovaca* **20:** 157–62.

NICOL, S. C., 1978. Non-shivering thermogenesis in the Potoroo, *Potorous tridactylus* (Kerr). *Comp. Biochem. Physiol.* **59C:** 33–7.

PENNYCUIK, P. R., 1967. A comparison of the effects of a variety of factors on the metabolic rate of the mouse. *Aust. J. Exp. Biol. Med.* **45:** 331–46.

ROWLATT, U., MROSOVSKY, N. AND ENGLISH, A., 1971. A comparative survey of brown fat in the neck and axilla of mammals at birth. *Biol. Neonate* **17:** 53–83.

RUBNER, M., 1902. "Die Gesetze des Energieverbrauches bei der Ernahrung". Leipzig: Deutiche.

WALLIS, R. L., 1976. Torpor in the dasyurid marsupial *Antechinus stuartii*. *Comp. Biochem. Physiol.* **53A:** 319–22.

WALLIS, R. L., 1978. Responses to low temperature in small marsupial mammals. *J. Thermal. Biol.* **4:** 105–11.

RESUMEN

28. La adaptacion al frío de un pequeño marsupial dasiúrido: *Antechinus stuartii*. Por Werner Reynolds y A. J. Hulbert.

Se comparó la estrategia de la adaptación al frío de la rata marsupial dasiúrida *Antechinus stuartii* con la de la rata euteria *Mus musculus*. De cada especie se compuso un grupo adaptado al frío (6 semanas a 5°C) y otro adaptado al calor (6 semanas a 25°C). Los *Antechinus* adaptados al frío mantuvieron la temperatura corporal, a 5°C, por un incremento en la producción de calor. Los *Antechinus adaptados* al calor fueron incapaces de mantener una constante temperatura corporal en el frío. Los *Mus* adaptados al frío tambien fueron más capaces de combatir el frío por un incremento en la producción de calor que los adaptados al calor. Esto fue debido al desarrollo de una termogénesis por medio de noradrenalina, no por escalofríos. La crecida producción de calor en los *Antechinus* adaptados al frío *no* fue debida a termogénesis por acción de la noradrenalina ni por escalofrío. Tanto en *Mus* como en *Antechinus* durante la adaptación al frio no hubo cambio ni en la conducción del cuerpo ni en la aislación por la piel. Las ratas marsupiales fueron más exitosas en la adaptación al frio que las *Mus;* mostraron una estrategia similar en la adaptación al frio que el clásico tipo de los euterios, aunque pareció que usaron un diferente mecanismo de producción térmica.

29. ADAPTATION TO LOW ENVIRONMENTAL TEMPERATURES IN THE CARNIVOROUS MARSUPIALS

ROBERT L. WALLIS[1]

Various responses of the Australian carnivorous marsupials (mainly the dasyurids) to low environmental temperatures are described and compared with those of other mammals. These adaptations to cold, which are not necessarily mutually exclusive, may include the development of non-shivering thermogenesis, seasonal changes in insulation, nest-sharing, fat storage and torpor.

Pages 285–91 *in* CARNIVOROUS MARSUPIALS, ed. by M. Archer, February, 1982. Roy. Zool. Soc. New South Wales: Sydney, Australia.

TEMPERATURE is one of the most important aspects of an animal's environment, whether it be the ambient temperature (Ta) of the animal's surroundings or the internal temperatures of its tissues or body fluids. There has been an evolutionary tendency in the higher animals to some sort of control or moderation of temperature extremes to which the animal's internal environment are exposed by modifications in behaviour, physiology and in many cases anatomy. There are advantages to an animal in maintaining a high, stable temperature of the internal environment by the regulation of the generation of internal heat and its retention; that such selective advantages exist are evidenced by the acquisition of endothermy and homeothermy in at least two phylogenetically disparate groups — the birds and mammals. Even between the three divisions of mammals there are quantitative differences in such characteristics as resting body temperature (Tb) and basal metabolic rate (Augee and Ealey 1968, Dawson and Hulbert 1970).

Endothermy is an energetically expensive process, particularly in the cold when often food is in short supply and metabolic demands are high. It is not surprising then, that mammals and birds have evolved a variety of adaptations to minimize the effect of low Ta's. These include behavioural changes such as seasonal migrations (Hart 1964), changes in the patterns of daily activity (Jansky and Hanak 1959) and nest–sharing or huddling (Prychodko 1958, Baudinette 1972).

Some workers have proposed that long–term adaptations to cold may involve selection for large body size and relatively short appendages. Such "climatic rules" are based on the presumption that larger animals (or those with short appendages) have relatively less of their body surface exposed to the environment, and hence would be expected to lose less body heat. There are, however, many other factors besides heat economy which influence selection of body size and shape and the validity of such "rules" has been questioned (Scholander 1958, Mayr 1965, McNab 1974). Other structural adaptations to low Ta involve seasonal increases in insulation by changes in thickness or texture of fur or the acquisition of more subcutaneous fat.

Torpor is another adaptation to cold observed in many mammals and birds and is any state of reduced Tb in which the animal is capable of spontaneously increasing its Tb to that level routinely observed in thermoneutrality. Because of this capacity for spontaneous arousal, and because Tb is regulated in many torpid endotherms (albeit at a lower level), torpor is considered an advanced adaptation and not an indication of physiological primitiveness or poor thermoregulation (Kalter and Folk 1979).

Increased heat production is a more energetically expensive means of coping with cold and may involve shivering or the acquisition of non-shivering thermogenesis (NST) (a longer term adaptation). The heat generated by NST adds to that made by shivering and exercise and is closely associated with the action of noradrenaline (Hsieh and Carlson 1957). Although common in new–born and hibernating eutherian mammals (Hayward 1971), it also develops in many eutherians in response to cold acclimation (Jansky 1973).

In this paper I aim to examine the various adaptations to low environmental temperatures employed by the Australian carnivorous marsupials (mainly the dasyurids) and to compare them with the strategies of coping with cold found in other mammals.

DIFFERENCES IN BASAL METABOLIC RATES

Basal metabolic rates (BMR) of some cold–adapted mammals are higher than those of their warm–adapted counterparts (Hart and Jansky 1973, Shield 1972); conversely the desert bandicoot *Macrotis* has a significantly lower BMR than that of two forest–dwelling species (Hulbert and Dawson 1974).

Reynolds and Hulbert (at this symposium) have similarly reported that cold acclimated *Antechinus stuartii* have significantly higher BMRs than warm acclimated animals.

[1] Department of Environmental Studies, Rusden State College, Clayton, Victoria, Australia, 3168.

However, a survey of Australian carnivorous marsupials by McNab (1978) indicated that no significant differences exist between BMR of species from deserts, grasslands, or woodlands. Although Ta as an influence on BMR cannot be entirely ruled out, McNab (1978) believes that diet is more important than climate in affecting BMR in mammals. In relation to this, he found that the numbat (*Myrmecobius fasciatus*), like eutherian termite and ant–eaters, has a BMR 50% less than that predicted for equivalent sized eutherians by Kleiber's (1961) equation.

CHANGES IN INSULATION

Two species of *Antechinus* have thicker fur and lower conductances in winter than summer. In *A. stuartii,* weight–specific conductances (C) of summer animals differ significantly ($p < 0.05$) from those of winter animals (Table 1). In some torpid *A. stuartii,* very low C values have been recorded. This is probably attributable to the low heat loss accompanying the near spherical posture adopted during sleeping, with head, tail and limbs tightly tucked under the thorax and abdomen. Regional heterothermy and superficial hypothermia is also possible in torpid animals; this would further reduce C. Some eutherians similarly have lower C values when torpid than when normothermic (Tucker 1965).

Table 1. Values of weight specific thermal conductance (C in cm^3 O_2/g.hr.°C) for 17 winter–captured, 4 torpid and 7 summer–captured *Antechinus stuartii.* (Taken from Wallis 1977).

	C ± SD	(range, n)
Winter, normothermic animals	0.16 ± 0.04	(0.091 – 0.24,17)
Winter, torpid animals	0.12 ± 0.05	(0.078 – 0.19,4)
Summer animals	0.24 ± 0.03	(0.20 – 0.27,7)

Changes in C with season have also been described in *A. swainsonii.* These changes were particularly noticeable in males where the mean C during summer (February) were 41.8% greater than during winter (August) (Gotts 1967).

BEHAVIOURAL ADAPTATIONS

Morton (1978a) has described nest–sharing in the dasyurid *Sminthopsis crassicaudata* in a field study at Werribee, Victoria. He found that nest–sharing was most frequent (> 70% of individuals in nests occurred in groups) during winter (May–July) when Ta was minimal, but diminished after July with the onset of breeding, even though Ta's were still low. During November–March, nest–sharing was minimal (e.g. 4% in November). *S. crassicaudata* even shared nests with a potential prey species, *Mus musculus,* although groups of *S. crassicaudata* were more common than "mixed" species groups. In the field *S. crassicaudata* was found to be primarily insectivorous; hence the energy saved through nest–sharing must outweigh that which would be gained from consuming the mice. The relationship between the two species during winter was termed "thermoregulatory mutualism" (Morton 1978a).

CAUDAL FAT STORAGE

A number of small dasyurids store fat in their tails. Such species include five species of *Sminthopsis,* three *Antechinus, Planigale gilesi* and *Dasycercus cristicauda* (Morton 1980). Morton (1978b) observed fat storage in *S. crassicaudata* was maximal during autumn, but concluded that caudal fat stores provided only limited protection against energy shortage and were more of a short-term adaptation to reduced arthropod food supply.

TORPOR

A range of depths of torpor exists in mammales. In hibernation, Tb is reduced below 10°C for several days (Hudson 1973), there is a loss of sensibility or power of motion (Kayser 1961) associated with a marked decline in oxygen consumption rate (OCR) and a depression in the activity of nervous system resembling coma (Swan *et al.* 1969). Hibernators are capable of rewarming without the application of exogenous heat. Hibernators are found in the following mammalian orders — Chiroptera, Rodentia, Primates, Marsupialia (Kalter and Folk 1979). At the other extreme is shallow, daily torpor in which Tb may be reduced by a few degrees for a few hours a day (Hudson 1978). In such a state, typical locomotor behaviours often remain, making it difficult to distinguish from a state of normorthermia.

Although hibernation has been described in four species of pigmy possum (Burramyidae), no dasyurid has yet been found to hibernate. Indeed, the most striking feature of torpor in dasyurids is the reasonable degree of uniformity of responses in terms of minimum Tb recorded (all > 14°C), maximum duration (< 10 hr) and arousal rates (0.33-0.50°Cmin) (Table 2). An apparent exception to such a pattern (*Dasycercus cristicauda*) listed in Wallis (1979) in fact conforms to the typical dasyurid pattern of daily torpor with evening arousals (Kennedy, pers. comm.).

The purpose of torpor is energy conservation; in an *A. stuartii* in which oxygen consumption fluctuated between 0.74 and 7–8 cm^3 O_2/g. hr over 25 hr, the energy saving in comparison with a normothermic animal was 59% (Wallis 1976). Most studies of torpor in Australian marsupials have been with laboratory animals due to difficulties in examining torpid animals in the field without disturbing them. In the only such field study, Morton (1978a) found 24 *S. crassicaudata* torpid in April to October, representing only 5.8% of the total number of animals captured over this period. Morton (1978a) concluded that the principal adaptation to cold and

Table 2. TORPOR IN SELECTED MARSUPIALS.

	Mass g	Ta range °C	Minimum Tb recorded °C	Maximum duration of torpor hr	Maximum arousal rate °C/min.	Spontaneous starvation induced Sp;St	Source
DASYURIDAE							
Sminthopsis crassicaudata	12 – 22	11 – 20.5	17(i)	10	0.50	St.	Godfrey (1968)
Sminthopsis crassicaudata	12 – 20	9 – 17	18.1	–	–	St.	Morton (1978a)
Sminthopsis macroura	14 – 30	15 – 24	18(i)	9	0.35	Sp.	Godfrey (1968)
Planigale ingrami	5.8 – 9.3	14	>14(ii)	4(iii)	–	Sp.	Fleay (1965), Dawson & Wolfers (1978)
Planigale tenuirostris	6.6 – 7.3	14	>14(ii)	4(iii)	–	Sp.	Dawson & Wolfers (1978)
Planigale gilesi	9.4 – 10.0	14	>14.8(iv)	4(iii)	–	St.	Dawson & Wolfers (1978)
Planigale maculata	9.7 – 16.4	17 – 32	19.6	1.5	0.50	Sp.	Morton & Lee (1978)
Antechinus stuartii	15 – 30	13 – 25	21	7.5	0.33	St.	Wallis (1976)
MYRMECOBIIDAE							
Myrmecobius fasciatus	–	–	–	–	–	Sp.	Seventy & Raymond (1973)

(i), skin surface temperature; (ii), no actual Tb's given, but authors note Tb's were close to Ta which dropped to near 14°C in winter; (iii), estimate; (iv), calculated from data in Dawson & Wolfers (1978), using M = C/Tb–Ta, where M is OCR in cm^3O_2/g.hr. and C in cm^3O_2/g.hr.°C.

food shortages was nest–sharing and that torpor was used only when foraging conditions were poor.

In most dasyurids torpor occurs seasonally in response to food shortages (Wallis 1976, Morton 1978a), but in three species of *Planigale* and in *S. macroura* it occurs spontaneously (Morton and Lee 1978, Dawson and Wolfers 1978, Godfrey 1968). Such variations in the degree of physiological preparation necessary for torpor also occurs in eutherians (Hudson and Bartholomew 1964). However, a higher incidence of torpidity in summer (aestivation) has yet to be demonstrated in a marsupial. In some dasyurids, torpor is a moderate Ta phenomenon and below Ta = 15°C normothermia is maintained (Kennedy and MacFarlane 1971, Wallis 1976).

Some eutherians are more likely to become torpid during the day when inactive (Kalter and Folk 1979); this also occurs in *A. stuartii, D. cristicauda, S. crassicaudata* and *P. maculata* (Wallis 1976, Kennedy and MacFarlane 1971, Morton 1978a, Morton and Lee 1978).

In other respects torpor in dasyurids is similar to that in burramyids and eutherians. Thus: the sleeping posture is adopted with head and tail tucked under the abdomen; breathing rates are low and apnea occurs (but not in *P. maculata;* Morton and Lee 1978); heart rate is slow although no consistent electrical abnormalities have been recorded in naturally torpid *A. stuartii,* in contrast to skipped beats apparent in a surgically induced hypothermic, summer animal (Wallis 1977); oxygen consumption is low; depth and duration of torpor is variable and may be related to the animal's condition and/or the amount of food it has recently consumed; Q_{10}'s for entry into torpor are similar; and 8–12 fold increases in oxygen consumption rate occur on arousal from torpor.

In *P. maculata,* entry into torpor is accompanied by a smooth decline in oxygen consumption rate (Morton and Lee 1978); in *A. stuartii*

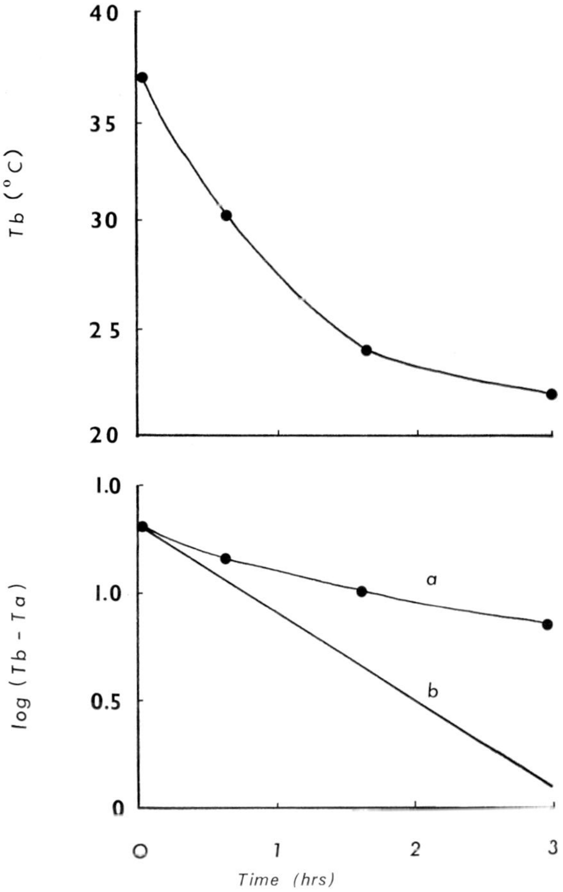

Fig. 1. Changes in Tb and $\log_{10}$ (Tb–Ta) with time for *A. stuartii* entering torpor at Ta=20°C. Curve (a) plots data obtained during experiment; graph (b) represents the predicted response of an *A. stuartii* cooling passively. (Taken from Wallis 1977).

both metabolic rate and Tb decline smoothly or occasionally decrease in a stepwise fashion to sequentially lower levels (Wallis 1976). The latter fluctuations in Tb and oxygen consumption rate before the final decline to low plateaux have been described in torpid hispid pocket mice by Wang and Hudson (1970) who interpreted them as representing bouts of shivering which serve as a break in the decline of Tb. Even in cases of smooth entry into torpor, *A. stuartii* cool more slowly than predicted if they were to cool passively (Fig. 1). Such controlled entry into torpor and the regulation of Tb above a critical minimum of 21°C has been described in many eutherians, and Hudson (1967) has concluded from this that torpor is not simply reversion to poikilothermy.

Maximum rates of warm-up of marsupials arousing from torpor are lower than values of equivalent sized eutherians (Heldmaier 1978, Wallis 1979). Table 3 indicates that warm-up rates of dasyurids are lower than those predicted for similar sized eutherians arousing from torpor. Slower rates of arousal have also been recorded in the didelphid *Marmosa* sp. and some burramyids (Morrison and McNab 1962, Hickman and Hickman 1960, Fleming 1976). It should be noted, however, that the ability to predict maximum rates of arousal in mammals from body mass data is contentious. Wang and Hudson (1971; 85) believe "that the differences in arousal rates of mammals, unlike birds, probably reflects different thermogenic abilities rather than a simple surface–volume relationship". Nevertheless, warm–up rates of marsupials are lower than those of eutherians of equivalent size (by comparison of species pairs). The difference is probably related to the lower rate of maximum sustained heat production in marsupials (Dawson 1977), and their apparent lack of brown adipose tissue (Wallis 1979).

Gaertner *et al.* (1973) and Hudson (1978) have described a shallow torpor in rodents where Tb decline is not great. Shallow torpor has also been recorded in *A. stuartii* in which Tb declined to 28–32°C but behaviour was undifferentiated from normothermic animals. In Fig. 2 it is apparent that such reductions in Tb occurred most frequently at Ta = 15–25°C (Wallis, 1977). Short bursts of shallow torpor were also indicated in *A. stuartii* kept at this Ta range during winter, and in which heart rate and breathing rates were monitored over several hours. Some of the *S. macroura* studied by Godfrey (1968) had Tb's of about 30°C but there were no accompanying data on heart rate, breathing rate or activity to suggest if this constituted a shallow torpor.

Similar drops in Tb (to 28.2°C) have been described in the larger Dusky Antechinus *A. swainsonii* (Gotts 1976, Ingham 1977). However, such low Tb's were generally found to occur in warm–acclimated (or "summer") animals exposed to cold (e.g. Ta = 5°C) and were not thought indicative of "torpor" as heart rate and breathing rate remained high (Ingham 1977).

Torpor has also been reported in the Numbat *Myrmecobius fasciatus*. Although no quantitative data are available, Serventy and Raymond (1973: 2235) note "numbats taken from hollow logs on cold winter mornings were often still and cold, seemingly dead. After a few hours lying in the sun the 'dead' animals would come to life and move away".

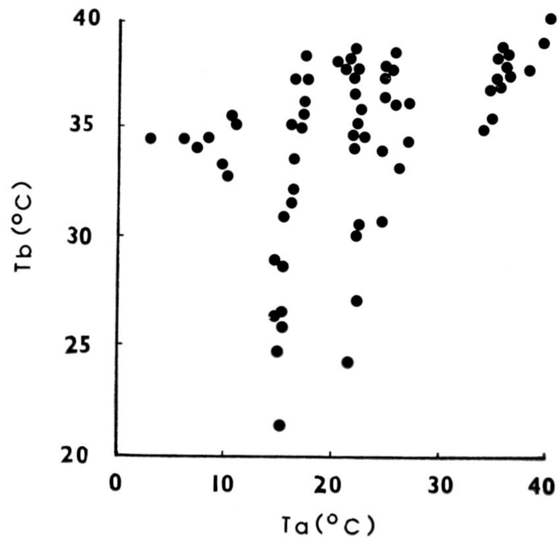

Fig. 2. Variation in Tb with Ta for resting *A. stuartii* (n=25). (Taken from Wallis 1977).

NON–SHIVERING THERMOGENESIS (NST)

A method to estimate NST involves comparison of working and resting oxygen consumption rates at various Ta's. Since the heat from NST adds to that derived from exercise, animals with well-developed NST have higher metabolic rates in the cold when exercising than at rest (Jansky 1973). Mammals with poorly developed NST have similar metabolic rates when exercising and resting.

This method has been applied to two species of dasyurid. In *A. swainsonii,* NST is acquired in the field only during winter by males (Gotts 1976). This results in winter animals having much lower lethal Tb's than summer animals. No development of NST or seasonal change in lower lethal Tb occurs in females. This finding reflects the greater activity (and exposure to cold) of the male during the mating period, June/July. Cold acclimated (8°C) juveniles also acquire NST. In such animals the curves for working and resting oxygen consumption rate plotted against Ta are parallel, but in winter–captured adult males, the curves converge as Ta decreases (Gotts 1976), indicating that some shivering occurs at low Ta's. Ingham (1977) observed that intensity of shivering at Ta = 5°C was much higher in "summer" than in "winter" animals.

Winter-captured *Antechinus stuartii* also demonstrate NST when they are forced to exercise (Wallis 1977). The working and resting oxygen consumption curves are not parallel (Fig. 3) indicating some shivering may occur at Ta <25°C but that its intensity increases as Ta decreases. Shivering in *A. stuartii* has been observed and recorded on EMG tracing at Ta <17°C but not at higher temperatures (Wallis 1979). It is conceivable, however, that resting

Table 3. Maximum rates of warm–up in dasyurid marsupials arousing from torpor.

	Mass	Maximum warm up rates Observed	Predicted*	%	Source
	g	°C/min		difference	
Planigale maculata	12	0.50	0.89	78	Morton & Lee (1978)
Sminthopsis crassicaudata	13.5	0.70	0.87	24	Godfrey (1968)
Sminthopsis macroura	20	0.35	0.70	100	Godfrey (1968)
Antechinus stuartii	25	0.33	0.58	76	Wallis (1976)

* Based on the equation of Heinrich & Bartholomew (1971) °C/min. $= 3.22\ BW(g)^{-0.51}$

animals are not shivering at their maximum capacity at 0°C and that exercise adds further heat production in the cold; i.e., it may not be necessary to assume acquisition of NST in this species. The degree of apparant NST in winter *Antechinus* species measured as the increase in OCR of exercising animals over resting animals at 0°C is similar to that in cold–acclimated, non-hibernating eutherians (Wallis 1979), but less than that found in eutherian hibernators. Such a difference may be due to the apparent absence of brown adipose tissue in marsupials studied to date.

NST is present in the macropod *Potorous tridactylus,* but the maximal dose of noradrenalin needed to effect increases in metabolic rates was much higher for the marsupial than for similar sized eutherians (Nicol 1978). Current investigations into noradrenalin influenced NST in *A. stuartii* have indicated a similar insensitivity to doses of NA that would kill mice (Wallis 1979). Reynolds and Hulbert (this symposium) have also observed NST in this species which is not induced by noradrenalin.

LARGER DASYURIDS IN THE COLD

Large mammals are generally better equipped than small ones to cope with cold due to lower relative surface areas for heat loss. *Dasyurus geoffroii* and *Sarcophilus harrisii* are both homeothermic in the cold (Arnold and Shield 1970, Guiler and Heddle 1974), a result which confirms the contention that marsupials are no less efficient homeotherms than eutherians and that they should not be considered more primitive in their thermophysiology.

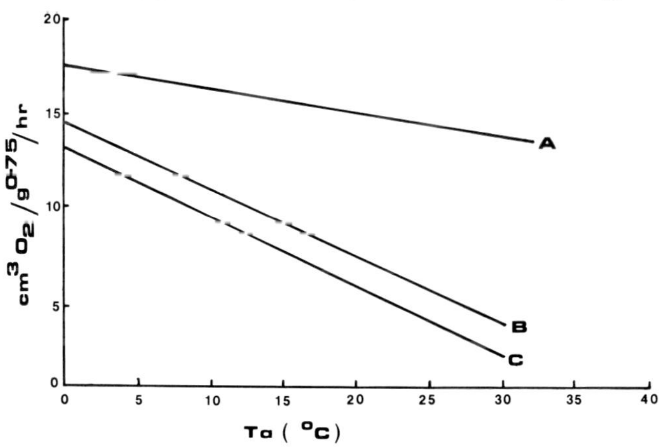

Fig. 3. Variation in OCR (cm³ O_2/g 0.75/hr) with Ta (°C) for resting and active *A. stuartii.* OCR of exercising *A. stuartii* is represented by line A. OCR of resting *A. stuartii* in which Tb>35°C are represented by line B; those in which Tb>32°C by line C. (Taken from Wallis 1977).

SUMMARY

Australia's carnivorous marsupials have a variety of adaptations to cold. The larger ones, by dint of good insulation and lower relative heat loss, can remain homeothermic at low Ta's, but smaller marsupials must generate much heat to remain normothermic. They may do this by metabolising stored fat (particularly obvious in the fat–tailed marsupials) and/or possibly by NST which might not be mediated by noradrenalin. Alternatively, small marsupials may temporarily suspend their homeothermy and become torpid. Torpor in dasyurids contrasts with that in pigmy possums and many eutherians in that it is a daily phenomenon, Tb is not reduced to very low levels (>14°C) and in some species does not occur at Ta <15°C. Seasonal changes in insulation and nest–sharing are further means of reducing heat loss at low Ta's.

ACKNOWLEDGEMENTS

Drs. S. Morton, T. Grant and M. Augee read and constructively criticised a draft of this paper.

REFERENCES

ARNOLD, J. AND SHIELD, J., 1970. Oxygen consumption and body temperature of the Chuditch *(Dasyurus geoffroii). J. Zool., Lond.* **160:** 391–404.

AUGEE, M. L., 1979. Monotremes and the evolution of homeothermy. *In* "Montreme biology" ed. by M. L. Augee. Royal Zoological Society of New South Wales: Sydney.

AUGEE, M. L. AND EALEY, E. H. M., 1968. Torpor in the Echidna, *Tachyglossus aculeatus. J. Mammal.* **49:** 446–54.

BAUDINETTE, R. V., 1972. The impact of social aggregation on the respiratory physiology of Australian hopping mice. *Comp. Biochem. Physiol.* **41:** 35–8.

DAWSON, T. J., 1977. Kangaroos. *Scient. Am.* **237**(2): 78–89.

DAWSON, T. J. AND HULBERT, A. J., 1970. Standard metabolism, body temperature, and surface areas of Australian marsupials. *Am. J. Physiol.* **218:** 1233–8.

DAWSON, T. J. AND WOLFERS, J. M., 1978. Metabolism, thermoregulation and torpor in shrew sized marsupials of the genus *Planigale. Comp. Biochem. Physiol.* **59A:** 305–9.

FLEAY, D., 1965. Australia's 'needle–in–a–haystack' marsupial. *Victorian Nat.* **82:** 195–204.

FLEMING, M., 1976. "Thermoregulation and torpor in *Burramys parvus*". Hons. thesis: Australian National University, Canberra.

GAERTNER, R. A., HART, J. S. AND ROY, O. Z., 1973. Seasonal spontaneous torpor in the White–footed Mouse, *Peromycus leucopus. Comp. Biochem. Physiol.* **45A:** 169–81.

GODFREY, G. K., 1968. Body-temperature and torpor in *Sminthopsis crassicaudata* and *S. laprinata* (Marsupialia: Dasuridae). *J. Zool. Lond.* **156:** 499–511.

GOTTS, D., 1976. "Energetics in *Antechinus Swainsonii*". M.Sc. thesis: Monash University, Melbourne.

GUILER, E. R. AND HEDDLE, R. W. L., 1974. Body temperatures in the Tasmanian Devil, *Sarcophilus harrisii* (Marsupialia: Dasyuridae). *Comp. Biochem. Physiol.* **47A:** 981–9.

HART, J. S., 1964. Geography and season: mammals and birds. *In* "Handbook of physiology, Section 4: Adaptation to environment", ed. by D. B. Dill. American Physiological Society: Washington D.C.

HART, J. S. AND HEROUX, O., 1955. Exercise and temperature regulation in lemmings and rabbits. *Can. J. Biochem. Physiol.* **33:** 428–35.

HART, J. S. AND JANSKY, L., 1963. Thermogenesis due to exercise and cold in warm and cold–acclimated rats. *Can. J. Biochem. Physiol.* **41:** 629–34.

HAYWARD, J. S., 1971. Non–shivering thermogenesis in hibernating mammals. Pp. 119–34 *in* "Non–shivering thermogenesis. Proceedings of the symposium", ed. by L. Jansky. Swets and Zeitlinger: Amsterdam.

HEINRICH, B. AND BARTHOLOMEW, G. A., 1971. An analysis of pre–flight warm–up in the sphinx moth, *Manduca sexta. J. exp. Biol.* **55:** 223–39.

HELDMAIER, G., 1978. Rewarming rates from torpor in mammals and birds. *J. therm. Biol.* **3:** 100.

HICKMAN, V. V. AND HICKMAN, J. L., 1960. Notes on the habits of the Tasmanian dormouse phalangers *Cercaertus nanus* (Demarest) and *Eudromicia lepida* (Thomas). *Proc. Zool. Soc. Lond.* **135:** 365–74.

HSIEH, A. C. H. AND CARLSON, L. D., 1957. Role of adrenalin and noradrenalin in the chemical regulation of heat production. *Am. J. Physiol* **190:** 243–6.

HUDSON, J. W., 1967. Variations in the patterns of torpidity of small homeotherns. Pp. 30–46 *in* "Mammalian hibernation III", ed. by K. C. Fisher. Elsevier: New York.

HUDSON, J. W., 1973. Torpidy in mammals. Pp. 97–165 *in* "Comparative physiology of thermoregulation" Vol. III, ed by G. C. Whittow. Academic Press: New York.

HUDSON, J. W., 1978. Shallow, daily torpor: a thermoregulatory adaptation. *J. therm. Biol.* **3:** 87.

HUDSON, J. W. AND BARTHOLOMEW, G. A., 1964. Terrestrial animals in the dry heat: estivators. *In* "Handbook of physiology Section 4: Adaptation to environment", ed. by D. B. Dill. American Physiological Society: Washington, D.C.

HULBERT, A. J. AND DAWSON, T. J., 1974. Standard metabolism and body temperature of perameloid marsupials from different environment. *Comp. Biochem. Physiol.* **47A:** 583–90.

INGHAM, P., 1977. "The physiology of *Antechinus swainsonii*". Biology 308 project report: Rusden State College, Melbourne.

JANSKY, L., 1973. Non–shivering thermogenesis and its thermoregulatory significance. *Biol. Rev.* **48:** 85–132.

JANSKY, L. AND HANAK, V., 1959. Studien uber Kleinsäuger–populationen in Südböhmen. II. Aktoitat der Spitzmause under natürlichen Bedingungen. *Saugetierkundliche Mitterlungen* 8: 55–63.

KALTER, V. G. AND FOLK, G. E., 1979. Humoral induction of mammalian hibernation. *Comp. Biochem. Physiol.* **63A:** 7–13.

KAYSER, C., 1961. "The physiology of natural hibernation". Pergamon: Oxford.

KENNEDY, P. M. AND MACFARLANE, W. V., 1971. Oxygen consumption and water turnover of the fat–tailed marsupials *Dasycercus cristicauda* and *Sminthopsis crassicaudata. Comp. Biochem. Physiol.* **40A:** 723–32.

KLEIBER, M., 1961. "The fire of life". Wiley: New York.

MAYR, E., 1965. Geographical character gradients and climatic adaptations. *Evolution* **10:** 105–8.

MCNAB, B. K., 1974. The energetics of endotherms. *Ohio J. Sci.* **74:** 370–80.

MCNAB, B. K., 1978. The comparative energetics of neo–tropical marsupials. *J. Comp. Physiol.* **125:** 115–28.

MORRISON, P. R. AND MCNAB, B. K., 1962. Daily torpor in a Brasilian murine opossum *(Marmosa). Comp. Biochem. Physiol.* **6:** 57–68.

MORTON, S. R., 1978a. Torpor and nest–sharing in free–living *Sminthopsis crassicaudata* (Marsupialia) and *Mus musculus* (Rodentia) *J. Mammal.* **59:** 569–75.

MORTON, S. R., 1978b. An ecological study of *Sminthopsis crassicaudata* (Marsupial: Dasyuridae) III. Reproduction and life history. *Aust. Wildl. Res.* **5:** 183–211.

MORTON, S. R., 1980. Ecological correlates of caudal fat storage in small mammals. *Aust. Mammal.* **3:** 81–6.

MORTON, S. R. AND LEE, A. K., 1978. Thermoregulation and metabolism in *Planigale maculata* (Marsupialia: Dasyuridae). *J. Therm. Biol.* **3:** 117–20.

NICOL, S., 1978. Non–shivering thermogenesis in the potoroo, *Potorous tridactylus* (Kerr). *Comp. Biochem. Physiol.* **59C:** 33–7.

PRYCHODKO, W., 1958. Effect of aggregation of laboratory mice *(Mus musculus)* on food intake at different temperatures. *Ecology* **39:** 500–3.

SCHOLANDER, P. F., 1958. Climatic rules. *Evolution* **10:** 339–40.

SERVENTY, V. AND RAYMOND, R., 1973. Torpidty in desert animals. *Australia's Wildlife Heritage* **14:** 2230–40.

SHIELD, J., 1972. Acclimation and energy metabolism of the dingo, *Canis dingo* and the coyote, *Canis latrans. J. Zool., Lond.* **168:** 483–501.

SWAN, H., JENKINS, D. AND KNOX, K., 1959. Metabolic torpor in *Protopterus aethiopicus*: an anti–metabolic agent from the brain. *Am. Nat.* **103:** 247–58.

TUCKER, V. A., 1965. The relation between the torpor cycle and heat exchange in the California pocket mouse, *Perognathus californicus. J. Cell. Comp. Physiol.* **65:** 405–14.

WALLIS, R. L., 1976. Torpor in the dasyurid marsupial *Antechinus stuartii. Comp. Biochem. Physiol.* **53A:** 319–22.

WALLIS, R. L., 1977. "Response to low temperature in *Antechinus stuartii* (Marsupialia: Dasyuridae)." M.Sc. thesis: Monash University, Melbourne.

WALLIS, R. L., 1979. Responses to low temperature in small marsupial mammals. *J. Therm. Biol.* **4:** 105–11.

WANG, L. C. AND HUDSON, J. W., 1970. Some physiological aspects of temperature regulation in the normothermic and torpid hispid pocket mouse, *Perognathus hispidus. Comp. Biochem. Physiol.* **32:**

WANG, L. C. AND HUDSON, J. W., 1971. Temperature regulation in normothermic and hibernating eastern chipmunk, *Tamias striatus. Comp. Biochem. Physiol.* **38A:** 59–90.

RESUMEN

29. Adaptacioness a bajas temperaturas ambientales de varios marsupiales carnívoros. Por Robert L. Wallis.

Varias respuestas de marsupiales carnívoros ausralia-otros mamíferos. Estas adaptaciones al frío, que nos (principalmente dasiúridos) a bajas temperaturas ambientales son descriptas y comparadas con las de necesariamente no son mutuamente excluyentes, pueden incluir el desarrollo de termogénesis distinta al tiritón, cambios estacionales del pelaje, compartimiento del nido, acumulación de grasa y letargo.

30. COMMUNICATION IN THE DASYURIDAE (MARSUPIALIA): A REVIEW

D. B. CROFT[1]

Communicatory behaviour is broadly defined to include any behaviour pattern (signal) that appears to transmit information about the sender's physiological or behavioural state to a receiver. Olfactory and auditory sensory channels are principally used by dasyurids for communication. Most species use olfactory investigation to establish the sex and sexual state of an individual, and possibly its identity in a social context. Both passive and active marking are common by faecal and urine deposition, or sternal and cloacal rubbing, respectively. Most marking seems to be associated with the establishment of a familiar home range area. Auditory communication is significant in most species. Most calls can be characterized within a squeak–churr–hiss–click continuum Calls are used in contexts such as alarm, agonistic interactions, courtship and mate attraction, nest and food defence, and the maintenance of proximity between mother and young. Few visual displays occur, the most common is an open–mouth threat that has been observed for all species studied. Tactile behaviour is often involved in courtship and copulation; for example, touching, neck gripping, palpation with the fore– or hindlimbs, and head rubbing.

The repertoire size for communicatory acts is around 10–20 different behaviour patterns for those species studied in greatest detail. This value is comparable with that for eutherian mammals of similar size and ecological niche. The behaviour of many species is poorly known but the communicatory pattern of dasyurids, in general, indicates a relatively asocial existence outside periods of mating behaviour and maternal care.

Pages 291–309 *in* CARNIVOROUS MARSUPIALS, ed. by M. Archer, February, 1982. Roy. Zool. Soc. New South Wales: Sydney, Australia.

THE FAMILY Dasyuridae has radiated widely in the insectivore–small carnivore niche throughout Australia and New Guinea. Yet, of the 49 species recognized (Kirsch and Calaby 1977), very few have been the subjects of detailed ethological studies. The behaviour of only the following species is well known: *Sminthopsis crassicaudata* (Ewer 1968a), *Dasyuroides byrnei* (Aslin 1974), *Sarcophilus harrisii* (Eisenberg *et al.* 1975, Golani 1976, Buchmann and Guiler 1977), *Planigale maculata* (Van Dyck 1979) and to a lesser extent, *Antechinomys laniger* (Happold 1972) and *Antechinus stuartii* (Marlow 1961, Braithwaite 1974, Settle and Croft 1982). This is perhaps not surprising as ethological studies are difficult for a group where individuals are small, often very fast moving, and for the most part nocturnal. Thus the social behaviour of even the most extensively studied species is known almost entirely from captive individuals.

Two other factors contribute to our ignorance of dasyurid social behaviour. First, most dasyurids are almost certainly solitary species and so intraspecific behavioural interactions occupy only limited periods of an individual's lifetime: e.g. in the mother–young unit, among members of the same litter, during courtship and mating behaviour, and in isolated encounters between conspecifics. Second, marsupials are often considered to be primitive mammals, lacking a corpus callosum and having a smaller brain size than eutherian counterparts (Andrew 1962, Jerrison 1973), and therefore their behavioural repertoire is assumed to be smaller or more stereotyped than eutherian mammals. However, if this is true, it should stimulate rather than impede research on marsupials, especially in a discipline such as ethology in which the comparative method has been so often stressed (Lorenz 1971, Tinbergen 1969). The work of Eisenberg amply illustrates the rewards of including ethological studies of marsupials in a general synthesis of the evolution of behaviour in mammals (Eisenberg and Golani 1977, Eisenberg 1977).

Animal communication has long intrigued ethologists and many broad comparative studies have been made (e.g. Sebeok 1977). Yet a definition of communication that encompasses all the diverse behaviour patterns shown by animals has proved elusive (Lewis and Gower 1980). The essential elements of a communication system are a sender who transmits information in a shared code to one or more receivers, who then act on it or merely store the information received (Smith 1977). Ethological studies have tended to concentrate on those instances where the behaviour of the sender alters the conditional probability pattern of the behaviour of a conspecific receiver (Wilson 1975). Natural selection is considered to have acted both on the production and reception of signals.

[1] School of Zoology, University of New South Wales, P.O. Box 1, Kensington, New South Wales, Australia, 2033.

In this review, I have attempted to cover a broad literature on dasyurid behaviour. This ranges from many fine accounts by Fleay of the natural history of various species, to detailed ethological studies as mentioned above. I have also included some unpublished material arising from observations of dasyurid species held in captivity at the School of Zoology, University of N.S.W. Thus I have necessarily adopted a broad definition of communication to include any behaviour pattern (signal) that appears to transmit information about the sender's physiological or behavioural state to a receiver. In this instance, I have not excluded interspecific communication such as threat behaviour to a potential predator. I have relied on the author's interpretation of the functional significance of the behaviour described even when a communicatory function is not proven in a strict sense; i.e. no attempt has been made to analyse the probability pattern of the receiver's behaviour (response) given the sender's behaviour (signal). Hutson's (1982) and Golani's (1976) studies are the only ones where this has been attempted.

SENSORY SYSTEMS

The brains of marsupial and placental mammals differ very little in both their structure and function (Johnson 1977). Thus there is considerable convergence in brain evolution in these two mammalian lines. Sensory systems in marsupial brains are well–developed, particularly the olfactory bulbs which are prominent in all marsupials. Marsupials are macrosommatic with extensive areas of olfactory epithelium and a well–developed vomeronasal organ (Broom 1896, Negus 1958). The auditory system of marsupials is similar to that of placental mammals, especially nocturnal species (e.g. *Rattus,* Johnson 1977). It is likely that olfactory and auditory systems are more important than the visual system in communication since dasyurids are generally nocturnal. In addition, vibrissae around the mouth region are usually long and suggest a well-developed tactile sense in this region (e.g. *Dasycercus cristicauda,* Jones 1924). Dasyurids also possess a well–developed manipulative ability with the forepaws which is used not only in prey catching and eating (e.g. Eisenberg and Leyhausen 1972) but also in grasping and pulling as forms of tactile behaviour in social interactions. Some genera such as *Sminthopsis* have large "nocturnal" eyes whereas others such as *Antechinus* have relatively small eyes and so the importance of visual displays in communication is likely to vary between genera.

Communication in the principal sensory modalities, olfaction and audition, are treated in separate sections. Tactile behaviour which is often difficult to observe and interpret functionally, and visual displays are reviewed in the broader context of selected patterns of social behaviour: agonistic and sexual interactions. Some comments are also made on the development of social behaviour through play.

OLFACTORY COMMUNICATION

Olfactory communication in mammals has recently been reviewed by Eisenberg and Kleiman (1972), Johnson (1973) and Stoddart (1976). Chemical messages are usually pheromones i.e., a chemical or mixture of chemicals, released into the external environment by an animal, that causes a specific behavioural reaction by a receiving conspecific (Shorey 1976). However, in mammals the chemical nature of secretions and the behavioural responses to them are often extremely complex (Beauchamp *et al.* 1976) and do not always meet the specific criteria for a pheromone (Kirschenblatt 1962). Thus Brown (1979) suggests that "social odours" is a more appropriate term for the chemical messages of mammals. These odours are either released from specialized glandular areas or, in some instances, represent specific metabolites in the animal's waste products. Chemical messages are deposited actively, for example by the animal deliberately rubbing a glandular area against the substrate, or passively through non-specific liberation of odours as the animal moves through its environment. Scent marking has the advantage that the message can be transferred to another individual or object and continue to serve as a signal in the absence of the emitting animal. Its disadvantage is that an animal cannot switch rapidly between one message and another as chemical communication is characterized by slowness of transmission and fade–out (Wilson 1975).

Mammalian scent marks serve a number of functions (Johnson 1973). The most important are probably: (i) a warning or deterrent to other conspecifics that a territory is occupied, (ii) a sex attractant or stimulant, (iii) a mechanism for labelling the home range for an individual's use in orientation or to maintain a sense of familiarity with an area, and (iv) an indicator of individual identity especially in regard to sexual status, age and perhaps dominance.

Various forms of scent marking behaviour are common in dasyurids (Table 1). All the species studied use cloacal dragging as a form of chemical deposition. Rubbing the chin and sternum are also common forms of marking behaviour. It is usually assumed that glandular areas are associated with these parts of the body and often this assumption is supported by the presence of secretion on marked objects. Russell (1981) has reviewed the distribution of scent glands in marsupials. In dasyurids, the most complete information is available for *Sminthopsis crassicaudata.* This species has ear, paracloacal, circum–anal and eccrine interdigital glands

Table 1. Modes of chemical deposition used by dasyurids in olfactory communication*

SPECIES	urine dribble	cloacal drag	chin rub	cheek rub	sternal rub	sandbathing ventral rub	sandbathing roll	saliva distribution bite objects	saliva distribution mouth wipe	"deliberate" defaecation
Planigale gilesi[1]										+
Planigale maculata[2]		+	+							+
Planigale tenuirostris[3]		+			+	—	—	+	+	+
Ningaui sp.[4]		+	+					—	—	+
Antechinus flavipes[5]	+	+			+	+				
Antechinus stuartii[6,7]		+	+		+			+		
Sminthopsis crassicaudata[5,8]	—	+			+					
Sminthopsis macroura[2]	+	+				+	+	+		+
Sminthopsis murina[1]									+	
Antechinomys laniger[9,10]		+	+							+
Dasycercus cristicauda[5,11]		+				+	+			
Dasyuroides byrnei[11,12]	+	+	+					+		+
Dasyurus viverrinus[10]	+				+	+				+
Sarcophilus harrisii[5,10]	+	+			+					

[1]Whitford (pers. commun.), [2]Van Dyck (1979), [3]Andrew & Settle (1982), [4]Fanning (1982), [5]Ewer (1968b), [6]Rigby (1972), [7]Braithwaite (1974), [8]Ewer (1968a), [9]Happold (1972), [10]Eisenberg & Golani (1977), [11]Sorenson 1970), [12]Aslin (1974). * (+) means observed; (—) stated not to occur; no sign means not observed to date.

(Green 1963). In addition, Cuttle (1978) found labial and sternal glands in *Phascogale tapoatafa*. Sternal glands have also been indicated in the marking behaviour of other species (*Planigale maculata*, Van Dyck 1979; *Dasyuroides byrnei*, Aslin 1974). In the males of both *Antechinus stuartii* (Woolley 1966) and *Planigale tenuirostris* (Andrew and Settle 1982), the sternal gland becomes especially prominent and active during the breeding season.

Urine and faeces are used in olfactory communication by some species (Table 1). Ewer (1968a) distinguished "deliberate" from "casual" defaecation and micturation and found both to occur in *Sminthopsis crassicaudata*. The same is true for *Planigale maculata* (Van Dyck 1979). Eisenberg and Golani (1977) have used the term "urine dribble" to identify "deliberate" micturation (see Table 1). In *Dasyuroides byrnei*, the pattern of defaecation differs between sexes (Aslin 1974). Females usually deposit faeces in a limited area whereas males deposit faeces over much of the cage, especially on novel objects.

Saliva may also contain some chemical message(s). Some species chew objects while others use a mouth wipe (Table 1). Both *Sminthopsis crassicaudata* (Green 1963) and *Phascogale tapoatafa* (Cuttle 1978) have glandular areas in and around the mouth which probably contribute to such chemical messages. The high frequency of mouth–to–mouth investigation in social encounters for *Sminthopsis crassicaudata* (Ewer 1968a) and *Sarcophilus harrisii* (Eisenberg and Golani 1977) support the hypothesis that saliva and/or some secretion from the mouth region conveys olfactory information. Eisenberg and Golani (1977) also contend that face–washing (common to all dasyurids studied) is a form of self–marking behaviour as well as a maintenance activity in removing foreign particles from the muzzle, vibrissae and eyes. Russell (1981) points out that face-washing is probably important in spreading mouth and ear secretions over the head of many species. The odours perhaps form the basis for individual identification as indicated in *Sminthopsis crassicaudata* (Ewer 1968a).

In species that sandbathe, a form of ventral rubbing is usually incorporated with side rubbing and rolling. Ewer (1968a) has argued that in *Sminthopsis crassicaudata* marking during sandbathing is of secondary importance to its function in cleaning the pelage. However, sandbathing is the most frequent marking behaviour in *Dasyuroides brynei* (Aslin 1974). The sternal area is pressed against the substrate and the animal vigorously pushes its chin and ventral surface along the sand. This form of behaviour is shown in encounters between individuals, especially when they are of the same sex, and so is most likely to be a specific form of marking behaviour.

Confining animals in captivity places a number of constraints on their use of space, especially in relation to other conspecifics. The principal function of marking behaviour in dasyurids appears to be a mechanism for labelling the animal's immediate environment (Table 2). For example, *Sminthopsis crassicaudata* (Ewer 1968a), *Antechinus flavipes* (Ewer 1968b) and at least the males of *Dasyuroides byrnei* (Aslin 1974) mark the whole cage area with glandular secretions (especially from the cloaca) and/or faeces and urine. However, in captivity neither *Sminthopsis crassicaudata* or *Planigale tenuirostris* (Andrew and Settle 1982) defend such an area nor do the scent marks necessarily deter another individual from using a marked area. Most species will respond to a novel object by marking it when it is placed in the home cage (Table 2). For example, *Sminthopsis macroura* will mark a novel object with cloacal secretions, saliva, faeces and urine (Van Dyck 1979).

Table 2. Functions of marking behaviour in captive dasyurids.

Species	Label familiar object	Label novel object	Label home cage area	Aggressive threat
Planigale maculata[1]			+	
Planigale tenuirostris[2]	+	+		
Ningaui sp.[3]		+	+	
Antechinus flavipes[4]			+	
Antechinus stuartii[5,6]	+	+		+
Sminthopsis crassicaudata[7]	+	+	+	
Sminthopsis macroura[1]		+		
Dasyuroides byrnei[8]		+	+	
Sarcophilus harrisii[9]				+

[1]Van Dyck (1979), [2]Andrew & Settle (1982), [3]Fanning (1982), [4]Ewer (1968b), [5]Rigby (1972), [6]Braithwaite (1974), [7]Ewer (1968a), [8]Aslin (1974), [9]Buchmann & Guiler (1977).

Planigale maculata will mark in response to a novel odour from a conspecific (Van Dyck 1979).

Some sex differences in marking behaviour have been noted. For example, only male *Planigale maculata* use sternal marking (Van Dyck 1979); female *Dasyuroides byrnei* sandbathe more than males but 'cloacal drag' less (Aslin 1974); male *Antechinus stuartii* mark more than females (Rigby 1972); and in *Sminthopsis crassicaudata,* only males mark with the cloaca in male–female encounters (Ewer 1968a). The reproductive state of the female influences her defaecation pattern in *Planigale maculata* (Van Dyck 1979) and *Sminthopsis crassicaudata* (Ewer 1968a). In both species, "deliberate" defaecation occurs principally when the female is pregnant or rearing young.

The dominance status of an individual can also influence marking behaviour. For example dominant individuals in *Antechinus stuartii* mark more than subordinants (Braithwaite 1974). In agonistic encounters between *Sarcophilus harrisii,* both members of a dyad mark by cloacal dragging in the initial stages of an interaction but once one individual establishes dominance the frequency of marking behaviour declines progressively in the subordinant (Buchmann and Guiler 1977). These authors suggest that cloacal dragging may reinforce vocal and visual display used in agonistic behaviour.

Olfactory investigation is an important component of dasyurid social behaviour (Table 3). Mouth sniffing, naso–nasal sniffing and touching, and cloacal sniffing are frequent behaviours. Ewer (1968a) suggests that mouth sniffing is important in individual recognition for *Sminthopsis crassicaudata* and so it possibly serves a similar function in the numerous other species in which it has been observed. Cloacal sniffing is generally confined to male–female interactions; the male sniffing the female's cloaca. Thus this behaviour is probably important in assessing the sexual state of the female either from glandular secretions in this area or the presence of steroid metabolites in the urine, or both. Male *Sarcophilus harrisii* and *Dasyuroides byrnei* also sniff the female's faeces and urine. Mothers frequently sniff their youngs' mouths, bodies or cloacas. Perhaps olfactory stimuli from the young are important in maintaining the female's maternal state. Finally, many species closely investigate and sniff novel objects or odours introduced into their cage (e.g. Fig. 1) and usually follow this investigatory behaviour with marking (Table 1, 2).

AUDITORY COMMUNICATION

Communication by sound production has obvious advantages for a nocturnal animal. Sounds can be broadcast day and night, in virtually all weather conditions and tend to flow around environmental obstacles. They require only moderate energy to transmit yet a wide range of signals can be produced and sent over long distances, and the sender can rapidly change from one message to another. The most common auditory signal in mammals is a vocalization which results from expulsion of air from the lungs causing a passive vibration of the glottis lips (Andrew 1963). The fundamental frequency so produced may be further modified during passage through the nasopharynx or buccal

Table 3. Olfactory investigation by dasyurids.

	mouth sniff	naso–nasal sniff	body sniff	scent sniff	sniff novel object	cloacal sniff
Planigale gilesi[1]	+	+				+
Planigale ingrami[2]	+					+
Planigale maculata[3]	+			+	+	+
Planigale tenuirostris[4]		+	+	+	+	+
Antechinus stuartii[5,6]		+	+			+
Sminthopsis crassicaudata[7-8]	+				+	+
Sminthopsis macroura[9]	+					
Sminthopsis murina[1]	+	+				+
Antechinomys laniger[10]	+					
Dasycercus cristicauda[10,11]			+			+
Dasyuroides byrnei[7,12]	+	+		+	+	+
Dasyurus geoffroii[13]		+	+			
Dasyurus maculatus[9]		+				
Sarcophilus harrisii[10,14]	+			+		+

[1]Whitford (pers. commun.), [2]McKay (1974), [3]Van Dyck (1979), [4]Andrew & Settle (1982), [5]Rigby (1972), [6]Braithwaite (1974), [7]Russell & Pearce (1971), [8]Ewer (1968a), [9]Settle (unpubl. obs.), [10]Eisenberg & Golani (1977), [11]Sorenson (1970), [12]Aslin (1974), [13]Archer (1974), [14]Eisenberg *et al.* (1975).

Table 4. Auditory signals of dasyurids given in alarm, agonistic and defensive contexts*

Species	Alarm	Agonistic (offensive/defensive) increasing arousal	Nest defence	Response to human intruder	Food defence
Planigale maculata[1,2]		fitt–zitt (♂), ssstt (♀)/tit (x3) (♂)	churr	sss. . .	rhitt (x3)
Planigale tenuirostris[3]		ca (x3)/chh (x3)	ca (x3)		chh (x3)
Ningaui sp.[4]	chatter	huff/hiss/ch (x3)	chh	hiss	
Antechinus flavipes[5,6]		churr	hiss		
Antechinus godmani[7]		tsst–tsst			
Antechinus stuartii[8]		huff/chee			
Antechinus swainsonii[9]		siss–siss			
Sminthopsis crassicaudata[10]		squeaks	squeaks		
Sminthopsis macroura[5]			hiss		
Sminthopsis murina[1]		ss (x3)/sh (x3)			s-s-s-s
Sminthopsis virginiae[11]		tzzzz			
Antechinomys laniger[12]		churr (x10)			hiss
Phascogale tapoatafa[13]	**tail rattle** **foot tap**				
Phascogale calura[15]	**foot tap**	soft hiss/loud hiss		churr	
Dasycercus cristicauda[16]					churr
Dasyuroides byrnei[5,17]		pant–hiss/churr, chatter/snort		churr	chatter
Dasyurus albopunctatus[5,18]		deep grunt	hiss–click/ churr		
Dasyurus geoffroii[15]		errutt			
Dasyurus hallucatus[19]				screech	
Dasyurus maculatus[20,21,22]	huff	hiss/sshah/scream,		hiss	
Dasyurus viverrinus[20,23]		cough/er–shish–a		hiss	hiss
Sarcophilus harrisii[5]	bark	growl/hiss/snort/**stamp**/ moan/whine, whine–growl/shriek			bark–**stamp**, snort–bark

* Mechanically produced sounds are in bold.

[1]Van Dyck (1979), [2]Fleay (1965), [3]Andrew (1977), [4]Fanning (1982), [5]Eisenberg *et al.* (1975), [6]Fleay (1949a), [7]Whitford (pers. commun.), [8]Braithwaite (1974), [9]Fleay (1932a), [10]Ewer (1968a), [11]Taplin (1980), [12]Happold (1972), [23]Fleay (1950), [14]Wakefield (1961), [15]Archer (pers. commun.), [16]Fleay (1961), [17]George (1964, [19]Fleay (1962), [20]Fleay (1932b), [21]Fleay (1948), [22]Settle (1978), [23]Fleay (1935a), [24]Settle & Croft (1982a), [25]Fleay (1934).

cavity. Sounds can also be produced by beating on the substrate. For example, foot–stamping by *Sarcophilus harrisii* (Eisenberg *et al.* 1975) and foot–tapping and tail–rattling by *Phascogale tapoatafa* (Fleay 1950, Wakefield 1961) are used in auditory communication.

The vocal behaviour of dasyurids is the best known aspect of their communicatory behaviour. Information ranges from onomatopoeic descriptions of vocalizations in the majority of species to detailed sonographic analysis of call types by Happold (1972), Aslin (1974), Eisenberg *et al.* (1975), Hutson (1982) and Fanning (1982). Eisenberg and coworkers (1975) surveyed the vocal behaviour of eight species of dasyurids among a larger sample of marsupial species. They recognized four main syllable types in the vocalizations of marsupials: (i) total syllables in which energy is organized into narrow frequency bands (e.g. squeak, chirp or moan), (ii) noisy syllables in which no discrete energy band occurs (e.g. hiss, scream), (iii) mixed syllables which combine the properties of the first two syllable types (e.g. growl, churr), and (iv) clicks which have little harmonic structure and are extremely brief ($<$ 20 msec) (e.g. click, cluck). Variation of the energy distribution, temporal patterning and duration of these fundamental syllable types provides a wide repertoire of call types.

Sarcophilus harrisii has the largest call repertoire yet found in dasyurids. Its eight vocalizations and two mechanically produced sounds cover virtually the whole spectrum of marsupial calls recorded by Eisenberg *et al.* (1975). However, as the behaviour of other species becomes better known repertoires of at least five calls are probably characteristic of dasyurids (Table 4). The function of a call is usually inferred from the context in which it is given but additionally some changes in vocalizations represent differences in arousal, especially in agonistic contexts.

Since few researchers have analyzed vocalizations using a sonograph, I have adopted the original onomatopoeic approximation given by the author in order to review auditory communication. However, Eisenberg *et al.* (1975) point out that some differences apparent in sonograms are not discriminated by the human ear. This is especially true of calls at ultrasonic frequencies above the human hearing range. Some calls (e.g. hisses, clicks and snorts) exceed audible frequencies (Happold 1972, Eisenberg *et al.* 1975) but no pure ultrasonic sounds (i.e. those with no audible frequencies) have as yet been recorded from dasyurids unlike many eutherian insectivores (Poduschka 1977).

Vocalizations are most commonly associated with agonistic encounters between conspecifics and defensive situations such as nest defence, food defence, threat to a human captor (Table 4). These vocalizations are usually noisy sounds such as hisses, huffs, grunts, growls and screams. The larger species produce extremely loud vocalizations; for example, the shriek of *Sarcophilus harrisii* (Eisenberg *et al.* 1975), and the "abrupt, piercing 'circular saw' screams" of *Dasyurus maculatus* (Fleay 1948). Male and female *Planigale maculata* apparently produce different calls but more usually differences in intensity (and call type) are distinguished, either in terms of defensive or offensive threat (e.g. *Planigale tenuirostris, Antechinus stuartii*) or arousal level (e.g. *Sarcophilus harrisii, Dasyuroides byrnei*).

Total calls tend to be associated with mother–young interactions (Table 5). The young of most species vocalize when displaced from their mother, giving calls such as sibilant hisses, chits, cha's or squeaks. These vocalizations normally result in retrieval behaviour by the mother (e.g. *Antechinus stuartii,* Settle and Croft 1982) and so are probably important in the maintenance of maternal care. In some species, mothers call to the young (e.g. *Dasyurus viverrinus, D. maculatus*). Happold (1972) made a particularly interesting observation that female *Antechinomys laniger* sometimes call their young out of the nestbox.

A mixture of calls are generally found in the context of courtship and mating behaviour. In the initial stages of courtship, some agonistic behaviour usually occurs. However, as yet few species have been observed to give calls specific to the courtship context. Male *Sminthopsis crassicaudata* and *Dasyurus albopunctatus* and *D. geoffroii* do give characteristic calls while following the female during courtship and *Sarcophilus harrisii* make a "huff" or "clap" sound (Table 5).

More significant is the increasing evidence in a wide range of dasyurids that the female attracts the male by calling (olfactory stimuli are also probably involved). Female *Dasyurus maculatus* give prolonged stanzas of a characteristic "cp–cp–" call during their oestrus period (Settle 1978, pers. obs.). Smaller species (*Planigale, Ningaui, Sminthopsis, Antechinomys*) often duet if males and females are housed in neighbouring cages during the females' oestrous. Usually the female calls, and the male answers with a similar call type followed by rapid searching movements about his cage. Happold (1972) found that duetting in *Antechinomys laniger* continued through the early stages of the youngs' development after birth but near the period when the young first detached from the teats the female's call changed from a polysyllabic to a monosyllabic one to which the male no longer responded. Female *Ningaui* also continue calling into early pregnancy (Fanning

Table 5. Auditory signals of dasyurids given in maternal, courtship and mating contexts and while exploring a novel environment*

SPECIES	♀ response to displaced young	Juvenile response to displacement	Courtship	Mate attraction	In novel environ./stress
Planigale gilesi[5]				clicks ♀, clicks ♂	
Planigale maculata[1,2]	wheeze	wheeze		stitt–stitt ♀, tsz–tsz ♂	twitter
Planigale tenuirostris[3]		chh–chh			twitter
Ningaui sp.[4]	soft tsss	sibilant ss–ss–ss	tsss ♂	tsitt x4 ♀, tsitt x1 ♂	twitter
Antechinus godmani[5]					twitter
Antechinus stuartii[24]		sibilant sisss			
Sminthopsis crassicaudata[10]		squeak	di–di ♂		
Sminthopsis macroura[5]		squeak			
Sminthopsis murina[1]				tsst ♀, ts–ts ♂	
Sminthopsis virginiae[11]		squeak			
Antechinomys laniger[12]	siss x1	sibilant ss–ss		bisyllabic ♀, trisyllabic ♂	
Phascogale tapoatafa[25]		cha–cha/tchit–tchit			
Dasycercus cristicauda[16]		cha–cha			
Dasyuroides byrnei[17]		grate			
Dasyurus albopunctatus[5]			cluck ♂		click
Dasyurus geoffroii[15]			cluck ♂		cluck
Dasyurus hallucatus[19]		chit–chitt/stitt–stitt			
Dasyurus maculatus[20,21,22]	chh–chh–chh	echh–echh–echh		cp–cp–cp ♀	
Dasyurus viverrinus[20,23]	er–chuck	ch–cha/shish–shish			
Sarcophilus harrisii[5]			huff/clap		

* References as listed under Table 4.

1982). *Ningaui* and *Sminthopsis murina* males, by contrast, continue to call throughout the breeding season (Fanning 1982, Whitford pers. commun.).

Some vocalizations have also been recorded during mating. Male *Ningaui* make a repeated soft "sst" call while copulating with the female once she has ceased struggling (Fanning 1982). Female *Dasyurus maculatus* make a brief "huff" sound as the male thrusts, possibly as a result of his pressure on the diaphragm during the thrust (pers. obs.).

Finally, *Planigale tenuirostris, P. maculata* and *Ningaui* produce a soft twittering sound while exploring a novel environment. The same is probably true of most *Antechinus* species (Whitford, pers. commun.). *Dasyurus albopunctatus* apparently "clicks" in a similar context.

AGONISTIC BEHAVIOUR

Agonistic behaviour is a composite term to cover aspects of threat, attack, defence and submission. The complete course of agonistic encounters between dasyurids are not reviewed but rather I have concentrated only on the postures and motor patterns that may communicate aggressive threat or submission to an opponent. Fighting behaviour in dasyurids usually includes chases, bites, wrestling, and in *Sarcophilus harrisii* oponents locking their jaws together (Buchmann and Guiler 1977, Eisenberg and Golani 1977). The postures and motor patterns used in threat and submission can be interpreted as visual displays, i.e. behavioural patterns that have become adapted in evolution to serve a communicatory function (Moynihan 1970), and so are usually stereotyped within a species. However, Golani (1976) and Eisenberg and Golani (1977) question the concept of a motor display patterns in social interactions. The motor patterns used by *Sarcophilus harrisii* in social encounters are interpreted in terms of the formation of "joints" between interactors (e.g. head to head, nose to mouth) and much of the motor activity in an encounter is related to the formation and maintenance of such "joints". Each interactor alters its own postures and attempts to manipulate its "opponent" to achieve a favourable position (e.g. in courtship the male attempts to gain a position from which he can grip the female's neck and achieve mounting). This methodology for analysing interaction patterns has only recently been derived and only *Sarcophilus harrisii* among the Dasyuridae studied in this way.

All dasyurid species so far studied use an open–mouth threat posture, usually accompanied by a vocalization (see Table 4), in agonstic encounters (Table 6). The common form is head forward to the opponent, mouth open displaying the teeth, ears laid back and eyes narrowed (Fig. 2). Variations occur in some species (e.g. *Planigale tenuirostris*) between an offensive and defensive threat. In the latter case, *P. tenuirostris* open the mouth fully, draw the head and body back away from the opponent, the ears and vibrissae are forward, and commonly one forepaw is raised (Andrew 1977). Morton (1978) suggests that most open–mouth threats in *Sminthopsis crassicaudata* are defensive, and the raised forepaw is used to ward off an attacker (Ewer 1968a). In contrast, Aslin (1974) interpreted the raised forepaw in *Dasyuroides byrnei* as offensive in preparation to strike.

Fig. 1. Ningaui sp. sniffing novel object.

Many species adopt an upright posture in threat and some authors have suggested that the display of the lighter ventrum has a communicatory function, given that dasyurids are largely nocturnal. Additionally, *Sarcophilus harrisii* adopts a lateral posture, often "shouldering" against its opponent. *Dasyuroides byrnei* sometimes twitches its tail (which has a conspicuous brush) when approaching an opponent and *Planigale maculata* may show this behaviour after a fight. Both Aslin (1974) and Van Dyck (1979) interpreted "tail–twitching" as a displacement activity reflecting a conflict between approach and retreat. However, the elevated tail of *Dasyuroides byrnei* is an important part of the threat display (Hutson 1982).

Submissive postures have been described for some species (Table 6). Usually the submissive animal adopts a low crouch, head lowered and turned sideways away from the opponent. Morrison (1975) describes *Planigale maculata*

Table 6. Communicatory behaviour observed in agonistic encounters of dasyurids.

SPECIES	THREAT		POSTURE					SUBMISSIVE
	open–mouth	vocal	upright	face	lateral	tail twitch	paw up	
Planigale gilesi[1]	+	+						head/body turn, vocal
Planigale maculata[2,3]	+	+				+	defence	low crouch, head turn
Planigale tenuirostris[4]	+	+	+	+				head turn
Ningaui sp.[5]	+	+						
Antechinus flavipes[6,7,8]	+	+	+					
Antechinus godmani[1]	+	+						
Antechinus stuartii[9,10]	+	+		+				belly up
Antechinus swainsonii[11]	+	+						
Sminthopsis crassicaudata[8,12]	+	+	+				defence	head turn/belly up
Sminthopsis macroura[7]	+	+						
Sminthopsis murina[13]	+	+						
Antechinomys laniger	+	+						
Dasycercus cristicauda[15]	+		+	+				
Dasyuroides byrnei[16]	+	+		+		+	offence	head low/jaws closed
Dasyurus geoffroii[17]	+	+	+					
Dasyurus hallucatus[18]	+	+						
Dasyurus maculatus[19,20]	+	+		+				
Dasyurus viverrinus[21]	+	+						
Sarcophilus harrisii[7,8,22]	+	+	+		+			

[1]Whitford (pers. commun.), [2]Van Dyck (1979), [3]Morrison (1975), [4]Andrew & Settle (1982), [5]Fanning (1982), [6]Fleay (1949b), [7]Eisenberg *et al.* (1975), [8]Eisenberg & Golani (1977), [9]Rigby (1972), [10]Braithwaite (1974), [11]Fleay (1932b), [12]Ewer (1968a), [13]Settle (unpubl. obs.), [14]Happold (1972), [15]Sorenson (1970), [16]Aslin (1974), [17]Archer (1974), [19]Fleay (1962), [19]Fleay (1940), [20]Fleay (1948a), [21]Fleay (1935a), [22]Fleay (1948b).

Fig. 2. Open–mouth threat of a female *Dasyurus maculatus* (Photo G. Settle).

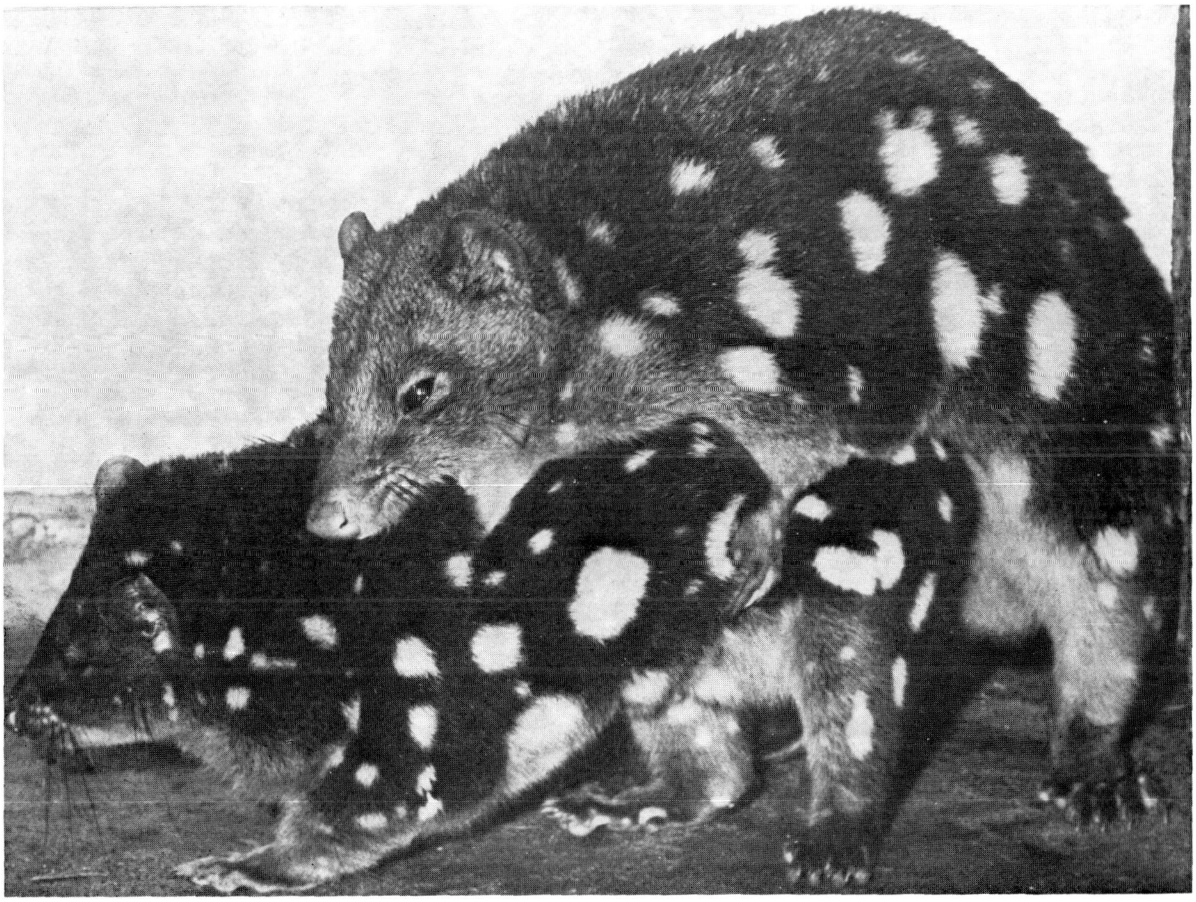

Fig. 3. Male *Dasyurus maculatus* gripping neck of female during copulation (Photo G. Settle).

as turning the body, but not the head, away laterally to the opponent. A full "belly–up" posture has been described for *Antechinus stuartii* and *Sminthopsis crassicaudata* but it is uncommon. Ewer (1968a) interprets "head (or throat) turning" as an intention movement derived from the full "belly–up" posture.

Fig. 4. Play–fighting in juvenile *Dasyurus maculatus* (Photo G. Settle).

SEXUAL BEHAVIOUR

Tactile behaviour, with olfactory investigation and some vocal behaviour, is especially important in courtship and copulation in dasyurids (Table 7). Communication by touch is maximally developed in those interactions, such as mating, that bring animals into closest bodily contact (Smith 1977). Patterns of touching behaviour that have been adapted for a communicatory function are common in mammals but nonetheless difficult to observe and interpret as some touching behaviour is certainly not stereotyped or may be secondary in function (e.g. in olfactory investigation and marking).

Courtship in dasyurids is usually accompanied by agonistic behaviour and, at times, prolonged chasing of the female by the male. Oestrus females rarely immediately accept the male's mounting attempts. *Dasyurus maculatus* is perhaps exceptional as calling females will readily accept a male unless, apparently, it is a cage–mate (Settle 1978). Some olfactory investigation of the female's mouth and cloaca precedes mounting attempts in some species but in others (e.g. *Ningaui*) this behaviour is absent (Table 7). Grasping at the female's body is common during mating chases but with the exception of "paws–on" behaviour in *Sminthopsis crassicaudata* (Ewer 1968a), this behaviour is not stereotyped and is unlikely to be specifically communicatory. Only *Sminthopsis crassicaudata* are known to allogroom during courtship although male *Dasyurus maculatus* lick the neck of the female during copulation (pers. obs.). With the exception of *Sminthopsis crassicaudata,* male dasyurids grip the neck of the female with their teeth prior to mounting (Fig. 3c). Males of some *Antechinus* species and *Sarcophilus harrisii* also drag the female about the cage using a

Table 7. Behaviour of the male during courtship in dasyurids.

SPECIES	Cloaca sniff	Mouth sniff	Follow	grasp body	"paw–on"	neck grip	neck–grip and drag	chin rub	vocal
Planigale maculata[1]	+		+	+		+			
Ningaui sp.[2]			+	+		+			♂
Antechinus flavipes[3]			+			+	+	+	♂
Antechinus stuartii[4]			+			+	+		
Sminthopsis crassicaudata[5]	+	+	+	+	+			grooms	
Sminthopsis virginiae[6]						+			
Dasycercus cristicauda[7]						+			
Dasyuroides byrnei[8]	+	+	+			+	+		
Dasyurus albopunctatus[9]									♂
Dasyurus geoffroii[10]	+		+			+			♂
Dasyurus maculatus[11]	+		+	+		+			
Dasyurus viverrinus[12]						+			
Sarcophilus harrisii[13]			+	+		+	+		♂

[1]Van Dyck (1979), [2]Fanning (1982), [3]Marlow (1961), [4]Ewer (1968b), [5]Ewer (1968a), [6]Taplin (1980), [7]Sorenson (1970) [8]Aslin (1974), [9]Eisenberg *et al.* (1975), [10]Archer (1974), [11]Croft (unpubl. obs.), [12]Fleay (1940), [13]Eisenberg & Golani (1977).

neckgrip; *Ningaui* males, by contrast, may be dragged by the female while they hold the neck grip. Females of the *Dasyurus geoffroii* (Archer 1974) and *D. maculatus* (Settle 1978) develop a swollen neck region during oestrous which serve multiple functions; a visual stimulus to the male, assist the male in gaining a neck grip, protect the neck from the male's bite.

the female's action. *Ningaui* however, seems to be exceptional in that the male also vocalizes in concert with this behaviour. Vocalizations by the males of other dasyurid species have only been recorded in the courtship phase of mating behaviour. Finally, "falling on the side" is characteristic of the smaller dasyurids during copulation. In *Mus musculus* (McGill 1962), "falling

Table 8. Behaviour of the male during copulation in dasyurids*

SPECIES	Clasps ♀'s body	Palpates c forelimbs	Palpates c hindlimbs	Rubs chin on ♀'s nape	Fall on side	Shakes ♀'s head	Vocal
Planigale maculata[1]	+	+	+		+	+	
Ningaui sp.[2]	+	+	+	+	+		♂
Antechinus flavipes[3]	+	+		+			
Antechinus stuartii[4]	+	+	+		+		
Sminthopsis crassicaudata[5]	+	+		+	+		
Sminthopsis virginiae[6]	+	+		+		+	
Dasycercus cristicauda[7]	+	+					
Dasyuroides byrnei[8]	+		+	noses			
Dasyurus maculatus[11]	+	+		licks		+	♀
Sarcophilus harrisii[11]	+	+					

* References as listed under Table 7.

The pattern of copulatory behaviour is similar in all dasyurid species studied so far (Table 8). Copulation in dasyurids is generally prolonged, lasting several hours or more (Eisenberg 1977). Males typically clasp the female's torso anterior to the hindlimbs, tucking their forelimbs in between the female's thighs. Most species palpate the female's abdomen during copulation, especially when thrusting. Ewer (1968a) describes these as jerky movements in *Sminthopsis crassicaudata* and suggests that they facilitate intromission by the male. *Dasyurus maculatus* males also stroke the female's flanks with their forepaws when maintaining a neck grip (pers. obs.). Unlike *Dasyurus maculatus* it is more common for males to tread, stroke or scratch the female's posterior with the hindlimbs. Treading movements with the feet are especially characteristic of *Planigale* species (Van Dyck 1979, pers. obs.). Another common behaviour pattern in small dasyurids is for the male to lay his head on the female's nape and frequently rub his head to and fro, stroking the female's nape and head. In *Dasyuroides byrnei* (Aslin 1974) and *Ningaui* (Fanning 1982), the male also occasionally thrusts his nose into the female's nape. In contrast, male *Planigale maculata* and *Dasyurus maculatus* sometimes shake the female's head laterally while gripping the neck. Ewer (1968a) interpreted the "head rubbing" behaviour of the male *Sminthopsis crassicaudata* as pacifying the female following her struggles. However, in *Ningaui* this behaviour usually follows thrusting by the male, and is not necessarily correlated with

on the side" follows the male's ejaculation but, unlike *Mus*, dasyurids do not necessarily disengage after this behaviour. In some instances in dasyurids, this behaviour results from the female's struggles but in others it perhaps indicates the male's ejaculation.

The neck grip is not only used to facilitate the male's initial mounting in courtship but is often maintained intermittently during copulation, especially in the initial stages when the female struggles. In *Dasyurus maculatus* (pers. obs.) the male in fact grips the female's neck whilst thrusting and releases the grip on her abdomen with his forepaw. Following thrusting the positions are reversed and the male frequently licks the female's neck. In other species (e.g. *Ningaui, Dasyuroides byrnei*), the male relinquishes the neck grip once the female becomes quiescent and follows with "head–rubbing" or "head–thrusting" behaviours.

PLAY BEHAVIOUR

Hunsaker and Shupe (1977) briefly reviewed play behaviour in marsupials as part of a discussion of the behaviour of New World Marsupials. They concluded that play behaviour is little developed in marsupials, "probably due to their simplistic behavioural activities". However, in the Dasyuridae, social play is extensive in the juveniles of all the larger species (Table 9). Fleay has provided numerous brief accounts of play behaviour in, particularly, *Dasyurus*. For example, describing young *Dasyurus viverrinus* he writes ". . . These young dasyures made a bright and interesting spectacle as they indulged in all the vivacious antics of the play–age. Frequently after nightfall they were brought into a drawing–room and from the moment of liberation they careered over the floor, about the couches and up the curtains, and wrestled with one another for hours" (Fleay 1935: 13).

The formulation of a precise definition of play has long been a contentious issue in ethology (e.g. Beckoff 1976, Fagen 1974, Symons 1978). The motivation of play is somewhat obscure but it generally occurs in contexts in which the causal factors of adult motor patterns seem not to be involved. The motor patterns themselves appear to serve no immediate adaptive function, or at least not the function ascribed to adult patterns of the same or similar form. Two distinct categories of play are generally recognized: social and manipulative play with objects. Object play may function to provide the animal with a diversity of sensory experience and to facilitate the aquisition of motor skills. Social play may aid the acquisition of social skills and perhaps have a role in the formation of social bonds. Thus social play is the more likely behaviour to be involved in the development of communicatory behaviour.

The predominant form of social play in dasyurids is play–fighting. Juveniles chase and wrestle with each other in long bouts of play (Fig. 4). Symons (1978) proposes that play–fighting in immature animals functions as practice for the development of fighting skills and strategies. He predicts that play–fighting will be most frequent in those species in which males compete for females and so it should be more frequent between males than females. Dasyurids, in general, are sexually dimorphic in body size, the males being the larger sex. If this dimorphism is the result of sexual selection for male competitive abilities, as seems likely, then play–fighting in dasyurids may well serve the function proposed by Symons.

However, it is still necessary to explain why social play is apparently absent in the smaller dasyurids. Two possibilities should be considered in future research: (i) species in which play is common are more social than those in which it it is absent (ii) play is related to the development of predatory skills as well as social skills and so it is more important in the larger dasyurids which take larger and/or more noxious prey. Little evidence exists at present to discriminate between these possibilities. *Sarcophilus harrisii* have overlapping home ranges, and possibly feed communally in closed groups at large carcasses (Buchmann and Guiler 1977). But *Sminthopsis crassicaudata* also have overlapping home ranges and will share nests at certain times of the year and yet are considered asocial (Morton 1978) and do not apparently play (Ewer 1968a). However, predatory behaviour in *Dasyurus viverrinus, Dasyuroides byrnei* and *Sarcophilus harrisii* does appear to be more complex than certain smaller dasyurids (*Antechinomys laniger, Sminthopsis macroura, Antechinus flavipes*) in that larger species "stalk" prey using an ambush technique (Eisenberg and Leyhausen 1972).

COMMUNICATION AND SOCIAL ORGANIZATION

Communication serves an essential co-ordinating function in the social activities of animals (Brown 1975, Wilson 1975). In general, the more complex the social organization of a species the larger its repertoire of communicatory acts (Wilson 1975). Thus asocial species are expected to have a smaller repertoire of communicatory behaviour and such behaviour is probably more stereotyped than in social species. Experience in a social group enables an animal to better adapt its social behaviour to individual differences of group members and such behaviour may therefore be more plastic.

Little is known about the social organization of dasyurids but most species are thought to be

Table 9. Play behaviour of some dasyurids.

SPECIES	
Dasycercus cristicauda	Stalkings and simulated attacks (Fleay 1961), play–fighting (Michener 1969), contagious locomotor play, 'fighting' play (Ewer 1968; 1969).
Dasyuroides byrnei	Locomotor play, simple social play (Aslin 1974), extensive play behaviour (Collins 1973).
Dasyurus hallucatus	Lomomotor play, wrestling, adults playful (Fleay 1962).
Dasyurus maculatus	Locomotor play, wrestling (Fleay 1940), play with objects (Troughton 1954), chasing, stalking, wrestling (Settle 1978).
Dasyurus viverrinus	Locomotor play, solitary and wrestling play, adult females playful (Fleay 1932b, 1935a; Nelson & Smith 1971).
Phascogale tapoatafa	'Hide and seek' (Fleay 1934).
Sarcophilus harrisii	Locomotor play, mock attack, wrestling (Fleay 1935b), manipulative play, chasing, fighting (Hediger 1958), 'frolicsome' (Ewer 1968b).
Sminthopsis crassicaudata	No genuine play (Ewer 1968a).
Antechinomys laniger	No play (Happold 1972).

asocial; individuals living a solitary existence for most of their lifetime. Communicatory behaviour such as the "mate attraction" call of males and females in some species is interpreted as a consequence of a solitary habit (Van Dyck 1979). Species such as *Sarcophilus harrisii* with overlapping home ranges and possible group formation (Buchmann and Guiler 1977), *Sminthopsis crassicaudata* with overlapping ranges and nest sharing facilitating contact between adults (Morton 1978), and *Dasyuroides byrnei* which at least huddle in captivity when alarmed (Aslin 1974) apparently have no such call. However, *Planigale gilesi* will also nest communally in winter (Andrew and Settle 1982) and probably have overlapping ranges (Read 1982) yet males and females call in the breeding season (Whitford, pers. commun.). Adult male and female *Sminthopsis murina* also share nests outside the breeding season and call (Whitford, pers. commun.). Nonetheless, *Ningaui* (Fanning 1982) and *Sminthopsis virginiae* (Taplin 1980) are relatively intolerant when housed in captivity and so support Van Dyck's hypothesis. However, inferences from the behaviour of various species in captivity are probably of little value. It will be necessary to study the social relationships of males and females through non-breeding and breeding periods in the wild before the relationships between social organization and "mate attraction" calling can be determined. Those species in which males and females rarely have any social contact prior to breeding may well prove to be the ones which call in the breeding season.

In captivity, a number of species show a propensity to form dominance relationships, at least in contrived encounters (Tables 10). For example, cage mates in *Sarcophilus harrisii* establish stable dominance relationships after a series of agonistic encounters (Buchmann and Guiler 1977). The male is hierarchically dominant over the female at the onset of the breeding season and will attack males (Guiler (1970). The male also often actively confines the female to a den for 10 to 12 days after mating (Roberts 1915, Fleay 1935b), but the female attains dominance over the male during pregnancy and fighting may interrupt birth (Guiler 1971). Postparturient females are hostile to males (Roberts 1915, Fleay 1935b) and presumably other females as both will eat young left in an unguarded den (Guiler 1971). However, studies in captivity, such as those listed in Table 10, tell us very little about the occurrence of stable dominance hierarchies in the wild. We can only conclude that individuals differ in their competitive abilities, aggressiveness increases during the breeding season, and lactating females tend to defend their young against conspecific intruders, possibly because of cannibalistic tendencies in dasyurids (Woolley 1971). Although some species show submissive postures (Table 6), it is most common for one opponent to retreat by "escape–leaping" (e.g. Aslin 1974). Thus without the physical constraints of a captive environment, most agonistic encounters in the wild probably conclude with the retreat of one opponent rather than submissive behaviour which would facilitate continued proximity and reflect more stable relationships between individuals.

Marking behaviour in captivity indicates that most species probably label their home ranges with scent marks. Whether these marks deter other neighbouring individuals from entering a home range is unknown. Ewer (1968a) found no evidence of territorial behaviour in captive *Sminthopsis crassicaudata* and this result is sup-

Table 10. Dominance relationships in some dasyurids.

SPECIES	Dominance relationship observed	Source
Planigale tenuirostris	Males form dominance relationships in paired encounters in a "neutral" arena.	Andrew (1977) Andrew & Settle (1982)
Antechinus stuartii	Near linear heirarchy formed as assessed by outcome of agonistic interactions. Females (3) generally dominant to males (2).	Wood (1970)
	Dominance relationships formed in dyadic encounters in a "neutral" arena	Braithwaite (1974)
Sminthopsis crassicaudata	Dominance established between two males competing for an oestrus female.	Ewer (1968a)
Antechinomys laniger	One male dominant over litter–mates at food.	Happold (1972)
Dasyuroides byrnei	Female–male–female hierarchy as assessed by outcome of agonistic encounters between three captive individuals.	Sorenson (1970)
	Dominance relationships formed between males in paired encounters in a "neutral" arena. No clear–cut relationships formed in male–female or female–female encounters.	Aslin (1974)
Sarcophilus harrisii	Stable dominance relationships formed between two cage–mates.	Buchmann & Guiler (1977)
	Male dominant to female outside breeding season. Pregnant and lactating females dominant to males.	Roberts (1915), Fleay (1935b)

ported by Morton's (1978) field observations. Van Dyck (1979) supports the idea that marking behaviour in *Planigale maculata* principally functions as a method of creating a familiar habitat for the individual. However, *Antechinus stuartii* mark after victories in agonistic encounters with the scent possibly serving a threat function (Braithwaite 1974) although home ranges overlap considerably (Wood 1970). Aslin (1974) concludes that sand–bathing in *Dasyuroides byrnei* may serve to mark territories which exclude conspecifics of the same sex. Happold (1972), on very little evidence, also suggests that *Antechinomys laniger* are territorial using a vocalization, and perhaps cloacal marking, to exclude conspecifics. Field studies and studies of captive individuals in large pens will be necessary to discover whether scent marks serve a deterrent function at territorial boundaries.

The form of sexual behaviour in dasyurids suggests that pair–bonds are usually not formed. Most matings are preceded by agonistic behaviour and chasing, and males usually take a neck grip to restrain the female. Little coordination in the male and females' behaviour is evident until the female becomes quiescent after mounting is achieved. Some copulations lead to injuries to the female's neck region (Fleay 1940, Marlow 1961). Considerable tactile stimulation of the female seems necessary before the male achieves intromission. There are some similarities in the mating behaviour of dasyurids and certain eutherian mammals. For example, a neck grip is also common in the mating behaviour of relatively asocial insectivores (Poduschka 1977) and carnivores (Ewer 1973). Male Tenrecs (e.g. *Setiferes*) also show "stimulant scratching" at the female's posterior during mating (Poduschka 1977).

Van Dyck (1979) has argued that the highly aggressive nature of the female during mating in *Planigale maculata* reflects an element of "female choice" in the mating system of this species and other "socially intolerant" dasyurids. He suggests that only the largest and fittest males will be able to overcome the female's aggression and so there is selection against smaller and weaker males. However, in that the species is sexually dimorphic with males larger than females, the male starts with a relative advantage. This latter point would seem to argue against Van Dyck's hypothesis. In mammalian species in which males grow to a larger size than females it is usually argued that this is the result of sexual selection for male competitive abilities (Brown 1975). Thus males fight other males in competition for a mate rather than males fight females in order to mate. It is more probable that sexual dimorphism in dasyurids has evolved as the result of sexual selection for male competitive abilities rather than female choice. Male *Antechinus stuartii* become significantly more aggressive in the breeding season whereas the aggressiveness of females does not change (Braithwaite 1974). Cohesive behaviour in male *Sminthopsis crassicaudata* declines markedly in the breeding season although the mean but not the range of scores for aggressive behaviour also declines slightly (Morton 1978). Further, Read (1982) found several male *Planigale gilesi* in pit traps close to a copulating pair and Whitford (pers. commun.) observed a similar aggregation of males around a female while pit–trapping for *Sminthopsis murina* during the breeding season. These results suggest that males compete for females during the breeding season.

The aggressiveness of females to males during mating in captivity may be accentuated by the small cages in which they are housed and the relatively brief periods in which they are in contact. Nonetheless a certain degree of aggressiveness by the female is not surprising if the sexes are normally segregated. The female is likely to be aggressive towards an intruding individual, particularly during early oestrous and repeated brief contacts between males and females prior to mating may gradually overcome this initial aggression. Such a phenomenon is well–known in territorial birds (e.g. Hinde 1953, Marler 1956).

Finally, in common with other nocturnal mammals, auditory communication is well–developed in dasyurids. Vocalizations occur in the context of all social interactions — mating behaviour, agonistic behaviour, maternal care. In some species such as *Sarcophilus harrisii* the vocal repertoire is highly developed for a non–primate mammal. Communication at a distance, as achieved by vocal behaviour, perhaps again reflects the lack of cohesive group formation in dasyurids but other factors such as nocturnal activity and, for some species, a dense habitat will also be important in the development of auditory communication. In contrast to insectivores (Poduschka 1977) and rodents (Sales and Pye 1974) pure ultrasonic calls are not given by dasyurids. Vocalizing at ultrasonic frequencies is thought to be an adaptation in rodents to avoid predation during social encounters at close proximity because ultrasounds are both difficult to detect and locate at a distance (Sales and Pye 1974). Some insectivores may use ultrasounds in echo–location (Poduschka 1977). Whether selective pressures to develop ultrasonic communication are less for dasyurids than for these other mammalian groups is unknown. Murid rodents in Australia vocalize at ultrasonic frequencies (Watts 1975, 1976) but perhaps this is a phylogenetically old trait in rodents. In any case, Watts' studies indicate that ultrasonic vocalizations are largely produced only during early development.

In summary, dasyurids have a well–developed repertoire of communicatory behaviour. Like other nocturnal mammals, olfactory and auditory communication are emphasized and visual displays are limited. Tactile behaviour seems to be important in close social encounters. Rather than being primitive and restricted in their behavioural repertoire in comparison with eutherian groups such as the insectivores, the communicatory behaviour of dasyurids seems to be equally well–developed and reflects the ecological niche they occupy rather than their phylogenetic status. As one example, Poduschka's (1977) review of communication in the Soricidae (shrews) indicates that visual communication is similarly poorly developed. Gould (1969) identified several call types from shrews (e.g. chirps, clicks, twittering, "putt", buzzes, chirp buzzes and putter twitter) which are comparable both in form and repertoiral size to dasyurid calls. Chemical communication varies between species of shrews but is apparently well–developed in some such as *Crocidura oliveri* and *Suncus murinus* (Poduschka 1977) and is similar in form and function to dasyurid behaviour. Similarities in the tactile behaviour of some insectivores and dasyurids during mating have already been noted.

ACKNOWLEDGEMENTS

I wish to thank Dick Whitford for sharing many of his observations on dasyurid behaviour with me. The free use of extensive reference collections held by the late Graham Settle and Dr. Michael Archer greatly facilitated the writing of this review. I wish to acknowledge my debt to the writings of Dr. John Eisenberg and his colleagues which provided an extremely useful background on which to build this review. My sincere thanks to the referees, Dr. Eleanor Russell and Mr. Steven Van Dyck for their valuable comments on the initial draft of this paper. I am especially grateful to Dr. Russell for allowing me to read a draft of her forthcoming paper on chemical communication in marsupials and monotremes.

REFERENCES

ANDREW, D. L., 1977. "The behaviour of *Planigale tenuirostris* (Dasyuridae: Marsupialia), with comparative notes on other species". Hons. thesis: University of New South Wales Sydney.

ANDREW, D. L. AND SETTLE, G. A., 1982. Observations on the behaviour of *Planigale* (Dasyuridae, Marsupialia), with particular reference to Narrow–nosed Planigale *(Planigale tenuirostris)*. Pp.311–24 *in* "Carnivorous marsupials" ed. by M. Archer. Roy. Zool. Soc. New South Wales: Sydney.

ANDREW, R. J., 1962. Evolution of intelligence and vocal mimicking. Science **137:** 585–9.

ANDREW, R. J. 1963. The origin and evolution of the calls and facial expressions of primates. *Behaviour* **20:** 1–109.

ARCHER, M., 1974. Some aspects of reproductive behaviour and the male erectile organs of *Dasyurus geoffroii* and *D. hallucatus* (Dasyuridae: Marsupialia). *Mem. Qd Mus.* **17:** 63–7.

ASLIN, H., 1974. The behaviour of *Dasyuroides byrnei* (Marsupialia) in captivity. *Z. f. Tierpsychol.* **35:** 187–208.

BEAUCHAMP, G. K., DOTY, R. L., MOULTIN, D. B. AND MUGFORD, R. A., 1976. The pheromone concept in mammalian chemical communication: a critique Pp.144–60 *in* "Mammalian olfaction, reproductive processes and behaviour", ed. by R. L. Doty. Academic Press: New York.

BECKOFF, M. 1976. Animal play: problems and perspectives. *In* "Perspectives in ethology. Vol. 2". ed. by P. P. G. Bateson and P. H. Klopfer. Pp.165–88. Plenum: New York.

BRAITHWAITE, R. W., 1974. Behavioural changes associated with the population cycle of *Antechinus stuartii* (Marsupialia). *Aust. J. Zool.* **22:** 45–62.

BROOM, R., 1896. On the comparative anatomy of the organ of Jackobson in marsupials. *Proc. Linn. Soc. N.S.W.* **21:** 591–63.

BROWN, J. L., 1975. "The evolution of behaviour". Norton: New York.

BROWN, R. E., 1976. Mammalian social odours: a critical review. *Adv. Study Behav.* **10:** 103–62.

BUCHMANN, O. L. K. AND GUILER, E. R., 1977. Behaviour and ecology of the Tasmanian Devil, *Sarcophilus harrisii*. Pp.155–68 *in* "The Biology of Marsupials", ed. by B. Stonehouse and D. Gilmore. MacMillan: London.

COLLINS, L. R., 1973. "The monotremes and marsupials". Smithsonian Press: Washington.

CUTTLE, P. 1978. The behaviour of the Tuan, *Phascogale tapoatafa*. Unpubl. M.Sc. thesis: Monash University, Melbourne.

EISENBERG, J. F., 1977. The evolution of the reproductive unit in the Class Mammalia. Pp.39–71 *in* "Reproductive behaviour and evolution Vo. I". ed. by J. S. Rosenblatt and B. Komisaruk. Plenum: New York.

EISENBERG, J. F. AND GOLANI, I., 1977. Communication in Metatheria. Pp.575–599 *in* "How animals communicate" ed. by T. A. Sebeok. Indiana University Press: Bloomington.

EISENBERG, J. F. AND KLEIMAN, D. G., 1972. Olfactory communication in mammals. *Ann. Rev. Ecol Syst.* **3:** 1–32.

EISENBERG, J. F. AND LEYHAUSEN, P., 1972. The phylogenesis of predatory behaviour in mammals. *Z.f. Tierpsychol.* **30:** 59–93.

EISENBERG, J. F., COLLINS, L. R. AND WEMMER, C., 1975. Communication in the Tasmanian Devil *(Sarcophilus harrisii)* and a survey of auditory communication in the Marsupialia. *Z.f. Tierpsychol.* **37:** 379–99.

EWER, R. F., 1968a. A preliminary survey of the behaviour in captivity of the dasyurid marsupial, *Sminthopsis crassicaudata* (Gould). *Z.f. Tierpsychol.* **25:** 319–65.

EWER, R. F., 1968b. "Ethology of mammals". Logos Press: London.

EWER, R. F., 1969. Some observations on the killing and eating of prey by two dasyurid marsupials: the Mulgara, *Dasycercus cristicauda,* and the Tasmanian Devil, *Sarcophilus harrisii. Z.f. Tierpsychol.* **26:** 23–38.

EWER, R. F., 1973. "The carnivores". Weidenfield and Nicholson: London.

FAGEN, R., 1974. Selective and evolutionary aspects of animal play. *Amer. Nat.* **108:** 850–8.

FANNING, D. F., 1982. Reproduction and development in *Ningaui* sp. (Dasyuridae, Marsupialia) from the Northern Territory. Pp.23–37 *in* "Carnivorous marsupials", ed. by M. Archer. Roy. Zool. Soc. New South Wales: Sydney.

FLEAY, D., 1932a. The rare dasyures (native cats). *Vict. Nat.* **49:** 63–8.

FLEAY, D., 1932b. Swainson's Phascogale (the "bush mouse"). *Vict. Nat.* **49:** 132–4.

FLEAY, D., 1934. The Brush–tailed Phascogale. First record of breeding habits. *Vict. Nat.* **51:** 88–100.

FLEAY, D., 1935. Breeding of *Dasyurus viverrinus* and general observations on the species. *J. Mammal.* **16:** 10–16.

FLEAY, D., 1935b. Notes on breeding of Tasmanian Devils. *Vict. Nat.* **52:** 100–5.

FLEAY, D., 1940. Breeding of the Tiger Cat. *Vict. Nat.* **56:** 159–63.

FLEAY, D., 1948a. Australia's marsupial "Tiger Cat". *Animal Kingdom* **51:** 36–41.

FLEAY, D., 1948b. The marsupial Devil of Tasmania. *Animal Kingdom* **51:** 86–91.

FLEAY, D., 1949a. The Yellow–footed Marsupial Mouse. *Vict. Nat.* **65:** 273–7.

FLEAY, D., 1949b. The Brush–tailed Tuan. A story of a fruitless vigil. *Wildlife* **11**(2): 118–21.

FLEAY, D., 1950. Experiences with Australia's Brush–tailed Tuan. *Animal Kingdom* **53:** 152–7.

FLEAY, D., 1961. Breeding the Mulgara. *Vict. Nat.* **78:** 288–93.

FLEAY, D., 1962. The Northern Quoll, *Satanellus hallucatus*. *Vict. Nat.* **78:** 188–293.

FLEAY, D., 1965. Australia's "needle–in–a–haystack" marsupial. *Vict. Nat.* **82:** 195–204.

GEORGE, G. G., 1964. Notes on a New Guinea Quoll. *Vict. Nat.* **81:** 254–6.

GOLANI, I. 1976. Homeostatic motor processes in mammalian interactions: a choreography of display. Pp.69–134 *in* "Perspectives in ethology. Vol. 2", ed by P. P. G. Bateson and P. H. Klopfer. Plenum: New York.

GOULD, E., 1969. Communication in three genera of shrews (Soricidae): *Suncus*, *Blarina* and *Cryptotis*. *Commun. in Behav. Biol.* **3A:** 11–31.

GUILER, E. R., 1970. Observations on the Tasmanian Devil, *Sarcophilus harrisii* (Marsupialia: Dasyuridae). II Reproduction, breeding and growth of pouch young. *Aust. J. Zool.* **18:** 63–70.

GUILER, E. R., 1971. The Tasmanian Devil, *Sarcophilus harrisii* in captivity. *Int. Zool. Yb.* **11:** 32–3.

GREEN, L. M. A., 1963. Distribution and comparative histology of cutaneous glands in certain marsupials. *Aust. J. Zool.* **11:** 250–72.

HAPPOLD, M., 1972. Maternal and juvenile behaviour in the marsupial Jerboa, *Antechinomys spenceri* (Dasyuridae). *Aust. Mammal.* **1:** 27–37.

HEDIGER, H., 1958. Verhalten de Beuteltiere Marsupialia. *Handbuch der Zoolögie* **8:** 1–27.

HINDE, R. A., 1953. The conflict between drives in the courtship and copulation of the chaffinch. *Behaviour* **5:** 1–31.

HUNSAKER II, D. AND SHUPE, D., 1977. Behaviour of New World marsupials. Pp.279–347 *in* "The biology of marsupials" ed. by D. Hunsaker II. Academic Press: New York.

HUTSON, G. D., 1982. An analysis of offensive and defensive threat displays in *Dasyuroides byrnei* (Dasyuridae, Marsupialia). Pp.345–63 *in* "Carnivorous marsupials", ed. by M. Archer. Roy. Zool. Soc. New South Wales: Sydney.

JERRISON, H. J., 1973. "Evolution of the brain and intelligence". Academic Press: New York.

JOHNSON, R. P., 1973. Scent marking in mammals. *Anim. Behav.* **21:** 521–35.

JOHNSON, J. I. JR., 1977. Central nervous system of marsupials. Pp.157–278 *in* "The biology of marsupials", ed. by D. Hunsaker II. Academic Press: New York.

JONES, F. WOOD, 1924. The external characters of pouch embryos of marsupials. No. 6 — *Dasycercus cristicauda*. *Trans. Roy. Soc. S. Aust.* **48:** 136–41.

KIRSCH, J. A. W. AND CALABY, J. H., 1977. The species of living marsupials: an annotated list. Pp.9–26 *in* "The biology of marsupials", ed. by B. Stonehouse and D. Gilmore. Macmillan: London.

KIRSCHENBLATT, J., 1962. Terminology of some biologically active substances and validity of the term 'pheromones'. *Nature* **195:** 916–7.

LEWIS, D. B. AND GOWER, D. M., 1980. "Biology of communication". Blackie: Glasgow.

LORENZ, K. 1971. Methods of approach to the problems of behaviour. Pp.246–80 *in* "Studies in animal and human behaviour. Vol. II". Methuen: London.

MARLER, P., 1956. Behaviour of the chaffinch *(Fringilla coelebs)*. *Behaviour, suppl.* **5:** 1 184.

MARLOW, B. J., 1961. Reproductive behaviour of *Antechinus flavipes* (Waterhouse) Marsupialia, and the development of pouch young. *Aust. J. Zool.* **9:** 203–18.

MCGILL, T. E., 1962. Sexual behaviour of three inbred strains of mice. *Behaviour* **19:** 341–50.

MCKAY, G. M., 1974. *Planigale ingrami* in captivity. *Koolewong* **3:** 4–5.

MICHENER, G. R., 1969. Notes on the breeding and young of the Crest–tailed Marsupial Mouse, *Dasycercus cristicauda*. *J. Mammal.* **50:** 633–5.

MORRISON, R. G. B., 1975. Emergence of the Pigmy Antechinus. *Aust. Nat. Hist.* **18:** 164–7

MORTON, S. R., 1978. An ecological study of *Sminthopsis crassicaudata* (Marsupialia: Dasyuridae). II. Behaviour and social organization. *Aust. Wildl. Res.* **5:** 163–82.

MOYNIHAN, M. H., 1970. Control, suppression, decay, disappearance and replacement of displays. *J. Theoret. Biol.* **29:** 85–112.

NEGUS, V. E., 1958. "Comparative anatomy and physiology of the nose and paranasal sinuses". Livingstone: London.

NELSON, J. E. AND SMITH, G., 1971. Notes on growth rates in native cats of the family Dasyuridae. *Int. Zool. Yb.* **11:** 37–41.

PODUSCHKA, W., 1977. Insectivore communication. Pp.600–633 *in* "How animals communicate", ed. by T. A. Sebeok. Indiana University Press: Bloomington.

READ, D., 1981. Observations of the movements of two arid zone planigales (Dasyuridae, Marsupialia). Pp.227–31 *in* "Carnivorous marsupials" ed. by M. Archer. Roy. Zool. Soc. New South Wales: Sydney.

RIGBY, R. G., 1972. A study of the behaviour of caged *Antechinus stuartii*. *Z.f. Tierpsychol.* **31:** 15–25.

ROBERTS, M. G., 1915. The keeping and breeding of Tasmanian Devils *(Sarcophilus harrisii)*. *Proc. Zool. Soc. Lond.* **1915:** 575–81.

RUSSELL, E. M., 1981. Scent glands and olfaction in the social behaviour of marsupials and monotremes. *In* "Social odours in mammals", ed. by D. McDonald. (In press).

RUSSELL, E. M. AND PEARCE, G. A., 1971. Exploration of novel objects by marsupials. *Behaviour* **40:** 312–22.

SALES, G. AND PYE, D., 1974. "Ultrasonic communication by animals". Chapman and Hall: London.

SEBEOK, T. A., 1977. "How animals communicate". Indiana University Press: Bloomington.

SETTLE, G. A., 1978. The quiddity of Tiger Quolls. *Aust. Nat. Hist.* **19:** 164–9.

SETTLE, G. A. AND CROFT, D. B., 1982a. Maternal behaviour of *Antechinus stuartii* (Dasyuridae, Marsupialia) in captivity. Pp.365–81 *in* "Carnivorous marsupials", ed. by M. Archer, Roy. Zool. Soc. New South Wales: Sydney.

SHOREY, H. H., 1976. "Animal communication by pheromones". Academic Press: New York.

SMITH, W. J., 1977. "The behaviour of communicating". Harvard University Press: Cambridge, Mass.

SORENSON, M. W., 1970. Observations on the behaviour of *Dasycercus cristicauda* and *Dasyuroides byrnei* in captivity. *J. Mammal.* **51:** 123–30.

STODDART, D. M. 1976. "Mammalian odours and pheromones". Edward Arnold: London.

SYMONS, D., 1978. "Play and agression: a study of Rhesus monkeys". Columbia University Press: New York.

TAPLIN, L. E., 1980. Observations on the reproductive biology of *Sminthopsis virginiae* (Tarragon) (Marsupialia: Dasyuridae). *Aust. Zool.* **20:** 407–18.

TINBERGEN, N., 1969. Ethology. Pp.238–268 *in* "Scientific thought 1900–1960", ed. by R. Harre. Clarendon Press: Oxford.

TROUGHTON, E., 1954. The marsupial "Tiger Cat". Birth and growth in captivity. *Aust. Mus. Mag.* **11:** 200–2.

VAN DYCK, S., 1979. Behaviour in captive individuals of the dasyurid marsupial *Planigale maculata* (Gould 1851). *Mem. Qd. Mus.* **19:** 413–31.

WAKEFIELD, N. A., 1961. Notes on the Tuan. *Vic. Nat.* **78:** 232–5.

WATTS, C. H. S., 1975. Vocalizations of Australian mice (Rodentia: *Notomys*). *J. Zool., Lond.* **177:**

WATTS, C. H. S., 1976. Vocalizations of the Plains Rat *Pseudomys australis* Gray (Rodentia: Muridae). *Aust. J. Zool.* **24:** 95–103.

WILSON, E. O., 1975. "Sociobiology: the new synthesis". Harvard University Press: Cambridge, Mass.

WOOD, D. H., 1970. An ecological study of *Antechinus stuartii* (Marsupialia) in a southeast Queensland rainforest. *Aust. J. Zool.* **18:** 185–207.

WOOLLEY, P., 1966. Reproduction in *Antechinus* spp. and other dasyurid marsupials. *Symp. Zool. Soc., Lond.* **15:** 281–94.

WOOLLEY, P., 1971. Breeding patterns, and the breeding and laboratory maintenance of dasyurid marsupials. *Exp. Animals, suppl.* **22:** 161–72.

RESUMEN

30. La communicación entre los Dasyuridae una revisión. Por D. B. Croft.

El comportamiento de la comunicación es generalmente definido como incluyendo cualquier tipo de conducta (señal) que aparente trasmitir información sobre el estado fisiológico o emotivo de un emisor a un receptor. La mayoría de las especies usan la investigación olfativa para establecer el sexo y el estado sexual de un individuo, y posiblemente su identidad en el contexto social. El marcado pasivo es comúnmente cumplido por la deposición de las heces o de la orina. El activo por la fricción de la región esternal o la cloacla. La mayoria del marcado parece estar relacionado con el establecimiento de un área territorial familiar. La communicación auditiva es significativa en la mayoría de las especies. La mayoría de los llamados suelen caer dentro de un continuo "squeak–churr–hiss–click". Los llamados son usados como alarmas, interacción agonística, atracción por cortejo y apareo, defensa del nido o la comida, y el mantenimiento de la proximidad entre la madre y el hijo. Pocos despliegues visuales ocurren, el más común es por una amenazante boca abierta observado en todas las especies estudiadas. La conducta tacttl a menudo se cumple durante el cortejo y la copulación, como por ejemplo tocando, asiéndose del cuello, palpando con los miembros anteriores o posteriores y sobando la cabeza.

La variedad del repertorio de actos de comunicación in cluye aproximadamente de 10 a 20 distintos tipos de comportamiento para aquellas especies estudiadas con mayor detalle. Este valor es comparable al de los mamíferos euterios de tamaño y nicho ecológico similares. El comportamiento de muchas especies es pobremente conocido, pero el tipo de comunicación en los dasiúridos en general indica que, fuera de los períodos de celo y cuidado maternal, su existencia es relativamente asocial.

31. OBSERVATIONS ON THE BEHAVIOUR OF SPECIES OF *Planigale* (DASYURIDAE, MARSUPIALIA) WITH PARTICULAR REFERENCE TO THE NARROW-NOSED PLANIGALE *(Planigale tenuirostris)*

D. L. ANDREW[1] and G. A. SETTLE

Observations on nine *Planigale tenuirostris* (8♂, 1♀) and one *Planigale gilesi* (♂) maintained under reverse day/night and normal lighting conditions are described. Some casual observations on *Planigale maculata* (typical form) and *Planigale ingrami* are also included. Animals were housed in groups (up to 3) or individually in enclosures designed to simulate their natural environment. Video recording and direct methods of observation were employed. Neither *P. tenuirostris* or *P. gilesi* can be described as strictly nocturnal since both display variable degrees of diurnal activity throughout the year. Sunbasking is characteristic of diurnal activity in *P. tenuirostris, P. gilesi* and *P. ingrami*. During cooler months planigales will sometimes become torpid. When provided with suitable substrate, *P. tenuirostris* and *P. gilesi* excavate short tunnels leading to sleeping chambers but do not line these with nest material. Alternative sleeping sites were within or under grass tussocks and hollow logs. Audition and olfaction appear well developed in species of *Planigale* while vision is developed more for contrast than visual acuity. Planigales display many of the postures characteristic of the Dasyuridae and their sitting pose appears to be typical of species of this genus. *Planigale tenuirostris* and *P. gilesi* can be distinguished by their modes of locomotion. Animals defecate and micturate more frequently on solid objects and in specific locations, than they do randomly about their enclosure. Cloacal marking was also observed. Individuals tend to follow specific pathways about enclosures but cage areas utilized primarily by specific animals are not defended. Agonistic interactions develop mostly over food or nest sites but rarely result in serious fighting. A number of distinct vocalizations are produced by *P. tenuirostris* and young emit a hissing call when separated from the mother. Planigales will attack prey equal in body size to themselves and exhibit considerable flexibility in their mode of attack, dependent primarily on prey type, but always administer a killing bite to the head.

Pages 311–24 *in* CARNIVOROUS MARSUPIALS, ed. by M. Archer, February, 1982. Roy. Zool. Soc. New South Wales: Sydney, Australia.

IN HIS REVISION of the genus *Planigale*, Archer (1976) recognises five species: *P. ingrami, P. tenuirostris, P. gilesi, P. maculata* and *P. novaeguineae*. In this paper we shall only deal with the first four species, in particular, *P. tenuirostris*. All *Planigale* species are small body size, *P. ingrami* and *P. tenuirostris* being minute (6 – 10 g.). Until recent years there has been little detailed information available on these species, the smallest of the Dasyuridae. Van Dyck (1979) comprehensively examined behaviour in captive *Planigale maculata* and Fleay (1965, 1967) made general observation on this species (not *P. ingrami*, Archer 1976). Observations on some aspects of reproduction have been recorded for *P. maculata* by Morrison (1975) and for *P. maculata sinualis* by Aslin (1975) and Taylor *et al.* (1982). McKay (1974) commented on the behaviour of captive *P. ingrami* and feeding habits of a Northern Territory planigale of uncertain identity (Archer 1976) were reported by Van Deusen (1969). Some information is available on habitat preferences of *P. maculata* (Marlow 1958, Fleay 1965, Van Dyck 1979), *P. maculata sinualis* (Davies 1960, Parker 1973, Taylor *et al.* 1982), *P. gilesi* (Aitken 1972, Denny 1975) and *P. ingrami* (Troughton 1967, Parker 1973).

ORIGIN OF ANIMALS STUDIED

Individuals of *Planigale ingrami* (2 ♂♂, 2 ♀♀), *P. maculata sinualis* (3 ♂♂, 2 ♀♀) held by Taronga Zoo, and *P. gilesi* (1 ♂♂) and *P. tenuirostris* (2 ♂♂) (Fig. 1) held by the University of New South Wales were observed during these studies. Female *P. gilesi* and male *P. maculata* (typical form) were not observed. A female of the typical form of *P. maculata* (Fig. 2) was captured (by G.S.) near Woodenbong, New South Wales, in August, 1976 (Fig. 3).

Fig. 1. Planigale tenuirostris (♂) grooming forepaws. (Photo. G. A. Settle).

[1] School of Zoology, University of New South Wales, P.O. Box 1, Kensington, New South Wales, Australia, 2033. Graham Settle died in 1979.

Fig. 2. Typical form of *Planigale maculata* (♀) (Photo. G. A. Settle).

Vegetation covering these lowlands is primarily chenopod shrubs (e.g. *Kochia, Bassia, Atriplex*) and perennial tussock grasses (e.g. *Astrebla*). The composition of these communities is influenced by both landform and soil type. The alluvial plains are covered by a mosaic of areas with and without perennial vegetation, reflecting run-off and run-on control by the microtopography. Bladder salt bush (*Atriplex vesicaria*) occurs above subsoils impeding downward water movement and tussock grasslands predominate on the heavier textured soils (Burrell 1973).

Fig. 3. The capture site of *Planigale maculata* (Fig. 2) on the Mt. Lindsay Highway, 4 km northeast of Woodenbong, New South Wales in August, 1976. The specimen was found under a sheet of iron by the roadside. (Photo. G. A. Settle).

Fig. 4. The Fowlers Gap Creek flood plain is extensively gilgaied (foreground) with a vegetation cover of chenopod shrubs and perennial tussock grasses. The tree line of gums indicates the creek's position. (Photo. G. A. Settle).

A further nine *P. tenuirostris* (Fig. 1) were captured using drift fences with cylindrical pits (25 x 45 cm) placed every 4 m, on the Fowlers Gap Creek flood plain (Fig. 4), approximately 110 km N. of Broken Hill, New South Wales during January and February, 1977. A single juvenile male *P. gilesi* (Fig. 5) was also taken but not recognised as a distinct species until a few weeks later. Only two *P. tenuirostris* captured were adults, a lactating female (released) and an adult male. The others (6♂, 1♀) were uniformly sized, half-grown juveniles.

DISTRIBUTION OF SPECIES OF PLANIGALE IN AUSTRALIA

No polytypic dasyurid genera have all species occupying just one type of habitat. Figures 6a and 6b show the locations from which modern specimens of the four Australian *Planigale* species have been captured (Parker 1973, Archer 1976, Morton and Martin 1979, Bird and Sinclair 1980, Smith and Medlin 1980), together with certain features of land form, soil type and rainfall (modified after Moore 1975). Three species, *P. gilesi, P. ingrami* and *P. tenuirostris,* exhibit very close alliance with the grey, brown and

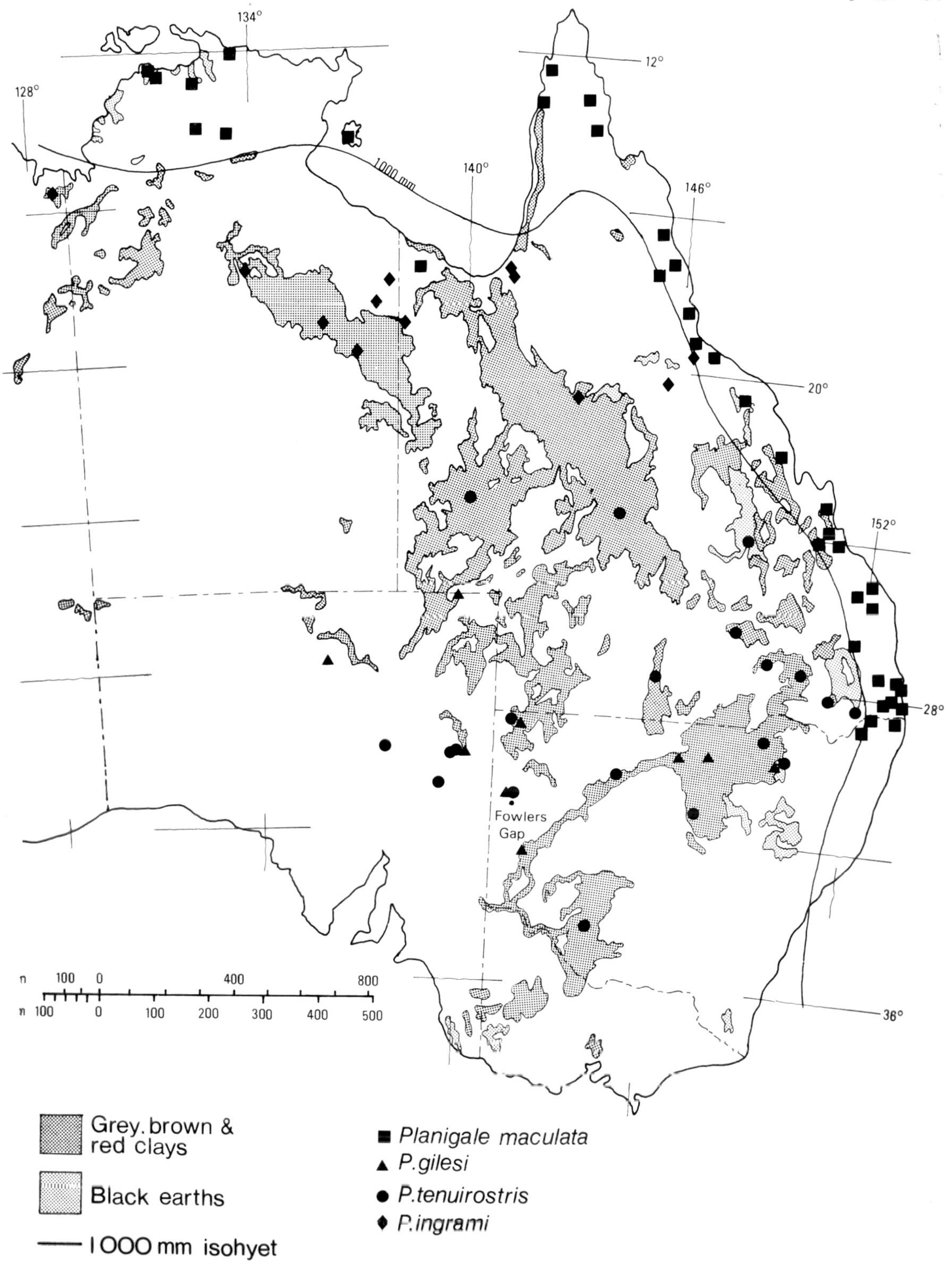

Fig. 6 (a) The distribution of modern specimens of the four Australian species of *Planigale* (after Parker 1973, Archer 1976, Denny 1975, Morton and Martin 1979, Bird and Sinclair 1980, Smith and Medlin 1982) in relation to the occurrence of the grey, brown and red clays and black earths, collectively referred to as the cracking clay or gilgai soils (after Moore 1975). The 1000 mm rainfall isohyet is also shown.

Fig. 5. Planigale gilesi (♂) occurs sympatrically with *P. tenuirostris* at Fowlers Gap, New South Wales. (Photo. G. A. Settle).

red clays, and the black earths, all of which are characterized by their deep cracking and gilgai formations (Fig. 6a).

Small localized pockets of these soils eg. at Fowlers Gap, are not indicated on a map of this scale (1 : 12,000,000), but all points have been plotted on more detailed soil maps (1 : 2,000,000) (C.S.I.R.O., Atlas of Australian soils) which further substantiates the correlation between soil type and the distribution of these three species of *Planigale*. The two soil groups occur primarily on broad riverine plains and gently rolling uplands (relief to 30 metres) and it is apparent that many of the points which do not correspond to either soil group lie on these landforms (Fig. 6b). *Planigale gilesi* shows particular restriction to riverine flood plains and overflows.

Two distinct forms of *Planigale maculata* are recognised by Archer (1976): the typical form (Fig. 2) occurring in eastern Queensland and northeastern New South Wales, and the *sinualis* form in the Northern Territory. The distribution of both forms is broadly delimited by 1000 mm isohyet, and neither soil type, nor land form, appears restrictive to the distribution of typical *P. maculata* (Figs. 6a and 6b). The *sinualis* form appears to be somewhat restricted by landform and exhibits some preference for cracking clay soils, but the number of capture localities is low and further data are required. Archer (1976) suggests a third form may be represented by a few specimens from Aurukun Mission in the Gulf of Carpentaria, a site receiving only slightly less than 1000 mm of rain annually.

BEHAVIOUR OF PLANIGALE IN CAPTIVITY

Most of the observations reported here were conducted under controlled day–night cycle (twilight transitions were included in the lighting regime) at Taronga Zoo, Sydney. Specimens of *P. tenuirostris* were housed in pit–type cages (base area 53 x 152 cm) furnished with red sand, hollow branches, dried grasses and small shrubs collected from their natural habitat at Fowlers Gap. Small aquaria (60 x 30 x 40 cm) similarily 'decorated' were used for some detailed close–range observations e.g. grooming and predatory sequences. Both video and direct methods of observations were used.

Locomotor Activity: In the field, all animals were captured at night when the minimum overnight temperature was above 19.8°C. Large numbers of terrestrial invertebrates such as spiders, scorpions, centipedes, cockroaches, beetles and crickets, were also captured in the pit traps on these warm nights.

It is likely that *P. tenuirostris* and *P. gilesi* spend a significant proportion of time below the surface of the ground. The capture site is characterized by well–developed micro–relief gilgai formations (Fig. 4) in the deep, yellowish–red, neutral, loaming sands which comprise the alluvial soils along the back plains of creeks (Corbett 1973).

Figure 7 shows the cumulative number of captive *P. tenuirostris* (1 ♀♀, 5 ♂♂, 3 per cage) active at the begining of each 15 minute period of the 24 hour day/night cycle on six separate occasions (i.e. no consecutive days). Activity increased markedly after 'sunset', especially following simulated twilight when more animals were active than during any other period of the day or night. The number of animals active generally declined towards midnight increasing again during the early hours of morning. Some individuals remained active or emerged again soon after sunrise, continuing their activity until about 10:00 hours when a timing switch activated bright lights to simulate direct solar radiation. Low levels of activity were recorded for the remaining daylight. The single female was generally less active than the five males, often emerging only when the two males sharing her enclosure were not active. Food was provided

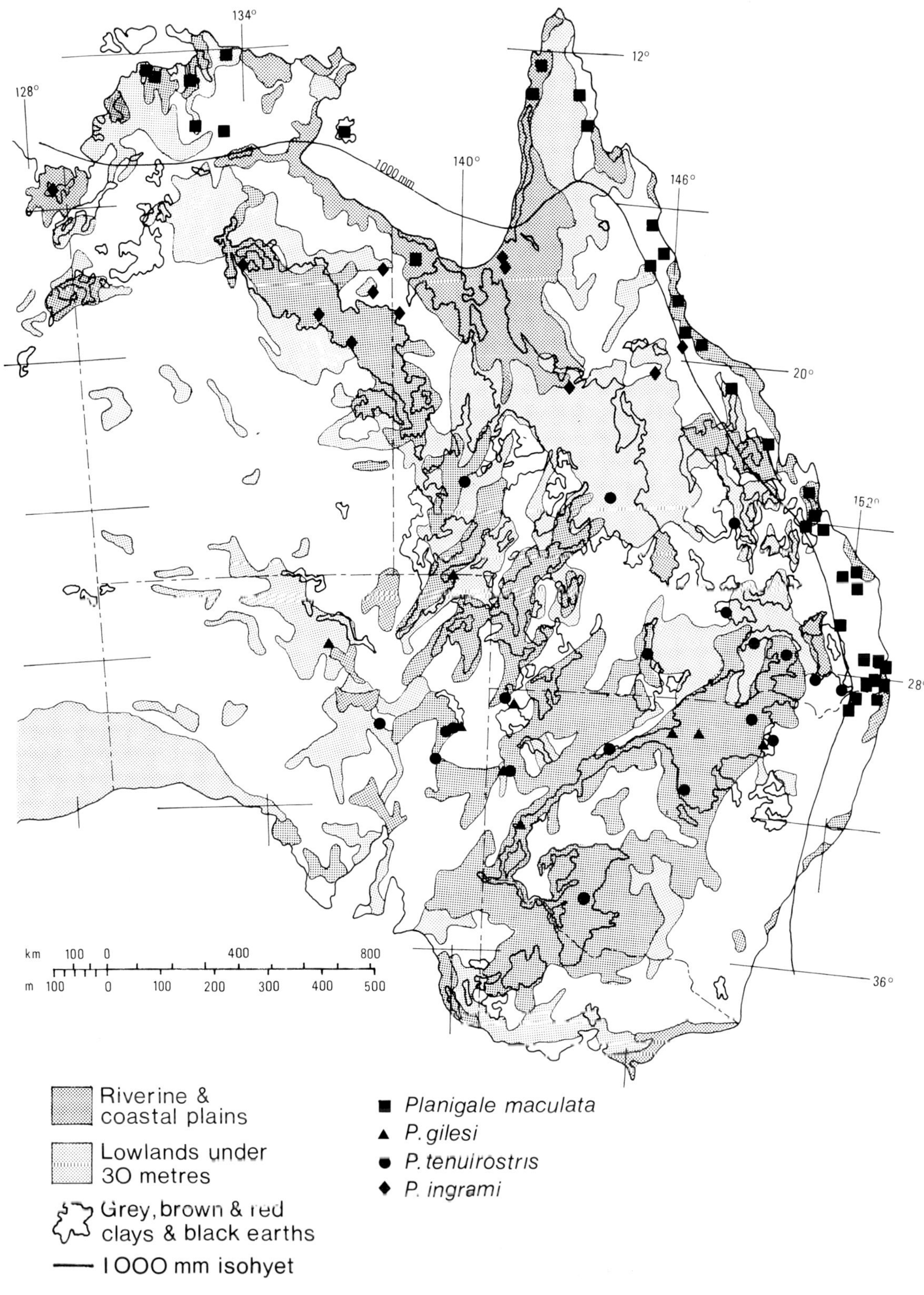

Fig. 6 (b) The distribution of modern specimens of the four Australian species of *Planigale* in relation to low lying landforms. The cracking clay soils and 1000 mm isohyet are also shown.

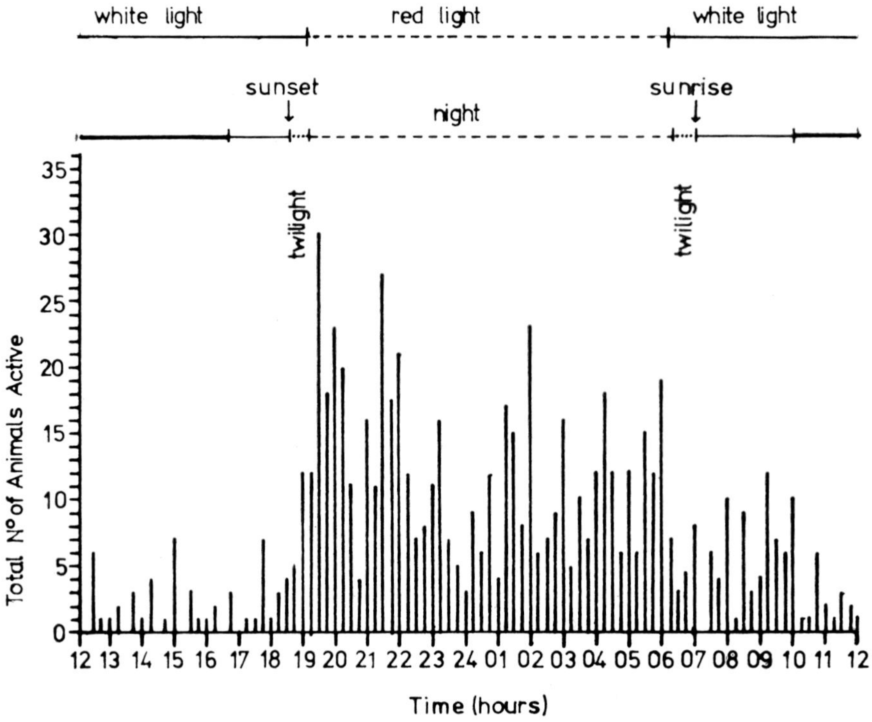

Fig. 7. Diurnal activity pattern of *Planigale tenuirostris*. The graph shows the number of individuals (N=5♂, 1♀; 3 per cage) active at the beginning of each 15 minute period of the 24 hour day/night cycle cumulated from data obtained on six different occasions. Animals were housed under reverse day/night conditions and the above graph is a guide to the lighting used.

ad libitum and possibly affected the total amount of activity by decreasing the necessary time required to find food.

Planigale tenuirostris (1♀♀, 2♂♂), housed under natural lighting conditions and exposed to direct sunlight, displayed variable amounts of diurnal activity throughout the year and between individuals. Some became active during daylight to defecate, micturate, feed or sunbask. Others were intermittently active for much of the day, especially during the morning, climbing and running about the enclosures. The proportion of diurnal activity increased progressively throughout the cooler winter months (overnight minima often 10°C or less) when animals frequently became torpid overnight and often remained so until 11:00 hours, eventually emerging to feed or sunbask. In summer months, the planigales again became predominantly nocturnal in their activity rythms.

The single male *P. gilesi* displayed higher levels of diurnal activity, whilst the sole female *P. maculata* (typical form) was rarely active during daylight (this specimen has been in captivity for over two years).

Gaits, Postures, Grooming and Comfort Movements: Generally, all species of *Planigale* adopt similar positions during their daily activities. No differences are apparent in posture or locomotory patterns between *P. tenuirostris* and *P. ingrami*.

When sitting, both species splay hind feet forwards and outwards at approximately 45° and extend the tail out immediately behind forming a tripod–like configuration to support the weight of the body. In this position, the spine is markedly hunched in the upper thoracic region (Fig 8a). *Planigale gilesi* places its feet more directly under the body and the spinal hump is less conspicuous.

In the sleeping position the forelimbs are extended and placed inside the hind limbs and the head is lowered and tucked under, lying on the lower abdomen (Fig. 8b). *Planigale gilesi* curls the body tightly so the head lies directly on the hind feet (Fig. 8c) while the tail is sometimes employed to wrap tightly around the sparsely furred ears and feet.

Following bouts of activity, planigales frequently adopt a crouched inactive position. The animal crouches low to the substrate and draws its spine into a convex arc, its head rests on the substrate with ears and vibrissae relaxed and eyes half closed. It will, however, still respond to external stimuli, particularly sound, opening its eyes after any disturbance.

During semi–rapid locomotion, *P. tenuirostris* employs a sinuous scurrying run, where the body is held close to the substrate and the fore and hind legs move in a splayed arc–like manner to produce a smooth flowing motion. At higher speeds, a quadrapedal bounding gait is used. *Planigale gilesi* also uses this scurrying run but additionally will move in a series of short leaps and bounds, with its tail moving up and down in co-ordination with each individual movement. Unlike *P. tenuirostris*, *P. gilesi* jumps freely from one object to another while running, easily spanning distances of up to 30 cm without interrupting its gait.

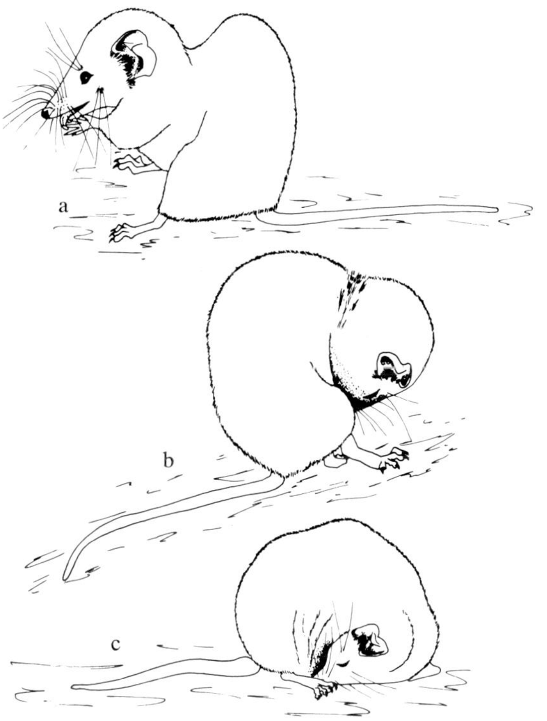

Fig 8. (a) Sitting position of *P. tenuirostris* (see also Fig. 1); note the extreme thoracic hunch of the spine. (b) Sleeping position of *P. tenuirostris*. (c) Sleeping position of *P. gilesi*.

When alerted, Planigales will adopt one of three investigatory postures. The most commonly observed is the raised paw posture with one fore foot raised, the tail held out horizontally and not contacting the substrate, and the ears vibrissae held erect (Fig. 9a). The bipedal stance is similar to the normal sitting position but less crouched. If the source of interest requires further investigation, an extended upright stance may be adopted. In this posture only the toes of the hind feet and the tail contact the substrate and the head is raised and moved through a small arc (Fig. 9b).

Stretching occurs most frequently after arousal, sunbasking or grooming. One fore foot is extended at a time, pulling the body out, and progressively extending the back and hindlimbs. Yawning generally follows stretching and occasionally grooming and is sometimes accompanied by a sharp "chh" vocalization. Tail wriggling was observed when planigales were sitting or sunbasking.

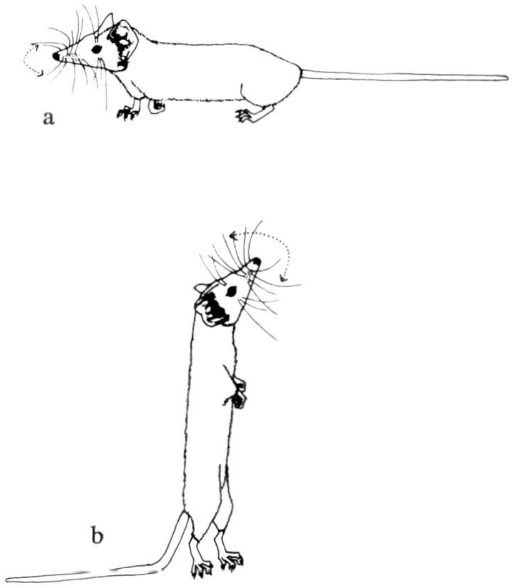

Fig. 9. Investigatory positions of *P. tenuirostris* (a) Raised paw posture; (b) Extended upright posture.

Planigales employed three grooming methods. (1) Rubbing with licked forepaws is used for the snout, face and ears. Initially the forepaws move synchronously beginning with the snout but begin to alternate as they reach the ears. (2) Licking, nibbling and combing with forepaws is used for grooming most of the body and is combined with (3) scratching and licking of the hindfoot.

SUNBASKING: Sunbasking is characteristic of diurnal activity of *P. ingrami*, *P. gilesi* and *P. tenuirostris*, as in many other dasyurids, and is especially frequent during cooler months. Animals usually orientate facing the light (heat) source with forelimbs outstretched and their trunks flattened against the substrate. Juvenile *P. tenuirostris*, ca.39 days (eyes just open), adopted a basking posture almost immediately when exposed to sunlight. *Planigale tenuirostris* will also curl up in grass clumps with a sunny aspect.

SANDBATHING: Sandbathing occurs in some dasyurids such as *Sminthopsis crassicaudata* (Ewer 1969) and *Dasyuroides byrnei* (Aslin 1974) and in particular those from dry regions. It is rare behaviour among species from areas of high rainfall (e.g. it does not occur in species of *Dasyurus*, *Antechinus stuartii*, *A. flavipes*, *A. swainsonii* or *Phascogale tapoatafa*). It does however, occasionally occur in *Sminthopsis murina*. Sandbathing was not observed in any species of *Planigale*.

BURROWING AND NEST BUILDING: Substrates similar to those in natural habitats and suitable for observations of digging behaviour were produced by packing moistened sand–clay mixtures into the bottoms of aquaria and allowing them to dry.

While neither *P. tenuirostris* nor *P. gilesi* are prolific burrowers, both species excavated short tunnels leading into sleeping chambers. These tunnels were gradually extended and finally resulted in small networks. Digging in a 'new' environment invariably commenced on the under-surfaces of solid objects (e.g. pieces of mulga root embedded in the soil), with a very narrow, vertically flattened elliptical entrance. This led into a passage which usually followed horizontally or downwards inclining under–surfaces of objects. Most burrows turned downwards when the edge of these objects was reached or when the lower surface of the object turned upwards. Artificially produced 'cracks' in the soil were also used as starting points for tunnels, but generally animals used hollow logs, bark and grass clumps as sleeping sites. The single female *P. maculata* was never observed to construct burrows.

None of the burrows or hollow logs utilized by *P. gilesi* and *P. tenuirostris* (including a female rearing five young) as sleeping sites were lined with nest material. Small spherical depressions were formed in the bases of grass tussocks (*Astrebla* sp.) but these appeared to result from the animal curling up within these areas, rather than from deliberate construction.

In contrast, the female *P. maculata,* during an oestrus cycle, did construct a cup shaped nest from grass and leaves under a piece of bark. Subsequently one of the authors (D.A.) observed *P. maculata* (N=2) construct nests from *Eucalyptus* sp. leaves in nest boxes and hollow logs (diameter 4 cm). These observations agree with those of Fleay (1965) and Van Dyck (1979).

SENSORY PERCEPTION: As in most nocturnal mammals, the vision of *P. tenuirostris* appears well–developed for contrast but not for visual acuity. Animals frequently passed within 1–2 cm of stationary prey (e.g. grasshoppers) without apparently detecting their presence, whereas moving prey rarely escaped 'attention'.

Hearing is acute in all planigales and high frequency sounds appear most easily detected. Unidentified sounds led to cessation of movement, pricking of the ears and orientation of the head towards the source. Olfaction is also well–developed, and seems important in the detection and identification of prey, other individuals and inert objects. Animals invariably responded to changes in environmental stimuli by sniffing, frequently standing bipedally with the nose pointing upwards (Fig. 9b), or extending the head forwards and sniffing in a small arc (Fig. 9a).

COMMUNICATION AND UTILIZATION OF CAGE AREA. (a) Vocalizations: *Planigale tenuirostris* employs several distinctive calls, often emitted at considerable volume and intensity, seemingly inconsistent with their minute size. They are most frequently associated with agonistic interactions between individuals and at least five audible calls were distinguished. (1) During a chase the agressor commonly produces stanzas of "chh, chh, chh", made with the mouth closed. This call is also employed by individuals carrying captured prey to a retreat and appears to have an offensive context. (2) Some animals also utter a constant chattering sound while consuming their prey. (3) Animals within refuges produce a sharp staccato "ca ca ca" towards intruders, who usually then retreat. This call is also produced by individuals interlocked in wrestling fights and appears to have a defensive context. (4) *P. tenuirostris* can be induced to produce a high pitched twittering call, similar to that of a small bird, by either removing all cover from the enclosure or placing it into the cage of another individual. The animal often holds its ears and vibrassae back and performs some type of displacement behaviour, e.g. grooming. (5) Young planigales, approximately 39 days old and barely 2 cm in length, produce continuous stanzas of "chh chh chh" when they become separated from the mother.

(b) Olfactory Signals—Marking: Although individuals (1♀♀, 5♂♂, 3 per cage) utilized some cage areas more commonly than others, except for sleeping sites, these were not actively defended. Planigales deposit faeces and urine on solid objects they frequently cross or in the corners of enclosures. Cloacal dragging is employed, particularly following eliminative acts and may continue for several centimetres. During the breeding season the sternal gland develops in male *P. tenuirostris.* Its nature and the extent of its use remains undertermined, although it seems likely that it plays a similar role for this species as it does for *P. maculata* (Van Dyck 1979) and other marsupials. Mouthing does not appear to be important in *P. tenuirostris.*

In the four species of *Planigale* studied, all individuals tended to follow specific pathways within their enclosures. Where more than one individual was present in a cage, some areas remained more or less exclusive to the movements of particular individuals and pathway sharing occurred in other areas. Sleeping sites were frequently shared by conspecifics (up to three), particularly during colder months, and interspecific sharing of sleeping sites was observed for *P. tenuirostris* and *P. gilesi* (both males).

Fig. 10. (a, above) The agressive animal *(P. tenuirostris)* approaches with its head raised and the defending animal adopts a sideways defence posture. (b, below) An open mouthed threat frequently causes the defending animal to flee. (Photo. G. A. Settle).

SOCIAL BEHAVIOUR: Agonistic encounters between established cage mates are infrequent and cage areas utilized primarily by specific individuals are not defended. Agonistic behaviour precipitates primarily from disputes over food or refuges and is expressed as threats, chases and wrestling fights, but these are rarely of serious consequence, injuries being only superficial in nature, e.g. the removal of tufts of fur.

Several sterotyped postures can be identified in agonistic encounters. The aggressor frequently approaches with its head raised and the defending animal adopts a sideways defence posture with head lowered (Fig. 10a). An open–mouthed threat by an aggressor generally produces a fleeing response in the defending animal (Fig. 10b) but, if it fails to flee, contact fighting can result. An aggressor can be repelled by the defending animal rearing to a threat posture with one or both forepaws raised (Figs. 11a and 11b). *Planigale tenuirostris* approaches opponents from the front or side, whereas *P. maculata* uses a slow 'deliberate' walk to approach opponents from behind, with eyes almost closed during the final stages of the approach.

Bites are delivered to the back, flanks and necks of opponents and are the most common form of contact. Wrestling bouts are brief and most frequently follow prolonged holding bites by aggressors, preventing their opponent's escape. Interlocked face to face wrestling occasionally occurs but is of short duration.

Acts which might be termed 'amicable' are infrequent in planigales. Naso–nasal contacts are more frequent than naso–cloacal contacts. The introduction of a strange conspecific increases the number of these acts and all following encounters involve some amount of olfactory assessment, even between previously familiarized animals. As mentioned above communal nesting is common during cooler months when animals usually huddle together.

PARENTAL BEHAVIOUR: A female *P. tenuirostris* gave birth to and reared five male young to ca 39 days in the presence of two adult males without apparent male interference with the young. The female retrieved the young by picking them up in her mouth one by one and carrying them back to the nest. She also transported them to new nesting sites in this manner. During this period, the mother displayed increased levels of aggression towards her adult male companions, if they came near to the nest, and defended the young against human interference, rearing back on her hind feet and lunging open–mouthed towards intruding fingers or noses. The size of these offending 'bodies' was not inhibitory, for she gripped them between her jaws whenever possible. Unfortunately, we decided to place her in another cage to allow us to observe activity within the nest and she rejected the young a few days later.

PREDATORY BEHAVIOUR: All species of *Planigale* exhibited a remarkable degree of flexibility in the structure of predatory sequences. The comments outlined here refer to *P. tenuirostris* (primarily) and *P. gilesi*. Variations in the method of attack appear largely dependent upon the type of prey. Only two features were common to all sequences: (1) identification of prey was by olfaction, or perhaps taste where moving prey was rapidly captured and identified; and (2) the killing bite was oriented towards the prey's head. Furthermore, species of *Planigale* will attempt to capture and kill some potential prey species as large or larger than themselves in size (see Fleay 1965).

Small, slow moving prey (e.g. mealworms, caterpillars) were generally detected by olfaction. These were sometimes patted several times with the forepaws before being picked up in the mouth, held in the forepaws, and a killing bite effected to the head. In contrast, Van Dyck (1979) reports that *P. maculata* does not employ a killing bite when dealing with mealworms. Stationary, but potentially highly mobile, small prey (e.g. small moths, cockroaches, crickets) were first approached slowly, then suddenly grasped with the forepaws with a simultaneous killing bite being delivered to the head. Fluttering or struggling of prey sometimes evoked head–shaking by the predator.

Moving prey (e.g. cockroaches, flying or crawling moths, small lizards) were visually detected. If the prey moved slowly planigales sometimes watched them for short periods before approaching. The method of approach was variable; rapidly moving prey were usually rushed (aerial capture was observed for flying insects), as were some slower moving prey. In the latter case, planigales sometimes approached prey obliquely, moving adjacent to cover until they were only a short distance away and then rushing forward to effect capture.

A form of ambush was also observed. Planigales sometimes remained motionless if prey moved towards them, rushing forward only when the prey was close or changed direction.

Perhaps the most interesting method of prey capture by species of *Planigale* is that used for large invertebrates, particularly large grasshoppers. Planigales rarely rush these prey. Instead, they make several slow approaches and withdrawals before capture is attempted. Initial lunges are usually oriented towards the head but if unsuccessful are redirected to the base of a hind leg which is gripped between the jaws

Fig. 11. (a, above) An agressive approach *(P. tenuirostris)* can be repelled (b, below) by defending animal rearing to a threat posture with one or both forepaws raised. (Photo. G. A. Settle).

and held. The limb is usually dismembered by the kicking of the grasshoppers themselves as they attempt to escape. An identical sequence follows for the remaining hind limb, and occasionally for other limbs as well. This mode of attack effectively renders the insect incapable of escape. *Planigale maculata* has also been observed to use this technique when dealing with large grasshoppers, but apparently attacks are generally ineffective (Van Dyck 1979). Mauling sometimes develops during struggles with large prey. However, the sequence is completed with one or more directed killing bites to the head.

Dead or immobilized prey were frequently carried to cover before being consumed. Where possible, a bounding–type gait (with a horizontal component reciprocally dependent upon the body weight of the prey) was used to reach cover. The animal invariably commenced feeding at the anterior end. Limbs and wings of prey were discarded as they were encountered. Planigales sometimes abandoned large grasshoppers they had killed, apparently being unable to break through the exoskeleton.

One of us (G.S.) has also observed *P. tenuirostris* and a female *P. maculata* 'attacking' the base of some types of flowers. Petals were chewed and the skin of a grape was eaten by one *P. tenuirostris*. There is, therefore, a possibility that these dasyurids may utilize nectar or other plant products as a food source during appropriate seasons.

DISCUSSION

We suggest that the observed interspecific differences in locomotary patterns between species of *Planigale* are associated with habitat structure and the manner in which the habitat is utilized. Sinuous body movement occurs in three species (*P. ingrami, P. gilesi* and *P. tenuirostris*) found in association with soils having deep cracking characteristics. Descriptions by other authors of habitat structure at localities where these species have been captured refer to the banks and flood plains of perennial watercourses, dry swamps (or adjacent areas when full) and the drainage areas leading to them. Typical plant association at these sites are tussocky grass, chenopod shrubs, and sedge–bullrush complexes around bores or other permanent water bodies. The covering vegetation layer is thin, particularly during dry seasons when plants die off and blow away (almost completely in some regions such as around Fowlers Gap) leaving a broken vegetative network interspaced with open ground. Under these conditions a sinuous form of lomotion will be advantageous; the oblique movement of the limbs permits extreme flattening of the body during passage through narrow cracks in the soil, and a bounding gait is used on the surface to cross open expanses of ground between clumps of vegetation. The more frequent adoption of a bounding gait and readiness to leap between objects may indicate that *P. gilesi* utilizes the above ground habitat more than subterranean cracks and gilgais which are more likely inhabited by *P. tenuirostris*. This probably provides some ecological separation between these species where they occur sympatrically (e.g. Fowlers Gap, New South Wales and Moolawatana Station, South Australia (Bird and Sinclair 1980)).

The variety of habitats occupied by *P. maculata* is considerable in comparison with the habitat specificity exhibited by the species discussed above. The typical form is predominantly found in sclerophyll forest (Marlow 1958), more open timber country, flooded marshes and rocky areas. The *sinualis* form occupies flood plains during the dry season and *Pandanus* and *Melaleuca* fringe areas bordering flood plains during the wet season (Davies 1960, Taylor *et al.* 1982), and hummock grass beneath *Acacia* on sandstone boulders in sand (D. Kitchener, pers. comm. cited by Archer 1976). Thus habitat selection by *P. maculata maculata* appears primarily dependent upon the availability of surface cover, presumably for protection against predators, and not upon the cracking characteristics of the substrate. This does not, however, appear to be true for the *sinualis* form. Their restriction to regions of high rainfall correlates with substantial vegetative growth. Rocky sites are alternatively occupied where vegetation is less dense. Sinuous locomotory movements and a bounding gait at higher speeds would not appear to be advantageous to small species inhabiting such environments, and these forms of locomotion are noticeably absent in *P. maculata maculata*. Additionally, Van Dyck (1979) noted that, following a noise or disturbance, *P. maculata* 'froze' remaining completely motionless for up to two minutes. This behaviour was not noted in *P. tenuirostris* or *P. gilesi*. In contrast, when startled, they immediately dashed for cover.

Archer's (1976) suggested taxonomic affinities between *P. maculata* and *P. novaeguineae* are supported by the structural similarity of the habitats of both species (see Tate and Archbold 1941, Menzies 1972 and Ziegler 1972 for descriptions of *P. novaeguineae* habitat; also pers. comm. with M. McCoy) and, consequently, we predict that sinuous body movements and a bounding gait do not characterize the locomotory patterns of the New Guinean species.

Burrowing is important to *P. tenuirostris* and *P. gilesi* with both species occupying habitats where ground cover is sparce. (*P. ingrami* was not observed in this regard). It does not appear important to *P. maculata* which occupies habitats with comparatively dense ground cover.

Planigale tenuirostris and *P. gilesi* clearly are not purely nocturnal since both species exhibit considerable diurnal activity in cooler months. These observations are in agreement with those of other authors. For *P. gilesi,* Aitkin (1972) notes that they "... exhibit short bursts of activity spread throughout each 24–hour period ...". Martin Denny (pers. comm.), in trapping studies (Elliot box traps) throughout winter months at Mount King, New South Wales, found that the majority of his *P. gilesi* were caught during daylight hours.

The activity patterns of captive *P. tenuirostris* suggest that in its natural habitat this species is also likely to become increasingly diurnal during winter months. Future long–term pitfall trapping programmes (Elliot traps rarely succeed in capturing *P. tenuirostris*) may confirm this opinion. Both species probably become active opportunistically, following the activity patterns of their insect prey.

In contrast, *P. maculata* rarely displayed diurnal activity, an observation supported by Van Dyck (1979) who notes that "The Planigales rarely extended their activities beyond sunrise. Most animals returned to nests at least one or two hours before sunrise".

Unlike the arid habitat of *P. tenuirostris* and *P. gilesi,* where diurnal temperature fluctuations are great and overnight minimum temperatures frequently drop below 0°C in winter, *P. maculata* occupies areas experiencing more temperate coastal climes, with few environmental extremes. Consequently the insect prey of *P. maculata* probably remains active at night throughout winter.

Data presented here which relate to social behaviour in species of *Planigale* are regarded as heuristically useful only in so far as they indicate a need for further research before any clear picture of social organization emerges.

In *P. tenuirostris* and *P. gilesi,* a considerable degree of mutual tolerance is exhibited, with correspondingly low levels of agonistic encounters between familiar individuals. Communal nesting is a frequent occurrence during cooler months (see also Archer 1976). Animals tend to follow specific pathways within their enclosures and marking behaviour, using deposits of urine and faeces on particular objects, is common. In enclosures containing more than one individual, some areas remained more or less exclusive to the movements of particular individuals but these areas were not defended and pathway sharing occurred. Asociality is not indicated by the absence of amicable, expressive or complex vocal acts similar to those which have evolved in some eutherians. The absence of this behaviour appears to be a feature common to nearly all the dasyurids.

Nest construction was only undertaken by a female *P. maculata* during an oestrus cycle and did not occur in a female *P. tenuirostris* with young. Other planigales used unlined chambers in sand or logs as refuges, and 'nests' in grass tussocks which were formed by repeated occupation rather than by active construction. Other authors (Heinsohn 1970, Archer 1976) have reported the construction of nests, but details are scant and there is a need for further study.

In agonistic encounters, *P. maculata* (both forms) approached the rear of opponents (as noted by Morrison 1975), whereas in other species frontal or lateral approaches were used. Male *P. tenuirostris* did not interfere with the well–being of a female or her young. Maternal aggression towards interspecific intruders is extreme in *P. tenuirostris* (this study) and *P. maculata* (Fleay 1965), but two conspecific males housed in the same enclosure as the female *P. tenuirostris* were not attacked.

The predatory sequence in *P. tenuirostris* and *P. gilesi* is both complex and environmentally labile. Stable components of the sequence appear limited to three behaviours: (1) grasping of the prey with the forepaws; (2) shaking of the head in apparent response to prey struggling; and (3) a well–directed killing bite oriented at the head of the prey. Other components show a remarkable degree of flexibility considering the very small size of these marsupials. Both the manner of approach and attack vary with the size and type of prey, the rate, direction and form of locomotion used by prey, and the physical structure of the environment. Indeed both the complexity and flexibility of the predatory sequence in *P. tenuirostris* and *P. gilesi* is greater than those described in many other dasyurids of larger size, such as *Sminthopsis crassicaudata* (Ewer 1968), *Dasycercus cristicauda* and *Sarcophilus harrisii* (Ewer 1969), *Dasyuroides byrnei* (Aslin 1974, Hutson 1975), and *Antechinomys laniger, Sminthopsis murina* (Settle, unpubl. obs.).

In conclusion, the behaviour of species of *Planigale* is complex and most certainly does not fit the stereotyped image of marsupials as 'lesser advanced mammals' incapable of behavioural complexity.

ACKNOWLEDGEMENTS

We wish to thank Dr. P. Crowcroft for the provision of space at Taronga Zoo and Mr. D. Thomas, Mr. L. Moyes, Mr. L. Naylor, Mr. W. Zirkler and Mr. G. Bingham for the maintenance of the study animals and for permitting us to observe planigales held by the Zoo.

We also wish to thank Mr. D. Read (University of New South Wales) who checked trap lines at Fowlers Gap in our absence, and Mr. D. Whitford who maintained planigales at the University of New South Wales.

We gratefully acknowledge the valuable comments of Dr. Michael Archer, Ms. Joan Dixon, Dr. Eleanor Russell and Dr. David Croft on the first draft of this paper.

REFERENCES

AITKEN, P. F., 1972. *Planigale gilesi* (Marsupialia, Dasyuridae); a new species from the interior of southeastern Australia. *Rec. Sth. Aust. Mus.* **16**(10): 1–14.

ANDREW, D. L., 1977. The behaviour of *Planigale tenuirostris* (Dasyuridae:Marsupialia). Hons. thesis: University of New South Wales, Sydney.

ARCHER, M., 1976. Revision of the Marsupial genus *Planigale* Troughton (Dasyuridae). *Mem. Qd. Mus.* **17**(3): 341–65.

ASLIN, H. J., 1974. The behaviour of *Dasyuroides byrnei* (Marsupialia) in captivity. *Z. Tierpsychol.* **35:** 187–208.

ASLIN, H. J., 1975. Reproduction in *Antechinus maculatus* Gould (Dasyuridae). *Aust. Wildl. Res.* **2:** 77–80.

BIRD, P. L. AND SINCLAIR, R. G., 1980. Preliminary results of small mammal surveys in northern South Australia (Abstract). *Aust. Mammal. Soc. Bull.* **6:** 26.

BURRELL, J. P., 1973. Vegetation of Fowlers Gap Station. *Fowlers Gap Arid Zone Res. Ser. No.* **4**.

C.S.I.R.O., 1961. Atlas of Australian soils, sheets 1–10, 1961, Melbourne University Press: Melbourne.

CORBETT, J. R., 1973. Soils on Fowlers Gap Station. *Fowlers Gap Arid Zone Res. Ser. No.* **4**.

DAVIES, S. J. J. F., 1960. A note on two small mammals of the Darwin area. *J. Roy Soc. W. Aust.* **43:** 63–74.

DENNY, M. J. S., 1975. Mammals of Sturt National Park, Tibooburra, New South Wales. *Aust. Zool.* **18:** 179–95.

EWER, R. F., 1968. A preliminary survey of the behaviour in captivity of the dasyurid marsupial *Sminthopsis crassicaudata* (Gould). *Z. Tierpsychol.* **25:** 319–65.

EWER, R. F., 1969. Some observations on the killing and eating of prey by two dasyurid marsupials: the Mulgara, *Dasycercus cristicauda* and the Tasmanian Devil, *Sarcophilus harrisii*. *Z. Tierpsychol.* **26:** 23–38.

FLEAY, D., 1965. Australia's "needle–in–a–haystack" marsupial. *Vict. Nat.* **82:** 195–204.

HEINSOHN, G. F., 1970. World's smallest marsupial. The flat headed marsupial mouse. *Animals* **13:** 220–2.

HUTSON, G., 1975. Sequences of prey catching behaviour in the Brush–tailed Marsupial Rat *(Dasyuroides byrnei)*. *Z. Tierpsychol.* **39:** 39–60.

HUNSAKER, D., (ed.) 1977. "The biology of marsupials". Academic Press: New York.

MARLOW, B. J., 1958. A survey of the marsupials of New South Wales. *C.S.I.R.O. Wildl. Res.* **3:** 71–114.

MCKAY, G. M., 1974. *Planigale ingrami* in captivity. *Koolewong* **3:** 4–5.

MENZIES, J. I., 1972. The relative abundance of *Planigale novaeguineae* and other small mammals in the south Papuan savannas. *Mammalia* **36:** 400–5.

MORRISON, R. G. B., 1975. Emergence of the pygmy Antechinus. *Aust. Nat. Hist.* **18**(5): 164–7.

MORTON, S. R. AND MARTIN, A. A., 1979. Feeding ecology of the Barn Owl, *Tyto alba,* in arid southern Australia. *Aust. Wildl. Res.* **6:** 191–204.

MOORE, R. M., 1975. "Australian grasslands". Aust. Nat. Univ. Press: Canberra.

PARKER, S. A., 1973. An annotated checklist of the native land mammals of the Northern Territory. *Rec. S. Aust. Mus.* **16:** 1–57.

SMITH, M. J. AND MEDLIN, G. C., 1982. Dasyurids in the northern Flinders Ranges, South Australia, before pastoral development. *In* "Carnivorous marsupials", ed. by M. Archer. Roy. Zool. Soc. New South Wales: Sydney.

TATE, G. M. H. AND ARCHBOLD, R., 1941. New rodents and marsupials from New Guinea. *Amer. Mus. Novit.,* **1101:** 1–9.

TAYLOR, J. M., CALABY, J. H. AND REDHEAD, T. D., 1982. Breeding in wild populations of *Planigale maculata sinualis* (Dasyuridae, Marsupialia). Pp.83–87 *in* "Carnivorous marsupials", ed. by M. Archer. Roy. Zool. Soc. New South Wales: Sydney.

TROUGHTON, E., 1967. 'Furred animals of Australia'. 9th ed. Angus and Robertson: Sydney.

VAN DEUSEN, H. M., 1969. Feeding habits of *Planigale*. *J. Mammal.* **50:** 616–8.

VAN DYCK, S., 1979. Behaviour in captive individuals of the dasyurid marsupial *Planigale maculata* (Gould 1851). *Mem. Qd Mus.* **19:** 413–29.

ZIEGLER, A. C., 1972. Additional specimens of *Planigale novaeguineae* (Dasyuridae: Marsupialia) from Territory of Papua. *Aust. Mammal.* **1:** 43–5.

RESUMEN

31. Observación del comportamiento de *Planigale* (Dasyuridae, Marsupialia), con particular referencia al planigale de hocico estrecho (*Planigale tenuirostris*). Por D. L. Andrew y G. A. Settle.

En su revisión del género *Planigale,* Archer (1976) reconoce cinco especies: *P. ingrami, P. tenuirostris, P. gilesi, P. maculata* y *P. novaeguineae.* En este trabajo nosotros trataremos solamente las primeras cuatro especies, en particular *P. tenuirostris.* Todas las especies de *Planigale* son de pequeño tamaño corporal, P. *ingrami* y *P. tenuirostris* diminutas (6–10 gramos). Hasta hace pocos años hubo poca información detallada accesible sobre estas especies, las más pequeñas de todos los Dasyuridae. Van Dyck (1979) examinó comprensivamente el comportamiento en cautividad de *Planigale maculata* y Fleay (1965, 1967) hizo observaciones generales sobre esta especie (nó *P. ingrami,* Archer 1976). Observaciones sobre ciertos aspectos de la reproducción habían sido registrados por Morrison (1975) para *P. maculata* y para *P. maculata sinualis* por Aslin (1975) y por Taylor *et al.* (1980). McKay (1974) comentó sobre el comportamiento en cautividad de *P. ingrami.* Van Deusen (1969) informó sobre los hábitos alimentarios de un planigale del Northern Territory de incierta identidad (Archer 1976). Es asequible alguna información sobre las preferencias de hábitat de *P. maculata* (Marlow 1958, Fleay 1965, Van Dyck 1979), *P. maculata sinualis* (Davies 1960, Parker 1973, Taylor *et al.* 1980), P. *gilesi* (Aitken 1972, Denny 1975) y *P. ingrami* (Troughton 1967, Parker 1973).

En este trabajo se describe el comportamiento en cautividad de *P. tenuirostris,* y algunas observaciones adicionales de otras especies de *Planigale.* Los resultados proveen una vasta categorización del comportamiento de *Planigale* como punto de partida de investigaciones futuras.

32. A PRELIMINARY REPORT ON ASPECTS OF THE BEHAVIOUR OF THE DASYURID MARSUPIAL *Phascogale tapoatafa*

PETER CUTTLE[1]

Phascogale tapoatafa is a nocturnal, arboreal, carnivorous marsupial. Activity is confined to the dark phase of the daily light cycle, and occurs in bouts interspersed with periods of inactivity. There appears to be some degree of social facilitation in the timing and duration of activity bouts. *Phascogale tapoatafa* shows a range of prey–catching movements adapted to the stimulus properties of different prey items. Invertebrate prey are killed with a series of bites which destroys the segmented nervous system. Vertebrates are killed with a crushing bite to the skull. The paws are used extensively to restrain and manipulate prey prior to killing. *Phascogale tapoatafa* exhibit anatomical, behavioural and developmental adaptations to their arboreal niche. These behaviour patterns are described and discussed both in terms of adaptation to the total environment and in relation to knowledge of other dasyurids.

Pages 325–32 *in* CARNIVOROUS MARSUPIALS, ed. by M. Archer, February, 1982. Roy. Zool. Soc. New South Wales: Sydney, Australia.

THE BRUSH-TAILED PHASCOGALE, *Phascogale tapoatafa*, is a small, nocturnal, arboreal, carnivorous marsupial. Each of these attributes is important in the interaction of the genome with the environment, which manifests itself as the behaviour of the species. Little work has been done on the behaviour of *P. tapoatafa*, either in the wild or in captivity, other than by Fleay (1923, 1934) whose anecdotes are mainly concerned with reproduction. However, the behaviour of several other species of dasyurids has been studied more closely. Of particular interest are studies of *Sminthopsis crassicaudata* (Ewer 1968), *Dasycercus cristicauda* (Ewer 1969a, Sorenson 1970), *Dasyuroides byrnei* (Sorenson 1970, Aslin 1974) and *Antechinus stuartii* (Rigby 1972). Further studies concerning prey–catching behaviour of dasyurids (Ewer 1969b, Eisenberg and Leyhausen 1972, Hutson 1973) and activity cycles in dasyurids (Crowcroft and Godfrey 1968, Wood 1970) are also of interest. These studies reveal the overall homogeneity of the Dasyuridae, insofar as behaviour is concerned, and are of direct relevance in interpreting the observed behaviour of *P. tapoatafa*.

This study was conducted to obtain a conspectus of various aspects of the behaviour in captivity of *P. tapoatafa*, particularly those behaviour patterns related to the nocturnal and arboreal habit and the carnivorous diet of this species.

MATERIALS AND METHODS

Animals were collected at Christmas Hills, Victoria (37°35'S 145°20'E) in habitat described elsewhere in this volume (Cuttle 1982). Twenty–six animals, either wild–caught or captive–reared offspring of females captured pregnant, were observed in captivity over a period of three years. Animals were housed in indoor brick enclosures, approximately 3.6 m x 3.0 m x 3.0 m high with an observation window in one corner, and furnished with tree trunks and branches in which wooden nest boxes were provided. Animals were maintained on a reversed lighting cycle controlled by an automatic time switch. The lighting cycle was altered to simulate the natural light cycle throughout the year. "Day" lighting was provided by two 40 watt fluorescent lights and one 15 watt incandescent light. "Night" lighting was provided by the 15 watt incandescent light alone. No attempt was made to simulate dawn and dusk.

Observations on general behaviour, prey–catching behaviour and activity rhythms were made on groups of animals housed in the indoor enclosures. Pencil–and–paper records were supplemented with audiotape recordings made during observation. Further observations were made with a remote–controlled video–recording system fitted with a low–light camera mounted in the observation window. A monitor was located in a different room to minimise disturbance to the animals. Behaviour was recorded under "night" lighting conditions and later replayed for closer analysis. For closer analysis of prey–catching behaviour, animals were isolated in encounter boxes approximately 100 cm x 50 cm x 50 cm high with a sloping glass front. Prey was introduced either through the top of the box or through one end. Sequences were filmed at 64 frames per second with a Bolex H16 Reflex camera, with the minimum lighting necessary to expose 200 ASA film at f1.9. Film was later analysed frame–by–frame with an L.W.

[1] Department of Zoology, Monash University, Clayton, Victoria, Australia, 3168. Present Address: Natureland Zoological Gardens, P.O. Box 109, Coolangatta, Queensland, Australia, 4225.

Optics Co. Photo–optical Data Analyzer. Successive frames were traced from a back–projection screen for comparison of postures and movements. In addition, video–recording was used to examine aspects of prey–catching behaviour in encounter boxes.

Information on daily activity rhythms was recorded automatically with a microwave movement detector (MMD) mounted on the ceiling of the enclosures and coupled to a multi–channel event–recorder. The MMD emits a microwave pulse and responds to a change in the reflected signal caused by movement of the animals in the enclosure. The setting of the MMD could be altered so that movements of a chosen magnitude triggered a response while smaller movements were ignored. Other data on behaviour of the animals were entered in parallel with the activity data by a manual keyboard connected to the same event–recorder.

RESULTS

Daily Activity Cycle: *Phascogale tapoatafa* has previously been shown to be nocturnal (Fleay 1923, 1934). However the manner in which the species partitions its activities during the night has not previously been studied. The activity of two animals housed together, recorded over nine consecutive daily lighting cycles, shows that almost all of the activity was recorded during the dark phase, with only sporadic periods of very low score during the light phase (Fig. 1a). On all nine experimental days, activity commenced at about the beginning of the dark phase and ceased shortly after the end of the dark phase. Dark phase activity comprised "bouts" of high score, alternating with periods of very low or zero score. The pattern of bouts and gaps was similar from day to day, but the timing and duration of bouts of activity varied (Fig. 1b).

Direct observation on a group of six animals sharing an indoor enclosure revealed that there were short periods when all animals were active together, while there were periods around the middle of the dark phase when all were inactive inside the nest boxes (Fig. 2). Peaks in the number of animals active follow the pattern of peaks

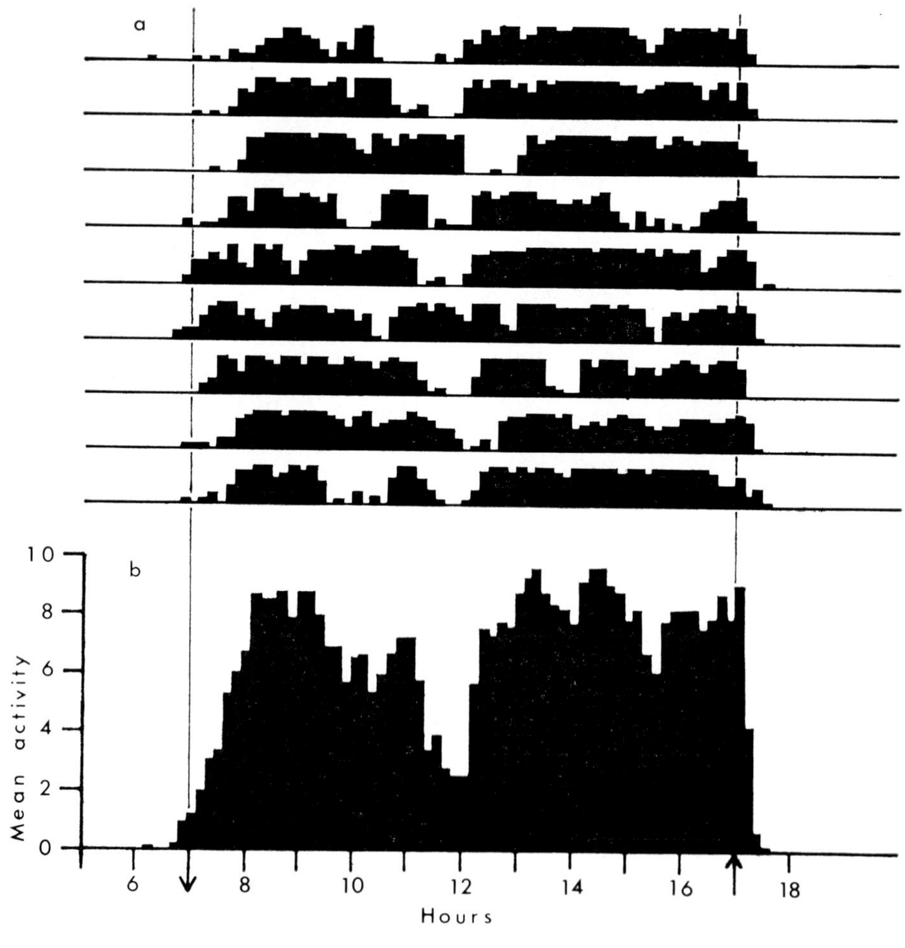

Fig. 1. Activity scores for two *P. tapoatafa* housed together over nine consecutive dark phases. (a) MMD record. Unit of time is ten minutes. For each unit the score represents the number of minutes in which movement was recorded (range 0–10). (b) Mean activity score. Arrow pointing down = lights off. Arrow pointing up = lights on.

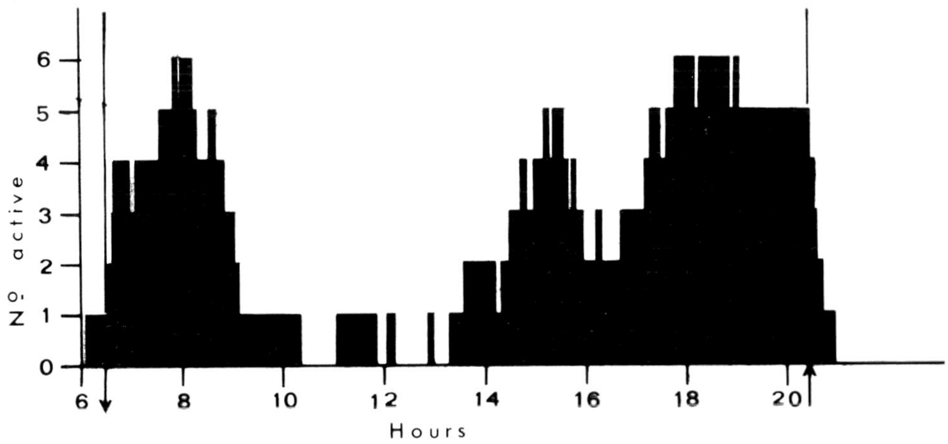

Fig. 2. Number of *P. tapoatafa* active in a group of six during a single dark phase.

of activity recorded with the MMD (compare Figs. 1b and 2). There appeared to be some degree of social facilitation in the timing and duration of activity bouts. Animals sharing a single nest box emerged in rapid succession at the beginning of a bout of activity. Individual animals briefly entered nest boxes alone many times during an activity bout, and the composition of a group of active animals changed frequently. However, the end of an activity bout was marked by the return in quick succession of most animals to the nest box. The presence of other animals outside the nest box may delay the return of an individual to the nest box. Individual animals occasionally remained active alone outside the nest box for varying periods. There may be consistent individual differences between animals, in both the timing and the duration of bouts of activity.

Manual recording of the separate times of activity of two animals housed together, in parallel with MMD recording, revealed a difference in activity pattern between the two animals (Fig. 3). However, the periods of highest MMD activity score corresponded with the periods of joint activity. Both animals showed brief periods of eating during each activity bout. Drinking followed most occurrences of eating when prepared dry food was available. Both animals visited the nest box during each period of activity.

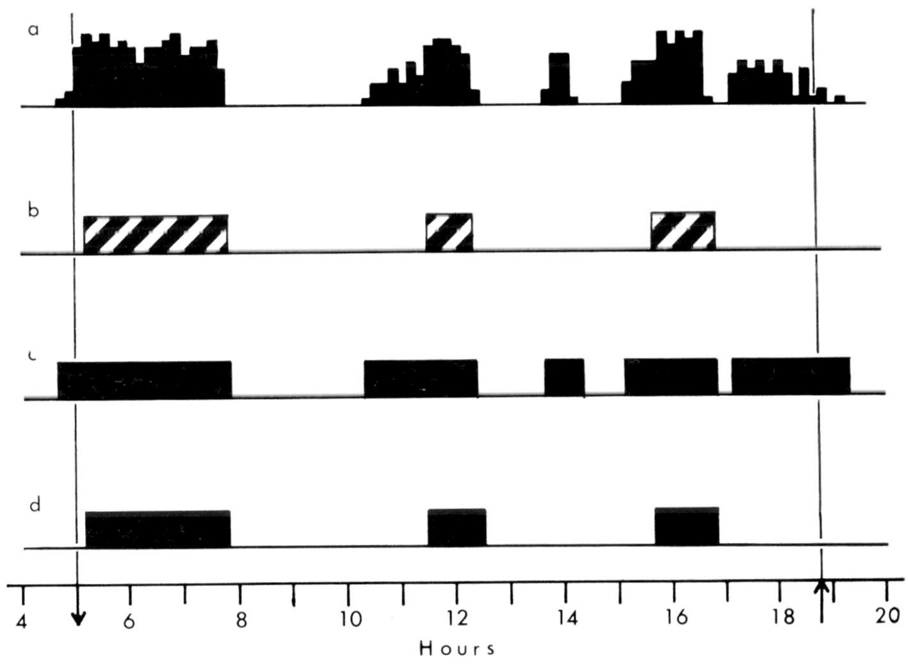

Fig. 3. Activity records for two *P. tapoatafa* housed together over a single dark phase. (a) MMD record. (b) Times when both animals were active. (c), (d) Times of activity of male and female respectively.

The food bowl in an enclosure containing six animals was placed on a hinged tray, the free end of which was supported by a spring balance. A clock was placed beside the scale of the balance, and a time–lapse Super 8 mm film (2 frames per minute) was made of the clock and scale. This sampled both the times of presence of an animal at the food dish — indicated by gross movement of the scale pointer — and the steady decrease in the mass of food remaining, which was proportional to the decrease in the steady–state scale reading (Fig. 4). Periods of activity at the food tray did not always result in an appreciable decrease in the scale reading. However, periods when the mass of food remaining decreased most rapidly corresponded with periods when greatest activity was recorded with the time-lapse camera. These periods fell within periods of activity recorded with the MMD, but did not correspond closely with periods of most intense overall activity. When food is clumped, as in a bowl, then foraging movements are accordingly reduced. Animals were also observed to pause to consume all classes of prey.

often allowing the first to escape. However prey that had been captured were killed before others were chased. Alternatively, small prey that could be eaten quickly were completely devoured before others were pursued. Several prey were thus captured and killed in quick succession. This behaviour may be seen to be advantageous in capturing prey which are found clumped, rather than distributed uniformly or randomly throughout the environment. After eating each prey item, the animal often returned to the site of capture and searched for further prey. Animals could often be induced to kill two mice before commencing to eat either, even though none was ever observed to eat all of one mouse at a single feeding. Remains of a kill were often left where they lay, but were consumed later.

Responses to Various Prey Items: Olfactory stimuli appeared to be important in the location and indentification of small, slow–moving or inactive prey such as the larvae and pupae of the mealworm beetle *Tenebrio molitor*. The prey were simply picked up in one paw and passed to the mouth to be eaten. Although the

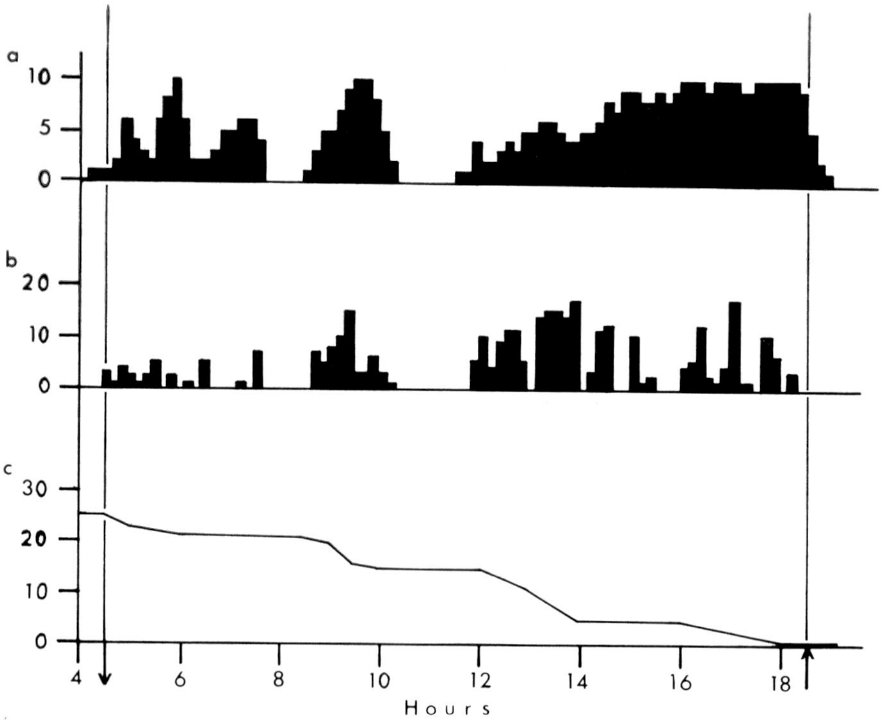

Fig. 4. Feeding activity records for six *P. tapoatafa* housed together during a single dark phase. (a) MMD record. (b) Time–lapse film record. Unit of time is ten minutes. Score for each unit represents the number of frames (2 f.p.m.) on which activity at the food bowl is recorded. (c) Time–lapse film record of scale reading (arbitrary scale). Decrease in scale reading is proportional to amount of food eaten.

Prey–catching Behaviour: Captive *P. tapoatafa* showed a variable response to superabundant prey. An animal was easily distracted from chasing one prey item to chase another,

mouth was close to the paw at the time of grasping, prey were not taken directly with the mouth. Prey were usually held in one paw to be chewed (Fig. 5a).

Fig. 5. Prey–catching movements of *P. tapoatafa* traced from 16mm film. (a) Grasping mealworm prior to eating. (b) Grasping and biting at scorpion. (c) Rapid series of bites along the length of a centipede. (d) Grasping and pinning movement with a mouse.

Phascogale tapoatafa showed mainly visual orientation to living cockroaches *Periplaneta americana*. Movement of the prey elicited an immedicate pinning movement of the forepaws. If the cockroach escaped, it was pursued with repeated attempts to restrain it with the forepaws. When restrained, the cockroach was passed to the mouth with both paws, and a rapid series of bites was delivered to its body until it ceased to struggle. The animal appeared to push the spiny legs of the cockroach away from its mouth, resulting in many bites being delivered to the anterior of the insect. Deep–frozen cockroaches were readily accepted as prey, but as they did not move or struggle they did not elicit the pinning movement of the forepaws or the rapid series of bites. The animal did only what was necessary to procure the prey.

Scorpions did not flee when attacked, but turned to face the attacker, attempting periodically to cling with the pedipalps and strike with the sting. Thus there was no flight to elicit a pinning movement of the paws. However, olfactory stimuli were apparently enough to release a grasping movement which was accompanied by a bite to any convenient part of the prey within reach of the mouth (Fig. 5b). A rapid series of bites was then delivered to the body (= prosoma plus mesosoma), the paws grasping the pedipalps or the tail. Biting was repeated until the whole body was crushed and all struggling had ceased. Scorpions sometimes managed to cling to the vibrissae or fur, and sting the Phascogale on the snout or inside the mouth. This resulted in a sharp retraction of the head and renewed efforts to restrain the scorpion. Occasionally a defensive movement by a scorpion led to temporary withdrawal of the Phascogale from the scene, but prey were never abandoned. The prey–catching sequence commenced again almost immediately. Thus the defensive behaviour of scorpions elicited prey–catching movements which did not appear in sequences with cockroaches or mealworms.

Centipedes differed from scorpions as prey in two important ways. Firstly, they were highly mobile and showed a flight response rather than defence, thus eliciting following and grasping movements. Secondly, their long thin shape elicited an exaggerated form of the rapid series of bites, as the body was drawn back and forth through the mouth, causing bites to fall along the entire length of the body of the centipede (Fig. 5c). One centipede, removed after having been killed, was found to have puncture wounds in several segments, while other segments had been damaged internally. Thus the series of bites appears to be a response to the segmented nervous system of invertebrate prey.

Vertebrate prey elicited generally the same set of prey–catching responses as invertebrate prey, up to and including the initial grasping movement. Some subsequent responses were specific to vertebrate prey. Chicks *Gallus domesticus* were treated as if they were novel or intimidating prey. Typically, the Phascogale closed on the chick with a series of approach and withdrawal movements, interspersed with circling movements and pauses. A sudden movement of a chick precipitated an immediate and rapid retreat for a short distance. With repeated approaches, and with greater experience of chicks, the animal merely retracted the head in response to incipient attack by the chick. Chicks were nosed and grasped before the first bite was applied opportunistically to any convenient part of the chick. A chick was often carried to a more secluded spot before the bite was adjusted towards the skull. A killing bite was then applied to the skull, and this bite was adjusted and re-applied as necessary until struggling ceased.

Phascogale tapoatafa were routinely fed laboratory mice *Mus musculus* as part of their diet, and prey–catching techniques for this prey were similar to those for chicks. However, as the mouse was more active in its defence and escape behaviour, it offered a greater variety of prey stimuli. Consequently the prey–catching response of the Phascogale was more variable for the mouse than for the chick. Orientation to mice was primarily visual, in response to the movement of the mouse. Animals did not orientate readily to a stationary mouse, nor were they easily able to locate a stationary mouse at any distance.

The approach to a mouse was more rapid and direct than to a chick, possibly because of greater familiarity with mice as prey. However, if a mouse suddenly adopted a defensive posture, or lashed with its tail, the Phascogale rapidly withdrew a short distance or retracted its head. If the mouse fled it was immediately chased. A mouse that darted aside and suddenly "froze" was temporarily lost to view, indicating the importance of movement in prey–catching behaviour. Pinning and grasping movements of the paws followed approach (Fig. 5d). A killing bite was immediately delivered to the skull if the prey did not struggle, but wrestling and repeated opportunistically directed biting were used to subdue a struggling mouse. Wrestling continued until either the mouse managed to break away or a killing bite was delivered to the mouse's skull. Re–adjustment of this bite occurred if the mouse continued to struggle. Neither chicks nor mice were shaken at any time.

ARBOREALITY: The hindlimb of *P. tapoatafa* shows a great deal of rotatory mobility, probably rivalling that of the Sciuridae of North America. This is strongly adaptive for the arboreal habit, as it allows the animal to cling at any orientation on a tree surface at any angle. In captivity, Phascogales spent long periods clinging head–downwards on the underside of sloping branches. The reason for this was not clear. The long thin digits of the manus are adapted both for clinging with the recurved claws and for extracting small prey items from crevices in or under the bark of trees. Prey is captured or extracted with the forepaws while the hind limbs maintain a grip on the tree surface. The tail is sometimes used as a third support. The large conspicuous brush–like tail appears not to be used as a counter–balance during locomotion either in the trees or on the ground. Rather, its movement is governed by pelvic rotation during spinal flexion and extension, and is opposite to that expected of a counter–balance. One female, tailless since infancy, was at no disadvantage either in the trees or on the ground.

Phascogale tapoatafa exhibits a behaviour pattern which appears to be unique amongst the Dasyuridae. This is a rapping of the forefeet on the bark or trunk of a tree, often advancing a short distance with each rap. This behaviour is shown in response to a threatening stimulus and appears to have some communicative value, as the rapping of one animal is often taken up by another.

Phascogale tapoatafa is the only dasyurid known to nest regularly in tree hollows at any height above ground. This has implications for the development of the young. Locomotor ability apears early in development, and the young are able to climb unaided before the eyes open, well before they emerge from the nest unaided. The slow lateral sequence walk used by Phascogales in the trees in an extremely stable gait, with always at least three feet in contact with the surface. The time of first separation from the nipple is a time of potential litter mortality, as young that become separated from the mother away from the nest may fall to the ground and not be recovered (Cuttle 1982). In *P. tapoatafa,* the young remain firmly attached to the nipples until they are suddenly all left in the nest at once. The young are not carried by the mother but remain in the nest until ready to emerge unaided.

DISCUSSION

Whilst the activity rhythms of putative prey items of *P. tapoatafa* are completely unknown, it is likely that some of the wide range of prey species will be available at any time. In becoming active periodically throughout the night, and feeding during each activity bout, *P. tapoatafa* can sample the available prey repeatedly during the night. In addition, by becoming inactive inside the insulating nest mass when the stomach becomes full, rather than remaining outside, the animals suffer less heat loss and are thus able to conserve energy. Presumably this strategy would be of greatest advantage during winter when environmental demands on energy are greatest.

By synchronizing their activity periods, the occupants of a single shared nest are able to take greatest advantage of this strategy. Periodic visits to the nest during activity periods would reveal the state of occupancy of the nest. A nest with several animals already in it may be very attractive, since several animals crowded together are better able to conserve metabolic energy. An empty, or near–empty nest may be less attractive for the relatively lower energy saving that may be made. It may additionally indicate that, since most animals remain active, food is presently sufficiently abundant to warrant the expense of metabolic energy. Phascogales are attracted to the sound of another's chewing, which leads, in captivity, to frequent food theft. In the wild, with synchronization of activity, this may lead to the efficient cropping of locally abundant food resources, as more food would be caught per unit searching effort.

The actual sequence of prey–catching behaviour shown by *P. tapoatafa* in any prey encounter is a function of the stimulus properties of the prey. Some features of the sequence are common to all prey situations, while others are seen only in response to specific behaviours of certain prey.

A movement of the paws contacting the prey occurs before the first bite in any prey encounter. Primitively, this serves to pick up prey and hold it for chewing. Mobility of prey elicits a pinning movement, while unpleasant stimuli such as spines, clinging or stinging elicit a pushing away action of the paws. With small prey that struggle, the grip is released and reapplied in synchrony with a repeated bite, until all struggling ceases. Long–bodied prey may be passed back and forth through the mouth by the paws. Thus the "paws–on" movement for invertebrate prey becomes more complex as the prey become harder to catch, restrain, handle or immobilise. The implications of arboreality extend into the area of prey–catching. *P. tapoatafa* makes greater use of its paws in prey–catching than does any other dasyurid so far studied. This compensates somewhat for the reduced use of movements such as shaking the prey by this species. Shaking of prey is a regular feature of the prey–catching behaviour of more terrestrial dasyurids (Ewer 1969b, Eisenberg and Leyhausen 1972, Hutson 1973, van Dyck and Ogilvie 1977), but could conceivably lead to the loss of prey from an elevated position in a tree.

With active vertebrate prey, the paw movements become highly modified to cope with the extra difficulty of capture. Movement of a mouse elicits a pinning movement; attempted escape is countered by grasping; with continued struggle, a wrestling movement involving all four paws is used. Thus the use of the paws becomes more complex as the potential for escape or retaliatory behaviour of the prey becomes more complex. The complexity of the response does not reflect the phyletic relationships of the prey, or even the relative size of the prey; the paw movements in response to the cockroach are more complex than the movements in response to the chick!

The complexity of the biting response also varies with the prey stimuli. No special bite is used with small innocuous prey. Active invertebrate prey are immobilised with a rapid series of bites to all parts of the body, which effectively destroys the segmented nervous system. Struggling vertebrate prey are weakened by repeated opportunistic biting before a killing bite is delivered to the thin part of the skull. The central nervous system is thus extensively damaged, killing the prey outright or rendering it immobile.

Other small dasyurids show a similar set of prey–catching responses to various forms of prey. *S. crassicaudata* shows simple responses to innocuous prey and a bite–and–release response to noxious arthropods (Ewer 1968). *D. byrnei* and *D. cristicauda* also show more active responses to active and noxious prey than to innocuous prey (Sorenson 1970). *D. cristicauda* regularly shakes prey, presumably as a means of disorientating or immobilising the prey (Ewer 1969b).

All dasyurids so far studied exhibit a killing bite delivered to the skull of vertebrate prey. However, species differ in the method of immobilising vertebrate prey. *D. byrnei* make regular use of a bite–shake–release movement to immobilise a mouse before placement of a killing bite (Hutson 1973). Bite-shake is used only by young or inexperienced *D. cristicauda* (Ewer 1969b). *Sarcophilus harrisii* does not use its paws at all, but bites directly on the head of a rat (Ewer, *op. cit.*).

The killing bite of dasyurids is functionally similar to the "cervical bite" of the Carnivora (Eisenberg and Leyhausen 1972), in that both are anteriorly placed and both serve to kill the prey. However, the canine teeth of dasyurids are not adapted for separating the cervical vertebrae and lacerating the hind–brain, as are those of the Carnivora. Instead, the skull is crushed, causing massive damage to the parietal region of the brain.

ACKNOWLEDGEMENTS

I thank J. E. Nelson for his supervision of the study, and A. Lill, D. Croft and E. Russell for their valued comments on a draft of this paper. I am deeply grateful to Mr. and Mrs. W. R Maughan, of Natureland Zoological Gardens, Coolangatta, Queensland for generous financial support throughout the study. Jo-ann Hyndman kindly typed the manuscript.

REFERENCES

Aslin, H., 1974 The behaviour of *Dasyuroides byrnei* (Marsupialia) in captivity. *Z. Tierpsychol.* **35:** 187–208.

Crowcroft, P. and Godfrey, G. K., 1968. The daily cycle of activity in two species of *Sminthopsis* (Marsupialia: Dasyuridae). *J. Anim. Ecol.* **37:** 63–73.

Cuttle, P., 1982. Life–history strategy of the dasyurid marsupial *Phascogale tapoatafa. In* "Carnivorous marsupials", ed. by M. Archer. Roy. Zool. Soc. New South Wales: Sydney.

Eisenberg, J. F. and Leyhausen, P., 1972. The phylogenesis of predatory behaviour in mammals. *Z. Tierpsychol.* **30:** 59–93.

Ewer, R. F., 1968. A preliminary survey of the behaviour in captivity of the dasyurid marsupial, *Sminthopsis crassicaudata* (Gould). *Z. Tierpsychol.* **25:** 319–65.

Ewer, R. F., 1969a. More about the Mulgara. *Wildl. Aust.* **6:** 78–9.

Ewer, R. F., 1969b. Some observations on the killing and eating of prey by two dasyurid marsupials: the Mulgara, *Dasycercus cristicauda,* and the Tasmanian Devil, *Sarcophilus harrisii. Z. Tierpsychol.* **26:** 23–38.

Fleay, D., 1923. The Brush–tailed Phascogale. *Victorian Nat.* **46:** 135–6.

Fleay, D., 1934. The Brush–tailed Phascogale: first record of breeding habits. *Victorian Nat.* **51:** 88–100.

Hutson, G. D., 1973. "The causation of grooming, prey–catching and agonistic behaviour in *Dasyuroides byrnei* (Marsupialia: Dasyuridae)". Ph.D. thesis: Monash University, Melbourne.

Rigby, R. G., 1972. A study of the behaviour of caged *Antechinus stuartii. Z. Tierpsychol.* **31:** 15–25.

Sorenson, M. W., 1970 Observations on the behaviour of *Dasycercus cristicauda* and *Dasyuroides byrnei* in captivity. *J. Mammal.* **51:** 123–30.

Van Dyck, S. and Ogilvie, P., 1977. *Antechinus swainsonii* (Waterhouse 1840), the Dusky Marsupial Mouse, an addition to the mammal fauna of Queensland. *Mem. Qd Mus.* **18:** 69–73.

Wood, D. H., 1970. An ecological study of *Antechinus stuartii* (Marsupialia) in a south–east Queensland rainforest. *Aust. J. Zool.* **18:** 185–207.

RESUMEN

32. Informe preliminar de aspectos del comportamiento del marsupial dasiúrido *Phascogale tapoatafa.* Por Peter Cuttle.

Phascogale tapoatafa es un marsupial carnívoro, nocturno y arborícola. Su actividad está confinada a la fase oscura del ciclo solar diario, y ocurre en turnos alternados con períodos de inactividad. Parece haber cierto grado de facilidad social según el tiempo y la duración de los períodos de actividad. *Phascogale tapoatafa* muestra una variedad de movimientos específicos en respuesta a las propiedades estimulantes de las diferentes presas. Los invertebrados son matados por una serie de mordiscos que les destruyen el segmentado sistema nervioso. Los vertebrados son matados por un mordisco que les aplasta el cráneo. Las garras son usadas intensamente para retener y manipular la presa antes de matarla. *Phascogale tapoatafa* exhibe adaptaciones anatómicas, de comportamiento y adaptaciones ontogenéticas a su nicho arbóreo. Esos tipos de comportamiento son descriptos y discutidos, tanto en términos de adaptación al ambiente en su totalidad como en relación al conocimiento del de otros dasiúridos.

33. TRAPPABILITY OF *Antechinus stuartii* (DASYURIDAE, MARSUPIALIA) AT PETROI, NORTHEASTERN NEW SOUTH WALES

H. L. STATHAM[1]

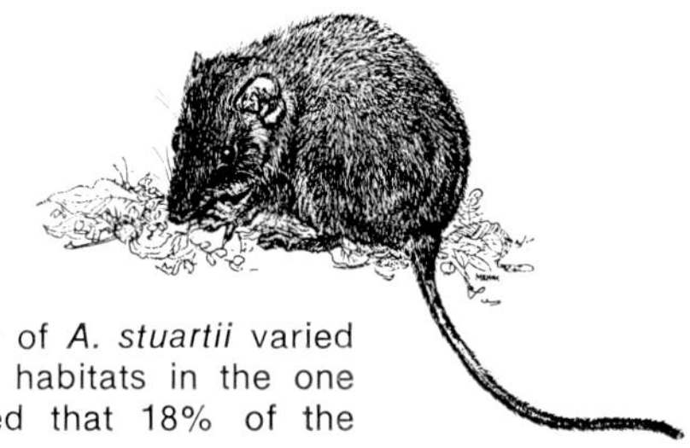

Antechinus stuartii were live trapped monthly on large grids (9 ha) in tall woodland, wet sclerophyll forest, cool temperate rainforest and logged cool temperate rainforest from March, 1974, until September, 1976. The number of traps per trap station was adjusted to ensure at least one spare trap per station. During the study tracking strips were established in the cool temperate rainforest grid to investigate the problem of trappability.

Analyses of the capture–recapture data showed that the trappability of *A. stuartii* varied seasonally in each habitat for both males and females and between habitats in the one year. Additionally, the tracking strips used in the rainforest showed that 18% of the individuals in that habitat were not being captured.

Hence the equal probability of capture assumption, implicit in most population estimation models, was invalid. The interpretation of population estimates determined by the capture–recapture method then needs to be approached with considerable caution.

Pages 333–44 *in* CARNIVOROUS MARSUPIALS, ed. by M. Archer, February, 1982. Roy. Zool. Soc. New South Wales: Sydney, Austtralia.

MANY METHODS of estimating population numbers from capture–recapture data are available. These range from the Peterson method (Lincoln Index) introduced by Peterson in 1889 (Le Cren 1965) to the more sophisticated models of Seber (1965), Jolly (1965) and Manly and Par (1968).

Smith *et al.* (1975) reviewed some general models for capture–recapture studies and listed the following common assumptions:

i) The animals do not lose their marks.

ii) The captures are correctly recorded as marked or not marked (actual mark is recorded in most studies).

iii) Marking does not affect the probility of survival.

iv) The population is either open or closed, therefore —
   a) there is no gain or loss of members during sampling, or
   b) there is recruitment and immigration but death and emigration affect marked and unmarked animals equally (in this case survival rates are assumed to be constant, i.e., not correlated with age), or
   c) knowledge is available from other sources which permits an allowance to be made for migration, birth and death prior to the analysis of the data.

v) The population is randomly sampled so that either —
   a) every animal has the same probability of capture, or
   b) if there exist strata within the population which have different probabilities of capture then the marked animals are proportionally distributed within each of these strata.

All capture–recapture models require that assumptions, i, ii and iii hold. Those for closed populations, e.g. Peterson method, additionally require assumptions iv(a) and v(a). The Jolly–Seber stochastic model requires assumptions iv(b) and c(a) while the Manly–Parr modification requires assumptions iv(c) and v(a) (Manly 1970). Finally, a population study with differing trap behaviour between different strata of animals, such as between males, juvenile females and parous females, require assumption v(b).

The assumption most frequently violated in capture–recapture studies is number v, that of equal catchability. This is so whether strata within the population are considered separately (v(b)) or not (v(a)). This factor has been reported as a major weakness in capture–recapture studies (e.g. Tanaka 1956, Tanton 1965, Balph 1968, How 1972 and Fletcher 1974). The situation is complicated because changes in trappability cannot readily be detected by statistical methods (Roff 1973) or compensated for statistical refinements (Hanson 1967 cited by Smith *et al.* 1975). Field methods designed to reduce heterogeneity of trapping, for example shifting traps during the sampling period, have been recommended by Eberhardt (1969) and Woodland (personal communication), but were not used in this study. Leonard (1972) and Freeland (1972) both used a system of setting randomized traps within regularly spaced squares,

[1] School of Natural Resources, University of New England, Armidale, New South Wales, Australia. Present address: Forestry Commission, Tasmania; C/- Department of Agriculture, 41A Cameron Street, Launceston, Tasmania, Australia, 7250.

but neither discussed the merits of their system. Statistical methods of analysing data which violate assumption v have been given by Janion *et al.* (1968), Eberhardt (1969), Marten (1970) and Wilbur and Landwehr (1974).

The purpose of the present study was to test the assumption of the equal catchability. Two approaches were adopted. Firstly, capture–recapture data were examined to see if there were differences in the recapture frequencies for different sexes, seasons, years or habitats. Secondly, by using a method with assumptions independent of those of the capture–recapture model, an attempt was made to detect the presence of, and extricate the relative number of, non–captured animals in one trapping grid.

While a number of non–trapping techniques are available for the detection of mammals, few are suitable for population estimation of small mammals. Density estimation based on direct observation was not possible due to the nocturnal and cryptic habits of *A. stuartii*. Indirect observations using tracking boards (Lord *et al.* 1970), dropping board (Emlen *et al.* 1957), smoked–paper tracking (Sheppe 1965, 1967), runway surveys (Pearson 1959) and sand transects (Bider, Sarrazin and Bider 1973, Doucet and Bider 1974) have been used to detect the relative but not the actual numbers of small mammals in an area.

METHODS

A full description of the study area, habitats and trapping procedure has been given elsewhere (Fletcher 1977, Statham and Harden 1982). Briefly, *A. stuartii* were trapped for 31 months in tall woodland, wet sclerophyll forest and logged and unlogged cool temperate rainforest and 8 months in dry sclerophyll forest. All sites were at Petroi, 70 km east of Armidale, N.S.W., and trapping took place between March 1974 and September 1976.

The method used to detect untrapped individuals in the present study was an adaptation of that of Bider (1968). Results could be quantified by the relative frequency of tracks left by captured, toe-cut animals and uncaptured animale without toe–cuts.

In November, 1975, small quantities of bulldust, river sand, sandy loam and termite mound material were collected on Petroi, sun dried, sieved and levelled for testing as a small mammal tracking medium. Using several house mice (*Mus musculus*) as test animals, bulldust was found to reveal the clearest tracks.

In December a supply of bulldust derived from sheared muddy greywackes was collected, sieved through a 320 μm sieve and stored. A reference collection of photographs of tracks made by the most abundant small mammal species on Petroi was prepared.

The rainforest grid was used for this tracking study for two reasons. Firstly trapping records suggested a lower trappability of animals there than in the sclerophyll areas. Secondly, the paucity of vegetation near ground level made the operation simpler and caused less less disturbance than in the other habitats.

Eight strips, each 10 m x 0.5 m were raked to remove litter. Each strip was then mattocked to a depth of 15 cm. All clods of dirt were broken up and roots were removed. A strip 30 cm wide, raised slightly above the surrounding ground, was smoothed then compacted by hand with a plasterer's trowel. Litter was redistributed to the edge of the strip. A sloping roof of clear plastic was placed over the strip to prevent damage due to rain and litter fall. To prepare the actual tracking surface, a 1 cm thick layer of bulldust was spread over the compacted soil and levelled with the plaster's trowel. Finally, a fine layer of bulldust was sieved over the strip.

Four strips were operational in December 1975, and were read for four consecutive nights that month. Abnormally high rainfall over summer and autumn (>500 mm/month) prevented further tracking until May, 1976, when the other four strips were completed and tracking recommenced. The strips were serviced on each of the remaining field trips — June, July, August and September.

All *A. stuartii* captured in the trapping grids were identified by an ear marking system. In 1976 all juvenile *A. stuartii* captured in the rainforest grid had two outside toes on their front left foot removed. Parous females from the previous generation had a different toe removed. Then, when *A. stuartii* tracks on the bulldust were read, it was generally possible to record whether they had been made by captured or uncaptured animals.

RESULTS

1. Population Size and Structure: The numbers of *A. stuartii* captured in each habitat each generation, together with population structure and density estimates have been reported elsewhere (Statham and Harden 1982). Juvenile *A. stuartii* first entered the traps in December or January, about three months after their birth in September. There were usually more males than juvenile females known to be alive, with exceptions in logged cool temperate rainforest during 1974 and in tall woodland in 1975. Although there were large numbers of females known to be alive in each habitat during the breeding season of 1974, few individuals were captured then but were captured parous later in the year. This trend for females not to enter traps at breeding was observed, less noticeably, during each year of the study (Fletcher 1977).

Table 1. Summary of capture frequency and female *A. stuartii* in each habitat and for each generation.

Category		Habitat				
		W.Sc.	L.Rf.	Rf.	T.Wdl.	D.Sc.
Generation 1, March–September, 1974						
Male:	No. of trapping nights.	39.0	39.0	39.0	39.0	
	No. of individuals	55.0	30.0	40.0	46.0	
	No. of captures	314.0	125.0	166.0	196.0	
	Mean no. of captures/individual	5.7	4.2	4.2	4.3	
	Mean no. males/night	8.1	3.2	4.3	5.0	
Female:	No. of individuals	54.0	52.0	39.0	30.0	
	No. of captures	270.0	143.0	127.0	138.0	
	Mean no. captures/individual	5.0	2.8	3.3	4.6	
	Mean no. females/night	6.9	3.7	3.3	3.5	
Generation 2, October, 1974–September 1975						
Male:	No. of trapping nights	48.0	47.0	47.0	48.0	
	No. of individuals	63.0	44.0	54.0	36.0	
	No. of captures	456.0	153.0	271.0	205.0	
	Mean no. captures/individual	7.2	3.5	5.0	5.7	
	Mean no. males/night	9.5	3.3	5.8	4.3	
Female:	No. of individuals	55.0	49.0	38.0	38.0	
	No. of captures	311.0	174.0	144.0	193.0	
	Mean no. captures/individual	5.7	3.6	3.8	5.1	
	Mean no. females/night	6.5	3.7	3.1	4.0	
Generation 3, October, 1975–September, 1976						
Male:	No. of trapping nights	36.0	32.0	32.0	36.0	24.0
	No. of individuals	68.0	36.0	34.0	29.0	33.0
	No. of captures	393.0	138.0	131.0	105.0	188.0
	Mean no. captures/individual	5.8	3.8	3.9	3.6	5.7
	Mean no. males/night	10.9	4.3	4.1	2.9	7.8
Female:	No. of individuals	49.0	22.0	23.0	21.0	18.0
	No. of captures	193.0	36.0	74.0	72.0	62.0
	Mean no. captures/individual	3.9	1.6	3.2	3.4	3.4
	Mean no. females/night	5.4	1.1	4.1	2.0	2.6

ABBREVIATIONS: W. Sc. = Wet sclerophyll forest; L. Rf. = Logged cool temperate rainforest; Rf. = Cool temperate rainforest; T. Wdl. = Tall woodland; D. Sc. = Dry sclerophyll forest.

2. FREQUENCY OF CAPTURE: The frequency of capture data for each generation of males and females in each habitat are summarised in Table 1. For these analyses juvenile and parous females have been pooled. During each generation the mean numbers of captures per individual and per night were greatest in wet sclerophyll forest for both males and females. Individual males were captured more often than females during each generation in each habitat except generation 2 in logged rainforest. Also the number of males captured per night was greater than the number of females in all habitats except logged rainforest during generations 1 and 2. This was so despite the fact that for generations 2 and 3, data from October, November and December were used in these calculations, during which time only parous females were captured.

To compare the frequency distributions of captures of male and female *A. stuartii* in the five habitats the histograms shown in Figure 1 were drawn and the data were analysed using the χ^2–test. The results for tall woodland, wet sclerophyll, logged and unlogged rainforest were obtained by summing data from each generation. Thus, the few females captured during two generations are counted twice in the figure and analyses. Results for dry sclerophyll are for the 1976 generation only. Results of tests are given in Appendix 1 and only distributions different at the five per cent level are considered here.

Between 1974 and 1976 the frequency distribution of male captures in wet sclerophyll was significantly different from those in logged and unlogged rainforest. This was caused by a greater proportion of males being captured more often in wet sclerophyll than in either of the rainforest areas. In 1976, the frequency distribution of male captures in sclerophyll habitats (tall woodland, wet and dry sclerophyll combined) was significantly different from that in the two rainforest habitats. Once again, this was caused by a greater proportion of males being captured more often in the sclerophyll areas than in the rainforest areas.

From 1974 to 1976 a greater proporation of females in the sclerophyll areas (wet sclerophyll and tall woodland) were captured more frequently than those in rainforest areas causing significant differences between the frequency distributions. The 1976 frequency distribution of female captures in dry sclerophyll did not differ significantly from that in any other habitat.

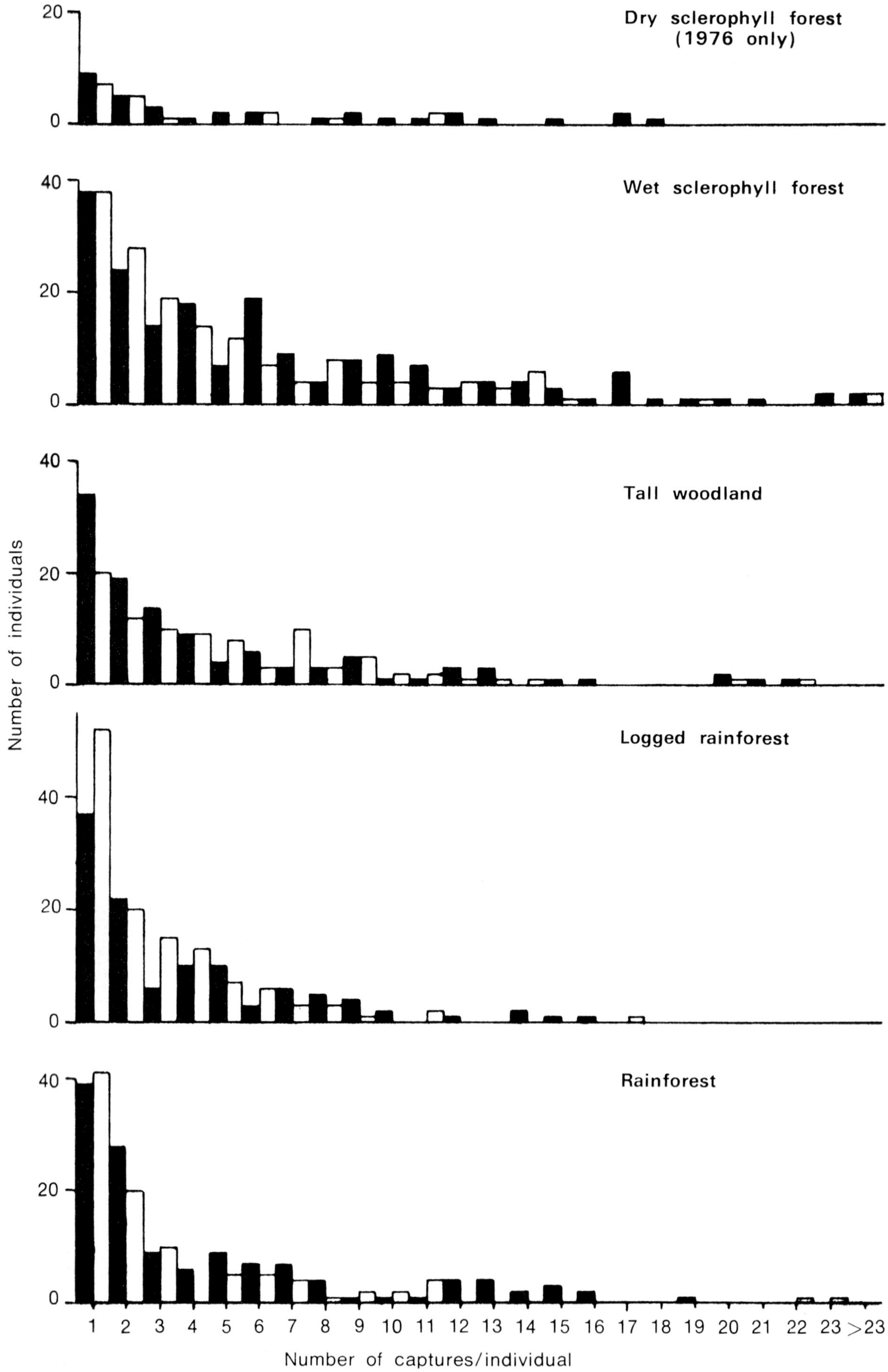

Fig. 1. A stuartii frequency of capture for males (solid black) and females (open) for all generations in each habitat. Values were obtained by combining data from each generation.

When data from the two rainforest areas were combined it was found that the frequency distributions of the two sexes varied, with a large proportion of males being captured more frequently than females. No significant differences between sexes were found in the other habitats.

3. EFFECT OF RAIN: During the study it was thought that rain influenced the numbers of *A. stuartii* captured, such that more animals were captured on wet nights than on dry nights. Graphs were drawn of several month's data, for the number of *A. stuartii* captured in wet sclerophyll against 24-hourly rainfall. These showed no apparent relationship. It was therefore decided to test the difference between the number of individuals captured on wet and dry nights. This comparison was made for tall woodland, wet sclerophyll and logged and unlogged rainforest, using the analysis of variance. The classification based on unweighted analysis of cell means (Snedecor and Cochran 1967: 475) was used because of the unequal numbers and proportions represented in the data. Results from each of the three years were pooled into four three–monthly groups. A trapping night was counted wet if 0.5 mm or more of rain were recorded at the daily rainguage check.

There were no significant differences between the number of *A. stuartii* captured on wet or dry nights in any of the four habitats (Analysis of Varience tables are given in Appendix 2).

4. EFFECT OF NIGHT TRAPPING IN WINTER: Known to be alive estimates of males, juvenile females and parous females in dry sclerophyll between May and August, 1976, based on trap clearing at 2300 hours and 0630 hours, are given in Table 2.

During the 1976 winter more animals were known to be alive in dry sclerophyll when results from both night and morning trap clearing regimes were considered than when only those from night trap clearing were used. One additional animal captured in the morning was male, one was a parous female and the remainder were juvenile females. Thus, in 3 nights of trapping, in July, there were four different juvenile females captured by 2300 hours and an additional five individuals captured at the 0630 hours trap clearing. Although no similar tests were car-

Table 2. A stuartii night trapping, the number of males, juvenile females and parous females known to be alive in dry sclerophyll during winter of 1976 based on night and morning trap clearing regimes.

Month	Category						Total	
	Males		Juvenile females		Parous females			
	2300 (hrs)	0630 (hrs)	2300 (hrs)	0630 (hrs)	2300 (hrs)	0630 (hrs)	2300 (hrs)	0630 (hrs)
May	16	16	6	8	2	2	24	26
June	17	17	6	7	1	2	24	26
July	17	17	4	9	1	2	22	28
August	12	13	4	6	1	1	17	20

Table 3. A. stuartii trap deaths, number and percentage of males and females found dead in traps, March 1974–September 1976

Trapping grid	Sex	Trap deaths							
		Generation 1		Generation 2		Generation 3		Total	
		Number	%	Number	%	Number	%	Number	%
W.Sc.	males	8	15	8	13	9	13	25	13
	females	1	2	6	11	8	16	15	13
L. Rf.	males	1	3	4	9	2	6	7	6
	females	1	2	3	6	1	5	5	5
Rf.	males	4	10	9	17	17	50	30	23
	females	5	13	–	–	1	4	6	8
T. Wdl.	males	4	9	6	17	5	17	15	14
	females	1	3	2	5	5	24	8	11
D.Sc.	males	–	–	–	–	4	11	4	11
	females	–	–	–	–	2	9	2	9
Total	males	17	10	27	14	37	18	81	14
	females	8	5	11	6	17	12	36	9
	males and females	25	7	38	10	48	14	117	12

ABBREVIATIONS: W.Sc. = Wet sclerophyll forest; L. Rf. = Logged cool temperate rainforest; Rf. = Cool temperate rainforest; T. Wdl. = Tall woodland; D. Sc. = Dry sclerophyll forest.

ried out in the other habitats, these results suggest that a significant proportion of juvenile females enter traps after midnight. Thus, night trap clearing carried out during winter months in wet sclerophyll, tall woodland, logged and unlogged rainforest might have seriously underestimated juvenile females but probably had little effect on males. The sample size for parous females was too small to determine whether or not this stratum might have been underestimated by night trapping.

5. TRAP MORTALITY: A total of 117 *A. stuartii* (12% of individuals captured) died in traps between March, 1974 and September, 1976. The number and percentage of male and female trap deaths for each generation in each habitat are given in Table 3.

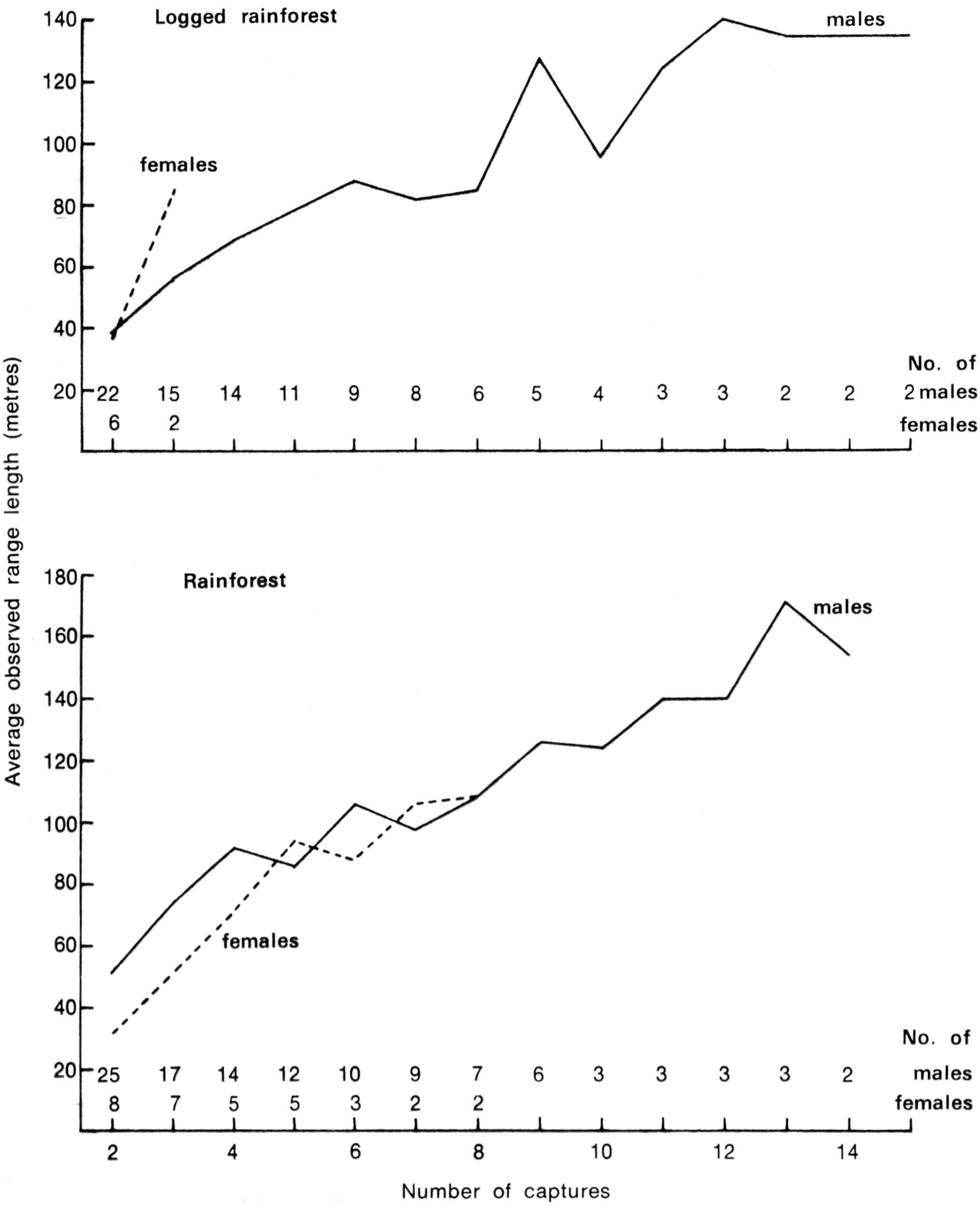

Fig. 2. A. stuartii movements, average O.R.L. at consecutive captures for males and females captured in ach trapping grid during 1976.

Fig. 2 continued.

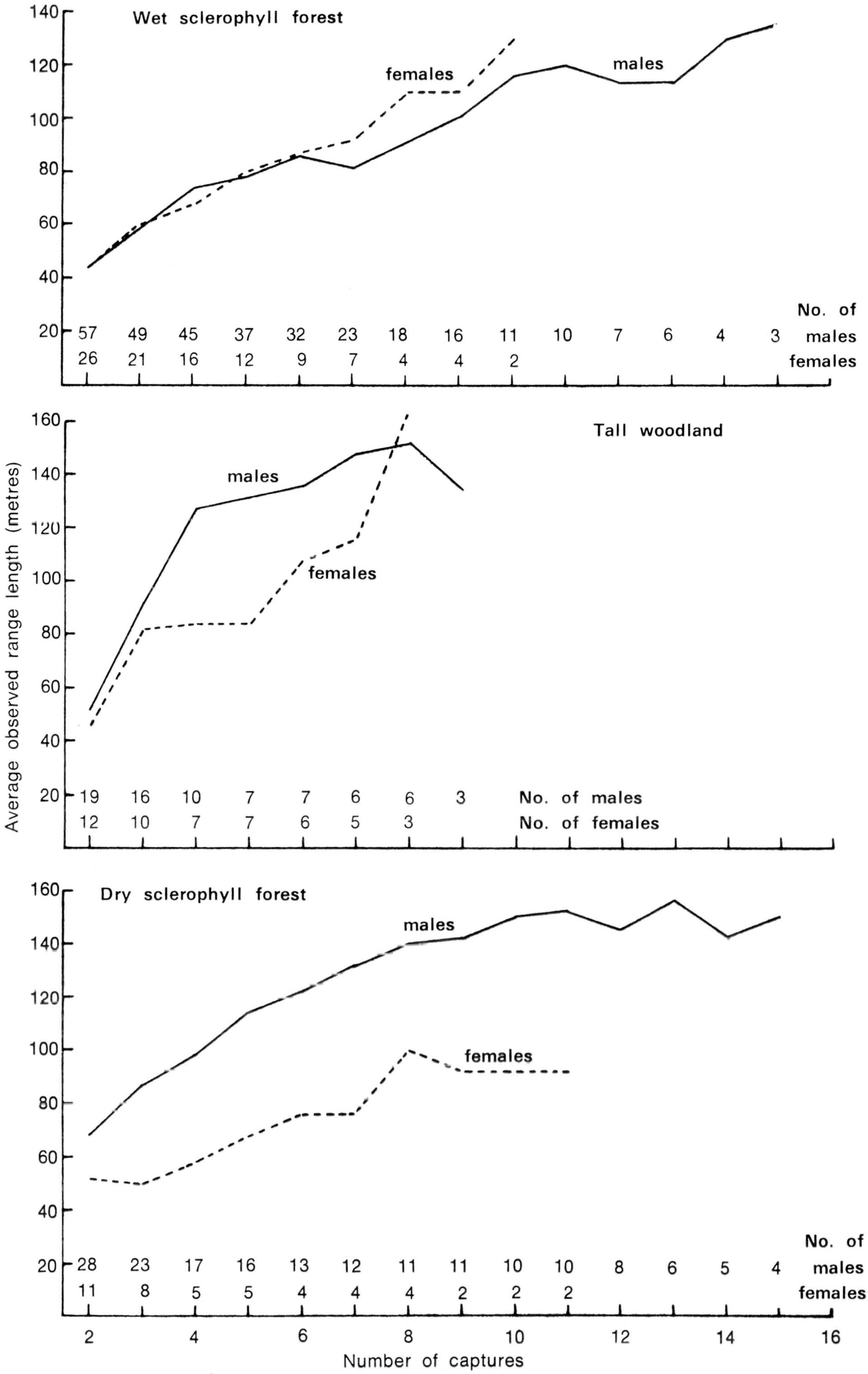

The percentage trap mortality was usually greater in males than in females. Exceptions occurred in rainforest during generation 1 and in wet sclerophyll and tall woodland during generation 3. Thirty–six (44%) of the dead males, including all 17 dead in rainforest in 1976, were animals captured in August and September at the time of the male die–off. Excluding these deaths, trap mortality was eight per cent.

Two types of nematode worms were found inside the peritoneum and in the gut of a large proportion of dead *A. stuartii*. Specimens of both types were examined by officers of the Armidale District Veterinary Laboratory. One was identified as *Toxocara mystax* while specimens of the other were too broken up to allow specific identification.

6. Trap Spacing and Grid Size: The 1976 capture records were used to determine the average observed range length (O.R.L.) at consecutive captures for male and female *A. stuartii* in each habitat. These data are shown graphically in Figure 2 and are summarised in Table 4. An individual's O.R.L. is defined as the maximum distance between its sites of capture.

Table 4. A. stuartii movements, average O.R.L. (metres) for males and females in each habitat during 1976.

	Habitat				
Category	L.Rf.	Rf.	W.Sc.	T.Wdl.	D.Sc.
Males	135	154	136 ↑	130	150
Females	86 ↑	106 ↑	130 ↑	160 ↑	92

ABBREVIATIONS: ↑, curve still increasing at last point; W.Sc. = Wet sclerophyll forest; L. Rf. = Logged cool temperate rainforest; Rf. = Cool temperate rainforest; T. Wdl. = Tall woodland; D. Sc. = Dry sclerophyll forest.

For males captured in all habitats except wet sclerophyll and for females captured in dry sclerophyll the estimated O.R.L. had levelled off by the time all 1976 capture data had been considered, suggesting that the true O.R.L. had been reached. In the remaining five categories, insufficient captures had been made to allow accurate O.R.L. determination. The average O.R.L. was greater for males than for females in the dry sclerophyll logged and unlogged rainforest. In the latter two habitats female O.R.L. estimates may be considerably underestimated because of the scarcity of individuals captured more than a few times.

For both males and females the O.R.L. was approximately 2–3 times greater than the minimum distance between traps (50 metres). Also the O.R.L. was not more than half the length of the square grid (300 m metres).

7. Tracking as a Test of the Capture–Recapture Technique: Table 5 shows the results of *A. stuartii* tracking on bulldust strips in the rainforest between December, 1975 and September, 1976.

Since all juvenile *A. stuartii* captured in the rainforest in 1976 had two toes removed and all parous females had one toe removed, the percentage of tracks made by toe-cut animals was a measure of the proportion of animals in the population which had been captured in the trapping programme.

In August 1976, tracks made by 21 *A. stuartii* were recorded and of the 17 tracks deciphered, 82 per cent were made by marked animals. Thus 18 per cent of the animals detected by tracking had not previously been captured. These data are supported by the fact that six new *A. stuartii* were captured in August and 15 in September in that habitat (Statham and Harden 1982). The total number of *A. stuartii* tracks recorded in months other than August was too small to allow accurate quantification of the ratio of marked population to total population. Thus calculation of the proportion of marked animals in the other habitats was not considered meaningful.

Table 5. A. stuartii tracking results, number of tracks of toe–cut and non–toe–cut animals recorded on ten–metre bulldust strips in the rainforest.

Month	Length of strips examined (metres)	Number of tracks			Total tracks	% of tracks deciphered	% of deciphered tracks made by toe–cut animals
		by toe–cut animals	by non–toe–cut animals	undeciphered			
Dec. '75	160	–	–	–	–	–	–
May '76	80	0	1	0	1	100	0
June '76	230	1	2	2	5	60	33
July '76	400	0	1	1	2	50	0
Aug. '76	640	14	3	4	21	81	82
Sept. '76	240	3	0	2	5	60	100

DISCUSSION

The results have illustrated several weaknesses in the capture–recapture model which are often overlooked by zoologists. For example, the trapping behaviour of *A. stuartii* differed between months, between habitas and sometimes between sexes.

Annual peak captures and K.T.B.A. estimates were composed of large numbers of males and juvenile females rather than parous females. They usually occurred in autumn or early winter.

Similar trends in other studies have led to the erroneous suggestion that *A. stuartii* populations peak in autumn (e.g. Wood 1970). From a knowledge of the biology of this species, it seems obvious that the maximum number of independent animals would occur in summer, immediately after weaning. Thus the peak in captures and population estimates in autumn reflect trap behaviour rather than a true population peak. It is suggested, however, that these months of peak captures would be the best time to attempt one–off estimates of *A. stuartii* numbers.

At most times there were several *A. stuartii* K.T.B.A. but not actually captured. The fact that one of these animals was seen near its usual capture sites and that most were subsequently recaptured at the same trap site or one adjacent, strongly suggests that these animals were present when not captured, rather than that they had moved away. Individuals cannot have been excluded from traps because there was always one spare trap per station. Although *A. stuartii* preference for clean or dirty traps was not tested, individuals were found in traps previously occupied by *R. fuscipes* and vice versa.

Individual male *A. stuartii* in wet sclerophyll were captured more often than those in logged or unlogged rainforest. Females were captured more often in wet sclerophyll and tall woodland than in either rainforest areas. Males were generally captured more often than females, although the only significant differences was when data from the rainforest areas were pooled. Thus the *A. stuartii* populations in the different habitats were not randomly sampled· so that assumption v(a) (see Introduction) was violated. Since it is known that individual females rarely entered traps during the breeding season and one male (mentioned above) was observed at night but not captured, it seems that assumption v(b) (see Introduction) was also violated. Determination of total population numbers within each habitat is therefore suspect. Clearly, comparisons between habitats and with other studies, at this stage, are limited to consideration of trappable populations rather than total populations.

In this study, no significant differences were found between the number of *A. stuartii* captured on wet and dry nights. However more detailed data such as the time of rainfall, cloud cover etc. may be more informative.

The practice of clearing traps at night was found to reduce the number of juvenile females captured in the dry sclerophyll and was inefficient because traps had to be revisited the following day to be reopened. Weighed against these disadvantages, there were two factors strongly in favour of night trapping. Firstly, one person could run both day and night trap rounds without assistance, and secondly, with the insulation provided by dry leaves, night trapping reduced *A. stuartii* trap mortality to 12 per cent, which was 50 per cent less than it had been in 1973 (Fletcher 1974). Thus it is proposed that the most sound winter trapping regime would be to leave traps open continuosuly during each trapping period, clearing them twice daily, once around midnight and once at first light. Then hopefully, all trappable animals would be captured and trap mortality would remain low.

The average O.R.L. estimated for male and female *A. stuartii* in each habitat ranged approximately from 100 metres to 150 metres. Since individuals moved these distances between traps, it is probable that, in this study, all trappable animals were able to locate traps spaced at 50 metres. The practice of having at least one empty trap at each station would have eliminated competition for traps at any particular station. At the same time, the nine hectares covered by each of these grids seemed sufficiently large an area to encompass most movements of a large proportion of animals captured. It is therefore considered that these nine hectare grids, of seven rows of seven trap stations, at 50 metre spacing, gave an efficient balance between trap spacing, overall grid size and the use of as few traps as possible (up to 130 per trapping grid).

Usually, more individual males were captured per generation than females. This may have been a characteristic of the populations or it may have been a consequence of the capture–recapture programme as suggested by the night and morning trap clearing carried out in the dry sclerophyll.

The peak in *A. stuartii* captures which occurred in rainforest during September 1976 was caused by 15 new males, representing 27% of the population in that habitat. A similar influx into the traps occurred in this habitat in September of 1975 (Statham and Harden 1982) and in 1973 (Fletcher 1974). It occassionally occurred, less markedly, in the other habitats The reason for this influx is not certain. Wood (1970) reported an increase in male O.R.L. at about the time of breeding. This could be interpreted to mean that new captures were caused by animals from outside the grid moving in. The concept of a mass migration of males into the rainforest at the time of breeding seems unlikely and is not supported by the small increase in number of males observed occasionally at this time in other habitats. If the phenomenon were solely due to individual males moving greater distances than at other times, less animals previously captured over the grid should have been captured with more new animals captured at peripheral rather than at interior traps. Neither of these changes was observed. It is therefore suggested that the large influx of new males into the traps was caused by a change in the trapping behaviour of males already present at the rainforest grid. The bulldust tracking data, with 18 per cent of *A. stuartii* tracks made in August being left by animals not previously captured supports this idea.

Results from the rainforest bulldust tracking strips provided an independent means of testing the capture–recapture programme in that habitat and confirmed the suspicion that not all *A. stuartii* in the rainforest were being captured. However, extrapolation of these data to other habitats is uncertain due to the variations in *A. stuartii* trap behaviour between habitats. It seems important that this tracking should be extended to include more tracking strips in this trapping grid and a similar number of strips in each of the other grids. Only by doing this can an accurate determination of total population numbers in each habitat be made. The different trap behaviour discussed for *A. stuartii* in sclerophyll areas suggests that the trappable populations in each of these habitats might have represented a higher proportion of the total population than in rainforest.

ACKNOWLEDGEMENTS

This research was part of a study on habitat relationships among small mammals at Petroi, undertaken in fulfilment of the requirements for the degree of Master of Natural Resources in the University of New England. I would like to thank Mr. R. Harden, N.P.W.S., N.S.W., other members of his N.P.W.S. team and numerous students for their assistance with field work. Sincere thanks are extended to Mr. & Mrs. Joe Davies for permission to work and camp on Petroi and for their kind hospitality. Dr. V. Bofinger who helped with statistical analyses and Mr. S. Williams programmed the animal capture data. I am grateful to the N.P.W.S. for assistance with travel and other aspects of the study. For two years of the work I was supported by a Commonwealth Postgraduate Research Award. Mr. R. Harden, Dr. B. Fox, Dr. D. Croft and Mr. S. Van Dyck all gave useful comments on the manuscript.

REFERENCES

Balph, D. F., 1968. Behavioural response of unconfined Unita Ground Squirrels to trapping. *J. Wildl. Manage.* **32:** 778–94.

Bider, J. R., 1968. Animal activity in uncontrolled terrestrial communities as determined by a sand transect technique. *Ecol. Monogr.* **38:** 269–308.

Doucet, G. J. and Bider, J. R., 1974. The effects of weather on the activity of the Masked Shrew. *J. Mammal.* **55:** 348–63.

Eberhardt, L. L., 1969. Population estimates from recapture frequencies. *J. Wildl. Manage.* **33:** 28–39.

Emlen, J. T., Hine, R. L., Fuller, W. A. and Alfonso, P., 1957. Dropping boards for population studies of small mammals. *J. Wildl. Manage.* **21:** 300–14.

Fletcher, H. L., 1974. "A study of the distribution and abundance of two small mammal species in four different habitats." Dip. Nat. Res. Dissertation: University of New England, Armidale.

Fletcher, H. L., 1977. "Habitat relationships among small mammals at Petroi, northeastern N.S.W." M. Nat. thesis: University of New England, Armidale.

Freeland, W. J., 1972. "A rainforest and its rodents." M.Sc. thesis: University of Queensland, Brisbane.

How, R. A. "The ecology and management of *Trichosurus* species (Marsupialia) in N.S.W." Ph.D. thesis: University of New England, Armidale.

Janion, S. M., Ryszkowski, L. and Wierzbowska, T. 1968. Estimate of number of rodents with variable probability of capture. *Acta Theriol.* **13:** 285–94.

Jolly, G. M., 1965. Explicit estimates from capture–recapture data with both death and immmigration–stochastic model. *Biometrika.* **52:** 225–47.

Le Cren, E. D., 1965. A note on the history of mark–recapture population estimates. *J. Anim. Ecol.* **34:** 453–54.

Leonard, B. V., 1972. "The effect of fire upon selected small mammals and leaf litter fauna in sclerophyll forest in southern Australia". M.Sc. thesis Monash University, Melbourne.

Lord, R. D., Vilches, A. M., Maiztegui, J. I. and Soldini, C. A., 1970. The tracking board: A relative census technique for studying rodents. *J. Mammal.* **51:** 828–29.

Manly, B. F. J., 1970. A simulation study of animal population estimation using the capture–recapture method. *J. Appl. Ecol.* **7:** 13–39

Manly, B. F. J. and Parr, M. J., 1968. A new method of estimating population size, survivorship and birth rate from capture–recapture data. *Trans. Soc. Brit. Ent.* **18:** 81–9.

Marten, G. G., 1970. A regression method for mark–recapture estimation of population with unequal catchability. *Ecology.* **51:** 291–95.

Pearson, O. P., 1959. A traffic survey of *Microtus–Reithrodontomys* runways. *J. Mammal.* **40:** 169–80.

Roff, D. A., 1973. An examination of some statistical tests used in the analysis of mark–recapture data. *Oecolgia.* **12:** 35–54.

Sarrazin, J. P. R. and Bider, J. F., 1973. Activity, a neglected parameter in population estimates — the development of new technique. *J. Mammal.* **54:** 369–82.

Seber, G. A. F., 1965. A note on multiple recapture census. *Biometrika.* **52:** 249–59.

Sheppe, W., 1965. Characteristics and uses of *Peromyscus* tracking data. *Ecology.* **46:** 630–34.

Sheppe, W., 1967. The effect of live trapping on the movement of *Peromyscus. Am. Midl. Nat.* **78:** 471–86.

SMITH M. H., GARDNER, R. H., GENTRY, J. B., KAUFMAN, D. W. AND O'FARRELL, M. H., 1975. Chapter 2: Density estimations of small mammal population. *In* "Small mammals: Their productivity and population dynamics" ed. by F. D. Golley, K. Petrusewicz and L. Ryszkowski. IBP 5. Cambridge University Press: Cambridge.

SNEDECOR, G. W. AND COCHRAN, W. G., 1967. "Statistical methods". 6th Edition. Iowa State University Press: Iowa.

STATHAM, H. L. AND HARDEN, R. H., 1982. Habitat preferences of *Antechinus stuartii* (Dasyuridae, Marsupialia) at Petroi, northeastern N.S.W. *In* "Carnivorous marsupials", ed. by M. Archer. Roy. Zool. Soc. New South Wales: Sydney, Australia.

TANAKA, R., 1956. On differential response to live traps of marked and unmarked small mammals. *Annotnes Zool. Jap.* **29:** 44–51.

TANTON, M. T., 1965. Problems of live trapping and population estimation for the wood mouse, *Apodemus slyvaticus* (L) *J. Anim. Ecol.* **34:** 1–22.

WILBUR, H. M. AND LANDWEHR, J. M., 1974. The estimation of population size with equal and unequal risk of capture. *Ecology.* **55:** 1339–48.

WOOD, D. H., 1970. An ecological study of *Antechinus stuartii* (Marsupialia) in a southeast Queensland rainforest. *Aust. J. Zool.* **18:** 185–207.

RESUMEN

15. Probabilidad de entrampado de *Antechinus stuartii* (Marsupialia) en Petroi, Noreste de New South Wales. Por H. L. Statham.

Especímenes de *Antechinus stuartii* fueron mensualmente entrampados desde Marzo de 1974 hasta Septimbre de 1976 a lo largo de grandes reticulados (9 ha) en bosque alto, bosque esclerófilo húmedo, foresta lluviosa templado–fría y en foresta lluviosa templado–fría talada. El número de trampas por estación de trampeo fue ajustado de manera de asegurar al menos una por estación. Durante el estudio se establecieron carriles de fajas en bosque lluvioso templado–frío con el objeto de investigar las probabilidades de entrampamiento.

El análisis de los datos de captura–recaptura mostró que el entrampado de *A. stuartii* varió estacionalmente en cada hábitat "tanto para los machos como para las hembras y entre los hábitats dentro del mismo áno. Adicionalmente, los carriles de fajas usados en el bosque lluvioso mostraron que el 18% de los individuos de hábitat no habían sido capturados.

Así, la supuesta igual probabilidad de captura, implíen la mayor parte de los modelos de estinación de una población, quedo invalidada. La interpretación de la tasa poblacional determinada por el método de captura–recaptura debe ser tomado con mucha cautela.

APPENDIX 1

A. stuartii, results of χ^2–tests comparing frequency distributions of the number of captures for males and females in all habitats.

Comparison	df	χ^2	Sig.⁺
Males 1974–76			
W.Sc × L.Rf. × Rf. × T.Wdl.	21	36.0	**
(W.Sc. + T.Wdl.) × (L.Rf. + Rf.)	7	19.6	**
W.Sc. × L.Rf.	7	19.6	**
W.Sc. × Rf.	7	16.8	*
W.Sc. × T.Wdl.	7	11.3	n.s.
L.Rf. × Rf.	7	5.3	n.s.
L.Rf. × T.Wdl.	7	8.0	n.s.
Rf. × T.Wdl.	7	6.3	n.s.
Males 1976			
(W.Sc, T.Wdl. + D.Sc.) × (L.Rf. + Rf.)	5	12.1	*
W.Sc × T.Wdl. × D.Sc.	4	9.4	–
D.Sc × W.Sc.	4	4.2	n.s.
D.Sc. × L.Rf.	3	2.4	n.s.
D.Sc. × R.f.	3	1.8	n.s.
D.Sc × T.Wdl.	2	1.3	n.s.
Females 1974–76			
W.Sc. × L.Rf. × Rf. × T.Wdl.	15	37.1	**
(W.Sc. + T.Wdl.) × (L.Rf. + Rf.)	5	28.2	***
W.Sc. × L.Rf.	5	20.4	***
W.Sc. × Rf.	5	13.7	*
W.Sc. × T.Wdl.	5	4.6	n.s.
L.Rf. × Rf.	5	5.0	n.s.
L.Rf. × T.Wdl.	5	17.3	**
Rf. × T.Wdl.	5	13.2	*
Females 1976			
(W.Sc. + T.Wdl. + D.Sc.) × (L.Rf. + Rf.)	3	3.2	n.s.
D.Sc. × W.Sc.	2	3.5	n.s.
D.Sc. × L.Rf.	1	2.3	n.s.
D.Sc. × Rf.	2	0.2	n.s.
D.Sc. × T.Wdl.	1	1.9	n.s.
Males × Females 1974–76			
W.Sc. + L.Rf. +Rf. +T.Wdl.	13	22.1	–
W.Sc.	5	11.0	–
L.Rf.	5	9.2	n.s.
Rf.	5	4.9	n.s.
T.Wdl.	5	5.5	n.s.
W.Sc. + T.Wdl.	5	5.2	n.s.
L.Rf. + Rf.	5	13.9	*
Males × Females 1976 – D.Sc.	2	2.9	n.s.
Males all habitats 1974 × 75 ×76	20	25.8	n.s.
Females all habitats – 1974 × 75 × 76	14	16.6	n.s.

Symbols: ⁺ levels of significance; n.s. = $P>0.05$; * = $P<0.05$; ** = $P<0.01$; *** = <0.001.

APPENDIX 2

Analysis of variance, comparisons of the number of animals captured on wet and dry nights, using the method of unweighted analysis of cell means (Snedecor and Cochran 1967 : 475).

	Source	df	ms	sig.⁺
a) Wet sclerophyll forest, n_h = 12.2	Wet/dry	1	11.69	n.s.
	Months	3	184.21	*
	Interaction	3	11.46	n.s.
	Within cells	97	5.52	
b) Logged cool temperate rainforest, n_h = 9.06	Wet/dry	1	0.13	n.s.
	Months	3	14.78	–
	Interaction	3	2.02	n.s.
	Within cells	86	2.28	
c) Cool temperate rainforest, n_h = 10.55	Wet/dry	1	0.01	n.s.
	Months	3	28.46	*
	Interaction	3	1.46	n.s.
	Within cells	93	1.96	
d) Tall woodland, n_h = 8.51	Wet/dry	1	0.66	n.s.
	Months	3	34.68	–
	Interaction	3	4.62	n.s.
	Within cells	92		

Symbols: ⁺ levels of significance as below appendix 1.

34. AN ANALYSIS OF OFFENSIVE AND DEFENSIVE THREAT DISPLAYS IN *Dasyuroides byrnei* (DASYURIDAE, MARSUPIALIA)

GEOFFREY D. HUTSON[1]

A series of male–male and female–female agonistic encounters of *Dasyuroides byrnei* were observed and analysed. The causation of the threat displays, pant–hiss and chatter, was examined by analysis of transitions with other behaviour patterns and the time intervals between them. Pant–hiss was associated with a high tendency to attack and chatter with a high tendency to flee. Analysis of the effect of the displays on the behaviour of the opponent showed that pant–hiss was given in response to non–agonistic behaviour whereas chatter was a response to overt aggression. Thus, pant–hiss appears to be an offensive threat display and is an initial warning to an opponent to withdraw. Chatter appears to be a defensive threat display and indicates an intention to fight back if attacked.

Analysis of the context of the displays and the structure of the sound component revealed a surprising amount of variability in the displays. The rate of note repetition and harshness of the calls appeared to reflect the underlying motivations to flee and attack.

A series of playback experiments showed that the different vocalizations had different effects on the behaviour of an opponent and suggested that the variation in motivation evident in the calls was communicated to an opponent. Experimental manipulation of a visual component of the displays, dyeing the tail completely black, reduced fighting behaviour, and indicated that the tail was an effective visual threat signal.

The applicability of the conflict theory to the causation of the displays is discussed.

Pages 345–63 *in* CARNIVOROUS MARSUPIALS, ed. by M. Archer, February, 1982. Roy. Zool. Soc. New South Wales: Sydney.

THE TRADITIONAL interpretation of the causation of vertebrate threat displays is that they are the result of an underlying conflict between the tendency to attack and the tendency to flee from an opponent. This conflict theory was initially developed in birds (e.g. Moynihan 1955, Tinbergen 1959, and reviewed by Blurton Jones 1968) but has also been applied to mammals. For example, Leyhausen (1956) has described how the threat posture of the domestic cat changes with different absolute and relative strengths of the tendencies to attack (aggression) and flee (fear). Similarly, Grant and Mackintosh (1963) have shown an ambivalence among the tendencies to flee, attack and mate for four species of rodents. However, as Ewer (1968 a) has pointed out, the proof of a conflict theory for the causation of mammalian threat displays is largely negative; where there is no threat there is no conflict.

Andrew (1972) has criticised this approach on the grounds that use of major intervening variables (such as attack and flee) in accounting for displays tends to obscure an understanding of the relationships between specific responses. Also, brain stimulation experiments in the cat and the Oppossum (*Didelphis virginiana*) failed to confirm the conflict theory, because threat was elicited only by stimulation at a single site in the brain (Brown and Hunsperger 1963, Roberts, Steinburg and Means 1967).

However, despite questioning of the motivational basis of threat displays their function in communication has generally been considered to be obvious. The threat displays presumably reflect changes in underlying motivation of the animal which are then communicated to the opponent. In an offensive threat the message is, "If you do not retreat, I will attack you", whereas in defensive threat the message is, "I am not about to attack, but if you attack me, I will retaliate". The function of the threat is to deter the opponent; to drive him away in the first case, to prevent him making an attack in the second (Ewer 1968a).

The aim of this study of *Dasyuroides byrnei* (Spencer 1896) was firstly, to determine the immediate causation of threat displays used in agonistic interactions, and especially the applicability of the conflict theory of causation; and secondly, to determine the function of the displays in communication. Dasyurid marsupials are especially interesting from this point of view because they appear to have very restricted threat display repertoires in comparison with the large variability shown by other carnivorous mammals, such as Cats, Dogs and Wolves.

[1] Department of Zoology, Monash University, Clayton, Victoria, Australia, 3168. Present address: School of Agriculture and Forestry, University of Melbourne, Parkville, Victoria, Australia, 3052.

Encounters were staged in a neutral observation cage. Causation of the threat displays was examined by analysis of transitions with other behaviour patterns and the time intervals between them. The function of the displays was assessed from their effect on the behaviour of the opponent. The predictions of these analyses were then tested experimentally by playback of the auditory component of the displays and by modification of a visual component — dyeing tails completely black.

METHODS

LABORATORY MAINTENANCE: A breeding colony of *Dasyuroides* was established in the laboratory with animals trapped on Sandringham Station, Queensland, (24°03'S 139°03'E). Animals were housed individually in wooden boxes (45 x 45 x 60 cm) with hinged wire mesh tops. The floor of these boxes was covered with sand and a cardboard tube (30 x 8 cm) was supplied together with shredded paper or dried grass for nesting material. Animals were fed daily on tinned dog food and twice weekly on live day–old chickens. Water was initially available *ad libitum* but was discontinued when it was observed that some animals did not drink at all.

ENCOUNTERS: Both male–male (n=12 animals) and female–female (n=8 animals) encounters were conducted using only wild caught animals. Encounters were staged in a neutral 90 x 60 x 45 cm observation cage. The floor of the cage was covered in sand and a cardboard tube (10 x 12 cm) was provided as a refuge in each half of the cage. Animals were paired randomly to give a "round–robin" sequence such that each animal was paired once with every other one. Animals were transferred from their home cage to the observation cage in a cotton bag. Each pair was then allowed one hour to become accustomed to the cage before a partition was removed and the encounter allowed to proceed for 15 minutes.

No one animal was used on successive days so that the minimum interval between one encounter and the next, for any individual, was 48 hours. After each encounter the walls of the observation cage were wiped down with a damp cloth and the sand was sifted through a fine grid to remove faecal pellets. Any sand that was damp with urine was removed and replaced by fresh sand which was thoroughly mixed with the old. Thus although scents were not entirely removed it was hoped that there was some equivalence between the residual scent environments of separate encounters.

Sequences of behaviour were recorded using an Esterline Angus 20 channel Event Recorder connected to a push–button keyboard. Fourteen categories of behaviour were recorded. Similar categories have been described by Braithwaite (1974) in *Antechinus stuartii*.

(i) APPROACH: Locomotion towards an opponent. The *approach* to an opponent may be direct or indirect, slow or rapid, or composed of a number of pauses. The body posture is variable, but is often extended, with the head and tail held in a horizontal plane or slightly lower. In close proximity to an opponent, the body is often further extended, resulting in both hindlegs being drawn back behind the body. The ears are slightly folded and the hair on the shoulders and terminal half of the tail is erected. One forepaw is often raised (one–paw–up), a posture which Ewer (1968 b) has reported for many other dasyurid marsupials. Occasionally an *approach* was accompanied by tail–switching, which has been described by Aslin (1974). The animal dragged its hindquarters on the substrate and switched its tail vigorously from side to side.

(ii) INVESTIGATION: Sniffing usually begins as a naso–nasal contact and extends backwards along the jaw angle to the eye, ear and neck. Occasionally, with the more posterior orientations the fur is nosed and one paw is raised and sometimes placed onto the opponent.

(iii) ATTACK: Two types of *attack* occur, but they have been included within one category.

(a) *Lunge*: One animal will lunge at the other with the mouth open. Usually only the head is moved and the bite is directed at the side of the head or neck. The movement rarely involves actual delivery of a bite.

(b) *Attack leap*: This movement is similar to the lunge in that the mouth is open, but here the whole body is involved as the animal leaps off the ground at an opponent. Again the bite is directed to the side of the head or neck and is often delivered.

(iv) PANT–HISS: A low intensity vocalization usually uttered with the mouth open. The sound is produced on expiration and is often repeated at regular intervals. Aslin (1974) refers to this vocalization as "hissing".

The *pant–hisses* comprising a series are not usually uttered at the same intensity, but either increase or decrease in intensity with repetition. The body posture which accompanies the sound has a characteristic form (Fig. 1a). The head is usually held above the horizontal plane of the body and the ears forward. The tail is usually lowered slightly and the long black hairs of the terminal half are erected.

(v) CHATTER: A rapid series of loud, harsh sounds uttered with the mouth open. There is usually little variation in the intensity of the

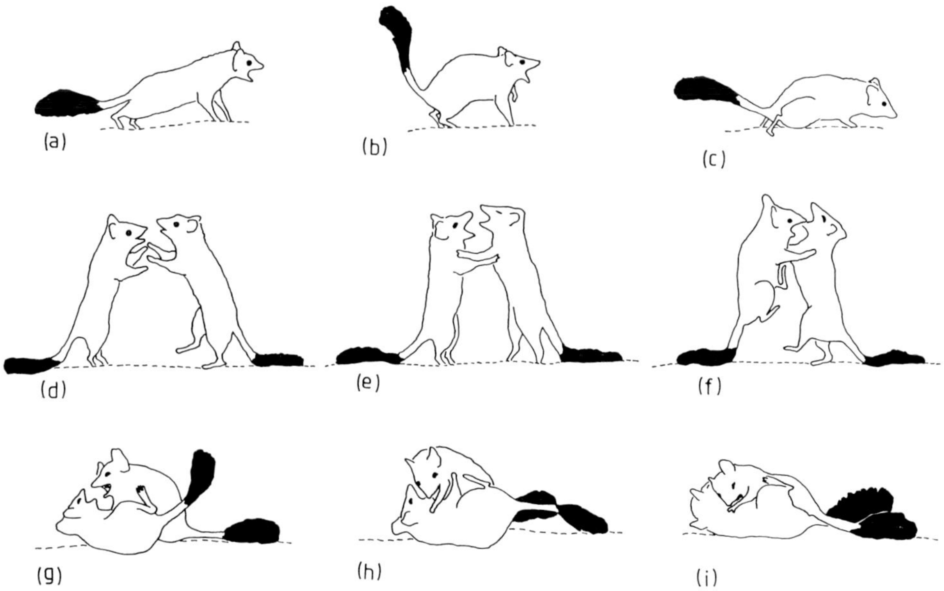

Fig. 1. Agonistic behaviour of *Dasyuroides*: (a) *pant–hiss*, (b) *chatter*, (c) *cloacal dragging*, (d)–(f) sequence in the upright phase of *fighting*, (g)–(i) sequence in the wrestling phase of *fighting*. Drawings were taken from 16mm cine film.

sounds with repetition. The body posture which accompanies *chatter* is sometimes similar in form to the *pant–hiss* posture. However, the head is often lowered slightly and the ears folded. The tail position also varies but it is usually held above the horizontal plane, and often is held vertically (Figure 1b).

(vi) WITHDRAWAL: Two types of *withdrawal* occur, but these are the ends of a continuum in which the first type grades over into the second. The first is a sudden backwards movement of the head, which is then sometimes turned away from the opponent. The second type of withdrawal, more frequently observed, is locomotion away from an opponent.

(vii) RETREAT: Locomotion away from an opponent to a refuge. Thus *withdrawal* may become a *retreat*.

(viii) FLEEING: Rapid locomotion away from an opponent, distinguished from *withdrawal* by erratic leaping and bounding movements.

(ix) CHASE: The active pursuit of an opponent, frequently accompanied by a *sniffing* vocalization which is usually uttered with the mouth closed. The sound appears to result from the rapid expiration of air through the nostrils and may rise and fall in intensity with repetition.

(x) FIGHTING: Two distinct and occasionally alternating phases of *fighting* occur: an upright phase and a wrestling phase. In the upright phase both animals rear up on their hindlegs, grasping one another with their paws, and attempting to kick out with one or both hindlegs at the abdomen. The mouth is usually open and the tail held on the ground to aid balance (Fig. 1d–f). Eventually, one animal drags the other down, leading to the wrestling phase. Both animals now lie on their sides and kick and push the other's abdomen with their hindlegs. The paws are used to maintain a grip around the neck region and to pull the opponent into a position for a bite to the side of the neck (Fig. 1g–h). The tails are not used in an interlocking fashion. When both animals have a neck grip the hindlegs are used more to push the opponent away, whilst at the same time maintaining the bite. The end result is that the animals may roll over and over a number of times until one breaks away. The bite to the side of the neck is used to maintain a grip on the opponent and is not used as a killing bite as in prey–catching behaviour (Hutson 1975). Harrison Matthews (1964) has described similar forms of ritualized fighting in diverse mammalian species. Occasionally, the adoption of a distinct "belly–up" posture precedes the end of a bout of *fighting*. One animal lies motionless on its back displaying the conspicuous white throat and ventrum. In an evenly matched encounter a fight often lasts up to 5 minutes or more and both animals may lose large quantities of fur pulled from the side of the neck.

(xi) NON-AGONISTIC BEHAVIOUR: Three behaviour patterns which are not specifically directed towards an opponent have been included within this category. *Exploration* involves the investigation of the surrounding environment by "sniffing" and "nosing" and is characterized by three distinctive postures — an "extended" body posture, "one–paw–up" and an upright bipedal stance. *Digging* is performed primarily with the forepaws, but sand is also kicked back with the hindlegs. *Sandbathing* involves the rubbing of the sternum and ventrum along the substrate as the animal pushes forward with the hindlegs. The flanks and sides of the head and snout are occasionally rubbed along the substrate in a similar way.

(xii) GROOMING: All the grooming movements described by Hutson (1976) were recorded in the encounters, but *grooming* was more usually confined to brief bouts of face–wiping or hindleg scratching.

(xiii) ELIMINATION: Both micturition and defaecation have been included within this category. These behaviour patterns are characterized by a crouched posture in which the head and ventrum are held close to the substrate and the tail is raised vertically.

(xiv) CLOACAL DRAGGING: This behaviour often consists of only a brief depression of the cloaca to the substrate. More commonly the depression is accompanied by a forward shuffling movement as the animal moves forward sitting on its haunches. The tail is usually slightly raised (Fig. 1c).

TRANSITION ANALYSIS: Transition analysis is a common method of analysing sequences of behaviour. The frequency with which one behaviour pattern is preceded or followed by another is summed into a matrix and then the total observed transitions or pairings between particular behaviour patterns are tested for significance. It is generally considered that the occurrence of two behaviour patterns close together in time indicates that they share common causal factors, which may be internal or external to the animals. Where internal causation is of interest it is assumed that the underlying motivation will change more slowly than the overt behaviour.

A problem with transition analysis of encounter data is that associations between behaviour patterns within an individual may be due to changes in the behaviour of an opponent rather than reflecting motivational changes. To overcome this difficulty I have scored each transition from one behaviour pattern to another according to whether there was no change in the behaviour of the opponent, or whether the opponent changed his behaviour after the first behaviour pattern of the transition commenced and before the second commenced. These transitions were then summed separately for individual animals over all encounters. The relative importance of within individual effects and between individual effects on the frequency of occurrence of the transitions was tested using the Wilcoxon matched–pairs test. There are three possible results from this test : (a) a transition is significantly more common when there is a change in the behaviour of the opponent; (b) a transition is significantly more common when there is no change in behaviour of the opponent; or (c) there is no significant difference in frequency of occurrence of the transition when there is no change or a change in the opponent's behaviour. Thus (a) will indicate that internal causal factors may be responsible for the transition, (b) will indicate an opponent or interaction effect on the transition, and (c) will indicate that both effects are of equal (or no) importance.

The advantages of this method are that it circumvents many of the problems of transition analysis pointed out by Slater (1973). Each individual contributes equally to the analysis; there is no need to compare transition frequencies with a random model using χ^2 or an equivalent test (e.g. Rechten and Fernald 1979); and the problem of repetitions of the same behaviour pattern distorting a transition matrix is avoided. The main disadvantage of the method is that it provides only a relative and not an absolute measure of the importance of within and between–individual effects. A further disadvantage, which is common to transition analysis methods, is that it gives no temporal information. Thus a transsition could show a strong within–individual effect because it is of such short duration that there is no time for the opponent to change his behaviour. It is therefore important to supplement transition analysis with analysis of the time intervals (gaps) between behaviour patterns.

COMMUNICATION ANALYSIS: Smith (1965) has pointed out that the function of a signal is to provide information on the state of the central nervous system of the communicator. Thus, the information content of a particular signal can be inferred from the effect it has on the behaviour of the recipient.

Transition analysis has therefore also been used in this study to examine the effect of the behaviour of one individual on the behaviour of another. Thus the interaction A1 – B2 – C1 – D2 would provide the three transitions A – B, B – C, and C – D. In this analysis the transitions for each individual in the first series of encounters have been summed separately and I have only considered transitions involving the threat displays, *pant–hiss* and *chatter*. A *no response* category was included in addition to the 14 behaviour categories and indicated that an animal was stationary in a crouched or sitting posture. The observed transition frequencies were compared with expected values calculated from the overall transition matrix. The expected value is the cross-product of row and column totals divided by the total number of transitions. Slater and Ollason's

(1972) technique for comparing observed and expected values was used. An index was calculated for each animal by subtracting the expected from the observed value. In a random model the mean value of this index should be zero. To test for significant departures from zero, the standard error of the index for each transition was calculated and multiplied by Student's t (for $p = 0.05$, and n–1 degrees of freedom). If the mean was greater than this figure it indicated a significant departure from the random model.

CONTEXT AND STRUCTURAL VARIATION OF VOCALIZATIONS: A second series of 28 male-male encounters was staged in a circular observation arena about 1m in diameter. Eight males were paired randomly in a "round–robin" sequence, and no individual was used on successive days. Five of these animals had been used in the first series of encounters, one wild caught animal had not been used previously, nor were two animals which had been raised in the laboratory. The arena was cleaned after each encounter. Sounds produced during these encounters were recorded on magnetic tape using a Nagra IIIB tape recorder and Beyer M88 microphone with a Grampian 60 cm diameter parabolic reflector mounted on a tripod. Recordings were made at a tape speed of 3.75 inches per second. Descriptions of the context of the sounds (i.e. the vocalizing animal's orientation, behaviour and distance from the opponent) were simultaneously recorded on a cassette recorder and later transcribed.

Sounds recorded were later numbered individually by speaking directly onto the tape, using a two–channel Tandberg tape recorder. The sounds were then played at normal or half speed from the Tandberg recorder into a Kay Electric Company Sonagraph (Model 6061A). The "high–shape" and "narrow band" filter settings were used throughout. Although the parabolic reflector acts as a high–pass filter cutting off at 1 kHz (Little 1964) a comparison of sonograms of calls recorded with and without the reflector showed no discernible differences. Precise measurements of the frequency and duration of calls were made from the sonograms using a grid marked on a clear plastic overlay. In the analysis of chatter vocalizations all notes separated by a period of less than one second were considered to be within a call and gaps of more than 1 second were defined as between call intervals.

EXPERIMENTAL PLAYBACK OF VOCALIZATIONS: Vocalizations selected for playback were copied onto a tape loop and played back at the recording speed of 3.75 in/sec. on a Uher 4000 L portable tape recorder. A Wharfedale speaker was attached to one end of an observation cage outside a 6 cm diameter circular hole. An 18 X 12 cm mirror was mounted inside the cage directly beneath the speaker hole. Eight animals (5 males, 3 females) were tested. For each trial one animal was placed in one half of the cage (distal to the speaker/mirror) and allowed one hour to become accustomed to the cage. A partition was then removed and recording commenced immediately the animal had approached to within 45 cm of the speaker. Each playback trial consisted of a 2 minute test period during which the sound was continuously played back, and a 2 minute control period during which there was no sound playback. The order of presentation of test and control periods was alternated so that for each sound, half of the animals were presented with the control period first, and half were presented with the experimental period first. Sequences of behaviour were recorded with an Esterline Angus Event Recorder. The behaviours recorded were *approach, investigation, withdrawal, retreat, grooming non–agonistic behaviour, elimination* and *cloacal dragging*. A mirror was used in addition to the sound playback to provide a fixed reference point for both visual and auditory localization. Behaviours such as *approach, investigation, withdrawal* and *retreat* could then be defined with respect to this reference point and so a more meaningful comparison with the encounter was possible. Differences between test and control periods were compared with Wilcoxon's matched–pairs test.

VISUAL MODIFICATION OF THE THREAT DISPLAYS: The same eight females which were used in the initial series of encounters were selected for this experiment. The tails of four of these animals were dyed to the base with "Durafur" black dye. Animals were then randomly paired to give a "round–robin" sequence where each dyed animal was paired with each un–dyed one. A total of 16 encounters between dyed and un–dyed animals was run. As a control series of encounters, the 16 duplicate encounters between the same animals in the initial encounter series were used. Differences between experimental and control encounters were compared by a Wilcoxon matched–pairs test.

RESULTS

TRANSITION ANALYSIS: Sixty–six male–male encounters and 28 female–female encounters were analysed. The results are given in Table 1. A comparison of the frequency of the two types of significant transition indicates that within – individual effects are more pronounced in transitions with *chatter* and between individual effects more pronounced in transitions with *pant–hiss*. This suggests that *chatter* might reflect a stronger motivational basis, whereas *pant–hiss* is more

Table 1. Results of Wilcoxon tests for within-individual transitions. + indicates the transition is significantly more common when there is a change in the behaviour of the opponent; O indicates the transition is significantly more common when there is no change in the behaviour of the opponent; NS indicates there is no significant difference in frequency of occurrence of the transition when there is a change or no change in the opponent's behaviour.

Transition	Preceding or following behaviour pattern													
	AP	IN	AT	PH	CH	WI	RE	FL	CS	FI	NA	GM	EL	CD
Males														
Pant-hiss precedes:	NS	NS	NS	+	+	NS	NS	NS	NS	NS	NS	NS	NS	NS
Pant-hiss follows:	O	NS	NS	+	+	NS	NS	NS	NS	NS	+	NS	NS	NS
Chatter precedes:	NS	NS	NS	+	NS	NS	NS	O	NS	NS	NS	NS	NS	NS
Chatter follows:	NS	NS	NS	+	NS	NS	NS	O	NS	O	NS	NS	NS	NS
Females														
Pant-hiss precedes:	NS	NS	O	+	NS	NS	+	NS	NS	NS	+	NS	NS	NS
Pant-hiss follows:	NS	NS	+	+	NS	NS	+	NS	NS	NS	+	NS	NS	NS
Chatter precedes:	NE	NS	NS	NS	NS	NS	NS	O	NS	O	NS	NS	NS	NS
Chatter follows:	NS	NS	NS	NS	NS	NS	NS	O	NS	O	NS	NS	NS	NS

Abbreviations: AP – *Approach*, IN – *Investigation*, AT – *Attack*, PH – *Pant-hiss*, CH – *Chatter*, WI – *Withdrawal*, RE – *Retreat*, FL – *Fleeing*, CS – *Chase*, FI – *Fighting*, NA – *Non-agonistic behaviour*, GM – *Grooming*, EL – *Elimination*, CD – *Cloacal dragging*.

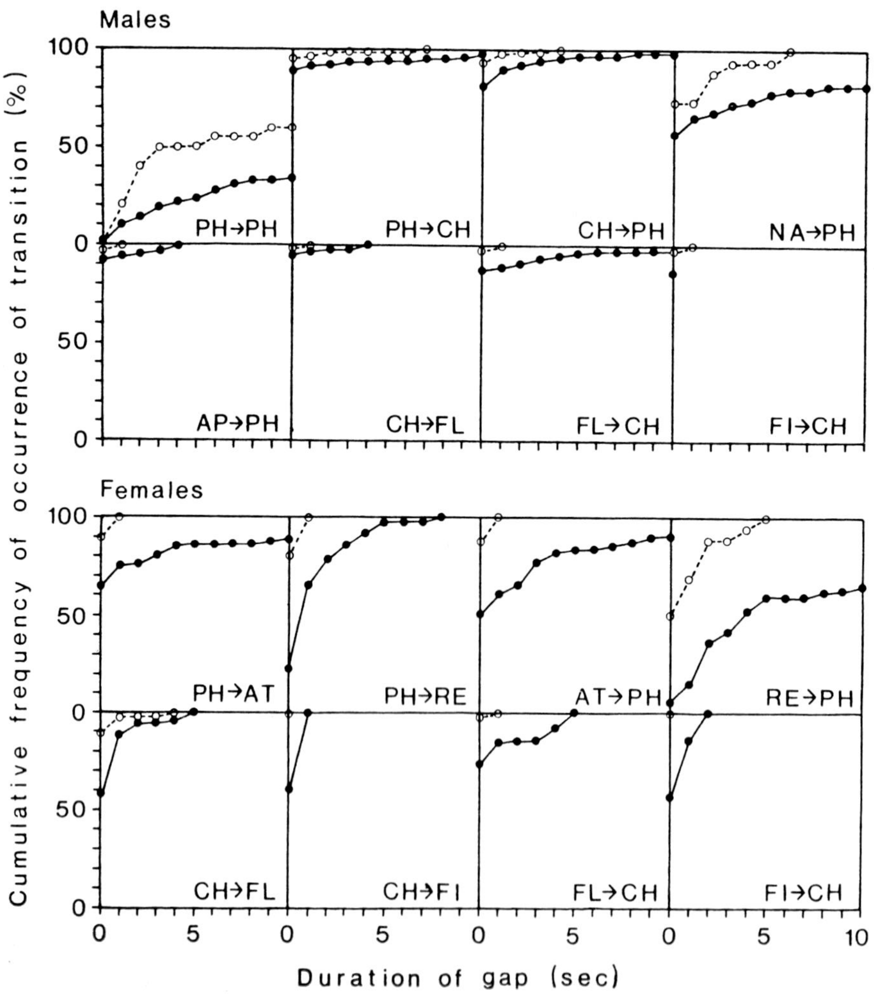

Fig. 2. Cumulative frequency distributions of gaps between significant transitions. (o- - -o) Transitions followed by no change in behaviour of the opponent; (●——●) transitions followed by a change in behaviour of the opponent. Abbreviations: AP *Approach*, AT – *Attack*, PH – *Pant-hiss*, CH – *Chatter*, RE – *Retreat*, FL – *Fleeing*, FI – *Fighting*, NA – *Non-agonistic behaviour*. There were insufficient data to plot graphs for PH → NA and PH → PH transitions in females

sensitive to changes in the behaviour of the opponent.

Pant–hiss appears to be an offensive display. In males *pant–hiss* followed *approach* and in females *pant–hiss* preceded *attack* significantly often when there was no change in the opponent's behaviour. This suggests that *pant–hiss* is associated with an aggressive (i.e. attacking) motivation. All other significant transitions were related to change in the opponent's behaviour, e.g. transitions with *non–agonistic behaviour* in males and females and transitions with *retreat* in females. *Pant–hiss* was also often repeated in reponse to an opponent's change in behaviour, by both males and females.

In contrast to *pant–hiss, chatter* appears to be a defensive display. In both males and females *chatter* was strongly associated with *fighting* and *fleeing* when there was no change in the opponent's behaviour. Thus, the underlying motivation may be to flee, but if attacked the animal will fight back.

The most marked difference between male and female encounters appears to be the greater frequency of transitions between *pant–hiss* and *chatter* in males, in response to a change in the opponent's behaviour.

GAPS BETWEEN TRANSITIONS: There are two possible explanations for a transition between two behaviour patterns occurring in response to no change in the opponent's behaviour. The first is that the transition takes place so quickly that there is insufficient time for the opponent to react, and the second is that there is sufficient time for the opponent to respond, but the opponent does not do so (Halliday 1975). To examine this possibility I have pooled data from individual animals and plotted cumulative frequency distributions of the gaps between all the significant transitions shown in Table 1. Thus for any given duration the graphs will indicate the proportion of transitions completed (Fig. 2). In all cases, transitions in response to no change in the opponent's behaviour were completed more quickly, i.e. there were shorter gaps between the two behaviour patterns, than transitions in response to a change in the opponent's behaviour. This indicates that transitions which are significantly common when there is no change in behaviour occur too quickly for the opponent to respond. Inspection of the graphs for these transitions (e.g. ♀♀ PH → AT, ♂♂ AP → PH) shows that the maximum gap is usually 1 sec or less. The graphs also illustrate a difference between the two displays. Gaps between transitions involving *chatter* are generally much shorter than between transitions involving *pant–hiss*.

DISPLAY DURATION: In addition to the gap between a transition, the duration of the display itself could determine whether an opponent has sufficient time to react to the transition. In Fig. 3 I have plotted cumulative frequency distributions of display duration according to whether the display was the first or second behaviour pattern in the transition. The data from individual animals have been pooled for all transitions involving a display. The effect of an opponent changing its behaviour can be examined by comparing the graphs for no change or a change in

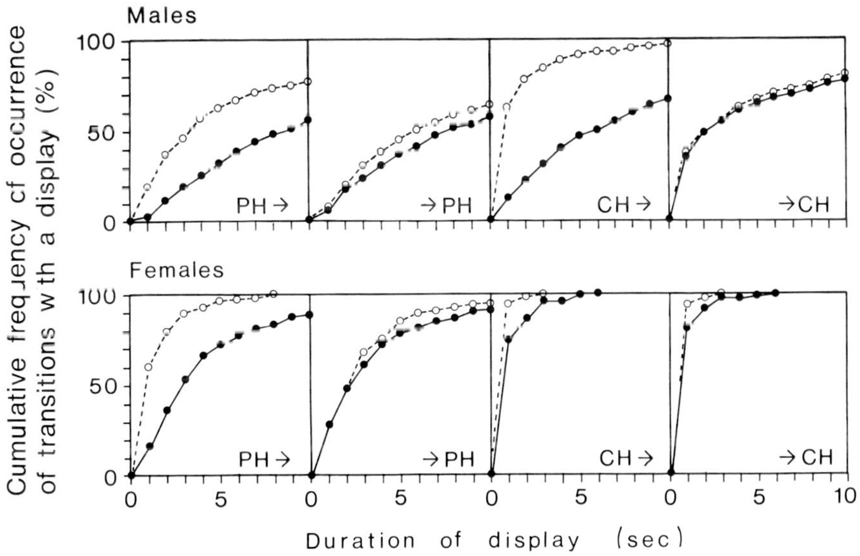

Fig. 3. Cumulative frequency distributions of display duration for all transitions involving a display. (o- - -o) Transitions followed by no change in behaviour of the opponent; (●——●) transitions followed by a change in behaviour of the opponent.

the opponent's behaviour. When the display was the second behaviour in a transition there was little difference between the two graphs. Thus, a change in the opponent's behaviour does not affect display duration. This result would be predicted by a simple interactive model of communication between two animals where the duration of a behaviour will depend on what the opponent does next rather than what he has done previously. However, when a display is the first behaviour in a transition a change in the opponent's behaviour does affect display duration. Displays are shorter when there is no change and longer when there is a change in the opponent's behaviour. When considered together with the results of the gap analysis this suggests that transitions which are significantly more common in response to no change in an opponent's behaviour occur too quickly for the opponent to react and therefore presumably are the result of internal motivational factors. In contrast, in transitions which are more common in response to a change in behaviour the display may be prolonged or there may be longer gaps which give the opponent a greater opportunity to react.

The main difference between male and female encounters revealed by this analysis is that males tend to have longer displays. Variation in dislay duration will be considered in more detail later.

Effect of Displays on Behaviour of the Opponent: The results of the communication analysis are given in Table 2. *Pant–hiss* was positively associated with *approach, withdrawal, non–agnostic behaviour, elimination* and *cloacal dragging,* and negatively with *no response, chatter, fleeing, chase,* and *fighting.* Thus *pant–hiss* is given in response to non–agonistic rather than agonistic behaviour of the opponent, which suggests that it may function as a warning to an opponent to keep away. *Chatter,* on the other hand, was strongly associated with *chase* and *fighting* in males and females, and in males also with *approach, attack, withdrawal, retreat and cloacal dragging.* The fewer significant transitions in females is probably due to the smaller sample size. Thus *chatter* appears to be a response to overt aggression (especially *attack, chase* and *fighting*) of an opponent. This analysis supports the interpretation of display motivation based on transition analysis of the individual transitions. *Pant–hiss* appears to be an offensive threat display and functions as an initial warning to an opponent to withdraw. *Chatter* appears to be a defensive threat display given in response to an attacking animal and presumably functions to deter further attacks. The significant transitions with *withdrawal* and *retreat* in males indicate that the display is often effective in this function.

Table 2. Results of transition analysis of between–individual transitions. + indicates the transition is significantly more common than expected in comparison with a random model; O indicates there is no significant difference in the frequency of the transition; — indicates the transition is significantly less common than expected on a random basis.

Transition		Preceding or following behaviour pattern of opponent														
		O	AP	IN	AT	PH	CH	WI	RE	FL	CS	FI	NA	GM	EL	CD
Males	*Pant–hiss* precedes:	O	+	O	O	O	—	O	O	—	—	—	+	O	+	O
	Pant–hiss follows:	—	O	O	O	O	—	+	O	—	—	—	+	O	O	+
Females	*Pant–hiss* precedes:	—	+	O	O	O	O	+	O	—	—	O	O	O	O	+
	Pant–hiss follows:	—	+	O	O	O	O	+	O	—	O	—	—	O	O	+
Males	*Chatter* precedes:	—	O	—	+	—	—	+	+	—	+	+	O	O	O	+
	Chatter follows:	—	+	—	+	—	—	O	O	—	+	+	O	O	—	+
Females	*Chatter* precedes:	O	O	O	O	O	O	O	O	O	+	+	O	O	O	O
	Chatter follows:	O	O	O	O	O	O	O	O	O	+	+	O	—	O	O

Abbreviations: O – *No response,* AP – *Approach,* IN – *Investigation,* AT – *Attack,* PH – *Pant–hiss,* CH – *Chatter,* WI – *Withdrawal,* RE – *Retreat,* FL – *Fleeing,* CS – *Chase,* FI – *Fighting,* NA – *Non–agonistic behaviour,* GM – *Grooming,* EL – *Elimination,* CD – *Cloacal dragging.*

Table 3. Frequency of occurrence of displays (per cent) in different contexts.

	Stationary	Moving towards opponent	Moving away from opponent	*Fleeing*	*Fighting*
Pant–hiss	77.0 ± 6.5	15.0 ± 5.8	8.0 ± 2.4	0	0
Chatter	42.4 ± 13.8	2.1 ± 0.6	2.0 ± 0.7	37.2 ± 9.3	16.5 ± 5.9

Mean values are shown ± SE. N = 5 animals for *pant–hiss,* N = 6 for *chatter.* At least 30 displays were recorded from each animal.

Table 4. Frequency of occurrence (per cent) of static displays according to orientation to the opponent.

	Display orientation		
	Towards opponent	Away from opponent	To either side of opponent
Pant–hiss	73.8 ± 4.4	10.2 ± 4.0	16.0 ± 6.9
Chatter	88.1 ± 8.9	1.8 ± 0.7	10.2 ± 8.9

Mean values are shown ± SE. N = 5 for *pant–hiss*, N= 4 for *chatter*.

Table 5. Frequency of occurrence (per cent) of static displays according to distance of displaying animal from the opponent.

	Distance from opponent (cm)					
	0 – 10	10 – 20	20 –30	30 – 40	40 – 50	50 – 100
Pant–hiss	34.3 ± 8.8	8.3 ± 2.3	2.2 ± 0.9	12.2 ± 3.4	6.4 ± 2.0	36.6 ± 8.0
Chatter	71.5 ± 8.8	7.9 ± 2.4	1.0 ± 0.6	6.3 ± 2.3	2.5 ± 1.1	10.9 ± 5.2

Mean values are shown ± SE. N = 5 for *pant–hiss*, N= 4 for *chatter*.

CONTEXT OF THE DISPLAYS: In addition to the motivational basis of a display the context in which it is performed could affect the information transmitted to an opponent. Most *pant–hiss* displays were performed when stationary whereas *chatter* was common in three different situations –stationary, *fleeing* and *fighting* (Table 3). Both displays had a similar static orientation and were directed towards, rather than to either side or away from an opponent (Table 4). The distance of a displaying animal from an opponent may also form an important part of the context of static displays. *Pant–hiss* was frequently performed at varying distances from an opponent, but especially in close proximity (< 10 cm) and also at some distance (50 – 100 cm) from an opponent. *Chatter* occurred most frequently in close proximity to an opponent (Table 5).

Thus, *pant–hiss* occurred in a more fixed context than *chatter,* and was usually a static display directed towards an opponent over a wide range of distances. *Chatter* occurred commonly in three contexts – as a static display directed towards an opponent in close proximity, during *fighting* and during *fleeing* away from an opponent.

Three types of vocalization were recorded from both males and females during the agonistic encounters of *Dasyuroides* : *sniffing, pant–hiss* and *chatter*. *Sniffing* has not been analysed in detail as it appeared to have no function in communication.

SNIFFING: This sound resulted from the rapid expiration of air through the nostrils and was composed of a series of notes repeated at regular intervals (Fig. 4). The duration of an individual note varied from 0.05 to 0.11 sec and occupied a frequency range from 0 to 15 kHz. More usually the maximum frequency of a note was about 10 kHz, with most of the energy between 1.5 and 8 kHz. The interval between notes varied from 0.13 to 0.20 sec. The notes were usually repeated at a rate of about 5 per second, and might rise and fall in intensity with repetition.

Ewer (1968b) has described a similar low intensity "di–di–di–di" sound when a male *Antechinus* or *Sminthopsis* is chasing a female and has suggested that it is a reflection of the male's highly excited condition and without signal value. In *Dasyuroides*, the sound has been recorded from both sexes during the *chase* of an opponent, and occasionally during *fighting*. The sound appears to have no effect on the behaviour of the opponent and may be associated with a high level of arousal.

Table 6. Individual variation in duration and frequency range of *pant–hiss* vocalizations.

Animal number	Number of calls	Mean duration (sec)	Mean frequency range (kHz)
3	20	0.51 ± 0.02	6.8 ± 0.3
4	56	0.58 ± 0.01	7.7 ± 0.2
6	162	0.52 ± 0.01	4.8 ± 0.3
9	41	0.50 ± 0.02	6.4 ± 0.3
11	116	0.46 ± 0.01	6.5 ± 0.1

Means are shown ± SE. Duration: $F = 8.31$, df =4, 390, $P < 0.001$; Frequency range: $F = 22.88$, df = 4, 390, $P < 0.001$.

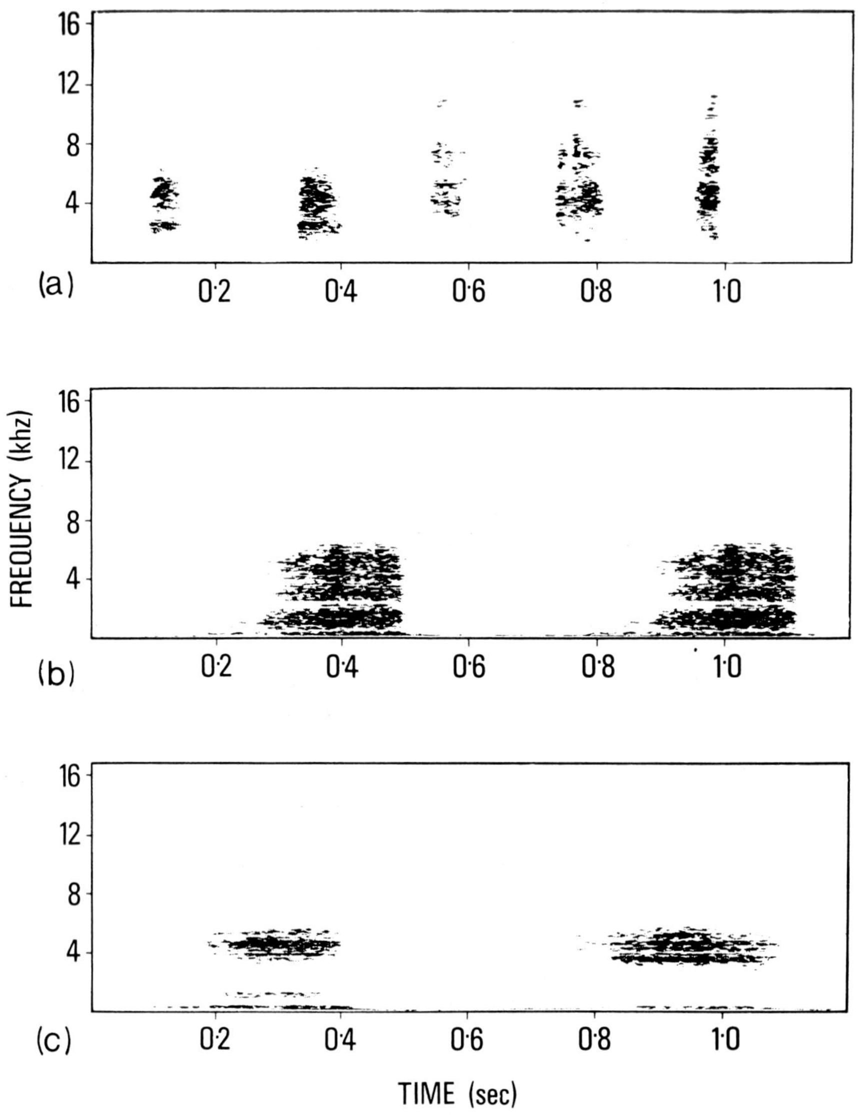

Fig. 4. Sonograms of (a) *sniffing* recorded during a *chase*, (b) *pant–hiss* recorded in close proximity (<15 cm) to an opponent, and (c) *pant–hiss* recorded some distance away (>15 cm) from an opponent.

PANT–HISS: *Pant–hiss* is a low intensity breathing sound usually uttered with the mouth open, but sometimes with the mouth closed (Fig. 4). The duration of the sound varied from 0.10 to 0.95 sec and the frequency range from 0 to 11 kHz. There was significant variation between individuals in both the duration and frequency range of the sound (Table 6). The average duration of a single *pant–hiss* was about 0.5 sec and most of the energy was between 0.5 and 7.0 kHz, with peaks between 0.4 and 0.7 kHz, 1.0 and 2.5 kHz, and 5.0 and 6.5 kHz. During low intensity calling the higher frequency components of the sound were often omitted and only the low frequency band (0.4 to 0.7 kHz) was visible on the sonograms. During high intensity calling a fourth band between 3.0 and 4.0 kHz could be discerned. The *pant–hiss* vocalization was often repeated at regular intervals and might change in intensity with successive renditions. Also, the rate of repetition was highest when an opponent was in close proximity (< 15 cm), but as the opponent moved away the duration of the intervals between successive *pant–hisses* increased. In a sample of 103 calls from ♂ 6 where the interval between successive calls could be measured from sonograms, the mean interval when the opponent was within 15 cm was 0.28±0.02 sec, compared with an interval of 0.52±0.03 sec when the opponent was more than 15 cm away ($F = 44.3$, d.f. = 1,101, $p<0.001$). The structure of the call may also change according to the proximity of the opponent. Table 7 shows the effect of opponent distance on duration and frequency

Table 7. Structural variation in *pant-hiss* vocalizations.

Animal number	Distance to opponent (cm)	Note duration (sec)	P	Frequency range (kHz)	P
6	<15	0.55 ± 0.02	<0.001	5.5 ± 0.3	<0.01
	>15	0.47 ± 0.01		3.9 ± 0.4	
9	<15	0.56 ± 0.02	<0.001	6.9 ± 0.3	<0.05
	>15	0.41 ± 0.03		5.5 ± 0.5	
11	<15	0.54 ± 0.03	<0.05	6.2 ± 0.7	NS
	>15	0.45 ± 0.13		6.5 ± 0.1	

range of *pant–hiss* calls in three animals. Insufficient calls were recorded from ♂3 and ♂4 to be included in this analysis. In close proximity to an opponent the sound was of longer duration and covered a wider frequency range than when the opponent was more than 15 cm away.

CHATTER: This vocalization is a rapid series of loud, harsh sounds uttered with the mouth open. The individual notes of the call might be repeated at regular or irregular intervals, but were usually uttered at the same level of intensity. *Chatter* was recorded frequently in three separate contexts (Fig. 5) : during *fleeing, fighting* and when directed towards an opponent in close proximity (<15 cm).

There was significant individual variation in both the duration and the frequency range of *chatter* (Table 8). The call ranged in frequency from 0 to 18 kHz, but more usually this range was restricted to 0 to 14 kHz, with most of the

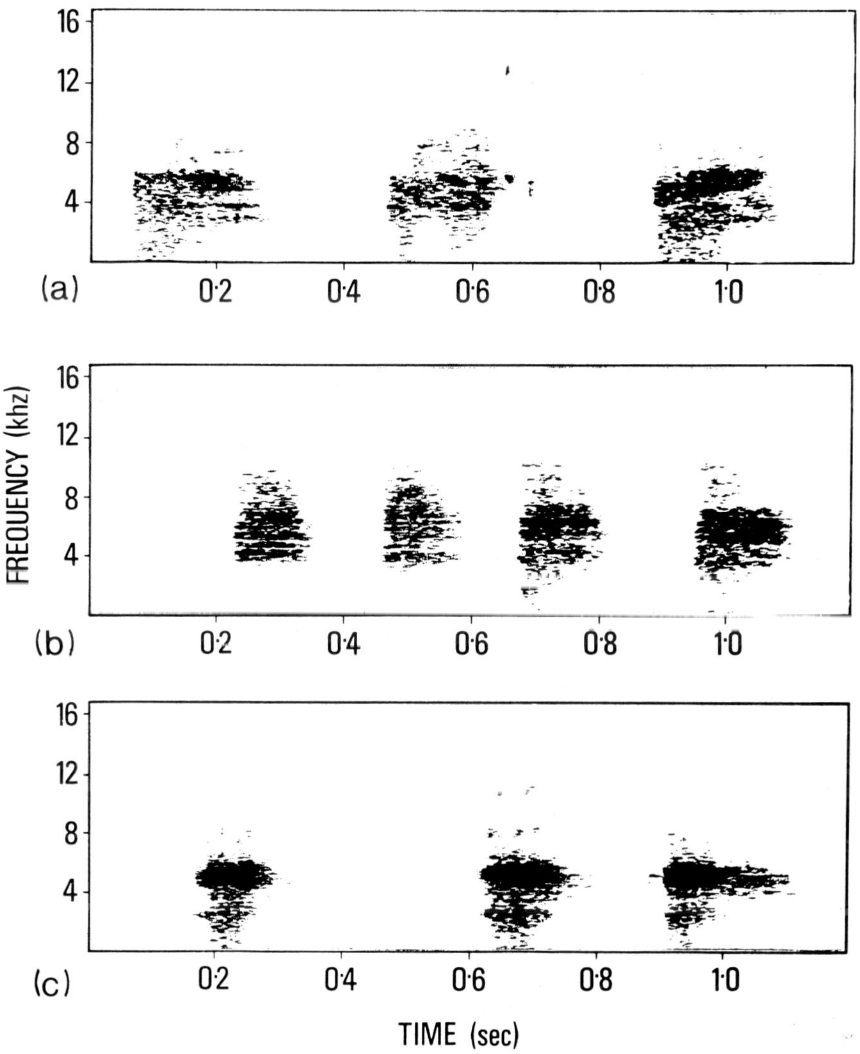

Fig. 5. Sonograms of (a) *chatter* recorded when directed towards an opponent in close proximity, (b) *chatter* recorded during *fighting,* and (c) *chatter* recorded during *fleeing.*

Table 8. Individual variation in duration and frequency range of *chatter* vocalizations.

Animal number	Number of calls	Mean duration (sec)	Mean frequency range (kHz)
1	75	0.66 ± 0.13	12.9 ± 0.3
3	131	2.46 ± 0.31	13.7 ± 0.1
6	56	2.37 ± 0.33	12.8 ± 0.3
11	25	1.40 ± 0.38	11.8 ± 0.6
12	143	0.65 ± 0.05	12.7 ± 0.2
13	66	1.22 ± 0.15	13.6 ± 0.2

Means are shown ± SE. Duration: $F = 13.81$, $df = 5,490$, $P < 0.001$; Frequency range: $F = 6.28$, $df = 5,490$, $P < 0.001$.

Table 9. Structural variation in *chatter* vocalizations.

Context	Number of notes per call	Note duration (sec)	Interval duration (sec)	Frequency range (kHz)
Towards (4)	6.2 ± 0.7	0.16 ± 0.01	0.12 ± 0.02	13.0 ± 0.5
Fighting (5)	5.1 ± 0.6	0.15 ± 0.00	0.09 ± 0.01	14.2 ± 0.5
Fleeing (6)	3.5 ± 0.6	0.13 ± 0.01	0.13 ± 0.01	12.6 ± 0.3
P	<0.05	<0.025	<0.10	<0.05

Mean values are given ± SE. Sample sizes are shown in parentheses.

energy between 4 and 8 kHz. The variation in call duration was primarily the result of a variable number of notes per call.

Although *chatter* is an extremely variable vocalization the structure of the call appears to depend on the context in which *chatter* occurs. A mean value was calculated for each animal for each call parameter measured and the means for the three contexts compared by analysis of variance (Table 9). *Chatter* which was directed towards an opponent in close proximity was composed of a large number of notes of long duration which were also separated by intervals of long duration. The notes occupied a wide frequency range. *Chatter* which occurred during fighting had a structure of a large number of fairly long notes with a wide frequency range, but very short intervals separated these notes. During *fleeing,* there were a small number of notes per call and the notes were of short duration and were separated by fairly long intervals. However, in this context, the call occupied a narrow range of frequencies.

When considered together with the variation in *pant–hiss* these results suggest that the structure of the vocalizations can be characterized by two variables – rate of note repetition and harshness (the frequency range and duration of individual notes). These two variables appear to vary independently and may be good indicators of underlying motivation, with harshness related to an attacking tendency and rate of calling related to a fleeing tendency.

INTERMEDIATE VOCALIZATIONS: The *pant–hiss* and *chatter* vocalizations were usually quite discrete sounds. However, occasionally the two intergraded to form an intermediate sound. This intergradation was rare and has only been recorded when the sound was directed towards an opponent in close proximity. The intermediate sound was not a stable, discrete vocalization, and usually took the form of a single note which was then commonly followed by *Chatter.* Some examples of these intermediate sounds are shown in Fig. 6. In each case the sound resembled a *pant–hiss* in that it was of fairly long duration (0.20 – 0.40 sec), but the low frequencies and discrete energy bands characteristic of *pant–hiss* were absent. Most of the energy of the sound was between 2.0 and 6.0 kHz, which closely resembled the intensity emphasis in the following *chatter* call.

RESPONSES TO EXPERIMENTAL PLAYBACK OF VOCALIZATIONS: Analysis of the context and structure of the displays has revealed a surprising amount of variability in the displays. To test whether the information contained in these displays is communicated to an opponent, the effects of different vocalizations on behaviour were compared in a series of playback experiments. Six sounds were used : (1) *sniffing,* (2) *pant–hiss* recorded in close proximity (<15 cm) to an opponent, (3) *pant–hiss* recorded more than 15 cm away from an opponent, (4) *chatter* recorded in close proximity, (5) *chatter* recorded during *fleeing,* and (6) *chatter* recorded during fighting. Playback of these sounds in a controlled situation eliminates much of the variation in response due to the context of the call and so permits a finer comparison of the effects of signal variation.

Apart from two occasions when an animal attempted to bite at the speaker, no behaviour patterns other than those listed in Tables 10 and 11 were observed during the playbacks. Thus, there was a conspicuous absence of many of the agonistic behaviours which followed the displays during the interactions. Some of these behaviours (e.g. *chase, fighting*) obviously cannot occur in the playback situation. The absence of *attack* and the displays themselves may be a result of the incomplete stimulus situation. The visual stimulus provided by a mirror image does not closely resemble the posture normally used in a display. However, even with this limitation, the results obtained indicate the different effects of the sounds.

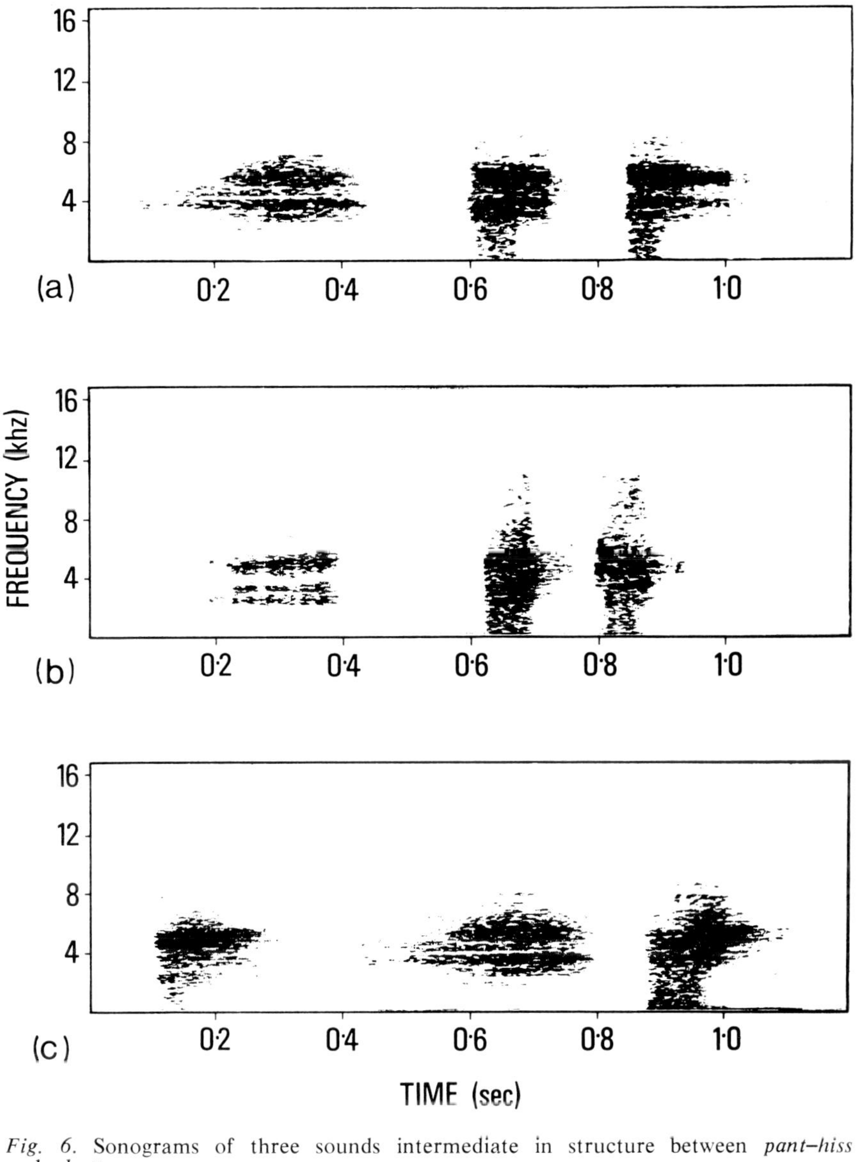

Fig. 6. Sonograms of three sounds intermediate in structure between *pant–hiss* and *chatter.*

All sounds except *sniffing* and *chatter* in close proximity resulted in a significant increase in the frequency of *withdrawal* during the playback period (Table 10). For total duration of the behaviours (Table 11), there was a significant increase in the time spent *withdrawing* in response to both the *pant–hiss* calls and in response to *chatter* recorded during *fleeing* and *fighting*. Also there was a significant decrease in the total duration of *non–agonistic behaviour* during the test periods for *pant–hiss* recorded in close proximity and *chatter* recorded during *fleeing* and *fighting*.

Table 10. Responses to playback: median frequency of occurrence of different behaviour patterns during test and control periods. Ranges are given in parentheses.

	Sniffing			Pant-hiss (Towards >15 cm)			Pant-hiss (Towards <15 cm)			Chatter (Towards <15 cm)			Chatter (During fleeing)			Chatter (During fighting)		
	Test	Control	P	Test	Control	P	Test	Control	P	Test	Control	P	Test	Control	P	Test	Control	P
Approach	4 (0-6)	2 (0-6)	NS	9 (1-14)	6.5 (0-11)	NS	3.5 (0-10)	2 (0-4)	NS	2.5 (0-7)	1.5 (0-7)	NS	2 (0-3)	1 (0-5)	NS	4.5 (0-11)	4 (1-6)	NS
Investigation	3.5 (0-10)	3 (0-5)	NS	2.5 (0-10)	1 (0-7)	NS	3.5 (0-9)	2 (0-4)	NS	1 (0-2)	1.5 (0-5)	NS	1 (0-3)	1.5 (0-6)	NS	3 (0-7)	2.5 (0-5)	NS
Withdrawal	1 (0-4)	1 (0-4)	NS	8 (1-13)	2.5 (0-9)	<0.02	2 (1-9)	0 (0-2)	<0.01	1 (0-5)	0 (0-5)	NS	1.5 (0-4)	0.5 (0-2)	<0.05	4.5 (0-11)	1 (0-2)	<0.02
Retreat	0 (0-2)	0 (0-1)	NS	1 (0-5)	0.5 (0-5)	NS	1 (0-2)	0 (0-1)	NS	1.5 (0-7)	0 (0-5)	NS	1.5 (0-3)	0 (0-1)	NS	1.5 (0-3)	0 (0-3)	NS
Grooming	0 (0-2)	0 (0-5)	NS	0 (0-1)	0 (0-2)	NS	0 (0-3)	0.5 (0-4)	NS	0 (0-2)	0 (0-2)	NS	1 (0-1)	1.5 (0-8)	NS	0 (0-3)	0.5 (0-3)	NS
Non-agonistic behaviour	5 (1-8)	5 (0-8)	NS	2 (0-6)	2.5 (0-8)	NS	5.5 (0-9)	5.5 (0-8)	NS	4 (0-8)	4.5 (1-10)	NS	4.5 (2-10)	5.5 (1-9)	NS	4 (2-10)	5 (4-7)	NS
Elimination	0 (0-0)	0 (0-2)	NS	0 (0-1)	0 (0-0)	NS	0 (0-0)	0 (0-0)	NS	0 (0-1)	0 (0-1)	NS	0 (0-0)	0 (0-1)	NS	0 (0-0)	0 (0-0)	NS
Cloacal dragging	0 (0-1)	0 (0-4)	NS	0 (0-1)	0 (0-2)	NS	0 (0-0)	0 (0-5)	NS	0 (0-1)	0 (0-5)	NS	0 (0-2)	0 (0-2)	NS	0 (0-2)	0 (0-1)	NS

Table 11. Responses to playback: median total duration of different behaviour patterns during test and control periods. Ranges are given in parentheses.

	Sniffing			Pant-hiss (Towards >15 cm)			Pant-hiss (Towards <15 cm)			Chatter (Towards <15 cm)			Chatter (During fleeing)			Chatter (During fighting)		
	Test	Control	P	Test	Control	P	Test	Control	P	Test	Control	P	Test	Control	P	Test	Control	P
Approach	6 (0-13)	2.5 (0-15)	NS	30 (2-46)	13.5 (0-67)	NS	4.5 (0-21)	3.5 (0-7)	NS	4.5 (0-17)	3 (0-10)	NS	4.5 (0-7)	2 (0-10)	NS	15.5 (0-31)	9 (3-20)	NS
Investigation	13 (0-44)	9 (0-28)	NS	18 (0-56)	4.5 (0-54)	NS	8.5 (0-45)	5.5 (0-9)	NS	0.5 (0-6)	3 (0-21)	NS	1 (0-11)	4.5 (0-33)	NS	10 (0-26)	19.5 (0-42)	NS
Withdrawal	1 (0-6)	1 (0-6)	NS	13 (1-23)	3 (0-22)	<0.01	2 (1-8)	0 (0-1)	<0.01	2 (0-5)	0 (0-10)	NS	1.5 (0-3)	0.5 (0-1)	<0.05	10 (0-18)	1 (0-4)	<0.02
Grooming	0 (0-4)	0 (0-18)	NS	0 (0-1)	0 (0-28)	NS	0 (0-3)	0.5 (0-6)	NS	0 (0-1)	0 (0-2)	NS	1 (0-1)	2 (0-31)	NS	0 (0-2)	0.5 (0-5)	NS
Non-agonistic behaviour	72 (5-112)	84 (0-108)	NS	6.5 (0-55)	11.5 (0-115)	NS	33.5 (0-83)	79.5 (0-100)	<0.05	24 (0-114)	63.5 (0-117)	NS	40 (4-99)	69.5 (8-112)	<0.01	18.5 (4-71)	67 (42-112)	<0.01
Elimination	0 (0-0)	0 (0-2)	NS	0 (0-1)	0 (0-0)	NS	0 (0-0)	0 (0-0)	NS	0 (0-2)	0 (0-1)	NS	0 (0-0)	0 (0-3)	NS	0 (0-3)	0 (0-0)	NS
Cloacal dragging	0 (0-1)	0 (0-3)	NS	0 (0-1)	0 (0-2)	NS	0 (0-0)	0 (0-3)	NS	0 (0-1)	0 (0-3)	NS	0 (0-1)	0 (0-3)	NS	0 (0-3)	0 (0-3)	NS

Playback of *sniffing* resulted in no significant differences between the test and control periods and suggests that this vocalization does not have an obvious communicatory function. In contrast, all the *pant–hiss* and *chatter* calls, except *chatter* in close proximity, increased the frequency of *withdrawal,* which suggests that these vocalizations do in fact function as threat signals. Further, the effects of the playback of these sounds indicate that the context of the call has an important role in determining subsequent behaviour. Thus *pant–hiss* recorded in close proximity was more likely to inhibit the performance of *non–agonistic behaviour* than *pant–hiss* recorded more than 15 cm away. This suggests that the structural differences of *pant–hiss* uttered in close proximity (greater harshness, increased repetition rate) may reflect a more effective type of threat. Similarly, although all *chatter* calls were followed by an increase in *withdrawal,* the difference between test and control periods was not signficicant for *chatter* in close proximity. Also, only *chatter* uttered during *fleeing* or *fighting* tended to inhibit the performance of *non–agonistic behaviour* during playback. The structure of these calls may also reflect differences in effectiveness as threat signals. *Chatter* uttered during *fighting* is harsher and the notes repeated more rapidly than *chatter* directed towards an opponent. Thus for both *pant–hiss* and *chatter* the rate of note repetition and harshness of the call are greatest in contexts where the calls have the most marked effect.

RESPONSES TO A VISUAL MODIFICATION OF THE THREAT DISPLAYS: The body posture or form which accompanies a display is often as characteristic of it as a vocalization. In *Dasyuroides,* one such easily recognized component of the display postures is the position of the tail. During *pant–hiss* the tail was usually lowered slightly, whereas during *chatter* it was often held in a vertical position. In both displays the long black hairs on the terminal half of the tail were erected to form a conspicuous black brush. In order to test the effect on behaviour of a visual modification to the form of the displays, a series of encounters was run in which the tails of half of the animals were dyed completely black. This procedure should intensify the visual signalling properties of the black tail if indeed it does function in communication. The results are shown in Tables 12 and 13. The most striking result was the dramatic reduction in frequency and duration of *fighting* during the experimental series of encounters. Few other behaviour patterns were affected. In dyed animals there was a reduction in the frequency and duration of *fleeing* and an increase in the frequency of *retreat.* In the undyed animals there was a reduction in the frequency of *grooming* and an increase in the duration of *elimination.*

These results suggest that the black tail of *Dasyuroides* does function as a visual threat signal, and the effect of intensifying this signal is to reduce overt *fighting* behaviour of an opponent. The reduction in *fighting* behaviour appeared to be caused by undyed animals avoiding their opponent, and may have been the result of fortuitous sighting of the black tail during *non–agonistic behaviour* such as exploring.

Table 12. Effect of dyed tails on frequency of occurrence of different behaviour patterns during experimental and control encounters. Median values are shown with ranges in parentheses.

		Behaviour pattern													
		AP	IN	AT	PH	CH	WI	RE	FL	CS	FI	NA	GM	EL	CD
Dyed animals	Control	6 (0-35)	0 (0-3)	6.5 (0-19)	11 (0-41)	2.5 (0-44)	6.5 (1-22)	4.5 (0-16)	1.5 (0-20)	0 (0-3)	1.5 (0-8)	9 (1-21)	5 (0-15)	1 (0-4)	0.5 (0-8)
	Experimental	8.5 (0-56)	0 (0-2)	7.5 (3-22)	18 (2-46)	0 (0-58)	7 (0-34)	10 (1-35)	0.5 (0-3)	0 (0-11)	0 (0-4)	6.5 (0-54)	3 (0-14)	0 (0-2)	1 (0-6)
	P	NS	NS	NS	NS	NS	NS	<0.05	<0.05	NS	<0.01	NS	NS	NS	NS
Un-dyed animals	Control	35 (10-70)	0 (0-6)	3 (0-22)	1 (0-13)	0 (0-1)	25.5 (5-56)	9.5 (0-43)	0 (0-18)	0.5 (0-10)	1.5 (0-8)	30 (0-44)	5 (1-22)	1 (0-3)	4.5 (0-15)
	Experimental	36.5 (3-55)	0 (0-1)	3 (0-14)	1 (0-28)	0 (0-1)	28 (5-48)	10.5 (0-17)	0 (0-10)	0.5 (0-4)	0 (0-4)	33.5 (8-55)	3.5 (0-17)	1 (0-5)	7 (0-14)
	P	NS	NS	NS	NS	NS	NS	NS	NS	NS	<0.01	NS	<0.01	NS	NS

Abbreviations: AP - Approach, IN - Investigation, AT - Attack, PH - Pant-hiss, CH-Chatter, WI - Withdrawal, RE - Retreat, FL - Fleeing, CS - Chase, FI - Fighting, NA - Non-agonistic behaviour, GM - Grooming, EL - Elimination, CD - Cloacal dragging.

Fig. 13. Effect of dyed tails on duration (sec.) of different behaviour patterns during experimental and control encounters. Median values are shown with ranges in parentheses.

		Behaviour pattern												
		AP	IN	AT	PH	CH	WI	FL	CS	FI	NA	GM	EL	CD
Dyed animals	Control	21 (0-91)	0 (0-5)	3.5 (0-12)	31.5 (0-266)	1 (0-32)	14 (1-45)	3 (0-148)	0 (0-14)	1 (0-39)	76 (2-548)	7.5 (0-175)	3 (0-9)	0.5 (0-15)
	Experimental	21 (0-104)	0 (0-5)	4 (1-11)	78 (7-250)	0 (0-43)	12 (0-46)	0.5 (0-11)	0 (0-19)	0 (0-4)	39.5 (0-506)	3.5 (0-36)	0 (0-15)	1 (0-7)
	P	NS	NS	NS	NS	NS	NS	<0.05	NS	<0.01	NS	NS	NS	NS
Un-dyed animals	Control	68 (21-167)	0 (0-7)	2.5 (1-10)	2 (0-162)	0 (0-1)	55.5 (14-104)	0 (0-36)	0.5 (0-106)	1 (0-39)	341.5 (68-615)	8 (1-42)	3 (0-11)	6.5 (0-29)
	Experimental	76.5 (3-114)	0 (0-3)	1 (0-6)	3 (0-193)	0 (0-3)	40.5 (5-77)	0 (0-7)	0.5 (0-13)	0 (0-4)	39.9 (9-609)	4.5 (0-33)	8 (0-21)	8 (0-21)
	P	NS	NS	NS	NS	NS	NS	NS	NS	<0.01	NS	NS	<0.01	NS

Abbreviations: AP - Approach, IN - Investigation, AT - Attack, PH - Pant-hiss, CH - Chatter, WI - Withdrawal, FL - Fleeing, CS - Chase, FI - Fighting, NA - Non-agonistic behaviour, GM - Grooming, EL - Elimination, CD - Cloacal dragging.

Grant, Mackintosh and Lerwill (1970) found that when the black marking on the chest of the golden hamster was intensified with dye, the marked animals showed more aggression and the unmarked animals showed more flight behaviour. In *Dasyuroides,* there was no significant change in the frequency of *fleeing* by unmarked animals, although the reduced amount of *fleeing* and increased frequency of *retreat* by dyed animals may have been a response to less aggression by unmarked animals.

DISCUSSION

Pant–hiss is clearly an offensive threat display which indicates an intention to attack and *chatter* a defensive threat display which indicates an intention to fight back if attacked. Analysis of within individual transitions suggests that the motivational basis of these displays may be the simultaneous presence of two incompatible types of behaviour – *attacking* and *fleeing. Pant–hiss* reflected a high tendency to attack and a low tendency to flee, and *chatter* was associated with a low tendency to attack and a high tendency to flee. This interpretation fits in well with the conflict theory for the causation of threat displays, which has been applied to a diverse range of mammalian species, e.g. the rat (Grant 1963), the cat (Leyhausen 1956), Grant's gazelle (Walther 1965) and various primates (e.g. Hinde and Rowell 1962, van Hooff 1967).

Hinde (1970) has warned against an over–interpretation by implying that all displays, let alone all signals used in intra–specific communication, depend on conflict. In some instances, all aspects of a display are correlated and there is no need to postulate two conflicting tendencies (e.g. Simpson 1968). Other authors (Wallis 1962, Stout and Brass 1969, Stout 1975) have interpreted display as variations along a continuous scale or dimension, without the need to refer to conflicting tendencies. However, in *Dasyuroides* the conflict theory is of help in examining the relationships of the displays to a fleeing tendency, rather than their resulting from say a low level of attack. Thus, although *chatter* is associated with a high tendency to flee, it commonly occurs in situations where an animal will also *attack,* e.g. during *fighting* and defence of a refuge.

Andrew (1956, 1972) has suggested that in descriptions of conflict situations it is more useful to specify which responses are incompatible, rather than an incompatibility between groups of responses or major variables such as an attacking or fleeing tendency. Thus various *Emberiza* spp. flick their tails when there is a conflict between flying towards a food dish at which a feared superior is feeding and also when moving through twigs, when the birds have simultaneous tendencies to move by flight and sidling. In both cases the essential incompatibility is between two locomoting movements, one at least of which involves flight. The conflict theory has also been criticised from a physiological view point. Brown and Hunsperger (1963) elicited threat by direct electrical stimulation of one area in the brain of cats, and, finding no areas in which attack could be elicited without threat, concluded that a conflict theory for causation of threat displays was inadequate. Brown and Hunsperger's paper provoked considerable argument (Rowell 1964, Brown 1964), especially with regard to the interpretation of the term "tendency", and illustrates both the confusion and dangers which exist in describing behaviour at one level and testing or interpreting it at another.

Caryl (1979) has argued against the conflict theory on theoretical grounds. Using a model based on games theory he suggests that information about aggressiveness should not be encoded in a display. This view contrasts dramatically with the traditional interpretation that aggressive displays have evolved to transmit information about the probability of attack or escape. However, a basic assumption of the model, that the contestants are equally matched, is seldom met in mammalian encounters, and so the predictions of this model are of very limited use in assessing display causation.

Despite these various objections to the conflict theory it appears to be the most appropriate model for the causation of threat displays in *Dasyuroides*. Although it is difficult to specify the exact nature of the conflict without experimental manipulation of the conflicting tendencies (e.g. Blurton Jones 1968), the displays appear to be caused by the interaction of the actual tendencies to attack and to flee rather than a conflict between non–specific locomotory movements either towards or away from an opponent. Often a display is so well adapted for its signal function that the original motivational relationships are no longer clear. The behaviour patterns are elaborated upon and ritualized (e.g. the "typical intensity" of Morris 1957) during the course of evolution so that relationships with their original causal factors are not obvious. Lorenz (1961) has demonstrated these changes in the displays of many duck species. However, in *Dasyuroides* the displays appear to have retained a close relationship to their causal factors, the tendencies to attack and to flee, and the variation in motivation evident in the calls appears to be communicated to an opponent. Playback of both *pant–hiss* and *chatter* increased the occurrence of *withdrawal,* whereas playback of the *sniffing* vocalization, associated with *chasing* (but no characteristic static form), had no effect. Further, intensification of one of the visual components of these displays reduced overt *fighting*. This clearly indicates that the displays contain information about the probability of attack or escape which is transmitted to the opponent.

Playback of the calls suggested that their effectiveness in communication was determined by the rate of note repetition and harshness. These two variables appeared to reflect the underlying motivations to attack and to flee, with increased harshness indicating a greater tendency to attack, and decreased rate of calling indicating a greater tendency to flee. Thus an animal which *chatters* during *fighting* is more likely to *attack* than an animal which *chatters* directly towards an opponent or during *fleeing*. Similarly, an animal which utters a *pant–hiss* in close proximity to an opponent is more likely to *attack* than an animal which calls some distance away from an opponent. The predictions of this analysis, that calls which convey the most information about the probability of attack should have the greatest effect on an opponent, were confirmed by the playback experiments. The only exception was *chatter* recorded in close proximity, which did not significantly increase the frequency of *withdrawal* in the test period.

Thus in *Dasyuroides* the context of a display appears to be closely related to its causation. Such a relationship may help to explain the occurrence of a display in what seems an inappropriate situation. An animal which has just *attacked* an opponent is unlikely to *withdraw* in response to the *chatter* from the *fleeing* animal. However, the *chatter* in this situation may indicate more precisely than the actual *fleeing* behaviour the intention of the *fleeing* animal.

Marler (1967) has pointed out that a basic function of many animal signals is to help individuals locate or avoid each other, and that even this elementary function can have profound implications for signal structure. The *chatter* call of *Dasyuroides* possesses all the characteristics which Marler (1960) describes as facilitating the precise localization of a source of sound. The call is composed of both high frequency components for localization by means of differences in intensity and low frequency components for localization by means of phase differences. The rapid repetition of individual notes with an abrupt onset and termination facilitates localization by time differences. In contrast, the *pant–hiss* call is less easily located. Although the individual notes may be repeated at regular intercals, each note has a gradual onset and termination which hinders location by comparison of time differences. The call also has fewer high frequency components for localization by comparison of intensity differences. Thus, on the basis of strustural differences of the auditory components the displays, *chatter* is ideally suited for communication at a distance, whereas *pant–hiss* would be more effective at close range. For the visual components of the displays, however, there is no need for special structural characteristics facilitating localization since perception will be direct.

The interpretation of a signal by the recipient animal may be influenced also by its context. For example, Smith (1963) has shown that the kitter vocalization of the Eastern Kingbird may have different meanings according to whether the calling bird is approaching a perch, a potential perch, a mate or another bird. In *Dasyuroides pant–hiss* was characteristically a static display which was orientated towards an opponent over a fairly wide distance range. On the other hand *chatter* occurred in three separate contexts, with the structure of the call directly related to the context. However, care must be exercised in interpreting these contexts with regard to the natural situation. The limitations of the encounter situation are such that the maximum distance between two animals was about 100 cm. In fact, *pant–hiss* was clearly audible to the human ear up to about 5 m, and *chatter* up to about 25 m. Within this wider frame of reference, the structural variations in *pant–hiss* can be considered as occurring in close range, whereas *chatter* has the potential for communication over long distances. Thus the *pant–hiss* call would be enhanced by visual signals such as changes in the gape of the mouth, position of the ears and the degree of piloerection and eye closure. However, *chatter*, which is more suited to communication at a distance, shows greater variation in both harshness and rate of calling. Thus subtle changes in motivation may be reflected more in the auditory component than the visual components of this display. In *Dasyuroides* both displays are quite distinct, and vocalizations intermediate between the two are rare. Within each display however there is a fairly continuous variation which indicates a range of motivation (Smith 1963) and appears to correspond to Moynihan's (1955) "band of motivation" and Morris's (1957) "typical intensity". In contrast, many of the higher primates have a large number of highly graded signals (e.g. the Rhesus monkey, Rowell 1962) which Marler (1967) has suggested are ideally suited for close range communication in diurnal primates living in close–knit social groups. However in nocturnal mammals such as *Dasyuroides* selection may have favoured unambiguous signals where the exact intention (e.g. attack or flee) of the signaller is easy to predict and there is no danger of misinterpretation. Thus the threat display repertoire has been restricted to an offensive and defensive display, but there is a great deal of variation within each display.

ACKNOWLEDGEMENTS

This work forms part of my PhD thesis submitted to Monash University and is dedicated to my wife, Betsa, who has helped and supported me during many years of preparation. I would like to thank Dr. John Nelson for his encouragement and supervision. Drs. John Eisenberg and David Croft provided useful criticisms of a draft of this manuscript. The study was carried out with the assistance of a Commonwealth Postgraduate Research Award.

REFERENCES

ANDREW, R. J., 1956. Some remarks on behaviour in conflict situations, with special reference to *Emberiza* spp. *Brit. J. Anim. Behav.* **4:** 41–5.

ANDREW, R. J., 1972. Information potentially available in mammalian displays. Pp.179–206 *in* "Non-verbal communication" ed. by R. A. Hinde. Cambridge University Press: Cambridge.

ASLIN, H., 1974. The behaviour of *Dasyuroides byrnei* (Marsupialia) in captivity. *Z. Tierpsychol.* **35:** 187–208.

BLURTON JONES, N. G., 1968. Observations and experiments on causation of threat displays of the great tit *(Parus major). Anim. Behav. Monogr.* **1:** 75–158.

BRAITHWAITE, R. W., 1974. Behavioural changes associated with the population cycle of *Antechinus stuartii* (Marsupialia). *Aust. J. Zool.* **22:** 45–62.

BROWN, J. L., 1964. Goals and terminology in ethological motivation research. *Anim. Behav.* **12:** 538–41.

BROWN, J. L. AND HUNSPERGER, R. W., 1963. Neuroethology and the motivation of agonistic behaviour. *Anim. Behav.* **11:** 439–48.

CARYL, P. G., 1979. Communication by agonistic displays: what can games theory contribute to ethology? *Behaviour* **68:** 136–69.

EWER, R. F., 1968a. "Ethology of mammals". Logos Press: London.

EWER, R. F., 1968b. A preliminary survey of the behaviour in captivity of the dasyurid marsupial *Sminthopsis crassicaudata* (Gould). *Z. Tierpsychol.* **25:** 319–65.

GRANT, E. C., 1963. An analysis of the social behaviour of the male laboratory rat. *Behaviour* **21:** 260–81.

GRANT, E. C. AND MACKINTOSH, J. H., 1963. A comparison of the social postures of some common laboratory rodents. *Behaviour* **21:** 246–59.

GRANT, E. C., MACKINTOSH, J. H. AND LERWILL, C. J., 1970. The effect of a visual stimulus on the agonistic behaviour of the Golden Hamster. *Z. Tierpsychol.* **57:** 73–81.

HALLIDAY, T. R., 1975. An observational and experimental study of sexual behaviour in the Smooth Newt, *Triturus vulgarus* (Amphibia: Salamandridae). *Anim. Behav.* **23:** 291–322.

HARRISON MATTHEWS, L., 1964. Overt fighting in mammals. Pp.23–32 *in* "The natural history of aggression" ed. by J. D. Carthy and F. J. Ebling. Academic Press: London.

HINDE, R. A., 1970. Animal behaviour. A synthesis of ethology and comparative psychology. 2nd ed. McGraw–Hill: New York.

HINDE, R. A. AND ROWELL, T. E., 1962. Communication by postures and facial expressions in the Rhesus Monkey (*Macaca mulatta*). *Proc. Zool. Soc. Lond.* **138:** 1–21.

HOOFF, J. A. R. A. M., VAN, 1967. The facial displays of the catarrhine monkeys and apes. Pp.7–68 *in* "Primate ethology" ed. by D. Morris. Weidenfeld and Nicholson: London.

HUTSON, G. D., 1975. Sequences of prey-catching behaviour in the Brush–tailed Marsupial Rat *(Dasyuroides byrnei). Z. Tierpsychol.* **39:** 39–60.

HUTSON, G. D., 1976. Grooming behaviour and birth in the dasyurid marsupial *Dasyuroides byrnei. Aust. J. Zool.* **24:** 277–82.

LEYHAUSEN, P. 1956. Verhaltensstudien an Katzen. *Z. Tierpsychol., Beiheft* **2:** 1–120.

LITTLE, R. S., 1964. Acoustic properties of parabolic reflectors. *Bioacoust. Bull.* **4:** 1–3.

LORENZ, K., 1961. Phylogenetische Anpassung und adaptive Modifikation des Verhaltens. *Z. Tierpsychol.* **18:** 139–87.

MARLER, P., 1960. Bird songs and mate selection. Pp.348–67 *in* "Animal sounds and communication" ed. by W. E. Lanyon and W. N. Tavolga. American Institute Biological Sciences: Washington.

MARLER, P., 1967. Animal communication signals. *Science* **157:** 769–74.

MORRIS, D., 1957. "Typical intensity" and its relation to the problem of ritualisation. *Behaviour* **11:** 1–12.

MOYNIHAN, M., 1955. Some aspects of reproductive behaviour in the black–headed gull (*Larus ridibundus ridibundus* L) and related species. *Behaviour Suppl.* **4:** 1–201.

RECHTEN, C. AND FERNALD, R. D., 1979. A sampled randomization test for examining single cells of behavioural transition matrices. *Behaviour* **69:** 217–27.

ROBERTS, W. W., STEINBERG, M. L. AND MEANS, L. W., 1967. Hypothalamic mechanisms for sexual, aggressive and other motivational behaviours in the Opossum, *Didelphis virginiana. J. Comp. Physiol. Psychol.* **64:** 1–15.

ROWELL, C. H. F., 1964. Comments on a recent discussion of some ethological terms. *Anim. Behav.* **12:** 535–7.

ROWELL, T. E., 1962. Agonistic noises of the Rhesus Monkey *(Macaca mulatta). Symp. Zool. Soc. Lond.* **8:** 91–6.

SIMPSON, M. J. A., 1968. The display of the Siamese Fighting Fish, *Betta splendens. Anim. Behav. Monogr.* **1:** 1–73.

SLATER, P. J. B., 1973. Describing sequences of behaviour. Pp.131–53 *in* "Perspectives in ethology" ed. by P. P. G. Bateson and P. H. Klopfer. Plenum Press: London.

SLATER, P. J. B. AND OLLASON, J. C., 1972. The temporal pattern of behaviour in isolated male Zebra Finches : transition analysis. *Behaviour* **42:** 248–69.

SMITH, W. J., 1963. Vocal communication of information in birds. *Amer. Nat.* **97:** 117–25.

SMITH, W. J., 1965. Message, meaning, and context in ethology. *Amer. Nat.* **99:** 405–9.

STOUT, J. F., 1975. Aggressive communication by *Larus glaucescens* III. Description of the displays related to territorial protection. *Behaviour* **55:** 181–208.

STOUT, J. F. AND BRASS, M. E., 1969. Aggressive communication by *Larus glaucescens.* Part II. Visual communication. *Behaviour* **34:** 42–54.

TINBERGEN, N. 1959. Comparative studies of the behaviour of gulls (Laridae) : a progress report. *Behaviour* **15:** 1–70.

WALLIS, D. I., 1962. Aggressive behaviour in the ant, *Formica fusca. Anim. Behav.* **10:** 267–74.

WALTHER, F. R., 1965. Verhaltensstudien an der Grantgazelle (*Gazella granti Brooke,* 1872) in Ngorongoro–Krater. *Z. Tierpsychol.* **22:** 167–208.

RESUMEN

34. Un análisis del despliegue ofensivo y de la amenaza defensiva de *Dasyuroides byrnei*. Por Geoffrey D. Hutson.

Se observaron y analizaron una serie de disputados encuentros de machos–machos y hembras–hembras de la especie *Dasyuroides byrnei.* Las causas del despliegue de amenazas, jadeo siseante y rechinamiento fueron examinadas por el análisis de las transiciones respecto de otros modelos de comportamiento y de los intervalos del tiempo existente entre ellos. El jadeo siseante está asociado a una alta predisposición al ataque y el rechinamiento auna alta predisposición a la huída. El análisis del efecto del despliegue de esos comportamientos sobre el del oponente mostró que el jadeo siseante fue dado en respuesta a una actitud no agresiva, mientras que el rechinamiento fue en respuesta a una abierta agresión. De tal manera, el jadeo siseante parece ser un despliegue de amenazante ofensiva y es una advertencia inicial a un oponente en retirada. El rechinamiento parece ser una amenaza defensiva e indica la intención de luchar si es atacado.

El análisis del contexto de los despliegues y de la estructura del sonido componente reveló una sorprendente cantidad de variantes de los despliegues. El ritmo de la repetición de los sonidos y la aspereza de los llamados parecieron reflejar las subyacentes motivaciones a la fuga y al ataque.

Una serie de experimentos mostraron que las diferentes vocalizaciones tuvieron diferentes efectos sobre el comportamiento de un oponente y sugieren que variación en la evidente motivaciòn de los llamados es comunicada al oponente. La manipulación evidente de un componente visual del desplieque, como el teñido totalmente negro de la cola, redujo al comportamiento de lucha, e indicó por ello que la cola fue una efectiva señal visual de amenaza.

Se discute la applicabilidad de la teoría del conflicto en la causa de los despliegues del comportamiento.

35. MATERNAL BEHAVIOUR OF *Antechinus stuartii* (DASYURIDAE, MARSUPIALIA) IN CAPTIVITY

G. A. SETTLE and D. B. CROFT[1]

Pregnant female *Antechinus stuartii* were obtained from the wild. Maternal behaviour, after parturition, was observed in the nestbox and its near vicinity from video recordings. *Antechinus* mothers care for their young by providing protection and support in a pouch and later a nest, warmth, grooming and nutrition. Construction of a nest, however, was independent of the female's reproductive state and so not directly related to the birth or development of young. Grooming of the young as a proportion of the total time mothers spent grooming in the nest increases with the development of the young, especially during the period when the young's eyes open and they detach from the nipples. Grooming in the nest then declines as the young begin to make excursions outside the nest. While the young are still attached to the teats (30–50 days post–partum), pouch grooming is sequentially related to grooming of the young. When the young are detached and more mobile (60–80 days pp.) pouch grooming becomes an integral part of the mother's self–grooming behaviour.

Once the young detach from the teat but while their motor co-ordination is poor, the female maintains proximity with her young using "pull–under" retrieval behaviour. As the young's activities become more independent and better co–ordinated, the young are primarily responsible for maintaining proximity in the mother–young dyad. While young are left in the nest during the mother's excusions outside, the mother uses a 'tumbling' behaviour to dislodge young from the pouch.

Litters of less than three young are rarely reared to independence. Nonetheless, litter size does not have a significant effect on the rate of development of the young. The importance of the suckling stimulus in the maintenance of maternal behaviour is discussed.

Pages 365–81 *in* CARNIVOROUS MARSUPIALS, ed. by M. Archer, February, 1982. Roy. Zool. Soc. New South Wales: Sydney, Australia.

SOME FORM of parental behaviour is essential to the survival of the young of mammals. The young are fed on milk secreted by the mammary glands of the female and so the mother usually plays the role of the young's principal caretaker. Maternal care, however, is generally not just confined to the provision of milk but often includes other behavioural patterns. For example, the mother provides protection, she cleans and keeps the young warm, and in some cases plays an active role in the development of the young's behaviour in its physical and social environment (Ewer 1968a).

Maternal behaviour in marsupials is particularly interesting because the young are born in an almost embryonic state after a short gestation period, and are then supported in a pouch where they obtain milk, warmth and protection. This reproductive pattern fosters a close association between the mother and young during the young's pouch life. In this period maternal care extends additionally to the maintenance of the hygiene of the pouch and young by the mother's licking and grooming behaviour (Aslin 1974, Russell and Giles 1974). When the young first vacates the pouch, it is either left in a nest, clings to the mother and is carried about by her, or follows at foot. During this stage, the mother–young relationship often encompasses a more varied pattern of behavioural interactions such as retrieval behaviour (Ewer 1968b).

Fleay (1932, 1934, 1949, 1961, 1965) has provided accounts of maternal behaviour in a number of dasyurid species but this behaviour has been studied in detail for only a few species (*Sminthopsis crassicaudata,* Ewer 1968b; *Dasyuroides byrnei,* Aslin 1974; *Antechinomys laniger,* Happold 1972; *Dasyurus maculatus,* Settle 1978). Little is known, however, about maternal behaviour of *Antechinus stuartii* even though reproduction in this species has been extensively studied (Marlow 1961, Woolley 1966, Braithwaite 1973). Therefore, maternal behaviour of captive *Antechinus stuartii* was studied by focussing on the mother and young's behaviour in the proximity of the nest site. Nest building by the female, her grooming and suckling of the young, and the spatiotemporal patterning of the mother and young's behaviour were investigated.

METHODS

Observations were focused on maternal behaviour in the nest and its immediate vicinity as the physical constraints of this site are similar in the wild and in captivity. Pregnant females were obtained from the wild (Sydney region) and, for most of the observations reported here, only those producing six or more young were retained. Large litters were reduced to a standard size of six young. The maximum litter size ing in the Sydney region is eight but litters of six are the most common in the wild.

[1] School of Zoology, University of New South Wales, P.O. Box 1, Kensington, New South Wales, Australia, 2033. Graham Settle died in 1979.

Females and their young were housed indoors under natural lighting penetrating through windows. Temperature was maintained at around 22°C by an air–conditioner. The subjects were caged in a box (60 x 46 x 62cm) made from 4mm 6–ply wood. The box had a wire mesh back and hinged glass door at the front. A nest box was attached to the outer surface of one side with a 5 cm diameter entrance hole, 40 cm from the cage floor. Access to the nest box was provided by a branch leading from the cage floor and terminating against the cage wall just below the nest box entrance. The nest box was designed with a viewing window in one side, a port for the introduction of a flash for still photography, and variable red illumination for observations using a video camera (Fig. 1).

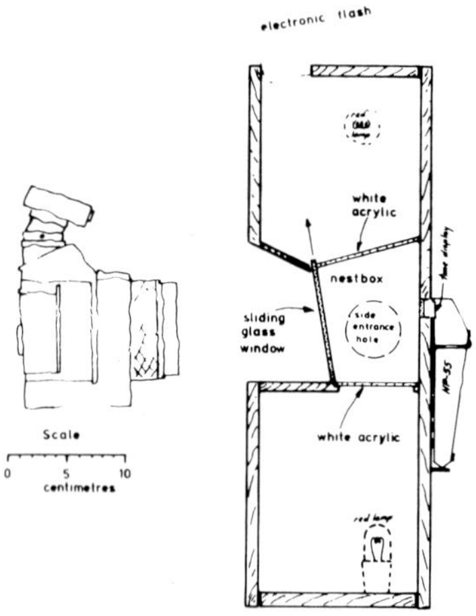

Fig. 1. Cage–end view of nestbox showing details of construction (vertical section), placement of 35mm photographic equipment (or video camera) and timer. The frontface of the nestbox was 19cm wide. The two lamps were 15 watt, 240 volt red pilot lamps controlled by separate dimmer circuits.

Cage floors were covered with clean sand or soil and quantities of dry *Eucalyptus* leaves randomly scattered over the surface. A variety of cover objects were suplied but orientated so as not to obscure the subjects activities from the observer.

Antechinuses were fed a mincemeat–vitamin–calcium–cheese mix with supplementary live insect food (e.g. meal worm larvae, cockroaches, African plague locusts). Water was supplied *ad libitum*. Specific methods used in the study of various aspects of the maternal behaviour of *Antechinus stuartii* will be given in the relevent sections following.

RESULTS

NEST BUILDING: *Antechinus stuartii* build nests within hollow logs and rock cavities in the wild. The size of the completed structure depends largely on the size of the cavity selected as a nest site. For example, 16 nests collected in the Muogamarra Nature Reserve (Sydney) ranged in size from 10 x 12 x 8cm to 30 x 30 x 15cm. These nests consisted mostly of *Eucalyptus* and *Angophora* leaves, especially broad lanceolate ones (i.e. 10–15mm wide, 4cm long).

In captivity, both males and females will readily build nests in a nesting place with a restricted entrance.

CONSTRUCTION OF THE NEST BY THE FEMALE: Gathering material for the construction of a new nest commences quite suddenly, usually following normal maintenance activities. Much nest building took place between midnight and dawn. Females display very high levels of activity during the initial gathering and construction period sometimes filling nestboxes with as many as 260 leaves on the first evening. Most females completed nests in less than three consecutive nights.

Leaves in good condition are selected from the leaf litter, picked up in the mouth and carried one at a time to the nestbox where they are dropped forming a small pile. Subsequent visits to the nestbox scatter the leaves about the floor but females spend little time rearranging these during early stages of nest construction.

Gathering becomes sporadic and interspersed with bouts of leaf manipulation and other building activities within the nestbox as the quantity of gathered material increases. Construction of the nest is achieved through a combination of leaf placement and compression of nesting material by body movements. The mouth and forpaws are jointly used for manipulation and placement of leaves. Turning and tumbling movements flatten leaves against the walls, compressing the outer layers and enlarging the central cavity. These body movements also seem to function as a test for leaf placement.

During the early stages of nest construction animals frequently burrow into the walls of the nest and re-emerge at some other point, whereupon they set about repairing the damage caused These 'structural adjustments' to the nest frequently damage leaves, as does the considerable force often applied during placement. Thus a substantial amount of broken material is often present in the outer layers of the nest by the time it is completed. Leaf fragments and leaves with V-shaped wedges torn from them (where gripped between the teeth) are typical signs of forced placement.

The presence of small twigs in nest walls is not a result of intentional gathering. Occasionally branches bearing three or four leaves are dragged into the nest and the leaves removed by bites to the petiole. The remaining twigs become part of the nest wall as they are covered with further layers of leaves.

In the final stages of nest–building female *Antechinus stuartii,* in particular those approaching parturition or with neonates, often gather leaves direct from foliage. These are exised by biting through the petiole, carried to the nest and used as a nest lining. Dry leaves are rarely pliable but green leaves are readily curled to provide an exceedingly smooth lining to the nest chamber. Furthermore, as the green leaves dry they do not only retain their curvature but also improve the rigidity of the inner walls of the nest. Presumably green leaves are rare components of the outer walls because of their moisture content and the associated risk of fungal attack.

GATHERING BEHAVIOUR OF THE FEMALE: Experiments were performed to ascertain whether the nest building behaviour of pregnant females or those with young differed from that of females without young. Four female *Antechinus stuartii,* were provided with liberal quantities of nesting material (leaf litter from *Eucalyptus* forest scattered over the floors of cages) throughout the breeding season. Each morning gathered material was removed from the nestboxes, and a record made of the number of pieces gathered. These data were collected from six weeks prior to mating until the young were first observed gathering leaves (ca. 100 days post–partum).

In a preliminary experiment, material gathered was marked with a felt pen and returned to the cage. The results of this experiment indicated that females select a high proportion of leaves used previously for nest construction (Table 1). Such selection could bias the quantities of leaves gathered, and repeated gathering of the same material caused excessive damage and fragmentation of leaves, making counting difficult. Therefore, in the main experiment, gathered leaves were discarded and the supply in cages replenished with fresh leaf litter.

Table 1. Repetitive gathering of the same leaves by female *Antechinus stuartii.* Leaves in nests were marked with a felt–tipped pen and returned to the cage floor where they formed less than 20% of the available nesting material.

Female	Night 1 No. leaves marked	Night 1 No. regathered	Night 2 Cumulative No. leaves marked	Night 2 No. regathered
AS102	127	83	198	76
AS111	56	49	114	68
AS124	105	66	137	89
AS128	182	103	219	116
Totals	470	301	668	349
% marked leaves regathered		64%		52%

Data for the number of leaves gathered daily by the four females were grouped into successive five–day intervals from 60 days before to 100 days after parturition, and a median calculated (Fig. 2). Around 200 leaves (horizontal dashed line in Fig. 2) represented a nest constructed with well compacted walls. Nests of 150 leaves (or more) were completed in the sense that they filled nestboxes, and only compacting the walls of the nest permitted the female to add more leaves. Even 100 leaves was sufficient to provide a good 'first–night' nest. Most of the median values lie above 100 leaves gathered per night, except when animals mated (ca. 30 days before parturition) and just before and after parturition. At these times all females gathered less than 100 per night. These were the only periods when

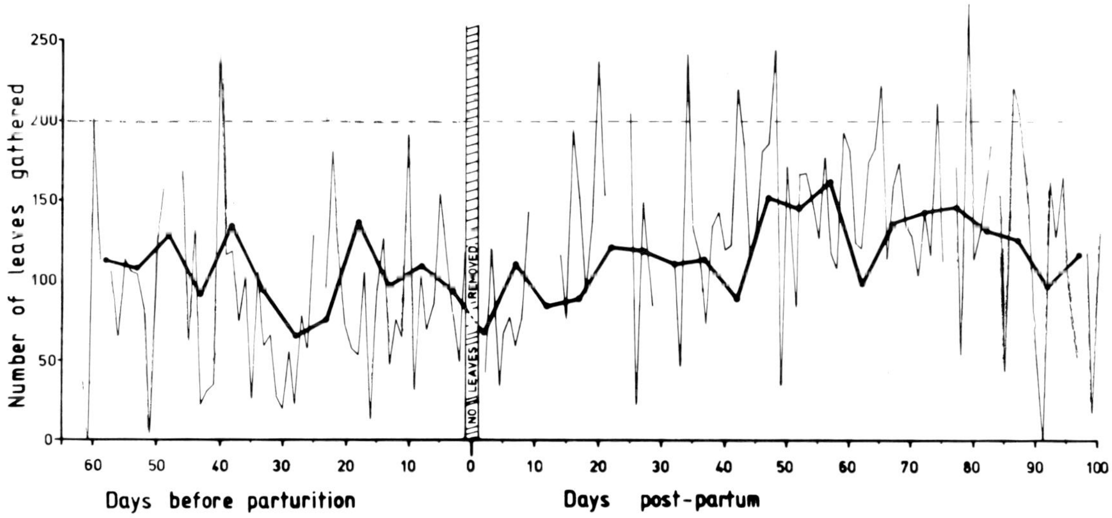

Fig. 2. Gathering of nest–building material by female *Antechinus stuartii* (N = 4) from 60 days before until 100 days after parturition. The median number of leaves gathered in successive 5–day intervals (thick line) and the daily results for a single female (thin line) are shown. The broken line at 200 leaves indicates a full nestbox.

the reproductive state of the female influenced her nest building behaviour. Females were prolific nest builders regardless of whether they were virgin, pregnant or lactating. However, the general pattern of gathering behaviour was highly sporadic; ranges extend from near zero to the capacity of the nestbox over the whole sampling period.

GROOMING: Licking, saliva spreading with the forepaws, scratching and nibbling and combing of the fur with the teeth are the principal components of grooming behaviour. Most grooming of the young by the mother is licking, especially around the cloacal region. Changes in the duration of grooming behaviour and its temporal patterning as the young developed were investigated.

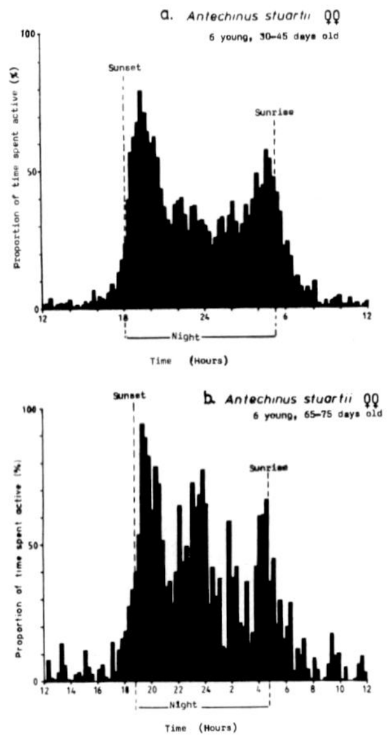

Fig. 3. Diel activity patterns of female *Antechinus stuartii* (N = 6) with litters of six young between (a) 30 and 45 days old, and (b) 65 and 75 days old. Results are expressed as the mean proportion (%) of time in successive 15–min. intervals spent active. Each female was observed on two occasions. Mean scores for each individual were calculated and used to obtain population means.

CHANGES IN THE AMOUNT OF TIME THE MOTHER SPENDS SELF–GROOMING AND GROOMING THE YOUNG: A post–sunset peak occurs in the activity pattern of females with young (Fig. 3). Observations of grooming behaviour were made from one hour videotape records of this period. The duration of Gm (self–grooming young) and Gy (grooming young) grooming bouts by the mother were timed to the nearest 5–second unit. In cases where both Gm and Gy occurred in the same bout, the two categories were treated separately. Brief pauses (< 5sec) in bouts of Gm or Gy were included but not pauses between Gm and Gy subbouts in a bout of Gm+Gy. The scoring method provides an approximation of duration of bouts, not an abesolute measure. Duration scores for for Gm and Gy were cumulated for each observation session from 20 to 109 days post–partum (pp) and analysed in intervals of five days. A total of 3000 grooming bouts by nine *Antechinus stuartii* mothers were scored.

The intervals chosen for analysis include the following stages in the development of the young: i; 20–38 days pp—young always in pouch; ii, from ca. 39 days pp—young still carried in pouch but able to relinquish the teats; iii, from ca 53 days pp—young first left in the nest on a regular basis; iv, 60–69 days pp—eyes of young closed to eyes open but young not leaving nestbox; v, 70–79 days pp—progressive development of excursions from the nestbox by young; vi, 80–110 days pp—young moving freely in and out of the nestbox and eating increasing amounts of solid food.

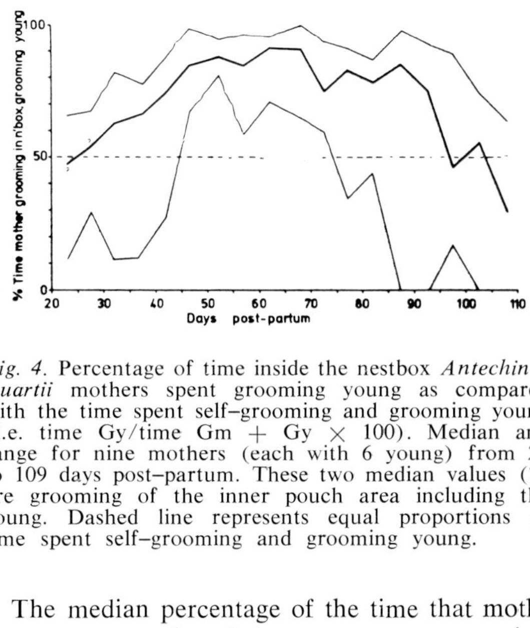

Fig. 4. Percentage of time inside the nestbox *Antechinus stuartii* mothers spent grooming young as compared with the time spent self–grooming and grooming young (i.e. time Gy/time Gm + Gy × 100). Median and range for nine mothers (each with 6 young) from 20 to 109 days post–partum. These two median values (?) are grooming of the inner pouch area including the young. Dashed line represents equal proportions of time spent self–grooming and grooming young.

The median percentage of the time that mothers spent grooming their young as a proportion of the total time they spent grooming when inside the nestbox was calculated (Fig. 4). Mothers spent more time grooming their young than self–grooming in general. This is not only apparent from the median but also from the position of the median between upper and lower range limits, i.e. the distribution within intervals is skewed towards upper values in most cases.

The grooming behaviour of the mother, however, changes during the young's development. Mothers began to groom their young more than themselves from 25 days pp. By 54 days pp mothers spent much much more time in Gy than Gm although there was no sudden increase in grooming of the young associated with this period when they were first left in the nestbox. Mothers always spent more time grooming young than self–grooming between 45 and 74 days pp.

From 70 to 95 days pp there was a tendency for mothers to spend a slightly lower proportion of their time grooming the young. Over the same period the lower range limit of percent time grooming the young fell to zero, i.e. during some sampling periods some mothers did not groom their young at all by 85 days pp. Young between 75 and 95 days old often pestered and groomed their mothers, and were sometimes groomed as a result. Much of the maternal grooming in the interval 80 to 95 days pp was in response to these interactions between mother and young. Beyond 95 days pp mothers began spending an increasing greater proportion of time grooming themselves than grooming their young.

MATERNAL GROOMING SEQUENCES: The temporal investment of the female in grooming her young changes with the developmental stage of the young. To look at whether the way in which she grooms the young also changes, the structure of grooming sequences was investigated. In this analysis, the self–grooming of the mother (Gm) was sub–divided into two categories: pouch grooming (Gp) and other acts of self–grooming (i.e. Gm — Gp). Pouch–grooming is the grooming of the pouch area and its margins without contact with any young in the pouch. It is an important part of maternal care in marsupials e.g. in macropods just before parturition (Sharman and Calaby 1964) and following birth when young are in the pouch (Russell and Giles 1974a), and in *Dasyuroides byrnei* at parturition (Hutson 1976) and during the first 30 days after birth (Aslin 1974). Since pauses (Pau) were also viewed as important, they formed a fourth discrete category.

Grooming sequences were analysed at two developmental stages of the young: i, 30–50 days pp when young were still attached to the teats and were not left in the nest during the mother's excursions; ii, 60–80 days pp which is from just before the eyes open until the young are moving freely in and out of the nestbox. In comparison with (i), these young are physically much larger and considerably more mobile. The period between the two age classes served to partition the data and reduce overlap of patterns.

Grooming bouts by nine mothers were analysed and transition frequencies between grooming of the young, pouch grooming, other acts of

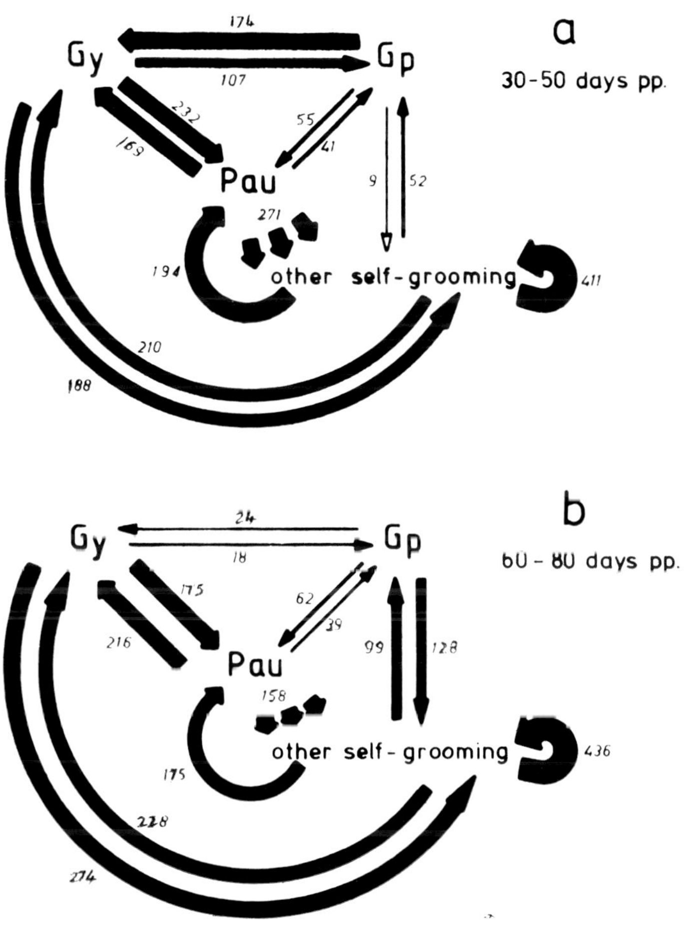

Fig. 5. Flow diagrams of grooming sequences by *Antechinus stuartii* mothers (N = 9). The transition frequencies between grooming of the young (Gy), pouch grooming (Gp), other acts of self–grooming, and pauses (Pau) are indicated. (a) Results for 213 bouts in the interval 30 to 50 days pp; (b) for 196 bouts in the interval 60 to 80 days pp.

self–grooming, and pauses calculated (Fig. 5). A marked change between the two sampling periods in the primary association of pouch grooming with other grooming categories occurred. when the young were in the later stages of pouch life, 30 to 50 days pp, pouch grooming tended to be more closely associated with grooming of the young than with other self–grooming acts by the mother (Fig. 5a). However, this link must be interpreted with some caution because Gp and Gy acts tend to overlap and cycle backwards and fowards once pouch grooming is initiated. It was often difficult to tell whether the mother was grooming the heads and faces of young or around the nipples. Some overlap of these two categories occurs, especially when young and pouch are contacted in a single lick. Perhaps a more important aspect is that pouch grooming betwen 30 and 50 days pp tended to be initiated after pauses and other self–grooming acts. It then cycled with grooming of the young, and was followed by further grooming of the young (usually in the perineal region) rather than returning to self–grooming.

When the young were older (60–80 days pp), pouch–grooming was largely an integral part of self–grooming (Fig. 5b). At this time mothers rarely groomed their pouches while a large proportion of the young were attached to the teats. There was no indication that the observed change was due to anything more than the physical obstruction to pouch–grooming as a consequence of suckling by the older and physically larger infants.

Pauses in grooming bouts were important and seemed to function not only as rest periods but operationally as periods during which mothers redirected their grooming activities. Transfers direct from self–grooming (including pouch grooming) to grooming the young were less often associated with the behaviour of the young (e.g. onset of activity, grappling at the mother's face) than were those that took place via pauses. During pauses mothers were more responsive to the behaviour of the young. Alternatively, when the mother was grooming parts of her body other than the pouch and was disturbed by the young, she often paused before grooming them. Probably most transitions direct to grooming of the young are associated with the obvious care–giving functions of cleaning, whereas transitions via pauses become increasingly associated with social interactions between mother and young, especially after the eyes of the young are open. This is supported by the change in sign of the net transfer from Pau to Gy, and Gy to 'other self–grooming' between the two age intervals, i.e.

		30–50 days	60–80 days
Gy ➤	Gm Gp	–22	+46
Pau ➤	Gy	–63	+40

where net transfer is the difference in transition frequency between the first and second behaviour, and the second and first behaviour.

Retrieving Behaviour: The maintenance of proximity between mother and young is essential to many components of maternal care (e.g. suckling, provision of warmth, protection). Female *Antechinus stuartii* have a type 1 pouch (Woolley 1974) i.e. the pouch does not enclose the young in a fold of skin, and so the young, especially once they detach from the teat, can become displaced. The female, herself, often actively displaces young from the pouch preceding excursions from the nest (see next section). Thus retrieving behaviour has developed in female *Antechinus stuartii* which enables young, with as yet poorly developed locomotory abilities, to regain the pouch.

Fig. 6. "Pull-under" retrieving behaviour of a 45–day–old *Antechinus stuartii* infant by its mother.

Retrieval by the Mother: A form of behaviour called 'pull–under' retrieving was used by females as the sole method of gathering young dispersed about the nestbox (Fig. 6). Here the female pulls the young under her with her forepaws as she moves forward, pausing to enable the young to grasp her ventrum and usually to attach to a teat. For example, often when mothers awoke and found some of the young scattered about the nestbox they reached out and pulled them under their ventrum. At times mothers were so industrious in grooming their young that odd ones were dislodged from teats and tumbled out amongst their sibs. If this happened to young whose eyes were still closed they were often unable to re–orient towards their mother and crawled about making short sibilant distress cries until retrieved by their mother. However, distress calls were not necessary to elicit retrieval. The opposite was often true as females frequently retrieved silent young wondering about the nestbox shortly after their eyes had opened and cotinued to do so even though the infant was trying to crawl away and then uttering distress calls. However, it is not known whether young *Antechinus* call at frequencies above those audible to human.

When young are first able to relinquish the teat (but long before they are able to be left in the nest i.e. 40–45 days old), they have another means of attracting the mother's attention. Unable to crawl about effectively these young perform a flicking movement of the body if they become separated from the mother. The young lies prostrate for a second or so, arches its spine and then suddenly flicks it downwards and its head bobs up. In a box with no nesting material these 'convulsing' young are usually seen and retrieved. In a normal nest they possibly become hidden by leaves and so 'convulsing' perhaps attracts the attention of the mother in the absence of distress calls.

Antechinus stuartii mothers were never observed to use their mouths to pick up or carry an infant of any age. *Sminthopsis crassicaudata* (Ewers 1968b) and *Antchinomys laniger* (Happold 1972) similarly do not use mouth transport to retrive young. But *Dasyuroides byrnei* (Aslin 1974) and *Dasyurus maculatus* (Settle 1978) mothers sometimes pick up young (eyes closed) who have wandered out of the nest, and carry them in their mouth back to the nest.

she entered the nest (and thereby approached the young), or happened to pass by them and the young came to her.

'SELF–RETRIEVAL' BY THE YOUNG: Another type of retrieving in dasyurids involves the 'clinging–response' and/or following behaviour of the young but occurs in a different context and is distinct from active retrieval by the mother. This was called 'self–retrieval'.

Self–retrieval is the independent movement of the young towards their mother which is associated with an attempt, successful or otherwise, to maintain bodily contact with her. Self–retrieval differs from Ewer's (1968b) 'clinging–response' in that the young do not necessarily effect contact with the mother and it also excludes instances of bodily contact where the retrieving function seems to be absent. Some examples illustrate this difference:

i. mother and young sitting adjacent, mother suddenly bounds away, young bounds after her and is apparently trying to catch her, mother stop at nest entrance, young stops a few centimetres away.

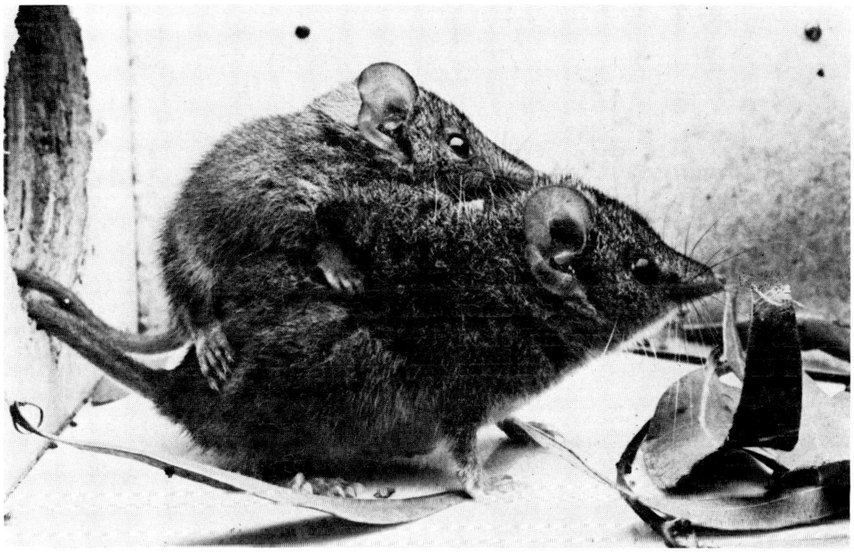

Fig. 7. Juvenile *Antechinus stuartii* clinging to its mother's back.

Dasyurids utilize a form a retrieval which most other workers have referred to as 'the clinging–response', 'clinging on by' etc. the young (e.g. Ewer 1968b). Accounts of dasyurid maternal behaviour often describe mothers approaching young not old enough to leave the nest. The young then attach to a teat or cling to the mother's flanks without receiving any further attention from her (Fig. 7). *Antechinus stuartii* mothers were not observed to actively retrieve young by using their 'clinging–response' alone. This behaviour occurred, however, as a result of some passive approaches by the mother: e.g. when

ii. mother feeding in corner of cage, then grooms and sits, young exploring cage some distance away from mother; mother bounds towards nestbox (or some other point), young bounds towards mother and when it succeeds in clinging to the fur on her back, mother with young on back proceeds to destination where either young climbs off or remains clinging.

iii. young approaches mother during general exploration, (i.e. undirected movements) climbs on her back, gets off again and moves away without mother moving more than a few centimetres.

Both (i) and (ii) are acts of self–retrieval, (iii) is not. Self–retrieval also occurs when young are startled or threatened by sound, novel object etc. and flee to the mother who is not necessarily moving. In this case the young invariably cluster about trying to climb on the mother. Ewer (1968b; 356–7) possibly did not distinguish a category of self–retrieval in *Sminthopsis crassicaudata* both because she did not have the opportunity to observe events within the maternal nest and because *S. crassicaudata* young do not leave the nest until 62 days old, 12 days after their eyes have opened. By contrast *Antechinus stuartii* show progressive development of excursions from the nest when about 70 days old (i.e. 6 or 7 days after the eyes open).

In *Antechinus stuartii* self–retrieval behaviour developed gradually. There was a phase of development beginning just before eye–opening (ie. from about 60 days old) during which the efforts of the young to contact their mother, when she first entered the nestbox following an excursion, became more vigorous. When a mother entered the nestbox she usually did so quickly, contacted two or three young and the remainder crawled about until contact with another sib or the mother. If another sib was contacted first it was crawled over or under until the mother was reached. At this age the only cues available were the scents and vocalizations of the young and mother, and disturbances due to increased activity in the nestbox following the mother's return. These more strenuous efforts to contact the mother apeared to develop from the contact behaviour exhibited by younger offspring, i.e. their movement towards the mother to feed and huddle etc. However, once contact was made (the young usually received grooming or other interactions from the mother) the young often left the mother soon after and resumed activities (or sleep) elsewhere in the nestbox. Thus contact behaviour seems dependent on the nurturative role of the mother at this stage of development.

Following eye–opening, self–retrieval increased markedly presumably because the young were able to use vision to follow the movements of the mother. This coincided with the rapid development of locomotory skills and by 70 days young often moved to the entrance of the nestbox where they sat looking out. At this stage when the mother returned to the nestbox the young were displaced into the nestbox as she entered and self–retrieval took place inside the nestbox.

A few days later when young sat or explored just outside the nestbox, mothers often passed them when leaving and entering the nestbox. If mothers passed slowly on the way back into the nest, which they did infrequently, young tended to follow them less often than when mothers passed by quickly. Young also followed their mothers into nestboxes more than they followed them out. When young followed mothers out, they rarely ventured further than outside the nestbox entrance even though mothers usually moved away from the nest quite rapidly. At this stage of development, both the rapidity and direction of the mother's movements are probably factors regulating the self–retrieval behaviour of the young. The young tended to follow rapid movements of the mother more so as mothers returned to the nest .

By 80 days pp young were venturing to all parts of the cage. At this stage self–retrieving was less frequent and only took place when the mother moved rapidly away from the young. The direction in which the mother moved was less important.

SEPARATING THE YOUNG FROM THE TEAT: Young dasyurids who are old enough to be left in the nest but still attached to the teats immediately before an excursion by the mother, must either be carried with her, independently relinquish the teats, or induced to let go. Since young are usually left in the nest, it is probable that the costs of carrying the young (e.g. impaired locomotory ability of the female increasing the risk of predation and reducing foraging efficiency) outweigh the costs of leaving them in the nest (e.g. hypothermia, predation). Thus not only is retrieval of the young a component of maternal care but also separation of young from the mother is also important.

The nature of excursions by mothers away from their litters undergoes considerable change during the first 90 days of the youngs' lives. Prior to 45 days pp all young are attached to the mother's nipples during her excursions. Mothers begin to leave young behind in the nest over the next ten days. Generally, all young are left in the nest during the mother's excursions from 54 days pp until shortly after their eyes have opened (ca. 63 days pp). During this phase the mean duration of excursions by mothers remains virtually constant but there is a progressive and marked increase in their frequency (Fig. 8). There is a reversal of this trend once the eyes of the young have opened and the frequency of the mother's excursions begins to decline, while the mean duration of her excursions begins to increase (Fig. 8). The total time spent away from their litters by mothers increases continuously from 45 to 84 days pp.

Passive Teat Relinquishment: On many occasions when the mother stood up (particularly if she had rested for some time) her young came away from the pouch freely with their mouths closed. Often it was clear that these

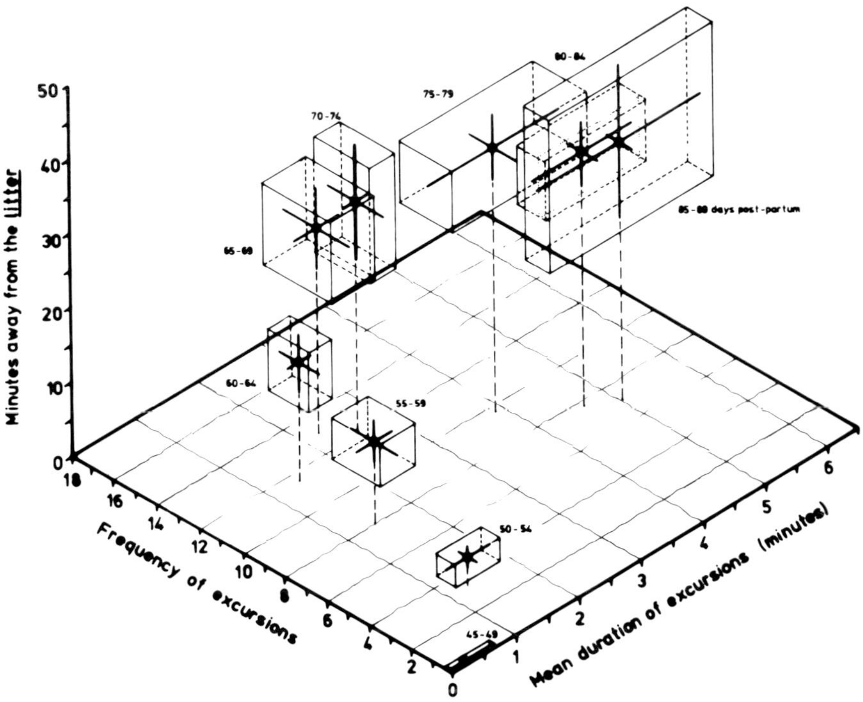

Fig. 8. The time *Antechinus stuartii* mothers (N = 9) with six young spent away from their litters during the post–sunset activity peak (Fig. 3) as a function of excursion frequency and duration, and the age of the young.

young had remained in the suckling position but had let go of the teats and were at rest when their mother arose. The time over which the 'rest after suckling' position was maintained became shorter with increasing age of the young and was replaced with general huddling.

Young were sometimes dragged upwards with the body of the mother as she rose but then fell away from the pouch with their mouths open. These young rarely showed more than occasional twitching movements of the hindlimbs beforehand, suggesting that they had been resting and retained only a loose grip on the teats.

Induced Teat relinquishment: Young attached to the teats and showing signs of activity usually remained attached to the teats when the mother rose. However, prior activity was not a necessary condition and many times resting young began to grapple at the region of the mother's pouch as soon as she began to move. Even young free of the teats were often able to re–attach before their mother had moved more than a few centimetres. Young were particularly quick at re–attaching and their skill in doing so improved as their locomotory powers developed. There were three main methods used by mothers to induce their young to relinquish the teats. Combinations of the three were also common.

i. Nudging and grooming the young: Mothers nudged at the young with their snouts and often licked them about the head. When the young appeared to have a firm grip on the teats, grooming by mothers was often prolonged and sometimes led to extensive bouts. Grooming of the young generally lasted until the young became quiescent and mothers often shuffled about in small semi–circles as they groomed. These shuffling movements tended to be associated with pre–excursionary grooming. If the young let go when groomed or nudged by their mother, she left the nestbox. This behaviour, however, was the least effective in inducing teat relinquishment.

ii. Shaking of the mother's body or hindquarters: Gentle body shaking by mothers was observed a few times, but not the vigorous shaking of, for example, *Dasycercus cristicauda* (Fleay 1961). Again this behaviour was relatively ineffective in inducing teat relinquishment.

iii. Disorientation of the young: To induce teat relinquishment and disorient the young, female *Antechinus stuartii* used a behaviour pattern called 'tumbling'. When the mother was about to tumble, she usually sat facing the nestbox exit with her young grouped in front of her, lent forward over the young, tucking her head beneath them, and quickly performed a complete somersault. Any young attached to the mother's teats were spun around with her, opening their mouths on the way round. When upright again the mother made a rapid dash for the nestbox exit, leaving her young in disarray.

Tumbling was performed obliquely or by circling the young, especially during the first week the young were left in the nest as a litter

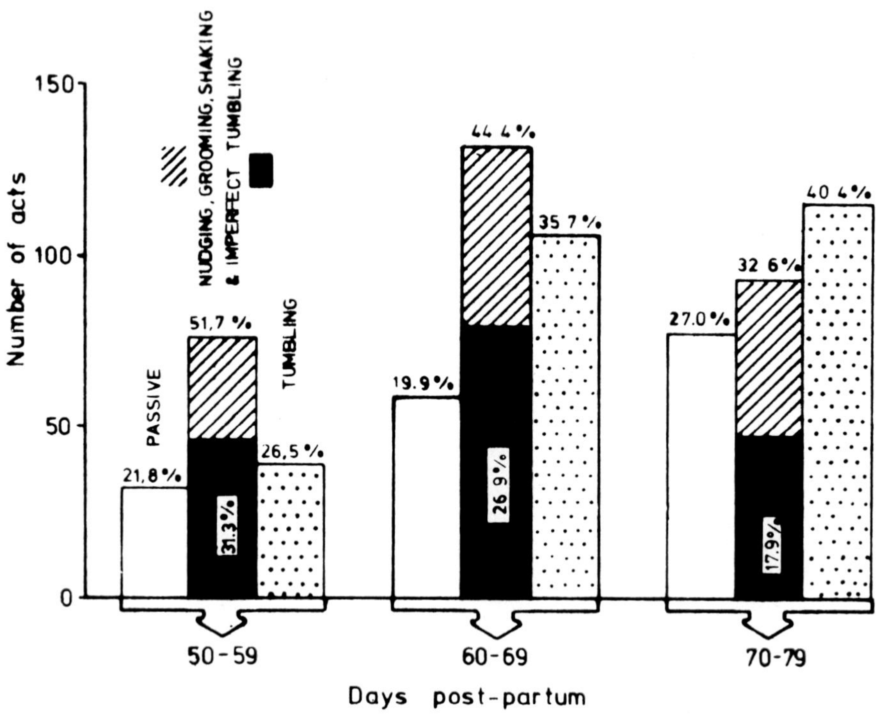

Fig. 9. Methods used by *Antechinus stuartii* mothers to leave young of three age–groups in the nestbox. Data pooled for 13 mothers. Given are the number of acts and the percentage composition for each age group. The term "passive" refers to instances when mothers were not encumbered with young and did not perform any specific act involving the young immediately before leaving. "Tumbling" refers to tumbles in a vertical plane. Other methods are combined with "imperfect tumbling", but imperfect tumbles are distinguished within that category.

(i.e. ca. 53 to 60 days pp). The relative frequency of 'imperfect' tumbles as compared to tumbles in a vertical plane fell with increasing age of the young (Fig. 9). As their motor skills increased with age, young were able to follow their mother's movements through imperfect tumbles. Mothers overcame this by performing an increasing proportion of perfect tumbles as their young grew (Fig. 9) and these continued to disorient and separate young from the teats. Although mothers maintained the success rate of tumbling by performing tumbles more precisely as their young grew, 'imperfect' tumbling was still used in situations where it was effective (e.g. when the young were drowsy). That tumbling was used as required by the situation is reflected by the proportionate increase with the age of the young in the number of times mothers were able to leave nestboxts unimpeded by the young (% passive, Fig. 9), i.e. young were more often free of the teats when older. Thus, tumbling by mothers varied with changes in the young's behaviour. Tumbling was not performed when it was not necessary. If a tumble failed the excursion was aborted, the mother withdrew into the nestbox, groomed the young, and tumbled again with greater speed and precision.

Apart from dislodging young from the teats, tumbling served another function which became more important with increasing age of the young. Any young which were close to the mother but not attached to her teats were also disoriented by tumbling while those further away were often not disturbed at all (e.g. Fig. 10). Young whose eyes had opened only a day or so before (i.e. ca. 65 days pp) were often surprisingly quick at recontacting mothers who tumbled but were slow leaving the nestbox. By 70 days pp young were very quick indeed. On a number of occasions mothers tumbled when young were grappling at their pouches and flanks even though none were attached to the teats. The tendency of young to follow their mother was greatest when they were close to her. By tumbling and disorienting nearby young, mothers thwarted their attempts to follow.

Overall, tumbling was the primary manner by which *Antechinus stuartii* mothers prevented young from riding with them outside the nestbox. When the young began making excursions away from the nest without the mother (ca. 77 days pp) tumbling declined rapidly.

SUCKLING: This section discusses suckling behaviour from two perspectives: (i) its possible influence in the control of maternal behaviour, and (ii) its nutritive function for the young.

A number of studies of marsupials indicate that the suckling stimulus provided by the young is important for initiation and maintenance of lactation. In *Trichosurus vulpecula* and *Macropus rufus* lactation is initiated when neonates are fostered to non–lactating parous and virgin surrogates (Sharman 1962, Sharman and Calaby 1964). More specifically, the mammary glands of macropodids are known to be able to produce a different type of milk depending on the age of the young (Bailey and Lemon 1966). Lactation is probably maintained by a combination of suckling stimulation and the quantity of milk suckled although these two variables are difficult to separate. Control mechanisms, including the suggested existence of endocrine systems regulating milk composition (Tyndale–Biscoe 1973) have not been studied in detail. However, Renfree (1979) has shown that milk secretion in *Macropus eugenii* is unimpaired by denervation of the mammary gland; and furthermore that oxytocin may be a facultative rather than obligatory hormone for milk discharge. Russell (1973) suggested that a combination of sucking stimulation and stimulation due to the physical presence of young could initiate and maintain maternal behaviour in marsupials. In the Macropodidae and those other marsupial species (e.g. *Trichosurus vulpecula*) where the norm is the birth of a single embryon, the stimulus provided by that young is sufficient. In some species where the litter contains several young the stimulation provided may either be additive, i.e. dependent upon the number of young successfully attaching to the teats, or there may be a threshold effect.

INFANT MORTALITY, SUB-MAXIMAL LITTER SIZE AND MATERNAL BEHAVIOUR: In *Didelphis virginiania,* where the average litter size is about seven (Collins 1973, Hunsaker 1977), lactation is not maintained when only a single young is present (Reynolds 1952). *Antechinus stuartii* females similarly do not rear single young. The minimum surviving litter size is the current study was two and Marlow (1961) reported a minimum of three young per litter for this species. Three young were maintained with no apparent changes in maternal response provided that the litter size had been stable since parturition. Only one out of five litters containing two young were

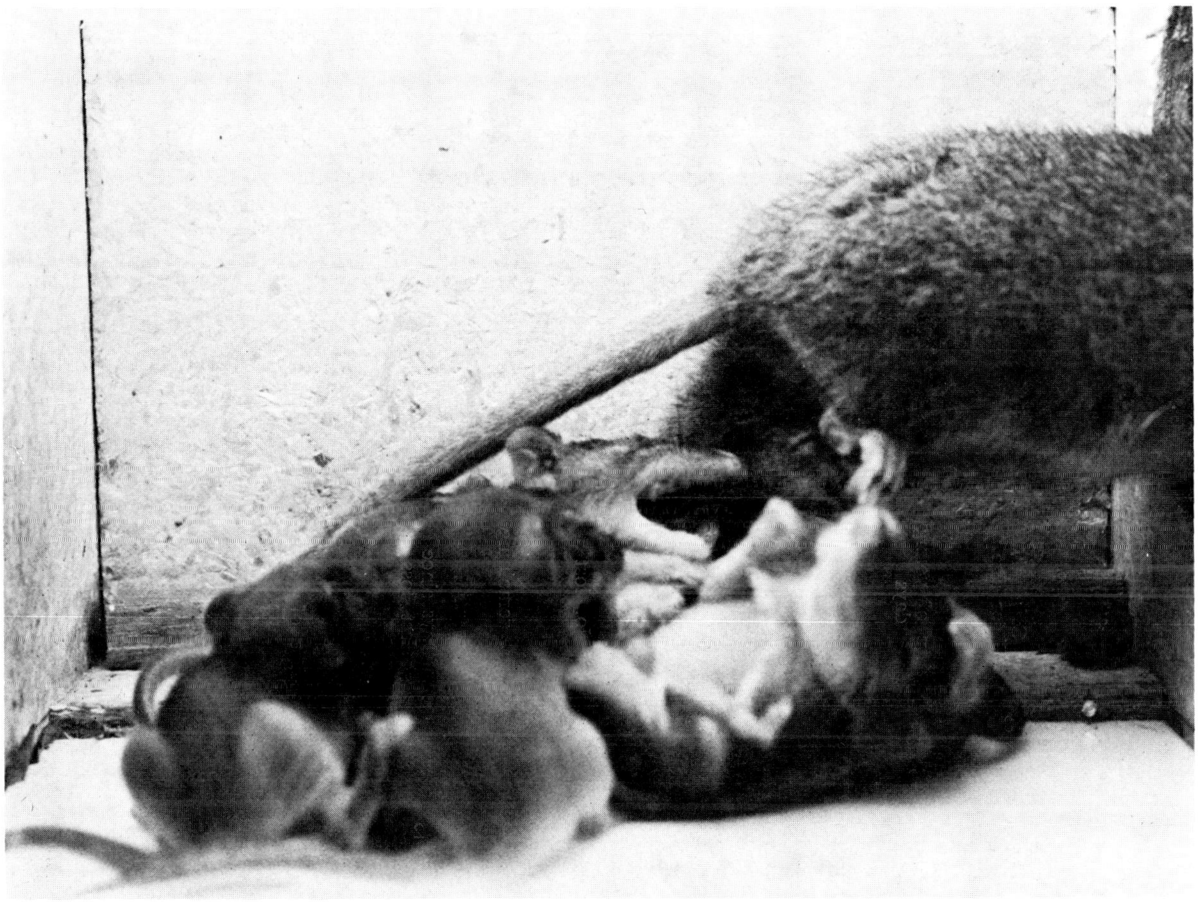

Fig. 10. The effect of "tumbling" by an *Antechinus stuartii* mother. Mother makes for nestbox entrance after "tumbling". Two of the 64–day–old young were attached to the teats, one is about to let go, the other's mouth is wide open. Another young who was close to the mother lies on its back disoriented, two more who were further away watch but are otherwise undisturbed. (A sixth young is obscured).

reared to independence by their mother (AS-108). In the other four litters the young disappeared when less than 30 days old. In the instance where two young were reared by AS108, two only were attached to the teats following parturition and their development was normal (Table 3).

Survival rates for litters containing six or more young were generally very high. However, if a mother ate some of her young after they were more than two weeks old she usually continued to do so at intermittent intervals over a period of a few weeks until none were left. The initial stimulus for eating young less than 60 days pp was probably the distrubance of the mother. Poor diets are another source of infanticide but it is well-known that stress disrupts normal patterns of maternal care (e.g. in tree shrews, D'Souza and Martin 1974). Infant mortality was higher during this study in those litters not housed in cages with observation nestboxes. This was attributed to the greater amount of disturbance to these mothers and young in order to examine their progress.

These findings indicate that the stimulation provided by the young in maintaining the maternal state of the female must be maintained above a minimum or threshold level. When the number of pouch young falls below a particular level, thought to be three for *Antechinus stuartii,* the stimulation provided is insufficient to maintain maternal care-giving and the young are usually eaten if still attached to the teats. Once the young are old enough to be left in the nest they are often deserted rather than eaten but mortality of these older infants due to the actions of the mother is low. Perhaps their large size increases the stimulus input to the mother per young. The ability of the mammary glands to respond correctly to the age of the suckling young in other marsupial species (Bailey and Lemon 1966) suggests that this is probably so.

Once mothers have eaten some of the small young they probably continue to do so because of the reduced stimulation provided by the survivors. No female (with original litter size > 2) ate all her young at one time. The common trend was for one or two to disappear, then some days later more and so on. This step-wise mortality suggests that it takes some days before the effects of the decreased stimulus from the young is expressed in the mother's behaviour. Moreover, the interval between infant losses usually shortens as the number of remaining young decreases (Table 2).

In *Antechinus stuartii* a mammary gland does not instantaneously regress following removal of the young but first becomes turgid and then diminishes in size over a week or so. In *Trichosurus vulpecula* there is a general effect on the suckling stimulus which causes the mammary glands to regress more slowly than post-parturient non-lactating females (Sharman 1962). Such effects imply some involvement of endocrine systems. Lamb (1975) cites numerous examples of lag times between hormone injections to rats and resultant changes in maternal behaviour. In no cases did hormone injection whether specific substances or 'cocktail' mixtures of them, reduce the latency to onset of maternal behaviour to less than 40 hours. We might expect a more protracted waning of maternal care to accompany gradually declining hormone levels. Russell and Giles (1974) suggested that hormones are probably

Table 2. Losses of young in litters of *Antechinus stuartii.*

		1st loss			2nd loss				3rd loss			
Serial No.	Original litter size	Age of young days	No. lost	No. remaining	Age of young days	No. lost	No. remaining	Days since last loss	Age of young days	No. lost	No. remaining	Days since last loss
AS128	8	31	3	5	42	2	3	11	48	3	0	5
751M	6	34	3	3	39	2	1	5	<40	1	0	<1
761KL	6	12	1	5	23	3	2	11	27	2	0	4
AS104	5	9	2	3	16	1	2	7	20	2	0	4
AS107	5	15	3	2	17	2	0	2				
757N	5	21	2	3	28	3	0	7				
753W	5	42	1	4	48	2	2	6	51	2	0	3
782C	4	3	1	3	46	1	2	8	49	2	0	3
AS7	2	2	2	0								
765KV	2	3	2	0								
767KV	2	2	2	0								
763KL	2	1	2	0								

Table 3. Litter size and age at which certain events occcurred in the development of *Antechinus stuartii* young. The serial number identifies the mother and adjacent is the year in which the young were born. C and AS prefixes to serial numbers are for animals captured in Muogamarra Nature Reserve, Cowan; suffixes give the capture localities of other animals (C = Muogamarra, H = Hornsby, KV = Kangaroo Valley, N = Narrabeen, SL = Smiths Lake, W = Wollombi). Please refer to text for explanation of development stages.

Serial No.	Year 19..	No. of young	left in nest	eyes open	investigate nest entrance	out of nest	away from nest
C3	'73	8	53	62	63	69	77
AS105	'74	8	53	64	66	72	79
751SL	'75	8	49	64	65	70	79
AS102	'74	7	54	64	67	71	80
757N	'75	7	50	63	66	71	76
AS101	'74	6	55	63	64	70	78
AS111	'74	6*	51	64	66	70	79
751C	'75	6	54	62	64	69	76
752C	'75	6	54	64	64	71	76
753C	'75	6*	56	64	68	72	80
754C	'75	6*	53	65	67	72	81
751W	'75	6*	54	64	64	70	75
752W	'75	6	51	63	66	71	74
752H	'75	6*	53	63	66	71	77
751N	'75	6*	53	64	65	71	77
761KV	'76	6*	54	63	65	70	80
762KV	'76	6	50	61	63	69	75
763KV	'76	6	55	62	64	69	78
762SL	'76	6	53	62	64	70	75
C5	'73	5	54	63	64	69	75
AS124	'74	5	54	64	65	69	78
751H	'75	5	51	63	64	70	78
C6	'73	4	53	62	63	71	77
AS112	'74	3	56	63	64	71	79
AS121	'74	3	53	61	64	69	74
764H	'76	3	55	63	67	70	78
761SL	'76	3	55	63	65	69	75
AS108	'74	2	54	62	63	70	76

* Litter size reduced to six soon after parturition.

important in initiating pouch cleaning in *Macropus eugenii* in addition to stimuli provided directly by the young. In three of the female *Antechinus stuartii* that lost their young during this study, pouch cleaning remained high for a week but then declined to levels similar to those of females in a non-reproductive state.

There is probably no simple explanation of the observed step-wise maternal infacticide of *Antechinus stuartii* particularly in the absence of any information on the hormonal and neural control of lactation and maternal behaviour in this species. There is a delay between infant mortality, mammary gland regressions and further acts of infanticide. The latency to further infanticide possibly involves changes in endocrine secretion and a lag between these changes and expression through the mother's behaviour.

RATES OF DEVELOPMENT OF THE YOUNG: Marlow (1961) provides growth rates for *Antechinus stuartii* young. In view of the proclivity of this species to eat young when disturbed (e.g. Woolley 1966), disturbance to animals was minimized in this study and so no attempt was made to weigh the young. General observations suggested that young within small litters were rather larger than those in large litters during the period from about 30 to 60 days pp. There was no obvious difference in the size of young between litters of three to eight members.

Table 3 gives the ages at which certain events occurred during development of litters of *Antechinus stuartii* (N = 28). Births were recorded within 24 hours. The five stages of development recorded were defined as follows: i; left in the nest: the first night on which the mother left all her young in the nest; ii; eyes open: both eyes of the young are just open, i.e. open as a narrow slit. A further 24 hours is required for the eyes of the young to open fully; iii; investigate nest entrance: the first day on which young sat at the nest entrance and looked out but did not move outside the nestbox; iv; out of nest: young moving in and out of the nest but never more than a few centimetres away from the entrance to the nestbox; v; away from the nest: young making excursions at least as far as the cage floor.

The data in Table 3 are insufficient to treat all litter sizes separately. The data were therefore cumulated into three groups: (i) litters of > 5 young, (ii) litters of 5 and 6 young, and (iii) litters of > 6 young. The rate development in these small, medium and large litters was then investigated.

There were no significant differences in development rates between the three litter sizes (Table 4). However, the age at which the young were first left in the nest yields a lower probability than do the other events. The ω^2 estimate for left in the nest indicates that around 12% of the variance was due to between litter size differences. Analysis is of variance for the number of days from 'left in the nest' to 'eyes open' between small, medium and large sized litters yields:

Total SS	110.71
Total SS	35.75
Error SS	74.96
F (2, 25)	5.96
P	0.04
ω^2	0.24

The result is significant at the 5% level. The ω^2 estimate has doubled. What this analysis indicates is that there are differences between small, medium and large litters in the interval 'left in the nest' to 'eyes open' but these are probably masked overall by variability among all litters in the rate of development. In other words as the young grow older it becomes more and more difficult to detect variations in development due to litter size. However, it is concluded that litter size does not affect the rate of development of the young. Nutrition of the young, therefore seems to be less dependent on litter size than the maintenance of the maternal drive.

DISCUSSION

Many mammals utilize a den or nest as a site for rearing altricial young until the youngs' motor skills have developed sufficiently to allow them to range further afield (Ewer 1968a). As a consequence the nest site becomes a centre of activity for the family unit as a whole, restricting not only the movement in and the relationship of the young with the physical environment but also placing strictures upon the movements of the parents themselves. Construction of a nest is common in many dasyurids; for example, *Dasyuroides byrnei* (Aslin 1974), *Sminthopsis crassicaudata* (Ewer 1968b), *Dasyurus viverrinus* (Fleay 1932). A nest provides protection both from the environment and predation, and during reproduction a stable site in which to rear the young. This latter function is especially important where immature young are not always carried with the female during foraging excursions as is the case with dasyurids. Both male and female *Antechinus stuartii* build nests (Rigby 1972) but, as the current study has shown, the nest building behaviour of the female is independent of her reproductive state and the development

Table 4. The effect of litter size on rates of development of *Antechinus stuartii young*. Data from Table cumulated into 3 litter-size classes; i.e. > 6, 5 and 6, < 5 young per litter. Mean ($\bar{x}$), variance (S^2) and range of values (Ra) within each litter class. Results of analysis of variance between classes. Probabilities are given to two decimal places. ω^2 (est.), i.e. proportion of variance in each development stage accounted for by litter size, calculated by the methods of Hays (1973).

litter size		left in nest	eyes open	investigate nest entrance	out of nest	away from nest
	$\bar{x}$	51.8	63.4	65.4	70.6	78.2
> 6	S^2	3.76	0.64	1.84	1.04	2.16
(N=5)	Ra	49-54	62-64	63-67	69-72	76-80
	$\bar{x}$	53.24	63.12	64.82	70.18	77.18
6 and 5	S^2	2.53	1.04	1.56	0.97	4.03
(N=17)	Ra	50-56	61-65	64-68	69-72	74-81
	$\bar{x}$	54.33	62.33	64.33	70.0	76.5
< 5	S^2	1.22	0.56	1.89	0.67	2.92
(N=6)	Ra	53-56	61-63	63-67	69-71	74-79
	$\bar{x}$	53.21	63	64.82	70.21	77.21
Overall	S	3.10	1.0	1.79	0.95	3.74
	Total SS	86.71	28.0	50.11	26.71	104.71
Analysis	Treat SS	17.52	3.70	3.10	1.04	7.94
of	Error SS	69.19	24.29	47.0	25.67	96.77
Variance	$F_{(2,25)}$	3.17	1.90	0.83	0.51	1.03
	P	0.21 ns	0.46 ns	0.79 ns	0.90 ns	0.73 ns
	ω^2	0.12	0.05	0	0	0

stage of her young. Thus nest building is not specifically linked to the maternal state of the female even though nest construction is advantageous in rearing the young. This result is in contrast to *Dasyuroides byrnei,* for Aslin (1974) noted an increase in nest-building behaviour in the period 0 to 30 days after birth. Ewer (1968b) also found that maternal *Sminthopsis crassicaudata* have an increasing tendency to carry bedding into the nest after birth of the young.

Over most of the youngs' development, female *Antechinus stuartii* spend more time grooming their young than themselves while in the nestbox. This is particularly true from 45 — 74 days after birth. Stimulation from the young is probably important in the initiation of this allgrooming behaviour (cf. Russell and Giles 1974). From 39 days pp, young are able to relinquish the teat and become more active until by 70 days pp they are making excursions from the nestbox. Thus the development stage in which the female grooms her young most, encompasses a period when the young are increasingly active but still largely dependent on the female for their wellbeing. Grooming serves to maintain the hygiene of the young and the nest. Females frequently groom the cloacal region of the young, stimulating the elimination of waste products which the female then consumes. Thus fouling of the nest site and perhaps the build up of a strong odour that could possibly attract a predator are prevented. In addition, grooming could conceivably be important in stimulating the somato–sensory system of the young and aiding its development. The female probably also monitors the responses of the young to her grooming as feedback on their developmental progress. *Antechinus stuartii* mothers spent an exceptionally long time grooming young that were weak and lethargic (and subsequently died).

Analysis of generalized sequences of maternal grooming show that the organization of this behaviour changes with increasing age of the young. The grooming sequences were highly variable but perhaps this variability is of no functional significance provided that the care–giving role of the mother is maintained. Ewer (1968b) predicted for *Sminthopsis crassicaudata* that analysis of the mother's grooming behaviour in the nest would show an association between self-grooming and grooming the young as demonstrated here for *Antechinus stuartii*. The finding that grooming sequences of *Antechinus stuartii* were labile in structure is not exceptional among marsupials. In the Tammar *Macropus eugenii* a high proportion of pouch cleaning (82%) is closely associated with other grooming activities, but pouch cleaning and other grooming activities are not ordered into a regular sequence (Russell and Giles 1974). Since *Antechinus stuartii* mothers interrupt grooming themselves to groom their offspring this suggests that the causal factors involved are hierarchically organized. If so the mechanisms are complex for the actions of the mother were not only to stimuli from the young but also lack of stimuli (i.e. movement and non–movement of the young).

Once the young begin to relinquish the teats (i.e. about 39 days pp), contact between mother and young must be maintained if maternal care is to continue. In the nest the mother uses "pull–under" retrieving behaviour to gather young dispersed about the nest, especially when the young make distress calls or "convulsing" movements. This form of retrieval behaviour is also used by female *Dasycercus cristicauda* (Ewer 1968b) and *Dasuroides byrnei* (Aslin 1974) and probably other dasyurid species. Once the young become more independent in their movements, and particularly when they begin to make excursions outside the nestbox, the young take a much greater role in maintaining proximity with their mother and perform "self–retrieval" behaviour. This switch in the relative roles of the mother and young in maintaining proximity as the youngs' locomotor abilities improve also occurs in macropodid marsupials (Russell 1973, Croft 1981).

Table 5. A summary of methods used by dasyurid mothers to free young from the teats before leaving the nestbox.

Species	Nudging and grooming	Body shaking	Rapid locomotory movements	Tumbling	Source
Antechinomys laniger	+	–	–	–	Happold 1972
Antechinus flavipes	+	–	+	++	(see also Fleay 1949)
Antechinus stuartii	+	(+)	+	++	pers. obs.
Antechinus swainsonii	+	–	+	+(?)	pers. obs.
Dasycercus cristicauda	?	++	–	–	Fleay 1961
Dasyuroides byrnei	+	+	+	–	Aslin 1974
Dasyurus maculatus	++	–	–	–	Settle 1978
Planigale tenuirostris	+	–	–	–	pers. obs.
Sminthopsis murina	+	once only	+	–	pers. obs.

During the period when the mother leaves her young in the nest to make foraging excursions outside, "tumbling" is performed to induce young to relinquish the teats. Observations on *Antechinus flavipes* and a single *Antechinus swainsonii* mother were made, and similar behaviour was performed by these species. The behaviour has not been reported for other dasyurid species (Table 5); for these shaking or rapid locomotory movements are more common. Nudging and grooming the young to induce teat relinguishment is the behaviour pattern that is common to most species studied (Table 5). The rather extreme form of behaviour (tumbling) used by *Antechinus* mothers suggests little communication between mother and young in relation to teat relinguishment. This is probably a consequence of the relatively brief period (54–63 days pp) in which mothers leave all their young in the nest while foraging outside.

The stimulation the mother receives from her youngs' suckling behaviour appears to be important in the maintenance of her maternal behaviour. Successful rearing of less than three young is extremely rare in *Antechinus stuartii* and loss of young often leads to acts of infanticide and progressive loss of the whole litter. It is suggested that the stimulation the female receives from each suckling young is additive and a threshhold exists below which females cease to maintain a litter. Among the Dasyuridae, only *Sarcophilus harrisii* are known to rear a single young (Green 1967, Guiler 1970) but 80% of litters contain greater numbers (Guiler 1970). Settle (1978) reported a case of the successful rearing of a single young by a *Dasyurus maculatus* mother but this was probably exceptional. Godfrey (1969) noted that litters of *Sminthopsis larapinta* containing only one or two young were never reared.

Although maternal care depends on the female maintaining the full complement of the young from birth, litters of different sizes are produced (or at least attach to the teats after birth) and reared successfully. The development rates of young in small (2–5), medium (5–6) or large (6–8) litters was independent of litter size. Thus either the nutrition the female provides, and probably other maternal care, are adjusted to the litter size, or development rates are relatively fixed provided adequate maternal care is given. Probably both factors are involved.

The failure of female *Antechinus stuartii* to rear small litters (< 3) or litters in which mortality occurs at early stages is surprising for a species that is monoestrus and the males die after a single breeding season (Wood 1970, Braithwaite 1973). However, females can breed again in a second and sometimes a third year (Lee Bradley and Braithwaite 1977) and so there is some insurance against the loss of a litter. Superfoetation may also assure that the number of young successfully attaching to the teats at birth exceeds the minimum viable litter size. In other respects, the maternal behaviour of the female is well developed and comparable to other Dasyurid species that have been studied. The female is highly responsive to stimuli presented by her young and her behaviour in many respects is tuned to their developmental stage.

ACKNOWLEDGEMENTS

Dr. Eleanor Russell was instrumental in the inception and formulation of this study. Her advice and assistance throughout the project is gratefully acknowledged. Thanks are given to Cathy Settle for her assistance in the maintenance of captive antechinuses. Advice and information kindly provided by Mr. David Fleay is gratefully acknowledged. Dr. J. F. Eisenberg and Dr. E. Russell provided useful criticism of a draft of this paper. During the course of this study, and until his death, the senior author was a PhD candidate supervised by Dr. D. Croft.

REFERENCES

Aslin, H., 1974. The behaviour of *Dasyuroides byrnei* (Marsupialia) in captivity. *Z. f. Tierpsychol.* **35:** 183–208.

Bailey, L. F. and Lemon, M., 1966. Specific milk proteins associated with the resumption of development by the quiescent blastocyst of the lactating Red Kangaroo. *J. Reprod. Fertil.* **11:** 473–75.

Braithwaite, R. W., 1973. "An ecological study of *Antechinus stuartii* (Marsupialia: Dasyuridae)". M.Sc. thesis: University of Queensland, Brisbane.

Collins, L. R., 1973. "Monotremes and marsupials". Smithsonian Institution Press: Washington.

Croft, D. B., 1981. Behaviour of Red Kangaroos, *Macropus rufus* (Desmarest, 1822) in northwestern New South Wales. *Aust. Mammal.* **4:** 5–58.

D'Souza, F. and Martin, R. D., 1974. Maternal behaviour and the effects of stress in tree shrews. *Nature* **251:** 309–11.

Ewer, R. F., 1968a. "Ethology of mammals". Logos Press: London.

Ewer, R. F., 1968b. A preliminary survey of the behaviour in captivity of the dasyurid marsupial, *Sminthopsis crassicaudata* (Gould). *Z. f. Tierpsychol.* **25:** 319–65.

Fleay, D., 1932. Swainson's Phascogale (the "bush mouse"). *Vic. Nat.* **49:** 132–4.

Fleay, D., 1934. The Brush-tailed Phascogale. First record of breeding habits. *Vict. Nat.* **51:** 88–100.

Fleay, D., 1949. The Yellow-footed Marsupial Mouse. *Vic. Nat.* **65:** 273–7.

Fleay, D., 1961. Breeding the Mulgara. *Vict. Nat.* **78:** 160–7.

Fleay, D., 1965. Australia's 'needle-in-a-haystack' marsupial. *Vict. Nat.* **82:** 195–204.

Godfrey, G. K., 1969. Reproduction in a laboratory colony of the marsupial mouse *Sminthopsis larapinta* (Marsupialia: Dasyuridae). *Aust. J. Zool.* **17:** 637–54.

GREEN, R. H., 1967. Notes on the Devil *(Sarcophilus harrisii)* and the Quoll *(Dasyurus viverrinus)* in northeastern Tasmania. *Rec. Queen Victoria Mus. (Tas.)* **27:** 1–13.

GUILER, E. R., 1970. Observations on the Tasmanian Devil *Sarcophilus harrisii* (Marsupialia: Dasyuridae) II. Reproduction, breeding and growth of the young. *Aust. J. Zool.* **18:** 63–70.

HAPPOLD, M., 1972. Maternal and juvenile behaviour in the marsupial Jerboa, *Antechinomys spenceri* (Dasyuridae). *Aust. Mammal.* **1:** 27–37.

HAYS, W. L., 1973. "Statistics for the Social Sciences". 2nd Ed. Holt, Reinhart and Winston: New York.

HUNSAKER, II, D., 1977. Ecology of New World marsupials. Pp.95–156 *in* "The biology of marsupials", ed. by D. Hunsaker II. Academic Press: New York.

HUTSON, G. D., 1976. Grooming behaviour and birth in the dasyurid marsupial *Dasyuroides byrnei. Aust. J. Zool.* **24:** 277–82.

LAMB, M. E., 1975. Physiological methanisms in the control of maternal behaviour in rats: A review. *Psychol. Bull.* **82:** 104–19.

LEE, A. K., BRADLEY, A. J. AND BRAITHWAITE, R. W., 1977. Corticosteroid levels and male mortality in *Antechinus stuartii.* Pp.209–20 *in* "The biology of marsupials", ed. by B. Stonehouse and D. Gilmore. Macmillan Press: London.

MARLOW, B. J., 1961. Reproductive behaviour of *Antechinus flavipes* (Waterhouse) Marsupialia, and the development of pouch young. *Aust. J. Zool.* **9:** 203–18.

RENFREE, M. B., 1979. Initiation of development of diapausing embryo by mammary denervation during lactation in a marsupial. *Nature* **278:** 549–51.

REYNOLDS, H. C., 1952. Studies on reproduction in the opossum *(Didelphis virginiana). Univ. Calif. Berkeley, Publ. Zool.* **52:** 223–84.

RIGBY, R. G., 1972. A study of the behaviour of caged *Antechinus stuartii. Z. f. Tierpsychol.* **31:** 15–25.

RUSSELL, E. M., 1973. Mother–young relations and early behavioural development in the marsupials *Macropus eugenii* and *Megaleia rufa. Z. f. Tierpsychol.* **33:** 163–203.

RUSSELL, E. M. AND GILES, D. C., 1974. The effects of young in the pouch on pouch cleaning in the Tammar Wallaby, *Macropus eugenii* Desmarest (Marsupialia). *Behaviour* **51:** 19–37.

SETTLE, G. A., 1978. The quiddity of Tiger Quolls. *Aust. Nat. Hist.* **19:** 164–9.

SHARMAN, G. B., 1962. The initiation and maintenance of lactation in the marsupial *Trichosurus vulpecula. J. Endocrinol.* **25:** 375–85.

SHARMAN, G. B. AND CALABY, J. H., 1964. Reproductive behaviour in the Red Kangaroo *Megaleia rufa,* in captivity. *CSIRO Wildl. Res.* **9:** 58–85.

TYNDALE–BISCOE, H., 1973. "Life of marsupials". Edward Arnold: London.

WOOD, D. H., 1970. An ecological study of *Antechinus stuartii* (Marsupialia) in a southeast Queensland rainforest. *Aust. J. Zool.* **19:** 347–53.

WOOLLEY, P., 1966. Reproduction in *Antechinus* spp. and other dasyurid marsupials. *Symp. Zool. Soc. Lond.* **15:** 281–94.

WOOLLEY, P., 1974. The pouch of *Planigale subtilissima* and other dasyurid marsupials. *J. Roy. Soc. of W.A.* **57:** 11–5.

RESUMEN

35. Comportamiento maternal de *Antechinus stuartii* en cautividad. Por G. A. Settle y D. B. Croft.

De los ambientes naturales se capturaron hembras preñadas de *Antechinus stuartii.* Se. observó el cuidado maternal, después del parto, en las cajas de cautividad y en sus vecindades por medio de registros de video. Los cuidados maternales de los hijos de *Antechinus* son la provisión de protección y sostén en el marsupio y ulteriormente en un nido, calor, acicalamiento y nutrición. Sin embargo, la construcción del nido fue independiente del estado de preñez de la hembra, y por eso no relacionado directamente al nacimiento o desarrollo de los jóvenes. El acicalamiento del joven crece proporcionalmente con el desarrollo de éste, especialmente durante el período en que los jóvenes abren sus ojos y se destetan. Luego el acicalamiento en el nido decrece desde que el joven comienza a hacer excursiones fuera del nido. Cuando el joven se ha destetado y es más movil (60–80 dias pp.) el acicalamiento deviene en propio de la madre.

Una vez que el joven se ha destetado y mientras su corrdinación motora es pobre la hembra mantiene la proximidad a su hijo usando un ocmportamiento de "pull–under". En cuanto las actividades del joven devienen más independientes y mejor coordinadas, los jóvenes son primariamente responsables de mantener su proximidad dentro del círculo madre–cría. A fin de que los jóvenes queden en el nido mientras la madre hace su excursión, ésta hace uso del particular comportamiento de "brincar" o "tumbarse" para desalojarlos del marsupio.

Cuando las pariciones son de menos de tres hijos raramente éstos son criados hasta el final. Sin embargo, el tamaño de la parición no tiene un significativo efecto en el desarrollo del joven. Se discute la impor tancia del estímulo de la succión del lactante en el mantenimiento del comportamiento maternal.

36. THE DEVELOPMENT OF EXPLORATORY BEHAVIOUR IN *Antechinus stuartii* (DASYURIDAE, MARSUPIALIA) YOUNG IN CAPTIVITY

G. A. SETTLE and D. B. CROFT[1]

Exploratory behaviour of juvenile *Antechinus stuartii* develops progressively from the nestbox into various parts of the cage once the youngs' eyes have opened. Females and their litters of six young were studied. The youngs' activity increases markedly once their eyes open and continues to increase until observations were terminated at 100 days post-partum. Exploratory behaviour similarly increases as the young mature. Initially it is confined to the nestbox, entrance, then just outside the nestbox until finally the whole cage is explored. As each new environment is encountered, exploration in the now familiar areas tends to decline although the total time invested in exploratory behaviour continues to increase. The relationship between the mother's activities and the development of exploratory behaviour in the young is discussed. Some comparisons are made on the development of exploratory behaviour in marsupial and eutherian young.

Pages 383–96 *in* CARNIVOROUS MARSUPIALS, ed. by M. Archer, February, 1982. Roy. Zool. Soc. New South Wales: Sydney, Australia.

EXPLORATION is a broad category of behaviours which assist an animal to become familiar with its environment. This behaviour is not the only way of gaining knowledge of the environment or the stimulus qualities of some object, but it does contain a clear statement of choice within the animal, i.e. it may choose to explore or not explore. Shillito (1963) divides exploratory behaviour into two components. Firstly investigation which is behaviour orientated to unfamiliar objects. And secondly reconnaissance which is exploratory behaviour in a familiar environment, possibly to renew or revise information about it. Information about the environment is primary gained by a process of latent learning i.e. learning without obvious reinforcement (Alcock 1975). Like other broad categories of behaviour (e.g. play) more precise definitions are difficult to formulate (Hinde 1970, Barnett and Cowan 1976).

A great many of the authors who have dealt with exploratory behaviour have been concerned with an animal's response to novel objects or new places under artificially contrived circumstances which bear little if any resemblance to a normal mode of existence. This is especially true of the many studies on movement patterns and elimination of laboratory rodents in an "open field" (Barnett 1976). Some authors (Russell and Pearce 1971, Barnett and Cowan 1976) have attempted to place their observations into a more comprehensive framework by outlining differences in sensory capabilities, modes of life and the possible contributions made by a caged environment to the observed reactivity of subjects.

Exploration may be associated with many factors in the environment (e.g. Berlyne 1960, Glickman and Hartz 1964, Hinde 1970, Bateson 1973, Humphrey 1973). Although novelty is conducive to exploratory behaviour in most circumstances, this is not always so. Familiarity with the surrounding environment is important (Russell 1967), and non–novel objects are sometimes investigated for just as long as novel ones (Fentress 1967 and unpublished observations cited by Fentress 1976).

Hinde (1970: 143) points out that during exploration the animal may attend to all changes in the environment or focus on particular ones; or both may occur. With eye–opening, the ability to the young to locate the source of stimulus is markedly increased. At this time a great many new behavioural patterns develop as vision accelerates the maturation of an integrated control of behaviour no longer solely dependent upon thermotactile and olfactory input (Hinde 1970, Roseenblatt 1976). In examining the development of exploratory behaviour in *Antechinus stuartii* we therefore include not only the exploratory movements of the young but also orienting responses, of which visual exploration of the environment forms an important part. Berlyne (1960:80) would refer to much of this behaviour as 'diversive exploration.'

Exploratory behaviour cannot be divorced from the general pattern of an animal's activities as it forms a specific subset of such activities. We consider first the activity patterns of young *Antechinus stuartii* from 30 to 100 days pp, both in relation to changes with age and changes in the mother's behaviour. Then, against this background the development of exploratory behaviour is investigated.

---

[1] School of Zoology, University of New South Wales, P.O. Box 1, Kensington, New South Wales, Australia, 2033. Graham Settle died in 1979.

METHODS

SUBJECTS AND HOUSING: Pregnant female *Antechinus stuartii* were obtained from the wild. Only those with litters of six young were observed; larger litters were reduced to six within a day of birth.

The general housing and maintenance of subjects is given in Settle and Croft (1982).

MEASUREMENT OF ACTIVITY PATTERNS IN DEVELOPING YOUNG: Unless otherwise stated, the behaviour of the young was scored from a video record of a standard one-hour sampling period during the mother's post-sunset activity peak (see figure 3 in Settle and Croft 1982). The data for Fig. 3 (this paper) were scored as follows: An estimate of the time spent active by one young was calculated using the equation:

$$\% \text{ scored time active per young} = \frac{\Sigma \text{ Ym}}{6 \times \text{m}} \times 100$$

where Ym is the number of young active during each scored minute, 6 is the number of young in each litter, and M is the number of minutes for which scores were made. Data were not collected unless at least half the young could be seen, unobscured by other individuals. Young outside the nestbox, either on the mother's back or independent of her (i.e. not attached to the teats) were scored as active. Insufficient data were collected for young less than 40 days old and have been omitted. The data are grouped into 5–day class intervals for the age of the young and the median solid, mean broken and range of values are given for each interval.

A slightly different method was used for gathering data for Table 1. The data were collected for states of the mother; (i) mother absent, i.e. outside the nestbox, (ii) mother present inside the nestbox and active (iii) mother present and inactive. Observations were made on 14 *Antechinus stuartii* females with litters of six. So as not to bias scores towards a particular litter a maximum of ten minutes was permitted for each of the three above categories on any night for the same litter. Some observations were made from video–records (as above) but most of the data comes from observations made later at night direct from the video monitor.

Results are expressed in the same manner as for Fig. 3; however, the scoring method was more stringent. The activity of the young was estimated every 30 seconds. All six young had to be visible and the mother's state had to remain unchanged for at least two minutes before scoring commenced. The latter prerequisite was necessary to reduce the effect of the mother's previous state on the activity of her young. For example a mother may groom her young and they become active as a result, the mother then curls up and goes to sleep but the young may take some minutes to become inactive again. In practice very few scores were made when the mother was asleep, especially when the young were less than 70 days old because mothers almost always curl around or over the young when sleeping. Inactive mothers most frequently crouched or lay by their young. A second criterion for scoring was that the mothers should maintain the same state for at least two minutes after scoring commenced.

In table 1, the estimate marked with an asterisk (*) refers to four occasions on which young less than 50 days old were left in the nest. One young was left behind on each of the three occasions and two on one other. They were soon retrieved by the mother and remained active for the short period they were alone. The data refer only to the young left in the nest and the prerequisite two–minute delay before scoring was waived.

DEFINITION AND MEASUREMENT OF THE YOUNGS' EXPLORATORY BEHAVIOUR: Quantitative data were scored from video-recordings

Table 1. The percentage of 3–second scoring periods spent active per young by litters of *Antechinus stuartii* when the mother was absent, present and active, present and inactive. Number in italics give the number of scoring periods on which estimates are based. Please see text for further details, including special explanation of the figure marked *.

CIRCUMSTANCE

Age–Class of Young (days)	♀♀ absent		♀♀ present–active		♀♀ present–inactive	
	% activity per young	No. 30 sec. periods scored	% activity per young	No. 30 sec. periods scored	% activity per young	No. 30 sec. periods scored
30–39	–	–	30.9	*61*	14.9	*87*
40–49	100*	*6*	53.5	*82*	27.0	*135*
50–59	22.3	*154*	44.1	*175*	18.0	*196*
60–69	38.9	*974*	72.0	*534*	26.4	*238*
70–79	44.1	*1137*	77.9	*816*	25.1	*629*
80–89	60.9	*1979*	72.8	*780*	42.4	*534*
90–99	55.1	*1587*	63.7	*941*	53.6	*492*

made during standard one–hour sampling periods. These were supplemented by observations made at other times to ensure that the samples were representative of events. The eyes of the young in each litter opened at a mean of 63.1 days (± 0.42 SE) and the age of each litter was adjusted so that eye–opening occurred on day 63 for each litter. This procedure helps to correct for differences in the rate of development between litters (see Settle and Croft 1982), and the term *standardized age* is used to indicate this. Figure 1 shows the standardized age at which video–samples were made for nine litters of *Antechinus stuartii* whose mothers are identified by the serial numbers on the left of the figure.

young were freely moving about their cages and beyond that time it is unlikely that the amount of exploration by these captive animals is a valid assessment of events in the field. Four categories of exploration were used.

i) Exploration of the Interior of the Nestbox, (NB): Young moved about inside the nestbox as if searching for something. They investigated objects (debris, bits of food, corners of the nest, edges of the glass observation panel, the floor, the roof etc., and the inner lip of the entrance hole), other animals and sometimes places which had no particular features (e.g. a uniform part of a wall).

Fig. 1. The standardized age at which video–samples were made of the behaviour of nine litters of *Antechinus stuartii.* Serial numbers of mothers are given out on the left.

Before eye–opening the movements of young about the nestbox may be broadly grouped into three categories: i) movements towards other sibs or the mother, or both; ii) movements away from others to rest or sleep in some other part of the nestbox, and iii) general movements about the nestbox which do not seem to be associated with either i) or ii). In the last category movements could be associated with exploration or exercise, or a combination of both. However, they may be related to other events or functions, such as the development of co-ordinated locomotor skills. Occasionally young of this age displayed interest in some object and 'nosed' it several times before moving on. In short, although there is some indication that young do begin to explore before their eyes open, no quantitative estimate of the extent to which they do so can be reliably determined.

We shall consider in some detail the development of exploratory behaviour in the standardized age period between eye-opening (63 days) and 90 days old. By the end of this period all

ii) Exploration Associated with Nestbox Entrance, (E): Young climbed into the nestbox entrance or attempted to (Fig. 2a), and inspected the entrance itself or gazed into the cage. The latter undoubtedly includes components of visual, auditory and olfactory stimulation. Frequently orientating of the head in a new direction indicated that the attentional process of that animal had shifted to a new source of stimulation. When in the nestbox entrance the young spent very little time orientated towards the nestbox interior. As they grew in size, the young were able to look out through the entrance without climbing into it (Fig. 2b).

iii) Exploration Associated with the Area just Outside the Nestbox Entrance, (OE): In each cage a branch, obliquely angled from the cage floor to the entrance of the elevated nestbox, was the only pathway by which young could move between the nestbox and the general cage environment. Before they began to explore the whole cage, young often moved out of the nestbox entrance and began to explore the

Fig. 2. (a) Juvenile *Antechinus stuartii* investigating in the nestbox entrance. (b) Juvenile *Antechinus stuartii* investigating the area just outside the nestbox entrance.

upper end of the branch, or simply crouched on it and looked about the cage (Fig. 2b). Later the young ventured further and further down the branches and spent a long time investigating irregularities in the surface. The time in which young spent exploring the upper one–third of these branches, excluding activies such as grooming and feeding in which they often engaged, was scored.

iv) Exploration Away from the Nest-box, (AN): Young exploring any part of the cage not covered by any of the above categories, i.e. general exploration of the cage environment.

The feature common to each of these categories is the location of the exploring individual rather than the location of the source of stimulation. Manipulative solitary play (including 'tail–pulling') was regarded as investigative and is therefore included in scores.

As an index of exploratory behaviour a score of the number of young who spent more than 15 seconds of each 30–second interval exploring was cumulated for each observation session and each of the four categories. This result was divided by six (the number of young in each litter) and expressed as a percentage of time. Data were then grouped into two–day intervals and the mean percentage of time per young per litter calculated. The results in the following figures are expressed in this form. Note that the range of values given indicates the range of mean values per young between litters, and not for individual young.

RESULTS

ACTIVITY RATES IN DEVELOPING YOUNG AND THE EFFECT OF THE MOTHER'S PRESENCE: The activity of young *Antechinus stuartii* generally increased as they matured from 40 to 100 days old (Fig. 3). The figure should be interpreted with some caution. The data were gathered during the post–sunset activity peaks of the mothers and activity of the mother tends to increase the activity of the young (see below). Also, for young less than 54 days old (i.e. carried attached to the teats of the mother during her excursions) scoring was practical only when the mother was inside the nestbox. Even then sampling time was reduced still further because resting mothers normally curl around or hunch over their young. The latter observational difficulties diminished as the young grew in size (and hence were more obvious) and as their mother spent more time away from the nestbox. Possible effects of sampling bias are indicated on Figure 3 by question marks. However, the results clearly show that the young *Antechinus stuartii,* once left in the nest, become more active as they get older.

If the presence and activity of the mother is taken into account (Table 1), then *Antechinus stuartii* young 50–100 days old are more active when the mother was present and active than when she is absent. Although we can test this difference (Wilcoxon matched pairs, P(α) = 0.05), to do so is probably not justifiable. Beyond ca. 70 days the young were easier to see and increasingly active. The mother was visibly disturbed by the activities of the young when she attempted to rest; in turn she spent more time active (Fig. 4) and by her own activity tended to keep the young active when near them. When the mother was away these older young often rested and in general the activity of the young was more spasmodic than that of the mother. From around 80 days post–partum onwards the results probably reflect artifacts of the captive confinement;

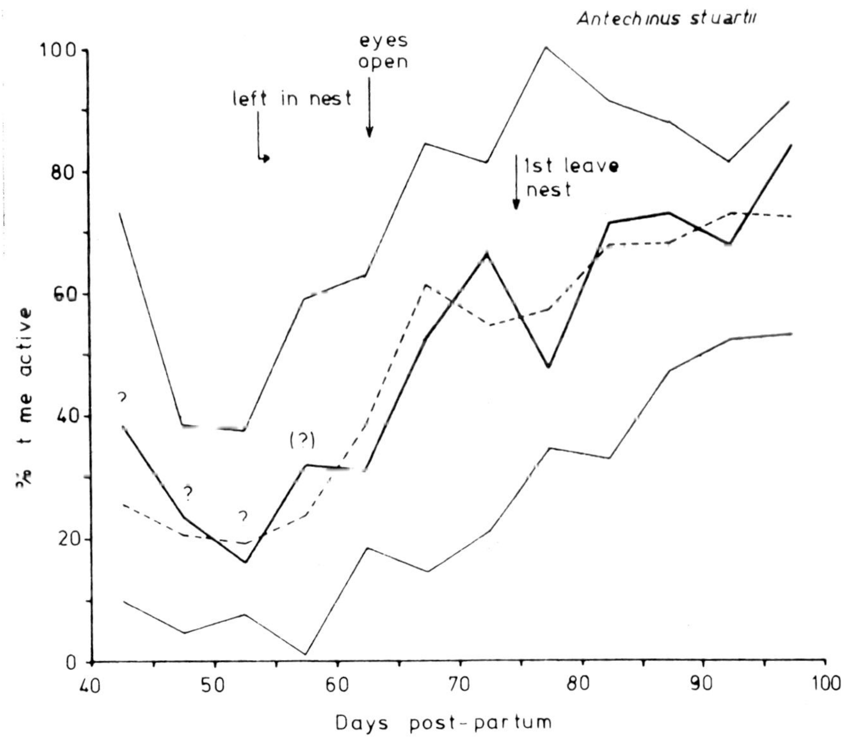

Fig. 3. The percentage of time *Antechinus stuartii* young spent active as they matured from day 40 pp to day 100 pp. (See text for further details).

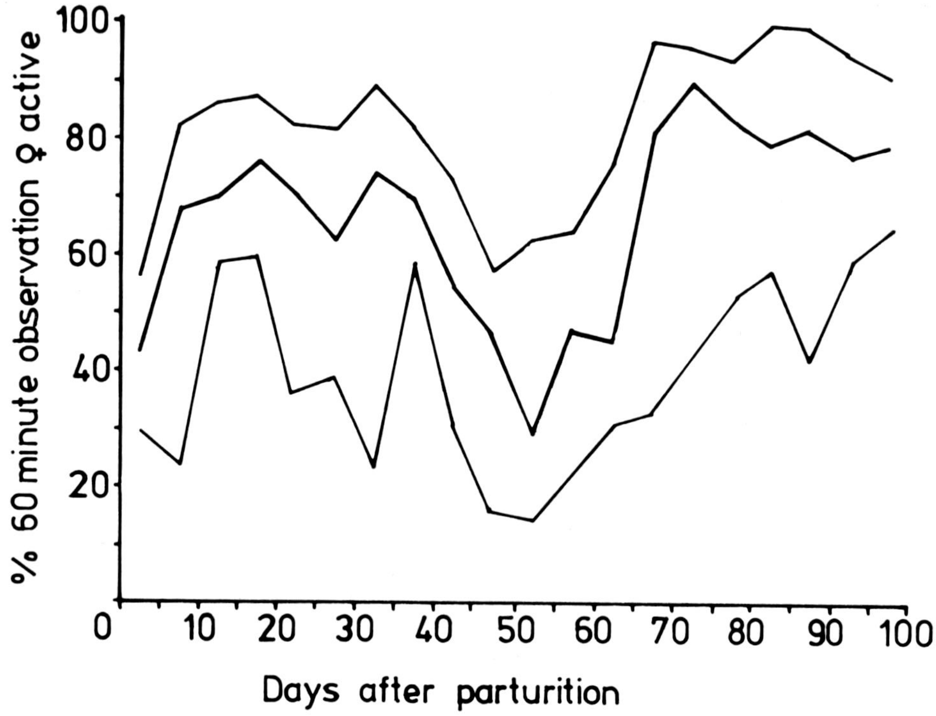

Fig. 4. The percentage of time *Antechinus stuartii* females spent active from parturition until their young were 100 days pp. Mean bounded by range is shown.

by analogy a tendency towards 'perpetual motion' caused by the inability of the individuals concerned to spatially separate themselves during periods of activity.

Young in the presence of active mothers were more active than those whose mothers were inactive (Table 1, Wilcoxon matched pairs, P(a) = 0.01). The basis for testing these data is more sound since there is only a single variable in the state of the mother (her activity); where above there were two (proximity and activity).

If we compare the activity of the mother (Fig. 4) with that of the young (Fig. 3) bearing in mind the effects of the mother's activity and that both sets of data come from the same series of observations, there are some interesting parallels and divergences. Both young and mother show depressed levels of activity about the time the young are first left in the nest (ca. 54 days pp.). For the young this is true whether the sampling period is defined in terms of the mother's activity cycle (Fig. 3) or in terms of the mother's activity state (Fig. 5). It is also true for the young independent of the activity state of the mother (Fig. 5). The transitionary periods between different states and levels of activity of the mother tend to compensate for one another. For example, if a mother groomed her young and then rested it took some time before the young became quiescent, thus averaging their activity between her two states. Thus there is little difference between the activity of ca. 54 day old young when their mother was present and inactive, and their activity when she was away from the nestbox.

The activity of both young and mother increases beyong 55 days pp. This is most obvious during periods when the mother is relatively active (Fig. 3 and 4 and combination of a and b in Fig. 5) but is delayed if we consider only times during which the mother is inactive (Fig. 5c).

When the young first begin to make short excursions from the nest (from ca. 75 days pp.) there was a transient decrease in the proportion of time they spent active. The decrease is more apparent in the median of scores than in the mean (Fig. 3). In other words the young's activity tended to come in bursts, they were either very active or resting. In Fig. 5 these effects are masked by the sampling method, cumulation of data into large age–class intervals of the young, and greater number of subject litters. The latter is important because all litters do not develop at precisely the same rate. If an event like first excursion from the nest causes brief transient changes in behaviour, between litter variations in the timing of the event will average out the data.

Sudden bursts of activity by *Antechinus stuartii* were observed. Although quite active inside the nestbox after their eyes had opened, their movements were less energetic than once they had started to venture from the nest. The young would then wake, emerge from the nestbox, clamber about branches for several minutes, re-

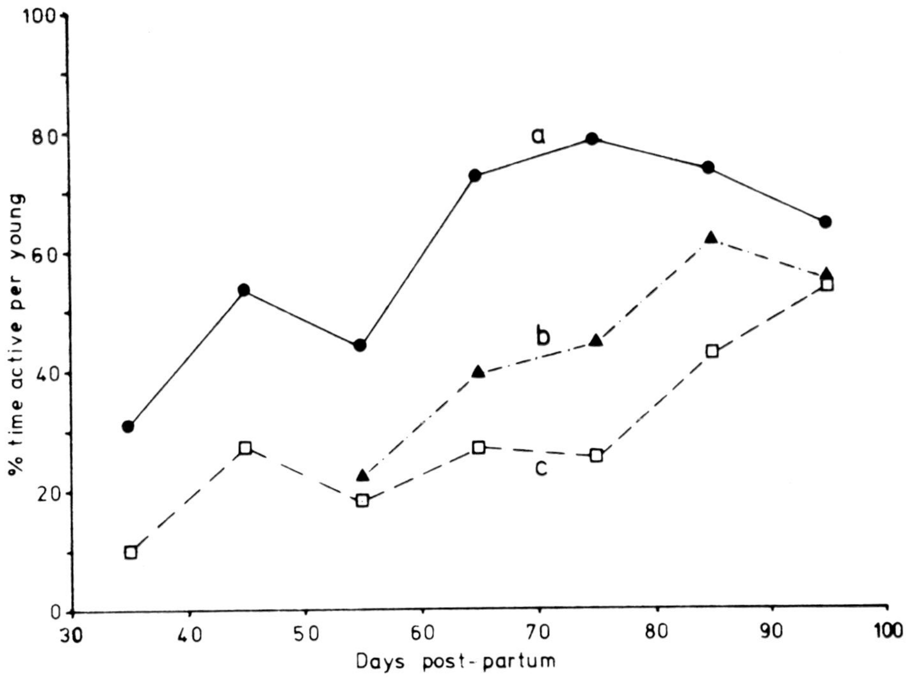

Fig. 5. The percentage of time *Antechinus stuartii* young (35 to 95 days pp) spent active when their mother was (a) present in the nestbox and active, (b) absent from the nestbox and active, (c) present in the nestbox but inactive.

enter the nestbox and fall asleep (see the following section). There are likely to be physiological correlates here associated with the training and exercise of the young's muscles, as Fagen (1976) suggested in his review of exercise, play and physical training in animals. Possibly these initial bursts of high activity by young *Antechinus stuartii* led to muscle fatigue and hence, periodic resting.

As the young *Antechinus stuartii* approach 100 days old their activity becomes more independent of the movements of the mother (Fig. 5). At the same age small groups of two or three young are often encountered in the field well away from the nest site. General observations suggest that by around 100 days pp. the young have well–developed activity rhythms similar to those of the mother. Thus, although we still find that the level of the young's activity during the post–sunset activity peak is still climbing after 90 days pp. (Fig. 3) and by 100 days pp. has reached or exceeded that of the mother's (Fig. 4), scores of activity collected later in the evening are considerably lower (Fig. 5).

TOTAL EXPLORATORY BEHAVIOUR: From the time when the eyes of the young opened (63 days) there was a progressive increase in the total time they spent exploring up until 84 days (Fig. 6). The last point on the figure suggests that exploration may continue to increase beyond an age of 90 days but this was not so for these sampling periods. Data collected during standard sampling periods for young older than 90 days fluctuate in a random manner between 30 and 55 percent (mean values). However, observations in larger cages and in the field suggest that *Antechinus stuartii* young do spend increasing amounts of time exploring beyond 90 days old. Probably the restricted environment within a cage does not provide sufficient environmental complexity to stimulate further increases in exploration at this age. These older young were certainly stimulated to explore for longer uninterrupted periods if, for example, branches were replaced in their cages with new ones.

On Fig. 6, the first observed occurrences of three major stages in the development of young's excursions from the nest are indicated. The horizontal bars indicate the spread of days over which these events were first observed in the nine litters to which the data refer. The arrows show the mean days of first occurrence. The results suggest that these events are associated with more rapid increases in the time spent exploring and are followed by small plateaus.

Exploratory Behaviour in Various Parts of the Cage: The changes with age in the percentage of time the young spent engaged in exploratory activities when in different locations in the cage are shown in Fig. 7 for: (a) inside the nestbox and (b) at the entrance to the nestbox, (c) just outside the nestbox entrance and (d) well away from the nestbox. Note that the ordinate axes are not drawn to the same scale in each graph.

(i) Exploration confined to the nestbox and its entrance: The young began to explore the confines of the nestbox as soon as their eyes

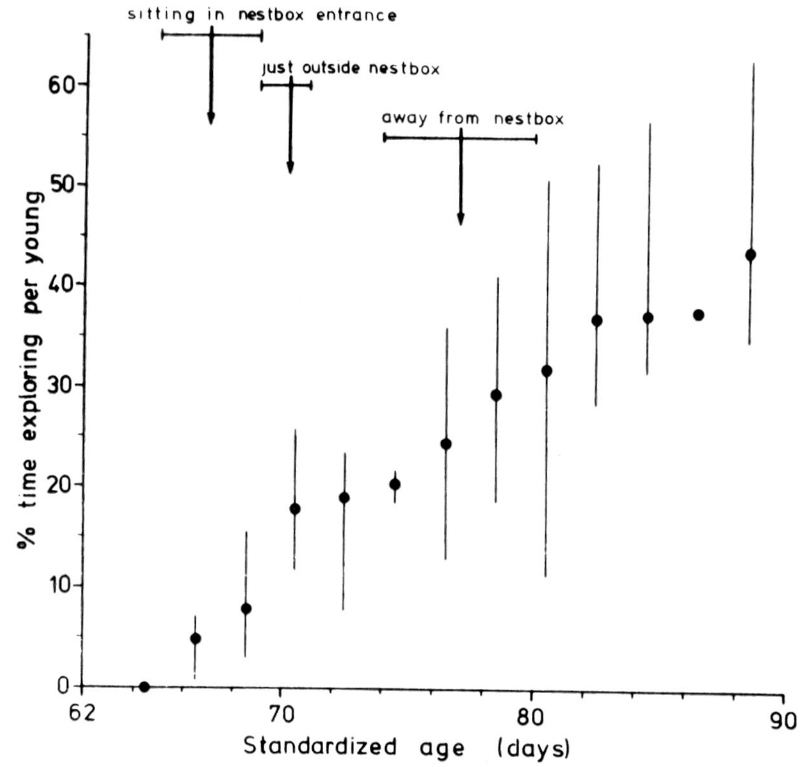

Fig. 6. The change with age in the percentage of time young *Antechinus stuartii* spent exploring from 66 to 90 days pp.

were open (Fig. 7), and became increasingly attentive to the movements of the mother out of the nestbox. The first observed explorations of the nestbox entrance took place immediately after the mother left the nest when an infant attempted to follow her. Figure 2a shows a young looking out through the nestbox entrance on the same day the eyes of all the young in this litter opened. The stance of this infant is typical of young between 63 and 66 days old

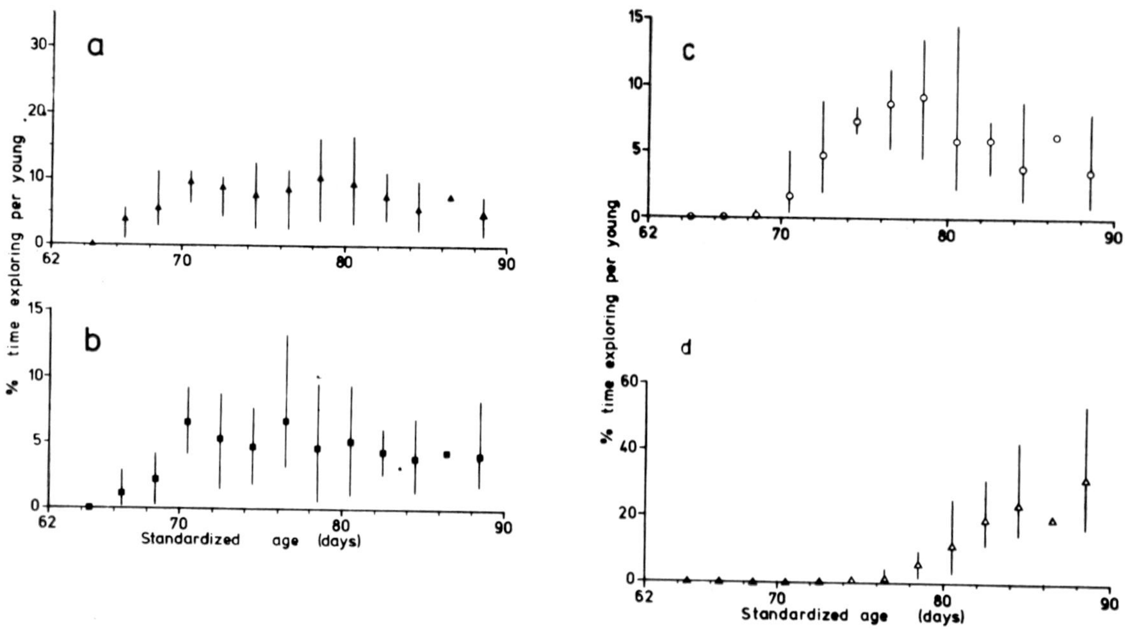

Fig. 7. The change with age in the percentage of time *Antechinus stuartii* young (from 66 to 90 days pp) spent exploring (a) inside the nestbox, (b) at the nestbox entrance, (c) just outside the nestbox entrance, and (d) well away from the nestbox.

when they lacked the co-ordinated locomotory movements and strength necessary to climb into and balance in the entrance hole provided.

When young moved into the nestbox entrance they too were watched by other sibs (e.g. Fig. 2a). After these initial events the young began to explore at the nestbox entrance at times during the mother's excursions other than immediately after she had left the nest. The motor skills of the young improved rapidly and they began to follow others into the entrance. By the time they were 68 days old they often climbed into the entrance when the mother was inside the nestbox and active, but rarely did so when she was inactive. Time spent exploring both the nestbox and its entrance continued to increase (Figure 7a, b). The young became more responsive to the movements of other young from the nestbox into the entrance and of the mother either into the entrance or out through it. This led to bouts of exploratory activity at the nestbox entrance. Sometimes a whole litter clustered around the entrance, all trying to climb in even though there was only room for three or four. At other times when the young were active inside the nestbox, none occupied the entrance. As a consequence of the design of the nestboxes these bouts tended to enhance scores for exploration inside the nestbox while concomitantly depressing scores for exploration at the entrance.

(ii) Exploration and excursions outside the nestbox: There was a marked increase in the amount of visual inspection of objects and events (e.g. movements of the mother or prey) outside the nestbox during the last day or so before the young began to emerge from the entrance (ca. 70 days old). As a consequence the young spent more time exploring in the entrance (Fig. 7b).

The first excursions from the entrance onto the branch just outside took place in one of three ways. The most common method was for one young to be pushed out of the entrance by young climbing into the entrance from inside the nestbox. These young usually investigated a few features of the branch (sniff) and then forced their way back into the entrance of the nestbox. These 'compulsory' excursions were soon followed by voluntary ones, sometimes only a few minutes later. Early voluntary excursions always took place when at least one other sibling, usually more (rarely the mother, who was usually well away from the nestbox when these events occurred), was sitting in the entrance or on the branch just outside it. The first infant to emerge usually 'hesitated' for some time, sniffing the branch or touching it with one forepaw and withdrawing several times before moving out onto it. If no others moved out to join this innovative individual, it soon returned to the nestbox. When one moved out there was an increasing trend for others to follow with less hesitation. These first groups remained out for longer than did a single young.

A few early emergences from the nestbox took place when young followed their mothers out of the entrance or tried to jump onto mothers who were just outside the entrance. These were clearly attempts to contact mothers and climb onto their backs. If unsuccessful the young made their way back into the nestbox. Mothers did not wait for their young to catch up nor did young attempt to contact mothers more than a few centimetres away from the entrance to the nestbox. The manner in which mothers moved into and out of nestboxes seems important here. If they paused just outside for more than a few seconds after emerging from the nestbox they usually re–entered, either remaining inside, or emerging again and moving swiftly down the branch to the cage floor. The rapidity of this movement away from the nestbox is probably the reason why the young do not follow.

The proportion of time spent exploring just outside the nestbox entrance increased after 70 days (Fig. 7c) and by 74 days the young were moving from inside the nestbox to the branch outside quite freely. By this age the amount of time exploring inside the nestbox and at the entrance had stabilized and overall increases in time spent exploring were due to increased exploration outside the nestbox. Movement of the young out of the nestbox whilst the mother was inside was rare at these times.

From around 75 days onwards the young spent more time exploring just outside the nestbox and began to do so when their mother was inside. As with exploration at the entrance, there was a limit to the number of young who could gather on the top of the branch and individuals were often displaced to positions further down. It was at these times when young first began to explore farther from the nestbox. Individuals initiated short exploratory excursions down the branch leading to the floor. They were often followed or joined by one or two others as they moved away from the nestbox. Time spent exploring just outside the nestbox reaches a peak between 76 and 79 days (Fig. 7c) and then falls as the young begin to make more frequent and long excursions away from the nest. From this time on increases in the total time exploring are due to increases in eploration away from the nestbox (Fig. 7d).

Time spent exploring outside the nestbox is a fair indicator of total time spent outside up until about 85 days of age. From then on other activities (eating, grooming, etc.) become more frequent and the young become more independent from other sibs during exploratory forays into the cage. The first excursions from the nestbox were usually made as small groups but later the same evening individuals frequently moved about independent of others. This seems to have been a by-product of the small size of cages as groups persist in the wild to about 160 days of age.

DISCUSSION

The overall proportion of time *Antechinus stuartii* young spend active and engaged in exploratory activities increases as a function of age after their eyes open. The results show that exploration does not continue to increase in previously explored parts of the caged environment once the young begin to explore a new part of the environment. Instead, overall increases in the proportion of time spent exploring are due to increases in the time the young spend exploring the new part of the environment. When the young begin to explore a new part of the environment the time they spend exploring at now familiar places decreases with age. Exploration just outside the nestbox decreased markedly when the young began to make excursions further from the nest.

Each of the exploration categories refers to a subsequent stage in the development of the young corrected at the day on which the eyes open for earlier differences in developmental rates; the first observed occurrence of each subsequent stage did not overlap with any other. Therefore, the development of excursions from the nest has at least four discrete stages.

However, we cannot assume that the increasing exploration at the nestbox entrance by young *Antechinus stuartii* aged between 63 and 69 days was due to more than a general increase in activity after their eyes opened. Comparison of the percentage of time the young spent exploring at the nestbox entrance (Fig. 7b) with the percentage of time the same young spent active in any manner during the same observation periods (Fig. 3) reveals that the rate of increase in activity is the greater of the two. In addition, there were no differences between the rate of increase of exploration inside the nestbox (Fig. 7a) and that at the entrance (Fig. 7b) during this period of development. Thus exploratory behaviour seems to develop at a slower rate than general activity at this developmental stage.

The novelty of newly encountered parts of the environment may be sufficient to explain why the young explore when they move into that part of the environment but does not suggest why the young had not advanced into that sector in a more gradual way, or at an earlier age, or why they paused for some time before exploring further. The development of perceptual and motor skills are clearly constraining factors when young *Antechinus stuartii* are small (as in many other animals, e.g. in moose: Altmann 1963, in langurs: Jay 1963, in kangaroos: Russel 1973, in lions: Schenkel 1966), but in the last few days before the young begin to explore outside the nest these skills do not appear to be limiting.

The motor skills of 73–day old *Antechinus stuartii* (i.e. those who had spent time outside the nestbox but not away from it) were sufficient to enable them to run up the branch leading to the nestbox after being placed on the cage floor. Only occasionally did the young of this age have difficulty in finding the correct and only way back to the nestbox. By contrast young about 68 days old (i.e. with little or no time spent outside the nestbox) rarely chose the correct path back. These younger infants certainly knew where the nest was located (and indicated this by moving across the floor in the direction of the nestbox which they were then unable to reach) but they apparently possessed an incomplete knowledge of the cage environment. They were, however, quite able to climb the branch to the nestbox when placed upon it.

These observations suggest that the young learnt a great deal about the structural organization of the cage environment before they had moved about it. The most plausible means for the young to gather this knowledge is by visual exploration of the cage while sitting in the entrance to the nestbox or on the branch immediately outside it. It is unlikely that these young individuals were unable to perceive the environment in a meaningful way.

If one repositioned the upper end of the branch leading to the nestbox so that it terminated against a blank part of the wall and placed another familiar branch so that it then led to the nestbox, an adult female placed on the cage floor would try to return to the nestbox via the first branch several times before she used the new one. This suggests that *Antechinus* rely upon a knowledge of known pathways rather than perceiving small changes in the environment.

As indicated above, 74 day–old young possessed both a knowledge of the cage environment and motor skills sufficient to find their way back to the nest yet they rarely ventured further from the nest than just outside the entrance. Familiarity with the environement is known to be an important factor influencing the exploratory behaviour of animals (Hutt, 1966, Russell 1967, Sale 1969, Hinde 1970, and Barnett and Cowan 1976), and seems to be important in limiting the movement of young *Antechinus stuartii* away from the nest. Moreover, from the above observation of 74 day–old infants, it is probable that visually acquired knowledge of the environment is the complementary component of familiarity but this will lead to decreasing novelty and the young animals will seek further changes in stimulus input. By doing so they encounter new and unfamiliar segments of the environment. If the dicrepancy between the perceived stimuli originating in new environment and those of the previous one is small exploration may continue but large changes may elicit fear. To this extent fear and exploration are overlapping categories of behaviour (Hinde 1970).

However, it is not known whether fear was a general factor inhibiting exploration of the cage environment by *Antechinus stuartii* young. If fear was important, then it would seem most likely to be associated with the complexity of the cage environment rather than the stimulus qualities of particular objects. But the natural environment is far more complex than that provided in the cages used in this study. Even when the young *Antechinus* retreated from newly encountered stimuli, e.g. sand on the floor of the cage, they did not retreat for long. Fearfulness of new sectors of the environment could influence the overall state of the young and delay entry. However, it will be difficult to identify such components because if they are operative under general conditions then they will also be masked by the normal pattern of behavioural development.

Perhaps one of the more imporant factors regulating the stage–wise development of the environmental exploration in *Antechinus stuartii* young is the tendency of the young to cluster together near the nest. Attachment to the nest site and attachment to other sibs, either alone or in combination, are probably involved in this behaviour.

Attachment to the nest site is clearly strong. Both captive and hand–caught free–living *Antechinus stuartii* young about 74 days old, when liberated nearby, rapidly return to nests from which all other animals have been removed. When mothers and their young are moved to new cages of identical construction to the old ones the young wander away from the nestbox, but when the old nestboxes are placed in the same new cages this does not happen.

Attachment to sibs may also slow progression to the next stage of environmental exploration. Young rarely moved into new places unless others were nearby, especially when they first began to explore just outside the nestbox and later when they first began to make excursions farther into the cage. This will tend to promote exploration by groups (see below) but will hinder exploration of new places by individuals if others do not follow.

Thus far we have primarily dealt with casual factors which may operate to slow the progression of the young to the next stage of environmental exploration. Both external and internal factors are viewed as important. In an operational sense these factors acted in two ways: i) to impede progression to the next encountered part of the environment (e.g. motor skills, attachment to the nest site, unfamiliarity with the new environment), and ii) to retain the attentional process of the young at the current stage of environmental exploration by providiing sufficient stimulation at that place (e.g. the novelty of the current location, whether in terms of stimuli directly associated with that place or as a consequence of stimuli received at that place from another place; attachment to sibs). However, it was clear that some factors did not have discrete boundaries and could operate in both positive and negative ways (e.g. familiarity, unfamiliarity, novelty and fear). We shall now consider some factors promoting progression from one stage of environmental exploration to the next.

It was clear that infants first climbed into the nestbox entrance not to explore but in an attempt to maintain contact with the mother. Exploration at the nestbox entrance thus began as a consequence of the infant's presence there, presumably in response to a new set of stimuli once attention was no longer focused on the mother. Attachment to the mother did not contribute in a direct manner to either the first exploratory forays by the young from the nestbox to the branch outside, or to the first ones away from the proximity of the nestbox. Mothers leaving nests were frequently followed by young as far from the nest as the young ventured during the current stage of environmental exploration when not accompanied by the mother. Following of the mother increased in the period 64 to 78 days post–partum and declined thereafter. Whilst this may be conducive to further exploration by young following the mother to that location, it was apparent that the tendency to follow the mother back into the nestbox was greater than the tendency to follow her out. However, this does not mean that following of the mother is unimportant to the development of environmental exploration in the young.

Settle and Croft (1982) described the changes with age of the young in the temporal parameters of the mother's excursions away from them. After 64 days pp. the duration of the mother's excursions away from the young increased. Further, there was a trend towards an increasing frequency of the mother's excursions until about 70 days post–partum after which excursion frequency gradually declined. It seems unlikely that these changes in the behaviour of the mother correspond by chance to the mean age at which the eyes of the young opened and the mean age at which they first began to venture from the nestbox respectively.

Therefore, a combination of the form of the mother's excursions away from the young and following of the mother by the young produce what is effectively a casual factor which acts in an indirect manner to promote the transition from one stage of the environmental exploration to the next (i.e. NB → E → OE → AN). Any increase in the time the young spend exploring in the entrance to the nestbox, on the branch just outside the entrance, will shorten the period of time between stages of environmental exploration because it will tend to cancel out most of the factors considered above as acting to prolong the period of time at a given stage. But if (as indicated above) young follow mothers back into the nestbox more readily than they follow them out, how does following of the mother increase the time the young spend at each of the above locations?

The following behaviour of *Antechinus stuartii* young and the form of the mother's excursions away from the young (i.e. few, many, short or comparatively long) may combine to change the ratio of the time the young spend in the nestbox. For example, when the young are between 66 and 71 days old the following events occur.

Frequent excursions of short duration by the mother will increase the amount of time that young spend in the entrance (and hence the amount of time exploring there) if the young have already retired to the nestbox when the mother returns. This was often true at the beginning of stage E; young remained in the nestbox entrance for only brief periods before moving back inside the nestbox.

As the young grew older they spent an increasing amount of time in the entrance. Although the mean duration of the mother's ex cursions increased only slightly, mother's began to make a number of longer excursions. The brief excursions by the mother at the end of stage E ought to lead to an increase in the time the young spend inside the nestbox because they tend to follow the mother in when she returns. Although the young may independently return to the entrance after following their mother into the nestbox (i.e. before the mother leaves again). As the young become older they both follow the mother in and then independently return to the entrance more often; especially just before they begin to venture from the nestbox. Thus the high frequency of the mother's excursions does not intefere with exploration at the entrance by these older young.

In summary, it appears that a large number of short excursions by the mother may enhance exploration at the nestbox entrance by the young in the period just after they begin to explore there. Later, when the young spend more time in the entrance and move into it more readily other factors, e.g. following of other sibs, are likely to be more important.

During the next stage of environmental exploration (i.e. when the young explore just outside the nestbox entrance but go no further, OE) a similar sequence of events takes place. Initially the mother's excursions are frequent and of short duration, and thus encourage movement of the young into and out of the nestbox, increasing familiarity of the young with the process of leaving and entering. As in the case above, the young at first spend only brief periods at this newly encountered part of the environment, and they often retreated to the nestbox entrance when the mother returned from short excursions. As the young became increasingly familiar with the region just outside the nestbox and moved out more freely, they spent more time there. Over the same time period the mother's excursions became longer and less frequent. The increase in duration of her excursions by the time the young were ready to move further from the nest was substantial but total time spent away was similar. This allowed the young to spend longer periods outside the nestbox exploring before their mother returned, since they still followed her back into the nestbox.

Within each of the two stages considered above (E, OE), the following behaviour of the young and the form of the mother's excursions away from them seem jointly to promote the development of environmental explorations in two ways. Firstly, in the period immediately after the young begin to make exploratory forays into a new part of the environment, the time they spend at the part of the environment may be augmented. This may accelerate the process of exploration at and familiarization with that part of the environment. Secondly, following of the mother into and out of the nest may assist them to become more familiar with rapid movement from one point to another, since the movements of the mother are usually rapid. Hence the combined effects seem to operate rather differently in the early and late part of each stage. Neither the form of the mother's excursion nor following by young foster exploration directly. Both may do so by influencing factors directly causal to exploration, e.g. familiarity factors, time spent at the place in question, free movement of the young between that place and another familiar place. There are numerous other more directly related reasons why these two areas of the behaviour of mother and young should be so organized, especially in the field. Changes in the form of the mother's excursions, for example, allow her to find increasing quantities of food necessary to meet the increasing nutritional requirements of the young. Following behaviour of the young may lead to receipt of maternal care and, once the young start to move out of the nest predator avoidance. The purpose of dwelling here on what may be considered 'minor factors' overall, is to illustrate some of the more subtle features of the system we are dealing with. This is important because it demonstrates that the individual behaviour of mother and young tends to mesh together in areas which at first level of examination appear unrelated.

Few other studies, especially for marsupials, have looked specifically at the development of exploratory behaviour in a young mammal. Adult dasyurids usually create a familiar home environment with marking behaviour and will investigate and often mark novel objects placed into such an area (Croft 1982). The form and latency of their investigatory behaviour depends on how intimidatory the novel object is (e.g. *Sminthopsis crassicaudata*. Ewer 1968). In a comparison of responses to novel objects by one or two species from four marsupial families, Russell and Pearce (1971) concluded that of the more carnivorous species *Dasyuroids byrnei* displayed more object investigation than did the herbivorous and insectivorous species. In a broader study, Glickman and Sroges (1966) looked at object investigation by various vertebrate orders. They concluded that within an order, species with complex feeding strategies and low risk of predation show more exploratory behaviour than species which are herbivorous and risk high predation. Between orders, the two marsupials studied (*Didelphis virginiana, Macropus rufus*) were ranked with insectivores and edentates as showing relatively few responses to novel objects as compared to rodents, carnivores and primates.

On current knowledge, it is difficult to place the results of these studies and our observations on *Antechinus stuartii* into a developmental perspective. Marsupials potentially are weaned from their mother after exposure to the external environment over a longer developmental period than eutherians. In macropodids, young frequently investigate the immediate environment of the mother while carried in her pouch with her head out (Russell 1974, Croft 1981a, b). Possibly the same is true for species such as dasyurids that spend the later part of their development in a nest. However, this early exposure to the environment from the pouch may be less significant as *Antechinus stuartii* young have relinquished the teats before overt exploratory activities occur. Eye opening correlates with the onset of exploratory activity and it increases progressively thereafter. In the Tiny Musk Shrew *Crocidura bicolor,* eye opening occurs 12 days pp., investigatory behaviour of objects commences at day 13, and exploration at day 17 when the young first leave the nest (Dippenaar 1979). Thus the development of exploratory behaviour may be similar in both marsupial and altricial eutherian young reared in nests. Eye-opening is the common starting point of this developmental sequence. The same is not true, however, for a macropid young leaving the pouch for the first time and the precocial young of, for instance, an ungulate leaving the womb in parturition.

ACKNOWLEDGEMENTS

Dr. Eleanor Russell was instrumental in much of the conception and development of this study. Her help and advice is gratefully acknowledged especially her valuable criticisms of an earlier draft of this paper. Ms. Debbie Andrew also constructively criticised a draft of this paper. Thanks are given to Cathy Settle for her help in maintaining the Antechinuses. During the course of this study, and until his death, the senior author was a PhD student of Dr. D. B. Croft.

REFERENCES

Alcock, J., 1975. "Animal behaviour: an evolutionary approach". Sinauer: Massachusetts.

Altmann, M., 1963. Naturalistic studies of maternal care in Moose and Elk. Pp.233–253 *in* "Maternal behaviour of mammals", ed. by H. L. Rheingold. Wiley: New York.

Barnett, S. A., 1976. "The rat: a study in behaviour". Revised ed. ANU Press: Canberra.

Barnett, S. A. and Cowan, P. E., 1976. Activity, exploration, curiosity and fear: an ethological study. *Interdiscip. Sci. Rev.* **1:** 41–62.

Bateson, P. P. G., 1973. Preference for familiarity and novelty: a model for the simultaneous development of both. *J. Theoret. Biol.* **41:** 249–59.

Berlyne, D. E., 1960. "Conflict, arousal and curiosity". McGraw–Hill: New York.

Croft, D. B., 1981a. Social behaviour of the Euro, *Macropus robustus* (Gould) in the Australian arid zone. *Aust. Wildl. Res.* **8:** 13–49.

Croft, D. B., 1981b. Behaviour of Red Kangaroos, *Macropus rufus* (Desmarest, 1822) in northwestern New South Wales. *Aust. Mammal.* **4:** 5–58.

Croft, D. B., 1982. Communication in the Dasyuridae: a review. Pp.293–309 *in* "Carnivorous marsupials", ed. by M. Archer. Roy. Zool. Soc. New South Wales: Sydney.

Dippenaar, N. J., 1979. Notes on the early post–natal development and behaviour of the Tiny Musk Shrew, *Crocidura bicolor.* Bocage, 1889 (Insectivora: Soricidae). *Mammalia* **43:** 83–91.

Ewer, R. F., 1968. A preliminary survey of the behaviour in captivity of the dasyurid marsupial, *Sminthopsis crassicaudata* (Gould). *Z. f. Tierpsychol.* **25:** 319–65.

Fentress, J. C., 1967. Observations on the behavioural development of a hand–reared male Timber Wolf. *Amer. Zool.* **7:** 338–51

Fentress, J. C., 1976. Dynamic boundaries of patterned behaviour: interaction and self-organization. Pp.33–169 *in* "Growing points in ethology", ed. by P. P. G. Bateson and R. A. Hinde. Cambridge University Press: London.

Glickman, S. E. and Hartz, K. E., 1964. Exploratory behaviour in several species of rodents. *J. Comp. Physiol. Psychol.* **58:** 101–4.

Glickman, S. E. and Sroges, R. W., 1966. Curiosity in zoo animals. *Behaviour* **26:** 151–88.

Hinde, R. A., 1970. "Animal behaviour: a synthesis of ethology and comparative psychology". 2nd ed. McGraw–Hill: New York.

Humphrey, N. K., 1973. 'Interest' and 'pleasure': two determinations of a monkey's visual preferences. *Perception* **1:** 395–416.

Hutt, C., 1966. Exploration and play in children. *Symp. Zool. Soc., Lond.* **18:** 61–81.

Jay, P. C., 1963. Mother–infant relations in langurs. Pp.202–304 *in* "Maternal behaviour in mammals", ed. by H. L. Rheingold. Wiley: New York.

Rosenblatt, J. S., 1976. Stages in the early behavioural development of altricial young of selected species of non-primate mammals. Pp.345–383 *in* "Growing points in ethology", ed. by P. P. G. Bateson and R. A. Hinde. Cambridge University Press: London.

Russell, E. M., 1967. The effect of experience of surroundings on the response of *Lebistes reticulatus* to a strange object. *Anim. Behav.* **15:** 586–94.

Russell, E. M., 1973. Mother–young relations and early behavioural development in the Marsupials, *Macropus eugenii* and *Megaleia rufa*. *Z. f. Tierpsychol.* **33:** 163–203.

Russell, E. M., 1974. The biology of Kangaroos (Marsupialia–Macropodidae). *Mammal. Rev.* **4:** 1–59.

Russell, E. M. and Pearce, G. A., 1971. Exploration of novel objects by marsupials. *Behaviour.* **40:** 312–22.

Sale, P. F., 1969. A suggested mechanisms for habitat selection by the juvenile Manini *Acanthusus triostegus sandvicensis* Streets. *Behaviour.* **35:** 27–44.

Schenkel, R., 1966. Play, exploration and territoriality in the wild Lion. *Symp. Zool. Soc. Lond.* **18:** 145–54.

Settle, G. A. and Croft, D. B., 1982. Maternal behaviour of *Antechinus stuartii* (Dasyuridae, Marsupialia) in captivity. Pp.365–81 *in* "Carnivorous marsupials", ed. by M. Archer. Roy. Zool. Soc. New South Wales Sydney. This symposium.

Shillito, E. E., 1963. Exploratory behaviour in the short–tailed Vole, *Microtus agrestis*. *Behaviour.* **21:** 145–54.

RESUMEN

36. El desarrollo del comportamiento exploratorio en jóvenes de *Antechinus stuartii* (Dasyuridae, Marsupalia) en cautividad. Por G. A. Settle y D. B. Croft.

El comportamiento exploratorio de jóvenes de *Antechinus stuartii* se dasarrolla progresivamente desde la caja–nido hacia varias partes de la jaula una vez que han abierto los ojos. La actividad de los jóvenes crece marcadamente una vez que han abierto los ojos y continúa aumentando hasta los 100 días que duraron las observaciones posteriores al parto. De la misma manera el comportamiento exploratorio se incrementa con la madurez del joven. Inicialmente está confinado a la caja-nido, luego a su entrada, más adelante hasta afuera de ella hasta que finalmente es explorada toda la jaula. A medida que cada nuevo ambiente es registrado la exploración en esa nueva área tiende a declinar, aunque el tiempo total dedicado a la exploración continúa incrementándose. Se discute la relación entre la activdad de la madre y el desarrollo del comportamiento exploratorio en los jóvenes. Se hacen algunas comparaciones del desarrollo del comportamiento exploratorio en los jóvenes marsupiales con el de los euterios.

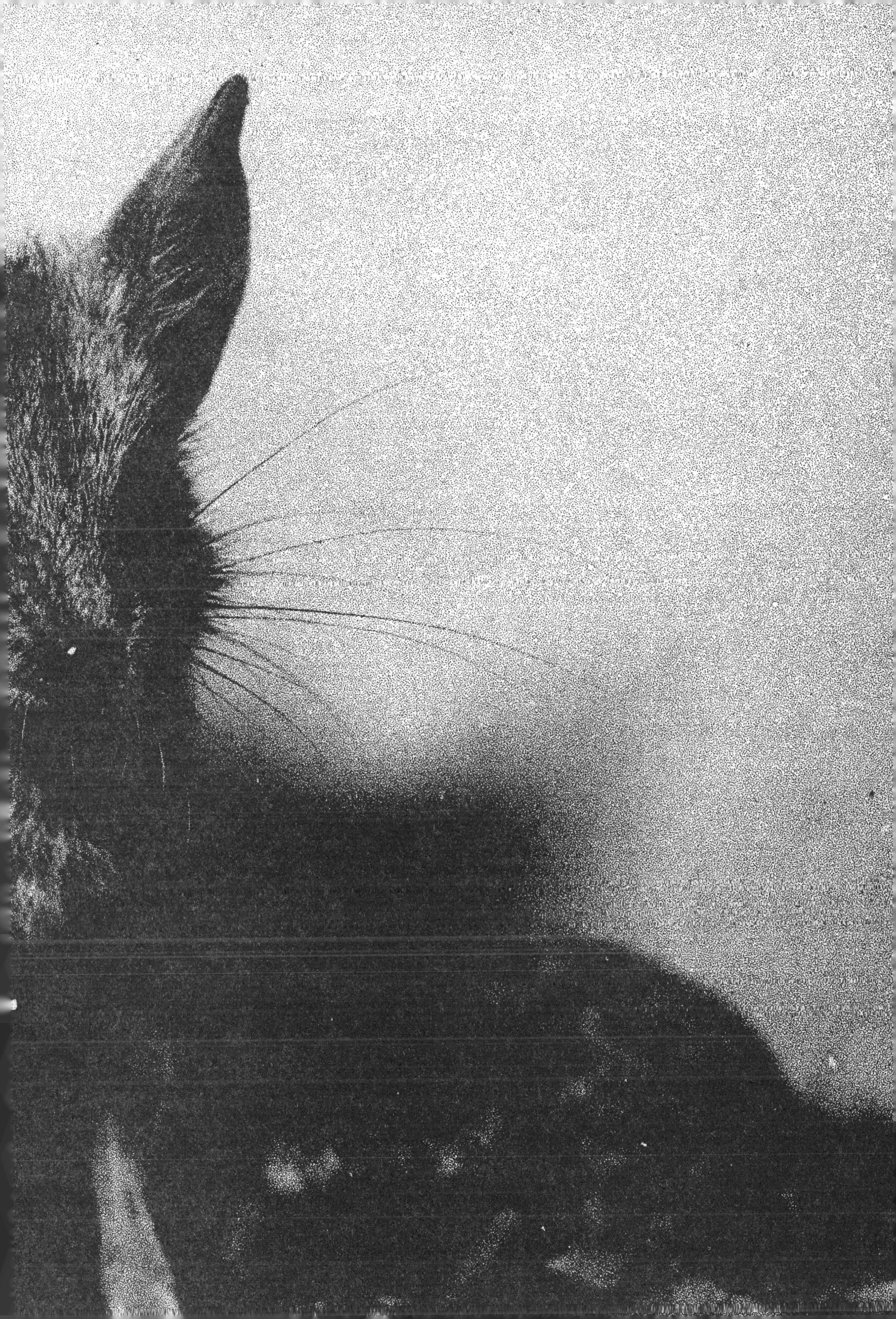

CARNIVOROUS MARSUPIALS

CARNIVOROUS MARSUPIALS
Volume 2

Edited by

Michael Archer

Published by

Royal Zoological Society of New South Wales

P.O. Box 20, Mosman, 2088, N.S.W., Australia

ISBN—0 9599951 3 7
National Library Canberra

Published February, 1982

PRINTED IN AUSTRALIA BY
SURREY BEATTY & SONS PTY. LIMITED
43 Rickard Road, Chipping Norton, 2170, N.S.W.

CONTENTS

Volume 1

PHYSIOLOGY

BEHAVIOUR

Volume 2

PALAEONTOLOGY

MORPHOLOGY, CYTOLOGY, GENETICS, ELECTROPHORESIS AND PHYLOGENETIC RELATIONSHIPS

CARNIVOROUS MARSUPIALS

Volume 2

37. REVIEW OF THE DASYURID (MARSUPIALIA) FOSSIL RECORD, INTEGRATION OF DATA BEARING ON PHYLOGENETIC INTERPRETATION, AND SUPRAGENERIC CLASSIFICATION

MICHAEL ARCHER[1]

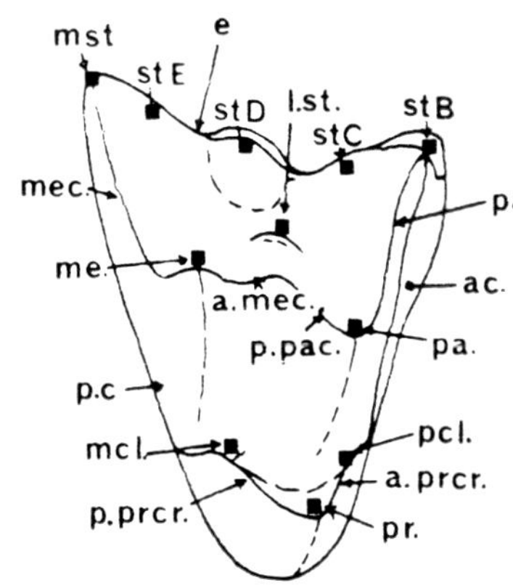

All extinct species of dasyurid are critically reviewed. New material is described for the middle Miocene *Ankotarinja tirarensis* and *Keeuna woodburnei. Dasylurinja kokuminola* n. gen. and sp. is described from middle Miocene sediments of South Australia. *Dasyuroides achilpatna* n. sp. is described from Plio–Pleistocene sediments of New South Wales. *Sminthopsis floravillensis* n. sp. is described from Plio–Pleistocene sediments of northwestern Queensland. Species of *Antechinus, Planigale, Sminthopsis, Ankotarinja* and *Dasyurus* are recognized in various Tertiary and Plio–Pleistocene faunas. The phylogenetic and palaeoecological significance of each of these fossil dasyurids is considered.

The palaeontological data are integrated with that from other studies in order to interpret the broad aspects of dasyurid evolution. Nine groups of dasyurids are recognized: 1) the species of *Ankotarinja;* 2) the species of *Keeuna;* 3) the species of *Dasylurinja;* 4) the species of *Murexia;* 5) the species of *Neophascogale* and *Phascolosorex;* 6) the species of *Phascogale* and *Antechinus;* 7) the species of *Planigale;* 8) the species of *Wakamatha, Ningaui* and *Sminthopsis;* and 9) the species of all other genera. The interrelationships of the species and genera within these groups is also considered, particularly those within the ninth group.

Dasyurid intrafamilial classification is reviewed and revised. Nine subfamilial level groups are recognized but only six are given taxonomic distinction: Murexinae for group 4; Phascolosorexinae for group 5; Phascogalinae Gill for group 6; Planigalinae for group 7; Sminthopsinae for group 8; and Dasyurinae (Goldfuss) for group 9. Within the Dasyurinae, two tribes are recognized: Parantechini and Dasyurini (Goldfuss).

Pages 397–443 *in* CARNIVOROUS MARSUPIALS, ed. by M. Archer, February, 1982. Roy. Zool. Soc. New South Wales: Sydney, Australia.

In 1964, Ride reviewed (Ride 1964b) the status and signficance of the two fossil dasyurids known at the time: *Glaucodon ballaratensis* and *Sarcophilus laniarius*. In 1971, Bartholomai added *Dasyurus dunmalli*. Since that time, partly because of the use of screenwashing techniques, partly because of the increase in the number of Australian palaeontologists, and partly because of the rapid growth in the number of Tertiary deposits known, a further seven genera and at least twelve species have been discovered.

It is my intention here to review the status and probable phylogenetic significance of all fossil dasyurids described to date, as well as several which are described for the first time in the sections that follow. Some reveal little beyond aspects of the radiation of the modern generic groups to which they belong. Other seemingly annectant forms suggest relationships that extend beyond Australia.

For the purposes of this review, the fossil dasyurids will be treated under four headings: Miocene dasyurids; Pliocene dasyurids; Plio-Pleistocene dasyurids: late Quaternary dasyurids. The Plio-Pleistocene heading covers forms which are uncertainly dated or known to lie close to the Pliocene–Pleistocene boundary (about 1.8 to 3.0 my, depending on authority).

Abbreviations for specimens in particular institutional collections are as follows: AMNH, modern specimens in the Archbold Collections of the American Museum of Natural History; AM M, modern specimens in the Australian Museum, Sydney; AM F, fossil specimens in the Australian Museum, Sydney; C, modern specimens in the National Museum of Victoria, Melbourne; J, modern specimens in the Queensland Museum, Brisbane; NMV P, fossil specimens in the National Museum of Victoria; SAM P, fossil specimens in the South Australian Museum, Adelaide; QM F, fossil specimens in the Queensland Museum, Brisbane; UCMP, fossil specimens in the collections of the University of California, Berkeley; UCR, fossil specimens in the University of California, Riverside.

Tooth number follows the nomenclatural system established elsewhere (Archer 1978a), such that the maximum dasyurid dental formula is as follows: $dI^4/_3$; $I^4/_3$; $dC^1/_1$; $C^1/_1$; $P^3/_3$; $M^5/_5$. The M1 of this system is the deciduous

[1] School of Zoology, University of New South Wales, P.O. Box 1, Kensington, New South Wales, Australia, 2033.

tooth previously called dP4 by some authors (e.g. Ride 1964b, Archer 1976b) and dP3 by others (e.g. Stirton 1957).

MIOCENE DASYURIDS

ANKOTARINJA TIRARENSIS Archer, 1976b (Figs 1 and 2)

This middle Miocene Ngapakaldi taxon from the Etadunna Formation of South Australia has been described (Archer 1976b) as a possible dasyurid with marked similarities to some didelphoids such as species of *Alphadon*. It has little bearing on interpreting modern dasyurid radiations other than to indicate that among modern dasyurids the most similar, and hence structurally ancestral, Australian forms are the species of the genera *Murexia, Neophascogale* and *Phascolosorex*. *A. tirarensis* is much too late in time to be an ancestor of all dasyurids but is regarded to be a descendant of ancestral dasyurids that retained a large number of didelphid character states.

Since the original description in 1976 of this middle Miocene taxon, five more specimens have been obtained from the type locality (Tedford Locality) at Lake Palankarinna. Four of these duplicate previously known parts of the dentition (UCR17001, LM_5; UCR15332, a dentary fragment with RM_2; UCR16998, LM^3; and AM F62531 a right dentary fragment with M_5) and add nothing new to the concept of *A. tirarensis*.

The fifth new specimen, a right dentary fragment retaining P_{1-3}, was sorted out of the Tedford Locality concentrate about two years later than the specimens described in 1976, and it is now clear that it is part of the dentary of the holotype SAM P18190 (Figs 1 and 2). The fit between the two specimens is perfect and suggests that the dentary was damaged during collection. The whole dentary is still missing the RM_2, the tooth overlying the original break. This addition to the holotype provides our first insight into the antemolar region of this primitive dasyurid. All of the premolars are shorter in height than M_3 (and M_2 of other specimens). The P_2 is just taller–crowned than P_3 which is just taller–crowned than P_1. The P_3 is just narrower than the subequally wide P_1 and P_2. The crowns of all three premolars have the protocone placed anteriorly, away from which descend anterior and posterior crests. There is an incipient tiny cusp at the upturned posterior end of each posterior crest. There are narrow lingual cingula and narrow posterobuccal cingula on at least P_{2-3}. The posterobuccal corner of P_1 may be worn. The P_1 anterior root occurs immediately buccal to the posterior half of the C_1 alveolus. The anterior part of the P_1 crown was therefore probably displaced, lying adjacent to the posterobuccal flank of the C_1. All premolars contact anteroposteriorly and are slightly displaced such that the anterior end of P_3 rests against the posterolingual edge of P_2, and the anterior end of P_2 rests against posterobuccal end of P_1. The posterior edge of P_3 would have just contacted the anterior cingulum of M_2. There are two foramina present in the anterior dentary fragment, one below the posterior root of P_1 and a larger foramen below what would have been the middle of M_2. The gently inclined symphysis extends posteriorly as far as the posterior root of P_2.

Of all Australian taxa used for this comparison (Archer 1976b), taking into consideration the new information about premolar morphology, the closest similarity may be seen to some species of *Murexia* and *Antechinus*. Of particular significance is the size of P_3 which, although slightly smaller than P_2, is clearly not reduced to anything like the extent it is in most other dasyurids, including species of the otherwise similar genera *Neophascogale* and *Phascolosorex*. Considerable similarity may be seen between the premolars of *Murexia rothschildi*, the only obvious differences being that the premolars in *M. rothschildi* are slightly wider posteriorly and narrower anteriorly. This similarity between *A. tirarensis* and species of *Murexia* is in keeping with the conclusions previously reached (Archer 1976b) that *A. tirarensis* was essentially a very primitive type of dasyurid. *Murexia* species have generally been interpreted to be among the most primitive of living dasyurids (Tate 1947 and Archer 1976c).

KEEUNA WOODBURNEI Archer, 1976b (Figs 2 and 3)

This dasyurid, described elsewhere (Archer 1976b), also from the Ngapakaldi local fauna of the Etadunna Formation, more closely resembles species of modern dasyurid genera, such as those of *Antechinus*, than does *Ankotarinja tirarensis* to which it also shows considerable similarity. Like *A. tirarensis*, it is of little value in understanding dasyurid radiations and it has been interpreted as a more modern type of dasyurid representing a lineage that did not survive into modern times. Like *A. tirarensis*, it shows more didelphid characters than do living dasyurids.

Since the original description, four additional specimens evidently referable to *Keeuna woodburnei*, have been recovered from the type locality (Tedford Locality) at Lake Palankarinna. One of these (UCR17004), an RM^4, is in all essentials the same as a previously described LM^4 (part of QM F7333). The second specimen (UCR15269) is a trigonid of an RM_3 and indistinguishable from a previously described LM_3 (UCR15274). The remaining two specimens, UCR15324 and UCR16999, appear

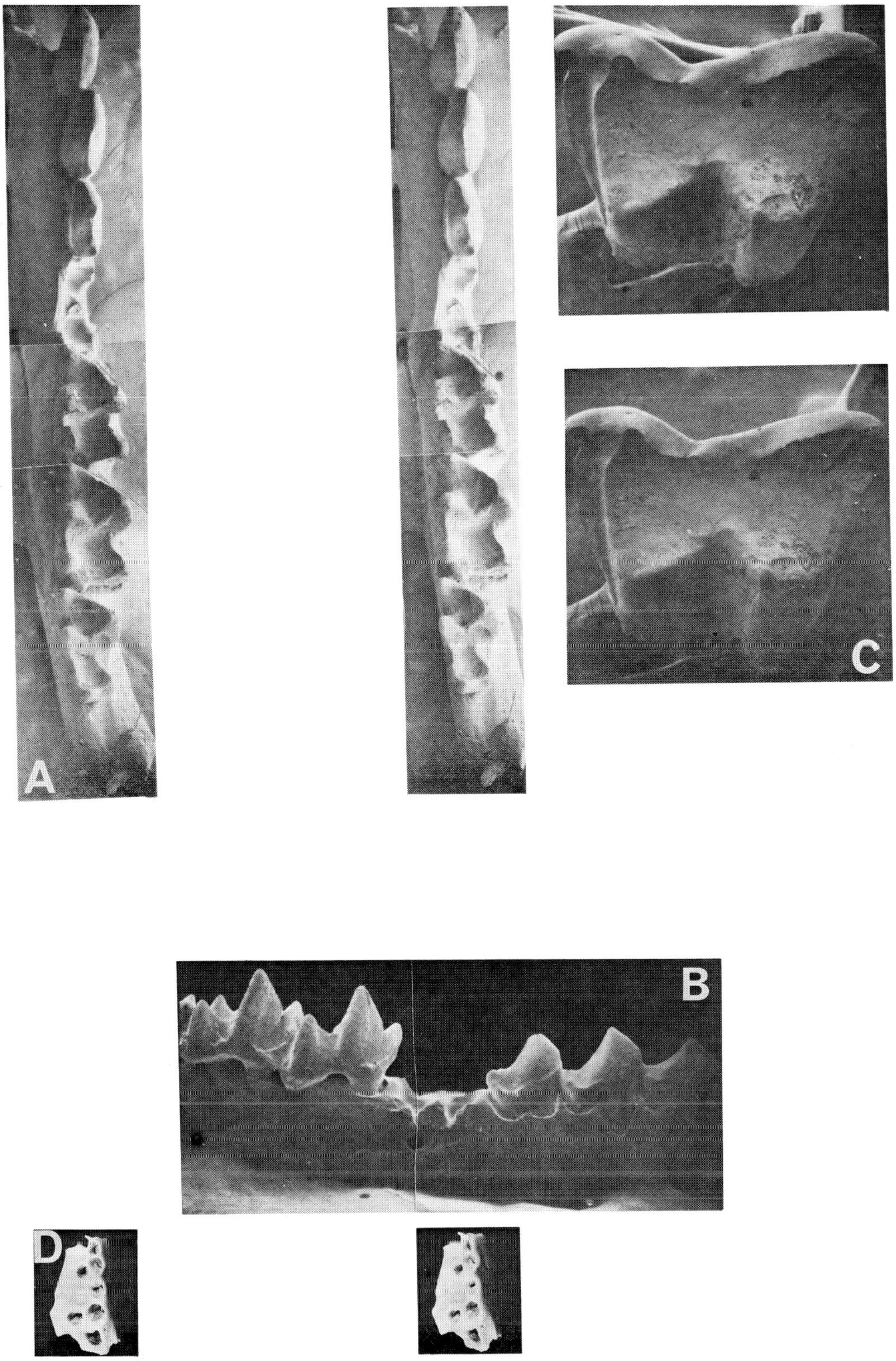

Fig 1. Ankotarinja species. Measurements for 1A–C are given in Fig. 2. A, occlusal view of the holotype of *Ankotarinja tirarensis*, from the Ngapakaldi local fauna. The original holotype is shown with the new addition of the premolar region. B, the holotype in buccal view. C, *Ankotarinja* sp., a left molar fragment, from the Kutjamarpu local fauna. D, dasyurid maxillary fragment, QM F10640, Ngapakaldi local fauna X2.

to be an LM^2 and an RM^2 respectively. In the original description, M^2 was unknown. The description of this tooth that follows is based on UCR16999, the least worn of the two teeth. Reference is made to UCR15324 only when it appears to differ from UCR16999.

The metacone of M^2 is higher than the paracone, which is in turn higher than the protocone. There is a distinct rise and thickening in the preprotocrista representative of a diminutive protoconule. Similarly, a diminutive metaconule is present on the postprotocrista. With wear (as in UCR15324), the positions of these conules are still visable. There is a well–developed stylar cusp B, small stylar cusp A (=parastyle), larger stylar cusp D, and a small stylar cusp E between stylar cusp D and the metastylar corner of the tooth. Between stylar cusp D and B, buccal to the point of confluence of the postparacrista and premetacrista, there are two contiguous small cusps collectively representative of stylar cusp C. In UCR15324, there is only one cusp in the position of stylar cusp C, and in mass it about equals the combined mass of the two contiguous cusps in UCR16999. The preparacrista is about equal in length to the postparacrista, and runs to the anterolingual base of stylar cusp B. The premetacrista is just longer than the postparacrista and a little less than half the length of the postmetacrista. The preprotocrista is continuous with a well–developed anterior cingulum which in turn runs to the parastyle. The postprotocrista runs buccally only as far as the base of the metacone. Although unclear in UCR 16999 in UCR15324 there is a slight posterior cingulum on the base of the crown from below the level of the apex of the metacone to halfway to the metastylar corner of the tooth. This posterior cingulum is not connected to any other crests. A continuous fine crest connects stylar cusps C, D and E to the metastylar corner of the tooth. Stylar cusps A, B and C are also connected but only just near their bases. With wear, as in UCR15324, stylar cusps A, B, C and E are connected by a continuous crest. Also with wear, a semilunar crest of enamel develops just buccal to the confluence of the postparacrista and premetacrista. In the relatively unworn tooth UCR16999, this semilunar crest has not been developed. It was noted previously (Archer 1976b) to occur in the worn LM^3 and LM^4 of QM F7333. Clearly it is a feature only of worn molars. The protoconal basin is deep and enclosed posteriorly by a short cristid extending from the base of the metacone to the metaconule.

Compared with the corresponding structures in the more posterior molars previously described, in M^2 stylar cusp D is slightly larger than it is in M^3, and about twice as large as it is in M^4. The preparacrista of M^2 is about half the length of that crest in M^4. The postparacrista is subequal in length in M^2 and M^4. The postmetacrista of M^2 is about subequal to or just shorter in length than that crest in M^4. The whole crown of M^2 is longer than it is wide, a condition reversed in M^4.

In comparison with the M^2 of other dasyurids, the closest resemblance is to species of *Antechinus, Murexia,* and *Ankotarinja,* all three being relatively plesiomorphic genera. But there are also resemblances to didelphids, such as species of *Marmosa,* just as there are between *Ankotarinja tiranensis* and didelphids. The M^2 of *Keeuna woodburnei* gives a clearer indication of the plesiomorphic nature of this taxon within the Dasyuridae than does M^3 or M^4, but it does not indicate a close relationship to the species of any particular plesiomorphic genus of dasyurids.

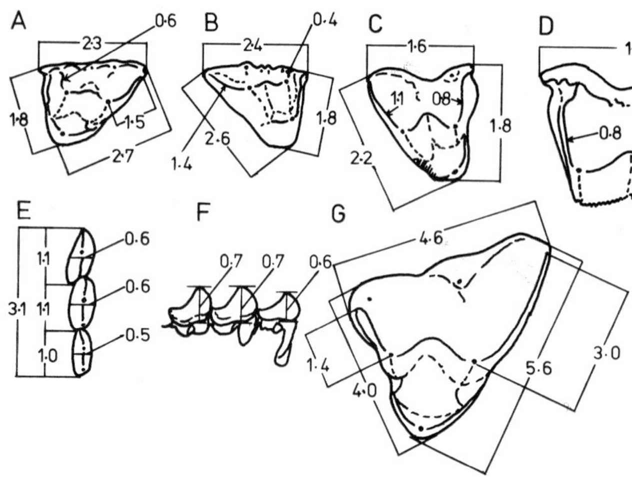

Fig. 2. Measurements (mm) of fossil dasyurids. A, UCR15324, LM^2 of *Keeuna woodburnei* from the Ngapakaldi local fauna. B, UCR16999, RM^2 of *K. woodburnei.* C, QM F10621, right upper molar of the Riversleigh dasyurid. D, AM F62527, *Ankotarinja* sp. from the Kutjamarpu local fauna. E–F, RP_{1-3} of the holotype, *Ankotarinja tirarensis* from the Ngapakaldi local fauna, occlusal and buccal views. G, NMV P28889, LM^3 of *Dasyurus hallucatus* from the Fisherman's Cliff local fauna.

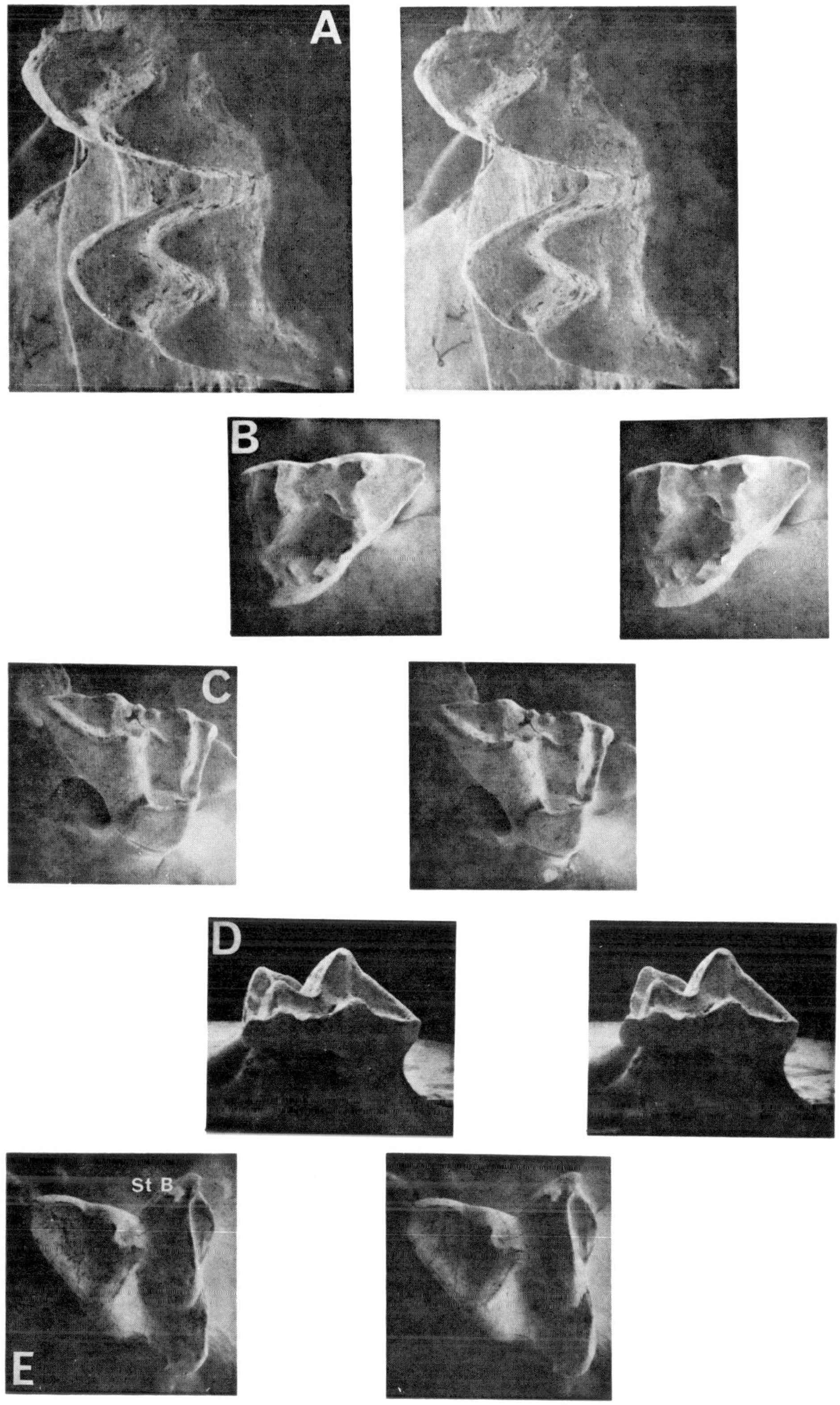

Fig. 3. A, QM F7333, left maxillary fragment with M^{3-4}, *Keeuna woodburnei* from the Ngapakaldi local fauna. Measurements of this specimen are given in Archer (1976b). B, UCR15324, LM^2 of *K. woodburnei.* C–D, UCR16999, RM^2 of *K. woodburnei,* occlusal and buccal views. E, QM F10621, the Riversleigh right upper molar, with stylar cusp B indicated (St B). Measurements are given in Fig. 2.

Other Etadunna Dasyurids from the Ngapakaldi Local Fauna of South Australia

Stirton, Tedford and Miller (1961:35) note another Etadunna dasyurid which they regard as an animal that ". . . may not be far removed from the ancestry of *Thylacinus*.". Elsewhere (Archer 1976b and c) it has been suggested that this animal shows close affinities to dasyurids and not thylacinids. Premolar size gradient increases posteriorly and M_2 lacks a metaconid. These characters are not unique to thylacinids and occur in some dasyurids. Total loss of the M_2 metaconid occurs in some specimens of *Dasyurus maculatus* (e.g. J9809) and the same premolar gradient occurs in species of *Murexia, Sminthopsis* and *Antechinomys*. The very reduced M_2 paraconid in the Etadunna form (visible in photographs kindly made available by Dr. W. A. Clemens and Mr. C. Campbell) is a common feature in dasyurids with a reduced M_2 metaconid and contrasts with the condition found in thylacinids.

A left maxillary fragment (AM F62528) from Tedford Locality, Lake Palankarinna, represents yet another middle Miocene dasyurid from South Australia. It retains alveoli for a possible M^{2-4}. It is smaller in size than material referred by Stirton, Tedford & Miller (1961) to the as yet unnamed taxon just discussed above, but is decidedly larger than *Keeuna woodburnei* and *Ankotarinja tirarensis*. Little about this fourth taxon can be deduced except that its molar size gradient was very steep. No dental material referable to this taxon has been recovered.

A Small Dasyurid from the Middle Miocene Riversleigh Local Fauna of Northwestern Queensland

(Figs 2 and 3)

On 26 October, 1976, Mr. Henk Godthelp, a member of the 1976 Ray E. Lemley Expedition to North Queensland, spotted a tiny bat bone in sandy limestone on Riversleigh Station, in north Queensland. Details of the stratigraphy of this site will be published elsewhere in connection with description of the local fauna. Between the collections made that year and others made in 1977, we obtained a total of one tonne of limestone matrix. Acid processing has so far resulted in several hundred specimens of a small microchiropteran similar to species of hipposiderids, two specimens of a larger microchiropteran, a fragment of a peramelid molar, a small macropodid premolar, and an upper molar of a very small dasyurid. Acid processing continues.

The dasyurid upper molar, QM F10621, is probably an RM^2 or possibly an RM^3. It has a very large metacone, smaller paracone, and a low protocone. The stylar cusps are reduced with only a medium–sized stylar cusp D, and a small stylar cusp B. The preparacrista is only about three quarters the length of the large postmetacrista just longer than the prematacrista, and almost twice the length of the postmetacrista. The postmetacrista is very well developed and clearly was a very important shearing crest. The preprotocrista terminates on the anterior edge of the crown below the paracone. The postprotocrista is damaged but clearly did not extend buccally beyond the base of the metacone. There is no posterior cingulum. Although the anterior cingulum is confined to the anterobuccal corner of the tooth, it is very well developed. Stylar cusp B is just posterior to the buccal end of the preparacrista and in this regard somewhat unusual. In most dasyurids the preparacrista runs directly to the stylar cusp B. Similarly unusual is the fact that the anterobuccal cingulum does not connect directly or indirectly to stylar cusp B. Such a normal connection in dasyurids is here prohibited by the preprotocrista which runs uninterrupted straight out to the parastylar corner of the tooth. Overall, the tooth is not compressed anteroposteriorly as are molars of some dasyurids such as species of *Sminthopsis* and *Ningaui,* but then neither is it as elongate as the instars of many other genera, including the more plesiomorphic forms such as species of *Murexia, Neophascogale,* and *Phascolosorex*. The ectoloph is deeply indented immediately anterior to stylar cusp D.

Comparisons throughout the family have failed to find a close match with any particular modern species. The greatest similarity is to some species of *Antechinus,* such as *A. macdonnellensis*. These share the apomorphic features of a well–developed postmetacrista, comparably reduced stylar shelf morphology, deep ectoloph indentation, lack of a posterior cingulum, and restricted anterobuccal cingulum. However they lack some features of the Riversleigh tooth such as the technically just isolated position of stylar cusp B, and the lack of connection between the anterobuccal cingulum and stylar cusp B. Nevertheless, this tooth presents a more "modern" appearance than the Etadunna dasyurids discussed above. This impression comes from its specialized stylar shelf area, enlarged postmetacrista (probably carnivorous adaptions), loss of a posterior and anteromesial cingulum, deeply indented ectoloph and short anteroposterior length (features commonly related in some small dasyurids to elongation of the postmetacrista).

Dasylurinja kokuminola n. gen. and sp. from the Middle Miocene Namba Formation, Lake Yanda, South Australia

In the course of naming the middle Miocene fluvio–lacustrine Namba Formation, Callen and Tedford (1976) note the presence of vertebrates from Namba outcrops at Lake Yanda, approximately 63 km northeast of Frome Downs Station homestead, in the Lake Frome area of South

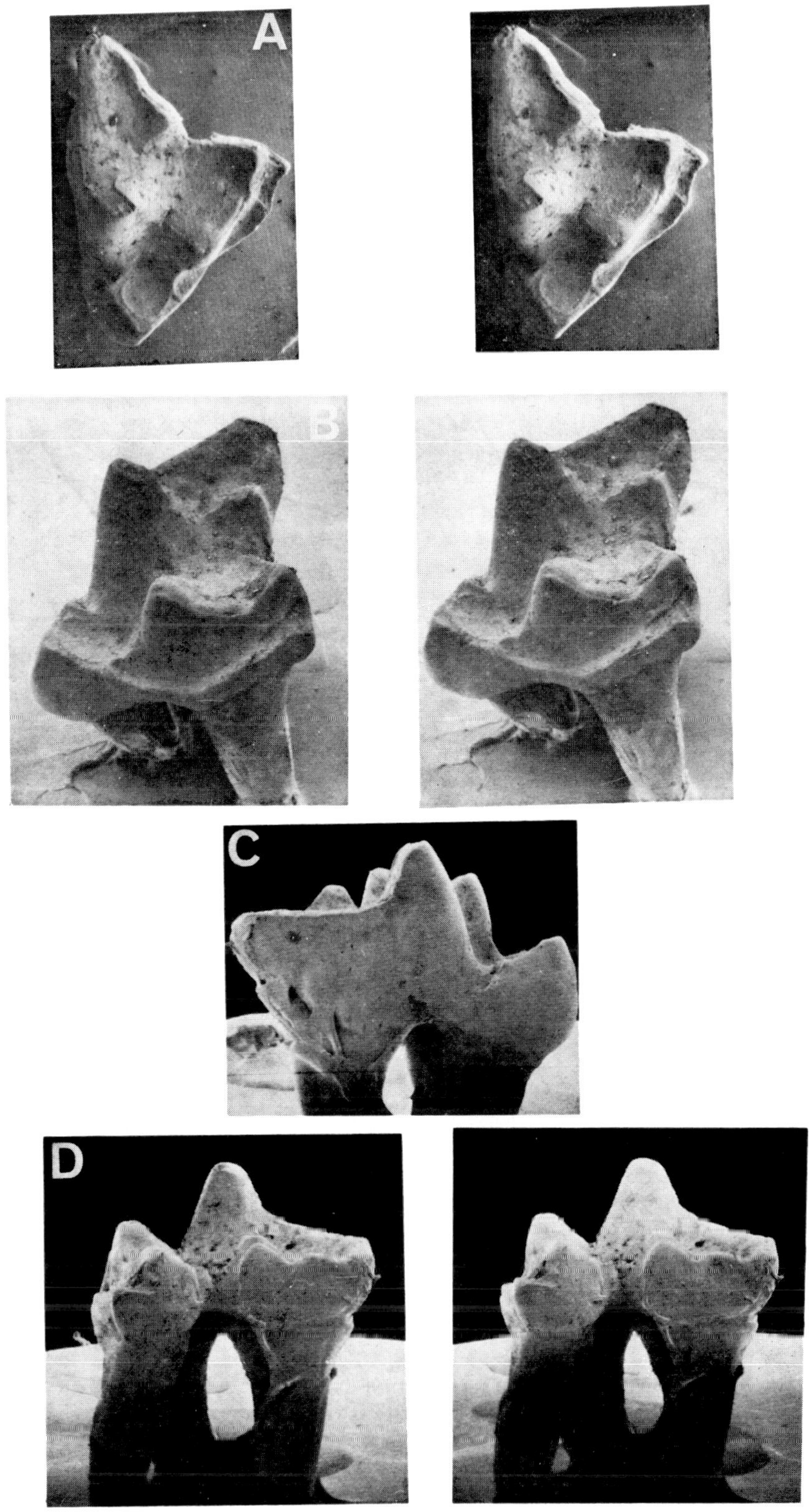

Fig. 4. The holotype of *Dasylurinja kokuminola*, a right upper molar, from the Yanda local fauna. Measurements are given in Fig. 5. A, occlusal view. B, anterior oblique view. C, posterior view. D, buccal view.

Australia. In June, 1973, R. H. Tedford, R. T. Wells, M. P. Plane, R. Brown, N. Pledge and M. Archer recovered additional vertebrate remains from Lake Yanda. These were briefly noted in Tedford *et al.* (1977) together with a stratigraphic section of the fossil locality (fig. 2D in Tedford *et al.* 1977).

In January, 1980, a field party from the University of New South Wales screenwashed surficial eroded sediments from the same fossil locality at Lake Yanda and obtained yet additional vertebrate remains, including the dasyurid described below. It is unrepresented among the previously collected Yanda taxa, most of which will be described elsewhere by Tedford.

After making comparisons of this Yanda specimen with all described dasyurids, it is evident that it can be referred to none of the known genera. Because it is so distinct morphologically, I do not hesitate to distinguish it with a new generic and specific name. Further work is planned at Lake Yanda and hopefully additional elements referable to this unique taxon will be found.

Dasylurinja kokuminola n. gen. and n. sp.
(Figs 4 and 5)

Holotype: SAM P23623, an isolated right upper fourth molar, collected 17 January, 1980 by M. Archer, T. H. Rich, and students and staff of the University of New South Wales, School of Zoology.

Type Locality: Western edge of Lake Yanda, approximately 63 km northeast of Frome Downs Station Homestead, Frome Downs Station, South Australia.

Stratigraphic horizon and age: Tedford *et al.* (1977) regard the fossiliferous horizon at Lake Yanda to probably be in the lower member of the Namba Formation. The actual material sieved by the 1980 field party was the upper 5 cm of sediments from the western edge of the lake. Bones were not seen within the deeply weathered surficial clays and presumably have been concentrated at the point of collection by slope wash from the visably fossiliferous arenaceous clays forming very gentle knolls at the western edge of the lake. Tedford *et al.* (1977) interpret the Lake Yanda fauna to be part of the Pinpa local fauna and thus middle Miocene in age[1].

Etymology: *Dasy* refers to the Dasyuridae and means hairy, *lurinja* means "a stranger" in the language of a tribe of central Australian Aboriginals (Kempe 1891); the allusion is to a strange

[1] Some doubt about this faunal identification must remain because a dolphin fragment was just recovered from the same sediments (Rich and Archer, in prep.), a beast known so far only from the Ericmas local fauna (also middle Miocene in age).

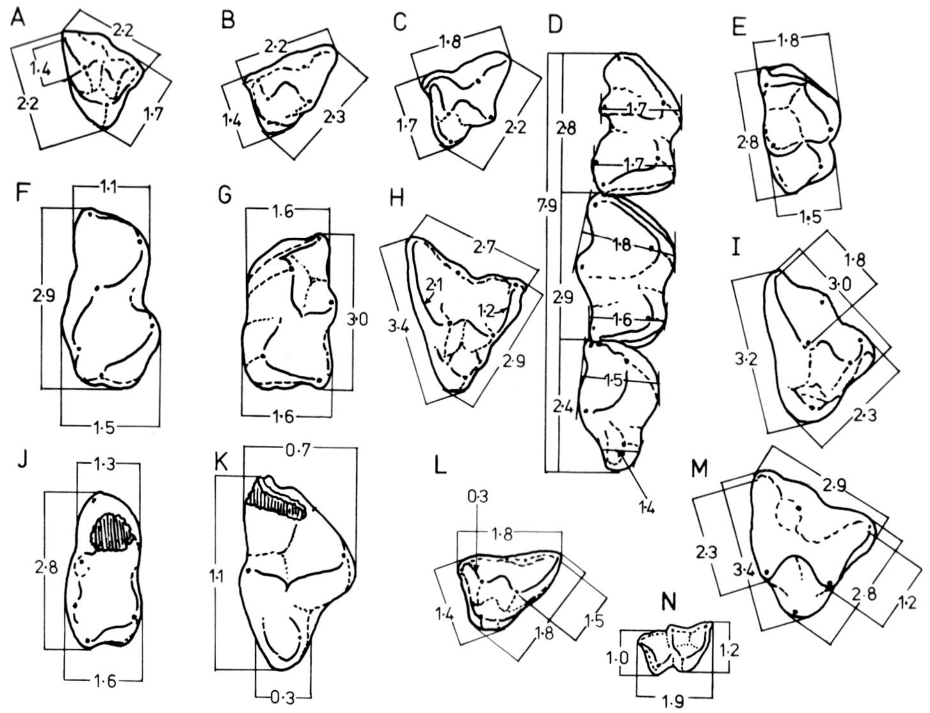

Fig. 5. Measurements (mm) of fossil dasyurids. A, QM F10626, an RM² of *Antechinus* sp. from the Floraville local fauna. B, QM F10624, LM² of *Sminthopsis floravillensis*. C, QM F10625, LM⁴ of *S. floravillensis*. D. NMV P28888, RM_{3-5} of *Dasyuroides achilpatna*. E, NMV P38343, RM_X of *D. achilpatna*. F, NMV P38429, RM_2 of *D. achilpatna*. G, NMV P38334, LM_3 of the holotype of *D. achilpatna*. H, SAM P23623, the holotype of *Dasylurinja kokuminola*. I, NMV P38331, RM^2 of *Dasyuroides achilpatna*. J, UCMP P63640, RM_2 of *Myoictis* sp., from the Awe local fauna. K, QM F10620, RM_5 of *Planigale* sp. from the Bluff Downs local fauna. L, NMV P160482, LM^2 of *Antechinus* sp. from the Hamilton local fauna. M, NMV P38442, RM^4 of Dasyuroides achilpatna. N, QM F10627, RM?, of *Sminthopsis floravillensis.*

dasyurid or hairy stranger. The specific name comes from the Aboriginal words *kokuma,* "to bite", and *inola,* "a spider" (Kempe 1891); the allusion is to the probable ability of this dasyurid to tackle the large terrstrial megalomorph spiders, i.e. a competent spider–bite.

Generic diagnosis: The upper fourth molar of this dasyurid differs from all others in the following combination of characters: The lack of a posterior cingulum; the presence of a complete anterior cingulum; presence of stylar cusps B, D, and E; an anterposteriorly compressed photocone; a postmetacrista that is almost twice the length of the preparacrista; and a poorly developed metaconule and protoconule; it most closely resembles the M^4 of some species of *Dasyurus* (*Satanellus*) but differs from these (besides in being much smaller) in its retention of stylar cusp E, smaller conules, and in the morphology of the parastylar area of the crown.

Description: The fourth upper molar: The metacone is much taller than (just under twice the height of) the paracone, which is just taller than the subequal stylar cusp B, and stylar cusp D, which are taller than the protocone. Stylar cusp E is subequal in height to the parastyle (= stylar cusp A). There is a continuous well–developed anterior cingulum connecting the preprotocrista to the parastylar corner of the tooth. There is a slight indenture for the metastylar corner of M^3 near the anterobuccal end of the anterior cingulum. The anterior cingulum extends only to the parastylar corner of the tooth. It does not contact stylar cusp B. There is no trace of a posterior cingulum. The preparacrista is greater than twice the length of the postparacrista. The almost vertical premetacrista is about twice the length of the postparacrista. The postmetacrista is just less than twice the length of the preparacrista. Stylar cusp B is the anterior end of a buccal stylar shelf crest. The crest has a fissure at a point directly buccal to the position of the intersection of the postparacrista and premetacrista, and coincident with the point of maximum indenture in the ectoloph. On either side of this fissure, the buccal stylar crest shows slight swellings which may be homologues of a bicusped stylar cusp C such as occurs in some molars of *Neophascogale* and *Phascolosorex* (Archer 1976c). Stylar cusp D is just posterior to this fissure. It is connected by a low crest to stylar cusp E, being separated only by another but less prominent fissure. There is a poorly developed protoconule on the preprotocrista which has a short buccal crest connecting to the lingual base of the paracone. Similarly, a poorly–developed metaconule has a short buccal crest linking the postprotocrista to the lingual base of the metacone. The crown length (2.7 mm) is just shorter than the anterior crown width (2.9 mm) as measured along a line extended through stylar cusp B and the protocone. The posterior crown width (3.5 mm) is considerably longer, reflecting the elongate postmetacrista. Other crown measurements are as follows: preparacrista, 0.9 mm; postparacrista, 0.4 mm; premetacrista, 0.9 mm; postmetacrista, 2.0 mm; metacone–stylar cusp D, 1.1 mm; metacone–paracone, 1.1 mm; and ectoflexus (point of maximum buccal concavity) – postparacrista–premetacrista junction, 0.9 mm.

Discussion: This highly distinctive dasyurid stands out among other Miocene dasyurids in its decidedly carnivorous specialization of the elongate postmetacrista. It resembles the species *Satanellus* (see below under the discussion of *Dasyurus dunmalli,* and the discussion, for comments on the generic status of the various species of *Dasyurus* and *Satanellus*) but differs from these as noted in the diagnosis. Although most of the species of *Dasyurus* and *Satanellus* are more specialized than *Dasylurinja kokuminola,* in some details of upper molar morphology such as the relatively less elongate postmetacrista and relatively less reduced paracone, *Satanellus hallucatus* is less specialized than *D. kokuminola.* Assuming that all species of *Satanellus* and *Dasyurus* are monophyletic, *D. kokuminola* is then prohibited from being ancestral to this group unless the more primitive features of *S. hallucatus* have secondarily been acquired from a more specialized ancestor. If *D. kokuminola* is not ancestral to the species of *Satanellus,* then it is probable that acquisition of the new features in which it is similar to species of *Satanellus* and *Dasyurus* has been independently achieved. This is not surprising because they appear to be adaptations for a carnivorous habit. Already convergence in these characters has been demonstrated in other lineages (e.g. between thylacinids and dasyurids such as *Sarcophilus,* Archer 1976c).

Assuming that the similarity between *D. kokuminola* and species of *Satanellus* and *Dasyurus* is the result of convergence, the next most similar group of dasyurids is that which contains only the less specialized *Keeuna woodburnei.* The Yanda dasyurid and *K. woodburnei* are similar in many features including a complete anterior but absent posterior cingulum. However, they differ in that *D. kokuminola* lacks the specialized reduction of stylar cusps B and D found in *K. woodburnei* but has a specialized metacone and postmetacrista morphology (noted above) which *K. woodburnei* lacks. Therefore it is only reasonable to conclude that neither is likely to have been the others ancestor or descendent. *Dasylurinja kokuminola* seems to represent yet another distinctive middle Miocene lineage that is not clearly related to any of the other known dasyurid lineages. In its specialized features, however, it parallels the carnivorous adaptions shown by other dasyurid lineages. In particular, the enlarged metacone and postmetacrista are adaptions for increasing the overall shearing components of the molar row, a feature that characterizes the more specialized carnivorous members of other dasyurid and thylacinid lineages (Bensley 1903; Archer 1976c).

It may be predicted on the basis of the occlusal counterparts that the lower molars of *Dasylurinja kokuminola* will be found to have a proportionately enlarged paracristid, reduced metaconid, and possibly an anteroposteriorly foreshortened talonid.

Dasyurids from the Kutjamarpu Local Fauna of the Wipijiri Formation, South Australia

In 1971, a joint expedition including the University of California, the South Australian Museum, and the author made extensive collections of vertebrates representing the Kutjamarpu local fauna, from the Leaf Locality in the Wipijiri Formation of Lake Ngapakaldi, South Australia. The Kutjamarpu dasyurids have since been under study by Dr. C. R. Campbell and, in part, formed the basis for his Ph.D. thesis (Campbell 1976). Dr. Campbell has kindly allowed me to see an epoxy cast of the left dentary that is the proposed holotype of an as yet unnamed Kutjamarpu taxa, as well as photographs of many of the other Kutjamarpu specimens. In addition, collections made at the Leaf Locality by a joint Queensland Museum, National Museum of Victoria, Bureau of Mineral Resources, and South Australian Museum expedition have produced a broken upper molar (AM F62527) of a taxon evidently unrepresented in the UCMP collection studied by Campbell, and apparently referable to *Ankotarinja*. Because the taxa studied by Campbell are not yet published, it would be premature to present detailed analyses about these here. I will confine my remarks to the general summary of Dr. Campbell's thesis published in *Dissertation Abstracts International* B (Science and Engineering 37 (9) 1977).

Campbell (1976, as cited in *Dissertation Abstracts International,* ibid: 4375–B) recognizes two new monotypic Kutjamarpu dasyurid genera. In addition, he refers other isolated teeth to species of the living dasyurid genus *Antechinus*. Considering the monotypic genera, he suggests that one is a ". . . suitable ancestor for *Dasyurus maculatus* as well as the late Pliocene animal *Glaucodon ballaratensis* . . ." The second is ". . . very similar to *Myoictis* of New Guinea and *Dasycercus* and *Dasyuroides* of Australia. It could be ancestral to all or any one of these genera". Of the teeth referable to *Antechinus,* he states that they ". . . compare very closely to those in living species and no formal taxonomic designations are made."

Kutjamarpu Dasyurid Genus No. 1

The epoxy cast available for study is of the proposed holotype for the *Dasyurus maculatus*–like form, referred to here as Kutjamarpu dasyurid genus No. 1. It reveals two kinds of character states. The first, and most numerous, are plesiomorphic character states (Archer 1976c). These include a relatively large P_3, elongate, narrow and uncrowded premolars, a well–developed talonid on M_5 with three cusps, unspecialized trigonids (except possibly on M_2 which is broken), a talonid on M_3 which is wider than the trigonid, a carnassial notch in the metacristid M_4 that is lingual to the midpoint of the crest, the lack of complete buccal cingula on the molars, hypocristids which do not contact the entoconids, and a cristid obliqua of M_4 that contacts the trigonid buccal to the point below the carnassial notch in the metacristid. In species of *Dasyurus* and *Satanellus* examined in this study (*D. maculatus, D. geoffroii, D. viverrinus, D. dunmalli, S. hallucatus* and *S. albopunctatus*) many of these plesiomorphic dasyurid features are also present. However, no special relationship is indicated between the Kutjamarpu taxon and species of *Dasyurus* on the basis of plesiomorphic character states. The other type of character states involved are the specialized or apomorphic ones. The ones shared with species of *Dasyurus* include a possibly (in the Kutjamarpu form) reduced paraconid on M_2 (which occurs in the extreme in species of *Dasyurus*) a metaconid on M_5 which is taller than the paraconid, metacristids which are not at right angles to the long axis of the tooth row, conical entoconids, and low talonids. Similarity in these apomorphic states to species of *Dasyurus* are not exclusive and occur also with the majority of other dasyurids. Pending an opportunity to examine the Kutjamarpu material in more detail, I cannot see why any clear affinities based on synapomorphic character states to any other known dasyurid genus. However, it is noteworthy that it possesses two plesiomorphic character states (the large M_5 talonid and the relatively large P_3) that, among living dasyurids, only occur together in species surviving in the lush areas of New Guinea.

Kutjamarpu Genus No. 2: *Ankotarinja* sp. (Figs 1 and 2)

The single Kutjamarpu dasyurid tooth fragment (AM F62527 Figs 1 and 2) collected by the joint QM, SAM, NMV and BMR expedition, from the Leaf Locality, represents an LM^3 or LM^4. It is missing the protocone. The paracone is lower than the metacone. There is a small stylar cusp B, subequal and posteriorly positioned stylar cusp D, and smaller central stylar cusp C. All three stylar cusps are united by a low longitudinal buccal crest. There is an anterior cingu-

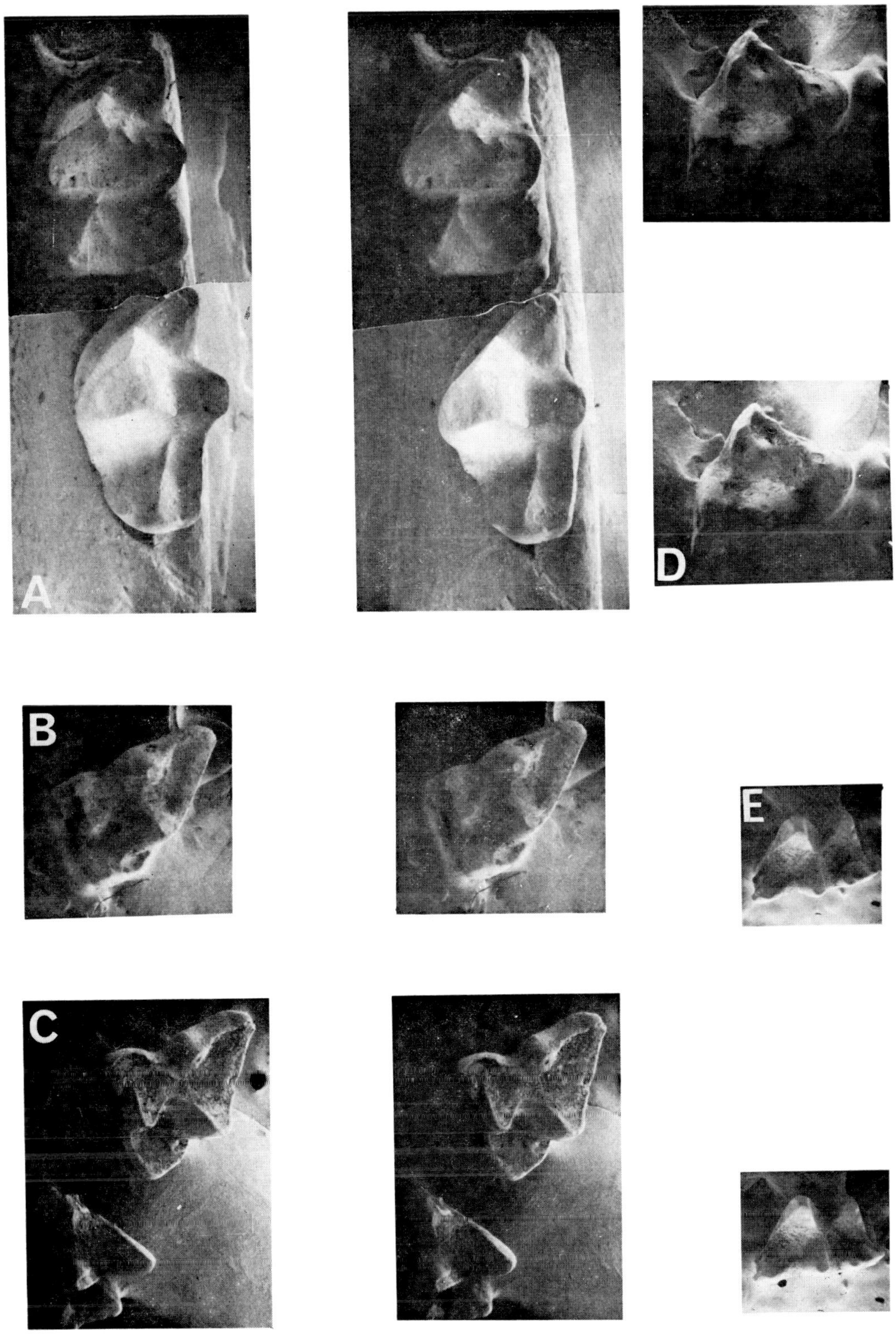

Fig. 6. Fossil dasyurids. A, part of the holotype of *Wakamatha tasselli*, showing the LM_{4-5}. Measurements are given in Archer and Rich (1979). B, QM F10624, LM^2, *Sminthopsis floravillensis* from the Floraville local fauna. C, QM F10625, LM^4 of *S. floravillensis*. D, QM F10626, RM^2, *Antechinus* sp. from the Floraville local fauna. E, QM F10627, $RM?_4$, *S. floravillensis*. B–E, measurements given in Fig. 5.

lum extending from the parastylar corner of the tooth to the broken edge of enamel below the tip of the paracone. It was probably continuous with the preprotocrista. There is no evidence of a posterior cingulum. The postparacrista and premetacrista are subequal and shorter than the preparacrista which is about three–quarters the length of the postmetacrista.

All of these character states are also those of the slightly older *Ankotarinja tiranensis* from the Ngapakaldi local fauna. The only significant differences between the Kutjamarpu tooth and the Ngapakaldi teeth (e.g. UCR 16998) are 1) the slightly larger size, 2) the slightly longer postmetacrista, and 3) the deeper ectoflexus in the ectoloph of the Kutjamarpu tooth. These differences suggest the possibility that the Ngapakaldi species might have been ancestral to the Kutjamarpu form.

This sharing of closely related species between the Ngapakaldi and Kutjamarpu faunas corroborates a growing body of data that suggests the two faunas are very similar in age, with the Kutjamarpu fauna appearing to be the younger of the two. The Kutjamarpu fauna comes from the Wipijiri Formation which was deposited in stream channels developed in the Etadunna clays. Although there is as yet no way of ascertaining the actual difference in age between the two faunas, on the basis of faunal similarity, it seems improbable that the difference would be more than a couple of million years.

WAKAMATHA TASSELLI Archer and Rich, an Enigmatic Dasyurid of Possible Miocene Age from South Australia (Figs 6 and 7)

In 1974 a field party of the National Museum of Victoria found a left dentary fragment (NMV P33253) of a smal dasyurid in the bottom sediments of Billeroo Creek, Frome Downs Station, South Australia. This has been described by Archer and Rich (1979) as *Wakamatha tasselli.* Subsequent investigations have failed to determine the probable stratigraphic level from which it came. There are exposures of both the middle Miocene Namba Formation and the Pleistocene Eurinilla Formations in the banks of Billeroo Creek so its age is at present uncertain.

In morphology, *Wakamatha tasselli* resembles many other Miocene dasyurids and species of *Murexia* in having a large P_3 (as indicated by the alveoli). However, it also resembles some modern Australian dasyurids in its possession of a reduced M_5 talonid and transverse hypocristid that contacts the entoconid (Fig. 6). Although among modern dasyurids the last character state is unique to *Sminthopsis crassicaudata,* and might therefore suggest a special relationship to this modern species, it is the only distinctive character state that the two species share. Further, details of the character state in the two species indicate that the overt similarity is the result of convergence (Archer and Rich 1979).

The phyogenetic affinities of *Wakamatha tasselli* are unclear. Despite the particular convergence on *Sminthopsis crassicaudata,* there are other general similarities to all species of *Sminthopsis.* These include the unreduced size of P_3 (although it was evidently larger in *W. tasselli* than in any known species of *Sminthopsis*), the transverse orientation of the metacristids, and the transverse orientation of the hypocristids. Of known dasyurids, the closest forms to species of *Sminthopsis* are those of *Antechinomys* (a subgenus of *Sminthopsis, fide* Archer (1981a) and *Ningaui* (Archer 1975). Tentatively I would propose that *Wakamatha tasselli* is probably the sister group of a combined *Sminthopsis–Antechinomys–Ningaui* group.

PLIOCENE DASYURIDS

Antechinus sp. from the Early Pliocene Hamilton Fauna of Victoria

Turnbull and Lundelius (1970) have described a diverse taxonomic assemblage from a fossil soil on the Grange Burn, Victoria. The fossiliferous soil is immediately overlain by a basalt that yielded a K/Ar date of 4.35 ± 0.1 million years (Turnbull *et al.* 1965). By screen–washing the fossil soil, they recovered many small mammals, but no dasyurids. In 1977, T. H. Rich began reworking the Hamilton site and extended the excavations. Two years later, he started using a finer screen size in his washing boxes. Sorting of this finer concentrate revealed two small dasyurid molars. It also revealed ten teeth of a tiny dasyurid–like peramelid. Both of these taxa will be dealt with in more detail elsewhere (Archer and Rich in preparation). Here I will only make preliminary comments about the two dasyurid molars.

Antechinus sp. (Figs 5 and 8)

An isolated LM^2 (NMV P160482) represents a very small dasyurid. Comparison of its morphology with didelphids and dasyurids reveals that the closest similarity is to some of the species of *Antechinus.* I have checked the morphology of M^2 in all species of *Antechinus* and found that they may be divided into several groups. One of these *Antechinus* groups is only represented (as far as I am aware) by taxonomically unallocated specimens from Mt. Wilhelm in Papua, New Guinea (e.g. AMNH1909-01), and specimens of *A. mayeri* from Papua, New Guinea. Their combination of M^2 character states includes an only slightly swollen base to stylar cusp D. Their other features are mostly plesiomorphic dasyurid features (Archer 1976c) and certainly plesiomorphic *Antechinus* features.

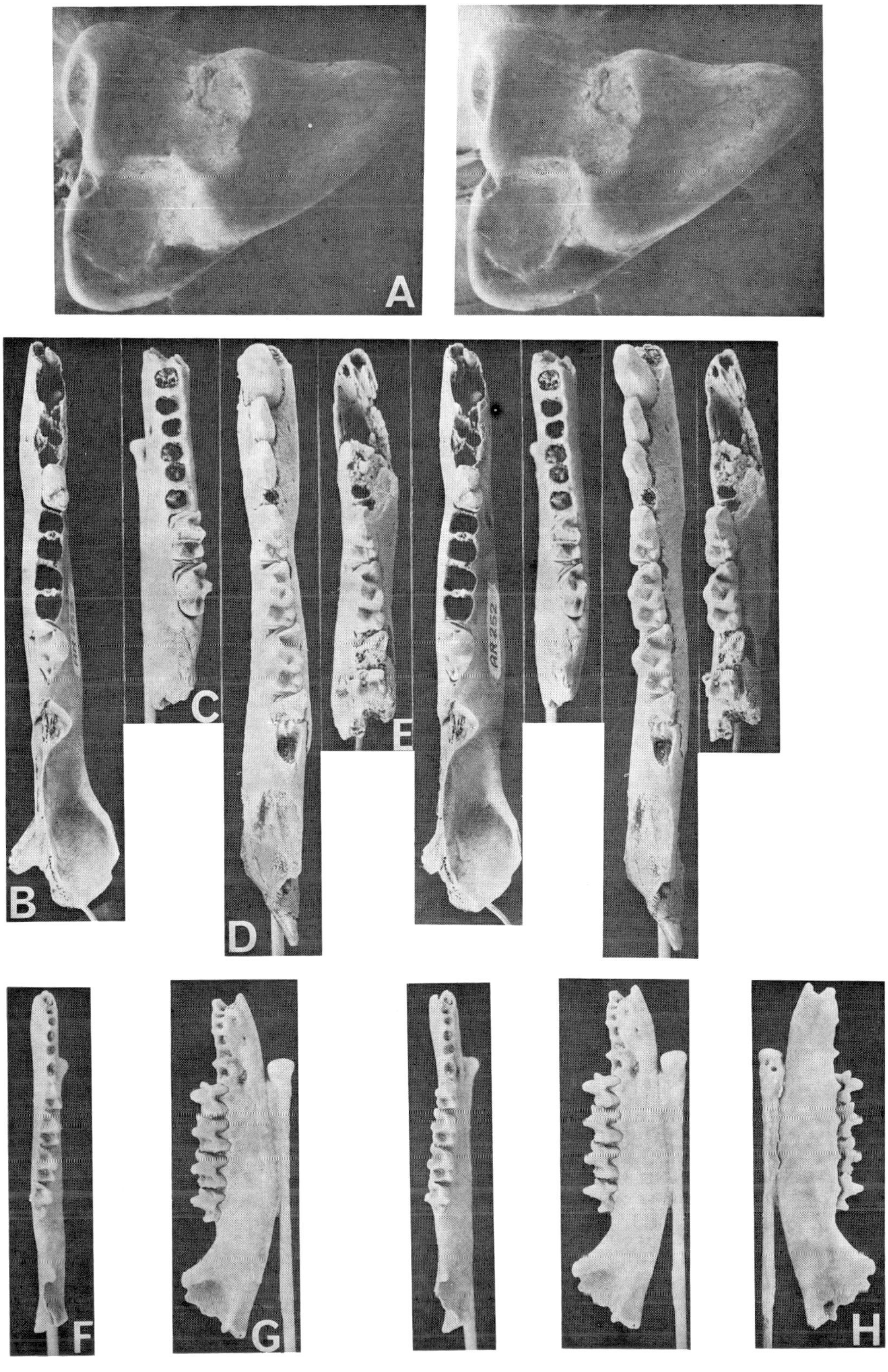

Fig. 7. A, NMV P28889, LM^3 of *Dasyurus hallucatus* from the Fisherman's Cliff local fauna. B, the holotype of *Glaucodon ballaratensis*, X1. C, the holotype of *Wakamatha tasselli* from an unknown stratigraphic provinence, X2. D, QM F742, left dentary of *Dasyurus dunmalli* from the Chinchilla local fauna showing an alveolus for P_3, X2. E, AM F62530, left dentary fragment of *D. dunmalli* from the Bow local fauna, showing the alveolus for P_3 and the posterior alveolus for P_2, X2. F–H, holotype of *Sminthopsis floravillensis* from the Floraville local fauna of northwestern Queensland. F, occlusal view of the right dentary, X2. G, buccal view, X2. H, lingual view, X2. Measurements in Figs. 5 and 9.

In all of these features, including size, the Hamilton *Antechinus* tooth is most similar to *A. mayeri* and the Mt. Wilhelm *Antechinus*. It only differs in having a slightly better developed posterior cingulum. Well developed posterior cingula are also probably plesiomorphic dasyurid features, and its greater development of this structure, the Hamilton tooth is probably more plesiomorphic than modern species of *Antechinus*. The significance of the unusual stylar cusp D morphology is more difficult to interpret. It could be a plesiomorphic *Antechinus* feature, the more swollen condition being a synapomorphic state of the remaining species of *Antechinus*. This seems unlikely because species of otherwise essentially plesiomorphic genus *Murexia* resemble Australian and other New Guinean species of *Antechinus* in this feature, not *A. mayeri* or the Mt. Wilhelm *Antechinus*. It seems best to interpret *A. mayeri*, the Hamilton and Mt. Wilhelm *Antechinus* species as representatives of an essentially primitive branch of the genus that had, however, developed a specialized stylar cusp B morphology.

As it now stands, this early Pliocene Hamilton *Antechinus* is Australia's oldest representative of the genus. Older specimens recorded by Campbell (1976), from the mid to late Miocene Kutjamarpu local fauna, have been tentatively referred by him to two species of *Antechinus*. I have not had the chance to examine this material.

The second Hamilton dasyurid tooth (NMV P160483) is a broken isolated right lower molar. The posterolingual half of the talonid is missing. It is slightly larger than the M_4 of the Mt. Wilhelm specimen noted above, and therefore may have represented a slightly larger dasyurid than the other Hamilton tooth. However, in its overall morphology, it is also similar to the Mt Wilhelm *Antechinus* and may represent a larger individual of the same Hamilton species.

Planigale sp. from the Early Pliocene Bluff Downs Fauna of North Queensland

On 3 May, 1975, sediments of the early Pliocene Allingham Formation, from Barrett Locality (located between the positions of sections 1 and 2 along the Type Section shown in Figure 1A of Archer and Wade 1976), were screen washed in Allingham Creek. When the Bluff Downs local fauna of the Allingham Formation was first described by Archer (in Archer and Wade 1976), no dasyurids were known. Subsequent sorting of the 1975 Barrett Locality wash was concentrate by Mr. Henk Godthelp produced by a tiny dasyurid dentary fragment.

Planigale sp.
(Figs 5 and 9)

A right dentary fragment (QM F10620) retaining M_5 and the alveoli for $M_{2\text{-}3}$ represents a tiny species of *Planigale*. In size, it is larger than *P. ingrami*, smaller than *P. maculata* and *P. novaeguineae*, and approximately the size of *P. tenuirostris* and *P. gilesi*. Molar morphology in species of *Planigale* is not clearly species specific (Archer 1976a) and I am at a loss to more specifically determine the affinities of the Bluff Downs *Planigale*. However, it exhibits a slightly better developed and broader talonid on M_5 than occurs in *P. ingrami, P. tenuirostris* or *P. gilesi*. The cristid obliqua is subequal in height to the lingual talonid crest and thereby encloses a distinct talonid basin. In *P. ingrami* the lingual crest is poorly developed and the talonid slopes away on both sides of the cristid obliqua. This feature of the talonid morphology of the Bluff Downs *Planigale* is plesiomorphic within the genus. The trigonid is similar in morphology to that of other *Planigale* species. The paraconid is damaged.

Because of the very incomplete condition of the Allingham dentary, little can be determined of its phylogenetic significance. It represents the first known Tertiary and therefore also oldest known occurrence of a species of the genus.

Dasyurus dunmalli Bartholomai from the Pliocene Chinchilla Local Fauna of Queensland and the Bow Local Fauna of New South Wales

Dasyurus dunmalli was originally described by Bartholomai (1971) on the basis of specimens from the early to mid Pliocene Chinchilla Sand Formation of Southeastern Queensland. A specimen (AM F62530) evidently referable to *D. dunmalli* was recovered in 1978 by students of the University of New South Wales (Fig. 7). The Bow local fauna is interpreted by Stillbeck (in press) to be early to mid Pliocene in age on the basis of its overall similarities to the Chinchilla local fauna. Differences in the Bow Specimen are noted below.

Bartholomai (1971) describes *Dasyurus dunmalli* as being generally similar to the modern *D. viverrinus* but differing in possessing a single–rooted peg–like P_3. He notes (1971: 22) that "Apart from the presence of the minute additional third premolar . . . no other consistent morphological differences are evident between this species and . . . *D. viverrinus*". In dasyurids it is almost always the case that when P_3 is reduced, P^3 is noticeably larger than P_3 (as in all species of *Antechinus*). If this relationship applies to *D. dunmalli*, P^3 (as yet unknown) may be expected to be relatively larger than P_3, perhaps about the size of P^3 in *Phascolosorex doriae*.

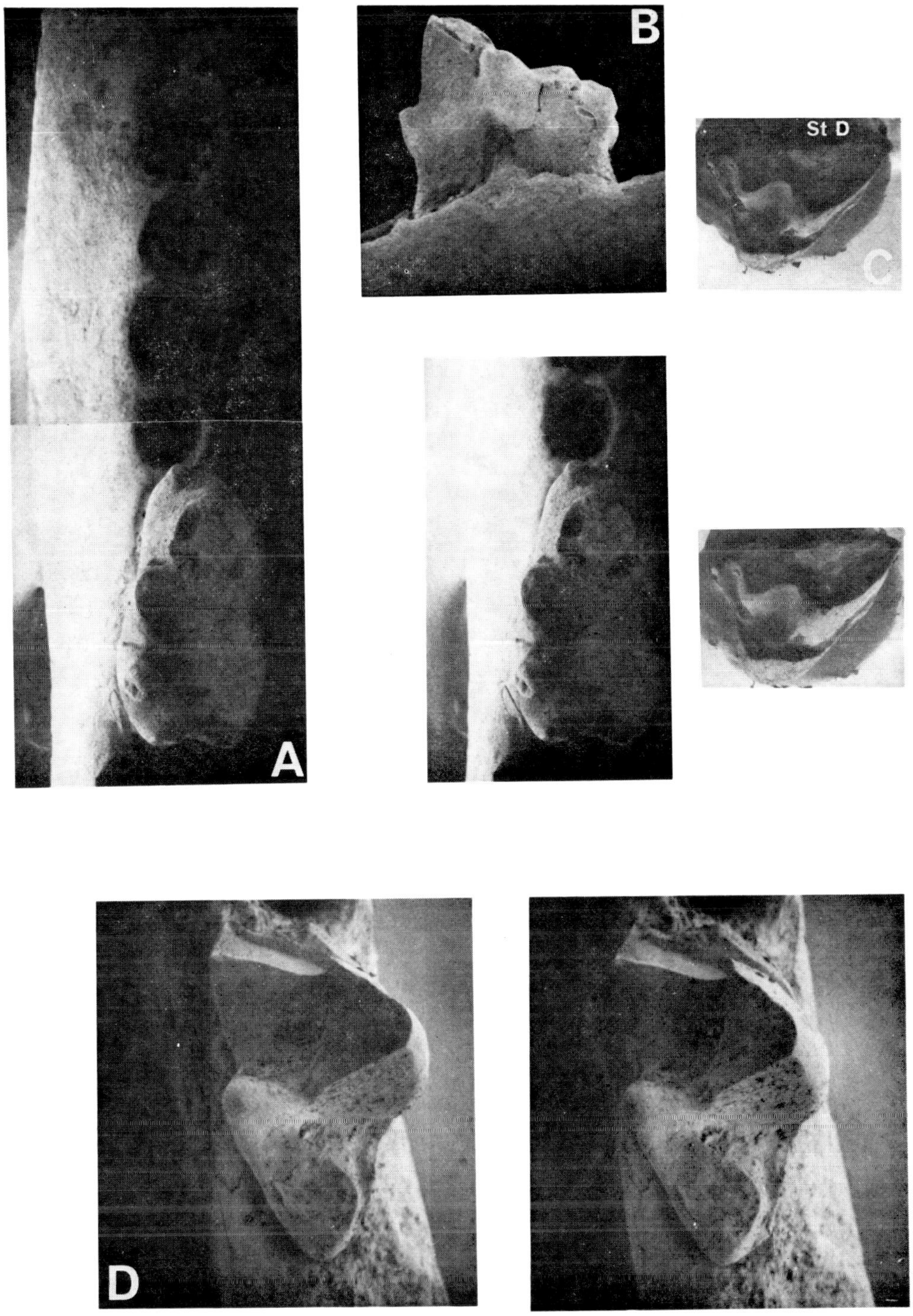

Fig. 8. A–B, UCMP63640, a right dentary fragment with M_2, *Myoictis* sp. from the Awe local fauna. The alveolus directly anterior to the RM_2 is interpreted to represent a single–rooted P_3. C, NMV P160482, an isolated LM^2, *Antechinus* sp. from the Hamilton local fauna. The position of the narrow stylar cusp D is indicated. D, QM F10620, a fragmentary dentary with RM_5, *Planigale* sp. from the Bluff Downs local fauna. Measurements are given in Fig. 5.

Further, the relatively narrow talonid of the M_2 of *D. dunmalli* suggests that paraconid and stylar cusp B may have both been present on the as yet unknown M^2. In modern species of *Dasyurus*, the relative width of the talonid (relative to the trigonid) decreases from M_2 to M_4 in proportion to the distance between the paracone and stylar cusp B of the upper molars. The talonid of M_2 of *D. dunmalli* does not appear to be wide enough to reflect a fused or single cusp represening the paracone and stylar cusp B of M^2 such as is typical of all modern species of the genera *Dasyurus, Dasycercus* and *Dasyuroides*.

The presence of P_3 in *Dasyurus dunmalli* raises a question about the possible convergent loss of P_3 in other species of *Dasyurus*. As noted by Bensley (1903) and Archer (1976c), molar morphology of *D. hallucatus* is structurally ancestral within the genus. And yet, except in its retention of P_3 *D. dunmalli* resembles the relatively specialized *D. viverrinus*.

If the presence of P_3 has much value in interpretations phylogeny, there may be two interpretations of the significance of *D. dunmalli.* If *D. hallucatus* is structurally ancestral to modern *Dasyurus* the placement of *dunmalli* in *Dasyurus* is possibly unjustified, its resemblance in molar morphology to *D. viverrinus* being the result of convergence. Alternately, *D. dunmalli* could be regarded as structurally ancestral to all modern *Dasyurus* except *hallucatus* which should not be placed in *Dasyurus*. The second alternative has some merit in terms of molar morphology. It is also indirectly supported by other authors (e.g. Pocock 1926, Tate 1947, Troughton 1967) who have allocated the modern Quolls into four genera, *Satanellus* (containing *hallucatus* and *albopunctatus*), *Dasyurus* (containing *maculatus*), *Dasyurus* (containing *geoffroii*) and *Dasyurinus* (containing *viverrinus*). However, no author has suggested that *hallucatus* and *albopunctatus* are the only Quolls of separate generic classification and taking external characters into consideration, this would seem an untenable alternative. The first alternative that *D. dunmalli* is convergent on *D. viverrinus* seems equally untenable because the specializations that characterise the larger species of *Dasyurus*. are complex. They include a posterior shift in the point of maximum metacrista–paracristid shear, relative increase in metacrista to paracrista length, less–transverse orientation of the metacristids, and reduction of talonid size.

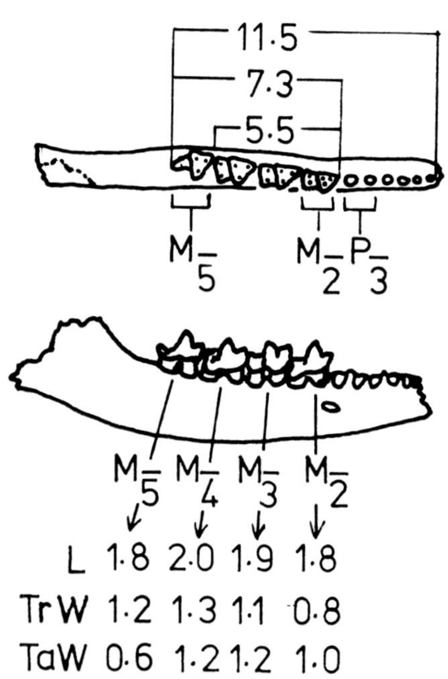

Fig. 9. Measurements of the holotype of *Sminthopsis floravillensis*. The length from M_5 to the anterior edge of the alveoli for P_1 = 11.5 mm. Abbreviations are: L = crown length; TrW = width of the trigonid through the protoconid; TaW = width of the talonid through the hypoconid. All measurements are in millimetres.

On the other hand, the presence of P_3 may have little value in interpreting phylogeny. This is certainly suggested by loss of P_3 in only one of five species of *Planigale* (Archer 1976a) as well as its independent loss in several other dasyurid lineages including species of *Dasycercus* and *Dasyuroides, Antechinus rosamondae* and *A. macdonnellensis*. Gross reduction foreshadowing loss of P_3 occurs in yet other independent lineages such as species of *Myoictis* and *Antechinus apicalis*. If *D. dunmalli* is in fact most closely related to other large *Dasyurus* species, and *D. hallucatus* is in all respects except P_3 ancestral to all other *Dasyurus,* it is only necessary to postulate loss of P_3 twice within the genus: once from the ancestral *hallucatus–albopunctatus* stock (subsequent to the differentiation of the other ancestral forms of *Dasyurus*); and once from a common ancestor of the species *geoffroii, viverrinus* and *maculatus*. The common ancestor of the latter species could well have resembled *D. dunmalli* in molar morphology. The exact position of *maculatus* in this scheme is problematical. It appears to be structurally ancestral to *Glaucodon ballaratensis* and *Sarcophilus harrisii,* a lineage which extends back at least into the Pliocene. Differentiation of *D. dunmalli* from other large *Dasyurus* may have preceeded the development of ancestral *maculatus* and been a late Tertiary event.

Reference of the Bow Quoll (AM F62530, Fig. 7) to *Dasyurus dunmalli* needs some qualification. It is a left dentary fragment preserving M_2, M_3, the talonid of M_4, and damaged alveoli for $I_{1\text{-}3}$, C_1 P_1, P_2 and P_3. It differs from the holotype and paratype of *Dasyurus dunmalli* in having slightly smaller entoconids, a crowded premolar row such that the posterior root of P_1 lies buccal to the anterior root of P_2, and a lingual displacement of the single–rooted P_3 such that it lies anterolingual to the anterior root of M_2. The crowded premolar row cannot be due to age because the Bow specimen, although not representative of an old adult, shows comparable tooth wear to that of the Chinchilla paratype QM F742, which does not show crowding. In all other adult specimens of species of *Dasyurus* known to me, the premolar rows are uncrowded. Crowding may occur in juveniles before the dentary elongates (e.g. *Dasyurus viverrinus,* J20381; Archer 1976c, fig. 8A).

It is tempting to see the Bow Quoll as representative of a stage in the evolution of the species of *Dasyurus*. If selection on an as yet unknown, 3-premolared ancestral Quoll favoured reduction of the premolar row length, a probable resultant condition would be a crowded premolar row and a reduced P3. This stage is represented by the Bow Quoll. In other Quolls two trends are evident: complete loss of P_3 and, perhaps as a result, secondary reduction of premolar crowding. These latter character states are those of the living species. The Chinchilla Quoll represents a stage in which the premolars are less crowded than the Bow Quoll, but in which the diminutive P_3 is anterior (not anterolingual) to the M_2.

PLIO-PLEISTOCENE DASYURIDS

Dasyurids from Plio–Pleistocene Deposits Along the Leichhardt River, Northwestern Queensland

In 1976 and 1977, Dr. Ray E. Lemley sponsored Queensland Museum expeditions to North Queensland in search of fossil marsupials. One of the localities discovered occurs in dissected riverine sediments exposed between the western bank of the Leichhardt River and the parallel north–south Floraville Station track, south of the Floraville Station homestead. Details of the stratigraphy will be published elsewhere as part of the report on the whole fauna. The local fauna containing the dasyurids reported below is called here the Floraville local fauna.

The age of the Floraville local fauna is in doubt. It includes no undoubted Pleistocene species such as *Diprotodon optatum*. On the other hand, there are also no undoubted Pliocene species. Nothing about the sediments enables an age assessment. They are surficial and underlain by other undated fluviatile sediments. The significance of the apparent lack of Pliocene or Pleistocene index fossils is difficult to interpret because of our almost total lack of knowledge about late Pliocene to Pleistocene faunas from anywhere in northern Australia. Considering the regional endemism of modern Australian faunas, it would not be surprising if a fossil taxon characteristic of a certain period in southern Australia was absent at the same time in northern Australia. The detailed analysis in preparation of the Floraville local fauna, and other as yet unreported local faunas from north Queensland, will help to clarify this matter.

Sminthopsis floravillensis n. sp.
(Figs. 5, 6, 7 and 9)

Holotype: QM F10628, a right dentary with $M_{2\text{-}5}$ and avleoli for C_1, and $P_{1\text{-}3}$, collected by M. Archer 10 October, 1977.

Type locality: Site 5C (Archer's field notes for 30 September, 1977 to 29 October, 1977: p.89 for discovery of the dentary; p.110 for photographs of Site 5C; and p.111 for a stratigraphic section of Site 5C), in dissected riverine sands and gravels, south of Floraville Homestead, between the north–south Floraville Station track and the western bank of the Leichhardt River, Floraville Station, northwestern Queensland.

Paratypes: From Site 5C: (Isolated teeth) QM F10627, LM_4; QM F10629, LM_5; QM F10630, $RM^{?3}$; QM F10624, LM^2; QM F10631, $LM^{?3}$; QM F10636, $LM^{?3}$; QM F10625, $LM^{?4}$; QM F10626, $RM^{?2}$; QM F10632, $RM^{?2}$; QM F10633, right dentary fragment with very worn $M_{4\text{-}5}$ and fragment of M_3; (from Site 5B) QM F10634, LM^2; and QM F10635, $LM^{?4}$.

Diagnosis: *Sminthopsis floravillensis* differs from all species of *Sminthopsis* in having (on the basis of alveoli) a much larger P_3 that considerably exceeds P_2 in length; it also differs from *S. hirtipes, S. ooldea, S. butleri,* and *S. granulipes* in being larger, and in having better-developed entoconids on $M_{2\text{-}4}$; it also differs from *S. ooldea* in lacking short wide premolars and reduced talonids; it also differs from *S. psammophila* and *S. douglasi* in being smaller; it also differs from *S. crassicaudata* in being larger and in lacking hypocristid–entoconid contact; it also differs from *S. macroura* in being larger and in having entoconids that are smaller; it also differs from *S. virginiae* and *S. longicaudata* in being larger.

Description (based on the holotype and paratypes): Second upper molar (based on QM F10624, QM F10626, QM F10632): The metacone is taller than stylar cusp D which is much taller than the paracone which is taller than stylar cusp B which is taller than the protocone. The preparacrista connects directly to stylar cusp B. No crest links stylar cusp B to stylar cusp D. A low crest links stylar cusp D to the metastylar corner of the tooth. A low crest links stylar cusp B to the parastylar corner of the tooth. There is a short anterobuccal cingulum that does not extend as far lingually as the base of the paracone. The preprotocrista does not extend bucally beyond the base of the paracone. There is thus no continuous anterior cingulum. The postprotocrista extends posterobuccally to the posterolingual base of the metacone. There is thus no posterior cingulum. There is a low anterlingually directed crest that extends from the base of the metacone to a point on the postprotocrista just posterior to the protocone in the position of a metaconule. The preparacrista is subequal to the postparacrista and shorter than the premetacrista which is about half the length of the postmetacrista. The basal crown length (2.2mm) is longer than the basal anterior width between stylar cusp B and the protocone (1.7mm). The ectoloph has only a slight inflexion between stylar cusps B and D.

Third upper molar (QM F10631, QM F10636, QM F10630): This tooth is as described above for M^2 except as follows: Stylar cusp D is only slightly taller than the paracone. The anterobuccal cingulum is shorter than it is in M^2, but the parastylar corner of the tooth projects further in a buccal direction. The point of flexion in the ecotoloph is deeply indented into the crown. The preparacrista is longer than the postparacrista, and subequal to the premetacrista. The crown length (1.9mm) is equal to the anterior crown width.

Fourth upper molar (QM F10625; QM F10634): This tooth is as described above for M^3 except as follows. There is less parastylar development. The paracone is taller than stylar cusp D. The preparacrista is markedly longer than the postparacrista and the premetacrista. The preparacrista is approximately three–quarters the length of the postmetacrista. The crown length (1.8mm) is shorter than the anterior crown width (2.3mm).

Meristic gradients from M^2 to M^4. The preparacrista increases in length posteriorly. The premetacrista decreases in length posteriorly. Crown length decreases posteriorly. Stylar cusp D size and parastylar development decreases posteriorly. The degree of flexion of the ectoloph increases posteriorly. Relative paracone height increases posteriorly.

The dentary (QM F10628, the holotype; and QM F10633): The most complete dentary at hand is that of the holotype. It preserves the region from C_1 to the masseteric fossa below the ascending ramus. A mental foramen occurs in a depression below a point between the posterior root of P_3 and the anterior root of M_2. The mandibular foramen is preserved at the posterior edge of the dentary fragment. The symphysis extends posteriorly to a point below the anterior root for P_3. The maximum depth of the dentary occurs below the posterior root of M_4.

The canine and premolar alveoli (QM F10628): The posterior end of the canine alveolus lies lingual to the posterior root of P_1. The anterior alveolus of P_1 is not preserved. The alveolar lengths are shown in Fig. 10. Clearly P_3 was longer than P_2, and (on the basis of alveolar lengths, 1.7mm) subequal to M_2. The anterior and posterior roots of both P_3 (and P_2) were subequal in size to each other. The premolars probably contacted one another anteroposteriorly because there are no diastemata between the alveoli.

The lower molars of the holotype and QM F10633 are slightly and very worn respectively. The isolated RM_4, QM F10627, is little worn For this reason, M_4 will be described first. The other teeth will then be described only in so far as they differ from M_4.

Fourth lower molar (QM F10627; QM F10633; QM F10628); The trigonid is just wider than the talonid, and conspicuously longer. The metacristid and hypocristid are transverse to the long axis of the dentary. The protoconid is taller than the metaconid which is taller than the paraconid. The unworn hypoconid was probably (it is worn) shorter than the paraconid and subequal in height to the entoconid which is taller than the hypoconulid. The carnassial notch in the metacristid occurs just lingual to the midpoint of the crest. The paracristid (although worn) was probably lower than the metacristid. The cristid obliqua intersects the posterior wall of the trigonid at a point just buccal to the carnassial notch of the metacristid. The hypocristid extends from the hypoconid directly to the posterolingual corner of the tooth without even approximating the entoconid. There is a slight suggestion of a very short vertical crest on the posterior flank of the entoconid. Otherwise the entoconid is essentially conical without obvious crests. The posterior cingulum is well–developed and extends from the posterobuccal corner of the base of the hypoconid to the hypoconulid. A short cingulum bridges the anterobuccal basal corner of the hypoconid with the posterobuccal basal corner of the protoconid. The anterior cingulum, although not well–preserved in any specimen, appears to have been as well–developed as the posterior cingulum. It extended from the anterobuccal base of the protoconid to the notch for the hypoconulid of M_3. There is an anterior parastylid projection that extends anteriorly from the paraconid and forms the lingual wall of the hypoconulid notch.

Second lower molar (QM F10628): This tooth is comparable to M_4, except as follows. The talonid is wider than the trigonid. The paracristid extends in a more anterior direction than it does in M_3. The metacristid is shorter. The cristid obliqua intersects the trigonid in a more buccal position. The posterolingual corner of the tooth is subrounded. The paraconid is relatively low on the trigonid.

Third lower molar (QM F10628; QM F10633): This tooth is closely comparable in morphology to M_4 and only differs in the following ways. The trigonid is just narrower than the talonid. The cristid obliqua intersects the trigonid at a slightly more buccal position.

The fifth molar (QM F10628; QM F10633; QM F10629): The trigonid of this tooth is comparable to the trigonid of M_4. The talonid differs as follows. It is laterally compressed and reduced in complexity. The hypoconid, hypoconulid and entoconid are barely differentiated on the continuous U–shaped cristid obliqua–hypocristid–lingual talonid crest. There is, however, a shallow talonid basin enclosed by the talonid crests. The cristid obliqua intersects the trigonid at a position lingual to the carnassial notch of the metacristid. There is a reasonable possibility that one of the M_5s here provisionally referred to *Sminthopsis floravillensis* in fact represents another taxon. There is a difference in the relative heights of buccal and linqual talonid crests on the M_5s that suggests the presence of more than one similar–sized species. The trigonids, however, all suggest *Sminthopsis* morphology and, if a second species is represented, it does not appear to be referable to any of the other Floraville dasyurids noted below.

Meristic tooth gradients from M_2 to M_5 (QM F10628; QM F10633): The paraconid increases in height posteriorly. The trigonid widens posteriorly from M_2 to M_4, then decreases in width in M_5. The paracristids and metacristids lengthen from M_2 to M_4. The length of the paracristid in M_5 is subequal to that of M_4, but the M_5 metadristid is shorter than that crest in M_4. The cristid obliqua intersects the trigonid in a progressively more lingual position from M_2 to M_5.

Discussion: This species is regarded to be a species of *Sminthopsis* for three reasons: First, the metacristids and hypocristids are oriented transversely with respect to the long axis of the dentary. Second the P_3 is very large, a condition restricted to some species of *Murexia, Sminthopsis, Antechinomys,* and *Ningaui.* Third, in all of its minor features such as entoconid morphology, short and wide molars, lack of posterior cingula on the upper molars and so on (Archer 1976a), it most closely resembles species of *Sminthopsis* rather than species of any other genus.

On the basis of the size of the premolar alveoli of the holotype, *Sminthopsis floravillensis* had a proportionately larger P_3 than any other species of the genus. A large P_3 has been interpreted to be primitive for dasyurids in general (Archer 1976c). This would then suggest that this extinct species was, at least in this feature, more primitive than any other *Sminthopsis.* Its other features, such as the presence of an entoconid and its distance form the hypocristid, have also been regarded (Archer 1981a) as primitive within the genus.

Wakamatha tasselli, interpreted above to represent a sister group of the combined *Sminthopsis–Antechinomys–Ningaui* group, is similar in evidently having had a large P_3 and entoconids, as well as several other features typical of all species of *Sminthopsis* such as the transverse metacristids and hypocristids.

Right dentary fragment (QM F10633) preserves remnants of M_{3-5}. The trigonid and talonid of M_3 are worn so extensively that they were no longer in contact. The trigonid of M_4 is also heavily worn and almost smooth. The talonids of M_4 and M_5 similarily are very heavily worn. No wild–caught specimen known to me of a "semelparous" dasyurid species shows such extensive wear. However, comparable wear could develop in a dasyurid that survived for several years.

Antechinus sp.
(Fig. 6)

An isolated RM^2 (QM F10626) represents a small, probably new species of *Antechinus* in the Floraville local fauna. It is unlike the M^2 of all species of *Antechinus* in its possession of the following combination of character states. It has a continuous well–developed anterior cingulum and a very well-developed post–protocrista that appears to have terminated on the posterior base of the metacone. Stylar cusp D is very large, being only just shorter than the metacone, and closely approximated to the metacone. Its large size produces a very conspicuous swelling on the posterobuccal side of the crown. In contrast, stylar cusp B, although clearly distinct from the paracone, is small and produces only a very slight buccal swelling. The paracone is very low relative to the metacone and subequal in height to stylar cusp B.

This constellation of characters occurs in no other dasyurid known to me. Species of *Phascogale* and *Antechinus mayeri* are similar in postprotocrista construction but not in paracone–stylar cusp B-metacone–stylar cusp D construction. *Antechinus macdonnellensis* is also similar in some respects but its stylar cusp B closely approximates the paracone and also produces a conspicuous buccal swelling. The postmetacrista, and hence tooth length of M^2 in *A. macdonnellensis* are also proportionately longer. Some *Antechinus bellus* (e.g. AM M11304) are similar but differ in their lesser development of the postprotocrista, slightly less swollen stylar cusp D and slightly more swollen stylar cusp B. *Antechinus stuartii* is similar in having an almost comparable stylar shelf, paracone and metacone construction. It differs only in having a less well–developed postprotocrista.

Although I consider that this tooth represents a distinct species of *Antechinus,* I hesitate to establish a new *Antechinus* when only a single tooth is known. For the same reasons, its phylogenetic significance is difficult to interpret without additional information about premolar and lower molar morphology.

Dasyurus sp.

An isolated LM^2 (QM F10637), a fragment of a right lower molar (QM F10638), and a left edentulous dentary fragment with alveoli for M_3 and M_4 (QM F10639) represent a medium-sized species of *Dasyurus* on the basis of the generic character states reviewed elsewhere (Archer 1976c). The M^2 is too large and bulbous to be referable to *D. hallucatus,* the only species presently known to occur in northern Australia. Although the New Guinean *D. albopunctatus* has more bulbous molars (eg. AMNH109410), it is smaller than the Floraville species. The Floraville beast is smaller than *D. maculatus* although it is similar in the development of a bulbous–based crown. The closest morphological and size approximation is to *D. geoffroii* and *D. viverrinus. D. geoffroii* was known to occur in northern central Australia and has recently been found alive in New Guinea (Waitheman 1980). It clearly then must have also occurred in northern Australia. *D. viverrinus* has never been known to occur outside of eastern and southeastern Australia.

It is also possible that the Floraville *Dasyurus* could represent *D. dunmalli* (see above). Unfortunately the parts preserved of *D. dunmalli* and the Floraville *Dasyurus* do not usefully correspond. *D. dunmalli* is in the same size range and is similar in molar morphology to *D. geoffroii* and *D. viverrinus* (see above). Not enough of the Floraville *Dasyurus* is preserved to enable it to be either more precisely identified or phylogenetically assessed.

MYOICTIS SP. FROM THE PLIOCENE AWE LOCAL FAUNA OF PAPUA, NEW GUINEA

Plane (1967) records a specimen (UCMP 63640) of a small dasyurid from the Otibanda Formation of Papua, New Guinea. This dasyurid and other animal remains from the same formation are described by Plane as parts of the Awe local fauna which was determined by K/A dates on assocated tuffs to be between 3.9 and 7.6 million years old (Plane 1972). Page and McDougall (1972) suggest, however, that dates older than about 3.1 my for these tuffs may be unreliable.

Plane (1967) compares the Awe dasyurid specimen with specimens of *Antechinus apicalis, A. flavipes, A. minimus, A. swainsonii, A. bellus,* and *Planigale maculata.* He concludes that it belongs in the *Antechinus* group and that it is more similar to *A. apicalis* (as *Parantechinus*) than any of the other species. We (Plane and I, in preparation) have now made wider taxonomic comparisons and consider that the closest affinities of the Awe dasyurid are with species of the endemic New Guinean genus *Myoictis.* A preliminary description and abbreviated discussion of this, the only New Guinean Tertiary dasyurid, are given below. Plane and I will present a more detailed analysis elsewhere.

Myoictis sp.
(Figs 5 and 8)

Description: This Awe dasyurid is represented by a right dentary fragment containing M_2, alveoli for P_{1-3}, and broken alveoli for C_1 and M_3. The M_2 is worn on the posterior face of the trigonid and talonid. The relatively unworn paraconid is extremely reduced and low on the anterior face of the trigonid. No metacristid is present and there is only a barely visable trace of a paracristid. The metaconid occurs on the lingual edge of the crown, is worn to a level lower than the worn and broken protoconid, and is higher than the paraconid. The metacristid is obliterated by wear. The protoconid is massive and extends from the buccal edge of the trigonid to the lingual edge of the trigonid between the anterior paraconid and the posterolingual metaconid. All of the unworn faces of the protoconid are gently convex. The occlusal face of the hypoconid is worn. The unworn posterior face is gently concave. The unworn buccal surface is gently convex. The cristid obliqua is very worn and low. The hypocristid directly connects the hypoconulid and hypoconid, and then extends posterolingually beyond the hypoconulid to meet the posterolingual edge of the posterior cingulum. The entoconid is separated from the hypoconulid and metaconid by slight rifts; but all three cusps are united at their bases on the talonid. The entoconid is better–developed than the paraconid and hypoconulid, but is smaller than the other cusps. The crown is slightly swollen at the crown–root junction. Cingular development occurs only along the posterior face, on the buccal edge of the valley between the hypoconid and the protoconid, and on the anterior face of the trigonid buccal to the paraconid. A large wear facet is present on the posterior face of the hypoconid and is formed by the anterolingual face of the M^2 metacone. The posterior face of the trigonid and the whole occlussal surface of the talonid is worn, the result of occlusion with the protocone of M_2. The crown is fractured transversely across the anterior edge of the talonid. The tips of the protoconid and entoconid are broken.

Five alveoli are present anterior to M^2 and posterior to the canine alveolus. They are interpreted here (and by Plane 1967) to represent a two–rooted P_1, a two–rooted P_2 and a single–rooted P_3. On this basis, the posterior roots of P_1 and P_2 were larger than the anterior roots, P_2 was slightly larger than P_1, and P_3 was presumably peg–like. The premolars were all evidently shorter anteroposteriorly than M_2. The premolars would also have been in contact anteroposteriorly. The anterior root of P_1 almost contacts the basal edge of the canine. This alveolar proximity presumably meant that there was a slight buccal deflection of the anterior root of P_1. The canine alveolus is large, indicating that the transverse diameter of the C_1 was almost half again as large as any premolar. The alveolus is directed posterolingually and suggests that the canine was recurved and probably projected bucally beyond the buccal margin of the postcanine teeth.

We (Plane and I) compared the Awe specimen with all dasyurids and found the greatest agreement in morphology with species of *Myoictis*. We then compared the Awe specimen with all the nominate forms (whether in the synonomy of other names or not) referable to species of *Myoictis*. Details of these comparisons will be presented elsewhere. Overall, the poorest resemblance is to the two forms referable to *Myoictis wallacei*, all of which have a two–rooted P_3. Resemblance is better to the six forms referable to *M. melas*, all of which have a single–rooted P_3 or no P_3 at all.

The Awe *Myoictis* is, however, unique in one respect: the well–developed posterobuccal swelling below the hypoconid. This feature is otherwise unknown in any species of the genus. It has the effect of closing up the buccal embrasure between M_2 and M_3. When better Awe material becomes available, it will probably be appropriate to recognize the Awe *Myoictis* as a distinct species.

Glaucodon ballaratensis Stirton
From the Plio–Pleistocene of Victoria
(Fig. 7)

Stirton (1957), Ride (1964b), and Archer (1976c) review aspects of *Glaucodon ballaratensis* from Smeaton, Victoria. Ride (1964b) demonstrates its structurally intermediate position between *Dasyurus maculatus* and *Sarcophilus harrisii*.

Its age has been in doubt. Gill (1957) interprets the sediments from which *Glaucodon ballaratensis* came to be late Pliocene or early Pleistocene in age. Stirton (1957) suggests that they might be late Miocene or older in age. A K–Ar date of 2.1 my on basalt from the West Berry Consols Mine No. 1 Bore is reported by Aziz–ur–Rahman and McDougall (1972) who suggest that the sediments from which *Glaucodon ballaratensis* came are younger than this basalt.

Small Dasyurids from the Plio–Pleistocene Fishermans Cliff Local Fauna of New South Wales

The Moorna Formation, containing the Fisherman's Cliff local fauna of southwestern New South Wales, is interpreted by Marshall (1973) to be Pliocene or Pleistocene in age. On the basis of recent and more extensive studies (Crabb 1978), the Fisherman's Cliff local fauna is regarded to be more probably early Pleistocene in age.

This fauna contains at least six species of dasyurids: *Sarcophilus* sp.; a large species of *Dasyurus; Dasyurus hallucatus; Dasyuroides* n. sp.; *Sminthopsis* sp.; and a species probably referable to *Antechinus*.

Sarcophilus moornaensis Crabb

Marshall (1973) refers an isolated upper molar and a fragmentary left dentary with $M_{2\text{-}5}$ to *Glaucodon* cf. *G. ballaratensis*. He suggests that the Moorna material differs from the holotype of *G. ballaratensis* in having a more robust mandible. Nevertheless, he concludes 1973: 156) that there is an ". . . absence of notable morphological differences . . ." and ". . . it appears probable that they are conspecific." From a comparison of Marshall's *Illustrations*, and a cast of the holotype of *Glaucodon ballaratensis*, it appears that the Moorna animal has a narrower, less well–developed talonid, a smaller metaconid on M_5, and a more longitudinal metacristid orientation. Superficially at least, these features appear to be in the direction of (but fall well short of) the condition seen in *Sarcophilus harrisii*, and away from that of *G. ballaratensis*. Overt (but perhaps not phylogenetically significant) similarity to *S. harrisii* is evident in the proportion of tooth–row–length to dentary–length, and in the relatively robust molars. The Moorna specimen differs from *Sarcophilus harrisii* in the same characters that appear to differentiate it from *G. ballaratensis*, but in the opposite direction: the talonid of M_2 is markedly better–developed; the metaconid of M_5 is larger and the metacristids are more transversely orientated.

The isolated Moorna upper molar illustrated by Marshall (1973), is referable to the same species of Moorna dasyurid. *Glaucodon ballaratensis* is unrepresented by upper teeth so comparison of the isolated Moorna left M^3 can only be made with *Sarcophilus harrisii* and *Dasyurus maculatus*. It is structurally much closer to *S. harrisii* in the approximation of stylar cusp B and stylar cusp D to the paracone and metacone respectively, the reduction in size of stylar cusp D, elongation of the metacrista, and relatively robust and swollen crown base. It differs from *S. harrisii* in that these stylar cusps are relatively more distinct and, although joined by a horizontal crest to the metacone, stylar cusp D is still markedly better–developed in the Moorna tooth than it is in *S. harrisii*.

Elsewhere in this volume (Crabb 1982), this Moorna species has been described as *Sarcophilus moornaensis* Crabb. Subsequent to Crabb's work, the Moorna material is again being restudied by K. Alpin and J. Hope within the context of a broader Australian–wide analysis of

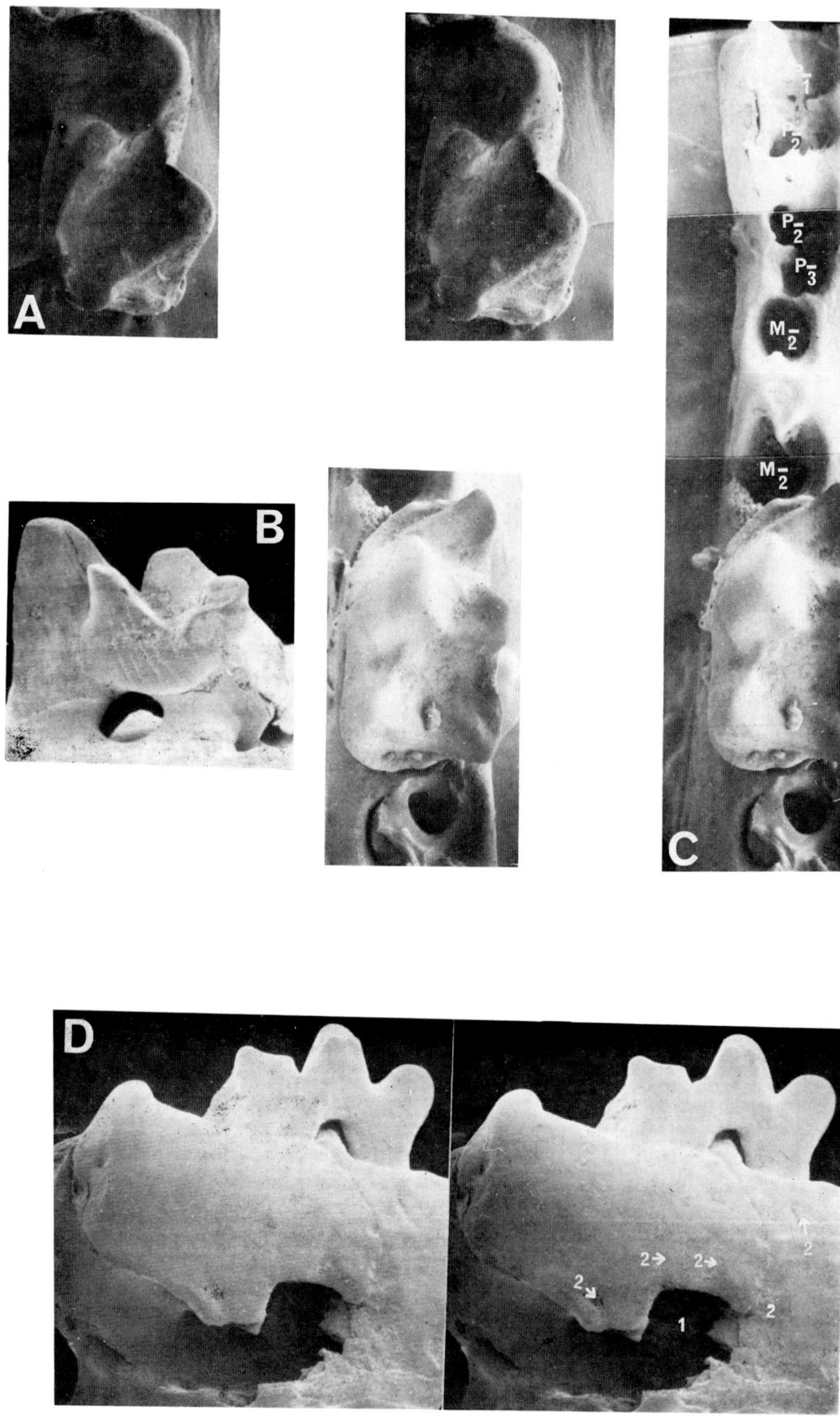

Fig. 10. Dasyuroides achilpatna. Measurements are given in Fig. 5. A, NMV P38429, RM_2, occlusal view. B, NMV P38429, lingual view to show the relatively large metaconid and virtually absent paraconid. C, the holotype, with LM_3, alveoli for P_1 and P_2, and the single alveolus for P_3. D, the holotype, lingual view of the posterior part of the dentary fragment, showing the hole (1) and depressed fracture interpreted to be the result of puncture by a canine of a species of *Dasyurus*, and associated tooth scratch marks (2) in the same area.

the species of *Sarcophilus*. Their conclusions will be published elsewhere.

Dasyuroides achilpatna n. sp.
(Figs 5, 10, 11 and 12)

Holotype: NMV P38334, a left dentary fragment with M_3 and alveoli for P_1, P_2, P_3, M_2, and M_4.

Type locality: Fisherman's Cliff, part of "Site 13" (see Gill 1973, figs. 30–31 for a section of Fisherman's Cliff; and his pp.40–42 for a description of the Moorna Formation at Fisherman's Cliff), Moorna Formation, Moorna Station, southwestern New South Wales.

Age: Late Pliocene or early Pleistocene (see above).

Paratypes (all NMV, specimens): P38373, RM_2; P38281, RM_2; P38419, LM_5; P38331, RM^2; P38442, RM^3; P38283, RM^5; P38372, LM^2; P38418, LM^3; P38399, LM^4; P28886, right dentary fragment with alveoli for C_1-M_2; P28888, right dentary fragment with $M_{3\text{-}5}$; all paratypes were collected from the type locality.

Diagnosis: *Dasyuroides achilpatna* differs from the only other species in the genus, *D. byrnei,* and from its otherwise closest relative *Dasycercus cristicauda,* in possession of a single–rooted P_3, evidently broader and proportionately shorter premolar crowns, larger metaconid on M_2, and a better–developed (than *D. cristicauda*), anterior cingulum on M^2. Differences from other otherwise related dasyurids are discussed below.

Description (based on the holotypes and paratypes): Second upper molar (NMV38331; NMV P38372): The metacone is taller than stylar cusp D, which is taller than the subequal (or inseparable) stylar cusp B–paracone, which is taller than the protocone. There is a complete although short anterior cingulum formed by the merger of the anterbobuccal cingulum and the preprotocrista. There is no posterior cingulum. The postprotocrista extends posterobuccally and then swings buccally to merge with the posterior base of the metacone. The preprotocrista appears to have been either nonexistant (with confluence of stylar cusp B and the paracone) or very short (both M^2s are slightly worn). The preparacrista is not preserved. The metacone is hooked up directly with stylar cusp D. If there was a preparacrista (and I assume there was on the basis of unworn molars of dasyurid species with similar M^2 morphology), it must have been a reduced crest. The crest hooking up the metacone to stylar cusp D is not the premetacrista; rather it is the breeched edge of enamel lying between the metacone and stylar cusp D. The postmetacrista is about twice as long as the metacone–stylar cusp D crest. The postprotocrista extends posterobuccally, parallel to the ectoloph. It bypasses the lingual base of the metacone but then turns abruptly bucally to intersect and terminate against the posterobuccal side of the base of the metacone. A distinct vertical crest occurs on the anterobuccal flank of stylar cusp B. It merges with and becomes the anterobuccal end of the anterior cingulum. A low crest connects stylar cusps B and D. Another low crest connects stylar cusp D to the metastylar corner of the tooth. Stylar cusp B produces a slight basal convexity on the buccal edge of the crown, but one that is longer than the bulge produced by stylar cusp B. The crown length (P38331, 2.9 mm; P38372, 2.9mm) is considerably longer than the anterior width (P38331, 2.4mm; P38-372, 2.1mm).

Third upper molar (NMV P38418; NMV P38442): This tooth is similar to the M^2 except as follows. The crest lengths differ (see meristic gradients below). The anterior cingulum is not complete. A preparacrista separates stylar cusp B from the paracone. The metacone – stylar cusp D distance is about three–quarters the length of the postmetacrista. There is no crest connecting the metacone to the stylar cusp D. There is a distinct premetacrista that is confluent with the postparacrista. The notch in the ectoloph between stylar cusps B and D, is pronounced. The tooth length (NMV P38418, 2.9mm) is approximately equal to its anterior width (NMV P38418, approximately 2.9 mm).

The fourth upper molar (NMV P38499, with a damaged metastylar corner): This tooth is very similar to M^3 but differs as follows. The crest lengths differ (see meristic gradients below). The anterobuccal cingulum is better–developed. The postparacrista and premetacrista are subequal in length. The ectoflexus is more sharply indented between stylar cusps B and D. Other possible differences are obscured by damage. The anterior width (3.0mm) can be determined, but not the crown length.

The fifth upper molar (NMV P38283): This tooth is very short (1.2mm) but as wide (2.9 mm) as the other posterior molars. The paracone is taller than the subequal stylar cusp B and the very reduced protocone. There are no other cusps. The almost horizontal preparacrista is very long. The steeply inclined postparacrista is about one third of its length. The pre– and postprotocrista are very short. The anterior cingulum is well–developed but confined to the buccal third of the crown. It connects directly with stylar cusp B. There is no posterior cingulum.

Meristic gradients: The preparacristae increase in length from M^2 to M^5. The postmetacristae increase in length at least from M^2 to M^3. Stylar cusp D becomes smaller and more distant from the metacone from M^2 to M^3. The protocone becomes shorter (anteroposteriorly) from M^2

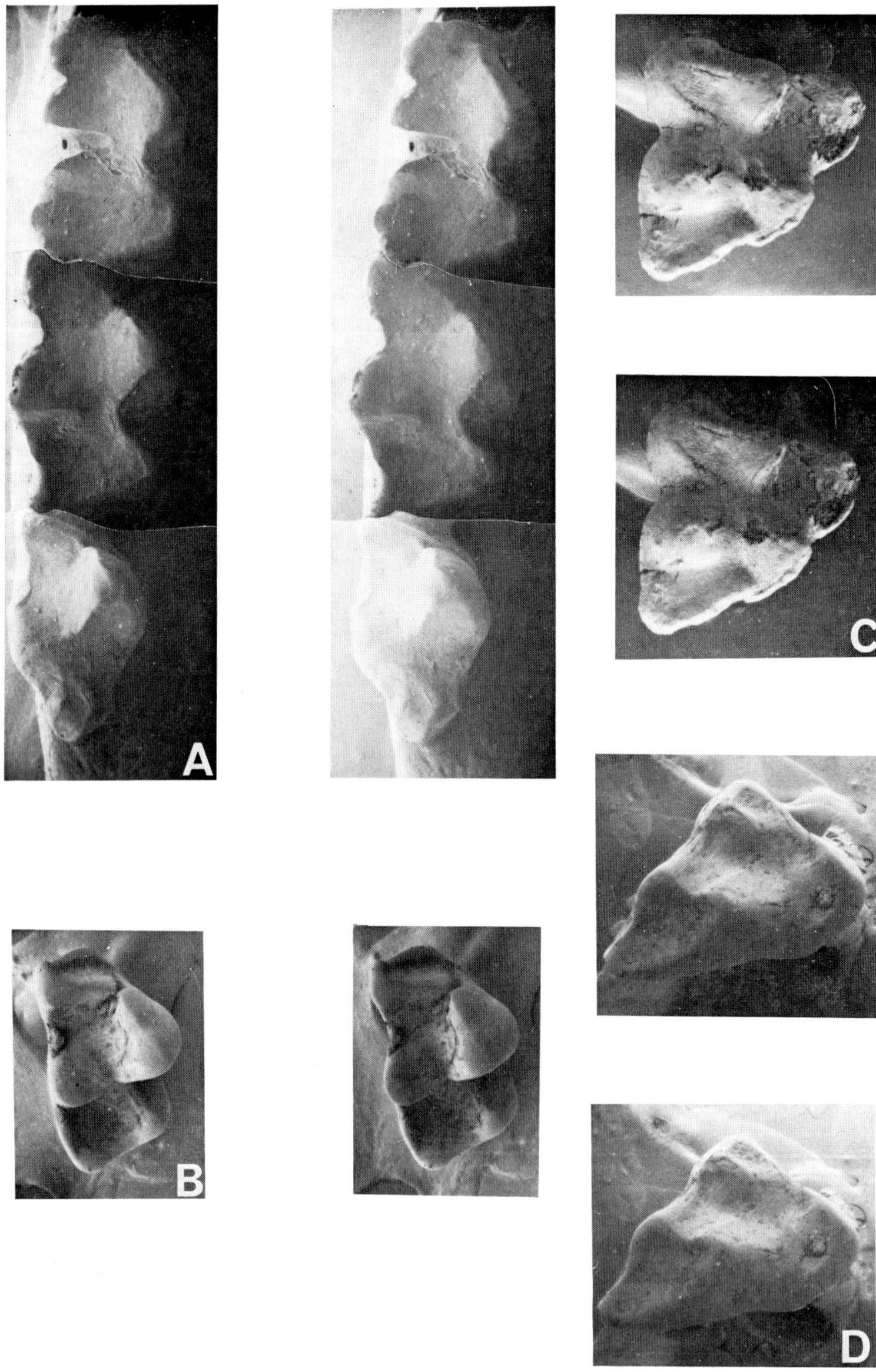

Fig. 11. Dasyuroides achilpatna from the Fisherman's Cliff local fauna. Measurements are given in Fig. 5. A, NMV P28888, right dentary fragment with worn M_{3-5}. B, NMV P38343, RM_X. C, NMV P38442, $RM^{?3}$. D, NMV P38331, RM^2.

to M^5. The ectoflexus becomes more marked from M^2 to M^4.

The dentary (NMV P28886; NMV P38334): The only two dentary fragments that preserve information about the premolars are, in all essentials, the same. Both show a single well–developed alveolus immediately anterior to that for the anterior root of M^2. This represents a single-rooted P_3. Four alveoli anterior to that for P_3 represent a small (relative to M_2) P_2 and an even smaller P_1. The alveoli are compressed antero-posteriorly and strongly suggest that the premolar row was crowded. The anterior root of P_1 in NMV P38334 is also displaced bucally, suggesting that the anterior end of the crown was slightly buccally displaced. All premolar alveoli, except that for P_3 and the anterior root of P_1, are wider than long, and suggest that the premolar crowns were broad. The combined length of the alveoli for P_2 (NMV P38334, 1.5mm; NMV P28886, 1.9mm), compared with the alveolar length of M_2) NMV P38334, 2.8mm; NMV P28886, 2.8mm), indicates that the premolars were relatively short (ratios of LP_2/LM_2 = 0.54 and 0.68; compare the similar ratios in four *Dasyuroides byrnei* of 0.69, 0.71, 0.72, and 0.80). On the basis of the canine alveolus, this tooth was large, and certainly wider than any premolar. The symphysis extends back to the level of the anterior root of M_2.

A feature of damage should be noted here. There are clear indications on the lingual side of NMV P38334 of tooth marks, some of which have left depressed fractures below the M_2. The rounded bone edges, other gouge marks, and size suggest that this specimen, (the holotype) may at one time have been part of a larger predator's dropping. The size of the tooth marks suggest the predator was most probably a species of *Dasyurus,* species of which are also present in the Moorna local fauna.

The second lower molar (NMV P38281; NMV P 38429; NMV P38373): The trigonid is narrower (all specimens 1.3mm) than the talonid (all specimens 1.5mm). The protoconid is much taller than the basal hypconulid, which is just taller than the extremely reduced (to absent) entoconid. There is no evidence whatsoever of a paraconid on two of the specimens, but on NMV P38281 there is a slight wear facette at the extreme anterior end of the crown that may have represented a very tiny paraconid. The para cristid is steeply inclined and extends antero-lingually (much more anteriorly than lingually) to the anterobasal tip of the crown. The metacristid is even more steeply inclined and extends posterobuccally, at right angles to the paracristid, to the reduced metaconid. Between the metaconid and protoconid, but closer to the metaconid, there is a carnassial notch in the metacristid. The short anterobucally directed cristid obliqua connects the hypoconid to the posterior side of the protoconid at a point below the apex of the protoconid. The cristid obliqua contacts the trigonid via an almost vertical ridge that occurs on the posterior flank of the protoconid. The intersection of these two crests is marked by a steeply inclined carnassial notch. The hypocristid extends posterolingually to the hypoconulid without approximating the vestigial entoconid. The entoconid is a slight rise in a low crest that surrounds the lingual side of the talonid. There is a continuous very narrow buccal cingulum in NMV P38373. In NMV P38429 it is not clear that the buccal cingulum actually extended beneath the protoconid. The buccal cingulum in all specimens merges with the narrow anterior cingulum that extends across the anterobuccal face of the trigonid to contact the anterior end of the paracristid. The posterior cingulum extends from beneath the hypoconid to the hypoconulid. A broadly concave wear facette represents the impression of the anterior cingulum of M_3.

The third lower molar (NMV P38334): The trigonid width (1.5mm) is slightly less than the talonid width (1.6mm). The protoconid is taller than the metaconid, which is taller than the paraconid, which is taller than the subequal hypoconid, entoconid and hypoconulid. The paracristid has a carnassial notch about midway along its length. The metacristid is not transverse but rather orientated posterolingually. There is a carnassial notch about midway along its length. The cristid obliqua is joined to the trigonid in the same manner as it is in M_2. The hypocristid parallels the metacristed in orientation and runs directly to the hypoconulid. Although worn, it appears that the entoconid is poorly differentiated from the crest forming the lingual margin of the talonid. It occurs about midway between the trigonid and the hypoconulid. The anterior cingulum is well–developed and extends from the anterobuccal base of the protoconid. There is a well–developed hypoconulid notch near its lingual end. The posterior cingulum extends from the posterior base of the hypoconid to the hypoconulid. Clearly a buccal cingulum occurred at least around the base of the talonid. It is not clear whether there was a buccal cingulum around the base of the trigonid, as there was in M_2, because this area of the M_3 seems to be worn.

The fourth lower molar: Two isolated teeth are reasonable candidates to represent M_4. The first, NMV P38343, is a right lower molar whose talonid is more reduced in width than the talonid of M_3, but much less so than it is in M_5. On the basis of meristic changes down the tooth row in *Dasyuroides byrnei,* this tooth would be a reasonable serial successor to the M_3 of *D. achilpatna.* The second tooth, NMV P28887, has a markedly reduced posterolingual corner of the talonid. Further, its anterior and posterior cingula are greatly reduced. These features seem to be too extreme for a serial successor to the M_3. However, the tooth is about the right size for an M_4, its talonid width is narrower than that of M_3 but wider than that of M_5, and it has a very reduced entoconid. Possibly it is abnormal or an extreme variant of the M_4. Whatever the case, it will not be included in the description of the M_4 of *D. achilpatna* given here.

Based on NMV P38343, the M_1 is very similar to the M_3, except as follows. The entoconid is barely distinguishable along the lingual crest of the talonid. The trigonid width (1.8mm) is greater than the talonid width (1.5mm). The posterior cingulum appears to be less well developed, but this may in part be due to wear. Similarly, the anterobuccal end of the anterior cingulum is either more poorly developed or worn. The cristid obliqua intersects the trigonid in a more lingual position than it does in M_3.

The fifth lower molar (NMV P38419): The trigonid width (1.4mm) is greatly in access of the talonid width (0.7mm). The trigonid construction is very similar to that of M_4 except that the paraconid is only just, if at all, shorter than the metaconid. Also the carnassial notch of the metacristid occurs in a more lingual position. The talonid is extremely reduced. The low lingual talonid crest is present but no entoconid is present. There is one poorly–differentiated cusp on the posterior edge of the talonid. This is probably the hypoconid. A vestigial cristid obliqua extends from the cusp to the rear of the trigonid at a position directly below the carnassial notch in the metacristid. A very shallow talonid basin is defined by the buccal (cristid obliqua) and lingual talonid crests.

Meristic changes along the lower molar row: The paracristid rotates trasversely from M_2 to M_3. The metacristid lengthens from M_2 to M_3. The carnassial notch in the metacristid occurs in a progressively more lingual position from M_2 to M_5. The talonid decreases in width (in proportion to the trigonid) from M_2 to M_5. The trigonid width increases from M_2 to M_4, and then decreases in M_5. The entoconid is tiny on M_2 and M_3, miniscule on M_4 and missing on M_5.

Origin of the species name: *Achilpa* is a central Australian Aboriginal name (Spencer and Gillen 1899) for the Western Quoll (*Dasyurus geoffroii*); *atna* is a central Australian Aboriginal word meaning "dung" (Kempe 1891). The allusion is to the fact that the holotype of *Dasyuroides achilpatna* bears tooth marks of a size consistant with interpretation that it was chewed by a species of *Dasyurus* (which are also in the Fisherman's Cliff local fauna). Further, as judged by the reduced size of the holotype, its rounded edges, and what appears to be evidence for chemical erosion, it was quite probably thoroughly processed by a gastronomically discerning Moorna Quoll.

Discussion: This Moorna dasyurid was regarded by Marshall (1973: 156) to closely compare ". . . in size and molar morphology with *Dasyuroides byrnei* (C458) and *Dasycercus cristicauda* (C655) . . . This species is regarded as a possible ancestral form of *Dasyuroides* or *Dasycercus* (or both)."

The singular features of this Moorna species include its tiny to miniscule entoconids, single–rooted P_3, crowded premolars, vestigial to absent paraconid on M_2, non–transverse metacristids and hypocristids, united or very tightly approximated paracone and stylar cusp B on M_2, and lack of any posterior cingula on the upper molars. These features prohibit reference of this species to the following genera: *Dasyurus, Sarcophilus, Ankotarinja, Keeuna, Phascogale, Planigale, Sminthopsis, Wakamatha, Antechinomys, Ningaui, Phascolosorex, Murexia,* and *Myoictis.* This leaves some species of *Antechinus* (sensu lato), *Dasycercus*, and *Dasyuroides.* Within *Antechinus,* only *Antechinus apicalis, A. macdonnellensis, A. rosamondae,* and *A. bilarni* have all or most of these features. I have made detailed comparison with all of these species as well as with those of *Dasycercus* and *Dasyuroides,* and conclude that the Moorna species is clearly most similar to *Dasyuroides byrnei.* Nevertheless, there is a striking similarity in molar morphology between these species of *Antechinus, Dasycercus* and *Dasyuroides,* and this is suggestive of a more broadly based relationship (see below).

The Moorna species differs from *Antechinus apicalis* (which, incidentally, has a single–rooted P_3) in having a more closely approximated paracone and stylar cusp B, a narrow talonid on M_4, in being considerably larger, in lacking any trace of a posterior cingulum on the upper molars, having proportionately longer preparacrista on M^3 and M^4, in having a longer M^5, and in being larger. It differs from *Antechinus macdonnellensis* in being much larger, in having a tiny to miniscule entoconid on a lingual talonid crest (although entoconid development varies within *A. macdonnellensis*), a longer (anteroposteriorly) protocone on M^3–M^4, a proportionately longer buccal half of $M^{2\text{-}4}$, a more transversely reduced talonid on M_4, and a larger metaconid on M_2. It differs from *A. rosamondae* in having a shorter (anteroposteriorly) and wider M^5, a single-rooted P_3 (*A. rosamondae* has no P_3), narrower hypocristids, longer molars, and in being much larger. It differs from *A. bilarni* in being much larger, in having much better–developed postprotocristae, and in having proportionately longer upper molars.

The reasons for choosing to refer the Moorna species to *Dasyuroides* rather than *Dasycercus* are less easy to defend. There are a few consistant differences between the two modern species in those parts of the dentition that are preserved in the Moorna specimens. The most useful involve aspects of the morphology of M_2 and M^5. The M_2 of *Dasyuroides byrnei* is relatively (to *Dasycercus cristicauda*) longer and narrower. The occlusal angle between the metacristid and paracristid is clearly greater than 90°. *In Dasycercus cristricauda* it is normally about 90° (in one specimen from the Nullarbor, however, the talonid angle was comparable to that seen in *D. byrnei*). The talonid is relatively swollen in *Dasycercus cristicauda* such that the lingual basal margin is strongly convex. There is also often a better–developed anterobuccal basal swelling or cingulum on the M_2 of *Dasycercus cristicauda*. The M^5 of *Dasyuroides byrnei* normally has an anterobuccal cingulum which *Dasycercus cristicauda* appears to lack. In all of these features, the Moorna species most closely resembles *Dasyuroides byrnei,* differing only in the character states indicated in the diagnosis.

In its retention of P_3 and larger metaconid on M_2, the Moorna species is clearly more plesiomorphic than either *D. byrnei* or *Dasycercus cristicauda*. However, in its wider premolars and greater reduction of the entoconid, it is also more specialized than either living species. Marshall's (1973) contention that this species could be ancestral to one or both of the living species is not supported by the autamorphic features shown by this Moorna dasyurid. An alternative hypothesis (Fig.12) would be that *D. achilpatna* represents an autapomorphically specialized (entoconid reduction and wide premolars) species that was the primitive sister group of the living *D. byrnei*. This would also imply an independent reduction of the metaconid on M_2 and the loss of P_3 in *Dasyuroides byrnei* and *Dasycercus cristicauda*. Both of these developments have been independently achieved in other dasyurid lineages such as *Antechinus* (*A. rosamondae*), *Dasyurus* (independently in *D. hallucatus* and the other living species of *Dasyurus,* see above), *Myoictis* (*M. melas*), and *Planigale* (*P. gilesi*).

The autapomorphic increase in premolar crowding in *Dasyuroides achilpatna* is a comparable evolutionary event to that discussed above in relation to *Dasyurus dunmalli*. The phylogentic stage in premolar reduction where P_3, although vestigial, is still present, is commonly associated with anterposterior crowding of the tooth row. Following phylogenetic loss of P_3 (as in modern *Dasyuroides*), the premolar row becomes less crowded.

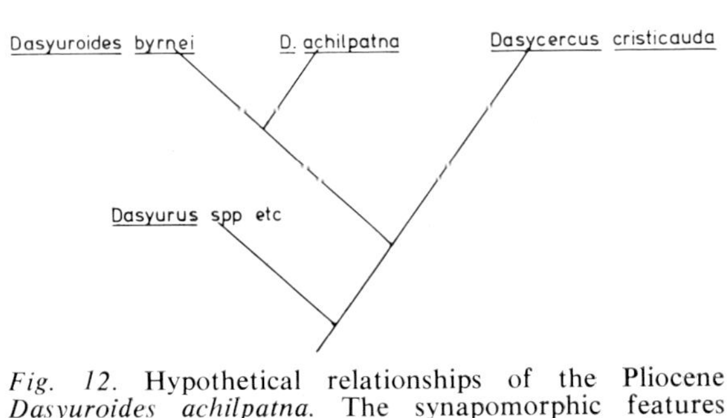

Fig. 12. Hypothetical relationships of the Pliocene *Dasyuroides achilpatna*. The synapomorphic features taken as the basis for its sister–group relationship to the modern *Dasyuroides byrnei* are noted in the text. *Dasyurus* spp. etc. refers to the rest of the Tribe Dasyurini as defined below in the text and shown in Fig. 13 on branch 9.2.

The presence of a species of *Dasyuroides* in the Moorna formation suggests that the environment of the Fisherman's Cliff local fauna was at least becoming arid. Reduction of entoconids, shortening of the premolar row, and reduction of the M_5 talonid are features commonly found in modern arid-adapted dasyurids, but rare in coastal non–arid–adapted species (Archer 1976c, 1981a).

Dasyurus hallucatus
(Fig. 7)

Marshall (1973) recognizes *Dasyurus* cf. *D. hallucatus* in the Fisherman's Cliff local fauna on the basis of NMV P28889. He regards this tooth to be a LM^4. I agree that the tooth is very similar to *D. hallucatus* but consider it to be an LM^3. The preparacrista is far too short for the tooth to be an LM^4. Further, I would refer it without hesitation to *D. hallucatus* because, in so far as the morphology of this tooth is concerned, it cannot be distinguished from the M^3 of *D. hallucatus*.

It is clear that this southern New South Wales occurrence of *D. hallucatus* is far south of its present known limit (Gympie in Queensland). The zoogeographic significance of this is unclear in part because we do not yet know what factors limit the southern spread of this species, and in part because some considerable doubt must remain about the specific identity. It is one thing to say the M^3 is indistinguishable from the modern species but quite another to know that the tooth came from an animal whose remaining morphology and ecological requirements were the same as the modern species.

A Larger Species of *Dasyurus*

A larger species of *Dasyurus* is represented by several teeth including: NMV P28890, a broken LM^4; NMV P38438, LM_3; NMV P29428, RM_4; NMV P38428, RM_4. The size of this species is just smaller than any specimens of *D. geoffroii* available to me (south-western Western Australian *D.g. fortis* and Papuan *D.g. geoffroii*). It is also close to the size of *D. viverrinus* and *D. dunmalli*. Molar morphology of the larger species of *Dasyurus* (particularly *D. geoffroii*) appears to be remarkably variable within species. For this reason, I do not consider that it would be a useful thing to try to determine the affinity of the isolated Moorna teeth more closely.

Marshall (1973) considers that one of the Moorna teeth, NMV P29428, a RM_4, agrees more closely in size and morphology with *D. geoffroii* than *D. viverrinus*. The reasons for his conclusions are: 1) the tooth is narrow relative to its length; 2) the talonid is narrower than the trigonid; and 3) the protoconid is set slightly anteriad of the metaconid. Elsewhere (Archer 1978b) it has been shown that these features are not consistant in the two species and do not serve to diagnose either species on the basis of isolated teeth.

The occurrence of more than one species of *Dasyurus* in a single modern local fauna is known from northern Australia (*D. geoffroii* and *D. hallucatus*, the former being now extinct in northern Australia), New Guinea (*D. albopunctatus* and *D. geoffroii*), and eastern Australia (*D. maculatus* and either *D. geoffroii*, which is extinct now in eastern Australia, *D. viverrinus*, or *D. hallucatus*). *D. geoffroii* would have occurred in the Moorna area prior to European settlement. *D. viverrinus* and, to a lesser extent, *D. maculatus* are confined to the more coastal regions of eastern Australia.

Two isolated teeth are probably referable to this larger species but allocation is less certain. These include NMV P38437, an RM_2, and NMV P38430, a fragment of an LM^3. The RM_2 is small and it is conceivable, although unlikely, that it represents a large individual of a *D. hallucatus*–sized species. The LM^3 fragment similarly could be a fragment of a large (for *D. hallucatus*) M^3 or a small (for a *D. geoffroii*–sized species) M^4.

Sminthopsis sp., *cf. S. macroura*

Two teeth (NMV P38311, RM^3; NMV P38498, RM_4) appear to represent a species of *Sminthopsis* comparable in size to the Common Dunnart, *Sminthopsis murina*. The lower molar exhibits a worn but well–developed entoconid in the plesiomorphic position for the genus. The upper molar is short (anteroposteriorly) and wide, without any of the more specialized features of some of the species within the genus (Archer 1981a). Stylar cusp D of the Moorna species is larger than that found in *S. floravillensis* (see above). The entoconid size and morphology are comparable to that found in *S. virginiae, S. douglasi* and *S. macroura*. Of these three similar species, only the smaller *S. macroura* is comparable in size. *Sminthopsis virginiae* also normally has a complete anterior cingulum on the upper molars, a feature lacking in the Moorna M^3. However, with only two isolated teeth, it is not possible to confidently allocate the Moorna species to a particular living species of the genus.

Sminthopsis macroura is a decidedly arid-adapted species based on its current central Australian distribution (Archer 1981a). However, because of the somewhat doubtful basis for suggesting close affinity of the Moorna species with *S. macroura,* constraint should be exercised in postulating palaeoecological implications based on this affinity.

Antechinus sp.

An isolated worn RM^3 (NMV P38272) appears to represent a species of *Antechinus* (*sensu latu*) that has only a very rudimentary posterior cingulum, a poorly-developed but just complete anterior cingulum, and a well-developed postmetacrista that is just less than twice the length of the preparacrista. This morphology is similar to that of more specilaized species such as *A. bilarni* rather than the more plesiomorphic species such as *A. stuartii*. However, with only a single isolated molar, there is little point in making more exhaustive comparisons.

LATE QUATERNARY DASYURIDS

A critical review of late Pleistocene dasyurids is much beyond the scope of the present work. There have been many studies of late pleistocene and Holocene dasyurids from cave deposits in Australia. Many of these are based on an adequate study of the morphology of living dasyurids. Examples include the valuable contributions of Wakefield (e.g. Wakefield 1972, Smith (1972), Hope *et al.* (1977), Baynes *et al.* (1976), and (forgive me) Archer (1978b). Many other thorough studies are cited in their reference lists. A few studies have been based on less rigorous analyses of modern dasyurid morphology and/or intraspecific variation. This is regretable for two reasons. First, it has sometimes led subsequent researchers to base their studies on unsound taxonomic or zoogeographic foundations. Second, it often results in the need for a subsequent worker to spend a lot of time sorting out muddles that might not have occurred had the original worker bothered to examine the morphology of all relevant taxa and attempted to determine intraspecific variation for all possibly relevant modern species before making identifications or naming new species.

It is surprising to me that with rare exception, most dasyurids known from late Pleistocene deposits in Australia appear, when rigorously analysed, to be referable to living species (e.g. Dawson, this volume). There are four possible exceptions. One is a large species of *Antechinus* represented by a dentary (QM F8445) from Russenden Cave (Archer 1978b). Stephen Van Dyck is at present examining this problem in the context of a broadly based and thorough analysis of dental morphology and variation in related living species of *Antechinus*. The second is a species of *Planigale* from the Madura Cave fauna (Lundelius and Turnbull 1973). Although most of the features of this animal that were originally interpreted by Lundelius and Turnbull

(1973) to be indicative of specific distinction have been shown (Archer 1976a) to be variably present in living species of the genus, it is distinctive (although at what taxonomic level is debatable) in the length of its P_3. The third is represented by a dentary (SAM P161182) referable to a species of *Sminthopsis* from the Victoria Cave fauna. Tentatively it has been referred by Smith (1972) to *S. murina*. However, she notes it is much smaller (relative to other *S. murina*) in size. Modern *S. murina* is clearly in need of closer examination. The isozymes of various "*S. murina*" studied by Baverstock *et al.* (this volume) suggest there may be more than one species involved. The Victoria Cave dentary may be related to this problem. The fourth is *Dasyurus mordax* Owen, 1877, from an unspecified cave in Australia. The holotype (a left dentary with P_2 – M_4) and only known specimen is missing. Owen's almost useless illustration (his plate 10, fig. 10) suggests a beast unlike any other species of *Dasyurus*. Taken as drawn, the left P_2 and M_2 resemble those of other species of *Dasyurus*. However, $M_{3\text{-}4}$ are clearly right (not left) teeth in the same dentary as the left P_2 and M_2. Either it is a hopelessly abnormal individual, Owen's delineator (or Owen) was out of sorts the morning he drew the specimen, or the specimen was faked. Whatever the case, the holotype is now missing (Mahoney and Ride 1975). Owen did not describe the specimen other than to say (in the legend for plate V on page 107): "Fig. 10. Inner side view of the mandible and teeth of *Dasyurus mordax*.".

Apart from these four possibly distinctive late Pleistocene forms, all other late Pleistocene records appear to represent modern species. Often they add considerably to our knowledge about formerly wider ranges (e.g. Archer 1978b). Examples include *Myrmecobius fasciatus* from the southern part of the Nullarbor Plain (R. Thompson pers. comm.), *Dasycercus cristicauda* from the coastal heath of Western Australia (Lundelius 1960), a species of *Planigale* from the southern Nullarbor (Lundelius and Turnbull 1973), and *Sminthopsis longicaudata* and *Antechinus rosamondae* from Cape Range, northwestern Australia (Kendrick and Porter 1973).

THE STATE OF THE DASYURID FOSSIL RECORD

The fossil record of dasyurids reviewed above is summarized in Table 1. It leaves a lot to be desired. Of a total of 20 genera and 70 species (living and extinct), only 5 genera and 14 species of non-extant dasyurids are known — that is only about 25% of the total number of genera and 20% of the total number of species known. Compare this with the record for kangaroos (macropodids, figures being based on Bartholomai 1978 and Ride 1970) of about 7 extinct genera and 54 extinct species in a total of 18 genera (living and extinct) and a total of about 94 species. The kangaroo fossil record then represents about 39% of known genera and 58% of known species. The relatively poor showing of fossil dasyurids is probably the result of two factors. Most of the smaller insectivorous dasyurids are easily overlooked by casual fossil collectors. Second, the larger dasyurids must have been less common than the smaller or comparable–sized herbivores upon which they fed.

Despite this paucity of fossil dasyurids, the known record (Table 1) reveals four important points. First, with two exceptions, middle Miocene dasyurids are all similar (but not closely related) in being largely plesiomorphic. They commonly reveal a mixture of dilelphid–like and modern highland New Guinean dasyurid–like features. This condition is clearly exemplified in *Ankotarinja tirarensis*. The first possible exception is the Riversleigh dasyurid which has some of the apomorphic features of species of modern *Antechinus*–like dasyurids. The second exception is *Dasylurinja kokuminola* which although plesiomorphic in many features is also apomorphic in its enlarged metacone and postmetacrista.

Second, taxa undoubtedly referable to modern genera are unknown in the record before early Pliocene time. Species of *Antechinus*, *Planigale* and *Dasyurus* are known from early Pliocene deposits. Nevertheless, it is highly likely that species of some of the more plesiomorphic genera, such as *Antechinus*, will be found in Miocene faunas. Further, the early Pliocene species of *Planigale* is only doubtfully distinguishable from some of the modern species of the genus and may thus be expected to at least have had late Miocene congeneric ancestors.

Third, of the early Pliocene dasyurids known, only one, the Hamilton *Antechinus*, appears to be related to a particular modern species, the Mt. Wilhelm *Antechinus*. This modern dasyurid is an inhabitant of a high–rainfall, heavily vegetated environment. To the extent that the Hamilton dasyurid is related to this Mt. Wilhelm animal, we may be justified in hypothesising a wet sclerophyll to rainforest habitat for that area of Victoria in the early Pliocene. This supports the ecological implications of other taxa such as *Phalanger* sp. and *Dorcopsis* sp. in the Hamilton fauna (Turnbull and Lundelius 1970).

Fourth, the Plio-Pleistocene species assemblages from southern New South Wales (Fisherman's Cliff fauna) and northwestern Queensland (Floraville fauna) are the first to include dasyurids either clearly related to modern arid–adapted species or which show suites of arid-adapted character–states.

Table 1 **Fossil Dasyurids, Their Age And The Faunas In Which They Occur**

Age	Epoch	South and Western Australia	Victoria	New South Wales	Queensland	Papua
10,000	↑ Recent					
1 my	↑ Pleistocene ↓	Western Australia Madura Fauna *Planigale* sp. South Australia Victoria Cave Fauna *Sminthopsis* sp.	Cave and Lake Deposits	Cave and Lake Deposits	Russenden Cave Fauna *Antechinus* sp.	Cave Deposits
2 my	1.8 my ↑ Opinions vary ↓	South Australia Kanunka Fauna (Various ?Recent Species)	Smeaton Fauna *Glaucodon ballaratensis*	Fisherman's Cliff Fauna *Sarcophilus moornaensis* *Antechinus* sp. *Dasyurus hallucatus* *D.* sp. *Sminthopsis* sp. *Dasyuroides achilpatna*	Floraville Fauna *Sminthopsis floravillensis* *Antechinus* sp. *Dasyurus* sp.	Awe Fauna *Myoictis* sp.
3 my	3.0 my ↑					
4 my	Pliocene ↓				Chinchilla Fauna *Dasyurus dunmalli*	
5 my			Hamilton Fauna *Antechinus* sp.	Bow Fauna *Dasyurus dunmalli*	Bluff Downs Fauna *Planigale* sp.	
8 my	↑	South Australia Horizon & Fauna Unknown *Wakamatha tasselli*				
10 my	Miocene					
13 my	↓	South Australia Kutjamarpu Fauna *Ankotarinja* sp. Other unnamed forms				
16 my		South Australia Ngapakaldi Fauna *Ankotarinja tirarensis* ***Keeuna woodburnei*** Other Unnamed Forms South Australia Yanda Fauna *Dasylurinja kokuminola*			Riversleigh Fauna *Antechinus*-like Form	
	To 23 5 my					

These aspects of the albeit poor fossil dasyurid record appear to broadly corroborate a hypothesis expanded elsewhere (Archer 1981b) about marsupial evolution in Australia since middle Miocene time. The middle Miocene forms suggest a lush well–vegetated central Australia. Gradually the forests opened up to give way to the spreading grasslands. By early Pliocene time, this process had reduced the wetter environments to isolated areas in eastern Australia and New Guinea. By Plio-Pleistocene time, some of the dasyurids present in inland areas of Australia (e.g. the species of *Dasyuroides* in the Fisherman's Cliff fauna) were developing arid–adapted character–states, presumably in response to continued deterioration of the continental climate. While this was going on, New Guinea had been rising into higher, cooler, wetter altitudes ever since the Australian block began to crumple up against southeast Asia. In rising, it preserved the wet sclerophyll to rainforest habitats (the Tumbunan region of Schodde and Calaby 1972) that were being lost across the Australian mainland. Some of the modern dasyurids, such as species of *Murexia,* in these highland New Guinean refuges are the most similar, among all modern dasyurids, to the middle Miocene forms found in central Australia. This similarity is based on a sharing of a large number of plesiomorphic, rather than derived, features and is thus not indicative of close relationship.

THE IMPLICATIONS OF THE FOSSIL DASYURIDS FOR PHYLOGENETIC INTERPRETATION

The middle Miocene *Ankotarinja tirarensis* provides evidence for the phylogenetic relationship between Cretaceous American didelphid–like taxa, such as species of *Alphadon,* and modern dasyurids (Archer 1976c). *Keeuna woodburnei,* also shows, but to a lesser extent, many didelphid like features. These Australian taxa do not appear to be clearly related to any modern dasyurid lineage.

The middle Miocene Riversleigh dasyurid suggests a considerable antiquity for the lineage leading to species such as *Antechinus bilarni* and *A. macdonnellensis.* It also suggests that the synapomorphic specializations within this lineage (the *Pseudantechinus* species group of Fig. 16) are not arid–adapted features because of their presence in the Riversleigh dasyurid.

Within the *Antechinus* suprageneric group (Fig. 14), the early Pliocene Hamilton dasyurid clearly represents a very primitive species. If it weren't for the apparent survival of a similar form on Mt. Wilhelm, it would be our only glimpse of this peculiar branch of the *Antechinus* phylogenetic tree. Hopefully closer attention will now be focused on the Mt. Wilhelm species because of its seemingly distinctive phylogenetic position among the living species of the genus.

The Fisherman's Cliff *Dasyuroides achilpatna* focuses attention on the difference between the cladistic and phenetic methods of phylogenetic analysis and the resultant classifications. The fossil form is clearly most similar to *Dasyuroides byrnei* (notably in plesiomorphic morphology of M_2). And yet, *Dasyuroides byrnei* and *Dasycercus cristicauda* share certain synapomorphic features (loss of P_3 and reduced metaconid on M_2) which are not present in the Fisherman's Cliff dasyurid. If *Dasyuroides byrnei* and *Dasycercus cristicauda* are to be distinguished at the generic level, then the logic of the cladistic system would demand a new taxonomic rank for the Fisherman's Cliff dasyurid at or above the generic level. I have not done this, although I am in agreement with the logic of the cladistic (phylogenetic systematic) system. By referring the Moorna species to *Dasyuroides,* it would appear that I have used a phenetic approach in classification and accordingly grouped the Moorna species with *Dasyuroides byrnei,* on the basis of overall similarity. This is not the case. The real reason is that I have doubts about the assumption that *Dasyuroides byrnei* and *Dasycercus cristicauda* should be separated at a generic level. They appear to be remarkably synapomorphic in cranial and dental morphology (Archer 1976c) and could, with little injury to the standard concept of dasyurid intrageneric variation (e.g. that which occurs in *Pseudantechinus* or *Sminthopsis*), be regarded as congeneric. In this context, no purpose would be served in erecting yet a third genus for what is an undoubted member of this possibly congeneric group (forgive me Willi). Debates about what particular degree of divergence is necessary before a particular species should be distinguished at the generic level are of much less interest than determination of the phylogentic relationships of the species.

Sarcophilus moornaensis may be structurally (and possibly temporally) intermediate between *Glaucodon ballaratensis* and *Sarcophilus harrisii.* Alternatively, it may be more closely related to *Glaucodon ballaratensis.* The age of *G. ballaratensis* is in doubt, but it could be older than *S. moornaensis.* Certainly there are no undoubted records of *S. harrisii* older than either *S. moornaensis* or *Glaucodon ballaratensis.*

The early Pliocene *Dasyurus dunmalli* appears to provide evidence for polyphyletic loss of P_2 in species commonly referred to *Dasyurus* (*sensu latu,* e.g. Ride 1970). This evidence for polyphyly, combined with data from other tissue systems noted below, contributes to the rationale of using (as does Tate 1947) Pocock's genus *Satanellus* for *hallucatus* and *albopunctatus.*

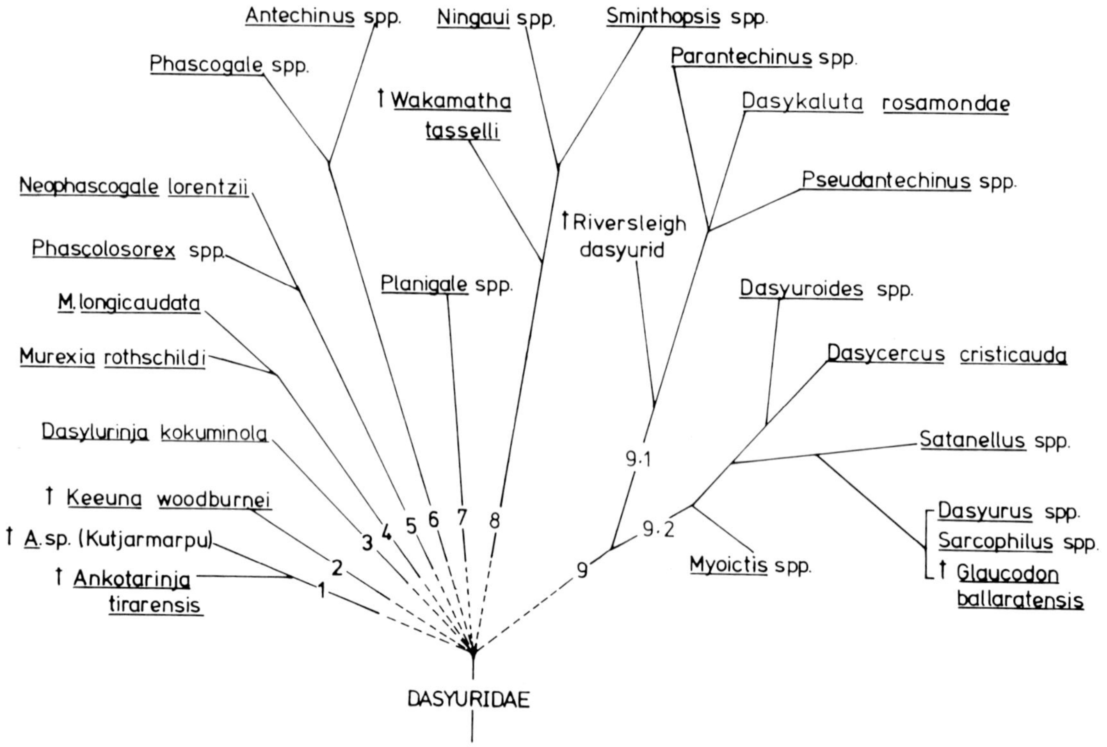

Fig. 13. A composite dendrogram of the nine intrafamilial dasyurid subgroups discussed in the text and portrayed in Figs. 14–16. The subgroups are accorded subfamilial rank in the consideration of dasyurid classification. There is as yet no clear basis for recognition of special relationships interrelating any of these nine subgroups.

The Awe species of *Myoictis* tells us little about the phylogenetic history of the genus. Its P_3 morphology, although apomorphic within the family, closely resembles that of *M. melas* and is thus plesiomorphic within the genus.

AN ATTEMPT TO SYNTHESIZE DATA BEARING ON INTERPRETATION OF DASYURID EVOLUTION

There appear to be nine groups of dasyurid species (Fig. 13). The boundaries of these groups are defined by the lack of clear synapomorphic character-states. They will be considered in the sequence in which they occur in Fig. 13.

The first group: This most plesiomorphic of the dasyurid groups contains only the middle Miocene species of *Ankotarinja*. The probable origins and radiations of the group have been discussed elsewhere (Archer 1976c) and above. They are at least structurally annectant between didelphid–like Cretaceous marsupials and other living dasyurids.

The second group: This contains only *Keeuna woodburnei*. It is also, but less so than *A. tirarensis,* a plesiomorphic species. It appears to closely relate to none of the other dasyurid groups.

The third group: This contains only *Dasylurinja kokuminola* which shows no clear special relationship to any other group although it appears to converge in aspects of molar morphology on species of *Dasyurus.*

The fourth group: This contains the species of *Murexia*. Besides external anatomy and gross dental and cranial morphology, there have been no tissue systems examined. This remains one of the glaring omissions in any attempt to interpret dasyurid phylogeny. I have pointed out elsewhere (Archer 1976c) that species of *Murexia* appear to be among the most plesiomorphic of all living dasyurids. In many of their plesiomorphic character–states they resemble species of *Ankotarinja* and even some of *Antechinus. Murexia longicaudata parva* is so similar in dental and cranial morphology to plesiomorphic species of *Antechinus* that the generic boundary can only be maintained by resorting to external morphology.

Tate (1947) considers thylacinids to be most closely related to species of *Murexia.* Elsewhere I have suggested that there is little support for this view (Archer 1976c).

The fifth group: The species of *Neophascogale* and *Phascolosorex* should probably be regarded as congeneric. As a group, they are among the most phylogenetically enigmatic of the dasyurids. As in the case of the species of *Murexia,* they have been glaringly unrepresented in the comparative enzyme and serological studies.

Dental morphology suggests that these species are either very autapomorphically specialized dasyurids, or they represent another major phylogenetic branch, hierarchically equivalent taxonomically to the rest of the dasyurids. At present, lacking any guidelines from other tissue systems, I have conservatively segregated them off as just another subgroup of dasyurids.

Earlier (1976c) I noted several cranial and dental features shared by species of *Sminthopsis* and those of *Neophascogale* and *Phascolosorex*. These include palatine vacuities, a bifid I_3 (present in *S. granulipes*), squamosal–frontal contact, an I^4 which is longer– crowned than I^2, and the somewhat posteriorly narrowed nasals. None of these shared features are undoubted synapomorphs (Archer 1976c). Nevertheless, I am inclined to pay attention to any sharing of these unusual dasyurid features, even though all it may mean is that both the *Neophascogale–Phascolosorex* and the *Sminthopsis–Ningaui* groups are plesiomorphic and independent groups within the family. In a revision of *Sminthopsis* (Archer 1981a), I have suggested that the species of *Neophascogale* and *Phascolosorex* may be the sister group of the *Ningaui–Sminthopsis* species group. We clearly need enzyme and/or serological data to resolve the various possibilities raised by studies of gross morphology.

The morphology of the upper molar stylar shelf (particularly of M^2) and of the incisors, sets species of *Neophascogale* and *Phascolosorex* apart from other dasyurids. Curiously, some of the unusual stylar shelf morphology is reminiscent of that found in bandicoots. When I proposed earlier (Archer 1976c) that the peramelid upper molar pattern was foreshadowed in some of the Cretaceous didelphids, I considered the stylar shelf arrangement in species *Phascolosorex* and *Neophascogale* (Archer 1976c) to be secondarily developed from a more standard

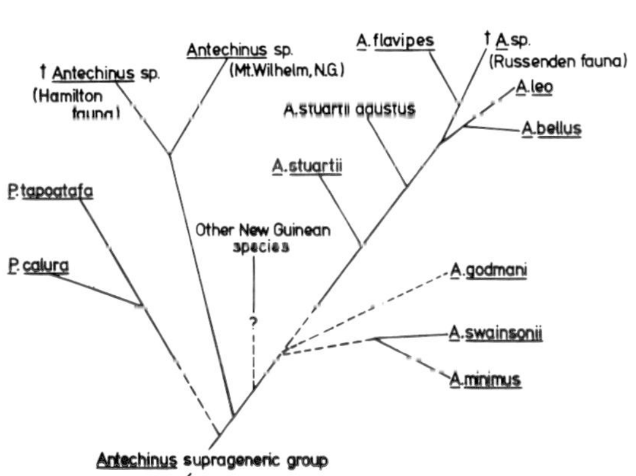

Fig. 14. Dendrogram of the hypothetical relationships of the species of *Phascogale* and *Antechinus* (sens. strict.). The bases for these hypotheses are discussed in the text. Most arise from the enzyme, dental and cranial data. A crucifix indicates a fossil taxon. Dashed lines indicate greater uncertainty. The question indicates uncertainty because the New Guinean species of *Antechinus* are very poorly known.

dasyurid pattern. Now I am not so sure. Again, we need more data from other tissue systems.

The sixth group (Fig. 14): This group contains the species of *Phascogale* and *Antechinus* (*sensu strictu*). All of the species tested group on the basis of enzymes, serology, and dental and cranial morphology. The autapomorphic species of *Phascogale* clearly belong here. The smallest of these, *P. calura,* is very *Antechinus*–like.

I have suggested above that, of the rest of this group, the living Mt. Wilhelm and Pliocene Hamilton antechinuses form a separate very plesiomorphic cluster. Tests of this must await examination of enzymes or other non–gross–morphological features of the Mt. Wilhelm animal. I do not know that this animal is conspecific with *A. wilhelmina;* in fact, I doubt that it is.

The other New Guinean species (the number varies according to authority; Tate 1947 lists three — *A. melanurus, A. mayeri* and *A. wilhelmina*) have not been the subject of enzyme, serological or phallic studies, nor even recent morphological studies. Like other New Guinean species, they remain glaringly unexamined.

Comparative enzyme studies suggest that *Antechinus stuartii adustus* from north Queensland and *A. s. stuartii* differ as much from one another as *A. flavipes* does from *A. bellus,* or as *A. swainsonii* does from *A. minimus.* Stephen Van Dyck (pers. comm.) suggests on the basis of gross dental morphology, of the species *A. stuartii, A. leo, A. bellus* and *A. flavipes, A. stuartii,* is the most plesiomorphic. Of the two types of *stuartii. A. s. adustus* is more closely related on the basis of enzymes to the *A. flavipes–leo–bellus* cluster.

The enzymes also most closely relate *Antechinus leo* to *A. bellus,* rather than to *A. flavipes* or *A. godmani* the two species with which specimens of *A. leo* were frequently confused (Van Dyck 1980). This corroborates Van Dyck's (1980) decision that *A. leo* is a distinct species.

Antechinus sp., from late Pleistocene deposits in Russenden Cave, southeastern Queensland (Archer 1978b), is allied most closely to *A. flavipes,* but solely of course on dental morphology. Little can be done to test this hypothesis unless the species still exists. Stephen Van Dyck is currently reviewing the status of the Russenden *Antechinus.*

The interrelations of *Antechinus godmani* are in doubt. Dental morphology at least would suggest that it is more similar to *A. flavipes* than to the *A. swainsonii–A. minimus* cluster, but only because *A. swainsonii* and *A. minimus* have as a group, autopomorphic features that make them dissimilar to other antechinuses. On the other hand, *A. godmani, A. swainsonii* and *A. minimus* are externally similar, with small eyes, and long pointed heads. They are also similar ecologically, being essentially small ground–living crepuscular dasyurids. The enzyme data do not help to resolve these alternatives although there is a very slight suggestion of closer affinity between *A. godmani* and the *A. swainsonii* – *A. minimus* group. Very tentatively, I have accepted the ecological and external morphological data over the dental morphological data as a guide to the relationships of *A. godmani.*

The *Antechinus swainsonii* – *A. minimus* species cluster is supported by data from all tissue systems examined. If, as a group, it is related to *A. godmani,* the distribution of these three species could be interpreted to be the result of speciation, following disruption of the formerly continuous east coastal population of a common ancestor. Of these three species, *A. godmani* is by far the most plesiomorphic. Of the closely related *A. swainsonii* and *A. minimus, A. minimus* is by far the more plesiomorphic. *A. godmani* is restricted to the rainforests of the Atherton Tableland. *Antechinus minimus* occurs in Tasmania, Victoria, and southeastern South Australia. *Antechinus swainsonii* occurs in the same regional areas as *A. minimus* but also in eastern New South Wales and southeastern Queensland. Because *A. minimus* is the more plesiomorphic of the two, it is probably phylogenetically closer than *A. swainsonii* to the common ancestor shared by these species with *A. godmani.* A population of this common ancestor may have become isolated in Tasmania following an interglacial rise in sea–level. During isolation it evolved into what we now regard as *A. swainsonii,* and flooded back onto the mainland during the subsequent glacial low stand of sea–level. It eventually managed to get as far north as the rainforests of the Lamington Plateau in southeastern Queensland (Van Dyck and Ogilvy 1977). At the same time *A. minimus* developed from the mainland populations of the common *swainsonii–minimus* ancestor isolated (by disruption of a formerly continuous east coastal habitat) in southeastern Australia. When Tasmania was again connected to the mainland, *A. minimus* would have been sufficiently different from *A. swainsonii* to enable it to invade Tasmania. Loss of suitable rainforests between Atherton in northeastern Queensland and the Lamington Plateau in southeastern Queensland could have produced the large northern distjunction in distribution of the species of this group of antechinuses.

The seventh group: The species of *Planigale* form the seventh group (Fig. 13). There is a slight suggestion on the basis of the enzyme data that species of *Planigale* may be more closely related to species of the eighth group (including species of *Sminthopsis* and *Antechinomys*). However, the enzyme data here are inadequate to allow confidence about such a hypothetical intergroup relationship. There is nothing else about the morphology of species of *Planigale* known to me that would support such a relationship.

On the basis of morphological similarity, the species of *Planigale* have been previously grouped (e.g. Tate 1947, Archer 1976c) with the antechinuses. However, the reasons for the gross morphological similarity between species of *Planigale* and *Antechinus* are mainly plesiomorphic dasyurid features such as the posteriorly widened nasals, a straight supratragus in the ear, post–interdigital pads on the hind feet, lack of palatine vacuities, posterior cingula on the upper molars, and so on. The possible exception to this is reduction of P_3, but this apormophic state occurs independently in many dasyurid groups (Archer 1976c) and cannot serve as an adequate basis for linking species of *Planigale* and *Antechinus.*

Within the *Planigale* group, species relationships are becoming clear. Dental and cranial morphology suggests (Archer 1976a and 1976c) that *P. maculata* is the most plesiomorphic species. Enzyme relationships (involving only *P. maculata, P. gilesi,* and *P. tenuirostris*) clearly support this interpretation. *Planigale gilesi* also must represent a very distinct part of the *Planigale* radiation because it completely lacks P^3, a tooth not even reduced in the other species. However, the enzymes suggest that it groups closer to *P. tenuirostris* than *P. maculata.* This implies that the dental autapomorphy of *P. gilesi* is not, in this context anyway, phylogenetically important. It must have occurred after differentiation of *P. tenuirostris* (probably with *P. ingrami*) stock from the *P. maculata* (probably with *P. novaeguineae*) stock.

The eighth group: The species of *Sminthopsis, Wakamatha* and *Ningaui* form the eighth group (Fig. 15). *Wakamatha tasselli* is only known from fossils and therefore nothing can be added here to what was said above.

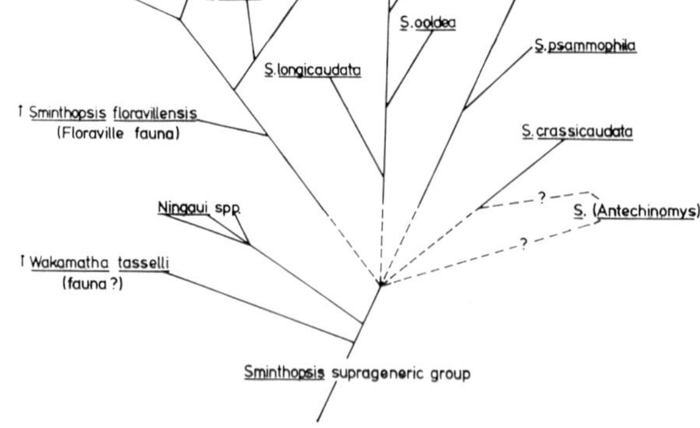

Fig. 15. Dendrogram of the hypothetical relationships of the species of *Wakamatha, Ningaui* and *Sminthopsis.* Symbols are as given in Fig. 13. The species of *Sminthopsis* distribute onto three or four branches whose broader interrelationships are unclear.

The species of *Ningaui* are clearly the most immediate sister group of *Sminthopsis*. Relationships within *Ningaui* have yet to be examined. At this time it is only possible to say that there are clearly more species than the original two I described in 1975, and possibly as many as four. Work with the species of this genus continues.

Within *Sminthopsis* itself (Fig. 15), there are unresolved problems in the interpretaion of species relationships. In a revision of the genus (Archer 1981a), I attempted a cladistic analysis of the 13 species then known (including *S.* (*Antechinomys*) *laniger*). In preparation of Fig. 15, I have had the additional information about *Sminthopsis floravillensis* (described above) and enzyme data for *S. crassicaudata, S. virginiae, S. macroura, S. ooldea, S. leucopus, S. murina,* and *S.* (*Antechinomys*) *laniger*. Morphology and the enzyme data are in agreement about: 1) the uniquences of *S. crassicaudata;* 2) the close relationship of *S. virginae* and *S. macroura;* 3) the specific distinction of *S. ooldea,* and also its close relationship to *S. murina;* 4) the specific distinction of *S. leucopus,* and also its close relationship to *S. murina;* and 5) the placement of *Antechinomys laniger* within *Sminthopsis* (i.e. taxonomically distinct at or below the subgeneric level). The enzyme data differed in suggesting that: 1) *S. crassicaudata* was not the sister species of *S.* (*A.*) *laniger,* but rather the sister group of all the other *Sminthopsis* tested; 2) *S. leucopus* was the sister species of *S. murina,* not of an *S. murina* – *S. ooldea* species group; 3) *S. murina* consists of more than one species; 4) and in suggesting that the *S. virginae* – *S. macroura* species group, is the sister group of all other *Sminthopsis* species except *S. crassicaudata.* Points of difference numbers 1 and 4 are rather fundamental. My tentative solution given in Fig. 15 is to portray the interrelationships of the major *Sminthopsis* species groups (and there is a basic agreement between the morphological and enzyme data about included members in these groups) as an unresolved five-branched polychotomy. Because I am inclined to firmly believe that the numerous morphological synapomorphies shown by *S. crassicaudata* and *Antechinomys laniger* indicate a special relationship, I have not been inclined here to give in to the enzyme data. Other problems of course include the unavailability of many of the more interesting species, such as *S. granulipes, S. longicaudata,* and *S. psammophila* for enzyme analysis. Hopefully the opportunity to test these will arise.

The phylogenetic position of *S. floravillensis,* shown in Fig. 15 as the sister group of an *S. virginiae* species group (for reasons described above), does not clash with relationships suggested by enzyme data. It does imply, however, that its specific differentiation predated that of the members of the *S. virginiae* species group. This is, therefore, a testable hypothesis.

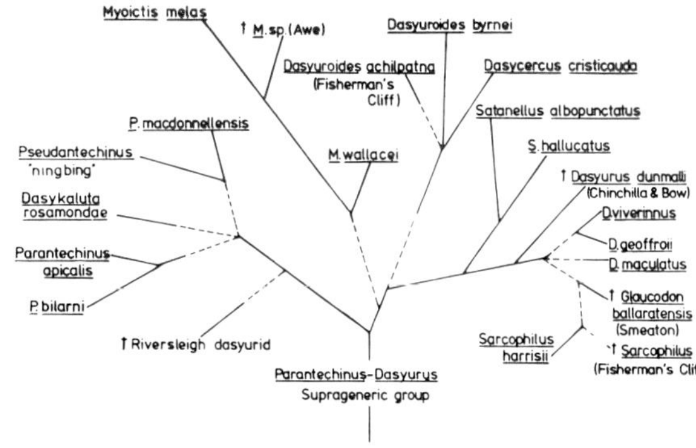

Fig. 16. A dendrogram of the hypothetical relationships of the species of *Parantechinus, Pseudantechinus, Myoictis, Dasyuroides, Dasycercus, Satanellus, Dasyurus, Glaucodon, Sarcophilus,* and related fossils. The basis for these hypotheses are discussed in the text. Symbols are as in Fig. 13. Trichotomies indicate conflict in interpretation of particular dichotomies.

The ninth group: This group (Fig. 16) is by far the most complex. It appears to consist of two subgroups: the *Parantechinus* suprageneric group; and the *Dasyurus* suprageneric group.

The *Parantechinus* suprageneric group is loosely coherent. The species contained clearly are not referable to *Antechinus,* the genus within which they have commonly been placed (e.g. Ride 1970). However, membership in *Antechinus* followed by exclusion from this genus is clearly not an adequate basis for recognizing monophyly.

Tate (1947) considers that the degree of reduction of P3 indicates that *Antechinus macdonnellensis, A. apicalis* and *A. mimulus* belong in his dasyurine subfamily, as opposed to the other species of *Antechinus* which are part of his phascogaline subfamily. For *A. macdonnellensis* and *mimulus* he proposes the generic name *Pseudantechinus*. For *A. apicalis* he proposes the generic name *Parantechinus*. Beyond the generic level revision, he suggests *P. apicalis* is the sister group of *Neophascogale lorentzii,* and that the *Pseudantechinus* species group is the sister group of a *Dasycercus–Myoictis* suprageneric group.

In describing *Antechinus rosamondae* as a species completely lacking P3, Ride (1964a) suggests that he could either follow Tate's (1947) lead and erect a third monotypic genus, or he could regard all of the taxa Tate considered, as well as *rosamondae,* to be species of *Antechinus*. In opting for the latter, he suggests that there is enough variation in P3 reduction in *A. macdonnellensis* alone to cast doubt on the wisdom of maintaining, let alone proliferating, the number of genera in this group of dasyurids.

Woolley (1977 and 1981) presents an important body of phallic data relevant to the problem of interpreting the relationships of this group of dasyurids. She considers that *A rosamondae, A. macdonnellensis, A. bilarni, A. apicalis,* and the "ningbing" *Antechinus* stand well apart from the taxa listed above as part of group six. Of these species, *A. rosamondae* appears to her to be primitive in penile morphology. *A. macdonnellensis* and the "ningbing" *Antechinus* have two accessory corpora cavernosa, while *A. bilarni* and *A. apicalis* have three.

Interpretation of the enzyme data (Baverstock *et al.* 1981) for the *Parantechinus* suprageneric group backs up most of the interpretations of cranial, dental, and penile data. It differs (for example from the hypothesis presented in Archer 1976c) in the suggestion that the *Dasyuroides–Dasycercus* suprageneric group appears to be more closely related to the *Dasyurus–Sarcophilus* (and presumably *Glaucodon*) suprageneric group than it is to the *A. bilarni–A. rosamondae–A. macdonnellensis* group. It is unfortunate that *A. apicalis* and the *"ningbing" Antechinus* were not available for the enzyme study. Within the *A. macdonnellensis – A. bilarni – A. rosamondae* group, the enzyme data suggest very large interspecific differences. Monophyly is indicated by only a 3% similarity in enzymes between *A. rosamondae* and *A. bilarni*, and again by only a 3% similarity between the hypothetical common ancestor of these two species and *A. macdonnellensis*. Richardson (pers. comm.) suggests that percentage differences of less than 5% are probably meaningless or, at best, poor indications of real genetic distance. Considering also the magnitude of the enzyme differences between these three species, it seems best to regard the interrelationships of *A. bilarni, A. macdonnellensis* and *A. rosamondae* as an unresolved monophyletic trichotomy, at least in terms of the enzyme data.

After examining details of dental morphology, I originally came to the conclusion (Archer

Table 2. Cranial, dental and external character states in the *Parantechinus* suprageneric group and selected outgroups that appear to differentiate species and genera within the *Parantechinus* suprageneric group[1].

Character	The Parantechinus suprageneric group: rosamondae	macdonnellensis	"ningbing"	bilarni	apicalis	Dasyuroides byrnei	Murexia longicaudata	Plesiomorphic within Parantechinus gp.
1. I^{2-4}	*2 >4	*2 ≤ 4	2 < 4	2 < 4	*2 ≃ 4	2 < 4	2 < 4	2 < 4
2. P3 root no.	*P3 absent	*P^3, 0-1R; P_3, 0-1R	*P^3, 0-2R; P_3, 0-1R	P^3, 2R; P_3, 1R	*P^3, 1R; P_3, 1R	*P^3, 1R; P_3, absent	P^3, 2R; P_3, 2R	P^3, 2R; P_3, 1R
3. St.B on M^2	*low + tiny	*? gone	*low + tiny	high + separate	high + separate	*? gone	high + separate	high + separate
4. M^5 width	M^5 just < M^4	*variable	*M^5 much < M^4	*M^5 much < M^4	variable	*M^5 much < M^4	M^5 ≃ M^4	M^5 just < M^4
5. M_2 paraconid	tiny	*variable	tiny	tiny	tiny	tiny	low	tiny
6. M_2 metaconid	*small to nr absent	*mod. to absent	small	small	small	small to tiny	tall	small
7. M_4 entoconid	*absent	*tiny to absent	tiny	tiny	tiny	small	tall	tiny
8. M_5 talonid	*extremely reduced	*very reduced	*extremely reduced	reduced	reduced	reduced	barely reduced	reduced
9. M^4 protocone	*compressed	uncompressed	*slightly compressed	uncompressed	uncompressed	uncompressed	uncompressed	uncompressed
10. Nasal shape	*rapid flare	*prox. flare	*nr Sminthopsis	prox. flare	?	*as Sminthopsis	prox. flare	prox. flare?
11. Sqam.-frontal	not	*no, but some close	*yes, but some close	no	no	not	not	not
12. Palatine vac.	*none	*tiny to none	*none	*tiny to none	small	small	none	small
13. Middle ear	*enormous; mast. large	*enormous; mast. varies	moderate; no mast.	moderate; no mast.	moderate, no mast.	*enormous; mast. large	small	moderate no mast.
14. Skull	deep	*slightly flat	*flat	*flat	deep	deep	deep	deep
15. Caudal incrass.	*yes	*yes	*yes	no	no	no	no	no

[1] Also tabulated is the most plesiomorphic species *(D. byrnei)* of the nearest sister group, and a generally plesiomorphic living dasyurid *(M. longicaudata).* The last column lists the state interpreted to be plesiomorphic within the *Parantechinus* suprageneric group on the basis of outgroup comparison. An asterisk (*) indicates an apomorphic character state (although its presence within the taxon may be variable). Character 1 refers to the crown length of the upper incisors. Character 2 is a measure of P3 reduction as judged by root number. Character 10 refers to the degree of posterior flaring of the nasal bones (unflared in species of *Sminthopsis*). Character 11 refers to the approximation or contact of the squamosal and frontal bones on the side of the braincase (Archer 1976d). Character 13 generally refers to the degree of development of the alisphenoid and periotic hypotympanic sinuses, while mast. refers to the mastoid epitympanic sinus (Archer 1976d). Character 14 refers to skull height at the level of basisphenoid. Character 15 refers to the ability to develop a fattened tail.

1976c) that *A. bilarni*, *A. rosamondae* and *A. macdonnellensis* are related on the basis of synapomorphic features, and together are the sister group of *A. apicalis*. Following a concept of monophyly for *Antechinus* (*sensu latu*), I placed this group as the apomorphic sister group of the *A. flavipes* species group (I now regard this particular conclusion to be naive). As a whole, this group appeared to be the sister group of a *Dasycercus* – *Dasyuroides* suprageneric group. It further appeared that this combination group was the sister group of the *Dasyurus* – *Sarcophilus* – *Glaucodon* suprageneric group.

The cranial and dental characters need reconsideration for two reasons: There is now data for the "ningbing" *Antechinus* (a taxon which will be formerly described elsewhere), and I must now be doubtful about the placement of *rosamondae* as the specialized sister group of the *macdonnellensis* species group. It may represent an albeit autapomorphically specialized basal stock within the whole of the suprageneric *Parantechinus* group. This is the conclusion Woolley (1981) reaches on the basis of penile structure. Because it conflicts with other morphological data, I will briefly consider the problem here.

The character–state distribution of fifteen cranial, dental and external features interpreted (Archer 1976c) to be phylogenetically distinctive as well as interspecifically distinctive within the *Parantechinus* suprageneric group have been listed in Table 2. Some difficulty in interpreting the polarity of the character–state morphoclines comes with the choice of which particular group should serve as the outgroup. I have used both *Dasyuroides byrnei* (the more plesiomorphic member of the nearest sister group) and *Murexia longicaudata* (a generally plesiomorphic living dasyurid), because both species within the nearest suprageneric sister group (*Dasyuroides byrnei* and *Dasycercus cristicaudata*) are also highly specialized autapomorphic taxa. In fact, 5 of the 15 character states of *Dasyuroides byrnei* are most probably convergent apomorphys independently developed in some of the *Parantechinus* suprageneric group species (Archer 1976c). This would include such character states as reduction of P3, otherwise *bilarni* must have redeveloped a 2–rooted P^3 and revedeloped an ancestrally lost P_3 – a possible but unlikely event. Convergent suppression of P3 within the *Dasyuroides* and the *Parantechinus* suprageneric group species seems more likely.

It is apparent, from a simple parsimonious assessment than non–penile character states do not appear to support the concept of relative primitiveness of *rosamondae*. If *rosamondae* is the primitive sister group to the other species, it must have convergently developed at least half of the apomorphic character states it possesses. On the basis of the same characters, *bilarni* and *apicalis* appear to be the most plesiomorphic. I have portrayed this and the penile morphology data in Fig. 17, as a cladogram based on the data given in Tables 2 and 3. Table 4 summarizes the distribution of the potentially synapomorphic non–penile character states among the five species involved.

In spite of the appeal of parismony, I have presented in (Fig. 17) the fundamental breakdown of the five *Parantechinus* suprageneric group species as an unresolved trichotomy. The reason is simply that I am uncertain about how to resolve the most probable phylogenetic hypothesis based on the two sets of conflicting data. If penile morphology is ignored, *rosamondae* seems to group best (with the least convergence) as the plesiomorphic sister group of the *macdonnellensis* – "ningbing" species group. However, it lacks the accessory corpora cavernosa present in the other four species and must, if it is the sister group of the *macdonnellensis* – "ningbing" species group, have lost these structures. It is important to note here that there is already evidence for convergence in some aspects of the penile morphology. A penile appendage occurs in *macdonnellensis* and *apicalis*, but not in "ningbing" or *bilarni*. Yet there are two accessory corpora cavernosa in "ningbing" and *macdonnellensis*, but three in *bilarni* and *apicalis*. Clearly, either the number of accessory corpora cavernosa or the presence of a penile appendage is the result of convergence. Balanced against this is the already acknowledged convergence noted above in some cranial and dental characters within dasyurid Group nine. It is also clear from Table 2 and Fig. 17, that despite the more parisimonius placement of *rosamondae* as the plesiomorphic sister group of the "ningbing"–*macdon-*

Table 3. The character states interpretable for the hypothetical taxonomic ancestors (HTA) of the species group indicated within the *Parantechinus* suprageneric group. Abbreviations: a, *apicalis;* b, *bilarni;* m, *macdonnellensis;* n, *"ningbing";* r, *rosamondae*. The numbers 1–15 refer to the characters listed in Table 2.

HTA	HTA1 (r+b+a+n+m)	HTA2 (m+n)	HTA3 (r+n+m)	HTA4 (b+a)
1	P	P	P	P
2	P	P	P	P
3	P	A	A	P
4	P	P	P	P
5	P	P	P	P
6	P	P	P	P
7	P	P	P	P
8	P	A	A	P
9	P	P	P	P
10	P	P	P	P
11	P	A	P	P
12	P	A	A	P
13	P	P	P	P
14	P	A	P	P
15	P	A	A	P
	15P	6A+9P	4A+11P	15P

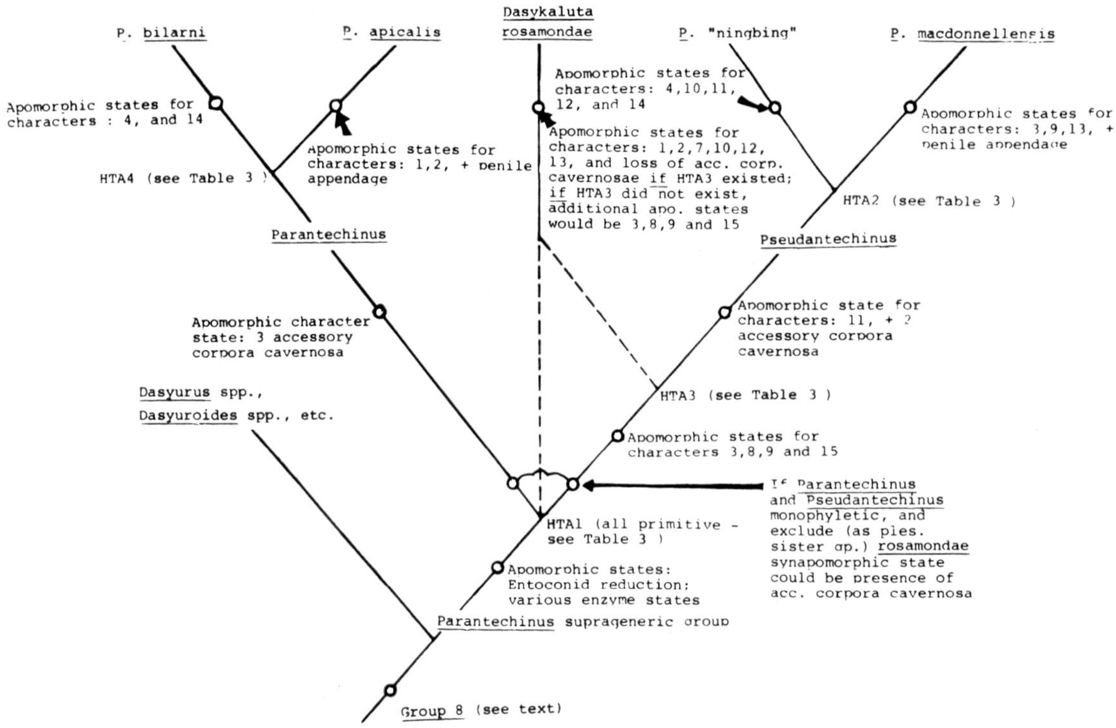

Fig. 17. Cladogram of hypothetical phylogenetic relationships of the species *bilarni, apicalis, rosamondae,* "ningbing" and *macdonnellensis.* The major unresolved problem concerns the position of *rosamondae.* Tate's generic names *Parantechinus* and *Pseudantechinus* are available for two of the supraspecific groups recognized. The characters upon which this analysis is based are given in Table 2. The character–states of the hypothetical taxonomic ancestors (HTA) are listed in Table 3.

nellensis species group, this position still implies the convergent development of 6 of the 15 dental and cranial character states. Clearly more data, from other character systems, are required before this trichotomy can be confidently resolved.

Above this level, the penile and other morphological data are in better agreement. The union of "ningbing" and *macdonnellensis* is backed by the presence in both of 2 accessory corpora cavernosa as well as 8 synapomorphic dental and cranial features (Table 4). Rationalizing the *bilarni – apicalis* union shown in Figs. 16–17 is more difficult. Both species are almost entirely plesiomorphic. However, both stand out in their possession of 3 accessory corpora cavernosa, a feature which serves to tie these two otherwise essentially plesiomorphic species together.

Accordingly, I propose that *bilarni* and *apicalis* both be referred to Tate's (1947) genus *Parantechinus,* and that the "ningbing" *Antechinus* and *macdonnellensis* be referred to Tate's genus *Pseudantechinus.* The fifth species, *rosamondae,* should be referred to a monotypic genus. Either it is the sister group of the whole *Parantechinus – Pseudantechinus* species groups, or the sister group of the *Pseudantechinus* species group. Either way, it lies outside all other gen-

Table 4. Shared apomorphic character–states involving dental, cranial and external morphology (see Table 1). The number in parentheses represents the number of these apomorphic features that are uniquely shared by the particular species pair considered.

Taxa	*rosamondae*	*macdonnellensis*	*"ningbing"*	*bilarni*	*apicalis*
rosamondae	11				
macdonnellensis	9(1)	13			
"ningbing"	7(0)	8(1)	10		
bilarni	1(0)	3(0)	3(0)	3	
apicalis	2(0)	2(0)	1(0)	0	2

eric boundaries. The enzyme data similarly suggests that *rosamondae* differs from *Pseudantechinus macdonnellensis* at about the 50% level and from *Parantechinus bilarni* at about the 40% level. Finally, because of the mosaic nature of the character state distributions, *rosamondae* can not unambiguously be allied with one or the other, or without the *Parantechinus* or *Pseudantechinus* species groups. For *rosamondae,* I propose the geenric name *Dasykaluta. Dasy* refers to dasyurid and means hairy. *Kaluta* is the name given to *rosamondae* by the peoples of the Nyamal linguistic group of the Pilbarra region including the area involving Woodstock Station, the type locality for the species. Mr. R. A. Johnston, Ranger for Woodstock Reserve, determined the Aboriginal name by showing photographs of a live individual to Aboriginals at Yandearra, Western Australia. They distinguished this animal from all other small dasyurids known from the area. The name would be prounced Dásé kálútà, and is here considered to be feminine.

The *Dasyurus* suprageneric group (the other part of group eight) is itself divisable into two groups. One of these groups, the species of *Myoictis,* is placed here because the penis has accessory corpora cavernosa, and there is reduction or loss of P_3. Tate (1947) groups the species of *Myoictis* specifically with those of *Dasycercus* and *Dasyuroides*. In terms of dental morphology this is plausible. However, a detailed gross morphological analysis is needed, as are of course serological and enzyme data. This is another New Guinean group about which almost nothing is known. The second subgroup of the *Dasyurus* suprageneric group contains the species of all the remaining dasyurid genera (Fig. 17). These are all highly apomorphic and most are very specialized carnivores. The species of *Dasycercus* and *Dasyuroides* are the most plesiomorphic within this group. The significance of the extinct species *Dasyuroides achilpatna* has been discussed above in regard to its taxonomic implications.

The phylogenetic implications of *Dasyurus dunmalli* have also been discussed above. It is the reason that I have used Pocock's genus *Satanellus* for *hallucatus* and *albopunctatus.* The enzyme data support this decision by demonstration that *S. hallucatus* (presumably with *S. albopunctatus,* although this species was not tested) is the sister group of the combined *Sarcophilus – Dasyurus – Glaucodon* suprageneric group (species tested included *harrisii, geoffroii, viverinnus,* and *maculatus*). The extinct *Glaucodon ballaratensis* was of course not tested, but its close morphological relationship to species of *Sarcophilus* has not been doubted. Crabb (1982) has discussed the phylogenetic relationships of the species of *Sarcophilus.*

Moeller (1973) has suggested that *Sarcophilus harrisii* should be separated from other dasyurids at the taxonomic level of tribe because of its distinct skeletal and dental morphology. This decision is difficult to corroborate because the seemingly large structural gap between *Sarcophilus harrisii* and other dasyurids is rapidly being filled by structurally intermediate forms. Already we have seen that *Sarcophilus moornaensis* is considerably more plesiomorphic than *S. harrisii,* and *Glaucodon ballaratensis* is at least structurally annectant between species of *Sarcophilus* and those of *Dasyurus.*

SUPRAGENERIC CLASSIFICATION WITHIN THE DASYURIDAE

Previous Concepts

We have come a long way in our appreciation of the diversity of marsupials and in our recognition of this diversity in classification. In 1811 Illiger managed to jam all of the marsupials, from Oppossums to wombats, into one family. In a more recent classification (Kirsch 1977), marsupials are regarded as a superorder containing at least 3 orders and 28 families. The increase in taxonomic breadth based on Australian marsupials is largely the result of recognition of diversity within the order Diprotodonta. The Australian carnivorous marsupials have been more conservately dealt with, and it is with only these that I am concerned here. Although thylacoleonids are also carnivorous Australian marsupials, their classification will not be considered here. Myrmecobiids, notoryctids and thylacinids are considered here because they have, at one time or another, been included within the Dasyuridae.

Goldfuss (1820) was the first to regard the Australian carnivorous marsupials as warranting taxonomic distinction above the generic level. Accordingly, he distinguished them at the tribal level as the Dasyurini.

Waterhouse (1838) treated most of this group as a distinct family, the Dasyuridae, and separated *Myrmecobius* off as the sole member of another family, the Myrmecobiidae.

In 1838, Bonoparte suggested that *Thylacinus* should represent a distinct family, the Thylacinidae. He also recognized, within the Dasyuridae, the subfamily Phascogalinae. In 1850, Bonaparte recognized within the Dasyuridae, the single division of the Dasyurina, and within the Peramelidae, the three divisions Phascogalina, Myrmecobiina, and Peramelina. He continued to recognize the Thylacine as a separate family, Thylacinidae, containing the single division Thylacinina.

Owen (1841) chose to split the Australian carnivorous marsupials in quite a different way, recognizing two tribes: The Sarcophaga (containing *Thylacinus, Dasyurus,* and *Phascogale*) and the Entomophaga (including *Myrmecobius*).

These tribal names reflected the supposed commonality in dietary preference of the members of each group.

Gill (1872) recognized the subfamilies Sarcophilinae (including *Sarcophilus*), and Dasyurinae (including *Dasyurus*) as well as the Phacogalinae of Bonaparte, all three being placed within the Dasyuridae. *Myrmecobius* was still regarded to represent a monotypic family. He placed the Myrmecobiidae together with Dasyuridae in a unique suborder, the Dasyuromorphia, which is the greatest taxonomic distinction so far afforded to this group.

In a burst of zealous lumping, Thomas (1888) placed *Thylacinus* and *Myrmecobius* back within the single family Dasyuridae, and then separated the family into two subfamilies, the Myrmecobiinae containing *Myrmecobius*, and the Dasyurinae containing all the rest.

In a most important early work, Bensley (1903) divided the Dasyuridae into three subfamilies, the monotypic *Myrmecobiinae*, the Dasyurini (containing all other genera except *Thylacinus*), and the monotypic Thylacininae. However, he expressed considerable doubt about the dasyurid nature of the Thylacine, regarding it to be very like the South American borhyaenids. Bensley was also among the first to hypothesize (but not taxonomically formalize), through the medium of a dendrogram (Fig. 18a), suprageneric groups involving all of the genera recognized at the time. He recognized at least six hierarchically arranged sets of genera within the Dasyuridae, including the highest infrafamilial grouping (which he clearly regarded as dubious) between *Thylacinus* and all other dasyurids.

Iredale and Troughton (1934), in a checklist of Australian mammals, once again recognized the monotypic family Myrmecobiidae, as well as three divisions within the Dasyuridae, the Thylacininae (containing *Thylacinus*), the Dasyurinae (containing *Dasyurus, Satanellus, Dasyurops* and *Sarcophilus*), and the Phascogalinae (containing all the rest).

Winge (as indicated in the 1941 translation) placed all of the above forms in two tribes within the Dasyuridae: the Thylacinini (containing *Thylacinus*); and the Dasyurini (containing all forms considered above as well as *Myrmecobius*). He also was the first to place *Notoryctes* (previously placed by other authors in a monotypic family) within the Dasyuridae, as a third Tribe, the Notoryctini.

Simpson (1945), in his most influential classification of mammals, placed *Notoryctes* in the Notoryctidae and, together with the Dasyuridae, placed both in his superfamily Dasyuroidae. He recognized four subfamilies within the Dasyuridae, the monotypic Myrmecobiinae, the monotypic Thylacininae, the Dasyurinae (containing *Dasyurus, Dasyurops, Satanellus, Dasyurinus,* and *Sarcophilus*), and the Phascogalinae (containing all other dasyurids).

A most important advance in classification was provided by Tate (1947). In the first major species–level revision of the family since Thomas (1888), he divided the Dasyuridae up into two subfamilies, the Phascogalinae and the Dasyurinae. He regarded *Myrmecobius* to represent a monotypic family. Although his two subfamily concept for the Dasyuridae was not new, he had a very atypical view about how the various genera should be allocated to the subfamilies. His Phascogalinae included all forms with relatively unreduced upper third premolars (*Murexia, Thylacinus, Sminthopsis, Antechinomys, Antechinus* (*sens. strict.*), *Planigale,* and *Phascogale*). All others, including many of the smaller forms such as *Pseudantechinus* and *Parantechinus,* were placed within the Dasyurinae. Further, within the subfamilies, he hypothesized (Fig. 18B), but did not formalize taxonomically, 21 suprageneric groups.

Ride (1964b) expanded taxonomic freedom within marsupial classification by recognition of four orders. One of these, his order Marsupicarnivora, contains the superfamily Dasyuroidea comprising two families, the monotypic Thylacinidae and the Dasyuridae. The latter contains the monotypic Myrmecobiinae and the Dasyurinae (containing all other dasyurid genera). He did not allocate *Notoryctes* to any particular order. In 1970, Ride used external (as well as internal) characteristics to facilitate recognition of nine (ten if *Myrmecobius* is included) suprageneric dasyurid groups (his Groups 23-31). These groups were not intended to necessarily represent either hypotheses about phylogenetic relationships or taxonomic units.

Previously (Archer 1976c,d) I presented phenograms (Figs. 18 C and D) summarizing what appeared at the time to be suprageneric groupings based on dental and basicranial morphology. Thylacines appeared to be most similar to either didelphids (based on basicranial evidence) or borhyaenids (based on dental evidence). In both studies Thylacines were not regarded to be related to dasyurids. I did not recognize any of the suprageneric groups as formal taxonomic units because these groups were based on study of single character systems.

Kirsch (1977) has provided the most recent comprehensive classification of marsupials, based largely on his studies of comparative serology. He utilizes Gill's (1872) suborder Dasyuromorphia to contain the superfamily Dasyuroidea which in turn contains the monotypic family Myrmecobiidae, and the Dasyuridae. He places the family Thylacinidae within the superfamily Borhyaenoidea, following my earlier suggestions

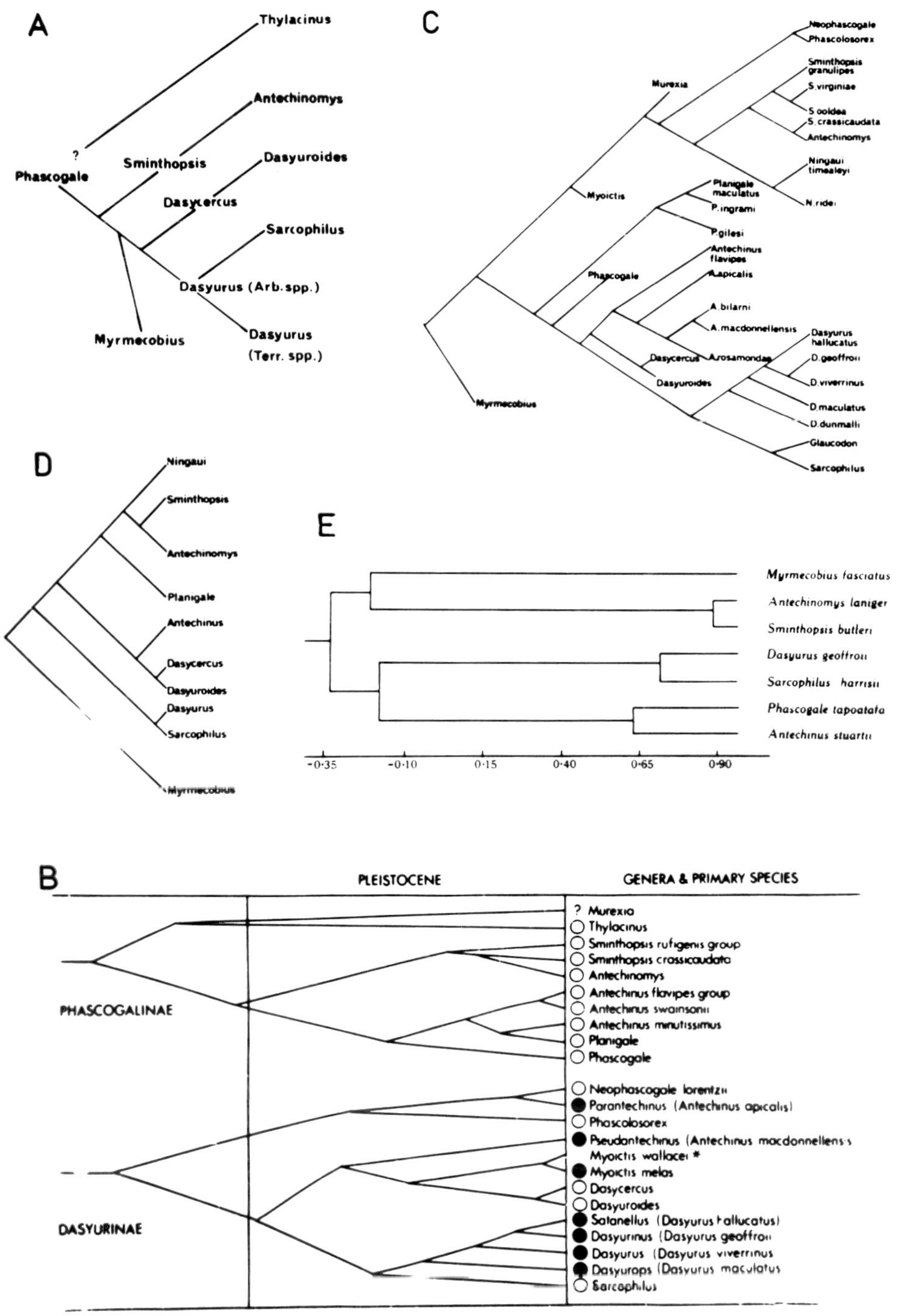

Fig. 18. Dendrograms of infrafamilial dasyurid relationships based on previous studies. A, the dendrogram of Bensley 1903, based on dental and pedal morphology, in which *Thylacinus* was regarded as dubiously dasyurid. B, dendrogram based on Tate 1947, involving dental and cranial morphology. The solid circles indicate species that possess accessory erectile tissue in the penis (Woolley and Webb 1977). C, a dendrogram of dental similarity (Archer 1976c). D, a dendrogram of basicranial similarity (Archer 1976d). E, the dendrogram of Kirsch (1977) summarizing serological affinity of some of the dasyurids tested.

(Archer 1976c,d) about the non–dasyurid affinities of Thylacines. He also portrayed, as a phenogram (Fig. 18E), six suprageneric groups involving *Myrmecobius* and the dasyurid, that he had tested serologically.

Through all of this, there have been two basic kinds of approaches to suprageneric classification within the Dasyuridae. The first and most common has been a fundamental dichotomy approach in which characters assumed to have fundamental phylogenetic significance have been used in an *ex cathedral* manner to divide the whole group, from family level down, into two to four major sections. Most of these classifications have only resulted in recognition of subfamily level taxa. The second approach to dasyurid classification (such as that used by Bensley 1903, Tate 1947, and Kirsch 1977) first involves species–level character analyses which are then followed by clustering based on affinity. These studies have resulted in hypotheses of up to seven hierarchical levels of relationship between the generic and familial levels. This latter approach has been more useful simply because it provides a richer tapestry of phylogenetic hypotheses to test. To this background, we have now added the phenogram of family–wide enzyme relationships (Baverstock *et al.* 1982), and new cranial and dental studies (Kirsch and Archer 1982). Other studies of relationship based on species–level analyses have dealt only with subgroups of the Dasyuridae, such as those based on DNA (Timms *et al.* 1982) and penile morphology (Woolley 1982) of taxa previously referred to *Antechinus*.

Towards a Synergistic Classification of the Dasyuridae

One of the important conclusions that emerged from the 1980 Royal Zoological Society Australian Carnivorous Marsupial Symposium was the surprising lack of disagreement about intrafamilial relationships within the Dasyuridae. This is apparent when cranial, dental, penile morphology, serological, enzyme, DNA, and paleontological data are considered. However, some of these tissue systems have not been checked as comprehensively within the family as have enzymes or dental morphology. This is why any attempt to derive a single classification from all of the data must be synergistic. Further, despite the overall agreement about most phylogenetic relationships, there is still conflicting evidence about the higher level group relationships. For this reason, a single dasyurid classification must also be syncretic if it is based, as it should be, on diverse data sets.

The suprageneric hierarchical arrangement given in Table 5 is based on Fig. 13. This in turn is based on comparative data from all of the character systems that have been analyzed. Nine primary groups are recognized at the subfamilial level.

SYSTEMATICS

Family Dasyuridae Goldfuss, 1820

Groups One, Two and Three

The first three groups are all extinct and based respectively on the species of *Ankotarinja, Keeuna,* and *Dasylurinja.* Although I consider that they are all distinct at the subfamilial level, I prefer not to erect three monotypic subfamilies without knowing something at least about their basicranial anatomy. There is a slight but important possibility that one or more of them may prove to have some special relationship to the Thylacinidae, the Didelphidae, the Microbiotheriidae, or the Pediomyidae. At present, much too little is known about these three extinct Australian groups to enable them to be confidently placed within or adjacent to any particular living group of dasyurids.

Group Four: Subfamily Murexinae *nov.*

This group contains only the species of *Murexia.* It is the most plesiomorphic of the living dasyurid groups. Nothing yet is known of the enzyme or serological relationships of this group within the family. Nevertheless, its dental and basicranial morphology leave little doubt that it is distinct from other dasyurids at a fundamental level (Archer 1976c,d). Tate (1947) even considered that it was the primary sister group of the Thylacinidae, a view which I do not share (see above). Nevertheless, Tate's view indicates his agreement about its distinctive phylogenetic position.

Group Five: Subfamily Phascolosorexinae *nov.*

This group may be as fundamentally distinct within the Dasyuridae as is the Murexinae. Its membership consists of the species of *Phascolo-*

Table 5. Proposed suprageneric classification within the family Dasyuridae. Group numbers refer to those illustrated in Fig. 14. The generic composition of each group is discussed in the text.

Group No.	Representative Genera	Subfamily	Tribe
Group 1	*Ankotarinja*	Unnamed SF.	monotypic SF
Group 2	*Keeuna*	Unnamed SF.	monotypic SF
Group 3	*Dasylurinja*	Unnamed SF.	monotypic SF
Group 4	*Murexia*	Murexinae *nov.*	monotypic SF
Group 5	*Phascolosorex, Neophascogale*	Phascolosorexinae *nov.*	monotypic SF
Group 6	*Phascogale, Antechinus*	Phascogalinae Gill	monotypic SF
Group 7	*Planigale*	Planigalinae *nov.*	monotypic SF
Group 8	*Sminthopsis, Wakamatha,* etc.	Sminthopsinae *nov.*	monotypic SF
Group 9	*Parantechinus, Dasyurus,* etc.	Dasyurinae (Goldfuss)	
Gp. 9.1	*Parantechinus, Pseudantechinus,* etc.		T. Parantechini *nov.*
Gp. 9.2	*Dasyuroides, Dasyurus,* etc.		T. Dasyurini (Goldfuss)

sorex and *Neophascogale,* all of which exhibit strikingly distinctive dental and basicranial features (Archer 1976 c,d, and above), some of which (e.g. the very large metacone on M^5) are even more plesiomorphic than the corresponding states in species of *Murexia.* Others (e.g. the complex stylar shelf) may be more specialized than those of any of the other dasyurids. Like the murexines, none have been the subject of enzyme or serological studies. As noted above, there is a remote possibility that the phascolosorexines will prove to have a special but distant relationship to sminthopsines.

Group Six: Subfamily Phascogalinae Gill, 1872

Gill's (1872) concept of this group was undoubtedly much broader than that of the present work. He presumably meant it to incorporate all small dasyurids although he did not actually allocate specific genera to any of his three dasyurid subfamilies, the Phascogalinae, Dasyurinae, and Sarcophilinae. In the sense used here, Phascogalinae includes all species of *Antechinus* (*sens. strict.,* see above) and *Phascogale.* After murexines, the phascogalines are the most pleisomorphic of living dasyurids. However, they also show enlargement of I1, reduction of M^5 size and complexity, and slight reduction of posterior premolar size. Other features have been noted elsewhere (Archer 1976c).

Group Seven: Subfamily Planigalinae *nov.*

The species of *Planigale* are placed in a monotypic subfamily simply because they cannot be convincingly allied to any other particular group of living or extinct dasyurids. The very slight suggestion, on the basis solely of the enzyme data, of a relationship with species of the Sminthopsinae (the Eighth Group) is much too tenuous to serve as a basis for lumping these groups into one subfamily.

Group Eight: Subfamily Sminthopsinae *nov.*

The interrelatedness of the genera included in this subfamily has only recently become clear. It is based on studies of serology, enzymes, dental and cranial morphology, and plaeontology.

Antechinomys laniger (there is evidently only one species in *Antechinomys,* Archer 1977) is part of the *Sminthopsis* radiation on the basis of all tissue systems examined. Clearly these, as a group, are also most closely related to the species of *Ningaui* (Archer 1975 and Baverstock *et al.* 1982). The extinct *Wakamatha tasselli* also undoubtedly belongs here, but whether it is part of the *Sminthopsis* radiation per se, or the sister group of the *Sminthopsis* – *Ningaui* suprageneric group, is not clear.

Group Nine: Subfamily Dasyurinae (Goldfuss, 1820)

The interrelationships of the species included in this subfamily are demonstrated by serology, enzyme studies, penile morphology, and sketetal and dental morphology. It is the most diverse subfamily within the Dasyuridae and future studies may indicate that it is illogically confined within the dasyurid hierarchy by classification at subfamilial level. On the other hand, its diversity is largely made up of arid–adapted forms whose phyolgenetic history is unlikely to have been greater than 10 to 15 million years old.

Tribe Parantechini *nov.*

This group contains the species of *Parantechinus, Dasykaluta* and *Pseudantechinus,* as well as the unnamed extinct Riversleigh dasyurid. The interrelationships of this suprageneric group have been discussed above.

Tribe Dasyurini (Goldfuss, 1820)

The species of *Myoicitis,* none of which have been examined serologically or studied in terms of comparative enzymes, are hypothesized to be related to this group on the basis of dental and penile morphology. This agrees also with the conclusions of Tate (1947), although Tate considered the species of *Myoictis* to also have a special relationship to the *Dasyuroides* – *Dasycercus* suprageneric group within the Dasyurini. This specific relationship is not supported by penile morphology, but deserves further consideration.

The *Dasyuroides* – *Dasycercus* group relationship to the *Satanellus* – *Dasyurus* – *Sarcophilus* – *Glaucodon* group is supported by dental morphology, serological and enzyme data. Assuming that it is correctly placed within the Dasyurini, it is also the plesiomorphic sister group of the *Dasyurus* – *Satanellus* – *Sarcophilus* – *Glaucodon* suprageneric group.

Within the *Dasyurus* – *Satanellus* – *Sarcophilus* – *Glaucodon* suprageneric group. *Satanellus* spp. are plesiomorphic and *Sarcophilus* spp. highly autapomorphic. Discussion above in connection with the fossil dasyurines elaborates hypotheses about intergeneric relationships within this group.

CONCLUSION

Seventeen years after Ride's (1964) monumental review of fossil marsupials and marsupial classification, we are still groping in the (albeit not so total) darkness about dasyurid phylogeny and classification. Much has been learned in the meantime, but most of this learning serves only to show how much more we have yet to discover. The newly described middle Miocene fossil material serves more to indicate a previously unsuspected breadth for the dasyurid radiation than it does to provide ancestors for known groups. Not until we reach the early Pliocene do we begin to see the threads of the lineages of modern–type dasyurids in the fossil record.

Studies of tissue systems including dentitions, skulls, male reproductive organs, blood, enzymes, and of DNA demonstrate that all of the living dasyurids are monophyletic. Further, Australian (not New Guinean) dasyurids appear to reflect a relatively recent evolutionary burst leading to polyphyletic development of arid–adapted forms. This polyphyly of arid–adapted forms (e.g. species of *Dasyuroides, Ningaui,* and *Parantechinus*) is evidenced by the existence in each case of phylogenetically closer non–arid–adapted forms (e.g. *Satanellus* spp., plesiomorphic *Sminthopsis* spp., and *Myoictis* spp. respectively). This suggests that the development of widespread aridity in Australia post–dated the time of the last phylogenetic divergence of these dasyurid lineages. Polyphyletic arid–adaptation appears to have even occurred within the genus *Sminthopsis,* making the same point, perhaps more effectively. The result of this adaptive response within the family has been the overlay of superficially similar morphological features on phylogenetically diverse dasyurids. Resolution of interrelationships at the subfamilial level on the basis of just morphology is, as a result, often very difficult.

However, the difficulties of determining higher level taxonomic relationships within the family are aggravated most by the lack of enzyme, serological, or data other than cranial and dental morphology, for any of the species of the four endemic New Guinean dasyurid genera, *Neophascogale, Phascolosorex, Murexia* and *Myoictis.* At least two of these appear, on the basis of cranial and dental morphology, to represent fundamentally distinctive subfamilies of their own. Tests of these hypotheses, using other tissue systems, are certainly among the most important steps that should be taken in clarifying phylogenetic relationships within the family. Additionally, and more to the point of this review, many more and much older fossil dasyurids must be recovered before any realistic idea of the breadth and history of the family's evolution is obtained.

ACKNOWLEDGEMENTS

A great many people over the last ten years have helped, indirectly as well as directly, to make this study possible.

For assistance in the field (and all of it given without an arm being twisted), I am most grateful to (in alphabetical order): Ms E. Archer, Mr J. Barrett, Mr L. Barrett, Ms T. Barrett, Dr A. Bartholomai, Mr R. Brown, Mr F. Bussat, Dr R. Callen, Mr E. Camp, Ms G. Clayton, Mr B. Cooke, Ms J. Covacevich, Ms L. Dawson, Dr T. Dawson, Mr A. Elliot, Dr B. Fox, Mr H. Godthelp, Ms S. Hand, Dr G. Hardy, Mr W. Head, Dr. G. Heinsohn, Ms J. Hundt (now Taylor), Mr D. Joff, Mr R. Kahout, Mr A. Kowanko, Mr P. Lawson, Mr C. Limpus, Dr G. Maynes, Mr M. Plane, Mr N. Pledge, Ms P. Rainbird, Dr P. Rich, Dr T. Rich, Mr I. Stewart, students of James Cook University, students of Kelvin Grove College of Advanced Education, students of the University of New South Wales (notably Messrs A. Smith, M. Ellis, K. Patterson, M. Watson, J. Tapsell, M. Graham, S. Byrnes, C. Hart, Ms B. Jones, Ms W. Davies, Ms C. Farrelly, Ms M. McDonald and many others), Mr S. Van Dyck, Dr M. Wade, and Dr M. Woodburne.

For financial support, I am grateful to The University of New South Wales and the Queensland Museum. Dr Ray E. Lemley also kindly financed expeditions through the Queensland Museum to Bluff Downs Station (the Allingham fauna) and Riversleigh Station (the Riversleigh fauna). Australian Research Grant Committee support (grant number E76/15122) was the basis for field work in the Pinpa, Etadunna and Wipijiri formations of South Australia. Australian Research Grant Committee Support, was responsible for enabling the concentrate from Riversleigh to be sorted (grant number E7915864), by Ms G. Clayton. Additionally, Australian Research Grant Committee support (great number D176/16079) enabled collection and research on many of the living species of *Antechinus* — like dasyurids, including the species of *Parantechinus, Pseudantechinus* and *Dasykaluta.*

For permission to work on station properties, and often for totally unexpected displays of hospitality and help, I most particularly wish to thank Mr and Mrs F. Naughton (Saxby Downs and Riversleigh Stn.), Mr and Mrs J. Nelson (Riversleigh Stn.), Mr B. and Mrs C. Oldfield Etadunna Stn.), Mr B. Napier and Mr Baverstock (Frome Downs Stn.), The Merriwa Shire Council, the Postmistress at Bow, The Chinchilla Shire Council, Mr and Mrs Camp (Floraville Stn.), Ms S. Blacklock (Southwick Stn., nr Bluff Downs), and Mr A. McDougal and Mr W. Angliss (Bluff Downs Stn.).

Access to specimens or casts in other collections was kindly provided by Dr W. Clemens, Dr C. Campbell and Dr B. Waters (University of California at Berkeley), Dr M. Woodburne (University of California at Riverside), Mr M. Plane (Bureau of Mineral Resources, Canberra), Dr T. Rich (National Museum of Victoria), Mr N. Pledge (South Australian Museum), Drs D. Kitchener, D. Merrilees, and A. Baynes (Western Australian Museum), Drs A. Bartholomai and R. Molner (Queensland Museum), and Dr A.

Ritchie and Mr B. Marlow (Australian Museum).

Aspects of the theoretical research and sections of this manuscript have at various stages of completion been constructively criticised by Mrs W. Ride, A. Bartholomai, P. Baverstock, J. Kirsch, M. Woodburne, T. Rich, G. Simpson, O. Reig, Messrs M. Plane and S. Van Dyck. Access to unpublished data was also kindly provided by Dr P. Woolley, Dr P. Baverstock, Mr S. Van Dyck, Dr C. Campbell, and Dr P. Crabb.

Dr W. Ride, Ms S. Meagher, Mr R. Johnston, Dr B. Wright, and the Aboriginal people of Yandearra, Western Australia, all helped to determine the appropriate generic name, *Dasykaluta,* for *rosamondae* Ride.

The mind-boggling task of sorting through tons of fossil concentrate for the tiny teeth described here was the lot of (again in alphabetical order): Ms E. Archer, Ms K. Brown, Ms G. Clayton, Mr H. Godthelp, Ms J. Hundt (now Taylor), Mr D. Joff, Mr R. Kahout, Ms P. Rainbird, Mr S. Van Dyck and Ms K. Watkins. Other able sorters in Victoria, including Mr T. Flannery, helped recover the Hamilton teeth. To these and all the other doers of similar related and difficult tasks, my sincerest thanks.

The excellent scanning electron microscope photographs were taken by Dr M. Dickson (University of New South Wales Electron Microscope Unit), and the excellent light photographs by Mr L. Meier. The graphic illustrations were drafted by Mrs J. Taylor. The manuscript was typed by Ms D. Thomas.

REFERENCES

ARCHER, M., 1975. *Ningaui,* a new genus of tiny dasyurids (Marsupialia) and two new species from arid Western Australia, *N. timealeyi* and *N. ridei. Mem. Queensland Mus.* **17:** 237–49.

ARCHER, M., 1976a. Revision of the marsupial genus *Planigale* Troughton (Dasyuridae). *Mem. Queensland Mus.* **17:** 341–65.

ARCHER, M., 1976b. Miocene marsupicarnivores (Marsupialia) from central South Australia, *Ankotarinja tirarensis gen et sp. nov., Keeuna woodburnei gen. et sp. nov.,* and their significance in terms of early radiations. *Trans. R. Soc. South Australia.* **100:** 53–73.

ARCHER, M., 1976c. The dasyurid dentition and its relationships to that of didelphids, thylacinids, borhyaenids (Marsupicarnivora) and peramelids (Peramelina: Marsupialia). *Aust. J. Zool. Suppl. Ser.* **39:** 1–34

ARCHER, M., 1976d. The basicranial region of marsupicarnivores (Marsupialia), inter–relationships of carnivorous marsupials, and affinities of the insectivorous marsupial peramelids. *J. Linn. Soc. Lond. (Zool.).* **59:** 217–332.

ARCHER, M., 1977. Revision of the dasyurid marsupial genus *Antechinomys* Krefft. *Mem. Queensland Mus.* **18:** 17–29.

ARCHER M., 1978a. The nature of the molar-premolar boundary in marsupials and reinterpretation of the homology of marsupial cheekteeth. *Mem. Queensland Mus.* **18:** 157–64.

ARCHER, M., 1978b. Quaternary vertebrate faunas from the Texas Caves of southeastern Queensland. *Mem. Queensland Mus.* **19:** 61–109.

ARCHER, M., 1981a. Revision of the dasyurid marsupial genus *Sminthopsis* Thomas. *Bull. Amer. Mus. Nat. Hist.*

ARCHER, M., 1981b. Origins and radiations of Australian mammals. *In* "Ecological biogeography" ed. by A. Keast. Junk.

ARCHER, M. AND RICH, T. H., 1979. *Wakamatha tasselli gen et sp. nov.,* a fossil dasyurid (Marsupialia) from South Australia convergent on modern *Sminthopsis. Mem. Queensland Mus.* **19:** 309–17.

ARCHER, M. AND WADE, M., 1976. Results of the Ray E. Lemley Expeditions, Part I. The Allingham Formation and a new Pliocene vertebrate fauna from northern Queensland. *Mem. Queensland Mus.* **17:** 379–97.

AZIZ–UR–RAHMEN AND MCDOUGALL, I., 1972. Potassium–argon ages on the New Volcanics of Victoria. *J. Proc. Soc. Victoria* **85:** 61–9.

BARTHOLOMAI, A., 1971. *Dasyurus dunmalli,* a new species of fossil marsupial (Dasyuridae) in the upper Cainozoic deposits of Queensland. *Mem. Queensland Mus.* **16:** 19–26.

BARTHOLOMAI, A., 1978. The fossil kangaroos. *Aust. Mammal.* **2:** 15–22.

BAVERSTOCK, P. R., ARCHER, M., ADAMS, M. AND RICHARDSON, B. J., 1982. Genetic relationships between 32 species of Australian dasyurid marsupials. *In "Carnivorous marsupials"* ed. by M. Archer. Roy. Zool. Soc. New South Wales: Sydney.

BAYNES, A., MERRILEES, D. AND PORTER, J., 1976. Mammal remains from the upper levels of a late Pleistocene deposit in Devils Lair Western Australia. *J. Proc. R. Soc. Western Australia* **58:** 97–126.

BENSLEY, B. A., 1903. On the evolution of the Australian Marsupialia; with remarks on the relationships of the marsupials in general. *Trans. Linn. Soc. Lond. (Zool.).* **9:** 83–217.

BONAPARTE, C. L., 1938. Synopsis vertabratorum systematis. *Nuovi Ann. Sci. Nat., Bologna* **2:** 105–33.

BONAPARTE, C. L., 1850. "Conspectus systematicum. Mastazoolgiae. Edito altera reformata . . . Classis I. Mammalia." Lugduni Batavorum, apud E. J. Brill, Academic Typographum.

CALAN, R. AND TEDFORD, R. H., 1976. New late Cainozoic rock units and depositional environments, Lake Frome area, South Australia. *Trans. R. Soc. South Australia* **100:** 125–67.

CAMPBELL, C. R., 1976. "Tertiary Dasyuridae and Peramelidae (Marsupialia) from the Tirari Desert, South Australia". Unpublished Ph.D. thesis: University of California, Berkeley.

CRABB, P. L., 1978. "Fossil mammals of the lower Pleistocene Moorna Sands, Southwest New South Wales, Australia." Unpublished Ph.D. thesis: Monash University, Melbourne.

CRABB, P., 1982. Pleistocene dasyurids from southwestern New South Wales. *In* "Carnivorous marsupials" ed. by M. Archer. Roy. Zool. Soc. New South Wales: Sydney.

GILL, E. D., 1957. The stratigraphical occurrence and palaeoecology of some Australian Tertiary marsupials. *Mem. Nat. Mus. Victoria* **21:** 135–203.

GILL, E. D., 1973. Geology and geomorphology of the Murray River region between Mildura and Renmark, Australia. *Mem. natn. Mus. Victoria* **34:** 1–97.

GILL, T., 1872. Arrangement of the families of mammals with analytical tables. *Smithson. Misc. Collect.* **2:** i–vi, and 1–98.

GOLDFUSS, G. A., 1820. "Handbuch der Zoologie". J. L. Schrag: Nurnberg.

Illiger, C., 1811. "Prodromus sysematis mammalium et avium additis terminis zoogeographicus utriudque Classis". C. Salfeld: Berlin.

Kempe, H., 1891. A grammer and vocabulary of the language spoken by the Aborigines of MacDonnell Ranges, South Australia. *Trans. and Proc. and Report of the Roy. Soc. South Australia* **14:** 1–54.

Kendrick, G. and Porter, J. K., 1973. Remains of a Thylacine (Marsupialia: Dasyuroidea) and other fauna from caves in the Cape Range, Western Australia. *J. Proc. R. Soc. Western Australia* **56:** 116–22.

Kirsch, J. A. W., 1968. Prodromus of the comparative serology of Marsupialia. *Nature (Lond.)* **217:** 418–20.

Kirsch, J. A. W., 1977. The comparative serology of Marsupialia, and a classification of marsupials. *Aust. J. Zool Suppl. Ser. No.* **52**.

Kirsch, J. A. W. and Archer, M., 1982. Polythetic cladistics.

Lundelius, E. L., 1960. Post Pleistocene faunal succession in Western Australia and climatic interpretation. *Proc. 21st Internat. Geol. Cong., 1960, pt. 4, sec.* **4:** 142–53.

Lundelius, E. L. and Turnbull, W. D., 1973. The mammalian fauna of Madura Cave, Western Australia, Part 1. *Fieldianna, Geol.* **31:** 1–35.

Mahoney, J. A. and Ride, W. D. L., 1975. Index to the genera and species of fossil Mammalia described from Australia and New Guinea between 1838 and 1968. *Spec. Publs. Western Australian Mus.* **6:** 1–250.

Marshall, L. G., 1973. Fossil vertebrate faunas from the Lake Victoria region, S. W. New South Wales, Australia. Mem. natn. Mus. Victoria **34:** 151–81.

Moeller, H., 1973. Zur Kenntnis der Schadelgestalt grosser Raubbeutler (Dasyuridae Waterhouse, 1838) eine allometrische Formanalyse. *Zool. Jb. Anat. Bd.* **91:** 257–303.

Owen, R. 1841. Outlines of a classification of the Marsupialia. *Proc. Zool. Soc. Lond.* **7:** 5–19.

Owen, R., 1877. "Researches on the fossil remains of the extinct mammals of Australia: with a notice of extinct marsupials of England". Vol. 1: 107; Vol. 2, pl. 5, fig. 10. J. Erxleben: London.

Page, R. W. and McDougal, I., 1972. Ages of mineralization of gold and porphyry copper deposits in the New Guinean Highlands. *Econ. Geol.* **67:** 1034–48.

Plane, M. D., 1967. Stratigraphy and vertebrate fauna of the Otibanda Formation, New Guinea. Bull. Bur. *Miner. Resour. J. Aust Geol. Geophys. Aust.* **86:** 1–64.

Plane, M. D., 1972. A New Guinea fossil macropodid (Marsupialia) from the marine Pliocene of Victoria, Australia. *Mem. Nat. Mus. Victoria* **33:** 33–6.

Pocock, R. I., 1926. The external characters of *Thylacinus, Sarcophilus,* and some related marsupials. *Proc. Zool. Soc. London* **1926:** 1037–84.

Ride, W. D. L., 1964a. *Antechinus rosamondae,* a new species of dasyurid marsupial from the Pilbara District of Western Australia; with remarks on the classification of *Antechinus. Western Australian Nat.* **9:** 58–65.

Ride, W. D. L., 1964b. A review of Australian fossil marsupials. *J. Proc. R. Soc. Western Australia.* **47:** 97–131.

Ride, W. D. L., 1970. "A guide to the native mammals of Australia." Oxford Univ. Press: Melbourne.

Schodde, R. and Calaby, J. H., 1972. The biogeography of the Australo–Papuan bird and mammal faunas in relation to Torres Strait. Pp.257–300 *in* "Bridge and barrier: the natural history of Torres Strait" ed. by D. Walker. Australian National University, Dept. of Biogeography and Geomorphology, Publication B6/3 (1972).

Simpson, G. G., 1945. The principles of classification and a classification of mammals. *Bull. Amer. Mus. Nat. Hist.* **85**.

Smith, M., 1972. Small fossil vertebrates from Victoria Cave, Naracoorte, South Australia. II. Peramelidae, Thylacinidae and Dasyuridae (Marsupialia). *Trans. R. Soc. South Australia* **96:** 125–37.

Spencer, B. and Gillen, F. J., 1899. "The native tribes of central Australia". Macmillan and Co. Ltd: London.

Skillbeck, C. G., 1981. A preliminary report on the late Cainozoic geology and fossil fauna of Bow, New South Wales. *J. Proc. Linn. Soc. New South Wales* **104:** 171–81.

Stirton, R. A., 1957. Tertiary marsupials from Victoria, Australia. *Mem. natn. Mus. Victoria* **21:** 121–34.

Stirton, R. A., Tedford, R. H. and Miller, A. H., 1961. Cenozic stratigraphy and vertebrate paleontology of the Tirari Desert, South Australia. *Rec. South Australian Mus.* **14:** 19–61.

Tate, G. H. H., 1947. Results of the Archbold Expeditions No. 56. On the anatomy and classification of the Dasyuridae (Marsupialia). *Bull. Amer. Mus. Nat. Hist.* **88:** 97–156.

Tedford, R. H., Archer, M., Bartholomai, A., Plane, M. D., Pledge, N. S., Rich, P. V.and Wells, R. T., 1977. The discovery of Miocene vertebrates, Lake Frome area, South Australia. *Bur. Miner. Resour. J. Aust. Geol. Geophys.* **2:** 53–7.

Thomas, O., 1888. "Catalogue of the Marsupialia and Monotremata in the collection of the British Museum (Natural History)." British Museum (Natural History): London.

Timms, M., Westerman, M.and During, B., 1982. Studies on the DNA of some *Antechinus* species. *In* "Carnivorous marsupials" ed. by M. Archer. Roy. Zool. Soc. New South Wales: Sydney.

Troughton, E. LeG., 1967. "Furred animals of Australia". Ninth ed. Angus and Robertson Ltd.: Sydney.

Turnbull, W. D. and Lundelius, E. L., 1970. The Hamilton fauna. A late Pliocene mammalian fauna from the Grange Burn, Victoria, Australia. *Fieldianna: Geology* **19:** 1–163.

Turnbull, W. D., Lundelius, E. L. and McDougall, I., 1965. A potassium/argon dated Pliocene marsupial fauna from Victoria, Australia. *Nature* **206:** 816.

Van Dyck, S., 1980. The Cinnamon Antechinus, *Antechinus leo* (Marsupialia: Dasyuridae), a new species from the vine–forests of Cape York Peninsula. *Australian Mammal.* **3:** 5–17.

Van Dyck, S. and Ogilvie, P., 1977. *Antechinus swainsonii* (Waterhouse, 1840), the Dusky Marsupial Mouse, an addition to the mammal fauna of Queensland. *Mem. Queensland Mus.* **81:** 69–73.

Wakefield, N. A., 1972. Palaeoecology of fossil mammal assemblages from some Australian caves. *Proc. R. Soc. Victoria.* **85:** 1–26.

Waterhouse, G. R., 1846. "A natural history of the Mammalia". Vol. 1. *Hippolyte Bailliere:* London.

Waithman, J., 1979. A report on a collection of mammals from Southwest Papua, 1972–1973. *Australian Zool.* **20:** 313–26.

Winge, H., 1941. "The interrelationships of the mammalian genera Vol. 1: Monotremata, Marsupialia, Insectivora, Chiroptera, Edentata. C. A. Reitzels Forlag: Copenhagen.

Woolley, P. A., 1982. Phallic morphology of the Australian species of *Antechinus* (Marsupialia: Dasyuridae): A new taxonomic tool. *In* "Carnivorous marsupials", ed. by M. Archer. Roy. Zool. Soc. New South Wales: Sydney.

WOOLLEY, P. A. AND WEBB, S. J., 1977. The penis of dasyurid marsupials. Pp.307–23 *in* "The biology of marsupials" ed. by Stonehouse and D. Gilmore. MacMillan Press Ltd.: London.

RESUMEN

37. Reanálisis registro de los dasiúridos australianos (Marsupialia), integración de los datos conducentes a la interpretación filogenética y clasificación supragenérica. Por Michael Archer.

Todas las especies extinguidas de dasiúridos son críticamente revisadas. Se describe nuevo material del Mioceno medio de las especies *Ankotarinja tirarensis* y *Keeuna woodburnei*. *Dasylurinja kokuminola* n. gen. y sp. es descripta de sedimentos del Mioceno medio de South Australia. *Dasyuroides achilpatna* n.sp. es descripta de sedimentos del Plioceno–Pleistoceno del Noroeste de Queensland. Se reconocen especies de *Antechinus, Planigale, Sminthopsis, Ankotarinja* y *Dasyurus* de varias faunas terciarias y plio–pleistocenas. La signifficación filogenética y paleontológica de cada uno de estos dasiuridos fósiles es considerada.

Los datos paleontológicos son integrados con aquellos de otros estudios de manera de interpretar los aspectos mayores de la evolución de los dasiúridos. Son reconocidos nueve grupos de dasiúridos: 1) las especies de *Ankotarinja;* 2) las especies de *Keeuna;* 3) las especies de *Dasylurinja;* 4) las especies de *Murexia;* 5) las especies de *Neophascogale* y *Phascolosorex;* 6) las especies de *Phascogale* y *Antechinus;* 7) las especies de *Planigale;* 8) las especies de *Wakamatha, Ningaui* y *Sminthopsis;* y 9) las especies de todas los otros géneros. Tambien se consideran las relaciones de las especies y géneros de estos grupos, particularmente aquellas del grupo noveno.

Se revisa y revee la classificación intrafamiliar de los Dasyuridae Se reconocen nueve grupos de nivel subfamiliar, pero sólo seis reciben distincion taxonomica: Murexinae para el grupo 4; Phascolosorexinae para el grupo 5; Phascogalinae Gill para el grupo 6; Planigalinae para el grupo 7; Sminthopsinae para el grupo 8; y Dasyurinae (Goldfuss) para el grupo 9. Se reconocen dos tribus dentro de los Dasyurinae: Parantechini y Dasyurini (Goldfuss).

38. A REVIEW OF MIOCENE THYLACINIDS (THYLACINIDAE, MARSUPIALIA), THE PHYLOGENETIC POSITION OF THE THYLACINIDAE AND THE PROBLEM OF APRIORISMS IN CHARACTER ANALYSIS

MICHAEL ARCHER[1]

The Tertiary thylacinid record is reviewed. Only the Miocene Alcoota *Thylacinus potens* Woodburne is represented by enough material to permit useful phylogenetic analysis. It displays a complex of features which lead to the following conclusions: it descended from a tribosphenic form that had at least stylar cusps D and E, cusps absent in all known borhyaenids but present, for example, in some dasyurids and didelphids; it had a complete posterolateral palatal foramen, a primitive feature lacking in all known dasyurids; lower molar morphology in this Miocene species is as specialized as the modern species and gives no clear indications of ancestral morphology; *T. potens* and the modern species *T. cynocephalus* probably represent collateral branches of the thylacinid family tree, rather than an ancestor–descendant lineage.

The relationships of thylacinids are reconsidered using all available data: dental morphology; basicranial morphology; post-cranial (non–pedal) morphology; pedal morphology; serological data, and palaeontological data. Of four phylogenetic sister–group hypotheses considered (thylacinid–dasyuroid, t–borhyaenoid, t–didelphid, and t–pediomyid), the most probable is a dasyuroid–thylacinid sister–group hypothesis if the serological and pedal morphology data are given more weight than data from the other character systems. The molecular clock separation date (7 million years) for thylacinids from dasyurids appears to be in error. Thylacinid fossils are known from middle Miocene deposits. A second molecular clock date of 24 million years for the most recent thylacinid–myrmecobiid common ancestor does not conflict irreconcilably with the data from other tissue systems or the palaeontological data.

The most recent support for the thylacinid–dasyuroid sister–group hypothesis is that of Marshall (1977). However, the character analyses upon which his conclusions are based appear to be an inadequate basis for these conclusions. Problems of apriorism are aggravated by his application of the weighting scheme of Hecht (1976) and Hecht and Edwards (1976). It is suggested that this scheme for assessing the phylogenetic significance of apomorphies is, contrary to the intention of its authors, subject to arbitrariness.

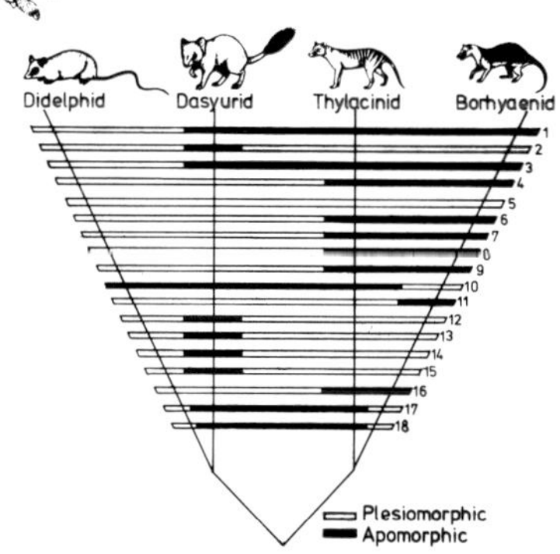

Pages 445–76 *in* CARNIVOROUS MARSUPIALS, ed. by M. Archer, Roy. Zool. Soc. New South Wales: Sydney, Australia.

FEW DEBATES about marsupial phylogeny have exceeded, in long–term interest, the question of the relationships of the Tasmanian Thylacine. The matter was raised by Bensley (1903) who suggested that the Tasmanian Thylacine might be a foreign element in the Australian fauna, noting that it seemed to show a striking similarity to the South American carnivorous borhyaenids. His opinion was seconded by Sinclair (1906:444) who notes that "The reality of a former land connection between the Australian region and South America . . . is plainly indicated . . . and it is only by such a connection that the distribution of the Thylacinidae can be explained". He considers, in his monograph on the Santa Cruzian (middle Miocene) borhyaenids, that these South American carnivorous marsupials were, in fact, thylacinids. There followed a long, but sporadic debate about the relationships of the thylacinids. Opposition to the views of Bensley and Sinclair came most vigorously from Simpson (1941:5) who considers that the similarities between thylacinids and borhyaenids are the result of parallelism: "As for *Thylacinus*, there seems to be abundant, conclusive evidence that it is merely a specialized dasyurid". Wood (1924) and Archer (1976a,b) support and amplify Bensley's view; Simpson (1948), Tate (1947), and Marshall (1977) support Simpson's (1941) view. Until recently, other authors have taken sides but contributed little to a comprehensive understanding of group relationships (most are noted in Simpson 1941, Archer 1976a and Marshall 1977).

[1] School of Zoology, University of New South Wales, P.O. Box 1, Kensington, New South Wales, Australia, 2033.

New data presented by Sarich *et al.* (1982) and Szalay (1982) appear to tip the balance in favour of Simpson's (1941) view. However, there are still problems about the precise nature of the relationships indicated by the immunological data of Sarich *et al.* (1982).

The Tertiary fossil record of thylacinids is very poor. Late Tertiary thylacines are noted by Plane (1976), from the Miocene Awe local fauna in New Guinea, and by De Vis (1893) from the Pliocene Chinchilla local fauna of Queensland. These have been critically reviewed by Dawson (1982) who concludes that, on the basis of specimens known, they are not separable from the modern *Thylacinus cynocephalus.* The specific status of the Awe thylacine, known from an isolated premolar, cannot be confidently assessed because of inadequate representation. *Thylacinus potens,* the only named Miocene thylacine, is considered by Woodburne (1967 : 39) to be a thylacine very similar to modern Tasmanian Thylacine and "... nearly as distinct from any known dasyurid as are the Pleistocene and Recent members of the genus." He concludes that "If *Thylacinus,* or the Thylacinidae, diverged from the Dasyuridae, as seems probable, the time of divergence was surely no later than middle Tertiary and was probably much earlier." In the course of preparing the present review, I have had the opportunity to re-examine the holotype of *Thylacinus potens* and some of its features that appear to bear on questions about thylacinid evolution have been critically examined. The Miocene Kutjamarpu thylacine is only known from a premolar.

Dental terminology used follows that established elsewhere (Archer 1978) such that the complete thylacinid dental formula is as follows: $dI\frac{1-4}{1-3}$; $I\frac{1-4}{1-3}$; $dC\frac{1}{1}$; $C\frac{1}{1}$; $P\frac{1-3}{1-3}$; $M\frac{1-5}{1-5}$. The $M\frac{1}{1}$ is vestigial and resorbed from the juvenile dentition before it erupts. The holotype of *Thylacinus potens* (CPC6746) is in the palaeontological collections of the Bureau of Mineral Resources, Canberra. Paratypes of *T. potens* and the Kutjamarpu thylacine specimen are in the paleontological collections of the University of California at Berkeley.

GROUPS OF CARNIVOROUS MARSUPIALS RELEVANT TO HYPOTHESES ABOUT THYLACINID RELATIONSHIPS

There are at least nineteen families of carnivorous (including insectivorous and insectivorous–omnivorous) marsupials (most are noted in Kirsch 1977). There have been at least twenty-five ancestor–descendant and/or sister–group

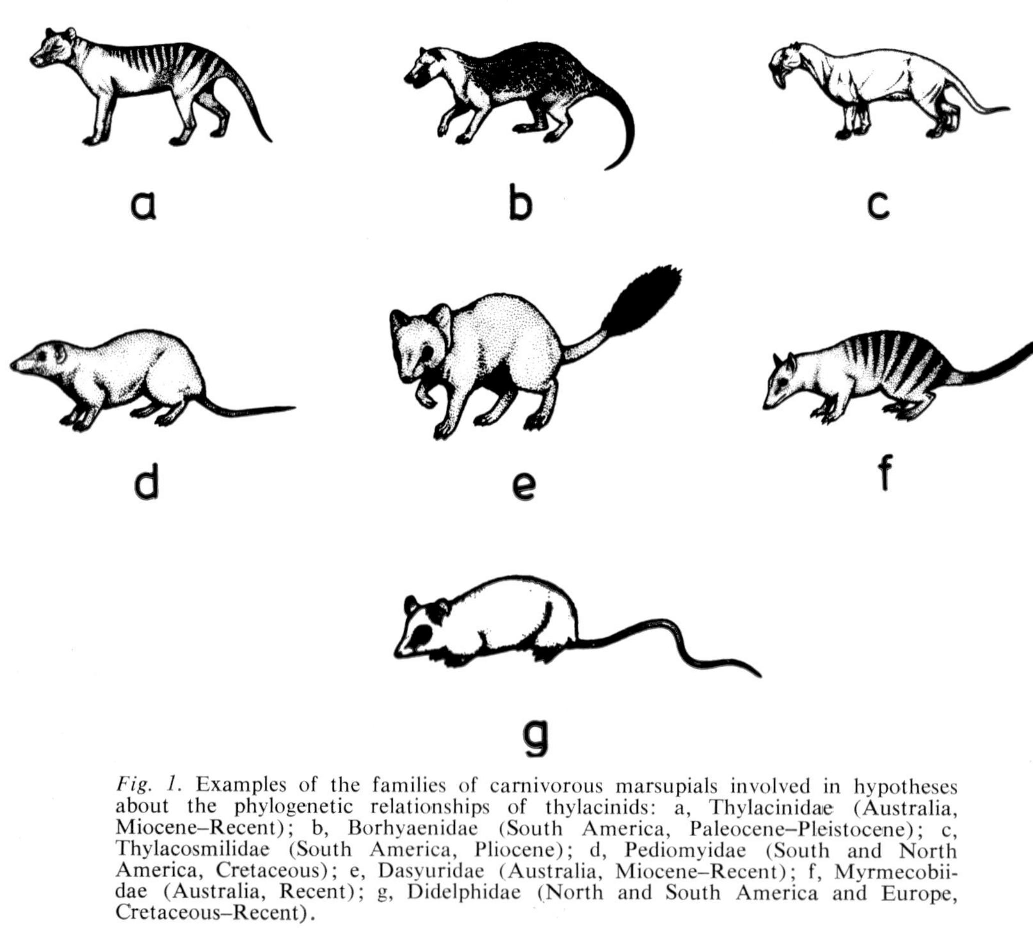

Fig. 1. Examples of the families of carnivorous marsupials involved in hypotheses about the phylogenetic relationships of thylacinids: a, Thylacinidae (Australia, Miocene–Recent); b, Borhyaenidae (South America, Paleocene–Pleistocene); c, Thylacosmilidae (South America, Pliocene); d, Pediomyidae (South and North America, Cretaceous); e, Dasyuridae (Australia, Miocene–Recent); f, Myrmecobiidae (Australia, Recent); g, Didelphidae (North and South America and Europe, Cretaceous–Recent).

hypotheses of intergroup relationships involving these families. Here I will only consider the seven families and four main hypotheses about intergroup relationships that are relevant to interpreting the relationships of thylacinids.

1. *The Thylacinidae* (Fig. la): These are the Thylacines or Tasmanian "wolves". There are only two distinctive species known (Dawson 1982), from Australia, New Guinea and Tasmania. *Thylacinus cynocephalus* (including *spelaeus* and *rostralis*) is the modern, Pleistocene and probably Pliocene species. *Thylacinus potens* occurs in the Miocene Alcoota fauna. Both species are morphologically very similar. Apomorphic character states of this family include (Archer 1976a) reduction of incisor number to 4/3, marked reduction in size of M1, reduced stylar shelf, approximation of the paracone and metacone on the upper molars, reduction of the protocone, loss of the metaconid, enlargement of the preparacristid and gross reduction of the epipubic bones.

2. *The Borhyaenidae* (Fig. 1b): This family of extinct cat to lion–sized South American carnivores includes approximately twenty-seven genera and forty–two species extending from the Paleocene to the Pliocene (Marshall 1978). Apomorphic character states include the same as those noted above for thylacinids as well as nasal–lacrimal contact and apparent reduction of the mastoid bone on the posterior surface of the skull. Other distinctive features of this group are noted below in the discussion of characters.

No one has recently doubted Simpson's view (e.g. as discussed in Simpson 1948) that borhyaenoids (see below) evolved from didelphids. Several late Paleocene and Eocene South American taxa (such as *Eobrasilia coutoi* and species of *Patene*) are structurally intermediate in morphology. However these particular annectant taxa also occur contemporaneously with more specialized borhyaenids and indicate that the probable time of origin of the borhyaenoids was the late Cretaceous or early Paleocene (Marshall 1976).

3. *The Thylacosmilidae* (Fig. 1c): These are large, extinct, South American marsupial saber–toothed carnivores. There are four genera and five species (Marshall 1976). Apomorphic features include loss of incisors, enormous enlargement of the upper canines and corresponding development of flange on the lower jaw, loss of protocones, and possibly the "enclosed" external auditory meatus. These and other character states are listed, but not analyzed, as part of a diagnosis of the family by Marshall (1976).

No one has hypothesized that thylacosmilids gave rise to any other group of marsupials, nor that they are related to any other family than the Borhyaenidae.

2–3. *The Borhyaenoidea*: This superfamily contains the Borhyaenidae and Thylacosmilidae (Marshall 1978). The thylacosmilids are assumed to have evolved from as yet unknown borhyaenoids sometime during the early Tertiary. Scott (1937) suggests that a species of the borhyaenid genus *Arminiheringia* (early Eocene to early Oligocene) might have been ancestral to thylacosmilids, a view also held by Marshall (1978). Borhyaenoids have been involved in discussions or proposals of ancestor–descendant and/or sister–group hypotheses involving didelphids (e.g. Simpson 1941), thylacinids (e.g. Sinclair 1906), dasyurids (e.g. Cabrera 1972) and pediomyids (e.g. Marshall 1977).

Features of this superfamily include a small lunar which contacts a large magnum, a reduced or lost transverse canal, a lacrimal bone that extends onto the rostrum and usually has a large tuberosity developed above the lacrimal canal which opens within the orbit. Marshall (1976 and 1978) gives these and other character–states in a cladistically unanalyzed list of superfamily features.

4. *The Pediomyidae* (Fig. 1d): There are possibly two genera, *Pediomys* and *Aquiladelphis,* and at least nine species. All occurred in the Cretaceous and are known only from North and South America. Clemens (1966) raised the group to family status. Prior to that (e.g. Simpson 1945), they were regarded as a subfamily within the Didelphidae. Apomorphic character–states include (Clemens 1966) a reduced stylar shelf on the upper molars with a very reduced or absent stylar cusp B, and a cristid obliqua that intersects the trigonid on the posterior flank of the protoconid.

McKenna (pers. comm. 1976) has noted the possibility that pediomyids may not even be marsupials because many of the apomorphic features of the group are in fact common features of placental mammals. Their possession of four adult molars has been the principal reason for their being regarded to be marsupials. This cannot be an apomorphic character unique to marsupials in the strict sense unless marsupials are considered to include ancestral placentals.

Nevertheless, assuming that they are marsupials, their origin has reasonably been interpreted to be didelphid (Clemens 1966) at least in part because of structurally intermediate species such as *Pediomys elegans* and *P. clemensi* which preserve remnants of the stylar shelf and in this regard resemble species of *Alphadon,* characteristic Cretaceous didelphids. Russell (1976) suggests that the reduction of stylar cusp B in pediomyids might indicate that they were ancestral to dasyurids. Marshall (1977) also raised this general possibility, but more specifically in regard to the ancestry of thylacinids and/or borhyaenids.

5. *The Dasyuridae* (Fig. 1e): These are the Quolls (sometimes called Native–cats) and their smaller relatives in Australia, New Guinea and Tasmania. They include forms ranging in size from a shrew to a bulldog. At present, seventeen genera and about sixty species are recognized including three Miocene genera (*Ankotarinja, Dasylurinja* and *Keeuna*) and one Pliocene genus (*Glaucodon*) (Archer 1982). The group is morphologically diverse including such extremes as the tiny didelphid–like *Ankotarinja tirarensis* and the massive bulldog–like *Sarcophilus harrisii.* Apomorphic character–states include (Archer 1976a,b) enlargement and posterior shif of stylar cusp D on the upper molars, reduction (relative to plesiomorphic didelphids) of the size of M1, development of a hypoconulid notch in the anterior cingulum, reduction (relative to plesiomorphic didelphids) of the size of P_3, loss of a complete posterolateral palatal foramen and development of a large (relative to the plesiomorphic didelphid condition such as occurs in species of *Philander* or *Marmosa*) alisphenoid tympanic wing.

6. *The Myrmecobiidae* (Fig. 1f): This monotypic family encompasses only *Myrmecobius fasciatus,* the Australian Numbat. There are no pre–Pleistocene records. Numerous autapomorphic states of the only species have been documented by Archer and Kirsch (1977).

No one has recently doubted that myrmecobiids evolved from dasyurids and are also the sister–group of an, as yet unrecognized, subgroup of dasyurids. This dichotomy probably occurred prior to the onset of widespread aridity in middle Miocene times. Details of the construction of the floor of the middle ear (Archer 1976b) suggest that acquisition of the enlarged alisphenoid hypotympanic sinus in myrmecobiids occurred in parallel with the similar but different enlargement in dasyurids.

5–6. *The Dasyuroidea*: The Myrmecobiidae and Dasyuridae comprise the Dasyuroidea. Synapomorphic features of this superfamily include the reduced incisor number (4/3 from the didelphid 5/4), loss of the foramen ovale and its replacement by a foramen pseudovale.

There has been a surprising lack of disagreement about the phylogenetic origins of the dasyuroids. Clemens (1966) concludes that Cretaceous didelphines, such as species of *Alphadon,* could be ancestral to dasyuroids. Archer (1982) reviews a middle Miocene Australian taxon (*Ankotarinja tirarensis*) that is structurally intermediate between Cretaceous didelphines and modern dasyurids. The only recent hypothesis that in any way differs from this is that of Russell (1976) who suggests that pediomyids may have been ancestral to dasyurids because he considers that stylar cusp B is small in dasyurids. However, in the oldest named dasyuroids (Archer 1976c), and most living structurally primitive species such as those of *Murexia,* stylar cusp B is as large as it is in most didelphines. Reduction of stylar cusp B does indeed occur in some dasyurids, such as species of *Dasyurus, Dasycercus, Dasyuroides, Pseudantechinus* and *Dasykaluta,* but these are all highly specialized forms at least in terms of their dentitions (Archer 1976a and 1982). Reduction of stylar cusp B in these dasyurids and in pediomyids would appear to be the result of convergence.

There is some difference of opinion about the descendents of dasyuroids. Although no modern researcher has expressed doubts about myrmecobiids being specialized dasyurids, there is no consensus about the role played by dasyurids in the origin of the Australian bandicoots and diprotodont marsupials. Elsewhere (Archer 1976a), I have suggested that dasyuroids were a sterile branch, and that perameloids were the base for the diprotodont radiation. Others (e.g. Bensley 1903, Ride 1964) have suggested the reasonableness of the hypothesis that dasyuroids gave rise to perameloids which in turn gave rise to the diprotodonts. Kirsch (1977) remains open–minded about the possibilities, but points out that at least in terms of serological affinity, dasyuroids group with perameloids (bandicoots) while the latter show no special affinities with the diprotodontans.

Other ancestor–descendant or sister–group hypotheses have involved dasyuroids with thylacinids (e.g. Simpson 1941), borhyaenoids (e.g. Cabrera 1927) and the Australian marsupial moles or notoryctids (e.g. Stirling 1891).

7. *The Didelphidae* (Fig. 1g): As presently treated (e.g. Kirsch 1977a), this diverse family of South American, North American and formerly European opposums contains at least 43 genera including the plesiomorphic genus *Alphadon* and the highly apomorphic genus *Chironectes,* the only species of which is aquatic. The number of species is well over twice the number of genera but depends on the authority consulted. Over half the genera known are represented by only extinct species. I do not know what, if any, synapomorphic character states this family possesses. The only character states primitively possessed by all groups currently placed in this taxon are essentially synapomorphic features of the Marsupialia (and hence plesiomorphic within this group) such as a large stylar shelf, a well–developed stylar cusp C, twinned hypoconulid and entoconid on the lower molars, and complete reduction of true tooth replacement in the post–canine dentition.

The relationships of the Didelphidae are particularly difficult to interpret, precisely because it is essentially a plesiomorphic (and possibly paraphyletic?) group. As the probable stem groups(s) for the remainder of the marsupial radiation, there may well be many sister–group pairs or at least structural ancestors for other families contained within the Didelphidae. Elsewhere (Archer 1976a) it has been demonstrated that the oldest known dasyuroids were morphologically most similar to some Cretaceous didelphines. Other proposals for ancestor–descendant and/or sister–group hypotheses involving the Didelphidae have related it to the Borhyaenidae (e.g. Simpson 1941), Thylacinidae (e.g. Archer 1976b), Pediomyidae (e.g. Clemens 1966), Peramelidae (Archer 1976a and b), and so on. Most of the remaining families of South American marsupials have at one time or another been postulated to have descended from ancestral didelphids.

If the Didelphidae is also considered to include *Holoclemensia texana,* then the origins of the family may be in effect those of the Marsupialia and lie somewhere in or near the Aegialodontidae. Familial allocation of *H. texana,* however, seems highly debatable.

Didelphoidea: Under this very broad taxonomic umbrella are placed (e.g. by Kirsch 1977a and others) the Didelphidae, Stagodontidae, Pediomyidae, Microbiotheriidae, Sparassocynidae, Caroloameghiniidae and Glasbiidae. In the present paper, reference to didelphoids may be taken to include all or any one of these families. Many of these families contain taxa characterized by highly distinctive apomorphic features. But, as a whole, this superfamily is probably, like the Didelphidae itself, paraphyletic and is more reasonably regarded as a grade of marsupial evolution. Studies in progress by John Kirsch *et al.* with didelphoids will no doubt help to clarify the boundaries of monophyletic groups within this superfamily.

HYPOTHESES ABOUT THE RELATIONSHIPS OF THYLACINIDS

There have been four sister–group hypotheses proposed involving thylacinids: 1, thylacinids and dasyuroids; 2, thylacinids and didelphoids; 3, thylacinids and borhyaenoids; and 4, thylacinids and pediomyids. Thre are of course also ancestor–descendant hypotheses.

Simpson (1941), Archer (1976a and b), and Marshall (1977) have reviewed most of these. The principal contenders are:

1, thylacinids and borhyaenoids have a common ancestor that they do not share with any other group;

2, thylacinids are derived from a particular group of borhyaenoids or vice versa (Sinclair 1906, Wood 1924, Archer 1976a);

3, thylacinids are derived from dasyuroids and borhyaenoids from didelphids (e.g. Simpson 1941, Marshall 1977);

4, an unspecified "dasyuroid" (in Cabrera's 1927 sense including dasyuroids, thylacinids and borhyaenoids) gave rise independently to dasyurids, thylacinids and borhyaenids;

5, dasyuroids, thylacinids, and/or borhyaenoids descended from pediomyids (Russell 1976 and/or Marshall 1977).

In considering basicranial structure, I previously (Archer 1976b) concluded that the thylacinid pattern could be derived from that of didelphids but not dasyuroids or borhyaenoids. Borhyaenoid basicranial structure could be derived from that of didelphids, but was unlike that of thylacinids. In terms of dentition (Archer 1967a), while thylacinids and borhyaenoids share unique derived features (the bases for earlier proposals of a thylacinid–borhyaenid relationship by Sinclair 1906, Bensley 1903 and Wood 1924), thylacinids and dasyuroids share none. The most parsimonious interpretation seemed to be (Archer 1976a and b) that thylacinids evolved from a borhyaenoid that had a generalized (i.e. didelphid–like) basicranial structure but a thylacinid–borhyaenoid–like dentition. Synapomorphic character states of such a hypothetical borhyaenoid–thylacinid taxon would include, as well as many dental features, reduction or loss of the epipubic bones. The current alternative hypothesis, most recently considered by Marshall (1977), is that thylacinids are specialized dasyuroids that have converged on borhyaenoids.

Marshall's (1977) analysis purports to be cladistic. My earlier analyses (Archer 1976a and b) were not cladistic and interpretations of affinity were mainly based on overall similarity in dental and basicranial structure. I now prefer to initially analyse character–states cladistically. Because Marshall (1977) reached a conclusion about thylacinid affinity that was different from my own, I have attempted here to re-evaluate the character–state transformations of what still appear to be the potentially most significant characters considered in my earlier works, as well as those considered by Marshall (1977). The results of this analysis seem to indicate that the differences in our conclusions based on cranial and dental morphology only result from: 1, the choice of characters (he considered 14; I have considered the same 14 as well as 93 other potentially useful characters); 2, his assumptions about or interpretations of the polarity of morphoclines involving these characters; and 3, his weighting system (an application of Hecht 1976 and Hecht and Edwards 1976).

ASSESSMENT OF CHARACTER–STATES AND THE PROBLEM OF WEIGHTING

In order to evaluate the hypothetical relationships of the various groups of marsupial carnivores noted above, it is necessary to consider taxonomic distribution as well as the significance of the various character–states involved. To do this I will consider character–state morphoclines and what importance (or weighting) each shared (synapomorphic) state should receive. This involves first a critique of the procedures advocated by Hecht (1976) and Hecht and Edwards (1976), as well as those used by Marshall (1977). In the discussions of character–states, ancestor–descendant as well as cladistic relationships will be considered. Justification for considering ancestor–descendant as well as sister–group hypotheses has been reviewed by Szalay (1977).

Hecht and Edwards (1976) and Hecht (1976) have proposed a general system for determinating the information value of character–states in phylogenetic analysis. This adds a second level of bias to character analysis subsequent to the one produced by the original choice of characters.

The system proposed by Hecht (1976) and Hecht and Edwards (1976) has been used by Marshall (1977) to analyze my earlier dental and basicranial (Archer 1976a,b) analyses. For this reason I have found it necessary to consider why application by Marshall of the weighting scheme of Hecht and Edwards led him to a conclusion so different from my own.

I consider that the system of Hecht and Edwards also has several intrinsic problems and does not, as they suggest it does (Hecht and Edwards 1976: 655), provide a means for non–arbitrarily allocating weights to characters or character–states.

They recognize five levels of ascending significance (I–V) for homologous character–states which they summarize as follows (Hecht and Edwards 1976: 656): "(I) Shared and derived character states which are the result of loss. Such loss characters are the zero state because there is no developmental evidence. (II) Shared and derived character states which are the result of simplification or reduction of complex structures as indicated by comparative or developmental anatomy. (III) Shared and derived character states which are the result of growth and developmental processes dependent on size, age, and hormonal and other physiological relationships. Characters related to allometric functions are the best example. (IV) Shared and derived character states which are highly integrated functionally and are subject to directional selection. (V) Shared and derived character states which are unique and innovative in structure (the most important type of information)."

Hecht (1976: 344) further suggests that the actual value of a group V character–state synapomorphy is probably a thousand–fold or greater than the value of a group I character–state synapomorphy. He suggests that an even more realistic evaluation of the different groups would be expressed by incrementing powers of 10 such that group I=1, II=10, III=100, IV=1,000, and V=10,000.

Considered as an internally logical system, I do not understand why a group III character–state should receive more weight than a group II character. Surely if the state of character is truely determined for example by allometry, this character complex is no more useful as a synapomorphy than is a group II synapomorphy. Hecht and Edwards (1976) note that group III character states are probably very susceptible to parallel evolution. For this reason alone group III character–state synapomorphies would appear to be less potentially significant than a group II synapomorphy for which the ontogeny of development is known. For the same reasons, group IV synapomorphies, precisely because they are highly integrated functionally, should merit very little more weight than the group III synapomorphies. Practical problems also arise (as they have in Marshall's work) in recognition of a group IV character synapomorphy. If it can be demonstrated (as I have done below) that the features of such a supposedly integrated functional unit are known to vary independently in closely related forms then each feature should be treated as an independent apomorphy. In that case, the combined character weights would probably be very different than a collective group IV weight.

A more serious problem concerns the practice of deciding to which particular weighting group a character–state or unique character belongs. An example will help make the point. Many herbivorous marsupials have gaps called diastemata in the tooth row between the C^1 and P^3, and between the I_1 and P_3. This is clearly a loss character. The gaps are occupied by additional teeth in plesiomorphic groups. But consider also that Ardan, Kemp and Ride (1961) have convincingly demonstrated that these diastemata are very important acquisitions in herbivorous mammals. Instead of having to constantly keep their heads down at ground level while obtaining food, they can bite off long grass stems allowing the excess to protrude from the side of the mouth through the diastema. This reduces the risks of predation while they continue to feed the grass into the molar region using the tongue. So it is a loss or reduction character (teeth lost), weighting group I or II, but it is also an acquisition character, weighting group V (we know how it evolved, i.e. by selection for elimination of the missing teeth). Similarly, and of more immediate concern, Marshall (1977) considers that a broad nasal–lacrimal contact is a group V autapomorph for all borhyaenids. I have demonstrated below that this is not the case, for several reasons, but here I only want to focus on the question of his decision to regard this as a group V character. It *is* a group V character but only if we describe the character as "apomorphic acquisition of a nasal–lacrimal contact". However, we can alternatively and equally correctly describe it as "apomorphic loss of the plesiomorphic frontal–maxillary contact". Both descriptions are accurate, but one demands a weight of 10,000, the other a weight of no more than 10.

The general point I would make here is that commonly character loss will be accompanied by some associated character acquisition. The choice of which of the two opposite character–states is used is unfortunately susceptible to the apriorisms of the investigator. Further, dismissal of the significance of a character–state as a loss character could be justified but obscures the potential value of an immediately related acquisition character–state. The unique apomorphic reduction of epipubic bones in thylacinids and borhyaenids is technically a low weight character because it involves reduction, but discounting the character for this reason obscures its value as a signal of some as yet unrecognized simultaneously acquired apomorphic character–state such as new muscle insertions, pouch support mechanisms, and so on.

The Hecht and Edwards (1977) system, in its attempt to provide a non–arbitrary guide to weighting apomorphic characters in phylogenetic analysis, has this basic weakness: It cannot be applied without involving the bias of the investigator. Theoretically, it should be useful. But practically, it is unreliable in the hands of anyone who uses it.

This does not mean that I prefer a numerical analysis of the characters. As Hecht rightly points out, such an approach also involves a weighting bias that is dependent on the number of characters recognized. Although some attempt to assess the significance of apomorphic character–states is undoubtedly desirable, I do not think that phylogenetic issues can satisfactorily be settled by counting points that have been allocated in such an arbitrary manner, something that Marshall (1977) has tried albeit ineffectually to do using the system of Hecht and Edwards (1976). In the analysis that follows, I do not propose to follow any rigorous guide for allocating weights to character–states. All I will try consistantly to do is lay the same information before you that I have available to me, and then make assessments about the significance and relationships based on this information. If these assessments involve more than this information as they sometimes do, they are speculations. And although perhaps I should apologise for speculation, it is too much fun to avoid.

Apart from the problem of deciding how or if to allocate weights to character–states, there is the even more fundamental problem, also exemplified in Marshall's (1977) analysis, of what features or attributes constitute distinct characters.

It is often contended by systematists who question the possibility of common origin of two overtly similar groups, that the similarity (plesiomorphic and apomorphic) may be the result of total convergence or parallelism, a response to similar selection pressures. Features involved in this type of evolution do not warrant equal individual phylogenetic weighting compared with characters that are supposedly evolving independently. Taken in the simplistic extreme, all similar features found in thylacinids and borhyaenids that can be shown to be adaptations to a common selection pressure, could be treated as one character. For example, Marshall (1977 : 423) when tallying up his interpretations of character weights, says ". . . characters 11 and 12 received special attention. Because a functional (group IV) relationship exists between metaconid loss and stylar shelf reduction (e.g. both are associated with improved carnassialization), *both* characters are given a combined maximum weight of 4." The main flaw in Marshall's analysis here is to assume that the two characters are functionally tied to one another, which, as I have demonstrated elsewhere (e.g. Archer 1976a), is simply not the case. Apart from the evidence given below that they do not always vary concordantly (which is part of the definition of "character" given by Hecht 1976 : 339), and hence demonstrate their separate genetic control, they do not contact one another — the stylar shelf overhangs the lower molars and is not controlled by demands for occlusion with any parts of the lower molars, let alone specifically the metaconids. Marshall (1977) appears *a priori* to assume these characters vary concordantly in spite of the fact that there is more than enough evidence to show that the characters are not functionally or phylogenetically tied.

As a general guide, it seems reasonable to suggest that if, in related taxa, it is possible to find a feature evolving independently, that feature should be treated as a separate character in each of the related taxa. I have done this in the character analysis below.

MIOCENE THYLACINIDS: A BRIEF REVIEW OF THE MATERIAL

Pliocene and Pleistocene thylacinids have all been referred to the modern species *Thylacinus cynocephalus* (see Dawson 1982 for a review). There are only two Miocene faunas known to contain thylacinids: The Alcoota local fauna of the Northern Territory which contains *Thylacinus potens* Woodburne; and the Kutjumarpu local fauna of South Australia which contains an unnamed thylacinid based on an isolated premolar.

Thylacinus potens Woodburne
(Figs 2 and 3B)

Woodburne (1967 : 39) concludes that *Thylacinus potens* is very similar to Quaternary thylacines but in its possession of ". . . small palatal fenestrae, the presence of a stylar cusp anterior to the metastyle of M^3, the more symmetrical arrangement of the parts of the upper molars and the strong cleft in the labial outline of the upper molars . . .", these character–states in the Miocene species ". . . could be interpreted as primitive features relative to a remote dasyurid ancestry." In order to consider the possible phylogenetic significance of these and other character–states, I have re-examined the holotype. Comments based on this re-examination are presented in the character analysis.

Material: Woodburne (1967) provides an excellent description of the holotype (CPC6746), a damaged rostrum including a palate and damaged dentition. It was collected from Paine Quarry, Alcoota Homestead, in the Northern

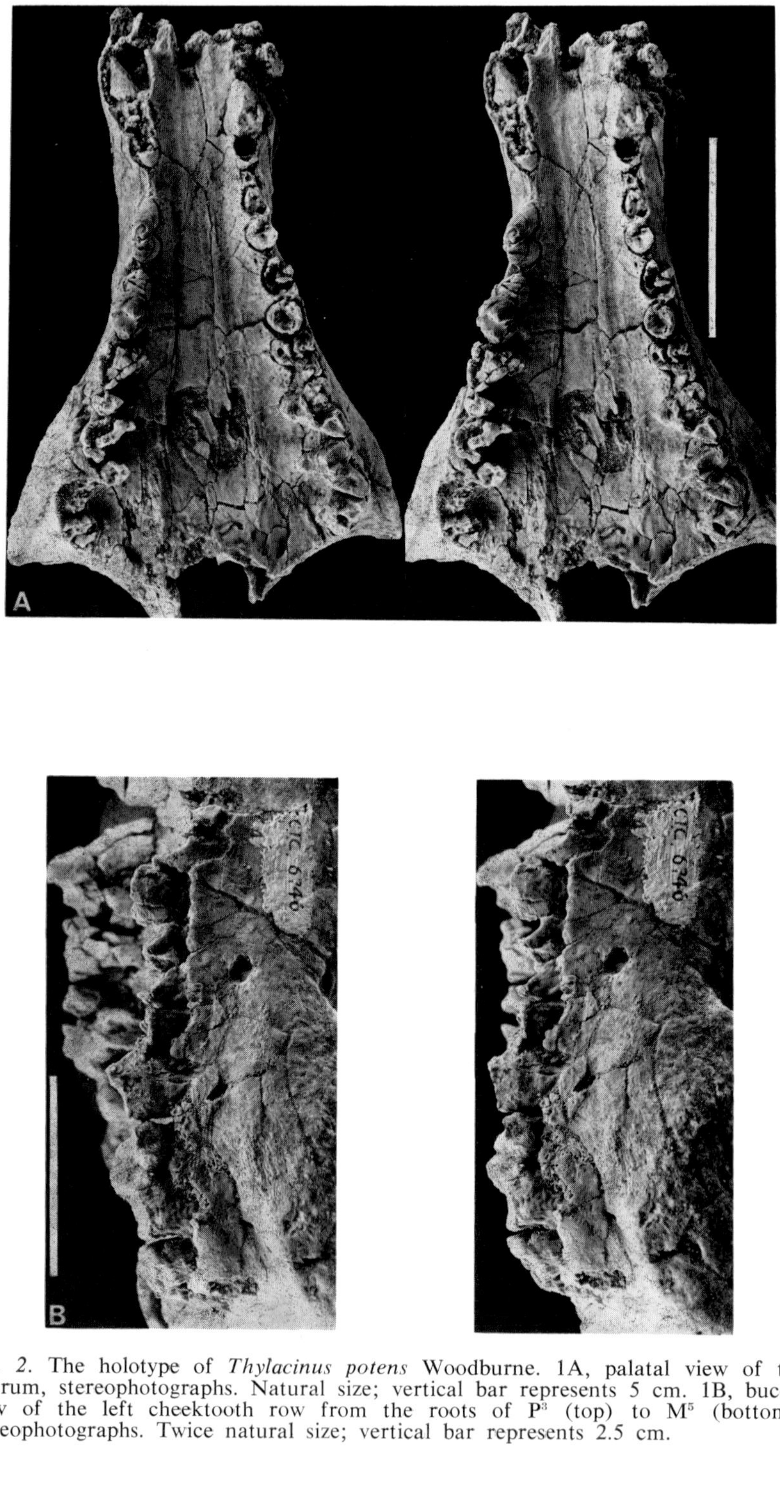

Fig. 2. The holotype of *Thylacinus potens* Woodburne. 1A, palatal view of the rostrum, stereophotographs. Natural size; vertical bar represents 5 cm. 1B, buccal view of the left cheektooth row from the roots of P^3 (top) to M^5 (bottom), stereophotographs. Twice natural size; vertical bar represents 2.5 cm.

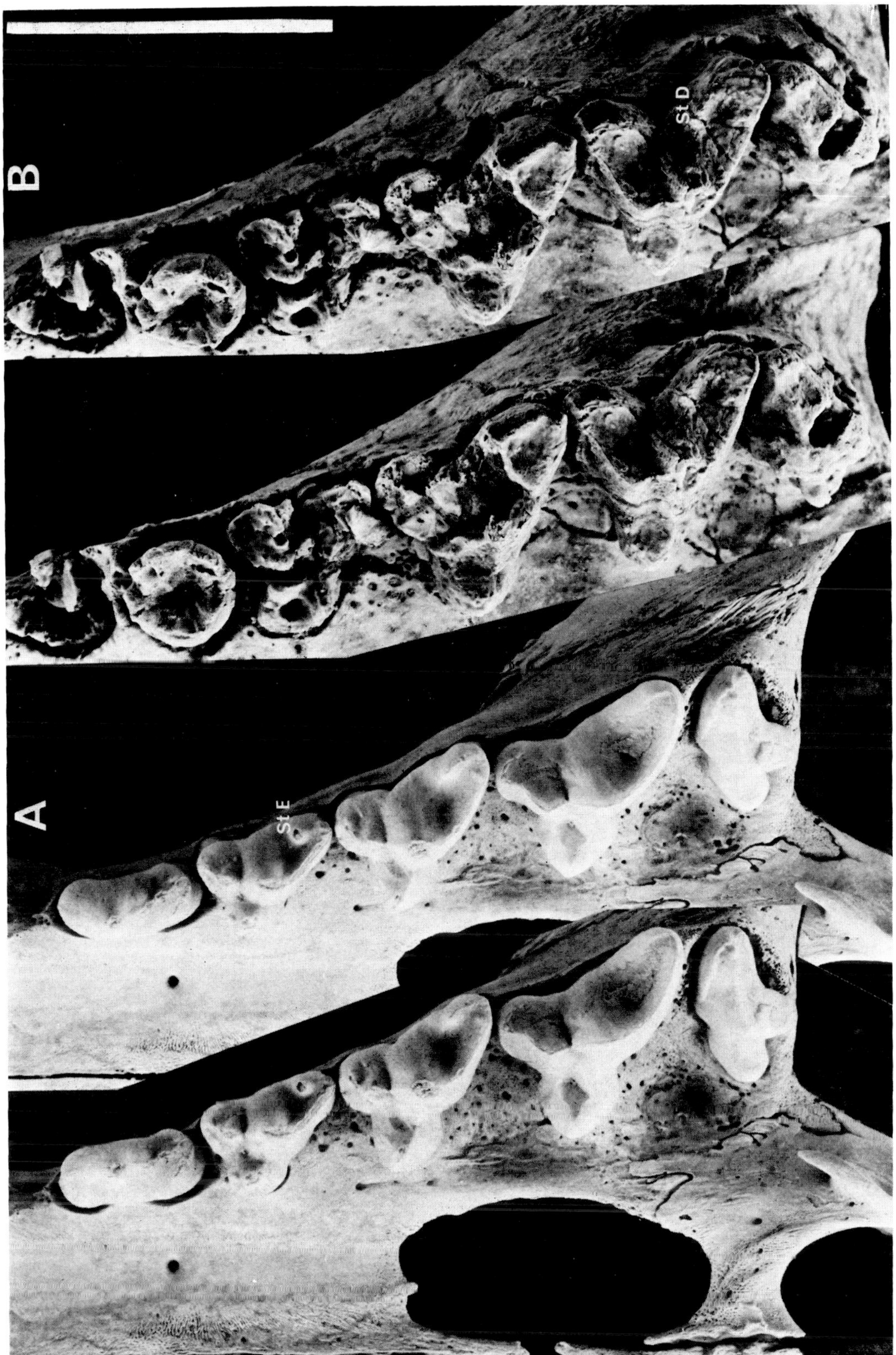

Fig. 3. Thylacinid molar morphology. A, *Thylacinus cynocephalus*, AM M217, left P^3–M^5. B, *Thylacinus potens*, holotype, roots for P^3–M^2, and M^{3-5}. Abbreviations: ST.E, stylar cusp E; ST.D, stylar cusp D. Twice natural size; vertical bar represents 2.5 cm.

Territory. Dentary fragments are also known (University of California, Berkeley, paleontological collections) but these are much less complete than the rostrum. A subsequent trip to Alcoota in 1975 (the Ray E. Lemley Expedition to Alcoota) recovered a few additional mandibular fragments and a small fragment of the premolar region of a maxilla of a second skull. These new specimens (Queensland Museum) add little to the concept of the species as understood by Woodburne (1967).

Age: The Alcoota local fauna, from the Waite Formation, is interpreted (M.O. Woodburne, pers. comm., and Woodburne *et al.,* in preparation) to be late Miocene in age (about 7 million years old). The fauna is dated on the basis of comparison with other dated faunas. On this basis, it is older than the early Pliocene local faunas, such as the Bluff Downs, Hamilton, Chinchilla and Bow faunas but younger than the suite of middle Miocene local faunas (dated at about 15 million years old) such as the Ngapakaldi, Pinpa, Tarkarooloo, and Ericmas faunas. Unfortunately there is as yet no means for absolutely dating the Alcoota local fauna.

The Kutjamarpu Thylacinid

An isolated premolar, recovered in 1971 from the Leaf Locality of the Wipijiri Formation, Lake Ngapakaldi, Etadunna Station, South Australia, appears to represent a thylacinid (Campbell, pers. comm.). The fauna from the Leaf Locality is known as the Kutjamarpu local fauna (Stirton, Tedford and Woodburne 1967) and is interpreted (Woodburne, pers. comm.) to be about 12 million years old. The Kutjamarpu thylacine is therefore the oldest known thylacinid.

Very little can be determined about this tooth that is phylogenetically significant. Several subsequent collecting trips to the Leaf Locality have failed to turn up more complete specimens.

CHARACTER ANALYSIS

In earlier reviews (Archer 1976a,b), I considered 62 basicranial characters and 45 dental characters. Marshall (1977 : 414) states that "The affinities of borhyaenoids, dasyuroids, didelphoids, and thylacinids are based on an array of characters which, although not originally stated [by Archer, I presume] as shared derived characters, were used as such to diagnose the various groups. These characters are . . ." and he lists only 14 characters. In the present analysis, I have again considered these 14 (one, the structure of the manus and pes, was not a character I previously used). In addition, I have reconsidered fifteen dental and basicranial characters. These additional characters include sixteen selected from among those I earlier considered because they appear to reveal relevant information about thylacinid relationships. Others not considered in detail here, but dealt with previously, have not been selected either because they involve autapomorphic states in each group, or because the character–states vary so widely within different families that their phylogenetic significance is obscure.

Dental Characters

1. *Incisor number*: All dasyuroids, thylacinids and some borhyaenids (prothylacynines, hathlyacynines) have 4/3. At least one didelphid (*contra* Marshall 1977) also has less than 5/4 (Paula Couto 1952 : 14). Early reports of higher marsupial numbers than 5/4 (Marshall 1977) have almost certainly been based on misinterpretation of ontogenetic data (Kirkpatrick 1978). Lower numbers than 4/3 are known in some borhyaenids and thylacosmilids. Reduction from the interpreted ancestral marsupial number of 5/4 has occurred in all of the marsupial superfamilies. It is not possible to determine if any of these reductions are synapomorphic or all achieved independently. It was concluded by Simpson (1941) to be, information–wise, a neutral character. I agree with Simpson and did not use it in my earlier analyses.

2. *Premolar number*: Marshall (1977) notes that all borhyaenids, didelphids, thylacinids and most dasyuroids have three premolars. Thylacosmilids and some dasyurids have two. The number of premolars has not been an important character in discussions of the phylogenetic relationships of thylacinids, borhyaenids, and dasyurids, but it is discussed by Marshall (1977) who concludes that it has no direct bearing on the problem of borhyaenid–thylacinid affinity.

3. *Lower premolar gradient*: The gross size (crown length and height) gradient of the lower premolars may be phyletically significant (Archer 1976a). In borhyaenoids, thylacinids, and most didelphoids, P_1 is smaller than P_2 which is smaller than P_3 (presumably the plesiomorphic state based on its wide taxonomic distribution and ubiquitous occurrence among late Cretaceous didelphids, pediomyids and stagodontids). Thylacosmilids represent an extreme development with loss of P_1 (Riggs 1934). In dasyurids P_3 is either smaller than or, less commonly, subequal in size to P_2. This character–state distribution suggests dasyurids are autapomorphic in this regard, although not uniquely so. One Cainozoic didelphid, *Zygolestes paranensis,* also exhibits reduction of P_3 (Reig 1957).

4. *The extent of development of M1 (=dP4 of Thomas 1888)*: In didelphoids, dasyuroids, and various other Australian groups, the extent of development of the M1 varies. In some species it is large and molariform (considered to be the plesiomorphic state), at least in non–Diprotodonta (see Clemens 1966; Archer 1976a, 1978), and in others it is reduced. In *Thyla-*

cinus cynocephalus it is minute. In borhyaenoids, a small M1 is known to occur in three species, all referable to the single subfamily Borhyaeninae (Marshall 1976b).

Marshall (1977 : 416) states that "Size reduction of the dP4 [=M1 of this paper] and its tendency toward suppression in dasyurids and thylacinids is regarded [by him] as a synapomorphy for these groups, and contrasts markedly with the molariform symplesiomorphic condition in didelphids and borhyaenids." In the first place, some dasyurids have a small but molariform M1, such as species of *Murexia* (Tate 1947) and *Antechinomys* (Archer 1976a), and these do not "contrast markedly" with those of didelphids and borhyaenids. The M1 in myrmecobiids, the sister–group of dasyurids, is not markedly reduced and is decidedly molariform, thereby strongly suggesting that this is the primitive dasyuroid condition. Variation in didelphids is also considerable. Although a little-reduced and molariform M1 is undoubtedly primitive among didelphids, in species of at least two distinct didelphoid groups (e.g. *Marmosa robinsoni* and *Dromiciops australis*), the M1 is reduced more than it is in those dasyurids cited above. Similarly, despite Marshall's (1977) contention that the M1 in borhyaenids is molariform and markedly better–developed than it is in dasyurids, it is not. Marshall (1976b : 753) describes M^1 in a specimen referable to *cf. Pharsophorus* sp. as a tooth that ". . . basically resembles a diminutive M^1 [$=M^2$ of this paper] although the proportions are considerably different." His description (p. 752) of an M_1 of *Borhyaena tuberata* as a tooth with a ". . . large central cusp and a smaller cuspule both anterior and posterior to it", suggests that it is almost premolariform. There is no mention of a talonid. Sinclair (1906 : 378, pl. 59 and fig. 6) illustrates the M_1 of a species of *Cladosictis*, and again it is a reduced, almost premolariform tooth. The lack of evidence for an M1 in other borhyaenid subfamilies, although negative evidence, is at least suggestive that caution should be exercised in generalizing, as Marshall does, about their presence in the family as a whole.

Polyphyletic reduction in size and molariformity of M1 in borhyaenids, didelphids, dasyurids and thylacinids seems the most reasonable conclusion. There is no evidence that reduction of M1, when it occurs in dasyurids and thylacinids, is a synapomorphy such as is asserted by Marshall (1977). Size and molariformity of the M1 therefore do not seem to be of any clear phyletic significance above the family level.

5–11. *Molar structure*: Bensley (1903), Sinclair (1906), Wood (1924), Simpson (1941) and Archer (1976a), among others, have noted the close similarity in molar morphology between thylacinids and borhyaenids. Among the six marsupial groups known to have developed a strictly carnivorous (as opposed to insectivorous) dentition, including the borhyaenoids, thylacinids, dasyurids (the lineage leading to species of *Sarcophilus*), didelphids (the lineage leading to species of *Sparassocynus*), thylacoleonids and stagodontids, only borhyaenids and thylacinids share such similar apomorphic molar morphology. The others are almost as structurally different from each other as they are from the borhyaenid–thylacinid–type of dentition.

Marshall (1977 : 420) states that "The supposed existence of this common hypothetical dental borhyaenid–thylacinid ancestor is, however, based totally on loss (group I) or reduction (group II) characters which represent, in part, functional complex (group IV) characters. Reduction or loss of the paracone, metaconid, and stylar shelf are all features correlated with carnassialization. The changes in these characters represent but a single functional complex . . . The changes must have occurred in unison in order to maintain mechanical efficiency during occlusion, and separation of the complex into many characters gives to each undue weighting . . . These identical changes have occurred in all cases involving carnassialization of dentitions in mammalian groups (Reig and Simpson, 1972 : 534)."

On all points, Marshall (1977) has oversimplified the case, or is simply in error. Consider the question of the so–called *single* functional complex. It is common to find many of these characters varying independently, such as: the species of the pediomyid genus *Pediomys*, where the stylar shelves are reduced, but not the metaconids or paracones (Clemens 1966); the stagodontid *Didelphodon vorax* where the stylar shelves and paracones are very well-developed but the metaconids are reduced; some specimens of *Dasyurus* which have reduced metaconids but unreduced stylar shelves and paracones (Archer 1976a); the didelphid *Hondadelphys* which has vestigial metaconids but large stylar shelves (Marshall 1976c); the didelphid *Sparassocynus bahiai* which has very reduced almost vestigial paracones but distinct metaconids (Reig and Simpson 1972); in *Coona pattersoni*, a didelphid, the paraconids rather than the metaconids are reduced. In these and other examples there is ample evidence to suggest that many of the features noted by Marshall are not immutably linked, and therefore there is good reason to consider them as individual characters. On the other hand there are also several functional complexes in molar morphology, such as protocone and talonid width.

Finally, concerning Marshall's (1977) suggestion that the carnassial specializations common to borhyaenids and thylacinids are based totally on loss or reduction characters, this is simply untrue. The enormous development of the postmetacristae, the paracristids, and the anteroposterior shift in orientation of the paracristids and postmetacristae are clearly not characters of this sort.

For unstated reasons of his own, Marshall (1977) ignored most of my earlier (1976a) character analyses of molar structure and discussed only two: metaconid development and stylar cusp development. These are considered below, as well as other molar characters that might be potentially significant in interpreting relationships.

5. *Metaconids*: All marsupial carnivores (including the very specialized species of *Sarcophilus, Didelphodon, Sparassocynus* and *Thylacoleo*) possess metaconids except thylacinids and most borhyaenids. Two of the oldest known (late Paleocene to early Eocene) borhyaenids, *Patene simpsoni* and *Arminiheringia auceta,* as well as some later forms, had small metaconids. At least one other early (early Eocene) borhyaenid, *Angelocabrerus daptes,* and most later forms lacked this cusp. Paulo Couto (1952 : 417) says of the late Paleocene ?*Arminiheringia* sp. that there is a ". . . minute, almost vestigial metaconid, the latter compressed against the base of the posterointernal edge of the protoconid." Therefore, extreme metaconid reduction was well underway by late Paleocene time in one of two known lineages, and complete metaconid loss had occurred in at least one borhyaenid (*Angelocabrerus daptes*) by early Eocene time. The oldest known thylacinid, the Miocene *Thylacinus potens,* shows no trace of metaconids.

Although Marshall's (1977 : 417) conclusion about relationships based on this character is really in the nature of an ancestor–descendant statement, it seems appropriate to briefly consider it here before going on to other characters. He says "The important point is that metaconid loss in *Thylacinus* and some borhyaenids represents, at best, a character–state acquired independently and in parallel. *Thylacinus* could not have been derived from a borhyaenid lacking a metaconid," and ". . . Apomorphic cranial characters in borhyaenids . . . place time constraints on a hypothetical borhyaenid–thylacinid divergence, indicating that such a divergence could not have occurred before the earliest Paleocene or latest Cretaceous." The cranial characters interpreted by Marshall to be apomorphic (presumably he means here autapomorphic in borhyaenids such as the nasal–lacrimal contact) appear to be incorrectly interpreted (see below). His conclusion is simply unwarranted. Australia, Antarctica and South America were evidently connected by at least island arcs until late middle Eocene time (e.g. Raven and Axelrod 1973, Deighton *et al.* 1976) and hence until after early Eocene time when borhyaenids lacking metaconids (e.g. *Angelocabrerus daptes*) had already developed.

6. *The stylar shelf*: Thylacinids, borhyaenids and pediomyids show reduction of the stylar shelves and/or stylar cusps. They are well developed in all didelphids and most dasyurids. *Thylacinus potens* has a slightly better-developed stylar shelf on M^4 than does *T. cynocephalus.* Woodburne (1967 : 28) notes that the M^4 (= his M^3) of both a paratype (UCMP 66971) and the holotype of *Thylacinus potens* ". . . have a suggestion of a small stylar cusp posterolabial to the metacone and posterior to the cleft between the parastylar and metastylar wings." A comparable stylar cusp is suggested in a similar position on the M^3 of *T. potens.* Posterior stylar cusps also occur on the $M^{2\text{-}3}$ (but evidently not M^4) of some modern and Quaternary thylacines (Archer 1971), but it is not at all clear to me that the stylar cusp on at least M^3 of *T. potens* and *T. cynocephalus* are homologous. In *T. cynocephalus,* this cusp occurs in a position suitable for stylar cusp E in dasyurids and some didelphids, close to the metastylar corner of the crown (Fig. 3A). However, in *T. potens,* the poorly–defined stylar cusp occurs in a more anterior position closer to the metacone, which is more in keeping with interpretation as a vestigial stylar cusp D (Fig. 3B). In *T. potens* there is then no cusp in the position of stylar cusp E.

There are several possible interpretations of this stylar cusp data in terms of intrageneric evolution. First, perhaps it is merely an expression of the variation in both species of *Thylacinus.* Although this is possible, no individual of *T. cynocephalus* known to me shows a remnant of a stylar cusp in the position of stylar cusp D. Second, perhaps the position of the stylar cusp has shifted with time, stylar cusp D of *T. potens* moving posteriorly in *T. cynocephalus* to the position normally occupied by stylar cusp E. This is possible and might have been correlated with the greater posterior development of the postmetacrista and metastylar corner of the crown in $M^{3\text{-}4}$ of *T. cynocephalus* (see below). Third, perhaps the stylar cusp differences indicate that *T. potens* and *T. cynocephalus* are representatives of collateral phylogenetic branches of the family tree rather than points along a single lineage. If the stylar cusp differences between the two species are significant, this is a more probable interpretation of the relationship of the two species than is a patrilineal hypothesis involving first loss of stylar cusp E in *T. potens* followed by its secondary redevelopment in *T. cynocephalus.* Following the hypothesis of collateral branches, one descendant (*T. potens*) of the common ancestor could have lost stylar cusp E, and the other (*T. cynocephalus*) lost stylar cusp D, both cusps being present in a common ancestor. With just these data, I cannot confidently decide whether an ancestor–descendant or collateral lineage hypothesis is more probable. However, in view of details of lower molar morphology (the talonids of *T. potens* are more reduced and the trigonids wider than those structures in *T. cynocephalus,* Woodburne 1967), a hypothesis of collateral lineages is apparently the most parsimonious.

Whatever the significance of the stylar cusps for determining the relationships of the two species, the presence of a stylar cusp in the position of D in *Thylacinus potens,* on the face of it, indicates a more plesiomorphic condition than that which characterizes *T. cynocephalus*. Most tribosphenic marsupials have a well–developed stylar cusp D on M^{2-3} and commonly on M^{4}. This is true for most dasyurids and many didelphids. Presence of a remnant of this cusp in *T. potens* strongly suggests derivation of thylacinids from a tribosphenic marsupial that a normal stylar cusp assemblage, rather than from an already specialized tribosphenic form. Although a plesiomorphic character state, this would appear to be a dental feature (the only one) that suggests *phylogenetic* relationship of thylacinids to some group *other* than borhyaenids, all of which, as far as I am aware, lacked stylar cusp D.

7. *The postmetacrista and paracristid*: In borhyaenids, thylacinids, thylacosmilids, and some dasyurids (*Sarcophilus harrisii*), the postmetacrista and paracristids are apomorphically proportionately enlarged. In forms I have seen, these two crests are correlated in development and form a functional complex. As Bensley (1903) points out, this is an adaptation to carnivority in these marsupial groups, as is the marked anteroposterior shift in orientation of these crests that also occurs in all of these groups. In the dasyurid *Sarcophilus harrisii,* this specialized condition can be seen as one end in a structural morphocline beginning with small dasyurids, then *Dasyurus hallucatus, D. maculatus, Glaucodon ballaratensis,* and *S. harrisii* (Archer 1976a). Similarly, the structure of the postmetacristae and paracristids in thylacosmilids and borhyaenids can be seen as structural end points in morphoclines starting with didelphids, such as species of *Alphadon* and *Marmosa,* then species of *Patene,* then other borhyaenids. The postmetacrista of M^{4} in *Thyalcinus potens* is slightly less elongate than it is in *T. cynocephalus,* although it is proportionately much more elongate in both species than in any dasyurid (except *Sarcophilus harrisii*) or didelphid.

Overall, with the exception of thylacinids, complete morphoclines in the structure of the paracristids and metacristae occur within each of the groups in which they are also markedly specialized. This suggests that the marked specialization of the crests in borhyaenoids and dasyurids is convergent. However, the same conclusion cannot be drawn for thylacinids. Their specialization is as likely to be a synapomorph with that found in one of the other two groups as it is to be a third independant development.

8. *Paracone height*: Paracone height in marsupials is plesiomorphically lower than metacone height (e.g. Slaughter 1971). But in some marsupial groups, the paracone has undergone further reduction. In notoryctids, the paracone appears to be entirely suppressed. Simpson (1928) documents interpreted phylogenetic changes in paracone height in fossil North American and European didelphids. Elsewhere (Archer 1976a) I have suggested that reduction of paracone height relative to metacone height in dasyurids has occurred in two ways: by absolute enlargement of the metacone such as has occurred in *Sarcophilus harrisii;* and by absolute reduction of the paracone which seems to have occurred in some of the smallest dasyurids such as *Sminthopsis ooldea* and species of *Planigale* and *Ningaui*.

Marked paracone reduction occurs in some dasyurids (noted above), most borhyaenoids (*Patene* spp. are exceptions and representatives of the forms Simpson regards to be structurally intermediate in terms of molar structure between didelphids and borhyaenids), and all thylacinids and sparassocynids. Structurally annectant taxa link these specialized dasyurids and borhyaenoids to less specialised taxa, suggesting that the apomorphic state of paracone reduction is convergent in these groups. However, no structurally annectant taxa are known within the Thylacinidae or Sparassocynidae that preclude the possibility that the apomorphic state in these groups could be a synapomorphy with one of the other groups.

9. *M^{4} preparacrista and postmetacrista*: Woodburne (1967) points out that the postmetacrista of M^{4} in *Thylacinus potens* is less well–developed than it is in *T. cynocephalus*. The molar outline in *T. potens* is thereby more symmetrical around a transverse axis, from the protocone through the paracone–metacone conjunction. He also notes that the ectoflexus (buccal cleft) is much better–developed in *T. potens* than it is in *T. cynocephalus*. He regards that both of these features indicate the more primitive condition of *T. potens* relative to the structure in dasyurids.

The M^{4} of *Thylacinus potens* (Fig. 3) differs from that in *T. cynocephalus* in three additional features. The first is the presence of what appears to be a vestigial stylar cusp D (discussed above). The second is a proportionately more elongate preparacrista. The third is the less obtuse internal angle formed by the preparacrista and the postmetacrista.

These upper molar features are most reasonably interpreted to indicate the relatively plesiomorphic structure of the M^1 of *T. potens* in comparison with that of *T. cynocephalus*. Woodburne (1967) suggests that this plesiomorphy reflects an ancestral affinity with dasyurids, and it is certainly the case that the plesiomorphic character–states for these characters can be found in various dasyurids. However, it is also the case that they are not confined to dasyurids. They variously occur in some didelphids, primitive borhyaenids, pediomyids and so on, because they are essentially primitive marsupial features.

These plesiomorphic M^1 features in *Thylacinus potens* are, however, relevant to discussions of thylacinid origins, not because they point unequivocably to dasyurids as thylacinid ancestors, but because they suggest that development of the specialized upper molar morphology seen in species of *Thylacinus* could have occurred relatively late, perhaps even as late as the early Miocene. This must be balanced against the evidence of lower molar morphology which, judging from Woodburne's (1967) figures and description, are as or more specialized (wider trigonids and more reduced talonids) than those of *T. cynocephalus*. They completely lack metaconids, metacristids and entoconids, have very reduced talonids, and have enlarged paracristids and protoconids. Because the lower molars are more specialized (i.e. in their loss of metaconids and metacristids) than the upper molars in *T. cynocephalus,* perhaps we would expect to see signs of greater plesiomorphy in the lower molars of *T. potens,* which we evidently do not. Considering just lower molar morphology, Woodburne's (1967 : 39) conclusion that the time of divergence of thylacinids from their non–thylacinid ancestor ". . . was surely no later than middle Tertiary and was probably much earlier" is supported.

Resolution of the differences in interpretation about the phyletic distinction of *Thylacinus potens* based on first upper and then lower molar morphology should not be based on a simple quantitative tallying of points. Clearly we have no idea how long it took to develop the specialised thylacinid lower molar morphology. However, if the age of *T. potens* is correctly interpreted (Woodburne pers. comm., see above), and *T. potens* shows all of the complex lower molar apomorphic character–states present in *T. cynocephalus,* it is reasonable to assume that phyletic isolation of thylacinids from their non–thylacinid ancestor (whatever it was) probably took place greater than 15 million years ago, or at any rate no later than middle Miocene time.

10. *Protocone and talonid widths*: The protocone occludes with the talonid and, in all forms known to me, protocone and talonid width are correlated. This appears to be an independent functional complex. In thylacinids, borhyaenoids and some dasyurids (*Sarcophilus harrisii*), the two features are markedly reduced. Paula Couto (1961) illustrates lower molars referred to the borhyaenid *Patene simpsoni* that have talonids structurally intermediate between those of many didelphids and other borhyaenids. Similarly, the talonid and protocone structure of some dasyurid species such as *Dasyurus maculatus* and *Glaucodon ballaratensis* are structurally intermediate between the less apomorphic states of other dasyurids and that seen in *Sarcophilus harrisii*. These again suggest that acquisitions of the apomorphic state in dasyurids and borhyaenids have been independent events. Again, this cannot be determined to be the case for thylacinids.

11. *The cristid obliqua and metacristid*: Ride (1964) points out that in the apomorphic dasyurid *Sarcophilus harrisii,* the metacristid preserves its plesiomorphic connection with the albeit reduced metaconid. He further suggests that this is not the case in thylacinids where the crest he interprets to be the homologue of the metacristid contacts the hypoconid. I previously (Archer 1976a) supported Ride's conclusion, but now have some doubts. It seems equally plausible that the very poorly–defined crest extending posteriorly from the protoconid to the hypoconid is in fact part of the plesiomorphic cristid obliqua, and that the metacristid has disappeared. In plesiomorphic didelphids and dasyurids there is often a minor crest on the posterior flank of the protoconid that contacts the anterior end of the cristid obliqua. I do not know what state this character has in borhyaenids and thylacosmilids because early illustrations are generally not given in enough detail. Certainly the metacristid is reduced in borhyaenids, but the nature and variation of this reduction is not clear.

Cranial Characters

12. *Palatal vacuities*: Most marsupials commonly have three sets of palatal vacuities or fenestrae adjacent to the palatal midline: the premaxillary vacuities formed between the premaxillary and maxillary bones; the maxillary vacuities, formed between the maxillary and palatine bones; and the palatine vacuities, formed within the palatine bones (Archer 1975). Marshall (1977:415) has incorrectly interpreted my statements in citing my work (Archer 1976a:28) as follows: "Maxillary fenestrae [palatine vacuities] are very small . . .". My intent *was* maxillary fenestrae (or as I now prefer, maxillary vacuities), *not* "palatine vacuities" as Marshall inserted.

Marshall (1977) regards "palatal vacuities" (I presume he means here only maxillary and palatine vacuities) to generally be a plesiomorphic character for the Marsupialia (the presumed apomorphic state being a solid palate), and lack of them in borhyaenoids to be one important reason for concluding that thylacinids could not have been derived from borhyaenoids. However, the matter is not quite so simple. Ontogenetically, marsupial posterior palatal vacuities, in at least marsupials so far examined (see Parker 1886:-270; Archer 1976b and 1981), develop secondarily and enlarge proportionately at the expense of the palate as the pouch–young develops. Fenestration is secondary, although widespread. Small adult maxillary vacuities are present in most didelphoids and most dasyuroids. Maxillary vacuities are lacking in borhyaenoids, sparassocynids, some didelphids (species of *Caluromys, Glironia, Caluromysiops* and *Thylatheridium*), and some dasyuroids (e.g., absent in *Myrmecobius fasciatus* and extremely tiny in species of *Planigale*). This provides an interesting contrast in conclusions about polarity based first only on application of the principle of using common occurrence to determine primitiveness (widespread fenestration) compared with an awareness of ontogenetic development (secondary fenestration). The apomorphic state then is for a fenestrated palate.

Although possible, it is not likely that palatal fenestration in all groups in which it occurs is a synapomorphic state. This is because morphoclines in this character exist in most groups. In thylacinids for example, the most plesiomorphic species, *Thylacinus potens*, has maxillary vacuities that are markedly smaller than those of other species. Similarly, among peramelids, the most plesiomorphic taxa (species of *Microperoryctes, Peroryctes* and *Rhynchomeles*) have very small palatal vacuities while highly apomorphic species (e.g. *Chaeropus ecaudatus*) have large palatal vacuities. Further, I have previously (Archer 1976b) suggested that there appears to be a correlation, at least in Australian marsupials, between degree of palatal fenestration and aridity. As a generality, taxa with the largest palatal vacuities occur in the most arid environments. This could be interpreted to suggest that there has been convergent evolution of large palatal vacuities in Australian families in response to the deterioration of the Australian climate, a process that began in the Miocene.

Lack of palatal fenestration in the borhyaenids and several other groups may therefore be an indicator of their common occupation of non–arid habitats rather than a phyletically significant apomorphy.

Another possibility in regard to lack of maxillary vacuities in some groups is that this could result from secondary closure of an evacuated palate, perhaps in response to climatic amelioration. Simpson (1947:3) for example clearly assumes that borhyaenids descended from didelphids that had maxillary vacuities. "As far as known, definitely referable borhyaenids have no vacuities, although it cannot be doubted that they were derived from forms with vacuities and vacuities may have occurred in their most primitive forms". The ignorance about the structure and possible similarity of the palates in truely primitive early borhyaenids and thylacinids remains. Simpson (1947:6) also notes that "The small, single pair of vacuities in *Eobrasilia* [an early Eocene marsupial considered by Simpson (1947) to probably be a didelphid, but one structurally intermediate in several characters between primitive borhyaenids and didelphids] is, indeed, as near the specialized borhyaenid condition (no vacuities) as the typical didelphid condition (large vacuities, and often two pairs). Again, this region is not known in truly primitive early borhyaenids, which might be expected to resemble *Eobrasilia* here." Remarkably, Marshall (1977:70) independently concludes of *Eobrasilia* that "The small single pair of palatal vacuities is as near the specialized borhyaenid condition (no vacuities) as the typical didelphid condition (large vacuities, and often two pairs). Neither of these characters are known in primitive borhyaenids which might be expected to resemble *Eobrasilia*". (This statement in itself appears to be a most remarkable case of convergence).

If early borhyaenids resembled *Eobrasilia* in the state of this character as Simpson suggests, there would be no significant difference between this early borhyaenid condition and that of *Thylacinus potens.*

13. *Nasal–lacrimal contact*: Marshall (1977) claims incorrectly that the nasal–lacrimal contact is a unique character–state in borhyaenids. It is also present in some vombatids (wombats), wynyardiids (Spencer 1901), and phalangerids (cuscuses and brush–tailed possums). A nasal–lacrimal–frontal–maxillary point contact is evidently present in at least one didelphid (Simpson 1947) and is almost achieved in a number of other diprotodont groups. A frontal–maxillary contact occurs in other marsupials. Marshall (1977:415) therefore illogically contends that "Because this condition (nasal–lacrimal contact) is unknown in any other marsupial group it represents an important apomorphy for the family Borhyaenidae."

This character–state is clearly not unique to borhyaenids although it is probable that it has developed independently in some vombatids, wynyardiids, phalangerids and borhyaenids. But this also cautions that it is possible that it could have arisen by parallel evolution more than once within the Borhyaenidae. Some borhyaenids did not have a distinct and typically broad contact. Simpson (1947:3) states that "In *Arminiherringia,* however, naso–lacrimal contact is short. This part of the face is not known in other very early or in the most primitive borhyaenids. They probably were near the didelphid condition in this region, and it is distinctly possible that primitive borhyaenids here resembled *Eobrasilia*" where the nasal, lacrimal, frontal and maxillary bones evidently contact at a point. Further, Simpson (1947:3) states that "*Borhyaena* from the middle Miocene typically has a well-established naso–lacrimal contact, but in Princeton No. 15120 this is probably no longer than in *Eobrasilia*". It is therefore variably developed within at least some of the later borhyaenids.

Marshall (1977) notes that thylacosmilids, the "saber–toothed" borhyaenoid sister–group of the borhyaenids, lacked a nasal–lacrimal contact, and not surprisingly, interprets this lack as secondary loss (if loss of this sate is acceptable to Marshall here, why not in ancestral thylacinids?). It seems equally probable that thylacosmilids and borhyaenids have descended from borhyaenoids that lacked a nasal–lacrimal contact, Marshall (1976) notes that the oldest known thylacosmilids (Pliocene) were already highly specialized and indicate that the group had a long as yet unknown history.

In cladistic analysis, morphotype construction would indicate that the common borhyaenoid ancestor of the Borhyaenidae and Thylacosmilidae lacked a nasal–lacrimal contact (assuming of course that Marshall is correct in assuming they are sister–groups of each other), a conclusion that would be unavoidable when outgroup comparison is made with the primitive condition in didelphoids, the presumed sister–group of the borhyaenoids.

There is yet a further problem here about Marshall's (1977) methodology and use of Hecht (1976) and Hecht and Edwards (1976) weighting catagories. Marshall (1977 : 415) interprets the nasal–lacrimal contact in borhyaenids to be a group V character, a most significant indicator of relationship (or in this case lack of relationship to thylacinids). As noted above, this type of character assessment seems to me to be subject to interpretation, which is precisely what Hecht and Edwards (1976 : 655) say it should not be: "The weight or the importance of a character is not an arbitrary process dependent on the whim of the investigator." Just how significant is apomorphic acquisition of a contact between two bones that were plesiomorphically separated by two others? Ontogenetically, all that may be required is for growth of the frontal and/or maxillary to cease or slow relative to the constant growth of either the frontal or the lacrimal, which would then contact. In other words, this could be thought of as a loss or reduction character–state of the lowest (group I or II) value in interpreting relationship (which according to Hecht 1976, would be realistically regarded as 1,000 – 10,000 times less significant than a group V character–state).

In some dasyurids (Archer 1976c) and some phascolarctids (Ride 1957), it is evident that contact between alternative and opposite bones in the skull can vary within species. Obviously according to Simpson (1947) it is even variable within some borhyaenids. This does not mean that this type of character is useless in phylogenetic interpretation, only that use of high weight for this type of character in phylogenetic analysis is probably unwarranted.

Nasal–lacrimal contact in borhyaenids clearly does not prohibit a sister–group relationship between borhyaenids and any other marsupial group. Neither does it prohibit borhyaenoids from being the stem group for thylacinids.

14. *Transverse canal*: Marshall (1977 : 415) states that "Didelphoids, dasyuroids, thylacinids, and most other marsupial groups have large transverse canals perforating the basisphenoid" and suggests that this is a plesiomorphic character state for marsupials. Transverse canals are absent in some borhyaenids. In fact, loss of transverse canals has occurred, presumably independently, in dasyurids (in some it is variably present) (Archer 1975, 1976b,c), several didelphids (e.g. species of *Caluromys, Philander,* some *Lutreolina, Hondadelphys* and others), and in the species of five of the six borhyaenoid genera that are represented by suitable basicranial material (there are thirty–one borhyaenoid genera known). It can only be concluded that the presence of a transverse canal in thylacinids (the condition is only known in the modern species), and its absence in some borhyaenids (again, the condition in any of the pre-late Oligocene borhyaenids is not known), is of little phyletic significance. Reduction or loss of this canal has evidently occurred independently several times.

15. *Foramen ovale*: The mandibular branch of the trigeminal nerve exists from the basicranium via a foramen pseudovale developed between the periotic and the alisphenoid in all dasyuroids. In all thylacinids, borhyaenoids (except species of *Borhyaena*) and most didelphoids, this nerve leaves the endocranium via a true foramen ovale which penetrates the alisphenoid. Marshall (1977 : 416) concludes that "Presence of a foramen pseudovale in dasyuroids is regarded as a group I or II character, but this has no direct bearing on the problem of borhyaenid–thylacinid affinity." In regard to cladistic analysis, I agree, assuming that the presence of a foramen ovale is a plesiomorphic character–state (MacIntyre 1967). It should also be noted here that a very small foramen, topographically analogous to the foramen pseudovale of dasyuroids, occurs in almost all groups of carnivorous marsupials (Archer 1976b) including many of those that also possess a foramen ovale.

Although perhaps uninformative cladistically, the ubiquitous loss of the foramen ovale in dasyuroids does appear to have bearing on certain ancestor–descendent hypotheses (see below).

16. *Position of the infraorbital canal in Thylacinus potens*: Woodburne (1967) suggests that the infraorbital canal opened out onto the face in *Thylacinus potens* farther anteriad with respect to the dention that it does in the modern *T. cynocephalus.* I suggested (Archer 1976: 28) that this might indicate that the orbit could be relatively farther forward in *T. potens* thereby approaching the borhyaenid condition. Alternatively, it might indicate that the infraorbital foramen opened onto the external surface of the maxilla without making contact with the jugal, a condition that also occurs in at least some borhyaenids but not *Thylacinus cynocephalus.* Marshall (1977 : 416) suggests that uncertainty about these character–states makes them useless for phylogenetic analysis. After reconsideration, I agree. However, it should be pointed out that whichever condition is actually the case, it distinguishes *T. potens* from *T. cynocephalus* but not all borhyaenids.

The plesiomorphic condition would be for the infraorbital foramen to open through the side of the maxilla without contact with the jugal, as it does in all didelphids, dasyurids and borhyaenids that I have examined. In this regard. *Thylacinus cynocephalus,* but perhaps not *T. potens,* may be autapomorphic in its state of jugal–infraorbital foramen contact.

17. *Posterolateral palatal foramen*: Woodburne (1967) notes that *Thylacinus potens* has a posterolateral palatal foramen (Fig. 2A). It is well–developed and completely enclosed by the thick posterior edge of the palatine. Didelphoids, myrmecobiids, perameloids and at least some borhyaenoids also have a closed posterolateral palatal foramen. In sparassocynids, dasyurids and the modern *Thylacinus cynocephalus,* it is incomplete posteriorly. Marshall (1977 : 416) cites Simpson (1941 : 4) who notes that the foramen is absent in *T. cynocephalus.* From this, and my statement (Archer 1976a : 28) that the foramen is incomplete in *T. cynocephalus,* he concludes it is a highly variable character and of low reliability. My statement and Simpson's amount to the same thing — there is no complete posterolateral foramen in *T. cynocephalus.* It is not, as Marshall contends, a highly variable character in the modern species. Although his consequent dismissal of this character as phylogenetically significant is unwarranted, it is clear that because presence of a complete posterolateral palatal foramen is the plesiomorphic condition in marsupials, presence of this feature in two marsupial groups is not an indication of their common descent. However, presence of this plesiomorphic character–state is important in the matter of determining the character–state of the non-thylacinid ancestor of the thylacinids. This will be discussed below in connection with the significance of certain similarly plesiomorphic features of the thylacinid basicranium.

18. *The squamosal epitympanic sinus*: Marshall (1977 : 417) states that ". . . a well developed squamosal epitympanic sinus, which is covered by the pars flaccida of the tympanic membrane, is present in all known Australian marsupials . . . while it is absent in all known American forms (Van der Klaauw 1931)." I have concluded previously (Archer 1976b) that several American and Australian marsupial groups contain taxa with poorly–developed squamosal epitympanic sinuses. Several Australian forms completely lack them. For example, among didelphids I have examined, *Monodelphis domestica* has a well–developed posterolateral squamosal flange forming a structurally simple squamosal epitympanic sinus; *Lutreolina crassicaudata* also has a shallow depression developed in the squamosal which is homologous with this sinus; *Caluromys* sp. has a similar but even better–developed squamosal depression; specimens of species of *Metacheirus* and *Chironectes* have comparable simple squamosal epitympanic sinuses. *Borhyaena excavata* has a distinct squamosal epitympanic sinus although it is not a simple depression such as occurs in most other groups. Other borhyaenids that I have examined (only three other genera) appear to lack such a sinus. Some peramelids such as *Peroryctes longicauda* and *Rhynchomeles prattorum,* which are also regarded as plesiomorphic in general among perameloids, lack a squamosal sinus, depression, or posterolateral flange. In this regard, although they are an Australian group, they show less development of this sinus than do the didelphids just noted. Among dasyurids, species of *Murexia,* regarded by Tate (1947) and Archer (1976a) to be overall plesiomorphic, have very poorly–developed squamosal epitympanic sinuses, only just better–developed than those of some of the didelphids noted above. There is a common misconception that dasyurids all have large epitympanic and hypotympanic sinuses. Enlarged sinuses do occur in some taxa such as species of *Dasyuroides* and *Dasycercus,* but these extreme developments occur in taxa that are otherwise apomorphic dentally and that are adapted to the relatively recently developed arid habitats in central Australia.

The significance of this is that in Australian groups, the taxa regarded to be overall the structurally most plesiomorphic, also lack or at best have poorly–developed squamosal epitympanic sinuses. Some didelphids and at least one borhyaenid (very few are known from skulls) also have poorly–developed squamosal epitympanic sinuses. I conclude from this that there is no evidence to support Marshall's contention that the occurrence of these sinuses is ". . . a synapomorphy not only for dasyurids and thylacinids, but for all Australian marsupials" (Marshall 1977 : 417). The evidence in fact suggests that acquisition of this sinus by Australian and perhaps other American groups was by inheritance from didelphoids in which it was developed to much the same degree as it is in several living didelphids. Subsequent enlargement of this sinus in thylacinids and those dasyuroids in which it is well–developed, could have been the result of parallel evolution in response to the development of late Cainozoic aridity in Australia. The complete absence of this sinus in otherwise structurally primitive perameloids suggests that the appearance of this sinus in arid–adapted perameloids may even have been the result of convergent rather than parallel evolution, in the sense of Hecht and Edwards (1976).

Marshall (1977) considers that this "synapomorphy" warrants a group V weighting value. But because his conclusion that it is a synapomorphy is unjustifiable, so is his conclusion that it is phyletically very significant.

19. *Epitympanic recess*: Only thylacinids, some didelphids and some borhyaenids are known to have a sulcus for a vessel that passes around the posterior wall of the epitympanic recess (Archer 1976b). Such a sulcus is absent in dasyuroids. The condition in other South American groups is unknown. It is not clear what the plesiomorphic state of this character is.

20. *Periotic hypotympanic sinus*: The floor of the middle ear, in marsupials that have enclosed this area with bone, is variously composed of the alisphenoid, petrosal and/or mastoid parts of the periotic, squamosal, paraoccipital and/or even the ectotympanic (Archer 1976b). Only microbiotheriids, dasyuroids, some peramelids (none of the primitive forms), some didelphids (e.g. species of *Caluromys*), and possibly notoryctids develop a tympanic wing from the petrosal part of the periotic. This floors specifically the periotic hypotympanic sinus. It is not ubiquitously well–developed in any group except the Dasyuridae. It is entirely absent in all borhyaenids, thylacosmilids, thylacinids, sparassocynids, caenolestoids and the vast majority of didelphids. From this distribution I would conclude that lack of development is primitive. Further, because this feature is virtually absent in myrmecobiids while it is ubiquitously present in dasyurids, development of the apomorphic state in dasyuroids probably occurred independently and was not a feature of ancestral dasyuroids. In this aspect of middle ear structure, primitive dasyuroids probably resembled the majority of living didelphids (except species of *Caluromys*).

21. *Mastoid epitympanic sinus*: In some groups, the mastoid part of the periotic contributes a tympanic wing that floors a mastoid epitympanic sinus. This wing is well–developed only in some dasyuroids but is sometimes present in rudimentary form in didelphids and peramelids. The borhyaenids are unique (Archer 1976b) in the apparent (certain in species of *Borhyaena*) suppression of the mastoid contribution to the posterior surface of the skull (see below), but in regard to their lack of mastoid tympanic wing, they are evidently simply plesiomorphic. Although in myrmecobiids the mastoid tympanic wing directly contacts the alisphenoid tympanic wing, an autapomorphic condition (see below), this does not preclude the possibility that development of the mastoid tympanic wing in dasyurids and myrmecobiids is a synapomorphic feature.

22. *Mastoid contribution to the rear of the skull*: In all carnivorous marsupials (except at least some borhyaenoids) known to me, the mastoid portion of the periotic bone contributes to the posterior external surface of the skull. In at least some borhyaenids, the mastoid apparently fails to contribute (Archer 1976b). This failure may be a feature common to all borhyaenoids (although very few are presently known from skulls) and if so, autapomorphic for the group.

23. *Mastoid–alisphenoid tympanic wing contact*: Only myrmecobiids among marsupial carnivores develop a contact between the alisphenoid and mastoid tympanic wings. This is an autapomorphic condition and results from the failure of myrmecobiids to develop an intervening periotic tympanic wing, such as occurs in all dasyurids (Archer and Kirsch 1978).

24. *Structure of the stylomastoid foramen*: The stylomastoid foramen conducts the facial nerve through the ventral portion of the periotic. In some didelphids (e.g. *Marmosa, Monodelphis* and *Didelphis* spp.), perameloids, at least some borhyaenids, thylacinids and possibly myrmecobiids, this foramen lacks an anterior bony boundary. In other didelphids (e.g. *Caluromys* sp.) it is entire, being enclosed anteriorly by a process of the squamosal. In most dasyurids (e.g. *Dasyurus* spp.), this foramen is also complete anteriorly, but the primary anterior bony contribution comes from the periotic, the squamosal forming a bar that secondarily overlaps the periotic. In other dasyurids (e.g. some species of *Sminthopsis* and *Planigale*), the squamosal contribution is minimal. *Sarcophilus harrisii* is autapomorphic in its elaboration of a periotic for passage of the facial nerve.

Presumably the character state of an incomplete anterior rim is plesiomorphic because several groups have developed an anterior rim using different bones.

25. *Sulcus or canal connecting the facial nerve canal with the postglenoid canal*: This sulcus is present in some didelphids, borhyaenids (those in which the condition can be examined), and thylacinids. In didelphids this linking canal is entirely enclosed and passes through the periotic bone. In borhyaenids and thylacinids it is an open sulcus. No sulcus appears to link the postglenoid and facial nerve canals in dasyuroids or other carnivorous marsupials examined (Archer 1976b). It is not clear from this distribution which state of character is plesiomorphic.

26. *Internal jugular canal*: The internal jugular canal in dasyurids carries the internal jugular vein (Archer 1976b), not a branch of the internal carotid as Patterson (1965) and Marshall (1978) imply is the widespread condition in marsupials. Development of this canal differs in various marsupial families. It is short and poorly–enclosed dorsally in didelphids and, evidently, borhyaenids. It is similarly short but enclosed for a short distance in thylacinids. In dasyuroids it is also well–enclosed but very long, extending forward to the basisphenoid (Archer 1976b). A short open canal is evidently the plesiomorphic condition, dasyuroids being the most apomorphic in this regard.

27. *Ectotympanic shape*: Familial variations in this bone in marsupial carnivores have been described elsewhere (Archer 1976b). The ectotympanic of didelphids and juvenile thylacinids are comparable in being simple and uncomplicated, the condition interpreted here to be plesiomorphic. The condition in dasyuroids is a function of degree of auditory sinus development. In sparassocynids, it is modified in a way comparable to some apomorphic dasyurids. In some borhyaenids (*Cladosictis* spp.) and adult thylacinids, the ectotympanic appears highly modified albeit in dissimilar ways. Nevertheless, the borhyaenid, adult thylacinid and dasyurid conditions can be seen as simple independent modifications of the plesiomorphic condition such as is seen among didelphids.

POSTCRANIAL CHARACTERS

28. *Epipubic bones*: All marsupials have well–developed epipubic bones except thylacinids and the borhyaenoids. In the thylacinids they are cartilaginous and vestigial. Ossified epipubic bones are lost in at least one borhyaenid (*Cladosictis lustratus*) and not known to occur in any other (Sinclair 1906). There can be no doubt that presence of these bones is plesiomorphic because they are present in mammal–like reptiles, prototherians and even some eutherians (Kielan–Jawarowska 1975). It has even been suggested that the os penis of eutherians is the homologue of these bones, in part because it ossifies from two bilateral centres. Marshall (1977) concludes that reduction (or loss) of ossified epipubic bones in thylacinids and borhyaenids is a loss character (weighting group I) and hence provides minimal information about relationship. I think his conclusion is much too simplistic. We again have the problem of application of the weighting scheme of Hecht (1976) and Hecht and Edwards (1976). It may be strictly correct to relegate reduction of these bones to a low weight category following the tenets of Hecht and Edwards. But loss of these bones must indicate that their function has been changed or replaced by development of some other structure(s). It is not a case of simple loss. Muscle attachments must be changed, abdominal support has been rearranged, ontogentic changes must occur in development of all associated structures, perhaps even the pouch or pouch–area support system has undergone change. The mere fact that epipubic bones have been so rigorously maintained (despite the possibility of loss as demonstrated by thylacinids and borhyaenids) in all other marsupials, including many larger and smaller forms, must indicate that thylacinids and borhyaenids made some significant change(s) in abdominal anatomy. Our osseus signal of such change is this loss or gross reduction of the epipubic bones.

It should be possible to determiine at least some of the acquisition character–states accompanying reduction of the epipubic bones in thylacinids. Once these are known, corresponding checks on borhyaenids might reveal osseus alterations that fall on the ecquisition side of Hecht's weighting scheme. In the meantime I would suggest that it is simplistic to conclude, as Marshall (1977) does, that among marsupials this highly unique apomorphic reduction of the epipubic bones in these two groups is an unimportant character–state.

30. *The number of digits on the hind foot*: In *Thylacinus cynocephalus* and at least some borhyaenids (e.g., *Borhyaena* spp and *Prothylacinus patagonicus*) the hallux is lost. Presence of the hallux is plesiomorphic in marsupials. Reduction of the hallux has also occurred (evidently independently) in the argyrolagoids (e.g., *Argyrolagus* spp.), in the dasyurids (e.g., *Antechinomys laniger*, *Dasyuroides byrnei* and some species of *Dasyurus*), in the peramelids (e.g., *Chaeropus castonotis*) and the macropodoids (all except *Hypsiprymnodon moschatus*). For these reasons, I have not previously used this character–state as an indicator of relationship between thylacinids and borhyaenids.

31. *Morphology of the tarsal bones*: Tarsal bones have been recently studied by Szalay (1982) who concludes that thylacinids share derived tarsal character–states with Australian marsupials, but not borhyaenoids. Szalay consid-

ers that this tarsal morphology is not the result of convergence and hence is more important as an indicator of phylogeny than is data from other character systems such as teeth and sera. He does not, however, discuss which of the features of the tarsal bones are functionally independent or demonstrably different characters.

Molecular Characters

Sarich *et al.* (1982) obtained albumin from dried Thylacine (*Thylacinus cynocephalus*) skin and muscle in the Australian Museum mammal collection. They found, using microcompliment fixation and radioimmunoassay, that Thylacine albumin was indistinguishable from that of a species of *Dasyurus* and *Dasyuroides byrnei,* but clearly different from the albumin of *Myrmecobius fasciatus.*

They also calculated the immunological distance between these and other marsupials based on the albumin data. This distance is interpreted to be approximately proportional to the time elapsed since the existance of the most recent common ancestor. One unit (on a scale of 0 to 180 units, the latter representing the greatest distance yet measured between mammalian albumins) is hypothesized to represent the equivalent of 600,000 years since separation. On this basis, they conclude that the Thylacine shared a common ancestor with the species of *Dasyuroides* and *Dasyurus* about 7 million years ago and a common ancestor with peramelids about 48 million years ago.

Their conclusion about sister-group hypotheses (they only considered as alternatives the thylacinid–dasyurid and thylacinid–borhyaenid sister–group hypotheses) is that the thylacinid–dasyurid sister–group hypothesis is supported. Although they could not examine immunological data for either borhyaenids or pediomyids, the immunological proximity of the thylacinids to dasyurids appeared to them to preclude support for the borhyaenid–thylacinid sister–group hypothesis.

There are a few problems that need examination here. First, there are (see above and below) evidently thylacines considerably older than the 7 million year old date determined by the immunological data to have been the time of the most recent common ancestor of thylacines and species of *Dasyurus* and *Dasyuroides.* Further, the Alcoota thylacine, itself determined to be about 7 million years old, is such a well–developed thylacinid, that its describer regarded it to be only distinct from modern species at the specific level (Woodburne 1967). This is not to cast doubt on the branching sequence implied by the albumin data, just doubt on the assumption about the relation between real time and immunological distance time. The geological record does not provide a precise age for the known Tertiary thylacines, but I would be amazed if thylacines did in fact descend from a common ancestor shared with *Dasyuroides–Dasyurus*–type dasyurids within the last 7 million years. It would involve the assumption that the middle Miocene Kutjamarpu animal discussed above was not a thylacine as well as the assumption that the age of the undoubted Alcoota thylacine has been incorrectly interpreted such that it was about five rather than eight million years old.

Of greater curiosity is the implication that, if the thylacinid lineage became isolated only 7 million years ago from the lineage leading to species of *Dasyurus* and *Dasyuroides,* the *Dasyurus* and *Dasyuroides* species lineage must have developed in the same short space of time. The common ancestor of those two groups (the thylacinids and the dasyurids) must have been plesiomorphic to both. And herein lies a fundamental problem. In terms of dentition, a plesiomorphic common ancestor would have to resemble species of *Murexia,* or possibly some of the primitive species of *Antechinus.* These are among the most primitive known living dasyurids (Archer 1982). It would also have had a didelphid–like premolar gradient in contrast to all known dasyurids. In terms of cranial features, a plesiomorphic common ancestor would have had a didelphid–like palate with a complete posterolateral palatal foramen and didelphid-like basicranium with a foramen ovale, character states also unknown in any dasyurid *all* of which exhibit synapomorphic states for these characters. In fact, *Mymecobius fasciatus* exhibits the synapomorphic character–state of loss of the foramen ovale, although the albumin immunological distance between the Thylacine and *M. fasciatus* is interpreted to represent 24 million years.

There are two alternative ways of resolving these problems, neither of which is very satisfactory. The first is to conclude that all of the Thylacine's plesiomorphic morphological character–states are the result of a reversal in evolution having recently developed form a dasyurid with the dasyurid synapomorphic states. The second alternative is to conclude that either the immunological distance itself is incorrect or that the assumption about the nature of the relationship between this distance and real time is incorrect. I cannot judge the probability of this second alternative but it must be considered.

SUMMARY OF CLADISTIC INTERPRETATION OF THE MORPHOLOGICAL CHARACTERS

Table 1 summarizes the distribution of apomorphic (but not autapomorphic) states for the families discussed above. Using the cladistic in-

Table 1. Apomorphic (but not autapomorphic) morphological character–states[1] for thylacinids, borhyaenids, thylacosmilids, dasyurids, myrmecobiids, pediomyids and didelphids. Characters are as follows: 1, Upper incisor number reduced from 5; 2, P_3 smaller or just equal in size to P_2; 3, M1 reduced in size and molariformity; 4, Metaconids very reduced; 5, Stylar shelf and/or cusps very reduced; 6, Postmetacristae and paracristids markedly enlarged; 7, Paracones markedly reduced; 8, Protocones and talonids very reduced; 9, Metacristids very reduced in length in caranassial teeth; 10, Maxillary palatal vacuities present; 11, Transverse canal extremely reduced or absent; 12, Trigeminal nerve exits from cranium via a foramen pseudovale; 13, Posterolateral palatal foramen incomplete posteriorly; 14, Periotic hypotympanic sinus developed; 15, Mastoid part of periotic provides a tympanic wing to floor part of the middle ear; 16, Epipubic (marsupial) bones reduced to cartilage or lost; 17–18, Foot structure (*fide* Szalay).

Apomorphic states of morphological characters	Thylacinidae	Borhyaenidae	Thylacosmilidae	Pediomyidae	Myrmecobiidae	Dasyuridae	Didelphidae	"Ancestor"
1.	A	A	A	?P	A	A	P	P
2.	P	P	P	P	P	A	P	P
3.	A	A	?A	P	?P	A	P	P
4.	A	A	A	P	?P	P	P	P
5.	P	A	A	A	?P	P	P	P
6.	A	A	A	P	?P	P	P	P
7.	A	A	A	P	?P	P	P	P
8.	A	A	A	P	A	P	P	P
9.	A	A	A	P	P	P	P	P
10.	A	P	P	?P	P	A	A	P
11.	P	A	A	?P	P	P	P	P
12.	P	P	P	?P	A	A	P	P
13.	P	P	?P	?P	A	A	P	P
14.	P	P	P	?P	A	A	P	P
15.	P	P	P	?P	A	A	P	P
16.	A	A	?A	?P	P	P	P	P
17.	A	P	?P	P	?A	A	P	P
18.	A	P	?P	P	?A	A	P	P

[1] The condition of some characters has not been determined by direct observation for some groups, generally because that part of the anatomy is not yet known, is obscure in terms of homology, or else inadequately described. In these cases, "?P" or "?A" indicate the state hypothesized (plesiomorphic or apomorphic) for the purposes of the Wagner analysis. These hypotheses are guesses based on what little data there is, or else on the assumption of general plesiomorphy in otherwise essentially plesiomorphic taxa. In this tabulation the structurally annectant and extinct *Patene* spp. (between didelphids and borhyaenids), *Ankotarinja* sp. (between didelphids and dasyurids), and *Eobrasilia* sp. (between didelphids and borhyaenids) have not been considered because of their doubtful familial position and because they are known from only very incomplete material. They are discussed in the text. Autapomorphic character–states for individual families have not been included. These are discussed in the text.

terpretations in an unweighted manner, the number of derived (but not necessarily synapomorphic) character–states present in both thylacinids and borhyaenids is eight. The number in common to thylacinids and dasyurids is possibly as many as six. The number in common to Thylacinids and the didelphids is one. The six apomorphic states in common to Thylacinids and dasyurids include: No. 1, reduced incisor number (which thylacinids also share with borhyaenoids and several other groups); no. 3, the reduced size of M1 (also shared with several other groups including borhyaenoids, and in any case evidently not apomorphic character of dasyuroids because the M1 is not markedly reduced in myrmecobiids); no. 10, maxillary palatal vacuities (which may or may not be apomorphic in thylacinids, see discussion above); nos. 17–18, structure of the tarsal bones (Szalay 1982), although the actual number of functionally independent characters here is unclear; and albumin morphology. I can see no compelling reason for regarding any of the first three to be indicative of a hypothetical Thylacinidae–Dasyuridae (or–oidea) group because they are states also exhibited by groups in most other families. Only nos. 17–18 and the albumin results appear to be phylogenetically significant. Of the eight apomorphic states in common to the Thylacinids and borhyaenids, five (nos. 1, 3, 7 and 8) are the same low level of reliability because they are widespread among other families. The remaining four (nos. 4, 6, 9 and 16) are apomorphs shared only by thylacinids and borhyaenoids. Further, no. 16, reduced or absent marsupial bones, is a unique apomorphic character–state among all marsupials being shared only by thylacinids and borhyaenids (and possibly thylacosmilids, although pelvic remains for thylacosmilids are unknown).

Figure 4 illustrates one of the potential cladograms supported by the analysis of morphological characters for the families of marsupials considered in this study, and the distribution of apomorphic character–states. Twelve dichotomously branching cladograms involving the Didelphidae, Borhyaenidae, Thylacinidae and Dasyuridae are shown in Fig. 5. Pediomyids are not considered here because, as can be seen from Table 1, there are evidently no potential synapomorphies linking pediomyids to thylacinids (loss of stylar cusps in thylacinids is interpreted to be a convergent feature — see the character analysis). Of these, only cladograms 5C, D, and E are supported by unique synapomorphies for each branch (Table 1). The remaining nine are unsupported by this or any other character analysis and most have not been the focus of controversy. In all of the cladograms illustrated, I have made the assumption (supported by the character analysis) that of the four families considered here, the most plesiomorphic is the Didelphidae.

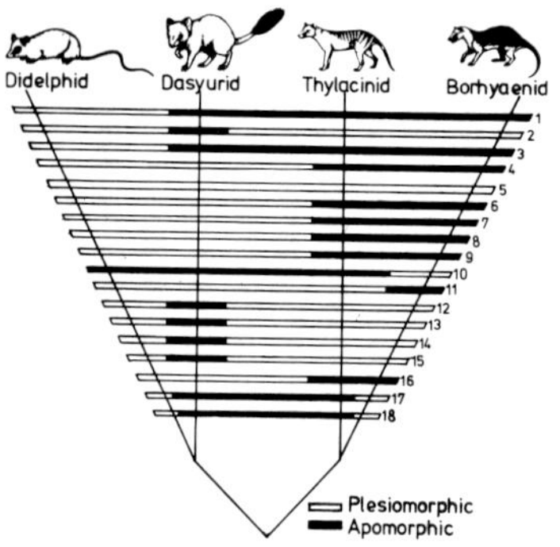

Fig. 4. One of several plausible cladograms supported by the distribution of character–states for morphological characters 1–18 (see Fig. 5 and Table 1).

Marshall (1977) considers five of these possible cladograms (those shown here in Fig. 5 as 5C, 5D, 5E, 5H and 5J), and considers that his interpretation of his data lends most support to either Fig. 5C or 5J (his renditions of these phyletic arrangements are reconstructed here such that the more plesiomorphic branches are to the left).

Marshall (1977) considers that the supposed dasyurid–thylacinid synapomorphies supporting 5J include a reduced M1 and an enlarged squamosal epitympanic sinus. Accordingly he interprets as convergence (or parallelism) the similar molar morphology (which involves more than only the two characters mentioned by Marshall, see above), and reduction or loss of the epipubic bones in thylacinids and borhyaenids. I have given reasons above for why I regard his *reasons* for these conclusions to be unacceptable. This cladogram (5J) also involves a plesiomorphic group (didelphid–borhyaenid). By his own acknowledgement (his fig. 3), there are no synapomorphic characters recognized that unite his hypothetical borhyaenid–didelphid group. In passing, it should be noted that most of the supposed autapomorphic character–states listed by Marshall, although unimportant in cladistic analysis, are also unacceptably interpreted by him to be plesiomorphic to all members of the groups hypothesized in Fig. 5J (see above).

Cladogram 5C is supported by Marshall (1977) and is one of the three supported by synapomorphies in this study (along with 5D and 5E). Of these three, the most support, on the basis of characters 1–18, occurs for 5E and the least for 5C.

The molecular data (not analyzed cladistically), could support any one of these three hypothese. Although Sarich *et al.* (1982) conclude that their data is support for a thylacinid–dasyurid synapomorphic group (i.e. they would support either Fig. 5C or 5D), they cannot, using molecular data alone, exclude Fig. 5E.

So on the face of it, cladistic analysis of morphological characters 1–18, and the molecular data, provide support for cladograms 5C, 5D, and 5E.

WAGNER ANALYSIS OF TABLE 1

As another means of clustering taxa and portraying conclusions about distribution of apomorphic states for the morphological characters given in Table 1, I used a program written by R. H. Crozier (University of New South Wales) called *PHYLOG* that uses the algorithm proposed by Farris (1970) for constructing Wagner networks. A hypothetical "Ancestor" was added to the taxonomic groups and attributed only plesiomorphic character states in order to provide a root for the Wagner trees. Out of a run of 100 trees, 85 had the shortest maximum matrix distance, 25. Of these, there were only two essentially different trees (Fig. 6A–B). Several general points can be made about these rooted Wagner trees. First, Thylacinidae consistantly groups with the Borhyaenoidea and, as a combined group, these were always at least four units from any HTU ("hypothetical taxonomic unit") occurring at the common base of two branches leading to OTUs ("operational taxonomic units", i.e. the taxa and "Ancestor" actually coded) over the shortest total matrix distance of 25. Second, Thylacinidae never appeared closer than eight (8–15) units to the common ancestral HTU for the Dasyuroidea but was never more than five units from the Borhyaenoidea. Third, the Dasyuroidea always appeared as the sister-group of a combined Thylacinidae–Borhyaenoidea sister–group. Fourth, the Pediomyidae never

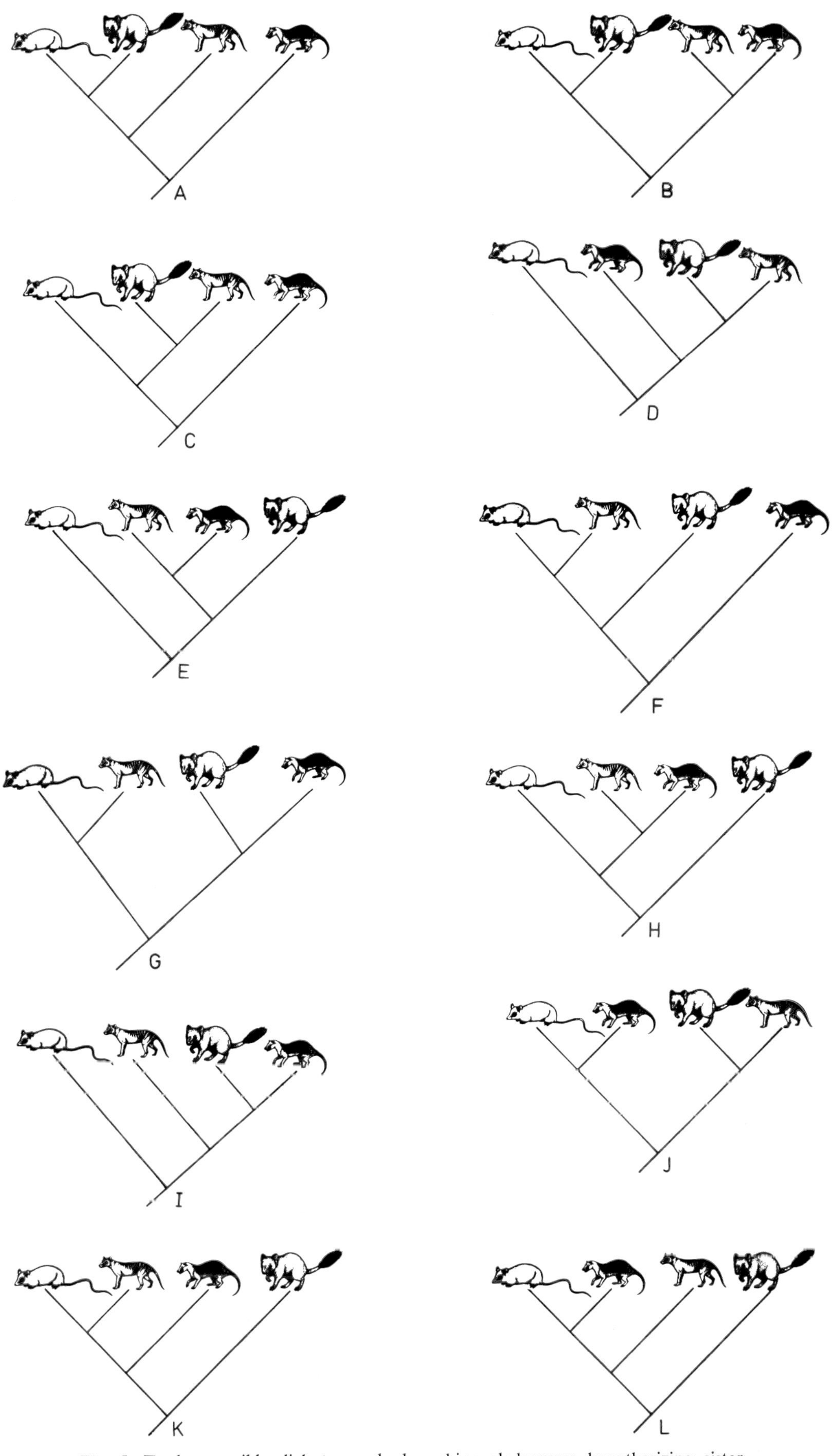

Fig. 5. Twelve possible dichotomously branching cladograms hypothesizing sister–group relationships of the families Didelphidae, Borhyaenidae, Thylacinidae and Dasyuridae.

grouped with the dasyuroids, thylacinids or borhyaenoids. Instead, along with the didelphids, it grouped either in a basic trichotomy, or fell out as the sister–group of all other descendents of "Ancestor".

The groupings provided by this Wagner analysis come as no real surprise in view of the morphological character–state distributions given in Table 1, and the character analyses discussed above. However, the Wagner trees do provide an interesting quantification of this data. Undoubtedly, if autapomorphic character–states were considered, the inter–taxa distances would be considerably different, but the differences would always be increases. For example, the Myrmecobiidae would probably extend much farther from "Ancestor" but remain most closely associated to the Dasyuridae. Similarly, although the Thylacosmilidae and Borhyaenidae coded identically, if autapomorphic states (e.g. the nasal–lacrimal contact) were involved, the Borhyaenidae would appear at a distance from the Thylacosmilidae. Although these distances are interesting as a measure of phyletic divergence rather than just branching sequences, it is impossible for all autapomorphic states to be recognized in taxa such as these. It is too tempting to recognize only the obvious autapomorphic states rather than the enormous number of others that form the total uniqueness of each group (e.g. body odour, ear shape, colour of the nasal membrane, and so on *ad infinitum*). Admittedly, I have also used distinctive synapomorphic states in my analysis and undoubtedly ignored many others of which I am blissfully unaware. Nevertheless, adding a second set of incomplete data in the form of autapomorphic states would do little to clarify the overall picture and nothing to clarify the branching sequence of the Wagner trees.

Again, we should note here that molecular data, although support for clustering Thylacinidae and Dasyuridae, cannot provide intrinsic data to falsify either of the two basic Wagner trees resulting from analysis of the morphological data.

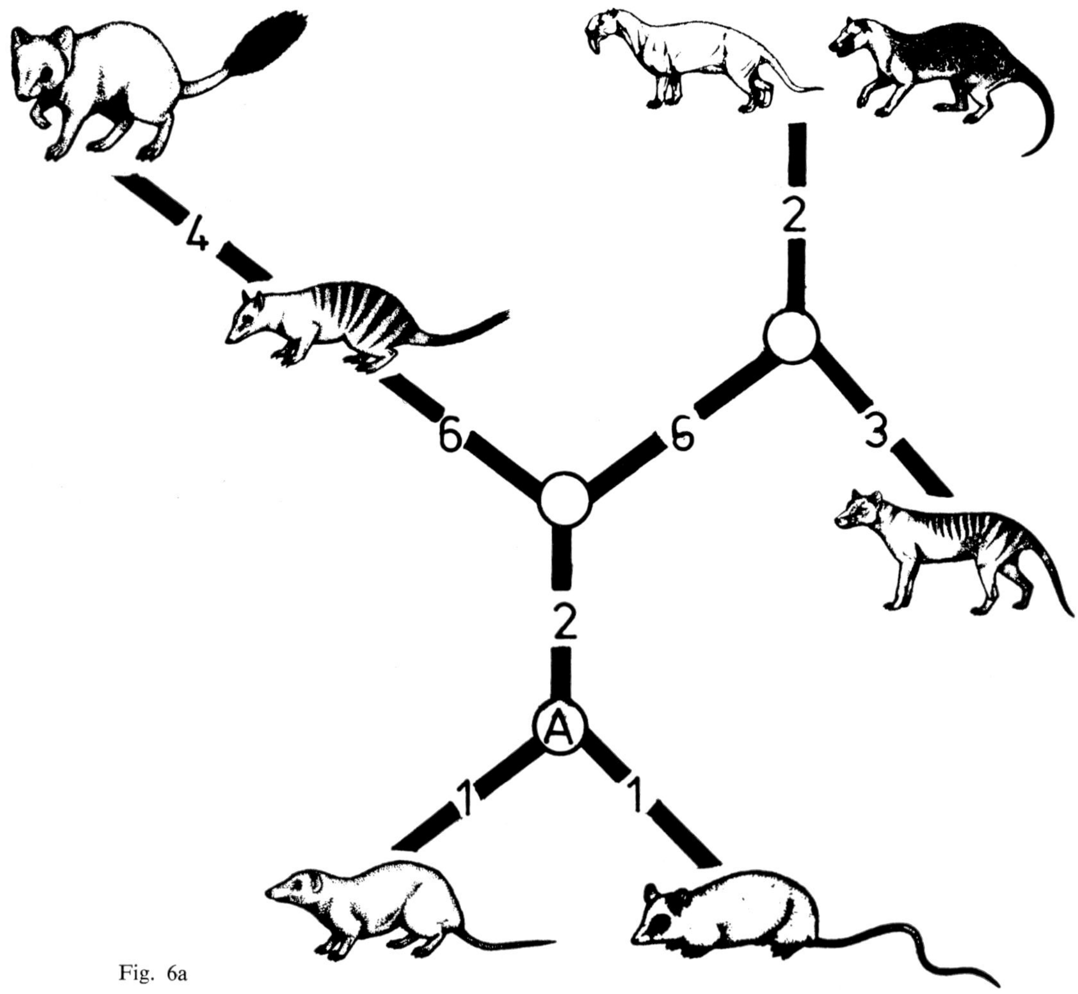

Fig. 6a

Fig. 6. The two essentially different types of equal–shortest trees produced by the Wagner analysis of the data summarized in Table 1. Open circles are hypothetical taxonomic units. Matrix distances are given as numbers between the taxa. A hypothetical ancestor, coded for all plesiomorphic states, is represented by "A".

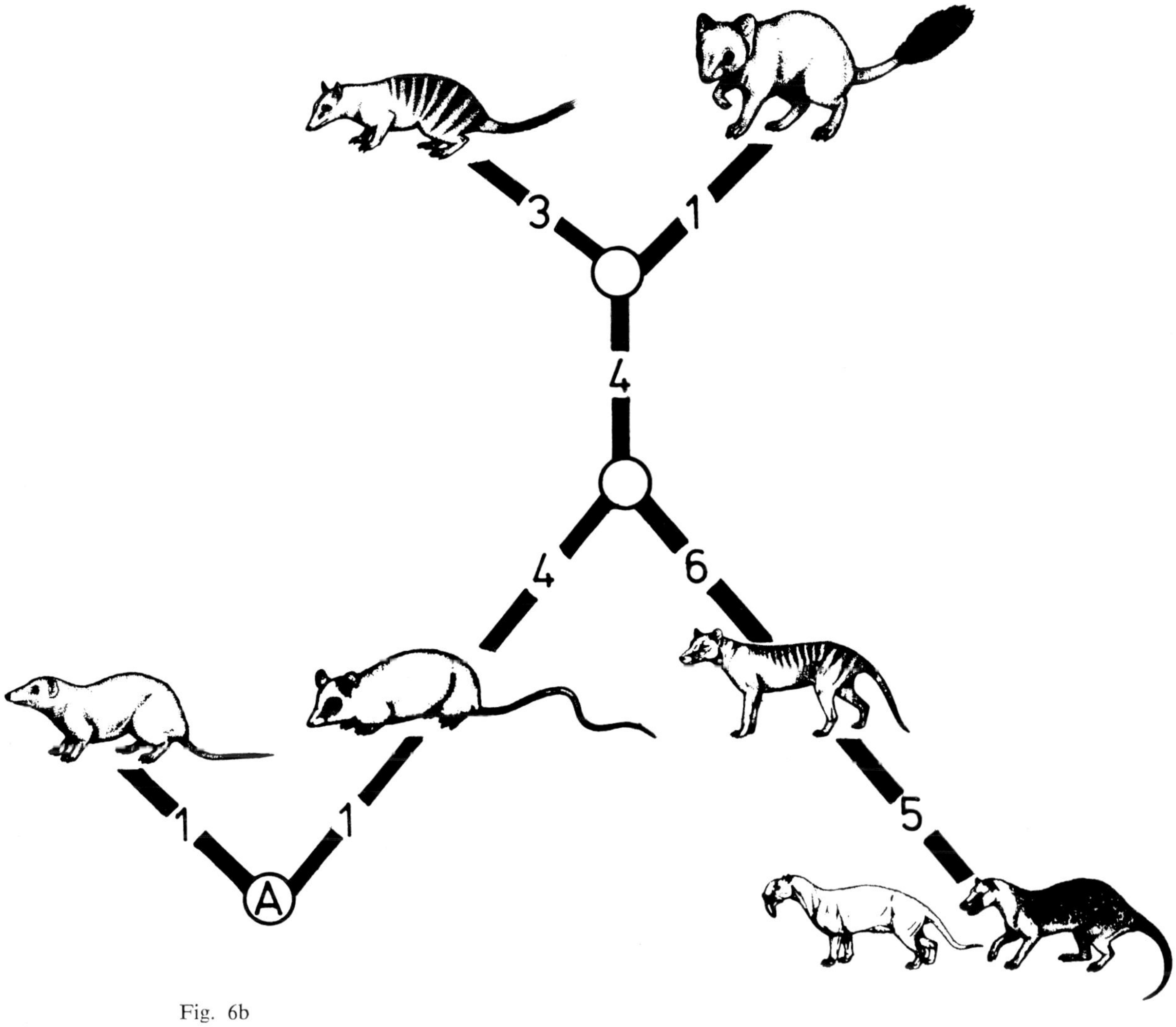

Fig. 6b

DISCUSSION OF THE CLADISTIC INTERPRETATION OF RELATIONSHIP BASED ON MORPHOLOGICAL DATA

The conclusions of Marshall (1977) about the cladistic relationships of the Borhyaenoidea, Thylacinidae, Dasyuroidea (Marshall used this taxon concept at the family level, regarding Myrmecobiidae to be a subfamily within the Dasyuridae) and Didelphidae cannot be supported on the bases he proposed. Even following his own character analyses, Marshall concludes (p. 442–3) that "If an unweighted system is used, then it would be logical and in the interest of parsimony to conclude that thylacinids are more closely related to borhyaenids . . . than to dasyurids . . .". After critical re-examination of the morphological characters used by Marshall (1977), there are found to be no cranial or dental morphological synapomorphies shared exclusively by dasyurids and thylacinids. However, there are tarsal characters (Szalay 1982) that appear to unite these groups, at least within the broader context of the whole of Australian marsupials. There is also one unique character–state shared by thylacinids and borhyaenids (reduced epipubics). Considering morphological character–state synapomorphies as a whole, the number shared by the Thylacinidae and Borhyaenidae is greater than the number shared by the Thylacinidae and Dasyuridae, and many more than the number that might serve to ally either the Borhyaenoidea or Thylacinidae to any of the other marsupial groups considered in this analysis.

If we were to apply the weighting scheme of Hecht and Edwards (1976) to this analysis, as Marshall (1977) did to his, the "weight" of evidence could (depending on how and why the weights were applied) reduce to insignificance the probability that the Thylacinidae is the sister–group of the Dasyuridae. This has not been done here precisely because application of this weighting scheme by Marshall serves to show the scheme's most important weakness: its potential arbitrariness and susceptability to use in supporting apriorisms. Having shown that there are important practical problems in applying this weighting scheme without recourse to apriorism, what are the alternatives? The most obvious would be a parsimonious assessment of relationship using a cladistic analysis of characters whose transformation states have been reasonably demonstrated to be independent of those of other characters. Wiley (1975: 236) suggests that the testing of phylogenetic hypotheses " . . . must be done under the rules of parsimony, not because nature is parsimonious, but because only parsimonious hypotheses can be defended by the investigator without resorting to authoritariarism or apiorism". With just the morphological information available, it could be concluded that the most parsimonious cladistic hypothesis would be one that suggests the Borhyaenoidea and Thylacinidae are sister–groups of each other.

If we ignore the most parsimonious conclusion based on the analysis of the gross morphology, and conclude instead that the Thylacinidae is the sister–group of the Dasyuroidea, we must simultaneously assume a remarkable amount of convergent (or parallel) evolution in order to account for the otherwise striking resemblance of thylacinids to borhyaenids. And it cannot be contended that most of the characters involved in this convergence (or parallelism) are varying concordantly (see above) because all can be shown to vary independently within and outside the Marsupialia. Lovetrop (1975: 514) suggests that "The probability of a phylogenetic classification is inversely proportional to the number of cases of convergence it implies . . .". Highly complex characters or character–states, by virtue of their complexity, are not likely to have arisen more than once. Even Marshall (1976c: 416) recognizes the improbability of hypotheses involving marked convergence or parallelism) when considering the affinities of *Hondadelphys*: "A third possibility is that the characters shared in common by these species are the result of convergent evolution; but considering the large number of characters involved this seems unlikely". And yet, comparable examples of remarkable convergence are known such as the similarity of marsupial moles (notoryctids) and placental moles (e.g. chrysoclorids). For this reason, the thylacinid–dasyuroid sister–group hypothesis should not be abandoned on the basis of parisimony alone.

CONCLUSIONS

A: The Cladistic Relationships of Thylacinids

Taking into consideration the data from all tissue systems analysed at the time, I previously concluded (Archer 1976a) that a special thylacinid–borhyaenoid relationship was the most probable of all those considered. One of the least probable seemed to be a close thylacinid–dasyurid relationship. Now albumin and tarsal morphology have provided additional data and it is time to reconsider the hypothesis. A summary of all evidence presented above supporting various sister–group hypotheses is presented in Table 2.

Cranial morphology provides no clear synapomorphs in support of any of the four sister–group hypotheses involving thylacinids. However, gross cranial morphology suggests (Archer 1976b) that the thylacinid basicranium is most similar to that of didelphids in that it lacks dasyurid and borhyaenid autapomorphic states and, as a result, is essentially plesiomorphic.

Postcranial apomorphic character–states supporting the borhyaenid–thylacinid hypothesis include the gross reduction of the epipubic bones and loss of the hallux, although the latter is considered here of dubious value.

Tarsal character–states supporting the thylacinid–dasyurid hypothesis are those discussed by Szalay (1982). He has not distinguished how many of the apomorphic features represent independently evolving characters which is the reason for my uncertainty about the number of apomorphic states involved. Szalay is also mainly concerned with demonstrating that thylacinid tarsal bones resemble those of other Australian

Table 2. A summary of support for four sister–group (SG) hypotheses involving thylacinids (T). The number indicates the number of apomorphic (and potentially synapomorphic) character–states involved. A question mark indicates that data based on the particular tissue system cannot be tested for the group involved.

SG Hypothesis	T–dasyurid	T–borhyaenid	T–didelphid	T–pediomyid
1. Cranial morphology	1	0	1	?
2. Postcranial (non–tarsal) morphology	0	2	0	?
3. Tarsal morphology	2?	0	0	0?
4. Dental morphology	2	7	0	0
5. Albumin	1	?	0	?

groups rather than those of borhyaenids. Hence he has not noted specific reasons for supporting a thylacinid–dasyurid sister–group hypothesis.

Dental morphology provides one of the main data sets supporting the thylacinid–borhyaenid hypothesis, with seven apomorphic character–states. The two apomorphic dental character–states that support a thylacinid–dasyurid hypothesis also support a thylacinid–borhyaenid hypothesis. There are therefore no unique dental character–states that support the thylacinid–dasyurid sister–group hypothesis. Despite this seemingly strong support for a borhyaenid–thylacinid sister–group hypothesis, I have no illusions (persistant ones anyway) about the immunity of dental characters to convergence. Here I just want to record the apparent evidence of the dental morphology.

Albumin immunological distance provides support only for the thylacinid–dasyurid sister–group hypothesis although it cannot at present be used to test sister–group hypotheses involving borhyaenids or pediomyids.

Taken on a simple quantitative basis, without weighting the data from one character system over another, we would have to conclude that a thylacinid–borhyaenid sister–group hypothesis was at least as probable as a thylacinid–dasyurid sister–group hypothesis. However, phylogenetic debate is rarely resolved simply by counting shared apomorphic states. Dental morphology is known to have converged in other groups (e.g. the marsupial and placental moles noted above) and convergence is possible here. Of course there is no basis for assuming that dental convergence has in fact occurred between thylacinids and borhyaenids unless the thylacinid–borhyaenid sister–group hypothesis contradicts another arising from analysis of a more reliable character system. The problem is to rank the various character systems according to their ability to discriminate between convergence and relationship. Of the other non–dental character systems examined, the albumin data does seem (e.g. Maxon, Sarich and Wilson 1975) to be a reliable indicator of phyletic relationship. In groups so far studied, albumin appears to have evolved divergently at a reasonably constant rate, despite seemingly gross departures in the rates of regulatory evolution (Wilson 1976), and convergence has evidently not been detected in the albumins of clearly unrelated groups.

In the present instance, there must still be some doubt about aspects of the interpretation (Sarich *et al.* 1982) of the relationships of thylacinids on the basis of just the albumin data. These reasons for doubt are given above. As a result, I hope comparable analyses of other enzymes from the same museum Thylacine skins will be tested to see if the results support the conclusions based on dried albumin. Perhaps the effects on reactivity of using dried as opposed to fresh albumin should also be checked. Although this would seem like an obvious thing to do, Sarich *et al.* (1982) evidently did not consider it necessary.

Because tarsal morphology also contradicts a thylacinid–borhyaenid sister–group hypothesis, I am now convinced that the thylacinid–borhyaenid dental similarity, although marked, is probably convergent. So also must be the common reduction of epipubic bones and loss of the hallux. This reversal of my 1976 opinion is based not on a simple recounting of potential synapomorphic character–states, but on the conclusion that immunological data are probably more reliable indicators of phylogenetic history than others involved in this particular debate. Because tarsal morphology supports the albumin immunological data, I regard both of these character systems to probably be more reliable indicators of relationships in these groups than dental morphology. If, on the other hand, tarsal morphology were the only character system whose data refuted a thylacinid–borhyaenid sister–group, I would not abandon the thylacinid–borhyaenid sister–group hypothesis in favour of some other quite so readily. I am not convinced that tarsal morphology is as demonstrably free of convergence as albumins. The conflict of the phylogenetic relationships of the South American opossum *Dromiciops australis* highlights the need for caution. Tarsal morphology suggests (Szalay 1982) that *D. australis* is actually a part of the Australian marsupial radiation. Serological reactivity, on the other hand, suggests (Kirsch 1982) that *D. australis* is part of the American marsupial radiation and (although a microbiotheriid) closest to didelphids. On the face of it, this appears to be a rather fundamentally different view and at least makes me cautious about unhesitatingly accepting Szalay's assumption that the tarsal morphology is not as likely as dental morphology to be the result of convergence.

Having concluded that it is probable that thylacinids are most closely related to dasyurids, it is appropriate to consider what the detailed nature of this cladistic relationship is. That is, are thylacines the sister–group of a particular subfamilial group of dasyurids, of dasyurids as a whole, or of dasyuroids (i.e. dasyurids and myrmecobiids). The albumin data suggests that the Thylacine groups closely with *Dasyurus* and *Dasyuroides* and Sarich *et al.* (1982) suggest that ". . . as the thylacine is closer to *Dasyurus* and *Dasyuroides* than it is to *Myrmecobius* it must be included in the Dasyuridae." This is of course logical only if *Myrmecobius fasciatus* is regarded to be a dasyurid. Elsewhere (Archer and Kirsch 1977) the case has been made for regarding *M. fasciatus* to be the sole representative of a distinct family, the Myrmecobiidae, within the superfamily Dasyuroidea. Nevertheless, preserving the spirit of the message of Sarich *et al.*, they are suggesting that the level of phylogenetic distinction of the Thylacine, based on its albumin reactivity, is very close to the two dasyurids with which it was compared.

The second view, that thylacinids represent the sister–group of dasyurids (not including myrmecobiids) as a whole, is similar to the first if *Myrmecobius fasciatus* is regarded to represent a distinct family. The albumin data could be seen as consistent with this view because it has not yet been demonstrated that Thylacine albumin is closer to that of species of *Dasyurus* and *Dasyuroides* than it is to that of some other dasyurid. Many modern workers regard Thylacines to be distinct at the family level (e.g. Kirsch 1977).

Both of these views about the relationships of Thylacines appear to be inconsistent with the morphological data. Thylacines have plesiomorphic character–states that both myrmecobiids and dasyurids have lost (Archer 1976a,b). These include the P_3 which is much larger than P_2, and retention of a foramen ovale (see above). Similarly, thylacinids primitively had a complete posterolateral palatal foramen whereas dasyurids apomorphically have lost this structure. For these to be retained plesiomorphic features in thylacinids, the whole radiation of the synapomorphic myrmecobiids and dasyurids must have occurred subsequent to the 7 million years postulated on the basis of the albumin data to have been the date of the thylacine–dasyurid dichotomy. This is possible, but I think enormously improbable. Dasyurids of a modern sort are known from the fossil record at least as early as middle Miocene time (e.g. the Riversleigh dasyurid). Further, myrmecobiids, which are considered on the basis of the albumin data to have shared a common ancestor with thylacinids 24 million years ago, have the dasyurid apomorphic character–states noted above. If these characters are synapomorphs in myrmecobiids and dasyurids, and if thylacinids are correctly interpreted to be plesiomorphic in these features, thylacinids must have branched off from the common dasyurid–myrmecobiid ancestral stock prior to acquisition of the synapomorhpic states

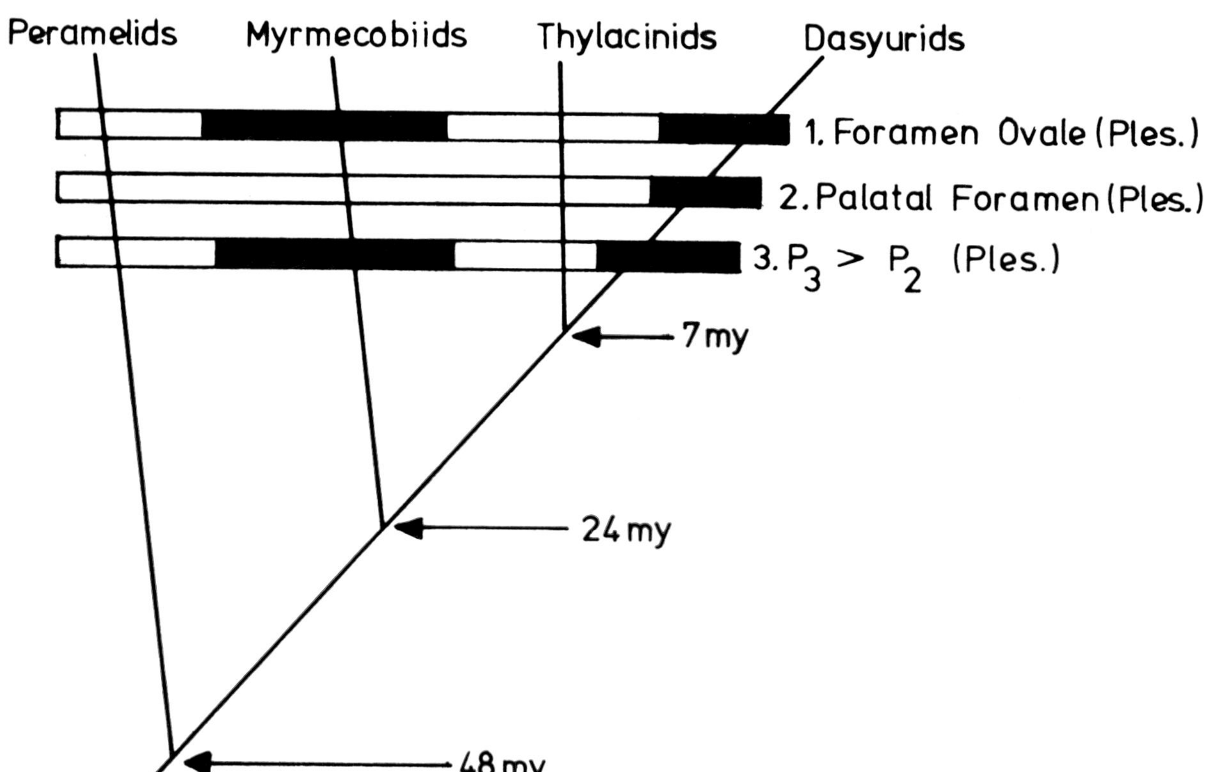

Fig. 7. Distribution of the states for three characters that appear not to support a thylacinid–dasyurid sister–group hypothesis superimposed on a hypothetical branching sequence arising from the albumin data of Sarich *et al.* (1982). The interpreted plesiomorphic condition is indicated by a blank space, the apomorphic condition by a solid bar. The character analyses are given in the text. The million year figures are those given by Sarich *et al.* (1982) for the most recent common ancestor based on the albumin molecular clock.

of these characters. This would mean that thylacinids could not have a 7 million year old common ancestor with dasyurids unless they are convergently plesiomorphic on other primitive marsupials, or dasyurids are convergently apomorphic on myrmecobiids. Both these alternatives are possible, but again, I think highly improbable. Figure 7 illustrates this apparent inconsistency.

The apomorphic character–states are the principle synapomorphies of a supposedly monophyletic dasyurid–myrmecobiid taxon. Similarly, redevelopment of a foramen ovale and postero–lateral palatal foramen in thylacinids seems at least unlikely, particularly in view of the obvious evidence (see above for *T. potens*) for loss of the posterolateral palatal foramen within the thylacinid radiation for what would then have been a second time after its redevelopment from an ancestor in which this foramen had already been apomorphically suppressed.

The third of the three possible kinds of special relationship considered above seems to me to be the most probable: thylacinids form the plesiomorphic sister–group of the whole of the dasyuroid radiation which includes only dasyurids and myrmecobiids. This would imply that the 7 million year old albumin divergence date calculated for the thylacinid–dasyurid dichotomy is in error, but perhaps not the 24 million year old thylacinid–myrmecobiid albumin divergence date. This third phylogenetic hypothesis is supported by the cranial and palaeontological data, the latter providing evidence for greater thylacinid antiquity than is indicated by the immunological data.

B. Autapomorphic Character States and Ancestor–Descendant (Patristic) Considerations

It is commonly held that autapomorphic character states have little if any significance in cladistic analysis because acquisition of these states can occur after a phyletic dichotomy and thus have no bearing on that event or the sister–group relationship of the resultant taxa. However, if at least one of the autapomorphic character–states of a taxon, say Taxon B, is not theoretically required to be present in the earliest member of that taxon, then Taxon B cannot exclusively be defined systematically *or* phylogenetically. This is simply because Taxon B will then *include* one member of an ancestral taxon, say Taxon A, that possesses only the autapomorph(s) of Taxon A. This is the nature of paraphyly: the ancestor of every taxon is a *member* of some other taxon. An alternative approach would be to regard all taxa as groups containing only members that possess (or have secondarily lost) one or more of the autapomorphic character–states used to recognize or define that taxon.

This simple distinction seems important to me because it appears to be at the heart of the rationale for making hypotheses about ancestor–descendant relationships. For example, in the present context, I am tempted to suggest that thylacinids could not have been derived from dasyurids because *all* dasyurids known (about 60 species) possess autapomorphic states (e.g. foramen pseudovale, loss of the posterolateral palatal foramina, and the P_3 which is smaller than, or very rarely subequal to, P_2) for characters represented in thylacinids only by plesiomorphic states. This sort of statement is, however, only meaningful if these higher level taxa can be mutually exclusively defined. If we require of any species that to qualify as a member of one of these families it must possess at least one autapomorphic character–state of the family in which we would place it, then the hypothesis that thylacinids (with these plesiomorphic character–states) did not descend from dasyurids (which possess the apomorphic character–states) becomes true by definition and distribution of character–states. If the Thylacinidae was in fact derived from a lineage of marsupials before that *same* lineage acquired one or more of the autapomorphic character–states of the Dasyuridae, that lineage, at the point in time when it sired the ancestral thylacinid, would have been didelphoid.

This is more than just semantics. It is a heuristic statement. Given that the albumin data accurately indicate that the Thylacine's greatest affinity lies with dasyurids, and that a common ancestor of thylacinids and dasyurids would have been structurally didelphoid in morphology (except perhaps in tarsal morphology), it indicates the existence of an as yet unknown group of Australian marsupials that was didelphoid in derivation, largely didelphoid in morphology, and yet serologically synapomorphic with living dasyurids — i.e. a group ancestral to living dasyurids.

In other words, the dasyuroid radiation must include at least two essential groups. The first is a group containing all of the living dasyurids and their immediate ancestors. The second is this as yet unknown plesiomorphic group, descendants of which may have included the thylacinids. In this patristic context, the myrmecobiids must have descended from dasyuroids more plesiomorphic than the living dasyurids (because they have a complete posterolateral palatal foramen) but more apomorphic (because myrmecobiids possess a foramen pseudovale) than the older as yet unknown dasyuroids which were ancestral to thylacinids on the one hand and the group of living dasyurids on the other (Fig. 8).

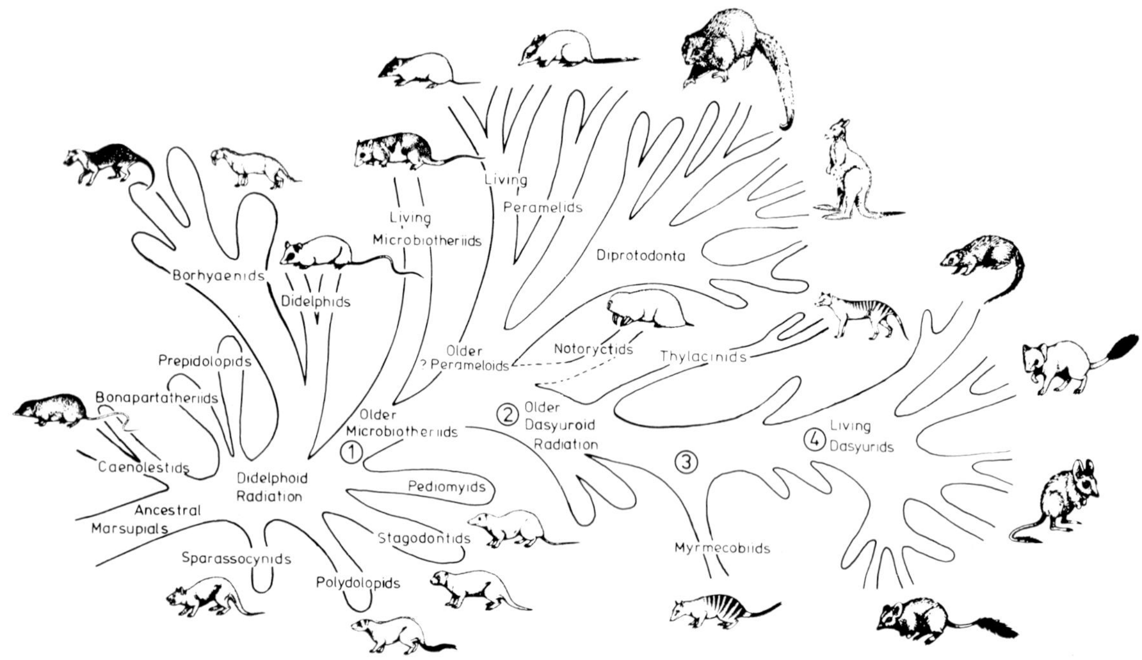

Fig. 8. An attempt to resolve conflicts in the data, from all character systems so far examined, about the phylogenetic relationships of the Thylacinidae within the Marsupialia. The circled numbers represent points in time for acquisition of certain features as follows: 1, the apomorphic tarsal structure reported by Szalay (1982); 2, the albumin affinity shared by thylacinids and dasyuroids (Sarich *et al.* 1982); 3, apomorphic acquisition of the foramen pseudovale and the premolar gradient $P_2 > P_3$ (Archer 1976a,b); 4, apomorphic loss of the posterior bar of the posterolateral palatal foramen (Archer 1976b). The existence of the as yet unknown "older dasyuroid radiation" is a hypothesis based on the syncretic resolution of the albumin and gross morphological data.

C. From Syncretism : A Homage to Convergence

If just one conclusion is drawn from this study, it should be that hypotheses about phylogenetic relationships of a group of extant organisms, particularly if the group has a poor fossil record, will almost certainly be doomed to early obsolescence unless the data from many tissue systems have been taken into account. No single tissue system has yet been shown to invariably clearly reflect *Real Phylogeny.* In fact, such a demonstration is probably impossible because it involves an *a priori* assumption that *Real Phylogeny* is actually determinable.

Some tissue systems are undoubtedly more reliable indicators of phylogeny than others, but all appear to be subject to either convergence, glitches in evolutionary rates of divergence (which, if time dependency is assumed, may lead to errors in interpreting branching sequences), or limitations in applicability.

For the moment, based on all available information, it would seem that the most probable interpretation about thylacinid relationships is that they are the plesiomorphic sister-group of the Dasyuroidea and have convergently developed a borhyaenid–like dentition. Because this syncretic hypothesis is based on the data from five rather than the previous three tissue systems, it might last a bit longer than the last! Hopefully, however, it will be tested and refined long before this. Much more new data needs to be gathered, particularly for the immunological characteristics of other proteins, as a basis for specifically testing the albumin results, as well as this new general hypothesis about the relationships of thylacines.

ACKNOWLEDGEMENTS

Many colleagues have contributed to the development and clarification of the ideas given here. Notable among them were Drs John Kirsch, David Ride, Ross Crozier, Tom Rich, Fred Szalay, Barry Richardson, Dick Tedford, George Simpson, Osvaldo Reig and Rosendo Pascual. The manuscript was constructively criticised by Dr. Ralph Molnar and Dr. Gaven Young. The holotype of *Thylacinus potens* was made available for study by Mr Mike Plane (Bureau of Mineral Resources). Ms Jenny Taylor helped produce the line figures and Mr. Leo Meier the photographs. Ms Jessica Johansen and Ms Deanne Thomas typed drafts of the manuscript.

POSTSCRIPT

Since this paper was written, two matters of relevance have occurred. Lowenstein, Sarich and Richardson (1982) have presented another discussion of the results they presented in "Carnivorous marsupials" (Sarich, Lowenstein and Richardson 1982), and an older middle Miocene thylacine has turned up. The new Miocene thylacine is based on a lower molar from the middle Miocene Riversleigh local fauna of northwestern Queensland, recovered by G. Clayton as part of a research project sponsored by the Australian Research Grants Committee. The age of this fauna, obtained from the Carl Creek Limestone, has most recently been concluded to between 12 and 15 million years old (Woodburne *et al.*, in prep.). It is distinctly thylacinid in morphology and corroborates my doubts (expressed above) about the 7 million year figure given by Sarich, Lowenstein and Richardson (1982) as the time when thylacines diverged from dasyurids. The Thylacine–dasyurid data and conclusions presented by Lowenstein, Sarich and Richardson are similar to those presented by Sarich *et al.* (1982). However, they have revised their estimate of Thylacine–dasyurid divergence to ". . . some 6 10 myr ago . . ." (p 411). Although, based on the palaeontological data, this is a revision in the right direction, the molecular date still seems much too young.

REFERENCES

ARCHER, M., 1971. A re-evaluation of the Fromm's Landing thylacine tooth. *Proc. R. Soc. Vict.* **84:** 229–34.

ARCHER, M., 1975. *Ningaui,* a new genus of tiny dasyurids (Marsupialia) and two new species, *N. timealeyi* and *N. ridei,* from arid Western Australia. *Mem. Qd Mus.* **17:** 237–49.

ARCHER, M., 1976a. The dasyurid dentition and its relationship to that of didelphids, thylacinids, borhyaenids (Marsupicarnivora) and peramelids (Peramelina: Marsupialia). *Aust. J. Zool. Suppl. Ser.* No. **39:** 34pp.

ARCHER, M., 1976b. The basicranial region of marsupicarnivores (Marsupialia), inter-relationships of carnivorous marsupials, and affinities of the insectivorous marsupial peramelids. *J. Linn. Soc. London (Zool.)* **59:** 217–322.

ARCHER, M., 1976c. Revision of the marsupial genus *Planigale* Troughton (Dasyuridae). *Mem. Qd Mus.* **17:** 341–65.

ARCHER, M., 1978a. The nature of the molar–premolar boundary in marsupials and a reinterpretation of the homolgy of marsupial cheekteeth. *Mem. Qd Mus.* **18:** 157–64.

ARCHER, M., 1982. Review of the Australian dasyurid (Marsupial) fossil record, integration of data bearing on phylogenetic interpretation, and suprageneric classification. Pp.397–443 *in* "Carnivorous marsupials", ed by M. Archer. Roy. Zool. Soc. New South Wales: Sydney, Australia.

ARCHER, M. AND KIRSCH, J. A. W., 1977. The case for the Thylacomyidae and Myrmecobiidae, Gill, 1872, or why are marsupial families so extended? *Proc. Linn. Soc. New South Wales* **102:** 18–25.

ARDEN, G. M., KEMP, F. H. AND RIDE, W. D. L., 1958. A radiographic analysis of mastication and swallowing in the domestic Rabbit, *Oryctolagus cuniculus* (L). *Proc. Zool. Soc. Lond.* **130:** 257–74.

BENSLEY, B. A., 1903. On the evolution of the Australian Marsupialia; with remarks on the relationships of the marsupials in general. *Trans. Linn. Soc. Lond. (Zool.)* **9:** 83–217.

CABRERA, A., 1927. Datos para el conocimiento de los dasyuroides fósiles Argentinos. *Rev. Mus. La Plata* **30:** 271–315.

CLEMENS, W. A. JR., 1966. Fossil mammals of the type Lance Formation Wyoming. *Univ. Calif. Publ. Geol. Sci.* **62:** 1–122.

DAWSON, L., 1982. The taxonomic status of fossil thylacines (*Thylacinus*: Thylacinidae, Marsupialia) from late Quaternary deposits in eastern Australia. Pp.527–36 *in* "Carnivorous marsupials", ed. by M. Archer. Roy. Zool. Soc. New South Wales: Sydney, Australia.

DEIGHTON, I., FALVEY, D. A. AND TAYLOR, D. J., 1976. Depositional environments and geotectonic framework: southern Australian continental margin. *A.P.E.A. Journal* **1976:** 25–36.

DE VIS, C. W., 1973. A thylacine of the earlier Nototherian Period in Queensland. *Proc. Linn. Soc. New South Wales* **1893:** v (Abstract).

HECHT, M. K., 1976. Phylogenetic inference and methodology as applied to the vertebrate record. Pp.335–63 *in* "Evolutionary biology", Vol. 9, ed. by T. Dobzhansky, M. Hecht and W. Steere. Appleton–Century–Crofts: New York.

HECHT, M. K. AND EDWARDS, J. L., 1976. The determination of parallel or monophyletic relationships: the proteid salamanders — a test case. *Amer. Nat.* **110:** 653–77.

KIELAN–JAWOROWSKA, Z., 1975. Possible occurrence of marsupial bones in Cretaceous eutherian mammals. *Nature* **255:** 698–9.

KIRKPATRICK, T. H., 1978. The development of the dentition of *Macropus giganteus* (Shaw). *Aust. Mammal.* **2:** 29–80.

KIRSCH, J. A. W., 1977a. The classification of marsupials with special reference to karyotypes and serum proteins. Pp.1–50 *in* "The biology of marsupials", ed. by D. Hunsaker. Academic Press: New York.

KIRSCH, J. A. W., 1977b. The comparative serology of Marsupialia, and a classification of marsupials. *Aust. J. Zool.*, Suppl. Ser. No. **52:** 152pp.

KIRSCH, J. A. W., 1982. The builder and his bricks. Pp.587–94 *in* "Carnivorous marsupials", ed. by M. Archer. Roy. Zool. Soc. New South Wales: Sydney, Australia.

LOVETROP, S., 1975. On the phylogenetic classification. *Acta Zool. Cracoviensia* **20:** 499–523.

MARSHALL, L. G., 1977. Cladistic analysis of borhyaenoid, dasyuroid, and thylacinid (Marsupialia: Mammalia) affinity. *Syst. Zool.* **26:** 410–25.

MARSHALL, L. G., 1978. Evolution of the Borhyaenidae, extinct South American predaceous marsupials. *Univ. Calif. Publ. Geol. Sci.* **117:** 1–89.

MAXSON, L. R., SARICH, V. M. AND WILSON, A. C., 1976. Continental drift and the use of albumin as an evolutionary clock. *Nature* **225:** 397–400.

PAULA COUTO, C. DE., 1952a. Fossil mammals from the beginning of the Cenozoic in Brazil. Marsupialia: Didelphidae. *Amer. Mus. Novit.* **1567:** 1–26.

PAULA COUTO, C. DE., 1961. Marsupiais fósseis do Paleoceno do Brazil. *An. Acad. brasil. Cienc.* **33:** 321–33.

PARKER, W. K., 1886. On the structure and development of the skull in the Mammalia. Part II, Edentata. Part III, Insetivora. *Phil. Trans. Roy. Soc. Lond.* **177:** 1–275.

PLANE, M. D., 1976. The occurrence of *Thylacinus* in Tertiary rocks from Papua New Guinea. *J. Aust. Geol. Geophys.* **1**(1).

RIGGS, E. S., 1934. A new marsupial saber-tooth from the Pliocene of Argentina and its relationships to other South American predaceous marsupials. *Trans. Amer. Phil. Soc. n.s.* **24:** 1–31.

REIG, O. A., 1957. Sobre la posición sistemática de *Zygolestes paranensis* Amegh. y de *Zygolestes entrerrianus* Amegh., con una recónsideración sobre la edad y la correlación del Mesopotamiense. *Holmbergia* **5:** 209–26.

REIG, O. A. AND SIMPSON, G. G., 1972. *Sparassocynus* (Marsupialia, Didelphidae), a peculiar mammal from the late Cenozoic of Argentina. *J. Zool., Lond.* **167:** 511–39.

RIDE, W. D. L., 1957. *Protemnodon parma* (Waterhouse) and the classification of related wallabies *(Protemnodon, Thylogale* and *Setonix). Proc. Zool. Soc. Lond.* **128:** 327–46.

RIDE, W. D. L., 1964. A review of Australian fossil marsupials. *Proc. R. Soc. West. Aust.* **47:** 97–131.

RUSSELL, L. S., 1976. The palaeogeographic significance of the polyprotodont marsupials. Abstract, 25th International Geological Congress, Sydney, p.333–4.

SARICH, V., LOWENSTEIN, J. M. AND RICHARDSON, B. J., 1982. Phylogenetic relationships of *Thylacinus* as reflected in comparative serology. Pp.707–9 *in* "Carnivorous marsupials", ed. by M. Archer. Roy. Zool. Soc. New South Wales: Sydney, Australia.

SIMPSON, G. G., 1941. The affinities of the Borhyaenidae. *Amer. Mus. Novit.* **1118:** 1–6.

SIMPSON, G. G., 1945. The principles of classification and a classification of mammals. *Bull. Amer. Mus. Nat. Hist.* **85:** 1–350.

SIMPSON, G. G., 1947. A new Eocene marsupial from Brazil. *Amer. Mus. Novit.* **1357:** 1–7.

SIMPSON, G. G., 1948. The beginning of the age of mammals in South America; Part 1. Introduction, Systematics: Marsupialia, Edentata, Condylarthra, Litopterna and Notoprogonia. *Bull. Amer. Mus. Nat. Hist.* **91:** 1–232.

SINCLAIR, W. J., 1906. Mammalia of the Santa Cruz beds. *Rept. Princeton Univ. Exped. Patagonia* **4:** 333–460.

SLAUGHTER, R., 1971. Mid–Cretaceous (Albian) therians of the Butler Farm local fauna, Texas. *In* "Early mammals", ed. by D. M. Kermack and K. A. Kermack. Academic Press: London.

SPENCER, B., 1901. A description of *Wynyardia bassiana*, a fossil marsupial from the Tertiary beds of Table Cape, Tasmania. *Proc. Zool. Soc. Lond.* **1900:** 776–94.

STIRLING, R., 1891. Description of a new genus and species of Marsupialia *'Notoryctes typhlops'. Trans. R. Soc. S. Aust.* **14:** 154–87.

STIRTON, R. A., TEDFORD, R. H. AND WOODBURNE, M. O., 1967. A new Tertiary formation and fauna from the Tirrari Desert, South Australia. *Rec. S. Aust. Mus.* **15:** 427–62.

SZALAY, F. S., 1977. Ancestors, descendants, sister groups and testing of phylogenetic hypotheses. *Syst. Zool.* **26:** 12–8.

SZALAY, F. S., 1982. The polyphyly of the Marsupicarnivora, and another look at the phylogeny and classification of the marsupials. Pp.621–40 *in* "Carnivorous marsupials", ed. by M. Archer. Roy. Zool. Soc. New South Wales: Sydney, Australia.

TATE, G. H. H., 1947. On the anatomy and classification of the Dasyuridae (Marsupialia). *Bull. Am. Mus. Nat. Hist.* **88:** 97–156.

WILEY, E. O., 1975. Karl R. Popper, systematics, and classification: a reply to Walter Bock and other evolutionary taxonomists. *Syst. Zool.* **24:** 233–43.

WILSON, A. C., 1976. Gene regulation in evolution. Chapter 13. *In* "Molecular evolution", ed. by F. J. Ayala. Sinauer Associates Inc.: Sunderland.

WOOD, H. E., 1924. The position of the "Sparassodonts": with notes on the relationships and history of the Marsupialia. *Bull. Amer. Mus. Nat. Hist.* **51:** 77–101.

WOODBURNE, M. O., 1967. The Alcoota fauna, central Australia: an integrated palaeontological and geological study. *Bur. Min. Resour. Aust. Bull.* **87:** 1–187.

RESUMEN

38. Una revisión de los tilacínidos (Thylacinidae, Marsupialia) del Mioceno, la posición filogenética de los Thylacinidae y el problema de apriorismos en el análisis de los caracteres. Por Michael Archer.

Se revisa el registro del Terciario de los tilacínidos. Sólo la especie *Thylacinus potens* Woodburne del Mioceno de Alcoota está representada por suficiente material como para permitir un análisis filogenético útil; muestra un complejo de caracteres que condujeron a las siguientes conclusiones: descendió de una forma tribosfénica que tuvo al menos las cúspides estilares *D* y *E*, ausentes en todos los boriénidos conocidos pero presente, por ejemplo, en algunos dasiúridos y didélfidos; tuvo un complejo foramen palatal posterolateral, carácter primitivo ausente en todos los dasiúridos conocidos; la morfología de los molares superiores de esta especie del Mioceno es tan especializada como la de la especie moderna y no da indicaciones claras de que sea ancestral; *T. potens* y la actual especie *T. cynocephalus* probablemente representan ramas colaterales del árbol familiar de los tilacínidos, más que una línea antecesor–descendiente.

Son reconsideradas las relaciones de los tilacínidos usando todos los datos asequibles: morfología dentaria, basicraniana, poscraniana (no pedal), pedal, datos serológicos y paleontológicos. De las cuatro hipótesis de grupos hermanoss consideradas (tilacínidos–dasiúridos, t–boriénidos, t–didélfidos y t–pediómidos) la más probable es la hipótesis de grupos hermanos dasiurideos–tilacínidos, si de mayor peso a los datos serológí cos y a la morphología pedal que a aquellos de otros sistemas de caracteres. La fecha de bifurcación de los tilacínidos y dasiúridos (7 millones de años) estimada por el "reloi" molecular parece ser errónea. Tilacínidos son conocidos de depósitos del Mioceno medio. Una segunde fecha crono–molecular de 24 milliones de años para el más reciente antecesor comú de los tilacínidos y mirmecóbidos no contradice irreconciliablemente los datos de otros sistemas de tejidos o los datos paleontológicos.

El más reciente apoyo a la hipótesis de que los tilacínidos y dasiúridos son grupos hermanos es el provisto por Marshall (1977). Sin embargo, el análisis de los caracteres en que se basaron esas conclusiones parecen ser inadecuados. Los problemas de apriorismo están agravados por su aplicación del esquema de evaluación de Hecht (1976) y Hecht y Edwards (1976). Se sugiere que este esquema para determinar la significación filogenética de las apomorfías está sujeto a arbitrariedad, contrariando la intencion de sus autores.

39. REVISION OF MARSUPIAL LIONS OF THE GENUS *Thylacoleo* GERVAIS (THYLACOLEONIDAE, MARSUPIALIA) AND THYLACOLEONID EVOLUTION IN THE LATE CAINOZOIC

M. ARCHER[1] and L. DAWSON[1]

Thylacoleo Gervais is regarded to be a senior synonym of *Plectodon* Krefft, *Schizodon* Stutchbury and *Thylacopardus* Owen, and *Thylacoleo carnifex* is regarded to be a senior synonym of *Mylodon australis* Krefft, *Thylacopardus australis* Owen, *Thylacoleo oweni* McCoy, and *Thylacoleo robustus* Krefft. *Thylacoleo crassidentatus* Bartholomai and *Thylacoleo hilli* Pledge are considered to be specifically distinct. New material referable to *T. crassidentatus* is described from the lower Pliocene Allingham Formation of Queensland and the Pliocene sediments at Bow in New South Wales. The Allingham material reveals much about the late Cainozoic morphological differentiation of the species of the genus. Although the Bow material reveals little that is new about *T. crassidentatus* another Bow specimen appears to be referable to the tiny species *T.* sp., *cf. T. hilli*. This is the first example of two thylacoleonid species occurring in the same local fauna and gives some insights into the "marsupial lion niche" in Australia. Pliocene–Pleistocene evolution of the species of the genus is considered. Cladistic and evolutionary approaches to the phylogenetic analysis of thylacoleonids are contrasted.

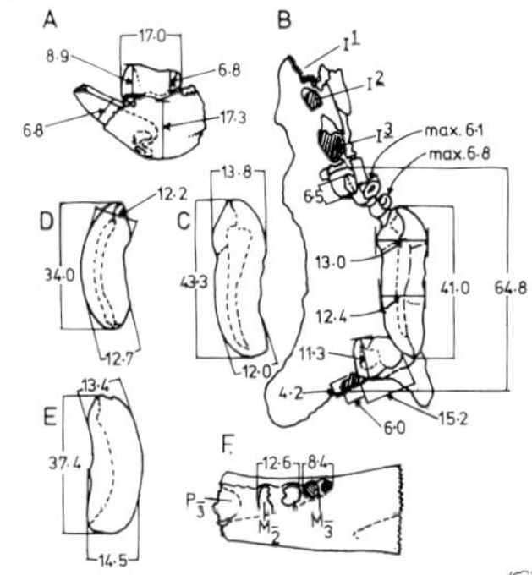

Pages 477–94 *in* CARNIVOROUS MARSUPIALS, ed. by M. Archer, February, 1982. Roy. Zool. Soc. New South Wales: Sydney, Australia.

MAHONEY and Ride (1975) have listed, with annotations, four generic and six specific names applied to thylacoleonids (marsupial lions) between 1852 and 1962: *Plectodon* Krefft, 1870; *Schizodon* Stutchbury, 1853; *Thylacoleo* Gervais, 1852; *Thylacopardus*, Owen, 1888; *Mylodon australis* Krefft, 1870; *Thylacopardus australis* Owen, 1888; *Thylacoleo carnifex* Owen, 1858; *Thylacoleo crassidentatus* Bartholomai, 1962; *Thylacoleo oweni* McCoy 1876; and *Thylacoleo robustus* Krefft, 1872. To these now may be added *Wakaleo* Clemens and Plane, 1974; *Wakaleo oldfieldi* Clemens and Plane, 1974; *Wakaleo vanderleuri* Clemens and Plane, 1974; *Wakaleo alcootaensis* Archer and Rich, 1982 and *Thylacoleo hilli* Pledge, 1977.

Wellington Caves, New South Wales, is the type locality of five of the above–named taxa. The Wellington deposits have also yielded a relatively large sample of Pleistocene thylacoleonid specimens. It is our intent first to consider variation in the Wellington Caves sample; second, to use insights gained from this analysis, with other data, to re–evaluate the specific distinction of all taxa referable to *Thylacoleo* Gervais; third, to draw attention to new material, possibly referable to the tiny species *Thylacoleo hilli,* recently obtained from Pliocene sediments in New South Wales, and to thylacoleonid material referable to *T. crassidentatus,* recently obtained from Pliocene sediments in Queensland; and finally to consider the significance of the new Pliocene material to interpretation of late Tertiary thylacoleonid evolution.

PLEISTOCENE THYLACOLEONIDS FROM WELLINGTON CAVES, NEW SOUTH WALES

MATERIALS AND METHODS

The Australian Museum Wellington Caves collection includes 5 nearly complete skulls, 13 maxillary fragments containing premolar teeth, and 37 mandibular specimens with P_3, including 10 nearly complete rami.

Institutional abbreviations for specimen numbers are as follows: AM F or AM MF, Australian Museum, Sydney; QM F, Queensland Museum, Brisbane; NMV P, National Museum of Victoria; SAM P, South Australian Museum, Adelaide; WAM, Western Australian Museum, Perth; BM No., British Museum (Natural History), London.

Comparison of the Wellington Caves material was made with AM F16457, a mandibular syntype of *Thylacoleo carnifex,* and AM F16457, a mandibular syntype of *Thylacoleo oweni* from the Darling Downs, Queensland; AM F4662, a cast of BM 39271, a cranium which is also a syntype of *Thylacoleo oweni* from the Darling

[1] School of Zoology, University of New South Wales, P.O. Box 1, Kensington, New South Wales, Australia, 2033.

Downs, Queensland; and with a cast of the premaxillary region and cranium of BM M16591, a syntype of *T. carnifex* from Lake Colungulac, Victoria. Five skulls and 3 mandibles from Naracoote Caves, South Australia, were also measured (SAM P16751, P16730, P16731, P16720, P20217, P16719, P16722, and one unregistered skull). BM M16591, a mutilated cranium from Lake Colongulac, has been selected as the lectotype of *Thylacoleo carnifex* Owen.

Measurements have been made as follows. A) Dental measurements (the mean, SD, and CV were calculated for each parameter): length P_3 – M_3 = the greatest anteroposterior dimension taken on the enamel; width P_3 = width at widest part of enamel in middle of tooth; width M_2 = maximum anterior width of the tooth; depth I_1 = depth measured in a dorsoventral plane at the base of enamel; length P^3 = maximum length at base of enamel on buccal side; width P^3 = maximum anterior width at base of enamel; and width M^2 = maximum width on enamel. B) Skull and mandible measurements: L1 = length of mandible from anterior edge of symphysis at base of I_1 to posterior surface of condyle in mid–line; L2 = length of mandible from base of condylar articular surface in midline to mental foramen; D1 = depth of mandible at posterior edge of M_2; D2 = depth of mandible from lower anterior edge of enamel of P_3 to lower border of mandible; palate width = distance between anterior edges of enamel of P^2 at alveolus; basicranial length = length from anterior edge of premaxilla at midline to lower edge of foramen magnum at midline; length of rostrum = length from anterior edge of premaxilla to posterior edge of M^2 along midline; and zygomatic width = the maximum external width across the zygomatic arches. C) Premolar teeth showed a variable degree of wear, described as follows: Wear Stage I = unworn, enamel not breached; II = early wear, enamel breached on anterior edge only of upper teeth, and just breached in lower teeth; III = medium wear, enamel breached to expose a narrow strip of dentine along full length of crown; and IV = late wear, wide exposure of dentine along full length of tooth.

Dental terminology follows Archer (1978). The maximum known dental formula for species of *Thylacoleo* is therefore considered to be: $I^{1\text{-}3}/_1$, $C^1/_0$, $P^{1\text{-}3}/_{1\text{-}3}$, $M^{2\text{-}3}/_{2\text{-}3}$. The deciduous cheektooth, regarded by Archer (1978) to be M_1, has not been reported in any thylacoleonid, despite the knowledge of at least one very tiny pouch–young specimen (Finch, pers. comm.). The true homology of the vestigal single–rooted teeth regarded here to be $P_{1\text{-}2}$ is in doubt. Conceivably they could also represent $I_{2\text{-}3}$, or I_3 and C_1, or C_1 and P_1.

Results and Discussion of Analysis of the Pleistocene Samples

A full description of the skull of the Pleistocene *Thylacoleo carnifex,* based on material from the fluviatile deposits of the eastern Darling Downs, Queensland, has been given by Woods (1956). No significant morphological differences (other than size) from this description have been found in any skull from Wellington Caves.

A summary of the dental measurements and statistical analysis of the data for the Wellington maxillary and mandibular teeth is given, with comparative data, for a sample from Naracoorte Caves, South Australia, in Table 1. Low variability of tooth size within the Wellington Caves sample is indicated by the relatively low values of CV, consistent with those usually found in populations of a single species (Simpson *et al.* 1960). From Table 1, it can been seen that the values for the length of $P^3/_3$ of Naracoorte Caves specimens fall within the observed range for the Wellington Caves sample, and also have a low CV value. Similarly, the sample means of

Table 1. Summary of measurements of teeth of *Thylacoleo carnifex* from Wellington Caves and Naracoorte Caves.

	WELLINGTON CAVES					NARACOORTE CAVES				
Character	N	$\overline{X}$	O.R.	S	CV	N	$\overline{X}$	O.R.	S	CV
LP_3	35	39.2	35.3 – 44.0	1.91	4.9	6	39.1*	37.0 – 42.0	1.69	4.3
WP_3	34	14.0	11.5 – 15.0	0.73	5.2	6	39.1*	37.0 – 42.0	1.69	4.3
LM_2	22	14.2	12.0 – 15.9	1.04	7.3					
WM_2	21	10.0	9.0 – 11.0	0.45	4.5					
$LP_3 + LM_2$	20	51.8	49.3 – 57.5	2.45	4.7					
depth I_1	12	18.3	17.0 – 19.3	0.90	4.9					
LP^3	16	49.8	46.9 – 55.5	2.46	4.9	5	50.7*	47.7 – 53.3	2.06	4.1
WP^3	18	14.7	13.2 – 15.8	0.66	4.5	5	15.3*	13.8 – 17.5	1.52	9.9
$LP^3 + LM^3$	8	47.2	42.5 – 51.5	2.89	6.1					
LM^2	4	14.2	13.8 – 15.0	0.53	4.0					
WM^2	4	9.3	8.8 – 9.8	0.46	5.3					

* $P > 0.05$ for difference between means of the two populations.

the two populations do not differ significantly. The apparent relative uniformity of premolar tooth size agrees with that found by Finch and Freedman (1982).

Measurements of the P3, skull and mandible, with estimated wear stages, are given in Tables 2 and 3 for specimens from Wellington Caves; Naracoorte Caves; L. Colongulac, Victoria; and the eastern Darling Downs, Queensland. The relative uniformity of tooth size noted above contrasts with the large differences observed in skulls and mandibles. For example, the basicranial length of AM F4659 is approximately 40% greater than that of AM MF9, while premolar length in the two specimens is approximately equal. Allometric growth differences caused by age or sex have been considered as a possible reason for this variation.

The age of a mammalian skull is commonly reflected by the stage of dental eruption, degree of tooth wear, degree of fusion of cranial sutures and degree of development of muscle attachment crests on the outside of the cranium. However, the Wellington Caves skulls appear to obfuscate these simple correlations. The largest available skull (AM F18666) has unworn premolar teeth (Wear Stage I) and weak development of the temporal ridges and sagittal crest indicating it is a young animal. In contrast, the smallest skull (AM MF9) has premolar teeth that are slightly worn (Wear Stage II) and stronger temporal ridges that meet posteriorly to form the sagittal crest (Fig. 1). In other words, this smaller skull appears to represent an older individual. To investigate this apparent incongruency, a size index was obtained by dividing the values of particular cranial and mandibular parameters (for

Table 2. Some mandibular dimensions and values for Size Index (I) for *Thylacoleo carnifex* from Wellington Caves and other Pleistocene fossil deposits.

Specimen Number	Location	Wear Stage	length P_3	L1	D1	D2	SIZE INDEX $\frac{L1}{LP_3}$	SIZE INDEX $\frac{D1}{LP_3}$
(AM) F44664	Wellington Caves	I	39.5	–	47.0	43.5	–	1.19
MF734	”	I	43.0	–	34.8	38.5	–	0.80
MF 21	”	I	37.8	140.0	34.5	34.0	3.70	0.91
F18884	”	I	37.3	–	32.0	38.0	–	0.86
F51286	”	I	44.0	134.0	36.5	39.8	3.05	0.83
F16625	”	I	37.5	–	40.5	41.0	–	1.08
F10835	”	I	39.2	–	41.0	–	–	1.05
MF 83	”	I	38.5	–	31.0	–	–	0.80
F16627	”	I	39.8	–	37.0	38.0	–	0.92
F62525	”	II	42.2	–	44.0	47.5	–	1.04
F37692	”	II	39.0	–	48.4	55.5	–	1.24
F61231	”	II	42.0	–	50.3	51.5	–	1.20
F16623	”	II	39.5	–	40.3	42.0(a)	–	1.00
F61230	”	II	37.1	–	44.9	48.0	–	1.21
F16617	”	II	39.5	–	42.0	38.0	–	1.06
F19370	”	II	41.5	–	44.8	45.5	–	1.08
F51288	”	III	39.0	170.0(a)	43.0	48.3	4.36	1.10
F16458	”	III	37.5	–	44.1	50.5	–	1.18
F10836	”	III	36.0	–	44.5	48.5	–	1.24
MF116	”	III	38.0	–	45.0	46.5	–	1.20
F61228	”	III–IV	40.2	170.0	47.0	56.0	4.23	1.17
F37619	”	IV	41.0	–	48.0	54.5	–	1.17
F61229	”	IV	38.0	–	53.0	60.0	–	1.39
F51287	”	IV	40.2	180.0	46.0	56.5	4.48	1.14
MF733	”	IV	39.0	–	53.0	54.5	–	1.36
F62524	”	IV	37.2	196.0	52.7	57.0	5.27	1.42
MF 80	”	IV	39.0		47.6	55.0	–	1.20
F16618	”	IV	38.5	–	–	45.0	–	–
(SAM) P20217	Naracoorte Caves	II–III	42.0	180.0	52.0	50.0	4.29	1.24
P16722	”	III	37.0	–	43.0	–	–	1.16
P16719	”	IV	39.0	175.0	50.0	59.0	4.49	1.28
(AM) F16456	Eastern Darling Downs, Qld	II	41.0	–	48.3	–	–	1.18
F16457	”	IV	39.6	–	52.6	–	–	1.35

(a) approximate measurement.

Table 3. Some skull dimensions and values for Size Index (I) for *Thylacoleo carnifex* from Wellington Caves and other Pleistocene fossil deposits.

Specimen Number	Location	Wear Stage	Length P^3	Palate Width	Basicran. Length	Rostral Length	Zygomatic Width	(I) Palat W / LP^3	(I) Basicran L / LP^3
(AM)									
F4660	Wellington Caves	I	49.4	73.6	192.0	87.0	180.0	1.49	3.89
F18666	”	I	52.0(a)	81.5	215.0	95.0(a)	–	1.57	4.28
F4661	”	I	49.5	70.0	190.0(a)	–	160.0	1.41	3.84
MF 9	”	II	50.0	64.3	163.0	78.0	155.0	1.29	3.26
F4659	”	III	50.4	77.5	225.0	96.2	–	1.54	4.46
(AM)									
F4662	Darling Downs, Qld	IV	53.5(a)	92.5	235.0	100.0(a)	–	1.73	4.39
(SAM)									
P16730	Victoria Cave Naracoorte, S.A.	II	47.7	72.5	195.0	82.0	185.0	1.52	4.09
P16751	”	II	53.3	86.0	230.0(a)	–	210.0(a)	1.61	4.32
P16720	”	III	51.5	86.5	207.0	91.0	190.0	1.68	4.07
P16731	”	IV	51.0	84.0	205.0	85.5	200.0	1.64	4.02
P ?	”	IV	50.0	79.0	200.0	83.5	–	1.58	4.00
(BM)									
M16591 &	L. Colongulac Vict.	II*	56.8	–	254.0()**	–	–	–	4.47
(NMV									
P1902	”								

(a) Approximate measurement; * Wear stage estimated from plate 56 of Gill (1973); ** Basicranial Length taken as 25.4 cm from Gill (1973).

example, basicranial length or mandibular depth) by the premolar length of the individual from which they were obtained. The resultant index, I, of each specimen is presented in Tables 2 and 3. These indices were then related to premolar wear stage as shown in Fig. 2.

In Fig. 2D, the mean values and observed ranges of 27 values of I for the parameter mandibular depth (D1) have been plotted against premolar wear stage. The slope of the line joining the mean values at each wear stage indicates a positive correlation between mandibular depth and the degree of premolar wear. This supports the common observation for mammals that tooth wear increases with age. However, in the case of parameters where fewer data are available, this trend is less evident. In parts A, B and C of Fig. 2, the values of I are represented by a vertical bar for each specimen. Figures 2A and 2B do not show a trend of increasing size index values with increase in premolar wear for the parameters palate width and basicranial length. A slight increase in values of I with increasing tooth wear is evident for the parameter mandibular length (Fig. 2 C).

These results do not support a hypothesis that the variations observed in skull size are due to increase in size with age in the Pleistocene Marsupial Lion. It is more probable that the differences in skull size represent large differences in individual size regardless of the age of the animal.

Sexual dimorphism is frequently responsible for size differences within populations of modern marsupials. Ride (1964) demonstrated this clearly for the other large Australian carnivorous species *Thylacinus cynocephalus*. Sexual dimorphism is generally reflected in bimodality of tooth size as well as total body size. However, histograms of thylacoleonid premolar lengths (Fig. 3) do not indicate strong bimodality even in the reasonably large samples ($N = 35$ for P^3) from Wellington Caves. In reasonably large samples an equal representation of sexes might be expected. Clearly, additional comparative data for skull and postcranial skeletons of *Thylacoleo carnifex* is required to test our hypothesis about extreme sexual dimorphism. At the moment, this explanation for the observed variation in skull and mandible size in the Wellington Caves sample seems to be more probable than a hypothesis that there are two species with similar sized teeth in the single population.

The Lake Colongulac skull, the lectotype of

Thylacoleo carnifex, has a larger P^3 than any other specimen measured. However, the estimated size index (I = 4.5, based on basicranial length) indicates that, although this is a very large individual, the skull length has the same relationship to premolar length as at least one skull (AM F4659) from Wellington Caves. It also agrees closely with skulls from Naracoorte Caves in proportional size. It apparently represents a very large but young animal. In the absence of other morphological differences, it is considered here to represent an individual conspecific with those from the Wellington and Naracoorte Caves populations.

From the measurements made in this study, the teeth of all specimens referable to species of *Thylacoleo* have been shown to vary little in size and morphology both within and between Pleistocene populations. We conclude that they are all therefore referable to *Thylacoleo carnifex* Owen, the earliest available name for the species (Mahoney and Ride 1975). The data presented here do however suggest that other parts of the skeleton of *T. carnifex* may not be so conservative in size or proportion. It is possible, even probable, that these differences are the result of a sexual dimorphism that is not reflected in tooth size.

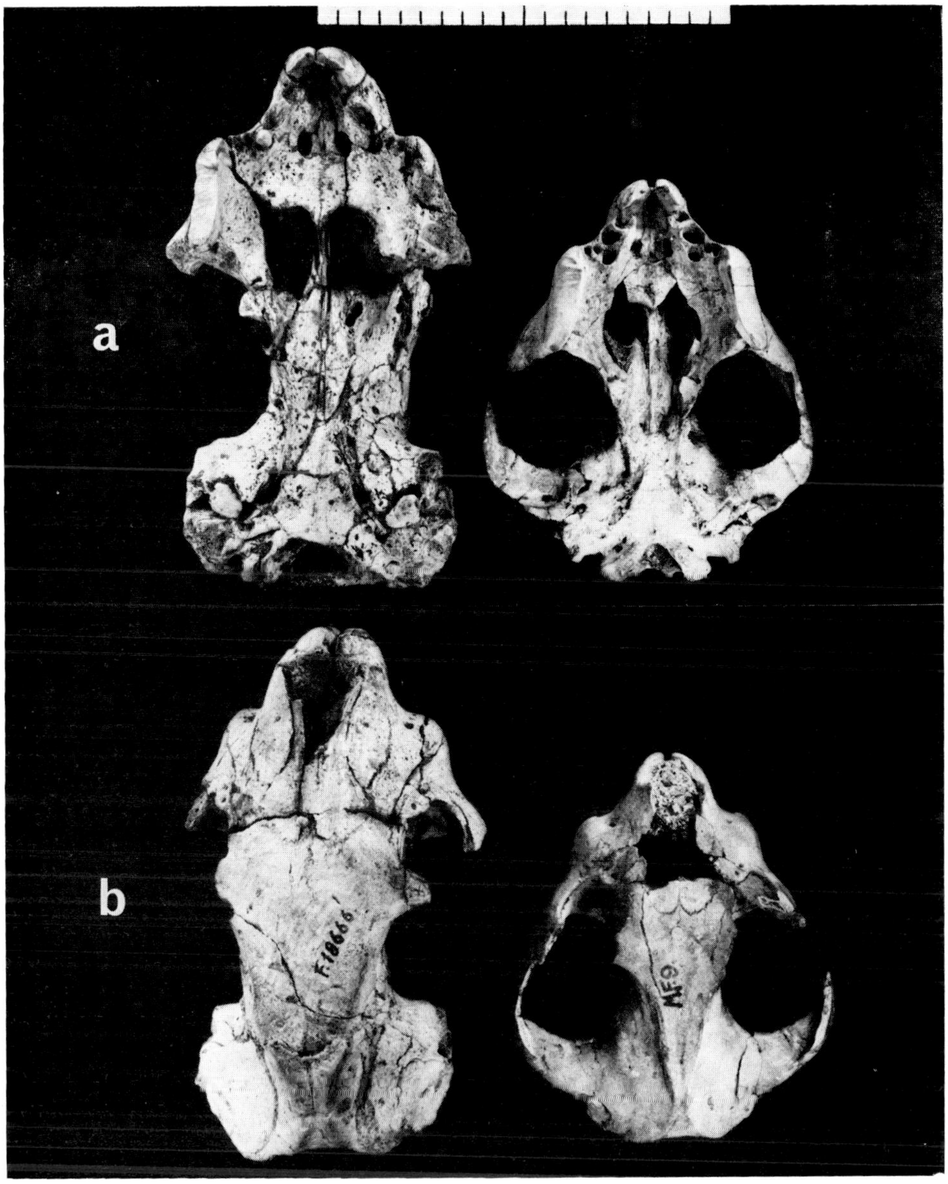

Fig. 1. Extreme size variation in skulls of *Thylacoleo carnifex* from Wellington Caves; AM MF9 (left) and AM F18666 (right): a) ventral view, b) dorsal view. Scale in cm.

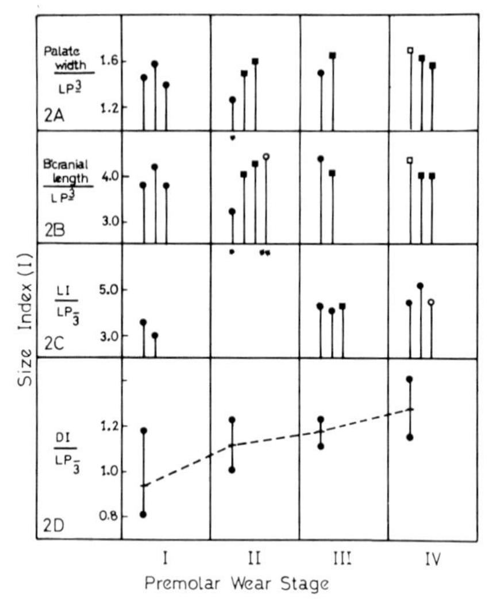

Fig. 2. The relationship between Size Index and premolar wear stage in *Thylacoleo carnifex*. Data from Table 2 and Table 3. In Fig. 2A, 2B and 2C, vertical lines represent values of three size indices of individual specimens of *Thylacoleo carnifex* from various Pleistocene localities. * AM MF9; ** Lectotype *T. carnifex* BM M16591. •— Wellington Caves; o— Lake Colongulac; □— Darling Downs; ■— Naracoorte Caves. In Fig. 2D vertical bars represent the observed range of size indices for Wellington Caves specimens of *T. carnifex*. Dotted line joins the mean values for size index obtained for specimens at each wear stage.

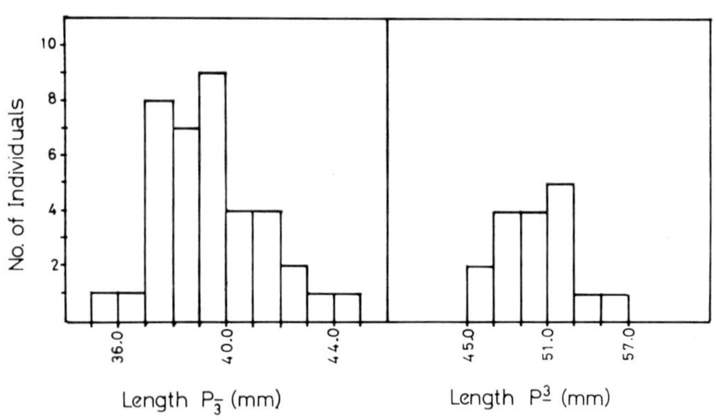

Fig. 3. Histograms showing distribution of premolar lengths in *Thylacoleo carnifex* from Wellington Caves, New South Wales.

THE REFERENCE OF CERTAIN GENERIC AND SPECIFIC NAMES TO Thylacoleo carnifex OWEN

Schizodon STUTCHBURY, 1853

No species was named for *Schizodon*. The specimen described by Stutchbury was a right mandibular ramus from King Creek, Clifton, Darling Downs, Queensland. It subsequently became a syntype of *Thylacoleo carnifex* Owen, 1858, and *Thylacoleo oweni* McCoy, 1876 (Mahoney and Ride 1975). The generic name, on the basis of this action of Owen and McCoy, is clearly a junior synonym of *Thylacoleo* Gervais.

Plectodon KREFFT, 1870

This genus was described (without an associated species name) on the basis of the tip of an isolated upper incisor, from Wellington Caves, mistakenly identified by Krefft as a lower incisor of a genus allied to *Thylacoleo*. Krefft (1870) notes, in the original description, that it has a) a dental fold on the inner side, b) the outer surface marked by a shallow groove in the middle, and c) enamel with worm-eaten appearance.

Examination of a cast, AM L1366, of BM 42638, a specimen of *Plectodon* sent to Owen by Krefft, clearly indicates that the specimen is actually the unworn tip of a left upper thylacoleonid incisor. The mistaken identification of the homology of this tooth by Krefft (1870) led him to then make comparisons with other forms that are now clearly spurious. No specimen available in the Australian Museum collection has I^1 unworn and *in situ* in a skull. However, the specimen agrees very closely in size, morphology and wear stage with an isolated RI^1 (AM F53-475b), referable to *Thylacoleo carnifex* from Wellington Caves. *Plectodon* Krefft is therefore regarded here to be a junior synonym of *Thylacoleo* Gervias.

Mylodon (?) *australis* KREFFT, 1870

An isolated distal phalanx (AM F7232) from Wellington Caves breccia, bearing a large ungual crest and ungual process, was considered by Krefft (1870:8) to be " . . . not unlike the South American genus *Mylodon*". He tentatively assigned it to that genus as *Mylodon* (?) *australis*. The recent discovery and reconstruction of the manus of *Thylacoleo carnifex* from Naracoorte Caves (Wells and Nichol 1977), a cast of which (AM F59745) was available for comparison, leaves no doubt that AM F7232 is in fact a distal phalanx of the first digit of the right manus of *T. carnifex*. The Wellington phalanx agrees completely in size and morphology with that of the Naracoorte Caves specimen. Therefore, the species *Mylodon australis* is a junior synonym of *Thylacoleo carnifex* Owen.

Thylacoleo oweni McCOY, 1876

Specimens of *Thylacoleo*, described by Owen in 1859, 1866 and 1871 from localities in "New South Wales" (i.e. N.S.W. and Queensland, see Mahoney and Ride 1975), were considered by McCoy (1876) to differ specifically from specimens available to him from Lake Colongulac, Victoria, which included the cranial syntype of *Thylacoleo carnifex* Owen. From the thirteen syntypes listed by Mahoney and Ride (1975) for *Thylacoleo oweni*, we have selected a cranium, BM 39271, as the lectotype. The four main features considered by McCoy (1876) to distinguish *Thylacoleo carnifex* from *T. oweni* are critically reviewed as follows:

1) *The position of the premaxillo–maxillary suture*: McCoy (1876) observed that this suture crossed the middle of the socket of the fourth tooth (C^1) in the Victorian specimen. He compared this condition with the observation of Owen (1871) that this suture crossed the socket of the third upper tooth in specimens from the Darling Downs. In the present study, this suture has been observed to vary in position, particularly buccal to the tooth sockets, where it characteristically descends to approach, but usually not meet, the alveolus of I^3 near its mid point (e.g. AM F4660, AM MF9, AM F18666), then abruptly turns posteriad and crosses the alveolus of C^1 variably near its anterior edge (AM F18-666) or as far posteriorly as its mid–line (AM F4660 and NMV P1902). The exact course of the suture is often difficult to determine in damaged fossil specimens. It is clear that this is a variable character. The condition seen in the Victorian syntype of *Thylacoleo carnifex* is also found in specimens from the Wellington Caves and the eastern Darling Downs, and therefore is not of taxonomic value in distinguishing a Victorian species.

2) *The palate width between the second and third incisor teeth*: McCoy (1876) considered the anterior palate of the Victorian species (as represented by NMV P1902) to be narrower than that of the specimens referable to *Thylacoleo oweni*. Some comparative measurements for the distance between the inner edges of the alveoli of the left and right I_2 are given in Table 4. The data indicate that the anterior palate width of the Victorian specimen falls within the observed range for specimens from Queensland and New South Wales.

3) *Size of the second and third upper incisors*: these teeth were considered by McCoy (1876) to be larger in the Victorian species. Comparative data for the maximum crown length of I^3 at its occlusal edge are given in Table 4. The data indicate that the length of I^3 of the Victorian specimen NMV P1902 falls at the top of the observed range for length I^3 of Wellington Caves specimens, not inconsistent with interpretation as a large individual of the same species.

4) *Depth of the mandible*: McCoy (1876) noted that a mandibular specimen from Lake Colongulac, Victoria, differed from a "New South Wales specimen" figured by Owen (1871, plate XII) in that the relative depth of the mandible at the posterior edge of M_3 (measurement A) was greater than the mandibular depth below the anterior edge of the crown of P_3 (measurement B). These measurements correspond with the parameters D1 and D2 as defined above and data for Wellington Caves specimens are given in Table 3. Two specimens from Wellington Caves had D1 > D2 (i.e. A > B), thus resembling McCoy's specimen. As noted above, mandibular depth and proportions appear to be subject to considerable allometric variation caused by growth with age.

Thus the characters selected by McCoy (1876) appear to have no taxonomic value in distinguishing a Victorian species from a New South Wales and Queensland species. The data presented in Table 3 have indicated that the P^3 of the Lake Colongulac skull exceeds any other specimen in length. However, the size index $\frac{\text{Basicranial L}}{\text{LP}^3}$ for this skull (Table 3, Fig. 2)

Table 4. Some comparative measurements (mm) of incisor and canine teeth of *Thylacoleo carnifex*.

Specimen number	Location	Distance between P^{2-2}	Max. crown length I^3	Max. crown length C^1	Max. crown width C^1
NMV P1902	L. Colongulac	20.6	16.4	(Rt) 10.4	7.2 (a)
,,	,,	–	–	(L) 8.9*	6.6
AM F18666	Wellington C.	20.3	–	–	–
AM F4660	,,	21.3	–	–	–
AM MF 9	,,	17.0	–	–	–
AM F4662	,,	27.2	–	–	–
AM F16481	,,	–	16.5	–	–
AM F16641	,,	–	15.8	–	–
AM F16473	,,	–	15.2	–	–
AM F53479	,,	–	10.2	–	–
AM F16640	,,	–	14.1	–	–
AM F16642	,,	–	13.7	–	–
AM F16479	,,	–	–	6.7	6.2
AM F16483	,,	–	–	8.9	6.2
AM F53481a	,,	–	–	8.2	5.5
AM F53481b	,,	–	–	9.2	6.0
AM F53481c	,,	–	–	7.3	5.4
AM F53481d	,,	–	–	9.5	6.0
AM F16475	,,	–	–	10.0	6.9

* Tooth may be damaged; (a) approximate measurement

suggests that it represents a large individual proportionally similar to most skulls from Wellington Caves. The name *Thylacoleo oweni* McCoy is therefore regarded by us to be a junior synonym of *Thylacoleo carnifex* Owen.

Thylacoleo robustus KREFFT, 1872

An isolated right upper canine tooth (AM F16479) from the Wellington Caves, was considered by Krefft (1872) to represent a species larger than *Thylacoleo carnifex*. Considerable variability in crown size of canine teeth of *Thylacoleo carnifex* is indicated by the data given in Table 4. The dimensions of AM F16479, the holotype of *T. robustus*, fall within the observed range of the total sample. The high variability of these teeth may be a reflection of sexual dimorphism or, more likely, of their reduced functional importance. Root size also varies greatly in the available specimens. The data indicate that *Thylacoleo robustus* Krefft is a junior synonym of *T. carnifex* Owen.

Prochaerus celer DE VIS, 1886

There were eight isolated syntypes for this most heterogeneous of De Vis's taxonomic creations (De Vis 1886, 1887). These include QM F757, a damaged tooth representing a P^3 of *Zygomaturus trilobus*, and QM F756, a left I_1 of a thylacoleonid. Longman (1916) refers QM F756 to *Thylacoleo carnifex* (and with this decision we are in agreement). Mahoney and Ride (1975) regard QM F757 to probably represent a diprotodontid (and with this we also agree). Because QM F756 is the only syntype that can at present be confidently identified, we propose that it become the lectotype, and all other syntypes become paralectotypes. Because the lectotype appears indistinguishable from the left lower incisor of *Thylacoleo carnifex*, we therefore regard *Prochaerus celer* De Vis to be a junior synonym of *Thylacoleo carnifex* Owen.

Thylacopardus australis OWEN, 1888

Owen (1888) cites the presence of second upper molar in the holotype of this species (AM MF9) as warranting generic distinction. The present study supports Anderson (1929: 48) who could find no evidence for a second molar in this skull, and who considered the "molariform prominence" of Owen to represent the " . . . truncated end of the bar which bounds the posterior palatal vacuity . . . ". Generic distinction on the basis of Owen's criteria is therefore unwarranted. The extremely small size of the skull is attributed by Anderson (1929) to immaturity. He notes however, that it is similar to other Wellington Caves specimens in premolar size.

The data presented above (Table 3, Fig. 2) of this paper indicate that this skull differs considerably in proportion from others of similar dental wear stage. As indicated above the degree of development of the sagittal crest and temporal ridges also suggests that this was not a very young animal. While it could be concluded that the proportional differences of the holotype justify its specific distinction, such a decision appears unwarranted at present. Because of its otherwise close morphological agreement with the remaining sample of *Thylacoleo carnifex* noted above, we therefore support Anderson (1929) in placing *Thylacopardus australis* Owen in the synonymy of *Thylacoleo carnifex* Owen.

ALLINGHAM AND BOW THYLACOLEONIDS REFERABLE TO *Thylacoleo crassidentatus* BARTHOLOMAI

Archer (in Archer and Wade 1976) notes the occurrence of a species of *Thylacoleo* in the Bluff Downs local fauna from the Pliocene Allingham Formation of north Queensland. At the time of description, only one fragment (QM F7808) of a left P^3 represented any part of the upper dentition of the Allingham thylacoleonid. Mainly on the basis of a right dentary retaining part of P_3 and all of M_2 (QM F7762), the Allingham form appeared to be specifically unique. Its similarity to *T. crassidentatus* (from the Pliocene Chinchilla local fauna of the Chinchilla Sand, southeastern Queensland) was noted but it was also considered to differ from the Chinchilla form in having a proportionately much shorter P_3. Archer and Bartholomai (1978) note a skull fragment (Fig. 4) of *Thylacoleo* sp. from the Bluff Downs fauna that was collected by J. Barratt in 1975. Although this also appears similar to *T. crassidentatus*, there are some differences. Additional Bluff Downs thylacoleonid specimens that have not previously been noted include an isolated left I^1 (QM F10642) and an isolated left P^3 (QM F10623, Fig. 4). Considering the whole Bluff Downs sample, it is probable that this thylacoleonid is referable to *T. crassidentatus* even though it differs markedly in several features.

The Bow local fauna of northeastern New South Wales has been the subject of a recent study by C. G. Skilbeck. Collections were made by J. A. Mahoney, C. G. Skilbeck, E. W. McDonald, A. Ritchie, L. Dawson and M. Archer. A preliminary faunal summary is presented in Skilbeck (1980). Work continues at the type locality of the Bow fauna and has resulted in, among other things, discovery of a very tiny thylacoleonid (see below). There are three additional thylacoleonid specimens from Bow including AM F59594, an isolated left P_3, AM F59-569, an isolated right P_3, and AM F59593, a left dentary fragment with alveoli for M_3, M_2 and part of P_3 (Fig. 4). Of all known thylacoleonids, these most closely resemble *Thylacoleo crassi-*

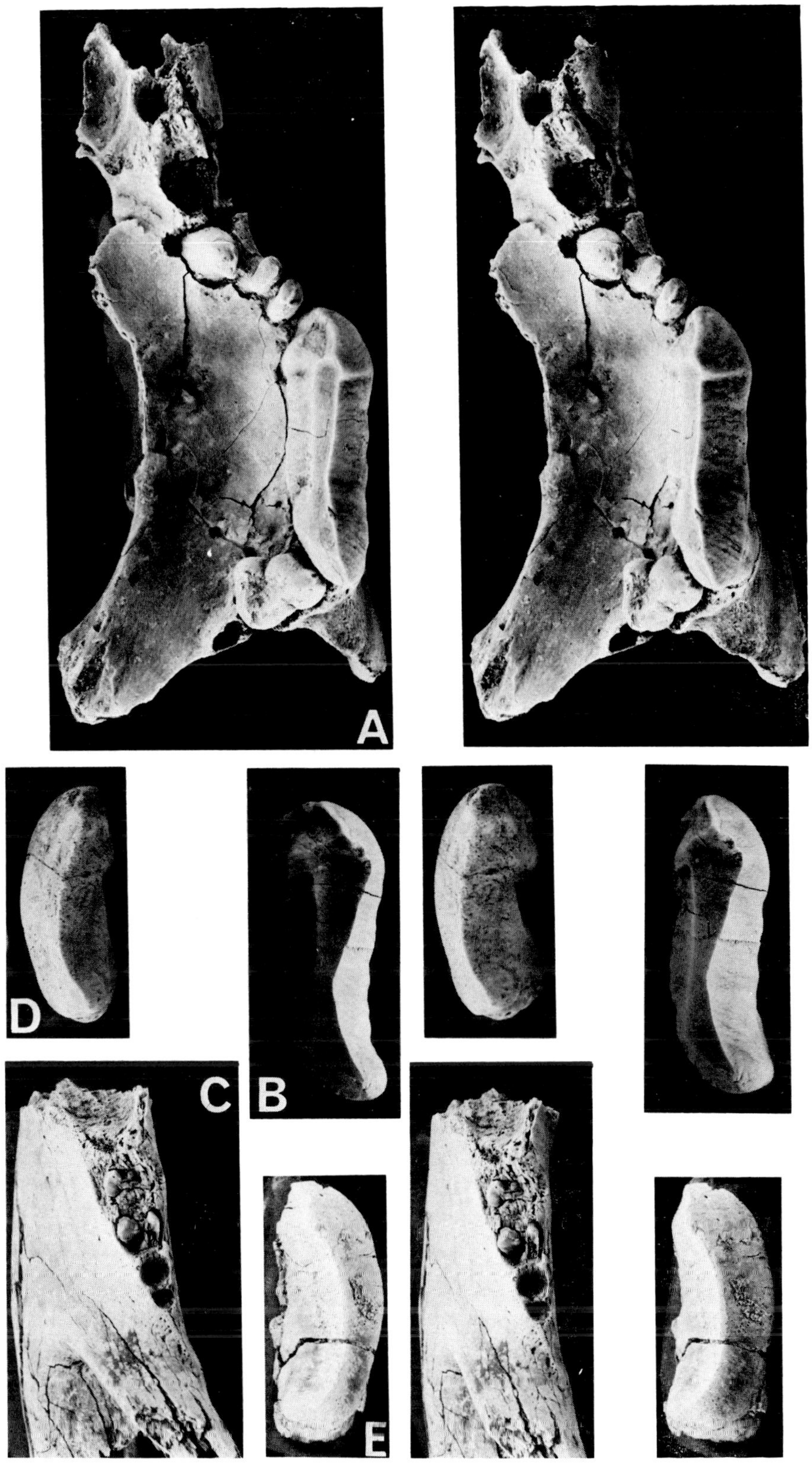

Fig. 4. Thylacoleo crassidentatus, X 1. A, QM F10622, left skull fragment with alveoli for I^{1-3}, M^3 and C^1–M^2, from the Bluff Downs local fauna. The M^3 is incipiently two–rooted. B, QM F10623, isolated left P^3, also from the Bluff Downs local fauna. C, AM F59593, left dentary fragment with alveoli for P_3, a two–rooted M_3, and broken roots for M_2, from the Bow local fauna. D, AM F59594, left P_3, from the Bow fauna. E, AM F59569, RP_3, also from the Bow fauna.

dentatus. The Bow local fauna, based mainly on its close faunal similarities with the Chinchilla local fauna, is probably early to middle Pliocene in age.

After coming to the conclusion that these Allingham and Bow thylacoleonids are referable to *T. crassidentatus,* it is clear that these new specimens enable us to expand the concept of that species.

Thylacoleo crassidentatus BARTHOLOMAI, 1962

Holotype: QM F3565, a partial right dentary with the root for I_1, $P_{1\text{-}3}$, M_2 and M_3 from the Chinchilla Sand of southeastern Queensland.

Referred material: Bartholomai (1962) lists seven additional mandibular fragments or lower teeth and three maxillary fragments from the Chinchilla fauna; noted above are a further six Bluff Downs specimens including a partial skull fragment and three Bow specimens.

Description of the Bluff Downs skull fragment QM F10622 (Figs. 4,5): The skull fragment preserves parts of the left premaxilla, jugal, most of the left maxilla, the left C^1–M^2 and alveoli for the left $I^{1\text{-}3}$ and left M^3. This specimen is particularly important for several reasons. First, it demonstrates that *T. crassidentatus,* in the two appropriately preserved maxillary fragments (the

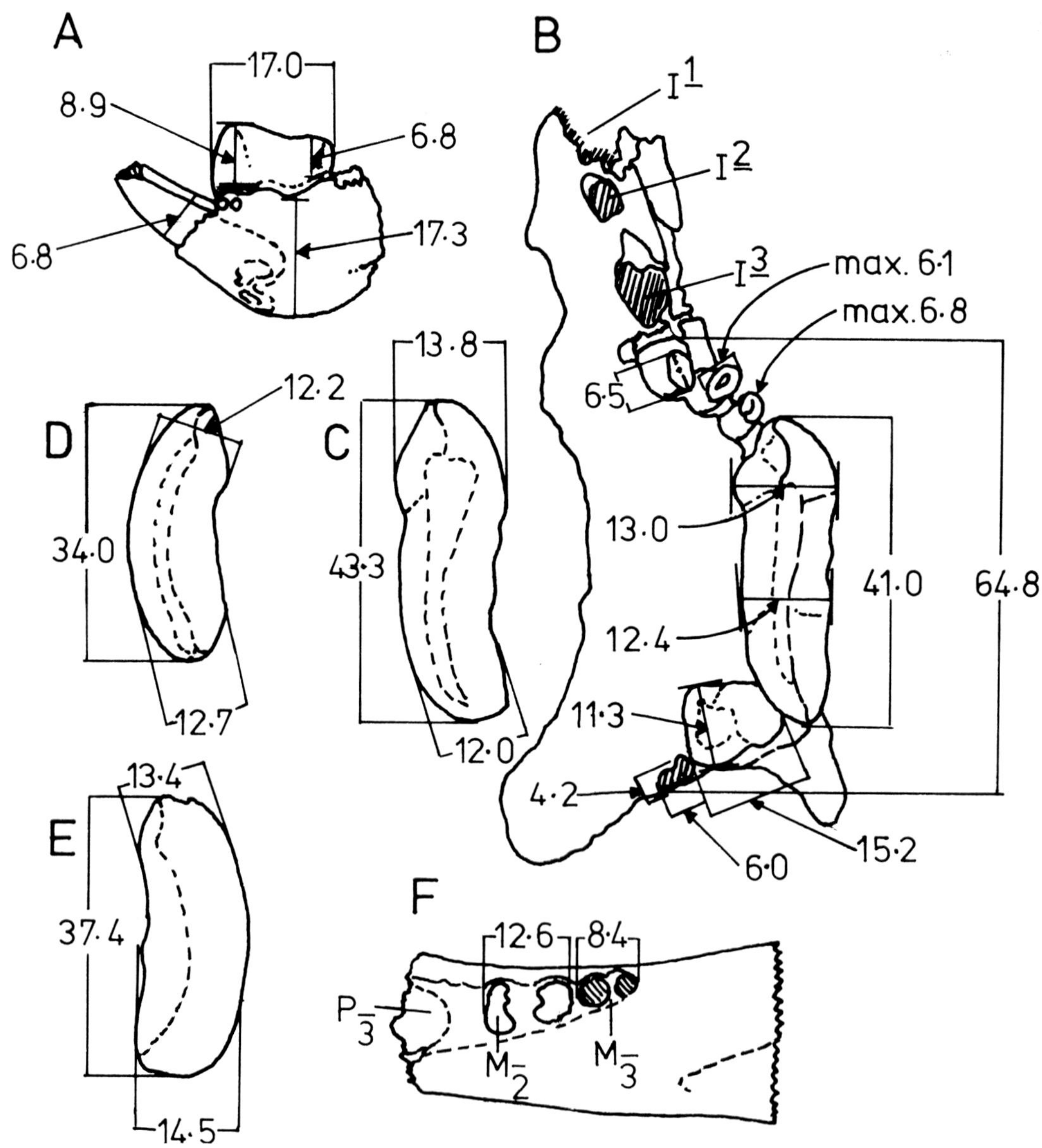

Fig. 5. Measurements (mm) of fossil thylacoleonids. A, AM F63584, right dentary fragment of *Thylacoleo* sp. cf. *T. hilli.* The height of the crown is measured below the anterior and posterior cusps. The depth of the dentary is measured at a point below lowest point in the P_3 longitudinal crest. B, QM F10622, left skull fragment of the Bluff Downs *T. crassidentatus.* The crown lengths of C^1–P^2 are maximum lengths no matter how the crown is oriented. C, QM F10623, another left P^3 of the Bluff Downs *T. crassidentatus.* D, AM F59594, left P_3 of the Bow *T. crassidentatus.* E, AM F59569, right P_3 of the Bow *T. crassidentatus.* F, AM F59593, a left dentary fragment of the Bow *T. crassidentatus* showing alveoli for a rather large M_3.

other being the Chinchilla specimen AM F2954), had a well–formed, incipiently two–rooted M^3, a tooth phylogenetically lost in the Pleistocene *T. carnifex*. Second, the M^2 is proportionately very large (16.3mm in maximum length) in comparison with the length of P^3 (41.4mm). The ratio of the maximum crown length of P^3 over the crown length of M^2 is 2.5 in the Bluff Downs specimen as opposed, for example, to 3.6 and 3.7 in two specimens of *T. carnifex* from Wellington Caves. Third, the M^2 contacts the P^3 near its extreme posterior end. In every specimen of *T. carnifex* examined, the P^3 extends for a considerable distance posterior to this point of intersection. In the Bluff Downs specimen, the P^3 extends approximately 3.0mm beyond this point of contact (Fig. 5). In two specimens of *T. carnifex* measured, the P^3 extends 10.5mm and 13.0mm beyond this same point. The reliability of this last interspecific difference is, however, doubted because of the structure of the isolated Bluff Downs left P^3 (QM F10623). It is somewhat longer (max. length 42.8mm) and may represent a male. On its lingual face, near the posterior end, are three wear facettes. The first and longest is clearly the occlusal counterpart of P^3. The second, at a steep angle to the first, is evidently the occlusal counterpart of the trigonid of M_2. The third is a small, oval, very shallow facette at the junction of the first two facettes. If it represents the interdental wear facette produced by the adjacent M^2, this point of contact on P^3 is followed posteriorly by about 9.2mm of P^3 crown, a length not significantly different from that which occurs in some specimens of *T. carnifex*.

Other Bluff Downs specimens: The dentary (QM F7762) has been figured and briefly discussed elsewhere (Archer *in* Archer and Wade 1976). Its most conspicious feature is the apparently small size of the broken P_3. The anterior end of the crown is missing. The alveolar length of the tooth is only 30.7mm as opposed to approximately 37.0mm in the holotype of *T. crassidentatus* and 36.4, 37.4, 42.6, and 43.4mm in four specimens of *T. carnifex* (Wellington Caves). Considered as a ratio, the P_3 alveolar length over the alveolar length of M^2, the Bluff Downs dentary is 2.1, the holotype of *T. crassidentus* is 2.9, and the four specimens of *T. carnifex* (Wellington Caves) are 3.3, 3.1, 3.5 and 3.3 respectively.

The morphology of P_3 was singled out by Bartholomai (1962: 35) to differ markedly in *T. crassidentatus* and *T. carnifex*. The posterior end of P_3 was said to be much broader and the "... longitudinal convexity much stronger ..." in *T. crassidentatus* than in *T. carnifex*. However, the breadth of the posterior end of P_3 in the holotype of *T. crassidentatus* is not noticeably greater than it is in many Wellington Cave specimens (e.g. AM F18884) of *T. carnifex*. The relative longitudinal convexity of P_3 is also not always clearly greater in *T. crassidentatus*. However, the holotype of *T. crassidentatus* and the isolated P_3 from Bow (see below) have the anterior–most vertical crest, which descends from the anterior cusp, orientated anterolingually with respect to the long axis of the tooth. In all *T. carnifex* specimens checked, this crest extends anteriorly with only a slight lingual turn. Other features of the P_3 of *T. crassidentatus* which may appear different on the basis of small samples, appear to be variable in large samples of *T. carnifex*. This includes the extent of mid–crown thickening, the width of the posterior part of the tooth, and the extent of vertical transverse crest development leading away from the main cusp. Variation in crown length in *T. carnifex* has been discussed above. It may in part be sexually as well geographically based.

The I^1 (QM F10642) appears to belong to a larger individual than does the skull fragment (QM F10622) because it will not fit into the I^1 alveolus of that specimen.

The Bow specimens: The dentary fragment (AM F59593) retains all alveoli for $M_{2\text{-}3}$ and part of the posterior alveolus for P_3. The alveoli for M_3 indicate that it was distinctly two–rooted and proportionately large. A ratio of alveolar lengths, M_2/M_3, is 1.6 for the Bow dentary, 2.1 for the Bluff Downs dentary, and 2.3, 3.3, and 3.4 for three dentaries of *T. carnifex* (Wellington Caves).

The two Bow P_3s differ from each other in massiveness, AM F59569 being the larger of the two. Otherwise they are essentially similar in morphology and closely resemble P_3 of the holotype of *T. crassidentatus*.

Comments: There is temptation to interpret the differences between the Pliocene *T. crassidentatus* and the Pleistocene *T. carnifex* as continuous change with time along a series of simple correlated morphoclines. In *T. carnifex*, the P^3 has lengthened posteriorly, the M^2 has become reduced in size, the M^3 has been lost, the P^3 has lengthened (presumably posteriorly) and the M_2 has been reduced in size. All of these changes could be interpreted as adaptations to the increasing specialization of the carnassial tooth. Reduction of the non–carnassial teeth (M2 and M^3) is a feature of progressive carnassialization in other mammalian lineages such as felids. Assuming this happened only once in late Tertiary thylacoleonids, it could provide a guide to the age determination of undated thylacoleonids.

Thylacoleo SP., cf. T. hilli PLEDGE FROM THE PLIOCENE BOW LOCAL FAUNA

In 1979, students and staff of the School of Zoology, University of New South Wales, de-

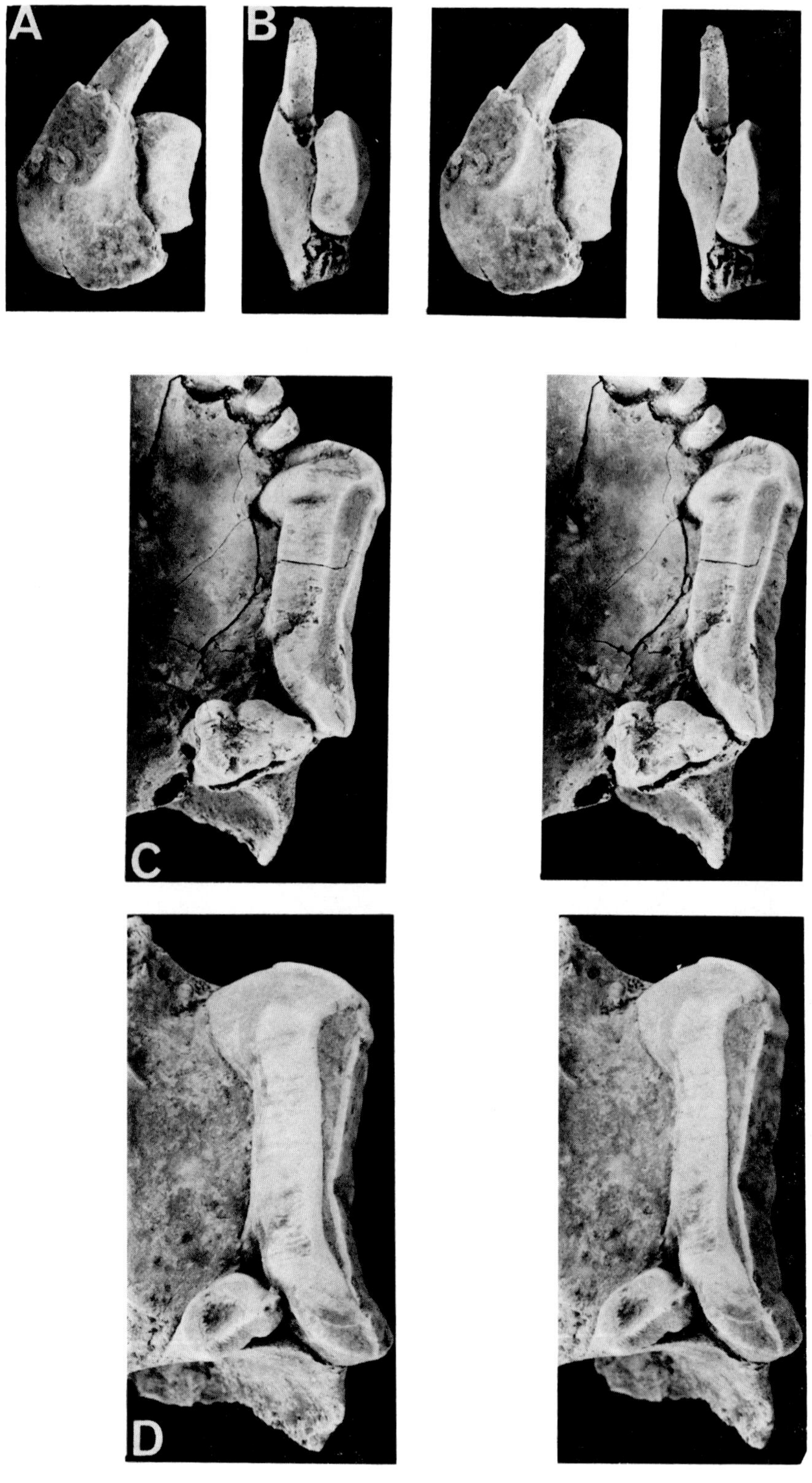

Fig. 6. Thylacoleo species, X 1. A, AM F63584, right dentary fragment with I_1 and P_3, and alveoli for P_{1-2} and M_3, *Thylacoleo* sp. *cf. T. hilli* from the Bow local fauna. B, AM F63584, occlusal view. C, QM F10622, left maxilla with P^1–M^2 and alveoli for M^3, *T. crassidentatus* from the Bluff Downs local fauna. D, AM F62526, left maxilla with P^3–M^2, *T. carnifex* from Wellington Caves, showing the proportionate elongation of the posterior part of P^3 past the M^2, the relatively reduced M^2, and the anterobuccal development of M^2.

veloped a quarry operation at the Bow fossil site. Washing of the quarry spoils by G. Clayton, S. Churchill and J. Hundt produced additional specimens including the partial dentary of a tiny thylacoleonid similar to *Thylacoleo hilli* Pledge. The only previously known material of *T. hilli* is the holotype, an isolated left P^3.

Thylacoleo sp., cf. *T. hilli* Pledge

Holotype of Thylacoleo hilli Pledge: SAM P17621, an isolated left P^3 from Town Cave (Y2), Curramulka, York Peninsula, South Australia.

Referred specimen: Besides the holotype, the only other possibly referable specimen is AM F 63584 (Figs 5, 6), the Bow right dentary fragment of a young adult marsupial lion. It preserves most of the right I_1, all of P_3, alveoli for P_{1-2} and partial alveoli for M_2. It was collected by the staff and students of the University of New South Wales during October, 1979, from immediately north of the Bow Post Office, on the side of the Bow road cutting, in northern New South Wales. The grid reference is 323033 on the 1:250,000 Geograpical Sheet (no. S1–56–1) for Singleton, New South Wales.

Stratigraphic unit and horizon of the Bow specimen: The dentary was obtained by screen–washing sediment obtained from an unnamed sandy, riverine conglomerate, loosely cemented by fine silt and clay, and overlying olivine flood basalts dated (elsewhere) at 34 to 42 million years (Skilbeck 1980). The riverine conglomerates are overlain by approximately 0.5 to 1.0m of blacksoil. The sediment from which the dentary was obtained was quarried from between 1.0 and 2.0m below the ground surface and therefore between 0.0 and 1.0m below the blacksoil–riverine conglomerate contact.

Probable age: The age of the Bow fauna is probably Pliocene based on similarities to that from the Chinchilla Sand (see Archer and Wade 1976).

Description of the Bow dentary (Fig. 6): The dentary fragment extends from the alveolus of I_1 to the posterior alveolus for M_2. All edges are worn and indicate that the specimen was damaged prior to its final deposition and recovery. There is a small mental foramen just anterior to the bulge for the anterior root of the P_3, and lateral to the I_1, a position similar to that found in all other thylacoleonids. The very short diastema between I_1 and P_3 is taken up by the alveoli for P_{1-2}. From the alveolar margins, it is clear that both P_1 and P_2 were present, as in other species of *Thylacoleo,* but unlike the condition in *Wakaleo oldfieldi* (Clemens and Plane 1974). The P_1 was wider than P_2. Both teeth were either very short–rooted or perhaps had not had time to develop longer roots, because both alveoli are extremely shallow and floored by the upper surface of I_1. A similar condition occurs in *T. carnifex* and probably *T. crassidentatus.*

The I_1, although much smaller, is essentially similar to the I_1 of *T. carnifex.* The only apparent difference is that in the Bow I_1, in occlusal view, the medial and posterior (lateral or buccal) sides of the tooth are more–or–less parallel. In *T. carnifex,* they tend to diverge posteriorly (towards the I_1 root). Reflecting this difference, but the ratio of the maximum width (transverse) of I_1 to the maximum width of P_3 in the Bow dentary is 0.59 (4.6/7.8) while that in a specimen of *T. carnifex* is 0.76 (11.5/15.2), indicating the proportionately narrower I_1 of the Bow dentary. In comparison with the I_1 of *Wakaleo oldfieldi* (Clemens and Plane 1974), the Bow I_1 is more similar to that of species of *Thylacoleo* in being wider in cross–section at its dorsal (occlusal) side, and in having the posterodorsal margin of its alveolus posterior to the anterior edge of the anterior root of P_3.

The P_3 of the Bow specimen is also essentially similar to that tooth in other species of *Thylacoleo.* It differs from all other species of *Thylacoleo* in being very small. It differs from all thylacoleonids in that the anterior vertical crest bifurcates near its base, with one ventral crest extending posterolingually and the other continuing down the anterior edge of the P_3 crown. As in other thylacoleonids (although there is some intraspecific variation in large samples of *T. carnifex*), there is a prominant vertical lingual crest extending down from the principal anterior cusp and a distinct basal swelling in a medial position along the lingual side of the crown. This latter swelling is more pronounced in *Wakaleo oldfieldi* than it is in any species of *Thylacoleo* and even extends well up the lingual side of the crown towards the longitudinal crest. The longitudinal crest of the Bow P_3 shows slight occlusal wear. The crest has a marked inflection near the posterior third of the crown where it swings in a posterolingual direction. The acuity of this inflection varies somewhat in specimens of *T. carnifex,* being more even and gentle in *T. carnifex* than it is in *T. crassidentatus.* It is most acute in *Wakaleo oldfieldi,* but only just more so than it is in the Bow P_3. There is some intraspecific variation in this feature in large samples of *T. carnifex,* ranging from a completely smooth curve (e.g. AM MF21) to a less–even curve (e.g. BM No. 39995). The posterior end of the longitudinal crest swings sharply linguad and then descends, merging evenly with the base of the crown. This is the condition in other species of *Thylacoleo* but not *Wakaleo.* In the latter, this crest swings anterolingually, terminating abruptly on the posterolingual side of the crown. There is also a more poorly–developed posterobuccal vertical crest on the Bow P_3 that descends from the longitudinal crest. This is better–developed in the Bow P_3 than on the P_3s of other species of *Thylacoleo*. There does not seem to be any crest in this position on the only known P_3 of *Wakaleo oldfieldi*. In lateral view, the anterior cusp of the Bow P_3 is higher than the posterior cusp but slightly less so than occurs in other species of *Thylacoleo*. In *Wakaleo oldfieldi*, the posterior cusp is only just lower than the anterior cusp. The main crest descending from the anterior cusp extends anteriorly, as it does in *T. carnifex*, rather than as it does in *T. crassidentatus* or *Wakaleo oldfieldi* where it swings anterolingually. The alveolus for M_2 reveals little of use in interpreting M_2 crown structure, except that the anterior root was clearly well–developed. Compared with a comparable preserved specimen of *T. carnifex* (AM F18884), the anterior alveolus of M_2 in the Bow specimen is proportionately wider.

The basis for referring the Bow dentary to Thylacoleo sp., cf. T. hilli Pledge: The holotype and only known specimen of *T. hilli* Pledge is an isolated left P^3. The Bow specimen is a dentary. On the face of it, the length of the Bow P_3 (16.6 mm) compared with the length of the P^3 of *T. hilli* (24.4mm) would seem to suggest they were not conspecific. Similarly suggestive of specific distinction, is the ratio of P_3 over P^3 lengths. The ratio of the Bow P_3 length over the *T. hilli* P^3 length is 0.68. This compares with a ratio of the mean of Wellington Cave *T. carnifex* P_3 lengths (39.5mm, $n = 33$) over the mean Wellington Cave *T. carnifex* P^3 lengths (44.6mm, $n = 24$) of 0.89. However, if we take the ratio of the smallest Wellington Cave *T. carnifex* P_3 length (36.2mm) over the longest Wellington Cave *T. carnifex* P^3 length (56.4mm), the value is 0.64 which is clearly not very different from the Bow/*T. hilli* ratio of 0.68. For this reason, and because we cannot make direct structural comparisons between the Bow dentary and corresponding specimens of *T. hilli*, we have decided to take a conservative approach in regarding the Bow dentary to be possibly, though clearly not certainly, referable to *T. hilli*.

Comments: The Bow specimen of *T.* sp. cf. *T. hilli* demonstrates at least three important points. First, it indicates that more than one species of thylacoleonid can exist in a single local fauna. In all other faunas, only one species, either of *Thylacoleo* or of *Wakaleo*, occurs. In the Bow local fauna, both *T. crassidentatus* and *T.* sp. cf. *T. hilli* occur. This apparent sympatry makes it clear that the "marsupial lion niche" can be partitioned up in much the same way as has been done by felids on other continents. In the Bow case, sympatry was probably facilitated by the great size difference between the two species.

Second, it is curious that, although greatly different in size, the two sympatric species are essentially similar in dental morphology. These thylacoleonids must have been utilizing a food

Table 5. Character–states of features that appear to be useful in phylogenetic interpretation of thylacoleonid evolution. The characters (1–12) are abbreviated as follows: 3–6, molar size proportionate to P3; 7, posterobuccal crest on P_3; 9–10, premolar length proportionate to M2. Alternative apomorphic character–states for characters 1–4 and 9–11 may be grades along continuous (but as yet incompletely documented) morphoclines. Question marks are an indication of uncertainty either because the condition has not been thoroughly considered (only character 9 for *T. crassidentatus*, C) or because the relevant part of the dentition is unknown. *Thylacoleo crassidentatus* (BD) refers to specimens from the Allingham Formation (Bluff Downs local fauna), and *T. crassicaudatus* (C) refers to Chinchilla specimens. *Thylacoleo carnifex* (B) refers to the Balladonia specimens (noted by Finch 1982), and *T. carnifex* (T) refers to all other typical specimens referable to *T. carnifex*. * Under this name we have combined data based on both the holotype of *T. hilli* as well as on the Bow specimen of *T.* sp. *cf. T. hilli.*

Character	Plesiomorphic states (P)	Apomorphic states (A)	*Wakaleo oldfieldi*	*W. vandeleuri*	*W. alcootaensis*	*Thylacoleo hilli**	*T. crassidentatus* (BD)	*T. crassidentatus* (C)	*T. carnifex* (T)	*T. carnifex* (B)
1. Premolar number	P^{1-3}/P_{1-3}	reduction	A	A	A	P	P	P	P	P
2. M_4	present	loss	P	P	P	A	A	A	A	A
3. M_3 (proportionate)	large	reduced to loss (1–2)	P	P	?P	?A1	A1	A1	A1	A2
4. M^3 (proportionate)	large	reduced to loss (1–3)	?P	?P	P	?A1	A2	A2	A3	A3
5. M_2 (proportionate)	large	reduced (1–2)	P	P	?P	?A1	A2	A2	A2	A2
6. M^2 (proportionate)	large	reduced (1–3)	?P	?P	P	?A1	A2	A2	A3	A3
7. Post. Buc. crest on P_3	absent	presence	P	P	?P	A	P	P	P	P
8. P_3 anterior crest	simple	bifurcated	P	P	?P	A	P	P	P	P
9. P_3 crown length (prop.)	short	long (1–4)	P	P	?A1	A2	A3	?A3	A4	A4
10. P^3 crown length (prop.)	short	long (1–4)	?P	?P	A1	?A2	A3	A4	A4	A4
11. Absolute size P3	small	larger (1–6)	P	A1	A2	A3	A5	A5	A6	A4
12. I_1 width	narrow	wide (1–2)	P	?P	?P	A1	A2	A2	A2	A2

resource that failed to lead to selection for significantly different tooth shape despite this size difference. Dasyurid carnivores of comparably different sizes have quite different tooth morphologies. This is not true, however, for many herbivorous macropodids of comparable sizes. This matter of thylacoleonid diets is open to wild (albeit enjoyable) speculation, but at present no clear solution.

Third, the Bow dentary clearly indicates that all thylacoleonids in the size range of species of *Wakaleo* need not be species of *Wakaleo*. The presence of $P_{1\text{-}2}$ and the *Thylacoleo*–like details of P_3 morphology, appear to clearly ally *T.* sp. *cf. T. hilli* with *Thylacoleo* rather than *Wakaleo*. This at least corroborates the decision of Pledge (1977) to refer *hilli* to *Thylacoleo*.

THYLACOLEONID EVOLUTION IN THE LATE CAINOZOIC: SOME TENTATIVE SPECULATION

We have tabulated twelve characters that appear to be significant in interpretation of thylacoleonid phylogeny (Table 5). Interpretation of the polarity of the character–states has been determined by outgroup comparison with diprotodontans as a whole. Some apomorphic states are graded (e.g. A1 to A6) according to the degree of development of the apomorphic condition. In one instance (character 11) an apparent reversal of apomorphic states appears to have occurred such that the Balladonia thylacoleonid (Finch 1982 and Finch and Freedman 1982) became secondarily smaller in size than its ancestor.

Figure 7 presents this data in the form of a cladogram with synapomorphic character states indicated on the various branches. In Fig. 8, this same data is presented in the form of a phylogenetic tree with the taxa placed according to their interpreted ages.

Clearly two major temporally overlapping lineages occur, the species of *Wakaleo* and those of *Thylacoleo*. *Wakaleo* species have reduced P1–2 and are in this regard more specialized than species of *Thylacoleo*. They have been discussed elsewhere (Archer and Rich 1982).

Within *Thylacoleo*, there appears to be a fundamental dichotomy between the more plesiomorphic *T. hilli* (holotype and the Bow speci-

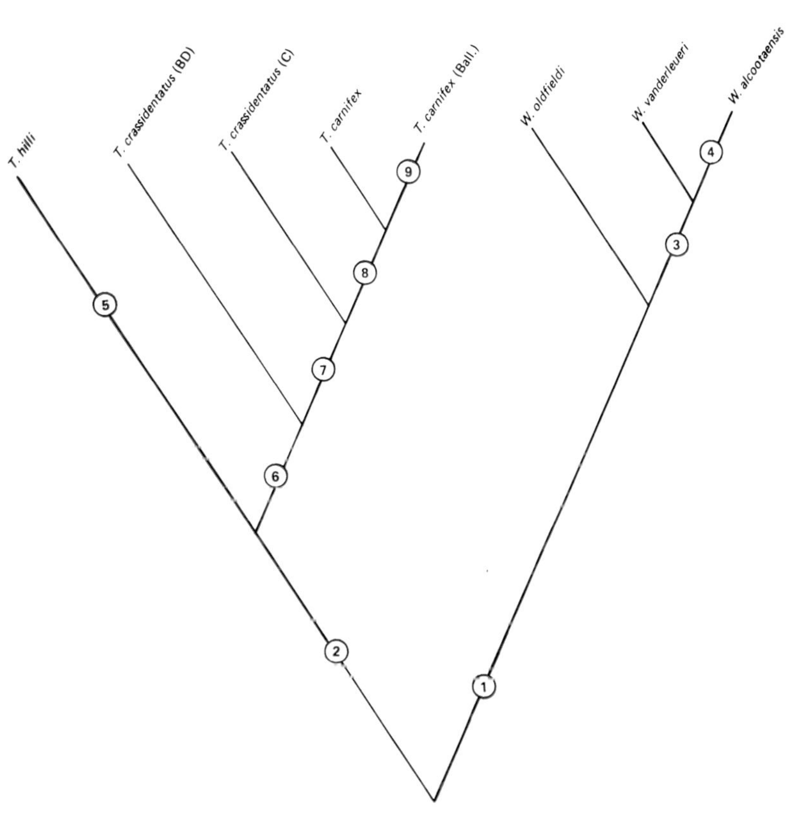

Fig. 7. Cladogram of hypothetical thylacoleonid phylogeny. Apomorphic character–states (Table 4): *Branch 1, states* 1A; *Branch 2,* states 2A, 3A, 4A, 5A, 6A, 12A; *Branch 3,* 11A1; *Branch 4,* 9?A1, 10A1, 11A2; *Branch 5,* 4?A1, 5?A1, 7A, 8A, 9A2, 10?A2, 11A3, 12A1; *Branch 6,* 4A2, 5A2, 6A2, 9A3, 10A3, 11?A4, 12A2; *Branch 7,* 10A4; *Branch 8,* 4A3, 6A3, 9A4; *Branch 9,* 3A2, 11A4. Abbreviations: BD, Bluff Downs; C, Chinchilla; Ball., Balladonia. The taxon *T. hilli* includes (in this analysis) the holotype as well as the Bow specimen of *T.* sp. *cf. T. hilli.*

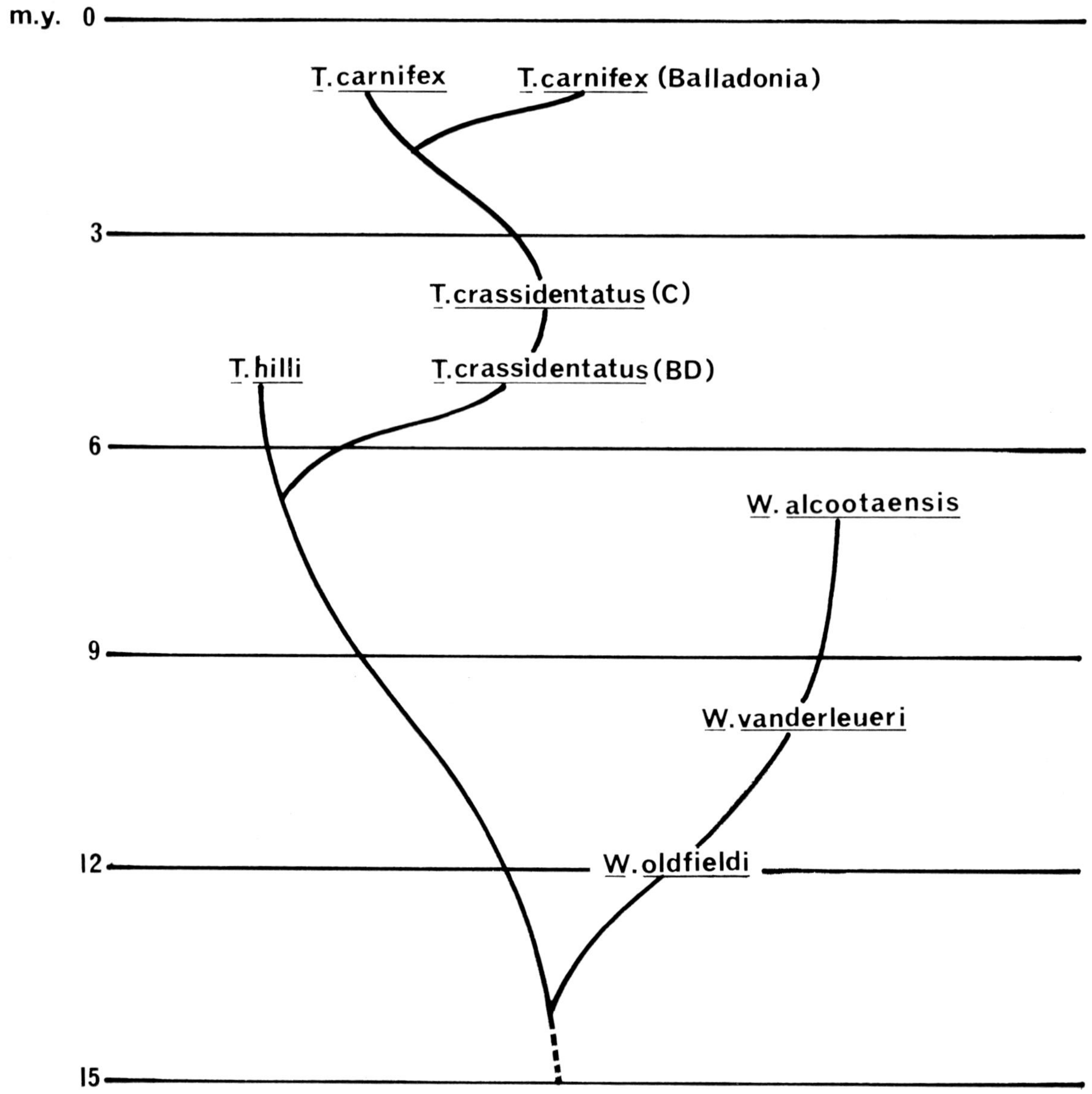

Fig. 8. Patristic hypothesis relating known thylacoleonids. The different placement of *T. crassidentatus* (BD) in Figs. 7 and 8 reflects the concept that ancestors as such cannot be recognized in phylogenetic systematics (cladistic proceedure), even though there is nothing as yet known about the Bluff Downs *crassidentatus* that would prohibit it as an ancestor of the Chinchilla *crassidentatus*. Abbreviations as in Fig. 7.

men) and all other species. It is of course possible that *T. hilli* represents a secondary reversion in absolute size from a *T. crassidentatus*–like ancestor. We doubt this because 1) small size charactersizes the other major branch of thylacoleonids (*Wakaleo* spp.) and is therefore probably a plesiomorphic character for the family; 2) *T.* sp. *cf. T. hilli* possesses a relatively less–specialized I_1 than do the other species of *Thylacoleo;* and 3) it appears to have also had a proportionately smaller P_3 in contrast to the other species of *Thylacoleo*. Within the remaining species, the Bluff Downs *Thylacoleo crassidentatus* appears to be the most plesiomorphic, retaining the least specialized P^3 and least reduced M^2. The Chinchilla (and possibly Bow) *T. crassidentatus* exhibits a somewhat more elongate P^3. *T. carnifex* is more specialized in its marked enlargement of P^3, reduction of molar size and loss of M^3. Within *T. carnifex,* the Balladonia population stands out in its apparent reduction of overall size.

There is little doubt in our minds that when more comes to be known of late Tertiary thylacoleonid evolution, the picture we have presented above will rapidly become outdated. As recently as 1978, Archer and Bartholomai (1978: 11) suggested that, considering what was then known about the fossil record of thylacoleonids, " . . . evolution has been almost orthogenetic in the continuous development of a proportionately longer premolar blade and reduced molar row. On these grounds, the recently described and undated *Thylacoleo hilli* with its very short premolar, could be interpreted as probably pre–Pliocene in age". The little Bow specimen has pulled the plug on that all too simple view: Australian palaeontology lurches ahead.

ACKNOWLEDGEMENTS

Dr. Ray E. Lemley financed and entirely made possible the expeditions to north Queensland where the new Bluff Downs *Thylacoleo crassidentatus* material was obtained. Mr. I. M. Tiley, Mr. Boughton and son, Mr G. Colton, Mr L. Cameron, Mrs. P. Wicks, Mr. L. Bailey, Mr. T. Houlahan, Mr. and Mrs. K. Creagan and other residents in the Merriwa and Bow areas extended invaluable official help. The Merriwa Shire Council was particularly helpful in making the trips to the fossil site possible. Messrs J. A. Mahoney and C. G. Skilbeck (University of Sydney), and Dr. A. Ritchie (Australian Museum) helped make many of the early collections at Bow. Dr. T. J. Dawson and family, Ms. E. Archer, Ms. D. Andrews, Dr. G. Maynes, Dr. B. Fox, Dr. M. Fox, Dr. G. Hardy, Ms. M. Hardy and about sixty enthusiastic students and friends (all staff, students or associates of the University of New South Wales) were involved in collecting the bulk of the Bow fossil material, including the unique dentary of *Thylacoleo* sp. *cf. T. hilli*. Dr. A. Ritchie, Dr. T. H. Rich, Dr. D. Merrilees and Dr. R. T. Wells kindly allowed access to the specimens in collections of which they are in charge. Lyn Dawson was in receipt of a Commonwealth Post Graduate Research Award and a Grant from the Trustees of the Australian Museum.

Mr. L. Meier (U.N.S.W.) took the light photographs, Dr. P. W. Keller took the X-ray photograph that enabled us to determine that the Merriwa dentary of *T. hilli* was adult. Ms. J. Hundt (now Taylor) (U.N.S.W.) and Dr. T. H. Rich (National Museum of Victoria) drafted the figures. Ms. J. Johansen typed the manuscript. Ms. E. Finch, Dr. T. Rich, Dr. R. Wells and Mr. M. Plane kindly read and criticised drafts of the manuscript.

REFERENCES

ANDERSON, C., 1929. Palaentological notes No. 1. *Macropus titan* Owen and *Thylacoleo carnifex* Owen. *Rec. Aust. Mus.* **17:** 35–49.

ARCHER, M., 1978. The nature of the molar–premolar boundary in marsupials and a reinterpretation of the homology of marsupial cheekteeth. *Mem. Qd Mus.* **18:** 157–64.

ARCHER, M. AND BARTHOLOMAI, A., 1978. Tertiary mammals of Australia: a synoptic review. *Alcheringa* **2:** 1–19.

ARCHER, M. AND RICH, T. H., 1982. Results of the Ray E. Lemley Expeditions. *Wakaleo alcootaensis* n. sp. (Thylacoleonidae: Marsupialia), a new marsupial lion from the Miocene of the Northern Territory, Australia, with a consideration of early radiations in the family. Pp.495–502 *in* "Carnivorous marsupials", ed. by M. Archer. Roy. Zool. Soc. New South Wales: Sydney.

ARCHER, M. AND WADE, M., 1976. Results of the Ray E. Lemley Expeditions, Part I. The Allingham Formation and a new Pliocene vertebrate fauna from northern Queensland. *Mem. Qd Mus.* **17:** 379–97.

BARTHOLOMAI, A., 1962. A new species of *Thylacoleo* and notes on some caudal vertebrae of *Palorchestes azael*. *Mem. Qd Mus.* **14:** 33–40.

CLEMENS, W. A. AND PLANE, M., 1974. Mid–Tertiary Thylacoleonidae (Marsupialia, Mammalia). *J. Paleo.* **48:** 652–60.

DE VIS, C. W. (IN ANON.), 1886. A post–Pliocene artiodactyle. *The Brisbane Courier, No. 8837, vol.* **41**, *May 12, 1886*: p.6, cols. 4–5. (See Mahoney and Ride 1975 for additional details of publication and republication in *The Daily Observer (Brisbane), No. 879, vol.* **3**, *May 12th, 1886:* p.2, cols. 5–6).

FINCH, M. E., 1982. The discovery and interpretation of *Thylacoleo carnifex* (Thylacoleonidae, Marsupialia). Pp.537–51 *in* "Carnivorous marsupials", ed. by M. Archer. Roy. Zool. Soc. New South Wales: Sydney.

FINCH, M. E. AND FREEDMAN, L., 1982. An odontometric study of *Thylacoleo* (Thylacoleonidae, Marsupialia). Pp.533–61 *in* "Carnivorous marsupials", ed. by M. Archer. Roy. Zool. Soc. New South Wales: Sydney.

GILL, E. D., 1973. Antipodal distribution of the holotype bones of *Thylacoleo carnifex* Owen (Marsupialia). *Sci. Rep., Tohoku Univ., 2nd Ser. (Geol.), spec. vol.* **6:** 497–99.

KREFFT, G., 1870. In [New South Wales Parliamentary Paper] *Wellington Caves (correspondence relative to the exploration of).* Pp.1–12.

KREFFT, G., 1872. Natural History. The phalanger tribe (continued). *The Sydney Mail and New South Wales Advertiser,* No. 621, Vol. 13, May 25th, 1872, p.649, cols. 2–4, figs. 1–14.

LONGMAN, H. A., 1916. The supposed artiodactyle Queensland fossils. *Proc. R. Soc. Qd* **28:** 83–7.

MAHONEY, J. A. AND RIDE, W. D. L., 1975. Index to the genera and species of fossil Mammalia described from Australia and New Guinea between 1838 and 1968 (including citations of type species and primary type specimens). *Spec. Publs. West. Aust. Mus.* **6.**

McCOY, F., 1876. *Thylacoleo carnifex* (Ow). *Prodromus of the Palaeontology of Victoria; or, figures and descriptions of Victorian organic remains.* Decade **3:** 7–12, pl.21. Melbourne, Geological Survey of Victoria.

OWEN, R., 1859. On the fossil mammals of Australia — Part I. Description of a mutilated skull of a large marsupial carnivore (*Thylacoleo carnifex,* Owen), from a calcareous conglomerate stratum, eighty miles southwest of Melbourne, Victoria. *Phil. Trans. R. Soc.* **149:** 309–22, pls.11–15.

OWEN, R., 1866. On the fossil mammals of Australia — Part II. Description of an almost entire skull of the *Thylacoleo carnifex,* Owen, from a freshwater deposit, Darling Downs, Queensland. *Phil. Trans. R. Soc.* **156:** 73–82, pls.2–4.

OWEN, R., 1871. On the fossil mammals of Australia — Part IV. Dentition and mandible of *Thylacoleo carnifex,* with remarks on the arguments for its herbivority. *Phil. Trans. R. Soc.* **161:** 213–66, pls.11–14.

OWEN, R. [*in* Anon.], 1888. On *Thylacopardus australis,* Owen. *Nature, Lond.* **39:** 215.

PLEDGE, N. S., 1977. A new species of *Thylacoleo* (Marsupialia: Thylacoleonidae) with notes on the occurrences and distribution of Thylacoleonidae in South Australia. *Rec. S. Aust. Mus.* **17:** 277–83.

RIDE, W. D. L., 1964. A review of Australian fossil marsupials. *J. Proc. Roy. Soc. West. Aust.* **47:** 97–131.

SKILBECK, C. G., 1980. A preliminary report on the late Cainozoic geology and fossil fauna of Bow, New South Wales. *Proc. Linn. Soc. New South Wales* **104:** 171–81.

SIMPSON, G. G., ROE, A. AND LEWONTIN, R. C., 1960. "Quantitative zoology". Rev. ed. Harcourt: New York.

WELLS, R. T. AND NICHOL, B., 1977. On the manus and pes of *Thylacoleo carnifex* Owen (Marsupialia). *Trans. Roy. Soc. S. Aust.* **101:** 139–146.

WOODS, J. T., 1956. The skull of *Thylacoleo carnifex. Mem. Qd Mus.* **13:** 125–40.

RESUMEN

39. Revisión de los "leones marsupiales" del género *Thylacoleo* Gervais (Thylacoleonidae: Marsupialia) y evolución de los Thylacoleonidae en el Cenezoico tardío. Por M. Archer y L. Dawson.

Se considera a *Thylacoleo* Gervais como el sinónimo más antiguo de *Plectodon* Krefft, *Schizodon* Stutchbury y *Thylacopardus* Owen. *Thylacoleo carnifex* es considerado como el más antiguo sinónimo de *Mylodon australis* Krefft, *Thylacopardus australis* Owen, *Thylacoleo oweni* McCoy y *Thylacoleo robustus* Krefft. Se considera a *Thylacoleo crassidentatus* Bartholomai y *Thylacoleo hilli* Pledge como especies distintas. Se describe nuevo material referible a *T. crassidentatus* del Plioceno inferior de la Formación Allingham de Queensland y de sedimentos del Plioceno de Bow en New South Wales. El material de Allingham revela mucha información sobre la diferenciación morfológica de las especies del género durante el Cenozoico tardío. Aunque el material de Bow revela poco nuevo acerca de *T. crassidentatus,* otro espécimen de Bow parece ser referible a la *T.* sp. cf. *T. hilli.* Es éste el primer caso en que dos especies de Thylacoleonidae aparecen en la misma fauna local, por lo que da cierta compresión sobre el "nicho del león marsupial" en Australia. La evolución de las especies del género durante el Plioceno–Pleistoceno es también considerada. Se confrontan los métodos cladístico y evolutivo del análisis filogenético de los tilacoleónidos.

40. RESULTS OF THE RAY E. LEMLEY EXPEDITIONS. *Wakaleo alcootaensis* n. sp. (THYLACOLEONIDAE, MARSUPIALIA), A NEW MARSUPIAL LION FROM THE MIOCENE OF THE NORTHERN TERRITORY WITH A CONSIDERATION OF EARLY RADIATION IN THE FAMILY

M. ARCHER[1] and T. H. RICH[2]

Wakaleo alcootaensis, a new species of marsupial lion from the Miocene Alcoota local fauna, Northern Territory, is described. It provides the first data about the upper dentition of species in the genus. The P^{1-2} are absent (lost phylogenetically), M^2 is tritubercular, M^3 was smaller than M^2, and there may have been an M^4. Relative to species of *Thylacoleo,* the molars are large and P^3 less well–developed. The *Thylacoleo* lineage must have been at least coeval with the *Wakaleo* lineage.

Evidence from the structure of M^2 suggests that the tooth has, within the scope of thylacoleonid evolution, suppressed an ancestral hypocone, become secondarily tribosphenic, rotated clockwise and developed a new cusp anterobuccal to the protocone to become again quadritubercular. It has evidently changed its function from a meat–cutting tooth (*Wakaleo* spp.) to a spatulate honing instrument opposing the thegotic facette of the M_2 trigonid (*Thylacoleo* spp.).

Pages 495–502 *in* CARNIVOROUS MARSUPIALS, ed. by M. Archer, February, 1982. Roy. Zool. Soc. New South Wales: Sydney, Australia.

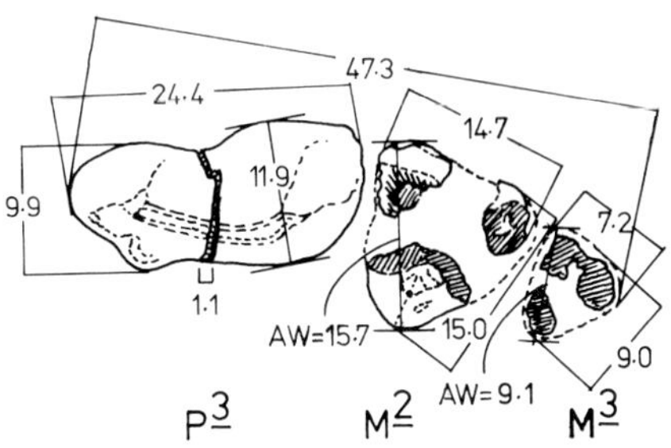

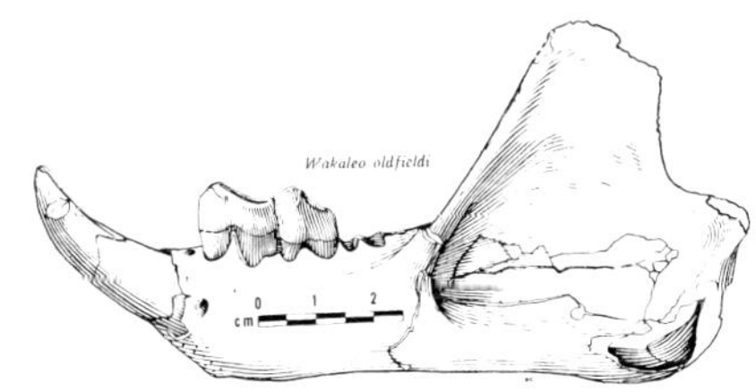

IN 1967, a Bureau of Mineral Resources field party, working in the Miocene Camfield Beds of the Northern Territory (Bullock Creek local fauna), recovered a fragment of a right dentary that retained M_2. In 1971, another field party consisting in part of palaeontologists from the University of California and the South Australian Museum, working in the Miocene Wipijiri Formation (Kutjamarpu local fauna) of South Australia, recovered an almost complete left dentary (the dentary shown next to the abstract of this paper) retaining P_3 and M_2, as well as an isolated partial RP_3 and an RM_2. These specimens have been described by Clemens and Plane (1974) as new species of the genus *Wakaleo* (*W. vanderleueri* and *W. oldfieldi* respectively). All currently described specimens referable to species of the thylacoleonid genus *Wakaleo* contain only portions of the lower dentitions (Clemens and Plane 1974).

On 4 June, 1974, we recovered, or rather almost destroyed while vigorously blocking out a large area of bone bed, a fragmentary maxilla of a small thylacoleonid from the Miocene Waite Formation (Alcoota local fauna) of the Northern Territory. This Alcoota specimen forms the subject of the present paper. It preserves the first known portions of the upper dentition of a species of *Wakaleo.*

Dental terminology follows Archer (1978). Accordingly, the maximum dental formula for known thylacoleonids is:

I^{1-3}; I_1; C^1; P^{1-3}; P_{1-3}; M^{2-4}; M_{2-4}.

The M1 has not been observed in any thylacoleonid specimen. As a result of information presented in the present paper and by Clemens and Plane (1974), it is evident that species of *Wakaleo* have lost P^{1-2} and P_{1-2}. Species of *Thylacoleo,* the only other known genus of thylacoleonids, have lost M^4 and (in some species) M^3, and M_4.

Measurements are given in Fig. 1. Museum abbreviations are as follows: NTM P1, palaeontological collections of the Northern Territory Museum, Darwin.

SYSTEMATICS

FAMILY THYLACOLEONIDAE

Genus *Wakaleo* Clemens & Plane, 1974

Wakaleo alcootaensis n. sp. (Figs 1-4)

Holotype: NTM P1, a left maxillary fragment with P^3, broken M^2, and alveoli (with roots)

1 School of Zoology, University of New South Wales, P.O. Box 1, Kensington, New South Wales, Australia, 2033.
2 National Museum of Victoria, Russell Street, Melbourne, Australia, 3000.

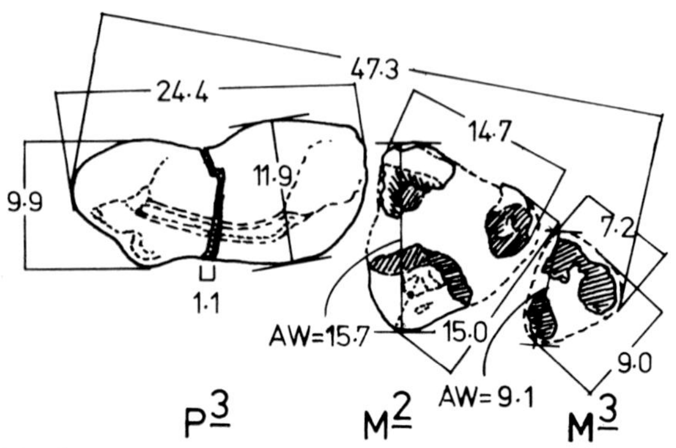

Fig 1. Measurements (mm) of the teeth of the holotype of *Wakaleo alcootaensis*. The P^3 is fractured transversely and the two halves separated by a 1.1 mm gap. The true length of P^3 is therefore 23.3 mm. Approximate crown outlines are dashed in around the remnants of M^2 and M^3. No trace of an M^4 remains, although this tooth was probably present. AW = minimum anterior width, from the lingual margin of the protocone remnant to the buccal margin of the anterobuccal corner of the tooth.

for M^3, collected by Mr. Archer and T. and P. Rich, 4 June, 1974.

Type locality: Three metres southwest of Paine Quarry (the quarry name and location are discussed in Woodburne 1967), 6.4 km southwest of Alcoota Homestead and 3.4 km southwest of the junction of the Waite and Ongeva Creeks, Alcoota Station, Northern Territory, Australia.

Age: The Alcoota local fauna is interpreted to be approximately 8 million years old (Woodburne *et al.* pers. comm., Archer and Bartholomai 1978), and thus possibly Cheltenhamian (late Miocene).

Diagnosis: *Wakaleo alcootaensis* differs from all species of the related genus *Thylacoleo* in lacking P^{1-2}, in having P^3 to M^2 anteroposteriorly oriented in a gently arcuate row, in having a discrete anterolingual cusp on P^3, in having a proportionately very large triangular M^2, and in probably retaining M^4. It differs from other species of *Wakaleo* in being much larger, the P^3 of *W. alcootaensis* being approximately twice the length of P_3 of *W. oldfieldi*.

Description: The maxillary fragment retains the base of the zygomatic arch which arises above M^2, in a more posterior position than occurs in species of *Thylacoleo*. The palate slopes steeply away from the tooth row, as in species of *Thylacoleo*. There is no evidence of any alveoli for P^1 or P^2, despite preservation of at least 7 mm of relatively undamaged maxilla anterior to P^3. In species of *Thylacoleo*, the posterior edge of the alveolus for P^2 occurs within 2 mm of the P^3 alveolus. The palate continues posteriorly behind the alveoli for M^3, but because of damage, it is uncertain whether M^4 was present. Nevertheless, it seems highly probable that it was considering the large size of M^3 and the width and length of palatal area available behind M^3. An unfortunately accurate swing with Archer's Marsh pick at the moment of discovery severed us from certainty about this point. Several tiny perforations lingual to P^3 may be palatal foramina or could be artifacts of preservation.

The P^3 is conspicuously wider posteriorly than anteriorly, in marked contrast to the P^3 in species of *Thylacoleo*. At opposite ends of the longitudinal crest are two principal cusps. The anterior cusp is slightly higher than the posterior which is much better-developed than the posterior cusps of some specimens of species of *Thylacoleo*. However, it seems probable that this posterior cusp in *Wakaleo alcootaensis* is not homologous with the posterior cusp in P^3 of species of *Thylacoleo*. From the anterior cusp, a pronounced crest ascends more–or–less vertically, in a posterolateral direction, to merge with the base of the crown. In contrast to P^3 of most species of *Thylacoleo,* there is no crest ascending the lingual flank of the crown from the anterior cusp such as occurs in many (but not all) specimens of *T. carnifex* and *T. crassidentatus*. There is, however, a distinct anterolingual cusp at the base of the crown above the main anterior cusp. This cusp is crescentic, its concave side pointing posteroventrally, its anterior end terminating half way up the lingual side of the crown, and its posterior end merging with the basal swelling of the crown. It sustains wear that would have been produced by the anterior cusp of P_3. The posterior end of P^3 may be damaged, although this is uncertain because the resultant crown surface is smooth and regular. Whether due to damage or not, a crest extends posterolaterally up the buccal side of the crown from the posterior cusp and merges with the base of the crown. On the lingual side, although there is not a crest as such, there is a distinct, vertically oriented, elongate swelling extending from the posterior cusp to the crown. The longitudinal crest is sinuous with inflections at the anterior cusp, the mid–point of the crest, the posterior cusp, and midway between the posterior cusp and the posterior edge of the crown. The short vertical crest that extends anterior to the anterior cusp has a small cuspule associated with its anterodorsal end on the anterior base of the crown. The part of the longitudinal crest posterior to the posterior cusp swings buccally to the base of the crown. This postero–buccal crest, assuming it is not an artefact resulting from a piece of enamel having spalled from the posterobuccal corner of the tooth, has no homologue in species of *Thylacoleo*. On the buccal side of the crown, between the anterolateral vertical crest and the posterolateral vertical crest, there are two minor swellings along the base of the crown. Two roots are present. The posterior root is the larger, supporting 56% of the crown length. In specimens of *T. carnifex,* the much

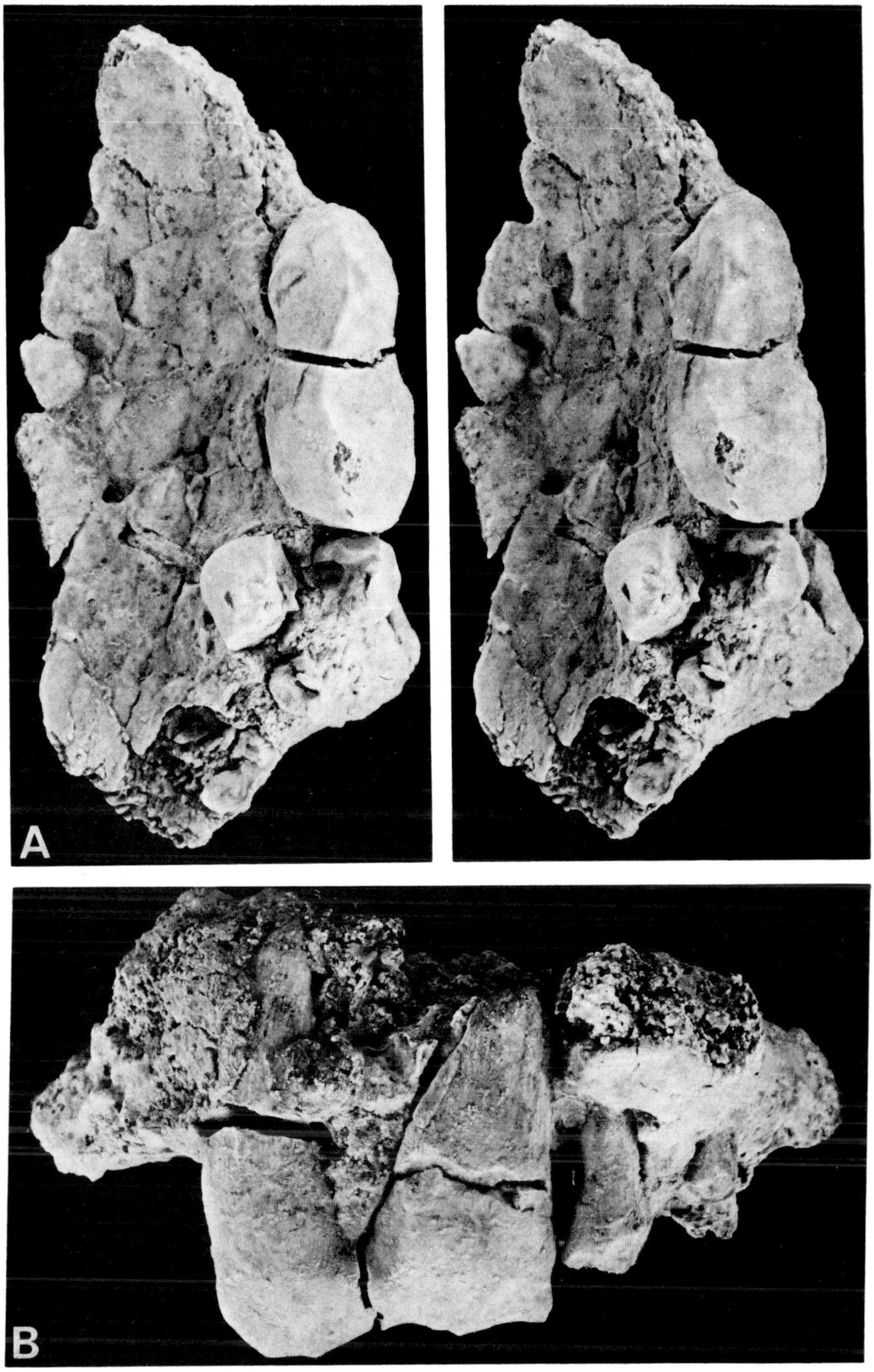

Fig. 2. The holotype of *Wakaleo alcootaensis*, X2. A, occlusal view. B, buccal view.

larger posterior root supports about 70% of the crown length and in *T. crassidentatus,* about 60% of the crown. The wear pattern on P^3 is comparable to that seen in specimens of *Thylacoleo* spp. except that the anterolingual basal cusp is distinct with a discrete wear facette.

The large M^2 is badly damaged. It is represented by three triangularly distributed roots with the lingual root still supporting a significant remnant of the crown. This portion of the crown indicates that the protocone was low, with a poorly defined postprotocrista. There were crenulations in the protoconal basin. A tiny fossette occurs near the apex of the low crown, just posterior to the protocone and on the lingual edge of the postprotocrista. The base of the crown extends out beyond the protocone, and thus must have appeared to be swollen at its base. The basal swelling and protoconal basin crenulations are the only clear similarities to M^2 of species of *Thylacoleo,* and in particular to *T. crassidentatus* from Bluff Downs (see the illustrations in Archer and Dawson 1982). The anterobuccal (paracone) root is surmounted by a portion of crown. If the shape is not the result of damage, the base of the crown extended beyond the base of the paracone, leaving a buccally concave surface between the missing paracone and the base of the crown. Alternatively, this buccal concavity may be the result of spalling. The intact crown was evidently basically triangular with its three sides roughly subequal, and all much longer than the width of P^3.

The relatively much reduced M^2 of *T. carnifex* is normally quadrilateral in shape, with an anterobuccal parastyle that contacts the lingual flank of P^3. In *T. crassidentatus* the M^2 retains vestiges of a triangular outline, there being only a small lingual swelling between the parastyle and the protocone. The M^2 in *W. alcootaensis* also differs from that tooth in species of *Thylacoleo* in occurring immediately posterior to P^3. In species of *Thylacoleo,* M^2 occurs either posterolingual to P^3 (*T. crassidentatus*) or lingual to P^3 (*T. carnifex*).

Only roots of M^3 remain. There was a distinct lingual protoconal root and on the buccal side, two incipiently differentiated roots, one anterior and one posterior, fused by a narrow strap of dentine on the buccal side. This tooth also

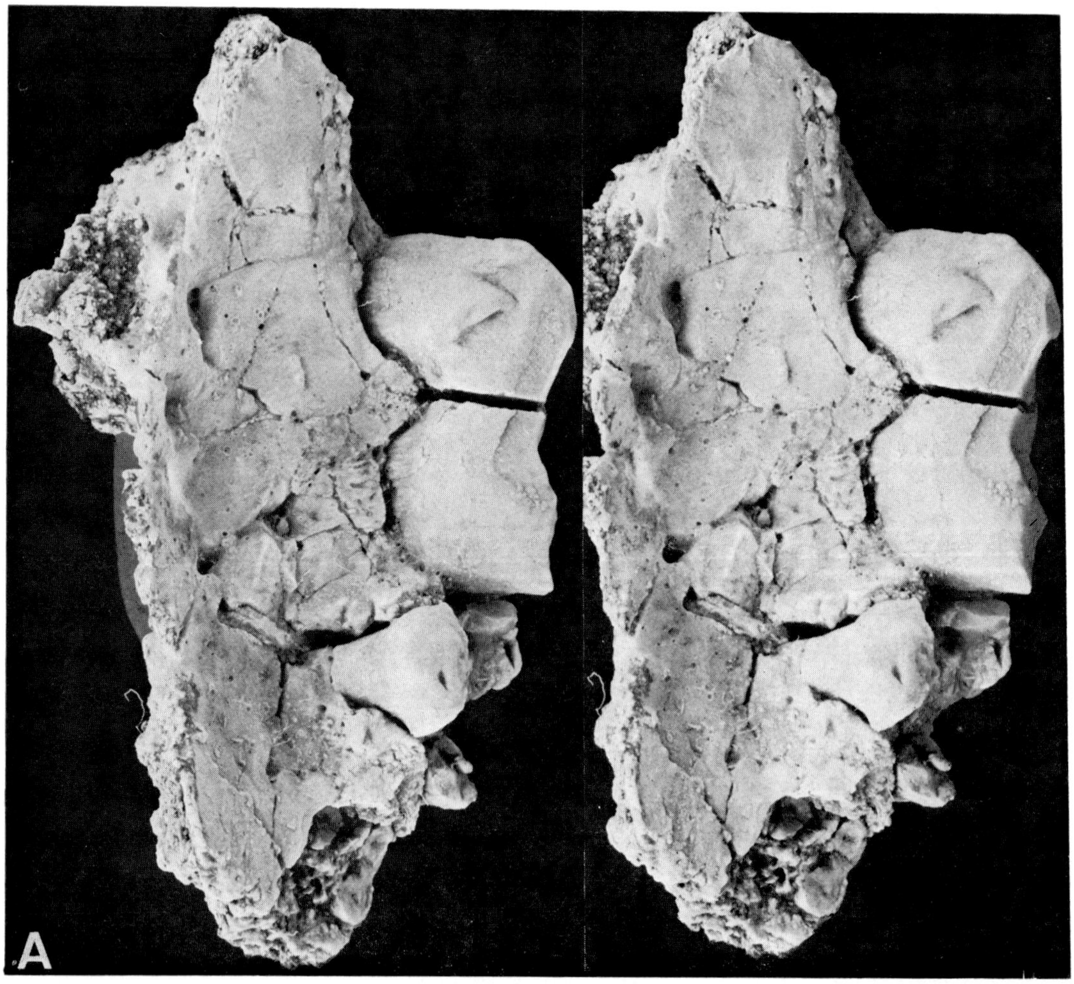

Fig. 3. Left maxilla of *Wakaleo alcootaensis,* holotype, occlusal oblique view. X2.

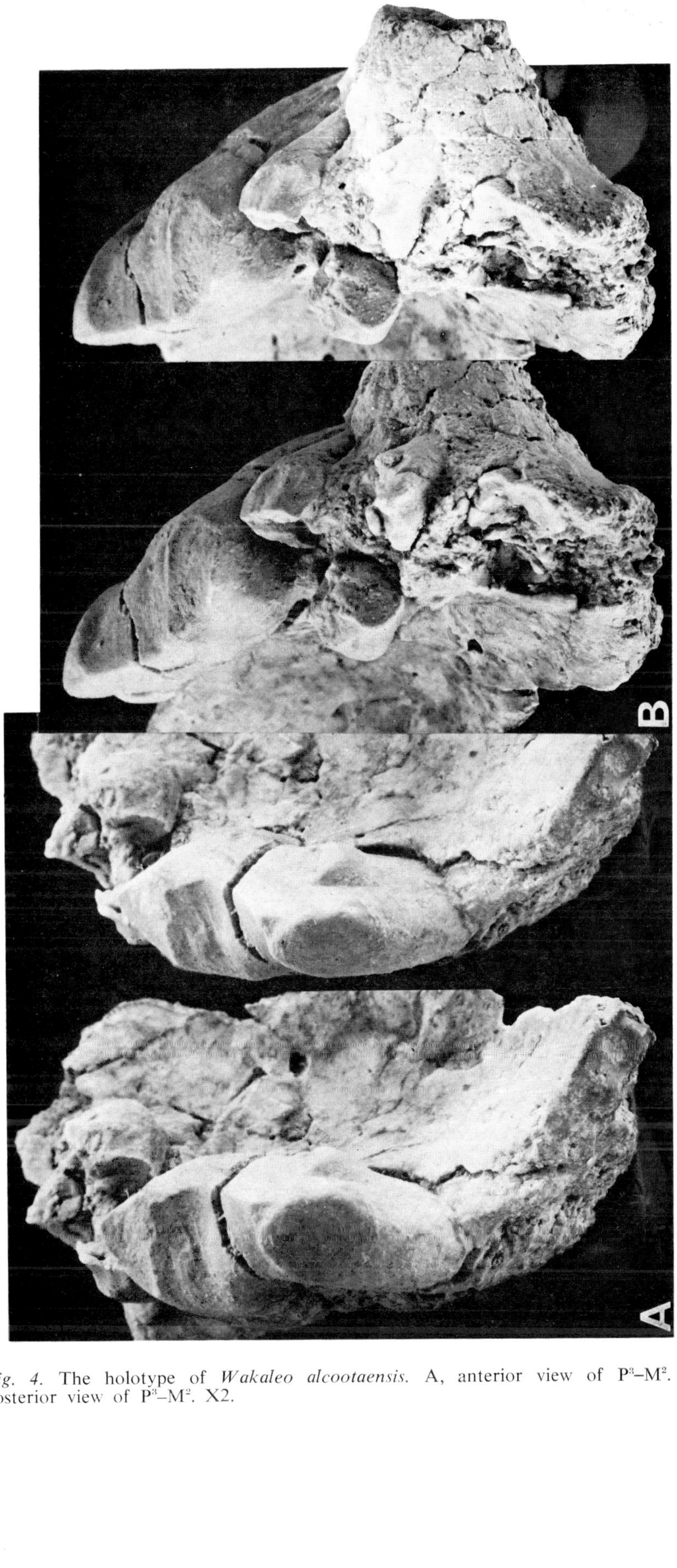

Fig. 4. The holotype of *Wakaleo alcootaensis.* A, anterior view of P^3–M^2. B, posterior view of P^3–M^2. X2.

occurred immediately posterior to M^2. In the only other thylacoleonid (*T. crassidentatus*) known to possess M^3, M^2 is proportionately much smaller, and although it has the same relative position with respect to M^2 because of the greater curvature of the whole molar row in species of *Thylacoleo*, M^3 actually lies posterolingual to M^2.

Presence of a possible M^4 is discussed above in respect to the palatal area available. No actual remnant of such a tooth remains. If present, it would have occurred immediately posterior to M^3.

The whole tooth curves gently, being convex buccally. This even curvature is due partly to the short length of P^3, and partly due to the sequential alignment of the teeth. In species of *Thylacoleo*, the molar row occurs at a steep angle to the relatively much more elongate P^3.

DISCUSSION AND A CONSIDERATION OF THYLACOLEONID ORIGINS

Referral of the Alcoota species to *Wakaleo* is not justified on the basis of the relatively (with respect to *Thylacoleo*) well–developed molars, short premolar, and non–rotated molar row of *alcootaensis*, because these are most probably plesiomorphic conditions within the Thylacoleonidae. On the other hand, loss of the anterior premolars is an apomorphic character state shared with the species of *Wakaleo* (Clemens & Plane 1974). Although common loss is probably more likely to be the result of convergence than is acquisition, it stands nevertheless as a synapomorphic character state uniting *alcootaensis* with the species of *Wakaleo*. Because upper dentitions are not yet known for *Wakaleo oldfieldi* or for *W. vanderleueri*, it seems preferable at present to refer *alcootaensis* to *Wakaleo* rather than to split it away from these species as a monotypic genus. Accordingly, we suggest that a synapomorphic and diagnostic character state for species of *Wakaleo* is the absence of both upper and lower first and second premolars.

Wakaleo alcootaensis provides insight into the probable evolution of the dentition of thylacoleonids as firstly, it corroborates the suggestion of Clemens and Plane (1974) that species of *Wakaleo* differ from *Thylacoleo* spp. in having lost at least one of the anterior premolars. Although this suggests that species of *Wakaleo* could not be ancestral to species of *Thylacoleo*, they point out that because there is a known instance of reappearance of a lost tooth (Kurten 1963, M_2 in a lineage of felids), a species of *Wakaleo* might still have been directly ancestral to species of *Thylacoleo*, all of which have P_2. With this caution in mind, it is nevertheless most parsimonious to suggest that the *Wakaleo* lineage is a specialised collateral branch of thylacoleonids not directly ancestral to the *Thylacoleo* lineage. As suggested by Clemens and Plane (1974), however, species of *Wakaleo* may otherwise indicate a grade of structural development that was involved in the evolution of the species of *Thylacoleo*. A further implication is that the lineage leading to *Thylacoleo* spp. must have been at least coeval with that leading to *Wakaleo* spp. assuming that *Wakaleo* is defined in part by reduction of the anterior premolars (Fig. 5).

The Alcoota species also gives an indication of how the molar orientation characteristic of species of *Thylacoleo* developed (Fig. 6). In *W. alcootaensis* M^2 was evidently triangular with the anterobuccal cusp abutting the posterior surface of P^3. From indications within *Thylacoleo* (Archer and Dawson 1982), it seems that P^3 elongation has occurred mainly by development of the posterior end of the crown, such that P^3 overrides the anterobuccal corner of M^2. In the Pliocene Bluff Downs (north Queensland) specimens of *T. crassidentatus*, M^2 is subtriangular, but the paracone (the anterobuccal cusp of the M^2 in *W. alcootaensis*) is oriented buccally and abuts P^3 on the posterolingual edge of the P^3 crown. The protocone–metacone side of the tooth faces lingually rather than posterolingually as it does in *W. alcootaensis*. In the Pleistocene *T. carnifex*, the quadrangular M^2 has been bypassed by the posterior development of P^3, and its anterobuccal cusp abuts against the lingual flank of P^3. The M^2 is also orientated such that the convex protocone–metacone side of the crown faces lingually. The reasons for this apparent phylogentic rotation of M^2 may have been related to the progressive enlargement of the P^3, enlargement of the trigonid height of M_2 and reduction of talonid width of M_2. In species of *Wakaleo*, P^3 crown height and M_2 trigonid height are relatively low. Phylogenetic increase in the height of the anterior end of P^3 and the trigonid of M_2 in species of *Thylacoleo*, combined with the posterior bypass of M^2 by P^3, evidently left M^2 in a better position to assist

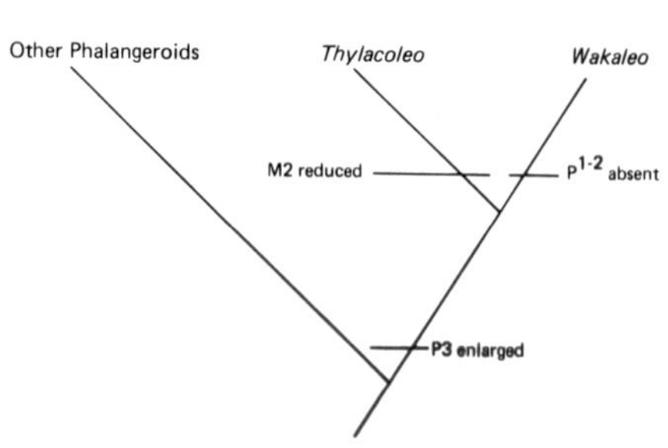

Fig. 5. Cladogram depicting three key apomorphic developments leading to the lineages of thylacoleonids. Evidence for the loss of P^{1-2} in the *Wakaleo* lineage is based solely on *W. alcootaensis* although reduction to loss of P_{1-2} in the other species of *Wakaleo* probably reflects similar apomorphic reduction in the upper premolars of these species.

with the honing of the posterobuccal face of the tall trigonid of M_2. The talonid of M_2 also became reduced in width, presumably in conjunction with a reduction in need for crushing elements in the dentition. This progressive reduction evidently occurred on the buccal side of the M_2 talonid, and may have provided a selection pressure for M^2 to rotate lingually in order to maintain occlusion.

The most striking thing about the molars of *W. alcootaensis,* besides their large size, is their evidently triangular outline. It is commonly held that thylacoleonids descended from some diprotodontan group, almost all of which have quadritubercular upper molars. Ride (1964:118), for example, suggests that thylacoleonids are "Obviously derived from large phalangers . . ." and ". . . represent the supreme diprotodont adaptation to the rapacious habit, a habit which is not strange to some of their smaller relatives". Similarly, Archer (1976) suggests that thylacoleonids might have been derived from phalangerids. All known phalangerids, as well as all other diprotodontans (including all of the possums, kangaroos, diprotodontids, koalas, and ektopodontids) have upper molars exhibiting hypocone development. From the shape and orientation of the roots and details of crown structure adjacent to the protocone, we infer that *Wakaleo alcootaensis* either had no hypocone development or at best only very slight hypocone development. We are then faced with two interpretations. Either this presumably tritubercular tooth was more primitive in structure than that of any other diprotodontan and not a modification of a quadritubercular ancestor, *or* it is secondarily tritubercular, having suppressed an ancestral hypocone. We favour the latter interpretation. There is nothing else about thylacoleonids that would lead to a conclusion that they are structurally the most primitive diprotodontans. Elsewhere (Archer 1976) it has been argued that the most primitive diprotodontans were selenodont forms similar to ancestral phascolarctids or petaurids. It seems more likely that thylacoleonids descended from quadritubercular ancestors, perhaps some group such as burramyids that were predisposed to eating meat or at least insects and under some selective pressure to redevelop tritubercular sectorial structures in their molars.

The *Wakaleo* structural grade may then have involved tritubercular upper molars combined with a long sectorial (probably carnassial) premolar. To get from this *Wakaleo* grade to the *Thylacoleo* grade, the molars must have been reduced in size and the premolar enlarged. Evidently faced with two types of potentially carnivorous adaptations available in the *Wakaleo* grade dentition — a sectorial premolar and tritubercular upper molars — selection in the *Thylacoleo*

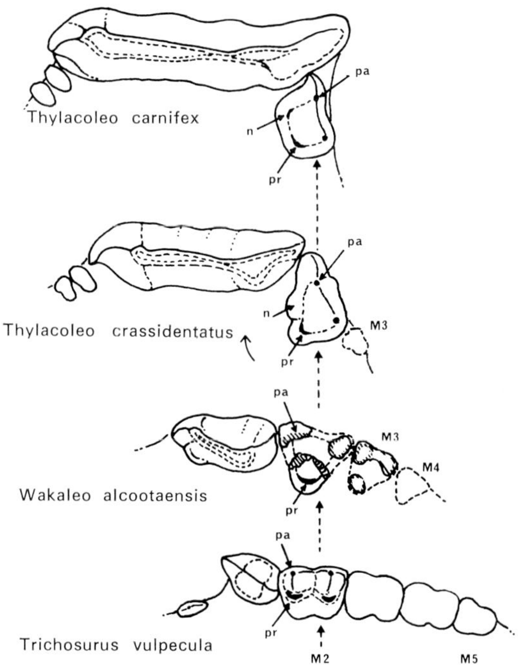

Fig. 6. A schematic portrayal of the hypothetical changes in structure and orientation of M^2 in thylacoleonids. In a presumed quadritubercular ancestor (morphologically probably similar to a burramyid or phalangerid — *Trichosurus* is shown), the protocone (pr) and hypocone were well–developed lingual cusps, and the paracone (pa) was anterior in position. In the Miocene *Wakaleo alcootaensis,* the hypocone has evidently been suppressed, perhaps in response to selection for the redevelopment of a tribosphenic molar, a more suitable molar morphology for a carnivore. The protocone is still lingual. In the Pliocene *Thylacoleo crassidentatus,* M^2 has evidently rotated clockwise (in the direction of the curved arrow) in response to the now massive enlargement of P^3. In response to probable selection for maintenance of occlusion with P^3, and also in response to the need for a structure on M^2 to hone the major thegotic facette of the M_2 trigonid, a new cusp (n) buccal to the anterolingual protocone appears to have developed in *T. crassidentatus.* This new anterobuccal cusp is even better–developed in the Pleistocene *T. carnifex* where it is commonly as large as the protocone.

lineage favoured gross specialisation of the sectorial premolar at the expense of the number and size of the upper molars. As noted above, this evolution also appears to have resulted in the rotation of M^2. In *Thylacoleo crassidentatus* (the Bluff Downs material), the essential triangular shape of M^2 is still clear, but here an asymmetry has developed. The paracone–metacone side and the paracone–protocone side have lengthened relative to the protocone–metacone side. Additionally, a basal swelling between the protocone and paracone heralds the first development of a new fourth corner for M^2. In *T. carnifex,* where M^2 is relatively even smaller, and the protocone–metacone side of the tooth even shorter relative to the paracone–metacone side, this fourth cusp between the protocone and the paracone is greatly enlarged and the whole tooth is once again quadritubercular. But most curiously, this squaring up has not been the result of addition of a hypocone such as was probably the method of squaring up used by other diprotodontans and perameloids.

If this speculative hypothesis is correct, M^2 of thylacoleonids has had a most interesting phylogenetic history. To recount, we hypothesize that: 1) it began as a tribospheric tooth in the ancestors of diprotodontans; 2) it acquired a hypocone (probably a modified metaconule; Archer 1976, Bensley 1903) and became quadritubercular in ancestral diprotodontans; 3) it suppressed the hypocone to become tribosphenic in ancestral thylacoleonids (the *Wakaleo* grade of development); and 4) it developed a new fourth cusp between the protocone and paracone to again acquire a quadritubercular structure (the *Thylacoleo* grade of development). The selective reason for M^2 regaining a quadritubercular shape is, we think, in part related to the need to function as a flat broad honing surface for the M_2 trigonid. This function is improved by anterior development of the crown in the region between the protocone and paracone. It also would help to enclose the embrasure between the lingual flank of P^3 and the anterior flank of M^2.

ACKNOWLEDGEMENTS

Our sincerest thanks are extended to Dr. Ray E. Lemley who financed and accompanied the 1974 Queensland Museum trip to Alcoota Station at which time the holotype was obtained. Tom and Wendy Webb of Alcoota Station also extended many courtesies during our visit. Most of the success of the expedition depended on them. Dr. Colin Jack-Hinton, Director of the Northern Territory Museum and Art Gallery, kindly permitted collection of material at Alcoota. Mr. Carl Roth, also of the Northern Territory Museum and Art Gallery, accompanied the expedition and provided (with his wife Bobby) indispensible logistical help during trips to Alice Springs. Dr. Pat Rich (now of Monash University) and Mr. Errol Beutel (Queensland Museum) each spent several weeks recovering the enormous blocks of Alcoota bone–bed. Mr. Rudy Kahout (Queensland (Museum), Mr. Henk Godthelp (then of the Queensland Museum) and other Queensland Museum staff all shared in the albeit delightful labours of preparing the Alcoota material.

Dr. R. T. Wells and Ms. E. Finch provided useful criticism of a draft of this paper.

REFERENCES

Archer, M., 1976. Phascolarctid origins and potential of the selenodont molar in the evolution of diprotodont marsupials. *Mem. Qd Mus.* **17:** 367–71.

Archer, M., 1978. The nature of the molar–premolar boundary in marsupials and reinterpretation of the homology of marsupial cheekteeth. *Mem. Qd Mus.* **18:** 157–64.

Archer, M. and Bartholomai, A., 1978. Tertiary mammals of Australia: a synoptic review. *Alcheringa* **2:** 1–19.

Archer, M. and Dawson, L., 1982. Revision of marsupial lions of the genus *Thylacoleo* Gervais (Thylacoleonidae, Marsupialia) and thylacoleonid evolution in the late Cainozoic. Pp.477–94 *in* "Carnivorous marsupials", ed. by M. Archer. Roy. Zool. Soc. New South Wales: Sydney.

Bensley, B. A., 1903. On the evolution of the Australian Marsupialia with remarks on the relationships of the marsupials in general. *Trans. Linn. Soc. Lond.* **9:** 83–217.

Clemens, W. A. and Plane, M., 1974. Mid-Tertiary Thylacoleonidae (Marsupialia, Mammalia). *J. Paleo.* **48:** 652–60.

Kurten, B., 1963. Return of a lost structure in the evolution of the felid dentition. *Soc. Sci. Fennica Commentations Biol.* **26:** 1–10.

Ride, W. D. L., 1964. A review of Australian fossil marsupials. *J. Proc. R. Soc. West. Aust.* **47:** 97–131.

RESUMEN

38. Resultados de las expediciones Ray E. Lemley. *Wakaleo alcootaensis* n.sp. (Marsupialia: Thylacoleonidae), un nuevo león marsupial del Mioceno del Northern Territory, con consideraciones sobre la temprana radiación de la familia. Por M. Archer y T. H. Rich.

Se describe *Wakaleo alcootaensis*, una nueva especie de león marsupial de la fauna local Alcoota, del Mioceno del Northern Territory. Provee los primeros datos sobre la dentición superior de una del géero. Los $P^{1\text{-}2}$ están ausentes (perdidos filogenéticamente), el M^2 es tribubercular, el M^3 mas pequeño que el M^2, y pudo haber habido un M^4. En relación a las especies de *Thylacoleo*, los molares son más grandes y el P^3 menos hipertrofiado. La linea de *Thylacoleo* debe haber sido al menos coeva de la linea de *Wakaleo*.

La evidencia obtenida de la estrucura del M^2 sugiere que este diente, dentro de los límites de la evolución de los Thylacoleonidae, ha perdido el ancestral hipocono, ha rotado hacia la derecha y desarrollado una nueva cúspide anterobucal al protocono, deviniendo así nuevamente caudritubercular. Evidentemente ha cambiado su función de un diente carnicero (*Wakaleo* spp.) a otro espatulado, actuante como instrumento de afilación de la opuesta tegósica del trigó del M_2 (*Thylacoleo* spp.).

41. DASYURIDS (MARSUPIALIA) IN LATE QUATERNARY COMMUNITIES IN SOUTHWESTERN AUSTRALIA

ALEXANDER BAYNES[1]

In Hastings Cave the paramount factor controlling net relative abundances of small mammal species accumulated during the Holocene was substrate areas in the edaphic zones surrounding this coastal plain site. Of the abundant dasyurids, *Sminthopsis granulipes* was restricted to the quartz sand zone community; *S. murina* probably occurred initially in the laterite gravel sand zone community, and later invaded the developing calcareous sand zone (which may have carried more than one mammal subcommunity); and *Dasycercus cristicauda* probably occurred mainly in the calcareous sand zone. *Phascogale calura* was lost from the Hastings Cave area in the early to mid Holocene, and *Sarcophilus harrisii* probably in the late Holocene.

The Devils Lair deposit contains 8 species of mammal present in the extreme south-west of Australia during the last glacial which became locally extinct by the late Holocene. Most, if not all, had presumably invaded the area since the last interglacial. No small dasyurids are included in this group.

Several formerly abundant rodent species have become extinct in the Jurien district around Hastings Cave since European man colonized Australia. The same two native species now occupy the small and large rodent niches in all the communities. In contrast the original niche relations of *S. granulipes* and *S. murina* have probably not altered. However, *Antechinus apicalis, A. flavipes* and *Dasycercus cristicauda* have probably become locally extinct in European times.

The Dasyuridae is the only large group of non-flying Australian mammals in which no species is known to have become totally (as opposed to locally) extinct in European times. Some have, however, suffered drastic range reduction.

The above points are consistent with the dasyurids' conservative population dynamics and generally unspecialized ecological strategies.

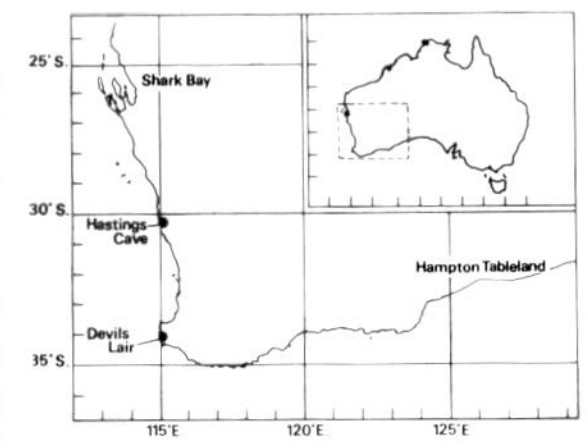

Pages 503–10 *in* CARNIVOROUS MARSUPIALS, ed. by M. Archer, February, 1982. Roy. Zool. Soc. New South Wales: Sydney, Australia.

A NUMBER of palaeoecological studies have recently been carried out on remains of mammals from late Quaternary cave deposits in southwestern Australia. The information which these provide on dasyurids in the source communities is surveyed in this paper. Although the paper concentrates on dasyurids (particularly the Abstract which, in the nature of symposia, was submitted before the final draft of the paper was completed), it has been found necessary to consider other small mammals in describing the communities. The most important of these are the rodents, which are abundant elements in most Australian mammal subcommunities, especially in heaths. The palaeoecological interpretation in the principal study reported, is based mainly upon changes in rodent abundances. Rodents provide the most useful basis for comparson with dasyurids.

Southwestern Australia is defined as that part of the continent which lies southwest of the boundary of the arid zone as defined by Slatyer and Perry (1969). This boundary line runs between Shark Bay and the western end of the Hampton Tableland (see Fig. 1), with a slight convexity on the southern side.

HASTINGS CAVE

The fossil mammals in Hastings Cave (at latitude 30°15′S., longitude 115°05′E. Fig. 1) were first investigated (as from "Drover's Cave") by Lundelius (1960), and have since been more intensively studied by Baynes (1979).

Hastings Cave is an inclined fissure cave (Bastian 1964), typical of those which occur in the Tamala Limestone of the northern Swan Coastal Plain. A large volume of water–transported sandy sediment has entered through the wide entrance and accumulated in one main chamber. The cave lies about mid way across the coastal plain, some 5 km from the nearest point on the modern coast at Jurien Bay.

The modern habitats of the Jurien district are mostly heaths. They occur in zones parallel to the coast, principally controlled by edaphic factors which depend upon the underlying geomorphic systems. At the western edge of the coastal plain is a calcareous sand zone, corresponding to the Quindalup Dune System of McArthur and Bettenay (1960), of beach ridges and dunes of beach sand which are either still mobile or have become stabilized by vegetation. Between this System and the Spearwood Dune System to the

[1] Zoology Department, University of Western Australia, Nedlands, Western Australia, Australia, 6009. Present address: Western Australian Museum, Francis Street, Perth, Western Australia, Australia, 6000.

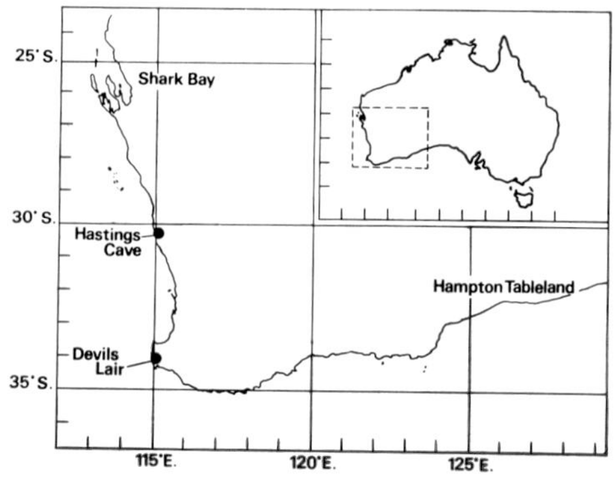

Fig. 1. Locality map showing the principal cave sites mentioned.

east are winter salt lakes with mainly clay substrates. Spearwood System substrates consist of low outcrops of Tamala Limestone and quartz sand which increases in depth to the east until it merges into the local Bassendean Dune System composed entirely of quartz sand (Lowry 1974). Together these two Systems constitute a quartz sand zone. Hastings Cave lies in the western half of this zone. To the east the ground rises into the dissected Gairdner Range. Lateritic gravel sands form the principal substrate in the range; others include Mesozoic sandstone screes, laterite pavements, colluvial sands and clays.

During the final cycle of glacioeustatic sea–level lowering in the last glacial the western coastal plain was extended westward, probably to the edge of the continental shelf. The calcareous components of surface sands would have been leached away to leave a quartz sand substrate across most of the plain (Clarke and Dortch 1977). As sea–level began to rise again after the glacial maximum, the sea coast moved eastward gradually reducing the area of quartz sand substrate. After the maximum marine transgression, when the sea reached close to its present level at about 6,000 B.P. (Thom and Chappell 1975), the coast began to move westward again as the beach ridges of calcareous sand built up. The area of quartz sand substrate was further reduced after this date when large mobile calcareous Quindalup dunes were blown onto the top of the Spearwood System.

The mammal fossil results which I obtained from a new excavation in Hastings Cave were similar in fauna and species' relative abundances to those of Lundelius (1960). I sorted and identified mammals from 10 layers selected from the full depth of the excavation, each of which was radiocarbon dated. The deposit I excavated spans the Holocene, from about 11,400 B.P. to modern. The mammal remains are a death assemblage. A strong bias to small mammals, and the form of remains, indicates that owls (*Tyto alba*) were the principal predators, with minor contributions from mammalian predators. This suite of predators would have accumulated remains of mammals occurring within a radius from Hastings Cave of about 20 km (the estimated range of the owls).

Most of the well represented mammal species show simple trends in relative abundance over the time span of the deposit; and six species present in older levels can be demonstrated to be absent from layers younger than mid Holocene. In the summary of results below all relative abundance figures (%) are rounded approximate values.

Steady decreases over the time span of the deposit are shown by *Pseudomys albocinereus* (60% – 30%) and *Notomys* sp. (9% – 3%); *perameles bougainville* declines from 1½% to 0 after the mid Holocene; and *Sminthopsis granulipes* shows a general, though variable, decrease over time (3% – 1½%. The greatest increases are shown by *Rattus fuscipes* and *Rattus tunneyi* which are present in older layers but increase rapidly after the mid Holocene to each exceed 11% in the youngest layers; *Pseudomys shortridgei* shows a general increase (5% – 12%); *Isodon obesulus* shows a steady increase (1% – 3%); and *Sminthopsis murina* a general, though variable, increase (2% – 5%). The relative abundances of two species differ markedly from either of the general trends. *Pseudomys occidentalis,* the second most abundant species in the deposit, rises from 12% in the older layers to exceed 20% between about 5,500 B.P. and 4,000 B.P. but falls to only 10% in the late Holocene. *Dasycercus cristicauda* does not exceed ½% until the late Holocene when it rises to 2%. In addition to *P. bougainville,* the species which can be shown to be absent from layers younger than mid Holocene are *Phascogale calura, Leporillus apicalis, Lagorchestes hirsutus, Lagostrophus fasciatus* and *Petrogale penicillata.*

As discussed in another section below, many of the mammal species which occurred in the Jurien district when European man first colonized Australia are now locally extinct. These changes preclude direct investigation of the original local mammal subcommunities. However, information on habitat requirements of several of the species, obtained from modern populations throughout southwestern Australia, provides some insight into the processes which occurred in the Jurien district over the Holocene. A pattern emerges of a relationship between substrate preferences and common trends in relative abundance. Of the species which decrease, *Pseudomys albocinereus* and *Sminthopsis granulipes* occur principally on quartz sands, and *Notomys mitchellii* (the likely species of *Notomys*) is also usually found on quartz sand substrates. Of those which increase, both *Rattus* species can occur in high populations in calcareous coastal Quindalup dunes; and *Sminthopsis murina* and *Isoodon obesulus* occur in wide ranges of habitats which include this System; nothing is known of the habitats of *Pseudomys shortridgei* in coastal south–western Australia. Of the anomalous species, the habitats in southwestern Australia of *Dasycercus cristicauda* are unknown, but *Pseudomys occidentalis* is typically recorded from substrates including lateritic gravel (McKenzie *et al.* 1973, Kitchener and Chapman 1976).

The changes in relative abundances through the Holocene of several mammals in Hastings Cave are therefore consistent with the interpreted changes in the areas of the edaphic zones within predator range (Baynes 1979). Thus, the relative abundances of the quartz sand species were highest in the early Holocene when this substrate was most extensive, and declined steadily as it was covered first by the advancing sea and then by calcareous sands. The *Rattus* species did not increase until the calcareous sand habitats had developed. The relative abundance of *Pseudomys occidentalis* rose to a maximum shortly after the date of the maximum marine transgression and then declined, corresponding to the proportion of the predator search area around Hastings Cave made up by the lateritic gravel sand substrates of the Gairdner Range. The absolute area of this zone did not, of course, change.

Change in the relative areas of the edaphic zones within predator range of Hastings Cave was probably the paramount factor controlling changes in relative abundances of mammals in the deposit.

The unique pattern of relative abundance of *Pseudomys occidentalis* suggests that this species was also unique among abundant species in being restricted to the laterite sand zone. As such it was independent of the paramount limiting factor. This provides a potential means of inferring the community groupings of the more abundant mammals in the Jurien district. Since the area of this edaphic zone did not change, the sample abundances of *P. occidentalis* may be treated as constants, to which the contemporary abundances of other species may be related as ratios. The changes in these ratios through time (Figs 2 and 3) show clear patterns.

The strongest common pattern of ratios (Fig. 2) is that shown by quartz sand mammals, including *Sminthopsis granulipes*. These probably constituted the common small mammals of this zone community on the coastal plain, and the deep sand colluvium of the Range. The plunge in the ratios of these species (in the case of *Perameles bougainville* to zero), at about the time of the maximum marine transgression, may reflect a crisis in the productivity of this community on the coastal plain, as a result of changed proximity to the ocean, as well as a decrease in its extent. On the basis of the sea–level curve of Chappell (1976), this transgression brought the sea shore to within 10 km of Hastings Cave for the first time since about 30,000 B.P. This may have had an adverse effect upon vegetation not adapted to salt laden winds etc. Such a crisis may have been a factor in the local demise of *P. bougainville*.

The later increases in ratios of the quartz sand community mammals (Fig. 2), despite continuing reduction in the area of that zone, may have resulted from one or all of three factors. As the calcareous zone substrates differentiated under subaerial weathering, and the primary succession in vegetation proceeded, they may have been gradually rendered suitable for the quartz zone mammals to invade. Or, the productivity of the vegetation may have recovered, either with changes in association dominants, or because the sea coast withdrew westward with the development of the Quindalup System.

Mammals whose ratios rise to a maximum after the maximum of the marine transgression are plotted in Fig. 3. The parallel changes in ratio of *Pseudomys shortridgei* and *Rattus* spp. after about 5,000 B.P. suggest that *P. shortridgei* also occurred in the developing calcareous sand zone. However, there were substantial numbers of *P. shortridgei* (more than 5% relative abundance) present before the calcareous sand zone developed; and the ratio of the species does not plunge around 6,000 B.P., as those of quartz sand mammals do (Fig. 2). This suggests that the early *P. shortridgei* population was independent of habitat changes on the coastal plain. It probably occurred in the laterite sand zone. The generally similar plots for *Isoodon obesulus* and *Sminthopsis murina* (Fig. 3a) are consistent with these species also being members of this community which invaded the developing calcareous sand zone.

The inferred abundant small mammals of each zone are shown in Table 1. Those of the quartz sand and laterite sand zone communities are consistent with simple ecological separation by resource partitioning, in the sense of Schoener (1974). Primary separation is along the major habitat niche axis. Each of these communities includes one small rodent species, one larger rodent, one small insectivore and one large insectivore. However, relations within the assemblage of abundant small mammals from the calcareous sand zone are not so clear, because all the rodents are about the same size. All three may have occurred in one community: the substrates may be more productive because they are less leached, so that the habitats they bear can be

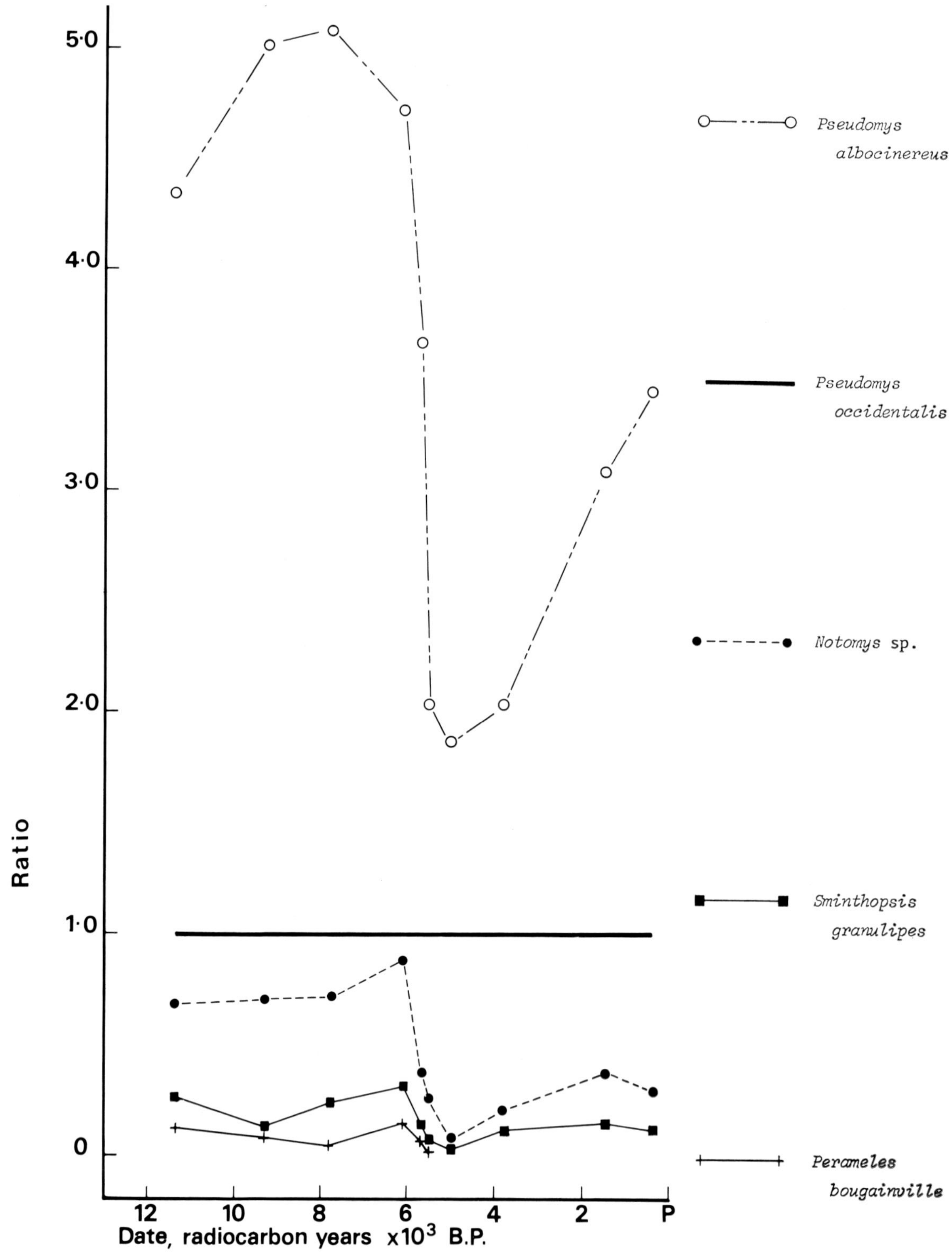

Fig. 2. Abundances of common quartz sand mammals in Hastings Cave, expressed as ratios of contemporary abundances of *Pseudomys occidentalis*, plotted against radiocarbon dates of samples.

successfully exploited by less specialized species than those of the other zones. Communities containing up to three large rodents are known in heaths in eastern Australia (Braithwaite *et al.* 1978). Alternatively, the calcareous zone may have included more than one mammal subcommunity. Different rodent species may have been dominant in communities at different primary successional stages on the dunes. *Sminthopsis murina* and *Isoodon obesulus* appear to have been the small and large insectivores common to all. The late date of the rise in ratio of *Dasycercus cristicauda* (Fig. 3b) suggests that this small carnivore may also have exploited later stages in the primary succession in the calcareous sand zone.

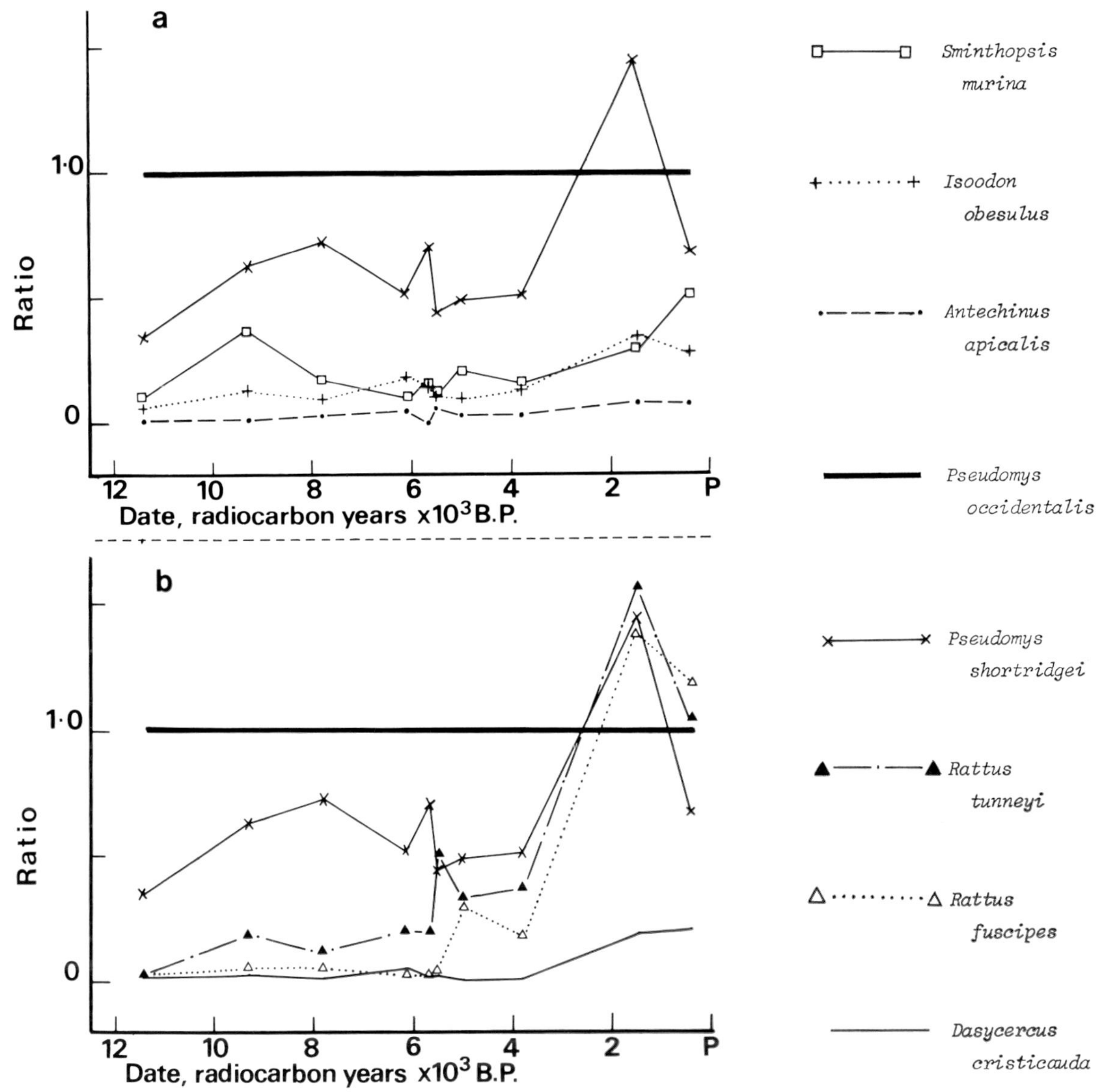

Fig. 3. Abundances of other (non quartz sand) substrate common mammals in Hastings Cave, expressed as ratios of contemporary abundances of *Pseudomys occidentalis*, plotted against radiocarbon dates of samples. (Figure in two parts for clarity).

Because change in edaphic zone areas was the paramount factor controlling relative abundances of fossil mammals in Hastings Cave, it would have swamped any effects of climate, and climatic interpretation of mammal relative abundances in this deposit is meaningless. On the other hand the six mammals which are absent from sediments young than mid Holocene shared a common pattern in original (i.e. immediately pre–European) distributions. All were absent from the forest block and occurred in the drier parts of southwestern Australia. Four also had

Table 1. Inferred assemblages of abundant small mammals in the three principal edaphic zones of the Jurien district around Hastings Cave.

Calcareous sand zone	Quartz sand zone	Laterite sand zone
Pseudomys shortridgei	*Pseudomys albocinereus*	*Pseudomys occidentalis*
Rattus fuscipes	*Notomys* sp.	*Pseudomys shortridgei*
Rattus tunneyi		
Sminthopsis murina	*Sminthopsis granulipes*	*Sminthopsis murina*
Isoodon obesulus	*Perameles bougainville*	*Isoodon obesulus*

extensive ranges in the arid zone: in particular *Phascogale calura* (see below). Their disappearance is therefore consistent with an increase in winter rainfall. *P. calura* probably occurred in white gum woodland on clays in the Range, but a local factor, as suggested above for *Perameles bougainville,* may have affected all mammals lost from the coastal plain.

The date of the inferred climatic change is, however, uncertain because of possible upward derivation of specimens in the Hastings Cave deposit, suggested by stratigraphy. It may have occurred in the mid Holocene, or as early as Termination I (i.e. the Pleistocene – Holocene boundary).

CAVE DEPOSITS IN THE EXTREME SOUTHWEST OF AUSTRALIA

Recent studies of cave deposits in the extreme southwest of Australia (including Devils Lair), have shown that eight mammal species have become locally extinct and one has invaded the area since Termination I. The only dasyurid among these is *Sarcophilus harrisii,* which may have persisted until as recently as 400 years B.P. (Archer and Baynes 1973); there, as elsewhere, it probably became extinct as a result of the arrival in mainland Australia of the dog (e.g. Milham and Thompson 1976). The other seven species lost, plus one which made a brief appearance before the last glacial maximum, and the single invader, have all probably changed in presence as a result of edaphic or climatic factors, or a combination of both (Baynes *et al.* 1976, Balme *et al.* 1978, Porter 1979). None of the species occurs in forest habitats in southwestern Australia, and all were probably restricted to the corridors of shrubby habitats between the coast and the forest block.

Indirect evidence (e.g. Bé and Duplessy 1976) suggests that the climate of southwestern Australia was as wet in the last interglacial as in the Holocene, and sea-level was at least as high (Chappell 1976). It is therefore unlikely that the species persisted in the extreme southwest through the last interglacial, and reasonable to assume that most, if not all, invaded, or reinvaded the area along the coastal corridors during the last glacial.

One of the invaders was *Pseudomys albocinereus. Sminthopsis granulipes* is shown by the Hastings Cave study (above) to have had very similar habitat requirements on the northern Swan Coastal Plain. Also, the original range limits of *S. granulipes,* like those of *P. albocinereus* and *Notomys mitchellii* (the species most likely to be the *Notomys* sp. found in Devils Lair) lie close to the ends of the two coastal corridors. Yet *S. granulipes* has not been detected in any of the cave deposits in the extreme southwest. The original range limits of *Antechinus apicalis* lie even closer to the ends of the coastal corridors, though less is known of the habitat requirements of this species. Neither of these dasyurids appears to have taken the opportunity to expand its distribution into the extreme southwest during the last glacial, which was seized by eight species belonging to a range of other taxa.

DASYURIDS IN EUROPEAN AUSTRALIA

The dasyurids also differ from other taxa in their response to the colonization of Australia by European man. In the relatively well studied Jurien district around Hastings Cave large changes in mammal fauna have occurred during this time (Table 2). The absence from the fossil fauna of *Eptesicus regulus, Chalinolobus gouldii, C. morio, Tarsipes spencerae, and Tachyglossus aculeatus* is probably the result of sampling effects; vespertilionid bats and *T. aculeatus* were probably never taken by the owls, and *T. spencerae* was not detected because only tooth–bearing structures of the mammals were identified. The present day fauna, based on collecting by

Table 2. Mammal faunas of the Jurien district: original, recorded as fossils in the surface sediments of Hastings Cave (radiocarbon dated at about 400 years B.P. to modern), and present day, recorded in the living fauna.

	Original	Present day
Pseudomys albocinereus	+	+
Pseudomys occidentalis	+	·
Pseudomys shortridgei	+	·
Pseudomys praeconis	+	·
Notomys sp.	+	·
Rattus fuscipes	+	+
Rattus tunneyi	+	·
Mus musculus	+	+
Oryctolagus cuniculus	·	+
Macroderma gigas	+	·
Nyctophilus geoffroyi	+	+
Eptesicus regulus	·	+
Chalinolobus gouldii	·	+
Chalinolobus morio	·	+
Canis familiaris	+	·
Vulpes vulpes	·	+
Felis catus	·	+
Dasyurus geoffroii	+	·
Dasycercus cristicauda	+	·
Antechinus apicalis	+	·
Antechinus flavipes	+	·
Sminthopsis crassicaudata	·	+
Sminthopsis granulipes	+	+
Sminthopsis murina	+	+
Isoodon obesulus	+	·
Tarsipes spencerae	·	+
Bettongia penicillata	+	·
Macropus irma	+	+
Macropus fuliginosus	+	+
Tachyglossus aculeatus	·	+

+ = positive record.

Chapman and Kitchener (1977) and myself, may also be incomplete. However, about nine species which were abundant or moderately common in the Hastings Cave deposit appear to have become locally extinct. These include *Antechinus apicalis, A. flavipes,* and *Dasycercus cristicauda.*

Extinction in the Jurien district has apparently been greater among the rodents than the dasyurids. The two native species *Pseudomys albocinereus* and *Rattus fuscipes* now fill the small and large rodent niches in all three edaphic zones. The introduced *Mus musculus* now occurs in almost all habitats in the Jurien district (more than *P. albocinereus* and *R. fuscipes*). Its niche relations in this district have not been investigated, but the continued presence of the smallest native rodent, *P. albocinereus*, suggests that usurpation of niche space by *Mus musculus* may not have been only, or even mainly, at the expense of locally pre–existing rodents (cf. Posamentier and Recher 1974). In contrast, available evidence (mainly from Chapman and Kitchener 1977) suggests that *Sminthopsis granulipes* and *S. murina* still occur in their original edaphic zones. This probably indicates that the niche relations of these small insectivores have not so far changed. However, *S. crassicaudata* has invaded the district, probably very recently and changes may occur in the future.

Apart from bats, the Dasyuridae (as currently understood) is the only large group of Australian mammals in which no species is known to have become totally (as opposed to locally) extinct since European man colonized Australia. Some species have, however, suffered drastic range reductions; examples which occur in southwestern Australia are *Antechinus apicalis, Dasyurus geoffroii,* and *Phascogale calura.*

Fossil *Antechinus apicalis* occur in surface deposits in limestone caves along the west coast of Australia from Shark Bay (e.g. Western Australian Museum vertebrate palaeontological collection No. 71.7.84 from near False Entrance Well, at latitude 26°24'S.) to Yanchep at latitude 31° 33'S. (e.g. specimen 73.1.146). This suggests that the species originally occurred all along this coast. It is also known from the south coast of Western Australia between longitudes of about 118°E. and 121°E. It may have originally been widespread in the area between the southwestern boundary of the arid zone (as defined in Slatyer and Perry 1969) and the forest block. There are also early records from South Australia and Queensland, but these need confirmation (Morcombe 1967). Probably extant populations are only known at two localities on the south coast of Western Australia (Woolley 1977).

Dasyurus geoffroii was originally widely distributed in arid and semi–arid Australia (Ride 1968). It is now probably restricted to an area in southwestern Australia south of about latitude 31°S. and west of about longitude 124°S.

The original range limits of *Phascogale calura* are fairly well defined by published records: it occurred north to about 20°S. in central Australia (Parker 1973), and western Australia (Burbidge and Fuller 1979); a fossil from a cave in Cape Range in northwestern Australia (Kendrick and Porter 1974) is of unknown age but may also represent the original range; it is present in surface cave deposits at Shark Bay (e.g. 71.7.31); it occurrs in central southwestern Australia; it is present in Hampton Tableland (see Fig. 1) cave deposits (Lundelius 1957); it was first described from South Australia; and it was recorded from near the junction of the Murray and Darling Rivers (Wakefield 1966). However, the details of its distribution within these limits are not yet known. Extant populations are only known in central southwestern Australia.

CONCLUSIONS

The few studies of long term change discussed above suggest that dasyurids are generally less ecologically enterprising than rodents, the other principal group of small mammals in most Australian communities, but have withstood the impact of European man upon Australia better than the rodents. These points probably reflect more conservative population dynamics and less specialized ecological strategies in the dasyurids.

ACKNOWLEDGEMENTS

The Hastings Cave study formed the principal subject of a Ph.D project supervised first by Dr. W. D. L. Ride and then by Prof. A. R. Main. Costs of fieldwork and radiocarbon dates were met by a Royal Society Leverhulme scholarship, and an A. R. G. C. grant to Dr. Ride. I thank B. J. Fox and J. A. Mahoney for referees' comments, and A. K. Lee for comments in discussion of the paper at the symposium. The figures are adapted from originals drawn by Ms. P. Bradbury. I thank the W.A. Museum for the facilities used in the preparation of the paper. I am also most grateful to the Royal Zoological Society of New South Wales for assistance with travel to the symposium.

REFERENCES

ARCHER, M. AND BAYNES, A., 1973. Prehistoric mammal faunas from two small caves in the extreme south–west of Western Australia. *J. R. Soc. West. Aust.* **55:** 80–9.

BALME, J. M., MERRILEES, D. AND PORTER, J. K., 1978. Late Quaternary mammal remains, spanning about 30,000 years, from excavations in Devil's Lair, Western Australia. *J. R. Soc. West. Aust.* **61:** 33–65.

BASTIAN, L., 1964. Morphology and development of caves in the southwest of Western Australia. *Helictite* **2:** 105–18.

BAYNES, A., 1979. "The analysis of a late Quaternary mammal fauna from Hastings Cave, Jurien, Western Australia". Ph.D thesis: University of Western Australia, Perth.

BAYNES, A., MERRILEES, D. AND PORTER, J. K., 1976. Mammal remains from the upper levels of a late Pleistocene deposit in Devil's Lair, Western Australia. *J. R. Soc. West. Aust.* **58:** 97–126.

BE, A. W. H. AND DUPLESSY, J. C., 1976. Subtropical Convergence fluctuations and Quaternary climates in the middle latitudes of the Indian Ocean. *Science, N.Y.* **194:** 419–22.

BRAITHWAITE, R. W., COCKBURN, A. AND LEE, A. K., 1978. Resource partitioning by small mammals in lowland heath communities of south–eastern Australia. *Aust. J. Ecol.* **3:** 423–45.

BURBIDGE, A. A. AND FULLER, P. J., 1979. Mammals of the Warburton Region, Western Australia. *Rec. West. Aust. Mus.* **8:** 57–73.

CHAPMAN, A, AND KITCHENER, D. J., 1977. Mammals of Cockleshell Gully Reserve and adjacent areas. *Rec. West. Aust. Mus. Suppl. No.* **4:** 15–25.

CHAPPELL, J. M. A., 1976. Aspects of late Quaternary palaeogeography of Australian – east Indonesian region. Pp. 11–22 *in* "The origin of the Australians" ed. by R. L. Kirk and A. G. Thorne. Australian Institute of Aboriginal Studies: Canberra.

CLARKE, J. AND DORTCH, C. E., 1977. A 10,000 year BP radiocarbon date for archaeological finds within a soil of the Spearwood Dune System, Mosman Park, W.A. *Search* **8:** 36–8.

KENDRICK, G. W. AND PORTER, J. K., 1974. Remains of a Thylacine (Marsupialia: Dasyuroidea) and other fauna from caves in the Cape Range, Western Australia. *J. R. Soc. West. Aust.* **56:** 116–22.

KITCHENER, D. J. AND CHAPMAN, A., 1976. Mammals of the Tarin Rock and North Tarin Rock Reserves. *Rec. West. Aust. Mus. Suppl. No.* **2:** 61–7.

LOWRY, D. C., 1974. "Dongara – Hill River, SH 50–5 and SH 50–9, geological sheet. 1: 250,000 map and explanatory notes". Australian Government Printing Service: Canberra.

LUNDELIUS, E. L., 1957. Additions to knowledge of the ranges of Western Australian mammals. *West. Aust. Nat.* **5:** 173–82.

LUNDELIUS, E. L., 1960. Post Pleistocene faunal succession in Western Australia and its climatic interpretation. *Int. geol. Congr. XXI (IV)*: 142–53.

MCARTHUR, W. M. AND BETTENAY, E., 1960. The development and distribution of the soils of the Swan Coastal Plain, Western Australia. *Soil Publ. C.S.-I.R.O., Aust. No.* **16:** 1–55.

MCKENZIE, N. L., BURBIDGE, A. A. AND MARCHANT, N. G., 1973. Results of a biological survey of a proposed wildlife sanctuary at Dragon Rocks, near Hyden, Western Australia. *Rep. Fish. Fauna Dep. West. Aust. No.* **12:** 1–27.

MILHAM, P. AND THOMPSON, P., 1976. Relative antiquity of human occupation and extinct fauna at Madura Cave, southeastern Western Australia. *Mankind* **10:** 175–80.

MORCOMBE, M. K., 1967. The rediscovery after 83 years of the Dibbler *Antechinus apicalis* (Marsupialia, Dasyuridae). *West. Aust. Nat.* **10:** 103–11.

PARKER, S. A., 1973. An annotated checklist of the native land mammals of the Northern Territory. *Rec. S. Aust. Mus.* **16 (11):** 1–57.

PORTER, J. K., 1979. Vertebrate remains from a stratified Holocene deposit in Skull Cave, Western Australia. *J. R. Soc. West. Aust.* **61:** 109–17.

POSAMENTIER, H. AND RECHER, H. F., 1974. The status of *Pseudomys novaehollandiae* (the New Holland Mouse). *Aust. Zool.* **18:** 66–71.

RIDE, W. D. L., 1968. On the past, present, and future of Australian mammals. *Aust. J. Sci.* **31:** 1–11.

SCHOENER, T. W., 1974. Resource partitioning in ecological communities. *Science, N.Y.* **185:** 27–39.

SLATYER, R. O. AND PERRY, R. A. (eds.), 1969. "Arid lands of Australia". Australian National University Press: Canberra.

THOM, B. G. AND CHAPPELL, J., 1975. Holocene sea levels relative to Australia. *Search* **6:** 90–3.

WAKEFIELD, N. A., 1966. Mammals of the Blandowski expedition to northwestern Victoria, 1856–57. *Proc. R. Soc. Vict.* **79:** 371–91.

WOOLLEY, P., 1977. In search of the Dibbler, *Antechinus apicalis* (Marsupialia: Dasyuridae). *J. R. Soc. West. Aust.* **59:** 111–7.

RESUMEN

41. Los dasiúridos de las comunidades del Cuaternario tardío del Sudoeste de Australia. Por Alexander Baynes.

En Hastings Cave el factor más importante que ha controlado las relativas abundancias de especies de pequeños mamíferos acumulados durante el Holoceno fue la calidad del substrato de las zonas edáficas que rodean este sitio de la llanura costera. De los abundantes dasiúridos, *Sminthopsis granulipes* estuvo restringido a la comunidad de la zona de arena cuarzosa; *S. murina* probablemente vivió incialmente en la comunidad de la zona de laterita arenosa–conglomerádica, y más tarde invadió la zona areno–calcarea en desarrollo (la que pudo haber tenido más de una subcomunidad de mamíferos); y *Dasycercus cristicauda* probablemente vivió principalmente en la zona areno–calcárea. *Phascogale calura* desapareció del área de la Hastings Cave en el Holoceno temprano a medio, y *Sarcophilus harrisii* probablemente en el Holoceno tardío.

El depósito de Devils Lair contiene 8 especies de mamíferos presentes en el extremo sudoeste de Australia durante el último glacial, las que localmente devinieron extintas por el Holoceno tardío. La mayoría, si no todos, presumiblemente habían invadido el área desde el último interglacial. Ningún pequeño dasiúrido está incluído en este grupo.

Varias especies de roedores originalmente abundantes se extinguieron en el distrito de Jurien de los alrededores de Hastings Cave desde que los europeos colonizaron Australia. Las mismas dos especies nativasocupan ahora los pequeños y grandes nichos de roedores en todas las comunidades. Por el contrario las relaciones originales de los nichos de *S. granulipes* y *S. murina* probablemente no se alteraron. Sin embargo, *Antechinus apicalis*, *A. flavipes* y *Dasycercus cristicauda* probablemente devinieron localmente extintos en los tiempos de los europeos.

Los Dasyuridae son el único gran grupo de mamíferos australianos no voladores conocido de los que ninguna especie devino totalmente (en oposición a localmente) extinta en los tiempos europeos. Sin embargo, algunas sufrieron drástica reducción en su distribución.

Los puntos mencionados precedentemente son consistentes con dinámica conservadora de sus poblaciones y con sus generalmente inespecializadas estrategias ecológicas.

42. PLEISTOCENE DASYURIDS (MARSUPIALIA) FROM SOUTH-WESTERN NEW SOUTH WALES

P. L. CRABB[1]

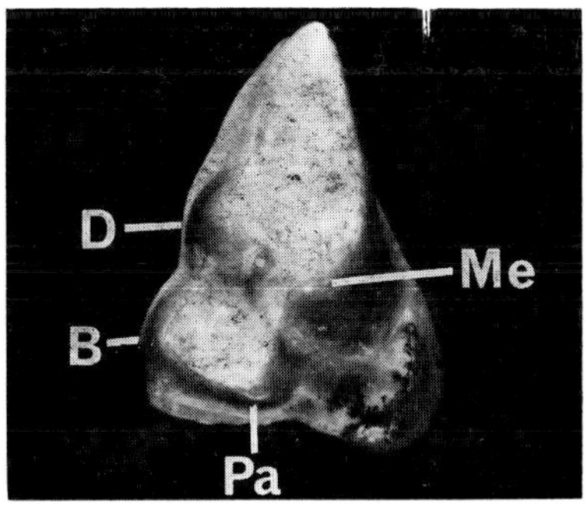

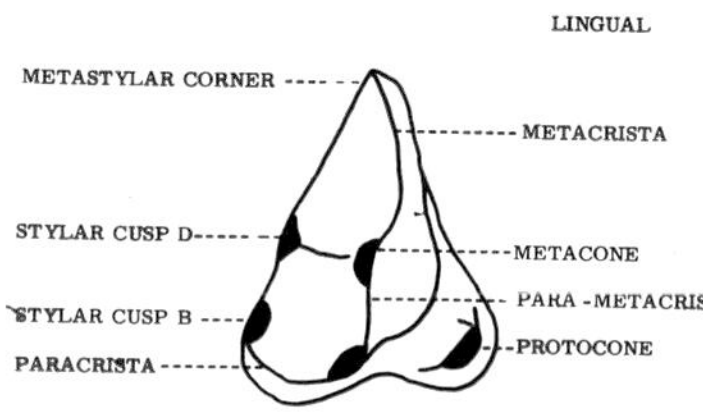

Sarcophilus moornaensis n.sp. is described from the early Pleistocene Moorna Sands. The single upper molar and incomplete lower molars and mandible of *S. mooraensis* are similar to the molars and mandibles of *S. harrisii* except for the development of the stylar cusps of the upper molars and the metaconids and talonids of the lower molars. *S. moornaensis* is probably in the direct evolutionary line leading to *S. harrisii* and possibly shares a common ancestor with *Glaucodon ballaratensis*.

Pages 511–16 *in* CARNIVOROUS MARSUPIALS, ed. by M. Archer, February, 1982. Roy. Zool. Soc. New South Wales: Sydney.

VERTEBRATE fossils were initially collected from the Pleistocene Moorna Sand at Fishermans Cliff, Moorna Station, southwestern New South Wales, by E. D. Gill and K. N. G. Simpson (Gill 1973). Further collections were made by parties from the Zoology Department, Monash University, in 1971 (Marshall 1973) and 1972.

A new species of *Sarcophilus* is described from a single upper molar and an incomplete left mandible. Several isolated teeth are referred to other dasyurid taxa. These specimens are from the Fishermans Cliff local fauna (Marshall 1973). All the specimens are deposited in the National Museum of Victoria (NMV). The terminology for the molars and teeth of dasyurids follows Archer (1974, 1975, 1978).

SYSTEMATICS

SUPERORDER MARSUPIALIA

ORDER POLYPROTODONTIA

FAMILY DASYURIDAE

Sarcophilus moornaensis n.sp.

Holotype: NMV P28684 an isolated unworn left M^3 or M^4, roots absent (Figs 1 and 3).

Paratype: NMV P28268 an incomplete left mandible with damaged M_2, M_3, M_4 and M_5, and the alveoli of the incisors (I_1 to I_3), the canine and the premolars (P_2 and P_3) (Figs 2 and 4).

Etymology: Moorna is named for the station where the Fishermans Cliff local fauna was found.

Locality: Fishermans Cliff is on the northern bank of the Murray River 5 kilometres up stream from Moorna Station Homestead, which is 20 kilometres west of Wentworth, southwestern New South Wales.

Age: An early Pleistocene age of the Moorna Sands has been deduced from the geological setting. The Moorna Sands are interbedded with the early Pleistocene Blanchtown Clay (Firman 1972). Gill (1973) and Marshall (1973) suggested a Pliocene and Plio–Pleistocene age respectively for this deposit. Although the hypothesis of Pliocene age cannot be justified, the degree of similarity of the mammal fauna to the extant fauna indicates the possibility that the fauna could be middle Pleistocene in age. However, the phylogenetic histories of most early and middle Pleistocene mammals and their evolutionary relationships to extant species are unknown. Until these have been clarified, the precise age of the Fishermans Cliff local fauna must remain undetermined. Tentatively, based on stratigraphy, I suggest it is early Pleistocene in age.

Differential diagnosis: Differs from *Sarcophilus harrisii* in its smaller size. The single upper molar (NMV P28684) also has a stronger stylar cusp D which is separated by a relatively longer and deeper intercusp valley from the metacone. Stylar cusp B is equal in development to that of the paracone and is connected anteriorly by the paracrista. In NMV P28268, the metaconids of M_4 and M_5 are high cusps, extending approximately three quarters of the height of the protoconid. These specimens are essentially similar in form and relative proportions to the equivalent elements in *S. harrisii*. In these latter features, and in the more robust mandible of NMV P28268, *S. moornaensis* differs from *Glaucodon ballaratensis*.

Description: The upper molar (NMV P28684) is triangular in occlusal outline (Fig. 3). The three primary cusps, the paracone, protocone and metacone, are distinct. Stylar cusp D is triangular

[1] Zoology Department, Monash University, Clayton, Victoria, Australia, 3168. Present address: Zoology Department, La Trobe University, Bundoora, Victoria, 3083.

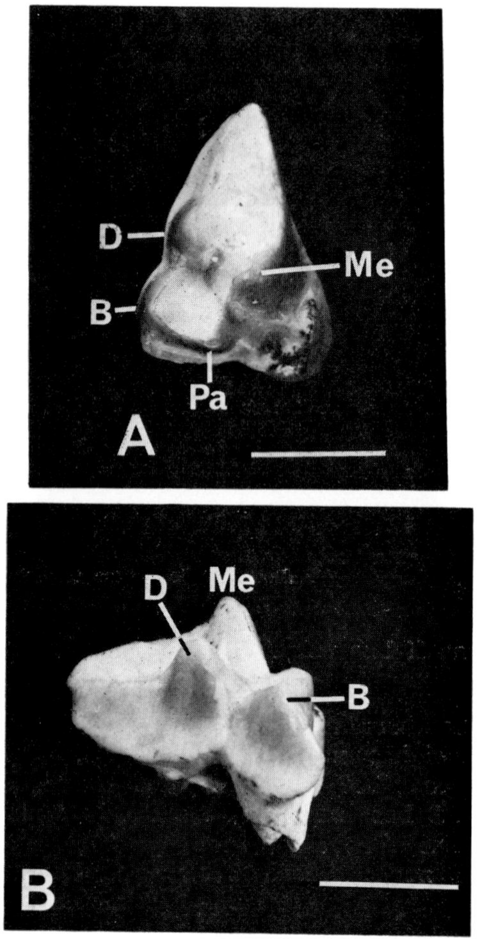

Fig. 1 Upper molar of *S. moornaensis* n.sp. (NMV P28684). A. Occlusal view showing the broad separation of stylar cusp D and the metacone and stylar cusps B and the paracone. B. Labial view showing the strong development of stylar cusps B and D (B–stylar cusp B: D–stylar cusp D: Me–metacone: Pa–paracone). Scale = 10mm.

in shape from the occlusal aspect and is slightly convex labially with a small crest extending medially to the base of the labial edge of the metacone. There are broad valleys anterior and posterior to this crest. The posterior valley opens out onto the metastylar corner of the molar. Stylar cusp B and the paracone are of equal height. They are well separated and are linked anteriorly by the paracrista.

The protocone is a crescentic cusp positioned low on the crown lingual to the paracone and linked to the base of the metacone by a small crest. The paracone is anterior to the metacone and is linked to stylar cusp B by the paracrista and to the metacone by the para–metacone crest. The metacrista extends from the apex of the high central metacone to the metastylar corner of the molar. The parametacone crest and the metacrista form an irregular, longitudinal shearing crest the length of the molar. A slight anterior cingulum is present anterior to the paracone and stylar cusp B, and extends lingually to the protocone.

The labial side of the molar is vertical, reflecting the strong development of stylar cusps B and D, and their broad separation from the paracone and the metacone respectively (Fig. 1A).

The mandible (NMV P28268) is incomplete (Figs 2 and 4). The ascending process and the posterior part of the horizontal ramus are broken off. The incisors, canine and premolars are lost. The three incisor alveoli are eroded but it can be seen that they extend anterior and ventral to the large elliptical canine alveolus. The four premolar alveoli are crowded so that the anterior root of P_2 is labial to the posterior edge of the canine and the anterior root of P_3 is labial to the posterior root of P_2, indicating that the long axis of the premolars were oblique to the long axis of the mandible.

The crown of the first molar is eroded and most of the enamel is absent. This molar is narrow laterally and it has a complete, crescentic talonid with no cusps. The peaks of the cusps are eroded in the second molar, but there is a slight metaconid adpressed to the protoconid (Fig. 2C). In M_3 the talonid is rectangular but it is damaged so that the talonid cusps are difficult to delimit. The crown of the third molar is eroded, the protoconid is high and a distinct metaconid is adpressed labially for three quarters of its height (Fig. 2C). The talonid of M_4 is strongly developed and the talonid cusps are small but distinct. The crown of the fourth molar is eroded and fractured but it retains a distinct small metaconid and a paraconid. The talonid is narrow and triangular with no cusps evident(Fig. 2C).

In comparison with *Sarcophilus harrisii,* the molars of *S. moornaensis* differ in the development and position of the upper molar stylar cusps, and in the relatively greater development of the metaconids and talonids in the lower molars. In *S. harrisii,* stylar cusp D is relatively more compressed laterally and more closely adpressed to the labial edge of the metacone. The apex of stylar cusp D is approximately 2.2 mm from the edge of the metacone. In contrast, in *S. moornaensis,* this distance is 3.2 mm. Stylar cusp B in *S. harrisii* is larger than the paracone, not equally developed as it is in *S. moornaensis,* and these two cusps are further apart in the latter species. In *S. harrisii,* the labial sides of the upper molars are more inclined lingually, reflecting the relatively reduced development of stylar cusp D and the relatively more lingual position of stylar cusps D and B compared with the condition in *S. moornaensis* where the labial edge of the upper molar is vertical.

The shape of the mandible of *S. harrisii* is similar to that of *S. moornaensis,* but the talonids of M_3 and M_4 are not as elongate and the talonid cusps are not as distinct. Similarly, the metaconids are not as strongly developed and the trigonids are less compressed laterally in *S. harrisii*. The incisor, canine and premolar alveoli are similarly developed in both species and the molar and mandibular measurements taken are comparable (Table 1).

Sarcophilus laniarius is larger than *S. moornaensis*. The stylar cusps of the upper molars of species of *Dasyurus* are more distinct relative to those in *S. moornaensis* and the large metaconids are not as adpressed to the protoconid. The talonids aand talonid cusps are relatively larger and more distinct in species of *Dasyurus* than they are in *S. moornaensis*. The molars and the mandible of *D. maculatus* and *D. affinis* are considerably smaller than the molars and mandible of *S. moornaensis*.

DISCUSSION

Marshall (1973) referred specimens NMV P28684 and NMV P28268 to *Glaucodon* (Stirton 1957) and compared them to *G. ballaratensis* the single species of this genus. He stated (p.156) that there are no " . . . notable morphological differences . . ." between these specimens and *G. ballaratensis*. I consider that the Fishermans Cliff specimens have a distinctive sarcophiline form. Although the metaconids are similarly distinct and the talonids are as relatively well–developed as in *G. ballaratensis*, the Fishermans Cliff specimens are significantly larger and more robust (Table 1). In comparison, the more gracile molars and mandible of *G. ballaratensis* are similar in form to the molars and mandibles of species of *Dasyurus*.

Tate (1947) noted that in *Sarcophilus harrisii*, the teeth have become very massive compared with the original dasyurine structure, that there has been an extreme reduction of all the talonids and an almost total obsolesence of the outer styles of the upper molars and of the metaconids. These developments constitute a morpho-

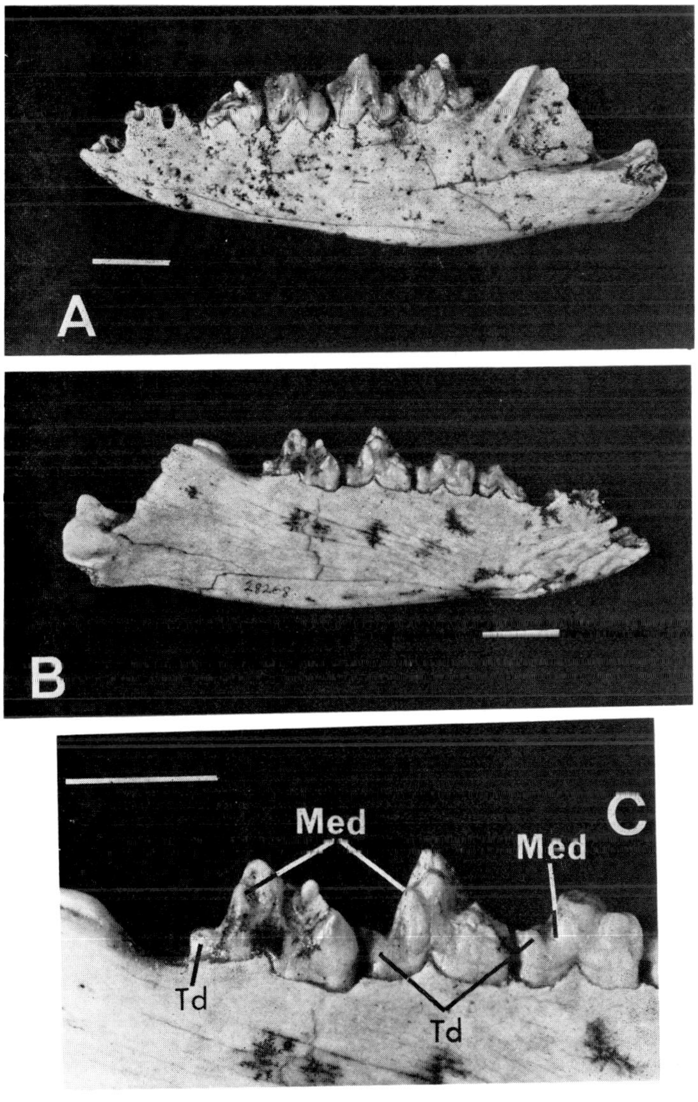

Fig. 2. Mandible of *S. moornaensis* n.sp. (NMV P28268). A. Labial view. B. Lingual view. C. Lingual view of M_2, M_3 and M_4 showing the relatively strong development of the metaconids and talonids. (Med – Metaconid: Td – talonid). Scale = 10mm.

logical trend from species of *Dasyurus* through *Sarcophilus harrisii*. Within this morphological trend, *S. moornaensis* and *G. ballaratensis* occupy a position between species of *Dasyurus* (as a whole) and *S. harrisii* in that the stylar cusps are distinct, as in species of *Dasyurus*, and not as adpressed lingually as in *S. harrisii*. Further, the metaconids and talonids are relatively smaller than in species of *Dasyurus* and relatively larger than in *S. harrisii*.

Ride (1964) considered that *G. ballaratensis* was structurally ancestral to *Sarcophilus harrisii*, and in some characters was close to species of *Dasyurus*. Archer and Bartholomi (1978:5) commented that "A structural lineage starting with a hypothetical fossil form similar to *Dasyurus maculatus*, to *Glaucodon ballaratensis*, to the annectant extinct species of *Sarcophilus* and ending with *S. harrisii* is one of the best documented marsupial structural phylogenies". The annectant extinct species is *S. moornaensis*.

I consider that the molars of *S. moornaensis* are similar to the molars of *G. ballaratensis* and, in terms of a structural lineage, they are as "primitive" as the molars of *G. ballaratensis*. Thus, if *S. moornaensis* is structurally ancestral to *S. harrisii*, I consider that *S. moornaensis* and *Glaucodon ballaratensis* may have shared a common ancestor, the more slender *G. ballaratensis* retaining the relative proportions of the species of *Dasyurus* while *Sarcophilus harrisii* became more robust. Both lineages reduced the labial edge of the molars along with a general crowding of the anterior teeth and an increase in overall size. Thus, from a hypothetical fossil form similar to *Dasyurus maculatus*, there are two structural lineages, one leading through *Sarcophilus moornaensis* to the late Pleistocene and extant sarcophilines, and the other leading to *Glaucodon ballaratensis*. If this hypothesis is correct we may expect to find in the fossil record species which have reduced stylar cusps, metaconids and talonids as seen in the younger sarcophilines, but which retain the gracile form of *Glaucodon ballaratensis*.

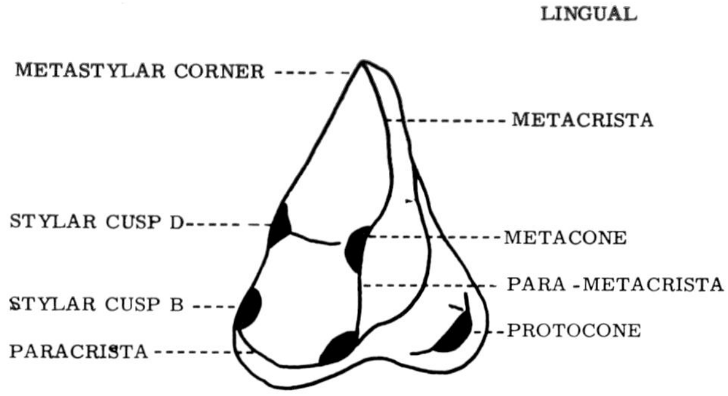

Fig. 3. Diagram of the upper molar of *S. moornaensis* n.sp. (NMV P28684) showing the terminology used in the text.

OTHER MOORNA DASYURIDS

Dasyurus sp. indet.

Material. NMV P28889 ?M^3, NMV P38430 $M^?$, NMV P28890 $M^?$, NMV P38437 M_2, NMV P29428 M_3 or M_4, NMV P38438 ?M_4, NMV P38428 M_3 or M_4. All the isolated specimens listed fall within the morphological and size ranges of *D. hallucatus*, *D. viverrinus* and *D. geoffroyi*. Marshall (1973) referred specimen NMV P28889 to *D. cf. D. hallucatus* on overall size and the proportionally smaller metastylar ridge. I consider that the broad overlap in the dimensions and the morphology of the extant species of *Dasyurus* precludes reference of these isolated molars to particular species.

cf. Dasyuroides–Dasycercus gen. indet.

Material. NMV P38372 M^2, NMV P38442 ?M^3, NMV P38418 M^3 or M^4, NMV P38331 M^3 or M^4, NMV P38283 M^5, NMV P38373 M_2,

Table 1. The dimensions of *S. moornaensis* compared with *S. harrisii* and *G. ballaratensis* (Stirton 1957). The mandible measurements in particular show the robust nature of *Sarcophilus* compared with *Glaucodon*.

CHARACTER	*Sarcophilus moornaensis*	*Sarcophilus harrisii*		*Glaucodon ballaratensis*
Upper molar	NMV P28684	M^3	M^4	–
length — labial	10	12.3	11.7	–
width — anterior	9	9.7	9.7	–
MANDIBLE	NMV P28268	MANDIBLE		NMV P16136
Length of molar row	36.7	40.2		34.6
Distance between C and M_2	10.9	11.3		10.3
Length of M_2	8.0	8.8		7.6
Width of M_2	5.2	6.3		4.2
Length of M_3	8.5	9.6		–
Width of M_3	5.6	6.3		–
Length of M_4	10.0	11.3		–
Width of M_4	6.4	6.5		–
Length of M_5	11.0	11.2		10.0
Width of M_5	6.3	6.7		5.8
Depth of mandible below P_2 between roots	14.6	18.5		10.1
Depth of mandible below M_5 between roots	20.0	20.6		17.3
Thickness of mandible below M_5	8.0	10.3		6.7

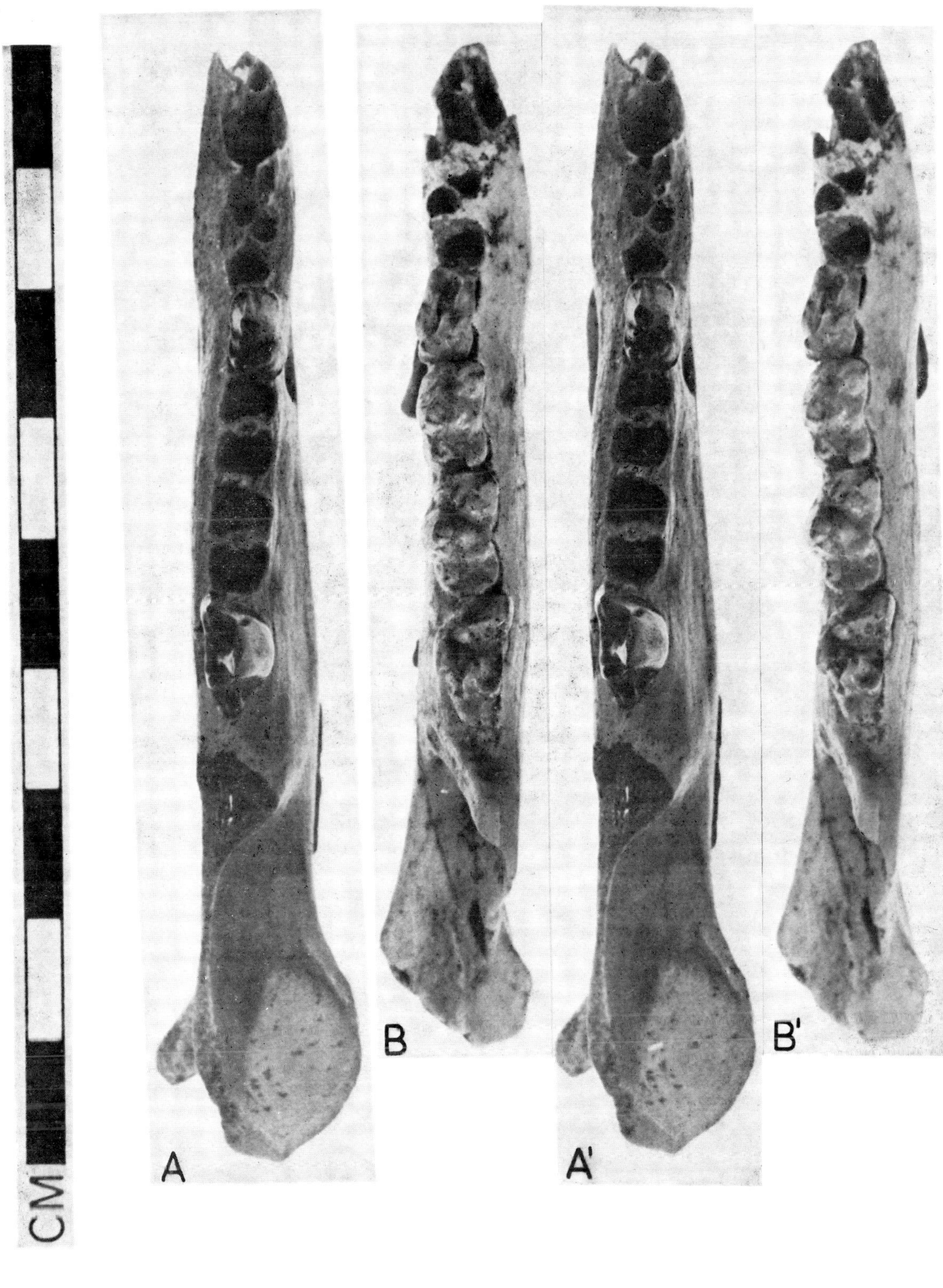

Fig. 4. Stereopairs of the holotype of *Glaucodon ballaratensis* (A–A') and the mandible of *Sarcophilus moornaensis* n.sp. (B–B').

NMV P38429 M_2, NMV P38281 M_2, NMV P28887 ?M_4, NMV P38343 ?M_4, NMV P38343 ?M_4, NMV P38370 M_4, NMV P38334 left mandible with alveoli of C, P_{1-3}, M_2, molar M_3, and alveolus of M_4, NMV P28888 right mandible with alveolus of M_2, molars M_{3-5}, NMV P28886 right edentulous mandible with alveoli of the C, P_{1-3}, M_2. These specimens are compared with the monotypic genera *Dasyuroides* and *Dasycercus* because these two taxa cannot be differentiated on the characters of isolated molars or incomplete mandibles.

Marshall (1973) referred specimens NMV P28886 and NMV P28888 to a small unnamed species close in size and molar morphology to *Dasyuroides byrnei* and *Dasycercus cristicauda.* He considered that this new species was a possible ancestral form of the two above species because of the presence of a single alveolus, possibly representing a rudimentary P_3 (e.g., in NMV P28886). However, Tate (1947) states that the last premolar (P_3 of Archer 1978 = P_4 of Tate 1947) in *Dasycercus cristicauda* is reduced to a single root and is absent in *Dasyuroides byrnei.* Thus specimens NMV P28886, NMV P28888 and NMV P38334, fall within the morphological variation of *D. cristicauda* and they do not appear to be separable from *D. byrnei.*

CONCLUSION

The dasyurid specimens from the Fishermans Cliff local fauna are isolated molars or incomplete mandibles. These specimens have some molar characters which are suggestive of a slightly different and temporally earlier grade of dental morphology than the extant grade of dental morphology (Marshall 1973). However, except for the distinctive development of the stylar cusps of the upper molar and the metaconids of the lower molars of *S. moornaensis,* the dental characters cannot be used to differentiate the fossil specimens from extant species.

ACKNOWLEDGEMENTS

This research was supported by a Monash University Post–graduate Scholarship and the Zoology Department, Monash University. I gratefully acknowledge the assistance during this research of Jim Warren, Brian Roberts, Max Hart, Peter Domelow, Vince Salanitrie, Thea McConnell, fellow students of the Monash Zoology Department, and in particular Gordon Sanson and Ken Simpson (Burwood College, Melbourne). I also thank Anne Warren, Tom Rich and Gordon Sanson for reading the manuscript, D. Walsh for the photographs in Figs. 1 and 3, Jeanette Hope and Ken Alpin for the photographs used in Fig. 4, and Mrs. R. McLauchlan and Mrs. B. Stubbs for typing the final draft.

REFERENCES

Archer, M., 1974. The development of the premolar and molar crowns of *Antechinus flavipes* (Marsupialia, Dasyuridae) and the significance of cusp ontogeny in mammalian teeth. *J. Proc. R. Soc. West. Aust.* **57:** 118–25.

Archer, M., 1975. *Ningaui,* a new genus of tiny dasyurids (Marsupialia) and two new species, *N. timealeyi* and *N. ridei,* from arid Western Australia. *Mem Qd Mus.* **17:** 237–49.

Archer, M., 1978. The nature of the molar–premolar boundary in marsupials and an interpretation of the homology of marsupial cheekteeth. *Mem. Qd Mus.* **18:** 157– 64.

Archer, M. and Bartholomai, A., 1978. Tertiary mammals of Australia: a synoptic review. *Alcheringa* **2:** 1–20.

Crabb, P. L., 1978. Fossil mammals of the Lower Pleistocene Moorna Sands, southwest New South Wales, Australia. Ph.D. Thesis. Zoology Dept., Monash University: Melbourne.

Firman, J. B., 1972. Regional stratigraphy of surficial deposits in the Murray Basin and Gambier Embayment. *Rept. Bk. No. 71/1,* Department of Mines, South Australia.

Gill, E. D., 1973. Geology and geomorphology of the Murray River region between Mildura and Renmark, Australia. *Mem. natn Mus., Melb.* **34:** 1–98.

Marshall, L. G., 1973. Fossil vertebrate faunas from the Lake Victoria region, S.W. New South Wales, Australia. *Mem. natn Mus., Melb.* **34:** 151–72.

Ride, W. D. L., 1964. A review of Australian fossil marsupials. *J. Proc. R. Soc. West. Aust.* **47:** 97–131.

Stirton, R. A., 1957. Tertiary marsupials from Victoria, Australia. *Mem. natn Mus., Melb.* **21:** 21–134.

Tate, G. H. H., 1947. On the anatomy and classification of the Dasyuridae (Marsupialia). *Bull. Amer. Mus. nat. Hist.* **88:** 101–55.

RESUMEN

42. Dasiúridos (Marsupialia) Pleistocénicos del sudoeste de New South Wales. Por P. L. Crabb.

Se describe *Sarcophilus moornaensis* n.sp. del Pleistoceno temprano de Moorna Sands. El único molar superior y los incompletos molares inferiores y mandíbula de *S. moornaensis* son similares a los mismos de *S. harrisii,* excepto por el desarrollo de las cúspides estilares del molar superior y los metacónidos y talónidos de los molares inferiores. Probablemente *S. moornaensis* está en la línea evolutiva directa que lleva a *S. harrisii,* y posiblemente comparte un antecesor común con *Glaucodon ballaratensis.*

43. TAXONOMIC STATUS OF FOSSIL DEVILS (*Sarcophilus*, DASYURIDAE, MARSUPIALIA) FROM LATE QUATERNARY EASTERN AUSTRALIAN LOCALITIES

LYNDALL DAWSON[1]

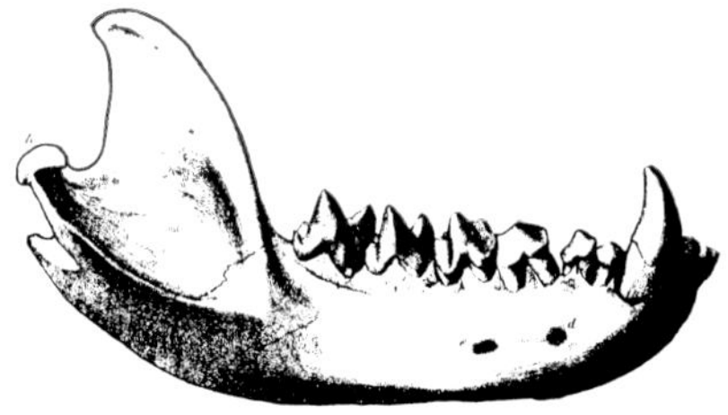

Dental measurements of fossil devils from the Pleistocene deposits of the Wellington Caves, New South Wales are compared with those of modern *S. harrisii* and found to differ in dental proportions, supporting their distinction as *Sarcophilus laniarius* (Owen). The data indicate that one species only, *S. laniarius,* is represented in the Wellington Caves fauna. Similarly large devils are shown to occur in other Late Pleistocene deposits of eastern Australia, the species surviving at least until 23,000 years BP at Clogg's Cave, Buchan, Victoria. However, a form of devil significantly smaller than modern *S. harrisii* also occurred in late Pleistocene deposits of western New South Wales. Two alternative hypotheses have been considered that might account for the relationships of Pleistocene and modern forms.

Pages 517–25 *in* CARNIVOROUS MARSUPIALS, ed. by M. Archer, February, 1982. Roy. Zool. Soc. New South Wales: Sydney, Australia.

DURING the past 150 years, there has been much controversy about the age and mode of deposition of the richly fossil–bearing sediments of the Wellington Caves, New South Wales (see Lane and Richards 1963 for a review). While some doubt remains about the age of the Wellington Caves themselves (Colditz 1943, Frank 1971, 1972, Francis 1973), the geological evidence presented by Frank (1971, 1972) suggests that these caves were filled with fossil–bearing sediments during the Pleistocene period and that most entrances have been closed (until disturbance in historic times) since approximately 15,000 B.P. An aim of the present study of marsupial fossil bones from the Wellington Caves has been to determine whether the fauna also supports this geological hypothesis. Because of the probable unreliability of C^{14} dating of the bones themselves, and because of the absence of datable charcoal in the deposits, the present study has been based on comparison of the Wellington Caves species with populations of equivalent modern species (where appropriate) and with the species from other Quaternary and Late Tertiary fossil sites throughout Australia.

A large dasyurid was recognised by Sir Richard Owen in the specimens sent to England by Sir Thomas Mitchell from the Wellington Caves or "Wellington Valley" of New South Wales (Mahoney and Ride 1975: 32–4). He named it *Dasyurus laniarius* Owen, 1838 (Owen *in* Mitchell 1838). Owen (1877) later recognised that *D. laniarius* belonged in the genus *Sarcophilus* but retained the name *laniarius* for specimens from the Wellington Caves which he considered to be larger than the living species. However, he also reported (Owen 1877: 105) that material ". . . from the Breccia Cavern . . ." also included specimens of *S. harrisii* (= *S. ursinus* of Owen) on the basis of size. Subsequently all fossil devils have been referred to either as *S. harrisii* (or *ursinus*) or to *S. laniarius,* the former name being applied to "small" specimens, the latter to "large" ones. Subjective judgement has been the only guide to such nomenclature because no comprehensive dental measurements have been available for either species. Ride (1964) recommended that the distinction between the large and small forms continue to be recognised at the species level, largely on the basis of Gill's (1953: 87) statement that the mandibular measurements of *S. laniarius* were ". . . from 15 to 50 percent larger than the average extant *Sarcophilus.*"

Although now extinct on the mainland, devil remains have been found in Pleistocene and Recent deposits of all states of mainland Australia. Recent dated occurrences are 3,120 ± 100 years B.P. from Padypadiy, Northern Territory (Calaby and White 1967), and 430 ± 160 years B.P. from Cave 1, Turner Brook, southwestern Western Australia (Archer and Baynes 1972).

It was the purpose of this study to establish whether recognition of a larger species of devil (i.e. *S. laniarius*) is justified on the basis of the characters of fossil jaws and teeth from the Wellington Caves and, if so, to establish whether more than one species is present in that fauna.

[1] School of Zoology, University of New South Wales, P.O. Box 1, Kensington, New South Wales, Australia, 2033.

Fossil specimens from various areas in eastern Australia have been included to provide comparative data for faunal correlation. The results of this study have also thrown some light on the supposed phenomenon of late Pleistocene dwarfing as described by Marshall and Corruccini (1978).

Institutional abbreviations used are as follows: AM, Australian Museum; BM, British Museum (Natural History); QM, Queensland Museum; NMV, National Museum of Victoria; SAM, South Australian Museum; SPS/ANU, School of Pacific Studies, Australian National University; UCMP, University of California Museum of Paleontology.

MATERIALS

The Wellington Caves Fossil Sample

The sample of fossil specimens referable to *Sarcophilus* included in this study from the Australian Museum collection from Wellington Caves comprised 81 mandibular fragments or lower teeth, and 66 maxillary fragments or upper teeth as follows (Australian Museum numbers):

Mandibular fragments: F58559, F58672, F58638, F58641, F58653, F58652, F58628, F58646, F58632, F58631, F58552, F58553, F58558, F58557, F58556, F58651, F58541, F58652, F58653, F58622, F58555, F58542, F58623, F58562, F58643, F58657, F58621, F58661, F58636, F58547, F58635, F58546, F58615, F58620, F58647, F58633, F58549, F58544, F58626, F58560, F18890, F58644, F58624, F58639, F58655, F18869, F58640, F58629, F58630, F58550, F58618, F58642, F58614, F58616, F58656, F58649, F58637, F58634, F58617, F58660, F58613, F58645, F58692, F58648, F58554, F57925, F58551, F58670, F57926, F58654, F58561, F58650, F58659, F16561, F58619, F58548, F58543, F16511b, F16588, F16535.

Maxillary fragments: F16564, F58663, F58666, F57930, F58568, F58579, F57927, F58580, F58575, F58567, F58612, F58599, F58569, F58667, F58572, F16505a, F16505b, F58577, F16565, F18870, F58664, F58611, F58597, F58594, F58590, F58500, F58578, F58592, F58564, F58573, F58608, F58591, F58571, F58566, F58570, F58576, F58589, F58606, F58609, F16566, F58595, F58610, F16567, F58598, F58583, F58602, F58585, F58605, F58603, F58607, F58604, F58593, F58574, F58581, F58582, F58565, F58601, F58586, F58668, F58584, F58587, F58662, F57928, MF434, F31045, F16540, F58665/F58588.

All Wellington Caves specimens are highly fragmented. Specimens were sorted into left and right sides and then checked for the possibility that any two or more represented one individual on the basis of tooth size, wear and preservation. Where matching specimens were recognised, only the most complete side was measured for statistical purposes, unless the other side retained a character unrepresented on the more complete side. Loose teeth were included if their position in the tooth row could be determined, but data from excessively worn teeth were excluded. Two specimens in the Australian Museum collection, labelled as being from the Wellington Caves (F58588, F58665), are atypical in preservation in being less mineralized, and have been treated separately from the rest of the sample. Casts of the maxillary syntypes, BM M10799 and BM M10798 were also examined.

The Sample of Modern *Sarcophilus harrisii*

The jaws and teeth of 20 skulls of *S. harrisii* from the collection of modern mammals in the Australian Museum were measured. All were collected in Tasmania between 1866 and 1932. No data on sex were available.

Other Comparative Material

Specimens of fossil devils were measured from the collections of the Queensland Museum, the National Museum of Victoria, and the South Australian Museum. Some specimens were loaned by J. Hope, School of Pacific Studies, ANU, Canberra. The comparative material is listed in Tables 1 and 2.

METHODS

The following measurements were made with a Helios dial caliper to 0.1 mm as follows. Upper molars: length (L) = greatest length at the base of the enamel on the buccal side of the tooth. Width (W) = greatest anterior width at the base of the enamel. Lower molars: length (L) = greatest length along the longitudinal axis of the tooth: width (W) = greatest width across the base of the enamel below the apex of the protoconid taken with the calipers held perpendicular to the base of the tooth. The combined (inclusive) length of the alveoli of M^{2-4} and M_{2-5} were also measured.

The mean (X), standard deviation (SD), observed range (OR) and coefficient of variation (CV) were calculated for each character of the fossil Wellington Caves and modern Tasmanian sample. Variances were compared using the F–test and the means for each character were compared using student's t–test by the method (assuming heteroscadisticity) of Sokal and Rohlf (1969). The t–test was also used to compare data from individual specimens with the sample means. The coefficient of differences (CD) was calculated for dental measurements using the

Table 1. Dimensions (mm) of the upper teeth of fossil devils from eastern Australian localities.

NUMBER	LOCATION	LP^2	WP^2	LP^3	WP^3	LM^2	WM^2	LM^3	WM^3	LM^4	WM^4	LM^5	LM^{2-4}
SAM, unregd.	Victoria Cave, S.A.	5.0	5.4	7.5	6.5	13.5(a)	–	–	–	15.0	11.4	–	–
QM F1872	Marmor, Qld.	–	–	–	–	12.4	9.9	14.4	11.0	15.2	10.2	8.9	–
SAM P13911	Lake Tandou, N.S.W.	–	–	5.8	5.5	10.2	9.0	11.3	9.5	10.8	9.0	–	32.2
QM F1693	Marmor, Qld.	–	–	–	–	13.5	10.9	–	–	–	–	–	–
SAM P17941	Dempsey's Lake, S.A.	–	–	5.4	4.5	9.8	7.8(a)	11.0	8.0(a)	11.8	9.0	6.8	–
SAM DL1	(unregd.) Dempsey's Lake, S.A.	–	–	–	–	–	–	10.0(a)	9.0(a)	–	–	–	–
SAM DL2	" " " "	–	–	–	–	–	–	–	–	11.5(a)	8.5	–	–
SAM P12647	Mair's Cave, S.A.	4.0	4.2	5.7	4.8	10.0	8.2	11.2	9.2	–	–	–	–
AM F58040	Lake Garnpung, S.A.	4.0	4.4	5.4	5.0	10.0	8.5	11.0	9.0	10.8(a)	8.5	7.0	30.0
SPS/ANU "B"	Clogg's Cave, Buchan, Vic.	–	–	–	–	12.9	10.1	–	–	–	–	–	–
SPS/ANU K6/9	Seton, Kangaroo Is., S.A.	–	–	–	–	–	–	12.0	10.9	–	–	–	–
SPS/ANU K7/10	" " " "	–	–	–	–	11.0	7.5	–	–	–	–	–	–
QM F14/728	E. Darling Downs, Qld.	–	–	–	–	12.4	9.6	14.8	10.4	14.1	10.7	8.7	–
QM F3709	Cement Mills, Gore, Qld.	–	–	–	–	10.9	9.5	12.5	10.8	–	–	–	–
QM F3710	" " " "	–	–	–	–	–	–	14.2	10.9	–	–	–	–
NMV P48011	Lancefield Swamp, Vic.	–	–	–	–	–	–	14.5	11.6	–	–	–	–
NMV P7432	Queenscliff, Vic.	4.5	4.8	6.3	5.6	11.2	8.8	12.2	9.5	12.8	9.3	7.3	34.5

a) Approximate measurement.

Table 2. Dimensions (mm) of the lower teeth of fossil devils from eastern Australian localities.

NUMBER	LOCATION	LP_3	WP_3	LM_2	WM_2	LM_3	WM_3	LM_4	WM_4	LM_5	WM_5	LM_{2-5}
SPS/ANU K6/6	Seton, Kangaroo Is, S.A.	–	–	–	–	–	–	11.0(a)	6.5(a)	11.0	6.5	39.5
SAM P20454	Victoria Cave, S.A.	7.9	7.2	10.7	7.4	12.2	7.8	14.2	7.4	–	–	–
SAM P13532	" " "	–	–	8.9	6.5	–	–	–	–	–	–	–
SAM P13536	" " "	–	–	–	–	–	–	–	–	12.3	6.8	–
SAM P13723	Haystall Cave, S.A.	7.5	6.5	11.5	6.8	13.1	6.9	14.8	7.4	14.0	7.5	49.7
SAM P18657	Hensche's Cave, S.A.	7.0	6.3	10.6	7.2	12.7	7.9	15.0	8.2	14.4	7.7	–
AM F60138–9	Lake Tandou, N.S.W.	5.8	4.7	8.0	5.9	9.9	6.3	11.7	6.3	–	–	–
AM F60136–7	Kangaroo Lake, N.S.W.	7.4	7.2	10.2	7.0	13.0	7.5	14.5	7.5	–	–	–
SAM P57989	Wombeyan Quarry, N.S.W.	7.5	7.0	10.2	8.5	12.5	8.1	14.6	8.1	15.3	8.8	50.0
UCMP 46193	Kalaminurina, S.A.	–	–	9.2	6.3	11.1	6.7	12.7	6.6	12.8	6.8	44.2
AM MF973	Bingara, N.S.W.	–	–	9.7	–	10.9	7.4	13.5	7.0	–	–	–
QM F721	E. Darling Downs, Qld.	–	–	10.2	6.6	12.5	7.2	15.2	7.6	14.2	7.4	–
SAM P17941	Dempsey's Lake, S.A.	–	–	–	–	–	–	–	–	10.7	6.7	–
QM F724	E. Darling Downs, Gowrie, Qld.	–	–	10.7	6.7	–	–	13.9	7.4	14.3	7.8	–
QM F14/722	" " " , Pilton, "	–	–	10.3	6.5	11.4	7.1	13.8	7.3	13.6	7.3	–
QM F3725	Cement Mills, Gore, Qld.	–	–	–	–	14.8	–	15.2	8.6	16.4	8.5	–
QM F3706	" " " "	–	–	–	–	11.8	7.9	15.0	8.1	–	–	–
QM F3704	" " " "	–	–	9.5	6.6	11.6	7.4	13.8	7.0	13.3	7.0	–
QM F2771	" " " "	–	–	–	–	–	–	13.1	7.7	13.5	8.2	–
QM F3705	" " " "	–	–	10.6	7.4	12.4	7.7	14.3	7.3	13.0	7.2	–
NMV P1857	Gisborne, Vic.	6.3	5.3	8.5	5.9	10.0	6.5	12.3	6.7	11.0	6.0	40.3
NMV P7432	Queenscliff, Vic.	6.5	5.8	9.0	5.9	10.5	6.8	12.6	7.0	12.3	6.9	–

(a) Approximate measurement.

formula $CD = \frac{\overline{X}_a - \overline{X}_b}{S_a + S_b}$ where $\overline{X}_a$ and $\overline{X}_b$ are the sample means and S_a and S_b are the respective standard deviations (Mayr *et al.* 1953).

Variations in the proportional size of teeth were illustrated using log ratio diagrams as exemplified by Simpson *et al.* (1960) in which the formula for the log difference, d = log X — log standard, where X represents the value of the specimen or population mean being tested. In Fig. 1 the mean value for the Wellington Caves sample are plotted using the means of the modern Tasmanian sample as standard. The observed range of values for both samples are also shown.

Bivariate comparisons between various dental characters were plotted graphically, and 95 percent confidence ellipses for the resultant clusters

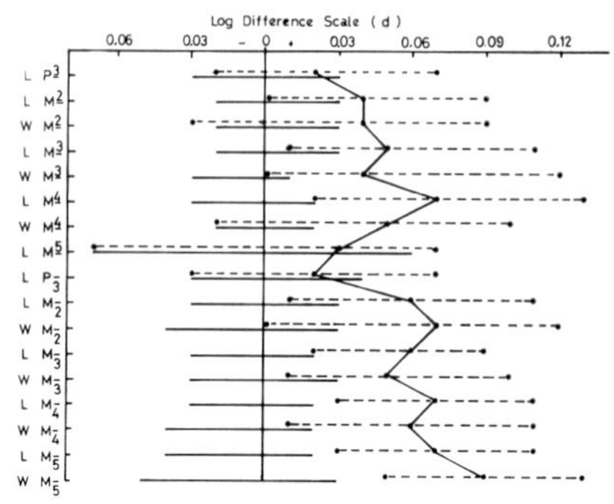

Fig. 1. Log difference diagram showing proportional relationships of tooth length (L) and width (W) in fossil *Sarcophilus laniarius* from Wellington Caves, compared with a sample of modern *S. harrisii* from Tasmania as standard. Log difference, d = Lox x − Log Standard. Solid vertical lines join mean values of d for each sample. Data is from Table 3.

were determined using a Munro Calculator programmed with the formula of Sokal and Rohlf (1969). Comparisons were made of length and width of premolar and molar teeth, and of length of $P^3/_3$ with length $M^2/_2$, and of length M_4 with length M_5. In a given population it would be expected that 95% of the members would fall within the ellipse for the characters being compared. The method provides a graphic way of estimating the similarity of two populations. Corresponding values for fossil specimens from other localities were also plotted on these graphs, thus obtaining a direct visual comparison of individual specimens with population parameters of the reference samples. Examples of 95% confidence ellipses showing discrimination between the Tasmanian and Wellington Caves samples for bivariate comparisons are shown in Fig. 2.

Dental terminology follows Archer (1978a). Accordingly, the adult dental formula for *Sarcophilus harrisii* is

$$I\frac{1\text{-}4}{1\text{-}3},\ C\frac{1}{1},\ P\frac{2\text{-}3}{2\text{-}3},\ M\frac{2\text{-}5}{2\text{-}5}.$$

RESULTS

Dental and mandibular measurements of the Wellington Caves specimens and the sample of *Sarcophilus harrisii* from Tasmania are summarised with the results of statistical analysis in Table 3. Variances of some parameters were found to differ significantly at the 5% level or less, while the mean dimensions for each character differ significantly between the two populations at the 1% level, or less. The mean values for CV are slightly higher for the Wellington Caves sample (mean CV = 5.4) than for the modern Tasmanian sample (mean CV = 4.1), but both values fall within the range acceptable for a population of one species (Simpson *et al.* 1960, Gingerich 1974). The coefficient of difference (CD) was calculated as an indication of

Table 3. Summary and statistical comparisons of dental dimensions (mm) of modern *Sarcophilus harrisii* from Tasmania and *S. laniarius* from Wellington Caves.

Parameter		Wellington Caves					S. harrisii, Tasmania					C.D.	F–test Variance		t–test Mean	
		N	Mean (mm)	O.R.	S (mm)	C.V.	N	Mean (mm)	O.R.	S (mm)	C.V.		F	P*	t's**	P
P^3	L	27	6.7	6.0– 7.5	0.34	5.0	20	6.4	6.0– 6.8	0.24	3.7	0.52	2.01	NS	3.545	$>0.01<0.001$
	W	27	6.1	5.0– 7.0	0.45	7.4	20	5.8	5.2– 6.3	0.30	5.1	0.40	2.25	0.05	2.739	>0.01
M^2	L	53	12.3	11.3–13.4	0.51	4.1	20	11.2	10.7–12.0	0.37	3.3	1.25	1.90	NS	10.147	>0.001
	W	53	10.0	8.3–10.9	0.49	4.9	20	9.1	8.6– 9.7	0.25	2.8	1.22	3.84	0.001	10.286	>0.001
M^3	L	53	13.7	12.5–15.7	0.60	4.4	20	12.1	11.5–13.0	0.43	3.5	1.55	1.95	NS	12.63	>0.001
	W	53	11.0	9.8–13.2	0.60	5.5	20	9.9	9.3–10.2	0.42	4.2	1.08	2.04	0.05	8.804	>0.001
M^4	L	41	14.1	12.6–16.0	0.67	4.7	18	11.9	11.0–12.4	0.41	3.5	2.04	2.67	0.05	15.446	>0.001
	W	40	11.0	9.0–13.0	0.73	6.6	18	9.8	9.3–10.3	0.33	3.4	1.13	4.89	0.001	8.622	>0.001
M^5	W	14	9.0	7.0– 9.7	0.70	7.8	18	8.2	6.8– 9.4	0.76	9.3	0.55	0.85	NS	3.089	$>0.01<0.001$
P_3	L	17	7.0	6.3– 7.8	0.38	5.5	20	6.7	6.2– 7.3	0.31	4.6	0.43	0.78	NS	2.601	>0.02
	W	17	6.3	5.6– 6.8	0.35	5.5	20	5.8	5.1– 6.1	0.28	4.9	0.79	1.56	NS	4.740	>0.001
M_2	L	52	9.9	8.9–11.2	0.50	5.1	20	8.7	8.2– 9.4	0.30	3.4	1.50	2.78	0.01	12.438	>0.001
	W	51	6.9	6.0– 8.0	0.57	8.3	20	6.1	5.5– 6.6	0.30	4.9	0.92	3.61	0.05	7.673	>0.001
M_3	L	56	11.6	10.6–12.5	0.49	4.2	20	10.2	9.6–10.8	0.35	3.4	1.67	1.96	NS	13.720	>0.001
	W	54	7.2	6.5– 8.0	0.34	4.7	20	6.4	6.0– 6.8	0.20	3.1	1.48	2.89	0.01	12.432	>0.001
M_4	L	49	13.9	12.5–15.0	0.65	4.7	19	11.7	11.0–12.3	0.37	3.1	2.16	3.09	0.01	17.487	>0.001
	W	50	7.4	6.5– 8.3	0.39	5.3	19	6.4	5.8– 6.7	0.28	4.4	1.49	1.94	NS	11.811	>0.001
M_5	L	33	13.9	12.7–15.1	0.61	4.4	18	11.8	10.7–12.3	0.39	3.3	2.10	2.45	0.05	14.952	>0.001
	W	32	7.6	7.0– 8.5	0.38	5.0	18	6.3	5.5– 6.8	0.34	5.4	1.81	1.25	NS	12.432	>0.001
$LM^{2\text{-}4}$		15	38.4	35.5–41.0	1.46	3.8	16	34.2	32.5–36.5	0.98	2.9	1.72	2.22	NS	9.342	>0.001
$LM_{2\text{-}5}$		5	46.4	44.5–48.0	1.30	2.9	17	42.6	40.5–45.0	1.26	3.0	1.48	1.06	NS	5.786	>0.001
Mean CV (dental)						5.4					4.1					

* NS = $P>0.05$
** t-test assuming heteroscedasticity, Sokal and Rohlf (1969).

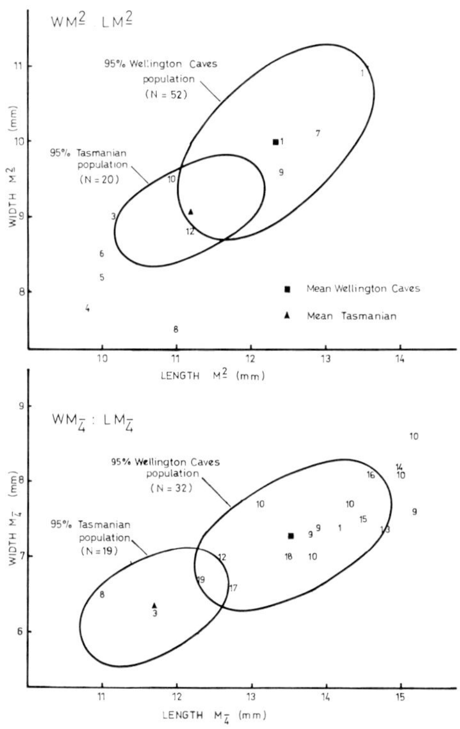

Fig. 2. Examples of 95% confidence ellipses for bivariate comparisons of tooth length (L) and width (W) for samples of devils from the Wellington Caves fossil deposits and from the modern Tasmanian population. Individual specimens from other fossil deposits in eastern Australia are also plotted. Numbers indicate the approximate coordinates of specimens other than those from the Wellington Caves deposits, and represent the following localities and specimens:— 1. Victoria Cave, Naracoorte, S.A. (SAM P20454, unregistered skull SAM); 2. Marmor, Qld. (QM F1872); 3. Lake Tandou, N.S.W. (SAM P13911, AM F60138–9); 4. Dempsey's Lake, S.A. (SAM P17941); 5. Mair's Cave, S.A. (SAM P12647); 6. Lake Garnpung, N.S.W. (AM F58040); 7. Clogg's Cave, Buchan, Vic. (SPS/ANU "B"); 8. Seton Rock Shelter, Kangaroo Isl., S.A. (SPS/ANU K7/10, SPS/ANU K6/6); 9. Eastern Darling Downs, Qld. (QM F14/728, QM F721, QM F724, QM F14/722); 10. Cement Mills, Gore, Qld. (QM F3709, QM F3706, QM F3704, QM F2771, QM F3705); 11. Lancefield Swamp, Vic. (NMV P48011); 12. Queenscliff, Vic. (NMV P7432); 13. Haystall Cave, S.A. (SAM P13723); 14. Hensche's Cave, S.A. (SAM P18657); 15. Kangaroo Lake, N.S.W. (AM F60136–7); 16. Wombevan Quarry, N.S.W. (SAM P57989); 17. Kalaminurina, S.A. (UCMP16193 cast); 18. Bingara, N.S.W. (AM MF973); 19. Gisborne Cave, Vic. (NMV P7432). Measurements for these specimens are given in Tables 1 and 2.

the degree of difference present in each character between the two populations. The results are shown in Table 3. CD values of 1.5 or greater, suggested by Ride (1964) as the requirement for subspecific separation, were found for most dimensions of the lower molars, but for only two dimensions of the upper teeth (i.e. LM^3 and LM^4).

The data indicate an "average" percent difference in size between the two populations (determined using the method of Marshall and Corruccini 1978) of approximately 14%.

Differences in dental proportions between the two populations are also depicted in the log ratio diagram (Fig. 1). This indicates that there is less difference in length between anterior teeth than posterior molars, the posterior molars becoming progressively longer in the Wellington Caves sample. In the width of M^2 and M_2 there is little difference in proportion between the two populations, being only slightly wider in the Wellington Caves sample. The more posterior molars are proportionately narrower compared to their length in the Wellington Caves sample, except for M_5, which is proportionately wider than in *S. harrisii*. The great variability in the size of M^5 in *S. harrisii* may reflect a reduction in the functional importance of this tooth.

Dental measurements of fossil specimens from many mainland localities were compared with those of the Wellington Caves and Tasmanian samples by plotting the data on bivariate scattergrams examples of which are shown in Fig. 2. Students t–tests were used to compare the measurements of each specimen with the mean values of the Tasmanian population of *S. harrisii*. It can be seen (Table 4) that while some fossil specimens are significantly larger than modern *S. harrisii*, several are significantly smaller. While most data for the specimens that are significantly larger than *S. harrisii* fall within the 95% confidence ellipses for the Wellington Caves sample, the data for specimens that are significantly smaller than *S. harrisii* fall outside (below) the 95% confidence ellipses for that species in most of the characters tested.

Table 5 indicates that specimens corresponding in dental size and proportion with the Wellington Caves sample occur in association with other large extinct marsupial species and in deposits with a minimum age greater than 16,000 years B.P. However, Table 5 also indicates that fossil devils that are significantly smaller than *S. harrisii* have been found in deposits dated approximately from 16,000 to 30,000 years BP, although for some deposits an age of greater than 30,000 years is possible, for example Dempsey's Lake, South Australia (D. Williams, pers. comm. 1980). This small form occurs in association with at least some large extinct marsupial species in all deposits in which it has been recorded so far.

Mainland fossil specimens measured here, and known to be younger than approximately 10,000 B.P., were found to be close in size to the upper range of modern. *S. harrisii*. However, Calaby and White (1967: 474) note that the specimen from Padypadiy, Northern Territory associated with a c^{14} date of 3,120 — 100 B.P. is ". . . appreciably smaller than modern Tasmanian examples". The 10,000 year old specimen from Laura, north Queensland, may also be smaller than modern *S. harrisii* (Horton 1977).

Table 4. Significance levels of t-tests comparing the sample mean of *Sarcophilus harrisii* (Tasmania) with single fossil specimens.

	Lake Tandou				Dempsey's Lake	Lake Garnpung	Mair's Cave, S.A.	Seton K.I.	Wellington Caves	Gore Qld.	Bingara N.S.W.	Darling Downs	Wombeyan N.S.W.	Victoria Cave, S.A.	Kangaroo L. N.S.W.
LP^3				2.0%	0.1%	0.1%	1.0%	–	0.1%					0.1%	
LM^2				1.0%	0.1%	1.0%	1.0%	NS	0.1%					0.1%	
WM^2				NS	0.1%	5.0%	1.0%	0.1%	–						
LM^3				NS	0.1%	0.1%	5.0%	NS	0.1%						
WM^3				NS	0.1%	5.0%	NS	5.0%	–						
LM^4				2.0%	NS	2.0%			NS					0.1%	
WM^4				5.0%	2.0%	0.1%			–					0.1%	
LP_3	1.0%												1.0%	1.0%	5.0%
LM_2	5.0%	NS								0.1%	0.1%	0.1%	0.1%	1.0%	0.1%
WM_2	2.0%	NS								0.1%	–	1.0%	0.1%	1.0%	1.0%
LM_3	NS									0.1%	NS	0.1%	0.1%	0.1%	0.1%
WM_3	NS									0.1%	0.1%	0.1%	0.1%	1.0%	0.1%
LM_4	NS							NS		0.1%	0.1%	0.1%	0.1%	0.1%	0.1%
WM_4	NS							–		0.1%	5.0%	0.1%	–	5.0%	0.1%
LM_5			NS		1.0%			5.0%		0.1%	–	0.1%	–	2.0%	–
WM_5			NS		NS			NS		0.1%	–	1.0%	–	0.1%	–
	** F60138–9	** P13532	** P13536	** P13911	** P17941	** F58040	** P12647	** K7/10 K6/9	** F58588 F58565	* Mean values	* MF973	* F721	* P57989	* Unregd. skull +P20454	* F60136–7

** Significantly smaller. * Significantly larger.

Table 5. Size correlation of specimens of *Sarcophilus* with the Wellington Caves and modern Tasmanian populations. The table includes only specimens measured by this author, or for which dental measurements are available in the literature.

Location	Correlation	Age	Reference
Texas Caves, Qld.	1,E		Archer (1978)
Marmor, Qld.	1,E		Longman (1925)
Gore Cement Mills, Qld.	1,E		Bartholomai (1977)
Darling Downs, Qld.	1,E		
Darling Downs, Qld.	1,E		
Bingara, N.S.W.	1,E		
Wombeyan Caves, N.S.W.	1,E		J. Hope, pers. comm. (1980)
Kangaroo Lake, N.S.W.	1,–		
Keilor, Dry Creek Fauna, Vic.	1,E	> 18,000 BP	Marshall (1974)
Clogg's Cave, Buchan, Vic.	1,E	22,980±2,000 BP	Flood (1973)
Lancefield Swamp, Vic.	1,E	26,000±500 BP	Gillespie *et al.* (1978)
Victoria Cave, Naracoorte, S.A.	1,E		R. Wells, pers. comm.
Haystall Cave, S.A.	1,E		
Hensches Cave, S.A.	1,E		
Kalaminurina, S.A.	1,E		Bartholomai & Marshall (1973)
Laura, Nth. Qld.	2,M	10,000 BP app.	Horton (1977)
L. Menindee, N.S.W.	2,E	18–26,000 BP	Tedford (1967)
L. Garnpung, N.S.W.	2,E	> 12,000 BP	Hope (1978)
L. Tandou, N.S.W.	2,3,E	> 20,000 BP	Hope, pers. comm. (1980)
Dempsey's Lake, S.A.	2,E	> 30,000 BP	Wells (1978) D. Williams, pers. comm. (1980)
Mair's Cave, S.A.	2,E		
Kangaroo Isl., S.A.	3,E	16,000±100 BP	Hope *et al.* (1977)
Queenscliff, Vic.	3,M	Holocene	Bartholomai & Marshall (1973)
Gisborne Cave, Vic.	3,M	Holocene	Gill (1964)

Key: 1. Agrees in size with Wellington Caves population; 2. Significantly smaller than *S. harrisii*, Tasmania; 3. Agrees in size with *S. harrisii*, Tasmania; E. Associated with extinct megafauna; M. Associated with modern species only.

DISCUSSION

This study indicates that the sample of teeth and jaws referable to species of *Sarcophilus* from the Wellington Caves is interpretable as representative of a single species that is also significantly larger than modern *S. harrisii.* Values obtained for the coefficient of difference (CD) between the fossil and modern populations were mostly between 1.0 and 2.0 with a mean of 1.3, a magnitude which would barely justify distinction at the subspecific level according to the interpretation of this index as used in previous studies of fossil marsupials (Ride 1964, Bartholomai 1975). However, the Wellington Caves fossils appear to differ in dental proportions as well as size from *S. harrisii,* and to such an extent that it is possible to justify recognition of a second species, *Sarcophilus laniarius* (Owen) The occurrence of similarly large devils in the Pleistocene deposits of the Darling Downs, Cement Mills, Bingara and Victoria Cave, Naracoorte (Table 5) supports a Pleistocene age for the Wellington Caves sample of this species.

Comparison of other fossil specimens from throughout Australia with the Wellington Caves and modern samples throws some doubt on the hypothesis of a simple dwarfing of late Pleistocene *S. laniarius* to produce *S. harrisii,* such as is suggested by Marshall (1973) and Marshall and Corruccini (1978). Rather, the contemporary occurrence of two species of devil is suggested by the presence of *S. laniarius* in the deposits of Lancefield Swamp (26,000 BP) and Cloggs Cave, Buchan (23,000 BP), while a form significantly smaller than modern *S. harrisii* occurs in deposits at Lake Garnpung, Lake Tandou and Dempsey's Lake, which are also older than 20,000 years (J. Hope and W. Williams pers. comm. 1980, and references in Table 5). These two forms of devil, occurring in the Late Pleistocene (approximately 25,000 years ago) differed from each other in size to a degree greater than that by which *S. laniarius* from the Wellington Caves differs from the modern species.

Although temporal overlap of two species has been shown here to have been possible, their sympatric contemporaneous occurrence has not been demonstrated. Two specimens in the Australian Museum collection, F58588 and F58665, maxillary fragments from the right and left hand, respectively, of a single individual, labelled as coming from the Wellington Caves, are significantly smaller than *S. harrisii* (Table 4, Fig. 2). However, they are preserved in a way that is quite different from that typical of other specimens from Wellington Caves, being light and unmineralised. They may therefore be mislabelled, or have come from a more recent deposit in the caves. These specimens clearly illustrate the distinction between the two forms of devil. However, their unknown stratigraphic provenance and different preservation prohibit them from being regarded as evidence for sympatry.

Archer (1978) has suggested that *S. harrisii* and *S. laniarius* occur together in the fauna from Cement Mills, Queensland, described by Bartholomai (1977). Of this sample of 9 mandibular and 4 maxillary specimens, Archer (1978) attributed only two mandibular fragments, QM F3734 and QM 3725 to *S. laniarius* suggesting that the rest of the specimens represented *S. harrisii.* Some of the specimens from Cement Mills have been measured in this study and the data presented in Tables 1 and 2. When data for the mandibular fragments are plotted on the bivariate scattergrams as shown in Fig. 2, all fall above the 95% confidence limits of *S. harrisii,* and all except QM F3725 fall within, or very near the 95% confidence limits of the Wellington Caves population. Values for the exceptional specimens fall *above* those of the Wellington Caves population.

The situation is not quite as clear for maxillary specimens from the Cement Mills sample. The coordinates for length and width of M^2 and M^3 of QM F3709 fall within, or close to, the zone of overlap between the 95% confidence ellipses of the Wellington Caves and the modern sample. The dimensions of QM F3710 fall clearly within the ellipse of the Wellington Caves sample only. However, the dimensions given by Bartholomai (1977) of a single M^2, QM F3728, fall within the *S. harrisii* ellipse only, indicating that this species may be present in the Cement Mills sample. The data indicate that most specimens in the Cement Mills fauna could have been drawn from a population having dental dimensions similar to those of the Wellington Caves sample. The data also indicate, however, that individuals both smaller and larger than those in the Wellington Caves sample are present. Unfortunately, there is no stratigraphic information associated with the Cement Mills sample, and for this reason this data cannot be considered sufficient evidence to support the sympatric occurrence of two species. It is possible that more than one fauna is represented by the Cement Mills material, as noted by Archer (1978).

Another possible case of sympatry is suggested by specimens from Lake Tandou, western New South Wales (Merrilees 1973). This deposit is currently being studied by Jeanette Hope who suggested (pers. comm.) that the two forms of devil present come from different stratigraphic horizons, and that the larger form is, in fact, much older than the smaller.

In conclusion, data from the Wellington Caves population has substantiated the distinction of a a large fossil species of devil, for which the name *S. laniarius is* appropriate. This large species typically occurs in Pleistocene faunas and appears to have survived as recently as 23,000 years BP at Clogg's Cave, Buchan. However, the demonstration of a small late Pleistocene species, distinct at least from *S. laniarius* has complicated the interpretation of the ancestry of the modern species, *S. harrisii.* While many phylogenetic pathways are possible, the two most parsimonious hypotheses that can be erected on the present data are presented in Fig. 3. The first is that modern *S. harrisii* is descended from the small species existing on the mainland at approximately 25,000 years BP. Although this small species is clearly distinct from *S. laniarius,* its distinction from *S. harrisii* has yet to be demonstrated. This hypothesis requires that some increase in size occurred in the Tasmanian population and in recent specimens from Victoria.

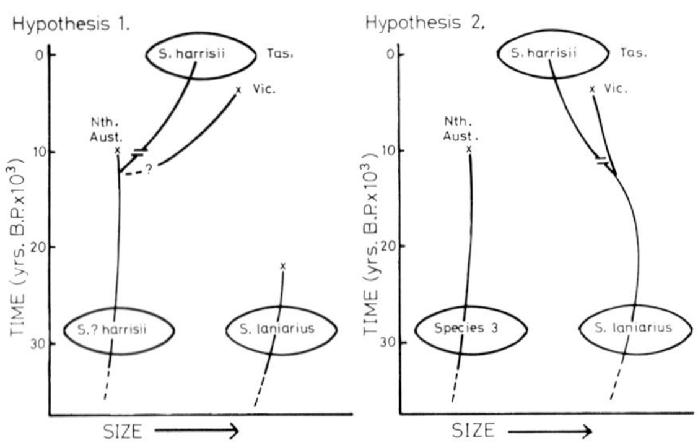

Fig. 3. Two alternative hypotheses regarding the interrelationships of populations of devils, based on data presented in this paper.

The second hypothesis, similar to that of Marshall and Corruccini (1978), proposes that modern *S. harrisii* is a dwarfed descendant of *S. laniarius*. However, this hypothesis requires recognition of a small species (Species 3) that was distinct from *S. harrisii.* This distinction cannot be justified on available data, either of a morphological kind, or on a basis of a statistically significant size difference.

Before either hypothesis can be adequately tested, more data must be accumulated about the geographical and stratigraphic distribution of small late Pleistocene devils.

ACKNOWLEDGEMENTS

Work on the Wellington Caves fossils has been undertaken while the author was in receipt of a Commonwealth Postgraduate Research Award and has also been supported by a Postgraduate Grant from the Australian Museum and a further Grant-in-aid from the Trustees of the Australian Museum, Sydney. I wish to thank Dr. Alex Ritchie of the Australian Museum for his support and permission to study in that Institution. Dr. A. Bartholomai, Queensland Museum; Dr. N. Pledge, South Australian Museum; and Dr. T. Rich, Victorian Museum also allowed me to use their collections. Dr. J. Hope, Dr. R. Wells and Dr. M. Archer kindly lent specimens for this study. Permission was obtained from the Library of The Australian National University to quote unpublished data from the Ph.D thesis of R. Frank (1972). Dr. Michael Archer gave much valuable encouragement and advice as have many other members of the School of Zoology, University of New South Wales. Mr. M. Plane and Mr. J. Mahoney provided useful criticism of a draft of this paper.

REFERENCES

Archer, M., 1978a. The nature of the molar–premolar boundary in marsupials and reinterpretation of the homology of marsupial cheekteeth. *Mem. Qd Mus.* **18:** 157–64.

Archer, M., 1978b. Quaternary vertebrate faunas from the Texas Caves of southeastern Queensland. *Mem. Qd Mus.* **19:** 61–109.

Archer, M. and Baynes, A., 1972. Prehistoric mammal faunas from two small caves in the extreme southwest of Western Australia. *J. Roy. Soc. West. Aust.* **55:** 80–9.

Bartholomai, A., 1975. The genus *Macropus* Shaw (Marsupialia: Macropodidae) in the upper Cainozoic deposits of Queensland. *Mem. Qd Mus.* **17:** 195–235.

Bartholomai, A., 1977. The fossil vertebrate fauna from Pleistocene deposits at Cement Mills, Gore, southeastern Queensland. *Mem. Qd Mus.* **18:** 41–51.

Bartholomai, A. and Marshall, L. G., 1972. The identity of the supposed dasyurid marsupial *Sarcophilus prior* De Vis 1883, with comments on other reported "Pliocene" occurrences of *Sarcophilus. Mem. Qd Mus.* **16:** 396–74.

Calaby, J. H. and White, C., 1967. The Tasmanian Devil (*Sarcophilus harrisii*) in northern Australia in Recent time. *Aust. J. Sci.* **29:** 473–75.

Colditz, M. J., 1942. Physiography of the Wellington District, N.S.W. *J. Roy. Soc. N.S.W.* **76:** 235–51.

Francis, G., 1973. Evolution of the Wellington Caves landscape. *Helictite* **11:** 79–91.

Frank, R., 1972. "Sedimentological and morphological study of selected cave systems in eastern New South Wales". Unpublished Ph.D. thesis: A.N.U., Canberra.

Flood, J., 1973. Pleistocene human occupation and extinct fauna in Clogg's Cave, Buchan, southeastern Australia. *Nature (London)* **246** (5431): 303.

Gill, E. D., 1953. Distribution of the Tasmanian Devil, the Tasmanian Wolf, and the Dingo in southeastern Australia in Quaternary times. *Vic. Nat.* **70:** 86–90.

Gill, E. D., 1964. Appendix in Mahoney, J. A., 1964. The taxonomic status of *Dasyurus affinis* McCoy, 1865 (Dasyuridae) and *Hypsiprymnus trisulcatus* McCoy, 1865 (Macropodidae), two marsupials from a Holocene cave near Gisborne, Victoria. *Proc. R. Soc. Vic.* **77:** 525–33.

Gillespie, R., Horton, D. R., Ladd, P., Macumber, P. G., Rich, T. H., Thorne, R. and Wright, R. V. S., 1978. Lancefield Swamp and the extinction of the Australian megafauna. *Science* **200:** 1044–8.

GINGERICH, P. D., 1974. Size variability of the teeth of living mammals and the diagnosis of closely related sympatric fossil species. *J. Paleontology* **48:** 895–903.

HOPE, J. H., 1978. Pleistocene mammal extinctions: the problem of Mungo and Menindee, New South Wales. *Alcheringa* **2:** 65–82.

HOPE, J. H., LAMPERT, R. J., EDMONSON, E., SMITH, M. J. AND VAN TETS, G. F., 1977. Late Pleistocene faunal remains from Seton Rock Shelter, Kangaroo Island, South Australia. *Jour. Biogeography* **4:** 363–85.

HORTON, D. R., 1977. A 10,000 year old *Sarcophilus* from Cape York. *Search* **10:** 374–5.

LANE, E. A. AND RICHARDS, A. M., 1963. The discovery, exploration and scientific investigation of Wellington Caves, New South Wales. *Helictite* **2:** 3–52.

LONGMAN, H., 1925. Fossil marsupials from Marmor. *Mem. Qd Mus.* **8:** 109.

MAHONEY, J. A. AND RIDE, W. D. L., 1975. Index to the genera and species of fossil Mammalia described from Australia and New Guinea between 1838 and 1968. *West. Aust. Mus. Spec. Pub.* **6**.

MARSHALL, L. G., 1973. Fossil vertebrate faunas from the Lake Victoria region of southwestern New South Wales. *Mem. Nat. Mus. Vic.* **34:** 151–71.

MARSHALL, L. G., 1974. Late Pleistocene mammals from the Keilor Cranium site, southern Victoria, Australia. *Mem. Nat. Vic.* **35:** 63–86.

MARSHALL, L. G. AND CORRUCINI, R. S., 1978. Variability, evolutionary rates and allometry in dwarfing lineages. *Paleobiology* **4:** 101–19.

MAYR, E., LINDSAY, E. G. AND USINGER, R. L., 1953. "Methods and principles of systematic zoology". McGraw Hill: New York.

MERRILEES, D., 1973. Fossiliferous deposits at Lake Tandou, New South Wales, Australia. *Mem. Nat. Mus. Vic.* **34:** 177–82.

MITCHELL, T. L., 1838. "Three expeditions into the interior of eastern Australia, with descriptions of the recently explored region of Australia Felix, and of the present colony of New South Wales". T. and W. Boone: London. Vol. 2.

OWEN, R., 1877. "Researches on the fossil remains of the extinct mammals of Australia; with a notice on the extinct marsupials of England". London.

RIDE, W. D. L., 1964. A review of Australian fossil marsupials. *J. Proc. Roy. Soc. West. Aust.* **47:** 97–131.

SIMPSON, G. G., ROE, A. AND LEWONTIN, R. C., 1960. "Quantitative zoology". Rev. ed. Harcourt: New York.

SOKAL, R. R. AND ROHLF, F. J., 1969. "Biometry – The principles and practice of statistics in biological research". Freeman and Co.: San Francisco.

TEDFORD, R. H., 1967. The fossil Macropodidae from Lake Menindee, New South Wales. *Univ. Calif. Publs. Geol. Sci.* **64:** 1–156.

WELLS, R. T., 1978. Fossil mammals in the reconstruction of Quaternary environments with examples from the Australian fauna. Pp.103–24 *in* "Biology and Quaternary environments" ed. by D. Walker and J. C. Guppy. Australian Academy of Science: Canberra.

RESUMEN

43. Status taxonómico de los "deviles" fósiles (Sarcophilus : Dasyuridae, Marsupialia) del Cuaternario tardío de localidades del Este de Australia. Por Lyndall Dawson.

Las medidas dentarias de los "deviles" fósiles de los depósitos de las cuevas de Wellington, New South Wales, son comparadas con las del actual *S. harrisii;* sus diferentes proporciones apoyan su distinción como *Sarcophilus laniarius* (Owen). Los datos indican que en la fauna de las Wellington Caves está representala sólo una especie, *S. laniarius.* Se demuestra que "deviles" de tamaño similar aparecen en otros depósitos del Pleistoceno tardío del Este de Australia, cuyas especies sobrevivieron al menos hasta 23.000 años BP en la Cueva Clogg, Buchan, Victoria. Sin embargo, una forma de "devil" significativamente más pequeña que la actual *S. harrisii* aparece también en depósitos del Pleistoceno tardío del Oeste de New South Wales. Se han considerado dos hipótesis alternativas que podrían contar en las relaciones de las formas pleistocenas y actual.

44. TAXONOMIC STATUS OF FOSSIL THYLACINES (*THYLACINUS*, THYLACINIDAE, MARSUPIALIA) FROM LATE QUATERNARY DEPOSITS IN EASTERN AUSTRALIA

LYNDALL DAWSON[1]

Fossil thylacines from Wellington Caves, New South Wales, in the Australian Museum, Sydney, have been examined in an effort to clarify the taxonomic status of the fossil species *Thylacinus spelaeus* Owen, 1845. Comparison has been made with a sample of modern *T. cynocephalus* from Tasmania, and with fossil specimens from locations throughout eastern Australia. Although the Wellington Caves thylacines are shown to be significantly larger than the Tasmanian thylacine, these differences in size alone are not considered sufficient to warrant the specific distinction of *T. spelaeus*. *Thylacinus rostralis* De Vis, 1893 and *Thylacinus major* Owen, 1877, are considered to be junior synonyms of *T. cynocephalus*. Much variation in size is demonstrated for fossils from Pleistocene and Holocene localities, some specimens being significantly smaller than modern *T. cynocephalus*.

Pages 527–36 *in* CARNIVOROUS MARSUPIALS, ed. by M. Archer, February, 1982. Roy. Zool. Soc. New South Wales: Sydney, Australia.

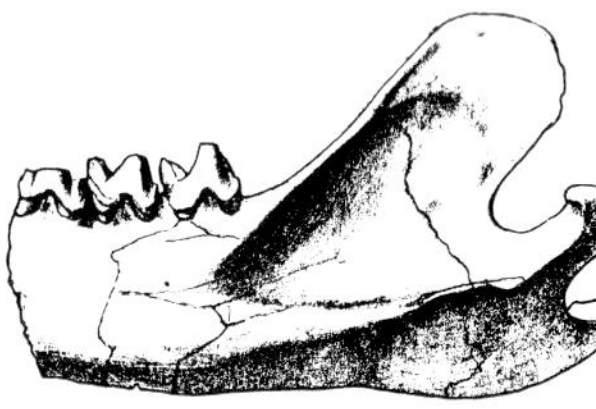

THE AIMS of the present study of fossil marsupials from the Wellington Caves deposits in New South Wales have been set out previously (Dawson 1982, this volume). In this paper the taxonomic status and variability of fossil thylacine specimens will be assessed.

Owen (1845) described the first fossil thylacine specimens from material probably collected in the Wellington Caves by Count de Strzelecki (Mahoney and Ride 1975: 40). At that time he also correctly referred to a species of *Thylacinus* the two specimens (M10800 and M10801 in the collections of the British Museum (Natural History) he had previously (1838) described as *Dasyurus laniarius* Owen. Although he considered that some of this fossil material was indistinguishable in size and morphology from the living *Thylacinus cynocephalus,* he assigned two partial mandibular rami (including M10801) and an isolated M_4 to *Thylacinus spelaeus,* a species which he considered to be similar to but larger than *T. cynocephalus*. Two of the syntypes of *Thylacinus spelaeus* have since been lost and M10801, the anterior part of a mandibular ramus from Wellington Caves is the single remaining type specimen (Mahoney and Ride 1975 : 39–40). Lydekker (1887) referred the few fossil specimens available to him to *T. spelaeus* (four from Wellington Caves and two from Pleistocene deposits in Queensland). Lydekker considered the smaller specimens to represent females of *T. spelaeus*.

Owen (1877) later named *Thylacinus major* based on specimens from Wellington Caves. The validity of this species has been in doubt since the time of De Vis (1894) and will be discussed in this paper.

On the basis of Owen's (1845) description, De Vis (1893, 1894) interpreted *T. spelaeus* to be morphologically almost identical with *T. cynocephalus,* differing only in its larger size. De Vis (1893) named a second large fossil species from Queensland, *Thylacinus rostralis,* considering it to differ from the living species and from *T. spelaeus* in some cranial and dental proportions.

A Tertiary species, *Thylacinus potens* Woodburne, has been described from the late Miocene Alcoota local fauna (Woodburne 1967). This is the earliest record of an undoubted thylacine. A single premolar from the middle to late Miocene Kutjamarpu local fauna (Archer and Bartholomai 1978) may represent an older thylacine. *Thylacinus potens* has been shown (Woodburne 1967) to be distinct from modern *T. cynocephalus* and *T. rostralis* on the basis of several morphological features of the palate and molar teeth as well as its larger size. A premolar fragment of an unknown species is also known from the Pliocene Awe fauna of Papua New Guinea (Plane 1976).

Some preliminary studies on the validity of fossil thylacine taxa have been reported by Ride (1964). He compared dental and mandibular characters of thylacines from Western Australia with a small sample (max. N = 4) of "eastern cave–fossil *Thylacinus*" (including the existing

[1] School of Zoology, University of New South Wales, P.O. Box 1, Kensington, New South Wales, Australia, 2033.

syntypical specimen of *T. spelaeus*) and with a large sample (N = approx. 50) of modern *T. cynocephalus*. However, statistical tests did not justify even sub-specific distinction for the Western Australian sample. Ride (1964) did, however, recommend retention of *T. rostralis* as an available name for eastern Australian thylacines.

The aim of the present study is to investigate the nomenclatural status of the eastern Australian fossil material. To do this it is necessary to 1) measure variability of a large sample of Wellington Caves thylacines, this being the probable type locality for the earliest name proposed for a larger fossil species, *T. spelaeus* (Owen 1845); 2) to compare the Wellington Caves sample with modern *T. cynocephalus;* and 3) to investigate the status of the names *Thylacinus rostralis* De Vis and *Thylacinus major* Owen.

Institutional abbreviations used are as follows: AM, Australian Museum, Sydney; QM, Queensland Museum, Brisbane; NMV, National Museum of Victoria, Melbourne; SAM, South Australian Museum; BM, British Museum of Natural History; SPS/ANU, School of Pacific Studies, Australian National University, Canberra.

Dental nomenclature follows Archer (1978). Accordingly, the adult thylacine dental formula is

$$I\frac{1\text{-}4}{1\text{-}3};\ C\frac{1}{1};\ P\frac{1\text{-}3}{1\text{-}3};\ M\frac{2\text{-}5}{2\text{-}5}.$$

MATERIALS

The Wellington Caves Sample

Referred Specimens from the Australian Museum: Mandibular fragments: F16533, F57850, F57855, F16550, F57868, F57848, F57846, F57875, F57847, MF308, F57849, F57859, F57856, F16588, F31745, F57898, F57905, F57907, F57896, F57899, F57897, F16552, F57861, F57854, F57862, F57850, F57876, MF119, F57901, F57894. Maxillary fragments: F50818, F50823, F50821, F50825 -6, F50824, F57873, F57871, F57877, F50827, F50819, F50822, F57869, F50820, F57882, F31731, F57880, F57878.

The sample of fossil thylacines included in this study from the Australian Museum collection from Wellington Caves comprises 17 maxillary and 31 mandibular fragments. The material is very fragmentary and no entire skulls or mandibles are represented. Broken or severely worn teeth were not included in the statistical treatment, nor were isolated teeth because it is difficult to always be certain of the correct designation of an isolated tooth. Specimens were sorted into left and right sides and checked for the possibility that any two or more represented one individual, by comparing tooth size, wear and preservation. No matching specimens were recognised. A cast of a syntype of *T. spelaeus,* BM M10801, a mandibular fragment, was available.

The Sample of Modern *Thylacinus cynocephalus*

Twenty entire skulls were available in the Australian Museum. All were collected in Tasmania between 1866 and 1925, but bear no further locality, age or sex data. The age of individuals was roughly estimated from tooth eruption and wear stages. Thus four specimens were regarded to be juveniles (M^5 was not fully erupted), 10 were regarded to be "young adults" (all teeth were erupted but lightly worn), and 5 were older adults (worn teeth).

Other Comparative Material

Fossil thylacine specimens were measured from the collections of the Queensland Museum and the South Australian Museum. Specimens from western New South Wales and Papua New Guinea were made available by J. Hope (School of Pacific Studies, ANU, Canberra). A skull from Yarrangobilly Caves, New South Wales was loaned by J. Calaby (C.S.I.R.O. Canberra). Specimens from Jenolan Caves and Bingara,

Table 1. Dimensions (mm) of the skull and upper teeth of fossil thylacines from eastern Australian localities.

NUMBER	LOCATION	LP^1	LP^2	LP^3	LM^2	WM^2	LM^3	WM^3	LM^4	WM^4	LM^5	WM^5	L $M^{1\text{-}3}$	Basi-cran. L.	Zygo-matic W.
SAM P20450	Victoria Cave, Naracoorte, S.A.	6.5	7.4	—	11.5	8.4	13.8	10.7	—	—	10.6	6.4	39.0	—	—
SAM P20448	" " " "	8.0	—	—	—	—	15.0	11.5	—	—	—	—	—	—	—
SAM P16754	" " " "	6.5	8.3	—	11.5	8.4	—	—	—	—	11.5	6.4	—	—	—
SAM P14511	Mair's Cave, S.A.	—	—	13.5	12.5	—	15.0	12.5	17.2	13.6	—	—	42.6	—	—
SAM (unregd.)	" " "	8.3	—	11.2	10.8	7.8	—	—	—	—	—	—	35.5	—	—
SAM P13827	" " "	6.6	—	9.6	10.5	7.5	11.8	10.0	—	—	11.0	6.0	35.0	170(a)	—
SAM (unregd.)	" " "	6.2	9.2	11.3	11.9	9.0	13.2	10.4	14.6	11.4	11.5	5.8	39.7	210.0	118.0
—	Yarrongobilly Caves, N.S.W.	—	9.2	11.8	12.0	8.8	14.7	11.6	15.6	12.4	12.2	6.2	—	241.0	170.0
QM F730	Ellangowan, Qld.	—	—	11.6	12.1	8.3	14.0	11.5	16.0	13.0	9.0	—	—	220.0	—
AM F413	Jenolan Caves, N.S.W.	5.7	8.0	11.8	11.0	8.0	12.3	9.6	15.5	12.5	10.4	6.8	—	—	—

(a) Approximate measurement.

New South Wales were available from the Australian Museum collections. The comparative material used is listed in Tables 1 and 2, which also give the dimensions of these specimens.

METHODS

MEASUREMENTS

Dental measurements (mm), made with Helios dial calipers, are defined as follows: Tooth length — The greatest anteroposterior dimension of the enamel crown; Tooth width — (1) Upper molars, the maximum anterior width at the base of the enamel across the paracone — protocone, (2) Lower molars, the width across the widest part of the protoconid. It should be noted that these measurements differ from those used to compare modern and fossil thylacine teeth by Ride (1964) and Lowry (1972). They both measured tooth length only, a measure defined by them, for upper molars, as the diagonal length through protocone to metacone. The measurement of lower molar width used here is not maximum tooth width. The mean values obtained in this study for the widths of lower molars of modern *T. cynocephalus* are very similar to those obtained by Woodburne (1967) for his character, W1 (anterior width of lower molars), of a modern Tasmanian sample, although this width is defined differently.

Other measurements taken were the inclusive length of M_{2-5}, measured at the alveolar level; depth of the mandible beneath P_3 and beneath M_5; the inclusive length from P_3 to the canine at the alveolar level ($P_3 - C$); basicranial length; and the maximum width across the zygomatic arches. Spacing of premolar teeth was measured as follows:

$$\text{Ratio a/b} = \frac{\text{the diastemal length between alveoli of } P^1 \;\&\; P^2}{\text{the diastemal length between alveoli of } P^2 \;\&\; {}^3;}$$

$$\text{Ratio a'/b'} = \frac{\text{distance between the apices of } P^1 \text{ and } P^2}{\text{distance between the apices of } P^2 \text{ and } P^3;}$$

$$\text{Ratio c/d} = \frac{\text{diastemal length between the alveoli of } P^1 \;\&\; P^2}{\text{diastemal length between the alveoli of } P^2 \;\&\; P^3;}$$

$$\text{Ratio c'/d'} = \frac{\text{distance between the apices of } P^1 \text{ and } P^2}{\text{distance between the apices of } P^2 \text{ and } P^3.}$$

STATISTICAL ANALYSIS

The data was analysed as described in Dawson (1982), for a similar study of fossil species of *Sarcophilus*.

LOG DIFFERENCE DIAGRAM

In the log difference diagram (Fig. 1) (Simpson *et al.* 1960) the mean values of the modern sample of *Thylacinus cynocephalus* from Tasmania have been used as the standard for comparison of the proportional differences of the Wellington Caves sample. The observed ranges of both samples are shown.

RESULTS

Dental, cranial and mandibular measurements of the sample of *Thylacinus cynocephalus* from Tasmania, and of the Wellington Caves sample are summarised with the results of statistical analysis in Tables 3 and 4. The means of the dental measurements for the Wellington Caves sample are significantly larger (at the 5% level or lower) than those of the modern Tasmanian

Table 2. Dimensions (mm) of the mandible and lower teeth of fossil thylacines from eastern Australian localities.

NUMBER	LOCATION	LP_3	LM_2	WM_2	LM_3	WM_3	LM_4	WM_4	LM_5	WM_5	Alv. L M_{2-5}	Mand. depth below M_5	Mand. depth below P_3
SAM P20451	Victoria Cave, Naracoorte, S.A.	—	11.0	—	13.2		15.8	[illegible]	17.6	8.2	58.5	35.6	—
SAM P20453	" " " "	11.3	10.0	4.8	12.3	5.8	—	—	—	—	—	28.8	—
SAM P20452	" " " "	11.2	10.5	5.2	13.0	6.3	15.0	6.9	—	—	—	—	—
SAM P16750	" " " "	12.4	10.0(a)	—	13.5	6.1	—	—	—	—	—	28.1	—
SAM (unregd.)	" " " "	11.3	9.6	4.4	—	—	14.6	6.0	16.5	7.8	53.0	32.2	—
SAM (unregd.)	Mair's Cave, S.A.	10.2(a)	8.0(a)	—	—	—	—	—	13.0(a)	—	43.0	18.2	—
SAM (unregd.)	" " "	9.5	8.2	3.9			—	—	—	—	42.0	18.8	—
SAM (unregd.)	" " "	12.5	10.3	5.1	13.0	6.0	14.9	6.5	16.6	7.8	55.5	35.4	—
SAM P13827	" " "	10.0	8.0	4.6	11.0(a)	5.3	12.5	5.5	15.0	7.0	46.0	26.5	—
SAM P13728	Hensche's Cave, S.A.	12.0	—	—	12.3	6.0	14.6	—	17.3	7.8	55.0	30.0	—
SPS/ANU (unregd.)	Clogg's Cave, Buchan, Vic.	—	—	—	—	—	—	—	—	9.1	—	—	—
SPS/ANU MM5	Lake Mungo, N.S.W.	—	—	—	—	—	11.6	4.9	13.6	6.4	—	—	—
SPS/ANU —	Lake Menindee, N.S.W.	—	—	—	11.5	5.5	13.8(a)	5.5(a)	14.5	7.0	44.0	—	—
ANU, NCA/B/3	Nombe, Papua–New Guinea	9.2	8.2	4.0	10.5	5.5	11.6	6.0	—	—	43.9	—	—
QM F1737	Marmor Quarry, Qld.	—	—	—	—	—	16.5	7.9	—	—	—	—	—
QM F730	Ellangowan, Qld.	12.0	10.5	5.7	13.6	6.4(a)	14.3	7.4	16.9	7.8(a)	—	—	38.9
AM MF413	Jenolan Caves, N.S.W.	11.6	9.2	4.4	11.3	5.6	13.3	6.9	15.5	8.1	51.0	32.0	25.0
AM MF1185	Bingara, N.S.W.	—	—	—	—	—	—	—	—	—	—	—	21.6

(a) Approximate measurement.

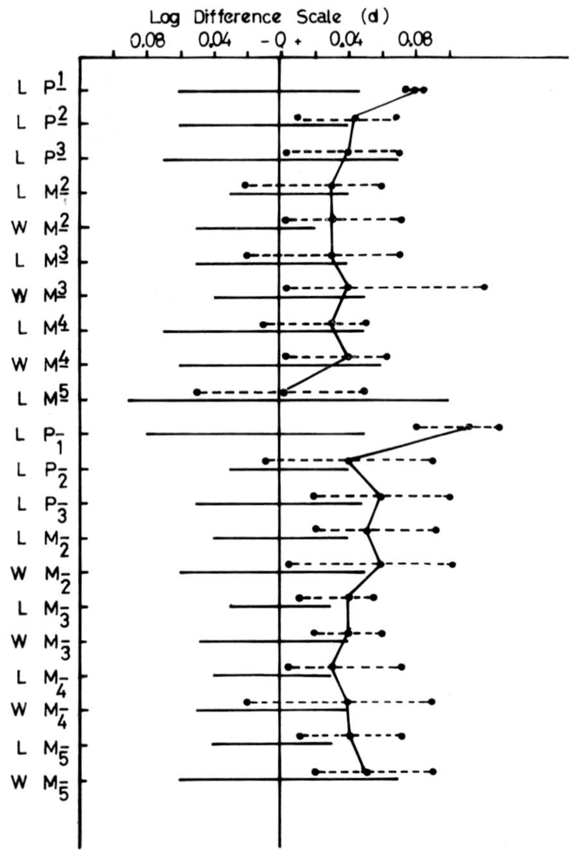

Fig. 1. Log difference diagram showing proportional relationships of tooth length (L) and width (W) in fossil *Thylacinus cynocephalus* from Wellington Caves, compared with a sample of modern *T. cynocephalus* from Tasmania as standard. Log difference, d = log x–log standard. Solid vertical lines join mean values of d for each sample. Horizontal lines represent the observed range of values for each sample. Data from Table 3.

sample for all characters except the length and width of M^5. The Tasmanian sample exhibited a slightly higher mean value for CV (mean CV = 6.8) than the fossil sample (mean CV = 6.0). Comparably high CV values for a sample of modern *T. cynocephalus* were noted by Woodburne (1967) (mean CV of 30 measurements = 6.8) and Ride (1964) (mean CV of 5 dental characters = 6.0). Both authors attributed these somewhat high values to sexual dimorphism. Thus a mean value of CV = 6.0 for the Wellington Caves sample is consistent with the variation expected for a population of a single species. Because of the fragmentary nature of the specimens from Wellington Caves, few comparisons could be made of cranial or mandibular dimensions other than tooth size. The data suggest that mandibular length (as indicated by $M_{2\text{-}5}$ and length P_3 – C) was not significantly greater in the Wellington Caves specimens than in the modern Tasmanian sample. However, data presented in Table 4 indicate that there was little overlap in the observed ranges of the values for mandibular depth below P_3 and below M_5, these being greater in the fossil population. There also was no overlap in observed range between the two populations for LP^1 and LP_1.

The coefficient of difference (CD) has been calculated using the formula CD $\frac{Xa - Xb}{sa - sb}$ where Xa and Xb are the sample means and sa and sb are the respective standard deviations (Mayr *et al.* 1953) as a guide to estimating the degree of size difference between the populations. Values for CD are given in Table 3 and were generally much lower than the value of 1.5 suggested by Ride (1964) to indicate subspecific difference.

The log difference diagram (Fig. 1) illustrates the large amount of overlap in size between the two populations tested, and indicates that they do not differ significantly in most dental proportions. The data indicates that $P^1/_1$ was slightly larger proportionately in the fossil sample than in the modern population.

The lectotype of *T. spelaeus* (BM M10801) is a fragment of a right mandibular ramus with incisors, the canine, and P_1, P_2 and P_3. The premolar lengths of this specimen are LP_1 = 7.7mm, LP_2 = 9.8mm and LP_3 = 11.4mm. Length of P_1 falls within the observed range of that tooth for the rest of the Wellington Caves sample, but not for the modern Tasmanian sample while LP_2 and LP_3 fall within the observed range of the Tasmanian population.

Dental measurements for the lectotype of *T. rostralis* (QM F730) are as follows: LP^3 = 11.6mm, LM^2 = 12.1mm, WM^2 = 8.3mm, LM^3 = 14.0mm, WM^3 = 11.5mm, LM^4 = 16.0mm, WM^4 = 13.0mm, basicranial length = 220mms; LP_3 = 12.0mm, LM_2 =10.5mm, WM_2 = 5.7 mm, LM_3 = 13.6mm, WM_3 = 6.4 (a) mm, LM^4 = 14.3mm, WM^4 = 7.4mm, LM^5 = 16.9 mm; these values do not differ significantly from the mean values for the Wellington Caves population.

Comparison of these values with the mean values for the modern Tasmanian sample showed that some teeth were significantly larger (Table 5). The basicranial length of QM F730 falls within the O.R. of the modern Tasmanian sample.

De Vis (1894) and Woodburne (1967) suggested that the spacing of the premolar teeth may be of taxonomic importance in distinguishing *T. rostralis* from *T. cynocephalus*. Variation in this character within the Tasmanian sample has therefore been investigated. The results, given in Table 4, indicate that although the diastemal lengths between the alveoli of the premolar teeth is rather variable in the Tasmanian population, the apices of the teeth maintain a rather more constant ratio for the distances between $P^1/_2$ and $P^2/_2$, and between $P^2/_1$ and $P^3/_3$.

Table 3. Summary and statistical comparisons of dental measurements (mm) of modern *Thylacinus cynocephalus* from Tasmania and of fossils from Wellington Caves.

	Wellington Caves					*T. cynocephalus*–Tasmania						Variance		student's t*	
Character	N	$\overline{X}$	O.R.	S	CV	N	$\overline{X}$	O.R.	S	CV	CD	F	P	t	P
LP^1	2	7.5	7.4– 7.5	0.07	—	18	6.2	5.4– 7.0	0.46	7.3	2.45	43.18	NS*	10.90	>0.001
LP^2	4	9.3	8.6–10.0	0.62	7.1	20	8.4	7.3– 9.3	0.62	7.4	0.73	—	—	2.65	>0.02<0.01
LP^3	6	11.5	10.7–12.4	0.66	6.0	18	10.6	9.1–12.2	0.79	7.4	0.62	1.43	NS	2.74	>0.02<0.01
LM^2	14	11.7	10.5–12.5	0.57	4.9	18	11.0	10.2–12.0	0.53	4.8	0.55	1.16	NS*	3.55	>0.01<0.001
WM^2	14	8.6	7.9– 9.5	0.43	5.0	18	8.0	7.2– 8.4	0.36	4.5	0.75	1.43	NS	4.45	>0.001
LM^3	15	14.1	12.8–15.5	0.77	5.4	20	13.3	11.9–14.5	0.72	5.4	0.54	1.14	NS	3.13	>0.01
WM^3	15	11.1	10.2–13.4	0.80	7.3	20	10.2	9.2–11.4	0.51	5.0	0.75	2.46	NS	3.81	>0.001
LM^4	8	15.9	14.8–16.8	0.69	4.3	20	15.1	12.8–16.9	1.06	7.0	0.45	2.36	NS	2.23	>0.05<0.02
WM^4	10	12.9	11.7–13.7	0.85	6.6	18	11.8	10.2–13.3	0.90	7.7	0.68	1.12	NS	3.21	>0.01
LM^5	7	6.5	6.0– 7.5	0.55	8.4	14	6.8	5.5– 8.8	0.96	14.1	0.20	3.05	NS	0.91	NS
WM^5	6	12.0	11.0–13.6	1.00	8.3	14	12.2	10.0–15.5	1.53	12.5	0.08	2.34	NS	0.35	NS
LP_1	6	7.8	7.2– 8.2	0.40	5.4	18	6.0	5.0– 6.7	0.49	8.3	2.02	1.50	NS	9.00	>0.001
LP_2	7	10.0	8.9–11.4	0.84	8.7	16	9.1	8.5–10.0	0.46	5.0	0.69	3.33	NS	2.67	>0.02<0.01
LP_3	5	12.2	11.1–13.5	1.11	9.5	17	10.6	9.5–11.9	0.66	6.2	0.91	2.83	NS	3.06	>0.01<0.001
LM_2	9	10.7	10.0–11.8	0.57	5.5	20	9.6	8.7–10.6	0.61	6.3	0.92	1.22	NS	4.69	>0.001
WM_2	9	5.1	4.4– 5.5	0.34	6.7	20	4.4	3.8– 4.9	0.29	6.7	1.11	1.37	NS	5.36	>0.001
LM_3	8	13.1	12.4–13.5	0.34	2.6	20	12.0	11.1–12.9	0.64	5.4	1.12	3.54	0.05	5.88	>0.001
WM_3	8	6.3	5.9– 6.6	0.25	4.0	20	5.7	5.1– 6.2	0.31	5.4	1.05	1.63	NS	5.34	>0.001
LM_4	13	15.2	14.2–16.4	0.57	3.7	20	14.1	12.8–15.0	0.55	3.9	0.97	1.04	NS	5.49	>0.001
WM_4	13	7.5	6.6– 8.5	0.48	6.4	20	6.9	6.1– 7.5	0.39	5.7	0.70	1.60	NS	3.77	>0.001
LM_5	10	17.4	16.0–18.4	0.73	4.2	17	15.7	14.2–17.0	0.90	5.7	0.99	1.28	NS	5.35	>0.001
WM_5	7	8.5	7.9– 9.3	0.52	6.1	17	7.6	6.6– 8.9	0.54	7.1	0.86	1.04	NS	3.81	>0.01<0.001
LM_{2-5}	1	—	57(a)	—	—	18	52.0	47.0–58.0	2.82	4.7	—	—	—	—	—
LP_3-C	3	53.0	48.0–56.0	4.36	—	20	48.0	37.0–58.5	5.53	11.5	—	—	—	—	—
	Mean CV = 6.0					Mean CV = 6.8									

* Method assuming heteroscedasticity (Sokal and Rohlf 1969)
NS = P>0.05

Table 4. Summary of some cranial and mandibular measurements (mm) of *Thylacinus cynocephalus* from Tasmania, and fossil specimens.

Character	Tasmania				Wellington Caves				Yarrongo-billy Caves	QM* F730	AM** MF413
	N	$\overline{X}$	O.R.	S	N	$\overline{X}$	O.R.	S			
Basicranial Length	11	206.6	170.0–240.0	23.4	—	—	—	—	241.0	220.0	—
Zygomatic Width	11	124.8	98.0–150.0	16.3	—	—	—	—	170.0	—	—
Ratio a/b (a)	11	1.2	0.9– 1.6	0.26	9	1.2	0.9– 1.3	0.34	—(b)	—	0.7
Ratio a'/b' (a)	11	1.0	0.8– 1.2	0.10	2	—	0.8– 1.2	—	—	—	0.7
Mand. d. below P_3	20	20.0	16.0– 23.7	2.78	9	25.9	21.0–33.0	3.62	—	30.0	25.0
Mand. d. below M_5	18	28.5	23.0– 40.0	4.41	3	39.2	37.0–43.0	3.33	—	43.0	32.0
Ratio c/d (a)	11	1.0	0.6– 2.2	0.45	11	0.9	0.4– 2.3	0.55	—	2.0	0.3
Ratio c'/d' (a)	11	0.8	0.7– 0.9	0.09	1	—	0.8	—	—	1.2	0.6

* Lectotype of *T. rostralis*
** Jenolan Caves, N.S.W.
(a) Ratios are defined in text; Section 3.1.
(b) Abnormal due to loss of teeth during life.

For example, in the modern specimen, AM S410, the value for ratio c/d = 2.2 while c'/d' = 0.9. The 11 modern skulls examined for this character included juveniles, young adults and aged individuals, and no pattern of variation with age was observed. It is suggested here that the need for good occlusion between upper and lower premolars determines the spacing of the apices of the teeth and that only this character is likely to be of taxonomic significance. The apical spacing of premolars could only be measured in a few specimens from Wellington Caves because of high incidence of broken teeth. However, the ratios obtained agreed closely with those obtained for the modern Tasmanian sample.

The lectotype of *T. rostralis* lacks P^2 which has apparently been lost during the life of the animal because the alveolus is partly filled with bone. Thus apical spacing of the upper premolars cannot be measured. While P_2 is closer to P_3 in this mandible than any specimen of the modern sample, it is considered here that this variation may represent an abnormality resulting from the lack of need to occlude with the missing P^2. Larger samples of undamaged jaws showing this variation are needed to establish any possible taxonomic distinction. The premolar teeth of the only known partial skull of *Thylacinus potens* are broken off and, therefore, apical spacing of these teeth cannot be measured.

Dental measurements of fossil thylacine specimens from many other mainland deposits have been compared with those from Wellington Caves and with the modern Tasmanian sample. The origin of these specimens and their dental dimensions are given in Tables 1 and 2. Table 6 presents a summary of this comparative information. Student's t-tests were used to compare some specimens with the population means of the modern Tasmanian sample (Table 5). Most specimens agreed in size with the modern Tasmanian sample, or fell within the observed range of Wellington Caves sample. However, the specimens from Lake Mungo and from Kiowa, New Guinea, were significantly smaller than the Tasmanian population. Other thylacines reported in the literature to be significantly smaller than modern *T. cynocephalus* are from Fromm's

Table 5. Significance levels of t–tests comparing the sample means of *Thylacinus cynocephalus* from Tasmania with some single fossil specimens.
* Significantly larger; ** Significantly smaller.

	E. Darling Downs	Yarrongo-billy Cave	Jenolan Caves	Lake Mungo	Lake Menindee	Nombe N.G.
LP^3	NS	NS	NS			
LM^2	5.0%	5.0%	NS			
WM^2	NS	2.0%	NS			
LM^3	NS	5.0%	NS			
WM^3	2.0%	1.0%	NS			
LM^4	NS	NS	NS			
WM^4	NS	NS	NS			
LM^5	NS	NS	NS			
WM^5	—		NS			
LP_3	NS		NS			2.0%
LM_2	NS		NS			5.0%
WM_2	0.1%		NS			NS
LM_3	5.0%		NS		NS	5.0%
WM_3	5.0%		NS		NS	NS
LM_4	5.0%		NS	0.1%	NS	0.1%
WM_4	NS		NS	0.1%	NS	5.0%
LM_5	NS		NS	5.0%	NS	—
WM_5	NS		NS	5.0%	NS	—
	* QM F730	* Unregd. specimen J. Calaby, CSIRO	AM MF413	** MM5 Unregd. Coll. Hope '78	Unregd. Coll. Hope '78	** ANU/NCA/B/3

Landing, South Australia (Macintosh and Mahoney 1964), and from the Eucla Division of Western Australia (Lowry 1972).

DISCUSSION

Comparison of the Wellington Caves thylacine samples with a sample of *T. cynocephalus* from Tasmania has revealed neither morphological differences, nor any consistent proportional differences, with the exception of a slightly larger $P^1/_1$ in the fossil sample. Although dental dimensions of the Wellington Caves thylacines have been shown to be significantly larger than those of the modern Thylacine, the low C.D. values indicate that the magnitude of this difference is not sufficient to warrant subspecific distinction on the basis of size alone. It is considered here that the observed difference between the two populations in the single character, proportional size of $P^1/_1$, does not constitute sufficient evidence for taxonomic distinction at the species level. Thus synonomy of *T. spelaeus* with *T. cynocephalus,* as previously suggested by Lydekker (1887) and Ride (1964), is supported.

Thylacinus major Owen is based on specimens from Wellington Caves. Mahoney and Ride (1975) have discussed the uncertainty of Owen's intention regarding this name, and the composite and restored nature of the illustrations of the syntypes. In describing the species, Owen (1877: 106) stated that " . . . the upper canine is proportionately larger than the lower one than it is in *T. cynocephalus*" and also suggested its " . . . superiority of size . . .". One syntype of *Thylacinus major* is a partial upper canine, BM 42549, figured by Owen (1877, pl. 5, Fig. 7). This specimen has been selected as the lectotype of *T. major*. It is considered here that size of the canine teeth of thylacines is of very little taxonomic value at the species level, because the crowns are subject to heavy wear, while the roots grow considerably with the age of the animal. Comparative measurements on isolated specimens are, therefore, almost impossible to make. For this reason it is considered here that BM 42549 is inadequate to serve on its own as a basis for a distinct thylacine taxon.

The other syntype listed in Mahoney and Ride (1975), BM 42548, is an edentulous manibular fragment. Mahoney and Ride (1975) have established that the figure of that syntype (Owen 1877, pl. 5, Fig. 8) is a composite drawing, the three teeth being added, the last two of which (i.e. M_4 and M_5) being molars of *Sarcophilus*. While that mandibular fragment belongs to a thylacine, without the teeth it does

Table 6. Size correlation of fossil thylacine specimens with the Wellington Caves and modern Tasmanian populations. Table includes only specimens measured by this author, or for which dental measurements are available in the literature.

Location	Correlation	Age	Reference
Marmor, Qld	1, E		
Gore Cement Mills, Qld	1, E		
Ellangowan, D. Dns, Qld	1, E		
Chinchilla Sand, Qld (as *T. rostralis*)	1, E	Pliocene	Ride (1964)
Bingara, N.S.W.	3, E	Pleistocene	Marcus (1976)
Lake Menindee, N.S.W.	3, E	>18–26,000 BP	Tedford (1967)
Lake Mungo, N.S.W.	2, M	>19,000 BP	Hope (1978)
Clogg's Cave, Buchan, Vic	1, E	17–23,000 BP	Flood (1973)
Jenolan Caves, N.S.W.	3,		
Yarrangobilly Caves, N.S.W.	1, —		
Victoria Cave, Naracoorte, S.A.	1, E	Pleistocene	Smith (1972)
Mair's Cave, S.A.	1,2, E		
Henschke's Cave, S.A.	1, E		
Fromm's Landing, S.A.	2, M	4,000 BP	Macintosh and Mahoney (1964)
Kiowa, New Guinea	2, M	9,920±200 BP	Van Deusen (1963), Bulmer (1964)
Nombe, New Guinea	2, M	"human occupation levels"	Plane (1976)
Thylacine Hole, Eucla Div. W.A.	2, —	4,650±153 BP	Lowry and Merrilees (1969), Merrilees (1970), Lowry (1972)
Murra–el–elevyn Cave Eucla Div. W.A.	2, —	3,280±90 BP	Partridge (1967), Lowry (1972)

KEY: 1. Significantly larger than Tasmanian population. 2. Significantly smaller than Tasmanian population. 3. Agrees in size with Tasmanian population. E. Associated with extinct megafauna. M. Associated with modern species only.

not contain taxonomic characters adequate to establish a new species. Because neither the lectotype, BM 42549, nor BM 42548 contains characters of taxonomic value at the specific level it is proposed here that the name *T. major* should be regarded as a junior synonym of *T. cynocephalus*. The synonymy of *T. major* and *T. spelaeus* has been previously accepted by De Vis (1894), Ride (1964) and Woodburne (1967).

The description of *T. rostralis* was first published in an abstract (De Vis 1893: v). It was based on " . . . a number of fragmentary portions of the cranium (which) have for some time been in the Queensland Museum . . . (and) . . . the major part of the left side of an adult skull . . . ". These specimens are individually listed and described in De Vis (1894) and it was clearly De Vis's intention that they should be syntypes of the new species. They include four specimens from locations on the eastern Darling Downs, Queensland: QM F730, a partial skull from Ellangowan; Cat. No. 12,027, a right maxillary fragment from Pilton; Cat. No. 12,028, a left maxillary fragment, and Cat No. 12,030, a left mandibular fragment, both from Gowrie. Two specimens from Chinchilla on the western Darling Downs are also listed: Cat No. 12,029, a right mandibular fragment; and Cat. No. 12,-031, at left maxillary fragment (catalogue numbers refer to old numbers of the Queensland Museum). It would appear that Mahoney and Ride (1975) were mistaken in citing QM F730 as the holotype, because no holotype was indicated in the original description (De Vis 1893). Ride (1964) also appears to have misinterpreted De Vis (1894) in stating that both *T. rostralis* and *T. cynocephalus* occurred in the Chinchilla Sand, since De Vis (1894) clearly referred the Chinchilla specimens to *T. rostralis* only. The partial skull, QM F730, from Ellangowan, which is the most complete of the syntypes, is selected here as the lectotype of *Thylacinus rostralis* De Vis, 1893. The other specimens then become paralectotypes.

Thylacinus rostralis was described by De Vis (1893, 1894) as a larger species than *T. cynocephalus,* also differing in proportion from that species as well as from *T. spelaeus*. The teeth of the lectotype of *T. rostralis* have been shown here (Table 5) to be larger for at least some dimensions, than the mean size of the modern Tasmanian Thylacine. Neverthless, they are comparable in size to the fossil thylacines from Wellington Caves. Ride (1964: 108) has also presented measurements of *T. rostralis,* represented by the partial skull (QM F730) and " . . . the only other specimens from the same beds", i.e. QM F3742 from King's Creek, Clifton, and BM 35973 from Gowrie. All dental measurements given by him for these specimens fell within the observed range given for *T. cynocephalus* (i.e. eastern Cave–fossils plus the Tasmanian sample), with the single exception of the longer P_3 of QM F3742. Some mandibular dimensions given by Ride (1964) exceeded those of modern *T. cynocephalus* and supported his suggestion that *T. rostralis* might possibly represent a larger Pleistocene form than *T. cynocephalus* (and consequently than *T. spelaeus*) He stated further (Ride 1964 : 108) that "A value of 18.6 for a single isolated lower molar . . . from the Pliocene Chinchilla Sand also suggests this form may be possibly distinct. [? from *T. cynocephalus*] as well.". However, the length given, 18.6mm, fell within Ride's (1964) observed range for the length of M_4 ($= M_5$) of eastern Cave–fossils. The skull QM F730 is of an aged individual, possibly accounting for its more massive mandibular dimensions. The data presented here, and those of Ride (1964), therefore suggest that specific distinction (as *T. rostralis*) for all or part of the Queensland sample cannot be justified on the basis of size alone.

Most of the proportional differences noted by De Vis (1894) are in such plastic characters as relative muzzle length, basicranial length, size of the premaxillary bone, size of the occipital condyle and mandibular depth. These differences, noted by De Vis (1894) after comparison of two skulls only, are probably attributable to allometric variations observed to occur in the modern Tasmanian population as a result of differences in age or sex. The dental proportions of the Queensland skull agree closely with those of the Wellington Caves and Tasmanian samples as illustrated in the log difference diagrams presented in Fig. 1. The spacing of the premolars, considered by Woodburne (1967) as possibly having taxonomic signficance has been investigated in this study. The results have shown that intraspecific variation could account for the observed differences between skull QM F730 and the Wellington Caves and Tasmanian samples. It is concluded, from this study, therefore, that *T. rostralis* should also be placed in the synonymy of *T. cynocephalus*.

Lowry (1972) expanded Ride's assessment of the status of small fossil thylacines from Western Australia. She found that, although the dental characters she tested were significantly smaller than those of modern *T. cynocephalus* from Tasmania, using the 75% rule of Mayr (1969), even sub–specific distinction was not justified on the basis of size alone. However, extreme variability was evident in the Western Australian Quaternary sample which came from a wide geographic area. Most specimens also lacked stratigraphic data.

The data presented in Table 6 indicate that while most deposits from eastern Australia, known to be Pleistocene in age, contain large thylacines, thylacines significantly smaller than the modern Tasmanian population also occur in deposits of likely Pleistocene age in South Australia and western New South Wales. An isolated M_5 larger than that tooth in the modern population, is known from Clogg's Cave, Buchan, New South Wales (Table 2), in sediments dated at 17 – 23,000 BP (Flood 1973). The sample of thylacines in the South Australian Museum from Mair's Cave (Tables 1 and 2) is of interest because it appears to contain specimens which are smaller, larger and within the observed range of the modern Tasmanian sample for dental size. However, this deposit is unstudied and the exact provenance of the specimens is unknown.

In summary, the data available for fossil thylacines do not show a clear distinction between large and small forms in the Late Quaternary although some size differences are apparent. It is possible that a high degree of sexual dimorphism such as is known to occur in the modern Tasmanian population (Ride 1964), could obscure possible specific differences in geographically or temporally separate populations. Study of more complete cranial and postcranial material might reveal morphological or proportional differences which would justify the specific distinction of some fossil populations. Tentatively, it is concluded that only one variable species, *Thylacinus cynocephalus,* can be recognised on the basis of the data available at present in the Pliocene, Pleistocene and modern faunas of eastern Australia, and New Guinea.

ACKNOWLEDGEMENTS

Work on the Wellington Caves fossils has been undertaken while the author was in receipt of a Commonwealth Postgraduate Research Award and has also been supported by a Postgraduate Grant from the Australian Museum and a further Grant–in–aid from the Trustees of the Australian Museum, Sydney. I wish to thank Dr. Alex Ritchie of the Australian Museum for his support and permission to study in that institution. Dr. A. Bartholomai, Queensland Museum; Dr. N. Pledge, South Australian Museum; and Dr. T. Rich, Victorian Museum also allowed me access to the collections of those institutions. Dr. J. Hope, Dr. R. Wells and Dr. J. Calaby kindly lent specimens for this study. I wish to thank Dr. Michael Archer, particularly, for his support and advice, and acknowledge help from many members of the School of Zoology, University of New South Wales. Dr. A. Bartholomai and Mr. J. Mahoney provided useful criticism of a draft of this paper.

REFERENCES

ARCHER, M., 1978. The nature of the molar–premolar boundary in marsupials and reinterpretation of the homology of marsupial cheekteeth. *Mem. Qd Mus.* **18:** 157–64.

ARCHER, M. AND BARTHOLOMAI, A., 1978. Tertiary mammals of Australia: a synoptic review. *Alcheringa* **2:** 1–19.

BARTHOLOMAI, A., 1977. The fossil vertebrate fauna from Pleistocene deposits at Cement Mills, Gore, southeast Queensland. *Mem. Qd Mus.* **18:** 41–51.

BULMER, S., 1964. Radiocarbon dates from New Guinea. *J. Polynesian Soc.* **73:** 327–8.

DAWSON, L., 1982. Taxonomic status of fossil devils (*Sarcophilus,* Dasyuridae, Marsupialia) from Late Quaternary eastern Australian fossil localities. Pp.527–36 *in* "Carnivorous marsupials", ed. by M. Archer. Roy. Zool. Soc. New South Wales: Sydney.

DE VIS, C. W., (In Anon, 1893). A thylacine of the earlier nototherian period in Queensland. *Abstr. Proc. Linn. Soc. N.S.W.* for Nov. 29th, 1893.

DE VIS, C. W., 1894. A thylacine of the earlier nototherian period in Queensland. *Proc. Linn. Soc. N.S.W.* **8:** 443–7.

FLOOD, J., 1973. Pleistocene human occupation and extinct fauna in Clogg's Cave, Buchan, southeastern Australia. *Nature, Lond.* **246** (5431): 303.

HOPE, J. H., 1978. Pleistocene mammal extinctions: the problem of Mungo and Menindee, New South Wales. *Alcheringa* **2:** 65–82.

LONGMAN, H., 1925. Fossil marsupials from Marmor. *Mem. Qd Mus.* **8:** 109.

LOWRY, J. W., 1972. The taxonomic status of small fossil Thylacines from Western Australia. *J. R. Soc. West. Aust.* **55:** 19–29.

LOWRY, J. W. AND MERRILEES, D., 1969. Age of the dessicated carcase of a thylacine (Marsupialia, Dasyuroidea) from Thylacine Hole, Nullarbor Region, Western Australia. *Helictite* **7:** 15–6.

LYDEKKER, R., 1887. Catalogue of the fossil Mammalia in the British Museum (Natural History) Cromwell Rd., S.W. Part 5. Containing the group Tillodontia, the Orders Sirenia, Cetacea, Edentata, Marsupialia, Monotremata, and supplement. Trustees of the British Museum (Natural History): London.

MACINTOSH, N. W. G. AND MAHONEY, J. A., 1964. A 4,000 year old Thylacine tooth (Dasyuridae) from Shelter 2, Fromm's Landing. An appendix to Mulvaney, D. J., Lawton, G. H. and Twidale, C. R., 1964. Archeological excavation of rock shelter No. 6, Fromm's Landing, South Australia, *Proc. R. Soc. Vict.* **77:** 479–516.

MAHONEY, J. A. AND RIDE, W. D. L., 1975. Index to the genera and species of fossil Mammalia described from Australia and New Guinea between 1838 and 1968. *West. Aust. Mus. Special Publication No.* 6.

MARCUS, L. F., 1976. The Bingara fauna. A Pleistocene vertebrate fauna from Murchison Country, New South Wales, Australia. *Univ. Calif. Pubs. Geol. Sciences.* **114:** 1–145.

MAYR, E., 1969. "Principles of systematic zoology" McGraw Hill: New York.

MAYR, E., LINDSAY, E. G. AND USINGER, R. L., 1953. "Methods and principles of systematic zoology". McGraw Hill: New York.

MERRILEES, D., 1970. A check on the radiocarbon dating of dessicated Thylacine (Marsupial "wolf") and Dog tissue from Thylacine Hole, Nullarbor region, Western Australia. *Helicite* **8:** 39–42.

OWEN, R., 1845. "Descriptive and illustrative catalogue of the fossil organic remains of Mammalia and Aves contained in the Museum of the Royal College of Surgeons of England". Richard and John E. Taylor: London.

OWEN, R., 1877. "Researches on the fossil remains of the extinct mammals of Australia; with a notice of the extinct marsupials of England". J. Erxleben: London. Vol. 1 (text), Vol. 2 (plates).

PARTRIDGE, J., 1967. A 3,300 year old Thylacine (Marsupialia, Thylacinidae) from the Nullarbor Plain, Western Australia. *J. R. Soc. West. Aust.* **50:** 57–9.

PLANE, M., 1976. The occurrence of thylacines in tertiary rocks from Papua New Guinea. *J. Aust. Geol. Geophys.,* **1.**

RIDE, W. D. L., 1964. A review of Australian fossil marsupials. *J. R. Soc. West. Aus.* **47:** 97–131.

SMITH, M., 1972. Small fossil vertebrates from Victoria Cave, Naracoorte, South Australia. II Peramelidae, Thylacinidae and Dasyuridae (Marsupialia). *Trans. R. Soc. S. Aust.* **96:** 125–37.

SOKAL, R. R. AND ROHLF, F. J., 1969. "Biometry. The principles and practice of statistics in biological research". Freeman and Co.: San Fransisco.

TEDFORD, R. H., 1967. The fossil Macropodidae from Lake Menindee, New South Wales. *Univ. Calif. Publs. Geol. Sci.* **64:** 1–156.

VAN DEUSEN, H. M., 1963. First New Guinea record of *Thylacinus*. *J. Mammal.* **44:** 279–80.

WOODBURNE, M. O., 1967. The Alcoota fauna, central Australia. *Bull. Bur. Min. Res. Geol. and Geophys.* **87:** 1–174.

RESUMEN

44. Status taxonómico de los tilacinos fósiles (*Thylacinus*: Thylacinidae, Marsupialia) de los depósitos del Cuaternario tardío del Este de Australia. Por Lyndall Dawson.

Se examinaron los tilacinos fósiles de Wellington Caves, New South Wales, del Australian Museum, Sydney, con el objeto de aclarar el status taxonómico de la especie fósil *Thylacinus spelaeus* Owen, 1845. Se hicieron comparaciones con muestras del actual *T. cynocephalus* de Tasmania, y con especímenes fósiles de localidades de todo el Este de Australia. Aunque los tilacinos de las Wellington Caves mostraron ser de talla significativamente mayor que el tilacina de Tasmania, estas diferencias en tamaño solamente no son consideradas suficientes para garantizar la separación específica de *T. spelaeus*. *Thylacinus rostralis* De Vis 1893 y *Thylacinus major* Owen, 1877 son considerados como sinónimos de *T. cynocephalus*. Los fósiles de localidades pleistocenas y holocenas demostraron mucha variación en el tamaño, siendo algunos especímenes significativamente más pequeños que el actual *T. cynocephalus*.

45. THE DISCOVERY AND INTERPRETATION OF *Thylacoleo carnifex* (THYLACOLEONIDAE, MARSUPIALIA)

M. E. FINCH[1]

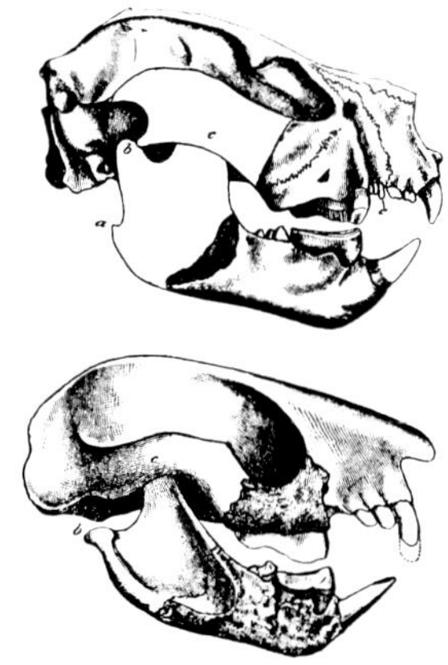

The history of the discovery of *Thylacoleo carnifex* (Thylacoleonidae: Marsupialia) from the early 1830s up to the present is outline '. Much of the early skull material was fragmentary, making reconstruction of the head difficult. The dentition is divided into two functional units and the other teeth have been reduced. The upper and lower anterior incisors meet at their tips in a "caniniform unit", while the third upper and lower premolars are huge trenchant slicing blades. On the basis of this unusual dentition, various authors have ascribed herbivorous, omnivorous and carnivorous habits to the marsupial lion. Reconstruction of adductor jaw musculature and estimation of relative muscle weights supports the carnivore hypothesis. Preliminary work on the strongly clawed limbs suggests feline-like locomotion.

Pages 537–51 *in* CARNIVOROUS MARSUPIALS, ed. by M. Archer, February, 1982. Roy. Zool. Soc. New South Wales: Sydney. Australia.

THYLACOLEO CARNIFEX, widely known as the marsupial lion, was a heavily–built animal about the size of a large dog (Fig. 1). Its broad head had but a short snout, forward–looking eyes and was equipped with a most remarkable dentition. Expressed as a formula, I 3/1, C 1/0, P 3/3, M 1/2, the dentition appears basically phalangerid, except for the reduction in molar size and number. Viewed in the skull, the large caniniform upper first incisors, the semi-procumbent, laniariform lower first incisors and the huge upper and lower trenchant third premolars (P3/3) immediately attract attention. The upper canine (C^1) and the first two premolar teeth (P1/1, P2/2) are reduced and often considered functionless. Such a unique dentition has naturally given rise to conflicting views on the phylogenetic position of the animal and engendered heated controversy on its feeding habits. Skull features and particularly the dentition provided virtually the only evidence for the early debates as, until comparatively recently, the postcranial skeleton was only represented by a few isolated limb bones which had been discovered in the vicinity of *Thylacoleo* skulls.

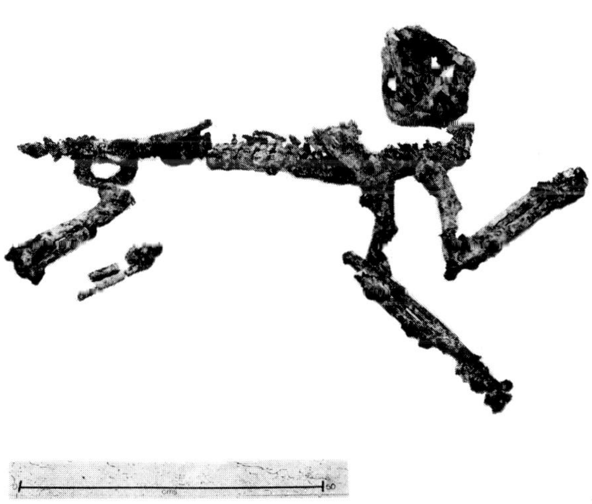

Fig. 1. The Moree *Thyacoleo carnifex* adult (F 52398–1).

In the present paper, the discovery and interpretation of the material of *Thylacoleo carnifex* is outlined. Based on an analysis of the known, and also some new, material, a functional interpretation of the feeding habits and locomotion of the marsupial lion is presented.

EARLY DISCOVERIES AND DESCRIPTIONS

The first intimation of the existence of *Thylacoleo* came from material collected from the Wellington Valley area, New South Wales (Fig. 2), in the early 1830s by Major (later Sir) Thomas Mitchell. Incisor and premolar teeth, now known to be of this animal, were illustrated in Mitchell's "Three Expeditions into the Interior of Eastern Australia" of 1838 but were not mentioned in the Zoological Appendix to volume 2 contributed by Professor Richard Owen, the eminent English palaeontologist. He was of the opinion (1877, vol. 1: 107) that the premolar, the "large, two–fanged sectorial tooth", was so unique that it was impossible to determine the affinities of the animal without study of more complete material.

Owen subsequently received fragments of left and right maxillae, each bearing P^3 and M^1, and a portion of a left dentary with P_3 intact but with only the roots of M_1 and M_2. These had been found in the bed of a tributary of the Condamine River, Queensland, but no date of discovery was recorded. Lydekker's 1887 "Cata-

[1] Department of Zoology, University of Western Australia, Nedlands, Western Australia, 6009.

Fig. 2. Thylacoleo and *Wakaleo* sites: 1. Mammoth Cave; 2. Balladonia; 3. Bullock Creek; 4. Alcoota; 5. Bluff Downs; 6. Spring Creek; 7. Chinchilla; 8. Dalby; 9. King River; 10. Gore; 11. Moree; 12. Bingara; 13. Wellington Caves; 14. Geurie; 15. Lake Victoria; 16. Lake Colongulac; 17. Lake Colac; 18. Buchan; 19. Strathdownie; 20. Duck Ponds; 21. Montagu and Scotchtown; 22. Florentine Valley; 23. Lake Palankarinna; 24. Lake Ngapakaldi; 25. Naracoorte; 26. Curramulka; 27. Normanville; 28. Mount Gambier; 29. Buckalowie; 30. Bundey; 31. Port Pirie.

logue of the Fossil Mammalia in the British Museum (Natural History)" lists M3653 and M3654, the right and left maxillary fragments of *Thylacoleo* from Queensland, as having been figured by Owen in 1877 and presented by him to the Museum.

In his summary of his research on fossil mammals of Australia (1877), Owen described the Condamine fragments and commented on the orientation of the 'working surfaces' of the upper and lower sectorials. He noted the vertical striae visible on them and concluded that these teeth were shearing blades used in the slicing of meat. However, he could find no evidence of the marsupial or placental relationship of this carnivore.

Study of specimens M3653 and M3654 by the present author has verified the presence of a well–marked ridge near the posterior end of the wear facet of each upper carnassial and a second, more pronounced ridge at the inter–rootial level on the occlusal surface of M3654. The latter ridge is not developed in other specimens available for study and the former is usually only a low boundary between the wear facets caused by the lower carnassial and the adjacent lower first molar. Owen pointed out that the ridges were comparable with those discernible in the sectorial teeth of *Felis spelaea* and *Hyaena spelaea* but on present evidence it is now considered highly probable that the Condamine animal had suffered deformity of, or damage to, its lower teeth during life, resulting in an unusual wear pattern on the upper premolar.

In 1845, fragmentary skull material of the same animal was gathered by W. Adeney from his property at Lake Colongulac, eighty miles southwest of Melbourne, Victoria, and a year later it was forwarded by Dr. H. Hobson of Melbourne to Owen at the British Museum. The material included a cranium with an incomplete posterolateral surface and lacking zygomatic arches (Fig. 3A). There was also a maxillary fragment, bearing the carnassial premolar and the single molar, which proved to be part of the broken skull. Owen realized that any valid interpretation of the affinities and functions of this unique material could only be based on wide–ranging comparisons of *Thylacoleo* with other mammals. Such studies were not complete until 1858, at which time he presented his findings to the Royal Society. In this account he was able also to include information on a cast of part of a mandible (Fig. 3B) which had been sent to him by Samuel Stutchbury, the Geological Surveyor of New South Wales. This cast comprised most of the ramus and bore the carnassial premolar (P_3), M_1 and the alveolus of M_2, and it clearly indicated the nature of the symphysis. It is notable that this first account of *Thylacoleo* was based on fossil fragments from two widely separated localities, the skull from Victoria and the mandible from King Creek, Darling Downs, Queensland (Fig. 2). The definitive description of the species was not published until 1859, only an abstract and an article 'Odon-

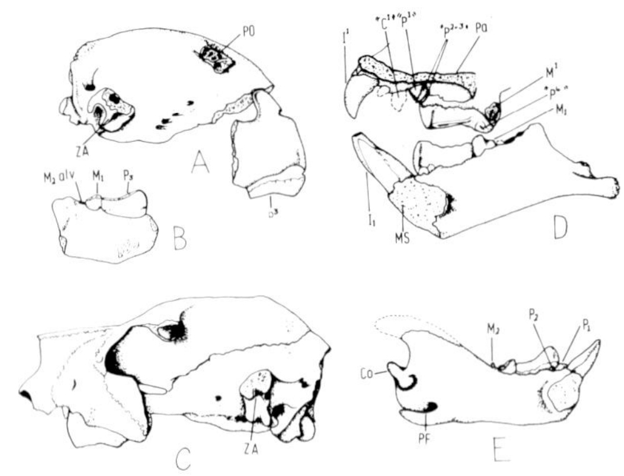

Fig. 3. Thylacoleo carnifex material described by Owen (modified from Owen). A = lateral aspect of skull fragments from Lake Colongulac, Victoria (1859); B = lingual aspect of the mandibular fragment from King Creek, Queensland (1859); C = lateral aspect of skull from Darling Downs, Queensland (1866) showing the complete rostral region with C^1 and P^3; D = sagittal section through upper and lower jaws viewed from the lingual aspect (1871). I^3, C^1, P^1, P^2, P^3 were labelled "C^1", "P^1", "P^2", "P^3" and "P^4" respectively by Owen; E = lingual aspect of mandible from Wellington Caves, New South Wales (1883). alv = alveolus; MS = mandibular symphysis; Pa = palate; PF = pterygoid fossa; PO = postorbital bar (broken); ZA = squamosal pillar of zygomatic arch.

tology' in the Encyclopaedia Britannica appearing in 1858.

An earlier reference to *Thylacoleo* is found in Gervais' "Zoologie et Paléontologie Françaises" published in parts between 1848 and 1852, but no dates are given for the individual parts. It appears that Gervais visited Owen in London in 1848 and was shown the Lake Colongulac material in the course of a discussion on the affinities of *Hyaenodon*, a placental which was considered by many at that time to be a marsupial. Gervais stated clearly in his book that Owen had named the Australian material *Thylacoleo* but he gave no description of it. This reference to *Thylacoleo* was obviously unknown to Stutchbury who, in a technical report (1853) to the Colonial Secretary, had given the name *Schizodon* to the original of the mandibular cast which he had sent to Owen after comparing the carnassial with that figured by Mitchell (1838). Owen (1859) pointed out that even if *Schizodon* Stutchbury had priority over *Thylacoleo*, it was a homonym of *Schizodon* Agassiz and *Schizodon* Waterhouse, names given to a fish in 1829 and to a mammal in 1842, respectively.

In his 1859 paper Owen presented a concise and detailed description of the Lake Colongulac and the Darling Downs material, although he gave Hodgson's Creek rather than King Creek as the source of the mandible. In this paper, he used the species name *carnifex* but unfortunately did not designate a holotype which has led Mahoney and Ride (1975) to record as syntypes both the cranium and the mandible on which Owen's description was based. On the basis of detailed comparisons between *Thylacoleo* and the marsupial carnivores *Dasyurus* (*Sarcophilus*) *ursinus* and *Thylacinus harrisii*, and also with *Felis spelaea*, the eutherian cave lion, no doubt remained in Owen's mind as to the marsupial affinities of *Thylacoleo*. He noted its large posterior palatal vacuities, the placement of the various basicranial foramina, the participation (albeit a minor one) of the alisphenoid in the auditory bulla while the tympanic bone was largely free, the extension of the lacrymal bone on to the face, the extraorbital position of the lacrymal foramen and the relatively small size of the brain. Within the Marsupialia he suggested that *Thylacoleo* was most closely related to the dasyurids and to *Dasyurus ursinus* in particular "although the interval be still very great between them" (Owen 1859: 321).

Seven years later Owen (1866) was able to amplify his description, having received more skull fragments from Darling Downs which had been sent by Edward Hill of Sydney. These, when assembled, made a more complete specimen (Fig. 3C) than the Lake Colongulac skull. On the upper jaw, the alveoli for the pre-carnassial teeth were all present and the canine itself was in position. Thus Owen determined that *Thylacoleo* must have possessed a large, tusk-like I^1 and two smaller incisors on the premaxilla while the maxilla must have carried the small blunt canine tooth followed by one or two small premolars placed medially to the anterior root of the carnassial. He also noted the double wear facet on the medially-directed occlusal surface of P^3 (Fig. 4), which indicated a slicing action against both P_3 and M_1. Owen drew attention to the fact that the glenoid cavity was not semi-cylindrical in shape as in typical eutherian carnivores. Although transversely elongated and with processes limiting lateral excursion of the condyle, its articulating surface was slightly convex ventrally and extended forwards for about 2 cm from the postglenoid process.

Information on the mandible had been provided by a photograph and a drawing of Wellington Caves material sent by Mr. Gerard Krefft, Curator at the Australian Museum, Sydney. From these, Owen (1866) determined that there was only one large tooth at the anterior margin of the dentary, although its form could not be ascertained as the crown was missing. Only two small alveoli lay between this tooth and the lower carnassial. On the basis of this pictorial information, Owen postulated that the thylacoleonid affinities lay with the diprotodontid marsupials rather than with the carnivorous dasyurids. He gave the dental formula of the animal as $I\frac{3-3}{?}$, $C\frac{1}{?}$, $P\frac{2-2}{2-2}$, $M\frac{1-1}{2-2} = 24$ denoting his uncertainty as to whether a second incisor or a canine lay behind I_1 and believing that the small anterior premolar was two-rooted. Owen reiterated his belief in the carnivority of *Thylacoleo*, writing that it "exemplifies the simplest and most effective dental machinery for predatory life and carnivorous diet known in the Mammalian class. It is the extreme modification, to this end, of the Diprotodont type of *Marsupialia*" (Owen 1866: 81).

By 1871 further Darling Downs fossil material (Fig. 3D) had reached Owen and from this he described I^1, C^1, the two small premolars anterior to the upper carnassial and a lower incisor in which the base of the crown remained. Unfortunately the premaxillary-maxillary suture was unclear and appeared to run through the alveolus of I^3 which Owen therefore labelled C^1, while the actual canine became P^1. Krefft's photographs had served as bases for illustrations of isolated teeth in this paper and, in the following year, Krefft (1872b) pointed out that his pictures of I^2, I^3 and C^1 had been labelled P^1, C^1 and P^2 respectively by Owen. The confusion was soon clarified. Flower in 1868 had published a figure of the skull in which the canine was correctly labelled and in 1876 McCoy gave a detailed description of the path of the premaxillary-maxillary suture in new material from

Lake Colongulac. The suture was shown to approach the alveolus of I^3 but then, at the margin of the jaw, it turned back to run through the alveolus of the canine tooth.

Both Owen (1871) and Krefft (1872b) noted the serrated enamel on the lateral edge of each I_1. Krefft also drew attention to the restriction of the enamel to the tips of the crowns on the interdental surfaces of these two teeth. At the same time, he found wear facets in this region indicating that there was movement between the two lower incisors and that, consequently, the mandibular symphysis could not have been fused. Each of the large upper first incisors proved to have a wear facet on its lingual surface demonstrating that the lower incisors passed behind the upper ones at jaw closure. The crowns of I^2 and I^3 also showed evidence of wear at their tips due to contact with I_1.

The next instalment of the saga of *Thylacoleo carnifex* came in 1883 when Owen (1883a) enthusiastically acknowledged receipt of an upper and a lower jaw from Wellington Valley, each of which carried the full complement of teeth. The lower jaw (Fig. 3E) exhibited an undamaged, transversely–elongated condyle and a deeply inflected mandibular angle but unfortunately the coronoid process was broken. The broad, backwardly–directed outline of the latter was finally determined in 1887 from a cast of a complete dentary forwarded to Owen by G. P. Ramsay, Krefft's successor. In this, the last of his concise and thought–provoking papers on *Thylacoleo,* Owen restated his opinion that the marsupial lion was a diprotodontid which had become adapted for a carnivorous habit.

None of the descriptions mentioned so far deals with two almost unique marsupial lion features. The orbit is delineated posteriorly by a slender bar which seems to have been lost from many fossil specimens. This postorbital bar is also found in *Thylacosmilus* but is lacking in other marsupials. It was illustrated without comment in a skull reconstruction in Lydekker's catalogue (1887, fig. 28) but attention was drawn to the feature by Anderson (1929) and by Gill (1954). The second feature, one peculiar to *Thylacoleo,* was also commented upon by Anderson. Its sphenopalatine foramen, through which the orbit and nasal cavity are in communication, is considerably larger, relatively, than in any other land mammal. No explanation for this has been advanced.

Woods in 1956 published a comprehensive description of the skull of the marsupial lion based on the Darling Downs material held in the Queensland Museum. From one skull he was able to prepare an endocranial cast and he compared this with the one described by Gervais as early as 1869. The brain appears to most closely resemble that of the phascolomids amongst marsupials. The ear region (including the stapes) and the temporomandibular joint, on the other hand, show great similarity to those of the phascolarctids. Woods also drew attention to the fact that no deciduous teeth had been found in any specimen.

Up to the present time, skull and dental material of *Thylacoleo carnifex* have been discovered in Pleistocene deposits in all Australian states (Fig. 2). The Western Australian and Tasmanian specimens are few in number but appear to have belonged to animals of smaller size than those found in the eastern states. This size discrepancy has been investigated by Finch and Freedman (1982).

POSTCRANIAL MATERIAL

So far only the skull has been considered. Owen described a metacarpal (1859), a radius, ulna and terminal phalanx (1883a) and a pelvic girdle (1883b) while De Vis gave an account of a femur (1887) and a radius (1900) and Longman (1925) ascribed a calcaneum to *Thylacoleo.* Krefft found ungual phalanges, one of which he described (1870a) as *Mylodon australis* although Owen (1871, 1877) was inclined to believe them to be thylacolean. However, although many of these bones had been found at the same localities as *Thylacoleo* skulls, there was no clear proof that they were actually parts of a marsupial lion.

The only published account of postcranial bones which were unequivocably referable to *Thylacoleo* was that of Wells and Nichol (1977). They described the manus in both articulated and disarticulated condition and were of the opinion that the forelimb stance was digitigrade. According to these authors the pollex would have shown considerable mobility and appeared to be pseudo–opposable to the pisiform. The pes was incomplete but its structure implied a plantigrade stance and a hallux that diverged from the other digits.

THE MOREE THYLACOLEO

In 1966 an almost complete skeleton of a marsupial lion (Fig. 1) was unearthed from a loam pit near Moree in northern New South Wales (Fig. 2). The right scapula and right ilium were adherent to vertebrae and the groups of vertebrae fitted together to make a complete column except for the terminal caudal units. Limb bones were broken but again could be reassembled to form a complete forelimb and hindlimb which lacked only the pes. The adult specimen was accompanied by a juvenile, represented by a large skull fragment and a partial lower jaw. Associated with these was the crushed skull of a pouch–young *Thylacoleo*. This material is presently being used in studies on mastication, feeding habits and locomotion (Finch and Freedman 1982, Finch 1981).

THYLACOLEONID SPECIMENS ASCRIBED TO DIFFERENT GENERA AND SPECIES

The following genera and species have been described at different times since 1859 and work by Archer and Dawson (1982) has clarified the standing of some of these taxa.

Plectodon Krefft, 1870: Krefft (1870a) published a short description of material found in Wellington Caves which included some badly fragmented lower incisor teeth said to belong to an animal "allied to *Thylacoleo*" which he named *Plectodon*. No type was designated and no details of its discovery locality were given. A specimen was sent to Owen but no species name was ever published, even though Krefft made further reference to *Plectodon* in later publications (1870b, 1882) and gave an illustration of the incisor (1882). Archer and Dawson (1982) regard *Plectodon* as a junior synonym of *Thylacoleo* Gervais.

Mylodon? australis Krefft, 1870: Amongst some of the Wellington Cave collections Krefft found a large, hooded ungual phalanx resembling that of a South American sloth, *Mylodon*. He assigned the bone to the genus *Mylodon?* and made it the type of *Mylodon? australis*. A short description was published (Krefft 1870a) but it was Owen (1871, 1877) who was the first to illustrate the specimen from a series of photographs of two differently shaped sorts of phalanges sent to him by Krefft. The eminent British palaeontologist figured both sorts of phalanges and, by rejecting all other marsupials known to have lived in the discovery locality at that time as being of the wrong size, suggested that *Thylacoleo* must have been the possessor of the curved ungual phalanges. He discounted the possibility that *Mylodon* had ever lived in Australia. Etheridge (1918) investigated the confused situation which had arisen from Owen's referral of two different sorts of phalanges to *Thylacoleo* and the listing by Lydekker (1887) of a cast of a third variety of phalanx as also being thylacoleonid. Etheridge decided that Krefft's large, hooded ungual bone was probably referable to *Thylacoleo*. Of the two unsheathed phalanges, the highly arched and compressed form approximated most nearly to that of the great flying phalanger while the longer, less compressed claws were Koala–like. In their recent study, Archer and Dawson (1982) showed that *Thylacoleo carnifex* is a senior synonym of *Mylodon australis*.

Thylacoleo oweni McCoy, 1876: The problems inherent in having to base the taxonomy of fossil species on a few incomplete specimens are highlighted by the studies of McCoy (1876). In 1875 McCoy had been given a partial skull and a mandible collected by Adeney from the same locality at Lake Colongulac as the material sent to Owen thirty years before. The skull appeared to be the missing part of Owen's specimen and carried all precarnassial teeth *in situ*. McCoy's detailed comparison between the Victorian material (including the newly acquired lower jaw) and that from Darling Downs and Wellington Caves (Owen 1859, 1866, 1871) led him to distinguish the New South Wales *Thylacoleo* as a new species, *T. oweni*.

The basis for the new species given by McCoy was as follows:

a) The longer convex margin of I^3 was posterior in *T. carnifex* but anterior in *T. oweni*. However, Owen had based his description of *T. carnifex* on isolated teeth and had obviously reversed I^3.
b) The premaxillary–maxillary suture traversed the alveolus for the canine tooth in *T. carnifex* but was said to meet the alveolus anterior to that of the canine (i.e. the tooth originally designated as I^3) in Owen's 1871 account.
c) The dorsoventral depth of the mandibular ramus was measured in front of the carnassial and again behind M_2. The Victorian material showed the anterior distance to be the greater while the *T. oweni* mandibles were deeper at the posterior end.
d) The palate of *T. carnifex* was narrower across the premaxillary region and 6 mm longer anteroposteriorly than was the case in *T. oweni*.
e) I^2, I^3 and C^1 were larger in the Victorian than in the New South Wales specimens.

Of these supposed differences, the first two points (a and b) have been shown to be invalid by Owen (1877) while the remaining differences appear to be within the range of variability shown by *T. carnifex*. Archer and Dawson (1982) regard *Thylacoleo oweni* as a junior synonym of *Thylacoleo carnifex*.

Thylacoleo robustus Krefft, 1872: Krefft (1872a) gave the species name *Thylacoleo robustus* to a right upper canine tooth found in Wellington Caves. His only justification for erecting a new species was the fact that the homologous tooth of *T. carnifex* was said to be considerably smaller than that of the new species. This species has been relegated to junior synonymy with *Thylacoleo carnifex* (Archer and Dawson 1982).

Thylacopardus australis Owen, 1888: A collection of fossils from a newly discovered cave in Wellington Valley, New South Wales, provided Owen with a skull which lacked a lower jaw and was smaller than his other specimens of

Thylacoleo. He named it *Thylacopardus australis*, the name first appearing in print in *Nature* (1888). Owen presented his full description of this skull to the Royal Society of London in 1888, but the paper itself was only published by title (1889 or 1888). Two features were used to distinguish this new genus from *Thylacoleo*, its small size and the presence of a small molar or "molariform prominence" on the posterior border of the palate behind M^1.

In his unpublished manuscript (quoted by C. Anderson 1929) Owen wrote that "the dentinal structure" of the molariform prominence had been demonstrated, thus proving the existence of a second upper molar. Anderson studied the manuscript and the type specimen (which was then housed in the Geological and Mining Museum, Sydney) and came to the conclusion (1929) that the "M^2" was but the broken edge of the post palatine bar and showed no evidence of dentine. He believed that the animal was *Thylacoleo carnifex* but could only have been a juvenile due to its size and unworn teeth. Anderson stressed that although Owen's measurements of *Thylacopardus* and *Thylacoleo* skull bones showed a considerable discrepancy in size, the lengths of the upper carnassials were almost identical. It is thus unlikely that *Thylacopardus australis* constitutes a valid genus and species and Archer and Dawson (1982) consider it a junior synonym of *T. carnifex*.

Thylacoleo crassidentatus Bartholomai, 1962: Amongst the *Thylacoleo* material from the Darling Downs area housed in the Queensland Museum is a small collection from Chinchilla Sand at Chinchilla, northwestern Darling Downs. Woods (1956) considered these specimens to be sufficiently distinct to warrant further investigation and proposed Pliocene age for them. All known *Thylacoleo carnifex* specimens are considered to be Pleistocene in age. The acquisition of a partial right mandibular ramus from Chinchilla provided Bartholomai (1962) with these necessary material for the erection of a new species, *T. crassidentatus*. He lists the main distinguishing features of this species in comparison to *T. carnifex* as:

a) P_3 is broader posteriorly and more convex buccally than the P_3 of *T. carnifex*.
b) M_1 is wider over both the anterior and posterior roots than in *T. carnifex*. The posterior portion of the crown has a wide median ridged area and a shallow posterolabial fossette. M_2 is also more robust and one specimen gives an indication of an alveolus for M_3.
c) The mandibular ramus is broader and less concave medially.
d) I_1 is more recumbent, the angle between the tooth and the base of the mandible being smaller.

The maxillary material was fragmentary and exhibited no marked difference from *T. carnifex* although one specimen may have possessed an M^2.

Further material referable to this species has been described from the Lower Pliocene Allingham Formation of Northern Queensland (Archer and Wade 1976) and at Bow in New South Wales (Archer and Dawson 1982).

Balladonia specimens: Glauert (1912) and Merrilees (1968) both record *Thylacoleo* from caves near Balladonia, Western Australia (Fig. 2). Merrilees points out that Glauert did not differentiate between the various caves around Balladonia but Merrilees himself specifies Wonberna Cave as the source of *Thylacoleo* "probably not *carnifex*". This animal was considerably smaller than *T. carnifex* and may be a different species (Finch and Freedman 1982). The age of the deposit is not known.

Thylacoleo hilli Pledge, 1977: Pledge (1977) has described an isolated upper left carnassial of *Thylacoleo* found in 1956 in a cave at Curramulka, South Australia (Fig. 2). This specimen is of unknown age but is presumably Late Miocene or Early Pliocene. It is remarkable for its small size (Finch and Freedman 1982), being approximately half the length of the P^3 of *T. carnifex*. For this reason it has been placed in a new species, *Thylacoleo hilli*. Archer and Dawson (1982) report the finding of material referable to this species at Bow, New South Wales.

Wakaleo: Miocene deposits in South Australia and Northern Territory have yielded thylacoleonids which Clemens and Plane (1974) consider to be quite distinct from *Thylacoleo*. These animals retained three lower molars, had a comparatively short carnassial tooth and possessed only one precarnassial premolar. The latter is a feature not apparent in *Thylacoleo* species. Such differences, they believed, warranted the creation of a new genus, *Wakaleo*.

The species name, *W. oldfieldi*, was given to the South Australian Lake Ngapakaldi specimen which consisted of a left dentary bearing the incisor, P_3 and M_1. Present also were alveoli for P_1 (or P_2), M_2 and M_3. Because of facets developed on an isolated M_2, it was surmised that there were two upper molars which were set posterior to, and not lingual to, the carnassial tooth.

The Bullock Creek, Northern Territory (Fig. 2), fossil was a mandibular fragment carrying a single molar tooth, M_1, alveoli for M_2 and M_3, and a partial alveolus for I_1. The dentary was slightly larger than that of *W. oldfieldi*, and, taking into account the spatial separation and a probable age difference between the two specimens, Clemens and Plane (1974) felt justified in erecting a second species, *W. vanderleuri*.

A third species, *W. alcootaensis*, from the Miocene Alcoota local fauna of the Northern Territory (Archer and Rich 1982), appears to be the largest of the *Wakaleo* species so far discovered. This skull fragment retained tribosphenic molars, a feature implying secondary modification of the ancestral quadritubercular molars for a carnivorous diet.

The oldest known thylacoleonid: Clemens and Plane (1974) mention an edentulous skull fragment from the mid-Miocene Ngapakaldi Fauna at present under study by Clemens. It was possibly a thylacoleonid, exhibiting alveoli for an enlarged P^3 and four upper molars.

INTERPRETATION OF FEEDING HABITS

The unusual skull and dental characters of the marsupial lion, along with the paucity of postcranial material, gave rise to a heated and protracted controversy on its feeding habits.

Owen, in all of his papers, entertained no other possibility than that of a carnivorous habit and described *Thylacoleo* as "one of the fellest and most destructive of predatory beasts" (1859). Broom (1898) and Woods (1956) agreed with Owen while Glauert (1912) was more inclined to believe that *Thylacoleo* was a scavenger. The finding of broken and incised bones led De Vis (1883), W. Anderson (1889), Zeitz (1907) and Spencer and Walcott (1911) to postulate that the carnassial teeth of the marsupial lion were used in bone crushing. Cope (1882) on the other hand suggested crocodile eggs or carrion as its main food. Bensley (1903) pointed out that the phalangerid stock had been of an omnivorous–herbivorous habit in which there was a trend towards reduction of the canine teeth. He believed that *Thylacoleo* had evolved from such ancestors towards carnivority, which would explain many of its dental peculiarities.

On the other hand, several palaeontologists favoured a herbivorous diet for the marsupial lion. Flower (1868), Dawkins (1864), Krefft (1866) and Lydekker (1894) all wrote of plant food but, puzzled by the reduction of the molars, added that fruits or even an occasional bird or mammal might have been consumed. C. Anderson (1929) suggested that cycad pith or fruits of the Cucurbitaceae might have sustained the animal, while Gill (1954) gave a critical discussion of the problem but put forward no idea for a possible food source.

In the light of such a puzzling array of arguments it is necessary to reassess the morphological features which provide evidence for feeding habits and masticatory movements in *Thylacoleo*. A more comprehensive description of *Thylacoleo* morphology is given in Finch (1981).

THE SKULL
(Fig. 3)

A brachycephalic skull, such as is found in *Thylacoleo*, is characteristic of many carnivores. It is accompanied by a shortening of the jaws, thus bringing the canine (or caniniform) teeth closer to the condyle and allowing greater force to be exerted on them by the adductor musculature. At the same time, the reduction in snout size allows both eyes to be directed anteriorly, giving binocular vision and leading to the greater visual acuity vital to a hunter.

The zygomatic arch is deep and diverges widely from the cranium, providing an extensive bar for muscle attachment and at the same time allowing space for a large temporalis muscle. The size of the latter is also indicated by the well-developed sagittal and lambdoidal crests from which part of the temporalis originates. The postorbital bar, an uncommon structure in carnivores, is found also in the marsupial sabre-tooth, *Thylacosmilus*, and may well serve as an attachment area for temporalis fibres and the aponeurosis which blankets the whole muscle. At the same time, it protects and stabilises the eyes from pressure by the contracting jaw muscles. The bar also acts as a buttress supporting the anterior end of the jugal bone which is subject to compression and tension forces at the maxillo-jugal suture in piercing by the incisors and slicing by the carnassials respectively (Buckland-Wright 1978). No clear boss or bony projection from the anteroventral border of the zygomatic arch is visible, implying that the superficial masseter muscle was comparatively weak.

THE MANDIBLE
(Fig. 3D)

The condyle is a laterally-elongated bar about 45 mm wide. Its dorsal surface is slightly flattened and it rests in a glenoid fossa, the articulating surface of which is somewhat convex ventrally. A well-defined postglenoid process prevents backward movement of the condyle but there is no preglenoid barrier. The condition is thus similar to that of the phalangers, in particular that of the omnivorous *Phalanger maculatus* and does not resemble the typical eutherian carnivore condition in which only orthal movement is possible.

The condyle is low on the dentary, being on a level with the tooth row. The condylar neck is thick and short while the coronoid process is broad and high. This leads one to suggest a forceful, scissors-like jaw closure. The thick condylar neck resists force, and the low articulation elongates the effort lever arm of the temporalis muscle by increasing the distance between the fulcrum and the centre of the insertion area of the muscle (Scapino 1972).

The masseteric fossa is wide and deep, giving insertion space for a moderately large deep masseter muscle and the pterygoid fossa is similarly well developed for insertion of the internal pterygoid muscle. The latter muscle is rarely of large size in a typical carnivore. There is no fusion at the mandibular symphysis which would appear to detract from the strength of the jaw during a forceful bite.

The Teeth

Of the three pairs of upper incisors, the first pair is large, curved and caniniform, converging at the tips and showing wear facets on their adjacent surfaces. An oblique groove on the posterior surface of each has been made by the lower incisors. I^2 is small but I^3 has a large crown which is inclined anteriorly and lingually.

The small blunt upper canine tooth is located in the palate posterior to I^3 and bears no resemblance to the homologous tooth in most carnivores. The first two upper premolars are also small and are set at the anterior border of P^3.

The third upper premolar is a huge trenchant tooth which may be up to 57 mm long and has a lingually–directed occlusal surface. As this tooth is worn down, the buccal side of the root elongates faster than the lingual surface and thus the crown 'rotates' lingually to maintain close contact with P_3. The edge of the maxilla appears to grow downwards at the same time which increases the vaulting of the palate (Fig. 4). Such a rotation has been reported by Mellett (1969) in the upper carnassial teeth (M^1 and M^2) of *Hyaenodon* but does not seem to occur in fissipede carnivores.

The single upper molar is set obliquely on the lingual side of the posterior root of P^3. It shows an anteromedial wear surface commensurate with its function as a stop for the lower first molar.

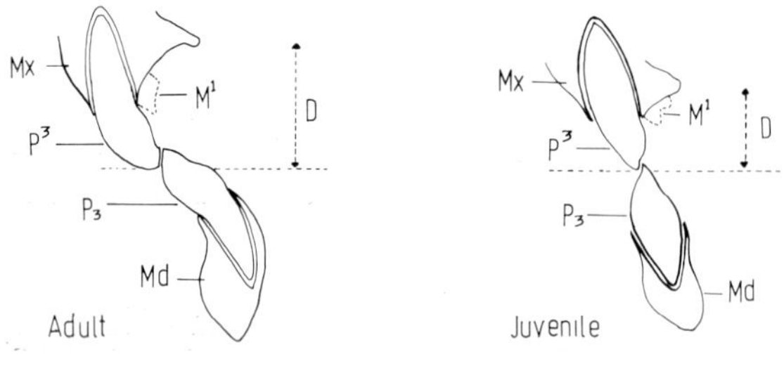

Fig. 4. Transverse section through the maxilla and mandible of a juvenile and an adult *T. carnifex* in the region of P3. D = height of palatal vault at mid line; Md = mandible; Mx = maxilla; M^1 = position of upper molar (lingual to the posterior root of P^3).

In the lower jaw, the single pair of procumbent incisors lie with their tips behind I^1 when at rest. Wear occurs at the tip and on the outer border of the crown and an interdental facet is located on the contiguous incisor surfaces.

As in the top jaw there are two small premolars at the anterolingual border of the lower carnassial. The latter tooth is inclined buccally to meet the occlusal surface of P^3 (Fig. 4) and grew in this direction during life as appears to have been the case in *Hyaenodon* (Mellett 1969). The lower sectorial surface is completed by a facet on the trigonid of M_1 so that both P_3 and M_1 shear past P_3. M_2 is very small and rarely persists in fossil material.

The picture which may be built up from *Thylacoleo* skull morphology is thus predominantly that of an animal exhibiting adaptations to carnivority parallel to those seen in eutherians except in the form of the condyle and glenoid fossa, the reduction of canine teeth, the procumbent I_1 and the enlarged I^1.

Masticatory Musculature

Reconstruction of the masticatory musculature of *Thylacoleo* using a dissection of the similar phalangerid, *Phalanger maculatus,* as a guide (Finch 1981) showed that the temporalis muscle contributed about 56% of the total adductor musculature, while the masseter complex (superficial and deep masseters and zygomaticomandibularis) made up 34% and the pterygoids only 10%. Turnbull (1970) gives figures for a wide range of mammals which he groups into four main categories: A, generalised or unspecialised type; B, carnivore–shear or 'scissors' type; C, ungulate–grinding or 'mill' type; and D, rodent–gnawing or 'forward–shift' type. The range of relative weights of the three muscle groups for the four categories (A–D), *Phalanger* and *Thylacoleo*, are plotted in Fig. 5. There is only a difference of degree between categories A and B as, in each, the temporalis is dominant and the pterygoid is the smallest of the adductor muscle groups. The forward–shift type (D) differs from the grinding type (C) in that the dominant masseter complex is considerably larger and the temporalis and pterygoid tend to be more reduced than in type C.

Phalanger appears intermediate between types C and D. Inspection of its skull shows a glenoid fossa which is not closed anteriorly and has an articulating surface which is convex ventrally. This implies a forward shift mechanism to permit occlusion of the incisors as, at rest, the tip of I_1 lies behind I^1. To attain this forward position,

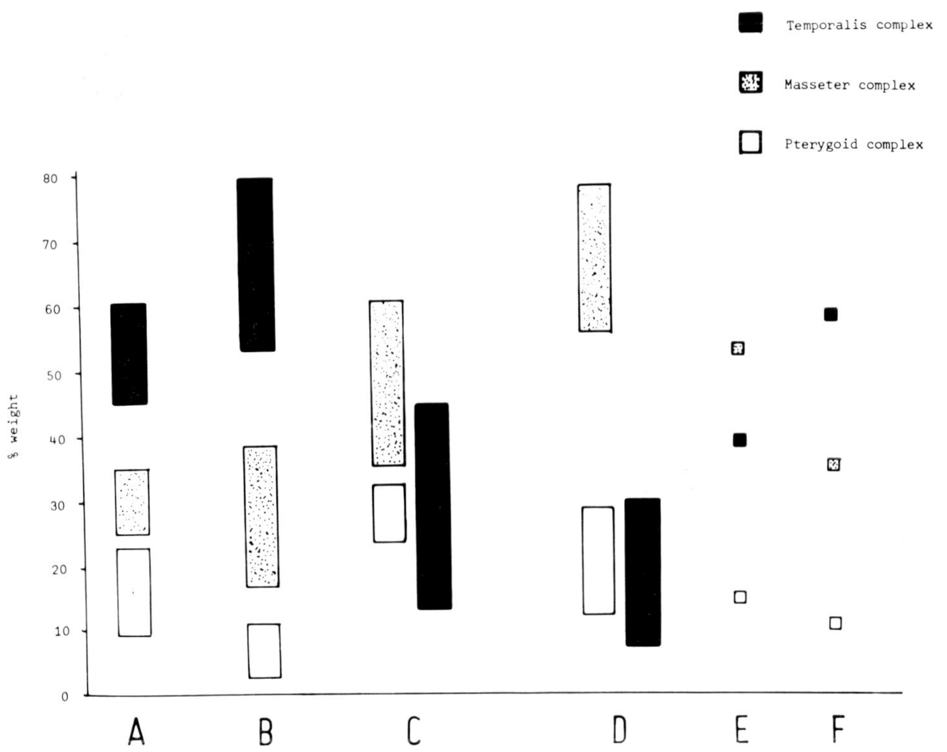

Fig. 5. Diagram showing the ranges of percentage weights of the temporalis, masseter and pterygoid muscle groups in generalised (A), carnivore–shear (B), grinding (C) and forward–shift (D) mammals. (Weights taken from Turnbull 1970). E. and F are the percentage weights of these muscles in *Phalanger* and *Thylacoleo* respectively.

the condyle is moved anteriorly on to the convexity of the glenoid fossa which allows the incisor tips to meet. At the same time, this prevents occlusion of the cheek teeth when the incisors are in use.

Thylacoleo, in contrast to *Phalanger*, shows a dominant temporalis and a small pterygoid which places the animal clearly in the carnivore-shear group (B). Van Valen (1969: 114) states his belief that the masseter complex outweighs the temporalis in the marsupial lion but gives no supporting data.

In order to gain further insight into mastication in *Thylacoleo*, the probable areas of origin and insertion of the adductor muscles were outlined on an accurate drawing of the skull and the midpoints of the two attachments of each muscle were joined. This gave the direction of pull of the 'central' fibre of the muscle. Taking this into account, along with the distance of the carnassial tooth from the condyle (resistance lever arm) and the length of the effort lever arm (fulcrum to centre of muscle insertion area), it is possible to calculate the useful power (E) of a muscle. This is done from Turnbull's formula (1970):

$$E = M \times F_L \times F_\times \times r$$

where M is the percentage weight of the muscle and r is the ratio of the effort lever arm to the resistance lever arm. F_L and $F_\times$ are correction factors which compensate for the deviation, in the parasagittal and transverse planes, of fibre direction from the plane of jaw closure. A muscle orientated at 90° to the effort lever arm in the parasagittal plane is considered capable of exertiing its full force to close the jaw and thus $F_L =$ 1. However, any deviation from 90° will lower its efficiency until a muscle lying parallel to the lever arm has $F_L = 0$. Similarly, $F_\times = 1$ if the muscle, viewed from the posterior aspect of the skull, is vertical in its orientation (lying parallel to the coronoid process). $F_\times$ tends towards zero as the muscle tends towards a horizontal orientation.

Of the total useful power (E) generated by the adductor muscles of *Thylacoleo*, the temporalis exerts 62%, the superficial masseter 10%, the deep masseter 11%, the zygomaticomandibularis 14% and the pterygoid 3% (Finch 1981). When these figures are compared to the percentage weights, only the zygomaticomandibularis and pterygoid exert less force than one would expect from their relative mass. Study of the probable fibre direction (Fig. 6) reveals the reason. These two muscles approach a horizontal rather than a vertical direction. Clearly their functions are not purely adductor. Scapino (1965) suggests that in the dog, where the mandibular symphysis is not fused, these two muscles regulate the angle at which the lower carnassial approaches the upper one by tilting the dentary to give the

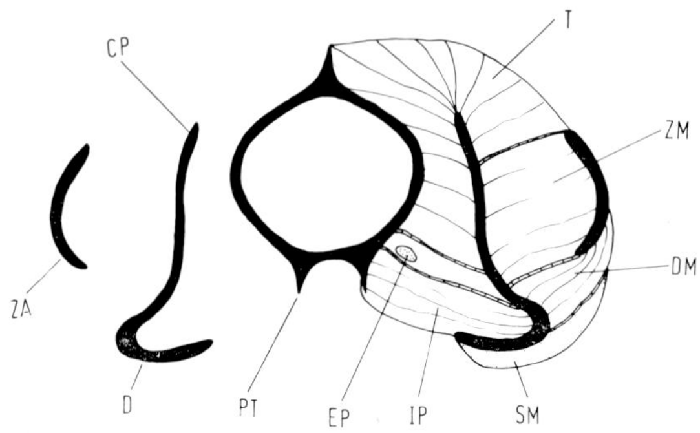

Fig. 6. Diagrammatic transverse section of a *Thylacoleo* skull in the region of the coronoid process (CP) showing the disposition of the adductor jaw muscles. Bone is black. D = dentary; DM = deep masseter muscle; EP = external pterygoid muscle; IP = internal pterygoid muscle; PT = pterygoid; SM = superficial masseter muscle; T = temporalis muscle; ZA = zygomatic arch; ZM = zygomaticomandibularis muscle.

most efficient shear. This would also appear to be the case in *Thylacoleo* in which the mandibular symphysis remains unfused. Further, only limited lateral excursion of the condyle is possible at the glenoid fossa in *Thylacoleo* and, although the carnassial teeth continue to grow as they are worn down, accurate placement of the shearing surfaces requires some adjustment to the position of the lower blade.

Feeding Habits

From a consideration of skull morphology and reconstructed musculature, it is suggested that *Thylacoleo* could exert a powerful slicing action with the carnassial teeth and a strong, piercing stab with the four convergent first incisors. Wear facets indicate occlusion between I_1 and the three upper incisors which would permit the biting or tearing of smaller pieces of food. Wear on the posterolateral surface of the canine can only have been caused by contact with the anterior buttress of P_3 when the jaw was in the forward position. The location of M^1 and a facet on the posterolateral wall of M_1 support Woods' (1956) contention that the upper molar was a stop for M_1 as the jaw moved back into its resting position.

The lack of a molar grinding battery indicates that the marsupial lion could not have subsisted on plant material. The extreme size of the carnassial teeth and the concomitant carnivore adaptations point to a flesh–eating habit. However, the animal has retained the ancestral forward–shift mechanism along with the reduction of the canines, two features which are not carnivore characteristics. A comparable case is that of the giant panda, *Ailuropoda melanoleuca* which has been shown by Davis (1964) to have evolved in the opposite direction. From an omnivore–carnivore stock, it has become a specialised herbivore while still retaining certain carnivore traits.

Wells (1982) is in general agreement with the hypothesis that *Thylacoleo* was a carnivore but envisages the animal as a bone–crusher which could pivot its mandible laterally on one condyle thus enabling it to occlude the I_1 with the three ipsilateral upper incisors.

INTERPRETATION OF LOCOMOTORY ADAPTATIONS

The Moree *Thylacoleo* specimen (Fig. 1) includes both of the forelimbs and one hind–limb which lacks the pes. It was the first nearly complete specimen to have been discovered and provides the basis for limb comparisons with those animals whose locomotory habits are known.

Howell (1944), Hildebrand (1954, 1961), Smith and Savage (1956) and Savage (1977) have discussed various indices of intra– and inter–limb lengths which are characteristic of different mammalian locomotory patterns. The morphological trends which become apparent are the retention of the primitive condition of a slightly shorter fore- than hindlimb in quadrupedal walking animals, for the elongation of the distal limb elements in cursorial forms (e.g. ungulates) and for the development of short but stout legs in fossorial and swimming animals. It must be remembered, however, that the forelimb may serve functions other than locomotion, such as food manipulation, which may modify the adaptation of the limb.

In this preliminary investigation, measurements were taken of the length of the vertebral column (taken as atlas – last lumbar vertebra), humerus, radius, metacarpal 3, femur and tibia. Attention was confined to the Marsupialia although the eutherian lion was included for comparison. Skeletal material of the following taxa was made available by Dr. W. D. L. Ride from the Western Australian Museum collections. Figures in parentheses give the number of specimens measured and, where possible, the measurements of both right and left limbs of each animal were obtained.

Order Marsupicarnivora

Family Dasyuridae: (*Antechinus flavipes* (1), *Dasyurus geoffroii* (1), *Sarcophilus harrisii* (1))

Order Peramelina

Families Thylacomyidae and Peramelidae: (*Macrotis lagotis* (1), *Isoodon obesulus* (2))

Order Diprotodonta

Families Petauridae and Phalangeridae: (*Pseudocheirus occidentalis* (2), *Trichosurus vulpecula* (4))

Family Vombatidae: (*Lasiorhinus latifrons* (2), *Vombatus hirsutus* (1))

Family Phascolarctidae: (*Phascolarctos cinereus* (2))

Families Macropodidae and Potoroidae (*Macropus* sp. (1), *Bettongia lesueur* (1))

The most useful comparisons which can be made in the case of the Moree *Thylacoleo* are: (a) humerus + radius + metacarpal 3 / length of vertebral column (atlas to last lumbar) x 100; (b) length of forelimb / length of hindlimb x 100; (c) length of tibia / length of femur x 100; (d) length of radius / length of humerus x 100. The widely used ratio of hindlimb to vertebral column length could not be calculated owing to the absence of the pes in the Moree specimen.

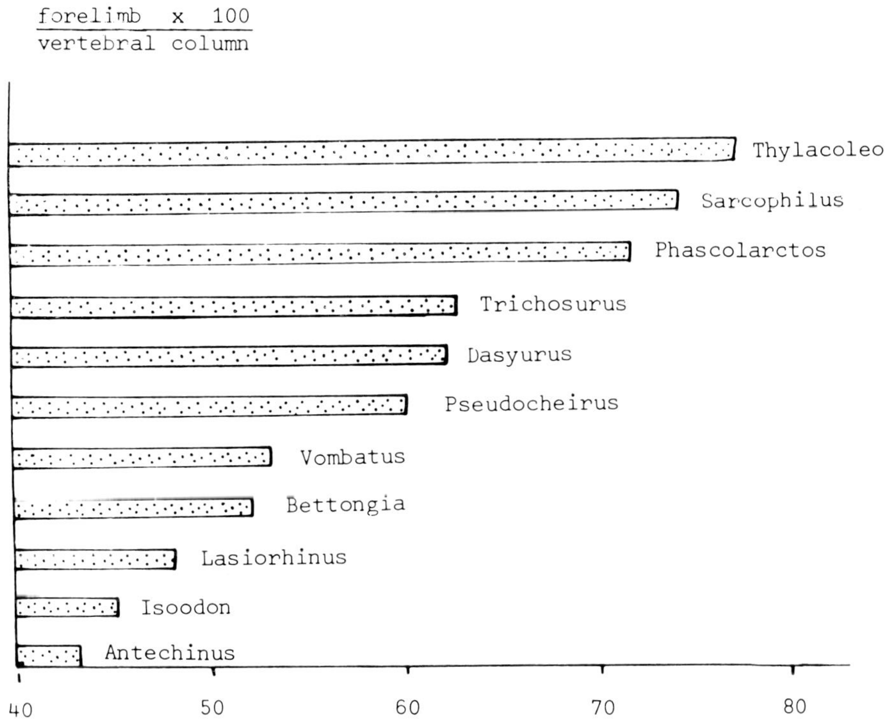

Fig. 7. Ratio of forelimb to vertebral column length (atlas to last lumbar) in a series of marsupials.

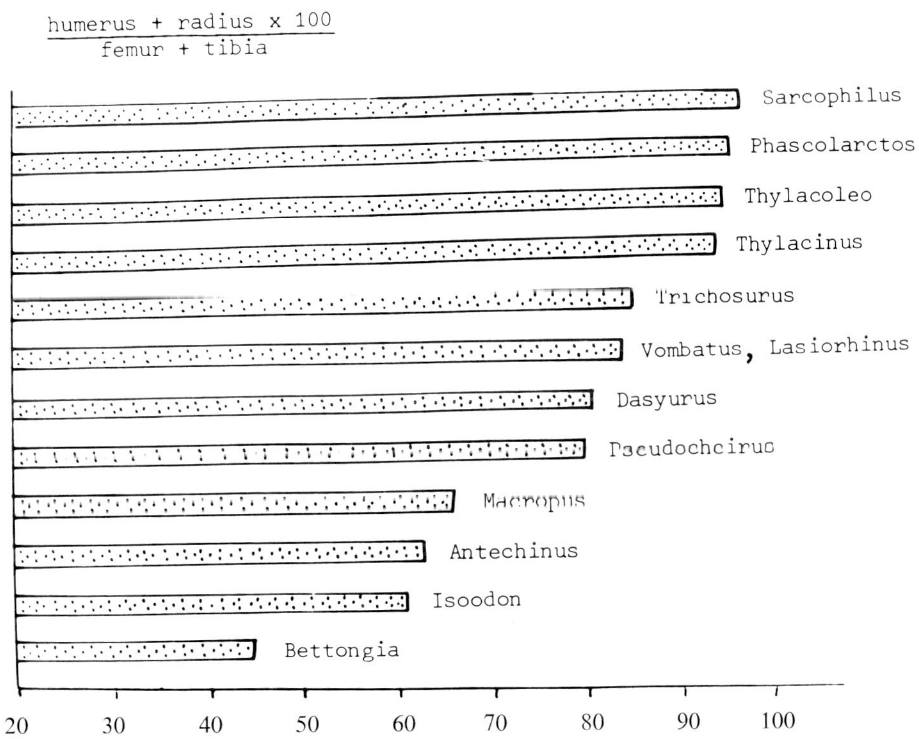

Fig. 8. Intermembral index in a series of marsupials. The value for *Thylacinus* is taken from Howell (1944).

Ratio (a), forelimb/vertebral column, differentiates cursorial and climbing animals from the fossorial and saltatorial forms. Animals of the first two groups tend to show comparatively long forelimbs in comparison with the length of the vertebral column. Fig. 7 demonstrates that *Thylacoleo's* forelimbs are considerably longer than those of the arboreal *Trichosurus* and *Pseudocheirus* and the smaller cursorial marsupial cat, *Dasyurus*. The ratios for the fossorial wombats, *Vombatus* and *Lasiorhinus* and the largely saltatorial *Bettongia, Isoodon* and *Antechinus* are low, ranging from 43 to 53.

Ratio (b), the intermembral index, similarly allows the separation of cursorial and arboreal from saltatorial animals. Fig. 8 shows this distinction clearly, but it is notable that wombats tend to have limb ratios similar to those of the climbers. *Thylacoleo* is again placed with the larger cursorial carnivorous forms.

difference between *Thylacoleo* and *Sarcophilus* is usually about 100 in unspecialised animals, ris- approximately equal length.

The final ratio, (d), that of radius/humerus is usually about 100 in unspecialised animals, rising to 140 in saltatorial forms, to 120 in swift cursorial animals and falling to below 100 in fossorial and walking forms. The eutherian lion has a humero–radial index of only about 90 (Howell 1944) in spite of its ability to run swiftly for short periods. Amongst the marsupials (Fig. 10), *Thylacoleo* shows a much higher index (115) than its eutherian counterpart and it is placed among animals showing climbing adaptations.

In the four figures (7–10), the position of the arboreal koala, *Phascolarctos,* is somewhat puzzling. It appears close to *Thylacoleo* in every case and, in Figs. 8 and 9 in particular, there is considerable divergence in its indices from those of

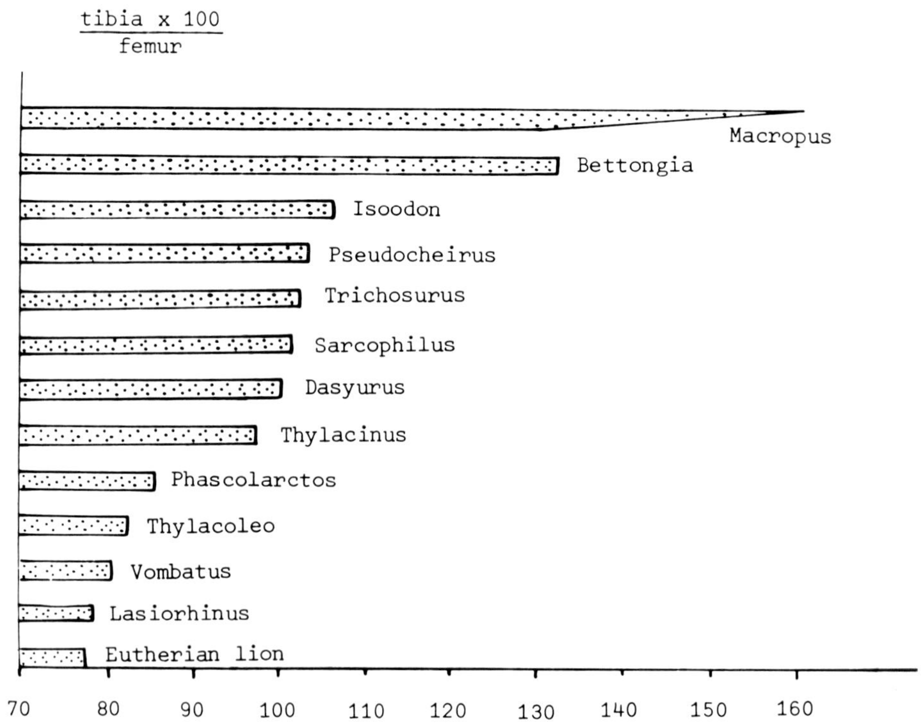

Fig. 9. Femoro–tibial index in marsupials and the eutherian Lion. Values for *Macropus, Thylacinus* and the lion are taken from Howell (1944). The *Macropus* ratios (130–160) cover a range of species of this genus.

Within the hindlimb, the tibia/femur index (c) has been found by Howell (1944) to be high in saltatorial animals and in swift runners, but low in walking forms such as cattle, lions and hyaenas. In Fig. 9, *Thylacoleo* is ranged alongside the eutherian Lion and the wombats. Clearly the marsupial Lion cannot be a swift cursorial animal and its femoro–tibial index (82) does not approach that of the arboreal possums, *Trichosurus* and *Pseudocheirus* (102–103). The the other arboreal marsupials, the possums. However, Archer (1976) postulates that the Phascolarctidae and the closely related Vombatidae arose directly from ancestral diprotodonts with selenodont teeth and that the phalangerids are a quite independent evolutionary line from the basic stock. Work on comparative serology by Kirsch (1977) supports this hypothesis. If this is correct, the divergence between the adaptations of Koalas and possums is not unexpected.

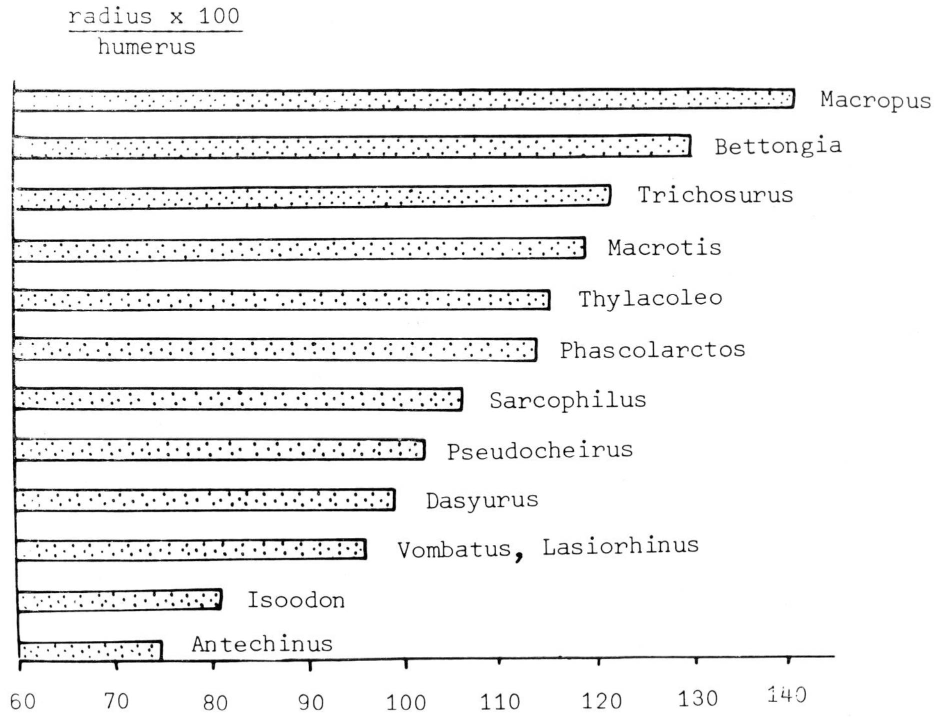

Fig. 10. Humero–radial index in marsupials.

On the basis of the limb proportions studied, *Thylacoleo* is likely to have been a cursorial marsupial which was not capable of swift locomotion. It showed no marked climbing adaptations (Figs 7, 8 and 9) although the relatively long radius (Fig. 10) places it among the arboreal and the saltatorial forms whose forelimb is commonly used for manipulative and food gathering purposes as well as for locomotion.

This preliminary investigation will be supplemented by studies on the limb girdles and the vertebral column but at present it points to *Thylacoleo* being a heavy–bodied animal normally employing a deliberate walk but which might have been capable of short bursts of speed. The unusually long forelimb with its huge claw on the first digit is likely to have been used in reaching out to drag food towards the mouth and would indicate a predatory rather than a scavenging habit. The pseudo–opposability of the powerful first digit suggests a scansorial habit to Wells and Nichol (1977) and, considering the phalangerid affinities of the group, this may have been the case. *Thylacoleo*, the 'pouched lion', may well have been more aptly named Thylacopardus or 'pouched leopard' had the complete skeleton been available for study almost a hundred years ago.

ACKNOWLEDGEMENTS

Sincere appreciation is due to Dr. W. D. L. Ride, then Director of the Western Australian Museum, without whose expert guidance this study could not have been carried out. He introduced the author to *Thylacoleo* and continued to instil encouragement as the initial problems unfolded. Dr. L. Freedman of the University of Western Australia has given much of his time and considerable expertise in fruitful discussions on the subject, while Dr. R. J. G. Savage offered hospitality and stimulating ideas during the author's study period at Bristol University. To Dr. A. Ritchie is owed a considerable debt of gratitude for permission to work on a specimen so important to the Australian Museum's outstanding display series. Ms. S. McConnell kindly undertook the photography, while Ms. Z. Lewis gave helpful advice on the diagrams.

The manuscript was read by Dr. L. Freedman, and Mr. J. Mahoney (University of Sydney) and Dr. W. D. L. Ride to whom the author is indebted for their helpful and constructive comments.

REFERENCES

ANDERSON, C., 1929. Palaeontological Notes No. 1. *Macropus titan* Owen and *Thylacoleo carnifex* Owen. *Rec. Aust. Mus.* **17:** 35–49.

ANDERSON, W., 1889–90. On Post-Tertiary ossiferous clays near Myall Creek, Bingera. *Rec. Geol. Surv. N.S.W. Vol.* **1:** 116–26.

ARCHER, M., 1976. Phascolarctid origins and the potential of the selenodont molar in the evolution of diprotodont marsupials. *Mem. Qd Mus.* 17: 367–71.

ARCHER, M. AND DAWSON, L., 1982. Revision of marsupial lions of the genus *Thylacoleo* Gervais (Thylacoleonidae : Marsupialia) and thylacoleonid evolution in the late Cainozoic. Pp.477–94 *in* "Carnivorous marsupials", ed. by M. Archer. Roy. Zool. Soc. N.S.W.: Sydney.

ARCHER, M. AND RICH, T. H., 1982. Results of the Ray E. Lemley Expeditions. *Wakaleo alcootaensis* n.sp. (Thylacoleonidae: Marsupialia), a new marsupial lion from the Miocene of the Northern Territory, with a consideration of the early radiation of the family. Pp.495–502 *in* "Carnivorous marsupials", ed. by M. Archer. Roy. Zool. Soc. N.S.W.: Sydney.

ARCHER, M. AND WADE, M., 1976. Results of the Ray E. Lemley Expeditions. Part 1 The Allingham Formation and a new Pliocene vertebrate fauna from northern Queensland. *Mem. Qd Mus.* **17:** 379–97.

BARTHOLOMAI, A., 1962. A new species of *Thylacoleo* and notes on some caudal vertebrae of *Palorchestes azael*. *Mem. Qd Mus.* **14:** 33–40.

BENSLEY, B. A., 1903. On the evolution of the Australian Marsupialia; with remarks on relationships of the marsupials in general. *Trans. Linn. Soc. Lond.* (*Zool.*) (2), IX: 83–217.

BROOM, R., 1898. On the affinities and habits of *Thylacoleo*. *Proc. Linn. Soc. N.S.W.* (2) **22:** 57–74.

BUCKLAND–WRIGHT, J. C., 1978. Bone structure and the patterns of force transmission in the cat skull (*Felis catus*). *J. Morph.* **155:** 35–62.

CLEMENS, W. A. AND PLANE, M., 1974. Mid–Tertiary Thylacoleonidae (Marsupialia, Mammalia). *J. Paleontol.* **48:** 652–60.

COPE, E. D., 1882. The ancestry and habits of *Thylacoleo*. *Amer. Nat.* **1882:** 520–2.

DAVIS, D. D., 1964. The Giant Panda. A morphological study of evolutionary mechanisms. *Fieldiana; Zool., mem. ser.* **3:** 1–339.

DAWKINS, W. B., 1864. On the Rhaetic beds and White Lias of western and central Somerset and on the discovery of a new fossil mammal in the Grey Marlstones beneath the bone bed. *Q. J. Geol. Soc. Lond.* XX: 396–412.

DE VIS, C. W., 1883. On tooth-marked bones of extinct marsupials. *Proc. Linn. Soc. N.S.W.* **8:** 187–90.

DE VIS, C. W., 1887. On a femur probably of *Thylacoleo*. *Proc. R. S. Qd* **3:** 122–8.

DE VIS, C. W., 1900. Bones and diet of *Thylacoleo*. *Ann. Qd Mus.* **5:** 7–11.

ETHERIDGE, R., 1918. The ungual phalanges termed *Mylodon australis* by Krefft, spelaean animal vel *Thylacoleo* by Owen, and *Thylacoleo* by Lydekker. *Ann. Mag. Nat. Hist., ser.* 9, **2:** 307–18.

FINCH, M. E., 1981. "*Thylacoleo*, an Australian fossil phalangerid marsupial". Unpublished thesis.

FINCH, M. E. AND FREEDMAN, L., 1982. An odontometric study of *Thylacoleo* (Thylacoleonidae, Marsupialia). Pp.533–61 *in* "Carnivorous marsupials", ed. by M. Archer. Roy. Zool. Soc. N.S.W.: Sydney.

FLOWER, W., 1868. On the affinities and probable habits of the extinct Australian marsupial *Thylacoleo carnifex* Owen. *Quart. J. Geol. Soc. Lond.* XXIV: 307–19.

GERVAIS, P., 1869. Mémoire sur les formes cérébrales propres aux marsupiaux. *Nouv. Arch. Mus. Hist. Nat. Paris* **5:** 229–51.

GERVAIS, P., 1848–52. Zoologie et paléontologie francaises (animaux vértebrés) ou nouvelles recherches sur les animaux vivantes et fossiles de la France. Arthus Bertrand: Paris. Tome I, 271 pp. Tome II, 146 pp.

GILL, E. D., 1954. Ecology and distribution of *Thylacoleo*. *Vict. Nat.* **71:** 18–35.

GLAUERT, L., 1912. Fossil marsupial remains from Balladonia in the Eucla Division. *Rec. W. Aust. Mus.* **1:** 47–60.

HILDEBRAND, M., 1954. Comparative morphology of the body skeleton in recent Canidae. *Univ. Calif. Publ. Zool.* **52:** 399–470.

HILDEBRAND, M., 1961. Body proportions of didelphid (and some other) marsupials, with emphasis on variability. *Am. J. Anat.* **109:** 239–49.

HOWELL, A. B., 1944. Speed in Animals. Univ. of Chicago Press: Chicago.

KIRSCH, J. A. W., 1977. The comparative serology of Marsupialia, and a classification of marsupials. *Aust. J. Zool., Suppl. Ser., No.* **52:** 1–152.

KREFFT, G., 1866. On the dentition of *Thylacoleo carnifex* (Ow.). *Ann. Mag. nat. Hist., ser.* 3 **18:** 148–9, Pl. 11.

KREFFT, G., 1870a. Wellington Caves (Correspondence relative to exploration of). Pp.1–12 *in* "New South Wales Parliamentary Paper". [Extracts concerning *Plectodon* and *Mylodon? australis* republished in Mahoney and Ride 1975, p.199].

KREFFT, G., 1870b. Guide to the Australian fossil remains, exhibited by the Trustees of the Australian Museum, and arranged and named by Gerard Krefft, F.L.S., Curator and Secretary. The Trustees of the Australian Museum: Sydney.

KREFFT, G., 1872a. Natural history. The phalanger tribe – (continued). The Sydney Mail and New South Wales Advertiser, No. 621, Vol *13*, May 25th 1972. [Extract concerning *Thylacoleo* republished in Mahoney and Ride, 1975, pp 200–202].

KREFFT, G., 1872b. A Cuvierian principle in palaeontology, tested by evidences of an extinct leonine marsupial (*Thylacoleo carnifex*), by Professor Owen, F.R.S., D.C.L., Foreign Associate of the Institute of France. *Ann. Mag. nat. Hist.* (4) **10:** 169–82.

KREFFT, G., 1882. Exploration of the caves and rivers of New South Wales. Pp.1–52 *in* "New South Wales Parliamentary Paper, 1882".

LONGMAN, H. A., 1925. Fossil marsupials from Marmor. *Mem. Qd Mus.* **8:** 109–10.

LYDEKKER, R., 1887. Catalogue of the fossil Mammalia in the British Museum (Natural History). Part 5. London, the Trustees of the British Museum (Natural History), XXXV, 345 pp.

LYDEKKER, R., 1894. "A handbook to the Marsupialia and Monotremata". W. H. Allen: London.

MAHONEY, J. A. AND RIDE, W. D. L., 1975. Index to the genera and species of fossil Mammalia described from Australia and New Guinea between 1838 and 1968. *Spec. Publs. West. Aust. Mus.* **6:** 1–250.

MCCOY, F., 1876. *Thylacoleo carnifex* (Ow.). Prodromus of the palaeontology of Victoria; or, Figures and descriptions of Victorian organic remains. *Decade* **3:** 7–12, Melbourne, Geological Society of Victoria.

MELLETT, J. S., 1969. Carnassial rotation in a fossil carnivore. *Amer. Midl. Nat.* **82:** 287–9.

MERRILEES, D., 1968. Man the destroyer: late Quaternary changes in the Australian marsupial fauna. *J. Roy. Soc. W. Aust.* **51:** 1–24.

MITCHELL, T. L., 1838. Three expeditions into the interior of eastern Australia, with descriptions of the recently explored region of Australia Felix, and of the present colony of New South Wales. T. and W. Boone: London. Vol. 1 and Vol. 2.

OWEN, R., 1858. "Odontology", in the Encyclopaedia Britannica, or dictionary of arts, sciences and general literature (8th ed.), **16:** 407–84.

OWEN, R., 1859. On the fossil mammals of Australia. Part 1. Description of a mutilated skull of the large marsupial carnivore (*Thylacoleo carnifex* Owen), from a calcareous conglomerate stratum, eighty miles S.W. of Melbourne, Victoria. *Phil. Trans. R. Soc.* **149:** 309–22.

OWEN, R., 1866. On the fossil mammals of Australia. Part II. Description of an almost entire skull of the *Thylacoleo carnifex,* Owen, from a freshwater deposit, Darling Downs, Queensland. *Phil. Trans. R. Soc.* **156:** 73–82.

OWEN, R., 1871. On the fossil mammals of Australia. Part IV. Dentition and mandible of *Thylacoleo carnifex,* with remarks on the arguments for its herbivority. *Phil. Trans. R. Soc.* **161:** 213–66.

OWEN, R., 1877. Researches on the fossil remains of the extinct mammals of Australia; with a notice of the extinct marsupials of England. J. Erxleben: London. Vol. 1 (text) and Vol. 2 (plates).

OWEN, R., 1883a. On the affinities of *Thylacoleo. Phil. Trans. R. Soc.* **174:** 575–82.

OWEN, R., 1883b. Pelvic characters of *Thylacoleo carnifex. Phil. Trans. R. Soc.* **1883:** 639–43.

OWEN, R., 1887–1888. Additional evidence of the affinities of the extinct marsupial quadruped *Thylacoleo carnifex* (Owen). *Phil. Trans. R. Soc.* **1887 (1888:)** 1–3.

OWEN, R., (In Anon.) (1888). On *Thylacopardus australis,* Owen. *Nature, Lond.* **39:** 215.

PLEDGE, N., 1977. A new species of *Thylacoleo* (Marsupialia, Thylacoleonidae), with notes on the occurrence and the distribution of Thylacoleonidae in South Australia. *Rec. S. Aust. Mus.* **17:** 277–83.

SAVAGE, R. J. G., 1977. Evolution in carnivorous mammals. *Palaeontol.* **20:** 237–71.

SCAPINO, R. P., 1965. The third joint of the canine jaw. *J. Morph.* **116:** 23–50.

SCAPINO, R. P., 1972. Adaptive radiation of mammalian jaws. *In* "Morphology of the maxillo–mandibular apparatus", ed. by G. Schumacher, Georg Thieme. Leipzig.

SMITH, J. M. AND SAVAGE R. J. G., 1956. Some locomotory adaptions in mammals. *J. Linn. Soc. Zool.* **42:** 603–22.

SPENCER, W. B. AND WALCOTT, R. H., 1911. The origins of cuts on bones of Australian extinct marsupials. *Proc. Roy. Soc. Vict. Vol.* XXIV: 92–123.

STUTCHBURY, S., 1850–1853 (MS date). Dairy of a geological and mineralogical survey of the colony of New South Wales, during the years 1850, 1851, 1852, 1853 by Samuel Stutchbury, F. G. S. etc., geologist to the Colonial Government, Pp.1–863. Mitchell Library Manuscripts A2639. [Entries on *Schizdon* published by Mahoney and Ride 1975, Pp.239–240].

STUTCHBURY, S., 1953. New South Wales. Geological and Mineralogical Surveys. Eleventh tri–monthly report upon the geological and mineralogical structure of New South Wales, 9–11.

TURNBULL, W. D., 1970. Mammalian masticatory apparatus. *Fideldiana: Geology* **18:** 149–356.

VAN VALEN, L., 1969. Evolution of dental growth and adaptation in mammalian carnivores. *Evol.* **23:** 96–117.

WELLS, R. T., HORTON, D. R. AND ROGERS, P., 1982. *Thylacoleo carnifex* Owen (Thylacoleonidae, Marsupialia): marsupial carnivore? Pp.574–86 *in* "Carnivorous marsupials", ed. by M. Archer. Roy. Zool. Soc. N.S.W.: Sydney.

WELLS, R. T. AND NICHOL, B., 1977. On the manus and pes of *Thylacoleo carnifex* Owen (Marsupialia). *Trans. R. Soc. S. Aust.* **101:** 139–46.

WOODS, J. T., 1956. The skull of *Thylacoleo carnifex. Mem. Qd Mus.* **13:** 125–40.

ZEITZ, A., 1907. Exhibit of *Thylacoleo carnifex* and kangaroo bones found at Salt Creek, Normanville. Abst. of Proc., *Trans. R. Soc. S. Aust.* **31:** 317.

RESUMEN

45. Historia e interpretación de *Thylacoleo carnifex* (Thylacoleonidae, Marsupialia). Por M. E. Finch.

Se presenta una reseña de la historia del descubrimiento de *Thylacoleo carnifex* (Thylacoleonidae : Marsupialia) desde comienzos de 1830 hasta la actualidad. La mayoría del primer material craneano fue fragmentario, por lo que la reconstrucción de la cabeza fue difícil. La dentición está dividida en dos unidades funcionales y los ostros dientes están reducidos. Los incisivos anteriores, superiores e inferiores, convergen por sus extremos en una "unidad caniniforme", mientras que los premolares terceros superior e inferior son enormes hojas cortantes. Sobre la base de esta desusada dentición, varios autores han atribuido al león marsupial hábitos herbívoros, omnívoros y carnívoros. La reconstrucción de la musculatura mandibular adductora y la estimación del relativo peso de los músculos apoyan la hipótesis carnívora. El trabajo preliminar sobre sus patas fuertemente armadas de garras sugiere una locomoción semejante a los felinos.

46. AN ODONTOMETRIC STUDY OF THE SPECIES OF *Thylacoleo* (THYLACOLEONIDAE, MARSUPIALIA)

M. E. FINCH[1] and L. FREEDMAN[2]

Primarily using the dimensions of P3, samples of *Thylacoleo* from various parts of Australia have been compared, and their inter–relationships determined, by univariate and multivariate analyses. *T. hilli* and *T. crassidentatus* are distinguishable from each other and from *T. carnifex*, the former by its much smaller size, the latter by the size relationship of P^3 to P_3. Within *T. carnifex*, the Victorian P3 dimensions are the largest of the eastern states samples which, however, do not vary greatly. The Western Australian and Tasmanian samples mainly differ from those in the eastern states on Penrose's 'size' and 'shape' criteria, respectively. These results are in agreement with the specific separation of *T. carnifex, T. hilli* and *T. crassidentatus*. The Western Australian and, more especially, the Tasmanian populations of *T. carnifex* may merit taxonomic recognition but confirmation by other features and more material is required.

Pages 553–61 *in* CARNIVOROUS MARSUPIALS, ed. by M. Archer, February, 1982. Roy. Zool. Soc. New South Wales: Sydney, Australia.

THERE is now a substantial amount of material which has been referred to the genus *Thylacoleo* from all parts of continental Australia and also from Tasmania. In addition to the type species, *Thylacoleo carnifex* Owen, 1858, two other valid species, *T. crassidentatus* Bartholomai, 1962, and *T. hilli* Pledge, 1977, have been described (reviewed by Archer and Dawson, 1982). A closely related thylacoleonid genus, *Wakaleo*, includes two species, *W. oldfieldi* Clemens and Plane, 1974 and *W. vanderleuri* Clemens and Plane, 1974. The available thylacoleonid skull and postcranial material is too limited and fragmentary to be of much value in elucidating differences within and between these taxa. However, a significant number of teeth of *Thylacoleo*, especially P3, are now available.

This study investigates relationships within the genus *Thylacoleo* by univariate and multivariate analysis of the dimensions of P3.

MATERIALS AND METHODS

A large number of specimens was measured by one of us (M.E.F.) and all measurements available in the literature were also utilised. In addition, measurements were kindly made available by Dr. M. Archer, School of Zoology, University of New South Wales, Mr. N. Pledge, South Australian Museum and Dr. R. Wells, School of Biological Sciences, Flinders University of South Australia. All measurements, together with details of source and locality, are in Finch (1981); and a map of the localities is included in Finch (1982).

For each tooth included in this study maximum length and/or maximum breadth measurements were available. Where any tooth was measured by more than one person and the figures differed slightly the mean was used. It is possible that some P_3 breadth measurements from Queensland, which should have been grouped with *T. crassidentatus*, may have been included in *T. carnifex*. Only those specimens actually referred by Bartholomai (1962) to *T. crassidentatus* are listed with that species. Any others with the locality label "Chinchilla" are included in *T. carnifex* on the assumption that Bartholomai would have studied all such material and listed all specimens referable to his new species.

A statistical summary of the available odontometric data of *Thylacoleo* is presented in Tables 1 and 2. The dental nomenclature used in this paper for this genus is: I 3/1, C 1/0, P 3/3, M 1/2. Most of the teeth can be seen in the lateral view of the *Thylacoleo* skull (Fig. 1). For the maximum length and maximum breadth

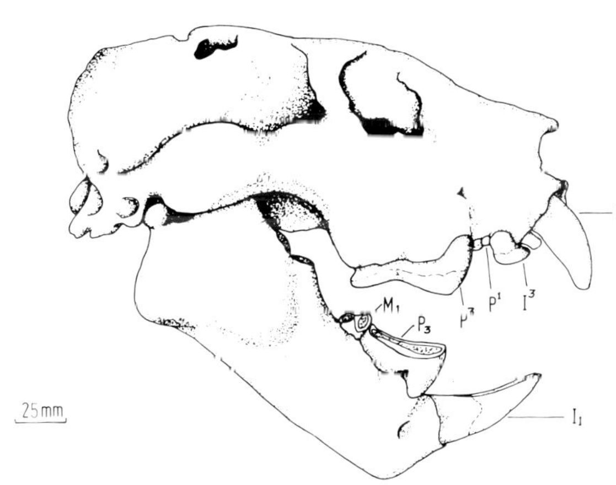

Fig. 1. Lateral view of the skull of *T. carnifex* based on the Moree specimen (F 52398–1). The teeth not visible are: C^1 (medial to P^1), M^1 (medial to posterior root of P^3), P_1 and P_2 (medial to anterior root of P_3), M_2 alveolus lies posterior to M_1.

[1] Department of Zoology, University of Western Australia, Nedlands, Western Australia, 6009.
[2] Department of Anatomy and Human Biology, University of Western Australia, Nedlands, Western Australia, 6009.

of each tooth, the number (N), mean value ($\overline{X}$), observed range (OR), standard deviation (s) and coefficient of variation (V) are listed. Material referred to *T. carnifex* was analysed by individual states, and also for all localities together. In most cases, the distance between the localities within each state is far less than that between states, making this a useful subdivision. Possible exceptions are Moree and Bingara in New South Wales which are slightly closer to the Queensland than to the main New South Wales localities, Wellington Caves and Geurie. Published dimensions of *Wakaleo* have been included in Table 2 for comparison with *Thylacoleo*.

Comparisons between the *Thylacoleo* samples using P3 mean lengths and breadths have been made by the following methods:

1) Histograms to examine for sexual dimorphism (Fig. 2 – P3 lengths).

2) Ratio diagram (Simpson, Roe and Lewontin 1960) of mean lengths, using Victoria (the largest) as the standard (Fig. 3). Because of the paucity of data for *T. hilli* and *T. crassidentatus*, the breadth values were not similarly analysed.

3) Plots of mean length against mean breadth for P^3 (Fig. 4) and P_3 (Fig. 5).

4) Student's 't' tests (Table 3).

5) Penrose's mean square distance, 'size' and 'shape' (Table 4 and Fig. 6).

Penrose's multivariate method of assessing morphometric distance between pairs of samples is based on the differences ($d_1, d_2, \ldots\ldots d_n$) between the mean values of each character studied, expressed in pooled standard deviation units (Penrose 1954). The mean square distance (C_H^2) is calculated for each pair of samples as follows:

$C_H^2 = \frac{\Sigma(d^2)}{m}$, where m is the number of characters examined. This generalised distance can be split into 'size' (C_Q^2) and 'shape' (C_Z^2) components. 'Size' and 'shape' distances are determined by the following formulae:

$$C_Q^2 = \frac{(\Sigma d)^2}{m^2}; \text{ and } \frac{m-1}{m} C_Z^2 = C_H^2 - C_Q^2$$

RESULTS

Examination of Tables 1 and 2 shows that only for P3 are there sufficient numbers of specimens for analysis. Further, the high coefficients of variation reflect difficulties in measuring small

Tooth	Locality	Length					Breadth				
		N	X	O R	s	V%	N	X	O R	s	V%
I^1	*T. carnifex*										
	NSW	5	17.64	15.0-21.0	2.24	12.7	5	11.74	10.2-13.0	1.13	9.4
	WA	2	14.65	14.4-14.9	0.35	-	2	10.35	10.2-10.5	0.21	-
	All localities	7	16.79	14.4-21.0	2.35	14.0	7	11.34	10.2-13.0	1.15	10.1
I^2	*T. carnifex*										
	NSW	3	8.77	8.5- 9.0	0.25	2.9	3	8.43	8.3- 8.5	0.12	1.4
	Qld	1	10.6	-	-	-	-	-	-	-	-
	All localities	4	9.23	8.5-10.6	0.94	10.2	3	8.43	8.3- 8.5	0.12	1.4
I^3	*T. carnifex*										
	NSW	10	14.76	12.6-17.0	1.20	8.1	10	8.73	7.5-10.6	1.07	12.3
C^1	*T. carnifex*										
	NSW	8	8.58	7.3-10.6	1.05	12.2	7	7.11	6.3- 8.5	0.76	10.7
	Qld	1	6.4	-	-	-	-	-	-	-	-
	All localities	9	8.33	6.4-10.6	1.22	14.7	7	7.11	6.3- 8.5	0.76	10.7
P^1	*T. carnifex*										
	NSW	4	6.05	5.3- 7.5	1.0	16.5	3	5.53	5.0- 6.0	0.50	0.1
P^2	*T. carnifex*										
	NSW	4	7.13	5.3- 8.2	1.33	18.6	3	4.97	4.4- 5.5	0.55	11.1
	Qld	1	6.2	-	-	-	1	5.4	-	-	-
	All localities	5	6.94	5.3- 8.2	1.22	17.6	4	5.08	4.4- 5.5	0.50	9.8
P^3	*T. carnifex*										
	Qld	11	52.49	50.0-55.4	1.93	3.7	6	15.32	14.7-16.0	0.51	3.3
	NSW	38	51.13	45.6-56.4	2.47	4.8	31	15.15	13.6-16.5	0.64	4.2
	Vic	3	54.60	51.3-56.8	2.91	5.3	3	16.17	15.1-17.4	1.16	7.2
	SA	58	51.41	46.5-57.9	2.42	4.7	58	15.21	12.7-17.1	0.95	6.2
	WA	2	42.7	42.1-43.3	0.85	-	2	12.5	12.4-12.6	0.14	-
	Tas	1	40.0	-	-	-	1	14.5	-	-	-
	All localities	113	51.26	40.0-57.9	2.89	5.6	101	15.17	12.4-17.4	0.92	6.1
	T. crassidentatus	1	53.5	-	-	-	-	-	-	-	-
	T. hilli	1	24.4	-	-	-	-	-	-	-	-
M^1	*T. carnifex*										
	NSW	8	8.90	7.3-10.5	1.02	11.5	8	13.95	12.9-15.5	0.88	6.3
	WA	1	6.8	-	-	-	1	11.4	-	-	-
	All localities	9	8.67	6.8-10.5	1.19	13.7	9	13.67	11.4-15.5	1.19	8.7

Table 1. Statistical summary (in mm) of upper teeth of *Thylacoleo* spp. NSW = New South Wales, Qld = Queensland, SA = South Australia, Tas = Tasmania, Vic = Victoria, WA = Western Australia.

Tooth	Locality	Length N	Length $\bar{X}$	Length O R	Length s	Length V%	Breadth N	Breadth $\bar{X}$	Breadth O R	Breadth s	Breadth V%
I_1	*T. carnifex*										
	NSW	8	19.9	18.1-22.8	1.77	8.9	8	12.13	10.7-13.2	0.84	6.9
	WA	2	15.9	15.2-16.6	0.99	-	2	10.0	10.0-10.0	-	-
	All localities	10	19.1	15.2-22.8	2.32	12.2	10	11.7	10.0-13.2	1.17	10.0
P_1	*T. carnifex*										
	NSW	1	11.0	-	-	-	1	7.0	-	-	-
P_2	*T. carnifex*	-	-	-	-	-	-	-	-	-	-
P_3	*T. carnifex*										
	Qld	15	40.18	35.2-42.0	1.53	3.8	9	13.61	11.0-15.5	1.33	9.8
	NSW	62	39.18	34.5-44.4	2.26	5.8	59	13.78	11.3-15.2	0.79	5.7
	Vic	12	40.98	36.3-44.3	2.28	5.6	11	14.69	13.5-16.1	0.72	4.9
	SA	82	39.56	36.3-43.4	1.44	3.7	80	14.04	11.9-16.4	1.03	7.4
	WA	4	33.15	30.4-36.0	2.30	6.9	3	11.50	11.4-11.6	0.10	0.9
	Tas	1	32.50	-	-	-	1	13.70	-	-	-
	All localities	176	39.39	30.4-44.4	2.18	5.5	163	13.92	11.0-16.4	1.01	7.3
	T. crassidentatus	5	37.44	35.9-41.4	2.28	6.1	4	13.86	13.5-14.3	0.39	2.8
	T. hilli	-	18.8 (calculated)	-	-	-	-	-	-	-	-
	Wakaleo										
	W. oldfieldi	1	12.4	-	-	-	1	7.8	-	-	-
	W. vanderleuri	1	13.5-15.5 (range of estimates)	-	-	-	-	-	-	-	-
M_1	*T. carnifex*										
	Qld	9	13.27	10.5-15.1	1.39	10.5	2	9.05	7.8-10.3	1.77	-
	NSW	21	13.20	10.0-15.0	1.03	7.8	14	10.24	9.5-11.5	0.61	6.0
	WA	1	11.50	-	-	-	1	7.90	-	-	-
	Tas	1	11.50	-	-	-	-	-	-	-	-
	All localities	32	13.11	10.0-15.1	1.17	8.9	17	10.12	7.8-11.5	0.80	7.9
	T. crassidentatus	4	14.80	13.3-15.5	1.04	-	4	10.90	10.0-11.5	0.67	-
	Wakaleo										
	W. oldfieldi	2	8.9	7.5-10.4	-	-	2	7.5	6.3- 8.7	-	-
	W. vanderleuri	1	11.7	-	-	-	1	7.9	-	-	-

Table 2. Statistical summary (in mm) of lower teeth of *Thylacoleo* and *Wakaleo* spp. Abbreviations as in Table 1.

and/or irregularly shaped teeth. For P3 the coefficients of variation suggest reasonably homogeneous samples (less than 10% – Simpson, Roe and Lewontin 1960). Unfortunately, only an upper and lower P3 are available from Tasmania and, for Western Australia (P3) and Victoria (P^3), the samples are very small (2–4). However, most other samples are of reasonable size and some (e.g. South Australia and New South Wales) are quite large.

Before attempting to analyse the data it was felt necessary to examine the *T. carnifex* material for sexual dimorphism as this has been demonstrated in some marsupial genera, including *Thylacinus* (Ride 1964), *Perameles* (Freedman and Joffe 1967) and *Macropus* (Bartholomai 1971). For this purpose P3 lengths and breadths were plotted as histograms but only those for length have been included (Fig. 2). The data appear to be normally distributed except that the 'tails' appear long on the lower dimension side due to the small size of the few available specimens from Tasmania and Western Australia. P_3 shows no sign of bimodality. In P^3 there is a 'dip' in the distribution at the 51 mm mark and a smaller 'dip' at 55 mm. Histograms of P3 breadths are very similar to those

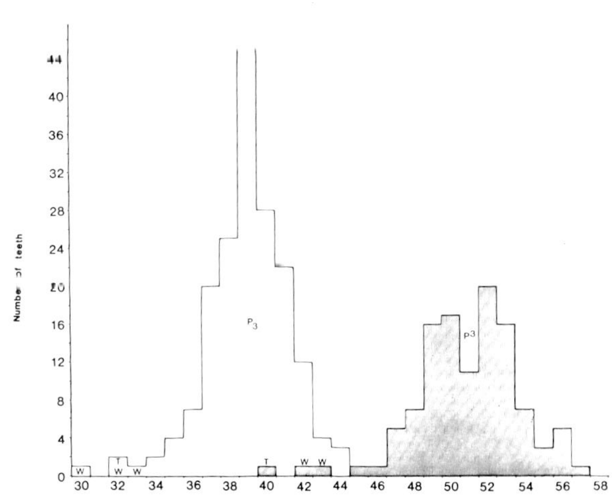

Fig. 2. Histograms of P3 lengths for *T. carnifex*. T = Tasmanian and W = Western Australian specimens.

of length but the 'central dip' is not as marked. From these distributions, particularly in view of the considerably larger size of the P_3 sample, it was felt reasonable to conclude that no obvious metrical sexual dimorphism could be demonstrated for P3.

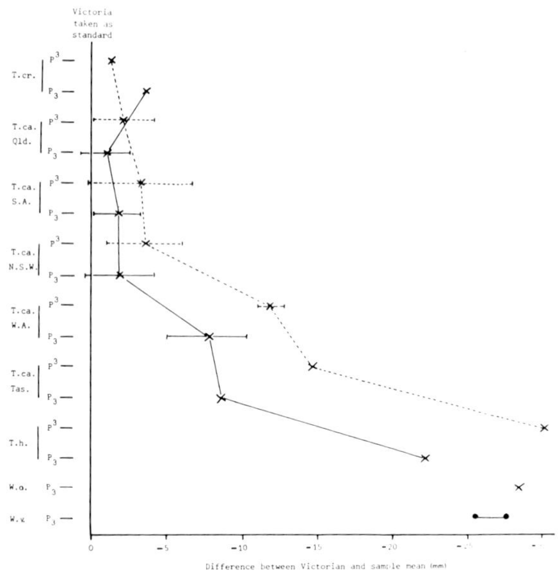

Fig. 3. Ratio diagram of divergences of mean P3 lengths of thylacoleonid samples using the Victorian sample as a standard. The samples compared are listed on the vertical scale; the horizontal scale represents the differences (in mm). T.cr. = *T. crassidentatus*, T.ca. = *T. carnifex*, T.h. = *T. hilli*, W.o. = *Wakaleo oldfieldi*, W.v. = *W. vanderleuri*. (—— P^3, - - - P_3, ●——● range of estimates, |——| plus/minus 1 s.d.).

The divergences of the various *T. carnifex* sample mean lengths of P3 from those of Victoria (the largest) were next examined (Fig. 3). From this analysis it can be seen that the Queensland, South Australian and New South Wales samples are of similar mean lengths. In each case both mean measurements, but especially the P^3 lengths, are slightly less than the Victorian sample mean. The Western Australian and Tasmanian values are similar to each other but less than those of the eastern states samples. Of the other two species of *Thylacoleo,* the *T. hilli* lengths are considerably shorter than any of those of the *T. carnifex* samples. The other species, *T. crassidentatus,* although similar in P^3 and P_3 mean lengths to the eastern states samples, is the only instance in which P_3 mean length differs by more than P^3 length from the Victorian values. The *Wakaleo* P_3 values, inincluded for comparison only, are one third of those for *T. carnifex* and *T. crassidentatus,* and are even shorter than those of *T. hilli.*

Plots of mean length against mean breadth for P^3 (Fig. 4) and P_3 (Fig. 5) reiterate the close metrical similarity of the *T. carnifex* samples from New South Wales, South Australia and Queensland. However, the considerably larger

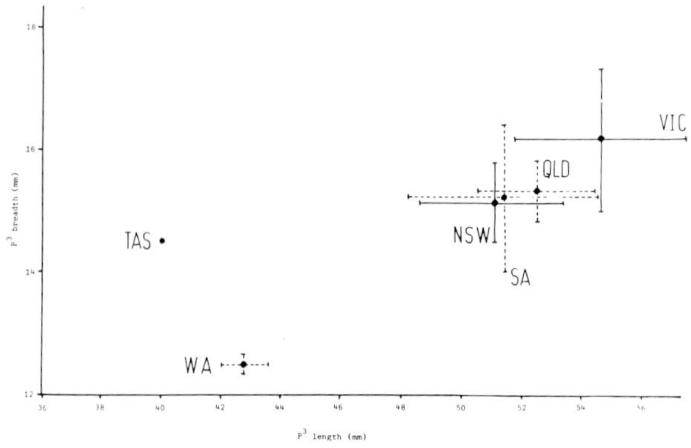

Fig. 4. Plot of mean length against breadth of P^3 for *T. carnifex* samples. Lines represent one standard deviation on either side of mean.

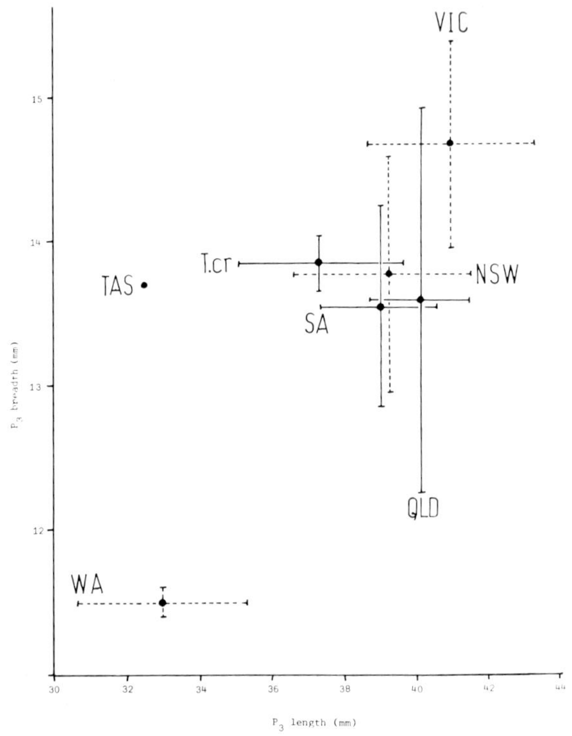

Fig. 5. Plot of mean length against breadth of P_3 for *T. carnifex* and *T. crassidentatus.* Lines represent one standard deviation on either side of mean.

mean breadth of both P^3 and P_3, and the greater mean length of P^3 of the Victorian sample is highlighted. The *T. carnifex* sample from Tasmania is considerably shorter in length than those from other parts of Australia, whilst that from Western Australia is both much shorter and much narrower.

Student's 't' tests were performed on all available pairs of samples for both length and breadth of P3 (Table 3). In the *T. carnifex* tests, the most important results were the significant differences, mostly at the 1% level, between the sample from Western Australia and those from all other parts of continental Australia. A similar result was obtained in the Victorian–New

			N.S.W.	Qld	S.A.	W.A.	Tas	T. crassidentatus
Vic.	P^3	L	+++	-	+	++*	+*	-*
		B	++	-	-	+*	-*	0
	P_3	L	++	-	+++	+++	+++	-
		B	+++	+	+	+++	-	-
N.S.W.	P^3	L		-	-	+++	+++	-
		B		-	-	+++	-	0
	P_3	L		-	-	+++	+++	-
		B		-	-	+++	-	-
Qld	P^3	L			-	+++	-	-
		B			+++	+++	-	0
	P_3	L			-	+++	+++	+++
		B			-	++	-	-
S.A.	P^3	L				+	+++	-
		B				+++	-	0
	P_3	L				+++	+++	+++
		B				+++	-	-
W.A.	P^3	L					0	0
		B					0	0
	P_3	L					-*	-
		B					+++*	+++
Tas.	P^3	L						0
		B						0
	P_3	L						-*
		B						-*

Table 3. Student's 't' tests for P3 length (L) and breadth (B) between samples of *T. carnifex* and *T. crassidentatus*. + = significant at 5% level, ++ = significant at 2% level, +++ = significant at 1% level, — = not significant, O = no measurement available, * = numbers *very* small.

	Vic	N.S.W.	Qld	S.A.	W.A.	Tas	T. crassidentatus
Vic		1.37 (96.5%)	0.79 (91.9%)	1.01 (93.5%)	18.62 (98.2%)	12.20 (74.4%)	1.22 (76.3%)
N.S.W.	1.42		0.08 (44.4%)	0.03 (75.0%)	9.88 (98.3%)	5.38 (58.9%)	0.00 (0.0%)
Qld	0.86	0.18		0.01 (7.1%)	11.75 (95.8%)	6.79 (57.7%)	0.07 (8.3%)
S.A.	1.08	0.04	0.14		10.97 (98.7%)	6.20 (62.8%)	0.02 (2.7%)
W.A.	18.96	10.05	12.27	11.11		0.68 (21.0%)	10.01 (90.5%)
Tas	16.40	9.13	11.76	9.87	3.24		8.28 (61.3%)
T. crassidentatus	1.60	0.64	0.84	0.73	11.06	13.50	

Size Distance

Mean Square Distance

Table 4. Penrose's mean square distance and 'size' distance between samples of *T. carnifex* and *T. crassidentatus*, using P3 length and breadth. Figures in parentheses show 'size' distance as a percentage of mean square distance.

South Wales 't' tests, and 7 of the possible 18 tests between Tasmanian and the other Australian *T. carnifex* samples were significant at the 1% level. However, some of the tests where the samples are small may be biased towards the null hypothesis. Because of the small sample sizes, or lack of available measurements, it was not possible to do a comprehensive set of tests between *T. carnifex* and the other two species. However, for *T. crassidentatus,* only 3 of the 16 possible tests were significant and, for *T. hilli,* of only 3 tests feasible, all were significant.

The results of the Penrose multivariate analysis, using the mean lengths and breadths of P3, are included in Table 4. In the intra–*T. carnifex* comparisons for mean square distance, the Western Australian and Tasmanian samples stand out as being very different to those from the eastern states; they also differ from each other more than from the eastern states samples. Of the eastern states comparisons the morphometrically most divergent sample is Victoria, while South Australia and New South Wales are the most similar. In the *T. crassidentatus* distances, those from Tasmania and Western Australia are again the greatest, but distances to the eastern states samples are small.

Examination of the eastern states 'size' components of the mean square distances (Table 4) reveals that, for the Victorian sample, the 'size' factor is by far the most important. 'Size' is again the most important component for all of the comparisons with Western Australia, except the one with Tasmania, where 'shape' contributes about 79% of the mean square distance. In the Tasmanian morphometric distances 'shape' plays a considerably greater part, although 'size' is still the major contributor. 'Shape' is the dominant factor in the Queensland–South Australia distance. The *T. crassidentatus* morphometric distance is mainly 'size' with Victoria and Western Australia, but almost totally 'shape' with New South Wales, Queensland and South Australia; in the Tasmanian comparison 'size' contributes 61%.

The plot of Penrose's 'size' against 'shape' in Fig. 6, reveals a most interesting set of relationships. Area A indicates great similarity between samples and includes all of the inter–relationships between the eastern states samples and between them and *T. crassidentatus*. Area B is high on the 'shape' axis but also includes some 'size' component; it covers all of the relationships involving Tasmania. Area C is mainly displaced on the 'size' axis and includes all of the Western Australian sample pairs, except that with Tasmania.

DISCUSSION

Three important considerations underlie the use of the carnassial teeth (P3) in these comparisons. Thylacoleonid third premolars are sufficiently unique to ensure that no misidentifications have occurred and their very uniqueness has stimulated collectors to gather all available specimens. The samples are thus relatively large ones. The second reason lies in the apparent non–replacement of this tooth during growth.

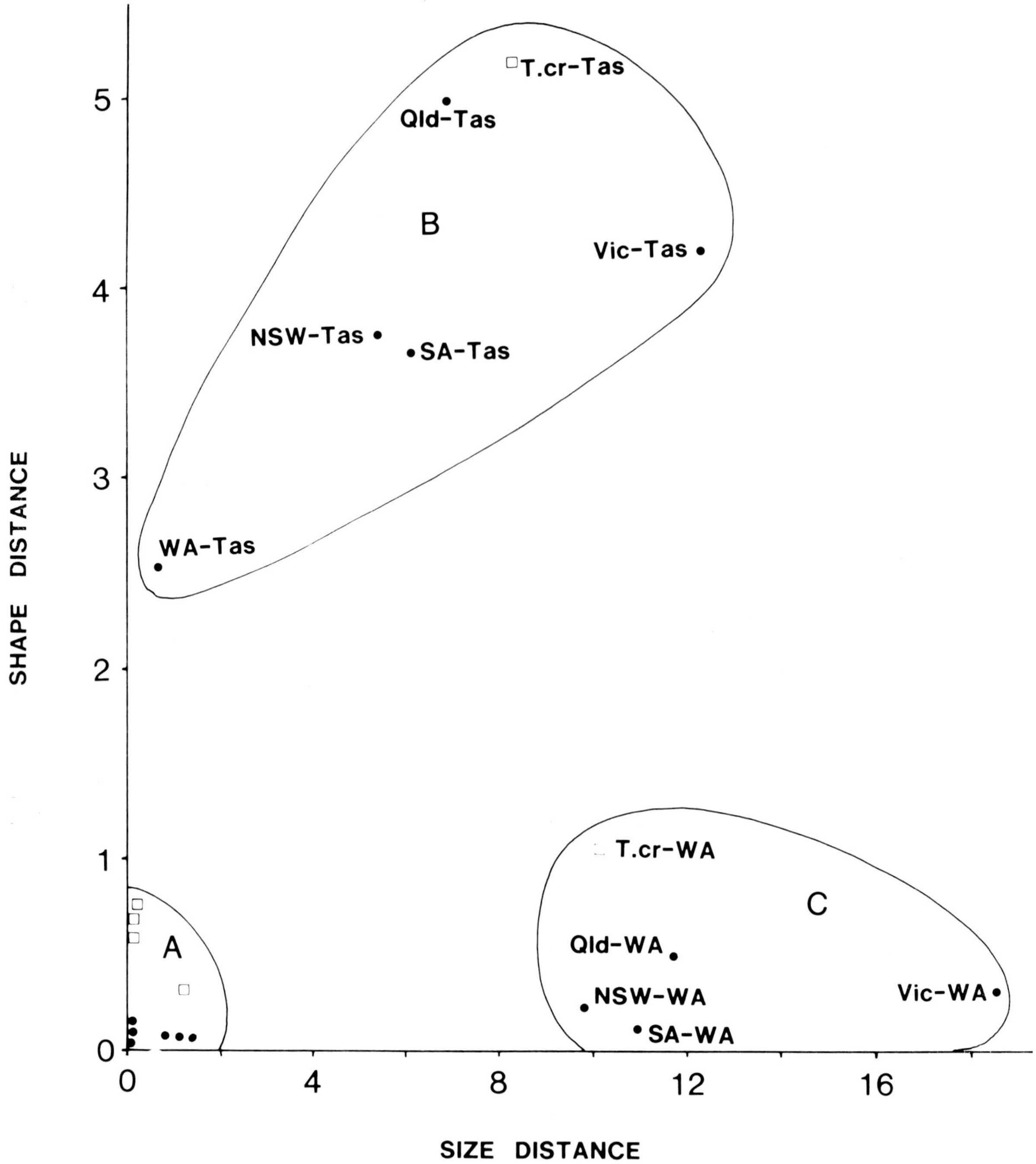

Fig. 6. Plot of Penrose's 'shape' against 'size' for *T. carnifex* and *T. crassidentatus*. Pairs involving *T. crassidentatus* are indicated by □.

Juvenile individuals may therefore be used without fear of biasing the sample by the inclusion of deciduous teeth. Thirdly, an elongated, trenchant tooth with a comparatively simple outline is much easier to measure than a small tooth or one with more complex margins.

The measurements used on P3 were maximum length and maximum breadth. It may be speculated that, in very old animals, the constant abrasion of the P^3 lingual surface against the P_3 buccal surface during mastication would appreciably diminish breadth measurements. However, the P^3 crown is angled lingually and that of P_3 is inclined buccally, considerably reducing the amount of wear on the contiguous side walls of these teeth. (Fig. 1). Thus, the facet which is clearly visible on the lateral wall of P_3 of the figured adult does not extend to the anterior one third of the tooth, nor to the cementoenamel junction which is the widest part of the crown and therefore the site of measurement. As the maximum length measurement is at the base of the crown it too is unaffected by wear.

To place much reliance on tooth size as a criterion for differentiating interspecific or even intraspecific populations might seem to be over–weighting what could merely be relatively minor variations. However, according to Gould (1975) and Gingerich (1979), tooth size shows a high correlation with the size of the animal, and body size is an important feature of an animal's adaptation to its environment (McNab 1971, Van Valen 1973). Gould used the area of the whole postcanine tooth row when demonstrating allometric relationships between tooth and body size as he was dealing with herbivores and omnivores in which all grinding surfaces would have been involved in food treatment. In *Thylacoleo* the carnassials are the main postcanine teeth involved in food processing (Fig. 1), and hence they should be valid indicators of body size. M_1 actually shares masticatory functions with P_3 but so few fossil mandibles with associated P_3 and M_1 are available that it is not feasible to limit anlysis to those few specimens. In this connection we may note that Gingerich (1979) has plotted length against breadth of the lower carnassial tooth (M_1) of Michigan canids and from this one tooth was able to easily distinguish between the wolf and coyote which have the same general morphology but differ in body size.

Neither the histograms (Fig. 2) nor the coefficients of variation (Tables 1 and 2) suggested size sexual dimorphism as a probable major factor in the analysis of P3 in *Thylacoleo.* At though the number of teeth tested for sexual dimorphism was large (P^3,113 and P_3, 176), it is possible that the intersample variation later found may have obscured minor sex size differences. However, histograms of New South Wales P^3 ($N = 38$) and P_3 ($N = 62$) gave no evidence of bimodality for breadth or for P_3 length; P^3 length showed a minor depression at the peak of the curve.

The overall conclusion from the various analyses made on the P3 dimensions is that the variation found between the Queensland, New South Wales and South Australian samples of *T. carnifex* is minor, probably random, and that no obvious overall clinal pattern emerges. The Victorian *T. carnifex* sample mean values are larger than each of those of the other eastern states and, in the case of the 't' test comparisons with New South Wales, all four tests were significant at the 2% level (Table 3). This difference could be of consequence.

Although recent discoveries of Tertiary thylacoleonids have permitted the formulation of a tentative phylogeny of the group (Archer and Bartholomai 1978, Archer and Rich 1982 and Archer and Dawson 1982) any attempt to trace the radiation or migration of the Pleistocene *T. carnifex* within Australia must be purely conjectural. However, if Wellman's (1979) theory of the slow but continuous uplift of the highland chain running parallel to the coast from Victoria to Queensland is correct, the Murray–Darling river drainage area to the north and west of this chain must have been in existence for about 90 million years. Therefore it is not surprising that the marsupial lions found in Queensland, New South Wales and South Australia within this drainage basin appear very similar. This geographical barrier could also account for the larger size of the Victorian animals which lived on the southern slopes of the range.

With regard to the Western Australian *Thylacoleo* specimens, all analyses revealed their P3 values to be considerably smaller than all those of the eastern states samples (Figs. 4 and 5). Of the relevant 't' tests most were significant at the 1% level (Table 3).

The presently described Western Australian thylacoleonids come from a region in the south of the State (Balladonia) from which other fossil marsupials of diminished size have been described e.g. Thylacines (Ride 1964, Lowry 1972). The Balladonia marsupial lions also seem to be from a population of smaller *T. carnifex* than those found in other parts of the continent. During the latter part of the Pleistocene there was increasing aridity over much of the continent with the hot dry conditions extending to the Bight and including Balladonia (Wyrwoll 1979). Under such conditions populations of a particular species may be reduced in size (Bergmann's rule). Lowry (1972) made a detailed statistical comparison of the teeth of the small fossil thylacines from the Eucla Division of Western Australia with those of modern *Thylacinus cynocephalus.* She came to the conclusion, in agreement with that of Ride (1964), that the smaller fossil animals were conspecific with the modern form. However, sexual dimorphism may have been a complicating factor in her analysis. The much smaller size of Western Australian *Thylacoleo* suggests that this population might merit subspecific separation, a suggestion supported by the small size of M1 (Tables 1 and 2). However, until additional and more complete specimens are available, and also information as to their dating, it does not seem appropriate to formally propose a taxonomic distinction at this time.

The Tasmanian *Thylacoleo* material is considered to be from "the end of the Pleistocene, possibly as late as 10,000 years ago" (Murray and Goede, 1977). The specimens were found to differ from the mainland samples both in 'shape' and in 'size' (Table 4, Fig. 6). Although of similar size to the Western Australian *T. carnifex*, they were shown to differ strongly from the sample in 'shape'. The reduced 'size' and different 'shape' of P3, and also the small size of M_1 (Table 2), point to the possibility of at least subspecific distinction. The considerable difference of the Tasmanian *T. carnifex* sample from those of the mainland could reflect the intermittent separation of the 'island' from the mainland during the Pleistocene (Jones 1977). However, as with the Western Australian sample, more dental and skull material is necessary before proposing for the Tasmanian sample separate taxonomic status to the eastern states forms.

Of the two other species of *Thylacoleo* studied, *T. hilli* is only represented by the length dimensions of single upper and lower carnassials, and the latter is a calculated measurement. However, as both of these measurements are less than half of the respective mean values for the *T. carnifex* samples it would seem that its specific separation is validated. The proposed Miocene–Pliocene dating (Pledge 1977) is not unreasonable.

The third species of *Thylacoleo, T. crassidentatus,* is also possibly of Pliocene age (Bartholomai 1962), but the sample of P_3, unlike that of *T. hilli,* is of more reasonable size (Table 2). On overall size the lower carnassial of *T. crassidentatus* is very similar to that of the eastern states samples (Fig. 5) but its shape is different to that of all the other P_3 specimens (Fig. 3). In addition, the relative lengths of the upper and lower carnassials are different to those of both *T. carnifex* and *T. hilli.* The P_3/P^3 ratios as percentages are: *T. carnifex,* 76.7% (range, 75.0% – 77.6% for the mainland samples and 81.2% for Tasmania); *T. hilli,* 77.0%; whilst for *T. crassidentatus* the ratio is only 70.0%. These differences appear to firmly support the specific distinctness of *T. crassidentatus,* despite its apparent agreement with the eastern states samples in the Penrose 'size' – 'shape' plot (Fig. 6).

ACKNOWLEDGEMENTS

We gratefully acknowledge the assistance given by Dr. M. Archer, School of Zoology, University of New South Wales, Mr. N. Pledge, South Australian Museum, and Dr. R. Wells, School of Biological Sciences, Flinders University of South Australia, in obtaining measurements of teeth. Dr. A. Ritchie gave permission for the loan of *Thylacoleo* teeth from the Australian Museum, Sydney and his kindness in allowing one of us (M.E.F.) to retain the Moree marsupial lion material for study has made this work possible. Use of the facilities of the Queensland Museum, Brisbane, was generously permitted by Dr. A. Bartholomai, while the co-operation of Dr. D. Merrilees and Dr. A. Baynes of the Western Australian Museum, Perth, in making available the Balladonia material is much appreciated.

REFERENCES

Archer, M. and Bartholomai, A., 1978. Tertiary mammals of Australia. *Alcheringa* **2:** 1–19.

Archer, M. and Dawson, L., 1982. Revision of marsupial lions of the genus *Thylacoleo* Gervais (Thylacoleonidae, Marsupialia) and thylacoleonid evolution in the late Cainozoic. Pp.477–94 *in* "Carnivorous marsupials", ed. by M. Archer. Roy. Zool. Soc. N.S.W.: Sydney.

Archer, M. and Rich, T. H., 1982. Results of the Ray E. Lemley expeditions. *Wakaleo alcootaensis* n.sp. (Thylacoleonidae, Marsupialia), a new marsupial lion from the Miocene of the Northern Territory, with a consideration of the early radiation of the family. Pp.495–502 *in* "Carnivorous marsupials", ed. by M. Archer Roy. Zool. Soc. N.S.W.: Sydney.

Bartholomai, A., 1962. A new species of *Thylacoleo* and notes on some caudal vertebrae of *Palorchestes azael. Mem. Qd Mus.* **14:** 33–40.

Bartholomai, A., 1971. Morphology and variation of the cheek teeth in *Macropus giganteus* (Shaw) and *Macropus agilis* (Gould). *Mem. Qd. Mus.* **16:** 1–18.

Clemens, W. A. and Plane, M., 1974. Mid-Tertiary Thylacoleonidae (Marsupialia, Mammalia). *J. Paleontol.* **48:** 652–60.

Finch, M. E., 1981. *Thylacoleo,* an Australian fossil phalangerid marsupial. Unpublished Thesis.

Finch, M. E., 1982. The discovery and interpretation of *Thylacoleo carnifex* (Thylacoleonidae, Marsupialia). Pp.537–51 *in "Carnivorous marsupials",* ed. by M. Archer Roy. Zool. Soc. N.S.W.: Sydney.

Freedman, L. and Joffe, A. D., 1967. Skull and tooth variation in the genus *Perameles.* Part 2: Metrical features of *P. nasuta.* Part 3: Metrical features of *P. gunnii* and *P. bougainville. Rec. Austr. Mus.* **27:** 183–95, 197–212.

Gingerich, P. D. 1979. The stratophenetic approach to phylogeny and reconstruction in vertebrate paleontology. Pp.41–77 *in* "Phylogenetic analysis and paleontology", ed. by J. Cracraft and N. Eldredge. Columbia University Press: N.Y.

Gould, S. J., 1975. On the scaling of tooth size in Mammals. *Amer. Zool.* **15:** 351–62.

Jones, R., 1977. Man as an element of a continental fauna: The case of the sundering of the Bassian Bridge. Pp.317–86 *in* "Sunda and Sahul", ed. by J. Allen, J. Goulson and R. Jones. Academic Press.

Lowry, J. W. J., 1972. The taxonomic status of small fossil thylacines (Marsupialia, Thylacinidae) from Western Australia. *J. R. Soc. West. Aust.* **55:** 19–26.

McNab, B. K., 1971. On the ecological significance of Bergmann's rule. Ecology **52:** 845–54.

Murray, P. and Goede, A., 1977. Pleistocene vertebrate remains from a cave near Montagu, N.W. Tasmania *Rec. Q. Vict. Mus.* **60:** 1–29.

Penrose, L. S., 1954. Distance, size and shape. *Ann. Eugen.* **18:** 337–43.

PLEDGE, N. S., 1977. A new species of *Thylacoleo* (Marsupialia: Thylacoleonidae) with notes on the occurrences of Thylacoleonidae in South Australia. *Rec. South Aust. Mus.* **17:** 277–83.

RIDE, W. D. L., 1964. A review of Australian fossil marsupials. *J. R. Soc. West. Aust.* **47:** 97–131.

SIMPSON, G. G., ROE, A. AND LEWONTIN, R., 1960. "Quantitative zoology". Rev. Ed. Harcourt, Brace and Co: New York.

VAN VALEN, L., 1973. Evolution of dental growth and adaptation in mammalian carnivores. *Evolution* **23:** 96–117.

WELLMAN, P., 1979. On the Cainozoic uplift of the southeastern Australian highland. *J. Geol. Soc. Aus.* **26:** 1–9.

WYRWOLL, K.–H., 1979. Late Quaternary climates of Western Australia: evidence and mechanisms. *J. R. Soc. West. Aust.* **62:** 129–42.

RESUMEN

46. Un estudio odontométrico de *Thylacoleo* (Thylacoleonidae, Marsupialia). Por M. E. Finch y L. Freedman.

Por el uso primario de las dimensiones del P3 se compararon especímenes de *Thylacoleo* de varias partes de Australia, determinándose sus interrelaciones por análisis uni y multivariados. *T. hilli* y *T. crassidentatus* son distinguibles entre sí y de *T. carnifex*, el primero por su tamaño mucho menor, el último por la relación del tamaño del P^3 respecto a P_3. Dentro de *T. carnifex*, las dimensiones del P3 de lot de Victoria son las mayores de las muestras de los estados del Este, los que — sin embargo — no varían grandemente. Las muestras de Western Australia y de Tasmania difieren fundamentalmente de aquellas de los estados del Este en los criterios del "tamano" y "forma" de Penrose, respectivamente. Estos resultados están de acuerdo con la separación especifica de *T. carnifex, T. hilli* y *T. crassidentatus*. Las poblaciones de Western Australia y, más específicamente, las de Tasmania de *T. carnifex* podrían merecer reconocimiento taxonómico, pero se requiere confirmación por otros caracteres y por más material.

47. DASYURIDS OF THE NORTHERN FLINDERS RANGES BEFORE PASTORAL DEVELOPMENT

M. J. SMITH[1] and G. C. MEDLIN[2]

Bones of small mammals were collected from deposits in caves in Chambers Gorge, Big Moro Gorge and Top John Gorge in the northern Flinders Ranges, South Australia. These deposits had formed by breakdown of owl pellets, intact pellets being present on the surface. The low abundance of *Mus musculus* suggested that little of the bone accumulated after European settlement of the area. Eight dasyurid species were identified. *Sminthopsis crassicaudata, S. macroura* and *Planigale tenuirostris* occur near the area now. Range extensions are noted for *Antechinomys laniger, Phascogale calura, Dasycercus cristicauda* and *Dasyuroides byrnei*. *Dasyurus* sp. *cf. D. geoffroii* was found also.

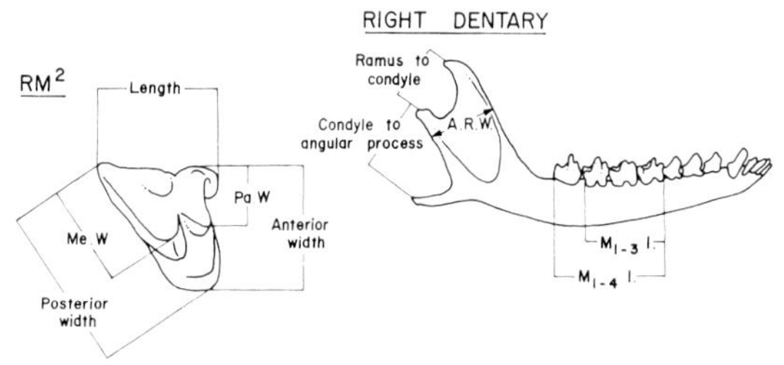

Pages 563–72 *in* CARNIVOROUS MARSUPIALS, ed. by M. Archer, February, 1982. Roy. Zool. Soc. New South Wales: Sydney, Australia.

THE NORTHERN Flinders Ranges and the saltbush plains to the west and east were stocked with sheep and cattle in the 1850's. The early pastoralists were required to run on their land not less than 100 sheep or 16 cattle to the square mile (0.386 sheep or 0.06 cattle to the hectare) to retain their leases. Such a rate of stocking is at least double that which the land has since been found to sustain. Severe overstocking soon degraded the vegetation and areas within a day's travel of the few water sources were almost bared. The serious situation was exacerbated by unrelieved drought from 1863 to 1866, and was rendered irreversible by the rabbit invasion in the 1880's. The utilization and destruction of the saltbush–bluebush vegetation had happened before it was considered of interest to survey the flora or fauna (Mincham 1971).

It is fortunate, then, that Barn Owls, *Tyto alba,* roosted in caves and gorges of the ranges and there, regurgitating the bones of their prey, left a record of the vertebrate fauna. A preliminary report of the species represented by bones in the caves of Chambers Gorge listed 16 native mammals, many of them previously unrecorded from the region (Smith 1977). Since that report, many more species have been identified, including five dasyurids additional to the three found previously.

MATERIALS AND METHODS

The northern Flinders Ranges are drained to the east by a series of creeks that flow onto the Frome plains. Some of these creeks have cut deep gorges into the Proterozoic Pound Quartzite, three such gorges being Chambers, Big Moro and Top John, the gorge of Big John Creek (Fig. 1). The walls of the gorges are marked with fissures, crevices and shallow caves. Although the creeks seldom flow, some permanent pools remain.

The climate is warm and arid. Rainfall records from stations near the three gorges show that, on average, rain falls on less than 33 days a

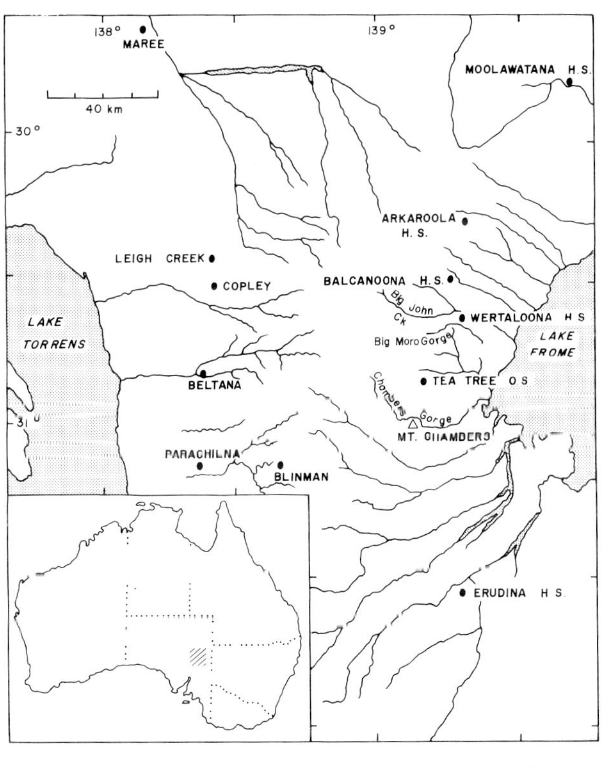

Fig. 1. Location in Australia of the northern Flinders Ranges (hatched) and location within that area of some places referred to in the text.

[1] Department of Zoology, University of Adelaide, Box 498 G.P.O., Adelaide, South Australia, Australia, 5001.
[2] Mawson High School, Colton Avenue, Hove, South Australia, Australia, 5048.

year (Table 1). More rain falls in summer than in winter. Summer temperatures are high. At Balcanoona the average maximum January temperature for 12 years (1945–56) was 33.6°C and the extreme maximum 43.3°C. At Leigh Creek, with a 27–year temperature record the average maximum January temperature (1952–1978) was 35.1°C and the extreme maximum 47.9°C. Winters are mild, the average minimum July temperature being 5.3°C at Balcanoona and 4.7°C at Leigh Creek. The extreme minimum was –2.5°C at both recording stations. Frost occurred at Leigh Creek on an average of 13.5 days a year (data from Bureau of Meteorology, Adelaide).

Chambers, 433 m high, the distance from D cave to the Frome plains is less than 3 km. Big Moro Gorge and Top John Gorge have been explored less thoroughly and collections made from only one cave in each. Neither cave is registered. They are known to us as N and M caves, respectively. The distance from these caves to the plains is even less than from D cave of Chambers Gorge.

Material was collected in bulk mainly from the surface. In D cave, material was excavated to a depth of 30 cm. No stratification was seen.

Taxonomy follows Ride (1970) with modifications from Kirsch and Calaby (1977). The

Table 1. Mean and median monthly rainfalls (mm) and mean rain days for three recording stations on the eastern border of the northern Flinders Ranges. Data from Bureau of Meteorology, Adelaide.

Recording Station		Number of years	Jan	Feb	Mar	Apr	May	June	July	Aug	Sept	Oct	Nov	Dec	Annual
Balcanoona	Mean	26	36	18	37	26	19	11	10	14	9	16	16	15	227
	Median	26	16	11	7	10	3	2	9	9	4	3	3	10	158
	Rain days	26	4	2	2	3	3	1	3	3	3	3	2	3	32
Arrowie	Mean	70	16	14	13	9	13	14	7	7	10	9	12	16	140
	Median	69	2	4	3	0	4	9	1	4	4	6	6	4	127
	Rain days	53	1	1	1	1	1	1	1	1	1	1	1	1	12
Wertaloona	Mean	65	20	21	19	11	13	12	8	7	10	12	10	13	156
	Median	62	6	9	1	2	5	3	2	3	1	8	4	5	133
	Rain days	17	2	2	2	2	2	1	2	2	2	2	2	2	23

The vegetation near the gorges varies from river red gum, *Eucalyptus camaldulensis,* association along the watercourses, to wattle associations on the hills. As well as *E. camaldulensis,* the watercourse vegetation includes *Melaleuca glomerata, M. linophylla* and *Acacia salicina.* On the hills, *A. rivalis* and *A. aneura* occur together with *Eremophila* spp. and *Cassia* spp. Stands of *Casuarina cristata* grow on the shallow, calcareous soils of the Mt Chambers Range fault block system. On the adjacent plains the vegetation is saltbush, *Atriplex* spp., bluebush, *Maireana* spp., with patches of mulga and casuarina. The herb–level plants are *Bassia* sp., grasses and many ephemerals.

Many crevices and caves in the eastern 3 or 4 km of Chambers Gorge contain deposits of bone on the floor or on ledges, and samples from them have been examined. The deep accumulations were all found in crevices in limestone or associated rock types. The largest and most diverse deposit is in a fissure in the cliff face bounding the southern side of a tributary of Mt Chambers Creek. It is known to us as D cave and is registered as 5F38. In it material deposited in an upper section has fallen down a crack into a lower section, and some has spilled out of the cave onto the rock below. Although the gorge is cut through hills and passes by Mt

tooth nomenclature of Archer (1978) has been noted and nomenclature used here is that of Archer (1976a). Measurements (Fig.2) were made to 0.01 mm with needlepoint, dial reading calipers. Representative specimens are lodged in the South Australian Museum (S.A.M.).

RESULTS

Each deposit consisted of an accumulation of bone, fur, disintergrating owl pellets and some intact pellets. Although a few nearly complete skulls were recovered, mostly from D cave, nearly all the material was fragmentary.

Marsupials constituted a small proportion of the terrestrial mammals (Table 2). Three species of bat (Chiroptera) were present in low abundance, and bird and reptile bones were found. Amongst the marsupials dasyurids were more abundant than bandicoots, phalangerids or macropodids. The native rodents comprised thirteen species, including two species of *Leporillus*, five species of *Notomys*, five species of *Pseudomys* and *Rattus villosissimus*.

Introduced mammals were represented by *Mus musculus* and by one fragment of Rabbit. Although *M. musculus* accounted for only 1.4% of the individuals recovered from the loose ma-

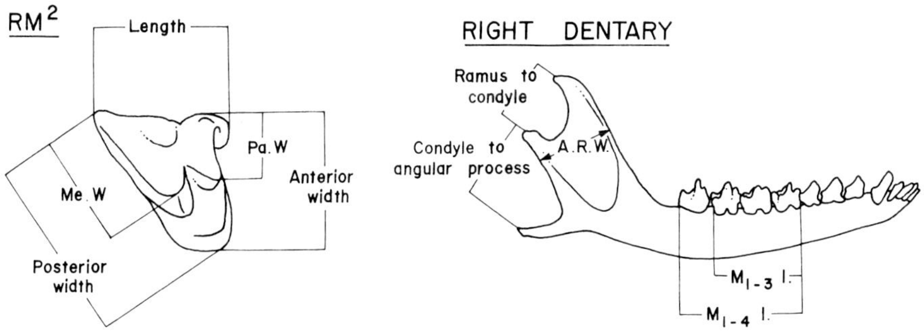

Fig. 2. Method of measuring teeth and dentaries. A.R.W., width of ascending ramus; Me.W., width from metacone to metastylar corner of tooth; Pa.W., width from paracone to parastylar corner of tooth.

terial, it was commonly found when whole pellets were broken up. In whole pellets, *M. musculus* made up 26.2% of the 168 individuals found and was second in abundance to *Pseudomys hermannsburgensis* with 32.1%. *M. musculus* was found together with *A. laniger* and *S. macroura* in pellets but has not yet been found with any rare or extinct species.

were always associated with short, wide molar teeth (Fig. 3). The length of M^{1-3} (Table 3) was similar to that of specimens measured by Archer (1977) and to specimens in the South Australian Museum (Table 3). The lower molar row and the dentary shape conformed with Archer's (1977) description and differed from *Sminthopsis* species in the deposit (Table 3).

Table 2. Composition of the mammalian fauna in caves in three Flinders Ranges gorges. Specimens recovered from intact owl pellets are not included.

	Chambers Gorge D Cave, upper level		Big Moro Gorge		Top John Gorge	
	Minimum number	%	Minimum number	%	Minimum number	%
Native rodents	2892	93.2	424	93.0	33	91.7
Mus musculus	44	1.4	1	0.2	0	—
Bandicoots	86	2.7	6	1.3	0	—
Macropodids and phalangerids	5	0.2	1	0.2	2	5.6
Dasyurids	110	3.5	24	5.3	1	2.8
Total	3137		456		36	

Antechinomys laniger (Gould, 1956)

This species was listed previously (Smith 1977) as *A. spenceri. A. spenceri* is now considered to be a junior synonym of *A. laniger* (Archer 1977).

Complete dentaries are readily identified by the narrow, parallel–sided ascending ramus and masseteric fossa, high condyle and fine, curved angular process. The tip of the angular process is much farther removed from the condyle than is the tip of the coronoid process. The talonids of M_1 –M_3 have only one lingual cuspid, perhaps an entoconid, which is connected to the lingual end of the hypocristid. The upper molars are relatively wide and the interdental fenestrae are large and numerous (Archer 1977).

Antechinomys laniger maxillae from D cave were characterised by the large fenestrae in the intermolar palate, and by a distinct indentation of the ectoloph of M^1 –M^3 between stylar cusp D and the metastylar corner. These characters

Antechinomys laniger was relatively abundant in Chambers Gorge D cave, e.g. P 22882, but occurred less frequently in Big Moro Gorge and was not found in Top John Gorge (Table 4).

The *spenceri* form of *A. laniger* was abundant in central Australia when Finlayson made his surveys in 1932–35 and 1953–56. In the Everard and Musgrave Ranges he found it more plentiful than any of the *Sminthopsis* species. He recorded it also from Oodnadatta and Charlotte Waters and from localities farther north (Finlayson 1961) but he did not find it in the Lake Eyre basin. Specimens in the South Australian Museum were taken at Oodnadatta and as far south as Cootra East (Fig. 4). It has a wide range in Western Australia (Ride 1970). The *laniger* form of *A. laniger* occurred in the far northwest of Victoria in 1857 (Wakefield 1966) and its range extends into central and western New South Wales and Queensland (Archer 1977).

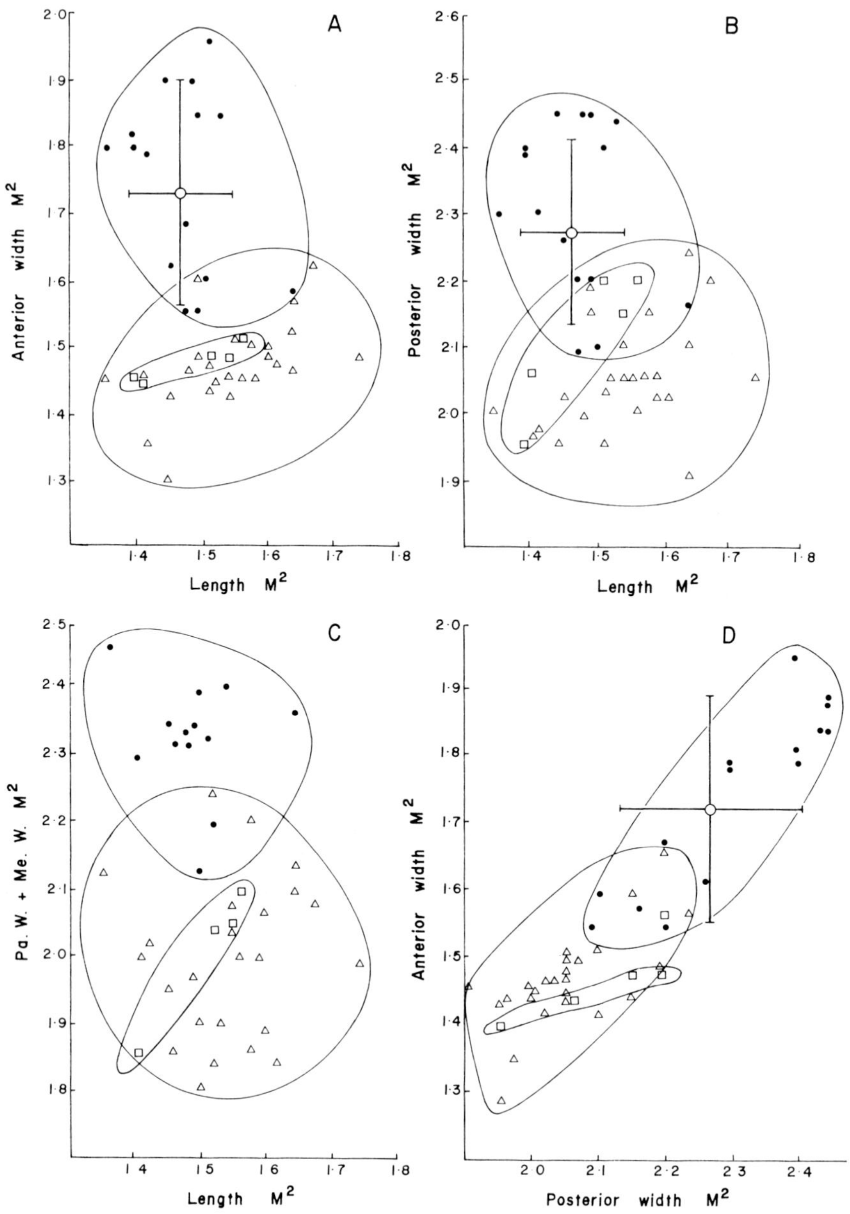

Fig. 3. Bivariate graphs showing the differences in proportions of M^2 of specimens assigned to *Antechinomys laniger*, and those assigned to *Sminthopsis* spp. △ *S. macroura*, □ *S. crassicaudata*, ● *A. laniger*, all from D cave, Chambers Gorge (see text for criteria for identification). ○ Mean of ten modern specimens of *A. laniger* of *spenceri* form in the S.A.M. Bars extend one standard deviation on either side of mean. Boundaries of clusters of points are drawn by eye and have no mathematical significance.

Table 3. Measurements in mm of *Antechinomys laniger* and *Sminthopsis macroura* from D cave, compared with 10 modern specimens of each species.

	Antechinomys laniger				*Sminthopsis macroura*			
	Cave deposit			Modern	Cave deposit			Modern
	n	Range	Mean s.e.	Mean s.e.	n	Range	Mean s.e.	Mean s.e.
M^{1-3} length, alveolar	12	4.30–4.79	4.55±.05	4.88±.08	12	4.35–4.82	4.60±.05	4.59±.04
P^4 width	4	0.49–0.60	0.56±.02	0.57±.02	10	0.45–0.52	0.49±.01	0.51±.02
M_{1-3} length, alveolar	33	4.06–4.77	4.41±.03	4.63±.09	31	4.06–4.51	4.33±.02	4.25±.05
M_{1-4} length, alveolar	31	5.60–6.35	5.97±.04	6.2 ±.11	31	5.43–6.10	5.78±.03	5.67±.06
Ascending ramus width	30	3.30–4.20	3.70±.05	3.42±.08	27	3.78–5.00	4.42±.06	4.26±.09
Ramus to condyle	26	2.39–3.74	2.90±.06	3.03±.12	22	3.15–3.90	3.48±.04	3.83±.09
Condyle to angular process	25	4.06–6.25	5.50±.11	5.46±.27	15	3.60–4.92	4.40±.09	4.37±.11
A.R.W./Condyle to ang. proc.	24	0.60–0.85	0.68±.01	0.64±.04	15	0.91–1.10	1.01±.02	0.98±.02

Table 4. Relative abundance of dasyurid species in the three deposits.

	D cave, Chambers Gorge		Big Moro Gorge		Top John Gorge	
	Minimum number of individuals	% of dasyurids	Minimum number of individuals	% of dasyurids	Minimum number of individuals	% of dasyurids
Antechinomys laniger	32	29.1	5	20.8	0	—
Sminthopsis macroura	36	32.7	13	54.2	1	100
Sminthopsis crassicaudata	2	1.8	0	—	0	—
Planigale tenuirostris	3	2.7	0	—	0	—
Phascogale calura	3	2.7	1	4.2	0	—
Dasycercus cristicauda	28	25.5	3	12.5	0	—
Dasyuroides byrnei	4	3.6	0	—	0	—
Dasyurus geoffroii	2	1.8	2	8.3	0	—

Chambers Gorge is about 270 km southeast of Stuarts Creek, the nearest known locality, and so it is surprising that *A. laniger* is the second most abundant dasyurid in the cave bone accumulation. Similarly, *A. laniger* was the most abundant dasyurid in a deposit from Madura Cave on the edge of the Hampton Tableland, Western Australia (Lundelius and Turnbull 1975), although it has not been recorded alive in that area (Ride 1970).

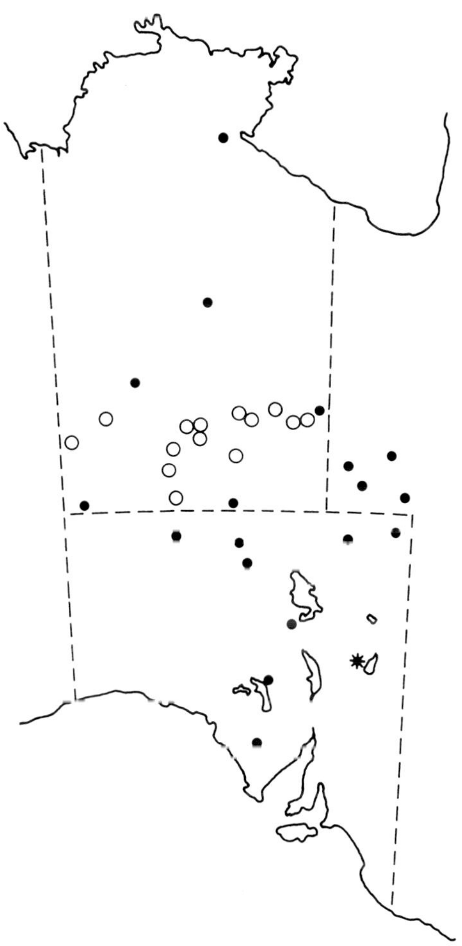

Fig. 4. Distribution of *Antechinomys laniger* in South Australia, the Northern Territory and part of Queensland. • locality records of specimens in S.A.M. o additional records from Parker (1973). * location of Chambers Gorge.

The *spenceri* form of *A. laniger* lives in various habitats including sandhills, mulga and gidgee woodland, Mitchell grass plain, gibbers and sparsely vegetated gravelly plain (Archer 1977, Parker 1973).

Sminthopsis macroura (Gould, 1845)

Maxillae of *Sminthopsis* species can be distinguished from *Antechinomys* by their narrower teeth, from *Planigale* species by the absence of posterior cingula on M^{1-3} (Archer 1976b) and from *Ningaui* species by the stronger development of paracones on M^{1-3} (Archer 1975). Dentaries of *Sminthopsis* species differ from those of *Antechinomys* in having a broad masseteric fossa with the anterior and posterior borders diverging, so that the distance from tip of ascending ramus to condyle nearly equals the distance from condyle to tip of angular process (Lundelius and Turnbull 1975). In *Planigale* species P_4 is reduced or absent whereas it is well developed in *Sminthopsis* species. Most *Sminthopsis* species have larger dentaries and longer M_{1-4} than *Ningaui* species, *S ooldea* being an exception.

In D cave material, maxillae of *Sminthopsis* species differed from *Antechinomys laniger* in having narrower teeth, stylar cusp D less well-developed and a straight ectoloph border between stylar cusp D and metastylar corner of M^{1-3}. Most of these maxillae had the interdental palate finely pitted as in South Australian specimens of *S. macroura* and were considered to be conspecific with *S. macroura.* Eight maxillae differed in having small interdental fenestrae and these were considered to be possibly *S. crassicaudata* (see below).

Sminthopsis macroura is one of the group of *Sminthopsis* species having well-developed, high entoconids on M_{1-3}. Within this group it is smaller than *S. virginiae* and differs from *S. crassicaudata* which is the only species in which the hypocristid lingual end contacts the entoconid base (Archer 1976a). Dentaries from D cave that were ascribed to *S. macroura* were relatively robust and comparable in size and shape to modern specimens (Table 3). The M_{1-3}

all showed a tall, sharp, columnar entoconid. The hypocristid did not contact the entoconid and was separated from it by a valley.

The species, represented by P 22883 from D cave, Chambers Gorge, was the most abundant dasyurid in the cave deposits. Specimen P 191-18a, considered by Smith (1977) to be *S. crassicaudata,* is *S. macroura.*

There are several records of *S. macroura.* from the plains between the Flinders Ranges and Lake Frome, and from south of Lake Frome (Fig. 5). A skull of *S. macroura* was found in a recently deposited owl pellet collected in 1975 near Erudina homestead about 60 km south of Chambers Gorge (Morton and Martin 1979).

At Fowlers Gap in western New South Wales, *S. macroura* occurred mainly in denser vegetation along the margins of the creeks and *S. crassicaudata* inhabited the open alluvial plains (Morton and Martin 1979). In northern South Australia similar habitat preference was observed, with *S. macroura* being found amongst the herbs and grass tussocks and *S. crassicaudata* on the open stony ground (Aitken, pers. comm.).

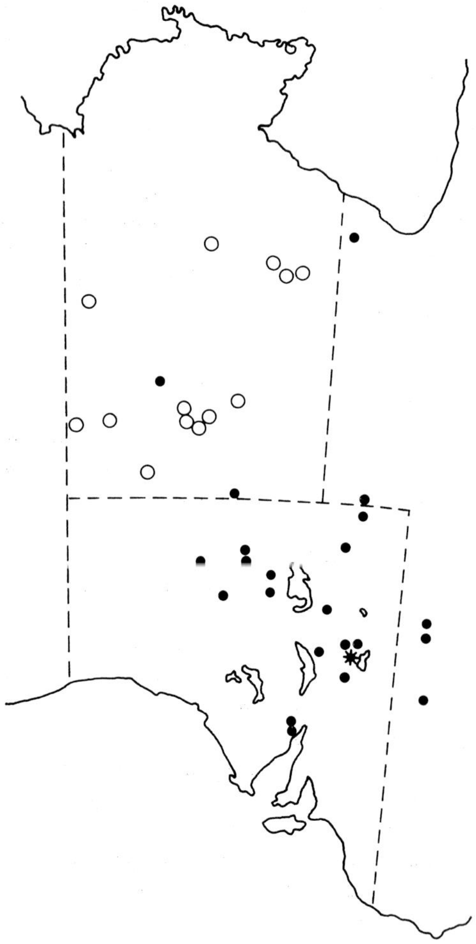

Fig. 5. Distribution of *Sminthopsis macroura* in South Australia, the Northern Territory and parts of N.S.W. and Queensland. Symbols as for Fig. 4.

Sminthopsis crassicaudata (Gould, 1844)

Dentaries of *S. crassicaudata* can be distinguished from other species of *Sminthopsis* by the presence on M_{1-3} of a well–developed entoconid, the base of which is contacted by the lingual end of the hypocristid (Archer 1976a). No such clearcut character is available for distinguishing upper molars. In all *S. macroura* in the South Australian Museum that were collected south of Lake Eyre and west of Lake Torrens the interdental palate is finely pitted whereas in *S. crassicaudata* the interdental palate is often fenestrated. Hence *Sminthopsis* specimens with fenestrae between the upper molars were tentatively referred to *S. crassicaudata*

Teeth in the tentatively assigned maxillae fell within the range of variation of *S. macroura* (Fig. 3). In dentaries, the length M_{1-4} was 5.42 and 6.34 mm in two specimens and the length M_{1-3} was 4.04 and 4.91.

Sminthopsis crassicaudata (e.g. P 22884 from D cave) was rare in the cave deposits, being represented by three dentaries (1 L, 2R), and perhaps by 8 maxillae (7 left, 1 right).

This species is widespread in South Australia and specimens in the South Australian Museum attest to its presence on the Frome Plains. Both *S. crassicaudata* and *S. macroura* were found in recently deposited barn owl pellets from several localities in arid Australia in 1975 (Morton and Martin 1979).

Planigale tenuirostris Troughton, 1928

Maxillae of *Planigale* species are characterized by small size, reduced paracone and metacone on M^{1-3}, and posterior cingula present on M^{1-3}. Dentaries are characterized by the single–rooted or absent P_4 and relatively short talonids on the first three molars. *Ningaui* species have P_4 two–rooted. *Planigale tenuirostris* has three premolars above and below and in this respect differs from *P. gilesi* with only 2 upper and 2 lower premolars on each side. *Planigale tenuirostris* is smaller than *P. maculata* but similar in size to *P. ingrami* from which it can be distinguished by differences in some cranial ratios (Archer 1976b).

Dentaries found in D cave were similar in size to *P. tenuirostris,* with M_{1-4} 4.11, 4.42 and 3.88 mm. They were assigned to *P. tenuirostris* rather than *P. ingrami* because *P. ingrami* has been found only in northern Australia (Archer 1976b).

The three dentaries (1 left, 2 right, e.g. P 22-885) from D cave were the only specimens found.

The distribution of *P. tenuirostris* is in northern and northwestern New South Wales and southern Queensland (Archer 1976b). On the southwestern fringe of this distribution, the first specimen recorded from South Australia was a mummified body at the bottom of a disused stone tank at Mulga Creek Well near the north-eastern tip of the Flinders Ranges (Aitken 1971). More recently several specimens have been trapped near Moolawatana homestead about 130 km north of Chambers Gorge (S.A.M. records).

At Fowlers Gap in western New South Wales *P. tenuirostris* lived on the open alluvial plains as did *S. crassicaudata* (Morton and Martin 1979).

Phascogale calura Gould, 1844

In species of *Phascogale* as in those of *Antechinus* the upper molars are narrower relative to length than in *Sminthopsis* species. In lower molars the metacristids extend posterolingually from the protoconid whereas in *Sminthopsis* species the metacristids are transverse (Archer 1976a). The P_4 is reduced relative to P_3 in *Phascogale* although P^4 is not reduced relative to P^3. This condition occurs also in *A. flavipes* but not in *A. apicalis* and *A. macdonnellensis* where P^4 also is reduced or in *A. rosamondae* where $P^4/_4$ are absent (Archer 1976a).

Phascogale tapoatafa is larger than *P. calura*, the former having the M^{1-3} length about 9 mm compared with about 6 mm for the latter. Fragments of *P. calura* skulls and dentaries could be confused with specimens of *A. macdonnellensis* having $P^4/_4$ relatively well developed. They can be distinguished by the prominent labial cingulum on upper and lower premolars of *P. calura*, a labial cingulum being barely discernible in *A. macdonnellensis* premolars. Most specimens of *A. macdonnellensis* lack stylar cusp B on M^1, a condition typical of dasyurids with P^4 strongly reduced or absent (Archer 1976a). Even in specimens such as Western Australian Museum M14740 with P^4 higher than P^3 and stylar cusp B present on M^1, it is a very small cusp, whereas in *P. calura* it is well developed. In the dentaries, M_1 of *P. calura* has a large cuspid internal to the protoconid whereas in *A. macdonnellensis* a cuspid in this position is very small.

Fragments of *P. calura* from the cave deposits conformed with recent specimens from Western Australia (Table 5) and were larger than 6 specimens of *A. macdonnellensis*. The species was rare, being represented by one right maxilla and three left dentaries from D cave and one right maxilla (P 22886) from Big Moro Gorge.

Many specimens of *P. calura* have been collected from southwest Australia, where the species still occurs (Kitchener and Chapman 1977). One specimen is known from northern Western Australia, viz. SAM M1511 from Well No. 44 on the Canning Stock Route (21°01'S, 126°07'E) taken in 1931 (Fig. 6). *Phascogale calura* was collected at Alice Springs, Barrow Creek and Tennant Creek at the end of last century (Parker 1973) in the Macdonnell Ranges in 1900 (SAM M3823) and at Gol Gol Creek near Mildura in 1856 or 1857 (Wakefield 1966). Bones of this species have been found in several caves in Western Australia (Archer 1974, Kendrick and Porter 1973, Lundelius 1963, Baynes, pers. comm.).

In Western Australia specimens of *P. calura* were trapped in several different plant associations, all having trees or tall shrubs (Kitchener and Chapman 1977).

Dasycercus cristicauda (Krefft, 1867)

The upper molars are relatively narrow and P^4 is single-rooted and minute. The dentary is robust, with only two premolars. The M_1 anterior cingulum is short, and the M_1 metaconid is separate from the protoconid. *Dasyuroides byrnei* has a similar dentition but can be distinguished by its large size, much better developed premolar cingula and M_1 morphology.

The cave deposit specimens resembled modern specimens in structure and size (Table 6). The P^4 was a tiny spicule located lingual to the anteroexternal corner of M^1.

The species was found frequently in Chambers Gorge and Big Moro Gorge. It was represented not only by maxillae and dentaries (e.g. P 22887 from D cave) but also by nearly complete skulls from D cave.

Its distribution is from central Northern Territory south through western South Australia to

Table 5 Dimensions of *Phascogale calura* fragments from Chambers Gorge and Big Moro Gorge, compared with seven modern specimens, for which range is given, followed by mean ± standard error.

	Cave deposit specimens	Modern specimens	
P^3 width	0.67	0.55–0.66	0.60 ± .02
M^{1-3} length, alveolar	5.60, 5.79	5.56–6.05	5.88 ± .07
M^2 length, crown	2.06, 1.90	1.92–2.16	2.03 ± .03
M^2 anterior width	2.01, 2.06	1.82–2.09	1.96 ± .03
M^2 posterior width	2.57, 2.71	2.39–2.58	2.50 ± .03
P_3 width	0.68	0.62–0.80	0.71 ± .03
M_{1-3} length, alveolar	5.24, 5.50, 5.77	5.37–5.82	5.60 ± .06
M_{1-4} length, alveolar	7.01, 7.21, 7.48	7.14–7.60	7.37 ± .07

Western Australia. It has also been found in southwestern Queensland and in northeastern South Australia between Lake Eyre and the Queensland border at Mulka and Lake Eyre South. Chambers Gorge is about 250 km southeast of Lake Eyre South.

Dasycercus cristicauda lives in sand-ridge areas (Parker 1973).

The known distribution is around the junction of South Australia, Queensland and the Northern Territory, between Charlotte Waters, Killalpaninna, Birdsville and Coorabulka. Chambers Gorge is about 400 km south of Killalpaninna.

In its present distribution *D. byrnei* appears to be restricted to gibber plain (Watts and Aslin 1974).

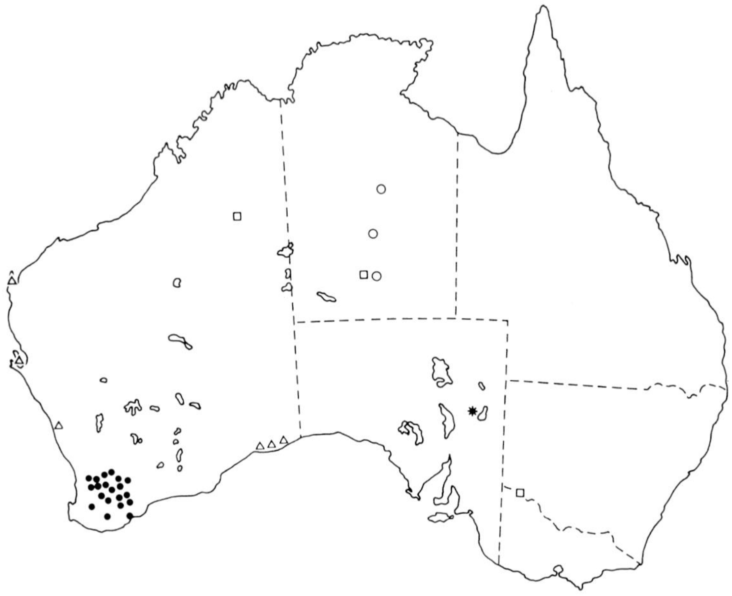

Fig. 6. Distribution of *Phascogale calura*. ● locality records of specimens in Western Australian Museum. O records from Parker (1973), □ additional locality records of whole specimens (see text). △ records from bones in cave deposits. * location of Chambers Gorge.

Dasyuroides byrnei Spencer, 1896

The upper fourth molar is very small and the lower is absent. The teeth are larger than in *D. cristicauda*. Buccal and lingual cingula on $P^3/_3$ are complete in *D. byrnei* but are scarcely distinguishable on *D. cristicauda* (Archer 1976a). The M_1 anterior cingulum is well–developed and the M_1 metaconid small relative to the protoconid.

In the two cave deposit specimens with P^4 preserved, the tiny tooth was located in the buccal line of the tooth–row.

Although this species (e.g. P 22888) was rare, one of the specimens representing it was a nearly complete skull from D cave.

Dasyurus sp., *cf. D. geoffroii* Gould, 1841

No maxillae of *Dasyurus* were found in the cave deposits. The dentaries (e.g. P 22889 from Big Moro Gorge) found were similar in size to those of both *D. viverrinus* and *D. geoffroii* and their assignment to the latter species was on geographical location.

The distribution of *D. geoffroii* in South Australia may have covered most of the State although there are few specimens to attest to this. Finlayson (1961) believed that it was once distributed over a large area of Central Australia. *Dasyurus viverrinus* occurred mainly in the southeast of South Australia, extending into the Mt Lofty Ranges as far north as Quorn where a specimen was taken in 1909 (SAM M3852).

Table 6. Dimensions of *Dasycercus cristicauda* from Chambers Gorge, compared with ten modern specimens from South Australia and southern Northern Territory. Range is given, followed by mean and standard error, and for cave specimens the number of specimens in parentheses.

	Chambers Gorge specimens		Modern specimens	
M^{1-3} length, alveolar	6.93–8.17,	7.47 ± .07 (19)	7.57–8.26,	7.98 ± .08
M^2 length, crown	2.44–2.91,	2.67 ± .04 (18)	2.61–2.89,	2.82 ± .03
M^2 anterior width	2.22–2.76,	2.45 ± .03 (18)	2.41–2.77,	2.60 ± .04
M^2 posterior width	3.05–3.78,	3.29 ± .04 (18)	3.03–3.72,	3.43 ± .06
M_{1-3} length, alveolar	6.66–7.95,	7.21 ± .05 (29)	7.31–8.09,	7.77 ± .09
M_{1-4} length, alveolar	9.12–10.48,	9.62 ± 0.8 (22)	10.00–10.92,	10.48 ± .11

DISCUSSION

The age of the deposits is not known. The low abundance of *Mus musculus*, except in intact pellets, indicates that most of the bones accumulated before about 1850, when *M. musculus* reached the Flinders Ranges (Aitken 1969). On the other hand, lack of deterioration of the bone suggests that it is not very old. No significant morphological differences were detected between the cave specimens and modern specimens of each species.

The dasyurid component of the fauna from Chambers Gorge not only extends the known range of four species but also provides a unique faunal composition. The most nearly similar dasyurid species assemblages have been found in central Australia south of 18°S and north of 29°S latitude. *Dasyurus geoffroii* was once widespread over the whole area but had become rare by 1961 (Finlayson 1961). *Antechinomys laniger, S. crassicaudata* and *S. macroura* have been found throughout and *D. cristicauda* in all but the northeast of the area (Parker 1973; Watts and Aslin 1974; specimens in South Australian Museum). The three remaining dasyurids from Chambers Gorge do not form part of this widespread dasyurid species assemblage. *Phascogale calura* has been recorded only from Alice Springs, Barrow Creek and Tennant Creek within central Australia (Parker 1973) whilst *D. byrnei* is found in the Lake Eyre basin but occurs only marginally in the Northern Territory, at Charlotte Waters (Parker 1973) and *P. tenuirostris* has not been recorded at all from the area (Archer 1976). On the other hand, *Antechinus macdonnellensis* is widespread in central Australia but has not been recorded from Chambers Gorge.

The environment near Chambers Gorge, particularly on the nearby Frome Plains, provides habitats similar to the known habitats of species for which range extensions have been recorded. *Phascogale calura* in the western part of its range lives in sclerophyll woodland but habitat information is not available for the rare specimens from central and southeastern Australia. *Dasyuroides byrnei* inhabits gibber–sandhill associations, *D. cristicauda* typically is found in sand–ridge areas but has been trapped in other habitats and *A. laniger*, and *D. geoffroii* have been found in a variety of habitats. Hence the greater diversity of dasyurids in the Flinders Ranges area in the past does not necessarily suggest climatic change.

Until the deposits are dated a direct relationship between introduction of sheep and cattle and decline of native species cannot be demonstrated. However the comments of the few travellers in the area who did notice the fauna show that they saw small mammals that no longer can be found there (Aitken 1969). It seems likely that the main reduction in the native mammal population closely followed the arrival of the hoofed mammals.

ACKNOWLEDGEMENTS

D. Lowe collected the material from Big Moro Gorge and Top John Gorge. Dr. Darrel Kitchener, Western Australian Museum and Mr Peter Aitken, South Australian Museum allowed access to their collections and provided information on localities and habitats. Ruth Altmann prepared the figures.

REFERENCES

AITKEN, P. F., 1969. The mammals of the Flinders Ranges. Pp.255–356 *in* "The natural history of the Flinders Ranges", ed. by D. W. P. Corbett. Libraries Board of South Australia: Adelaide.

AITKEN, P. F., 1971. *Planigale tenuirostris* Troughton the Narrow Nosed Planigale, an addition to the mammal fauna of South Australia. *S. Aust. Nat.* **46:** 18.

ARCHER, M., 1974. New information about the Quaternary distribution of the Thylacine (Marsupialia, Thylacinidae) in Australia. *J. R. Soc. W. Aust.* **57:** 43–50.

ARCHER, M., 1975. *Ningaui*, a new genus of tiny dasyurids (Marsupialia) and two new species, *N. timealeyi* and *N. ridei*, from arid Western Australia. *Mem. Qd Mus.* **17:** 237–49.

ARCHER, M., 1976a. The dasyurid dentition and its relationships to that of didelphids, thylacinids, borhyaenids (Marsupicarnivora) and peramelids (Peramelina: Marsupialia). *Aust. J. Zool. Suppl. Series* No. **39:** 1–34.

ARCHER, M., 1976b. Revision of the marsupial genus *Planigale* Troughton (Dasyuridae). *Mem. Qd Mus.* **17:** 341–65.

ARCHER, M., 1977. Revision of the dasyurid marsupial genus *Antechinomys* Krefft. *Mem. Qd Mus.* **18:** 17–29.

ARCHER, M., 1978. The nature of the molar–premolar boundary in marsupials and a reinterpretation of the homology of marsupial cheekteeth. *Mem. Qd Mus.* **18:** 157–64.

FINLAYSON, H. H., 1961. On central Australian mammals. Part IV — The distribution and status of central Australian species. *Rec. S. Aust. Mus.* **14:** 141–91.

KENDRICK, G. W. AND PORTER, J. K., 1973. Remains of a Thylacine (Marsupialia: Dasyuroidea) and other fauna from caves in the Cape Range, Western Australia. *J. R. Soc. W. Aust.* **56:** 116–22.

KIRSCH, J. A. W. AND CALABY, J. H., 1977. The species of living marsupials: an annotated list. Pp.9–26 *in* "The biology of marsupials" ed. B. Stonehouse and D. Gilmore. Macmillan: London.

KITCHENER, D. J. AND CHAPMAN, A., 1977. Mammals of Bendering and West Bendering Nature Reserves. *Rec. West. Aust. Mus. Suppl. No.* **5:** 17–29.

LUNDELIUS, E. L., 1963. Vertebrate remains from the Nullarbor Caves, Western Australia. *J. R. Soc. W. Aust.* **46:** 75–80.

LUNDELIUS, E. L. AND TURNBULL, W. D., 1975. The mammalian fauna of Madura Cave, Western Australia Part II. *Fieldiana Geology* **31** (2): 37–117.

MINCHAM, H., 1971. "The story of the Flinders Ranges". Rigby: Adelaide.

Morton, S. and Martin, A. A., 1979. Feeding ecology of the Barn Owl, *Tyto alba,* in arid southern Australia. *Aust. Wildl. Res.* **6:** 191–204.

Parker, S., 1973. An annotated checklist of the native land mammals of the Northern Territory. *Rec. S. Aust. Mus.* **16** (11): 1–57.

Ride, W. D. L., 1970. "A guide to the native mammals of Australia". Oxford University Press: Melbourne.

Smith, M. J., 1977. Remains of mammals including *Notomys longicaudatus* (Gould) (Rodentia: Muridae) in owl pellets from the Flinders Ranges, S.A. *Aust. Wildl. Res.* **4:** 159–70.

Wakefield, N. A., 1966. Mammals recorded for the mallee, Victoria. *Proc. Roy. Soc. Vict.* **79:** 627–36.

Watts, C. H. S. and Aslin, H. J. 1974. Notes on the small mammals of northeastern South Australia and southwestern Queensland. *Trans. R. Soc. S. Aust.* **9:** 61–9.

RESUMEN

47. Los dasiúridos de las Flinders Ranges del norte antes del desarrollo pastoril. Por Meredith J. Smith y Graham C. Medlin.

Se coleccionaron huesos de pequeños mamíferos de cuevas de Chambers Gorge, Big Moro Gorge y Top John Gorge del norte de las Flinders Ranges, South Australia. Estos depósitos se formaron por la rotura de regurgitaciones de lechuzas; regurgitaciones intactas están presentes en la superficie. La baja frecuencia de *Mus musculus* sugiere que poco de la acumulación ocurrió después del establecimiento de los europeos en el área. Se identificaron ocho especies de dasiúridos. Todavía pueblan las cercanías del área *Sminthopsis crassicaudata, S. macroura* y *Planigale tenuirostris.* Se anotan las distribuciones de *Antechinomys laniger, Phascogale calura, Dasycercus cristicauda* y *Dasyuroides byrnei. Dasyurus* c.f. *D. geoffroii* fué registrada tambien.

48. *Thylacoleo carnifex* OWEN (THYLACOLEONIDAE): MARSUPIAL CARNIVORE?

R. T. WELLS[1], D. R. HORTON[2] and P. ROGERS[3]

A study of theogotic wear on the teeth combined with a reconstruction of the jaw mechanics indicates that *Thylacoleo carnifex* used its lower incisors for stabbing and its third premolars for shearing. These movements appear to have been achieved with an essentially phalangerid jaw musculature. An analysis of the attritional wear on teeth indicates a greater similarity to carnivores. Empirical support for this hypothesis was gained by constructing a facsimile thylacoleonid and "feeding" it different diets. It is concluded that *Thylacoleo carnifex* was a carnivore adapted to climbing. A Leopard-like niche is suggested.

Pages 573–86 *in* CARNIVOROUS MARSUPIALS, ed. by M. Archer, February, 1982. Roy. Zool. Soc. New South Wales: Sydney, Australia.

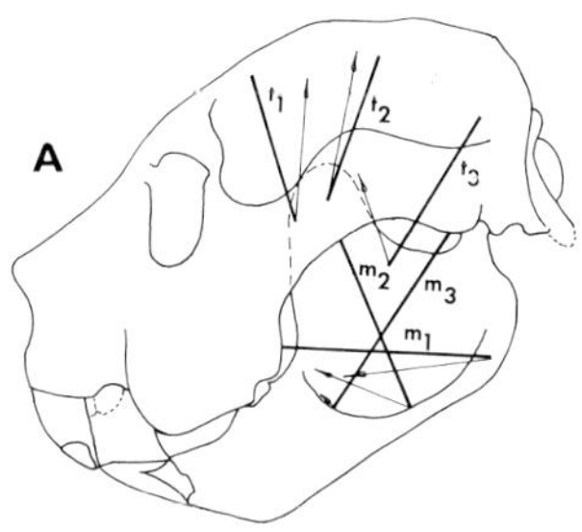

THE EATING HABITS of *Thylacoleo* have been a matter for debate since the first description of the species. Owen was convinced that the animal was carnivorous, but a number of authors, including Flower (1868) challenged this view.

The argument for a herbivorous diet was demolished by Broom (1898), who pointed out that it was largely based on the idea that 'diprotodonts can't be carnivores'. He also suggested that it was impossible to consider an animal which lacked grinding molars to be herbivorous. He dismissed the dietary hypotheses involving soft, rich food (e.g. ripe fruit) which attempted to explain the lack of molars.

That Broom did not succeed in ending the *Thylacoleo* debate was largely the fault of De Vis, who previously described *Thylacoleo* as "a carnivore resembling in its habit of feeding, the hyaena rather than the lion" (De Vis 1883). This statement changed the nature of the debate from carnivore versus herbivore to bone crusher versus herbivore. The 'bone crusher' hypothesis was so indefensible anatomically and was based on such a poor piece of research, that it could be eliminated from consideration, thus leaving the herbivore hypothesis as the only viable one.

De Vis began his analysis by stating that there were large numbers of "scored, cracked, chopped and crushed bones" to be found in the Darling Downs fossil sites. These bones De Vis attributed to the activities of the dingo. He then described a kangaroo jaw with two deep cuts which he said had been caused by *Thylacoleo*. De Vis then made a remarkable deduction — "though it might have attempted the fracture of the victims jaw for the sake of its investment of flesh, as the cat breaks and swallows the bones of her prey, yet the first idea suggested by the specimen that it was one of the many bones crunched for their own sakes, is probably the true one".

At the end of the paper came an extraordinary reversal. De Vis described a new large species of *Sarcophilus, S. prior*. Now the cut marks which were originally apparently caused by dingoes, and had then been clearly caused by *Thylacoleo* ("no such teeth save those of the thylacoleo are known"), were attributed to the new species which "must have been fully competent to the work whereof so many traces have been left". Bartholomai and Marshall (1973) re-examined the specimen on which "*Sarcophilus prior*" was based and decided that it came from a wombat.

Spencer and Walcott (1912) seemed to compensate for the De Vis paper when they examined some cut–marked bones from Victoria and showed experimentally that the marks matched those which *Thylacoleo* teeth could produce. To see why Spencer and Walcott's work did not settle the question we must understand what they set out to achieve and also look at the nature of their data.

There had been, before 1912, considerable debate as to how long man had been in Australia. The most promising line of evidence giving man a long antiquity was the attempt to prove human association with the extinct megafauna. The only basis for this was the presence of "cut marks" on the bones of megafauna. A number of such bones had been found including most notably the "Colongulac Bone" and the "Buchan Bone". Spencer and Walcott were trying to find out whether such marks really were caused by stone tools or whether they were tooth marks. Because the alternative hypotheses were man or animal, they did not carry out tests to settle the question of *which* animal was involved.

Another reason for not testing which animal had caused the marks was the nature of their data. They were dealing almost entirely with bone fragments — i.e. there were almost no

[1] School of Biological Sciences, Flinders University of South Australia, Bedford Park, South Australia 5042.
[2] Australian Institute of Aboriginal Studies, Acton House, Acton, Australian Capital Territory, A.C.T. 2601.
[3] School of Medicine, Flinders University of South Australia, Bedford Park, South Australia 5042.

opposing marks present which would have provided proof that the marks were tooth marks and left Spencer and Walcott free to concentrate on the nature of those teeth. In fact they did not conclusively prove that the marks *were* tooth marks; their arguments against a human origin were that Aborigines were not known to have cut bones in this way. There were also methodological problems. Spencer and Walcott concentrated most of their attention on the "cut" ends of bones, showing that the shape of these agreed with the shape of *Thylacoleo* teeth. They paid little attention to the numerous smaller marks, and provided very few measurements against which the claims of *Thylacoleo* (as distinct from *Thylacinus*) could be evaluated.

As Spencer and Walcott failed to provide convincing evidence that *Thylacoleo* was carnivorous, the way was left for a return to the herbivore hypothesis by an attack on De Vis. This was precisely what Anderson (1929) did. De Vis had created a glaring weakness in the carnivore hypothesis with his depiction of *Thylacoleo* as a 'marsupial hyaena'. Anderson was, quite rightly, scornful — "nor does the hypothesis that it was a sort of marsupial hyaena commend itself, although several writers have described fossil bones which exhibit what they suppose to be tooth marks of *Thylacoleo*. An animal adapted for a bone diet has broad strong teeth with several cusps, and a well developed cingulum on the molars and premolars, the purpose of which is to protect the gums from injury by splinters of bone. The teeth of *Thylacoleo* are not of this type; nor is such a cingulm to be found on them".

Logically, all that Anderson had done was to disprove the hypothesis that *Thylacoleo* was a "bone cruncher". As far as he was concerned if *Thylacoleo* was not a bone crusher it must be a herbivore. By adopting this position Anderson had painted himself into a corner, because he was also aware of Broom's arguments against the herbivore hypothesis. He reluctantly settled for the 'soft, rich food' compromise (in this case the fruit of native curcurbits or cycad nuts) but pointed out "there are difficulties in accepting the view that *Thylacoleo* could obtain an adequate food supply from these sources".

The debate now seemed to have reached a stalemate. Anderson had apparently disproved the carnivore theory, but the herbivore theory had an inherent contradiction. The lack of molar teeth meant that the food had to be easily broken up and readily digestible, but if this were the case what was the point of the extraordinary premolars?

Gill (1954) attempted to resolve part of this contradiction. Pointing out first that cucumbers were rare or absent in just that part of Australia where *Thylacoleo* was found, he then turned his attention to the premolars. He suggested that they were no different in kind from the sectorial premolars of such macropodids as *Bettongia* and *Sthenurus* which were browsing animals. Gill saw in the premolars of *Thylacoleo* a likeness to "the short, curved lopping blades near the handles of hedge clippers" (note that in the plate illustrating this, a left maxilla has been inadvertently associated with a right mandible). He continued — "If these [the premolars] served an herbivorous function, then they must have been for mincing and dicing the plant food ingested insted of grinding it as the molars usually do . . . they could mince leaves, twigs and fine branches; they could handle with ease food that would be far too coarse for most herbivores".

It is difficult to see how the premolars could have been used to 'mince and dice' the food (the first cut through a 'fine branch' is easy to visualise but not the numerous succeeding ones). The other objection to this model is the lack of molars. All the browsers with sectorial premolars have extensive molar arrays adapted to even finer cutting and crushing of plant material. Food "too coarse for most herbivores" would require considerable grinding to break down woody tissue — 'mincing and dicing' would be inadequate to break down cell walls.

Gill (1951, 1952) also discussed Spencer and Wallcott's work and provides us with another insight into why the carnivore hypothesis was rejected. He claimed that the 'Colongulac Bone' could not be the work of *Thylacoleo* because the angle of the cuts was wrong and because "the two wedges are in a direct line and not slightly offset as they would be if due to the bite of an animal". Gill also (quite correctly) criticized Spencer and Wallcott's work on the 'cut ends' of bones – "it is my opinion that the long clean surfaces on some of the bones from Pejark Marsh, for instance, are a result of natural fracture after the bones have dried. Similar surfaces are found in bones belonging to horizons later than the extinction of *Thylacoleo*".

Gill added another sentence – "The incisions on some bones are due to rats" – which gave the impression that all of the marks and cuts considered by Spencer and Walcott had now been accounted for. This was not the case, and the reference to rats was misleading. Spencer and Walcott knew what rodent gnawing looked like – "The side of the bone opposite to these cuts has been gnawed for its whole length by some small rodent, such as a rat, the marks of both upper and lower incisors being perfectly distinct". The marks they were concerned with were not caused by rats — "vary in shape from a straight line to a curve similar to that of a boomerang. The short, straight, chisel–like incisions are, as a rule, terminated abruptly at one end . . . in one or two cases they terminate in a slightly marked bifurcation . . . In addition to these there are more definite cuts, of gashes, of a V or wedge–shaped form in cross section".

Spencer and Walcott had not proved that these marks could *only* have been made by *Thylacoleo* (although they reproduced the bifurcation experimentally) and neither Gill nor Anderson attempted to explain them.

The cut marks described by De Vis and Spencer and Walcott created a problem – how could (and why should) *Thylacoleo* set out to break up bones without teeth adapted to the purpose? This was quite clearly a contradiction and Gill and Anderson were quite right to reject a carnivore hypothesis based on such a premise. The reason for the development of this premise is that De Vis and Spencer and Walcott had developed their hypothesis on the basis of the cut marks (rather than from the question of what did *Thylacoleo* eat). The cut marks therefore became the focus of attention and assumed great significance – if *Thylacoleo* had made such marks they must have been made *intentionally*. The correct interpretation was that the marks indicated that *Thylacoleo* had been *involved* with a particular carcass. On that basis one could then ask questions about the *nature* of that involvement. The recovery of a large series of well preserved *Thylacoleo* skulls, associated limb elements and assorted vertebrae from the Victoria Fossil Cave at Naracoorte in South Australia (Wells 1974) provided an ideal opportunity to reconsider the niche of this animal. We present here some new evidence relevant to this discussion. A description of the unworn and worn teeth, an analysis of jaw movement, a reconstruction of the jaw musculature and a scanning electron microscope study of attritional wear on the enamel.

DENTITION

The accepted dental formula for *Thylacoleo carnifex* is $I\frac{1.2.3.}{1}\ C\frac{1}{0}\ Pm\frac{1.2.3.}{1.2.3.}\ M\frac{1}{2}$ (Woods 1956) (Fig. 1). The position and alignment of two alveoli between I_1 and Pm_3 suggest the presence of either $I_{2.3}$ or $Pm_{1.2}$ in juvenile specimens. The unworn crown of C^1 in pouch young specimens is identical in form to that of the unworn I^1 (e.g., S.A. Museum specimen P20458). However, the position of the premaxillary/maxillary suture at the anterior border of the alveolus confirms the identification of the tooth as a canine leaving the animal with a typically diprotodont dentition indicative of a phalangerid ancestry.

T. carnifex differs from all other phalangerids in the development of a caniniform I_1 and a large sectorial third premolar. The molar tooth row is reduced to a single medially directed, generally double though occasionally triple rooted M^1 and an M_1 which abuts Pm_3 to form a continuation of the lower sectorial blade. In juvenile specimens (eg., S.A.M. P16714) a very tiny single rooted button like tooth M_2 is often present, only the alveolus remaining in adult specimens examined.

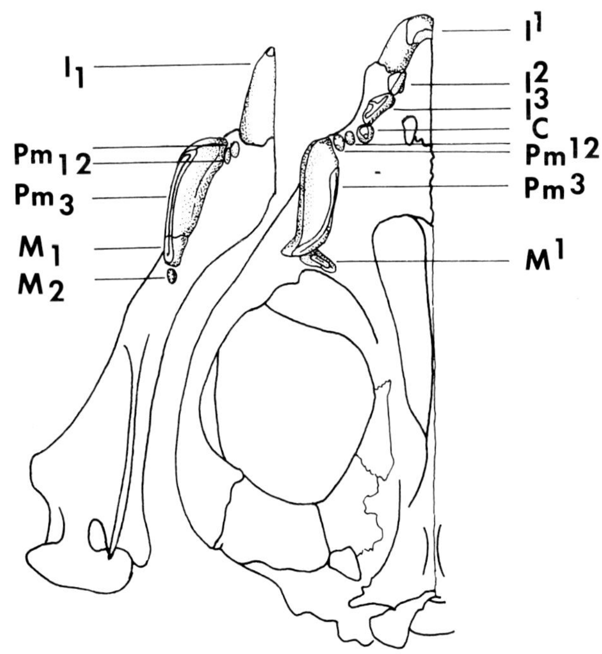

Fig. 1. Occlusal view of dentition of *Thylacoleo carnifex*.

I^1: The convex antero–labial aspect of the upper incisor is caniniform in unworn specimens; the antero–medial and postero–lateral borders bearing narrow mildly serrate ridges. The posterior aspect is convex in transverse section, a small facet occurs medially at the contact with the opposite I^1. The crown is completely enveloped by enamel. With age the tip of the crown and the enamel coating of the posterior and medial surfaces is worn away producing, in old animals, an incisiform tooth with a small convex enamel cap confined to the antero–lateral aspect, a rounded convex occlusal outline and a deeply grooved postero–medial surface of dentine (Fig. 2).

I_1: The lower incisor remains caniniform throughout life. The tooth is triangular in transverse section; medial and posterior surfaces at right angles to each other with a convex antero–lateral surface. This tooth is the "bayonet–like lower laniary" of Owen (1871), (Fig. 3). It bears serrate ridges on the anterior and posterior labial edges and a double serrate ridge on the postero–medial edge. In unworn specimens enamel completely invests the crown, in older animals wear against the posterior surface of the I^1 produces an oblique triangular occlusal outline similar to that of the upper canines of felids and sabertooths, and as with the latter (Turnbull 1978) the beaded or serrate ridges giving it "super-added modifications strengthening and perfecting it for its work as a piercer, holder and lacerator" (Owen 1871). In older animals (eg., S.A.M. P 16566) I_1 sometimes bears a small facet on its antero–lateral aspect adjacent to the crown tip and apparently produced by contact with the medial aspect $I^{2,3}$.

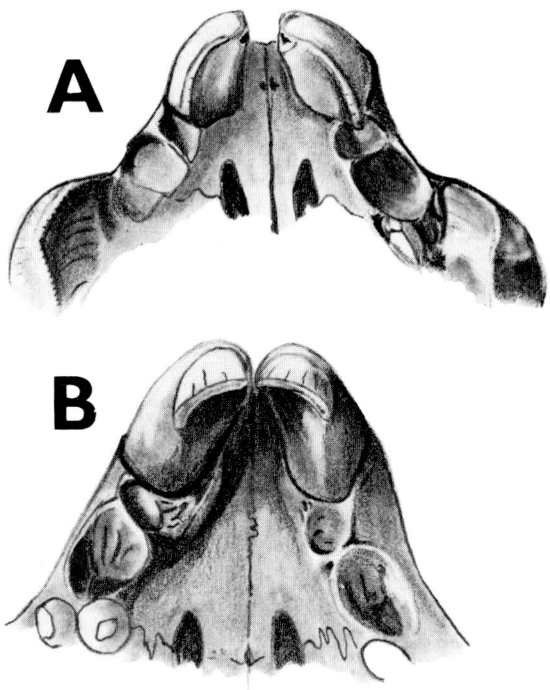

Fig. 2. I^1 occlusal view Juvenile (A), crown enveloped by enamel and bearing mildly serrate ridges; adult (B), eruption and wear reduces enamel to convex antero–lateral cap.

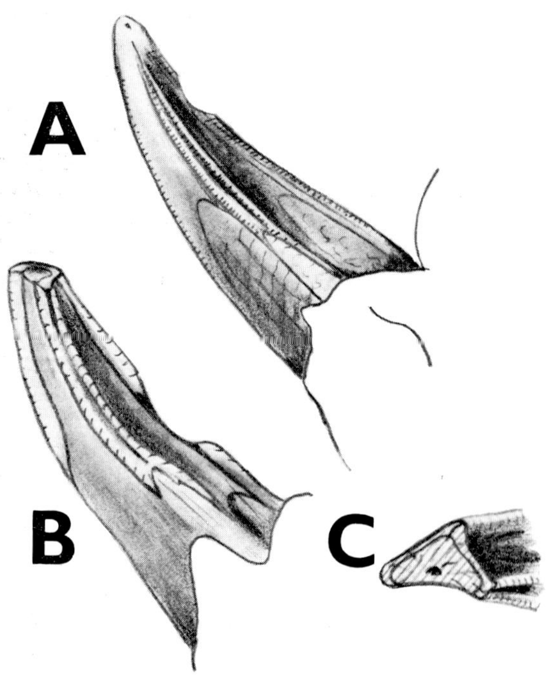

Fig. 3. I_1 postero–medial aspect. Juvenile (A) crown capped by enamel with borders forming serrate ridges. Adult (B), serrate ridges retained. (C) Occlusal view of adult I_1 showing distribution of enamel and parallel thegosing striae.

$I^{2,3}$: These teeth have fallen from their alveoli in all the Naracoorte specimens. All these isolated teeth are in an advanced state of wear and the original cusp pattern has been lost. I^2 is the smaller, it is subtriangular in occlusal view, low crowned with a single relatively long tapering root. I^3 is a subrectangular low crowned tooth approx. twice the size of I^2. It has a large recurved single root. A median enamel ridge descends down the labial face of the crown. Both teeth have well developed antero–posterior thegosing facets produced by contact with the antero–lateral aspect of I_1.

C^1: In its unworn form the crown of the canine is typically caniniform bearing antero–medial and vestigial postero–lateral ridges and convex labial surface. Unlike the incisors it is low crowned and soon develops a flattened occlusal surface from contact with the anterior most tip of the Pm_3.

$Pm^{1,2}$: These shallow rooted peg–like teeth have an oval occlusal outline with a much suppressed median ridge orientated parallel with the antero–lingual facet of Pm^3. These teeth, unlike the canine, show no evidence of wear regardless of age. They appear to be truly vestigial.

$Pm^3/_3$: These remarkable sectorial teeth (Fig. 4) have ben described and discussed in detail by many workers (eg., Krefft 1866, Owen 1871, Murchison 1868, Broom 1898, Gill 1954). For the purpose of this analysis some characteristics are worth restating.

(1) The single antero–posterior cutting ridge of the crown in both upper and lower teeth descends anteriorly to the alveolus and in unworn specimens is finely serrate along its entire length.

(2) The crest of the crown rises anteriorly and is concave in both lateral and occlusal views, the

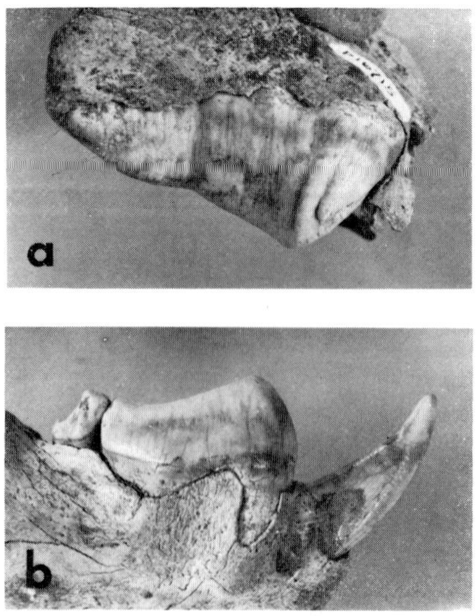

Fig. 4. Unworn juvenile Pm^3 (a) and I_1Pm_3 (b). Note the serrate borders to the crowns of each tooth.

upper bowed lingually, the lower labially. The anterior root, in keeping with the form of the tooth, is the larger of the two.

(3) The movement of the lower premolar to the lingual side of the upper wears an oblique facet along the cutting edges. The angle of the occlusal plane relative to the palate varies greatly between individuals from 14° to 48° Gill (1954) and this study.

(4) The longitudinal axis of these premolars is aligned with the saggital plane and not with the axis of the jaw.

M^1 This tooth, like the M^1 in the felids, is rotated medially and acts as a guide for the talonid of the M_1 directing and arresting the jaw in its medial movement during occlusion of the sectorial premolars.

FUNCTIONAL ANALYSIS OF THE DENTITION

Functionally the *Thylacoleo* dentition comprises stabbing incisors and shearing premolars. Wear on the convergent upper incisors produces a blunted semi–circular occlusal outline with enamel restricted to the labial surface (Fig. 2). Wear on the convergent lower incisors maintains a sharp triangular occlusal outline bounded by enamel with serrate ridges descending from the crown (Fig. 3). The lower incisors thus retain a stabbing or piercing function with the uppers acting as both an anvil against which food is restrained and as a stone against which the lowers are thegosed. Engagement of the incisors results in disengagement of the premolars.

The 3rd premolars are shearing blades. Occlusion commences posteriorly and advances anteriorly. The concave outline results initially in a single point of shear advancing anteriorly, but as the teeth occlude, the high anterior portions of the crowns engage to produce a second point of shear advancing posteriorly, thereby trapping food between the converging blades and preventing it from spilling forward. This buttressing effect of the high anterior portion of the crown is reflected in the larger anterior root of the third premolars. The curved palatal aspect of these teeth also serves to maintain tension on the food during shearing. A further refinement is seen in the serrate ridges which descend anteriorly from the crown of the third premolars. During the latter stages of occlusion when food is forced against the anterior of the crown, these ridges begin to diverge thereby maintaining tension on the food during the final cutting phase.

Wear on the premolars eventually breaches the enamel and in old individuals this produces broad, flat facets of dentine surrounded by enamel. Nonetheless, the vertical ribbing of the enamel on the lingual side of the lowers and labial side of the uppers ensures maintenance of micro-serrations along the cutting edge and hence effective shearing capacity of the teeth.

JAW MECHANICS AND SKULL ARCHITECTURE

The masticatory apparatus is important in determining the architecture of the skull. The dentition in *Thylacoleo* is adapted for stabbing or piercing food with the incisors, and shearing or slicing food with the premolars. Use of the incisors requires a simple hinge–like articulation with a powerful closing action over a wide range of angles, whereas the shearing premolars only begin to function with the jaws almost closed. Furthermore shearing requires continuous unilateral pressure on the blades and hence some lateral movement at the hinge and a corresponding medial shift of the mandible as the teeth occlude (Fig. 5).

The jaw musculatures of *Trichosurus vulpecula* (3 specimens) and *Pseudocheirus peregrinus* (1 specimen) were dissected by the method of Bock and Shear 1972. Sites of origin and insertion of individual muscles were compared with similar muscle scars on the skulls and jaws of *Thylacoleo*.

Three major muscle groups were examined, the temporalis, masseter and pterygoids. A zygomatico mandibularis was not evident in either extant species; the fibre groups of the temporalis form a continuous array in both species. In adult *Thylacoleo* there appear to be three distinctly sculptured areas on the lateral aspect of

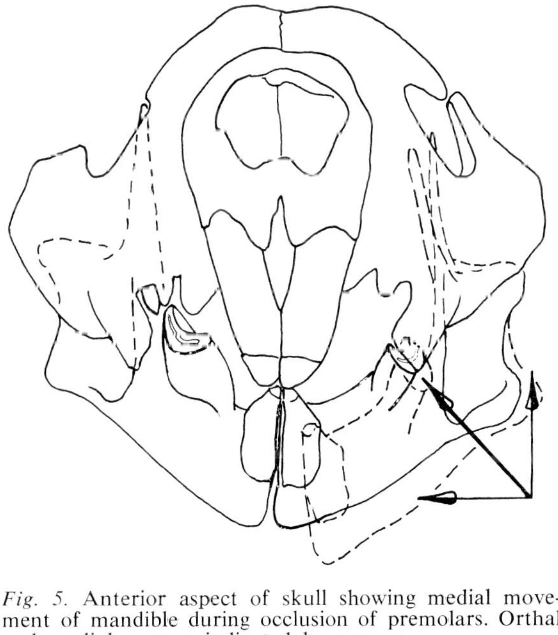

Fig. 5. Anterior aspect of skull showing medial movement of mandible during occlusion of premolars. Orthal and medial vectors indicated by arrows.

the cranium and at least two on the medial aspect of the coronoid process, these, along with a strong post–orbital bar, suggest a powerful anterior fibre group in the temporalis and the possibility of three fibre groupings (t_1, t_2, t_3, Fig. 6).

The line of action of each muscle or fibre group has been approximated in *Thylacoleo* by joining the mid–point of origin with the mid–point of insertion as deduced from the muscle scars. No attempt has been made to account for multipenniform fibre groups, e.g., as in the deeper portions of the masseter. To determine moments of each muscle about the jaw hinge the force exerted along the line of action of the muscle has been resolved into two components, one from the hinge to the mid–point of insertion, and the other at right angles to it. For the purpose of this vector analysis, all forces along the line of action of each muscle have been given equal magnitude (Fig. 6).

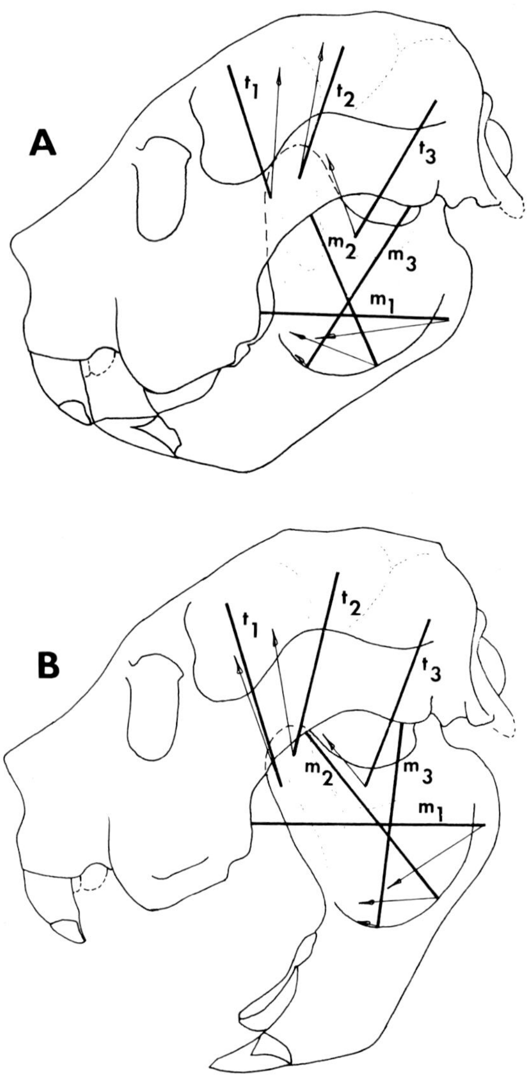

Fig. 6. Skull of *Thylacoleo* with jaw closed (A) and open (B). Lines of action of superficial, medial and deep masseter m_1, m_2, m_3 and temporalis fibre groups t_1, t_2, t_3 are indicated. Arrows indicate respective turning movements about condyle. For explanation see text.

From Fig. 6, the turning moments of the superficial (m_1) and medial masseter (m_2) and the temporalis fibre groups (t_1, t_2, t_3) are of similar magnitude for all positions of the jaw. The deep masseter (m_3) however, exerts only a small turning moment about the hinge, its main function would be to exert a strong backward pull on the mandible, thereby maintaining the condyle tightly in the fossa of the occluding side during the early phase of shearing. This would seem unnecessary as the shearing of tough material would tend to displace the jaw posteriorly on the occluding side (Fig. 7). However, contraction of the deep masseter on the balancing side would be of greater significance as it would aid the medial shift of the jaw during occlusion.

This leads to a discussion of the fossa and the pterygoid musculature. In felids where the cheek tooth row consists of shearing carnassials, the symphysis of the jaw is tightly bound and the whole jaw must slide laterally to engage the carnassials on a given side. *Thylacoleo* has a cheek tooth row which consists functionally of a single giant 'carnassial' yet it retains a phalangerid–like glenoid fossa strongly buttressed medially and posteriorly. The condyle of the jaw is sub–cylindrical for most of its lateral aspect but is somewhat hemispherical on its medial aspect.

When the jaw is positioned centrally and open, the lower incisors are readied for stabbing. Reaction to the thrust is directed posteriorly into

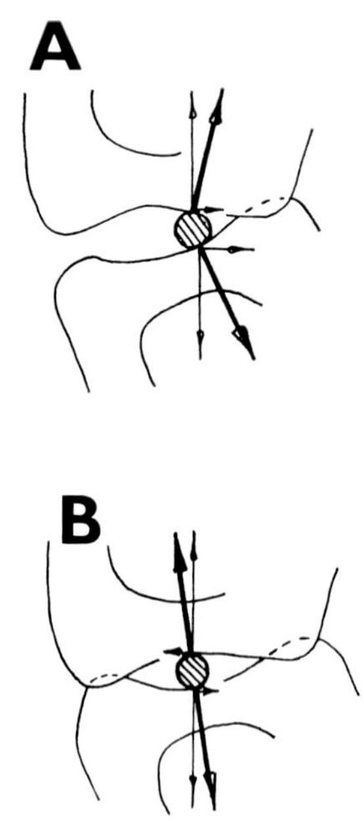

Fig. 7. A and B depict subsequent phases of biting by the premolars upon a resistant substrate. The reaction forces (heavy arrows) are resolved into components (light arrows) perpendicular and parallel to a line joining the axis of the condyle to the point of application of the bite. The resultant effect in A would be to displace the food anteriorly to position B. The resultant effect at B would be to displace jaw posteriorly. This movement would be resisted by the glenoid fossa.

the buttressed rim of the fossa. To bring the 'carnassials' into play, the jaw must be rotated medially towards the occluding side (Fig. 8) by contraction of the balancing side pterygoid complex. As the bite progresses, lateral pressure could be provided by a labial rotation of the mandible about its longitudinal axis and be produced by the pterygoids on the occluding side This may account for the unusually well–developed insertion in the superficial pterygoid (Fig. 9). This motion would undoubtedly be used for both thegosis and shearing food. Thegosis of the lower premolar produces a planar occlusal surface while the upper tends to attain a slighly rounded lingual profile· With continued thegosing, the occlusal angle gradually decreases, becoming almost horizontal in old individuals. The final phase of shearing in old animals is largely one of a medial shift of the mandible.

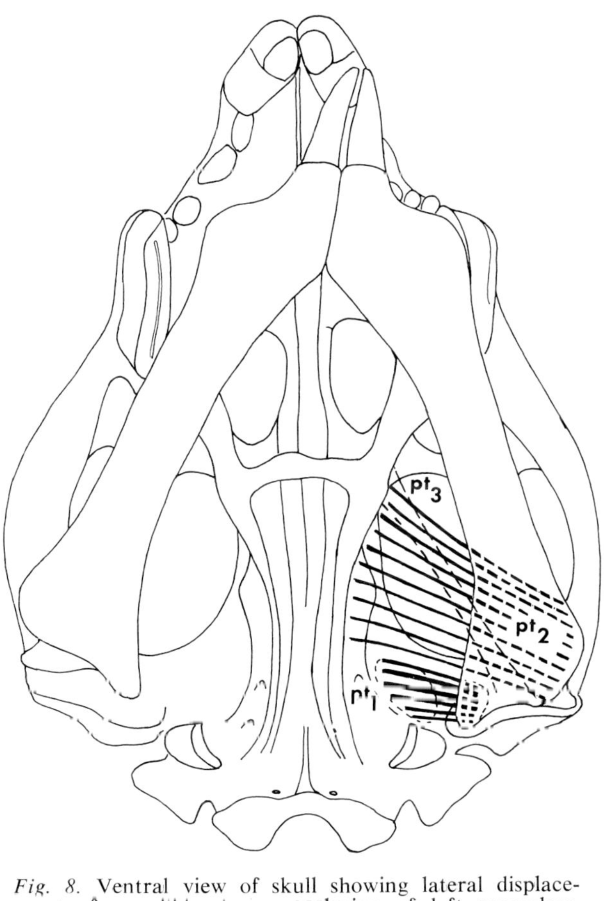

Fig. 8. Ventral view of skull showing lateral displacement of mandible during occlusion of left premolars. Right mandible has been drawn in position it would assume if symphysis was fused. Note displacement of right condyle.

These shearing movements necessitate a flexible symphysis, otherwise lateral rotation of the ramus and maintenance of lateral pressure during occlusion would lead to dislocation of the condyle on the balancing side (Fig. 9). Among the Potoroinae, the rat kangaroos, *Bettongia* spp., have developed a large, fluted shearing premolar, yet retain the molar tooth row. In these species the deep layer of the masseter has invaded the mandible, presumably as a means of increasing leverage at the premolar (Abbie 1939). This has not occurred in *Thylacoleo* as reduction of the molar tooth row has led to a foreshortening of the jaw greatly increasing the power available at the premolars. Thus in *Thylacoleo* we find the essentially phalangerid jaw mechanics retained in association with a sectorial dentition.

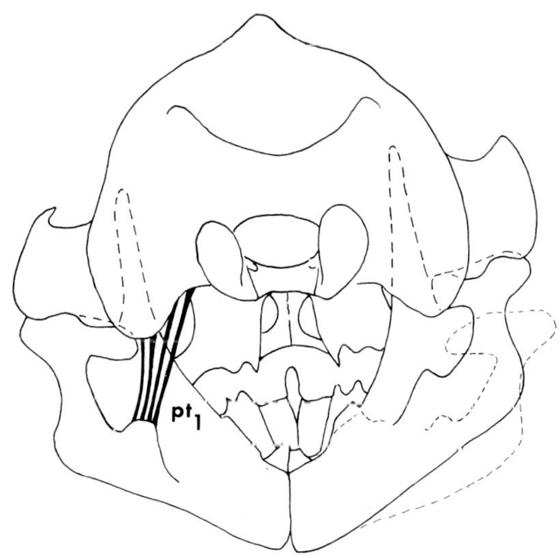

Fig. 9. Posterior view of skull with left premolars in occlusion. Dotted outline indicates displacement of right condyle if symphysis was fused. The proposed line of action of the superficial pterygoid (pt. 1) from its origin medially of the fossa to its insertion on the prominent angular process is indicated.

The question then arises as to whether both craniomandibular joints are acting together during unilateral shearing. Greaves (1978) has proposed a geometrical model to explain the position of the grinding tooth row in the mandible of selenodont artiodactyls. This model is based on the assumption that the balancing side articulation contributes significantly to maintenance of maximum force along the occluding tooth row. Fig. 10 indicates the position of the shearing blade in *Thylacoleo* mandible relative to Greaves' zone of maximum force, i.e. BCD. Where the line CB represents the most posterior position of the resultant of the muscle forces that close the jaws, while D represents the most anterior position. The efficiency of the lever system will decrease anteriorly of D and posteriorly of CB· Given that Greaves' model withstands empirical testing as evidenced by Hylander's study of Galagos and Macaques (1979), it would suggest that *Thylacoleo* jaw geometry has been selected for maximising power at the shearing blades, utilising both balancing and occluding side musculature and placing the fulcrum at the balancing condyle. Such an arrangement is in accord with the proposed function of the deep masseter on the balancing side. The insertion of the strong roots of I_1 behind and medially of the anterior root of Pm_3 also allows greater power to be exerted at the tips of the caniniform teeth.

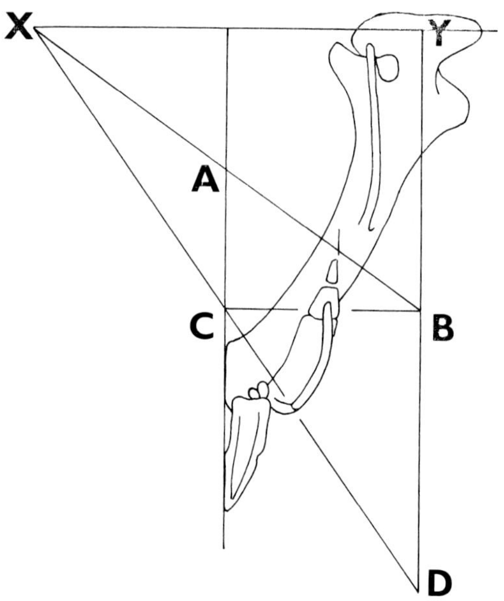

Fig. 10. Using the balancing side condyle (X), as a fulcrum the cheek tooth row falls within the zone CBD where maximum bite forces are predicted q.v. Greaves (1978).

THE DIET

Historical arguments regarding the diet of *Thylacoleo* have already been discussed. The details of the dentition suggests a carnivorous habit. Further supportive evidence can be gained from an SEM study of food wear on non–thegosed areas of enamel from the cheek teeth of mammals with known dietary preferences. Numbers of teeth from several species were examined under the light microscope and a representative tooth chosen from each for examination under the scanning micoscope. Fig. 11 provides a comparison of attritional striae attributed to the passage of food over the enamel. Striae from herbivores (left column) are markedly finer than those of known carnivores (right column), while *Thylacoleo* is clearly similar to the latter group (Fig. 12). The widths of the abrasion striae were measured directly from photographs and corrected for enlargement size. There is an order of magnitude difference between the mean size of striae of herbivores and carnivores (Table 1). Striae on the teeth of *Thylacoleo*, although similar to known carnivores, are on the average 5X larger.

To further test this hypothesis, facsimile *Thylacoleo* third premolars were ground and filed from mild steel and lapped on 600 grit wet and dry emery paper to leave the occluding surface with fine horizontal 'thegosing' marks. The teeth were inserted in a machine (Fig. 12) which replicated the geometry and simulated the reconstructed jaw movements of *Thylacoleo.* Light machine oil substituted for saliva and the machine was 'fed' a number of different foods. A simulated herbivore diet included fresh leaves and petioles of assorted *Eucalyptus* species, while a carnivorous diet included limb muscles and hide as well as portions of the rib cage and hide of specimens of *Macropus fuliginosus* and *M. eugenii.* After each trial a small insert was removed from the lower premolar and the abrasion striae examined under the SEM. The pattern that emerged was similar to that observed on the enamel of real molars (Fig. 12, Table 1). Plant material produced fine striae, possibly due to silica phytoliths within the leaves, while meat and hide or meat and bone produced larger striae attributed to either sand grains within the pelt or bone fragments respectively.

Thus the weight of evidence indicates a greater similarity to the carnivores than the herbivores.

DISCUSSION

Ewer (1967) produced an interesting hypothesis concerning hyaena evolution – "The modern felids have a general purpose dentition : their carnassial shear is well developed, and their premolars are capable of a fair degree of bone crushing. Primary carnivores of this type would not seem to provide a very encouraging situation for a potential scavenger. The sabertooths, however, are very different. Here the cheek teeth are specialized to an extreme degree for the one function of slicing. The carnassial shear is highly efficient, but the premolars are so reduced that the animal must have been unable to deal with more than the very smallest bones. In the presence of a primary predator of this type, the niche for bone crushing specialists stands invitingly wide open . . . Hyaenas and sabertooths would form an ideal combination". It is tempting to simply substitute *Thylacoleo* for sabertooth and Tasmanian Devil for hyaena in the above paragraph, but we need to examine the situation in more detail.

The important difference between the sabertooth and *Thylacoleo* dentitions is of course the enormously enlarged canines in the sabertooth. These were quite clearly killing weapons, probably used for piercing vertebrae. There seems to be an impression that most large carnivores kill in this way. For example Ewer (1973) quotes Leyhausen (1965) to the effect that "the typical felid kill results from one of the canine teeth entering between two neck vertebrae, forcing them apart and breaking the spinal cord . . . the laterally compressed canines are admirably

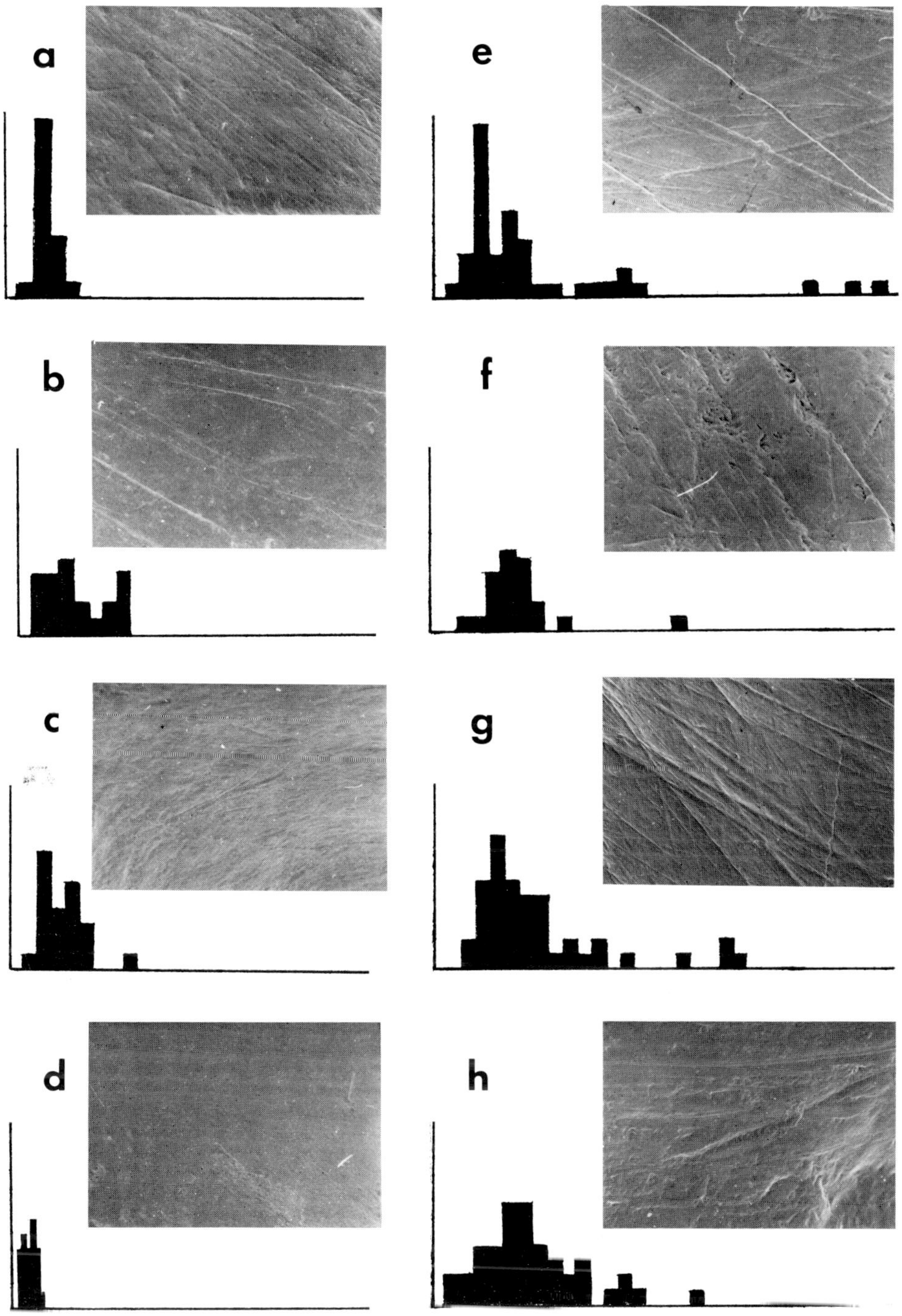

Fig. 11. A comparison of attritional striae on the enamel of known herbivores left column compared with that of known carnivores right column. All specimens X 207. Histograms of striae distributions all at same scale qv. Table 1. a, *Macropus giganteus;* b, *Macropus rufus;* c, *Trichosurus vulpecula;* d, *Phascolarctos cinereus;* e, *Canis familiaris;* f, *Felis catus;* g, *Thylacinus cynocephalus;* h, *Sarcophilus harrisii.*

adapted to act in this way and much less so to bite directly on bone".

The lack of such teeth has been used as a major argument against *Thylacoleo* being carnivorous. Anderson (1929) says "Thylacoleo was but poorly equipped for killing and devouring animals . . . The median incisors of the upper jaw were blunt tusks rather than piercing implements, they early developed signs of wear, and there was no interlocking of upper and lower incisors to enable the animals to get a firm and deadly grip on its prey". Gill (1954) adds "a living animal caught in such incisors could spin about as if on a pivot".

Most large carnivores seem to kill only rarely, if at all, in the way in which Leyhausen has described. Lions for example normally kill prey animals by suffocation. Schaller (1972) reports "Unlike small prey which is usually killed by a bite in the back of the neck, wildebeest, zebra and others are usually grabbed by the throat. The teeth of lions are not built for crushing bones, nor are they particularly long, making it difficult for them to reach a vulnerable part by penetrating the layers of muscle, skin, and bone at the nape. On the other hand, the throat is unprotected, and a lion may grasp it for as long as 8, 10, and 13 minutes to mention the three longest periods noted, until prey ceases to move. The throat of zebra and other thin–skinned prey is often lacerated by the canines but damage is usually limited to deep punctures. However, the skin on the throat of wildebeest is up to 1 cm thick and in many instances the canines

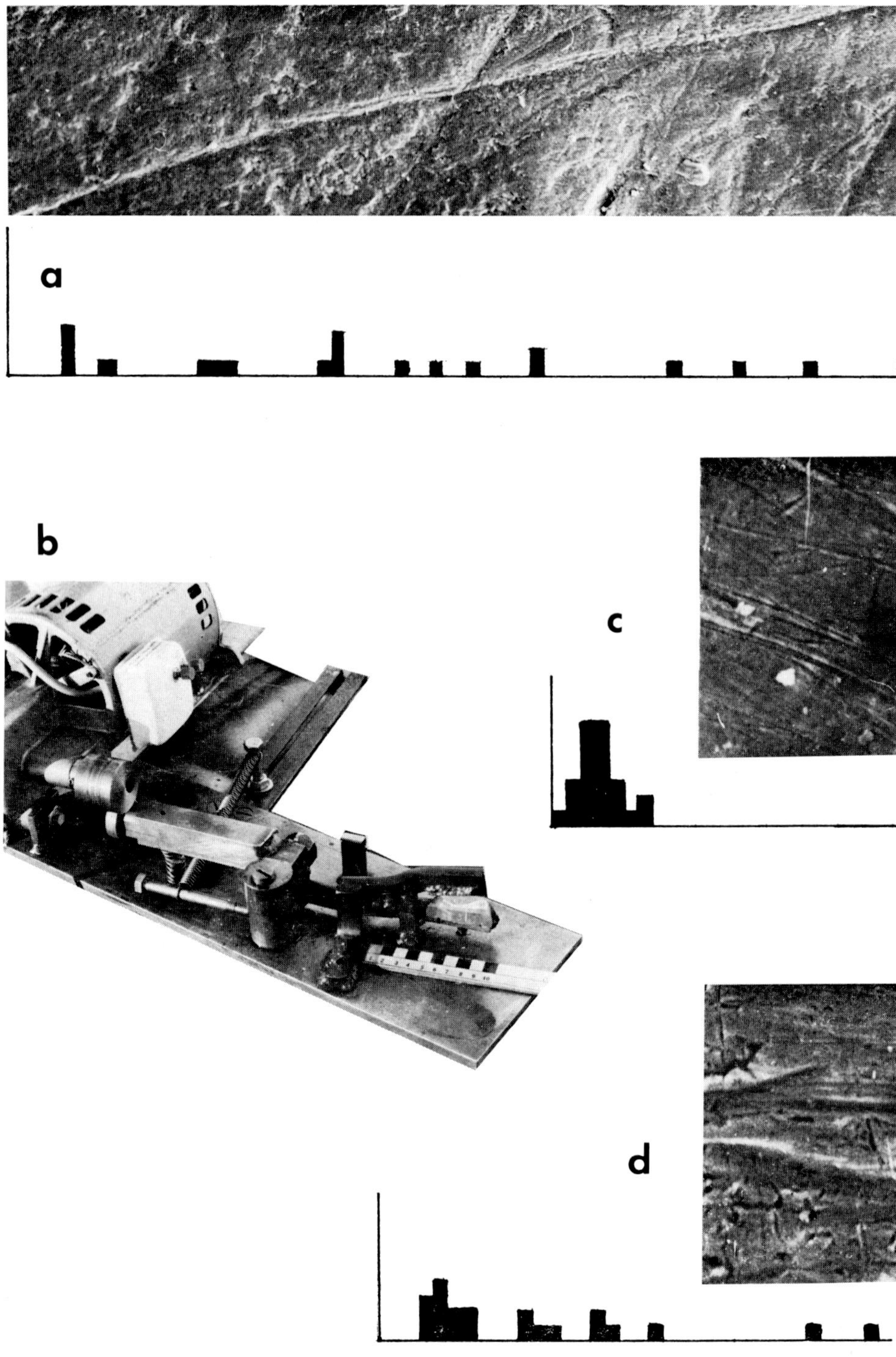

Fig. 12. a, attritional striae on the enamel of *T. carnifex* Pm^3 X 207; b, facsimile thylacoleonid premolars mounted in machine to replicate reconstructed chewing movements; c, attritional striae on facsimile Pm^3 from chewing eucalypt leaves X 400; d, attritional striae on facsimile Pm^3 from chewing meat, bone and hide X 400; Histograms for (a), (c) and (d) at same scale as in Fig. 11.

Table 1. Average width of attritional striae on non–thegosed areas of enamel from cheekteeth of carnivores and herbivores and facsimile thylacoleonid.

Niche	Species	$\bar{x}$ width of attrition striae mm
?	*Thylacoleo carnifex*	0.0096 ± 0.0065 S.D. (n = 22)
CARNIVORES	*Thylacinus cynocephalus*	0.0014 ± .0008 (n = 43)
	Sarcophilus harrisii	0.0012 ± .0006 (n = 41)
	Canis familiaris	0.0012 ± .0007 (n = 21)
	Felis catus	0.0015 ± .0015 (n = 24)
	Facsimile thylacoleonid	0.0020 ± 0.0015 S.D. (n = 21)
HERBIVORES	*Macropus giganteus*	0.0005 ± 0.0001 S.D. (n = 17)
	Macropus rufus	0.0008 ± 0.0004 (n = 22)
	Trichosurus vulpecula	0.0005 ± 0.0001 (n = 23)
	Phascolarctos cinereus	0.00015 ± 0.00006 (n = 29)
	Facsimile thylacoleonid	0.0006 ± 0.0002 (n = 24)

fail to penetrate it; four slight indentations, made by the tips of the canines, may be the only evidence of the throat hold. Death in this and other large species appears to be due to strangulation. Another method of killing is suffocation; a lion places its mouth over the muzzle of its prey and holds it shut until the animals dies".

The same pattern is true for the cheetah. Schaller (1972) notes – "The cheetah tries to grab the throat of its quarry, usually remaining behind the animal . . . although the skin was often punctured by the canines and occasional chewing movements were made by the cheetah, death in most instances seemed to be due to strangulation". Other authors confirm this. Kruuk and Turner (1967) describe Cheetah kills resulting from "a very fast chase after which the victim is knocked over with a forepaw. The victim is killed by suffocation through a bite in the throat or around the muzzle". Eaton (1970) observed that "Teeth did not puncture major vessels and no necks were broken or damaged . . . loss of blood was not appreciable, and apparently not the cause of death. Punctures of the preys skin were slight and not always present . . . Death . . . is the result of strangulation".

The Cheetah's skull shows morphological features associated with this killing pattern (Ewer 1973) – "The canines are small by felid standards and not very much flattened". Ewer also notes that the Cheetah has a very small gape.

It is clear from the above that specialized "killing teeth" are not necessary for a carnivore, and that *Thylacoleo* could have killed by strangulation or suffocation. It is of interest that most of the large carnivores use the front paws to knock down the prey and to hold it down while strangling it. The large claw on the manus of *Thylacoleo* (Wells and Nichol 1977) and long forelimb (Finch 1971) would be admirably suited to this function.

Granted that *Thylacoleo* could kill other animals, could it then remove flesh from the carcass? Anderson (1929) thought not – "Nor did *Thylacoleo* possess a convenient provision for removing the flesh of its victim. Typical placental and marsupial carnivores are provided with transverse incisor rows above and below, which enable them to tear the flesh from their prey, but *Thylacoleo* had no teeth adapted for this purpose". It could be noted (immediately) that on this basis a sabretooth could not be carnivorous either. But let us see how modern carnivores eat.

Schaller (1967) notes that "The tiger's carnassial teeth are highly efficient cutting tools, parting the skin of its victim like scissors; with a combination of cutting, pulling and tearing, the cat rapidly bolts down the meat, skin and viscera". Schaller (1972) observed Lions "cutting skin, fascia and other tough parts with their carnassials, tearing chunks of muscle and viscera with canines, and pulling off small pieces with their incisors". It seems that *Thylacoleo* could have coped with a carcass quite well by using its premolars, with some help from the incisors.

Finally there is the question of tooth marks on bones. Many of the large carnivores, Lions, Tigers and Leopards, often eat a large number of the bones in a carcass. However as noted earlier, this is unlikely to have been the case with sabretooths, nor is it the case with Cheetahs. Ewer (1973) suggests — "The cheek teeth are extremely narrow and blade–like, adapted almost exclusively for meat slicing, which suggests that the Cheetah must be incapable of dealing with bones of any size". Eaton (1970) notes "the Leopards carnassials have cusps on the inner edges which are an adaptation for crushing. The Cheetah lacks these cusps". Brain (1970) analysed the carcasses left by Cheetahs and confirms these hypotheses — "damage to bovid skeletons in the 50–150 pound range was minimal. It was restricted to fragile parts such as ribs, vertebral processes and scapulae".

A small number of fossil bones bearing cut marks attributable to *Thylacoleo* have been reported from Queensland (De Vis 1883) and Victoria (Spencer and Walcott 1911) and most recently from Lancefield in Victoria (Horton *et al.* 1979). Although the evidence from Victoria sites is persuasive it is interesting that remains of *Thylacoleo* have not been found in association with incised bones. Gill (1954) has noted that sites such as Buchan Caves (Victoria) and Scotchtown Cave (Tasmania) where *Thylacoleo* remains are relatively common do not yield incised bones. This has also been our experience with the Victoria Cave (South Australia). That incised bones and the remains of *Thylacoleo* have not been found together may reflect the mode of accumulation of the fossils rather than the absence of an interaction between the species. More significantly one of us (Horton) reports that only 0.35% (45) bones, mainly foot elements, from Lancefield bear cut marks that could be attributed to *Thylacoleo*. Such evidence is not incompatible with a basically flesh–eating habit.

Finch (1971) in a study of the skeleton of *Thylacoleo* noted that the fore and hind limbs are of approximately equal length and are long in comparison with the length of the vertebral column, characters which we have confirmed from the Naracoorte specimens. Finch pointed out the similarities to climbing forms such as the Koala (*Phascolarctos*) and the Brush–tailed Possum (*Trichosurus*) but implied that the absence or mobile digits on the forepaw would exclude an arboreal niche. Wells and Nichol (1977) in a detailed study of the manus and pes demonstrated the grasping ability of both as ideal adaptions for climbing.

Assigning extinct animals to niches on the basis of a functional analysis of their morphology can only ever be an historical interpretation. It is made more convincing if it correlates well with modern functional analogues, but even here one could envisage difficulties, e.g., given only the skeleton of a tree kangaroo one might be hesitant in concluding that the saltorial nature of the hind limbs was an adaptation to climbing. Bearing these difficulties in mind we advance the following interpretation for this paradoxical animal namely

(i) that *Thylacoleo carnifex* was a flesh eating marsupial of diprotodontid ancestry;

(ii) that its limb adaptations were for climbing and grasping rather than running;

(iii) that the absence of interlocking canines as a means of holding prey was compensated by the powerful grasping mechanism and large clawed first digit of the manus;

(iv) that it killed its prey by holding it against the upper incisors and stabbing with the lowers possibly leading to strangulation, suffocation or severence of the vertebral column of the prey;

(v) that it used its large shearing premolars to divest its prey of flesh.

This interpretation leads to the question as to whether it preyed on arboreal animals. *Thylacoleo* was equivalent in size to a Leopard and in the absence of any brachiating adaptations it seems unlikely that it pursued arboreal species which can move with ease through the highest branches. What then could be the prey of this relatively large extinct marsupial? *Thylacoleo* remains are often found in association with extinct leaf–eating kangaroos of the genus *Sthenurus*. Studies of the skeletons of these kangaroos (Wells in prep. and Wells and Tedford in prep.) indicate that they were ground dwelling species. They would form suitable prey for an animal the size of *Thylacoleo*. It is our speculative hypothesis that *Thylacoleo* preyed on these kangaroos and that it hauled its prey into trees as does the leopard, thereby removing it from the reach of scavenging forms such as the Tasmanian Devil (*Sarcophilus*). Such an hypothesis is the simplest which accords with all the available evidence and provides a niche for a carnivore which avoids competition with the Thylacine *Thylacinus*, the only other large Australian marsupial exhibiting predominantly flesh–shearing cheek teeth.

ACKNOWLEDGEMENTS

This work was supported by a grant from the Australian Research Grants Committee. We extend our thanks to John Warman for his help in the construction of the facsimile thylacoleonid and Ms. Annette Ward for typing the manuscript. Jack Mahoney and Eileen Finch provided useful criticism of a draft of this manuscript.

REFERENCES

Abbie, A. A., 1939. A masticatory adaption peculiar to some diprotodont marsupials. *Proc. Zool. Soc. Lond. B* **109** (2): 261–79.

ANDERSON, C., 1929. Palaeontological notes No. 1. *Macropus titan* Owen and *Thylacoleo carnifex* Owen. *Rec. Aust. Mus.* **17:** 35–45.

BARTHOLOMAI, A. AND MARSHALL, L. G., 1973. The identity of the supposed dasyurid marsupial, *Sarcophilus prior* De Vis, 1883, with comments on other reported 'Pliocene' occurrences of *Sarcophilus*. *Mem. Qd Mus.* **16:** 369–74.

BOCK, W. J. AND SHEAR, CH. R., 1972. A staining method for gross dissection of vertebrate muscles. *Anat. Anz. Bol.* **130, S:** 222–7.

BRAIN, C. K., 1970. New finds at Swartkrans Australopithecine site. *Nature* **225:** 1112–9.

BROOM, R., 1898. On the affinities and habits of *Thylacoleo*. *Proc. Linn. Soc. N.S.W.* **23:** 57–74.

DE VIS, C. W., 1883. on tooth-marked bones of extinct marsupials. *Proc. Linn. Soc. N.S.W.* **8:** 187–90.

EATON, R. L., 1970. The predatory sequence, with emphasis on killing behaviour and its ontogeny in the Cheetah (*Acinonyx jubatus* Schreber). *A. Tierpsychol.* **27:** 492–504.

EWER, R. F., 1967. The fossil hyaenids of Africa — reappraisal. Pp.109–23 *in* "Background to evolution in Africa", ed. by W. W. Bishop and J. D. Clark. University of Chicago Press: Chicago.

EWER, R. F., 1973. "The carnivores". Weidenfeld and Nicholson: London.

FINCH, E., 1971. *Thylacoleo*, marsupial lion or marsupial sloth? *Aust. Nat. Hist.* **17:** 7–11.

FLOWER, W. H., 1868. On the affinities and probable habitats of the extinct Australian marsupial *Thylacoleo carnifex* Owen. *Quart. J. Geol. Soc.* **24:** 307–19.

GILL, E. D., 1951. New evidence from Victoria relative to the antiquity of Australian Aborigines. *Aust. J. Sci.* **14:** 69–73.

GILL, E.. D., 1952. *Thylacoleo* and incised bones. *Aust. J. Sci.* **14:** 201.

GILL, E. D., 1954. Ecology and distribution of the extinct giant marsupial, *"Thylacoleo"*. *Vic. Nat.* **71:** 18–35.

GREAVES, W. S., 1978. The jaw lever system in ungulates: a new model *J. Zool., Lond.* **184:** 271–85.

HORTON, D. R., WELLS, R. T. AND WRIGHT, R. V. S., 1979. *Thylacoleo*, marsupial carnivore. Abstract only. *Aust. Mamm. Soc. Bull.* **6** (1): 25.

HYLANDER, W. L., 1979. Mandibular formation in *Galago crassicaudatus* and *Macaca fascicularis*: An *in vivo* approach to stress analysis of the mandible. *J. Morph.* **159:** 253–96.

KREFFT, G., 1866. On the dentition of *Thylacoleo carnifex*. *Ann. Mag. Nat. Hist.* **3** (18): 148 9.

KRUUK, H. AND TURNER, M., 1967. Comparative notes on predation by Lion, Leopard, Cheetah and Wild Dog in the Serengeti area, east Africa. *Mammalia* **31:** 1–27.

LEYHAUSEN, P., 1965. Uber die Funktion der Relativen Stimmungshierarchie. *Z. Tierpsychol.* **22:** 412–94.

MURCHISON C., 1868. "Palaeontological memoirs and notes of the late Hugh Falconer A.M.M.D." London.

OWEN, R., 1871. On the fossil mammals of Australia. 4. Dentition and mandible of *Thylacoleo carnifex* with remarks on the arguments for herbivority. *Phil. Trans. Roy. Soc. Lond.* **161:** 213–66.

SCHALLER, G. B., 1967. "The deer and the tiger". University of Chicago Press: Chicago.

SCHALLER, G. B., 1972. "The Serengeti Lion". University of Chicago Press: Chicago.

SPENCER, B. AND WALCOTT, R. H., 1912. The origin of cuts on bones of Australian extinct marsupials. *Proc. Roy. Soc. Vic.* **24:** 92–133.

TURNBULL, W. D., 1978. Another look at dental specialisations in the extinct sabre–toothed marsupial, *Thylacosmilus*, compared with its placental counterparts. *In* "Development, function and evolution of teeth" ed. by P. M. Butler and K. A. Joysey. Academic Press: New York.

WELLS, R. T., 1974. Reconstructing the past. Excavations in fossil caves. *Aust. Nat. Hist.* **18:** 208–11.

WELLS, R. T. AND NICHOL, B., 1977. On the manus and pes of *Thylacoleo carnifex* Owen (Marsupialia). *Trans. R. Soc. S.A.* **101:** 139–46.

WOODS, J. T., 1956. The skull of *Thylacoleo carnifex*. *Mem. Qd Mus.* **13:** 125–40.

RESUMEN

48. *Thylacoleo carnifex* Owen (Thylacoleonidae): un marsupial carnivoro? Por R. T. Wells, D. R. Horton y P. Rogers.

El estudio del desgaste tegótico de los dientes combinado con la reconstrucción de los mecanismos mandibulares indican que *Thylacoleo carnifex* usó sus incisisivos inferiores para apuñalar y sus terceros premolares para cortar. Estos movimientos parecen haberse adquirido con una musculatura mandibular esencialmente falangérida. Un análisis de las facetas del desgaste de estos dientes muestra una mayor similitud con las de los carnívoros que con los herbívoros. El apoyo empírico a esta hipótesis se obtuvo mediante la construcción de un facsímil de tilacoleónido y "alimentándolo" con diferentes dietas. Se concluyó que *Thylacoleo carnifex* fue un carnívoro adaptado a trepar. Se sugiere un nicho semejante al del leopardo.

49. THE BUILDER AND THE BRICKS: NOTES TOWARD A PHILOSOPHY OF CHARACTERS

JOHN A. W. KIRSCH[1]

It is common to find, when attempting to reconstruct phylogeny, that different characters give different arrangements; this is true even when logical mistakes (such as basing groups on shared primitive rather than shared derived features) are corrected. Discrepancies among character–phylogenies are usually the result of convergence, parallelism, or reversals of state (collectively termed *homoplasy*), and one solution to the problem they raise is the democratic one of accepting the indications of the majority — a solution said to be based on parsimony, but really the method of minimum pathlength. Adopting the shortest tree is equivalent to ignoring homoplasy and thus a formal contradiction of true parsimony, which is the philosophical position that the most general, all–encompassing explanation is to be preferred to several specific ones. Since homoplastic resemblances normally represent adaptive similarities arrived at by independent evolutionary routes, a more likely solution to the problem of homoplasy would be to *weight* characters negatively on the basis of their sensitivity to environmental constraints. One such scheme is suggested by the Aristotelean conceit of multiple causality: anatomical characters are to be trusted least since, being adaptive, they serve the final cause of organisms (their reproductive fitness). On the other hand, an impressive body of evidence indicates close tracking of divergence in biochemical characters (specifically, structural genes and their products) with time, making them ideal for use in the reconstruction of phylogeny. Physiological processes and the program of development occupy the middle ground of, respectively, efficient and formal causes, and ontogenetic analysis in particular may be useful in determining the place of individual characters in an hierarchical weighting scheme.

Pages 587–94 *in* CARNIVOROUS MARSUPIALS, ed. by M. Archer, February, 1982. Roy. Zool. Soc. New South Wales: Sydney, Australia.

THESE DAYS the word "relationships" connotes mainly one aspect of phylogeny, cladistics, or the sequence of branchings that produced the natural groups of organisms we recognize through taxonomic procedures. There is considerable agreement in these volumes ("Carnivorous marsupials") about the groupings of marsupial carnivores, but there are also minor and major discrepancies. One point that nearly all authors agree upon is that *Antechinus* (*sens.*, e.g., Ride 1970) is polyphyletic; they do not all agree as to how the separate sections interrelate or stand with respect to other genera. And the issue of bandicoot affinities remains unresolved by these new studies.

No one, of course, expects perfect congruence among taxonomic characters, but there does seem to be a consensus that progress in systematics ought to lead to general agreement. Nevertheless, taxonomy cannot be pure and simply a numbers game: some judgment as to the relative worth, or weight, of characters must be made at some point. That judgment is not easy, but one can start by attempting to understand what it means when characters do *not* agree in their indications for phylogeny. I shall describe some of the reasons, and suggest what we have to do to make our arrangements "come out right," using mostly examples from papers presented in this symposium, alluded to only by the authors' names.

I begin by making some general points within the current outlines of systematic theory, and then proceed to consider a possible context in which some problems may be understood and perhaps resolved. The title should be taken seriously: these are a series of notes for ideas that cannot be fully worked out in this space; many important aspects of character–analysis, such as the subtraction of fabricational noise (Seilacher 1973), have not even been considered. Further, this paper obviously owes much to discussions that are common in the systematic literature, and because of that I have not always referenced each point thoroughly.

RECONSTRUCTION OF PHYLOGENY: LOGICAL CONSIDERATIONS

The first, and most obvious, source of disagreement among character–sets is a logical one relating to the fundamental principle of cladistic analysis, that only shared, derived character–states have meaning for documenting branching points (Hennig 1966). No one would place the monotremes in a clade with reptiles just because they lay eggs: laying eggs is a primitive feature that apparently had not changed early in the evolution of mammals. Equally, few would doubt that since marsupials and eutherians bear living young these two taxa are each other's nearest relatives within Mammalia (in fact, some have disagreed, but that disagreement illustrates a different point, considered below). All therians share a derived character–state and are therefore deemed to be products of a recent branching, attested to by the transition in state of the character "mode of reproduction" from "egg–laying" to "live–bearing". The character–state shared by monotremes and various kinds of reptiles, amphibians, and fishes demonstrates no

[1] Museum of Comparative Zoology, Harvard University, Cambridge, Massachusetts, U.S.A. 02138.

special relationship among these groups: they could have diverged at various times, and reproductive mode alone provides no marker of these events.

If we accept that shared primitiveness cannot be the evidence for a special branching relationship, many of the differences between groupings based on distinct character–sets are resolved. For example, there is no inconsistency between Harding's and Renfree *et al.*'s demonstration that the sperm of *Tarsipes* are very like the dasyurids' and the more common view that *Tarsipes* is related to other diprotodontans, with which it shares a number of derived states such as diprotodonty, syndactyly, and a superficial thymus. Presumably, *Tarsipes* simply retains the primitive sperm morphology that must have characterized the earliest diprotodontans.

Of course, agreeing that shared, derived states are the only proper basis for inferring cladistic events just raises the question of how to make the distinction between primitive and advanced. That is a question that is easier to answer in the abstract than in particular cases: one author may argue strenuously that a character–state is derived which another holds to be primitive. The chief criteria used to determine the direction of change are "common is primitive" and "outgroup comparison". The first simply means that the character–state found in the widest range of taxa is likely to be the primitive condition, while the second criterion is that the state held in common by some members of the group and a taxon known to be phylogenetically distinct from it is probably ancestral. These two criteria may sometimes intergrade (Estabrook 1977), encompass other variants (Stevens 1980), or be in conflict. In the case of reproductive mode, "common is primitive" would clearly be misleading, since most mammals are live–bearers; egg–laying is regarded as primitive, because several outgroups with respect to mammals display the state. On the other hand, for the larger taxon Vertebrata the principle of commonality is sufficient, applicability always being relative to the taxonomic level being considered. Determination of the direction of change in marsupial karyotypes shows where both criteria are used and yet may conflict (e.g., Sharman 1961, 1973; Hayman and Martin 1974): while $2_n = 14$ is clearly the most widespread number among marsupials, their outgroup, Eutheria, has a much higher average number. The question is therefore whether the wide taxonomic distribution of $2_n = 14$ is a better approximation to the starting–point in marsupial chromosome evolution than the closer match between $2_n = 22$ and the modal value of the outgroup. Whichever number is chosen, the decision affects our view of marsupial phylogeny, since it determines whether we are more likely to consider $2_n = 14$ or $2_n = 22$ derived and hence a basis for special cladistic groupings.

CONVERGENCE: THE CENTRAL PROBLEM

The Pervasiveness of Homoplasy

The reproduction example above was chosen advisedly, because there has been some argument about whether marsupials and eutherians can be convincingly related just on this basis (Lillegraven 1969, Sharman 1970, Tyndale–Biscoe 1973). This is not because anyone doubts that the character–state "live–bearing" is derived, but rather because it seems possible that this condition could be reached independently in the two therian groups; after all, vivipary has evolved separately in fish, sharks, snakes, and lizards as well. Mistaking states evolved convergently or in parallel for true homologies is a common reason for disagreement about phylogeny; at least for simple characters, there is also the possibility of reversion to a previous condition. Parallelism, convergence, and reversion together are now generally called *homoplasy*. Many discrepancies among the cladistic hypotheses proposed herein, such as the difference of opinion between Szalay and myself about the position of *Dromiciops* (Kirsch 1977), or that between Sarich *et al.* and Archer (1976) on the place of the Thylacine, must ultimately be due to homoplasy in at least one set of data. The question is, which set of data is misinterpreted? Or are both homoplastic?

Escape through Parsimony

Since one goal of character–analysis is achieving concordance among disparate data–sets, the procedure of most taxonomists has been, as it were, to flip the coin again. If two sets are in conflict but a third agrees with one or the other of them, then the set agreed with is said to be confirmed, while the discrepant set is falsified. The usual interpretation of dentition places *Dromiciops* with other didelphoids rather than with Australian marsupials, as its feet do; serological data indicate a real, though distant, alliance with Didelphidae, confirming the dental characters. This scorecard approach to resolving questions of relationship is generally referred to as "parsimony". Obviously, the more sets of data in agreement the stronger one's confidence in the confirmed set.

Matters are not always so simple, of course: Archer's list of derived characters shared by thylacinids and borhyaenids is a long one. If Archer and the earlier anatomists who reached conclusions similar to his are wrong, then the "lesson of the thylacine" is just that convergence can be extremely close, involving very many characters. Part of the difficulty in applying the criterion of parsimony is the classic one of deciding what "a" character is: are the numerous dental characters listed and examined by Archer one or many? Doubtless some features of dentition imply others (the numbers of upper molars rarely differ by more than one from the lowers, and are usually complementary in shape for obvious functional reasons). Whether the details of molar structure are quite separate characters from, or rather necessarily implied by, the shapes of other kinds of teeth is arguable. Yet the whole point of taxonomy is to discover correlations among features, so that it might be counter–productive to discount a data–set just for this reason.

Problems with the parsimony criterion – and thus of algorithms for dealing simultaneously with many characters – do not stop there. First, there is the argument that confirming instances are no proof of an hypothesis anyway: the proper business of science is to try to disconfirm, or falsify, hypotheses, so that amassing data which are mutually supportive tells you nothing. The modern proponent of that view is Popper (1968), but it strikes me as inherently paradoxical when applied to taxonomy. A probabalistic approach to taxonomic characters, as in biology generally, is clearly more appropriate (e.g., Harper 1979).

There is a more serious problem with naive character–counting. It is usually maintained that parsimony is a starting–point, or the only concievable methodological principle (Novacek 1980), but using the simplest tree to detect homoplasy is in effect a statement that most of the time convergence does not occur. On the contrary, much homoplasy seems to be present in every large data–set, particularly those that have directly to do with gross anatomy.

More importantly, the equation of parsimony with character–compatibility is a trivialization of Ockman's argument that a unitary explanation for a wide range of independent phenomena is to be preferred over several that account for each set of events separately. Classifications are a means of summarizing (or rather indexing, Hull 1970) the distribution of characters over a set of taxa. They are descriptive of associations of features, and it is that correlation which gives classifications some "predictive" value: knowing that two species are congeners, for example, means that they probably have a long list of characters in common. This sense of prediction is quite different from what is meant by "cause and effect" in physics or most other areas of biology. Predictions in classification thus are primarily descriptive statements, and have little explanatory power. To treat classifications as hypotheses in the usual sense vulgarizes what the latter are meant to do.

One comes a lot closer to Ockham's meaning in the mutual testing of classificatory statements and biogeographic hypotheses, since the critical thing becomes exactly that the same causal explanation should apply to a number of potentially independent taxonomic and geographic relationships. As Ashlock pointed out (1974, summarizing earlier discussions of Hennig and others), one of the most attractive "uses" of cladistics is to establish a phyletic framework which can be compared with the history of the places where those organisms live. Much of the excitement among biologists about plate tectonics comes from just such comparisons. Whereas Wegener (1924) used biological relationships to support his theory, the facts of continental drift are now used to validate phylogenies. If South America and Australia have been disjunct for forty million years or more, we might expect a degree of divergence between the geographic representatives of taxa common to both continents reflecting this separation in time and place. Special relationships – such as the mooted thylacinid–borhyaenid one – would obviously require an exception to passive separation and is correspondingly unparsimonious in the true sense. In addition, the causal connection between geological history and phylogeny receives further support from the repetition of the pattern of relationships for many taxa. Using serological comparisons, Maxson *et al.* (1975) showed that tree frogs in Australia and South America have much the same temporal and phyletic relatedness to each other as marsupials. There is no sense in which one can say the frog and marsupial albumins are part of the same data–set!

The contrast between even this simple application of true parsimony and the vulgar parsimony implied by constructing the shortest trees for possibly correlated characters is important. Moreover, using historical biogeography to test various taxonomic characters frequently indicates that some sorts of data are "better" than others (the Wagner trees of marsupicarnivores shown in my paper with Archer are inconsistent with serology, for example), and suggests that certain kinds of information are *intrinsically* more likely to give reliable phyletic information. It is thus also necessary to recognize degrees of reliability and to *weight* characters accordingly.

Escape through Weighting

Pattern and process: The useful and truly parsimonious combination of earth history with taxonomic characters suggests that satisfactory reconstruction of phylogeny requires considering the processes that give rise to patterns of character distribution among taxa.

Yet there is a general feeling among taxonomists (articulated by Eldredge and Cracraft 1980) that the proper business of taxonomists is to consider only the patterns – or at least that establishing these must come first. On the contrary, an understanding of process can, as I have argued elsewhere (Johnson *et al.* 1980), alter the very perception of pattern. The most obvious examples come from ontogenetic studies such as where states previously thought primitive turn out to represent the retention of early parts of an ontogenetic series into adult life. As Nelson (1978) has rightly concluded, these retained juvenile features are then properly considered advanced, which clearly makes a substantial difference to the arrangement of organisms in a phyletic sequence. Although working out the relationship of ontogeny and phylogeny has been part of phylogenetic speculation at least since Darwin, the contemporary formalisms for using ontogeny in classification have been developed by Gould (1977) and implemented by several authors (Albrecht *et al.* 1979).

Structural vs. regulatory change: Awareness that process influences relationships is one basis of the distinction between structural and so-called regulatory gene changes, popularized by Wilson and his colleagues (Wilson *et al.* 1974). Their explanation was designed to explain the disjunction between schemes based on characters which mark the passage of time in a regular way (proteins) and those which seem to evolve by fits and starts (anatomy). The property required of regularly evolving characters is that they be insensitive to environmental demands, or selectively neutral; the arguments that this is at least approximately true for structural genes and their protein products are now familiar, deriving largely from Kimura's (1968) demonstration that mutation and fixation rates should be equal for neutral characters. If that is so, then similarity in selectively neutral characters (or, more exactly, their underlying genetic coding) should be inversely proportional to the lengths of time of separation from common ancestors. That is a generality that has been shown to be approximately true over long periods of evolutionary time (Wilson *et al.* 1977). Short–term exceptions occur for stochastic reasons: change in nucleotide sequences or proteins is often likened to radioactivity decay, and given fewer "nuclei" involved in single gene changes, short–term irregularities should occasion no surprise. But it is often overlooked that the possibility of mistaking branching *sequences* (as opposed to mis–estimating branch *lengths*) is a real one over the short term (Romero–Herrera *et al.* 1978).

Clearly anatomical characters do not evolve at steady rates. The structural vs. regulatory paradigm holds that large–scale anatomical differences may be caused by relatively small (and sometimes rapid) genetic change or even gene-rearrangement. Wilson *et al.* (1974) argue that the great anatomical and chromosomal diversity of mammals in comparison with frogs is explained in this way, although their albumins appear to have changed at the same rate.

Relative rates of genetic and regulatory change in mammals need not follow this pattern. The Dasyuridae provide a striking counter–example, since their chromosome numbers and banding patterns, and distribution as well as amounts of repetitive DNA (one candidate for a regulatory role), differ very little over the entire family, while electrophoretic distances (a measure of structural change) are as great as in other families of mammals.

No easy dichotomy between rates for the two types of taxonomic characters is likely to be universally applicable, but the conclusion that anatomical characters are unlikely to provide a clock as useful as the biochemical one is almost self–evidently true. The strategy of cladistics is designed to overcome this limitation of morphology by avoiding the question of degrees of resemblance and making the issue in phyletics one of branching *sequences* rather than of evolutionary *rates*. Even recognizing that there are rate differences among characters does not solve the problem of homoplasy, for which a theory of the likelihood that various sorts of characters will converge is needed.

Multiple causality: A key to developing such a theory lies in the recognition that phenomena may be understood or explained in terms of several distinct types of causes. For Aristotle there were four – the material, efficient, formal and final. His literally homely analogy was of a house, its material cause being the bricks of which it is made. The house's efficient cause is the labor that went into its construction — the assembly of the bricks — while its formal cause is the plan which specified how those bricks were to be put together. The final cause of a house, however, is its purpose as a structure to live in.

Modern science tends to focus on efficient causes, which are accessible to experiment and falsifiable hypotheses, although study of formal causes is often implicit, especially in molecular biology. Material causes are denigrated as "mere description", and final causes as vitalistic. But consideration of taxonomic characters in Aristotelean terms can show why some are more trustworthy than others for phylogenetic reconstruction, and thus unsprings the trap of vulgar parsimony.

The neutralist argument is clearly a call for study of the "bricks." In principle, there seems to be no reason why all houses could not be built of the same materials, but, as Williams (1966) points out, it is just the inability of natural selection to reduce the mutation rates to zero that ultimately accounts for change. The limited variation observed in proteins and their underlying genetic codes results from a truly random process (thermal noise), amplified to the phenotypic level. The reason for studying biochemistry, then, is not that it is virtuous to be "close to the genome" (some biochemical characters are not: see Fig. 1), but that the processes which are responsible for the anatomical characters operate ultimately on the molecular "building blocks" of proteins.

But the universality of biochemical pathways is another reason why the changes in biochemicals may be relatively slow and steady. Smith and Morowitz (1982) point out that early establishment of common biosynthetic and physiological pathways, as well as of the genetic code, constrained many subsequent changes. Any mutation that interferes at this fundamental level – deep, as it were, in the hierarchy of historical changes – must be compatible with every process dependent upon it.

Reliable phylogenetic indicators will therefore be those related to such fundamental characters, and the reason that macromolecules track the passage of time so well is not only because their variants are selectively neutral, but because so many possible variants are impermissible in the context of hierarchically deep processes. Some proteins, like the "families" of cytochromes *c*, may in fact change to the point where functional considerations influence judgments of homology more than structure (Dickerson 1980).

Physiological processes in this sense can be considered the efficient cause of an organism, but not all such processes of assembly and functioning are hierarchically deep. The hierarchical depth of a feature is probably best appreciated through ontogenetic analysis, and it is now almost trite to point out that many of the differences among species, and even perhaps the origins of major groups, may be explained by alterations in the timing of events in development, by allometric growth, by the monstrous results of ontogenetic variants, or other epigenetic mechanisms (Gould 1977, Rachootin and Thomson 1981). The unfolding program of ontogeny is equivalent to Aristotle's formal cause, and since Gould has even tied heterochronic changes to *r* and *K* selection, Van Valen's aphorism (1980 and earlier papers) that evolution is the control of development through ecology takes on particular meaning.

What is often forgotten, however, is that phyletic relationships are often assumed prior to ontogenetic analysis: the pattern is explained, but not established, by the developmental argument. Nevertheless, I would state again that understanding the reciprocal relationship between pattern and process need not be entirely circular, but mutually illuminating.

The fourth component of causation involves questions of purpose and is traditionally the most disreputable phase of biological study. But science, especially biology, is pointless without the direction given by teleology. That does not all imply that answers to questions of purpose need be transcendental: the purpose of an organism is to reproduce, and all of its adaptations can be analyzed from this point of view. Teleology is a program for research, and the study of physiology, for example, is inconceivable without a functional assumption (McCune, unpublished ms.).

DISCUSSION

The materials of which organisms are made are put together according to a plan in order to increase their fitness. Thus may the Aristotelean approach to character–weighting be summarized, for the usefulness of characters in reconstructing phylogeny is proportional to the extent to which they reflect the formal, efficient, or final causes on the one hand (the builder) or the material cause on the other (the bricks). Because gross anatomy is the level at which the organism most directly interacts with its environment, morphology is primarily the study of adaptation. Since the demands of the environment are almost boringly repetitive, it is at the anatomical level that we might expect to encounter homoplasy most often. Viewing organisms through a haze of ecological prejudice will almost always subvert the phyletic goal, and ecology is thus the enemy of phylogenetic taxonomy. To the extent that functional analysis of anatomy also documents adaptive response, it too, may fall short of giving the true history of organisms.

The study of characters reflecting purpose is likely to generate branching sequences far removed from the true cladistics of the group, and if we fail to recognise the the ubiquity of convergence we additionally commit the fundamental logical fallacy of imputing the same cause (in this case, the "same" evolutionary pathway) to similar effects. On the other hand, recognition in specific cases that only similar ends have been reached (the wings of bats and those of birds) demonstrates the distinctness of the pathways followed. But by definition "convergence" supposes that we can recognize apparent similarities as superficial, incomplete. Were convergence total, we would never know it to have occurred. Hence all our examples of the phenomenon are bound to seem obvious, and the problem less real than it in fact is.

The long controversy about the regularity of, or lack of it, in protein evolution arose in part because macromolecules do differ among themselves, as do anatomical characters. But they differ only because organisms are imperfect in maintaining the exact same genotype through the generations. The imperfection, combined with selective neutrality, is the basis of the evolutionary clock. If we desire, first of all, to obtain a record of phyletic branching, then clearly biochemistry has primacy of place in this endeavour.

That is not to say that biochemical compounds are *never* subject to selection – indeed, the paradox is precisely that the constraints may be so tight on their functioning, and the relationship between structure and operation so intimate, that virtually no alteration is possible without physiological disaster, as is true for some haemoglobins. It has even been suggested that the long–term regularities in rate can actually be accounted for by selection, as, by the Red Queen (Van Valen 1974). Under this hypothesis each organism is imagined to be on an adaptive treadmill, racing just to keep up with every other. The multiplicity, constancy, and descriptively random character of selection pressures is such that the alterations in protein structure would — being so continuous — mimic the apparently stochastic changes predicted by the neutrality hypothesis. On the other hand, apparent lack of morphological changes in some organisms over long time spans might be underlain by a flurry of genotypic alterations all producing or maintaining the same anatomy, as is implied by the unexpected degree of the genetic variation in *Limulus* (Selander *et al.* 1970).

Notwithstanding the examples of direct and obvious selection on proteins, were selection on protein structure the rule, short–term departures from regularity would be considerably greater than they are. However, the equation of particular types of causes with kinds of characters is not perfect: the important point is that biochemical features are *more* material than are anatomical ones. Rather, the most obvious limitation of biochemistry is that its study is largely restricted to living organisms.

Another important limitation of material analysis is that the structure of the genome is turning out to be much more complex than we had imagined only a few years ago (Brown 1981). Far from being a linear sequence of nucleotides corresponding to a sequence of amino acids, with occasional punctuation marks, genes are interrupted, have their parts scattered, change position, can be transferred between species, or may even originate by viral insertion; nuclear DNA may include enormous amounts of apparently non–functional or possibly parasitic components. The reductionist hope of twenty years ago that we might learn everything there was to know about an organism by sequencing its genome (which is now close to being technologically feasible) seems pious and naive: to understand how eucaryotic organisms are put together we have to understand much more about how the genome is organised and operates in ontogeny.

But that, of course, is just the point: mechanisms for reorganising the genotype are both many and swift to evolve; the complex way the coding system works is part of formal and efficient causality and is not necessarily inconsistent with the time–dependent nature of change in the materials which are shuffled about. Kohne's scheme (1970) for the origin of repeated DNA sequences by amplification of unique ones may be too simple, but the evidence (largely work of Sibley in prep.) is now good that comparison of large numbers of unique DNAs from many species gives an accurate measure of divergence times.

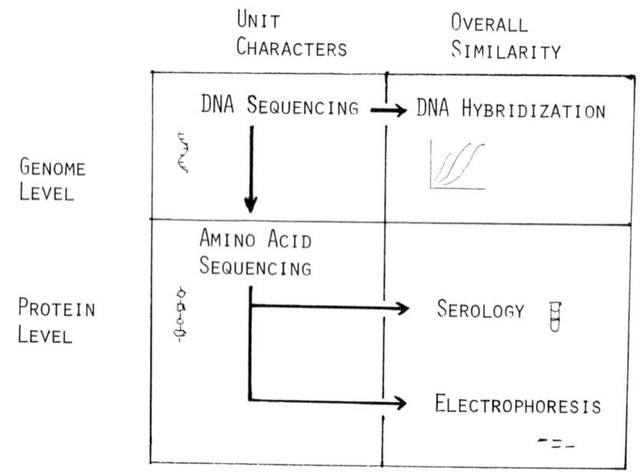

Fig. 1. Modes and levels in the study of structural genes and their products. The various kinds of biochemical investigation are arranged according to whether they take place at the level of genes or their products (rows), and whether the results are expressed in terms of unit characters or overall similarity indices (columns). Electrophoretic data are considered to be single–number measures because only one property of the whole protein — charge — is assessed.

The limits of biochemistry for phylogenetic reconstruction are instead technological and interpretive. Many of the methods of biochemical systematics produce information several steps removed from nucleotide sequences (Fig. 1). This is especially true for electrophoresis, which measures only overall charge and is therefore highly sensitive to mistaken homology; but it is also true for serological resemblances and even amino–acid sequences. Recent DNA sequencing shows that our guesses about the underlying codes for some proteins were incorrect since the degeneracy of the code was not sufficiently taken into account (Jukes 1980).

There is nonetheless a remarkably close tracking between nucleotide sequence, amino–acid sequence, serological resemblances, and, at least at low categorical levels, electromorph similarities. Nevertheless, mistaken homologies and variations in rates of change do adversely affect reconstruction of true branching sequences. There is much similarity between Sarich's marsupial phylogeny and my own, despite technical differences in our serological methods, except for a rather striking discrepancy within Diprotodonta. Sarich *et al.* show the kangaroos well separated from all other diprotodontans, while the majority of my experiments separate kangaroos, together with the various phalangers, from other diprotodontans. The methodological difference is that Sarich's work involves only one protein while my results are based on whole sera. In this case at least the reason for the discrepancy may be that albumins can evolve somewhat faster in some marsupial lineages than in others, as admitted by Maxson *et al.* (1975). Since nephelometric comparisons of whole sera probably measure mainly albumin similarities, it is interesting that Wagner distance analysis of my quantitative data (from Kirsch 1977) gave the same relationships among diprotodontans as did the microcomplement fixation tests on albumin alone. Similarly, the close affinity of the Thylacine with a few dasyurids, which is a remarkable feature of Sarich's results, may be exceptional. It is also worth pointing out that ordering of divergences where branch–lengths are short (as they are in this region of the tree) may be unduly influenced by experimental error, real homoplasies, or stochastic events (Kirsch 1969).

This last point raises the question of how biochemical data should be interpreted. It is probable that their interpretation should not be straightforwardly phenetic. Most clustering algorithms begin with the assumption that the most similar taxa are closely related and distant ones less so. The assumption inevitably constrains the tree–structure. Yet Wagner distance analysis may, because of variation in rates of change in different branches, yield a tree where the most similar taxa are not necessarily the closest cladistically (Bottjer unpublished, Farris 1972). It is not even clear that a simple Euclidean interpretation of distances is entirely appropriate in dealing with immunological data. Many interpretive problems are reduced when the data can be manipulated as individual characters, as can nucleotide or amino–acid sequences.

CONCLUSIONS

The common experience of taxonomists is that characters disagree. So long as we adopt the method of vulgar parsimony we are at the mercy of assuming the equivalence of characters. One escape is to recognise an hierarchical relationship among the kinds of characters themselves with respect to their trustworthiness for inferring the true phylogeny, depending on which of the underlying types of causes most strongly constrains their evolution. A distinction between the builder (efficient, formal, and final causes) and the bricks (the material basis of evolution) helps to organise characters into such a scheme: those most sensitive to adaptation, reflecting the "purpose" of organisms, are the record of changing ecological needs and organisms' responses to them, not necessarily to the passage of time. The results may easily be repetitive.

Biochemical characters, in the sense of structural genes or their products and to a lesser extent metabolic processes, offer perhaps the best hope for a direct and unequivocal reconstruction of historical relationships, since they represent respectively the material and the hierarchically deepest efficient causes of organisms both phyletically and ontogenetically. In turn, ontogenetic analysis helps to establish the hierarchical depth of efficient causes by establishing the formal cause of organisms and their evolution.

Yet even material features may reflect adaptation, so that there are limits on the phyletic significance of normally time–dependent changes in the material bases of organisms. Besides, one may argue that the goals of classification are not limited to reconstructing branching sequences. Even within the narrow confines of historical reconstruction, the bare cladogram is only a starting–point for documenting changes in the features of organisms: the sequence of such changes, and the reasons for them, should be interesting to taxonomists along with other evolutionary biologists. A home, after all, is more interesting than the bricks of which it is built.

ACKNOWLEDGEMENTS

Any worthy thoughts in this paper, I owe to many discussions and arguments with students and colleagues over an extended period; any peculiarities of interpretation or logic, however, are my own: Temple Smith in particular will surely deplore my dissection of causality, while Stan Rachootin will find it naive. (And Aristòtle would be – perhaps was – the first to point out the dangers of arguing from any analogy, especially his own). Nevertheless, I am grateful to all my associates for at least pretending interest in my ideas, and to David Ride for convincing me long ago that having them was worthwhile. Peter Bottjer has kept me more honest than I might otherwise have been, and was patient and resourceful in typing the several versions of the manuscript. Finally, I thank Michael Archer for requesting that I write this paper, and for being so patient in waiting for me to do so. He and Dave Briscoe also read the paper and provided constructive criticism.

REFERENCES

Albrecht, P., Gould, S. J., Oster, G. G. and Wake, D. B., 1979. Size and shape in ontogeny and phylogeny. *Paleobiology.* **5:** 296–317.

Archer, M. A., 1976. The dasyurid dentition and its relationships to that of didelphids, thylacinids, borhyaenids (Marsupicarnivora) and peramelids (Peramelina: Marsupialia). *Aust. J. Zool. Suppl. Ser.* **39:** 1–34.

Ashlock, P. D., 1974. The uses of cladistics. *Ann. Rev. Ecol. Syst.* **5:** 81–99.

Brown, D. D., 1981. Gene expression in eukaryotes. *Science.* **211:** 667–74.

Dickerson, R. E., 1980. Cytochrome c and the evolution of energy metabolism. *Sci. Amer.* **242:** 136–53.

Eldredge, N. and Cracraft, J., 1980. "Phylogenetic patterns and the evolutionary process: method and theory in comparative biology". Columbia Univ. Press: New York.

ESTABROOK, G. G., 1977. Does common equal primitive? *Syst. Bot.* **2:** 36–42.

FARRIS, J. S., 1972. Estimating phylogenetic trees from distance matrices. *Am. Nat.* **106:** 645–68.

GOULD, S. J., 1977. "Ontogeny and phylogeny". The Belknap Press of Harvard Uni. Press: Cambridge.

HARPER, C. W., 1979. A Bayesian probability view of phylogenetic systematics. *Syst. Zool.* **28:** 547–53.

HAYMAN, D. L. AND MARTIN, P. G., 1974. Animal cytogenetics 4, Chordata 4, Mammalia I: Monotremata and Marsupialia. Gebrüder–Borntraeger: Berlin–Stuttgart.

HENNIG, W., 1966. "Phylogenetic systematics". University of Illinois Press: Urbana.

HULL, D., 1970. Contemporary systematic philosophies. *Ann. Rev. Ecol. Syst.* **1:** 19–54.

JOHNSON, J. I., SWITZER, R. C. AND KIRSCH, J. A. W., 1980. Phylogeny through brain traits relation of lateral olfactory tract fibers to the accessory olfactory formation as a palimpsest of mammalian descent. *Brain, Behavior, and Evolution.* **71:** 339–63.

JUKES, T. H., 1980. Silent nucleotide substitutions and the molecular evolutionary clock. *Science* **210:** 973–8.

KOHNE, D. E., 1970. Evolution of higher–organism DNA. *Quart. Rev. Biophys.* **33:** 327–75.

KIMURA, M., 1968. Evolutionary rate at the molecular level. *Nature, (London).* **217:** 624–6.

KIRSCH, J. A. W., 1969. Serological data and phylogentic inference: the problem of rates of change. *Syst. Zool.* **18:** 296–311.

KIRSCH, J. A. W., 1977. The comparative serology of Marsupialia, and a classification of marsupials. *Aust. J. Zool. Suppl. Ser.* **52:** 1–152.

LILLERGRAVEN, J. A., 1969. Latest Cretaceous mammals of upper part of Edmonton Formation of Alberta, Canada, and review of marsupial–placental dichotomy in mammalian evolution. *University of Kansas Palaeontological Contributions.* **50** (Vertebrata 12): 1–122.

MAXSON, L. R., SARICH, V. M. AND WILSON, A. C., 1975. Continental drift and the use of albumin as an evolutionary clock. *Nature (London)* **225:** 397–400.

NELSON, G., 1978. Ontogeny, phylogeny, and the biogenetic law. *Syst. Zool.* **27:** 324–45.

NOVACEK, M. J., 1980. Cranioskeletal features in tupaiids and selected Eutheria as phylogenetic evidence. Pp.35–93 *in* "Comparative biology and evolutionary relationships of tree shrews", ed. by W. P. Luckett. Plenum Press: New York.

POPPER, K. R., 1968. "Conjectures and refutations: the growth of scientific knowledge". Harper and Row: New York.

RACHOOTIN, S. AND THOMSON, K. S., 1981. Epigenetics paleontology, and evolution. Pp.181–93 *in* "Evolution today, Proceedings of the Second International Congress of Systematic and Evolutionary Biology", ed. by G. G. E. Scudder and J. L. Reveal. Hunt Institute for Botanical Documentation, Carnegie–Mellon University: Pittsburgh.

ROMERO–HERRERA, A. E., LEHMANN, H. F. R. S., JOYSEY, K. A. AND FRIDAY, A. E., 1978. On the evolution of myoglobin. *Philosophical Transactions of the Royal Soc. London. B. Biol. Sci.* **283:** 61–163.

RIDE, W. D. L., 1970. "A guide to the native mammals of Australia". Oxford Univ. Press: Melbourne.

SEILACHER, A., 1973. Fabricational noise in adaptive morphology. *Syst. Zool.* **22:** 451–65.

SELANDER, R. K., YANG, S. Y., LEWONTIN, R. C. AND JOHNSON, W. E., 1970. Genetic variation in the horseshoe crab (*Limulus polyphemus*), a phylogenetic "relic". *Evolution.* **24:** 402–14.

SHARMAN, G. B., 1961. The mitotic chromosomes of marsupials and their bearing on taxonomy and phylogeny. *Aust. J. Zool.* **9:** 38–60.

SHARMAN, G. B., 1970. Reproductive physiology of marsupials. *Science.* **167:** 1221–8.

SHARMAN, G. B., 1973. Marsupial taxonomy and phylogeny. *Aust. Mammal.* **1:** 137–54.

SMITH, T. F. AND MOROWITZ, H. J., 1982. Between history and physics. *J. Mol. Evol.* (in press).

STEVENS, P. F., 1980. Evolutionary polarity of character states. *Ann. Rev. Ecol. Syst.* **11:** 333–58.

TYNDALE–BISCOE, H., 1973. "Life of marsupials". American Elsevier Publishing Company, Inc.: New York.

VAN VALEN, L. M., 1974. Molecular evolution as predicted by natural selection. *J. Mol. Evol.* **3:** 89–101.

VAN VALEN, L. M., 1980. Evolution as a zero–sum game for energy. *Evolutionary Theory* **4:** 289–300.

WEGENER, A., 1924. "The origin of continents and oceans". Methuen: London.

WILSON, A. C., CARLSON, S. S. AND WHITE, T. J., 1977. Biochemical evolution. *Ann. Rev. Biochem.* **46:** 573–639.

WILSON, A. C., MAXSON, L. R. AND SARICH, V. M., 1974. Two types of molecular evolution. Evidence from studies of interspecific hybridization. *Proc. Nat. Acad. Sci. U.S.A.* **71:** 2843–7.

WILLIAMS, G. C., 1966. "Adaptation and natural selection a critique of some current evolutionary thought". Princeton University Press: Princeton.

RESUMEN

49. El constructor y los ladrillos: notas conducentes a una filosofía de los caracteres. Por J. A. W. Kirsch.

En el intento de reconstruir filegenias es común encontrar que a diferentes caracteres se les dan diferentes ordenaciones; esto es cierto aún cuando lógicos errores son corregidos (como el basar grupos en caracteres primitivos compartidos en lugar de derivados compartidos). Las discrepancias entre las filogenias basadas en los caracteres son usualmente el resultado de convergencia, paralelismo o inversiones del estado (colectivamente denominado *homoplasia*); una solución al problema que ellos promueven es la aceptación democrática de las indicaciones de la mayoría — una solución que se dice basada en la economía pero que realmente es el método de la distancia más corta. El adoptar el árbol más corto equivale a ignorar la homoplasia y por ello una formal contradicción de la verdadera parsimonia, que es la posición filosófica según la cual la explicación más general, más integradora, es preferible a varias específicas. Dado que las semejanzas homoplásticas normalmente representan similitudes adaptativas logradas por rutas evolutivas independientes, una mejor solución al problema de la homoplasia podría ser el *peso* de los caracteres negativamente sobre la base de su sensibilidad a apremios ambientales. Uno de tales esquemas es sugerido por la aristotélica concepción de la múltiple causalidad: los caracteres anatómicos son de fiar menos desde que, siendo adaptativos, sirven a la causa final de los organismos (su aptitud reproductiva). Por otro lado, un impresionante cuerpo de evidencias indica que en el tiempo se produce un ajustado trazo de divergencia en los caracteres bioquímicos (especialmente los genes estructurales y sus productos), haciéndolo ideal para su uso en la reconstrucción de la filegenia. Los procesos fisiológicos y el plan del desarrollo ocupan el campo intermedio de, respectivamente, las causas eficientes y formales; el análisis ontogenético en particular puede ser útil para determinar la posición de los caracteres individuales en un esquema de evaluación jerárquica.

50. POLYTHETIC CLADISTICS, OR, WHEN PARSIMONY'S NOT ENOUGH: THE RELATIONSHIPS OF CARNIVOROUS MARSUPIALS

JOHN A. W. KIRSCH[1] AND
MICHAEL ARCHER[2]

Wagner tree analysis was carried out on 71 species of carnivorous marsupials, using two non–overlapping data–sets, one of which (26 characters) had proved useful for grouping opossums, while the other (28 characters) had been developed for dasyuroids; of the 54 characters, 37 were dental. When both geographic groups were analyzed separately, both data–sets gave rather similar groupings of dasyuroids, especially at intrageneric levels (for example, the polyphyly of *Antechinus* and *Dasyurus,* suspected on other grounds, was supported); but the relationships of opossums were rather different under the two regimes of coding. When both geographic groups were examined together, opossum relations were further distorted, with *Lestodelphys* and others intercalated with dasyurids; the latter were otherwise generally depicted as an isolated portion of one branch of a technically paraphyletic Didelphoidea. The dasyurid–oriented characters separated thylacines from borhyaenids so long as the fossil antecedents of the latter were included among the taxa examined; but opossum characters of both data–sets in combination placed thylacinids with borhyaenids, as did dental (but not non–dental) features from either or both character–sets. Bandicoots were closer to opossums on the dasyurid characters, but appeared derived from New Guinean dasyurids when opossum codings were used. None of the characters were cladistically perfect — that is, there was considerable convergence in state both within and between the geographic units of marsupials — and most groups were based on singular combinations of changed characters which individually occurred in other clades. In some cases, no single character–state was held by all members of a cladistic group, thus approximating the "polythetic" criterion for membership of the numericists.

Pages 595–619 *in* CARNIVOROUS MARSUPIALS, ed. by M. Archer, February, 1982. Roy. Zool. Soc. New South Wales: Sydney, Australia.

THE PROGRESS of dasyurid systematics during the past quarter–century, like that of marsupial taxonomy in general, has followed what may be a unique pathway for mammals: several broad, promising, and quite new hypotheses of relationship were developed not from the usual morphological data, but from karyotypes and biochemistry; and for a time it was the anatomists whose ideas were tested against these schemes, not, as is usually the case, that the serology or chromsomes had to be reconciled with bones and teeth. In fact, anatomical data were often in disagreement with the newer information (*cf.* Kirsch and Archer 1976).

That situation has changed, particularly as regards the internal classification of Dasyuroidea, which proved intractable to ordering by cytogenetics (Hayman and Martin 1974) and were only subjected to limited serological study. Hence arguments about their relationships remained based on morphology. Now, however, the more discriminating methods of enzyme analysis, banding studies, and DNA characterization are being applied successfully to intrageneric and species–level questions, and are proving to be just as useful as close quantitative comparisons of tooth or penis morphology (see the various papers in volume II of "Carnivorous marsupials").

But perhaps not more so, since the renewed enthusiasm for classical analysis has fed on a restructuring in the way taxonomic characters are used. We refer, of course, to cladistic methods, which rest on the rather obvious principle that only shared, derived characters can verify the special relationship of any putative phyletic group (Hennig 1966). With a more disciplined application of this old idea, many of the apparent discrepancies among the "old" morphological classifications of marsupials and the "new" biochemical or chromosomal ones have disappeared (Kirsch 1977). In the resolution, it has become apparent that one reason chemical characters can give such a direct approximation to phyletics is that structural gene products seem to evolve divergently at roughly constant rates (Wilson, Carlson and White 1977). When they do not, the location and extent of variations are generally obvious in a reasonably complete data–set (see Sarich, Lowenstein and Richardson 1982). There is now considerable empirical and theoretical support for the notion of a "molecular clock", and it is clear that cladistically interpreted characters which do not evolve time-

[1] Museum of Comparative Zoology, Harvard University, Cambridge, Massachusetts, U.S.A., 02138.
[2] School of Zoology, University of New South Wales, P.O. Box 1, Kensington, New South Wales, Australia, 2033.

Table 1:

Characters \ Taxa	Ancestor	*Lestodelphys halli*	*Sparassocynus* spp.	*Glasbius intricatus*	*Patene simpsonii*	*Sipalocyon gracilis**	*Lycopsis longirostris**	*Arminiheringia* spp.	*Borhyaena tuberata**	*Didelphodon vorax*	*Caenolestes obscurus*	*Microperoryctes murina**	*Peroryctes longicaudata*	*Perameles nasuta*	*Echymipera clara*	*Isoodon macrourus*	*Chaeropus ecaudatus*	*Rhynchomeles prattorum*	*Macrotis lagotis*	*Notoryctes typhlops*	*Dromiciops australis*	*Metachirus nudicaudatus*	*Didelphis virginiana*	*Lutreolina crassicaudata*	*Monodelphis brevicaudata*	*Chironectes minimus*	*Pediomys elegans*	*Glironia venusta*	*Alphadon rhaister*	*Thylophorops chapalmalensis**	*Hondadelphys fieldsi*	*Philander opossum*	*Caluromys*
1. Upper incisor number.	0.0	0.0	0.0	0.0	1.0	1.0	1.0	2.0	2.0	1.0	1.0	0.0	0.0								0.0	0.0	0.0	0.0	0.0	0.0	0.0	0.0	0.0	0.0	0.0	0.0	0.
2. Upper incisor shape.	0.0	0.0	0.0	0.0	1.0	1.0	0.0	0.0	0.0	0.0	1.0	1.0	1.0								1.0	1.0	0.0	0.0	1.0	1.0	0.0	1.0	0.0	0.0	0.0	0.0	0.
3. Relative sizes, $I^{2 \& 4}$.	0.0	1.0	1.0	0.0	0.0	0.0	0.0	0.0	1.0	0.0	1.0	1.0	1.0								1.0	0.0	0.0	1.0	0.0	0.0	0.0	0.0	0.0	0.0	0.0	1.0	1.
4. Relative sizes, $P^{2 \& 3}$.	0.0	0.0	0.0	0.0	0.0	0.0	0.0	0.0	0.0	0.0	0.0	0.0	1.0								0.0	0.0	0.0	0.0	0.0	0.0	0.0	0.0	0.0	0.0	0.0	0.0	1
5. Relative sizes, $P_{2 \& 3}$.	0.0	0.0	0.0	1.0	0.0	0.0	0.0	0.0	0.0	0.0	2.0	0.0	1.0								0.0	2.0	2.0	2.0	1.0	2.0	0.0	1.0	1.0	2.0	0.0	2.0	2.
6. Number of cusps on M^1.	0.0	0.0	0.0	0.0	0.0	0.0	0.0	0.0	0.0	0.0	1.0	1.0	1.0								0.0	0.0	0.0	0.0	0.0	0.0	0.0	0.0	0.0	0.0	0.0	0.0	0
7. Size of metacone, M^5.	0.0	1.0	2.0	0.0	0.0	2.0	2.0	2.0	2.0	0.0	2.0	1.0	1.0								2.0	0.0	0.0	1.0	2.0	0.0	0.0	0.0	0.0	0.0	0.0	1.0	0
8. Posterior cingulum, M^x.	0.0	1.0	1.0	0.0	1.0	1.0	1.0	1.0	1.0	1.0	1.0	1.0	1.0								1.0	1.0	1.0	1.0	1.0	1.0	0.0	1.0	0.0	1.0	1.0	1.0	1.
9. Size of paracone, M^3.	0.0	1.0	1.0	0.0	0.0	1.0	1.0	1.0	1.0	1.0	0.0	0.0	0.0								0.0	0.0	0.0	1.0	0.0	1.0	0.0	0.0	0.0	1.0	0.0	0.0	0
10. Size of metacone, $M^{3 \& 4}$.	0.0	0.0	1.0	0.0	1.0	1.0	1.0	1.0	1.0	1.0	0.0	0.0	0.0								0.0	0.0	0.0	1.0	0.0	0.0	0.0	0.0	0.0	1.0	0.0	0.0	0
11. Size of protocone, M^4.	0.0	1.0	1.0	0.0	1.0	1.0	1.0	1.0	1.0	0.0	0.0	0.0	0.0								0.0	0.0	0.0	1.0	0.0	0.0	0.0	0.0	0.0	0.0	0.0	0.0	0
12. Stylocone D on M^3.	0.0	0.0	1.0	0.0	1.0	1.0	1.0	1.0	1.0	0.0	1.0	0.0	0.0								1.0	0.0	0.0	0.0	0.0	0.0	0.0	1.0	0.0	0.0	1.0	0.0	1
13. Position of stylocone B, M^2.	0.0	0.0	0.0	0.0	0.0	1.0	1.0	1.0	1.0	0.0	1.0	0.0	0.0								0.0	0.0	0.0	0.0	0.0	0.0	1.0	0.0	0.0	0.0	0.0	0.0	0
14. Size of paraconid, M_2.	0.0	0.0	1.0	0.0	0.0	1.0	0.0	0.0	0.0	0.0	0.0	0.0	0.0								0.0	0.0	0.0	0.0	0.0	0.0	0.0	0.0	0.0	0.0	0.0	0.0	0
15. Size of metaconid, M_4.	0.0	0.0	0.0	0.0	1.0	1.0	1.0	1.0	1.0	0.0	0.0	0.0	0.0								0.0	0.0	0.0	0.0	0.0	0.0	0.0	0.0	0.0	0.0	1.0	0.0	0
16. Shape of metaconid, M_4.	0.0	1.0	1.0	1.0	1.0	0.0	0.0	0.0	0.0	1.0	1.0	1.0	1.0								1.0	1.0	1.0	1.0	1.0	1.0	1.0	0.0	1.0	1.0	1.0	1.0	
17. Entoconid size, M_4.	0.0	1.0	2.0	0.0	1.0	2.0	2.0	2.0	2.0	2.0	0.0	0.0	0.0								0.0	0.0	0.0	0.0	1.0	0.0	0.0	0.0	0.0	0.0	0.0	0.0	
18. Shape of entoconid, M_4.	0.0	0.0	1.0	0.0	0.0	0.0	0.0	0.0	1.0	0.0	0.0	0.0	0.0								0.0	0.0	0.0	0.0	0.0	0.0	0.0	0.0	0.0	0.0	0.0	0.0	
19. Number of cusps, talonid of M_5.	0.0	1.0	1.0	1.0	1.0	1.0	1.0	1.0	1.0	0.0	1.0	1.0	1.0								1.0	0.0	0.0	1.0	1.0	0.0	0.0	0.0	0.0	0.0	0.0	0.0	
20 Size of talonid, M_4.	0.0	1.0	1.0	0.0	1.0	1.0	1.0	1.0	1.0	0.0	0.0	0.0	0.0								0.0	0.0	0.0	1.0	0.0	0.0	0.0	0.0	0.0	0.0	0.0	0.0	
21 Maxillary vacuities.	0.0	1.0	0.0	0.0	0.0	0.0	0.0	0.0	0.0	0.0	2.0	2.0	2.0								2.0	1.0	1.0	1.0	1.0	1.0	1.0	1.0	1.0	1.0	1.0	1.0	
22. Posterior palatal foramina.	0.0	0.0	1.0	0.0	0.0	0.0	0.0	0.0	0.0	0.0	0.0	0.0	0.0								0.0	0.0	0.0	0.0	0.0	0.0	0.0	0.0	0.0	0.0	0.0	0.0	
23. Foramen ovale.	0.0	0.0	0.0	0.0	0.0	0.0	0.0	0.0	1.0	0.0	1.0	0.0	0.0								0.0	0.0	0.0	0.0	1.0	0.0	0.0	0.0	0.0	0.0	0.0	0.0	
24. Transverse canal.	0.0	0.0	1.0	0.0	1.0	1.0	0.0	0.0	1.0	0.0	0.0	0.0	0.0								0.0	1.0	0.0	0.0	0.0	0.0	0.0	0.0	0.0	0.0	1.0	1.0	
25. Squamosal epitympanic recess.	0.0	0.0	0.0	0.0	0.0	0.0	0.0	0.0	0.0	0.0	0.0	0.0	0.0								0.0	0.0	0.0	0.0	0.0	1.0	0.0	0.0	0.0	0.0	0.0	0.0	
26. Periotic tympanic wing.	0.0	1.0	0.0	0.0	0.0	0.0	0.0	0.0	0.0	1.0	0.0	0.0	0.0								2.0	0.0	0.0	0.0	0.0	0.0	0.0	0.0	0.0	0.0	0.0	0.0	
27. Alisphenoid.	0.0	0.0	1.0	0.0	0.0	0.0	0.0	0.0	0.0	0.0	0.0	0.0	0.0								1.0	0.0	0.0	0.0	0.0	0.0	0.0	0.0	0.0	0.0	0.0	0.0	
28. Squamosal-frontal relation.	0.0	0.0	0.0	0.0	0.0	0.0	0.0	0.0	1.0	0.0	0.0	1.0	1.0								0.0	0.0	0.0	0.0	0.0	0.0	0.0	0.0	0.0	0.0	0.0	0.0	
29. Presence/absence of M conules	0.0	2.0	1.0	0.0	0.0	0.0	0.0	2.0	2.0	2.0	1.0	1.0	1.0	1.0	1.0	1.0	1.0	1.0	2.0	2.0	1.0	1.0	1.0	1.0	1.0	1.0	0.0	0.0	0.0	1.0	0.0	1.0	
30. Relative length $M^{3 \& 4}$.	0.0	0.0	1.0	0.0	1.0	0.0	1.0	1.0	1.0	1.0	0.0	0.0	0.0	1.0	1.0	1.0	0.0	0.0	0.0	0.0	0.0	1.0	1.0	1.0	1.0	1.0	0.0	0.0	1.0	1.0	1.0	1.0	
31. Outer border M^4.	0.0	1.0	0.0	1.0	0.0	0.0	0.0	0.0	0.0	1.0	0.0	0.0	0.0	0.0	0.0	0.0	1.0	1.0	0.5	1.0	0.0	0.0	0.0	0.0	1.0	0.0	1.0	0.0	0.0	0.0	0.0	0.0	
32. Relative size & position of cusps on M^x.	0.0	2.0	2.0	0.0	2.0	2.0	2.0	2.0	2.0	1.0	0.0	1.0	1.0	1.0	1.0	0.5	0.5	1.0	0.0	3.0	0.0	1.0	1.0	2.0	2.0	1.0	0.0	0.0	0.0	2.0	1.0	1.0	
33. Stylar shelf.	0.0	0.5	2.0	0.0	1.0	2.0	2.0	2.0	2.0	0.5	2.0	1.0	1.0	1.0	1.0	1.0	1.0	1.0	1.0	2.0	0.0	0.0	0.0	0.0	0.0	0.0	1.0	1.0	0.0	0.0	1.0	0.0	
34. Paracrista & stylocone of M^x.	0.0	1.0	0.0	1.0	1.0	0.0	0.0	1.0	0.0	0.0	1.0	0.0	0.0	0.0	0.0	0.0	0.0	0.0	0.5	0.0	0.0	0.0	0.0	0.0	0.0	1.0	1.0	1.0	0.0	0.0	0.0	0.0	
35. Metastylar spur & metacrista of M^x.	0.0	1.0	2.0	0.0	0.0	1.0	2.0	2.0	2.0	0.0	0.0	0.0	0.5	1.0	0.5	1.0	1.0	0.0	0.5	1.0	0.0	1.0	1.0	2.0	1.0	1.0	0.0	0.0	0.0	2.0	1.0	1.0	
36. Shape of protocones of M^x.	0.0	1.0	2.0	0.0	1.0	2.0	1.0	2.0	2.0	0.0	0.0	0.0	0.0	0.0	0.0	0.0	0.0	0.0	0.0	2.0	0.0	1.0	1.0	1.0	1.0	1.0	0.0	0.0	0.0	2.0	0.0	1.0	
37. Relative size & shape of $M^{4 \& 5}$.	0.0	2.0	2.0	2.0	2.0	2.0	2.0	2.0	2.0	0.0	1.5	2.0	2.0	2.0	2.0	2.0	2.0	2.0	2.0	3.0	0.0	0.0	2.0	2.0	2.0	2.0	2.0	0.0	0.0	2.0	2.0	0.0	
38. Relative size & shape of $M_{5 \& 4}$.	0.0	0.0	0.0	2.0	1.0	0.0	0.0	0.0	0.0	0.0	2.0	1.0	1.0	0.0	0.0	0.0	0.0	0.0	0.0	0.0	0.0	1.0	0.0	0.0	1.0	0.0	1.0	0.0	0.0	0.0	0.0	0.0	
39. Relative size of M_x talonid & trigonid.	0.0	1.0	1.0	1.0	0.5	1.0	1.0	1.0	1.0	0.0	0.0	0.5	0.5	0.5	1.0	0.5	0.5	1.0	0.5	2.0	0.0	1.0	1.0	1.0	1.0	1.0	0.0	0.0	0.0	1.0	1.0	1.0	
40. Size & shape of M_5 talonid.	0.0	1.0	1.0	0.0	1.0	1.0	1.0	1.0	1.0	1.0	0.0	1.0	1.0	1.0	1.0	1.0	1.0	1.0	1.0	2.0	1.0	0.0	0.0	1.0	1.0	0.0	1.0	0.0	0.0	1.0	1.0	0.0	
41. Relative sizes, meta- and paraconid of M_x.	0.0	0.0	1.0	0.0	1.0	2.0	2.0	2.0	2.0	2.0	0.0	0.0	0.0	0.0	0.0	0.0	0.5	0.0	0.0	0.0	1.0	1.0	1.0	2.0	1.0	1.0	0.0	1.0	1.0	2.0	2.0	1.0	
42. Development of protoconid.	0.0	1.0	1.0	2.0	1.0	1.0	1.0	1.0	1.0	2.0	0.0	1.0	1.0	1.0	1.0	2.0	2.0	1.0	1.0	1.0	0.0	0.0	0.0	1.0	2.0	0.0	1.0	0.0	0.0	1.0	1.0	0.0	
43. Upper incisor shape.	0.0	0.0	0.0	0.0	0.0	1.0	0.0	0.0	0.0	0.0	1.0	1.0	1.0	1.0	1.0	1.0	1.0	1.0	1.0	0.0	1.0	0.0	0.0	0.0	0.0	0.0	0.0	0.0	0.0	0.0	0.0	0.0	
44. Size of P^x.	0.0	0.0	0.0	0.0	0.0	0.0	0.0	0.0	0.0	0.0	0.0	0.0	0.0	0.0	0.0	0.0	0.0	0.0	0.0	0.0	0.0	0.0	0.0	0.0	0.0	0.0	0.0	1.0	0.0	0.0	0.0	0.0	
45. Shape of P^x.	0.0	0.0	0.0	0.0	0.0	0.0	0.0	0.0	0.0	1.0	0.0	0.0	0.0	0.0	0.0	1.0	1.0	0.0	0.0	0.0	0.0	0.0	0.0	0.0	0.0	1.0	0.0	0.0	0.0	0.0	0.0	1.0	
46. Body size I.	0.0	0.0	0.0	0.0	0.0	1.0	2.0	2.0	2.0	1.0	0.0	0.0	0.0	1.0	1.0	1.0	0.0	1.0	1.0	0.0	0.0	1.0	1.0	1.0	0.0	1.0	0.0	0.0	1.0	1.0	1.0	1.0	
47. Body size II.	0.0	1.0	0.0	0.0	0.0	0.0	0.0	0.0	0.0	0.0	1.0	1.0	0.0	0.0	0.0	0.0	0.0	0.0	0.0	1.0	1.0	0.0	0.0	0.0	1.0	0.0	0.0	0.0	0.0	0.0	0.0	0.0	
48. Palatal vacuities.	0.0	0.0	1.0	1.0	1.0	1.0	1.0	1.0	1.0	1.0	0.0	0.0	0.0	0.0	0.0	0.0	0.0	0.0	0.0	0.0	0.0	0.0	0.0	0.0	0.0	0.0	0.0	1.0	0.0	0.0	1.0	0.0	
49. Otic region I.	0.0	4.0	3.0	0.0	0.0	0.0	0.0	0.0	0.0	1.0	1.0	1.0	0.0	0.0	0.0	0.0	0.0	0.0	2.0	3.0	0.0	0.0	0.0	0.0	1.0	0.0	0.0	1.0	0.0	0.0	0.0	0.0	
50. Otic region II.	0.0	0.0	0.0	0.0	0.0	0.0	0.0	0.0	0.0	0.0	0.0	1.0	1.0	1.0	1.0	1.0	1.0	1.0	0.0	0.0	0.0	0.0	0.0	0.0	0.0	0.0	0.0	0.0	0.0	0.0	0.0	0.0	
51. Size of paroccipital process.	0.0	0.5	0.0	0.5	0.5	0.5	0.5	0.5	0.5	0.5	0.5	1.0	1.0	0.5	0.5	0.0	0.0	0.5	0.0	1.0	1.0	0.0	0.0	0.0	0.0	0.0	0.5	0.0	0.0	0.0	0.5	0.0	
52. Size of ectotympanic process.	0.0	2.0	0.0	0.0	0.0	0.0	0.0	0.0	0.0	0.0	1.0	1.0	1.0	1.0	1.0	2.0	1.0	1.0	1.0	2.0	0.0	0.0	0.0	0.0	1.0	0.0	0.0	0.0	0.0	0.0	0.0	0.0	
53. Condition of pars mastoidea.	0.0	0.0	0.0	0.0	0.0	0.0	0.0	0.0	0.0	0.0	0.5	0.0	0.0	0.0	0.0	0.0	0.0	0.0	0.0	1.0	1.0	0.0	0.0	0.0	0.0	0.0	0.0	0.0	0.0	0.0	0.0	0.0	
54. Size of sinus epitympanicus.	0.0	0.0	1.0	0.0	0.0	0.0	0.0	0.0	0.0	0.0	0.0	0.0	0.0	1.0	0.0	1.0	1.0	0.0	1.0	0.0	0.0	0.0	0.0	0.0	0.0	0.0	0.0	0.0	0.0	0.0	0.0	0.0	

Table 1. Data used in computing the Wagner trees discussed herein; character–states are defined in Table 2. Characters 1–28 are the "dasyurid codings" devised by Archer (1976), while 29–54 are after Reig, Kirsch and Marshall (in preparation). Underlined scores were inferred from incomplete or ambiguous material, and blanks indicate that scoring was not done on those taxa. All computations were performed on data read into the program in the order shown, by columns from left to right. Starred taxa are ones for which the values are averages for the entire genus in the opossum codings.

Peratherium cuvieri	*Murexia longicaudata*	*Myoictis melas*	*Phascolosorex doriae*	*Phascolosorex dorsalis*	*Neophascogale lorentzii*	*Ankotaringia tirarensis*	*Keeuna woodburnei*	*Sminthopsis murina*	*Sminthopsis ooldea*	*Sminthopsis crassicaudata*	*Sminthopsis virginiae*	*Sminthopsis granulipes*	*Antechinomys laniger*	*Ningaui* spp.	*Planigale tenuirostris*	*Planigale maculata*	*Planigale gilesi*	*Antechinus stuartii*
0.0	1.0	1.0		1.0		1.0	1.0	1.0							1.0			1.0
0.0	0.0	0.0		1.0		0.0	0.0	0.0							0.0			0.0
0.0	0.0	0.0		0.0		0.0	0.0	0.0							1.0			1.0
0.0	0.0	1.0		1.0		0.0	0.0	0.0							0.0			0.0
0.0	1.0	2.0		2.0		1.0	1.0	1.0							2.0			2.0
0.0	1.0	1.0		1.0		1.0	1.0	0.0							1.0			1.0
1.0	1.0	2.0		1.0		1.0	1.0	2.0							2.0			2.0
1.0	0.0	0.0		0.0		1.0	1.0	1.0							0.0			0.0
0.0	0.0	0.0		0.0		0.0	0.0	0.0							1.0			0.0
0.0	0.0	0.0		0.0		0.0	0.0	0.0							0.0			0.0
0.0	0.0	0.0		0.0		0.0	0.0	0.0							1.0			0.0
1.0	0.0	0.0		0.0		0.0	0.0	0.0							0.0			0.0
0.0	0.0	0.0		0.0		0.0	0.0	0.0							1.0			0.0
0.0	0.0	1.0		1.0		0.0	0.0	0.0							1.0			1.0
0.0	0.0	0.0		0.0		0.0	0.0	0.0							0.0			0.0
0.0	0.0	0.0		0.0		0.0	0.0	1.0							0.0			0.0
0.0	1.0	0.0		0.0		0.0	0.0	2.0							2.0			0.0
0.0	0.0	0.0		0.0		0.0	0.0	1.0							1.0			0.0
0.0	1.0	0.0		0.0		0.0	0.0	1.0							1.0			0.0
0.0	0.0	0.0		0.0		0.0	0.0	0.0							1.0			0.0
1.0	1.0	1.0		1.0		1.0	1.0	2.0							1.0			2.0
0.0	1.0	1.0		1.0		1.0	1.0	1.0							1.0			1.0
0.0	1.0	1.0		1.0		1.0	1.0	1.0							1.0			1.0
0.0	0.0	0.0		0.0		0.0	0.0	0.0							1.0			0.0
0.0	1.0	1.0		1.0		1.0	1.0	1.0							1.0			1.0
0.0	1.0	1.0		1.0		1.0	1.0	1.0							1.0			1.0
0.0	1.0	1.0		1.0		1.0	1.0	1.0							1.0			1.0
0.0	0.0	0.0		1.0		0.0	0.0	1.0							0.0			0.0
0.0	0.0	0.5	1.0	0.5	1.0	0.0	1.0	1.5	2.0	1.0	1.5	2.0	2.0	2.0	1.5	2.0	1.0	1.5
0.0	0.0	0.0	0.0	0.0	0.0	0.0	0.0	0.0	0.0	0.0	0.0	0.0	0.0	0.0	0.0	0.0	0.0	0.0
1.0	1.0	1.0	1.0	1.0	1.0	1.0	1.0	1.0	1.0	1.0	1.0	1.0	1.0	1.0	1.0	1.0	1.0	1.0
0.0	1.0	1.0	1.0	1.0	1.0	1.0	1.0	1.0	1.0	1.0	1.0	1.0	1.0	2.0	2.0	2.0	2.0	1.0
0.0	0.5	1.0	1.0	1.0	1.0	0.5	0.5	1.0	1.0	1.0	1.0	1.0	1.0	1.0	2.0	2.0	2.0	1.0
0.0	0.0	0.0	0.0	0.0	0.0	0.0	0.0	0.0	0.0	0.0	0.0	0.0	0.0	0.0	0.0	0.0	0.0	0.0
1.0	1.0	1.0	1.0	1.0	0.0	1.0	1.0	1.5	1.5	1.5	1.5	1.5	1.5	1.5	1.5	1.5	1.5	0.5
0.0	0.0	0.0	0.0	0.0	0.0	0.0	0.0	1.0	1.0	0.5	1.0	1.0	1.0	1.0	1.0	1.0	1.0	0.0
2.0	2.0	2.0	2.0	2.0	2.0	2.0	1.5	2.0	2.0	2.0	2.0	2.0	2.0	3.0	2.0	2.0	2.0	2.0
0.0	1.0	1.0	1.0	1.0	1.0	1.0	1.0	1.0	1.0	1.0	1.0	1.0	1.0	1.0	1.0	1.0	1.0	1.0
1.0	1.0	1.0	1.0	1.0	1.0	1.0	1.0	1.0	1.0	1.0	1.0	1.0	1.0	1.0	1.0	1.0	1.0	1.0
1.0	1.0	1.0	1.0	1.0	1.0	1.0	0.0	1.0	1.0	1.0	1.0	1.0	1.0	1.0	1.0	1.0	1.0	1.0
0.0	0.0	0.0	0.0	0.0	0.0	0.0	0.0	0.0	0.0	0.0	0.0	0.0	0.0	0.0	0.5	1.0	1.0	0.0
1.0	1.0	1.0	1.0	1.0	1.0	1.0	1.0	1.0	1.0	1.0	1.0	1.0	1.0	1.0	1.0	1.0	1.0	1.0
0.0	0.0	0.0	1.0	1.0	1.0	0.0	0.0	0.0	0.0	0.0	0.0	1.0	0.0	0.0	0.0	0.0	0.0	0.0
0.0	0.0	0.0	0.0	0.0	0.0	0.0	0.0	0.0	0.0	0.0	0.0	0.0	0.0	0.0	0.0	0.0	0.0	0.0
0.0	0.0	0.0	0.0	0.0	0.0	0.0	0.0	0.0	0.0	0.0	0.0	0.0	0.0	0.0	1.0	1.0	0.0	0.0
0.0	0.0	0.0	0.0	0.0	0.0	0.0	0.0	0.0	0.0	0.0	0.0	0.0	0.0	0.0	0.0	0.0	0.0	0.0
0.0	1.0	1.0	1.0	1.0	0.0	1.0	1.0	1.0	1.0	1.0	1.0	1.0	1.0	1.0	1.0	1.0	1.0	1.0
1.0	0.0	0.0	0.0	0.0	0.0	0.0	0.0	0.0	0.0	0.0	0.0	0.0	0.0	0.0	0.0	0.0	0.0	0.0
0.0	2.0	2.0	2.0	2.0	2.0	2.0	2.0	2.0	2.0	2.0	2.0	2.0	2.0	2.0	2.0	2.0	2.0	2.0
0.0	0.0	0.0	0.0	0.0	0.0	0.0	0.0	0.0	0.0	0.0	0.0	0.0	0.0	0.0	0.0	0.0	0.0	0.0
0.5	0.5	0.5	0.0	0.0	0.0	0.0	0.0	0.0	0.0	0.0	0.0	0.0	0.5	0.5	0.5	0.0	0.5	0.0
0.0	2.0	2.0	1.0	2.0	3.0	2.0	2.0	2.0	2.0	2.0	2.0	2.0	1.0	[illegible]	2.0	2.0	2.0	2.0
0.0	0.0	0.0	0.0	0.0	0.0	0.0	0.0	0.0	0.0	0.0	0.0	0.0	1.0	1.0	1.0	1.0	1.0	0.5
0.0	0.0	0.0	0.0	0.0	0.0	0.0	0.0	0.0	0.0	0.0	0.0	0.0	0.0	0.0	0.0	0.0	0.0	0.0

Parantechinus bilarni	*Pseudantechinus macdonnellensis*	*Dasykaluta rosamondae*	*Parantechinus apicalis*	*Antechinus swainsonii*	*Antechinus leo*	*Dasyuroides byrnei*	*Dasycercus cristicauda*	*Satanellus hallucatus*	*Dasyurus maculatus*	*Dasyurus viverrinus*	*Dasyurus geoffroii*	*Phascogale tapoatafa*	*Sarcophilus harrisii*	*Glaucodon ballaratensis*	*Wakamatha tasselli*	*Myrmecobius fasciatus*	*Thylacinus cynocephalus*	*Thylacinus potens*
1.0	1.0	1.0				1.0		1.0	1.0				1.0		1.0	1.0	1.0	1.0
0.0	0.0	0.0				0.0		0.0	0.0				0.0		0.0	0.0	0.0	0.0
0.0	0.0	1.0				0.0		0.0	0.0				0.0		0.0	1.0	0.0	0.0
1.0	1.0	1.0				1.0		1.0	1.0				1.0		0.0	1.0	0.0	0.0
2.0	2.0	2.0				2.0		2.0	2.0				2.0		0.0	2.0	0.0	0.0
1.0	1.0	1.0				1.0		1.0	1.0				1.0		0.0	0.0	1.0	1.0
2.0	2.0	2.0				2.0		1.0	2.0				2.0		2.0	1.0	1.0	1.0
1.0	1.0	1.0				1.0		1.0	1.0				1.0		1.0	0.0	1.0	1.0
0.0	0.0	0.0				0.0		0.0	0.0				1.0		0.0	0.0	1.0	1.0
0.0	0.0	0.0				0.0		0.0	1.0				1.0		0.0	0.0	1.0	1.0
0.0	0.0	0.0				0.0		0.0	0.0				1.0		0.0	0.0	1.0	1.0
0.0	0.0	0.0				0.0		0.0	0.0				1.0		0.0	0.0	1.0	0.0
0.0	1.0	0.0				1.0		1.0	1.0				1.0		0.0	0.0	0.0	0.0
1.0	1.0	1.0				1.0		1.0	1.0				1.0		0.0	0.0	0.0	0.0
0.0	0.0	0.0				0.0		0.0	0.0				1.0		0.0	0.0	1.0	1.0
0.0	0.0	0.0				0.0		0.0	0.0				0.0		1.0	0.0	0.0	0.0
2.0	2.0	2.0				1.0		0.0	1.0				2.0		0.0	0.0	2.0	2.0
0.0	0.0	0.0				0.0		0.0	0.0				1.0		1.0	1.0	1.0	1.0
1.0	1.0	1.0				1.0		1.0	0.0				1.0		1.0	0.0	1.0	1.0
0.0	0.0	0.0				1.0		0.0	1.0				1.0		0.0	0.0	1.0	1.0
1.0	1.0	1.0				2.0		2.0	1.0				1.0		2.0	0.0	2.0	1.0
1.0	1.0	1.0				1.0		1.0	1.0				1.0		1.0	0.0	1.0	0.0
1.0	1.0	1.0				1.0		1.0	1.0				0.0		1.0	1.0	0.0	0.0
0.0	0.0	0.0				0.0		0.0	1.0				0.0		0.0	0.0	0.0	0.0
1.0	2.0	2.0				2.0		1.0	1.0				2.0		1.0	1.0	1.0	1.0
1.0	2.0	2.0				2.0		2.0	1.0				2.0		1.0	0.0	1.0	1.0
1.0	1.0	1.0				1.0		1.0	1.0				1.0		1.0	1.0	0.0	0.0
0.0	0.0	0.0				0.0		0.0	0.0				0.0		1.0	0.0	1.0	1.0
1.0	0.5	2.0	2.0	1.0	0.5	1.0	2.0	0.0	0.0	0.0	0.0	0.5	1.0	0.0	1.5	0.5	2.0	2.0
0.0	0.0	0.0	0.0	0.0	0.0	0.0	0.0	0.0	1.0	0.0	0.0	0.0	0.0	0.0	0.0	0.0	1.0	0.5
1.0	1.0	1.0	1.0	1.0	1.0	1.0	1.0	1.0	0.0	0.0	0.0	1.0	0.0	0.0	1.0	0.0	0.0	0.0
1.0	1.0	1.0	1.0	1.0	1.0	1.0	1.0	1.0	1.0	1.0	1.0	1.0	2.0	1.5	1.0	0.0	2.0	2.0
1.0	1.0	1.0	1.0	1.0	1.0	1.0	1.0	1.0	1.0	1.0	1.0	1.0	2.0	2.0	1.0	2.0	2.0	2.0
0.0	0.0	0.0	0.0	0.0	0.0	0.0	0.0	0.0	0.0	0.0	0.0	0.0	0.0	0.0	0.0	1.0	0.0	0.0
1.0	1.0	1.0	1.0	1.0	1.0	1.5	1.5	1.0	2.0	1.5	2.0	1.0	2.0	2.0	1.5	1.0	2.0	2.0
1.0	1.0	1.0	1.0	0.0	0.0	0.0	1.0	0.0	0.0	0.0	0.0	0.0	2.0	1.5	1.0	0.0	2.0	2.0
3.0	3.0	3.0	3.0	2.0	3.0	3.0	3.0	2.0	2.0	3.0	3.0	3.0	1.5	3.0	2.0	0.0	3.0	3.0
1.0	1.0	1.0	1.0	1.0	1.0	1.0	1.0	1.0	1.0	1.0	1.0	1.0	0.0	0.5	0.0	0.0	0.0	0.0
1.0	1.0	1.0	1.0	1.0	1.0	1.0	1.0	1.0	1.0	1.0	1.0	1.0	1.0	1.0	1.0	0.5	1.0	1.0
1.0	1.0	1.0	1.0	1.0	1.0	1.0	1.0	1.0	1.0	1.0	1.0	1.0	1.0	1.0	1.0	1.0	1.0	1.0
0.5	1.0	0.5	1.0	0.0	0.0	0.0	0.5	0.0	1.0	0.0	0.0	0.0	2.0	1.5	0.0	1.0	2.0	2.0
1.0	1.0	1.0	1.0	1.0	1.0	1.0	1.0	1.0	1.0	1.0	1.0	1.0	1.0	1.0	1.0	0.0	1.0	1.0
0.0	0.0	0.0	0.0	1.0	0.0	0.0	0.0	0.0	0.0	0.0	0.0	0.0	0.0	0.0	0.0	0.0	0.0	0.0
0.0	0.0	0.0	0.0	0.0	0.0	0.0	0.0	0.0	0.0	0.0	0.0	0.0	0.0	0.0	0.0	0.0	0.0	0.0
0.0	0.0	0.0	0.0	0.0	0.0	0.0	0.0	0.0	0.0	0.0	0.0	1.0	1.0	0.5	0.0	0.0	0.0	0.0
0.0	0.0	0.0	0.0	0.0	0.0	0.0	0.0	0.0	1.0	1.0	1.0	0.0	1.0	1.0	0.0	0.0	2.0	2.0
1.0	1.0	1.0	1.0	1.0	1.0	1.0	1.0	0.0	0.0	0.0	0.0	0.0	0.0	0.0	0.0	1.0	0.0	0.0
0.0	0.0	0.0	0.0	0.0	0.0	0.0	0.0	0.0	0.0	0.0	0.0	0.0	0.0	0.0	0.0	1.0	0.0	0.0
2.0	2.0	3.0	2.0	2.0	2.0	3.0	3.0	2.0	2.0	2.0	2.0	2.0	2.0	2.0	2.0	4.0	0.0	0.0
0.0	0.0	0.0	0.0	0.0	0.0	0.0	0.0	0.0	0.0	0.0	0.0	0.0	0.0	0.0	0.0	0.0	1.0	1.0
0.0	0.0	1.0	0.0	0.0	0.0	1.0	1.0	0.0	0.0	0.0	0.0	0.0	0.0	0.0	0.0	0.5	0.0	0.0
2.0	2.0	2.0	2.0	2.0	2.0	2.0	2.0	2.0	2.0	2.0	2.0	2.0	3.0	2.5	2.0	2.0	3.0	3.0
1.0	1.0	1.0	0.0	0.0	0.0	1.0	1.0	0.0	0.0	0.0	0.0	0.0	0.0	0.0	0.0	0.0	0.0	0.0
0.0	0.0	0.0	0.0	0.0	0.0	0.0	0.0	0.0	0.0	0.0	0.0	0.0	0.0	0.0	0.0	0.0	0.0	0.0

dependently must (if they are good ones) give a phylogeny that is congruent with a biochemical one.

Yet the clock is manifestly erratic, and its variations may be sufficient to give incorrect branching patterns as well as markedly different branch–lengths over equal periods of time, especially when rather close relationships are involved (Kirsch 1969). That seems to be true whether amino–acid sequences, serological resemblences, or genetic differences are examined (see papers in Ayala 1976), and the occurence of such mistaken branchings enhances the argument for continually comparing diverse data–sets for congruence and compatibility. Moreover, the mistaken associations implied by some biochemical comparisons have left many cladists suspicious of the utility of such methods altogether. At times, their skepticism borders on a rejection of any phenetic comparison simply because it *is* phenetic (Hennig 1966).

In order to allay such doubts and to correct the inevitable errors, then, it is important to demonstrate in specific cases the general (and particular) correspondence between biochemistry and anatomy. This demonstration has recently been attempted by Reig, Kirsch and Marshall (in preparation) for the relationships of American marsupials. Using a data–set comprising 31 dental and other osteological characters supplemented by inferences drawn from reproduction, behaviour, and paleontology, they have constructed a cladogram of Didelphoidea that is remarkably consistent (in a technical sense to be defined later) and closely congruent to the separate indications of chromosomes and serology. When convergence occurs, it is readily detectable and can be rationalized as such. The high degree of compatibility among the anatomical characters themselves thus seems to argue that these were a good choice for adumbrating opossum relationships, and their overall correspondence with biochemistry and cytogenetics suggests that the "parsimony", or minimum pathlength, approach to phylogenetic reconstruction is a sound one.

It was under the influence of the apparent success of the opossum study that we decided to attempt a similar comparison of data–sets for the dasyurids and their near relatives. We hoped to demonstrate that the now abundant (if still regrettably incomplete) data on numerous aspects of their morphology and chemistry were similarly in accord. And we hoped to extend the inferences to unknown areas of dasyurid taxonomy, setting the terms for future study. As the following will show, we were only partly successful, but the results are perhaps less interesting for what they say about dasyurid relationships than for what they indicate about the fundamental constraints imposed by character–choice and the validity of the parsimony criterion itself.

MATERIALS AND METHODS

Characters and Taxa

Since Reig *et al.'s* characters seemed to work so well for opossums, we began by scoring a number of dasyurids and their collateral relatives according to that scheme. Some of Reig *et al.'s* taxa, and additional American marsupials scored by Archer, were included in the subsequent analyses. Since the initial results were not entirely satisfactory – that is, did not agree with the prejudices of both of us! – we also coded 46 Australian and American taxa for characters that had previously seemed apt for discriminating and relating various marsupial carnivores (Archer 1976 and other papers). Both sets of scorings are given in Table 1, where characters 1–28 are the "dasyurid codings" and 29–54 are the "opossum codings"; the polarities and hence ordering of states are described in Table 2, but the bases for decisions about these matters are given elsewhere (in Reig *et al.* and Archer's papers). Note that only two characters are common to both schemes of coding. These are 2 and 43, shape of upper incisors, and 22 and 48, palatal vacuities. However, 2 and 43 have distinct primitive conditions and the various taxa are scored relative to that beginning state; thus the scores are not necessarily identical for the same taxa in these two features. And while the scorings *are* the same for characters 22 and 48, since a distinction has been made between maxillary and posterior palatal vacuities in the dasyurid codings but not the opossum subset, we have included both characters in the experiments which involved the entire data–set. Hence there is really no overlap between the opossum and dasyurid codings. As for correlation among the characters, we have considered this to be operationally absent for this set since some variation over the set of taxa occured independently for any pair of characters we considered. There are no characters and very few character–states unique to one taxon. The set of taxa scored totalled seventy–one for the opossum codings, and 46 of these were also coded for dasyurid characters. An hypothetical all–primitive *Ancestor* was included in all computer runs for reasons explained below.

Wagner Analysis

A besetting sin of much cladistics is over–reliance on a few characters, common because the simultaneous reconciliation of the indications of many features is difficult to perform intuitively. Yet consideration of a large number of characters is required by the parsimony approach, and

Table 2. List and description of the characters used to generate the data of Table 1. Characters 1–28 are based on previous studies of dasyurids, and 29–54 are taken from the set employed by Reig, Kirsch and Marshall (in preparation).

# and name of character	Primitive condition	Advanced state(s)	Comments
1. Upper incisor number	Five, 0.0	Four, 1.0; fewer than four, 2.0	
2. Upper incisor shape	Round, 0.0	Spatulate, 1.0	Cf. character 43
3. Relative sizes, I^2 and I^4	I^2 less than I^4, 0.0	I^2 equal or greater than I^4, 1.0	
4. Relative sizes, P^2 and P^3	P^3 greater than P^2, 0.0	P^3 less than P^2, 1.0	
5. Relative sizes, P_2 and P_3	P_3 greater than P_2, 0.0	P_3 less than P_2, 1.0	
6. Number of cusps on M^1	Three cusps, 0.0	Less than three, 1.0	
7. Size of metacone on M^5	Large metacone, 0.0	Small, 1.0; none, 2.0	
8. Posterior cingulum, M^X	Has cingulum, 0.0	Lacks cingulum, 1.0	
9. Size of paracone, M^3	Paracone unreduced, 0.0	Paracone reduced, 1.0	
10. Size of metacone, $M^{3\ and\ 4}$	M^4 about same as M^3, 0.0	M^4 metacone larger, 1.0	
11. Size of protocone, M^4	Protocone unreduced, 0.0	Protocone reduced, 1.0	
12. Stylocone D on M^3	Stylocone present, 0.0	Stylocone absent, 1.0	
13. Position of stylocone B on M^2	Distinct from paracone, 0.0	Near paracone, 1.0	
14. Size of paraconid on M_2	Paraconid normal, 0.0	Paraconid reduced, 1.0	
15. Size of metaconid on M_4	Metaconid normal, 0.0	Metaconid reduced, 1.0	
16. Shape of metaconid, M_4	Metaconid not transverse, 0.0	Metaconid transverse, 1.0	
17. Entoconid size, M_4	Entoconid tall, 0.0	Small, 1.0; very reduced, 2.0	
18. Shape of entoconid, M_4	Entoconid bladelike, 0.0	Entoconid round, 1.0	
19. Number of cusps, talonid of M_5	Talonid with 3 cusps, 0.0	With fewer than 3, 1.0	
20. Size of talonid, M_4	Talonid normal, 0.0	Talonid reduced, 1.0	
21. Maxillary vacuities	Vacuities absent, 0.0	Small, 1.0; large, 2.0	
22. Posterior palatal foramina	Foramina present, 0.0	Foramina incomplete, 1.0	Cf. character 48
23. Foramen ovale	Foramen ovale present, 0.0	Foramen ovale absent, 1.0	
24. Transverse canal	Canal present, 0.0	Canal absent, 1.0	
25. Squamosal epitympanic recess	Recess absent, 0.0	Small, 1.0; large, 2.0	
26. Periotic tympanic wing	Wing absent or tiny, 0.0	Wing small, 1.0; large, 2.0	
27. Alisphenoid	Aisphenoid isolated, 0.0	Touches periotic or mastoid, 1.0	
28. Squamosal/frontal contact	No contact, 0.0	Contact, 1.0	
29. Presence/absence of conules on molars	Both para– and metaconules present, 0.0	Para– absent, metaconule present, 1.0; both absent, 2.0	Intermediates signify reductions
30. Relative lengths, $M^{3\ and\ 4}$	M^3 longer or subequal to M^4, 0.0	M^3 shorter than M^4, 1.0	Intermediates occur
31. Outer border of M^4	Subequal or longer than anterior border, 0.0	Noticeably shorter than anterior border, 1.0	Intermediates occur
32. Relative size and position of cusps on M^X	With subequal, separated para– and metacones, 0.0	With separated paracone smaller, 1.0; with approximated paracone smaller, 2.0; with only metacone present, 3.0	Intermediates indicate variability between molars
33. Stylar shelf	With usually five enlarged cusps, 0.0	Perhaps five but some small, 0.5; shelf narrow with small cusps, 1.0; cusps all very reduced, 2.0	

TABLE 2 CONTINUED

# and name of character	Primitive condition	Advanced state(s)	Comments
34. Paracrista and stylocone, M^X	Paracrista united with stylocone, 0.0	Not united, 1.0	Borderline indicated as 0.5
35. Metastylar spur and metacrista of M^X	Spur and metacrista moderately developed, latter not longer than postparacrista, 0.0	Both well developed, metacrista longer than post-paracrista, 1.0; both highly developed, espec. in M^4, 2.0	Intermediates occur
36. Shape of protocones, M^X	Bulbous, not compressed anteroposteriorly, 0.0	Moderately compressed, 1.0; highly compressed, 2.0	Intermediates occur
37. Relative size and shape of $M^{4 \text{ and } 5}$	M^5 subtriangular, wide as M^4, not compressed anteroposteriorly, para–cone subequal to metacone, 0.0	M^5 as in 0.0 but much reduced relatve to M^4, 1.0; with M^5 very reduced, 1.5; M^5 wider or as wide as M^4, compressed anteroposteriorly, metacone reduced relative to paracone or absent, 2.0; M^5 also narrower than M^4, 3.0	
38. Relative size and shape, $M_{5 \text{ and } 4}$	M_5 longer or subequal to M_4, 0.0	M_5 just shorter than M_4, 0.5; or moderately shorter and narrower than M_4, 1.0; M_5 markedly shorter and narrower than M_4, 2.0	
39. Relative sizes of M_X talonid and trigonid	Talonid longer than trigonid, 0.0	About equal, 0.5; talonid shorter, 1.0; talonid absent, 2.0	
40. Size and shape of M_5 talonid	Normal, 0.0	Reduced and narrower than that of M_4, 1.0; talonid missing, 2.0	
41. Relative sizes, meta– and paraconid of M_X	Metaconid larger than paraconid, 0.0	About equal, 1.0; paraconid larger than metaconid, 2.0	Intermediates occur
42. Development of protoconid	Moderately developed, 0.0	Much higher than paraconid, 1.0; both conids high, 2.0	
43. Upper incisor shape	Normal, 0.0	Spatulate, 1.0	Cf. character 2: initial states not the same
44. Size of P^X	Normal, 0.0	First P^X reduced, 1.0	
45. Shape of P^X	Normal, 0.0	$P^{2 \text{ and } 3}$ bulbous, 1.0	
46. Body size I.	Medium, 0.0	Large, 1.0; very large, 2.0	Character is two–part to show branching (cf. 47)
47. Body size II.	Medium, 0.0	Small, 1.0	
48. Palatal vacuities	Vacuities present, 0.0	Vacuities absent, 1.0	Cf. characters 20 and 21: combined here
49. Otic region I.	With alisphenoid and periocticum not or little expanded, 0.0	With both moderately expanded and ectotympanic contributing floor of tympanic cavity, 1.0; with alisphenoid and perioticum closing floor, 2.0; with alisphenoid, pars petrosal, and pars mastoidea united to form complete bulla, 3.0; with enlarged alisphenoid forming majority of bulla, 4.0	Character is two–part to show branching (cf. 50)
50. Otic region II.	As in state 1.0 of character 49, 0.0	With process of increasing length, 1.0, 2.0 or 3.0	
51. Size of paroccipital process	Process long, 0.0	Very small but not absent, 0.5; absent or altered in form, 1.0	
52. Size of ectotympanic process	Without process, 0.0	With process of increasing length, 1.0, 2.0 or 3.0	Intermediates occur
53. Condition of pars mastoidea	Normal, 0.0	Just swollen, 0.5; expanded, 1.0	
54. Size of sinus epitympanicus	Sinus absent, 0.0	Sinus present, 1.0	

accordingly, a number of computer programs have been devised for reconstructing the evolution of several characters across a set of many taxa (Sneath and Sokal 1973). A readily available, and especially appropriate, one is the Wagner tree method developed by Farris (Kluge and Farris 1969, and later papers). A simple example will illustrate both the necessity for such an algorithm and the features which make it particularly attractive to us.

Probably most marsupial biologists regard syndactyly as a significant feature uniting bandicoots and diprotodontans, and separating both from dasyuroids: it is a clearly derived state of the pes. However, diprotodontans differ from perameloids and dasyuroids in having procumbent lower incisors. In terms of just these two, two–state characters, we could imagine bandicoots and diprotodontans to be related through an ancestor that was syndactyl and polyprotodont; note that bandicoots are, in terms of *just* these two characters, identical to that ancestor, so that the perameloids and diprotodontans may be said to differ from each other by only one "evolutionary step". That common ancestor could in turn have been derived from something like a dasyuroid (primitive in both features), and again a single evolutionary change is all that separates them. Wagner tree analysis is simply a formalization of this logic, and similarly begins with three taxa and constructs the character–states of the taxon that would uniquely connect them; sometimes no known taxon possesses the necessary states, and then an hypothetical form is interposed along the connecting network. It is when additional taxa are added to a simple three–taxon network that it becomes obvious why allowance for convergence through the use of many characters, and hence the use of a computer, must be admitted. For example, consider the caenolestoids, which are diprotodont but also have didactyl feet. It is unclear how they should join the string of Australian taxa: if between diprotodontans and their most recent ancestor, then the syndactyl foot must have been lost and re-evolved; if nearer the dasyuroids, then diprotodonty must have evolved twice. (In fact, the computer program takes the latter course, as shown).

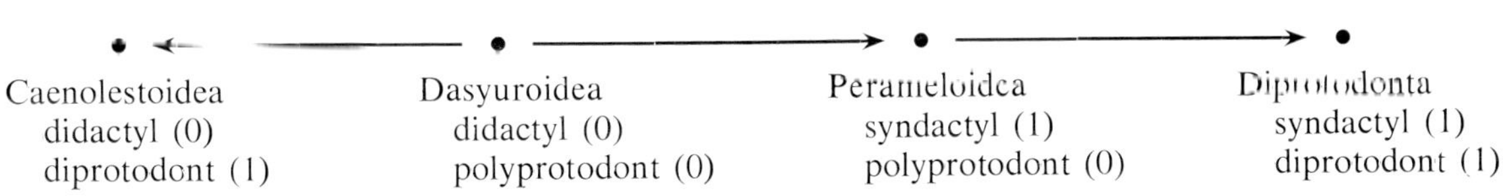

In either case, some "extra" step, or change in state, is required; however, it is a reasonable conclusion that one or the other character evolved the derived state twice, and one reason we like the Wagner approach is because it allows for just such parallel or convergent changes and reversals in state.

When many characters are considered, the number of such parallelisms, convergences, and reversals – collectively known as "homoplasy" – usually rises; since the whole point of the Wagner algorithm is to minimize homoplasy, it attempts to build the tree that requires the fewest homoplastic changes – the most parsimonious or shortest network. Achieving that result is, unfortunately, a mathematical problem without a formal solution, so the Wagner (or any other pathlength–minimizing) program can only be considered heuristic; it follows that the trees should be viewed as investigative tools rather than final answers – this is especially true since unless the taxa are extremely well–separated and obviously clumped, even the order in which the data are entered may affect the branching pattern. Nevertheless, by keeping track of the history of characters throughout the networks, various trees based on different character–subsets can be compared for their *consistency*, or relative freedom from homoplasy, formally defined for any character as the ratio of its *range* (the minimal number of steps that would occur were its evolutionary path unique, which is one less than its number of states) to its *length* on the tree (the actual number of changes, including homoplastic ones). The consistency of the entire tree is calculated analogously, by adding the ranges of all characters and dividing by the sum of the lengths of all segments of the network.

Once a tree has been generated, there are further experiments that can be performed. Since a record is kept of the history of each character, those which are most homoplastic could simply be eliminated and a new analysis performed on the more consistent ones; alternatively, the consistent characters may be weighted to increase their contribution to the form of the tree in a subsequent analysis (Farris 1969); or convergent characters might be recoded as separate characters to define the problem away (in the example above, the condition of caenolestoid incisors might be recognized as a new character with a derived state, "pseudo–diprotodonty", not possessed by any of the Australian taxa). As noted above, different trees can sometimes be obtained simply by altering the order of taxa or by adding or removing them; once again, some of these alterations result from limitations in any heuristic algorithm and should not be considered special faults of the Wagner program – and, moreover, those changes may sometimes have real biological meaning.

Table 3. Partial listing of Wagner Tree experiments performed on the data of Table 1. Consistency ratios (Range/Length; see text for explanation) are given for each tree, and the figure number is shown when illustrated. All trees were rooted on an hypothetical *Ancestor* primitive in every character, added to each set of taxa. Symbols: Ø means that calculation was not performed because the data were not complete; * indicates that midpoint rooting was also performed, but in every case the consistencies were the same, so only rooting on *Ancestor* is shown.

Characters / Taxa	Either Australian or American			Both Australian and American				
	21 Australian	46 Australian	25 American	29 of both	29 of both (tree fitted)	30 of both	46 of both	71 of both
1 – 28 (all dasyurid codings)	34/91=.37 (Fig. 1)	Ø	35/96=.35 (Fig. 2)	35/132=.27	35/170=.21	35/132=.27* (Fig. 9)	35/181=.19* (Fig. 8)	Ø
1 – 20 (dasyurid dental)	23/59=.39	Ø	24/71=.34	24/84=.29	24/124=.19	24/86=.28	24/125=.19 (Fig. 10)	Ø
21 – 28 (dasyurid non–dental)	11/20=.55	Ø	10/16=.63	11/31=.35	11/46=.24	11/31=.35	11/35=.31	Ø
29 – 54 (all opossum codings)	38/79=.48 (Fig. 3)	43/119=.36 (Fig. 4)	40/119.5=.33 (Fig. 5)	41/125.5=.33	41/161=.25	42/131.5=.32* (Fig. 12)	42/183=.23* (Fig. 11)	45/230=.20 (Fig. 14)
29 – 45 (opossum dental)	24/45=.53	28/76.5=.37	26/82=.32	26/69=.38	26/113.5=.23	27/71=.38	27/107=.25 (Fig. 13)	30/141.5=.21
46 – 54 (opossum non–dental)	14/22=.64	15/27=.56	14/21=.67	15/33=.45	15/47.5=.32	15/35=.43	15/36.5=.41	15/42=.36
1 – 54 (all characters)	72/177.5=.41 (Fig. 6)	Ø	74/235.5=.31 (Fig. 7)	76/275.5=.28	76/331=.23	77/281=.27* (Fig. 16)	77/393.5=.20* (Fig. 15)	Ø
1 – 20; 29 – 45 (all dental)	47/110=.43	Ø	50/172=.29	50/175=.29	50/237.5=.21	51/177.5=.29	51/272=.19 (Fig. 17)	Ø
21 – 28; 46 – 54 (all non–dental)	25/50=.50	Ø	24/45.5=.53	26/73=.36	26/93.5=.28	26/78.5=.33	26/88=.30 (Fig. 18)	Ø

An option of the program package we used ("Wagner 78", available from J. S. Farris, Department of Biological Sciences, State University of New York at Stony Brook, Long Island 11790) is that the characters may be fitted to a predetermined tree, a useful technique for finding out how well new data fit one's preconceived notions of phylogeny.

One important limitation of the Wagner method is that the direction of change is not necessarily indicated – the base, or *root,* of the tree must be inferred. If one has strong confidence in the polarities of individual characters, the root will simply be the taxon with the largest number of primitive character–states, or an hypothetical all–primitive "ancestor" may be included in the analysis as the root taxon (our general procedure). Alternatively, the directions of change can be inferred by including an "outgroup", or taxon known to be phyletically more distinct from all members of the group under study than they are from each other; the point of linkage will then indicate the starting–point for evolution within that group (we have included several putative outgroups in most of our trees). Or (and we have done this in a few cases) the tree can be rooted at the midpoint between the two most distant taxa, a decision tantamount to assuming regular rates of change which may give a quite different orientation to the tree from rooting based on all–primitive ancestral taxon or an outgroup.

Trees Generated

Table 3 lists most of the analyses which we performed on the data of Table 1. All data were entered in the order of taxa shown in that table, by columns, from left to right. As indicated in Table 3, we systematically partitioned both the data–set and set of taxa, computing trees for just American or just Australian marsupials, or for both together, using only dasyurid codings, only opossum codings, or both, and utilizing all characters within the two subsets or just dental or non–dental features; and we varied the number of taxa to some extent (necessarily in the case of Australian taxa, since not all were coded for dasyurid characters; the 47–taxon trees including members of both groups were taxa common to both subsets of the characters*). In addition, the 29 taxa included in our set which were shown on the tree of dental similarities in Archer (1976) were separately analyzed and the various subsets of characters fitted to Archer's tree.

All of the trees listed in Table 3 were computed using the method of rooting based on an all–primitive *Ancestor* added to the set of taxa. In addition, trees corresponding to Figs 8, 9, 11, 12, 15 and 16 were recomputed and rooted at the midpoint. The consistencies of these additional trees did not differ from those rooted on *Ancestor,* so they are not listed separately. For those listed, the summed ranges of characters (numerators) and tree-lengths (denomenators) are given for each computation as well as the quotient. Note that the summed ranges may differ for two trees computed on the same character subset if the number of taxa differs. This is because one or more character–states may not be represented in a reduced set of taxa. Strictly, consistencies are only comparable for equal numbers of taxa computed on the same number of characters and states, as for example between the values in the two columns headed "29 of both".

In the trees illustrated here, we have followed the convention that *Ancestor* appears at the right or upper righthand corner, and successively more derived taxa are toward the left. When (as in the example above) a real taxon is found to be exactly the same in all its character–states as the hypothetical immediate ancestor uniting it and other taxa, that real taxon is placed on the node which represents its ancestor. More generally, then, the lengths of lines connecting nodes are proportional to the number of evolutionary steps (sums of state–changes, in increments of one–half unit) separating any two taxa; that number is sometimes zero. For the sake of clarity, we have presented our trees in the form of phenograms and included a scale for each that can be used to measure the pathlengths, which are the usually horizontal lines separating linkage levels; vertical distances have no absolute meaning. It is useful, too, to remember that branches emanating from any node may be rotated around that point.

Not all of the trees generated in this study are illustrated or will be discussed in detail, nor are all the trees not listed in Table 3 considered. Early in the study we experimented with even smaller subsets of the data – comprised of the characters shown to be "consistent" in the initial analyses – but these were so small that little discrimination among taxa was possible, and the attempt to build additive or nearly additive trees was abandoned. For the same reason, the amount of computation necessary to weight characters successively according to their consistencies (Farris 1969) did not seem likely to be rewarded.

RESULTS

Although we began this study by constructing large trees for all taxa based on the opossum codings and progressed to more restrictive trees

* That is not strictly true, as *Phascolosorex doriae* was inadvertently substituted for *P. dorsalis* in the computations based on opossum codings. However, these taxa differ by only 0.5 units in the opossum codings (see Figs 4 and 14; *P. doriae* was not scored on the dasyurid characters), and so restitution of the correct taxon only shortens the line leading to *P. doriae* by this amount, without changing its cladistic association. In that respect, then, the relationship of *Phascolosorex* is the same for both species.

and to others based on dasyurid codings, it seems best to begin the presentation with trees most likely to agree with readers' intuitions and then demonstrate how more elaborate or differently organized Wagner networks depart from expectation. Thus, the first set of trees we will discuss are the several resulting from analyses of just American or just Australian taxa, using either subset of the characters. Clearly, the richness of relationships shown in any of our trees cannot be discussed exhaustively, so we will focus throughout on rather few points, mainly the relationships of the thylacinids, peramelids, the numbat, and several taxa suspected on other grounds of being polyphyletic. Taxa are referred to by just their generic names where no confusion is likely.

AUSTRALIAN OR AMERICAN TAXA ALONE

Dasyurid codings, Figs 1 *and* 2: The first tree depicts the arrangement of the 21 dasyurids scored for both subsets of the characters and is based on the dasyurid codings. Readers of this volume will immediately recognize some familiar patterns, in that the two dasyures are cladistically separated, as is *Antechinus stuartii* from its congeners. The fossils *Ankotarinja tirarensis* and *Keeuna woodburnei* are, with the species of the New Guinean genus *Murexia,* close to another fossil, *Wakamatha tasselli,* and to species of *Sminthopsis;* the Numbat is quite distinct from all other dasyuroids and the two bandicoots even more so. One surprise, however, is that the two thylacines are close to *Sarcophilus harrisii* which in turn is close to the species of *Planigale*; all form a sister–group to the Tiger Quoll, which is, as noted, well–separated from *Satanellus hallucatus.* The significance of these associations will become apparent in trees which include American marsupials, but the cluster of "large carnivores" is quite consistent throughout most of the trees, and it is generally true that groupings at lower categorical levels are much the same whether based on dasyurid or opossum codings. As a consequence of the thylacines' close relationship with the Devil, *Thylacinus potens* is the more derived of the two; but its "derived" features are in fact secondary reversions to primitiveness in characters 12 and 22. This tree will be considered in more detail in the Discussion.

The tree of Fig. 2 shows how the same characters relate the opossums, and these groupings do not make a great deal of taxonomic sense (see the discussion in Kirsch 1977), except in that, as in the first tree, there is an obvious trend toward increasing carnivory; otherwise, the 2n=14 and 2n=22 opossums are intermixed, the caluromyines are not a group, and *Dromiciops australis* (which should be a sister–group to all other didelphids) is derived from within an unlikely clade; *Caenolestes obscurus,* the sole diprotodont included in our experiments, is apparently related to *D. australis,* but this is a grouping by exclusion: the linkage is the last to occur in the successive addition of taxa, and no characters truly descriptive of the caenolestid were included in either subset of the data. Yet

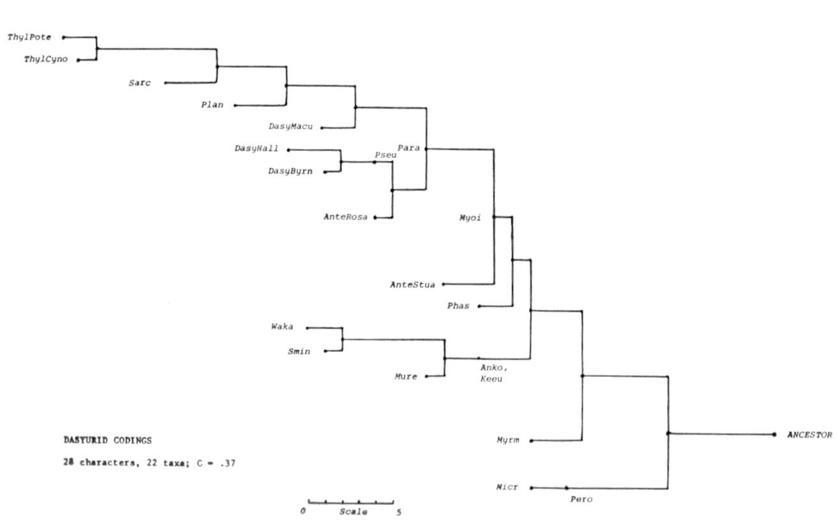

Fig. 1. Wagner tree of 22 Australian taxa computed on the basis of characters 1–28 (dasyurid codings), with overall consistency of .37. In this and all subsequent illustrations taxa are identified by the first four letters of their genera, and, if confusion is possible, the first four of their specific epithets as well. Errors in labeling or deviations from this rule are noted for individual trees. Horizontal lines between linkage levels are drawn in proportion to evolutionary distances, and the scale may be used to determine the exact number of evolutionary steps separating nodes; distances are in increments of 0.5 unit. Since the all-primitive *Ancestor* is at right, taxa to the left are more derived. Arrows on some figures indicate the approximate position of the tree's midpoint if the appropriate computation was carried out. DasyHall = *Satanellus hallucatus;* and AnteRosa = *Dasykaluta rosamondae.*

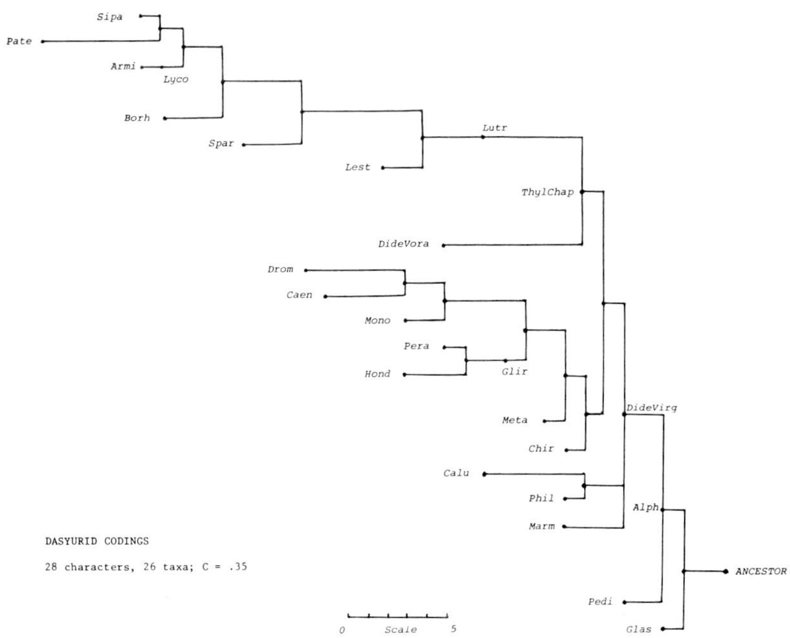

Fig. 2. Wagner tree of 26 American taxa computed on the basis of characters 1–28 (dasyurid codings), with overall consistency of .35. Conventions as for Fig. 1.

this tree does show the Cretaceous species of *Alphadon,* often cited as ancestral to opossums (Clemens 1979), near the base along with two other North American Cretaceous marsupials. Like the first and most subsequent trees, there is a notable linearity to this one: apparently most of our characters tend to discriminate marsupials as more or less carnivorous mammals; but that after all, is largely what the evolution of marsupicarnivores is about.

Although the results are not shown, these groups of taxa were reanalyzed using just the dental (1 – 20) or non–dental (21–28) characters of the subset. Most of the relationships were altered in ways similar to those shown in taxonomically more inclusive trees, discussed later on; with the exception of the intercalation of bandicoots and the Numbat into the Dasyuridae, most of the lower–level relationships were preserved. Non–dental characters gave little discrimination but did separate the thylacines from the Devil.

Opossum codings, Figs 3, 4 and 5: Trees 3 and 5 show the same sets of taxa as in the first two trees, but related on the basis of the opossum codings; thus, the opossums (Fig. 5) group much as in Reig *et al.*'s trees, except that those authors did not include most of the fossils depicted here: the main effect of their inclusion is to dismember the "large opossum" (2n=22) group so that some members tend toward or are included in the greatly derived carnivore branch (again at the top, or terminus, of the tree). But the derivation of caluromyines (here in association with *Caenolestes obscurus*) and of *Dromiciops australis* as major branches, and the placement of species of *Marmosa, Lestodelphys,* and *Monodelphis* together, are results in accord with expectation.

On the other hand, the cladistics of the Australian marsupials (Fig. 3) are altered somewhat from those shown in Fig. 1: the bandicoots are now derived from within the Dasyuridae, and the thylacines (with the Devil) are an early–derived branch (significantly, the next to split off are species of *Planigale*). But on the whole the differences between Figs 1 and 3 are fewer than those between Figs 2 and 5. Certainly,associations at the generic or specific levels are altered very little for the dasyurids using the different subsets, though, in terms of the opossum codings, *T. potens* is the more primitive thylacine and *Satanellus hallucatus* is the nearest cladistic relative of *D. maculatus;* but the *Antechinus* species are still clearly diphyletic, at least, with *Dasyuroides byrnei* close to them. In other words, though the two character subsets seem to say different things about opossums, they indicate much the same about dasyurid relationships. That being so, it is worth considering Fig. 4, a tree showing all 46 of the dasyurids which were coded according to the opossum characters. There are few surprises here, since for the most part the additional taxa belong to apparently secure clades already established among the smaller set. However, the addition of species of *Ningaui* and other *Planigale* removes *Planigale* from its proximity to the Thylacine–Devil clade and places them near

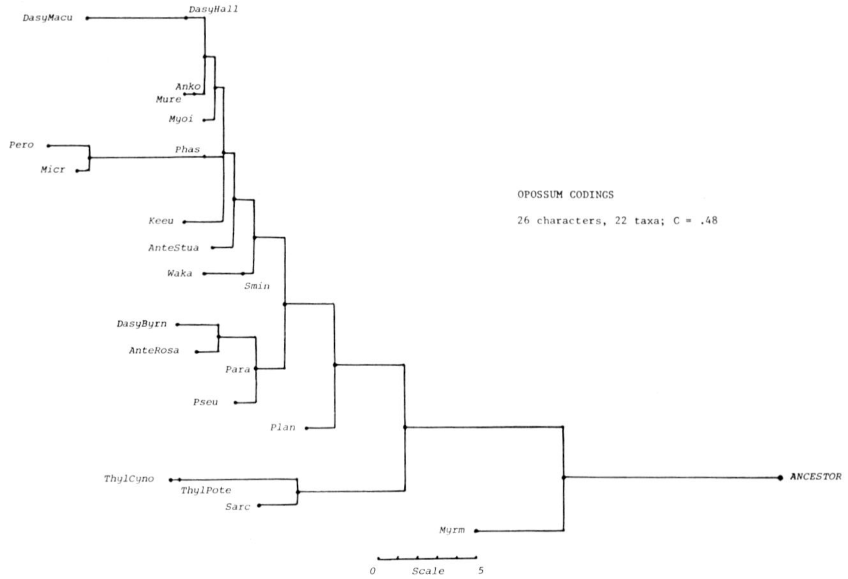

Fig. 3. Wagner tree of 22 Australian taxa computed on the basis of characters 29–54 (opossum codings), with overall consistency of .48. Conventions as for Fig. 1.

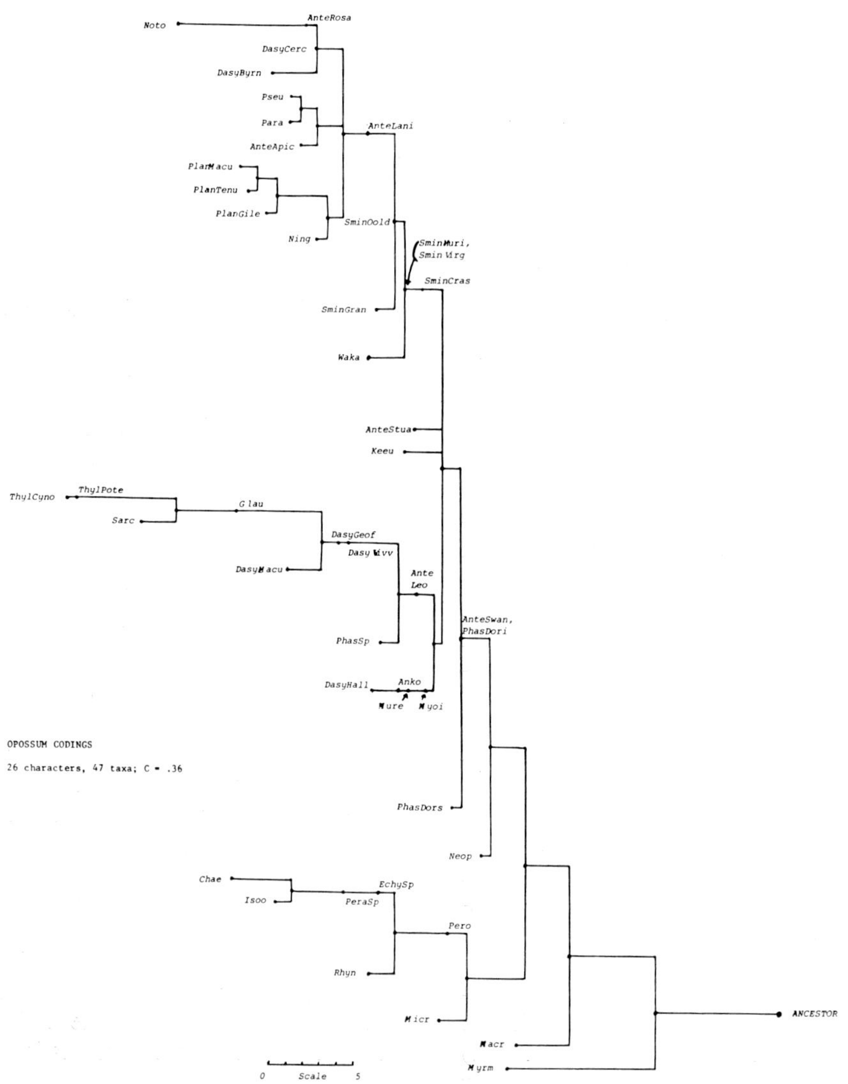

Fig. 4. Wagner tree of 47 Australian taxa computed on the basis of characters 29–54 (opossum codings), with overall consistency of .36. Conventions as for Fig. 1, except that *AnteSwan* is *Antechinus swainsonii, EchySp* is *Echymipera clara, PeraSp* is *Perameles nasuta,* and *PhasSp* is *Phascogale tapoatafa.*

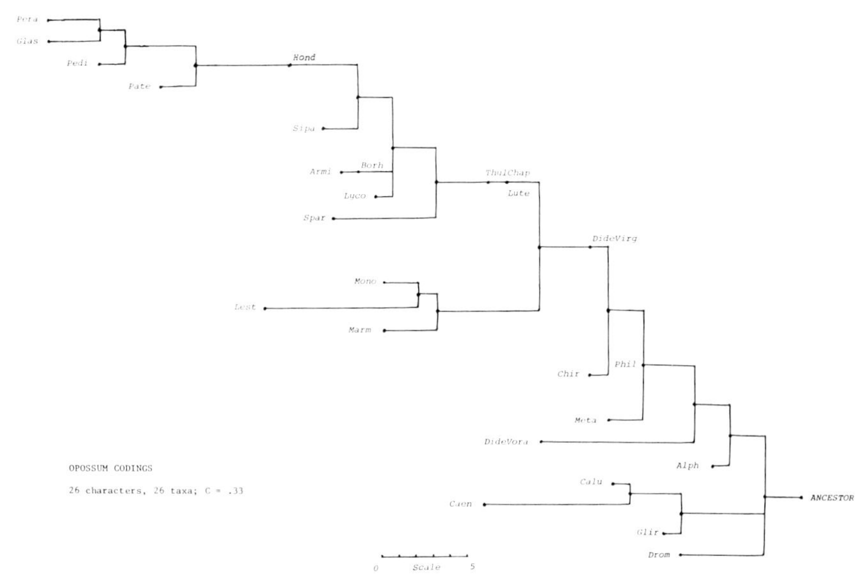

Fig. 5. Wagner tree of 26 American taxa computed on the basis of characters 29–54 (opossum codings), with overall consistency of .33. Conventions as for Fig. 1.

to the "true" species of *Antechinus*. However, that genus seems even more diverse here than before, since *A. leo* is near the base of the "large carnivore" branch while *A. swainsonii* is more primitive. The bandicoots form a believable sequence except that the species of *Macrotis* stand for a separate lineage in contrast to all other bandicoots plus the dasyuroids (which includes *Notorycytes typhlops*, grouped by exclusion with some *Antechinus* and *Dasycercus cristicauda*). We emphasize again that these relationships are, as it were, from the point of view of the opossums; it is unfortunate that we did not extend the coding by dasyurid characters to the additional Australian taxa represented here.

Analysis on dental characters alone again caused few noteworthy changes, and the nine non–dental characters produced trees with very little discrimination, although bandicoots were not a single clade and tended to be clustered, along with New Guinean dasyurids, near the base of the trees; opossum relationships were, on the basis of non–dental features, very close to expectation.

Both codings combined, Figs 6 and 7. Some indication of the relative influence of the two subsets of data on relationships is given by trees generated using all 54 characters together. The branchings among Australian carnivores (Fig. 6) are very similar to those shown in the first tree, except that *Wakamatha tasselli* and species of *Sminthopsis* are placed very near the Thylacine–Devil–planigale clade instead of near the other fossil and New Guinean taxa; the two dasyures are here sister–taxa, but the polyphyly of *Antechinus* remains evident, as does the early derivation of *Myrmecobius fasciatus* and the still–earlier derivation of the two bandicoots. As expected, then, combination of the two subsets reinforces most groups of Australian marsupials.

The effect of the dasyurid codings on the opossums (Fig. 7) is also relatively minor at fine levels of relationship, but there is a major depar-

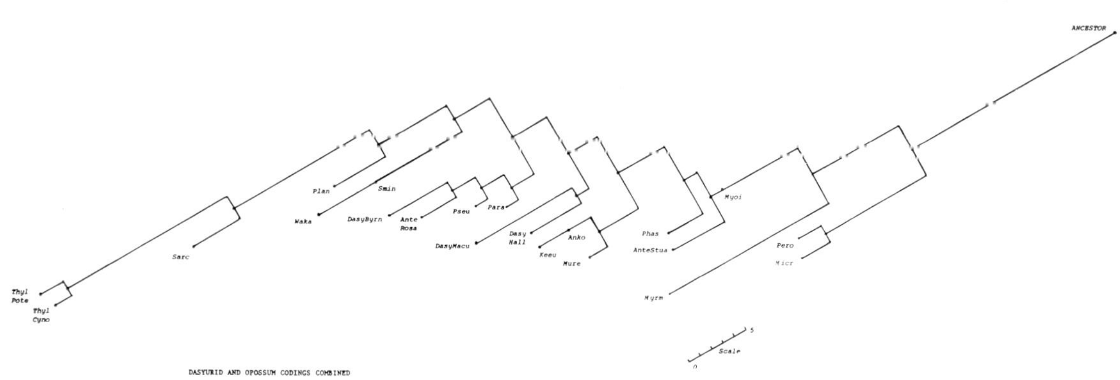

Fig. 6. Wagner tree of 22 Australian taxa computed on the basis of characters 1–54 (dasyurid and opossum codings combined), with overall consistency of .41. Conventions as for Fig. 1.

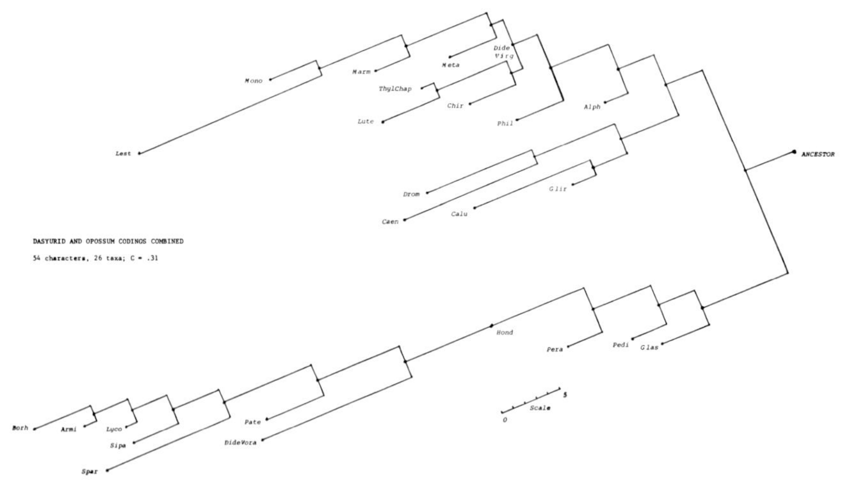

Fig. 7. Wagner tree of 26 American taxa computed on the basis of characters 1–54 (dasyurid and opossum codings combined), with overall consistency of .31. Conventions as for Fig. 1.

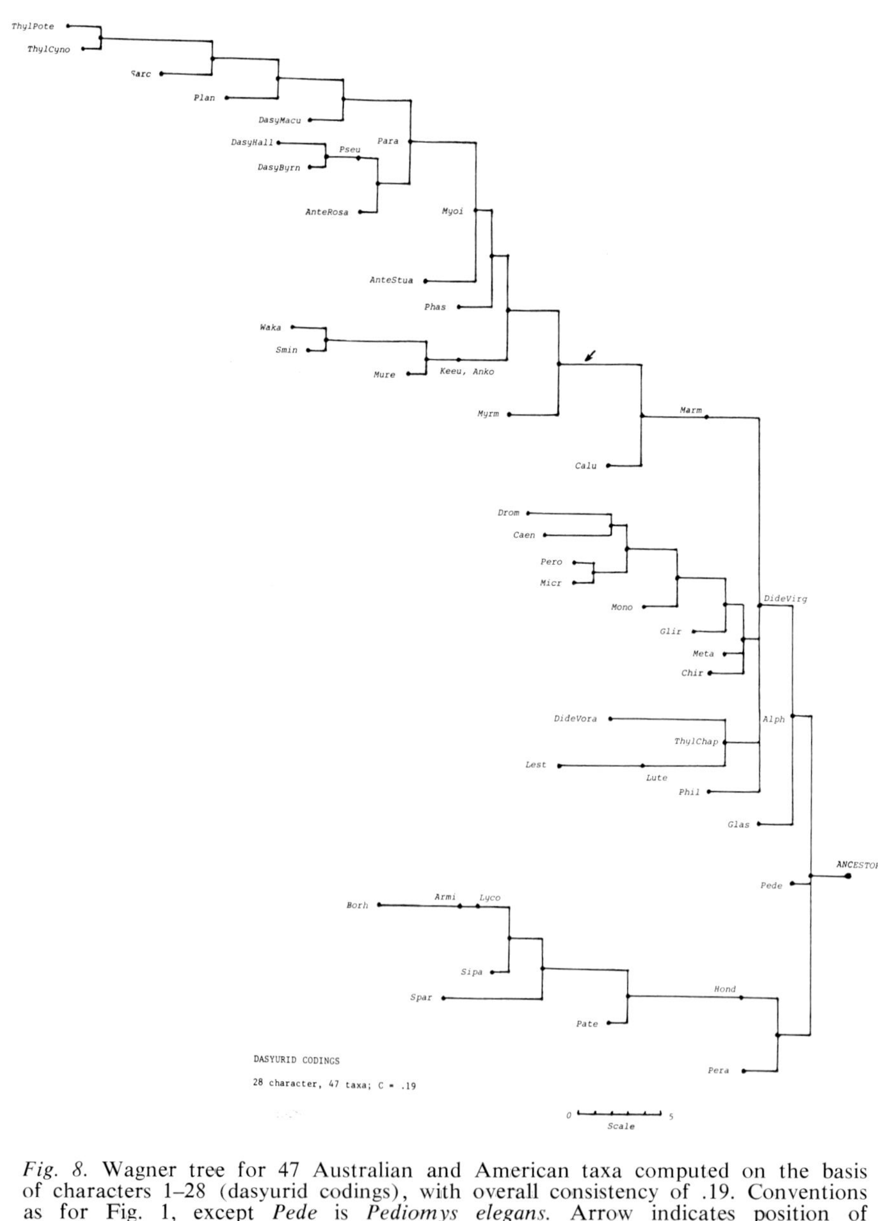

Fig. 8. Wagner tree for 47 Australian and American taxa computed on the basis of characters 1–28 (dasyurid codings), with overall consistency of .19. Conventions as for Fig. 1, except *Pede* is *Pediomys elegans*. Arrow indicates position of midpoint.

ture from either of the two previously illustrated trees involving opossums in that the fossil carnivores are all placed in a single branch, the result of a fundamental dichotomy among opossums (there are two exceptions — species of *Thylaphorops*, manifest relatives of *Lutreolina crassicaudata*, and species of *Alphadon*, the putative ancestor of *Didelphis*–like forms). One noteworthy feature of the "fossil" clade is the close relationship among *Sparassocynus* spp. (usually considered to be didelphids; Reig and Simpson 1972) and successively derived borhyaenids. The living opossums are themselves dichotomous, with the caluromyines plus *Dromiciops australis* and *Caenoloestes obscurus* opposed to the "large opossum" branch. *Alphadon rhaister* is near the base of the tree.

Again, partitioning of the characters into dental or non–dental does not have marked effects, except that bandicoots are dentally less distinct from dasyuroids; and the close association of caluromyines and *Dromiciops australis* is lost when only non–dental features are used.

Australian and American Taxa Together

Dasyurid codings, Figs 8, 9, and 10: These and the remainder of the trees illustrated include a combination of Australian and American marsupials and thus explore the relationships between the two geographic moieties. The first question we asked concerned the thylacines and borhyaenids, so the most striking feature of Fig. 8, a tree based on Archer's dasyurid characters, is that the thylacines (with, as usual, species of *Planigale* and *Sarcophilus*) are the most derived taxa of all, while the borhyaenids and other fossil American carnivores form a branch at the first dichotomy after *Ancestor;* in fact, midpoint rooting of this tree, at a point just to the right of the node before *Myrmecobius fasciatus,* places the thylacines and *Borhyaena* spp. equidistant from the base. Interestingly, the separation between them is not maintained when a smaller subset of taxa (31; see Fig. 9) or the same set of 47 but analyzed on just the 20 dental characters (Fig. 10) are used. With the smaller set of taxa, *Borhyaena* spp. and *Sparassocynus* spp., among the most derived of their clade in Fig. 8, join *Thylacinus* spp. at the terminus of the dasyurid branch of the tree (midpoint rooting of the tree in Fig. 9 again places the base near the divergence of the Numbat); the significance of this alteration in position of American carnivores will be discussed below. Similarly, when only dental characters are considered (Fig. 10), many of the most derived Australian marsupials are re–placed in a branch with the borhyaenids and other American marsupials of markedly carnivorous aspect. However, in the latter case it cannot be said that the tree is clearly partitioned into American and Australian groups, whereas when all characters are used (for either 31 or 47 taxa), the Australian marsupials are nearly all concentrated in a single, very derived clade (which is, in fact, one branch of the fundamental dichotomy produced when the tree of Fig. 8 is rooted at the midpoint): they appear, in a way, as an extension of the "opossum theme" of increasing carnivory. The only major exception to this generalization is that the bandicoot group is placed with *Dromiciops australis* and *Caenolestes obscurus*. With fewer taxa, the position of *Myrmecobius fasciatus* is more entrenched within Dasyuridae, as it is when only dental characters are employed (Fig. 10). The eight non–dental characters give little discrimination (the trees are not shown), but the thylacines and borhyaenids never grouped. Otherwise, the lower–category taxa have relationships in these trees not dissimilar to those of earlier figures. In summary, this character–subset is quite good at maintaining separation of Australian from American marsupials, although, again, the former are merely a clade within a subgrouping

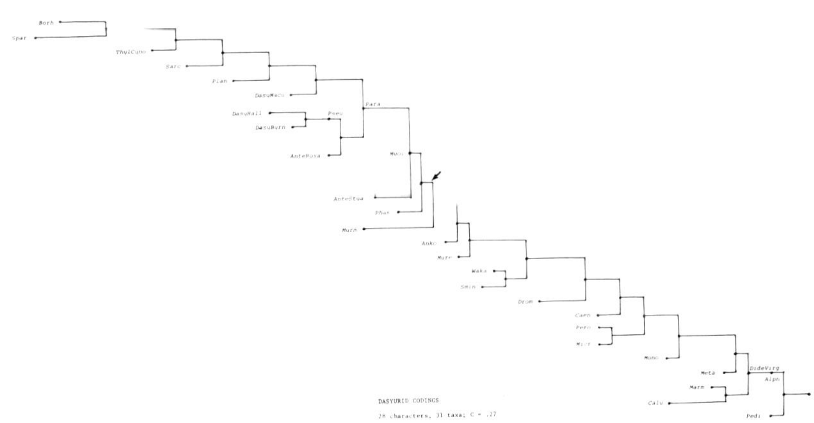

Fig. 9. Wagner tree of 31 Australian and American taxa computed on the basis of characters 1–28 (dasyurid codings), with overall consistency of .27. Conventions as for Fig. 1. Arrow indicates position of midpoint.

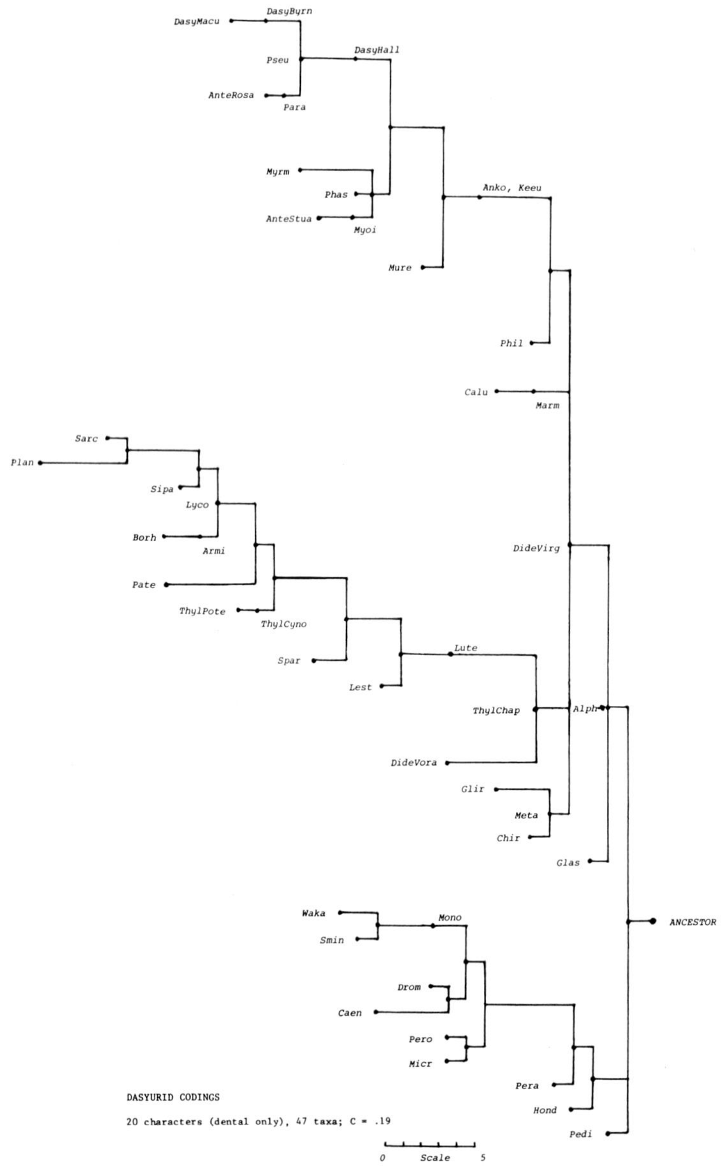

Fig. 10. Wagner tree of 47 Australian and American taxa computed on the basis of characters 1–20 (dasyurid dental), with overall consistency of .19. Conventions as for Fig. 1.

of the opossums, rather than the expected alternative branch of a fundamental dichotomy within the marsupicarnivores.

Opossum codings, Figs 11–14: The opossum codings, in contrast, separate the dasyuroids much less cleanly. Moreover, the thylacines and *Sarcophilus harrisii,* but not always *Planigale* spp., are associated with the borhyaenids throughout this series of trees. The relationships of both geographic groups are further distorted by the intercalation of American marsupials among the Dasyuridae: thus, *Caenolestes obscurus* and the bandicoots are near the termini of the trees calculated from all characters (Fig. 11 for 47 taxa, and Fig. 12 for 31), but not when only dental ones were employed (Fig. 13); midpoint rooting, about the point of divergence of *Didelphis virginiana* in the 47–taxon tree and of *Wakamatha tasselli* in the 31–taxon dendrogram, did not separate American from Australian marsupials. *Lestodelphys halli,* when included, is very close to *W. tasselli* and species of *Sminthopsis;* yet *Myrmecobius fasciatus* placed near the base of the tree with caluromyines and *Dromiciops australis* when 47 or 72 taxa (Fig. 14) were analyzed, but when only 31 taxa were examined the Numbat appeared along with *Caluromys* spp. to be very derived (that is, they were placed with Australian marsupials). These last remarks hold for all characters or for only the dental ones (Fig. 13).

The particularities of intra– and many intergeneric relationships of the dasyuroids are much as in Figs 3, 4, and 6, but differ for opossums

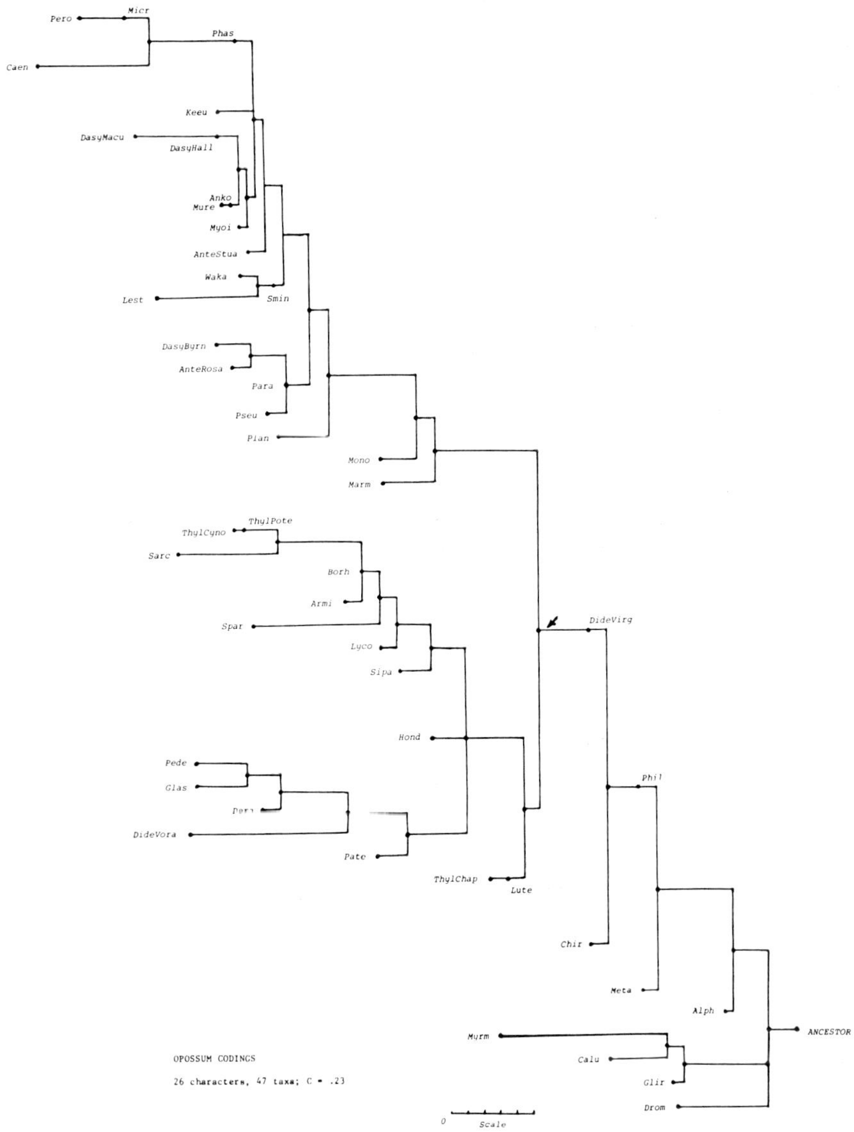

Fig. 11. Wagner tree of 47 Australian and American taxa computed on the basis of characters 29–54 (opossum codings), with overall consistency of .23. Conventions as for Fig. 1, except that *Pede* is *Pediomys elegans.* Arrow indicates position of midpoint.

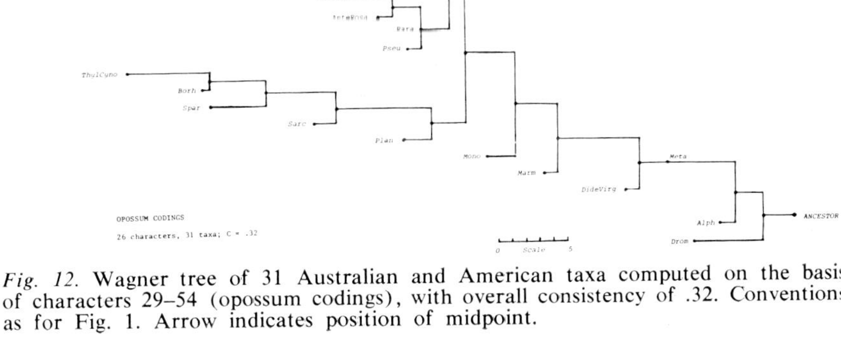

Fig. 12. Wagner tree of 31 Australian and American taxa computed on the basis of characters 29–54 (opossum codings), with overall consistency of .32. Conventions as for Fig. 1. Arrow indicates position of midpoint.

(*cf.* Figs 5 and 7) in a few respects relatable to the intercalation of *Lestodelphys halli* and other opossums with dasyuroids: in a sense the relationships of opossums are "twisted" out of shape as one branch of their tree is forced to graft with one of the dasyurids', so it is fair to say that the opossum codings are less successful than the dasyurid ones in keeping the

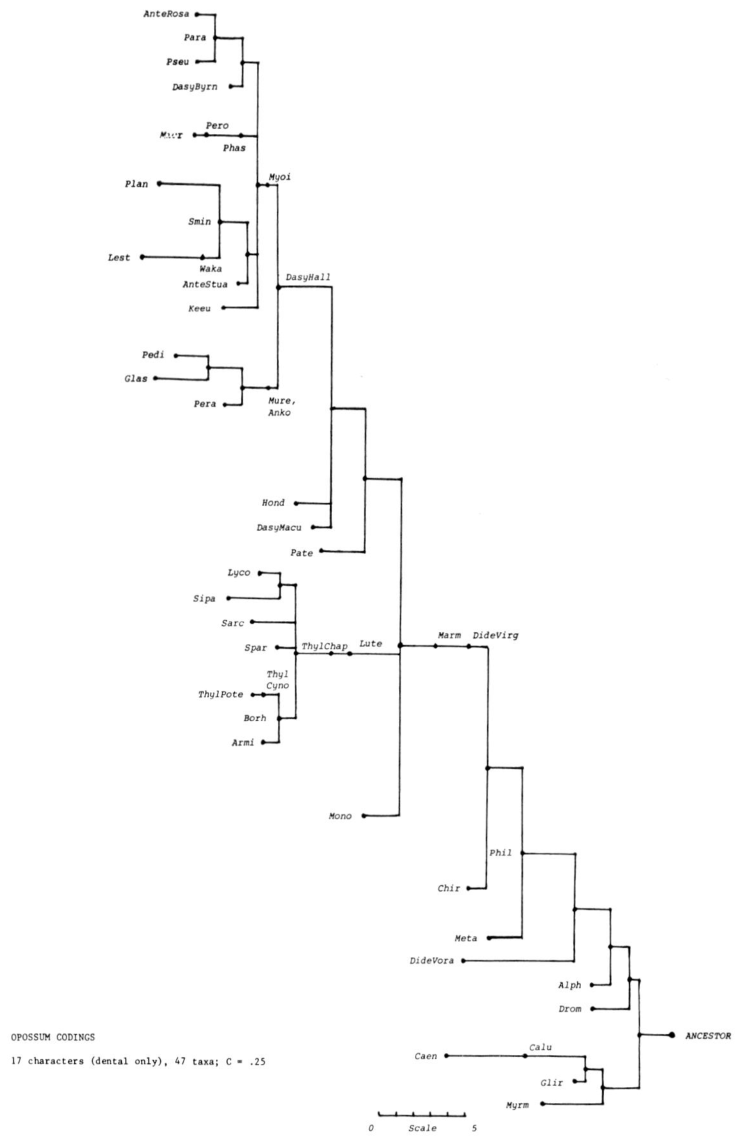

Fig. 13. Wagner tree of 47 Australian and American taxa computed on the basis of characters 29–45 (opossum dental), with overall consistency of .25. Conventions as for Fig. 1.

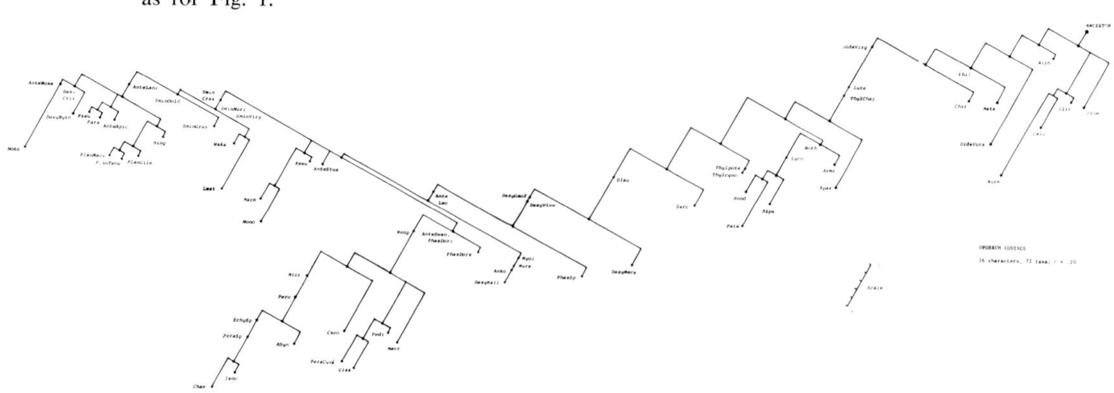

Fig. 14. Wagner tree of 72 Australian and American taxa computed on the basis of characters 29–54 (opossum codings), with overall consistency of .20. Conventions as for Fig. 1, except that *AnteSwan* is *Antechinus swainsonii, EchySp* is *Echymipera clara, PeraSp* is *Perameles nasuta, PhasSp* is *Phascogale tapoatafa,* and *SminGras* is *Sminthopsis granulipes.*

American and Australian marsupials apart. Again, use of dental characters alone does not alter these statements.

Curiously, the thylacines and borhyaenids were separated on non–dental features when 47 or more taxa were studied, but linked when the reduced set of 31 taxa was examined in terms of these nine characters.

Both codings combined, Figs 15 – 18: The last set of trees shown are constructed from the entire data–set or the combined dental or non–dental portions of it, and, as do the trees for Australian and American marsupials alone based on the combined subsets, they strengthen most of the points made heretofore. The opossum codings are apparently decisive in placing the thylacines and the Devil, but not *Planigale* spp., in the borhyaenid clade in the 47–taxon tree based on all 54 characters (Fig. 15); midpoint rooting neatly separates the upper and lower branches, and this tree is thus the only one to have *Ancestor* at the base for both all–primitive and midpoint rooting methods. In a 31–taxon tree (Fig. 16), however, where midpoint rooting puts the base just to the left of the divergence of *Wakamatha tasselli* and *Sminthopsis* spp., both *Borhyaena tuberata* and *Sparassocynus* spp. lie along the lineage of *Thylacinus* spp., *Sarcophilus harrisii,* and *Planigale* spp., and all of the large American carnivores are part of that lineage on the basis of dental characters (Fig. 17); the reason for their separation in trees calculated from just dasyurid characters is evident when only non–dental characters are used (Fig. 18), for then the American carnivores are again quite cleanly separated in the resulting tree. At the same time, the intra– and intergeneric relationships are not much altered from previous trees whatever portion of the data–set is employed.

TREE–FITTING EXPERIMENTS

A set of 29 taxa common to both subsets of characters and to the phenogram of dental similarity in Archer (1976) were analyzed un-

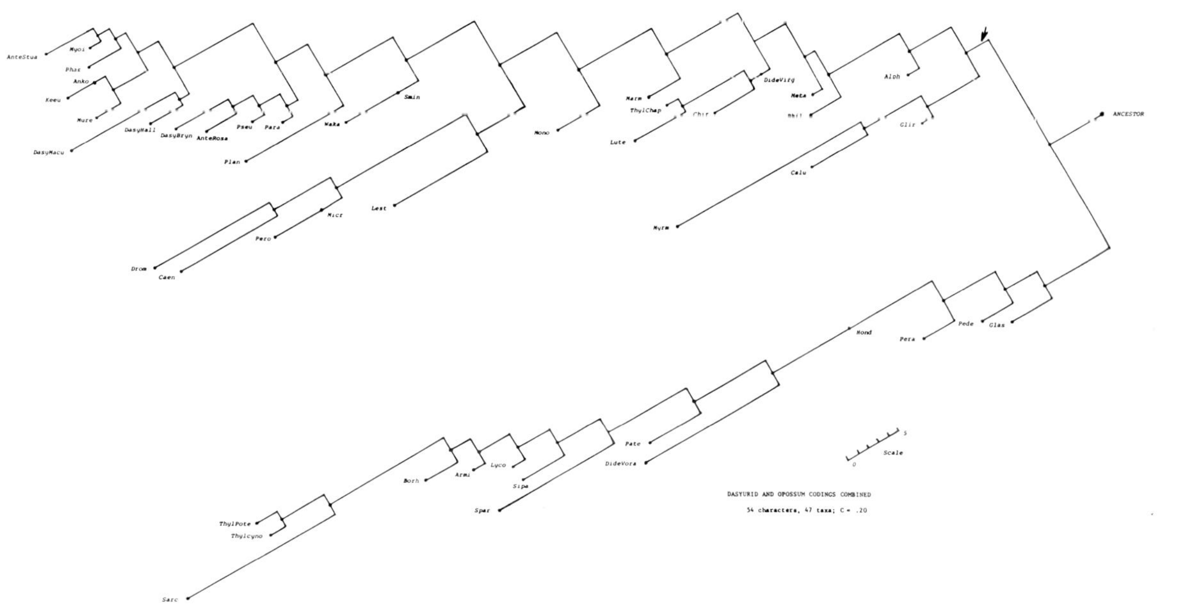

Fig. 15. Wagner tree of 47 Australian and American taxa computed on the basis of characters 1–54 (dasyurid and opossum codings combined), with overall consistency of .20. Conventions as for Fig. 1, except that *DasyBryn* is *Dasyuroides byrnei* and *Pede* is *Pediomys elegans.* Arrow indicates position of midpoint.

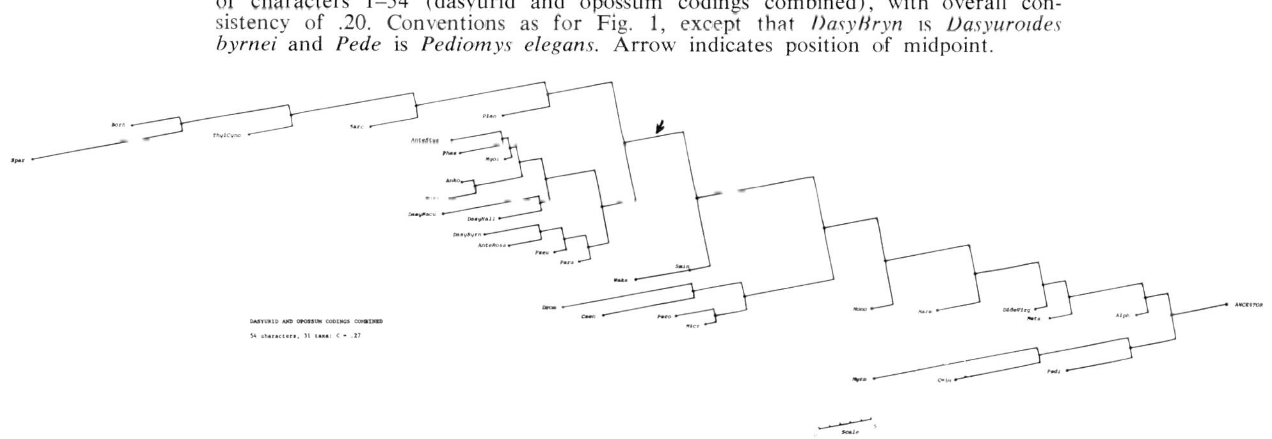

Fig. 16. Wagner tree of 31 Australian and American taxa computed on the basis of characters 1–54 (dasyurid and opossum codings combined), with overall consistency of .27. Conventions as for Fig. 1. Arrow indicates position of midpoint.

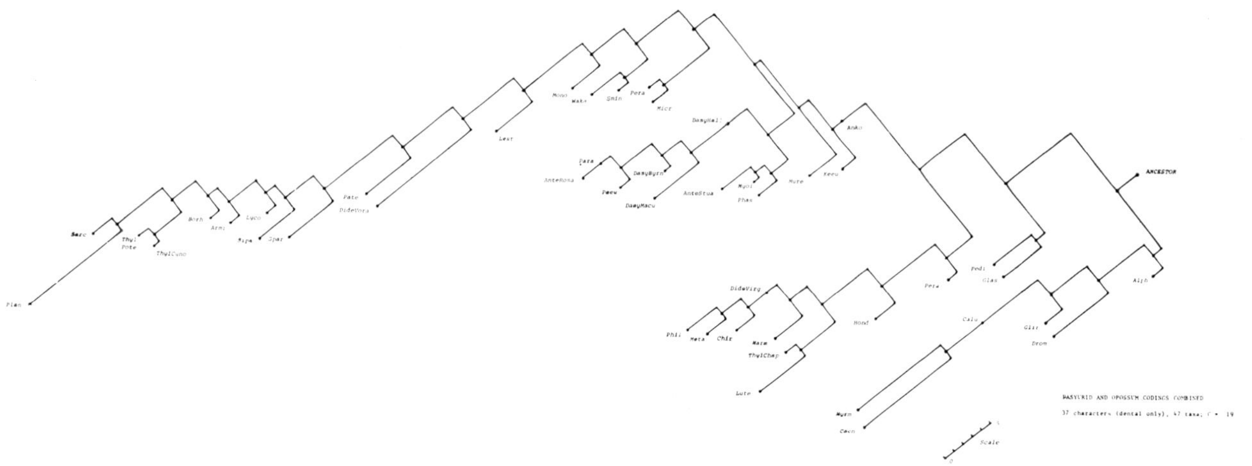

Fig. 17. Wagner tree of 47 Australian and American taxa computed on the basis of characters 1–20 and 29–45 (dasyurid dental and opossum dental combined), with overall consistency of .19. Conventions as for Fig. 1, except that the *Pera* linked with *Micr* is *Peroryctes longicaudata.*

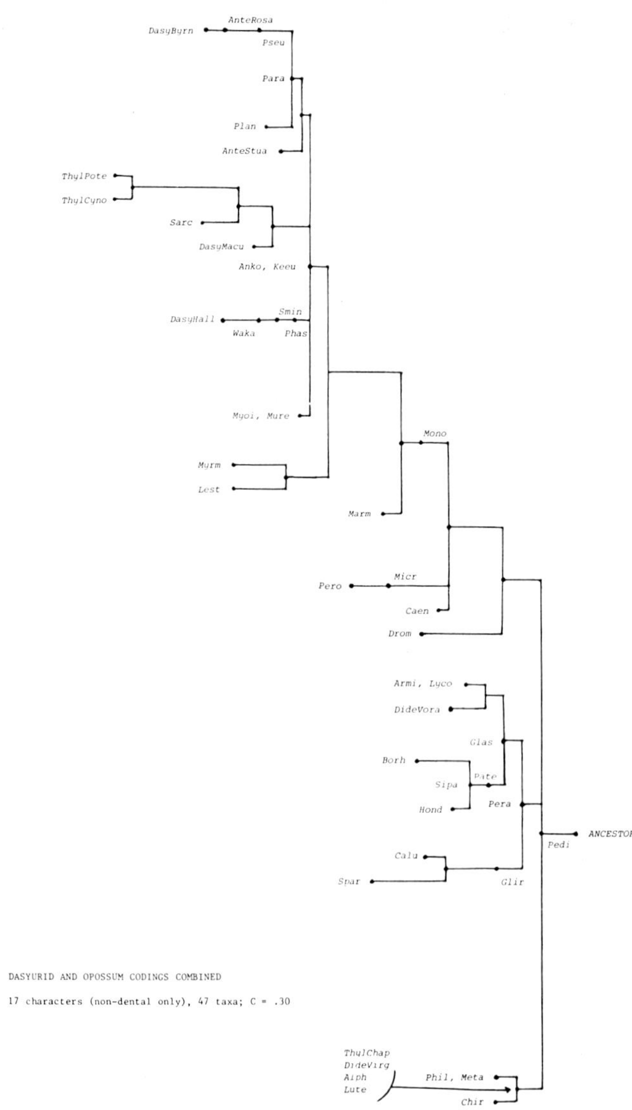

Fig. 18. Wagner tree of 47 Australian and American taxa computed on the basis of characters 21–28 and 46–54 (dasyurid non-dental and opossum non-dental combined), with overall consistency of .30. Conventions as for Fig. 1.

der all the usual regimes of characters; in each case these characters were also fitted to Archer's tree (*Antechinus stuartii* was substituted for *A. flavipes* and *Ancestor* was added to the set), a chief feature of which is the placement of the thylacines with borhyaenids on the American half of the basic dichotomy he recognized in Marsupicarnivora. Another feature of Archer's dendrogram is that the bandicoots are shown branching off between the thylacinid–borhyaenid and didelphid branches. None of these trees are illustrated here, since the Wagner trees were quite similar to the foregoing, and Archer's scheme is well–known. The consistencies of all trees are given in Table 3, however, where it can be seen that the Wagner–generated trees are always shorter than the fit of the same characters on Archer's diagram. Since Archer's tree was ostensibly based on just dental characters, it is surprising that no marked or regular improvement in fit results from using only these features. Some of these trees also figured in related experiments by Smith and Kirsch (1982) which will be noted in the Discussion.

DISCUSSION

As we have pointed out repeatedly in discussing individual trees, the relationships of congeners and even of some groups of genera, particularly among the Dasyuridae, are quite consistent throughout our several manipulations of the data–set. Thus, the long–suspected polyphyly of *Antechinus* is supported by Wagner tree analysis, as it is by several other studies reported in this volume; the peramelid bandicoots usually group well; *Satanellus hallucatus* is generally separated from other dasyures, again as suggested by enzyme analysis; and the overall groupings of opossums (into large, small, caluromyine, and highly carnivorous) are stable at least under the subset of opossum codings. There is, therefore, little need to discuss many of the taxonomic indications of these experiments: as one would hope if science were the justification of intuition, the numerical analysis of characters previously held (in less formal studies) to give these groupings does indeed give similar terminal structure.

Yet the larger questions of opossum and dasyuroid relationships are not answered so simply and clearly by these trees. The broader groupings of genera, and some special relationships of particularly interesting taxa, are not only equivocal but variable between trees. The contrast between stability at lower levels and equivocation at higher would not, perhaps, be much of a surprise to a pheneticist: the large number of character–states likely to be shared by congeners means that no clustering technique can fail to recognize the implied relationship, while the more tenuous affinities of distant relatives render them extremely sensitive to the compromises necessary in forming groups of higher–category taxa. Even a program which traces the successively changing links between taxa, as the Wagner method does, is limited in this way when dealing with small differences or similarities.

But the aims of pheneticists are humble: they only wish to recognize and interpret affinities in a statistical framework, while cladists seek the truth of phylogeny. Since we, too, are interested in the course of marsupicarnivore evolution, the departures of our trees from expectation, their instability, and the reasons that we have obtained unexpected or divergent results all require some comment. The book–keeping operations of the Wagner program show that the characters, much less the taxa, do not evolve simply, and to a large extent what we have learned about character–evolution transcends the original aims of the study. We begin the discussion, however, with a review of the most remarkable phylogenetic results.

The most important of these concern the relationships of the thylacines and borhyaenids, which Archer argued show a suite of shared, derived characters difficult to view as convergent (Archer 1982). It was thus surprising that, on Archer's own characters, the thylacines and American large carnivores occupied the extremes of the tree (Fig. 8), and, moreover, that *Thylacinus* spp. are not representative of one branch of a basic dichotomy within Dasyuroidea. That is true at least when all of characters 1 – 28 are used; when the subset is partitioned into dental and non–dental features, it becomes apparent that it is the latter which effect the separation, for the thylacines and borhyaenids *do* link on the basis of dental similarities (Fig. 10), as they do when opossum characters (29 – 54) are used to generate a tree. That is very much what Archer said in 1976, in stressing the dental but not basicranial similarities of thylacinids and borhyaenids.

Borhyaena tuberata and *Thylacinus* spp. also derive from the same branch of the tree (an Australian one) when a reduced set of taxa are considered (e.g., Fig. 9), whether all or only dental, but usually not non–dental, characters are analyzed; this result seems to illustrate the point of Simpson (1941) and others that any apparent phyletic association of *Thylacinus* spp. and the borhyaenids is one between the derived, terminal members of their respective lineages: again, when the presumed antecedents of the borhyaenids are included, the American carnivores form a smooth and progressively derived sequence from their common ancestor in a (presumed) didelphoid. (In fairness, however, it must be pointed out that many of the non-dental characters are only inferred for these taxa). Since many of the antecedants of *Thylacinus* spp. are apparently still extant, it is then obvious that the derived characters of advanced borhyaenids and thylacinids were evolved in parallel. The Wagner program (and probably any such algorithm) must inevitably pair *Borhyaena tuberata* with *Thylacinus* spp. when only these endpoints of the two series are included, since no distinction between "true synapomorphies" and convergences was made in the coding; the tree of 47 taxa provides a basis for making that distinction, and illustrates the importance of including as many *taxa* (as well as characters) as possible in any cladistic analysis. In view of the disparagement of the fossil record by some cladists, we think our demonstration is important: no clearer illustration of the vital significance of fossils to the cladistic enterprise could be wanted.

At the same time, the characters used must be appropriate. Since *Thylacinus* spp. are transferred to the borhyaenid branch when opossum codings are used (Fig. 11, and others), the characters which sometimes keep them close to dasyuroids must be peculiar to the dasyurid subset. Moreover, if *Thylacinus* spp. are borhyaenids, then so is *Sarcophilus harrissi*, for this is one relationship among dasyuroid genera that is not altered by any coding regime; often too, these genera are associated with other dasyurids like the planigales and some species of *Antechinus*.

The position of the bandicoots is a problem scarcely less contentious than the thylacine question, here connected with the relationships of such "aberrant" or unusual taxa as species of *Caenolestes*, *Dromiciops* and *Myrmecobius*. On the basis of the dasyurid codings, the peramelids tended to have a separate derivation from American marsupials, as Archer (1976) said they do, usually grouping with *Caenolestes* and *Dromiciops*, and deriving from forms like *Monodelphis brevicaudata* (Fig. 8); this was true whether the characters were dental (Fig. 10) or non–dental. Opossum characters, on the other hand, indicated that the bandicoots are derived from within the Australian marsupials, again in association with *Caenolestes obscurus* (and *Dromiciops australis* as well when only 31 taxa were considered); but while that relationship held for non–dental characters as well as for the whole subset of opossum codings, it was not true for dental characters alone: then, the peramelids were well separated from species of *Caenolestes* and *Dromiciops* and located at the terminus of the tree (*cf.* Figs 11, 12 and 13). When both dasyurid and opossum subsets were used together, the combined dental characters also separated both *Dromiciops australis* and *Caenolestes obscurus* from the peramelids (Fig. 17) – in other words, the indications from opossum dental characters overrode the dasyurid ones. There are strong indications that, where it was examined, *Macrotis lagotis* is distinct from the other bandicoots (Fig. 14). The taxonomic isolation of this genus (Archer and Kirsch 1977) is thus supported by Wagner analysis.

The involvement of *Caenolestes obscurus* with bandicoots should not, however, be taken too seriously, despite Osgood's (1921) arguments in favour of such a relationship, because this genus and *Myrmecobius fasciatus* tended to join the trees at a late cycle of clustering, showing that they were not really very similar to any other taxon in the tree; but this was not usually true for bandicoots, thylacinids or borhyaenids. Since the caenolestid and Numbat showed no special similarities to anything, one would expect their specific linkage points to depend on the number and type of taxa included in any particular run, and this was true: thus, while *Myrmecobius fasciatus* tended to be associated with American marsupials on the basis of opossum characters (either all 26 or dental only, but not non-dental), and with Australian marsupials on any part or all of the dasyurid codings, its position varied depending on whether 31 or 47 taxa were examined (*cf.* Figs 11 and 12, for example); *Caenolestes obscurus* linked more consistently with the bandicoots, wherever they might be. And finally, *Notoryctes typhlops*, also a late joiner, was always near to dasyurids like *Dasykaluta rosamondae* (Figs 4 and 14), although *Lestodelphys halli* was sometimes linked with the mole as well. Of course, all the experiments involving *Notoryctes typhlops* utilized opossum codings only.

The problem posed by aberrant taxa such as these illustrates the dilemma of cladist seeking to fix their phyletic origin: on the one hand the many unique character–states of divergent taxa contribute nothing to establishment of branching patterns, and must be discounted or excluded altogether (we included very few "autapomorphies" in our data-set); on the other, the features common to the several "normal" taxa do not adequately represent the special nature of aberrants, and any derived states shared by them may be too few to fix the positions of the divergent taxa with any confidence.

Still another "big" question is whether these 54 characters adequately indicate the supposed phyletic separation of Australian and American marsupials, the special relationship of thylacines aside. If so, one might expect the two groups to fall separately on the two branches of an initial dichotomy, but this was clearly not the case for any of those trees shown here, although many of the trees (especially those based on dasyurid codings) did maintain a special, derived position for the dasyuroids from one section of a paraphyletic American taxon. Midpoint rooting of the tree in Fig. 8 did effect a separation of American from most Australian forms, but that was the only case; and this technique of fixing a root, besides assuming rather equal rates of morphological change, implied parallel change in the major branches rather than progressive derivation. (More than that, since in all but one case of midpoint rooting, the *Ancestor* was placed near the terminus of the opossum branch, trees rooted in the middle implied progressive *reversion* of characters among some American marsupials). But it is certainly true that there is considerable convergence or parallelism between and within the several families of marsupials: the somewhat lower stability of opossum relationships under different coding regimes, the intercalation of some didelphoids and dasyurids, and the erratic positioning of what should be the nearest outgroup (*Caenolestes obscurus;* but see the discussion of late–joiners above) all suggest that it is the opossums which converge on dasyurids, and not the other way round.

Whether that is a distinction without a difference or not (or is simply an expected result of the presumably longer evolutionary history of didelphoids), convergence is certainly a major feature of this data–set, which raises the question of why any of these experiments "worked" at all. Again, the question may be rhetorical for the lower categorical levels, where similarities often involve so many characters; but it is probable that in shifting attention from one genus to another a taxonomist makes quite subtle decisions about what is and is not "the same", and these distinctions certainly cannot be encoded very simply, nor made *a priori.* Of course, inconsistency means precisely that supposedly homologous characters, or at least character–states, are only analogous; those characters (or states) could be recorded separately in the convergent taxa to eliminate the inconsistency. But, besides being somewhat tautologous, this cannot be done without guidance from the tree–structure (consider again the thylacinid–borhyaenid case), and it is only possible to perform recoding if *some* fully consistent features are present to give shape to the trees to indicate which characters are most subject to convergence. Since few of our characters were particularly "good" in the sense of being consistent, this is one of the reasons we did not indulge in recoding or character–weighting; but the main reason is that the very shapes of our trees and the bases for nearly all cladistic groupings come from the homoplasy, not the consistency. That leads to one of the more interesting observations to be drawn from these experiments: the groups are founded not on uniquely derived states, but on singular combinations of changed characters which may individually occur in other clades. To demonstrate that this is so, and to explore some of the theoretical implications, we will consider the tree shown in Fig. 1 in some detail.

That tree depicts the relationships of 21 Australian taxa (plus *Ancestor*) and is based on just the 28 dasyurid–oriented characters. The overall consistency of the tree is .37, which compares favourably with most others calculated from as many characters on as many taxa. We chose this tree for discussion because we believe there will be general agreement about the "truth" of most of its features. In addition, over half of the individual characters have consistencies of .50 or above, which is mainly an effect of the few taxa included; in contrast, the tree of Fig. 8, which shows 47 taxa and was calculated from the same 28 characters, has an overall consistency of only .19 and all characters are below .50.

To begin with, there are only four cladistically "perfect" characters determining the tree of Fig. 1, the two–state characters 1, 9, 11 and 15; but these are all "loss" or reduction characters, a type fashionably considered unreliable — in fact, the consistencies of all four drop to .33 or below in calculating Fig. 8, where character 1 becomes three–state to accommodate the further reduction of incisor number in species of *Arminiheringia* and *Borhyaena.* However, in Fig. 1 that character cleanly separates the dasyuroids from the two bandicoots, and the latter are moreover united by six synapomorphies representing as many evolutionary steps. These are all advanced states of characters 2, 8, 16, 19, 21 and 28 (that of 21 being two steps removed from the primitive condition). Yet not a single one of the six is a *uniquely* derived condition: the "best" characters are 2 and 16, which have changed only twice over the tree; character 2 is shared in its derived condition with *Phascolosorex doriae,* as is 28, and the identical condition of character 28 as well as of 16 and 21 are held in common by bandicoots and the group comprised of *Wakamatha tasselli* and *Sminthopsis* spp., these three shared states being the highest number of matches between bandicoots and any other taxon in this tree. Thus, while this *combination* of states may be unique to the peramelids, no state of any one feature is peculiar to them; in a large enough tree it seems possible that even the combination might occur more than once, if it were arrived at by alternative evolutionary pathways.

Yet it is not even true that cladistic groups on a Wagner tree are necessarily defined by such a combination, which in a sense could be considered a "higher–order" synapomorphy: if we consider the clade comprised of species of *Wakamatha, Sminthopsis* and *Murexia,* whose common ancestor possessed characters 17 and 19 in their singly derived conditions, we find that the synapomorphies of that group are not present in all members of it. Thus character 17 changes to the third state in species of *Sminthopsis* and reverts to the primitive condition in *Wakamatha tasselli.*

In the case of larger groupings it may even be that not even a single defining state is present in all taxa. Consider the assemblage that includes the thylacines and *Sarcophilus harrisii,* united with each other by the "perfect" character 15 (and some not-so-perfect ones) and with *Planigale* spp. by the two remaining true synapomorphies, and extending back to a common ancestor which was a taxon apparently corresponding to *Parantechinus bilarni* in all of its character-states. The supposed synapomorphies for this group are characters 8, 17, and 19, the first and last in derived condition and 17 in the doubly derived state (2.0). Examination of the states of the nine terminal taxa, however, shows that not all of them possess the defining characters in the conditions of their common ancestor, as was true for the *Wakamatha–Sminthopsis–Murexia* clade; but it is further the case that none of the three characters passes through the branching sequence without at least one further alteration, meaning that the group has no defining feature present in all members, a formal contradiction of the synapomorphy concept. To be specific, *Planigale* spp. have 17 and 19 but not 8 in the required state; *Dasyurus maculatus* has only 8 in the derived condition; *Satanellus hallucatus* has 8 and 19 but not 17 in the "proper" state; and *Dasyuroides byrnei* similarly lacks the needed state of character 17. The group is thus, in the terminology of a rival taxonomic methodology, polythetic (Sneath and Sokal 1973): that is, while a set of derived states may be unique to the group, possession of no one of these is required to qualify for membership in it. Unlike in the numericists' notion of a polythetic group, however, members need not even possess a certain number of the traits. The reason the Wagner tree character–distributions can depart so much from true polythetics is of course because the network traces a chain of derivation: additional synapomorphies of more restrictive groupings are necessarily acquired within the tree, insuring that taxa may keep membership in more inclusive groups even without retaining previously gained states; and so *D. maculatus,* which holds only one of the three character–states in common with *Parantechinus bilarni* (and that a state also present in *Ankotaringa tirarensis* and most of its descendants, as well as in the bandicoots) maintains its association with the other derivatives of *Parantechinus* spp. because of the "new" synapomorphies it shares with species of *Planigale, Sarcophilus* and the thylacines (though these number only four, and one character undergoes a further change in the Devil).

All of this suggests that if it is phylogenetic truth that one is seeking, then the parsimony approach may not be the best one, since the low consistency of so many characters and the somewhat polythetic nature of the groups formed shows that homoplasy is too integral to the trees' structure for its minimization to have much real biological meaning. Some final experiments should be mentioned here to give precision and weight to that declaration. Smith and Kirsch 1982) subjected trees similar to those listed in Table 3 to their information metric, a method that shows how well a particular hierarchical structure predicts or recovers the information contained in a data–set. As noted in Results, the fit of any group of characters from Table 1 to Archer's (1976) dendrogram was poorer than their fit on trees generated by the Wagner program itself. Yet the information content of Archer's tree was *better* than was the corresponding Wagner tree (in one case, by 27% though Archer's was 31% longer). To reiterate, this is true even though Archer's diagram is a much less *parsimonious* representation of the evolutionary pathways implied by the characters.

Moreover, it is clear that even slightly different ways of looking at the characteristics of the same organs — mostly the teeth in the present study — may give quite different dendrograms, implying a cladistic equivalent of the numericists' failed "non–specificity hypothesis." And it is also apparent that none of the trees are particularly good minimizations of homoplasy, whatever the value of so doing; taken together with the experiments just noted above, which call into question the usefulness of parsimony methods anyway, the worth of the whole effort of recent workers to escape convergence in taxonomic analyses is thereby seriously questioned.

Yet for all their deviation from the neat diagrams of classical cladistics, our trees do have purely taxonomic implications that are not so very much at odds with expectation. Again, these may be mostly at the lower categorical levels, but such relationships are precisely the ones of interest to the majority of biologists, and it is gratifying to note the concordance of so many disparate approaches to problems at *any* level. If the broader questions of marsupicarnivore history are still unanswered, we can at least enter the next quarter-century of marsupial systematics with some confidence that the group which seemed so impenetrable to our understanding in the sixties, Dasyuridae, is at last beginning to yield some of its genealogical secrets.

ACKNOWLEDGEMENTS

We thank O. A. Reig and Larry Marshall for many discussions of opossum characters, and Steve Farris for provision of the Wagner 78 program. Kirsch's part in this study was supported by a grant from the Milton Fund of Harvard University, and Archer's by funds from the Australian Research Grants Committee and other sources. Alfonso Coleman performed incredible feats of photographic reduction in retaining the clarity of our trees, and Peter Bottjer heroically assembled the table of data and typed the final drafts.

REFERENCES

ARCHER, M., 1976. The dasyurid dentition and its relationships to that of didelphids, thylacinids, borhyaenids (Marsupicarnivora) and peramelids (Peramelina: Marsupialia). *Aust. J. Zool. Suppl. Ser.* **39:** 1–34.

ARCHER, M., 1982. Review of the dasyurid (Marsupialia) fossil record, integration of data bearing on phylogenetic interpretation, and suprageneric classification. Pp.397–443 *in* "Carnivorous marsupials" ed. by M. Archer. Roy. Zool. Soc. New South Wales: Sydney.

ARCHER, M. AND KIRSCH, J. A. W., 1977. The case for the Thylacomyidae and Myrmecobiidae, Gill, 1872, or why are marsupial families so extended? *Proc. Linn. Soc. New South Wales* **102:** 18–25.

AYALA, F. J. (ed.), 1976. "Molecular evolution". Sinauer Associates, Inc: Sunderland.

CLEMENS, W. A., 1979. Marsupialia. Pp.192–220 *in* "Mesozoic mammals" ed. by J. A. Lillegraven, S. Kielan–Jaworowska and W. A. Clemens. University of California Press: Berkeley.

FARRIS, J. S., 1969. A successive approximations approach to character weighting. *Syst. Zool.* **18:** 374–85.

HAYMAN, D. L. AND MARTIN, P. S., 1974. Mammalia I.: Monotremata and Marsupialia. *Animal Cytogenetics,* Vol. 4: Chordata 4. Gebrüder–Borntraeger: Berlin–Stuttgart.

HENNIG, W., 1966. "Phylogenetic systematics". University of Illinois Press: Urbana.

KIRSCH, J. A. W., 1968. Prodomus of the comparative serology of Marsupialia. *Nature (London)* **217:** 418–20.

KIRSCH, J. A. W., 1969. Serological data and phylogenetic inference: the problem of rates of change. *Syst. Zool.* **18:** 296–311.

KIRSCH, J. A. W., 1977. The comparative serology of Marsupialia, and a classification of marsupials. *Aust. J. Zool. Suppl. Ser.* **52:** 1–152.

KLUGE, A. G. AND FARRIS, J. S., 1969. Quantitative phyletics and the evolution of anurans. *Syst. Zool.* **18.** 1–32.

OSGOOD, W. H., 1921. A monographic study of the American marsupial, *Caenolestes. Publ. Field Museum Nat. Hist. Zool. Series* **14:** 1–156.

REIG, O. A., KIRSCH J. A. W. AND MARSHALL, L. G. Relationships of living and Neocenozoic didelphoid marsupials. (In preparation).

REIG, O. A. AND SIMPSON, G. G., 1972. *Sparassocynus* (Marsupialia, Didelphidae), a peculiar mammal from the late Cenozoic of Argentina. *J. Zool. Lond.* **167:** 511–39.

SARICH, V., LOWENSTEIN, J. M. AND RICHARDSON, B. J., 1982. Phylogenetic relationships of the Thylacine (*Thylacinus cynocephalus,* Marsupialia) as reflected in comparative serology. Pp.707–9 *in* "Carnivorous marsupials", ed. by M. Archer. Roy. Zool. Soc. New South Wales: Sydney.

SIMPSON, G. G., 1941. The affinities of the Borhyaenidae. *Novitates No.* **1118:** 1–6.

SMITH, T. F. AND KIRSCH, J. A. W., 1982. On assessment of the information content of classifications. *Syst. Zool.,* under review.

SNEATH, P. H. A. AND SOKAL R. R., 1973. "Numerical taxonomy". W. H. Freeman and Company: San Francisco.

WILSON, A. C., CARLSON, S. S. AND WHITE, T. J., 1977. Biochemical evolution. *Ann. Rev. Biochem.* **46:** 573–639.

RESUMEN

50. El cladismo politético, o cuando el método de la parsimonia no es suficiente: las relaciones de los marsupiales carnívoros. Por John A. W. Kirsch y Michael Archer.

El análisis filético de Wagner fué aplicado a 71 especies de marsupiales carnívoros, usando dos series de datos qu no se superponen; una de ellas (26 caracteres) probó ser útil para agrupar las zarigüeyas, mientras que la otra (28 caracteres) fué usada para los dasiúridos; de los 54 caracteres, 37 fueron dentales. Cuando ambos grupos geográficos fueron analizados separadamente, ambas series de datos dieron similares agrupamientos de los dasiúridos, especialmente de los niveles intragenéricos (por ejemplo, fue probado el polifiletismo de *Antechinus* y *Dasyurus,* sostenido con otros fundamentos); pero las relaciones de las zarigüeyas fueron bastante diferentes según los dos regímenes de codificación. Cuando se examinaron conjuntamente ambos grupos geográficos, las relaciones de las zarigüeyas fueron muy alteradas, con *Lestodelphys* y otros intercalados con los Dasyuridae; además, los últimos aparecieron como una porción aislada de una rama de Didelphoidea técnicamente parafilética. Los caracteres pro–dasiúridos separaron los tilacinos de los boriénidos tanto como se incluyeron los antecedentes fósiles de los últimos entre los taxa examinados; pero los caracteres de las zarigüeyas o ambas series de datos en combinación ubicaron los tilacínidos con los boriénidos, como lo hacen los rasgos dentarios (aunque no los no–dentarios) de cada una de las series de caracteres o de ambas. Según los caracteres dasiúridos los "bandicoots" están mas próximos a las zarigüeyas, pero aparecen como derivados de los dasiúridos de Nueva Guinea cuando se usan los códigos de las zarigüeyas. Ninguno de los caracteres fueron cladísticamente perfectos, esto es, hubo considerable convergencia en el estado tanto dentro como entre las unidades geográficas de los marsupiales; la mayoría de los grupos estuvieron basados en singulares combinaciones de caracteres cambiados que individualmente existían en otros clados. En algunos casos, no se mantpuvo el mismo estado del carácter en todos los miembros de un grupo cladístico, a proximándose así al criterio "politético" de los taxonomistas numéricos.

51. A NEW APPRAISAL OF MARSUPIAL PHYLOGENY AND CLASSIFICATION

FREDERICK S. SZALAY[1]

Detailed character analysis of the tarsal morphology of nearly all living marsupial genera and representatives of the fossil record, along with a survey of dental and cranial morphology, resulted in a new phylogenetic hypothesis and a new evolutionary classification of the Metatheria. The monophyly of the Andean Patagonian *Dromiciops australis* and all Australasian marsupials is supported and it is expressed by the cohort Australidelphia. The cohort Ameridelphia, including the order Didelphida is divided into the suborders Didelphiformes (Pediomyidae and Didelphidae) and Borhyaeniformes (Borhyaenoidea, Caenolestoidea, Argyrolagoidea). The australidelphian radiation is divided into: (1) order Dromiciopsia, forms with polyprotodont dentition and diadactylous grasping pes with the complexly derived australidelphian tarsal morphology; (2) order Dasyurida, derivable from dromiciopsians; and (3) order Syndactyla with their homologously derived syndactylous pes, also originating from a dromiciopsian ancestry. Several problematical groups can be allocated based on the tarsal evidence. Thylacinids show no special, homologously derived, shared similarities with any of the Borhyaeniformes. Their most recent ties are with other australidelphians, particularly within the Dasyurida. Bensley's allocation of the Notoryctidae to the Syndactyla is confirmed. The distal tarsus, the presence of similar sized metatarsals II and III, a large hallux, and the polyprotodont anterior dentition suggest derivation from the ancestral syndactylan suborder, the Syndactyliformes.

Pages 621–40 *in* CARNIVOROUS MARSUPIALS, ed. by M. Archer, February, 1982. Roy. Zool. Soc. New South Wales: Sydney, Australia.

AN ONGOING appraisal of cranial, dental, pedal, and particularly tarsal character clines in fossil and living groups of marsupials made it apparent to me that the stimulating and important phylogenetic hypotheses and classifications based on them presented by past students cannot accommodate the known evidence bearing on metatherian phylogeny. It is with some reluctance but with firm convictions resting on the significance of newly analyzed character clines that I offer for scrutiny and testing a new synthesis of marsupial phylogeny, and an evolutionary (not cladistic) classification to express the phylogenetic and adaptational history of these animals.

The outlines of metatherian relationships which are still debated today have been originally conceived around the concepts of the polyprotodont–diprotodont dichotomy (see especially Archer 1975, 1976, Clemens 1979, Kirsch 1977, Marshall 1977, Ride 1964, Simpson 1970, Tedford 1974, and numerous others). Primarily based on the benchmark studies of Bensley (1903), the phylogenetic hypotheses subsequently proposed and the classifications adopted considered these two groups as basic.

In addition to the characters derived from the retention or reduction of the incisors, the dentally based divisions of the metatherians had to take into consideration the modifications of syndactyly in some Australasian groups in contrast to the diadactylous primitive condition in the pes. Until recently, when various biochemical techniques (specifically serology; see Kirsch 1977) have been brought to bear on hypotheses of phylogeny, the characters utilized to erect phylogenetic hypotheses and classifications were primarily the dentition and the skull, with the occasional restraints observed due to the syndactylous condition of the pes in a number of groups. In order to clarify some ambiguities of the serological data, however, doubts were eventually cast on the monophyletic nature of the complexly derived syndactylous foot by Kirsch (1977).

My work, the preliminary results of which are presented here, is only a beginning to investigate new character complexes in addition to those traditionally pursued. Yet, *in conjunction* with the known aspects of cranial hard anatomy and the superficial morphology of the pes, a survey of the tarsal complex of marsupials resulted in some startling insights into metatherian phyletics. A detailed and extensive survey of metatherian tarsal morphology was started, over two years ago, and I hoped to shed some light on what I considered contradictory hypotheses in the published literature based on the assumed morphocline polarities of dental, cranial and some sero logical features. My initial attempts were to understand the basic differences and their mechanical significance in the various groups before I formulated polarity hypotheses of the reasonably complex tarsal morphology. Throughout this work my efforts centered on the weighted, but combined utilization of hard morphology to generate, as nearly as possible, a realistic

[1] Hunter College, City University of New York, New York City, New York, U.S.A., 10021; and the American Museum of Natural History, New York, New York, U.S.A., 10024.

taxon phylogeny hypothesis rather than another character phylogeny, based on the tarsal morphology.

New phylogenetic hypotheses should be based on newly conceived syntheses of all known morphocline polarities, or on a newly arranged relative recency of origin of known, well researched morphocline polarities, or a combination of both. Invariably investigation of a new character complex necessitates the reconsideration of the time of relative origin of all characters, even the ones well known previously. All of this phylogenetic analysis, whether implicitly or explicitly, deals with notions of transformation of homologous character–states. I realise that this synthesis is an interim answer. I am convinced, however, that the addition of several well surveyed, intraspecifically and intragenerically stable tarsal characters of considerable complexity means progress towards marsupial taxon phylogeny when considered along with other morphoclines.

Much of the evidence relied on in this paper is thoroughly discussed and amply documented in a monograph in preparation. Detailed reasons are given there for the emphasis in postcranial comparisons on the astragalus and calcaneus. Briefly, these bones, in addition to reflecting a very complex species specific mechanical relationship between the crus and the more distal tarsus and cheiridia, are also the most commonly represented elements of the skeleton in fossil mammals. The latter reason alone would make their use especially justified. Most importantly, however, the mechanically complex tarsus, invariably moulded around the requirements of substrate preference and locomotor mechanics, is also powerfully constrained by the genetic mechanisms inherited in any given group (see pp.315–320 in Szalay 1977b). Simply summarized, in addition to the species specific solutions evident on the bones of the proximal tarsus, these "habitus" characters are apparent only as an aspect of the inherited, ancestral morphological matrix, the "heritage" of this character complex.

METHODS

I would like to state briefly that throughout this study, and some others (Szalay 1976, 1977b, etc.), my approach to phylogenetic analysis is based on the search for homologously shared and derived characters and not on estimated morphological distance. This practice, clearly enunciated by Hennig (1950), practiced, however, as early as 1910 by Gregory, should not be confused with what has recently become known as the cladistic methodology. I prefer to refer to the methodology employed here, clearly outlined by Bock (1977a), as a phylogenetic one, or phylogenetics. The reasons for this superficially minor, but in reality profound, distinction are deeply rooted in what different systematists conceive the role of time, ancestral stages, and evolutionary biology to be in systematics.

As a result of Hennig's justifiably major impact, the school of cladistic analysis, *sensu stricto,* has quickly evolved, centered primarily at the American Museum of Natural History. Students of this discipline, primarily in the pages of *Systematic Zoology,* have increasingly come to define and narrow a methodological approach to focus on the search for sister–group, or cladistic relationships. The method of cladistic analysis is based on axioms of logic and various dictums which became progressively removed from evolutionary theory, from functional and adaptional analysis of the characters, from concepts of character transformation through phyletic evolution, and ancestor–descendent relationships. Practitioners of the neocladistic method reject the aid of the fossil record and stratigraphic zonation of organisms to phylogenetic analysis. Attempts to deny the temporal and spatial constraints on phylogeny, the concept of ancestry, and consequently of transformation from one condition to another, along with several misapplications of Popper's philosophy, have led to unconvincing views of evolution itself.

Cladograms order all evolutionary relationships, artificially, as sister–group (i.e. cladistic) ties. Phylogenetic hypotheses on the other hand, and their graphic expressions, the phylogenetic trees, include all information expressable in cladograms in addition to temporal information and transformational hypotheses. To use philosophical jargon, the information content of a phylogenetic hypothesis is greater than that of its cladogram expression, consequently opportunities for its "falsification" and "verification" are more ample. It is, simply, a more precise and more heuristic type of hypothesis. The arguments offered that ancestor–descendent relationships cannot be "falsified" are specious. An ancestor–descendent hypothesis depends on a greater number of constraints than a cladistic one, and therefore, in addition to its greater information content, it can be more easily contradicted (see Szalay 1977a; Naylor, pers. comm.).

But the most severe criticism that I can offer of the post–Hennigean neocladistic school lies in its effect on character analysis, the very heart of phylogenetics. There is a general disregard among cladistic practitioners for the analysis of the function and biological role (*sensu* Bock and von Whalert 1965) of features. The various rules (e.g., outgroup comparison) are supposed to lead to the rigidly preferred polarity hypothesis once the data for a morphocline is gathered. One of the numerous reasons, however, why systematists should attempt to understand developmentally, functionally, and adaptationally the characters studied relates to the significance of "non–functional" attributes, or epiphenomena of mechanically significant modifications, and their usefulness as "phylogenetic markers". Before one can appreciate and make use of "neutral" characters in phylogenetic analysis, one must understand the entire character complex sensitive to change and the manner in which it did change in all the available species in which it is displayed. Those aspects of the underlying morphological matrix which canalize change, emanating from the heritage morphology, are of exceptional value to the phylogeneticist. Epiphenomena of some important mechanical alteration, or the reworking of the mechanics of a whole character complex, often mark a community of ancestry, or channel further changes recognizably even when the form–function complex of the ancestor is drastically modified or obliterated. One of the many examples illustrating this phenomenon is the elaborate "re-evolution" of spatulate, incisiform "incisors" of the advanced indrioids of the Malagasy lemuriform radiations from their tooth-combed ancestry (see Szalay and Delson 1979), and the subsequently channelled occlusal relationships of the caniniform teeth.

Needless to say, in depth functional analysis of characters results in heightened perception of a character complex, recognition of characters where there were none noticed before, and a subsequent increase in the available data base for the systematist (Szalay 1981). In this study, one of the most important newly recognized morphocline polarity hypotheses, that relating to the modifications of the australidelphian lower ankle joint in dasyuridans, was conceived by using a simple functional transformational model.

It appears then that adherence to the rigid rules of a cladistic approach in character analysis often results in glib and simplistic solutions of morphocline polarities, results which become clearly inappropriate after more detailed functional and adaptational analyses of a character complex are conducted in a group of species (see especially Bock's 1981 comprehensive account of functional adaptive analysis in phylogenetics).

Terminology for tarsal morphology. There is no brief, readily usable terminology comparable to that employed for the details of molar teeth. I have therefore attempted to develop a functional–descriptive system which has its base firmly in the conventional anatomical designations for the bones and articulating facets. This system with its patterns of abbreviations hopes to accomplish the following: 1) maintain its connection with descriptive anatomical terminology employed (almost never uniformly, however) in anatomy texts; 2) treat the morphology of the bones in a functionally significant framework; 3) use clearly recognizable, decipherable abbreviations which can be broken down into alphabet–like units, resulting in savings of space in the text and ease of reading; 4) give an immediate clue to the student whether the unit referred to is a) a whole synovial joint between two bones, b) a particular facet of a joint on a particular bone, or c) a bony process, tuberosity, etc. Table 1, in addition to a glossary of the abbreviations, contains further notes and explanations on the terminology and its use. Fig. 1 illustrates the terminology as it pertains to the astragalus and calcaneus.

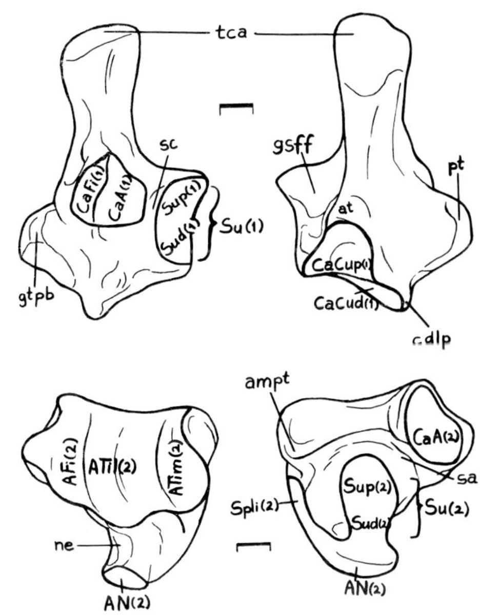

Fig. 1. The nomenclature used in the text for the morphology of the calcaneus and astragalus. For explanations of the system of abbreviations see Table 1. Above — dorsal and plantar views of the calcaneus, whereas below — the dorsal and plantar views of the astragalus are depicted. The tarsals are of *Didelphis patagonicus*.

DISCUSSION

Cohort Ameridelphia, new

One of the several major problems which has emerged from this investigation centers on the diagnosis, definition, and identifiability of various groups of the order Didelphida, usually referred to in the past as didelphoids. Even the most careful and exacting studies and diagnoses (Clemens 1966, Simpson 1974, Crochet 1978) have come to reflect the fact that dental criteria alone, at least those proposed so far, are usually constrained to use primitive metatherian characters, thus often lacking the power of resolution to allocate a particular taxon to its monophyletic group. Thus, for example, the traditional dental concepts of Didelphidae and primitive Borhyaenidae have never been satisfactorily resolved on dental grounds alone (see discussions in Simpson 1970).

Table 1. The following abbreviations are used in the text and in Fig. 1. The tarsal bones are designated with numbers in parentheses; calcaneus (1), astragalus (2), cuboid (3), navicular (4), ectocuneiform (5), mesocuneiform (6), and entocuneiform (7). The bones of the crus are designated as: fibula (8), tibia (9). In the representation of a particular facet the abbreviation cites the two bones in contact, whether the portion of the facet is lateral, medial, or other, then, in parentheses, the number indicating the bone on which the facet is found. For example, ATil (2) stands for the lateral articular facet for the tibia on the astragalus, or, the lateral tibioastragular facet. The numerical designation of the bones is preferred to changing the sequence of abbreviations for facets of the same joint.

Specific joints are abbreviated by the combination of the first letters in capitals of the names of those units which name the joint, and the capital letter J for joint. For example CCJ and UAJ denote the calcaneocuboid and upper ankle joints, respectively.

Certain conditions or diagnostic modifications, often repeated in the text, are also abbreviated. For example, CLAJP refers to the continuous lower ankle joint pattern found in australidelphians. Abbreviations entirely in lower case designate bony processes, tuberosities, grooves, etc.

A, astragalus; ACJ, astragalocalcaneal joint (= lower ankle joint); ampt, astragalar medial plantar tuberosity; AFi, astragalofibular facet; AN, astragalonavicular facet; ANJ, astragalonavicular joint; *at,* anterior plantartubercle; ATil, lateral astragalotibial facet; ATim, medial astragalotibial facet; *c,* calcaneum; CaA, calcaneoastragalar facet; CaCu, calcaneocuboid facet; CaCup, proximal calcaneocuboid facet; CaCud, distal calcaneocuboid facet; CaFi, calcaneofibular facet; CAJ, cuboidoastragalar joint; CCJ, calcaneocuboid joint; *cdlp,* calcaneal distolateral process; CLAJP, continuous lower ankle joint pattern; Cu, cuboid; CuA, cuboidoastragalar facet; d, distal; Ec, ectocuneiform; EMIJ, entocuneiform–metatarsal I joint; En, entocuneiform; f, facet; Fi, fibula; *gsff,* groove on sustenaculum for tendon of M. flexor (digitorum) fibularis; *gtpb,* groove for tendon of M. peroneus brevis; l, lateral; LAJ, lower ankle joint (= CAJ); m, medial; Mc, mesocuneiform; N, navicular; NCJ, naviculocuboid joint; ne, neck of astragalus; p, proximal; *pt,* peroneal tubercle (or process); *sa,* sulcus astragali; *sc,* sulcus calcanei; SLAJP, separate lower ankle joint pattern; Spli, spring (calcaneonavicular) ligament facet; Su, sustenacular facet.

The use of tarsal characters clearly does not offer a panacea for the allocation of the growing number of dental taxa of the Cretaceous and early Paleogene didelphoids and borhyaeniforms. The recognition and delineation of diagnostically derived characters of the living radiation of the Didelphidae, however, and a similar diagnosis of the tarsal morphology of several genera of fossil borhyaeniforms facilitates the evaluation of dental taxa in the fossil record. The presence of certain types of distal tarsal elements in various early Tertiary quarries sheds light on the possible affinities of the taxa identified by dentitions from the same localities. Based on primitive dental features many metatherian dentitions have been allocated to the Didelphidae (e.g. *Alphadon*). Unless these allocations are based on demonstrably homologously shared and derived characters, the practice has limited phylogenetic significance.

In spite of some light at the end of the taxonomic tunnel in regards to a crisper diagnosis of the primitive marsupial groups, there is as yet no satisfactory solution concerning the use of family or subfamily rankings of the various didelphoids; this being a reflection of our ignorance of their phylogeny. My reasons for using a broadly defined, but well subdivided concept of Didelphoidea rests on reasons rooted in evolutionary classification. Although their internal phylogeny is poorly understood, the didelphoids are cranially, dentally, and postcranially the most representative of the archaic marsupials, as this has long been recognized. Of this assemblage only the Didelphidae can be clearly delineated cranially and postcranially as a truly monophyletic group. The Pediomyidae is expanded to be a temporarily useful horizontal category whose various subfamilies, particularly the pediomyines, were probably more primitive than either the known Didelphidae or Borhyaeniformes. In terms of phenetic divergence based on teeth, however, the separation of stegodontines, glasbiines, or peradectines from each other on the family level is not warranted. It appears, however, that studies like those of Crochet (1978, 1979a, 1979b) are beginning to provide some of the answers to problems of early didelphoid evolution.

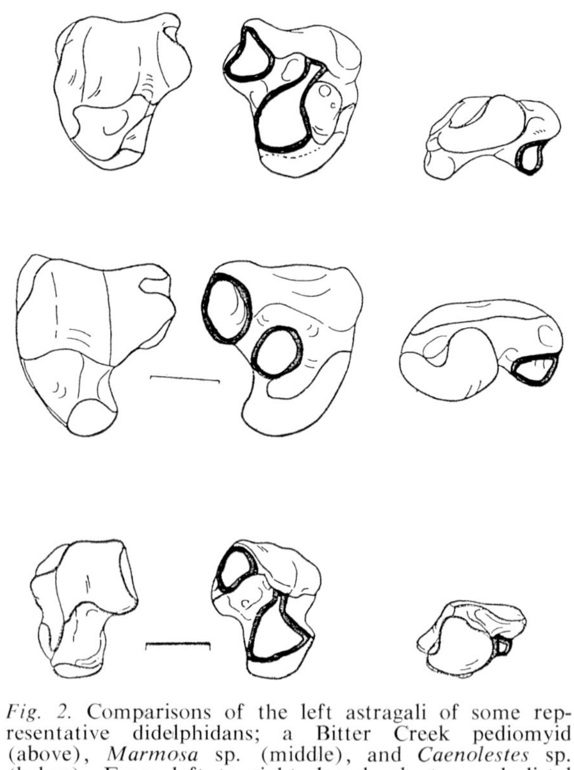

Fig. 2. Comparisons of the left astragali of some representative didelphidans; a Bitter Creek pediomyid (above), *Marmosa* sp. (middle), and *Caenolestes* sp. (below). From left to right dorsal, plantar and distal views, respectively. Note the primitive SLAJP on all three (heavy black outline). Scales represent one mm.

The tarsal evidence attests that no member of the Didelphidae which share the diagnostic features of the living taxa and their most recent common ancestor gave rise to the Borhyaeniformes. Furthermore, no Mesozoic or early Paleogene genus can be convincingly allocated to the

Didelphidae, *sensu stricto*. The use of the horizontal and paraphyletic concept Didelphoidea, however, should allow the continued future reassessment of the ancestor–descendant and branching relationships of the various groups within the superfamily a) without depriving the classification of the adaptational information or b) by adding unnecessary nomenclatorial confusion as the minor differences in sister–group relationships among the didelphoids may become apparent. Such a paraphyletic taxon adds to classificatory stability without stifling research or sacrificing adaptational information for the non specialist[1].

The branching relationships of the genera within the Pediomyidae are not understood as this is clearly reflected in the most recent, careful studies (Clemens 1966, 1977). It would, however, behoove students of mammal faunas, who assiduously recover and describe teeth from wash concentrates to do the same equally carefully for postcranials, and attempt to allocate and identify these as well. (No part of an organism can be said, *a priori,* to possess those characters which are the most recently acquired and homologously shared with another taxon, and are therefore indicative of their phylogenetics most accurately. Neither dentitions, skulls, molecules, or tarsal remains have the exclusive privilege to mark the most recent shifts of phylogeny. Skeletal morphology has the advantage, whenever fossils are available, however, to potentially maintain connection between extinct and extant taxa.)

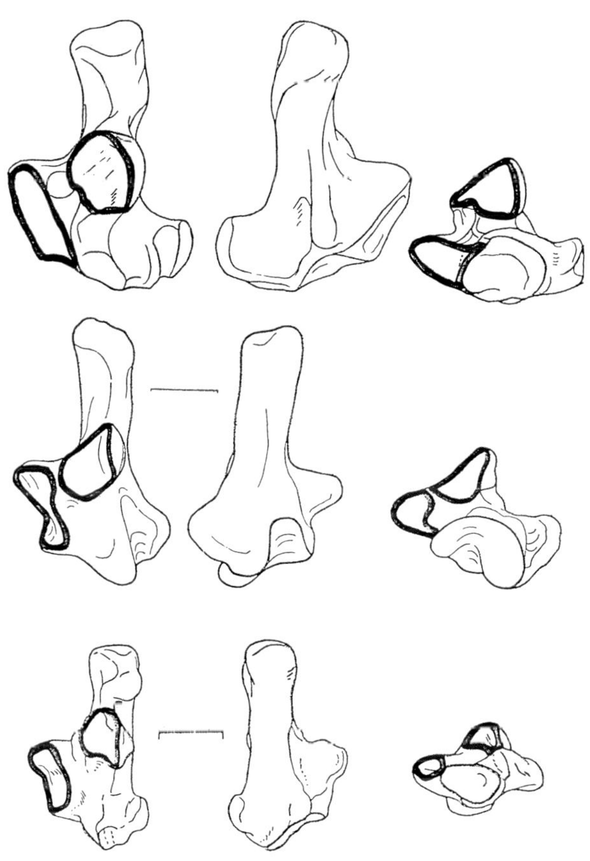

Fig. 3. Comparisons of the left calcanea of some representative didelphidans; a Bitter Creek pediomyid (above), *Marmosa* sp. (middle), and *Caenolestes* sp. (below). From left to right dorsal, plantar and distal views, respectively. Note the primitive SLAJP on all three (heavy black outline). Scales represent one mm.

Based on comparisons of didelphoids, *Dromiciops* and *Microbiotherium,* the most recent relationships of the stem australidelphians may have been either with unknown stocks of primitive didelphids or with peradictine or pediomyine pediomyids, a most unsatisfactorily ambiguous hypothesis. The dentition of *Dromiciops* does not share the distinctly enlarged metacone of many didelphids, or their characteristic straight and long paraconid–protoconid crest. The relatively poorly known pediomyid tarsus precludes a definite choice between a pediomyid or didelphid source for the Australidelphia. Lack of any suggestion in australidelphians for the ancestral presence of the didelphid calcaneocuboid pivot–like depression (CaCup) on the calcaneus or the projection on the cuboid tends to indicate a source more primitive than the Didelphidae.

PEDIOMYIDAE: A sample of proximal tarsals from the Lance late Cretaceous, and the Paleogene Itabori, Tiffany, Bitter Creek, Four Mile, Green River, Bridger, and Tepee Trail faunas has profound bearing on the concepts of various family definitions within the Didelphida, on what is primitive or derived in metatherian tarsal morphology, and subsequently, on hypotheses of marsupial phylogeny. The fossil record, together with morphology, is utilized to propose morphocline polarity hypotheses (*fide* Simpson 1975, Szalay 1976, Bock 1977a).

The tarsal morphology from the faunas listed above is not didelphid–like, lacking the derived modifications of the most recent common ancestor of the Didelphidae. The underlying ancestral morphological matrix of the proximal tarsals of metatherians from the faunas noted suggests a primitive pattern which is best considered antecedent to a) the didelphids, b) borhyaenoids, and c) caenolestoids (there is no known tarsal morphology for the polydolopids). At this stage of our understanding, then the Pediomyidae (Simpson 1927a) appears to have the morphological characteristics which were antecedents to all other didelphidans (but see Clemens 1979).

The general conformation of the proximal tarsus, as best examplified by specimens from the Lance and Bitter Creek faunas, of some pediomyids (in a broad sense) is the best estimate at

[1] Crochet's (1978, 1979a) extensive studies of European marsupials have shed new light on didelphoid differentiation. It is clear from a comparison of the Cretaceous forms, including such forms as *Alphadon,* and the diversity of Paleogene genera that none of these, based on dental characters, can be satisfactorily allocated to the Didelphidae, *sensu stricto.* It appears that both dentally and tarsally the didelphids are more derived than pediomyids and that their ancestry stems from some hitherto unknown pediomyids. This is not the concensus, however, which is based on dental morphology, as reflected in such recent studies as those of Clemens (1979) and Fox (1971, 1979a, 1979b).

present for the general ancestral metatherian pattern. Although the age spread is great for these specimens, the morphological pattern underlying the differences in the taxa assures a reliable assessment for their common ancestor. Their morphology, along with other metatherian tarsal remains, is being described and documented in greater detail elsewhere (see Figs 4–9).

The proximal tarsus in known taxa judged to be pediomyid is quite similar, as is that of didelphids, in overall morphology to such primitive eutherians as the late Cretaceous leptictimorph *Procerberus* and the even more primitive tarsus of the condylarth from the same deposits, *Protungulatum* (see Szalay 1977b). As in these, the metatherian pediomyid tuber calcaneum is relatively long, and the peroneal process is primitively well developed. The LAJ makes contact on two well developed and separated facets, the CaA and Su, and the maintenance of the distinct groove between these is in fact one of the diagnostic primitive therian features of this joint. Unlike in primitive eutherians the distal portion of the sustentacular facet (Sud) is always well developed, giving the didelphidan astralagus and calcaneus relatively larger surfaces to slide and rotate. This aspect, along with the morphology and mechanics of the CCJ of the metatherian morphotype, is likely to be related to the habitually inverted mobile foot of a primarily arboreal animal. In addition to the primitively (in therians) long tuber, the early Eocene specimens of pediomyids and Itaborai Paleocene borhyaenids have remarkably broad transverse dimensions from peroneal process to the medial margin of the sustentaculum.

As in multituberculates, the CaA (1) is relatively very large compared to Su (1), more so than in didelphids. CaA (1) is gently rounded, the axis for this curved plane running primarily transversely. CaFi (1) for the fibular contact is narrow, and like in multituberculates, the CaCu (1) faces quite medially, more so than in didelphids. CaCu(1) is similar in outline to 150° of a circle and it is nearly uniformly depressed towards the center of the facet. This depressed ovoid receives the cuboid and facilitates a great deal of free rotation. The deep and striking plantar depression on CaCu(1) of didelphids (even found in the highly modified tarsus of *Chironectes*) for the projecting proximal male body of the cuboid is entirely lacking not only in pediomyids but even in the most primitive of borhyaeniforms, the early Tertiary borhyaenids.

A relatively large *gtpb* on the dorsal and distal end of the peroneal process characterizes the primitive metatherian condition.

The general conformation of the pediomyid astragalus is not dissimilar to that of didelphids. When the UAJ, LAJ, and ANJ are carefully examined and compared, however, important differences emerge. The relative size of the AFi(2) compared to the size of the total tibial articular contact [ATil(2) and ATim(2)] is much smaller than in didelphids or in australidelphians. The ATim(2) forms a more acute angle with ATil(2) and consequently the medial profile of the astragalus is considerably higher than that in didelphids or in primitive australidelphians. These pediomyid characters become even more accentuated in borhyaenids. As in primitive eutherians or in didelphids the CaA(2) and Sup(2) are well separated by a sulcus. The Sup(2) is usually relatively larger than the homologous facet in didelphids. As in other didelphidans, this facet is on the lateral side of the plantar surface, and it is usually continuous with the ANa(2). This latter articulation with the navicular is orientated transversely across the distal end of the astragulus, and the facet lacks the characteristic dorsal twist on the lateral side of the didelphids.

To summarize the diagnostic features of the pediomyid tarsus, the following attributes may be noted:

1) The UAJ is characterized by relatively narrow fibular contact, the CaFi(1) and AFi(2) being narrower relatively than in didelphids. The ATim(2) joins ATil(2) with a more acute angle, in contrast to the more gently curving transition found in didelphids.

2) The LAJ facets form the primitive SLAJP, the facets being clearly separated as in other didelphidans.

3) The CCJ is reminiscent of a ball and socket articulation. Unlike in didelphids, it completely lacks the pivot-like modification of the proximal segment of the joint. The depression homologous with CaCup(1) on the didelphid calcaneus cannot be identified on any Cretaceous–Paleogene marsupial calcaneus.

4) The ANJ lacks the extreme dorsal twist of the lateral half of the ANa(2) seen in didelphids, or the dorsoplantar alignment of this facet and its extreme broadening seen in terrestrial caenolestids and other borhyaeniforms. The ANa(2) and the Spli(2), which are continuous with each other, run essentially on the neck and head of the astragalus.

DIDELPHIDAE: The concept of the Didelphidae, espoused by me, attempts to single out the diagnostic presence of those derived characters, or a derivation from them, which were possessed by the last common ancestor of the extant radiation. In light of the tarsal and dental morphology surveyed, this can be satisfactorily accomplished. Taking into account characters other than the dentition for the family diagnosis, it appears that a number of dental attributes should not serve to identify didelphids. The 5/4 incisor formula, as well as the stylar cusp of the upper molars are best considered primitive metatherian characters, or perhaps even primitive therian features if the latter concept includes only the metatherians and eutherians. Identifying dentitions, therefore, as belonging to didelphines in the Mesozoic (e.g. *Alphadon*) has little phylogenetic relevance without corroborative evidence from the shared and derived characters of the animals. Cusps A and B are the therian parastyle and stylocone, respectively, whereas cusp E is the therian metastyle. Cusps C and D occur not only in those taxa like *Alphadon* which Clemens (1966) referred to as the Cretaceous "didelphines" (not based on the stylar cusp criteria). These cusps are also recognizably developed in the otherwise more primitive dentitions of *Pediomys*. These points, I believe, largely nullify even the dental grounds on the bases of which certain dental taxa have been assigned to the Didelphinae.

The CaA(1) is usually large, widest laterally and narrowed medially towards Su(1). It is usually sharply convex and its long axis orientation compared to the long axis of the calcaneus is usually dependent on the substrate preference of the species. The CaFi(1) is usually well developed. The CCJ is diagnostic, and that alone can differentiate major groups of metatherians. The concave and somewhat crescent shaped CaCud(1) faces mostly distally whereas the continuous CaCup(1) is a distinctive concavity for the cuboid pivot, facing plantad. The cuboid facets form a ledge–like platform [CaCud(3)] and the didelphid pivot [CaCup(3)] seems to support on its margin great masses of tendinous connection between the cuboid and calcaneus. A similarly immense cuboid projection can also be found in fossil and living dermopterans (Szalay and Drawhorn 1980). The didelphid CCJ is slightly distad and plantad to the LAJ.

The combined upper ankle joint facet of the astragalus is diagnostically broad transversely. The AFi(2) is large and prominent and forms an angle with ATil(2). The contact between these facets is not sharp, but rather gentle and gradual. Whereas AFi(2) is convex and rounded, ATil(2) is gently concave. ATim(2) joins ATil(2) approximately at the same angle as does the AFi(2) laterally. Mobility of the entire crus on the combined dorsal surface of the body of the astragalus is not particularly restricted. Fore and aft mobility is slightly stabilized by the shallow groove on the ATil(2).

To summarize the diagnostic characteristics of the didelphid proximal tarsus, the following attributes of the ancestral morphological matrix of the family may be noted:

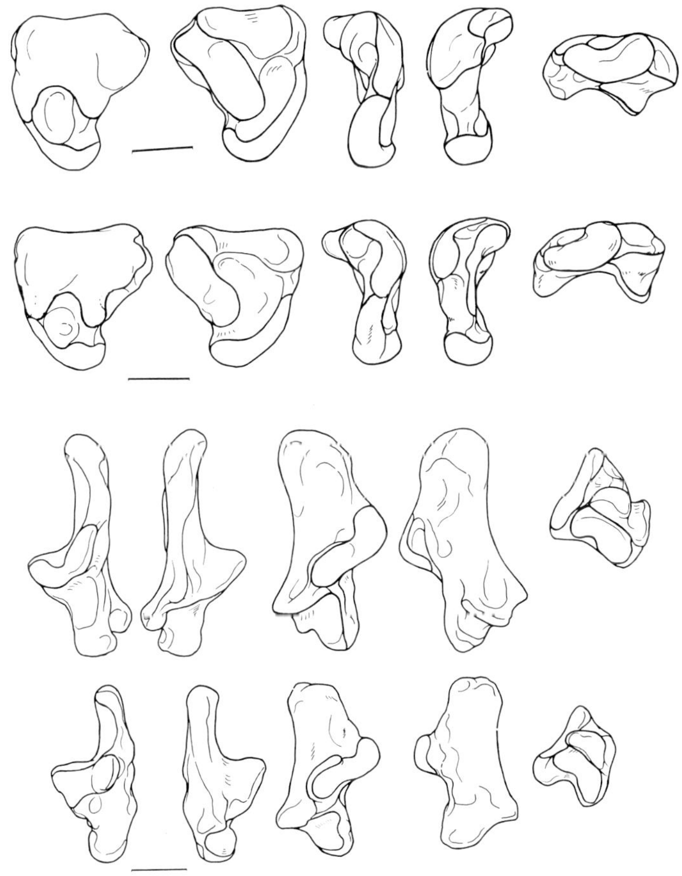

Fig. 4. Comparisons of the left astragali (above) and calcanea (below) of *Dromiciops* and a representative phalengeroid, *Distoechurus*. Note the detailed and derived similarities shared between the two taxa. From left to right, dorsal, plantar, medial, lateral and distal views, respectively. Scales represent one mm.

1) The UAJ is broad and gently curving, facilitating mediolateral and fore and aft movements, the male surfaces consisting of well developed CaFi(1), a transversely extensive AFi(2), and a broad ATi(2).

2) As in pediomyids, the LAJ facets are well separated.

3) The CCJ in all living taxa is characterized by a well developed cuboid pivot which clearly separates didelphids from both the pediomyids and borhyaeniforms, and from the australidelphians.

4) The ANJ is approximately transversely oriented in relation to the distal view of the astragalus. The AN(2) is laterally thickened and the lateralmost area is twisted dorsally.

BORHYAENIFORMES: The first major group, the Borhyaenoidae, is cranially and dentally, as well as in size range, a very heterogeneous assemblage (see especially Marshall 1976). As far as it is known, however, the tarsal morphology of the early Paleocene forms or those of the advanced *Thylacosmilus* bear the stamp of a common morphotype specialization. My present acquaintance with the tarsal morphology of the group is based on a collection of South American Paleocene (Itaborai) footbones, identified on the basis of pedal material of *Prothylacinus patagonicus* (PU157000), "*Sipalocyon*" ("*Amphiproviverra manzaniana*" PU151-54), *Cladosictis patagonica* (PU15702, 15701), and *Thylacosmilus* (CNHM P 14344). Although the genera span two subfamilies of borhyaenids (Hathlyacyninae and Prothylacininae) and thylacosmilids, the characteristic proximal tarsal adaptations are uniformly present.

The calcaneus retains some of its primitive, pediomyid–like, rather than didelphid–like, morphology. The most significant difference concerns the morphology of the CaFi(1). From the primitive condition it is expanded laterally into a prominent rounded facet which seems to have allowed a great deal of mobility between the fibula and the calcaneus of the upper ankle joint. In borhyaenoids, as in caenolestids, the fibular contact with the tarsus is largely through the calcaneus, the AFi(2) being discernably reduced. Primitively the peroneal process is large as in the pediomyids.

The borhyaenoid astragalus, as that of caenolestids, is more drastically modified from the ancestral condition than the calcaneus. The mechanical reasons for this are in the primary resculpturing of the upper ankle joint. The tibia comes to greatly and characteristically restrict the UAJ articulation medially. Extremes of habitual eversion as a result of a powerful push–off phase probably during ground running result in selectional forces which reposition the head of the astragalus and alter the direction of habitual and close packed contact between the latter and the navicular.

In addition to the above modifications, an equally characteristic special similarity shared between borhyaenoids and caenolestids is the conformation of the robust long *ampt*. This overhanging protuberance on the medioproximal corner of the plantar surface of the astragalus is coupled with the presence of an attenuated Sup (2) which runs, ribbon–like, above the *ampt* in both groups. The facet and the process suggests a powerful locking mechanism of the LAJ. This preliminary explanation will be further elaborated elsewhere.

The UAJ is medially constricted by a greatly hypertropheid tibial medial malleolus for which the ATim(2) extends far forward onto the neck of the astragulus particularly medially. This facet is sharply angled from the ATil(2) and it has a long and deep groove for the guidance and restriction of the tibial malleolus. The ATim(2) facet is gently rounded with a slight groove running fore and aft laterally. The medial part of this facet extends far forward onto the medial side of the neck. As noted above, the AFi(2) is relatively small and joins ATil(2) without forming an angle with it.

The CCJ of borhyaenoids and caenolestids can be compared with that of the didelphids as these are known not only by the calcaneus but also by the entire pedal skeleton. Although all borhyaeniforms (present sense) have a discernible lateral shelf on the distal surface of the CaCu(2), these facets have no resemblance to the didelphid condition with its diagnostic and striking cuboid pivot.

The extensive CaFi(1), the distally prominent peroneal process, the basically similar, virtually identical, UAJ and ANJ and the prominent *ampt* along with the articular morphology and mechanics of the LAJ appear to form a complex shared from a common ancestry between borhyaenoids and caenolestids. A specialization which postdates the acquisition of the borhyaeniform character complex can be found in "*Sipalocyon*" and *Thylacosmilus*. In these, and probably other taxa, a derived astragalo–cuboid contact is established (see Fig. 6) for the necessity to distribute the forces transmitted through the crus of cursorially specialized animals. Thylacinids have widely separated astragali and cuboids, and they share none of the tarsal specializations of borhyaeniforms.

The ANJ in borhyaenoids, as in caenolestids, is largely reoriented in comparison with the dorsal contours of the LAJ. The surfaces which are in close packed contact during the extremes of eversion [as the dorsolateral portion of the AN (2)], as during the push–off phase of a step or leap, are greatly expanded. The inferior and medial portion of the AN(2) is effectively under the neck, and, as it becomes continuous with the Spli (2), it runs medially. This joint, like in all terrestrial cursorial forms (e.g., canids, ungulates, macropodids, etc.), but with its own characteristic epigenetic constraints, strengthens the tarsus by close packing the ANJ during the extremes of eversion.

Of the second major subdivision of the Borhyaeniformes, the Caenolestoidea, only the Caenolestidae is known by tarsal remains. Their inclusion in this suborder, however, based on the morphology of the tarsus, seems to be firmly based. All the important shared and derived similarities between borhyaenoids and caenolestids have been noted. In the summary of the borhyaeniform modifications of the distal tarsus I will note some of the differences between the two major subdivisions, but to emphasize their close derived similarity the summary below hopes to give an account of the borhyaeniform morphotype.

The Argyrolagidae were monographed and greatly clarified by Simpson (1970). The specific details of the distal tarsals are not clear from the illustrations, however, and I have not as yet had the opportunity to study the material. The illustrations (Fig. 16F in particular, in Simpson 1970) of the astragalus show a striking proximodistal enlargement of the LAJ medially. This

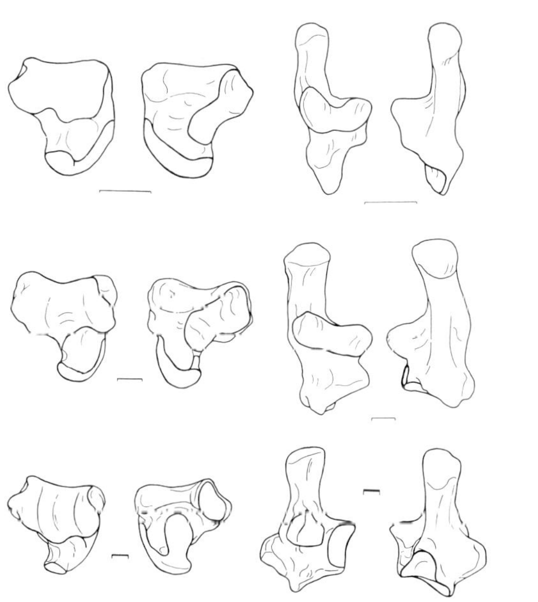

Fig. 5. Comparisons of the right astragali and calcanea of a representative didelphid (*Didelphis*, below), a dasyuridan (*Neophascogale*, middle), and a syndactylan (*Cercartetus*, above) to show the diagnostic differences of the australidelphian pattern of the distal tarsus from that seen in primitive ameridelphian didelphids. Note the general overall conformation of morphology between *Neophascogale* and *Cercartetus*, in particular the morphology of the LAJ. From left to right dorsal, and plantar views, respectively. Scales represent one mm.

character is one of the distinct specializations of the borhyaeniform morphotype, and it does not occur in other metatherian groups except the Borhyaeniformes. More detailed analysis cannot be made until the material is examined.

The following characters of the tarsus were probably present in the last common ancestor of the Borhyaeniformes, present sense.

1) The UAJ is characterized by a mortise–like articulation of the crus medially and the proximal tarsals. An unusually developed ATim (2) on the medial side of the astragalus extends distally onto the neck, receiving in a depression the medial tibial malleolus. Continuous onto the neck of the astragalus is the mediodistal extension of the ATil(2). The AFi(2) is transversely very narrow. Unlike in caenolestids, the CaFi(1), although equally large, is on the dorsal surface of the calcaneus in borhyaenoids, rather than on the lateral side.

2) The most characteristic aspect of the LAJ lies in its ability to lock. The greatly enlarged *ampt* and the extended Sup(2) prevent extremes of foot inversion in borhyaeniforms. Of the well separated LAJ facets Su(2) is wide and triangular and broadly continuous laterally with the AN(2).

3) The CCJ appears only slightly modified from a pediomyid–like condition, and there are no indications in any borhyaeniform for a didelphid–like CCJ being the ancestral source. Lateral to the largely convex CaCu(3) a small shelf, and the corresponding part of the CaCu(1), ensure restricted mediolateral translation in this joint. The morphology and mechanics of the LAJ appear to be identical in the smallest caenolestids or the largest *Thylacosmilus*.

4) The ANJ is a highly modified part of the borhyaeniform tarsus. The dorsal (lateral in non–borhyaeniforms) portion of the AN(2) is greatly expanded, and the more plantar (previously medial) portion quickly narrows and is continuous with the Spli(2), the latter condition as in most other metatherians. The ANJ is effectively dorso–plantar in orientation, the AN(2) running continuously from the dorsal tip of the head ventrally, slightly medially, and posteriorly. The dorsally expanded border of the AN(2) is nearly parallel with the combined AFi(2)–ATi(2) surfaces.

Cohort Australidelphia, new

The tarsal morphology of *Dromiciops australis* from Chile and all marsupials of Australasia possess a number of complex derived characters, so far discovered only in their tarsus, which I believe had evolved from a common ancestor after the differentiation of that putative ancestor from a didelphoid stock. These homologously

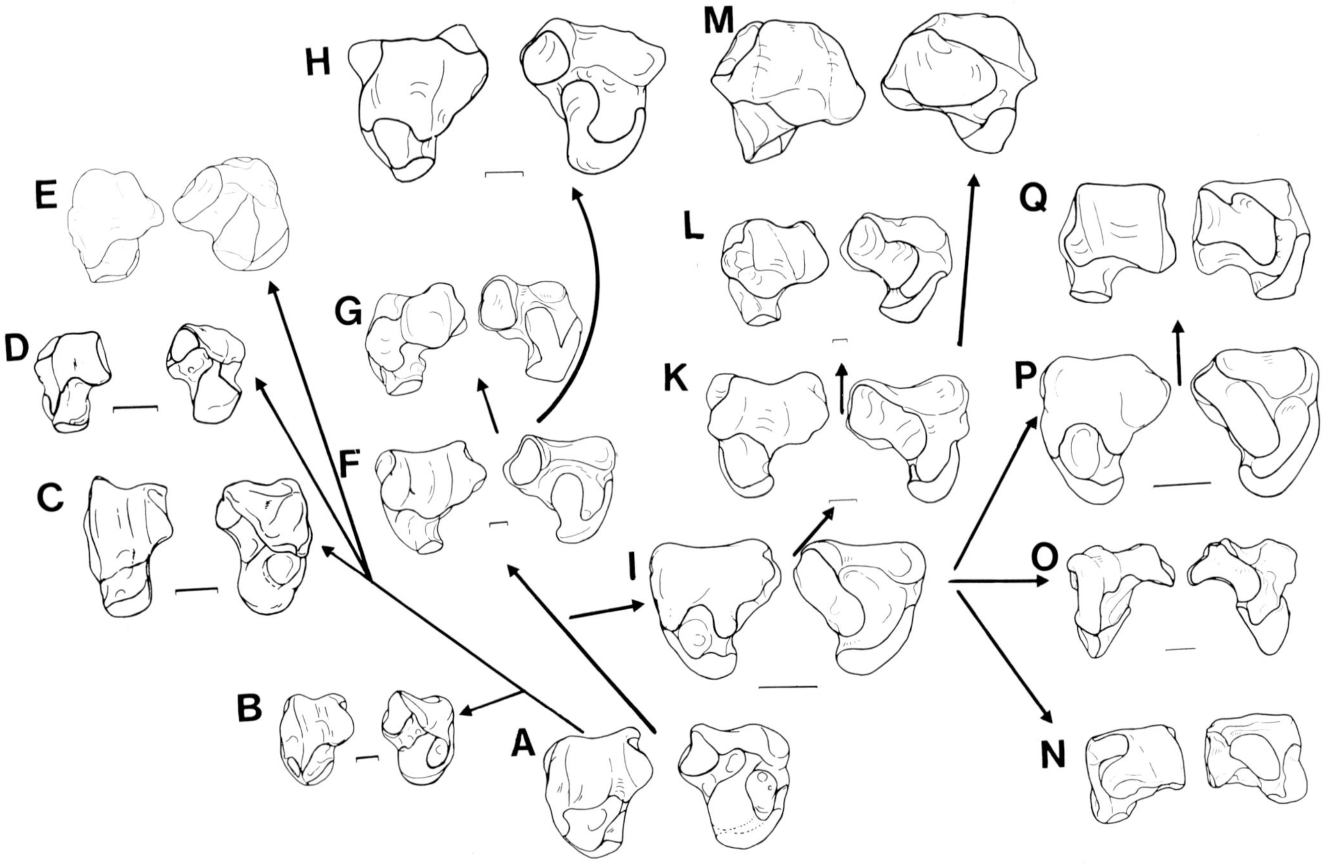

Fig. 6. Hypothetical transformation series to show the morphological changes of the UAJ and LAJ on the left astragalus in major groups of the Metatheria. The series (each representative consisting of a dorsal and plantar view), which is made up of a large number of contemporaneous taxa, is not intended as a phylogenetic summary of taxon relationships. It illustrates, however, the espoused polarity hypotheses of the joint complexes as reflected on the astragalus. For details see under the appropriate places in text. A. Bitter Creek (early Eocene) pediomyid; B. Bridger (middle Eocene) ?borhyaeniform; C. Tepee Trail (late Eocene) borhy aeniform; D. *Caenolestes;* E. *Thylacosmilus* (late Miocene–early Pliocene); F. *Didelphis;* G. *Chironectes;* H. *Glironia;* I. *Dromiciops;* K. *Neophascogale;* L. *Myrmecobius;* M. *Thylacinus;* N. *Perameles;* O. *Notoryctes;* P. *Distoechurus;* Q. *Hypsiprymnodon.* Note the differences between ameridelphians (A–H) and australidelphians (I–Q), the former sharing the primitive SLAJP, whereas all the latter display the CLAJP.

shared and derived characters, an intricate character complex, represent one of the diagnostic features of the common ancestor of the Australidelphia.

Rather than attempting to provide details (but see representative illustrations) for all major groups of australidelphians (a task beyond the ambitions of this communication), I will give an account of the australidelphian characters of the tarsus and comment on the evolution of the most significant attributes (see Figs 4–9).

The fundamental similarity possessed by all australidelphians is centered on the modifications in the LAJ of their morphotype. In the polyprotodont and diadactylus *Dromiciops,* with a strongly grasping arboreal type foot (in spite of its reported terrestrial habits), in dasyuridans, with their reduced hallux and largely terrestrial substrate preference, in numbats, in *Notoryctes,* in terrestrial bandicoots and macropodids, and in phalangers, the morphology of the LAJ articulations is based on the same underlying pattern. The difference from the primitive ameridelphian pattern (Figs 3–5) can be easily discerned. In all australidelphians (with some clearly derived exceptions among the most advanced macropodids) the LAJ facets are diagnostically continuous (CLAJP) and primitively prominent Sud is lost. The significance of this fundamental, derived, similarity cannot be overemphasized. It is an unvarying feature of the amazingly heterogeneous locomotor adaptations of the Australasian marsupials and *Dromiciops.* Irrespective of the specific form–function complexes found in the diversely adapted species, this common heritage is the derived common heritage of the cohort. When, in a few species, distal calcaneal astragalar contact is reestablished, it is still invariably accompanied by a CLAJP, and from the different location and conformation of this reevolved contact, its secondary, nonhomologous nature is obvious. Thus, for example, in *Myrmecobius* and *Thylacinus*, the presence of a "Sud" is relocated in a position relatively different from that in ameridelphians.

Noting similarities, however, does not carry the same weight in phylogenetics as the added attempts, combined with data gathering, of explaining the nature of the original mechanical significance of a character complex and, if possible, the subsequent independence from adaptational significance. I will, therefore, first outline a preliminary hypothesis for the evolution of the australidelphian tarsal complex which appears to explain the transformation, then suggest reasons why the dasyuridan condition is derived from the primitive australidelphian pattern, still preserved by *Dromiciops* and pygmy possums.

Table 2. An abridged evolutionary classification of the Metatheria, based on the phylogenetic hypothesis outlined in this paper and depicted in Fig. 10. Whenever new names or new concepts are used, "new" appears after the name.

- Class Mammalia
  - Infraclass Metatheria
    - Cohort Ameridelphia, new
      - Order Didelphida, new
        - Suborder Didelphiformes, new
        - Suborder Borhyaeniformes, new
          - Superfamily Borhyanoidea
          - Superfamily Caenolestoidea
          - Superfamily Argyrolagoidea
    - Cohort Australidelphia, new
      - Order Dromiciopsia, new
      - Order Dasyurida, new
      - Order Syndactyla
        - Suborder Syndactyliformes, new
          - Superfamily unknown
          - Superfamily Notoryctoidea
        - Suborder Perameliformes, new
        - Suborder Phalangeriformes, new
          - Superfamily Phalangeroidea
          - Superfamily Vombatoidea
          - Superfamily Diprotodontoidea
          - Superfamily Macropodoidea

The first major question which arises concerns the nature of the transformation and the mechanical and adaptive reasons for it. The explanation requires an understanding of the tarsal mechanics in the groups which have the morphological complex, and those which, as putative ancestral stages, never acquired it. The simultaneous modifications of the LAJ and CCJ is suggested by the habitual orientation of the foot in arboreal phalangeroids, large and small. Even when on level substrate (personal observations on *Acrobates, Burramys, Cercartetus, Gymnobelideus, Pseudocheirus, Phalanger, Trichosurus,* etc.) these animals keep the hallux pointed forward, in the direction of progression, whereas the remaining toes are pointed laterally. On the arboreal substrate, however, the hallux usually rests on the top branch, pointing forward, the remaining grasping toes, while pointing laterally and downward, encircle the branch in a highly inverted position. The forces habitually incurred by phalangeroids on a branch are such, that were it not for some osseous modification of the tarsus, the proximal tarsus would tend to be forced, along the transverse tarsal joint, downward. The evolution of the enormous *cdlp,* therefore, along with the resculpturing of the entire CCJ, has probably occured in the primitive australidelphian to counteract these forces encountered in a habitually inverted foot. In addition to the relief that the modifications brought to those soft structures of the tarsus, the australidelphian CCJ had also come to channel movements of the pes on the proximal tarsals. Flexing and extending the fore pes while the hallux and the heel of the foot was resting on the branch has come to be accomplished on a LAJ which has become transversely constricted and, as noted above, continuous. This LAJ transformation was probably partly due to the obliteration of the Sud articulation as a result of the reformed CCJ. The articulation of the cuboid with the calcaneus has taken up the place previously occupied by the Sud (see especially the medial view of the calcaneus of *Dromiciops* and *Distoechurus,* Fig. 4).

Although the wedged CCJ of the primitive australidelphian is not a universal australidelphian character, as is the CLAJP, its heritage influence is very strong on the CCJ of australidelphian groups. Once terrestrial progression evolved (several times) from such morphological base, the restrictions of the primitive australidelphian CCJ had to be modified to allow the fore and aft orientation of a level or slightly everted pes. These modifications were carried out differently in the dasyuridan, perameliform, and macropodoid ancestors, among other australidelphians.

Although dental morphology has been a main stay of a metatherian phylogenetics, there remain some hitherto unutilized features which are likely to be synapomorphies of *Dromiciops* and other australidelphians, particularly the polyprotodont dasyurids and peramelids. These features are further modified in the stem phalangeriforms. The relatively tiny, almost peg like incisors of Didelphidae are so uniformly prevalent among the living representatives that its presence in the morphotype of the family was almost certain. The condition may have been primitive in the ancestry of the protoaustralidelphian. This tiny incisor conformation is in sharp contrast to the australidelphians noted, in which the incisors are mesiodistally expanded. For example, in juxtaposition to a length of incisor row/length of C–M_4 ratio of 0.17 in *Marmosa* or in other didelphid genera, in *Dromiciops* the ratio is 0.42, similar proportions holding true for dasyurids, after adjustments are made for the one less incisor. Thus, the ratio of 0.28 in *Antechinus flavipes,* for example, is 0.35 if an additional incisor is considered. Incisor hypertrophy was likely to have been a derived character of the australidelphian ancestor.

Both *Dromiciops* and most dasyurids (possibly all), unlike didelphids, have strikingly developed distal heels on their lower incisors. *Dromiciops* and the small, very likely primitive, dasyurids have mandibular condyles which are both relatively small and short transversely. In these forms the molar trigonids are much less open than in the didelphids, which also possess relatively large and very transverse mandibular condyles.

Given the characters discussed above, we may now briefly take stock of the protoaustralidelphian characters which diagnosed this species from its putative didelphidan ancestor. This first australidelphian was polyprotodont, with at least a 5/4 incisor formula, with hypertrophied incisors, the lower incisors with expanded distal heels. It was an evolved pedal grasper, diadactylous, with a sellar EMIJ, in contrast to the grooved condition of this articulation in ameridelphians. Its diagnostically derived tarsal morphology is detailed below.

The position of the Dasyurida is of special importance in any scheme which maintains that their substrate preference and pedal adaptations are derived from arboreal ancestry. A crucial difference in past assessments of marsupial phylogeny and that proposed here is reflected in a remark by Ride (1964: 102) in which he considered "... that Bensley's (1903: 163) view that the terrestrial foot of dasyuroid marsupials is secondarily derived from arboreal one is no more than an assumption". This remark, of course, is only a counterstatement to Bensley's assertions which was neither preceded nor subsequently followed up by research in the relevant character-complex, the osteology of the pes, by either Ride or other students. Such statements, or similar tacit assumptions, however, continued to influence phylogeneticists working on metatherian evolution. Inuendos, assumptions, or explicit statements as to the overall primitiveness of dasyurids (almost invariably aspects of their dentition however) within the Metatheria abound in the literature.

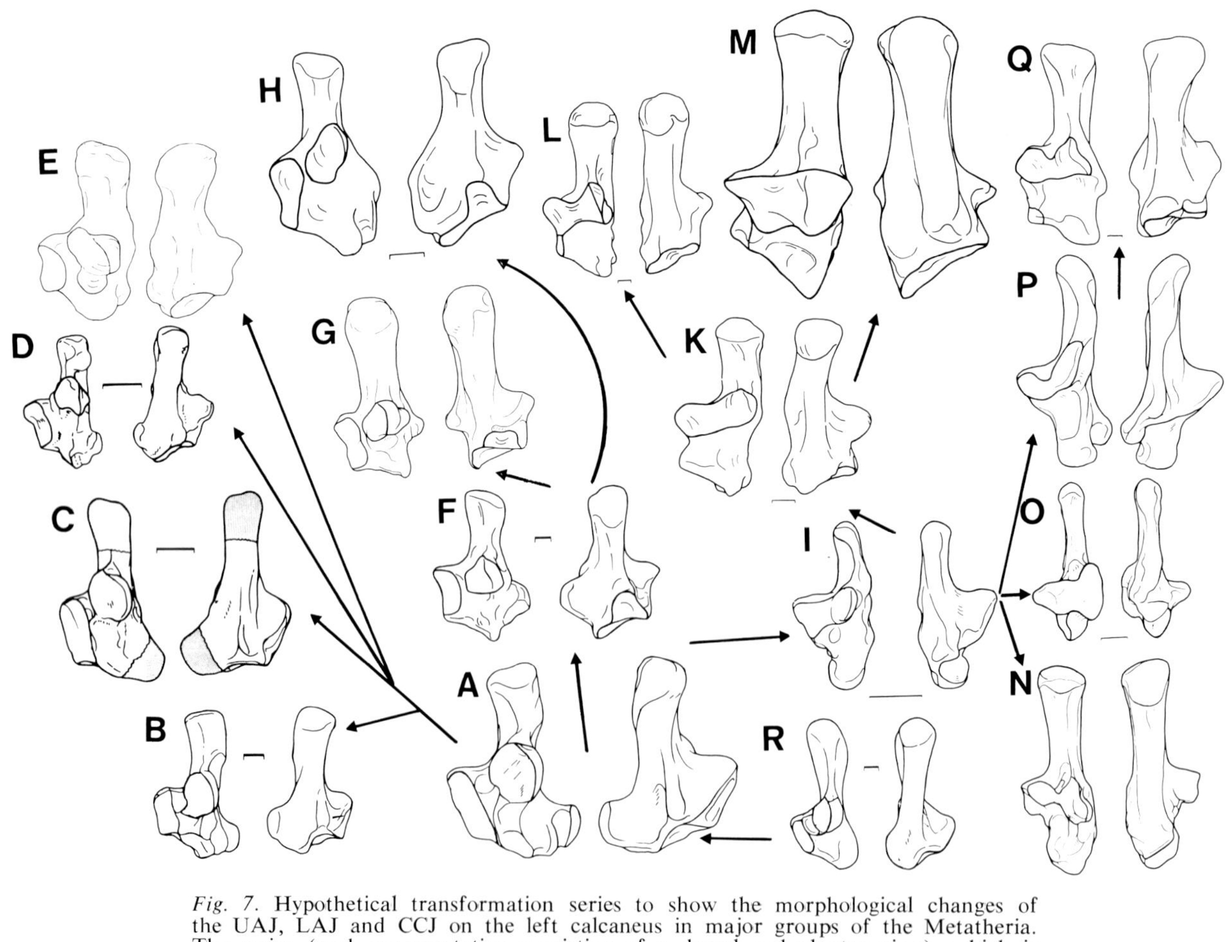

Fig. 7. Hypothetical transformation series to show the morphological changes of the UAJ, LAJ and CCJ on the left calcaneus in major groups of the Metatheria. The series (each representative consisting of a dorsal and plantar view), which is made up of a large number of contemporaneous taxa, is not intended as a phylogentic summary of taxon relationships. It illustrates, however, the espoused polarity hypotheses of the joint complexes as reflected on the calcaneus. For details see under the appropriate places in text. A to Q same as in Fig. 6. R. Lance Cretaceous pediomyid. Note the differences between the didelphid CCJ and that of other metatherians, and also the differences between ameridelphians (A–H) and australidelphians (I–Q), the former showing the primitive therian SLAJP, whereas all the latter display the CLAJP.

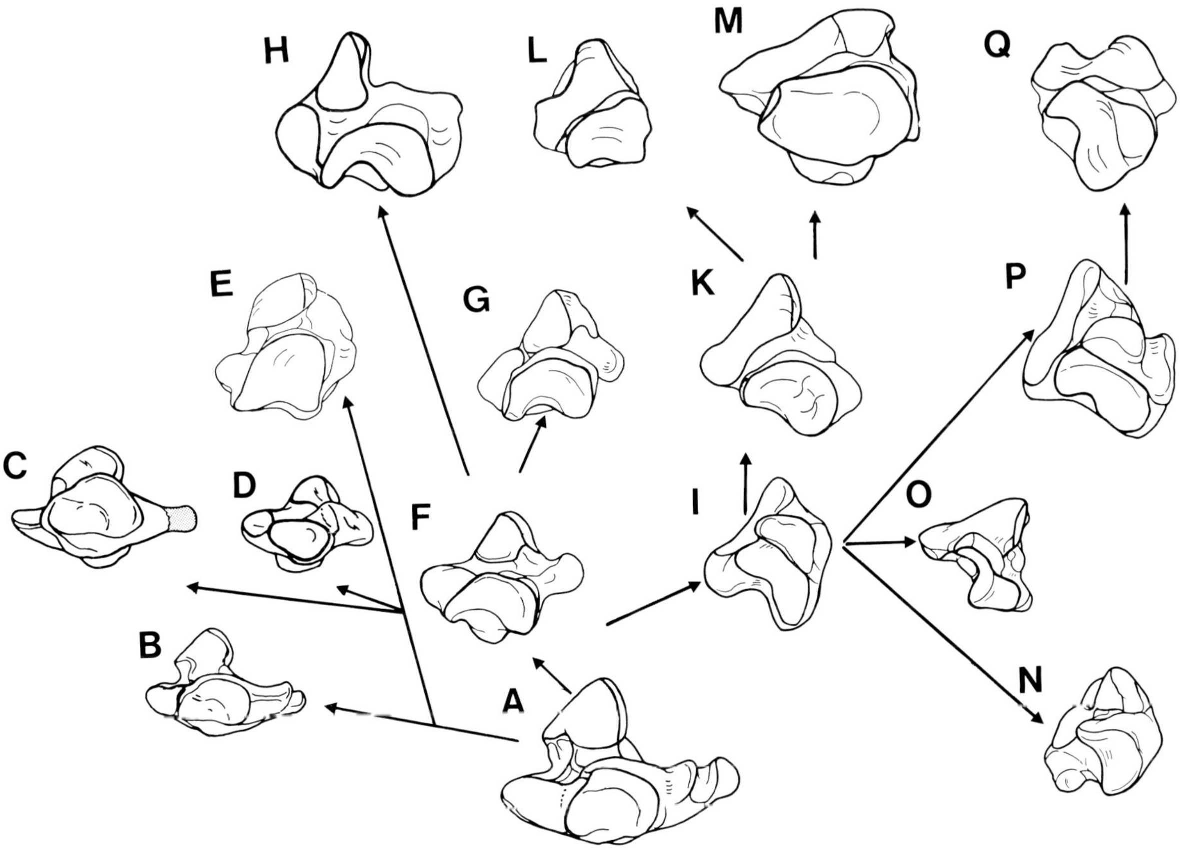

Fig. 8. Hypothetical transformation series to show the morphological changes of the CCJ on the left calcaneus in distal view. The series, which is made up of a large number of contemporaneous taxa, is not intended as a phylogenetic summary of taxon relationships. It illustrates, however, the espoused polarity hypothesis of the CCJ as reflected on the calcaneus. For details see under appropriate places in text. A to Q same as in Fig. 6. Note differences between didelphids (F–H) and other ameridelphians.

Despite the prevalent notion that Bensley's original ideas were incorrect, all factual information and their explanation by the least contradicted hypotheses point to the view that the ancestral dasyuridan, a terrestrial species, modified the ancestral australidelphian tarsal pattern discussed above. To accomplish rapid terrestrial locomotion on a level substrate the ancestral dasyuridan CCJ had to be freed from the constraints placed on it by the wedge shaped angle–iron type articulation which has evolved to stabilize a permanently inverted foot. But stating this is merely assuming it. The first part of a somewhat satisfactory answer lies in the reason why the Sud was obliterated in the dasyuridan morphotype. A look, either in *Dromiciops* or in arboreal phalangeriforms, at the space displaced by the CCJ at the expense of the distal portion of the LAJ is highly suggestive that the Sud was obliterated concurrently with the evolution of the australidelphian CCJ–LAJ complex. That the space on the distal dorstal surface of the dasyuridan calcaneus is present where previously the Sud was located is due, according to this view, to the modifications of the dasyuridan CCJ to free that joint from the previously useful restraint that it acquired in the first, habitually grasping–inverting australidelphians. This occurred by reconverting the CCJ to a transversely mobile complex, but not unlike the primitive metatherian pattern. It is revealing, however, that the orientation of the CCJ to the CLAJP, when viewed distally, is identical to that in the primitive australidelphian pattern (compare *Neophascogale* to *Dromiciops* or *Cercartetus*, Figs 4–5). The similarity of the CCJ in *Antechinus melanurus* to *Dromiciops* and phalangers is especially striking.

The second reason why the facts indicate that dasyuridans evolved from an arboreal, and specifically australidelphian stock rests on the morphology of the EMIJ of the dasyuridans. Why would a group of animals, which have all but eliminated the first digit, evolve a mobile, sellar shaped joint? The morphology of the EMIJ in didelphids and known borhyaeniforms (pediomyids are unknown in that respect) have two prominent ridges on the proximal surface of the first metatarsal which ensures that the adduction and abduction of this digit on the entocuneiform is firmly guided, and somewhat restricted. Why would the EMIJ morphology of those dasyurids which still retain this articulation, as *Neophascogale,* for example, have the same sellar articulation as that of *Dromiciops* and phalangeriforms, an articulation so well suited for great mobility and extremes of abduction? The answer, simply, or the most parsimonious hypothesis, given that other aspects of the biology of these animals are not contradictory, is that the dasyuridan morphotype inherited that condition from the dromiciopsian ancestor, as it also inherited its characteristic CLAJP.

The character of the EMIJ of the dasyuridan morphotype is perhaps the best independent test so far, after the morphology of the LAJ, of the proposed australidelphian ties of the Dasyurida. This feature is of especial significance, as the mobile EMIJ of *Dromiciops* and phalangeriforms is also possessed by a group of primarily terrestrial taxa, the dasyuridans, with their virtual loss of all hallucial biological roles. In spite of this rudimentation of the hallux, the form–function attributes display a stasis from their arboreal phase of evolutionary history, specifically, from a dromiciopsian phase of that history.

The dasyuridan cuboid, as seen in *Neophascogale* for example, when compared with that of didelphids and australidelphians (*Dromiciops* and phalangeriforms), at first glance lacks the extreme modifications of the latter. Yet the characteristic indentations for the ectocuneiform on the didelphid cuboid is absent from dasyurids. On the other hand, the broadly elevated and extremely convex CaCu(3), although clearly not as in the australidelphians mentioned, is much more similar to the latter than that of its didelphid homologue.

Based on all that has been stated above, it is clear that the nature of comparative evidence of the tarsus in the Thylacinidae and representatives of the Borhyaenoidea indicates that thylacinids are not borhyaeniforms (see also Simpson 1941, Marshall 1977). Borhyaeniforms and thylacinids unmistakably show in their tarsal morphology the characteristic constraints of their respective ancestors, and therefore, also indicate the most likely hypothesis that thylacinids are derived from other australidelphians. Showing and emphasizing in detail that the complex tarsal

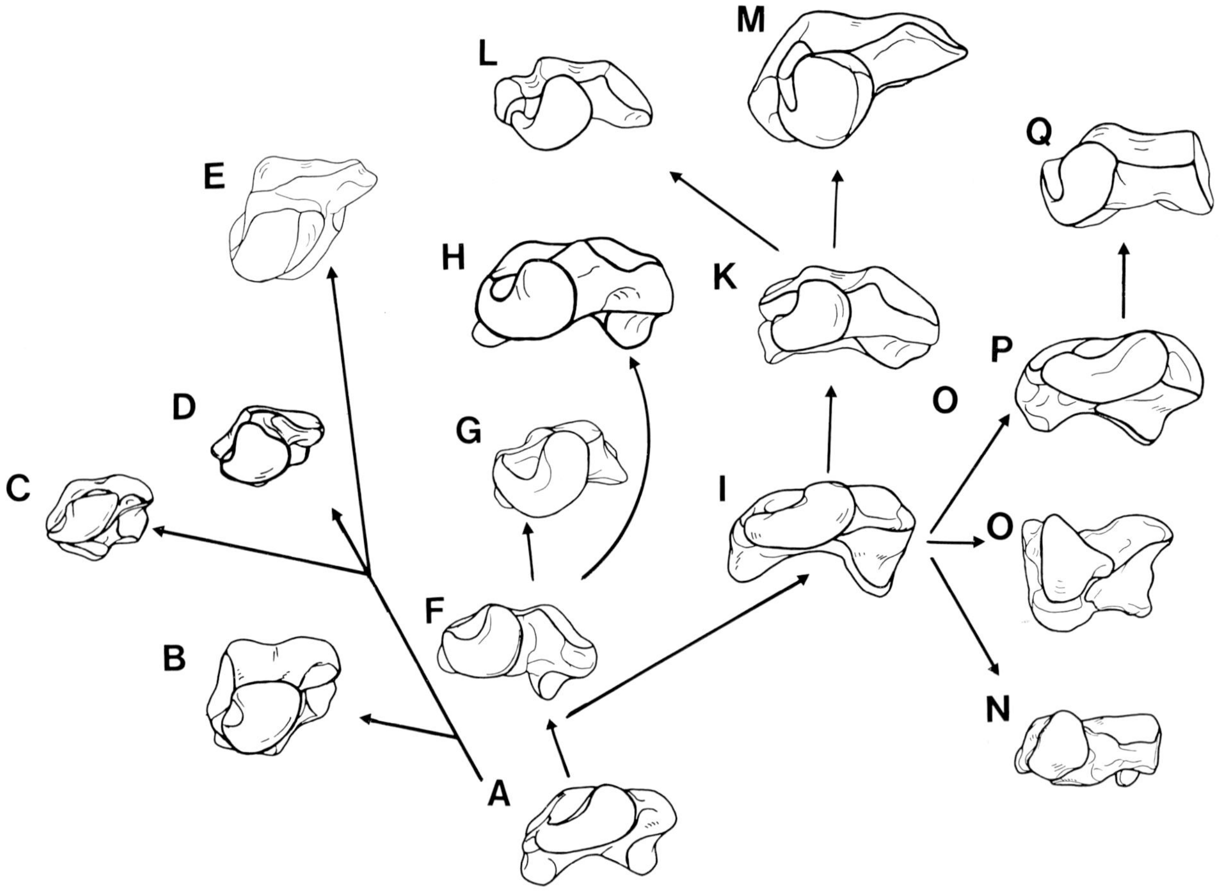

Fig. 9. Hypothetical transformation series to show the morphological changes of the ANJ on the left astragalus in distal view. The series, which is made up of a large number of contemporaneous taxa, is not intended as a phylogenetic summary of taxon relationships. It illustrates, however, the espoused polarity hypothesis of the ANJ as reflected on the astragalus. For details see under appropriate places in text. A to Q same as in Fig. 6. Note the differences between the borhyaeniform modifications (C–E) and those in other metatherians, whether terrestrial or not.

morphology of these groups employs different morphological solutions, constraints inherited from divergent ancestries, will hopefully further the acceptance of the idea that the ancestral australidelphian, a derived, arboreal species of the order Dromiciopsia, was a rather distinctive form compared to known ameridelphians.

Before I outline in detail some of the known diagnostic modifications at specific points in the proposed metatherian phylogeny the following osteological features of the foot should be noted which appeared to have characterized the ancestor of the Australidelphia.

1) The UAJ was transversly broad, mobile not unlike in arboreal didelphids or pediomyids. Fibular contact with the calcaneus was absent or minimal. Both ATil(2) and ATim(2) extended onto the neck of the astragalus.

2) The LAJ facets have become one continuous articulation; CaA became confluent with Sup, whereas Sud was lost, and contact between Sud (2) and AN(2) was broken.

3) The CCJ was transformed into a laterally extremely rigid calcaneocuboid contact, whereby the cuboid firmly abutts into the angled space provided by the distolaterally lengthened calcaneus, reminiscent of the inner surface of an angle iron. Consequent to the probably simultaneous remodelling of the LAJ and CCJ, the calcaneus, profoundly reshaped, lost most of the peroneal process distally.

4) The ANJ was mobile, facilitating considerable movement; its orientation was primarily transverse in comparison to the distal outline of the astragalus.

5) The EMIJ was remodeled, taking on a sellar shape, extending the range and mobility of the hypertrophied first digit of the foot.

6) Superficial comparisons suggest that the tendon M. flexor (digitorum) fibularis (= M. flexor hallucis longus) has greatly increased in its relative size in the australidelphian morphotype.

EPILOGUE AND CLASSIFICATION

A summary of the metatherian relationships proposed here is shown in Fig. 10. Arrows numbered 1–12 depict the derivations and approximate points of origin of some of the diagnostic character combinations which were present in a clade at the time shown. The classifactory scheme from which the taxa are selected is basically an evolutionary one with a strong phylogenetic emphasis on monophyletic groups, with the appropriate use of paraphyletic categories when the evidence is equivocal (e.g. Pediomyidae, Syndactyliformes). The following list corresponds to the horizontal arrows.

1) Presence of primitive ameridelphian features. Primitive polyprotodont dentition with at least a 5/4 incisor formula, transversely broad sectorial molars with stylar cusps; foot probably with large grasping hallux without claw; tarsal morphology the most primitive known (see details above): a) SLAJP between proximal tarsals, b) ball and socket type CCJ, c) transversely orientated ANJ.

2) Polyprotodont dentition in ancestral borhyaeniform, with incisor formula probably reduced to 4/3, sectorial molars, and foot probably with reduced hallux. Tarsal morphology, modified for a primarily terrestrial locomotor repertoire (see details above), is characterized by: a) a mortise–like UAJ medially and a locking LAJ, b) CCJ slightly modified to enhance mediolateral stability, c) ANJ modified to ensure greatest resistance to deformation and to ensure stability at the end of extreme eversion and the push–off phase of the stride.

3) Diprotodont modification of the lower incisors. Posterior premolar not plagiaulacoid. All tarsal characters under 2) are present in the Caenolestidae but tarsus is unknown in the other two probably caenolestoid taxa.

4) Plagiaulacoid condition of the posterior premolars. A group known primarily dentally.

5) Extensive modifications of the hindlimb for jumping. Forelegs reduced. Tubular bony snouts anterior to the incisors (2/2). Teeth are rootless. Affinities of the Argyrolagidae are unclear; detailed appraisal of the tarsus is yet to be carried out. The visibly enlarged medial portion of the upper ankle joint (see Simpson 1970, fig. 16F) suggests borhyaeniform rather than didelphiform ties.

6) Primitive polyprotodont dentition; very small, peg–like incisors, transversely shortened upper molars and mesiodistally attenuated protolophid; very tight mesial twining of the entoconid and hypoconulid. Diadactylous foot with large grasping hallux. Tarsal morphology characterized by: a) broad and gently curving UAJ, b) transversely extensive AFi(2), c) SLAJP between proximal trasals, d) extensive, projecting cuboid pivot, e) EMIJ articulation with double ridged pattern on proximal end of metatarsal I.

7) Acquisition of the australidelphian morphological matrix. Retention of the primitive polyprotodont dentition with 5/4 incisor formula, hypertrophied incisors, and sectorial cheek teeth. Diadactylous foot with hypertrophied grasping hallux. Tarsal morphology profoundly modified and characterized by: a) LAJ with CLAJP and without the Sud; b) CCJ with the angle iron type modification, c) calcaneofibular contact absent or greatly reduced, d) EMIJ modified into sellar pattern on both the articulating surfaces of the entocuneiform and metatarsal I.

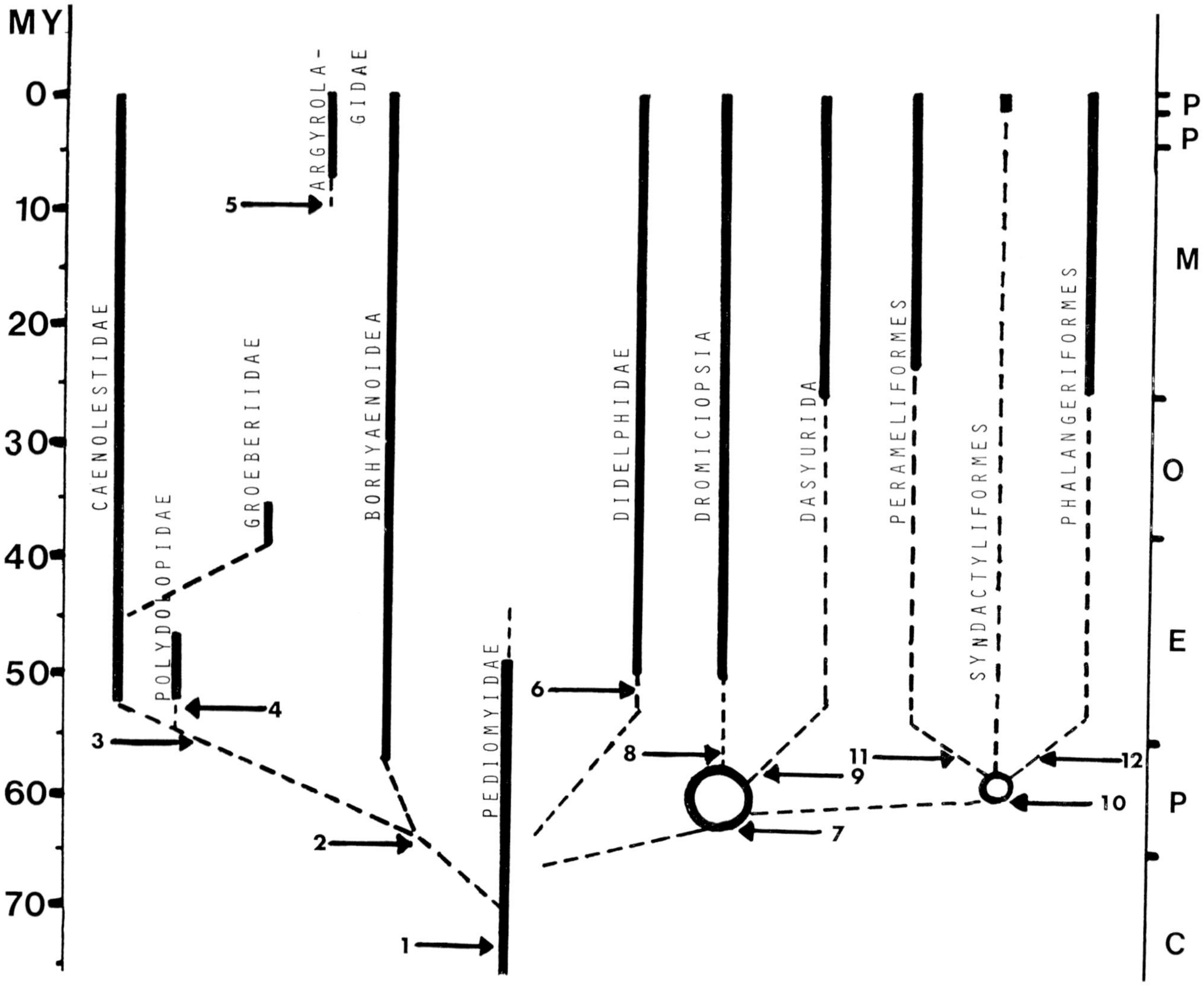

Fig. 10. A hypothesis of metatherian relationships. The arrows indicate the derivation and approximate point of origin of some of the traits which were the bases for the hypothesis. No attempt was made to depict the diagnostic features of all family or higher level taxa. Those shown, however, cover all of the Metatheria. The continuous thick vertical lines represent the known stratigraphic range of the taxa; vertical broken lines represent their estimated total duration; slanted broken lines show derivation from an ancestral taxon. Circles represent unsolved branching or ancestor-descendant relationships. On the left border subdivisions in ten million year intervals are indicated; on the right border the subdivisions correspond to the Upper Cretaceous and the epochs of the Cenozoic. For the list of characters depicted by the arrows see the corresponding part of the text in the Epilogue and Classification under the appropriate numbers.

8) All the characters postulated to be present at 7). The most recent common ancestor of *Dromiciops* and *Microbiotherium* developed composite bullae, highly inflated, formed anteriorly by the tympanic process of the alisphenoid and posteriorly by the petrosal. This condition was not present in the ancestral dromiciopsian.

9) Acquisition of the dasyuridan character complex is based on the modification of the carpal and tarsal morphology of a dromiciopsian stock for a primarily terrestrial existence. Polyprotodont dentition with a reduced 4/3 incisor formula but with the hypertrophied australidelphian incisors, and a diadactylous foot with a greatly reduced hallux. Tarsal morphology retains many primitive australidelphian features; a) LAJ with CLAJP and lost Sud, b) CCJ freed from the constraints of the ancestral australidelphian pattern, c) EMIJ sellar in configuration as in primitive australidelphians in spite of the rudimentation of the hallux.

10) Acquisition of the primitive syndactylan complex by a form with the primitive polyprotodont dentition with a 5/4 incisor formula and molar teeth more squared than in either known dasyuridans or dromiciopsians. Major postcranial modification involves the syndactylous foot for specific arboreal locomotion and subsequently as a grooming adaption. Tarsal morphology remains unchanged from primitive australidelphian pat-

tern represented by the tarsus of *Dromiciops*. The character combination noted was present in early members of the hypothetical Syndactyliformes. It is this stock, as Bensley (1903) suggested, which probably also gave rise to the Notoryctidae.

11) Acquisition of the character combination of perameliform ancestor is by the remodelling of the syndactylous grasping foot into a terrestrially adapted structure. Emphasis of the digit IV and reduction of the syndactylous II–III, and V by a polyprotodont ancestral peramelid. Tarsal morphology is taxon specifically remodelled to suit cursorial demands, while still retaining the CLAJP and traces of the australidelphian CCJ pattern.

12) Phalangeriform morphotype develops fascicular aberrans, reduces the number of lower incisors, and emphasizes the first pair of diprotodont pattern, but retains australidelphian pes with the syndactylous grasping feature of the ancestral syndactyliform.

As I present a new classification of the Metatheria, I would like to comment briefly on this subject. The most recent and valuable discussions on marsupial classifications may be found in Simpson (1970) and Kirsch (1977). I will elsewhere, in a monographic study in preparation discuss fully the necessity and the *raison d'être* for the classification offered here.

It has been too often stated in discussions of metatherian classifications that a particular arrangement is based on either the skull, dentition, feet or serology. This approach is wrong as it clearly attempts to reflect character phylogeny rather than the phylogeny of the taxa. Although classification can heuristically remain evolutionary, it should nevertheless be based on the most probable taxon phylogeny available (*fide* Simpson 1945, 1961, Bock 1977b). It is also of paramount importance that classifications should become increasingly phylogenetic (not cladistic). Most phylogenetic studies are somewhat rooted in previous classifications and it is for that reason that the utmost importance should be attached to defining taxa crisply and as unambiguously as possible, based on all of their known derived character–states. As soon as I began this research it became obvious to me that definitions of marsupial groups, particularly suprafamilial taxa, have been defined on one or another set of characters, rather than attempting, even osteologically, a reasonably complete characterization. Clearly, as the paper above shows, the tarsus, a stable and complex area, never before was utilized to detect the presence of derived character complexes. If no attempt is made to ferret out those features which suggest the most recent phylogenetic shifts, then no unambiguous groupings are possible, and subsequently there is little hope for approaching a realistic phylogenetic hypothesis which best accounts for all known character clines.

Past suprafamilial classifications, particularly those presented during the last two decades, tended to elevate previously known to be distinct groups to ranks more equivalent to eutherian taxa. Thus, even Ride's (1964) radical but welcome departure from previous classificatory practices gave few well researched character clines for a substantive phylogenetic hypothesis which is usually required for more conservative classifications [see Simpson's (1970) thorough discussion of this]. The concepts Paucituberculata or Peramelina, for example, emphasize the distinctness but not the phylogenetic ties of these taxa. The Marsupicarnivora, however, was a new concept which attempted to express affinity in a phylogenetic sense. It stimulated thought, research, and as all worthwhile scientific ideas, to paraphrase Popper, carried its seeds of obsolescence. As other students have also come to realize, the Marsupicarnivora is based on both primitive metatherian characters and convergent specializations of distinct groups of carnivorous marsupials. The Marsupicarnivora is not a horizontal group in the highly acceptable sense of modern evolutionary classification (for a clearly presented account of the latter see Simpson 1961 and Bock 1977a, 1977b). It should be emphasized that both the Borhyaenoidea and Dasyuroidea appear to be more recently related to the Caenolestoidea (Paucituberculata of Ride 1964) and the Syndactyla, respectively, than to the Didelphoidea. Here lies the reason for considering the Marsupicarnivora polyphyletic. Following the arguments offered in this paper, the concept Marsupicarnivora (a group informally envisaged by Simpson 1948) is based on the primitive morphology of the Metatheria in two independently attained, primarily terrestrial, radiations of carnivorous marsupials, and therefore this taxon cannot account for the derived character complexes of borhyaenids and dasyurids which tie them to two different higher taxa.

In his thorough discussion of argyrolagid affinities Simpson (1970) reviewed various classifactory arrangements of higher categories within the Metatheria and added, without discussing it, what he called an alternative suggestion. Based on the geographical distribution of marsupials he listed the suborder Hesperometatheria (including the Didelphoidea and Caenolestoidea) and the suborder Eometatheria (including the Dasyuroidea, Perameloidea, and Phalangeroidea). Concerning the origins and classification of the Argyrolagidae he remarked in a footnote (p. 43): "And therefore reference to the Hesperometatheria, if my only half–serious earlier suggestion for ordinal or subordinal classification were adopted".

Simpson's groups were based on zoogeography (hence perhaps his remark "half–serious"), he considered no character clines or any aspect of morphology ("Those groups (i.e. the Didelphoidea and Dasyuroidea) were barely or perhaps not distinguishable in their most primitive forms" (Simpson 1970: 43)), and his Eometatheria did not, and would not have, included taxa from the Americas. Simpson's subordinal concepts are not based on demonstrably shared key adaptations, and therefore they cannot be considered evolutionary. The proposed cohort Ameridelphia does not include all "hesperometatherians", and the cohort Australidelphia includes not only "eometatherians" but also "hesperometatherians". As the Hesperometatheria and Eometatheria are not equivalent groupings to the Ameridelphia and Australidelphia, the former nomina were not employed in this paper.

The essential spirit of Ride's (1964) classficatory scheme in which he championed the taxonomic expression of marsupial diversity is followed here. The classification proposed in this paper recognises four orders based on a combination of the importance of phylogenetic ties as reflected by shared and derived characters and on adaptive diversity. As far as I am aware, there are no polyphyletic taxa in this classification.

OVERVIEW AND SUMMARY

During this study a number of points of major significance became apparent. Subsequent to these discoveries of similarities and formulation of polarity hypotheses, taking into account the cranial, dental, and pedal character clines known in the Metatheria, a number of phylogenetic interpretations have emerged.

1. The living didelphid marsupials possess an underlying homologous similarity in their tarsus which can be reliably assessed as diagnostically derived when compared to a marsupial morphotype or the tarsal remains of the Cretaceous–Paleogene didelphoid Pediomyidae.

2. The diagnostic modifications of the borhyaenoid tarsus bears neither primitive nor special shared and derived similarity to that of any Australasian group, neither dasyurid nor thylacinid. It is probably pediomyid derived.

3. The tarsal morphology of the Caenolestidae is especially similar to that of the highly characteristic borhyaenoid lineages which were present both in North and South America during the Paleogene.

4. All Australasian marsupials, both polyprotodont and diprotodont, possess a strongly similar morphological matrix on their tarsal morphology which sets them aside from the didelphidan taxa. This australidelphian pattern is shared even by the most distinctly and divergently adapted species. The australidelphian heritage pattern is almost certainly shared and derived when contrasted with the homologous ameridelphian complex from which it has evolved.

5. Uniquely, among the living South American marsupials, the montoypic Chilean *Dromiciops* shares the entire underlying tarsal morphology with australidelphian pygmy possums and other phalangeriforms. It shares the canalizing australidelphian morphological pattern of the tarsus with all australidelphians. In spite of the fact that *Dromiciops australis* is primitively polyprotodont and diadactylous, its derived tarsal morphology securely ties it with the Australidelphia.

6. As ascertained by Segall (1969), the South American fossil *Microbiotherium* shares a complex derived similarity in its basicranium with *Dromiciops*, probably homologously. Neither dental nor tarsal characters support Reig's (1955) inclusion of *Caluromys, Caluromysiops,* and *Glironia*, with either *Microbiotherium* (not known in its tarsal morphology) or *Dromiciops* in the same family. The former taxa are clearly didelphid in their known derived characters, their tarsus.

7. The syndactylous condition of the foot, as proposed by other students in the past, appears to be a monophyletic, uniquely shared and derived character complex. There is no known or documented evidence for the hypothesis that it has evolved more than once. Syndactyly probably postdates an adaptive modification in a dromiciopsian species in which the second and third toes became reduced in order to avoid injurious contact with the greatly enlarged and highly mobile big toe when grasping and encircling small terminal branches. The hypothesis of Jones (1925) that syndactyly is a grooming adaptation is probably correct, but it is probably based on a preadapted condition which is the result of a specific grasping adaptation early in an australidelphian lineage leading to the common ancestor of the Syndactyla.

8. The combined evidence of dental and tarsal morphology strengthens Bensley's suggestion that notoryctids are syndactylans, derived from stem syndactyliforms.

These points, along with the reevaluation of the pertinent dental, cranial, and pedal evidence known, where reconsidered and synthesised into a new metatherian phylogeny especially on the supraordinal level. In addition, an evolutionary classification is offered which, while being more phylogenetic than previous ones, still employs paraphyletic taxa (e.g. the Pediomyidae), it does not attempt to express all known sister–group relationships in taxonomic dichotomies, or all relationships by the use of an inordinate number of new taxonomic categories with the subsequent necessity for numerous new zoological nomina (see comments and discussion on cladistic classificatory practices in Bock 1977b, and Szalay 1977b).

ACKNOWLEDGEMENTS

I wish to extend my deep appreciation to the Department of Zoology, Monash University, Victoria, particularly its chairman, Dr. James Warren, for generously providing space and facilities to carry out part of the research reported in this paper. A John Simon Guggenheim Foundation Fellowship for 1980 enabled me to devote my time to marsupial research in Australia. A CUNY Doctoral Faculty Grant defrayed the expenses for illustrations.

Ms. Anita Cleary, throughout the past year has patiently and with great talent drawn the outline figures of the tarsals in this paper. I am responsible for the details of the various joint facets on the drawings.

For their trust and help I am indebted to Drs. Walter J. Bock, Clark F. Howell, W. Patrick Luckett and George Gaylord Simpson.

For permission to study and borrow specimens I thank the following colleagues and their institutions. Drs. Sidney Anderson, Malcolm C. McKenna, and Richard H. Tedford of the American Museum of Natural History; Dr. Mary Dawson of the Carnegie Museum; Dr. William D. Turnbull of the Chicago Natural History Museum; Dr. Robert M. West of the Milwaukee Public Museum; Drs. William A. Clemens, Jr., Donald E. Savage, and Ms. Barbara Waters of the University of California (Berkeley) Museum of Paleontology; Dr. Peter Robinson of the University of Colorado.

An early draft of this paper was constructively criticised by Drs G. B. Sharman, M. Archer and an anonymous referee.

REFERENCES

Archer, M., 1975. The basicranial region of marsupicarnivores (Marsupialia), interrelationships of carnivorous marsupials, and affinities of the insectivorous marsupial peramelids. *J. Linn. Soc. (Lond.)* **59:** 217–322.

Archer, M., 1976. The dasyurid dentition and its relationships to that of didelphids, thylacinids, borhyaenids (Marsupicarnivora) and peramelids (Peramelina: Marsupialia). *Aust. J. Zool., Suppl. Ser.* **39:** 1–34.

Bensley, A., 1903. On the evolution of the Australian Marsupialia; with remarks on the relationships of the marsupials in general. *Trans. Linn. Soc. Lond. Zool.* **2:** 83–217.

Bock, W. J., 1977a. Adaptation and comparative method. Pp.57–82. *in* "Major patterns in vertebrate evolution" ed. by M. K. Hecht, P. C. Goody, and B. M. Hecht. Plenum Press: New York.

Bock, W. J., 1977b. Foundations and methods of evolutionary classification. Pp.851–96. *in* "Major patterns in vertebrate evolution" ed. by M. K. Hecht, P. C. Goody, and B. M. Hecht. Plenum Press: New York.

Bock, W. J., 1981. Functional–adaptive analysis in evolutionary classification. *American Zoologist.*

Bock, W. J. and von Wahlert, 1965. Adaptation and the form function complex. *Evolution,* **91:** 269–99.

Clemens, W. A. Jr., 1966. Fossil mammals of the type Lance Formation, Wyoming. Part II. Marsupialia. *Univ. Calif. Publ. Geol. Sci.* **62:** 1–222.

Clemens, W. A. Jr., 1977. Phylogeny of the marsupials. Pp.51–67. *in* "The biology of marsupials" ed. by B. Stonehouse and P. Gilmore. Macmillan Press.

Clemens, W. A. Jr., 1979. Marsupialia. Pp.192–220 *in* "Mesozoic mammals", ed. by J. A. Lillegraven, Z. Kielan–Jaworowska, and W. A. Clemens, Jr. University of California Press: Berkeley.

Crochet, J. Y., 1978. "Les marsupiaux du Tertiare d'Europe". Thèse Univ. Sci. et Techn: Languedoc, 2 vols. 360 p.

Crochet, J. Y., 1979a. Diversité systématique des Didelphidae (Marsupialia) Européens Tertiares. *Geobios* **12** (3): 365–78.

Crochet, J. Y., 1979b. Données nouvelles sur l'histoire paleogeographique des Didelphidae (Marsupialia). *C. R. Acad. Sc. Paris.* **288** (19): 1457–60, ser. D.

Fox, R. C., 1971. Marsupial mammals from the early Campanian Milk River Formation, Alberta, Canada. *in* "Early mammals" ed. by D. M. Kermack and K. A. Kermack. *Zool. J. Linn. Soc.* **50:** (Suppl. 1). 145–64.

Fox, R. C., 1979a. Mammals from the Upper Cretaceous Oldman Formation, Alberta. I. *Alphadon* Simpson. *Canad. Journ. Earth Sci.* **16:** 91–102.

Fox, R. C., 1979b. Mammals from Upper Cretaceous Oldman Formation Alberta. II. *Pediomys* Marsh (Marsupialia). *Canad. Journ. of Earth Sci.* **16:** 103–13.

Gregory, W. K., 1910. The orders of mammals. *Bull. Amer. Mus. Nat. Hist.* **3:** 1–524.

Hennig, W., 1950. "Grundzuge einer Theorie der phylogenetischen Systematik". Berlin: Deutscher Zentralverlag.

Jones, F., Wood, 1925. The mammalian toilet and some considerations arising from it. *Proc. Roy. Soc. Tasmania.* Pp.14–62.

Jones, F., Wood, 1923–1925. "The mammals of South Australia. Parts I–II". Government Printer: Adelaide.

Kirsch, J. A. W., 1977. The comparative serology of Marsupialia, and a classification of marsupials. *Aus. J. Zool., Suppl. Ser.* **52:** 1–152.

Marshall, L. G., 1976. Evolution of the Borhyaenidae, extinct South American predaceous marsupials. *U. Calif. Publ., Geo. Sci.* **117:** 1–89.

Marshall, L. G., 1977. Cladistic analysis of borhyaenoid, dasyuroid, didelphoid, and thylacinid (Marsupialia: Mammalia) affinity. *Syst. Zool.* **26:** (4): 410–25.

Reig, O., 1955. Noticia preliminar sobre la presencia de microbiotherinos viventes en la fauna Sudamericana. *Invest. Zool. Chil.* **2:** 121–30

Ride, W. D. L., 1964. A review of Australian fossil marsupials. *J. Proc. R. Soc. West Aust* **47** (4). 97–131.

Segall, W., 1969. The middle ear region of *Dromiciops. Acta Anat.* **72:** 489–501.

Simpson, G. G., 1972a. Mammalian fauna of the Hell Creek Formation of Montana. *Amer. Mus. Novitates* **267:** 1–7.

Simpson, G. G., 1941. The affinities of the Borhyaenidae. *Amer. Mus. Novit.* **1118:** 1–6.

Simpson, G. G., 1945. The principles of classification and a classification of mammals. *Bull. Amer. Mus. Nat. Hist.* **85:** vi–xvi, 1–350.

SIMPSON, G. G., 1948. The beginning of the age of mammals in South America. *Bull. Amer. Mus. Nat. Hist.* **91:** 1–232.

SIMPSON, G. G., 1961. "Principles of animal taxonomy". Columbia University Press: New York.

SIMPSON, G. G., 1970. The Argyrolagidae, extinct South American marsupials. *Bull. Mus. Comp. Zool.* **39** (1) 1–86.

SIMPSON, G. G., 1974. Notes on Didelphidae (Mammalia, Marsupialia) from the Huayquerian (Pliocene) of Argentina. *Amer. Mus. Novit.* **2259:** 1–15.

SIMPSON, G. G., 1975. Recent advances in methods of phylogenetic inference. Pp.3–19. *in* "Phylogeny of the primates: a multidisciplinary approach" ed. by W. P. Luckett and F. S. Szalay. Plenum Press New York.

SZALAY, F. S., 1976. Systematics of the Omomyidae (Tarsiiformes, Primates): taxonomy, phylogeny, and adaptations. *Bull. Amer. Mus. Nat. Hist.* **156:** 157–450).

SZALAY, F. S., 1977a. Ancestors, descendants, sister-groups, and testing of phylogenetic hypotheses. *Syst. Zool.* **26:** 12–18.

SZALAY, F. S., 1977b. Phylogenetic relationships and a classification of the eutherian Mammalia. Pp.317–74. *in* "Major patterns of vertebrate evolution" ed. by M. K. Hecht, P. C. Goody and B. M. Hecht. Plenum Press: New York.

SZALAY, F. S., 1981. Functional analysis and the practice of the phylogenetic method as reflected by some mammalian studies. *Amer. Zool.* **21:** 37–45.

SZALAY, F. S. AND DELSON, E., 1979. "Evolutionary history of the primates". Academic Press: New York. i–xiv, 1–580.

SZALAY, F. S. AND DRAWHORN, 1980. Evolution and diversification of the Archonta in an arboreal milieu. *In* "Tree Shrews". ed. by W. P. Luckett. Plenum Press: New York.

TEDFORD, R. H., 1974. Marsupials and the new paleogeography. *In* "Paleogeographic provinces and provinciality", ed. by C. A. Ross. *Soc. Econ. Pal. Min., Spec. Publ.* **21:** 109–26.

RESUMEN

51. Una nueva apreciación de la filogenia y clasificación de los marsupiales. Por Frederick S. Szalay.

Un detallado anàlisis del carácter de la morfolgía del tarso de casi todos los marsupiales vivientes y representativos del registro fósil, junto a un examen de la morfología dentaria y craneana, resultó en una nueva hipótesis filogenética y en una nueva clasificación evolutiva de los Metatheria. Se sostiene la monofilia de la especie patagónica *Dromiciops australis* y de todos los marsupiales australoasiáticos, la que es expresada por la cohorte Australidelphia. La cohorte Ameridelphia, incluyendo el orden Didelphida, es dividida en los subórdenes Didelphiformes (Pediomyidae y Didelphidae) y Borhyaeniformes (Borhyaenoidea, Caenolestoidea, Argyrolagoidea). La radiación australididelfiana es dividida en: 1) orden Dromiciopsia, formas con dentición poliprotodonta y patas con aprehensión diadactilida, con la morfología tarsal australidelfiana complejamente derivada; 2) orden Dasyurida, derivada de la dromi ciopsiana; y 3) orden Syndactyla con sus homólogamente derivados pies síndactilios, también originado de un antecesor dromiciopsiano. Varios grupos problemáticos pueden ser ubicados basados en la evidencia de los tarsos. Los tilacínidos no muestran especiales simifitudes compartidas, homólogamente derivadas, con ninguno de los Borhyaeniformes. Sus más recientes vinculaciones son con otros austradidelfianos, particularmente dentro de los Dasyurida. La ubicación de Bensley de los Notoryctidae en los Syndactyla es confirmada. El tarso distal, la presencia de metatarsales II y III de similares tamaños, un gran hallux, y la anterior dentición poliprotodonta sugieren derivación del ancestral suborden sindactilio, los Syndactyliformes.

52. GENETIC RELATIONSHIPS AMONG 32 SPECIES OF AUSTRALIAN DASYURID MARSUPIALS

P. R. BAVERSTOCK[1], M. ARCHER[2], M. ADAMS[1] AND B. J. RICHARDSON[3]

Isozyme electrophoresis of 32 loci in 90 individuals of 31 species of dasyurid marsupials was used to determine genetic relationships between and within genera. Cladistic analysis of the data revealed the following: 1) A phylogenetic scheme, based upon genetic data, largely follows the scheme based upon classical criteria. 2) The genus *Antechinus* is polyphyletic, with *A. rosamondae, A. bilarni,* and *A. macdonnellensis* forming a group separate from "true" *Antechinus.* 3) *A. bilarni* and *A. macdonnellensis* are nevertheless genetically very distinct from each other. 4) *Dasyuroides byrnei* and *D. cristicaudata* form a monophyletic group near *Dasyurus/Sarcophilus.* 5) *Ningaui* from Ayers Rock is specifically distinct from *Ningaui* from the west coast of S.A. 6) *Sminthopsis murina* from coastal N.S.W. is specifically distinct from *S. murina* from western S.A.

Pages 641–50 *in* CARNIVOROUS MARSUPIALS, ed. by M. Archer, February, 1982. Roy. Zool. Soc. New South Wales: Sydney, Australia.

ISOZYME electrophoresis can be a powerful taxonomic tool. It has been used, for example, to analyse sibling species complexes (e.g. Kirsch and Poole 1967, Adams *et al.* 1981), to determine specific status of allopatric populations (e.g. Baverstock *et al.* 1977), or for determining genetic relationships between species (Avise 1974). When analysed correctly and due account taken of differential rates of evolution, genetic relationships can in turn be used to infer cladistic relationships (Richardson *et al.* 1973, Mikevich and Johnson 1976, Baverstock *et al.* 1979).

The marsupial family Dasyuridae consists of about 50 recognised species, of which about 39 species in 10 genera occur in Australia (Kirsch and Calaby 1977, Archer 1978). Despite the fact that the group is currently the centre of extensive taxonomic revision (e.g. Archer 1975, 1976a, 1977, 1979, 1981 and 1982, and Van Dyck 1980), many problem areas remain, not the least of which are the generic relationships of the species *Dasyuroides byrnei* and *Dasycercus cristicauda* and the possible polyphyletic origin of the species commonly placed (e.g. Ride 1970) within *Antechinus.*

Kirsch (1977) conducted a series of serological studies of the family aimed primarily at elucidating intergeneric relationships. However, species of several genera (*Planigale* and *Ningaui*) were unavailable at the time for serological analysis. Moreover, as Kirsch (1977) admits, some of the serological techniques he used have since been shown to be rather imprecise and to have since been improved upon by more precise and quantitative techniques.

Chromosomal studies of dasyurids have so far contributed nothing to our understanding of relationships within the group. All species studied by conventional staining have a diploid number of 14 (Hayman and Martin 1974), and G-banding studies of seven genera (Rofe 1979) have confirmed the suspected conservative nature of chromosomal evolution in the family. Only the genus *Ningaui* shows evidence of intrageneric chromosomal diversity that might prove to be taxonomically useful (Baverstock and Gelder, unpublished).

We have initiated an extensive electrophoretic study of the dasyurids which is incomplete since some species have not yet become available to us, while for other species we have specimens from only one locality. Nevertheless, the results obtained to date reveal some interesting interspecific and intergeneric relationships and indicate areas where additional data are required.

The present study is based on 90 specimens of 32 species, an average of about three individuals per species. While some may be alarmed at such small numbers, it has been amply demonstrated that they are sufficient for determining genetic relationship using isozyme electrophoresis (Sarich 1977, Nei 1978, Gorman and Renzi 1979, Baverstock *et al.* 1981). Indeed it is far more important to score as many loci as possible. Of the loci screened, 32 gave staining of sufficient intensity and resolution in all but one species (*Dasyurus geoffroii*) to be scored confidently.

MATERIALS AND METHODS

Animals were bled by cardiac puncture under halothane anaesthesia; plasma was stored at –80°C and erythrocytes at –17°C in ethylene glycol (VandeBerg and Johnston 1977). Where possible, liver tissue was obtained by biopsy under halothane anaesthesia (a technique that proved satisfactory even for animals as small as *Planigale gilesi* and *P. tenuirostris*) and stored at –80°C.

[1] Laboratory of Animal Services, Institute of Medical and Veterinary Science, Frome Road, Adelaide, South Australia, Australia, 5001.
[2] School of Zoology, University of New South Wales, P.O. Box 1, Kensington, New South Wales, Australia, 2033.
[3] Bureau of Flora and Fauna, Department of Home Affairs and Environment, P.O. Box 1252, Canberra, A.C.T., Australia, 2601.

Table 1. Proteins stained for, number of loci, and E.C. Number.

Enzyme/Protein	Number of Loci	E.C. Number
Acid phosphatase	1	3.1.3.2
Adenosine deaminase	1	3.5.4.4
Alcohol dehydrogenase	1	1.1.1.1
Albumin	1	–
Adenylate kinase	1	2.7.4.3
Diaphorase	1	1.6.2.2
Enolase	1	4.2.1.11
Glyceraldehyde–3–phosphate dehydrogenase	1	1.2.1.12
Glyoxalase I	1	4.4.1.5
Glutamate–oxaloacetate transaminase	2	2.6.1.1
Glycollate oxidase	1	1.1.3.1
α Glycerophosphate dehydrogenase	1	1.1.1.8
Glucose–6–phosphate dehydrogenase	1	1.1.1.49
Glucose–phosphate isomerase	1	5.3.1.9
Haemoglobin	1	–
Isocitrate dehydrogenase	2	1.1.1.42
Lactate dehydrogenase	1	1.1.1.27
Malate dehydrogenase	2	1.1.1.37
Mannose–phosphate isomerase	1	5.3.1.8
Nucleoside phosphorylase	1	2.4.2.1
Peptidases	2	3.4.11 or 13
6–Phosphogluconate dehydrogenase	1	1.1.1.44
Phosphoglycerate kinase	1	2.7.2.3
Phosphoglucomutase	1	2.7.5.1
Pyruvate kinase	1	2.7.1.40
Superoxide dismutase	1	1.15.1.1
Triose–phosphate isomerase	1	5.3.1.1
Transferrin	1	–
	32	

The 32 loci scored are shown in Table 1; note that this list excludes esterases, since they cannot be scored with confidence in dasyurids. Electrophoretic and staining procedures have been described elsewhere (Baverstock *et al.* 1980).

The data were assembled into a species–locus matrix and analysed in three ways.

1) *Phenetic analysis.* Such an analysis clusters species according to their gross genetic similarities. Of the various clustering procedures available, we used the Average Linkage Cluster method (Cormack 1971). Such an analysis is useful for assessing the genetic diversity within genera and between allopatric populations of species, but cannot be used to infer cladistic relationships since it takes no account of differential rates of evolution at the genetic level.

2) *Wagner method.* The Wagner method does not assume constancy of rate of evolution but does assume maximum parsimony. It is therefore more likely to yield the correct branching sequence between taxa. We used the Distance Wagner Procedure of Farris (1972).

3) *Hennig analysis.* This technique determines the order of dichotomy within a group by producing taxa that have derived states in common, since primitive character–states held in common are not phylogenetically informative (Hennig 1966). In the present study, we assumed that the primary dichotomy was between *Sminthopsis, Antechinomys* and *Ningaui* on the one hand, and all other genera on the other, and used each as a sister–group to the other. The Hennig analysis was done by one of us (B. J. R.) without prior knowledge of results from either the Wagner analysis or the phenetic analysis.

RESULTS

GENETIC SIMILARITIES (Fig. 1)

a) *Family Level.* The dasyurids are about as diverse genetically as would be expected for a family of mammals, the most divergent groups being different at about 76% of loci.

b) *Generic Level.* The genera *Sminthopsis* and *Dasyurus* are genetically diverse, with a maximum of about 55% of loci showing differences. The genus *Antechinus* as currently recognised (e.g. by Ride 1970) is genetically almost as diverse as the whole family, the first branch occurring at 68% difference. However, the "true" *Antechinus* (*bellus, flavipes, godmani, leo, minimus, stuartii* and *swainsonii*) form a reasonably tight grouping at 30% difference. The genus *Planigale* is also a tight group at 17% difference.

c) *Species Level.* Within "true" *Antechinus*, good species such as *A. swainsonii* and *A. minimus* differ at 9% loci, and *A. bellus* and *A. leo* at 13% of loci. Species that have been somewhat confused in the past, such as *A. godmani, A. flavipes* and *A. leo* (Van Dyck 1980), differ from each other at over 20% of loci and are therefore quite distinct. *A. flavipes* groups with *A. stuartii*, but differs from it at 16% of loci.

ii) *S. macroura* from Sandringham, S.W. Qd did not differ from *S. macroura* from Mulgaria near Lake Torrens in S.A.

iii) *Ningaui* sp. from Ayers Rock, N.T. differed from *Ningaui* sp. from the west coast of S.A. at 28% of loci.

iv) *Planigale maculata* from Townsville, Qd did not differ from *P. maculata* from Humpty Doo, N.T.

v) *Antechinus flavipes* from Mt. Spec, Qd did not differ from *A. flavipes* from Adelaide, S.A., nor from specimens from Vic. and N.S.W.

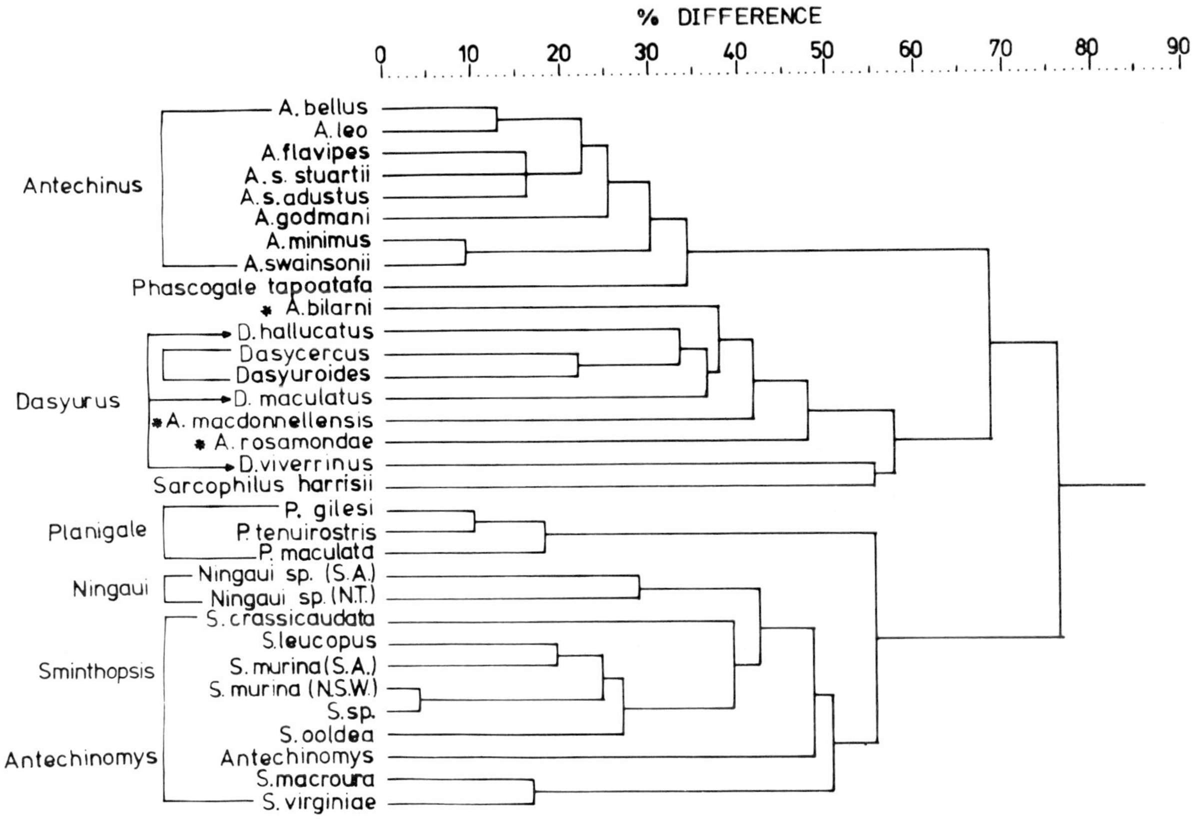

Fig. 1. Phenogram of genetic similarities based upon 32 loci in 32 taxa of Australian dasyurids established by the Average Linkage Cluster method.

Antechinus bilarni and *A. macdonnellensis*, two species that have been confused in the past, are genetically very different (41% difference).

Within *Sminthopsis*, *S. ooldea* and *S. leucopus* are distinct from *S. murina*. However, *S.* sp. 1 (a taxon which Archer (unpubl.) placed near but possibly distinct from *S. murina*) is genetically very similar to *S. murina* from N.S.W.

d) *Population Level.* For seven of the species studied we had representatives from widely differing geographic localities. They showed the following differences.

i) *Sminthopsis murina* from coastal N.S.W. differed from *S. murina* from the west coast of S.A. (a distance of some 2,000km) at 26% of loci.

vi) *A. s. stuartii* from south coastal N.S.W. differed from *A. s. adustus* from Ravenshoe, Qd at 15% of loci.

vii) *A. macdonnellensis* from Woodstock in the Kimberleys W.A. did not differ from *A. macdonnellensis* from Macdonnell Ranges, central Australia.

Wagner Tree

Inspection of the Wagner Tree (Fig. 2) reveals the following relationships:

i) The genus *Antechinus* as presently recognised is almost certainly polyphyletic, with *A. bilarni, A. macdonellensis* and *A. rosamon-*

dae standing apart from the "true" *Antechinus*. Moreover, *A. bilarni, A. macdonellensis* and *A. rosamondae* form a monophyletic assemblage.

ii) Within "true" *Antechinus*, three lineages are apparent:

a) *A. swainsonii* and *A. minimus.*

b) *A. godmani.*

c) *A. stuartii, A. bellus, A. leo* and *A. flavipes.*

dae falling close to *Dasyuroides byrnei, Dasycercus cristicauda, Dasyurus* spp. and *Sarcophilus harrisii.*

ii) *Phascogale tapoatafa* is monophyletic with the "true" *Antechinus* spp.

iii) *Planigale* spp. are distinct from all these genera.

iv) *Sminthopsis* spp. are monophyletic with respect to *Ningaui* spp.

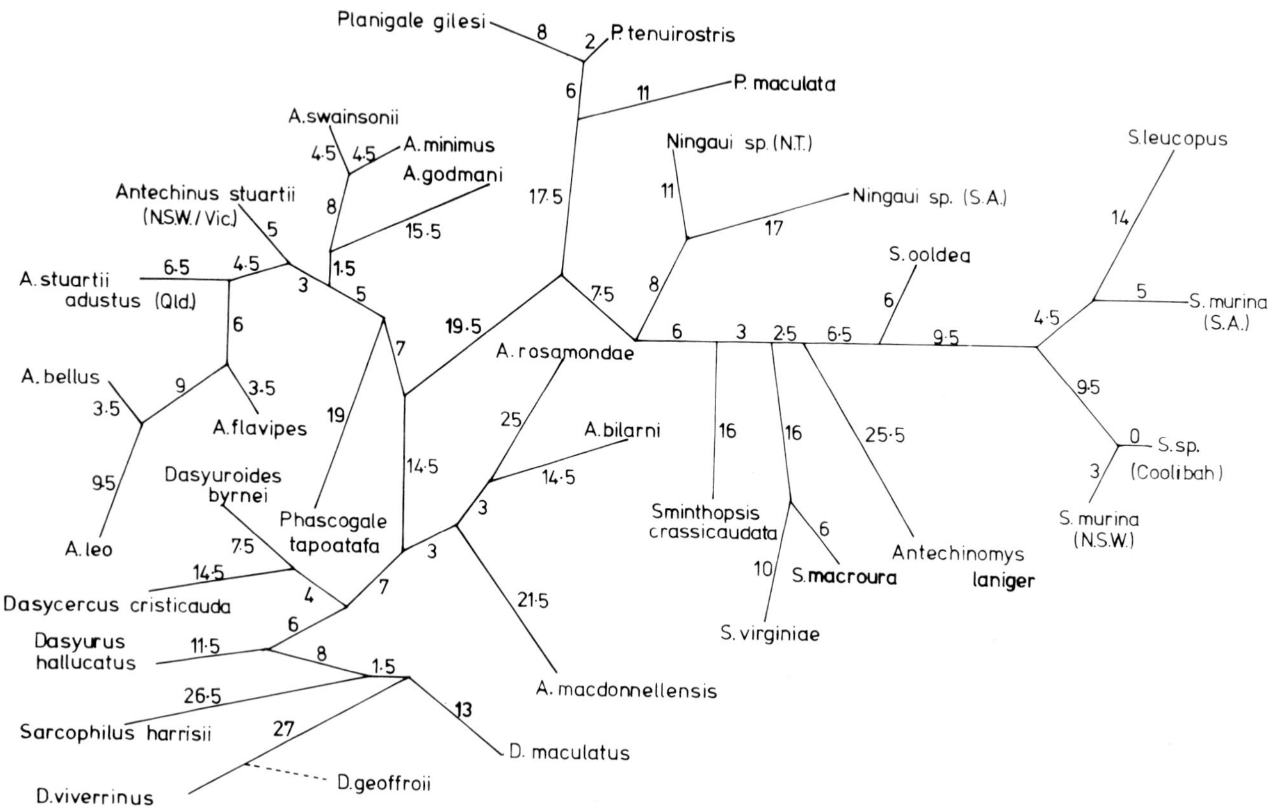

Fig. 2. Wagner Tree showing relationships between 32 taxa of Australian dasyurids. Branch lengths shown indicate percentage difference. Thus a distance of 10 indicates 3.2 loci.

iii) *Phascogale tapoatafa* is close to "true" *Antechinus* spp. but is outside that genus.

iv) *Dasyuroides byrnei* and *Dasycercus cristicauda* form a monophyletic group.

v) *Sarcophilus harrisii* occurs within the *Dasyurus* species group.

vi) *Planigale* spp. are distinct from both the *Antechinus* lineage and the *Sminthopsis* species lineages.

vii) *Antechinomys laniger* and *Ningaui* spp. are all near *Sminthopsis.*

Hennig Analysis (Fig. 3a,b)

The Hennig analysis shows quite clearly the following groupings) i.e. those based on at least two derived character–states in common):

i) *Antechinus* spp. are polyphyletic, with *A. bilarni, A. macdonnellensis* and *A. rosamon-*

Weaker arguments (i.e. where divergent states in common exceed convergent states in common by only one) can be made for the following:

i) Within "true" *Antechinus, A. minimus* and *A. swainsonii* form a monophyletic group, as do *A. bellus* and *A. flavipes.*

ii) *A. bilarni* is closer to *A. rosamondae* than either is to *A. macdonnellensis.*

iii) *Dasyuroides byrnei* and *Dasycercus cristicauda* form a monophyletic group.

iv) *Dasyurus* is a monophyletic genus.

v) The *S. murina* species group is monophyletic.

vi) *S. macroura* and *S. virginiae* form a monophyletic group.

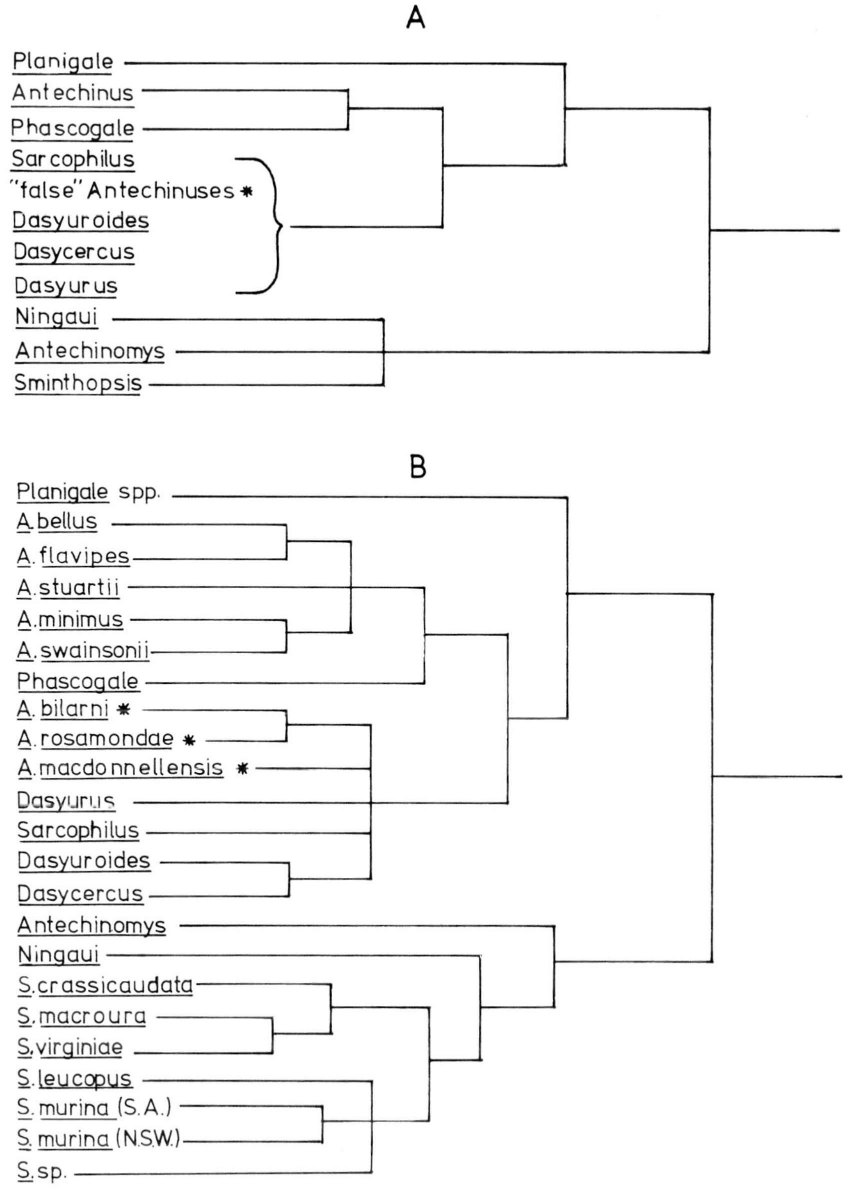

Fig. 3. Cladistic relationships between Australian dasyurids based upon derived character–states (synapomorphies) in common (Hennig Analysis) based upon a given primary dichotomy between species of *Ningaui, Antechinomys* and *Sminthopsis* on the one hand and those of all other genera on the other. 3A. Dichotomies where at least two synopomorphies indicate each dichotomy. 3B. Dichotomies where at least one synapomorphy indicates each dichotomy.

DISCUSSION

Population Differentiation

The primary aim of the present study was to determine the genetic relationships between species and between genera. For this reason we concentrated upon obtaining specimens of as many species as possible rather than obtaining specimens from as many localities as possible. If, however, geographically divergent representatives of a supposedly single species were available, they were included in the study. As a result, evidence for large differences between widely separated allopatric populations emerged in three cases:

i) *Sminthopsis murina* from the west coast of S.A. differed from *S. murina* from coastal N.S.W. at 26% of loci (lower 90% limit 15%).

ii) *Ningaui* sp. from Ayers Rock differed from *Ningaui* sp. from western S.A. at 28% of loci (lower 90% limit 18%).

iii) *Antechinus stuartii stuartii* from south coastal N.S.W. and *A.s. adustus* from north Qd differed at 15% of loci (lower 90% limit 8%).

Among rodents, populations that differ electrophoretically at more than 15% of loci are usually different species (Baverstock *et al.* 1977). While the same need not necessarily be true for dasyurids, there seems no reason to expect them to be any different. Moreover, the present study has shown that otherwise distinct species of dasyurids may differ by as little as 10% of loci.

Therefore the results suggest that *Sminthopsis "murina"* consists of at least two allopatric species and that this taxon should be studied electrophoretically in more detail.

The *Ningaui* specimens from the west coast of S.A. and the *Ningaui* specimens from Ayers Rock almost certainly belong to different species. The only two species of the genus described so far are *N. ridei* and *N. timealeyi* (Archer 1975), both from Western Australia, although Archer (1978) has indicated that more species may be represented among other as yet unassessed samples from the N.T., S.A., Qd, Vic. and N.S.W. Unfortunately, we did not have specimens of the two Western Australian species for direct electrophoretic comparison. Nevertheless it is clear that this genus should be thoroughly investigated with electrophoretic techniques.

Although *Antechinus stuartii stuartii* and *A. s. adustus* show genetic differentiation, the level of difference found is insufficient by itself to justify their recognition as full species.

There is no evidence of genetic differentiation between *Sminthopsis macroura* from Qd and S.A., *Planigale maculata* from Qd and N.T., *Antechinus flavipes* from Qd, N.S.W., Vic. and S.A., or *A. macdonnellensis* from the Kimberleys and central Australia.

Species Differentiation

Antechinus godmani, A. leo, A. flavipes and *A. stuartii* were considered by Tate (1947) to be one species, *A. flavipes*. Wakefield and Warnecke (1967) resurrected *A. godmani* and *A. stuartii*, a view amply supported by the present data. More recently, *A. leo* has been distinguished from *A. godmani* (Van Dyck 1980), to which it is otherwise morphologically somewhat similar. Again the genetic data show clear separation of these two species.

Antechinus bilarni was considered conspecific with *A. macdonnellensis* by Ride (1970). Kirsch and Calaby (1977), however, treated *A. bilarni* as a separate species. Genetically, *A. bilarni* differs greatly from *A. macdonnellensis, viz* at 41% of loci.

Until recently (e.g. Ride 1970), *S. ooldea* has been treated as no more than a subspecies of *S. murina*, as Troughton (1965) originally proposed. However, Archer (1981) considers it to be a full species. The genetic data supported the latter conclusion. There has also been some doubt about the specific distinctness of *S. leucopus* from *S. murina*, specific distinction being supported by Archer (1981) but not Troughton (1965). The present study shows that *S. leucopus* from Tasmania is quite distinct from *S. murina*.

Cladistic Relationships between Species and Genera

Based upon the genetic relationships, the Wagner Tree and the Hennig analysis, we have constructed a branching sequence (Fig. 4) which we believe is most compatible with the data. We would stress that this branching sequence is based entirely upon *genetic* data and takes no account whatever of other data. Moreover, it should be pointed out that, while electrophoretic comparison is clearly a tool of choice for discerning *close* relationships, the problem of homology would seem to be greater with increasing distances (e.g. Maxson and Maxson 1979). Therefore, the confidence that can be placed in our suggested cladogram will decrease with increasing genetic distances.

The genetic data suggest at least three major groups among Australian dasyurids tested — the *Sminthopsis* suprageneric group (*Sminthopsis, Antechinomys* and *Ningaui*), the *Planigale* group (containing only *Planigale*), and the *Antechinus* suprageneric group (*Antechinus, Phascogale, Dasyuroides, Dasycercus, Dasyurus* and *Sarcophilus*).

i) *Sminthopsis* Suprageneric Group

This lineage consists of the species of *Sminthopsis, Ningaui* and *Antechinomys*. The genetic data suggest that *Sminthopsis* may be paraphyletic with *Antechinomys*. Certainly the relationship is a close one. Ningaui, however, appears to be quite distinct from the *Sminthopsis* species we examined. Archer (1975, 1977) also recognises the probable close ıelationships of these genera on morphological criteria. Kirsch (1977), on serological grounds, recognises the close relationship between *A. laniger* and species of *Sminthopsis*, but *Ningaui* was not availabe to him.

Unfortunately, a number of species of *Sminthopsis* were not available to us (*S. hirtipes, S. granulipes, S. longicaudata, S. psammophila, S. butleri* and *S. douglasi*). For the species we had, we found three groupings:

1. *S. murina* group–*S. murina* (coastal N.S.W.), *S.* sp. 1 (near *murina*), *S. murina* (S.A.), *S. leucopus* and *S. ooldea*.
2. *S. virginiae* and *S. macroura*.
3. *S. crassicaudata*.

The *S. virginiae* and *S. murina* groups were already predicted on the basis of morphological criteria (Archer 1976 and 1980).

Fig. 4. Proposed cladistic relationships between 33 Australian dasyurids based entirely upon electrophoretic data. Broken lines indicate areas where the branching sequence is uncertain.

ii) *Planigale* Group

This consists only of the species of *Planigale*. It has been suggested that *Planigale* is a derivative of an *Antechinus*-like dasyurid (e.g. Troughton 1967, Ride 1970). However, Archer (1976a and b) has pointed out that *Planigale* has characters of both *Ningaui* and *Antechinus*, and was therefore less certain whether it belonged near the *Antechinus* group or the *Sminthopsis* group. The genetic data fails to resolve the question and for this reason we have left *Planigale* as a separate group based only on the genetic data. Unfortunately, there are no serological data because species of *Planigale* were not available to Kirsch (1977).

Within *Planigale*, *P. gilesi* and *P. tenuirostris* are genetically and cladistically closer to each other than either is to *P. maculata*. Archer (1976a) recognises these as three separate groups *viz P. maculata* with *P. novaeguineae*, *P. ingrami* with *P. tenuirostris*, and *P. gilesi* alone. Because we lacked *P. ingrami* and *P. novaeguineae*, we were unable to test these groupings.

Until quite recently, *P. maculata* was placed (although often with reservations) within *Antechinus* (e.g. Ride 1970). The genetic data show it to be very distinct from all species of *Antechinus* and very close to *P. gilesi* and *P. tenuirostris*, strongly supporting Archer's (1976a) reference of *maculata* to the genus *Planigale*.

iii) *Antechinus* Suprageneric Group

This suprageneric group is genetically more heterogeneous than either the *Sminthopsis* or *Planigale* lineage, and consists of more genera.

The first dichotomy gives the *Antechinus* group (i.e. "true" *Antechinus* species along with species of *Phascogale*) and the remaining species. The second dichotomy gives the "false" antechinuses on the one hand and species of *Dasyuroides*, *Dasyurus* and *Sarcophilus* on the other. The third dichotomy gives the species of *Dasyuroides* and *Dasycercus* in one group and those of *Dasyurus* and *Sarcophilus* in the other.

Kirsch (1977) found a similar relationship for the taxa he examined. Thus, serologically, species of *Dasyurus* group close to *Sarcophilus harrisii, Phascogale* species group close to *Antechinus stuartii,* and these four generic assemblages cluster before any clusters with the *Sminthopsis/Antechinomys* suprageneric group. His data for *Dasyuroides byrnei* were incomplete, but linked it with species of *Dasyurus* and *Sarcophilus*. His data for *Dasycercus cristicauda* were even more incomplete, but seemed to exclude it from this group. The genetic data appear to contradict this latter conclusion based on serology.

Kirsch (1977) also found that *Antechinus rosamondae* was serologically closer to *Dasyurus* then to species of true *Antechinus* (e.g. *A. flavipes*), but that *A. macdonnellensis* grouped closer to the species of true *Antechinus*. Certainly *A. macdonnellensis* and *A. rosamondae,* although apparently in the same group, are not genetically close, but there is no genetic evidence that *A. macdonnellensis* links with the species of true *Antechinus*.

True *Antechinus* Species

Within the true *Antechinus,* three groups emerge: *A. swainsonii* with *A. minimus; A. bellus* with *A. leo; A. flavipes* with *A. stuartii;* and *A. godmani* alone. The distinctness of *A. swainsonii* and *A. minimus* as a single group has been anticipated on morphological grounds (e.g. Archer 1976b). The degree of distinctness of *A. godmani,* however, has not been recognized.

False *Antechinus* Species

This group contains *Antechinus macdonnellensis, A. bilarni* and *A. rosamondae.* On the basis of genetic data, it is cladistically closer to species of *Dasyurus* than to those of "true" *Antechinus*. Recognition of at least one distinct generic rank for the species of this group is warranted. *A. macdonnellensis, A. bilarni* and *A. rosamondae* are genetically more distinct from each other than say, *Dasycercus cristicauda* and *Dasyuroides byrnei.* For this reason, there may be grounds for recognising three separate genera. Other species possibly referable to this group but unavailable for study are *A. apicalis* and *A.* "ningbing" (an as yet undescribed species).

Three New Guinean species of *Antechinus* (*A. melanurus, A. naso* and *A. wilhelmina*) were also not available to us for this study, and it will be of interest to determine within which group (the true antechinuses or false antechinuses) each of these species belong.

Dasyuroides and *Dasycercus* Species

Interpretation of interrelationships of the species of these two genera has been in doubt. The genetic data show that they are a monophyletic group, possibly close to the species of *Dasyurus* and *Sarcophilus*.

Dasyurus and *Sarcophilus* Species

Dasyurus viverrinus and *D. geoffroii* appear to be sister–species, a view supported by Kirsch's (1977) serological data as well as external, dental and basicranial morphology. Serologically, *D. hallucatus* is evidently closer to this pair than is *D. maculatus* (Kirsch 1977). The genetic data suggest the opposite, that *D. maculatus* is closer to this pair.

Dasyurus species may be paraphyletic with respect to *Sarcophilus harrisii,* although the data supporting this view are weak. If it is the case, this may be support for recognition of *Satanellus* for *hallucatus*. This view is also supported by palaeontological data (see Archer 1982).

CONCLUSION

There seems little doubt that the results of the present study have contributed significantly to our understanding of evolutionary relationships within Australian dasyurids. Principal conclusions include: 1, The clear polyphyletic nature of *Antechinus* (*sensu latu*); 2, The relationship of *Dasyuroides byrnei* and *Dasycercus cristicauda*; 3, The confirmation of the generic placement of *Planigale maculata;* 4, Clarification of the status and relationships of *Sminthopsis ooldea, S. leucopus, Antechinus godmani* and *A. leo;* 5, The indications of genetic diversity within *S. murina* and within *Ningaui.*

Archer (1982), after considering the results reported here, those reported by others such as Woolley (1982) and those of his own studies on morphology, has proposed that *bilarni* and *apicalis* be placed in the genus *Parantechinus* Tate, *macdonnellensis* and the "ningbing" Antechinus in *Pseudantechinus* Tate, and *rosamondae* in the monotypic genus *Dasykaluta* Archer. Our results, to the extent that they can do so, support this proposal.

It should be borne in mind that approximately 9 species of Australian dasyurids were not available to us, nor were any of the 11 New Guinean species. Moreover, the indications of genetic heterogeneity within *S. murina* and *Ningaui* make it desirable to study more material of these groups. Finally, while small numbers of individuals of each taxon are sufficient for assessing the genetic relationships between taxa, for widespread taxa, it is highly desirable to consider specimens from several parts of the range to determine the possible specific status of geographically distant or isolated forms. For this reason, we would like to use this opportunity to request blood and/or liver from any animal whose relationships may be in doubt. These tissues can be sampled from live animals with only a small risk of death of the animal.

ACKNOWLEDGEMENTS

We thank Drs. J. A. W. Kirsch and D. Briscoe for critical comments on the manuscript. Drs. M. Nagy and J. Reide provided technical assistance. For assistance with collecting, supplying or loaning animals or supplying tissues we thank Dr. H. Aslin, Dr. R. Baker, R. Baverstock, Dr. R. Begg, Dr. P. Bird, Mr. C. Moritz, Dr. K. Johnson, Dr. D. Ling, Dr. S. Nichol, Mr. D. Read, Mr. D. Rofe, Dr. R. Sinclair, Dr. C. Tidemann, Mr. S. Van Dyck, Dr. H. Wapstra, Dr. C. Watts, Mr. R. Whitford, Dr. P. Woolley, Dr. P. Johnson, Dr. J. Winter and Ms. J. Brislow. Financial support for the project was provided by a University of New South Wales Biological Sciences Faculty Grant to M. Archer. Ms. D. Thomas, Ms. J. Johansen and Ms. G. O'connor typed drafts of the manuscript.

REFERENCES

Adams, M., Baverstock, P. R., Tidemann, C. R. and Woodside, D. P., 1981. Large genetic differences between three sibling species of Australian *Eptesicus* (Microchiroptera). *Heredity* (in press).

Archer, M., 1975. *Ningaui*, a new genus of tiny dasyurids (Marsupialia) and two new species, *N. timealeyi* and *N. ridei*, from arid Western Australia. *Mem. Qd Mus.* **17:** 287–49.

Archer, M., 1976a. Revision of the marsupial genus *Planigale* Troughton (Dasyuridae). *Mem. Qd Mus.* **17:** 341–465.

Archer, M., 1976b. The dasyurid dentition and its relationships to that of didelphids, thylacinids, borhyaenids (Marsupicarnivora), and peramelids (Peramelina, Marsupialia). *Aust. J. Zool., Suppl. Ser.* **39:** 1–34.

Archer, M., 1977. Revision of the dasyurid marsupial genus *Antechinomys* Krefft. *Mem. Qd Mus.* **18:** 17–29.

Archer, M., 1978. The status of the Australian dasyurids, thylacinids and myrmecobiids. Pp.29–43 *in* "The status of endangered Australasian wildlife" ed. by M. J. Tyler. Royal Zoological Society of South Australia Inc.: Adelaide.

Archer, M., 1980. Two new species of *Sminthopsis* Thomas (Dasyuridae: Marsupialia) from northern Australia, *S. butleri* and *S. douglasi*. *Aust. Zool.* **20:** 327–45.

Archer, M., 1981. Results of the Archbold Expeditions. No. 104. Systematic revision of the marsupial dasyurid genus *Sminthopsis* Thomas. *Bull. Amer. Mus. Nat. Hist.* **168:** 61–224.

Archer, M., 1982. Review of the dasyurid (Marsupialia) fossil record, integration of data bearing on phylogenetic interpretation, and suprageneric classification. Pp. 397–443 *in* "Carnivorous marsupials", ed. by M. Archer. Roy. Zool. Soc. New South Wales: Sydney.

Avise, J. C., 1974. Systematic value of electrophoretic data. *Syst. Zool.* **23:** 465–81.

Baverstock, P. R., Watts, C. H. S. and Cole, S. R., 1977. Electrophoretic comparisons between allopatric populations of five Australian pseudomyine rodents (Muridae). *Aust. J. Biol. Sci.* **30:** 471–85.

Baverstock, P. R., Cole, S. R., Richardson, B. J. and Watts, C. H. S., 1979. Electrophoresis and cladistics. *Syst. Zool.* **28:** 214–19.

Baverstock, P. R., Watts, C. H. S., Adams, M. and Gelder, M., 1980. Chromosomal and electrophoretic studies of Australian *Melomys* (Rodentia: Muridae). *Aust. J. Zool.* **28:** 553–74.

Baverstock, P. R., Watts, C. H. S., Adams, M. and Cole, S. R., 1981. Genetical relationships among Australian rodents (Muridae). *Aust. J. Zool.* **29:** 289–303.

Cormack, R. M., 1971. A review of classification. *Roy. Stat. Soc. J., Ser. A.* **134:** 321–53.

Farris, J. S., 1972. Estimating phylogenetic trees from distance matrices. *Amer. Nat.* **106:** 645–68.

Gorman, G. C. and Renzi, J. Jr., 1979. Genetic distance and heterozygosity estimates in electrophoretic studies: effects of sample size. *Copeia* **1979:** 242–9.

Hayman, D. L. and Martin, P. G., 1974. Mammalia I: Monotremata and Marsupialia, Vol. 4: Chordata 4. *In* "Animal cytogenetics" ed. by B. John. Gebruder Borntraeyer: Berlin–Stuttgart.

Hennig, W., 1966. "Phylogenetic systematics". University of Illinois Press: Urbana.

Kirsch, J. A. W., 1977. The comparative serology of Marsupialia, and a classification of marsupials. *Aust. J. Zool., Suppl. Ser. A.* **52:** 1–152.

Kirsch, J. A. W. and Poole, W. E., 1967. Serological evidence for speciation in the Grey Kangaroo, *Macropus giganteus* Shaw, 1790: (Marsupialia: Macropodidae). *Nature* **215:** 1097–8.

Maxson, L. R. and Maxson, R. D., 1979. Comparative albumin and biochemical evolution in plethodontid salamanders. *Evolution* **33:** 1057–62.

Mikevich, M. F. and Johnson, M. S., 1976. Congruence between morphological and allozyme data in evolutionary inference and character evolution. *Syst. Zool.* **25:** 260–70.

Nei, M., 1978. Estimation of average heterozygosity and genetic distance from a small number of individuals. *Genetics* **89:** 583–90.

Richardson, B. J., Johnston, P. G., Clarke, P. and Sharman, G. B., 1973. An evaluation of electrophoresis as a taxonomic method using comparative data from the Macropodidae (Marsupialia). *Biochem. Sys.* **1:** 203–9.

Ride, W. D. L., 1970. "A guide to the native mammals of Australia". Oxford University Press: Melbourne.

Rofe, R. H., 1979. G–banding and chromosomal evolution. Unpublished PhD thesis: University of Adelaide, Adelaide.

Sarich, V. M., 1977. Electrophoresis in evolutionary studies: Rates, sample sizes, and the neutrality hypothesis. *Nature* **265:** 24–8.

TATE, G. H. H., 1947. On the anatomy and classification of the Dasyuridae (Marsupialia). *Bull. Amer. Mus. Nat. Hist.* **88:** 97–156.

TROUGHTON, E. LEG., 1965. A review of the marsupial genus *Sminthopsis* (Phascogalinae) and diagnoses of new forms. *Proc. Linn. Soc. N.S.W.* **89:** 309–21.

VANDEBERG, J. L. AND JOHNSTON, P. G., 1977. A simple technique for long-term storage of erythrocytes for enzyme electrophoresis. *Biochem. Genet.* **15:** 213–5.

VAN DYCK, S., 1980. The Cinnamon Antechinus, *Antechinus leo* (Marsupialia: Dasyuridae), a new species from the vine-forests of Cape York Peninsula. *Aust. Mammal.* **3:** 5–17.

WAKEFIELD, N. A. AND WARNEKE, R. M., 1967. Some revision in *Antechinus* (Marsupialia) — 2. *Vic. Nat.* **84:** 69–99.

RESUMEN

52. Las relaciones genéticas entre los Dasyuridae reveladas por la electroforesis de la isozima. Por P. R. Baverstock, M. Archer, M. Adams, y B. J. Richardson.

Para determinar las relaciones genéticas entre y dentro de los géneros marsupiales dasyuridos se usó la electroforesis de la isozima de 32 loci en 90 individous de 31 especies. El análisis cladístico de los datos reveló lo siguiente: 1) Un esquema filogenético basado en los datos genéticos concuerda considerablemente con el esquema basado en los criterios clásicos. 2) El género *Antechinus* es polifilético, con *A. rosamondae*, *A. bilarni* y *A. macdonnellensis* integrando un grupo separado del "verdadero" *Antechinus*. 3) Sin embargo, *A. bilarni* y *A. macdonnellensis* son entre sí genéticamente muy distintas. 4) *Dasyuroides byrnei* y *D. cristicaudata* forman un grupo monofilético cercano a *Dasyurus*/*Sarcophilus*. 5) *Ningaui* de Ayers Rock específicamente distinta de *Ningaui* de la costa oeste de South Australia. 6) *Sminthopsis murina* de las costas de N.S.W. es específicamente distinta de *S. murina* de la costa oeste de South Australia.

53. STEROID BINDING PROTEINS IN THE PLASMA OF DASYURID MARSUPIALS

A. J. BRADLEY[1]

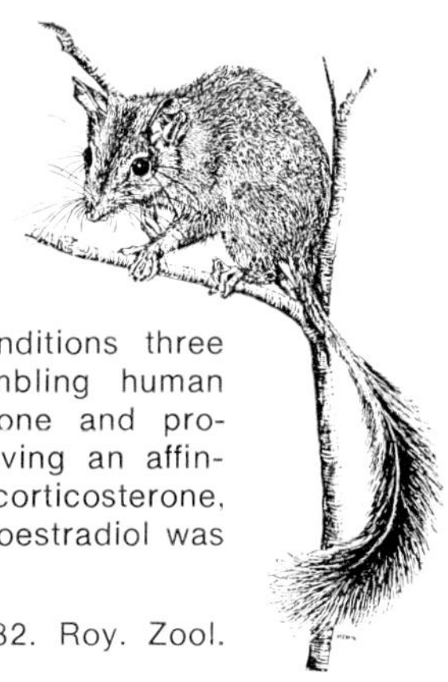

The binding of cortisol, progesterone, testosterone, oestradiol and aldosterone was examined in 29 marsupial species by equilibrium dialysis and/or steady state polyacrylamide gel electrophoresis. The association constants ($K\alpha$ 36°C) and binding capacities ($[\Sigma\rho]$) for cortisol and testosterone were determined by micro–dialysis to be from 2.7 to 9.6 x 10^{7} M^{-1} and 0.3 to 3.7 x 10^{-7} M respectively for cortisol and 1.2 to 8.3 x 10^{8} M^{-1} and 0.3 to 4.0 x 10^{-7} M for testosterone. High affinity binding for cortisol was detected in all 29 species studied, but testosterone was bound with high affinity in only 10 species. By polyacrylamide gel electrophoresis under steady state conditions three electrophoretically distinct steroid transport proteins were detected, one resembling human corticosteroid binding globulin (CBG) having an affinity for cortisol, corticosterone and progesterone, a second resembling human sex hormone binding globulin (SHBG) having an affinity for testosterone, and albumin which had an affinity for testosterone, cortisol, corticosterone, progesterone, aldosterone and oestradiol. High affinity binding for aldosterone and oestradiol was not detected.

Pages 651–7 *in* CARNIVOROUS MARSUPIALS, ed. by M. Archer, February, 1982. Roy. Zool. Soc. New South Wales: Sydney, Australia.

IT IS COMMONLY accepted that all classes of steroid hormones are transported in the plasma in the form of complexes with binding proteins (Westphal 1971). Since many of these steroid protein complexes render the steroid hormone biologically inactive, it may be appreciated that simple measurements of the total quantity of a steroid in the blood are an inadequate measure of the hormone stimulation of a target tissue. The importance of measurements other than total quantity has been aptly demonstrated for a number of steroid hormones under laboratory conditions (Slaunwhite, Lockie, Back and Sandberg 1962; Sandberg, Rosenthal, Schneider and Slaunwhite 1966; Oppenheimer 1968) and for a field population of marsupials (Bradley, McDonald and Lee 1980).

The biological role of the steroid transport proteins has been variously reported as the direction of hormone signals toward organs having protein permeable vascular beds (Keller, Richardson and Yates 1969), the protection of globulin bound steroids from hepatic catabolism and the damping out of rapid fluctuations in active unbound steroid concentration (Tait and Burstein 1964; Baird, Horton, Longcope and Tait 1969). It is considered that all steroid binding protein, including those with low association constants, have a part to play in regulating the metabolism of steroid hormones (Idler and Truscott 1972).

While steroid hormone binding proteins resembling the corticosteroid–binding globulin (CBG) and sex hormone binding globulin (SHBG) reported in human plasma (Daughaday 1958; Mercier–Bodard, Alfsen and Baulieu 1970) have also been found both in other eutherian mammals and non–mammalian vertebrates (Seal and Doe 1965, 1966; Westphal 1967, 1971; Idler 1972; Corvol and Bardin 1973; Sernia 1978), there is a paucity of information on these proteins in marsupials. Seal and Doe (1965), in an investigation of species distribution of CBG, reported finding glucocorticoid binding in the plasma of the Opossum, *Didelphis marsupialis,* and the Red Kangaroo, *Macropus rufus.* Other reports of CBG like systems in marsupials include the Brush–tailed Possum, *Trichosurus vulpecula* (Khin Aye Than and McDonald, 1976), the Brown Antechinus, *Antechinus stuartii* (Bradley, McDonald and Lee 1976, 1980), the Quokka, *Setonix brachyurus* (McDonald and Bradshaw 1977), and the Opossum, *Didelphis marsupialis* (Cook, Sutterlin, Graber and Nalbandov 1977). These authors also reported the absence of SHBG in the Opossum plasma. More recently, plasma from 29 different marsupial species was examined by Sernia, Bradley and McDonald (1979). They found that all species studied that CBG activity but that less than half of the species had SHBG activity.

Prior to this study the only other report of SHBG in marsupial plasma was for a kangaroo of unstated species (Wenn, Kamberti, Keyvanjah and Johannes 1977).

This study was undertaken to extend the available information on marsupial steroid hormone binding proteins.

MATERIALS AND METHODS

Blood Samples

Blood samples were taken from either laboratory held animals or animals captured in the

[1] Department of Zoology, University of Western Australia, Nedlands, Western Australia, Australia, 6009.

field, by orbital sinus puncture in the case of the smaller marsupials, or from either the tail vein or via cardiac puncture in larger marsupials. Blood was immediately centrifuged and the plasma was removed, frozen and stored at –20°C until required. Before use endogenous steroids were removed by incubation of plasma with charcoal (Norit A, 50 mg/ml) for 40 min at 37°C.

Equilibrium Dialysis

The association constant (K_a) and binding capacity ($[\Sigma\rho]$) for cortisol and testosterone were measured by the microdialysis method of Englund, Huberman, Jovin and Kornberg (1969) using the modifications described by Sernia, Bradley and McDonald (1979). K_a and $[\Sigma\rho]$ were calculated from Scatchard plots (Scatchard 1949, Tait and Burstein 1964) after correction of the data for low affinity binding (Sandberg, Rosenthal, Schneider and Slaunwhite 1966).

Steady–State Polyacrylamide Gel Electrophoresis

The technique of Ritzen, French, Weddington, Nayfeh and Hansson (1974) was used since it makes possible the detection of both high and low affinity binding proteins.

Charcoal treated plasma was diluted 10–fold with 10% glycerol in phosphosaline buffer (0.05 M pH 7.4) and 50 μl samples were incubated with labelled steroid (3, 5, 10 nM) for 2 h at 4°C. Labelled steroid was incorporated into 7% polyacrylamide gels onto which the preincubated sample was transferred. Unlabelled steroid (50 times the concentration of labelled steroid) was incorporated into gels on which were run samples previously incubated with unlabelled steroid as described by Cook, Hunter and Kelly (1977). Gels were run at 4°C at 1mA per gel for 3–4 h using a small quantity of bromophenol blue in the upper buffer chamber which marked the migration of both albumin and the free dye front. Protein mobilities (Rm) in the gels were expressed relative to the mobility of the free dye front.

Gels were sliced and counted as described in Serina *et al.* (1980).

Materials

The Radiochemical Centre, Amersham, U.K., supplied [1, 2 – ^{3}H] progesterone (sp. act. 47 Ci/m mol), [1 – 2 – ^{3}H] testosterone (sp. act. 58 Ci/m mol), [1, 2 –^{3}H] cortisol (sp. act. 49 Ci/m mol), [6, 7 – ^{3}H] oestradiol (sp. act. 41 Ci/m mol) and [1, 2, 6, 7 – ^{3}H] aldosterone (sp. act 93 Ci/m mol). All of these radioactive steroids were purified before use. All other chemicals were of A R grade. A toluene scintillation fluid was used for non–aqueous samples and Instagel (Packard, Australia) for aqueous samples. Radioactivity was measured in Packard Tricarb 3320 scintillation counter and quenching was measured by external standardisation.

RESULTS

Equilibrium Dialysis

Dialysis of plasma proteins using cortisol, testosterone, oestradiol and aldosterone was carried out in 29 species which included 16 species of dasyurids. High affinity binding of cortisol was detected in all species studied. The association constant (K_a) for cortisol binding fell between 2.7 x 10^7M^{-1} in *Antechinus flavipes* and 9.6 x 10^7M^{-1} in *Dasyurus maculatus*. The binding capacity ($[\Sigma\rho]$) fell between 0.3 x 10^{-7}M for *Sminthopsis leucopus* and *Antechinus stuartii*, and 3.7 x 10^{-7}M for *Macropus eugenii*. In at least two species, *A. stuartii* and *A. flavipes*, there were seasonal variations in binding capacity for cortisol, the variation in *A. stuartii* spanning almost the whole range mentioned above. Only about one–third of the species examined exhibit high affinity binding for testosterone (SHBG). High affinity binding was not detected in dasyurids, while only about a quarter of the diprotodontans were found to lack such binding proteins. The range of association constants was from 1.2 x 10^8 (*Macropus fuliginosis*) to 8.3 x 10^8M^{-1} (*Pseudocheirus peregrinus*) and binding capacity ranged from 0.3 x 10^{-7}M (*Macropus rufus*) to 4 x 10^{-7}M (*Trichosurus arnhemensis*). High affinity binding for oestradiol or aldosterone was not detected in any of the species studied.

Steady–State Polyacrylamide Gel Electrophoresis

Plasma from 18 species which had been preincubated at 4°C with radioactive and/or unlabelled steroid was run as described. The results of the steady–state polyacrylamide gel electrophoresis are shown in Fig. 1 and Table 2.

The Rm values for CBG were between 0.45 in *Antechinus swainsonii* and 0.62 in *Sarcophilus harrisii* (Table 2). Results using this method confirm the presence of CBG in all the species studied in addition to indicating the presence of multiple CBGs in the plasma of *Macropus irma* and *Setonix brachyurus*. In these two animals cortisol was not distributed equally between the two CBGs; the CBG with the higher Rm carried most of the steroid (Fig. 1). The saturable nature (i.e. the low binding capacity) of the CBG binding was shown by the displacement of cortisol and corticosterone binding from the CBG by an excess of unlabelled steroid in all species studied. At high steroid concentrations, albumin was the most significant steroid binding protein.

Table 1. Binding capacity ([Σρ]) and association constant (Kα, 36°) for the binding of cortisol and testosterone in the plasma of 29 marsupial species.

Species	Cortisol		Testosterone	
	Kα x 10⁻⁷ M⁻¹	[Σρ] x 10⁷ M	Kα x 10⁻⁸ M⁻¹	[Σρ] x 10⁷ M
Order: Polyprotodonta				
Dasyuridae				
Antechinomys spenceri (♂)	7.8	1.2	A	
Antechinus flavipes	2.7	0.5 – 3.6 (Seasonal variation)	A	
Antechinus rosamondae (♂)	4.2	1.5	A	
Antechinus stuartii	7.3	0.3 – 3.6 (Seasonal variation)	A	
Antechinus swainsonii	4.7	2.4	A	
Dasyuroides byrnei	8.2	1.3	A	
Dasyurus geoffroii (♀)	5.2	2.7	A	
Dasyurus maculatus (♂)	9.6	0.6	A	
Phascogale tapoatafa	6.2	1.0	A	
Planigale maculata (♂)	6.8	1.4	A	
Sarcophilus harrisii	8.7	1.2	A	
Sminthopsis crassicaudata	7.1	1.1	A	
Sminthopsis leucopus (♂)	3.8	0.3	A	
Sminthopsis macroura (♀)	7.5	1.8	A	
Sminthopsis murina	8.3	0.9	A	
Sminthopsis ooldea (♂)	5.7	0.9	A	
Order: Diprotodonta				
Macropodidae				
Bettongia penicillata (♀)	6.2	1.9	4.6	1.1
Macropus eugenii	6.5	3.7	A	
Macropuss fuliginosus (♂)	7.2	1.2	1.2	2.8
Macropus irma (♂)	5.3	1.6	A	
Macropus rufus (♂)	5.0	1.1	6.9	0.3
Setonix brachyurus	4.8	1.2	2.4	0.3
Thylogale billardierii (♀)	4.3	1.5	5.4	0.6
Wallabia bicolor (♂)	5.7	1.0	4.7	1.4
Wallabia rufogriseus (♀)	6.0	1.6	1.6	0.4
Phalangeridae				
Trichosurus arnhemensis (♀)	3.2	1.5	0.6	4.0
Trichosurus vulpecula	4.8	1.2	2.4	1.3
Petauridae				
Pseudocheirus peregrinus (♂)	5.2	1.2	8.3	0.4
Vombatidae				
Vombatus ursinus (♂)	3.8	1.5	A	

A. High affinity binding absent. Both sexes were examined except where indicated. In no case was high affinity binding of either oestradiol or aldosterone detected.

Progesterone bound to CBG in all species studied (Table 2), but again it could be displaced in the presence of excess unlabelled steroid. SHBG was detected in plasma from two of the three macropodid species studied. *Bettongia penicillata* (Rm = 0.31) and *Setonix brachyurus* (Rm = 0.31), but could not be detected in *Macropus irma* (Table 2 and Fig. 1). In species which lacked SHBG, competitive displacement of cortisol from CBG was investigated in the presence of 50–fold excess of unlabelled testosterone; in all cases there was little or no displacement of cortisol from CBG binding sites (Fig. 1). In *Dasyurus geoffroii* where significant binding of testosterone to CBG was detected (Table 2) in the absence of cortisol, the animal was suffering from a mammary tumor and may have had an abnormally high blood protein concentration.

Plasma albumin was found to have Rm values between 0.67 in *Antechinus rosamondae* and 0.79 in *Phascogale tapoatafa*. This protein bound all of the steroids examined.

DISCUSSION

These results confirmed the previously reported presence of CBG and SHBG activity in the plasma of a number of marsupials, and also support the conclusion of Corvol and Bardin (1973) that there appears to be a more limited species distribution of SHBG than of CBG. It is apparent that the binding affinity, steroid specificity and electrophoretic mobility of marsupial CBG, SHBG and albumin are quite similar to those found in eutherian mammals, but

Table 2. Electrophoretic mobility of marsupial albumin, CBG and SHBG in various species under steady-state conditions.

Species	Relative Mobility		
	Albumin	CBG	SHBG
Order: Polyprotodonta			
Dasyuridae			
Antechinus flavipes	0.76	0.55	A
Antechinus rosamondae	0.67	0.50	A
Antechinus stuartii	0.77	0.51	A
Antechinus swainsonii	0.70	0.45	A
Dasyuroides byrnei	0.76	0.59	A
Dasyurus geoffroii	0.71	0.51**	A
Dasyurus maculatus	0.70	0.52	A
Phascogale tapoatafa	0.79	0.55	A
Planigale maculata	0.68	0.47	A
Sarcophilus harrisii	0.77	0.62	A
Sminthopsis crassicaudata	0.76	0.57	A
Sminthopsis leucopus	0.68	0.52	A
Sminthopsis macroura	0.76	0.47	A
Sminthopsis murina	0.70	0.56	A
Sminthopsis ooldea	0.68	0.50	A
Order: Diprotodonta			
Macropodidae			
Bettongia penicillata	0.72	0.59	0.31
Macropus irma	0.70	0.49, 0.61*	A
Setonix brachyurus	0.73	0.47, 0.57*	0.31

SHBG, bound testosterone but not oestradiol. CBG, bound cortisol, corticosterone and progesterone in all species. Albumin, bound all steroids including aldosterone. *, CBG to which most steroid was bound. **, Significant binding of testosterone to CBG.

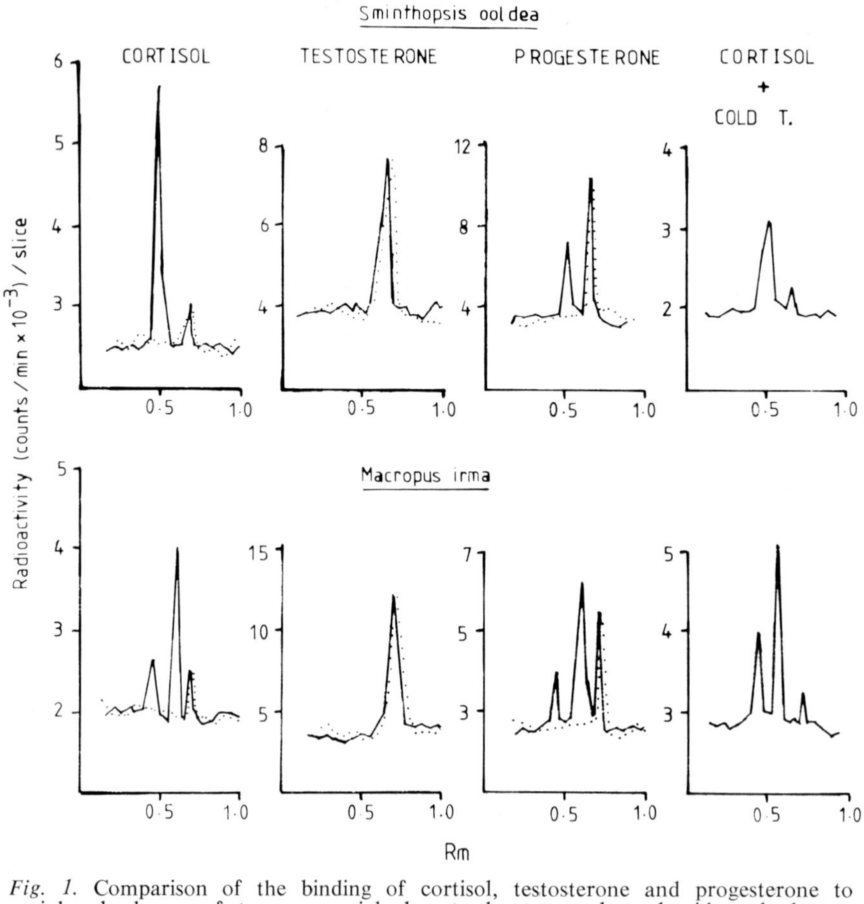

Fig. 1. Comparison of the binding of cortisol, testosterone and progesterone to peripheral plasma of two marsupials by steady-state polyacrylamide gel electrophoresis. Samples were incubated with 5 nM steroid and processed on gels containing 5 nM steroid (—). Duplicate samples, incubated and run in the same way, but with 250 nM unlabelled steroid included in the incubation medium and gel are represented by the dotted line (. . . .) where displacement occurred. Protein mobility (Rm) relative to the bromophenol blue dye front appears on the horizontal axis.

as pointed out by Sernia, Bradley and McDonald (1979), binding capacity for glucocorticoids is generally low, which may be related to the generally low plasma glucocorticoid concentrations in marsupials (Oddie, Blaine, Bradshaw, Coghlan, Denton, Nelson and Scoggins 1976; McDonald 1977; McDonald and Waring 1979). Values for Ka and [Σρ] are in agreement with previously published values (Khin Aye Than and McDonald 1975; McDonald and Bradshaw 1977; and Sernia *et al.* 1979). The double CBG system detected in *Macropus irma* is unusual, but not unique. Sernia *et al.* (1979) detected a similar system in four species from the families Peramelidae, Macropodidae and Phascolarctidae. Presumably the two proteins would have different affinity constants and binding capacities for bound steroids. In the case of *Macropus irma* and *Setonix brachyurus* the CBG with the lower Rm appears to be very readily saturable, so the apparent Ka value probably refers to the CBG with the higher Rm. The use in this study of steady–state polyacrylamide gel electrophoresis has overcome a problem encountered by Sernia *et al.* (1979) that SHBGs of low binding capacity could not be detected using normal polyacrylamide gel electrophoresis. They could detect testosterone binding to SHBG in *Setonix brachyurus* using micro–equilibrium dialysis but not using gel electrophoresis. In this study binding was detected for *Bettongia penicillata* and *Setonix brachyurus*.

A seasonal variation occurs in [Σρ] for plasma of both *A. stuartii* and *A. flavipes*. The striking consequences of a decrease in plasma CBG concentration ([Σρ]) coinciding with an elevated plasma glucocorticoid concentration have been described in a field population of *A. stuartii* (Bradley *et al.* 1980) and similar changes are known to occur in both *A. flavipes* (Bradley, unpublished) and *A. swainsonii* (McDonald and Lee, pers. comm.). Although the [Σρ] for *A. flavipes* does not fall to the same low value found in *A. stuartii*, the lower (Ka) value for *A. flavipes* would probably result in the exposure of the tissues of both species to similar biologically active glucocorticoid concentrations near the time of the male mortality. The evidence for the belief that CBG and albumin bound steroid is biologically inactive is discussed by Ballard (1979); however, no such studies have yet been carried out using marsupials.

An unusual feature of marsupial SHBG is that while binding testosterone it does not bind oestradiol, which is bound only to albumin. While this does imply that oestrogen is present in a biologically active form, it should be appreciated that albumin, at concentrations of about 4% in the blood, may become a significant competitor for oestrogen because of its high concentration.

Underestimates of quantity of oestrogen receptor and affinity for oestrogen binding in the presence of albumin have been reported by Wittliff, Hilf, Brooks, Savlov, Hall and Orlando (1972) and Clark and Peck (1979). Probably one of the most interesting features of reports so far on marsupial binding proteins, and of particular relevance to this Volume, is the total absence of SHBG from the plasma of dasyurids, with the interesting correlation that the only species so far described as having a total male postmating mortality belong to this family (Woolley 1966; Wood 1970; Lee, Bradley and Braithwaite 1977; and Bradley, McDonald and Lee 1980). Large increases in the plasma androgen concentration at the time of mating have been described in *Antechinus stuartii* (Moore 1974; Bradley *et al.* 1980), and are known to occur in both *A. flavipes* (Bradley, unpublished) and *A. swainsonii* (McDonald and Lee, pers. comm.), all species which show the postmating mortality, but not in *S. crassicaudata* (McDonald and Lee, pers comm.), which does not have this life history. Presumably most of the androgen is biologically active in these species, with elevation of the plasma concentrations of this steroid coinciding with changes in behavioural patterns, at least in *A. stuartii* (Braithwaite 1974; Lee Bradley and Braithwaite 1977).

For the binding of testosterone to SHBG the values of binding affinities for marsupials cover a similar range to the published values for the eutherian mammals; however, again low binding capacities are more common for the marsupials. The tendency toward low testosterone binding capacity is surprisingly not accompanied by low testosterone concentrations since in marsupials these have been found to fall within the range for the Eutheria (Cook, McDonald and Gibson 1977; Lincoln 1978; Bradley *et al.* 1980).

Many factors govern the biological activity of a steroid. The amount of steroid available for *in vivo* receptor binding depends not only on blood binding parameters, but also on the rate of metabolism and excretion of that hormone. These factors which influence the biological activity of a steroid should be carefully considered in any studies involving the endocrinology of marsupials.

ACKNOWLEDGEMENTS

Plasma samples were obtained with the assistance of Dr. I. R. McDonald, Department of Physiology, Monash University, Professor A. K. Lee and Mr. P. Cuttle, Department of Zoology, Monash University, Dr. P. Woolley, Department of Zoology, Latrobe University, Drs. P. Presidente and I. Beveridge, Veterinary Clinical

Centre, Werribee, Victoria, Dr. J. E. Kinnear, Department of Fisheries and Wildlife, W.A., and Dr. D. King, Agriculture Protection Board, W.A.

I thank Dr. C. Sernia, Department of Physiology, University of Queensland, for helpful discussion. An earlier draft of this paper was constructively criticised by Drs. D. W. Cooper and D. Bradshaw. Mrs. B. Curtis and Mrs. H. Greenhill typed the manuscript.

REFERENCES

BAIRD, D. T., HORTON, R., LONGCOPE, C. AND TAIT, J. F., 1969. Steroid dynamics under steady–state conditions. *Rec. Prog. Hormone Res.* **25:** 611–664.

BALLARD, P. L., 1979. Delivery and transport of glucocorticoids to target cells. Pp.25–48 *in* "Glucocorticoid hormone action" ed. by J. D. Baxter and G. G. Rousseau. Springer Verlag: Berlin.

BRADLEY, A. J., MCDONALD, I. R. AND LEE, A. K., 1976. Corticosteroid binding globulin and mortality in a dasyurid marsupial. *J. Endocrinol.* **70:** 323–324.

BRADLEY, A. J., MCDONALD, I. R. AND LEE, A. K., 1980. Stress and mortality in a small marsupial *Antechinus stuartii* Macleay. *Gen. Com. Endocr.* **40:** 188–200.

BRAITHWAITE, R. W., 1974. Behavioural changes associated with the population cycle of *Antechinus stuartii. Aust. J. Zool.* **22:** 45–62.

CLARK, J. H. AND PECK, E. J. Jr., 1979. "Female sex steroids". Springer Verlag: Berlin.

COOK, B., HUNTER, R. H. H. AND KELLY, A. S. L., 1977. Steroid–binding proteins in follicular fluid and peripheral plasma from pigs, cows and sheep. *J. Reprod. Fert.* **51:** 65–71.

COOK, B., MCDONALD, I. R. AND GIBSON, W. R., 1978. Prostatic function in the Brush–tailed Possum *Trichosurus vulpecula. J. Reprod. Fert.* **53:** 369–375.

COOK, B., SUTTERLIN, N. S., GRABER, J. W. AND NALBANDOV, A. V., 1977. Synthesis and action of gonadal steroids in the American Opossum, *Didelphis marsupialis.* Pp.253–54 *in* "Reproduction and evolution" ed. by J. H. Calaby and C. H. Tyndale–Biscoe. Australian Academy of Science: Canberra.

CORVOL, P. AND BARDIN, C. W., 1973. Species distribution of testosterone–binding globulin. *Biol. Reprod.* **8:** 277–282.

DAUGHADAY, W. H., 1958a. Binding of corticosteroid by plasma proteins. III. The binding of corticosteroid and related hormones by the human plasma and plasma protein fractions as measured by equilibrium dialysis. *J. Clin. Invest.* **37:** 511–518.

DAUGHADAY, W. H., 1958b. Binding of corticosteroid by plasma proteins. IV. The electrophoretic demonstration of corticosteroid binding globulin. *J. Clin. Invest.* **37:** 519–523.

ENGLUND, P. T., HUBERMAN, J. A., JOVIN, T. M. AND KORNBERG, A., 1969. Enzymatic synthesis of deoxyribonucleic acid. XXX. Binding of triphosphates to deoxyribonucleic polymerase. *J. Biol. Chem.* **244:** 3038–3044.

IDLER, D. R., 1972. "Steroids in non–mammalian vertebrates". Academic Press: New York.

IDLER, D. R. AND TRUSCOTT, B., 1972. Corticosteroids in fish. Pp.126–211 *in* "Steroids in non–mammalian vertebrates" ed. by D. R. Idler. Academic Press: New York.

KELLER, N., RICHARDSON AND YATES, F. E., 1969. Protein binding and the bioligical activity of corticosteroids. *In vivo* induction of hepatic and pancreatic alanine aminotransferase by corticosteroids in normal and estrogen treated rats. *Endrocinol.* **84:** 49–62.

KHIN AYE THAN AND MCDONALD, I. R., 1976. Corticosteroid binding by plasma protiens and the effects of estrogens in the marsupial Brush–tailed Possum *Trichosurus vulpecula. J. Endocrinol.* **68:** 31–41.

LEE, A. K., BRADLEY, A. J. AND BRAITHWAITE, R. W., 1977. Life history and male mortality in *Antechinus stuartii.* Pp.209–20 *in* "The biology of marsupials" ed. by D. Gilmore and B. Stonehouse. Macmillan Scientific PPress: London.

LINCOLN, G., 1978. Plasma testosterone profiles in macropodid marsupials. *J. Endocrinol.* **77:** 347–351.

MCDONALD, I. R., 1977. Adrenal function in marsupials. Pp.345–77 *in* "The biology of marsupials" ed. by D. Gilmore and B. Stonehouse. Macmillan Scientific Press: London.

MCDONALD, I. R. AND BRADSHAW, S. D., 1977. Plasma corticosteroids and the effect of adrenocorticotrophin in a macropodid marsupial (*Setonix brachyurus* Quoy and Gaimard). *J .Endrocrinol.* **75:** 409–18.

MCDONALD, I. R. AND WARING, H., 1979. Hormones of marsupials and monotremes. Pp.873–924 *in* "Hormones and evolution" ed. by E. J. W. Barrington. Academic Press: London.

MERCIER–BOARD, C., ALFSEN, *A.* AND BAULIEU, E.–E., 1970. Sex–steroid binding plasma protein. *Acta Endocrinol. Suppl.* **147:** 204–24.

MOORE, G. H., 1974. "Aetiology of the die–off of male *Antechinus stuartii"*. Ph.D. thesis, Australian National University, Canberra.

ODDIE, C. J., BLAINE, E. H., BRADSHAW, S. D., COGHLAN, J. P., DENTON, D. A., NELSON, J. F. AND SCOGGINS, B. A., 1976. Blood corticosteroids in Australian marsupials and placental mammals and one monotreme. *J. Endocrinol* **69:** 341–8.

OPPENHEIMER, J. H., 1968. Role of plasma proteins in the binding, distribution and metabolism of the thyroid hormones. *N. Eng. J. Med.* **278:** 1153–62.

RITZEN, E. M., FRENCH, F. S., WEDDINGTON, S. C., NAYFEH, N. AND HANSSON, V., 1974. Steroid binding in polyacrylamide gels – quantitation at steady state conditions. *J. Biol. Chem.* **249:** 6597–604.

SANDBERG, A. A., ROSENTHAL, H. E., SCHNEIDER, S. L. AND SLAUNWHITE, W. R. Jr., 1966. Protein–steroid interactions and their role in the transport and metabolism of steroids. Pp.1–61 *in* "Steroid dynamics" ed. by G. Pincus, T. Nakao and J. R. Tait. Academic Press: New York.

SCATCHARD, G., 1949. The attraction of proteins for small molecules and ions. *Ann. N.Y. Acad. Sci.* **51:** 660–72.

SEAL, U. S. AND DOE, R. P., 1965. Vertebrate distribution of corticosteroid–binding globulin and some endocrine effects of concentration. *Steroids* **5:** 827–41.

SEAL, U. S. AND DOE, R. P., 1966. Corticosteroid–binding globulin: biochemistry, physiology and phylogeny. Pp.63–90 *in* "Steroid dynamics" ed. by G. Pincus, N. Takao and J. F. Tait. Academic Press: New York.

SERNIA, C., 1978. Steroid binding proteins in the plasma of the Echidna *Tachyglossus aculeatus,* with comparative data for some marsupials and reptiles. *Aust. Zool.* **20:** 87–98.

SERNIA, C., BRADLEY, A. J. AND MCDONALD, I. R., 1979. High affinity binding of adrenocortical and gonadal steroids by plasma proteins of Australian marsupials. *Gen. Comp. Endocr.* **38:** 496–503.

SLAUNWHITE, W. R. Jr., LOCKIE, G. N., BACK, N. AND SANDBERG, A. A., 1962. Inactivity *in vivo* of transcortin–bound cortisol. *Science* **135:** 1062–3.

TAIT, J. F. AND BURSTEIN, S., 1964. *In vivo* studies of steroid dynamics in man. Pp.441–557 *in* "The hormones. Vol. V." ed. by C. Pincus, K. V. Thiman and E. B. Astwood. Academic Press: New York.

WENN, R. U., KAMBERTI, I. A., KEYVANJAH, M. AND JOHANNES, A., 1977. Distribution of testosterone–estradiol binding globulin (TeBG) in higher vertebrates. *Endokrinologie* **69:** 151–6.

WESTPHAL, U., 1967. Steroid protein interactions. XIII. Concentration and binding affinities of corticosteroid–binding globulins in sera of man, monkey, rat, rabbit and guinea–pig. *Arch. Biochem. Biophys.* **118:** 556–67.

WESTPHAL, U., 1971. "Steroid–protein interactions". Springer–Verlag: Berlin.

WITTLIFFE, J. L., HILF, R., BROOKS, W. F. JR., SAVLOV, E. D., HALL, T. C. AND ORLANDO, R. A., 1972. Specific estrogen–binding capacity of the cytoplasmis receptor in normal and neoplastic breast tissues of humans. *Cancer Res.* **32:** 1983–92.

WOOD, D. H., 1970. An ecological study of *Antechinus stuartii* (Marsupialia) in a south–east Queensland rain forest. *Aust. J. Zool.* **18:** 185–207.

WOOLLEY, P., 1966. Reproduction in *Antechinus* spp. and other dasyurid marsuplals. Pp.281–94 *in* "Com parative biology of reproduction in mammals", 15th Symposium of the Zoology Society, London. Academic Press: New York/London.

WOOLLEY, P., 1973. Breeding patterns and the breeding and laboratory maintenance of dasyurid marsupials. *Exp. Animals* **1:** 161–72.

RESUMEN

53. Union de esteroides a las proteinas del plasma de los marsupiales dasiuridos. Por A. J. Bradley.

Las uniones del cortisol, progesterona, testosterona, estradiol y aldosteron fueron examinadas en 29 especies de marsupiales por equilibrio de diálisis y/o por electroforesis del gel estable de poliacrilamida. Las constantes de asociación (Kα 36° C) y las capacidades de unión ([$\Sigma\rho$] del cortisol y la testosterona fueron determinadas por microdiálisis, demostrándo que van desde 2.7 a 9.6 x 10^7 M^{-1}, y desde 0.3 a 3.7 x 10^{-7} M para el cortisol, y de 1.2 a 8.3 x 10^8 M^{-1} y 0.3 a 4.0 x 10^{-7} M para la testosterona. En las 29 especies estudias das se detectó una alta afinidad de unión del cortisol, pero la alta afinidad de unión de la testosterona estuvo limitada a sólo 10 especies. Por electroforesis en gel de poliacrilamida bajo condiciones estables se detectaron tres proteinas de transporte de esteroides, electroforéticamente distintas, una parecida al corticosteroide humano unido a la globulina (CBG) que tiene afinidad por el cortisol, corticosterona y progesterona; una segunda parecida a la hormona sexual humana unida a la globulina (SHBG) que tiene afinidad por la testosterona y la albúmina, que mostró una afinidad por la testosterona, cortisol, corticosterona, progesterona, aldosterona y estradiol. No se detectó una alta afinidad de unión en la aldosterona y el estradiol.

54. SPERM ULTRASTRUCTURE, SPERMIOGENESIS AND EPIDIDYMAL SPERM MATURATION IN DASYURID MARSUPIALS: PHYLOGENETIC IMPLICATIONS

H. R. HARDING[1], P. A. WOOLLEY[2], C. D. SHOREY[3], AND F. N. CARRICK[4]

The distinctive features of sperm ultrastructure, spermiogenesis and epididymal sperm maturation are described for 13 dasyurid species from 6 genera. An extreme consistency in sperm form and development between these species is evident. Comparisons are made with sperm from other marsupial families. The most notable is that some unusual features of the dasyurid sperm flagellum are shared by *Tarsipes spenserae* and the peramelids *(Perameles nasuta, Isoodon macrourus)*. Further features are shared by the dasyurids and *Tarsipes* but not the peramelids. The similarities between dasyurid, *Tarsipes* and peramelid sperm are suggested to represent phylogenetic links rather than convergence, and on this basis comments are made on current marsupial phylogenetic trees and on a possible ancestral sperm form for the Australian marsupials.

Pages 659–73 *in* CARNIVOROUS MARSUPIALS, ed. by M. Archer, February, 1982. Roy. Zool. Soc. New South Wales: Sydney, Australia.

DETAILED DESCRIPTIONS of dasyurid sperm ultrastructure, spermiogenesis and epididymal sperm maturation have yet to be published. Ultrastructural observations have been reported only in brief abstracts (Cleland 1965, Hughes 1976, Harding *et al.* 1977, 1979b), or incorporated into broader comparative studies (Harding *et al.* 1979a, Sapsford *et al.* 1967, 1969, 1970).

Light microscopy is necessarily limited to gross structures, and apart from the early observations and lengthy descriptions of Fürst (1887) on sperm development in *Phascogale albipes* (= *Sminthopsis murina*), it has likewise only resulted in brief reports forming part of broader comparative studies (Bishop and Austin 1957, Bishop and Walton 1960, Hughes 1965). The unpublished thesis of one of the authors (Harding 1977) probably provides the most comprehensive description of dasyurid sperm to date.

Despite the lack of detail in the literature, there are indications that dasyurid sperm possess a number of unusual features which make them of considerable interest for comparative structure–function studies of sperm, and for our continuing speculation on marsupial phylogeny. For example : Hughes (1965) found dasyurid sperm to be among the largest in mammals and consequently suggested (pers. comm.) their utility for manipulation at the intracellular level, for chemical dissection and for analysis of the components. In addition, Harding (1977) has noted that the flattened nature of the dasyurid sperm flagellum provides differentiated external axes of orientation which, coupled with their large size, should make these sperm particularly suitable for studying the relationship of internal flagellar structure to motility. Cleland (1965) first drew attention to the unusual arrangement of the intraflagellar components in dasyurid sperm, and their similarity in this respect with the peramelids.

Hughes (1965) stressed the uniformity in sperm morphology within the Dasyuridae at the light microscope level. His observations were based on examination of *Dasyuroides byrnei* and *Sarcophilus harrisii.* In addition he reviewed the previous reports on *Dasyurops maculatus* (= *Dasyurus maculatus*) (Bishop and Austin 1957, Bishop and Walton 1960), *Phascogale albipes* (= *Sminthopsis murina*) (Fürst 1887) and *Antechinus flavipes flavipes, A.f. leucogaster, A. swainsonii, A. stuartii, Sminthopsis crassicaudata* (Wolley, unpublished observations). This uniformity among dasyurid sperm has been confirmed at the ultrastructural level with observations on *Antechinus* and *Sminthopsis* (Cleland 1965), the 'marsupial mouse' (Sapsford *et al.* 1967, 1969, 1970) (= *Antechinus flavipes leucogaster,* Cleland, pers. comm.), *Sarcophilus harrisii* and *Dasyurus viverrinus* (Hughes 1976), *D. maculatus, Dasycercus cristicauda, Antechinus rosamondae, A. macdonnellensis* and *Planigale tenuirostris* (Harding *et al.* 1977, 1979a,b).

We have extended our observations on spermiogenesis, epididymal sperm maturation and mature sperm ultrastructure, from the latter 6 to 13 dasyurid species. In this paper we do not de-

[1] Department of General Studies, University of New South Wales, P.O. Box 1, Kensington, New South Wales, Australia, 2033.
[2] Department of Zoology, La Trobe University, Bundoora, Victoria, Australia, 3083.
[3] Department of Histology and Embryology, University of Sydney, Broadway, New South Wales, Australia, 2007.
[4] Department of Veterinary Anatomy, University of Queensland, St Lucia, Brisbane, Queensland, Australia, 4006.

scribe these aspects of dasyurid sperm in detail, as this will be done separately (Harding, Woolley, Shorey and Carrick, in preparation) nor do we emphasise those features of dasyurid sperm which also serve to characterise marsupial from eutherian sperm, as this has been discussed previously (Harding *et al.* 1979a). Instead, we intend to concentrate on those features of dasyurid sperm structure and development which have comparative interest at both the intra- and interfamilial level of the Marsupialia. In this way we hope that a contribution may be made to the continuing debate on marsupial phylogeny, through the incorporation of information from sperm studies with that based on the traditional skeletal criteria and the more recently used serological and cytological criteria.

MATERIALS AND METHODS

Testicular material and epididymal sperm samples have been examined from the following 13 dasyurid species (the generic usage follows Archer 1982): *Antechinus bellus, Parantechinus bilarni, Antechinus leo, Pseudantechinus macdonnellensis, Dasykaluta rosamondae, Antechinus stuartii, Dasycercus cristicauda, Dasyuroides byrnei, Dasyurus maculatus, D. viverrinus, Planigale ingrami, P. tenuirostris, Sminthopsis crassicaudata.*

To obtain the testicular material the tunic of the testis was cut to expose the seminiferous tubules and small samples of the tubules were cut and placed for 30–60 minutes in Karnovsky's fixative with picric acid added. Further processing was as described previously (Harding *et al.* 1976a). Epididymal samples were obtained from 5 regions : those taken from the proximal, central and distal regions of the epididymis are referred to as caput, corpus and cauda epididymidal samples respectively, whilst intermediate material from between the former two and the latter two regions is referred to as caput–corpus and corpus–cauda respectively. The samples were cut from appropriately located segments of the epididymis using fine scissors, and then transferred to a drop of fixative (Karnovsky's with added picric acid) on a piece of dental wax. Multiple parallel cuts (1 mm or intermediate material from between the former less apart) were made in the epididymal segments with a new razor blade whilst the samples were under fixative, and the process was repeated making a second series of cuts at right angles to the first. The cut epididymal tissue was then transferred to a specimen tube containing a large volume of the fixative for 30–60 minutes. Further processing was as for the testicular samples. Specimens were examined using a Philips EM 200, EM 201, EM 300 or a JEOL 100CX.

OBSERVATIONS AND DISCUSSION

In a previous paper (Harding *et al.* 1979a) we outlined some unusual features of mature sperm structure and epididymal sperm maturation in the Dasyuridae. Examination of an additional 7 species has shown these features to be common to all 13 dasyurid species we have examined to date, thus giving further support to the earlier indications that dasyurids exhibit a high degree of uniformity in their sperm structure.

In order to ensure a reasonably complete picture of dasyurid sperm characteristics in this paper as a basis for interfamily comparisons, we reiterate our previous observations, but in this instance we support them with evidence from our examination of the additional species, and also describe some unusual, and previously unreported, features of spermiogenesis.

Mature Epididymal Sperm

With the exception of sperm of the Honey Possum, *Tarsipes spenserae* (Scarlett 1978), and the Thylacine, *Thylacinus cynocephalus* (Woolley, unpublished observations), dasyurid sperm are the largest found in marsupials examined to date. Hughes (1965, 1976, pers. comm.) reported that flagellar length of 192–242μm in the 3 dasyurid species he studied was approached only in the peramelids (165–194μm), whilst macropod, phalangerid and petaurid sperm had considerably shorter flagella ($\approx$100μm). The nucleus is similarly very long in the dasyurids, *Tarsipes spenserae* and *Thylacinus cynocephalus,* and comparison of its length/width ratio with that of other marsupials, shows it to be of slender proportions (see Table 1).

Except in the vombatoids, the nucleus overlies the proximal midpiece by varying amounts in mature sperm of all marsupials (see Harding *et al.* 1979a, for explanation of this, and comparison with the eutherian situation). A consequence of the extreme length of the nucleus in dasyurids is that the overlap of the proximal mitochondria is greatest in this group (Fig. 1a), and in order to accommodate the mitochondria and also ensure a streamlined form for the mature spermatozoon, the distal nucleus comprises only a very thin curved dorsal plate that closely covers the underlying mitochondria (Fig. 3c). This is matched most closely in *Tarsipes spenserae* (Harding *et al.* 1981) and *Potorous tridactylus;* species in which the nucleus is equally long (see Table 1). Apart from their large size, dasyurid sperm are most notable for the flattened nature of the midpiece and for their unusual intraflagellar organisation. Although flattening of the flagellum is common in the principal piece in marsupial sperm, only in the dasyurids and *Tarsipes spenserae* (Harding *et al.* 1981) is the midpiece also flattened (Figs 2a, b). In terms of intraflagellar organisation, the dasyurids, *Tarsipes*

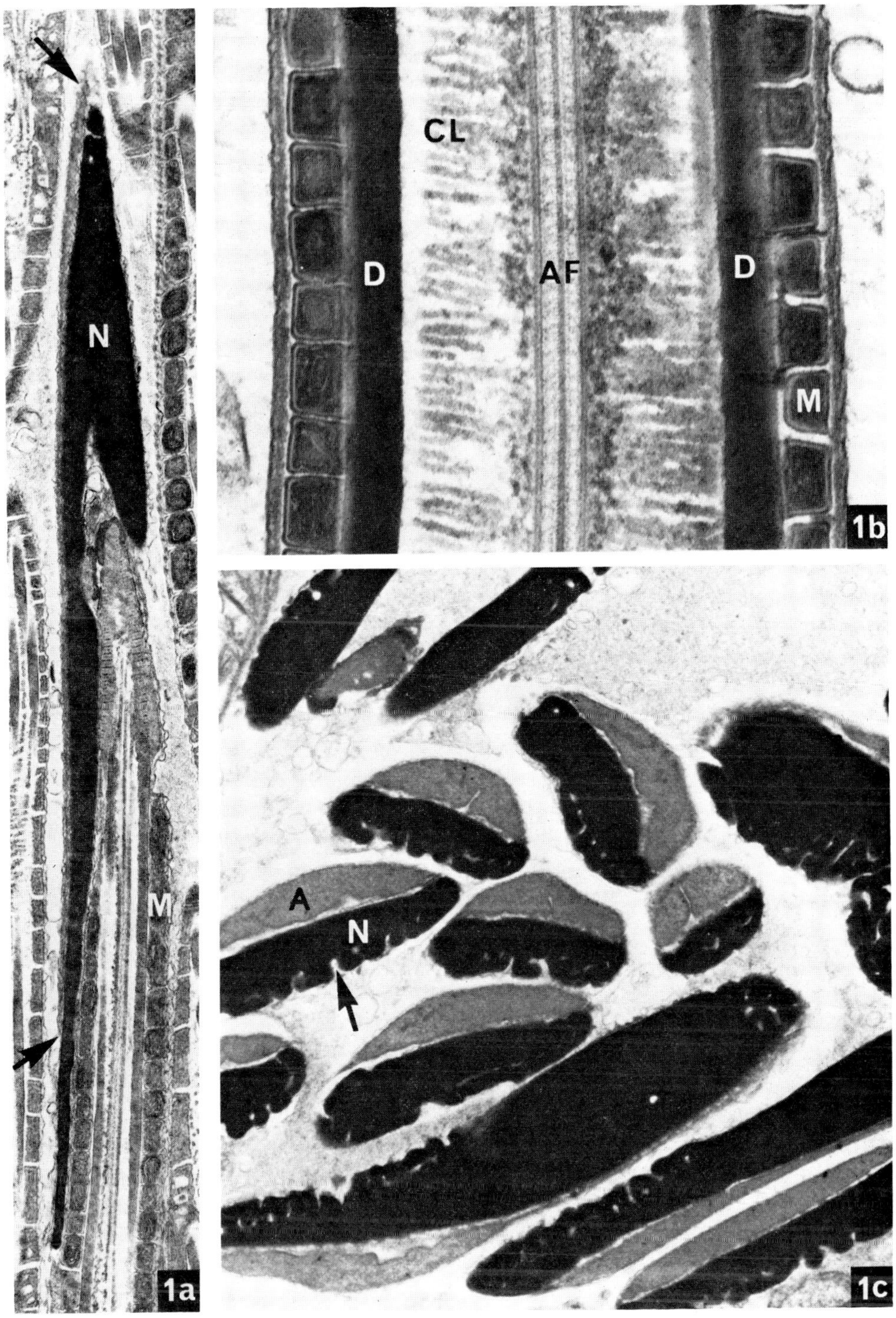

Fig. 1a. Longitudinal section through the head and proximal midpiece of a corpus–cauda epididymidal sperm of *Dasyurus viverrinus*. Arrows indicate the anterior and posterior margins of the elongated acrosome. Note that the distal parachromatin nuclear appendage is not evident (compare with Figure 4). M mitochochondria; N nucleus. X 14,000.

Fig. 1b. Longitudinal section through part of the midpiece of a cauda epididymidal sperm of *Dasykaluta rosamondae*. The dense outer fibres (D) are widely separated from the axial filament complex (AF) and the two are connected by connecting laminae (CL) which are not continuous longitudinally. M mitochondria. X 44,000.

Fig. 1c. Series of transverse and oblique sections through the heads of cauda epididymidal sperm of *Planigale tenuirostris*. The periphery of the nucleus (N) is indented over parts of ventral surface (arrowed). A acrosome. X 17,500.

Table 1. Approximate dimensions of the nucleus in fifteen marsupial species.

Family/species	Mean length of nucleus (μm) ± S.E.M.	Mean width of nucleus (μm) ± S.E.M.	Length width ratio
PHALANGERIDAE[1]			
Trichosurus vulpecula	3.7±0.16	1.6±0.06	2.3
PETAURIDAE[1]			
Petaurus norfolcensis	4.1±0.22	2.3±0.11	1.8
Petaurus breviceps	5.5±0.13	1.9±0.21	2.8
Pseudocheirus peregrinus	5.0±0.14	3.2±0.11	1.6
MACROPODIDAE[1]			
Macropus robustus	4.5±0.01	1.5±0.03	3.0
Macropus parma	5.3±0.27	1.4±0.03	3.8
Megaleia rufa	3.3±0.14	2.0±0.09	1.7
Potorous tridactylus	9.1±0.15	1.9±0.09	4.8
DASYURIDAE[1]			
Dasykaluta rosamondae	9.8±0.54	1.6±0.05	6.2
Dasyurus viverrinus	11.0±0.9	1.9±0.08	5.9
Dasycercus cristicauda	9.5	1.9±0.04	5.1
Planigale tenuirostris	9.3	2.0±0.14	4.6
PERAMELIDAE[1]			
Isoodon macrourus	4.5	2.2±0.05	2.0
THYLACINIDAE[2]			
Thylacinus cynocephalus	10.6±0.07	2.3±0.04	4.7
TARSIPEDIDAE[3]			
Tarsipes spenserae	10.6±0.14	3.0±0.1	3.5

1. From Harding (1977). Measurements taken from electron micrographs.
2. From Woolley (unpublished observations). Whole sperm measured by light microscopy.
3. From Scarlett (1978). Whole sperm measured by light microscopy.

and the peramelids (Cleland and Rothschild 1959, Harding *et al.* 1979a) are characterised by an extreme radial displacement of their dense outer fibres from the corresponding axoneme doublets (Figs 1a, 2a,b). In these groups the dense outer fibres closely overlie the doublets in the proximal regions of the midpiece (Fig. 3c), but the two are widely separated over most of the length of the latter and of the proximal principal piece (Fig. 5). This radial separation narrows and certain dense outer fibres terminate in more distal regions of the tail (Fig. 5). Narrow fibrous bands connect the displaced dense outer fibres to their corresponding doublets, and appear in transverse sections as single bands in the dasyurids (Figs 2a,b) and *Tarsipes* (Harding *et al.* 1981) but double in the peramelids (Cleland and Rothschild 1959, Harding 1977). A characteristic of the dasyurids and *Tarsipes* but not the peramelids, is a precocious displacement of dense outer fibre in the most proximal regions of the midpiece where the remaining fibres are still closely associated with their underlying doublets (Fig. 3c).

Typically the dense outer fibres closely overlie the axoneme doublets in all regions of the flagellum in other marsupials examined to date: *Trichosurus vulpecula, Pseudocheirus peregrinus, Petaurus breviceps, P. norfolcensis, Macropus robustus, M. parma, Megaleia rufa, Phascolarctos cinereus* (Harding 1977, Harding *et al.* 1979a), *Caluromys philander* (Olson and Hamilton 1976). However, a very slight displacement of the fibres is evident in *Potorous tridactylus* (Harding *et al.* 1979a) and *Didelphis virginiana* (Olson *et al.* 1977). In view of this it is perhaps worth emphasising the resemblance of *P. tridactylus* to the dasyurids in sperm head length and proportions, which was mentioned earlier.

A consequence of the separation of the dense outer fibres and axoneme in the dasyurids, and to a lesser extent the peramelids, is that the inner surface of the mitochondrial sheath is 'sculptured' in order to accommodate the dense outer fibres (Figs 2a,b). This is not a pronounced feature in *Tarsipes* where the mitochondrial sheath appears to be of relatively even thickness in transverse section and to have a fairly smooth inner surface.

Additional features of the flagellum link these three groups. The diameter of the mid– and principal piece is almost identical at their juncture in these groups, such that the plasma membrane smoothly covers the annulus (Fig. 2d). In contrast, the plasma membrane is inturned at the annulus in other marsupials and the diameter of the distal midpiece typically exceeds that of the proximal principal piece (Harding 1977, Harding *et al.* 1979a). The form and arrangement of the mitochondria in the dasyurids is likewise very distinctive. In longitudinal sections the mitochondria appear as relatively small

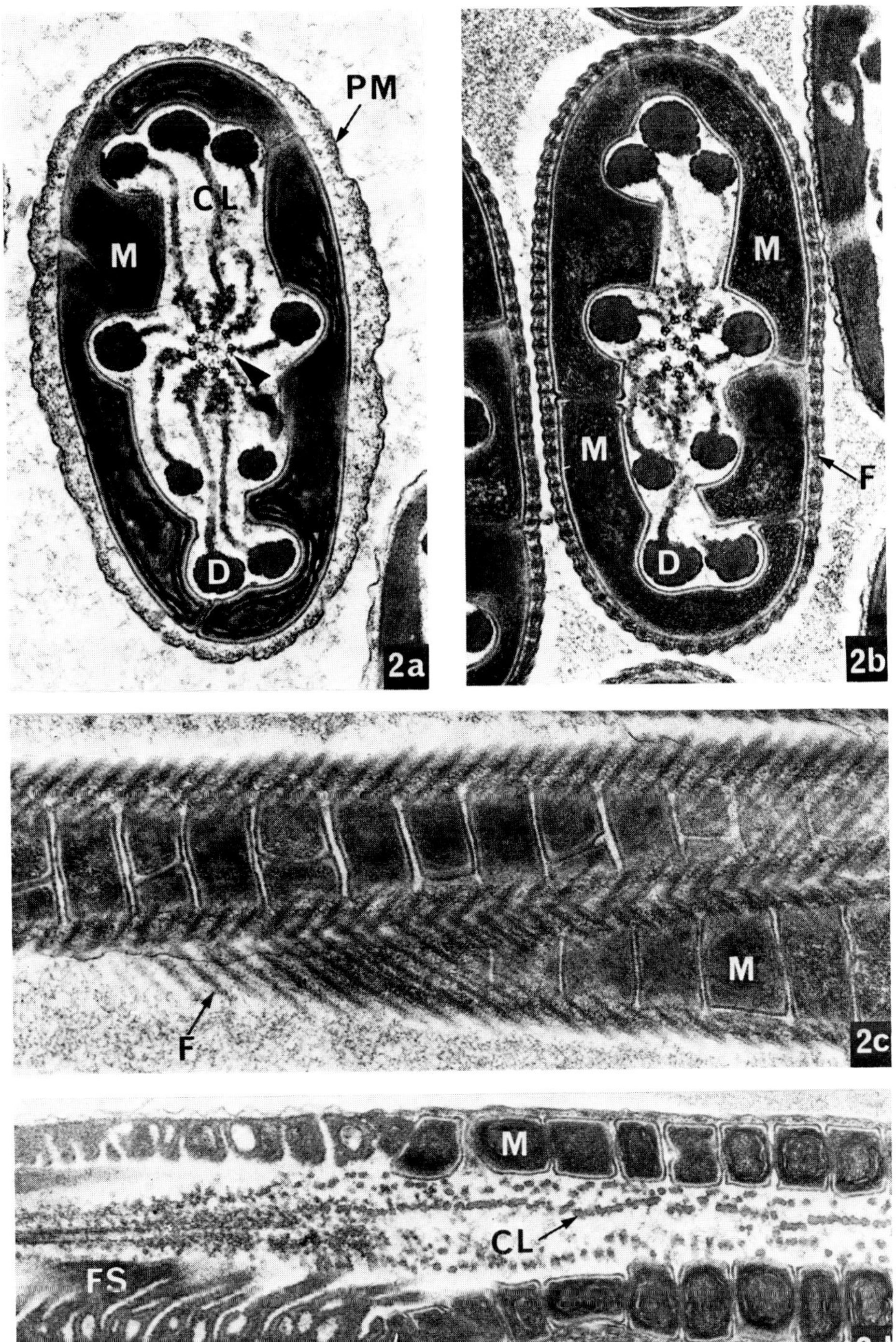

Fig. 2a. Transverse section through the midpiece of a caput–corpus epididymidal sperm of *Dasyurus viverrinus*. The midpiece is markedly flattened and the inner surface of the mitochondria (M) is sculptured to accommodate the dense outer fibres (D) which are radially displaced from the axoneme (arrowed). The fibrous bands (CL) connecting the dense outer fibres to the axoneme appear as single bands. The midpiece fibre network is not yet developed under the plasma membrane (PM). X 29,000.

Fig. 2b. Transverse section through the midpiece of a cauda epididymidal sperm of *Dasyurus viverrinus*. The same comments and labels as given for 2a apply, except the midpiece fibre network (F) is now developed. X 31,500.

Fig. 2c. Superficial longitudinal sections through the midpiece of corpus–cauda epididymidal sperm of *Dasyurus viverrinus*, showing the regular helical arrangement of the midpiece fibre network (F). Note the close fit of the regularly arranged 'brick–like' mitochondria (M). X 31,000.

Fig. 2d. Longitudinal section to show the similar diameter of the mid and principal pieces at the annulus (arrowed) of a corpus–cauda epididymidal sperm of *Dasyurus viverrinus*. CL oblique sections of the connecting laminae; FS, fibrous sheath of principal piece; M mitochondria of midpiece. X 24,500.

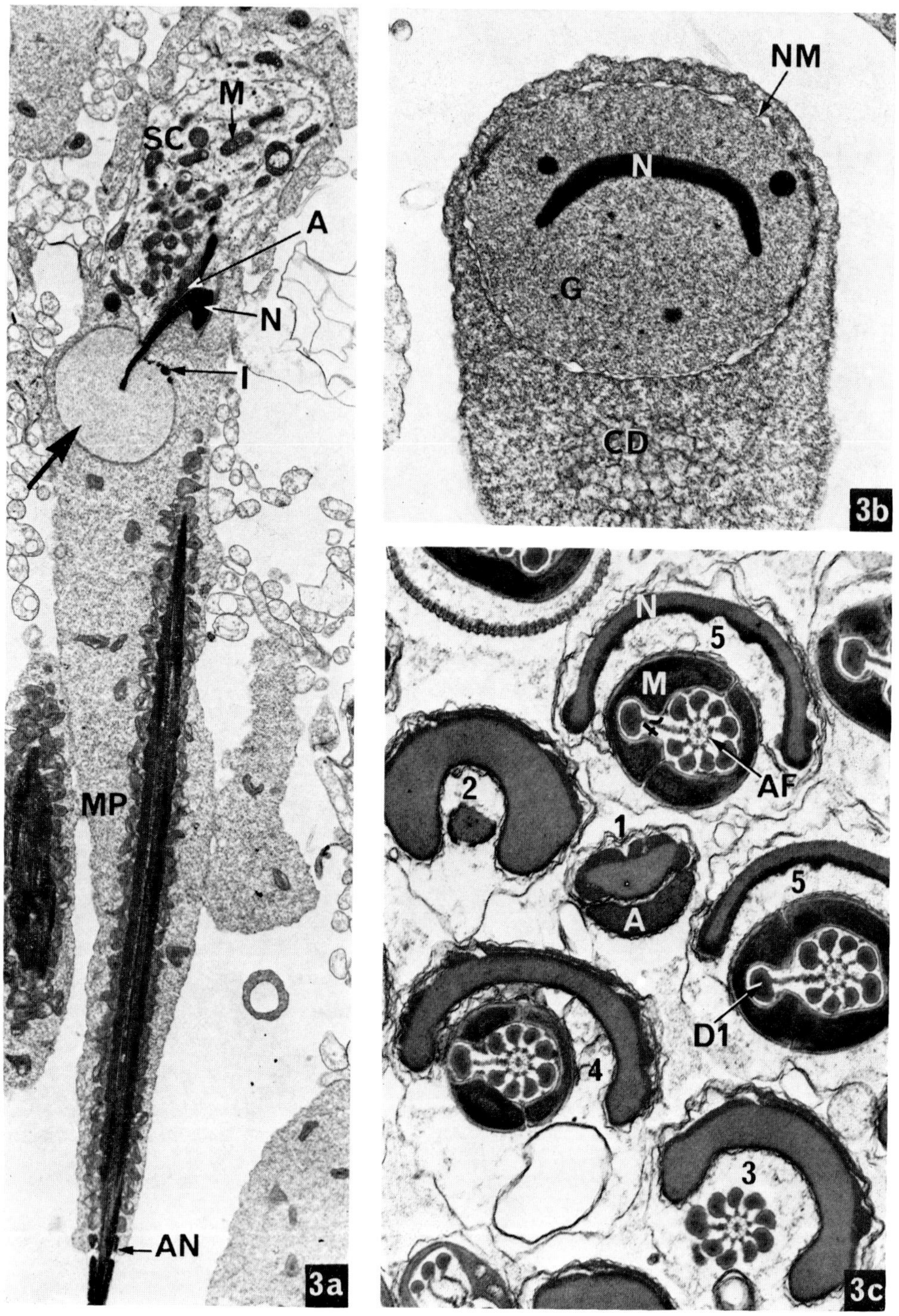

Fig. 3a. Longitudinal section of a late spermatid from the testis of *Antechinus leo.* Note the aggregation of mitochondria (M) over the developing acrosome (A) and the distal ballooning (arrowed) of the nucleus (N). Electron dense inclusions (I) are evident in this nuclear appendage. AN annulus; MP midpiece; SC Sertoli cell cytoplasm. X 4,000.
Fig. 3b. Transverse section through the distal region of the nucleus in a caput epididymidal sperm of *Sminthopsis crassicaudata.* Note that the granular parachromatin mass (G) which surrounds the electron dense nuclear material (N) is part of the nucleus and is bounded by the nuclear membrane (NM). CD cytoplasmic droplet. X 19,500.
Fig. 3c. Transverse sections through various regions of the head and underlying midpiece of sperm from the corpus–cauda epididymidis of *Antechinus leo.* The sections are numbered 1 to 5; 1 corresponding to the most proximal and 5 to the most distal sections of the nucleus. The nucleus (N) in distal regions of the sperm head comprises a thin curved plate covering the underlying mitochondria (M) of the midpiece. Note the precocious displacement of dense outer fibre I (D1) whilst other fibres still lie close to the axoneme (AF). Differentiation of parts of the periphery of the nucleus is evident in all sections, with the peripheral region being of greater electron density. Indentations of this differentiated peripheral layer are evident in section I. A acrosome. X 22,500.

'brick–like' structures which are very regularly arranged and close fitting (Figs. 1b, 2c, d) and the close fit is equally evident in transverse sections (Figs. 2a, b). The mitochondria in *Tarsipes* almost exactly match the dasyurid arrangement, and the peramelids whilst lacking the precise regularity and extremely close fit, also show some resemblance to the latter. In contrast, mitochondria are more rounded in profile in the other marsupials examined, and their arrangement lacks the dasyurid precision (Harding *et al.* 1979a).

The nucleus in dasyurid sperm is, as mentioned above, long and proportionally slender, whilst the acrosome forms a thin layer over approximately the anterior four–fifths of the dorsal nuclear surface (Fig. 1a). An extensive acrosomal coverage of the latter along its long axis is also found in *Petaurus breviceps, P. norfolcensis, Pseudocheirus peregrinus* (Harding *et al.* 1979a), *Tarsipes spenserae* (Harding *et al.* 1981), and the American species *Caluromys philander* (Olson and Hamilton 1976) and *Didelphis virginiana* (Olson *et al.* 1977), whilst in the macropodids, peramelids and *Trichosurus vulpecula* the acrosome is restricted to approximately the anterior one–third of the dorsal nuclear surface (Harding *et al.* 1979a).

An unusual modification of certain regions of the nuclear periphery is undoubtedly the most interesting feature of the dasyurid sperm head. Peripheral regions of the nucleus (particularly on the ventral surface) frequently form a demarcated layer of differing electron density to the remainder. Usually this layer is also indented so that the nuclear surface appears in sections as a series of globular projections (Fig. 1c). With the exception of the peramelids (Harding *et al.* 1979a, Sapford *et al.* 1969), such modifications of the nucleus have not been reported in other mammals. The ventral flanges of the peramelid nucleus frequently contain electron lucent areas giving a pitted or vesiculated appearance, which is reminiscent of, but not identical to, the 'globular projections' in the dasyurids. As discussed previously (Harding *et al.* 1979a) it is hard to imagine what the role of these modifications might be.

Only two features of the principal piece in dasyurid sperm are worthy of note here as departing from the usual marsupial form. The first, which has been described above, is the radial displacement of the dense outer fibres from the axoneme in all but the distal regions of the principal piece. The second is the absence of a principal piece fibre network (Fig. 5), namely, an irregular array of fibrous material which is found between the plasma membrane and fibrous sheath in at least the proximal half of the principal piece in the American species, *Caluromys philander* (Phillips 1970a) and *Didelphis virginiana* (Holstein 1965, Phillips 1974), and all Australian species examined to date except for the peramelids and possibly the Koala in which it may occur in a rudimentary form (Harding *et al.* 1979a). In the peramelids, although a fibre network is lacking, a granular accessory sheath surrounds the fibrous sheath in the proximal principal piece (Sapsford *et al.* 1970). No such accessory sheath is evident in the dasyurids.

SPERMIOGENESIS

In most respects spermiogenesis in the dasyurids follows the pattern reported for other marsupials (Sapsford *et al.* 1967, 1969, Phillips 1970b, Rattner 1972, Harding 1977, Harding *et al.* 1976a). However, two features are sufficiently different to be worthy of special mention.

From at least the stage of nuclear flattening (see Sapsford *et al.* 1967, 1969, for stages of spermiogenesis), a striking feature in dasyurids is an aggregation of mitochondria in the Sertoli cell cytoplasm overlying the developing acrosome (Fig. 3a). This 'mitochondria–rich' body of cytoplasm is enclosed by an almost continuous cisterna of the endoplasmic reticulum. The spatial relationship of the mitochondria with the developing acrosome suggests some functional connection between the two. However, if this is the case, it becomes necessary to question why such mitochondrial support for acrosome development is required in the dasyurids but not in other marsupials.

Late spermatids in dasyurids are notable for the presence of a spherical mass of granular material which surrounds the attenuated extremity of the nucleus distal to the posterior margin of the acrosome (Fig. 3a). This granular mass appears to be continuous with the condensed nuclear chromatin, and hence to lie within a distal ballooning of the nuclear membrane (Fig. 3b). Sapsford *et al.* (1969) have used the term parachromatin, for relatively electron lucent areas within the otherwise electron dense nucleus in *Perameles nasuta*, and this is probably the most suitable term to use for the granular mass in the dasyurids. Spherical electron dense inclusions are frequently observed within this parachromatin in the dasyurids, particularly in the portion lying ventral to the condensed nuclear chromatin (Figs. 3a,b).

Sapsford *et al.* (1969) reported that "In the ram and marsupial mouse the distal extremity of the nucleus of the relatively mature spermatid is ballooned to form a sac–like appendage which is devoid of condensed chromatin and is bounded by a nuclear envelope containing many nuclear pores" (p.238). They make no mention of the contents of this sac–like appendage. In addition, a similar ballooning of the distal nuclear

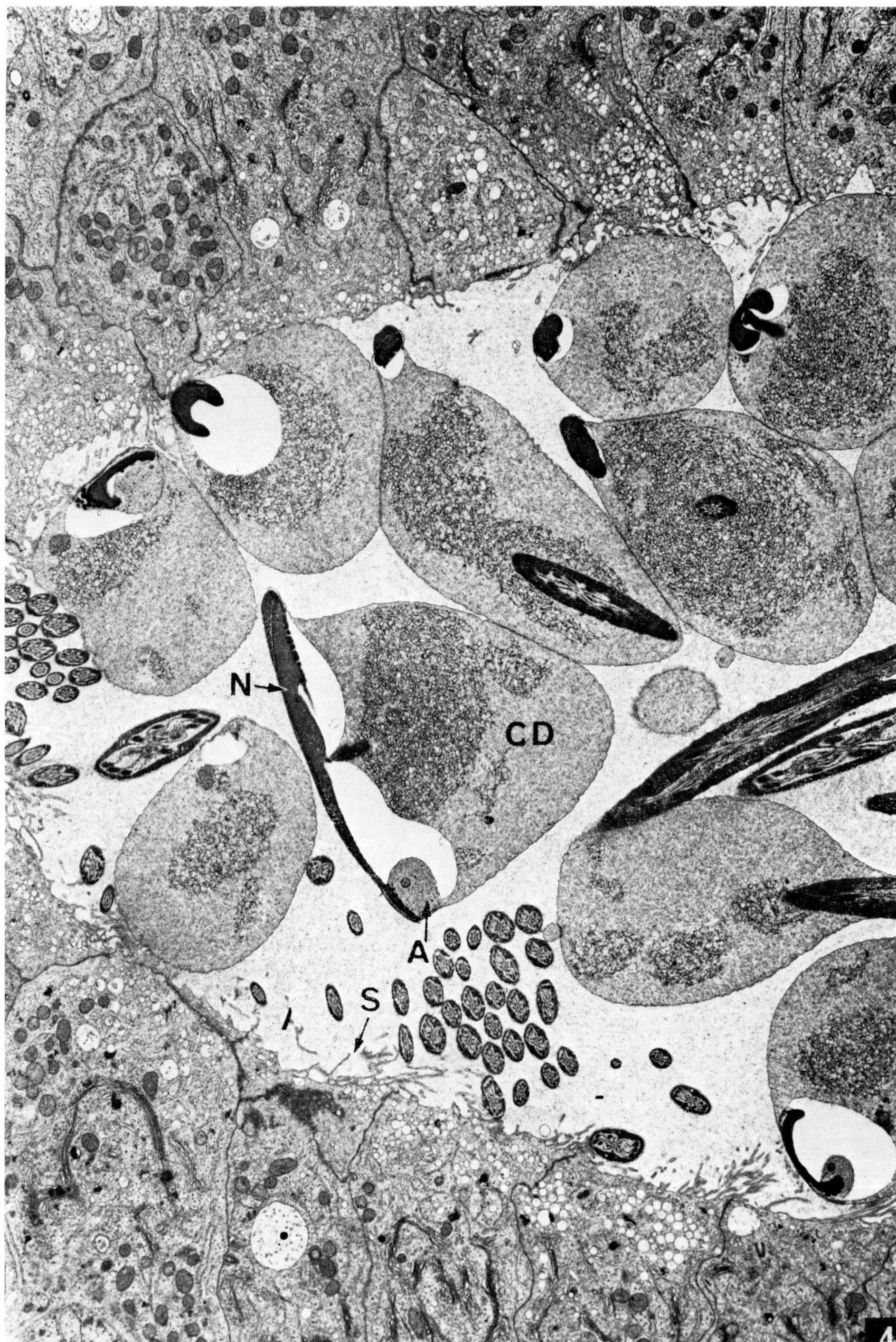

Fig. 4. Section through the caput epididymidis of *Dasykaluta rosamondae*. Long stereocilia (S) line the epididymal epithelium. Longitudial sections through immature sperm show the nucleus (N) perpendicular to the long axis of the sperm, and the presence of the cytoplasmic droplet (CD). The distal nuclear appendage (A) is still present. X 4,000.

envelope enclosing an electron lucent area is evident in *P. nasuta* (see fig. 5, plate 13, Sapsford *et al.* 1969). Sapsford *et al.* infer some functional relationship between areas of parachromatin and nuclear pores, presumably concerned with disposal of dispensable nuclear substances. However, whether such a relationship occurs in dasyurid sperm remains to be determined and thus further investigation will be required before we can assume that the nuclear appendages perform a similar function in these three groups (the ram, dasyurids and *P. nasuta*). The parachromatin appendage of the dasyurid nucleus is still present in caput epididymidal sperm (Fig. 4), but not in sperm from distal regions of the epididymis (Fig. 1a). Its fate during epididymal sperm maturation is not clear at present.

Epididymal Sperm Maturation

Striking morphological changes occur in the sperm as they pass through the epididymis in a number of marsupials. These changes include : an anteriorly directed contraction of the cytoplasmic droplet and its shedding from the proximal midpiece; rotation of the nucleus from a perpendicular to a parallel orientation with respect to the flagellum; an increase in the electron density of the mitochondrial matrix material; development of a helically arranged network surrounding the mitochondria and underlying the midpiece plasma membrane; development of randomly arranged flask–like invaginations of the plasma membrane between the parallel bands of the midpiece fibre network; a pronounced reorganisation of the acrosomal material that converts the acrosome from an attenuated cuplike structure into its definitive compact form closely adhering to the dorsal nuclear surface; the development of stacks of membranes (membranous whorls) that emanate from the neck region and project anteriorly and posteriorly to surround the distal region of the nucleus and proximal midpiece; pairing of sperm in the region of the acrosome in American species (Cummins 1976; Harding *et al.* 1975, 1976b, 1979a, Krause and Cutts 1979; Olson and Hamilton 1976; Olson *et al.* 1977; Temple–Smith and Bedford 1976). The occurrence of most of these maturation changes in marsupials is summarised in Table 2.

The maturation process in the Dasyuridae is perhaps most notable for the close similarity between species in the nature and timing of the observed changes. Overall it fits unremarkably into the generalised marsupial pattern, with the first four of the eight maturation changes listed above occurring in all the dasyurids examined for this study.

Loss of the cytoplasmic droplet from the proximal midpiece, and the rotation of the nu-

Table 2. Epididymal sperm maturation changes (modified from Harding *et al.* 1979a).

Family Species	Midpiece fibre network	Midpiece invagination	Acrosomal compaction	Rotation of head	Membranous whorls	Pairing
PHALANGERIDAE						
Trichosurus vulpecula	X	X	X	XX	X	—
MACROPODIDAE						
Macropus robustus	X	—	X	XX	?	—
Macropus parma	X	—	X	XX	X	—
Megaleia rufa	X	—	X	XX	X	—
Potorous tridactylus	X	—	X	XX	X	—
PETAURIDAE						
Petaurus breviceps	X	X	—	X	—	—
Petaurus norfolcensis	X	X	—	?	—	—
Pseudocheirus peregrinus	X	X	X	—	—	—
PHASCOLARCTIDAE						
Phascolarctos cinereus	?	—	—?	—	—	—
DASYURIDAE[1]	X	—	—	X	—	—
PERAMELIDAE						
Perameles nasuta	?	—	—	X	—	—
Isoodon macrourus	?	—	—	X	—	—
DIDELPHIDAE						
Didelphis virginiana	X?	—	—	X	—?	X
Caluromys philander	X?	—	—	X	—?	X

X = structure present or process occurs.

— = structure absent or process does not occur.

1 = Based on the 13 species examined for this study.

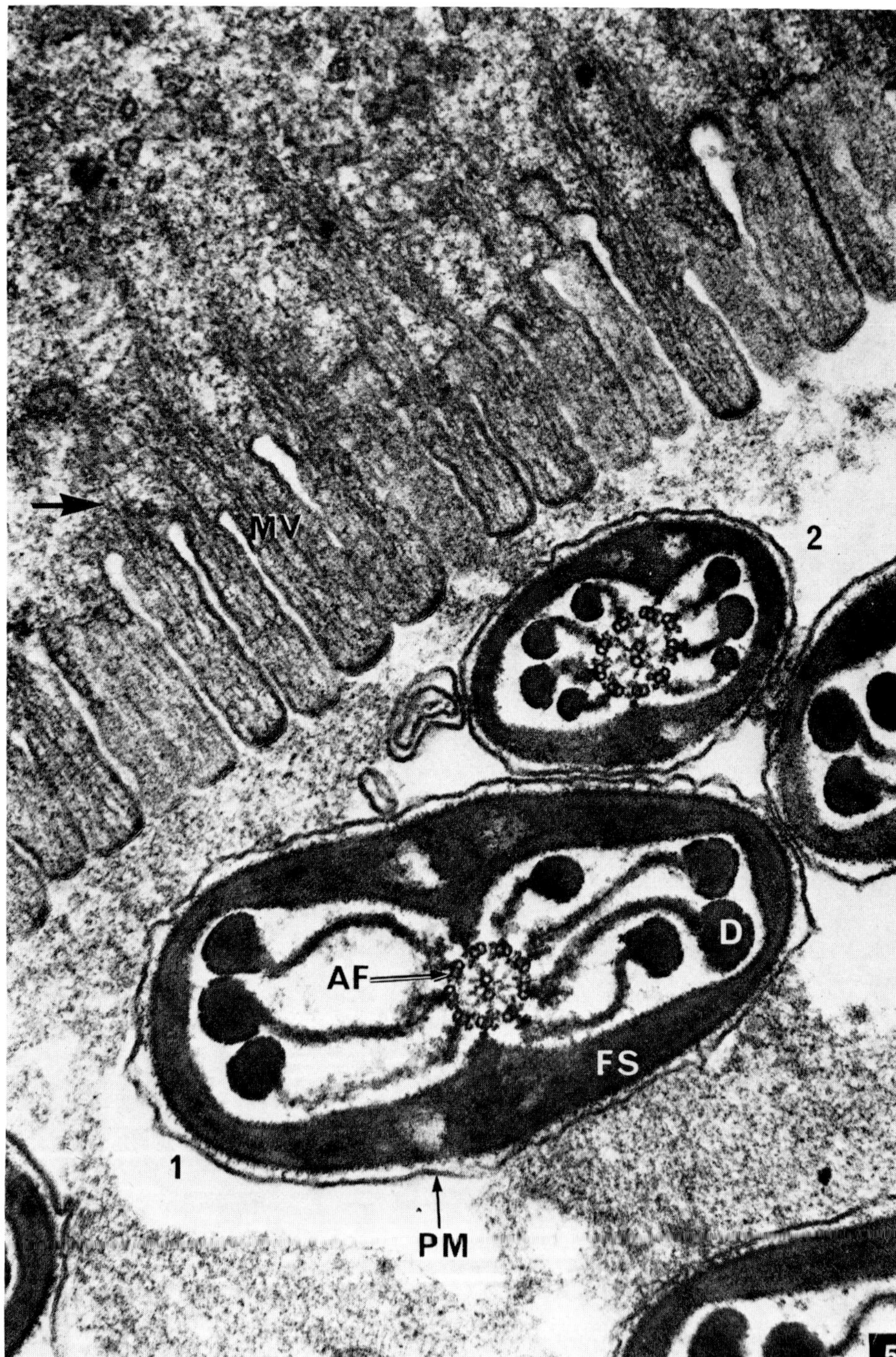

Fig. 5. Section through the corpus–cauda epididymidis of *Dasycercus cristicauda.* Microvilli (MV) line the epididymal epithelium and contain inner filaments (arrowed) which extend into the apical cytoplasm of the epithelium. Transverse sections of the principal piece are obvious in the lumen. A principal piece fibre network is not seen underlying the plasma membrane (PM). The dense outer fibres (D) are widely separated from the axoneme (AF) in section 1 but the gap has narrowed in the more distal section (2). FS fibrous sheath of principal piece. X 56,000.

cleus during epididymal sperm maturation, are features shared with virtually all marsupials examined to date, although minor variation in the region of the epididymis in which these changes occur is evident (Harding *et al.* 1979a). Similarly the occurence of maturation changes in the mitochondria appears to be a common feature in marsupials but this requires further investigation (Harding 1977). The development of a helically arranged midpiece fibre network is likewise the norm among marsupials (Figs 2a, b), although there are considerable differences between species in its nature and degree of development (Harding *et al.* 1979a). The network appears identical in all dasyurids examined and among other marsupials most closely resembles that found in *Trichosurus vulpecula, Petaurus breviceps, P. norfolcensis* and the macropodids. Some differences, between the network in the dasyurids and these diprotodonts are apparent, however, and have been discussed previously (Harding *et al.* 1979a). Perhaps the most interesting of these is that the helices are more tightly wound in the dasyurids, so that the separation between the centres of the fibres viewed in superficial longitudinal section (Fig. 2c) varies from 6nm – 80nm in the dasyurids but from 100nm – 130nm in the diprotodont species. We previously commented on the possible functional significance of this in noting that the narrower spacing of the fibres in the dasyurids appeared to match the nature of their mitochondrial sheath; namely, the mitochondria are more regularly arranged, narrowly spaced and compact in the dasyurids than in *T. vulpecula, P. breviceps, P. norfolcensis* or the macropodids (Harding *et al.* 1979a).

The extreme similarity between the dasyurids, *Tarsipes* and the peramelids in intraflagellar organisation, and at least between the dasyurids and *Tarsipes* in the nature of the mitochondrial sheath has been discussed above. Such close similarity between these groups does not apply for the midpiece fibre network however. Although a well developed network is found in *Tarsipes,* preliminary observations suggest that it differs from that in the dasyurids (Harding *et al.* 1981). A midpiece network is not well developed in the peramelids. Instead, a layer of granular material surrounds the mitochondrial sheath in caput epididymidal sperm in these species and this increases in thickness during epididymal transit. No sign of a fibre network is evident in transverse sections, but certain superficial longitudinal sections of the midpiece show a faint fibrous substructure (Harding *et al.* 1979a).

The above comments concern four maturation processes which have been shown to occur in the Dasyuridae. The final four changes of the eight listed above do not appear to occur in dasyurids. The absence of pairing of spermatozoa is not surprising since this phenomenon appears to typify American but not Australian marsupials. Sac–like invaginations of the midpiece plasma membrane develop only in some phalangeroids of the species examined so far, and thus their absence in the dasyurids aligns the latter with the majority of marsupial species.

Acrosomal changes of the type described for *Trichosurus vulpecula* (Harding *et al.* 1976a) do not occur in the dasyurids, but occasional small electron lucent areas are seen in the acrosomal material of some caput epididymidal sperm indicating that the acrosome may undergo very minor compaction during epididymal transit. The absence of the pronounced acrosomal maturation in the dasyurids concurs with our previous suggestions regarding the reasons for the occurrence of this process in certain marsupials but not in others (Harding *et al.* 1979a,b). It is worth noting here that of the two groups showing close links with the dasyurids in terms of sperm flagellar structure, acrosomal maturation changes are not evident in the peramelids, whilst minor changes occur in *Tarsipes* (Harding *et al.* 1981).

Perhaps the most interesting feature relating to epididymal sperm maturation in dasyurids concerns not the sperm, but the epididymal epithelium. In all the dasyurids we have examined to date, the luminal surface of the epididymal cells from the proximal half of the duct is lined with stereocilia (Fig. 4). In the corpus-cauda and more distal regions of the epididymis, stereocilia are no longer present and instead the luminal surface of the cells is covered with closely packed microvilli (Fig. 5). The latter resemble Fawcett's (1966) description of microvilli from the intestinal epithelium, particularly in their erect orientation, precisely ordered arrangement and the presence of filaments which run longitudinally in their interior and extend downward into the apical cytoplasm (Fig. 5). These features, as suggested by Fawcett (1966), are probably indicative of considerable structural stability in contrast to the much less rigid nature of the stereocilia from more proximal regions of the duct.

Microvilli of the type found in the dasyurids may be characteristic of this group. Among other Australian marsupials, the epididymal epithelium is lined throughout by sterocilia in *Trichosurus vulpecula, Petaurus breviceps, P. norfolcensis, Pseudocheirus peregrinus, Potorous tridactylus* and *Phascolarctos cinereus* (Harding 1977). We have not examined the epididymal epithelium in the peramelids, but in *Tarsipes spenserae* some similarity to the dasyurid condition is evident. Preliminary observations in this species show a change in the nature of the luminal border in the corpus–cauda and more distal regions of the epididymis, from the typical stereocilia which are found proximal to this, to structures which show the characteristics of microvilli, but which are longer and less closely packed than those in the dasyurids (Harding, Carrick and Shorey, unpub.obs).

The situation in American marsupials has not been reported fully enough for close comparison. Olson and Hamilton (1976) show microvilli lining the distal epididymal epithelium in *Caluromys philander,* but they are branched and rather irregularly arranged, and thus appear to differ considerably from the dasyurid microvilli. The lining of the proximal regions of the duct in this species is not described.

As far as we know, microvilli of the dasyurid type have not been found in eutherian epididymal cells. Thus the apparent uniqueness of the luminal border in the distal epididymis of the dasyurids (and possibly *Tarsipes*) raises the question of its functional significance, and this would seem an interesting topic for further study.

CONCLUDING REMARKS AND IMPLICATIONS FOR TAXONOMY

The major points which emerge from the above observations and discussion are as follows:

At least on the basis of the 13 species examined for this study, dasyurids display an extreme consistency in sperm form, such that separation of the species by qualitative examination of the sperm does not seem possible. Reliable separation may be possible for at least some species by quantitative techniques, and we are currently investigating this matter. A similar qualitative consistency in sperm form is evident in the two peramelids we have examined, *Perameles nasuta* and *Isoodon macrourus* (Harding 1977, Harding *et al.* 1979a), but since other peramelid species have not been studied it is not clear whether this consistency is maintained throughout the group. In contrast, sperm of the 4 macropodid species we have examined (*Macropus robustus, M. parma, Megaleia rufa* and *Potorous tridactylus*) can be separated quite easily by qualitative means, and the same is true of *Petaurus breviceps, P. norfolcensis* and *Pseudocheirus peregrinus* within the Petauridae (Harding 1977, Harding *et al.* 1979a).

At the interfamilial level, another major point is that dasyurids display a number of very unusual features in mature sperm ultrastructure and spermiogenesis. Close resemblance to dasyurid sperm is found only in the Western Australian Honey Possum *Tarsipes spenserae,* although a number of differences allow sperm of the two groups to be readily distinguished. The peramelids also share a number of features with these two groups: the pronounced radial displacement of the dense outer fibres from the axial filament complex; the almost identical diameter of the mid– and principal pieces at the annulus; and a lesser similarity is also evident in the shape and arrangement of the mitochondria in the midpiece and in the large size of the flagellum. Sperm of all other Australian and American marsupials examined so far differ quite markedly from those of the dasyurids.

The similarities between the dasyurids and peramelids in sperm structure support the long–held view of close relationship between these groups (e.g. Bensley 1903) and the more recent evidence of this from serology (Kirsch 1968). On the other hand *Tarsipes spenserae* is currently regarded as an aberrant marsupial species without apparent close affinities (Kirsch 1977); when it has been linked with other groups this has generally been with the phalangerids (e.g. Bensley 1903, Gregory 1910).

On the basis of sperm structure one would have to suppose either extreme convergence or close phylogenetic relationship between *Tarsipes* and the dasyurids. As discussed below, the latter does not necessarily imply that the two lineages had a recent divergence.

The very distinctive and highly unusual features of sperm structure shared by the dasyurids, *Tarsipes* and peramelids depart markedly from those of other mammals and do not suggest convergence. Although the ancient phylogenetic links between the peramelids and dasyurids can be argued convincingly on other grounds, the case against convergence for *Tarsipes* cannot be argued on the same grounds, and other relevant evidence bearing on this case is still required. For convergence to lead to such strikingly unusual features in sperm structure, especially without an apparent functional rationale, does not seem likely. The question of a functional rationale may, however, be contentious. It could be argued that the large size of the sperm flagellum in these groups requires structural modification (e.g. the radial displacement of the dense outer fibres) to facilitate motility. However, two points are relevant here. Firstly, it would seem unlikely that the 'solution' to this functional 'problem' should be so similar in the three groups since other 'solutions' to such a 'problem' have been suggested for other mammals (Fawcett 1970). However, it could be further argued that the marsupial prototype sperm allowed for only one type of 'solution', namely the very similar radial displacement of the dense outer fibres and its consequences in terms of mitochondrial arrangement etc. Secondly, the role of the various flagellar organelles in sperm motility is still far from clear. In the light of this, and the previous point, it seems more reasonable to assume that the similarities in sperm structure between these three groups indicate close phylogenetic relationship.

If one looks at current marsupial phylogenetic trees in terms of sperm structure, a number of comments can be made. As an example, Archer (1981) shows three lines diverging from a didelphoid origin to give rise to the Australian radiation – (1) borhyaenid/thylacinid, (2) myrmecobiid/dasyurid and (3) peramelid/notoryctid/tarsipedid/vombatoid/phalangeroid lines (see Fig. 6). Since the unusual flagellar features described above are known to occur in the modern species of the dasyurids of line 2, possibly the thylacines of line 1 (see Hughes 1965) and the early branching peramelids and *Tarsipes* of line 3, but not in the later branching groups of the third line, one can suggest the following:

(i) that since the unusual flagellar characteristics are shared by at least two of the three lines mentioned above, they were probably present in their didelphoid derived progenitors;

(ii) if the latter were representative of the American didelphoid radiation also, then we need to assume these characteristics were lost in all modern American descendants of the didelphoid line; and

(iii) were lost in all the later branching groups of the third line, namely the vombatoids and phalangeroids (at least on the basis of the macropodid, petaurid, phalangerid and vombatoid species examined so far).

Similar comments would apply to Kirsch's (1977) phylogeny.

If we accept the position of at least the dasyurids and peramelids in the current phylogenies (Archer 1981, Kirsch 1977), then we would argue for the existence of an ancestral sperm form for the major Australian radiation, which displayed the common characteristics of the dasyurids, *Tarsipes* and peramelids outlined above. The highly unusual and rather bizarre form of this sperm type in relation to flagellar structure has been stressed above, and thus our idea of an 'ancestral sperm form' is counter–intuitive in flagellar structure at least, in that it appears highly derived rather than generalised. It is entirely possible however, to design a sequence of stages through which phalangeroid[1] sperm could be derived from this 'ancestral form', and indeed we drew attention above to the long nucleus and minor displacement of the dense outer fibres from the axial filament complex in *Potorous tridactylus;* two features which could be regarded as 'links' with the dasyurid condition.

With respect to intercontinental relationships between the marsupial groups, the presence of sperm pairing in all the American species and none of the Australian suggests that the divergence between these groups took place before the radiation of the extant American families, namely the Didelphidae, Microbiotheriidae and Caenolestidae (see Fig. 6).

[1] The comments made here ignore the vombatoids, since one of the authors (Harding 1979) has suggested a rather different placement for this group on the basis of sperm structure, to that shown in most current phylogenies.

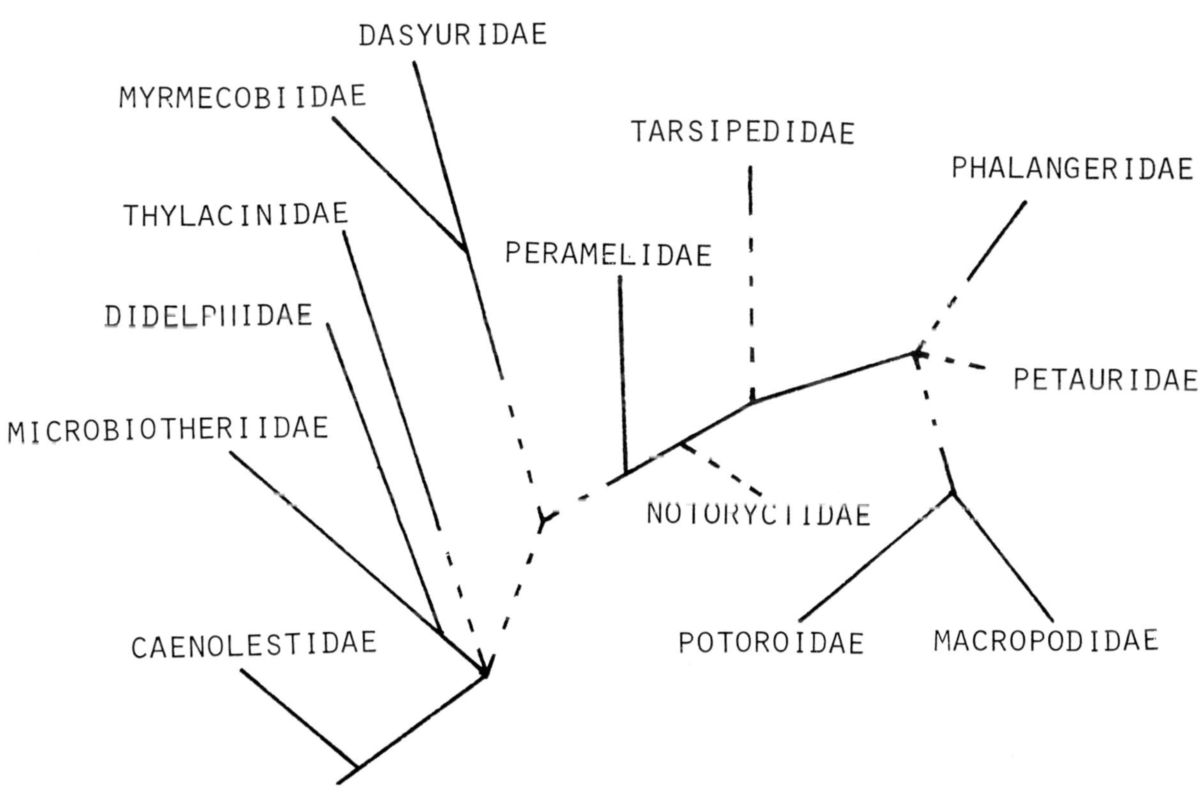

Fig. 6. Hypothetical phylogeny of marsupial families. Modified from Archer (1981) with some families excluded.

Finally, in line with the current phylogenies (Kirsch 1977, Archer 1981), the similarities between *Tarsipes*, the peramelids and the dasyurids detailed here need not imply a close modern affinity stemming from a recent divergence. Rather, the particularly close similarities between the dasyurids and *Tarsipes* may indicate conservatism with respect to sperm structure in these groups. Moreover, the consistency in sperm form within the Dasyuridae has been stressed above and it is interesting to note that the dasyurids appear to be a very 'compact' group by certain other criteria also. For example, Kirsch (1977) reported the serum proteins of all Dasyuridae to be very similar, whilst Martin and Hayman (1967) found that 6 dasyurid species they examined varied very little in their karyotypes. Perhaps the radiation of the 'modern' dasyurid form is relatively recent.

ACKNOWLEDGEMENTS

The authors would like to thank the following people for provision of animals: Dr. A. Burbidge for *Tarsipes spenserae;* Dr. P. G. Johnston for *Potorous tridactylus;* Dr. G. Lyne for *Isoodon macrourus* and *Perameles nasuta* and Dr. G. Maynes for *Macropus parma.* An earlier draft of this paper was constructively criticised by Drs. L. Hughes and G. B. Sharman.

Part of this study was supported by ARGC Grants: D1 7715525 to Drs. Shorey and Harding and D1 68 16887 to Dr. P. A. Woolley.

REFERENCES

Archer, M., 1981. A review of the origins and radiations of Australian mammals. Ch. 37. *In* "Ecological biogeography in Australia" ed. by A. Keast. Junk: The Hague.

Archer, M., 1982. Review of the dasyurid (Marsupialia) fossil record, integration of data bearing on phylogenetic interpretation, and suprageneric classification. Pp.397–443 *in* "Carnivorous marsupials", ed. by M. Archer. Roy. Zool. Soc. New South Wales: Sydney.

Bensley, A. B., 1903. On the evolution of the Australian Marsupialia: with remarks on the relationships of marsupials in general. *Trans. Linn. Soc. Lond. (Zool).* **9:** 83–217.

Bishop, M. W. H. and Austin, C. R., 1957. Mammalian spermatozoa. *Endeavour* **16:** 137–50.

Bishop, M. W. H. and Walton, A., 1960. Spermatogenesis and the structure of mammalian spermatoza. Pp.1–129 *in* "Marshall's physiology of reproduction" Vol. 1, Pt. 2 ed. by A. S. Parkes. 3rd ed. Longmans, Green and Co: London.

Cleland, K. W., 1965. Electron microscope structure of the dasyuroid sperm. *J. Anat., Lond.* **99:** 953.

Cleland, K. W. and Rothschild, Lord, 1959. The bandicoot spermatozoon: an electron microscope study of the tail. *Proc. roy. Soc. B.* **150:** 24–42.

Cummins, J. M., 1976. Epididymal maturation of spermatozoa in the marsupial *Trichosurus vulpecula:* changes in motility and gross morphology. *Aust. J. Zool.* **24:** 499–512.

Fawcett, D. W., 1966. "An atlas of fine structure: the cell, its organelles and inclusions". W. B. Saunders: Philadelphia.

Fawcett, D. W., 1970. A comparative view of sperm ultrastructure. *Biol. Reprod.* **2**, *suppl.* **2:** 90–127.

Furst, C. M., 1887. Uber die Entwicklung der Samenkörperchen bei den Beutelthieren. *Arch. mikr. Anat.* **30:** 336–65.

Gregory, W. K., 1910. The orders of mammals. *Bull. Amer. Mus. Nat. Hist.* **271:** 1–524.

Harding, H. R., 1977. "Reproduction in male marsupials: a critique, with additional observations on sperm development and structure". Ph.D. thesis: University of New South Wales, Sydney.

Harding, H. R., 1979. Affinities of the Koala on the basis of sperm ultrastructure. *Aust. Mamm. Soc. Bull.* **6:** 22–3.

Harding, H. R., Carrick, F. N. and Shorey, C. D., 1975. Ultrastructural changes in spermatozoa of the Brush-tailed Possum, *Trichosurus vulpecula* (Marsupialia), during epididymal transit. Part I: the flagellum. *Cell Tiss. Res.* **164:** 121–32.

Harding, H. R., Carrick, F. N. and Shorey, C. D., 1976a. Spermiogenesis in the Brush-tailed Possum, *Trichosurus vulpecula* (Marsupialia): The development of the acrosome. *Cell Tiss. Res.* **171:** 75–90.

Harding, H. R., Carrick, F. N. and Shorey, C. D., 1976b. Ultrastructural changes in spermatozoa of the Brush-tailed possum, *Trichosurus vulpecula* (Marsupialia), during epididymal transit. Part II: the acrosome. *Cell Tiss. Res.* **171:** 61–73.

Harding, H. R., Carrick, F. N. and Shorey, C. D., 1977. Spermatozoa of Australian marsupials: ultrastructure and epididymal development. Pp.151–2 *in* "Reproduction and evolution, fourth international symposium on comparative biology of reproduction" ed. by J. H. Calaby and C. H. Tyndale-Biscoe. Australian Academy of Science: Canberra.

Harding, H. R., Carrick, F. N. and Shorey, C. D. 1979a. Special features of sperm structure and function in marsupials. Pp.289–303 *in* "The spermatozoon" ed. by D. W. Fawcett and J. M. Bedford. Urban and Schwarzenberg Inc.: Baltimore-Munich.

Harding, H. R., Carrick, F. N. and Shorey, C. D., 1979b. A comparative study of acrosome development during spermiogenesis and epididymal sperm maturation in some Australian marsupials. *Procs 11th Ann. Conf. Aust. Soc. Reprod. Biol.* Pp.82.

Harding, H. R., Carrick, F. N. and Shorey, C. D., 1981. Marsupial phylogeny: new indications from sperm ultrastructure and development in *Tarsipes spenserae? Search* **12:** 45–7.

Holstein, A. F., 1965. Elektronenmikroskopische untersuchungen am Spermatozoon des Opossum (*Didelphys virginiana* Kerr). *Z. Zellforsch.* **65:** 904–14.

Hughes, R. L., 1965. Comparative morphology of spermatozoa from five marsupial families. *Aust. J. Zool.* **13:** 533–43.

Hughes, R. L., 1976. Reproduction in the male marsupial Devil *Sarcophilus harrisii* with particular reference to the ultrastructure of epididymal spermatozoa. Pp.908–10 *in* "Pathology of reproduction and A.I.". *Proc. VIIIth Int. Congr. Reprod. and A.I.* Vol. 4. Krakow, July, 1976.

KIRSCH, J. A. W., 1968. Prodromus of the comparative serology of Marsupialia. *Nature, Lond.* **217:** 418–20.

KIRSCH, J. A. W., 1977. The comparative serology of Marsupialia, and a classification of marsupials. *Aust. J. Zool. Suppl. Ser. No.* **52:** 1–152.

KRAUSE, W. J. AND CUTTS, J. H., 1979. Pairing of spermatozoa in the epididymyis of the Opossum *(Didelphis virginiana)*: a scanning electron microsopic study. *Arch. histol. jap.* **42:** 181–90.

MARTIN, P. G. AND HAYMAN, D. L., 1967. Quantitative comparisons between the karyotypes of Australian marsupials from three different superfamilies. *Chromosoma* **20:** 290–310.

OLSON, G. E. AND HAMILTON, D. W., 1976. Morphological changes in the midpiece of Wooly Opossum spermatozoa during epididymal transit. *Anat. Rec.* **186:** 387–404.

OLSON, G. E., LIFSICS, M., FAWCETT, D. W. AND HAMILTON, D. W., 1977. Structural specializations in the flagellar plasma membrane of opossum spermatozoa. *J. Ultrastruct. Res.* **59:** 207–21.

PHILLIPS, D. M., 1970a. Ultrastructure of spermatozoa of the Wooly Opossum, *Caluromys philander. J. Ultrastruct. Res.* **33:** 381–97.

PHILLIPS, D. M., 1970b. Development of spermatozoa in the Wooly Opossum with special reference to the shaping of the sperm head. *J. Ultrastruct. Res.* **33:** 369–80.

PHILLIPS, D. M., 1974. "Spermiogenesis". Academic Press: New York and London.

RATTNER, J. B., 1972. Nuclear shaping in marsupial spermatids. *J. Ultrastruct. Res.* **40:** 498–512.

SAPSFORD, C. S., RAE, C. A. AND CLELAND, K. W., 1967. Ultrastructural studies on spermatids and Sertoli cells during early spermiogenesis in the bandicoot *Perameles nasuta* Geoffroy (Marsupialia). *Aust. J. Zool.* **15:** 881–909.

SAFSFORD, C. S., RAE, C. A. CLELAND, K. W., 1969. Ultrastructural studies on maturing spermatids and on Sertoli cells in the bandicoot *Perameles nasuta* Geoffroy (Marsupialia). *Aust. J. Zool.* **17:** 195–292.

SAFSFORD, C. S., RAE, C. A. AND CLELAND, K. W., 1970. Ultrastructural studies on the development and form of the principal piece sheath of the bandicoot spermatozoon. *Aust. J. Zool.* **18:** 21–48.

SCARLETT, G., 1978. "Reproduction in the male Honey Possum, *Tarsipes spencerae* Gray (Marsupialia: Tarsipedidae)". Unpublished Hons. thesis: La Trobe Uuniversity, Melbourne.

TEMPLE–SMITH, P. AND BEDFORD, J. M., 1976. The features of sperm maturation in the epididymis of a marsupial, the Brushtailed Possum, *Trichosurus vulpecula. Amer. J. Anat.* **147:** 471–500.

RESUMEN

54. Ultraestructura del esperma, espermatogénesis y maduración del esperma epididimal en los marsupiales dasiúridos: implicancias filogenéticas. Por H. R. Harding, P. A. Woolley, C. D. Shorey y F. N. Carrick.

Se describen los rasgos distintivos de la ultraestructura del esperma, espermatogénesis y madurción del esperma epididimal de 13 especies de 6 géneros de dasiúridos. Es evidente la extrema consistencia de la forma del esperma con la embriogénesis de estas especies. Se hicieron comparaciones con el esperma de otras familias de marsupiales. La más notable es que ciertos rasgos inusuales del flagelo del esperma de los dasiúridos son compartidos por *Tarsipes spenserae* y los perámelidos *(Perameles nasuta, Isoodon macrourus)*. Otros rasgos son compartidos por los dasiúridos y *Tarsipes*, pero nó los perámelidos. Se sugiere que las similitudes entre el-esperma de los dasiúridos, *Tarsipes* y perámelidos representan relaciones filogenéticas más que convergencia; y sobre esta base se hacen comentarios sobre los árboles filogenéticos actuales de los marsupiales y sobre una posible forma de esperma ancestral para los marsupiales australianos.

55. THE THYLACINE (THYLACINIDAE, MARSUPIALIA): HOW GOOD A PURSUIT CARNIVORE?

ALLEN KEAST[1]

The thylacinid, *Thylacinus cynocephalus,* the major pursuit carnivore of the Australian region, is less convergent structurally on its Holarctic counterpart, the placental Wolf *(Canis lupus),* than superficial criteria suggest. When considered relative to a series of smaller and more generalized dasyurids *(Antechinus flavipes* and *Dasyurus maculatus),* it departs from the series only in its significantly longer neck and slightly longer forelimb. However, the point is made that the smaller dasyurids already have somewhat long limbs and hence, to a degree, the Thylacine, as a dasyuroid, could be said to have been pre–adapted to a running role. In the ratios of the three limb segments, both in hindlimb and forelimb, the dasyuroids as a group differ from arctoids as a group (see analysis of the canids by Hildebrandt 1952).

The smaller dasyurids, as a group, have fairly long skulls. This is also the case for the skull of the Thylacine and, as Möeller (1968) notes, there are also certain similarities in this regard to the Wolf.

Basically, although it is clear that the Thylacine was adequate for its role on the isolated Australian continent as a pursuit carnivore, and that in its structural adaptations for this role it was similar to the placental Wolf, in virtually no structural adaptation was it as specialized as the Wolf.

Pages 675–84 *in* CARNIVOROUS MARSUPIALS, ed. by M. Archer, February, 1982. Roy. Zool. Soc. New South Wales: Sydney, Australia.

THE MARSUPIAL THYLACINE (*Thylacinus cynocephalus*) and placental Wolf (*Canis lupus*) provide one of the more striking examples of convergent evolution amongst the larger mammals. Each is the major pursuit carnivore of its respective biogeographic region. They share remarkably similar 'doglike' body forms, large body–size, skull shape, elongate jaws and sectorial molars. Since there is a close correlation between hindlimb and forelimb structure and kind of locomotion in mammals (Taylor 1976), considerable convergence might also be expected here. Simpson (1965), in a discussion of convergent evolution, points out that it never involved the entire body but only those parts or organic systems adaptive to the particular role. The Thylacine retains the full set of basic marsupial characteristics (see the discussion of Smith 1982). Simpson has also written, moreover, that virtually every bone in the Thylacine body is identifiable as such and distinct from its counterpart in the Wolf.

The most interesting single question posed by the resemblance of the Thylacine and Wolf is how closely the former resembles the latter in its adaptations to the pursuit carnivore role. Exploration of this question forms the basis of the present contribution.

METHODS

The approach used is to compare the Thylacine and Wolf relative to each other and in terms of two small and generalized members of their respective dasyuroid and arctoid assemblages. The examples of the former are the Yellow–footed Antechinus (*Antechinus flavipes*), a small carnivore and insectivore that is mainly terrestrial, and the Tiger Quoll (*Dasyurus maculatus*) a cat–sized arboreal and terrestrial form that is a major predator of possums and other small mammals. Equivalents in the arctoid series used are the common Weasel (*Mustela frenata*), a rat–sized terrestrial predator, and the Martin (*Martes americana*), an arboreal animal approximately equivalent to the Tiger Quoll and a generalized predator of squirrels, and other small mammals, as well as partly a fruit eater. The use of mustelids as generalized members of the arctoid assemblage is not ideal because they are relatively short–limbed and long–bodied, but they are still the most appropriate for the present purpose.

The major adaptive structures (limbs, skull, axial skeleton) form the basis of the study. The two species are compared with smaller, generalized members of their groups as well as with each other (Fig. 1). For comparative purposes, the presacral region of the spine is regarded as the 'conservative' segment and neck and limb length are investigated relative to this. A longer neck places the biting jaws further in advance of the body. Longer limbs, and equalization of the forelimb and hindlimb, improve speed and functional efficiency.

Sixteen sets of measurements were taken on each skeleton. The length of each limb segment was taken as the distance between the articulat-

[1] Biology Department, Queen's University, Kingston, Ontario, Canada, K7L3N6.

Fig. 1. The Thylacine *(Thylacinus cynocephalus)* and Wolf *(Canis lupus)*, relative to small generalized members of their respective series including the Brush-tailed Phascogale *(Phascogale tapoatafa)*, the Weasel *(Mustela frenata)*, and intermediate sized forms such as the Tiger Quoll *(Dasyurus maculatus)* and the Martin *(Martes americanus)*.

ing surfaces. The third metacarpal and fourth metatarsal, as the longest members of their series, provided a measurement for the terminal limb element. Addition of the figures for the proximal, medial, and terminal elements, gave the figure for total limb length. Lengths of the spinal segments were obtained by placing their centra end to end.

Only fully adult members of each species were measured. Since emphasis is on species differences, the sexes were grouped. This also removed the problem of having to deal with unsexed or dubiously sexed animals, exclusion of which would have made the series too small. Even so, and despite bringing together material from the larger museums of the world, it was usually not possible to get more than 15 good skeletons of each species.

Sexual differences are considerable in some carnivores, especially the Weasel (see Hall and Kelson 1959). In the Wolf, females average only 4% smaller than the males in a range of measurements (Hildebrand 1952). Sexual differences in the Thylacine are not great and are due largely to the slightly larger male skull, with its correspondingly longer face (Smith 1982).

In the making of comparisons, the general methodology of Hildebrand (1952) is followed since this enables comparisons to be made with a range of canids in addition to the Wolf. Ratios are used to bring the various species to the same unit length (Mayr *et al.* 1953), and from these to identify major adaptive shifts.

The use of ratios in evolutionary studies has recently been the subject of controversy and discussion (e.g. Atchley *et al.* 1976, Dodson 1978, Albrecht 1978, Atchley and Anderson 1978). Thus, contrary to what is commonly believed, ratios do not remove the effect of size from the data and, particularly when compounded, large and systematic statistical changes occur with respect to the structure and underlying distribution of data. These stem primarily from changes in the coefficient of variation of the denominator variable (Atchley *et al.* 1976). Nevertheless, the use of ratios in growth studies is relevant because all characters increase in absolute size and are all positively intercorrelated (Dodson 1978). Atchley and Anderson (1978), in a further statement, discriminate between 'doing calculations on a series of numbers which happen to be ratios (p. 73), which is appropriate, and using ratios as a basis for carrying out statistical analyses, which is not.

RESULTS

The Limbs

Hindlimb/spine (presacral region) ratios for the various species are given in Fig. 2, and fore/hindlimb ratios in Fig. 3. The proportions of the individual limb components (femur–tibia–4th metatarsal, and humerus–radius–3rd metacarpel), expressed as percentages, are given in Table 1. In the case of the former, the horizontal line indicates the observed range, the vertical line the arithmetic mean, the black rectangle plus or minus twice the standard error of the mean, and the white rectangle the calculated range (i.e. plus or minus 3 times the standard deviation of the mean).

Relative to the presacral spine, the hindlimb of the Thylacine is no longer than that of the two smaller and more generalized dasyurids. In the mustelid–canid series there is a progressive increase in the ratio from the Weasel through the Martin to the Wolf. The mustelids are short–limbed and long–bodied whereas the smaller dasyurids are of more 'normal' proportions. The hindlimb of the Wolf is relatively longer than that of the Thylacine.

The forelimb/hindlimb ratio of the Thylacine is only slightly greater than in the two smaller dasyurids. That of the Wolf is markedly greater

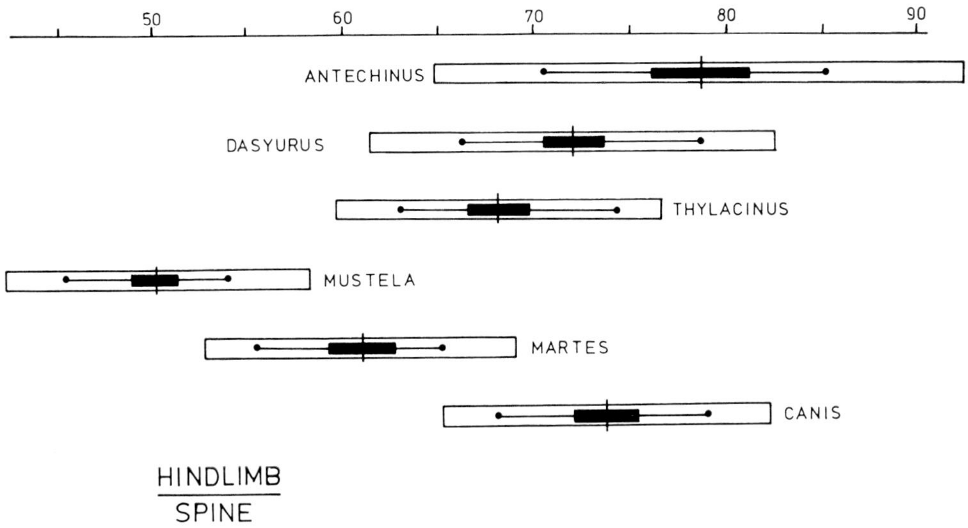

Fig. 2. Hindlimb/spine, the various species, expressed as percentages.

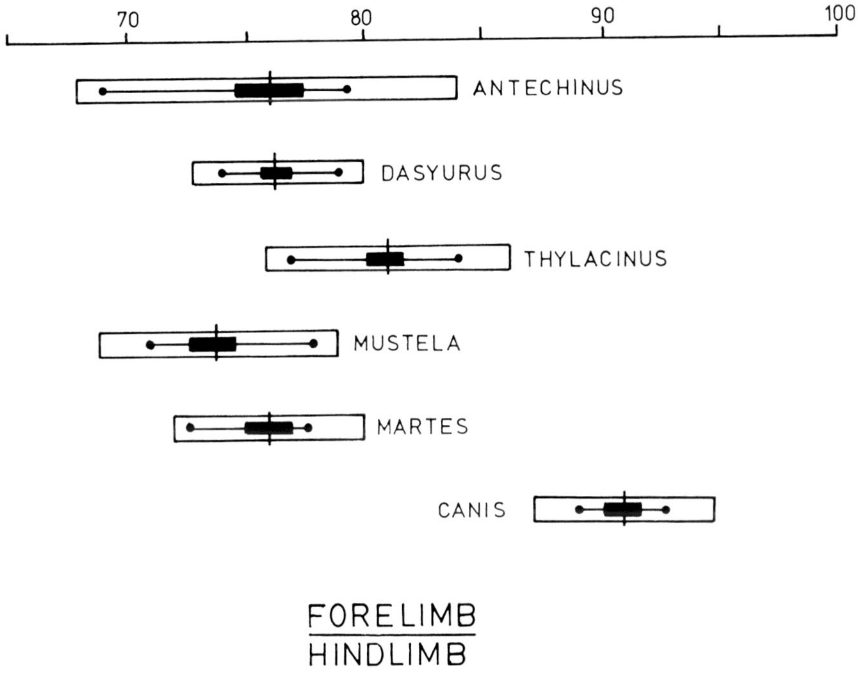

Fig. 3. Forelimb/hindlimb of the various species, expressed as percentages.

than those of the Martin and the Weasel and distinctly greater than that of the Thylacine which has 'generalized dasyurid' proportions.

In both these ratios, the Yellow-footed Antechinus shows the greatest individual variation (both in the observed and calculated range). The reason for this is not obvious. The Weasel and Martin do not display a greater variability than the Wolf, despite their great sexual size dimorphism. That is to say, the ratio of these characters is the same in the two sexes.

When the length of the components of each limb are expressed as percentages of the limb length, the metatarsal segment accounts for 15.7% in the Antechinus, 13.7% in the Tiger Quoll and 13.9% in the Thylacine; compared to 18.8%, 17.3% and 17.8% in the Weasel, the Martin and the Wolf respectively. The femur is relatively shorter in the mustelid–canid series than in the dasyuroids studied. In the forelimb, the metacarpal segment accounts for 12.3% of the length, 10.3% in the Tiger Quoll and 11.9% in the Thylacine; and 16.7%, and 15.9% and 17.4% in the Weasel, the Martin and the Wolf. The radius and humerus are of approximately equal length in the smaller marsupials, and are closely similar in the Thylacine. The humerus is

Table 1. Percentage lengths of the limb components in the hindlimb and forelimb. Standard error of the mean is given in brackets. Based on adult animals, males and females grouped.

	4th Metatarsal	Tibia	Femur	3rd Metacarpal	Radius	Humerus
DASYUROIDS						
Antechinus flavipes	15.7 (0.31)	43.0 (0.67)	41.3 (0.51)	12.3 (0.34)	45.2 (0.61)	42.5 (0.54)
Dasyurus maculatus	13.7 (0.43)	43.2 (0.11)	43.1 (0.17)	10.3 (0.18)	43.5 (0.22)	46.2 (0.26)
Thylacinus cynocephalus	14.0 (0.20)	42.9 (0.18)	43.1 (0.22)	11.7 (0.17)	43.8 (0.27)	44.5 (0.24)
CANOIDS						
Mustela frenata	18.8 (0.18)	42.4 (0.19)	38.8 (0.20)	16.7 (0.22)	34.5 (0.22)	48.8 (0.38)
Martes americana	17.3 (0.15)	42.9 (0.30)	39.8 (0.28)	15.9 (0.35)	37.5 (0.31)	46.6 (0.45)
Canis lupus (Arctic Canada)	17.8 (0.20)	42.5 (0.33)	39.7 (0.29)	17.4 (0.22)	42.3 (0.30)	40.3 (0.18)

significantly longer than the radius in the two mustelids. In the Wolf, however, the reverse is the case and the metacarpal and radius are of equal length.

Thus, in the forelimb of the three placentals, the terminal member is relatively long. This is also the case with the hindlimb except for the Wolf where it is disproportionately long. The Thylacine does not differ from the smaller dasyurids. In vertebrates, where an increase in limb length occurs, it is commonly the terminal member that demonstrates the greatest proportional increase (Dilger 1956).

These results gain an additional dimension when the ratios are compared as pairs of segments. The ratios of the radius to humerus and of the 3rd metacarpal to the total forelimb ratios (Figs 4, 5) are similar in the Thylacine and the smaller dasyurids. However, in the former, the Wolf has a significantly higher ratio than the other two placentals. The 3rd metacarpal to total forelimb ratio is uniformly higher in the placentals.

Tibia/femur ratios are little different within series but tend to be slightly higher in the placentals (Fig. 6). Fourth metatarsal to hindlimb ratios are uniformly higher in the mustelids and the canid (Fig. 7).

The Skull

In the skull width/length ratio (width at zygomatic arch to epicondylar length) the Thylacine is no different from the two smaller dasyurids (Fig. 8). In the Wolf, it is slightly higher than in the two smaller placentals. It corresponds well with the Tiger Quoll but slightly exceeds the Thylacine. Variability varies between the different species but no trend is discernible.

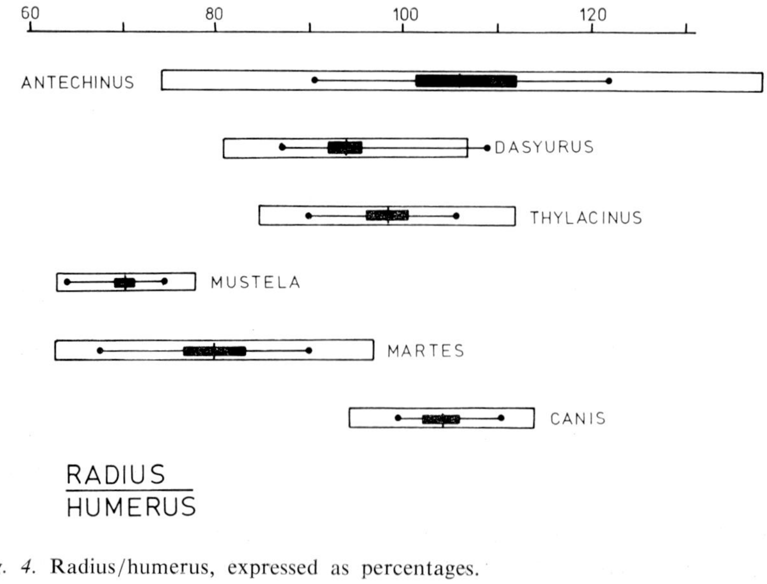

Fig. 4. Radius/humerus, expressed as percentages.

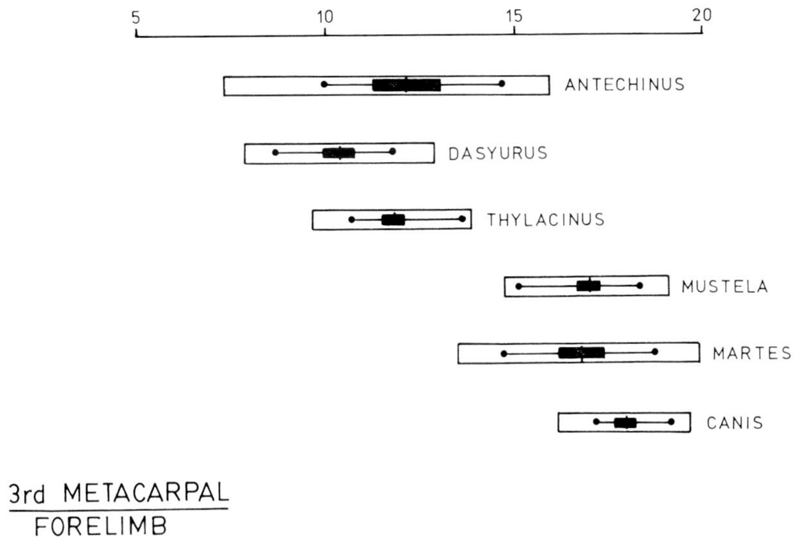

Fig. 5. Third metacarpal/forelimb, expressed as percentages.

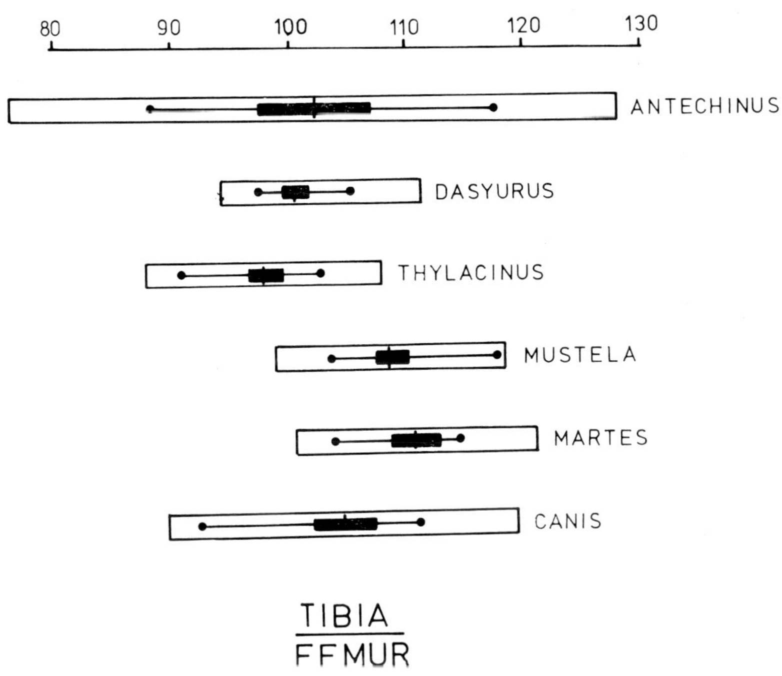

Fig. 6. Tibia/femur, expressed as percentages.

A high skull width/length ratio (along, of course, with depth) reflects massiveness and hence ability to hold and subdue prey.

Alternatively, an elongate skull, with forward position of the canines (the slashing and piercing members of the dental series) could be advantageous to a pursuit hunter. Here it would seem that the optimum skull length in the dasyuroids has been achieved in the Tiger Quoll. A more significant feature, functionally, is probably jaw length, which is in turn linked to the position of the articulation on the floor of the skull. The Thylacine has an elongate jaw (Smith 1982). (The Thylacine also has a long neck; see next section).

Major differences, of course, characterize the dentition of the placental and dasyuroid series. The wolf shows regional specialization of an advanced type. The rear molars are reduced to two, but one of these is a particularly large, flattened crushing element. Powerful carnassials in the cheek region fill the cutting and bone-breaking role. The Thylacine, by contrast, has a triangular (tribosphenic) molar series, with each tooth

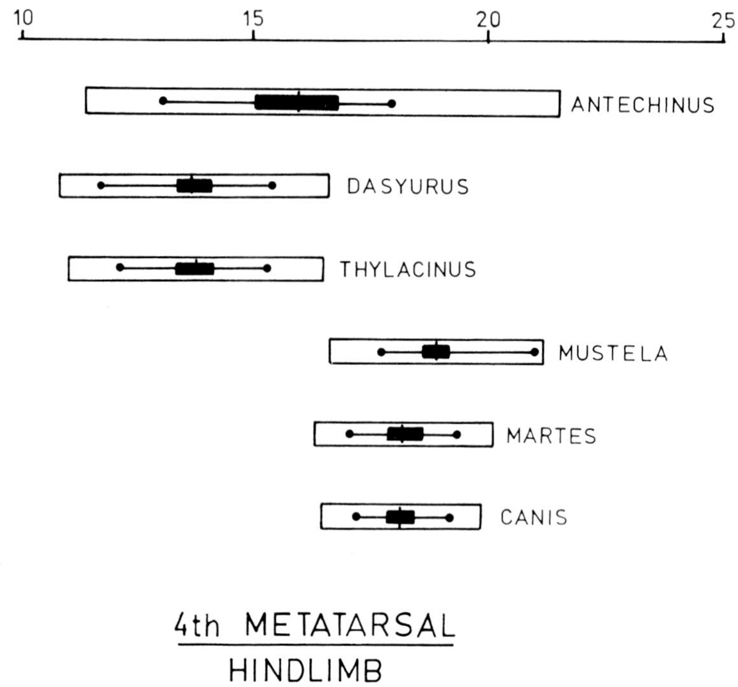

Fig. 7. Fourth metatarsal/hindlimb, expressed as percentages.

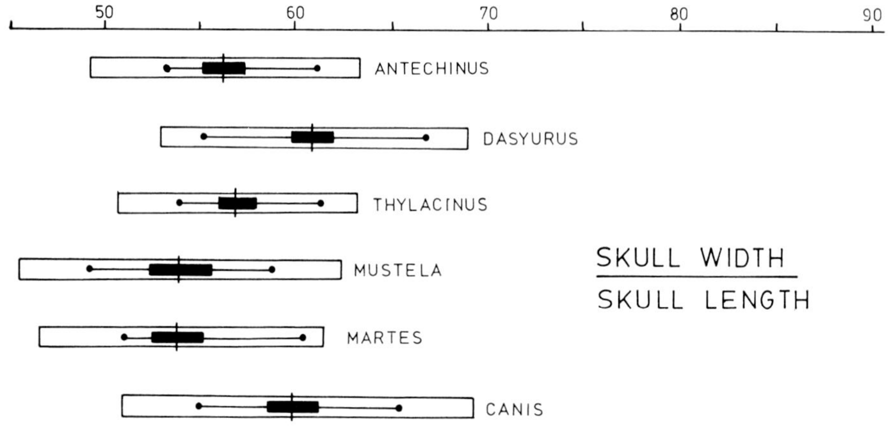

Fig. 8. Skull width/skull length, expressed as percentages.

having both a crushing and slicing function. Regional differentiation is relatively absent. Thus whilst the molar series in the Thylacine represents a more specialized condition than that of the smaller dasyurids, it shows less regional differentiation than the teeth of the Wolf. However, since the two systems are so different, it is difficult to equate them.

General Body Proportions

The basic body proportions of the Thylacine, its two smaller relatives, the Wolf and the two mustelids, are shown diagramatically in Fig. 9. They are based on mean lengths. To bring out differences in neck length, this section is here treated separately from the rest of the vertebral column and comparisons are made between it, the limb lengths, and the thoracic–lumber section of the vertebral column. The features are calculated as a percentage of the latter.

The necks of the Antechinus, the Tiger Quoll and the Thylacine form a series of increasing length. The neck of the Thylacine is 23% longer than that of the Tiger Quoll and 60% longer than in the Antechinus. The latter two are semi–arboreal, and terrestrial species, respectively, but the shorter necks can hardly be correlated with these habits. Rather, it would appear that the long neck of the Thylacine is adaptive for its particular role. The Wolf in turn has a longer neck than the two mustelids. Its neck, nevertheless, is 25% longer than that of the Thylacine.

As previously documented, the hindlimb/spine ratio in the Thylacine is no different from that in the two small dasyurids. In the Wolf, relative to the thoracic section of the spine, it is 30% longer than the Martin. It is also 30% longer than in the Thylacine. The Thylacine shows a very modest increase in forelimb length relative to the smaller dasyurids. In the Wolf, it is 56% longer than in the Martin and 30% longer than in the Thylacine.

THE ADAPTIVE ZONE OF THE THYLACINE RELATIVE TO THAT OF THE WOLF

As noted elsewhere in this Volume (Smith 1982), the way of life, hunting methods and diet of the Thylacine are imperfectly known. The general concensus is that it hunted in pairs, followed the scent or trotted after prey and then secured it with a final rush. However, as Smith points out, this account could be wisdom drawn by analogy with the Wolf. There is no evidence that it ranged far in the course of its hunts, although hunting routes across many miles of country have been inferred (Smith 1982). When pursued, its gait has been described as a 'shambling canter' (Troughton 1948). It has been said that the Thylacine occasionally engaged in a 'bipedal hop'. This has been refuted by some authorities (e.g. Ride 1970). A film made by the Government of Tasmania shows a Thylacine erect upon its hindlimbs (Smith 1982), but this of course does not provide evidence that it could progress using a bipedal hop.

The diet of the Thylacine was apparently small macropodids, echidnas, rats, birds and lizards. These are all items that would be better stalked that pursued.

DISCUSSION

The findings of Möeller (1968) that the Thylacine is, in effect, an 'overgrown' dasyurid and that its limbs and skull show little, if any advances over that of smaller members of the dasyurid series is confirmed. The present analysis brings out that it is different from the Wolf in virtually all structural features relating to the pursuit carnivore role. Only in its increased neck length does it show an adaptation that parallels that of the Wolf. The forelimb/hindlimb ratio shows a minor increase. This may be functionally quite significant in bringing the forelimb closer to the hindlimb in length and in so doing, leading to a more efficient stride.

It should be noted that dasyuroids as a group are all rather long–limbed. The hindlimb/spine ratio is about 68–78%, while in the Wolf it is only about 73%. The degree of difference from the Thylacine (68%) is not all that great. Compare this with the Weasel and the Martin that, like all mustelids, are long–bodied and short–limbed. The Wolf stands out from all the others in the length of the forelimb which is about 12% longer than in any of the other five species. This is obviously linked to the Wolf's ability to roam over great distances and run swiftly. It can be equated with long gait and probably a greater energetic efficiency.

The generally similar skull shape in the Wolf and the Thylacine is obvious. The Thylacine's skull width/length proportions are not very different from these features in the smaller dasyuroids and could be looked upon as the result of pre–adaptation. However, it also shows that the 'elongate' skulls of the Thylacine and the Wolf

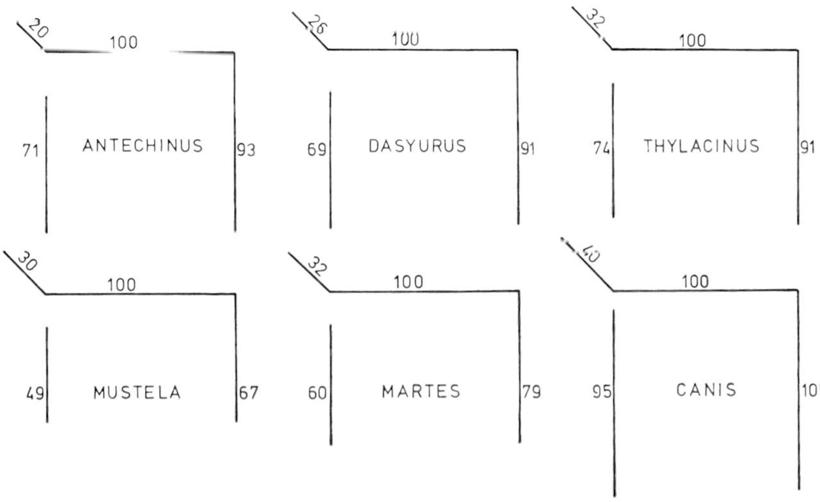

Fig. 9. Neck and individual limb lengths, expressed as a percentage of the trunk region of the vertebrae (thoracic and lumbar vertebrae added).

contrast strikingly with the foreshortened and massive ones of the scavanger and 'bone–crushing' members of their respective series: the Tasmanian Devil (*Sarcophilus harrisii*) amongst the dasyuroids; the Wolverine (*Gulo luscus*) amongst the arctoid carnivores; and also the Spotted Hyaena (*Crocuta crocuta*) within the African aeluroids (viverrids, cats and hyaenas).

The studies of Hildebrant (1952) show that some of the canids may differ fairly markedly from the Wolf in basic dimensions. This, of course, is to be expected in a series that contains forms as diversified and differently adapted as small, delicate foxes, the long–legged Neotropical Maned Wolf (*Chrysocyon brachyurus*), the short–legged Bush Dog (*Speothos venaticus*) and the African Hunting Dog (*Lycaon pictus*). Some of the differences between these species far exceed those between the Thylacine and the Wolf, and between the former and the smaller dasyurids. Thus the long–legged Maned Wolf has hindlimb/total spine ratio of 89–94%, and the short–legged Bush Dog of only 50–54%. Equivalent figures for the forelimb/hindlimb ratios are 99–106%, and 56–59%. The African Hunting Dog which, in ranging ability, stamina and speed, matches the Wolf, has a hindlimb/spine ratio of 65% and a forelimb/hindlimb one of about 73%. In the member of the canid series immediately below the Wolf in size, the Nearctic Coyote (*Canis latrans*), these two ratios are about 69–71% and 74–77% respectively. However, in foxes like the Grey Fox (*Urocyon cinereoargenteus*) they are about 57–59%, and 69–73% respectively. On the basis of these figures, the Thylacine does not come through too badly for the canid series as a whole: i.e. it is not very different from the generalized canids in limb proportions. It only differs from the more specialized pursuit carnivore members of the canid series.

Also of interest here is how the proportions of the different limb segments compare in the other canids studied by Hildebrand relative to the Wolf. In the case of the hindlimb, the tibia is invariably longer than the femur with the exception of the Bush Dog, where it is the same. The terminal member (longest metatarsal) ranges from 18–20% of the total (17.8% in the Bush Dog). In the forelimb, either the proximal or middle member may be longest, and the terminal member ranges from 16.6% to 19.2% of the whole. Thus there are differences between the canids and dasyuroids as groups. That is, in the three members of the latter analysed, the terminal member of the hindlimb ranges from 13.7–15.7% and in the forelimb, from 10.3–12.3%. The femur is shorter than the tibia and the humerus and radius are of equal length. In ratio of limb segments, the Thylacine does not break free from the basic dasyuroid mould.

The ratio length of cervical vertebrae to length of thoracics plus lumbars ranges from about 36% to about 49% in canids (Hildebrand) so that the 32% in the Thylacine, whilst high for dasyuroids, and equal to that in mustelids, lies below the limit of the canid range.

Möeller (1968) points out that the facial region of the skull of the Thylacine shows some similarity to the 'weaker' canids such as the Coyote and *Canis aureus*, and is more like these than the Wolf. However, he also finds that similarities between the Thylacine and the canids increase during the course of development. The length of the upper row of molars in the Thylacine and species of *Canis* become nearly identical, with occlusion of the teeth proceeding in the same manner, a 'pair of tongs' becoming a 'pair of scissors' (p. 338).

ACKNOWLEDGEMENTS

The writer would like to express his thanks to the authorities of the following museums for allowing him to measure the skeletons housed in their institutions: American Museum of Natural History, New York; Smithsonian Institution, Washington; Field Museum of Natural History, Chicago; Kansas State Museum, Lawrence; Transvaal Museum, Pretoria; British Museum, London; Royal Ontario Museum, Toronto; Australian Museum, Sydney; National Museum of Victoria, Melbourne; Tasmanian Museum, Hobart.

The work was carried out while the writer was a recipient of a research grant from the National Science and Engineering Research Council of Canada, the grant being administered by Queen's University, Kingston, Ontario.

I should like to acknowledge my indebtedness to these bodies and to the Trustees of the Australian Museum, Sydney, for the award of a Research Fellowship. Mike Archer and Ron Strahan provided useful criticism of a draft of this paper.

REFERENCES

Albrecht, G. H., 1978. Some comments on the use of ratios. *Syst. Zool.* **27:** 67–71.

Atchley, W. R., Gaskins, C. T. and Anderson, D., 1976. Statistical properties of ratios. I. Empirical results. *Syst. Zool.* **25:** 137–48.

Atchley, W. R. and Anderson, D., 1978. Ratios and the statistical analysis of biological data. *Syst. Zool.* **27:** 78–83.

Dilger, W. C., 1956. Adaptive modifications and ecological isolation mechanisms in the thrush genera *Catharus* and *Hylocichla*. *Wilson Bull.* **68:** 171–99.

Dodson, P., 1978. On the use of ratios in growth studies. *Syst. Zool.* **27:** 62–7.

Hall, E. R. and Kelson, K. R., 1959. "The mammals of North America". Ronald Press: New York.

HILDEBRAND, M., 1952. An analysis of body proportions in the Canidae. *Am. J. Anat.* **90:** 217–56.

MAYR, E., LINSLEY, E. G. AND USINGER, R. L., 1953. "Methods and principles of systematic zoology." McGraw–Hill: New York.

MOELLER, H., 1968. Zur Frage der Parallelerscheinungen bei Metatheria und der Eutheria. Vergleichende untersuchungen an Beutelwolf und Wolf. *Z. wiss. Zool.* **177:** 283–392.

RIDE, W. D. L., 1970. "A guide to the native mammals of Australia". Oxford Univ. Press: Melbourne.

SIMPSON, G. G., 1965. "The geography of evolution". Chilton Books: New York.

SMITH, M., 1982. Review of the Thylacine (Thylacinidae, Marsupialia). *In* "Carnivorous marsupials", ed. by M. Archer. Roy. Zool. Soc. New South Wales: Sydney.

TAYLOR, M. E., 1976. The functional anatomy of the hindlimb of some African Viverridae (Carnivora). *J. Morphol.* **148:** 227–49.

TROUGHTON, E. LEG., 1948. "Furred animals of Australia". Angus and Robertson: Sydney.

RESUMEN

55. Los Thylacine (Thylacinidae, Marsupialia): percecion carnivora? Por Allen Keast.

Los thylacinid *Thylacinus cynocephalus,* la mejor percecución carnivora de la región Australiana, es menos convergente estructural sobre holarctic contraparte, la placental Lobo *(Canis lupus),* la superficial criteria sugiere cuando considerada la relativa a una serie de pequeños y mas generalizado dasyurido (*Antechinus flavipes* y *Dasyurus maculatus*). Para partir desde las series en significante largo brazo. Como sea, el punto esta echo que los penqueños dasyuridos ya tengan largos miembros y des de aqui, para un grado, la Thylacine, como un dasyuroidos, pue de decir para haber estado preadactada para una corrida parte. En los radios de los tres miembros segmentos, cada uno en pierna y brazo, los dasyuridos como un grupo diferenciado desde arctoides como un grupo (ver analysis de los canidos por Hildebrandt 1952).

Los pequeños dasyuridos, como un grupo, tienen regularmente largas calaveras. Esto es tambien el caso por la calavera de la Thylacine, como Moeller (1968) notas, y esos son ciertos similares en esta observación a los Lobos. Basical, tambien es claro que los Thylacine estubo adecuado por partes sobre la continente Australiano como proposito carnivoro, y que es estructural adaptación por esta parte fue similal a la placental Lobo, en virtual no estructural adaptación fue como especializada como el Lobo.

56. SOME ASPECTS OF THE ECOLOGICAL AND GENETICAL STRUCTURE OF A NATURAL POPULATION OF *Antechinus stuartii* (DASYURIDAE, MARSUPIALIA)*

C. MORITZ[1]

A mark recapture study of the small dasyurid *Antechinus stuartii* on two grids at Cumberland Junction, Victoria, provides evidence for a high turnover in both males and females between successive generations. There is also evidence for considerable gene flow through the populations as at least 30% of females captured during the breeding season were immigrants. The absence of genetic differentiation at the transferrin locus between the grid populations is consistent with the occurrence of gene flow.

These observations are inconsistent with Braithwaite's (1973, 1979) proposal that *A. stuartii* population structure is characterised by localised demes and that the polygynous mating system results in limited gene flow.

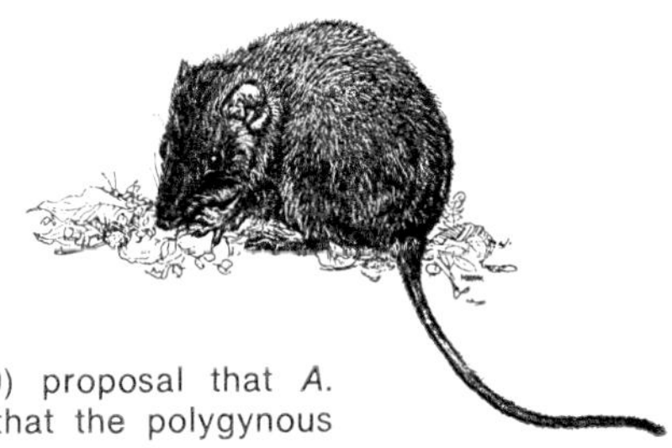

Pages 685–94 *in* CARNIVOROUS MARSUPIALS, ed. by M. Archer, February, 1982. Roy. Zool. Soc. New South Wales: Sydney, Australia.

THE SUBDIVISION of populations into small, partially isolated and internally panmictic demes has important biological consequences. By promoting random genetic drift, demic population structure has been proposed to allow for evolutionary trial and error and to increase the rate of evolution (Wright 1951, Wilson *et al.* 1975, Bush *et al.* 1977). However, random genetic drift is only likely to be a significant force where demes are small, the composition of demes is temporally stable and gene flow resulting from dispersal is minimal. (McCracken and Bradbury 1977). While there have been numerous studies of population structure of small placental mammals, small marsupials have been largely neglected in this regard.

Antechinus stuartii is a small dasyurid marsupial (♂♂ <70g, ♀♀ <35g) with a remarkable life history. At the end of the annual two week mating season complete male mortality occurs and also, few females survive to breed a second time (Woolley 1966, Wood 1970, Lee *et al.* 1977). On the basis of an intensive ecological and behavioural study, Braithwaite (1973, 1979) has suggested that the population structure of *A. stuartii* may consist of local demes centered in areas of favourable habitat. Furthermore Braithwaite proposed that behavioural interactions amongst males may largely restrict mating to a few dominant males.

This study attempts to examine population structure in *A. stuartii* by combining ecological and genetical techniques. Specifically, two questions are addressed:

(i) Is the composition of demes stable between generations?

(ii) Will dispersal lead to gene flow?

METHODS

HABITAT LOCATION AND DESCRIPTION: Following survey trapping of the Marysville region of the Great Dividing Range, Victoria (Fig. 1), the Cumberland Junction site was selected for a detailed study. This area has a high mean annual rainfall (2.64m) but for the duration of the study rainfall was approximately half the usual (Big River Records, Forestry Comm. Vic.).

The vegetation at Cumberland Junction was described by Gullan *et al.* (1979) as transitory between tall open forest ('wet sclerophyll') and open forest ('dry sclerophyll') and is dominated by messmate, *Eucalyptus obliqua*. Three species of small mammals were trapped during the study. These, in order of decreasing trap success were the Bush Rat, *Rattus fuscipes*, the target species *A. stuartii* and the Dusky Antechinus, *A. swainsonii*. The only potential predators seen were occasional Tiger Snakes, feral Cats and a Whistling Kite *Haliastur sphenurus*; however, no nocturnal observations were made.

FIELD TECHNIQUES: Live trapping at Cumberland Junction was conducted from February, 1979, to September, 1979, inclusive with a total of 5,350 trap nights. With the exception of line trapping conducted between February and

[1] Department of Genetics, Melbourne University, Parkville, Victoria, 3052. Present Address: Department of Population Biology, Research School of Biological Sciences, Australian National University, P.O. Box 475, Canberra City, A.C.T., Australia, 2601.

* Dedicated with deep affection to James Lord on the occasion of his 80th birthday.

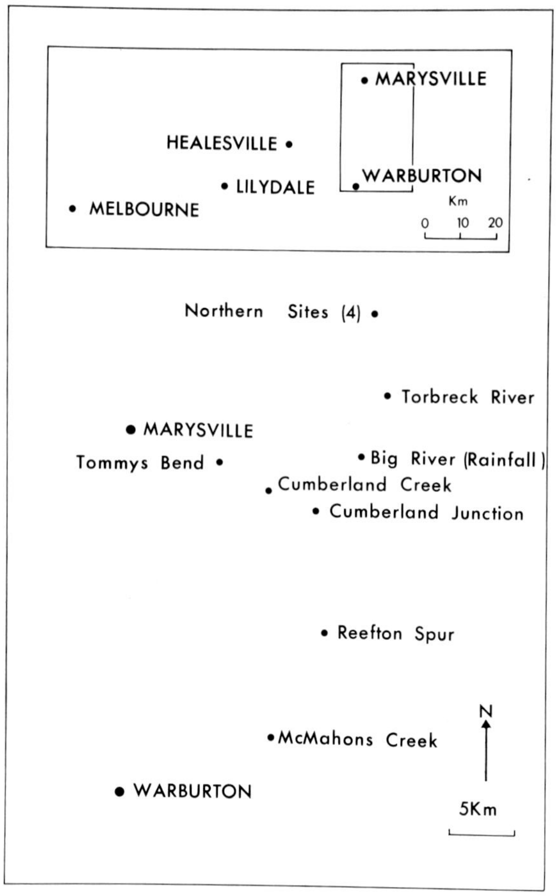

Fig. 1. Sampling localities.

March, the trapping methodology followed that of Myers (1974). Two grids, each a square grid matrix of 100 traps, were established in May. Traps were separated by 20m and the two grids by 100m. This intergrid distance was calculated so as to exclude routine movement of animals between grids but to allow detection of dispersal. No physical barriers to movement were evident between the grids.

One folding small mammal trap (Elliot Scientific Equipment Co., Upwey, Vic.) was set at each station on a grid for three consecutive nights per month from May to September and trapping on grid 2 followed that on grid 1. Traps were baited with a mixture of peanut butter, rolled oats and honey and precautions taken to minimise trap mortality.

At first capture *A. stuartii* were uniquely toe-clipped and a blood sample collected in a heparinised microhaematocrit tube (Clay Adams Inc., U.S.A.) by puncturing the sub–orbital sinus after light etherisation. In this and subsequent captures the identity, weight, reproductive condition and capture site were recorded. All animals were released at the point of capture.

Electrophoretic Procedures: Eletrophoresis was performed on commercially produced cellulose acetate gels ('Cellogel', Chemetron, Milan). The techniques for the preparation of samples, running and staining of gels are given by Baverstock *et al.* (1977, 1980). The transferrin bands were identified by the specific rivanol precipitation technique (Gartside 1972). All samples were analysed for transferrin phenotype within three weeks of collection with samples from previous months acting as mobility controls. The transferrin allozymes were designated fast if anodal and slow if cathodal with the superscripts F and S applied respectively.

RESULTS

Population Ecology: In the course of the study 41 male (144 captures) and 31 female (135 captures) *A. stuartii* were caught at Cumberland Junction. The total population size was estimated by the frequency of capture model (Caughley 1977) which relaxes the assumption of equal catchability. The fit of the theoretical distributions to the observed was examined with the Kolmogorov Smirnov test (Sokal and Rohlf 1969). Although it was not possible to arrive at a single estimate for each month, the negative binomial model consistently failed to describe the observed values and the number of animals caught was always less than the estimated total population size. Consequently the results presented and conclusions drawn in this study can only be applied to the trap–revealed population.

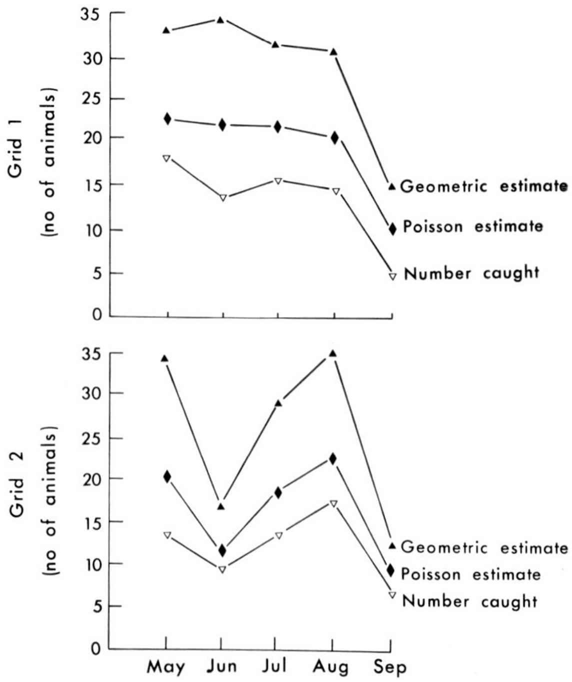

Fig. 2. Comparison of frequency of capture estimates of total population size and the number of animals caught.

A complete enumeration of this *A. stuartii* population may be precluded by a combination of arboreal activity (Wood 1970), torpor during winter (Wallis 1975) and variation in the behavioural response to traps (Roff 1973). Further analyses are based on the number of animals caught. On this basis the approximate density of the population fluctuated around three animals per hectare.

Observations of the reproductive condition of males and females indicated that the mating season spanned the August trapping period. At this time pronounced red swelling around the penis, pendulous testes and increased trap mortality was noticed in males. Females generally had well developed nipples and the pouch skin was flushed red. In the September trapping period all females trapped were either in advanced stages of pregnancy or had just given birth. Furthermore, three females gave birth while held and given a 26 to 35 day gestation period (Woolley 1973) this confirms that mating occurred during the August trapping period. The sex ratio remained even until September when no males were trapped. Only two second year females were caught and these have been included with the first year female cohort for further analyses.

Males were significantly heavier than females (for t–tests each month, all P <0.05) and steadily increased in weight towards the breeding season when the variance was minimal (Table 1). Conversely, excepting in August, females did not increase in weight.

been combined. Amongst both males and females, there were large numbers of new animals in the February to May period. In June and July, few new animals were caught. In August, the majority of males were new and conversely, no new females were detected at this time.

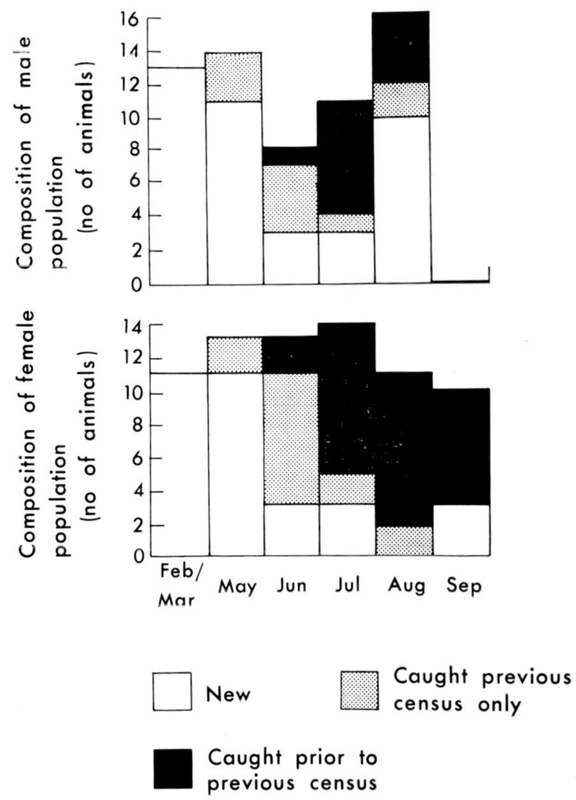

Fig. 3. Composition of *A. stuartii* population at Cumberland Junction during the study.

Table 1. Body weights (grams) of males and females at Cumberland Junction.

MONTH	MALE MEAN ± S.D. (Sample Size)	FEMALE MEAN ± S.D. (Sample Size)
FEB/MAR	18.5 ± 3.6 (17)	15.8 ± 1.9 (11)
MAY	20.5 ± 1.8 (17)	15.4 ± 1.3 (11)
JUNE	22.3 ± 2.3 (7)	15.3 ± 1.4 (11)
JULY	24.3 ± 3.7 (11)	15.6 ± 1.6 (13)
AUGUST	26.5 ± 1.7 (12)	17.6 ± 2.0 (9)

Population Composition and Turnover: Stability of demic composition in *A. stuartii* would not be represented by the same individuals being present between generations since all males and most females are dead before the second mating period (Wood 1970, Braithwaite 1979). Instead if mating was to be restricted to siblings then the composition of demes could be said to be temporally stable.

The number of new animals and those caught previously (defined as residents) for each sex in each month is shown in Fig. 3. The temporal pattern of population composition was the same for both grids and accordingly the data have

Of the new animals, those that were not recaptured in another month are defined as transients. This is supported by Wood's (1970) conclusion that individuals of *A. stuartii* not recaptured for several months were off-grid residents. As Fig. 4 shows, the level of transients of both sexes was high during February and March. In August all new males were transient by definition. The locations of transient male captures during the breeding season are of particular interest: only two of the thirteen captures were more than 20 m inside the grid area. The degree of population turnover is expressed as the proportion of native animals, i.e. those remaining in their natal site, present in the population. Since the grids were established and intensively trapped during May, the animals caught during and prior to May are taken to represent the native population. However, as the population at this time is likely to include animals that entered the population during the first dispersal phase (see discussion) this definition will underestimate population turnover. The results of the analysis are given in Table 2. The proportion of native males and females decreased between the February to May period and June. The level remained stable in the female population for the rest of the study; males however, show a further decrease (i.e. increased turnover) prior to the breeding season.

Movements and Home Range: The trapping programme, three consecutive nights per grid per month, was designed to facilitate the estimation of the magnitude of movements. The estimates used (Brant 1962) are the average distance between captures (Av. D) and the average maximum distance between captures (Av. M). The results are given in Table 3. Av. D was similar in both sexes ($t_{20} = 1.10$, $0.4 > P > 0.2$). Pairwise t–tests revealed a significant increase in

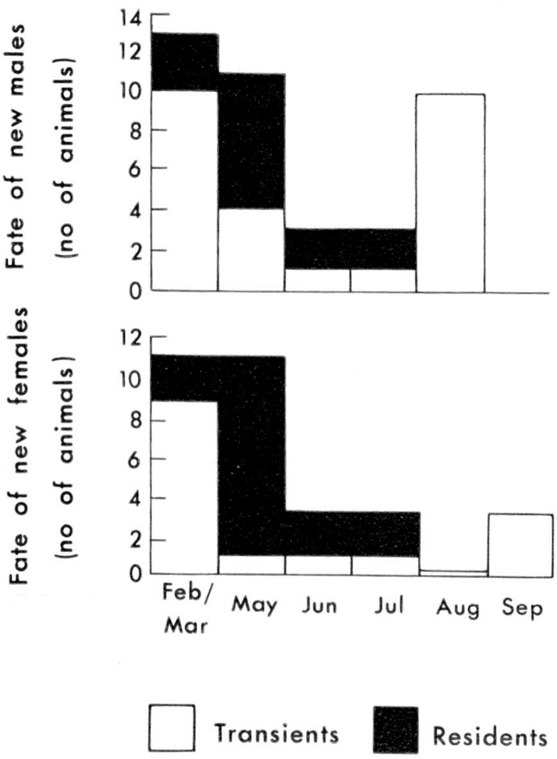

Fig. 4. Number of transients amongst new animals at Cumberland Junction during the study period.

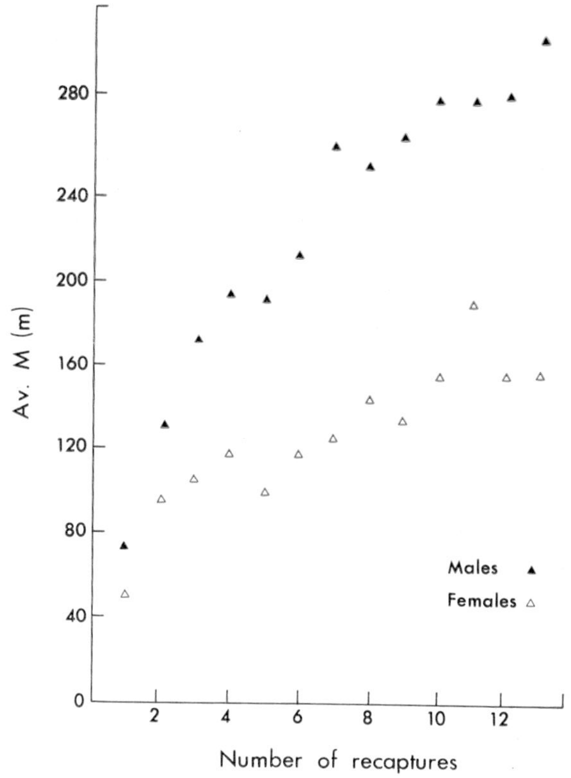

Fig. 5. Relationship between Av. M and the number of recaptures.

Table 2. Population turnover at Cumberland Junction.

		FEB–MAY	JUNE	JULY	AUG.	SEPT.
MALES	NATIVES* REMAINING	24	5	7	6	0
	TOTAL MALES IN POPULATION	24	7	11	16	0
	PROPORTION OF NATIVES IN POPULATION	1.00	0.71	0.64	0.38	
FEMALES	NATIVES REMAINING	21	9	8	8	5
	TOTAL FEMALES IN POPULATION	21	12	13	12	10
	PROPORTION OF NATIVES	1.00	0.75	0.62	0.67	0.50

* Trap deaths excluded.

Table 3. Summary of movement estimates for each sex. Numbers in brackets are sample sizes.

ESTIMATE (m)	MONTH				
	MAY	JUNE	JULY	AUG.	SEPT.
Av.D ± S.E.					
Male	40.56 ± 9.87 (10)	24.84 ± 6.97 (8)	61.85 ± 18.44 (8)	73.94 ± 11.49 (9)	
Female	25.96 ± 5.32 (8)	36.43 ± 6.86 (8)	71.6 ± 21.8 (8)	41.9 ± 10.97 (7)	43.44 ± 18.48 (6)
Av.M ± S.E.					
Male	94.3 ± 22.34 (11)	93.6 ± 25.9 (9)	123.9 ± 22.3 (8)	157.9 ± 38.0 (7)	
Female	53.34 ± 15.02 (8)	68.66 ± 15.74 (8)	77.77 ± 19.3 (9)	41.0 ± 13.0 (6)	65.11 ± 24.53 (6)

male movements towards the mating season (June c.f. August: t_{15} = 3.54, 0.01 > P > 0.001) whereas no such trend was evident in females. The apparent temporary increase in female movements in July approaches significance (May c.f. July: t_{14} = 0.1 > P >0.05). The absence of any clear change in population composition at this time indicates that this temporary increase is not associated with dispersal. Instead, this phenomenon, if real, may be due to more widespread foraging necessitated by unusually arid conditions in July (rainfall was 42% of the July mean). The mean maximum distance between captures (Av. M) is larger in males than females (t_{23} = 2.78, 0.025 > P > 0.01). The apparent increase in male Av. M towards the mating season is not significant. However, given the consistently large standard errors associated with the estimate, this is not surprising. Amongst males, Av. M continues to rise with increasing number of recaptures whereas female, Av. M increases only slightly after the eighth capture. (Fig. 5). These Av. M values are taken over the whole study period and hence represent an index of the lifetime home range.

The grid system employed failed to detect dispersal. While several individuals were captured on both grids this was attributable to their home ranges encompassing the adjacent borders of the two grids.

Home range overlap was assessed by comparing the observed distribution of individuals as revealed by trapping to a random distribution with the index of dispersion (Southwood 1978). If trap sites can be taken to represent home range loci then home range exclusion is revealed as even dispersion and extensive home range overlap as aggregation (Davies 1978, Metzgar 1979). This approach is compatible with the concept of a home range as an array of sites frequently visited (Leyhausen 1965, Jewell 1966, Ewers 1968). With exception of the males and females on grid 1 during the breeding season, the dispersion of individuals was random (Table 4).

POPULATION GENETICS: A preliminary electrophoretic survey revealed the transferrin (Trf) polymorphism which was subsequently used as a genetic marker. Three phenotypes were identified; single banded fast and slow homozygotes and a double banded heterozygote. A biallelic codominant autosomal system of inheritance is assumed.

The frequency of Trf alleles did not vary significantly over the area sampled (N = 240 G_6 = 2.714, P > 0.5). Smith's H statistic (Smith 1970) was used to test for deviations from Hardy–Weinberg equilibrium genotypic proportions (Table 5). The true value of the deviation is taken to lie within the 95% confidence intervals (i.e. ± 2 standard deviations) and a positive H value indicates a deficiency of heterozygotes. The only significant departures detected were an excess of heterozygotes at Torbreck River in February to March period and a defi-

Table 4. Summary of dispersion of individuals analyses. I_D is the index of dispersion (Southwood 1978) and approximates the χ^2 distribution ($\chi^2_{0.05[99]}$ = 123.23).

TEST			I_D	P
NON-BREEDING	GRID 1.	MALES	84.89	0.9 > P > 0.5
		FEMALES	84.55	0.9 > P > 0.5
	GRID 2.	MALES	107.62	0.5 > P > 0.1
		FEMALES	104.3	0.5 > P > 0.1
BREEDING	GRID 1.	MALES	124.74	0.05 > P > 0.025
		FEMALES	125.1	0.05 > P > 0.025
	GRID 2.	MALES	99.99	0.5 > P > 0.1
		FEMALES	88.28	0.9 > P > 0.5

ciency of heterozygotes amongst grid 2 residents in the May to September period.

For the Cumberland Junction population, G tests for independence (Sokal and Rohlf 1969) revealed no temporal variation on either grid for genotypic proportions of allele frequency. This allowed the pooling of data across months on each grid. Also, there was no association between trappability and genotype (Moritz, unpublished data). Two by two G tests with Yates correction (Sokal and Rohlf 1969) for comparisons of allele frequency between residents and transients, and between the residents of the two grids revealed no significant heterogeneity (Table 6). Similarly, no differences in genotypic composition between grids or with respect to status were apparent.

DISCUSSION

The life history pattern of *A. stuartii* observed in this study is concordant with the results of previous ecological studies (Fig. 6, Wood 1970, Braithwaite 1973, 1974, 1979, Lee *et al.* 1977, Braithwaite and Lee 1979). It incorporates two dispersal periods. The first of these affects both males and females and occurs at a time corresponding to the transition from communal to solitary organisation. The level of immigration associated with this dispersal phase cannot be estimated from this study as the increased trapping area associated with the replacement of live trapping by grids in May would be expected to reveal previously untrapped residents. Nevertheless, the failure to recapture 70% of animals caught prior to May implies considerable turnover.

The second dispersal phase involves only males and occurs during the breeding season. Statham has argued that the detection of new males at this time is due to increased trap susceptibility rather than immigration. At Cumberland Junction, *A. stuartii* male movements increased markedly during the breeding season

Table 5. Fit of observed genotypic distribution to the Hardy–Weinberg equilibrium expectations using Smith's H. (Smith 1970).

SITE	NUMBER OF ANIMALS	Trf^F/Trf^F	Trf^F/Trf^S	Trf^S/Trf^S	H ± 2 S.D.
FEBRUARY/MARCH					
Northern	10	0	2	8	–0.006 ± 0.058
Torbreck River	13	0	6	7	–0.1151 ± 0.100
Cumberland Jn.	30	3	9	18	0.0414 ± 0.069
Cumberland Creek + Tommys Bend	11	0	5	6	–0.028 ± 0.1084
MAY–SEPTEMBER					
Cumberland Jn.					
Residents Grid 1	21	3	5	13	0.081 ± 0.085
Grid 2	12	1	1	10	0.076 ± 0.074
Transients Grid 1	11	0	2	9	–0.004 ± 0.051
Grid 2	13	0	4	9	–0.019 ± 0.074
Total Grid 1	32	3	7	22	0.056 ± 0.058
Grid 2	25	1	5	19	0.023 ± 0.049
Reefton Spur	10	0	3	7	–0.017 ± 0.083
McMahons Creek	5	0	1	4	0.00

Table 6. Results of G tests for independence performed for the analysis of genetical structure at Cumberland Junction.

TEST	DEGREES OF FREEDOM	G	P
ALLELE FREQUENCY			
GRID–RESIDENTS–ALLELES	1	1.036*	0.1 <P<0.5
STATUS–ALLELES	1	0.938*	0.1 <P<0.5
GENOTYPIC COMPOSITION			
GRID–STATUS–GENOTYPE	7	9.052	0.1 <P<0.5
GRID–GENOTYPE	2	0.738	0.5 <P<0.9
STATUS–GENOTYPE	2	4.888	0.05<P<0.1
GRID–STATUS	1	1.946	0.1 <P<0.5

*G adjusted with Yates correction for small sample size.

and while it cannot be determined how far the transient males penetrated into the population, the predominance of peripheral captures indicates that these new males are in fact off–grid residents that have increased their movements. In contrast to the rainforest inhabiting population studied by Statham, the most plausible explanation for the appearance of new males in this population during the breeding season is immigration. The recapture data indicates a further 30% increase in male turnover resulting from this dispersal period.

The evidence presented above for considerable turnover of both males and females in this *A. stuartii* population clearly suggests that the composition of demes is unstable between generations. The critical question however, is whether this turnover actually leads to gene flow.

On the basis of contrived encounters between wild *A. stuartii,* Braithwaite (1973, 1974) suggests that during the period of solitary organisation males were territorial and females mutually avoided each other. However, defining a home as the area enclosed within the outermost capture loci, both Braithwaite (1973, 1979) and Wood (1970) report extensive home range overlap amongst males. This appears to contradict one of the essential features of territoriality which is the exclusion of rivals from the area (Brown and Orians 1970). This conflict is partially resolved by the approach taken in this study which indicated that extensive home range overlap does not occur amongst *A. stuartii* males.

Braithwaite has suggested that the main consequence of the male territoriality is to maintain the breeding success of dominant males. (Braithwaite 1973, 1979, Braithwaite and Lee 1979). In the highly heterogenous rainforest habitat where his study was conducted dominant males were found to occupy territories superior in prey abundance relative to those occupied by subordinates. Consequently the dominant males continued to increase in weight during the winter scarcity of prey items whereas subordinate males did not. He proposed that these heavy–dominant males were more active at the onset of the breeding season and so dominated mating. If male territoriality does indeed lead to a restriction of mating to dominant males as suggested by Braithwaite then social structuring in *A. stuartii* may limit gene flow despite the observed dispersal.

Braithwaite's hypothesis can be questioned on several grounds: (1) If valid it leads to the prediction that there should be a large variance in male body weight during the breeding season such as Braithwaite demonstrated at Mt. Glorious. Since the winter rainfall at Cumberland Junction was exceptionally low during the study period this effect on male body weight should be especially pronounced. In fact the variance in male weight at Cumberland Junction was lower during the breeding season than at any other time. A similar result in south-east Australia has been reported by Barnett *et al.* (1977). This evidence calls to question the generality of of Braithwaite's observations which may only apply to extremely heterogenous habitats in which the availability of prey items during winter is so reduced as to affect activity of animals in the mating season. (2) In the laboratory, females of *A. stuartii* have been observed to mate more than once (Woolley 1966) and there seems little reason to doubt that multiple mating occurs naturally. Since Braithwaite's hypothesis suggests that dominant males die first, multiple insemination would allow subordinate males to mate after the demise of dominant males. (3) If the prolonged copulation of *A. stuartii* (Marlow 1961, Woolley 1973) allows subordinate males to mate with other females or if the onset of oestrus is not perfectly synchronised then the mating success of subordinates would be increased. Clearly, these factors and the existence of multiple insemination in natural populations warrant further investigation.

Even if behavioural interactions between males precludes mating by subordinates, this is not likely to greatly effect gene flow. The first dispersal phase preceeds territory formation (Fig. 6) and consequently the animals forming territories are a mixture of immigrant and native animals. Braithwaite (1979) found no difference between transients and residents with respect to dominance. This is concordant with my observations: of the small number of new animals detected between May and September, most became residents.

The most substantial evidence for gene flow stems from the fact that immigrant females comprising at least 30% of the population, bred within the population. All of these females present during September were either pregnant or carrying young. Even in the unlikely event that gene flow did not result from male dispersal, successful reproduction by migrants of one sex is sufficient to produce appreciable gene flow (McCracken and Bradbury 1977, Greenwood *et al.* 1978).

No population subdivision was observed for the transferrin locus. This may be due to one or more of three causes. Firstly each grid population may represent a genetically heterogenous array of demes. Pooling data across these would obscure such variation. If this is the case then deficiency of heterozygotes due to the Wahlund effect (Crow and Kimura 1970) would be expected in the overall grid population. If the observed deficiency amongst grid 2 residents can be interpreted as a Wahlund effect then this

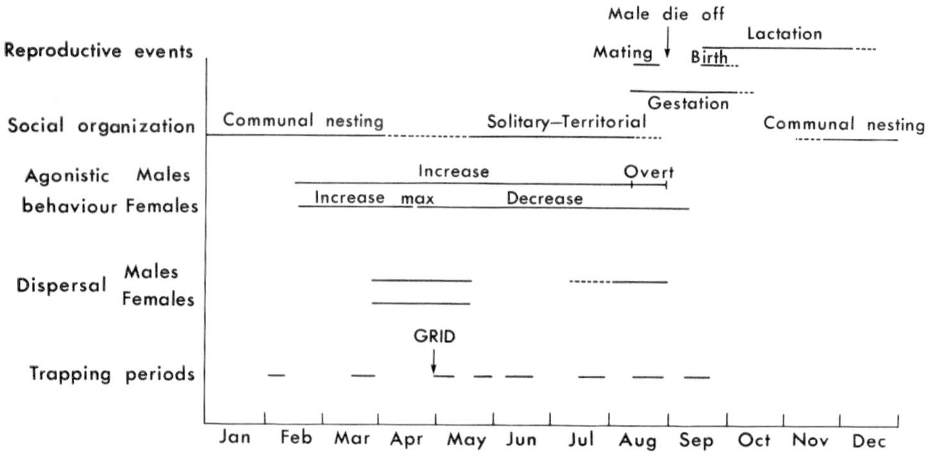

Fig. 6. Life history events in *A. stuartii* (from Braithwaite 1974; Lee *et al.* 1977; Braithwaite and Lee 1979 and this study).

may be an adequate explanation. The greatest heterozygote deficiency should occur prior to the first dispersal phase, however populations sampled at this time did not show this effect.

Secondly the lack of differentiation may be due to stabilising selection. Transferrin is polymorphic in many vertebrates (Manwell and Baker 1970) and differences in prenatal mortality between genotypes have been reported in inbred *Mus musculus* (Ashton and Dennis 1972) and the vole *Microtus ochrogaster* (Gaines *et al.* 1971). The degree to which selection has affected the pattern of variation at the transferrin locus in this population cannot be determined from this study.

An alternate explanation for the genetic homogeneity is that the two grid populations represent samples taken from the same deme within which appreciable gene flow occurs. This is consistent with the observed ecological structure which was characterised by high levels of population turnover, presumably resulting in gene flow.

When these ecological and genetical data are considered together, this *A. stuartii* population appears to conform more closely to the isolation by distance model than the island model of population structure (see Wright 1951). This hypothesis is currently being tested using electrophoretic techniques.

The generality of these results requires testing. However the density of this population is similar to that commonly found in southeastern Australia (Nagy *et al.* 1978) and if density adequately reflects habitat suitability then the outbreeding nature of *A. stuartii* may be general. High levels of population turnover and minimal genetic microdifferentiation have now been reported for a broad spectrum of small mammals (e.g. bats, McCracken and Bradbury 1977; deer mice, Fairbairn 1977; feral *Mus musculus* (cf. commensal populations) Bronson 1979; squirrels, Michener and Michener 1977, Schwartz and Armitage 1980). This indicates that models of small mammal population structure that propose extreme subdivision (e.g. Anderson 1970) are not valid for natural populations for many species (but see Daly 1980). In particular, models of mammalian evolution that assume a demic population structure (e.g. Wilson *et al.* 1975, Bush *et al.* 1977) may be unfounded. Clearly there is a need for further studies of small mammal populations that incorporate both ecological and genetical approaches.

ACKNOWLEDGEMENTS

This work was completed under the supervision of John Dearn who provided help in all aspects. I am also grateful to Tony Lee and Peter Baverstock for their assistance and to all those who accompanied me on field trips. The Fisheries and Wildlife department, Victoria, kindly allowed me to use their field station. The manuscript received constructive criticism from R. Crozier, B. Fox, J. Benzie, J. Daly, J. Dearn, B. John, J. Myers and J. Smith.

REFERENCES

Anderson, P. K., 1970. Ecological structure and gene flow in small mammals *Symp. Zool. Soc.* (*Lond.*) **26:** 299–325.

Ashton, G. C. and Dennis, M. N., 1971. Selection at the transferrin locus in mice. *Genetics* **67:** 253–65.

Barnett, J. L., How, R. A. and Humphries, W. F., 1977. Small mammal populations in pine and native forests in north-eastern New South Wales. *Aust. Wildl. Res.* **4** (3): 233–40.

Baverstock, P. R., Watts, C. H. S. and Cole, S. R., 1977. Electrophoretic comparisons between allopatric populations of five Australian pseudomyine rodents (Muridae). *Aust. J. Biol. Sci.* **30:** 471–85.

BAVERSTOCK, P. R., WATTS, C. H. S., ADAMS, M. AND GELDER, M., 1980. Chromosomal and electrophoretic studies of Australian *Melomys* (Rodentia: Muridae). *Aust. J. Zool.* **28:** 553–74.

BRAITHWAITE, R. W., 1973. "An ecological study of *Antechinus stuartii*". Unpublished M.Sc. thesis: Queensland University, Brisbane.

BRAITHWAITE, R. W., 1974. Behavioural changes associated with the population cycle of *Antechinus stuartii*. *Aust. J. Zool.* **22:** 45–62.

BRAITHWAITE, R. W., 1979. Social dominance and habitat utilisation in *Antechinus stuartii* (Marsupialia). *Aust. J. Zool.* **27:** 517–28.

BRAITHWAITE, R. W. AND LEE, A. K., 1979. A mammalian example of semelparity. *Am. Nat.* **113** (1): 151–5.

BRANT, D. H., 1962. Measures of the movements and population densities of small rodents. *Univ. Calif. Publs. Zool* **62** (2): 105–84.

BRONSON, F. H., 1979. The reproductive ecology of the House Mouse. *Quart. Rev. Biol.* **54:** 265–99.

BROWN, J. L. AND ORIANS, G. H., 1970. Spacing patiution in mammals. *Proc. Natl. Acad. Sci.* USA 239–62.

BUSH, G. L., CASE, S. M., WILSON, A. C. AND PATTON, J. L., 1977. Rapid speciation and chromosome evoiution in mammals. *Proc. Natl. Acad. Sci. USA* **74:** 3942–6.

CAUGHLEY, G., 1977. "Analysis of vertebrate populations". J. Wiley and Sons: New York, Toronto.

CROW, J. F. AND KIMURA, M., 1970. "An introduction to population genetics theory". Harper and Row: New York, Evanston and London.

DALY, J. C., 1981. The effects of social organisation and environmental diversity on determining the genetic structure of a population of the wild rabbit, *Oryctolagus cuniculus*. *Evolution* **35:** 689–706.

DAVIES, N. B., 1978. Ecological questions about territorial behaviour. Pp.317–50 *in* "Behavioural ecology" ed. by J. R. Krebs and N. B. Davies. Clackwell: Oxford.

EWER, R. F., 1968. "Ethology of mammals". Paul Elek (Scientific Books) Ltd.: London.

FAIRBAIRN, D. J., 1977. The spring decline in deer mice: death or dispersal. *Can. J. Zool.* **55:** 84–92.

GAINES, M. S., MYERS, J. AND KREBS, C. J., 1971. Experimental analysis of relative fitness in transferrin genotypes of *Microtus ochrogaster*. *Evol.* **25:** 443–53.

GARTSIDE, D. F., 1972. The *Littoria ewingi* complex in south–east Australia. III. Blood protein variation across a narrow hybrid zone between *L. ewingi* and *L. paraewingi*. *Aust. J. Zool.* **20:** 435–43.

GREENWOOD, P. J., HARVEY, P. H. AND PERRINS, C. M., 1978. Inbreeding and dispersal in the Great Tit. *Nature* **271:** 52–4.

GULLAN, P. K., PARKES, D. M., MORTON, A. G. AND BARTLEY, M. J., 1979. "Sites of environmental significance (botanical) in the Upper Yarra Valley and Dandenong Ranges". Ministry of Conservation publication: Victoria.

JEWELL, P. A., 1966. The concept of home range in mammals. *Symp. Zool. Soc. Lond.* **18:** 85–109.

LEE, A. K., BRADLEY, A. J. AND BRAITHWAITE, R. W., 1977. Corticosterold levels and male mortality in *Antechinus stuartii*. Pp.209–20 *in* 'The biology of marsupials' ed. by B. Stonehouse and D. Gilmore. Macmillan Scientific Press: London.

LEYHAUSEN, P., 1965. The communal organisation of solitary mammals. *Symp. Zool. Soc. (Lond.)* **14:** 249–63.

MANWELL, C. AND BAKER, C. M. ANN, 1970. "Molecular biology and the origin of the species". University of Washington Press: Seattle.

MARLOW, B. J., 1961. Reproductive behaviour of the marsupial mouse, *Antechinus flavipes* (Waterhouse) (Marsupialia) and the development of pouch young. *Aust. J. Zool.* **9:** 203–18.

MCCRACKEN, G. F. AND BRADBURY, J. W., 1977. Paternity and genetic heterogeneity in the polygynous bat, *Phyllostomus hastatus*. *Science* **198:** 303–6.

METZGAR, L. H., 1979. Dispersion patterns in a *Peromyscus* population. *J. Mamm.* **60** (1): 129–45.

MICHENER, G. R. AND MICHENER, D. J., 1977. Population structure and dispersal in Richardson's Ground Squirrels. *Ecology* **58:** 359–68.

MYERS, J. H., 1974. Genetic and social structure of feral House Mice populations on Grizzly Island, California. *Ecology* **55:** 747–59.

NAGY, K. A., SEYMOUR, R. S., LEE, A. K. AND BRAITHWAITE, R., 1978. Energy and water budgets in free living *Antechinus stuartii* (Marsupialia: Dasyuridae). *J. Mamm.* **59** (1): 60–8.

ROFF, D. A., 1973. An examination of some statistical tests used in the analysis of mark–recapture data. *Oecologia* **12:** 35–54.

SCHWARTZ, O. A. AND ARMITAGE, K. B., 1980. Genetic variation in social mammals: The marmot model. *Science* **207:** 665–7.

SMITH, C. A. B., 1970. A note on testing the Hardy Weinberg Law. *Ann. Hum. Genet. Lond.* **33:** 377–83.

SOKAL, R. R. AND ROHLF, F. J., 1969. "Biometry". W. H. Freeman and Co.: San Francisco.

STATHAM, H. L., 1982. Trappability of *Antechinus stuartii* (Dasyuridae, Marsupialia) at Petroi, northeastern New South Wales. Pp.333–44 *in* "Carnivorous marsupials" ed. by M. Archer. Roy. Zool. Soc. New South Wales: Sydney.

SOUTHWOOD, T. R. E., 1978. "Ecological methods". English Language Book Society and Chapman and Hall: London.

WALLIS, R. L., 1975. Torpor in the dasyurid marsupial *A. stuartii*. *Comp. Biochem. Physiol.* **53A:** 319–22.

WILSON, A. C., BUSH, G. L., CASE, C. M., AND KING, M. C., 1975. Social structuring of mammalian populations and rate of chromosomal evolution. *Proc. Natl. Acad. Sci. USA* **72:** 5061–5.

WOOD, D. H., 1970. An ecological study of *Antechinus stuartii* in a south–east Queensland rainforest. *Aust. J. Zool.* **18:** 185–207.

WOOLLEY, P., 1966. Reproduction in *Antechinus* spp. and other dasyurid marsupials. *Symp. Zool. Soc. Lond.* **15:** 281–94.

WOOLLEY, P., 1973. Breeding patterns, and the breeding and laboratory maintenance of dasyurid marsupials. *Exp. Animals* **22** *(Suppl.)*: 161–72.

WRIGHT, S., 1951. The genetic structure of populations. *Ann. Eugen.* **15:** 323–54.

RESUMEN

56. Algunos aspectos de la ecologica y genetica estructura de una natural populacion de *Antechinus stuartii* (Dasyuridae, Marsupialia). Por C. Moritz.

Una marca recapturada estudiada de el pequeño dasiuro *Antechinus stuartii* sobre dos acumuladas en Cumberland Juntion, Victoria, provienes evidencias para un alto numero en siertos masculinos y femeninos entre sucesivas generaciones. Tambien es evidente considerable gen fluir dentro la populacion por lo meños 30% de femeninos capturados dentro de la temporada reproducidora fueron emigrantes. La ausencia de genetica diferenciacion en la langosts transferencia entre la acumulada populacion esta consistente con la ocurrencia de gen fluir. Esas observaciones son inconsistentes con Braithwaite's (1973–1979) propuesta *stuartii* populacion estructura esta caracterisada por localidad demes y que la polygnous matutinos sistemas resultados en limitados gen fluir.

57. GEOGRAPHIC VARIATION IN THE EXTERNAL MORPHOLOGY OF *Sminthopsis crassicaudata* (DASYURIDAE, MARSUPIALIA)

S. R. MORTON[1] AND F. ALEXANDER[1]

Tail length and ear length of *Sminthopsis crassicaudata* increase northward through the species' range, while pes length and head and body length remain constant. Analysis of variance indicates that tail length is significantly related to latitude, but not to measures of moisture availability. Thus, changes in tail length, and probably ear length are consistent with Allen's rule. Because variation in these and other morphological characteristics is clinal, it is recommended that subspecific distinctions within *S. crassicaudata* be discarded.

Pages 695–8 *in* CARNIVOROUS MARSUPIALS, ed. by M. Archer, February, 1982. Roy. Zool. Soc. New South Wales: Sydney, Australia.

THE dasyurid marsupial *Sminthopsis crassicaudata* (Gould, 1844) is a small nocturnal insectivore (body weight 10–20 g) which is widespread in grassland and shrubland across southern Australia. It is characterized by an incrassated tail whose length does not exceed the head and body length, but shows considerable geographic variation. The variation in tail length, as well as in other external morphological characteristics such as ear length and foot–pad configuration, led Troughton (1964) to comment that the species displayed "bewildering variation" throughout its range. Later workers have remained satisfied that only one species is included within the taxon (Ride 1970, Archer 1981), but analysis of the patterns of variation has not been attempted. Tail length is generally least in individuals from southern populations and increases in more northerly populations (Spencer 1896, Thomas 1902, Finlayson 1933, Troughton 1964, Hope and Godfrey 1968, Morton 1978a,b, Archer 1981). This pattern is of interest because the tail is used as an organ of fat storage, and Morton (1978b) suggested that tail length may have increased in more arid northern regions to provide greater storage capacity. Alternatively, the tail may be shorter in southern *S. crassicaudata* in accordance with Allen's rule, i.e. that appendages become smaller in individuals inhabiting colder environments; thus, selection for fat storage may have been unimportant in determining tail length. In the present paper, variation in tail length is examined within the context of variation in other external chaarcteristics, and an attempt is made to distinguish between possible causes of geographic variation. Finally, subspecific distinctions with *S. crassicaudata* are examined in the light of this information.

MATERIALS AND METHODS

A total of 465 specimens of *S. crassicaudata* from the collections of the Australian Museum, the National Museum of Victoria, the Northern Territory Museum, the Queensland Museum, the South Australian Museum and the Western Australian Museum was examined. Measurements taken to the nearest mm from spirit specimens were head and body length, tail length, ear length, and pes length (*sine unguis*). Some measurements were copied from the labels of a series of skins in the South Australian Museum. Latitude and climatic data from the nearest meteorological station (average annual rainfall and average daily mean temperature; Bureau of Meteorology, 1966, 1969) were determined for the locality of each specimen. In addition, average evaporation at each locality was estimated from a map in Bureau of Meteorology (1968). The data were combined into climatic indices designed to provide measures of the availability of moisture to plants in the Australian arid zone (Gentilli 1971). The three dominant attributes of climate — annual rainfall, temperature, and evaporation — interact to produce varying amounts of moisture stress upon vegetation, and thus upon the invertebrates eaten by *S. crassicaudata.* Australian climatologists developed two indices in an effort to define such stress in terms of the readily available climatic factors (see Gentilli 1971 for a discussion): an index of aridity equal to rainfall X $(\text{temperature} + 10)^{-1}$; and a hydric index equal to rainfall X $(\text{evaporation})^{-0.7}$. Both indices were used in this study so as to broaden the search for possible influences of climate upon variation in *S. crassicaudata.*

Contour maps of tail length and ear length were created by dividing the range of *S. crassicaudata* into units of 2° latitude by 2° longitude and averaging the data within each unit; isopleths of mean values were then drawn by eye. To determine the relative effects of latitude and climate factors on changes in tail length, separate two–factor analyses of variance were conducted comparing the influence of latitude with

[1] School of Biological Sciences, Zoology Building, University of Sydney, New South Wales, Australia, 2006. Present address: Alligator Rivers Region Research Institute, Office of the Supervising Scientist, Jabiru, Northern Territory, Australia, 5796.

rainfall, index of aridity and hydric index. The latitudinal range of *S. crassicaudata* and each of the indices of moisture availability were divided into three equal spans, and a random sample of tail lengths was selected from each of the nine resultant cells for analysis.

RESULTS

Head and body length of *S. crassicaudata* is plotted against latitude in Fig. 1. A linear regression of the data upon which Fig. 1 is based has the formula

$$x = 0.1\ y + 70.5$$

where x = latitude and y = head and body length. The gradient of this regression does not differ significantly from zero (t-test, $0.2 > p > 0.1$); thus, head and body length does not vary with latitude. Similarly, pes length is constant throughout the species' range; the lowest and highest values recorded were 13 and 17 mm, and 95.5% of measurements were between 14 and 16 mm.

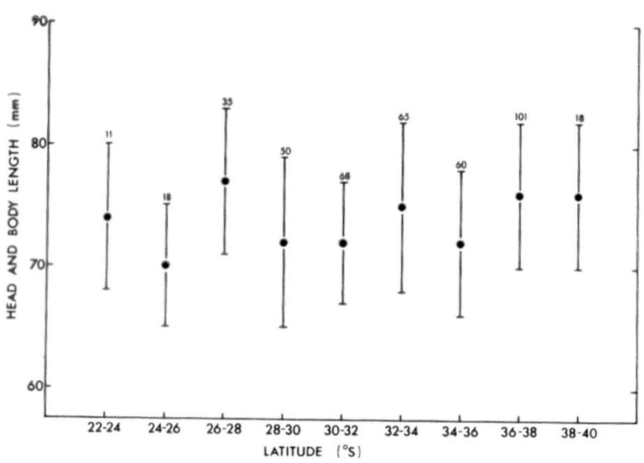

Fig. 1. Variation in head and body length of *S. crassicaudata* with latitude. Measurements of individuals from within bands of 2° latitude were pooled to give averages (●) and standard deviation (bars); numerals represent sample sizes.

In contrast, tail and ear lengths of *S. crassicaudata* vary considerably throughout the range with both increasing from south to north (Figs 2 and 3). Ear length was not analysed further, but tail length is negatively correlated with latitude (Fig. 4). A linear regression of the data upon which Fig. 4 is based has the formula

$$x = -1.7\ y + 110.9$$

where x = latitude and y = tail length. The gradient of this regression differs significantly from zero (t-test, $p < .001$). Because most of the range of *S. crassicaudata* lies within a region in which decreased latitude is strongly correlated with decreased moisture availability (see Gentilli 1971), however, similar regressions can be derived for tail length against rainfall, index of aridity and hydric index. Thus, two–factor analysis of variance was used to determine whether latitude or climatic factors concerned with moisture provided better correlations with tail length (Table 1; tail lengths transformed to natural logarithms to stabilize variances). Latitude had a significant relation to tail length in each test, and none of the indices of moisture availability had significant effects, either alone or in interaction with latitude. Thus, latitude is the most important factor influencing tail length.

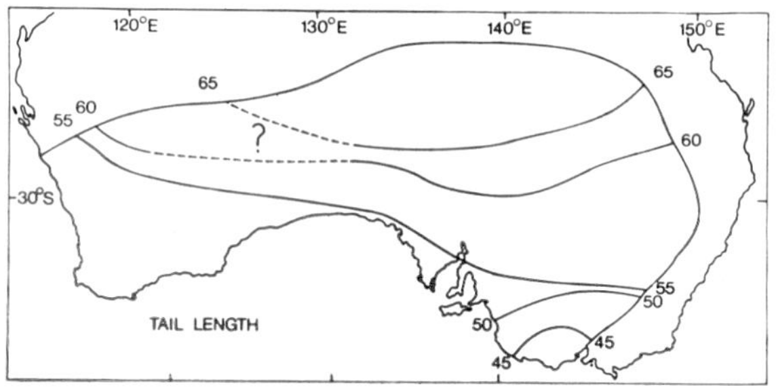

Fig. 2. The geographic range of *S. crassicaudata*, showing isopleths (in mm) of tail length.

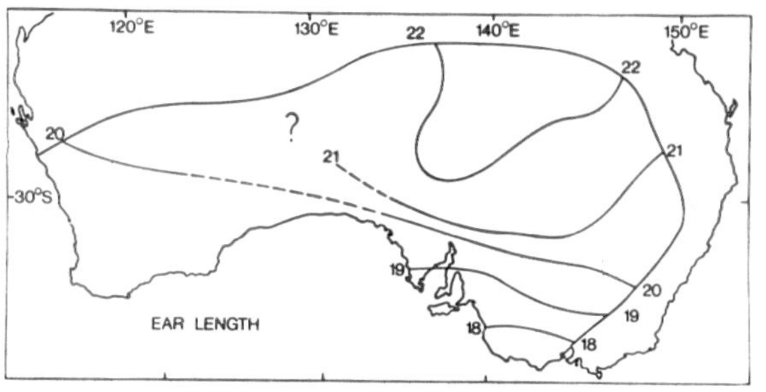

Fig. 3. The geographic range of *S. crassicaudata*, showing isopleths (in mm) of ear length.

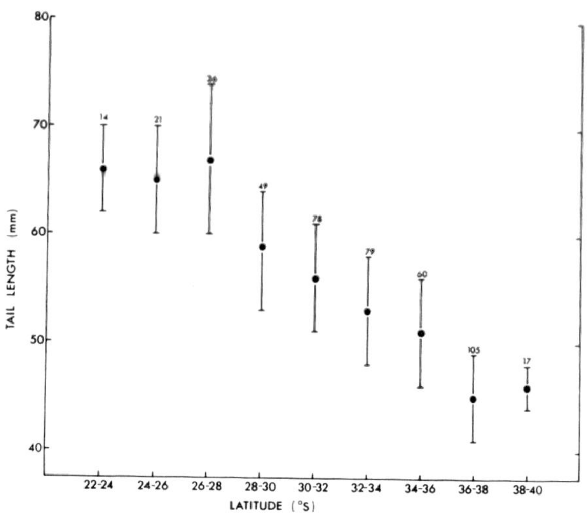

Fig. 4. Variation in tail length of *S. crassicaudata* with latitude. Measurements of individuals from within bands of 2° latitude were pooled to give averages (●) and standard deviations (bars); numerals represent sample sizes.

Table 1. Two–factor analysis of variance of tail lengths (log–transformed) of *S. crassicaudata.*

Source of variation	Sum of squares	d.f.	Mean square	F	Significance
	Latitude vs. Rainfall				
Latitude	.379	2	.190	21.13	$P < .001$
Rainfall	.003	2	.001	.14	NS
Interaction	.049	4	.012	1.36	NS
Residual	.485	54	.009	–	–
	Latitude vs. Index of Aridity				
Latitude	.281	2	.140	21.35	$P < .001$
Index of Aridity*	.012	1	.012	1.80	NS
Interaction	.029	2	.014	2.19	NS
Residual	.197	30	.007	–	–
	Latitude vs. Hydric Index				
Latitude	.521	2	.260	22.53	$P < .001$
Hydric Index	.054	2	.027	2.35	NS
Interaction	.072	4	.018	1.57	NS
Residual	.832	72	.012	–	–

* To gain satisfactory sample sizes, it was necessary to reduce the divisions of this index from three to two.

DISCUSSION

Causes of Geographic Variation

Body size remains constant throughout the range of *S. crassicaudata*, but tail and ear lengths decrease with increasing latitudes; this presumably represents a thermoregulatory response in accordance with Allen's rule. Thus, Morton's (1978b) suggestion that tail length has increased to enlarge fat stores in arid regions is not substantiated. Several authors have noted that the maximum incrassation of the tail in individuals from more arid regions is greater than that in moister regions (Finlayson 1933, Troughton 1964, Morton 1978b), and so it seems probable that the major response to increased aridity has been development of thicker tails rather than longer tails. More data are required to test this suggestion.

Morton (1978b) showed that variation in tail length was greater in populations of *S. crassicaudata* inhabiting more arid regions, and postulated that this heterogeneity reflected fluctuating food abundance and hence changing selective pressures for and against fat storage. In view of the lack of effect of aridity on tail length shown in the present study, such a situation is unlikely. Morton (1978b) also failed to demonstrate that significantly higher variation in tail length was not merely an allometric result of greater variation in body size within the arid zone population. Insufficient data from discrete populations were available in the present study to examine this problem further.

Taxonomic Implications

Thomas (1902) described the central desert representative of *S. crassicaudata* as a distinct subspecies, *S. c. centralis;* the type specimen came from Killalpaninna, Lake Eyre basin, South Australia. Finlayson (1933) documented the characteristics of this form more fully, and provisionally named a further form, *S. c. ferruginea,* based on specimens from the Macdonnell Ranges, Northern Territory. The characteristics distinguishing *S. c. centralis* from the typical subspecies were listed as longer tail, ear and limbs; paler colouration; and more coarsely granulated pads on the pes, tending towards a midline row of distinctly enlarged granules (Finlayson 1933; see Troughton 1964). Certain minor cranial differences were also noted. *S. c. ferruginea* was distinguished only in terms of its brilliant ferruginous colour, and Finlayson (1933) noted that dimensions of the specimens were comparable with those of specimens of *S. c. centralis.*

This study has shown that two of the distinctive characteristics of the central subspecies (longer tail and ear) are the result of clinal variation. The pes length is constant throughout the range, but even if lengths of other parts of the limbs increased towards the centre of the continent, it seems very likely that such variation would likewise be clinal. Similarly, populations of *S. crassicaudata* without enlarged granules on the hind foot–pads appear to occur only in the southeastern and southwestern corners of the range. Outside these regions, the foot–pads are more coarsely granulated, tending towards the development of a midline row of enlarged granules in the northern part of the range. Colour is tremendously variable and undoubtedly is a response to local edaphic factors (see Harrison 1975), pale individuals occur in many regions outside the Lake Eyre basin. Cranial characteristics also appear to display clinal variation (Archer 1981). Finally, nipple number does not change geographically in a regular manner (Morton 1978b).

Thus it is evident that the characters upon which *S. c. centralis* was based are the result of clinal variation or, in the case of colouration, adaptation to local edaphic conditions. Individuals with these characters occur not only in individuals from the centre of the arid zone, but also in western New South Wales, southern Queensland and inland Western Australia; hence, the subspecies cannot be geographically defined, and we recommend that the subspecies designation be discarded. Finlayson (1933) did not designate a type for his provisional taxon *S. c. ferruginea,* but it seems likely that he was referring to a series of brilliantly coloured *S. crassicaudata* presently held in the South Australian Museum (registration numbers M3490, M3491, M3494, M3495, and M3498). The locality for these individuals is given only as "Macdonnell Ranges". The specimens are not unusual in any characteristic apart from colour, and subspecific separation is likewise unwarranted. Accordingly, in the interests of introducing stability into zoological nomenclature, we propose that South Australian Museum no. M3495 (spirit preserved female with six young) be regarded as the Lectotype of *ferruginea,* and that *ferruginea* Finlayson be placed in the synonomy of *crassicaudata* Gould.

ACKNOWLEDGEMENTS

This study could not have been conducted without the cooperation and assistance of the curators who gave permission to examine specimens in their care: Ms. J. M. Dixon (National Museum of Victoria); Mr. B. J. Marlow (Australian Museum); Mr. P. F. Aitken (South Australian Museum); Dr. D. J. Kitchener (Western Australian Museum); and Dr. M. Archer (formerly of the Queensland Museum). Dr. A. A. Martin and Dr. M. Archer provided encouragement, Dr. G. J. Caughley and Mr. A. J. Press gave statistical advice and Mr. G. Ross aided in analysis. Dr. A. A. Martin, Dr. H. F. Recher and Dr. A. C. Hodson read and commented upon the manuscript. Mr. P. F. Aitken advised about the choice of the most appropriate specimen as lectotype for *ferruginea* Finlayson. The data were collected while S.R.M. held an Australian Commonwealth Postgraduate Research Award and grants from the M.A. Ingram Trust in the Department of Zoology, University of Melbourne. The analysis was conducted and the manuscript written during S.R.M.'s tenure as a postdoctoral Research Fellow in the University of Sydney.

REFERENCES

Archer, M., 1981. Results of the Archbold Expeditions No. 104. Systematic revision of the marsupial dasyurid genus *Sminthopsis* Thomas. *Bull. Amer. Mus. Nat. Hist.* **168:** 61–224.

Bureau of Meteorology, 1966. "Rainfall statistics, Australia". Government Printer: Melbourne.

Bureau of Meteorology, 1968. "Review of Australia's water resources – annual and monthly rainfall and evaporation". Government Printer: Melbourne.

Bureau of Meteorology, 1969. "Climatic averages, Australia". Government Printer: Melbourne.

Finlayson, H. H., 1933. On mammals from the Lake Eyre basin. Part 1. – The Dasyuridae. *Trans R. Soc. S. Aust.* **57:** 195–202.

Gentilli, J., 1971. The main climatological elements. Pp.119–88 *in* "Climates of Australia and New Zealand", ed. by J. Gentilli. Elsevier: Amsterdam.

Harrison, D. L., 1975. Desert coloration in rodents. Pp.269–76 *in* "Rodents in desert environments", ed. by I. Prakash and P. K. Ghosh. Junk: The Hage.

Hope, R. M. and Godfrey, G. K., 1968. Transferrin polymorphism in the Australian marsupial *Sminthopsis crassicaudata* (Gould). *Aust. J. Biol. Sci.* **21:** 587–91.

Morton, S. R., 1978a. An ecological study of *Sminthopsis crassicaudata* (Marsupialia: Dasyuridae) 1. Distribution, study areas and methods. *Aust. Wildl. Res.* **5:** 151–62.

Morton, S. R., 1978b. An ecological study of *Sminthopsis crassicaudata* (Marsupialia: Dasyuridae) III. Reproduction and life history. *Aust. Wildl. Res.* **5:** 183–211.

Ride, W. D. L., 1970. "A guide to the native mammals of Australia". Oxford University Press: Melbourne.

Spencer, B., 1896. Mammalia. Pp.1–52 *in* "Report of the Horn Expedition to central Australia, Part II — Zoology", ed. by B. Spencer. Dulau and Co.: London.

Thomas, O., 1902. Two new Australian small mammals. *Ann. Mag. Nat. Hist.* (7) **10:** 491–2.

Troughton, E. LeG., 1964. A review of the marsupial genus *Sminthopsis* (Phascogalinae) and diagnoses of new forms. *Proc. Linn. Soc. N.S.W.* **89:** 307–21.

RESUMEN

57. Variacion geografica de la morfologia externa de *Sminthopsis crassicaudata* (Dasyuridae, Marsupialia). Por S. R. Morton y F. Alexander.

La longitud de la cola y de las orejas de *Sminthopsis crassicaudata* se incrementa hacia el norte de su distribución, mientras que la longitud de los pies, de la cabeza y del cuerpo permanece constante. El análisis de la variación indica que la longitud de la cola está significativamente relacionada a la latitud, pero nó a la humedad aprovechable. Así, los cambios en la longitud de la cola, y probablemente los de las orejas, están de acuerdo a la regla de Allen. Como la variación de éstas y otras características es clinal, se recomienda que sean desconsideradas en las distinciones subespecíficas de *S. crassicaudata.*

58. ENCEPHALIZATION IN AUSTRALIAN MARSUPIALS

J. E. NELSON[1] and H. STEPHAN[2]

Data on brain weights (mg) and body weights (g) of 320 individuals of 89 species of Australian marsupials were analysed. The allometric equation, log brain weight = 1.938 + 0.592 log body weight, which was derived from data of all dasyurid species except those of *Planigale,* was used to calculate indices of encephalization for all marsupials and some eutherians. The indices for the *Planigale* spp. were distinctly lower than those for the rest of the dasyurids. Dasyurids, peramelids and burramyids have similar average indices which are lower than those of phalangerids, macropodids and vombatids. Within the Petauridae, *Pseudocheirus* spp. and *Schoinobates volans* have values among the lowest for marsupials, whereas the *Petaurus* spp., *Gymnobelideus leadbeateri* and *Dactylopsila trivirgata* have values among the highest. Brief comments are made on values within and between families and between marsupials and eutherians.

Pages 699–706 *in* CARNIVOROUS MARSUPIALS, ed. by M. Archer, February, 1982. Roy. Zool. Soc. New South Wales: Sydney, Australia.

MARSUPIALS have always been of great interest to mammalogists and to students of evolution because of their supposed primitiveness, and it is often implied, if not stated, that they represent a stage on the way to the evolution of 'true mammals'. Their brains have also aroused great interest, for the same reasons, and the differing brain structures within the order have been used as a basis for classification (Abbie 1941). Yet most of the statements about the brains of marsupials have been based on one or two brains from each of a small number of species, and no attempt has been made to determine if these are representative of the group, nor has any attempt been made to comprehensively examine brain structure across the full range of marsupials.

We present here an analysis of brain weight/body weight relationships so that a general idea of encephalization within marsupials, and between marsupials and eutherians can be obtained. Since this is only a preliminary report, the data on individual animals, as well as a more comprehensive examination of variation within species and within families, will be published later.

Since the brain weight (mg)/body weight (g) relationship is not linear (small species have relatively larger brains than large species), any assessment of encephalization must allow for differences in body weight. This is done by using the allometric equation (log brain weight = log b + a log body weight, which plots as a straight line with y intercept of log b and slope of a) and by the use of an index of encephalization.

The allometric equation provides a formula for calculating brain weight for any body weight according to a certain rate of increase (a) with increasing body weight, and the index of encephalization is a measure of the degree of increase or decrease of the real brain weight relative to the brain weight calculated from the allometric equation. Thus, in previous studies, Bauchot and Stephan (1966) and Stephan (1972), by using an allometric equation empirically derived from insectivorans, could examine the amount of increase or decrease in encephalization in any species compared to an insectivoran of the *same body weight.* The values of the indices of comparison thus depend upon the values of a (the slope) and log b (the y intercept).

In such comparisons, it thus becomes very important to exercise great care in choosing the slope (the y intercept is of less importance) as all subsequent comparisons depend on this value.

In this paper we examine which group gives the best slope for use as a basis for comparison of encephalization within marsupials. Using this value we examine the variations in encephalization within marsupials, and then relate these to those of eutherians.

MATERIAL AND METHODS

The data in this paper are derived from 320 specimens from 89 Australian species of marsupials. 275 of these were collected by John Nelson, 3 by Heinz Stephan, 8 by John Haight and 34 values were taken from the literature.

All brains were perfused with formalin, dissected out and weighed free of their membranes. The data in the literature were not collected in

[1] Zoology Department, Monash University, Clayton, Victoria, Australia, 3168; and Max-Planck-Institut fur Hirnforschung, Neurobiologische Abteilung, Frankfurt/M., Germany.

[2] Max-Planck-Institut fur Hirnforschung, Neurobiologische Abteilung, Frankfurt/M., Germany.

the same way, but we have used them when there was reasonable agreement with our data, or when we did not have any data (e.g. for species of *Myrmecobius*, *Thylacinus* and *Macrotis*).

Table 1 shows the number of individuals and the number of species within each family from which data were obtained. We have followed the classification used by Kirsch and Calaby (1977), but in addition we have subdivided the Dasyuridae, Petauridae and Macropodidae into groups suggested by our data. In Table 1 and Fig. 5, we have used the same numbering system for the families as Kirsch and Calaby used.

The log brain weights were plotted against log body weights, and regression line analysis (least squares) was used to determine the slope of the allometric equation for each of the families represented by more than one species. In the regression line analysis, the average brain weight and the average body weight for each species were used, rather than individual data, in order to minimise the effects of the differing sample size for each species. A canonical analysis was also made of the same groups using the brain weights and body weights of each individual.

An analysis of the above led to the selection of a reference line, and for any species, its distance measured vertically above or below this line is a measure of an increased or decreased level of encephalization above an animal (perhaps hypothetical) on this line and hence with the same body weight. The distance is thus an 'index of comparison'. All points on the reference line have distance values of 0. Since the antilog of 0 is 1, these points have an index value of 1 (or 100%) and so a doubling of the index will be reached at a distance of 0.301 (log of 2) from the reference line.

RESULTS

Table 1 lists the number of species (1) and the number of individuals (2) within each of the twelve Australian families of marsupials (and for subgroups within three of these families) for which we have data. Also listed are the slopes calculated by linear regression analysis (3), the correlation coefficients (4), and the ratio of the lowest to the highest body weight within each group (5).

The slope of the regression line for all Australian marsupials is 0.652. However, this does not necessarily mean that this slope is a good one to use when constructing a base line for comparison. In fact, the reverse is true since it is derived from species covering a wide range of adaptations and so tends to blur the differences between groups. Thus it may not be a true reflection of the slopes of the groups (families) providing the data. If it is imagined that the data in each group are along the rungs of a ladder whose sides are not at right angles to the rungs, then, for all these data, the regression line, which would be through the centre of the rungs, would give no indication of the stepwise distribution (grades) of the data. An indication of the grades within the data would be best obtained by selection of a regression line along one of the rungs; for, the regression line should have biological significance as well as mathematical validity.

Table 1. Brain weight/body weight relationships in various families or sub-groups of Australian marsupials: (1) number of species; (2) number of individuals; (3) slope of regression line; (4) correlation coefficient; and (5) ratio of lowest to highest body weight. The numbers in front of each family are the numbers given to that family by Kirsch and Calaby (1977).

	(1)	(2)	(3)	(4)	(5)
4 Dasyuridae	21	67	0.634	0.982	1 : 840
4A Dasyuridae (–4B)	19	62	0.592	0.996	1 : 437
4B *Planigale*–group	2	5	1.925		1 : 1.4
5 Myrmecobiidae	1	1	–	–	–
6 Thylacinidae	1	2	–	–	–
8 Peramelidae	5	18	0.254	0.697	1 : 3
9 Thylacomyidae	1	1	–	–	–
10 Phalangeridae	4	27	0.327	0.899	1 : 2
11 Burramyidae	4	7	0.606	0.987	1 : 5
12 Petauridae	12	38	0.264	0.678	1 : 12
12A *Pseudocheirus*–group	7	25	0.436	0.401	1 : 2
12B *Petaurus*–group	4	12	0.599	0.978	1 : 5
12C *Dactylopsila*	1	1	–	–	–
13 Macropodidae	36	142	0.537	0.974	1 : 76
13A Potoroinae	6	16	0.493	0.971	1 : 5
13B Macropodinae	30	126	0.566	0.964	1 : 32
14 Phascolarctidae	1	6	–	–	–
15 Vombatidae	2	9	–2.784	–	1 : 1
16 Tarsipedidae	1	2	–	–	–
Australian marsupials	89	320	0.652	0.987	1 : 4962

The slope for all marsupials (0.652) is higher than the slope for all of the groups or subgroups (see column 3 in Table 1) except the *Planigale*–group, which has only two species of about the same body weight and hence their slope has no real meaning (also no correlation coefficient). Thus, if this slope (0.652) is used when constructing a base line for comparison, then, within families and/or subgroups (i.e. along a rung), the larger species would have relatively lower indices of encephalization and the smaller ones would have relatively higher indices. Thus, the effect on the index of body size alone, would not have been optimally minimalised.

Ideally, the slope should be derived from species which cover a wide range of body weights and which are closely related (and so which might be expected to have the same general brain structure) and its value should be a true reflection of the trends within the groups.

The range of body weights is too low in peramelids, phalangerids, burramyids and vombatids, still low in petaurids, higher in macropodids and highest in dasyurids (see column 5 in Table 1). In the last two families, since the range of body weights is large, one would expect the value of the slope to have considerable stability. Also, since there are a large number of individuals and species involved in each, one would expect that this value would not alter significantly with new data. Dasyurids have the largest range in body weights, while the macropodids involve data for the largest number of species. The dasyurids are the more closely related group of species (Kirsch 1977).

Dasyurids have a slope (by linear regression analysis) of 0.634, macropodids have a slope of 0.537, and the average of these is 0.586. The two recognised subfamilies of the Macropodidae have very different slopes — the potoroines 0.493, and the macropodines 0.566. The potoroines have a small range of body weights and so their slope, in spite of the good correlation coefficient, is not significant. Within the Macropodidae, they are relatively small animals with relatively high indices of encephalization and so tend to flatten the slope for the family.

Within the Dasyuridae, the two very small *Planigale* species have much lower values of encephalization than the rest of the family, and when these are taken from the family, the slope becomes lower, 0.592 cf. 0.634. This slope of 0.592 is close to the average (0.586) of the two families with the best data (Dasyuridae and Macropodidae). It also is close to the canonical average of all marsupial groups (0.609), and has the best correlation coefficient (0.996). Therefore, for the basis of comparison, we have chosen the dasyurid regression line (without the *Planigale* species), since this group has the largest range of body weights, an even spread of species within this range (Fig. 1), a high correlation coefficient, and a slope, which is close

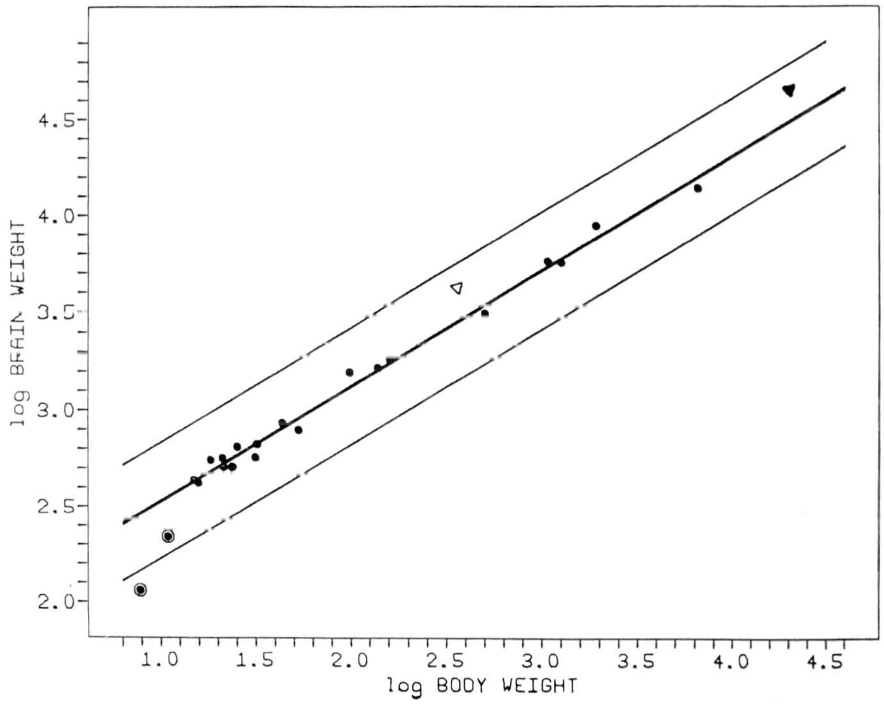

Fig. 1. Plot of average brain weight (mg) and average body weight (g) for the 21 species of dasyurids (●), as well as *Thylacinus* (solid triangle) and *Myrmecobius* (open triangle). The thickly drawn line in this and Figs 2 to 4 is the dasyurid (less *Planigale* spp.) regression line. The thinner line above indicates when a doubling of the index of comparison is reached and the one below indicates when a halving is reached.

to that obtained by the canonical averaging method (0.609), and which has a value that is reflected in several other groups.

The dasyurid (4A) regression line is thickly drawn in Figs 1-4. Its formula is log brain weight = 1.938 + 0.592 log body weight. The index of encephalization for any species is indicated by the distance from this line. In Figs 1-4, the thinner lines indicate when a doubling (above) or halving (below) of encephalization from this base line has been reached.

In Fig. 1, the various species of dasyurids have been plotted with those of *Myrmecobius* and *Thylacinus*. Figs 2-4 plot values for the other families or groups, with the dasyurid line superimposed. In Fig. 5, the indices of encephalization are summarized in graphic form: the 100 mark is the value of the dasyurid base line; and the range and mean are shown for each family or group.

Since the *Planigale* spp. were so divergent from the rest of the dasyurids, we have excluded them when we calculated the dasyurid regression line, which was used to calculate the indices of comparison. It is possible that our *Planigale* spp. are aberrant and that further data will show they are not as divergent as indicated. However, our data suggest it would be of considerable interest to examine *Planigale* and the possibly related *Ningaui* in some detail. It may not be insignificant that some of these flat–headed animals are the most shrew–like of the marsupial mice. The shrews have low indices and have been placed by Stephan (1967) in the basal Insectivora.

Within the Dasyuridae, the values for the major genera are: *Antechinus* 85 – 110, av. 97; *Sminthopsis* 90 – 114, av. 100, and *Dasyurus* 91 – 117, av. 103. It is of interest that in *Antechinus* and *Dasyurus*, the species with the

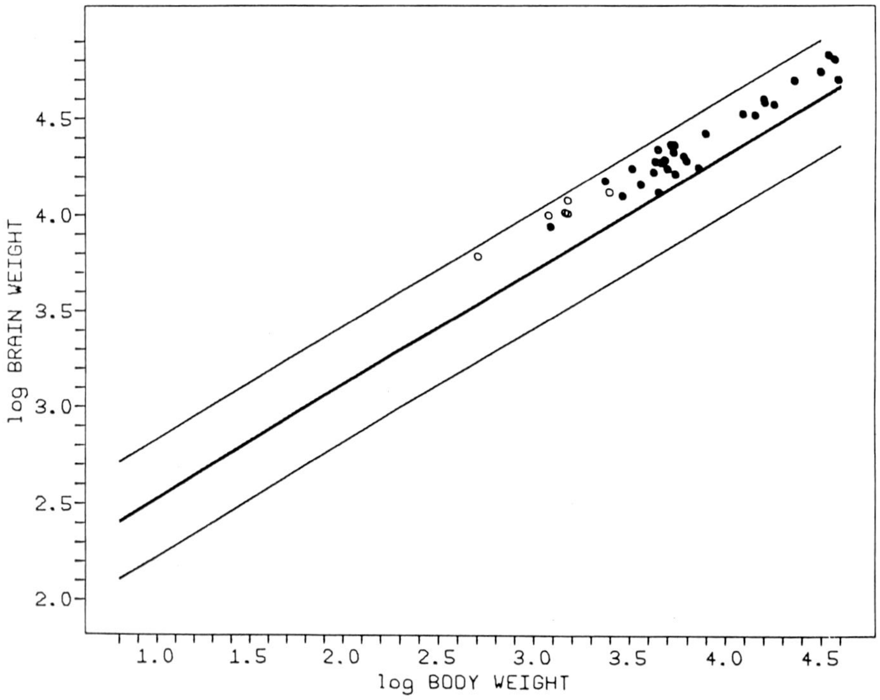

Fig. 2. Plot of average brain weight (mg) and average body weight (g) for the 36 species of macropodids. The potoroines are indicated by o, the macropodines by ●. The lines are the same as those drawn in Fig. 1.

DISCUSSION

Comparisons within the Dasyuridae

In Figs 1 and 5, it can be seen that the *Planigale* spp. are some distance from the rest of the dasyurids. *P. ingrami* has an index of comparison of 39 (n = 1) and *P. maculata* has an index of 61 (n = 4). The rest of the family have indices from 85 – 120 and, since this group was used as the basis for the indices of comparison, it has a mean of 100.

highest values (*A. stuartii* and *D. maculatus*) are the most arboreal within each genus, and so could be considered to occupy a more complex environment. *Sarcophilus harrisii* has a low value 88 (n = 6).

Comparisons with other Marsupials

Myrmecobiidae and Thylacinidae

Myrmecobius fasciatus has an index (149) which is well outside that of the Dasyuridae (see Figs 1 and 5). This value is based on data taken from the literature (Möeller 1973) and may not be accurate as it was based on measurements of cranial capacities (n = 2). However, the evidence here supports its separation from the Dasyuridae.

Thylacinus cynocephalus also has an index (148) which would place it outside of the Dasyuridae (see Figs 1 and 5). This index is also based on data taken from the literature: partly on brain weight/body weight data (Hrdlicka 1905, Warncke 1908), partly on cranial capacities (n = 30) (Möeller 1970).

Macropodidae

Macropodids have indices from 104 to 182 which is in agreement with Möeller (1973) that macropodids have higher values than dasyurids. As can be seen in Figs 2 and 5, all of the 36 species are above the dasyurid reference line. Potoroines tend to have higher indices (149 to 182, av. 166) than macropodines (104 to 176, av. 142). Within the Potoroinae the indices are: *Aepyprymnus* 149; *Bettongia* 155 – 182, av. 166; *Potorous* 175, *Hypsiprymnodon* 176. Within the Macropodinae the indices are: *Lagorchestes* 104; *Onychogalea* 105 – 115, av. 110; *Dendrolagus* 134; *Thylogale* 130 – 151, av. 143; *Wallabia* 150; *Macropus* 111 – 168, av. 146; and *Petrogale* 124 – 176, av. 148.

Although the range of the potoroines is within the range of the macropodines, their values are in the upper part of the macropodine range. This may be related to the fact that the diets of potoroines are in general less specialised than those of macropodines (some are partly insectivorous as well as herbivorous), or it may be that the group as a whole has become reduced in size during its evolution. The development of a reduced body size in domestic Dogs and in primate species appears to have produced forms with high indices of encephalization (Stephan 1972).

Petauridae and Phalangeridae

On the basis of our data, there are three groups within the Petauridae. The *Pseudocheirus*–group (including *Schoinobates volans*) with indices of 69 – 105, av. 83; the *Petaurus*–group (including *Gymnobelideus leadbeateri*) 139 – 171, av. 156; and *Dactylopsila trivirgata* 224 (see Figs 3 and 5).

We have only one specimen of *Dactylopsila trivirgata*. It may be that further specimens will bring its value closer to the *Petaurus*–group, but on our data, it is so far above that group (almost twice the range above the highest value of the group), that we have placed it apart.

There is no overlap between the *Pseudocheirus*–group, which has the second lowest average index of comparison, and the *Petaurus*–group which has the second highest average (see Fig. 5).

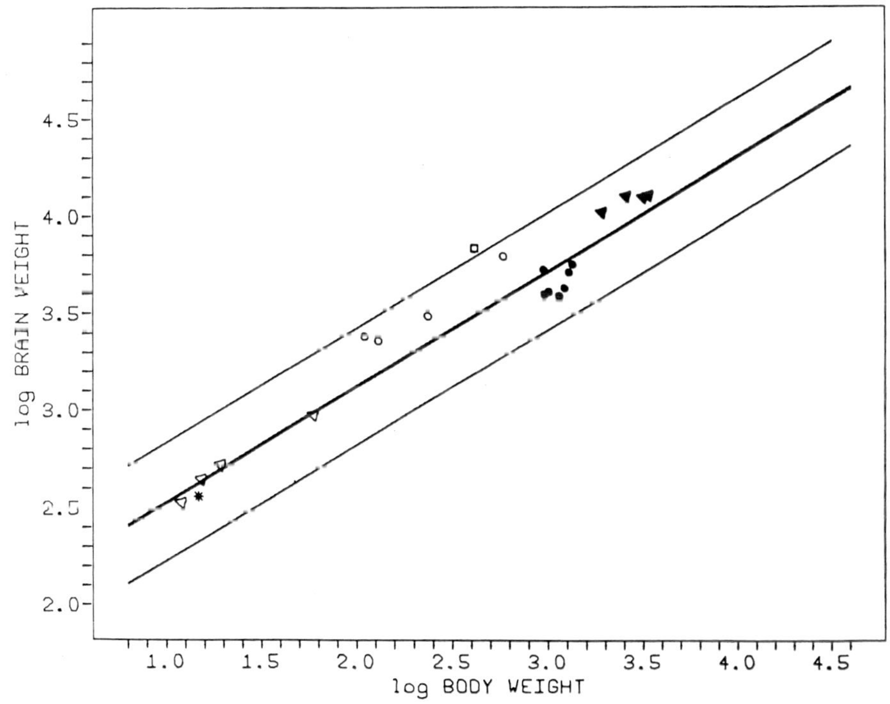

Fig. 3. Plot of average brain weight (mg) and average body weight (g) for the 4 species of burramyids (open triangle), the 12 species of petaurids (*Pseudocheirus*–group ●; *Petaurus*–group o; *Dactylopsila* □), the 4 species of phalangerids (solid triangle and the 1 species of tarsipedids (*). The lines are the same as drawn in Fig. 1.

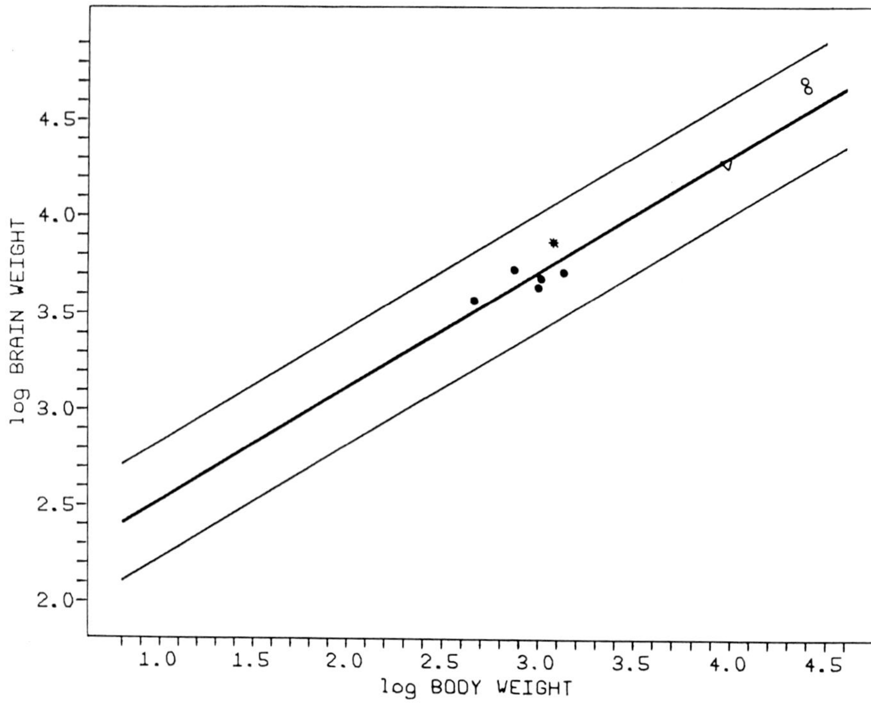

Fig. 4. Plot of average brain weight (mg) and average body weight (g) for the 5 species of peramelids (●), the 1 species of thylacomyid (*), the 1 species of phascolarctid open triangle), and the 2 species of vombatids (o). The lines are the same as drawn in Fig. 1.

The indices for phalangerids (120 – 140, av. 130) range between the *Pseudocheirus*–group and the *Petaurus*–group, with essentially no overlap.

Burramyidae, Tarsipedidae, Peramelidae, Thylacomyidae, Phascolarctidae and Vombatidae

The burramyids have a range of 89 – 105, av. 98; *Tarsipes* 85; peramelids 84 – 122, av. 99; *Phascolarctos cinereus* 97. All of these have values or averages close to the base of 100 (Figs 3, 4 and 5). The peramelids have a similar range and mean to those of dasyurids, but the burramyids have a much smaller range. *Macrotis lagotis* has a value (based on data taken from Möeller 1973) of 128, close to the maximum value found in the Peramelidae (Figs 4 and 5). Vombatids have relatively high indices: *Lasiorhinus* 131 and *Vombatus* 151 (Figs 4 and 5).

Didelphidae

The material for the didelphids is taken from the literature, where there is much confusion about the identity of species. We have lumped all *Didelphis* spp. to give an index of 75, and for didelphids, indices from 75 to 142, av. 116. They thus have a range extending above the dasyurids, but the number of species for which we found data is small (n = 6) and so the range and average are less reliable than the Australian data. *Caluromys* 137 and *Marmosa murina* 142 were outside the dasyurid range but not *Marmosa domina* 106, a *Marmosa* sp. 121, or *Philander opossum* 117.

Comparison with Eutherians

The brain weight and body weight data for eutherians were taken from Stephan *et al.* (1980). When indices for the insectivorans are calculated from the dasyurid reference line, the basal insectivorans (Stephan 1967) range from 47 to 71 (av. 60, n = 11). Insectivora without *Rhynchocyon stuhlmanni* (index 180) range from 71 to 140 (av. 96, n = 15) and with *R. stuhlmanni* the average is 101. Thus, the progressive insectivorans overlap the dasyurids (less *Planigale* spp.) but extend above them, while the basal insectivorans are below them and overlap the *Planigale* spp.

The calculated indices for the prosimians range from 155 – 474, av. 256 (n = 29). The only species within the marsupial range are within the Tupaiidae and Cheirogaleinae, together with a single species of *Lepilemur,* in the Lemurinae, whose species otherwise have high values.

Simians range from 278 to 811, av. 543 (n = 49) and *Homo sapiens* has an index of 2171, i.e. the human brain is about 22 times larger than a dasyurid of the same (hypothetical) body weight.

General

Our results support the current taxonomy of Australian marsupials except that some families, on our data, need reconsideration of species groupings below the family level. The brain is a very plastic organ and as Pirlot and Stephan (1970) have shown, brain weight/body weight relationships may sometimes reflect ecological adaptations (e.g. feeding strategies) rather than phylogenetic relationships and so our statements on groupings, and on the relationship of individual species (e.g. *Thylacinus cynocephalus*) to groups, are not necessarily taxonomic statements. Organisms, and not single characters, evolve. Hence, studies on single characters are of value only when individual species, or groups of species, show clear differences from the rest of their taxonomic group (e.g. the *Pseudocheirus*-group within the Petauridae; *Planigale* spp. and the Dasyuridae). It then has to be decided, by weighing up *all* known characters, if these differences are of phylogenetic importance.

In any system, if the size of the smallest components (e.g. neurons) is circumscribed, then the functional capacity of the system will depend upon its absolute size and on its structural differentiation. Thus, in general, brain size could be expected to be related to *potential functional capacity* and so the index of encephalization could be considered to be one important criterion of the grade of evolution.

If a group, as a whole, tends to have lower values than another group (e.g. the species of *Pseudocheirus* vs. those of *Petaurus;* marsupials vs. prosimians) then the former is at a lower grade of evolution. So to us, marsupials are above the grade of basal insectivorans, about the grade of progressive insectivorans and lower prosimians, but below higher prosimians and all simians.

ACKNOWLEDGEMENTS

We would like to thank all of the various state authorities who aided us in the collection of animals, the various individuals who have given us specimens, John Haight now at the School of Biological Sciences, A.N.U., for his data, Michael Stephan for computing and Helma Lehmann for typing. John Nelson was supported on various collecting trips by Monash University Funds especially the Deans Fund, Faculty of Science, and by the Ingram Trust. The manuscript was read and constructively criticised by Drs. J. A. W. Kirsch and L. Radinsky.

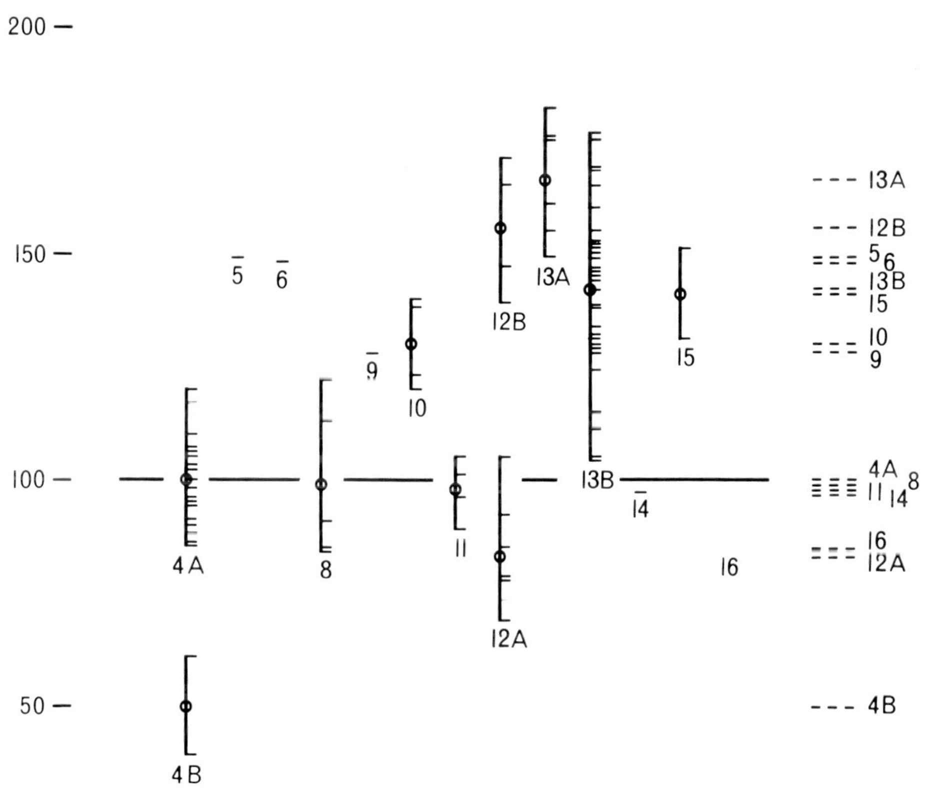

Fig. 5. Diagrammatic representation of the range (vertical bars) and mean (o) of the indices of encephalization (scaled on the vertical axis from 50 to 200) of the various families of Australian marsupials. The families are numbered as in Table 1. Each of the small horizontal bars represent the index for a species. The sequence of means is scaled on the right hand vertical axis.

REFERENCES

ABBIE, A. A., 1941. Marsupials and the evolution of mammals. *Aust. J. Sci.* **4:** 77–92.

BAUCHOT, R. AND STEPHAN, H., 1966. Données nouvelles sur l'encéphalisation des insectivores et des Prosimiens. *Mammalia* **30:** 160–96.

HRDLICKA, A., 1905. Brain weight in vertebrates. *Proc. Smithson. Misc. Coll.* **48:** 89–112.

KIRSCH, J. A. W., 1977. The comparative serology of Marsupialia, and a classification of marsupials. *Aust. J. Zool., Suppl. Ser.* **52:** 1–152.

KIRSCH, J. A. W. AND CALABY, J. H., 1977. The species of living marsupials: an annotated list. Pp.9–26 *in* "The biology of marsupials", ed. by B. Stonehouse and D. Gilmore. Macmillan Press: London.

MOELLER, H., 1970. Vergleichende Untersuchungen zum Evolutionsgrad der Gehrine grosser Raubbeutler *(Thylacinus, Sarcophilus* und *Dasyurus). Z. System–Evol.* **8:** 69–80.

MOELLER, H., 1973. Zur Evolutionshöhe des Marsupialiagehirns. *Zool. Jb. Anat.* **91:** 434–48.

PIRLOT, P. AND STEPHAN, H., 1970. Encephalization in Chiroptera. *Can. J. Zool.* **48:** 433–44.

STEPHAN, H., 1967. Zur Entwicklungshöhe de Insektivoren nach Merkmalen des Gehirns und die Definition der 'basalen Insektivoren'. *Zool. Anz.* **179:** 177–99.

STEPHAN, H., 1972. Evolution of primate brains: a comparative anatomical investigation. Pp.155–74 *in* "The functional and evolutionary biology of primates", ed. by R. Tuttle. Aldine/Atherton Inc.: Chicago.

STEPHAN, H., FRAHM, H. AND BARON, G., 1980. New and revised data on volumes of brain structures in insectivores and primates. *Folia primatol.* (in press).

WARNCKE, P., 1908. Mitteilung neuer Gehirn– und Körper– gewichtsbestimmungen bei Säugern, nebst Zusammenstellung der gesamten bisher beobachteten absoluten und relativen Gehirn– gewichte bei den verschiedenen Spezies. *J. Psychol. Neurol.* **13:** 355–403.

RESUMEN

58. Encefalización en marsupiales Australianos. Por John E. Nelson y Heinz Stephan.

Son analizados los datos del peso de los cerebros y del cuerpo de 320 individuos de 89 especies de marsupiales Australianos. La ecuación alométrica: log. del peso del cerebro =1.938 + log. del peso del cuerpo, derivida de datos provistos por todas las especies de Dasyuridae excepto aquellos de *Planigale,* fue usada para calcular los índices de encefalización de todos los marsupiales y algunos euterios. Los índices para las especies de *Planigale* fueron notoriamente más bajos que los del resto de los Dasyuridae. Los Dasyuridae, Peramelidae y Burramyidae tienen índices promedio similares y más inferiores que aquellos de los Phalangeridae, Macropodidae y Vombatidae. Dentro de los Petauridae, las especies de *Pseudocheirus* y *Schoinobates* tienen valores entre los más bajos de los marsupiales, mientras que aquellos de *Pseudocheirus* spp. *Gymnobelideus* y *Dactylopsila* están entre los más altos. Se hacen breves comentarios sobre los valores dentro y entre las familias, y entre los marsupiales y los euterios.

59. PHYLOGENTIC RELATIONSHIPS OF THE THYLACINE (*Thylacinus cynocephalus*, MARSUPIALIA) AS REFLECTED IN COMPARATIVE SEROLOGY

V. SARICH[1], J. M. LOWENSTEIN[2] AND B. J. RICHARDSON[3]

The Thylacine (*Thylacinus cynocephalus* Harris) has been considered to be a dasyurid marsupial, although it has been suggested that it may be more closely related to the extinct South American borhyaenid marsupials. Using microcomplement fixation procedures and albumen extracted from small pieces of untanned Thylacine skin, the genetic relationships of the Thylacine were measured as the immunological distances separating *Thylacinus* from various dasyurid genera. When these distances are compared with those separating other dasyurid genera from each other, it can be seen that *Thylacinus* belongs within the Dasyuridae as a distinct genus.

Pages 707–9 *in* CARNIVOROUS MARSUPIALS, ed. by M. Archer, February, 1982. Roy. Zool. Soc. New South Wales: Sydney, Australia.

THE THYLACINE (*Thylacinus cynocephalus* Harris) has been considered to be a dasyurid marsupial (Romer 1966). However several authors such as Bensley (1903), Wood (1923), and Archer (1976) have suggested that it might be more closely related to the Borhyaenidae, an extinct group of South American carnivorous marsupials. Clearly these two alternatives imply quite different phylogenies (Fig. 1). If the sister-group of *Thylacinus* is the Borhyaenidae then they must have separated before the ancestor of the Thylacine came to Australia and, as a result, the Thylacine would be genetically distinct from all living marsupial groups. If the Thylacine is a dasyurid then it will have close affinities with this group, a more distant relationship with *Myrmecobius fasciatus* and be only distantly related to the other superfamilies of marsupials.

Sarich (unpublished) has used quantitative microcomplement fixation to measure the immunological distances separating the albumins of various marsupial species and, as a result, it is possible to predict the immunological distances expected between thylacine albumin and that of other marsupials under the two hypotheses (Table 1).

The basic problem in testing these hypotheses is, of course, collecting a blood sample from a Thylacine, more particularly, a sample of Thylacine albumin. It occurred to us that there probably were a number of untanned Thylacine skins in museum collections around Australia and we felt that 2 square cm of skin should contain enough albumin as a contaminant to allow us to measure some immunological distances.

Small pieces of Thylacine skin were obtained from three different Thylacines through the kindness of the curators at the Australian Museum, the National Museum of Victoria and the Queen Victoria Museum in Launceston. As well, some dried muscle was collected from the femur of the Australian Museum specimen. The albumin was extracted from samples and tested using both microcomplement fixation and a radioimmunoassay procedure.

A method for studying immunological relationships by radioimmunoassay techniques is in the process of development. This has the advantage that it uses much less material than microcomplement fixation but at present, it has the disadvantage that the results are only qualitative. The method gives results equivalent to detecting spurs on an Ouchterlony Plate. The reaction observed when an antiserum is mixed with an homologous serum is the sum of a series of antibody/antigen reactions, each one affecting a different part of the albumin molecule. If the antiserum is first reacted with an heterologous albumin some of the various antigens present will be absorbed; the more similar the heterologous and homologous albumins the more of the antibody/antigen reaction will be used up, and only a few types of antigen will be left when the antiserum is then tested with homologous albumin. The radioimmunoassay system measures the amount of activity remaining in the antiserum after it is pretreated with a heterologous albumin. This is done by measuring the amount of radioactive globin that binds to the unreacted antiserum which has been collected using immobilised homologous albumin.

[1] Departments of Anthropology and Biochemistry, University of California, Berkeley, Carliforia, U.S.A., 94720.
[2] Department of Medicine (122 MRZ), University of California, San Francisco, California, U.S.A., 94143.
[3] Department of Population Biology, Australian National University, Canberra, A.C.T., Australia, 2600.

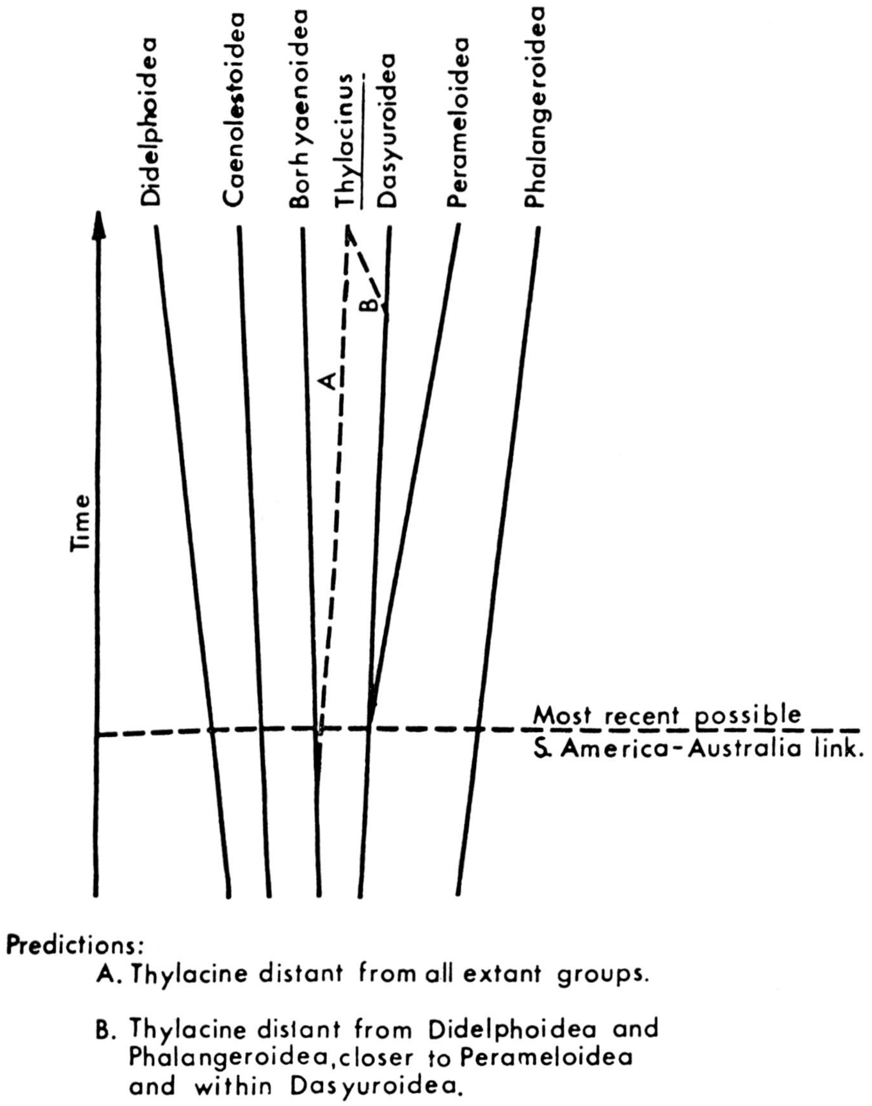

Fig. 1. Alternate phylogenetic relationships between the Thylacine and the various marsupial superfamilies.

Using this system, the Thylacine, *Dasyurus* sp. and *Dasyuroides byrnei* are indistinguishable while *Myremcobius fasciatus* has less than half of its antibody binding sites in common with the Thylacine.

Qualitative microcomplement fixation can be used to calculate the immunological distance between the albumins of different species. The immunological distance is measured as 100 x log of the relative concentration of antiserum (raised against a heterologous sera) required to produce a complement fixation curve whose peak is as high as that given by the homologous antigen (Sarich 1969). Immunological distances range

Table 1. Immunological distances expected between *Thylacinus cynocephalus* and other marsupials based on two alternative phylogenetic hypotheses.

Group Thylacine to:	Immunological Distance Predicted Hypothesis A	 Hypothesis B	 Observed	Time since Divergence
Didelphoidea	130	130	>>80	
Phalangeroidea	130	100	>80	
Perameloidea	130	90	80	48 m.y.
Myrmecobius fasciatus	130	35	40	24 m.y.
Dasyuroides byrnei	130	0–20	11	7 m.y.
Dasyurus sp.	130	0–20	11	7 m.y.

from 0 for reactions of immunological identity to 180 units for the maximum distance measured between mammalian albumins. As the immunological distance separating the albumins of different species has been shown to be approximately proportional to the time that has elapsed since the most recent common ancestor to the two species, it has been possible to relate immunological distance to time. One unit of distance is equivalent to 600,000 years since separation.

It can be seen (Table 1) that the measured immunological distances separating the species fit the pattern predicted by hypothesis B. The Thylacine shows close genetic relationships with *Dasyurus* sp. and *Dasyuroides byrnei,* it is further from *Myrmecobius fasciatus* and it is distant from the other superfamilies.

Clearly then, of the animals tested, the dasyurids constitute the closest living relatives of the Thylacine and, as the Thylacine is closer to species of *Dasyurus* and *Dasyuroides* than it is to *Myrmecobius fasciatus,* it must be included in the Dasyuridae. If the immunological distances are converted to time since the most recent common ancestor (Table 1), then the Thylacine diverged from *Dasyuroides byrnei* and *Dasyurus* sp. about 7 million years ago, from *Myrmecobius fasciatus* about 24 million years ago and from the peramelids about 48 million years ago. As all these divergences have occurred since Australia separated from Antarctica, the borhyaenid hypothesis cannot stand, even if the borhyaenids are considered close to the dasyurids, as the most recent common ancestor of the Thylacine and the borhyaenids has to be older than the link between the dasyurids and the peramelids. If the characters in common between the Thylacine and the borhyaenids are not paralellisms, then they must have been shared with the common ancestor of all dasyurids and the peramelids as well. The full analysis of the relationship of the Thylacine and the methodology used will be published when the work is completed.

ACKNOWLEDGEMENTS

A draft of this manuscript was read and constructively criticised by Drs. R. Crozier and M. Archer.

REFERENCES

Archer, M., 1976. The dasyurid dentition and its relationships to that of the didelphids, thylacinids, borhyaenids (Marsupicarnivora) and peramelids (Peramelina. Marsupialia). *Aust. J. Zool. Suppl. Ser. No.* **39:** 1–34.

Bensley, B. A., 1903. On the evolution of the Australian Marsupialia; with remarks on the relationships of the marsupials in general. *Trans. Linn. Soc. Lond. (Zool.)* **9:** 83–217.

Romer, A. S., 1966. "Vertebrate paleontology". University of Chicago Press: Chicago.

Sarich, V. M., 1969. Pinniped origins and the rate of evolution of carnivore albumins. *Syst. Zool.* **18:** 286–295.

Wood, H. E., 1924. The position of the "sparassodonts": with notes on the relationships and history of the Marsupialia. *Bull. Amer. Mus. Nat. Hist.* **51:** 77–101.

RESUMEN

59. Las relaciones filogenéticas de el Thylacine (*Thylacinus cynocephalus,* Marsupialia) manifestadas por la serologia comparada. Por V. Sarich, J. M. Lowenstein y B. J. Richardson.

El lobo marsupial de Tasmania (*Thylacinus cynocephalus*) ha sido considerado como un marsupial dasiúrido, aunque también se sugirió que podría estar más estrechamente relacionado a los extinguidos marsupiales Borhyaenidae de América del Sur. Por el uso de procedimientos de fijación de microcomplementos y albúmina extraída de pequénos trozos de piel del tilacino sin curtir, se midieron sus relaciones genéticas según las distancias immunológicas que lo separan de varios géneros de Dasyuridae. Cuando se comparan estas distancias con las que separan entre sí a otros dasiúridos puede advertirse que *Thylacinus* pertenece a los Dasyuridae como un género distinto.

60. KARYOTYPIC SIMILARITIES BETWEEN *Dromiciops australis* (MICROBIOTHERIIDAE, MARSUPIALIA) AND SOME AUSTRALIAN MARSUPIALS

G. B. SHARMAN[1]

Dromiciops australis (Family Microbiotheriidae) shares a common 2n = 14 karyotype with species of *Isoodon* (Peramelidae), *Cercartetus* (Burramyidae) and *Vombatus ursinus* (Vombatidae). Dasyurids also have 2n = 14 but usually have four chromosomes morphologically different from their equivalents in *Dromiciops australis* although species of *Ningaui* differ in only three chromosomes. The 2n = 14 chromosome complements of didelphids and caenolestids are different from each other, from those of *Dromiciops australis*, and from those of all members of the Australian marsupial radiation. The occurrence of similar karyotypes in *Dromiciops australis* and Australian forms representative of three families suggests that these, and the remaining members of the Australian radiation, had a common ancestor since they diverged from other marsupials.

Pages 711–14 *in* CARNIVOROUS MARSUPIALS, ed. by M. Archer, February, 1982. Roy. Zool. Soc. New South Wales: Sydney, Australia.

REIG, Fernandez and Spotorno (1972) described the 2n = 14 chromosome complement of a female *Dromiciops australis*, Hayman and Martin (1974) provided mean percent chromosome lengths for a further female specimen and Fernandez – Donoso, Barrios and Pinchera (1979) demonstrated nucleolar organizing regions by silver staining in two male specimens. These appear to be the only specimens of *D. australis* studied cytologically.

RESULTS AND DISCUSSION

The chromosomes of *Dromiciops* were compared with those of 2n = 14 didelphids, caenolestids, dasyurids and burramyids and with the *euchromatic* portions of the chromosomes of peramelids and vombatids (Fig. 1). *Dromiciops australis* chromosomes (Fig. 1d) are very similar to those of *Cercartetus concinnus* (Fig. 1e), *Isoodon macrourus* (Fig. 1f) and *Vombatus ursinus* (Fig. 1g) while they are quite unlike those of various species of *Marmosa* (Fig. 1a), *Caenolestes* (Fig. 1b) and *Dasyurus* (Fig. 1c).

The American carnivores (except *Dromiciops australis*) differ from each other and each differs from *Dasyurus* spp. especially in chromosome arm ratio (Fig. 1) but there are also slight differences in overall percentage length between chromosomes similarly numbered. Secondary constrictions (presumed nucleolar organizing regions) have not been reported on the chromosomes of didelphids but in *Caenolestes obscurus* and the dasyurids the smallest submetacentric pair (5, Fig. 1b, c) have satellited short arms with subterminal constrictions (Hayman, Kirsch, Martin and Waller 1971).

The arm ratios of the autosomes of species of *Dromiciops, Cercartetus, Isoodon* and *Vombatus* are remarkably alike (Fig. 1). Secondary constrictions occur on the no. 6 autosomal pair in *Isoodon* spp. and other peramelines and on the X chromosome of *Vombatus ursinus* but were not reported on any chromosome of *D. australis* (Reig, Gardner, Bianchi and Patton 1977). Examination of an earlier published *D. australis* karyotype suggests that near terminal secondary constrictions may be found on the second smallest autosomal pair, although no such claim was made by the authors (Reig *et al.* 1972). However, Fernandez–Donoso *et al.* (1979) report nucleolar organizing regions on the two smallest autosomes of *D. australis* (5 and 6, Fig. 1d) and on the C2 autosome of *Marmosa elegans* (6, Fig. 1a). Hayman and Martin (1969) stated that the satellited chromosomes of *Cercartetus nanus* were one of the two small autosomal pairs but Rofe (1979) did not report secondary constrictions on any chromosome of *Cercartetus concinnus*. Sharman (1961) reported secondary constrictions in a submedian position on the long arms of one of the larger chromosomal pairs in *Cercaërtus* (= *Cercartetus*) *lepidus* but in *Burramys parvus* the *X* chromosomes are satellited as in *Vombatus ursinus* (Carolyn E. Murtagh, unpublished observations). In spite of those differences the most obvious chromosome relationship between *D. australis* and some members of the Australian radiation is that expressed in Fig. 2.

Giemsa (G) banding studies have not been made on *Dromiciops australis* chromosomes. Rofe (1979) reported similar G banding karyotypes and similar morphologies in comparable autosomes of *Perameles nasuta, Isoodon obesulus, Cercartetus concinnus* and *Vombatus ursinus*. The G banded karyotype of *Ningaui* sp. differed in chromosomes 1, 2 and 4 and those of

[1] School of Biological Sciences, Macquarie University, North Ryde, New South Wales, Australia, 2113.

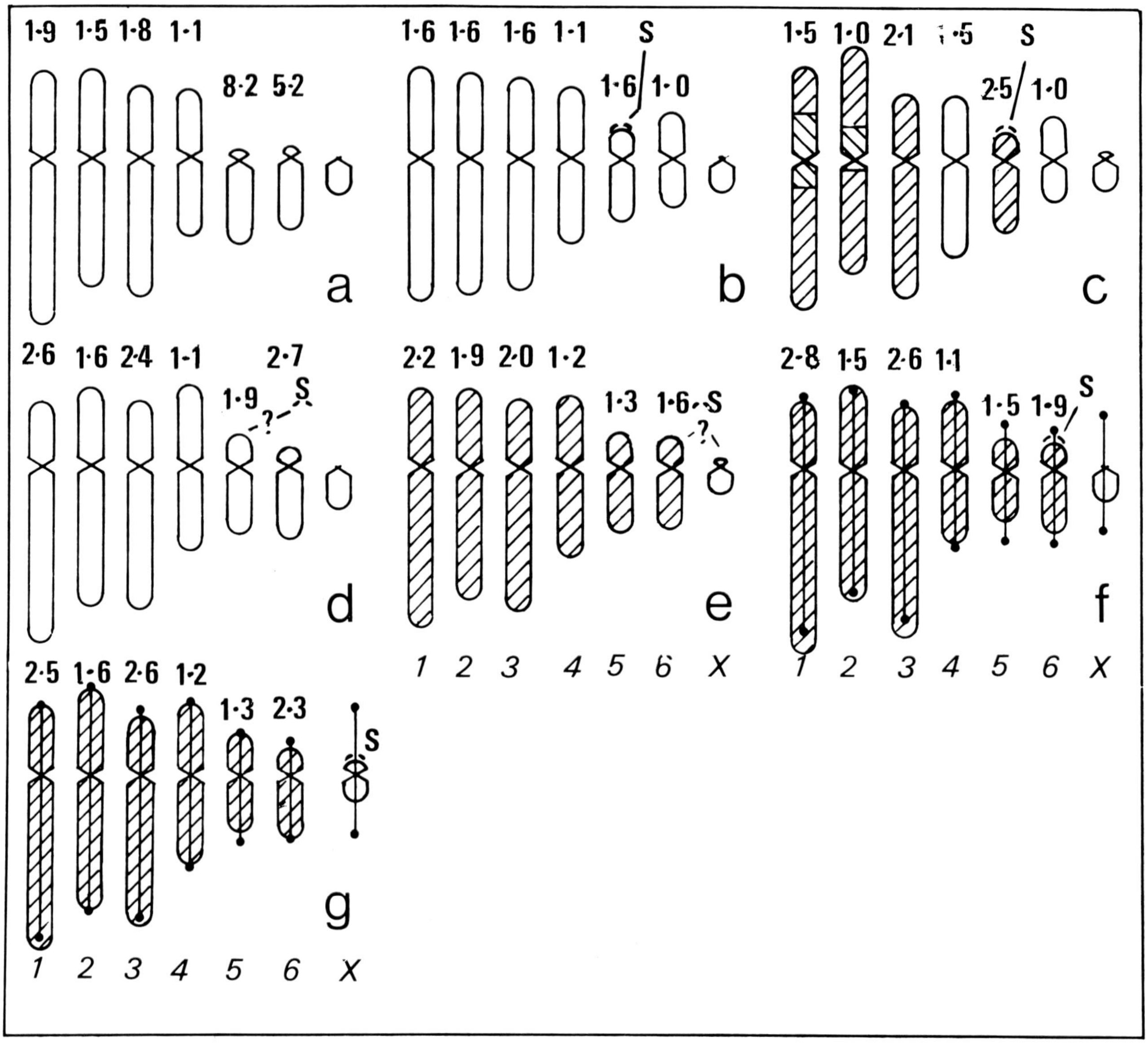

Fig. 1. Diagrams showing the relative lengths of autosomal arms (percentage length of haploid autosomal karyotype) in 2n = 14 marsupials. a. *Marmosa elegans*, b. Caenolestidae, c. *Dasyurus viverrinus*, d. *Dromiciops*, e. *Cercartetus* (3 spp.), f. *Isoodon macrourus*, g. *Vombatus ursinus*. In f and g, C banding material was neglected in calculating relative lengths and black discs at end of lines show relative lengths including C banding regions. X chromosomes shown as percentage length of haploid autosomal karyotype. Arm ratios of autosomes (r × length short arm = length long arm) are given above each chromosome — a, d from Reig *et al.* (1977) b, c, e based on measurements of Hayman and Martin (1974); cross hatched chromosomes have similar banding patterns to other cross hatched chromosomes of same number (Rofe 1979), banding pattern of *Isoodon obesulus* shown on autosomes of *I. macrourus*; S–satellites and secondary constrictions.

Sminthopsis crassicaudata and *Dasyurus viverrinus* differed in chromosomes 1, 2, 4 and 6 with respect to the G banded karyotypes of other 2n = 14 marsupials. The differences were considered to be due to pericentric inversions which changed the shapes of up to four of the six pairs of autosomes. The approximate limits of the postulated inversions in nos. 1 and 2 chromosomes are shown in Fig. 1. At present it is not possible to state whether or not the dasyurid karyotype might be just as readily derivable from the 2n = 14 karyotype of didelphids as comparisons between the published banding patterns of didelphids and Australian marsupials were not possible (Rofe 1979).

Obviously the standard didelphid karyotype is as different from that of *Dromiciops australis* as is that of the dasyurids (Fig. 1, see also Reig *et al.* 1977 for the didelphid – *D. australis* comparisons). Rofe (1979) attached no particular significance to the differing chromosomal locations of the secondary constrictions in dasyurids, peramelids and *Vombatus ursinus,* a point of interest in view of Hayman and Martin's (1969) contention that the location of the constrictions was of particular importance to their conclusion that 2n = 14 was the primitive number for all marsupials.

The placing of the dasyurids in an ordinal rating (Szalay 1982) of equivalent status to the remainder of the Australian radiation (i.e. the Syndactyla) is of interest in view of the demonstrated chromosomal differences between the 2n = 14 karyotypes of dasyurids and those of other 2n = 14 Australian marsupials. However, it is perhaps significant that *Ningaui* sp. although regarded as somewhat specialized (Archer 1975), differs from the Australidelphian 2n = 14 by but three, instead of four, changed chromosomes. The Australidelphian number and morphology may eventually be found amongst the dasyurids – perhaps in those New Guinea genera, regarded as primitive by Kirsch and Calaby (1977), which have not yet been studied cytologically.

Rofe (1979) presented strong evidence for derivation of the Australian marsupial radiation from a 2n = 14 ancestor with a karyotype similar to that of some burramyids, peramelids and *Vombatus ursinus.* The occurrence of a virtually identical karyotype in *Dromiciops australis* supports Szalay's contention (1982) of monophyly of *D. australis* and the Australian radiation. If the morphological similarity is eventually supported by the occurrence of similar banding patterns then the occurrence of an immediate *D. australis,* peramelid, burramyid, vombatid ancestor would appear to be beyond doubt. Presumably the heterochromatic (C banding) regions found on the autosomes, and especially on the X chromosomes, of the peramelids and vombatids (Fig. 1f, g) represent additions of repeated sequence DNA which have occurred since these forms diverged from a common ancestor. C banding regions are also a feature of the 2n = 16 to 2n = 22 chromosome complements of pseudocheirine phalangeroids (Murray, McKay and Sharman 1979; Murray, Sharman, McKay and Calaby 1980). The enigmatic *Notoryctes typhlops* is assigned by Szalay (1982)

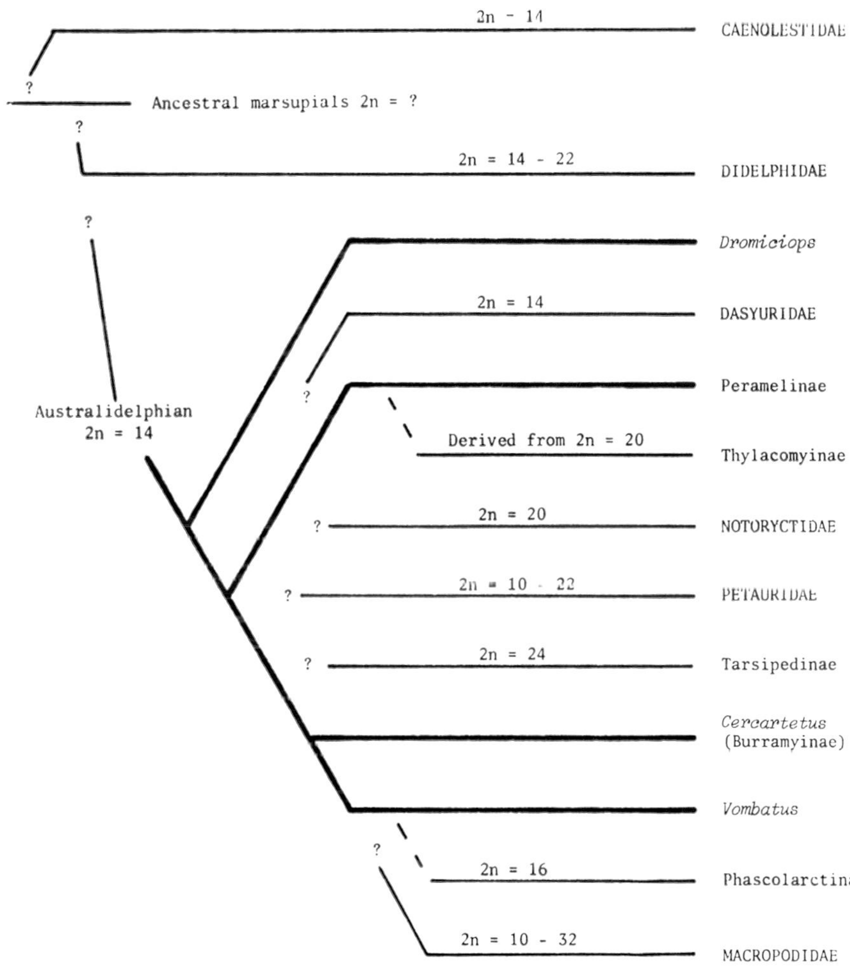

Fig. 2. The relationships of *Dromiciops* and certain members of the Australian marsupial radiation as expressed by chromosome number and morphology. The taxa connected by the heavy line share similar karyotypes. Families and subfamilies are as delineated by Szalay (this Volume).

to his (new) Phophalangeriformes on the basis of tarsal morphology. Calaby, Corbett, Sharman and Johnston (1974) drew attention to the cytological similarities of *N. typhlops* to some phalangeroids and Sharman (1973) pointed out that derivation of the Australian marsupial radiation from a 2n = 14 ancestor required increases in chromosome number to 2n = 20, 22 and even 32 in several different lines (see Fig. 2). Within the families Petauridae and Macropodidae (as in Szalay's classification) reduction from higher to lower numbers has occurred (Sharman 1973, Hayman and Martin 1974, Rofe 1979).

The monotremes have high chromosome numbers (Murtagh 1977) as have most primitive vertebrates and most eutherians so the 2n = 14 didelphids are presumably derived from ancestors which had a higher number of chromosomes. The 2n = 18 and 2n = 22 didelphids are thought to be derived from 2n = 14 ancestors (Reig *et al.* 1977, Kirsch 1977) so they represent one, or more likely two, further instances of chromosome number increase within the marsupial radiation. Given the antiquity of the marsupial group and the demonstrated occurrences of both increase and decrease in chromosome number in recent radiations it may never be possible to assign a primitive chromosome number and morphology to the group as a whole.

ACKNOWLEDGEMENTS

Mrs. Carolyn E. Murtagh provided the *Vombatus ursinus* preparations, G banded the chromosomes of *Vombatus* and *Isoodon* and gave freely of her time in discussion. Thanks are also extended to Dr. Ruth H. Rofe who loaned a copy of her Ph.D. thesis. Drs. R. Harding and M. Archer critically read a draft of this paper.

REFERENCES

Archer, M., 1975. *Ningaui* a new genus of tiny dasyurids (Marsupialia) and two new species, *N. timealeyi* and *N. ridei*, from arid Western Australia. *Mem. Qd Mus.* **17:** 237–49.

Calaby, J. H., Corbett, L. K., Sharman, G. B. and Johnston, P. G., 1974. The chromosomes and systematic position of the marsupial mole *Notoryctes typhlops*. *Aust. J. Biol. Sci.* **27:** 529–32.

Fernandez–Donoso, R., Barrios, S. and Pincheira, J., 1979. Position of the nucleolus within the nuclei of pachytene spermatocytes of *Dromiciops australis* and *Marmosa elegans* (Didelphoidea–Marsupialia). *Experientia* **35:** 1021–3.

Hayman, D. L., Kirsch, J. A. W., Martin, P. G. and Waller, P. F., 1971. Chromosomal and serological studies of the Caenolestidae and their implications for marsupial evolution. *Nature* **231:** 194–5.

Hayman, D. L. and Martin, P. G., 1969. Cytogenetics of marsupials. Pp.191–217 *in* "Comparative mammalian cytogenetics", ed. by K. Benirschke. Springer: New York.

Hayman, D. L. and Martin, P. G., 1974. Mammalia I: Monotremata and Marsupialia. *In* "Animal cytogenetics. Vol. 4: Chordata 4", ed. by B. John. Gebrüder Borntraeger: Berlin.

Kirsch, J. A. W., 1977. The comparative serology of Marsupialia and a classification of marsupials. *Aust. J. Zool. Suppl. Ser. No.* **52:** 1–152.

Kirsch, J. A. W. and Calaby, J. H., 1977. The species of living marsupials: an annotated list. Pp.9–26 *in* "The biology of marsupials", ed. by B. Stonehouse and D. Gilmore. MacMillan: London.

Murray, J. D., McKay, G. M. and Sharman, G. B., 1979. Studies on metatherian sex chromosomes. IX sex chromosomes of the Greater Glider (Marsupialia: Petauridae). *Aust. J. Biol. Sci.* **32:** 375–86.

Murray, J. D., Sharman, G. B., McKay, G. M. and Calaby, J. H., 1980. Karyotypes, constitutive heterochromatin and taxonomy of ringtail possums of the genus *Pseudocheirus* (Marsupialia: Petauridae). *Cytogenet. Cell Genet.* **27:** 73–81.

Murtagh, C. E., 1977. A unique cytogenetic system in monotremes. *Chromosoma* **65:** 37–57.

Reig, O. A., Fernandez, R. and Spotorno, A., 1972. Further occurrence of a karyotype of 2N = 14 chromosomes in two species of Chilean didelphoid marsupials. *Z. f. Säugetierkunde* **37:** 37–42.

Reig, O. A., Gardner, A. L., Bianchi, N. O. and Patton, J. L., 1977. The chromsomes of the Didelphidae (Marsupialia) and their evolutionary significance. *Biol. J. Linn. Soc.* **9:** 191–216.

Rofe, R., 1979. "G–banding and chromosomal evolution in marsupials". Ph.D. thesis: University of Adelaide, Adelaide.

Sharman, G. B., 1961. The mitotic chromosomes of marsupials and their bearing on taxonomy and phylogeny. *Aust. J. Zool.* **9:** 38–60.

Sharman, G. B., 1973. The chromosomes of non–eutherian mammals. Pp.485–530 *in* "Cytotaxonomy and vertebrate evolution", ed. by A. B. Chiarelli and E. Capanna. Academic Press: New York.

Szalay, F. S., 1982. A new appraisal of marsupial phylogeny and classification. Pp.621–40 *in* "Carnivorous marsupials", ed. by M. Archer. Roy. Zool. Soc. New South Wales: Sydney.

RESUMEN

60. Similitudes cariotípicas entre *Dromiciops australis* (Microbiotheriidae, Marsupialia) y algunos marsupiales australianos. Por G. B. Sharman.

Dromiciops australis (familia Microbiotheriidae) comparte un común cariotipo 2n = 14 con especies de *Isoodon* (Peramelidae), *Cercartetus* (Burramyidae) y con *Vombatus ursinus* (Vombatidae). los dasiúridos también tienen un cariotipo 2n = 14, aunque usualmente tienen cuatro cromosomas morfológicamente diferen tes de sus equivalentes en *Dromiciops australis*, aunque las especies de *Ningaui* difieren en sólo tres cromosomas. Los complementos cromosómicos 2n = 14 de los didélfidos y cenoléstidos son diferentes uno del otro, de los de *Dromiciops australis* y de aquellos de todos los miembros de la radiación marsupial australiana. La existencia de cariotipos similares en *Dromiciops australis* y en las formas australianas representativas de tres familias sugiere que éstas, y los restantes miembros de la radiación australiana, tuvieron un común antecesor cuando divergieron de los otros marsupiales.

61. STUDIES ON THE DNA OF SOME SPECIES OF *Antechinus* (DASYURIDAE, MARSUPIALIA)

M. TIMMS[1], M. WESTERMAN[1] AND B. DÜRING[1]

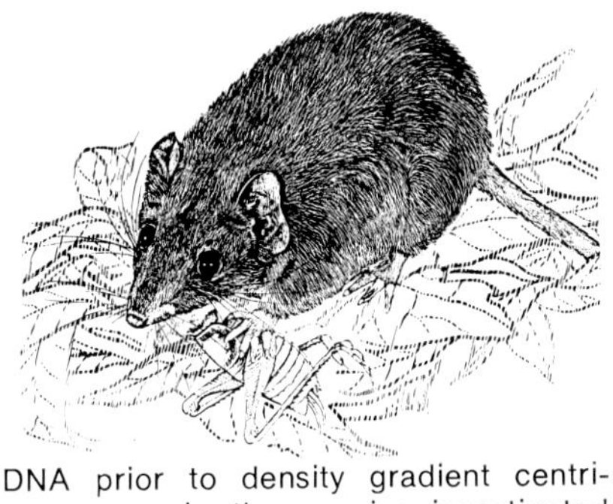

The marsupial genus *Antechinus* is characterized by the remarkable evolutionary stability of its karyotype including its G– and C–banding patterns. With one notable exception — the Dibbler, *A. apicalis,* the amounts of C–band positive material in the genomes are similar and very small. A study of the DNA from some species has shown that in general the amounts of highly repeated sequence DNA (Hr–DNA) in the genome are also small (about 1%). An obvious exception was *A. stuartii* with some 19% of its genome made up of Hr–DNA. Even where this class of DNA was small however, certain sequences contained in it were resolvable as cryptic satellites or could be seen to markedly skew the distribution of the DNA main-band when actinomycin–D, Ag^{+} or Hg^{2+} ions were bound to the DNA prior to density gradient centrifugation. A comparison of the distribution of some of these Hr–DNA sequences in the species investigated suggested the possible existence of two major species subgroups.

Pages 715–21 *in* CARNIVOROUS MARSUPIALS, ed. by M. Archer, February, 1982. Roy. Zool. Soc. New South Wales: Sydney, Australia.

THE GENOMES of most eukaryotes contain a class of DNA sequences repeated many times in tandem arrays (Britten and Kohne 1968), the repeating unit varying in size from as low as 2 base pairs (bp) in crabs (Josse *et al.* 1961) to over 1400 bp in calf (Kurnit *et al.* 1973). Although the *amounts* of particular sequences present in the genomes of related species may differ markedly, it would appear that those amplified in any one species come from a common "Library" (Salser *et al.* 1976). The repeats are often, but not invariably, associated with heterochromatin defined by C–banding techniques. Despite much speculation, little is known of the function(s) of highly repeated sequences. A number of authors have suggested that repeated sequence (satellite) DNA may have no adaptive function at all at the level of the organism, i.e. it is "selfish" DNA (Doolittle and Sapienza 1980, Orgel and Crick 1980). The apparent evolutionary stability of many sequences however, implies a function for at least some of them (Peacock and Lohe 1980), their stability probably reflecting the need for interactions of particular DNA sequences with specific proteins (Hsieh and Brutlag 1979). It has frequently been implied that repeated sequence DNA may play a (casual) role in the speciation process, either through a role in meiosis (Corneo 1976, Fry and Salser 1977) or in chromosomal rearrangements (Hatch *et al.* 1976). As has been forcibly pointed out recently, however, these speculations are frequently based on inadequate or erroneous interpretations of the existing data (see John and Miklos 1979, Miklos and John 1979, Miklos *et al.* 1980).

The genus *Antechinus* shows a remarkable degree of evolutionary conservation of the karyotype (see Young *et al.* this Volume). The present studies on sequence composition of the DNA from a number of species from this genus were initiated: 1, to see whether this conservatism of the karyotype was matched by a similar conservatism of highly repeated sequence DNA (Hr–DNA); 2, to see if there was a simple relationship between the amounts of Hr–DNA and heterochromatin in the genome and; 3, to see whether the DNA relationships throw any light on the phylogenetic relationships in this genus. This report outlines the results obtained to date.

MATERIALS AND METHODS

DNA from males of 10 species of *Antechinus* were used in the present study. These were *A. apicalis*[1], *A. bellus*[2], *A. bilarni*[2], *A. flavipes*, *A. godmani*[2], *A. leo*[2], *A. macdonnellensis*[2], *A. minimus*, *A. stuartii*, and *A. swainsonii.* DNA was extracted from various tissues using the method of Marmur (1961). The DNA was purified by centrifugation through a preparative neutral caesium chloride (CsCl) density gradient before use for further studies. Analytical centrifugations were carried out using an MSE Centriscan 75 analytical ultra centrifuge and traces were made of the gradients at equilibrium using the inbuilt U.V. scanning device.

Actinomycin D/CsCl: These gradients were prepared as per Brutlag *et al.* (1977).

Heavy metal/Cs_2SO_4: Silver or mercury ions were added to DNA previously dialysed into 5 mM sodium tetraborate solution until the ratio of $Ag^{+} : PO_4^{3-}$ was 0.3 or the ratio of $Hg^{2+} : PO_4^{3-}$

[1] DNA obtained from hypotetraploid female cells cultured by Dr. J. M. Graves.

[2] Material kindly supplied by Dr. Pat Woolley (Zoology Dept., La Trobe University).

[1] Department of Genetics and Human Variation, La Trobe University, Bundoora, Victoria, Australia, 3083.

was 0.1. Caesium sulphate was then added to give final densities of 1.525 g/cm^3 (for Ag^+) or 1.510 g/cm^3 (for Hg^{2+}).

Thermal denaturations: These were carried out as described in Narayan and Rees (1976).

Reassociation Kinetics: DNA samples were extensively dialysed into 0.12M KPB[3] or 0.75M KPB, cooled on ice, degassed with helium and sheared using a Technics 600 sonicator (4 x 1 min blasts using a 19 mm probe at stop 9 with 1 minute cooling intervals). The sonicated double–stranded DNA had an average size of 5 – 600 bp as determined by agarose gel electrophoresis using Hind III digested λ–phage DNA as size marker (Boehringer Mannheim). Reassociation kinetics of the sheared, degassed DNA were followed spectrophotometrically using the method of Britten *et al.* (1974). Low Cot points were obtained using 10 mm pathlength stoppered cuvettes and 0.12M KPB, higher Cot values were obtained using 2 mm pathlength stoppered cuvettes and 0.75M KPB. The estimate of percentage reassociation at a given Cot was usually the mean of three samples.

RESULTS

DNA Reassociation Kinetics

In order to estimate the proportion of the total DNA made up of highly repeated sequences (Hr–DNA), incubations were carried out until Cots of 10^1 or 10^2 were reached. The Hr–DNA fraction (reassociating before Cot 10^{-2}) varied between the eight species from about 1% of the genome in *A. godmani* and *A. macdonnellensis* to about 19% of the genome of *A. stuartii.* A moderately fast fraction (Mr–DNA) reassociates between Cot 10^{-2} and 10^2 and constitutes a further 20–30% of the genome for each species. As might be expected the Cot curves of the different species show a general similarity but the amount of repeated sequence DNA varies between the eight species studied. With the exception of *A. stuartii,* only a small portion of the genome (< 5%) is composed of very highly repeated sequences. In view of the general relationship between Hr–DNA and C–band positive material, the low percentages observed in the present study accord well with the amounts expected from the chromosome C–banding studies (Young *et. al.,* this Volume). When sheared DNA (500 bp) from *A. leo* was denatured and allowed to reassociate to a Cot of 10^{-1} in 0.12M KPB and the DNA immediately placed in neutral CsCl and spun to equilibrium, almost all of the DNA remains at a density characteristic of single–stranded DNA (near 1.717 g/cm^3), again indicating presence of very little well matched reassocciated Hr–DNA.

Thermal Denaturation

The melting profiles for the DNAs from different species, when measured under identical ionic conditions, give information on the base composition as well as on the heterogeneity of dispersion of the base pairs within the DNA. The melting profile for *A. leo* DNA is shown in Figure 1 and the Tms of all species studied are shown in Table 1. It can be seen that the melting curve for *A. leo* is generally steep and smooth, indicating a large measure of dispersion of AT and GC base pairs within the genome. The small "step" visible in the lower temperature part of the curve was consistently seen, not only in a number of replicate melts for *A. leo* but also in those of the other species. This would suggest the presence within the genome of the *Antechinus* species studied of a small component which is relatively much richer in A+T base pairs – a finding supported by observations of an A+T rich DNA fraction separable on Malachite green affinity columns (Westerman, unpublished data).

The Tms range from 66.6°C (*A. godmani*) to 70°C (*A. leo*) giving mole percent G+C values of 30.0 to 39.06 (see Table 1). There are thus no significant differences between any of the spe-

[3] 0.12M KPB = 0.06M KH_2PO_4, 0.06M K_2HPO_4, PPH 6.8.

Table 1. Mole–percent G+C values for the DNA from ten species of *Antechinus* (T_m and melting range in °C).

Species	Buoyant density in neutral CsCl	Percent G+C	T_m in 0.1 × SSC	Percent G+C	Melting range
A. apicalis	1.6934 g/cm^3	34.08	–	–	–
A. bellus	1.6951 g/cm^3	35.82	68.8	36.13	7.44
A. bilarni	1.6954 g/cm^3	36.12	69.2	37.11	7.93
A. flavipes	1.6942 g/cm^3	34.90	68.4	35.16	11.47
A. godmani	1.6949 g/cm^3	35.61	66.6	30.0	6.1
A. leo	1.6983 g/cm^3	39.08	70.0	39.06	8.78
A. macdonnellensis	1.6950 g/cm^3	35.71	67.3	32.23	6.1
A. minimus	1.6936 g/cm^3	34.29	67.5	32.96	6.34
A. stuartii	1.6946 g/cm^3	35.31	67.9	33.94	5.12
A. swainsonii	1.6945 g/cm^3	35.20	67.2	32.23	10.74

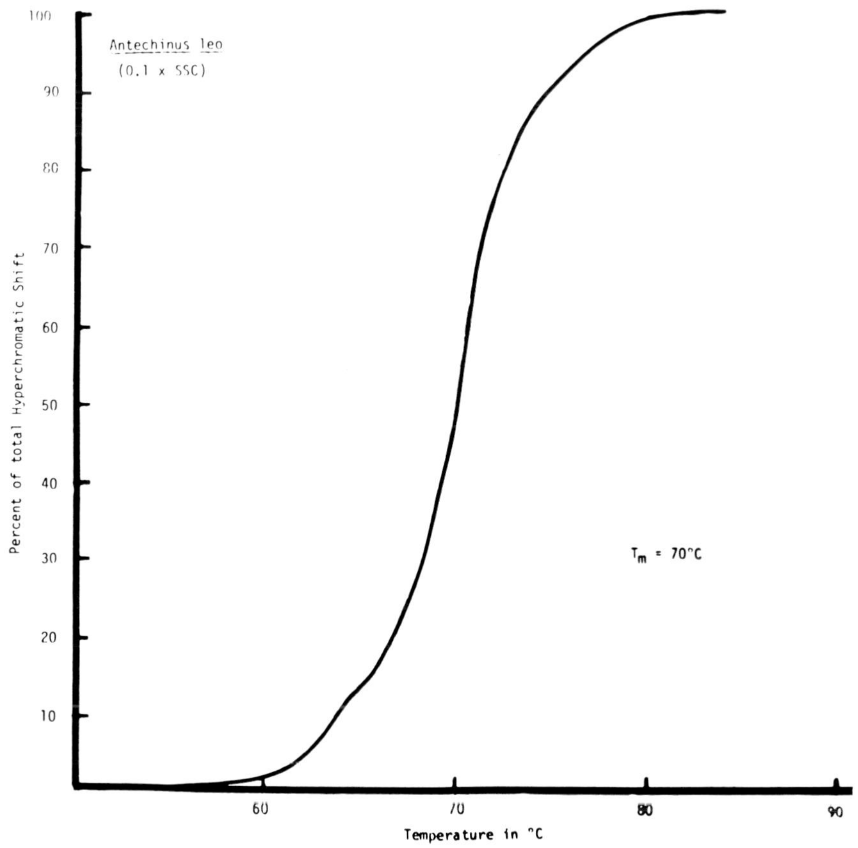

Fig. 1. The melting profile for the DNA of *A. leo*.

cies either in average base composition or in the base compositional heterogeneity value (2σ) which reflects the degree of dispersion of the base pairs.

Analytical Ultracentrifugation

DNA from 10 *Antechinus* species was centrifuged to equilibrium on neutral CsC1 gradients. No obvious buoyant density satellite DNAs were observed even in overloaded gradients. The main band densities calculated with reference to the marker DNA (Mandel and Marmur, 1968) ranged from 1.6934 g/cm^3 for *A. apicalis* to 1.6-938 g/cm^3 for *A. leo* (see Table 1). In some cases the DNA main bands were narrow and roughly symmetrical, in others they were clearly skewed towards the heavy side (e.g. *A. bellus, A. minimus, A. macdonnellensis, A. flavipes* and *A. stuartii*) indicating a GC rich component. The buoyant densities provided alternative estimates of GC ratios which correspond closely with those based on the Tms (see Table 1).

The antibiotic Actinomycin–D is known to bond preferentially to particular DNA sequences and can therefore be used to separate out repeated sequences in the genome whose average base composition is similar to bulk DNA (see Walker 1968, Peacock *et al.* 1973, Brutlag *et al.* 1977). Cryptic satellites were clearly resolved from the DNA of *A. leo* and *A. macdonnellensis* (see Fig. 2). In other cases although no distinct satellite was present the main bands were markedly skewed to either the heavy side (*A. godmani*) or to the light side of the gradient (*A. bellus, A. bilarni, A. macdonnellensis, A. swainsonii, A. minimus* and *A. stuartii*).

Satellites similar to those isolated from antibiotic–containing gradients can often be obtained from heavy metal/Cs_2SO_4 gradients. Under the experimental conditions obtaining Ag^+ ions bond preferentially to guanine rich regions (see Sissoeff *et al.* 1976). Cryptic satellites were resolved for *A. macdonnellensis, A. stuartii* and *A. godmani,* whilst prominent light shoulders were seen in *A. bilarni* (possibly two), *A. leo* and *A. godmani* (see Fig. 3). Skewness of the main band towards the heavy side of the gradient or distinct shoulders were seen in the profiles of *A. stuartii, A. swainsonii, A. minimus* and *A. flavipes.* Whilst some similarity between the results for Actinomycin–D/CsC1 and Ag^+ Cs_2SO_4 is to be expected, since both bind to GC–rich regions, differences will exist between them as the two binding agents are not expected to bind exactly the same base sequences.

Mercury/Cs_2SO_4 gradients were used since Hg^{2+} ions are known to bind preferentially to thymine bases (Sissoeff *et al.,* loc. cit.). Cryptic

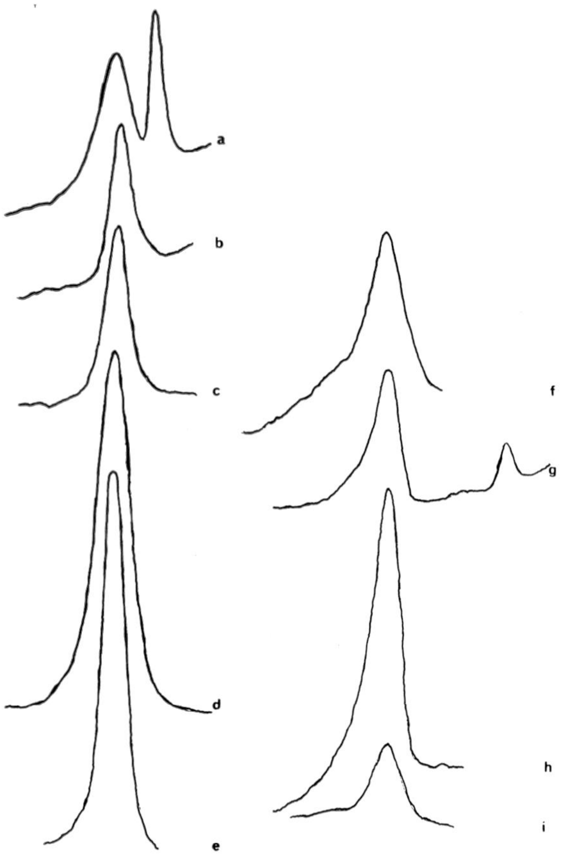

Fig. 2. Buoyant density profiles in the analytical ultracentrifuge of DNAs run in the presence of Actinomycin–D in Caesium Chloride. a. *A. leo;* b. *A. godmani;* c. *A. flavipes;* d. *A. stuartii;* e. *A. bellus;* f. *A. bilarni;* g. *A. macdonnellensis;* h. *A. swainsonii;* i. *A. minimus.*

satellites were resolved in the DNA of *A. stuartii* (heavy) and *A. bilarni* and *A. leo* (light). Once again skewness of the main band DNA or pronounced shoulders were seen on the heavy side for *A. stuartii, A. godmani, A. leo, A. bellus, A. swainsonii, A. minimus, A. bilarni* and *A. macdonnellensis* and on the light side for *A. leo* (see Fig. 4).

DISCUSSION

The usefulness of chromosome studies in the understanding of the phylogenetic relationships of species has been greatly enhanced by the development of G–, C– and Q–banding techniques which allow for greater linear differentiation of the chromosomes and hence more detailed comparisons between morphologically very similar karyotypes. The chromosomal approach is limited, however, in those cases where few obvious differences exist between the karyotypes of the related species as occurs in *Antechinus* (Young *et al.,* this Volume). Even where linear differentiation of chromosomes is very great as in Dipteran salivary gland chromosomes, it is possible that no chromosomal differences between species can be detected (Carson 1981, Wasserman 1981).

Close similarities in karyotypes of related species are known which mask significant differences in the amounts of DNA per diploid nucleus, e.g. 12.4 – 20.3 pg DNA in the genus *Lathyrus* (Narayan and Rees 1976) or 18–69 pg DNA in the genus *Plethodon* (Mizuno and Macgregor 1974). Narayan and Rees (loc. cit.) further demonstrated that in the genus *Lathyrus,* not all classes of DNA (Hr, Mr and unique) need be altered to the same extent but that in general quantitative changes in the the amount of DNA per nucleus usually involve mainly the repetitive fractions. In view of the apparent evolutionary stability of the satellites and Hr–DNA, these classes of DNA sequences have received considerable attention in the making of interspecific comparisons (Hennig and Walker 1970, Barnes *et al.* 1978, Gerlach *et al.* 1978, Marx *et al.* 1979). In most cases the distribution and conservation of components of the genome seem to reflect the more traditional phylogenies based on morphology (see Barnes *et. al.,* loc. cit. for references).

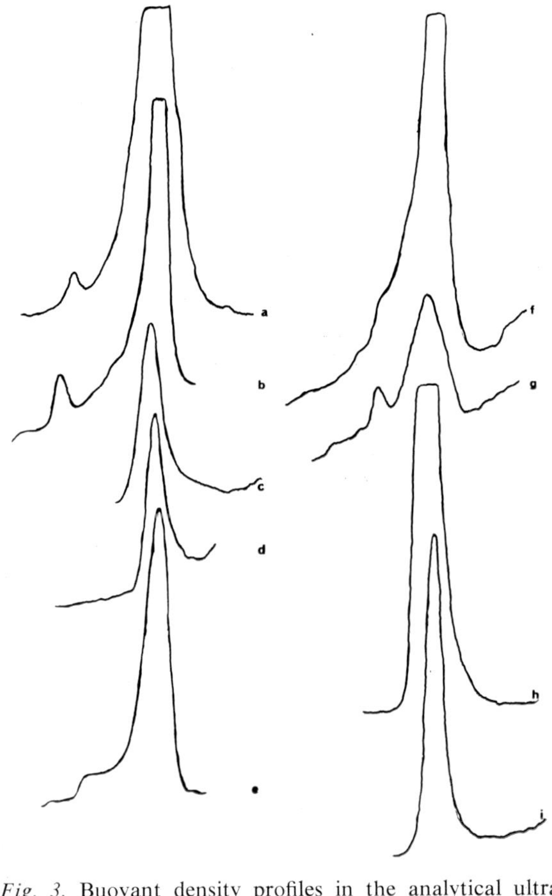

Fig. 3. Buoyant density profiles in the analytical ultracentrifuge of DNAs run in the presence of silver in Caesium Sulphate. a. *A. stuartii;* b. *A. godmani;* c. *A. swainsonii;* d. *A. minimus;* e. *A. leo;* f. *A. bilarni;* g. *A. macdonnellensis;* h. *A. bellus;* i. *A. flavipes.*

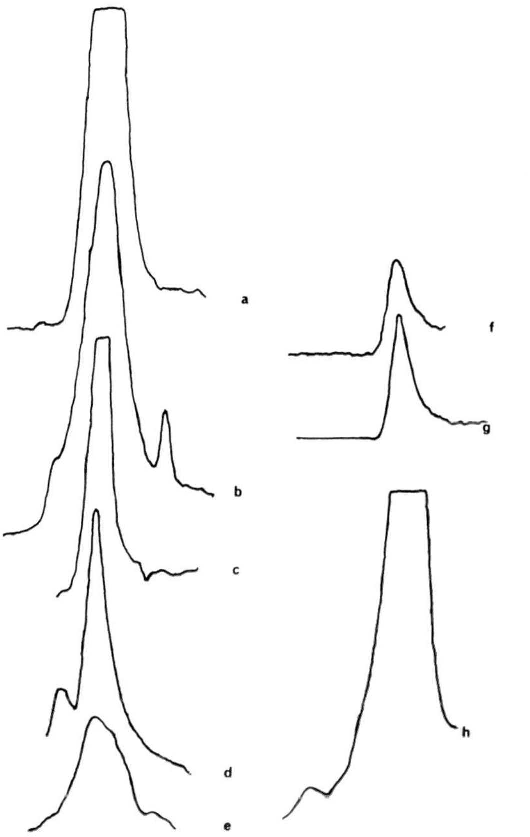

Fig. 4. Mercury/Cs_2SO_4 density gradient profiles for DNAs of a. *A. bellus;* b. *A. stuartii;* c. *A. flavipes;* d. *A. leo;* e. *A. godmani;* f. *A. swainsonii;* g. *A. minimus;* h. *A. bilarni.*

Little data is currently available on nuclear DNA contents for *Antechinus* species though Hayman and Martin (1974) showed that there was only about 1.2% variation in DNA amounts for seven species of Dasyurid. In the preliminary studies reported here we have seen that, with the exception of *A. stuartii,* the other *Antechinus* species investigated have only a small portion of their genome made up of highly repeated sequence DNA. This finding is consistent with the findings of Young *et al.* (this Volume) that most of these species have little C–band positive material. The relationship between Hr–DNA and constitutive heterochromatin in *Antechinus* is not straightforward, however. Thus *A. macdonnellensis,* which has substantial amounts of heterochromatin on the X–chromosome as well as on the autosomes, has only 1% of its genome present as Hr–DNA sequences. In contrast, *A. stuartii,* which has some 19% of its DNA present as highly repeated sequences, has relatively little heterochromatin as defined by C–banding. (For a detailed discussion of the relationship between Hr–DNA and constitutive heterochromatin see John and Miklos 1979).

Although no satellite DNA sequences were found in the DNA of *A. stuartii* using neutral CsCl density gradients, several cryptic satellites were revealed when particular DNA binding agents were used. Even when there was little Hr–DNA in the genome (e.g. *A. macdonnellensis*) cryptic satellites were still resolvable. In most cases, however, the differences in base composition of the Hr–DNA sequences were not sufficiently different from bulk DNA to allow resolution of distinct satellites even though the distribution profile of the DNA main band was markedly skewed to one side or the other by actinomycin–D, Ag+ of Hg^{2+} ions.

Although the work on *Antechinus* DNA is at an early stage, the results obtained so far suggest that there have been some changes in the amounts and types of Hr–DNA sequences accompanying speciation. On the basis of the studies using silver or mercury or actinomycin–D binding, an attempt has been made to ascertain the possible phylogenetic relationships of the species used. The dendrogram in Fig. 5 was obtained by unweighted pair group clustering of coefficients of DNA resemblance. *A. swainsonii* and *A. minimus* were indistinguishable from one another on the basis of the present experiments as were *A. bellus* and *A. flavipes.* These four species together with *A. stuartii* and *A. godmani* form one apparent major grouping. Within the other group, *A. bilarni* and *A. macdonnellensis* appear to be very similar in respect to Hr sequences in the DNA and both are apparently more closely related to *A. leo* than to the other species. Although necessarily provisional, with the exception of *A. leo* the relationships suggested in Fig. 5 largely agree with those proposed on the basis of penis structure (Woolley, this Volume) and enzyme polymorphisms (Baverstock *et al.,* this Volume). In view of the fact that the chromosomes of the Dibbler, *A. apicalis,* contain so much more constitutive heterochromatin than any other *Antechinus* species, it would be of great interest to see if the DNA from this species, contained proportionately

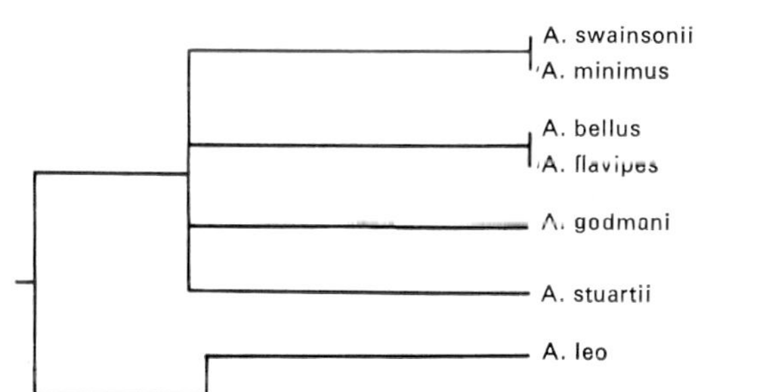

Fig. 5. A phenogram of nine species of *Antechinus* based on a U.P.G.M.A. cluster analysis of Euclidian distance measures using DNA differences.

more satellites or other highly repeated sequences and how it relates to the other species. Although it has been shown that the *Antechinus* species included in the current studies differ in their Hr–DNA sequences, this does not necessarily imply that the DNA changes which have occurred in the evolution of the genus have played any part in the speciation process.

ACKNOWLEDGEMENTS

An earlier draft of this paper was read and constructively criticised by Drs. R. Crozier and B. J. Richardson.

REFERENCES

Barnes, S. R., Webb, D. A. and Dover, G., 1978. The distribution of satellite and mainband DNA components in the *melanogaster* species sub–group of *Drosophila*. 1. Fractionation of DNA in actinomycin–D and distamycin–A density gradients. *Chromosoma* (*Ber.*) **67:** 341–63.

Baverstock, P., Archer, M., Adams, M. and Richardson, B. J., 1982. Genetic relationships among 32 species of Australian dasyurid marsupials. Pp641–50 *in* "Carnivorous marsupials", ed. by M. Archer. Roy. Zool. Soc. New South Wales: Sydney.

Britten, R. J., Graham, D. E., and Neufeld, B. R., 1974. Analysis of repeating DNA sequences by reassociation. *In* "Methods in enzymology", **29:** 363–418, ed. by L. Grossman and K. Moldave. Academic Press: New York.

Britten, R. J. and Kohne, D. E., 1968. Repeated sequences in DNA. *Science* **161:** 529–40.

Brutlag, R. J., Appels, R., Dennis, E. S. and Peacock, W. J., 1977. Highly repeated DNA in *Drosophila melanogaster*. *J. Mol. Biol.* **112:** 31–47.

Carson, H. L., 1980. Homosequential species of Hawaiian *Drosopila*. *In* "Proc. 7th Int. Chrom. Conf. Oxford" ed. by M. D. Bennett, G. M. Hewitt and M. Bobrow. George Allen and Unwin: London.

Corneo, G., 1976. Do satellite DNAs function as sterility barriers in eukaryotes? *Evol. Theory* **1:** 261–5.

Doolittle, W. F. and Sapienza, C., 1980. Selfish genes, the phenotype paradigm and genome evolution. *Nature* (*Lond.*) **284:** 601–4.

Fry, K. and Salser, W., 1977. Nucleotide sequences of HS–*a* satellite DNA from Kangaroo Rat *Dipodomys ordii* and characterization of similar sequences in other rodents. *Cell.* **124:** 1069–84.

Gerlach, W. L., Appels, R., Dennis, E. S. and Peacock, W. J., 1978. "Evolution and analysis of wheat genomes using highly repeated DNA sequences." Proc. 5th Int. Wheat Gen. Symposium.

Hatch, F. T., Bodmer, A. J., Mazrimas, J. A. and Moore, D. H., 1976. Satellite DNA and cytogenetic evolution. DNA quantity, satellite DNA and karyotypic variations in Kangaroo rats (genus *Dipodomys*). *Chromosoma (Berl.)* **58:** 155–68.

Hayman, D. L. and Martin, P. G., 1974. Monotremata and Marsupialia. Animal cytogenetics, Vol. 4: Chordata part 4. Mammalia I. Borntraeger: Stuttgart.

Hennig, W. and Walker, P. M. B., 1970. Variations in the DNA from two rodent families (Cricetidae and Muridae). *Nature* (*Lond.*) **225:** 915–9.

Hsieh, T. S. and Brutlag, D. L., 1979. A protein that preferentially binds *Drosophila* satellite DNA. *Proc. Nat. Acad. Sci.* **76:** 726–30.

John, B. and Miklos, G. L. G., 1979. Functional aspects of satellite DNA and heterochromatin. *Int. Rev. Cytol.* **58:** 1–114.

Josse, J., Kaiser, A. D. and Kornberg, A., 1961. Enzymatic synthesis of DNA VIII: Frequencies of nearest neighbour base sequences in deoxyribonucleic acid. *J. Biol. Chem.* **236:** 864–75.

Kurnit, D. M., Shafit, B. R. and Maio, J. J., 1973. Multi–satellite deoxyribonucleic acids in the calf and their relation to the sex chromosomes. *J. Mol. Biol.* **81:** 273–84.

Mandel, M. and Marmur, J., 1968. Use of ultraviolet absorbance–temperature profile for determining the gaunine plus cytosine content of DNA. Pp.195–206 *in* "Methods in enzymology", ed. by L. Grossman and K. Moldave. Academic Press: New York.

Marmur, J., 1961. A procedure for the isolation of DNA from micro–organisms. *J. Mol. Biol.* **3:** 208–18.

Marx, K. A., Purdom, I. F. and Jones, K. W., 1979. Primate repetitive DNAs; evidence for new satellite DNAs and similarities in non–satellite repetitive DNA sequence properties. *Chromosoma (Berl.)* **73:** 153–61.

Miklos, G. L. G. and John, B., 1979. Heterochromatin and satellite DNA in man: properties and prospects. *Amer. J. Hum. Gen.* **31:** 264–80.

Miklos, G. L. G., Willcocks, D. A. and Baverstock, P. R., 1980. Restriction endonuclease and molecular analyses of three rat genomes with special reference to chromosome rearrangements and speciation problems. *Chromosoma* (*Berl.*) **76:** 339–63.

Mizuno, S. and Macgregor, H. C., 1974. Chromosomes, DNA sequences and evolution in salamanders of the genus *Plethedon*. *Chromosoma* (*Berl.*) **48:** 239–96.

Narayan, R. K. J. and Rees, H., 1976. Nuclear DNA variation in *Lathyrus*. *Chromosoma* (*Berl.*) **54:** 141–54.

Orgel, L. E. and Crick, F. H. C., 1980. Selfish DNA: the ultimate parasite. *Nature* (*Lond.*) **284:** 604–8.

Peacock, W. J. and Lohe, A. R., 1981. Repeated sequences in the chromosomes of *Drosophila*. *In* "Insect cytogenetics" ed. by R. L. Blackman, G. M. Hewitt and M. Ashburner. Blackwells: Oxford.

Peacock, W. J., Brutlag, D. L. L., Goldring, E., Appels, R., Hinton, C. W. and Linsley, D. L., 1973. The organization of highly repeated sequences in *Drosophila melanogaster* chromosomes. *Cold Spring Harbor Symp. Quant. Biol.* **38:** 405–16.

Salser, W., Bowen, S., Brown, D., El Adli, R., Fedoroff, N., Fry, K., Heindell, H., Paddock, G., Poon, R., Wallace, B. and Whitcombe, P., 1976. Investigation of the organization of mammalian chromosomes at the DNA sequence level. *Fed. Proc. Fed. Am. Soc. Exp. Biol.* **35:** 23–35.

Sissoeff, O., Griovard, I. and Guille, E., 1976. Studies on metal ions – DNA interactions: specific behaviour of reiterative DNA sequences. *Prog. Biophys. Molec. Biol.* **31:** 165–99.

Walker, P. M. B., 1968. How different are the DNAs from related animals. *Nature* (*Lond.*) **219:** 228–35.

Wasserman, M., 1981. Evolution of the *repleta* group. *In* "The genetics and biology of *Drosophila*. Vol. 3" ed. by M. Ashburner, H. L. Carson and J. N. Thompson. Academic Press: London.

WOOLLEY, P. A., 1982. Phallic morphology of the Australian species of *Antechinus* (Dasyuridae, Marsupialia): a new taxonomic tool? Pp.767–81 in "Carnivorous marsupials", ed. by M. Archer. Roy. Zool Soc New South Wales: Sydney.

YOUNG, G. J., GRAVES, J. A. M., BARBIERI, I., WOOLLEY, P. A., COOPER, D. W. C. AND WESTERMAN, M., 1982. The chromosomes of dasyurids (Marsupialia). Pp.783–95 *in* "Carnivorous marsupials", ed. by M. Archer. Roy. Zool. Soc. New South Wales: Sydney.

RESUMEN

61. Estudios del DNA en algunas especies de *Antechinus* (Dasyuridae, Marsupialia). Por M. Timms, M., Westerman y B. Düring.

El género de marsupial *Antechinus* se caracteriza por la remarcable estabilidad de su cariotipo, incluyendo el tipo de las bandas G y C. Con una sola excepción notable, *A. apicalis* ("the Dibbler"), las cantidades de material positivo en la banda C de los genomas son similares y muy pequeñas. Un estudio del DNA de algunas especies mostró que en general las cantidades de la secuencia altamente repetida del DNA (Hr–DNA) del genoma son también pequeñas (alrededor del 1%). Una excepción obvia fue *A. stuartii,* con alrededor del 19% de su genoma compuesto por Hr–DNA. Aun cuando esta clase de DNA fue pequeña, ciertas secuencias contenidas en ellas fueron resolubles como satélites crípticos o considerados como remarcables modificado res de la distribución de la banda principal del DNA cuando iones de actinomicina–D, Ag ó Hg^{2+} son agregados al DNA antes de la centrifugación del gradiente de densidad. La comparación de la distribución de algunas de estas secuencias Hr–DNA en las especies investigadas sugirieron la posible existencia de dos subgrupos de especies principales.

62. THE RELATIONSHIPS OF *Antechinus stuartii* AND *A. flavipes* (DASYURIDAE, MARSUPIALIA) WITH SPECIAL REFERENCE TO QUEENSLAND

S. VAN DYCK[1]

Relationships between all forms of *Antechinus stuartii* and *A. flavipes* in Queensland are discussed in terms of external morphology, distribution, reproductive biology and cranial morphology. The significantly larger form *A. flavipes rubeculus* n. subsp. from northeastern Queensland is described, and the subspecific status of *A. stuartii adustus* is confirmed. *A. stuartii stuartii* is essentially restricted to subtropical vine–forest in southeastern Queensland. *A. s. adustus* is found only in high altitude vine–forests of northeastern Quensland. A regular pattern of high winter rainfall characterizes areas occupied by both forms. *A. flavipes flavipes* and *A. f. rubeculus* occupy a wide range of habitats and can tolerate lower, more irregular winter rainfall. In Queensland *A. s. stuartii* mates in the last two weeks of September, *A. s. adustus* mates probably in early July, *A. f. flavipes* mates in late September and *A. f. rubeculus* in June and July.

In cranial comparison with *A. f. flavipes, A. s. stuartii* is distinguished by the following: longer anterior and posterior palatal vacuities, longer inter–palatal vacuity length, narrower rostrum and zygomatic widths, wider interorbital width, longer premolar row, wider separation of transverse canal foramina, narrower snout–canine depth, larger entoconids and wider talonids. In comparison with *A. s. stuartii, A. s. adustus* is distinguished by: shorter basicranial length, smaller palatal vacuities, shorter premolar row, greater inside bullae width, shorter separation of transverse canal foramina and prominent posterolingually lobed upper premolars. *A. s. adustus* also differs significantly from the neotype of *A. s. stuartii* from Sydney.

Antechinus stuartii represents the more primitive antechinus while *A. flavipes* the more recently derived and more adaptable. The life histories and their consequences are discussed for both species throughout their range in eastern Australia.

Pages 723–66 *in* CARNIVOROUS MARSUPIALS, ed. by M. Archer, February, 1982. Roy. Zool. Soc. New South Wales: Sydney, Australia.

THE GENUS *ANTECHINUS* is comprised of a number of closely related forms of small semi–arboreal marsupicarnivores. Considerable controversy exists over the number of species and questions of interspecific relationships remain unresolved. While several species of antechinuses occupy temperate and arid areas throughout Australia, the greatest diversity is seen in the tropical and subtropical representatives of which five species and four subspecies occur in Queensland. Of these, much confusion has surrounded the species *A. stuartii* (Macleay 1841) and *A. flavipes* (Waterhouse 1838), two of the most abundant small marsupials in eastern coastal Australia.

The purpose of this paper is to examine the relationships of *A. flavipes* and *A. stuartii*, reassess the status of the two species and provide effective guidelines for their recognition. The results are largely drawn from newly available field observations and specimens collected on Queensland Museum mammal surveys in north eastern Queensland and Cape York Peninsula during 1976–78.

Skins and skulls from a total of 315 specimens were studied. Of forty–four cranial characters measured for the named taxa, thirteen were considered to demonstrate most systematic significance within the genus, while seven cranial characters in particular highlighted most trenchant differences between the two species. Three external body measurements were taken for samples of the taxa (Tables 1–7). Cranial measurements were taken with dial calipers to the nearest .1 mm (see Fig. 1). External body dimensions were measured to the nearest 1 mm and body weights to the nearest 1 g. Age variation was minimised by using for statistical analysis only those skulls which possessed fully erupted permanent premolars and no measurements were taken from sub–adult animals cap-during the four months prior to April in any particular year. Beside mensural characters, qualitative characters were evaluated; these included development of the entaconid, shape of talonid and quality of pelage. Pelage colour was compared by reference to Ridgway (1912). Terminology is shown in Fig. 1 and nomenclature follows Archer as follows: teeth (Archer 1974), basicranial structures (Archer 1976a) and tooth number (Archer 1978).

TAXONOMY OF Antechinus stuartii

In 1967 Wakefield and Warneke revised the genus *Antechinus* with reference to *A. stuartii*. In emphasizing the distinction of this species from *A. flavipes* they pointed out the error of Iredale and Troughton (1934) in applying the

[1] Queensland Museum, Gregory Terrace, Fortitude Valley, Queensland, Australia, 4006.

trinomial *A. f. flavipes* to the central eastern New South Wales population of *A. stuartii.* This error resulted in the same taxonomic misapplication being repeated in subsequent works such as Troughton (1941), Tate (1947), Horner and Taylor (1959), Marlow (1961) and Walker (1964). Wakefield and Warneke (1967) concluded that the type specimen of *A. stuartii* was lost and they designated a neotype for *A. s. stuartii* from Waterfall, National Park, New South Wales. With regard to the separation of races they concluded, firstly, that *A. stuartii*

Table 1. Absolute measurements for *Antechinus stuartii stuartii* in southeastern Queensland. Terminology and abbreviations used in TABLES 1–7. BL, basicranial length; ZW, zygomatic width; SWR–LC^1, skull width between right and left upper canines; SWR–LM^4, skull width between right and left upper fourth molars; APVL, anterior palatal vacuity length; PPVL, posterior palatal vacuity length; IPVL, inter–palatal vacuity length; IBW, inside bullae width; IOW, minimum interorbital width; TC, distance of separation of transverse canal foramina; SCD, snout–canine depth; HB, head-body length; TV, tail–vent length; W, weight (g); N, number of specimens in sample; $x \pm r$, sample mean ± one standard error; OR, observed range; S, standard deviation; CV, coefficient of variation; t, value of Students t–test; P, probability of exceeding the observed value of t, * information from Braithwaite (1973) for animals weighed in early September at Mt. Glorious (mean weight ± 2 standard errors).

	N	$\bar{x} \pm r$	OR	S	CV
BL	50♂	27.97 ± .13	25.5 - 30.5	0.95	3.4
	23♀	26.80 ± .17	24.9 - 28.7	0.80	3.0
ZW	51♂	16.65 ± .09	15.3 - 18.7	0.66	4.0
	25♀	15.97 ± .11	14.7 - 17.4	0.56	3.5
SWR-LC^1	49♂	4.96 ± .03	4.4 - 5.5	0.20	4.1
	25♀	4.85 ± .04	4.5 - 5.2	0.19	3.9
SWR-LM^4	50♂	12.96 ± .07	11.3 - 13.8	0.51	4.0
	25♀	12.18 ± .09	11.1 - 12.7	0.44	3.6
APVL	51♂	3.89 ± .06	3.0 - 4.9	0.45	11.6
	25♀	3.76 ± .10	3.0 - 4.7	0.49	13.2
PPVL	52♂	5.94 ± .05	5.3 - 6.7	0.35	5.9
	25♀	5.96 ± .06	5.5 - 6.5	0.31	5.1
IPVL	51♂	2.44 ± .08	1.1 - 3.7	0.56	23.0
	25♀	2.34 ± .10	1.1 - 3.1	0.51	21.9
$P^{\underline{1-2}}$	50♂	2.47 ± .02	2.1 - 2.7	0.13	5.1
	25♀	2.36 ± .02	2.1 - 2.6	0.12	4.9
$P^{\underline{1-3}}$	51♂	3.69 ± .02	3.4 - 4.0	0.16	4.3
	25♀	3.53 ± .02	3.3 - 3.7	0.10	2.9
IBW	50♂	4.18 ± .04	3.6 - 4.7	0.28	6.6
	25♀	4.07 ± .06	3.4 - 4.6	0.29	7.2
IOW	51♂	7.02 ± .03	6.5 - 7.7	0.20	2.9
	25♀	6.79 ± .04	6.4 - 7.2	0.22	3.3
TC	51♂	3.09 ± .03	2.6 - 3.7	0.24	7.9
	25♀	3.02 ± .05	2.6 - 3.7	0.26	8.4
SCD	49♂	4.89 ± .04	4.1 - 5.4	0.28	5.8
	25♀	4.48 ± .05	3.8 - 5.0	0.27	6.0
HB	115♂	116.59 ± .63	102 -136	6.78	5.8
	32♀	102.19 ± 1.02	94 -112	5.76	5.6
TV	114♂	96.54 ± .54	84 -106	5.82	6.0
	32♀	88.47 ± .91	64 -100	5.12	5.8
W*	11♂	60 ± 5.30	52 - 67		
	6♀	28 ± 2.44	24 - 32		

burrelli (Le Souef 1926) from Ebor, northeastern New South Wales was not a justifiable subspecies and secondly that there were insufficient specimens and distribution data to make an assessment of the status of *A. s. adustus* (Thomas 1923) from Ravenshoe, northeastern Queensland.

The present study clarifies and supports the subspecific status of *A. s. adustus*.

TAXONOMY OF Antechinus flavipes

A thorough resume of taxonomy of *A. flavipes* is presented in Wakefield and Warneke (1967). The only form they recognize subspecifically is *A. flavipes leucogaster* (Gray 1841 from southwestern Western Australia and they regard the form *A. flavipes rufogaster* (Gray 1841) as a subjective synonym of *A. f. flavipes*. The status of *A. flavipes* in northern

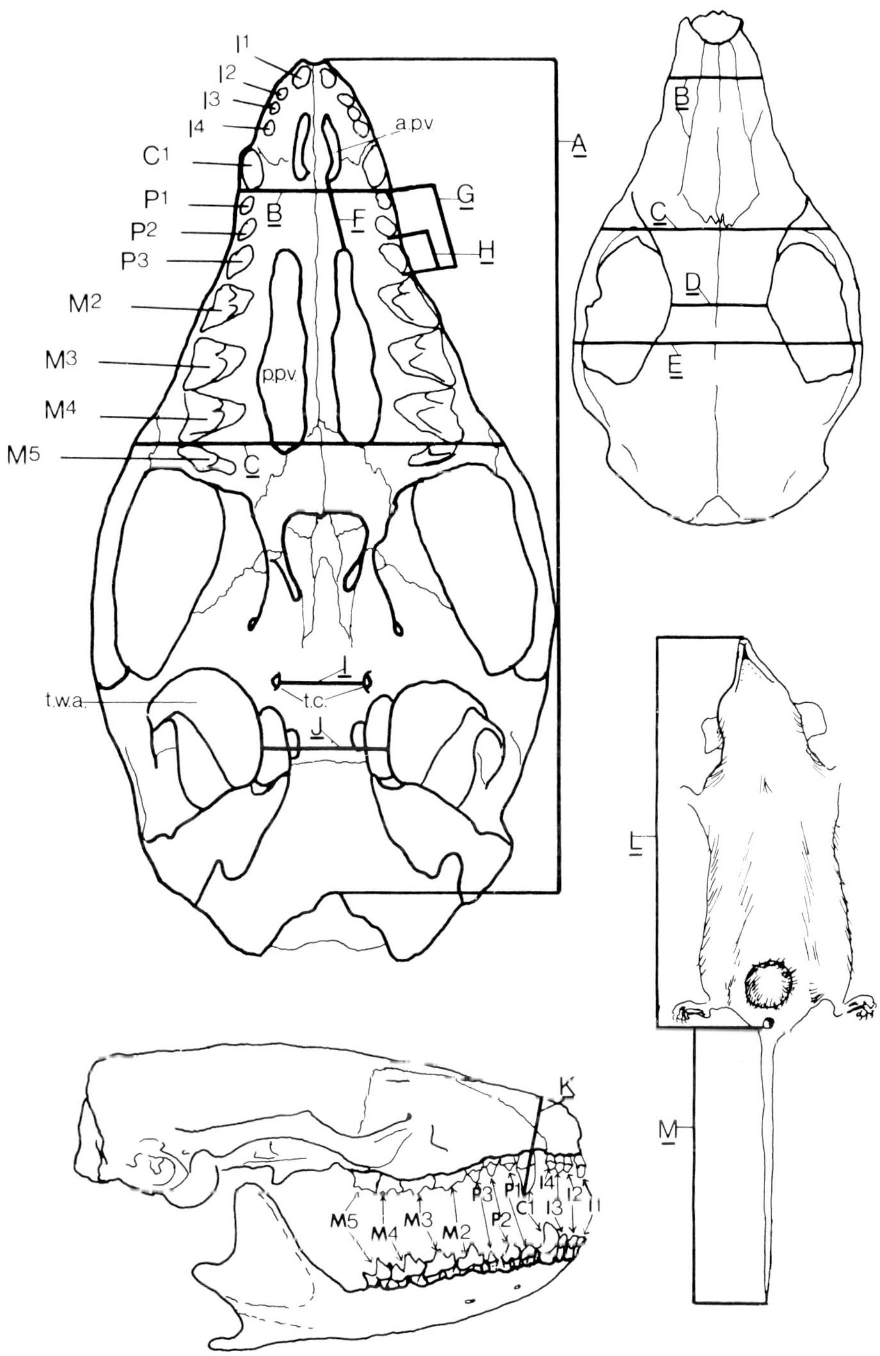

Fig. 1. Terminology and mensuration in the skull, dentary and body of *Antechinus*. A, basicranial length; B, skull width between right and left upper canines; C, skull width between right and left upper fourth molars; D, interorbital width; E, zygomatic width; F, inter-palatal vacuity length; G, length separating P^1 and P^3; H, length separating P^1 and P^2; I, length of separation of transverse canal foramina; J, inside bullae width; K, snout–canine depth; L, head–body length; M, tail–vent length; a.p.v., anterior palatal vacuity; p.p.v., posterior palatal vacuity; t.c., transverse canal foramina; t.w.a., tympanic wing of the alisphenoid.

Table 2. Absolute measurements for *Antechinus stuartii adustus.*

	N	$\bar{x} \pm r$	OR	S	CV
BL	14♂	27.06 ± .13	26.1 - 28.0	0.46	1.8
	14♀	26.01 ± .21	25.2 - 27.7	0.78	3.0
ZW	14♂	16.38 ± .10	15.8 - 17.0	0.39	2.4
	13♀	15.54 ± .15	15.0 - 16.2	0.54	3.5
SWR-LC[1]	14♂	4.99 ± .03	4.8 - 5.3	0.13	2.6
	15♀	4.80 ± .04	4.6 - 5.1	0.16	3.4
SWR-LM[4]	14♂	12.80 ± .10	12.1 - 13.4	0.36	2.8
	15♀	12.04 ± .17	11.2 - 13.8	0.66	5.5
APVL	14♂	2.71 ± .08	2.2 - 3.1	0.30	11.1
	15♀	2.51 ± .07	2.1 - 2.9	0.27	10.7
PPVL	13♂	4.84 ± .13	4.1 - 5.4	0.49	10.0
	13♀	4.74 ± .15	4.1 - 5.4	0.53	11.3
IPVL	14♂	4.62 ± .11	3.9 - 5.4	0.41	8.9
	15♀	4.50 ± .16	3.5 - 6.0	0.61	13.5
p^{1-2}	14♂	2.22 ± .04	2.0 - 2.4	0.14	6.2
	15♀	2.17 ± .02	2.0 - 2.3	0.10	4.5
p^{1-3}	14♂	3.40 ± .04	3.2 - 3.7	0.15	4.5
	15♀	3.35 ± .03	3.1 - 5.0	0.13	3.9
IBW	14♂	4.39 ± .05	4.0 - 4.7	0.20	4.6
	14♀	4.26 ± .05	3.9 - 4.5	0.19	4.6
IOW	14♂	6.86 ± .06	6.5 - 7.2	0.21	3.1
	14♀	6.68 ± .04	6.4 - 7.0	0.17	2.5
TC	14♂	2.81 ± .04	2.6 - 3.0	0.16	5.7
	14♀	2.86 ± .04	2.5 - 3.0	0.15	5.3
SCD	14♂	4.66 ± .06	4.2 - 5.0	0.23	5.0
	15♀	4.41 ± .08	4.1 - 5.1	0.30	6.7
HB	9♂	101.67 ± 2.91	94 -119	8.72	7.9
	10♀	93.00 ± .97	89 -100	3.05	3.3
TV	9♂	98.33 ± 1.26	91 -103	3.77	3.8
	10♀	90.10 ± .43	89 - 93	1.37	1.5
W	15♂	33.40 ± .86	30 - 42	3.31	10.0
	15♀	24.07 ± 1.00	21 - 34	3.86	16.11

Queensland has been characterized by conflicting taxonomies and lack of sufficient comparative material. The problems that Wakefield and Warneke encountered in claryifying the status of *A. s. adustus* they also faced with tropical *A. flavipes*. They recognized the distinct nature of the northern populations thus '(p. 71) . . . the relatively few specimens available from northern Australia have allowed no such clarification of that status of its tropical populations. The population of North–east Queensland probably warrants separate subspecific status, and that of Cape York certainly does.'

The Cape York Peninsula form from the Iron and McIlwraith Ranges has been referred to in the past as *Antechinus flavipes godmani* (Tate 1952, Horner and Taylor 1959, Degabriele 1976) and *A. godmani* (Marlow 1962). Wakefield and Warneke (1967) recognized the distinct nature of this form but refrained from describing it as new. This most northerly form has since been shown to differ from *A. flavipes* in colour, supratragal, dental and basicranial morphology (Van Dyck 1980). It is the only mammal endemic to Cape York Peninsula and is geographically isolated from the most northerly population of *A. flavipes* at Home Rule (15° 42′S 145° 14′E), by approximately 350 km to the south. It has been assigned full specific status as *Antechinus leo* (Van Dyck 1980).

The status of the remaining population from northern Queensland (Townsville 19° 16′S 146° 49′E to Cooktown 15° 28′S 145° 15′E) previously poorly represented in collections, is now clarified. Wakefield and Warneke (1967) were able to examine approximately 19 specimens from northeastern Queensland, and with this sample were unable to determine the significance of differences in this population when compared to southern *A. f. flavipes*. A larger series now exists in the collections of the Queensland Museum, the examination of which reveals many characters which differ significantly from those in southern populations.

For this reason subspecific recognition is given to this largest, most northerly, superbly red form of *A. flavipes*.

Antechinus flavipes rubeculus, n. subsp.

Holotype: Queensland Museum JM2676, adult male, puppet skin and extracted skull, with torso in spirit, collected 31 July 1976 by S. Van Dyck.

Locality and Habitat: Flaggy Ck Forestry barracks, Black Mountain Rd, Kuranda (15° 46′S 145° 38′E) northeastern Queensland. Mesophyll vine–forest on schist (Webb 1978).

Range. Coastal and Tablelands from Mt. Spec 18° 57′S 146° 11′E (approximately 60 km N.

Table 3. Absolute measurements for *Antechinus flavipes flavipes* in southeastern Queensland.

	N	$\bar{x} \pm r$	OR	S	CV
BL	28♂	28.63 ± .20	26.6 - 31.7	1.08	3.8
	18♀	26.46 ± .22	25.6 - 29.0	0.91	3.5
ZW	30♂	18.01 ± .16	16.1 - 19.5	0.89	4.9
	19♀	16.62 ± .12	15.6 - 17.7	0.53	3.2
SWR-LC[1]	30♂	5.82 ± .06	5.0 - 6.3	0.32	5.6
	23♀	5.20 ± .05	4.9 - 5.7	0.23	4.5
SWR-LM[4]	30♂	14.55 ± .09	13.4 - 15.4	0.52	3.6
	20♀	13.62 ± .15	12.6 - 15.3	0.69	5.1
APVL	27♂	2.64 ± .05	2.0 - 3.0	0.26	9.8
	23♀	2.41 ± .05	1.9 - 2.8	0.22	9.3
PPVL	30♂	4.52 ± .07	3.9 - 5.4	0.41	9.0
	23♀	4.23 ± .10	2.6 - 4.8	0.48	11.2
IPVL	30♂	5.34 ± .09	4.4 - 6.4	0.47	8.8
	23♀	5.01 ± .09	4.3 - 5.8	0.42	8.4
$P^{\underline{1-2}}$	30♂	2.05 ± .01	1.9 - 2.2	0.08	3.8
	23♀	2.03 ± .02	1.8 - 2.2	0.12	5.2
$P^{\underline{1-3}}$	30♂	3.33 ± .04	2.9 - 3.6	0.19	5.9
	23♀	3.19 ± .04	2.8 - 3.5	0.17	5.3
TBW	29♂	3.90 ± .05	3.4 - 4.5	0.28	7.3
	19♀	3.82 ± .08	3.2 - 4.7	0.35	9.2
IOW	30♂	6.52 ± .05	6.0 - 7.0	0.29	4.4
	21♀	6.37 ± .05	6.0 - 6.9	0.23	3.7
TC	29♂	2.58 ± .05	2.1 - 3.1	0.26	10.0
	19♀	2.45 ± .03	2.2 - 2.8	0.14	5.8
SCD	20♂	5.56 ± .08	4.8 - 6.4	0.41	7.4
	20♀	4.78 ± .06	4.2 - 5.2	0.27	5.7
HB	27♂	116.52 ± 2.60	93 -165	13.49	11.6
	23♀	102.83 ± 1.51	90 -122	7.26	7.1
TV	27♂	96.82 ± 5.20	80 -110	7.29	7.5
	23♀	84.65 ± 1.76	72 -107	8.43	10.0
W	8♂	47.50 ± 5.23	26 - 76	14.80	31.1
	4♀	28.00 ± 2.55	21 - 33	5.10	18.2

Townsville NEQ) to Home Rule 15° 42′S 145° 14′E (approximately 30 km S. Cooktown NEQ) – a total northerly spread of approximately 350 km (see Fig. 6). This form appears to be geographically isolated from southern populations. *Specimens Examined*: Mt. Spec 18° 57′S 146° 11′E, JM1200 ♂, J10148 ♀; Ingham 18° 29′S 146° 53′E, JM1918 ♂, JM1938 ♂, JM5890 ♂, JM1917 ♂, JM2213 ♂, JM2210 ♂, JM19410 ♀, JM590 ♀; Wallaman Falls 18° 35′S 145° 48′E, JM2350 ♀; Atherton 17° 16′S 145° 29′E, J5903 ♂, J5902 ♂, J19679 ♂, J1967A ♂, J193080 ♂, J19676B ♂, J19678 ♂, J5901 ♀; Mission Beach 17° 52′S 146° 06′E, JM1970 ♂; El Arish 17° 49′S 146° 00′E, J1490 ♀; Ravenshoe 17° 36′S 145° 29′E JM1940 ♀; Home Rule 15° 42′S 145° 14′E, JM680 ♂; Shipton's Flat 15° 48′S 145° 16′E, JM592 ♂; Kuranda 15° 46′S 145° 38′E, JM2211 ♂, JM1939 ♂; Kirrama 18° 09′S 145° 37′E, JM590; Thornton Peak 16° 10′S 145° 23′E, JM593 ♂; Herberton 17° 23′S 145° 23′E J9307 ♂; Mt Finnegan 15° 49′S 145° 17′E AMNH 154316 ♀, 154313 ♀; Speewah 16° 55′S 145° 38′E, AMNH 154245 ♀, 154243 ♂; Russell River 17° 22′S 145° 58′E, AMNH 154248 ♀.

Table 4. Absolute measurements for type and series of *Antechinus flavipes rubeculus*.

	JM2676	N	$\bar{x} \pm r$	OR	S	CV
BL	30.9	21♂	30.06 ± .37	26.6 - 32.9	1.72	5.7
		11♀	28.48 ± .27	27.3 - 29.7	0.90	3.2
ZW	19.4	22♂	18.80 ± .23	16.9 - 20.7	1.06	5.6
		11♀	17.77 ± .18	17.0 - 19.0	0.61	3.4
SWR-LC[1]	6.3	21♂	6.01 ± .09	5.1 - 6.9	0.41	6.9
		11♀	5.81 ± .08	5.4 - 6.2	0.25	4.4
SWR-LM[4]	15.9	21♂	14.90 ± .11	14.2 - 16.2	0.52	3.5
		11♀	14.37 ± .14	13.9 - 15.2	0.45	3.1
APVL	2.5	21♂	2.73 ± .07	2.3 - 3.2	0.30	11.1
		11♀	2.51 ± .08	2.1 - 2.8	0.27	10.6
PPVL	4.6	21♂	4.66 ± .11	3.6 - 5.6	0.51	10.9
		11♀	4.26 ± .10	3.5 - 4.6	0.32	7.4
IPVL	5.8	21♂	5.74 ± .14	4.8 - 7.4	0.63	11.1
		11♀	5.76 ± .12	5.1 - 6.4	0.39	6.7
P^{1-2}	2.1	21♂	2.15 ± .03	1.9 - 2.4	0.14	6.3
		11♀	2.12 ± .03	1.9 - 2.2	0.10	4.6
P^{1-3}	3.5	21♂	3.37 ± .05	3.0 - 3.8	0.24	7.1
		11♀	3.34 ± .05	3.1 - 3.6	0.17	5.2
IBW	4.4	20♂	4.26 ± .07	3.5 - 4.7	0.31	7.3
		11♀	4.38 ± .06	4.0 - 4.7	0.20	4.7
IOW	6.7	21♂	6.83 ± .07	6.3 - 7.5	0.34	4.9
		11♀	6.84 ± .06	6.4 - 7.2	0.21	3.1
TC	2.4	22♂	2.79 ± .05	2.4 - 3.2	0.24	8.4
		11♀	2.78 ± .04	2.5 - 2.9	0.12	4.5
SCD	5.6	21♂	5.81 ± .07	5.2 - 6.5	0.34	5.9
		11♀	5.34 ± .09	4.9 - 5.9	0.31	5.9
HB	144	22♂	125.68 ± 3.07	98 -151	14.39	11.4
		7♀	110.29 ± 3.16	98 -127	9.55	8.7
TV	110	22♂	103.68 ± 2.47	85 -131	11.56	11.2
		6♀	92.67 ± 2.81	86 -104	6.89	7.4
W	76	12♂	67.75 ± 3.75	35 - 79	12.98	19.2
		6♀	38.67 ± 3.90	30 - 52	9.54	24.7

Fig. 2. A, *Antechinus stuartii stuartii.* B, Habitat of *Antechinus stuartii stuartii*, Mt. Tamborine, southeastern Queensland.

Diagnosis: The largest form of A. flavipes, differing from *A. f. flavipes* and *A. f. leucogaster* in the following respects: in colour *A. f. rubeculus* exhibits the most extreme tendency toward redness; in body measurement and weight it is significantly larger and heavier than all other forms. In cranial morphology it is significantly larger than southern forms in the following measurements; basicranial length, zygomatic width, skull width at R–LC[1], inter–palatal vacuity length, P^{1-2}, inside bullae width, transverse canal foramina separation, skull–canine depth and interorbital width (Table 4).

Description: Pelage: colours for holotype as follows; dorsal crown fur (4 mm long) with basal 2 mm Slate–Gray, median 1 mm Buckthorn Brown, apical 1 mm Fuscous–Black. Medially thickened guard hairs (5 mm long) are interspersed through the fur and the head appears overall to be a grizzled Sepia. Fur of mid–back (9 mm long) with basal 5 mm Slate–Gray, median 3 mm Cinnamon–Brown. Fur of rump (7 mm long) with basal 4 mm Slate–Gray, median thickened 2 mm Hazel and apical 1 mm Black. Black–tipped guard hairs (10 mm long) are interspersed through the fur. The rump appears overall to be Hazel.

Eye–rings, lips and chin are Apricot Buff. Post auricular patches are Hazel.

Flanks, hindfeet and forefeet are Cinnamon–Rufous. The soft ventral fur (6mm long on belly and 3 mm on interramal region) is Slate–Gray on the basal half and Hazel on the apical half. The belly is thus a warm Hazel.

The tail is thickly haired dorsally with a mixture of Fuscous–Black and Hazel hairs (3 mm long) at the base. Toward the tip of the tail, hairs become predominantly Fuscous–Black (5 mm long at tip). The underside of the tail is thinly covered with Hazel hairs at the proximal end; the distal 48 mm, however, is thickly covered with Fuscous–Black Hairs (3–6 mm long) giving the tail a small ventral crest.

Body measurements: weight and body measurements of holotype, JM2676 and examined series appear in Table 4.

Vibrissae and foot pads: similar to *A. f. flavipes*.

Cranial morphology: *A. f. rubeculus* is significantly ($P < .001$) larger in nine of the thirteen characters considered here as systematically significant within the genus. Absolute measurements for the holotype JM2676 and examined series appear in Table 4. A comparative description is given under Cranial and Dental Morphology. The skull of the holotype JM2676 appears in Fig. 11.

Reproductive data are presented under Comparative Reproduction.

DISTRIBUTION

Antechinus stuartii stuartii. In southeastern Queensland the nominate form *A. s. stuartii* (Fig. 2A) is essentially restricted to subtropical vine–forests, and occurs from sea level (Tallebudgera Creek 28° 08′S 153° 26′E) to altitudes of 1000 m (Lamington Plateau 28° 16′S 153° 05′E). It is abundant in major tracts of vine-forest on or east of the Great Dividing Range south of Gympie 26° 11′S 152° 40′E (Fig. 3). Structural features of typical complex notophyll vine–forest inhabited by *A. s. stuartii* are described by Webb (1959) and include the abundance of notophylls and microphylls, lianas, vascular epiphytes, plant butresses, compound entire leaves and trunk–obscuring aroids such as *Pothos* (Fig. 2B). The uneven canopy of mixed evergreens and occasional deciduous emergents ranges in height from 21–45 m. In Queensland *A. s. stuartii* is rarely found in sclerophyll associations, where its presence is limited to humid, dense areas such as fern gullies and vine entanglements along watercourses. It is particularly abundant in the perimeter vegetation of vine–forests and favours dense tangled stands of invasive pioneer shrubs such as *Lantana camara,* wild raspberry *Rubus* spp., bleeding heart *Omolanthus* spp. and bracken fern *Pteridium esculentum.* According to Braithwaite (1973) who studied a population at Mt. Glorious 27° 20′S 152° 47′E (40 km northwest of Brisbane) *A. s. stuartii* avoids both mature forest with sparse ground cover and disturbed areas with cunjevoi (*Alocasia machorhiza*) and young palm growth. He found a greater preference for areas of tangled and fallen lianas and rotten logs. All areas where *A. s. stuartii* occurs in southeastern Queensland experience relatively high average annual rainfall of up to 2000 mm.

In contrast to those habitats preferred in Queensland, Wakefield and Warneke (1967) found that in Victoria and southern New South Wales *A. s. stuartii* was abundant in a wide variety of sclerophyll habitats receiving average annual rainfalls of up to 1500 mm and at altitudes up to approximately 1400 m. Trapping surveys conducted at Nadgee Wildlife Reserve, New South Wales by Newsome, McIlroy and Catling (1975) resulted in captures of *A. s. stuartii* in the following habitats: swamp, dune–swale, lowland sclerophyll forest, upland sclerophyll forest, and upland heath (full floristic descriptions appear in Newsome *et al.* (1975)).

Antechinus stuartii adustus. This small form (Fig. 4A) is restricted to dense tropical vine–forests (mostly complex mesophyll on basalt) from Paluma 19° 00′S 146° 12′E (60 km north of Townsville) to Mt. Spurgeon 16° 26′S 145° 12′E (near Mossman), a geographically isolated distribution with a north–south range of approximately 300 km (Fig. 3). This species does

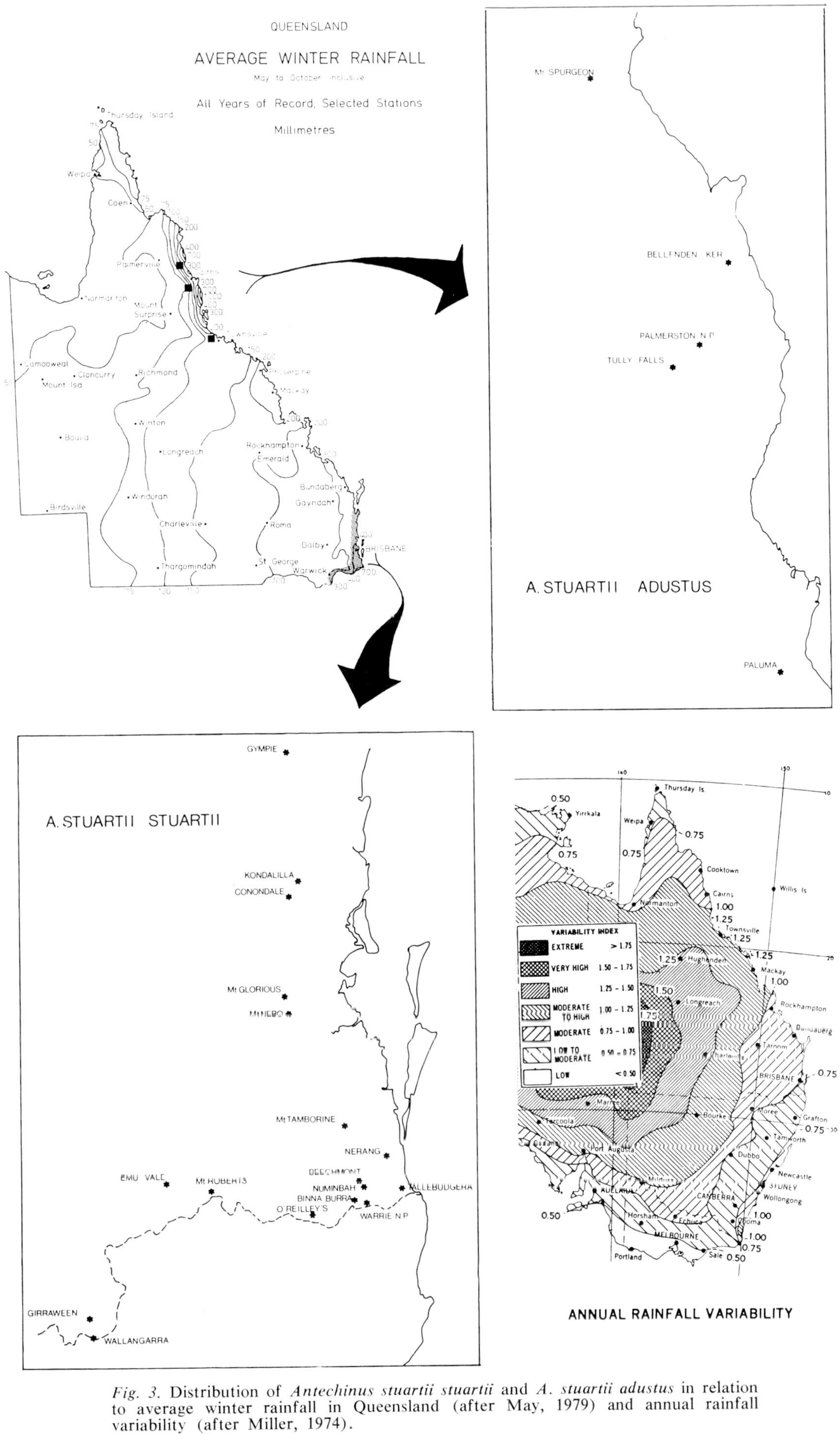

Fig. 3. Distribution of *Antechinus stuartii stuartii* and *A. stuartii adustus* in relation to average winter rainfall in Queensland (after May, 1979) and annual rainfall variability (after Miller, 1974).

not inhabit lowland vine–forest and is only found in very wet and cloudy uplands (Fig. 4B), at altitudes above approximately 800 m, in areas that receive highly seasonal rainfall of up to 1500 mm annually. Major populations occur at Paluma 19° 00′S 146° 12′E, Palmerston 17° 37′S 145° 40′E, Tully Falls 17° 45′S 145° 35′E, Bellenden Ker 17° 16′S 145° 51′E and Mt. Spurgeon 16° 26′S 145° 12′E (Fig. 3). This race appears to have similar habitat preferences to the southern *A. s. stuartii* and has been most frequently trapped along perimeter vegetation in entanglements of ferns and wild raspberry and around recently fallen and decaying logs. Around Ravenshoe it occurs sympatrically with *A. godmani* and *A. f. rubeculus* (pers. obs.).

Antechinus flavipes flavipes. In southeastern Queensland *A. f. flavipes* (Fig. 5A) occupies an extremely wide range of habitat types. Its distribution appears to be limited by two major factors: the presence of *A. s. stuartii,* and its ability to tolerate increasing aridity (Fig. 6). A third possible limiting factor hinted at by Dwyer, Hockings and Wilmer (1979) is that of the presence of heavy infestations of cane toads (*Bufo marinus*) in what appear to be suitable habitats for *A. flavipes.* It occurs both east and west of the Great Dividing Range in habitat types outside those vine–forests where *A. s. stuartii* is concentrated (Fig. 5B). At Cooloola 26° 12′S 153° 05′E and Beerwah 26° 51′E 152 57′E Dwyer *et al.* (1979) trapped *A. flavipes* in the following habitat types (see Dywer *et al.* (1979) for full floristic descriptions); scribbly gum – brush box, blackbutt –forest oak; rainforest; blackbutt – *Banksia aemula;* blackbutt – bloodwood; scribbly gum – bloodwood (burnt 1971 and 1973); scribbly gum – *B. aemula* (burnt 1971); *B. aemula* (burnt 1973); *B. aemula* – scribbly gum associations; heath and swamp; paper bark swamp, open forest and fringing rainforest; swamp oak – paperback (Kinaba Is.); and slash pine (30 yrs). At its most easterly limit, Frazer Island 25° 22′S 153° 07′E, *A. flavipes,* occupies complex notophyll vine–forest and adjacent wet sclerophyll forest. This area receives a yearly average rainfall of 1741 mm. At its western limit in Queensland, St. George 28° 03′S 148° 35′E, it occupies dry *Callitris* woodland which receives a yearly average rainfall of 511 mm. Throughout its southerly range it is common in all small relict vine–forest patches and dense vegetation bordering creeks and particularly in patches of *Lantana camara.* Vine-forests at Bunya Mts 26° 53′S 151° 37′E (near Kingaroy, southeastern Queensland) and at Gympie 26° 11′S 152° 40′E southeastern Queensland which are unoccupied by *A. s. stuartii* are inhabited by *A. flavipes.*

In mid–eastern Queensland (Eungella Range 20° 55′S 148° 30′E near Mackay), *A. flavipes* has been collected in vine–forests along Broken River. *Antechinus stuartii* is not present.

While populations of *A. flavipes* and *A. stuartii* in southeastern Queensland appear to be abruptly isolated from one another, it would be an over simplification to state categorically that the two species are allopatric. During May 1950 D. P. Vernon (pers. comm.) of the Queensland Museum collected both species in adjacent traps set 7 km NE Wallangarra 28° 55′S 151° 56′E, an area of dry sclerophyll woodland and abundant granite outcrops and bracken gullies. During June 1969 T. H. Kirkpatrick (pers. comm.) of the National Parks and Wildlife Service, Warwick, caught the two species in bracken fern adjacent to creeks through dry sclerophyll woodland at Emu Creek 28° 14′S 152° 15′E and also at Swan Ck 28° 09′S 152° 18′E (both localities near Warwick). In southern Queensland intermingling of the two species is extremely uncommon and the above occurrences are most reasonably explained by presuming that trapping either took place at or near the geographical interface of the two populations or that the area was surveyed at a time of population flux.

Antechinus flavipes rubeculus. In northeastern Queensland *A. f. rubeculus* (Fig. 7A) occupies lowland and highland tropical vine–forests (Fig. 7B) at altitudes from 20 m (Herbert R. 18° 29′S 145° 51′E) – 1200 m (Thornton Peak) see Fig. 6). In some highland areas (e.g. the vineforests around Ravenshoe) *A. f. rubeculus* occurs sympatrically with *A. s. adustus* and *A. godmani.* It also occurs outside vine–forest and specimens have been collected at altitudes from 600 m (Kirrama) – 1050 m (Herberton) in dry sclerophyll woodland dominated by casuarinas, bloodwood, stringybark, acacias and dense blady grass.

EXTERNAL MORPHOLOGY

Antechinus stuartii stuartii (Fig. 2A). The most obvious character of external morphology in *A. s. stuartii* is that of the overall uniformity of its fur colour, which dorsally is Snuff Brown (a light chocolate colour) and ventrally a lighter Buffy Brown. Most specimens show a slight warming of the brown from the mid dorsum to the rump but this is most obvious under the bright illumination. Both fore and hind feet of *A. s. stuartii* are a Drab colour (a light grey–brown). This species lacks distinct eye rings and post auricular patches and the tail is a uniform Drab (grey–brown). In some specimens the tail tip is slightly darker than the tail base but this dark brown cannot be confused with the Fuscous–Black tip of *A. flavipes.*

Antechinus stuartii adustus (Fig. 4A). Although this subspecies maintains the uniformity of colour seen in the nominate form, its overall

Fig. 4. A, *Antechinus stuartii adustus.* B, Habitat of *Antechinus stuartii adustus,* near Ravenshoe, northeastern Queensland.

Fig. 5. A, *Antechinus flavipes flavipes.* B, Habitat of *Antechinus flavipes flavipes,* near Dalby, southeastern Queensland.

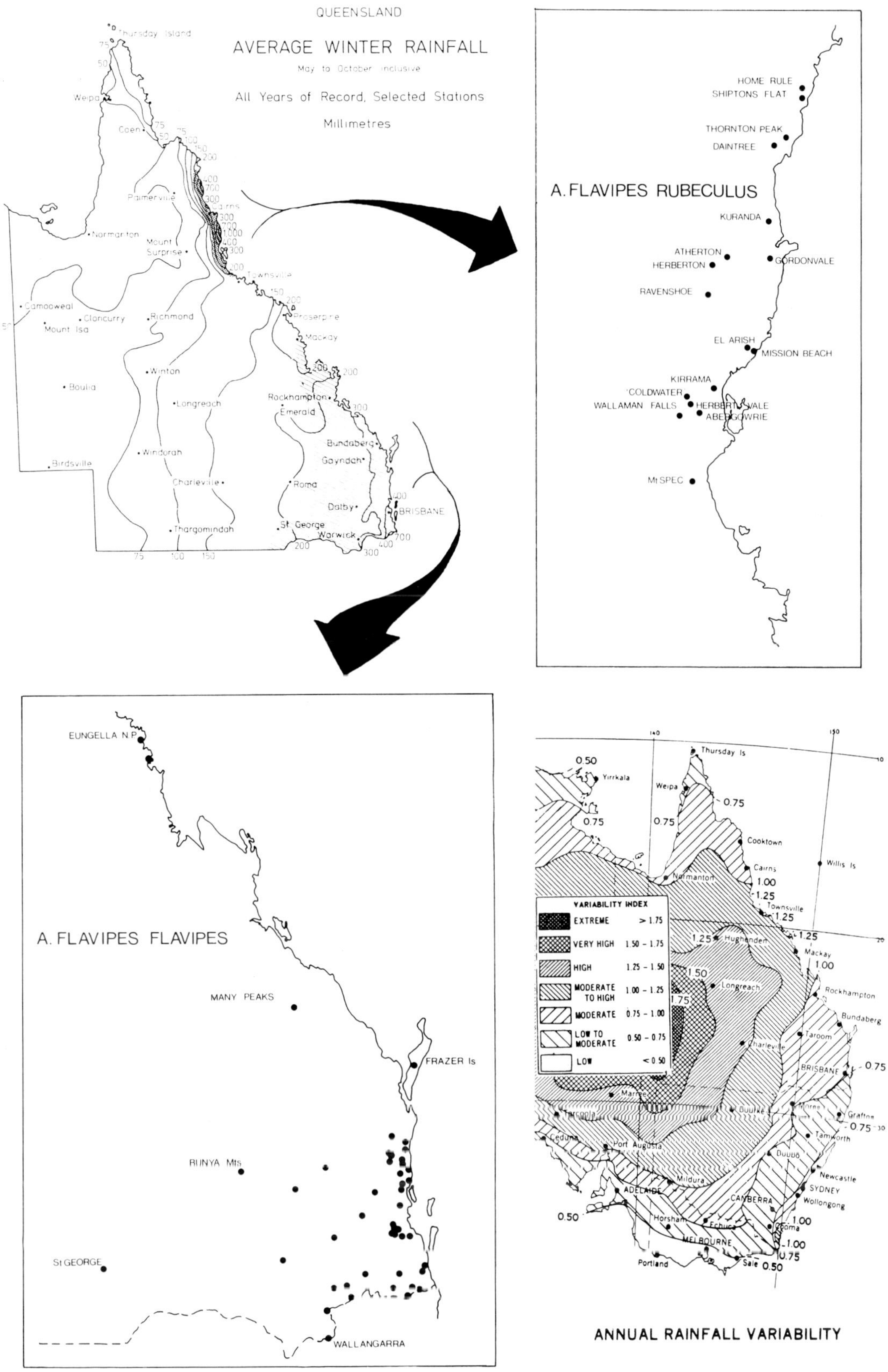

Fig. 6. Distribution of *Antechinus flavipes flavipes* and *A. flavipes rubeculus* in relation to average winter rainfall in Queensland (after May, 1979) and average rainfall variability (after Miller 1974). For *A. flavipes flavipes*, locality names have been deleted in area of heaviest density.

Fig. 7. A, *Antechinus flavipes rubeculus.* B, Habitat of *Antechinus flavipes rubeculus,* near Home Rule, northeastern Queensland.

appearance is much darker than the southeastern form, the depth of colour however is variable. Some specimens (e.g. JM809 from Mt. Spurgeon) are almost Fuscous (a gingery black) dorsally and Tawny ventrally, with Fuscous feet and tail.

Antechinus flavipes flavipes (Fig. 5A). The most striking characteristics of external morphology in *A. f. flavipes* are: (1) the obvious colour change which occurs dorsally from behind the Slate–Gray head to the more Tawny rump; (2) the Tawny sides of the abdomen, fore and hind feet; (3) the prominent Black tip of the tail; (4) the Deep Olive–Buff eye rings and Tawny post auricular patches.

Whilst the actual shade of fur colour varies slightly from one locality to another the difference between head and rump colour is prominent and consistent. Specimens from mideastern Queensland (Eungella Range 20° 55′S 148° 30′E near Mackay) are the drabbest, with all the characteristics of southeastern Queensland specimens toned down so that pelage colour appears almost *A. s. stuartii*–like. However, although eye rings and post auricular patches are barely discernible, and dorsal foot colouration is *stuartii*–like there is a noticeable head–rump colour difference and the tail colour changes from fawn at the base to darker brown at the tip.

Antechinus flavipes rubeculus (Fig. 7A). Of all races of *A. flavipes* the most supreme example of colour tendency toward redness is seen in this race from northeastern Queensland. Most specimens appear darkly coloured and gingery which accounts for many specimens (e.g. J5903, JM1490, J5902, JM197) having been previously misclassified as *A. godmani*. The overall darkened appearance (when compared to southeastern Queensland specimens) is brought about by a change in the colour of the median section of each dorsal hair (non guard). The Mouse Gray colour (found in southern populations) is replaced by Cinnamon–Rufous to Hazel so that the reddest examples of *A. f. rubeculus* (JM593 from Thornton Peak 16° 10′S 145° 23′E, J5902 from Atherton 17° 16′S 145° 29′E, JM688 from Home Rule 15° 42′S 145° 14′E) the head becomes almost Fuscous while the lower dorsum becomes Brick Red, abdomen sides are Mars Orange, fore and hind feet, eye rings and post auricular patches Tawny.

There are however, some adult specimens from northeastern Queensland (e.g. J9307 from Herberton 17° 23′S 145° 23′E, J10148 from Mt. Spec 18° 57′S 146° 11′E, JM591 from Atherton 17° 16′S 145° 29′E) which exhibit similar light colouring to those from southeastern Queensland. The faded appearance of many old skins and most puppets of *A. f. rubeculus* prepared from alcohol specimens suggests that the red colour seen in living specimens of this form is particularly vulnerable to bleaching. However a number of drably coloured specimens collected recently (1979) from Mt. Spec. by R. Whitford (pers. comm.) indicates that at least at the southern extremities of its range, lightly coloured geographic variants of *A. f. rubeculus* occur naturally.

EXTERNAL MEASUREMENTS AND WEIGHTS

Measurements (see Fig. 1) and weights of *A. stuartii* and *A. flavipes* appear in Tables 1–6, with population and species comparisons in Table 7. Except where otherwise stated no differentiation has been made between first and second year females.

Antechinus stuartii stuartii. In 1973 Braithwaite recorded details of specimens collected from Mt. Glorious 27° 21′S 152° 47′E (40 km northwest of Brisbane). Males trapped in September 1971 had an average head–body length of 127 ± 2.36 mm (O.R. = 120–131; N = 8) an average total length of 226 ± 4.66 mm (O.R. = 221–238; N = 8) and an average weight of 59 ± 4.58 g (O.R. = 48–68; N = 8). First–year females trapped in September had an average head–body length of 104 ± 5.34 mm (O.R. = 91–112, N = 6) an average total length of 194 ± 5.54 mm (O.R. = 180–201) and an average weight of 28 ± 2.44 g (O.R. 24–32; N = 6). Second–year females trapped in September had an average head–body length of 109 ± 4.62 mm (O.R. = 102–117; N = 5), an average total length of 198 ± 7.76 mm (O.R. = 186–211; N = 5) and an average weight of 28 ± 1.94 g (O.R. = 25–31; N = 6).

Braithwaite noted annual variation in average weights for adult males caught in September from 1963–66 (average weight 62 g (1963), 51 g (1964), 46 g (1965), 50 g (1966)). He considered that the overall drop in average annual rainfall on Mt. Glorious from 1964 (1016 mm) to 1966 (431.8 mm) was the major contributing cause for annual weight variation.

Adult males are approximately 2.03 times as heavy as adult females and approximately 1.22 times as long (head–body length). The heaviest specimen in the collections of the Queensland Museum is 72 g male J20265, collected 6 July 1970 from Tallebudgera Creek 28° 08′S 153° 26′E, West Burleigh, southeastern Queensland.

Comparison of these values with those of southern populations (New South Wales and Victoria) shows southeastern Queensland animals to be larger and heavier. A comparison with those external measurements presented by Wakefield and Warneke (1967) shows that

Table 5. Absolute measurements for combined sample of *A. flavipes flavipes* from southeastern Queensland and *A. flavipes rubeculus*.

	N	x̄ ± r	OR	S	CV
BL	49♂	29.24 ± .22	26.6 - 32.9	1.55	5.3
	29♀	27.23 ± .25	25.6 - 29.7	1.34	4.9
ZW	52♂	18.34 ± .14	16.1 - 20.7	1.03	5.6
	30♀	17.04 ± .14	15.6 - 19.0	0.79	4.6
SWR-LC[1]	51♂	5.90 ± .05	5.0 - 6.9	0.37	6.3
	34♀	5.40 ± .06	4.9 - 6.2	0.37	6.9
SWR-LM[4]	51♂	14.70 ± .08	13.4 - 16.2	0.54	3.7
	31♀	13.89 ± .13	12.6 - 15.3	0.71	5.1
APVL	48♂	2.67 ± .04	2.0 - 3.2	0.28	10.4
	34♀	2.44 ± .04	1.9 - 2.8	0.24	9.8
PPVL	51♂	4.58 ± .06	3.9 - 5.6	0.45	9.9
	34♀	4.24 ± .07	2.6 - 4.8	0.43	10.1
IPVL	51♂	5.51 ± .08	4.4 - 7.4	0.57	10.4
	34♀	5.26 ± .09	4.3 - 6.4	0.54	10.3
$p\underline{1-2}$	51♂	2.10 ± .02	1.9 - 2.4	0.12	5.8
	34♀	2.06 ± .02	1.8 - 2.2	0.11	5.4
$p\underline{1-3}$	51♂	3.35 ± .03	2.9 - 3.8	0.21	6.4
	34♀	3.24 ± .03	2.8 - 3.6	0.18	5.7
IBW	49♂	4.03 ± .05	3.4 - 4.7	0.34	8.3
	30♀	4.02 ± .07	3.2 - 4.7	0.41	10.2
IOW	51♂	6.64 ± .05	6.0 - 7.5	0.34	5.1
	32♀	6.53 ± .06	6.0 - 7.2	0.32	4.9
TC	51♂	2.66 ± .04	2.1 - 3.2	0.26	9.8
	30♀	2.57 ± .04	2.2 - 2.9	0.21	8.2
SCD	47♂	5.67 ± .06	4.8 - 6.5	0.40	7.0
	31♀	4.98 ± .07	4.2 - 5.9	0.39	7.9
HB	49♂	120.63 ± 2.07	93 - 165	14.50	12.0
	30♀	104.57 ± 1.52	90 - 127	8.32	8.0
TV	49♂	99.90 ± 1.42	80 - 151	9.96	10.0
	29♀	86.31 ± 1.61	72 - 107	8.67	10.0
W	20♂	56.20 ± 4.48	26 - 79	20.03	35.6
	10♀	34.40 ± 2.99	21 - 52	9.47	27.5

southeastern Queensland males are approximately 1.34 times longer (head–body) than males (N = 58) from Loch Valley 37° 48′S 146° 01′E Victoria, while southeastern Queensland females are approximately 1.24 times longer (head–body) than Loch Valley females (N = 13). Weight comparison with Victorian animals (7 ♂♂, 9 ♀♀) collected by L. Selwood at Kinglake 37° 32′S 145° 20′E and Narbethong 37° 34′S 145° 40′E from June to September 1976 shows that southeastern Queensland males and females are approximately 2.46 and 1.65 times heavier than their Victorian conspecifics respectively. Compared to individuals of *A. stuartii* (27 ♂♂, 21 ♀♀) at Murrambateman 34° 58′S 149° 02′E near Canberra ACT (C. Dickman pers. comm.) males from southeastern Queensland are approximately 1.28 times as long (head–body) and approximately 2.11 times as heavy as their Canberra conspecifics. The comparison with specimens from central coastal New South Wales shows that southeastern Queensland males are approximately 1.4 times longer (head–body) than New South Wales males (N = 24) and approximately 2.03 times as heavy (N = 8). Southeastern Queensland females are approximately 1.25 times as long (head–body) as New South Wales females (N =

17) and approximately 1.53 times as heavy (N = 7).

Southeastern Queensland *A. stuartii* have proportionately shorter tails than more southerly conspecifics. Comparisons between the southeastern Queensland, Loch Valley, Canberra and New South Wales series shows that the proportion of tail length (tail–vent) to head–body length for the southeastern Queensland animals is approximately 0.78 for males and 0.87 for females, while the Loch Valley, Victoria series the proportions are approximately for males 0.97 and females 0.98; for Canberra males 0.85 and females 0.86 and for the central New South Wales coast 0.89 for males and 0.87 for females.

Antechinus stuartii adustus. This form is smaller than any other antechinus found in Queensland. Males are approximately 0.87 times as long (head–body) and approximately 0.58 times as heavy as male *A. s. stuartii* from southeastern Queensland. Similarly, females of *A. s. adustus* are smaller, by approximately 0.86 times (head–body) and lighter by approximately 0.82 times than females from the southeast of Queensland. Male *A. s adustus* are longer (head–body) and heavier than female *adustus* by approximate factors of 1.19 and 1.38 times respectively. This subspecies, however, is larger and heavier than the nominate form in Victoria. Male *adustus* are larger and heavier by approximate factors of 1.16 and 1.39 times and female *adustus* by approximate factors of 1.09 and 1.36 times respectively. Compared to animals from mid–eastern New South Wales, *A. s. adustus* are again slightly longer and heavier, males by the approximate factors of 1.08 and 1.15 for length (head–body) and weight, and females by approximately 1.12 and 1.30 times. The comparison of specimens of *A. s. adustus* with the nominate form from southeastern Queensland shows both males and females of *adustus* as having proportionally longer tails than their southern-conspecifics. The proportion of tail length (tail–vent) to head–body length for *A. s. adustus* males is approximately 0.89 while that for southeastern Queensland males is 0.78. These figures correspond to the approximate vales of 0.96 and 0.87 for females. While the values for tail proportions for male *adustus* are similar to those from Canberra and New South Wales (mentioned previously), tails of female *adustus* are proportionately much longer than the tails of either New South Wales or Canberra females. The tail lengths of both male and female *adustus,* never-the-less are proportionately shorter than for either males or females from Victoria (Loch Valley).

Antechinus flavipes flavipes. In southeastern Queensland *A. f. flavipes* is similar in all measured body proportions to southeastern Queensland *A. s. stuartii* and is also similar to southern *A. flavipes* (see Wakefield and Warneke 1967). Dwyer *et al.* (1979) recorded weights of 830 specimens; September male weights ranged between 33.5–61.0 g and mid–August female weights ranged between 24–39 g. The heaviest animal they recorded was 74 g. male collected in mid–August. Adult males are approximately 1.13 times longer (head–body) and 1.7 times heavier than adult females.

Antechinus flavipes rubeculus. In northeastern Queensland *A. f. rubeculus* is similarly proportioned but essentially larger and much heavier than southern populations. Adult males are approximately 1.08 times longer (head–body) and 1.43 times heavier than southern males; adult females are approximately 1.07 times longer (head–body) and 1.38 times heavier than southern females. Males of *A. f. rubeculus* are approximately 1.14 times longer and 1.75 times heavier than females and are at least twice as heavy as *A. s. adustus* with which it occurs. sympatrically in some areas.

REPRODUCTIVE BIOLOGY

Antechinus stuartii stuartii. The unusual reproductive cycle of this form has received more attention than that of probably any other dasyurid. In 1961 Marlow noted the species' brief annual mating season. The sudden onset of senility in males after one breeding season and the ability of females to produce litters in two successive breeding seasons was later demonstrated by Woolley (1966). At Mt. Glorious, southeastern Queensland, mating occurs only in the last two weeks of September each year and is followed immediately by the mortality of all males (Wood 1970), Braithwaite 1973). This sudden demise in males is seen by Lee, Bradley and Braithwaite (1977) as aetiologically related to a stress response associated with the search for females, prolonged and repeated copulation and continuous agnostic encounters with rival males. The response involves a supression of immune system resulting from the increase of plasma corticosteroids (Barnett 1973) and thus opens the way to fatal pathogenic infection in association with gastric and duodenal haemorrhages, skin hydroxoproline (Barnett 1973) loss of body weight (Woolley 1966, Wood 1970, Braithwaite 1973) negative nitrogen balance (Woollard 1971) and neutrophilia and lymphopenia (Cheal, Lee and Barnett 1976). The evolution and consequences of "semelparity" in the males of *A. s. stuartii* and the mortality among females during the mating, gestation and lactation periods are discussed by Braithwaite and Lee (1979).

Females of *A. s. stuartii* undergo a gestation period lasting 25–26 days (Woolley 1966) and all births at Mt. Glorious occur in the last two weeks of October each year, after which young remain in the pouch for approximately 5 weeks and are weaned after a total of approximately three months lactation (Wood 1970). Young share a nest until approximately the end of July when territorial behaviour begins (Lee *et al.* 1977). Litter size at Mt. Glorious averages 7.5 young (Wood 1970). In more northerly districts mating occurs slight earlier than at Mt. Glorious. At the Conondale Range 26° 52′ 152° 40′E approximately 70 km N Brisbane, females of *A. s. stuartii* were captured 21, 22, 24, 25 October. All had pouches containing 7 or 8 young, each of which was between 1–1.5 cm long. Their ages were thus approximately 12–20 days (cf Marlow 1961), their birth having taken place early in October. In New South Wales (near Sydney) this species reproduces nearly 6 weeks earlier than animals in southeastern Queensland. Marlow (1961) found that mating took place in early August with births occurring in mid–September. In Canberra, Woolley (1966) observed mating of captive animals from July to September. In Victoria, Cheal, Lee and Barnett (1976) and Wakefield and Warneke (1967) note that this species mates in early August and that births occur in mid–September.

Nipple number appears to be regular in southeastern Queensland. Of thirty–two female specimens, representing localities over the entire range of *A. s. stuartii* in southeastern Queensland, all had pouch areas containing 8 nipples.

Antechinus stuartii adustus. Little is known regarding reproduction in this form. Several specimens (e.g. JM1907, JM1908, JM1911) collected near Malan 17° 37′S 145° 40′E, 2 August 1976 had well pouch areas. One of these females (JM1908) gave birth to 6 young 4 August 1976. It is probable that this race mates early in July and gives birth early in August. Of 10 specimens examined from 3 collecting localities all pouches contained 6 teats. Of thirteen males collected live between 30 June 1976 and 18 July 1976 none survived captivity beyond 29 July 1976. Preliminary post-mortem examinations revealed gastric and duodenal haemorrhages, hepatic necrosis, degeneration in proximal tubules of kidneys and anaemia. It seems probable that males of *A. s. adustus* experience a yearly die–off similar to that described for *A. s. stuartii* but which probably takes place at the end of July each year.

Antechinus flavipes flavipes. In 1966 Woolley demonstrated that *A. f. flavipes* from near Canberra and also southwestern Western Australia bred only once each year, with females only perhaps surviving to produce a second litter. This species did, however, breed earlier than *A. stuartii* and thus the two species remained reproductively isolated in some areas where they occurred sympatrically. Later, Inns (1976) described physiological and morphological conditions in captives of South Australian *A. flavipes.* His work demonstrated that wild individuals of this species followed a yearly reproductive cycle similar to that already described for *A. s. stuartii.* Features he notes in males include an increase in body weight, activity and metabolic rate until late June and an increase in testis development until the end of July after which there is a general decline in activity, body weight, scrotal width and pelage condition. At Beerwah 26° 51′S 152° 57′E southeastern Queensland *A. flavipes* mates late in September, males become moribund and die–off in late September and females give birth in late October (Dwyer *et al.* 1979). Most females caught at this locality were found to have pouch areas containing 10 teats, counts of eight or nine being uncommon. Of eighteen specimens examined from collections in the Queensland Museum, fifteen pouch areas contained 10 teats and three contained eight. At Mt. Nebo 27° 24′S 152° 47′E, 55 km south of Beerwah, a single female was collected by the author 20 September 1975 with eight pouch young all approximately 5 mm long (crown–rump). This indicates an early August mating – a period of approximately two months before the geographically nearest (Mt. Glorious 27° 21′S 152° 47′E) individuals of *A. stuartii.*

In southeastern Queensland (at Beerwah) *A. flavipes* may mate up to 3 months later than conspecifics in Canberra (cf Dwyer 1979, Woolley 1966) and Victoria (cf Wakefield and Warneke 1976).

Antechinus flavipes rubeculus. In northeastern Queensland matings and births may occur up to 4 months earlier than in southeastern Queensland (pers. obs.). One male (JM1918) was caught near Ingham 4 July 1979 while it was copulating under a log. A female (JM1941) captured near Ingham 5 July 1976 gave birth to eight young 8 July 1976; another female (JM1940) captured at Ravenshoe 15 July 1976 had 8 young each approximately 10.5 mm long (crown–rump) at capture. Another female captured by R. Whitford (pers. comm.) at Mt. Spec. 28 July 1979 gave birth to eight young 17 August 1979. It would appear from the scanty information available that *A. f. rubeculus* mates from mid June to mid July, a period similar to that discussed by Woolley (1966) for *A. f. flavipes* near Canberra, ACT. Of six specimens examined in the collections of the Queensland Museum, three pouch areas contained six nipples, one contained seven and two contained eight. In captivity males deteriorate in condition after the fashion described by Inns (1976). There are however a number of adult male specimens (eg. J19676, J9393, J19678, JM688) in the collections of the Queensland Museum which have been collected up until as late as 29 October in a particular year.

CRANIAL AND DENTAL MORPHOLOGY

ANTECHINUS STUARTII STUARTII AND ANTECHINUS FLAVIPES FLAVIPES

From a total of forty four measured characters, the following seven are considered as most significantly diagnostic in the comparison between *A. s. stuartii* and *A. f. flavipes* from Queensland. Characters 1–5 are displayed and compared for the two species in Fig. 12. Cranial measurements appear in Tables 1 and 3 and species are compared in Table 7. 1, Lengths of anterior and posterior palatal vacuities and the extent of their separation; 2, Rostral and interorbital widths; 3, Premolar row length; 4, Transverse canal foramina separation; 5, Snout–

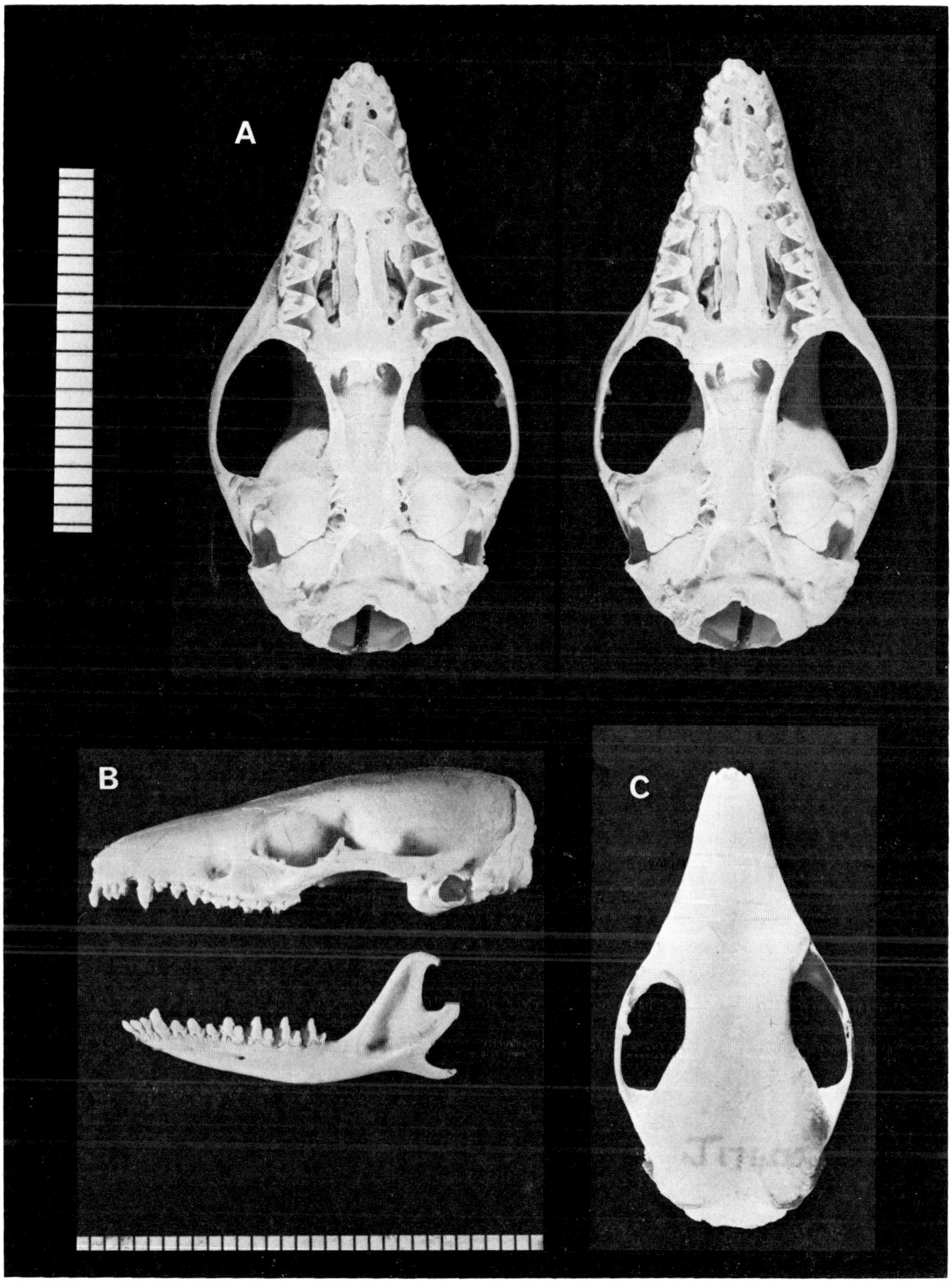

Fig. 8. A–C, *Antechinus stuartii stuartii.* J17400, adult male, Warwick, Queensland.

canine depth; 6, Entoconid development on lower molars and associated widening of talonid; 7, Degree of reduction of talonid of M_5.

Palatal Vacuities

In Queensland, *A. s. stuartii* has significantly ($P < .001$) larger anterior and posterior palatal vacuities than *A. f. flavipes*. As the extent of overlap between species (3.9% overlap in palatal vacuities, 3.1% overlap in posterior palatal vacuities) is minimal this is character alone is extremely useful for the identification of cranial material. Moreover, values for the distance of separation between anterior and posterior palatal vacuities show zero overlap between the two species. Comparisons of palatal vacuity lengths relative to basicranial lengths are made for both species in Figs. 13 and 14. *Antechinus stuartii stuartii* exhibits consistently small inter–palatal vacuity distances throughout its entire range in southeastern Queensland; conversely *A. f. flavipes* exhibits large inter–palatal vacuity distances

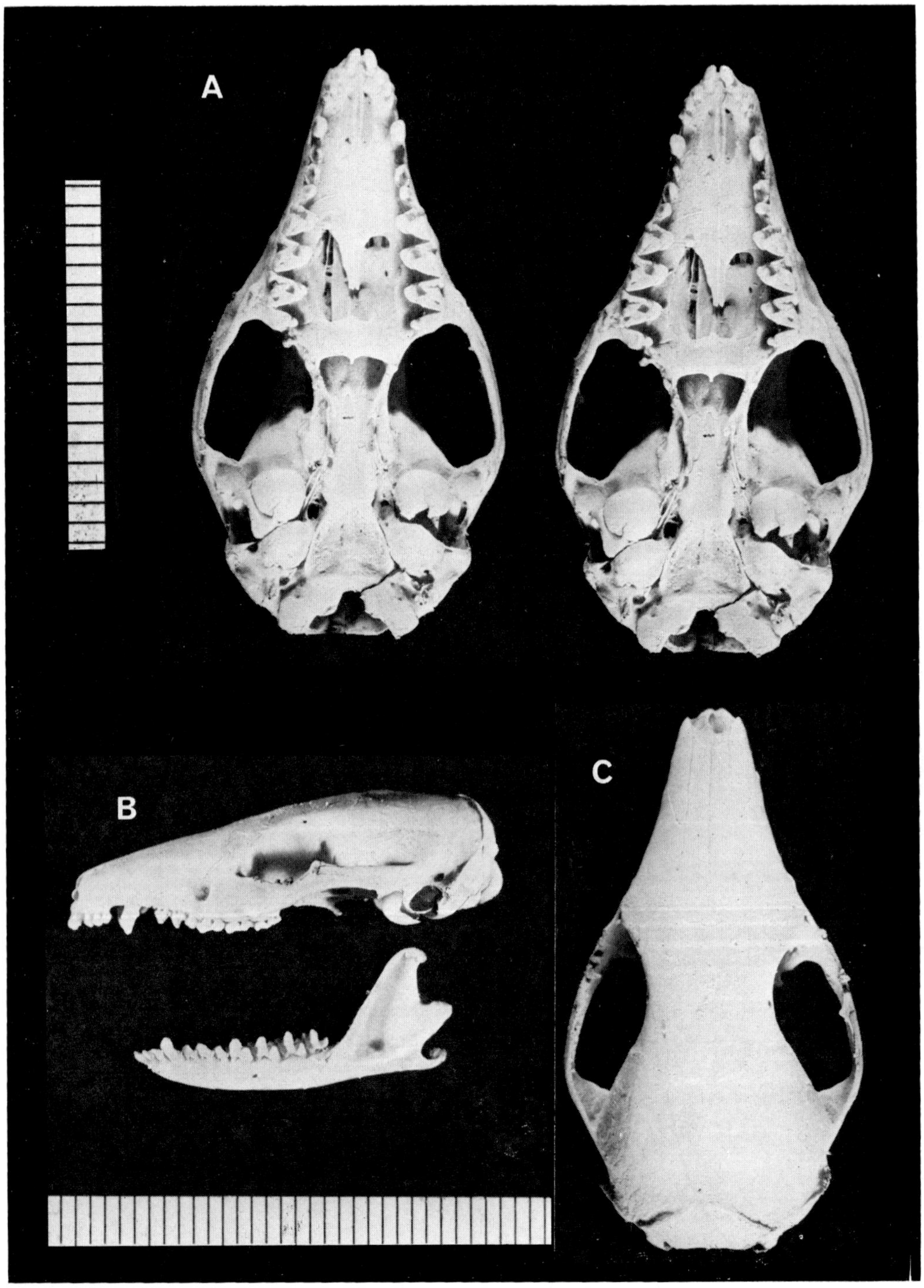

Fig. 9. A–C, *Antechinus stuartii adustus.* J17589, adult male, Mt. Spec, Queensland.

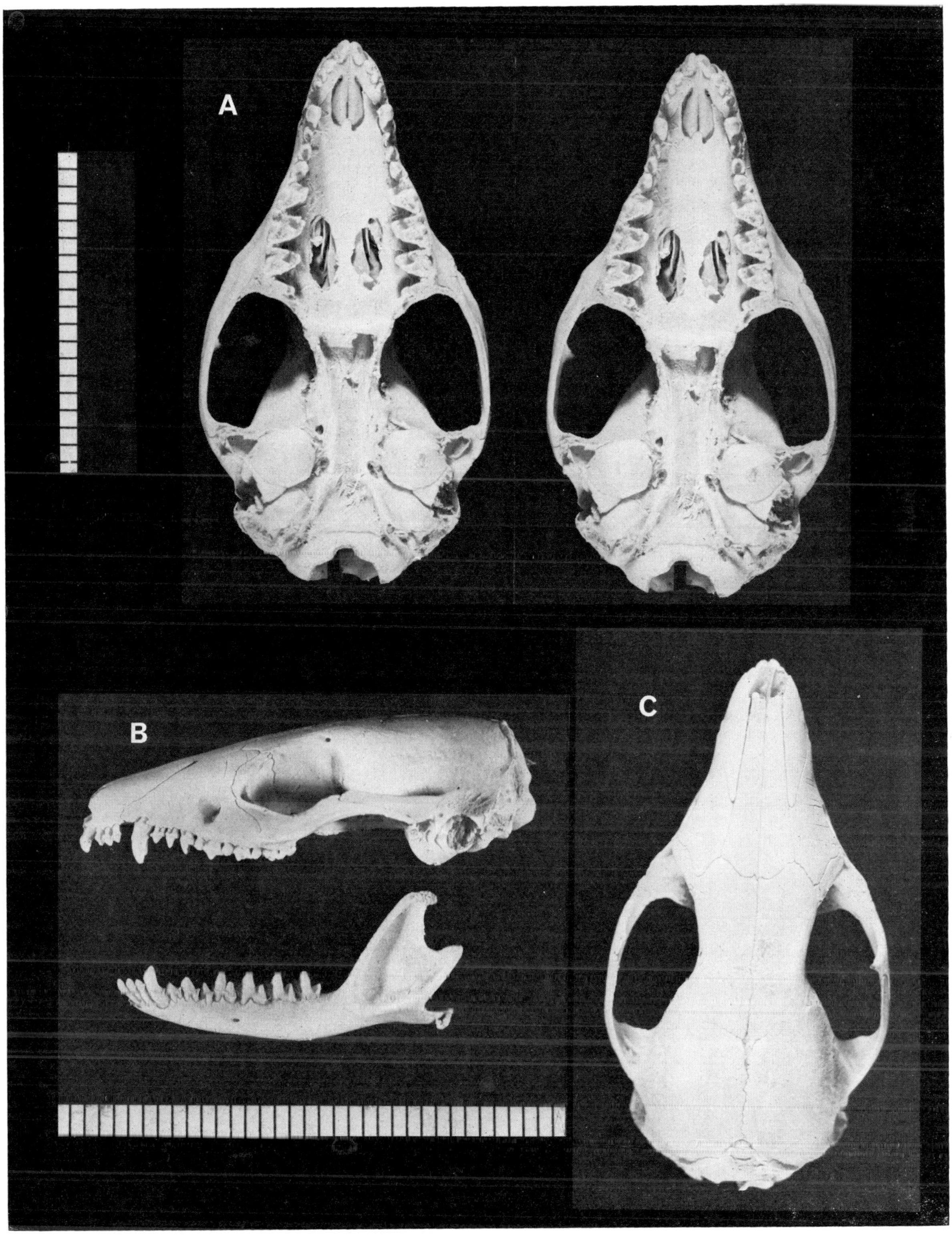

Fig. 10. A–C, *Antechinus flavipes flavipes*. J8968, adult male, Bunya Mts, Queensland.

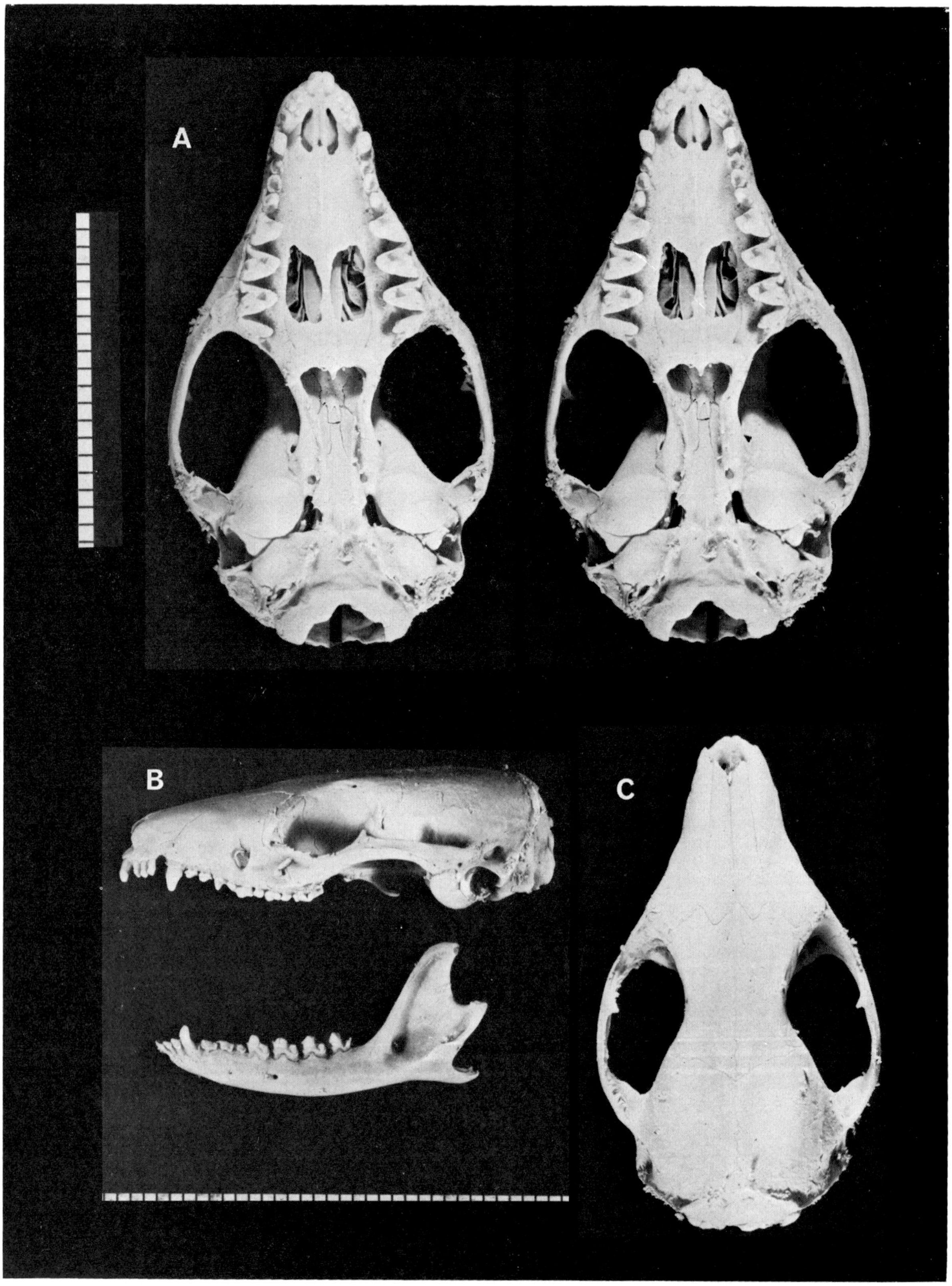

Fig. 11. A–C, *Antechinus flavipes rubeculus.* JM2676, adult male, Kuranda, Queensland.

Table 6. Absolute measurements for *Antechinus stuartii stuartii* in New South Wales.

	N	x̄ ± r	OR	S	CV
BL	20♂	26.24 ± .17	24.4 - 27.3	0.86	3.3
	17♀	25.22 ± .21	24.0 - 21.1	0.88	3.0
ZW	27♂	15.91 ± .13	14.3 - 16.9	0.66	4.2
	16♀	15.45 ± .14	14.7 - 16.4	0.57	3.7
SWR-LC¹	29♂	4.99 ± .04	4.6 - 5.5	0.22	4.4
	20♀	4.78 ± .09	4.0 - 5.6	0.39	8.2
SWR-LM⁴	26♂	12.95 ± .10	12.3 - 13.8	0.50	3.9
	18♀	12.57 ± .13	11.5 - 13.5	0.54	4.3
APVL	26♂	2.83 ± .05	2.3 - 3.5	0.26	9.2
	18♀	2.60 ± .06	2.1 - 3.2	0.27	10.2
PPVL	28♂	5.00 ± .09	4.0 - 6.1	0.50	10.1
	20♀	4.88 ± .17	3.5 - 6.1	0.20	15.3
IPVL	27♂	4.50 ± .08	3.0 - 5.1	0.44	9.8
	19♀	4.27 ± .11	3.2 - 4.9	0.47	11.1
P^{1-2}	28♂	2.49 ± .03	2.2 - 2.8	0.14	5.6
	20♀	2.30 ± .03	2.0 - 2.5	0.14	5.9
P^{1-3}	28♂	3.59 ± .03	3.2 - 3.8	0.17	4.8
	20♀	3.37 ± .04	2.9 - 3.7	0.19	5.8
IBW	27♂	3.93 ± .04	3.5 - 4.3	0.23	5.8
	19♀	3.84 ± .07	3.3 - 4.4	0.32	8.3
IOW	28♂	6.55 ± .05	6.2 - 7.0	0.28	4.3
	20♀	6.42 ± .07	6.1 - 7.0	0.31	4.8
TC	24♂	2.91 ± .05	2.5 - 3.6	0.26	8.8
	14♀	2.84 ± .05	2.5 - 3.1	0.21	7.3
SCD	25♂	4.83 ± .07	4.1 - 5.3	0.33	6.9
	20♀	4.31 ± .08	3.8 - 5.0	0.37	8.5
HB	22♂	102.41 ± 1.71	88 -114	8.03	7.8
	19♀	95.26 ± 1.60	83 -106	6.98	7.3
TV	24♂	91.46 ± 1.31	80 -100	6.43	7.0
	17♀	83.77 ± 2.01	70 - 99	8.30	9.9
W	8♂	29.00 ± 3.99	17 - 49	11.27	38.9
	7♀	18.57 + 1.82	18 - 28	4.83	2.6

throughout Queensland. Vacuity lengths, inter-palatal vacuity distances, degrees of overlap and percentages of palatal fenestration in the two species are compared in Fig. 15.

Rostral and Interorbital Widths

Antechinus s. stuartii has a significantly (P< .001) narrower rostrum, relative to basicranial length than *A. f. flavipes*. This condition is first noticeable as an increase in skull width in *A. f. flavipes* between the right and left upper canines (SWR–LC¹). The difference is accentuated in *A. f. flavipes* across the skull in the region between the right and left upper fourth molars (SWR–LM⁴). Skull breadths relative to basicranial lengths are compared for both species in Tables 1 and 3.

Although *A. f. flavipes* exhibits a significantly (P< .001) larger zygomatic width than *A. s. stuartii* there is a high degree of overlap and this character by itself cannot be used as a reliable determinant of species. The same is true of interorbital width comparison, where although *A. s. stuartii* exhibits significantly (P< .001) larger interorbital widths there is considerable overlap between the smaller values of *A. s. stuartii*

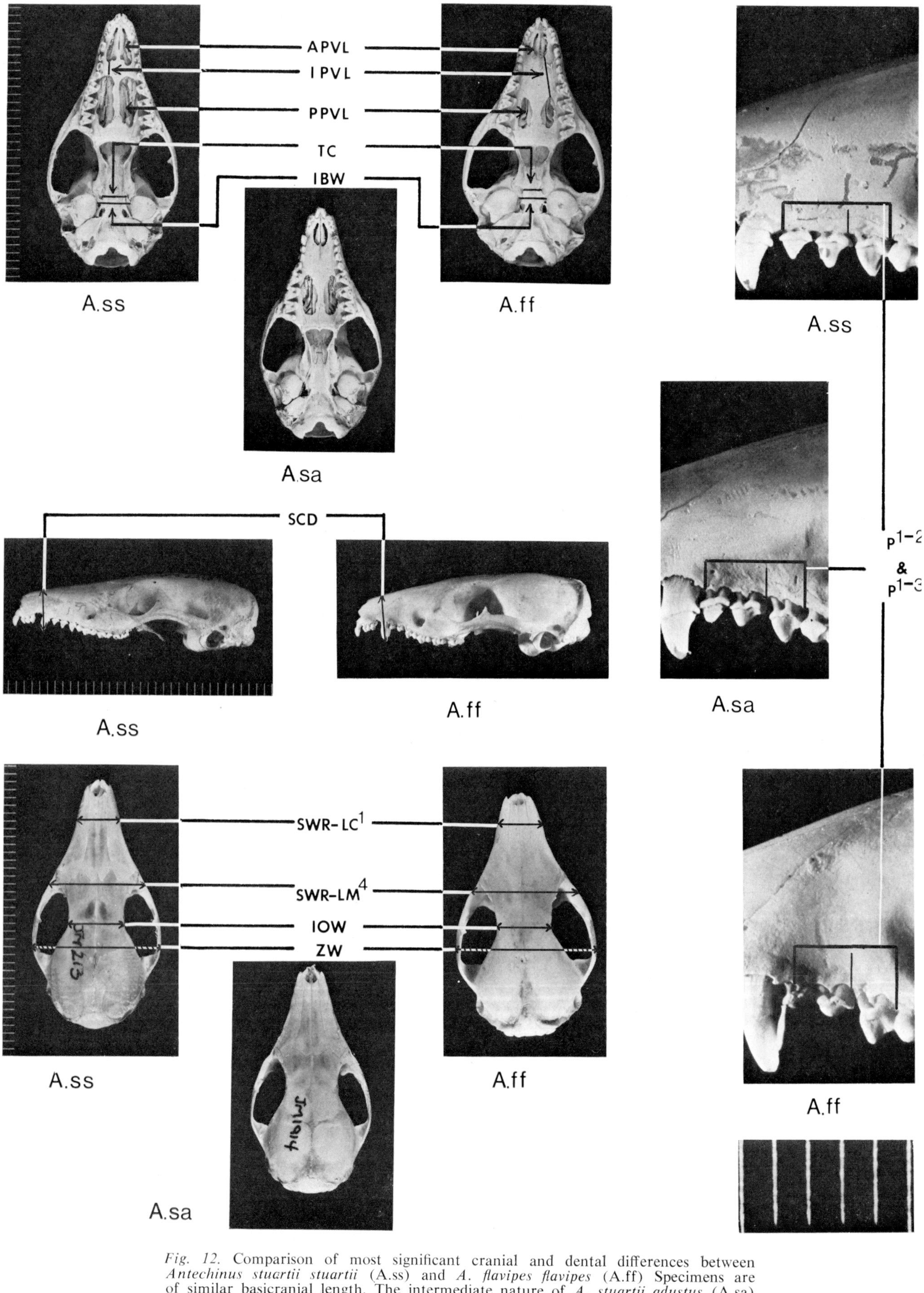

Fig. 12. Comparison of most significant cranial and dental differences between *Antechinus stuartii stuartii* (A.ss) and *A. flavipes flavipes* (A.ff) Specimens are of similar basicranial length. The intermediate nature of *A. stuartii adustus* (A.sa) is demonstrated. All scales in 1 mm calibrations. Abbreviations are as for Tables 1–7.

females and the larger of *A. f. flavipes* males. Interorbital and zygomatic widths are compared to relative basicranial lengths for both species in Fig. 17.

Entoconid Size

The size of the entoconid on M_4 is the most useful character used in the determination of species when only dentaries are present. Whilst $M_2 - M_4$ carry entoconids, the greatest variation in size, between species, is shown at M_4. *A. s. stuartii* shows a high bladelike, anteroposteriorly elongate, sharply pointed entoconid at M_4 (Fig. 21A) whilst that in *A. f. flavipes* is low, blunt and more compressed (Fig. 21C). This character is remarkably constant even under extreme cases of tooth wear; in some instances the crowns of lower canines, all premolars and part of M_2 may be completely absent due to excessive wear yet the entoconid on M_4 still exhibits those qualities necessary for accurate identification at the species level (Fig. 22). Associated with the more massive entoconid development in *A. stuartii* is a distinct widening of the talonid relative to the trigonid width. This is most apparent at M_3. Here the orientation of the talonid endolph relative to the long axis of the jaw is distinctly different in both species. In *A. stuartii* (Fig. 21E, F) that portion of the talonid endoloph which lies between the base of the metaconid and the base of the entoconid orients (anteriorly to posteriorly) lingually away from the long axis of the jaw. In *A. flavipes* (Fig. 21G, H) this portion of the endoloph orients bucally. For entoconid and talonid condition in *A. s. adustus* see Fig. 21B and F respectively and for *A. f. rubeculus* see Fig. 21D and H.

Premolar Row Length

Antechinus s. stuartii has a significantly ($P< .001$) longer upper premolar row (P^{1-3}) than *A. f. flavipes.* This character however is even more outstanding at the P^{1-2} length where for *A. s. stuartii* males $\bar{x} = 2.27 \pm .02$, females $\bar{x} = 2.36 \pm .02$ and for *A. f. flavipes* males $\bar{x} = 2.05 \pm .01$ and females $\bar{x} = 2.03 \pm .02$. Premolar row length is compared with relative basicranial lengths for both species in Fig. 18.

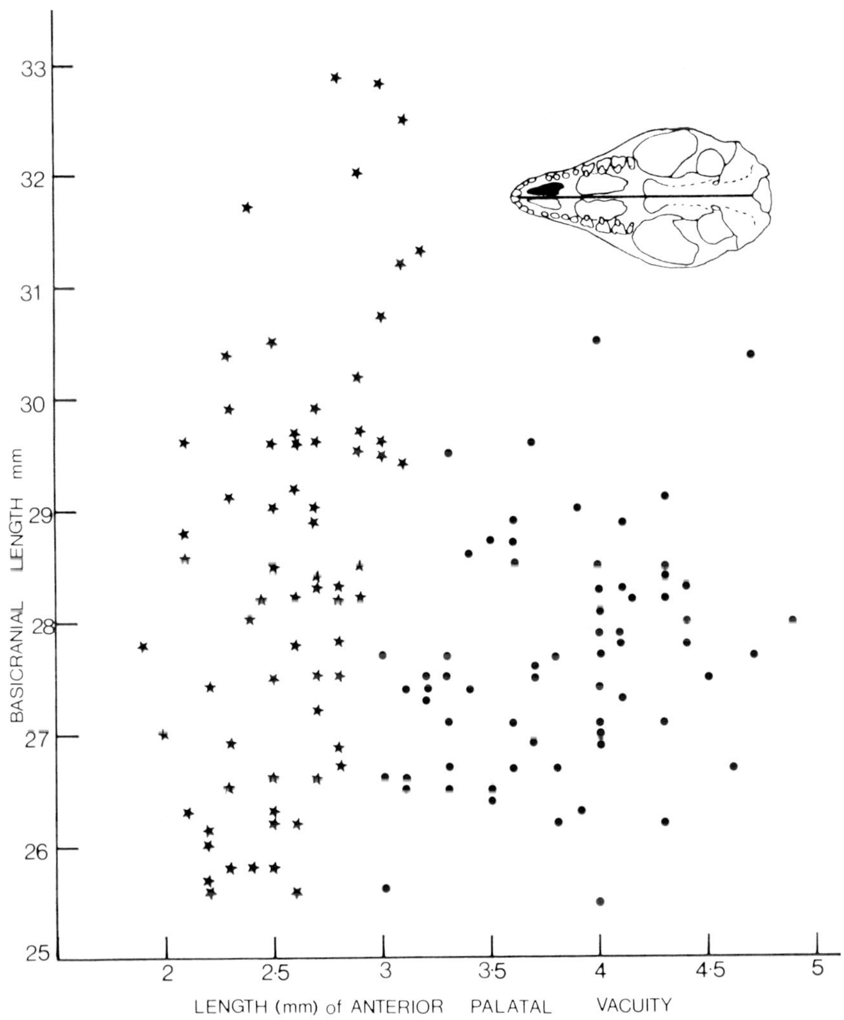

Fig. 13. Comparison of the relative size of anterior palatal vacuities in *Antechinus stuartii stuartii* (dots) and *A. flavipes flavipes* (stars).

Transverse Canal Foramina Separation

In *A. s. stuartii* the separation of the transverse canal foramina is significantly ($P < .001$) greater than in *A. f. flavipes*. Separation of the foramina are compared with relative basicranial lengths for both species in Fig. 19.

Snout–Canine Depth

Antechinus f. flavipes has a significantly ($P < .001$) greater snout to canine–tip depth than *A. s. stuartii*, the striking nature of this character, however, becomes blurred at the *A. f. flavipes* female – *A. s. stuartii* male interface. This character is made significant mainly, but not completely, by the longer and more massive nature of the upper canine in *A. f. flavipes*. Snout–canine depths for both species are compared with relative basicranial lengths in Fig. 20.

Talonid Reduction on M_5

The talonid of M_5 in *A. s. stuartii* is considerably more reduced and noticeably narrower than the corresponding talonid in *A. f. flavipes*.

ANTECHINUS STUARTII ADUSTUS

Cranial measurements appear in Table 2; comparisons with other forms in Table 7. *Antechinus s. adustus* differs from *A. s. stuartii* (in southeastern Queensland) in the following respects:

Basicranial Length

Antechinus s. adustus has a significantly ($P < .01$) shorter basicranial length than *A. s. stuartii*, yet individuals show no significant corresponding reduction in skull width to basicranial length i.e. zygomatic width, skull width at R–LC¹ and R–LM⁴ and interorbital width. This form is sexually dimorphic with males having significantly ($P < .001$) larger skulls (basicranial length) than females.

Palatal Vacuities

Antechinus s. adustus has significantly ($P < .001$) smaller anterior and posterior palatal vacuity lengths than *A. s. stuartii* (Fig. 15). The decreased palatal vacuity size leads to a corresponding significant ($P < .001$) increase in size of the interpalatal vacuity distance over *A. s. stuartii*. Lengths of anterior and posterior palatal vacuities relative to basicranial lengths in *A. s. adustus* are compared with the equivalent ranges for these characters found in *A. s. stuartii* and *A. f. flavipes* in Figs. 23 and 24.

Premolar Row

Antechinus s. adustus has a significantly (P .001) shorter premolar row than *A. s. stuartii* (Fig. 25A, B). This is so for both lengths P^{1-2} and P^{1-3}.

Bullae Separation

Antechinus s. adustus has a significantly (for males $P < .02$ for females $P < .05$) greater minimum distance separating the alisphenoid tympanic bullae than *A. s. stuartii;* consequently there is a reduction in alisphenoid tympanic bullae size in *A. s. adustus*.

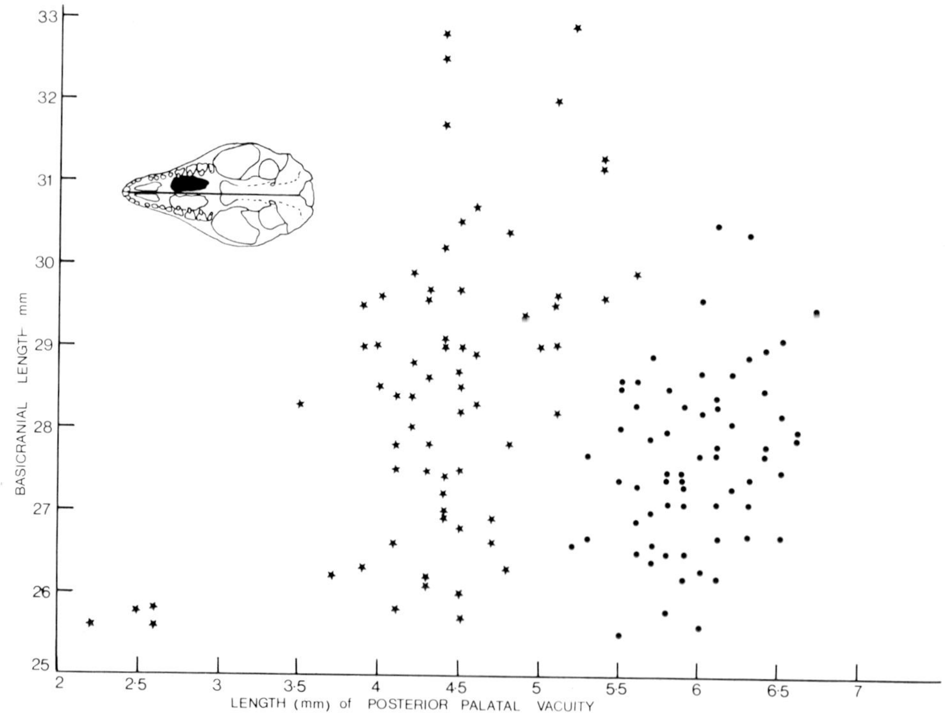

Fig. 14. Comparison of the relative size of posterior palatal vacuities in *Antechinus stuartii stuartii* (dots) and *A. flavipes flavipes* (stars).

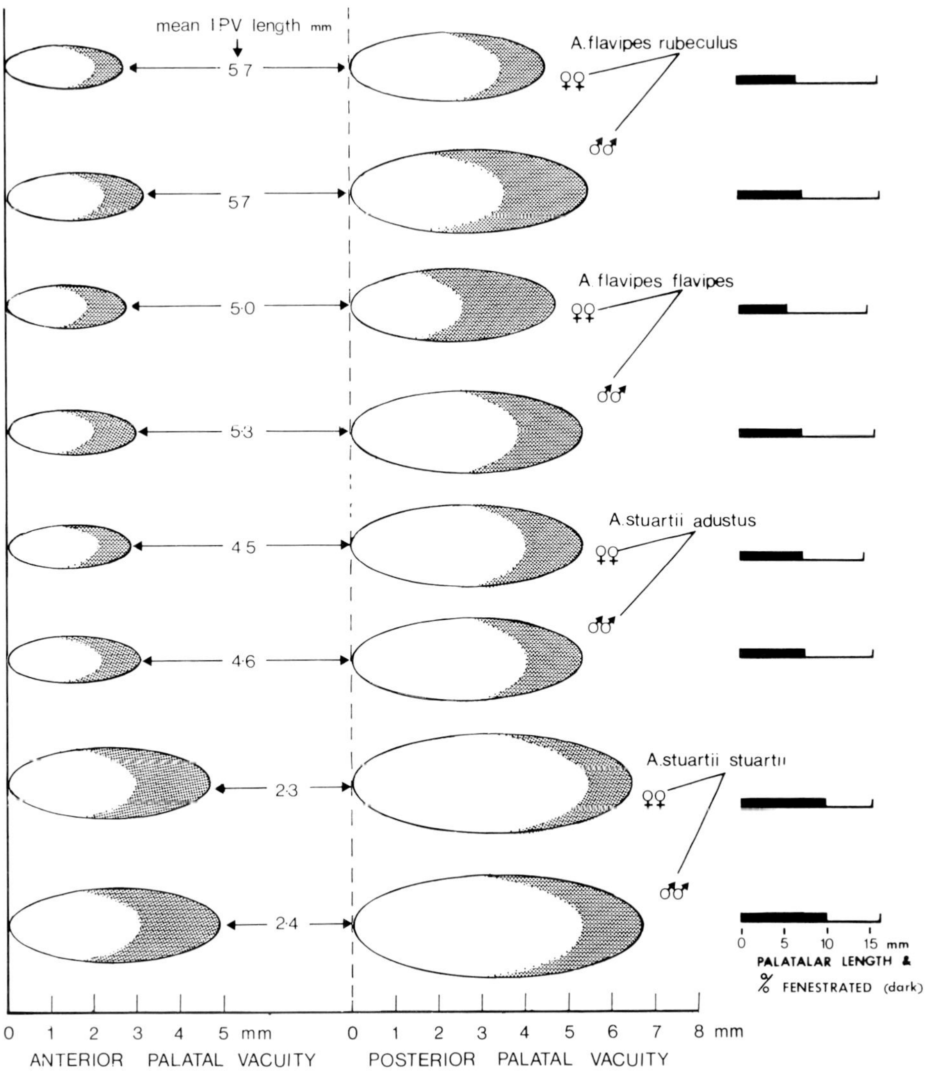

Fig. 15. Comparison of maximum and minimum lengths of anterior, posterior and the mean inter–palatal (IPV) vacuity lengths in Queensland antechinuses accompanied by a comparison (right) of palatal fenestration in the species. Shaded areas represent total variation in each group.

Transverse Canal Formation Separation

Antechinus s. adustus has a significantly (males P< .001, for females P< .05) shorter separation of the transverse canal foramina than *A. s. stuartii*.

Premolar Morphology

In contrast to *A. s. stuartii* from southeastern Queensland the upper premolars of *A. s. adustus* carry prominent postero–lingual lobes (Fig. 25C–H). This feature is most evident in P^2. In lower premolars, P_2 also is prominently lobed. The tendency therefore in *adustus* is toward more massive, stumpy, rounded premolars than the narrow bladelike condition seen in more southerly populations.

The Taxonomic Status of Antechinus stuartii adustus

Specimens *of A. s. adustus* from northeastern Queensland differ significantly from specimens of *A. s. stuartii* in southeastern Queensland (Table 1, 2 and 7). The differences however between *adustus* and the neotype of *A. s. stuartii* from Waterfall, Sydney, NSW, are apparent but less obvious. This lack of distinction applies similarly between central coastal *A. s. stuartii* in New South Wales and *adustus*. Nevertheless while the subspecific status of *adustus* is justified on the basis of pelage, cranial and dental uniqueness, these factors combined with its geographical isolation should accord it with full specific rank. Its status, however may be revealed more reliably through such channels as reproductive biology and biochemical comparison.

Those features which separate *adustus* from the neotype of *A. s. stuartii* and the central coastal series from New South Wales are set out below. Relevant cranial measurements appear in Tables 2, 6 and 7.

(1) The tendency in *adustus* toward more stumpy rounded premolars has led to a significant (P< .001) shortening in the pre-

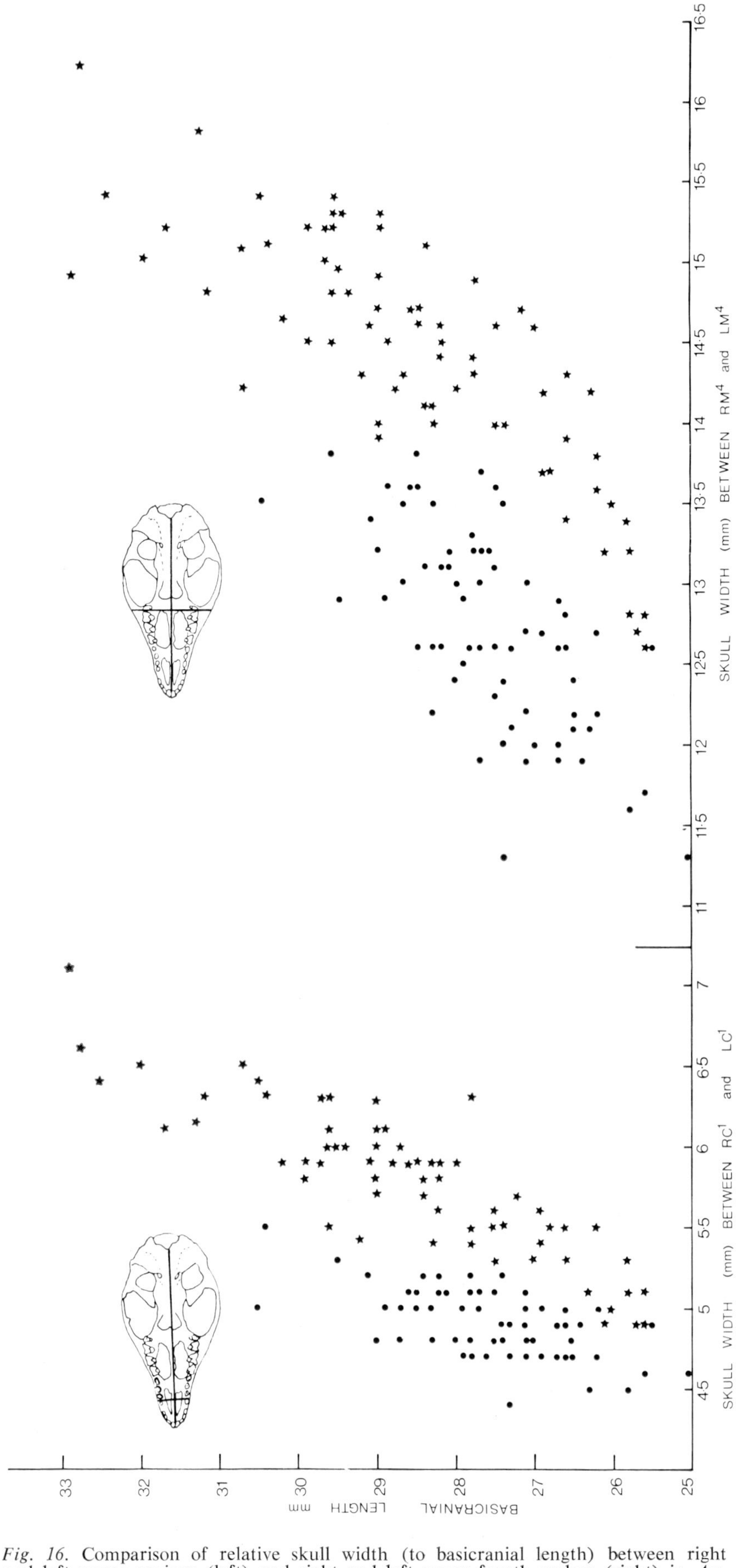

Fig. 16. Comparison of relative skull width (to basicranial length) between right and left upper canines (left) and right and left upper fourth molars (right) in *Antechinus stuartii stuartii* (dots) and *A. flavipes flavipes* (stars).

Table 7. Comparison of significant differences in species and subspecies using Student's t–test.

		A. s. stuartii (SEQ) / A. s. adustus (NEQ)		A. s. stuartii (SEQ) / A. s. stuartii (NSW)		A. s. adustus (NEQ) / A. s. stuartii (NSW)		A. s. stuartii (SEQ) / A. f. flavipes (SEQ)		A. f. flavipes (SEQ) / A. f. rubeculus (NEQ)	
		t	P<	t	P<	t	P<	t	P<	t	P<
BL	♂	3.46	.01	7.77	.001	3.31	.01	2.80	.01	3.56	.001
	♀	2.94	.01	5.92	.001	2.62	.02	1.27	.3	5.82	.001
ZW	♂	1.46	.2	4.71	.001	2.44	.05	7.85	.001	2.92	.01
	♀	2.27	.05	2.88	.01	1.78	.3	39.45	.001	5.42	.001
SWR-LC^1	♂	0.53	.6	0.62	.6	0	.9	14.72	.001	1.86	.05
	♀	0.85	.4	0.79	.5	0.19	.8	5.77	.001	7.04	.001
SWR-LM^4	♂	1.10	.3	0.82	.9	0.99	.5	13.40	.001	0.19	.9
	♀	0.81	.5	2.61	.02	2.54	.02	8.51	.001	3.23	.01
APVL	♂	9.24	.001	11.08	.001	1.32	.2	13.28	.001	1.11	.3
	♀	9.06	.001	9.08	.001	0.95	.4	12.13	.001	1.15	.3
PPVL	♂	9.32	.001	9.82	.001	0.96	.4	16.61	.001	1.08	.3
	♀	8.98	.001	13.48	.001	1.07	.3	14.95	.001	0.19	.9
IPVL	♂	13.57	.001	16.58	.001	0.85	.5	23.84	.001	2.60	.02
	♀	12.05	.001	12.86	.001	1.24	.3	19.70	.001	5.05	.001
P^{1-2}	♂	6.26	.001	0.63	.6	5.89	.001	15.95	.001	3.24	.01
	♀	5.15	.001	1.55	.2	3.05	.01	9.52	.001	2.15	.05
P^{1-3}	♂	6.08	.001	2.60	.02	3.54	.01	9.12	.001	0.66	.6
	♀	4.92	.001	3.63	.001	0.35	.8	8.53	.001	2.41	.05
IBW	♂	2.62	.02	3.97	.001	6.34	001	4.28	.001	5.63	001
	♀	2.19	.05	2.49	.02	4.27	001	2.59	.02	4.93	.001
IOW	♂	2.62	.02	8.68	.001	3.65	.001	7.74	.001	3.50	.01
	♀	1.72	.1	4.64	.001	5.06	.001	6.13	.001	5.65	.001
TC	♂	4.11	.001	2.97	.01	1.30	.3	8.86	.001	2.9	.01
	♀	2.11	.05	2.21	.05	1.29	.8	8.63	.001	6.57	.001
SCD	♂	2.69	.01	0.82	.5	1.60	.2	8.35	.001	2.24	.05
	♀	0.76	.5	1.78	.1	0.86	.3	3.70	.001	5.24	.001
HB	♂	2.47	.02	8.72	.001	2.54	.02	0.39	.9	2.29	.05
	♀	4.81	.001	3.84	.001	0.97	.4	0.36	.8	2.24	.05
TV	♂	0.91	.4	3.82	.001	3.00	.01	0.17	.9	2.53	.02
	♀	0.98	.4	2.45	.02	2.40	.05	2.09	.05	2.14	.05
W	♂	22.83	.001	8.88	.001	1.43	.2	2.77	.02	3.23	.01
	♀	2.41	.05	5.02	.001	4.53	.001	0	.9	2.02	.1

molar row which is most pronounced at P^{1-2}. For male *adustus* the average length (mm) of P^{1-2} is 2.22 ± .04, for male *A. s. stuartii* $\bar{x}$ = 2.49 ± .03 and for the neotype length P^{1-2} = 2.5. For female *adustus* $\bar{x}$ = 2.17 ± .02 and for female *A. s. stuartii* $\bar{x}$ = 2.30 ± .03.

(2) In *A. s. adustus* the alisphenoid tympanic bullae are more widely spaced (inside bullae width) than central New South Wales specimens. This character is directly correlated with a decrease in the size of the alisphenoid bullae in *adustus*.

(3) Interorbital width is significantly ($P < .001$) greater in individuals of *A. s. adustus* than in *A. s. stuartii*.

(4) There is considerable specialization in the incisors of *A. s. adustus*, I^{2-4} being considerably more massive than the more spatulate incisors of central New South Wales individuals, and the neotype. I^2 and I^3 are also higher crowned in *adustus* than in the nominate form.

(5) In premolar morphology, the prominent posterolingual lobes on P^{1-3} of *adustus*, which cause the premolars to appear round-

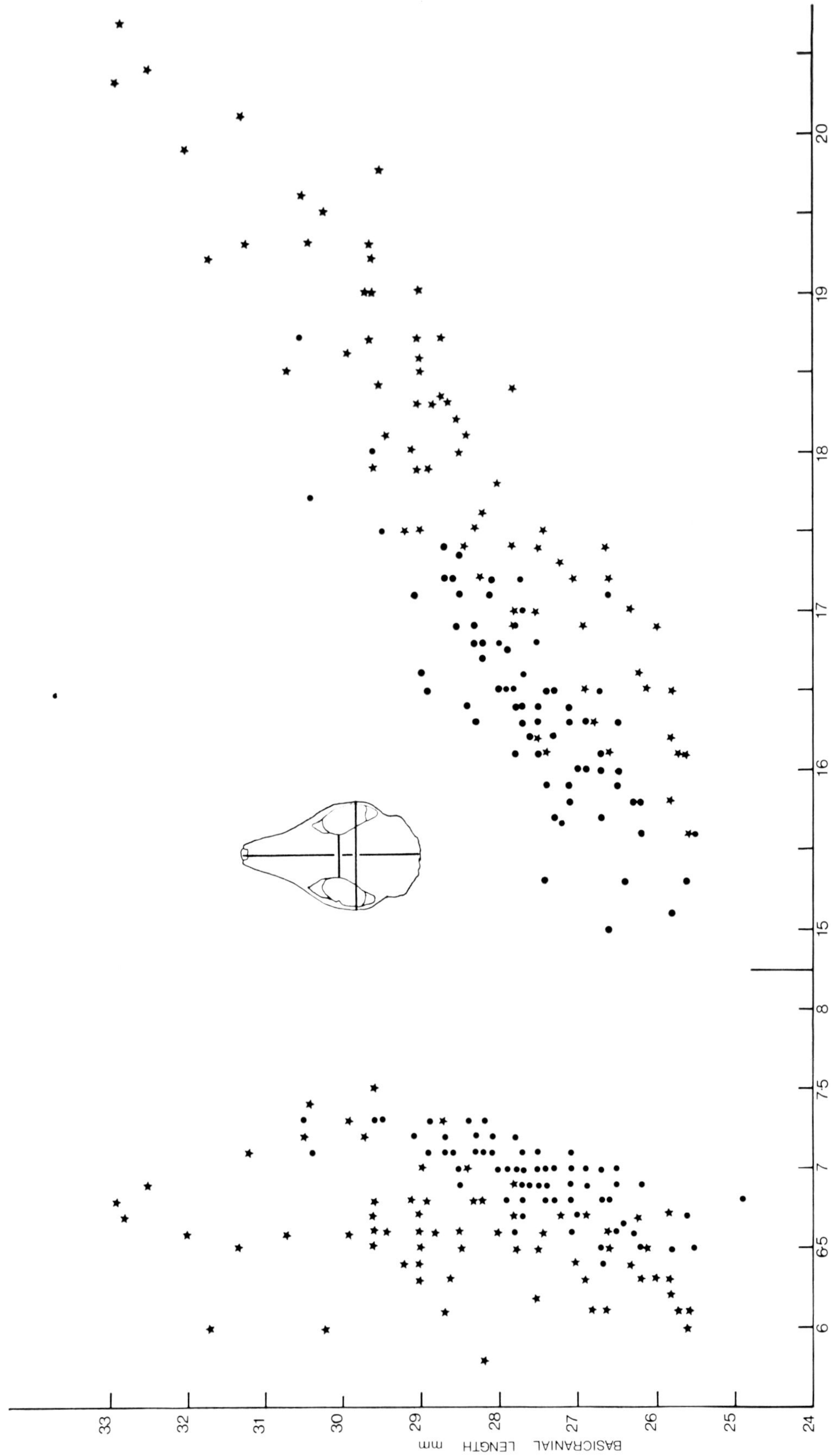

Fig. 17. Left, comparison of relative (to basicranial length) interorbital width and right, zygomatic width in *Antechinus stuartii stuartii* (dots) and *A. flavipes flavipes* (stars).

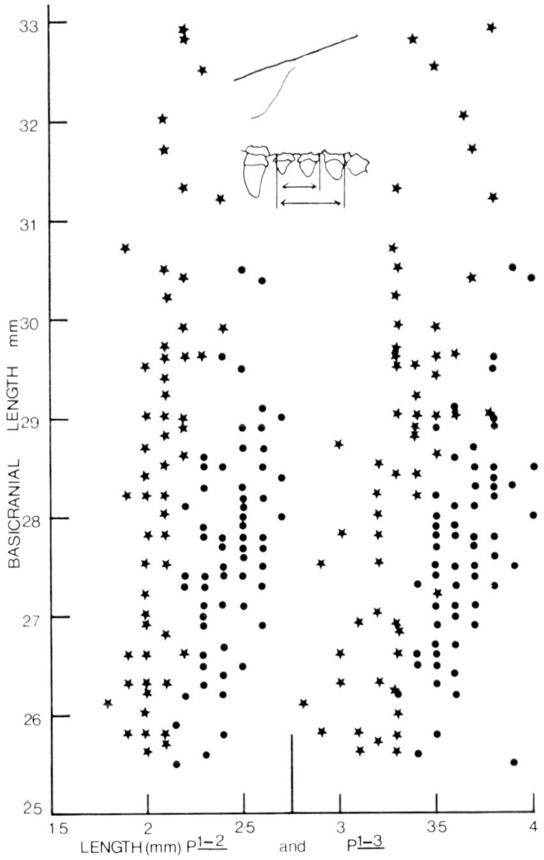

Fig. 18. Left, comparison of relative (to basicranial length) premolar row length, P^{1-2} and right, premolar row length, P^{2-3}, in *Antechinus stuartii stuartii* (dots) and *A. flavipes flavipes* (stars).

ed and stumpy are accentuated in comparison with the typical form and the neotype, because of the increased narrowing of premolar width in these southern forms.

ANTECHINUS FLAVIPES RUBECULUS

Cranial measurements appear in Table 4; comparisons with other forms in Table 7. *A. f. rubeculus* from northeastern Queensland differs from *A. f. flavipes* in southeastern Queensland in the following respects:

Basicranial Length

Antechinus f. rubeculus has a significantly (P< .001) greater basicranial length than *A. f. flavipes*. Sexual dimorphism is pronounced in this form with males having significantly (P< .01) larger skulls (basicranial length) than females.

Zygomatic Width

Antechinus f. rubeculus has a significantly (males P< .01, females P< .001) greater zygomatic width than *A. f. flavipes*.

Rostral Width

Antechinus f. rubeculus has a significantly (for males P< .05, for females P< .001) wider snout in the region R–LC¹ than *A. f. flavipes*. In males this condition becomes less significant (P< .9) at the region R–LM⁴, but is significant (P< .01) in females.

Palatal Vacuities

Although there is no significant difference between northern and southern populations with respect to sizes of anterior and posterior palatal vacuities, the significantly greater basicranial length of *A. f. rubeculus* results in a corresponding significant increase in the interpalatal vacuity distance (for males P< .02, for females P< .001).

Alisphenoid Tympanic Bullae

Antechinus f. rubeculus has a significantly (P< .001) wider separation of alisphenoid tympanic bullae (inside bullae width) than *A. f. flavipes*.

Transverse Canal Foramina

Antechinus f. rubeculus has a significantly (for males P< .01, for females P< .001) greater distance of separation between transverse canal foramina than *A. f. flavipes*.

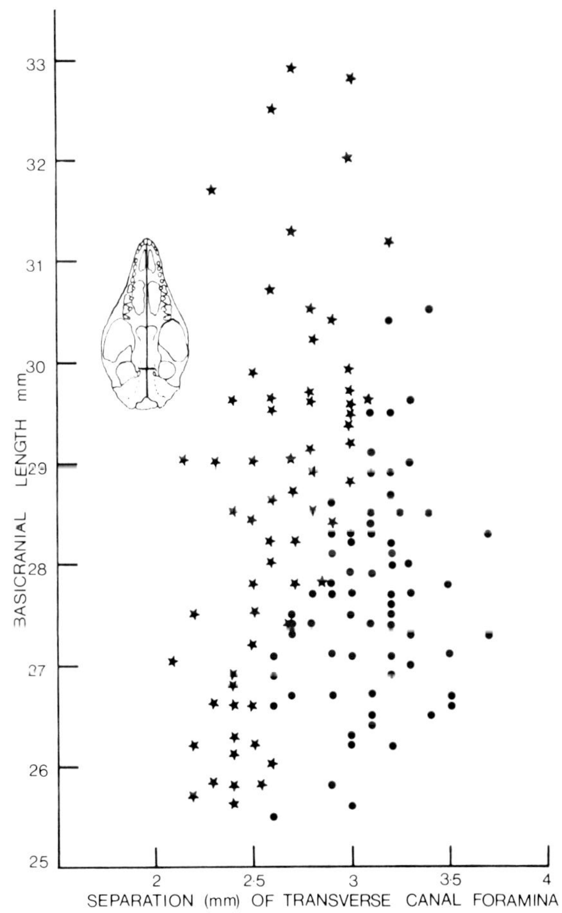

Fig. 19. Comparison of relative lengths of separation between the transverse canal foramina in *Antechinus stuartii stuartii* (dots) and *A. flavipes flavipes* (stars).

Skull–Canine Depth

Antechinus f. rubeculus has a significantly (for males $P < .05$, females $P < .001$) greater skull–canine depth than *A. f. flavipes.*

Interorbital Width

Antechinus f. rubeculus has a significantly (males $P < .01$, females $P < .001$) greater interorbital width than *A. f. flavipes.*

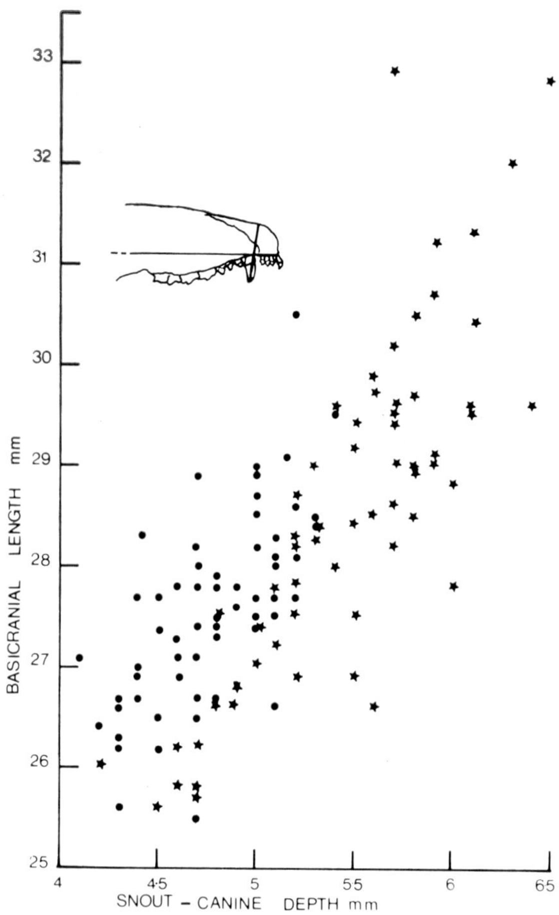

Fig. 20. Comparison of relative snout–canine depth in *Antechinus stuartii stuartii* (dots) and *A. flavipes flavipes* (stars).

AFFINITIES BETWEEN A. flavipes AND A. stuartii

The morphological diversity within *Antechinus*, as well as the presence of *Antechinus*–like fossil dasyurids in the middle Miocene Kutjamarpu fauna (Archer 1981a) suggests that phyletic lineages in these small dasyurids can be traced back well into the Tertiary, before the onset of continent–wide aridity. The relatively late development of aridity in Australia is supported by a number of vertebrate paleontologists (Stirton, Tedford and Woodburne 1968, Archer 1981a) and palynologists (Callen and Tedford 1976, Galloway and Kemp 1981). The early Miocene environment as described by Galloway and Kemp (1981) was relatively warm and wet, the dominant vegetation being rainforest along the wide coastal belts, with a high proportion of *Nothofagus* extending over much of the continent. The trend toward increasing aridity began in the mid–Miocene, 15 m.y. B.P. and had brought extreme aridity by approximately 17,000 years B.P., and with it reduction and fragmentation of rainforests and a lowering of sea levels. It is therefore proposed that the ancestral species of *Antechinus* was non arid–adapted and that those extant arid–adapted species evolved after the late Miocene climate deterioration. The marsupicarnivore *Ankotarinja tirarensis* from the mid–Miocene Etadunna Formation (Archer 1976b) and an *Antechinus*–like dentary from the Pliocene Otibanda Formation, New Guinea (Plane 1967) indicate that among modern dasyurids those from New Guinea rainforests such as *Murexia, Neophascogale* and *Phascolosorex* exhibit the most structurally ancestral characters. This supports the suggestion by Tate (1947) that the genus *Murexia* represents the least specialized of dasyurids, retaining more primitive features than any other group and for the purpose of this paper provides the framework for comparison within modern species of antechinuses.

Of the two dasyurids *A. stuartii* and *A. flavipes, A. stuartii* represents the more structurally primitive antechinus, while *flavipes* represents the more recent and successful in terms of radiation and adaptation.

The affinities between these two species are now examined using those character–states which have already been demonstrated to most readily separate the two species.

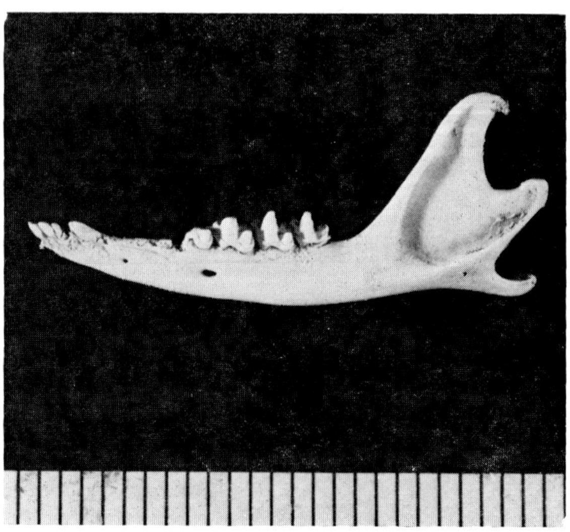

Fig. 22. Left dentary of aged male *Antechinus stuartii stuartii* showing tooth wear.

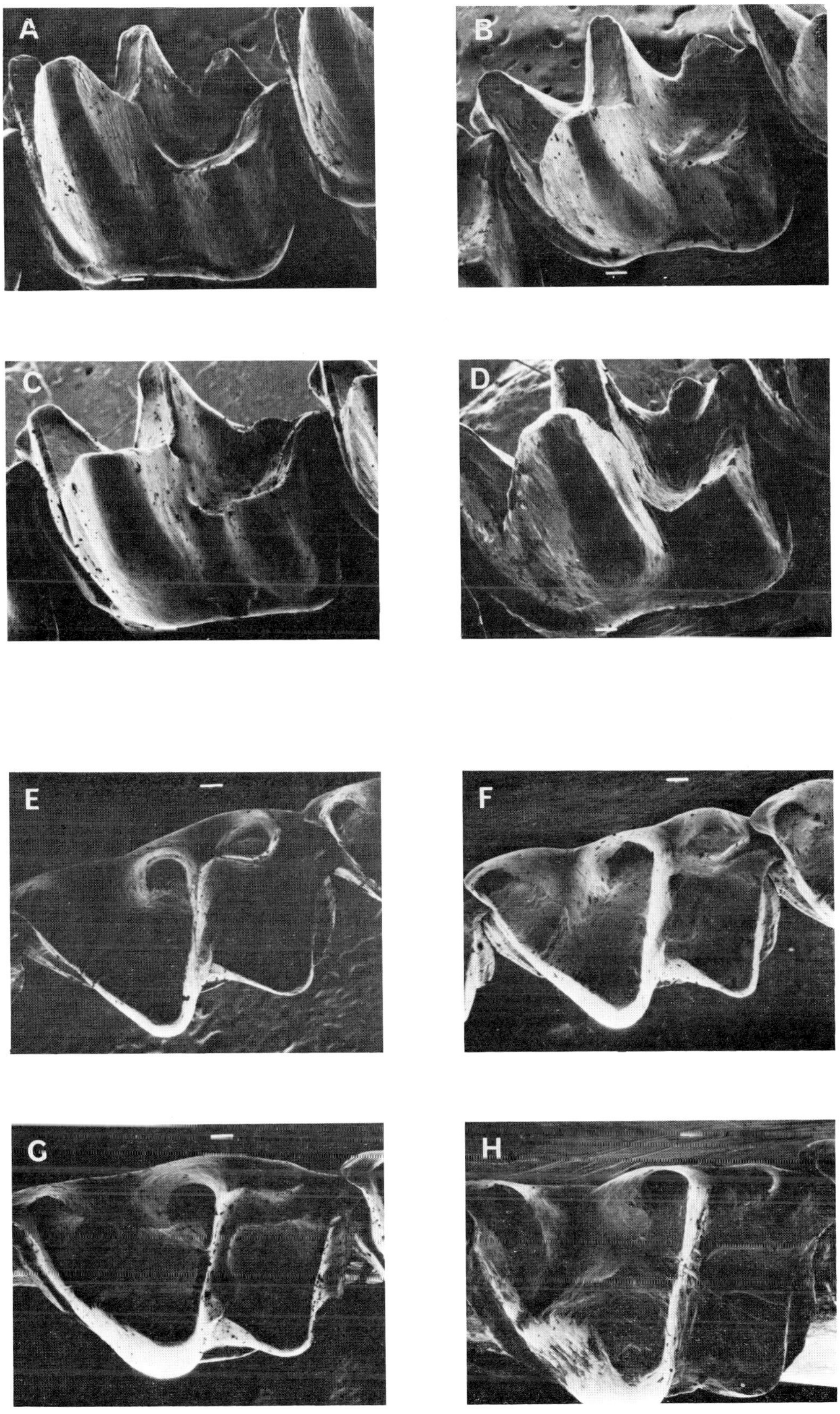

Fig. 21. A–D, entoconid development on M_4 in *Antechinus stuartii stuartii* (A) *A. stuartii adustus* (B), *A. flavipes flavipes* (C), *A. flavipes rubeculus* (D), E–H, talonid shape on M_3 in *A. stuartii stuartii* (E), *A. stuartii adustus* (F), *A. flavipes flavipes* (G), *A. flavipes rubeculus* (H).

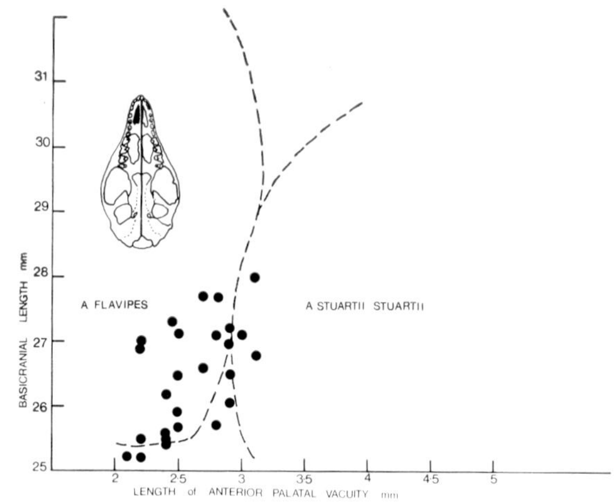

Fig. 23. Relative length of anterior palatal vacuities in *Antechinus stuartii adustus*. Dots are superimposed over the simplified pattern of distribution shown for *A. stuartii stuartii* and *A. flavipes flavipes* in Fig. 13.

Premolar Width and Premolar Row Length – Uncrowded and Narrow or Crowded and Wide

Archer (1976c) demonstrates that narrow premolars in uncrowded premolar rows represent the primitive premolar condition in didelphoids and hence dasyuroids. Elsewhere he notes (Archer 1981) that reduction of P3, widening of the premolars and crowding of the premolar tooth row are derived states within the genus *Sminthopsis* as well as within dasyurids as a whole. The crowded condition of the premolar row and in particular the crowding of P^{1-2} in *A. flavipes* has been demonstrated above. The more primitive condition of narrower, uncrowded premolars is seen in *A. stuartii*. Associated with the compression of the premolar row in *A. flavipes* there has been a posterior shifting of C^1 and a subsequent widening of the $I^4 - C^1$ diastema to accommodate a more massive C_1. The condition of more massive canines in *A. flavipes* is reflected in the greater snout–canine depth discussed above, and this combined with the crowded premolar row is almost certainly related to the tendency of *A. flavipes* to tackle vertebrate prey such as house mice and small birds which are usually ignored by *A. stuartii* (pers. obs.). Tate (1947) noted short muzzles and short palates accompanying short, crowded tooth rows in vertebrate–killing dasyurids. While the premolar row in *A. f. rubeculus* is proportionate to the southern nominate form, that of *A. s. adustus* is dissimilar to *A. s. stuartii* in that P^1 and P^2 of *adustus* carry prominent posterolingual lobes. The condition is most pronounced at P^2 (Fig. 25), and also pronounced at P_2. The premolars are therefore wider and more rounded than in *A. s. stuartii,* a condition which Archer (1976c) considers a derived state in didelphoids and dasyurids. Archer (1976c) directly correlates the development of shorter and wider premolars with extreme shortening of the premolar row. In *adustus* there is a significant proportional decrease in the distance P^{1-2} when compared with the nominate race. It is apparent that the most abrasive and damaging wear to *Antechinus* teeth occurs along the lower premolar row (Fig. 22). The nature of food, determined by geographic position probably determines the extent of wear. The small size of *adustus,* to some extent limits the

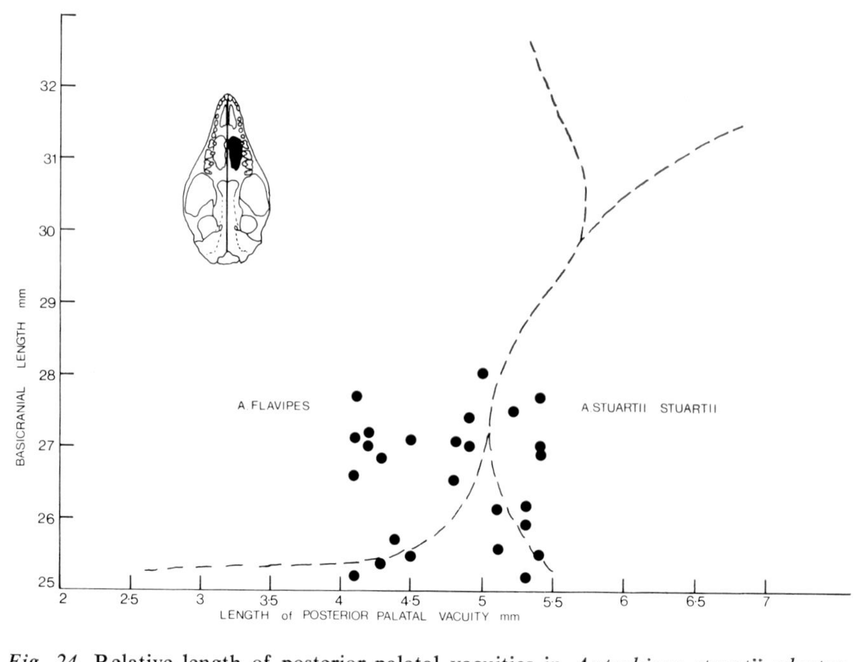

Fig. 24. Relative length of posterior palatal vacuities in *Antechinus stuartii adustus*. Dots are superimposed over the simplified pattern of distribution shown for *A. stuartii stuartii* and *A. flavipes flavipes* in Fig. 14.

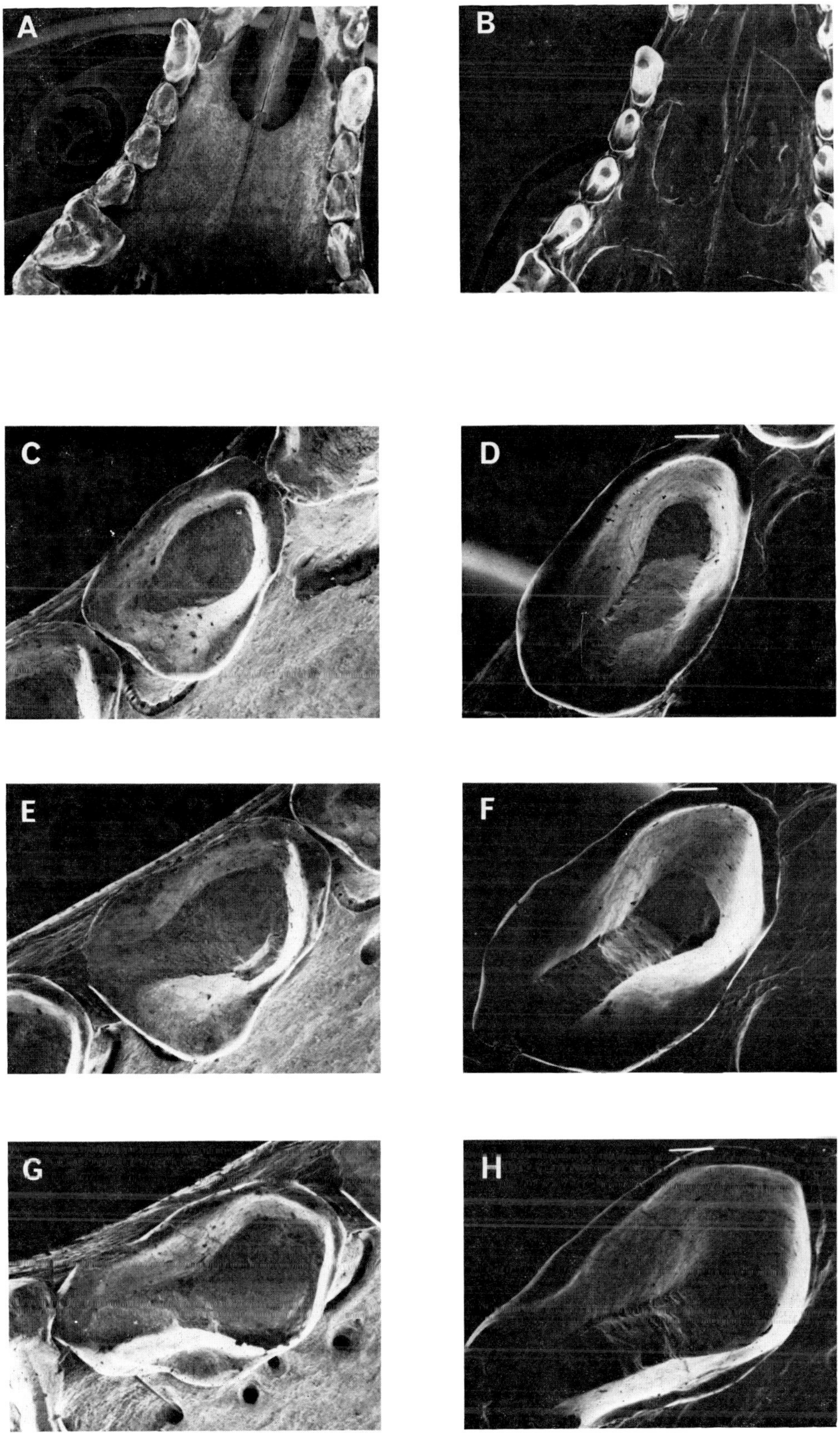

Fig. 25. A–B, premolar row in *Antechinus stuartii adustus* (A), and *A. stuartii stuartii* (B). C, P^1 in *A. stuartii adustus.* D, P^1 in *A. stuartii stuartii.* E, P^2 in *A. stuartii adustus,* F, P^3 in *A. stuartii stuartii.* G, P^3 in *A. stuartii adustus.* H, P^3 in *A. stuartii stuartii.*

types of food available to it and in this respect the widening of premolars may be a step toward retention of premolar constitution in response to an abrasive diet.

Rostral Width – Long and Narrow or Short and broad

A picture of the acuteness of wedging of the snout is given by comparing the rostral widths across right M^4 and left M^4 or right C^1 and left C^1 with basicranial length. Comparison with a number of didelphids and dasyurids shows that the primitive rostral condition in dasyurids is almost certainly long and narrow, such as the condition seen in the narrow snouted didelphids *Philander* and *Didelphis* and also in the dasyurids *A. stuartii* and *A. swainsonii* and contrasting to the brachycephalic condition seen in *Sarcophilus* and to a slightly lesser extent in *A. flavipes.* Lengthening of the forepalate and the rostrum have been associated with insect–killing habits and short muzzles in association with shortened cheek tooth row have been linked with vertebrate–killing (Tate 1947, Archer 1975, 1976c). The relatively long narrow snout in *A. stuartii* reflects the primitive dasyurid condition and that of *A. flavipes* again demonstrates the more derived nature of this species and reflects on its tendency to deal with larger or vertebrate prey. The relatively wider rostrum of *A. s. adustus* than *A. s. stuartii* in southeastern Queensland again suggests the more derived nature of this northern form.

Entoconid Development – Compressed and Elongate or Round and Relatively Isolated

Archer (1980) makes two major comments concerning entoconids in dasyurids: (1) that species occupying higher rainfall areas such as those of highland New Guinea have well developed entoconids while many Australian arid-adapted dasyurids lack them. (2) that the absence entoconids in Australian dasyurids is a derived condition.

Both *A. f. flavipes* and *A. f. rubeculus* exhibit extremely reduced, round and isolated entaconids. *A. s. stuartii* and *A. s. adustus* however, display the primitive condition of compressed elongate entoconids. The associated swelling of the talonid in M_3 of *A. s. stuartii* and *A. s. adustus* such that the width of the talonid is greater than that of the trigonid is a condition found in few other dasyurids. It is however characteristic of *Murexia longicaudata* and ancestral didelphids and is regarded by Archer (1976c) to represent the primitive condition. The derived condition of talonid width narrower than trigonid is evident in *A. flavipes.*

Interorbital Width – Pronounced or Slight Constriction

Tate (1947) considered *Murexia* as the least modified descendent of early Tertiary dasyurids. In his description of the 'hypothetical primitive dasyure' he depicted the animal's skull as 'slightly constricted' in the mid–brain area. Although didelphids and many dasyurids exhibit pronounced interorbital constriction, *Murexia,* and some species of *Antechinus* and *Planigale* show only slight constriction. While the primitive condition of the mid–brain region appears to have been markedly constricted, it appears that a reversal of the character occurred in primitive insectivorous dasyurids which expand ed the brain case at the expense of the development of the more massive temporalis muscles characteristic of those marsupicarnivores with pronounced interorbital constriction and carnivorous tendencies. In this respect *A. s. stuartii, A. s. adustus* and *Murexia* spp. exhibit the primitive dasyurid condition of slight constriction of the mid–brain. *Antechinus s. adustus* exhibits a tendency toward slightly smaller interorbital width relative to the nominate form in southeastern Queensland, a suggestion of this form's more recent evolution. *Antechinus flavipes* exhibits significantly smaller interorbital width than *A. s. stuartii* or *adustus,* a further indication of this species more recent derivation and adaptation toward managing larger or vertebrate prey.

Length of Anterior Palatal Vacuities – Long or Short

Archer (1981b) considers the primitive condition of premaxillary vacuities to be one of intermediate length (extending posteriorly to about the level of the canine) and Tate (1947) imagined the primitive dasyurid to have 'short anterior foramina'. The long condition of the anterior palatal vacuities in *Murexia* spp. and numerous New Guinea species of *Antechinus* leads to the supposition that the primitive dasyurid condition was characterized by long premaxillary vacuities extending posteriorly to a position midway between C^1 and P^1. *Antechinus s. stuartii* emphasizes this primitive condition, where in southeastern Queensland specimens the premaxillary vacuities extend posteriorly to a position midway between P^1 and P^2. The more derived *adustus* has relatively shorter premaxillary vacuities which extend to a position just short of the posterior limit of C^1 and the most derived state is seen in *A. flavipes* where the vacuities extend posteriorly to about the level of the canine.

Length of Posterior Palatal Vacuities – Long or Short

Although Archer (1981b) found a direct correlation existed between the degree of aridity of habitat and length of palatal vacuities in *Sminthopsis* such that those species occupying arid climates had larger vacuities and vice versa; the opposite is true for species of *Antechinus.* In fact *A. s. stuartii* from the vine–forests of southeastern Queensland has a greater proportion of palatal fenestration than any other dasyurid. This feature, however, varies interspecifically so that while vacuities are large in southeastern Queensland specimens of *A. s. stuartii,* a reduction in length occurs the further south that various specimens are examined. A definitive pronouncement on the primitive state of the character is therefore very speculative. The primitive character–state for didelphoids and hence dasyuroids, according to Archer (1981b) and Tate (1947) was of poor palatal fenestration. This view however overlooks the enlarged condition of posterior fenestra in species of adult *Murexia* and the more primitive of species of New Guinea *Antechinus* which suggest that for dasyurids at least, the ancestral condition was characterized by pronounced fenestration, a feature also apparent in *A. s. stuartii* in southeastern Queensland but reduced in the more derived *A. s. adustus* and extremely reduced in the relatively more arid adapted and derived *A. flavipes.*

Canine Size – Large or Small

Caniniform canines are clearly the primitive state in didelphoids and dasyuroids (Archer 1981b). While *A. stuartii* maintains the 'normal' primitive canine proportions, *A. flavipes* has enlarged upon this size to produce more massive upper and lower canines apparently adapting it to manage larger or vertebrate prey.

Alisphenoid Tympanic Wings (Bullae) – Enlarged or Small

Primitive dasyurids such as *Murexia* have very small alisphenoid tympanic bullae as also do most didelphids. This condition of small bullae is regarded by Archer (1981b) to be the primitive condition for dasyurids and is characteristic of both *A. s. stuartii* and *A. s. adustus* in comparison with the derived, more inflated condition seen in *A. flavipes.* Although Archer (1981b) finds a direct correlation in *Sminthopsis* between large bullae and arid habitat, the situation in *Antechinus* is not as clear. This is mainly due to the fact that in comparitively similar eastern vine–forest habitats the bullae of *A. flavipes* are still significantly larger than those of *A. s. stuartii.* This I regard as a relict condition in coastal *A. flavipes* and it is supported by a comparison with more primitive relatives of *A. flavipes. Antechinus leo* the sister species of *A. flavipes* and confined to vine–forests of Cape York Peninsula has significantly smaller bullae than the geographically closest population of *A. flavipes* (*rubeculus*) which inhabits both dry sclerophyll and vine–forest habitats (Van Dyck 1980). This suggests a reinvasion of more arid adapted *A. flavipes* into coastal vine–forest at a later date. The correlation of large bullae with 'arid adaptation' is further supported by the condition seen in *A. s. adustus* which is confined to wet, high altitude vine–forests and has relatively smaller bullae than either *A. s. stuartii* or *A. flavipes.*

DISCUSSION

According to Ziegler (1977) the ancestral stocks of *Murexia, Phascolosorex, Myoictis* and *Neophascogale* were very early (possibly early Miocene) Australian immigrants to northern New Guinea where the evolution of the endemic New Guinea antechinuses resulted from isolation from the northern Australian land mass in the late Pliocene. Although Zeigler (1977) sees no 'close relationship' between any extent Australian dasyurid genus and the four endemic New Guinea genera *Phascolosorex, Murexia, Myoictis* and *Neophascogale,* the cranial and dental morphology of Murexia indicates close affinity with the genus *Antechinus.* For example characters common to both *Murexia longicaudata* (the most primitive member of the genus) and *A. s. stuartii* include the following: separation of I^1 from I^2 by a small diastema; a gradient of decreasing crown height from I^2 – I^4; approximated P^2 and P^3 leaving a slight gap between P^1 and P^2; blade–like, uncrowded premolars with P^3 forming a continuous cutting crest with parastylar corner of M^2; complete buccal and lingual premolar cingula; M^5 reduced and without a metacone; M^4 paracone unreduced relative to metacone; M^4 paracone and metacone unapproximated; paracrista M^2 runs to stylar cusp B; a distinct stylar cusp B on M^2; a complete anterior cingulum developed on all upper molars; compressed elongate entoconids on M^{3-5}; talonid wider than trigonid; 3 cusps on the talonid of M_5; little reduction in the metaconid of M_2, presence of prominent paraconid on M_2; paraconid M_5 smaller in height than metaconid; metacristids oblique to the long axis of the tooth row; relatively long, narrow rostrum; slight interorbital constriction; long premaxillary and maxillary palatal vacuities; small alisphenoid tympanic bullae; widely separated transverse canal foramina.

The close affinity of these two genera suggests that the ancestors of Australian and New Guinea antechinuses were *Murexia*–like giving rise to an *A. stuartii*–like ancestor (possibly highland adapted) and an *A. leo*–like ancestor (possibly lowland adapted), both initially adapted to the areas within the moist forests dominating the continent during the early Miocene. It is possible that the increasing aridity during the mid Miocene and shrinking of the wet forests led to the evolution of *A. flavipes* and *A. bellus,* both more suited to dryer forest and grassland conditions than the *stuartii* and *leo* forms. Subsequent retreat of forests may have restricted *A.*

flavipes to the southwest of Western Australia, and concentrated it in eastern Queensland, New South Wales and Victoria. *Antechinus bellus* was restricted to Northern Territory and *A. leo* remained in the only significant area of lowland vine–forests in the continent – that in Cape York Peninsula. While *A. stuartii* was left occupying all remnants of eastern vine–forests its size decreased in the Atherton Tablelands possibly under the competitive influence of *A. godmani* and also decreased in those areas south of New South Wales–Queensland border where it ventured outside the optimum conditions offered by the warm, wet northern vine–forests. *Antechinus flavipes,* while it had never achieved true 'arid adaptation' had radiated and occupied many tolerable habitats outside the wet forests and then reinvaded all those wet habitats unoccupied by *A. stuartii* e.g. lowland vine-forests of northeastern Queensland, Bunya Mountains, Eungella Range and Frazer Island.

Antechinuses in Queensland utilize the vine–forest habitat in preference to all other habitats. *Antechinus s. stuartii* which presumably evolved in and is almost entirely restricted to vine–forests in southeastern Queensland apparently excludes *A. flavipes* from this habitat. The dependence of *A. s. stuartii* on vine–forest over other habitats in Queensland is undoubtedly related to the states highly seasonal summer rainfall pattern, the limits of its distribution being determined by the average winter ('dry season') rainfall. In southeastern Queensland *A. s. stuartii* is only found in those areas that receive at least approximately 350 mm of rain between the months May to October (Fig. 3). In highland vine–forests of northeastern Queensland where permanent water is assured over the dry season *A. s. adustus* is found in areas receiving at least 200 mm between May and October (Fig. 3). All such areas experience moderate to low annual rainfall variability. In all other areas winter rainfall is too low to support *A. s. stuartii* over the dry winter season.

The yearly influence of rainfall on *stuartii* living in vine–forest has been demonstrated by Braithwaite (1973). He showed that dramatic yearly fluctuations in mean body weights of *A. stuartii* at Mt. Glorious were related to variation in average annual rainfall; years of 'good' rainfall (e.g. 1000 mm p.a.) were associated with correspondingly high average body weight (eg. 62 g for weight of males in September), while periods of 'drought' (e.g. 430 mm p.a.) brought low average body weight (e.g. 46 g).

Water flux rates are relatively high in free living animals in Victoria (Nagy, Seymour, Lee and Braithwaite 1978) and such animals ingested free water far in excess of the expected daily turnover needed to maintain normal water balance. Even though the diet of arthropods alone In July appeared to supply well above the required water balance level, animals still drank free water to excess. While interpretation of the apparent excessive intake of free water may depend on other parameters such as temperature, diet and behaviour (Nagy *et al.* 1978) the significance of free water to *A. stuartii* is nevertheless highlighted by this study. On a broader scale it seems that the spread of *A. stuartii* out of the preferred vine–forest habitat and into the drier ecotones of coastal New South Wales and Victoria has only been made possible by the combined influence of the increased precipitation levels of the last 10,000 years (Galloway and Kemp 1981) and, more importantly, by the fact that annual rainfall in southern temperate climates of southern New South Wales and Victoria is more evenly spread, with significant rainfall occurring in all months of the year. The expansion of range of *A. stuartii* has had associated consequences. Body size and weight reduction is the most noticeable difference between the small individuals in Victoria and the larger southeastern Queensland individuals. While *A. flavipes* maintains a relatively robust size and weight throughout its continental range, the size of *A. s. stuartii* decreases with increasing latitude. At first appearances it is clear from the evenly spread or winter high rainfall patterns of southeastern New South Wales and Victoria, that water is not as critical to antechinuses in the temperate regions (except in the occasional times of drought or flood) as it is in tropical and subtropical Queensland. There are some data to support the interpretation that the low temperatures characteristic of southern winters influence *A. s. stuartii* by restricting growth and ultimate body size. Firstly the low temperatures of autumn and winter and the associated depression in insect abundance has been shown to cause a lull in growth in antechinuses (Wood 1970) as well as shrews (Shillito 1963) and field mice (Baker and Ranson 1933). In southeastern Queensland a period of low, zero or negative growth occurs through the cold months May to July in males of *A. stuartii* (Wood 1970). At Werribee, Victoria, a plateau or decrease in average monthly body weight occurs from May to June in the small insectivorous dasyurid *Sminthopsis crassicaudata* (Morton 1978). At Healesville, Victoria, where maximum and minimum air temperature for July average 11.8 and 3.9° C respectively, individuals of *A. stuartii* spend energy 4.5 times faster than their basal rate with about 85% of their energy allocated to body temperature regulation and basal maintenance, leaving approximately 100 Kcal/kg/day for activity and other expenditures (Nagy *et al.* 1978). At Mt. Glorious, southeastern Queensland, maximum and minimum July temperatures average 15.5 and 8.9° C respectively. Under these warmer conditions a greater energy surplus could be expected after the allocation of energy for body temperature regulation and basal maintenance. Comparison of growth rates of *A. stuartii* presented by Wood (1970) for southeastern Queensland, Barnett (1973) and Cheal *et al.* (1976) for a population at Powelltown, Victoria, reveals firstly that although in February animals have nearly the same average body weight as recorded by the following autumn months, animals in southeastern Queensland gain weight much faster than Victorian individuals. Secondly the plateau or winter slump in weights occurs at the two localities at the same time of the year (late May to late July). Thirdly, the last surge in weight gain before death occurs at Mt. Glorious from late July to mid September (a period of approximately 8 weeks) and in Victoria from late July to mid August (approximately 4 weeks).

This last consideration implies that it is impossible for Victorian individuals of *A. stuartii* to either match the weight gain of Queensland animals in the immediate pre–death surge (e.g. up to a 16 g gain in southeastern Queensland, 5 g in Victoria) or to reach an ultimate adult size equivalent to southeastern Queensland animals. The low temperature conditions of Victorian winters apparently preclude an earlier commencement of that last pre–death weight increase. June and July are periods when conditions are coldest, availability of food is at its lowest, periods of torpor are most frequent (Wallis 1976) and foraging conditions are poorest. (Wood (1970) suggested a sharp drop in survival rate of males in July 1965 following a heavy deluge of rain one night early that month).

Throughout Australia timing of reproduction varies with latitude, (Horner and Taylor 1959, Woolley 1966, Taylor and Horner 1970) and in *A. stuartii* also with altitude (pers. comms. of D. H. Wood, A. C. Robinson and A. K. Lee in Braithwaite 1979). While the breeding season in both *A. stuartii* and *A. flavipes* is undoubtedly geared so that young are being raised when their food supply is most plentiful (Baker 1938, Lack 1954, Morton 1978, Braithwaite 1979), those specific climatic conditions which are responsible for the 'ultimate' timing (Baker 1938) of mating are poorly understood.

The gross differences in breeding seasons of Queensland, New South Wales and Victorian *A. stuartii* and *A. flavipes* can be correlated with two principal climatic parameters which determine the seasonal abundance of insects for each area. In the tropics (e.g. Iron Range) where temperatures are constantly high, rarely falling below 20° C, insect flushes are determined by high rainfall. In the temperate regions (e.g. Melbourne) where annual rainfall is even and occurs significantly in all months, rarely falling below 50 mm per month, temperature determines insect abundance. In the subtropics (e.g. Brisbane) a blurring of the extremities of tropical and southern rainfall patterns and temperature variation occurs so that insect numbers are influenced to varying degrees by both.

At the tip of Cape York Peninsula, although temperatures increase in August, significant rainfall does not occur until October and November (Fig. 26). Insignificant rainfall occurs during the months May–September. *A. leo,* a close relative of *A. flavipes,* occupies the tropical vine–forests of the Iron and McIlwraith ranges, Cape York Peninsula (loc. 28, Fig. 26). The scanty information available suggests that this species mates early in October (Van Dyck 1980). Its young are therefore raised during the January–February–March period when insect numbers are highest.

In the Cairns, Atherton Tablelands district (loc. 32, Fig. 26) *A. f. rubeculus* mates from mid–June to mid–July. *A. s. adustus* (with which *A. f. rubeculus* occurs sympatrically in some areas) mates early in July. In these areas water is more abundant during the winter 'dry' season than in the far northern tropics and temperatures remain relatively high (minimum average July temperature Cairns 16.7° C, Atherton 10.0° C). Under these conditions insects are available throughout the winter and the antechinuses can reproduce earlier and gain maximum benefit from these insect flush occurring at a time when their young are becoming more independent (December, January, February). The success gained from the adoption of an early mating strategy by *A. f. rubeculus* under these optimum conditions is reflected in this form's relatively large body size.

In southeastern Queensland although average winter rainfall rarely falls below 350 mm and 200 mm in those areas where *A. s. stuartii* and *A. f. flavipes* respectively are found, average winter temperatures fall as low 4.7° C (Toowoomba). Rainfall does not increase significantly until October (for Brisbane mean monthly rainfall for September 48 mm, October 75 mm) and mean daily minimum temperatures increase significantly August (Brisbane 10.0° C) and September 12.7° C). The breeding season therefore coincides with a slightly later temperature and rainfall included insect boom.

In the temperate climates of southern New South Wales and Victoria, timing of reproduction in *A. stuartii* and *A. flavipes* appears to be mainly dependent on temperature increase, as rainfall is significant in all months. After the cold winter months, the rise in average monthly temperature during August brings a corresponding rise in insect abundance. Reproduction is therefore timed early.

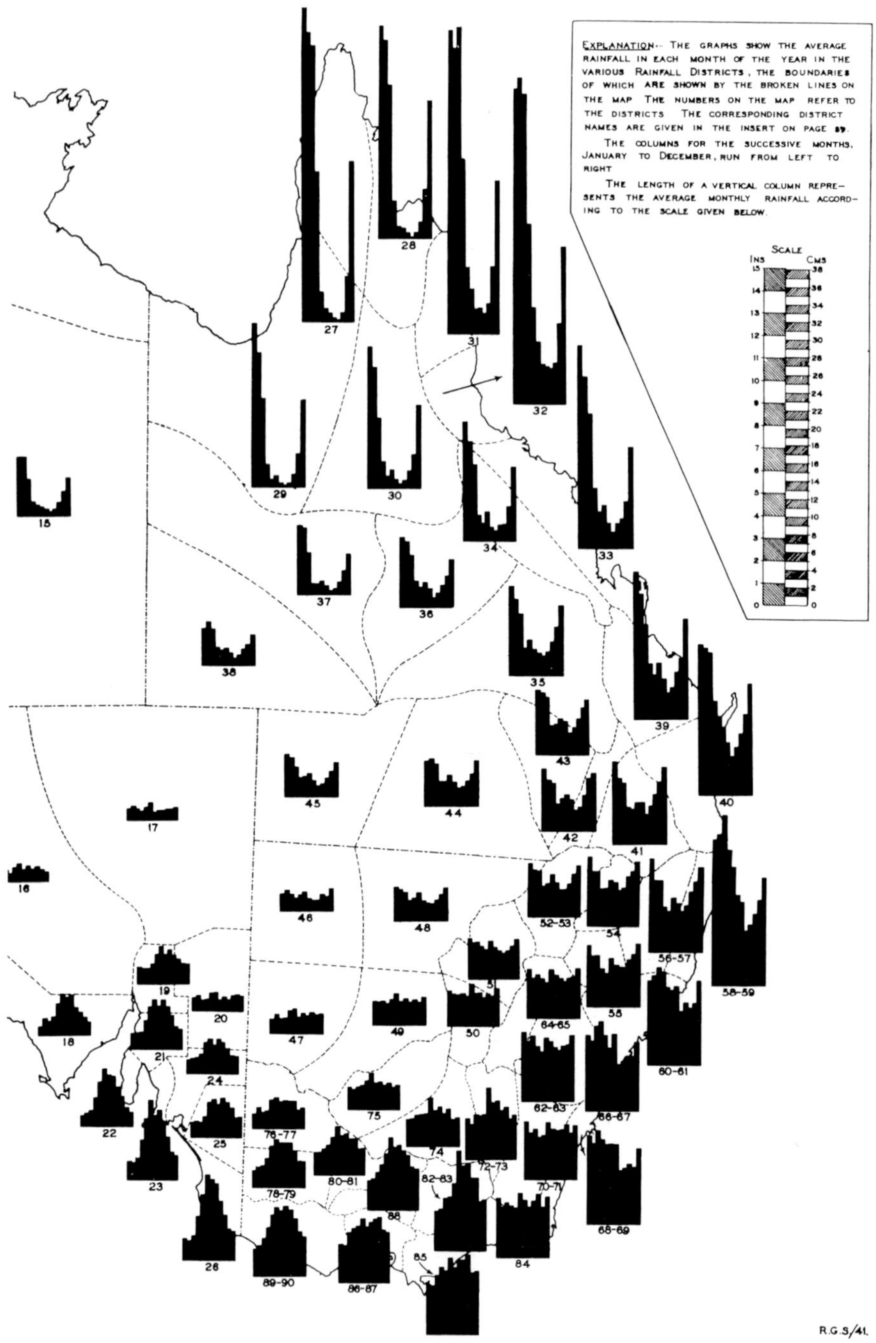

Fig. 26. Monthly distribution of rainfall over eastern Australia (after Archer 1966).

Data presented by Dwyer (1977) suggest that for two species of New Guinean antechinuses, *Antechinus melanurus* and *A. naso* inhabiting the Eastern Highlands, breeding may occur through the year with no suggestion of the synchronous reproductive pattern characteristic of Australian thin–tailed antechinuses.

This at first appears anomolous with the switching temperature–rainfall model I have already proposed to account for gross patterns of climatic reproductive activity shown in Australian antechinuses. However, an investigation of climatic patterns characteristic of the Eastern Highlands provides data which support the model and offer an explanation for such non–synchronous breeding behaviour.

The mere proximity of Cape York Peninsula to New Guinea may initially suggest that reproductive patterns seen in Cape York antechinuses should be repeated in New Guinean antechinuses. This is an unreasonable assumption, overlooking the fundamental climatological differences between the two land masses. "No rigorous climatic analysis is required to show that northern Australia and New Guinea occupy distinct climatic zones with only minimal overlap" (Nix and Kalma 1972: 60). The outmoded and inaccurate nature of this interpretation of New

Guinea as lying wholly in the tropics and experiencing a typical monsoonal climate is exposed and revised by Brookfield and Hart (1966) and thorough climatological analyses of the two areas are presented by Webster and Streten (1972) and Nix and Kalma (1972). Most specific information relating to the area where Dwyer (1977) collected his information is presented by Haantjens (1970).

Of greatest relevance to this discussion are the following general climatic trends characteristic of much of the New Guinea Highlands: a lack of distinct seasons, a moderate rainfall rather evenly distributed through the years, weak winds, high humidity, considerable fog and limited temperature range (Gressitt and Nadkarni 1978). According to Haantjens (1970), at altitudes of 1700 m and higher, the climate approximates the moist temperate or mesothermal climate of higher latitudes but differs from these in its very restricted annual temperature range. This contrasts to the highly seasonal monsoonal climate of Cape York Peninsula. Most importantly, in the Goroka–Mt. Hagen area, the climate is equable throughout the year in all its facets with little year to year variability. Mean monthly temperatures vary through the year by only 0.7° C. Rainfall becomes more seasonal in the east with the seasonal pattern similar to Port Moresby but with higher monthly totals. To the the west, seasonality decreases and approaches the almost complete lack of seasonality of the Southern Highlands District (McAlpine 1970). Even the "driest" parts of the Goroka–Mt. Hagen area are regions of substantial precipitation surplus with the driest months (May–August) registering average monthly falls between 31–84 mm (compared to 1–14 mm for the same months in Iron Range, Cape York Peninsula).

Much of the success of antechinuses in the highly seasonal areas of eastern and southern Australia is attributable to the phenomenon of breeding seasons varying through the species range to coincide with invertebrate food peaks which vary temporally around the coast. It is hardly surprising then that selection has favoured year–round breeding in some New Guinean species where year–round abundance of food is assured (D. Sands pers. comm.) by the strongly damped seasonal variation in temperature and the surplus precipitation that occurs throughout the year.

There is little geographical variation in the degree of sexual dimorphism in *A. stuartii* and *A. flavipes* in Queensland. Males are consistently significantly larger (head–body length) than females. Only in *A. f. rubeculus* is the sexual difference less. Selander (1966) reported that in birds, ecological and morphological polymorphism is frequently expressed in sexual dimorphism in size and structure of feeding apparatus and in differential foraging behaviour and niche utilization by the sexes. It is difficult to imagine that in such unspecialized insectivores as *A. stuartii* and *flavipes* the difference in body size unattended by other structural sexually dimorphic characters gives significant ecological separation between the sexes. Elsewhere (Van Dyck 1979) I have discussed a possible role of sexual dimorphism arising in response to the problems associated with reproduction in the apparently socially intolerant dasyurid *Planigale maculata*. In this species, it seems that the larger, aggressive, and most vocal males are selected by the uncooperative nature of oestrous females of that species. In *A. stuartii*, Braithwaite (1979) found that in males of the same age, dominance and weight were highly correlated. Concerning reproduction, he said (p. 525) "Hence the most aggressively active male may make the largest contribution to the gene pool, and contribute to strong selection for maintenance of the observed life history of this species".

Scudo's (1969) "genetically effective size" of a population turned out to be more strongly dependent on sexual dimorphism for environmental requirements and sex ratio than reproductive efficiency. Given Scudo's assumption, genetic and reproductive efficiency is maximised by a high proportion of short–lived small males in the population. In the case of "semelparous" antechinuses, although sexual selection has established a trait which makes a larger proportion of the environmental resources available to the females, the associated risk of decreasing the chances of the females being inseminated appears to be overcome by males being larger than their uncooperative mates. A compromise has therefore been reached where large agressive dominant males, which represent a low fraction of the population, give the highest reproductive efficiency possible and all males have much shorter life expectancy than females.

The greater adaptability of the more recently derived *A. flavipes* is evident from its distribution and morphology. In Queensland its range extends from the coast westward through a variety of habitat types into areas receiving low average annual rainfall with moderate to high annual rainfall variability (Fig. 6). Coincident with its movement into diverse and drier habitats, morphological changes have been selected which enable *A. flavipes* to deal with a wider range of food, in particular vertebrate prey. Its succes is further evidenced by the extent of its continental distribution and its reinvasion of vine–forested areas in northeastern Queensland. There is evidence to suggest that the more unpredictable nature of its wood and grassland habitat has selected females capable of producing more offspring so that average nipple number (10) is higher than that of the vine–forest dwelling *A. stuartii* (8).

Antechinus s. stuartii represents a more primitive *Antechinus* primarily adapted to an insectivorous diet and a vine–forest habitat. Its ability to extend its range outside vine–forests and into drier heaths and woodlands is limited by low winter rainfall in its northern range and by the limits to which it can tolerate the extended cold winters of the south. The reduction in body size in the south, appears to support the observation of Braithwaite (1979) that it is unlikely that the "life history" of *A. stuartii is* optimal in all situations where it is currently found.

In some areas of southeastern Australia, competitive interactions with more suitably adapted species have apparently caused local extinctions of *A. stuartii* and led to its replacement in some areas (Braithwaite 1979, Braithwaite and Gullan 1978, Braithwaite, Cockburn and Lee 1978). However, Braithwaite's (1979: 526) opinion that *A. stuartii* "appears to be trapped at an adaptive peak" and that its extinction is likely should its "life history cease to be more adaptive generally" is a misjudgment of the species life strategy on the basis of its performance in southeastern Australia. Certainly a false and dismal impression of any mammals' adaptive qualities is gained when the life history of a species is judged on its performance at the outer–extremeties of its range, where competitive interactions with more suitably adapted species threated its existence and climatic factors test the limits of the species tolerances (e.g. *A. stuartii* in southeastern Australia and *A. flavipes* in central eastern Australia).

The life histories of *A. stuartii* and *A. flavipes* are geared to exploit the highly predictable seasonal environments of Australia (Braithwaite and Lee 1979). The ability of these antechinuses to adapt and extend their ranges is demonstrated in the latitudinal and altitudinal adjustment of breeding seasons to compensate for seasonal variation in insect abundance. The variation in nipple number, from 6–10, in females of *A. stuartii* in Victorian localities (Wakefield and Warneke 1967) may also be an adaptation toward the more rigorous, less predictable climatic conditions of the south.

But contrary to the opinion of Braithwaite (1979) that "*A. stuartii* appears to be trapped at an adaptive peak . . . from which there seems little escape", I would contend that both *A. stuartii* and *A. flavipes* have developed homeostatic social systems for which there is no reason to escape.

ACKNOWLEDGEMENTS

I am especially grateful to Dr. M. Archer (University of New South Wales) who encouraged the study of Queensland antechinuses and who constructively criticized this work. Both he and his wife Elizabeth maintained many of my study animals in captivity. For their invaluable field assistance I thank Mr Jack Willows (Byron Bay), Mr Paul Filewood (Queensland Museum) and for their hospitality Mr. and Mrs. Angus Mackee (Ingham) and Mr. and Mrs. Rob Whiston (Gap Creek). Mr Bob Warneke (Fisheries and Wildlife Division, Victoria) generously offered details of specimens examined by him overseas, Mr. Basil Marlow (Australian Museum) provided free access to specimens and type material in his care and Dr. R. Braithwaite (Australian National Parks and Wildlife Service) permitted me to quote measurements from his unpublished thesis. I also thank Messrs A. Easton and R. Burgess (Queensland Museum) for the photographs, Mr. R. Raven (Queensland Museum) for advice on electron microscopy and Miss J. Hamilton for typing the manuscript. Mr. Bob Warneke and Dr. J. Calaby (C.S.I.R.O., Canberra) read and constructively criticized this work.

REFERENCES

Archer, K. M., 1966. "Year book, Australia". 52 Commonwealth Bureau of Census and Statistics: Canberra, Australia.

Archer, M., 1974. The development of cheek–teeth in *Antechinus flavipes* (Marsupialia, Dasyuridae). *J. R. Soc. West. Aust.* **57:** 54–63.

Archer, M., 1975. Abnormal dental development and its significance in dasyurids and other marsupials. *Mem. Qd Mus.* **17:** 251–65.

Archer, M., 1976a. The basicranial region of marsupicarnivores (Marsupialia), inter–relationships of carnivorous marsupials, and affinities of the insectivorous marsupial peramelids. *J. Linn Soc. Lond. Zool.* **59:** 217–322.

Archer, M., 1976b. Miocene marsupicarnivores (Marsupialia) from central South Australia, *Ankotarinja tirarensis* gen. et. sp. nov., *Keeuna woodburnei* gen. et sp. nov. and their significance in terms of early marsupial radiations. *Trans. R. Soc. S. Aust.* **100:** 53–73.

Archer, M., 1976c. The dasyurid dentition and its relationships to that of didelphids, thylacinids, borhyaenids (Marsupicarnivora) and peramelids (Peramelina: Marsupialia). *Aust. J. Zool., Suppl. Ser.* **39:** 1–34.

Archer, M., 1978. The nature of the molar–premolar boundary in marsupials and a reinterpretation of the homology of marsupial cheekteeth. *Mem. Qd Mus.* **18:** 157–64.

Archer, M., 1981a. A review of the origins and radiations of Australian mammals. Pp.1435–88 *in* "Ecological biogeography of Australia", ed. by A. Keast. *Monogr. Biol.* **41.** Dr. W. Junk: The Hague.

Archer, M., 1981b. Results of the Archbold Expeditions. No. 104. Systematic revision of the marsupial dasyurid genus *Sminthopsis* Thomas. *Bull. Amer. Mus. Nat. Hist.* **168:** 61–224.

Baker, J. R., 1938. The evolution of breeding seasons. Pp.161–77 *in* "Evolution, essays presented to E. S. Goodrich (Oxford)".

BAKER, J. R. AND RANSON, R. M., 1933. Factors affecting the breeding of the Field Mouse (*Microtus agrestis*). Part III. Locality. *Proc. R. Soc. Ser. B.* **113:** 486–95.

BARNETT, J. L., 1973. A stress response in *Antechinus stuartii* (Macleay) *Aust. J. Zool.* **21:** 501–13.

BRAITHWAITE, R. W., 1973. An ecological study of *Antechinus stuartii* (Marsupialia: Dasyuridae). M.Sc. thesis: University of Queensland, Brisbane.

BRAITHWAITE, R. W., 1979. Social dominance and habitat utilization in *Antechinus stuartii* (Marsupialia). *Aust. J. Zool.* **27:** 517–28.

BRAITHWAITE, R. W., COCKBURN, A. AND LEE, A. K., 1978. Resource partitioning by small mammals in lowland heath communities of south–eastern Australia. *Aust. J. Ecol.* **3:** 423–45.

BRAITHWAITE, R. W. AND GULLAN, P. K., 1978. Habitat selection by small mammals in Victorian heathland. *Aust. J. Ecol.* **3:** 109–27.

BRAITHWAITE, R. W. AND LEE, A. K., 1979. A mammalian example of semelparity. *Am. Nat.* **113:** 151–5.

BROOKFIELD, H. C. AND HART, D., 1966. "Rainfall in the tropical southwest Pacific". Australian National University, Research School of Pacific Studies, Dept. of Geography. Publ. G/3, Canberra.

CALLEN, R. A. AND TEDFORD, R. H., 1976. New late Cainozoic rock units and depositional environments, Lake Frome area South Australia. *Trans. R. Soc. Aust.* **100:** 125–68.

CHEAL, P. D., LEE, A. K. AND BARNETT, J. L., 1976. Changes in the haematology of *Antechinus stuartii* (Marsupialia) and their association with male mortality. *Aust. J. Zool.* **24**: 299–311.

DEGABRIELE, R., 1976. The forest mammals of Cape York Peninsula. *Habitat* **4** (4): 3–6

DWYER, P. D., 1977. Notes on *Antechinus* and *Cercartetus* (Marsupialia) in the New Guinea Highlands. *Proc. R. Soc. Qd* **88:** 69–73.

DWYER, P. D., HOCKINGS, M. AND WILMER, J., 1979. Mammals of Cooloola and Beerwah. *Proc. R. Soc. Qd.* **90:** 65–84.

GALLOWAY, R. W. AND KEMP, E. M., 1981. Late Cainozoic environments in Australia. Pp. 51–80 *in* "Ecological biogeography in Australia", ed. by A. Keast. *Monogr. Biol.* **41.** Dr. W. Junk: The Hague.

GRESSITT, J. L. AND NADKARNI, N., 1978. "Guide to Mt. Kaindi. Background to montane New Guinea ecology". Ecology Institute Handbook Number 5.

HAANTJENS, H. A., 1970. Lands of the Goroka–Mount Hagen area, Papua New Guinea. *Land Research Series No.* **27:** 7–12. CSIRO, Australia.

HORNER, B. E. AND TAYLOR, J. M., 1959. Results of the Archbold Expeditions. No. 80. Observations on the biology of the Yellow–footed Marsupial Mouse *Antechinus flavipes flavipes*. *Amer. Mus. Novit. No.* **1972**.

INNS, R. W., 1976. Some seasonal changes in *Antechinus flavipes* (Marsupialia. Dasyuridae). *Aust. J. Zool.* **24:** 523–31.

IREDALE, T. AND TROUGHTON, E. LEG., 1934. A check–list of the mammals recorded from Australia. *Mem. Aust. Mus.* **6:** 1–22.

LACK, D., 1954. "The natural regulation of animal numbers". Oxford University Press: London.

LEE, A. K., BRADLEY, A. AND BRAITHWAITE, R. W., 1977. Corticosteroid levels and male mortality in *Antechinus stuartii*. Pp.209–20 *in* "The biology of marsupials" ed. by B. Stonehouse and D. Gilmore. Macmillan: London.

LE SOUEF, A. S. AND BURRELL, H., 1926. "Wild animals of Australasia". Harrap and Co.: London.

MCALPINE, J. R., 1970. Climate of the Goroka–Mount Hagen area. *Land Research Series No.* **27:** 66–79. CSIRO, Australia.

MARLOW, B. J., 1961. Reproductive behaviour of the marsupial mouse *Antechinus flavipes flavipes* Waterhouse (Marsupialia) and the development of the pouch young. *Aust. J. Zool.* **9:** 203–18.

MARLOW, B. J., 1962. "Marsupials of Australia". Jacaranda Press: Brisbane.

MAY, O. M., 1979. "Queensland year book". 39. Australian Bureau of Statistics: Queensland Office.

MILLER, J. G., 1974. "Year book, Australia". 60. Australian Bureau of Statistics: Canberra Australia.

MORTON, S. R., 1978. An ecological study of *Sminthopsis crassicaudata* (Marsupialia: Dasyuridae) III. Reproduction and life history. *Aust. Wildl. Res.* **5:** 183–213.

NAGY, K. A., SEYMOUR, R. S., LEE, A. K. AND BRAITHWAITE, R., 1978. Energy and water budgets in free–living *Antechinus stuartii* (Marsupialia: Dasyuridae). *J. Mammal.* **59:** 60–8.

NEWSOME, A. E., MCILROY, J. AND CATLING, P., 1975. The effects of an extensive wildfire on populations of twenty ground vertebrates in south–east Australia. *In* "Managing terrestrial ecosystems" ed. by J. Kikkawa and H. A. Nix. *Proc. Ecol. Soc. Aust.* **9:** 1–335.

NIX, H. A. AND KALMA, J. D., 1972. Climate as a dominant control in the biogeography of northern Australia and New Guinea. Pp.61–91 *in* "Bridge and barrier: the natural and cultural history of Torres Strait", ed. by D. Walker. A.N.U.: Canberra.

PLANE, M. D., 1967. Stratigraphy and vertebrate fauna of the Otibanda Formation, New Guinea. *Bur. Min. Resour. Aust. Bull.* **86:** 1–64.

RIDGWAY, R., 1912. "Colour standards and colour nomenclature". Author's publication: Washington.

SCUDO, F. M., 1969. On the adaptive value of sexual dimorphism: II. Unisexuality. *Evolution* **23:** 36–49.

SELANDER, R. K., 1966. Sexual dimorphism and differential niche utilization in birds. *Condor* **68:** 113–51.

SHILLITO, J. F., 1963. Field observations on the growth reproduction and activity of a woodland population of the Common Shrew *Sorex araneus* L. *Proc. Zool. Soc. Lond.* **140:** 99–114.

STIRTON, R. A., TEDFORD, R. H. AND WOODBURNE, M. O., 1968. Australian Tertiary deposits containing terrestrial mammals. *Univ. Calif. Publ geol. Sci.* **77:** 1–30.

TATE, G. H. H., 1947. Results of the Archbold Expeditions. No. 56. On the anatomy and classification of the Dasyuridae (Marsupialia). *Bull. Amer. Mus. Nat. Hist.* **18:** 97–156.

TATE, G. H. H., 1952. Results of the Archbold Expeditions. No. 66. Mammals of Cape York Peninsula. *Bull. Amer. Mus. Nat. Hist.* **98:** 567–616.

TAYLOR, J. M. AND HORNER, B. E., 1970. Gonadal activity in the marsupial mouse, *Antechinus bellus*, with notes on other species in the genus (Marsupialia: Dasyuridae). *J. Mammal.* **51:** 659–88.

THOMAS, O., 1923. The Godman Exploration Fund: List of mammals from north Queensland collected by Mr. T. V. Sherrin. *Ann. Mag. Nat. Hist.* (Series 9), **11:** 174.

TROUGHTON, E., 1941. "Furred animals of Australia". Angus and Robertson: Sydney.

VAN DYCK, S., 1979. Behaviour in captive individuals of the dasyurid marsupial *Planigale maculata* Gould, 1851). *Mem. Qd Mus.* **19:** 413–29.

VAN DYCK, S., 1980. The Cinnamon Antechinus *Antechinus leo* (Marsupialia: Dasyuridae), a new species from the vine–forests of Cape York Peninsula. *Aust. Mammal.* **3:** 5–17.

WAKEFIELD, N. A. AND WARNEKE, R. M., 1967. Some revision in *Antechinus* (Marsupialia) –2. *Vic. Nat., Melb.* **84:** 69–99.

WALKER, E., 1964. "Mammals of the world: Vol. 1". Johns Hopkins University Press: Baltimore.

WALLIS, R. L., 1976. Torpor in the dasyurid marsupial *Antechinus stuartii*. *Comp. Biochem. Physiol.* **53A:** 319–22.

WEBB, L. J., 1959. A physiognomic classification of Australian rain forests. *J. Ecol.* **47:** 551–70.

WEBB, L. J., 1978. A general classification of Australian rainforest. *Australian Plants* **9:** 349–63.

WEBSTER, P. J. AND STRETEN, N. A., 1972. Aspects of the late Quaternary climate in tropical Australia. Pp.39–60 *in* "Bridge and barrier: the natural cultural history of Torres Strait" ed. by D. Walker. A.N.U.: Canberra.

WOOD, D. H., 1970. An ecological study of *Antechinus stuartii* (Marsupialia) in a south–east Queensland rain forest. *Aust. J. Zool.* **18:** 185–207.

WOOLLARD, P. 1971. Differential mortality of *Antechinus stuartii* (Macleay): nitrogen balance and somatic changes. *Aust. J. Zool.* **19:** 347–53.

WOOLLEY, P., 1966. Reproduction in *Antechinus* species and other dasyurid marsupials. Pp.281–94 *in* "Comparative biology of reproduction in mammals" ed. by I. W. Rowlands. *Symp. Zool. Soc. Lond.* **15**. 84.

ZIEGLER, A. C., 1977. Evolution of New Guinea's marsupial fauna in response to a forested environment. Ch. 7. *In* "The biology of marsupials" ed. by B. Stonehouse and D. Gilmore. Macmillan: Londdon.

RESUMEN

62. Las relaciones de *Antechinus stuartii* y *A. flavipes* (Dasyuridae, Marsupialia), con especial referencia a Queensland. Por Stephen Van Dyck.

Las relaciones entre todas las formas de *Antechinus stuartii* y *A. flavipes* de Queensland son discutidas en terminos de morfología externa, distribución, biología reproductiva, y morflogía craneana. Se describe la notoriamente más grande forma *A. flavipes rubeculus* n. subsp. del norte Queensland, y se confirma el status subespecífico de *A. stuartii adustus*. *A. stuartii stuartii* está esencialmente restringida a los bosques selváticos subtropicales del sudeste de Queensland. Las áreas ocupadas por ambas formas se caracterizan por un tipo regular de intensas lluvias invernales. *A. flavipes flavipes* y *A. f. rubeculus* ocupan una gran variedad de hábitats y pueden tolerar lluvias de invierno menos intensas y más irregulares. En Queensland *A. s. stuartii* se aparea durante los dos últimas semanas de Septiembre. *A. s. adustus* probablemente se aparea en la primera parte de Julio y *A. f. rubeculus* en Junio y Julio.

Las diferencias craneanas de *A. s. stuartii* con *A. flavipes* son: más largas vacuidades palatales anteriores y posteriores, mayor longitud de la vacuidad interpalatal, rostro y anchura bicigomática mas estrechos, diámetro interorbitario mayor, hilera de los premolares más larga, separación mayor de los foràmenes de los canales transversales, menor altura del hocico en los caninos, más grandes entocónidos y más amplios talónidos. *A. s. adustus* se diferencia de *A. s. stuartii* por los siguientes caracteres: longitud basicraneana menor, vacuidades palatales más pequeñas, más cortas hileras de los premolares, ancho anterior de las mayor, menor separación de los forámenes de los canales transversos premolares superiores prominentemente lobulados postero–lingualment. *A. s. adustus* también difiere significativamente del neotipo de *A. s. stuartii*, en Sydney.

A. stuartii representa el más primitivo antequino, mientras que *A. flavipes* el más recientemente derivado y más adapble. Se discuten las historias biológicas y sus consecuencias de ambas especies distribuídas en el Este de Australia.

63. PHALLIC MORPHOLOGY OF THE AUSTRALIAN SPECIES OF *Antechinus* (DASYURIDAE, MARSUPIALIA): A NEW TAXONOMIC TOOL?

P. A. WOOLLEY[1]

Among the twelve currently recognised species of *Antechinus* found in Australia differences between species in the morphology of the penis can be seen in 1) the form of the tip of the penis 2) the length of the urethral grooves 3) the median dorsal lobe 4) the levator muscles and their tendons and 5) the accessory structures derived from the corpora cavernosa. Two major groups can be recognised on differences in the first three features listed above. In Group 1, which contains seven species, *viz. A. bellus, A. flavipes, A. godmani, A. leo, A. stuartii, A. minimus* and *A. swainsonii* the tip of the penis is bifid and the urethral grooves are long. At the base of the cleft in the penis there is a median lobe which projects from the dorsal surface of the penis. In Group 2, which contains five species, *viz. A. rosamondae,* the Ningbing Antechinus, *A. macdonnellensis, A. bilarni* and *A. apicalis,* the tip is not bifid, the urethral grooves are short and there is no median dorsal lobe. The form of the levator muscles and their tendons is constant in Group 1 but variable in Group 2 species, which can also be further subdivided on the occurrence and form, of accessory structures. On the basis of these differences in phallic morphology it is concluded that the twelve species represent at least four, and possibly five, genera.

Pages 767–81 *in* CARNIVOROUS MARSUPIALS, ed. by M. Archer, February, 1982. Roy. Zool. Soc. New South Wales: Sydney, Australia.

AT THE PRESENT TIME fifteen species of *Antechinus* are recognised (Kirsch and Calaby 1977, Archer 1979, Van Dyck 1980). Three are found in New Guinea and twelve in Australia. The Australian species are *A. apicalis, A. bellus, A. bilarni, A. flavipes, A. godmani, A. leo, A. macdonnellensis, A. minimus, A. rosamondae, A. stuartii, A. swainsonii* and the Ningbing (= Ninbing of Archer 1979) Antechinus, an as yet undescribed species from the Kimberley region of Western Australia.

In a study of the penis of dasyurid marsupials Woolley and Webb (1977) found that two of the eight species of Australian *Antechinus* examined had a penis appendage. In so far as it could be determined from superficial examination the penis appendage in the two species (*A. apicalis* and *A. macdonnellensis*) was similar. Histological examination of the penis of one of them (*A. apicalis*) revealed that the appendage contained accessory erectile tissue derived from the corpora cavernosa.

Subsequent to the study by Woolley and Webb, the author has examined the penis of the four species of *Antechinus* not previously examined. In two of these (*A. bilarni* and the Ningbing Antechinus) an accessory structure was found. However, in these two species the structure differed in form, and in neither did it form a penis appendage of the type previously described by Woolley and Webb (1977). The observation of these and other differences in gross morphology led to a detailed study of the external and internal anatomy of the penis of the twelve species of *Antechinus* found in Australia.

The basic structure of the penis of dasyurid marsupials has been described in Woolley and Webb (1977) and the observations made in that paper on the gross morphology and internal anatomy of two species of *Antechinus* (*A. stuartii* and *A. apicalis*) have been extended in the present study.

MATERIALS AND METHODS

The distribution of the twelve species of *Antechinus* found in Australia is shown in Fig. 1. Live animals were collected from the field and maintained in the laboratory for various periods. Only males known to be reproductively mature were used for the study. They were killed and the penis removed from the body and straightened while fresh. Some penis specimens were pre-

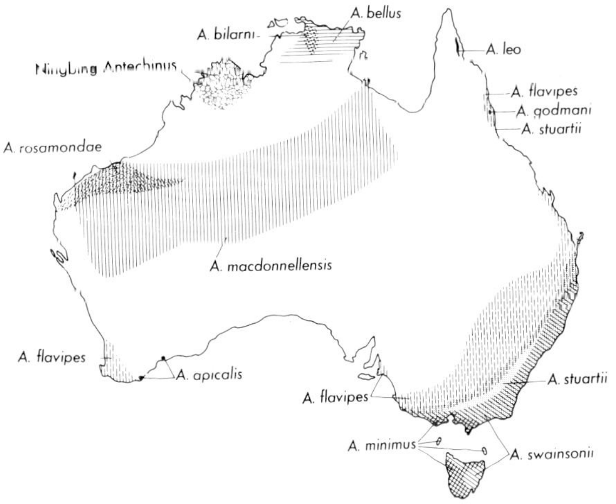

Fig. 1. Distribution of the twelve species of *Antechinus* found in Australia.

[1] Department of Zoology, La Trobe University, Bundoora, Victoria, Australia, 3083.

Table 1. The number of penis specimens dissected, and the number sectioned for each of the twelve species of *Antechinus*. Key to the Museums: QM = Queensland Museum, AM = Australian Museum, WAM = Western Australian Museum.

Species	Representative specimen	Number of penis specimens dissected	sectioned
A. apicalis	M15472 (WAM)	4	4
A. bellus	JM2028 (QM)	3	1
A. bilarni	JM2316 (QM)	7	2
A. flavipes	–	3	1
A. godmani	M11256 (AM)	2	1
A. leo	M11246 (AM)	2	1
A. macdonnellensis	JM1188 (QM)	6	3
A. minimus	M11257 (AM)	3	1
A. rosamondae	JM1442 (QM)	6	2
A. stuartii	–	8	4
A. swainsonii	M11250 (AM)	4	1
Ningbing Antechinus	JM2314 (QM)	4	3

served in 70% alcohol; others were fixed in aqueous Bouin's solution, embedded in paraffin wax and serial transverse sections cut at 10μm. All sections were mounted and stained with either haematoxlylin and eosin or Masson's trichrome stain. For each species at least one of the straightened penis specimens was photographed, in most cases while fresh. Table 1 shows the number of penis specimens dissected out and the number sectioned for each of the twelve species of *Antechinus*. Many of the donor animals have been lodged in Museums and the acquisition numbers of representative specimens are listed.

The external anatomy of the penis was studied in freshly dissected specimens, preserved specimens and photographic prints of the dorsal and ventral aspects of the penis. To facilitate comparison between species of the relative sizes of various parts of the penis the length of the penis from the union of the crura to the tip was standardised in the prints.

To assist in the analysis of the internal anatomy of the penis from the serial sections a Wild Drawing Microscope was used to project the sections to make outline drawings at 12.5x magnification. Drawings of sections approximately 500μm apart were made for each series and additional sections were drawn where necessary. These drawings were used to prepare plan diagrams of the form of the corpora cavernosa in the distal half of the penis.

EXTERNAL ANATOMY

The external anatomy of the penis of each of the twelve species is illustrated in Figure 2 by means of outline drawings prepared from the photographs of the dissected specimens. The ventral aspect of the entire penis and associated musculature, and the dorsal aspect of the distal portion is shown. The retractor muscles, which normally lie dorsal to the penis, have been shortened and drawn to the side where they insert onto the penis. The appendage on the penis of *A. macdonnellensis* and *A. apicalis* has been drawn to the side to show the underlying structures. The penis of *A. bilarni,* one of the species with an accessory structure which does not form an appendage, is also illustrated by means of photographs of the ventral and dorsal aspects of a freshly dissected specimen (Fig. 3). Photographs of the penis of *A. stuartii* and *A. apicalis* (a species with an appendage) appear in Woolley and Webb (1977).

Differences between species can be seen in Fig. 2 in: 1) the form of the tip of the penis 2) the length of the urethral grooves 3) the median dorsal lobe 4) the levator muscles and their tendons and 5) the accessory structures derived from the corpora cavernosa.

Two major groups can be recognised on differences in the first three features listed above. In Group 1, which contains seven species, *viz. A. bellus, A. flavipes, A. godmani, A. leo, A. stuartii, A. minimus* and *A. swainsonii* the tip of the penis is bifid and the urethral grooves are long. At the base of the cleft in the penis there is a median lobe which projects from the dorsal surface of the penis. The length of the lobe relative to the length of the bifid portion of the penis is much less in *A. leo* than in the other species. In Group 2, which contains five species, *viz. A. rosamondae,* the Ningbing Antechinus, *A. macdonnellensis, A. bilarni* and *A. apicalis,* the tip is not bifid, the urethral grooves are short and there is no median dorsal lobe.

Between these two groups differences can also be seen in the form of the levator muscles and their tendons. In all species in Group 1 the paired levator muscles are broad, relative to the width of the shaft of the penis, and long. The muscles do not fuse and the tendons, which do fuse, arise mesially. In Group 2 the form of the

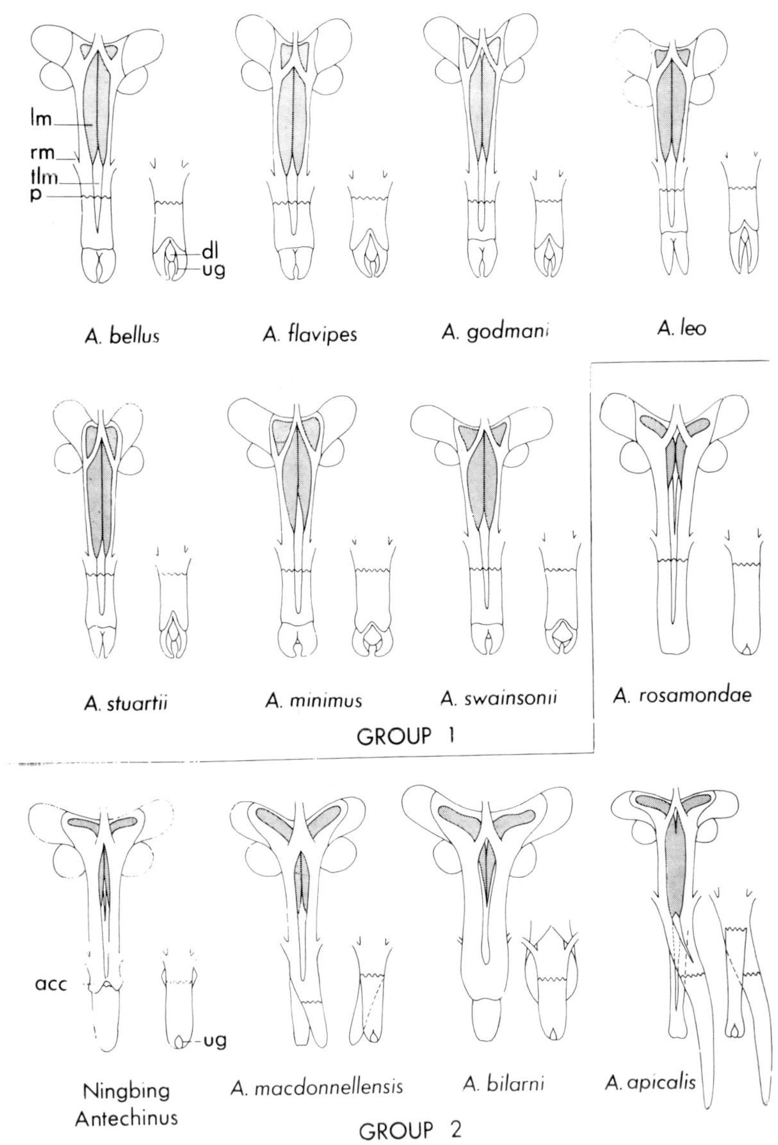

Fig. 2. External anatomy of the penis of the twelve species of *Antechinus* found in Australia. lm = levator muscle; tlm = tendon (s) of levator muscles; rm = insertion of retractor muscle; p = level of attachment of preputial skin; dl = dorsal lobe; ug = urethral groove, acc = accessory structure.

levator muscles and the origin of the tendons are variable. In four species (*A. rosamondae*, the Ningbing Antechinus, *A. macdonnellensis* and *A. bilarni*) the levator muscles do not fuse but they are relatively narrow and short. The tendons arise mesially in the first three species and laterally in the fourth. In the fifth species (*A. apicalis*) in this Group the levators fuse to form a single broad, long muscle mass from which two tendons arise mesially.

Within Group 1 two subgroups can be recognised on differences in the form of the median dorsal lobe. The lobe is narrow in five species *A. bellus, A. flavipes, A. godmani, A. leo* and *A. stuartii*) and broad in the other two (*A. minimus* and *A. swainsonii*). The terminal portions of the apices of the penis in species with a narrow lobe fold back onto the dorsal surface when the penis is retracted within the preputial sac, whereas in species with a broad lobe they curve towards the midline.

Within Group 2 one species (*A. rosamondae*) has no accessory structure; in the other four species an accessory structure is present but the form is different in each (see below). In two species (*A. macdonnellensis* and *A. apicalis*) it forms an appendage enclosed in a pocket of preputial skin.

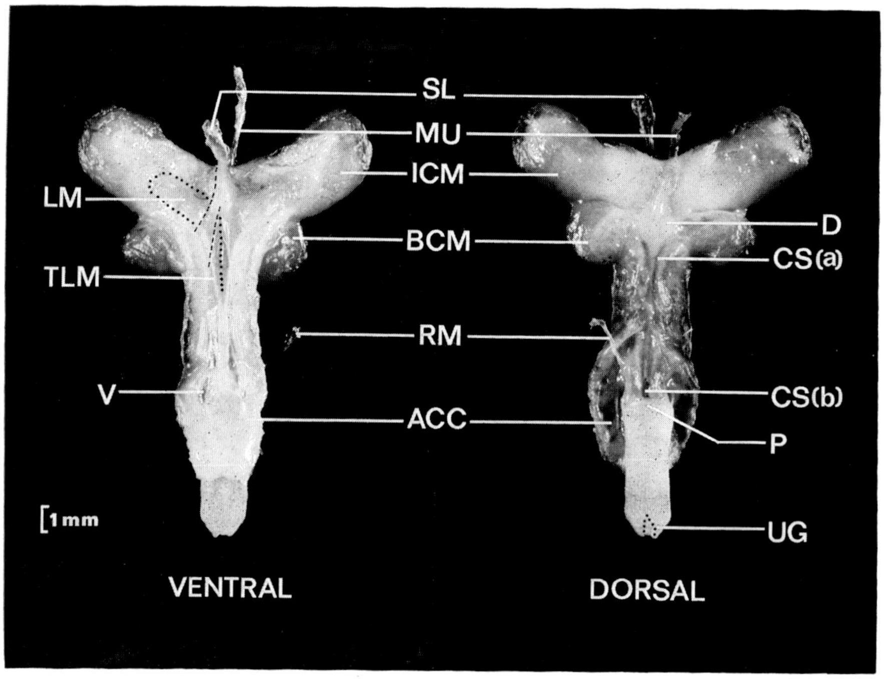

Fig. 3. Penis of *A. bilarni,* removed from body and straightened. SL = suspensory ligament; MU = membranous urethra; ICM = ischio–cavernosus muscle; BCM = bulbo–cavernosus muscle; RM = retractor muscle (removed on right); ACC = accessory corpora cavernosa; LM = levator muscle; TLM = tendon of levator muscle; V = vein draining corpus spongiosum; D = region of entry of ducts of bulbo–urethral glands into urethra; CS(a) = corpus spongiosum before dorsal union; CS (b) = corpus spongiosum after dorsal union; P = level of attachment of preputial skin; UG = urethral groove.

INTERNAL ANATOMY

The internal anatomy of the penis is illustrated for several species by means of a series of drawings of transverse sections at the levels shown in the accompanying outline drawing of the penis. The key to the structures illustrated is given in the first of the series of figures. The ventral surface of each section is shown uppermost because a ventral approach to the penis is made in dissection.

Group 1 Species

Figure 4 shows the structure of the penis of *A. bellus,* one of the species in Group 1 (see above). The crura of the corpora cavernosa (section i) unite medially to form a single mass (section ii) which extends distally on the ventro–lateral sides of the urethra to near the point of bifurcation of the penis, where it also divides and extends into each half of the tip (sections viii and ix). The cavernous tissue is surrounded by a thick tunica albuginea. At the base of the cleft in the penis the urethra opens dorsally into two urethral grooves which continue along each half of the tip of the penis (sections vii and viii). The bulbs of the corpus spongiosum (section ii) unite medially distal to the point of entry of the ducts of the bulbo-urethral glands into the urethra. They unite ventrally (see section iii) and then dorsally (section v) to form a thin layer of spongy tissue around the urethra. In the terminal free portion of the penis (distal to the level of attachment of the preputial skin) the corpus spongiosum extends around the sides of the corpus cavernosum to form the glans penis (sections vi–ix). The corpus spongiosum also extends into the median lobe (section viii) which forms at the base of the cleft in the penis. The corpus spongiosum is drained by two veins (section vi) which lie ventral to the corpus cavernosum. The veins unite anteriorally (section iii) to form the penis vein which passes through the penis vein muscle (section i). This muscle inserts onto the tunica albuginea of the crura. The ischio–cavernosus muscles, partially enclosing the crura of the penis and the levator penis muscles, which also insert onto the crura, can be seen in section i. The levator muscles converge (section ii) but do not fuse. They are broad, almost covering the ventral surface of the corpus cavernosum (sections iii and iv), and long, terminating at the level of insertion of the retractor penis muscles (section v). The tendons of the levator muscles arise mesially (section iii), shortly fuse (section iv) and insert distally onto the tunica albuginea of the corpus cavernosum (sections vi and vii). The bulbo–cavernosus muscles encase the bulbs of the penis (section ii).

Of the other six species in Group 1 the internal anatomy of the penis was found to be basically the same as that of *A. bellus*. The differences seen in the tip of the penis in species with a narrow median lobe and those with broad median lobe (see above) are illustrated with reference to *A. flavipes* (narrow lobe) in Fig. 5 and to *A. swainsonii* (broad lobe) in Fig. 6. The lobe commences to form between the urethral grooves on the dorsal surface of the penis and when fully formed it lies between the two halves of the tip. It extends to the ventral surface in *A. swainsonii* (sections v and vi) but not in *A. flavipes* (sections v and vi). At the widest point (section iv, Fig. 5 and 6) the lobe is much broader in *A. swainsonii* than in *A. flavipes*, and in *A. swainsonii* overlies the urethral grooves. In *A. flavipes* the urethral grooves lie on the dorsal surface of the two halves of the tip (sections ii–viii); in *A. swainsonii* they lie in a dorsal position proximally (sections ii–v) and on the inner surface of the apices distally (sections vi–viii). This apparent change in position of the urethral grooves may result from the curving of the apices towards the midline in species with a broad median lobe.

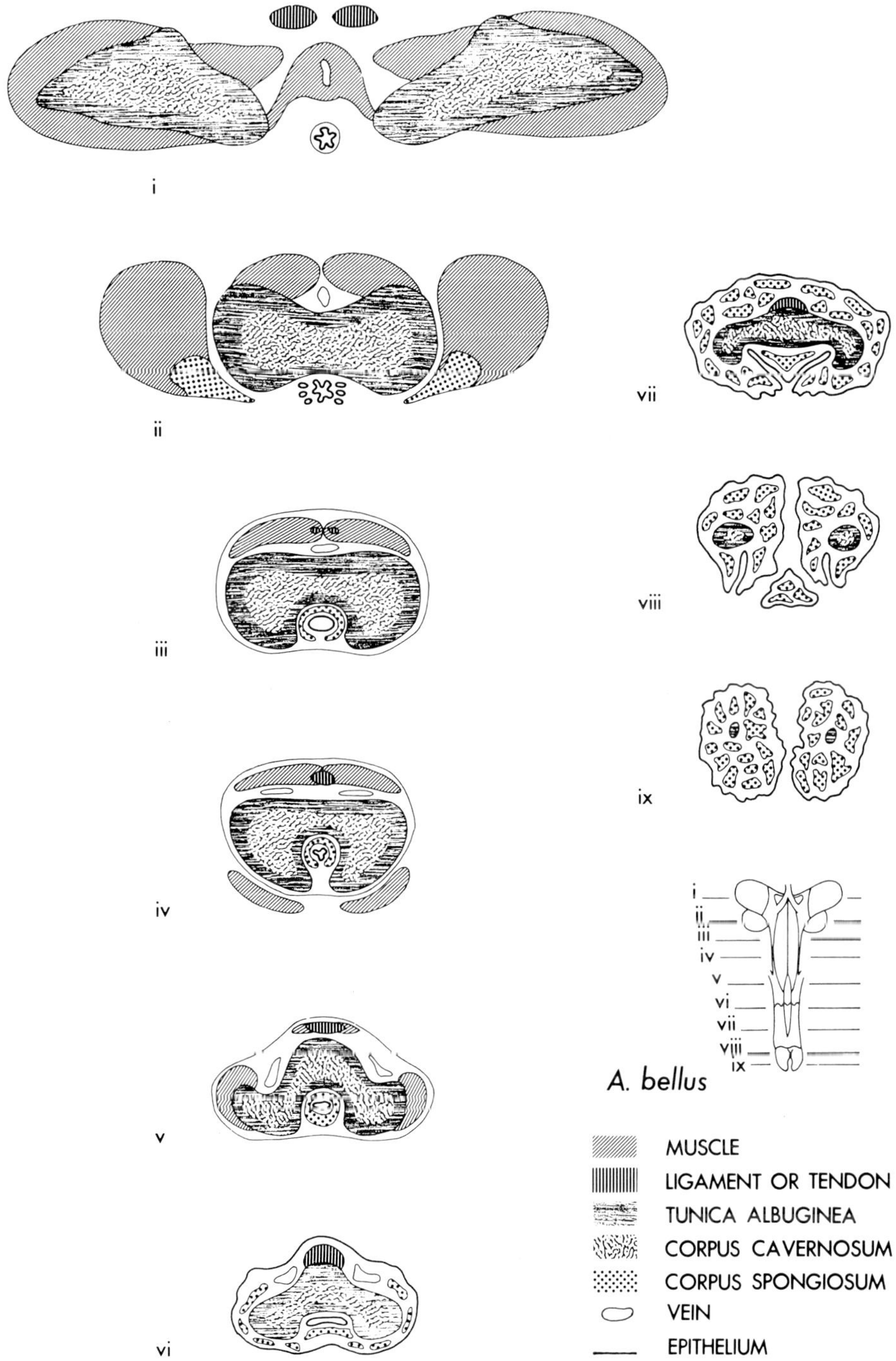

Fig. 4. Internal anatomy of the penis of *A. bellus*.

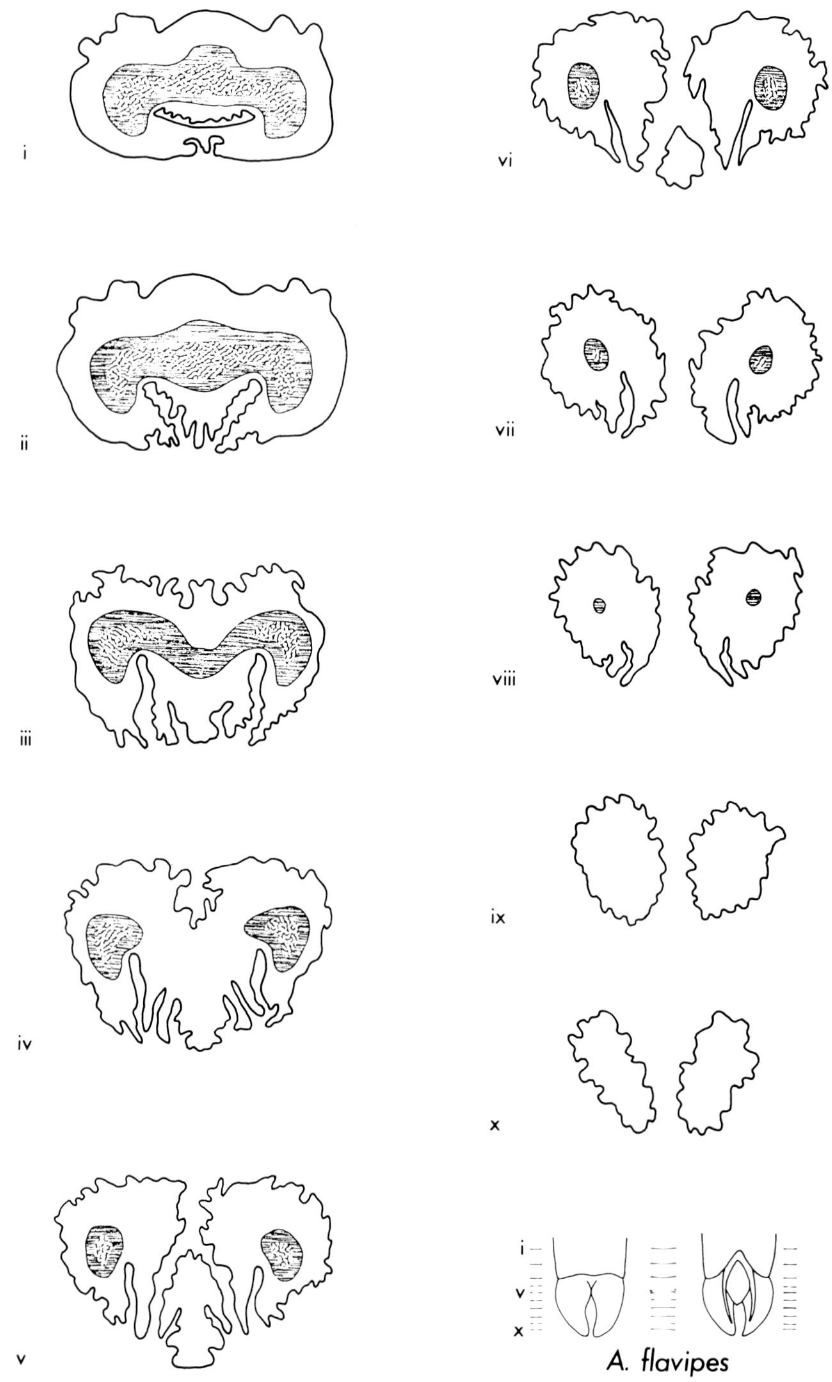

Fig. 5. Internal anatomy of the tip of the penis of *A. flavipes*. Corpus spongiosum not shown.

Group 2 Species

The structure of the penis in each of the species in group 2 is illustrated in Figs. 7, 8, 9, 10 and 11. The penis of *A. rosamondae* (Fig. 7) differs from that of *A. bellus* in the following features. The corpus cavernosum bifurcates only in the extreme tip of the penis (distal to section viii), which is not bifid. The urethral grooves are short. The levator muscles are narrow (section ii) and they lie in a deep depression in the ventral surface of the cavernosum (sections ii and iii). They are short, terminating (between sections iii and iv) far anterior to the insertion of the retractor muscles (section v). The tendons of the levator muscles, which arise mesially (section iii), remain separate for some distance before fusing (section v).

Accessory corpora cavernosa are not found in *A. rosamondae,* but are present in each of the other four species in group 2. In these four the anatomy of the penis in the region of the crura and bulbs is essentially the same as in *A. rosamondae* and has not been illustrated.

Figure 8 shows the internal anatomy of the penis of the Ningbing Antechinus. The corpus cavernosum forms dorsal and ventral branches (sections iv and v). The dorsal branch extends to the tip of the penis (sections v–x) and is bifid in the region of the short urethral grooves (section xi). The ventral branch subdivides (section vi) to form two accessory corpora cavernosa (section vii). The levator muscles are narrow and short, and they lie in a deep depression on the ventral surface of the corpus cavernosum (section i). The tendons of the levator muscles arise mesially (section i) and fuse between sections i and ii). The tendon passes between the accessory corpora cavernosa to insert onto the tunica albuginea of the dorsal branch of the corpus cavernosum (sections vi and vii). The pair of veins draining the corpus spongiosum pass lateral to the ventral branch of the corpus caver-

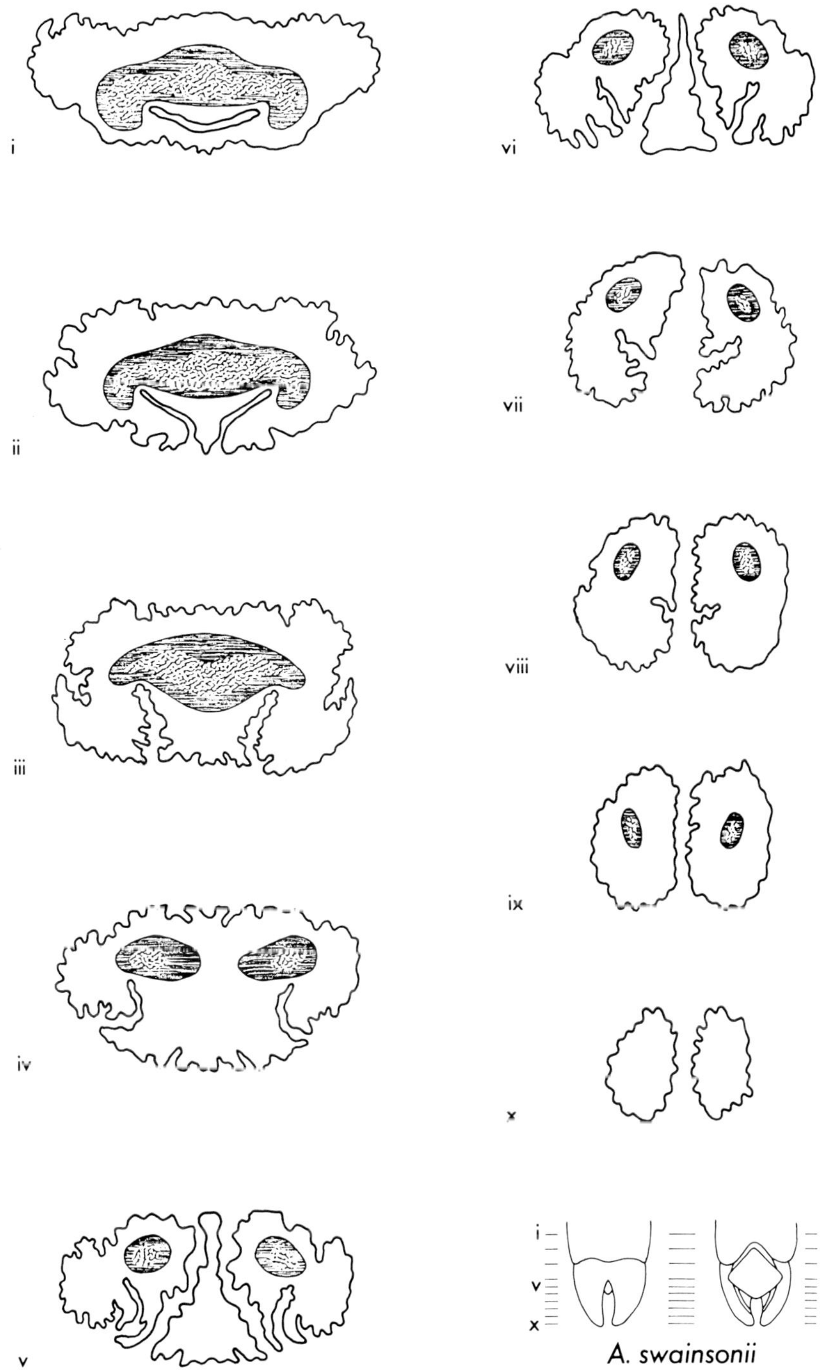

Fig. 6. Internal anatomy of the tip of the penis of *A. swainsonii*. Corpus spongiosum not shown.

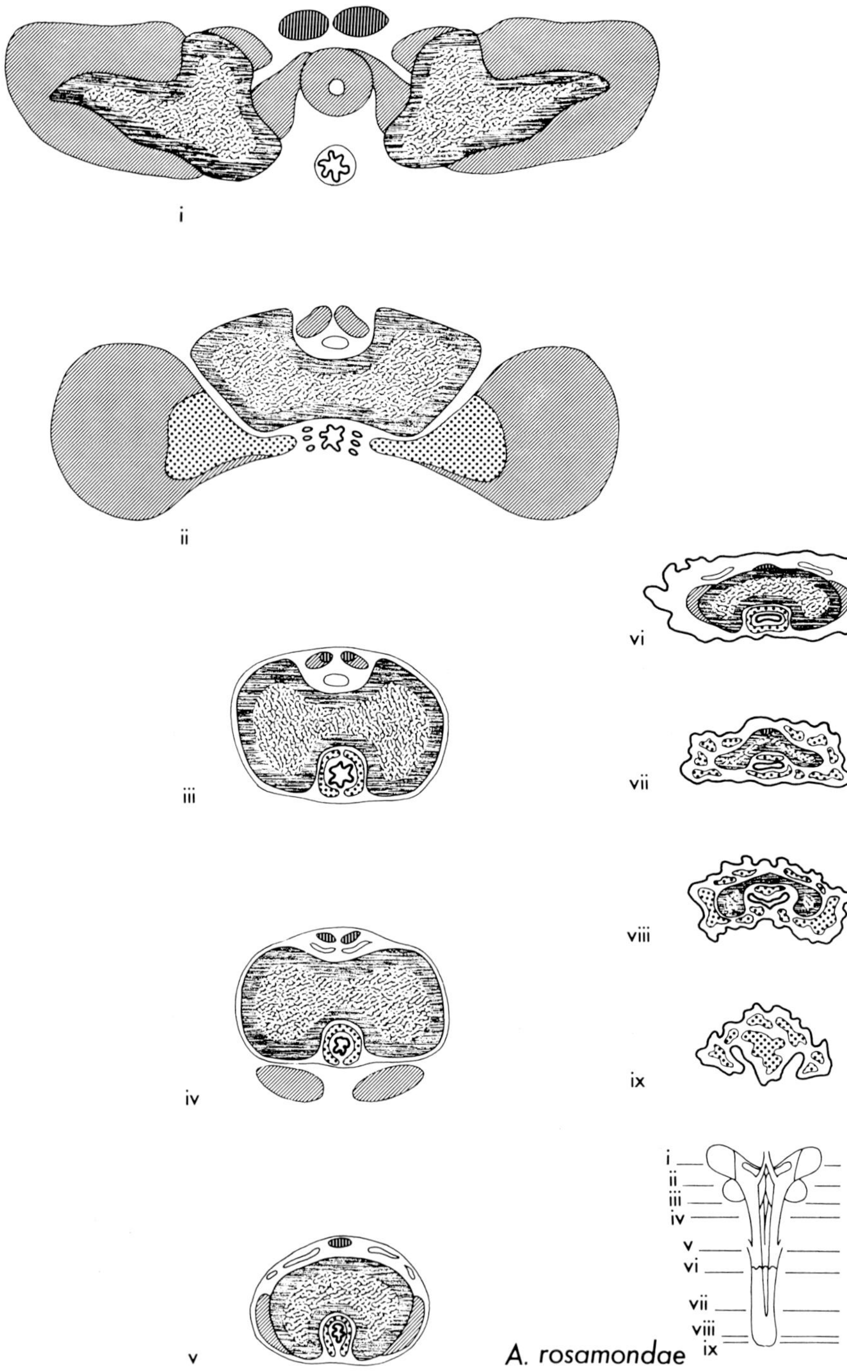

Fig. 7. Internal anatomy of the penis of *A. rosamondae.*

nosum (section v) and come to lie on the ventral surface (section iv).

The structure of the penis of *A. macdonnellensis* (Fig. 9) is similar to that of the Ningbing Antechinus in the following features. The corpus cavernosum divides to form dorsal and ventral branches (sections iv and v); the dorsal branch extends to the tip of the penis where it bifurcates (sections x and xi) and the ventral branch subdivides to form two accessory corpora cavernosa (section vii). The levator muscles are narrow, short, unfused and lie in a deep depression on the ventral surface of the corpus cavernosum; their tendons arise mesially (section i). The veins draining the corpus spongiosum pass lateral to the ventral branch of the corpus cavernosum (section v). It differs from the Ningbing Antechinus in that the accessory corpora cavernosa extend further distally to overlie the terminal free portion of the penis, and they lie in a pocket of preputial skin (sections ix–xi) thus forming an appendage to the penis. Also, the tendons of the levator muscles unite (section ii) soon after their origin and insert distally onto the tunica albuginea of the ventral branch of the corpus cavernosum before it subdivides (section vi).

Figure 10 shows the internal anatomy of the penis of *A. bilarni*. The corpus cavernosum forms dorsal and ventral branches (sections iii and iv). The dorsal branch extends to the tip of the penis, where it bifurcates. The ventral branch is much broader than the dorsal branch (section iv) and it can be seen in Fig. 3 that it wraps around the dorsal branch proximally. It subdivides to form three accessory corpora cavernosa (section vi). The two lateral accessory corpora cavernosa rejoin the median one (section vii). The accessory corpora cavernosa overlie part of the terminal free portion of the penis (section vi–viii and Fig. 3) but they do not lie in a pocket of preputial skin and do not form an appendage. The levator muscles are narrow and short, and lie in a depression on the ventral surface of the corpus cavernosum (section i). The tendons of the levator muscles arise laterally (section i and Fig. 3), fuse (section ii), then pass through the median accessory corpus cavernosum and insert onto the tunica albuginea of the dorsal branch of the corpus cavernosum (section v). The veins draining the corpus spongiosum pass between the lateral and median accessory corpora cavernosa (section v and Fig. 3).

In *A. apicalis* (Fig. 11) the corpus cavernosum divides to form dorsal and ventral branches (sections ii–iv) and the ventral branch subdivides to form three accessory corpora cavernosa (section v), as in *A. bilarni*. However, the lateral accessory corpora cavernosa (section v–ix) do not rejoin the median one, which extends beyond the tip of the penis (section xii). The accessory corpora cavernosa lie in a pocket of preputial skin and so form an appendage to the penis. The levator muscles unite to form a single broad muscle mass (section i) which does not cover the ventral surface of the corpus cavernosum as in Group 1 species, but which is much broader than in other Group 2 species. The levator muscles are also relatively longer than in other Group 2 species, and they terminate (between section vi and vii), distal to the insertion of the retractor muscles (sections iv and v). The retractor muscles are inserted onto the tunica albuginea of both the dorsal branch of the corpus cavernosum and the lateral accessory corpora cavernosa. Two tendons, one dorsal and one ventral arise mesially from the fused levator muscles (section vi). The dorsal tendon passes through the median accessory corpus cavernosum and inserts onto the tunica albuginea of the dorsal branch of the corpus cavernosum (sections v–x). The ventral tendon inserts onto the tunica albuginea of the median accessory corpus cavernosum. It is shorter than the dorsal tendon, terminating between sections viii and ix. The veins draining the corpus spongiosum pass between the lateral and median accessory corpora cavernosa (sections v and vi), as in *A. bilarni*, and they unite to form the penis vein proximal to section i.

THE CORPORA CAVERNOSA

To further illustrate the differences seen above in the form of the corpora cavernosa in the distal half of the penis they are drawn in plan view in Fig. 12. For comparison between species of the relative proportions and arrangement of the cavernous bodies the length of the terminal free portion of the penis (indicated by the level of attachment of the preputial skin) has been standardised by either photographic enlargement or reduction of the original plan diagrams prepared from the serial transverse sections.

In all species, the corpus cavernosum of the penis is bifid at the tip. The extent of the bifurcation is greater in Group 1 species, represented

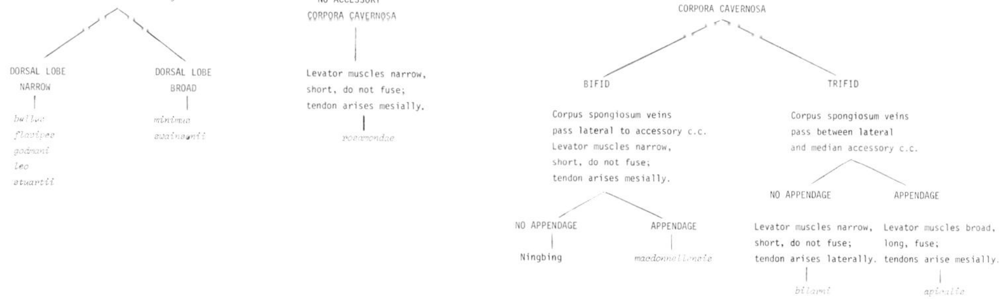

Table 2. Species groupings based on phallic morphology.

by *A. bellus,* in which the tip of the penis is also bifid and the urethral grooves are long, than in Group 2 species in which the tip of the penis is not bifid and the urethral grooves are relatively short.

In the four species of Group 2 in which the corpus cavernosum divides to form dorsal and ventral branches, the division occurs proximal to the level of attachment of the preputial skin.

The ventral branch in the Ningbing Antechinus and *A. macdonnellensis* is only a little broader than the dorsal branch and in both the ventral branch subdivides to form a bifid structure. In both species the veins draining the corpus spongiosum pass the lateral to these accessory corpora cavernosa to reach the ventral surface of the penis. Differences between the two species are seen in: 1) the level at which the subdivision of the ventral branch occurs (proximal to the level of attachment of the preputial skin to the corpus cavernosum of the penis in the Ningbing Antechinus and distal in *A. macdonnellensis*); 2) the site of insertion of the tendon of the levator muscle (onto the corpus cavernosum of the penis in the

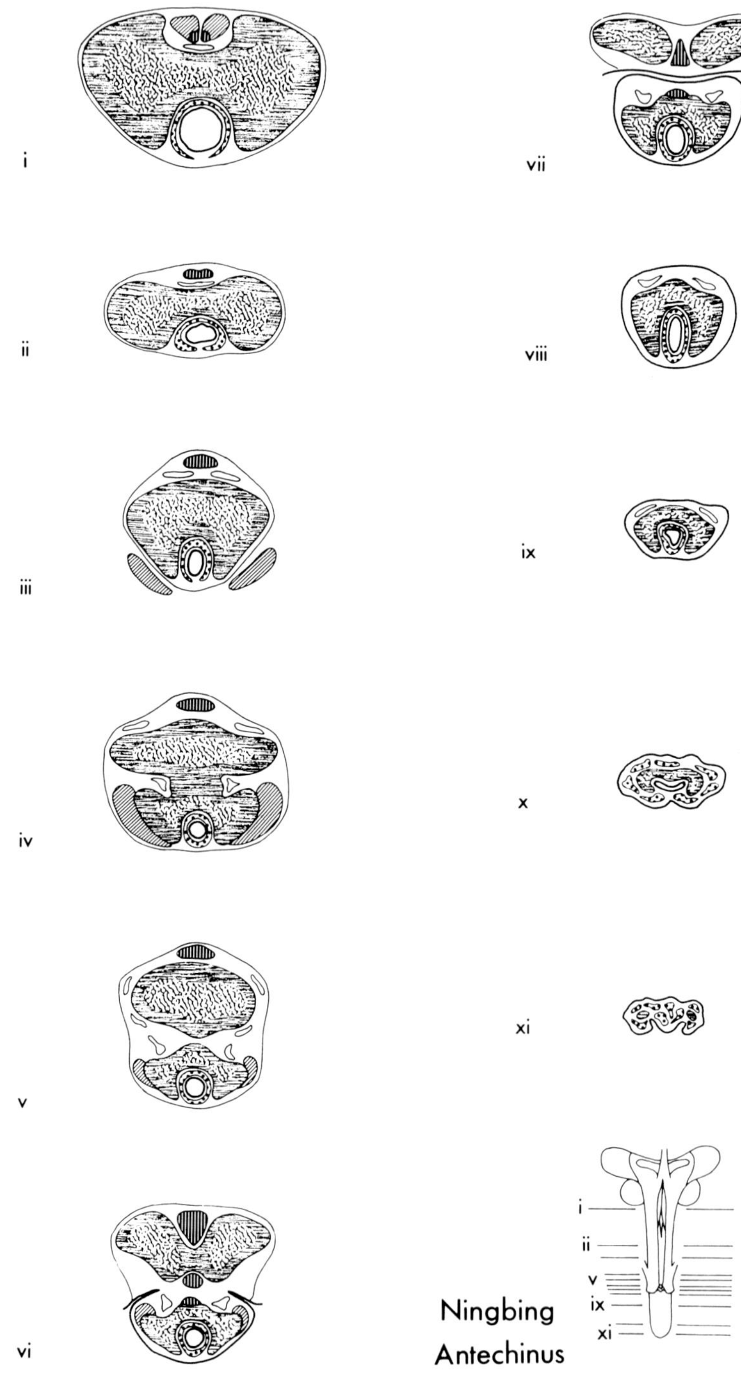

Fig. 8. Internal anatomy of the penis of the Ningbing Antechinus.

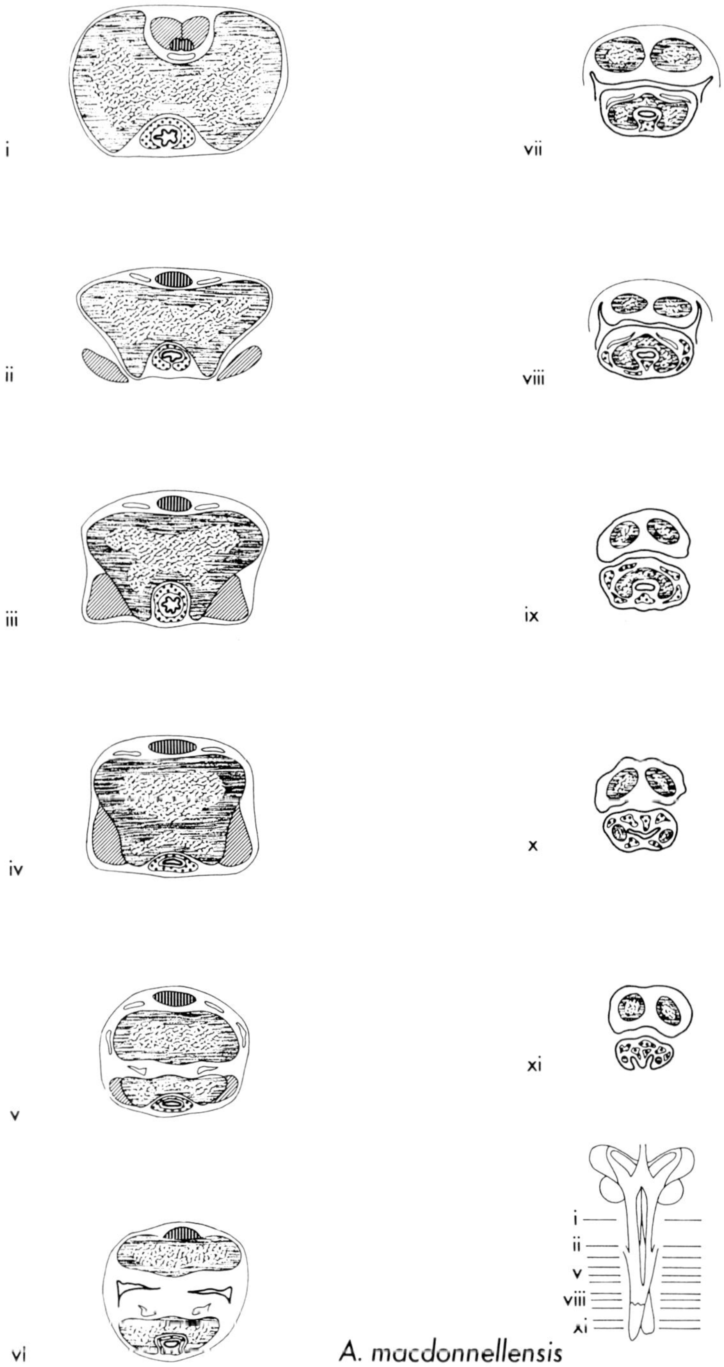

Fig. 9. Internal anatomy of the penis of *A. macdonnellensis.*

Ningbing Antechinus and onto the ventral branch of the corpus cavernosum in *A. macdonnellensis*); and 3) in the length of the accessory corpora cavernosa). They are much longer in *A. macdonnellensis* and form an appendage to the penis.

In both *A. bilarni* and *A. apicalis* the ventral branch of the corpus cavernosum is much broader than the dorsal and it subdivides to form a trifid structure. In both species the subdivision occurs proximal to the level of attachment of the preputial skin on the corpus cavernosum of the penis, the veins draining the corpus spongiosum pass between the lateral and median accessory corpora cavernosa to reach the ventral surface of the penis and the tendon of the levator muscle passes through the median accessory corpus cavernosum to insert onto the corpus cavernosum of the penis. Differences between the two species are seen in the accessory corpora cavernosa. In *A. bilarni* the lateral accessory corpora cavernosa rejoin the median one distally; in *A. apicalis* they remain separate. In addition, in *A. apicalis* the median accessory corpus cavernosum is much longer than in *A. bilarni* and the accessory structures form an appendage to the penis. *A. apicalis* also differs from *A. bilarni* in having a second tendon of the levator muscle which inserts onto the median accessory corpus cavernosum.

PHALLIC MORPHOLOGY AND SPECIES GROUPINGS

The grouping of species suggested by phallic morphology is summarised in Table 2. It is apparent, on the basis of phallic morphology, that the genus *Antechinus* as currently recognised is not a homogeneous group. Two major groups can be distinguished. Group 1 contains seven species which show considerable uniformity in the structure of the penis and on this basis appear to be closely related. They also have in common an unusual life history pattern (Lee *et al.* 1982) which is, however, exhibited by one of the species (*A. rosamondae*) in the second group. Group 2 contains five species, among which considerable variation is seen in the structure of the penis. With, perhaps, the exception

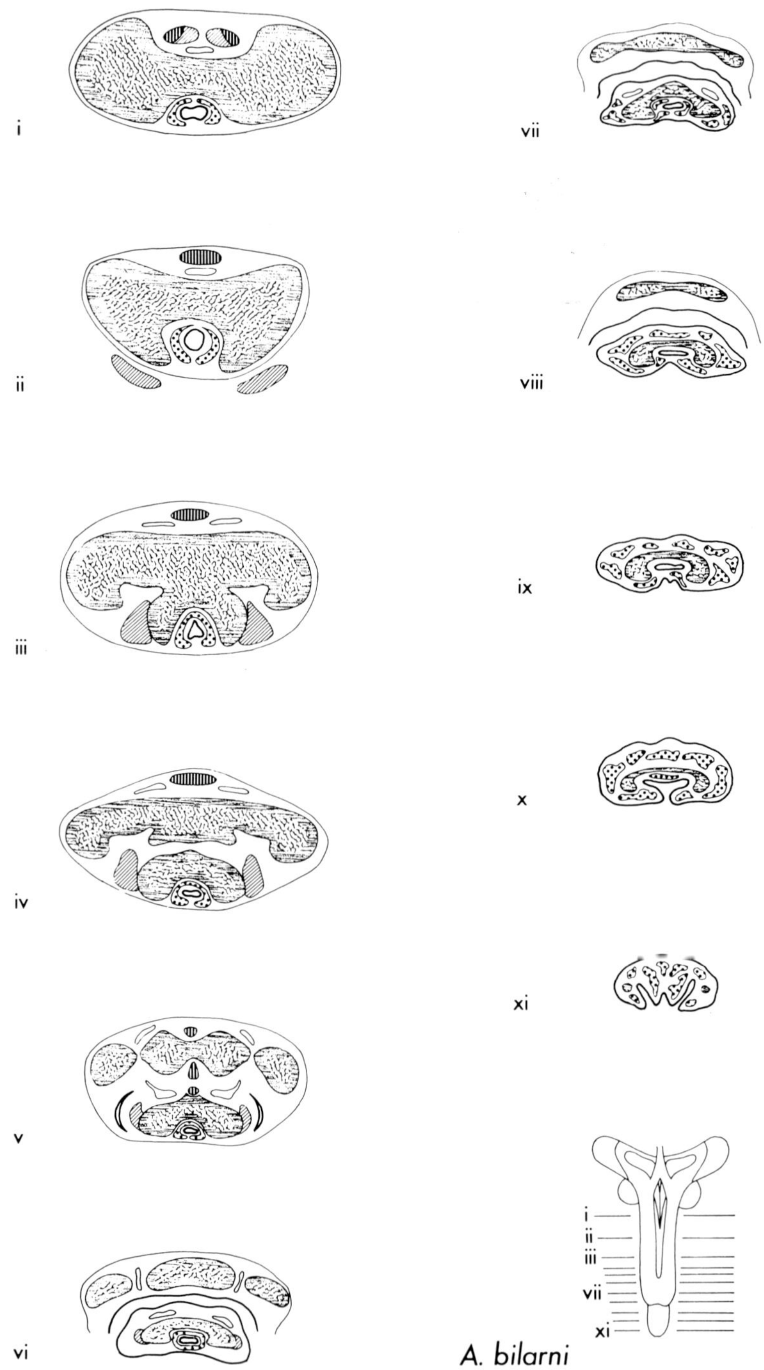

Fig. 10. Internal anatomy of the penis of *A. bilarni.*

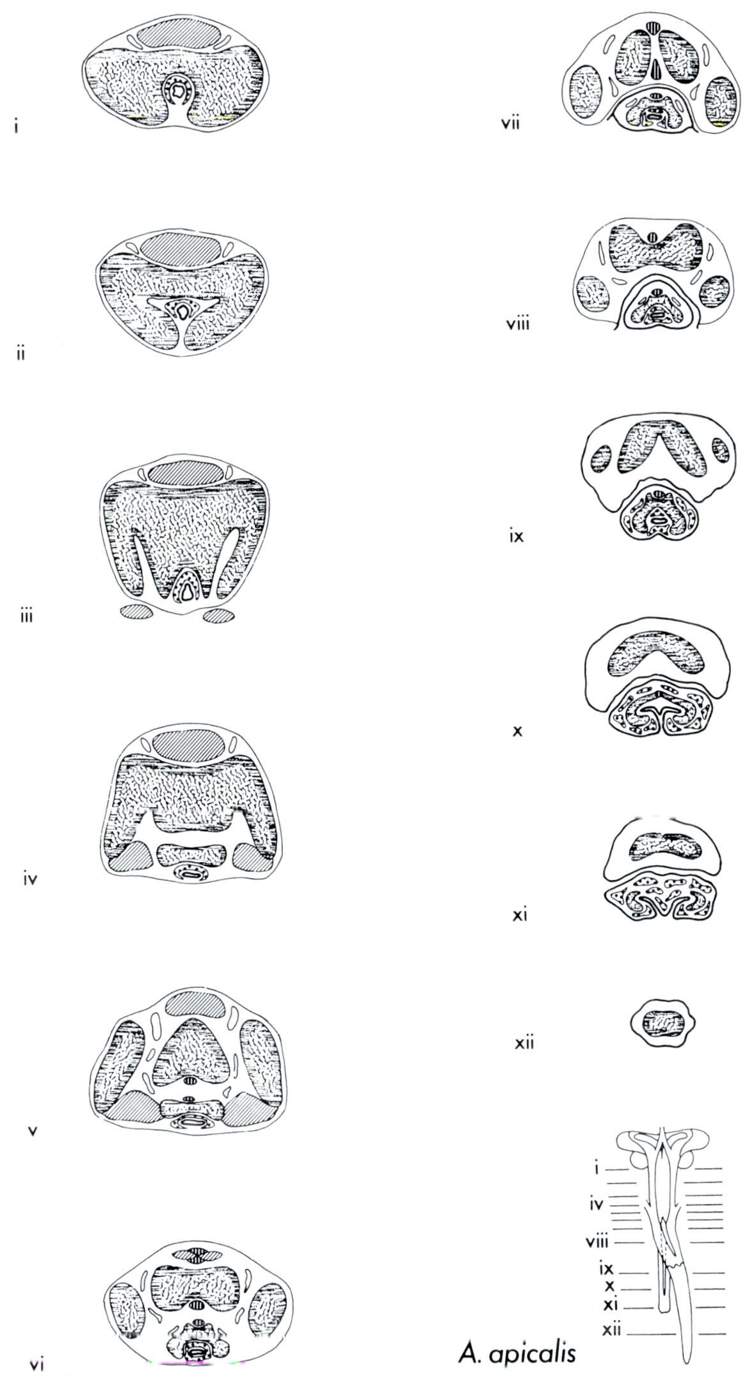

Fig. 11. Internal anatomy of the penis of *A. apicalis.*

of the Ningbing Antechinus and *A. macdonnellensis,* the species in Group 2 do not appear to be closely related on the basis of phallic morphology. One, *A. rosamondae,* lacks the accessory corpora cavernosa found in the other four, but otherwise shows greatest similarity to the two species with bifid accessory corpora cavernosa. Greater differences are seen between the two species with trifid accessory corpora cavernosa (*A. bilarni* and *A. apicalis*) than between the two with bifid accessory corpora cavernosa (the Ningbing Antechinus and *A. macdonnellensis*). The apparent similarity between the two species (*A. macdonnellensis* and *A. apicalis*) in which the accessory structures form an appendage to the penis is superficial.

It is of interest that Tate (1947) concluded, from a study of classic taxonomic characters, that the two species with a penis appendage were distinct from other species of *Antechinus* recognised at the time and he gave the two separate generic status (*A. macdonnellensis*= *Pseudantechinus macdonnellensis* and *A. apicalis*=

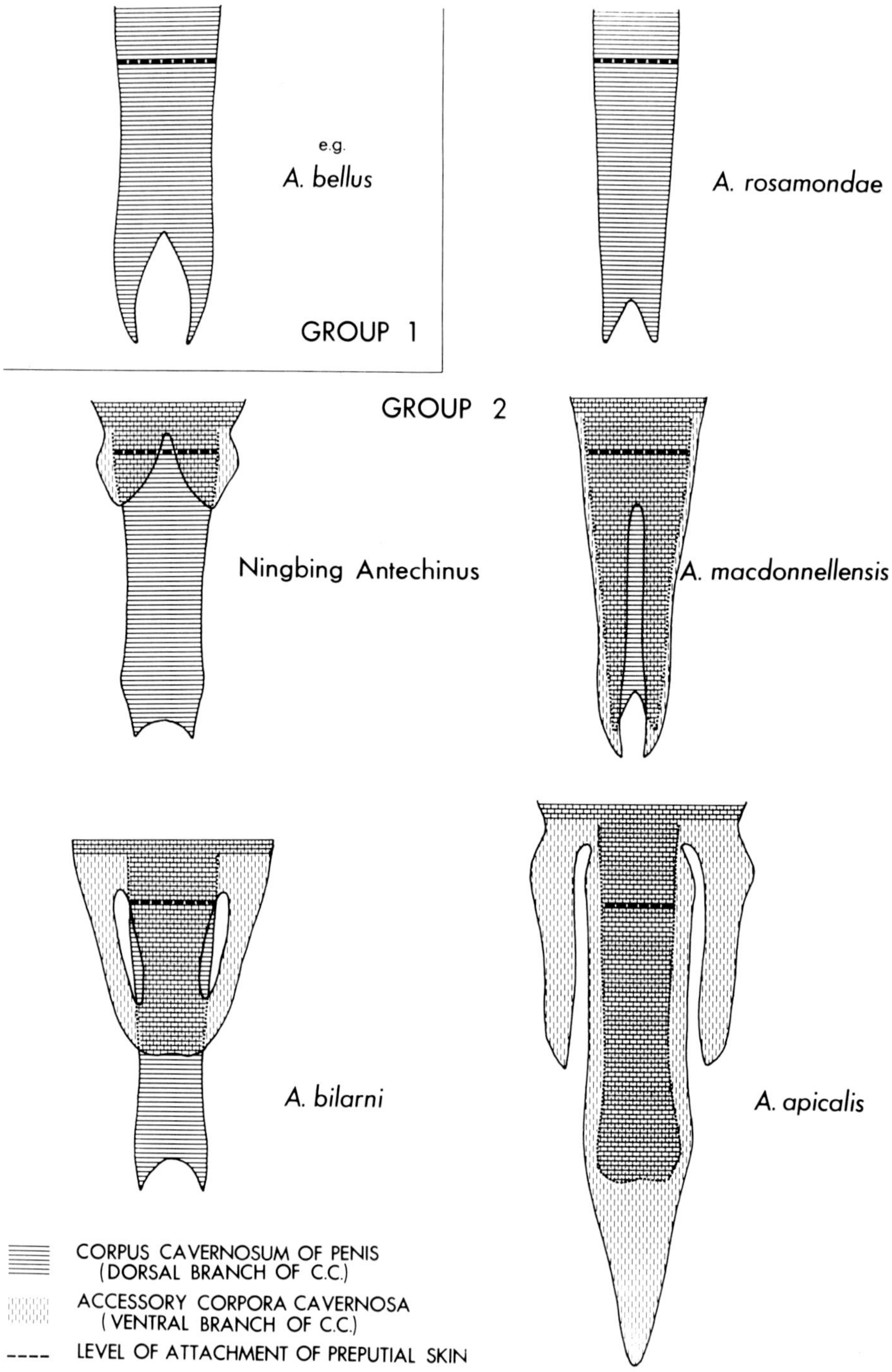

Fig. 12. Ventral plan view of the corpora cavernosa in the distal half of the penis of *A. bellus* (Group 1) and *A. rosamondae*, the Ningbing Antechinus, *A. macdonnellensis*, *A. bilarni* and *A. apicalis* (Group 2).

Parantechinus apicalis). In addition, Kirsch (1977) concluded from a serological study that the four species of *Antechinus* (*stuartii, flavipes, rosamondae* and *macdonnellensis*) studied by him possibly represented more than one genus. Baverstock *et al.* (1982) also, on the basis of serological studies, consider the genus *Antechinus* to be polyphyletic. These observations lend support to the conclusion, which can be drawn from the differences seen in phallic morphology, that the group represents at least four, and possibly five genera. The species in each genus are 1) *bellus, flavipes, godmani, leo, stuartii, minimus,* and *swainsonii* (*Antechinus?*), 2) *rosamondae,* 3) Ningbing and *macdonnellensis* (*Pseudantechinus?*) and 4) *bilarni* and *apicalis* (*Parantechinus?*). The differences seen between *bilarni* and *apicalis* may, however, be sufficient to warrant separate generic status for these two species. The overall relationships of the species of dasyurids are discussed by Archer (1982) where he allocates these species to four genera.

ACKNOWLEDGEMENTS

Financial assistance for this project was received from the Australian Research Grants Committee (Grant number D1 68/16887 to P. A. Woolley and M. Archer). The assistance given by Danny Greenstreet and Isabel Douglas in the analysis of sections is gratefully acknowledged. An earlier draft of this paper was read and constructively criticised by Drs J. H. Calaby and H. R. Harding.

REFERENCES

ARCHER, M., 1979. The status of Australian dasyurids, thylacinids and myrmecobiids. *In* "The status of endangered Australasian wildlife", ed. by M. J. Tyler. Royal Zoological Society of South Australia: Adelaide.

ARCHER, M., 1982. Review of the Australian dasyurid (Dasyuridae, Marsupialia) fossil record, integration of data bearing on phylogenetic interpretation, and suprageneric classification. Pp.397–443 *in* "Carnivorous marsupials", ed. by M. Archer. Roy. Zool. Soc. New South Wales: Sydney.

BAVERSTOCK, P. R., ARCHER, M., ADAMS, M. AND RICHARDSON, B. J., 1982. Genetic relationships among 32 species of Australian dasyurid marsupials. Pp.642–50 *in* "Carnivorous marsupials", ed. by M Archer. Roy. Zool. Soc. New South Wales: Sydney.

KIRSCH, J. A. W., 1977. The comparative serology of Marsupialia, and a classification of marsupials. *Aust. J. Zool., Suppl. Ser.,* **52:** 1–152.

KIRSCH, J. A. W. AND CALABY, J. H., 1977. The species of living marsupials: an annotated list. *In* "The biology of marsupials" ed. by B. Stonehouse and D. Gilmore. The Macmillan Press Ltd.: London and Basingstoke.

LEE, A. K., WOOLLEY, P. A. AND BRAITHWAITE, R. W., 1982. Life history strategies of dasyurid marsupials. Pp.1–11 *in* "Carnivorous marsupials", ed. by M. Archer. Roy. Zool. Soc. New South Wales: Sydney.

TATE, G. H. H., 1947. Results of the Archbold Expeditions. No. 56. On the anatomy and classification of the Dasyuridae (Marsupialia). *Bull. Amer. Mus. Nat. Hist.* **88:** 97–156.

VAN DYCK, S., 1980. The Cinnamon Antechinus, *Antechinus leo* (Marsupialia: Dasyuridae), a new species from the vine forests of Cape York Peninsula. *Aust. Mammal.* **3:** 5–17.

WOOLLEY, P. AND WEBB, S. J., 1977. The penis of dasyurid marsupials. *In* "The biology of marsupials" ed. by B. Stonehouse and D. Gilmore. The Macmillan Press Ltd: London and Basingstoke.

RESUMEN

63. La morfología fálica de las especies Australianas de *Antechinus* (Dasyuridae, Marsupialia): una nueva herramienta taxonómica? Por P. A. Woolley.

Entre las doce especies de *Antechinus* corrientemente reconocidas que se encuentran en Australia pueden distinguirse dos grupos basados en la forma del extremo del pene y en los músculos elevadores. En el Grupo 1, que incluye *A. stuartii, A flavipes, A. bellus, A. godmani, A. leo, A. swainsonii* y *A. minimus,* el extremo del pene es distalmente bífido y los músculos elevadores no se extienden ampliamente sobre la superficie ventral de la crura. En el Grupo 2, que incluye *A. rosamondae,* el "Ningbing Antechinus", *A. macdonnellensis, A. bilarni* y *A. apicalis,* el extremo del pene es romo y los músculos elevadores se extienden ampliamente sobre la superficie ventral de la crura. Dentro de cada grupo pueden reconocerse subgrupos. Son consideradas las relaciones sugeridas por la morfología fálica.

64. THE CHROMOSOMES OF DASYURIDS (MARSUPIALIA)

G. J. YOUNG[1,4], J. A. MARSHALL GRAVES[1]
I. BARBIERI[1], P. A. WOOLLEY[2], D. W. COOPER[3] and M. WESTERMAN[1]

Fibroblast cultures were derived from tissues of *Antechinus apicalis, A. bilarni, A. flavipes, A. leo,* two geographic variants of *A. macdonnellensis, A. minimus, A. rosamondae, A. stuartii, A. swainsonii,* the Ningbing Antechinus, *Planigale maculata* and *Dasyuroides byrnei.* Most cultures grew vigorously and apparently indefinitely, and retained the diploid karyotype.

Each species had 2n = 14 chromosomes. The autosomes of different species were almost indistinguishable by the criteria of gross morphology, arm lengths, or G–banding patterns. However, the amount and position of C–banding heterochromatin varied between the species. The X chromosomes of different species differed in centromere position and the size and position of interstitial C–banding heterochromatin, but most were of comparable length. An exception was the *A. apicalis* X chromosome, the short arm of which was comparable in size and banding pattern to those of entire X chromosomes in other species but which had in addition a long arm composed entirely of heterochromatin.

Because of the highly conserved karyotypes of dasyurids, the chromosomes of these marsupials are of limited use for taxonomic studies.

Pages 783–95 *in* CARNIVOROUS MARSUPIAILS, ed. by M. Archer, February, 1982. Roy. Zool. Soc. New South Wales: Sydney, Australia.

THE CLASSICAL description of species of the genus *Antechinus,* and of other small dasyurid species, has been based on morphological features (Wakefield and Warneke 1963, 1967, Archer 1975, 1976). However, several characters given as diagnostic of, for example, the genus *Antechinus* may individually be shared with other dasyurid species. The classical taxonomic criteria may fail to define accurately a species or genus, and new methods have been devised for comparing physical and biological properties of species.

In general, marsupials differ from other mammals by having fewer and larger chromosomes. Almost half of all known marsupial species have a diploid chromosome complement of 2n = 14, the 2n = 14 complement being found in species within the six large superfamilies (Caenolestoidea, Dasyuroidea, Didelphoidea, Perameloidea, Phalangeroidea and Vombatoidea) but not in the two monospecific Australian superfamilies (Notoryctoidea and Tarsipedoidea) (Hayman 1977).

A comparison of the gross morphology of the chromosomes of species with 2n = 14 complements from the six large superfamilies shows that the general form of the complement is remarkably constant (Hayman 1977). Hayman and Martin (1969, 1974) have argued that the occurrence of a similar 2n = 14 complement in several distinct lineages within the different superfamilies supports the hypothesis that a 2n = 14 complement represents the original marsupial chromosomal complement. The chromosomes of marsupial species with a complement other than 2n = 14 are thought to have arisen by centric fissions which produced uniarmed chromosomes from biarmed elements with a concomitant increase in the diploid number. Superimposed upon this basic mechanism for karyotypic change has been the alteration of chromosomal morphology in some lineages by the incorporation of pericentric inversions.

Hayman and Martin (1974) have reported that fifteen dasyurid species, representing eight genera and including two *Antechinus* species, all possess a diploid chromosome number of 2n = 14, the morphology of individual chromosomes being very similar between species. Recently, several techiques have been developed for delineating chromosomal banding patterns. These patterns are unique for every chromosome within a species and, in many instances, the banding patterns observed on specific chromosomes, or regions of chromosomes, from closely related species are similar (Miller 1977, for example). Also, the comparison of the band-patterns of specific chromosomal regions can reveal the evolutionary relationships of species. G–banding methods have been used to deduce the relationships between the chromosomes of *Macropus* species; although individual chromo-

[1] Department of Genetics and Human Variation, La Trobe University, Bundoora, Victoria, Australia, 3083.
[2] Department of Zoology, La Trobe University, Bundoora, Victoria, Australia, 3083.
[3] School of Biological Sciences, Macquarie University, North Ryde, New South Wales, Australia, 2113.
[4] Present address: Department of Medicine, Monash University, Alfred Hospital, Prahran, Victoria, Australia, 3181.

Table 1. Source of material cultured. The institutions in which the specimens are housed can be identified by the prefixes to the animal numbers: AM, Australian Museum; JM, Queensland Museum; WAM, Western Australian Museum. – specimens which have not been registered; * individual *A. swainsonii* cultures (three female, four male) derived from individual pouch young.

	Animal number	Sex	Collection site or Origin	Tissues cultured
Antechinus apicalis	WAM M15471	♀	Cheyne Beach, W.A.	Adipose, kidney, uterus
	WAM M15472	♂	Cheyne Beach, W.A.	Body wall, kidney, testis
A. bilarni	JM 1194	♂	Cannon Hill, N.T.	Body wall
A. flavipes	AM M11253	♀	"Double Tanks", near Coolabah, N.S.W.	Kidney, lung
A. leo	AM M11246	♂	McIlwraith Range, Qld.	Kidney, lung
A. macdonnellensis (Sedimentaries)	JM 1188	♂	Sedimentaries, N.W. of Ayers Rock, N.T.	Body wall
	JM 1189	♂	Sedimentaries, N.W. of Ayers Rock, N.T.	Body wall
	JM 1192	♂	Sedimentaries, N.W. of Ayers Rock, N.T.	Body wall
	JM 1439	♂	Sedimentaries, N.W. of Ayers Rock, N.T.	Body wall
A. macdonnellensis (Tanami)	JM 1190	♂	Tanami Desert, N.T.	Body wall
	JM 1191	♂	Tanami Desert, N.T.	Body wall
A. minimus	–	♀	Near mouth of Powlett River, Vic.	Body wall, kidney, lung
Ningbing Antechinus	JM 1481	♀	74.5 km. N. of Kununurra, W.A.	Body wall
	JM 1478	♂	77 km. N. of Kununurra, W.A., laboratory reared	Body wall
A. rosamondae	JM 1471	♀	Abydos Station, via Port Hedland, W.A.	Body wall, kidney
	JM 1472	♀	Young of JM 1471, laboratory reared	Body wall, kidney
	JM 1475	♂	Abydos Station, via Port Hedland, W.A.	Body wall, kidney
A. stuartii	–	♀	Kinglake, Vic.	Body wall, lung
	–	♂	Kinglake, Vic.	Body wall, lung
A. swainsonii	–*	♀, ♂	Cambarville, Vic.	Individual pouch young
Planigale maculata	JM 1486	♀	Cannon Hill, N.T.	Body wall, kidney
	JM 1487	♀	Cannon Hill, N.T.	Body wall, kidney
Dasyuroides byrnei	–	♂	La Trobe University colony	Heart, kidney, lung

somes are similar morphologically in several species, the G–banding patterns show that these chromosomes, previously considered to be homeologous, possess different arms or arm components (Rofe 1978).

In view of the reported similarity of the gross morphology of individual chromosomes of dasyurids (Hayman and Martin 1974) we have compared the gross morphology, arm lengths and banding patterns of the chromosomes of ten *Antechinus* species (including two geographic variants of *A. macdonnellensis*), *Planigale maculata* and *Dasyuroides byrnei.*

MATERIALS AND METHODS

Chromosome preparations were made from fibroblast cultures derived from tissues of adult animals of all species, and in addition from fibroblast cultures derived from pouch young of *A. stuartii* and *A. swainsonii* (Table 1). Fibroblast cultures were used because of the difficulties associated with both obtaining sufficient samples of either blood or bone marrow and preparing adequate chromosome preparations from these sources. Wherever possible, cells derived from both male and female specimens of each species were examined.

Fibroblast cultures were initiated, maintained in culture and stored in liquid nitrogen as described by Cooper *et al.* (1977). The culture medium was usually Ham's F10 (GIBCO), but some cell lines established from adult *A. rosamondae* tissue and from *A. swainsonii* pouch young were maintained successfully also in Dulbecco's Modified Eagle Medium (GIBCO). Both media were supplemented with either 10% or 20% foetal calf serum (GIBCO or CSL, Melbourne), 100 IU/ml penicillin (CSL), 50 μg/ml streptomycin sulphate (GLAXO), 100 μg/ml glutamine (Merck) and usually 100 μg/ml kanamycin (CSL). Fibroblasts were cultured in 100 ml glass prescription bottles, in the presence of 10% carbon dioxide, at a temperature of 35°C.

Chromosome preparations were made following metaphase arrest of cells by either 10^{-3}% (final concentration) colchicine (Calbiochem or CSL) or 0.1 μg/ml colcemid (GIBCO) for 1–2 hours. The cells were detached using a 0.1% solution of trypsin (Nutritional Biochemical Corporation), treated with warm hypotonic solution (0.075M KCl at 37°C for 12–17 min) and fixed for 1 hour at room temperature in 3:1 methanol: acetic acid. Following several changes of fixative (minimum of 3, 10 min each) the cells, resuspended in a small volume of fixative, were dropped on moist, alcohol cleaned microscope slides. The slides were air dried.

Standard karyotypes and chromosomal arm length measurements were obtained from large photographic prints of standard Giemsa–stained metaphase chromosomes using a map wheel. Between 10 and 20 photographs were measured for each species. Giemsa banding (G-banding) was performed by pretreating slides with 0.1% tryspin (NBC) for 1–10 min at either room temperature or 0°C (Seabright 1971), followed by staining with 3% Gurr's R66 Giemsa. In order to obtain C–banding of heterochromatic regions of the chromosomes, slides were pretreated with 5% barium hydroxide at 50°C for 1–2 min, incubated in 2 x SSC at 60°C for 1 hr, and Giemsa stained (Sumner 1972).

NOR–banding, which differentiates nucleolar organizer associated heterochromatin, was obtained by incubation of slides in 50% silver nitrate at 37°C, following a 30 min pretreatment with borate buffer (0.1M Na_2SO_4, 0.005M $Na_2B_4O_7$, pH 9.2) (G. C. Webb, personal communication). Chromosomes were photographed without counter staining.

Chromosome preparations were photographed on Kodak High Contrast copy film.

RESULTS

Cell Cultures

Fibroblast cultures derived from adult tissues became established more slowly than did those derived from pouch young. Once established, all cultures grew vigorously and most yielded permanent cell lines, which with two exceptions, retained the diploid karyotype and showed none of the usual signs of senescence over the period of the study. In this respect dasyurid cell lines appear to be unusual: most mammalian diploid cell lines have a limited life span (Hayflick 1965). The two exceptions were (1) a Ningbing Antechinus line which became predominantly tetraploid, and (2) an *A. rosamondae* line in which the majority of cells contained a 3/6 translocation involving the short arm of chromosome number 6. Banding studies showed that no, or very little, chromosomal material had been lost as a result of this translocation. Since this rearrangement did not occur in cell lines established from other tissues of the same animal, it is probably an artifact of culture. Oc casional spreads of *A. macdonnellensis* (Tanami population) contained an extra X chromosome; similar aneuploidy has frequently been observed in established *Sminthopsis crassicaudata* lines (J.A.M.G., unpublished results).

Gross Morphology of Chromosomes

The gross morphology of the chromosomes in the karyotypes of all species examined, with the exception of *A. apicalis,* is very similar (Fig. 1).

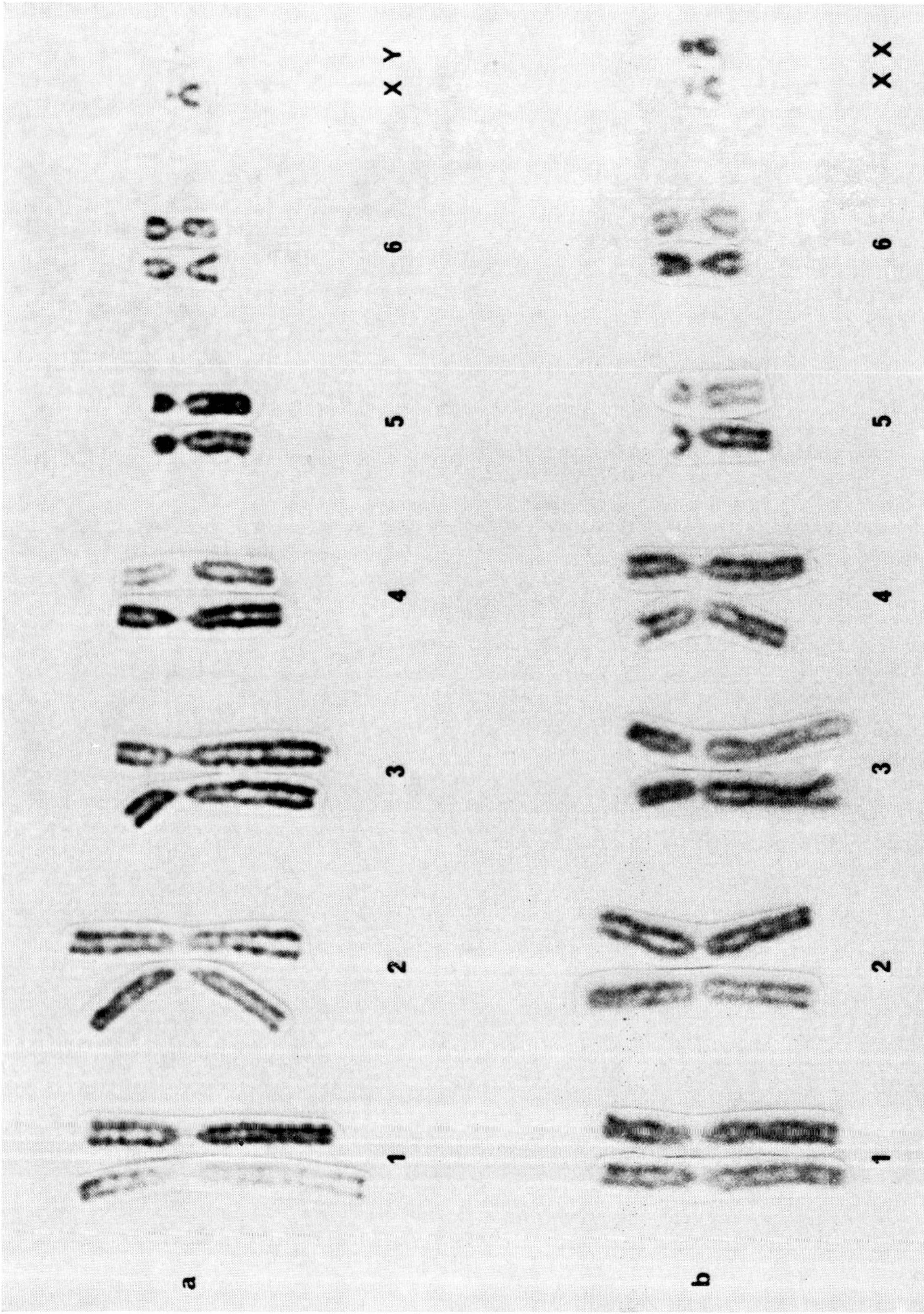

Fig. 1. Karyotypes of dasyurid marsupials. a) *Antechinus macdonnellensis* from the Sedimentaries, N.T., male. b) *A. rosamondae,* female.

All species possess a diploid chromosome number of 14, comprised of 12 autosomes plus the heteromorphic XX/XY sex chromosome pair. The autosomes, numbered 1 to 6 in decreasing size order (Hayman and Martin 1974) consist of four pairs of medium sized near–metacentric chromosomes (4), one pair of small sub–metacentric chromosomes bearing satellites on the short arm (probably the nucleolar organizing region) (5), and one pair of small metacentric chromosomes (6).

All species have a punctiform Y chromosome and a very small metacentric or acrocentric X chromosome, except *A. apicalis* which has a metacentric X chromosome similar in size to the smallest autosome.

Chromosomal Arm Lengths

The average length of the arms of each chromosome, expressed as the mean percent length relative to the total length of the autosomes, is shown in Table 2. For simplicity the standard errors associated with each calculation have been omitted from the Table; these were of the order of 0.5% except for several instances in which the standard errors were large, due in part at least to the small number of suitable metaphase figures in some preparations from which measurements could be obtained.

the length of the arms of this chromosome are difficult to obtain. The X chromosomes of all species examined are sub–metacentric, and in nine of the eleven species are of comparable size. The Ningbing Antechinus X chromosome is considerably smaller, and the *A. apicalis* X chromosome is considerably larger, than the X chromosomes of the other species (Fig. 2). The length of the short arm of the *apicalis* X chromosome is equivalent to, or is greater than, the length of the entire X chromosome of several of the other species examined.

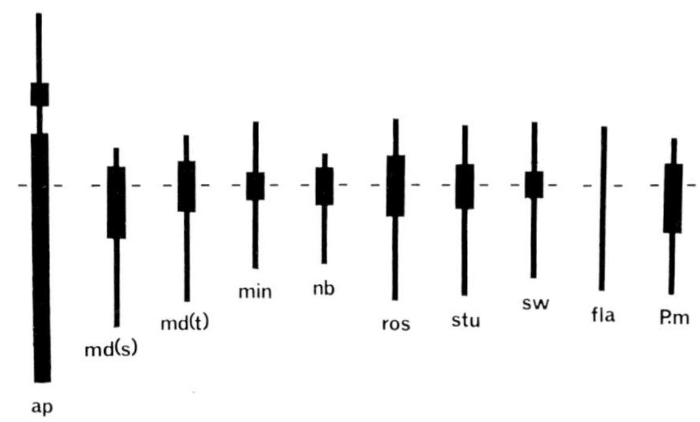

Fig. 2. Comparative lengths and C–banding heterochromatin patterns of dasyurid X chromosomes. (ap = *Antechinus apicalis;* md(s) = *A. macdonnellensis* (Sedimentaries population); md(t) = *A. macdonnellensis* (Tanami Desert population); min = *A. minimus;* nb = Ningbing Antechinus; ros = *A. rosamondae;* stu = *A. stuartii;* sw = *A. swainsonii;* fla = *A. flavipes;* P.m. = *Planigale maculata*). C-band data unavailable for *A. flavipes.*

Table 2. Mean percent length of chromosome arms of some dasyurids (calculated relative to total autosome length), (*A. mac.* (Sed) = *Antechinus macdonnellensis* population from the Sedimentaries, N.T. *A. mac* (Tan) = *A. macdonnellensis* population from the Tanami Desert, N.T.).

	1S	1L	2S	2L	3S	3L	4S	4L	5S	5L	6S	6L	XS	XL	TOTAL X	Y
Antechinus apicalis	9.6	13.3	9.3	12.1	6.2	14.2	6.4	9.4	3.4	6.8	4.1	5.5	3.7	4.3	8.0	0.6
A. bilarni	9.7	13.3	10.9	11.9	6.4	13.0	6.4	9.3	3.0	7.6	3.9	4.9	1.4	3.1	4.4	
A. flavipes	9.4	12.8	9.7	10.5	5.5	14.0	5.5	9.8	3.4	7.5	4.1	4.2	1.3	2.3	3.6	
A. leo	9.6	14.9	9.7	11.2	6.6	14.3	6.6	9.3	2.3	7.0	4.0	5.3				
A. mac (Sed)	10.0	14.7	10.4	11.4	6.1	13.9	6.3	9.5	2.9	6.9	3.8	4.0	0.8	3.1	3.9	
A. mac (Tan)	8.5	14.5	12.2	13.3	5.5	13.7	5.7	9.5	2.9	6.9	3.6	3.9	1.1	2.6	3.7	0.5
A. minimus	9.5	14.6	10.2	11.4	6.0	13.9	6.0	9.4	3.2	7.2	4.1	4.5	1.4	1.9	3.3	
Ningbing Antechinus	9.0	14.6	10.4	11.7	6.2	13.6	6.4	9.5	3.1	6.9	3.9	4.5	0.7	1.7	2.4	
A. rosamondae	9.4	13.9	10.2	11.3	6.3	13.7	6.3	9.5	3.0	7.2	4.1	4.8	1.5	2.4	4.0	
A. stuartii	9.0	14.3	10.4	11.0	6.0	14.0	6.1	9.5	3.8	7.2	4.2	4.5	1.3	2.4	3.7	
A. swainsonii	8.9	14.1	10.2	11.0	6.5	14.5	6.1	9.3	3.2	7.7	3.9	4.6	1.4	2.0	3.4	
Planigale maculata	9.2	14.8	10.2	11.2	5.8	14.6	6.2	10.1	2.6	7.0	3.7	4.4	1.0	2.5	3.5	0.5
Dasyuroides byrnei	9.4	13.5	10.3	11.1	6.8	12.6	5.6	9.7	4.3	7.6	4.0	5.3				

It can be seen from Table 2 that the lengths of the corresponding arms of the autosomes of the species examined are very similar; the few apparent differences between species are not statistically significant, because of the large standard errors.

The X chromosome of each of the species examined is the smallest chromosome of the complement (with the exception of the punctiform Y chromosome) and accurate measurements of

The relative position of the centromere of the X chromosome varies between species, but the resolution of G–banding patterns is not sufficient to permit determination of whether this variation is due to pericentric inversions. It can be seen from Figure 2 that the X chromosomes of *A. minimus* and *A. swainsonii* are indistinguishable with respect to both the position of the centromere and the total length, and that the X chromosomes of two other species, *A.*

flavipes and *A. stuartii,* also are virtually indistinguishable from each other. Although the X chromosomes of the *A. macdonnellensis* specimens obtained from the two different localities, Sedimentaries and Tanami, are very similar, the relative position of the centromere differs between them.

G–BAND PATTERNS

The G–band patterns of six species of *Antechinus* (*A. apicalis, A. minimus, A. rosamondae, A. stuartii, A. swainsonii* and the Ningbing Antechinus) and of *Planigale maculata* are shown in Fig. 3. The G–band patterns are consistent within species and there is very close correspondence of the banding patterns of individual chromosomes between the species examined. The clarity of the banding pattern is determined by several aspects of the experimental protocol, some of which may vary from preparation to preparation, and it was not possible to differentiate all of the minor bands of some species. However, it is clear that there are no differences in the presence, position and relative size of the major bands of the species examined. The distribution of minor bands was consistent between species in which minor bands were distinguishable (*A. rosamondae, A. swainsonii* and *Planigale maculata*). Preliminary examination of the G–band patterns of *A. leo, A. flavipes* and *Dasyuroides byrnei* revealed no major differences from those illustrated.

C–BANDING HETEROCHROMATIN

The chromosomes of seven *Antechinus* species (*A. apicalis, A. macdonnellensis, A. minimus, A. rosamondae, A. stuartii, A. swainsonii* and the Ningbing Antechinus) and *Planigale maculata* were examined for the presence of C–banding heterochromatin (Fig. 4). Several differences, in both the amount and the distribution of C–banding heterochromatin were detected between the species, these differences involving primarily the amount of centromeric heterochromatin on all chromosomes, the presence or absence of an interstitial heterochromatic band on the long arm of the chromosome number 3, and the number and position of C–bands on the short arm of chromosome number 5.

All of the chromosomes of all the species possess centromeric blocks of heterochromatin. The amount of centromeric heterochromatin was very uniformly distributed between chromosomes within species, but differed between species, being especially large in *A. apicalis* and relatively small in *A. stuartii* and *A. swainsonii.* Table 3 shows the estimated relative amounts of centromeric heterochromatin in the autosomes of the species examined as well as the variation between species in relation to the major non–centromeric autosomal heterochromatin. It can be seen that the size and distribution of the C–banding heterochromatin in the autosomes of *A. apicalis* is significantly different from that in the autosomes of the other species examined. The patterns of the distribution of heterochromatin in these other species varied somewhat between species, except that the patterns in *A. minimus* and *A. swainsonii* were very similar. The distribution of C–banding heterochromatin in *A. macdonnellensis* from the two localities is identical, but the size of the heterochromatic blocks in specimens from the Sedimentaries is consistently larger.

NOR–BANDING

The chromosomes of six *Antechinus* species (*A. apicalis, A. minimus, A. rosamondae, A. stuartii, A. swainsonii* and the Ningbing Antechinus) and *Planigale maculata* were examined by a silver–staining technique for the number and position of nucleolar organizer regions

Table 3. Distribution of C–banding heterochromatin in the autosomes of several dasyurids. (Abbreviations of specific names as in Fig. 2).

	ap	md(s)	md(t)	min	nb	ros	stu	sw	P.m
Relative amounts of centromeric heterochromatin	8+	+++	++	++	+++	+++	+	+	++
Interstitial (medial) band on 3L	+*	—	—	?	+	—	+	?	?
Telomeric bands									
a. autosome 1 only	—	—	—	—	+	—	—	—	—
b. autosomes 1 and 3	—	—	—	—	—	?	+	—	+
c. autosome 6	+	—	—	—	—	—	+	—	+
Nucleolar organizer associated (5S)									
a. terminal only	—	+	+	+	+	—	+	+	—
b. proximal only	—	—	—	—	—	+	—	—	—
c. terminal and proximal	+	—	—	—	—	—	—	—	+

* At least three bands resolved.

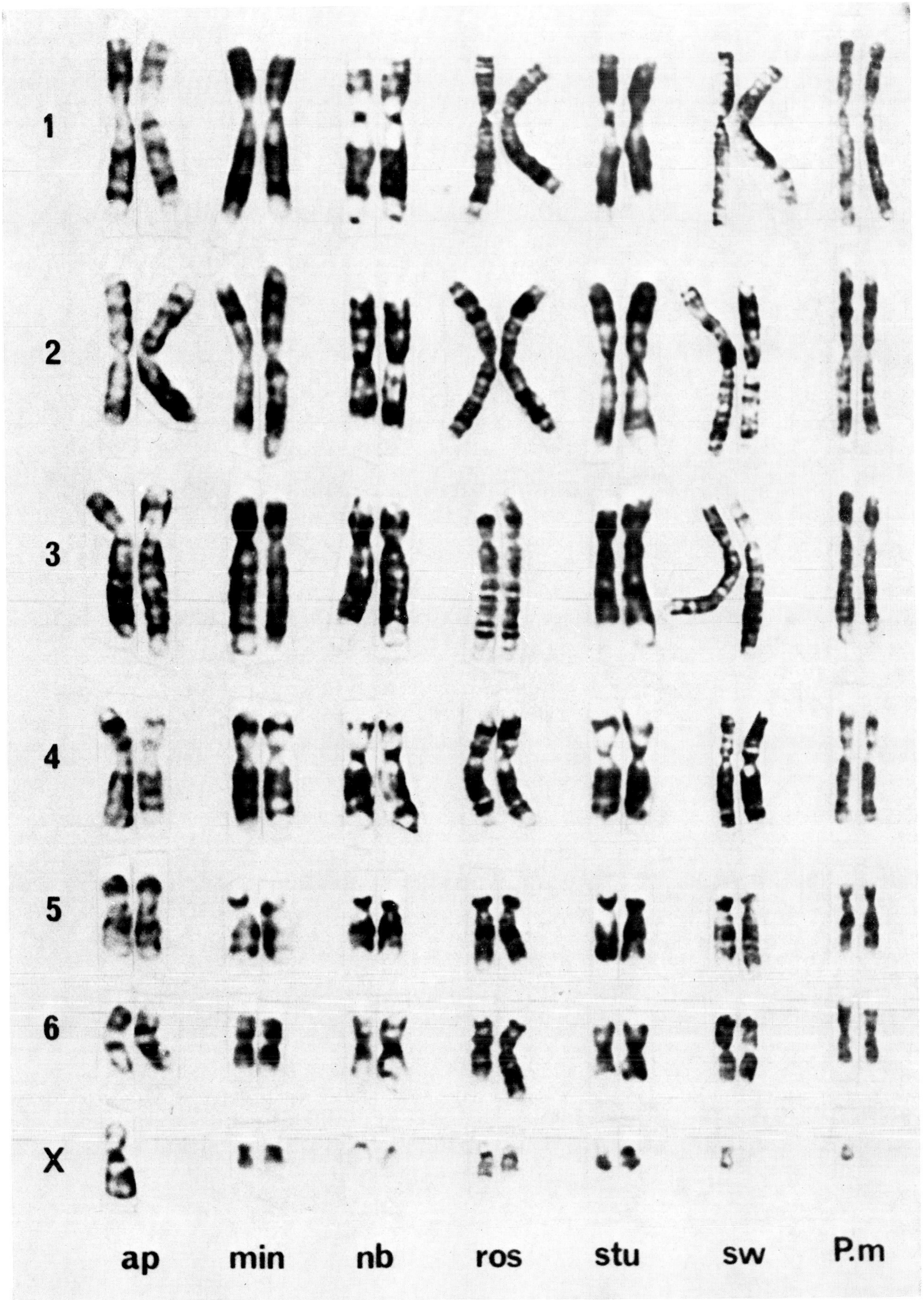

Fig. 3. G–banded karyotypes of some dasyurids. (Abbreviations of specific names as in Fig. 2). The Y chromosome has not been included in the karyotypes derived from males *(A. apicalis, A. swainsonii* and *P. maculata)* and the karyotype of *A. rosamondae* illustrates the 3/6 translocation which has been discussed in the text.

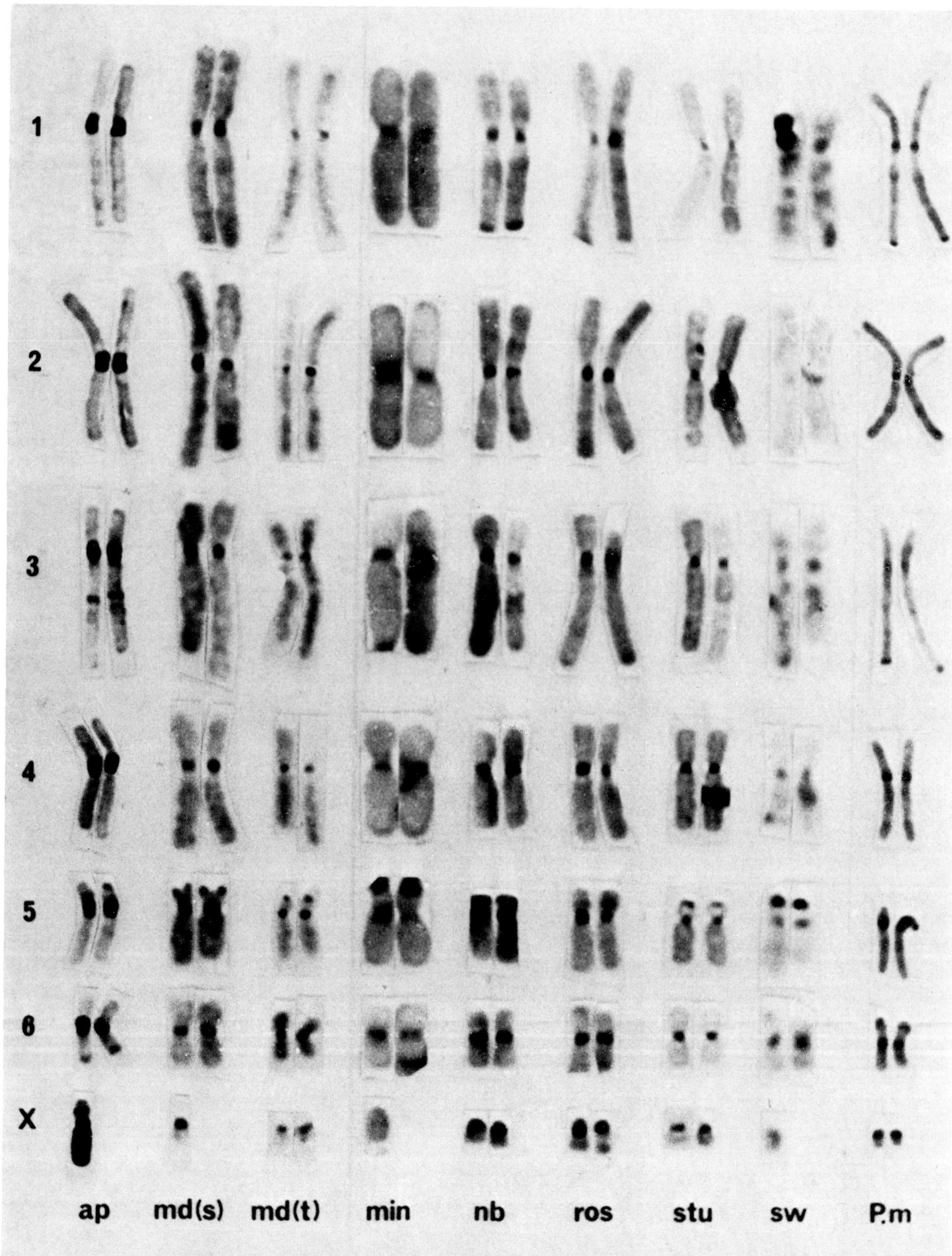

Fig. 4. C–banded karyotypes of some dasyurids (abbreviations of specific names as in Fig. 2). The Y chromosome has not been included in the karyotypes derived from males *(A. apicalis, A. macdonnellensis* from the Sedimentaries, *A. minimus* and *A. swainsonii).*

(NORs). In each of the species examined a single pair of NORs, situated at the distal end of the short arm of the chromosome number 5 pair, was detected. In some of the preparations examined the intensity of the silver–staining differed between the two homologues, and the NORs of *A. minimus* and *A. swainsonii* were more intensely stained than were the NORs of the other species (Fig. 5).

The relative size of the X chromosomes and the size and distribution of the centromeric C–banding heterochromatin of the X chromosomes of the species examined is shown in Fig. 2. It should be noted that while the arm length values were obtained from standard Giemsa–stained metaphase chromosome preparations, the size and distribution of heterochromatin bands were estimated from C–banded preparations. Hence the relationship between the position of the heterochromatin and the position of the centromere is, of necessity, subjective. It can be seen, however, that there is considerable variation between all of the species with respect to the distribution of C–banding heterochromatin, except that the distributions in *A. minimus* and *A. swainsonii* are very similar. The X chromosome of *A. macdonnellensis* from the Sedimen taries contains a larger heterochromatic block than does the X chromosome of *A. macdonnellensis* from the Tanami Desert.

DISCUSSION

Much valuable information on chromosome evolution in marsupials has been provided by earlier studies of gross chromosomal morphology (Sharman 1973, Hayman and Martin 1974, Hayman 1977, Reig *et al.* 1977). However, comparisons of chromosome size and centromere position may be inadequate to reveal even major differences in chromosome arrangement. The advent of chromosomal banding techniques makes it possible to identify homeologous portions of the genome and to compare both the amounts and the location of heterochromatin.

Dasyurids from eight different genera have been found to be karyotypically very conservative, showing little variation in either DNA content or chromosomal morphology (Hayman and Martin 1974, Hayman 1977). We have examined both the gross morphology and the specific banding patterns of the chromosomes of several

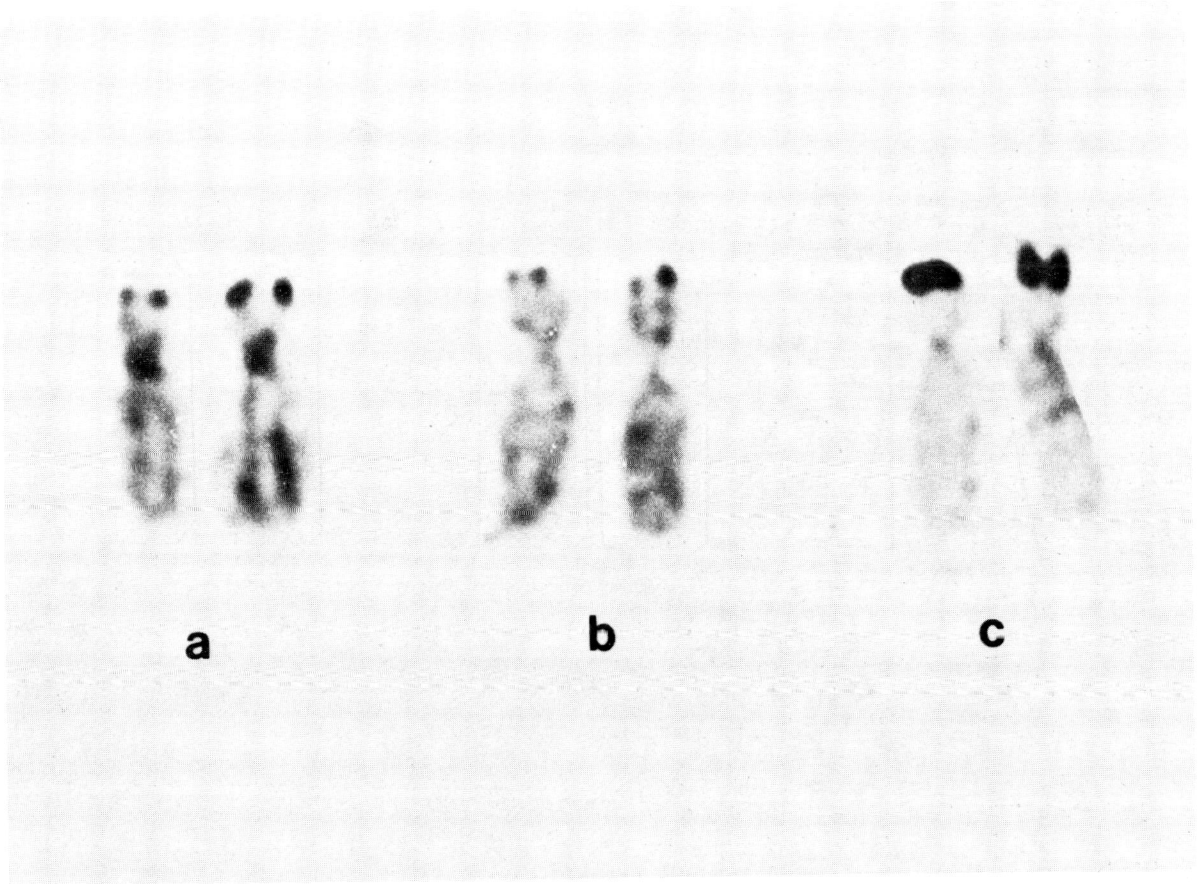

Fig. 5. NORs on chromosome number 5 of three *Antechinus* species. a, *A. minimus;* b, *A. rosamondae;* c, *A. swainsonii.* The chromosomes shown in a) were photographed from a preparation which had been C–banded prior to being destained and NOR–banded.

dasyurid species (mainly *Antechinus*) in order to determine the value of chromosomal banding patterns as a specific or generic diagnostic feature in dasyurids.

Although it is not clear which components of the DNA or nucleoprotein complex are detected by the various banding techniques (Comings 1978), banding patterns are highly repeatable and, in general, consistent within a species. Chromosomal banding patterns have been used to determine or confirm evolutionary relationships between species; for example, the genus *Rattus* (Yosida and Sagai 1973), the Australian rodents other than *Rattus* (Baverstock *et al.* 1977a), the primates (Finaz *et al.* 1977, Miller 1977) and the *Macropus* group (Rofe 1978); and between chromosome races of the same species (for example, the three races of the Australian gekko, *Phyllodactylus marmoratus*) (King and Rofe 1976).

The dasyurid species reported in this paper all possess a diploid chromosome number 2n = 14. The autosomes of those species from which the corresponding data were obtained are indistinguishable by the criteria of gross morphology, chromosomal arm lengths and G–banding patterns, but some minor differences in the size and centromere position of the X chromosome did occur.

Examination of C–banded chromosomes revealed several differences between the species (Fig. 4, Table 3) in the amount of centromeric heterochromatin (distributed uniformly between the chromosomes within a species), the amount and position of nucleolar organizer associated heterochromatin in the short arm of autosome number 5, and the number and position of interstitial bands. *A. apicalis* autosomes differed strikingly from those of the other species by the presence of very large centromeric heterochromatic regions. The X chromosomes of dasyurids are particularly interesting in that they make up the smallest proportion of the genome known for any mammal, yet they are thought to carry the same rigidly conserved set of genes (Ohno 1969). The length, centromere position, and G– and C–banding patterns of the X chromosomes were difficult to determine because the chromosomes are so small; nevertheless it is apparent that the X chromosomes are much more variable than the autosomes in this group. In many species, differences in length may be accounted for by differences in the amount of constitutive (C–banding) heterochromatin. This is particularly striking in the case of the atypically large, metacentric *A. apicalis* X chromosome, which is composed of a short arm, corresponding in size and banding pattern to the whole X chromosome of other species, plus an extra entirely heterochromatic long arm.

It is of interest that *A. minimus* and *A. swainsonii,* species which are deemed to be closely related by both classical taxonomic criteria (Wakefield and Warneke 1963) and phallic morphology (Woolley, this Volume), are virtually indistinguishable with respect to several karyotypic aspects (Fig. 2, Table 2, Table 3). Similarly, *A. flavipes* and *A. stuartii* are also considered to be closely related, both by the above criteria (Wakefield and Warneke 1967, Woolley, this Volume and by electrophoretic data (Cooper, James and Woolley, in preparation). Although G– and C–banding data are not available for *A. flavipes,* both the size and the centromere position of the X chromosome, which appear to be more variable between the species examined than are those of the autosomes, are virtually the same in *A. flavipes* and *A. stuartii.* Thus, there are several similarities between pairs of *Antechinus* species.

In contrast, we have detected several differences between the karyotypes of specimens of a single species, *A. macdonnellensis,* collected from different geographical localities. Both arms of chromosome number 2 of *A. macdonnellensis* specimens from the Tanami Desert may be longer than the equivalent arms of all of the other species examined (Table 2), but additional length measurements and the G–banding patterns of *A. macdonnellensis* (Tanami) are required to confirm this. The pattern of distribution of C–banding heterochromatin in the autosomes is the same for both populations (Table 3), but there is a greater amount of heterochromatin in both the autosomes and the X chromosomes of the population from the Sedimentaries. The total length of the X chromosomes in specimens collected from the two localities is very similar (Table 2), but the relative position of the centromere is different (Fig. 2). In addition, specimens collected from the two localities are distinguishable by their transferrin types (Cooper, James and Woolley, in preparation). Investigations of chromosomes and of serum proteins both indicate that the populations of *A. macdonnellensis* obtained from the Sedimentaries and from the Tanami Desert probably represent two races of the one species, and may be incipient species. It would be of great interest to examine both the karyotypic and the serological characteristics of specimens from other allopatric populations throughout the large geographic range of *A. macdonnellensis.*

In all mammalian species so far examined, the positions of nucleolar organizing regions (NORs) detected by the various silver–staining techniques correspond to those detected by *in situ* hybridization studies, although the silver–staining method does not always detect all of the NORs which are detected by *in situ* hybridization (Varley 1977). The intensity of silver–

staining, which may represent the degree of activity of the NOR (Goodpasture and Bloom 1975) has been reported to vary between homologues in some species (Varley 1977, Yosida 1978) but not in others (Yates *et al.* 1976). In view of this, and because of the fact that silver–stained NORs were observed in the present study on slides prepared two or three years before treatment, it is not clear what significance can be placed on the observation that the staining intensity of the NORs of several species differed between the homologues. However, it is of interest that the NORs of *A. minimus* and *A. swainsonii,* the two species examined which possess large blocks of C–banding heterochromatin at the distal end of the short arm of chromosome number five (Fig. 4), consistently stained more intensely than did the NORs of other species examined (Fig. 5). The significance of this terminal nucleolar organizer associated C–banding heterochromatin is not known.

C–banding heterochromatin is, in general, located in regions of the chromosomes rich in highly repeated simple–sequence (satellite) DNA (for review see Appels and Peacock 1978), and there is a correlation between the nuclear DNA content and the amount of C–banding heterochromatin in some marsupial species (Hayman 1977). All dasyurid species so far examined possess similar nuclear DNA content values (Hayman and Martin 1974), and although this report shows that the distribution of C–banding heterochromatin in the chromosomes of those species examined is quite varied, it is not clear if these variations are significant and suitable for specific or generic diagnosis. The nuclear DNA content value of *A. apicalis,* which contains large blocks of C–banding heterochromatin, has not been determined. Variation in the amount of C–banding heterochromatin in individual chromosomes within a species and between closely related species has been reported (Bradshaw and Hsu 1972, Dev *et al.* 1973, 1975, Fredga and Mandahl 1973, Pathak *et al.* 1973, Baverstock *et al.* 1977b, reviewed by John and Miklos 1979), and in several species the amount of C–banding heterochromatin in a chromosome complement bears no relation to the amount of highly repeated (satellite) DNA in the genome (reviewed by John and Miklos 1979).

Although a great deal is known about the structure and composition of highly repeated simple–sequence DNA and heterochromatin, the function of this material is still unknown. Several authors (for example, Gropp 1969, Yunis and Yasmineh 1971) have suggested that a major function is to facilitate chromosomal rearrangements which may lead to reproductive isolation and hence to speciation, but there is little experimental evidence to support this hypothesis (John and Miklos 1979, Miklos *et al.* 1980). Miklos *et al.* (1980) suggest that, instead, satellite DNA in the germline regulates meiotic recombination and therefore the genetic variability of a population. Alternatively, satellite DNA may confer no advantage at all on the organism, but may in fact be a class of "selfish" DNA, maintained by a drive mechanism (Doolittle and Sapienza 1980, Orgel and Crick 1980).

The role of chromosomes in speciation is not well understood. In most major vertebrate groups the rate of speciation is strongly correlated with the rate of evolutionary change in gene rearrangements (Wilson *et al.* 1975, Bush *et al.* 1977) and it has been suggested (Bush 1975, Wilson *et al.* 1975) that speciation and chromosomal evolution are casually related. The establishment of chromosomal rearrangement differences at the racial level may be a first step in speciation (Craddock 1975); intraspecific differentiation involving fixed cytological differences between geographically distinct forms has been reported for many species (Craddock 1975, King and Rofe 1976, reviewed by White 1973), although the evolutionary significance of such differentiation depends largely upon the nature of the rearrangement differences between the races. However speciation has occurred in the virtual absence of chromosomal change in several major vertebrate groups (Bickham and Baker 1976, Reig *et al.* 1977, and see Bush *et al.* 1975) suggesting (Reig *et al.* 1977) the formation and retention of close linkage of major genes which co–operate in a selectively advantageous way, i.e. supergenes (Sheppard 1953, Ford 1964). Thus the relative constancy of dasyurid karyotypes may reflect the establishment of supergenes early in the evolution of the family and retention, by epistatic selection, of chromosomal rearrangements advantageous for overall adaptations.

Thus, while there are several hypothetical roles for both chromosomes and heterochromatin in the processes of speciation and evolution, and while the amount of genetic variability within a species does appear to be related to the morphology and heterochromatin content of the chromosomes, the exact role of either the chromosomes or heterochromatin in speciation and evolution is still not understood. Although the morphology and heterochomatin content of the chromosomes of dasyurids is not diagnostic at either the specific or the generic level, seven of the species examined here are undoubtedly quite separate from each other. Cooper, James and Woolley (in preparation) have examined the electrophoretic mobility of 18 enzymes and proteins of *Antechinus apicalis, A. macdonnellensis, A. rosamondae, A. bilarni, A. stuartii* and the Ningbing Antechinus and have found no obvious groupings or relationships. On average, each pair of species differs for almost half the proteins and enzymes examined. The Nei genetic distances between any pair of species is about 0.75. The most natural interpretation of the enzyme data is that these species are the product of a single radiation. Combined with the karyotypic data presented here, they unquestionably show that protein evolution and speciation can occur in the absence of any major karyotypic changes.

ACKNOWLEDGEMENTS

The authors wish to thank Mr G. Singleton, Department of Genetics and Human Variation, for *A. minimus* tissues, Mrs. A. Monkman, Department of Genetics and Human Variation, for assistance with the preparation of photographs, and Mrs. M. Forrester, Department of Genetics and Human Variation, for typing the manuscript. The collection of animals was facilitated by A. R. G. C. grant number DI 68/16887 to P.A.W.

An earlier draft of this manuscript was constructively criticised by Professors O. Reig and M. White.

REFERENCES

Appels, R. and Peacock, W. J., 1978. The arrangement and evolution of highly repeated (satellite) DNA sequences with special reference to *Drosophila. Int. Rev. Cytol., Suppl.* **8:** 69–126.

Archer, M., 1975. *Ningaui,* a new genus of tiny dasyurids (Marsupialia) and two new species, *N. timealeyi* and *N. ridei,* from arid Western Australia. *Mem. Qd Mus.* **17:** 237–49.

Archer, M., 1976. Revision of the marsupial genus *Planigale* Troughton (Dasyuridae). *Mem. Qd Mus.* **17:** 341–65.

Baverstock, P. R., Watts, C. H. S. and Hogarth, J. T., 1977a. Chromosome evolution in Australian rodents. I. The Pseudomyinae, the Hydromyinae and the *Uromys/Melomys* group. *Chromosoma* (Berl.) **61:** 95–125.

Baverstock, P. R., Watts, C. H. S. and Hogarth, J. T., 1977b. Polymorphism of the X–chromosome, Y–chromosome and autosomes in the Australian hopping mice, *Notomys alexis, N. cervinus* and *N. fucus* (Rodentia, Muridae). *Chromosoma* (Berl.) **61:** 243–56.

Bickham, J. W. and Baker, R. J., 1976. Chromosome homology and evolution of emydid turtles. *Chromosoma* (Berl.) **54:** 201–19.

Bradshaw, W. N. and Hsu, T. C., 1972. Chromosomes of *Peromyscus* (Rodentia, Cricetidae) III. Polymorphism in *Peromyscus maniculatus. Cytogenetics.* **11:** 436–51.

Bush, G. L., 1975. Modes of animal speciation. *Ann. Rev. Ecol. Syst.* **6:** 339–64.

Bush, G. L., Case, S. M., Wilson, A. C. and Patton, J. L., 1977. Rapid speciation and chromosomal evolution in mammals. *Proc. Nat. Acad. Sci. USA* **47:** 3942–6.

Comings, D. E., 1978. Mechanisms of chromosome banding and implications for chromosomes structure. *Ann. Rev. Genet.* **12:** 25–46.

Cooper, D. W., Edwards, C., James, E., Sharman, G. B., VandeBerg, J. L. and Graves, J. A. M., 1977. Studies on metatherian sex chromosomes. VI. A third state of an X–linked gene: partial activity for the paternally derived *Pgk–A* allele in cultured fibroblasts of *Macropus giganteus* and *M. parryi. Aust. J. Biol. Sci.* **30:** 431–43.

Craddock, E. M., 1975. Intraspecific karyotypic differentiation in the Australian phasmatid *Didymuria violescenes* (Leach). I. The chromosome races and their structural and evolutionary relationships. *Chromosoma* (Berl.) **53:** 1–24.

Dev, V. G., Miller, D. A. and Miller, O. J., 1973. Chromosome markers in *Mus musculus* : strain differences in C–banding. *Genetics* **75:** 663–70.

Dev, V. G., Miller, D. A., Tantravahi, R., Schreck, Roderick, T. H., Erlanger, B. F. and Miller, O. J., 1975. Chromosome markers in *Mus musculus* : differences in C–banding between the subspecies *M. m. musculus* and *M. m. molissinus. Chromosoma* (Berl.) **53:** 335–44.

Doolittle, W. F. and Sapienza, C., 1980. Selfish genes, the phenotype paradigm and genome evolution. *Nature* **284:** 601–3.

Finaz, C., Nguyen Van Cong, Cochet, C., Frezal, J. and de Grouchy, J., 1977. Fifty–million–year evolution of chromosome 1 in the primates. Evidence from banding and gene mapping. *Cytogenet. Cell Genet.* **18:** 160–4.

Ford, E. B., 1964. "Ecological genetics". Methuen : London Pp.91–96.

Fredga, K. and Mandahl, N., 1973. Autosomal heterochromatin in some carnivores. *in* "Chromosome identification" *Nobel Symp.* **23:** 104–117.

Goodpasture, C. and Bloom, S. E., 1975. Visualization of nucleolar organizer regions in mammalian chromosomes using silver staining. *Chromosoma* (Berl.) **53:** 37–50.

Gropp, A. 1969. Cytologic mechanisms of karyotype evolution in insectivores. Pp.247–66 *in* "Comparative mammalian cytogenetics" ed. by K. Benirschke. Springer : Berlin, Heidelberg, New York.

Hayflick, L., 1965. The limited *in vitro* lifetime of human diploid cell strains. *Exptl. Cell Res.* **37:** 614–36.

Hayman, D. L., 1977. Chromosome number — constancy and variation. Pp.27–48 *in* "The biology of marsupials" ed. by B. Stonehouse and D. Gilmore. Macmillan : London.

Hayman, D. L. and Martin, P. G., 1974. "Animal Cytogenetics. Vol. 4: Chordata 4 Mammalia 1: Monotremata and Marsupialia". Gebrüder Borntraeger : Berlin–Stuttgart.

John, B. and Miklos, G. L. G., 1979. Functional aspects of satellite DNA and heterochromatin. *Int. Rev. Cytol.* **58:** 1–114.

King, M., and Rofe, R., 1976. Karyotypic variation in the Australian gekko *Phyllodactylus marmoratus* (Gray) (Gekkonidae: Reptilia) *Chromosoma* (Berl.) **54:** 75–87.

Miklos, G. L. G., Willcocks, D. A. and Baverstock, P. R., 1980. Restriction endonuclease and molecular analyses of three rat genomes with special reference to chromosome rearrangements and speciation problems. *Chromosoma* (Berl.) **76:** 339–63.

MILLER, D. A., 1977. Evolution of primate chromosomes. *Science* **198:** 1116–24.

OHNO, S., 1969. Evolution of sex chromosomes in mammals. *Ann. Rev. Genet.* **3:** 495–524.

ORGEL, L. E. AND CRICK, F. H. C., 1980. Selfish DNA: the ultimate parasite. *Nature* **284:** 604–7.

PATHAK, S., HSU, T. C. AND ARRIGHI, F. E., 1973. Chromosomes of *Peryomyscus* (Rodentia, Cricetidae) IV. The role of heterochromatin in karyotypic evolution. *Cytogenet. Cell. Genet.* **12:** 315–26.

REIG, O. A., GARDNER, A. L., BIANCHI, N. O. AND PATTON, J. L., 1977. The chromosomes of the Didelphidae (Marsupialia) and their evolutionary signicance. *Biol. J. Linn. Soc.* **9:** 191–216.

ROFE, R., 1978. G–banded chromosomes and the evolution of Macropodidae. *Aust. Mammal.* **2:** 53–63.

SEABRIGHT, M., 1971. A rapid banding technique for human chromosomes. *Lancet ii*: 971–72.

SHARMAN, G. B., 1973. The chromosomes of non-eutherian mammals. Pp. 485–530 *in* "Cytotaxonomy and vertebrate evolution" ed. by A. B. Chiarelli and E. Capanna. Academic Press : London, New York.

SHEPPARD, P. M., 1953. Polymorphism, linkage and the blood groups. *Amer. Nat.* **87:** 283–94.

SUMNER, A. T., 1972. A simple technique for demonstrating centromeric heterochromatin. *Exptl. Cell Res.* **75:** 304–6.

VARLEY, J. M., 1977. Patterns of silver staining of human chromosomes. *Chromosoma* (Berl.) **61:** 207–14.

WAKEFIELD, N. A. AND WARNEKE, R. M., 1963. Some revision of *Antechinus* (Marsupialia) –1. *Vict. Nat.* **80:** 194–219.

WAKEFIELD, N. A. AND WARNEKE, R. M., 1967. Some revision in *Antechinus* (Marsupialia) –2. *Vict. Nat.* **84:** 69–99.

WHITE, M. J. D., 1973. "Animal cytology and evolution". Third edition. Cambridge University Press : London.

WILSON, A. C., BUSH, G. L., CASE, S. M. AND KING, M. C., 1975. Social structuring of mammalian populations and rate of chromosomal evolution. *Proc. Natl. Acad. Sci. USA* **72:** 5061–5.

WOOLLEY, P. A., 1982. Phallic morphology of the Australian species of *Antechinus* (Marsupialia : Dasyuridae): A new taxonomic tool? Pp.767–81 *in* "Carnivorous marsupials", ed. by M. Archer. Roy. Zool. Soc. New South Wales: Sydney.

YATES, T. L., STOCK, A. D. AND SCHMIDLY, D. J., 1976. Chromosome banding patterns and the nucleolar organizer region of the Eastern Mole (*Scalopus aquaticus*). *Experientia* **32:** 1276–7.

YOSIDA, T. H., 1978. A preliminary note on silver–stained nucleolar organizer regions in the black and Norway rats. *Proc. Japan Acad.* **54** (Ser. *B*): 353–8.

YOSIDA, T. H. AND SAGAI, T., 1973. Similarity of Giemsa banding patterns of chromosomes in several species of the genus *Rattus*. *Chromosoma* (Berl.) **41:** 93–101.

YUNIS, J. J. AND YASMINEH, W. G., 1971. Heterochromatin, satellite DNA, and cell function. *Science* **174:** 1200–9.

RESUMEN

64. Los cromosomas de algunos dasiúridos (marsupialia). Por G. J. Young, J. A. Marshall Graves, I. Barbieri, P. A. Woolley, D. W. Cooper y M. Westerman.

Se hicieron cultivos de fibroblastos de tejidos de *Antechinus apicalis, A. bilarni, A. flavipes, A. leo,* dos variantes geográficas de *A. macdonnellensis, A. minimus, A. rosamondae, A. stuartii. A. swainsonii,* el "ningbing Antechinus", *Planigale maculata* y *Dasyuroides byrnei.* La mayor parte de los cultivos crecieron vigorosamente y aparentemente indifinidamente, y retuvieron el cariotipo diploide.

Cada especie 2n = 14 cromosomas. Los cromosomas de las diferentes especies son casi inindistinguibles por los criterios de su morfología general, longitudes de los brazos o tipo de la banda G. Sin embargo, la cantidad y posición de heterocramatina de la banda C varía según las especies. Los cromosomas X de diferentes especies difieren en la posición del centromero y en el tamaño y posición de la heterocromatina interistical de la banda C, aunque la mayoria son comparables en su longitud. Una excepción es el cromosoma X de *A. apicalis*, cuyo brazo corto es comparable en tamañ y tipo de bandeado al de aquellos cromosomas X de otras especies, aunque tienen en adición un largo brazo compuesto enteramente por heterocromatina.

Dado que el cariotipo de los dasiúridos es altamente conservador, los cromosomas de estos marsupiales tienen un limitado uso en los estudios taxonómicos.

INDEX TO TAXONOMIC NAMES

This index is provided to enable determination of the papers that mention particular taxa of mammals. Therefore, the numbers indicated refer to the *paper* numbers, not the page numbers. If a generic name has been noted in a paper without a species name (or its clear indication e.g. as would be the case for monotypic genera), the paper is indexed under the generic name only. Junior synonyms are indicated by an entry indicating the senior synonym where the paper numbers are given. Subspecies names are listed alphabetically as taxa within the generic sections and are followed by a reference to the species in which they are placed. Generic names are listed alphabetically and species (and subspecies) names are listed alphabetically within the generic sections.

ERRATA

The following errors in, or omissions from, articles in these Volumes should be noted. Spelling and other typographical errors, unless they impede the reader's understanding, have not been noted here. The first number in the code refers to the paper number, the second to the page number, the letter refers to the left (L) or right (R) column, the fourth number to the paragraph and the fifth to the line number.

7.66.Table 1. Under entry for *Macropus eugenii* and the reference Renfree and Tyndale–Biscoe (1978), the Secondary oocyte should be "+" (not "—") and the Fertilized uncleaved uterine egg should be "—" (not "+").

11.105.R.3.9–10. Replace lines 9–10 with "shows a threefold difference in mean body weight between *A. stuartii* and *A. swainsonii* during May".

15.154.L.1.11–13. Including the line starting with "terial were . . .", the rest of this paragraph should be replaced with "cent of the months sampled were Orthoptera and Larvae. Plant material did not show annual peaks in occurrence."

16.176.R.1. Between lines 3 and 4, insert "(50 cm). Log length was positively correlated with H.V.O. (50 cm) and".

20.226.L.6.3. Replace "170°S" with "17° 00′S".

24.256.L.3.21–22. The equation should be:

$$\frac{\text{Conductance}}{(\text{in W m}^{-2}{}^{0}\text{C}^{-1})} = \frac{\text{Heat production}}{(T_b - T_a)\ \text{Surface area}}$$

25.262.R.1.5. In the first equation, the phrase in parentheses should be "(where $\overline{V}o_2$ max is in units of ml O_2g^{-1})".

26.269.L.5.9. Line should read "mercial 1080 (0.2 mg/g of bait) replicated as closely".

37.406.R.1.40. Delete the word "why".

37.429.R.5.10. Line should read "of *stuartii, A. s. adustus* is more closely related".

37.433.L.2.11. Date cited for Woolley should be "1982".

37.433.L.Table 3. Legend. Line 5. Taxon represented by "n" should read "ningbing". Nowhere in this article or in these Volumes should this name be set in italics. It is informal and not intended to have any standing as an available taxonomic name.

37.441.L.2.4. For "Mrs." read "Drs." (for, in fact, I do not think there are any old women among them).

38.475.L.1. The date for the Lowenstein, Sarich and Richardson reference is 1981 (not 1982). The rest of the reference is: Albumin systematics of the extinct mammoth and Tasmanian Wolf. *Nature* **291:** 409–11.

38.474.R.2. Add "Particular thanks are extended to Ms Margy Oakden for the drawings of carnivorous marsupials used in this paper."

45.548.R.1.2. Line should be replaced with "is marked as the latter has a femur and tibia of".

46.559.L.1.10. Date cited should be "1969".

46.561.L. Van Valen reference date should be "1969".

47.565.L.2. Authorship of *Antechinomys laniger* should be "(Gould, 1856)".

51.628.L.5.2. For "Borhyaenoidae" read "Borhyaenoidea".

51.638.R.6.3. Line should read "known, were reconsidered and synthesized into".

53.651.R.1.23. Line should read "all species studied had CB6 activity but that".

54.659.R.2.12. For "Wolley" read "Woolley".

54.660.L.3. Delete line 20.

57.698.L.2.6. Line 6 should read "the centre of the arid zone,".

60.714.L.1.1. The fourth word should read "Syndactyliformes".

61.717.R.2.4. Line should read "the experimental conditions obtaining, Ag^{+} ions".

61.719.R.1.12. Line should read "actinomycin–D, Ag^{+} or Hg^{2+} ions."

62.726.L.1.7. Line should read "the northern populations thus (p.71) '. . . the".

62.740.L.3.5. Line should read "August 1976 had well–developed pouch areas. One of".

63.776.R.1.3. Delete "the".

64.793.R.2.8. Replace "casually" with "causally".

In addition, the following referees' names, inadvertently omitted from the acknowledgements sections of particular papers (an omission that was my fault), should be noted: For paper 2, Drs. B. Fox and M. Denny; 4, Drs. P. Woolley and B. Fox; 5, Drs. A. Lee and P. Dwyer; 7, Drs. L. Hughes and H. Tyndale–Biscoe; 9, Drs. P. Woolley and M. Denny and Mr. S. Van Dyck; 23, Mr. J. Mahoney and Dr. D. Rounsevell; 42, Dr. T. Rich and Mr. J. Mahoney; 46, Drs. R. Wells and T. Rich; 47, Dr. M. Archer and Mr. J. Mahoney.